Intermediate Algebra

Books in the Gustafson/Karr/Massey Series

Beginning Algebra, Ninth Edition

Beginning and Intermediate Algebra: An Integrated Approach, Sixth Edition

Intermediate Algebra, Ninth Edition

Intermediate Algebra

9th Edition

R. David Gustafson
Rock Valley College

Rosemary M. Karr
Collin College

Marilyn B. Massey
Collin College

Australia • Brazil • Japan • Korea • Mexico • Singapore • Spain • United Kingdom • United States

***Intermediate Algebra,* Ninth Edition**
R. David Gustafson, Rosemary M. Karr, Marilyn B. Massey

Publisher: Charlie Van Wagner
Acquisitions Editor: Marc Bove
Developmental Editor: Meaghan Banks
Assistant Editor: Shaun Williams
Editorial Assistant: Kyle O'Loughlin
Media Editors: Maureen Ross, Heleny Wong
Marketing Manager: Gordon Lee
Marketing Assistant: Angela Kim
Marketing Communications Manager: Katy Malatesta
Content Project Manager: Jennifer Risden
Creative Director: Rob Hugel
Art Director: Vernon Boes
Print Buyer: Karen Hunt
Rights Acquisitions Account Manager, Text: Timothy Sisler
Rights Acquisitions Account Manager, Image: Don Schlotman
Production Service: Chapter Two, Ellen Brownstein
Text Designer: Terri Wright
Photo Researcher: Meaghan Banks
Copy Editor: Ellen Brownstein
Illustrator: Lori Heckelman
Cover Designer: Terri Wright
Cover Image: Digital Archive Japan/DAJ/IPN
Compositor: Graphic World, Inc.

Library of Congress Control Number: 2009921040
ISBN-13: 978-0-495-83142-6
ISBN-10: 0-495-83142-5

Brooks/Cole
20 Davis Drive
Belmont, CA 94002-3098
USA

Cengage Learning is a leading provider of customized learning solutions with office locations around the globe, including Singapore, the United Kingdom, Australia, Mexico, Brazil, and Japan. Locate your local office at **www.cengage.com/global.**

Cengage Learning products are represented in Canada by Nelson Education, Ltd.

To learn more about Brooks/Cole, visit **www.cengage.com/brookscole**

Purchase any of our products at your local college store or at our preferred online store **www.ichapters.com.**

Printed in the United States of America
1 2 3 4 5 6 7 13 12 11 10 09

About the Authors

R. DAVID GUSTAFSON

R. David Gustafson is Professor Emeritus of Mathematics at Rock Valley College in Illinois and also has taught extensively at Rockford College and Beloit College. He is coauthor of several best-selling mathematics textbooks, including Gustafson/Frisk/Hughes, *College Algebra;* Gustafson/Karr/Massey, *Beginning Algebra, Intermediate Algebra, Beginning and Intermediate Algebra: A Combined Approach;* and the Tussy/Gustafson and Tussy/Gustafson/Koenig developmental mathematics series. His numerous professional honors include Rock Valley Teacher of the Year and Rockford's Outstanding Educator of the Year. He has been very active in AMATYC as a Midwest Vice-president and has been President of IMACC, AMATYC's Illinois affiliate. He earned a Master of Arts degree in Mathematics from Rockford College in Illinois, as well as a Master of Science degree from Northern Illinois University.

ROSEMARY M. KARR

Rosemary Karr graduated from Eastern Kentucky University (EKU) with a Bachelor's degree in Mathematics, attained her Master of Arts degree at EKU in Mathematics Education, and earned her Ph.D. from the University of North Texas. After two years of teaching high school mathematics, she joined the faculty at Eastern Kentucky University, where she earned tenure as Assistant Professor of Mathematics. A professor of mathematics at Collin College in Plano, Texas, since 1990, Professor Karr has written more than 10 solutions manuals, presented numerous papers, and been an active member in several educational associations (including President of the National Association for Developmental Education). She has been honored several times by Collin College, and has received such national recognitions as U.S. Professor of the Year (2007).

MARILYN B. MASSEY

Marilyn Massey teaches mathematics at Collin College in McKinney, Texas, where she joined the faculty in 1991. She has been President of the Texas Association for Developmental Education, featured on the list of Who's Who Among America's Teachers and received an Excellence in Teaching Award from the National Conference for College Teaching and Learning. Professor Massey has presented at numerous state and national conferences; her article "Service-Learning Projects in Data Interpretation" was one of two included from community college instructors for the Mathematical Association of America's publication, *Mathematics in Service to the Community.* She earned her Bachelor's degree from the University of North Texas and Master of Arts degree in Mathematics Education from the University of Texas at Dallas.

To
Harold and Monie,
Harder and Evelyn,

with love and affection
RDG

To my husband and best friend Fred,
for his unwavering support of my work
RMK

To Ron, for his unconditional love and support
MBM

Contents

Chapter 1 A Review of Basic Algebra 1

1.1 The Real Number System 2
1.2 Arithmetic and Properties of Real Numbers 14
1.3 Exponents 28
1.4 Scientific Notation 40
1.5 Solving Linear Equations in One Variable 48
1.6 Using Linear Equations to Solve Application Problems 61
1.7 More Applications of Linear Equations in One Variable 71
■ ***PROJECTS 79***
CHAPTER REVIEW 80
CHAPTER 1 TEST 89

Chapter 2 Graphs, Equations of Lines, and Functions 90

2.1 Graphing Linear Equations 91
2.2 Slope of a Line 106
2.3 Writing Equations of Lines 118
2.4 Introduction to Functions 131
2.5 Graphs of Other Functions 144
■ ***PROJECTS 156***
CHAPTER REVIEW 157
CHAPTER 2 TEST 163
CUMULATIVE REVIEW EXERCISES 164

Chapter 3 Systems of Linear Equations 165

3.1 Solving Systems of Two Linear Equations by Graphing 166
3.2 Solving Systems of Two Linear Equations by Substitution and Elimination (Addition) 176
3.3 Solving Systems of Three Linear Equations in Three Variables 190
3.4 Solving Systems of Linear Equations Using Matrices 200
3.5 Solving Systems of Linear Equations Using Determinants 209
■ ***PROJECTS 220***

CHAPTER REVIEW 221
CHAPTER 3 TEST 226

Chapter 4 Inequalities 228

4.1 Solving Linear Inequalities in One Variable 229
4.2 Solving Absolute-Value Equations and Inequalities in One Variable 240
4.3 Solving Linear Inequalities in One or Two Variables 253
4.4 Solving Systems of Linear and Quadratic Inequalities in Two Variables 261
4.5 Solving Systems Using Linear Programming 268
■ *PROJECTS 278*
CHAPTER REVIEW 279
CHAPTER 4 TEST 285
CUMULATIVE REVIEW EXERCISES 285

Chapter 5 Polynomials and Polynomial Functions 287

5.1 Polynomials and Polynomial Functions 288
5.2 Adding and Subtracting Polynomials 298
5.3 Multiplying Polynomials 303
5.4 The Greatest Common Factor and Factoring by Grouping 315
5.5 The Difference of Two Squares; the Sum and Difference of Two Cubes 323
5.6 Factoring Trinomials 332
5.7 Summary of Factoring Techniques 343
5.8 Solving Equations by Factoring 347
■ *PROJECTS 356*
CHAPTER REVIEW 357
CHAPTER 5 TEST 362

Chapter 6 Rational Expressions 363

6.1 Finding the Domain of Rational Functions and Simplifying Rational Expressions 364
6.2 Multiplying and Dividing Rational Expressions 375
6.3 Adding and Subtracting Rational Expressions 384
6.4 Simplifying Complex Fractions 394
6.5 Solving Equations Containing Rational Expressions 403
6.6 Dividing Polynomials 413
6.7 Synthetic Division 421
6.8 Proportion and Variation 428
■ *PROJECTS 442*
CHAPTER REVIEW 444
CHAPTER 6 TEST 450
CUMULATIVE REVIEW EXERCISES 451

Chapter 7 Radicals and Rational Exponents 453

7.1 Radical Expressions 454
7.2 Applications of the Pythagorean Theorem and the Distance Formula 467
7.3 Rational Exponents 474
7.4 Simplifying and Combining Radical Expressions 483
7.5 Multiplying and Dividing Radical Expressions 494
7.6 Radical Equations 504
7.7 Complex Numbers 513
■ ***PROJECTS 523***
CHAPTER REVIEW 524
CHAPTER 7 TEST 532

Chapter 8 Quadratic Functions, Inequalities, and Algebra of Functions 534

8.1 Solving Quadratic Equations Using the Square-Root Property and by Completing the Square 535
8.2 Solving Quadratic Equations by the Quadratic Formula 546
8.3 The Discriminant and Equations That Can Be Written in Quadratic Form 554
8.4 Graphs of Quadratic Functions 563
8.5 Quadratic and Other Nonlinear Inequalities 578
8.6 Algebra and Composition of Functions 589
8.7 Inverses of Functions 596
■ ***PROJECTS 606***
CHAPTER REVIEW 607
CHAPTER 8 TEST 615
CUMULATIVE REVIEW EXERCISES 616

Chapter 9 Exponential and Logarithmic Functions 618

9.1 Exponential Functions 619
9.2 Base-*e* Exponential Functions 631
9.3 Logarithmic Functions 639
9.4 Natural Logarithms 650
9.5 Properties of Logarithms 656
9.6 Exponential and Logarithmic Equations 666
■ ***PROJECTS 678***
CHAPTER REVIEW 679
CHAPTER 9 TEST 685

Chapter 10 Conic Sections and More Graphing 686

10.1 The Circle and the Parabola 687
10.2 The Ellipse 700

10.3 The Hyperbola 712
10.4 Systems Containing Second-Degree Equations 722
10.5 Piecewise-Defined Functions and the Greatest Integer Function 728
■ *PROJECTS 735*
CHAPTER REVIEW 735
CHAPTER 10 TEST 743
CUMULATIVE REVIEW EXERCISES 744

Chapter 11 Miscellaneous Topics 745

11.1 The Binomial Theorem 746
11.2 The *n*th Term of a Binomial Expansion 753
11.3 Arithmetic Sequences 756
11.4 Geometric Sequences 765
11.5 Infinite Geometric Sequences 773
11.6 Permutations and Combinations 777
11.7 Probability 788
■ *PROJECTS 794*
CHAPTER REVIEW 795
CHAPTER 11 TEST 800
CUMULATIVE REVIEW EXERCISES 801

Glossary G-1

Appendix I Symmetries of Graphs A-1

Appendix II Sample Final Examination A-7

Appendix III Tables A-11

Table A Powers and Roots A-11
Table B Base-10 Logarithms A-12
Table C Base-*e* Logarithms A-13

Appendix IV Answers to Selected Exercises A-15

Index I-1

Preface

TO THE INSTRUCTOR

This ninth edition of *Intermediate Algebra* is an exciting and innovative revision. The new edition reflects a thorough update, has new pedagogical features that make the text easier to read, and has an entirely new and fresh interior design. The revisions to this already successful text will further promote student achievement. This series is known for its integrated approach, for the clarity of its writing, for making algebra relevant and engaging, and for developing student skills. New coauthors Rosemary Karr and Marilyn Massey have joined David Gustafson, bringing more experience in, and contributions to, developmental education.

As before, our goal has been to write a book that

1. is enjoyable to read,
2. is easy to understand,
3. is relevant, and
4. will develop the necessary skills for success in future academic courses or on the job.

In this new edition, we have developed a learning plan that helps students transition to the next level in their coursework, teaching them the problem-solving strategies that will serve them well in their everyday lives. Most textbooks share the goals of clear writing, well-developed examples, and ample exercises, whereas the Gustafson/Karr/Massey series develops student success beyond the demands of traditional required coursework.

The ninth edition's learning tools have been developed with your students in mind and include several new features:

- *Learning Objectives* appear at the beginning of each section and provide a map to the content. Objectives are keyed to Guided Practice exercises, indicating which are satisfied by specific problems.
- The addition of the *Now Try This* exercises helps students develop a deeper conceptual comprehension of the material. These *Now Try This* exercises can also be used as questions for independent or group work, or for active classroom participation through in-class group discussions.
- The *Guided Practice* exercises are keyed both to Examples and to section Learning Objectives. Students working these problems are directed to the specific section in the text where they can find help on approaches to solve these problems.
- While *Guided Practice* offers students support on their homework, the *Additional Practice* sections reinforce and stretch their newly developed skills by having them solve problems independent of examples.

Through our collective teaching experience, we have developed an acute awareness of students' approach to homework. Consequently, we have designed the problem sets to extend student learning beyond the mimicking of a previous example.

NEW TO THIS EDITION

- New design
- New Learning Objectives
- New Vocabulary feature
- New Glossary
- New Now Try This feature
- New Guided Practice exercises
- Retooled Exercise Sets
- Redesigned Chapter Reviews
- New Basic Calculator Keystroke Guide

New Tabular Structure for Easier Feature Identification

The design now incorporates a tabular structure identifying features such as Objectives, Vocabulary, Getting Ready exercises, definitions and formulas, and the new Now Try This feature. Students are visually guided through the textbook, providing for increased readability.

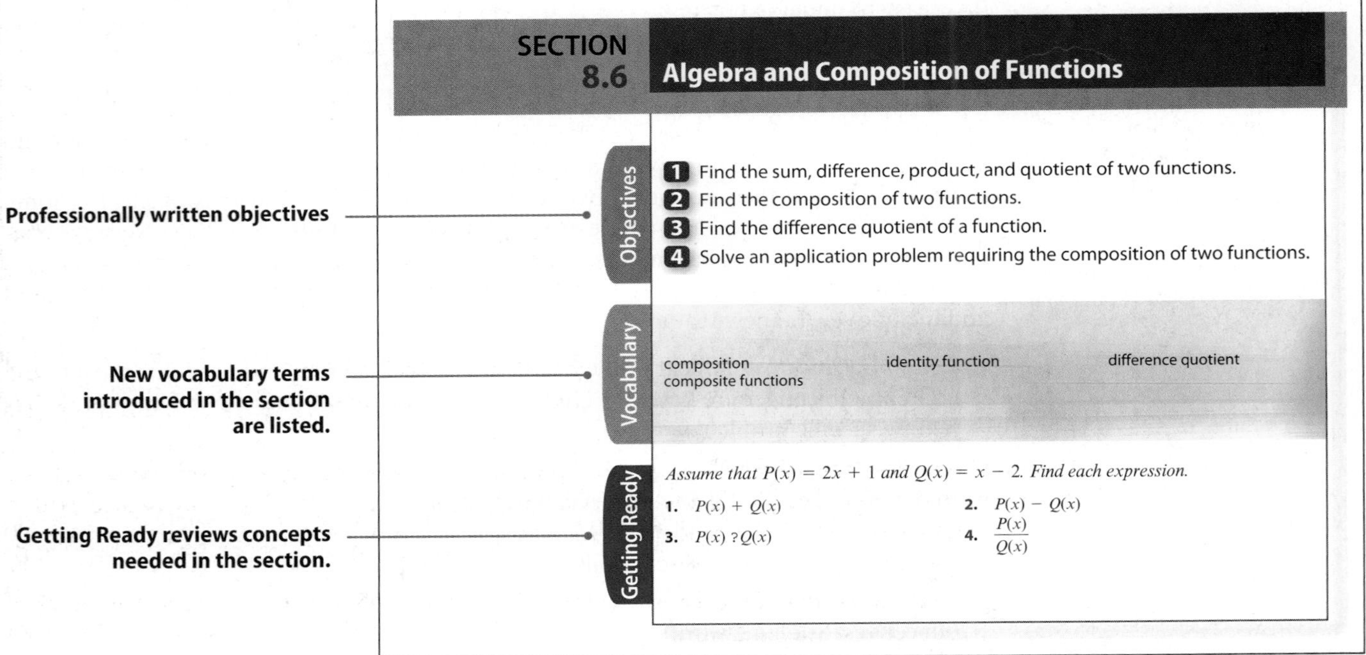

New Learning Objectives for Measurable Outcomes

Appearing at the beginning of each section, Learning Objectives are mapped to the appropriate content, as well as to relevant exercises in the Guided Practice section. Measurable objectives allow students to identify specific mathematical processes that may need additional reinforcement. For the instructor, homework can be more easily developed with problems keyed to objectives, thus facilitating the instructors' identification.

New Section Vocabulary Feature plus Glossary

In order to work mathematics, one must be able to speak the language. It is this philosophy that prompted us to strengthen the treatment of vocabulary. Not only are vocabulary words identified at the beginning of each section, these words are also bolded within the section. Exercises include questions on the vocabulary words, and a glossary has been included to facilitate the students' reference to these words.

New Now Try This Feature at the End of Each Section

The Something to Think About feature within the exercises already serves as an excellent transition tool, but we wanted to add transitional group-work exercises. Thus, each exercise set has been preceded with Now Try This problems intended to increase conceptual understanding through active classroom participation and involvement. To discourage a student from simply looking up the answer and trying to find a process that will produce that answer, answers to these problems will only be provided in the Annotated Instructor's Edition of the text.

Now Try This problems can be worked independently or in small groups and transition to the Exercise Sets, as well as to material in future sections. The problems will reinforce topics, digging a little deeper than the examples.

Now Try This exercises increase understanding through classroom participation.

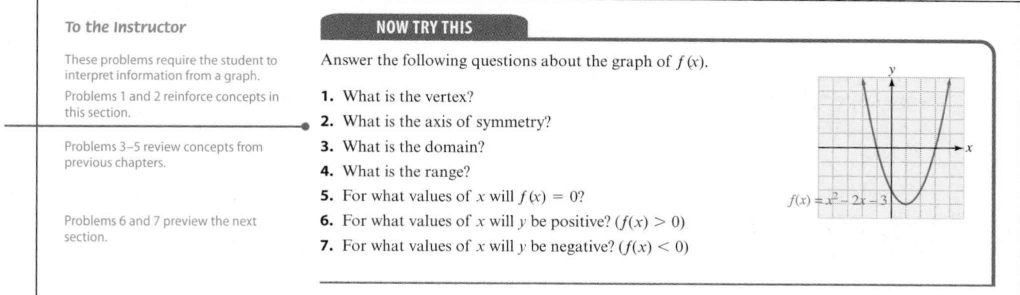

To the Instructor

These problems require the student to interpret information from a graph.

Problems 1 and 2 reinforce concepts in this section.

Problems 3–5 review concepts from previous chapters.

Problems 6 and 7 preview the next section.

NOW TRY THIS

Answer the following questions about the graph of $f(x)$.

1. What is the vertex?
2. What is the axis of symmetry?
3. What is the domain?
4. What is the range?
5. For what values of x will $f(x) = 0$?
6. For what values of x will y be positive? $(f(x) > 0)$
7. For what values of x will y be negative? $(f(x) < 0)$

New *Guided Practice* and Retooled Exercise Sets

The Exercise Sets for this ninth edition of *Intermediate Algebra* have been retooled to transition students through progressively more difficult homework problems. Students are initially asked to work quick, basic problems on their own, then proceed to working exercises keyed to examples, and finally complete application problems and critical thinking questions on their own.

Warm-Ups get students into the homework mindset, asking quick memory-testing questions. Review and Vocabulary and Concepts emphasize the main concepts taught in the section.

Guided Practice exercises are keyed to the objectives to increase student success by directing students to the concept covered in that group of exercises. Should a student encounter difficulties working a problem, a specific example within the objective is also cross-referenced.

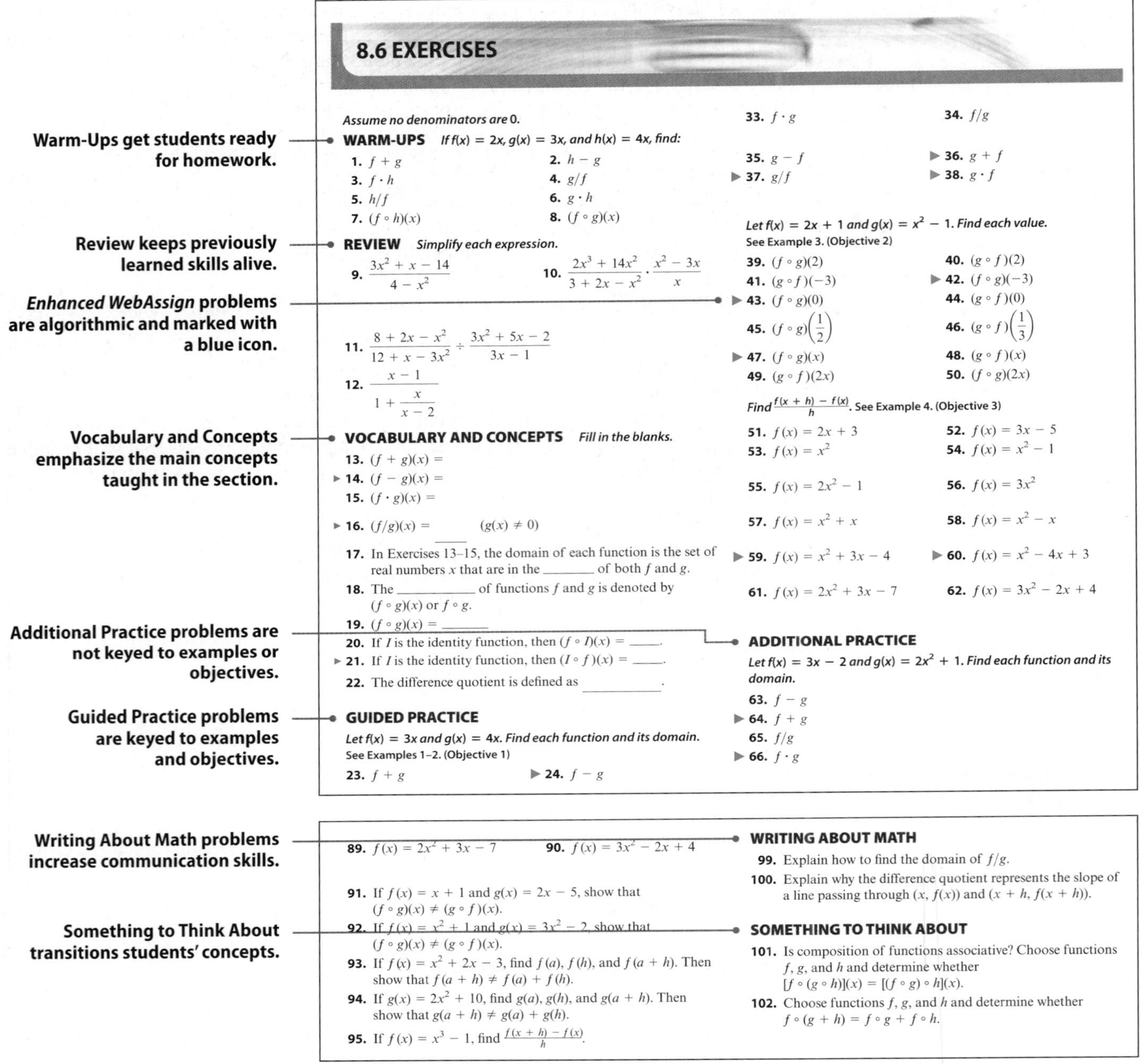

8.6 EXERCISES

Assume no denominators are 0.

WARM-UPS *If* $f(x) = 2x$, $g(x) = 3x$, *and* $h(x) = 4x$, *find:*

1. $f + g$
2. $h - g$
3. $f \cdot h$
4. g/f
5. h/f
6. $g \cdot h$
7. $(f \circ h)(x)$
8. $(f \circ g)(x)$

REVIEW *Simplify each expression.*

9. $\frac{3x^2 + x - 14}{4 - x^2}$
10. $\frac{2x^3 + 14x^2}{3 + 2x - x^2} \cdot \frac{x^2 - 3x}{x}$
11. $\frac{8 + 2x - x^2}{12 + x - 3x^2} \div \frac{3x^2 + 5x - 2}{3x - 1}$
12. $\frac{x - 1}{1 + \frac{x}{x - 2}}$

VOCABULARY AND CONCEPTS *Fill in the blanks.*

13. $(f + g)(x) =$
▸ **14.** $(f - g)(x) =$
15. $(f \cdot g)(x) =$
▸ **16.** $(f/g)(x) = \frac{}{\_\_\_\_}$ $(g(x) \neq 0)$
17. In Exercises 13–15, the domain of each function is the set of real numbers x that are in the ______ of both f and g.
18. The ______ of functions f and g is denoted by $(f \circ g)(x)$ or $f \circ g$.
19. $(f \circ g)(x) =$ ______
20. If I is the identity function, then $(f \circ I)(x) =$ ______.
▸ **21.** If I is the identity function, then $(I \circ f)(x) =$ ______.
22. The difference quotient is defined as ______.

GUIDED PRACTICE

Let $f(x) = 3x$ *and* $g(x) = 4x$. *Find each function and its domain.* See Examples 1–2. (Objective 1)

23. $f + g$
▶ **24.** $f - g$
33. $f \cdot g$
34. f/g
35. $g - f$
▶ **36.** $g + f$
▶ **37.** g/f
▶ **38.** $g \cdot f$

Let $f(x) = 2x + 1$ *and* $g(x) = x^2 - 1$. *Find each value.* See Example 3. (Objective 2)

39. $(f \circ g)(2)$
40. $(g \circ f)(2)$
41. $(g \circ f)(-3)$
▶ **42.** $(f \circ g)(-3)$
▶ **43.** $(f \circ g)(0)$
44. $(g \circ f)(0)$
45. $(f \circ g)\left(\frac{1}{2}\right)$
46. $(g \circ f)\left(\frac{1}{3}\right)$
▶ **47.** $(f \circ g)(x)$
48. $(g \circ f)(x)$
49. $(g \circ f)(2x)$
50. $(f \circ g)(2x)$

Find $\frac{f(x + h) - f(x)}{h}$. See Example 4. (Objective 3)

51. $f(x) = 2x + 3$
52. $f(x) = 3x - 5$
53. $f(x) = x^2$
54. $f(x) = x^2 - 1$
55. $f(x) = 2x^2 - 1$
56. $f(x) = 3x^2$
57. $f(x) = x^2 + x$
58. $f(x) = x^2 - x$
▶ **59.** $f(x) = x^2 + 3x - 4$
▶ **60.** $f(x) = x^2 - 4x + 3$
61. $f(x) = 2x^2 + 3x - 7$
62. $f(x) = 3x^2 - 2x + 4$

ADDITIONAL PRACTICE

Let $f(x) = 3x - 2$ *and* $g(x) = 2x^2 + 1$. *Find each function and its domain.*

63. $f - g$
▶ **64.** $f + g$
65. f/g
▶ **66.** $f \cdot g$

89. $f(x) = 2x^2 + 3x - 7$
90. $f(x) = 3x^2 - 2x + 4$
91. If $f(x) = x + 1$ and $g(x) = 2x - 5$, show that $(f \circ g)(x) \neq (g \circ f)(x)$.
92. If $f(x) = x^2 + 1$ and $g(x) = 3x^2 - 2$, show that $(f \circ g)(x) \neq (g \circ f)(x)$.
93. If $f(x) = x^2 + 2x - 3$, find $f(a)$, $f(h)$, and $f(a + h)$. Then show that $f(a + h) \neq f(a) + f(h)$.
94. If $g(x) = 2x^2 + 10$, find $g(a)$, $g(h)$, and $g(a + h)$. Then show that $g(a + h) \neq g(a) + g(h)$.
95. If $f(x) = x^3 - 1$, find $\frac{f(x + h) - f(x)}{h}$.

WRITING ABOUT MATH

99. Explain how to find the domain of f/g.
100. Explain why the difference quotient represents the slope of a line passing through $(x, f(x))$ and $(x + h, f(x + h))$.

SOMETHING TO THINK ABOUT

101. Is composition of functions associative? Choose functions f, g, and h and determine whether $[f \circ (g \circ h)](x) = [(f \circ g) \circ h](x)$.
102. Choose functions f, g, and h and determine whether $f \circ (g + h) = f \circ g + f \circ h$.

Additional Practice problems are mixed and not linked to objectives or examples, providing the student the opportunity to distinguish between problem types and select an appropriate problem-solving strategy. This will facilitate in the transition from a guided set to a format generally seen on exams.

Application problems ask students to apply their new skills to real-life situations. Writing About Math problems build students' mathematical communication skills. Something to Think About transitions students to a deeper comprehension of the section. These questions require students to take what they have learned in a section and use those concepts to work through a problem in a new way.

Many exercises, indicated in the text by a blue triangle, are available online through *Enhanced WebAssign.* These homework problems are algorithmic, ensuring that your students will learn mathematical processes, not just how to work with specific numbers.

Redesigned Chapter Review

For this edition of the text, we have combined the former Chapter Summaries and Chapter Reviews into a Chapter Review grid. The grid presents material cleanly and simply, giving students an efficient means of reviewing material.

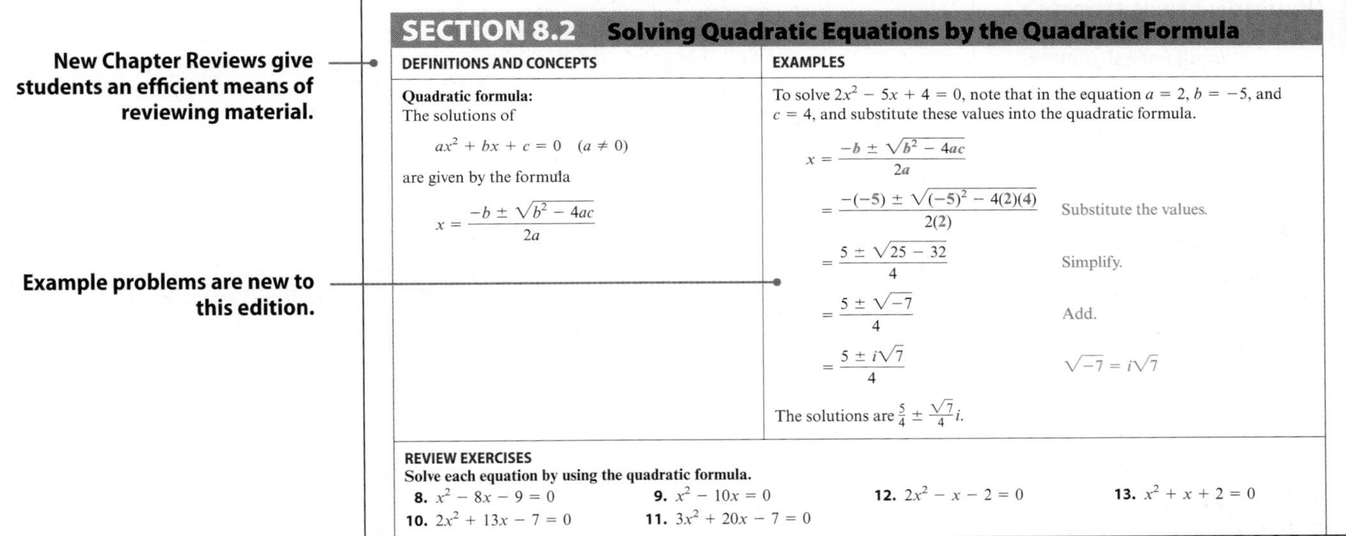

SECTION 8.2 Solving Quadratic Equations by the Quadratic Formula

DEFINITIONS AND CONCEPTS	EXAMPLES
Quadratic formula: The solutions of $ax^2 + bx + c = 0 \quad (a \neq 0)$ are given by the formula $x = \dfrac{-b \pm \sqrt{b^2 - 4ac}}{2a}$	To solve $2x^2 - 5x + 4 = 0$, note that in the equation $a = 2$, $b = -5$, and $c = 4$, and substitute these values into the quadratic formula. $x = \dfrac{-b \pm \sqrt{b^2 - 4ac}}{2a}$ $= \dfrac{-(-5) \pm \sqrt{(-5)^2 - 4(2)(4)}}{2(2)}$ Substitute the values. $= \dfrac{5 \pm \sqrt{25 - 32}}{4}$ Simplify. $= \dfrac{5 \pm \sqrt{-7}}{4}$ Add. $= \dfrac{5 \pm i\sqrt{7}}{4}$ $\sqrt{-7} = i\sqrt{7}$ The solutions are $\frac{5}{4} \pm \frac{\sqrt{7}}{4}i$.

REVIEW EXERCISES
Solve each equation by using the quadratic formula.

8. $x^2 - 8x - 9 = 0$ **9.** $x^2 - 10x = 0$ **12.** $2x^2 - x - 2 = 0$ **13.** $x^2 + x + 2 = 0$

10. $2x^2 + 13x - 7 = 0$ **11.** $3x^2 + 20x - 7 = 0$

New Basic Calculator Keystroke Guide

This tear-out card has been provided to assist students with their calculator functionality. It will serve as a quick reference for the TI-83 and TI-84 family of calculators, aiding students in building the technology skills needed for this course.

New tear-out guide serves as a quick reference to the TI-83/84 family of calculators.

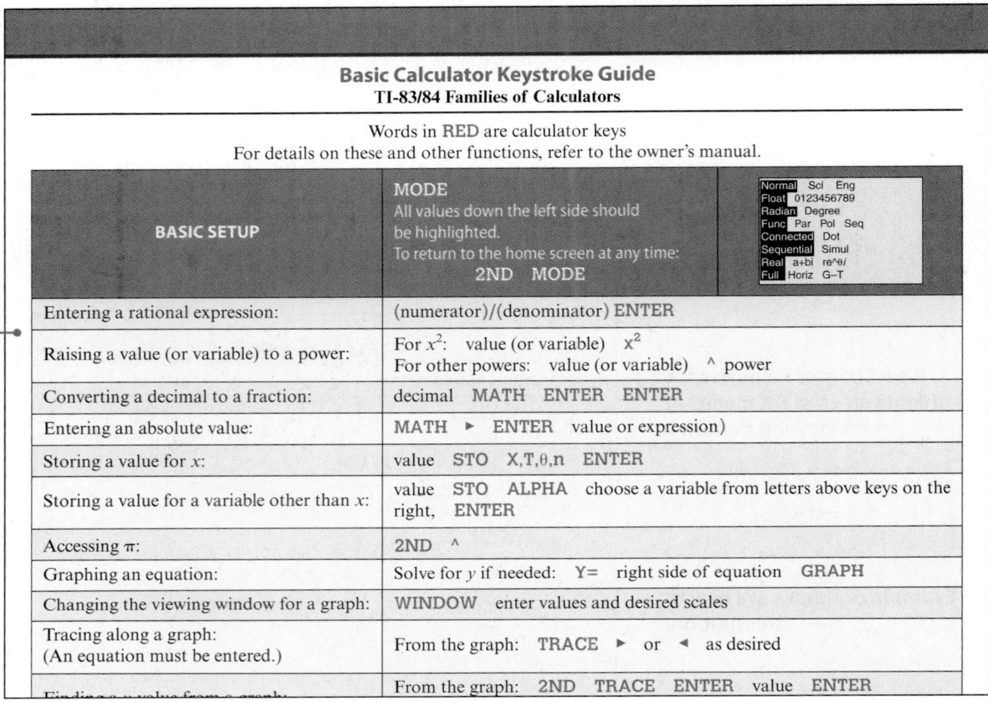

Basic Calculator Keystroke Guide
TI-83/84 Families of Calculators

Words in RED are calculator keys
For details on these and other functions, refer to the owner's manual.

BASIC SETUP	MODE All values down the left side should be highlighted. To return to the home screen at any time: 2ND MODE
Entering a rational expression:	(numerator)/(denominator) ENTER
Raising a value (or variable) to a power:	For x^2: value (or variable) x^2 For other powers: value (or variable) ^ power
Converting a decimal to a fraction:	decimal MATH ENTER ENTER
Entering an absolute value:	MATH ► ENTER value or expression)
Storing a value for x:	value STO X,T,θ,n ENTER
Storing a value for a variable other than x:	value STO ALPHA choose a variable from letters above keys on the right, ENTER
Accessing π:	2ND ^
Graphing an equation:	Solve for y if needed: Y= right side of equation GRAPH
Changing the viewing window for a graph:	WINDOW enter values and desired scales
Tracing along a graph: (An equation must be entered.)	From the graph: TRACE ► or ◄ as desired
Finding a y-value from a graph:	From the graph: 2ND TRACE ENTER value ENTER

TRUSTED FEATURES

- **Chapter Openers** showcase the variety of career paths available in the world of mathematics. We include a brief overview of each career, as well as job outlook statistics from the U.S. Department of Labor, including potential job growth and annual earnings potential.
- **Getting Ready** questions appear at the beginning of each section, linking past concepts to the upcoming material.
- **Comment** notations alert students to common errors as well as provide helpful and pertinent information about the concepts they are learning.
- **Accent on Technology** boxes teach students calculator skills to prepare them for using these tools in science and business classes, as well as for nonacademic purposes. Calculator examples are given in these boxes, and keystrokes are given for both scientific and graphing calculators. For instructors who do not use calculators in the classroom, the material on calculators is easily omitted without interrupting the flow of ideas.
- **Examples** are worked out in each chapter, highlighting the concept being discussed. We include Author Notes in many of the text's examples, giving students insight into the thought process one goes through when approaching a problem and working toward a solution. Most examples end with a Self Check problem, so that students may immediately apply concepts. Answers to each section's Self Checks are found at the end of that section.

- **Everyday Connections** boxes reveal the real-world power of mathematics. Each Everyday Connection invites students to see how the material covered in the chapter is relevant to their lives.
- **Perspectives** boxes highlight interesting facts from mathematics history or important mathematicians, past and present. These brief but interesting biographies connect students to discoveries of the past and their importance to the present.
- **Teaching Tips** are provided in the margins as interesting historical information, alternate approaches for teaching the material, and class activities.
- Chapter-ending **Projects** encourage in-depth exploration and a review of key concepts.
- **Chapter Tests** allow students to pinpoint their strengths and challenges with the material. Answers to all problems are included at the back of the book.
- **Cumulative Review** exercises follow the end of chapter material for every even-numbered chapter, and keep students' skills current before moving on to the next topic. Answers to all problems are included at the back of the book.

CONTENT CHANGES FOR THE NINTH EDITION

Although the Table of Contents is essentially the same as in the previous edition, we have added many more examples, Self Checks, and application problems throughout the text. Many sections have been rewritten to make the material easier for students to read. In the chapters, we have made the following specific changes.

Chapter 1: In Section 1.1, we have included more work on the intersection and union of intervals. In Sections 1.6 and 1.7, most solutions to application problems have been rewritten for better clarity.

Chapter 2: In Section 2.4, we include a discussion of relation and the domain and range of a relation before we discuss functions.

Chapter 3: Throughout the chapter, we now express the solution of a dependent system as a general ordered pair or an ordered triple.

Chapter 4: In Section 4.1, more emphasis is given to writing the solution set of an inequality in both set builder and interval notation.

Chapter 5: In this chapter, more emphasis is given to factoring by grouping.

Chapter 6: Throughout the chapter, we have given restrictions on the variable of all rational expressions to avoid any division by zero.

Chapter 7: Many sections have been reorganized and rewritten for improved clarity.

Chapter 8: In Section 8.5, we now solve quadratic and rational inequalities by constructing a sign chart, and also by plotting critical points and checking a test point in each interval.

Chapter 9: Many new exercises have been included in the Exercise Sets to provide more practice.

Chapter 10: This chapter has been largely rewritten to be easier for students to read.

Chapter 11: This chapter includes minor editing but remains essentially the same.

CALCULATORS

The use of calculators is assumed throughout the text. We believe that students should learn calculator skills in the mathematics classroom. They will then be prepared to use calculators in science and business classes and for nonacademic purposes. The directions within each exercise set indicate which exercises require the use of a calculator.

Since most intermediate algebra students now have graphing calculators, keystrokes are given for both scientific and graphing calculators. A new, removable Basic Calculator Keystroke Guide is bound into the book, as a resource for those students learning how to use a graphing calculator

ANCILLARIES FOR THE INSTRUCTOR

Print Ancillaries

Annotated Instructor's Edition (0-538-73662-3)
The *Annotated Instructor's Edition* provides the complete student text with answers next to each respective exercise. Those exercises that also appear in *Enhanced WebAssign* are clearly indicated.

Complete Solutions Manual (0-538-49520-0)
The *Complete Solutions Manual* provides worked-out solutions to all of the problems in the text.

Instructor's Resource Binder (0-538-73675-5)
New! Each section of the main text is discussed in uniquely designed Teaching Guides containing instruction tips, examples, activities, worksheets, overheads, and assessments, with answers provided.

Student Workbook (0-538-49543-X)
New! The *Student Workbook* contains all of the assessments, activities, and worksheets from the *Instructor's Resource Binder* for classroom discussions, in-class activities, and group work.

Electronic Ancillaries

Solution Builder (0-495-91103-8)
This online solutions manual allows instructors to create customizable solutions that they can print out to distribute or post as needed. This is a convenient and expedient way to deliver solutions to specific homework sets.

ENHANCED WebAssign *Enhanced WebAssign,* used by more than one million students at more than 1,100 institutions, allows you to assign, collect, grade, and record homework assignments via the web. This proven and reliable homework system includes thousands of algorithmically generated homework problems, an eBook, links to relevant textbook sections, video examples, problem-specific tutorials, and more. Contact your local representative for ordering details.

PowerLecture with ExamView® (0-538-49604-5)
This CD-ROM provides the instructor with dynamic media tools for teaching. Create, deliver, and customize tests (both print and online) in minutes with *ExamView® Computerized Testing Featuring Algorithmic Equations.* Easily build solution sets for homework or exams using *Solution Builder's* online solutions manual. Microsoft® PowerPoint®

lecture slides, figures from the book, and *Test Bank,* in electronic format, are also included on this CD-ROM.

Text Specific DVDs (0-538-73767-0)
These text-specific DVD sets, available at no charge to qualified adopters of the text, feature 10- 20-minute problem-solving lessons that cover each section of every chapter.

Instructor Website

ANCILLARIES FOR THE STUDENT

Print Ancillaries

Student Solutions Manual (0-538-49546-4)
The *Student Solutions Manual* provides worked-out solutions to the odd-numbered problems in the textbook.

Student Workbook (0-538-49543-X)
Get a head start! The *Student Workbook* contains all of the assessments, activities, and worksheets from the *Instructor's Resource Binder* for classroom discussions, in-class activities, and group work.

Electronic Ancillaries

ENHANCED WebAssign

Enhanced WebAssign, used by more than one million students at more than 1,100 institutions, allows you to do homework assignments and get extra help and practice via the web. This proven and reliable homework system includes thousands of algorithmically generated homework problems, an eBook, links to relevant textbook sections, video examples, problem-specific tutorials, and more.

Student Website

TO THE STUDENT

Congratulations! You now own a state-of-the-art textbook that has been written especially for you. We have tried to write a book that you can read and understand. The text includes carefully written narrative and an extensive number of worked examples with Self Checks. New Now Try This problems can be worked with your classmates, and Guided Practice exercises tell you exactly which example to use as a resource for each question. These are just a few of the many changes made to this text with your success in mind.

To get the most out of this course, you must read and study the textbook properly. We recommend that you work the examples on paper first, and then work the Self Checks. Only after you thoroughly understand the concepts taught in the examples should you attempt to work the exercises. A *Student Solutions Manual* is available, which contains the worked-out solutions to the odd-numbered exercises.

Since the material presented in *Intermediate Algebra, Ninth Edition,* will be of value to you in later years, we suggest that you keep this text. It will be a good source of reference in the future and will keep at your fingertips the material that you have learned here.

We wish you well.

Hints on Studying Algebra

The phrase "Practice makes perfect" is not quite true. It is "*Perfect* practice that makes perfect." For this reason, it is important that you learn how to study algebra to get the most out of this course.

Although we all learn differently, here are some hints on studying algebra that most students find useful.

Plan a Strategy for Success To get where you want to be, you need a goal and a plan. Your goal should be to pass this course with a grade of A or B. To earn one of these grades, you must have a plan to achieve it. A good plan involves several points:

- Getting ready for class,
- Attending class,
- Doing homework,
- Making use of the extensive extra help available, if your instructor has set up a course, and
- Having a strategy for taking tests.

Getting Ready for Class To get the most out of every class period, you will need to prepare for class. One of the best things you can do is to preview the material in the text that your instructor will be discussing in class. Perhaps you will not understand all of what you read, but you will be better able to understand your instructor when he or she discusses the material in class.

Do your work every day. If you get behind, you will become frustrated and discouraged. Make a promise that you will always prepare for class, and then keep that promise.

Attending Class The classroom experience is your opportunity to learn from your instructor and interact with your classmates. Make the most of it by attending every class. Sit near the front of the room where you can easily see and hear. Remember that it is your responsibility to follow the discussion, even though that takes concentration and hard work.

Pay attention to your instructor, and jot down the important things that he or she says. However, do not spend so much time taking notes that you fail to concentrate on what your instructor is explaining. Listening and understanding the big picture is much better than just copying solutions to problems.

Don't be afraid to ask questions when your instructor asks for them. Asking questions will make you an active participant in the class. This will help you pay attention and keep you alert and involved.

Doing Homework It requires practice to excel at tennis, master a musical instrument, or learn a foreign language. In the same way, it requires practice to learn mathematics. Since practice in mathematics is homework, homework is your opportunity to practice your skills and experiment with ideas.

It is important for you to pick a definite time to study and do homework. Set a formal schedule and stick to it. Try to study in a place that is comfortable and quiet. If you can, do some homework shortly after class, or at least before you forget what was discussed in class. This quick follow-up will help you remember the skills and concepts your instructor taught that day.

Each formal study session should include three parts:

1. Begin every study session with a review period. Look over previous chapters and see if you can do a few problems from previous sections. Keeping old skills alive will greatly reduce the amount of time you will need to prepare for tests.

2. After reviewing, read the assigned material. Resist the temptation of diving into the problems without reading and understanding the examples. Instead, work the examples and Self Checks with pencil and paper. Only after you completely understand the underlying principles behind them should you try to work the exercises.

Once you begin to work the exercises, check your answers with the printed answers in the back of the text. If one of your answers differs from the printed answer, see if the two can be reconciled. Sometimes, answers have more than one form. If you decide that your answer is incorrect, compare your work to the example in the text that most closely resembles the exercise, and try to find your mistake. If you cannot find an error, consult the *Student Solutions Manual.* If nothing works, mark the problem and ask about it in your next class meeting.

3. After completing the written assignment, preview the next section. This preview will be helpful when you hear that material discussed during the next class period.

You probably already know the general rule of thumb for college homework: two hours of practice for every hour you spend in class. If mathematics is hard for you, plan on spending even more time on homework.

To make doing homework more enjoyable, study with one or more friends. The interaction will clarify ideas and help you remember them. If you must study alone, a good study technique is to explain the material to yourself out loud.

Arranging for Special Help

Take advantage of any extra help that is available from your instructor. Often, your instructor can clear up difficulties in a short period of time. Find out whether your college has a free tutoring program. Peer tutors can often be of great help.

Taking Tests

Students often get nervous before taking a test because they are afraid that they will do poorly.

To build confidence in your ability to take tests, rework many of the problems in the exercise sets, work the exercises in the Chapter Reviews, and take the Chapter Tests. Check all answers with the answers printed at the back of the text.

Then guess what the instructor will ask, build your own tests, and work them. Once you know your instructor, you will be surprised at how good you can get at picking test questions. With this preparation, you will have some idea of what will be on the test, and you will have more confidence in your ability to do well.

When you take a test, work slowly and deliberately. Scan the test and work the easy problems first. Tackle the hardest problems last.

ACKNOWLEDGMENTS

We are grateful to the following people who reviewed the new edition of this series of texts. They all had valuable suggestions that have been incorporated into the texts.

Kent Aeschliman, Oakland Community College
Carol Anderson, Rock Valley College
Kristin Dillard, San Bernardino Valley College
Kirsten Dooley, Midlands Technical College
Joan Evans, Texas Southern University
Jeremiah Gilbert, San Bernardino Valley College
Joan Haig, University of Alaska Anchorage
Harvey Hanna, Ferris State University
Kathy Holster, South Plains College
Robert McCoy, University of Alaska Anchorage
John J. Morrell, Atlanta Metropolitan College
Gail Opalinski, University of Alaska Anchorage
John Squires, Cleveland State Community College
Mary Teel, University of North Texas

Additional Acknowledgments

We also thank the following people who reviewed previous editions.

Cynthia Broughtou, Arizona Western College
David Byrd, Enterprise State Junior College
Pablo Chalmeta, New River Community College
Michael F. Cullinan, Glendale Community College
Lou D'Alotto, York College-CUNY
Karen Driskell, Calhoun Community College
Hamidullah Farhat, Hampton University
Harold Farmer, Wallace Community College-Hanceville
Mark Fitch, University of Alaska, Anchorage
Mark Foster, Santa Monica College
Jonathan P. Hexter, Piedmont Virginia Community College
Dorothy K. Holtgrefe, Seminole Community College
Mike Judy, Fullerton College
Lynette King, Gadsden State Community College
Janet Mazzarella, Southwestern College
Donald J. McCarthy, Glendale Community College
Andrew P. McKintosh, Glendale Community College
Christian R. Miller, Glendale Community College
Feridoon Moinian, Cameron University
Brent Monte, Irvine Valley College
Daniel F. Mussa, Southern Illinois University
Joanne Peeples, El Paso Community College
Mary Ann Petruska, Pensacola Junior College
Linda Pulsinelli, Western Kentucky University
Kimberly Ricketts, Northwest-Shoals Community College
Janet Ritchie, SUNY-Old Westbury
Joanne Roth, Oakland Community College
Richard Rupp, Del Mar College
Rebecca Sellers, Jefferson State Community College
Kathy Spradlin, Liberty University
April D. Strom, Glendale Community College
Victoria Wacek, Missouri Western State College
Judy Wells, University of Southern Indiana
Hattie White, St. Phillip's College
George J. Witt, Glendale Community College
Margaret Yoder, Eastern Kentucky University

We are grateful to the staff at Cengage Learning, especially our publisher Charlie Van Wagner and our editor Marc Bove. We also thank Vernon Boes, Jennifer Risden, Meaghan Banks, and Heleny Wong.

We are indebted to Ellen Brownstein, our production service, to Jack Morrell, who read the entire manuscript and worked every problem, and to Mike Welden, who prepared the *Student Solutions Manual.* Finally, we thank Lori Heckelman for her fine artwork and Graphic World for their excellent typesetting.

R. David Gustafson
Rosemary M. Karr
Marilyn B. Massey

Index of Applications

Examples that are applications are shown with boldface numbers.
Exercises that are applications are shown with lightface numbers.

Architecture
Architecture, 355
Calculating clearance, 711
Designing an underpass, 711
Designing a patio, 764
Drafting, 440
Finding the dimensions of a truss, **352**
Gateway Arch, **695**
Landscape design, **707–708**
Width of a walkway, 699

Business
Analyzing ads, 27
Annual rate of depreciation, 130
Appreciation equations, 130
Arranging appointments, 787
Art, 130
Bookstores, 77
Break-point analysis, **184–185,** 189
Building construction, 441, 442
Buying boats, 267
Buying furniture, 267
Carpentry, 88, 302, 355, 473, 512
Cell phone growth, **625–627**
Choosing salary plans, 190
Computing profit, 553
Computing salaries, 70
Computing selling price, 77
Cost of carpet, **112–113**
Costs of a trucking company, 441
Customer service, 297
Cutting beams, 68
Cutting boards, 68
Demand equations, 105
Depreciation, **99,** 160, 512, 649
Depreciation equations, 130
Depreciation rates, **509–510**
Directory costs, 374, 375
Discount buying, 412
Draining a tank, 448
Drywalling a house, **407–408**
Earning money, 26
Finding the percent of markdown, **63–64**
Food service, 175
Framing a house, 374
Hourly wages, 104
House construction, 355
Housing, 143
Information access, 734
Inventories, 219, 277
Making bicycles, 188
Making clothing, 199
Making crafts, 277
Making fertilizer, 284
Making furniture, 78, 277
Making sporting goods, 260
Making statues, 199
Making water pumps, 189
Managing a beauty shop, 189
Managing a computer store, 189
Manufacturing footballs, 199
Manufacturing hammers, **195–196**
Marketing, 512
Maximizing revenue, 578
Merchandising, 187, 188
Metal fabrication, 553
Milling brass plates, 189
Mixing candies, 77, 188
Mixing nuts, **74,** 199
Office furniture, 343
Online research company, **107**
Online research service, **371–372**
Operating costs, 578
Ordering furnace equipment, 267
Percent of markdown, 87
Price increases, 412
Pricing, 421
Printer charges, 131
Printing books, 189
Printing stationery, **731**
Production, 277–278
Production problem, 278
Quality control, 793
Rate of decrease, 117
Rate of growth, 117
Reading proof, 411
Real estate, 130
Real estate listings, 130
Retailing, 174–175
Retail sales, **182**
Roofing a house, 412
Royalties, 734
Running a record company, 189
Running a small business, 189
Salvage value, 130, 631
Scheduling equipment, 240
Selling apples, 481
Selling calculators, 69
Selling clothes, 27
Selling DVD players, 143
Selling grass seed, 69
Selling hot dogs, 143
Selling shirts, 439
Selling tires, 143
Selling toys, 68
Selling trees, 260
Selling TVs, 314
Selling Vacuum Cleaners, **310**
Setting bus fares, 553
Shipping crates, 532
Shipping packages, 474
Siding a house, 448
Sorting records, 26

Storage, 343
Storing oil, 441
Supply equations, 105
Turkish rug for sale, **18**
Value of a car, 303
Value of a lathe, **125**
Value of two cars, 303
Workers' salaries, **20–21**
Work rates, 393

Education
Averaging grades, 27, 239
Calculating grades, 27
Choosing books, 787
Computing grades, 77
Cost and revenue, 175
History, 13
Picking committees, **783**
Planning a picnic, 787
Planning a work schedule, 239
Raising grades, 77
Rate of growth, 117
Saving for college, **627**
School enrollment, 578
Staffing, 440
Taking a test, 788

Electronics
Communications, 527
Electronics, 188, **320,** 402–403, 441, 466, 523
Ohm's Law, 60, 523
Power loss, 60
Radio frequencies, 190

Entertainment
Arranging an evening, 787
Balancing a seesaw, 70, 71, 88
Buying concert tickets, 261
Buying tickets, 105
Call letters, 787
Candy, 331
Checkers, 343
Concert receipts, 76
Concert tickets, 199
Darts, 331
Height of a rocket, **291–292**
Hobbies, 440
Juggling, 297
Model railroading, 440
Movie stunts, 331
Playing cards, **790, 791,** 793, 800, 802
Pricing concert tickets, 553
Production schedules, **274–275**
Roller coasters, 297
School plays, 76
Tension, 442
Theater seating, 208–209
TV coverage, 105
TV programming, **781**
Watching TV, **779**

Farming
Farming, 106, 188–189, 441
Fencing a field, 577
Fencing pastures, 69
Fencing pens, 70
Geometry problem, 188
Grooming horses, 27
Making whole milk, 77
Milk production, **74–75**
Planting corn, 798
Storing corn, 412

Finance
Amount of an annuity, **770**
Annuities, 772, 773
Bank accounts, 39
Comparing interest rates, 630
Comparing savings plans, 630
Comparison of compounding methods, 638, 639
Compound interest, 630, 677
Continuous compound interest, **633,** 638, 677
Declining savings, 772
Depreciation, 639
Determining a previous balance, 638
Determining the initial deposit, 638
Doubling money, 60, **628,** 655
Doubling time, **654**
Earning interest, 60, 104
Earning money, **256–257**
Financial planning, **72–73,** 278
Finding interest rates, 553
Frequency of compounding, 630
Inheriting money, 76
Installment loans, 764
Investing, 76, 89, 219, 260, 280, 451, 576
Investing money, 239, 728
Investment income, 188
Investment problem, 88
Managing a checkbook, 27
Maximizing income, **271–273**
Pension funds, 69
Piggy banks, 208
Portfolio analysis, **64**
Rule of seventy, 677
Saving money, **543–544,** 546, 764
Savings, 680
Savings accounts, 512
Savings growth, 772
Simple interest, **57**
Stock appreciation, 798
Stock averages, 27
Supplemental income, 76
Time for money to grow, 649
Tripling money, 655
Value of an IRA, 68

Geometry
Altitude of equilateral triangle, 503
Angles in a triangle, **65–66**
Angles of a polygon, 60
Area of a circle, 441
Area of an ellipse, 711
Area of a triangle, 383
Area of figures, 38
Area of many cubes, 473
Base of a triangle, 553
Circumference of a circle, 27
Complementary angles, 69
Complementary angles in a right triangle, 69
Curve fitting, **196–197,** 199, 200, 208
Dimensions of a rectangle, **550,** 553, 609
DVDs, **543**
Equilateral triangles, 69
Geometry, 88, 89, 188, 189, 198, 208, 302, 314, 354, 355, 420, 441, 472, 473, **489–491,** 529
Geometry problem, 728
Height of a triangle, 553
Heights of triangles, **56–57**
Hypotenuse of isosceles right triangle, 503
Interior angles, 764
Parallelograms, **186**
Perimeter of a rectangle, 393, 553
Perimeter of a square, 27, 355
Perimeter of a trapezoid, 26
Perimeter of a triangle, 26
Quadrilaterals, 69
Radius of a circle, 466
Side of a square, 553
Supplementary angles, 69
Supplementary angles and parallel lines, 70
Surface area of a cube, 473
Triangles, 70
Vertical angles, 69
Volume, 362
Volume of a cone, **436–437**
Volume of a pyramid, 362
Volume of a tank, 47
Volume of figures, 38–39
Volume of sphere, **36**
Width of a ring, 393

Home Management and Shopping
Arranging books, **779,** 787
Boat depreciation, 802
Building shelves, 78

Buying a computer, 239
Buying a TV and a VCR, 70
Buying a washer and dryer, 68
Buying books, 68
Buying compact discs, 239, 267
Buying furniture, 68
Buying roses, 70
Car depreciation, 105, 798
Car repairs, 131, 175
Choosing clothes, 788
Choosing housekeepers, 260
Cooking, 440
Depreciating a lawn mower, 130
Designing a swimming pool, 355
Dimensions of a window, 553
Dog runs, **66**
Draining a pool, 27, 412
Enclosing a swimming pool, 70–71
Family of five children, **792**
Filling a pond, 412
Filling a pool, 27, 374, 412
Finding dimensions, 69
Framing a picture, 323, 553
Framing pictures, 71
Gardening, 302, 355
Grocery shopping, **430–431**
House appreciation, 105, 772
Housecleaning, 412
House painting, 411
Ironing boards, 493
Landscaping, 175, **264–265**
Length of a rectangular garden, **551**
Making gutters, 297
Making Jell-O, 656
Maximizing area, **573–574,** 577
Mixing fuel, 440
Motorboat depreciation, 772
Nutritional planning, 198, 199
pH of pickles, 666
Preformed concrete, 362
Renting a rototiller, 239
Renting a truck, 239
Slope of a ladder, 116
Slope of a roof, 116
Storage capacity, 383
Value of a boat, 441
Value of a car, 104
Value of a house, 302, 303
Value of two houses, 303

Medicine
Alcohol absorption, 639
Body mass, 440, 474
Choosing a medical plan, 240
Clinical trials, **792**
Diet problem, 277
Diets, **273–274**
Epidemics, 639
FDA standards, **276**
Finding the variance, 577
Forensic medicine, 143, 656
Losing weight, 27
Medicine, 466, 512, 546, 639, 677, 793
Physical therapy, 189–190, 208
Pulse rates, 503
Recommended dosage, 440
Sizes of viruses, 27
Variance, **574**

Miscellaneous
Accidents, 546
Aquariums, 666
Backup generator, 793
Bouncing balls, 777
Cats and dogs, 208
Cell phone usage, 630
Chainsaw sculpting, 199
Choosing committees, 801
Choosing people, 799, 801
Combination locks, 787
Computers, 787
Cutting cables, 68
Cutting hair, **140**
Cutting ropes, 68
Dolphins, 143
Elevators, 239
Finding constant of variation, 441
Fine arts, 355
Focal length, 412–413
Forming a committee, 788, 802
Genealogy, 772
Hardware, 493
Height of a flagpole, 440
Height of a tree, **432–433,** 440
Ice cubes, 343
Inscribed squares, 772
Integer problem, 198, 354, 362, 553, 728
Lensmaker's formula, 413
Lifting cars, 70
Lining up, 787, 799, 802
Moving stones, 70
Number problem, 728
Organ pipes, 441
Packing a tennis racket, 473, 474
Palindromes, 787
Pendulums, 532, 545
Period of a pendulum, **461**
Phone numbers, 787
Photography, **500,** 503
Predicting heights and weights, **126–127**
Range of camber angles, 252
Reach of ladder, 473
Rodent control, 677
Signaling, 219
Standard deviation, **462–463**
Statistics, 466
Supporting a weight, 474
Wheelchair ramps, 117
Width of a river, 440
Winning a lottery, 788

Politics, Government, and the Military
Aircraft carriers, 323
Artillery, 728
Building a freeway, **470–471**
Building highways, **437**
City planning, **635**
Congress, **783–784**
Cost of electricity, 60, 375
Cost of water, 60
Crime prevention, 106
Displaying the flag, 27
Doubling time, **653**
Electric service, 473
Environmental cleanup, 374
Fighting fires, **468**
Figuring taxes, 260
Flags, 546
Forming committees, 799
Growth of a town, **769–770**
Highway design, 511–512, 699
Labor force, 554
Law enforcement, 466, 546
Legislation, 793
License plate numbering, **786**
Long-distance, **235–236**
Louisiana Purchase, 631
Making a ballot, 787
Making license plates, 787
Marine Corps, 190
Military science, 27
Monuments, 393
Paying taxes, 25
Pentagon's Defense Satellite Communications System, **438**
Police investigations, 577
Population decline, 639, 772
Population growth, **635–636,** 638, 655, **673–674,** 677, 772
Predicting burglaries, 131
Sending signals, **779, 780**
Space program, 554
Statistics, **236–237**
Taxes, 450
Telephone costs, 106
Telephone service, 473
Town population, 631
U.S. population, 681, 682
View from a submarine, 526
Water usage, 86, 578
Women serving in U.S. House of Representatives, **132**

World population growth, 638

Science
Alloys, 596
Alpha particles, 721
Angstroms per inch, 48
Artillery fire, 143
Astronomy, **44–45**
Atomic structure, **718–719**
Bacteria cultures, 630, 677
Bacterial growth, 677
Balancing levers, 71
Ballistics, 143, **353,** 355, **572–573,** 577, 609
Broadcast ranges, 699
Carbon-14 dating, **672–673,** 677, 684
Change in intensity, 666
Change in loudness, 666
Changing temperatures, 27
Chemistry, 554
Choosing sample size, 240
Controlling moths, 777
Conversion from degrees Celsius to degrees Fahrenheit, 143, 144
Converting temperatures, 60
Data analysis, 403
dB gain, 682
dB gain of an amplifier, 649
Diluting solutions, 78
Discharging a battery, 631
Distance to Alpha Centauri, 48
Distance to the Moon, 47
Distance to the Sun, 47
Earthquakes, 649, 682
Earth's atmosphere, 199
Electrostatic repulsion, 722
Engineering, **67,** 402, 413
Engineering designs, 70
Establishing equilibrium, 70
Falling objects, 441, 466, 545, 764, 798
Finding dB gain, **645**
Finding operating temperatures, 252
Finding temperature ranges, 252
Finding the gain of an amplifier, 649
Finding the hydrogen-ion concentration, **663**
Finding the pH of a solution, **663**
Flight to Pluto, 48
Force of gravity, 59
Free-falling objects, 639
Gas pressure, 441, 442
Generating power, 512
Generation time, **674–675**
Global warming, 117
Half-life, 677
Half-life of radon-22, **671–672**
Height of a bridge, **505–506**
Horizon distance, 512
Hurricanes, **93–94**
Hydrogen ion concentration, 666
Increasing concentration, 78
Lab experiments, 383
Lead decay, 677
Life of a comet, 48
Light intensity, **435–436**
Light year, 48
LORAN, 721
Mass of protons, 47
Measuring earthquakes, **646**
Meshing gears, 699
Mixing solutions, 77, 88, 89, **183–184,** 188, 481
Navigation, 175
Object Thrown Straight Up, **550**
Oceanography, 677
Path of a comet, 699
pH of grapefruit, 683
pH of solution, 666
Physics, 331
Projectiles, 699
Radioactive decay, 630, **636,** 638, 677, 681
Range of a comet, 48
Reading temperatures, 13
Research, 286
Satellite antennas, 699
Solar flares, 47
Solar heating, 71
Sonic boom, 721–722
Space shuttle orbit, **438**
Speed of light, 47
Speed of sound, 47
Springs, 252
Storm research, 77
Television translators, **692**
Temperature change, **594**
Temperature scales, **57,** 71
Thermodynamics, 60
Thorium decay, 677
Time of flight, 355
Tolerance of sheet steel, 252
Tritium decay, 677
Wavelengths, 47
Weather forecasting, 596
Weber-Fechner law, **664**

Sports
Area of a track, 711
Averaging weights, 27
Baseball, 472
Bowling, **471**
Buying golf clubs, 70
Cutting a rope, **62–63**
Cycling, 77
Diagonal of a baseball diamond, 466
Fitness equipment, 710–711
Football, **20,** 267
Mountain climbing, 27
NFL records, 199
Physical fitness, 117
Pool tables, 711
Rate of descent, **113**
Renting a jet ski, 734
Running a race, 77
Sailing, 473, 526
Ski runs, 440
Skydiving, 639
Targets, 503
Track and field, 267
U.S. sports participation, 106
Winter recreation, 355

Travel
Aviation, 252, 412
Boating, 239, 412
Braking distance, 297
Buying gasoline, 25
Car rentals, 70
Computing distance, 77
Computing time, 76
Computing travel time, 77
Direct variation, **434**
Distance, rate, and time, 383
Distance traveled, 104
Driving rates, 728
Driving to a convention, **408–409**
Finding distance, 441
Finding rates, 412, 521, 553
Flight paths, 440
Flying speed, 448
Gas consumption, 440
Gas mileage, **99**
Grade of a road, 116
Motion problem, 88
Planning a trip, 799
Renting trucks, 70
Riding a jet ski, 77
Riding a motorboat, 78
Riding in a taxi, 734
River cruise, **409–410**
Rowing a boat, 412
Sailing time, 451
Stowing baggage, 188
Stretching a vacation, 412
Taking a walk, 77
Time on the road, 412
Train travel, 412
Transportation, 403
Travel, 188, 576
Travel choices, 787
Travel times, **73,** 88, 374
Trip length, 448
Trucking, 77
Walking for charity, 412
Wind speed, 521
Winter travel, 420

©Shutterstock/sf2301420max

Careers and Mathematics

FARMERS, RANCHERS, AND AGRICULTURAL MANAGERS

American farmers and ranchers direct the activities of one of the world's largest and most productive agricultural sectors. They produce enough food to meet the needs of the United States and produce a surplus for export. Farmers, ranchers, and agricultural managers held nearly 1.3 million jobs in 2006. Modern farming requires increasingly complex scientific, business, and financial decisions. Therefore, people who were raised on farms must acquire the appropriate education.

Job Outlook:

The long-term trend toward consolidation of farms into fewer and larger farms is expected to continue over the 2006–2016 period. However, many small-scale farmers have developed successful market niches that involve direct contact with their customers, such as organic food production.

Since incomes of farmers vary greatly from year to year, many farmers have incomes from off-farm business activities.

Annual Earnings:

$39,832–$71,864

For More Information:

http://www.bls.gov/oco/ocos176.htm

For a Sample Application:

See Problem 45 in Section 1.6.

CHAPTER 1

A Review of Basic Algebra

1.1 The Real Number System
1.2 Arithmetic and Properties of Real Numbers
1.3 Exponents
1.4 Scientific Notation
1.5 Solving Linear Equations in One Variable
1.6 Using Linear Equations to Solve Application Problems
1.7 More Applications of Linear Equations in One Variable
- ***Projects***
CHAPTER REVIEW
CHAPTER TEST

In this chapter ▶

The language of mathematics is algebra. The word algebra comes from a book written by the Arabian mathematician Al-Khowarazmi around A.D. *800. Its title Ihm* ***al-jabr*** *wa'l muqabalah means restoration and reduction, a process then used to solve equations.*

SECTION 1.1

The Real Number System

Objectives

1. List any number in a set of real numbers that is composite, whole, irrational, natural, prime, rational, or an integer.
2. Insert a symbol $<$, $>$, or $=$ to define the relationship between two rational numbers.
3. Graph a real number, a set of real numbers, or an inequality on the number line.
4. Graph a compound inequality.
5. Find the absolute value of a real number.

Vocabulary

variables	prime numbers	real numbers
set	composite numbers	intervals
elements of a set	even integers	interval notation
natural numbers	odd integers	compound inequalities
whole numbers	roster method	union
integers	set-builder notation	intersection
subset	rational numbers	absolute value
coordinate	irrational numbers	

Getting Ready

1. Name some different kinds of numbers.
2. Is there a largest number?

In algebra, we will work with expressions that contain numbers and letters. The letters that stand for numbers are called **variables.** Some examples of algebraic expressions are

$$x + 45, \qquad \frac{8x}{3} - \frac{y}{2}, \qquad \frac{1}{2}bh, \qquad \frac{a}{b} + \frac{c}{d}, \qquad \text{and} \qquad 3x + 2y - 12$$

It is the use of variables that distinguishes algebra from arithmetic.

We begin the study of algebra by reviewing various subsets of the real numbers. We also will show how to graph these sets on a number line.

List any number in a set of real numbers that is composite, whole, irrational, natural, prime, rational, or an integer.

A **set** is a collection of objects. To denote a set, we can enclose a list of its **elements** with braces. For example,

$\{1, 2, 3, 4\}$ denotes the set with elements 1, 2, 3, and 4.

$\left\{\frac{1}{2}, \frac{2}{3}, \frac{3}{4}\right\}$ denotes the set with elements $\frac{1}{2}$, $\frac{2}{3}$, and $\frac{3}{4}$.

Three sets of numbers commonly used in algebra are as follows:

Natural Numbers The set of **natural numbers** includes the numbers we use for counting:

$\{1, 2, 3, 4, 5, 6, 7, 8, 9, \ldots\}$

Whole Numbers The set of **whole numbers** includes the natural numbers together with 0:

$\{0, 1, 2, 3, 4, 5, 6, 7, 8, 9, \ldots\}$

Integers The set of **integers** includes the natural numbers, 0, and the negatives of the natural numbers:

$\{\ldots, -5, -4, -3, -2, -1, 0, 1, 2, 3, 4, 5, \ldots\}$

Each group of three dots, called *ellipses,* indicates that the numbers continue forever.

In Figure 1-1, we graph each of these sets from -6 to 6 on a number line.

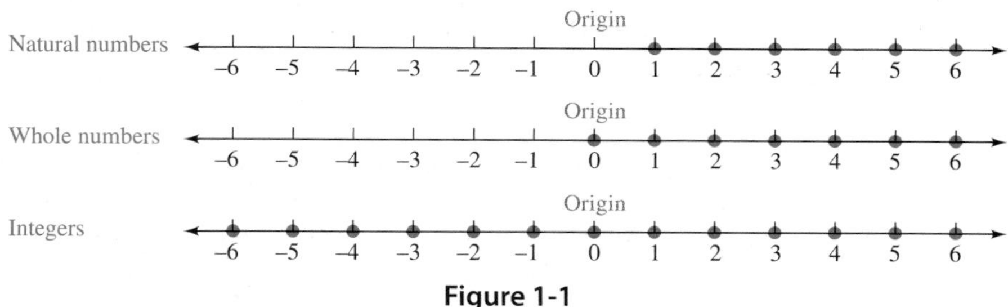

Figure 1-1

Since every natural number is also a whole number, we say that the set of natural numbers is a **subset** of the set of whole numbers. Note that the set of natural numbers and the set of whole numbers are also subsets of the integers.

To each number x in Figure 1-1, there corresponds a point on the number line called its *graph,* and to each point there corresponds a number x called its **coordinate.** Numbers to the left of 0 are *negative numbers*, and numbers to the right of 0 are *positive numbers*.

COMMENT 0 is neither positive nor negative.

There are two important subsets of the natural numbers.

Prime Numbers The **prime numbers** are the natural numbers greater than 1 that are divisible only by 1 and themselves.

$\{2, 3, 5, 7, 11, 13, 17, 19, 23, \ldots\}$

Composite Numbers The **composite numbers** are the natural numbers greater than 1 that are not prime.

$\{4, 6, 8, 9, 10, 12, 14, 15, 16, 18, 20, 21, \ldots\}$

Figure 1-2 shows the graphs of the primes and composites that are less than or equal to 14.

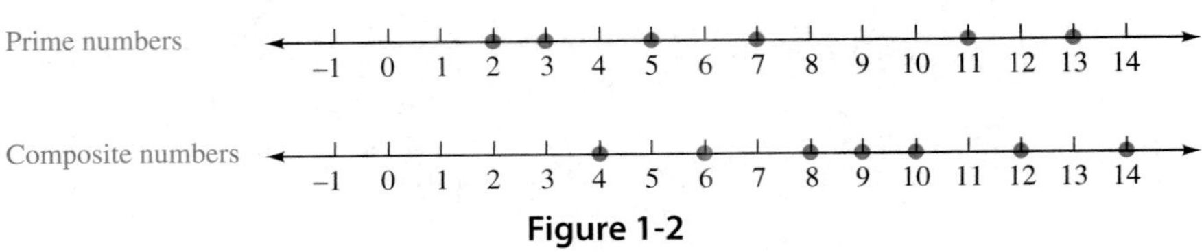

Figure 1-2

COMMENT 1 is the only natural number that is neither prime nor composite.

There are two important subsets of the integers.

Even Integers

The **even integers** are the integers that are exactly divisible by 2.

$\{. . ., -8, -6, -4, -2, 0, 2, 4, 6, 8, . . .\}$

Odd Integers

The **odd integers** are the integers that are not exactly divisible by 2.

$\{. . ., -9, -7, -5, -3, -1, 1, 3, 5, 7, 9, . . .\}$

Figure 1-3 shows the graphs of the even and odd integers from −6 to 6.

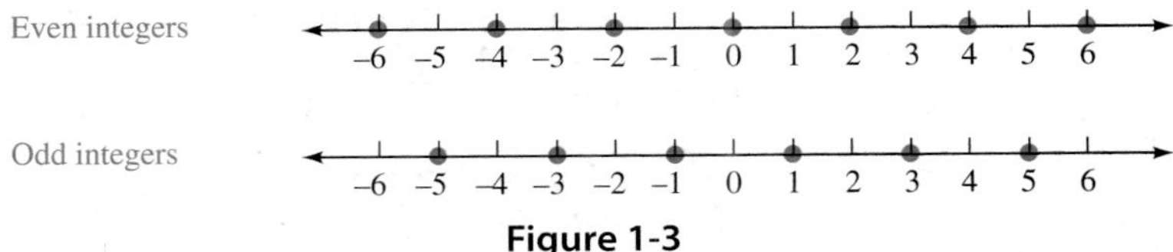

Figure 1-3

So far, we have written sets by listing their elements within braces. This method is called the **roster method.** A set also can be written in **set-builder notation.** In this method, we write a rule that describes which elements are in a set. For example, the set of even integers can be written in set-builder notation as follows:

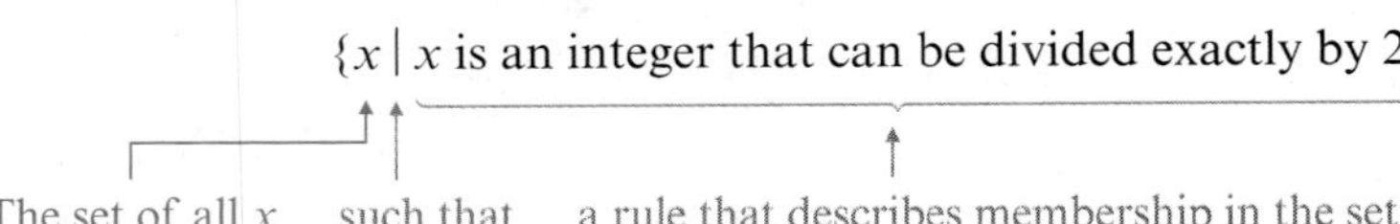

To find coordinates of more points on the number line, we need the *rational numbers.*

Rational Numbers

The set of **rational numbers** is

$$\left\{\frac{a}{b} \mid a \text{ is an integer and } b \text{ is a nonzero integer}\right\}$$

Each of the following numbers is an example of a rational number.

$$\frac{8}{5}, \quad \frac{2}{3}, \quad -\frac{44}{23}, \quad \frac{-315}{476}, \quad 0 = \frac{0}{7}, \quad \text{and} \quad 17 = \frac{17}{1}$$

Because each number has an integer numerator and a nonzero integer denominator

COMMENT Note that $\frac{0}{5} = 0$, because $5 \cdot 0 = 0$. However, $\frac{5}{0}$ is undefined, because there is no number that, when multiplied by 0, gives 5. The fraction $\frac{0}{0}$ is indeterminate, because all numbers, when multiplied by 0, give 0. Remember that the denominator of a fraction cannot be 0.

EXAMPLE 1 Explain why each number is a rational number: **a.** -7 **b.** 0.125 **c.** $-0.66. . . .$

Solution **a.** The integer -7 is a rational number, because it can be written as $\frac{-7}{1}$, where -7 and 1 are integers and the denominator is not 0. Note that all integers are rational numbers.

b. The decimal 0.125 is a rational number, because it can be written as $\frac{125}{1,000}$, where 125 and 1,000 are integers and the denominator is not 0. The fraction simplifies to $\frac{1}{8}$.

c. The decimal $-0.666. . .$ is a rational number, because it can be written as $\frac{-2}{3}$, where -2 and 3 are integers and the denominator is not 0.

SELF CHECK 1 Explain why each number is a rational number: **a.** 4 **b.** 0.5

The next example illustrates that a rational number can be written as a decimal that either terminates or repeats a block of digits.

EXAMPLE 2 Change each fraction to decimal form and determine whether the decimal terminates or repeats: **a.** $\frac{3}{4}$ **b.** $\frac{421}{990}$.

Solution **a.** To change $\frac{3}{4}$ to a decimal, we divide 3 by 4 to obtain 0.75. This is a terminating decimal.

b. To change $\frac{421}{990}$ to a decimal, we divide 421 by 990 to obtain 0.4252525. . . . This is a repeating decimal, because the block of digits 25 repeats forever. This decimal can be written as $0.4\overline{25}$, where the overbar indicates the repeating block of digits.

SELF CHECK 2 Change each fraction to decimal form and determine whether the decimal terminates or repeats:

a. $\frac{5}{11}$ **b.** $\frac{2}{5}$

The rational numbers provide coordinates for many points on the number line that lie between the integers (see Figure 1-4). Note that the integers are a subset of the rational numbers.

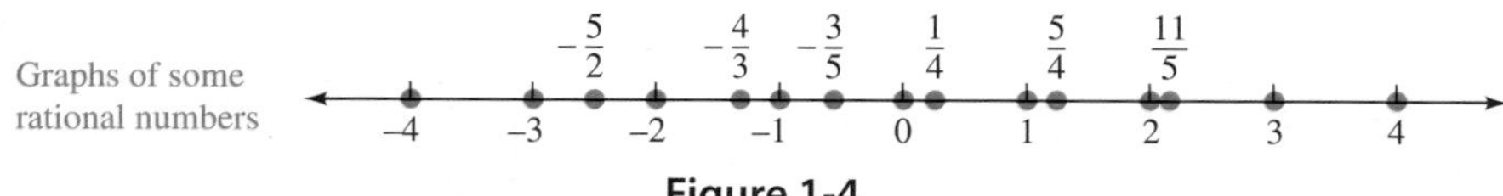

Figure 1-4

Points on the number line whose coordinates are nonterminating, nonrepeating decimals have coordinates that are called *irrational numbers.*

Irrational Numbers

The set of **irrational numbers** is

$$\{x \mid x \text{ is a nonterminating, nonrepeating decimal}\}$$

Some examples of irrational numbers are

$$0.313313331\ldots \qquad \sqrt{2} \approx 1.414213562\ldots \qquad \pi \approx 3.141592653\ldots$$

If we unite the set of rational numbers (the terminating and repeating decimals) and the set of irrational numbers (the nonterminating, nonrepeating decimals), we obtain the set of *real numbers.*

Real Numbers

The set of **real numbers** is

$\{x \mid x$ is a terminating decimal, a repeating decimal, or a nonterminating nonrepeating decimal$\}$

The number line in Figure 1-5 shows several points on the number line and their real-number coordinates. The points whose coordinates are real numbers fill up the number line.

François Vieta (Viête)
(1540–1603)
By using letters in place of unknown numbers, Vieta simplified the subject of algebra and brought its notation closer to the notation that we use today. One symbol he didn't use was the equal sign.

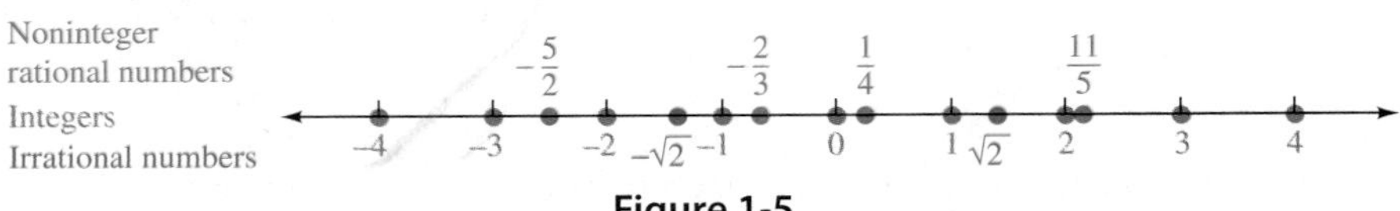

Figure 1-5

Figure 1-6 illustrates how various sets of numbers are interrelated.

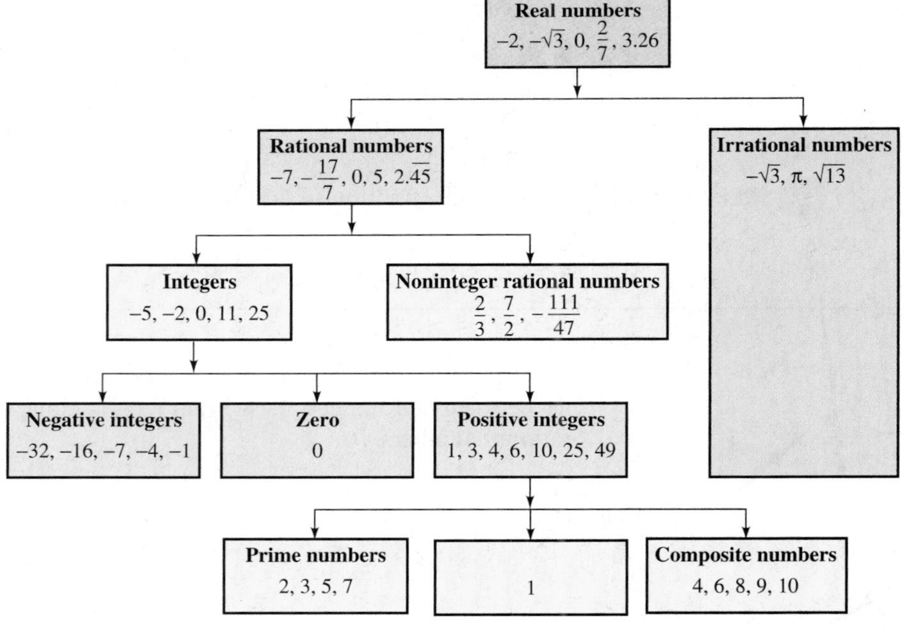

Figure 1-6

EXAMPLE 3 Which numbers in the set $\left\{-2, 0, \frac{5}{9}, 1.25, \sqrt{8}, 17\right\}$ are

a. natural numbers **b.** whole numbers **c.** integers
d. rational numbers **e.** irrational numbers **f.** real numbers?

Solution **a.** The only natural number is 17.

b. The whole numbers are 0 and 17.

c. The integers are -2, 0, and 17.

d. The rational numbers are $-2, 0, \frac{5}{9}$, 1.25, and 17. $\left(1.25\right.$ is rational, because 1.25 can be written in the form $\left.\frac{5}{4}.\right)$

e. The only irrational number is $\sqrt{8}$. It is irrational because $\sqrt{8} = 2.828427125\ldots$ and this decimal is a nonterminating, nonrepeating decimal.

f. All of the numbers are real numbers.

SELF CHECK 3 Which numbers in the set $\{-2, 0, 1.5, \sqrt{5}, 7\}$ are **a.** positive integers **b.** rational numbers?

2 Insert a symbol <, >, or = to define the relationship between two rational numbers.

To show that two quantities are not equal, we can use an inequality symbol.

Symbol	Read as	Examples
$\neq$	"is not equal to"	$6 \neq 9$ and $0.33 \neq \frac{3}{5}$
$<$	"is less than"	$22 < 40$ and $0.27 < 3.1$
$>$	"is greater than"	$19 > 5$ and $\frac{1}{2} > 0.3$
$\leq$	"is less than or equal to"	$1.8 \leq 3.5$ and $35 \leq 35$
$\geq$	"is greater than or equal to"	$25.2 \geq 23.7$ and $29 \geq 29$
$\approx$	"is approximately equal equal to"	$\sqrt{2} \approx 1.414$ and $\sqrt{3} \approx 1.732$

Table 1-1

We can always write an inequality with the inequality symbol pointing in the opposite direction. For example,

$$17 < 25 \quad \text{is equivalent to} \quad 25 > 17$$
$$5.3 \geq -2.9 \quad \text{is equivalent to} \quad -2.9 \leq 5.3$$

COMMENT Note that the inequality symbol always points to the smaller number.

On the number line, the coordinates of points get larger as we move from left to right, as shown in Figure 1-7. Thus, if a and b are the coordinates of two points, the one to the right is the greater.

This suggests the following general principle:

If $a > b$, point a lies to the right of point b on a number line.

If $a < c$, point a lies to the left of point c on a number line.

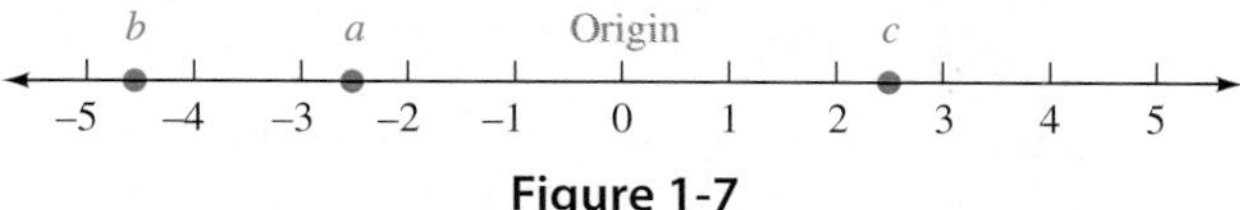

Figure 1-7

3 Graph a real number, a set of real numbers, or an inequality on the number line.

Graphs of some sets of real numbers are portions of a number line called **intervals.** For example, the graph shown in Figure 1-8(a) on the next page is the interval that includes all real numbers x that are greater than -5. Since these numbers make the inequality $x > -5$ true, we say that they satisfy the inequality. The parenthesis on the graph indicates that -5 is not included in the interval.

To express this interval in **interval notation,** we write $(-5, \infty)$. Once again, the parentheses indicate that the endpoints are not included.

The interval shown in Figure 1-8(b) is the graph of the inequality $x \leq 7$. It contains all real numbers that are less than or equal to 7. The bracket at 7 indicates that 7 is included in the interval. To express this interval in interval notation, we write $(-\infty, 7]$. The bracket indicates that 7 is in the interval.

COMMENT The symbol ∞ (infinity) is not a real number. It is used in Figure 1-8(a) to indicate that the graph extends infinitely far to the right. The $-\infty$ in Figure 1-8(b) indicates that the graph extends infinitely far to the left.

When using interval notation, $-\infty$ and ∞ will always be preceded or followed by a parenthesis.

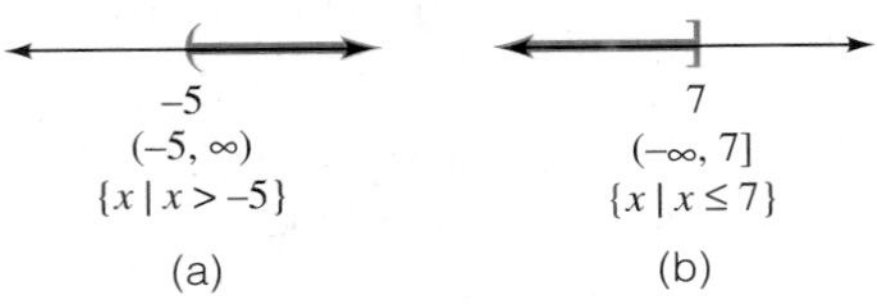

Figure 1-8

The graphs in Figure 1-8 also can be drawn using open and filled circles. An open circle indicates that an endpoint is not included, and a closed circle indicates that an endpoint is included.

–5
(–5, ∞)
{x | x > –5}

7
(–∞, 7]
{x | x ≤ 7}

EXAMPLE 4 Graph each set on the number line and then write the set in interval notation:
a. $\{x \mid x < 9\}$ **b.** $\{x \mid x \geq 6\}$.

Solution

a. $\{x \mid x < 9\}$ includes all real numbers that are less than 9. The graph is shown in Figure 1-9(a). This is the interval $(-\infty, 9)$.

b. $\{x \mid x \geq 6\}$ includes all real numbers that are greater than or equal to 6. The graph is shown in Figure 1-9(b). This is the interval $[6, \infty)$.

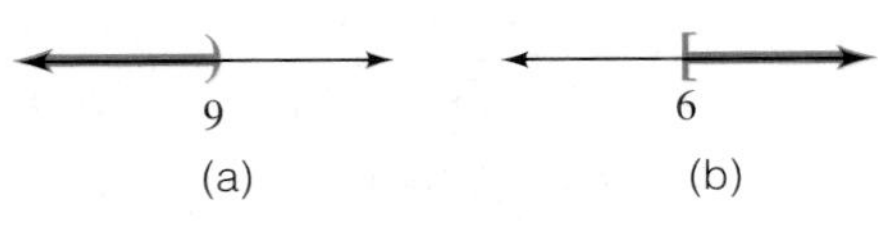

Figure 1-9

SELF CHECK 4 Graph $\{x \mid x \geq 5\}$ and then write the set in interval notation.

4 Graph a compound inequality.

COMMENT In a compound inequality involving *or,* the entire statement is true if either part is true.

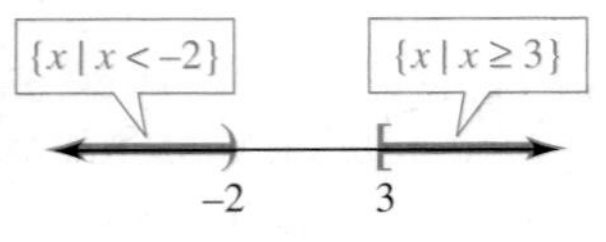

Figure 1-10

Expressions that involve more than one inequality are called **compound inequalities.** One type of compound inequality involves the word "or." For example, the graph of the set

$\{x \mid x < -2 \text{ or } x \geq 3\}$ Read as "the set of all real numbers x such that x is less than -2 or x is greater than or equal to 3."

is shown in Figure 1-10. This graph is called the **union** of two intervals and can be written in interval notation as

$(-\infty, -2) \cup [3, \infty)$ Read the symbol $\cup$ as "union."

EXAMPLE 5 Graph each set on the number line and write it in interval notation:
a. $\{x \mid x \leq -4 \text{ or } x > 5\}$ **b.** $\{x \mid x \geq -4 \text{ or } x > 5\}$.

Solution **a.** The set $\{x \mid x \leq -4 \text{ or } x > 5\}$ includes all real numbers less than or equal to -4 together with all real numbers greater than 5, as shown in Figure 1-11(a). This is the interval $(-\infty, -4] \cup (5, \infty)$.

b. The set $\{x \mid x \geq -4 \text{ or } x > 5\}$ includes all real numbers that are greater than or equal to -4 together with all real numbers that are greater than 5. This includes all numbers x that are greater than or equal to -4 or greater than 5. (See Figure 1-11(b).) This is the interval $[-4, \infty)$.

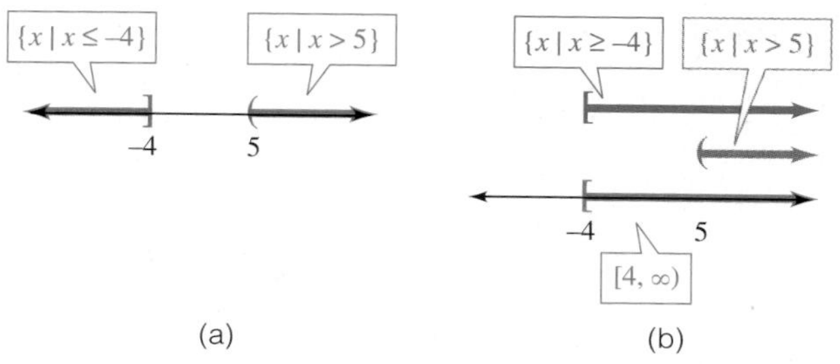

Figure 1-11

SELF CHECK 5 Graph $\{x \mid x < -1 \text{ or } x > 5\}$ and write it in interval notation.

Another type of compound inequality involves the word "and." For example, the graph of the set

$\{x \mid x \geq -5 \text{ and } x \leq 6\}$ Read as "the set of all real numbers x such that x is greater than or equal to -5 and x is less than or equal to 6."

is shown in red in Figure 1-12(a). It is obtained by finding the overlap (or **intersection**) of the graphs of $\{x \mid x \geq -5\}$ and $\{x \mid x \leq 6\}$. The overlap shown in red is the intersection of two intervals. The graph can be written in interval notation as

COMMENT In an intersection, the statement is true only when both statements are true.

$(-\infty, 6] \cap [-5, \infty)$ Read the symbol $\cap$ as "intersection."

From the graph, we see that the interval can be written more compactly as $[-5, 6]$.

To graph the set $\{x \mid x \geq -5 \text{ and } x > 6\}$, we find the overlap of the graphs of $\{x \mid x \geq -5\}$ and $\{x \mid x > 6\}$. This overlap is the interval $(6, \infty)$ as shown in Figure 1-12(b). In set-builder notation, the solution is $\{x \mid x > 6\}$.

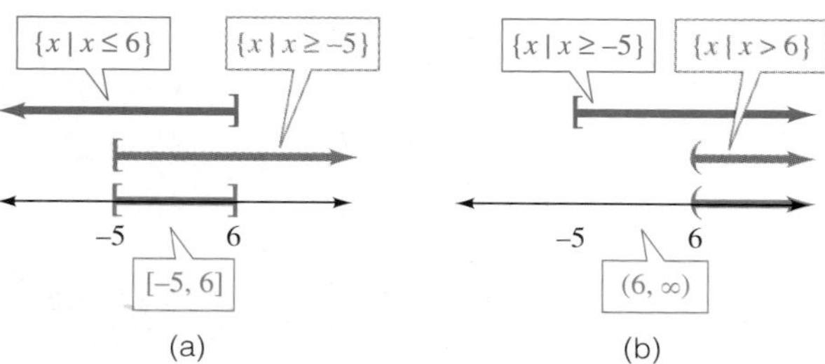

Figure 1-12

Two inequalities involving the word "and" often can be written as a single expression. For example, the compound inequality $x \geq -5$ and $x \leq 6$ previously discussed can be written as

$-5 \leq x$ and $x \leq 6$ From Figure 1-12(a), note that this includes all numbers from -5 to 6, including -5 and 6.

or as the single expression $-5 \leq x \leq 6$.

EXAMPLE 6 Graph each set on the number line and then write it in interval notation:
a. $\{x \mid -5 \leq x \leq 6\}$ **b.** $\{x \mid 2 < x \leq 8\}$ **c.** $\{x \mid -4 \leq x < 3\}$ **d.** $\{x \mid 0 < x < 7\}$

Solution

a. The set $\{x \mid -5 \leq x \leq 6\}$ includes all real numbers from -5 to 6, as shown in Figure 1-13(a). This is the interval $[-5, 6]$.

b. The set $\{x \mid 2 < x \leq 8\}$ includes all real numbers between 2 and 8, including 8, as shown in Figure 1-13(b). This is the interval $(2, 8]$.

c. The set $\{x \mid -4 \leq x < 3\}$ includes all real numbers between -4 and 3, including -4, as shown in Figure 1-13(c). This is the interval $[-4, 3)$.

d. The set $\{x \mid 0 < x < 7\}$ includes all real numbers between 0 and 7 as shown in Figure 1-13(d). This is the interval $(0, 7)$.

COMMENT In part **d.** $(0, 7)$ is an interval and not a point.

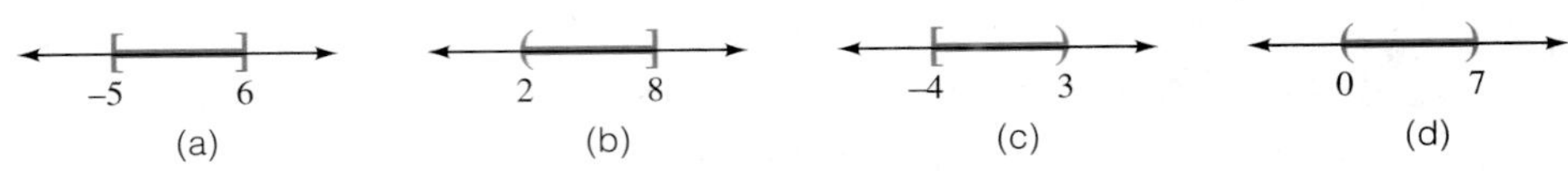

Figure 1-13

SELF CHECK 6 Graph $\{x \mid -6 < x \leq 10\}$ and then write it in interval notation.

The following table shows three ways to describe an interval.

Set notation	Graph	Interval notation
$\{x \mid x > a\}$	a	(a, ∞)
$\{x \mid x < a\}$	a	$(-\infty, a)$
$\{x \mid x \geq a\}$	a	$[a, \infty)$
$\{x \mid x \leq a\}$	a	$(-\infty, a]$
$\{x \mid a < x < b\}$	a b	(a, b)
$\{x \mid a \leq x < b\}$	a b	$[a, b)$
$\{x \mid a < x \leq b\}$	a b	$(a, b]$
$\{x \mid a \leq x \leq b\}$	a b	$[a, b]$
$\{x \mid x < a \text{ or } x > b\}$	a b	$(-\infty, a) \cup (b, \infty)$

Table 1-2

5 Find the absolute value of a real number.

The **absolute value** of a real number a, denoted as $|a|$, is the distance on a number line between 0 and the point with coordinate a. For example, the points shown in Figure 1-14 with coordinates of 3 and -3 both lie 3 units from 0. Thus, $|3| = |-3| = 3$.

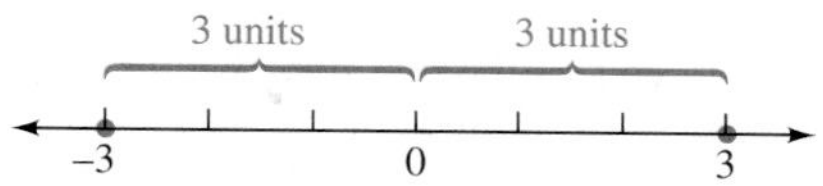

Figure 1-14

In general, for any real number a, $|a| = |-a|$.

The absolute value of a number can be defined more formally.

Absolute Value

For any real number x, $\begin{cases} \text{If } x \geq 0, \text{ then } |x| = x. \\ \text{If } x < 0, \text{ then } |x| = -x. \end{cases}$

If x is positive or 0, then x is its own absolute value. However, if x is negative, then $-x$ (which is a positive number) is the absolute value of x. Thus, $|x| \geq 0$ for all real numbers x.

EXAMPLE 7 Find each absolute value.

a. $|3| = 3$

b. $|-4| = 4$

c. $|0| = 0$

d. $-|-8| = -(8) = -8$ Note that $|-8| = 8$.

SELF CHECK 7 Find each absolute value. **a.** $|-9|$ **b.** $-|-12|$

SELF CHECK ANSWERS

1. a. $4 = \frac{4}{1}$ **b.** $0.5 = \frac{1}{2}$ **2. a.** $0.\overline{45}$, repeating decimal **b.** 0.4, terminating decimal **3. a.** 7 **b.** $-2, 0, 1.5, 7$ **4.**

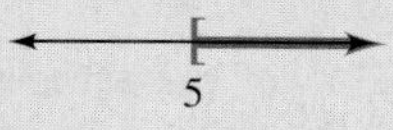

$[5, \infty)$ **5.**

$(-\infty, -1) \cup (5, \infty)$

6.

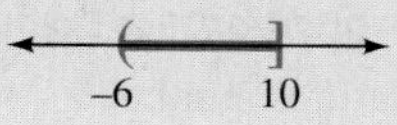

$(-6, 10]$ **7. a.** 9 **b.** -12

NOW TRY THIS

1. Which numbers in the set $\left\{-\frac{5}{8}, \sqrt{3}, 6.2222\ldots, 0, \sqrt{25}, 4, -9.01020304\ldots\right\}$ are

a. integers

b. irrational numbers

c. rational numbers

2. Describe each graph in set-builder and interval notation.

(a)

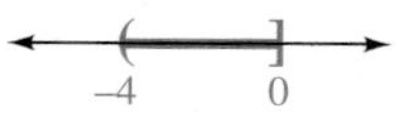

(b)

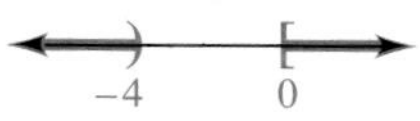

(c)

1.1 EXERCISES

WARM-UPS

1. List the first four prime numbers.

2. List the first five positive even integers.

3. Find $|-6|$.

4. Find $|10|$.

REVIEW

To simplify a fraction, factor the numerator and the denominator and divide out common factors. For example, $\frac{12}{18} = \frac{6 \cdot 2}{6 \cdot 3} = \frac{\not{6} \cdot 2}{\not{6} \cdot 3} = \frac{2}{3}$. Simplify each fraction.

5. $\frac{6}{8}$

6. $\frac{15}{20}$

7. $\frac{32}{40}$

8. $\frac{56}{72}$

To multiply fractions, multiply the numerators and multiply the denominators. To divide fractions, invert the divisor and multiply. Always simplify the result if possible.

9. $\frac{1}{4} \cdot \frac{3}{5}$

10. $\frac{3}{5} \cdot \frac{20}{27}$

11. $\frac{2}{3} \div \frac{3}{7}$

12. $\frac{3}{5} \div \frac{9}{15}$

To add (or subtract) fractions, write each fraction with a common denominator and add (or subtract) the numerators and keep the same denominator. Always simplify the result if possible.

13. $\frac{5}{9} + \frac{4}{9}$

14. $\frac{16}{7} - \frac{2}{7}$

15. $\frac{2}{3} + \frac{4}{5}$

16. $\frac{7}{9} - \frac{2}{5}$

VOCABULARY AND CONCEPTS *Fill in the blanks.*

17. A letter that represents a number is called a ________.

18. A ___ is a collection of objects called ________.

19. The numbers 1, 2, 3, 4, 5, 6, . . . form the set of ________ numbers.

20. The natural numbers together with 0 is called the set of ________ numbers.

21. The natural numbers, 0, and the negatives of the natural numbers form the set of ________.

22. The union of the rational and irrational numbers is called the set of _____ numbers.

23. An _____ integer can be divided exactly by 2.

24. An _____ integer cannot be divided exactly by 2.

25. A prime number is a ________ number that is larger than __ and can only be divided exactly by ______ and 1.

26. A ________ number is a natural number greater than __ that is not ______.

27. __ is neither positive nor negative.

28. The denominator of a fraction can never be __.

29. A repeating decimal represents a ________ number.

30. A nonrepeating, nonterminating decimal represents an ________ number.

31. The symbol ___ means "is less than."

32. The symbol ___ means "is greater than or equal to."

33. The symbol ___ means "is approximately equal to."

34. If x is negative, $|x| =$ ____.

35. $x \geq 5$ is called an ________ while "$x < -2$ or $x \geq 9$" is a ____________.

36. Integers are a ______ of the rational numbers.

37. The symbol $\cup$ means ______ and the symbol $\cap$ means ________.

38. An answer such as $\{0, 1, 2, 3, \ldots\}$ is in ________ notation, $(-2, \infty)$ is in ________ notation and $\{x \mid x > -2\}$ is in ________ notation.

GUIDED PRACTICE

Change each fraction into a decimal and classify the result as a terminating or a repeating decimal. **See Example 2. (Objective 1)**

39. $\frac{7}{8}$

40. $\frac{7}{3}$

41. $-\frac{11}{15}$

42. $-\frac{19}{16}$

List the elements in the set $\{-3, 0, \frac{2}{3}, 1, \sqrt{3}, 2, 9\}$ that satisfy the given condition. **See Examples 1–3. (Objective 1)**

43. natural number

44. whole number

45. integer

46. rational number

47. irrational number

48. real number

49. even natural number

50. odd integer

51. prime number

52. composite number

53. odd composite number

54. even prime number

Insert an $<$ or an $>$ symbol to make a true statement. **(Objective 2)**

55. 5 ▭ 9

56. 9 ▭ 0

57. −5 ▭ −10

58. −3 ▭ 10

59. $-7 \;\square\; 7$ **60.** $0 \;\square\; -5$

61. $6 \;\square\; -6$ **62.** $-6 \;\square\; -2$

Write each statement with the inequality symbol pointing in the opposite direction. **(Objective 2)**

63. $19 > 12$ **64.** $-3 \geq -5$

65. $-6 \leq -5$ **66.** $-10 < 13$

67. $5 \geq -3$ **68.** $0 \leq 12$

69. $-10 < 0$ **70.** $-4 > -8$

Graph each set on the number line. **See Example 4. (Objective 3)**

71. The set of prime numbers less than 8

0 1 2 3 4 5 6 7 8

72. The set of integers between -7 and 0

-7 -6 -5 -4 -3 -2 -1 0

73. The set of odd integers between 10 and 18

10 11 12 13 14 15 16 17 18

74. The set of composite numbers less than 10

0 1 2 3 4 5 6 7 8 9 10

75. $\{x \mid x > 3\}$ **76.** $\{x \mid x < 0\}$

77. $[-5, \infty)$ **78.** $(-\infty, 9]$

Graph each set on the number line. **See Examples 5–6. (Objective 4)**

79. $\{x \mid x < -3 \text{ or } x > 3\}$ **80.** $(\infty, -4] \cup (2, \infty)$

81. $(-\infty, -6] \cup [5, \infty)$ **82.** $\{x \mid x < -2 \text{ or } x \geq 3\}$

83. $\{x \mid 6 < x \leq 10\}$ **84.** $\{x \mid -3 \leq x \leq 8\}$

85. $\{x \mid -1 < x \leq 3\}$ **86.** $\{x \mid 2 < x < 5\}$

Write each expression without using absolute value symbols. Simplify the result when possible. **See Example 7. (Objective 5)**

87. $|20|$ **88.** $|-20|$

89. $-|-6|$ **90.** $-|8|$

91. $|-5| + |-2|$ **92.** $|12| + |-4|$

93. $|-5| \cdot |4|$ **94.** $|-6| \cdot |3|$

95. Find x if $|x| = 3$. **96.** Find x if $|x| = 7$.

97. What numbers x are equal to their own absolute values?

98. What numbers x when added to their own absolute values give a sum of 0?

ADDITIONAL PRACTICE *Graph each set on the number line.*

99. $\{x \mid x \leq 7\}$ **100.** $\{x \mid x \geq -2\}$

101. $[0, 5)$ **102.** $[-6, 9]$

103. $(6, \infty) \cup [9, \infty)$ **104.** $\{x \mid x > -8 \text{ and } x > 0\}$

APPLICATIONS

105. Reading temperatures On the thermometer in the illustration, graph each of the following temperature readings: 12°, 8°, 0°, -6°.

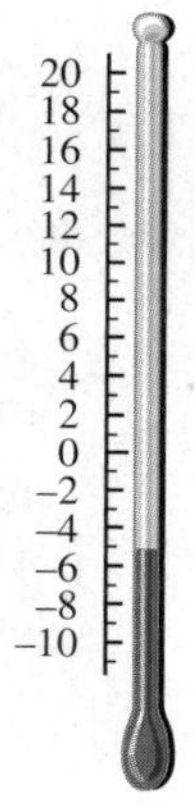

106. History On the number line in the illustration, the origin is the point B.C./A.D.

a. What happened in the year 1441?

b. What happened in the year 500 B.C.? (This would correspond to -500 on a number line.)

MAYA CIVILIZATION

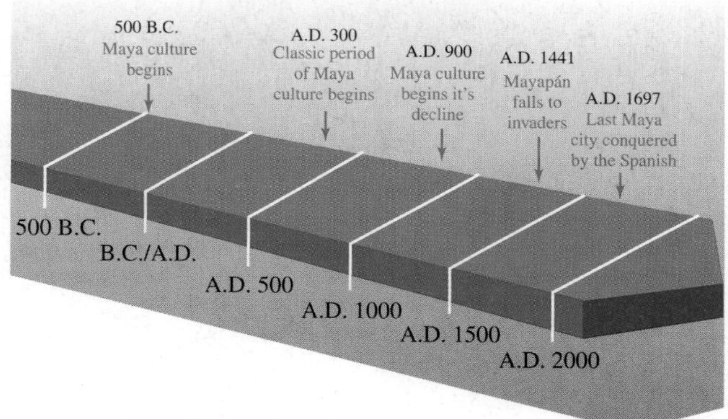

Based on data from People in Time and Place, Western Hemisphere *(Silver Burdett & Ginn Inc., 1991), p. 129*

WRITING ABOUT MATH

107. Explain why the integers are a subset of the rational numbers.

108. Explain why every integer is a rational number, but not every rational number is an integer.

109. Explain why the set of primes together with the set of composites is not the set of natural numbers.

110. Is the absolute value of a number always positive? Explain.

SOMETHING TO THINK ABOUT

111. How many integers have an absolute value that is less than 50?

112. How many odd integers have an absolute value between 20 and 40?

113. The **trichotomy property** of real numbers states that

If a and b are two real numbers, then

$$a < b \quad \text{or} \quad a = b \quad \text{or} \quad a > b$$

Explain why this is true.

114. Which of the following statements are always true?

a. $|a + b| = |a| + |b|$

b. $|a \cdot b| = |a| \cdot |b|$

c. $|a + b| \leq |a| + |b|$

SECTION 1.2 Arithmetic and Properties of Real Numbers

Objectives

1. Add real numbers.
2. Subtract real numbers.
3. Multiply real numbers.
4. Divide real numbers.
5. Evaluate a numeric expression following the rules for order of operations.
6. Find the mean, median, and mode when given a set of values.
7. Evaluate an algebraic expression when given values for its variables.
8. Identify the property of real numbers that justifies a given statement.

Vocabulary

sum	median	distributive property
difference	mode	additive identity
product	perimeter	multiplicative identity
quotient	circumference	additive inverse
grouping symbols	commutative properties	multiplicative inverse
mean	associative properties	

Getting Ready

Perform each operation.

1. $5 + 4$	**2.** $4 + 5$	**3.** $3 \cdot 4$	**4.** $4 \cdot 3$
5. $12 - 7$	**6.** $15 \div 3$	**7.** $21 \div 7$	**8.** $25 - 19$

We will begin this section by discussing how to add, subtract, multiply, and divide real numbers.

1 Add real numbers.

When two numbers are added, we call the result their **sum.** To find the sum of $+2$ and $+3$, we can use a number line and represent the numbers with arrows, as shown in Figure 1-15(a). Since the endpoint of the second arrow is at $+5$, we have $+2 + (+3) = +5$.

To add -2 and -3, we can draw arrows as shown in Figure 1-15(b). Since the endpoint of the second arrow is at -5, we have $(-2) + (-3) = -5$.

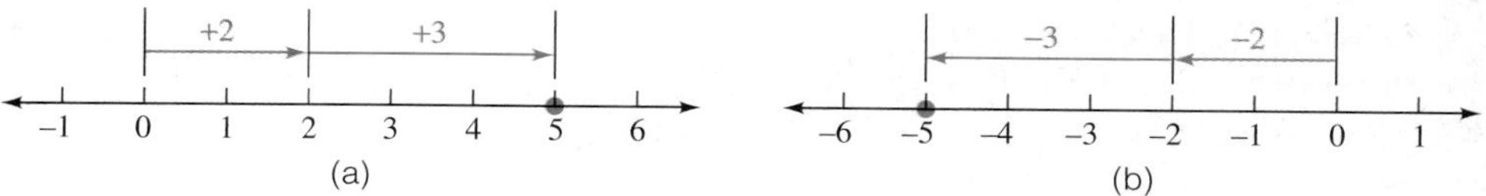

Figure 1-15

To add -6 and $+2$, we can draw arrows as shown in Figure 1-16(a). Since the endpoint of the second arrow is at -4, we have $(-6) + (+2) = -4$.

To add $+7$ and -4, we can draw arrows as shown in Figure 1-16(b). Since the endpoint of the final arrow is at $+3$, we have $(+7) + (-4) = +3$.

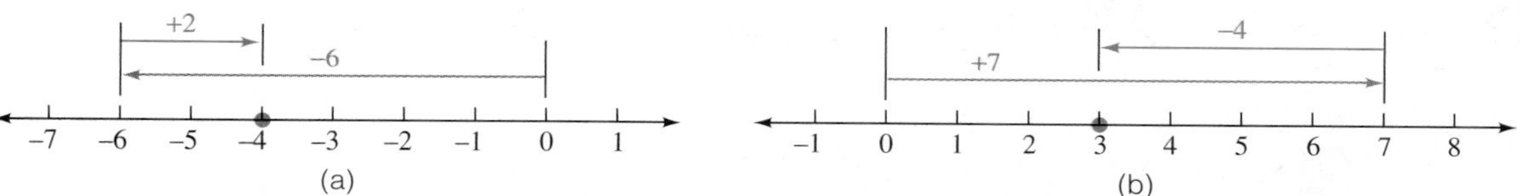

Figure 1-16

These examples suggest the following rules.

Adding Two Real Numbers

With like signs: Add the absolute values of the numbers and use the common sign.

With unlike signs: Subtract the absolute values of the numbers (the smaller from the larger) and use the sign of the number with the larger absolute value.

EXAMPLE 1 Add:

a. $+4 + (+6) = +10$ Add the absolute values and use the common sign: $4 + 6 = +10$.

b. $-5 + (-3) = -8$ Add the absolute values and use the common sign: $-(5 + 3) = -8$.

c. $+9 + (-5) = +4$ Subtract the absolute values and use a $+$ sign: $+(9 - 5) = +4$.

d. $-12 + (+5) = -7$ Subtract the absolute values and use a $-$ sign: $-(12 - 5) = -7$.

SELF CHECK 1 Add: **a.** $-7 + (-2)$ **b.** $-7 + 2$ **c.** $7 + 2$ **d.** $7 + (-2)$

2 Subtract real numbers.

When one number is subtracted from another number, we call the result their **difference.** To find the difference between two real numbers, we can change the subtraction into an equivalent addition. For example, the subtraction $7 - 4$ is equivalent to the addition $7 + (-4)$, because they have the same result:

$$7 - 4 = 3 \quad \text{and} \quad 7 + (-4) = 3$$

This suggests that to subtract two numbers, we can change the sign of the number being subtracted and add.

Subtracting Two Real Numbers

If a and b are real numbers, then $a - b = a + (-b)$.

EXAMPLE 2 Subtract.

a. $12 - 4 = 12 + (-4)$ Change the sign of 4 and add.
$= 8$

b. $-13 - 5 = -13 + (-5)$ Change the sign of 5 and add.
$= -18$

c. $-14 - (-6) = -14 + (+6)$ Change the sign of -6 and add.
$= -8$

SELF CHECK 2 Subtract. **a.** $-15 - 4$ **b.** $8 - 5$ **c.** $-12 - (-7)$

3 Multiply real numbers.

When two numbers are multiplied, we call the result their **product.** We can find the product of 5 and 4 by using 4 in an addition five times:

$$5(4) = \overbrace{4 + 4 + 4 + 4 + 4}^{\text{five addends of 4}} = 20$$

We can find the product of 5 and -4 by using -4 in an addition five times:

$$5(-4) = \overbrace{(-4) + (-4) + (-4) + (-4) + (-4)}^{\text{five addends of } -4} = -20$$

Since multiplication by a negative number can be defined as repeated subtraction, we can find the product of -5 and 4 by using 4 in a subtraction five times:

$$\begin{aligned} -5(4) &= -4 - 4 - 4 - 4 - 4 \\ &= -4 + (-4) + (-4) + (-4) + (-4) && \text{Change the sign of each 4 and add.} \\ &= -20 \end{aligned}$$

We can find the product of -5 and -4 by using -4 in a subtraction five times:

$$\begin{aligned} -5(-4) &= -(-4) - (-4) - (-4) - (-4) - (-4) \\ &= 4 + 4 + 4 + 4 + 4 && \text{Change the sign of each } -4 \text{ and add.} \\ &= 20 \end{aligned}$$

The products $5(4)$ and $-5(-4)$ both equal $+20$, and the products $5(-4)$ and $-5(4)$ both equal -20. These results suggest the first two of the following rules.

Multiplying Two Real Numbers

With like signs: Multiply their absolute values. The product is positive.

With unlike signs: Multiply their absolute values. The product is negative.

Multiplication by 0: If x is any real number, then $x \cdot 0 = 0 \cdot x = 0$.

EXAMPLE 3 Multiply.

a. $4(-7) = -28$ — Multiply the absolute values: $4 \cdot 7 = 28$. Since the signs are unlike, the product is negative.

b. $-5(-6) = +30$ — Multiply the absolute values: $5 \cdot 6 = 30$. Since the signs are alike, the product is positive.

c. $-7(6) = -42$ — Multiply the absolute values: $7 \cdot 6 = 42$. Since the signs are unlike, the product is negative.

d. $8(6) = +48$ — Multiply the absolute values: $8 \cdot 6 = 48$. Since the signs are alike, the product is positive.

SELF CHECK 3 Multiply. **a.** $(-6)(5)$ **b.** $(-4)(-8)$ **c.** $(17)(-2)$ **d.** $(12)(6)$

4 Divide real numbers

When two numbers are divided, we call the result their **quotient.** In the division $x \div y = q$ or $\frac{x}{y} = q \quad (y \neq 0)$, the quotient q is a number such that $y \cdot q = x$. We can use this relationship to find rules for dividing real numbers. We consider four divisions:

$$\frac{+10}{+2} = +5, \text{ because } +2(+5) = +10 \qquad \frac{-10}{-2} = +5, \text{ because } -2(+5) = -10$$

$$\frac{-10}{+2} = -5, \text{ because } +2(-5) = -10 \qquad \frac{+10}{-2} = -5, \text{ because } -2(-5) = +10$$

Dividing Two Real Numbers

With like signs: Divide their absolute values. The quotient is positive.

With unlike signs: Divide their absolute values. The quotient is negative.

Division by 0: Division by 0 is undefined.

COMMENT If $x \neq 0$, then $\frac{0}{x} = 0$. However, $\frac{x}{0}$ is undefined for any value of x.

These results suggest the first two rules for dividing real numbers.

EXAMPLE 4 Divide.

a. $\frac{36}{18} = +2$ — Divide the absolute values: $\frac{36}{18} = 2$. Since the signs are alike, the quotient is positive.

b. $\frac{-44}{11} = -4$ — Divide the absolute values: $\frac{44}{11} = 4$. Since the signs are unlike, the quotient is negative.

c. $\dfrac{27}{-9} = -3$ Divide the absolute values: $\frac{27}{9} = 3$. Since the signs are unlike, the quotient is negative.

d. $\dfrac{-64}{-8} = +8$ Divide the absolute values: $\frac{64}{8} = 8$. Since the signs are alike, the quotient is positive.

SELF CHECK 4 Divide. **a.** $\frac{55}{-5}$ **b.** $\frac{-72}{-6}$ **c.** $\frac{-100}{10}$ **d.** $\frac{50}{25}$

5 Evaluate a numeric expression following the rules for order of operations.

Suppose you are asked to contact a friend if you see a rug for sale while traveling in Turkey. After locating a nice one, you send the following message to your friend.

The next day, you receive this response.

The first statement in your friend's message says to buy the rug at any price. The second says not to buy it, because it is too expensive. The placement of the exclamation point makes these statements read differently, resulting in different interpretations.

When reading mathematical statements, the same kind of confusion is possible. To illustrate, we consider the expression $5 + 3 \cdot 7$, which contains the operations of addition and multiplication. To avoid the possibility of getting two different results when doing this type of calculation, we will agree to do the multiplications first.

Correct choice: Multiply first		*Incorrect choice: Add first*	
$5 + \mathbf{3 \cdot 7} = 5 + \mathbf{21}$	Multiply 3 and 7.	~~$\mathbf{5 + 3} \cdot 7 = \mathbf{8} \cdot 7$~~	Add 5 and 3.
$= 26$	Add 5 and 21.	~~$= 56$~~	Multiply 8 and 7.

Different results

To indicate that additions should be done before multiplications, we must use **grouping symbols** such as parentheses ( ), brackets [ ], or braces { }. In the expression $(5 + 3)7$, the parentheses indicate that the addition is to be done first:

$$(5 + 3)7 = 8 \cdot 7$$
$$= 56$$

To guarantee that calculations will have one correct result, we always will do calculations in the following order.

Rules for the Order of Operations for Expressions without Exponents

Use the following steps to perform all calculations within each pair of grouping symbols, working from the innermost pair to the outermost pair.

1. Perform all multiplications and divisions, working from left to right.
2. Perform all additions and subtractions, working from left to right.
3. Because a fraction bar is a grouping symbol, simplify the numerator and the denominator separately and simplify the fraction, whenever possible.

When all grouping symbols have been removed, repeat the rules above to finish the calculation.

EXAMPLE 5 Evaluate each expression.

a. $4 + 2 \cdot 3 = 4 + 6$ — Do the multiplication first.
$= 10$ — Then do the addition.

b. $2(3 + 4) = 2 \cdot 7$ — Because of the parentheses, do the addition first.
$= 14$ — Then do the multiplication.

c. $5(3 - 6) \div 3 + 1 = 5(-3) \div 3 + 1$ — Do the subtraction within parentheses.
$= -15 \div 3 + 1$ — Then do the multiplication: $5(-3) = -15$.
$= -5 + 1$ — Then do the division: $-15 \div 3 = -5$.
$= -4$ — Finally, do the addition.

d. $5[3 - 2(6 \div 3 + 1)] = 5[3 - 2(2 + 1)]$ — Do the division within parentheses: $6 \div 3 = 2$.
$= 5[3 - 2(3)]$ — Do the addition: $2 + 1 = 3$.
$= 5(3 - 6)$ — Do the multiplication: $2(3) = 6$.
$= 5(-3)$ — Do the subtraction: $3 - 6 = -3$.
$= -15$ — Do the multiplication.

e. $\dfrac{4 + 8(3 - 4)}{6 - 2(2)} = \dfrac{-4}{2}$ — Simplify the numerator and denominator separately.
$= -2$ — Do the division: $-4 \div 2 = -2$.

SELF CHECK 5 Evaluate. **a.** $5 + 3 \cdot 4$ **b.** $(5 + 3) \cdot 4$ **c.** $3(5 - 7) \div 6 + 3$
d. $\frac{5 - 2(4 - 6)}{9 - 2 \cdot 3}$

6 Find the mean, median, and mode when given a set of values.

Three measures of central tendency are commonly used in newspapers and magazines: the *mean,* the *median,* and the *mode.*

Mean The **mean** of several values is the sum of those values divided by the number of values.

$$\text{Mean} = \frac{\text{sum of the values}}{\text{number of values}}$$

EXAMPLE 6 **FOOTBALL** Figure 1-17 shows the gains and losses made by a running back on seven plays. Find the mean number of yards per carry.

Solution To find the mean number of yards per carry, we add the numbers and divide by 7.

$$\frac{-8 + (+2) + (-6) + (+6) + (+4) + (-7) + (-5)}{7} = \frac{-14}{7} = -2$$

The running back averaged -2 yards (or lost 2 yards) per carry.

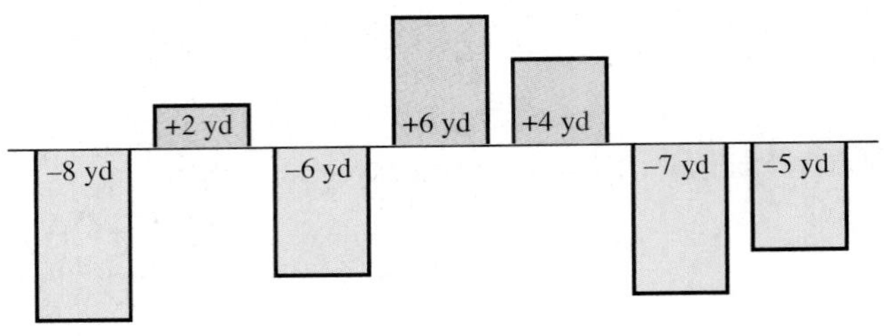

Figure 1-17

Median The **median** of several values is the middle value. To find the median,

1. Arrange the values in increasing order.
2. If there is an odd number of values, choose the middle value.
3. If there is an even number of values, find the mean of the middle two values.

Mode The **mode** of several values is the value that occurs most often.

EXAMPLE 7 Ten workers in a small business have monthly salaries of

\$2,500, \$1,750, \$2,415, \$3,240, \$2,790,
\$3,240, \$2,650, \$2,415, \$2,415, \$2,650

Find: **a.** the median **b.** the mode of the distribution.

Solution **a.** To find the median, we first arrange the salaries in increasing order:

\$1,750, \$2,415, \$2,415, \$2,415, \$2,500,
\$2,650, \$2,650, \$2,790, \$3,240, \$3,240

Because there is an even number of salaries, the median will be the mean of the middle two scores, \$2,500 and \$2,650.

$$\text{Median} = \frac{\$2{,}500 + \$2{,}650}{2} = \$2{,}575$$

b. Since the salary \$2,415 occurs most often, it is the mode.

If two different numbers in a distribution tie for occurring most often, the distribution is called *bimodal.*

COMMENT Although the mean is probably the most common measure of central tendency, the median and the mode are used frequently. For example, workers' salaries often are compared to the median salary. To say that the modal shoe size is 9 means that a shoe size of 9 occurs more often than any other size.

7 Evaluate an algebraic expression when given values for its variables.

Variables and numbers can be combined with the operations of arithmetic to produce *algebraic expressions.* To evaluate algebraic expressions, we substitute numbers for the variables and simplify.

EXAMPLE 8 If $x = 2$, $y = -3$, and $z = -5$, evaluate: **a.** $x + yz$ **b.** $\dfrac{xy + 3z}{y(z - x)}$.

Solution We substitute 2 for x, -3 for y, and -5 for z and simplify.

a.
$$\begin{aligned} x + yz &= 2 + (-3)(-5) \\ &= 2 + (15) \\ &= 17 \end{aligned}$$

b.
$$\begin{aligned} \frac{xy + 3z}{y(z - x)} &= \frac{2(-3) + 3(-5)}{-3(-5 - 2)} \\ &= \frac{-6 + (-15)}{-3(-7)} \\ &= \frac{-21}{21} \\ &= -1 \end{aligned}$$

SELF CHECK 8 If $x = 2$, $y = -5$, and $z = 3$, evaluate: **a.** $y - xz$ **b.** $\dfrac{xy - 2z}{xz + 2y}$.

Table 1-3 shows the formulas for the **perimeters** (the distance around a figure) of several geometric figures. The distance around a circle is called a **circumference.**

Figure	Name	Perimeter	Figure	Name	Perimeter/ circumference
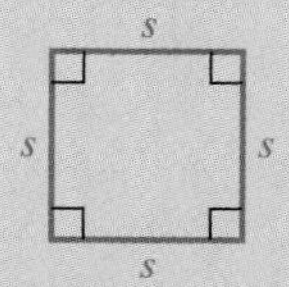	Square	$P = 4s$	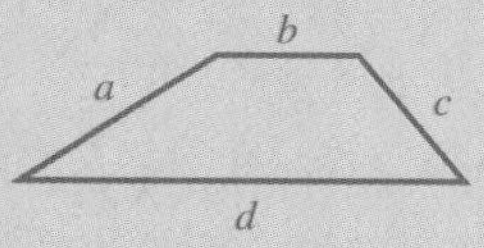	Trapezoid	$P = a + b + c + d$
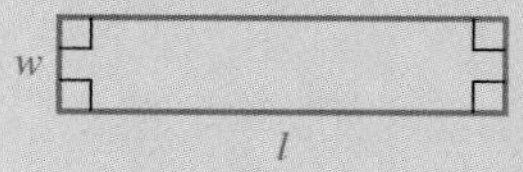	Rectangle	$P = 2l + 2w$	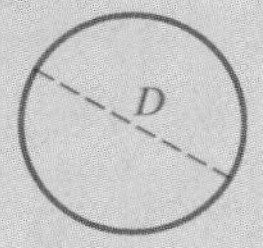	Circle	$C = \pi D$ $C = 2\pi r$ (π is approximately 3.1416)
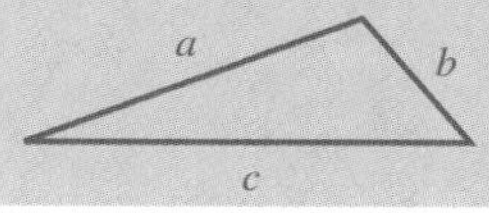	Triangle	$P = a + b + c$			

Table 1-3

EXAMPLE 9 Find the perimeter of the rectangle shown in Figure 1-18.

Solution We substitute 2.75 for l and 1.25 for w into the formula $P = 2l + 2w$ and simplify.

$$\begin{aligned} P &= 2l + 2w \\ P &= 2(2.75) + 2(1.25) \\ &= 5.50 + 2.50 \\ &= 8.00 \end{aligned}$$

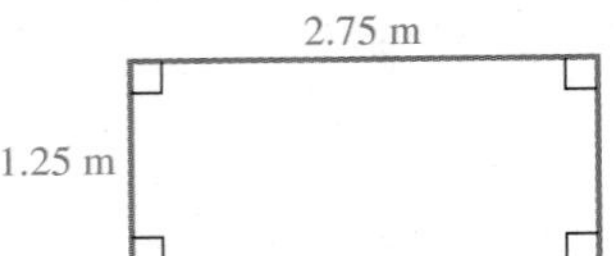

Figure 1-18

The perimeter is 8 meters.

SELF CHECK 9 Find the perimeter of a rectangle with a length of 8 meters and a width of 5 meters.

8 Identify the property of real numbers that justifies a given statement.

When we work with real numbers, we will use the following properties.

Properties of Real Numbers

If a, b, and c are real numbers, the following properties apply.

The commutative properties for addition and multiplication

$a + b = b + a \qquad ab = ba$

The associative properties for addition and multiplication

$(a + b) + c = a + (b + c) \qquad (ab)c = a(bc)$

The distributive property of multiplication over addition

$a(b + c) = ab + ac$

The *commutative properties* enable us to add or multiply two numbers in either order and obtain the same result. For example,

$2 + 3 = 5$ and $3 + 2 = 5$

$7 \cdot 9 = 63$ and $9 \cdot 7 = 63$

COMMENT Subtraction and division are not commutative, because doing these operations in different orders will give different results. For example,

$8 - 4 = 4$ but $4 - 8 = -4$

$8 \div 4 = 2$ but $4 \div 8 = \frac{1}{2}$

The *associative properties* enable us to group the numbers in a sum or a product in any way that we wish and still get the same result. For example,

$$\begin{aligned} (2 + 3) + 4 &= 5 + 4 \qquad & 2 + (3 + 4) &= 2 + 7 \\ &= 9 & &= 9 \end{aligned}$$

$$\begin{aligned} (2 \cdot 3) \cdot 4 &= 6 \cdot 4 \qquad & 2 \cdot (3 \cdot 4) &= 2 \cdot 12 \\ &= 24 & &= 24 \end{aligned}$$

COMMENT Subtraction and division are not associative, because different groupings give different results. For example,

$$(8 - 4) - 2 = 4 - 2 = 2 \quad \text{but} \quad 8 - (4 - 2) = 8 - 2 = 6$$
$$(8 \div 4) \div 2 = 2 \div 2 = 1 \quad \text{but} \quad 8 \div (4 \div 2) = 8 \div 2 = 4$$

The *distributive property* enables us to evaluate many expressions involving a multiplication over an addition. We can add first inside the parentheses and then multiply, or multiply over the addition first and then add.

$$2(3 + 7) = 2 \cdot 10 \qquad\qquad 2(3 + 7) = 2 \cdot 3 + 2 \cdot 7$$
$$= 20 \qquad\qquad\qquad = 6 + 14$$
$$= 20$$

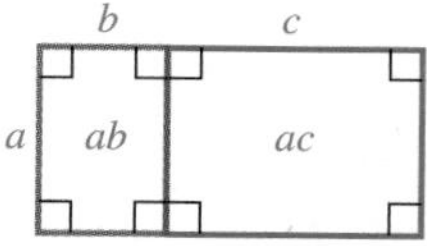

Figure 1-19

We can interpret the distributive property geometrically. Since the area of the largest rectangle in Figure 1-19 is the product of its width a and its length $b + c$, its area is $a(b + c)$. The areas of the two smaller rectangles are ab and ac. Since the area of the largest rectangle is equal to the sum of the areas of the smaller rectangles, we have $a(b + c) = ab + ac$.

A more general form of the distributive property is called the *extended distributive property.*

$$a(b + c + d + e + \cdots) = ab + ac + ad + ae + \cdots$$

EXAMPLE 10 Use the distributive property to write each expression without parentheses.
a. $2(x + 3)$ **b.** $2(x + y - 7)$.

Solution **a.** $2(x + 3) = 2x + 2 \cdot 3$
$= 2x + 6$

b. $2(x + y - 7) = 2x + 2y - 2 \cdot 7$
$= 2x + 2y - 14$

SELF CHECK 10 Remove parentheses. $-5(a - 2b + 3c)$.

The real numbers 0 and 1 have important special properties.

Properties of 0 and 1

Additive identity: The sum of 0 and any number is the number itself.

$$0 + a = a + 0 = a$$

Multiplication property of 0: The product of any number and 0 is 0.

$$a \cdot 0 = 0 \cdot a = 0$$

Multiplicative identity: The product of 1 and any number is the number itself.

$$1 \cdot a = a \cdot 1 = a$$

For example,

$$7 + 0 = 7, \quad 7(0) = 0, \quad 1(5) = 5, \quad \text{and} \quad (-7)1 = -7$$

If the sum of two numbers is 0, the numbers are called **additive inverses, negatives,** or **opposites** of each other. For example, 6 and -6 are negatives, because $6 + (-6) = 0$.

The Additive Inverse Property

For every real number a, there is a real number $-a$ such that

$$a + (-a) = -a + a = 0$$

The symbol $-(-6)$ means "the negative of negative 6." Because the sum of two numbers that are negatives is 0, we have

$$-6 + [-(-6)] = 0 \qquad \text{and} \qquad -6 + 6 = 0$$

Because -6 has only one additive inverse, it follows that $-(-6) = 6$. In general, the following rule applies.

The Double Negative Rule

If a represents any real number, then $-(-a) = a$.

If the product of two numbers is 1, the numbers are called **multiplicative inverses** or **reciprocals** of each other.

The Multiplicative Inverse Property

For every nonzero real number a, there exists a real number $\frac{1}{a}$ such that

$$a \cdot \frac{1}{a} = \frac{1}{a} \cdot a = 1 \quad (a \neq 0)$$

Some examples of reciprocals are

- 5 and $\frac{1}{5}$ are reciprocals, because $5\left(\frac{1}{5}\right) = 1$.
- $\frac{3}{2}$ and $\frac{2}{3}$ are reciprocals, because $\frac{3}{2}\left(\frac{2}{3}\right) = 1$.
- -0.25 and -4 are reciprocals, because $-0.25(-4) = 1$.

The reciprocal of 0 does not exist, because $\frac{1}{0}$ is undefined.

SELF CHECK ANSWERS

1. a. -9 **b.** -5 **c.** 9 **d.** 5 **2. a.** -19 **b.** 3 **c.** -5 **3. a.** -30 **b.** 32 **c.** -34 **d.** 72 **4. a.** -11 **b.** 12 **c.** -10 **d.** 2 **5. a.** 17 **b.** 32 **c.** 2 **d.** 3 **8. a.** -11 **b.** 4 **9.** 26 m **10.** $-5a + 10b - 15c$

NOW TRY THIS

1. Insert the proper signs and include any grouping symbols needed to make each expression equal to the given value.

a. 3 ▢ 5 = -2 **b.** -8 ▢ 4 = -4
c. 3 ▢ 5 ▢ 2 = 4 **d.** 3 ▢ 5 ▢ 2 = 9

2. Evaluate each of the following for $x = -2$, $y = 5$, and $z = -4$.

a. $x - y - z$ **b.** $x - y(-z)$

1.2 EXERCISES

WARM-UPS *Perform each operation.*

1. $+2 + (-4)$

2. $-5 - 2$

3. $7(-4)$

4. $(-7)(-4)$

5. $3 + (-3)(2)$

6. $\dfrac{-4 + (-2)}{-3 + 5}$

REVIEW *Graph each set on the number line.*

7. $\{x \mid x > 4\}$

8. $(-\infty, -5]$

9. $(2, 10]$

10. $\{x \mid -4 \le x \le 4\}$

11. Buying gasoline A man bought 32 gallons of gasoline at \$2.29 per gallon and 3 quarts of oil at \$1.35 per quart. The sales tax was included in the price of the gasoline, but 5% sales tax was added to the cost of the oil. Find the total cost.

12. Paying taxes On an adjusted income of \$57,760, a woman must pay taxes according to the schedule shown in the table. Compute the tax bill.

2008 Tax Rate Schedules			
Schedule X—If your filing status is Single			
If your taxable income is:		**The tax is:**	
Over—	**But not over—**		**Of the amount over—**
\$0	\$8,025	. . . . . 10%	\$0
8,025	32,550	**\$802.50 + 15%**	**8,025**
32,550	78,850	**\$4.481.25 + 25%**	**32,550**
78,850	164,550	**\$16,056.25 + 28%**	**78,850**
164,550	357,700	**\$40,052.25 + 33%**	**164,550**
357,700	. . . . .	**\$103,791.75 + 35%**	**357,700**

VOCABULARY AND CONCEPTS *Fill in the blanks.*

13. To add two numbers with like signs, we add their ________ values and keep the ________ sign.

14. To add two numbers with unlike signs, we ________ their absolute values and keep the sign of the number with the larger absolute value.

15. To subtract one number from another, we ________ the sign of the number that is being subtracted and ____.

16. The product of two real numbers with like signs is ________.

17. The quotient of two real numbers with unlike signs is ________.

18. The result of addition is a _____ and the result of subtraction is a ________.

19. The result of multiplication is a ________ and the result of division is a ________.

20. To simplify a numerical expression, we follow the ________________.

21. The denominator of a fraction can never be __.

22. The three measures of central tendency are the ______, ________, and ______.

23. Write the formula for the circumference of a circle. ________ or ________

24. Write the associative property of multiplication. ________________

25. Write the commutative property of addition. ____________

26. Write the distributive property of multiplication over addition. ________________

27. Multiplicative inverses are also called ___________.

28. What is the additive identity? __ What is the multiplicative identity? __

29. The additive inverse of -5 is __ and the multiplicative inverse of 7 is __.

30. $-(-a) =$ __

GUIDED PRACTICE

Add. **See Example 1. (Objective 1)**

31. $-3 + (-5)$

32. $2 + (+8)$

33. $-7 + 2$

34. $3 + (-5)$

35. $\dfrac{1}{2} + \left(-\dfrac{1}{3}\right)$

36. $-\dfrac{3}{4} + \left(-\dfrac{1}{5}\right)$

37. $-\dfrac{5}{9} + \left(\dfrac{3}{4}\right)$

38. $\dfrac{7}{8} + \left(-\dfrac{6}{11}\right)$

Subtract. **See Example 2. (Objective 2)**

39. $-3 - 4$

40. $-11 - (-17)$

41. $-33 - (-33)$

42. $14 - (-13)$

43. $\dfrac{1}{2} - \left(-\dfrac{3}{5}\right)$

44. $\dfrac{1}{26} - \dfrac{11}{13}$

45. $\dfrac{1}{3} - \dfrac{1}{2}$

46. $\dfrac{7}{8} - \left(-\dfrac{3}{4}\right)$

Multiply. **See Example 3. (Objective 3)**

47. $-2(6)$

48. $3(-5)$

49. $-3(-7)$

50. $-2(-5)$

51. $\left(-\frac{3}{5}\right)\left(\frac{10}{7}\right)$

52. $\left(-\frac{6}{7}\right)\left(-\frac{5}{12}\right)$

53. $\left(\frac{9}{8}\right)\left(-\frac{6}{7}\right)$

54. $\left(\frac{8}{3}\right)\left(-\frac{15}{16}\right)$

Divide. See Example 4. (Objective 4)

55. $\frac{-8}{4}$

56. $\frac{25}{-5}$

57. $\frac{-16}{-4}$

58. $\frac{-5}{-25}$

59. $\frac{3}{4} \div \left(-\frac{3}{8}\right)$

60. $-\frac{3}{5} \div \frac{7}{10}$

61. $-\frac{16}{5} \div \left(-\frac{10}{3}\right)$

62. $-\frac{5}{24} \div \frac{10}{3}$

Evaluate each expression following the order of operations. See Example 5. (Objective 5)

63. $3 + 4 \cdot 5$

64. $6 + 4 \cdot 7$

65. $2 - 3 \cdot 5$

66. $5 \cdot 3 - 6 \cdot 4$

67. $3 - 2 - 1$

68. $5 - 3 - 1$

69. $3 - (2 - 1)$

70. $5 - (3 - 1)$

71. $8 \div 4 \div 2$

72. $100 \div 10 \div 5$

73. $8 \div (4 \div 2)$

74. $100 \div (10 \div 5)$

Use the distribution: 7, 5, 9, 10, 8, 6, 6, 7, 9, 12, 9. You may use a calculator. See Examples 6–7. (Objective 6)

75. Find the mean.

76. Find the median.

77. Find the mode.

Use the distribution:
8, 12, 23, 12, 10, 16, 26, 12, 14, 8, 16, 23.

78. Find the median.

79. Find the mode.

80. Find the mean.

Let $a = 3, b = -2, c = -1$, and $d = 2$ and evaluate each expression. See Examples 8–9. (Objective 7)

81. $ab + cd$

82. $ad + bc$

83. $a(b + c)$

84. $d(b + a)$

85. $\frac{ad + c}{cd + b}$

86. $\frac{ab + d}{bd + a}$

87. $\frac{ac - bd}{cd - ad}$

88. $\frac{bc - ad}{bd + ac}$

Sorting records is a common task in data processing. A selection sort requires C comparisons to sort N records, where C and N are related by the formula $C = \frac{N(N-1)}{2}$. See Examples 8–9. (Objective 7)

89. How many comparisons are needed to sort 200 records?

90. How many comparisons are needed to sort 10,000 records?

91. Perimeter of a triangle Find the perimeter of a triangle with sides that are 23.5, 37.2, and 39.7 feet long.

92. Perimeter of a trapezoid Find the perimeter of a trapezoid with sides that are 43.27, 47.37, 50.21, and 52.93 centimeters long.

Determine which property of real numbers justifies each statement. See Example 10. (Objective 8)

93. $3 + 7 = 7 + 3$

94. $2 \cdot (9 \cdot 13) = (2 \cdot 9) \cdot 13$

95. $3(2 + 5) = 3 \cdot 2 + 3 \cdot 5$

96. $4 \cdot 3 = 3 \cdot 4$

97. $81 + 0 = 81$

98. $3(9 + 2) = 3 \cdot 9 + 3 \cdot 2$

99. $5 \cdot \frac{1}{5} = 1$

100. $3 + (9 + 0) = (9 + 0) + 3$

101. $a + (7 + 8) = (a + 7) + 8$

102. $1 \cdot 3 = 3$

103. $(2 \cdot 3) \cdot 4 = 4 \cdot (2 \cdot 3)$

104. $8 + (-8) = 0$

Use a calculator to verify each statement and then identify the property of real numbers that is being illustrated. (Objective 8)

105. $(37.9 + 25.2) + 14.3 = 37.9 + (25.2 + 14.3)$

106. $7.1(3.9 + 8.8) = 7.1 \cdot 3.9 + 7.1 \cdot 8.8$

107. $2.73(4.534 + 57.12) = 2.73 \cdot 4.534 + 2.73 \cdot 57.12$

108. $(6.789 + 345.1) + 27.347 = (345.1 + 6.789) + 27.347$

ADDITIONAL PRACTICE

Evaluate each expression.

109. $98 - 14 - 14$

110. $-43 - 19 + 59$

111. $20 - (-14) - 10$

112. $46 + (-21) - 31$

113. $-2(21)(-4)$

114. $-5(-12)(-3)$

115. $56 \div 7 \div 2$

116. $96 \div (-4) \div (-3)$

117. $2 + 6 \div 3 - 5$

118. $6 - 8 \div 4 - 2$

119. $(2 + 6) \div (3 - 5)$

120. $(6 - 8) \div (4 - 2)$

121. $\frac{3(8 + 4)}{2 \cdot 3 - 9}$

122. $\frac{5(4 - 1)}{3 \cdot 2 + 5 \cdot 3}$

123. $\frac{100(2 - 4)}{1{,}000 \div 10 \div 10}$

124. $\frac{8(3) - 4(6)}{5(3) + 3(-7)}$

Let $a = 3, b = -2, c = -1$, and $d = 2$ and evaluate each expression.

125. $2b - 5ac$

126. $4a(6c - b) + d$

127. $c - d(2 - a)$

128. $(a - c) - (b - d)$

APPLICATIONS

129. Earning money One day Scott earned $22.25 tutoring mathematics and $39.75 tutoring physics. How much did he earn that day?

130. Losing weight During an illness, Wendy lost 13.5 pounds. By the time she recovered, she had lost another 11.5 pounds. What integer represents her change in weight?

131. Changing temperatures The temperature rose 17° in the afternoon and then dropped 13° overnight. Find the overall change in temperature.

132. Displaying the flag Before the American flag is displayed at half-mast, it should first be raised to the top of the flagpole. How far has the flag in the illustration traveled?

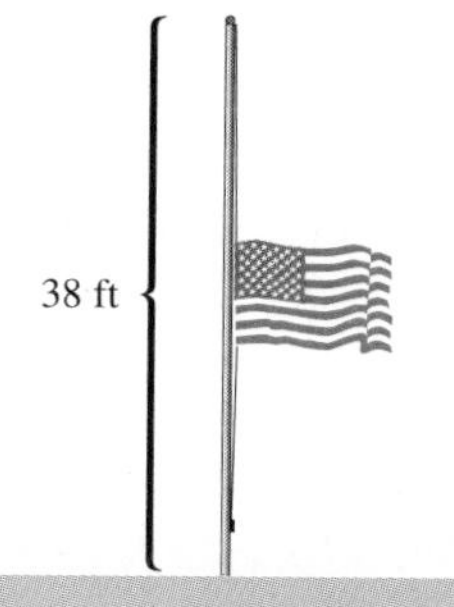

133. Changing temperatures If the temperature has been dropping 4° each hour, how much warmer was it 3 hours ago?

134. Mountain climbing A team of mountaineers climbed 2,347 feet one day and then came down 597 feet to a good spot to make camp. What signed number represents their net change in altitude?

135. Filling a pool The flow of water from a pipe is filling a pool at the rate of 23 gallons per minute. How much less water was in the pool 5 hours ago?

136. Draining a pool If a drain is emptying a pool at the rate of 12 gallons per minute, how much more water was in the pool 2 hours ago?

Use a calculator to help solve the following problems.

137. Military science An army retreated 2,300 meters. After regrouping, it moved forward 1,750 meters. The next day, it gained another 1,875 meters. What integer represents the army's net gain (or loss)?

138. Grooming horses John earned $8 an hour for grooming horses. After working for 8 hours, he had $94. How much did he have before he started work?

139. Managing a checkbook Sally started with $437.37 in a checking account. One month, she had deposits of $125.18, $137.26, and $145.56. That same month, she had withdrawals of $117.11, $183.49, and $122.89. Find her ending balance.

140. Stock averages The illustration shows the daily advances and declines of the Dow Jones average for one week. What integer represents the total gain or loss for the week?

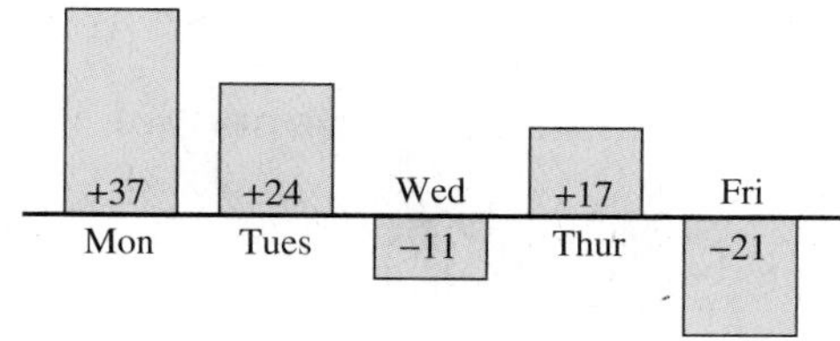

141. Selling clothes If a clerk had the sales shown in the table for one week, find the mean of daily sales.

Monday	$1,525
Tuesday	$ 785
Wednesday	$1,628
Thursday	$1,214
Friday	$ 917
Saturday	$1,197

142. Sizes of viruses The table gives the approximate lengths (in centimicrons) of the viruses that cause five common diseases. Find the mean length of the viruses.

Polio	2.5
Influenza	105.1
Pharyngitis	74.9
Chicken pox	137.4
Yellow fever	52.6

143. Calculating grades A student has scores of 75, 82, 87, 80, and 76 on five exams. Find his average (mean) score.

144. Averaging weights The offensive line of a football team has two guards, two tackles, and a center. If the guards weigh 298 and 287 pounds, the tackles 310 and 302 pounds, and the center 303 pounds, find the average (mean) weight of the offensive line.

145. Analyzing ads The businessman who ran the following ad earns $100,000 and employs four students who earn $10,000 each. Is the ad misleading?

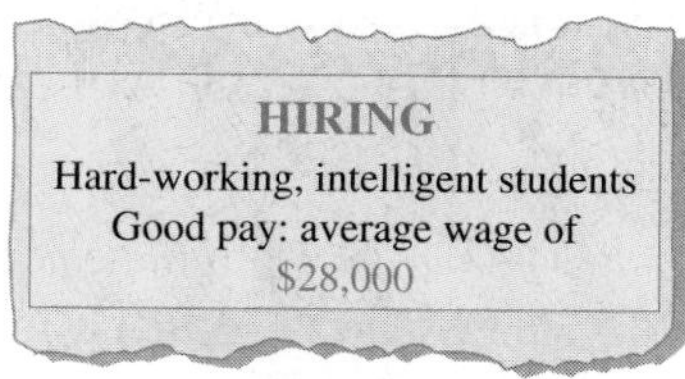

146. Averaging grades A student has grades of 78%, 85%, 88%, and 96%. There is one test left, and the student needs to average 90% to earn an A. Does he have a chance?

147. Perimeter of a square Find the perimeter of the square shown in the illustration.

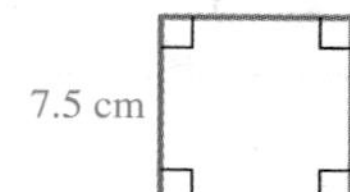

148. Circumference of a circle To the nearest hundredth, find the circumference of the circle shown in the illustration.

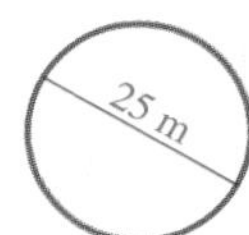

WRITING ABOUT MATH

149. The symmetric property of equality states that if $a = b$, then $b = a$. Explain why this property is often confused with the commutative properties. Why do you think this is so?

150. Explain why the mean of two numbers is halfway between the two numbers.

SOMETHING TO THINK ABOUT

151. Pick five numbers and find their mean. Add 7 to each of the numbers to get five new numbers and find their mean. What do you discover? Is this property always true?

152. Take the original five numbers in Exercise 151 and multiply each one by 7 to get five new numbers and find their mean. What do you discover? Is this property always true?

153. Give three applications in which the median would be the most appropriate average to use.

154. Give three applications in which the mode would be the most appropriate average to use.

SECTION 1.3 Exponents

Objectives

1. Identify the base and the exponent to simplify an exponential expression.
2. Simplify an expression by applying properties of exponents.
3. Evaluate an expression using the rules for the order of operations.
4. Apply the correct geometric formula to solve an application problem.

Vocabulary

factor	area	cone
power	volume	pyramid
base	sphere	cube
exponent	cylinder	rectangular solid

Getting Ready

Find each product.

1. $2 \cdot 2$

2. $3 \cdot 3 \cdot 3$

3. $(-4)(-4)(-4)$

4. $(-3)(-3)(-3)(-3)$

5. $\frac{1}{3} \cdot \frac{1}{3} \cdot \frac{1}{3}$

6. $-\left(\frac{2}{5} \cdot \frac{2}{5} \cdot \frac{2}{5} \cdot \frac{2}{5}\right)$

In this section, we will review exponents, a shortcut way of indicating repeated multiplication.

1 Identify the base and the exponent to simplify an exponential expression.

Exponents indicate repeated multiplication. For example,

$y^2 = y \cdot y$ Read y^2 as "y to the second power" or "y squared."
$z^3 = z \cdot z \cdot z$ Read z^3 as "z to the third power" or "z cubed."
$x^4 = x \cdot x \cdot x \cdot x$ Read x^4 as "x to the fourth power."

These examples suggest the following definition.

Natural-Number Exponents

If n is a natural number, then

$$x^n = \overbrace{x \cdot x \cdot x \cdot \cdots \cdot x}^{n \text{ factors of } x}$$

COMMENT A natural-number exponent tells how many times the base of an exponential expression is to be used as a factor in a product.

The exponential expression x^n is called a **power of x,** and we read it as "x to the nth power." In this expression, x is called the **base,** and n is called the **exponent.**

Base $\rightarrow x^n \leftarrow$ Exponent

EXAMPLE 1 Write each expression without exponents.

a. $2^5 = 2 \cdot 2 \cdot 2 \cdot 2 \cdot 2$
$= 32$

b. $(-2)^5 = (-2)(-2)(-2)(-2)(-2)$
$= -32$

c. $-4^4 = -(4^4)$
$= -(4 \cdot 4 \cdot 4 \cdot 4)$
$= -256$

d. $(-4)^4 = (-4)(-4)(-4)(-4)$
$= 256$

e. $\left(\frac{1}{2}a\right)^3 = \left(\frac{1}{2}a\right)\left(\frac{1}{2}a\right)\left(\frac{1}{2}a\right)$
$= \frac{1}{8}a^3$

f. $\left(-\frac{1}{5}b\right)^2 = \left(-\frac{1}{5}b\right)\left(-\frac{1}{5}b\right)$
$= \frac{1}{25}b^2$

SELF CHECK 1 Write each expression without exponents. **a.** 3^4 **b.** $(-5)^3$ **c.** $\left(-\frac{3}{4}a\right)^2$

COMMENT Note the difference between $-x^n$ and $(-x)^n$.

$$-x^n = -(\overbrace{x \cdot x \cdot x \cdot \cdots \cdot x}^{n \text{ factors of } x}) \qquad (-x)^n = \overbrace{(-x)(-x)(-x) \cdot \cdots \cdot (-x)}^{n \text{ factors of } (-x)}$$

Also, note the difference between ax^n and $(ax)^n$.

$$ax^n = a \cdot \overbrace{x \cdot x \cdot x \cdot \cdots \cdot x}^{n \text{ factors of } x} \qquad (ax)^n = \overbrace{(ax)(ax)(ax) \cdot \cdots \cdot (ax)}^{n \text{ factors of } (ax)}$$

2 Simplify an expression by applying properties of exponents.

Since x^5 means that x is to be used as a factor five times, and since x^3 means that x is to be used as a factor three times, $x^5 \cdot x^3$ means that x will be used as a factor eight times.

$$x^5x^3 = \underbrace{x \cdot x \cdot x \cdot x \cdot x}_{5 \text{ factors of } x} \cdot \underbrace{x \cdot x \cdot x}_{3 \text{ factors of } x} = \underbrace{x \cdot x \cdot x \cdot x \cdot x \cdot x \cdot x \cdot x}_{8 \text{ factors of } x}$$

In general,

$$x^m x^n = \underbrace{x \cdot x \cdot x \cdot x \cdot \cdots \cdot x}_{m \text{ factors of } x} \cdot \underbrace{x \cdot x \cdot \cdots \cdot x}_{n \text{ factors of } x} = \underbrace{x \cdot x \cdot x \cdot x \cdot x \cdot x \cdot x \cdot \cdots \cdot x}_{m+n \text{ factors of } x}$$

Thus, *to multiply exponential expressions with the same base, we keep the same base and add the exponents.*

The Product Rule of Exponents

If m and n are natural numbers, then

$$x^m x^n = x^{m+n}$$

COMMENT The product rule of exponents applies only to exponential expressions with the same base. The expression x^5y^3, for example, cannot be simplified, because the bases of the exponential expressions are different.

EXAMPLE 2 Simplify each expression.

a. $x^{11}x^5 = x^{11+5}$
$= x^{16}$

b. $a^5a^4a^3 = (a^5a^4)a^3$
$= a^9a^3$
$= a^{12}$

c. $a^2b^3a^3b^2 = a^2a^3b^3b^2$
$= a^5b^5$

d. $-8x^4\left(\frac{1}{4}x^3\right) = \left(-8 \cdot \frac{1}{4}\right)(x^4x^3)$
$= -2x^7$

SELF CHECK 2 Simplify each expression.

a. a^3a^5 **b.** $a^2b^3a^3b^4$ **c.** $-8a^4\left(-\frac{1}{2}a^2b\right)$

To find another property of exponents, we simplify $(x^4)^3$, which means x^4 cubed or $x^4 \cdot x^4 \cdot x^4$.

$$(x^4)^3 = x^4 \cdot x^4 \cdot x^4 = \underbrace{x \cdot x \cdot x \cdot x}_{x^4} \cdot \underbrace{x \cdot x \cdot x \cdot x}_{x^4} \cdot \underbrace{x \cdot x \cdot x \cdot x}_{x^4} = x^{12}$$

In general, we have

$$(x^m)^n = \underbrace{x^m \cdot x^m \cdot x^m \cdot \cdots \cdot x^m}_{n \text{ factors of } x^m} = \underbrace{x \cdot x \cdot x \cdot x \cdot x \cdot \cdots \cdot x}_{mn \text{ factors of } x} = x^{mn}$$

Thus, *to raise an exponential expression to a power, we keep the same base and multiply the exponents.*

To find a third property of exponents, we square $3x$ and get

$$(3x)^2 = (3x)(3x) = 3 \cdot 3 \cdot x \cdot x = 3^2x^2 = 9x^2$$

In general, we have

$$(xy)^n = \overbrace{(xy)(xy)(xy)\cdot\cdots\cdot(xy)}^{n \text{ factors of } xy} = \overbrace{xxx\cdot\cdots\cdot x}^{n \text{ factors of } x}\cdot\overbrace{yyy\cdot\cdots\cdot y}^{n \text{ factors of } y} = x^n y^n$$

To find a fourth property of exponents, we cube $\frac{x}{3}$ to get

$$\left(\frac{x}{3}\right)^3 = \frac{x}{3}\cdot\frac{x}{3}\cdot\frac{x}{3} = \frac{x\cdot x\cdot x}{3\cdot 3\cdot 3} = \frac{x^3}{3^3} = \frac{x^3}{27}$$

In general, we have

$$\left(\frac{x}{y}\right)^n = \overbrace{\left(\frac{x}{y}\right)\left(\frac{x}{y}\right)\left(\frac{x}{y}\right)\cdot\cdots\cdot\left(\frac{x}{y}\right)}^{n \text{ factors of } x/y} \quad (y \neq 0)$$

$$= \frac{\overbrace{xxx\cdot\cdots\cdot x}^{n \text{ factors of } x}}{\underbrace{yyy\cdot\cdots\cdot y}_{n \text{ factors of } y}}$$

Multiply the numerators and multiply the denominators.

$$= \frac{x^n}{y^n}$$

The previous three results are called the *power rules of exponents.*

The Power Rules of Exponents

If m and n are natural numbers, then

$$(x^m)^n = x^{mn} \qquad (xy)^n = x^n y^n \qquad \left(\frac{x}{y}\right)^n = \frac{x^n}{y^n} \quad (y \neq 0)$$

EXAMPLE 3 Simplify each expression.

a. $(3^2)^3 = 3^{2\cdot 3}$
$= 3^6$
$= 729$

b. $(x^{11})^5 = x^{11\cdot 5}$
$= x^{55}$

c. $(x^2x^3)^6 = (x^5)^6$
$= x^{30}$

d. $(x^2)^4(x^3)^2 = x^8x^6$
$= x^{14}$

SELF CHECK 3 Simplify each expression. **a.** $(a^5)^8$ **b.** $(a^4a^3)^3$ **c.** $(a^3)^3(a^2)^3$

EXAMPLE 4 Simplify each expression. Assume that no denominators are zero.

a. $(x^2y)^3 = (x^2)^3y^3$
$= x^6y^3$

b. $(x^3y^4)^4 = (x^3)^4(y^4)^4$
$= x^{12}y^{16}$

c. $\left(\frac{x}{y^2}\right)^4 = \frac{x^4}{(y^2)^4}$
$= \frac{x^4}{y^8}$

d. $\left(\frac{x^3}{y^4}\right)^2 = \frac{(x^3)^2}{(y^4)^2}$
$= \frac{x^6}{y^8}$

SELF CHECK 4 Simplify each expression. Assume that no denominators are zero.

a. $(a^4b^5)^2$ **b.** $\left(\dfrac{a^5}{b^7}\right)^3$

We can expand the rules for exponents hold for exponents of 0.

$$x^0x^n = x^{0+n} = x^n = 1x^n$$

Because $x^0x^n = 1x^n$, it follows that $x^0 = 1 \quad (x \neq 0)$.

Zero Exponents If $x \neq 0$, then $x^0 = 1$.

COMMENT 0^0 is undefined.

Because of the previous definition, any nonzero base raised to the 0 power is 1. For example, if no variables are zero, then

$$5^0 = 1, \qquad (-7)^0 = 1, \qquad (3ax^3)^0 = 1, \qquad \left(\tfrac{1}{2}x^5y^7z^9\right)^0 = 1$$

We can expand the rules for exponents to include negative integer exponents.

$$x^{-n}x^n = x^{-n+n} = x^0 = 1 \quad (x \neq 0)$$

Because $x^{-n} \cdot x^n = 1$ and $\frac{1}{x^n} \cdot x^n = 1$, we define x^{-n} to be the reciprocal of x^n.

Negative Exponents If n is an integer and $x \neq 0$, then

$$x^{-n} = \frac{1}{x^n} \quad \text{and} \quad \frac{1}{x^{-n}} = x^n$$

COMMENT By the definition of negative exponents, the base cannot be 0. Thus, an expression such as 0^{-5} is undefined.

Because of this definition, we can write expressions containing negative exponents as expressions without negative exponents. For example,

$$5^{-2} = \frac{1}{5^2} = \frac{1}{25} \qquad 10^{-3} = \frac{1}{10^3} = \frac{1}{1{,}000}$$

and if $x \neq 0$, we have

$$(2x)^{-3} = \frac{1}{(2x)^3} = \frac{1}{8x^3} \qquad 3x^{-1} = 3 \cdot \frac{1}{x} = \frac{3}{x}$$

EXAMPLE 5 Write each expression without negative exponents. Assume that $x \neq 0$.

a. $x^{-5}x^3 = x^{-5+3}$
$= x^{-2}$
$= \dfrac{1}{x^2}$

b. $(x^{-3})^{-2} = x^{(-3)(-2)}$
$= x^6$

SELF CHECK 5 Write each expression without negative exponents. Assume that $a \neq 0$.

a. $a^{-7}a^3$ **b.** $(a^{-5})^{-3}$

To develop a rule for dividing exponential expressions, we proceed as follows:

$$\frac{x^m}{x^n} = x^m\left(\frac{1}{x^n}\right) = x^m x^{-n} = x^{m+(-n)} = x^{m-n}$$

Thus, *to divide exponential expressions with the same nonzero base, we keep the same base and subtract the exponent in the denominator from the exponent in the numerator.*

The Quotient Rule

If m and n are integers, then

$$\frac{x^m}{x^n} = x^{m-n} \quad (x \neq 0)$$

EXAMPLE 6 Simplify each expression. Assume no variable is 0.

a. $\dfrac{a^5}{a^3} = a^{5-3}$
$= a^2$

b. $\dfrac{x^{-5}}{x^{11}} = x^{-5-11}$
$= x^{-16}$
$= \dfrac{1}{x^{16}}$

c. $\dfrac{x^4x^3}{x^{-5}} = \dfrac{x^7}{x^{-5}}$
$= x^{7-(-5)}$
$= x^{12}$

d. $\dfrac{(x^2)^3}{(x^3)^2} = \dfrac{x^6}{x^6}$
$= x^{6-6}$
$= x^0$
$= 1$

e. $\dfrac{x^2y^3}{xy^4} = x^{2-1}y^{3-4}$
$= xy^{-1}$
$= \dfrac{x}{y}$

f. $\left(\dfrac{a^{-2}b^3}{a^2a^3b^4}\right)^3 = \left(\dfrac{a^{-2}b^3}{a^5b^4}\right)^3$
$= (a^{-2-5}b^{3-4})^3$
$= (a^{-7}b^{-1})^3$
$= \left(\dfrac{1}{a^7b}\right)^3$
$= \dfrac{1}{a^{21}b^3}$

SELF CHECK 6 Simplify each expression. Assume that no variables are 0.

a. $\dfrac{(a^{-2})^3}{(a^2)^{-3}}$ **b.** $\left(\dfrac{a^{-2}b^5}{b^8}\right)^{-3}$

To illustrate one more property of exponents, we consider the simplification of $\left(\frac{2}{3}\right)^{-4}$.

$$\left(\frac{2}{3}\right)^{-4} = \frac{1}{\left(\frac{2}{3}\right)^4} = \frac{1}{\frac{2^4}{3^4}} = 1 \div \frac{2^4}{3^4} = 1 \cdot \frac{3^4}{2^4} = \frac{3^4}{2^4} = \left(\frac{3}{2}\right)^4$$

This example suggests that to raise a fraction to a negative power, we can invert the fractional base (take the reciprocal) and then raise it to a positive power.

Fractions to Negative Powers

If n is an integer, then

$$\left(\frac{x}{y}\right)^{-n} = \left(\frac{y}{x}\right)^{n} = \frac{y^n}{x^n} \quad (x \neq 0, y \neq 0)$$

EXAMPLE 7 Write each expression without using parentheses. Write answers without negative exponents.

a. $\left(\frac{3}{5}\right)^{-4} = \left(\frac{5}{3}\right)^{4}$
$= \frac{625}{81}$

b. $\left(\frac{y^2}{x^3}\right)^{-3} = \left(\frac{x^3}{y^2}\right)^{3}$
$= \frac{x^9}{y^6}$

c. $\left(\frac{2x^2}{3y^{-3}}\right)^{-4} = \left(\frac{3y^{-3}}{2x^2}\right)^{4}$
$= \frac{81y^{-12}}{16x^8}$
$= \frac{81}{16x^8} \cdot y^{-12}$
$= \frac{81}{16x^8} \cdot \frac{1}{y^{12}}$
$= \frac{81}{16x^8y^{12}}$

d. $\left(\frac{a^{-2}b^3}{a^2a^3b^4}\right)^{-3} = \left(\frac{a^2a^3b^4}{a^{-2}b^3}\right)^{3}$
$= \left(\frac{a^5b^4}{a^{-2}b^3}\right)^{3}$
$= (a^{5-(-2)}b^{4-3})^3$
$= (a^7b)^3$
$= a^{21}b^3$

SELF CHECK 7 Write $\left(\frac{3a^3}{2b^{-2}}\right)^{-5}$ without using parentheses.

We summarize the rules of exponents as follows.

Properties of Exponents

If there are no divisions by 0, then for all integers m and n,

$$x^mx^n = x^{m+n} \qquad (x^m)^n = x^{mn} \qquad (xy)^n = x^ny^n \qquad \left(\frac{x}{y}\right)^n = \frac{x^n}{y^n}$$

$$x^0 = 1 \quad (x \neq 0) \qquad x^{-n} = \frac{1}{x^n} \qquad \frac{x^m}{x^n} = x^{m-n} \qquad \left(\frac{x}{y}\right)^{-n} = \left(\frac{y}{x}\right)^{n} = \frac{y^n}{x^n}$$

The same rules apply to exponents that are variables.

EXAMPLE 8 Simplify each expression. Assume that $a \neq 0$ and $x \neq 0$.

a. $\frac{a^na}{a^2} = a^{n+1-2}$
$= a^{n-1}$

b. $\frac{x^3x^2}{x^n} = x^{3+2-n}$
$= x^{5-n}$

c. $\left(\dfrac{x^n}{x^2}\right)^2 = \dfrac{x^{2n}}{x^4}$
$= x^{2n-4}$

d. $\dfrac{a^n a^{-3}}{a^{-1}} = a^{n+(-3)-(-1)}$
$= a^{n-3+1}$
$= a^{n-2}$

SELF CHECK 8 Simplify each expression. Assume that $t \neq 0$.

a. $\dfrac{t^n t^2}{t^3}$ **b.** $\left(\dfrac{2t^n}{3t^3}\right)^3$

ACCENT ON TECHNOLOGY

Finding Powers

To find powers of numbers with many calculators, we use the y^x key. For example, to find 5.37^4, we enter these numbers and press these keys:

5.37 y^x 4 = Some calculators have an x^y key.

The display will read 831.5668016.

To use a graphing calculator, we enter these numbers and press these keys:

5.37 ∧ 4 ENTER

The display will read 5.37^4 831.5668016.

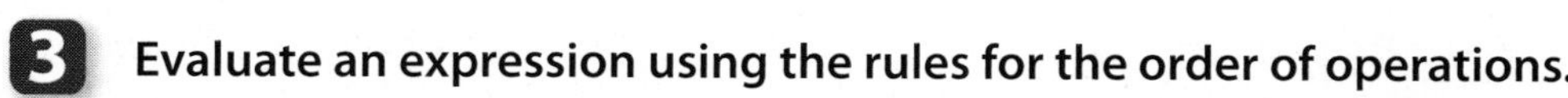

If these methods do not work, consult your owner's manual.

3 Evaluate an expression using the rules for the order of operations.

When simplifying expressions containing exponents, we find powers before performing additions and multiplications. Below we repeat the rules for order of operations, first discussed in the previous section, with a step included for exponents.

Rules for the Order of Operations

Use the following steps to perform all calculations within each pair of grouping symbols, working from the innermost pair to the outermost pair.

1. Find the values of any exponential expressions.
2. Perform all multiplications and divisions, working from left to right.
3. Perform all additions and subtractions, working from left to right.
4. Because a fraction bar is a grouping symbol, simplify the numerator and the denominator separately and simplify the fraction, whenever possible.

EXAMPLE 9 If $x = 2$ and $y = -3$, find the value of $3x + 2y^3$.

Solution

$3x + 2y^3 = 3(2) + 2(-3)^3$ Substitute 2 for x and -3 for y.
$= 3(2) + 2(-27)$ First find the power: $(-3)^3 = -27$.
$= 6 - 54$ Then do the multiplications.
$= -48$ Then do the subtraction.

SELF CHECK 9 Evaluate $-2a^2 - 3a$ if $a = -4$.

4 Apply the correct geometric formula to solve an application problem.

Table 1-4 shows the formulas used to compute the areas and volumes of many geometric figures. Remember that **area** is defined in square units and **volume** is defined in cubic units.

EXAMPLE 10 **VOLUME OF SPHERE** Find the volume of the sphere shown in Figure 1-20.

Solution The formula for the volume of a sphere is $V = \frac{4}{3}\pi r^3$. Since a radius is half as long as a diameter, the radius of the sphere is half of 20 centimeters, or 10 centimeters.

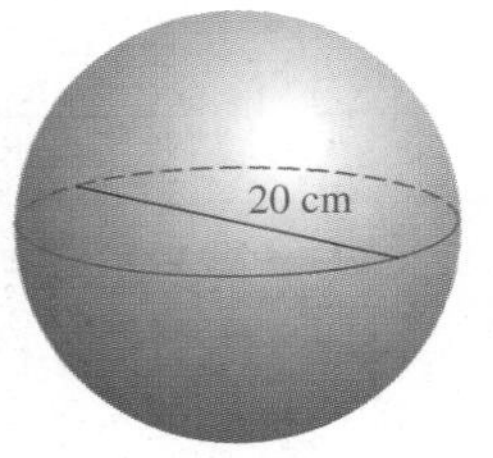

Figure 1-20

$$V = \frac{4}{3}\pi r^3$$

$$V = \frac{4}{3}\pi(10)^3 \quad \text{Substitute 10 for } r.$$

$$\approx 4188.790205 \quad \text{Use a calculator.}$$

To two decimal places, the volume is 4,188.79 cm^3.

SELF CHECK 10 Find the volume of a pyramid with a square base, 20 meters on each side, and a height of 21 meters.

Figure	Name	Area	Figure	Name	Volume
	Square	$A = s^2$		Cube	$V = s^3$
	Rectangle	$A = lw$		Rectangular solid	$V = lwh$
	Circle	$A = \pi r^2$		Sphere	$V = \frac{4}{3}\pi r^3$
	Triangle	$A = \frac{1}{2}bh$		Cylinder	$V = Bh$*
	Trapezoid	$A = \frac{1}{2}h(b_1 + b_2)$		Cone	$V = \frac{1}{3}Bh$*
				Pyramid	$V = \frac{1}{3}Bh$*

*B represents the area of the base.

Table 1-4

SELF CHECK ANSWERS **1. a.** 81 **b.** −125 **c.** $\frac{9}{16}a^2$ **2. a.** a^8 **b.** a^5b^7 **c.** $4a^6b$ **3. a.** a^{40} **b.** a^{21} **c.** a^{15} **4. a.** a^8b^{10} **b.** $\frac{a^{15}}{b^{21}}$ **5. a.** $\frac{1}{a^4}$ **b.** a^{15} **6. a.** 1 **b.** a^6b^9 **7.** $\frac{32}{243a^{15}b^{10}}$ **8. a.** t^{n-1} **b.** $\frac{8t^{3n-9}}{27}$ **9.** −20 **10.** 2,800 m^3

NOW TRY THIS

Let $a = -3$, $b = -2$, $c = 5$, and $d = -4$. Evaluate each expression.

1. $\dfrac{a-b}{c-d}$ **2.** $b^2 - 4ac$ **3.** $(c-a)^2 + (d-b)^2$

1.3 EXERCISES

WARM-UPS *Simplify each expression. Assume no variables are zero.*

1. 4^2
2. 3^3
3. x^2x^3
4. y^3y^4
5. 17^0
6. $(x^2)^3$
7. $(a^2b)^3$
8. $\left(\dfrac{b}{a^2}\right)^2$
9. 5^{-2}
10. $(x^{-2})^{-1}$
11. $\dfrac{x^5}{x^2}$
12. $\dfrac{x^2}{x^5}$

REVIEW *If a = 4, b = −2, and c = 5, find each value.*

13. $a + b + c$
14. $a - 2b - c$
15. $\dfrac{ab + 2c}{a + b}$
16. $\dfrac{ac - bc}{6ab + b}$

VOCABULARY AND CONCEPTS

Fill in the blanks.

17. In the exponential expression x^n, x is called the ____, and n is called the ________ or ______.

18. A natural-number exponent tells how many times the base is used as a ______.

19. $x^mx^n =$ _____
20. $(x^m)^n =$ ____
21. $(xy)^n =$ _____
22. $\left(\dfrac{x}{y}\right)^n = \dfrac{\_\_}{\_\_}$ $(y \neq 0)$
23. If $a \neq 0$, then $a^0 =$ __.
24. If $a \neq 0$, then $a^{-1} =$ __.
25. If $x \neq 0$, then $\dfrac{x^m}{x^n} =$ _____.
26. $\left(\dfrac{4}{5}\right)^{-3} = \left(\dfrac{}{\_}\right)^3$.

Write the formula to find each quantity.

27. Area of a square: _______
28. Area of a rectangle: _______
29. Area of a triangle: ________
30. Area of a trapezoid: _____________
31. Area of a circle: ________
32. Volume of a cube: _______
33. Volume of a rectangular solid: ________
34. Volume of a sphere: _________
35. Volume of a cylinder: ________
36. Volume of a cone: _________
37. Volume of a pyramid: _________
38. In Exercises 35–37, B represents the area of the _____ of a solid.

Identify the base and the exponent of each expression.

39. 5^3
40. -7^2
41. $-x^5$
42. $(-t)^4$
43. $2b^6$
44. $(3xy)^5$
45. $(-mn^2)^3$
46. $(-p^2q)^2$

GUIDED PRACTICE

Simplify each expression. See Example 1. (Objective 1)

47. 3^2
48. 3^4
49. -3^2
50. -3^4
51. $(-3)^2$
52. $(-3)^3$
53. $(-2x)^5$
54. $(-3a)^3$

Simplify each expression. Assume that no variables are zero. See Example 2. (Objective 2)

55. x^2x^3
56. y^3y^4

57. k^0k^7

58. x^8x^{11}

59. $x^2x^3x^5$

60. $y^3y^7y^2$

61. p^9pp^0

62. z^7z^0z

Simplify each expression. Assume that no variables are zero. See Example 3. (Objective 2)

63. $(-x)^2y^4x^3$

64. $-x^2y^7y^3x^{-2}$

65. $(x^4)^7$

66. $(y^7)^5$

67. $(a^2a^3)^4$

68. $(bb^2b^3)^4$

69. $(y^3y^{-1})^4$

70. $(z^{-2}z^5)^3$

Simplify each expression. Assume that no denominators are zero. See Example 4. (Objective 2)

71. $(-2x)^6$

72. $(-3y)^5$

73. $(x^3y^2)^4$

74. $(x^2y^5)^2$

75. $\left(\dfrac{a^3}{b^2}\right)^5$

76. $\left(\dfrac{a^2}{b^3}\right)^4$

77. $\left(\dfrac{1}{2}a^2b^5\right)^4$

78. $\left(-\dfrac{1}{3}mn^2\right)^6$

Simplify each expression. Assume that no variables are zero. (Objective 2)

79. 8^0

80. -9^0

81. $(-8)^0$

82. $(-9)^0$

83. $5x^0$

84. $-3x^0$

85. $7^0 + 7^2$

86. $5^2 - 5^0$

Simplify each expression. Assume that no variables are zero. See Example 5. (Objective 2)

87. 5^{-2}

88. 5^{-4}

89. -5^{-2}

90. -5^{-4}

91. $(-5)^{-2}$

92. $(-5)^{-4}$

93. $\dfrac{1}{a^{-4}}$

94. $\dfrac{3}{b^{-5}}$

Simplify each expression. Assume that no denominators are zero. See Example 6. (Objective 2)

95. $\dfrac{a^8}{a^3}$

96. $\dfrac{c^7}{c^2}$

97. $\dfrac{c^{12}c^5}{c^{10}}$

98. $\dfrac{a^{33}}{a^2a^3}$

99. $\dfrac{3m^5m^{-7}}{m^2m^{-5}}$

100. $\dfrac{(2a^{-2})^3}{a^3a^{-4}}$

101. $\dfrac{m^9m^{-2}}{(m^2)^3}$

102. $\dfrac{a^{10}a^{-3}}{a^5a^{-2}}$

Simplify each expression. Assume that no variables are zero. See Example 7. (Objective 2)

103. $\left(\dfrac{a^{-3}}{b^{-2}}\right)^{-2}$

104. $\left(\dfrac{k^{-3}}{k^{-4}}\right)^{-1}$

105. $\left(\dfrac{3x^5y^2}{6x^5y^{-2}}\right)^{-4}$

106. $\left(\dfrac{2a^3b^2}{3a^{-3}b^2}\right)^{-3}$

Simplify each expression. Assume that no variables are zero. See Example 8. (Objective 2)

107. $\dfrac{a^na^3}{a^4}$

108. $\dfrac{b^9b^7}{b^n}$

109. $\left(\dfrac{b^n}{b^3}\right)^3$

110. $\left(\dfrac{a^2}{a^n}\right)^4$

Evaluate each expression when $x = -2$ and $y = 3$. See Example 9. (Objective 3)

111. x^2y^3

112. x^3y^2

113. $\dfrac{x^{-3}}{y^3}$

114. $\dfrac{x^2}{y^{-3}}$

115. $(xy^2)^{-2}$

116. $-y^3x^{-2}$

117. $(-yx^{-1})^3$

118. $(-y)^3x^{-2}$

Find the area of each figure. Round all answers to the nearest unit. See Example 10. (Objective 4)

119.

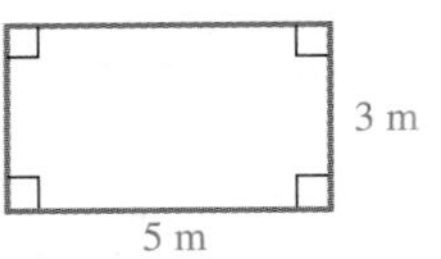

120.

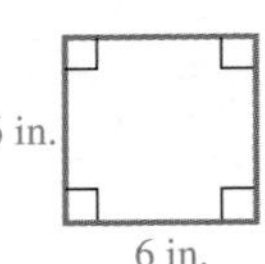

121.

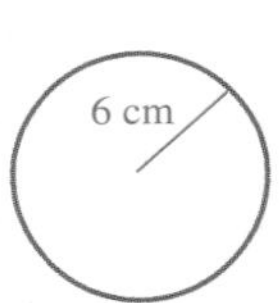

122.

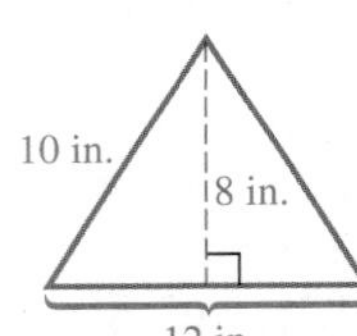

123.

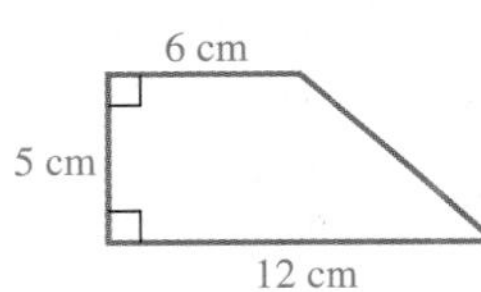

124.

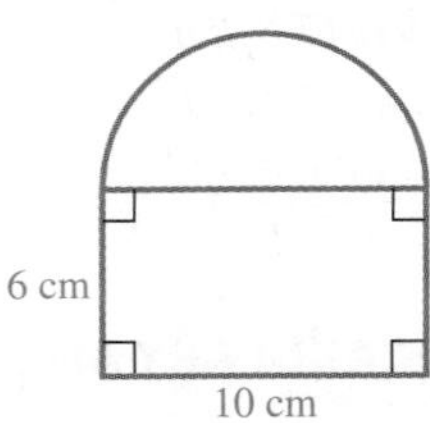

125.

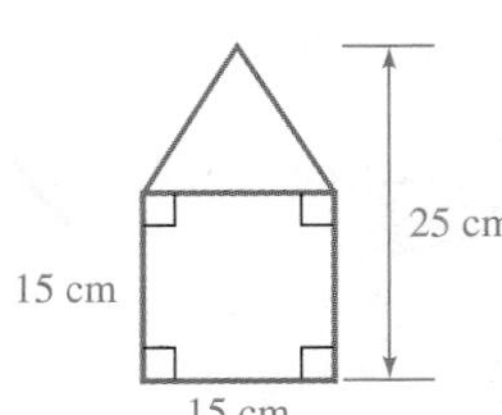

126.

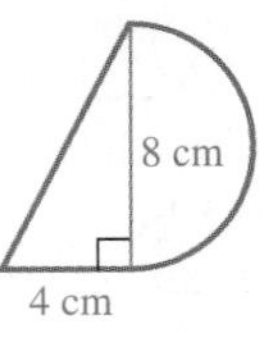

Find the volume of each figure. Round all answers to the nearest unit. See Example 10. (Objective 4)

127.

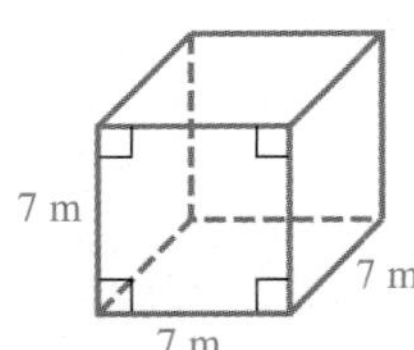

128.

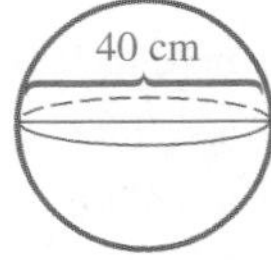

129.

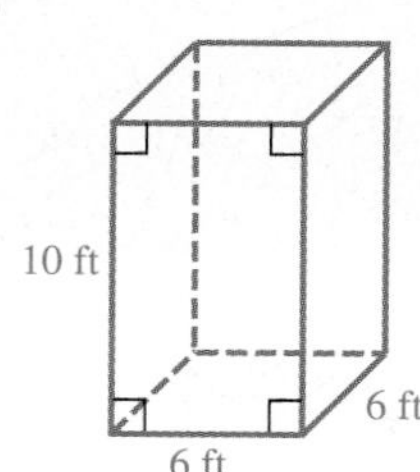

130.

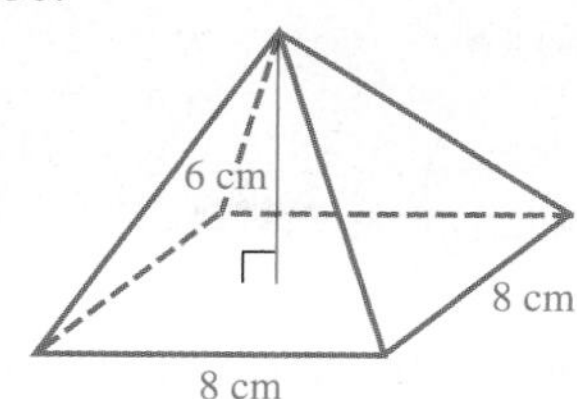

131.

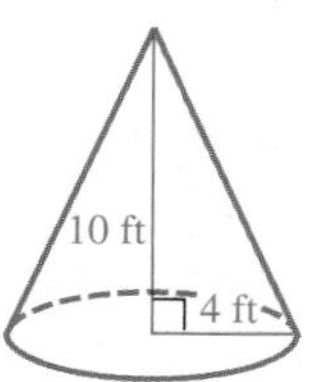

132.

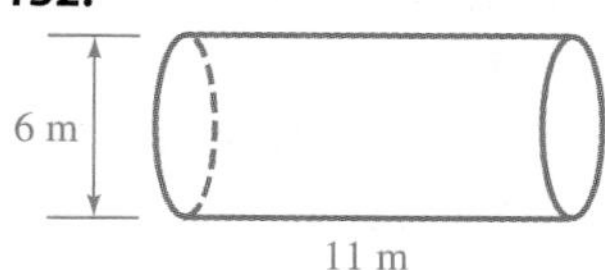

133.

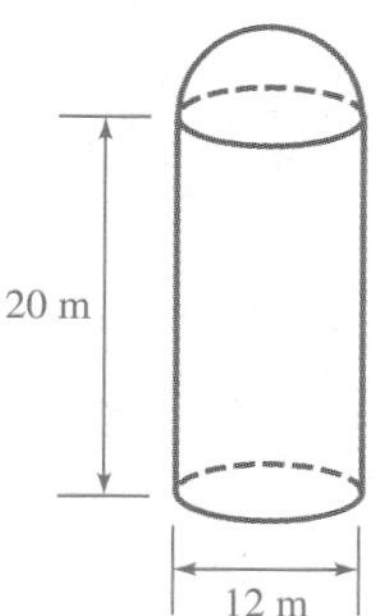

134.

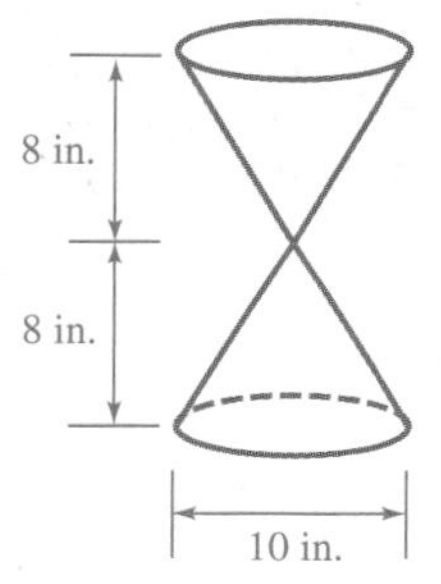

ADDITIONAL PRACTICE *Simplify each expression. Assume that no variables are zero.*

135. aba^3b^4

136. $x^2y^3x^3y^2$

137. $(-3p^2q^3)^5$

138. $(3x^3y^4)^3$

139. $(x^2)^3(x^3)^2$

140. $(z^{12})^2$

141. $(r^{-3}s)^3$

142. $(m^5n^2)^{-3}$

143. $(b^{-8})^9$

144. $(a^{-2})^4(a^3)^2$

145. $(-d^2)^3(d^{-3})^3$

146. $(c^3)^2(c^4)^{-2}$

147. $\left(\dfrac{3a^{-2}b^2}{17a^2b^3}\right)^0$

148. $\dfrac{a^0 + b^0}{2(a + b)^0}$

149. $\dfrac{(3x^2)^{-2}}{x^3x^{-4}x^0}$

150. $\dfrac{y^{-3}y^{-4}y^0}{(2y^{-2})^3}$

151. $\dfrac{a^{-n}a^2}{a^3}$

152. $\dfrac{a^na^{-2}}{a^4}$

153. $\dfrac{a^{-n}a^{-2}}{a^{-4}}$

154. $\dfrac{a^n}{a^{-3}a^5}$

155. $\left(\dfrac{4a^{-2}b}{3ab^{-3}}\right)^3$

156. $\left(\dfrac{2ab^{-3}}{3a^{-2}b^2}\right)^2$

157. $\left(\dfrac{-2a^4b}{a^{-3}b^2}\right)^{-3}$

158. $\left(\dfrac{-3x^4y^2}{-9x^5y^{-2}}\right)^{-2}$

Use a calculator to find each value.

159. 1.23^6

160. 0.537^4

161. -6.25^3

162. $(-25.1)^5$

Use a calculator to verify that each statement is true.

163. $(3.68)^0 = 1$

164. $(2.1)^4(2.1)^3 = (2.1)^7$

165. $(7.2)^2(2.7)^2 = [(7.2)(2.7)]^2$

166. $(3.7)^2 + (4.8)^2 \neq (3.7 + 4.8)^2$

167. $(3.2)^2(3.2)^{-2} = 1$

168. $[(5.9)^3]^2 = (5.9)^6$

169. $(7.23)^{-3} = \dfrac{1}{(7.23)^3}$

170. $\left(\dfrac{5.4}{2.7}\right)^{-4} = \left(\dfrac{2.7}{5.4}\right)^4$

APPLICATIONS *Use a calculator to find each value.*

171. Bank accounts The formula $A = P(1 + i)^n$ gives the amount A in an account when P dollars is the amount originally deposited (the principal), i is the annual interest rate, and n is the number of years. If \$5,000 is deposited in an account paying 11% compounded annually, how much will be in the account in 50 years?

172. Bank accounts The formula $P = A(1 + i)^{-n}$ gives the principal P that must be deposited at an annual rate i to grow to A dollars in n years. How much must be invested at 9% annual interest to have \$1 million in 50 years?

WRITING ABOUT MATH

173. Explain why a positive number raised to a negative power is positive.

174. Explain the rules that determine the order in which operations are performed.

175. In the definition of x^{-1}, x cannot be 0. Why not?

176. Explain why $(xyz)^2 = x^2y^2z^2$.

SOMETHING TO THINK ABOUT

177. Find the value: $2^{-1} + 3^{-1} - 4^{-1}$.

178. Simplify: $(3^{-1} + 4^{-1})^{-2}$.

179. Construct an example using numbers to show that $x^m + x^n \neq x^{m+n}$.

180. Construct an example using numbers to show that $x^m + y^m \neq (x + y)^m$.

SECTION 1.4 Scientific Notation

Objectives

1. Write a number in scientific notation given a number in standard notation.
2. Write a number in standard notation given a number in scientific notation.
3. Use scientific notation to simplify computations of very large or very small numbers.
4. Solve problems with numbers in scientific notation requiring answers rounded to the proper significant digit.

Vocabulary

scientific notation standard notation

Getting Ready

Evaluate each expression:

1. 10^1	**2.** 10^2	**3.** 10^3	**4.** 10^4
5. 10^{-2}	**6.** 10^{-4}	**7.** $4(10^3)$	**8.** $7(10^{-4})$

Very large and very small numbers occur often in science. For example, the speed of light is approximately 29,980,000,000 centimeters per second, and the mass of a hydrogen atom is approximately 0.000000000000000000000001673 gram.

Because these numbers contain a large number of zeros, they are hard to read and remember. In this section, we will discuss a notation that will enable us to express these numbers in a more compact form.

1 Write a number in scientific notation given a number in standard notation.

With exponents, we can write very large and very small numbers in a form called *scientific notation.*

Scientific Notation A number is written in **scientific notation** when it is written in the form $N \times 10^n$, where $1 \le |N| < 10$ and n is an integer.

EXAMPLE 1 Change 29,980,000,000 to scientific notation.

Solution The number 2.998 is between 1 and 10. To get 29,980,000,000, the decimal point in 2.998 must be moved ten places to the right. We can do this by multiplying 2.998 by 10^{10}.

$$29{,}980{,}000{,}000 = 2.998 \times 10^{10}$$

SELF CHECK 1 Change 150,000,000 to scientific notation.

EXAMPLE 2 Write 0.000000000000000000000001673 in scientific notation.

Solution The number 1.673 is between 1 and 10. To get 0.000000000000000000000001673, the decimal point in 1.673 must be moved twenty-four places to the left. We can do this by multiplying 1.673 by 10^{-24}.

$$0.000000000000000000000001673 = 1.673 \times 10^{-24}$$

SELF CHECK 2 Change 0.000025 to scientific notation.

EXAMPLE 3 Change -0.0013 to scientific notation.

Solution The absolute value of -1.3 is between 1 and 10. To get -0.0013, we move the decimal point in -1.3 three places to the left by multiplying by 10^{-3}.

$$-0.0013 = -1.3 \times 10^{-3}$$

SELF CHECK 3 Change $-45{,}700$ to scientific notation.

2 Write a number in standard notation given a number in scientific notation.

We can change a number written in scientific notation to **standard notation.** For example, to write 9.3×10^7 in standard notation, we multiply 9.3×10^7.

$$9.3 \times 10^7 = 9.3 \times 10{,}000{,}000 = 93{,}000{,}000$$

EXAMPLE 4 Change **a.** 3.7×10^5 and **b.** -1.1×10^{-3} to standard notation.

Solution **a.** Since multiplication by 10^5 moves the decimal point 5 places to the right,

$$3.7 \times 10^5 = 370{,}000$$

b. Since multiplication by 10^{-3} moves the decimal point 3 places to the left,

$$-1.1 \times 10^{-3} = -0.0011$$

SELF CHECK 4 Change each number to standard notation.
a. -9.6×10^4 **b.** 5.62×10^{-3}

Each of the following numbers is written in both scientific and standard notation. In each case, the exponent gives the number of places that the decimal point moves, and the sign of the exponent indicates the direction that it moves:

$5.32 \times 10^4 = 53200.$ — 4 places to the right

$6.45 \times 10^7 = 64500000.$ — 7 places to the right

$2.37 \times 10^{-4} = 0.000237$ — 4 places to the left

$9.234 \times 10^{-2} = 0.09234$ — 2 places to the left

$4.89 \times 10^0 = 4.89$ — No movement of the decimal point

COMMENT Numbers such as 47.2×10^3 and 0.063×10^{-2} appear to be written in scientific notation, because they are the product of a number and a power of 10. However, they are not in scientific notation, because 47.2 and 0.063 are not between 1 and 10.

EXAMPLE 5 Change **a.** 47.2×10^3 and **b.** 0.063×10^{-2} to scientific notation.

Solution Since the first factors are not between 1 and 10, neither number is in scientific notation. However, we can change them to scientific notation as follows:

a.
$$\begin{aligned} 47.2 \times 10^3 &= (4.72 \times 10^1) \times 10^3 && \text{Write 47.2 in scientific notation.} \\ &= 4.72 \times (10^1 \times 10^3) \\ &= 4.72 \times 10^4 \end{aligned}$$

b.
$$\begin{aligned} 0.063 \times 10^{-2} &= (6.3 \times 10^{-2}) \times 10^{-2} && \text{Write 0.063 in scientific notation.} \\ &= 6.3 \times (10^{-2} \times 10^{-2}) \\ &= 6.3 \times 10^{-4} \end{aligned}$$

SELF CHECK 5 Change each number to scientific notation.
a. 27.3×10^2 **b.** 0.0025×10^{-3}

PERSPECTIVE

The ancient Egyptians developed two systems of writing. In hieroglyphics, each symbol was a picture of an object. Because hieroglyphic writing usually was inscribed in stone, many examples still survive today. For daily life, Egyptians used hieratic writing. Similar to hieroglyphics, hieratic writing was done with ink on papyrus sheets.

One papyrus that survives, the Rhind Papyrus, was discovered in 1858 by a British archaeologist, Henry Rhind. Also known as the Ahmes Papyrus after its ancient author, it begins with a description of its contents: *Directions for Obtaining the Knowledge of All Dark Things.*

The Ahmes Papyrus

The Ahmes Papyrus and another, the Moscow Papyrus, together contain 110 mathematical problems and their solutions. Many of these were probably for education, because they represented situations that scribes, priests, and other government and temple administration workers were expected to be able to solve.

3 Use scientific notation to simplify computations of very large or very small numbers.

Scientific notation is useful when simplifying expressions containing very large or very small numbers.

EXAMPLE 6 Use scientific notation to simplify $\dfrac{(0.00000064)(24{,}000{,}000{,}000)}{(400{,}000{,}000)(0.0000000012)}$.

Solution After changing each number into scientific notation, we can do the arithmetic on the numbers and the exponential expressions separately.

$$\begin{aligned}\frac{(0.00000064)(24{,}000{,}000{,}000)}{(400{,}000{,}000)(0.0000000012)} &= \frac{(6.4 \times 10^{-7})(2.4 \times 10^{10})}{(4 \times 10^{8})(1.2 \times 10^{-9})}\\ &= \frac{(6.4)(2.4)}{(4)(1.2)} \times \frac{10^{-7}10^{10}}{10^{8}10^{-9}}\\ &= 3.2 \times 10^{4}\end{aligned}$$

In standard notation, the result is 32,000.

SELF CHECK 6 Simplify: $\dfrac{(320)(25{,}000)}{0.00004}$.

ACCENT ON TECHNOLOGY

Using Scientific Notation

Gottfried Wilhelm Leibniz (1646–1716)
Leibniz, a German philosopher and logician, is principally known as one of the inventors of calculus, along with Newton. He also developed the binary numeration system, which is basic to modern computers.

Scientific and graphing calculators often give answers in scientific notation. For example, if we use a calculator to find 301.2^8, the display will read

`6.77391496 19` Using a scientific calculator

`301.2 ^8`
`6.773914961E19` Using a graphing calculator

In either case, the answer is given in scientific notation and is to be interpreted as

$6.77391496 \times 10^{19}$

Numbers also can be entered into a calculator in scientific notation. For example, to enter 24,000,000,000 (which is 2.4×10^{10} in scientific notation), we enter these numbers and press these keys:

2.4 **EXP** 10
2.4 **EE** 10
} Whichever of these keys is on your calculator

To use a calculator to simplify

$$\frac{(24{,}000{,}000{,}000)(0.00000006495)}{0.00000004824}$$

we can enter each number in scientific notation. In scientific notation, the three numbers are

2.4×10^{10} $\quad 6.495 \times 10^{-8}$ $\quad 4.824 \times 10^{-8}$

(continued)

ACCENT ON TECHNOLOGY

Using Scientific Notation, continued

To use a scientific calculator to simplify the fraction, we enter these numbers and press these keys:

2.4 **EXP** 10 × 6.495 **EXP** 8 +/− ÷ 4.824 **EXP** 8 +/− =

The display will read **3.231343284** ^{10}. In standard notation, the answer is 32,313,432,840.

The steps are similar using a graphing calculator.

4 Solve problems with numbers in scientific notation requiring answers rounded to the proper significant digit.

If we measure the length of a rectangle and report the length to be 45 centimeters, we have rounded to the nearest centimeter. If we measure more carefully and find the length to be 45.2 centimeters, we have rounded to the nearest tenth of a centimeter. We say that the second measurement is more accurate than the first, because 45.2 has three *significant digits* but 45 has only two.

It is not always easy to know how many significant digits a number has. For example, 270 might be accurate to two or three significant digits. If 270 is rounded to the nearest ten, the number has two significant digits. If 270 is rounded to the nearest unit, it has three significant digits. This ambiguity does not occur when a number is written in scientific notation.

Finding Significant Digits

If a number M is written in scientific notation as $N \times 10^n$, where $1 \le |N| < 10$ and n is an integer, the number of significant digits in M is the same as the number of digits in N.

In a problem in which measurements are multiplied or divided, the final result should be rounded so that the answer has the same number of significant digits as the least accurate measurement.

EXAMPLE 7 **ASTRONOMY** Earth is approximately 93,000,000 miles from the Sun, and Jupiter is approximately 484,000,000 miles from the Sun. Assuming the alignment shown in Figure 1-21, how long would it take a spaceship traveling at 7,500 mph to fly from Earth to Jupiter?

Solution When the planets are aligned as shown in the figure, the distance between Earth and Jupiter is (484,000,000 − 93,000,000) miles or 391,000,000 miles. To find the length of time in hours for the trip, we divide the distance by the rate.

$$\frac{391{,}000{,}000 \text{ mi}}{7{,}500 \frac{\text{mi}}{\text{hr}}} = \frac{3.91 \times 10^8 \text{ mi}}{7.5 \times 10^3 \frac{\text{mi}}{\text{hr}}}$$

There are three significant digits in the numerator and two in the denominator.

$$\approx 0.5213333 \times 10^5 \text{ mi} \cdot \frac{\text{hr}}{\text{mi}}$$

$$\approx 52{,}133.33 \text{ hr}$$

Since there are 24 × 365 hours in a year, we can change this result from hours to years by dividing 52,133.33 by (24 × 365).

$$\frac{52{,}133.33 \text{ hr}}{(24 \times 365) \frac{\text{hr}}{\text{yr}}} \approx 5.951293379 \text{ hr} \cdot \frac{\text{yr}}{\text{hr}} \approx 5.951293379 \text{ yr}$$

Rounding to two significant digits, the trip will take about 6.0 years.

Figure 1-21

SELF CHECK 7 How long would it take if the spaceship could travel at 12,000 mph?

EVERYDAY CONNECTIONS CO_2 Emissions in Asia

	Carbon Dioxide (CO_2) Emissions				
	Total		Per capita		Cumulative 1800–2000
	(million metric tons) 1999	(percent change since 1990)	(metric tons per person) 1999	(percent change since 1990)	(million metric tons)
WORLD	23,172.2	8.9	3.9	(4.2)	1,017,359
ASIA (EXCL. MIDDLE EAST)	6,901.7	38.0	2.1	19.3	
Azerbaijan	33.2	. .	4.2	. .	2,300
Bangladesh	26.3	83.4	0.2	46.2	442
China	3,051.1	25.6	2.5	16.6	72,615
Georgia	5.3	. .	1.0	. .	380
India	903.8	52.9	0.9	31.9	20,275
Indonesia	244.9	76.9	1.2	56.0	4,872
Japan	1,158.5	10.5	9.1	7.7	36,577
Kazakhstan	114.5	. .	7.0	. .	8,264
Korea, Dem. People's Rep.	214.3	(1.2)	9.7	(10.8)	6,114
Korea, Rep.	410.4	75.5	8.8	62.2	7,120
Kyrgyzstan	4.7	. .	1.0	. .	440
Malaysia	101.3	90.4	4.6	55.7	1,832
Myanmar	9.0	122.2	0.2	90.0	257
Pakistan	92.2	48.9	0.7	17.9	1,952
Philippines	66.3	69.0	0.9	39.1	1,555
Singapore	53.2	53.1	13.6	17.9	1,690
Thailand	155.8	95.5	2.5	73.1	2,535
Turkmenistan	33.9	. .	7.3	. .	910
Uzbekistan	117.5	. .	4.8	. .	5,020
Viet Nam	36.6	103.7	0.5	74.1	1,061

Source: EarthTrends Data Tables: Climate and Atmosphere: http://earthtrends.wri.org

The table shows data for carbon dioxide (CO_2) emissions in Asia. Use the table to express each of the following using scientific notation.

1. Total CO_2 emissions in 1999 in India
2. Per capita CO_2 emissions in 1999 in Bangladesh
3. Cumulative CO_2 emissions in the world

SELF CHECK ANSWERS **1.** 1.5×10^8 **2.** 2.5×10^{-5} **3.** -4.57×10^4 **4. a.** $-96,000$ **b.** 0.00562 **5. a.** 2.73×10^3 **b.** 2.5×10^{-6} **6.** 200,000,000,000 **7.** about 3.7 years

NOW TRY THIS

1. An **angstrom** is a unit of length used to specify radiation wavelength. One angstrom is equal to one hundred-millionth of a centimeter.
   a. Write this number in standard notation.
   b. Write this number in scientific notation.
2. One micron is equal to 10,000 angstroms. Write this number in scientific notation.

1.4 EXERCISES

WARM-UPS

Give each numeral in scientific notation.

1. 352 **2.** 5,130
3. 0.002 **4.** 0.00025

Give each numeral in standard notation.

5. 3.5×10^2 **6.** 4.3×10^3
7. 2.7×10^{-1} **8.** 8.5×10^{-2}

REVIEW

Write each fraction as a terminating or a repeating decimal.

9. $\frac{3}{4}$ **10.** $\frac{4}{5}$
11. $\frac{13}{9}$ **12.** $\frac{14}{11}$

13. A man raises 3 to the second power, 4 to the third power, and 2 to the fourth power and then finds their sum. What number does he obtain?

14. If $a = -2$, $b = -3$, and $c = 4$, evaluate

$$\frac{5ab - 4ac - 2}{3bc + abc}$$

VOCABULARY AND CONCEPTS *Fill in the blanks.*

15. A number is written in ________ notation when it is written in the form $N \times$ ____, where $1 \le |N| < 10$ and n is an integer.

16. To change 6.31×10^4 to ________ notation, we move the decimal point in 6.31 ____ places to the right.

17. To change 6.31×10^{-4} to standard notation, we move the decimal point four places to the ____.

18. The number 6.7×10^3 ___ (insert $>$ or $<$) the number $6,700,000 \times 10^{-4}$.

GUIDED PRACTICE

Write each number in scientific notation. See Example 1. (Objective 1)

19. 3,900 **20.** 1,700
21. 17,600,000 **22.** 89,800,000

Write each number in scientific notation. See Example 2. (Objective 1)

23. 0.0078 **24.** 0.068
25. 0.0000096 **26.** 0.000046

Write each number in scientific notation. See Example 3. (Objective 1)

27. $-45,000$ **28.** $-547,000$
29. -0.00021 **30.** -0.00078

Write each number in standard notation. See Example 4. (Objective 2)

31. 2.7×10^2 **32.** 7.2×10^3
33. 7.96×10^5 **34.** 9.67×10^6
35. 3.23×10^{-3} **36.** 6.48×10^{-2}
37. 3.7×10^{-4} **38.** 4.12×10^{-5}

Write each number in standard notation. See Example 5. (Objective 2)

39. 23.65×10^6 **40.** 75.6×10^{-5}

41. 323×10^5 **42.** 689×10^6

Write each number in scientific notation and perform the operations. Give all answers in scientific notation. See Example 6. (Objective 3)

43. $\dfrac{(4{,}000)(30{,}000)}{(0.0006)}$

44. $\dfrac{(0.0006)(0.00007)}{21{,}000}$

45. $\dfrac{(640{,}000)(2{,}700{,}000)}{120{,}000}$

46. $\dfrac{(0.0000013)(0.000090)}{0.00039}$

47. $\dfrac{(0.006)(0.008)}{0.0012}$

48. $\dfrac{(600)(80{,}000)}{120{,}000}$

49. $\dfrac{(220{,}000)(0.000009)}{0.00033}$

50. $\dfrac{(0.00024)(96{,}000{,}000)}{640{,}000{,}000}$

ADDITIONAL PRACTICE *Write each number in scientific notation.*

51. $6{,}000 \times 10^{-7}$

52. 765×10^{-5}

53. 0.0527×10^{5}

54. 0.0298×10^{3}

55. 0.0317×10^{-2}

56. 0.0012×10^{-3}

57. 52.3×10^{0}

58. 867×10^{0}

Use a calculator to evaluate each expression.

59. $23{,}437^3$

60. 0.00034^4

61. $(63{,}480)(893{,}322)$

62. $(0.0000413)(0.0000049)^2$

63. $\dfrac{(320{,}000)^2(0.0009)}{12{,}000^2}$

64. $\dfrac{(0.000012)^2(49{,}000)^2}{0.021}$

65. $\dfrac{(69.4)^8(73.1)^2}{(0.0043)^3}$

66. $\dfrac{(0.0031)^4(0.0012)^5}{(0.0456)^{-7}}$

APPLICATIONS *Use scientific notation to find each answer. Round all answers to the proper number of significant digits.* See Example 7. (Objective 4)

67. Wavelengths Transmitters, vacuum tubes, and lights emit energy that can be modeled as a wave. List the wavelengths shown in the table in order, from shortest to longest.

Type	Use	Wavelength (m)
visible light	lighting	9.3×10^{-6}
infrared	photography	3.7×10^{-5}
x-ray	medical	2.3×10^{-11}
radio wave	communication	3.0×10^{2}
gamma ray	treating cancer	8.9×10^{-14}

68. Distance to the Sun The Sun is about 93 million miles from Earth. Find this distance in feet. (*Hint:* 1 mi = 5,280 ft.)

69. Speed of sound The speed of sound in air is 3.31×10^4 centimeters per second. Find the speed of sound in centimeters per hour.

70. Volume of a tank Find the volume of the tank shown in the illustration.

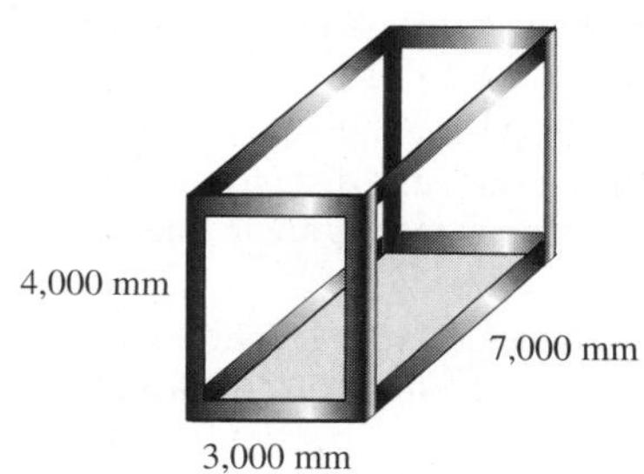

71. Mass of protons If the mass of 1 proton is 0.00000000000000000000000167248 gram, find the mass of 1 million protons.

72. Speed of light The speed of light in a vacuum is about 30,000,000,000 centimeters per second. Find the speed of light in mph. (*Hint:* 160,000 cm ≈ 1 mile.)

73. Distance to the Moon The Moon is about 239,000 miles from Earth. Find this distance in inches.

74. Distance to the Sun The Sun is about 149,700,000 kilometers from Earth. Find this distance in miles. (*Hint:* 1 km ≈ 0.6214 mile.)

75. Solar flares Solar flares often produce immense loops of glowing gas ejected from the Sun's surface. The flare in the illustration extends about 95,000 kilometers into space. Express this distance in miles. (*Hint:* 1 km ≈ 0.6214 mile.)

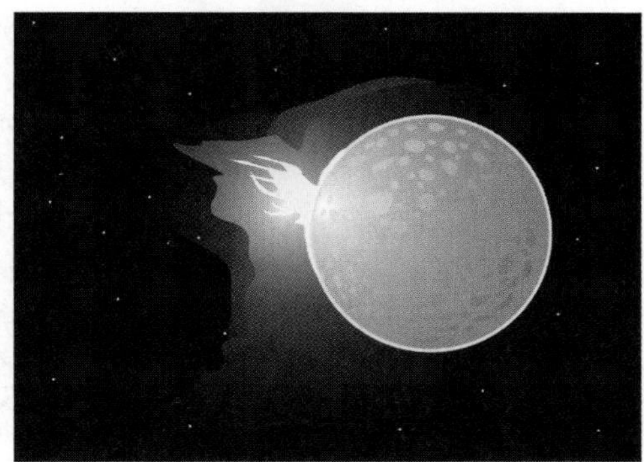

76. Distance to the Moon The Moon is about 378,196 kilometers from Earth. Express this distance in inches. (*Hint:* 1 km ≈ 0.6214 mile.)

77. **Angstroms per inch** One **angstrom** is exactly 0.0000001 millimeter, and one inch is 25.4 millimeters. Find the number of angstroms in one inch.

78. **Range of a comet** One **astronomical unit** (AU) is the distance from Earth to the Sun—about 9.3×10^7 miles. Halley's comet ranges from 0.6 to 18 AU from the Sun. Express this range in miles.

79. **Flight to Pluto** The planet Pluto is approximately 3,574,000,000 miles from Earth. If a spaceship can travel 18,000 mph, how long will it take to reach Pluto?

80. **Light year** Light travels about 300,000,000 meters per second. A **light year** is the distance that light can travel in one year. How many meters are in one light year?

81. **Distance to Alpha Centauri** Light travels about 186,000 miles per second. A **parsec** is 3.26 light years. The star Alpha Centauri is 1.3 parsecs from Earth. Express this distance in miles.

82. **Life of a comet** The mass of the comet shown in the illustration is about 10^{16} grams. When the comet is close to the Sun, matter evaporates at the rate of 10^7 grams per second. Calculate the life of the comet if it appears every 50 years and spends ten days close to the Sun.

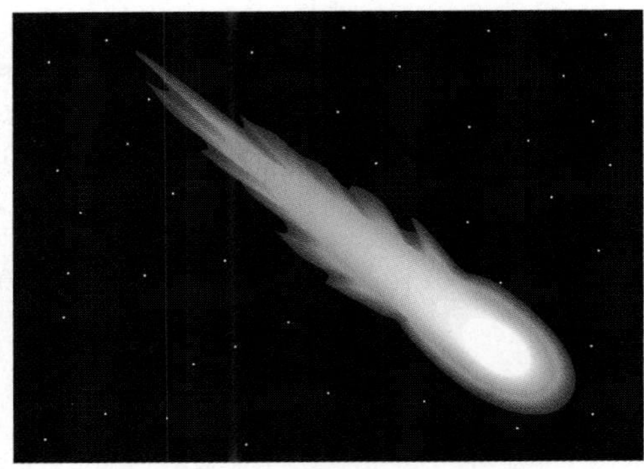

WRITING ABOUT MATH

83. Explain how to change a number from standard notation to scientific notation.

84. Explain how to change a number from scientific notation to standard notation.

SOMETHING TO THINK ABOUT

85. Find the highest power of 2 that can be evaluated with a scientific calculator.

86. Find the highest power of 7 that can be evaluated with a scientific calculator.

SECTION 1.5 Solving Linear Equations in One Variable

Objectives

1. Determine whether a number is a solution of an equation.
2. Solve a linear equation in one variable by applying the properties of equality.
3. Simplify an expression using the rules for order of operations and combining like terms.
4. Solve a linear equation in one variable requiring simplifying one or both sides.
5. Solve a linear equation in one variable that is an identity or a contradiction.
6. Solve a formula for an indicated variable.

Vocabulary

equation	algebraic terms	conditional equations
solution set	constants	identity
roots	numerical coefficients	contradiction
equivalent equations	like terms	
linear equations	similar terms	

Getting Ready

Fill in the blanks.

1. $\square + 3 = 5$ **2.** $8 - \square = 4$ **3.** $\frac{12}{\square} = 4$ **4.** $\square \cdot 5 = 30$

In this section, we show how to solve equations, one of the most important concepts in algebra. Then, we will apply these equation-solving techniques to solve formulas for various variables.

1 Determine whether a number is a solution of an equation.

An **equation** is a statement indicating that two quantities are equal. The equation $2 + 4 = 6$ is true, and the equation $2 + 4 = 7$ is false. If an equation has a variable (say, x) it can be either true or false, depending on the value of x. For example, if $x = 1$, the equation $7x - 3 = 4$ is true.

$$7(\mathbf{1}) - 3 = 4 \quad \text{Substitute 1 for } x.$$
$$7 - 3 = 4$$
$$4 = 4$$

However, the equation is false for all other values of x. Since 1 makes the equation true, we say that 1 *satisfies* the equation.

The set of numbers that satisfies an equation is called its **solution set.** The elements of the solution set are called **solutions** or **roots** of the equation. Finding the solution set of an equation is called *solving the equation.*

EXAMPLE 1 Determine whether 3 is a solution of $2x + 4 = 10$.

Solution We substitute 3 for x and see whether it satisfies the equation.

$$2x + 4 = 10$$
$$2(\mathbf{3}) + 4 \stackrel{?}{=} 10 \quad \text{Substitute 3 for } x.$$
$$6 + 4 \stackrel{?}{=} 10 \quad \text{First do the multiplication on the left side.}$$
$$10 = 10 \quad \text{Then do the addition.}$$

Since $10 = 10$, the number 3 satisfies the equation. It is a solution.

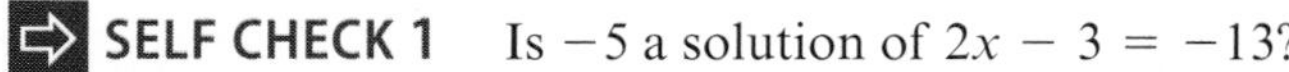

SELF CHECK 1 Is -5 a solution of $2x - 3 = -13$?

2 Solve a linear equation in one variable by applying the properties of equality.

To solve an equation we replace the equation with simpler ones, all having the same solution set. Such equations are called *equivalent equations.*

Equivalent Equations

Equations with the same solution set are called **equivalent equations.**

We continue to replace each resulting equation with an equivalent one until we have isolated the variable on one side of an equation. To isolate the variable, we can use the following properties:

Properties of Equality

If a, b, and c are real numbers and $a = b$, then

$$a + c = b + c \quad \text{and} \quad a - c = b - c$$

and if $c \neq 0$, then

$$ac = bc \quad \text{and} \quad \frac{a}{c} = \frac{b}{c}$$

In words, we can say: *If any quantity is added to (or subtracted from) both sides of an equation, a new equation is formed that is equivalent to the original equation.*

If both sides of an equation are multiplied (or divided) by the same nonzero quantity, a new equation is formed that is equivalent to the original equation.

The most basic equations that we will solve are *linear equations in one variable.*

Linear Equations

A **linear equation** in one variable (say, x) is any equation that can be written in the form

$$ax + c = 0 \quad (a \text{ and } c \text{ are real numbers and } a \neq 0)$$

EXAMPLE 2 Solve: $2x + 8 = 0$.

Solution To solve the equation, we will isolate x on the left side.

$$2x + 8 = 0$$

$$2x + 8 - \mathbf{8} = 0 - \mathbf{8} \qquad \text{To eliminate 8 from the left side, subtract 8 from both sides.}$$

$$2x = -8 \qquad \text{Simplify.}$$

$$\frac{2x}{\mathbf{2}} = \frac{-8}{\mathbf{2}} \qquad \text{To eliminate 2 from the left side, divide both sides by 2.}$$

$$x = -4 \qquad \text{Simplify.}$$

Check: We substitute -4 for x to verify that it satisfies the original equation.

$$2x + 8 = 0$$

$$2(-4) + 8 \stackrel{?}{=} 0 \qquad \text{Substitute } -4 \text{ for } x.$$

$$-8 + 8 \stackrel{?}{=} 0$$

$$0 = 0$$

Since -4 satisfies the original equation, it is the solution.

SELF CHECK 2 Solve $-3a + 15 = 0$.

EXAMPLE 3 Solve: $3(x - 2) = 20$.

Solution We isolate x on the left side.

$$3(x - 2) = 20$$

$$3x - 6 = 20 \qquad \text{Use the distributive property to remove parentheses.}$$

$$3x - 6 + 6 = 20 + 6 \qquad \text{To eliminate } -6 \text{ from the left side, add 6 to both sides.}$$

$$3x = 26 \qquad \text{Simplify.}$$

$$\frac{3x}{3} = \frac{26}{3} \qquad \text{To eliminate 3 from the left side, divide both sides by 3.}$$

$$x = \frac{26}{3} \qquad \text{Simplify.}$$

Check:

$$3(x - 2) = 20$$

$$3\left(\frac{26}{3} - 2\right) \stackrel{?}{=} 20 \qquad \text{Substitute } \tfrac{26}{3} \text{ for } x.$$

$$3\left(\frac{26}{3} - \frac{6}{3}\right) \stackrel{?}{=} 20 \qquad \text{Get a common denominator: } 2 = \tfrac{6}{3}.$$

$$3\left(\frac{20}{3}\right) \stackrel{?}{=} 20 \qquad \text{Combine the fractions: } \tfrac{26}{3} - \tfrac{6}{3} = \tfrac{20}{3}.$$

$$20 = 20 \qquad \text{Simplify.}$$

Since $\frac{26}{3}$ satisfies the equation, it is the solution.

SELF CHECK 3 Solve $-2(a + 3) = 18$.

3 Simplify an expression using the rules for order of operations and combining like terms.

To solve many equations, we will need to combine like terms. An **algebraic term** is either a number or the product of numbers (called **constants**) and variables. Some examples of terms are $3x$, $-7y$, y^2, and 8. The **numerical coefficients** of these terms are 3, -7, 1, and 8 (8 can be written as $8x^0$), respectively.

In algebraic expressions, terms are separated by + and − signs. For example, the expression $3x^2 + 2x - 4$ has three terms, and the expression $3x + 7y$ has two terms.

Terms with the same variables with the same exponents are called **like terms** or **similar terms:**

$5x$ and $6x$ are like terms.

$27x^2y^3$ and $-326x^2y^3$ are like terms.

$4x$ and $-17y$ are unlike terms. The terms have different variables.

$15x^2y$ and $6xy^2$ are unlike terms. The variables have different exponents.

By using the distributive law, we can combine like terms. For example,

$$5x + 6x = (5 + 6)x = 11x \quad \text{and} \quad 32y - 16y = (32 - 16)y = 16y$$

This suggests that to combine like terms, *we add or subtract their numerical coefficients and keep the same variables with the same exponents.*

EXAMPLE 4 Simplify: $2(x + 3) + 8(x - 2)$.

Solution To simplify the expression, we will use the distributive property to remove parentheses and then combine like terms.

$$\begin{aligned} 2(x + 3) + 8(x - 2) &= 2x + 2 \cdot 3 + 8x - 8 \cdot 2 && \text{Use the distributive property to remove parentheses.} \\ &= 2x + 6 + 8x - 16 && 2 \cdot 3 = 6 \text{ and } 8 \cdot 2 = 16. \\ &= 2x + 8x + 6 - 16 && \text{Use the commutative property of addition: } 6 + 8x = 8x + 6. \\ &= 10x - 10 && \text{Combine like terms.} \end{aligned}$$

SELF CHECK 4 Simplify: $-3(a + 5) + 5(a - 2)$.

COMMENT In this book, we will simplify expressions and solve equations. Recognizing when to do which is a skill that we will apply throughout this book.

When an expression does not contain an = sign, it can be simplified only by combining its like terms. Since an equation contains an = sign, it can be solved. Solving an equation means finding the values of the variable that make the equation true. Remember that

Expressions are to be simplified. *Equations are to be solved.*

4 Solve a linear equation in one variable requiring simplifying one or both sides.

EXAMPLE 5 Solve: $3(2x - 1) = 2x + 9$.

Solution

$$\begin{aligned} 3(2x - 1) &= 2x + 9 \\ 6x - 3 &= 2x + 9 && \text{Use the distributive property to remove parentheses.} \\ 6x - 3 + 3 &= 3x + 9 + 3 && \text{To eliminate } -3 \text{ from the left side, add 3 to both sides.} \\ 6x &= 2x + 12 && \text{Combine like terms.} \\ 6x - 2x &= 2x - 2x + 12 && \text{To eliminate } 2x \text{ from the right side, subtract } 2x \text{ from both sides.} \\ 4x &= 12 && \text{Combine like terms.} \\ x &= 3 && \text{To eliminate 4 from the left side, divide both sides by 4.} \end{aligned}$$

Check:

$$\begin{aligned} 3(2x - 1) &= 2x + 9 \\ 3(2 \cdot 3 - 1) &\stackrel{?}{=} 2 \cdot 3 + 9 && \text{Substitute 3 for } x. \\ 3(5) &\stackrel{?}{=} 6 + 9 \\ 15 &= 15 \end{aligned}$$

Since 3 satisfies the equation, it is the solution.

SELF CHECK 5 Solve $-4(3a + 4) = 2a - 4$.

To solve more complicated linear equations, we will follow these steps.

Solving Linear Equations

1. If the equation contains fractions, multiply both sides of the equation by a number that will eliminate the denominators.
2. Use the distributive property to remove all sets of parentheses and combine like terms on each side of the equation.
3. Use the addition and subtraction properties to get all variables on one side of the equation and all numbers on the other side. Combine like terms, if necessary.
4. Use the multiplication and division properties to make the coefficient of the variable equal to 1.
5. Check the result by replacing the variable with the possible solution and verifying that the number satisfies the equation.

EXAMPLE 6 Solve: $\frac{5}{3}(x - 3) = \frac{3}{2}(x - 2) + 2$.

Solution **Step 1:** Since 6 is the smallest number that can be divided by both 2 and 3, we multiply both sides of the equation by 6 to eliminate the fractions.

$$\frac{5}{3}(x - 3) = \frac{3}{2}(x - 2) + 2$$

$$6\left[\frac{5}{3}(x - 3)\right] = 6\left[\frac{3}{2}(x - 2) + 2\right] \quad \text{To eliminate the fractions, multiply both sides by 6.}$$

$$6 \cdot \frac{5}{3}(x - 3) = 6 \cdot \frac{3}{2}(x - 2) + 6 \cdot 2 \quad \text{Use the distributive property on the right side.}$$

$$10(x - 3) = 9(x - 2) + 12 \quad \text{Simplify.}$$

Step 2: We use the distributive property to remove parentheses and then combine like terms.

$$10x - 30 = 9x - 18 + 12$$

$$10x - 30 = 9x - 6$$

Step 3: We use the addition and subtraction properties by adding 30 to both sides and subtracting $9x$ from both sides.

$$10x - 30 \mathbf{- 9x + 30} = 9x - 6 \mathbf{- 9x + 30}$$

$$x = 24 \quad \text{Combine like terms.}$$

Since the coefficient of x in the above equation is 1, Step 4 is unnecessary.

Step 5: We check by substituting 24 for x in the original equation and simplifying:

$$\frac{5}{3}(x - 3) = \frac{3}{2}(x - 2) + 2$$

$$\frac{5}{3}(\mathbf{24} - 3) \stackrel{?}{=} \frac{3}{2}(\mathbf{24} - 2) + 2$$

$$\frac{5}{3}(21) \stackrel{?}{=} \frac{3}{2}(22) + 2$$

$$5(7) \stackrel{?}{=} 3(11) + 2$$
$$35 = 35$$

Since 24 satisfies the equation, it is the solution.

SELF CHECK 6 Solve $\frac{3}{2}(x - 2) = \frac{5}{3}(x - 3) + 2$.

EXAMPLE 7 Solve: $\dfrac{x + 2}{5} - 4x = \dfrac{8}{5} - \dfrac{x + 9}{2}$.

Solution

$$\frac{x + 2}{5} - 4x = \frac{8}{5} - \frac{x + 9}{2}$$

$$\mathbf{10}\left(\frac{x + 2}{5} - 4x\right) = \mathbf{10}\left(\frac{8}{5} - \frac{x + 9}{2}\right)$$ To eliminate the fractions, multiply both sides by 10.

$$2(x + 2) - 40x = 2(8) - 5(x + 9)$$ Remove parentheses.

$$2x + 4 - 40x = 16 - 5x - 45$$ Remove parentheses.

$$-38x + 4 = -5x - 29$$ Combine like terms.

$$-33x = -33$$ Add $5x$ and -4 to both sides.

$$\frac{-33x}{\mathbf{-33}} = \frac{-33}{\mathbf{-33}}$$ Divide both sides by -33.

$$x = 1$$ Simplify.

Check:

$$\frac{x + 2}{5} - 4x = \frac{8}{5} - \frac{x + 9}{2}$$

$$\frac{\mathbf{1} + 2}{5} - 4(\mathbf{1}) \stackrel{?}{=} \frac{8}{5} - \frac{\mathbf{1} + 9}{2}$$ Substitute 1 for x.

$$\frac{3}{5} - 4 \stackrel{?}{=} \frac{8}{5} - 5$$

$$\frac{3}{5} - \frac{20}{5} \stackrel{?}{=} \frac{8}{5} - \frac{25}{5}$$

$$-\frac{17}{5} = -\frac{17}{5}$$

Since 1 satisfies the equation, it is the solution.

SELF CHECK 7 Solve $\frac{a + 5}{5} + 2a = \frac{a}{2} - \frac{a + 14}{5}$.

EXAMPLE 8 Solve: $0.06x + 0.07(15{,}000 - x) = 990$.

Solution

$$0.06x + 0.07(15{,}000 - x) = 990$$

$$\mathbf{100}[0.06x + 0.07(15{,}000 - x)] = \mathbf{100}[990]$$ To remove the decimals, multiply both sides by 100.

$$6x + 7(15{,}000 - x) = 99{,}000$$ Use the distributive property to remove the brackets.

$$6x + 105{,}000 - 7x = 99{,}000$$ Use the distributive property to remove the parentheses.

$$-x = -6{,}000$$ Combine like terms and subtract 105,000 from both sides.

$$x = 6{,}000$$ Multiply both sides by -1.

Check:

$$0.06x + 0.07(15{,}000 - x) = 990$$
$$0.06(6{,}000) + 0.07(15{,}000 - 6{,}000) \stackrel{?}{=} 990$$
$$360 + 0.07(9{,}000) \stackrel{?}{=} 990$$
$$360 + 630 \stackrel{?}{=} 990$$
$$990 = 990$$

Since 6,000 satisfies the equation, the solution is 6,000.

SELF CHECK 8 Solve $0.05(8{,}000 - x) + 0.06x = 450$.

5 Solve a linear equation in one variable that is an identity or a contradiction.

The linear equations discussed so far are called **conditional equations.** These equations have exactly one solution. An **identity** is an equation that is satisfied by every number x for which both sides of the equation are defined. The solution of an identity is the set of *all real numbers* and is denoted by the symbol $\mathbb{R}$.

EXAMPLE 9 Solve: $2(x - 1) + 4 = 4(1 + x) - (2x + 2)$.

Solution

$$2(x - 1) + 4 = 4(1 + x) - (2x + 2)$$
$$2x - 2 + 4 = 4 + 4x - 2x - 2$$ Use the distributive property to remove parentheses.

$$2x + 2 = 2x + 2$$ Combine like terms.

The result $2x + 2 = 2x + 2$ is true for every value of x. Since every number x satisfies the equation, it is an identity. The solution set is $\mathbb{R}$.

SELF CHECK 9 Solve: $3(a + 4) + 5 = 2(a - 1) + a + 19$.

A **contradiction** is an equation that has no solution, as in the next example.

EXAMPLE 10 Solve: $\dfrac{x - 1}{3} + 4x = \dfrac{3}{2} + \dfrac{13x - 2}{3}$.

Solution

$$\frac{x - 1}{3} + 4x = \frac{3}{2} + \frac{13x - 2}{3}$$
$$6\left(\frac{x - 1}{3} + 4x\right) = 6\left(\frac{3}{2} + \frac{13x - 2}{3}\right)$$ To eliminate the fractions, multiply both sides by 6.

$2(x - 1) + 6(4x) = 3(3) + 2(13x - 2)$	Use the distributive property to remove parentheses.
$2x - 2 + 24x = 9 + 26x - 4$	Remove parentheses.
$26x - 2 = 26x + 5$	Combine like terms.
$-2 = 5$	Subtract $26x$ from both sides.

Since $-2 = 5$ is false, no number x can satisfy the equation. The solution set is the **empty set,** denoted as $\varnothing$.

SELF CHECK 10 Solve: $\frac{x+5}{5} = \frac{1}{5} + \frac{x}{5}$.

Every linear equation in one variable will be one of these types.

Type of equation	Examples		Solution set	
Conditional	$2x + 4 = 8$	$\frac{x}{2} - 4 = 12$	$\{2\}$	$\{32\}$
Identity	$x + x = 2x$	$2(x + 3) = 2x + 6$	$\mathbb{R}$	$\mathbb{R}$
Contradiction	$x = x - 1$	$2(x + 3) = 2x + 5$	$\varnothing$	$\varnothing$

Table 1-5

6 Solve a formula for an indicated variable.

Suppose we want to find the heights of several triangles whose areas and bases are known. It would be tedious to substitute values of A and b into the formula $A = \frac{1}{2}bh$ and then repeatedly solve the formula for h. It is easier to solve for h first and then substitute values for A and b and compute h directly.

To solve a formula for an indicated variable means to isolate that variable on one side of the equation.

EXAMPLE 11 Solve $A = \frac{1}{2}bh$ for h.

Solution

$$A = \frac{1}{2}bh$$

$2A = bh$	To eliminate the fraction, multiply both sides by 2.
$\frac{2A}{b} = h$	To isolate h, divide both sides by b.
$h = \frac{2A}{b}$	Write h on the left side.

SELF CHECK 11 Solve $A = \frac{1}{2}bh$ for b.

EXAMPLE 12 For simple interest, the formula $A = p + prt$ gives the amount of money in an account at the end of a specific time. A represents the amount, p the principal, r the rate of interest, and t the time. We can solve the formula for t as follows:

Solution

$$A = p + prt$$

$$A - p = prt \quad \text{To isolate the term involving } t \text{, subtract } p \text{ from both sides.}$$

$$\frac{A - p}{pr} = t \quad \text{To isolate } t \text{, divide both sides by } pr.$$

$$t = \frac{A - p}{pr} \quad \text{Write } t \text{ on the left side.}$$

SELF CHECK 12 Solve $A = p + prt$ for r.

EXAMPLE 13 The formula $F = \frac{9}{5}C + 32$ converts degrees Celsius to degrees Fahrenheit. Solve the formula for C.

Solution

$$F = \frac{9}{5}C + 32$$

$$F - 32 = \frac{9}{5}C \quad \text{To isolate the term involving } C \text{, subtract 32 from both sides.}$$

$$\frac{5}{9}(F - 32) = \frac{5}{9}\left(\frac{9}{5}C\right) \quad \text{To isolate } C \text{, multiply both sides by } \frac{5}{9}.$$

$$\frac{5}{9}(F - 32) = C \quad \frac{5}{9} \cdot \frac{9}{5} = 1$$

$$C = \frac{5}{9}(F - 32)$$

To convert degrees Fahrenheit to degrees Celsius, we can use the formula $C = \frac{5}{9}(F - 32)$.

SELF CHECK 13 Solve $S = \frac{180(n - 2)}{7}$ for n.

SELF CHECK ANSWERS

1. yes **2.** 5 **3.** -12 **4.** $2a - 25$ **5.** $-\frac{6}{7}$ **6.** 0 **7.** -2 **8.** 5,000 **9.** $\mathbb{R}$ **10.** $\varnothing$ **11.** $b = \frac{2A}{h}$ **12.** $r = \frac{A - p}{pt}$ **13.** $n = \frac{7S}{180} + 2$ or $n = \frac{7S + 360}{180}$

NOW TRY THIS

Solve each equation for the indicated variable.

1. $y - y_1 = m(x - x_1)$ for m

2. $A = p + prt$ for p

1.5 EXERCISES

WARM-UPS

Combine like terms.

1. $5x + 4x$

2. $7s^2 - 5s^2$

Determine whether each number is a solution of $2x + 5 = 13$.

3. 3

4. 4

5. 5

6. 6

Solve each equation.

7. $3x - 2 = 7$

8. $\frac{1}{2}x - 1 = 5$

9. $\frac{x-2}{3} = 1$

10. $\frac{x+3}{2} = 3$

REVIEW *Simplify each expression.*

11. $(-4)^3$

12. -3^3

13. $\left(\frac{x+y}{x-y}\right)^0$

14. $(x^2x^3)^4$

15. $\left(\frac{x^2x^5}{x^3}\right)^2$

16. $\left(\frac{x^4y^3}{x^5y}\right)^3$

17. $(2x)^{-3}$

18. $\left(\frac{x^2}{y^5}\right)^{-4}$

VOCABULARY AND CONCEPTS *Fill in the blanks.*

19. An ________ is a statement that two quantities are equal.

20. If a number is substituted for a variable in an equation and the equation is true, we say that the number ________ the equation. The numbers that satisfy an equation form its ________ set.

21. If two equations have the same solution set, they are called ________ equations.

22. The solutions of an equation are also called ______.

23. If a, b, and c are real numbers and $a = b$, then $a + c = b +$ __ and $a - c = b -$ __, using the addition and subtraction properties of equality.

24. If a, b, and c are real numbers and $a = b$, then $a \cdot c = b \cdot$ __ and $\frac{a}{c} = \frac{}{\_\_}$ $(c \neq 0)$, using the multiplication and division properties of equality.

25. A number or the product of numbers and variables is called an algebraic ______.

26. ______ terms (similar terms) are terms with the same variables and with the same exponents.

27. To combine like terms, add their ________________ and keep the same ________ and exponents.

28. An ________ is an equation that is true for all values of its variable and its solution set is denoted by $\mathbb{R}$. A contradiction is true for ____ values of its variable and its solution set is denoted by $\varnothing$.

GUIDED PRACTICE

Determine whether 5 is a solution of each equation. See Example 1. (Objective 1)

29. $3x + 2 = 17$

30. $7x - 2 = 33$

31. $\frac{3}{5}x - 5 = -2$

32. $\frac{2}{5}x + 12 = 8$

Solve each equation. See Example 2. (Objective 2)

33. $x + 6 = 8$

34. $y - 7 = 3$

35. $a - 5 = 20$

36. $b + 4 = 18$

37. $2u = 6$

38. $3v = 12$

39. $\frac{x}{4} = 7$

40. $\frac{x}{6} = 8$

Solve each equation. See Example 3. (Objective 2)

41. $3x + 1 = 3$

42. $8x - 2 = 13$

43. $2x + 1 = 13$

44. $2x - 4 = 16$

45. $3(x - 4) = -36$

46. $4(x + 6) = 84$

47. $3(r - 4) = -4$

48. $4(s - 5) = -3$

Determine whether the terms are like terms. If they are, combine them. See Example 4. (Objective 3)

49. $2x, 6x$

50. $-3x, 5y$

51. $-5xy, -7yz$

52. $-3t^2, 12t^2$

53. $3x^2, -5x^2$

54. $5y^2, 7xy$

55. $xy, 3xt$

56. $-4x, -5x$

Simplify each expression. See Example 4. (Objective 3)

57. $6(2x - 5) + 3(x + 1)$

58. $7(x + 3) + 4(3x - 8)$

59. $9(5x + 6) - 2(x + 5)$

60. $8(7x + 3) - 5(6x - 8)$

61. $8 + 2(x - 4) - (6x - 9)$

62. $9 - 3(6x + 2) - (4x + 3)$

63. $6x - 3(9 - 5x) + 8$

64. $-8x - 7(5 - 6x) - 9$

Solve each equation. See Example 5. (Objective 4)

65. $2(a - 5) - (3a + 1) = 0$

66. $8(3a - 5) - 4(2a + 3) = 12$

67. $3(y - 5) + 10 = 2(y + 4)$

68. $2(5x + 2) = 3(3x - 2)$

69. $9(x + 2) = -6(4 - x) + 18$

70. $3(x + 2) - 2 = -(5 + x) + x$

71. $-4p - 2(3p + 5) = -6p + 2(p + 2)$

72. $2q - 3(q - 5) = 5(q + 2) - 7$

Solve each equation. **See Example 6. (Objective 4)**

73. $\frac{x}{6} + 1 = \frac{x}{3}$

74. $\frac{3}{2}(y + 4) = \frac{20 - y}{2}$

75. $5 - \frac{x + 2}{3} = 7 - x$

76. $3x - \frac{2(x + 3)}{3} = 16 - \frac{x + 2}{2}$

77. $\frac{4x - 2}{2} = \frac{3x + 6}{3}$

78. $\frac{t + 4}{2} = \frac{2t - 3}{3}$

79. $\frac{a + 1}{3} + \frac{a - 1}{5} = \frac{2}{15}$

80. $\frac{2z + 3}{3} + \frac{3z - 4}{6} = \frac{z - 2}{2}$

Solve each equation. **See Example 7. (Objective 4)**

81. $\frac{1}{2}x - 4 = -1 + 2x$

82. $2x + 3 = \frac{2}{3}x - 1$

83. $\frac{x}{2} - \frac{x}{3} = 4$

84. $\frac{x}{2} + \frac{x}{3} = 10$

85. $\frac{5a}{2} - 12 = \frac{a}{3} + 1$

86. $\frac{5a}{6} - \frac{5}{2} = -\frac{1}{2} - \frac{a}{6}$

87. $\frac{a + 1}{4} + \frac{2a - 3}{4} = \frac{a}{2} - 2$

88. $\frac{y - 8}{5} + 2 = \frac{2}{5} - \frac{y}{3}$

Solve each equation. **See Example 8. (Objective 4)**

89. $0.45 = 16.95 - 0.25(75 - 3a)$

90. $3.2 + x = 0.25(x + 32)$

91. $0.09x + 0.14(10{,}000 - x) = 1{,}275$

92. $0.04(20) + 0.01l = 0.02(20 + l)$

Solve each equation and state whether the equation is an identity or a contradiction. Then give the solution set. **See Examples 9–10. (Objective 5)**

93. $4(2 - 3t) + 6t = -6t + 8$

94. $2(x - 3) = \frac{3}{2}(x - 4) + \frac{x}{2}$

95. $2x - 6 = -2x + 4(x - 2)$

96. $3(x - 4) + 6 = -2(x + 4) + 5x$

Solve each formula for the indicated variable. **See Example 11. (Objective 6)**

97. $A = lw$ for w

98. $p = 4s$ for s

99. $V = \frac{1}{3}Bh$ for B

100. $b = \frac{2A}{h}$ for A

101. $I = prt$ for t

102. $I = prt$ for r

103. $p = 2l + 2w$ for w

104. $p = 2l + 2w$ for l

Solve each formula for the indicated variable. **See Example 12. (Objective 6)**

105. $A = \frac{1}{2}h(B + b)$ for B

106. $A = \frac{1}{2}h(B + b)$ for b

107. $y = mx + b$ for x

108. $y = mx + b$ for m

Solve each formula for the indicated variable. **See Example 13. (Objective 6)**

109. $S = \frac{a - lr}{1 - r}$ for l

110. $C = \frac{5}{9}(F - 32)$ for F

111. $S = \frac{n(a + l)}{2}$ for l

112. $S = \frac{n(a + l)}{2}$ for n

ADDITIONAL PRACTICE

Solve each equation. If the equation is an identity or a contradiction, so indicate.

113. $3a - 22 = -2a - 7$

114. $a + 18 = 6a - 3$

115. $3(y - 4) - 6 = y$

116. $2x + (2x - 3) = 5$

117. $5(5 - a) = 37 - 2a$

118. $4a + 17 = 7(a + 2)$

119. $4(y + 1) = -2(4 - y)$

120. $5(r + 4) = -2(r - 3)$

121. $2(2x + 1) = 15 + 3x$

122. $-2(x + 5) = 30 - x$

123. $9.8 - 15z = -15.7$

124. $0.05a + 0.25 = 0.77$

125. $x(x - 3) = x^2 - 2x + 1 - (5 + x)$

126. $4 + 4(n + 2) = 3n - 2(n - 5)$

127. $4x - 2(3x + 2) = 2(x + 3)$

128. $y(y + 2) + 1 = y^2 + 2y + 1$

129. $l = a + (n - 1)d$ for n

130. $l = a + (n - 1)d$ for d

APPLICATIONS

131. Force of gravity The masses of the two objects in the illustration are m and M. The force of gravitation F between the masses is

$$F = \frac{GmM}{d^2}$$

where G is a constant and d is the distance between them. Solve for m.

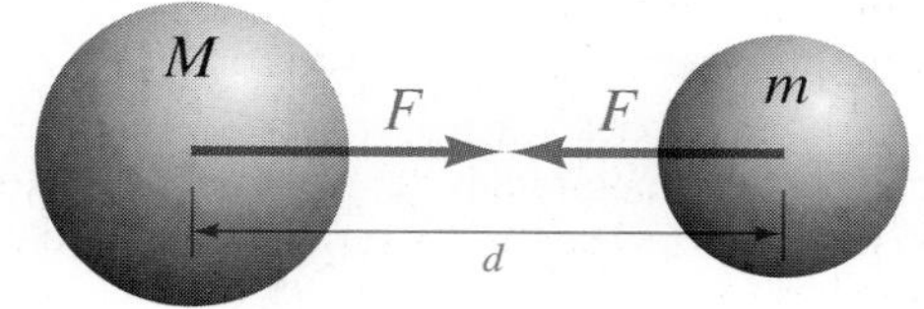

132. Thermodynamics The Gibbs free-energy formula is $G = U - TS + pV$. Solve for S.

133. Converting temperatures Solve the formula $F = \frac{9}{5}C + 32$ for C and find the Celsius temperatures that correspond to Fahrenheit temperatures of 32°, 70°, and 212°.

134. Doubling money A man intends to invest \$1,000 at simple interest. Solve the formula $A = p + prt$ for t and find how long it will take to double his money at the rates of 5%, 7%, and 10%.

135. Cost of electricity The cost of electricity in a certain city is given by the formula $C = 0.07n + 6.50$, where C is the cost and n is the number of kilowatt hours used. Solve for n and find the number of kwh used for costs of \$49.97, \$76.50, and \$125.

136. Cost of water A monthly water bill in a certain city is calculated by using the formula $n = \frac{5{,}000C - 17{,}500}{6}$, where n is the number of gallons used and C is the monthly cost. Solve for C and compute the bill for quantities used of 500, 1,200, and 2,500 gallons.

137. Ohm's Law The formula $E = IR$, called **Ohm's Law,** is used in electronics. Solve for R and then calculate the resistance R if the voltage E is 56 volts and the current I is 7 amperes. (Resistance has units of **ohms.**)

138. Earning interest An amount P, invested at a simple interest rate r, will grow to an amount A in t years, according to the formula $A = P(1 + rt)$. Solve for P. Suppose a man invested some money at 5.5%. If after 5 years, he had \$6,693.75 on deposit, what amount did he originally invest?

139. Angles of a polygon A regular polygon has n equal sides and n equal angles. The measure a of an interior angle is given by $a = 180°\left(1 - \frac{2}{n}\right)$. Solve for n. Find the number of sides of the regular polygon in the illustration if an interior angle is 135°.

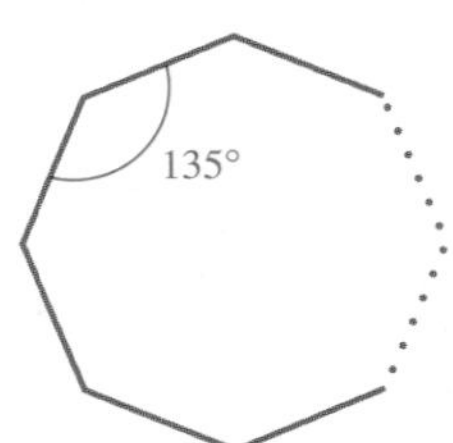

140. Power loss The illustration is the schematic diagram of a resistor connected to a voltage source of 60 volts. As a result, the resistor dissipates power in the form of heat. The power P lost when a voltage V is placed across a resistance R is given by the formula

$$P = \frac{E^2}{R}$$

Solve for R. If P is 4.8 watts and E is 60 volts, find R.

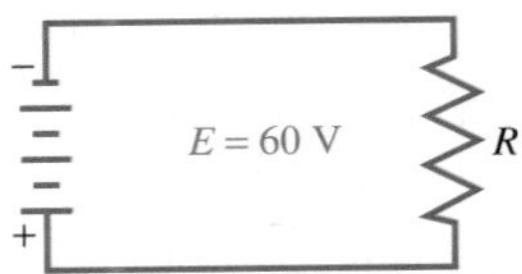

WRITING ABOUT MATH

141. Explain the difference between a conditional equation, an identity, and a contradiction.

142. Explain how you would solve an equation.

SOMETHING TO THINK ABOUT *Find the mistake.*

143.
$$\begin{aligned} 3(x - 2) + 4 &= 14 \\ 3x - 2 + 4 &= 14 \\ 3x + 2 &= 14 \\ 3x &= 12 \\ x &= 4 \end{aligned}$$

144.
$$\begin{aligned} A &= p + prt \\ A - p &= prt \\ A - p - pr &= t \\ t &= A - p - pr \end{aligned}$$

SECTION 1.6 Using Linear Equations to Solve Application Problems

Objectives

1. Solve a recreation problem by setting up and solving a linear equation in one variable.
2. Solve a business problem by setting up and solving a linear equation in one variable.
3. Solve a geometry problem by setting up and solving a linear equation in one variable.
4. Solve a lever problem by setting up and solving a linear equation in one variable.

Vocabulary

right angle
straight angle
acute angle
complementary angles
supplementary angles
right triangle
isosceles triangle
vertex angle
base angle
equilateral triangle
fulcrum

Getting Ready

Let $x = 18$.

1. What number is 2 more than x?
2. What number is 4 times x?
3. What number is 5 more than twice x?
4. What number is 2 less than one-half of x?

In this section, we will apply equation-solving skills to solve application problems. When we translate the words of a problem into mathematics, we are creating a *mathematical model* of the problem. To create these models, we can use Table 1-6 to translate certain words into mathematical operations.

Addition (+)	Subtraction (−)	Multiplication (·)	Division (÷)
added to	subtracted from	multiplied by	divided by
plus	difference	product	quotient
the sum of	less than	times	ratio
more than	less	of	half
increased by	decreased by	twice	

Table 1-6

We can then change English phrases into algebraic expressions, as in Table 1-7 on the next page.

English phrase	Algebraic expression
2 added to some number	$x + 2$
the difference between some number and 2	$x - 2$
5 times some number	$5x$
the product of 925 and some number	$925x$
5% of some number	$0.05x$
the sum of twice a number and 10	$2x + 10$
the quotient (or ratio) of 5 and some number	$\frac{5}{x}$
half of a number	$\frac{x}{2}$

Table 1-7

Once we know how to change phrases into algebraic expressions, we can solve many problems. The following list of steps provides a strategy for solving application problems.

Problem Solving

1. **Analyze the problem** and identify a variable by asking yourself "What am I asked to find?" Choose a variable to represent the quantity to be found and then express all other unknown quantities in the problem as expressions involving that variable.
2. **Form an equation** by expressing a quantity in two different ways. This may require reading the problem several times to understand the given facts. What information is given? Is there a formula that applies to this situation? Often a sketch, chart, or diagram will help you visualize the facts of the problem.
3. **Solve the equation** found in Step 2.
4. **State the conclusion** including proper units.
5. **Check the result.**

1 Solve a recreation problem by setting up and solving a linear equation in one variable.

EXAMPLE 1 **CUTTING A ROPE** A mountain climber wants to cut a rope 213 feet long into three pieces. If each piece is to be 2 feet longer than the previous one, where should he make the cuts?

Analyze the problem

We are asked to find the lengths of three pieces of rope. The information is given in terms of the length of the shortest piece. Therefore, we let x represent the length of the shortest piece and express the other lengths in terms of x. Then $x + 2$ represents the length of the second piece and $x + 2 + 2$, or $x + 4$, represents the length of the longest piece.

Form an equation

The climber knows that the sum of these three lengths can be expressed in two ways: as $x + (x + 2) + (x + 4)$ and as 213. (See Figure 1-22.)

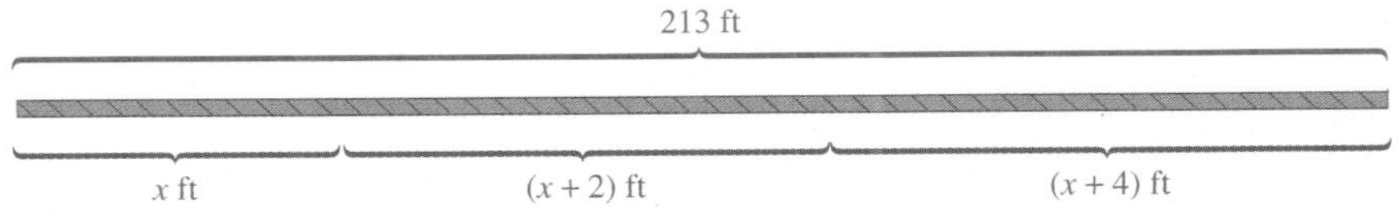

Figure 1-22

From Figure 1-22, we can set up the equation

The length of the first piece	plus	the length of the second piece	plus	the length of the third piece	equals	the total length of the rope.
x	$+$	$x + 2$	$+$	$x + 4$	$=$	213

Solve the equation

We can now solve the equation.

$$x + x + 2 + x + 4 = 213 \quad \text{This is the equation to solve.}$$
$$3x + 6 = 213 \quad \text{Combine like terms.}$$
$$3x = 207 \quad \text{Subtract 6 from both sides.}$$
$$x = 69 \quad \text{Divide both sides by 3. This is the length of the first piece.}$$
$$x + 2 = 71 \quad \text{This is the length of the second piece.}$$
$$x + 4 = 73 \quad \text{This is the length of the third piece.}$$

COMMENT Remember to include any units (feet, inches, pounds, etc.) when stating the conclusion to an application problem.

State the conclusion

He should make one cut 69 feet from one end. Then he should make a second cut 71 feet from one end of the remaining rope. This will leave a length of 73 feet.

Check the result

Each length is 2 feet longer than the previous length, and the sum of the lengths is 213 feet.

2 Solve a business problem by setting up and solving a linear equation in one variable.

When the regular price of merchandise is reduced, the amount of reduction is called the *markdown* (or *discount*).

Sale price = **regular price** − **markdown.**

Usually, the markdown is expressed as a percent of the regular price.

Markdown = **percent of markdown** · **regular price**

EXAMPLE 2 **FINDING THE PERCENT OF MARKDOWN** A home theater system is on sale for \$777. If the list price was \$925, find the percent of markdown.

Analyze the problem

We are asked to find the percent of markdown, so we let p represent the percent of markdown, expressed as a decimal. In this case, \$777 is the sale price, and \$925 is the regular price.

Form an equation

The amount of the markdown is found by multiplying the regular price by the *percent* of markdown. If we subtract this amount from the regular price, we will have the sale price. This can be expressed as the equation

Sale price	equals	regular price	minus	markdown
777	$=$	925	$-$	$p \cdot 925$

Solve the equation

We can now solve the equation.

$$777 = 925 - p \cdot 925 \quad \text{This is the equation to solve.}$$
$$777 = 925 - 925p$$

$$-148 = -925p \quad \text{Subtract 925 from both sides.}$$
$$0.16 = p \quad \text{Divide both sides by } -925.$$

State the conclusion: Because 0.16 equals 16%, the theater system is on sale at a 16% markdown.

Check the result: Since the markdown is 16% of \$925, or \$148, the sale price is \$925 − \$148, or \$777.

EXAMPLE 3 **PORTFOLIO ANALYSIS** A college foundation owns stock in IBC (selling at \$54 per share), GS (selling at \$65 per share), and ATB (selling at \$105 per share). The foundation owns equal shares of GS and IBC, but five times as many shares of ATB.

If this portfolio is worth \$450,800, how many shares of each type does the foundation own?

Analyze the problem: We must find how many shares of each type of stock is owned. The given information states that there are the same number of shares for both IBC and GS. If we will let x represent the number of shares of IBC, then x also represents the number of shares of GS. There are five times as many shares of ATB, so the number of shares of ATB will be represented by $5x$.

Form an equation: The value of the IBC stock plus the value of the GS stock plus the value of the ATB stock must equal \$450,800.

- Since each of the x shares of IBC stock is worth \$54, the value of this stock is \$$54x$.
- Since each share of x shares of GS stock is worth \$65, the value of this stock is \$$65x$.
- Since each share of the $(5x)$ shares of ATB stock is worth \$105, the value of this stock is \$$105(5x)$.

We set the sum of these values equal to \$450,800.

The value of IBC stock	plus	the value of GS stock	plus	the value of ATB stock	equals	the total value of the portfolio.
$54x$	$+$	$65x$	$+$	$105(5x)$	$=$	450,800

Solve the equation: We can now solve the equation.

$$54x + 65x + 105(5x) = 450{,}800 \quad \text{This is the equation to solve.}$$
$$54x + 65x + 525x = 450{,}800 \quad 105(5x) = 525x$$
$$644x = 450{,}800 \quad \text{Combine like terms.}$$
$$x = 700 \quad \text{Divide both sides by 644.}$$

State the conclusion: The foundation owns 700 shares of IBC, 700 shares of GS, and 5(700) or 3,500 shares of ABT.

Check the result: The value of 700 shares of IBC at \$54 per share is \$37,800. The value of 700 shares of GS at \$65 per share is \$45,500. The value of 3,500 shares of ATB at \$105 per share is \$367,500. The sum of these values is \$450,800.

3 Solve a geometry problem by setting up and solving a linear equation in one variable.

Figure 1-23 illustrates several geometric figures. A **right angle** is an angle whose measure is 90°. A **straight angle** is an angle whose measure is 180°. An **acute angle** is an angle whose measure is greater than 0° but less than 90°.

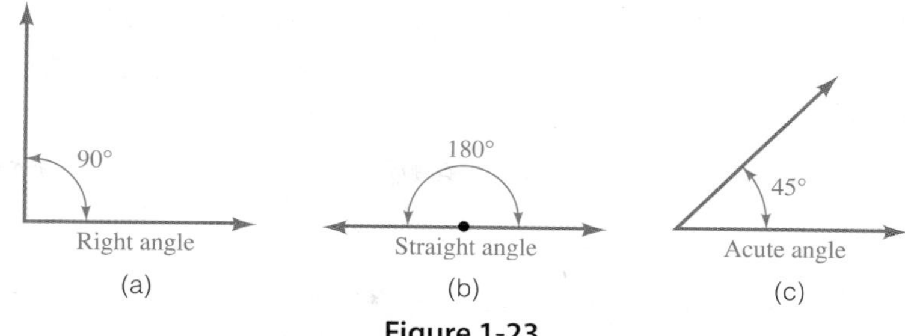

Figure 1-23

If the sum of two angles equals 90°, the angles are called **complementary angles,** and each angle is called the *complement* of the other. If the sum of two angles equals 180°, the angles are called **supplementary angles,** and each angle is called the *supplement* of the other.

A **right triangle** is a triangle with one right angle. In Figure 1-24(a), $\angle C$ (read as "angle C") is a right angle. An **isosceles triangle** is a triangle with two sides of equal measure that meet to form the **vertex angle.** The angles opposite the equal sides, called the **base angles,** are also equal. An **equilateral triangle** is a triangle with three equal sides and three equal angles.

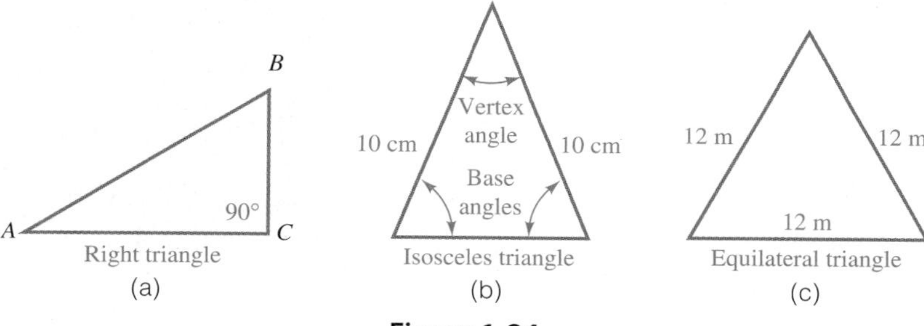

Figure 1-24

EXAMPLE 4 **ANGLES IN A TRIANGLE** If the vertex angle of the isosceles triangle shown in Figure 1-24(b) measures 64°, find the measure of each base angle.

Analyze the problem — We are asked to find the measure of each base angle. If we let x represent the measure of one base angle, the measure of the other is also x.

Form an equation — From geometry, we know that the sum of the measures of the three angles in any triangle is 180°. Thus, the sum of the angles in this triangle, $x + x + 64°$, is equal to 180°. We can form the equation

The measure of one base angle	plus	the measure of the other base angle	plus	the measure of the vertex angle	equals	180°.
x	$+$	x	$+$	64	$=$	180

Solve the equation — We now can solve the equation.

$$x + x + 64 = 180 \quad \text{This is the equation to solve.}$$
$$2x + 64 = 180 \quad \text{Combine like terms.}$$
$$2x = 116 \quad \text{Subtract 64 from both sides.}$$
$$x = 58 \quad \text{Divide both sides by 2.}$$

State the conclusion The measure of each base angle is 58°.

Check the result The sum of the measures of each base angle and the vertex angle is 180°:

$$58° + 58° + 64° = 180°$$

EXAMPLE 5 **DOG RUNS** A man has 28 meters of fencing to make a rectangular kennel. If the kennel is to be 6 meters longer than it is wide, find its dimensions.

Analyze the problem We are asked to find the dimensions of a rectangle. Therefore, we will need to find both the length and the width. If we let w represent the width of the rectangle in meters, the length will be represented by $w + 6$ meters.

Form an equation The perimeter P of a rectangle is the distance around it. Because this is a rectangle, opposite sides have the same measure (See Figure 1-25.) The perimeter can be expressed as either $w + w + 6 + w + w + 6 = P$ or as $2w + 2l = P$.

Figure 1-25

We can form the equation as follows.

Two widths	plus	two lengths	equals	the perimeter.
$2 \cdot w$	$+$	$2 \cdot (w + 6)$	$=$	28

Solve the equation We now can solve the equation.

$$2 \cdot w + 2 \cdot (w + 6) = 28 \quad \text{This is the equation to solve.}$$
$$2w + 2w + 12 = 28 \quad \text{Use the distributive property to remove parentheses.}$$
$$4w + 12 = 28 \quad \text{Combine like terms.}$$
$$4w = 16 \quad \text{Subtract 12 from both sides.}$$
$$w = 4 \quad \text{Divide both sides by 4.}$$
$$w + 6 = 10$$

State the conclusion The dimensions of the kennel are 4 meters by 10 meters.

Check the result If a kennel has a width of 4 meters and a length of 10 meters, its length is 6 meters longer than its width, and the perimeter is 2(4) meters + 2(10) meters = 28 meters.

4 Solve a lever problem by setting up and solving a linear equation in one variable.

EXAMPLE 6 **ENGINEERING** Design engineers must position two hydraulic cylinders as in Figure 1-26 to balance a 9,500-pound force at point A. The first cylinder at the end of the lever exerts a 3,500-pound force. Where should the engineers position the second cylinder, which is capable of exerting a 5,500-pound force?

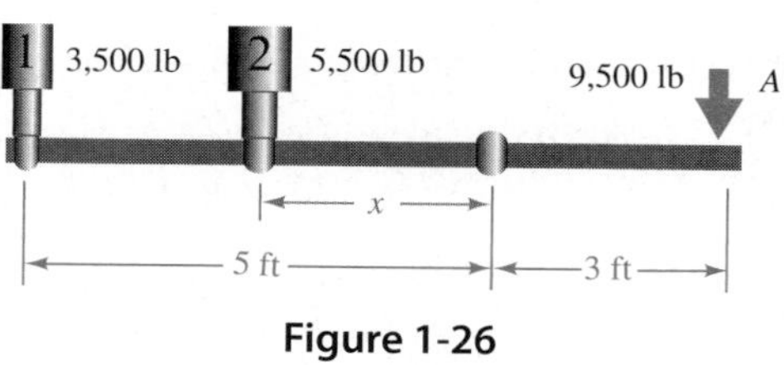

Figure 1-26

Analyze the problem: We must find the distance from the fulcrum that the second cylinder must be placed in order to balance the two sides. We will let x represent this distance.

Form an equation: From a simple principle in physics, the lever will be in balance when the force of the first cylinder multiplied by its distance from the pivot (also called the **fulcrum**) added to the force of the second cylinder multiplied by its distance from the fulcrum is equal to the 9,500 force multiplied by its distance from the fulcrum.

Using this information, we can form the equation

Force of cylinder 1 times the distance	plus	force of cylinder 2 times the distance	equals	force to be balanced times the distance.
$3{,}500 \cdot 5$	$+$	$5{,}500x$	$=$	$9{,}500 \cdot 3$

Solve the equation: We can now solve the equation.

$$3{,}500 \cdot 5 + 5{,}500x = 9{,}500 \cdot 3 \quad \text{This is the equation to solve.}$$
$$17{,}500 + 5{,}500x = 28{,}500 \quad \text{Multiply.}$$
$$5{,}500x = 11{,}000 \quad \text{Subtract 17,500 from both sides.}$$
$$x = 2 \quad \text{Divide both sides by 5,500.}$$

State the conclusion: The design must specify that the second cylinder be positioned 2 feet from the fulcrum.

Check the result: $3{,}500 \cdot 5 + 5{,}500 \cdot 2 = 17{,}500 + 11{,}000 = 28{,}500$
$9{,}500 \cdot 3 = 28{,}500$

NOW TRY THIS

Given the sum of two numbers is 92,

1. If one of the numbers is 27, what is the other?

2. If one of the numbers is -56, what is the other?

3. If one of the numbers is x, what expression represents the other number?

1.6 EXERCISES

WARM-UPS

Find each value.

1. 20% of 500

2. $33\frac{1}{3}\%$ of 600

If a stock costs $54, find the cost of

3. 5 shares

4. x shares

Find the area of the rectangle with the given dimensions.

5. 6 meters long, 4 meters wide

6. l meters long, $l - 5$ meters wide

REVIEW *Simplify each expression. Assume no variables are zero.*

7. $\left(\frac{3x^{-3}}{4x^2}\right)^{-4}$

8. $\left(\frac{r^{-3}s^2}{r^2r^3s^{-4}}\right)^{-5}$

9. $\frac{a^m a^3}{a^2}$

10. $\left(\frac{b^n}{b^3}\right)^3$

VOCABULARY AND CONCEPTS *Fill in the blanks.*

11. The expression ______ represents the phrase "4 more than 5 times x."

12. The expression ______ represents the phrase "6 percent of x."

13. The expression ____ represents the phrase "the value of x shares priced at $40 per share."

14. A right angle is an angle whose measure is ____. A straight angle measures _____.

15. An _____ angle measures greater than 0° but less than 90°.

16. If the sum of the measures of two angles equals 90°, the angles are called ____________ angles.

17. If the sum of the measures of two angles equals _____, the angles are called supplementary angles.

18. If a triangle has a right angle, it is called a _____ triangle.

19. If a triangle has two sides with equal measure, it is called an ________ triangle.

20. The equal sides of an isosceles triangle form the ______ angle and the remaining angles are called _________.

21. An _________ triangle has three equal sides and three equal angles.

22. The pivot point of a lever is called the _______.

APPLICATIONS

Set up an equation and solve each problem. See Example 1. (Objective 1)

23. Cutting ropes A 60-foot rope is cut into four pieces, with each successive piece twice as long as the previous one. Find the length of the longest piece.

24. Cutting cables A 186-foot cable is to be cut into four pieces. Find the length of each piece if each successive piece is 3 feet longer than the previous one.

25. Cutting boards The carpenter in the illustration saws a board into two pieces. He wants one piece to be 1 foot longer than twice the length of the shorter piece. Find the length of each piece.

26. Cutting beams A 30-foot steel beam is to be cut into two pieces. The longer piece is to be 2 feet more than three times as long as the shorter piece. Find the length of each piece.

Set up an equation and solve each problem. See Example 2. (Objective 2)

27. Buying a washer and dryer Find the percent of markdown of the sale in the following ad.

28. Buying furniture A bedroom set regularly sells for $983. If it is on sale for $737.25, what is the percent of markdown?

29. Buying books When the price of merchandise is increased, the amount of increase is called the *markup.* If a bookstore buys a used calculus book for $12 and sells it for $40, find the percent of markup.

30. Selling toys When the price of merchandise is increased, the amount of increase is called the *markup.* If the owner of a gift shop buys stuffed animals for $18 and sells them for $30, find the percent of markup.

Set up an equation and solve each problem. See Example 3. (Objective 2)

31. Value of an IRA In an Individual Retirement Account (IRA) valued at $53,900, a student has 500 shares of stock, some in Big Bank Corporation and some in Safe Savings and Loan. If Big Bank sells for $115 per share and Safe Savings sells for $97 per share, how many shares of each does the student own?

32. Pension funds A pension fund owns 12,000 shares in a stock mutual fund and a bond mutual fund. Currently, the stock fund sells for \$12 per share, and the bond fund sells for \$15 per share. How many shares of each does the pension fund own if the value of the securities is \$165,000?

33. Selling calculators Last month, a bookstore ran the following ad and sold 85 calculators, generating \$3,875 in sales. How many of each type of calculator did the bookstore sell?

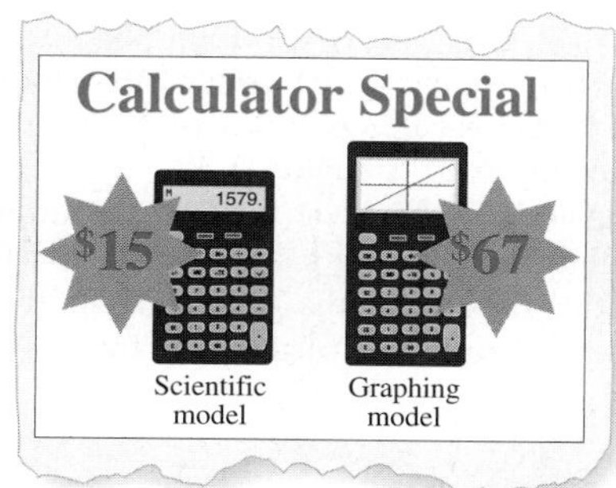

34. Selling grass seed A seed company sells two grades of grass seed. A 100-pound bag of a mixture of rye and bluegrass sells for \$245, and a 100-pound bag of bluegrass sells for \$347. How many bags of each are sold in a week when the receipts for 19 bags are \$5,369?

Set up an equation and solve each problem. See Example 4. (Objective 3)

35. Equilateral triangles Find the measure of each angle in an equilateral triangle.

36. Quadrilaterals The sum of the angles of any four-sided figure (called a *quadrilateral*) is 360°. The following quadrilateral has two equal base angles. Find x.

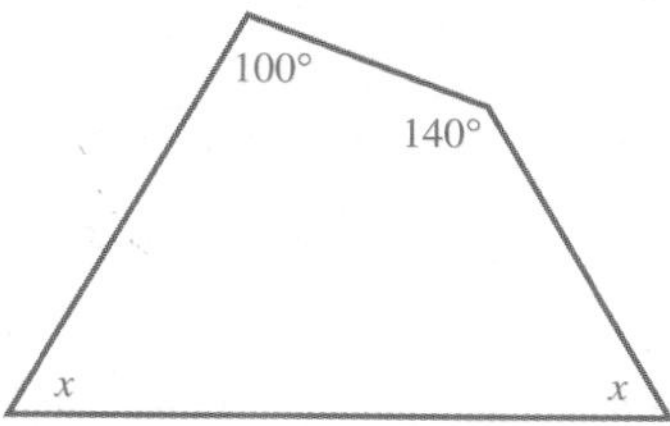

37. Vertical angles When two lines intersect as in the illustration, four angles are formed. Angles that are side-by-side, such as $\angle 1$ and $\angle 2$, are called **adjacent angles.** Angles that are nonadjacent, such as $\angle 1$ and $\angle 3$ or $\angle 2$ and $\angle 4$, are called **vertical angles.** From geometry, we know that if two lines intersect, vertical angles have the same measure. If $m(\angle 1) = (3x + 10)°$ and $m(\angle 3) = (5x - 10)°$, find x. Read $m(\angle 1)$ as "the measure of $\angle 1$."

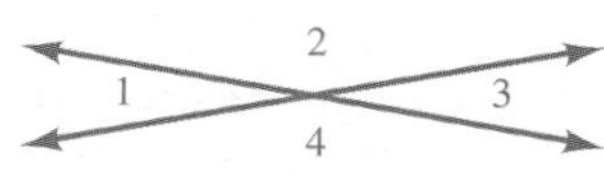

38. If $m(\angle 2) = (6x + 20)°$ in the previous illustration and $m(\angle 4) = (8x - 20)°$, find $m(\angle 1)$. (See Exercise 37.)

39. Supplementary angles If one of two supplementary angles is 35° larger than the other, find the measure of the smaller angle.

40. Supplementary angles Refer to the illustration and find x.

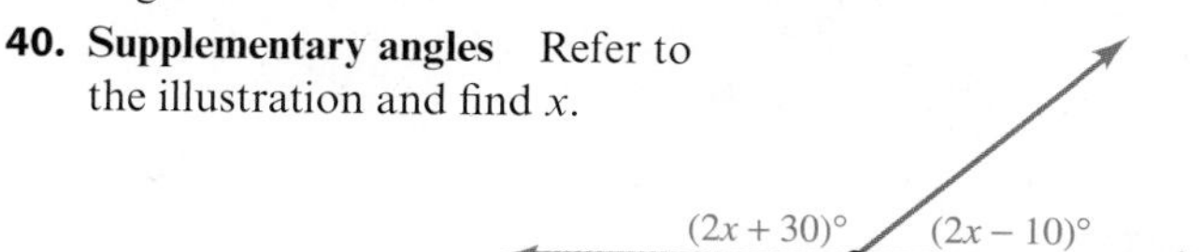

41. Complementary angles If one of two complementary angles is 22° greater than the other, find the measure of the larger angle.

42. Complementary angles in a right triangle Explain why the acute angles in the following right triangle are complementary. Then find the measure of angle A.

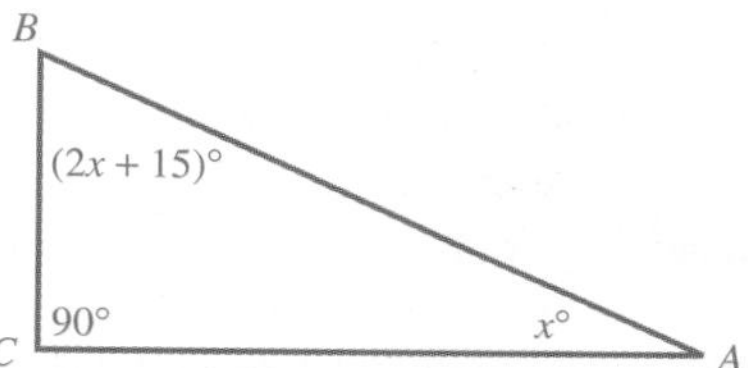

Set up an equation and solve each problem. See Example 5. (Objective 3)

43. Finding dimensions The rectangular garden shown in the illustration is twice as long as it is wide. Find its dimensions.

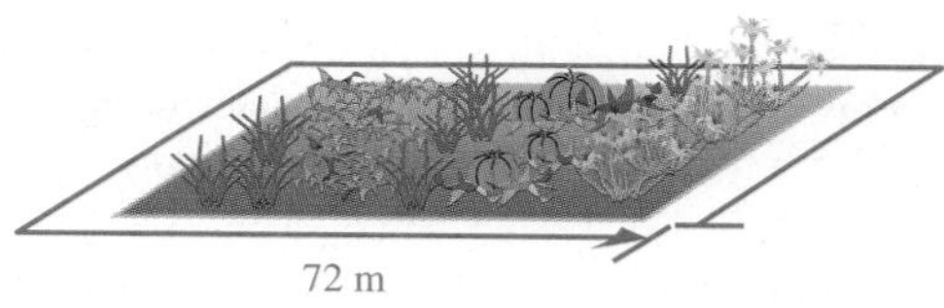

44. Finding dimensions The width of a rectangular swimming pool is one-third its length. If its perimeter is 96 meters, find the dimensions of the pool.

45. Fencing pastures A farmer has 624 feet of fencing to enclose the pasture shown in the illustration. Because a river runs along one side, fencing will be needed on only three sides. Find the dimensions of the pasture if its length is double its width.

46. **Fencing pens** A man has 150 feet of fencing to build the pen shown in the illustration. If one end is a square, find the outside dimensions.

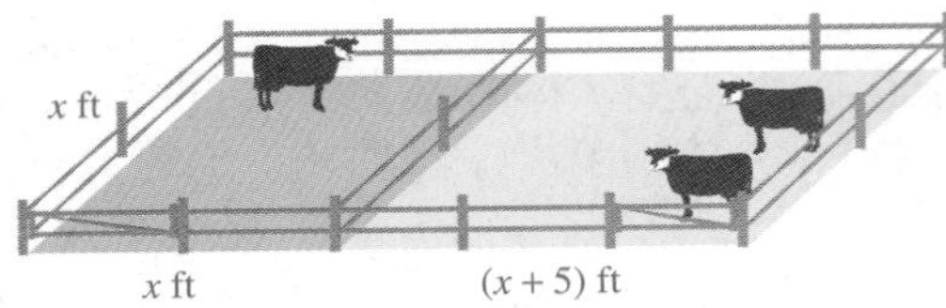

Set up an equation and solve each problem. Ignore the weight of the lever. **See Example 6. (Objective 4)**

47. **Balancing a seesaw** A seesaw is 20 feet long, and the fulcrum is in the center. If an 80-pound boy sits at one end, how far will the boy's 160-pound father have to sit from the fulcrum to balance the seesaw?

48. **Establishing equilibrium** Two forces—110 pounds and 88 pounds—are applied to opposite ends of an 18-foot lever. How far from the greater force must the fulcrum be placed so that the lever is balanced?

49. **Moving stones** A woman uses a 10-foot bar to lift a 210-pound stone. If she places another rock 3 feet from the stone to act as the fulcrum, how much force must she exert to move the stone?

50. **Lifting cars** A 350-pound football player brags that he can lift a 2,500-pound car. If he uses a 12-foot bar with the fulcrum placed 3 feet from the car, will he be able to lift the car?

ADDITIONAL APPLICATIONS ***Set up an equation and solve each problem.***

51. **Buying roses** A man with \$21.25 stops after work to order some roses for his wife's birthday. If each rose costs \$1.25 and there is a delivery charge of \$5, how many roses can he buy?

52. **Renting trucks** To move to Wisconsin, a man can rent a truck for \$49.95 per day plus 39¢ per mile. If he keeps the truck for one day, how many miles can he drive for a cost of \$147.45?

53. **Car rentals** While waiting for his car to be repaired, a man rents a car for \$12 per day plus 30¢ per mile. If he keeps the car for 2 days, how many miles can he drive for a total cost of \$42? How many miles can he drive for a total cost of \$60?

54. **Computing salaries** A student earns \$17 per day for delivering overnight packages. She is paid \$5 per day plus 60¢ for each package delivered. How many more deliveries must she make each day to increase her daily earnings to \$23?

55. **Buying a TV and a VCR** See the following ad. If the TV costs \$55 more than the VCR, how much does the TV cost?

56. **Buying golf clubs** The cost of a set of golf clubs is \$590. If the irons cost \$40 more than the woods, find the cost of the irons.

57. **Triangles** If the height of a triangle with a base of 8 inches is tripled, its area is increased by 96 square inches. Find the height of the triangle.

58. **Engineering designs** The width, w, of the flange in the following engineering drawing has not yet been determined. Find w so that the area of the 7-in.-by-12-in. rectangular portion is exactly one-half of the total area.

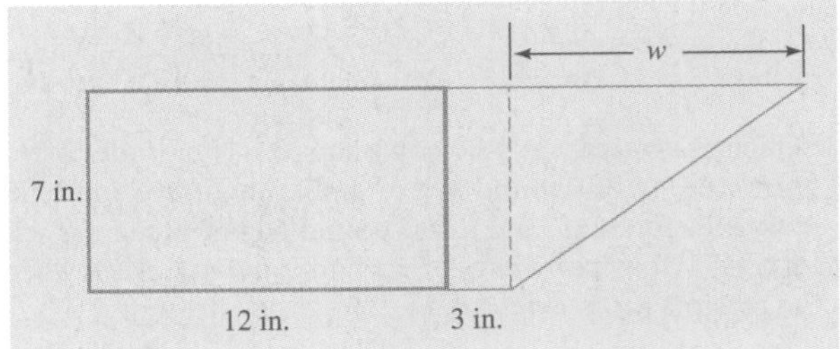

59. **Supplementary angles and parallel lines** In the illustration, lines r and s are cut by a third line l to form $\angle 1$ and $\angle 2$. When lines r and s are parallel, $\angle 1$ and $\angle 2$ are supplementary. If $m(\angle 1) = (x + 50)°$ and $m(\angle 2) = (2x - 20)°$ and lines r and s are parallel, find x. Read $m(\angle 1)$ as "the measure of $\angle 1$."

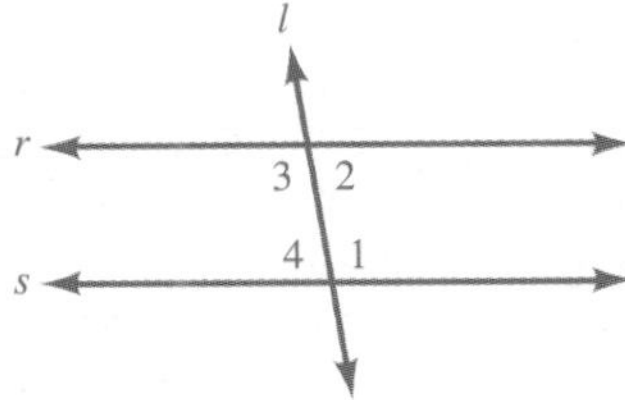

60. In the illustration, find $m(\angle 3)$. (*Hint:* See Exercise 59.)

61. **Enclosing a swimming pool** A woman wants to enclose the swimming pool shown in the illustration on the next page and have a walkway of uniform width all the way around. How wide will the walkway be if the woman uses 180 feet of fencing?

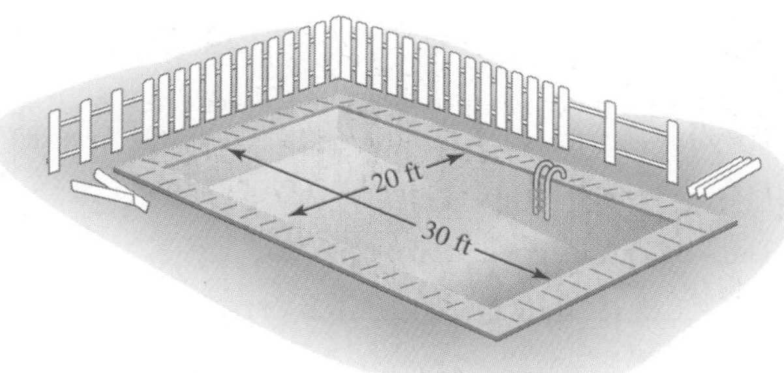

62. **Framing pictures** An artist wants to frame the following picture with a frame 2 inches wide. How wide will the framed picture be if the artist uses 70 inches of framing material?

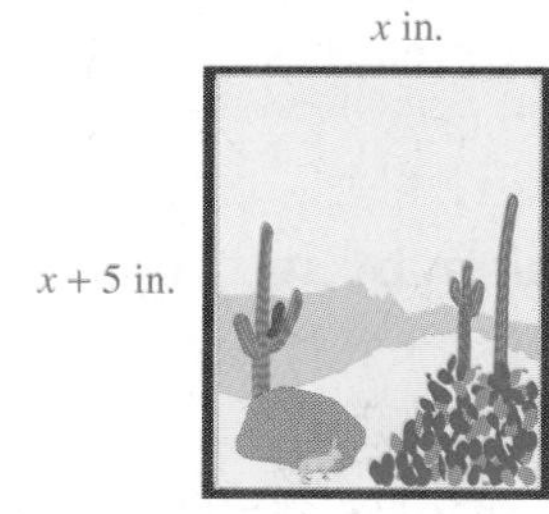

63. **Balancing levers** Forces are applied to a lever as indicated in the illustration. Find x, the distance of the smallest force from the fulcrum. Ignore the weight of the lever.

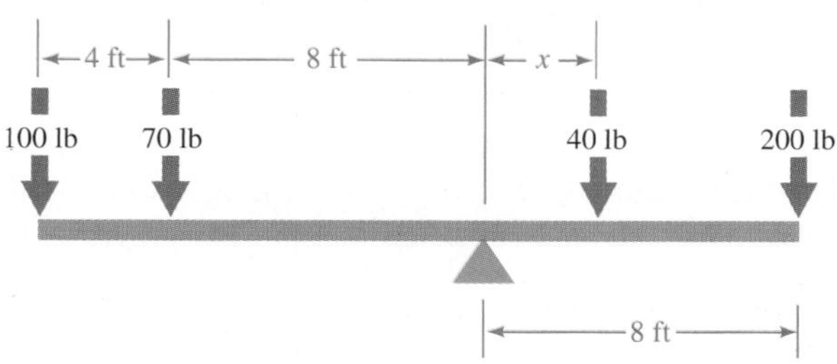

64. **Balancing seesaws** Jim and Bob sit at opposite ends of an 18-foot seesaw, with the fulcrum at its center. Jim weighs 160 pounds, and Bob weighs 200 pounds. When Kim sits 4 feet in front of Jim, the seesaw balances. How much does Kim weigh?

65. **Temperature scales** The Celsius and Fahrenheit temperature scales are related by the equation $C = \frac{5}{9}(F - 32)$. At what temperature will a Fahrenheit and a Celsius thermometer give the same reading?

66. **Solar heating** One solar panel in the illustration is to be 3 feet wider than the other, but to be equally efficient, they must have the same area. Find the width of each.

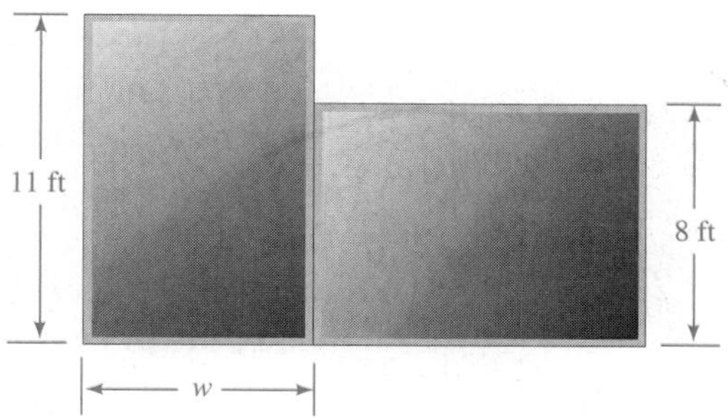

WRITING ABOUT MATH

67. Explain the steps for solving an applied problem.
68. Explain how to check the solution of an applied problem.

SOMETHING TO THINK ABOUT

69. Find the distance x required to balance the lever in the illustration.

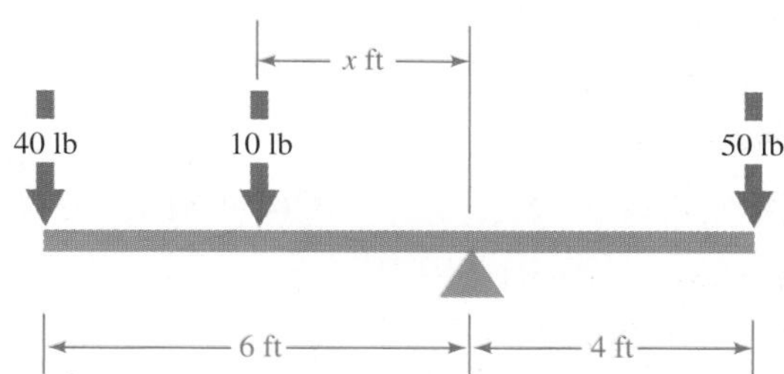

70. Interpret the answer to Exercise 69.

SECTION 1.7 More Applications of Linear Equations in One Variable

Objectives

1. Solve an investment problem by setting up and solving a linear equation in one variable.
2. Solve a uniform motion problem by setting up and solving a linear equation in one variable.
3. Solve a mixture problem by setting up and solving a linear equation in one variable.

Getting Ready

1. Find 8% of \$500.
2. Write an expression for 8% of \$$x$.
3. If \$9,000 of \$15,000 is invested at 7%, how much is left to invest at another rate?
4. If \$$x$ of \$15,000 is invested at 7%, write an expression for how much is left to be invested at another rate.
5. If a car travels 50 mph for 4 hours, how far will it go?
6. If coffee sells for \$4 per pound, how much will 5 pounds cost?

In this section, we will continue our investigation of problem solving by considering investment, motion, and mixture problems.

1 Solve an investment problem by setting up and solving a linear equation in one variable.

EXAMPLE 1 **FINANCIAL PLANNING** A professor has \$15,000 to invest for one year, some at 8% and the rest at 7% annual interest. If she will earn \$1,110 from these investments, how much did she invest at each rate?

Analyze the problem

We must find how much money the professor has invested in each of the two accounts. If we let x represent the amount she has invested in the account paying 8% interest, then the rest, or \$15,000 − x, is invested in the account paying 7% interest.

Form an equation

Simple interest is computed by the formula $i = prt$, where i is the interest earned, p is the principal, r is the annual interest rate, and t is the length of time the principal is invested. In this problem, $t = 1$ year. Thus, if \$$x$ is invested at 8% for one year, the interest earned is $i = \$x(8\%)(1) = \$0.08x$. If the remaining \$$(15{,}000 - x)$ is invested at 7%, the amount earned on that investment is \$$0.07(15{,}000 - x)$. See Figure 1-27. The sum of these amounts should equal the total earned interest of \$1,110.

	i	=	p	· r	· t
8% investment	$0.08x$		x	0.08	1
7% investment	$0.07(15{,}000 - x)$		$15{,}000 - x$	0.07	1

Figure 1-27

Using this information, we can form the equation

The interest earned at 8%	plus	the interest earned at 7%	equals	the total interest.
$0.08x$	+	$0.07(15{,}000 - x)$	=	1,110

Solve the equation

We can now solve the equation.

$$0.08x + 0.07(15{,}000 - x) = 1{,}110$$ This is the equation to solve.

$$8x + 7(15{,}000 - x) = 111{,}000$$ To eliminate the decimals, multiply both sides by 100.

$$8x + 105{,}000 - 7x = 111{,}000$$ Use the distributive property to remove parentheses.

$$x + 105{,}000 = 111{,}000$$ Combine like terms.

$$x = 6{,}000 \quad \text{Subtract 105,000 from both sides.}$$
$$15{,}000 - x = 9{,}000$$

State the conclusion She invested \$6,000 at 8% and \$9,000 at 7%.

Check the result The interest on \$6,000 is 0.08(\$6,000) = \$480. The interest earned on \$9,000 is 0.07(\$9,000) = \$630. The total interest is \$1,110.

2 Solve a uniform motion problem by setting up and solving a linear equation in one variable.

EXAMPLE 2 **TRAVEL TIMES** A car leaves Rockford traveling toward Wausau at the rate of 55 mph. At the same time, another car leaves Wausau traveling toward Rockford at the rate of 50 mph. How long will it take them to meet if the cities are 157.5 miles apart?

Analyze the problem Since we must find the time it takes for the two cars to meet, we will let t represent the time in hours.

Form an equation Uniform motion problems are based on the formula $d = rt$, where d is distance, r is rate, and t is time. In this case, the cars are traveling toward each other as shown in Figure 1-28(a) and we can organize the given information in a chart as shown in Figure 1-28(b).

We know that one car is traveling at 55 mph and that the other is going 50 mph. We also know that they travel for the same amount of time, t hours. Thus, the distance that the faster car travels is $55t$ miles, and the distance that the slower car travels is $50t$ miles. The sum of these distances equals 157.5 miles, the distance between the cities.

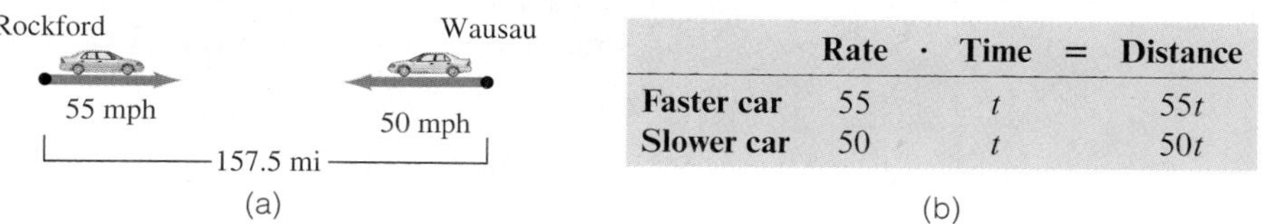

	Rate	· Time	= Distance
Faster car	55	t	$55t$
Slower car	50	t	$50t$

(b)

Figure 1-28

Using this information, we can form the equation

The distance the faster car goes	plus	the distance the slower car goes	equals	the distance between the cities.
$55t$	$+$	$50t$	$=$	157.5

Solve the equation We can now solve the equation.

$$55t + 50t = 157.5 \quad \text{This is the equation to solve.}$$
$$105t = 157.5 \quad \text{Combine like terms.}$$
$$t = 1.5 \quad \text{Divide both sides by 105.}$$

State the conclusion The two cars will meet in $1\frac{1}{2}$ hours.

Check the result The faster car travels 1.5(55) = 82.5 miles. The slower car travels 1.5(50) = 75 miles. The total distance traveled is 157.5 miles.

3 Solve a mixture problem by setting up and solving a linear equation in one variable.

EXAMPLE 3 **MIXING NUTS** The owner of a candy store notices that 20 pounds of gourmet cashews are getting stale. They did not sell because of their high price of \$12 per pound. The store owner decides to mix peanuts with the cashews to lower the price per pound. If peanuts sell for \$3 per pound, how many pounds of peanuts must be mixed with the cashews to make a mixture that could be sold for \$6 per pound?

Analyze the problem — We must find the number of pounds of peanuts needed to mix with the cashews, so we will let x represent the number of pounds of peanuts. We will be adding the x pounds of peanuts to the 20 pounds of cashews, so the total number of pounds of the mixture will be $20 + x$.

Form an equation — This problem is based on the formula $V = pn$, where V represents the value, p represents the price per pound, and n represents the number of pounds. We can enter the known information in a table as shown in Figure 1-29. The *value* of the cashews plus the *value* of the peanuts will be equal to the *value* of the mixture.

	Price	· Number of pounds	= Value
Cashews	12	20	240
Peanuts	3	x	$3x$
Mixture	6	$20 + x$	$6(20 + x)$

Figure 1-29

Using this information, we can form the equation

The value of the cashews	plus	the value of the peanuts	equals	the value of the mixture.
240	$+$	$3x$	$=$	$6(20 + x)$

Solve the equation — We can now solve the equation.

$$240 + 3x = 6(20 + x) \quad \text{This is the equation to solve.}$$
$$240 + 3x = 120 + 6x \quad \text{Use the distributive property to remove parentheses.}$$
$$120 = 3x \quad \text{Subtract } 3x \text{ and 120 from both sides.}$$
$$40 = x \quad \text{Divide both sides by 3.}$$

State the conclusion — The store owner should mix 40 pounds of peanuts with the 20 pounds of cashews.

Check the result — The cashews are valued at \$12(20) = \$240 and the peanuts are valued at \$3(40) = \$120, so the mixture is valued at \$6(60) = \$360.

The value of the cashews plus the value of the peanuts equals the value of the mixture.

EXAMPLE 4 **MILK PRODUCTION** A container is partially filled with 12 liters of whole milk containing 4% butterfat. How much 1% milk must be added to get a mixture that is 2% butterfat?

Analyze the problem — We will let x represent the number of liters of the 1% butterfat milk to be added. If there are already 12 liters of milk in the container, the total amount of milk will be $(12 + x)$ liters.

Form an equation Since the first container shown in Figure 1-30(a) contains 12 liters of 4% milk, it contains 0.04(12) liters of butterfat. To this container we will add the contents of the second container, which holds $0.01x$ liters of butterfat. The sum of these two amounts of butterfat $(0.04(12) + 0.01x)$ will be equal to the amount of butterfat in the third container, which is $0.02(12 + x)$ liters of butterfat. This information is presented in table form in Figure 1-30(b).

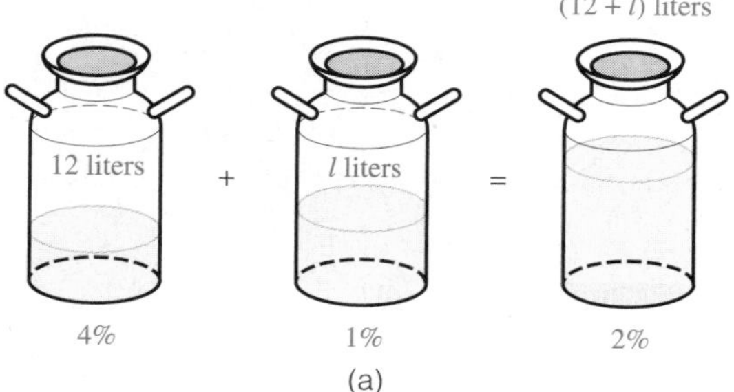

(a)

	Amount of butterfat	=	Amount of milk	·	Percent of butterfat
Whole milk	0.04(12)		12		0.04
1% milk	$0.01(x)$		x		0.01
2% milk	$0.02(12 + x)$		$12 + x$		0.02

(b)

Figure 1-30

Using this information, we can form the equation

The amount of butterfat in 12 liters of 4% milk	plus	the amount of butterfat in x liters of 1% milk	equals	the amount of butterfat in $(12 + x)$ liters of mixture.
0.04(12)	+	$0.01x$	=	$0.02(12 + x)$

Solve the equation We now solve the equation.

$$0.04(12) + 0.01x = 0.02(12 + x)$$ This is the equation to solve.

$$4(12) + 1x = 2(12 + x)$$ Multiply both sides by 100.

$$48 + x = 24 + 2x$$ Use the distributive property to remove parentheses.

$$24 = x$$ Subtract 24 and x from both sides.

State the conclusion Thus, 24 liters of 1% butterfat milk should be added to get a mixture that is 2% butterfat.

Check the result 12 liters of 4% milk contains 0.48 liters of butterfat and 24 liters of 1% milk contains 0.24 liters of butterfat. This gives a total of 36 liters of a mixture that contains 0.72 liters of butterfat. This is a 2% solution.

NOW TRY THIS

1. Let s represent the speed of a plane in still air. If it flies *into* the wind (headwind), its speed is reduced by the speed of the wind, and if it flies *with* the wind (tailwind), its speed is increased by the speed of the wind.
   a. Represent the speed of a plane flying into a 40 mph headwind.
   b. Represent the speed of a plane flying with a 40 mph tailwind.
   c. If a plane flies 480 miles in 4 hours into a 40 mph headwind, find its speed in still air.
   d. If a plane flies 480 miles in 2 hours with a 40 mph tailwind, find its speed in still air.

1.7 EXERCISES

WARM-UPS *Assume that all investments are for one year.*

1. How much interest will \$1,500 earn if invested at 6%?
2. Express the amount of interest \$$x$ will earn if invested at 5%.
3. If \$$x$ of \$30,000 is invested at 5%, how would you express the amount left to be invested at 6%?
4. If Brazil nuts are worth \$$x$ per pound, express how much 20 pounds will be worth.
5. If whole milk is 4% butterfat, how much butterfat is in 2 gallons?
6. If whole milk is 4% butterfat, express how much butterfat is in x gallons.

REVIEW *Solve each equation.*

7. $9x - 3 = 6x$
8. $7a + 2 = 12 - 4(a - 3)$
9. $\frac{8(y-5)}{3} = 2(y-4)$
10. $\frac{t-1}{3} = \frac{t+2}{6} + 2$

VOCABULARY AND CONCEPTS *Fill in the blanks.*

11. The formula for simple interest i is $i = prt$, where p is the ________, r is the ____, and t is the ____.
12. Uniform motion problems are based on the formula ______, where d is the ________, r is the ____ of speed, and t is the ____.
13. Dry mixture problems are based on the formula $V = pn$, where V is the ______, p is the ______ per unit, and n is the ________ of units.
14. The liquid mixture problem in Example 4 is based on the idea that the amount of ________ in the first container plus the amount of ________ in the second container is equal to the amount of ________ in the mixture.

APPLICATIONS

Set up an equation and solve each problem. See Example 1. (Objective 1)

15. **Investing** Lured by the following ad, a woman invested \$12,000, some in a money market account and the rest in a 5-year CD. How much was invested in each account if the income from both investments is \$680 per year?

First Republic Savings and Loan	
Account	**Rate**
NOW	2.5%
Savings	3.5%
Money market	4.0%
Checking	2.0%
5-year CD	6.0%

16. **Investing** A man invested \$14,000, some at 7% and some at 10% annual interest. The annual income from these investments was \$1,280. How much did he invest at each rate?
17. **Supplemental income** A teacher wants to earn \$1,500 per year in supplemental income from a cash gift of \$16,000. She puts \$6,000 in a credit union that pays 7% annual interest. What rate must she earn on the remainder to achieve her goal?
18. **Inheriting money** Paul split an inheritance between two investments, one paying 7% annual interest and the other 10%. He invested twice as much in the 10% investment as he did in the 7% investment. If his combined annual income from the two investments was \$4,050, how much did he inherit?
19. **Investing** Kyoko has some money to invest. If she invests \$3,000 more, she can qualify for an 11% investment. Otherwise, she can invest the money at 7.5% annual interest. If the 11% investment yields twice as much annual income as the 7.5% investment, how much does she have on hand to invest?
20. **Supplemental income** A bus driver wants to earn \$3,500 per year in supplemental income from an inheritance of \$40,000. If the driver invests \$10,000 in a mutual fund paying 8%, what rate must he earn on the remainder to achieve his goal?
21. **Concert receipts** For a jazz concert, student tickets are \$2 each, and adult tickets are \$4 each. If 200 tickets were sold and the total receipts are \$750, how many student tickets were sold?
22. **School plays** At a school play, 140 tickets are sold, with total receipts of \$290. If adult tickets cost \$2.50 each and student tickets cost \$1.50 each, how many adult tickets were sold?

Set up an equation and solve each problem. See Example 2. (Objective 2)

23. **Computing time** One car leaves Chicago headed for Cleveland, a distance of 343 miles. At the same time, a second car leaves Cleveland headed toward Chicago. If the first car averages 50 mph and the second car averages 48 mph, how long will it take the cars to meet?

24. Storm research During a storm, two teams of scientists leave a university at the same time to search for tornados. The first team travels east at 20 mph, and the second travels west at 25 mph, as shown in the illustration. If their radios have a range of 90 miles, how long will it be before they lose radio contact?

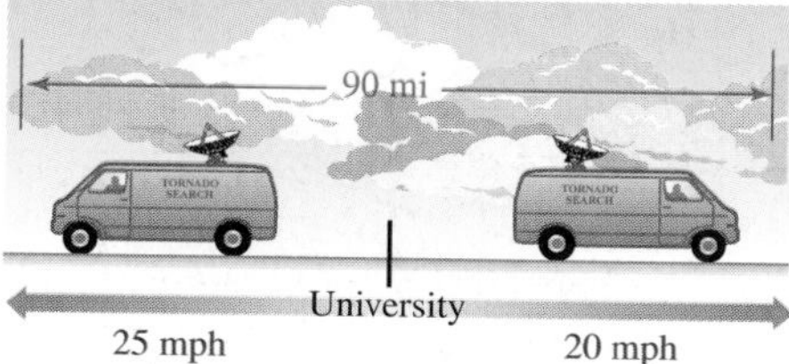

25. Cycling A cyclist leaves Las Vegas riding at the rate of 18 mph. See the illustration. One hour later, a car leaves Las Vegas going 45 mph in the same direction. How long will it take the car to overtake the cyclist?

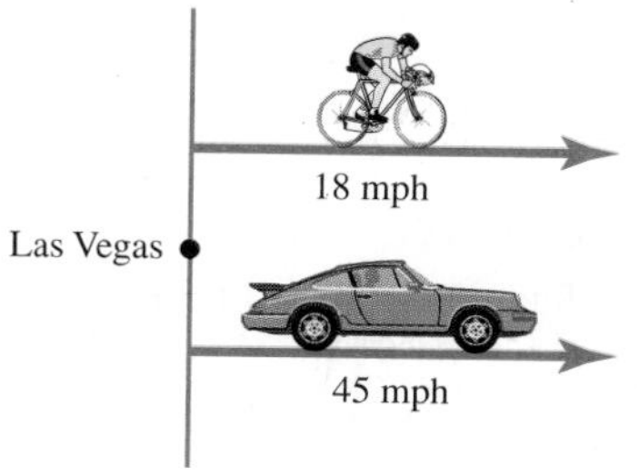

26. Computing distance At 2 P.M., two cars leave Eagle River, WI, one headed north and one headed south. If the car headed north averages 50 mph and the car headed south averages 60 mph, when will the cars be 165 miles apart?

27. Running a race Two runners leave the starting gate, one running 12 mph and the other 10 mph. If they maintain the pace, how long will it take for them to be one-quarter of a mile apart?

28. Trucking Two truck drivers leave a warehouse traveling in opposite directions, one driving at 50 mph and one driving at 56 mph. If the slower driver has a 2-hour head start, how long has she been on the road when the two drivers are 683 miles apart?

29. Taking a walk Sarah walked north at the rate of 3 mph and returned at the rate of 4 mph. How many miles did she walk if the round-trip took 3.5 hours?

30. Computing travel time Jamal traveled a distance of 400 miles in 8 hours. Part of the time, his rate of speed was 45 mph. The rest of the time, his rate of speed was 55 mph. How long did Jamal travel at each rate?

Set up an equation and solve each problem. **See Example 3. (Objective 3)**

31. Mixing candies The owner of a store wants to make a 30-pound mixture of two candies to sell for \$3 per pound. If one candy sells for \$2.95 per pound and the other for \$3.10 per pound, how many pounds of each should be used?

32. Computing selling price A mixture of candy is made to sell for \$3.89 per pound. If 32 pounds of candy, selling for \$3.80 per pound, are used along with 12 pounds of a more expensive candy, find the price per pound of the better candy.

33. Raising grades A student scores 70% on a test that contains 30 questions. To improve his score, the instructor agrees to let him work 15 additional questions. How many must he get right to raise his grade to 80%?

34. Raising grades On a second exam, the student in Exercise 33 earns a score of 60% on a 20-question test. This time, the instructor allows him to work 20 extra problems to improve his score. How many must he get right to raise his grade to 70%?

Set up an equation and solve each problem. **See Example 4. (Objective 3)**

35. Making whole milk Cream is approximately 22% butterfat. How many gallons of cream must be mixed with milk testing at 2% butterfat to get 20 gallons of milk containing 4% butterfat?

36. Mixing solutions How much acid must be added to 60 grams of a solution that is 65% acid to obtain a new solution that is 75% acid?

37. Computing grades Before the final, Estela had earned a total of 375 points on four tests. To receive an A in the course, she must have 90% of a possible total of 450 points. Find the lowest number of points that she can earn on the final exam and still receive an A.

38. Computing grades A student has earned a total of 435 points on five tests. To receive a B in the course, he must have 80% of a possible total of 600 points. Find the lowest number of points that the student can make on the final exam and still receive a B.

ADDITIONAL APPLICATIONS *Set up an equation and solve each problem.*

39. Bookstores A bookstore sells a history book for \$65. If the bookstore makes a profit of 30% on each book, what does the bookstore pay the publisher for each book? (*Hint:* The retail price = the wholesale price + the markup.)

40. Bookstores A bookstore sells a textbook for \$39.20. If the bookstore makes a profit of 40% on each sale, what does the bookstore pay the publisher for each book? (*Hint:* The retail price = the wholesale price + the markup.)

41. Riding a jet ski A jet ski can go 12 mph in still water. If a rider goes upstream for 3 hours against a current of 4 mph, how long will it take the rider to return?

42. Riding a motorboat The motorboat in the illustration can go 18 mph in still water. If a trip downstream takes 4 hours and the return trip takes 5 hours, find the speed of the current.

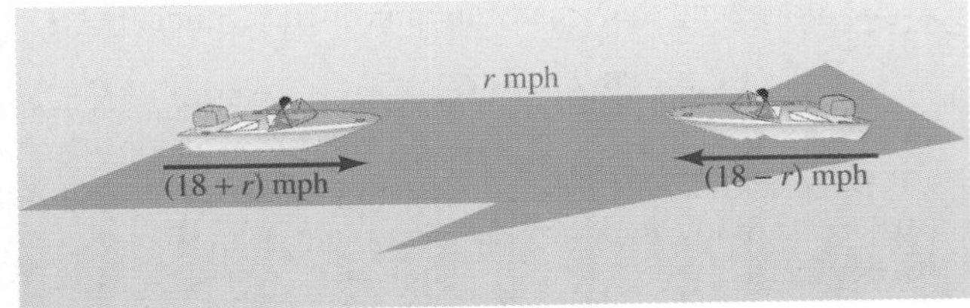

43. Diluting solutions In the illustration, how much water should be added to 20 ounces of a 15% solution of alcohol to dilute it to a 10% solution?

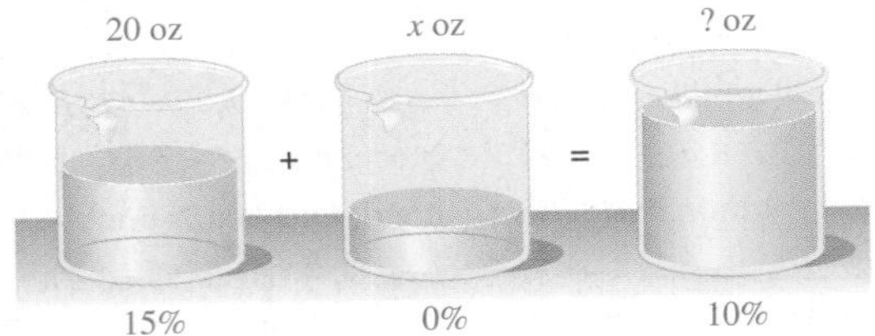

44. Increasing concentration How much water must be boiled away to increase the concentration of 300 gallons of a salt solution from 2% to 3%?

45. Making furniture A woodworker wants to put two partitions crosswise in the drawer shown in the illustration. (See the illustration.) He wants to place the partitions so that the spaces created increase by 3 inches from front to back. If the thickness of each partition is $\frac{1}{2}$ inch, how far from the front end should he place the first partition?

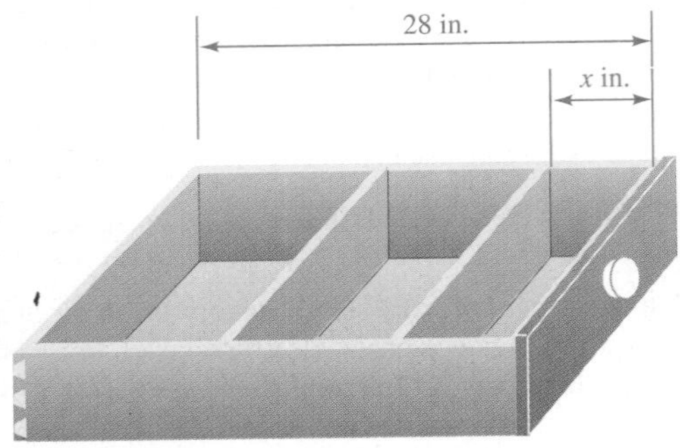

46. Building shelves A carpenter wants to put four shelves on an 8-foot wall so that the five spaces created decrease by 6 inches as we move up the wall. (See the illustration.) If the thickness of each shelf is $\frac{3}{4}$ inch, how far will the bottom shelf be from the floor?

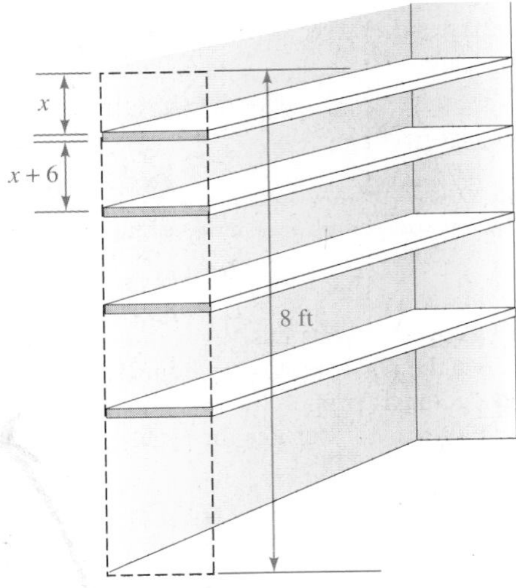

WRITING ABOUT MATH

47. What do you find most difficult in solving application problems, and why?

48. Which type of application problem do you find easiest, and why?

SOMETHING TO THINK ABOUT

49. Discuss the difficulties in solving this problem:

A man drives 100 miles at 30 mph. How fast should he drive on the return trip to average 60 mph for the entire trip?

50. What difficulties do you encounter when solving this problem?

Adult tickets cost $4, and student tickets cost $2. Sales of 71 tickets bring in $245. How many of each were sold?

PROJECTS

Project 1

Theresa manages the local outlet of Hook n' Slice, a sporting goods retailer. She has hired you to help solve some problems. Be sure to provide explanations of how you arrive at your solutions.

a. The company recently directed that all shipments of golf clubs be sold at a retail price of three times what they cost the company. A shipment of golf clubs arrived yesterday, and they have not yet been labeled. This morning, the main office of Hook n' Slice called to say that the clubs should be sold at 35% off their retail prices. Rather than label each club with its retail price, calculate the sale price, and then relabel the club, Theresa wonders how to go directly from the cost of the club to the sale price. That is, if C is the cost to the company of a certain club, what is the sale price for that club?

1. Develop a formula and answer this question for her.
2. Now try out your formula on a club that cost the company $26. Compare your answer with what you find by following the company procedure of first computing the retail price and then reducing that by 35%.

b. All golf bags have been sale-priced at 20% off for the past few weeks. Theresa has all of them in a large rack, marked with a sign that reads "20% off retail price." The company has now directed that this sale price be reduced by 35%. Rather than relabel every bag with its new sale price, Theresa would like to simply change the sign to say "_________% off retail price."

1. Determine what number Theresa should put in the blank.
2. Check your answer by computing the final sale price for a golf bag with a $100 retail price in two ways:
   - by using the percentage you told Theresa should go on the sign, and
   - by doing the two separate price reductions.
3. Will the sign read the same if the retail price is first reduced by 35%, and then the sale price reduced by 20%? Explain.

Project 2

The Parchdale City Planning Council is deciding whether or not the town should build an emergency reservoir for use in especially dry periods. The proposed site has room enough for a conical reservoir of diameter 141 feet and depth 85.3 feet.

As long as Parchdale's main water supply is working, the reservoir will be kept full. When a water emergency occurs, however, the reservoir will lose water to evaporation as well as supplying the town with water. The company that has designed the reservoir has given the town the following information.

If D equals the number of consecutive days the reservoir of volume V is used as Parchdale's water supply, then the total amount of water lost to evaporation after D days will be

$$0.1V\left(\frac{D-0.7}{D}\right)^2$$

So the volume of water left in the reservoir after it has been used for D days is

$$\text{Volume} = V - (\text{usage per day}) \cdot D - (\text{evaporation})$$

Parchdale uses about 57,500 cubic feet of water per day under emergency conditions. The majority of the council members think that building the reservoir is a good idea if it will supply the town with water for a week. Otherwise, they will vote against building the reservoir.

a. Calculate the volume of water the reservoir will hold. Remember to use significant digits correctly. Express the answer in scientific notation.

b. Show that the reservoir will supply Parchdale with water for a full week.

c. How much water per day could Parchdale use if the proposed reservoir had to supply the city with water for ten days?

Chapter 1 REVIEW

SECTION 1.1 The Real Number System

DEFINITIONS AND CONCEPTS	EXAMPLES
A **set** is a collection of objects. **Natural numbers:** $\{1, 2, 3, 4, 5, . . .\}$ **Whole numbers:** $\{0, 1, 2, 3, 4, 5, . . .\}$ **Integers:** $\{. . . , -3, -2, -1, 0, 1, 2, 3, . . .\}$ **Even integers:** $\{. . . , -6, -4, -2, 0, 2, 4, 6, . . .\}$ **Odd integers:** $\{. . . , -5, -3, -1, 1, 3, 5, . . .\}$ **Prime numbers:** $\{2, 3, 5, 7, 11, 13, . . .\}$ **Composite numbers:** $\{4, 6, 8, 9, 10, 12, . . .\}$ **Rational numbers:** numbers that can be written as $\left\{\frac{a}{b} \middle\vert a \text{ and } b \text{ are integers and } b \neq 0\right\}$ **Irrational numbers:** numbers that can be written as nonterminating, nonrepeating decimals **Real numbers:** numbers that can be written as decimals	Which numbers in the set $\left\{-5, 0, \frac{2}{3}, 1.75, \sqrt{49}, \pi, 6\right\}$ are **a.** natural numbers, **b.** whole numbers, **c.** integers, **d.** rational numbers, **e.** irrational numbers, **f.** real numbers, **g.** prime numbers, **h.** composite numbers, **i.** even integers, **j.** odd integers? **a.** $\sqrt{49}$, 6, $\sqrt{49}$ is a natural number since $\sqrt{49} = 7$. **b.** 0, $\sqrt{49}$, 6 **c.** -5, 0, $\sqrt{49}$, 6 **d.** $-5, 0, \frac{2}{3}, 1.75, \sqrt{49}, 6$ **e.** π **f.** $-5, 0, \frac{2}{3}, 1.75, \sqrt{49}, \pi, 6$ **g.** $\sqrt{49}$ **h.** 6 **i.** 0, 6 **j.** -5, $\sqrt{49}$

Graphing:

The following table shows three ways to describe an interval.

Set Notation	Graph	Interval Notation
$\{x \mid x > a\}$		(a, ∞)
$\{x \mid x < a\}$		$(-\infty, a)$
$\{x \mid x \geq a\}$		$[a, \infty)$
$\{x \mid x \leq a\}$		$(-\infty, a]$
$\{x \mid a < x < b\}$		(a, b)
$\{x \mid a \leq x < b\}$		$[a, b)$
$\{x \mid a < x \leq b\}$		$(a, b]$
$\{x \mid a \leq x \leq b\}$		$[a, b]$
$\{x \mid x < a$ or $x > b\}$		$(-\infty, a) \cup (b, \infty)$

The graph of $\{x \mid x < 7\}$ includes all real numbers that are less than 7 as shown in the figure below. This is the interval $(-\infty, 7)$.

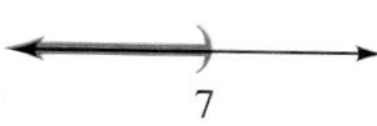

The graph of $\{x \mid x \geq -5\}$ includes all real numbers that are greater than or equal to -5 as shown in the figure below. This is the interval $[-5, \infty)$.

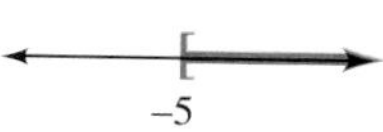

The set $\{x \mid 0 < x \leq 9\}$ includes all real numbers from 0 to 9, including 9, as shown in the figure below. This is the interval $(0, 9]$.

The set $\{x \mid -2 < x < 5\}$ includes all real numbers between -2 and 5, as shown in the figure below. This is the interval $(-2, 5)$.

Absolute value:

$$\begin{cases} \text{If } x \geq 0, \text{ then } |x| = x. \\ \text{If } x < 0, \text{ then } |x| = -x. \end{cases}$$

Evaluate: $-|-26|$.

$$-|\mathbf{-26}| = -(\mathbf{26}) = -26$$

REVIEW EXERCISES

List the numbers in $\left\{-4, -\frac{2}{3}, 0, 1, 2, \pi, 4\right\}$ that satisfy the given condition.

1. whole number
2. natural number
3. rational number
4. integer
5. irrational number
6. real number
7. negative number
8. positive number
9. prime number
10. composite number
11. even integer
12. odd integer
13. Graph the set of prime numbers between 20 and 30.

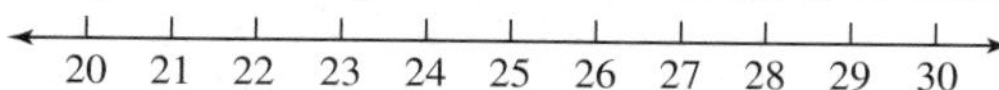

14. Graph the set of composite numbers between 5 and 13.

5 6 7 8 9 10 11 12 13

Graph each set on the number line.

15. $\{x \mid x \geq -4\}$
16. $\{x \mid -2 < x \leq 6\}$
17. $(-2, 3)$
18. $[2, 6]$
19. $\{x \mid x > 2\}$
20. $(-\infty, -1)$
21. $(-\infty, 0] \cup (2, \infty)$

Write each expression without absolute value symbols.

22. $|0|$
23. $|-1|$
24. $|8|$
25. $-|8|$

SECTION 1.2 Arithmetic and Properties of Real Numbers

DEFINITIONS AND CONCEPTS	EXAMPLES
Adding two real numbers: *With like signs:* Add their absolute values and keep the same sign. *With unlike signs:* Subtract their absolute values and keep the sign of the number with the greater absolute value.	 $(+1) + (+6) = +7 \qquad (-1) + (-6) = -7$ $(-1) + (+6) = +5 \qquad (+1) + (-6) = -5$
Subtracting two real numbers: $x - y$ is equivalent to $x + (-y)$.	$-8 - 2 = -8 + (-2)$ To subtract 2, add -2. $= -10$
Multiplying and dividing two real numbers: *With like signs:* Multiply (or divide) their absolute values. The sign is positive. *With unlike signs:* Multiply (or divide) their absolute values. The sign is negative. **Division with 0:** Zero divided by any number is 0. Division by zero is undefined.	$3(7) = 21 \qquad -3(-7) = 21$ $\frac{+6}{+2} = +3 \qquad \frac{-6}{-2} = +3$ $-3(7) = -21 \qquad 3(-7) = -21$ $\frac{+6}{-2} = -3 \qquad \frac{-6}{+2} = -3$ $\frac{0}{2} = 0$ because $(2)(0) = 0$. $\frac{2}{0}$ is undefined because no number multiplied by 0 gives 2.
Order of operations without exponents: Use the following steps to perform all calculations within each pair of grouping symbols, working from the innermost pair to the outermost pair. **1.** Perform all multiplications and divisions, working from left to right. **2.** Perform all additions and subtractions, working from left to right. **3.** Because a fraction bar is a grouping symbol, simplify the numerator and the denominator separately and then simplify the fraction, whenever possible. When all grouping symbols have been removed, repeat the rules above to finish the calculation.	Simplify $5(4 - 7) \div 5 + 1 = 5(-3) \div 5 + 1$ Do the subtraction within parentheses $= -15 \div 5 + 1$ Multiply. $= -3 + 1$ Divide. $= -2$ Add.
Evaluating algebraic expressions: To evaluate algebraic expressions, we substitute numbers for the variables and simplify.	If $x = 2$, $y = -3$, and $z = -5$, evaluate: $\frac{xy - 3z}{y(z + x)}$

	To evaluate the expression, we substitute 2 for x, -3 for y, and -5 for z and simplify. $\frac{xy - 3z}{y(z + x)} = \frac{2(-3) - 3(-5)}{-3(-5 + 2)}$ $= \frac{-6 + 15}{-3(-3)}$ $= \frac{9}{9}$ $= 1$
Commutative properties: $a + b = b + a$ $ab = ba$ **Associative properties:** $(a + b) + c = a + (b + c)$ $(ab)c = a(bc)$ **Distributive property of multiplication over addition:** $a(b + c) = ab + ac$	The commutative property of addition justifies the statement $x + 3 = 3 + x$. The commutative property of multiplication justifies the statement $x \cdot 3 = 3 \cdot x$. The associative property of addition justifies the statement $(x + 3) + 4 = x + (3 + 4)$. The associative property of multiplication justifies the statement $(x \cdot 3) \cdot 4 = x \cdot (3 \cdot 4)$. $7(2x - 8) = 7(2x) - 7(8) = 14x - 56$
Additive and multiplicative identities: $a + 0 = 0 + a = a$ $a \cdot 1 = 1 \cdot a = a$	$5 + 0 = 0 + 5 = 5$ $5 \cdot 1 = 1 \cdot 5 = 5$
Additive and multiplicative inverses: $a + (-a) = 0$ If $a \neq 0$, then $a\left(\frac{1}{a}\right) = \frac{1}{a} \cdot a = 1$	The additive inverse of -2 is 2 because $-2 + 2 = 0$. The multiplicative inverse of -2 is $-\frac{1}{2}$ because $-2\left(-\frac{1}{2}\right) = 1$.
Double negative rule: $-(-a) = a$	$-(-17) = 17$
Measures of central tendency: The **mean** of several values is the sum of those values divided by the number of values. $\text{Mean} = \frac{\text{sum of the values}}{\text{number of values}}$ The **median** of several values is the middle value. To find the median, **1.** Arrange the values in increasing order. **2.** If there is an odd number of values, choose the middle value. **3.** If there is an even number of values, find the mean of the middle two values.	Given the values 725, 650, 800, 500, and 725, find the **a.** mean $\frac{725 + 650 + 800 + 500 + 725}{5} = 680$ **b.** median 725 because it is the middle value

The **mode** of several values is the value that occurs most often.	**c.** mode 725 because it occurs more often than any other value

REVIEW EXERCISES

Perform the operations and simplify when possible.

26. $3 + (+5)$
27. $-6 + (-3)$
28. $-15 + (-13)$
29. $25 + 32$
30. $-2 + 5$
31. $3 + (-12)$
32. $8 + (-3)$
33. $7 + (-9)$
34. $-25 + 12$
35. $-30 + 35$
36. $-3 - 10$
37. $-8 - (-3)$
38. $27 - (-12)$
39. $38 - (-15)$
40. $(+5)(+7)$
41. $(-6)(-7)$
42. $\dfrac{-16}{-4}$
43. $\dfrac{-25}{-5}$
44. $4(-3)$
45. $-3(8)$
46. $\dfrac{-8}{2}$
47. $\dfrac{8}{-4}$

Simplify each expression.

48. $-4(3 - 6)$
49. $3[8 - (-1)]$
50. $-[4 - 2(6 - 4)]$
51. $3[-5 + 3(2 - 7)]$
52. $\dfrac{3 - 8}{10 - 5}$
53. $\dfrac{-32 - 8}{6 - 16}$

Consider the numbers 14, 12, 13, 14, 15, 20, 15, 17, 19, 15.

54. Find the mean.
55. Find the median.
56. Find the mode.
57. Can the mean, median, and mode of a group of numbers be the same?

Evaluate when $a = 5$, $b = -2$, $c = -3$, and $d = 2$.

58. $\dfrac{3a - 2b}{cd}$
59. $\dfrac{3b + 2d}{ac}$
60. $\dfrac{ab + cd}{c(b - d)}$
61. $\dfrac{ac - bd}{a(d + c)}$

Determine which property justifies each statement.

62. $3(4 + 2) = 3 \cdot 4 + 3 \cdot 2$
63. $3 + (x + 7) = (x + 7) + 3$
64. $3 + (x + 7) = (3 + x) + 7$
65. $3 + 0 = 3$
66. $3 + (-3) = 0$
67. $5(3) = 3(5)$
68. $3(xy) = (3x)y$
69. $3x \cdot 1 = 3x$
70. $a\left(\dfrac{1}{a}\right) = 1 \quad (a \neq 0)$
71. $-(-x) = x$

SECTION 1.3 Exponents

DEFINITIONS AND CONCEPTS	EXAMPLES
Properties of exponents: If no variable base is 0,	
$x^n = \overbrace{x \cdot x \cdot x \cdot \cdots \cdot x}^{n \text{ factors of } x}$	$x^5 = \overbrace{x \cdot x \cdot x \cdot x \cdot x}^{5 \text{ factors of } x}$
$x^m x^n = x^{m+n}$	$x^2 \cdot x^7 = x^{2+7} = x^9$
$(x^m)^n = x^{mn}$	$(x^2)^7 = x^{2\cdot 7} = x^{14}$
$(xy)^n = x^n y^n$	$(xy)^3 = x^3 y^3$
$\left(\dfrac{x}{y}\right)^n = \dfrac{x^n}{y^n}$	$\left(\dfrac{x}{y}\right)^3 = \dfrac{x^3}{y^3} \quad (y \neq 0)$
$x^0 = 1$	$(2x)^0 = 1 \quad (x \neq 0)$
$x^{-n} = \dfrac{1}{x^n}$	$x^{-3} = \dfrac{1}{x^3} \quad (x \neq 0)$
$\dfrac{x^m}{x^n} = x^{m-n}$	$\dfrac{x^7}{x^2} = x^{7-2} = x^5 \quad (x \neq 0)$
$\left(\dfrac{x}{y}\right)^{-n} = \left(\dfrac{y}{x}\right)^n = \dfrac{y^n}{x^n}$	$\left(\dfrac{5}{4}\right)^{-2} = \left(\dfrac{4}{5}\right)^2 = \dfrac{4^2}{5^2} = \dfrac{16}{25}$

Order of operations: Use the following steps to perform all calculations within each pair of grouping symbols, working from the innermost pair to the outermost pair. **1.** Find the values of any exponential expressions. **2.** Perform all multiplications and divisions, working from left to right. **3.** Perform all additions and subtractions, working from left to right. **4.** Because a fraction bar is a grouping symbol, simplify the numerator and the denominator separately and simplify the fraction, whenever possible. When all grouping symbols have been removed, repeat the rules above to finish the calculation.	To simplify $2^3 - 5(6 - 13) - 8^2$, we first perform the subtraction within the parentheses. $2^3 - 5(-7) - 8^2$ and then follow the steps in the rules for order of operations. $8 - 5(-7) - 64$ Evaluate exponential expressions. $= 8 + 35 - 64$ Perform the multiplication. $= 43 - 64$ Perform addition/subtraction left to right. $= -21$

REVIEW EXERCISES

Simplify each expression and write all answers without negative exponents.

72. 3^6

73. -2^6

74. $(-4)^3$

75. -5^{-4}

76. $(3x^4)(-2x^2)$

77. $(-x^5)(3x^3)$

78. $x^{-4}x^3$

79. $x^{-10}x^{12}$

80. $(3x^2)^3$

81. $(4x^4)^4$

82. $(-2x^2)^5$

83. $-(-3x^3)^5$

84. $(x^2)^{-5}$

85. $(x^{-4})^{-5}$

86. $(3x^{-3})^{-2}$

87. $(2x^{-4})^4$

88. $\frac{x^6}{x^4}$

89. $\frac{x^{12}}{x^7}$

90. $\frac{a^7}{a^{12}}$

91. $\frac{a^4}{a^7}$

92. $\frac{y^{-3}}{y^4}$

93. $\frac{y^5}{y^{-4}}$

94. $\frac{x^{-5}}{x^{-4}}$

95. $\frac{x^{-6}}{x^{-9}}$

Simplify each expression and write all answers without negative exponents.

96. $(3x^2y^3)^2$

97. $(-3a^3b^2)^{-4}$

98. $\left(\frac{3x^2}{4y^3}\right)^{-3}$

99. $\left(\frac{4y^{-2}}{5y^{-3}}\right)^3$

SECTION 1.4 Scientific Notation

DEFINITIONS AND CONCEPTS	EXAMPLES
Scientific notation: $N \times 10^n$, where $1 \le \lvert N \rvert < 10$ and n is an integer.	To change 352,000 to scientific notation, we note that 3.52 is between 1 and 10. To get 352,000, the decimal point in 3.52 must be moved five places to the right. We can do this by multiplying 3.52 by 10^5. $352{,}000 = 3.52 \times 10^5$ To write 0.0001593 in scientific notation, we note that 1.593 is between 1 and 10. To get 0.0001593, the decimal point in 1.593 must be moved four places to the left. We can do this by multiplying 1.593 by 10^{-4}. $0.0001593 = 1.593 \times 10^{-4}$
Significant digits: If a number M is written in scientific notation as $N \times 10^n$, where $1 \le \lvert N \rvert < 10$ and n is an integer, the number of significant digits in M is the same as the number of digits in N.	There are two significant digits in 23,000 and four significant digits in 126,200. The product $(23{,}000)(126{,}200) = 2.9026 \times 10^9$ should be rounded to two significant digits: 2.9×10^9.

REVIEW EXERCISES

Write each numeral in scientific notation.

100. 19,300,000,000

101. 0.0000000273

Write each numeral in standard notation.

102. 7.2×10^7

103. 8.3×10^{-9}

104. Water usage Each person in the United States uses approximately 1,640 gallons of water per day. If the population of the United States is 306 million people, how many gallons of water are used each day?

SECTION 1.5 Solving Linear Equations in One Variable

DEFINITIONS AND CONCEPTS	EXAMPLES
If a and b are real numbers and $a = b$, then $a + c = b + c$ $\quad a - c = b - c$ $ac = bc \quad (c \neq 0)$ $\quad \frac{a}{c} = \frac{b}{c} \quad (c \neq 0)$	If $a = b$, then $a + 3 = b + 3$ $\quad a - 5 = b - 5$ $4a = 4b$ $\quad \frac{a}{7} = \frac{b}{7}$
Solving linear equations: **1.** If the equation contains fractions, multiply both sides of the equation by a number that will eliminate the denominators. **2.** Use the distributive property to remove all sets of parentheses and combine like terms. **3.** Use the addition and subtraction properties to get all variables on one side of the equation and all numbers on the other side. Combine like terms, if necessary. **4.** Use the multiplication and division properties to make the coefficient of the variable equal to 1. **5.** Check the result.	To solve $\frac{x-2}{5} - x = \frac{8}{5} - x + 2$, we first eliminate the fractions by multiplying both sides by 5 and then proceed as follows: $5\left(\frac{x-2}{5} - x\right) = 5\left(\frac{8}{5} - x + 2\right)$ $x - 2 - 5x = 8 - 5x + 10$ Remove parentheses. $-4x - 2 = 18 - 5x$ Combine like terms. $x = 20$ Add $5x$ and 2 to both sides. Show that 20 satisfies the equation.
An **identity** is an equation that is satisfied by every number x for which both sides of the equation are defined. The solution of an identity is the set of *all real numbers* and is denoted by $\mathbb{R}$. A **contradiction** is an equation that has no solution. Its solution set is the **empty set,** $\varnothing$.	Show that the equation is an identity: $-5(2x + 3) + 8x - 9 = 6x - 8(x + 3)$ $-10x - 15 + 8x - 9 = 6x - 8x - 24$ Remove parentheses. $-2x - 24 = -2x - 24$ Combine like terms. Since the left side of the equation is the same as the right side, every number x will satisfy the equation. The equation is an identity and its solution set is all real numbers, $\mathbb{R}$. Show that the equation is a contradiction: $5(2x + 3) - 8x - 9 = 8x - 6(x + 3)$ $10x + 15 - 8x - 9 = 8x - 6x - 18$ Remove parentheses. $2x + 6 = 2x - 18$ Combine like terms. $6 = -18$ Subtract $2x$ from both sides. Because $6 = -18$ is a false statement, there is no number that will satisfy the equation. Therefore, it is a contradiction and the solution set is the empty set, $\varnothing$.

To solve a formula for an indicated variable means to isolate that variable on one side of the equation.

To solve $ax + by = c$ for y, we proceed as follows:

$$ax + by = c$$

$$by = c - ax \quad \text{Subtract } ax \text{ from both sides.}$$

$$\frac{by}{b} = \frac{c - ax}{b} \quad \text{Divide both sides by } b.$$

$$y = \frac{c - ax}{b} \quad \frac{b}{b} = 1$$

REVIEW EXERCISES

Solve each equation.

105. $5x + 12 = 37$

106. $-3x - 7 = 20$

107. $4(y - 1) = 28$

108. $3(x + 7) = 42$

109. $13(x - 9) - 2 = 7x - 5$

110. $\frac{8(x - 5)}{3} = 2(x - 4)$

111. $\frac{3y}{4} - 13 = -\frac{y}{3}$

112. $\frac{2y}{5} + 5 = \frac{14y}{10}$

113. $0.07a = 0.10 - 0.04(a - 3)$

114. $0.12x + 0.06(50{,}000 - x) = 4{,}080$

Solve for the indicated quantity.

115. $V = \frac{4}{3}\pi r^3$ for r^3

116. $V = \frac{1}{3}\pi r^2 h$ for h

117. $v = \frac{1}{6}ab(x + y)$ for x

118. $V = \pi h^2\left(r - \frac{h}{3}\right)$ for r

SECTION 1.6 Using Linear Equations to Solve Application Problems

DEFINITIONS AND CONCEPTS

To solve problems, use the following strategy:

1. *Analyze the problem* and identify a variable by asking yourself "What am I asked to find"? Choose a variable to represent the quantity to be found and then express all other unknown quantities in the problem as expressions involving that variable.
2. *Form an equation* by expressing a quantity in two different ways. This may require reading the problem several times to understand the given facts. What information is given? Is there a formula that applies to this situation? Often a sketch, chart, or diagram will help you visualize the facts of the problem.
3. *Solve the equation* found in Step 2.
4. *State the conclusion* including proper units.
5. *Check the result.*

EXAMPLES

A home theater system is on sale for $966. If the list price was $1,288, find the percent of markdown.

Analyze the problem. We must find the percent of markdown, so we let p represent the percent of markdown, expressed as a decimal. $966 is the sale price and $1,288 is the regular price.

Form an equation. The amount of the markdown is found by multiplying the regular price by the *percent* of markdown. If we subtract this amount from the regular price, we will have the sale price. This can be expressed as the equation

Sale price	equals	regular price	minus	markdown.
966	=	1,288	−	$p \cdot 1{,}288$

Solve the equation.

$$966 = 1{,}288 - p \cdot 1{,}288 \quad \text{This is the equation to solve.}$$

$$-322 = -1{,}288p \quad \text{Subtract 1,288 from both sides.}$$

$$0.25 = r \quad \text{Divide both sides by } -1{,}288.$$

State the conclusion. The theater system has been marked down 25%.

Check the result. A markdown of 25% from $1,288 is $322. The sale price is $1,288 − $322 or $966.

REVIEW EXERCISES

119. Carpentry A carpenter wants to cut a 20-foot rafter so that one piece is three times as long as the other. Where should he cut the board?

120. Geometry A rectangle is 4 meters longer than it is wide. If the perimeter of the rectangle is 28 meters, find its area.

121. Balancing a seesaw Sue weighs 48 pounds, and her father weighs 180 pounds. If Sue sits on one end of a 20-foot-long seesaw with the fulcrum in the middle, how far from the fulcrum should her father sit to balance the seesaw?

SECTION 1.7 More Applications of Linear Equations in One Variable

DEFINITIONS AND CONCEPTS	EXAMPLES
Use the strategy to solve application problems given in the Section 1.6 review.	A car leaves Dallas traveling toward Austin at the rate of 65 mph. At the same time, another car leaves Austin traveling toward Dallas at the rate of 55 mph. How long will it take them to meet if the cities are 192 miles apart? *Analyze the problem.* We must find the time it takes for the two cars to meet, so we will let t represent the time in hours. *Form an equation.* We will use the formula $d = rt$. We know that one car is traveling at 65 mph and that the other is going 55 mph. We also know that they travel for the same amount of time, t hours. Thus, the distance that the faster car travels is $65t$ miles, and the distance that the slower car travels is $55t$ miles. The sum of these distances must be 192 miles. We can form the equation $65t + 55t = 192$ *Solve the equation.* $65t + 55t = 192$ This is the equation to solve. $120t = 192$ Combine like terms. $t = 1.6$ Divide both sides by 120. *State the conclusion.* The two cars will meet in 1.6 hours or 1 hour 36 minutes. *Check the result.* The faster car travels $1.6(65) = 104$ miles. The slower car travels $1.6(55) = 88$ miles. The total distance traveled is 192 miles.

REVIEW EXERCISES

122. Investment problem Sally has $25,000 to invest. She invests some money at 10% interest and the rest at 9%. If her total annual income from these two investments is $2,430, how much does she invest at each rate?

123. Mixing solutions How much water must be added to 20 liters of a 12% alcohol solution to dilute it to an 8% solution?

124. Motion problem A car and a motorcycle both leave from the same point and travel in the same direction. (See the illustration.) The car travels at an average rate of 55 mph and the motorcycle at an average rate of 40 mph. How long will it take before the vehicles are 5 miles apart?

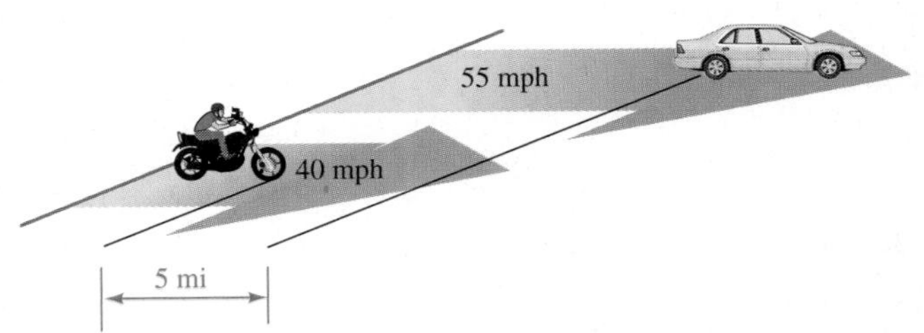

Chapter 1 TEST

Let $A = \{-2, 0, 1, \frac{6}{5}, 2, \sqrt{7}, 5\}$.

1. What numbers in A are natural numbers?
2. What numbers in A are irrational numbers?

Graph each set on the number line.

3. The set of odd integers between -4 and 6

−4 −3 −2 −1 0 1 2 3 4 5 6

4. The set of prime numbers less than 12

0 1 2 3 4 5 6 7 8 9 10 11 12

Graph each set on the number line.

5. $\{x \mid x > 4\}$
6. $[-3, \infty)$
7. $\{x \mid -2 \le x < 4\}$
8. $(-\infty, -1] \cup [2, \infty)$

Write each expression without using absolute value symbols.

9. $-|8|$
10. $|-5|$

Perform the operations.

11. $7 + (-5)$
12. $-5(-4)$
13. $\frac{12}{-3}$
14. $-4 - \frac{-15}{3}$

Consider the numbers $-2, 0, -2, 2, 3, -1, -1, 1, 1, 2$.

15. Find the mean.
16. Find the median.

Let $a = 2$, $b = -3$, *and* $c = 4$ *and evaluate each expression.*

17. ab
18. $a + bc$
19. $ab - bc$
20. $\frac{-3b + a}{ac - b}$

Determine which property of real numbers justifies each statement.

21. $3 + 5 = 5 + 3$
22. $a(b + c) = ab + ac$

Simplify. Write all answers without using negative exponents. Assume that no denominators are zero.

23. x^3x^5
24. $(2x^2y^3)^3$
25. $(m^{-4})^2$
26. $\left(\frac{m^2n^3}{m^4n^{-2}}\right)^{-2}$

Write each numeral in scientific notation.

27. 4,700,000
28. 0.00000023

Write each numeral in standard notation.

29. 6.53×10^5
30. 24.5×10^{-3}

Solve each equation.

31. $9(x + 4) + 4 = 4(x - 5)$
32. $\frac{2y + 3}{3} + \frac{3y - 4}{6} = \frac{y - 2}{2}$
33. $\frac{y - 1}{5} + 2 = \frac{2y - 3}{3}$
34. $400 + 1.5x = 500 + 1.25x$
35. Solve $P = L + \frac{s}{f}i$ for i.
36. **Geometry** A rectangle has a perimeter of 26 centimeters and is 5 centimeters longer than it is wide. Find its area.
37. **Investing** Jamie invests part of \$10,000 at 9% annual interest and the rest at 8%. If his annual income from these investments is \$860, how much does he invest at 8%?
38. **Mixing solutions** How many liters of water are needed to dilute 20 liters of a 5% salt solution to a 1% solution?

©Shutterstock.com/Lisa F. Young

CHAPTER 2

Graphs, Equations of Lines, and Functions

2.1 Graphing Linear Equations
2.2 Slope of a Line
2.3 Writing Equations of Lines
2.4 Introduction to Functions
2.5 Graphs of Other Functions
- ***Projects***

CHAPTER REVIEW
CHAPTER TEST
CUMULATIVE REVIEW EXERCISES

In this chapter ▶

It is often said that "A picture is worth a thousand words." In this chapter, we will show how numerical relationships can be described by using mathematical pictures called graphs.

Careers and Mathematics

ATMOSPHERIC SCIENTISTS—WEATHER FORECASTERS

Atmospheric science is the study of the atmosphere—the blanket of air covering the Earth. Atmospheric scientists, often called meteorologists, study the atmosphere's physical characteristics, motions, and processes. The best known application of this knowledge is in forecasting the weather. Atmospheric scientists held about 8,860 jobs in 2006. A bachelor's degree in meteorology is usually the minimum educational requirement for an entry-level position in the field.

Job Outlook:
Employment of atmospheric scientists is projected to increase about as fast as the average for all occupations through 2016. Opportunities will be greater in private industry than in the federal government.

Annual Earnings:
$55,530–$96,490

For More Information:
http://www.bls.gov/oco/ocos051.htm

For a Sample Application:
See Example 1 in Section 2.1.

SECTION
2.1
Graphing Linear Equations

Objectives

1 Plot an ordered pair on a coordinate plane and identify the coordinates of a point plotted on a coordinate plane.
2 Graph a linear equation in two variables by plotting points.
3 Identify the x-intercept and y-intercept given a linear equation and use them to graph the line.
4 Graph a horizontal line and a vertical line.
5 Apply the skills of graphing to application problems.
6 Find the midpoint of a line segment.

Vocabulary

rectangular coordinate system
x-axis
y-axis
origin
coordinate plane
quadrants
x-coordinate
y-coordinate
graph
ordered pair
linear equation
general form of a linear equation
y-intercept
x-intercept
intercept method
vertical line
horizontal line
midpoint

Getting Ready

Graph each set of numbers on the number line.

1. $\{-2, 1, 3\}$

−3 −2 −1 0 1 2 3 4

2. $\{x \mid x > -2\}$

3. $\{x \mid x \leq 3\}$

4. $(-3, 2)$

Many cities are laid out on a rectangular grid. For example, on the east side of Rockford, all streets run north and south and all avenues run east and west. (See Figure 2-1 on the next page.) If we agree to list the street numbers first, every location can be identified by using an ordered pair of numbers. If Jose lives on the corner of Third Street and Sixth Avenue, his location is given by the ordered pair (3, 6).

This is the street. → (3, 6) ← This is the avenue.

If Lisa has a location of (6, 3), we know that she lives on the corner of Sixth Street and Third Avenue. From the figure, we can see that

- Bob Anderson's location is (4, 1).
- Rosa Vang's location is (7, 5).
- The location of the store is (8, 2).

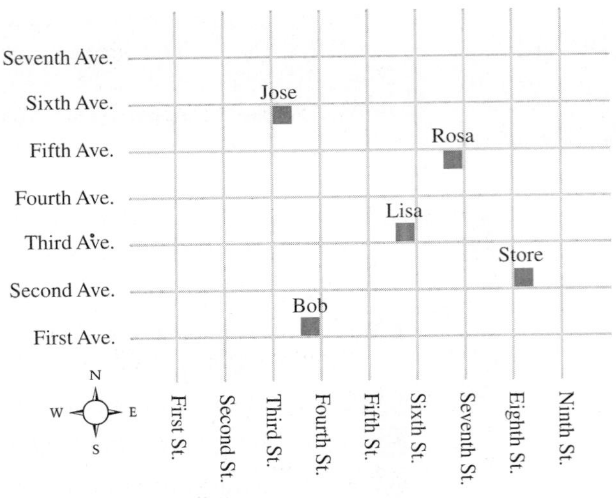

Figure 2-1

The idea of associating an ordered pair of numbers with points on a grid is attributed to the 17th-century French mathematician René Descartes. The grid often is called a **rectangular coordinate system,** or **Cartesian coordinate system,** after its inventor.

1 Plot an ordered pair on a coordinate plane and identify the coordinates of a point plotted on a coordinate plane.

A rectangular coordinate system (see Figure 2-2) is formed by two intersecting perpendicular number lines.

- The horizontal number line usually is called the ***x*-axis.**
- The vertical number line usually is called the ***y*-axis.**

The positive direction on the x-axis is to the right, and the positive direction on the y-axis is upward. If no scale is indicated on the axes, we assume that the axes are scaled in units of 1.

The point where the axes cross is called the **origin.** This is the 0 point on each axis. The two axes form a **coordinate plane** and divide it into four regions called **quadrants,** which are numbered as in Figure 2-2.

René Descartes (1596–1650)

Descartes is famous for his work in philosophy as well as in mathematics. His philosophy is expressed in the words "I think, therefore I am." He is best known in mathematics for his invention of a coordinate system and his work with conic sections.

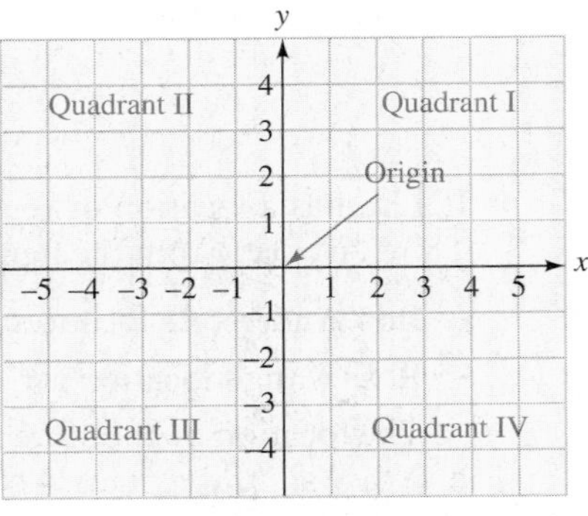

Figure 2-2

Every point in a coordinate plane can be identified by a pair of real numbers x and y, written as (x, y). The first number in the pair is the **x-coordinate,** and the second number is the **y-coordinate.** The numbers are called the *coordinates* of the point. Some examples of ordered pairs are $(-4, 6)$, $(2, 3)$, and $(6, -4)$.

$(-4, 6)$

In an ordered pair, the x-coordinate is listed first. The y-coordinate is listed second.

COMMENT Note that $(-4, 6)$ and $(2, 3)$ could define intervals or points depending on the context.

The process of locating a point in the coordinate plane is called *graphing* or *plotting* the point. In Figure 2-3(a), we show how to graph the point Q with coordinates of $(-4, 6)$. Since the x-coordinate is negative, we start at the origin and move 4 units to the left along the x-axis. Since the y-coordinate is positive, we then move up 6 units to locate point Q. Point Q is the **graph** of the point with coordinates $(-4, 6)$ and lies in quadrant II.

To plot the point $P(2, 3)$, we start at the origin, move 2 units to the right along the x-axis, and then move up 3 units to locate point P. Point P lies in quadrant I. To plot point $R(6, -4)$, we start at the origin and move 6 units to the right and then 4 units down. Point R lies in quadrant IV.

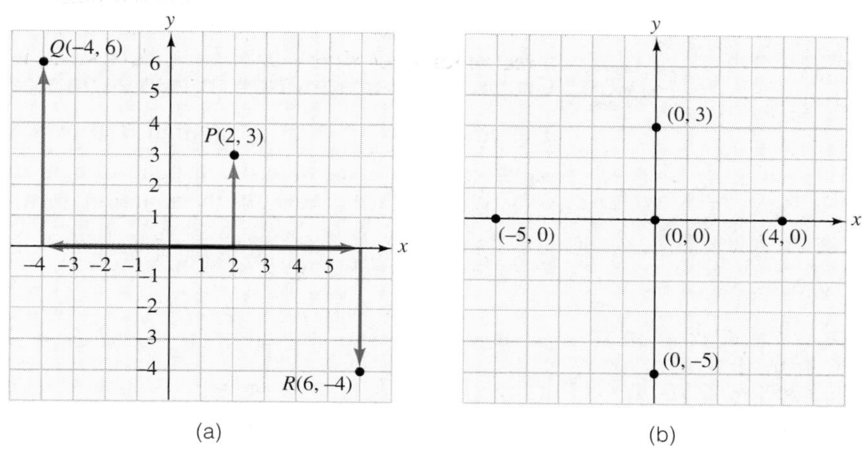

Figure 2-3

COMMENT Note that point Q with coordinates of $(-4, 6)$ is not the same as point R with coordinates $(6, -4)$. Since the order of the coordinates of a point is important, we call the pairs **ordered pairs.**

In Figure 2-3(b), we see that the points $(-5, 0)$, $(0, 0)$, and $(4, 0)$ all lie on the x-axis. In fact, every point with a y-coordinate of 0 will lie on the x-axis. We also see that the points $(0, -5)$, $(0, 0)$, and $(0, 3)$ all lie on the y-axis. In fact, every point with an x-coordinate of 0 will lie on the y-axis. The coordinates of the origin are $(0, 0)$.

EXAMPLE 1 **HURRICANES** The map shown in Figure 2-4 on the next page shows the path of a hurricane. The hurricane currently is located at 85° longitude and 25° latitude. If we agree to list longitude first, the location of the hurricane is given by the ordered pair (85, 25). Use the map to answer the following questions.

a. What are the coordinates of New Orleans?

b. What are the estimated coordinates of Houston?

c. What are the estimated coordinates of the landfall of the hurricane?

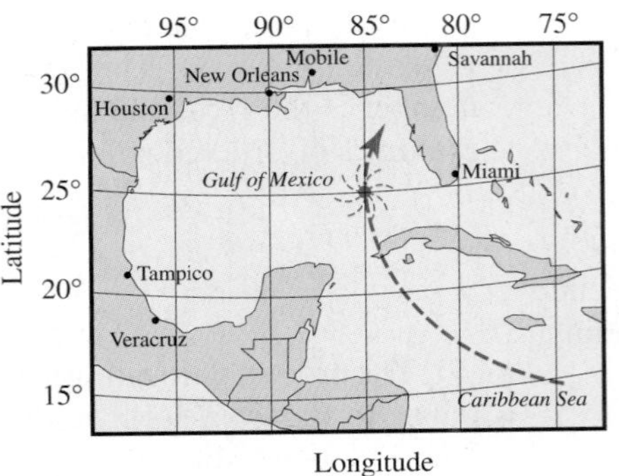

Figure 2-4

Solution

a. Since New Orleans is at 90° longitude and 30° latitude, its coordinates are (90, 30).

b. Since Houston has a longitude that is a little more than 95° and a latitude that is a little less than 30°, its estimated coordinates are (96, 29).

c. The hurricane appears to be headed for the point with coordinates of (83, 29).

2 Graph a linear equation in two variables by plotting points.

The equation $y = -\frac{1}{2}x + 4$ contains the two variables x and y. The solutions of this equation form a set of ordered pairs of real numbers. For example, the ordered pair $(-4, 6)$ is a solution, because the equation is satisfied when $x = -4$ and $y = 6$.

$$y = -\frac{1}{2}x + 4$$

$$6 = -\frac{1}{2}(-4) + 4 \quad \text{Substitute } -4 \text{ for } x \text{ and } 6 \text{ for } y.$$

$$6 = 2 + 4$$

$$6 = 6$$

$y = -\frac{1}{2}x + 4$

x	y	(x, y)
−4	6	(−4, 6)
−2	5	(−2, 5)
0	4	(0, 4)
2	3	(2, 3)
4	2	(4, 2)

Figure 2-5

This pair and some others that satisfy the equation are listed in the table shown in Figure 2-5.

The graph of the equation $y = -\frac{1}{2}x + 4$ is the graph of all points (x, y) on the rectangular coordinate system whose coordinates satisfy the equation.

EXAMPLE 2 Graph: $y = -\frac{1}{2}x + 4$.

Solution To graph the equation, we plot the ordered pairs listed in the table shown in Figure 2-5. These points appear to lie on a line, as shown in Figure 2-6. In fact, if we were to plot many more pairs that satisfied the equation, it would become obvious that the resulting points all will lie on the line. This line is the graph of the equation.

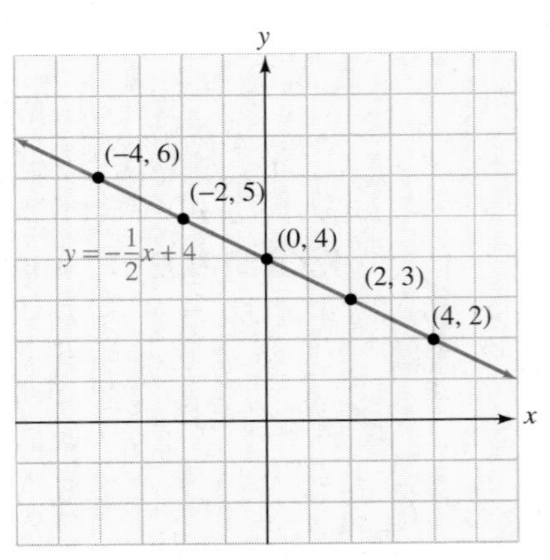

Figure 2-6

When we say that the graph of an equation is a line, we imply two things:

1. Every point with coordinates that satisfy the equation will lie on the line.
2. Any point on the line will have coordinates that satisfy the equation.

SELF CHECK 2 Graph: $y = 2x - 3$.

When the graph of an equation is a line, we call the equation a **linear equation.** Linear equations often are written in the form $Ax + By = C$, where A, B, and C stand for specific numbers (called constants) and x and y are variables. This form is called the **general form of the equation of a line.**

EXAMPLE 3 Graph: $3x + 2y = 12$.

Solution We can pick values for either x or y, substitute them into the equation, and solve for the other variable. For example, if $x = 2$,

$$3x + 2y = 12$$
$$3(2) + 2y = 12 \quad \text{Substitute 2 for } x.$$
$$6 + 2y = 12 \quad \text{Simplify.}$$
$$2y = 6 \quad \text{Subtract 6 from both sides.}$$
$$y = 3 \quad \text{Divide both sides by 2.}$$

The ordered pair (2, 3) satisfies the equation. If $y = 6$,

$$3x + 2y = 12$$
$$3x + 2(6) = 12 \quad \text{Substitute 6 for } y.$$
$$3x + 12 = 12 \quad \text{Simplify.}$$
$$3x = 0 \quad \text{Subtract 12 from both sides.}$$
$$x = 0 \quad \text{Divide both sides by 3.}$$

A second ordered pair that satisfies the equation is (0, 6).

These pairs and others that satisfy the equation lie on a line, as shown in Figure 2-7. The graph of the equation is the line shown in the figure.

$3x + 2y = 12$

x	y	(x, y)
−2	9	(−2, 9)
0	6	(0, 6)
2	3	(2, 3)
4	0	(4, 0)
6	−3	(6, −3)

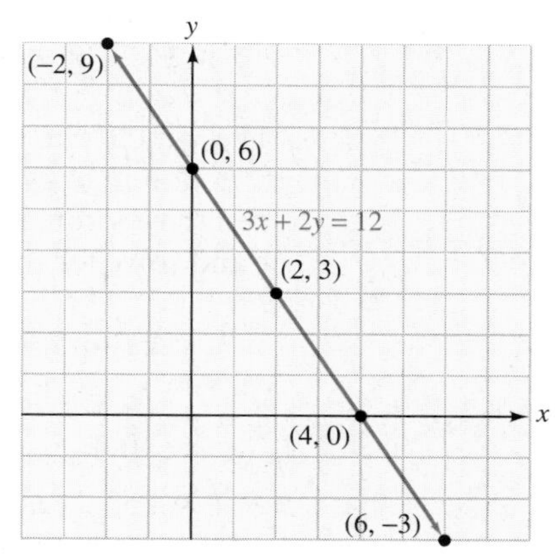

Figure 2-7

SELF CHECK 3 Graph: $3x - 2y = 6$.

3 Identify the x-intercept and y-intercept given a linear equation and use them to graph the line.

In Example 3, the graph intersects the y-axis at the point with coordinates (0, 6), called the ***y*-intercept,** and intersects the x-axis at the point with coordinates (4, 0), called the ***x*-intercept.** In general, we have the following definitions.

Intercepts of a Line

The ***y*-intercept** of a line is the point $(0, b)$ where the line intersects the y-axis. To find b, substitute 0 for x in the equation of the line and solve for y.

The ***x*-intercept** of a line is the point $(a, 0)$ where the line intersects the x-axis. To find a, substitute 0 for y in the equation of the line and solve for x.

EXAMPLE 4 Use the x- and y-intercepts to graph $2x + 5y = 10$.

Solution To find the y-intercept, we substitute 0 for x and solve for y:

$$\begin{aligned} 2x + 5y &= 10 \\ 2(\mathbf{0}) + 5y &= 10 && \text{Substitute 0 for } x. \\ 5y &= 10 && \text{Simplify.} \\ y &= 2 && \text{Divide both sides by 5.} \end{aligned}$$

The y-intercept is the point (0, 2). To find the x-intercept, we substitute 0 for y and solve for x:

$$\begin{aligned} 2x + 5y &= 10 \\ 2x + 5(\mathbf{0}) &= 10 && \text{Substitute 0 for } y. \\ 2x &= 10 && \text{Simplify.} \\ x &= 5 && \text{Divide both sides by 2.} \end{aligned}$$

The x-intercept is the point (5, 0).

Although two points are enough to draw a line, we plot a third point as a check. To find the coordinates of a third point, we can substitute any convenient number for x or y and solve for the other variable. If we let $x = -5$, we will obtain

$$\begin{aligned} 2x + 5y &= 10 \\ 2(\mathbf{-5}) + 5y &= 10 && \text{Substitute } -5 \text{ for } x. \\ -10 + 5y &= 10 && \text{Simplify.} \\ 5y &= 20 && \text{Add 10 to both sides.} \\ y &= 4 && \text{Divide both sides by 5.} \end{aligned}$$

The line also will pass through the point $(-5, 4)$.

A table of ordered pairs and the graph are shown in Figure 2-8.

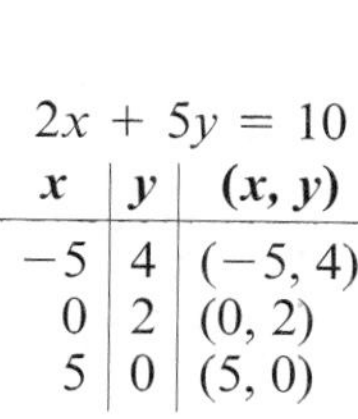

$2x + 5y = 10$

x	y	(x, y)
-5	4	$(-5, 4)$
0	2	$(0, 2)$
5	0	$(5, 0)$

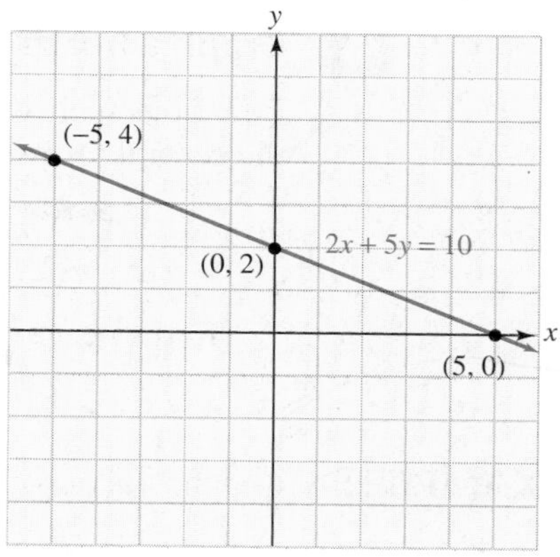

Figure 2-8

SELF CHECK 4 Find the x- and y-intercepts and graph $3x - y = -4$.

The method of graphing a line used in Example 4 is called the **intercept method.**

ACCENT ON TECHNOLOGY

Generating Tables of Solutions

Courtesy of Texas Instruments Incorporated

If an equation in x and y is solved for y, we can use a graphing calculator to generate a table of solutions. The instructions in this discussion are for a TI-83 Plus or TI-84 Plus graphing calculator. For specific details about other models, consult your owner's manual.

To construct a table of solutions for $2x + 5y = 10$, we first solve for y.

$$2x + 5y = 10$$

$$5y = -2x + 10 \quad \text{Subtract } 2x \text{ from both sides.}$$

$$y = -\frac{2}{5}x + 2 \quad \text{Divide both sides by 5 and simplify.}$$

To enter $y = -\frac{2}{5}x + 2$, we press **Y =** and enter $-(2/5)x + 2$, as shown in Figure 2-9(a).

To enter the x-values that are to appear in the table, we press **2ND** **WINDOW** (TBLSET) and enter the first value for x on the line labeled TblStart=. In Figure 2-9(b), -5 has been entered on this line. Other values for x that are to appear in the table are determined by setting an **increment value** on the line labeled ΔTbl=. Figure 2-9(b) shows that an increment of 1 was entered. This means that each x-value in the table will be 1 unit larger than the previous x-value.

The final step is to press the keys **2ND** **GRAPH** (TABLE). This displays a table of solutions, as shown in Figure 2-9(c).

Plot1 Plot2 Plot3
\Y1 = −(2/5)X + 2
\Y2 =
\Y3 =
\Y4 =
\Y5 =
\Y6 =
\Y7 =

(a)

TABLE SETUP
TblStart=−5
ΔTbl=1
Indpnt: Auto Ask
Depend: Auto Ask

(b)

X	Y1
−5	4
−4	3.6
−3	3.2
−2	2.8
−1	2.4
0	2
1	1.6

X=−5

(c)

Figure 2-9

4 Graph a horizontal line and a vertical line.

Equations such as $y = 3$ and $x = -2$ are linear equations, because they can be written in the general form $Ax + By = C$.

$$y = 3 \quad \text{is equivalent to} \quad 0x + 1y = 3$$
$$x = -2 \quad \text{is equivalent to} \quad 1x + 0y = -2$$

In the next example, we will discuss how to graph these equations.

EXAMPLE 5 Graph: **a.** $y = 3$ **b.** $x = -2$.

Solution **a.** Since the equation $y = 3$ does not contain x, the numbers chosen for x have no effect on y. The value of y is always 3. After plotting the pairs (x, y) shown in Figure 2-10, we see that the graph is a horizontal line with a y-intercept of $(0, 3)$. The line has no x-intercept.

b. Since the equation $x = -2$ does not contain y, y can be any number. After plotting the pairs (x, y) shown in Figure 2-10, we see that the graph is a vertical line with an x-intercept of $(-2, 0)$. The line has no y-intercept.

$y = 3$

x	y	(x, y)
−3	3	(−3, 3)
0	3	(0, 3)
2	3	(2, 3)
4	3	(4, 3)

$x = -2$

x	y	(x, y)
−2	−2	(−2, −2)
−2	0	(−2, 0)
−2	2	(−2, 2)
−2	6	(−2, 6)

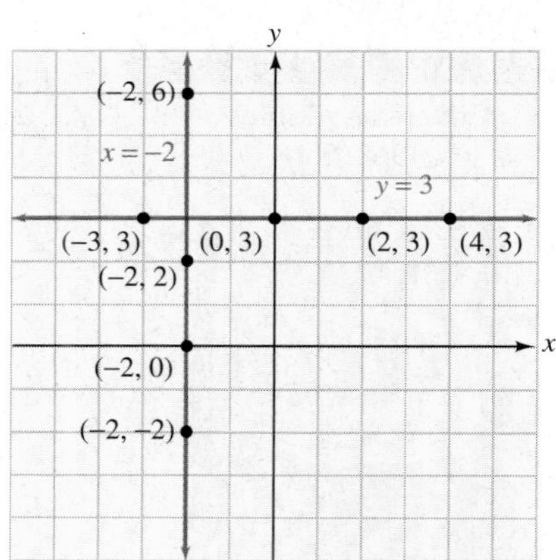

Figure 2-10

SELF CHECK 5 Graph $x = 4$ and $y = -3$ on one set of coordinate axes.

The results of Example 5 suggest the following facts.

Horizontal and Vertical Lines

If a and b are real numbers, then

The graph of $x = a$ is a **vertical line** with x-intercept at $(a, 0)$. If $a = 0$, the line is the y-axis.

The graph of $y = b$ is a **horizontal line** with y-intercept at $(0, b)$. If $b = 0$, the line is the x-axis.

5 Apply the skills of graphing to application problems.

EXAMPLE 6 **GAS MILEAGE** The following table gives the number of miles (y) that a bus can go on x gallons of gas. Plot the ordered pairs and estimate how far the bus can go on 9 gallons.

x	2	3	4	5	6
y	12	18	24	30	36

Solution Since the distances driven are rather large numbers, we plot the points on a coordinate system where the unit distance on the x-axis is larger than the unit distance on the y-axis. After plotting each ordered pair as in Figure 2-11, we see that the points lie on a line.

To estimate how far the bus can go on 9 gallons, we find 9 on the x-axis, move up to the graph, and then move to the left to locate a y-value of 54. Thus, the bus can go approximately 54 miles on 9 gallons of gas.

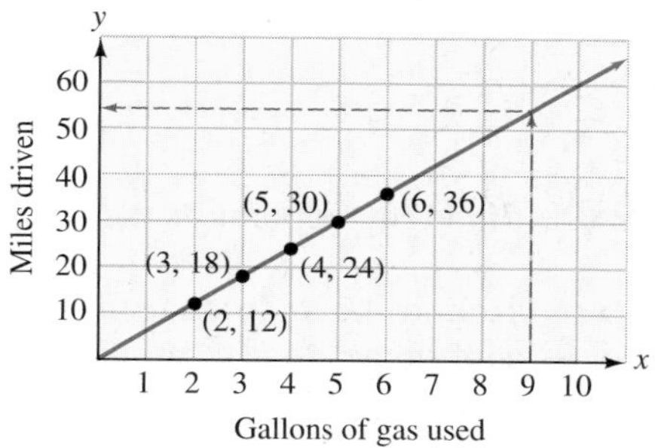

Figure 2-11

SELF CHECK 6 How far can the bus go on 10 gallons?

EXAMPLE 7 **DEPRECIATION** A copy machine purchased for \$6,750 is expected to depreciate according to the formula $y = -950x + 6{,}750$, where y is the value of the copier after x years. When will the copier have no value?

Solution The copier will have no value when its value (y) is 0. To find x when $y = 0$, we substitute 0 for y and solve for x.

$$y = -950x + 6{,}750$$

$$0 = -950x + 6{,}750$$

$$-6{,}750 = -950x \qquad \text{Subtract 6,750 from both sides.}$$

$$7.105263158 \approx x \qquad \text{Divide both sides by } -950.$$

The copier will have no monetary value in about 7.1 years.

SELF CHECK 7 When will the copier be worth \$2,000?

6 Find the midpoint of a line segment between two points.

To distinguish between the coordinates of two points on a line, we often use subscript notation. Point $P(x_1, y_1)$ is read as "point P with coordinates of x sub 1 and y sub 1." Point $Q(x_2, y_2)$ is read as "point Q with coordinates of x sub 2 and y sub 2."

If point M in Figure 2-12 lies midway between points $P(x_1, y_1)$ and $Q(x_2, y_2)$, point M is called the *midpoint* of segment PQ. To find the coordinates of M, we find the mean of the x-coordinates and the mean of the y-coordinates of P and Q.

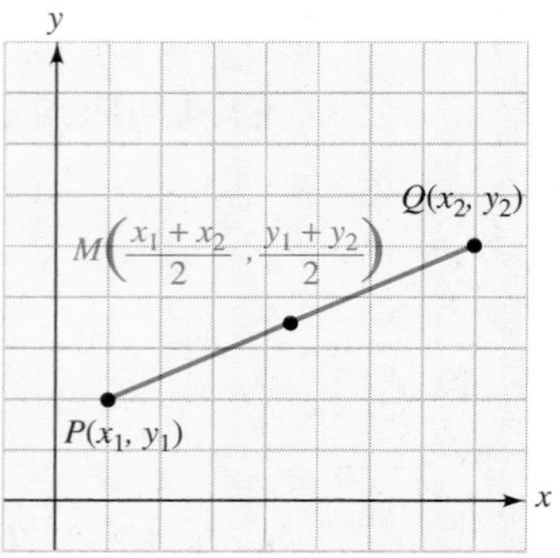

Figure 2-12

The Midpoint Formula

The **midpoint** of the line segment with endpoints at $P(x_1, y_1)$ and $Q(x_2, y_2)$ is the point M with coordinates of

$$\left(\frac{x_1 + x_2}{2}, \frac{y_1 + y_2}{2}\right)$$

EXAMPLE 8 Find the midpoint of the line segment joining $P(-2, 3)$ and $Q(7, -5)$.

Solution To find the midpoint, we find the mean of the x-coordinates and the mean of the y-coordinates to get

$$\frac{x_1 + x_2}{2} = \frac{-2 + 7}{2} = \frac{5}{2} \qquad \frac{y_1 + y_2}{2} = \frac{3 + (-5)}{2} = -1$$

The midpoint of segment PQ is the point $M\left(\frac{5}{2}, -1\right)$.

 **SELF CHECK 8** Find the midpoint of the segment joining $P(5, -4)$ and $Q(-3, 5)$.

ACCENT ON TECHNOLOGY

Graphing Lines

Courtesy of Texas Instruments Incorporated

We have graphed linear equations by finding ordered pairs, plotting points, and drawing a line through those points. Graphing is much easier if we use a graphing calculator.

Graphing calculators have a window to display graphs. To see the proper picture of a graph, we must decide on the minimum and maximum values for the x- and y-coordinates. A window with standard settings of

$$\text{Xmin} = -10 \quad \text{Xmax} = 10 \quad \text{Ymin} = -10 \quad \text{Ymax} = 10$$

will produce a graph in which the value of x is in the interval $[-10, 10]$ and y is in the interval $[-10, 10]$.

To graph $3x + 2y = 12$, we must first solve the equation for y.

$$
\begin{aligned}
3x + 2y &= 12 \\
2y &= -3x + 12 && \text{Subtract } 3x \text{ from both sides.} \\
y &= -\frac{3}{2}x + 6 && \text{Divide both sides by 2.}
\end{aligned}
$$

To graph the equation, press **Y =** and enter the right side of the equation. The display should look like

$$\backslash Y_1 = -(3/2)X + 6$$

We then press the **GRAPH** key to get the graph shown in Figure 2-13(a). To show more detail, we can draw the graph in a different window. A window with settings of $[-1, 5]$ for x and $[-2, 7]$ for y will give the graph shown in Figure 2-13(b).

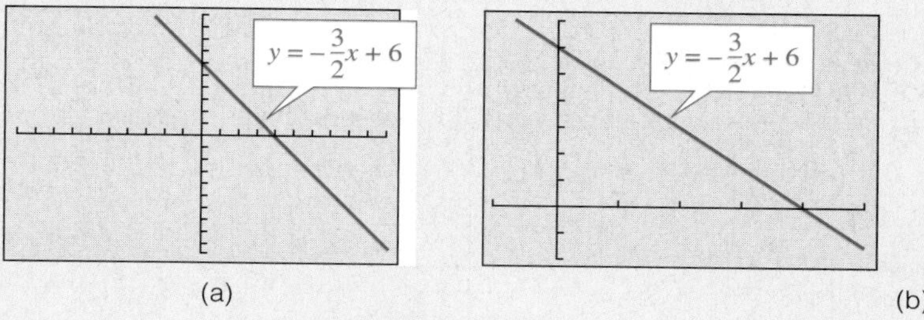

(a) (b)

Figure 2-13

We can use the trace command to find the coordinates of any point on a graph. For example, to find the x-intercept of the graph of $y = -\frac{3}{2}x + 6$, we graph the equation as in Figure 2-13(a) and press the **TRACE** key to get Figure 2-14(a). We then use the ▶ and ◀ keys to move the cursor along the line toward the x-intercept until we arrive at the point with the coordinates shown in Figure 2-14(b). To get better results, we can **ZOOM** in to get a magnified picture, trace again, and move the cursor to the point with the coordinates shown in Figure 2-14(c). Since the y-coordinate is nearly 0, this point is nearly the x-intercept. We can achieve better results with more zooms.

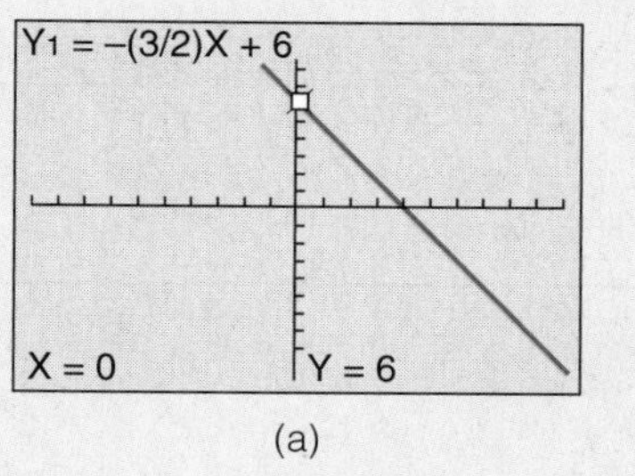

(a)

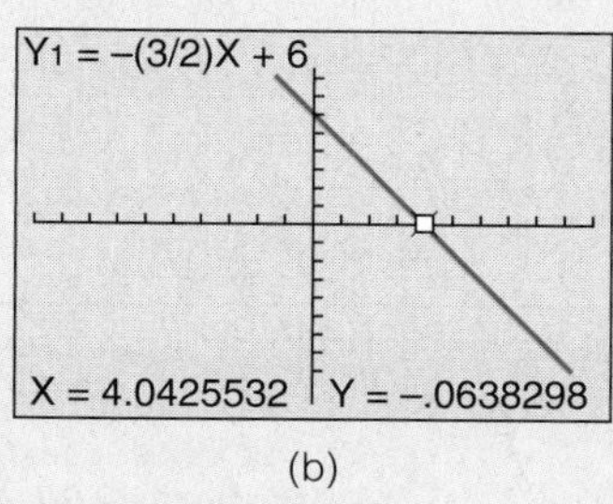

(b)

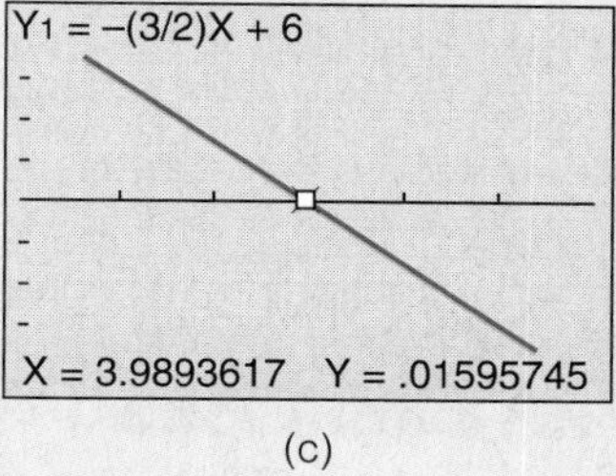

(c)

Figure 2-14

We can get the result more directly by using the zero command found in the CALC menu on the TI-83 or TI-84 family of graphing calculators. First, we graph the equation as in Figure 2-13(a). Then we press **2ND** **TRACE** (CALC) 2 (zero) to obtain Figure 2-15(a) on the next page. We then enter a left-bound guess (a number that lies to the left of the intercept) such as 2, press **ENTER**, enter a right-bound guess (a number that

(continued)

lies to the right of the intercept) such as 5, press **ENTER**, and press **ENTER** again to obtain Figure 2-15(b). From the figure, we see that the x-intercept is the point (4, 0).

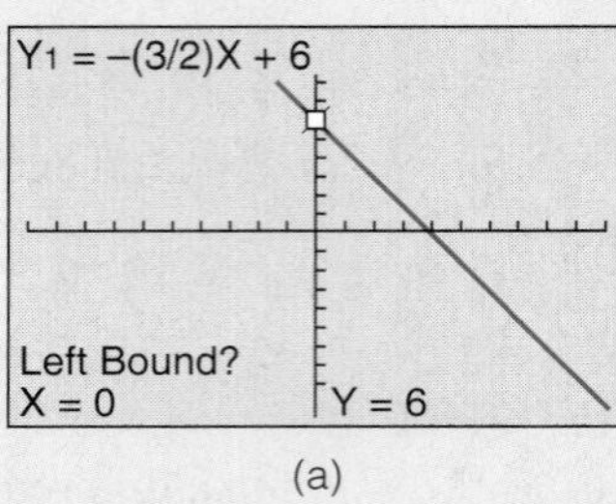

(a)

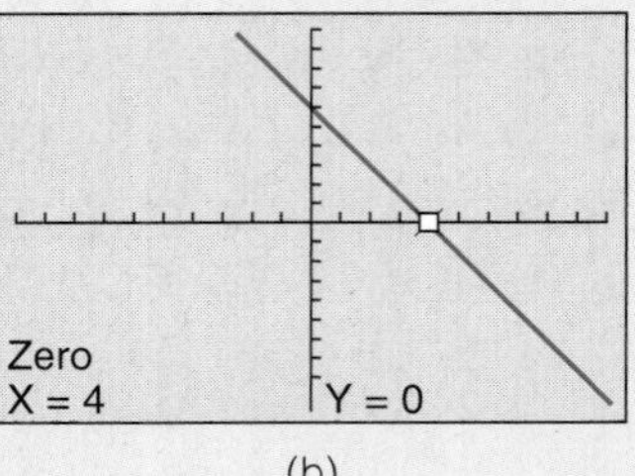

(b)

Figure 2-15

SELF CHECK ANSWERS

2.

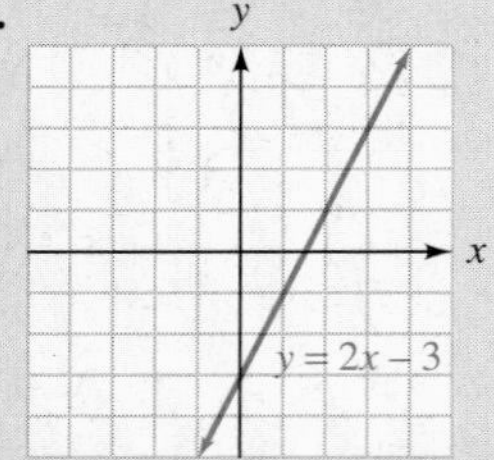

3.

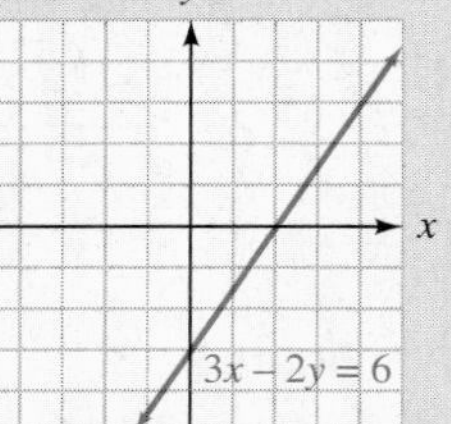

4.

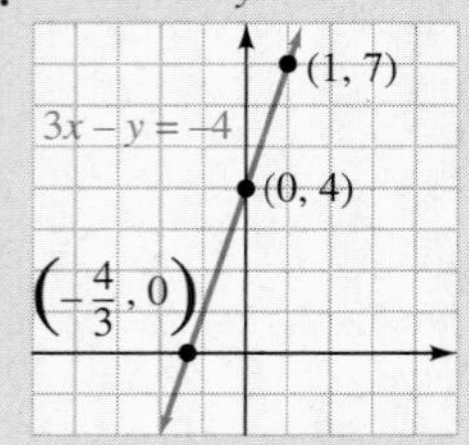

5.

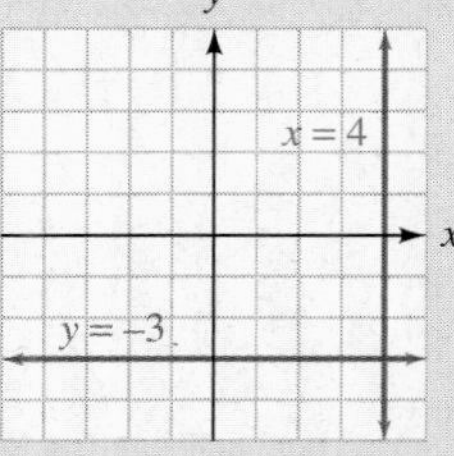

6. 60 miles **7.** 5 years **8.** $\left(1, \frac{1}{2}\right)$

NOW TRY THIS

1. Find the center of the circle with endpoints of its diameter at $(-5, 2)$ and $(-9, -7)$.

2. Find the midpoint of the segment joining the points with coordinates
a. $(p - 2, p)$ and $(4 - p, 5p - 2)$
b. $(p + 5, 3p - 1)$ and $(6p, p + 9)$

2.1 EXERCISES

WARM-UPS

Find the x- and y-intercepts of each line.

1. $x + y = 3$ **2.** $3x + y = 6$

3. $x + 4y = 8$ **4.** $3x - 4y = 12$

Determine whether the graphs of the equations are horizontal or vertical.

5. $x = -6$ **6.** $y = 8$

Find the midpoint of a line segment with endpoints at

7. $(2, 4), (6, 8)$ **8.** $(-4, 6), (4, -8)$

REVIEW

9. Evaluate: $-3 - 3(-5)$.

10. Evaluate: $(-5)^2 + (-5)$.

11. Simplify: $\dfrac{-3 + 5(2)}{9 + 5}$.

12. Simplify: $|-1 - 9|$.

13. Solve: $-4x + 7 = -21$.

14. Solve $P = 2l + 2w$ for w.

VOCABULARY AND CONCEPTS *Fill in the blanks.*

15. The pair of numbers $(6, -2)$ is called an ___________.

16. In the ordered pair $(-2, -9)$, -2 is called the __-coordinate and -9 is called the __-coordinate.

17. The point with coordinates $(0, 0)$ is called the ______.

18. The x- and y-axes divide the coordinate plane into four regions called __________.

19. Ordered pairs of numbers can be graphed on a ______________________________ system.

20. The process of locating the position of a point on a coordinate plane is called ________________ the point.

21. Any equation in the form $Ax + By = C$ is a ______________ and is said to be in ___________.

22. The point where a graph intersects the y-axis is called ___________.

23. The point where a graph intersects the ______ is called the x-intercept.

24. Using the x-intercept and y-intercept to graph an equation of a line is called the ______________.

25. The graph of any equation of the form $x = a$ is a _______ line.

26. The graph of any equation of the form $y = b$ is a _________ line.

27. The symbol x_1 is read as "x ______."

28. The midpoint of a line segment joining (a, b) and (c, d) is given by the formula _____________.

GUIDED PRACTICE

Plot each point on the rectangular coordinate system. (Objective 1)

29. $A(4, 3)$

30. $B(-2, 1)$

31. $C(3, -2)$

32. $D(-2, -3)$

33. $E(0, 5)$

34. $F(-4, 0)$

35. $G(2, 0)$

36. $H(0, 3)$

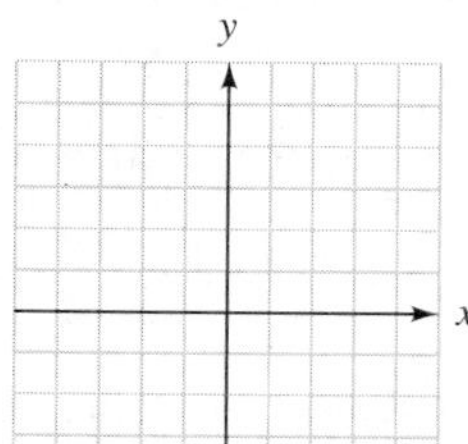

Give the coordinates of each point. See Example 1. (Objective 1)

37. A

38. B

39. C

40. D

41. E

42. F

43. G

44. H

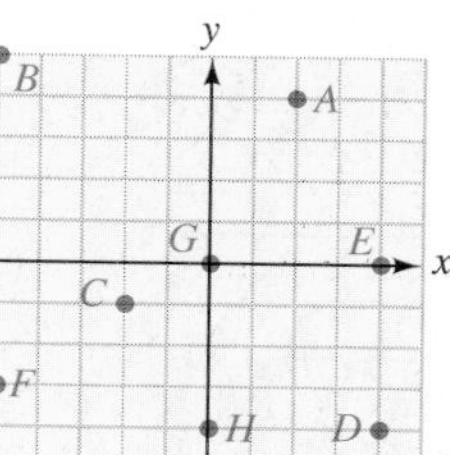

Complete each table and then graph by plotting the points. Check your work with a graphing calculator. See Example 2. (Objective 2)

45. $y = -x + 4$

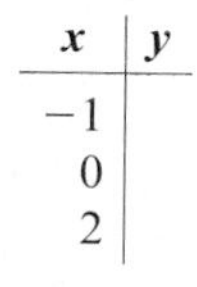

x	y
-1	
0	
2	

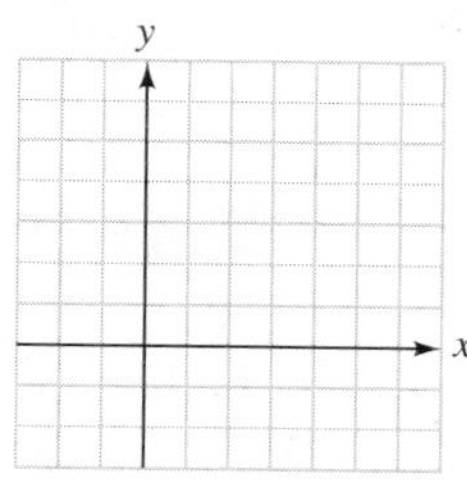

46. $y = x - 2$

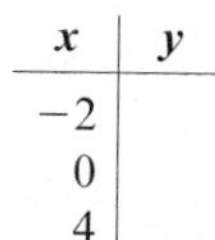

x	y
-2	
0	
4	

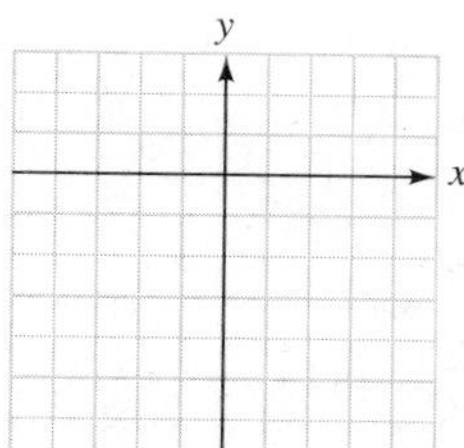

47. $y = 2x - 3$

x	y
-1	
0	
3	

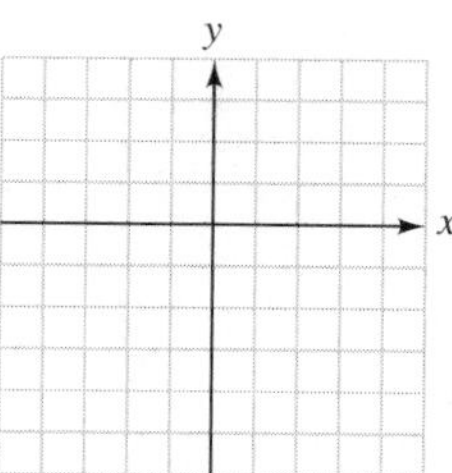

48. $y = -\dfrac{1}{2}x + \dfrac{5}{2}$

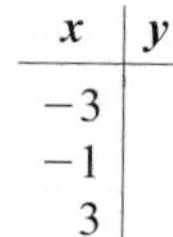

x	y
-3	
-1	
3	

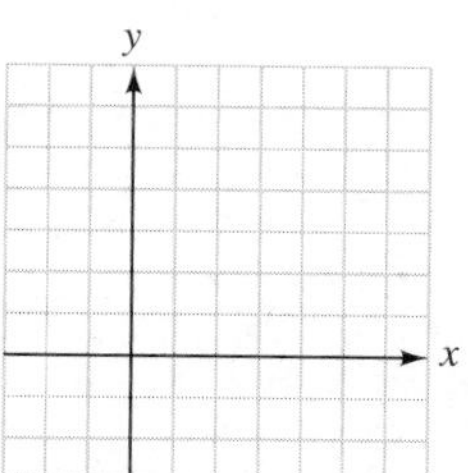

Graph each equation. Check your work with a graphing calculator. See Example 3. (Objective 2)

49. $3x + 4y = 12$

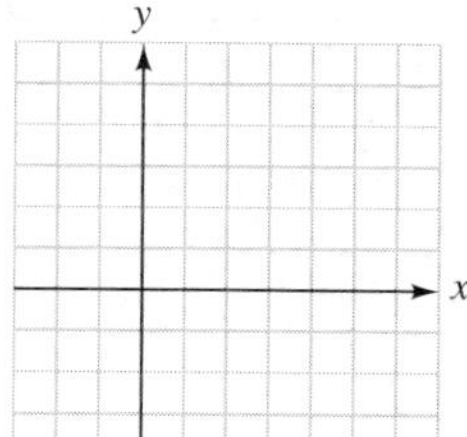

50. $4x - 3y = 12$

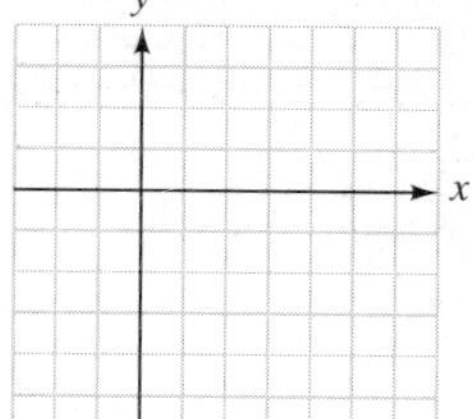

51. $6x - 3y = 12$

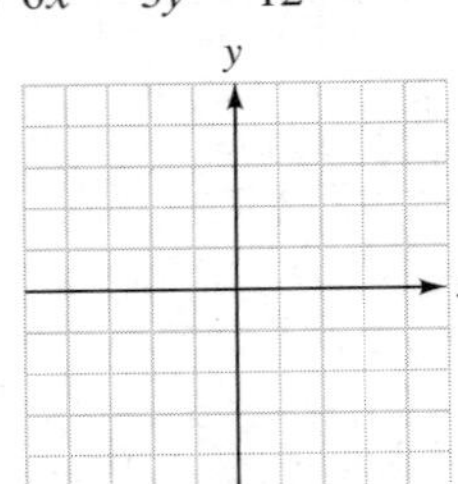

52. $3x + 6y = 12$

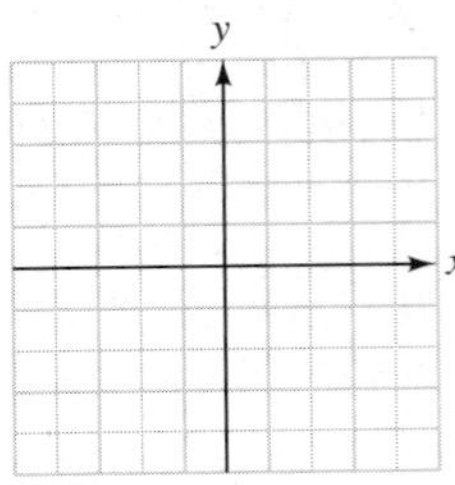

53. $y = -3x + 2$

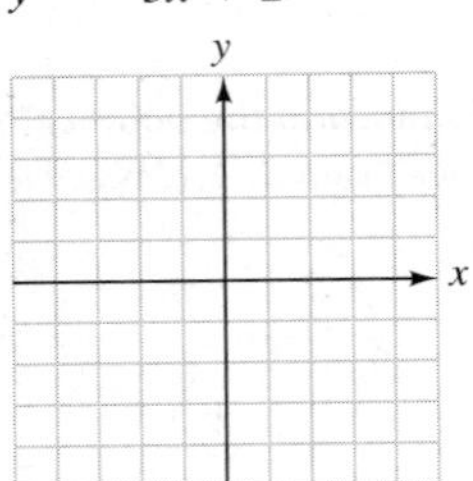

54. $y = 2x + 3$

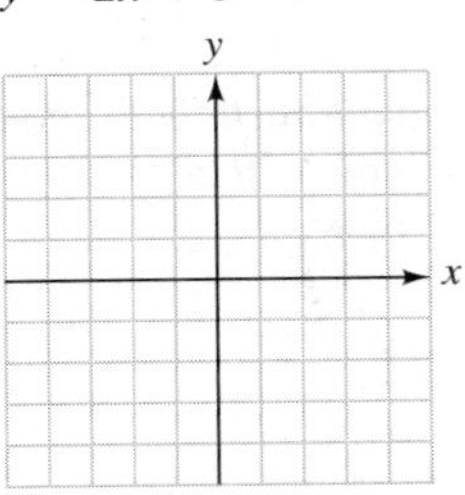

55. $y = \frac{3}{2}x$

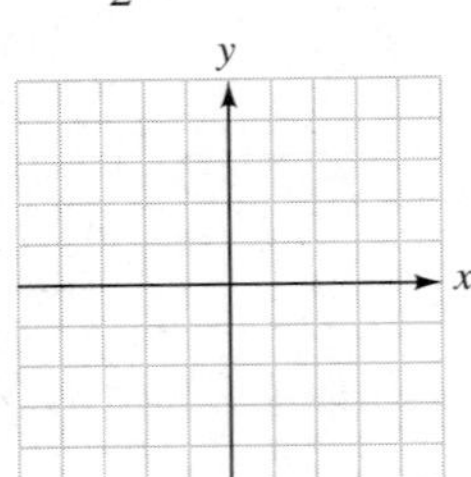

56. $y = -\frac{2}{3}x$

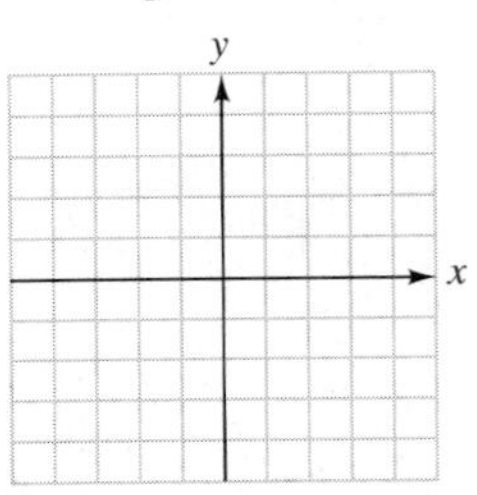

Graph each equation using the intercept method. Check your work with a graphing calculator. **See Example 4. (Objective 3)**

57. $3y = 6x - 9$

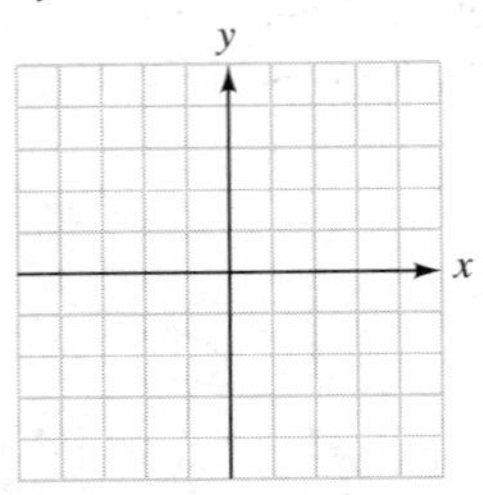

58. $2x = 4y - 10$

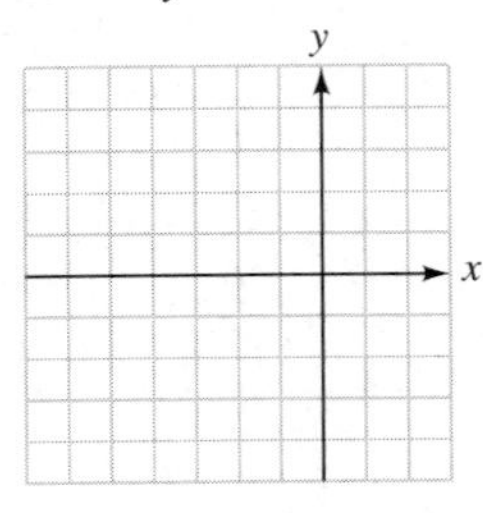

59. $3x + 4y - 8 = 0$

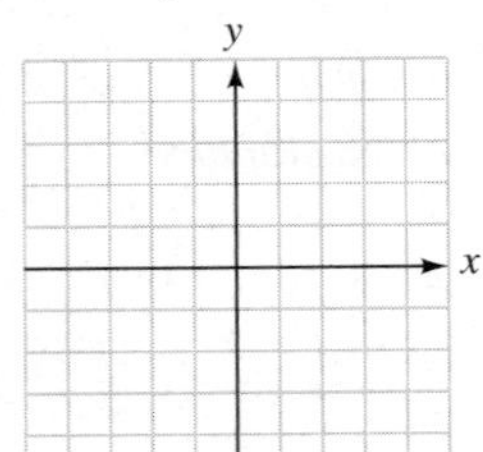

60. $-2y - 3x + 9 = 0$

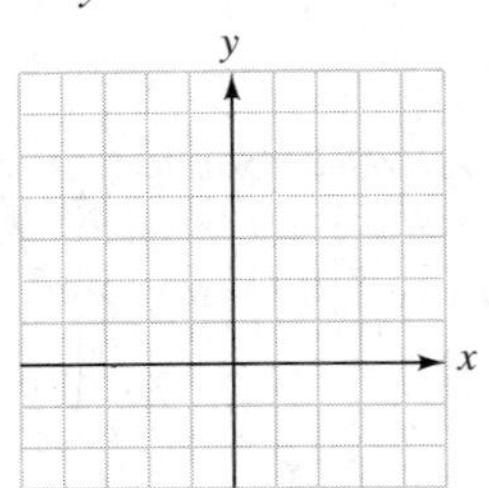

Graph each equation. **See Example 5. (Objective 4)**

61. $x = 3$

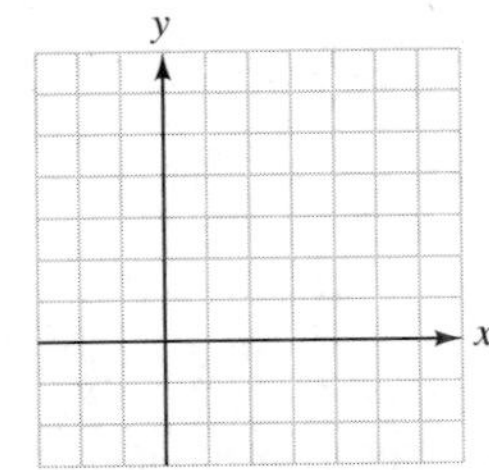

62. $y = -4$

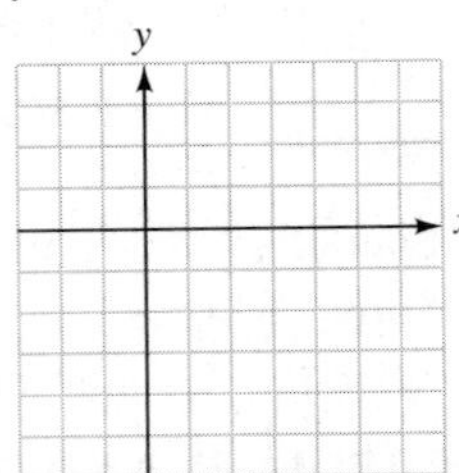

63. $-3y + 2 = 5$

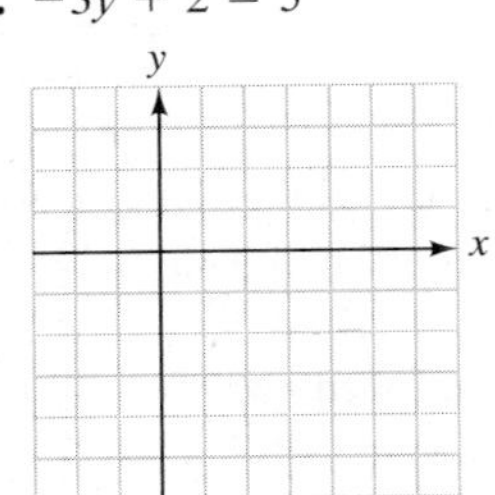

64. $-2x + 3 = 11$

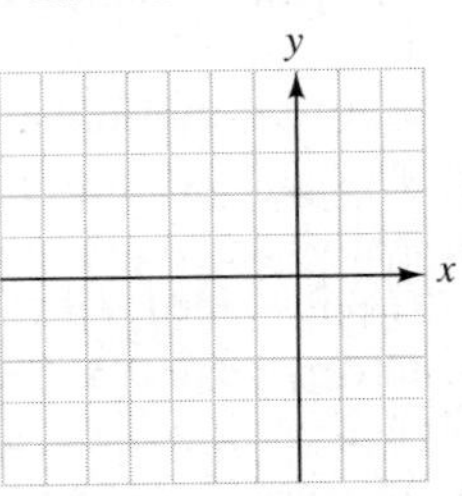

Plot the ordered pairs on a coordinate plane to find the answer. **See Example 6. (Objective 5)**

65. **Hourly wages** The table gives the amount y (in dollars) that a student can earn for working x hours. Plot the ordered pairs and estimate how much the student will earn for working 8 hours.

x	2	4	5	6
y	12	24	30	36

66. **Distance traveled** The table shows how far y (in miles) a biker can go in x hours. Plot the ordered pairs and estimate how far the biker can go in 8 hours.

x	2	4	5	6
y	30	60	75	90

67. **Value of a car** The table shows the value y (in dollars) of a car that is x years old. Plot the ordered pairs and estimate the value of the car when it is 4 years old.

x	0	1	3
y	15,000	12,000	6,000

68. **Earning interest** The table shows the amount y (in dollars) in a bank account drawing simple interest left on deposit for x years. Plot the ordered pairs and estimate the value of the account in 6 years.

x	0	1	4
y	1,000	1,050	1,200

Find the midpoint of segment PQ. See Example 8. (Objective 6)

69. $P(0, 0), Q(6, 8)$
70. $P(10, 12), Q(0, 0)$
71. $P(6, 8), Q(12, 16)$
72. $P(10, 4), Q(2, -2)$
73. $P(2, 4), Q(5, 8)$
74. $P(5, 9), Q(8, 13)$
75. $P(-2, -8), Q(3, 4)$
76. $P(-5, -2), Q(7, 3)$
77. $Q(-3, 5), P(-5, -5)$
78. $Q(2, -3), P(4, -8)$
79. **Finding the endpoint of a segment** If $M(-2, 3)$ is the midpoint of segment PQ and the coordinates of P are $(-8, 5)$, find the coordinates of Q.
80. **Finding the endpoint of a segment** If $M(6, -5)$ is the midpoint of segment PQ and the coordinates of Q are $(-5, -8)$, find the coordinates of P.

ADDITIONAL PRACTICE

Use a graphing calculator to find a table of solutions for each equation. Set the first value of x to be −2.5 and the increment value to be 0.5. Find the value of y that corresponds to x = −1. If an answer is not exact, round to the nearest thousandth.

81. $y = 2.5x + 1.5$
82. $y = 0.6x - 3.2$
83. $3.2x - 1.5y = 2.7$
84. $-1.7x + 3.7y = -2.8$

Use a graphing calculator to graph each equation, and then find the x-coordinate of the x-intercept to the nearest hundredth.

85. $y = 3.7x - 4.5$
86. $y = 0.6x + 1.25$
87. $1.5x - 3y = 7$
88. $0.3x + y = 7.5$

APPLICATIONS See Example 7. (Objective 5)

89. **House appreciation** A house purchased for \$125,000 is expected to appreciate according to the formula $y = 7{,}500x + 125{,}000$, where y is the value of the house after x years. Find the value of the house 5 years later.
90. **Car depreciation** A car purchased for \$17,000 is expected to depreciate according to the formula $y = -1{,}360x + 17{,}000$, where y is the value of the car after x years. When will the car have no monetary value?
91. **Demand equations** The number of TV sets that consumers buy depends on price. The higher the price, the fewer people will buy. The equation that relates price to the number of TVs sold at that price is called a **demand equation.** If the demand equation for a 17-inch TV is $p = -\frac{1}{10}q + 170$, where p is the price and q is the number of TVs sold at that price, how many TVs will be sold at a price of \$150?
92. **Supply equations** The number of TV sets that manufacturers produce depends on price. The higher the price, the more TVs manufacturers will produce. The equation that relates price to the number produced at that price is called a **supply equation.** If the supply equation for a 27-inch TV is $p = \frac{1}{10}q + 130$, where p is the price and q is the number produced for sale at that price, how many TVs will be produced if the price is \$150?
93. **TV coverage** In the illustration, a TV camera is located at point $(-3, 0)$. The camera is to follow the launch of a space shuttle. As the shuttle rises vertically, the camera can tilt back to a line of sight given by $y = 2x + 6$. Estimate how many miles the shuttle will remain in the camera's view.

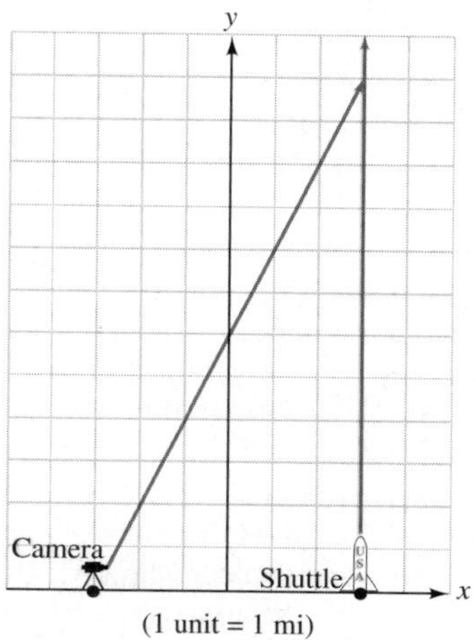

94. **Buying tickets** Children's tickets to the circus cost \$5 each plus a \$2 service fee for each block of tickets.
    a. Write a linear equation that gives the cost y for a student buying x tickets.
    b. Complete the table in the illustration and graph the equation.
    c. Use the graph to find the cost of buying 4 tickets.

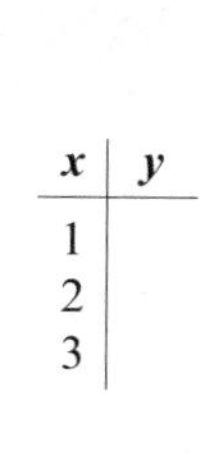

x	y
1	
2	
3	

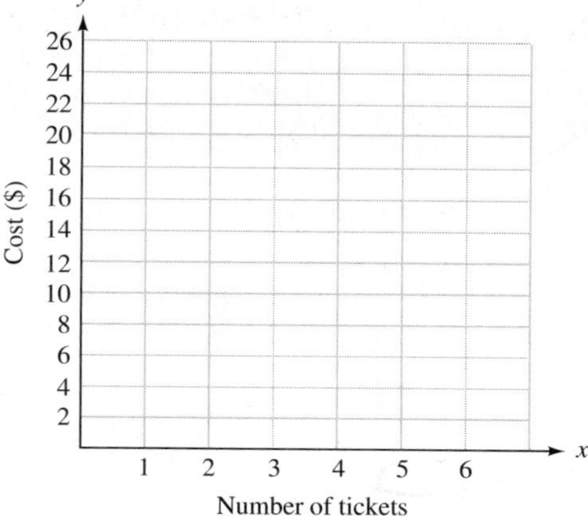

95. Telephone costs In one community, the monthly cost of local telephone service is \$5 per month, plus 25¢ per call.

a. Write a linear equation that gives the cost y for a person making x calls.

b. Complete the table in the illustration and graph the equation.

c. Use the graph to estimate the cost of service in a month when 20 calls were made.

x	y
4	
8	
12	
16	

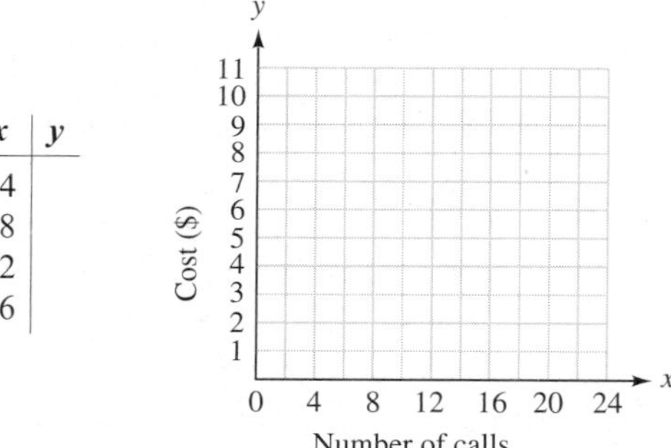

96. **Crime prevention** The number n of incidents of family violence requiring police response appears to be related to d, the money spent on crisis intervention, by the equation $n = 430 - 0.005d$. What expenditure would reduce the number of incidents to 350?

97. **U.S. sports participation** The equation $s = -0.9t + 65.5$ gives the approximate number of people 7 years of age and older who went swimming during a given year, where s is the annual number of swimmers (in millions) and t is the number of years since 1990. (*Source:* National Sporting Goods Association)

a. What information can be obtained from the s-intercept of the graph?

b. Estimate the number of swimmers in 2012.

98. **Farming** The equation

$$a = -3{,}700{,}000t + 983{,}000{,}000$$

gives the approximate number of acres a of farmland in the United States, t years after 1990. (*Source:* U.S. Department of Agriculture)

a. What information can be obtained from the a-intercept of the graph?

b. To the nearest million acres, estimate the number of acres of farmland in 2020.

WRITING ABOUT MATH

99. Explain how to graph a line using the intercept method.

100. Explain how to determine the quadrant in which the point $P(a, b)$ lies.

SOMETHING TO THINK ABOUT

101. If the line $y = ax + b$ passes only through quadrants I and II, what can be known about the constants a and b?

102. What are the coordinates of the three points that divide the line segment joining $P(a, b)$ and $Q(c, d)$ into four equal parts?

SECTION 2.2 Slope of a Line

Objectives

1. Find the slope of a line given a graph.
2. Find the slope of a line passing through two given points.
3. Find the slope of a line given its equation.
4. Identify the slope of a horizontal line and a vertical line.
5. Determine whether two lines are parallel, perpendicular, or neither.
6. Interpret slope in an application problem.

Vocabulary

slope
rise
run
perpendicular lines
negative reciprocals
parallel lines

Getting Ready

Simplify each expression.

1. $\dfrac{6-3}{8-5}$ 2. $\dfrac{10-4}{2-8}$ 3. $\dfrac{25-12}{9-(-5)}$ 4. $\dfrac{-9-(-6)}{-4-10}$

In Section 2.1, we graphed equations of lines. Later, we will show that we can graph a line if we know the coordinates of one point on the line and the slant (steepness) of the line. A measure of this slant is called the *slope* of the line.

1 Find the slope of a line given a graph.

A service offered by an online research company costs \$2 per month plus \$3 for each hour of connect time. The table in Figure 2-16(a) gives the cost y for certain numbers of hours x of connect time. If we construct a graph from these data, we get the line shown in Figure 2-16(b).

Grace Murray Hopper (1906–1992)

Grace Hopper graduated from Vassar College in 1928 and obtained a Master's degree from Yale in 1930. In 1943, she entered the U.S. Naval Reserve. While in the Navy, she became a programmer of the Mark I, the world's first large computer. She is credited for first using the word "bug" to refer to a computer problem. The first bug was actually a moth that flew into one of the relays of the Mark II. From then on, locating computer problems was called "debugging" the system.

	Hours of connect time					
x	0	1	2	3	4	5
y	2	5	8	11	14	17
	Cost					

(a)

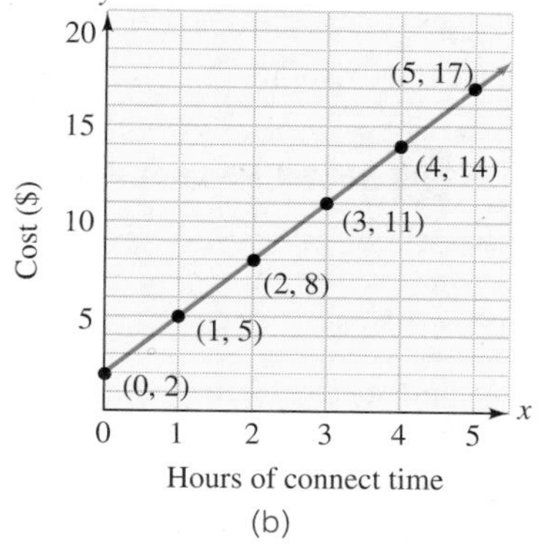

(b)

Figure 2-16

From the graph, we can see that if x changes from 0 to 1, y changes from 2 to 5. As x changes from 1 to 2, y changes from 5 to 8, and so on. The ratio of the change in y divided by the change in x is the constant 3.

$$\frac{\text{Change in } y}{\text{Change in } x} = \frac{5-2}{1-0} = \frac{8-5}{2-1} = \frac{11-8}{3-2} = \frac{14-11}{4-3} = \frac{17-14}{5-4} = \frac{3}{1} = 3$$

The ratio of the change in y divided by the change in x between any two points on any line is always a constant. This constant rate of change is called the **slope of the line.**

Slope of a Nonvertical Line

The slope m of the nonvertical line passing through points (x_1, y_1) and (x_2, y_2) is

$$m = \frac{\text{change in } y}{\text{change in } x} = \frac{y_2 - y_1}{x_2 - x_1}$$

2 Find the slope of a line passing through two given points.

EXAMPLE 1 Use the two points shown in Figure 2-17 to find the slope of the line passing through the points with coordinates $(-2, 4)$ and $(3, -4)$.

Solution We can let $(x_1, y_1) = (-2, 4)$ and $(x_2, y_2) = (3, -4)$. Then

$$\begin{aligned} m &= \frac{\text{change in } y}{\text{change in } x} \\ &= \frac{y_2 - y_1}{x_2 - x_1} \\ &= \frac{-4 - 4}{3 - (-2)} \\ &= \frac{-8}{5} \\ &= -\frac{8}{5} \end{aligned}$$

Substitute -4 for y_2, 4 for y_1, 3 for x_2, and -2 for x_1.

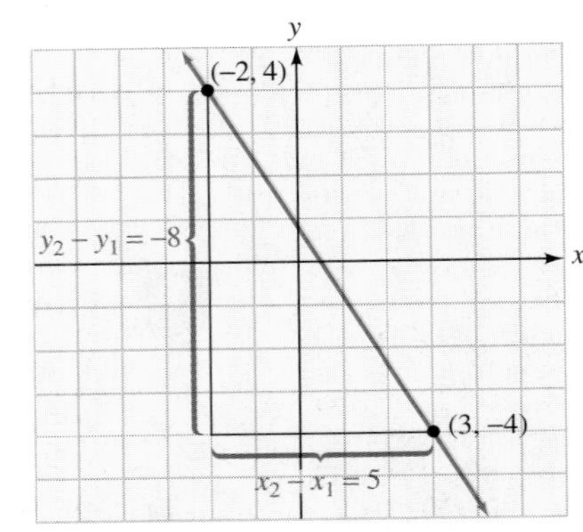

Figure 2-17

The slope of the line is $-\frac{8}{5}$. We would obtain the same result if we let $(x_1, y_1) = (3, -4)$ and $(x_2, y_2) = (-2, 4)$.

SELF CHECK 1 Find the slope of the line joining the points $(-3, 6)$ and $(4, -8)$.

COMMENT When calculating slope, always subtract the y-values and the x-values in the same order.

$$m = \frac{y_2 - y_1}{x_2 - x_1} \quad \text{or} \quad m = \frac{y_1 - y_2}{x_1 - x_2}$$

However, the following are not true:

$$m = \frac{y_2 - y_1}{x_1 - x_2} \quad \text{and} \quad m = \frac{y_1 - y_2}{x_2 - x_1}$$

The change in y (often denoted as Δy) is the **rise** of the line between two points. The change in x (often denoted as Δx) is the **run.** Using this terminology, we can define slope to be the ratio of the rise to the run:

$$m = \frac{\Delta y}{\Delta x} = \frac{\text{rise}}{\text{run}} \quad (\Delta x \neq 0)$$

3 Find the slope of a line given its equation.

To find the slope of a line from a given equation, we could graph the equation and count squares on the resulting line graph to determine the rise and the run. A better way is to find the x- and y-intercepts of the graph and use the slope formula.

EXAMPLE 2 Find the slope of the line determined by $3x - 4y = 12$.

Solution We first find the coordinates of two points on the line.

- If $x = 0$, then $y = -3$, and the point $(0, -3)$ is on the line.
- If $y = 0$, then $x = 4$, and the point $(4, 0)$ is on the line.

We then refer to Figure 2-18 and find the slope of the line between $(0, -3)$ and $(4, 0)$ by substituting 0 for y_2, -3 for y_1, 4 for x_2, and 0 for x_1 in the formula for slope.

$$m = \frac{\Delta y}{\Delta x}$$

$$= \frac{y_2 - y_1}{x_2 - x_1}$$

$$= \frac{0 - (-3)}{4 - 0}$$

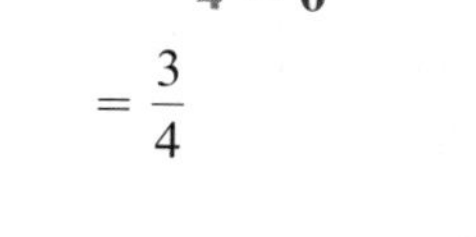

$$= \frac{3}{4}$$

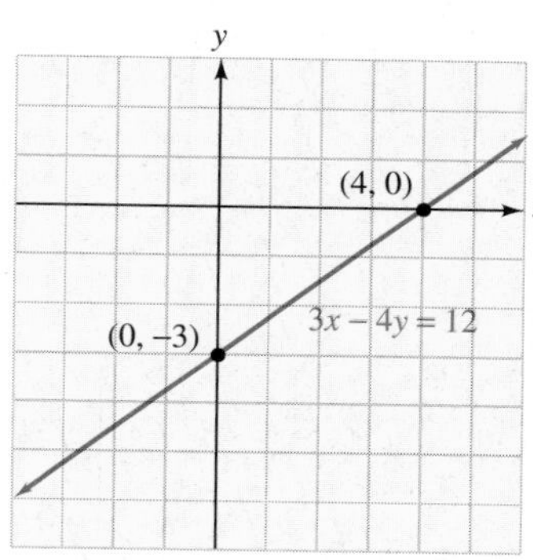

Figure 2-18

The slope of the line is $\frac{3}{4}$.

SELF CHECK 2 Find the slope of the line determined by $2x + 5y = 12$.

4 Identify the slope of a horizontal line and a vertical line.

If $P(x_1, y_1)$ and $Q(x_2, y_2)$ are points on the horizontal line shown in Figure 2-19(a) on the next page, then $y_1 = y_2$, and the numerator of the fraction

$$\frac{y_2 - y_1}{x_2 - x_1} \quad \text{On a horizontal line, } x_2 \neq x_1.$$

is 0. Thus, the value of the fraction is 0, and the slope of the horizontal line is 0.

If $P(x_1, y_1)$ and $Q(x_2, y_2)$ are two points on the vertical line shown in Figure 2-19(b) on the next page, then $x_1 = x_2$, and the denominator of the fraction

$$\frac{y_2 - y_1}{x_2 - x_1} \quad \text{On a vertical line, } y_2 \neq y_1.$$

is 0. Since the denominator cannot be 0, a vertical line has no defined slope.

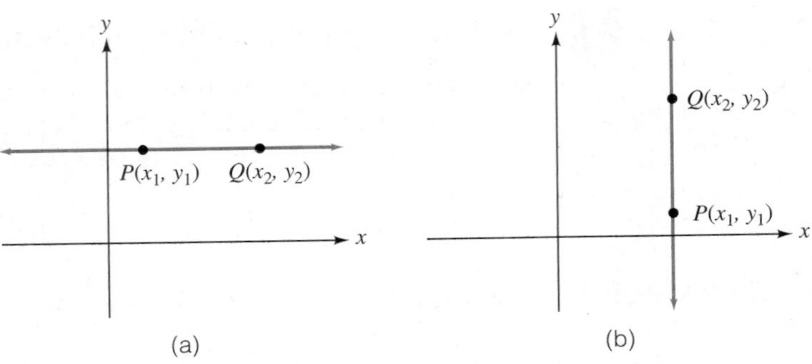

Figure 2-19

Slopes of Horizontal and Vertical Lines

All horizontal lines (lines with equations of the form $y = b$) have a slope of 0.

All vertical lines (lines with equations of the form $x = a$) have no defined slope.

If a line rises as we follow it from left to right, as in Figure 2-20(a), its slope is positive. If a line drops as we follow it from left to right, as in Figure 2-20(b), its slope is negative. If a line is horizontal, as in Figure 2-20(c), its slope is 0. If a line is vertical, as in Figure 2-20(d), it has no defined slope.

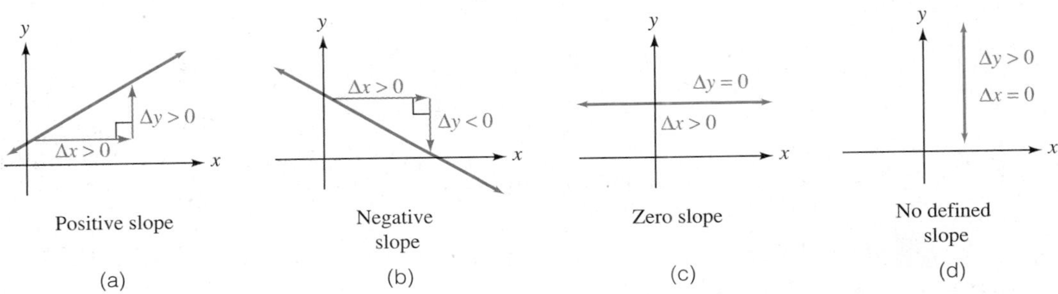

Figure 2-20

5 Determine whether two lines are parallel, perpendicular, or neither.

To see a relationship between parallel lines and their slopes, we refer to the parallel lines l_1 and l_2 shown in Figure 2-21, with slopes of m_1 and m_2, respectively. Because right triangles ABC and DEF are similar, it follows that

$$m_1 = \frac{\Delta y \text{ of } l_1}{\Delta x \text{ of } l_1}$$

$$= \frac{\Delta y \text{ of } l_2}{\Delta x \text{ of } l_2}$$

$$= m_2$$

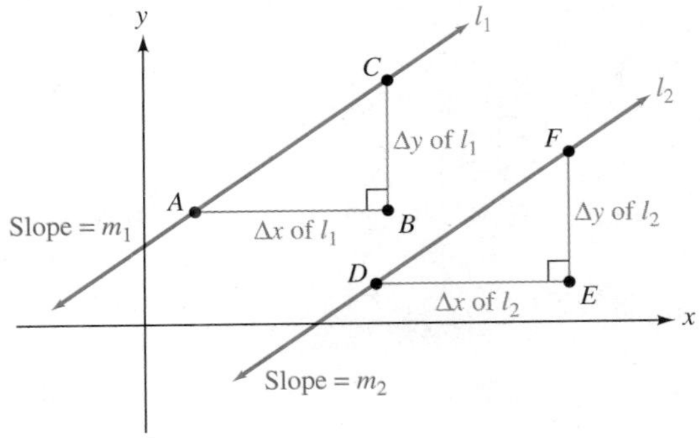

Figure 2-21

Thus, if two nonvertical lines are parallel, they have the same slope. It is also true that when two lines have the same slope, they are parallel.

Slopes of Parallel Lines

Nonvertical parallel lines have the same slope, and lines having the same slope are parallel.

Since vertical lines are parallel, lines with no defined slope are parallel.

EXAMPLE 3 The lines in Figure 2-22 are parallel. Find the slope of l_2.

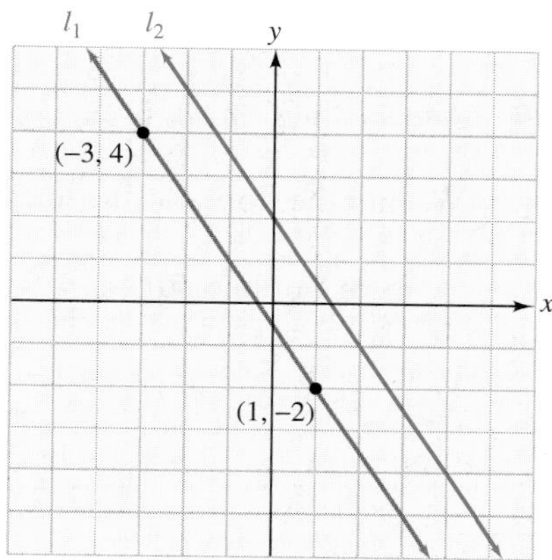

Figure 2-22

Solution From the figure, we can find the slope of line l_1. Since the lines are parallel, they will have equal slopes. Therefore, the slope of line l_2 will be equal to the slope of line l_1. We can use the points with coordinates $(-3, 4)$ and $(1, -2)$ on line l_1 to find the slope of l_1 as follows:

$$m = \frac{y_2 - y_1}{x_2 - x_1}$$

$$= \frac{-2 - 4}{1 - (-3)} \quad \text{Substitute } -2 \text{ for } y_2,\ 4 \text{ for } y_1,\ 1 \text{ for } x_2, \text{ and } -3 \text{ for } x_1.$$

$$= \frac{-6}{4}$$

$$= -\frac{3}{2}$$

The slope of l_1 is $-\frac{3}{2}$ and because the lines are parallel, the slope of l_2 is also $-\frac{3}{2}$.

SELF CHECK 3 Find the slope of any line parallel to a line with a slope of -3.

Two real numbers a and b are called **negative reciprocals** if $ab = -1$. For example,

$$-\frac{4}{3} \quad \text{and} \quad \frac{3}{4}$$

are negative reciprocals, because $-\frac{4}{3}\left(\frac{3}{4}\right) = -1$.

The following theorem relates perpendicular lines and their slopes.

Slopes of Perpendicular Lines

If two nonvertical lines are perpendicular, their slopes are negative reciprocals.

If the slopes of two lines are negative reciprocals, the lines are perpendicular.

Because a horizontal line is perpendicular to a vertical line, a line with a slope of 0 is perpendicular to a line with no defined slope.

EXAMPLE 4 Are the lines shown in Figure 2-23 perpendicular?

Solution We find the slopes of the lines and see whether they are negative reciprocals.

$$\text{Slope of line } OP = \frac{\Delta y}{\Delta x} = \frac{y_2 - y_1}{x_2 - x_1} = \frac{-4 - 0}{3 - 0} = -\frac{4}{3}$$

$$\text{Slope of line } PQ = \frac{\Delta y}{\Delta x} = \frac{y_2 - y_1}{x_2 - x_1} = \frac{4 - (-4)}{9 - 3} = \frac{8}{6} = \frac{4}{3}$$

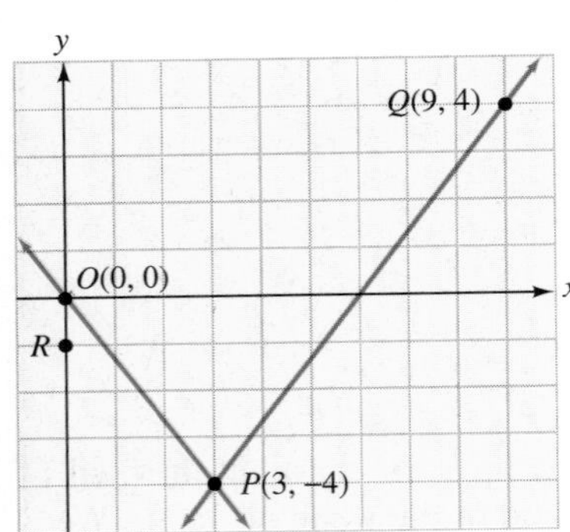

Figure 2-23

Since their slopes are not negative reciprocals, the lines are not perpendicular.

SELF CHECK 4 In Figure 2-23, is line PR perpendicular to line PQ?

6 Interpret slope in an application problem.

Many applied problems involve equations of lines and their slopes.

EXAMPLE 5 **COST OF CARPET** A store sells a carpet for \$25 per square yard, plus a \$20 delivery charge. The total cost c of n square yards is given by the following formula.

c	equals	cost per square yard	times	the number of square yards	plus	the delivery charge.
c	$=$	25	$\cdot$	n	$+$	20

Graph the equation $c = 25n + 20$ and interpret the slope of the line.

Solution We can graph the equation on a coordinate system with a vertical c-axis and a horizontal n-axis. Figure 2-24 shows a table of ordered pairs and the graph.

$c = 25n + 20$

n	c	(n, c)
10	270	(10, 270)
20	520	(20, 520)
30	770	(30, 770)
40	1,020	(40, 1,020)
50	1,270	(50, 1,270)

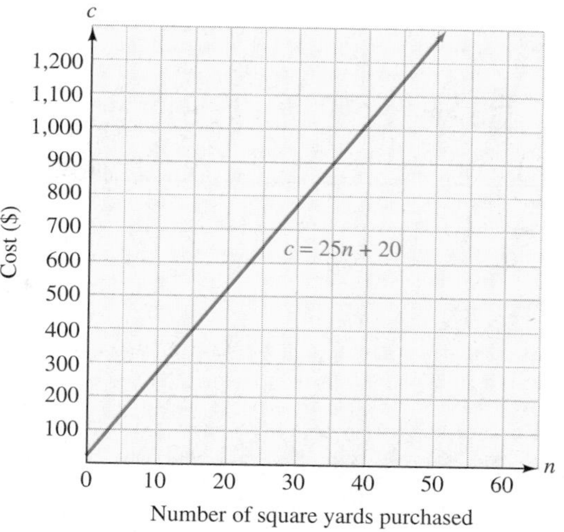

Figure 2-24

COMMENT Recall that any two points on a line can be used to find the slope. In this example, we chose (30, 770) and (50, 1,270), but we could have chosen any two points and obtained the same result.

If we choose the points (30, 770) and (50, 1,270) to find the slope, we have

$$
\begin{aligned}
m &= \frac{\Delta c}{\Delta n} \\
&= \frac{c_2 - c_1}{n_2 - n_1} \\
&= \frac{1{,}270 - 770}{50 - 30} && \text{Substitute 1,270 for } c_2\text{, 770 for } c_1\text{, 50 for } n_2\text{, and 30 for } n_1. \\
&= \frac{500}{20} \\
&= 25
\end{aligned}
$$

The slope of 25 is the cost of the carpet in dollars per square yard.

SELF CHECK 5 Interpret the y-intercept of the graph in Figure 2-24.

EXAMPLE 6 **RATE OF DESCENT** It takes a skier 25 minutes to complete the course shown in Figure 2-25. Find his average rate of descent in feet per minute.

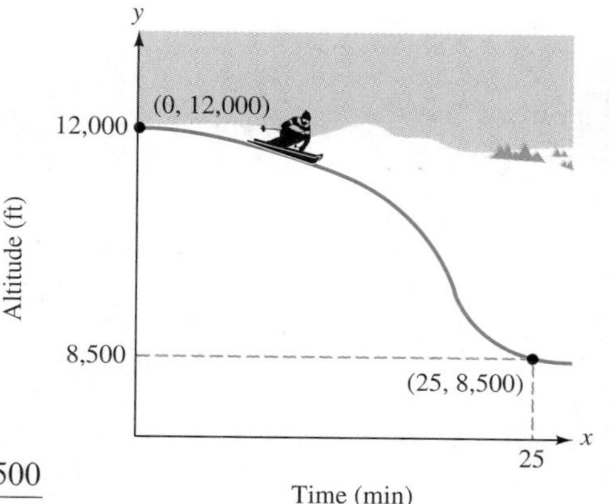

Figure 2-25

Solution To find the average rate of descent, we must find the ratio of the change in altitude to the change in time. To find this ratio, we calculate the slope of the line passing through the points with coordinates (0, 12,000) and (25, 8,500).

$$
\begin{aligned}
\text{Average rate of descent} &= \frac{12{,}000 - 8{,}500}{0 - 25} \\
&= \frac{3{,}500}{-25} \\
&= -140
\end{aligned}
$$

The slope is -140. Thus, the rate of change of descent is 140 ft/min.

SELF CHECK 6 Find the average rate of descent if the skier completes the course in 20 minutes.

EVERYDAY CONNECTIONS

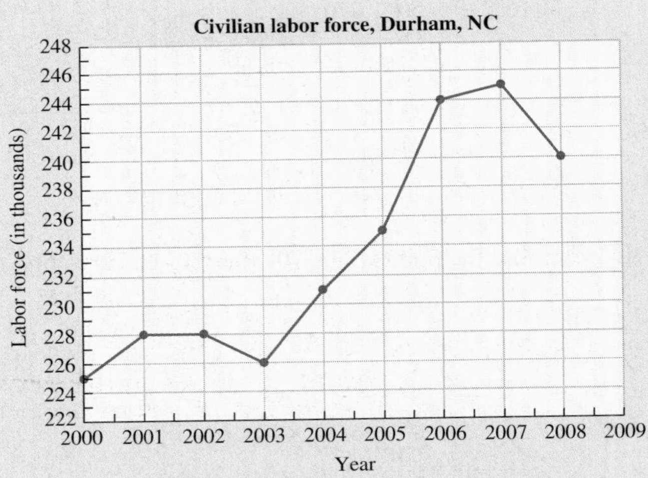

Source: http://www.bls.gov/eag/eag.nc_durham_msa.htm

We can approximate the **rate of growth (or decrease)** of a quantity during a given time interval by calculating the slope of the line segment that connects the endpoints of the graph on the given interval.

Use the data from the graph to compute the rate of growth (or decrease) of the following:

1. labor force from 2002 to 2003.
2. labor force from 2005 to 2006.
3. During which 1-year interval was a positive rate of growth of the labor force the smallest?

SELF CHECK ANSWERS

1. -2 **2.** $-\frac{2}{5}$ **3.** -3 **4.** no **5.** The y-coordinate of the y-intercept is the delivery charge. **6.** 175 ft/min

NOW TRY THIS

Find the slope of a line

1. parallel to $3x - 2 = 0$
2. perpendicular to $6x = 4$
3. perpendicular to $5y + 4 = 0$

2.2 EXERCISES

WARM-UPS *Find the slope of the line passing through*

1. (0, 0), (1, 3)
2. (0, 0), (3, 6)
3. Are lines with slopes of -2 and $-\frac{8}{4}$ parallel?
4. Find the negative reciprocal of -0.2.
5. Are lines with slopes of -2 and $\frac{1}{2}$ perpendicular?

REVIEW *Simplify each expression. Write all answers without negative exponents. Assume no variable is zero.*

6. $(x^3y^2)^3$
7. $\left(\frac{x^5}{x^3}\right)^3$
8. $(x^{-3}y^2)^{-4}$
9. $\left(\frac{x^{-6}}{y^3}\right)^{-4}$
10. $\left(\frac{3x^2y^3}{8}\right)^0$
11. $\left(\frac{x^3x^{-7}y^{-6}}{x^4y^{-3}y^{-2}}\right)^{-2}$

VOCABULARY AND CONCEPTS *Fill in the blanks.*

12. Slope is defined as the change in __ divided by the change in __.
13. A slope is a rate of ________.
14. The formula to compute slope is $m =$ ______.
15. The change in y (denoted as Δy) is the _____ of the line between two points.
16. The change in x (denoted as Δx) is the _____ of the line between two points.
17. The slope of a __________ line is 0.
18. The slope of a ________ line has no defined slope.
19. If a line rises as x increases, its slope is ________.
20. ________ lines have the same slope.
21. The slopes of nonvertical ____________ lines are negative __________.
22. A line with no defined slope and a line with a slope of __ are perpendicular.

GUIDED PRACTICE

Find the slope of the line that passes through the given points, if possible. See Example 1. (Objectives 1–2)

23.

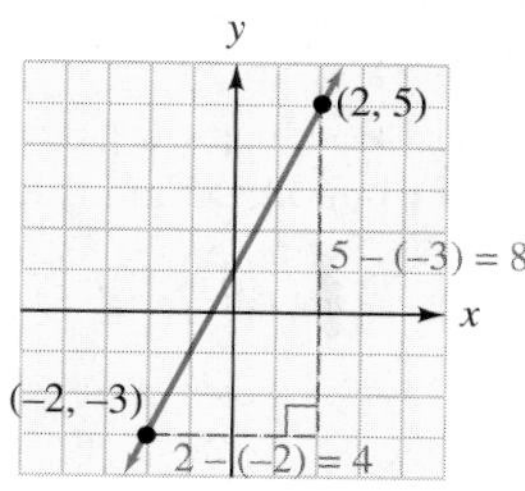

24.

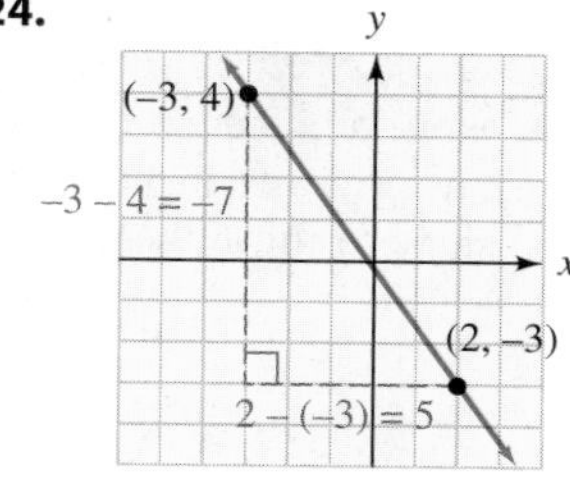

25. (0, 0), (3, 9)
26. (9, 6), (0, 0)
27. (−1, 8), (6, 1)
28. (−5, −8), (3, 8)
29. (3, −1), (−6, 2)
30. (0, −8), (−5, 0)
31. (7, 5), (−9, 5)
32. (2, −8), (3, −8)
33. (−7, −5), (−7, −2)
34. (3, −5), (3, 14)

Find the slope of the line determined by each equation. See Example 2. (Objective 3)

35. $3x + 2y = 12$
36. $2x - y = 6$
37. $3x = 4y - 2$
38. $x = y$
39. $y = \frac{x - 4}{2}$
40. $x = \frac{3 - y}{4}$
41. $4y = 3(y + 2)$
42. $x + y = \frac{2 - 3y}{3}$

Determine whether the slope of the line in each graph is positive, negative, 0, or not defined. (Objective 4)

43.

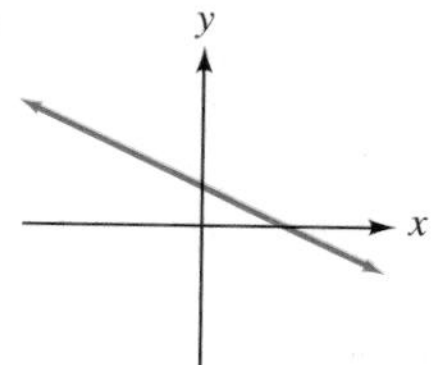

44.

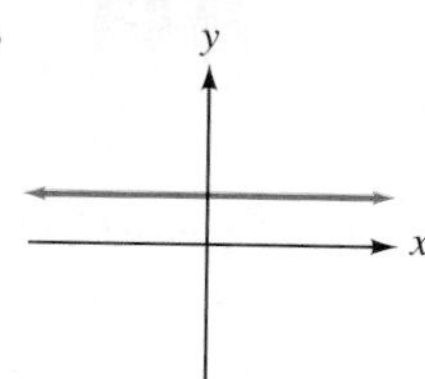

45.

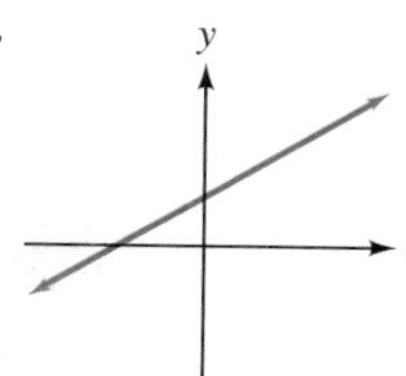

46.

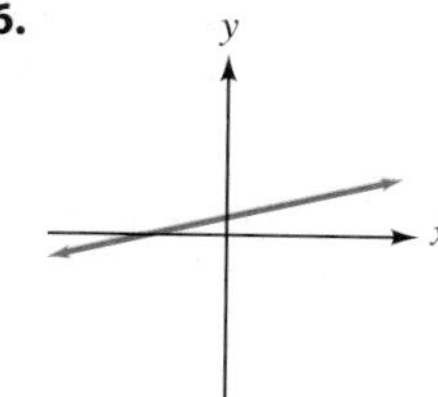

47.

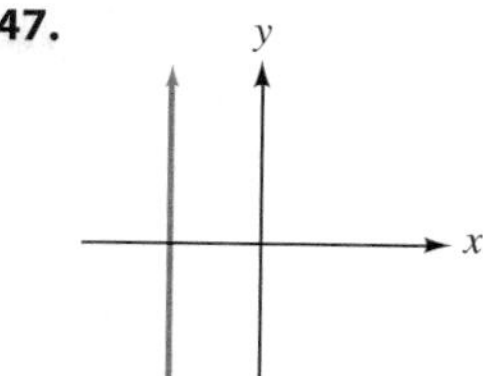

48.

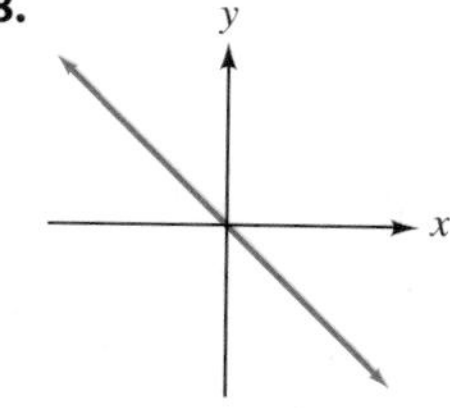

49.

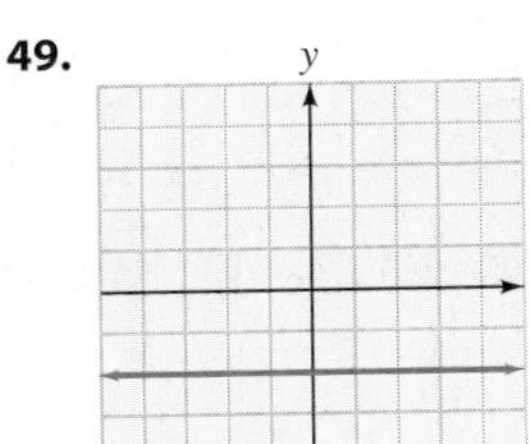

50.

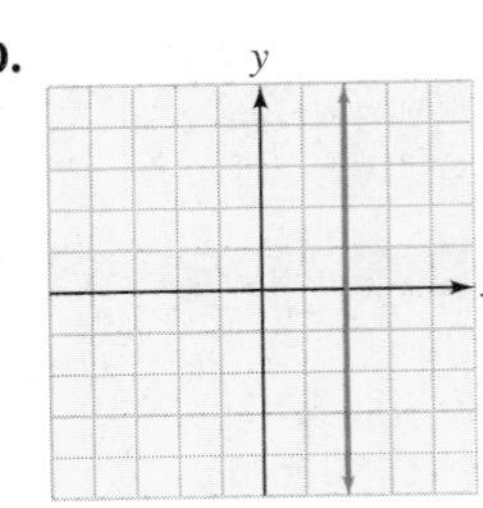

Determine whether the lines with the given slopes are parallel, perpendicular, or neither. **See Examples 3–4. (Objective 5)**

51. $m_1 = 3, m_2 = -\dfrac{1}{3}$

52. $m_1 = \dfrac{1}{4}, m_2 = 4$

53. $m_1 = 4, m_2 = 0.25$

54. $m_1 = -5, m_2 = \dfrac{1}{-0.2}$

Determine whether the line PQ is parallel or perpendicular or neither parallel nor perpendicular to a line with a slope of −2. **See Examples 3–4. (Objective 5)**

55. $P(3, 4), Q(4, 2)$

56. $P(6, 4), Q(8, 5)$

57. $P(-2, 1), Q(6, 5)$

58. $P(3, 4), Q(-3, -5)$

ADDITIONAL PRACTICE

59. Are the lines passing through the points (2.5, 3.7), (3.7, 2.5) and (1.7, −2.3), (2.3, −1.7) parallel, perpendicular, or neither parallel nor perpendicular?

60. Are the lines with slopes $m_1 = \frac{5.5}{2.7}$ and $m_2 = \left(\frac{2.7}{5.5}\right)^{-1}$ parallel, perpendicular, or neither parallel nor perpendicular ?

61. Are the lines with slopes $m_1 = \frac{3.2}{-9.1}$ and $m_2 = \frac{-9.1}{3.2}$ parallel, perpendicular, or neither parallel nor perpendicular?

62. Is the line passing through the points $P(5, 4)$ and $Q(6, 6)$ parallel, perpendicular, or neither parallel nor perpendicular to a line with a slope of −2?

63. Is the line passing through the points $P(-2, 3)$ and $Q(4, -9)$ parallel, perpendicular, or neither parallel nor perpendicular to a line with a slope of −2?

Find the slopes of lines PQ and PR and tell whether the points P, Q, and R lie on the same line. (Hint: Two lines with the same slope and a point in common must be the same line.)

64. $P(-2, 4), Q(4, 8), R(8, 12)$

65. $P(6, 10), Q(0, 6), R(3, 8)$

66. $P(-4, 10), Q(-6, 0), R(-1, 5)$

67. $P(-10, -13), Q(-8, -10), R(-12, -16)$

68. $P(-2, 4), Q(0, 8), R(2, 12)$

69. $P(8, -4), Q(0, -12), R(8, -20)$

70. Find the equation of the x-axis and its slope, if any.

71. Find the equation of the y-axis and its slope, if any.

72. Show that points with coordinates of (−3, 4), (4, 1), and (−1, −1) are the vertices of a right triangle.

73. Show that a triangle with vertices at (0, 0), (12, 0), and (13, 12) is not a right triangle.

74. A square has vertices at points $(a, 0)$, $(0, a)$, $(-a, 0)$, and $(0, -a)$, where $a \neq 0$. Show that its adjacent sides are perpendicular.

75. If a and b are not both 0, show that the points $(2b, a)$, (b, b), and $(a, 0)$ are the vertices of a right triangle.

76. Show that the points $(0, 0)$, $(0, a)$, (b, c), and $(b, a + c)$ are the vertices of a parallelogram. (*Hint:* Opposite sides of a parallelogram are parallel.)

77. If $b \neq 0$, show that the points $(0, 0)$, $(0, b)$, $(8, b + 2)$, and $(12, 3)$ are the vertices of a trapezoid. (*Hint:* A **trapezoid** is a four-sided figure with exactly two sides parallel.)

APPLICATIONS **See Examples 5–6. (Objective 6)**

78. Grade of a road Find the slope of the road. (*Hint:* 1 mi = 5,280 ft.)

79. Slope of a roof Find the slope of the roof.

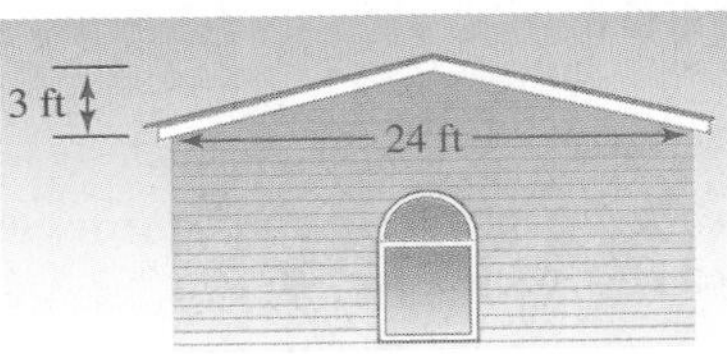

80. Slope of a ladder A ladder reaches 18 feet up the side of a building with its base 5 feet from the building. Find the slope of the ladder.

81. Physical fitness Find the slope of the treadmill for each setting listed in the table.

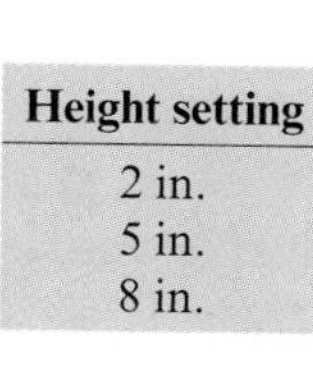

Height setting
2 in.
5 in.
8 in.

82. Global warming The following line graphs estimate the global temperature rise between the years of 1990 and 2040. Find the average rate of temperature change (the slope) of Model A: Status quo.

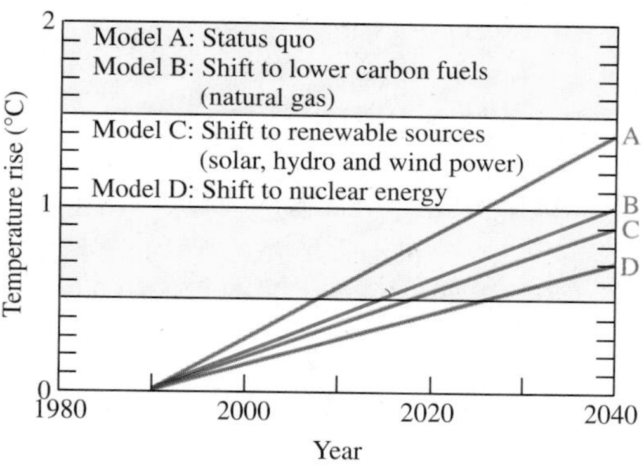

Based on data from The Blue Planet (Wiley, 1995)

83. Global warming Find the average rate of temperature change of Model D: Shift to nuclear energy.

84. Rate of growth When a college started an aviation program, the administration agreed to predict enrollments using a straight-line method. If the enrollment during the first year was 8, and the enrollment during the fifth year was 20, find the rate of growth per year (the slope of the line). (See the illustration.)

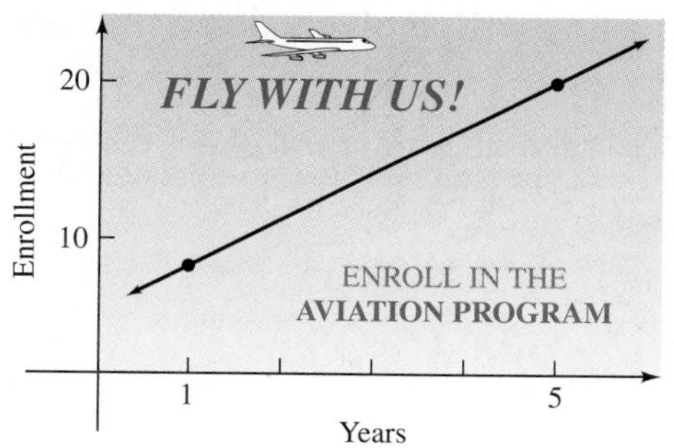

85. Wheelchair ramps The illustration shows two designs for a ramp to make a platform wheelchair accessible.

a. Find the slope of the ramp shown in design 1.

b. Find the slope of each part of the ramp shown in design 2.

c. Give one advantage and one disadvantage of each design.

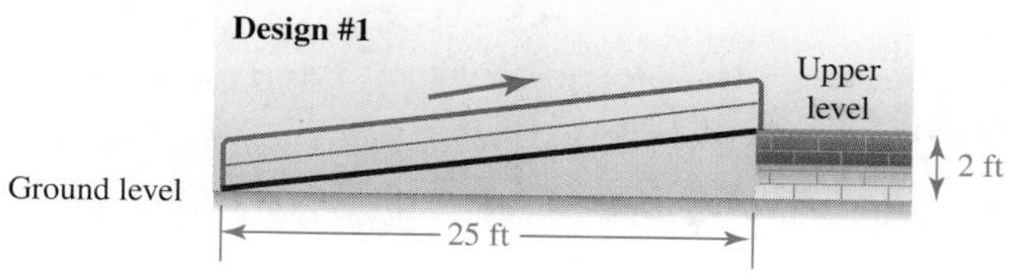

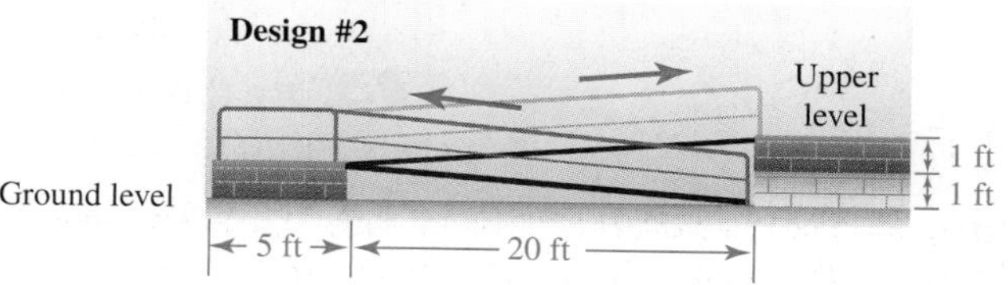

86. Rate of growth A small business predicts sales according to a straight-line method. If sales were \$85,000 in the first year and \$125,000 in the third year, find the rate of growth in sales per year (the slope of the line).

87. Rate of decrease The price of computer technology has been dropping for the past ten years. If a desktop PC cost \$5,700 ten years ago, and the same computing power cost \$1,499 two years ago, find the rate of decrease per year. (Assume a straight-line model.)

WRITING ABOUT MATH

88. Explain why a vertical line has no defined slope.

89. Explain how to determine from their slopes whether two lines are parallel, perpendicular, or neither.

SOMETHING TO THINK ABOUT

90. Find the slope of the line $Ax + By = C$. Follow the procedure of Example 2.

91. Follow Example 2 to find the slope of the line $y = mx + b$.

92. The points $(3, a)$, $(5, 7)$, and $(7, 10)$ lie on a line. Find a.

93. The line passing through points $(1, 3)$ and $(-2, 7)$ is perpendicular to the line passing through points $(4, b)$ and $(8, -1)$. Find b.

SECTION 2.3 Writing Equations of Lines

Objectives

1. Find the point-slope equation of a line with a given slope that passes through a given point.
2. Write the equation in slope-intercept form of a line that has a given slope and passes through a given point.
3. Graph a linear equation using the slope and *y*-intercept.
4. Determine whether two linear equations define lines that are parallel, perpendicular, or neither.
5. Write an equation of the line passing through a given point and parallel or perpendicular to a given line.
6. Use the general form of an equation of a line to determine whether two equations are parallel, perpendicular, or neither.
7. Write an equation of a line representing real-world data.

Vocabulary

point-slope form slope-intercept form general form

Getting Ready

Solve each equation.

1. $3 = \frac{x-2}{4}$ **2.** $-2 = 3(x+1)$

3. Solve $y - 2 = 3(x - 2)$ for y.

4. Solve $Ax + By + 3 = 0$ for x.

We now apply the concept of slope to write the equation of a line passing through two fixed points. We also will use slope as an aid in graphing lines.

1 Find the point-slope equation of a line with a given slope that passes through a given point.

Suppose that the line shown in Figure 2-26 has a slope of m and passes through the point (x_1, y_1). If (x, y) is a second point on the line, we have

$$m = \frac{y - y_1}{x - x_1}$$

or if we multiply both sides by $x - x_1$, we have

$$(1) \qquad y - y_1 = m(x - x_1)$$

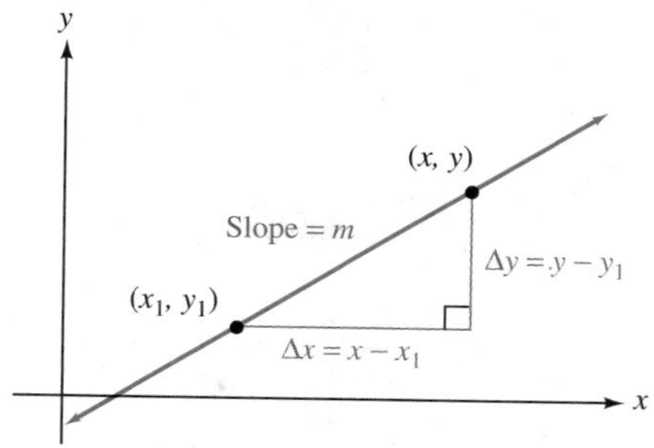

Figure 2-26

Because Equation 1 displays the coordinates of the point (x_1, y_1) on the line and the slope m of the line, it is called the **point-slope form** of the equation of a line.

Point-Slope Form

The **point-slope equation** of the line passing through $P(x_1, y_1)$ and with slope m is

$$y - y_1 = m(x - x_1)$$

EXAMPLE 1 Write the point-slope equation of the line with a slope of $-\frac{2}{3}$ and passing through $(-4, 5)$.

Solution We substitute $-\frac{2}{3}$ for m, -4 for x_1, and 5 for y_1 into the point-slope form and simplify.

$$y - y_1 = m(x - x_1)$$

$$y - 5 = -\frac{2}{3}[x - (-4)] \quad \text{Substitute } -\tfrac{2}{3} \text{ for } m, -4 \text{ for } x_1, \text{ and 5 for } y_1.$$

$$y - 5 = -\frac{2}{3}(x + 4) \quad -(-4) = 4$$

The point-slope equation of the line is $y - 5 = -\frac{2}{3}(x + 4)$.

SELF CHECK 1 Write the point-slope equation of the line with slope of $\frac{5}{4}$ and passing through $(0, 5)$.

EXAMPLE 2 Write the point-slope equation of the line passing through $(-5, 4)$ and $(8, -6)$. Then solve the equation for y.

Solution First we find the slope of the line.

$$m = \frac{y_2 - y_1}{x_2 - x_1}$$

$$= \frac{-6 - 4}{8 - (-5)} \quad \text{Substitute } -6 \text{ for } y_2, 4 \text{ for } y_1, 8 \text{ for } x_2, \text{ and } -5 \text{ for } x_1.$$

$$= -\frac{10}{13}$$

Because the line passes through both points, we can choose either one and substitute its coordinates into the point-slope form. If we choose $(-5, 4)$, we substitute -5 for x_1, 4 for y_1, and $-\frac{10}{13}$ for m and proceed as follows.

$$y - y_1 = m(x - x_1)$$

$$y - 4 = -\frac{10}{13}[x - (-5)] \quad \text{Substitute } -\tfrac{10}{13} \text{ for } m, -5 \text{ for } x_1, \text{ and 4 for } y_1.$$

$$y - 4 = -\frac{10}{13}(x + 5) \quad -(-5) = 5$$

To solve the equation for y, we proceed as follows:

$$y - 4 = -\frac{10}{13}x - \frac{50}{13} \quad \text{Remove parentheses.}$$

$$y = -\frac{10}{13}x + \frac{2}{13} \quad \text{Add 4 to both sides and simplify.}$$

The equation of the line is $y = -\frac{10}{13}x + \frac{2}{13}$.

SELF CHECK 2 Write the equation of the line passing through the points $(-2, 5)$ and $(4, -3)$ in point-slope form and then solve for y.

2 Write the equation in slope-intercept form of a line that has a given slope and passes through a given point.

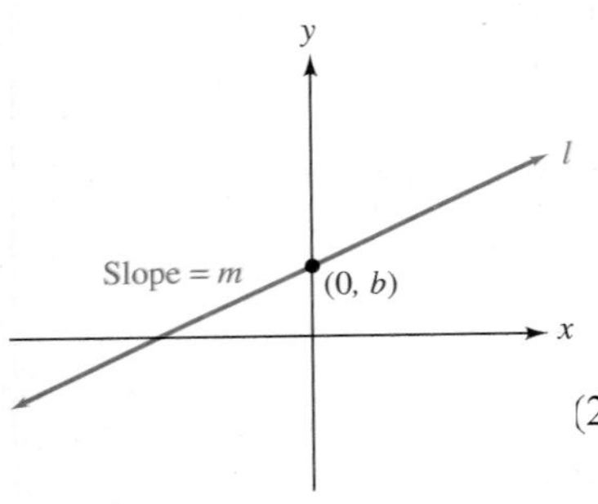

Figure 2-27

Since the y-intercept of the line shown in Figure 2-27 is the point $(0, b)$, we can write the equation of the line by substituting 0 for x_1 and b for y_1 in the point-slope form and then simplifying.

$$y - y_1 = m(x - x_1) \quad \text{This is the point-slope form of the equation of a line.}$$

$$y - b = m(x - 0) \quad \text{Substitute } b \text{ for } y_1 \text{ and 0 for } x_1.$$

$$y - b = mx \quad \text{Remove parentheses.}$$

$$(2) \quad y = mx + b \quad \text{Add } b \text{ to both sides.}$$

Because Equation 2 displays the slope m and the y-coordinate b of the y-intercept, it is called the **slope-intercept form** of the equation of a line.

Slope-Intercept Form The **slope-intercept equation** of a line with slope m and y-intercept $(0, b)$ is

$$y = mx + b$$

EXAMPLE 3 Use the slope-intercept form to write an equation of the line with slope 4 that passes through the point $(5, 9)$.

Solution Since we are given that $m = 4$ and that the ordered pair $(5, 9)$ satisfies the equation, we can substitute 5 for x, 9 for y, and 4 for m in the equation $y = mx + b$ and solve for b.

$$y = mx + b$$

$$9 = 4(5) + b \quad \text{Substitute 9 for } y, \text{ 4 for } m, \text{ and 5 for } x.$$

$$9 = 20 + b \quad \text{Simplify.}$$

$$-11 = b \quad \text{Subtract 20 from both sides.}$$

Because $m = 4$ and $b = -11$, the equation is $y = 4x - 11$.

SELF CHECK 3 Write the slope-intercept equation of the line with slope -2 that passes through the point $(-2, 8)$.

3 Graph a linear equation using the slope and y-intercept.

It is easy to graph a linear equation when it is written in slope-intercept form. For example, to graph $y = \frac{4}{3}x - 2$, we note that $b = -2$ so the y-intercept is $(0, b) = (0, -2)$. (See Figure 2-28.)

Because the slope of the line is $\frac{\Delta y}{\Delta x} = \frac{4}{3}$, we can locate another point on the line by starting at the point $(0, -2)$ and counting 3 units to the right and 4 units up. The line joining the two points is the graph of the equation.

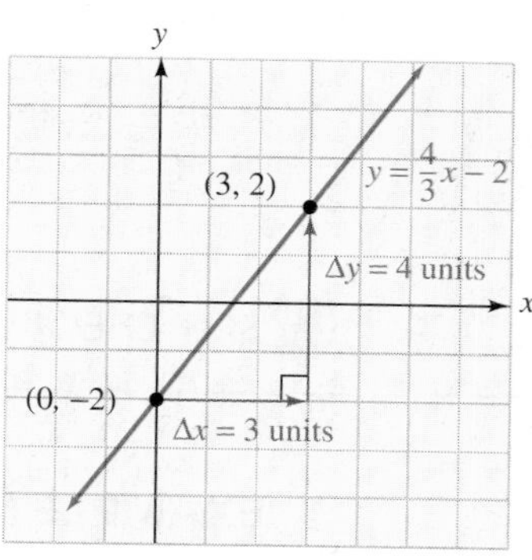

Figure 2-28

EXAMPLE 4 Find the slope and the y-intercept of the line with the equation $2(x - 3) = -3(y + 5)$. Then graph the line.

Solution We write the equation in the form $y = mx + b$ to find the slope m and the y-intercept $(0, b)$.

$$2(x - 3) = -3(y + 5)$$

$$2x - 6 = -3y - 15 \quad \text{Use the distributive property to remove parentheses.}$$

$$2x + 3y - 6 = -15 \quad \text{Add } 3y \text{ to both sides.}$$

$$3y - 6 = -2x - 15 \quad \text{Subtract } 2x \text{ from both sides.}$$

$$3y = -2x - 9 \quad \text{Add 6 to both sides.}$$

$$y = -\frac{2}{3}x - 3 \quad \text{Divide both sides by 3 and simplify.}$$

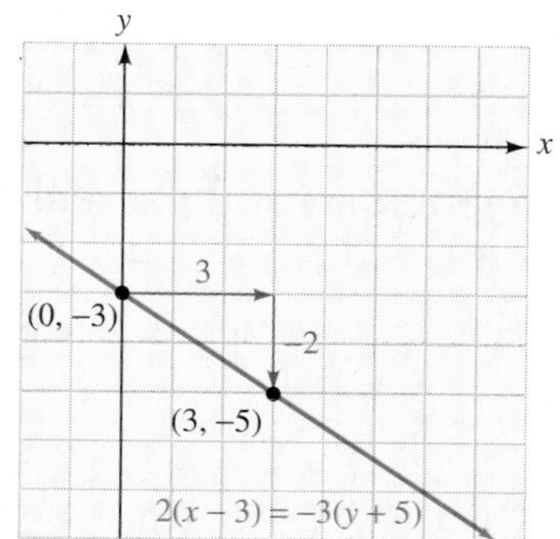

Figure 2-29

The slope is $-\frac{2}{3}$, and the y-intercept is $(0, -3)$. To draw the graph, we plot the y-intercept $(0, -3)$ and then locate a second point on the line by moving 3 units to the right and 2 units down. We draw a line through the two points to obtain the graph shown in Figure 2-29.

SELF CHECK 4 Find the slope and the y-intercept of the line with the equation $2(y - 1) = 3x + 2$ and graph the line.

4 Determine whether two linear equations define lines that are parallel, perpendicular, or neither.

EXAMPLE 5 Show that the lines represented by $4x + 8y = 10$ and $2x = 12 - 4y$ are parallel.

Solution In the previous section, we saw that distinct lines are parallel when their slopes are equal. To see whether this is true in this case, we can solve each equation for y.

$$4x + 8y = 10 \qquad\qquad 2x = 12 - 4y$$
$$8y = -4x + 10 \qquad\qquad 4y = -2x + 12$$
$$y = -\frac{1}{2}x + \frac{5}{4} \qquad\qquad y = -\frac{1}{2}x + 3$$

Because the values of $b\left(\frac{5}{4} \text{ and } 3\right)$ are different, the lines are distinct. Since the lines are distinct and have the same slope $\left(-\frac{1}{2}\right)$, they are parallel.

SELF CHECK 5 Are lines represented by $3x - 2y = 4$ and $6x = 4(y + 1)$ parallel?

EXAMPLE 6 Show that the lines represented by $4x + 8y = 10$ and $4x - 2y = 21$ are perpendicular.

Solution Since two lines are perpendicular when their slopes are negative reciprocals, we can solve each equation for y to see whether the slopes of their graphs are negative reciprocals.

$$4x + 8y = 10 \qquad\qquad 4x - 2y = 21$$
$$8y = -4x + 10 \qquad\qquad -2y = -4x + 21$$
$$y = -\frac{1}{2}x + \frac{5}{4} \qquad\qquad y = 2x - \frac{21}{2}$$

Since the slopes are $-\frac{1}{2}$ and 2 (which are negative reciprocals), the lines are perpendicular.

SELF CHECK 6 Are lines represented by $3x + 2y = 6$ and $2x - 3y = 6$ perpendicular?

5 Write an equation of the line passing through a given point and parallel or perpendicular to a given line.

We will now use the slope properties of parallel and perpendicular lines to write equations of lines.

EXAMPLE 7 Write the equation of the line passing through $(-2, 5)$ and parallel to the line $y = 8x - 3$.

Solution Since the equation is solved for y, the slope of the line given by $y = 8x - 3$ is the coefficient of x, which is 8. Since the desired equation is to have a graph that is parallel to the graph of $y = 8x - 3$, its slope also must be 8.

We substitute -2 for x_1, 5 for y_1, and 8 for m in the point-slope form and simplify.

$$\begin{aligned} y - y_1 &= m(x - x_1) \\ y - 5 &= 8[x - (-2)] && \text{Substitute 5 for } y_1\text{, 8 for } m\text{, and } -2 \text{ for } x_1. \\ y - 5 &= 8(x + 2) && -(-2) = 2 \\ y - 5 &= 8x + 16 && \text{Use the distributive property to remove parentheses.} \\ y &= 8x + 21 && \text{Add 5 to both sides.} \end{aligned}$$

The equation is $y = 8x + 21$.

SELF CHECK 7 Write the equation of the line that is parallel to the line $y = 8x - 3$ and passes through the origin.

EXAMPLE 8 Write the equation of the line passing through $(-2, 5)$ and perpendicular to the line $y = 8x - 3$.

Solution The slope of the given line is 8. Thus, the slope of the desired line must be $-\frac{1}{8}$, which is the negative reciprocal of 8.

We substitute -2 for x_1, 5 for y_1, and $-\frac{1}{8}$ for m into the point-slope form and simplify:

$$\begin{aligned} y - y_1 &= m(x - x_1) \\ y - 5 &= -\frac{1}{8}[x - (-2)] && \text{Substitute 5 for } y_1\text{, } -\tfrac{1}{8} \text{ for } m\text{, and } -2 \text{ for } x_1. \\ y - 5 &= -\frac{1}{8}(x + 2) && -(-2) = 2 \\ y - 5 &= -\frac{1}{8}x - \frac{1}{4} && \text{Remove parentheses.} \\ y &= -\frac{1}{8}x - \frac{1}{4} + 5 && \text{Add 5 to both sides.} \\ y &= -\frac{1}{8}x + \frac{19}{4} && \text{Combine terms: } -\tfrac{1}{4} + \tfrac{20}{4} = \tfrac{19}{4}. \end{aligned}$$

The equation is $y = -\frac{1}{8}x + \frac{19}{4}$.

SELF CHECK 8 Write the equation of the line that is perpendicular to the line $y = 8x - 3$ and passes through $(2, 4)$.

---

6 Use the general form of an equation of a line to determine whether two equations are parallel, perpendicular, or neither.

Recall that any linear equation that is written in the form $Ax + By = C$, where A, B, and C are constants, is said to be written in **general form.**

COMMENT When writing equations in general form, we usually clear the equation of fractions and make A positive. We also will make A, B, and C as small as possible. For example, the equation $6x + 12y = 24$ can be written as $x + 2y = 4$ by dividing both sides by 6.

Finding the Slope and y-Intercept from the General Form

If A, B, and C are real numbers and $B \neq 0$, the graph of the equation

$$Ax + By = C$$

is a nonvertical line with slope of $-\frac{A}{B}$ and a y-intercept of $\left(0, \frac{C}{B}\right)$.

You will be asked to justify the previous results in the exercises. You also will be asked to show that if $B = 0$, the equation $Ax + By = C$ represents a vertical line with x-intercept of $\left(\frac{C}{A}, 0\right)$.

EXAMPLE 9 Show that the lines represented by $4x + 3y = 7$ and $3x - 4y = 12$ are perpendicular.

Solution To show that the lines are perpendicular, we will show that their slopes are negative reciprocals. The first equation, $4x + 3y = 7$, is written in general form, with $A = 4$, $B = 3$, and $C = 7$. By the previous result, the slope of the line is

$$m_1 = -\frac{A}{B} = -\frac{4}{3} = -\frac{4}{3}$$

The second equation, $3x - 4y = 12$, also is written in general form, with $A = 3$, $B = -4$, and $C = 12$. The slope of this line is

$$m_2 = -\frac{A}{B} = -\frac{3}{-4} = \frac{3}{4}$$

Since the slopes are negative reciprocals, the lines are perpendicular.

SELF CHECK 9 Are the lines $4x + 3y = 7$ and $y = -\frac{4}{3}x + 2$ parallel?

We summarize the various forms for the equation of a line in Table 2-1.

Point-slope form of a linear equation	$y - y_1 = m(x - x_1)$ The slope is m, and the line passes through (x_1, y_1).
Slope-intercept form of a linear equation	$y = mx + b$ The slope is m, and the y-intercept is $(0, b)$.
General form of a linear equation	$Ax + By = C$ A and B cannot both be 0.
A horizontal line	$y = b$ The slope is 0, and the y-intercept is $(0, b)$.
A vertical line	$x = a$ There is no defined slope, and the x-intercept is $(a, 0)$.

Table 2-1

7 Write an equation of a line representing real-world data.

For tax purposes, many businesses use straight-line depreciation to find the declining value of aging equipment.

EXAMPLE 10 **VALUE OF A LATHE** The owner of a machine shop buys a lathe for \$1,970 and expects it to last for ten years. It can then be sold as scrap for an estimated **salvage value** of \$270. If y represents the value of the lathe after x years of use, and y and x are related by the equation of a line,

a. Find the equation of the line.

b. Find the value of the lathe after $2\frac{1}{2}$ years.

c. Find the economic meaning of the y-intercept of the line.

d. Find the economic meaning of the slope of the line.

Solution **a.** To find the equation of the line, we find its slope and use point-slope form to find its equation.

COMMENT When the size of data is large, we can insert a // (break) symbol on the x- or y-axis to indicate that the scale does not begin until the first value is listed.

When the lathe is new, its age x is 0, and its value y is \$1,970. When the lathe is 10 years old, $x = 10$ and its value is $y = \$270$. Since the line passes through the points (0, 1,970) and (10, 270), as shown in Figure 2-30, the slope of the line is

$$\begin{aligned} m &= \frac{y_2 - y_1}{x_2 - x_1} \\ &= \frac{\mathbf{270} - \mathbf{1{,}970}}{\mathbf{10} - \mathbf{0}} \\ &= \frac{-1{,}700}{10} \\ &= -170 \end{aligned}$$

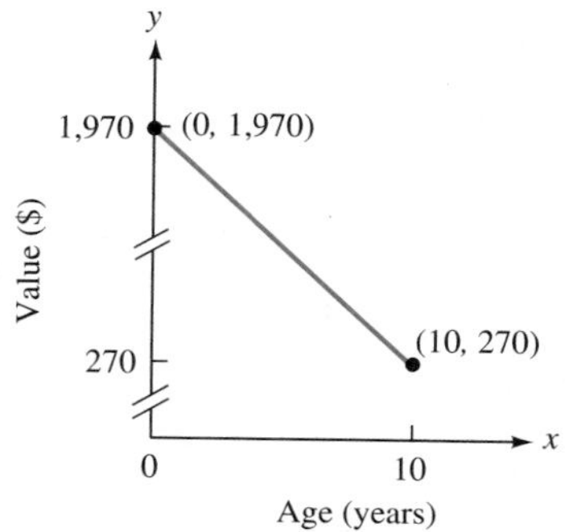

Figure 2-30

To find the equation of the line, we substitute -170 for m, 0 for x_1, and 1,970 for y_1 into the point-slope form and solve for y.

$$\begin{aligned} y - y_1 &= m(x - x_1) \\ y - \mathbf{1{,}970} &= \mathbf{-170}(x - \mathbf{0}) \\ \text{(3)} \qquad y &= -170x + 1{,}970 \end{aligned}$$

The current value y of the lathe is related to its age x by the equation $y = -170x + 1{,}970$.

b. To find the value of the lathe after $2\frac{1}{2}$ years, we substitute 2.5 for x in Equation 3 and solve for y.

$$\begin{aligned} y &= -170x + 1{,}970 \\ &= -170(\mathbf{2.5}) + 1{,}970 \\ &= -425 + 1{,}970 \\ &= 1{,}545 \end{aligned}$$

After $2\frac{1}{2}$ years, the lathe will be worth \$1,545.

c. The y-intercept of the graph is $(0, b)$, where b is the value of y when $x = 0$.

$$\begin{aligned} y &= -170x + 1{,}970 \\ y &= -170(\mathbf{0}) + 1{,}970 \\ y &= 1{,}970 \end{aligned}$$

Thus, b is the value of a 0-year-old lathe, which is the lathe's original cost, \$1,970.

d. Each year, the value of the lathe decreases by \$170, because the slope of the line is -170. The slope of the line is the **annual depreciation rate.**

In statistics, the process of using one variable to predict another is called *regression.* For example, if we know a man's height, we can make a good prediction about his weight, because taller men usually weigh more than shorter men.

Figure 2-31 shows the result of sampling ten men at random and finding their heights and weights. The graph of the ordered pairs (h, w) is called a **scattergram.**

Man	Height in inches	Weight in pounds
1	66	140
2	68	150
3	68	165
4	70	180
5	70	165
6	71	175
7	72	200
8	74	190
9	75	210
10	75	215

(a)

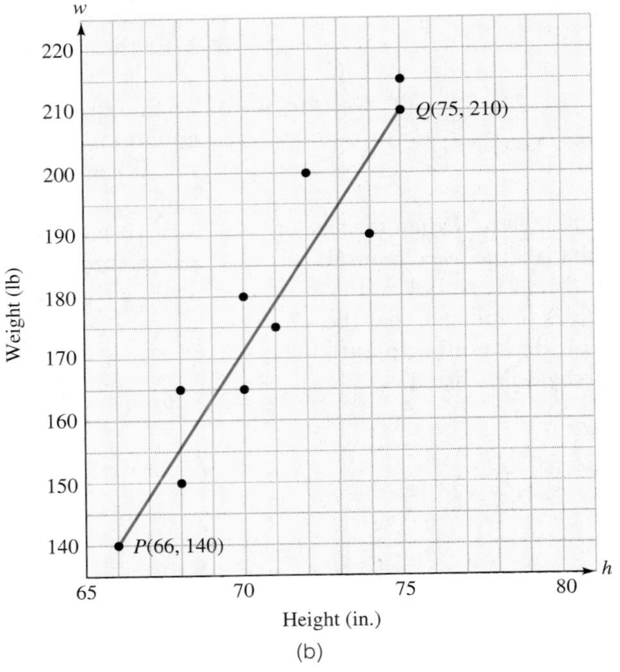

(b)

Figure 2-31

To write a prediction equation (sometimes called a *regression equation*), we must find the equation of the line that comes closer to all of the points in the scattergram than any other possible line. There are exact methods to find this equation, but we can only approximate it here.

To write an approximation of the regression equation, we place a straightedge on the scattergram shown in Figure 2-31 and draw the line joining two points that seems to best fit all the points. In the figure, line PQ is drawn, where point P has coordinates of (66, 140) and point Q has coordinates of (75, 210).

Our approximation of the regression equation will be the equation of the line passing through points P and Q. To find the equation of this line, we first find its slope.

$$\begin{aligned} m &= \frac{y_2 - y_1}{x_2 - x_1} \\ &= \frac{\mathbf{210} - \mathbf{140}}{\mathbf{75} - \mathbf{66}} \\ &= \frac{70}{9} \end{aligned}$$

We can then use point-slope form to find its equation.

$$\begin{aligned} y - y_1 &= m(x - x_1) \\ y - \mathbf{140} &= \frac{\mathbf{70}}{\mathbf{9}}(x - \mathbf{66}) \end{aligned}$$ Choose (66, 140) for (x_1, y_1).

$$y = \frac{70}{9}x - \frac{4{,}620}{9} + 140 \qquad \text{Remove parentheses and add 140 to both sides.}$$

$$(4) \qquad y = \frac{70}{9}x - \frac{1{,}120}{3} \qquad -\tfrac{4{,}620}{9} + 140 = -\tfrac{1{,}120}{3}$$

Our approximation of the regression equation is $y = \frac{70}{9}x - \frac{1{,}120}{3}$.

To predict the weight of a man who is 73 inches tall, for example, we substitute 73 for x in Equation 4 and simplify.

$$y = \frac{70}{9}x - \frac{1{,}120}{3}$$

$$y = \frac{70}{9}(73) - \frac{1{,}120}{3}$$

$$y \approx 194.4$$

We would predict that a 73-inch-tall man chosen at random will weigh about 194 pounds.

SELF CHECK ANSWERS

1. $y - 5 = \frac{5}{4}x$ **2.** $y = -\frac{4}{3}x + \frac{7}{3}$ **3.** $y = -2x + 4$

4. $m = \frac{3}{2}$, (0, 2)

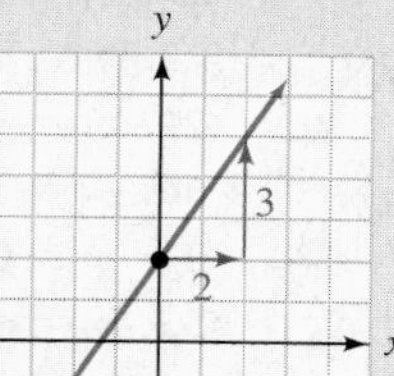

5. yes **6.** yes **7.** $y = 8x$ **8.** $y = -\frac{1}{8}x + \frac{17}{4}$ **9.** yes

NOW TRY THIS

1. Which of the following graphs could be the graph of $-px - q$?

a.

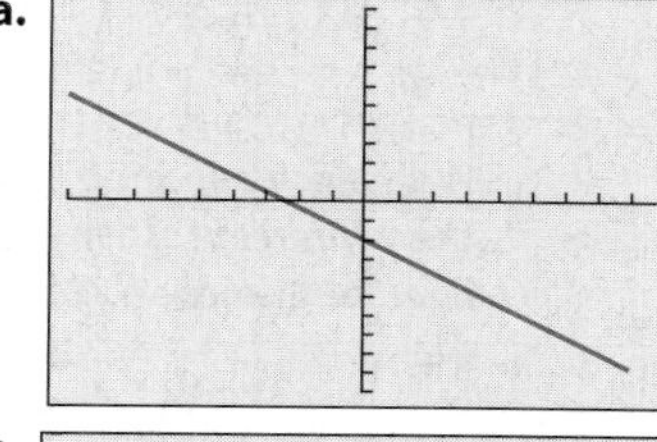

b.

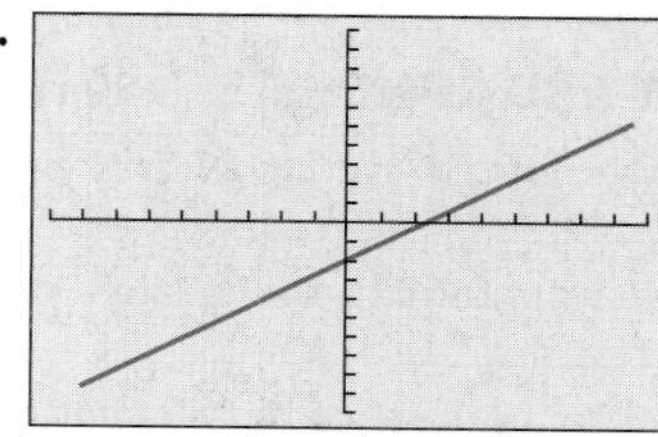

c.

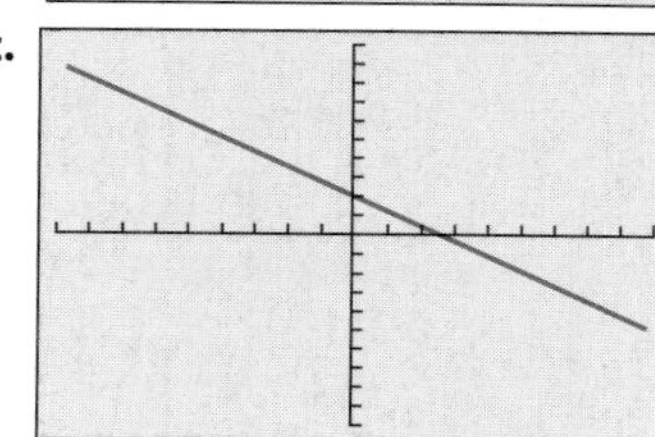

d.

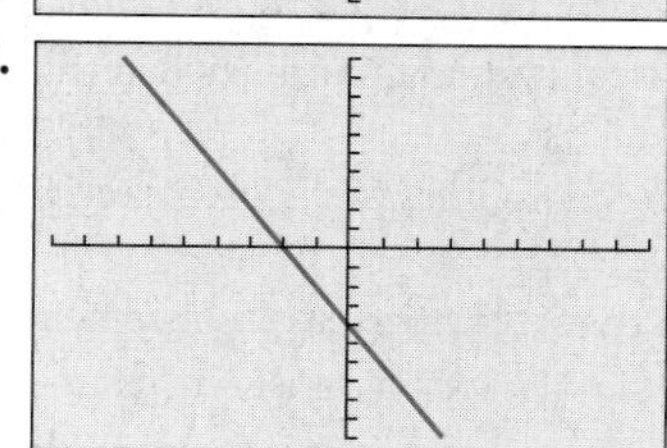

2. Write an equation of a line in slope-intercept form (if possible) with the given information.

a. parallel to $y = -3$ through (4, 6)

b. perpendicular to $x = 4$ through (0, 0)

c. parallel to $3x = 2y + 8$ through $(-4, -6)$

2.3 EXERCISES

WARM-UPS

Write the point-slope form of the equation of a line with $m = 2$, passing through the given point.

1. $(2, 3)$

2. $(-3, 8)$

Write the equation of a line with $m = -3$ and the given y-intercept.

3. $(0, 5)$

4. $(0, -7)$

Determine whether the lines are parallel, perpendicular, or neither.

5. $y = 3x - 4, y = 3x + 5$

6. $y = -3x + 7, x = 3y - 1$

REVIEW EXERCISES *Solve each equation.*

7. $3(x + 2) + x = 5x$

8. $12b + 6(3 - b) = b + 3$

9. $\frac{5(2 - x)}{3} - 1 = x + 5$

10. $\frac{r - 1}{3} = \frac{r + 2}{6} + 2$

11. Mixing alloys In 60 ounces of alloy for watch cases, there are 20 ounces of gold. How much copper must be added to the alloy so that a watch case weighing 4 ounces, made from the new alloy, will contain exactly 1 ounce of gold?

12. Mixing coffee To make a mixture of 80 pounds of coffee worth \$272, a grocer mixes coffee worth \$3.25 a pound with coffee worth \$3.85 a pound. How many pounds of the cheaper coffee should the grocer use?

VOCABULARY AND CONCEPTS *Fill in the blanks.*

13. The point-slope form of the equation of a line is ____________.

14. The slope-intercept form of the equation of a line is ____________.

15. The general form of the equation of a line is ____________.

16. Two nonvertical lines are parallel when they have the ______ slope.

17. If the slopes of two lines are negative reciprocals, the lines are ____________.

18. The process that recognizes that equipment loses value with age is called ____________.

GUIDED PRACTICE

Use point-slope form to write the equation of the line with the given properties. See Example 1. (Objective 1)

19. $m = 5$, passing through $(0, 7)$

20. $m = -8$, passing through $(0, -2)$

21. $m = -3$, passing through $(2, 0)$

22. $m = 4$, passing through $(-5, 0)$

Use point-slope form to write the equation of the line passing through the two given points and solve for y. See Example 2. (Objective 1)

23. $P(0, 0), Q(4, 4)$

24. $P(-5, -5), Q(0, 0)$

25. $P(3, 4), Q(0, -3)$

26. $P(4, 0), Q(6, -8)$

Use slope-intercept form to write the equation of the line with the given properties. See Example 3. (Objective 2)

27. $m = 3, b = 17$

28. $m = -2, b = 11$

29. $m = -7$, passing through $(7, 5)$

30. $m = 3$, passing through $(-2, -5)$

31. $m = 0$, passing through $(2, -4)$

32. $m = -7$, passing through the origin

33. Passing through $(6, 8)$ and $(2, 10)$

34. Passing through $(-4, 5)$ and $(2, -6)$

Find the slope and the y-intercept of the line determined by the given equation. See Example 4. (Objective 3)

35. $3x - 2y = 8$

36. $-2x + 4y = 12$

37. $-2(x + 3y) = 5$

38. $5(2x - 3y) = 4$

Write each equation in slope-intercept form to find the slope and the y-intercept. Then use the slope and y-intercept to graph the line. See Example 4. (Objective 3)

39. $y + 1 = x$

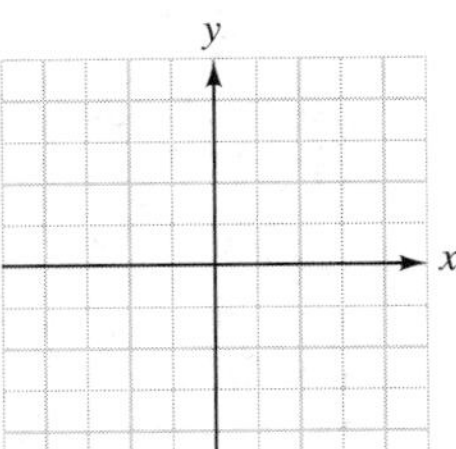

40. $x + y = 2$

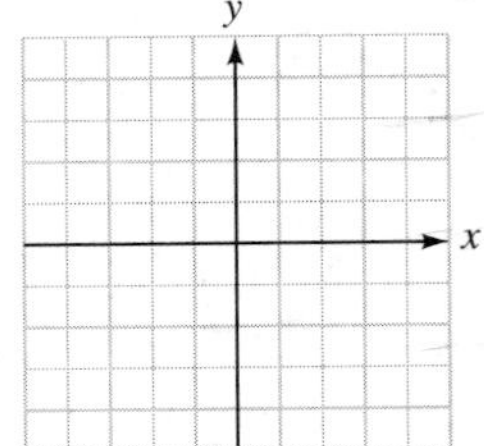

41. $x = \frac{3}{2}y - 3$

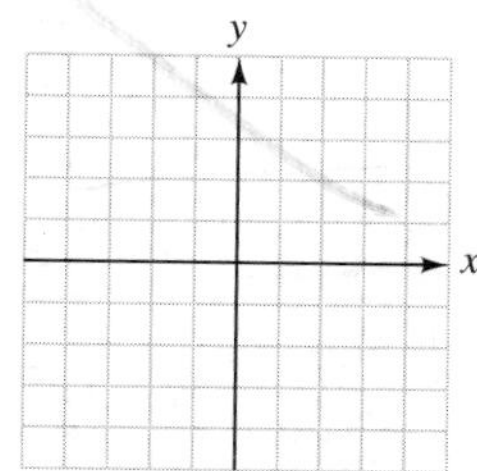

42. $x = -\frac{4}{5}y + 2$

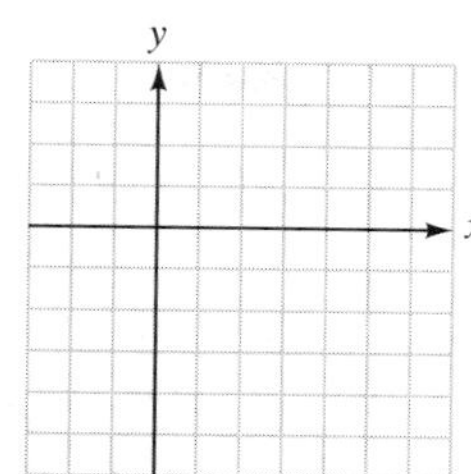

Determine whether the graphs of each pair of equations are parallel, perpendicular, or neither parallel nor perpendicular. See Examples 5–6. (Objective 4)

43. $y = 3x + 4,\ y = 3x - 7$

44. $y = 4x - 13,\ y = \frac{1}{4}x + 13$

45. $x + y = 2,\ y = x + 5$

46. $x = y + 2,\ y = x + 3$

47. $y = 3x + 7,\ 2y = 6x - 9$

48. $2x + 3y = 9,\ 3x - 2y = 5$

49. $x = 3y + 4,\ y = -3x + 7$

50. $3x + 6y = 1,\ y = \frac{1}{2}x$

51. $y = 3,\ x = 4$

52. $y = -3,\ y = -7$

53. $x = \frac{y - 2}{3},\ 3(y - 3) + x = 0$

54. $2y = 8,\ 3(2 + x) = 2(x + 2)$

Write the equation of the line that passes through the given point and is parallel or perpendicular to the given line. Write the answer in slope-intercept form. See Examples 7–8. (Objective 5)

55. $(0, 0)$, parallel to $y = 4x - 7$

56. $(0, 0)$, parallel to $x = -3y - 12$

57. $(2, 5)$, parallel to $4x - y = 7$

58. $(-6, 3)$, parallel to $y + 3x = -12$

59. $(0, 0)$, perpendicular to $y = 4x - 7$

60. $(0, 0)$, perpendicular to $x = -3y - 12$

61. $(2, 5)$, perpendicular to $4x - y = 7$

62. $(-6, 3)$, perpendicular to $y + 3x = -12$

63. $(4, -2)$, parallel to $x = \frac{5}{4}y - 2$

64. $(1, -5)$, parallel to $x = -\frac{3}{4}y + 5$

65. $(4, -2)$, perpendicular to $x = \frac{5}{4}y - 2$

66. $(1, -5)$, perpendicular to $x = -\frac{3}{4}y + 5$

Use the method of Example 9 to find whether the graphs determined by each pair of equations are parallel, perpendicular, or neither parallel nor perpendicular. See Example 9. (Objective 6)

67. $4x + 5y = 20,\ 5x - 4y = 20$

68. $9x - 12y = 17,\ 3x - 4y = 17$

69. $2x + 3y = 12,\ 6x + 9y = 32$

70. $5x + 6y = 30,\ 6x + 5y = 24$

ADDITIONAL PRACTICE

Find the slope and y-intercept and then graph the line.

71. $x = \frac{2y - 4}{7}$

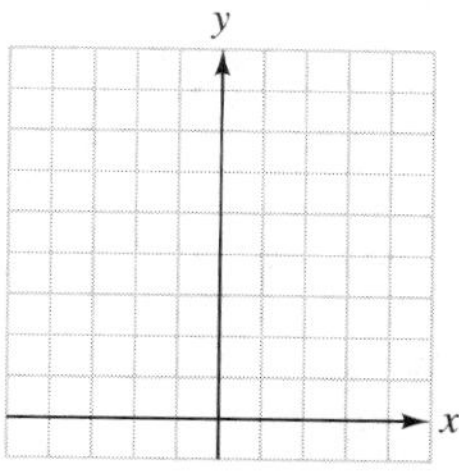

72. 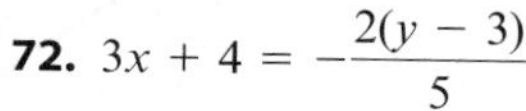 $3x + 4 = -\frac{2(y - 3)}{5}$

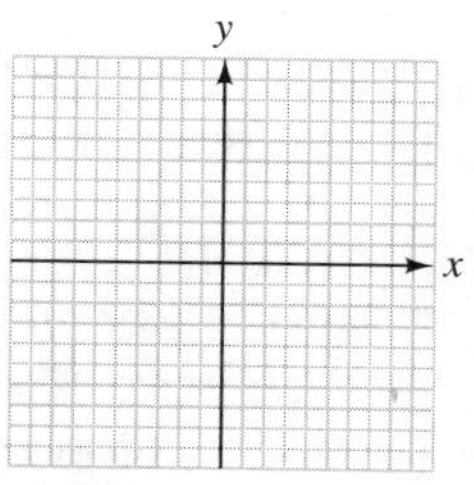

73. $3(y - 4) = -2(x - 3)$

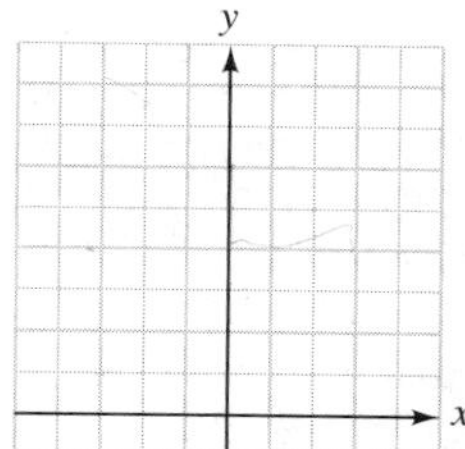

74. 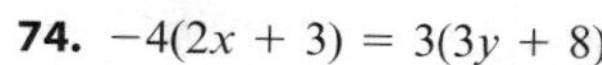 $-4(2x + 3) = 3(3y + 8)$

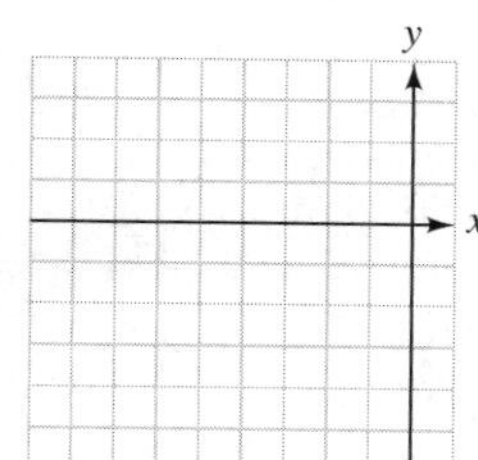

Use point-slope form to write the equation of each line and then write each answer in slope-intercept form.

75.

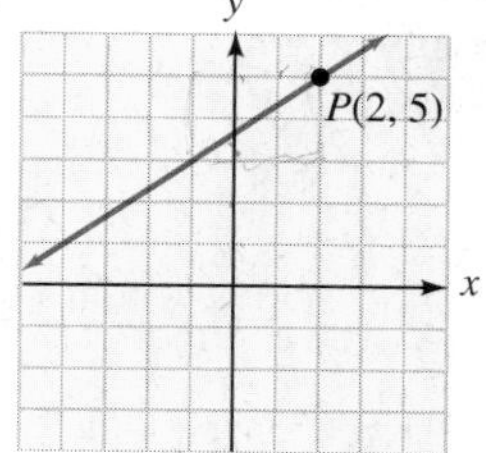

76.

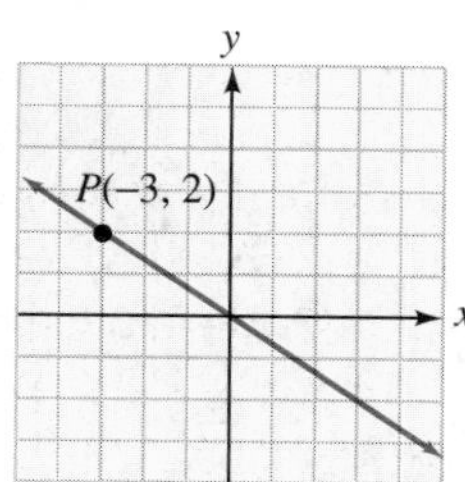

77.

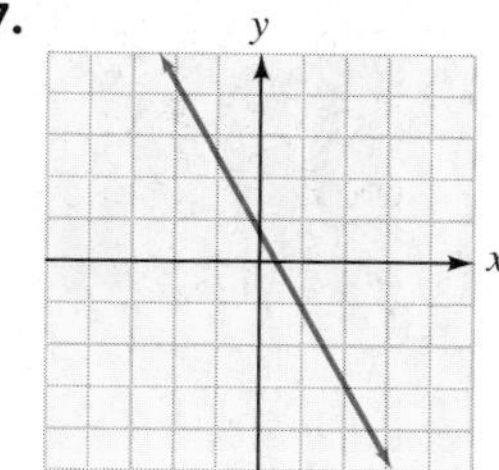

78.

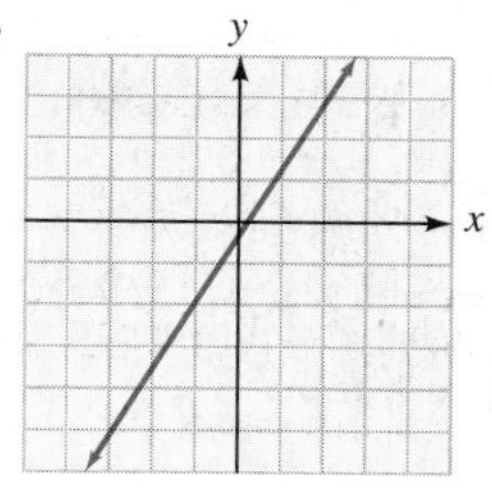

79. Find the equation of the line perpendicular to the line $y = 3$ and passing through the midpoint of the segment joining (2, 4) and (−6, 10).

80. Find the equation of the line parallel to the line $y = -8$ and passing through the midpoint of the segment joining (−4, 2) and (−2, 8).

81. Find the equation of the line parallel to the line $x = 3$ and passing through the midpoint of the segment joining (2, −4) and (8, 12).

82. Find the equation of the line perpendicular to the line $x = 3$ and passing through the midpoint of the segment joining (−2, 2) and (4, −8).

83. Solve $Ax + By = C$ for y and thereby show that the slope of its graph is $-\frac{A}{B}$ and its y-intercept is $\left(0, \frac{C}{B}\right)$.

84. Show that the x-intercept of the graph of $Ax + By = C$ is $\left(\frac{C}{A}, 0\right)$.

APPLICATIONS ***For problems involving depreciation or appreciation, assume straight-line depreciation or straight-line appreciation.*** **See Example 10. (Objective 7)**

85. Depreciation equations A truck was purchased for \$19,984. Its salvage value at the end of 8 years is expected to be \$1,600. Find the depreciation equation.

86. Depreciation equations A business purchased the computer shown. It will be depreciated over a 5-year period, when it will probably be worth \$200. Find the depreciation equation.

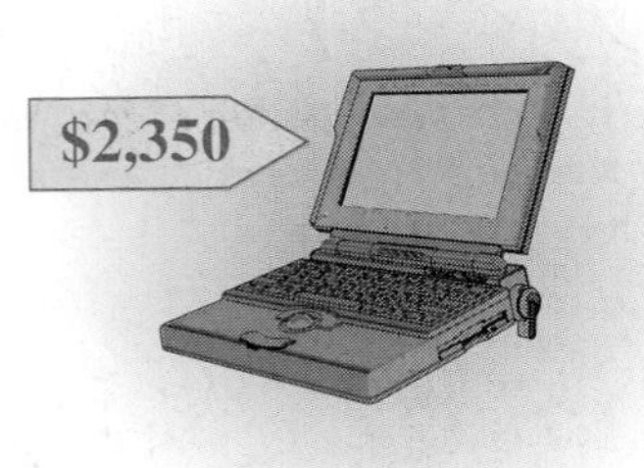

87. Art In 1987, the painting *Rising Sunflowers* by Vincent van Gogh sold for \$36,225,000. Suppose that an appraiser expected the painting to double in value in 20 years. Let x represent the time in years after 1987. Find the straight-line appreciation equation.

88. Real estate listings Use the information given in the following description of the property to write a straight-line appreciation equation for the house.

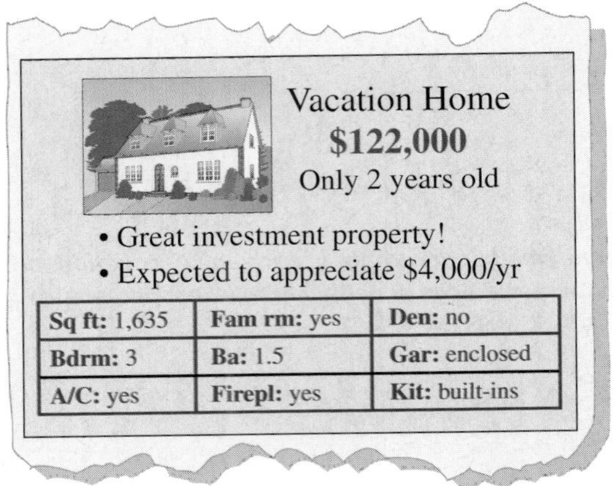

89. Appreciation equations A famous oil painting was purchased for \$250,000 and is expected to double in value in 5 years. Find the appreciation equation.

90. Appreciation equations A house purchased for \$142,000 is expected to double in value in 8 years. Find its appreciation equation.

91. Depreciation equations Find the depreciation equation for the TV in the want ad in the illustration.

For Sale: 3-year-old 65-inch TV, \$1,750 new. Asking \$800. Call 715-5588. Ask for Joe.

92. Depreciating a lawn mower A lawn mower cost \$450 when new and is expected to last 10 years. What will it be worth in $6\frac{1}{2}$ years?

93. Salvage value A copy machine that cost \$1,750 when new will be depreciated at the rate of \$180 per year. If the useful life of the copier is 7 years, find its salvage value.

94. Annual rate of depreciation A machine that cost \$47,600 when new will have a salvage value of \$500 after its useful life of 15 years. Find its annual rate of depreciation.

95. Real estate A vacation home is expected to appreciate about \$4,000 a year. If the home will be worth \$122,000 in 2 years, what will it be worth in 10 years?

96. Car repair A garage charges a fixed amount, plus an hourly rate, to service a car. Use the information in the table to find the hourly rate.

A-1 Car Repair Typical charges	
2 hours	\$143
5 hours	\$320

97. Printer charges A printer charges a fixed setup cost, plus \$15 for every 100 copies. If 300 copies cost \$75, how much will 1,000 copies cost?

98. Predicting burglaries A police department knows that city growth and the number of burglaries are related by a linear equation. City records show that 575 burglaries were reported in a year when the local population was 77,000, and 675 were reported in a year when the population was 87,000. How many burglaries can be expected when the population reaches 110,000?

WRITING ABOUT MATH

99. Explain how to find the equation of a line passing through two given points.

100. In straight-line depreciation, explain why the slope of the line is called the *rate of depreciation.*

SOMETHING TO THINK ABOUT *Investigate the properties of the slope and the y-intercept by experimenting with the following problems.*

101. Graph $y = mx + 2$ for several positive values of m. What do you notice?

102. Graph $y = mx + 2$ for several negative values of m. What do you notice?

103. Graph $y = 2x + b$ for several increasing positive values of b. What do you notice?

104. Graph $y = 2x + b$ for several decreasing negative values of b. What do you notice?

105. How will the graph of $y = \frac{1}{2}x + 5$ compare to the graph of $y = \frac{1}{2}x - 5$?

106. How will the graph of $y = \frac{1}{2}x - 5$ compare to the graph of $y = \frac{1}{2}x$?

107. If the graph of $y = ax + b$ passes through quadrants I, II, and IV, what can be known about the constants a and b?

108. The graph of $Ax + By = C$ passes only through quadrants I and IV. What is known about the constants A, B, and C?

SECTION 2.4 Introduction to Functions

Objectives

1. Find the domain and range of a relation and determine whether the relation is a function.
2. Find the domain and range from a graph and determine whether the graph represents a function.
3. Use function notation to evaluate a function at a given value.
4. Determine whether an equation represents a function.
5. Find the domain of a function given its equation.
6. Graph a linear function.

Vocabulary

relation	function	input
domain	vertical line test	dependent variable
range	output	independent variable

Getting Ready

If $y = \frac{3}{2}x - 2$, find the value of y for each value of x.

1. $x = 2$ **2.** $x = 6$ **3.** $x = -12$ **4.** $x = -\frac{1}{2}$

In this section, we will introduce *relations* and *functions*. We include these concepts in this chapter because they involve ordered pairs. The concept of *function* is one of the most important ideas in mathematics. The farther you go in mathematics, the more you will study functions.

1 Find the domain and range of a relation and determine whether the relation is a function.

Table 2-2 shows the number of women serving in the U.S. House of Representatives for several recent sessions of Congress.

Women in the U.S. House of Representatives					
Session of Congress	106th	107th	108th	109th	110th
Number of Female Representatives	56	59	59	68	71

Table 2-2

We can display the data in the table as sets of ordered pairs, where the *first component* represents the session of Congress and the *second component* represents the number of women serving in that session.

$$(106, 56) \qquad (107, 59) \qquad (108, 59) \qquad (109, 68) \qquad (110, 71)$$

Sets of ordered pairs like these are called **relations.** The set of all first components $\{106, 107, 108, 109, 110\}$ is called the **domain of the relation,** and the set of all second components $\{56, 59, 68, 71\}$ is called the **range of the relation.** Although 59 occurs twice as a second component, we list it only once in the range.

EXAMPLE 1 Find the domain and range of the relation $\{(3, 2), (5, -7), (-8, 2), (-9, -12)\}$.

Solution Since the set of first components is the domain, the domain is

$$\{3, 5, -8, -9\}$$

Since the set of second components is the range, the range is

$\{2, -7, -12\}$ The second component 2 is listed only once in the range.

SELF CHECK 1 Find the domain and range of the relation $\{(5, 6), (-12, 4), (8, 6), (5, 4)\}$.

When each first component in a relation determines exactly one second component, the relation is called a *function.*

Function A **function** is any set of ordered pairs (a relation) in which each first component determines exactly one second component.

EXAMPLE 2 Determine whether each relation defines a function.

a. $4 \rightarrow 3$, $6 \rightarrow 5$, $12 \rightarrow 9$

b.

x	y
8	2
1	4
8	6

c. $\{(-2, 3), (-1, 3), (0, 3), (1, 3)\}$

Solution **a.** The arrow diagram defines a function because each value in the first oval determines exactly one value in the second oval.

- $4 \rightarrow 3$ To the 4 in the first oval, there corresponds exactly one value, 3, in the second oval.
- $6 \rightarrow 5$ To the 6 in the first oval, there corresponds exactly one value, 5, in the second oval.
- $12 \rightarrow 9$ To the 12 in the first oval, there corresponds exactly one value, 9, in the second oval.

b. The table does not define a function, because the value 8 in the first column determines more than one value in the second column.

- In the first row, to the number 8, there corresponds the value 2.
- In the third row, to the number 8, there corresponds the value 6.

c. Since the first component of each ordered pair determines exactly one value for the second component, the set of ordered pairs is a function.

- $(-2, 3)$ To the number -2, there corresponds exactly one value, 3.
- $(-1, 3)$ To the number -1, there corresponds exactly one value, 3.
- $(0, 3)$ To the number 0, there corresponds exactly one value, 3.
- $(1, 3)$ To the number 1, there corresponds exactly one value, 3.

The results from parts (b) and (c) illustrate an important fact:

Two different ordered pairs of a function can have the same second component, but they cannot have the same first component.

SELF CHECK 2 Determine whether each relation is a function.

a. $0 \rightarrow 2$, $9 \rightarrow 3$, $9 \rightarrow 4$

b.

x	y
-1	-60
0	55
3	0

c. $\{(4, -1), (9, 2), (16, 15), (4, -4)\}$

2 Find the domain and range from a graph and determine whether the graph represents a function.

A **vertical line test** can be used to determine whether the graph of an equation represents a function. If every vertical line that intersects a graph does so exactly once, the graph represents a function, because every number x determines a single value of y. If any vertical line that intersects a graph does so more than once, the graph cannot represent a function, because some numbers x determines more than one value of y.

The graph in Figure 2-32(a) represents a function, because every vertical line that intersects the graph does so exactly once. The graph in Figure 2-32(b) does not represent a function, because some vertical lines intersect the graph more than once.

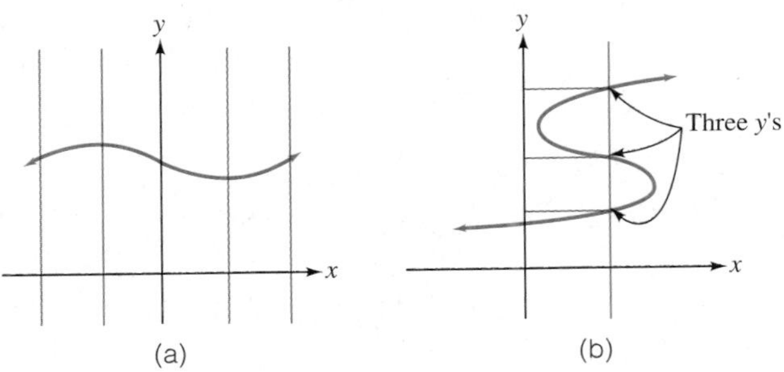

Figure 2-32

EXAMPLE 3 Find the domain and range of the relation determined by each graph and then tell whether the graph defines a function.

a.

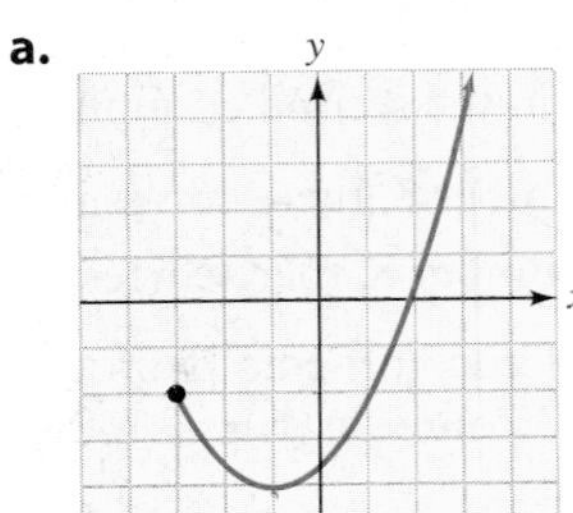

Figure 2-33

To find the domain, we look at the left-most point on the graph in Figure 2-33 and identify -3 as its x-coordinate. Since the graph continues forever to the right, there is no right-most point. Therefore, the domain is $[-3, \infty)$.

To find the range, we look for the lowest point on the graph and identify -4 as its y-coordinate. Since the graph continues upward forever, there is no highest point. Therefore, the range is $[-4, \infty)$.

Since every vertical line that intersects the graph will do so exactly once, the vertical line test indicates that the graph is a function.

b.

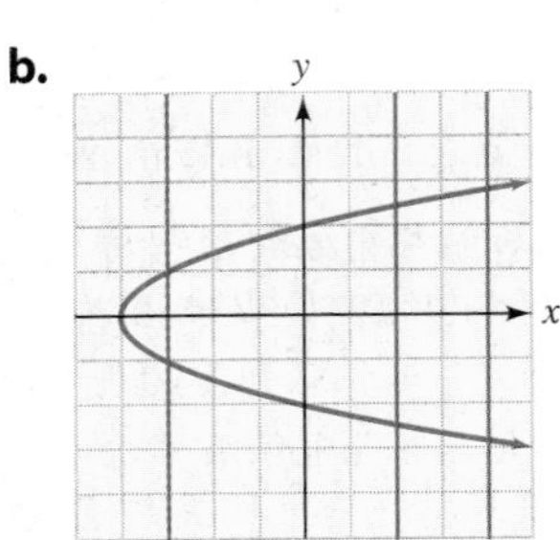

Figure 2-34

To find the domain, we note that the x-coordinate of the left-most point in Figure 2-34 is -4 and that there is no right-most point. Therefore, the domain is $[-4, \infty)$.

To find the range, we note that there is no lowest point or highest point. Therefore, the range is $(-\infty, \infty)$.

Since many vertical lines that intersect the graph will do so more than once, the vertical line test indicates that the graph is not a function.

SELF CHECK 3 Find the domain and range of the relation determined by the graph and then tell whether the graph defines a function.

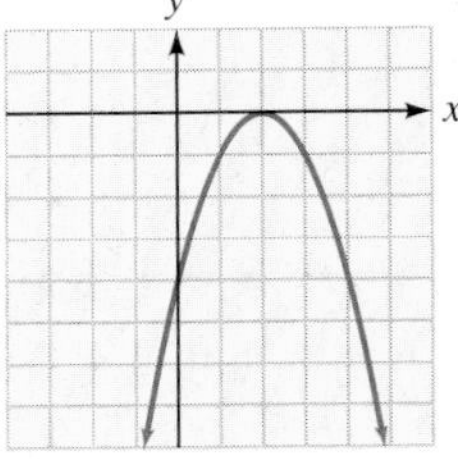

3 Use function notation to evaluate a function at a given value.

We can use a special notation to denote functions.

Function Notation The notation $y = f(x)$ denotes that the variable y is a function of x.

COMMENT The notation $f(x)$ does not mean "f times x."

The notation $y = f(x)$ is read as "y equals f of x." Note that y and $f(x)$ are two notations for the same quantity. Thus, the equations $y = 4x + 3$ and $f(x) = 4x + 3$ are equivalent.

The notation $y = f(x)$ provides a way of denoting the value of y (the *dependent variable*) that corresponds to some number x (the *independent variable*). For example, if $y = f(x)$, the value of y that is determined by $x = 3$ is denoted by $f(3)$.

EXAMPLE 4 Let $f(x) = 4x + 3$. Find: **a.** $f(3)$ **b.** $f(-1)$ **c.** $f(0)$ **d.** the value of x for which $f(x) = 7$.

Solution **a.** We replace x with 3.

$$\begin{aligned} f(x) &= 4x + 3 \\ f(3) &= 4(3) + 3 \\ &= 12 + 3 \\ &= 15 \end{aligned}$$

b. We replace x with -1.

$$\begin{aligned} f(x) &= 4x + 3 \\ f(-1) &= 4(-1) + 3 \\ &= -4 + 3 \\ &= -1 \end{aligned}$$

c. We replace x with 0.

$$\begin{aligned} f(x) &= 4x + 3 \\ f(0) &= 4(0) + 3 \\ &= 3 \end{aligned}$$

d. We replace $f(x)$ with 7.

$$\begin{aligned} f(x) &= 4x + 3 \\ 7 &= 4x + 3 \\ 4 &= 4x && \text{Subtract 3 from both sides.} \\ x &= 1 && \text{Divide each side by 4.} \end{aligned}$$

SELF CHECK 4 Let $f(x) = -2x - 1$. Find: **a.** $f(2)$ **b.** $f(-3)$ **c.** the value of x for which $f(x) = 5$.

To see why function notation is helpful, consider the following equivalent statements:

1. In the equation $y = 4x + 3$, find the value of y when x is 3.
2. If $f(x) = 4x + 3$, find $f(3)$.

Statement 2, which uses $f(x)$ notation, is much more concise.

We can think of a function as a machine that takes some input x and turns it into some output $f(x)$, as shown in Figure 2-35(a). The machine shown in Figure 2-35(b) turns the input number 2 into the output value -3 and turns the input number 6 into the output value -11. The set of numbers that we can put into the machine is the domain of the function, and the set of numbers that comes out is the range.

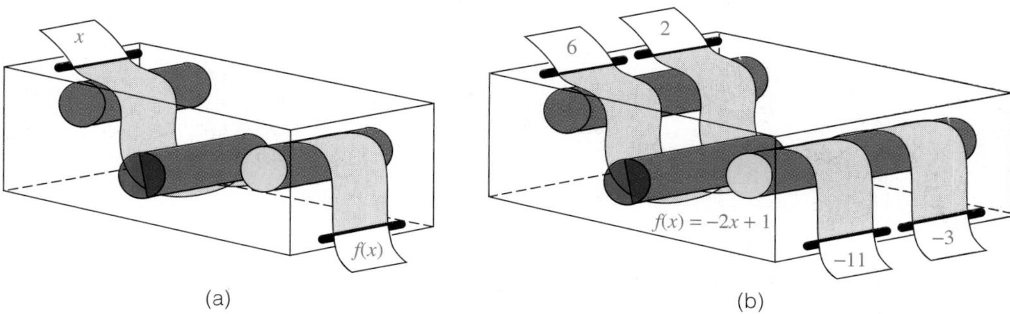

(a) (b)

Figure 2-35

The letter f used in the notation $y = f(x)$ represents the word *function.* However, other letters can be used to represent functions. The notations $y = g(x)$ and $y = h(x)$ also denote functions involving the independent variable x.

In Example 5, the equation $y = g(x) = x^2 - 2x$ determines a function, because every possible value of x gives a single value of $g(x)$.

EXAMPLE 5 Let $g(x) = x^2 - 2x$. Find: **a.** $g\left(\frac{2}{5}\right)$ **b.** $g(s)$ **c.** $g(s^2)$ **d.** $g(-t)$.

Solution **a.** We replace x with $\frac{2}{5}$.

$$\begin{aligned} g(x) &= x^2 - 2x \\ g\left(\frac{2}{5}\right) &= \left(\frac{2}{5}\right)^2 - 2\left(\frac{2}{5}\right) \\ &= \frac{4}{25} - \frac{4}{5} \\ &= -\frac{16}{25} \end{aligned}$$

b. We replace x with s.

$$\begin{aligned} g(x) &= x^2 - 2x \\ g(s) &= s^2 - 2s \\ &= s^2 - 2s \end{aligned}$$

c. We replace x with s^2.

$$\begin{aligned} g(x) &= x^2 - 2x \\ g(s^2) &= (s^2)^2 - 2s^2 \\ &= s^4 - 2s^2 \end{aligned}$$

d. We replace x with $-t$.

$$\begin{aligned} g(x) &= x^2 - 2x \\ g(-t) &= (-t)^2 - 2(-t) \\ &= t^2 + 2t \end{aligned}$$

SELF CHECK 5 Let $h(x) = -x^2 + 3$. Find: **a.** $h(2)$ **b.** $h(-a)$.

EXAMPLE 6 Let $f(x) = 4x - 1$. Find **a.** $f(3) + f(2)$ **b.** $f(a) - f(b)$.

Solution **a.** We find $f(3)$ and $f(2)$ separately.

$$\begin{aligned} f(x) &= 4x - 1 \\ f(3) &= 4(3) - 1 \\ &= 12 - 1 \\ &= 11 \end{aligned} \qquad \begin{aligned} f(x) &= 4x - 1 \\ f(2) &= 4(2) - 1 \\ &= 8 - 1 \\ &= 7 \end{aligned}$$

We then add the results to obtain $f(3) + f(2) = 11 + 7 = 18$.

b. We find $f(a)$ and $f(b)$ separately.

$$\begin{aligned} f(x) &= 4x - 1 \\ f(a) &= 4(a) - 1 \end{aligned} \qquad \begin{aligned} f(x) &= 4x - 1 \\ f(b) &= 4b - 1 \end{aligned}$$

We then subtract the results to obtain

$$\begin{aligned} f(a) - f(b) &= (4a - 1) - (4b - 1) \\ &= 4a - 1 - 4b + 1 \\ &= 4a - 4b \end{aligned}$$

SELF CHECK 6 Let $g(x) = -2x + 3$. Find: **a.** $g(-2) + g(3)$ **b.** $g(a) - g(b)$.

4 Determine whether an equation represents a function.

A function can be defined by an equation. For example, the equation $y = \frac{1}{2}x + 3$ sets up a rule in which each number x determines exactly one value of y. To find the value of y (called an **output value**) that corresponds to $x = 4$ (called an **input value**), we substitute 4 for x and simplify.

$$y = \frac{1}{2}x + 3$$

$$y = \frac{1}{2}(4) + 3 \quad \text{Substitute the input value of 4 for } x.$$

$$= 2 + 3$$

$$= 5$$

The ordered pair (4, 5) satisfies the equation and shows that a y-value of 5 corresponds to an x-value of 4. This ordered pair and others that satisfy the equation appear in the table shown in Figure 2-36. The graph of the equation also appears in the figure.

$y = \frac{1}{2}x + 3$

x	y	(x, y)	
−2	2	(−2, 2)	A y-value of 2 corresponds to an x-value of −2.
0	3	(0, 3)	A y-value of 3 corresponds to an x-value of 0.
2	4	(2, 4)	A y-value of 4 corresponds to an x-value of 2.
4	5	(4, 5)	A y-value of 5 corresponds to an x-value of 4.
6	6	(6, 6)	A y-value of 6 corresponds to an x-value of 6.
↑ Inputs	↑ Outputs		

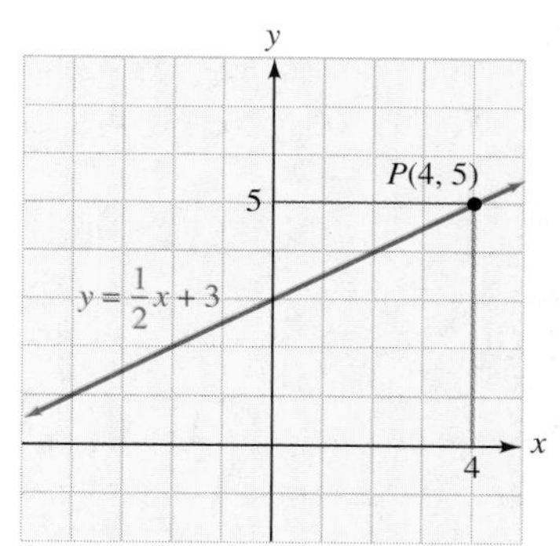

Figure 2-36

To see how the table in Figure 2-36 determines the correspondence, we find an input in the x-column and then read across to find the corresponding output in the y-column. For example, if we select 2 as an input value, we get 4 as an output value. Thus, a y-value of 4 corresponds to an x-value of 2.

To see how the graph in Figure 2-36 determines the correspondence, we draw a vertical and a horizontal line through any point (say, point P) on the graph, as shown in the figure. Because these lines intersect the x-axis at 4 and the y-axis at 5, the point $P(4, 5)$ associates 5 on the y-axis with 4 on the x-axis. This shows that a y-value of 5 corresponds to an x-value of 4.

In this example, the set of all inputs x is the set of real numbers. The set of all outputs y is also the set of real numbers.

When a correspondence is set up by an equation, a table, or a graph, in which only one y-value corresponds to each x-value, we say that *y is a function of x*. Since the value of y usually depends on the number x, we call y the **dependent variable** and x the **independent variable.**

y Is a Function of x

An equation, table, or graph in x and y in which each value of x (the *input*) determines exactly one value of y (the *output*) is a *function*. In this case, we say that *y is a function of x*.

The set of all input values x is the domain of the function, and the set of all output values y is the range.

EXAMPLE 7 Does $y = 2x - 3$ define y to be a function of x? If so, find its domain and range, and illustrate the function with a table and graph.

Solution For a function to exist, every value of x must determine exactly one value of y. To find y in the equation $y = 2x - 3$, we multiply x by 2 and then subtract 3. Since this arithmetic gives one result, each choice of x determines one value of y. Thus, the equation does define y to be a function of x.

Since the input x can be any real number, the domain of the function is the set of real numbers, denoted by the interval $(-\infty, \infty)$. Since the output y can be any real number, the range is also the set of real numbers, denoted as $(-\infty, \infty)$.

A table of values and the graph appear in Figure 2-37.

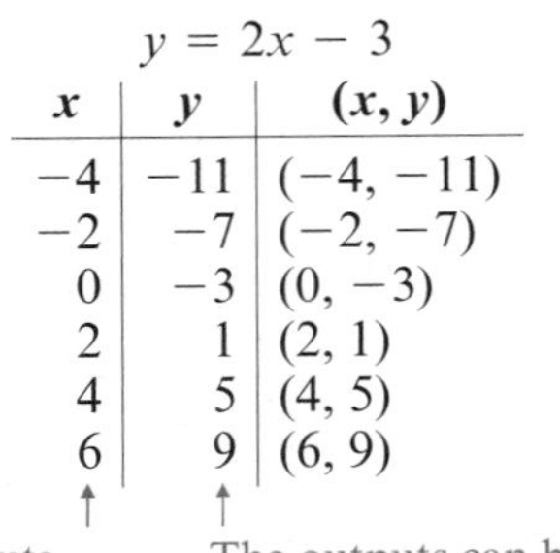

$y = 2x - 3$

x	y	(x, y)
−4	−11	(−4, −11)
−2	−7	(−2, −7)
0	−3	(0, −3)
2	1	(2, 1)
4	5	(4, 5)
6	9	(6, 9)

↑ The inputs can be any real number.

↑ The outputs can be any real number.

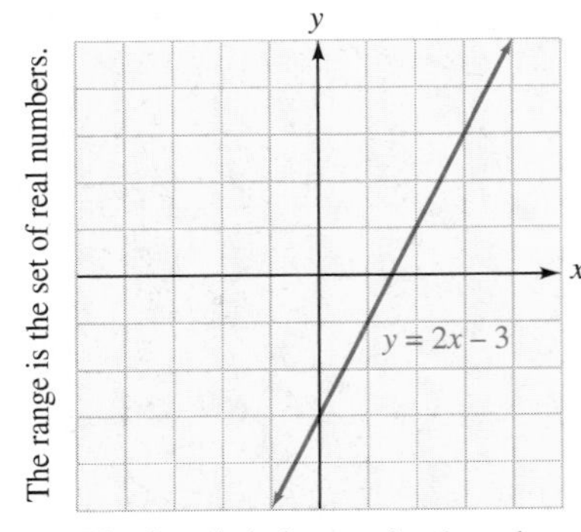

Figure 2-37

SELF CHECK 7 Does $y = -2x + 3$ define y to be a function of x?

EXAMPLE 8 Does $y^2 = x$ define y to be a function of x?

Solution For a function to exist, each value of x must determine exactly one value of y. If we let $x = 16$, for example, y could be either 4 or −4, because $4^2 = 16$ and $(-4)^2 = 16$. Since more than one value of y is determined when $x = 16$, the equation does not represent a function.

SELF CHECK 8 Does $|y| = x$ define y to be a function of x?

5 Find the domain of a function given its equation.

The domain of a function that is defined by an equation is the set of all numbers that are permissible replacements for its independent variable.

EXAMPLE 9 Find the domain of the functions defined by:

a. $f(x) = x^2 + 8x - 3$ **b.** $f(x) = \frac{1}{x - 2}$.

Solution **a.** Since any real number can be substituted for x in the function $f(x) = x^2 + 8x - 3$ to obtain a single value y, the domain is $(-\infty, \infty)$.

b. The number 2 cannot be substituted for x in the function $f(x) = \frac{1}{x - 2}$, because that would make the denominator 0. However, any real number, except 2, can be substituted for x to obtain a single value y. Therefore, the domain is the set of all real numbers except 2. This is the interval $(-\infty, 2) \cup (2, \infty)$.

SELF CHECK 9 Find the domain of the function defined by the equation $y = \frac{2}{x + 3}$.

6 Graph a linear function.

The **graph of a function** is the graph of the ordered pairs $(x, f(x))$ that define the function. For the graph of the function shown in Figure 2-38, the domain is shown on the x-axis, and the range is shown on the y-axis. For any x in the domain, there corresponds one value $y = f(x)$ in the range.

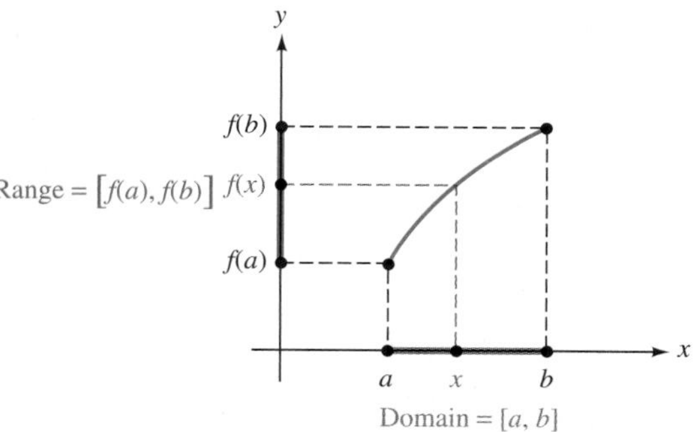

Figure 2-38

EXAMPLE 10 Graph the function $f(x) = -2x + 1$ and find its domain and range.

Solution We graph the equation as in Figure 2-39. Since every real number x on the x-axis determines a corresponding value of y, the domain is the interval $(-\infty, \infty)$ shown on the x-axis. Since the values of y can be any real number, the range is the interval $(-\infty, \infty)$ shown on the y-axis.

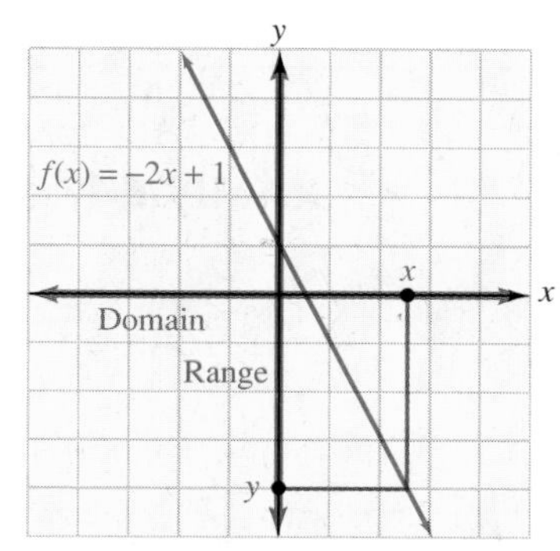

Figure 2-39

In Section 2.1, we graphed equations whose graphs were lines. These equations define basic functions, called *linear functions*.

Linear Functions

A **linear function** is a function defined by an equation that can be written in the form

$$f(x) = mx + b \qquad \text{or} \qquad y = mx + b$$

where m is the slope of the line graph and $(0, b)$ is the y-intercept.

EXAMPLE 11 Solve the equation $3x + 2y = 10$ for y to show that it defines a linear function. Then graph the function and find its domain and range.

Solution We solve the equation for y as follows:

$$3x + 2y = 10$$
$$2y = -3x + 10 \quad \text{Subtract } 3x \text{ from both sides.}$$
$$y = -\frac{3}{2}x + 5 \quad \text{Divide both sides by 2.}$$

Because the given equation can be written in the form $f(x) = mx + b$, it defines a linear function. The slope of its line graph is $-\frac{3}{2}$, and the y-intercept is $(0, 5)$. The graph appears in Figure 2-40. From the graph, we can see that both the domain and the range are the interval $(-\infty, \infty)$.

A special case of a linear function is the **constant function,** defined by the equation $f(x) = b$, where b is a constant. Its graph, domain, and range are shown in Figure 2-41.

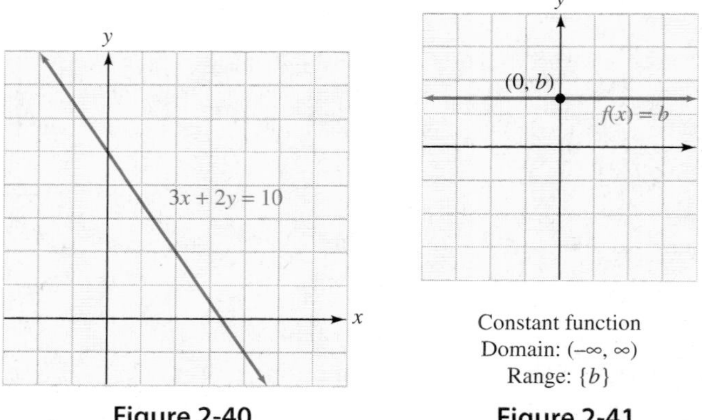

Figure 2-40

Figure 2-41

EXAMPLE 12 **CUTTING HAIR** A barber earns \$26 per day plus \$6.50 for each haircut she gives that day. Write a linear function that describes her daily income if she gives x haircuts each day.

Solution The barber earns \$6.50 per haircut, so if she serves x customers a day, her earnings for haircuts will be \$6.50x. To find her total daily income, we must add \$26 to \$6.50x. Thus, her total daily income $I(x)$ is described by the function

$$I(x) = 6.50x + 26$$

SELF CHECK 12 Write a linear function that describes her daily income if she gets a raise of \$0.50 per haircut.

SELF CHECK ANSWERS **1.** D: {5, −12, 8}; R: {6, 4} **2. a.** no **b.** yes **c.** no **3.** D: $(-\infty, \infty)$; R: $(-\infty, 0]$; yes **4. a.** −5 **b.** 5 **c.** −3 **5. a.** −1 **b.** $-a^2 + 3$ **6. a.** 4 **b.** $-2a + 2b$ **7.** yes **8.** no **9.** $(-\infty, -3) \cup (-3, \infty)$ **12.** $I(x) = 7x + 26$

NOW TRY THIS

1. Given $f(x) = 3x - 4$, find:
a. $f(x - 2)$ **b.** $f(x + h)$ **c.** $f(x + h) - f(x)$

2.4 EXERCISES

WARM-UPS

Determine whether each equation or inequality determines y to be a function of x.

1. $y = 2x + 1$
2. $y \geq 2x$
3. $y^2 = x$

If f(x) = 2x + 1, find:

4. $f(0)$
5. $f(1)$
6. $f(-2)$

REVIEW *Solve each equation.*

7. $\frac{y+2}{2} = 4(y + 2)$

8. $\frac{3z-1}{6} - \frac{3z+4}{3} = \frac{z+3}{2}$

9. $\frac{2a}{3} + \frac{1}{2} = \frac{6a-1}{6}$

10. $\frac{2x+3}{5} - \frac{3x-1}{3} = \frac{x-1}{15}$

VOCABULARY AND CONCEPTS

Fill in the blanks.

11. Any set of ordered pairs is called a ________.

12. A ________ is a correspondence between a set of input values and a set of output values, where each ______ value determines one ________ value.

13. In a function, the set of all input values is called the ________ of the function.

14. In a function, the set of all output values is called the ______ of the function.

15. The denominator of a fraction can never be __.

16. To decide whether a graph determines a function, use the ______________.

17. If any vertical line intersects a graph more than once, the graph ________ represent a function.

18. A linear function is any function that can be written in the form ______________.

19. In the function $f(x) = mx + b$, m is the ______ of its graph, and b is the y-coordinate of the __________.

Consider the function y = f(x) = 5x − 4. Fill in the blanks.

20. Any substitution for x is called an ______ value.

21. The value __ is called an output value.

22. The independent variable is __.

23. The dependent variable is __.

24. The notation $f(3)$ is the value of __ when $x = 3$.

GUIDED PRACTICE

Find the domain and range of each relation and determine if it is a function. See Examples 1–2. (Objective 1)

25. {(3, −2), (5, 0), (−4, −5), (0, 0)}

26. {(9, 2), (3, 3), (−6, −9), (2, 9)}

27. $\{(-2, 3), (6, 8), (-2, 5), (5, 4)\}$

28. $\{(3, -2), (5, 2), (4, 5), (3, 0)\}$

State the domain and range of the relation determined by each graph in interval notation and determine whether it represents a function. **See Example 3. (Objective 2)**

29.
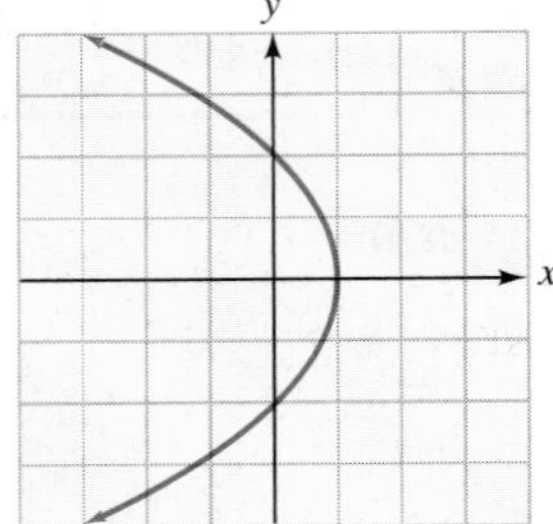

30.
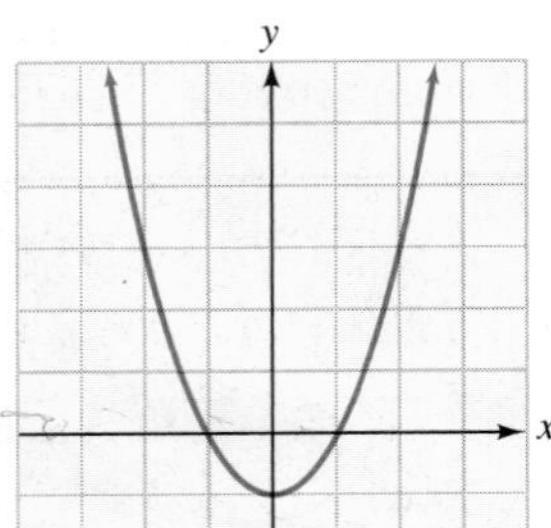

31.
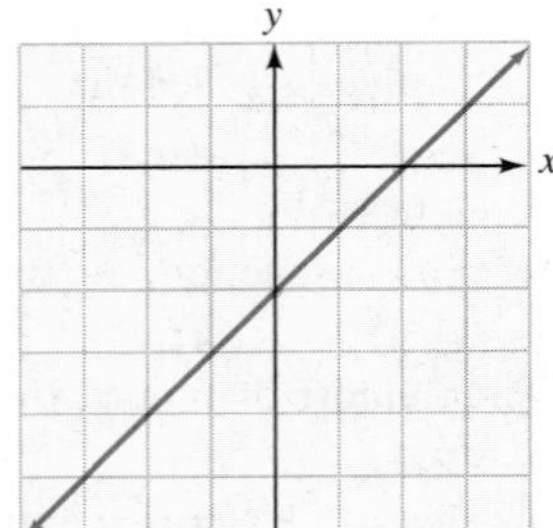

32.
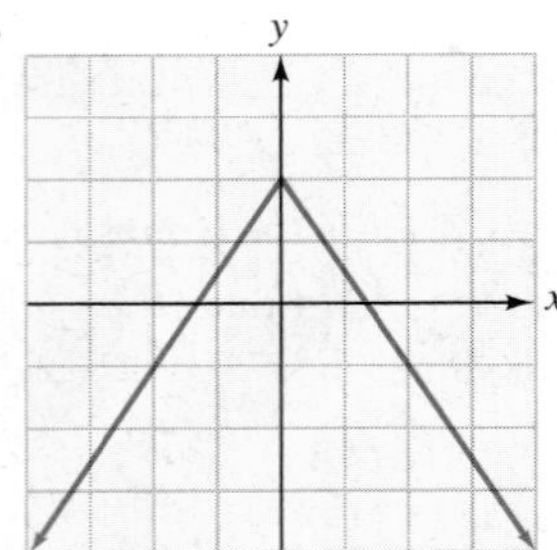

Find f(3), f(−1), and all values of x for which f(x) = 0. **See Example 4. (Objective 3)**

33. $f(x) = 3x$
34. $f(x) = -4x$
35. $f(x) = 2x - 3$
36. $f(x) = 3x - 5$

Find f(2) and f(3). **See Example 4. (Objective 3)**

37. $f(x) = x^2$
38. $f(x) = x^2 - 2$
39. $f(x) = x^3 - 1$
40. $f(x) = x^3$

Find f(2) and f(−2). **See Example 4. (Objective 3)**

41. $f(x) = |x| + 2$
42. $f(x) = |x| - 5$
43. $f(x) = x^2 - 2$
44. $f(x) = x^2 + 3$

Find g(w) and g(w + 1). **See Example 5. (Objective 3)**

45. $g(x) = 2x$
46. $g(x) = -3x$
47. $g(x) = 3x - 5$
48. $g(x) = 2x - 7$

Find each value given that f(x) = 2x + 1. **See Example 6. (Objective 3)**

49. $f(3) + f(2)$
50. $f(1) - f(-1)$
51. $f(b) - f(a)$
52. $f(b) + f(a)$

Determine whether the equation determines y to be a function of x. **See Examples 7–8. (Objective 4)**

53. $y = 2x + 3$
54. $y = -1$
55. $y = 2x^2$
56. $y^2 = x + 1$
57. $y = 3 + 7x^2$
58. $y^2 = 3 - 2x$
59. $x = |y|$
60. $y = |x|$

Find the domain of each function. **See Example 9. (Objective 5)**

61. $\{(-2, 3), (4, 5), (6, 7)\}$
62. $\{(0, 2), (1, 2), (3, 4)\}$
63. $f(x) = \dfrac{1}{x - 4}$
64. $f(x) = \dfrac{5}{x + 1}$
65. $f(x) = \dfrac{1}{x + 3}$
66. $f(x) = \dfrac{3}{x - 4}$
67. $f(x) = \dfrac{x}{x^2 + 2}$
68. $f(x) = \dfrac{x}{x - 3}$

Draw the graph of each linear function and give the domain and range. **See Examples 10–11. (Objective 6)**

69. $f(x) = 2x - 1$
70. $f(x) = -x + 2$

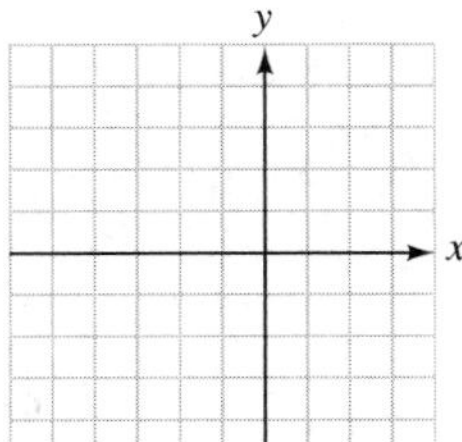
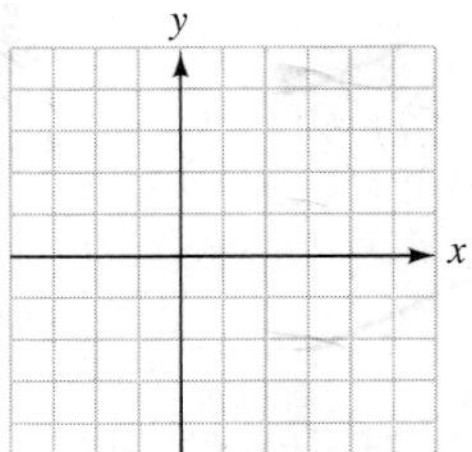

71. $2x - 3y = 6$

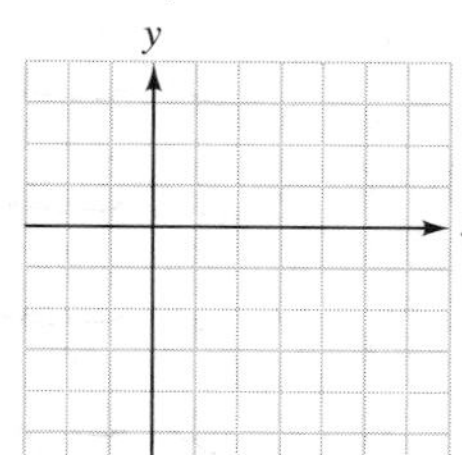

72. $3x + 2y = -6$

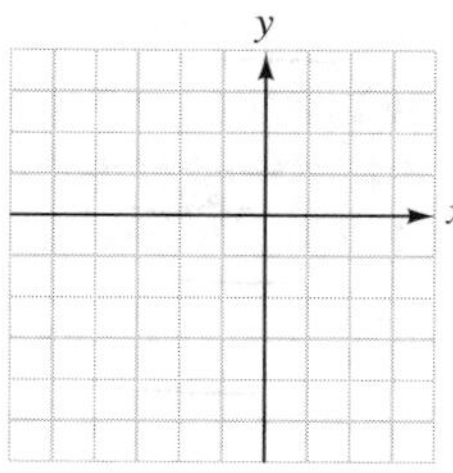

Determine whether each equation defines a linear function. See Examples 10–11. (Objective 6)

73. $y = 3x^2 + 2$

74. $y = \frac{x - 3}{2}$

75. $x = 3y - 4$

76. $x = \frac{8}{y}$

ADDITIONAL PRACTICE

For each function, find f(2) and f(3).

77. $f(x) = (x + 1)^2$

78. $f(x) = (x - 3)^2$

79. $f(x) = 2x^2 - x$

80. $f(x) = 5x^2 + 2x$

For each function, find f(3) and f(−1).

81. $f(x) = 7 + 5x$

82. $f(x) = 3 + 3x$

83. $f(x) = 9 - 2x$

84. $f(x) = 12 + 3x$

Find each value given that f(x) = 2x + 1.

85. $f(b) - 1$

86. $f(b) - f(1)$

87. $f(0) + f\left(-\frac{1}{2}\right)$

88. $f(a) + f(2a)$

APPLICATIONS ***For Problems 89–92, set up a linear equation.*** See Example 12. (Objective 6)

89. Selling DVD players An electronics firm manufactures portable DVD players, receiving \$120 for each unit it makes. If x represents the number of units produced, the income received is determined by the *revenue function* $R(x) = 120x$. The manufacturer has fixed costs of \$12,000 per month and variable costs of \$57.50 for each unit manufactured. Thus, the *cost function* is $C(x) = 57.50x + 12{,}000$. How many DVD players must the company sell for revenue to equal cost?

90. Selling tires A tire company manufactures premium tires, receiving \$130 for each tire it makes. If the manufacturer has fixed costs of \$15,512.50 per month and variable costs of \$93.50 for each tire manufactured, how many tires must the company sell for revenue to equal cost? (*Hint:* See Exercise 89.)

91. Selling hot dogs At a football game, a vendor sells hot dogs. He earns \$50 per game plus \$0.10 for each hot dog sold.

a. Write a linear function that describes the vendor's income if he sells h hot dogs.

b. Find his income if he sells 115 hot dogs.

92. Housing A housing contractor lists the following costs.

Fees and permits	\$14,000
Cost per square foot	\$102

a. Write a linear function that describes the cost of building a house with s square feet.

b. Find the cost to build a house having 1,800 square feet.

93. Ballistics A bullet shot straight up is s feet above the ground after t seconds, where $s = f(t) = -16t^2 + 256t$. Find the height of the bullet 3 seconds after it is shot.

94. Artillery fire A mortar shell is s feet above the ground after t seconds, where $s = f(t) = -16t^2 + 512t + 64$. Find the height of the shell 20 seconds after it is fired.

95. Dolphins See the illustration. The height h in feet reached by a dolphin t seconds after breaking the surface of the water is given by

$$h = -16t^2 + 32t$$

How far above the water will the dolphin be 1.5 seconds after a jump?

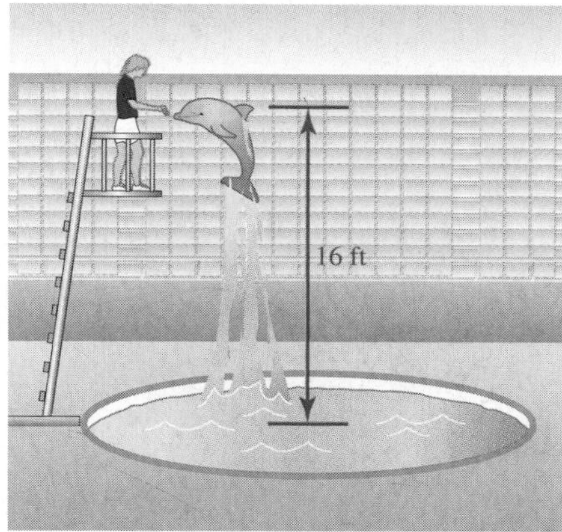

96. Forensic medicine The kinetic energy E of a moving object is given by $E = \frac{1}{2}mv^2$, where m is the mass of the object (in kilograms) and v is the object's velocity (in meters per second). Kinetic energy is measured in joules. Examining the damage done to a victim, a police pathologist estimates that the velocity of a club with a 3-kilogram mass was 6 meters per second. Find the kinetic energy of the club.

97. Conversion from degrees Celsius to degrees Fahrenheit The temperature in degrees Fahrenheit that is equivalent to a temperature in degrees Celsius is given by the function $F(C) = \frac{9}{5}C + 32$. Find the Fahrenheit temperature that is equivalent to 25°C.

98. Conversion from degrees Fahrenheit to degrees Celsius The temperature in degrees Celsius that is equivalent to a temperature in degrees Fahrenheit is given by the function $C(F) = \frac{5}{9}F - \frac{160}{9}$. Find the Celsius temperature that is equivalent to 14°F.

WRITING ABOUT MATH

99. Explain the concepts of function, domain, and range.

100. Explain why the constant function is a special case of a linear function.

SOMETHING TO THINK ABOUT *Let* $f(x) = 2x + 1$ *and* $g(x) = x^2$. *Assume that* $f(x) \neq 0$ *and* $g(x) \neq 0$.

101. Is $f(x) + g(x)$ equal to $g(x) + f(x)$?

102. Is $f(x) - g(x)$ equal to $g(x) - f(x)$?

SECTION 2.5 Graphs of Other Functions

Objectives

1. Graph the squaring, cubing, and absolute value functions.
2. Graph a vertical translation of the squaring, cubing, and absolute value functions.
3. Graph a horizontal translation of the squaring, cubing, and absolute value functions.
4. Graph a reflection of the squaring, cubing, and absolute value functions about the x-axis.

Vocabulary

squaring function	absolute value function	horizontal translations
parabola	vertical translations	reflection
cubing function		

Getting Ready

Give the slope and the y-intercept of each linear function.

1. $f(x) = 2x - 3$ **2.** $f(x) = -3x + 4$

Find the value of $f(x)$ *when* $x = 2$ *and* $x = -1$.

3. $f(x) = 5x - 4$ **4.** $f(x) = \frac{1}{2}x + 3$

In the previous section, we discussed *linear functions,* functions whose graphs are straight lines. We now extend the discussion to include nonlinear functions, functions whose graphs are not straight lines.

1 Graph the squaring, cubing, and absolute value functions.

If f is a function whose domain and range are sets of real numbers, its graph is the set of all points $(x, f(x))$ in the xy-plane. In other words, the graph of f is the graph of the equation $y = f(x)$. In this section, we will graph many basic functions. The first is $f(x) = x^2$ (or $y = x^2$), often called the **squaring function.** This function is a special case of a broader set of functions called *quadratic functions.* We will discuss these functions in detail in Chapter 8.

EXAMPLE 1 Graph the squaring function: $f(x) = x^2$.

Solution We substitute values for x in the equation and compute the corresponding values of $f(x)$. For example, if $x = -3$, we have

$$\begin{aligned} f(x) &= x^2 \\ f(-3) &= (-3)^2 \quad \text{Substitute } -3 \text{ for } x. \\ &= 9 \end{aligned}$$

The ordered pair $(-3, 9)$ satisfies the equation and will lie on the graph. We list this pair and others that satisfy the equation in the table shown in Figure 2-42. We plot the points and draw a smooth curve through them to get the graph, called a **parabola.**

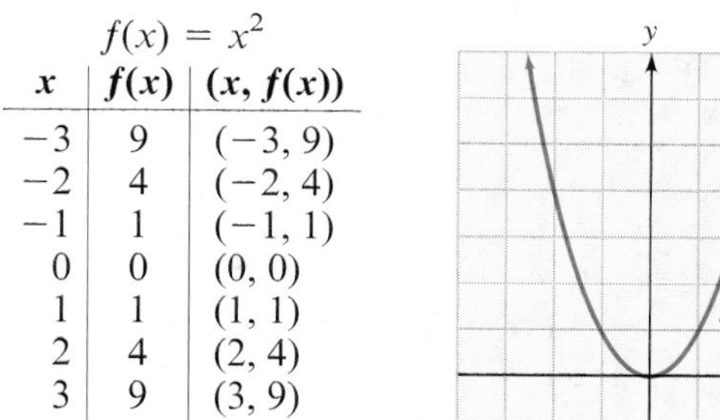

$f(x) = x^2$

x	$f(x)$	$(x, f(x))$
−3	9	(−3, 9)
−2	4	(−2, 4)
−1	1	(−1, 1)
0	0	(0, 0)
1	1	(1, 1)
2	4	(2, 4)
3	9	(3, 9)

Figure 2-42

From the graph, we see that x can be any real number. This indicates that the domain of the squaring function is the set of real numbers $\mathbb{R}$, which is the interval $(-\infty, \infty)$. We also can see that y is always positive or zero. This indicates that the range is the set of nonnegative real numbers, which is the interval $[0, \infty)$.

SELF CHECK 1 Graph $f(x) = x^2 - 2$ and compare the graph to the graph of $f(x) = x^2$.

The second basic function is $f(x) = x^3$ (or $y = x^3$), often called the **cubing function.**

EXAMPLE 2 Graph the cubing function: $f(x) = x^3$.

Solution We substitute values for x in the equation and compute the corresponding values of $f(x)$. For example, if $x = -2$, we have

$$\begin{aligned} f(x) &= x^3 \\ f(-2) &= (-2)^3 \quad \text{Substitute } -2 \text{ for } x. \\ &= -8 \end{aligned}$$

The ordered pair $(-2, -8)$ satisfies the equation and will lie on the graph. We list this pair and others that satisfy the equation in the table shown in Figure 2-43. We plot the points and draw a smooth curve through them to get the graph.

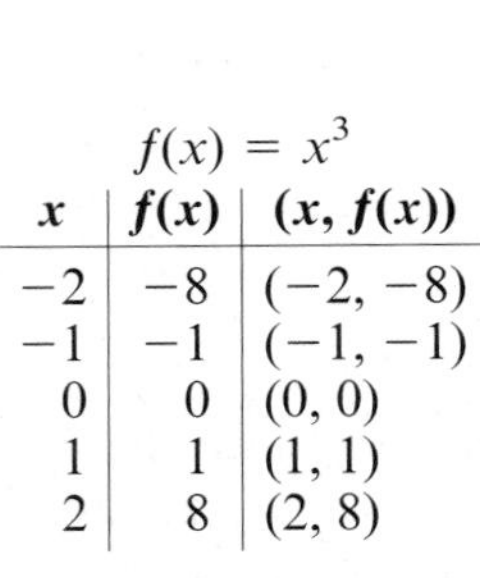

$f(x) = x^3$

x	$f(x)$	$(x, f(x))$
-2	-8	$(-2, -8)$
-1	-1	$(-1, -1)$
0	0	$(0, 0)$
1	1	$(1, 1)$
2	8	$(2, 8)$

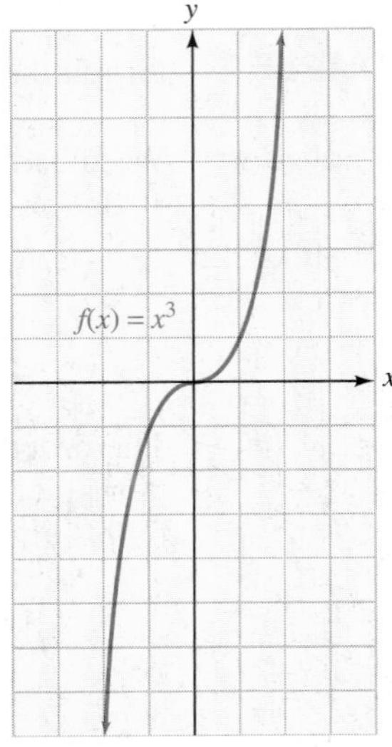

Figure 2-43

From the graph, we can see that x can be any real number. This indicates that the domain of the cubing function is the set of real numbers $\mathbb{R}$, which is the interval $(-\infty, \infty)$. We also can see that y can be any real number. This indicates that the range is the set of real numbers $\mathbb{R}$, which is the interval $(-\infty, \infty)$.

SELF CHECK 2 Graph $f(x) = x^3 + 1$ and compare the graph to the graph of $f(x) = x^3$.

The third basic function is $f(x) = |x|$ (or $y = |x|$), often called the **absolute value function.**

EXAMPLE 3 Graph the absolute value function: $f(x) = |x|$.

Solution We substitute values for x in the equation and compute the corresponding values of $f(x)$. For example, if $x = -3$, we have

$$f(x) = |x|$$
$$f(-3) = |-3| \quad \text{Substitute } -3 \text{ for } x.$$
$$= 3$$

The ordered pair $(-3, 3)$ satisfies the equation and will lie on the graph. We list this pair and others that satisfy the equation in the table shown in Figure 2-44. We plot the points and draw a V-shaped graph through them.

$f(x) = |x|$

x	$f(x)$	$(x, f(x))$
-3	3	$(-3, 3)$
-2	2	$(-2, 2)$
-1	1	$(-1, 1)$
0	0	$(0, 0)$
1	1	$(1, 1)$
2	2	$(2, 2)$
3	3	$(3, 3)$

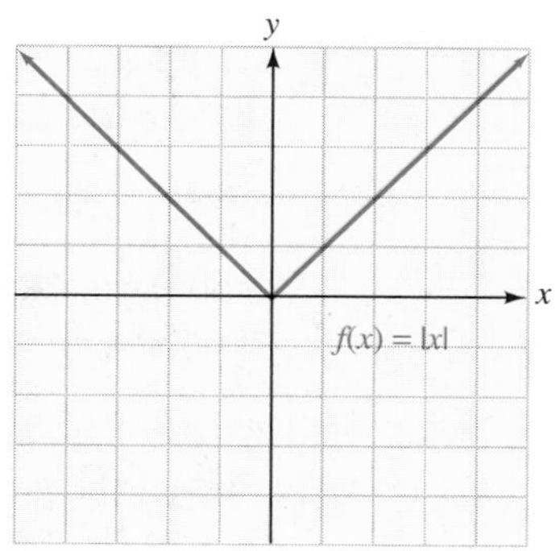

Figure 2-44

From the graph, we see that x can be any real number. This indicates that the domain of the absolute value function is the set of real numbers, ℝ, which is the interval $(-\infty, \infty)$. We also can see that y is always positive or zero. This indicates that the range is the set of nonnegative real numbers, which is the interval $[0, \infty)$.

SELF CHECK 3 Graph $f(x) = |x - 2|$ and compare the graph to the graph of $f(x) = |x|$.

ACCENT ON TECHNOLOGY

Graphing Functions

We can graph nonlinear functions with a graphing calculator. For example, to graph $f(x) = x^2$ in a standard window of $[-10, 10]$ for x and $[-10, 10]$ for y, we enter the function by entering x ^ 2 (or x x^2) and press the **GRAPH** key. We will obtain the graph shown in Figure 2-45(a).

To graph $f(x) = x^3$, we enter the function by entering x ^ 3 and press the **GRAPH** key to obtain the graph in Figure 2-45(b). To graph $f(x) = |x|$, we enter the function by selecting "abs" from the MATH NUM menu, typing x, and pressing the **GRAPH** key to obtain the graph in Figure 2-45(c).

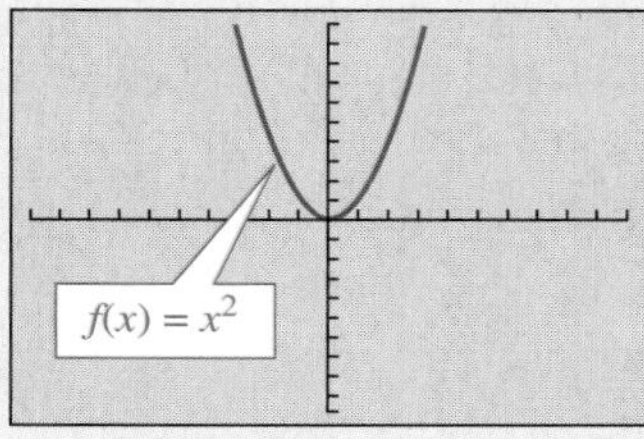

The squaring function

(a)

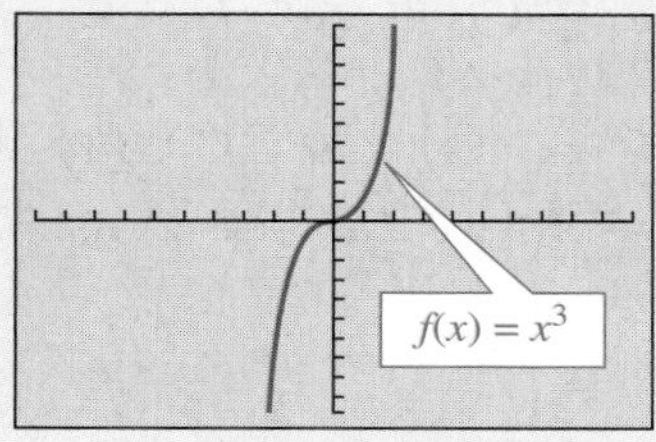

The cubing function

(b)

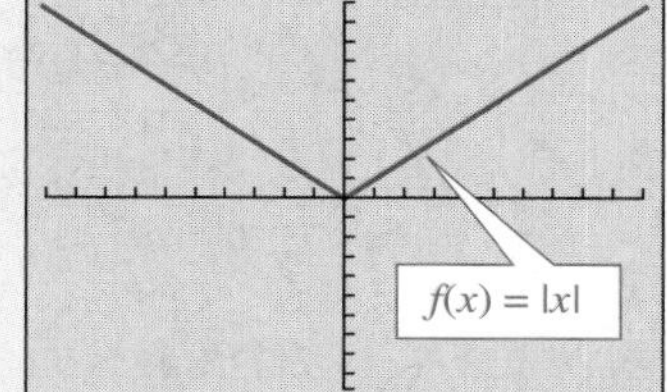

The absolute value function

(c)

Figure 2-45

When using a graphing calculator, we must be sure that the viewing window does not show a misleading graph. For example, if we graph $f(x) = |x|$ in the window $[0, 10]$ for x and $[0, 10]$ for y, we will obtain a misleading graph that looks like a line. (See Figure 2-46.) This is not true. The proper graph is the V-shaped graph shown in Figure 2-45(c).

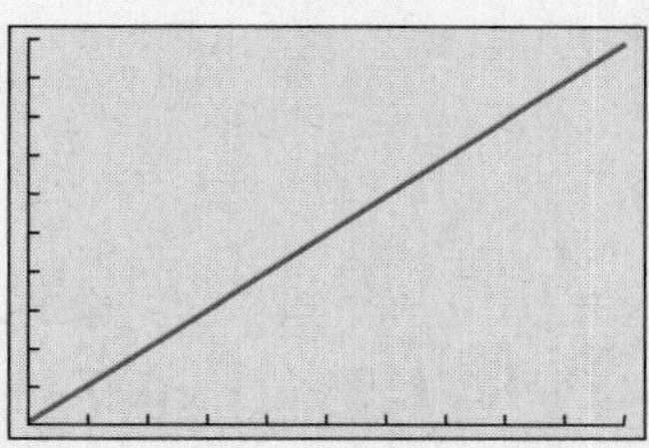

Figure 2-46

2 Graph a vertical translation of the squaring, cubing, and absolute value functions.

Examples 1–3 and their Self Checks suggest that the graphs of different functions may be identical except for their positions in the xy-plane. For example, Figure 2-47 shows the graph of $f(x) = x^2 + k$ for three different values of k. If $k = 0$, we get the graph of $f(x) = x^2$. If $k = 3$, we get the graph $f(x) = x^2 + 3$, which is identical to the graph of $f(x) = x^2$, except that it is shifted 3 units upward. If $k = -4$, we get the graph $f(x) = x^2 - 4$, which is identical to the graph of $f(x) = x^2$, except that it is shifted 4 units downward. These shifts are called **vertical translations.**

In general, we can make these observations.

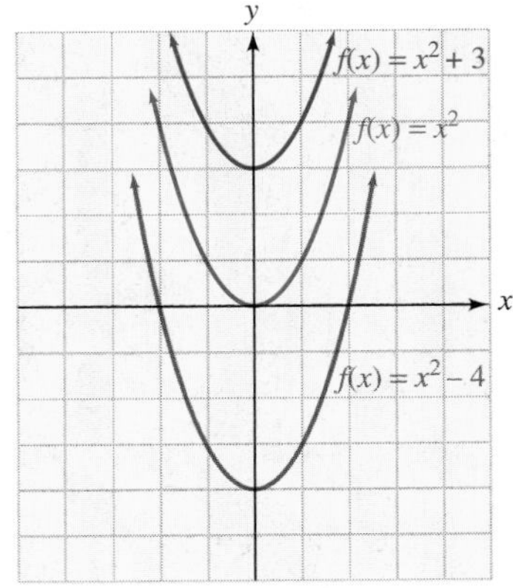

Figure 2-47

Vertical Translations

If f is a function and k is a positive number, then

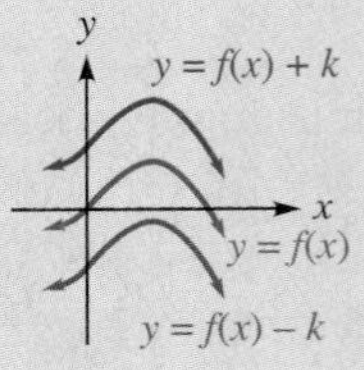

- The graph of $f(x) + k$ is identical to the graph of $f(x)$, except that it is translated k units upward.
- The graph of $f(x) - k$ is identical to the graph of $f(x)$, except that it is translated k units downward.

EXAMPLE 4 Graph: $f(x) = |x| + 2$.

Solution The graph of $f(x) = |x| + 2$ will be the same V-shaped graph as $f(x) = |x|$, except that it is shifted 2 units up. The graph appears in Figure 2-48.

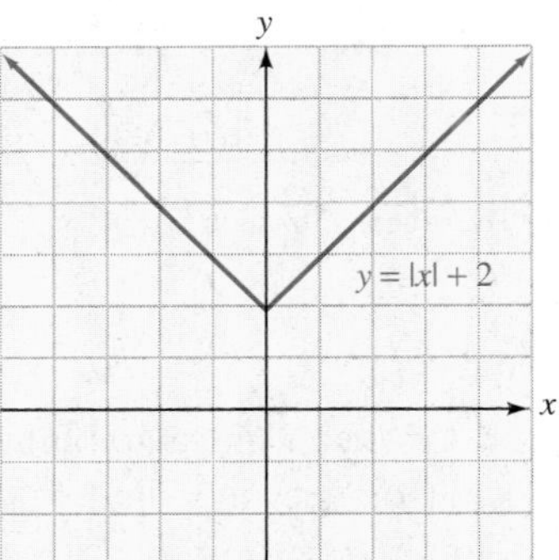

Figure 2-48

SELF CHECK 4 Graph: $f(x) = |x| - 3$.

3 Graph a horizontal translation of the squaring, cubing, and absolute value functions.

Figure 2-49 shows the graph of $f(x) = (x + h)^2$ for three different values of h. If $h = 0$, we get the graph of $f(x) = x^2$. The graph of $f(x) = (x - 3)^2$ is identical to the graph of $f(x) = x^2$, except that it is shifted 3 units to the right. The graph of $f(x) = (x + 2)^2$ is identical to the graph of $f(x) = x^2$, except that it is shifted 2 units to the left. These shifts are called **horizontal translations.**

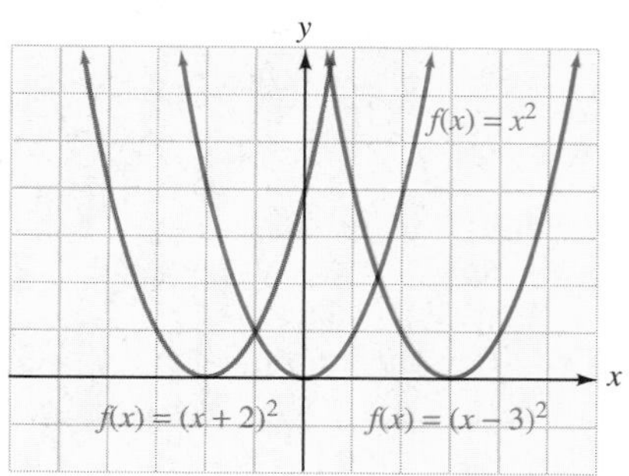

Figure 2-49

In general, we can make these observations.

Horizontal Translations

If f is a function and k is a positive number, then

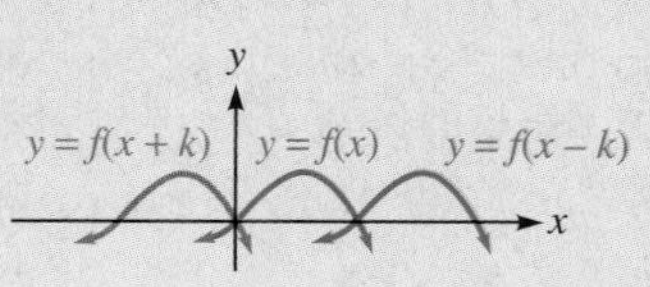

- The graph of $f(x - k)$ is identical to the graph of $f(x)$, except that it is translated k units to the right.
- The graph of $f(x + k)$ is identical to the graph of $f(x)$, except that it is translated k units to the left.

EXAMPLE 5 Graph: $f(x) = (x - 2)^2$.

Solution The graph of $f(x) = (x - 2)^2$ will be the same shape as the graph of $f(x) = x^2$, except that it is shifted 2 units to the right. The graph appears in Figure 2-50.

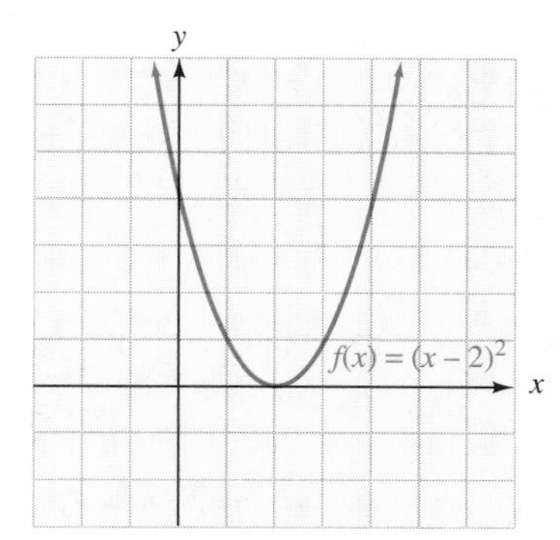

Figure 2-50

 **SELF CHECK 5** Graph: $f(x) = (x + 3)^3$.

EXAMPLE 6 Graph: $f(x) = (x - 3)^2 + 2$.

Solution We can graph this function by translating the graph of $f(x) = x^2$ to the right 3 units and then up 2 units, as shown in Figure 2-51.

COMMENT In Example 6, we could have translated the graph 2 units up and then 3 units to the right to obtain the same graph.

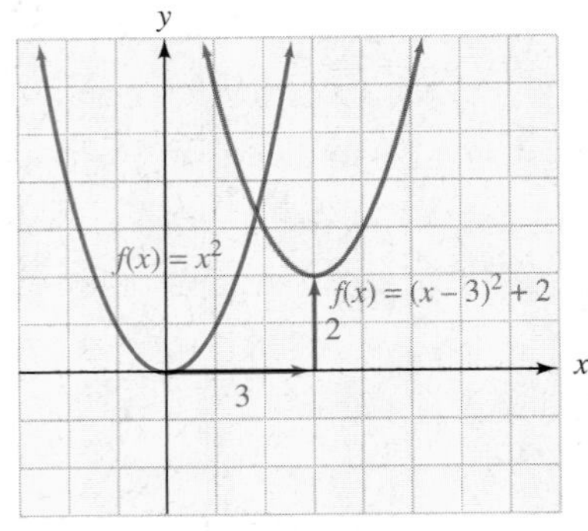

Figure 2-51

SELF CHECK 6 Graph: $f(x) = |x + 2| - 3$.

4 Graph a reflection of the squaring, cubing, and absolute value functions about the *x*-axis.

Figure 2-52 shows tables of solutions for $f(x) = x^2$ and for $f(x) = -x^2$. We note that for a given value of x, the corresponding y-values in the tables are *opposites.* When the values are graphed, we see that the (−) in $f(x) = -x^2$ has the effect of "flipping" the graph of $f(x) = x^2$ over the x-axis, so that the parabola opens downward. We say that the graph of $f(x) = -x^2$ is a **reflection** of the graph of $f(x) = x^2$ about the x-axis.

$f(x) = x^2$

x	$f(x)$	$(x, f(x))$
−2	4	(−2, 4)
−1	1	(−1, 1)
0	0	(0, 0)
1	1	(1, 1)
2	4	(2, 4)

$f(x) = -x^2$

x	$f(x)$	$(x, f(x))$
−2	−4	(−2, −4)
−1	−1	(−1, −1)
0	0	(0, 0)
1	−1	(1, −1)
2	−4	(2, −4)

Figure 2-52

EXAMPLE 7 Graph: $f(x) = -x^3$.

Solution To graph $f(x) = -x^3$, we use the graph of $f(x) = x^3$ from Example 2. First, we reflect the portion of the graph of $f(x) = x^3$ in quadrant I to quadrant IV, as shown in Figure 2-53. Then we reflect the portion of the graph of $f(x) = x^3$ that is in quadrant III to quadrant II.

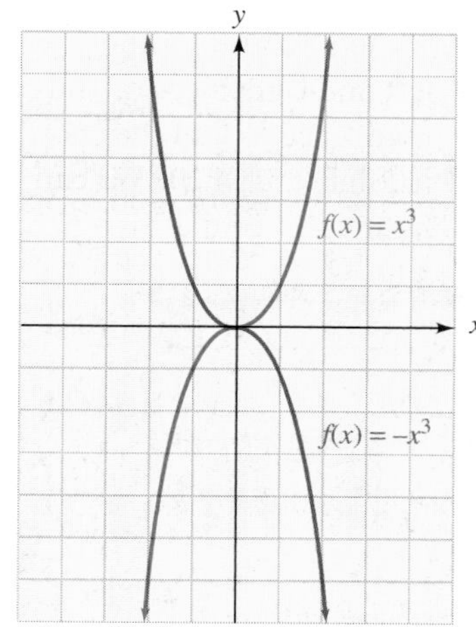

Figure 2-53

SELF CHECK 7 Graph: $f(x) = -|x|$.

Reflection of a Graph The graph of $y = -f(x)$ is the graph of $f(x)$ reflected about the x-axis.

PERSPECTIVE Graphs in Space

In an xy-coordinate system, graphs of equations containing the two variables x and y are lines or curves. Other equations have more than two variables, and graphing them often requires some ingenuity and perhaps the aid of a computer. Graphs of equations with the three variables x, y, and z are viewed in a three-dimensional coordinate system with three axes. The coordinates of points in a three-dimensional coordinate system are ordered triples (x, y, z). For example, the points $P(2, 3, 4)$ and $Q(-1, 2, 3)$ are plotted in Illustration 1.

Graphs of equations in three variables are not lines or curves, but flat planes or curved surfaces. Only the simplest of these equations can be conveniently graphed by hand; a computer provides the best images of others. The graph in Illustration 2 is called a **paraboloid;** it is the three-dimensional version of a parabola. Illustration 3 models a portion of the vibrating surface of a drum head.

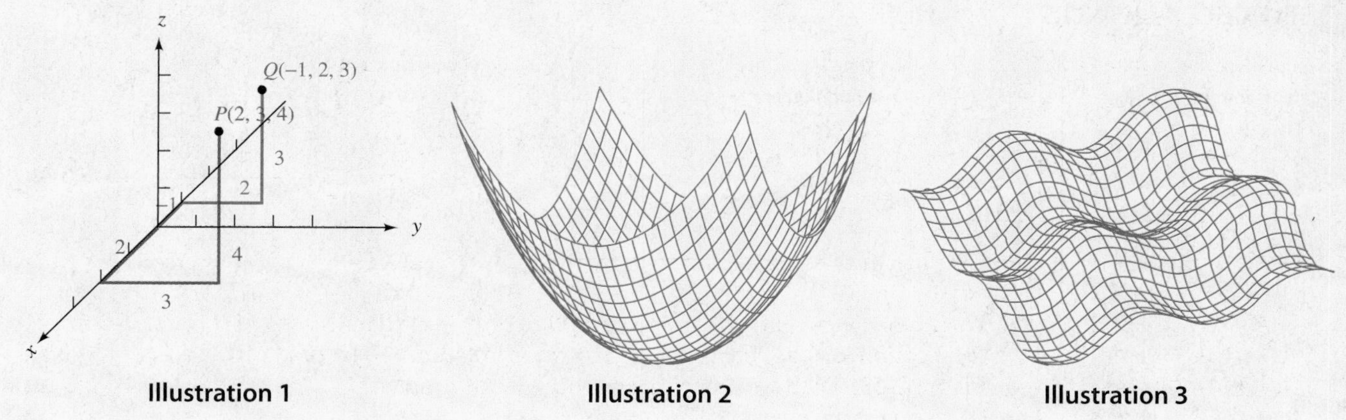

Illustration 1 Illustration 2 Illustration 3

ACCENT ON TECHNOLOGY

Solving Equations with Graphing Calculators

To solve the equation $2(x - 3) + 3 = 7$ with a graphing calculator, we can graph the left and right sides of the equation in the same window, as shown in Figure 2-54(a). We then trace to find the coordinates of the point where the two graphs intersect, as shown in Figure 2-54(b). We can then zoom and trace again to get Figure 2-54(c). From the figure, we see that $x = 5$.

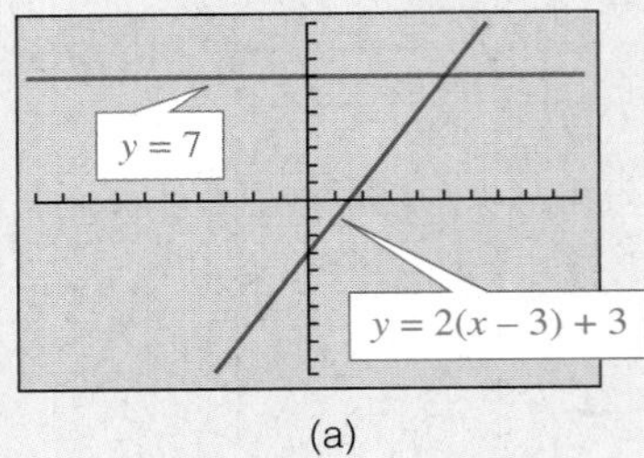

(a)

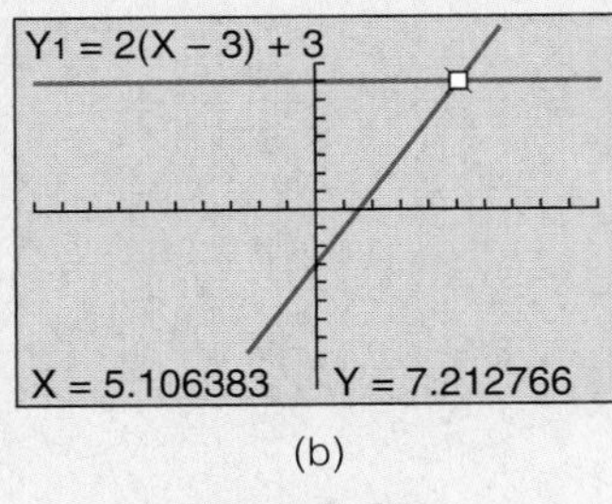

(b)

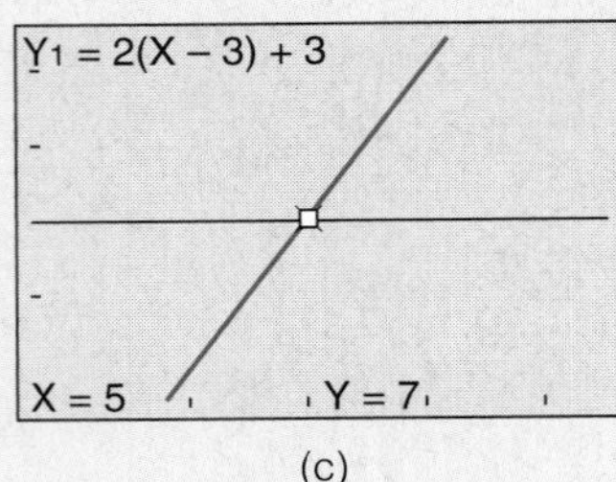

(c)

Figure 2-54

COMMENT The intersect method will work only if the x-coordinate of the intersection point of the graphs is visible on the screen. If it is not, you may need to adjust the viewing window until it is visible.

We also can solve the equation by using the intersect command found in the CALC menu on a graphing calculator. To use this method, we first clear the calculator of past work by selecting ZOOM and pressing 6 (Zstandard). We then graph the equations as shown in Figure 2-54(a), open the CALC menu, and select "5: intersect" to obtain Figure 2-55(a). We then select a point on the slanted line by pressing **ENTER**, and select a point on the horizontal line by pressing **ENTER**, and press **ENTER** again to obtain Figure 2-55(b). From the figure, we see that $x = 5$.

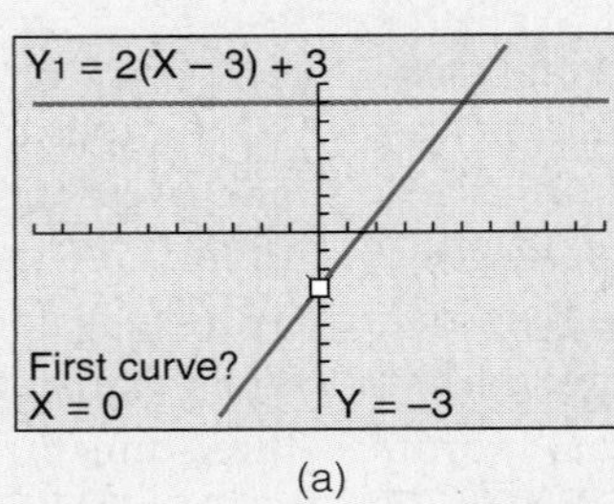

(a)

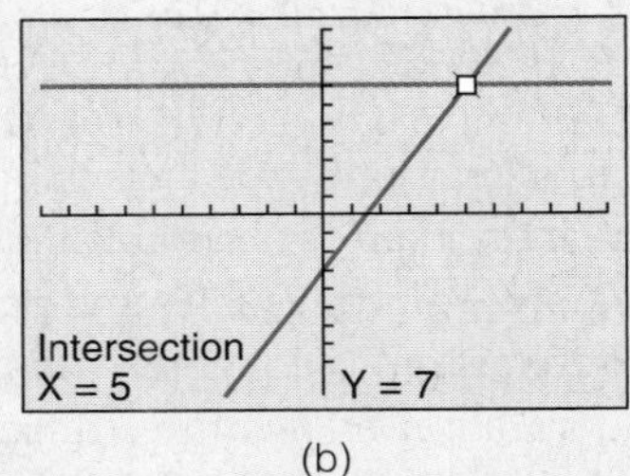

(b)

Figure 2-55

SELF CHECK ANSWERS

1. same shape, but 2 units lower

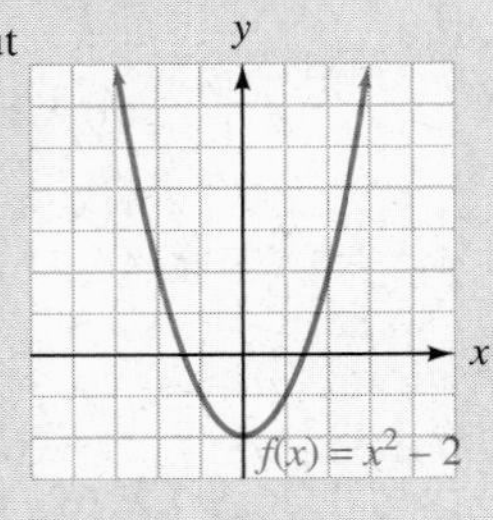

2. same shape, but 1 unit higher

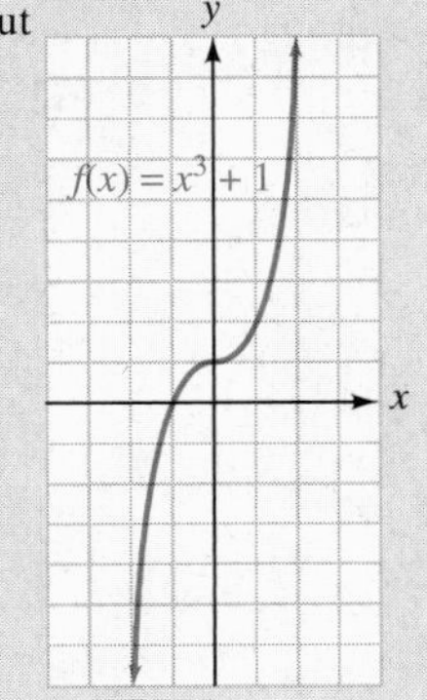

3. same shape, but 2 units to the right

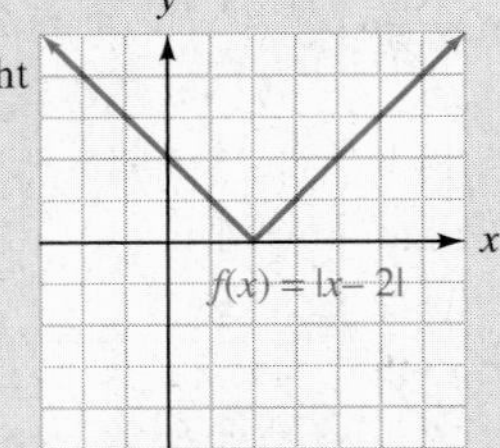

4.

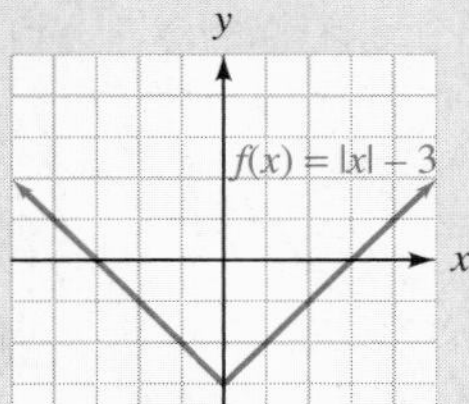

5.

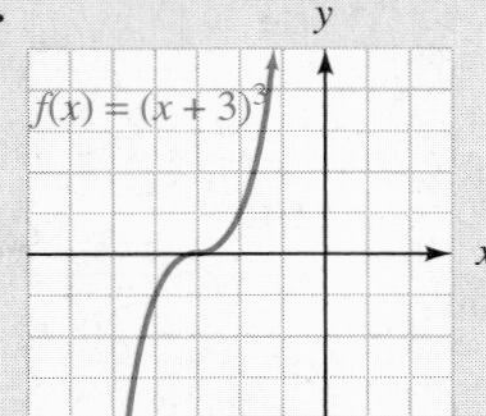

6.

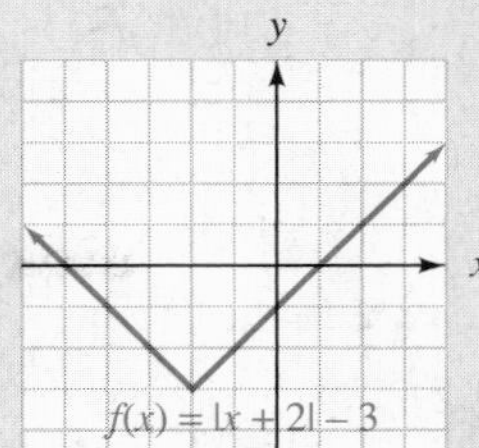

7.

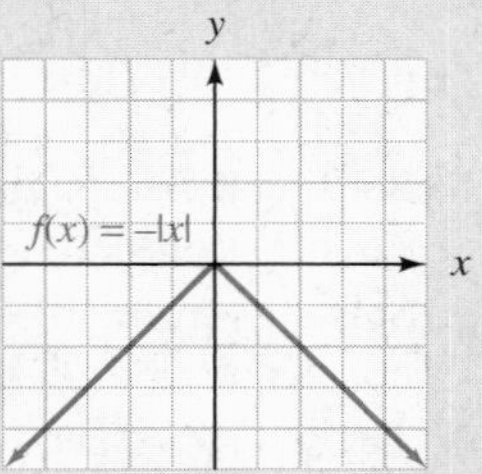

NOW TRY THIS

1. Given the graph of $f(x)$ below, sketch a graph of each translation or reflection.

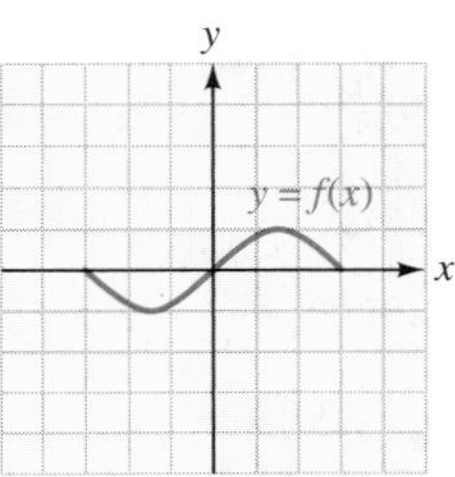

a. $f(x) + 2$

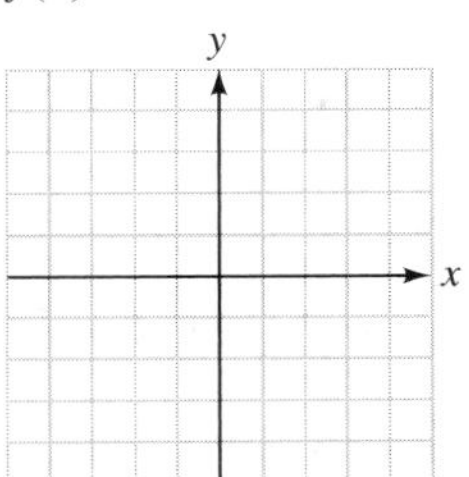

b. $-f(x)$

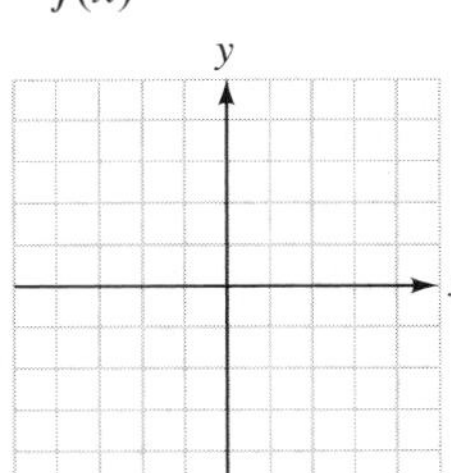

c. $f(x - 1)$

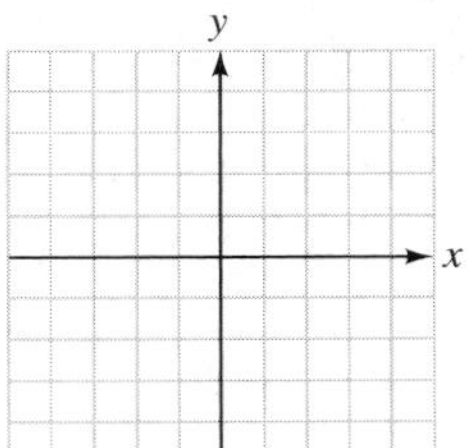

2.5 EXERCISES

WARM-UPS

1. Define the squaring function.
2. Define the cubing function.
3. Define the absolute value function.
4. Describe a parabola.
5. Describe the graph of $f(x) = |x| + 3$.
6. Describe the graph of $f(x) = x^3 - 4$.
7. What is meant by a reflection of a graph?
8. How does the graph of $y = x^2$ compare to the graph of $y = x^2 + 2$?

REVIEW

9. List the prime numbers between 40 and 50.
10. State the associative property of addition.
11. State the commutative property of multiplication.
12. What is the additive identity element?
13. What is the multiplicative identity element?
14. Find the multiplicative inverse of $\frac{5}{3}$.

VOCABULARY AND CONCEPTS *Fill in the blanks.*

15. The function $f(x) = x^2$ is called the ________ function. Its graph is a ________.

16. The function $f(x) = x^3$ is called the _______ function.

17. The function $f(x) = |x|$ is called the ____________ function.

18. Shifting the graph of an equation up or down is called a _______ translation.

19. Shifting the graph of an equation to the left or to the right is called a _________ translation.

20. The graph of $f(x) = x^2 + 5$ is the same as the graph of $f(x) = x^2$, except that it is shifted __ units ___.

21. The graph of $f(x) = x^3 - 2$ is the same as the graph of $f(x) = x^3$, except that it is shifted __ units _____.

22. The graph of $f(x) = (x - 5)^3$ is the same as the graph of $f(x) = x^3$, except that it is shifted __ units _________.

23. The graph of $f(x) = (x + 4)^3$ is the same as the graph of $f(x) = x^3$, except that it is shifted __ units ________.

24. The graph of $f(x) = (x - 3)^2 + 5$ is the same as the graph of $f(x) = x^2$, except that it is shifted 3 units ________ and 5 units ___.

25. The graph of $f(x) = -|x + 3|$ is the same as the graph of $f(x) = |x + 3|$, except that it is _______ about the x-axis.

26. To solve an equation with a graphing calculator, graph ____ sides of the equation and find the __________ of the point where the graphs intersect.

GUIDED PRACTICE

Graph each function by plotting points. Check your work with a graphing calculator. **See Examples 1–3. (Objective 1)**

27. $f(x) = x^2 - 3$

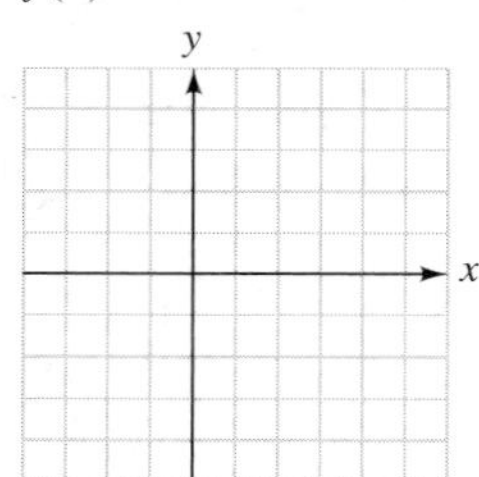

28. $f(x) = x^2 + 2$

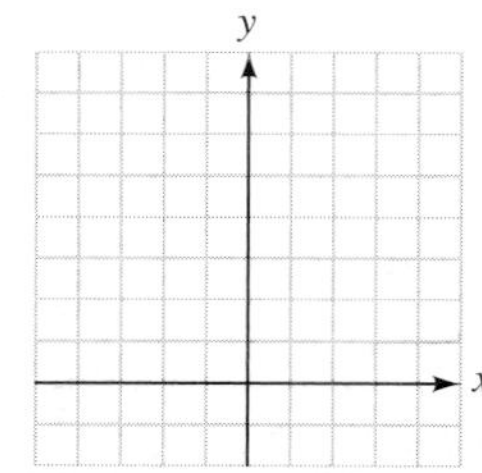

29. $f(x) = (x - 1)^3$

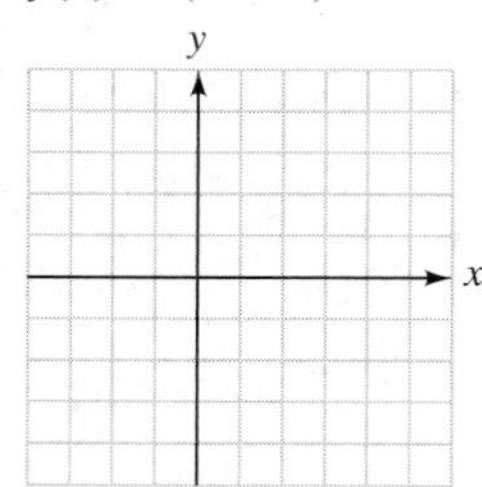

30. $f(x) = (x + 1)^3$

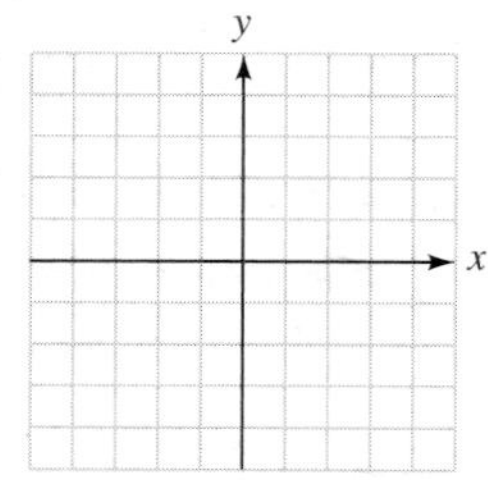

31. $f(x) = |x| - 2$

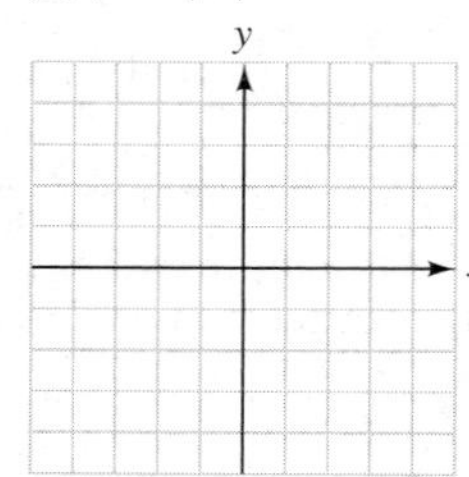

32. $f(x) = |x| + 1$

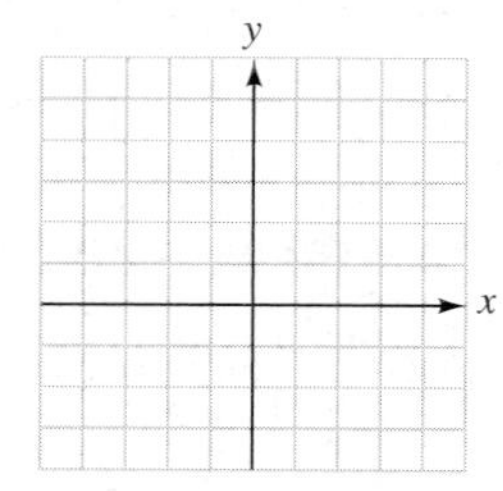

33. $f(x) = |x - 1|$

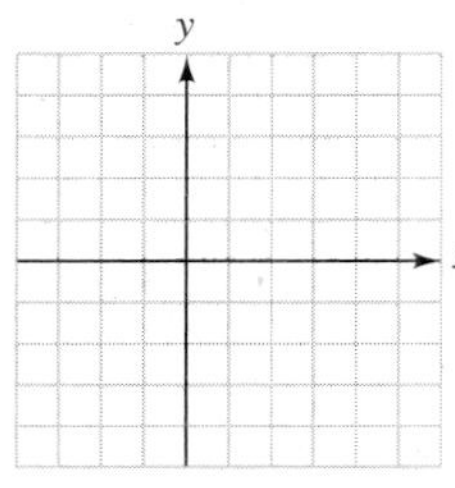

34. $f(x) = |x + 2|$

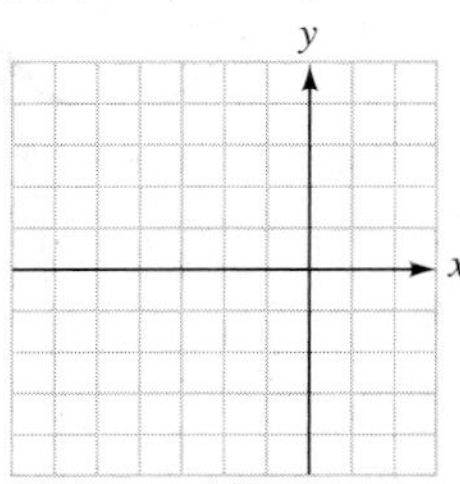

Graph each function using a vertical translation of the graph of $f(x) = x^2$, $f(x) = x^3$, or $f(x) = |x|$. **See Example 4. (Objective 2)**

35. $f(x) = x^2 - 5$

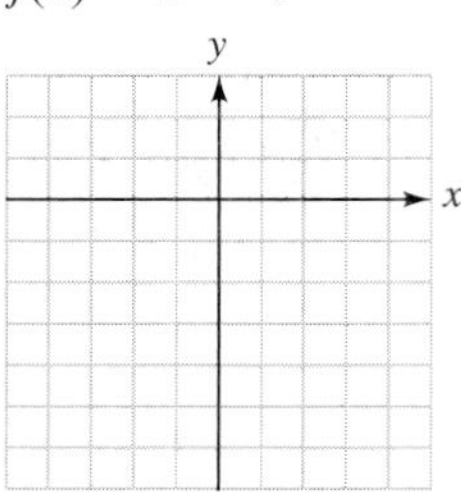

36. $f(x) = x^3 + 4$

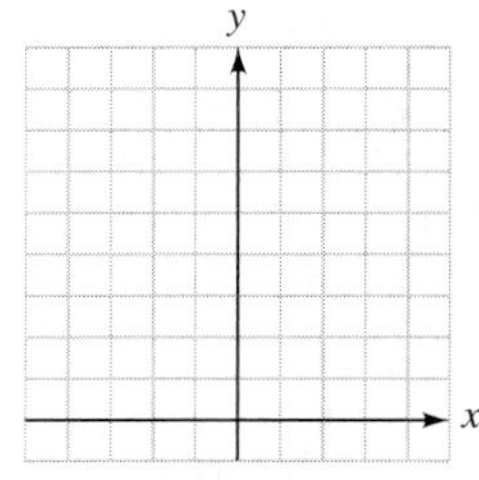

37. $f(x) = |x| + 6$

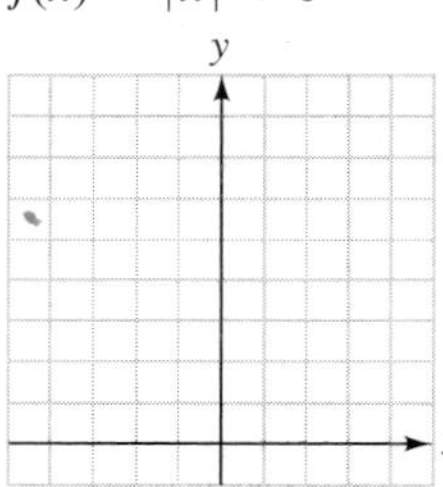

38. $f(x) = |x| - 4$

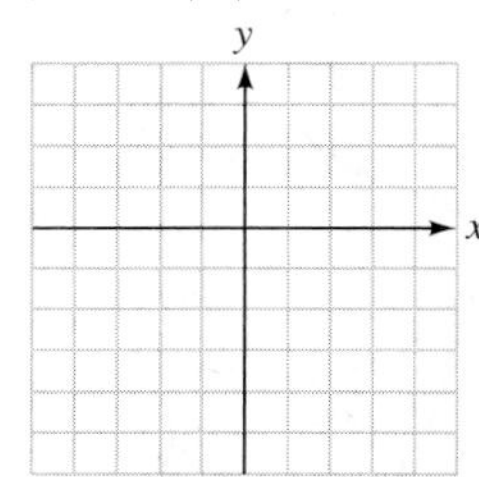

Graph each function using a horizontal translation of the graph of $f(x) = x^2$, $f(x) = x^3$, or $f(x) = |x|$. **See Example 5. (Objective 3)**

39. $f(x) = (x - 1)^3$

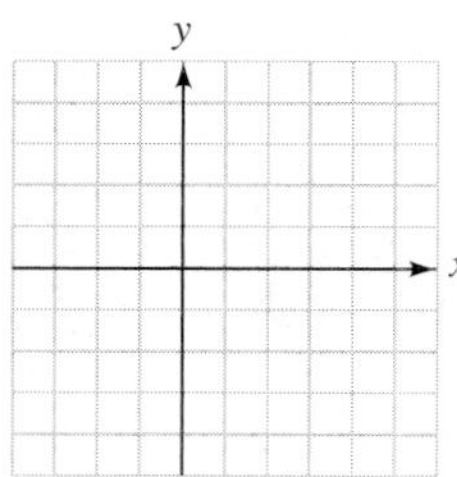

40. $f(x) = (x + 4)^2$

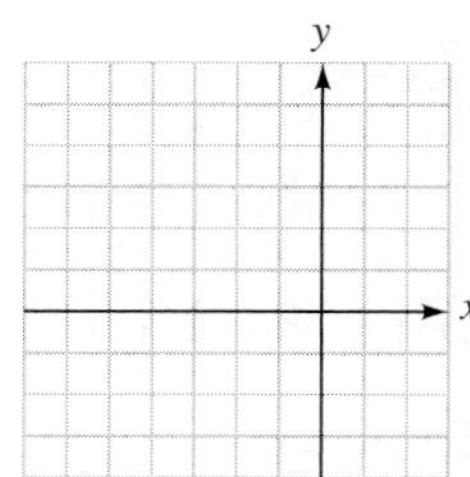

41. $f(x) = (x - 3)^2$

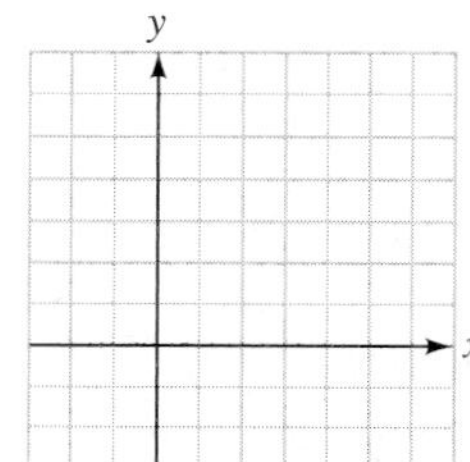

42. $f(x) = |x + 2|$

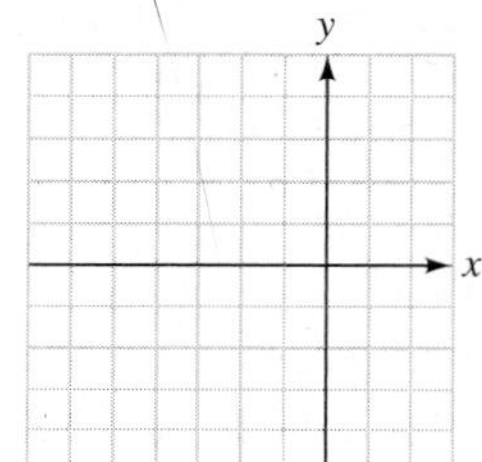

Graph each function using translations of the graph of $f(x) = x^2$, $f(x) = x^3$, or $f(x) = |x|$. See Example 6. (Objectives 2–3)

43. $f(x) = |x - 2| - 1$

44. $f(x) = (x + 2)^2 - 1$

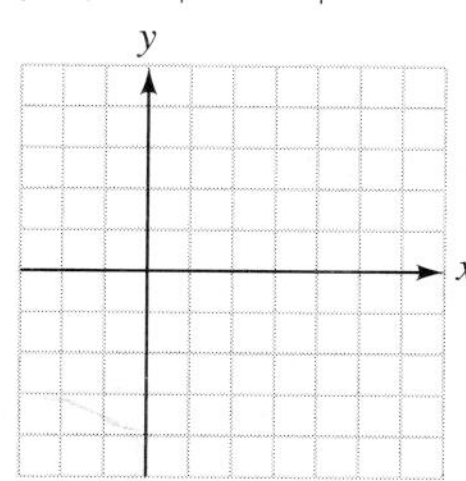

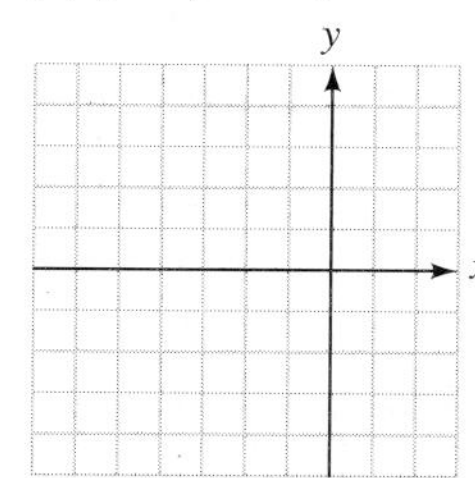

45. $f(x) = (x + 1)^3 - 2$

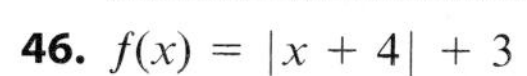

46. $f(x) = |x + 4| + 3$

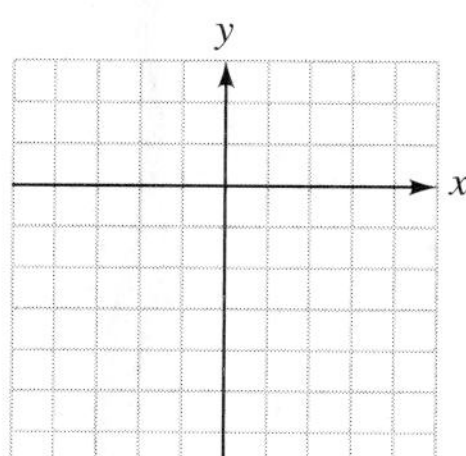

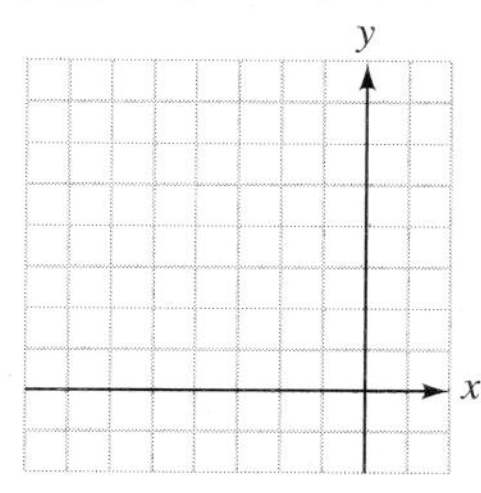

Graph each function using a reflection and a translation of the graph of $f(x) = x^2$, $f(x) = x^3$, or $f(x) = |x|$. See Example 7. (Objective 4)

47. $f(x) = -|x| + 1$

48. $f(x) = -x^3 - 2$

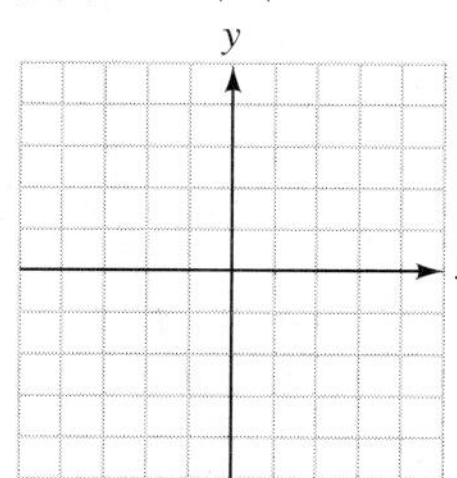

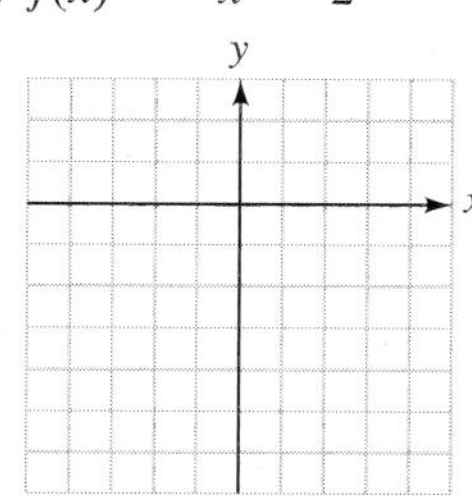

49. $f(x) = -(x - 1)^2$

50. $f(x) = -(x + 2)^2$

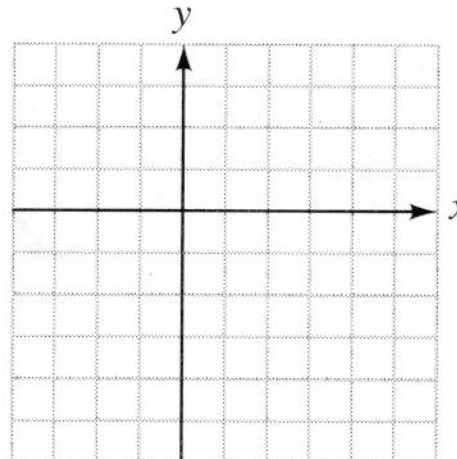

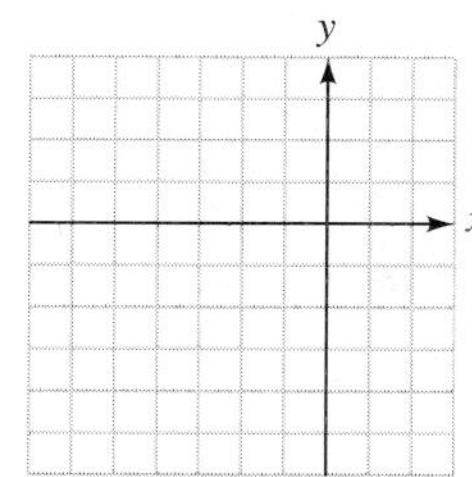

ADDITIONAL PRACTICE

Graph each function.

51. $f(x) = x^2 + 8$

52. $f(x) = x^3 - 8$

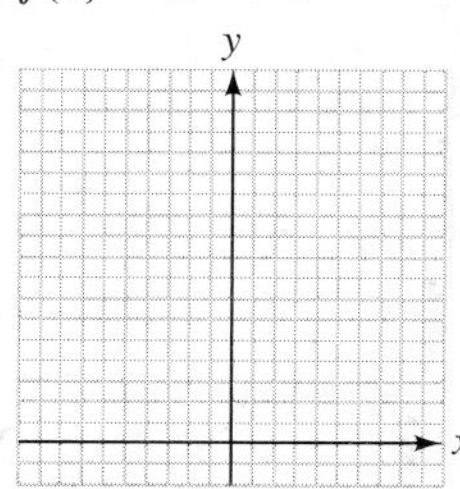

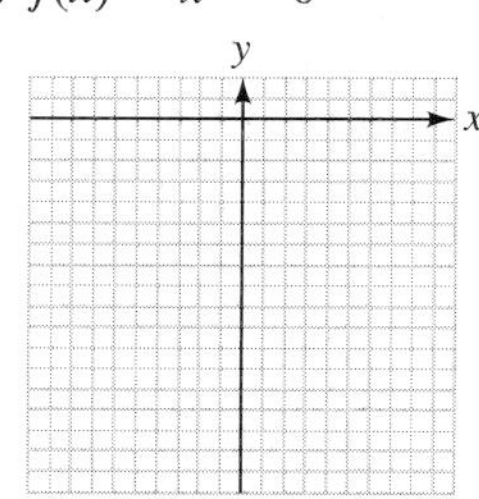

53. $f(x) = |x + 5|$

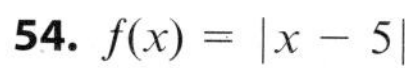

54. $f(x) = |x - 5|$

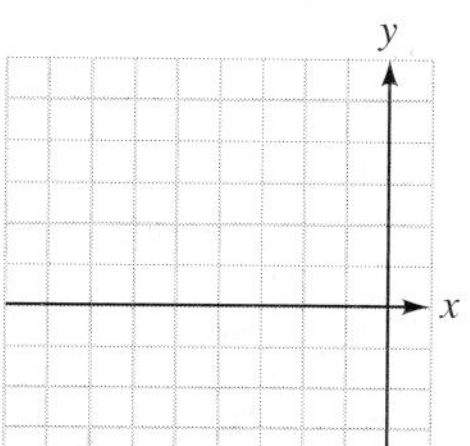

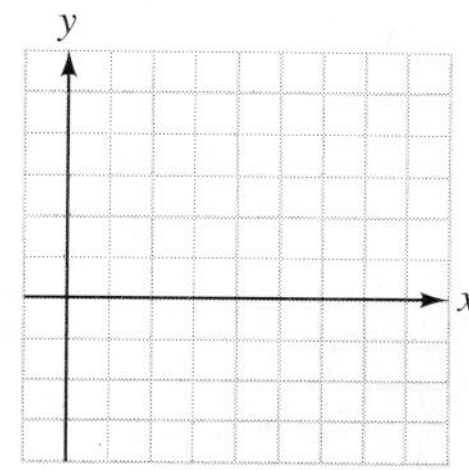

55. $f(x) = (x - 6)^2$

56. $f(x) = (x + 9)^2$

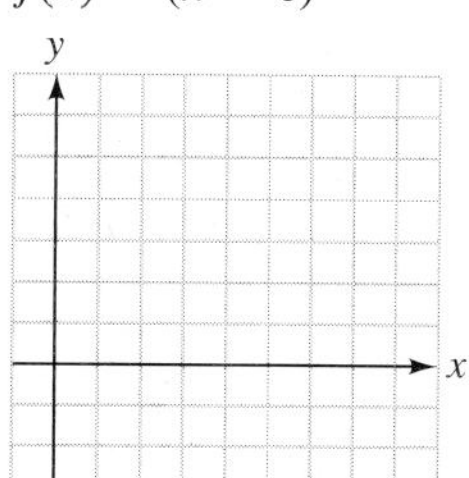

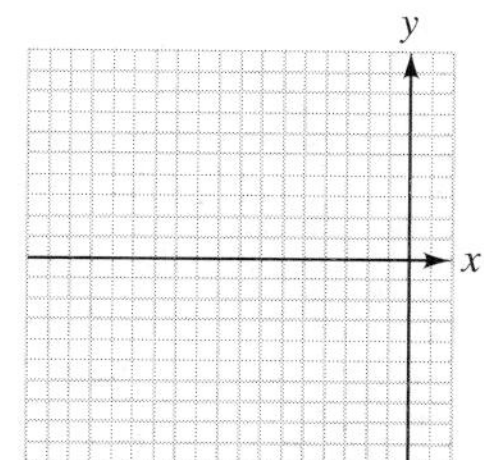

Use a graphing calculator to graph each function, using values of [−4, 4] for x and [−4, 4] for y. The graph is not what it appears to be. Pick a better viewing window and find a better graph.

57. $f(x) = x^3 + 8$

58. $f(x) = x^3 - 12$

Use a graphing calculator to solve each equation.

59. $3x + 6 = 0$

60. $7x - 21 = 0$

61. $4(x - 1) = 3x$

62. $4(x - 3) - x = x - 6$

63. $11x + 6(3 - x) = 3$

64. $2(x + 2) = 2(1 - x) + 10$

WRITING ABOUT MATH

65. Explain how to graph an equation by plotting points.

66. Explain why the correct choice of window settings is important when using a graphing calculator.

SOMETHING TO THINK ABOUT *Use a graphing calculator.*

67. Use a graphing calculator with settings of [−10, 10] for x and [−10, 10] for y to graph **a.** $y = -x^2$, **b.** $y = -x^2 + 1$, and **c.** $y = -x^2 + 2$. What do you notice?

68. Use a graphing calculator with settings of [−10, 10] for x and [−10, 10] for y to graph **a.** $y = -|x|$, **b.** $y = -|x| - 1$, and **c.** $y = -|x| - 2$. What do you notice?

69. Graph $y = (x + k)^2 + 1$ for several positive values of k. What do you notice?

70. Graph $y = (x + k)^2 + 1$ for several negative values of k. What do you notice?

71. Graph $y = (kx)^2 + 1$ for several values of k, where $k > 1$. What do you notice?

72. Graph $y = (kx)^2 + 1$ for several values of k, where $0 < k < 1$. What do you notice?

73. Graph $y = (kx)^2 + 1$ for several values of k, where $k < -1$. What do you notice?

74. Graph $y = (kx)^2 + 1$ for several values of k, where $-1 < k < 0$. What do you notice?

PROJECTS

Project 1

The Board of Administrators of a county has hired your consulting firm to plan a highway. The new highway is to be built in the outback section of the county.

The two main roads in the outback section are Highway N, running in a straight line north and south, and Highway E, running in a straight line east and west. These two highways meet at an intersection that the locals call Four Corners. The only other county road in the area is Slant Road, which runs in a straight line from northwest to southeast, cutting across Highway N north of Four Corners and Highway E east of Four Corners.

The county clerk is unable to find an official map of the area, but there is an old sketch made by the original road designer. It shows that if a rectangular coordinate system is set up using Highways N and E as the axes and Four Corners as the origin, then the equation of the line representing Slant Road is

$$2x + 3y = 12 \quad \text{(where the unit length is 1 mile)}$$

Given this information, the county wants you to do the following:

a. Update the current information by giving the coordinates of the intersections of Slant Road with Highway N and Highway E.

b. Plan a new highway, Country Drive, that will begin 1 mile north of Four Corners and run in a straight line in a generally northeasterly direction, intersecting Slant Road at right angles. The county wants to know the equation of the line representing Country Drive. You also should state the domain on which this equation is valid as a representation of Country Drive.

Project 2

You represent your branch of the Buy-from-Us Corporation. At the company's regional meeting, you must present your revenue and cost reports to the other branch representatives. But disaster strikes! The graphs you had planned to present, containing cost and revenue information for this year and last year, are unlabeled! You cannot immediately recognize which graphs represent costs, which represent revenues, and which represent which year. Without these graphs, your presentation will not be effective.

The only other information you have with you is in the notes you made for your talk. From these you glean the following financial data about your branch.

1. All cost and revenue figures on the graphs are rounded to the nearest \$50,000.
2. Costs for the fourth quarter of last year were \$400,000.
3. Revenue was not above \$400,000 for any quarter last year.
4. Last year, your branch lost money during the first quarter.
5. This year, your branch made money during three of the four quarters.
6. Profit during the second quarter of this year was \$150,000.

And, of course, you know that profit = revenue − cost.

With this information, you must match each of the graphs (Illustrations 1–4) with one of the following titles:

Costs, This Year	Costs, Last Year
Revenues, This Year	Revenues, Last Year

You should be sure to have sound reasons for your choices—reasons ensuring that no other arrangement of the titles will fit the data. The *last* thing you want to do is present incorrect information.

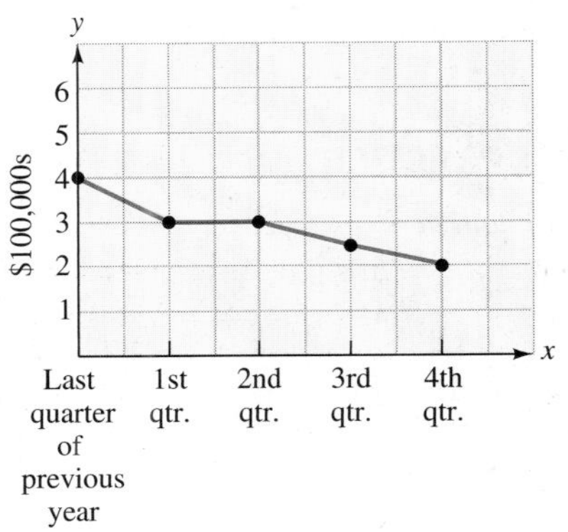

Illustration 1

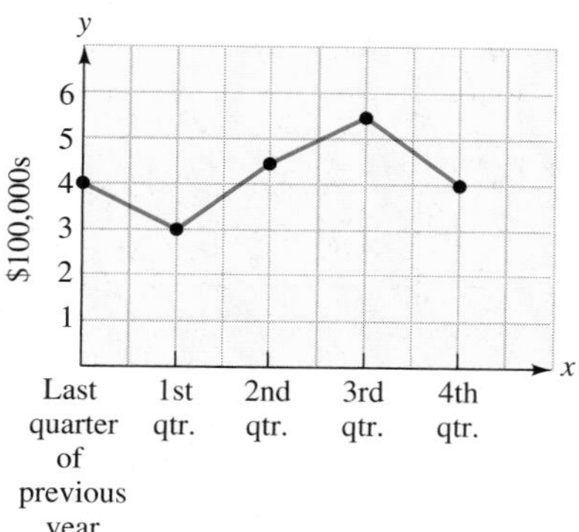

Illustration 2

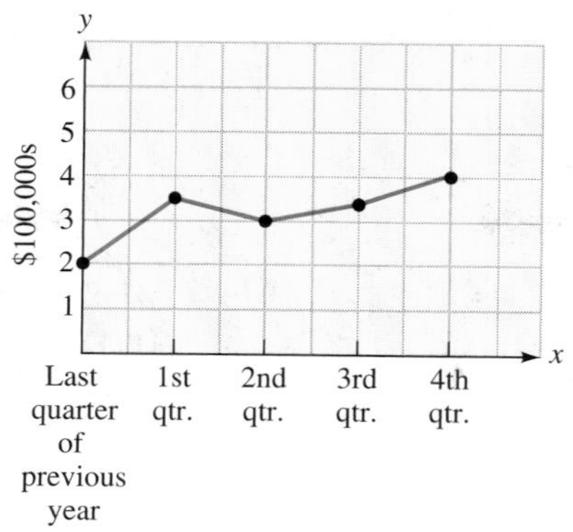

Illustration 3

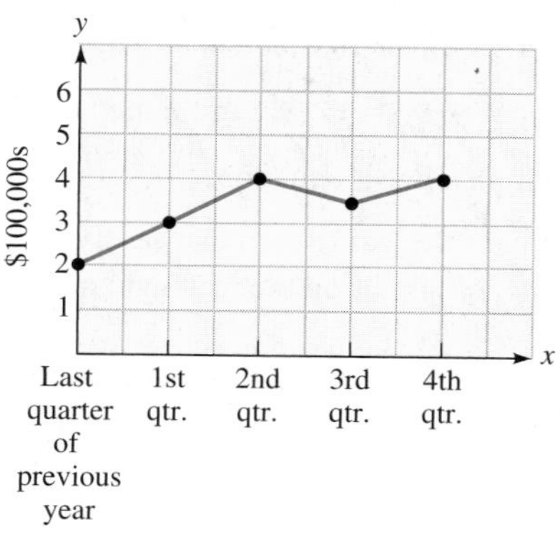

Illustration 4

Chapter 2 REVIEW

SECTION 2.1 Graphing Linear Equations

DEFINITIONS AND CONCEPTS	EXAMPLES
Any ordered pair of real numbers represents a point on the rectangular coordinate system.	On the coordinate axis, plot $(0, -2)$, $(2, -4)$, and $(-3, 5)$.
The point where the axes cross is called the *origin.* The four regions of a coordinate plane are called *quadrants.*	The origin is represented by the ordered pair $(0, 0)$. The point with coordinates $(5, 3)$ is in quadrant I. The point with coordinates $(-5, 3)$ is in quadrant II. The point with coordinates $(-5, -3)$ is in quadrant III. The point with coordinates $(5, -3)$ is in quadrant IV.
An ordered pair of real numbers is a **solution** of an equation in two variables if it satisfies the equation.	The ordered pair $(-1, 5)$ is a solution of the equation $x - 2y = -11$ because it satisfies the equation. $x - 2y = -11$ $(-1) - 2(5) \stackrel{?}{=} -11$ Substitute -1 for x and 5 for y. $-1 - 10 \stackrel{?}{=} -11$ $-11 = -11$ True. Since the results are equal, $(-1, 5)$ is a solution.

General form of an equation of a line:

$Ax + By = C$ (A and B are not both 0.)

To graph a linear equation,

1. Find three pairs (x, y) that satisfy the equation.
2. Plot each pair on the rectangular coordinate system.
3. Draw a line passing through the three points.

The equation $x + y = -2$ is written in general form. To graph it, we find three ordered pairs that satisfy the equation.

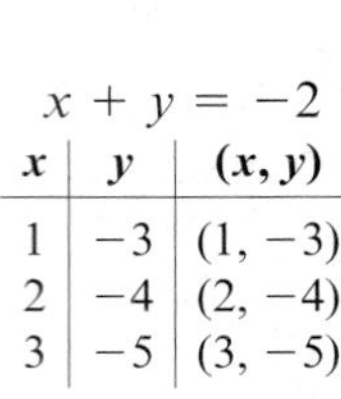

$x + y = -2$

x	y	(x, y)
1	-3	$(1, -3)$
2	-4	$(2, -4)$
3	-5	$(3, -5)$

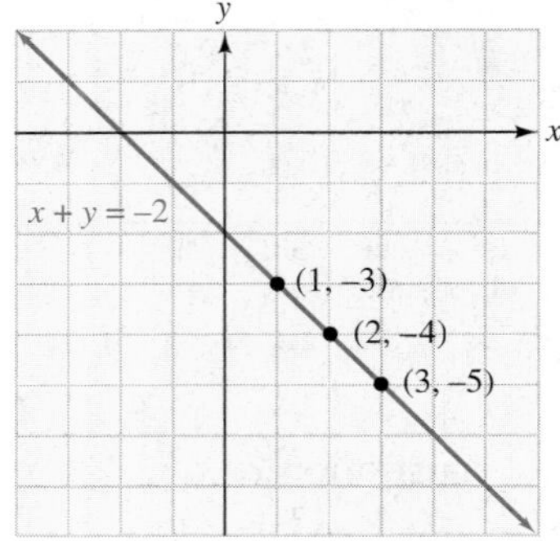

We then plot the points and draw a line passing through them.

The ***y*-intercept** of a line is the point where the line intersects the y-axis.

The y-intercept of the graph above is the point with coordinates $(0, -2)$.

The ***x*-intercept** of a line is the point where the line intersects the x-axis.

The x-intercept of the graph above is the point with coordinates $(-2, 0)$.

Graph of a vertical line:

$x = a$

x-intercept at $(a, 0)$

Graph of a horizontal line:

$y = b$

y-intercept at $(0, b)$

The graph of $y = -3$ is a horizontal line passing through $(0, -3)$.

The graph of $x = 2$ is a vertical line passing through $(2, 0)$.

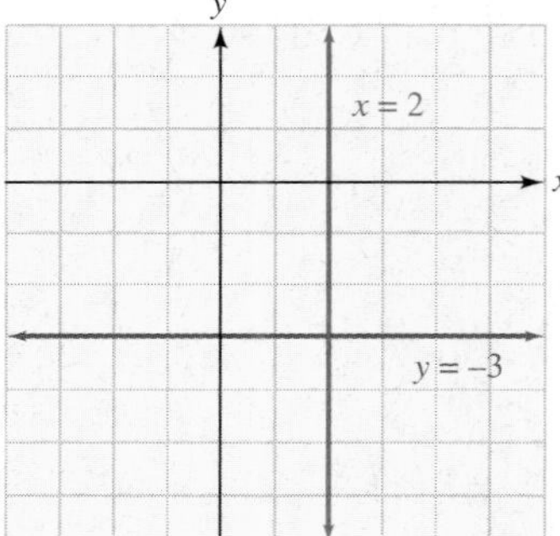

Midpoint formula:

If $P(x_1, y_1)$ and $Q(x_2, y_2)$, the midpoint M of segment PQ is

$$\left(\frac{x_1 + x_2}{2}, \frac{y_1 + y_2}{2}\right)$$

To find the midpoint of the line segment joining $P(3, -2)$ and $Q(2, -5)$, we find the mean of the x-coordinates and the mean of the y-coordinates to get

$$\frac{x_1 + x_2}{2} = \frac{3 + 2}{2} = \frac{5}{2}$$

$$\frac{y_1 + y_2}{2} = \frac{-2 + (-5)}{2} = -\frac{7}{2}$$

The midpoint of segment PQ is the point $M\left(\frac{5}{2}, -\frac{7}{2}\right)$.

REVIEW EXERCISES

1. Use the equation $2x - 3y = 12$ to complete the table.

x	y
-9	
	-8
-3	
	-4
3	
	0
	2

Graph each equation.

2. $x + y = 4$

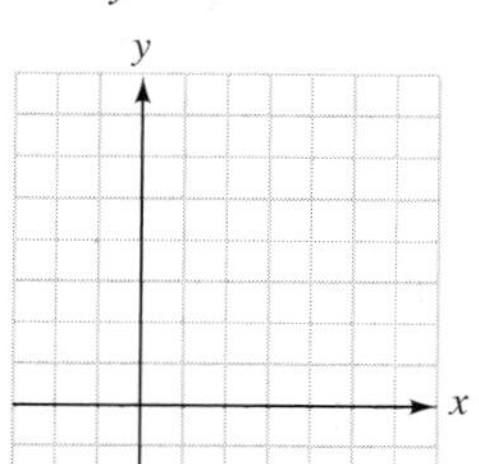

3. $2x - y = 8$

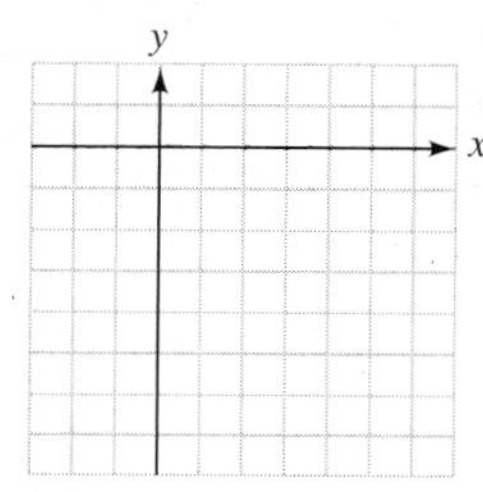

4. $y = 3x + 4$

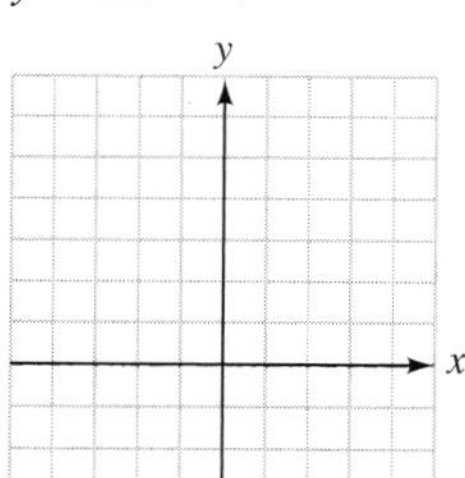

5. $x = 4 - 2y$

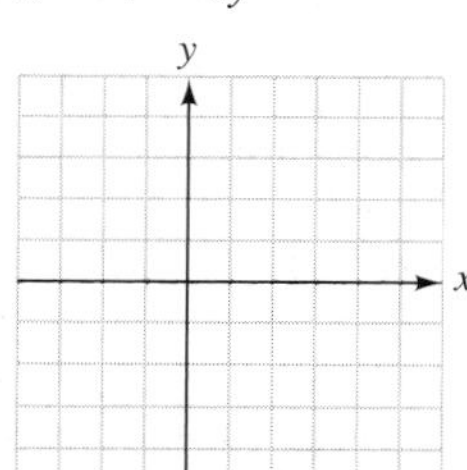

6. $y = 4$

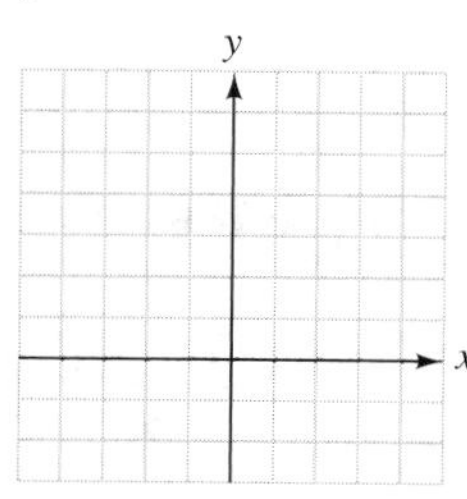

7. $x = -2$

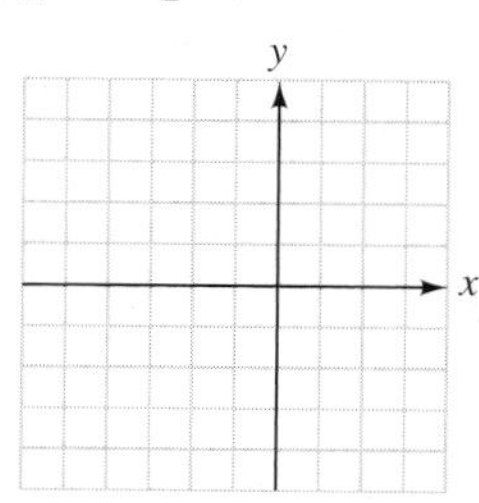

8. $2(x + 3) = x + 2$

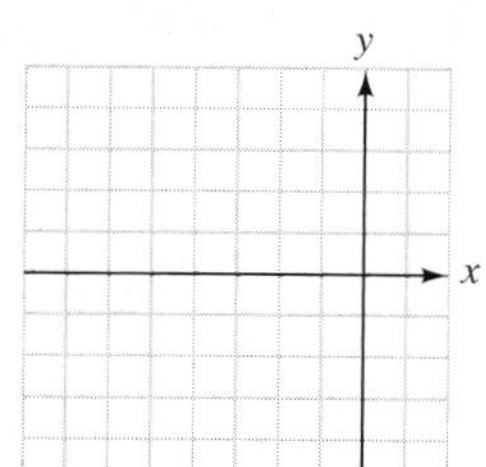

9. $3y = 2(y - 1)$

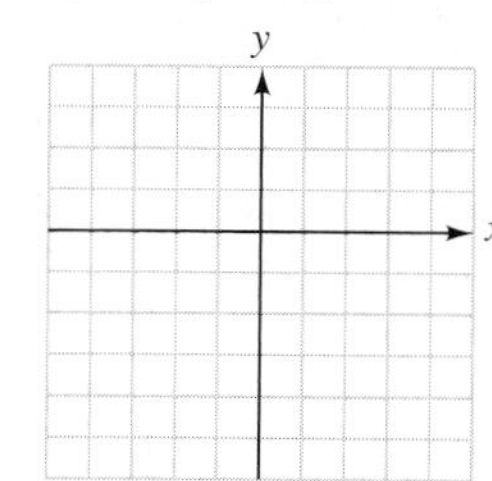

10. Find the midpoint of the line segment joining $P(-3, 5)$ and $Q(6, 11)$.

SECTION 2.2 Slope of a Line

DEFINITIONS AND CONCEPTS	EXAMPLES
Slope of a nonvertical line: If $x_2 \neq x_1$, $$m = \frac{\Delta y}{\Delta x} = \frac{y_2 - y_1}{x_2 - x_1}$$	The slope of a line passing through $(-1, 4)$ and $(5, -3)$ is given by $$m = \frac{y_2 - y_1}{x_2 - x_1} = \frac{-3 - 4}{5 - (-1)} = \frac{-7}{6}$$
Horizontal lines have a slope of 0.	$y = 5$ is a horizontal line with slope 0.
Vertical lines have no defined slope.	$x = 5$ is a vertical line with no defined slope.
Parallel lines have the same slope.	Two distinct lines with slopes $\frac{3}{2}$ and $\frac{3}{2}$ are parallel.
The slopes of two nonvertical perpendicular lines are negative reciprocals.	Two distinct lines with slopes 5 and $-\frac{1}{5}$ are perpendicular.

REVIEW EXERCISES

Find the slope of the line passing through points P and Q, if possible.

11. $P(2, 5)$ and $Q(5, 8)$

12. $P(-3, -2)$ and $Q(6, 12)$

13. $P(-3, 4)$ and $Q(-5, -6)$

14. $P(5, -4)$ and $Q(-6, -9)$

15. $P(-2, 4)$ and $Q(8, 4)$

16. $P(-5, -4)$ and $Q(-5, 8)$

Find the slope of the graph of each equation, if one exists.

17. $2x - 3y = 18$

18. $2x + y = 8$

19. $-2(x - 3) = 10$

20. $3y + 1 = 7$

Determine whether the lines with the given slopes are parallel, perpendicular, or neither.

21. $m_1 = 4, m_2 = -\frac{1}{4}$

22. $m_1 = 0.5, m_2 = \frac{1}{2}$

23. $m_1 = 0.5, m_2 = -\frac{1}{2}$

24. $m_1 = 5, m_2 = -0.2$

25. Sales growth If the sales of a new business were \$65,000 in its first year and \$130,000 in its fourth year, find the rate of growth in sales per year.

SECTION 2.3 Writing Equations of Lines

DEFINITIONS AND CONCEPTS	EXAMPLES
Equations of a line: *Point-slope form:* $y - y_1 = m(x - x_1)$	Write the point-slope form of a line passing through $(-3, 5)$ with slope $\frac{1}{2}$. $y - y_1 = m(x - x_1)$ This is point-slope form. $y - 5 = \frac{1}{2}[x - (-3)]$ Substitute. $y - 5 = \frac{1}{2}(x + 3)$
Slope-intercept form: $y = mx + b$	Write the slope-intercept form of a line with slope $\frac{1}{2}$ and y-intercept $(0, -4)$. $y = mx + b$ This is slope-intercept form. $y = \frac{1}{2}x - 4$ Substitute. $y = \frac{1}{2}x - 4$

REVIEW EXERCISES

Write the equation of the line with the given properties. Write the equation in general form.

26. Slope of 3; passing through $P(-8, 5)$

27. Passing through $(-2, 4)$ and $(6, -9)$

28. Passing through $(-3, -5)$; parallel to the graph of $3x - 2y = 7$

29. Passing through $(-3, -5)$; perpendicular to the graph of $3x - 2y = 7$

30. Depreciation A business purchased a copy machine for \$8,700 and will depreciate it on a straight-line basis over the next 5 years. At the end of its useful life, it will be sold as scrap for \$100. Find its depreciation equation.

SECTION 2.4 Introduction to Functions

DEFINITIONS AND CONCEPTS	EXAMPLES
A **relation** is any set of ordered pairs.	Find the domain and range of the relation $\{(2, -1), (6, 3), (2, 5)\}$ and determine whether it defines a function.
The **domain** is the set of first components of the ordered pairs.	**Domain:** $\{2, 6\}$ because 2 and 6 are the first components in the ordered pairs.
The **range** is the set of second components of the ordered pairs.	**Range:** $\{-1, 3, 5\}$ because -1, 3, and 5 are the second components in the ordered pairs.
A **function** is any set of ordered pairs (a relation) in which each first component determines exactly one second component.	The relation does not define a function because the first component 2 is paired with two different second components.

The **domain** of a function is the set of input values. The **range** is the set of output values.	Find the domain and range of the graph and determine whether the graph represents a function. **Domain:** Since the graph extends forever to the left, and stops at $x = 3$ on the right, the domain is $(-\infty, 3]$. **Range:** Since the graph extends forever downward and ends at $y = 4$, the range is $(-\infty, 4]$.
The **vertical line test** can be used to determine whether a graph represents a function.	Since every vertical line that intersects the graph will do so exactly once, the vertical line test indicates that the graph is a function.
$f(k)$ represents the value of $f(x)$ when $x = k$.	If $f(x) = -7x + 4$, find $f(2)$. $f(2) = -7(2) + 4$ Substitute 2 for x. $= -14 + 4$ $= -10$ In ordered pair form, we can write $(2, -10)$.
The domain of a function that is defined by an equation is the set of all numbers that are permissible replacements for its variable.	Find the domain of $f(x) = \frac{-5}{x+7}$. The number -7 cannot be substituted for x in the function $f(x) = \frac{-5}{x+7}$ because that would make the denominator 0. However, any real number, except -7, can be substituted for x to obtain a single value y. Therefore, the domain is the set of all real numbers except -7. This is the interval $(-\infty, -7) \cup (-7, \infty)$.
A **linear function** is a function defined by an equation that can be written in the form $f(x) = mx + b$ or $y = mx + b$ where m is the slope of the line graph and $(0, b)$ is the y-intercept.	The graph of $y = 4x - 1$ is a line with slope 4 and y-intercept $(0, -1)$. The domain is the set of real numbers $\mathbb{R}$, and the range is the set of real numbers $\mathbb{R}$.

REVIEW EXERCISES

Determine whether each equation determines y to be a function of x.

31. $y = 6x - 4$

32. $y = 4 - x$

33. $y^2 = x$

34. $|y| = x$

Assume that $f(x) = 3x + 2$ and $g(x) = x^2 - 4$ and find each value.

35. $f(-3)$

36. $g(8)$

37. $g(-2)$

38. $f(5)$

Find the domain of each function.

39. $f(x) = 4x - 1$

40. $f(x) = 3x - 10$

41. $f(x) = x^2 + 1$

42. $f(x) = \dfrac{4}{2 - x}$

43. $f(x) = \dfrac{7}{x - 3}$

44. $y = 7$

Use the vertical line test to determine whether each graph represents a function. If it does, give the domain and range.

45.

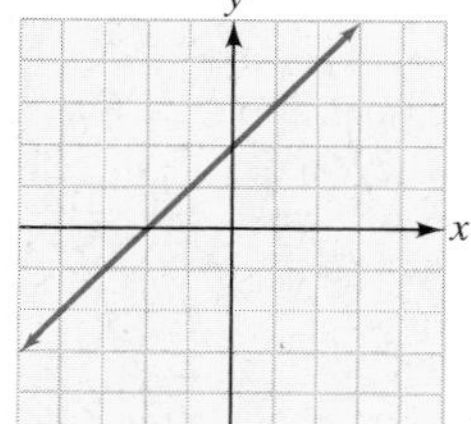

46.

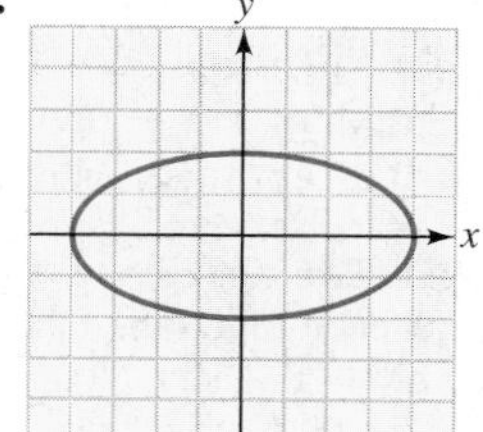

47.

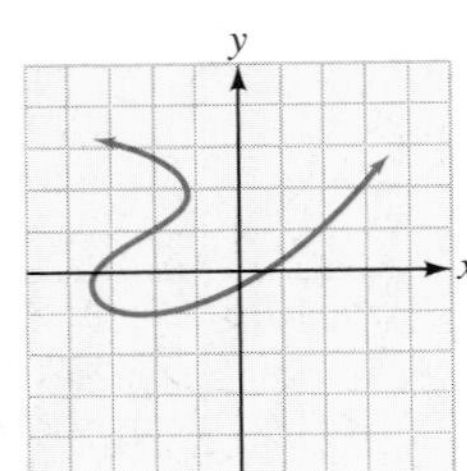

48.

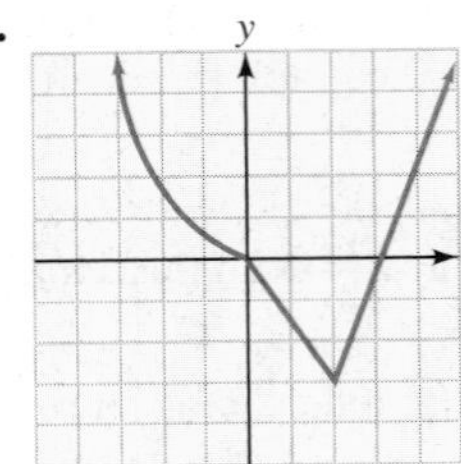

Determine whether each equation defines a linear function.

49. $y = 3x + 2$

50. $y = \dfrac{x + 5}{4}$

51. $4x - 3y = 12$

52. $y = x^2 - 25$

SECTION 2.5 Graphs of Other Functions

DEFINITIONS AND CONCEPTS	EXAMPLES
Vertical translations:	
If f is a function and k is a positive number, then	
• The graph of $f(x) + k$ is identical to the graph of $f(x)$, except that it is translated k units upward.	The graph of $f(x) = \|x\| + 4$ will be the same shape as $f(x) = \|x\|$, but shifted up 4 units.
• The graph of $f(x) - k$ is identical to the graph of $f(x)$, except that it is translated k units downward.	The graph of $f(x) = \|x\| - 4$ will be the same shape as $f(x) = \|x\|$, but shifted down 4 units.
Horizontal translations:	
If f is a function and k is a positive number, then	
• The graph of $f(x - k)$ is identical to the graph of $f(x)$, except that it is translated k units to the right.	The graph of $f(x) = (x - 3)^2$ will be the same shape as $f(x) = x^2$, but shifted 3 units to the right.
• The graph of $f(x + k)$ is identical to the graph of $f(x)$, except that it is translated k units to the left.	The graph of $f(x) = (x + 4)^2$ will be the same shape as $f(x) = x^2$, but shifted 4 units to the left.
Reflections:	
The graph of $y = -f(x)$ is identical to the graph of $f(x)$ except that it is reflected about the x-axis.	The graph of $f(x) = -x^3$ is identical to the graph of $f(x) = x^3$ except that it is reflected about the x-axis.

REVIEW EXERCISES

Graph each function.

53. $f(x) = x^2 - 3$

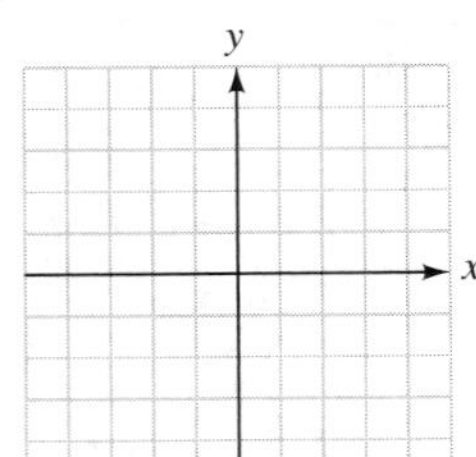

54. $f(x) = |x| - 4$

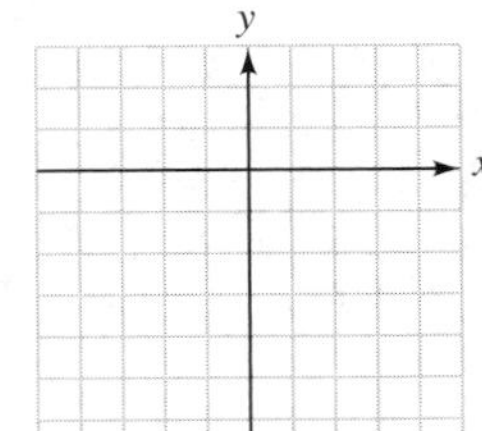

55. $f(x) = (x - 2)^3$

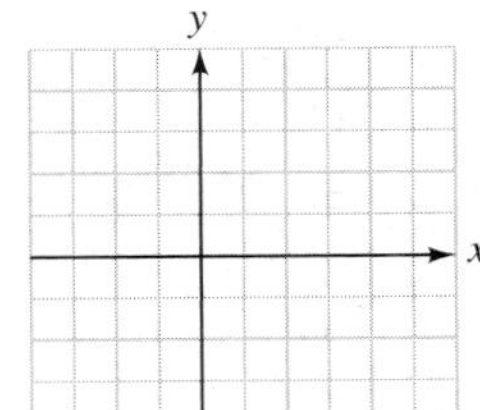

56. $f(x) = (x + 4)^2 - 3$

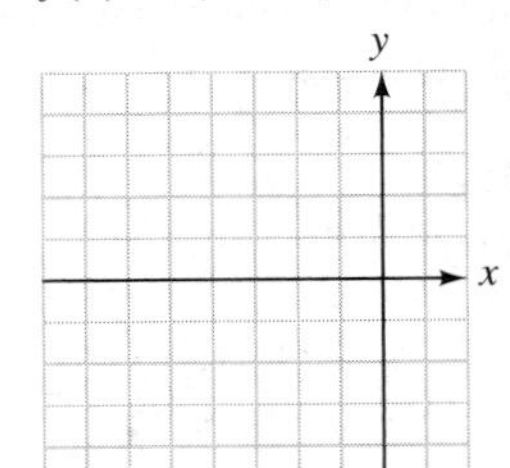

Use a graphing calculator to graph each function. Compare the results to the answers to Problems 53–56.

57. $f(x) = x^2 - 3$

58. $f(x) = |x| - 4$

59. $f(x) = (x - 2)^3$

60. $f(x) = (x + 4)^2 - 3$

61. Graph: $f(x) = -|x - 3|$.

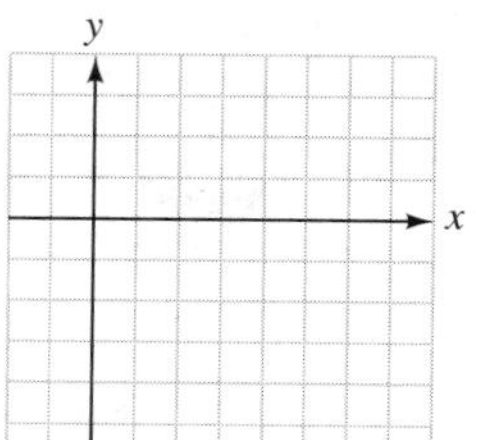

Chapter 2 TEST

1. Graph the equation $2x - 5y = 10$.

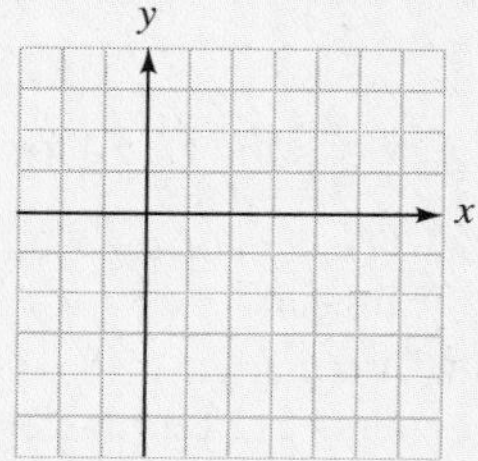

2. Find the x- and y-intercepts of the graph of $y = \frac{x - 3}{5}$.

3. Find the midpoint of the line segment.

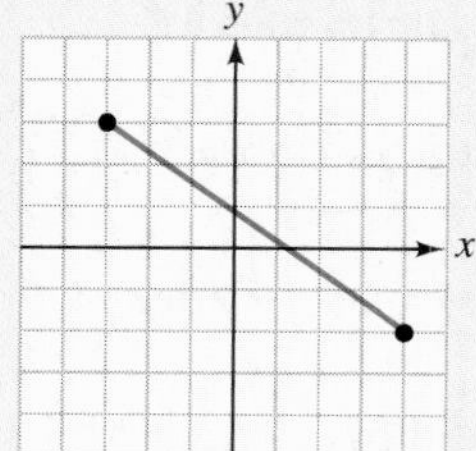

Find the slope of each line, if possible.

4. The line through $(-2, 4)$ and $(6, 8)$

5. The graph of $2x - 3y = 8$

6. The graph of $x = 12$

7. The graph of $y = 12$

8. Write the equation of the line with slope of $\frac{2}{3}$ that passes through $(4, -5)$. Give the answer in slope-intercept form.

9. Write the equation of the line that passes through $(-2, 6)$ and $(-4, -10)$. Give the answer in general form.

10. Find the slope and the y-intercept of the graph of $-2(x - 3) = 3(2y + 5)$.

11. Are the graphs of $4x - y = 12$ and $y = \frac{1}{4}x + 3$ parallel, perpendicular, or neither?

12. Determine whether the graphs of $y = -\frac{2}{3}x + 4$ and $2y = 3x - 3$ are parallel, perpendicular, or neither.

13. Write the equation of the line that passes through the origin and is parallel to the graph of $y = \frac{3}{2}x - 7$.

14. Write the equation of the line that passes through $P(-3, 6)$ and is perpendicular to the graph of $y = -\frac{2}{3}x - 7$.

15. Does $|y| = x$ define y to be a function of x?

16. Find the domain and range of the function $f(x) = |x|$.

17. Find the domain and range of the function $f(x) = x^3$.

Let $f(x) = 3x + 1$ and $g(x) = x^2 - 2$. Find each value.

18. $f(3)$

19. $g(0)$

20. $f(a)$

21. $g(-x)$

Determine whether each graph represents a function.

22.

23.

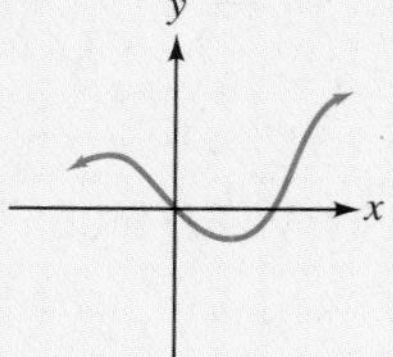

24. Graph: $f(x) = x^2 - 1$.

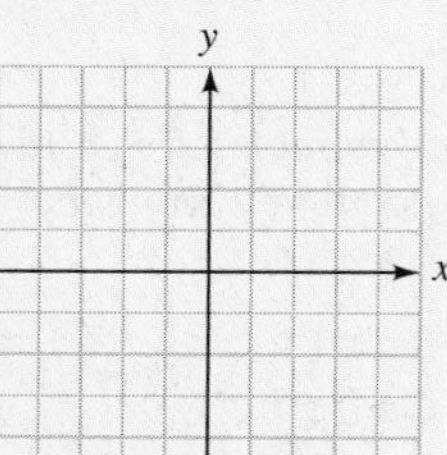

25. Graph: $f(x) = -|x + 2|$.

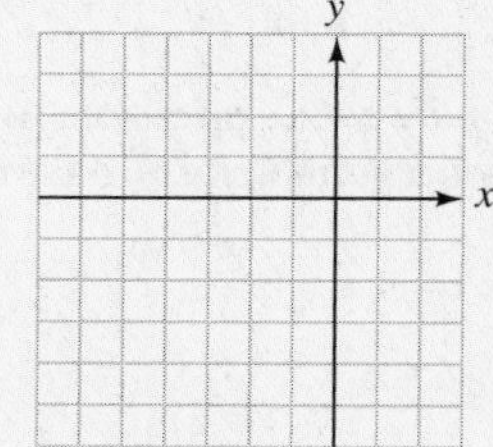

Cumulative Review Exercises

Determine which numbers in the set $\{-2, 0, 1, 2, \frac{13}{12}, 6, 7, \sqrt{5}, \pi\}$ *are in each category.*

1. Natural numbers
2. Whole numbers
3. Rational numbers
4. Irrational numbers
5. Negative numbers
6. Real numbers
7. Prime numbers
8. Composite numbers
9. Even numbers
10. Odd numbers

Graph each set on the number line.

11. $\{x \mid -2 < x \leq 5\}$
12. $[-5, 0) \cup [3, 6]$

Simplify each expression.

13. $-|5| + |-3|$
14. $\dfrac{|-5| + |-3|}{-|4|}$

Perform the operations.

15. $2 + 4 \cdot 5$
16. $\dfrac{8 - 4}{2 - 4}$
17. $20 \div (-10 \div 2)$
18. $\dfrac{6 + 3(6 + 4)}{2(3 - 9)}$

Evaluate each expression when $x = 2$ *and* $y = -3$.

19. $-x - 2y$
20. $\dfrac{x^2 - y^2}{2x + y}$

Determine the property of real numbers that justifies each statement.

21. $(a + b) + c = a + (b + c)$
22. $3(x + y) = 3x + 3y$
23. $(a + b) + c = c + (a + b)$
24. $(ab)c = a(bc)$

Simplify each expression. Assume that all variables are positive numbers and write all answers without negative exponents.

25. $(x^2y^3)^4$
26. $\dfrac{c^4c^8}{(c^5)^2}$
27. $\left(-\dfrac{a^3b^{-2}}{ab}\right)^{-1}$
28. $\left(\dfrac{-3a^3b^{-2}}{6a^{-2}b^3}\right)^0$
29. Change 0.00000497 to scientific notation.
30. Change 9.32×10^8 to standard notation.

Solve each equation.

31. $2x - 5 = 11$
32. $\dfrac{2x - 6}{3} = x + 7$
33. $4(y - 3) + 4 = -3(y + 5)$
34. $2x - \dfrac{3(x - 2)}{2} = 7 - \dfrac{x - 3}{3}$

Solve each formula for the indicated variable.

35. $S = \dfrac{n(a + l)}{2}$ for a
36. $A = \frac{1}{2}h(b_1 + b_2)$ for h
37. The sum of three consecutive even integers is 90. Find the integers.
38. A rectangle is three times as long as it is wide. If its perimeter is 112 centimeters, find its dimensions.
39. Determine whether the graph of $2x - 3y = 6$ defines a function.
40. Find the slope of the line passing through $P(-2, 5)$ and $Q(8, -9)$.
41. Write the equation of the line passing through $P(-2, 5)$ and $Q(8, -9)$.
42. Write the equation of the line that passes through $P(-2, 3)$ and is parallel to the graph of $3x + y = 8$.

Evaluate each expression, given that $f(x) = 3x^2 + 2$ *and* $g(x) = 2x - 1$.

43. $f(-1)$
44. $g(0)$
45. $g(t)$
46. $f(-r)$

Graph each equation and decide whether it is a function. If it is a function, give the domain and range.

47. $y = -x^2 + 1$
48. $y = \left|\frac{1}{2}x - 3\right|$

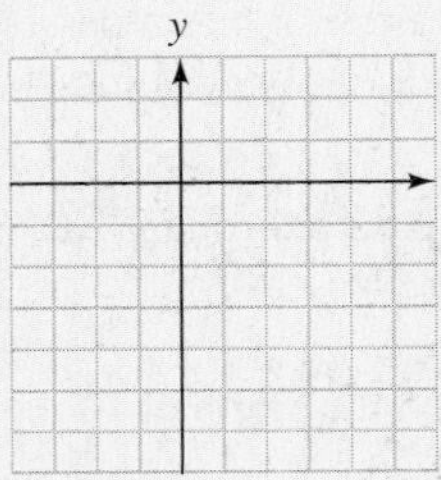

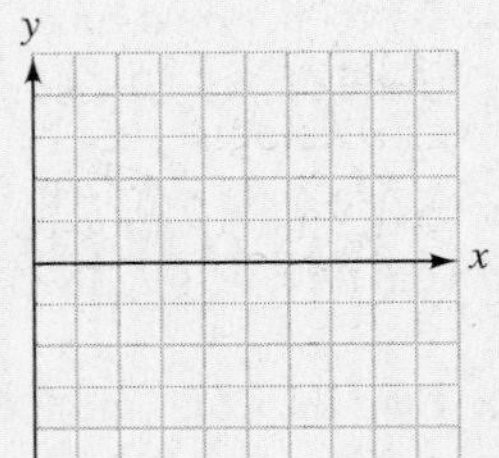

©Shutterstock.com/Nicholas Sutcliffe

CHAPTER 3

Systems of Linear Equations

3.1 Solving Systems of Two Linear Equations by Graphing

3.2 Solving Systems of Two Linear Equations by Substitution and Elimination (Addition)

3.3 Solving Systems of Three Linear Equations in Three Variables

3.4 Solving Systems of Linear Equations Using Matrices

3.5 Solving Systems of Linear Equations Using Determinants

■ *Projects*

CHAPTER REVIEW

CHAPTER TEST

Careers and Mathematics

COSMETOLOGISTS

Cosmetologists, often called hairdressers, help people look neat and well-groomed. They provide beauty services, such as shampooing, cutting, coloring, and styling hair.

Barbers and cosmetologists held about 825,000 jobs in 2006. Almost half of all barbers and cosmetologists are self-employed. All states require barbers and cosmetologists to be licensed, although the qualifications vary from state to state.

Job Outlook:
Overall employment of barbers and cosmetologists is projected to grow about as fast as the average for all occupations through 2016. Opportunities will be best for those licensed to provide a broad range of services.

Hourly Earnings:
\$8.71–\$14.25

For More Information:
http://www.bls.gov/oco/ocos169.htm

For a Sample Application:
See Problem 76 in Section 3.2.

In this chapter ▶

We have considered linear equations with the variables x and y. We found that each equation had infinitely many solutions (x, y), and that we can graph each equation on the rectangular coordinate system. In this chapter, we will discuss many systems of linear equations involving two or three equations.

SECTION 3.1 Solving Systems of Two Linear Equations by Graphing

Objectives

1. Solve a system of two linear equations by graphing.
2. Recognize that an inconsistent system has no solution.
3. Recognize that a dependent system has infinitely many solutions that can be expressed as a general ordered pair.
4. Solve an equation in one variable graphically.

Vocabulary

system of equations
consistent system
inconsistent system
independent equations
dependent equations
equivalent systems

Getting Ready

Let $y = -3x + 2$.

1. Find y when $x = 0$.
2. Find y when $x = 3$.
3. Find y when $x = -3$.
4. Find y when $x = -\frac{1}{3}$.
5. Find five pairs of numbers with a sum of 12.
6. Find five pairs of numbers with a difference of 3.

In the pair of equations

$$\begin{cases} x + 2y = 4 \\ 2x - y = 3 \end{cases} \quad \text{(called a system of equations)}$$

there are infinitely many ordered pairs (x, y) that satisfy the first equation and infinitely many ordered pairs (x, y) that satisfy the second equation. However, there is only one ordered pair (x, y) that satisfies both equations at the same time. The process of finding this ordered pair is called *solving the system.*

1 Solve a system of two linear equations by graphing.

We follow these steps to solve a system of two equations in two variables by graphing.

The Graphing Method

1. On a single set of coordinate axes, graph each equation.
2. Find the coordinates of the point where the graphs intersect. These coordinates give the solution of the system.
3. If the graphs have no point in common, the system has no solution.
4. If the graphs of the equations coincide (are the same), the system has infinitely many solutions.
5. Check the solution in both of the original equations.

When a **system of equations** has a solution (as in Example 1), the system is called a **consistent system.**

EXAMPLE 1 Solve the system by graphing: $\begin{cases} x + 2y = 4 \\ 2x - y = 3 \end{cases}$.

Solution We graph both equations on one set of coordinates axes, as shown in Figure 3-1.

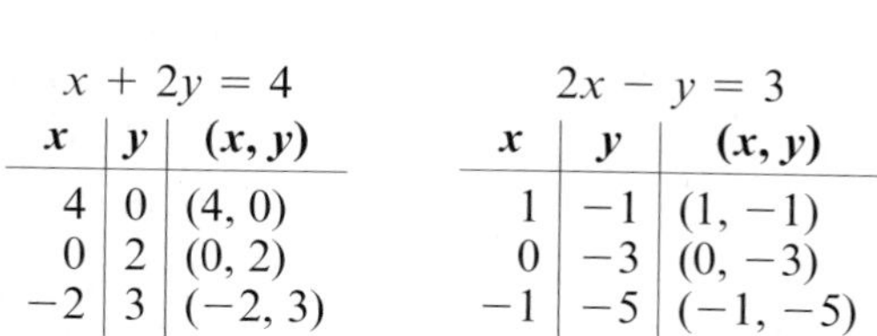

$x + 2y = 4$

x	y	(x, y)
4	0	(4, 0)
0	2	(0, 2)
−2	3	(−2, 3)

$2x - y = 3$

x	y	(x, y)
1	−1	(1, −1)
0	−3	(0, −3)
−1	−5	(−1, −5)

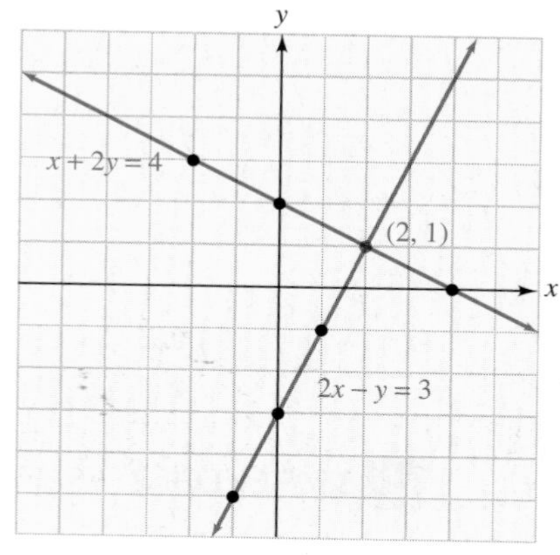

Figure 3-1

Although infinitely many ordered pairs (x, y) satisfy $x + 2y = 4$, and infinitely many ordered pairs (x, y) satisfy $2x - y = 3$, only the coordinates of the point where the graphs intersect satisfy both equations. Since the intersection point has coordinates of (2, 1), the solution is the ordered pair (2, 1), or $x = 2$ and $y = 1$.

When we check the solution, we substitute 2 for x and 1 for y in both equations and verify that (2, 1) satisfies each one.

SELF CHECK 1 Solve the system: $\begin{cases} 2x + y = 4 \\ x - 3y = -5 \end{cases}$.

2 Recognize that an inconsistent system has no solution.

When a system has no solution (as in Example 2), it is called an **inconsistent system.** Since there is no solution, the solution set is $\varnothing$.

EXAMPLE 2 Solve the system by graphing: $\begin{cases} 2x + 3y = 6 \\ 4x + 6y = 24 \end{cases}$.

Solution We graph both equations on one set of coordinate axes, as shown in Figure 3-2 on the next page. In this example, the graphs appear to be parallel. We can show that this is true by writing each equation in slope-intercept form.

$$\begin{aligned} 2x + 3y &= 6 \\ 3y &= -2x + 6 \\ y &= -\frac{2}{3}x + 2 \end{aligned} \qquad \begin{aligned} 4x + 6y &= 24 \\ 6y &= -4x + 24 \\ y &= -\frac{2}{3}x + 4 \end{aligned}$$

The slope of both lines is $-\frac{2}{3}$, but their y-intercepts, (2, 0) and (4, 0), are different. Since the slopes are equal and the y-intercepts are different, the lines are parallel.

Since the graphs are parallel lines, the lines do not intersect, and the system does not have a solution. The system is an *inconsistent system* and its solution set is $\varnothing$.

$2x + 3y = 6$

x	y	(x, y)
3	0	(3, 0)
0	2	(0, 2)
−3	4	(−3, 4)

$4x + 6y = 24$

x	y	(x, y)
6	0	(6, 0)
0	4	(0, 4)
−3	6	(−3, 6)

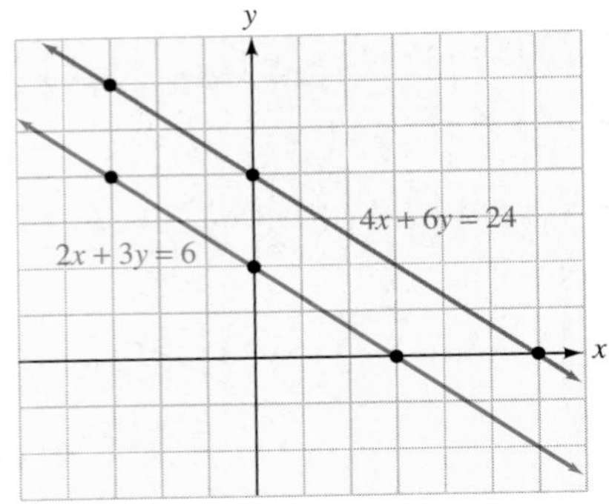

Figure 3-2

SELF CHECK 2 Solve the system: $\begin{cases} 2x - 3y = 6 \\ y = \frac{2}{3}x - 3 \end{cases}$.

3 Recognize that a dependent system has infinitely many solutions that can be expressed as a general ordered pair.

When the equations of a system have different graphs (as in Examples 1 and 2), the equations are called **independent equations.** Two equations with the same graph are called **dependent equations.**

EXAMPLE 3 Solve the system: $\begin{cases} 2y - x = 4 \\ 2x + 8 = 4y \end{cases}$.

Solution We graph each equation on one set of coordinate axes, as shown in Figure 3-3. Since the graphs are the same line (coincide), the system has infinitely many solutions. In fact, any ordered pair (x, y) that satisfies one equation satisfies the other also.

$2y - x = 4$

x	y	(x, y)
−4	0	(−4, 0)
0	2	(0, 2)

$2x + 8 = 4y$

x	y	(x, y)
−4	0	(−4, 0)
0	2	(0, 2)

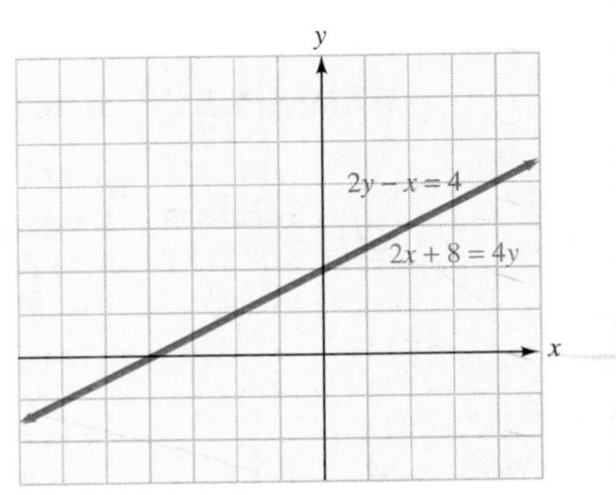

Figure 3-3

From the tables in Figure 3-3, we see that $(-4, 0)$ and $(0, 2)$ are two solutions.

To find a general ordered pair solution, we can solve either equation of the system to obtain $y = \frac{1}{2}x + 2$. Substituting $\frac{1}{2}x + 2$ for y in the ordered pair (x, y), we obtain the general solution $\left(x, \frac{1}{2}x + 2\right)$.

Because the equations have the same graph, they are *dependent equations.*

SELF CHECK 3 Solve the system: $\begin{cases} 2x - y = 4 \\ x = \frac{1}{2}y + 2 \end{cases}$.

We summarize the possibilities that can occur when we graph two equations, each with two variables.

If the lines are different and intersect, the equations are independent and the system is consistent. **One solution exists.**

If the lines are different and parallel, the equations are independent and the system is inconsistent. **No solution exists.**

If the lines coincide, the equations are dependent and the system is consistent. **Infinitely many solutions exist.**

If the equations in two systems are equivalent, the systems are called **equivalent systems.** In Example 4, we solve a more difficult system by writing it as a simpler equivalent system.

EXAMPLE 4 Solve the system: $\begin{cases} \frac{3}{2}x - y = \frac{5}{2} \\ x + \frac{1}{2}y = 4 \end{cases}$.

Solution We multiply both sides of $\frac{3}{2}x - y = \frac{5}{2}$ by 2 to eliminate the fractions and obtain the equation $3x - 2y = 5$. We multiply both sides of $x + \frac{1}{2}y = 4$ by 2 to eliminate the fraction and obtain the equation $2x + y = 8$. This will result in a new system

$$\begin{cases} 3x - 2y = 5 \\ 2x + y = 8 \end{cases}$$

which is equivalent to the original system and is easier to solve, since it has no fractions. If we graph each equation in the new system, as in Figure 3-4, we see that the coordinates of the point where the two lines intersect are $(3, 2)$. Verify that $x = 3$ and $y = 2$ satisfy each equation in the original system.

$3x - 2y = 5$		
x	y	(x, y)
0	$-\frac{5}{2}$	$\left(0, -\frac{5}{2}\right)$
$\frac{5}{3}$	0	$\left(\frac{5}{3}, 0\right)$

$2x + y = 8$		
x	y	(x, y)
4	0	(4, 0)
1	6	(1, 6)

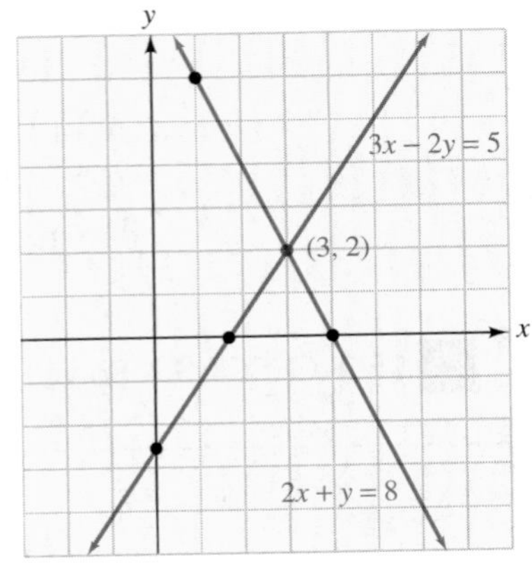

Figure 3-4

SELF CHECK 4 Solve the system: $\begin{cases} \frac{5}{2}x - y = 2 \\ x + \frac{1}{3}y = 3 \end{cases}$.

ACCENT ON TECHNOLOGY

Solving Systems by Graphing

To solve the system

$$\begin{cases} 3x + 2y = 12 \\ 2x - 3y = 12 \end{cases}$$

using a graphing calculator, we first solve each equation for y so we can enter them into a graphing calculator. After solving for y, we obtain the following equivalent system.

$$\begin{cases} y = -\frac{3}{2}x + 6 \\ y = \frac{2}{3}x - 4 \end{cases}$$

If we use window settings of $[-10, 10]$ for x and $[-10, 10]$ for y, the graphs will look like those in Figure 3-5(a). If we zoom in on the intersection point of the lines and trace, we will get an approximate solution like the one shown in Figure 3-5(b). To get better results, we can do more zooms.

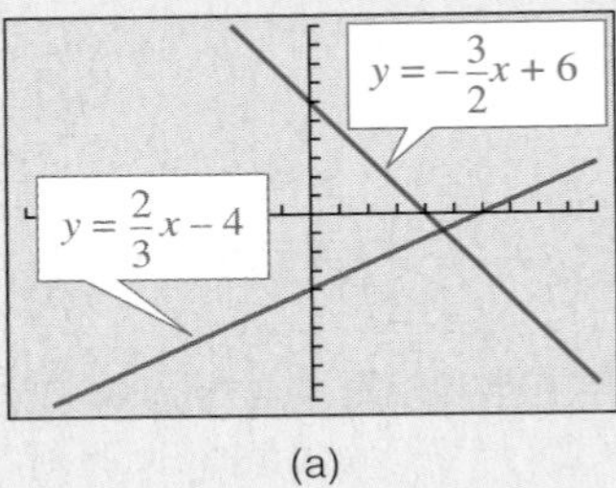

(a)

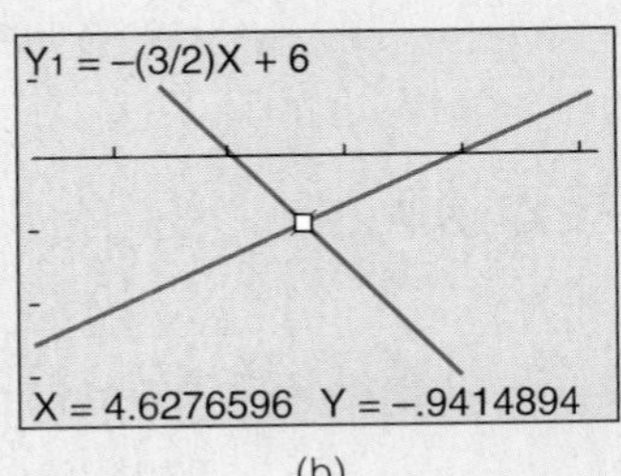

(b)

Figure 3-5

We also can solve this system using the INTERSECT command found in the CALC menu on a TI-83 Plus or TI-84 Plus graphing calculator. To use this method, we first draw the graph of each equation as shown in Figure 3-5(a), then open the CALC menu and select "5: intersect" to obtain Figure 3-6(a). Then we select a point on the first curve by pressing **ENTER**, select a point on the second curve by pressing **ENTER**, and press **ENTER** again to obtain Figure 3-6(b). From the figure, we see that the solution is approximately (4.6153846, −0.9230769).

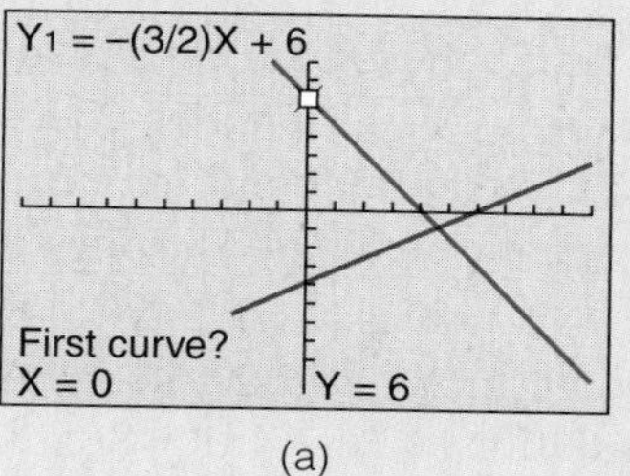

(a)

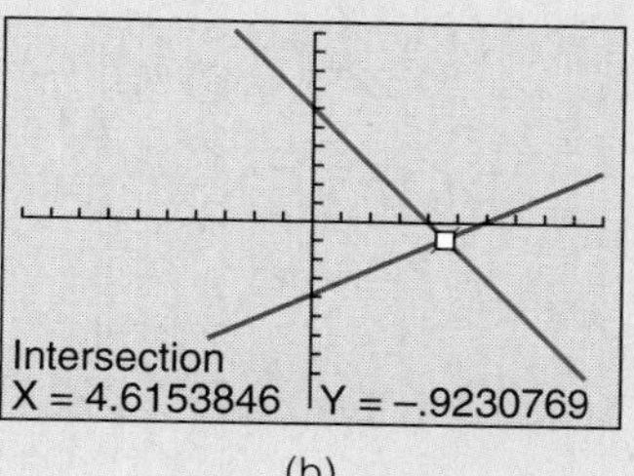

(b)

Figure 3-6

COMMENT Refer to the tear-out card for the method for converting decimal answers to fractions.

Verify that the exact solution is $x = \frac{60}{13}$ and $y = -\frac{12}{13}$.

4 Solve an equation in one variable graphically.

The graphing method discussed in this section can be used to solve equations in one variable.

EXAMPLE 5 Solve $2x + 4 = -2$ graphically.

Solution To solve $2x + 4 = -2$ graphically, we can set the left and right sides of the equation equal to y. The graphs of $y = 2x + 4$ and $y = -2$ are shown in Figure 3-7. To solve $2x + 4 = -2$, we need to find the value of x that makes $2x + 4$ equal to -2. The point of intersection of the graphs is $(-3, -2)$. This tells us that if x is -3, the expression $2x + 4$ equals -2. So the solution of $2x + 4 = -2$ is -3. Check this result.

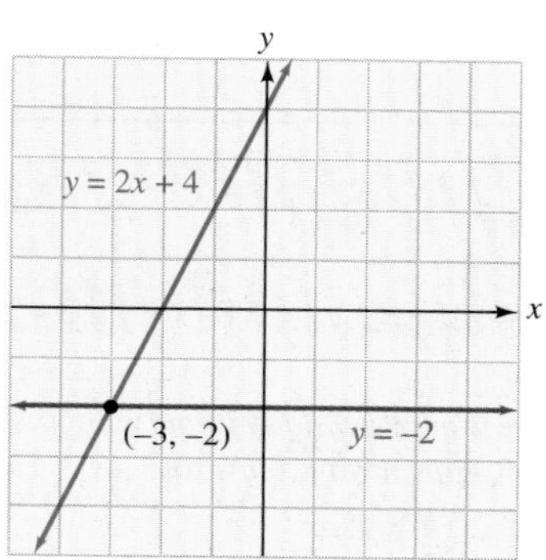

Figure 3-7

SELF CHECK 5 Solve $2x + 4 = 2$ graphically.

ACCENT ON TECHNOLOGY

Solving Equations Graphically

To solve $2(x - 3) + 3 = 7$ with a graphing calculator, we graph the left and right sides of the equation in the same window by entering

$$Y_1 = 2(x - 3) + 3$$
$$Y_2 = 7$$

Figure 3-8(a) shows the graphs, generated using settings of $[-10, 10]$ for x and for y.

The coordinates of the point of intersection of the graphs can be determined using the INTERSECT feature found on most graphing calculators. With this feature, the cursor automatically highlights the intersection point, and the x- and y-coordinates are displayed.

In Figure 3-8(b), we see that the point of intersection is (5, 7), which indicates that $x = 5$ is the solution of $2(x - 3) + 3 = 7$.

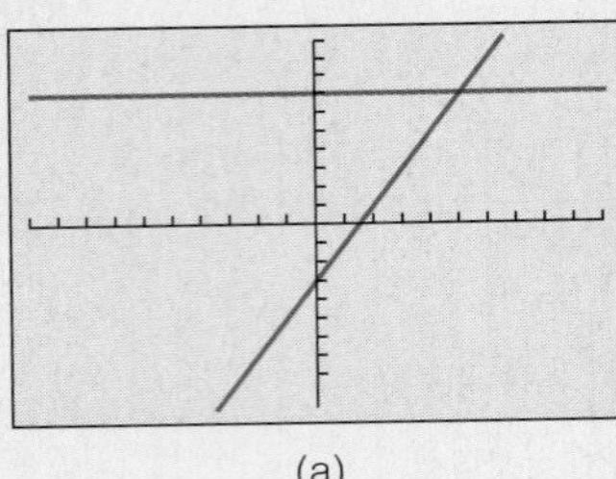
(a)

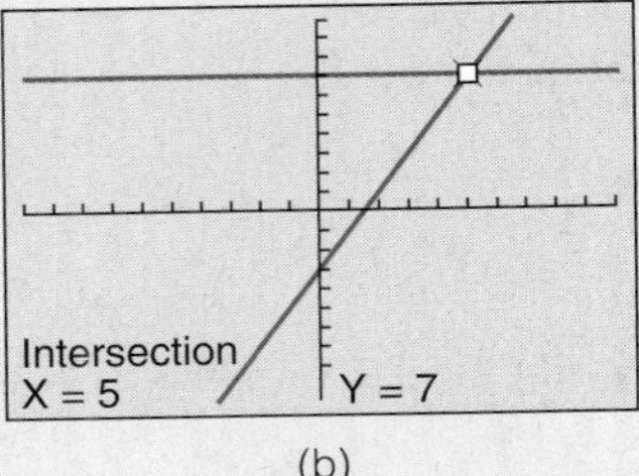

(b)

Figure 3-8

SELF CHECK ANSWERS **1.** (1, 2) **2.** ∅ **3.** $(x, 2x - 4)$ **4.** (2, 3) **5.** -1

NOW TRY THIS

1. Solve each system by graphing.

a. $\begin{cases} y = x^2 - 4 \\ y = 2x - 1 \end{cases}$ **b.** $\begin{cases} y = |x + 5| \\ y = -\frac{1}{3}x + 1 \end{cases}$ **c.** $\begin{cases} y = |x - 2| \\ y = x^2 \end{cases}$

3.1 EXERCISES

WARM-UPS *Determine whether the following systems will have one solution, no solution, or infinitely many solutions.*

1. $\begin{cases} y = 2x \\ y = 2x + 5 \end{cases}$

2. $\begin{cases} y = 2x \\ y = x + x \end{cases}$

3. $\begin{cases} y = 2x \\ y = -2x \end{cases}$

4. $\begin{cases} y = 2x + 1 \\ 2x = y \end{cases}$

REVIEW *Write each number in scientific notation.*

5. 93,000,000

6. 0.0000000236

7. 345×10^2

8. 752×10^{-5}

VOCABULARY AND CONCEPTS *Fill in the blanks.*

9. If two or more equations are considered at the same time, they are called a ______ of equations.

10. When a system of equations has one or more solutions, it is called a __________ system.

11. If a system has no solution, it is called an __________ system. Its solution set is $\varnothing$.

12. If two equations have different graphs, they are called __________ equations.

13. Two equations with the same graph are called __________ equations.

14. If the equations in two systems are equivalent, the systems are called __________ systems.

GUIDED PRACTICE

Determine whether the ordered pair is a solution of the system of equations. **(Objective 1)**

15. $(1, 2)$; $\begin{cases} y = 2x \\ y = \frac{1}{2}x + \frac{3}{2} \end{cases}$

16. $(-1, 2)$; $\begin{cases} y = 3x + 5 \\ y = x + 4 \end{cases}$

17. $(2, -3)$; $\begin{cases} y = \frac{1}{2}x - 2 \\ 3x + 2y = 0 \end{cases}$

18. $(-4, 3)$; $\begin{cases} 4x - y = -19 \\ 3x + 2y = -6 \end{cases}$

Solve each system by graphing. **See Example 1. (Objective 1)**

19. $\begin{cases} x + y = 6 \\ x - y = 2 \end{cases}$

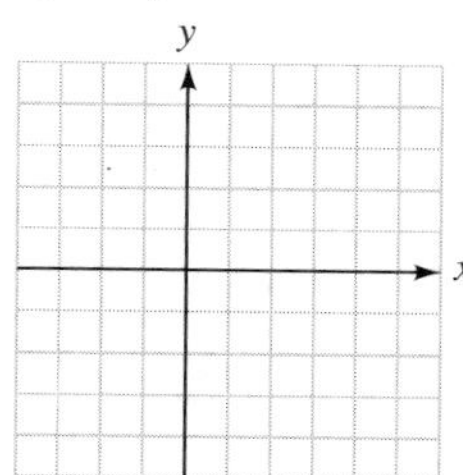

20. 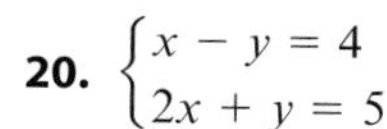
$\begin{cases} x - y = 4 \\ 2x + y = 5 \end{cases}$

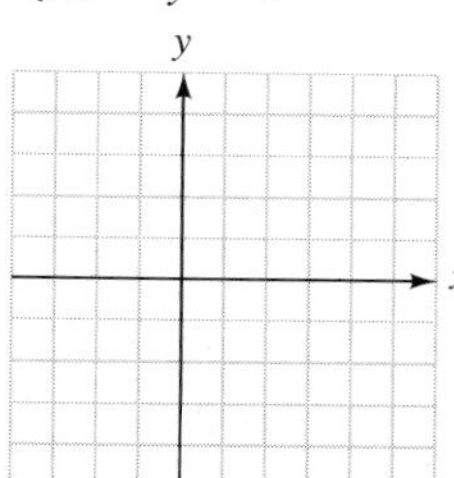

21. $\begin{cases} 2x + y = 1 \\ x - 2y = -7 \end{cases}$

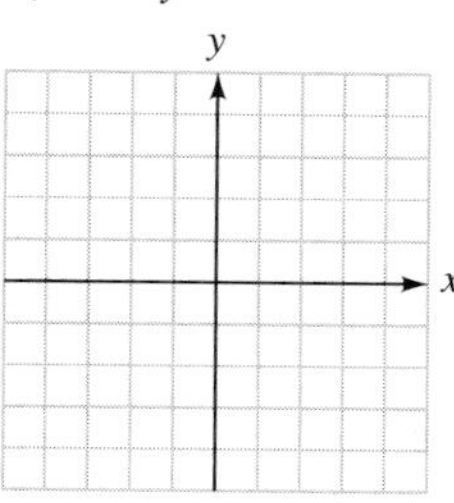

22. $\begin{cases} 3x - y = -3 \\ 2x + y = -7 \end{cases}$

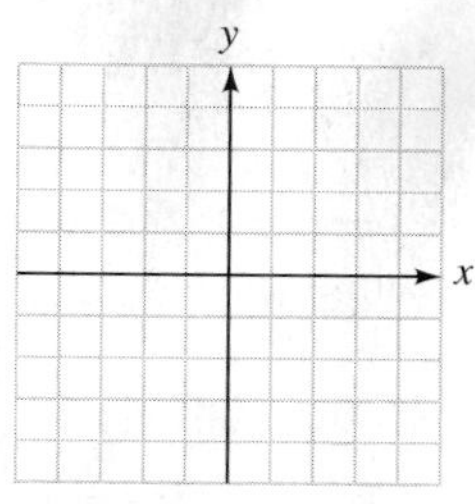

23. $\begin{cases} 2x + 3y = 0 \\ 2x + y = 4 \end{cases}$

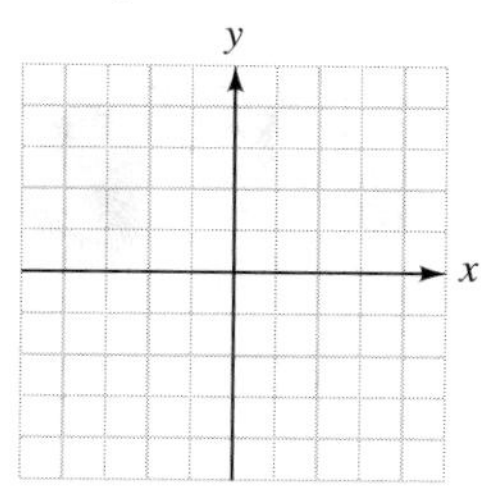

24. 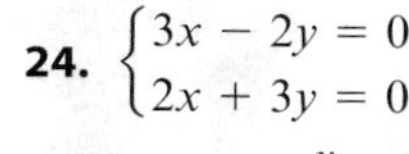
$\begin{cases} 3x - 2y = 0 \\ 2x + 3y = 0 \end{cases}$

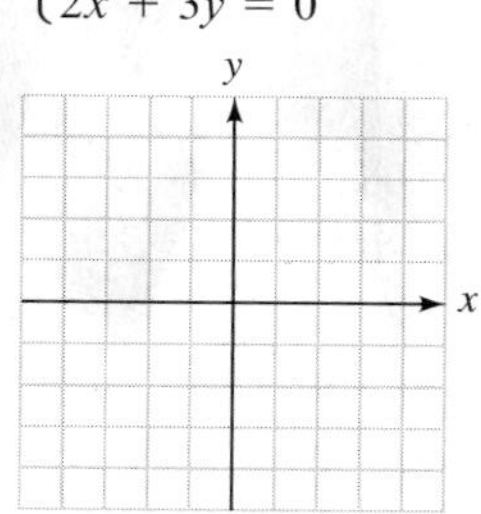

25. $\begin{cases} y = 3 \\ x = 2 \end{cases}$

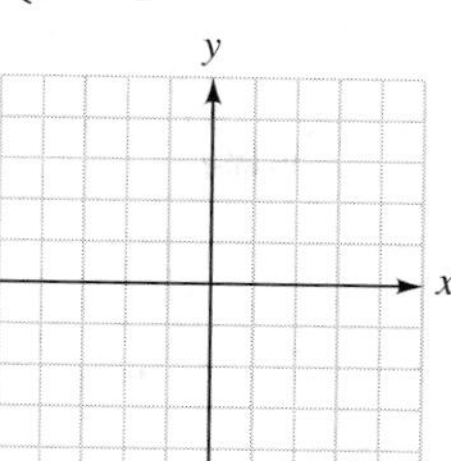

26. 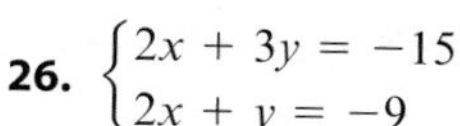
$\begin{cases} 2x + 3y = -15 \\ 2x + y = -9 \end{cases}$

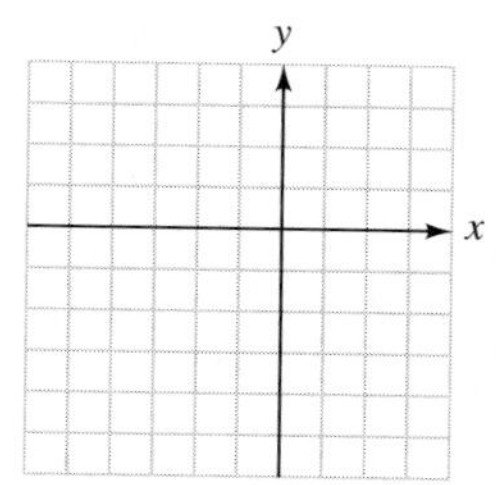

Solve each system by graphing. State whether the system is inconsistent or if the equations are dependent. Write each solution set. **See Examples 2–3. (Objectives 2–3)**

27. $\begin{cases} 3x = 5 - 2y \\ 3x + 2y = 7 \end{cases}$

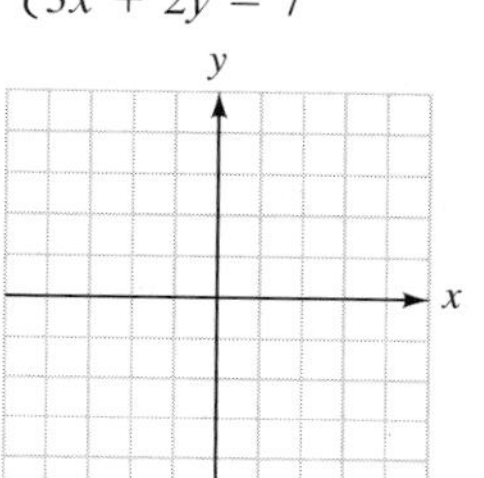

28. 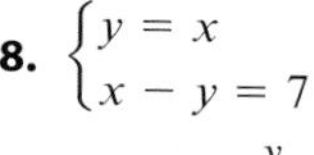
$\begin{cases} y = x \\ x - y = 7 \end{cases}$

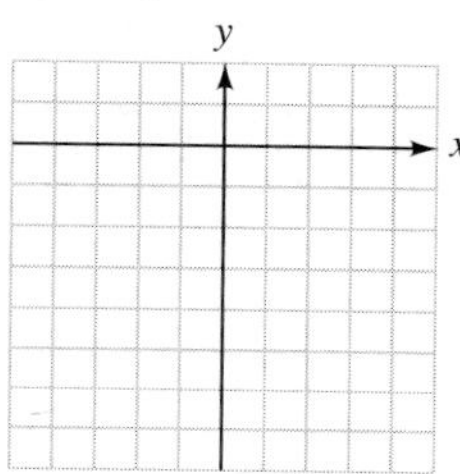

29. $\begin{cases} x = 3 - 2y \\ 2x + 4y = 6 \end{cases}$

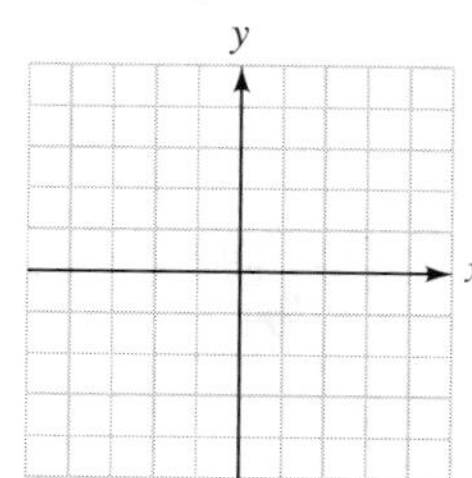

30. $\begin{cases} x = y \\ y - x = 0 \end{cases}$

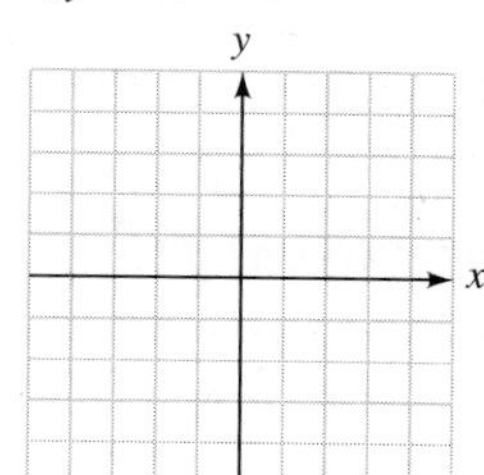

31. $\begin{cases} x = 2y - 8 \\ y = \frac{1}{2}x - 5 \end{cases}$

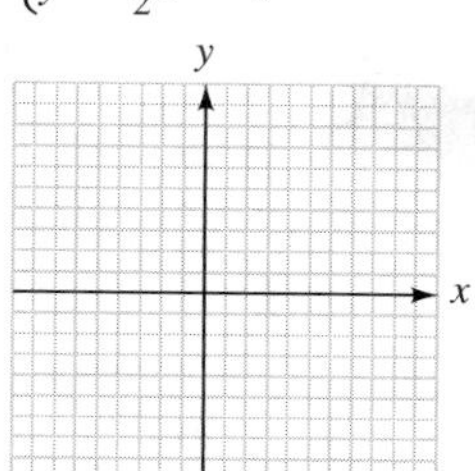

32. $\begin{cases} 6x + 3y = 9 \\ y + 2x = 3 \end{cases}$

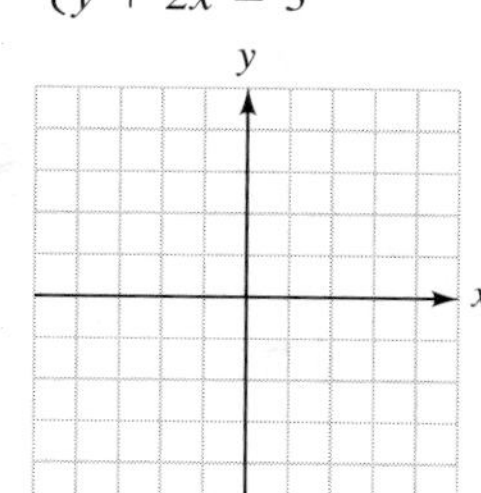

Solve each system by graphing. **See Example 4. (Objective 4)**

33. $\begin{cases} x = 2 \\ y = \dfrac{4 - x}{2} \end{cases}$

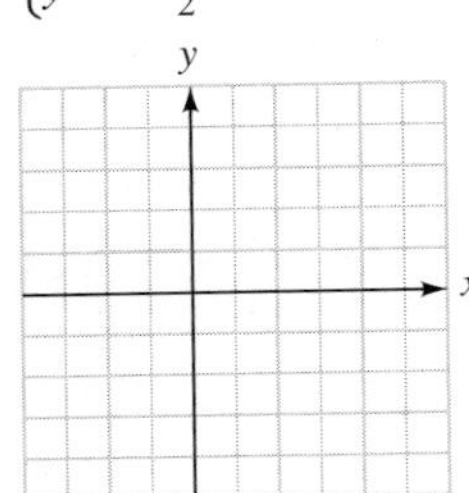

34. $\begin{cases} y = -2 \\ x = \dfrac{4 + 3y}{2} \end{cases}$

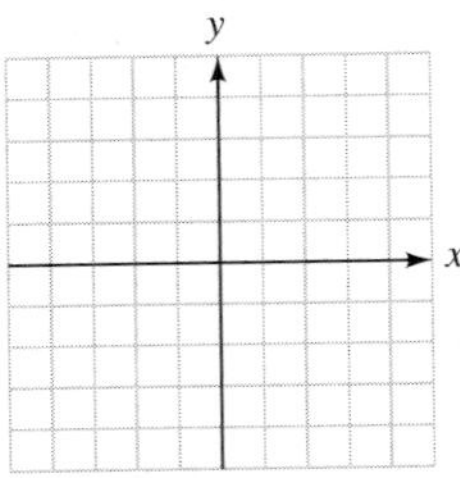

35. $\begin{cases} \frac{5}{2}x + y = \frac{1}{2} \\ 2x - \frac{3}{2}y = 5 \end{cases}$

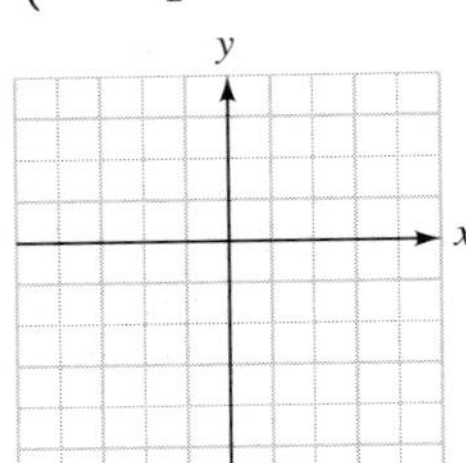

36. $\begin{cases} x = \dfrac{5y - 4}{2} \\ x - \frac{5}{3}y + \frac{1}{3} = 0 \end{cases}$

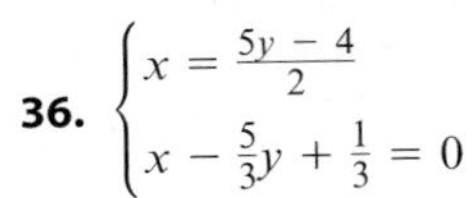

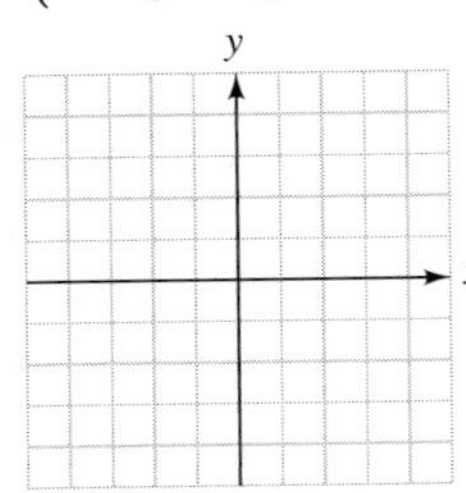

Solve each equation graphically. **See Example 4.**

37. $\frac{2}{3}b - 3 = -15$

38. $-7(a - 2) = 8$

39. $2(2x + 1) = 3x + 15$

40. $2(a - 5) = 3a + 1$

ADDITIONAL PRACTICE

Solve each system by graphing.

41. $\begin{cases} x = \dfrac{11 - 2y}{3} \\ y = \dfrac{11 - 6x}{4} \end{cases}$

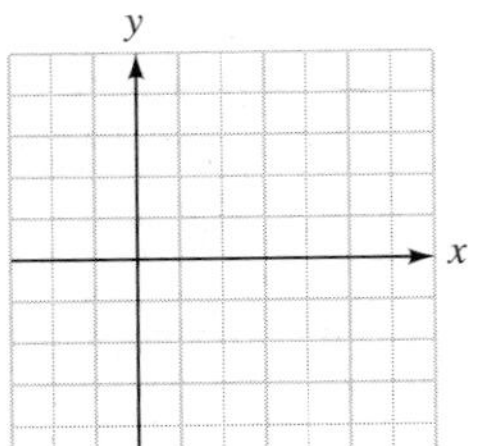

42. $\begin{cases} x = \dfrac{1 - 3y}{4} \\ y = \dfrac{12 + 3x}{2} \end{cases}$

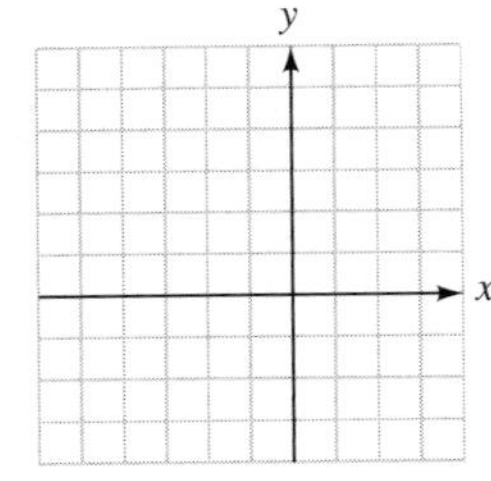

43. $\begin{cases} \frac{5}{2}x + 3y = 6 \\ y = \dfrac{24 - 10x}{12} \end{cases}$

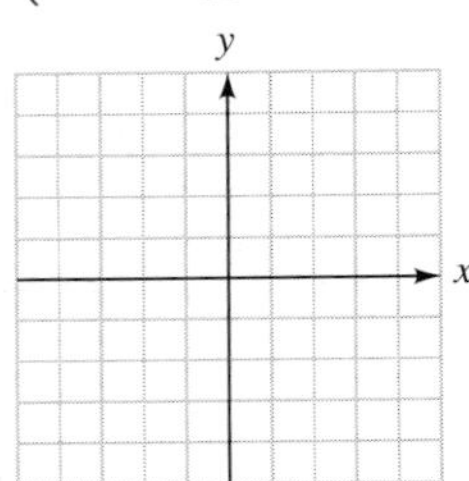

44. $\begin{cases} 2x = 5y - 11 \\ 3x = 2y \end{cases}$

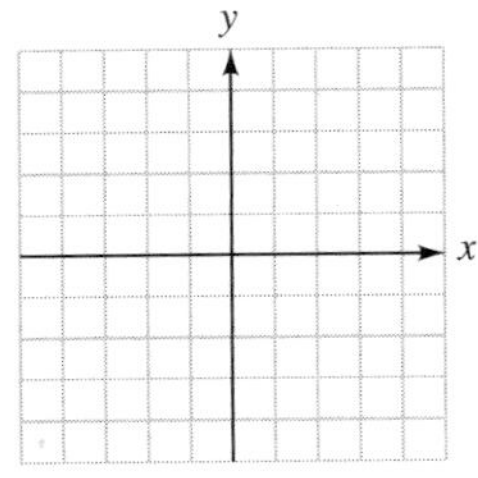

45. $\begin{cases} x - 3y = 9 \\ 3x - 27 = 9y \end{cases}$

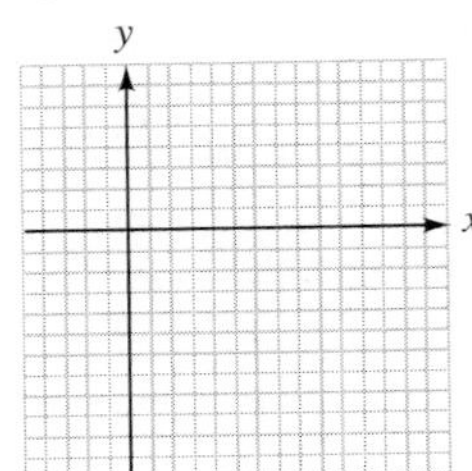

46. $\begin{cases} 8x - 2y = 7 \\ y = 4x + 3 \end{cases}$

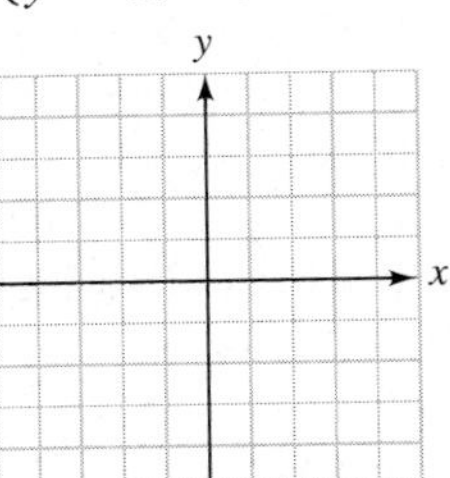

Use a graphing calculator to solve each system. Give the exact answer.

47. $\begin{cases} x = 13 - 4y \\ 3x = 4 + 2y \end{cases}$

48. $\begin{cases} 3x = 7 - 2y \\ 2x = 2 + 4y \end{cases}$

49. $\begin{cases} x = -\frac{3}{2}y \\ x = \frac{3}{2}y - 2 \end{cases}$

50. $\begin{cases} x = \dfrac{3y - 1}{4} \\ y = \dfrac{4 - 8x}{3} \end{cases}$

Use a graphing calculator to solve each system. Give all answers to the nearest hundredth.

51. $\begin{cases} y = 3.2x - 1.5 \\ y = -2.7x - 3.7 \end{cases}$

52. $\begin{cases} y = -0.45x + 5 \\ y = 5.55x - 13.7 \end{cases}$

53. $\begin{cases} 1.7x + 2.3y = 3.2 \\ y = 0.25x + 8.95 \end{cases}$

54. $\begin{cases} 2.75x = 12.9y - 3.79 \\ 7.1x - y = 35.76 \end{cases}$

APPLICATIONS

55. Retailing The cost of manufacturing one type of camera and the revenue from the sale of those cameras are shown in the illustration.

a. From the illustration, find the cost of manufacturing 15,000 cameras.

b. From the illustration, find the revenue obtained by selling 20,000 cameras.

c. For what number of cameras will the revenue equal the cost?

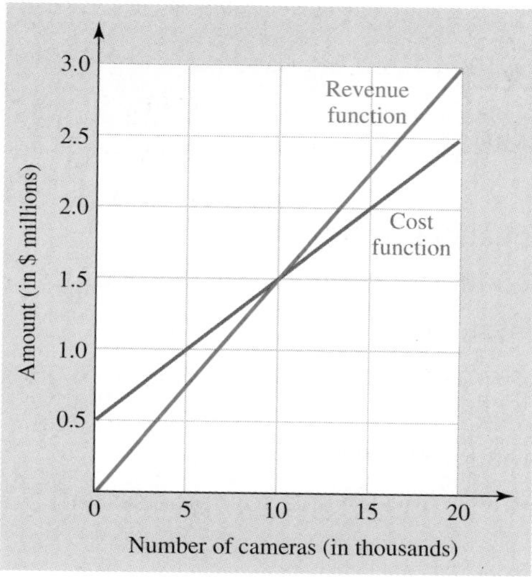

56. Food service

a. Estimate the point of intersection of the two graphs shown in the illustration. Express your answer in the form (year, number of meals).

b. What information about dining out does the point of intersection give?

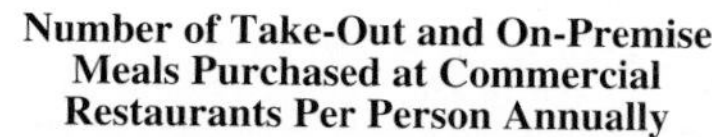

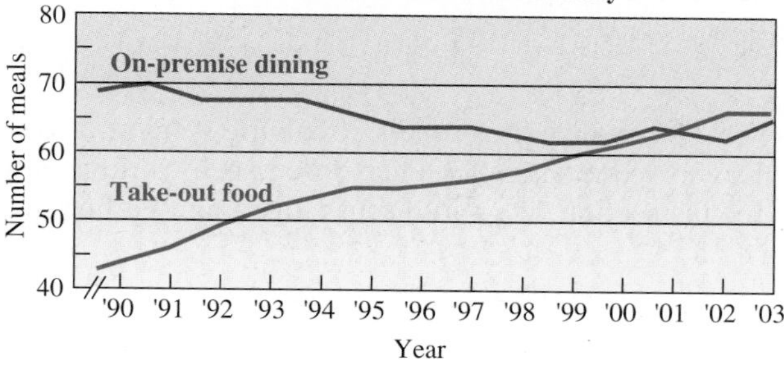

57. Navigation Two ships are sailing on the same coordinate system. One ship is following a course described by $2x + 3y = 6$, and the other is following a course described by $2x - 3y = 9$.

a. Is there a possibility of a collision?

b. Find the coordinates of the danger point.

c. Is a collision a certainty?

58. Navigation Two airplanes are flying at the same altitude and in the same coordinate system. One plane is following a course described by $y = \frac{2}{5}x - 2$, and the other is following a course described by $x = \frac{5y + 7}{2}$. Is there a possibility of a collision?

59. Car repairs Smith Chevrolet charges $50 per hour for labor on car repairs. Lopez Ford charges a diagnosis fee of $30 plus $40 per hour for labor. If the labor on an engine repair costs the same at either shop, how long does the repair take?

60. Landscaping A landscaper installs some trees and bushes at a bank. He installs 25 plants for a total cost of $1,500. How many trees ($t$) and how many bushes ($b$) did he install if each tree costs $100 and each bush costs $50?

61. Cost and revenue The function $C(x) = 200x + 400$ gives the cost for a college to offer x sections of an introductory class in CPR (cardiopulmonary resuscitation). The function $R(x) = 280x$ gives the amount of revenue the college brings in when offering x sections of CPR. Find the point where the cost equals the revenue by graphing each function on the same coordinate system.

62. In Exercise 61, how many sections does the college need to offer to make a profit on the CPR training course?

WRITING ABOUT MATH

63. Explain how to solve a system of two equations in two variables.

64. Can a system of two equations in two variables have exactly two solutions? Why or why not?

SOMETHING TO THINK ABOUT

65. Form an independent system of equations with a solution of $(-5, 2)$.

66. Form a dependent system of equations with a solution of $(-5, 2)$.

SECTION 3.2 Solving Systems of Two Linear Equations by Substitution and Elimination (Addition)

Objectives

1. Solve a system of two linear equations by substitution.
2. Solve a system of two linear equations by elimination (addition).
3. Identify an inconsistent linear system.
4. Identify a dependent linear system and express the answer as a general ordered pair.
5. Write a repeating decimal in fractional form.
6. Solve an application problem by setting up and solving a system of two linear equations.

Getting Ready

Remove parentheses.

1. $3(2x - 7)$ **2.** $-4(3x + 5)$

Substitute $x - 3$ *for* y *and remove parentheses.*

3. $3y$ **4.** $-2(y + 2)$

Add the left and right sides of the equations in each system.

5. $\begin{cases} 2x + 5y = 7 \\ 5x - 5y = 8 \end{cases}$ **6.** $\begin{cases} 3a - 4b = 12 \\ -3a - 5b = 15 \end{cases}$

The graphing method provides a way to visualize the process of solving systems of equations. However, it cannot be used to solve systems of higher order, such as three equations, each with three variables. In this section, we will discuss algebraic methods that enable us to solve such systems.

Solve a system of two linear equations by substitution.

To solve a system of two linear equations (each with two variables) by substitution, we use the following steps.

The Substitution Method

1. If necessary, solve one equation for one of its variables, preferably a variable with a coefficient of 1.
2. Substitute the resulting expression for the variable obtained in Step 1 into the other equation and solve that equation.
3. Find the value of the other variable by substituting the value of the variable found in Step 2 into any equation containing both variables.
4. State the solution.
5. Check the solution in both of the original equations.

EXAMPLE 1 Solve the system by substitution: $\begin{cases} 4x + y = 13 \\ -2x + 3y = -17 \end{cases}$.

Solution **Step 1:** We solve the first equation for y, because y has a coefficient of 1 and no fractions are introduced.

$$4x + y = 13$$

(1) $$y = -4x + 13 \quad \text{Subtract } 4x \text{ from both sides.}$$

Step 2: We then substitute $-4x + 13$ for y in the second equation of the system and solve for x.

$$\begin{aligned} -2x + 3y &= -17 && \\ -2x + 3(\mathbf{-4x + 13}) &= -17 && \text{Substitute } -4x + 13 \text{ for } y. \\ -2x - 12x + 39 &= -17 && \text{Use the distributive property to remove parentheses.} \\ -14x &= -56 && \text{Combine like terms and subtract 39 from both sides.} \\ x &= 4 && \text{Divide both sides by } -14. \end{aligned}$$

Step 3: To find y, we substitute 4 for x in Equation 1 and simplify:

$$\begin{aligned} y &= -4x + 13 \\ y &= -4(\mathbf{4}) + 13 \quad \text{Substitute 4 for } x. \\ &= -3 \end{aligned}$$

Step 4: The solution is $x = 4$ and $y = -3$, or $(4, -3)$. The graphs of these two equations would intersect at the point with coordinates $(4, -3)$.

Step 5: To verify that this solution satisfies both equations, we substitute $x = 4$ and $y = -3$ into each equation in the system and simplify.

$$\begin{aligned} 4x + y &= 13 \\ 4(\mathbf{4}) + (\mathbf{-3}) &\stackrel{?}{=} 13 \\ 16 - 3 &\stackrel{?}{=} 13 \\ 13 &= 13 \end{aligned} \qquad \begin{aligned} -2x + 3y &= -17 \\ -2(\mathbf{4}) + 3(\mathbf{-3}) &\stackrel{?}{=} -17 \\ -8 - 9 &\stackrel{?}{=} -17 \\ -17 &= -17 \end{aligned}$$

Since the ordered pair $(4, -3)$ satisfies both equations of the system, the solution checks.

SELF CHECK 1 Solve the system by substitution: $\begin{cases} x + 3y = 9 \\ 2x - y = -10 \end{cases}$.

EXAMPLE 2 Solve the system by substitution: $\begin{cases} \frac{4}{3}x + \frac{1}{2}y = -\frac{2}{3} \\ \frac{1}{2}x + \frac{2}{3}y = \frac{5}{3} \end{cases}$.

Solution First we find an equivalent system without fractions by multiplying both sides of each equation by 6.

(1) (2) $$\begin{cases} 8x + 3y = -4 \\ 3x + 4y = 10 \end{cases}$$

Because no variable in either equation has a coefficient of 1, it is impossible to avoid fractions when solving for a variable. We solve Equation 2 for x.

$$3x + 4y = 10$$

$$3x = -4y + 10 \quad \text{Subtract } 4y \text{ from both sides.}$$

$$(3) \qquad x = -\frac{4}{3}y + \frac{10}{3} \quad \text{Divide both sides by 3.}$$

We then substitute $-\frac{4}{3}y + \frac{10}{3}$ for x in Equation 1 and solve for y.

$$8x + 3y = -4$$

$$8\left(-\frac{4}{3}y + \frac{10}{3}\right) + 3y = -4 \quad \text{Substitute } -\tfrac{4}{3}y + \tfrac{10}{3} \text{ for } x.$$

$$-\frac{32}{3}y + \frac{80}{3} + 3y = -4 \quad \text{Use the distributive property to remove parentheses.}$$

$$-32y + 80 + 9y = -12 \quad \text{Multiply both sides by 3.}$$

$$-23y = -92 \quad \text{Combine like terms and subtract 80 from both sides.}$$

$$y = 4 \quad \text{Divide both sides by } -23.$$

We can find x by substituting 4 for y in Equation 3 and simplifying:

$$x = -\frac{4}{3}y + \frac{10}{3}$$

$$= -\frac{4}{3}(4) + \frac{10}{3} \quad \text{Substitute 4 for } y.$$

$$= -\frac{6}{3} \quad -\tfrac{16}{3} + \tfrac{10}{3} = -\tfrac{6}{3}$$

$$= -2$$

The solution is the ordered pair $(-2, 4)$. Verify that this solution checks.

SELF CHECK 2 Solve the system by substitution: $\begin{cases} \frac{2}{3}x + \frac{1}{2}y = 1 \\ \frac{1}{3}x - \frac{3}{2}y = 4 \end{cases}$.

2 Solve a system of two linear equations by elimination (addition).

In the elimination (addition) method, we combine the equations of the system in a way that will eliminate the terms involving one of the variables.

The Elimination (Addition) Method

1. Write both equations of the system in general form.
2. If necessary, multiply the terms of one or both of the equations by constants chosen to make the coefficients of one of the variables differ only in sign.
3. Add the equations and solve the resulting equation, if possible.
4. Substitute the value obtained in Step 3 into either of the original equations and solve for the remaining variable.
5. State the solution obtained in Steps 3 and 4.
6. Check the solution in both of the original equations.

EXAMPLE 3 Solve the system by elimination: $\begin{cases} 4x + y = 13 \\ -2x + 3y = -17 \end{cases}$.

Solution **Step 1:** This is the system in Example 1. Since both equations are already written in general form, Step 1 is unnecessary.

Step 2: To solve the system by elimination, we multiply the second equation by 2 to make the coefficients of x differ only in sign.

(1) $\begin{cases} 4x + y = 13 \\ -4x + 6y = -34 \end{cases}$

Step 3: When we add the equations, the terms involving x are eliminated because $4x$ and $-4x$ are additive inverses and their sum is 0. We will obtain

$$7y = -21$$

$$y = -3 \quad \text{Divide both sides by 7.}$$

Step 4: To find x, we substitute -3 for y in either of the original equations and solve for x. If we use Equation 1, we have

$$4x + y = 13$$

$$4x + (-3) = 13 \quad \text{Substitute } -3 \text{ for } y.$$

$$4x = 16 \quad \text{Add 3 to both sides.}$$

$$x = 4 \quad \text{Divide both sides by 4.}$$

Step 5: The solution is $x = 4$ and $y = -3$, or $(4, -3)$.

Step 6: The check was completed in Example 1.

SELF CHECK 3 Solve the system by elimination: $\begin{cases} 3x + 2y = 0 \\ 2x - y = -7 \end{cases}$.

EXAMPLE 4 Solve the system by elimination: $\begin{cases} \frac{4}{3}x + \frac{1}{2}y = -\frac{2}{3} \\ \frac{1}{2}x + \frac{2}{3}y = \frac{5}{3} \end{cases}$.

Solution This is the system in Example 2. To solve it by elimination, we find an equivalent system with no fractions by multiplying both sides of each equation by 6 to obtain

(1) $8x + 3y = -4$
(2) $3x + 4y = 10$

We can solve for x by eliminating the terms involving y. To do so, we multiply both sides of Equation 1 by 4 and both sides of Equation 2 by -3 to get

$$\begin{cases} 32x + 12y = -16 \\ -9x - 12y = -30 \end{cases}$$

When these equations are added, the y-terms are eliminated and we get the result

$$23x = -46$$

$$x = -2 \quad \text{Divide both sides by 23.}$$

To find y, we substitute -2 for x in either Equation 1 or Equation 2. If we substitute -2 for x in Equation 2, we get

$$\begin{aligned} 3x + 4y &= 10 \\ 3(-2) + 4y &= 10 &\quad& \text{Substitute } -2 \text{ for } x. \\ -6 + 4y &= 10 && \text{Simplify.} \\ 4y &= 16 && \text{Add 6 to both sides.} \\ y &= 4 && \text{Divide both sides by 4.} \end{aligned}$$

The solution is $(-2, 4)$.

SELF CHECK 4 Solve the system by elimination: $\begin{cases} \frac{4}{3}x + \frac{1}{2}y = -3 \\ \frac{1}{2}x + \frac{2}{3}y = -\frac{1}{6} \end{cases}$.

3 Identify an inconsistent linear system.

In the next example, there are no solutions.

EXAMPLE 5 Solve the system using any method: $\begin{cases} y = 2x + 4 \\ 8x - 4y = 7 \end{cases}$.

Solution Because the first equation is already solved for y, we use the substitution method.

$$\begin{aligned} 8x - 4y &= 7 \\ 8x - 4(2x + 4) &= 7 &\quad& \text{Substitute } 2x + 4 \text{ for } y. \end{aligned}$$

We then solve this equation for x:

$$\begin{aligned} 8x - 8x - 16 &= 7 &\quad& \text{Use the distributive property to remove parentheses.} \\ -16 &= 7 && \text{Combine like terms.} \end{aligned}$$

This impossible result shows that the equations in the system are independent, but that the system is inconsistent. Since the system has no solution, its solution set is $\varnothing$. If the equations of this system were graphed, the line graphs would be parallel.

SELF CHECK 5 Solve the system using any method: $\begin{cases} x = -\frac{5}{2}y + 5 \\ y = -\frac{2}{5}x + 5 \end{cases}$.

4 Identify a dependent linear system and express the answer as a general ordered pair.

In the next example, there are infinitely many solutions.

EXAMPLE 6 Solve the system using any method: $\begin{cases} 4x + 6y = 12 \\ -2x - 3y = -6 \end{cases}$.

Solution Since the equations are written in general form, we use the elimination (addition) method. To eliminate the x-terms when adding the equations, we multiply both sides of the second equation by 2 to obtain

$$\begin{cases} 4x + 6y = 12 \\ -4x - 6y = -12 \end{cases}$$

After adding the left and right sides, we get

$$0x + 0y = 0$$
$$0 = 0$$

Here, both the x- and y-terms are eliminated. The true statement $0 = 0$ shows that the equations in this system are dependent and that the system is consistent.

Note that the equations of the system are equivalent, because when the second equation is multiplied by -2, it becomes the first equation. The line graphs of these equations would coincide. Since any ordered pair that satisfies one of the equations also satisfies the other, there are infinitely many solutions.

To find a general solution, we can solve either equation of the system to obtain $y = -\frac{2}{3}x + 2$. Substituting $-\frac{2}{3}x + 2$ for y in the ordered pair (x, y), we obtain $\left(x, -\frac{2}{3}x + 2\right)$.

To find some solutions, we can substitute 0, 3, and 6 for x in the general solution to obtain $(0, 2)$, $(-3, 0)$, and $(6, -2)$.

Verify that each solution satisfies both equations.

SELF CHECK 6 Solve the system using any method: $\begin{cases} x = -\frac{5}{2}y + 5 \\ y = -\frac{2}{5}x + 2 \end{cases}$.

5 Write a repeating decimal in fractional form.

We have seen that to change a fraction into decimal form, we can divide its numerator by its denominator. The result is often a repeating decimal. By using systems of equations, we can change a repeating decimal into fractional form.

For example, to write $0.2\overline{54}$ as a fraction, we note that the decimal has a repeating block of two digits and then form an equation by setting x equal to the decimal.

(1) $x = 0.2545454\ldots$

We then form another equation by multiplying both sides of Equation 1 by 10^2.

(2) $100x = 25.4545454\ldots$ $10^2 = 100$

We can *subtract* each side of Equation 1 from the corresponding side of Equation 2 to obtain

$$\begin{aligned} 100x &= 25.4\ 54\ 54\ 54\ldots \\ x &= \ \ 0.2\ 54\ 54\ 54\ldots \\ \hline 99x &= 25.2 \end{aligned}$$

Finally, we solve $99x = 25.2$ for x and simplify the fraction.

$$x = \frac{25.2}{99} = \frac{25.2 \cdot \mathbf{10}}{99 \cdot \mathbf{10}} = \frac{252}{990} = \frac{18 \cdot 14}{18 \cdot 55} = \frac{14}{55}$$

We can use a calculator to verify that the decimal representation of $\frac{14}{55}$ is $0.2\overline{54}$.

The key step in the solution was multiplying both sides of Equation 1 by 10^2. If there had been n digits in the repeating block of the decimal, we would have multiplied both sides of Equation 1 by 10^n.

6 Solve an application problem by setting up and solving a system of two linear equations.

To solve problems using two variables, we follow the same problem-solving strategy discussed in Chapter 1, except that we use two variables and form two equations instead of one.

EXAMPLE 7 **RETAIL SALES** A store advertises two types of cell phones, one selling for \$67 and the other for \$100. If the receipts from the sale of 36 phones totaled \$2,940, how many of each type were sold?

Analyze the problem: We can let x represent the number of phones sold for \$67 and let y represent the number of phones sold for \$100.

Form two equations: Because a total of 36 phones were sold, we can form the equation

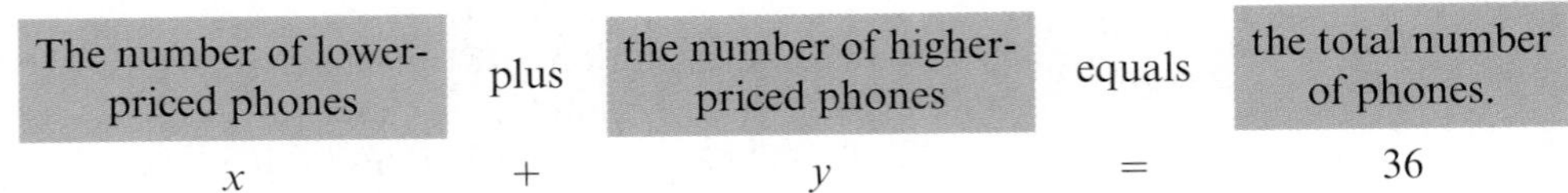

The number of lower-priced phones	plus	the number of higher-priced phones	equals	the total number of phones.
x	$+$	y	$=$	36

We know that the receipts for the sale of x of the \$67 phones will be \$67x and that the receipts for the sale of y of the \$100 phones will be \$100y. Since the sum of these receipts is \$2,940, the second equation is

The value of the lower-priced phones	plus	the value of the higher-priced phones	equals	the total receipts.
$67x$	$+$	$100y$	$=$	2,940

Solve the system: To find out how many of each type of phone were sold, we must solve the system

$$\begin{cases} x + y = 36 & (1) \\ 67x + 100y = 2{,}940 & (2) \end{cases}$$

We multiply both sides of Equation 1 by -100, add the resulting equation to Equation 2, and solve for x:

$$\begin{aligned} -100x - 100y &= -3{,}600 \\ 67x + 100y &= 2{,}940 \\ \hline -33x &= -660 \\ x &= 20 \quad \text{Divide both sides by } -33. \end{aligned}$$

To find y, we substitute 20 for x in Equation 1 and solve for y:

$$\begin{aligned} x + y &= 36 \\ \mathbf{20} + y &= 36 \quad \text{Substitute 20 for } x. \\ y &= 16 \quad \text{Subtract 20 from both sides.} \end{aligned}$$

State the conclusion: The store sold 20 of the \$67 phones and 16 of the \$100 phones.

Check the result: If 20 of one type were sold and 16 of the other type were sold, a total of 36 phones were sold.

Since the value of the lower-priced phones is 20(\$67) = \$1,340 and the value of the higher-priced phones is 16(\$100) = \$1,600, the total receipts are \$2,940.

EXAMPLE 8 **MIXING SOLUTIONS** How many ounces of a 5% saline solution and how many ounces of a 20% saline solution must be mixed together to obtain 50 ounces of a 15% saline solution?

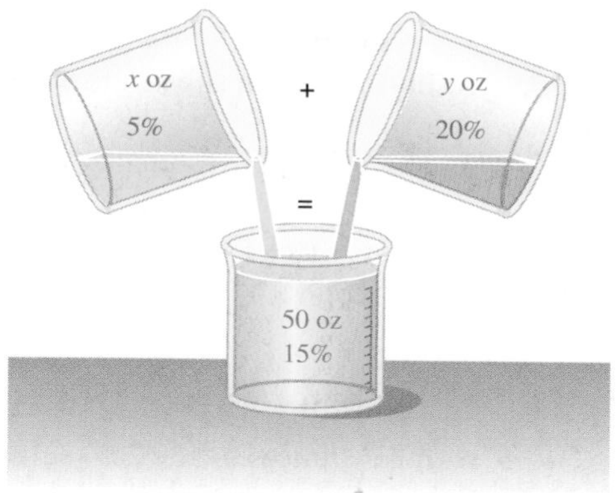

Figure 3-9

Analyze the problem: To find how many ounces of each type of saline solution should be mixed, we can let x represent the number of ounces of the 5% solution and y represent the number of ounces of the 20% solution.

Form two equations: Because a total of 50 ounces are needed, one of the equations will be

The number of ounces of 5% solution	plus	the number of ounces of 20% solution	equals	the total number of ounces in the mixture.
x	$+$	y	$=$	50

The amount of salt in the x ounces of 5% solution is $0.05x$, and the amount of salt in the y ounces of 20% solution is $0.20y$. The amount of salt in the 50 ounces of 15% solution will be $0.15(50)$. (See Figure 3-9.) This gives the equation

The salt in the 5% solution	plus	the salt in the 20% solution	equals	the salt in the mixture.
$0.05x$	$+$	$0.20y$	$=$	$0.15(50)$

Solve the system: To find out how many ounces of each are needed, we solve the following system:

$$\begin{cases} x + y = 50 & (1) \\ 0.05x + 0.20y = 7.5 & (2) \end{cases} \quad 0.15(50) = 7.5$$

To solve this system by substitution, we can solve Equation 1 for y

$$x + y = 50$$
$$(3) \quad y = 50 - x \quad \text{Subtract } x \text{ from both sides.}$$

and then substitute $50 - x$ for y in Equation 2.

$$0.05x + 0.20y = 7.5$$
$$0.05x + 0.20(\mathbf{50 - x}) = 7.5 \quad \text{Substitute } 50 - x \text{ for } y.$$
$$5x + 20(50 - x) = 750 \quad \text{Multiply both sides by 100.}$$
$$5x + 1{,}000 - 20x = 750 \quad \text{Use the distributive property to remove parentheses.}$$
$$-15x = -250 \quad \text{Combine like terms and subtract 1,000 from both sides.}$$

$$x = \frac{-250}{-15} \quad \text{Divide both sides by } -15.$$

$$x = \frac{50}{3} \quad \text{Simplify } \tfrac{-250}{-15}.$$

To find y, we can substitute $\frac{50}{3}$ for x in Equation 3:

$$\begin{aligned} y &= 50 - x \\ &= 50 - \frac{50}{3} \quad \text{Substitute } \tfrac{50}{3} \text{ for } x. \\ &= \frac{100}{3} \end{aligned}$$

State the conclusion — To obtain 50 ounces of a 15% solution, we must mix $16\frac{2}{3}$ ounces of the 5% solution with $33\frac{1}{3}$ ounces of the 20% solution.

Check the result — We note that $16\frac{2}{3}$ ounces of solution plus $33\frac{1}{3}$ ounces of solution equals the required 50 ounces of solution. We also note that 5% of $16\frac{2}{3} \approx 0.83$, and 20% of $33\frac{1}{3} \approx 6.67$, giving a total of 7.5, which is 15% of 50.

Running a machine involves both *setup costs* and *unit costs.* Setup costs include the cost of preparing a machine to do a certain job. Unit costs depend on the number of items to be manufactured, including costs of raw materials and labor.

Suppose that a certain machine has a setup cost of \$600 and a unit cost of \$3 per item. If x items will be manufactured using this machine, the cost will be

$$\text{Cost} = 600 + 3x \quad \text{Cost} = \text{setup cost} + \text{unit cost} \times \text{the number of items}$$

Furthermore, suppose that a larger and more efficient machine has a setup cost of \$800 and a unit cost of \$2 per item. The cost of manufacturing x items using this machine is

$$\text{Cost on larger machine} = 800 + 2x$$

The *break point* is the number of units x that need to be manufactured to make the cost the same using either machine. It can be found by setting the two costs equal to each other and solving for x.

$$\begin{aligned} 600 + 3x &= 800 + 2x \\ x &= 200 \quad \text{Subtract 600 and } 2x \text{ from both sides.} \end{aligned}$$

The break point is 200 units, because the cost using either machine is \$1,200 when $x = 200$.

$$\begin{aligned} \text{Cost on small machine} &= 600 + 3x \\ &= 600 + 3(\mathbf{200}) \\ &= 600 + 600 \\ &= 1{,}200 \end{aligned} \qquad \begin{aligned} \text{Cost on large machine} &= 800 + 2x \\ &= 800 + 2(\mathbf{200}) \\ &= 800 + 400 \\ &= 1{,}200 \end{aligned}$$

EXAMPLE 9 **BREAK-POINT ANALYSIS** One machine has a setup cost of \$400 and a unit cost of \$1.50, and another machine has a setup cost of \$500 and a unit cost of \$1.25. Find the break point.

Analyze the problem The cost of manufacturing x units using machine 1 is x times \$1.50, plus the setup cost of \$400. The cost of manufacturing x units using machine 2 is x times \$1.25, plus the setup cost of \$500. The break point occurs when the costs are equal.

Form two equations The cost C_1 using machine 1 is

The cost of using machine 1	equals	the cost of manufacturing x units at \$1.50 per unit	plus	the setup cost.
C_1	$=$	$1.5x$	$+$	400

The cost C_2 using machine 2 is

The cost of using machine 2	equals	the cost of manufacturing x units at \$1.25 per unit	plus	the setup cost.
C_2	$=$	$1.25x$	$+$	500

Solve the system To find the break point, we must solve the system $\begin{cases} C_1 = 1.5x + 400 \\ C_2 = 1.25x + 500 \end{cases}$. Since the break point occurs when $C_1 = C_2$, we can substitute $1.5x + 400$ for C_2 to get

$$
\begin{aligned}
1.5x + 400 &= 1.25x + 500 \\
1.5x &= 1.25x + 100 && \text{Subtract 400 from both sides.} \\
0.25x &= 100 && \text{Subtract } 1.25x \text{ from both sides.} \\
x &= 400 && \text{Divide both sides by 0.25.}
\end{aligned}
$$

State the conclusion The break point is 400 units.

Check the result For 400 units, the cost using machine 1 is $400 + 1.5(400) = 400 + 600 = 1{,}000$. The cost using machine 2 is $500 + 1.25(400) = 500 + 500 = 1{,}000$. Since the costs are equal, the break point is 400 units.

A *parallelogram* is a four-sided figure with its opposite sides parallel. (See Figure 3-10(a).) Here are some important facts about parallelograms.

1. Opposite sides of a parallelogram have the same measure.
2. Opposite angles of a parallelogram have the same measure.
3. Consecutive angles of a parallelogram are supplementary.
4. A diagonal of a parallelogram (see Figure 3-10(b)) divides the parallelogram into two *congruent triangles*—triangles with the same shape and same area.
5. In Figure 3-10(b), $\angle 1$ and $\angle 2$, and $\angle 3$ and $\angle 4$, are called pairs of *alternate interior angles.* When a diagonal intersects two parallel sides of a parallelogram, all pairs of alternate interior angles have the same measure.

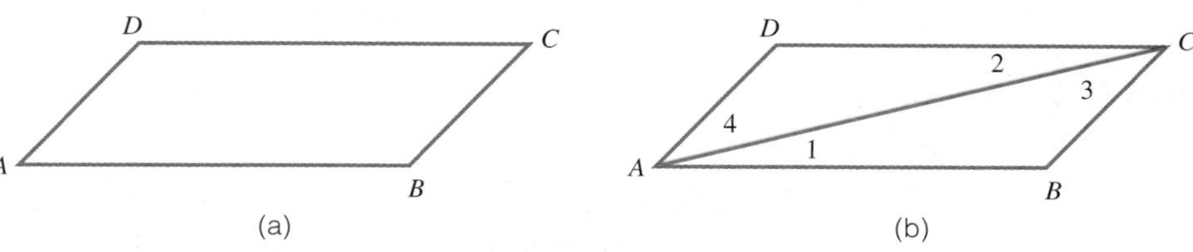

Figure 3-10

EXAMPLE 10 **PARALLELOGRAMS** Refer to the parallelogram shown in Figure 3-11 and find the values of x and y.

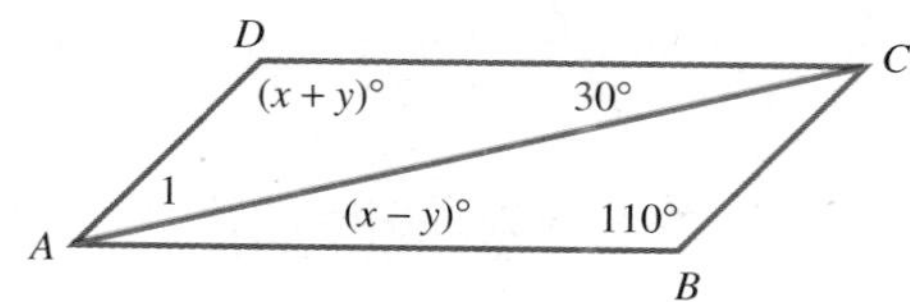

Figure 3-11

Solution Since diagonal AC intersects two parallel sides, the alternate interior angles that are formed have the same measure. Thus, $(x - y)° = 30°$. Since opposite angles of a parallelogram have the same measure, we know that $(x + y)° = 110°$. We can form the following system of equations and solve it by addition.

$$\begin{array}{l} (1) \\ (2) \end{array} \quad \begin{cases} x - y = 30 \\ x + y = 110 \end{cases}$$

$2x = 140$ Add Equations 1 and 2.

$x = 70$ Divide both sides by 2.

We can substitute 70 for x in Equation 2 and solve for y.

$x + y = 110$

$70 + y = 110$ Substitute 70 for x.

$y = 40$ Subtract 70 from both sides.

Thus, $x = 70$ and $y = 40$.

SELF CHECK ANSWERS **1.** $(-3, 4)$ **2.** $(3, -2)$ **3.** $(-2, 3)$ **4.** $(-3, 2)$ **5.** $\varnothing$ **6.** $\left(x, -\frac{2}{5}x + 2\right)$

NOW TRY THIS

1. Use substitution to solve the system $\begin{cases} t = r + 1 \\ r + s + t = 0. \\ r + s = -2 \end{cases}$

3.2 EXERCISES

WARM-UPS *Solve each system for x.*

1. $\begin{cases} y = 2x \\ x + y = 6 \end{cases}$

2. $\begin{cases} y = -x \\ 2x + y = 4 \end{cases}$

3. $\begin{cases} x - y = 6 \\ x + y = 2 \end{cases}$

4. $\begin{cases} x + y = 4 \\ 2x - y = 5 \end{cases}$

REVIEW *Simplify each expression. Write all answers without using negative exponents.*

5. $(a^2a^3)^2(a^4a^2)^2$

6. $\left(\dfrac{a^2b^3c^4d}{ab^2c^3d^4}\right)^{-3}$

7. $\left(\dfrac{-3x^3y^4}{x^{-5}y^3}\right)^{-4}$

8. $\dfrac{3t^0 - 4t^0 + 5}{5t^0 + 2t^0}$

VOCABULARY AND CONCEPTS *Fill in the blanks.*

9. Running a machine involves both ______ costs and ______ costs.

10. The ______ point is the number of units that need to be manufactured to make the cost the same on either of two machines.

11. A ____________ is a four-sided figure with both pairs of opposite sides parallel.

12. __________ sides of a parallelogram have the same length.

13. __________ angles of a parallelogram have the same measure.

14. __________ angles of a parallelogram are supplementary.

GUIDED PRACTICE

Solve each system by substitution. See Examples 1–2. (Objective 1)

15. $\begin{cases} y = x \\ x + y = 4 \end{cases}$

16. $\begin{cases} y = x + 2 \\ x + 2y = 16 \end{cases}$

17. $\begin{cases} x - y = 2 \\ 2x + y = 13 \end{cases}$

18. $\begin{cases} x - y = -4 \\ 3x - 2y = -5 \end{cases}$

19. $\begin{cases} x + 2y = 6 \\ 3x - y = -10 \end{cases}$

20. $\begin{cases} 2x - y = -21 \\ 4x + 5y = 7 \end{cases}$

21. $\begin{cases} \frac{x}{2} + \frac{y}{2} = 6 \\ \frac{x}{2} - \frac{y}{2} = -2 \end{cases}$

22. $\begin{cases} x = \frac{2}{3}y \\ y = 4x + 5 \end{cases}$

Solve each system by elimination (addition). See Examples 3–4. (Objective 2)

23. $\begin{cases} x - y = 3 \\ x + y = 7 \end{cases}$

24. $\begin{cases} x + y = 1 \\ x - y = 7 \end{cases}$

25. $\begin{cases} 2x + y = -10 \\ 2x - y = -6 \end{cases}$

26. $\begin{cases} x + 2y = -9 \\ x - 2y = -1 \end{cases}$

27. $\begin{cases} 5x - 2y = 19 \\ 3x + 4y = 1 \end{cases}$

28. $\begin{cases} 2y - 3x = -13 \\ 3x - 17 = 4y \end{cases}$

29. $\begin{cases} 4x + 6y = 5 \\ 8x - 9y = 3 \end{cases}$

30. $\begin{cases} 4x + 9y = 8 \\ 2x - 6y = -3 \end{cases}$

31. $\begin{cases} 2x + 3y = 8 \\ 3x - 2y = -1 \end{cases}$

32. $\begin{cases} \frac{5}{6}x + \frac{2}{3}y = \frac{7}{6} \\ \frac{10}{7}x - \frac{4}{9}y = \frac{17}{21} \end{cases}$

33. $\begin{cases} \frac{3x}{2} - \frac{2y}{3} = 0 \\ \frac{3x}{4} + \frac{4y}{3} = \frac{5}{2} \end{cases}$

34. $\begin{cases} \frac{3x}{5} + \frac{5y}{3} = 2 \\ \frac{6x}{5} - \frac{5y}{3} = 1 \end{cases}$

Solve each system using any method. See Examples 5–6. (Objective 3)

35. $\begin{cases} 3x = 2y - 4 \\ 6x - 4y = -4 \end{cases}$

36. $\begin{cases} 8x = 4y + 10 \\ 4x - 2y = 5 \end{cases}$

37. $\begin{cases} 8x - 4y = 16 \\ 2x - 4 = y \end{cases}$

38. $\begin{cases} x = \frac{3}{2}y + 5 \\ 2x - 3y = 8 \end{cases}$

39. $\begin{cases} y - 2x = 4 \\ 4x + 8 = 2y \end{cases}$

40. $\begin{cases} 3x - 6y = 18 \\ x = 2y + 3 \end{cases}$

41. $\begin{cases} 4(x + 2y) = 15 \\ x + 2y = 4 \end{cases}$

42. $\begin{cases} 9x = 3y + 12 \\ 4 = 3x - y \end{cases}$

Write each repeating decimal as a fraction. Simplify the answer when possible. (See Objective 5)

43. $0.\overline{3}$

44. $0.\overline{29}$

45. $-0.34\overline{89}$

46. $-2.3\overline{47}$

ADDITIONAL PRACTICE

Solve each system using any method.

47. $\begin{cases} 3x - 4y = 9 \\ x + 2y = 8 \end{cases}$

48. $\begin{cases} 3x - 2y = -10 \\ 6x + 5y = 25 \end{cases}$

49. $\begin{cases} 2x + 2y = -1 \\ 3x + 4y = 0 \end{cases}$

50. $\begin{cases} 5x + 3y = -7 \\ 3x - 3y = 7 \end{cases}$

51. $\begin{cases} \frac{x}{2} - \frac{y}{3} = -4 \\ \frac{x}{2} + \frac{y}{9} = 0 \end{cases}$

52. $\begin{cases} \frac{2}{5}x - \frac{1}{6}y = \frac{7}{10} \\ \frac{3}{4}x - \frac{2}{3}y = \frac{19}{8} \end{cases}$

53. $\begin{cases} \frac{3}{4}x + \frac{2}{3}y = 7 \\ \frac{3}{5}x - \frac{1}{2}y = 18 \end{cases}$

54. $\begin{cases} \frac{2}{3}x - \frac{1}{4}y = -8 \\ \frac{1}{2}x - \frac{3}{8}y = -9 \end{cases}$

Solve each system for x and y. Solve for $\frac{1}{x}$ and $\frac{1}{y}$ first.

55. $\begin{cases} \frac{1}{x} + \frac{1}{y} = \frac{5}{6} \\ \frac{1}{x} - \frac{1}{y} = \frac{1}{6} \end{cases}$

56. $\begin{cases} \frac{1}{x} + \frac{1}{y} = \frac{9}{20} \\ \frac{1}{x} - \frac{1}{y} = \frac{1}{20} \end{cases}$

57. $\begin{cases} \frac{1}{x} + \frac{2}{y} = -1 \\ \frac{2}{x} - \frac{1}{y} = -7 \end{cases}$

58. $\begin{cases} \frac{3}{x} - \frac{2}{y} = -30 \\ \frac{2}{x} - \frac{3}{y} = -30 \end{cases}$

APPLICATIONS

Set up a system of two linear equations to solve each problem. See Examples 7–10. (Objective 6)

59. Merchandising A pair of shoes and a sweater cost \$98. If the sweater cost \$16 more than the shoes, how much did the sweater cost?

60. **Merchandising** A sporting goods salesperson sells 2 fishing reels and 5 rods for $270. The next day, the salesperson sells 4 reels and 2 rods for $220. How much does each cost?

61. **Electronics** Two resistors in the voltage divider circuit in the illustration have a total resistance of 1,375 ohms. To provide the required voltage, R_1 must be 125 ohms greater than R_2. Find both resistances.

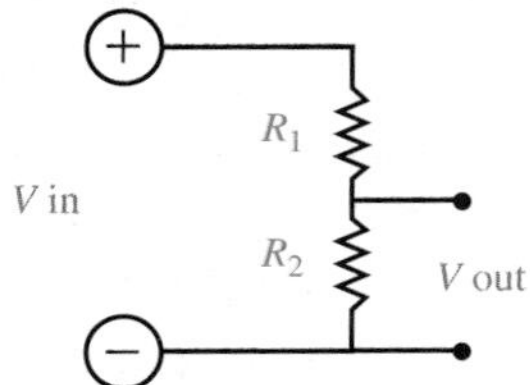

62. **Stowing baggage** A small aircraft can carry 950 pounds of baggage, distributed between two storage compartments. On one flight, the plane is fully loaded, with 150 pounds more baggage in one compartment than the other. How much is stowed in each compartment?

63. **Geometry problem** The rectangular field in the illustration is surrounded by 72 meters of fencing. If the field is partitioned as shown, a total of 88 meters of fencing is required. Find the dimensions of the field.

64. **Geometry** In a right triangle, one acute angle is 15° greater than two times the other acute angle. Find the difference between the angles.

65. **Investment income** Part of $8,000 was invested at 10% interest and the rest at 12%. If the annual income from these investments was $900, how much was invested at each rate?

66. **Investment income** Part of $12,000 was invested at 6% interest and the rest at 7.5%. If the annual income from these investments was $810, how much was invested at each rate?

67. **Mixing solutions** How many ounces of the two alcohol solutions in the illustration must be mixed to obtain 100 ounces of a 12.2% solution?

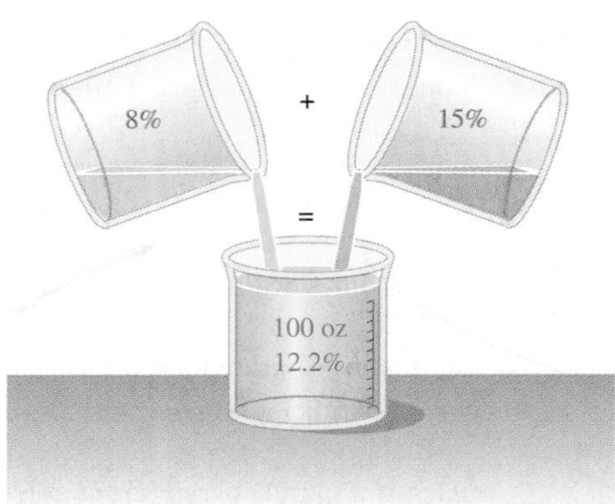

68. **Mixing candy** How many pounds each of candy shown in the illustration must be mixed to obtain 60 pounds of candy that is worth $3 per pound?

69. **Travel** A car travels 50 miles in the same time that a plane travels 180 miles. The speed of the plane is 143 mph faster than the speed of the car. Find the speed of the car.

70. **Travel** A car and a truck leave Rockford at the same time, heading in opposite directions. When they are 350 miles apart, the car has gone 70 miles farther than the truck. How far has the car traveled?

71. **Making bicycles** A bicycle manufacturer builds racing bikes and mountain bikes, with the per-unit manufacturing costs shown in the table. The company has budgeted $15,900 for labor and $13,075 for materials. How many bicycles of each type can be built?

Model	Cost of materials	Cost of labor
Racing	$55	$60
Mountain	$70	$90

72. **Farming** A farmer keeps some animals on a strict diet. Each animal is to receive 15 grams of protein and 7.5 grams of carbohydrates. The farmer uses two food mixes with nutrients shown in the illustration. How many grams of each mix should be used to provide the correct nutrients for each animal?

Mix	Protein	Carbohydrates
Mix A	12%	9%
Mix B	15%	5%

73. Milling brass plates Two machines can mill a brass plate. One machine has a setup cost of \$300 and a cost per plate of \$2. The other machine has a setup cost of \$500 and a cost per plate of \$1. Find the break point.

74. Printing books A printer has two presses. One has a setup cost of \$210 and can print the pages of a certain book for \$5.98. The other press has a setup cost of \$350 and can print the pages of the same book for \$5.95. Find the break point.

75. Managing a computer store The manager of a computer store knows that his fixed costs are \$8,925 per month and that his unit cost is \$850 for every computer sold. If he can sell all the computers he can get for \$1,275 each, how many computers must he sell each month to break even?

76. Managing a beauty shop A beauty shop specializing in permanents has fixed costs of \$2,101.20 per month. The owner estimates that the cost for each permanent is \$23.60. This cost covers labor, chemicals, and electricity. If her shop can give as many permanents as she wants at a price of \$44 each, how many must be given each month to break even?

77. Running a small business A person invests \$18,375 to set up a small business that produces a piece of computer software that will sell for \$29.95. If each piece can be produced for \$5.45, how many pieces must be sold to break even?

78. Running a record company Three people invest \$35,000 each to start a record company that will produce reissues of classic jazz. Each release will be a set of 3 CDs that will retail for \$15 per disc. If each set can be produced for \$18.95, how many sets must be sold for the investors to make a profit?

Break-point analysis *A paint manufacturer can choose between two processes for manufacturing house paint, with monthly costs shown in the table. Assume that the paint sells for \$18 per gallon.*

Process	Fixed costs	Unit cost (per gallon)
A	\$32,500	\$13
B	\$80,600	\$ 5

79. For process A, how many gallons must be sold for the manufacturer to break even?

80. For process B, how many gallons must be sold for the manufacturer to break even?

81. If expected sales are 6,000 gallons per month, which process should the company use?

82. If expected sales are 7,000 gallons per month, which process should the company use?

Making water pumps *A manufacturer of automobile water pumps is considering retooling for one of two manufacturing processes, with monthly fixed costs and unit costs as indicated in the table. Each water pump can be sold for \$50.*

Process	Fixed costs	Unit cost
A	\$12,390	\$29
B	\$20,460	\$17

83. For process A, how many water pumps must be sold for the manufacturer to break even?

84. For process B, how many water pumps must be sold for the manufacturer to break even?

85. If expected sales are 550 per month, which process should be used?

86. If expected sales are 600 per month, which process should be used?

87. If expected sales are 650 per month, which process should be used?

88. At what monthly sales level is process B better?

89. Geometry If two angles are supplementary, their sum is 180°. If the difference between two supplementary angles is 110°, find the measure of each angle.

90. Geometry If two angles are complementary, their sum is 90°. If one of two complementary angles is 16° greater than the other, find the measure of each angle.

91. Find x and y in the parallelogram.

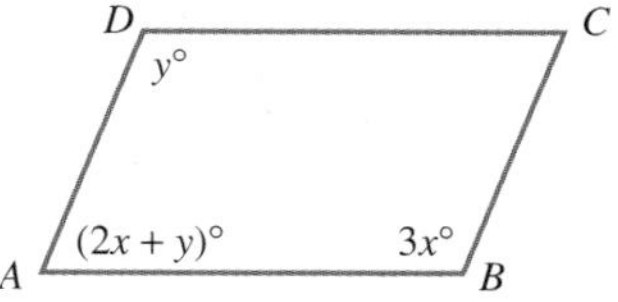

92. Find x and y in the parallelogram.

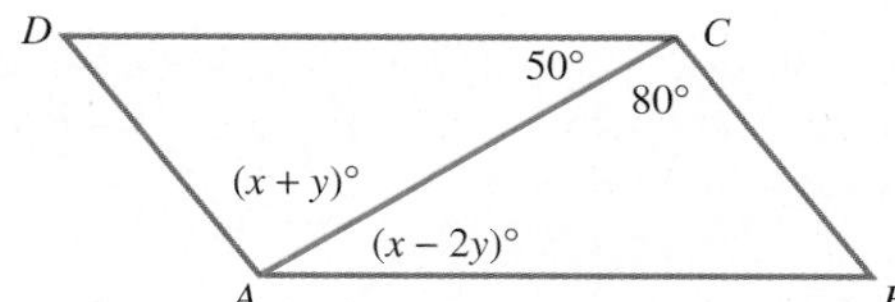

93. Physical therapy To rehabilitate her knee, an athlete does leg extensions. Her goal is to regain a full 90° range of motion in this exercise. Use the information in the illustration on the next page to determine her current range of motion in degrees.

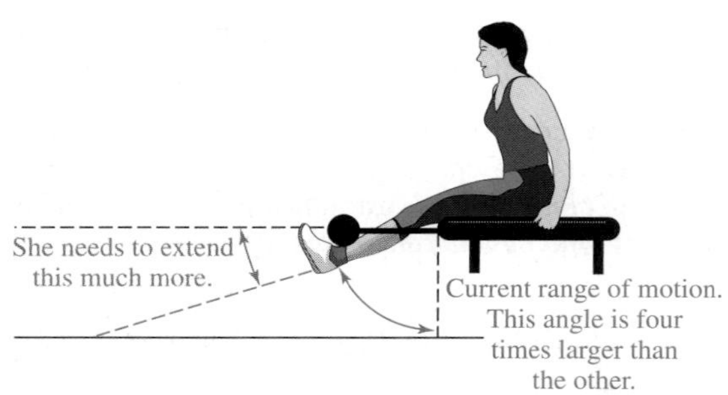

94. The Marine Corps The Marine Corps War Memorial in Arlington, Virginia, portrays the raising of the U.S. flag on Iwo Jima during World War II. Find the two angles shown in the illustration, if the measure of one of the angles is 15° less than twice the other.

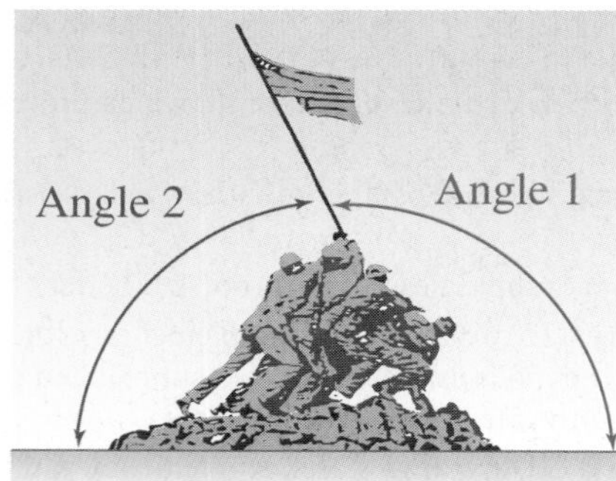

95. Radio frequencies In a radio, an inductor and a capacitor are used in a resonant circuit to select a wanted radio station at a frequency f and reject all others. The inductance L and the capacitance C determine the inductance reactance X_L and the capacitive reactance X_C of that circuit, where

$$X_L = 2\pi f L \quad \text{and} \quad X_C = \frac{1}{2\pi f C}$$

The radio station selected will be at the frequency f where $X_L = X_C$. Write a formula for f^2 in terms of L and C.

96. Choosing salary plans A sales clerk can choose from two salary options:

1. a straight 7% commission
2. \$150 + 2% commission

How much would the clerk have to sell for each plan to produce the same monthly paycheck?

WRITING ABOUT MATH

97. Which method would you use to solve the following system? Why?

$$\begin{cases} y = 3x + 1 \\ 3x + 2y = 12 \end{cases}$$

98. Which method would you use to solve the following system? Why?

$$\begin{cases} 2x + 4y = 9 \\ 3x - 5y = 20 \end{cases}$$

SOMETHING TO THINK ABOUT

99. Under what conditions will a system of two equations in two variables be inconsistent?

100. Under what conditions will the equations of a system of two equations in two variables be dependent?

SECTION 3.3 Solving Systems of Three Linear Equations in Three Variables

Objectives

1. Solve a system of three linear equations in three variables.
2. Identify an inconsistent system.
3. Identify a dependent system and express the solution as an ordered triple in terms of one of the variables.
4. Solve an application problem by setting up and solving a system of three linear equations in three variables.

Vocabulary

plane

Getting Ready

Determine whether the equation $x + 2y + 3z = 6$ is satisfied by the following values.

1. $(1, 1, 1)$ **2.** $(-2, 1, 2)$
3. $(2, -2, -1)$ **4.** $(2, 2, 0)$

We now extend the definition of a linear equation to include equations of the form $ax + by + cz = d$. The solution of a system of three linear equations with three variables is an ordered triple of numbers. For example, the solution of the system

$$\begin{cases} 2x + 3y + 4z = 20 \\ 3x + 4y + 2z = 17 \\ 3x + 2y + 3z = 16 \end{cases}$$

is the ordered triple (1, 2, 3), since each equation is satisfied if $x = 1$, $y = 2$, and $z = 3$.

$$\begin{aligned} 2x + 3y + 4z &= 20 \\ 2(1) + 3(2) + 4(3) &\stackrel{?}{=} 20 \\ 2 + 6 + 12 &\stackrel{?}{=} 20 \\ 20 &= 20 \end{aligned} \qquad \begin{aligned} 3x + 4y + 2z &= 17 \\ 3(1) + 4(2) + 2(3) &\stackrel{?}{=} 17 \\ 3 + 8 + 6 &\stackrel{?}{=} 17 \\ 17 &= 17 \end{aligned} \qquad \begin{aligned} 3x + 2y + 3z &= 16 \\ 3(1) + 2(2) + 3(3) &\stackrel{?}{=} 16 \\ 3 + 4 + 9 &\stackrel{?}{=} 16 \\ 16 &= 16 \end{aligned}$$

The graph of an equation of the form $ax + by + cz = d$ is a flat surface called a **plane.** A system of three linear equations in three variables is consistent or inconsistent, depending on how the three planes corresponding to the three equations intersect. Figure 3-12 illustrates some of the possibilities.

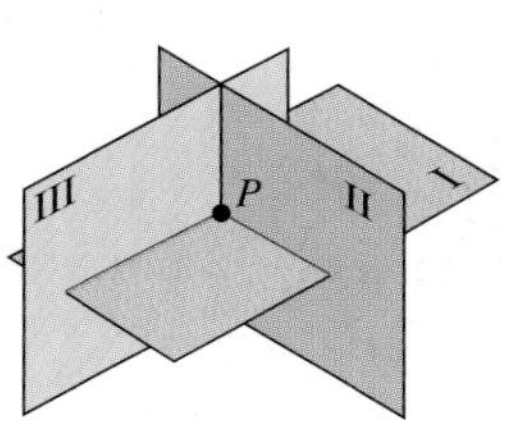

The three planes intersect at a single point P: One solution. A consistent system.

(a)

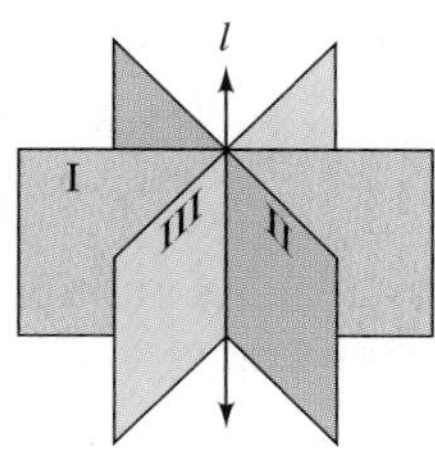

The three planes have a line l in common: Infinitely many solutions. A consistent system.

(b)

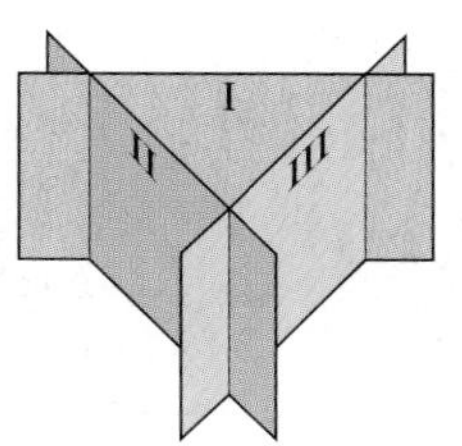

The three planes have no point in common: No solutions. An inconsistent system.

(c)

Figure 3-12

1 Solve a system of three linear equations in three variables.

To solve a system of three linear equations in three variables, we follow these steps.

Solving Three Equations in Three Variables

1. Pick any two equations and eliminate a variable.
2. Pick a different pair of equations and eliminate the same variable.
3. Solve the resulting pair of two equations in two variables.
4. To find the value of the third variable, substitute the values of the two variables found in Step 3 into any equation containing all three variables and solve the equation.
5. Check the solution in all three of the original equations.

EXAMPLE 1 Solve the system: $\begin{cases} 2x + y + 4z = 12 \\ x + 2y + 2z = 9 \\ 3x - 3y - 2z = 1 \end{cases}$.

Solution We are given the system

$$\begin{array}{ll} (1) & \\ (2) & \\ (3) & \end{array} \begin{cases} 2x + y + 4z = 12 \\ x + 2y + 2z = 9 \\ 3x - 3y - 2z = 1 \end{cases}$$

If we pick Equations 2 and 3 and add them, the variable z is eliminated:

$$\begin{array}{ll} (2) & x + 2y + 2z = 9 \\ (3) & \underline{3x - 3y - 2z = 1} \\ (4) & 4x - y = 10 \end{array}$$

We now pick a different pair of equations (Equations 1 and 3) and eliminate z again. If each side of Equation 3 is multiplied by 2 and the resulting equation is added to Equation 1, z is again eliminated:

$$\begin{array}{ll} (1) & 2x + y + 4z = 12 \\ & \underline{6x - 6y - 4z = 2} \\ (5) & 8x - 5y = 14 \end{array}$$

Equations 4 and 5 form a system of two equations in two variables:

$$\begin{array}{ll} (4) & \\ (5) & \end{array} \begin{cases} 4x - y = 10 \\ 8x - 5y = 14 \end{cases}$$

To solve this system, we multiply Equation 4 by -5 and add the resulting equation to Equation 5 to eliminate y:

$$\begin{array}{ll} & -20x + 5y = -50 \\ (5) & \underline{8x - 5y = 14} \\ & -12x = -36 \\ & x = 3 \quad \text{Divide both sides by } -12. \end{array}$$

To find y, we substitute 3 for x in any equation containing only x and y (such as Equation 5) and solve for y:

$$\begin{aligned}
(5) \qquad 8x - 5y &= 14 \\
8(3) - 5y &= 14 && \text{Substitute 3 for } x. \\
24 - 5y &= 14 && \text{Simplify.} \\
-5y &= -10 && \text{Subtract 24 from both sides.} \\
y &= 2 && \text{Divide both sides by } -5.
\end{aligned}$$

To find z, we substitute 3 for x and 2 for y in an equation containing x, y, and z (such as Equation 1) and solve for z:

$$\begin{aligned}
(1) \qquad 2x + y + 4z &= 12 \\
2(3) + 2 + 4z &= 12 && \text{Substitute 3 for } x \text{ and 2 for } y. \\
8 + 4z &= 12 && \text{Simplify.} \\
4z &= 4 && \text{Subtract 8 from both sides.} \\
z &= 1 && \text{Divide both sides by 4.}
\end{aligned}$$

The solution of the system is $(x, y, z) = (3, 2, 1)$. Verify that these values satisfy each equation in the original system.

SELF CHECK 1 Solve the system: $\begin{cases} 2x + y + 4z = 16 \\ x + 2y + 2z = 11 \\ 3x - 3y - 2z = -9 \end{cases}$.

2 Identify an inconsistent system.

The next example has no solution.

EXAMPLE 2 Solve the system: $\begin{cases} 2x + y - 3z = -3 \\ 3x - 2y + 4z = 2 \\ 4x + 2y - 6z = -7 \end{cases}$.

Solution We are given the system of equations

$$\begin{matrix} (1) \\ (2) \\ (3) \end{matrix} \quad \begin{cases} 2x + y - 3z = -3 \\ 3x - 2y + 4z = 2 \\ 4x + 2y - 6z = -7 \end{cases}$$

We can multiply Equation 1 by 2 and add the resulting equation to Equation 2 to eliminate y.

$$\begin{array}{ll}
 & 4x + 2y - 6z = -6 \\
(2) & \underline{3x - 2y + 4z = 2} \\
(4) & 7x \qquad\quad - 2z = -4
\end{array}$$

We now add Equations 2 and 3 to again eliminate y.

$$\begin{array}{ll}
(2) & 3x - 2y + 4z = 2 \\
(3) & \underline{4x + 2y - 6z = -7} \\
(5) & 7x \qquad\quad - 2z = -5
\end{array}$$

Equations 4 and 5 form the system

$$\begin{aligned} (4) \quad & \begin{cases} 7x - 2z = -4 \\ 7x - 2z = -5 \end{cases} \\ (5) \quad & \end{aligned}$$

Since $7x - 2z$ cannot equal both -4 and -5, this system is inconsistent; it has no solution. Its solution set is $\varnothing$.

SELF CHECK 2 Solve the system: $\begin{cases} 2x + y - 3z = 8 \\ 3x - 2y + 4z = 10 \\ 4x + 2y - 6z = -5 \end{cases}$.

3 Identify a dependent system and express the solution as an ordered triple in terms of one of the variables.

When the equations in a system of two equations in two variables were dependent, the system had infinitely many solutions. This is not always true for systems of three equations in three variables. In fact, a system can have dependent equations and still be inconsistent. Figure 3-13 illustrates the different possibilities.

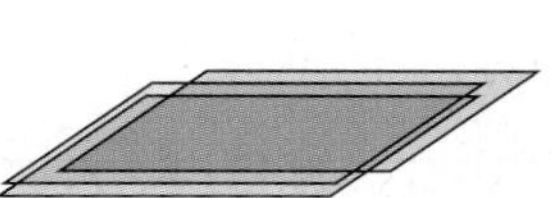

When three planes coincide, the equations are dependent, and there are infinitely many solutions.
(a)

When three planes intersect in a common line, the equations are dependent, and there are infinitely many solutions.
(b)

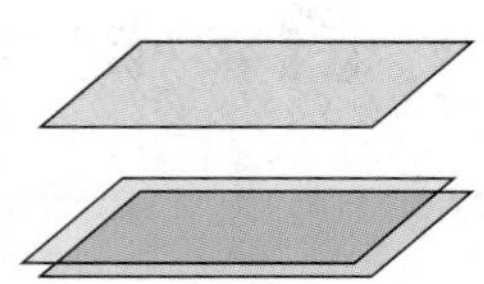

When two planes coincide and are parallel to a third plane, the system is inconsistent, and there are no solutions.
(c)

Figure 3-13

EXAMPLE 3 Solve the system: $\begin{cases} 3x - 2y + z = -1 \\ 2x + y - z = 5 \\ 5x - y = 4 \end{cases}$.

Solution We can add the first two equations to get

$$\begin{array}{rrcl} & 3x - 2y + z & = & -1 \\ & 2x + y - z & = & 5 \\ \hline (1) & 5x - y \phantom{{}-z} & = & 4 \end{array}$$

Since Equation 1 is the same as the third equation of the system, the equations of the system are dependent, and there will be infinitely many solutions. From a graphical perspective, the equations represent three planes that intersect in a common line, as shown in Figure 3-13(b).

To write the general solution to this system, we can solve Equation 1 for y to get

$$5x - y = 4$$

$$-y = -5x + 4 \quad \text{Subtract } 5x \text{ from both sides.}$$

$$y = 5x - 4 \quad \text{Multiply both sides by } -1.$$

We then can substitute $5x - 4$ for y in the first equation of the system and solve for z to get

$$3x - 2y + z = -1$$
$$3x - 2(5x - 4) + z = -1 \quad \text{Substitute } 5x - 4 \text{ for } y.$$
$$3x - 10x + 8 + z = -1 \quad \text{Use the distributive property to remove parentheses.}$$
$$-7x + 8 + z = -1 \quad \text{Combine like terms.}$$
$$z = 7x - 9 \quad \text{Add } 7x \text{ and } -8 \text{ to both sides.}$$

COMMENT The solution in Example 3 is called a *general* solution. If we had eliminated different variables, we could have expressed the general solution in terms of y or in terms of z.

Since we have found the values of y and z in terms of x, every solution to the system has the form $(x, 5x - 4, 7x - 9)$, where x can be any real number. For example,

If $x = 1$, a solution is $(1, 1, -2)$. $5(1) - 4 = 1$, and $7(1) - 9 = -2$.

If $x = 2$, a solution is $(2, 6, 5)$. $5(2) - 4 = 6$, and $7(2) - 9 = 5$.

If $x = 3$, a solution is $(3, 11, 12)$. $5(3) - 4 = 11$, and $7(3) - 9 = 12$.

This system has infinitely many solutions.

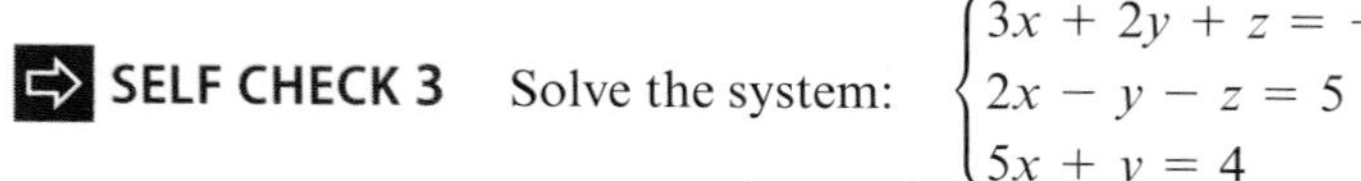

SELF CHECK 3 Solve the system: $\begin{cases} 3x + 2y + z = -1 \\ 2x - y - z = 5 \\ 5x + y = 4 \end{cases}$.

4 Solve an application problem by setting up and solving a system of three linear equations in three variables.

EXAMPLE 4 **MANUFACTURING HAMMERS** A company makes three types of hammers—good, better, and best. The cost of making each type of hammer is \$4, \$6, and \$7, respectively, and the hammers sell for \$6, \$9, and \$12. Each day, the cost of making 100 hammers is \$520, and the daily revenue from their sale is \$810. How many of each type are manufactured?

Analyze the problem We need to know the number of each type of hammer manufactured, so we will let x represent the number of good hammers, y represent the number of better hammers, and z represent the number of best hammers.

Form three equations Since x represents the number of good hammers, y represents the number of better hammers, and z represents the number of best hammers, we know that

The total number of hammers is $x + y + z$.

The cost of making good hammers is \$$4x$ (\$4 times x hammers).

The cost of making better hammers is \$$6y$ (\$6 times y hammers).

The cost of making best hammers is \$$7z$ (\$7 times z hammers).

The revenue received by selling good hammers is \$$6x$ (\$6 times x hammers).

The revenue received by selling better hammers is \$$9y$ (\$9 times y hammers).

The revenue received by selling best hammers is \$$12z$ (\$12 times z hammers).

The information leads to three equations:

The number of good hammers	plus	the number of better hammers	plus	the number of best hammers	equals	the total number of hammers.
x	$+$	y	$+$	z	$=$	100

The cost of making good hammers	plus	the cost of making better hammers	plus	the cost of making best hammers	equals	the total cost.
$4x$	$+$	$6y$	$+$	$7z$	$=$	520

The revenue from the good hammers	plus	the revenue from the better hammers	plus	the revenue from the best hammers	equals	the total revenue.
$6x$	$+$	$9y$	$+$	$12z$	$=$	810

Solve the system These three equations give the following system:

$$\begin{aligned} &(1)\quad \\ &(2)\quad \\ &(3)\quad \end{aligned}\begin{cases} x + y + z = 100 \\ 4x + 6y + 7z = 520 \\ 6x + 9y + 12z = 810 \end{cases}$$

that we can solve as follows:

If we multiply Equation 1 by -7 and add the result to Equation 2, we obtain

$$\begin{array}{rl} & -7x - 7y - 7z = -700 \\ & \underline{\;4x + 6y + 7z = 520\;} \\ (4) & -3x - y = -180 \end{array}$$

If we multiply Equation 1 by -12 and add the result to Equation 3, we obtain

$$\begin{array}{rl} & -12x - 12y - 12z = -1{,}200 \\ & \underline{\;6x + 9y + 12z = 810\;} \\ (5) & -6x - 3y = -390 \end{array}$$

If we multiply Equation 4 by -3 and add it to Equation 5, we obtain

$$\begin{aligned} 9x + 3y &= 540 \\ \underline{-6x - 3y} &\underline{= -390} \\ 3x &= 150 \\ x &= 50 \quad \text{Divide both sides by 3.} \end{aligned}$$

To find y, we substitute 50 for x in Equation 4:

$$\begin{aligned} -3x - y &= -180 \\ -3(50) - y &= -180 && \text{Substitute 50 for } x. \\ -y &= -30 && \text{Add 150 to both sides.} \\ y &= 30 && \text{Divide both sides by } -1. \end{aligned}$$

To find z, we substitute 50 for x and 30 for y in Equation 1:

$$\begin{aligned} x + y + z &= 100 \\ 50 + 30 + z &= 100 \\ z &= 20 && \text{Subtract 80 from both sides.} \end{aligned}$$

State the conclusion Each day, the company makes 50 good hammers, 30 better hammers, and 20 best hammers.

Check the result Check the solution in each equation in the original system.

EXAMPLE 5 **CURVE FITTING** The equation of the parabola shown in Figure 3-14 is of the form $y = ax^2 + bx + c$. Find the equation of the parabola.

Solution Since the parabola passes through the points shown in the figure, each pair of coordinates satisfies the equation $y = ax^2 + bx + c$. If we substitute the x- and y-values of each point into the equation and simplify, we obtain the following system.

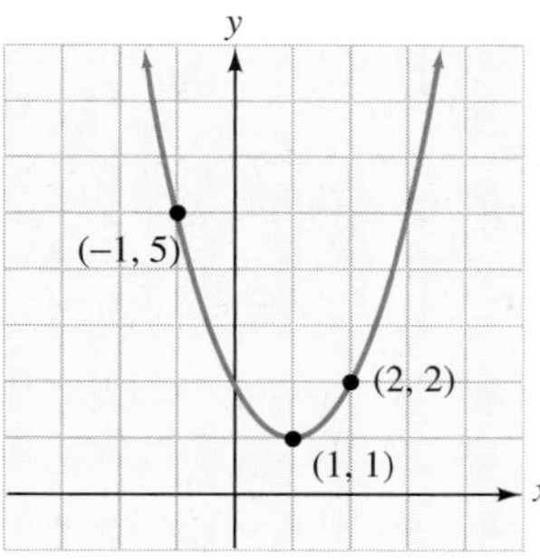

Figure 3-14

$$\begin{cases} a - b + c = 5 & (1) \\ a + b + c = 1 & (2) \\ 4a + 2b + c = 2 & (3) \end{cases}$$

If we add Equations 1 and 2, we obtain $2a + 2c = 6$. If we multiply Equation 1 by 2 and add the result to Equation 3, we get $6a + 3c = 12$. We can then divide both sides of $2a + 2c = 6$ by 2 and divide both sides of $6a + 3c = 12$ by 3 to get the system

$$\begin{cases} a + c = 3 & (4) \\ 2a + c = 4 & (5) \end{cases}$$

If we multiply Equation 4 by -1 and add the result to Equation 5, we get $a = 1$. To find c, we can substitute 1 for a in Equation 4 and find that $c = 2$. To find b, we can substitute 1 for a and 2 for c in Equation 2 and find that $b = -2$.

After we substitute these values of a, b, and c into the equation $y = ax^2 + bx + c$ we have the equation of the parabola.

$$y = ax^2 + bx + c$$
$$y = 1x^2 + (-2)x + 2$$
$$y = x^2 - 2x + 2$$

SELF CHECK ANSWERS **1.** (1, 2, 3) **2.** ∅ **3.** infinitely many solutions of the form $(x, 4 - 5x, -9 + 7x)$

NOW TRY THIS

1. The manager of a coffee bar wants to mix some Peruvian Organic coffee worth \$15 per pound with some Colombian coffee worth \$10 per pound and Indian Malabar coffee worth \$18 per pound to get 50 pounds of a blend that he can sell for \$17.50 per pound. He wants to use 10 fewer pounds of the Indian Malabar than Peruvian Organic. How many pounds of each should he use? (*Hint:* This problem is based on the formula $V = np$, where V represents value, n represents the number of pounds, and p represents the price per pound.)

3.3 EXERCISES

WARM-UPS *Is the ordered triple a solution of the system?*

1. (1, 1, 1); $\begin{cases} 2x + y - 3z = 0 \\ 3x - 2y + 4z = 5 \\ 4x + 2y - 6z = 0 \end{cases}$

2. (2, 0, 1); $\begin{cases} 3x + 2y - z = 5 \\ 2x - 3y + 2z = 4 \\ 4x - 2y + 3z = 10 \end{cases}$

REVIEW

Consider the line passing through **(−2, −4)** *and* **(3, 5).**

3. Find the slope of the line.

4. Write the equation of the line in general form.

Find each value if $f(x) = 2x^2 + 1$.

5. $f(0)$

6. $f(-2)$

7. $f(s)$

8. $f(2t)$

VOCABULARY AND CONCEPTS

Fill in the blanks.

9. The graph of the equation $2x + 3y + 4z = 5$ is a flat surface called a ______.

10. When three planes coincide, the equations of the system are ________, and there are ________ many solutions.

11. When three planes intersect in a line, the system will have ________ many solutions.

12. When three planes are parallel, the system will have ___ solutions.

Determine whether the ordered triple is a solution of the given system.

13. $(2, 1, 1)$, $\begin{cases} x - y + z = 2 \\ 2x + y - z = 4 \\ 2x - 3y + z = 2 \end{cases}$

14. $(-3, 2, -1)$, $\begin{cases} 2x + 2y + 3z = -1 \\ 3x + y - z = -6 \\ x + y + 2z = 1 \end{cases}$

GUIDED PRACTICE

Solve each system. See Example 1. (Objective 1)

15. $\begin{cases} x + y + z = 4 \\ 2x + y - z = 1 \\ 2x - 3y + z = 1 \end{cases}$

16. $\begin{cases} x + y + z = 4 \\ x - y + z = 2 \\ x - y - z = 0 \end{cases}$

17. $\begin{cases} 2x + 2y + 3z = 10 \\ 3x + y - z = 0 \\ x + y + 2z = 6 \end{cases}$

18. $\begin{cases} x - y + z = 4 \\ x + 2y - z = -1 \\ x + y - 3z = -2 \end{cases}$

19. $\begin{cases} 4x + 3z = 4 \\ 2y - 6z = -1 \\ 8x + 4y + 3z = 9 \end{cases}$

20. $\begin{cases} 2x + 3y + 2z = 1 \\ 2x - 3y + 2z = -1 \\ 4x + 3y - 2z = 4 \end{cases}$

Solve each system. See Example 2. (Objective 2)

21. $\begin{cases} 2a + 3b + c = 2 \\ 4a + 6b + 2c = 5 \\ a - 2b + c = 3 \end{cases}$

22. $\begin{cases} 2x + y - z = 1 \\ x + 2y + 2z = 2 \\ 4x + 5y + 3z = 3 \end{cases}$

Solve each system. See Example 3. (Objective 3)

23. $\begin{cases} x - 2y + 3z = 9 \\ -x + 3y = -4 \\ 2x - 5y + 3z = 13 \end{cases}$

24. $\begin{cases} 7x - 2y - z = 1 \\ 9x - 6y + z = 7 \\ x - 2y + z = 3 \end{cases}$

ADDITIONAL PRACTICE *Solve each system.*

25. $\begin{cases} a + b + 2c = 7 \\ a + 2b + c = 8 \\ 2a + b + c = 9 \end{cases}$

26. $\begin{cases} 3x - y - 2z = 12 \\ x + y + 6z = 8 \\ 2x - 2y - z = 11 \end{cases}$

27. $\begin{cases} x + \frac{1}{3}y + z = 13 \\ \frac{1}{2}x - y + \frac{1}{3}z = -2 \\ x + \frac{1}{2}y - \frac{1}{3}z = 2 \end{cases}$

28. $\begin{cases} x - \frac{1}{5}y - z = 9 \\ \frac{1}{4}x + \frac{1}{5}y - \frac{1}{2}z = 5 \\ 2x + y + \frac{1}{6}z = 12 \end{cases}$

29. $\begin{cases} x - 3y + z = 1 \\ 2x - y - 2z = 2 \\ x + 2y - 3z = -1 \end{cases}$

30. $\begin{cases} 2x + y - 3z = 5 \\ x - 2y + 4z = 9 \\ 4x + 2y - 6z = 1 \end{cases}$

31. $\begin{cases} 2x + 3y + 4z = 6 \\ 2x - 3y - 4z = -4 \\ 4x + 6y + 8z = 12 \end{cases}$

32. $\begin{cases} x - 3y + 4z = 2 \\ 2x + y + 2z = 3 \\ 4x - 5y + 10z = 7 \end{cases}$

APPLICATIONS

Set up a system of three linear equations to solve each problem. See Examples 4–5. (Objective 4)

33. Integer problem The sum of three integers is 18. The third integer is four times the second, and the second integer is 6 more than the first. Find the integers.

34. Integer problem The sum of three integers is 48. If the first integer is doubled, the sum is 60. If the second integer is doubled, the sum is 63. Find the integers.

35. Geometry The sum of the angles in any triangle is 180°. In triangle ABC, $\angle A$ is 100° less than the sum of $\angle B$ and $\angle C$, and $\angle C$ is 40° less than twice $\angle B$. Find the measure of each angle.

36. Geometry The sum of the angles of any four-sided figure is 360°. In the quadrilateral, $\angle A = \angle B$, $\angle C$ is 20° greater than $\angle A$, and $\angle D = 40°$. Find the measure of each angle.

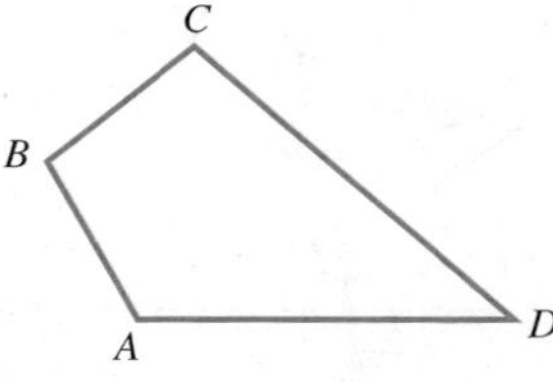

37. Nutritional planning One unit of each of three foods contains the nutrients shown in the table. How many units of each must be used to provide exactly 11 grams of fat, 6 grams of carbohydrates, and 10 grams of protein?

Food	Fat	Carbohydrates	Protein
A	1	1	2
B	2	1	1
C	2	1	2

38. Nutritional planning One unit of each of three foods contains the nutrients shown in the table. How many units of each must be used to provide exactly 14 grams of fat, 9 grams of carbohydrates, and 9 grams of protein?

Food	Fat	Carbohydrates	Protein
A	2	1	2
B	3	2	1
C	1	1	2

39. Making statues An artist makes three types of ceramic statues at a monthly cost of $650 for 180 statues. The manufacturing costs for the three types are $5, $4, and $3. If the statues sell for $20, $12, and $9, respectively, how many of each type should be made to produce $2,100 in monthly revenue?

40. Manufacturing footballs A factory manufactures three types of footballs at a monthly cost of $2,425 for 1,125 footballs. The manufacturing costs for the three types of footballs are $4, $3, and $2. These footballs sell for $16, $12, and $10, respectively. How many of each type are manufactured if the monthly profit is $9,275? (*Hint:* Profit = income − cost.)

41. Concert tickets Tickets for a concert cost $5, $3, and $2. Twice as many $5 tickets were sold as $2 tickets. The receipts for 750 tickets were $2,625. How many of each price ticket were sold?

42. Mixing nuts The owner of a candy store mixed some peanuts worth $3 per pound, some cashews worth $9 per pound, and some Brazil nuts worth $9 per pound to get 50 pounds of a mixture that would sell for $6 per pound. She used 15 fewer pounds of cashews than peanuts. How many pounds of each did she use?

43. Chainsaw sculpting A north woods sculptor carves three types of statues with a chainsaw. The times required for carving, sanding, and painting a totem pole, a bear, and a deer are shown in the table. How many of each should be produced to use all available labor hours?

	Totem pole	Bear	Deer	Time available
Carving	2 hours	2 hours	1 hour	14 hours
Sanding	1 hour	2 hours	2 hours	15 hours
Painting	3 hours	2 hours	2 hours	21 hours

44. Making clothing A clothing manufacturer makes coats, shirts, and slacks. The times required for cutting, sewing, and packaging each item are shown in the table. How many of each should be made to use all available labor hours?

	Coats	Shirts	Slacks	Time available
Cutting	20 min	15 min	10 min	115 hr
Sewing	60 min	30 min	24 min	280 hr
Packaging	5 min	12 min	6 min	65 hr

45. Earth's atmosphere Use the information in the graph to determine what percent of Earth's atmosphere is nitrogen, is oxygen, and is other gases.

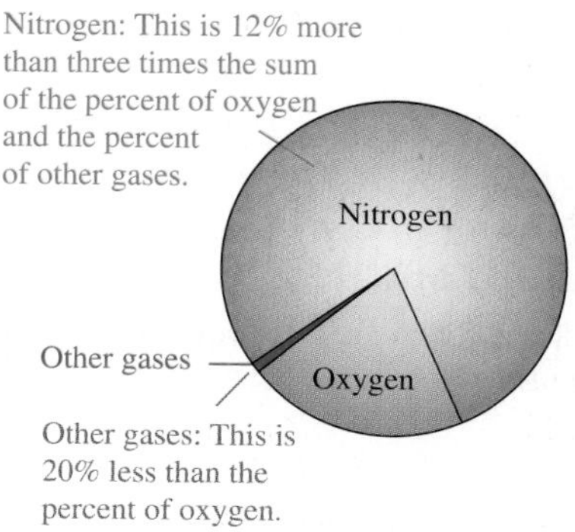

46. NFL records Jerry Rice, who played with the San Francisco 49ers and the Oakland Raiders, holds the all-time record for touchdown passes caught. Here are interesting facts about this feat.

- He caught 30 more TD passes from Steve Young than he did from Joe Montana.
- He caught 39 more TD passes from Joe Montana than he did from Rich Gannon.
- He caught a total of 156 TD passes from Young, Montana, and Gannon.

Determine the number of touchdown passes Rice has caught from Young, from Montana, and from Gannon as of 2003.

47. Curve fitting Find the equation of the parabola shown in the illustration.

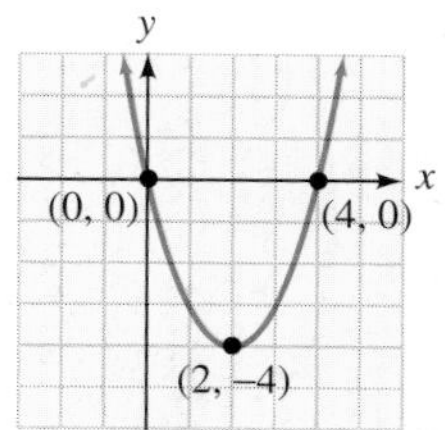

48. Curve fitting Find the equation of the parabola shown in the illustration.

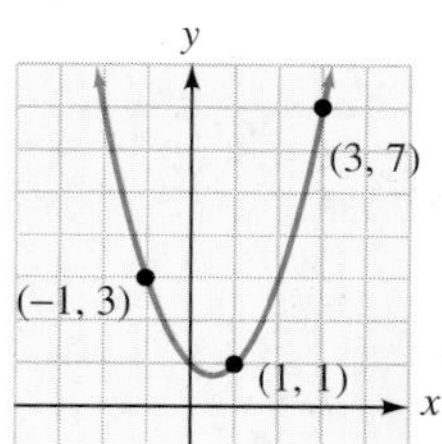

The equation of a circle is of the form $x^2 + y^2 + cx + dy + e = 0$.

49. Curve fitting Find the equation of the circle shown in the illustration.

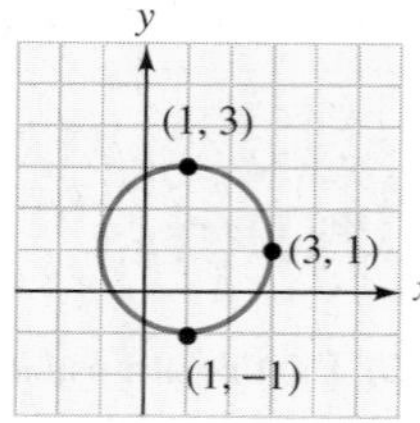

50. Curve fitting Find the equation of the circle shown in the illustration.

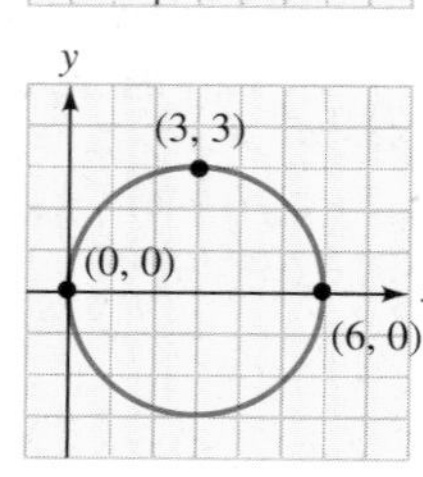

WRITING ABOUT MATH

51. What makes a system of three equations in three variables inconsistent?

52. What makes the equations of a system of three equations in three variables dependent?

SOMETHING TO THINK ABOUT

53. Solve the system:

$$\begin{cases} x + y + z + w = 3 \\ x - y - z - w = -1 \\ x + y - z - w = 1 \\ x + y - z + w = 3 \end{cases}$$

54. Solve the system:

$$\begin{cases} 2x + y + z + w = 3 \\ x - 2y - z + w = -3 \\ x - y - 2z - w = -3 \\ x + y - z + 2w = 4 \end{cases}$$

SECTION 3.4 Solving Systems of Linear Equations Using Matrices

Objectives

1. Solve a system of linear equations with the same number of equations as variables using Gaussian elimination.
2. Solve a system with more linear equations than variables using row operations on a matrix.
3. Solve a system with fewer linear equations than variables using row operations on a matrix.

Vocabulary

matrix	augmented matrix	triangular form of a matrix
element of a matrix	coefficient matrix	back substitution
square matrix	Gaussian elimination	

Getting Ready

Multiply the first row by 2 and add the result to the second row.

1. $\begin{matrix} 2 & 3 & 5 \\ 1 & 2 & 3 \end{matrix}$ **2.** $\begin{matrix} -1 & 0 & 4 \\ 2 & 3 & -1 \end{matrix}$

Multiply the first row by −1 and add the result to the second row.

3. $\begin{matrix} 2 & 3 & 5 \\ 1 & 2 & 3 \end{matrix}$ **4.** $\begin{matrix} -1 & 0 & 4 \\ 2 & 3 & -1 \end{matrix}$

In this section, we will discuss a streamlined method used for solving systems of linear equations. This method involves the use of *matrices.*

1 Solve a system of linear equations with the same number of equations as variables using Gaussian elimination.

Another method of solving systems of equations involves rectangular arrays of numbers called *matrices.*

Matrix A **matrix** is any rectangular array of numbers.

Arthur Cayley (1821–1895)
Cayley taught mathematics at Cambridge University. When he refused to take religious vows, he was fired and became a lawyer. After 14 years, he returned to mathematics and to Cambridge. Cayley was a major force in developing the theory of matrices.

Some examples of matrices are

$$A = \begin{bmatrix} 1 & 2 & 3 \\ 4 & 5 & 6 \end{bmatrix} \qquad B = \begin{bmatrix} 1 & 2 \\ 3 & 4 \\ 5 & 6 \end{bmatrix} \qquad C = \begin{bmatrix} 2 & 4 & 6 \\ 8 & 10 & 12 \\ 14 & 16 & 18 \end{bmatrix}$$

The numbers in each matrix are called **elements.** Because matrix A has two rows and three columns, it is called a 2×3 matrix (read "2 by 3" matrix). Matrix B is a 3×2 matrix, because the matrix has three rows and two columns. Matrix C is a 3×3 matrix (three rows and three columns).

Any matrix with the same number of rows and columns, like matrix C, is called a **square matrix.**

To show how to use matrices to solve systems of linear equations, we consider the system

$$\begin{cases} x - 2y - z = 6 \\ 2x + 2y - z = 1 \\ -x - y + 2z = 1 \end{cases}$$

which can be represented by the following matrix, called an **augmented matrix:**

$$\left[\begin{array}{rrr:r} 1 & -2 & -1 & 6 \\ 2 & 2 & -1 & 1 \\ -1 & -1 & 2 & 1 \end{array}\right]$$

The first three columns of the augmented matrix form a 3×3 matrix called a **coefficient matrix.** It is determined by the coefficients of x, y, and z in the equations of the system. The 3×1 matrix to the right of the dashed line is determined by the constants in the equations.

Coefficient matrix ***Column of constants***

$$\begin{bmatrix} 1 & -2 & 1 \\ 2 & 2 & -1 \\ -1 & -1 & 2 \end{bmatrix} \qquad \begin{bmatrix} 6 \\ 1 \\ 1 \end{bmatrix}$$

Each row of the augmented matrix represents one equation of the system:

$$\left[\begin{array}{rrr:r} 1 & -2 & -1 & 6 \\ 2 & 2 & -1 & 1 \\ -1 & -1 & 2 & 1 \end{array}\right] \begin{array}{c} \leftrightarrow \\ \leftrightarrow \\ \leftrightarrow \end{array} \begin{cases} x - 2y - z = 6 \\ 2x + 2y - z = 1 \\ -x - y + 2z = 1 \end{cases}$$

To solve a 3×3 system of equations by **Gaussian elimination,** we transform an augmented matrix into the following matrix that has all 0's below its main diagonal, which is formed by the elements a, e, and h.

$$\left[\begin{array}{ccc|c} a & b & c & d \\ 0 & e & f & g \\ 0 & 0 & h & i \end{array}\right] \quad (a, b, c, \ldots, i \text{ are real numbers})$$

We often can write a matrix in this form, called **triangular form,** by using the following operations.

Elementary Row Operations

1. Any two rows of a matrix can be interchanged.
2. Any row of a matrix can be multiplied by a nonzero constant.
3. Any row of a matrix can be changed by adding a nonzero constant multiple of another row to it.

- A type 1 row operation corresponds to interchanging two equations of a system.
- A type 2 row operation corresponds to multiplying both sides of an equation by a nonzero constant.
- A type 3 row operation corresponds to adding a nonzero multiple of one equation to another.

None of these operations will change the solution of the given system of equations.

After we have written the matrix in triangular form, we can solve the corresponding system of equations by a process called **back substitution,** as shown in Example 1.

EXAMPLE 1 Solve the system using matrices: $\begin{cases} x - 2y - z = 6 \\ 2x + 2y - z = 1 \\ -x - y + 2z = 1 \end{cases}$.

Solution We can represent the system with the following augmented matrix:

$$\left[\begin{array}{rrr|r} 1 & -2 & -1 & 6 \\ 2 & 2 & -1 & 1 \\ -1 & -1 & 2 & 1 \end{array}\right]$$

To get 0's under the 1 in the first column, we multiply row 1 of the augmented matrix by -2 and add it to row 2 to get a new row 2. We then multiply row 1 by 1 and add it to row 3 to get a new row 3.

$$\left[\begin{array}{rrr|r} 1 & -2 & -1 & 6 \\ 0 & 6 & 1 & -11 \\ 0 & -3 & 1 & 7 \end{array}\right]$$

To get a 0 under the 6 in the second column of the previous matrix, we multiply row 2 by $\frac{1}{2}$ and add it to row 3.

$$\left[\begin{array}{rrr|r} 1 & -2 & 1 & 6 \\ 0 & 6 & 1 & -11 \\ 0 & 0 & \frac{3}{2} & \frac{3}{2} \end{array}\right]$$

Finally, to clear the fraction in the third row, third column, we multiply row 3 by $\frac{2}{3}$, which is the reciprocal of $\frac{3}{2}$.

$$\left[\begin{array}{rrr:r} 1 & -2 & 1 & 6 \\ 0 & 6 & 1 & -11 \\ 0 & 0 & 1 & 1 \end{array}\right]$$

The final matrix represents the system of equations

$$\begin{array}{l} (1) \\ (2) \\ (3) \end{array} \quad \begin{cases} x - 2y - z = 6 \\ 0x + 6y + z = -11 \\ 0x + 0y + z = 1 \end{cases}$$

From Equation 3, we can see that $z = 1$. To find y, we substitute 1 for z in Equation 2 and solve for y.

$$\begin{array}{lrll} (2) & 6y + z &= -11 & \\ & 6y + \mathbf{1} &= -11 & \text{Substitute 1 for } z. \\ & 6y &= -12 & \text{Subtract 1 from both sides.} \\ & y &= -2 & \text{Divide both sides by 6.} \end{array}$$

Thus, $y = -2$. To find x, we substitute 1 for z and -2 for y in Equation 1 and solve for x:

$$\begin{array}{lrll} (1) & x - 2y - z &= 6 & \\ & x - 2(\mathbf{-2}) - \mathbf{1} &= 6 & \text{Substitute 1 for } z \text{ and } -2 \text{ for } y. \\ & x + 3 &= 6 & \text{Simplify.} \\ & x &= 3 & \text{Subtract 3 from both sides.} \end{array}$$

Thus, $x = 3$. The solution of the given system is $(3, -2, 1)$. Verify that this ordered triple satisfies each equation of the original system.

SELF CHECK 1 Solve the system using matrices: $\begin{cases} x - 2y - z = 2 \\ 2x + 2y - z = -5. \\ -x - y + 2z = 7 \end{cases}$

2 Solve a system with more linear equations than variables using row operations on a matrix.

We can use matrices to solve systems that have more equations than variables.

EXAMPLE 2 Solve the system using matrices: $\begin{cases} x + y = -1 \\ 2x - y = 7 \\ -x + 2y = -8 \end{cases}$.

Solution This system can be represented by the following augmented matrix:

$$\left[\begin{array}{rr:r} 1 & 1 & -1 \\ 2 & -1 & 7 \\ -1 & 2 & -8 \end{array}\right]$$

To get 0's under the 1 in the first column, we multiply row 1 by -2 and add it to row 2. Then we can multiply row 1 by 1 and add it to row 3.

$$\left[\begin{array}{rr|r} 1 & 1 & -1 \\ 0 & -3 & 9 \\ 0 & 3 & -9 \end{array}\right]$$

To get a 0 under the -3 in the second column, we can multiply row 2 by 1 and add it to row 3.

$$\left[\begin{array}{rr|r} 1 & 1 & -1 \\ 0 & -3 & 9 \\ 0 & 0 & 0 \end{array}\right]$$

Finally, to get a 1 in the second row, second column, we multiply row 2 by $-\frac{1}{3}$.

$$\left[\begin{array}{rr|r} 1 & 1 & -1 \\ 0 & 1 & -3 \\ 0 & 0 & 0 \end{array}\right]$$

The final matrix represents the system

$$\begin{cases} x + y = -1 \\ 0x + y = -3 \\ 0x + 0y = 0 \end{cases}$$

The third equation can be discarded, because $0x + 0y = 0$ for all x and y. From the second equation, we can read that $y = -3$. To find x, we substitute -3 for y in the first equation and solve for x:

$$\begin{aligned} x + y &= -1 \\ x + (-3) &= -1 && \text{Substitute } -3 \text{ for } y. \\ x &= 2 && \text{Add 3 to both sides.} \end{aligned}$$

The solution is $(2, -3)$. Verify that this solution satisfies all three equations of the original system.

SELF CHECK 2 Solve the system using matrices: $\begin{cases} x + y = 1 \\ 2x - y = 8 \\ -x + 2y = -7 \end{cases}$.

If the last row of the final matrix in Example 2 had been of the form $0x + 0y = k$, where $k \neq 0$, the system would not have a solution. No values of x and y could make the expression $0x + 0y$ equal to a nonzero constant k.

3 Solve a system with fewer linear equations than variables using row operations on a matrix.

We also can solve many systems that have more variables than equations.

EXAMPLE 3 Solve the system using matrices: $\begin{cases} x + y - 2z = -1 \\ 2x - y + z = -3 \end{cases}$.

Solution This system can be represented by the following augmented matrix.

$$\left[\begin{array}{rrr|r} 1 & 1 & -2 & -1 \\ 2 & -1 & 1 & -3 \end{array}\right]$$

To get a 0 under the 1 in the first column, we multiply row 1 by -2 and add it to row 2.

$$\left[\begin{array}{rrr|r} 1 & 1 & -2 & -1 \\ 0 & -3 & 5 & -1 \end{array}\right]$$

Then to get a 1 in the second row, second column, we multiply row 2 by $-\frac{1}{3}$.

$$\left[\begin{array}{rrr|r} 1 & 1 & -2 & -1 \\ 0 & 1 & -\frac{5}{3} & \frac{1}{3} \end{array}\right]$$

The final matrix represents the system

$$\begin{cases} x + y - 2z = -1 \\ y - \frac{5}{3}z = \frac{1}{3} \end{cases}$$

We add $\frac{5}{3}z$ to both sides of the second equation to obtain

$$y = \frac{1}{3} + \frac{5}{3}z$$

We have not found a specific value for y. However, we have found y in terms of z.

To find a value of x in terms of z, we substitute $\frac{1}{3} + \frac{5}{3}z$ for y in the first equation and simplify to get

$$x + y - 2z = -1$$

$$x + \frac{1}{3} + \frac{5}{3}z - 2z = -1 \quad \text{Substitute } \tfrac{1}{3} + \tfrac{5}{3}z \text{ for } y.$$

$$x + \frac{1}{3} - \frac{1}{3}z = -1 \quad \text{Combine like terms.}$$

$$x - \frac{1}{3}z = -\frac{4}{3} \quad \text{Subtract } \tfrac{1}{3} \text{ from both sides.}$$

$$x = -\frac{4}{3} + \frac{1}{3}z \quad \text{Add } \tfrac{1}{3}z \text{ to both sides.}$$

A solution of this system must have the form

$$\left(-\frac{4}{3} + \frac{1}{3}z, \frac{1}{3} + \frac{5}{3}z, z\right) \quad \text{This solution is a general solution of the system.}$$

for all values of z. This system has infinitely many solutions, a different one for each value of z. For example,

- If $z = 0$, the corresponding solution is $\left(-\frac{4}{3}, \frac{1}{3}, 0\right)$.
- If $z = 1$, the corresponding solution is $(-1, 2, 1)$.

Verify that both of these solutions satisfy each equation of the original system.

SELF CHECK 3 Solve the system using matrices: $\begin{cases} x + y - 2z = 11 \\ 2x - y + z = -2 \end{cases}$.

EVERYDAY CONNECTIONS Staffing

Matrices with the same number of rows and columns can be added. We simply add their corresponding elements. For example,

$$\begin{bmatrix} 2 & 3 & -4 \\ -1 & 2 & 5 \end{bmatrix} + \begin{bmatrix} 3 & -1 & 0 \\ 4 & 3 & 2 \end{bmatrix}$$
$$= \begin{bmatrix} 2+3 & 3+(-1) & -4+0 \\ -1+4 & 2+3 & 5+2 \end{bmatrix}$$
$$= \begin{bmatrix} 5 & 2 & -4 \\ 3 & 5 & 7 \end{bmatrix}$$

To multiply a matrix by a constant, we multiply each element of the matrix by the constant. For example,

$$5 \cdot \begin{bmatrix} 2 & 3 & -4 \\ -1 & 2 & 5 \end{bmatrix}$$
$$= \begin{bmatrix} 5 \cdot 2 & 5 \cdot 3 & 5 \cdot (-4) \\ 5 \cdot (-1) & 5 \cdot 2 & 5 \cdot 5 \end{bmatrix}$$
$$= \begin{bmatrix} 10 & 15 & -20 \\ -5 & 10 & 25 \end{bmatrix}$$

Since matrices provide a good way to store information in computers, they often are used in applied problems. For example, suppose there are 66 security officers employed at either the downtown office or the suburban office:

Downtown Office

	Male	Female
Day shift	12	18
Night shift	3	0

Suburban Office

	Male	Female
Day shift	14	12
Night shift	5	2

The information about the employees is contained in the following matrices.

$$D = \begin{bmatrix} 12 & 18 \\ 3 & 0 \end{bmatrix} \quad \text{and} \quad S = \begin{bmatrix} 14 & 12 \\ 5 & 2 \end{bmatrix}$$

The entry in the first row-first column in matrix D gives the information that 12 males work the day shift at the downtown office. Company management can add the matrices D and S to find corporate-wide totals:

$$D + S = \begin{bmatrix} 12 & 18 \\ 3 & 0 \end{bmatrix} + \begin{bmatrix} 14 & 12 \\ 5 & 2 \end{bmatrix}$$
$$= \begin{bmatrix} 26 & 30 \\ 8 & 2 \end{bmatrix}$$

We interpret the total to mean:

	Male	Female
Day shift	26	30
Night shift	8	2

If one-third of the force in each category at the downtown location retires, the downtown staff would be reduced to $\frac{2}{3}D$ people. We can compute $\frac{2}{3}D$ by multiplying each entry by $\frac{2}{3}$.

$$\frac{2}{3}D = \frac{2}{3}\begin{bmatrix} 12 & 18 \\ 3 & 0 \end{bmatrix}$$
$$= \begin{bmatrix} 8 & 12 \\ 2 & 0 \end{bmatrix}$$

After retirements, downtown staff would be

	Male	Female
Day shift	8	12
Night shift	2	0

SELF CHECK ANSWERS **1.** (1, −2, 3) **2.** (3, −2) **3.** infinitely many solutions of the form $\left(3 + \frac{1}{3}z, 8 + \frac{5}{3}z, z\right)$

NOW TRY THIS

1. A toy company builds authentic models of a compact car, a sedan, and a truck. The times required for preparation, assembly, and post-production are given below. Use matrices to help determine how many of each should be made in order to use all available labor hours.

	Compact	Sedan	Truck	Total labor hours
Preparation	1 hr	1 hr	2 hrs	60 hrs
Assembly	2 hrs	3 hrs	4 hrs	130 hrs
Post-production	2 hrs	2 hrs	3 hrs	100 hrs

3.4 EXERCISES

WARM-UPS

Consider the system $\begin{cases} 3x + 2y = 8 \\ 4x - 3y = 6 \end{cases}$.

1. Find the coefficient matrix.

2. Find the augmented matrix.

Determine whether each matrix is in triangular form.

3. $\begin{bmatrix} 4 & 1 & 5 \\ 0 & 2 & 7 \\ 0 & 0 & 4 \end{bmatrix}$

4. $\begin{bmatrix} 8 & 5 & 2 \\ 0 & 4 & 5 \\ 0 & 7 & 0 \end{bmatrix}$

REVIEW *Write each number in scientific notation.*

5. 93,000,000

6. 0.00045

7. 63×10^3

8. 0.33×10^3

VOCABULARY AND CONCEPTS

Fill in the blanks.

9. A ______ is a rectangular array of numbers.

10. The numbers in a matrix are called its ________.

11. A 3×4 matrix has __ rows and 4 ________.

12. A ______ matrix has the same number of rows as columns.

13. An __________ matrix of a system of equations includes the __________ matrix and the column of constants.

14. If a matrix has all 0's below its main diagonal, it is written in __________ form.

15. A ______ row operation corresponds to interchanging two equations in a system of equations.

16. A type 2 row operation corresponds to __________ both sides of an equation by a nonzero constant.

17. A type 3 row operation corresponds to adding a _______ multiple of one equation to another.

18. In the Gaussian method of solving systems of equations, we transform the _________ matrix into triangular form and finish the solution by using _____ substitution.

Use a row operation on the first matrix to find the missing number in the second matrix.

19. $\begin{bmatrix} 2 & 1 & 1 \\ 5 & 4 & 1 \end{bmatrix}$ $\begin{bmatrix} 2 & 1 & 1 \\ 3 & 3 & \blacksquare \end{bmatrix}$

20. $\begin{bmatrix} -1 & 3 & 2 \\ 1 & -2 & 3 \end{bmatrix}$ $\begin{bmatrix} -1 & 3 & 2 \\ \blacksquare & 1 & 5 \end{bmatrix}$

21. $\begin{bmatrix} 3 & -2 & 1 \\ -1 & 2 & 4 \end{bmatrix}$ $\begin{bmatrix} 3 & -2 & 1 \\ -2 & 4 & \blacksquare \end{bmatrix}$

22. $\begin{bmatrix} 2 & 1 & -3 \\ 2 & 6 & 1 \end{bmatrix}$ $\begin{bmatrix} 6 & 3 & \blacksquare \\ 2 & 6 & 1 \end{bmatrix}$

GUIDED PRACTICE

Use matrices to solve each system of equations. See Example 1. (Objective 1)

23. $\begin{cases} x + y = 2 \\ x - y = 0 \end{cases}$

24. $\begin{cases} x + y = 3 \\ x - y = -1 \end{cases}$

25. $\begin{cases} x + 2y = -4 \\ 2x + y = 1 \end{cases}$

26. $\begin{cases} 2x - 3y = 16 \\ -4x + y = -22 \end{cases}$

27. $\begin{cases} x + y + z = 6 \\ x + 2y + z = 8 \\ x + y + 2z = 9 \end{cases}$

28. $\begin{cases} x - y + z = 2 \\ x + 2y - z = 6 \\ 2x - y - z = 3 \end{cases}$

29. $\begin{cases} 2x + y + 3z = 3 \\ -2x - y + z = 5 \\ 4x - 2y + 2z = 2 \end{cases}$

30. $\begin{cases} 3x + 2y + z = 8 \\ 6x - y + 2z = 16 \\ -9x + y - z = -20 \end{cases}$

31. $\begin{cases} x + 2y + 2z = 2 \\ 2x + y - z = 1 \\ 4x + 5y + 3z = 3 \end{cases}$

32. $\begin{cases} x + 2y - z = 3 \\ 2x - y + 2z = 6 \\ x - 3y + 3z = 4 \end{cases}$

Use matrices to solve each system of equations. See Example 2. (Objective 2)

33. $\begin{cases} x + y = 3 \\ 3x - y = 1 \\ 2x + y = 4 \end{cases}$

34. $\begin{cases} x - y = -5 \\ 2x + 3y = 5 \\ x + y = 1 \end{cases}$

35. $\begin{cases} 2x - y = 4 \\ x + 3y = 2 \\ -x - 4y = -2 \end{cases}$

36. $\begin{cases} 3x - 2y = 5 \\ x + 2y = 7 \\ -3x - y = -11 \end{cases}$

37. $\begin{cases} 2x + y = 7 \\ x - y = 2 \\ -x + 3y = -2 \end{cases}$

38. $\begin{cases} 3x - y = 2 \\ -6x + 3y = 0 \\ -x + 2y = -4 \end{cases}$

Use matrices to solve each system of equations. Give a general solution. See Example 3. (Objective 3)

39. $\begin{cases} x + 2y + 3z = -2 \\ -x - y - 2z = 4 \end{cases}$

40. $\begin{cases} 2x - 4y + 3z = 6 \\ -4x + 6y + 4z = -6 \end{cases}$

41. $\begin{cases} x - y = 1 \\ y + z = 1 \\ x + z = 2 \end{cases}$

42. $\begin{cases} x + z = 1 \\ x + y = 2 \\ 2x + y + z = 3 \end{cases}$

ADDITIONAL PRACTICE

Use matrices to solve each system of equations.

43. $\begin{cases} 3x + 4y = -12 \\ 9x - 2y = 6 \end{cases}$

44. $\begin{cases} 5x - 4y = 10 \\ x - 7y = 2 \end{cases}$

45. $\begin{cases} 5a = 24 + 2b \\ 5b = 3a + 16 \end{cases}$

46. $\begin{cases} 3m = 2n + 16 \\ 2m = -5n - 2 \end{cases}$

47. $\begin{cases} 3a + b - 3c = 5 \\ a - 2b + 4c = 10 \\ a + b + c = 13 \end{cases}$

48. $\begin{cases} 2a + b - 3c = -1 \\ 3a - 2b - c = -5 \\ a - 3b - 2c = -12 \end{cases}$

49. $\begin{cases} x + 3y = 7 \\ x + y = 3 \\ 3x + y = 5 \end{cases}$

50. $\begin{cases} x + y = 3 \\ x - 2y = -3 \\ x - y = 1 \end{cases}$

51. $\begin{cases} 5x - 2y = 4 \\ 2x - 4y = -8 \end{cases}$

52. $\begin{cases} 2x - y = -1 \\ x - 2y = 1 \end{cases}$

53. $\begin{cases} 2x + y = -4 \\ 6x + 3y = 1 \end{cases}$

54. $\begin{cases} x - 5y = 7 \\ -2x + 10y = 9 \end{cases}$

55. $\begin{cases} 3x - 2y + 4z = 4 \\ x + y + z = 3 \\ 6x - 2y - 3z = 10 \end{cases}$

56. $\begin{cases} 2x + 3y - z = -8 \\ x - y - z = -2 \\ -4x + 3y + z = 6 \end{cases}$

57. $\begin{cases} 3x - y = 9 \\ -6x + 2y = -18 \end{cases}$

58. $\begin{cases} x - y = 1 \\ -3x + 3y = -3 \end{cases}$

59. $\begin{cases} x + y + z = 6 \\ x - y + z = 2 \end{cases}$

60. $\begin{cases} x - y = 0 \\ y + z = 3 \\ x + z = 3 \end{cases}$

61. $\begin{cases} x + 2y + z = 1 \\ 2x - y + 2z = 2 \\ 3x + y + 3z = 3 \end{cases}$

62. $\begin{cases} x - 2y + 3z = 9 \\ -x + 3y = -4 \\ 2x - 5y + 3z = 13 \end{cases}$

Remember these facts from geometry. Then solve each problem using two variables.

Two angles whose measures add up to 90° are complementary.
Two angles whose measures add up to 180° are supplementary.
The sum of the measures of the interior angles in a triangle is 180°.

63. Geometry One angle is 46° larger than its complement. Find the measure of each angle.

64. Geometry One angle is 28° larger than its supplement. Find the measure of each angle.

65. Geometry In the triangle below, $\angle B$ is 25° more than $\angle A$, and $\angle C$ is 5° less than twice $\angle A$. Find the measure of each angle in the triangle in the next column.

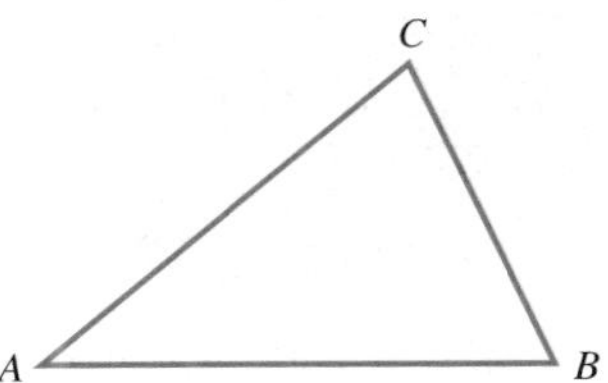

66. Geometry In the triangle below, $\angle A$ is 10° less than $\angle B$, and $\angle B$ is 10° less than $\angle C$. Find the measure of each angle in the triangle.

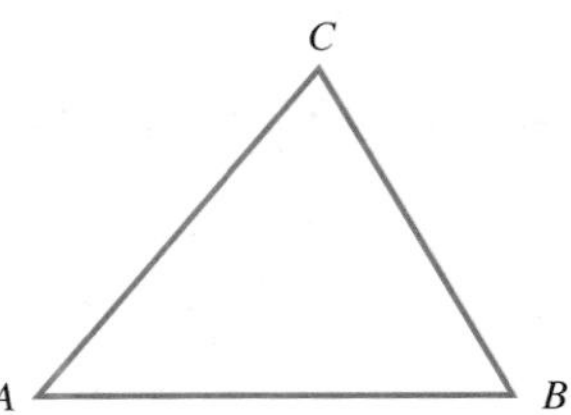

Remember that the equation of a parabola is of the form $y = ax^2 + bx + c$.

67. Curve fitting Find the equation of the parabola passing through the points (0, 1), (1, 2), and (−1, 4).

68. Curve fitting Find the equation of the parabola passing through the points (0, 1), (1, 1), and (−1, −1).

APPLICATIONS

69. Physical therapy After an elbow injury, a volleyball player has restricted movement of her arm. Her range of motion (the measure of $\angle 1$) is 28° less than the measure of $\angle 2$. Find the measure of each angle.

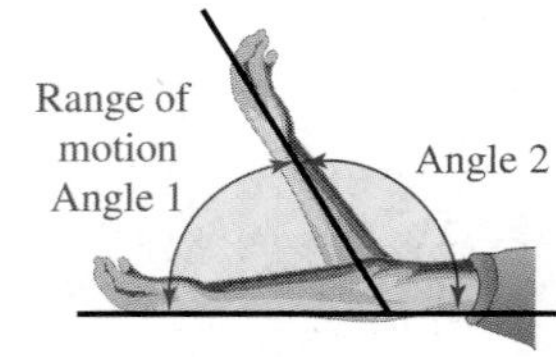

70. Cats and dogs In 2003, there were approximately 135 million dogs and cats in the U.S. If there were 15 million more cats than dogs, how many dogs and cats were there?

71. Piggy banks When a child breaks open her piggy bank, she finds a total of 64 coins, consisting of nickels, dimes, and quarters. The total value of the coins is \$6. If the nickels were dimes, and the dimes were nickels, the value of the coins would be \$5. How many nickels, dimes, and quarters were in the piggy bank?

72. Theater seating The illustration shows the cash receipts and the ticket prices from two sold-out performances of a play. Find the number of seats in each of the three sections of the 800-seat theater.

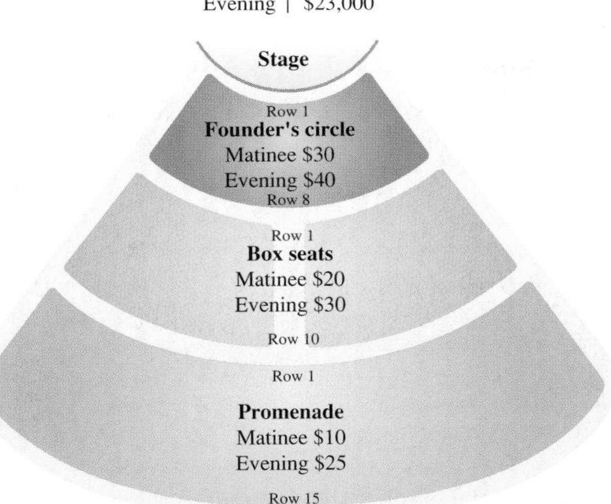

WRITING ABOUT MATH

73. Explain how to check the solution of a system of equations.

74. Explain how to perform a type 3 row operation.

SOMETHING TO THINK ABOUT

75. If the system represented by

$$\left[\begin{array}{ccc|c} 1 & 1 & 0 & 1 \\ 0 & 0 & 1 & 2 \\ 0 & 0 & 0 & k \end{array}\right]$$

has no solution, what do you know about k?

76. Is it possible for a system with fewer equations than variables to have no solution? Illustrate.

SECTION 3.5 Solving Systems of Linear Equations Using Determinants

Objectives

1. Find the determinant of a 2×2 and a 3×3 matrix without a calculator.
2. Solve a system of linear equations using Cramer's rule.

Vocabulary

determinant | minors | Cramer's rule

Getting Ready

Find each value.

1. $3(-4) - 2(5)$

2. $5(2) - 3(-4)$

3. $2(2 - 5) - 3(5 - 2) + 2(4 - 3)$

4. $-3(5 - 2) + 2(3 + 1) - 2(5 + 1)$

We now discuss a final method for solving systems of linear equations. This method involves *determinants,* an idea related to the concept of matrices.

1 Find the determinant of a 2 × 2 and a 3 × 3 matrix without a calculator.

If a matrix A has the same number of rows as columns, it is called a *square matrix.* To each square matrix A, there is associated a number called its *determinant,* represented by the symbol $|A|$.

Value of a 2 × 2 Determinant

If a, b, c, and d are real numbers, the **determinant** of the matrix $\begin{bmatrix} a & b \\ c & d \end{bmatrix}$ is

$$\begin{vmatrix} a & b \\ c & d \end{vmatrix} = ad - bc$$

COMMENT Note that if A is a matrix, $|A|$ represents the determinant of A. If A is a number, $|A|$ represents the absolute value of A.

The determinant of a 2 × 2 matrix is the number that is equal to the product of the numbers on the major diagonal

$$\begin{vmatrix} a & b \\ c & d \end{vmatrix}$$

minus the product of the numbers on the other diagonal

$$\begin{vmatrix} a & b \\ c & d \end{vmatrix}$$

EXAMPLE 1 Evaluate the determinants: **a.** $\begin{vmatrix} 3 & 2 \\ 6 & 9 \end{vmatrix}$ **b.** $\begin{vmatrix} -5 & \frac{1}{2} \\ -1 & 0 \end{vmatrix}$.

Solution

a. $\begin{vmatrix} 3 & 2 \\ 6 & 9 \end{vmatrix} = 3(9) - 2(6)$

$= 27 - 12$

$= 15$

b. $\begin{vmatrix} -5 & \frac{1}{2} \\ -1 & 0 \end{vmatrix} = -5(0) - \frac{1}{2}(-1)$

$= 0 + \frac{1}{2}$

$= \frac{1}{2}$

SELF CHECK 1 Evaluate: $\begin{vmatrix} 4 & -3 \\ 2 & 1 \end{vmatrix}$.

A 3 × 3 determinant can be evaluated by expanding by **minors.**

Value of a 3 × 3 Determinant

Minor of a_1 ↓ Minor of b_1 ↓ Minor of c_1 ↓

$$\begin{vmatrix} a_1 & b_1 & c_1 \\ a_2 & b_2 & c_2 \\ a_3 & b_3 & c_3 \end{vmatrix} = a_1 \begin{vmatrix} b_2 & c_2 \\ b_3 & c_3 \end{vmatrix} - b_1 \begin{vmatrix} a_2 & c_2 \\ a_3 & c_3 \end{vmatrix} + c_1 \begin{vmatrix} a_2 & b_2 \\ a_3 & b_3 \end{vmatrix}$$

To find the minor of a_1, we find the determinant formed by crossing out the elements of the matrix that are in the same row and column as a_1:

$$\begin{vmatrix} a_1 & b_1 & c_1 \\ a_2 & b_2 & c_2 \\ a_3 & b_3 & c_3 \end{vmatrix} \qquad \text{The minor of } a_1 \text{ is } \begin{vmatrix} b_2 & c_2 \\ b_3 & c_3 \end{vmatrix}.$$

To find the minor of b_1, we cross out the elements of the matrix that are in the same row and column as b_1:

$$\begin{vmatrix} a_1 & b_1 & c_1 \\ a_2 & b_2 & c_2 \\ a_3 & b_3 & c_3 \end{vmatrix} \qquad \text{The minor of } b_1 \text{ is } \begin{vmatrix} a_2 & c_2 \\ a_3 & c_3 \end{vmatrix}.$$

To find the minor of c_1, we cross out the elements of the matrix that are in the same row and column as c_1:

$$\begin{vmatrix} a_1 & b_1 & c_1 \\ a_2 & b_2 & c_2 \\ a_3 & b_3 & c_3 \end{vmatrix} \qquad \text{The minor of } c_1 \text{ is } \begin{vmatrix} a_2 & b_2 \\ a_3 & b_3 \end{vmatrix}.$$

EXAMPLE 2 Evaluate the determinant: $\begin{vmatrix} 1 & 3 & -2 \\ 2 & 1 & 3 \\ 1 & 2 & 3 \end{vmatrix}$.

Solution

$$\begin{vmatrix} 1 & 3 & -2 \\ 2 & 1 & 3 \\ 1 & 2 & 3 \end{vmatrix} = 1\underbrace{\begin{vmatrix} 1 & 3 \\ 2 & 3 \end{vmatrix}}_{\text{Minor of 1}} - 3\underbrace{\begin{vmatrix} 2 & 3 \\ 1 & 3 \end{vmatrix}}_{\text{Minor of 3}} + (-2)\underbrace{\begin{vmatrix} 2 & 1 \\ 1 & 2 \end{vmatrix}}_{\text{Minor of }-2}$$

$$= 1(3 - 6) - 3(6 - 3) - 2(4 - 1)$$

$$= -3 - 9 - 6$$

$$= -18$$

SELF CHECK 2 Evaluate: $\begin{vmatrix} 2 & -1 & 3 \\ 1 & 2 & -2 \\ 3 & 1 & 1 \end{vmatrix}$.

We can evaluate a 3×3 determinant by expanding it along any row or column. To determine the signs between the terms of the expansion of a 3×3 determinant, we use the following array of signs.

Array of Signs for a 3×3 Determinant

$$\begin{matrix} + & - & + \\ - & + & - \\ + & - & + \end{matrix}$$

The pattern above holds true for any square matrix.

PERSPECTIVE Lewis Carroll

One of the more amusing historical anecdotes concerning matrices and determinants involves the English mathematician Charles Dodgson, also known as Lewis Carroll. The anecdote describes how England's Queen Victoria so enjoyed reading Carroll's book *Alice in Wonderland* that she requested a copy of his next publication. To her great surprise, she received an autographed copy of a mathematics text titled *An Elementary Treatise on Determinants.* The story was repeated as fact so often that Carroll finally included an explicit disclaimer in his book *Symbolic Logic,* insisting that the incident never actually occurred.

Source: http://mathworld.wolfram.com/Determinant.html

Evaluate each determinant.

1. $\begin{vmatrix} 3 & 4 \\ 2 & 1 \end{vmatrix}$

2. $\begin{vmatrix} 3 & 4 & 2 \\ 1 & -1 & 5 \\ 1 & 2 & -2 \end{vmatrix}$

EXAMPLE 3 Evaluate $\begin{vmatrix} 1 & 3 & -2 \\ 2 & 1 & 3 \\ 1 & 2 & 3 \end{vmatrix}$ by expanding on the middle column.

Solution This is the determinant of Example 2. To expand it along the middle column, we use the signs of the middle column of the array of signs:

$$\begin{vmatrix} 1 & \mathbf{3} & -2 \\ 2 & \mathbf{1} & 3 \\ 1 & \mathbf{2} & 3 \end{vmatrix} = \underset{\text{Minor of 3}}{-3\begin{vmatrix} 2 & 3 \\ 1 & 3 \end{vmatrix}} + \underset{\text{Minor of 1}}{1\begin{vmatrix} 1 & -2 \\ 1 & 3 \end{vmatrix}} \underset{\text{Minor of 2}}{-2\begin{vmatrix} 1 & -2 \\ 2 & 3 \end{vmatrix}}$$

$$= -3(6 - 3) + 1[3 - (-2)] - 2[3 - (-4)]$$

$$= -3(3) + 1(5) - 2(7)$$

$$= -9 + 5 - 14$$

$$= -18$$

As expected, we get the same value as in Example 2.

SELF CHECK 3 Evaluate: $\begin{vmatrix} 2 & -1 & 3 \\ 1 & 2 & 2 \\ 3 & 1 & 1 \end{vmatrix}$.

2 Solve a system of linear equations using Cramer's rule.

The method of using determinants to solve systems of equations is called **Cramer's rule,** named after the 18th-century mathematician Gabriel Cramer. To develop Cramer's rule, we consider the system

$$\begin{cases} ax + by = e \\ cx + dy = f \end{cases}$$

where x and y are variables and a, b, c, d, e, and f are constants.

ACCENT ON TECHNOLOGY

Evaluating Determinants

To use a graphing calculator to evaluate the determinant in Example 3, we first enter the matrix by pressing the **MATRIX** key, selecting EDIT, and pressing the **ENTER** key. We then enter the dimensions and the elements of the matrix to get Figure 3-15(a). We then press **2ND** **QUIT** to clear the screen. We then press **MATRIX**, select MATH, and press 1 to get Figure 3-15(b). Next, press **MATRIX**, select NAMES, and press 1 to get Figure 3-15(c). To get the value of the determinant, we now press **ENTER** to get Figure 3-15(d), which shows that the value of the determinant is -18.

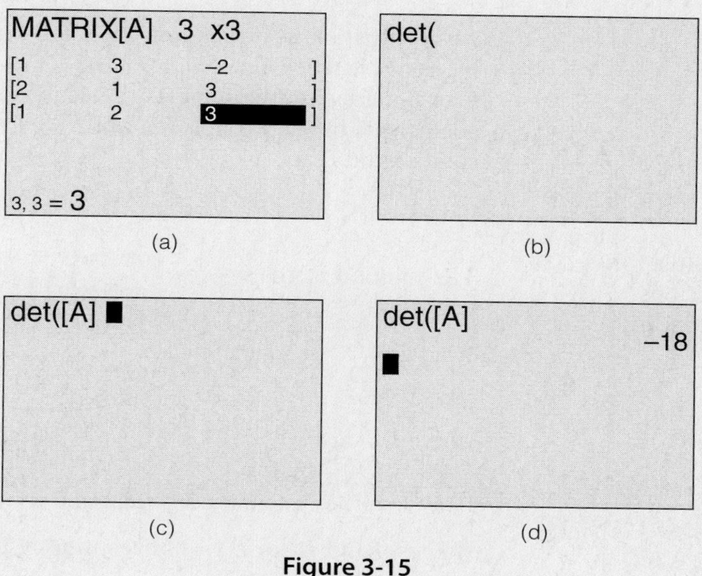

Figure 3-15

If we multiply both sides of the first equation by d and multiply both sides of the second equation by $-b$, we can add the equations and eliminate y:

$$\begin{array}{rcl} adx + bdy &=& ed \\ -bcx - bdy &=& -bf \\ \hline adx - bcx &=& ed - bf \end{array}$$

To solve for x, we use the distributive property to write $adx - bcx$ as $(ad - bc)x$ on the left side and divide each side by $ad - bc$:

$$(ad - bc)x = ed - bf$$

$$x = \frac{ed - bf}{ad - bc} \quad (ad - bc) \neq 0$$

We can find y in a similar manner. After eliminating the variable x, we get

$$y = \frac{af - ec}{ad - bc} \quad (ad - bc) \neq 0$$

Determinants provide an easy way of remembering these formulas. Note that the denominator for both x and y is

$$\begin{vmatrix} a & b \\ c & d \end{vmatrix} = ad - bc$$

Gabriel Cramer (1704–1752)

Although other mathematicians had worked with determinants, it was the work of Cramer that popularized them.

The numerators can be expressed as determinants also:

$$x = \frac{ed - bf}{ad - bc} = \frac{\begin{vmatrix} e & b \\ f & d \end{vmatrix}}{\begin{vmatrix} a & b \\ c & d \end{vmatrix}} \quad \text{and} \quad x = \frac{af - ec}{ad - bc} = \frac{\begin{vmatrix} a & e \\ c & f \end{vmatrix}}{\begin{vmatrix} a & b \\ c & d \end{vmatrix}}$$

If we compare these formulas with the original system

$$\begin{cases} ax + by = e \\ cx + dy = f \end{cases}$$

we note that in the expressions for x and y above, the denominator determinant is formed by using the coefficients a, b, c, and d of the variables in the equations. The numerator determinants are the same as the denominator determinant, except that the column of coefficients of the variable for which we are solving is replaced with the column of constants e and f.

Cramer's Rule for Two Equations in Two Variables

The solution of the system $\begin{cases} ax + by = e \\ cx + dy = f \end{cases}$ is given by

$$x = \frac{D_x}{D} \quad \text{and} \quad y = \frac{D_y}{D}$$

where $D = \begin{vmatrix} a & b \\ c & d \end{vmatrix}$, $D_x = \begin{vmatrix} \mathbf{e} & b \\ \mathbf{f} & d \end{vmatrix}$, and $D_y = \begin{vmatrix} a & \mathbf{e} \\ c & \mathbf{f} \end{vmatrix}$.

If $D \neq 0$, the system is consistent and the equations are independent.

If $D = 0$ and D_x or D_y is nonzero, the system is inconsistent.

If every determinant is 0, the system is consistent but the equations are dependent.

EXAMPLE 4 Use Cramer's rule to solve $\begin{cases} 4x - 3y = 6 \\ -2x + 5y = 4 \end{cases}$.

Solution The value of x is the quotient of two determinants. The denominator determinant is made up of the coefficients of x and y:

$$D = \begin{vmatrix} 4 & -3 \\ -2 & 5 \end{vmatrix}$$

To solve for x, we form the numerator determinant from the denominator determinant by replacing its first column (the coefficients of x) with the column of constants (6 and 4).

To solve for y, we form the numerator determinant from the denominator determinant by replacing the second column (the coefficients of y) with the column of constants (6 and 4).

To find the values of x and y, we evaluate each determinant:

$$x = \frac{D_x}{D} = \frac{\begin{vmatrix} 6 & -3 \\ 4 & 5 \end{vmatrix}}{\begin{vmatrix} 4 & -3 \\ -2 & 5 \end{vmatrix}} = \frac{6(5) - (-3)(4)}{4(5) - (-3)(-2)} = \frac{30 + 12}{20 - 6} = \frac{42}{14} = 3$$

$$y = \frac{D_y}{D} = \frac{\begin{vmatrix} 4 & 6 \\ -2 & 4 \end{vmatrix}}{\begin{vmatrix} 4 & -3 \\ -2 & 5 \end{vmatrix}} = \frac{4(4) - 6(-2)}{4(5) - (-3)(-2)} = \frac{16 + 12}{20 - 6} = \frac{28}{14} = 2$$

The solution of this system is (3, 2). Verify that $x = 3$ and $y = 2$ satisfy each equation in the given system.

SELF CHECK 4 Solve the system: $\begin{cases} 2x - 3y = -16 \\ 3x + 5y = 14 \end{cases}$.

EXAMPLE 5 Use Cramer's rule to solve $\begin{cases} 7x = 8 - 4y \\ 2y = 3 - \frac{7}{2}x \end{cases}$.

Solution We multiply both sides of the second equation by 2 to eliminate the fraction and write the system in the form

$$\begin{cases} 7x + 4y = 8 \\ 7x + 4y = 6 \end{cases}$$

When we attempt to use Cramer's rule to solve this system for x, we obtain

$$x = \frac{D_x}{D} = \frac{\begin{vmatrix} 8 & 4 \\ 6 & 4 \end{vmatrix}}{\begin{vmatrix} 7 & 4 \\ 7 & 4 \end{vmatrix}} = \frac{8}{0} \quad \text{which is undefined}$$

This system is inconsistent because the denominator determinant is 0 and the numerator determinant is not 0. Since this system has no solution, its solution set is $\varnothing$.

We can see directly from the system that it is inconsistent. For any values of x and y, it is impossible that 7 times x plus 4 times y could be both 8 and 6.

SELF CHECK 5 Solve the system: $\begin{cases} 3x = 8 - 4y \\ y = \frac{5}{2} - \frac{3}{4}x \end{cases}$.

EXAMPLE 6 Use Cramer's rule to solve $\begin{cases} y = -3x + 1 \\ 6x + 2y = 2 \end{cases}$.

Solution We first write the system in the form

$$\begin{cases} 3x + y = 1 \\ 6x + 2y = 2 \end{cases}$$

When we attempt to use Cramer's rule to solve this system for x, we obtain

$$x = \frac{D_x}{D} = \frac{\begin{vmatrix} 1 & 1 \\ 2 & 2 \end{vmatrix}}{\begin{vmatrix} 3 & 1 \\ 6 & 2 \end{vmatrix}} = \frac{0}{0} \quad \text{which is indeterminate.}$$

When we attempt to use Cramer's rule to solve this system for y, we obtain

$$y = \frac{D_y}{D} = \frac{\begin{vmatrix} 3 & 1 \\ 6 & 2 \end{vmatrix}}{\begin{vmatrix} 3 & 1 \\ 6 & 2 \end{vmatrix}} = \frac{0}{0} \quad \text{which is indeterminate.}$$

This system is consistent and its equations are dependent because every determinant is 0. Every solution of one equation is also a solution of the other. The solution is the general ordered pair $(x, -3x + 1)$.

SELF CHECK 6 Solve the system: $\begin{cases} y = 2x - 5 \\ 10x - 5y = 25 \end{cases}$.

Cramer's Rule for Three Equations in Three Variables

The solution of the system $\begin{cases} ax + by + cz = j \\ dx + ey + fz = k \\ gx + hy + iz = l \end{cases}$ is given by

$$x = \frac{D_x}{D}, \quad y = \frac{D_y}{D}, \quad \text{and} \quad z = \frac{D_z}{D}$$

where

$$D = \begin{vmatrix} a & b & c \\ d & e & f \\ g & h & i \end{vmatrix} \qquad D_x = \begin{vmatrix} \mathbf{j} & b & c \\ \mathbf{k} & e & f \\ \mathbf{l} & h & i \end{vmatrix}$$

$$D_y = \begin{vmatrix} a & \mathbf{j} & c \\ d & \mathbf{k} & f \\ g & \mathbf{l} & i \end{vmatrix} \qquad D_z = \begin{vmatrix} a & b & \mathbf{j} \\ d & e & \mathbf{k} \\ g & h & \mathbf{l} \end{vmatrix}$$

If $D \neq 0$, the system is consistent and the equations are independent.

If $D = 0$ and D_x or D_y or D_z is nonzero, the system is inconsistent.

If every determinant is 0, the system is consistent but the equations are dependent.

EXAMPLE 7 Use Cramer's rule to solve $\begin{cases} 2x + y + 4z = 12 \\ x + 2y + 2z = 9 \\ 3x - 3y - 2z = 1 \end{cases}$.

Solution The denominator determinant is the determinant formed by the coefficients of the variables. To form the numerator determinants, we substitute the column of constants for the coefficients of the variable to be found. We form the quotients for x, y, and z and evaluate the determinants:

$$x = \frac{D_x}{D} = \frac{\begin{vmatrix} 12 & 1 & 4 \\ 9 & 2 & 2 \\ 1 & -3 & -2 \end{vmatrix}}{\begin{vmatrix} 2 & 1 & 4 \\ 1 & 2 & 2 \\ 3 & -3 & -2 \end{vmatrix}} = \frac{12\begin{vmatrix} 2 & 2 \\ -3 & -2 \end{vmatrix} - 1\begin{vmatrix} 9 & 2 \\ 1 & -2 \end{vmatrix} + 4\begin{vmatrix} 9 & 2 \\ 1 & -3 \end{vmatrix}}{2\begin{vmatrix} 2 & 2 \\ -3 & -2 \end{vmatrix} - 1\begin{vmatrix} 1 & 2 \\ 3 & -2 \end{vmatrix} + 4\begin{vmatrix} 1 & 2 \\ 3 & -3 \end{vmatrix}} = \frac{12(2) - (-20) + 4(-29)}{2(2) - (-8) + 4(-9)} = \frac{-72}{-24} = 3$$

$$y = \frac{D_y}{D} = \frac{\begin{vmatrix} 2 & 12 & 4 \\ 1 & 9 & 2 \\ 3 & 1 & -2 \end{vmatrix}}{\begin{vmatrix} 2 & 1 & 4 \\ 1 & 2 & 2 \\ 3 & -3 & -2 \end{vmatrix}} = \frac{2\begin{vmatrix} 9 & 2 \\ 1 & -2 \end{vmatrix} - 12\begin{vmatrix} 1 & 2 \\ 3 & -2 \end{vmatrix} + 4\begin{vmatrix} 1 & 9 \\ 3 & 1 \end{vmatrix}}{-24} = \frac{2(-20) - 12(-8) + 4(-26)}{-24} = \frac{-48}{-24} = 2$$

$$z = \frac{D_z}{D} = \frac{\begin{vmatrix} 2 & 1 & 12 \\ 1 & 2 & 9 \\ 3 & -3 & 1 \end{vmatrix}}{\begin{vmatrix} 2 & 1 & 4 \\ 1 & 2 & 2 \\ 3 & -3 & -2 \end{vmatrix}} = \frac{2\begin{vmatrix} 2 & 9 \\ -3 & 1 \end{vmatrix} - 1\begin{vmatrix} 1 & 9 \\ 3 & 1 \end{vmatrix} + 12\begin{vmatrix} 1 & 2 \\ 3 & -3 \end{vmatrix}}{-24} = \frac{2(29) - 1(-26) + 12(-9)}{-24} = \frac{-24}{-24} = 1$$

The solution of this system is (3, 2, 1).

SELF CHECK 7 Solve the system: $\begin{cases} x + y + 2z = 6 \\ 2x - y + z = 9 \\ x + y - 2z = -6 \end{cases}$.

SELF CHECK ANSWERS **1.** 10 **2.** 0 **3.** -20 **4.** $(-2, 4)$ **5.** $\varnothing$ **6.** $(x, 2x - 5)$ **7.** $(2, -2, 3)$

NOW TRY THIS

Solve for x.

1. $\begin{vmatrix} x & 2 \\ x & 3 \end{vmatrix} = 4$

2. $\begin{vmatrix} 2x & 3 \\ -5x & -3 \end{vmatrix} = 10(x - 1)$

3. $\begin{vmatrix} x + 4 & 3 \\ 2x - 5 & 2 \end{vmatrix} = 2x + 5$

4. $\begin{vmatrix} 2 & x & -1 \\ 1 & 2x & 4 \\ -4 & x & 1 \end{vmatrix} = 30$

3.5 EXERCISES

WARM-UPS

Evaluate each determinant.

1. $\begin{vmatrix} 2 & 1 \\ 1 & 1 \end{vmatrix}$

2. $\begin{vmatrix} 0 & 2 \\ 1 & 1 \end{vmatrix}$

3. $\begin{vmatrix} 0 & 1 \\ 0 & 1 \end{vmatrix}$

When using Cramer's rule to solve the system $\begin{cases} x + 2y = 5 \\ 2x - y = 4 \end{cases}$,

4. Set up the denominator determinant for x.

5. Set up the numerator determinant for x.

6. Set up the numerator determinant for y.

REVIEW *Solve each equation.*

7. $3(x + 2) - (2 - x) = x - 5$

8. $\frac{3}{7}x = 2(x + 11)$

9. $\frac{5}{3}(5x + 6) - 10 = 0$

10. $5 - 3(2x - 1) = 2(4 + 3x) - 24$

VOCABULARY AND CONCEPTS *Fill in the blanks.*

11. A determinant is a ________ that is associated with a ________ matrix.

12. The value of $\begin{vmatrix} a & b \\ c & d \end{vmatrix}$ is ________.

13. The minor of b_1 in $\begin{vmatrix} a_1 & b_1 & c_1 \\ a_2 & b_2 & c_2 \\ a_3 & b_3 & c_3 \end{vmatrix}$ is ________.

14. We can evaluate a determinant by expanding it along any _____ or ________.

15. The method of solving a system of linear equations using determinants is called ____________.

16. The set up for the denominator determinant for the value of x in the system $\begin{cases} 3x + 4y = 7 \\ 2x - 3y = 5 \end{cases}$ is ________.

17. If $D \neq 0$, then the system is __________ and the equations are ____________.

18. If the denominator determinant for y in a system of equations is zero, the equations of the system are __________ or the system is ___________.

GUIDED PRACTICE

Evaluate each determinant. See Example 1. (Objective 1)

19. $\begin{vmatrix} 2 & 3 \\ -2 & 1 \end{vmatrix}$

20. $\begin{vmatrix} 3 & -2 \\ -2 & 4 \end{vmatrix}$

21. $\begin{vmatrix} -1 & 2 \\ 3 & -4 \end{vmatrix}$

22. $\begin{vmatrix} -1 & -2 \\ -3 & -4 \end{vmatrix}$

Evaluate each determinant. See Examples 2–3. (Objective 1)

23. $\begin{vmatrix} 1 & 0 & 1 \\ 0 & 1 & 0 \\ 1 & 1 & 1 \end{vmatrix}$

24. $\begin{vmatrix} 1 & 2 & 0 \\ 0 & 1 & 2 \\ 0 & 0 & 1 \end{vmatrix}$

25. $\begin{vmatrix} -1 & 2 & 1 \\ 2 & 1 & -3 \\ 1 & 1 & 1 \end{vmatrix}$

26. $\begin{vmatrix} 1 & 2 & 3 \\ 1 & 2 & 3 \\ 1 & 2 & 3 \end{vmatrix}$

27. $\begin{vmatrix} 1 & -2 & 3 \\ -2 & 1 & 1 \\ -3 & -2 & 1 \end{vmatrix}$

28. $\begin{vmatrix} 1 & 1 & 2 \\ 2 & 1 & -2 \\ 3 & 1 & 3 \end{vmatrix}$

29. $\begin{vmatrix} 1 & 2 & 3 \\ 4 & 5 & 6 \\ 7 & 8 & 9 \end{vmatrix}$

30. $\begin{vmatrix} 1 & 4 & 7 \\ 2 & 5 & 8 \\ 3 & 6 & 9 \end{vmatrix}$

Use Cramer's rule to solve each system, if possible. If the equations of the system are dependent, give a general solution. See Examples 4–6. (Objective 2)

31. $\begin{cases} 2x + y = 1 \\ x - 2y = -7 \end{cases}$

32. $\begin{cases} 3x - y = -3 \\ 2x + y = -7 \end{cases}$

33. $\begin{cases} x + y = 6 \\ x - y = 2 \end{cases}$

34. $\begin{cases} x - y = 4 \\ 2x + y = 5 \end{cases}$

35. $\begin{cases} 5x = 3y - 7 \\ y = \dfrac{5x - 7}{3} \end{cases}$

36. $\begin{cases} y = \dfrac{11 - 3x}{2} \\ x = \dfrac{11 - 4y}{6} \end{cases}$

37. $\begin{cases} 2x + 3y = 9 \\ y = -\dfrac{2}{3}x + 3 \end{cases}$

38. $\begin{cases} x = \dfrac{12 - 6y}{5} \\ y = \dfrac{24 - 10x}{12} \end{cases}$

Use Cramer's rule to solve each system, if possible. If the equations of the system are dependent, give a general solution. See Example 7. (Objective 2)

39. $\begin{cases} x + y + z = 4 \\ x + y - z = 0 \\ x - y + z = 2 \end{cases}$

40. $\begin{cases} x + y + z = 4 \\ x - y + z = 2 \\ x - y - z = 0 \end{cases}$

41. $\begin{cases} x + y + 2z = 7 \\ x + 2y + z = 8 \\ 2x + y + z = 9 \end{cases}$

42. $\begin{cases} x + 2y + 2z = 10 \\ 2x + y + 2z = 9 \\ 2x + 2y + z = 1 \end{cases}$

43. $\begin{cases} 2x + y - z = 1 \\ x + 2y + 2z = 2 \\ 4x + 5y + 3z = 3 \end{cases}$

44. $\begin{cases} 2x - y + 4z + 2 = 0 \\ 5x + 8y + 7z = -8 \\ x + 3y + z + 3 = 0 \end{cases}$

45. $\begin{cases} 2x + 3y + 4z = 6 \\ 2x - 3y - 4z = -4 \\ 4x + 6y + 8z = 12 \end{cases}$

46. $\begin{cases} x - 3y + 4z - 2 = 0 \\ 2x + y + 2z - 3 = 0 \\ 4x - 5y + 10z - 7 = 0 \end{cases}$

ADDITIONAL PRACTICE

Evaluate each determinant.

47. $\begin{vmatrix} x & y \\ y & x \end{vmatrix}$

48. $\begin{vmatrix} x + y & y - x \\ x & y \end{vmatrix}$

49. $\begin{vmatrix} a & 2a & -a \\ 2 & -1 & 3 \\ 1 & 2 & -3 \end{vmatrix}$

50. $\begin{vmatrix} 1 & 2b & -3 \\ 2 & -b & 2 \\ 1 & 3b & 1 \end{vmatrix}$

51. $\begin{vmatrix} 1 & a & b \\ 1 & 2a & 2b \\ 1 & 3a & 3b \end{vmatrix}$

52. $\begin{vmatrix} a & b & c \\ 0 & b & c \\ 0 & 0 & c \end{vmatrix}$

Use a graphing calculator to evaluate each determinant.

53. $\begin{bmatrix} 2 & -3 & 4 \\ -1 & 2 & 4 \\ 3 & -3 & 1 \end{bmatrix}$

54. $\begin{bmatrix} -3 & 2 & -5 \\ 3 & -2 & 6 \\ 1 & -3 & 4 \end{bmatrix}$

55. $\begin{bmatrix} 2 & 1 & -3 \\ -2 & 2 & 4 \\ 1 & -2 & 2 \end{bmatrix}$

56. $\begin{bmatrix} 4 & 2 & -3 \\ 2 & -5 & 6 \\ 2 & 5 & -2 \end{bmatrix}$

Use Cramer's rule to solve each system.

57. $\begin{cases} 2x + 3y = 0 \\ 4x - 6y = -4 \end{cases}$

58. $\begin{cases} 4x - 3y = -1 \\ 8x + 3y = 4 \end{cases}$

59. $\begin{cases} y = \frac{-2x + 1}{3} \\ 3x - 2y = 8 \end{cases}$

60. $\begin{cases} 2x + 3y = -1 \\ x = \frac{y - 9}{4} \end{cases}$

61. $\begin{cases} x = \frac{5y - 4}{2} \\ y = \frac{3x - 1}{5} \end{cases}$

62. $\begin{cases} y = \frac{1 - 5x}{2} \\ x = \frac{3y + 10}{4} \end{cases}$

63. $\begin{cases} 2x + y + z = 5 \\ x - 2y + 3z = 10 \\ x + y - 4z = -3 \end{cases}$

64. $\begin{cases} 3x + 2y - z = -8 \\ 2x - y + 7z = 10 \\ 2x + 2y - 3z = -10 \end{cases}$

65. $\begin{cases} 4x + 3z = 4 \\ 2y - 6z = -1 \\ 8x + 4y + 3z = 9 \end{cases}$

66. $\begin{cases} \frac{1}{2}x + y + z + \frac{3}{2} = 0 \\ x + \frac{1}{2}y + z - \frac{1}{2} = 0 \\ x + y + \frac{1}{2}z + \frac{1}{2} = 0 \end{cases}$

67. $\begin{cases} x + y = 1 \\ \frac{1}{2}y + z = \frac{5}{2} \\ x - z = -3 \end{cases}$

68. $\begin{cases} 3x + 4y + 14z = 7 \\ -\frac{1}{2}x - y + 2z = \frac{3}{2} \\ x + \frac{3}{2}y + \frac{5}{2}z = 1 \end{cases}$

Solve the equation.

69. $\begin{vmatrix} x & 1 \\ 3 & 2 \end{vmatrix} = 1$

70. $\begin{vmatrix} x & -x \\ 2 & -3 \end{vmatrix} = -5$

71. $\begin{vmatrix} x & -2 \\ 3 & 1 \end{vmatrix} = \begin{vmatrix} 4 & 2 \\ x & 3 \end{vmatrix}$

72. $\begin{vmatrix} x & 3 \\ x & 2 \end{vmatrix} = \begin{vmatrix} 3 & 2 \\ 1 & 1 \end{vmatrix}$

APPLICATIONS

73. Signaling A system of sending signals uses two flags held in various positions to represent letters of the alphabet. The illustration shows how the letter U is signaled. Find x and y, if y is to be 30° more than x.

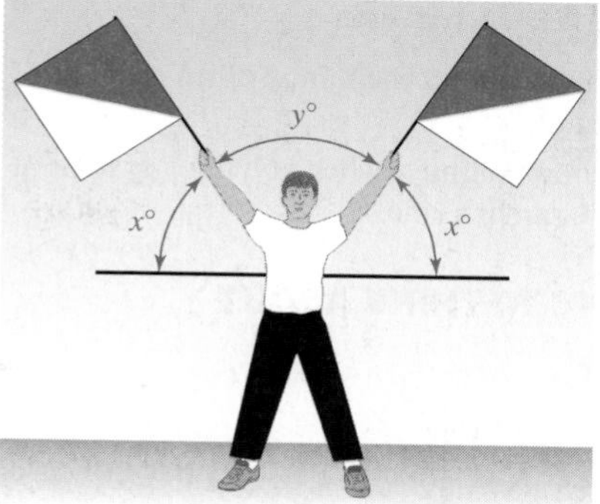

74. Inventories The table shows an end-of-the-year inventory report for a warehouse that supplies electronics stores. If the warehouse stocks two models of cordless telephones, one valued at \$67 and the other at \$100, how many of each model of phone did the warehouse have at the time of the inventory?

Item	Number	Merchandise value
Television	800	\$1,005,450
Radios	200	\$15,785
Cordless phones	360	\$29,400

75. Investing A student wants to average a 6.6% return by investing \$20,000 in the three stocks listed in the table. Because HiTech is considered to be a high-risk investment, he wants to invest three times as much in SaveTel and HiGas combined as he invests in HiTech. How much should he invest in each stock?

Stock	Rate of return
HiTech	10%
SaveTel	5%
HiGas	6%

76. Investing See the table. A woman wants to average a $7\frac{1}{3}$% return by investing \$30,000 in three certificates of deposit. She wants to invest five times as much in the 8% CD as in the 6% CD. How much should she invest in each CD?

Type of CD	Rate of return
12 month	6%
24 month	7%
36 month	8%

WRITING ABOUT MATH

77. Explain how to find the minor of an element of a determinant.

78. Explain how to find x when solving a system of linear equations by Cramer's rule.

SOMETHING TO THINK ABOUT

79. Show that

$$\begin{vmatrix} x & y & 1 \\ -2 & 3 & 1 \\ 3 & 5 & 1 \end{vmatrix} = 0$$

is the equation of the line passing through $(-2, 3)$ and $(3, 5)$.

80. Show that

$$\frac{1}{2}\begin{vmatrix} 0 & 0 & 1 \\ 3 & 0 & 1 \\ 0 & 4 & 1 \end{vmatrix}$$

is the area of the triangle with vertices at $(0, 0)$, $(3, 0)$, and $(0, 4)$.

Determinants with more than 3 rows and 3 columns can be evaluated by expanding them by minors. The sign array for a 4×4 determinant is

$$\begin{matrix} + & - & + & - \\ - & + & - & + \\ + & - & + & - \\ - & + & - & + \end{matrix}$$

Evaluate each determinant.

81. $\begin{vmatrix} 1 & 0 & 2 & 1 \\ 2 & 1 & 1 & 3 \\ 1 & 1 & 1 & 1 \\ 2 & 1 & 1 & 1 \end{vmatrix}$

82. $\begin{vmatrix} 1 & 2 & -1 & 1 \\ -2 & 1 & 3 & -1 \\ 0 & 1 & 1 & 2 \\ 2 & 0 & 3 & 1 \end{vmatrix}$

PROJECTS

Project 1

The number of units of a product that will be produced depends on the unit price of the product. As the unit price gets higher, the product will be produced in greater quantity, because the producer will make more money on each item. The *supply* of the product will grow, and we say that supply *is a function of* (or *depends on*) the unit price. Furthermore, as the price rises, fewer consumers will buy the product, and the *demand* will decrease. The demand for the product is also a function of the unit price.

In this project, we will assume that both supply and demand are *linear* functions of the unit price. Thus, the graph of supply (the y-coordinate) versus price (the x-coordinate) is a line with positive slope. The graph of the demand function is a line with negative slope. Because these two lines cannot be parallel, they must intersect. The price at which supply equals demand is called the *market price:* At this price, the same number of units of the product will be sold as are manufactured.

You work for Soda Pop Inc. and have the task of analyzing the sales figures for the past year. You have been provided with the following supply and demand functions. (Supply and demand are measured in cases per week; p, the price per case, is measured in dollars.)

The demand for soda is $D(p) = 19{,}000 - 2{,}200p$.

The supply of soda is $S(p) = 3{,}000 + 1{,}080p$.

Both functions are true for values of p from \$3.50 to \$5.75.

Graph both functions on the same set of coordinate axes, being sure to label each graph, and include any other important information. Then write a report for your supervisor that answers the following questions.

a. Explain why producers will be able to sell all of the soda they make when the price is \$3.50 per case. How much money will the producers take in from these sales?

b. How much money will producers take in from sales when the price is \$5.75 per case? How much soda will not be sold?

c. Find the market price for soda (to the nearest cent). How many cases per week will be sold at this price? How much money will the producers take in from sales at the market price?

d. Explain why prices always tend toward the market price. That is, explain why the unit price will rise if the demand is greater than the supply, and why the unit price will fall if supply is greater than demand.

Project 2

Goodstuff Produce Company has two large water canals that feed the irrigation ditches on its fruit farm. One of these canals runs directly north and south, and the other runs directly east and west. The canals cross at the center of the farm property (the origin) and divide the farm into four quadrants. The company is interested in digging some new irrigation ditches in a portion of the northeast quadrant. You have been hired to plan the layout of the new system.

Your design is to make use of ditch Z, which is already present. This ditch runs from a point 300 meters north of the origin to a point 400 meters east of the origin. The owners of Goodstuff want two new ditches dug.

- Ditch A is to begin at a point 100 meters north of the origin and follow a line that travels 3 meters north for every 7 meters it travels east until it intersects ditch Z.
- Ditch B is to run from the origin to ditch Z in such a way that it exactly bisects the area in the northeast quadrant that is south of both ditch Z and ditch A.

You are to provide the equations of the lines that the three ditches follow, as well as the exact location of the gates that will be installed where the ditches intersect one another. Be sure to provide explanations and organized work that will clearly display the desired information and assure the owners of Goodstuff that they will get exactly what they want.

Chapter 3 REVIEW

SECTION 3.1 Solving Systems of Two Linear Equations by Graphing

DEFINITIONS AND CONCEPTS	EXAMPLES
In a graph of two equations, each with two variables: If the lines are different and intersect, the equations are *independent* and the system is *consistent*. **One solution exists.**	Solve each system by graphing. **a.** $\begin{cases} x + y = 6 \\ x - y = 2 \end{cases}$ 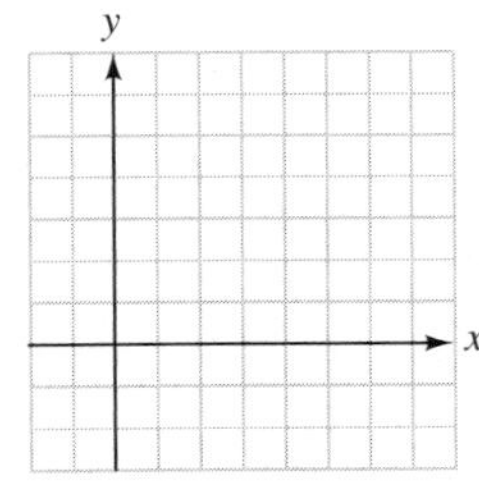 The solution is (4, 2). **b.** $\begin{cases} 2x + y = 6 \\ y = -2x - 3 \end{cases}$ 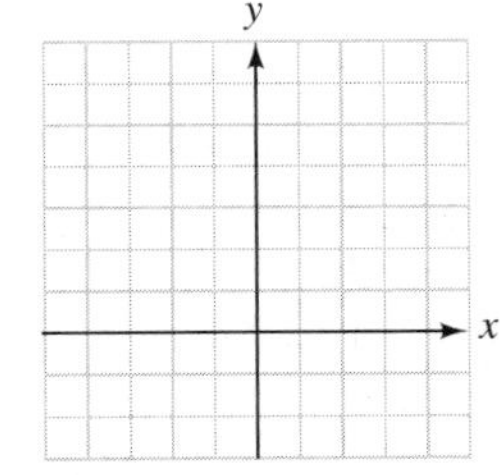 The solution set is $\varnothing$.
If the lines are different and parallel, the equations are *independent* and the system is *inconsistent.* **No solution exists.**	**c.** $\begin{cases} x - 3y = 6 \\ y = \frac{1}{3}x - 2 \end{cases}$ 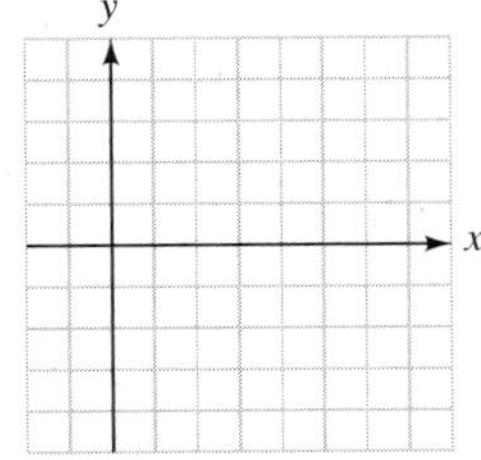
If the lines coincide (are the same), the equations are *dependent* and the system is *consistent.* **Infinitely many solutions exist.**	The equations of the system are dependent. A general solution is $\left(x, \frac{1}{3}x - 2\right)$.

REVIEW EXERCISES

Solve each system by the graphing method.

1. $\begin{cases} 2x + y = 11 \\ -x + 2y = 7 \end{cases}$

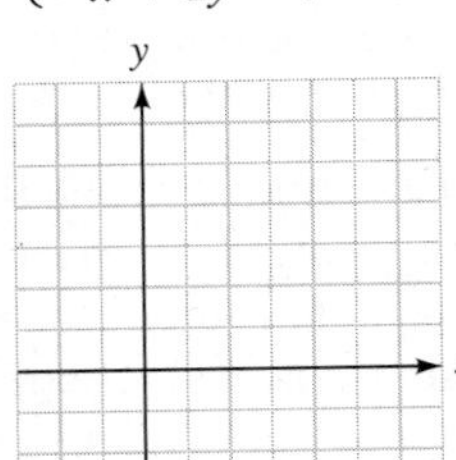

2. $\begin{cases} 3x + 2y = 0 \\ 2x - 3y = -13 \end{cases}$

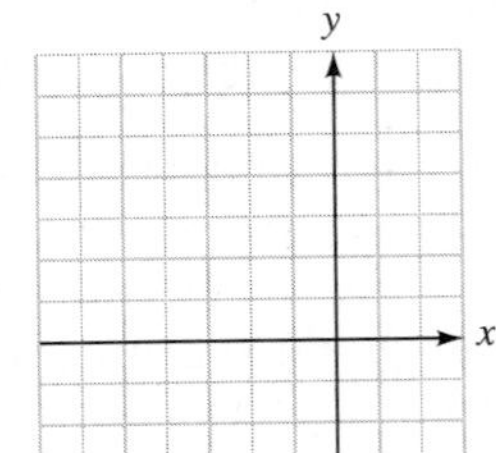

3. $\begin{cases} \frac{1}{2}x + \frac{1}{3}y = 2 \\ y = 6 - \frac{3}{2}x \end{cases}$

4. $\begin{cases} \frac{1}{3}x - \frac{1}{2}y = 1 \\ 6x - 9y = 2 \end{cases}$

SECTION 3.2 Solving Systems of Two Linear Equations by Substitution and Elimination (Addition)

DEFINITIONS AND CONCEPTS | **EXAMPLES**

To solve a system by *substitution,* solve one equation for a variable. Then substitute the expression found for that variable into the other equation. Then solve for the other variable to complete the ordered-pair solution.

Solve by substitution: $\begin{cases} x = 3y - 9 \\ 2x - y = 2 \end{cases}$.

Since the first equation is already solved for x, we will substitute its right side for x in the second equation.

$$2(3y - 9) - y = 2$$

$$6y - 18 - y = 2 \quad \text{Remove parentheses.}$$

$$5y - 18 = 2 \quad \text{Combine like terms.}$$

$$5y = 20 \quad \text{Add 18 to both sides.}$$

$$y = 4 \quad \text{Divide both sides by 5.}$$

To find x, we can substitute 4 for y in the first equation.

$$x = 3(4) - 9$$

$$x = 3$$

The solution is (3, 4).

To solve a system by *elimination* (addition), combine the equations of the system in a way that will eliminate one of the variables. The coefficients on one of the variables need to be additive inverses to eliminate that variable.

Solve by elimination: $\begin{cases} 2x - 3y = 8 \\ x + 2y = 4 \end{cases}$.

To eliminate x, we multiply the second equation by -2 and add the result to the first equation.

$$\begin{array}{r} 2x - 3y = 8 \\ \underline{-2x - 4y = -8} \\ -7y = 0 \end{array}$$

$$y = 0 \quad \text{Divide both sides by 7.}$$

To find x, we can substitute 0 for y in the first equation.

$$2x - 3(0) = 8 \quad \text{Substitute 0 for } y.$$

$$2x = 8 \quad \text{Simplify.}$$

$$x = 4 \quad \text{Divide both sides by 2.}$$

The solution is (4, 0).

REVIEW EXERCISES

Solve each system by substitution.

5. $\begin{cases} y = x + 4 \\ 2x + 3y = 7 \end{cases}$

6. $\begin{cases} y = 2x + 5 \\ 3x - 5y = -4 \end{cases}$

7. $\begin{cases} x + 2y = 11 \\ 2x - y = 2 \end{cases}$

8. $\begin{cases} 2x + 3y = -2 \\ 3x + 5y = -2 \end{cases}$

Solve each system by elimination.

9. $\begin{cases} x + y = -2 \\ 2x + 3y = -3 \end{cases}$

10. $\begin{cases} 3x + 2y = 1 \\ 2x - 3y = 5 \end{cases}$

11. $\begin{cases} x + \frac{1}{2}y = 7 \\ -2x = 3y - 6 \end{cases}$

12. $\begin{cases} y = \frac{x - 3}{2} \\ x = \frac{2y + 7}{2} \end{cases}$

SECTION 3.3 Solving Systems of Three Linear Equations in Three Variables

DEFINITIONS AND CONCEPTS	EXAMPLES
Strategy for solving three linear equations in three variables: **1.** Pick any two equations and eliminate a variable. **2.** Pick a different pair of equations and eliminate the same variable. **3.** Solve the resulting pair of equations in two variables. **4.** To find the value of the third variable, substitute the values of the two variables found in Step 3 into any equation containing all three variables and solve the equation. **5.** Check the solution in all three of the original equations.	To solve the system $\begin{cases} x + y + z = 4 \\ x - 2y - z = -9 \\ 2x - y + 2z = -1 \end{cases}$, we can add the first and second equations to obtain Equation 1: $x + y + z = 4$ $x - 2y - z = -9$ (1) $2x - y = -5$ We now multiply the second equation by 2 and add it to the third equation to obtain Equation 2: $2x - 4y - 2z = -18$ $2x - y + 2z = -1$ (2) $4x - 5y = -19$ To solve the system $\begin{cases} 2x - y = -5 \\ 4x - 5y = -19 \end{cases}$, formed by Equations 1 and 2, we can multiply the first equation by -2 and add the result to the second equation to eliminate x. $-4x + 2y = 10$ $4x - 5y = -19$ $-3y = -9$ $y = 3$ Divide both sides by -3. We can substitute 3 into either equation of the system to find x. $2x - y = -5$ This is the first equation of the system. $2x - 3 = -5$ Substitute 3 for y. $2x = -2$ Add 3 to both sides. $x = -1$ Divide both sides by 2. We now can substitute -1 for x and 3 for y into any of the equations in the original system and solve for z: $x + y + z = 4$ This is the first equation of the original system. $-1 + 3 + z = 4$ Substitute -1 for x and 3 for y. $2 + z = 4$ Simplify. $z = 2$ Subtract 2 from both sides. The solution is $(-1, 3, 2)$.

REVIEW EXERCISES

Solve each system.

13. $\begin{cases} x + y + z = 6 \\ x - y - z = -4 \\ -x + y - z = -2 \end{cases}$

14. $\begin{cases} 2x + 3y + z = -5 \\ -x + 2y - z = -6 \\ 3x + y + 2z = 4 \end{cases}$

SECTION 3.4 Solving Systems of Linear Equations Using Matrices

DEFINITIONS AND CONCEPTS	EXAMPLES
A **matrix** is any rectangular array of numbers. Systems of linear equations can be solved using matrices and the method of **Gaussian elimination** and back substitution.	To use matrices to solve the system $\begin{cases} x + y + z = 4 \\ 2x - y + 2z = -1 \\ x - 2y - z = -9 \end{cases}$ we can represent it with the following augmented matrix: $\left[\begin{array}{ccc\|c} 1 & 1 & 1 & 4 \\ 2 & -1 & 2 & -1 \\ 1 & -2 & -1 & -9 \end{array}\right]$ To get 0's under the 1 in the first column, we multiply row 1 of the augmented matrix by -2 and add it to row 2 to get a new row 2. We then multiply row 1 by -1 and add it to row 3 to get a new row 3. $\left[\begin{array}{ccc\|c} 1 & 1 & 1 & 4 \\ 0 & -3 & 0 & -9 \\ 0 & -3 & -2 & -13 \end{array}\right]$ To get a 0 under the -3 in the second column of the previous matrix, we multiply row 2 by -1 and add it to row 3. $\left[\begin{array}{ccc\|c} 1 & 1 & 1 & 4 \\ 0 & -3 & 0 & -9 \\ 0 & 0 & -2 & -4 \end{array}\right]$ Finally, to obtain a 1 in the third row, third column, we multiply row 3 by $-\frac{1}{2}$. $\left[\begin{array}{ccc\|c} 1 & 1 & 1 & 4 \\ 0 & -3 & 0 & -9 \\ 0 & 0 & 1 & 2 \end{array}\right]$ The final matrix represents the system $\begin{array}{l} (1) \\ (2) \\ (3) \end{array} \begin{cases} x + y + z = 4 \\ 0x - 3y + 0z = -9 \\ 0x + 0y + z = 2 \end{cases}$ From Equation 3, we see that $z = 2$. From Equation 2, we see that $y = 3$. To find x, we substitute 2 for z and 3 for y in Equation 1 and solve for x: (1) $x + y + z = 4$ $x + 3 + 2 = 4$ Substitute 2 for z and 3 for y. $x + 5 = 4$ Simplify. $x = -1$ Subtract 5 from both sides. Thus, $x = -1$. The solution of the given system is $(-1, 3, 2)$. Verify that this ordered triple satisfies each equation of the original system.

REVIEW EXERCISES

Solve each system by using matrices.

15. $\begin{cases} x + 2y = 4 \\ 2x - y = 3 \end{cases}$

16. $\begin{cases} x + y + z = 6 \\ 2x - y + z = 1 \\ 4x + y - z = 5 \end{cases}$

17. $\begin{cases} x + y = 3 \\ x - 2y = -3 \\ 2x + y = 4 \end{cases}$

18. $\begin{cases} x + 2y + z = 2 \\ 2x + 5y + 4z = 5 \end{cases}$

SECTION 3.5 Solving Systems of Linear Equations Using Determinants

DEFINITIONS AND CONCEPTS	EXAMPLES
A **determinant of a square matrix** is a number. $\begin{vmatrix} a & b \\ c & d \end{vmatrix} = ad - bc$ $\begin{vmatrix} a_1 & b_1 & c_1 \\ a_2 & b_2 & c_2 \\ a_3 & b_3 & c_3 \end{vmatrix}$ $= a_1\begin{vmatrix} b_2 & c_2 \\ b_3 & c_3 \end{vmatrix} - b_1\begin{vmatrix} a_2 & c_2 \\ a_3 & c_3 \end{vmatrix} + c_1\begin{vmatrix} a_2 & b_2 \\ a_3 & b_3 \end{vmatrix}$	Find the determinant $\begin{vmatrix} 8 & -3 \\ 2 & -1 \end{vmatrix}$. $\begin{vmatrix} 8 & -3 \\ 2 & -1 \end{vmatrix} = 8(-1) - (-3)(2)$ $= -8 + 6$ $= -2$ To evaluate the determinant $\begin{vmatrix} 1 & 3 & -2 \\ 1 & -2 & -1 \\ 2 & -1 & 3 \end{vmatrix}$, we can expand by minors: Minor of 1 ↓ Minor of 3 ↓ Minor of −2 ↓ $\begin{vmatrix} 1 & 3 & -2 \\ 1 & -2 & -1 \\ 2 & -1 & 3 \end{vmatrix} = 1\begin{vmatrix} -2 & -1 \\ -1 & 3 \end{vmatrix} - 3\begin{vmatrix} 1 & -1 \\ 2 & 3 \end{vmatrix} + (-2)\begin{vmatrix} 1 & -2 \\ 2 & -1 \end{vmatrix}$ $= 1(-6 - 1) - 3(3 + 2) - 2(-1 + 4)$ $= -7 - 15 - 6$ $= -28$
Cramer's rule for two equations in two variables: The solution of the system $\begin{cases} ax + by = e \\ cx + dy = f \end{cases}$ is given by $x = \frac{D_x}{D} = \frac{\begin{vmatrix} e & b \\ f & d \end{vmatrix}}{\begin{vmatrix} a & b \\ c & d \end{vmatrix}}$ and $y = \frac{D_y}{D} = \frac{\begin{vmatrix} a & e \\ c & f \end{vmatrix}}{\begin{vmatrix} a & b \\ c & d \end{vmatrix}}$ If $D \neq 0$, the system is consistent and the equations are independent. If $D = 0$ and D_x or D_y is nonzero, the system is inconsistent. If every determinant is 0, the system is consistent but the equations are dependent.	Solve using Cramer's rule: $\begin{cases} 2x - 4y = -14 \\ 3x + y = -7 \end{cases}$. $x = \frac{D_x}{D} = \frac{\begin{vmatrix} -14 & -4 \\ -7 & 1 \end{vmatrix}}{\begin{vmatrix} 2 & -4 \\ 3 & 1 \end{vmatrix}} = \frac{-14 - 28}{2 - (-12)} = \frac{-42}{14} = -3$ $y = \frac{D_y}{D} = \frac{\begin{vmatrix} 2 & -14 \\ 3 & -7 \end{vmatrix}}{\begin{vmatrix} 2 & -4 \\ 3 & 1 \end{vmatrix}} = \frac{-14 - (-42)}{2 - (-12)} = \frac{28}{14} = 2$ The solution is $(-3, 2)$.

Cramer's rule for three equations in three variables:

The solution of the system $\begin{cases} ax + by + cz = j \\ dx + ey + fz = k \\ gx + hy + iz = l \end{cases}$

is given by $x = \frac{D_x}{D}$, $y = \frac{D_y}{D}$, and $z = \frac{D_z}{D}$ where

$$D = \begin{vmatrix} a & b & c \\ d & e & f \\ g & h & i \end{vmatrix} \qquad D_x = \begin{vmatrix} j & b & c \\ k & e & f \\ l & h & i \end{vmatrix}$$

$$D_y = \begin{vmatrix} a & j & c \\ d & k & f \\ g & l & i \end{vmatrix} \qquad D_z = \begin{vmatrix} a & b & j \\ d & e & k \\ g & h & l \end{vmatrix}$$

If $D \neq 0$, the system is consistent and the equations are independent.

If $D = 0$ and D_x or D_y or D_z is nonzero, the system is inconsistent.

If every determinant is 0, the system is consistent but the equations are dependent.

To use Cramer's rule to solve $\begin{cases} x + y + z = 4 \\ 2x - y + 2z = -1 \\ x - 2y - z = -9 \end{cases}$, we can find D, D_x, D_y, and D_z and substitute these values into the formulas

$$x = \frac{D_x}{D}, \qquad y = \frac{D_y}{D}, \qquad \text{and} \qquad z = \frac{D_z}{D}$$

After forming and evaluating the determinants, we will obtain

$$D = 6, \qquad D_x = -6, \qquad D_y = 18, \qquad \text{and} \qquad D_z = 12$$

and we have

$$x = \frac{D_x}{D} = \frac{-6}{6} = -1,\ y = \frac{D_y}{D} = \frac{18}{6} = 3,\ z = \frac{D_z}{D} = \frac{12}{6} = 2$$

The solution of this system is $(-1, 3, 2)$.

REVIEW EXERCISES

Evaluate each determinant.

19. $\begin{vmatrix} 2 & 3 \\ -4 & 3 \end{vmatrix}$

20. $\begin{vmatrix} -3 & -4 \\ 5 & -6 \end{vmatrix}$

21. $\begin{vmatrix} -1 & 2 & -1 \\ 2 & -1 & 3 \\ 1 & -2 & 2 \end{vmatrix}$

22. $\begin{vmatrix} 3 & -2 & 2 \\ 1 & -2 & -2 \\ 2 & 1 & -1 \end{vmatrix}$

Use Cramer's rule to solve each system.

23. $\begin{cases} 3x + 4y = 10 \\ 2x - 3y = 1 \end{cases}$

24. $\begin{cases} 2x - 5y = -17 \\ 3x + 2y = 3 \end{cases}$

25. $\begin{cases} x + 2y + z = 0 \\ 2x + y + z = 3 \\ x + y + 2z = 5 \end{cases}$

26. $\begin{cases} 2x + 3y + z = 2 \\ x + 3y + 2z = 7 \\ x - y - z = -7 \end{cases}$

Chapter 3 TEST

1. Solve $\begin{cases} 2x + y = 5 \\ y = 2x - 3 \end{cases}$ by graphing.

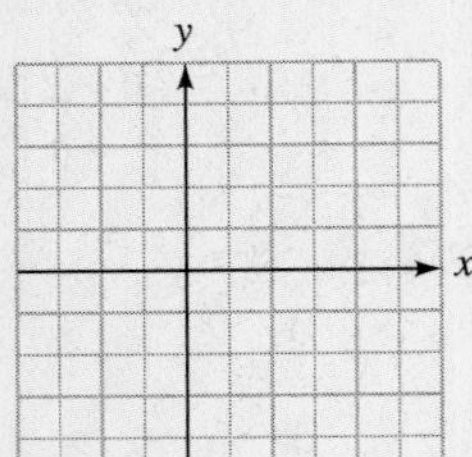

2. Use substitution to solve: $\begin{cases} 2x - 4y = 14 \\ x = -2y + 7 \end{cases}$.

3. Use elimination to solve: $\begin{cases} 2x + 3y = -5 \\ 3x - 2y = 12 \end{cases}$.

4. Use any method to solve: $\begin{cases} \frac{x}{2} - \frac{y}{4} = -4 \\ x + y = -2 \end{cases}$.

Consider the system $\begin{cases} 3(x + y) = x - 3 \\ -y = \frac{2x + 3}{3} \end{cases}$.

5. Are the equations of the system dependent or independent?

6. Is the system consistent or inconsistent?

Use an elementary row operation to find the missing number in the second matrix.

7. $\begin{bmatrix} 1 & 2 & -1 \\ 2 & -2 & 3 \end{bmatrix}, \begin{bmatrix} 1 & 2 & -1 \\ -1 & -8 & \blacksquare \end{bmatrix}$

8. $\begin{bmatrix} -1 & 3 & 6 \\ 3 & -2 & 4 \end{bmatrix}, \begin{bmatrix} -1 & 3 & 6 \\ 5 & -8 & \blacksquare \end{bmatrix}$

Consider the system $\begin{cases} x + y + z = 4 \\ x + y - z = 6 \\ 2x - 3y + z = -1 \end{cases}$.

9. Write the augmented matrix that represents the system.

10. Write the coefficient matrix that represents the system.

Use matrices to solve each system.

11. $\begin{cases} x + y = 4 \\ 2x - y = 2 \end{cases}$

12. $\begin{cases} x + y = 2 \\ x - y = -4 \\ 2x + y = 1 \end{cases}$

Evaluate each determinant.

13. $\begin{vmatrix} 2 & -3 \\ 4 & 5 \end{vmatrix}$

14. $\begin{vmatrix} -3 & -4 \\ -2 & 3 \end{vmatrix}$

15. $\begin{vmatrix} 1 & 2 & 0 \\ 2 & 0 & 3 \\ 1 & -2 & 2 \end{vmatrix}$

16. $\begin{vmatrix} 2 & -1 & 1 \\ 3 & 1 & 0 \\ 0 & 1 & 2 \end{vmatrix}$

Consider the system $\begin{cases} x - y = -6 \\ 3x + y = -6 \end{cases}$, *which is to be solved with Cramer's rule.*

17. When solving for x, what is the numerator determinant? **(Don't evaluate it.)**

18. When solving for y, what is the denominator determinant? **(Don't evaluate it.)**

19. Solve the system for x.

20. Solve the system for y.

Consider the system $\begin{cases} x + y + z = 4 \\ x + y - z = 6 \\ 2x - 3y + z = -1 \end{cases}$.

21. Solve for x.

22. Solve for z.

©Shutterstock.com/CREATISTA

Careers and Mathematics

AUTOMOTIVE SERVICE TECHNICIANS AND MECHANICS

Anyone whose car has broken down knows the importance of the jobs of automotive service technicians and mechanics. The ability to diagnose the source of a problem quickly and accurately requires good reasoning ability and a thorough knowledge of automobiles. Automotive service technicians and mechanics held about 773,000 jobs in 2006. Most training authorities strongly recommend that people seeking these jobs complete a formal training program in high school or in a post-secondary vocational school.

Job Outlook:
Employment of automotive service technicians and mechanics is expected to increase 14% between 2006 and 2016. Employment growth will continue to be concentrated in automobile dealerships and independent automotive repair shops.

Hourly Earnings:
\$11.96–\$21.56

For More Information:
http://www.bls.gov/oco/ocos181.htm

For a Sample Application:
See Problem 123 in Section 4.2.

CHAPTER 4
Inequalities

4.1 Solving Linear Inequalities in One Variable
4.2 Solving Absolute-Value Equations and Inequalities in One Variable
4.3 Solving Linear Inequalities in One or Two Variables
4.4 Solving Systems of Linear and Quadratic Inequalities in Two Variables
4.5 Solving Systems Using Linear Programming
- ***Projects***
CHAPTER REVIEW
CHAPTER TEST
CUMULATIVE REVIEW EXERCISES

In this chapter ▶

We previously have considered linear equations, statements that two quantities are equal. In this chapter, we will consider mathematical statements that indicate that two quantities are unequal.

SECTION 4.1

Solving Linear Inequalities in One Variable

Objectives

1. Solve a linear inequality in one variable and express the result as a graph and in set-builder and interval notation.
2. Solve a compound inequality in one variable involving "and," express the result as a graph, and express the solution in set-builder and interval notation.
3. Solve a compound inequality in one variable involving "or" and express the result as a graph and in set-builder and interval notation.
4. Solve an application problem by setting up and solving a linear inequality in one variable.

Vocabulary

inequality
linear inequality
interval notation
compound inequality

Getting Ready

Graph each set on the number line.

1. $\{x \mid x > -3\}$

2. $\{x \mid x < 4\}$

3. $\{x \mid x \leq 5\}$

4. $\{x \mid x \geq -1\}$

In this section, we will review the basic ideas of inequalities, first discussed in Chapter 1. After discussing many properties of inequalities, we will solve linear inequalities and application problems.

1 Solve a linear inequality in one variable and express the result as a graph and in set-builder and interval notation.

Inequalities are statements indicating that two quantities are unequal. Inequalities can be recognized because they contain one or more of the following symbols.

Inequality Symbols

Inequality	Read as	Example
$a \neq b$	"a is not equal to b."	$5 \neq 9$
$a < b$	"a is less than b."	$2 < 3$
$a > b$	"a is greater than b."	$7 > -5$
$a \leq b$	"a is less than or equal to b."	$-5 \leq 0$ or $-4 \leq -4$
$a \geq b$	"a is greater than or equal to b."	$9 \geq 6$ or $8 \geq 8$
$a \approx b$	"a is approximately equal to b."	$1.14 \approx 1.1$

By definition, $a < b$ means that "a is less than b," but it also means that $b > a$. If a is to the left of b on the number line, then $a < b$.

There are several basic properties of inequalities.

Trichotomy Property

For any real numbers a and b, exactly one of the following statements is true:

$$a < b, \qquad a = b, \qquad \text{or} \qquad a > b$$

The trichotomy property indicates that exactly one of the following statements is true about any two real numbers. Either

- the first is less than the second,
- the first is equal to the second, or
- the first is greater than the second.

Transitive Property

If a, b, and c are real numbers with $a < b$ and $b < c$, then $a < c$.

If a, b, and c are real numbers with $a > b$ and $b > c$, then $a > c$.

The first part of the transitive property indicates that:

If a first number is less than a second number and the second is less than a third, then the first number is less than the third.

The second part is similar, with the words "is greater than" substituted for "is less than."

Addition Property of Inequalities

Any real number can be added to (or subtracted from) both sides of an inequality to produce another inequality with the same direction as the original.

To illustrate the addition property, we add 4 to both sides of the inequality $3 < 12$ to get

$$3 + 4 < 12 + 4$$
$$7 < 16$$

We note that the $<$ symbol is unchanged (has the same direction).

Subtracting 4 from both sides of $3 < 12$ does not change the direction of the inequality either.

$$3 - 4 < 12 - 4$$
$$-1 < 8$$

Multiplication Property of Inequalities (Case 1)

If both sides of an inequality are multiplied (or divided) by a positive number, another inequality results with the same direction as the original inequality.

To illustrate the multiplication property, we multiply both sides of the inequality $-4 < 6$ by 2 to get

$$2(-4) < 2(6)$$
$$-8 < 12$$

The $<$ symbol is unchanged.

Dividing both sides by 2 does not change the direction of the inequality either.

$$\frac{-4}{2} < \frac{6}{2}$$
$$-2 < 3$$

Multiplication Property of Inequalities (Case 2)

If both sides of an inequality are multiplied (or divided) by a negative number, another inequality results, but with the opposite direction from the original inequality.

To illustrate the multiplication property, we multiply both sides of the inequality $-4 < 6$ by -2 to get

$$-4 < 6$$
$$-2(-4) > -2(6) \quad \text{Change } < \text{ to } >.$$
$$8 > -12$$

COMMENT Remember to change the direction of an inequality symbol every time you multiply or divide both sides by a negative number.

Here, the $<$ symbol changes to a $>$ symbol.

Dividing both sides by -2 also changes the direction of the inequality.

$$-4 < 6$$
$$\frac{-4}{-2} > \frac{6}{-2}$$
$$2 > -3$$

Linear Inequalities

A **linear inequality** in x is any inequality that can be expressed in one of the following forms (with $a \neq 0$).

$$ax + c < 0 \qquad ax + c > 0 \qquad ax + c \leq 0 \qquad ax + c \geq 0$$

We solve linear inequalities just as we solve linear equations, but with one exception. If we multiply or divide both sides by a negative number, we must change the direction of the inequality.

EXAMPLE 1 Solve each linear inequality and graph its solution set.

a. $3(2x - 9) < 9$ **b.** $-4(3x + 2) \leq 16$

Solution In each part, we use the same steps as for solving equations.

a. $3(2x - 9) < 9$

$6x - 27 < 9$ Use the distributive property to remove parentheses.

$$6x < 36 \quad \text{Add 27 to both sides.}$$
$$x < 6 \quad \text{Divide both sides by 6.}$$

The solution set is $\{x \mid x < 6\}$. The graph is shown in Figure 4-1(a). The parenthesis at 6 indicates that 6 is not included in the solution set.

b. $-4(3x + 2) \leq 16$

$-12x - 8 \leq 16$ Use the distributive property to remove parentheses.

$-12x \leq 24$ Add 8 to both sides.

$x \geq -2$ Divide both sides by -12 and reverse the $\leq$ symbol.

The solution set is $\{x \mid x \geq -2\}$. The graph is shown in Figure 4-1(b). The bracket at -2 indicates that -2 is included in the solution set.

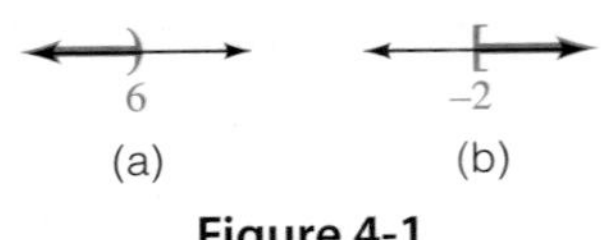

Figure 4-1

SELF CHECK 1 Solve the inequality and graph its solution set: $-3(2x - 4) \leq 24$.

Amalie Noether (1882–1935)

Albert Einstein described Noether as the most creative female mathematical genius since the beginning of higher education for women. Her work was in the area of abstract algebra. Although she received a doctoral degree in mathematics, she was denied a mathematics position in Germany because she was a woman.

In Chapter 1, we saw that **interval notation** is another way to express the solution set of an inequality in one variable. Recall that this notation uses parentheses and brackets to indicate whether endpoints of an interval are included or excluded from a solution set.

An interval is called a (an)

- **unbounded interval** if it extends forever in one or more directions. See Figure 4-2(a).
- **open interval** if it is bounded and has no endpoints. See Figure 4-2(b).
- **half-open interval** if it is bounded and has one endpoint. See Figure 4-2(c).
- **closed interval** if it is bounded and has two endpoints. See Figure 4-2(d).

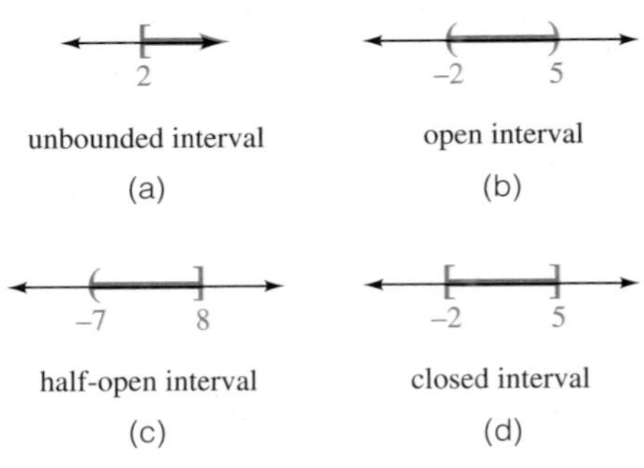

Figure 4-2

Table 4-1 shows the possible intervals that exist when a and b are real numbers.

Kind of interval	Set notation	Graph	Interval
Unbounded intervals	$\{x \mid x > a\}$	a	(a, ∞)
	$\{x \mid x < a\}$	a	$(-\infty, a)$
	$\{x \mid x \geq a\}$	a	$[a, \infty)$
	$\{x \mid x \leq a\}$	a	$(-\infty, a]$
Open interval	$\{x \mid a < x < b\}$	a b	(a, b)
Half-open intervals	$\{x \mid a \leq x < b\}$	a b	$[a, b)$
	$\{x \mid a < x \leq b\}$	a b	$(a, b]$
Closed interval	$\{x \mid a \leq x \leq b\}$	a b	$[a, b]$

Table 4-1

EXAMPLE 2 Solve each linear inequality and graph its solution set.

a. $5(x + 2) \geq 3x - 1$ **b.** $\frac{2}{3}(x + 2) > \frac{4}{5}(x - 3)$

Solution In both parts, we use the same steps as for solving linear equations.

a. $5(x + 2) \geq 3x - 1$

$5x + 10 \geq 3x - 1$ Use the distributive property to remove parentheses.

$2x + 10 \geq -1$ Subtract $3x$ from both sides.

$2x \geq -11$ Subtract 10 from both sides.

$x \geq -\frac{11}{2}$ Divide both sides by 2.

The solution set is $\left\{x \mid x \geq -\frac{11}{2}\right\}$, which is the interval $\left[-\frac{11}{2}, \infty\right)$. Its graph is shown in Figure 4-3(a) on the next page.

b. $\frac{2}{3}(x + 2) > \frac{4}{5}(x - 3)$

$15 \cdot \frac{2}{3}(x + 2) > 15 \cdot \frac{4}{5}(x - 3)$ Multiply both sides by 15, the LCD of 3 and 5.

$10(x + 2) > 12(x - 3)$ Simplify.

$10x + 20 > 12x - 36$ Use the distributive property to remove parentheses.

$-2x + 20 > -36$ Subtract $12x$ from both sides.

$-2x > -56$ Subtract 20 from both sides.

$x < 28$ Divide both sides by -2 and reverse the $>$ symbol.

The solution set is $\{x \mid x < 28\}$, which is the interval $(-\infty, 28)$. Its graph is shown in Figure 4-3(b).

−11/2 (a)   28 (b)

Figure 4-3

SELF CHECK 2 Solve the linear inequality and graph its solution set: $\frac{3}{2}(x + 2) < \frac{3}{5}(x - 3)$.

2 Solve a compound inequality in one variable involving "and," express the result as a graph, and express the solution in set-builder and interval notation.

To say that x is between -3 and 8, we write the inequality

$-3 < x < 8$ Read as "-3 is less than x and x is less than 8."

This inequality is called a **compound inequality,** because it is a combination of two inequalities:

$-3 < x$ and $x < 8$

The word *and* indicates that both inequalities are true at the same time.

Compound Inequalities The inequality $c < x < d$ is equivalent to $c < x$ and $x < d$.

EXAMPLE 3 Solve the inequality $-3 \le 2x + 5 < 7$ and graph its solution set.

Solution This inequality means that $2x + 5$ is between -3 and 7. We can solve it by isolating x between the inequality symbols.

$-3 \le 2x + 5 < 7$

$-8 \le 2x < 2$ Subtract 5 from all three parts.

$-4 \le x < 1$ Divide all three parts by 2.

The solution set is $\{x \mid -4 \le x < 1\}$, which is the interval $[-4, 1)$. The graph is shown in Figure 4-4.

−4 1

Figure 4-4

SELF CHECK 3 Solve $-5 \le 3x - 8 \le 7$ and graph its solution set.

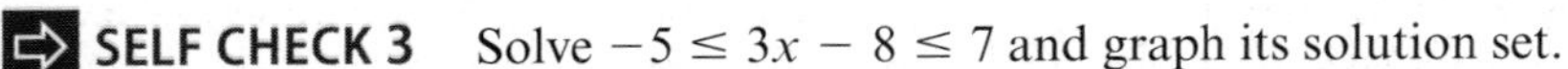

EXAMPLE 4 Solve: $x + 3 < 2x - 1 < 4x - 3$.

Solution Since it is impossible to isolate x between the inequality symbols, we solve each of the linear inequalities separately.

$x + 3 < 2x - 1$	and	$2x - 1 < 4x - 3$
$4 < x$		$2 < 2x$
$x > 4$		$1 < x$
		$x > 1$

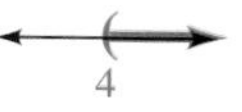
Figure 4-5

Only those x values where $x > 4$ and $x > 1$ are in the solution set. Since all numbers greater than 4 are also greater than 1, the solutions are the numbers x where $x > 4$. Thus, the solution set is the interval $(4, \infty)$, whose graph is shown in Figure 4-5.

SELF CHECK 4 Solve: $x + 2 < 3x + 1 < 5x + 3$.

3 Solve a compound inequality in one variable involving "or" and express the result as a graph and in set-builder and interval notation.

To say that x is less than -5 or greater than 10, we write the inequality

$x < -5$ or $x > 10$ Read as "x is less than -5 or x is greater than 10."

The word *or* indicates that only one of the inequalities needs to be true to make the entire statement true.

EXAMPLE 5 Solve the compound inequality $x \leq -3$ or $x \geq 8$.

Solution The word *or* in the statement $x \leq -3$ or $x \geq 8$ indicates that only one of the inequalities needs to be true to make the statement true. The graph of the inequality is shown in Figure 4-6. The solution set is $\{x \mid x \leq -3 \text{ or } x \geq 8\}$, which is the union of two intervals $(-\infty, -3] \cup [8, \infty)$.

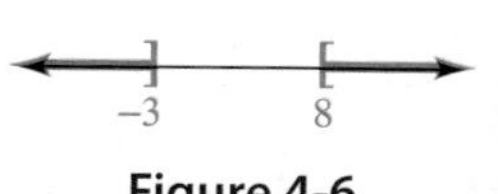
Figure 4-6

SELF CHECK 5 Solve: $x < -5$ or $x \geq 3$.

COMMENT In the statement $x \leq -3$ or $x \geq 8$, it is incorrect to string the inequalities together as $8 \leq x \leq -3$, because that would imply that $8 \leq -3$, which is false.

4 Solve an application problem by setting up and solving a linear inequality in one variable.

EXAMPLE 6 **LONG-DISTANCE** Suppose that a long-distance telephone call costs 36¢ for the first three minutes and 11¢ for each additional minute. For how many minutes can a person talk for less than $2?

Analyze the problem

We can let x represent the total number of minutes that the call can last.

Form an inequality

The cost of the call will be 36¢ for the first three minutes plus 11¢ times the number of additional minutes, where the number of additional minutes is $x - 3$ (the total number of minutes minus 3 minutes). The cost of the call is to be less than $2. With this information, we can form the inequality

The cost of the first three minutes	plus	the cost of the additional minutes	is less than	$2.
0.36	+	$0.11(x - 3)$	<	2

Solve the inequality We can solve the inequality as follows:

$$0.36 + 0.11(x - 3) < 2$$

$$36 + 11(x - 3) < 200$$ To eliminate the decimal points, multiply both sides by 100.

$$36 + 11x - 33 < 200$$ Use the distributive property to remove parentheses.

$$11x + 3 < 200$$ Combine like terms.

$$11x < 197$$ Subtract 3 from both sides.

$$x < 17.\overline{90}$$ Divide both sides by 11.

State the conclusion Since the phone company doesn't bill for part of a minute, the longest that the person can talk is 17 minutes.

Check the result If the call lasts 17 minutes, the customer will be billed \$0.36 + \$0.11(14) = \$1.90. If the call lasts 18 minutes, the customer will be billed \$0.36 + \$0.11(15) = \$2.01.

ACCENT ON TECHNOLOGY

Solving Linear Inequalities

We can solve linear inequalities with a graphing approach. For example, to solve the inequality $3(2x - 9) < 9$, we can graph $y = 3(2x - 9)$ and $y = 9$ using window settings of $[-10, 10]$ for x and $[-10, 10]$ for y to get Figure 4-7(a). We trace to see that the graph of $y = 3(2x - 9)$ is below the graph of $y = 9$ for x-values in the interval $(-\infty, 6)$. See Figure 4-7(b). This interval is the solution, because in this interval, $3(2x - 9) < 9$.

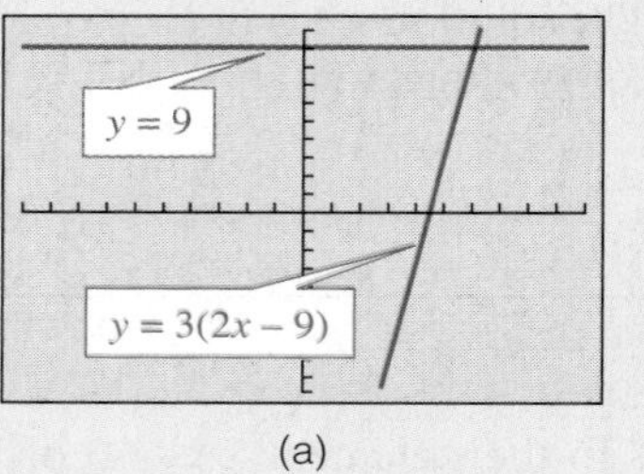

(a)

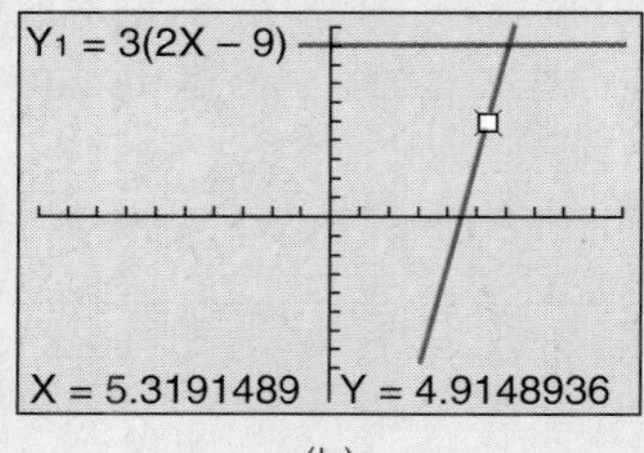

(b)

Figure 4-7

In statistics, researchers often estimate the mean of a population from the results of a random sample taken from the population.

EXAMPLE 7 **STATISTICS** A researcher wants to estimate the mean (average) real estate tax paid by homeowners living in Rockford. To do so, he selects a *random sample* of homeowners and computes the mean tax paid by the homeowners in that sample. How large must the sample be for the researcher to be 95% certain that the computed sample mean will be within \$35 of the true population mean—that is, within \$35 of the mean tax paid by all homeowners in the city? Assume that the standard deviation σ of all tax bills in the city is \$120.

Solution From statistics, the researcher has the formula

$$\frac{3.84\sigma^2}{N} < E^2$$

where σ^2 is the square of the standard deviation, E is the maximum acceptable error, and N is the sample size. The researcher substitutes 120 for σ and 35 for E in the previous formula and solves for N.

$$\frac{3.84(120)^2}{N} < 35^2$$

$$\frac{55,296}{N} < 1,225 \quad \text{Simplify.}$$

$$55,296 < 1,225N \quad \text{Multiply both sides by positive } N.$$

$$45.13959184 < N \quad \text{Divide both sides by 1,225.}$$

To be 95% certain that the sample mean will be within \$35 of the true population mean, the researcher must sample more than 45.13959184 homeowners. Thus, the sample must contain at least 46 homeowners.

SELF CHECK ANSWERS

1. $\{x \mid x \geq -2\}$ 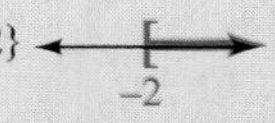

2. $\{x \mid x < -\frac{16}{3}\}$ or $\left(-\infty, -\frac{16}{3}\right)$ 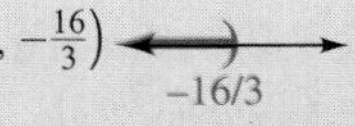

3. $\{x \mid 1 \leq x \leq 5\}$ or $[1, 5]$ 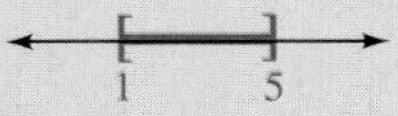

4. $\{x \mid x > \frac{1}{2}\}$ or $\left(\frac{1}{2}, \infty\right)$ 

5. $\{x \mid x < -5 \text{ or } x \geq 3\}$ or $(-\infty, -5) \cup [3, \infty)$ 

NOW TRY THIS

Solve the compound inequality.

1. $6x - 2(x + 1) < 10$ or $-5x < -10$

2. $4x \geq 2(x - 6)$ and $-5x \geq 30$

4.1 EXERCISES

WARM-UPS *Write the solution set of each inequality in both set-builder and interval notation.*

1. $2x < 4$

2. $3x \geq 9$

3. $-3x > 12$

4. $-2 < 2x \leq 8$

REVIEW *Simplify each expression.*

5. $\left(\dfrac{t^3t^5t^{-6}}{t^2t^{-4}}\right)^{-3}$

6. $\left(\dfrac{a^{-2}b^3a^5b^{-2}}{a^6b^{-5}}\right)^{-4}$

7. Small businesses A man invests \$1,200 in baking equipment to make pies. Each pie requires \$3.40 in ingredients. If the man can sell all the pies he can make for \$5.95 each, how many pies will he have to make to earn a profit?

8. Investing A woman invested \$15,000, part at 7% annual interest and the rest at 8%. If she earned \$2,200 in income over a two-year period, how much did she invest at 7%?

VOCABULARY AND CONCEPTS *Fill in the blanks.*

9. The symbol for "is not equal to" is ___.

10. The symbol for "is greater than" is ___.

11. The symbol for "is less than" is ___.

12. The symbol for "is less than or equal to" is ___.

13. The symbol for "is greater than or equal to" is ___.

14. If a and b are two numbers, then $a < b$, ______, or ______.

15. If $a < b$ and $b < c$, then ______.

16. If both sides of an inequality are multiplied by a ______ number, the direction of the inequality remains the same.

17. If both sides of an inequality are divided by a negative number, the direction of the inequality symbol must be ________.

18. $3x + 2 > 7$ is an example of a ______ inequality and $3 < 2x < 6$ is a __________ inequality.

19. The inequality $c < x < d$ is equivalent to ______ and ______.

20. The word "or" between two inequality statements indicates that only ____ of the inequalities needs to be true for the entire statement to be true.

21. The interval $(2, 5)$ is called an ______ interval.

22. The interval $[-5, 3]$ is called a _______ interval.

GUIDED PRACTICE

Solve each inequality. Express each result as a graph and in set-builder notation. **See Example 1. (Objective 1)**

23. $x + 4 < 5$

24. $x - 5 > 2$

25. $x + 2 \geq -3$

26. $x - 3 \leq 5$

27. $2x + 3 < 9$

28. $5x - 1 \geq 19$

29. $-3x - 1 \leq 5$

30. $-2x + 6 \geq 16$

Solve each inequality. Express each result as a graph and in interval notation. **See Example 2. (Objective 1)**

31. $\frac{1}{2}y + 2 > \frac{1}{3}y - 4$

32. $\frac{1}{4}x - \frac{1}{3} \leq x + 2$

33. $\frac{2}{3}x + \frac{3}{2}(x - 5) \leq x$

34. $\frac{5}{9}(x + 3) - \frac{4}{3}(x - 3) \geq x - 1$

Solve each inequality. Express each result as a graph and in interval notation. **See Example 3. (Objective 1)**

35. $-4 < 2x < 8$

36. $-3 \leq 3x < 12$

37. $-2 < -b + 3 < 5$

38. $2 < -t - 2 < 9$

39. $15 > 2x - 7 > 9$

40. $25 > 3x - 2 > 7$

41. $-6 < -3(x - 4) \leq 24$

42. $-4 \leq -2(x + 8) < 8$

Solve each inequality. Express each result as a graph and in interval notation. **See Example 4. (Objective 2)**

43. $x + 3 < 3x - 1 < 2x + 2$

44. $x - 1 \leq 2x + 4 \leq 3x - 1$

45. $x < -3$ and $x > 3$

46. $x + 2 < -\frac{1}{3}x < \frac{1}{2}x$

47. $5(x + 1) \leq 4(x + 3) < 3(x - 1)$

48. $-5(2 + x) < 4x + 1 < 3x$

49. $5(x - 2) \geq 0$ and $-3x < 9$

50. $4x \geq -x + 5 \geq 3x - 4$

Solve each inequality. Express each result as a graph and in interval notation. **See Example 5. (Objective 3)**

51. $3x + 2 < 8$ or $2x - 3 > 11$

52. $3x + 4 < -2$ or $3x + 4 > 10$

53. $-4(x + 2) \geq 12$ or $3x + 8 < 11$

54. $x < 3$ or $x > -3$

ADDITIONAL PRACTICE

Solve each inequality.

55. $5x - 3 > 7$

56. $7x - 9 < 5$

57. $8x + 30 > -2x$

58. $5x - 24 \leq 6$

59. $-3x + 14 \geq 20$

60. $-\frac{1}{2}x + 4 < 32$

61. $4(x + 5) < 12$

62. $-5(x - 2) \geq 15$

63. $3(z - 2) \leq 2(z + 7)$

64. $5(3 + z) > -3(z + 3)$

65. $-11(2 - b) < 4(2b + 2)$

66. $-9(h - 3) + 2h \leq 8(4 - h)$

67. $0.4x + 0.4 \leq 0.1x + 0.85$

68. $0.05 - 0.5x \geq -0.7 - 0.8x$

69. $0 \geq \frac{1}{2}x - 4 > 6$

70. $-6 \leq \frac{1}{3}a + 1 < 0$

71. $0 \leq \frac{4 - x}{3} \leq 2$

72. $-2 \leq \frac{5 - 3x}{2} \leq 2$

Use a graphing calculator to solve each inequality.

73. $2x + 3 < 5$

74. $3x - 2 > 4$

75. $5x + 2 \geq -18$

76. $3x - 4 \leq 20$

APPLICATIONS *Set up an inequality and solve each problem. Use a calculator to help solve each problem.*
See Examples 5–6. (Objective 4)

77. Renting a rototiller The cost of renting a rototiller is \$15.50 for the first hour and \$7.95 for each additional hour. How long can a person have the rototiller if the cost is to be less than \$50?

78. Renting a truck How long can a person rent the truck described in the ad if the cost is to be less than \$110?

79. Boating A speedboat is rated to carry 750 lb. If the driver weighs 205 lb and a passenger weighs 175 lb, how many children can safely ride along if they average 90 lb each?

80. Elevators An elevator is rated to carry 900 lb. How many boxes of books can the elevator safely carry if each box weighs 80 lb and the operator weighs 165 lb?

81. Investing money If a woman invests \$10,000 at 8% annual interest, how much more must she invest at 9% so that her annual income will exceed \$1,250?

82. Investing money If a man invests \$8,900 at 5.5% annual interest, how much more must he invest at 8.75% so that his annual income will be more than \$1,500?

83. Buying compact discs A student can afford to spend up to \$330 on a stereo system and some compact discs. If the stereo costs \$175 and the discs are \$8.50 each, find the greatest number of discs he can buy.

84. Buying a computer A student who can afford to spend up to \$2,000 sees the following ad. If she buys a computer, find the greatest number of DVDs that she can buy.

85. Averaging grades A student has scores of 70, 77, and 85 on three exams. What score is needed on a fourth exam to make an average of 80 or better?

86. Averaging grades A student has scores of 70, 79, 85, and 88 on four exams. What score does she need on the fifth exam to keep her average above 80?

87. Planning a work schedule Nguyen can earn \$5 an hour for working at the college library and \$9 an hour for construction work. To save time for study, he wants to limit his work to 20 hours a week but still earn more than \$125. How many hours can he work at the library?

88. Scheduling equipment An excavating company charges \$300 an hour for the use of a backhoe and \$500 an hour for the use of a bulldozer. (Part of an hour counts as a full hour.) The company employs one operator for 40 hours per week. If the company wants to take in at least \$18,500 each week, how many hours per week can it schedule the operator to use a backhoe?

89. Choosing a medical plan A college provides its employees with a choice of the two medical plans shown in the table. For what size hospital bills is Plan 2 better than Plan 1? (*Hint:* The cost to the employee includes both the deductible payment and the employee's insurance copayment.)

Plan 1	Plan 2
Employee pays \$100 Plan pays 70% of the rest	Employee pays \$200 Plan pays 80% of the rest

90. Choosing a medical plan To save costs, the college in Exercise 89 raised the employee deductible, as shown in the following table. For what size hospital bills is Plan 2 better than Plan 1? (*Hint:* The cost to the employee includes both the deductible payment and the employee's insurance copayment.)

Plan 1	Plan 2
Employee pays \$200 Plan pays 70% of the rest	Employee pays \$400 Plan pays 80% of the rest

91. Choosing sample size How large would the sample have to be for the researcher in Example 7 to be 95% certain that the true population mean would be within \$20 of the sample mean?

92. Choosing sample size How large would the sample have to be for the researcher in Example 7 to be 95% certain that the true population mean would be within \$10 of the sample mean?

WRITING ABOUT MATH

93. The techniques for solving linear equations and linear inequalities are similar, yet different. Explain.

94. Explain the concepts of *absolute* inequality and *conditional* inequality.

SOMETHING TO THINK ABOUT

95. If $x > -3$, must it be true that $x^2 > 9$?

96. If $x > 2$, must it be true that $x^2 > 4$?

97. Which of these relations is transitive?

a. $=$ **b.** $\leq$ **c.** $\not\geq$ **d.** $\neq$

98. The following solution is not correct. Why?

$$\frac{1}{3} > \frac{1}{x}$$

$$3x\left(\frac{1}{3}\right) > 3x\left(\frac{1}{x}\right) \quad \text{Multiply both sides by } 3x.$$

$$x > 3 \quad \text{Simplify.}$$

SECTION 4.2 Solving Absolute-Value Equations and Inequalities in One Variable

Objectives

1. Graph an absolute value function.
2. Solve an equation containing one absolute value term.
3. Solve an equation containing two absolute value terms.
4. Solve an inequality containing one absolute value term.

Getting Ready

Find each value. In Problems 3–4, assume that $x = -5$.

1. $-(-5)$ **2.** -0 **3.** $-(x - 5)$ **4.** $-(-x)$

Solve each inequality and give the result in interval notation.

5. $-4 < x + 1 < 5$ **6.** $x < -3$ or $x > 3$

In this section, we review the definition of absolute value and the graphs of absolute value functions. Then we will show how to solve equations and inequalities that contain absolute values.

1 Graph an absolute value function.

Recall the definition of the absolute value of x.

Absolute Value

For any real number x,

If $x \geq 0$, then $|x| = x$.

If $x < 0$, then $|x| = -x$.

This definition associates a nonnegative real number with any real number.

- If $x \geq 0$, then x (which is positive or 0) is its own absolute value.
- If $x < 0$, then $-x$ (which is positive) is the absolute value.

Either way, $|x|$ is positive or 0:

$|x| \geq 0$ for all real numbers x

EXAMPLE 1 Find each absolute value. **a.** $|9|$ **b.** $|-5|$ **c.** $|0|$ **d.** $|2 - \pi|$

Soltuion **a.** Since $9 \geq 0$, 9 is its own absolute value: $|9| = 9$.

b. Since $-5 < 0$, the negative of -5 is the absolute value:

$$|-5| = -(-5) = 5$$

c. Since $0 \geq 0$, 0 is its own absolute value: $|0| = 0$.

d. Since $\pi \approx 3.14$, it follows that $2 - \pi < 0$. Thus,

$$|2 - \pi| = -(2 - \pi) = \pi - 2$$

SELF CHECK 1 Find each absolute value. **a.** $|-3|$ **b.** $|\pi - 2|$

COMMENT The placement of a $-$ sign in an expression containing an absolute value symbol is important. For example, $|-19| = 19$, but $-|19| = -19$.

Earlier, we graphed the absolute value function $f(x) = |x|$ and considered many translations of its graph. The work in Example 2 reviews these concepts.

EXAMPLE 2 Graph the function $f(x) = |x - 1| + 3$.

Solution To graph this function, we translate the graph of $f(x) = |x|$. We move it 1 unit to the right and 3 units up, as shown in Figure 4-8 on the next page.

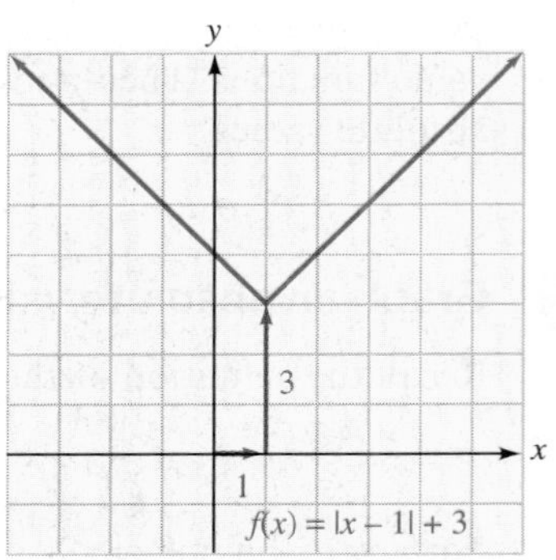

Figure 4-8

SELF CHECK 2 Graph: $f(x) = |x + 2| - 3$.

We also saw that the graph of $f(x) = -|x|$ is the same as the graph of $f(x) = |x|$ reflected about the x -axis, as shown in Figure 4-9.

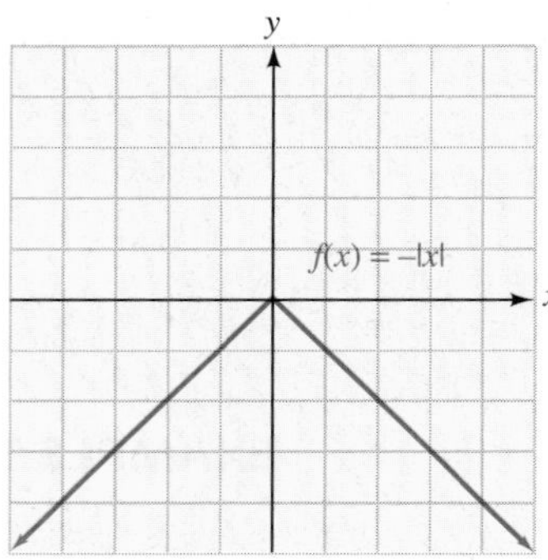

Figure 4-9

EXAMPLE 3 Graph the absolute value function $f(x) = -|x - 1| + 3$.

Solution To graph this function, we translate the graph of $f(x) = -|x|$. We move it 1 unit to the right and 3 units up, as shown in Figure 4-10.

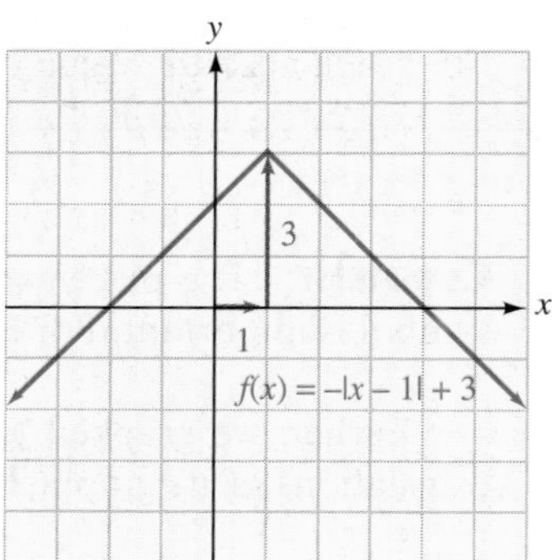

Figure 4-10

SELF CHECK 3 Graph: $f(x) = -|x + 2| - 3$.

2 Solve an equation containing one absolute value term.

In the equation $|x| = 5$, x can be either 5 or -5, because $|5| = 5$ and $|-5| = 5$. In the equation $|x| = 8$, x can be either 8 or -8. In general, the following is true.

Absolute Value Equations

If $k > 0$, then

$$|x| = k \quad \text{is equivalent to} \quad x = k \quad \text{or} \quad x = -k$$

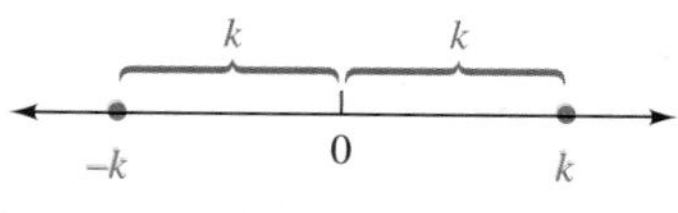

Figure 4-11

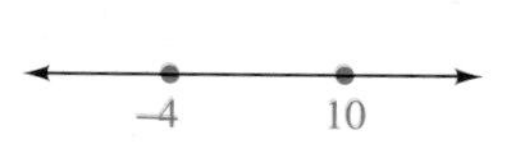

Figure 4-12

The absolute value of a number represents the distance on the number line from a point to the origin. The solutions of $|x| = k$ are the coordinates of the two points that lie exactly k units from the origin. See Figure 4-11.

The equation $|x - 3| = 7$ indicates that a point on the number line with a coordinate of $x - 3$ is 7 units from the origin. Thus, $x - 3$ can be either 7 or -7.

$$x - 3 = 7 \quad \text{or} \quad x - 3 = -7$$
$$x = 10 \quad \Big| \quad x = -4$$

The solutions of $|x - 3| = 7$ are 10 and -4, as shown in Figure 4-12.

COMMENT The equation $|x - 3| = 7$ also can be solved using the definition of absolute value.

If $x - 3 \geq 0$, then $|x - 3| = x - 3$, and we have

$$x - 3 = 7$$
$$x = 10$$

If $x - 3 < 0$, then $|x - 3| = -(x - 3)$, and we have

$$-(x - 3) = 7$$
$$-x + 3 = 7$$
$$-x = 4$$
$$x = -4$$

EXAMPLE 4 Solve the equation: $|3x - 2| = 5$.

Solution We can write $|3x - 2| = 5$ as

$$3x - 2 = 5 \quad \text{or} \quad 3x - 2 = -5$$

and solve each equation for x:

$$3x - 2 = 5 \quad \text{or} \quad 3x - 2 = -5$$
$$3x = 7 \quad \Big| \quad 3x = -3$$
$$x = \frac{7}{3} \quad \Big| \quad x = -1$$

Verify that both solutions check.

SELF CHECK 4 Solve: $|2x - 3| = 7$.

To solve more complicated equations containing a term involving an absolute value, we can isolate the absolute value on one side of the equation.

EXAMPLE 5 Solve the equation: $|5x + 3| - 7 = 4$.

Solution We first add 7 to both sides of the equation to isolate the absolute value on the left side.

$$|5x + 3| - 7 = 4$$

$$(1) \qquad |5x + 3| = 11 \qquad \text{Add 7 to both sides.}$$

We can now write Equation 1 as

$$5x + 3 = 11 \qquad \text{or} \qquad 5x + 3 = -11$$

and solve each equation for x.

$$\begin{array}{c|c} 5x + 3 = 11 \quad \text{or} & 5x + 3 = -11 \\ 5x = 8 & 5x = -14 \\ x = \frac{8}{5} & x = -\frac{14}{5} \end{array}$$

Verify that both solutions check.

SELF CHECK 5 Solve: $|7x - 2| + 4 = 5$.

EXAMPLE 6 Solve the equation: $\left|\frac{2}{3}x + 3\right| + 4 = 10$.

Solution We subtract 4 from both sides of the equation to isolate the absolute value on the left side.

$$\left|\frac{2}{3}x + 3\right| + 4 = 10$$

$$(2) \qquad \left|\frac{2}{3}x + 3\right| = 6 \qquad \text{Subtract 4 from both sides.}$$

We now can write Equation 2 as

$$\frac{2}{3}x + 3 = 6 \qquad \text{or} \qquad \frac{2}{3}x + 3 = -6$$

and solve each equation for x:

$$\begin{array}{c|c} \frac{2}{3}x + 3 = 6 \quad \text{or} & \frac{2}{3}x + 3 = -6 \\ \frac{2}{3}x = 3 & \frac{2}{3}x = -9 \\ 2x = 9 & 2x = -27 \\ x = \frac{9}{2} & x = -\frac{27}{2} \end{array}$$

Verify that both solutions check.

SELF CHECK 6 Solve: $\left|\frac{3}{5}x - 2\right| - 3 = 4$.

COMMENT Since the absolute value of a quantity cannot be negative, equations such as $\left|7x + \frac{1}{2}\right| = -4$ have no solution. Since there are no solutions, the solution set is empty. Recall that an **empty set** is denoted by the symbol $\varnothing$.

EXAMPLE 7 Solve the equation: $\left|\frac{1}{2}x - 5\right| - 4 = -4$.

Solution We first isolate the absolute value on the left side.

$$\left|\frac{1}{2}x - 5\right| - 4 = -4$$

$$\left|\frac{1}{2}x - 5\right| = 0 \qquad \text{Add 4 to both sides.}$$

Since 0 is the only number whose absolute value is 0, the binomial $\frac{1}{2}x - 5$ must be 0, and we have

$$\frac{1}{2}x - 5 = 0$$

$$\frac{1}{2}x = 5 \qquad \text{Add 5 to both sides.}$$

$$x = 10 \qquad \text{Multiply both sides by 2.}$$

Verify that 10 satisfies the original equation.

SELF CHECK 7 Solve: $\left|\frac{2}{3}x + 4\right| + 4 = 5$.

3 Solve an equation containing two absolute value terms.

The equation $|a| = |b|$ is true when $a = b$ or when $a = -b$. For example,

$$|3| = |3| \quad \text{or} \quad |3| = |-3|$$

$$3 = 3 \quad \Big| \quad 3 = 3$$

In general, the following statement is true.

Equations with Two Absolute Values

If a and b represent algebraic expressions, the equation $|a| = |b|$ is equivalent to the pair of equations

$$a = b \quad \text{or} \quad a = -b$$

EXAMPLE 8 Solve the equation: $|5x + 3| = |3x + 25|$.

Solution This equation is true when $5x + 3 = 3x + 25$, or when $5x + 3 = -(3x + 25)$. We solve each equation for x.

$5x + 3 = 3x + 25$	or	$5x + 3 = -(3x + 25)$
$2x = 22$		$5x + 3 = -3x - 25$
$x = 11$		$8x = -28$

$$x = -\frac{28}{8}$$

$$x = -\frac{7}{2}$$

Verify that both solutions check.

SELF CHECK 8 Solve: $|2x - 3| = |4x + 9|$.

4 Solve an inequality containing one absolute value term.

The inequality $|x| < 5$ indicates that a point with a coordinate of x is less than 5 units from the origin. (See Figure 4-13.) Thus, x is between -5 and 5, and

$$|x| < 5 \quad \text{is equivalent to} \quad -5 < x < 5$$

The solution set of the inequality $|x| < k \quad (k > 0)$ includes the coordinates of the points on the number line that are less than k units from the origin. (See Figure 4-14.)

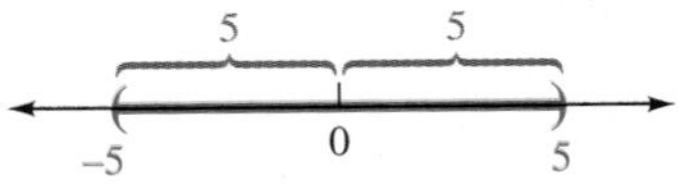

Figure 4-13

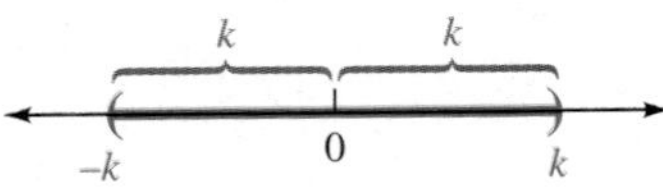

Figure 4-14

Solving $|x| < k$ and $|x| \leq k$

If $k > 0$, then

$$|x| < k \quad \text{is equivalent to} \quad -k < x < k$$

If $k \geq 0$, then

$$|x| \leq k \quad \text{is equivalent to} \quad -k \leq x \leq k$$

COMMENT Since $|x|$ is always positive or 0, the inequality $|x| < k$ has no solutions when $k < 0$. For example, the solution set of $|x| < -5$ is $\varnothing$.

EXAMPLE 9 Solve the inequality: $|2x - 3| < 9$.

Solution Since $|2x - 3| < 9$ is equivalent to $-9 < 2x - 3 < 9$, we proceed as follows:

$-9 < 2x - 3 < 9$

$-6 < 2x < 12$ Add 3 to all three parts.

$-3 < x < 6$ Divide all parts by 2.

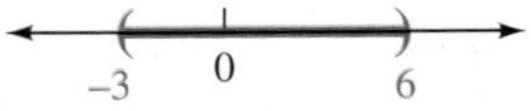

Any number between -3 and 6, not including either -3 or 6, is in the solution set. This is the interval $(-3, 6)$, whose graph is shown in Figure 4-15.

–3 0 6

Figure 4-15

SELF CHECK 9 Solve: $|3x + 2| < 4$.

COMMENT The inequality $|2x - 3| < 9$ also can be solved using the definition of absolute value.

If $2x - 3 \geq 0$, then $|2x - 3| = 2x - 3$, and we have

$$2x - 3 < 9$$
$$2x < 12$$
$$x < 6$$

If $2x - 3 < 0$, then $|2x - 3| = -(2x - 3)$, and we have

$$-(2x - 3) < 9$$
$$-2x + 3 < 9$$
$$-2x < 6$$
$$x > -3$$

These results indicate that $x < 6$ and $x > -3$, which can be written as $-3 < x < 6$.

EXAMPLE 10 Solve the inequality: $|2 - 3x| - 2 \leq 3$.

Solution We first add 2 to both sides of the inequality to obtain

$$|2 - 3x| \leq 5$$

Since $|2 - 3x| \leq 5$ is equivalent to $-5 \leq 2 - 3x \leq 5$, we proceed as follows:

$$-5 \leq 2 - 3x \leq 5$$

$$-7 \leq -3x \leq 3 \quad \text{Subtract 2 from all parts.}$$

$$\frac{7}{3} \geq x \geq -1 \quad \text{Divide all three parts by } -3 \text{ and reverse the directions of the inequality symbols.}$$

$$-1 \leq x \leq \frac{7}{3} \quad \text{Write an equivalent inequality with the inequality symbols pointing in the opposite direction.}$$

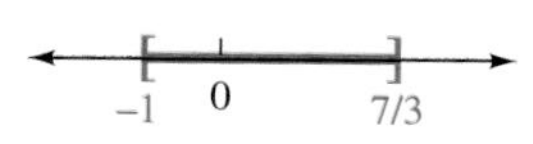

Figure 4-16

The solution set is the interval $\left[-1, \frac{7}{3}\right]$, whose graph is shown in Figure 4-16.

SELF CHECK 10 Solve: $|3 - 2x| + 3 \leq 8$.

The inequality $|x| > 5$ indicates that a point with a coordinate of x is more than 5 units from the origin. (See Figure 4-17.) Thus, $x < -5$ or $x > 5$.

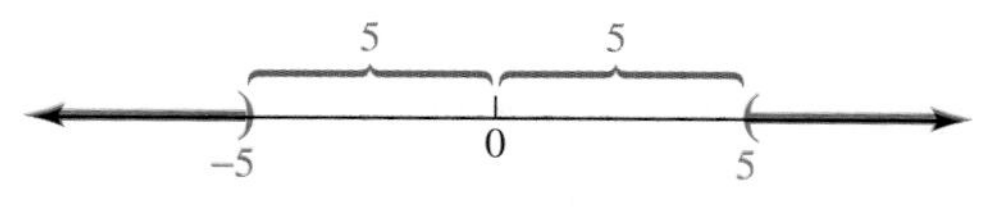

Figure 4-17

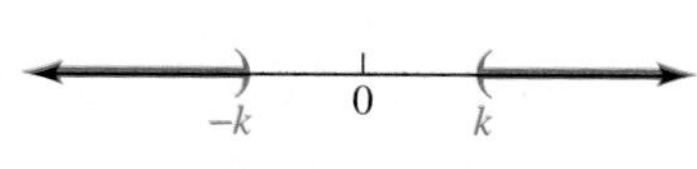

Figure 4-18

In general, the inequality $|x| > k$ can be interpreted to mean that a point with coordinate x is more than k units from the origin. (See Figure 4-18.)

Thus,

$$|x| > k \quad \text{is equivalent to} \quad x < -k \quad \text{or} \quad x > k$$

The *or* indicates an either/or situation. To be in the solution set, x needs to satisfy only one of the two conditions.

Solving $|x| > k$ and $|x| \geq k$

If $k \geq 0$, then

$$|x| > k \quad \text{is equivalent to} \quad x < -k \quad \text{or} \quad x > k$$
$$|x| \geq k \quad \text{is equivalent to} \quad x \leq -k \quad \text{or} \quad x \geq k$$

EXAMPLE 11 Solve the inequality: $|5x - 10| > 20$.

Solution We write the inequality as two separate inequalities and solve each one for x. Since $|5x - 10| > 20$ is equivalent to $5x - 10 < -20$ or $5x - 10 > 20$, we have

$$5x - 10 < -20 \quad \text{or} \quad 5x - 10 > 20$$
$$5x < -10 \quad\quad 5x > 30 \quad \text{Add 10 to both sides.}$$
$$x < -2 \quad\quad x > 6 \quad \text{Divide both sides by 5.}$$

Thus, x is either less than -2 or greater than 6:

$$x < -2 \quad \text{or} \quad x > 6$$

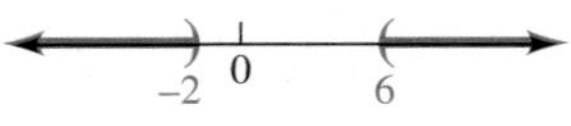

Figure 4-19

This is the interval $(-\infty, -2) \cup (6, \infty)$, whose graph appears in Figure 4-19.

SELF CHECK 11 Solve: $|3x + 4| > 13$.

COMMENT The inequality $|5x - 10| > 20$ can also be solved using the definition of absolute value.

If $5x - 10 \geq 0$, then $|5x - 10| = 5x - 10$, and we have

$$5x - 10 > 20$$
$$5x > 30$$
$$x > 6$$

or

If $5x - 10 < 0$, then $|5x - 10| = -(5x - 10)$, and we have

$$-(5x - 10) > 20$$
$$-5x + 10 > 20$$
$$-5x > 10$$
$$x < -2$$

Thus, $x < -2$ or $x > 6$.

EXAMPLE 12 Solve the inequality: $\left|\frac{3 - x}{5}\right| \geq 6$.

Solution We write the inequality as two separate inequalities:

$$\left|\frac{3 - x}{5}\right| \geq 6 \quad \text{is equivalent to} \quad \frac{3 - x}{5} \leq -6 \quad \text{or} \quad \frac{3 - x}{5} \geq 6$$

Then we solve each one for x:

$$\frac{3 - x}{5} \leq -6 \quad \text{or} \quad \frac{3 - x}{5} \geq 6$$
$$3 - x \leq -30 \quad\quad 3 - x \geq 30 \quad \text{Multiply both sides by 5.}$$
$$-x \leq -33 \quad\quad -x \geq 27 \quad \text{Subtract 3 from both sides.}$$
$$x \geq 33 \quad\quad x \leq -27 \quad \text{Divide both sides by } -1 \text{ and reverse the direction of the inequality symbol.}$$

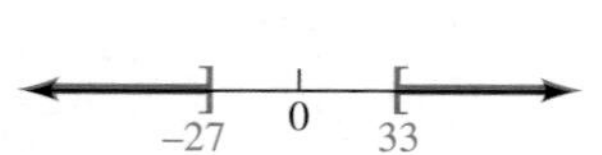

Figure 4-20

The solution set is $(-\infty, -27] \cup [33, \infty)$, whose graph appears in Figure 4-20.

SELF CHECK 12 Solve: $\left|\frac{3 - x}{6}\right| \geq 5$.

EXAMPLE 13 Solve the inequality: $\left|\frac{2}{3}x - 2\right| - 3 > 6.$

Solution We begin by adding 3 to both sides to isolate the absolute value on the left side. We then proceed as follows:

$$\left|\frac{2}{3}x - 2\right| - 3 > 6$$

$$\left|\frac{2}{3}x - 2\right| > 9 \qquad \text{Add 3 to both sides to isolate the absolute value.}$$

$$\frac{2}{3}x - 2 < -9 \quad \text{or} \quad \frac{2}{3}x - 2 > 9$$

$$\frac{2}{3}x < -7 \qquad \frac{2}{3}x > 11 \qquad \text{Add 2 to both sides.}$$

$$2x < -21 \qquad 2x > 33 \qquad \text{Multiply both sides by 3.}$$

$$x < -\frac{21}{2} \qquad x > \frac{33}{2} \qquad \text{Divide both sides by 2.}$$

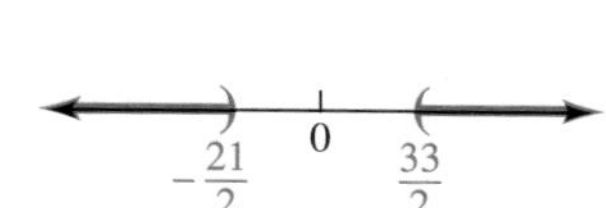

Figure 4-21

The solution set is $\left(-\infty, -\frac{21}{2}\right) \cup \left(\frac{33}{2}, \infty\right)$, whose graph appears in Figure 4-21.

SELF CHECK 13 Solve: $\left|\frac{3}{4}x + 2\right| - 1 > 3.$

EXAMPLE 14 Solve the inequality: $|3x - 5| \geq -2.$

Solution Since the absolute value of any number is nonnegative, and since any nonnegative number is greater than -2, the inequality is true for all x. The solution set is $(-\infty, \infty)$, whose graph appears in Figure 4-22.

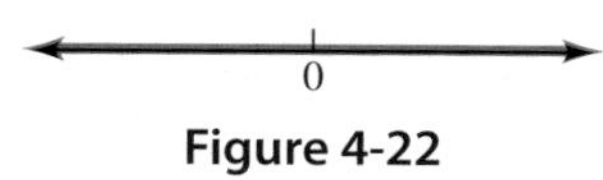

Figure 4-22

SELF CHECK 14 Find the solution set of $|3x - 5| \leq -2.$

ACCENT ON TECHNOLOGY

Solving Absolute Value Inequalities

We can solve many absolute value inequalities by a graphing method. For example, to solve $|2x - 3| < 9$ (shown in Example 9), we graph the equations $y = |2x - 3|$ and $y = 9$ on the same coordinate system. If we use window settings of $[-5, 15]$ for x and $[-5, 15]$ for y, we get the graph shown in Figure 4-23.

The inequality $|2x - 3| < 9$ will be true for all x-coordinates of points that lie on the graph of $y = |2x - 3|$ and below the graph of $y = 9$. By using the trace feature, we can see that these values of x are in the interval $(-3, 6)$.

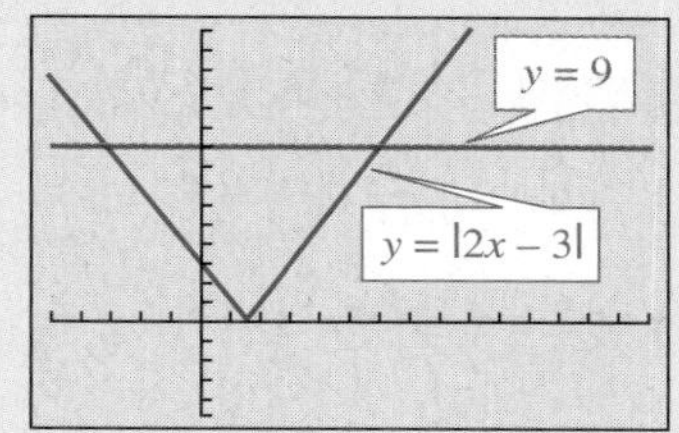

Figure 4-23

SELF CHECK ANSWERS

1. a. 3 **b.** $\pi - 2$ **2.**

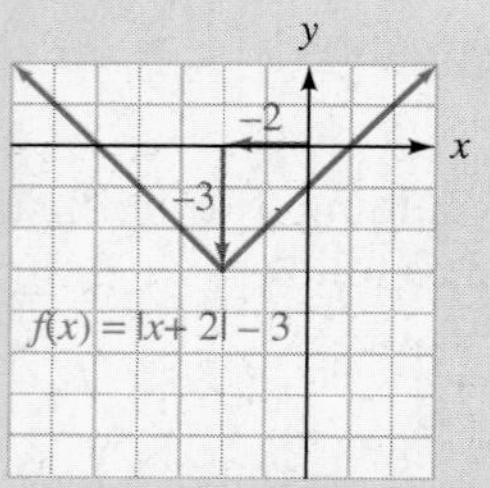

3.

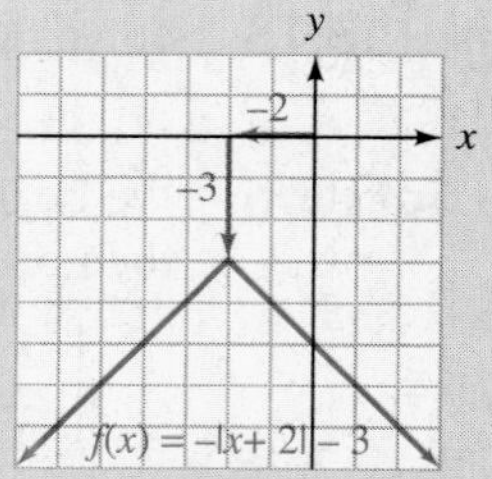

4. 5, −2 **5.** $\frac{3}{7}, \frac{1}{7}$

6. $15, -\frac{25}{3}$ **7.** $-\frac{15}{2}, -\frac{9}{2}$ **8.** −1, −6 **9.** $\left(-2, \frac{2}{3}\right)$

10. $[-1, 4]$

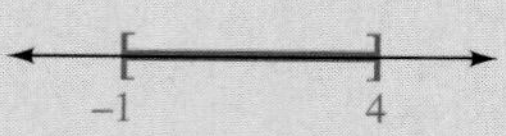

11. $\left(-\infty, -\frac{17}{3}\right) \cup (3, \infty)$

12. $(-\infty, -27] \cup [33, \infty)$

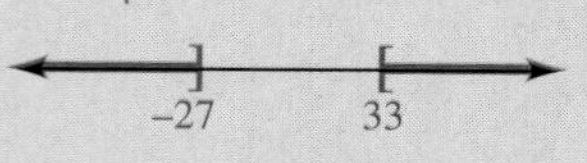

13. $(-\infty, -8) \cup \left(\frac{8}{3}, \infty\right)$

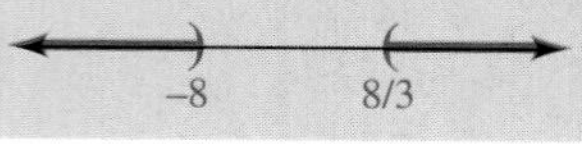

14. $\varnothing$

NOW TRY THIS

1. Express the following as an absolute value equation. $x = 2$ or $x = -2$
2. Express each of the following as an absolute value inequality.
   a. $-5 \leq x \leq 5$
   b. $3x - 5 < -7$ or $3x - 5 > 7$

4.2 EXERCISES

WARM-UPS

Find each absolute value.

1. $|-5|$

2. $-|5|$

3. $-|-6|$

4. $-|4|$

Solve each equation or inequality.

5. $|x| = 8$

6. $|x| = -5$

7. $|x| < 8$

8. $|x| > 8$

9. $|x| \geq 4$

10. $|x| \leq 7$

REVIEW *Solve each equation or formula.*

11. $3(2a - 1) = 2a$

12. $\frac{t}{6} - \frac{t}{3} = -1$

13. $\frac{5x}{2} - 1 = \frac{x}{3} + 12$

14. $4b - \frac{b+9}{2} = \frac{b+2}{5} - \frac{8}{5}$

15. $A = p + prt$ for t

16. $P = 2w + 2l$ for l

VOCABULARY AND CONCEPTS *Fill in the blanks.*

17. If $x \geq 0$, $|x| =$ __.

18. If $x < 0$, $|x| =$ ____.

19. $|x| \geq$ __ for all real numbers x.

20. The graph of the function $f(x) = |x - 2| - 4$ is the same as the graph of $f(x) = |x|$, except that it has been translated 2 units to the ______ and 4 units ______.

21. The graph of $f(x) = -|x|$ is the same as the graph of $f(x) = |x|$, except that it has been ________ about the x-axis.

22. If $k > 0$, then $|x| = k$ is equivalent to ______________.

23. $|a| = |b|$ is equivalent to ______________.

24. If $k > 0$, then $|x| < k$ is equivalent to ___________.

25. If $k \geq 0$, then $|x| \geq k$ is equivalent to ______________.

26. The equation $|x - 4| < -5$ has ____ solutions.

GUIDED PRACTICE

Find the value of each expression. See Example 1. (Objective 1)

27. $|8|$

28. $|-18|$

29. $-|2|$

30. $-|-20|$

31. $-|-30|$

32. $-|25|$

33. $|\pi - 4|$

34. $|2\pi - 4|$

Graph each absolute value function. See Examples 2–3. (Objective 1)

35. $f(x) = |x| - 2$

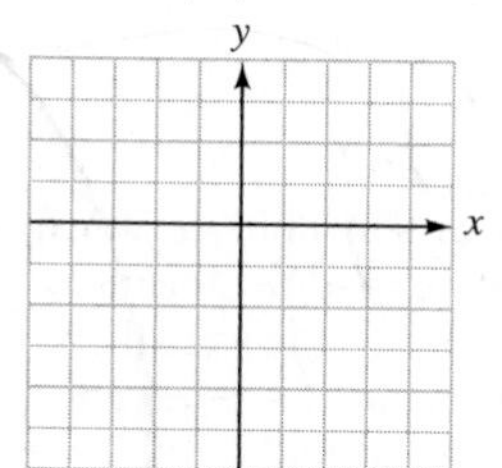

36. $f(x) = -|x| + 1$

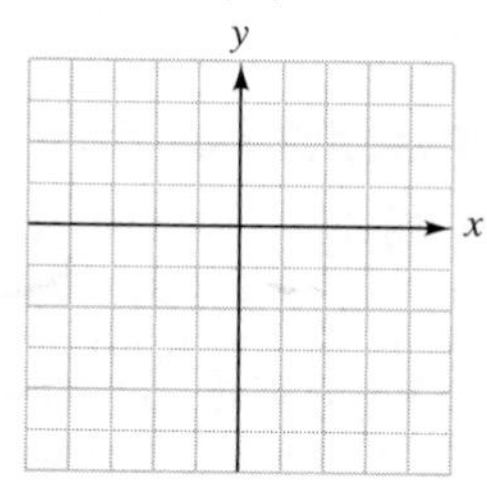

37. $f(x) = -|x + 4|$

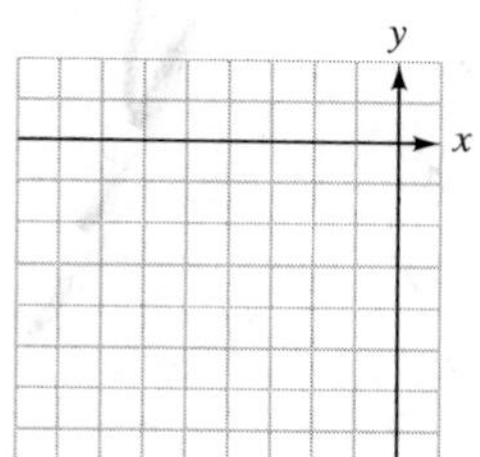

38. $f(x) = |x - 1| + 2$

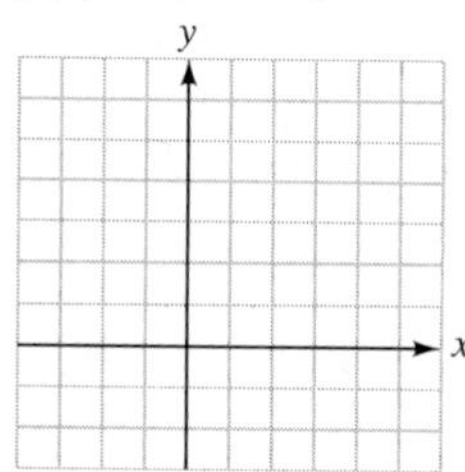

Solve each equation. See Examples 4–7. (Objective 2)

39. $|x| = 4$

40. $|x| = 9$

41. $|x - 3| = 6$

42. $|x + 4| = 8$

43. $|2x - 3| = 5$

44. $|4x - 4| = 20$

45. $\left|\frac{x}{2} - 1\right| = 3$

46. $\left|\frac{4x - 64}{4}\right| = 32$

47. $|3 - 4x| = 5$

48. $|8 - 5x| = 18$

49. $|3x + 24| = 0$

50. $|x - 21| = -8$

51. $\left|\frac{3x + 48}{3}\right| = 12$

52. $\left|\frac{x}{2} + 2\right| = 4$

53. $|2x + 1| - 3 = 12$

54. $|3x - 2| + 1 = 11$

Solve each equation. See Example 8. (Objective 3)

55. $|2x + 1| = |3x + 3|$

56. $|5x - 7| = |4x + 1|$

57. $|2 - x| = |3x + 2|$

58. $|4x + 3| = |9 - 2x|$

59. $\left|\frac{x}{2} + 2\right| = \left|\frac{x}{2} - 2\right|$

60. $|7x + 12| = |x - 6|$

61. $|3x + 7| = -|8x - 2|$

62. $-|17x + 13| = |3x - 14|$

Solve each inequality. Write the solution set in interval notation and graph it. See Examples 9–10. (Objective 4)

63. $|2x| < 8$

64. $|3x| < 27$

65. $|x + 9| \le 12$

66. $|x - 8| \le 12$

67. $|3x + 2| \le -3$

68. $|3x - 2| < 10$

69. $\left|\frac{1}{2}x - 3\right| - 4 < 2$

70. $\left|\frac{1}{6}x + 6\right| + 2 < 2$

71. $\left|\frac{x - 5}{10}\right| + 5 \le 5$

72. $\left|\frac{3}{5}x - 2\right| + 3 \le 3$

73. $\left|\frac{1}{7}x + 1\right| \le 0$

74. $|2x + 1| + 2 \le 2$

Solve each inequality. Write the solution set in interval notation and graph it. See Examples 11–14. (Objective 4)

75. $|5x| + 2 > 7$

76. $|7x| - 3 > 4$

77. $|3x + 2| > 14$

78. $|2x - 5| > 25$

79. $|4x + 3| > -5$

80. $|4x + 3| > 0$

81. $|2 - 3x| \ge 8$

82. $|-1 - 2x| > 5$

83. $|8x - 3| > 0$

84. $|7x + 2| > -8$

85. $\left|\frac{x-2}{3}\right| \leq 4$

86. $\left|\frac{x-2}{3}\right| > 4$

ADDITIONAL PRACTICE

Solve each equation or inequality. Graph the inequalities and state the solution in interval notation.

87. $|x+3| + 7 = 10$

88. $|2-x| + 3 = 5$

89. $3|2x+5| \geq 9$

90. $-2|3x-4| < 16$

91. $|4x-1| \leq 7$

92. $|5x-12| < -5$

93. $|3-2x| < 7$

94. $|4-3x| \leq 13$

95. $|3x-1| = |x+5|$

96. $|3x+1| = |x-5|$

97. $-|2x-3| < -7$

98. $-|3x+1| < -8$

99. $\left|x+\frac{1}{3}\right| = |x-3|$

100. $\left|x-\frac{1}{4}\right| = |x+4|$

101. $|5x-1| + 4 \leq 0$

102. $-|5x-1| + 2 < 0$

103. $|3x+1| + 2 < 6$

104. $|3x-2| + 2 \geq 0$

105. $\left|\frac{1}{3}x+7\right| + 5 > 6$

106. $\left|\frac{1}{5}x-5\right| + 4 > 4$

107. $|3x+2| = 16$

108. $|5x-3| = 22$

109. $|x-12| > 24$

110. $|x+5| \geq 7$

111. $\left|\frac{7}{2}x+3\right| = -5$

112. $|2x+10| = 0$

113. $\left|\frac{3}{5}x-4\right| - 2 = -2$

114. $\left|\frac{3}{4}x+2\right| + 4 = 4$

Write each inequality as an inequality using absolute values.

115. $-4 < x < 4$

116. $x < -4$ or $x > 4$

117. $x+3 < -6$ or $x+3 > 6$

118. $-5 \leq x-3 \leq 5$

APPLICATIONS

119. **Springs** The weight on the spring shown in the illustration oscillates up and down according to the formula $|d-5| \leq 1$, where d is the distance of the weight above the ground. Solve the formula for d and give the range of heights that the weight is above the ground.

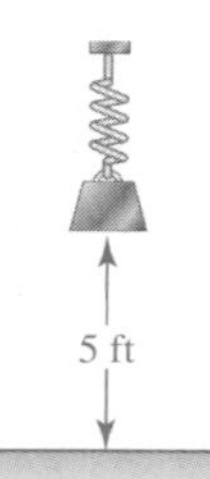

120. **Aviation** An airplane is cruising at 30,000 ft and has been instructed to maintain an altitude that is described by the inequality $|a - 30{,}000| \leq 1{,}500$. Solve the inequality for a and give the range of altitudes at which the plane can fly.

121. **Finding temperature ranges** The temperatures on a summer day satisfied the inequality $|t - 78°| \leq 8°$, where t is a temperature in degrees Fahrenheit. Express the range of temperatures as a compound inequality.

122. **Finding operating temperatures** A car CD player has an operating temperature of $|t - 40°| < 80°$, where t is a temperature in degrees Fahrenheit. Express this range of temperatures as a compound inequality.

123. **Range of camber angles** The specification for a car state that the camber angle c of its wheels should be $0.6° \pm 0.5°$. Express this range with an inequality containing absolute value symbols.

124. **Tolerance of sheet steel** A sheet of steel is to be 0.25 inch thick with a tolerance of 0.015 inch. Express this specification with an inequality containing absolute value symbols.

WRITING ABOUT MATH

125. Explain how to find the absolute value of a given number.

126. Explain why the equation $|x| + 5 = 0$ has no solution.

127. Explain the use of parentheses and brackets when graphing inequalities.

128. If $k > 0$, explain the differences between the solution sets of $|x| < k$ and $|x| > k$.

SOMETHING TO THINK ABOUT

129. For what values of k does $|x| + k = 0$ have exactly two solutions?

130. For what value of k does $|x| + k = 0$ have exactly one solution?

131. Under what conditions is $|x| + |y| > |x + y|$?

132. Under what conditions is $|x| + |y| = |x + y|$?

SECTION 4.3 Solving Linear Inequalities in One or Two Variables

Objectives

1. Graph a linear inequality in two variables.
2. Graph a compound linear inequality.
3. Solve an application problem requiring the use of a linear inequality in two variables.

Vocabulary

linear inequality
boundary line
half-plane
edge

Getting Ready

Do the coordinates of the points satisfy the equation $y = 5x + 2$?

1. $(0, 2)$ **2.** $\left(-\frac{2}{5}, 0\right)$ **3.** $(3, 18)$ **4.** $(-3, -13)$

In this section, we will show how to solve linear inequalities in the two variables that have two variables. We begin by discussing their graphs.

1 Graph a linear inequality in two variables.

Inequalities such as

$$2x + 3y < 6 \qquad \text{or} \qquad 3x - 4y \geq 9$$

are called *linear inequalities in two variables.* In general, we have the following definition:

Linear Inequalities

A **linear inequality** in x and y is any inequality that can be written in the form

$$Ax + By < C, \quad Ax + By > C, \quad Ax + By \leq C, \quad \text{or} \quad Ax + By \geq C$$

where A, B, and C are real numbers and A and B are not both 0.

The *graph of a linear inequality* in x and y is the graph of all ordered pairs (x, y) that satisfy the inequality.

The inequality $y > 3x + 2$ is an example of a linear inequality in two variables because it can be written in the form $-3x + y > 2$. To graph it, we first graph the related equation $y = 3x + 2$ as shown in Figure 4-24(a). This **boundary line,** sometimes called an **edge,** divides the coordinate plane into two **half-planes,** one on either side of the line.

To find which half-plane is the graph of $y > 3x + 2$, we can substitute the coordinates (0, 0) into the inequality and simplify.

$$y > 3x + 2$$
$$0 \overset{?}{>} 3(0) + 2 \quad \text{Substitute 0 for } x \text{ and 0 for } y.$$
$$0 \not> 2$$

Since the coordinates don't satisfy $y > 3x + 2$, the origin is not part of the graph. Thus, the half-plane on the other side of the dashed line is the graph, which is shown in Figure 4-24(b). We use a dashed line to indicate that the boundary line is not included in the graph.

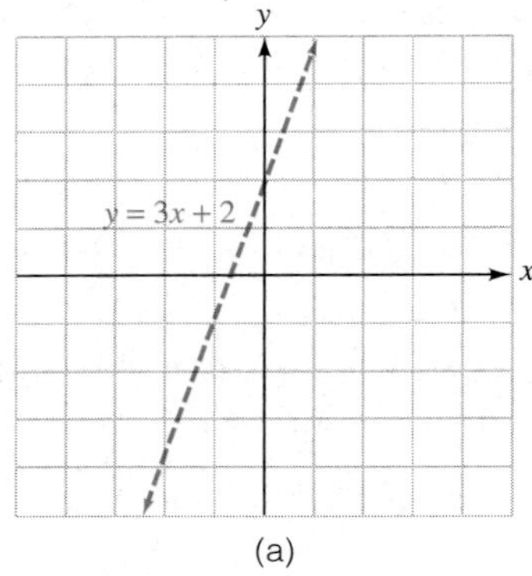

(a)

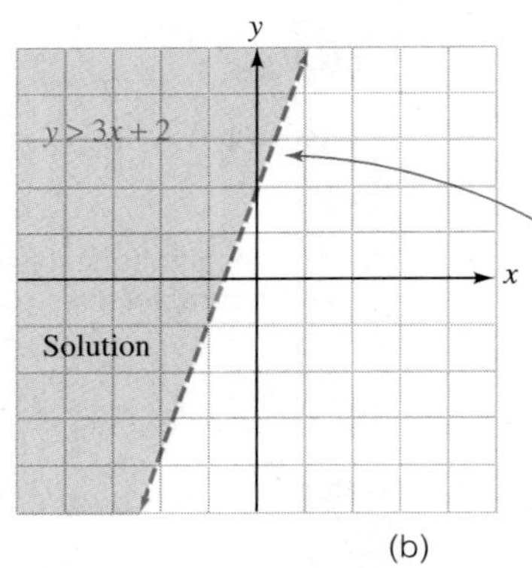

(b)

Figure 4-24

EXAMPLE 1 Graph the inequality: $2x - 3y \leq 6$.

Solution This inequality is the combination of the inequality $2x - 3y < 6$ and the equation $2x - 3y = 6$.

We start by graphing $2x - 3y = 6$ to find the boundary line. This time, we draw the solid line shown in Figure 4-25(a), because equality is included. To determine which half-plane represents $2x - 3y < 6$, we check to see whether the coordinates of the origin satisfy the inequality.

$$2x - 3y < 6$$
$$2(0) - 3(0) < 6 \quad \text{Substitute 0 for } x \text{ and 0 for } y.$$
$$0 < 6$$

Since the coordinates satisfy the inequality, the origin is in the half-plane that is the graph of $2x - 3y < 6$. The graph is shown in Figure 4-25(b).

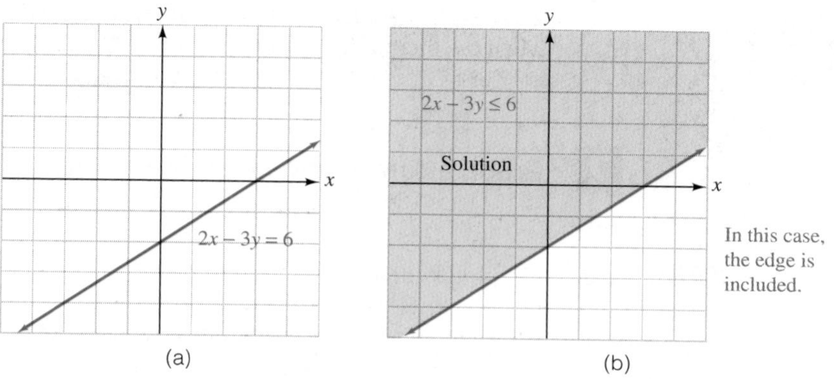

Figure 4-25

SELF CHECK 1 Graph: $3x - 2y \geq 6$.

EXAMPLE 2 Graph the inequality: $y < 2x$.

Solution We graph $y = 2x$, as shown in Figure 4-26(a). Because it is not part of the inequality, we draw the edge as a dashed line.

To determine which half-plane is the graph of $y < 2x$, we check to see whether the coordinates of some fixed point satisfy the inequality. We cannot use the origin as a test point, because the edge passes through the origin. However, we can choose a different point—say, (3, 1).

$y < 2x$

$1 \overset{?}{<} 2(3)$ Substitute 1 for y and 3 for x.

$1 < 6$

Since $1 < 6$ is a true inequality, the point (3, 1) satisfies the inequality and is in the graph, which is shown in Figure 4-26(b).

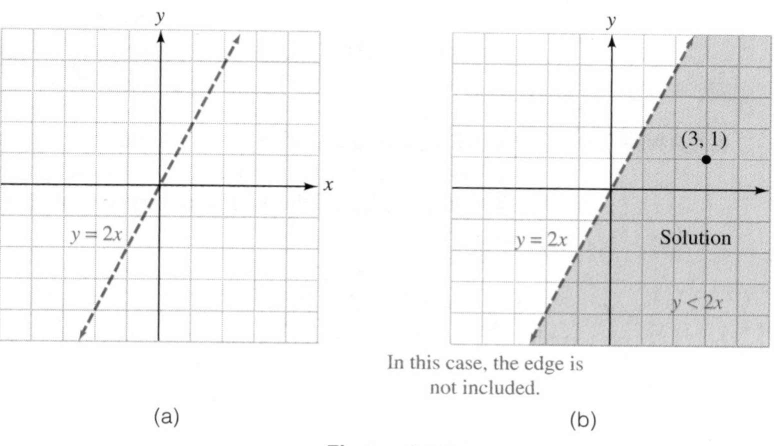

Figure 4-26

SELF CHECK 2 Graph: $y > 2x$.

COMMENT As the first two examples suggest, we draw a boundary line as a solid line when the inequality symbol is $\leq$ or $\geq$. We draw a dashed line when the inequality symbol is $<$ or $>$.

2 Graph a compound linear inequality.

EXAMPLE 3 Graph the inequality: $2 < x \leq 5$.

Solution The inequality $2 < x \leq 5$ is equivalent to the following two inequalities:

$$2 < x \quad \text{and} \quad x \leq 5$$

Its graph will contain all points in the plane that satisfy the inequalities $2 < x$ and $x \leq 5$ simultaneously. These points are in the shaded region of Figure 4-27.

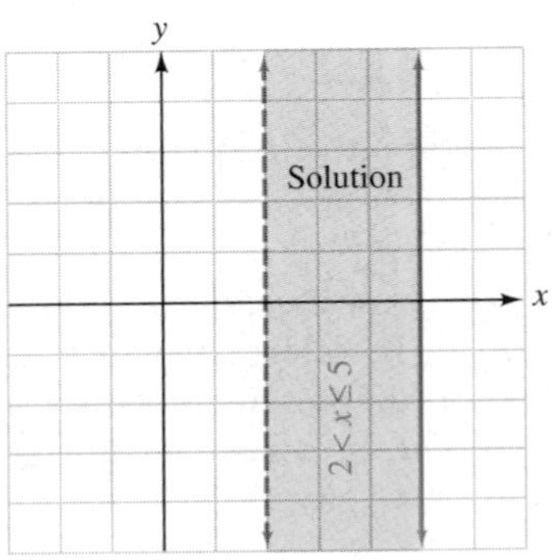

Figure 4-27

SELF CHECK 3 Graph: $-2 \leq x < 3$.

3 Solve an application problem requiring the use of a linear inequality in two variables.

EXAMPLE 4 **EARNING MONEY** Rick has two part-time jobs, one paying \$7 per hour and the other paying \$5 per hour. He must earn at least \$140 per week to pay his expenses while attending college. Write an inequality that shows the various ways he can schedule his time to achieve his goal.

Analyze the problem If we let x represent the number of hours per week he works on the first job, he will earn $\$7x$ per week on the first job. If we let y represent the number of hours per week he works on the second job, he will earn $\$5y$ per week on the second job.

Form an inequality To achieve his goal, the sum of these two incomes must be at least \$140.

The hourly rate on the first job	times	the hours worked on the first job	plus	the hourly rate on the second job	times	the hours worked on the second job	is greater than or equal to	\$140.
\$7	$\cdot$	x	$+$	\$5	$\cdot$	y	$\geq$	\$140

Solve the inequality

The graph of $7x + 5y \geq 140$ is shown in Figure 4-28. Any point in the shaded region indicates a way that he can schedule his time and earn \$140 or more per week. For example, if he works 10 hours on the first job and 15 hours on the second job, he will earn

$$\begin{aligned} \$7(10) + \$5(15) &= \$70 + \$75 \\ &= \$145 \end{aligned}$$

If he works 5 hours on the first job and 25 hours on the second job, he will earn

$$\begin{aligned} \$7(5) + \$5(25) &= \$35 + \$125 \\ &= \$160 \end{aligned}$$

Since Rick cannot work a negative number of hours, the graph has no meaning when x or y is negative, so only the first quadrant of the graph is shown.

COMMENT We could let x represent the number of hours he works on the second job and y represent the number of hours he works on the first job. The resulting graphs would be different, but the combination of hours would remain the same.

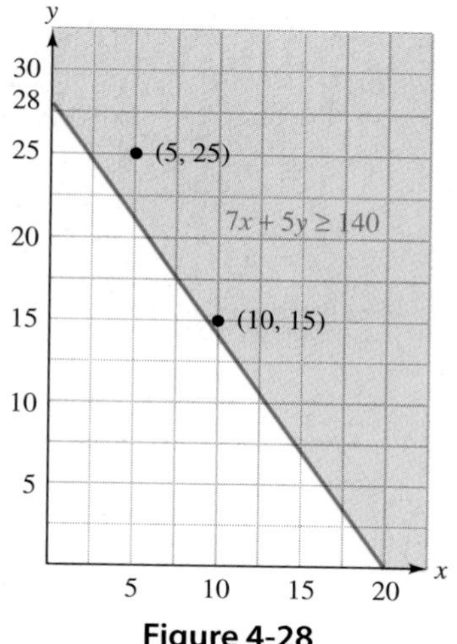

Figure 4-28

ACCENT ON TECHNOLOGY

Graphing Inequalities

Some calculators (such as the TI-83 Plus or the TI-84 Plus) have a graphing-style icon in the $y =$ editor. See Figure 4-29(a). Some of the different graphing styles are as follows.

\	line	A straight line or curved graph is shown.
◥	above	Shading covers the area above a graph.
◣	below	Shading covers the area below a graph.

We can change the icon by placing the cursor on it and pressing **ENTER**.

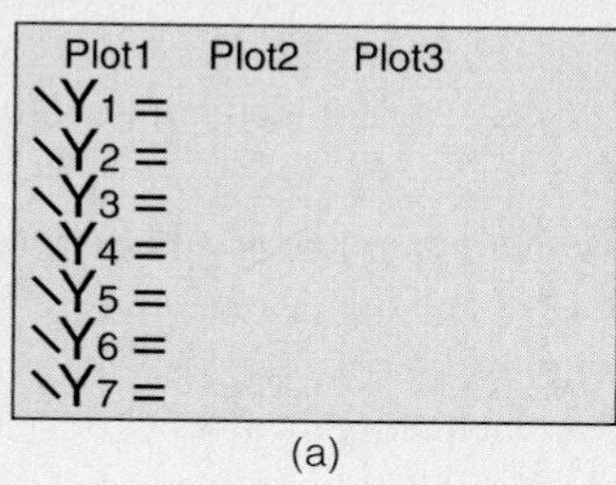

(a)

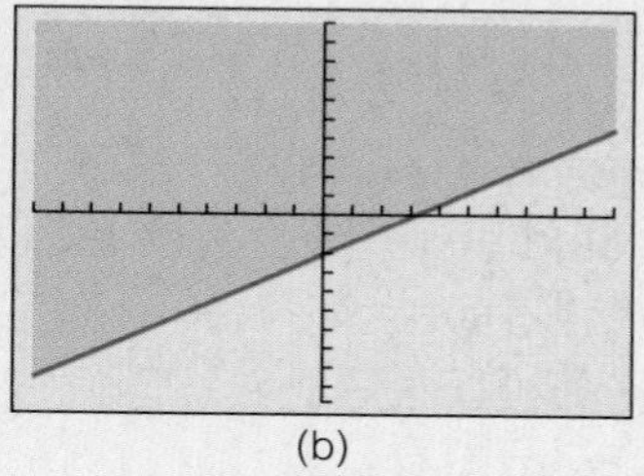

(b)

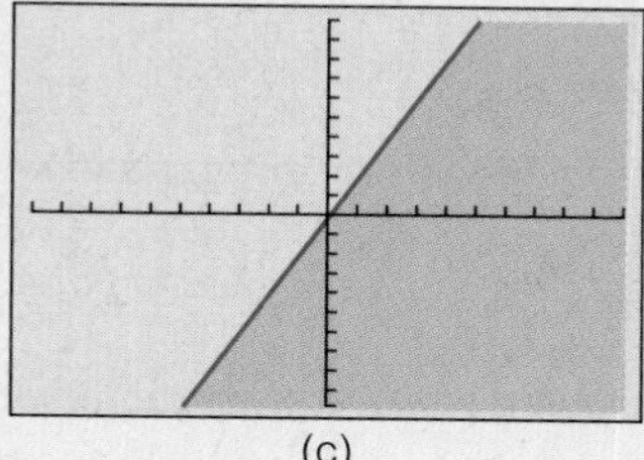

(c)

Figure 4-29

To graph the inequality of Example 1 using window settings of $x = [-10, 10]$ and $y = [-10, 10]$, we change the graphing-style icon to "above" (◥), enter the equation $2x - 3y = 6$ as $y = \frac{2}{3}x - 2$, and press **GRAPH** to get Figure 4-29(b).

To graph the inequality of Example 2 using window settings of $x = [-10, 10]$ and $y = [-10, 10]$, we change the graphing-style icon to "below" (◣), enter the equation $y = 2x$, and press **GRAPH** to get Figure 4-29(c).

If your calculator does not have a graphing-style icon, you can graph linear inequalities with a shade feature. To do so, consult your owner's manual.

Note that graphing calculators do not distinguish between solid and dashed lines to show whether or not the edge of a region is included within the graph.

SELF CHECK ANSWERS

1.

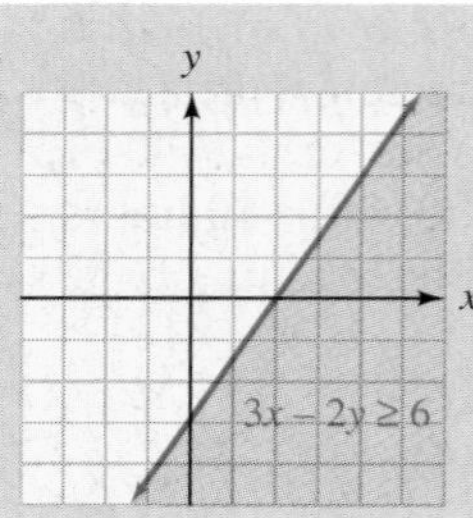

2.

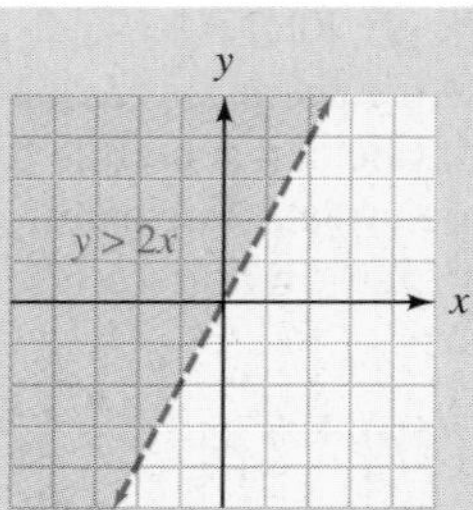

3.

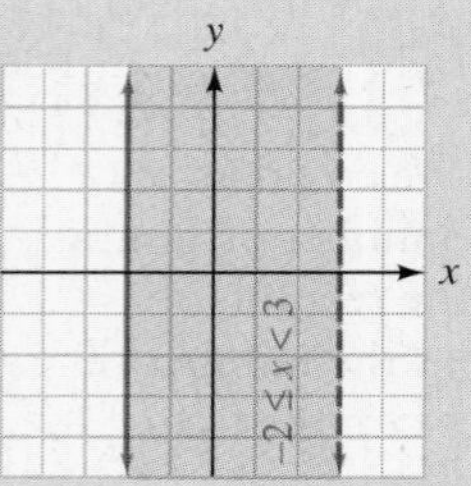

NOW TRY THIS

Solve each of the following by graphing.

1. $x \geq 2$ and $y < -1$

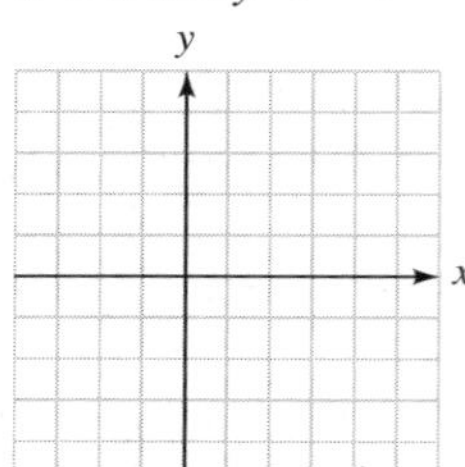

2. $x = 2$ and $y = -1$

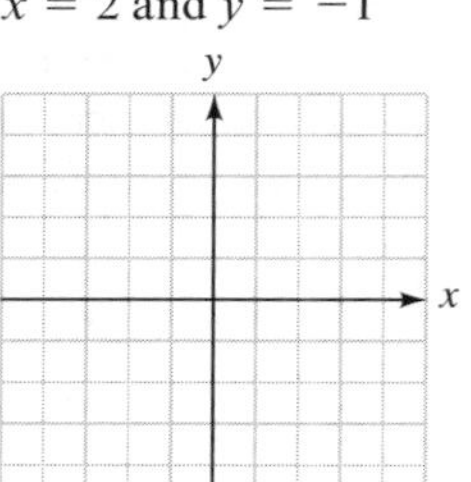

4.3 EXERCISES

WARM-UPS

Do points with the given coordinates satisfy $2x + 3y < 12$?

1. (0, 0)
2. (3, 2)
3. (2, 3)
4. (−1, 4)

Do points with the given coordinates satisfy $3x - 2y \geq 12$?

5. (0, 0)
6. (3, 2)
7. (2, −3)
8. (5, 1)

REVIEW *Solve each system.*

9. $\begin{cases} x + y = 4 \\ x - y = 2 \end{cases}$
10. $\begin{cases} 2x - y = -4 \\ x + 2y = 3 \end{cases}$
11. $\begin{cases} 3x + y = 3 \\ 2x - 3y = 13 \end{cases}$
12. $\begin{cases} 2x - 5y = 8 \\ 5x + 2y = -9 \end{cases}$

VOCABULARY AND CONCEPTS *Fill in the blanks.*

13. $3x + 2y < 12$ is an example of a ______ inequality.
14. Graphs of linear inequalities in two variables are __________.
15. The boundary line of a half-plane is called an _____.
16. If an inequality involves ≤ or ≥, the boundary line of its graph will be _____.
17. If an inequality involves < or >, the boundary line of its graph will be _______.
18. If $y < \frac{1}{2}x - 2$ and $y > \frac{1}{2}x - 2$ are false, then __________.

GUIDED PRACTICE

Graph each inequality. **See Examples 1–2. (Objective 1)**

19. $y > x + 1$

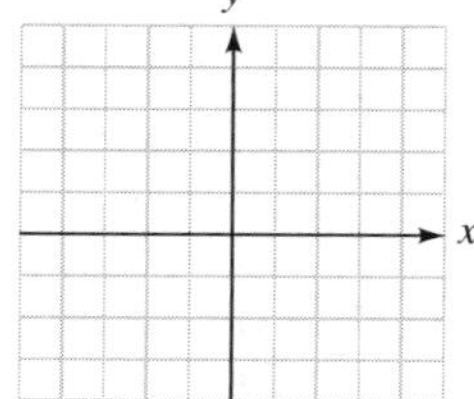

20. $y < 2x - 1$

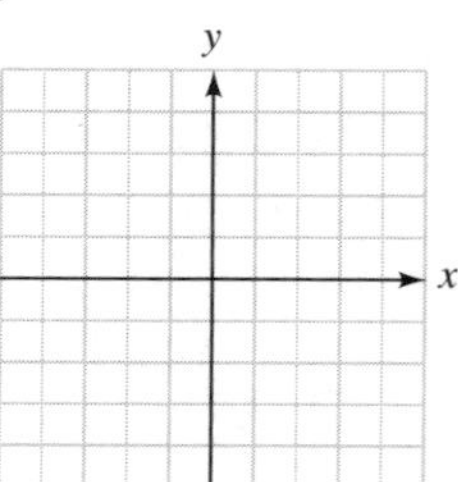

21. $y \geq x$

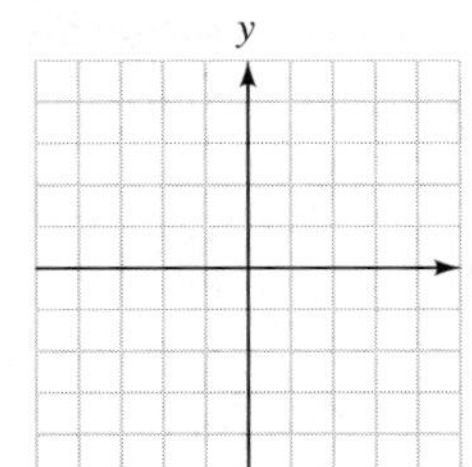

22. $y \leq 2x$

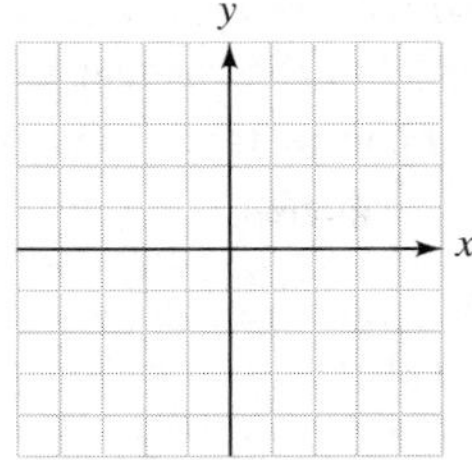

23. $x < 4$

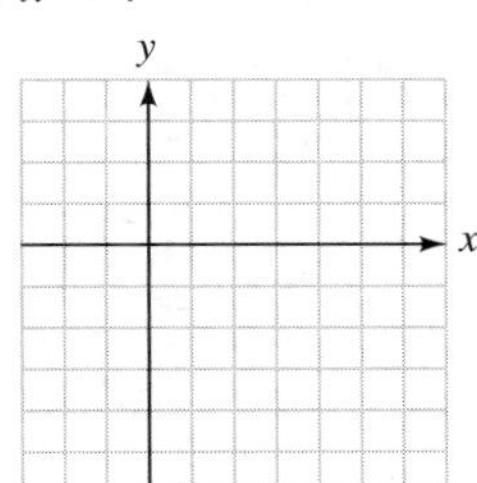

24. $x \geq -3$

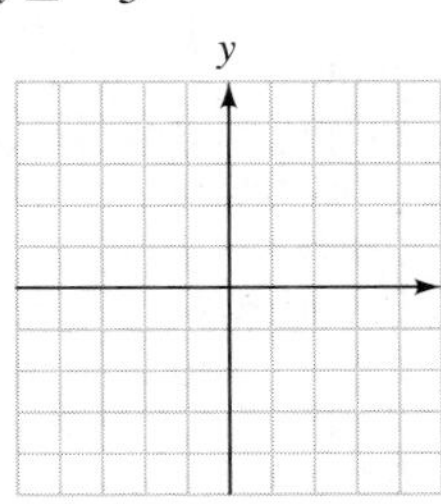

25. $y \geq -2$

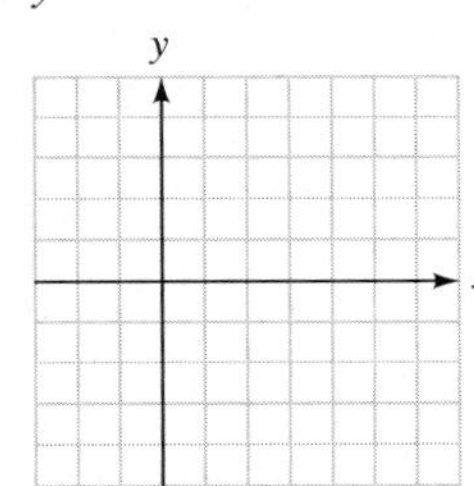

26. $y < 3$

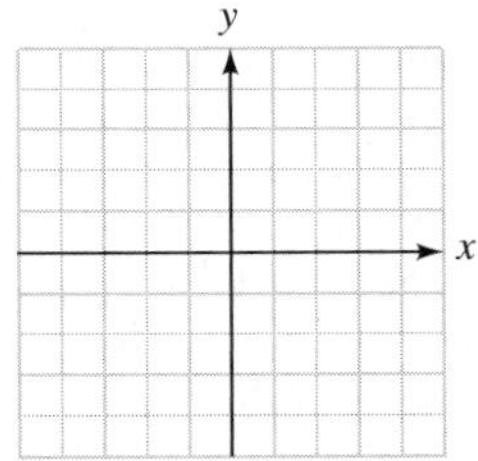

Graph each inequality. **See Example 3. (Objective 2)**

27. $-2 \leq x < 0$

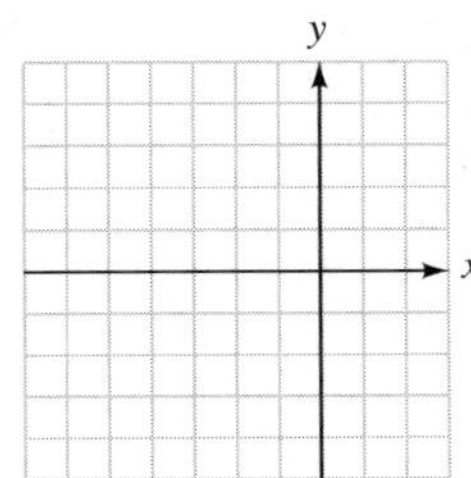

28. $-3 < y \leq -1$

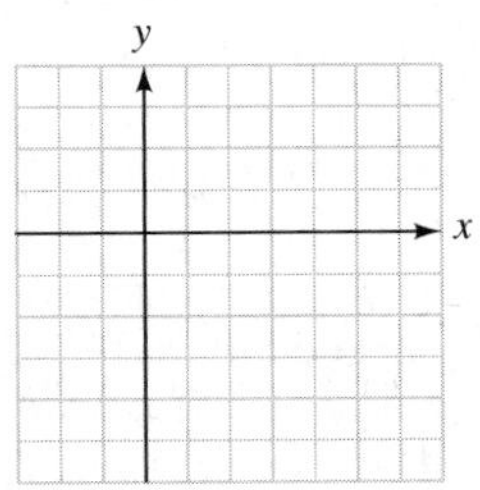

29. $y < -2$ or $y > 3$

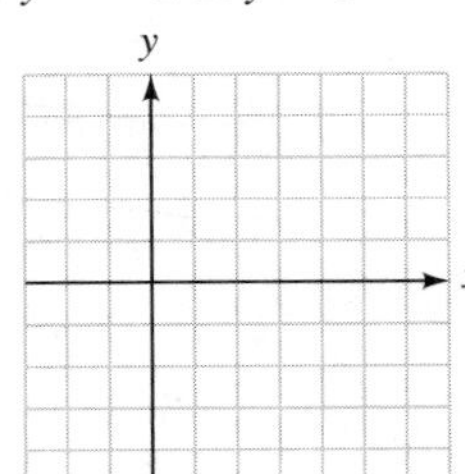

30. $-x \leq 1$ or $x \geq 2$

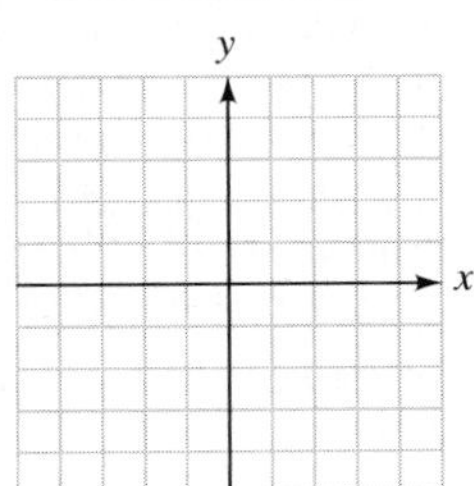

ADDITIONAL PRACTICE

Solve each inequality.

31. $y \geq 1 - \frac{3}{2}x$

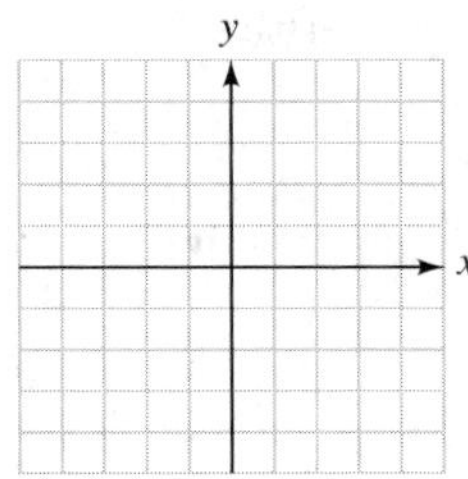

32. $y < \frac{1}{3}x - 1$

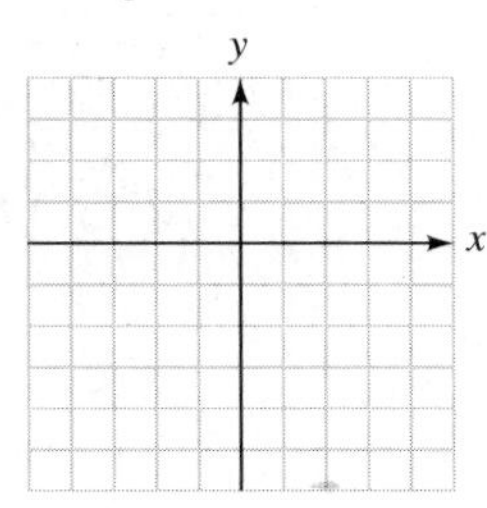

33. $0.5x + 0.5y \leq 2$

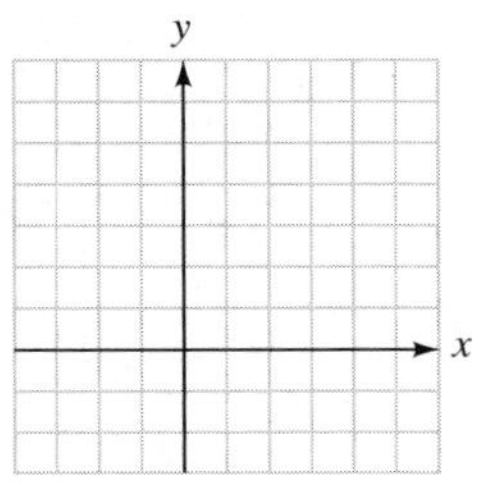

34. $0.5x + y > 1.5 + x$

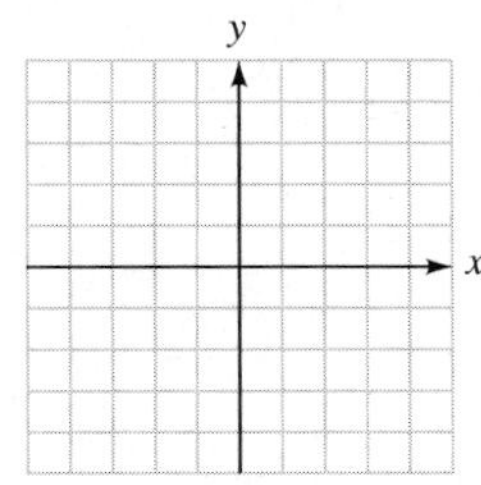

35. $2x + y \leq 6$

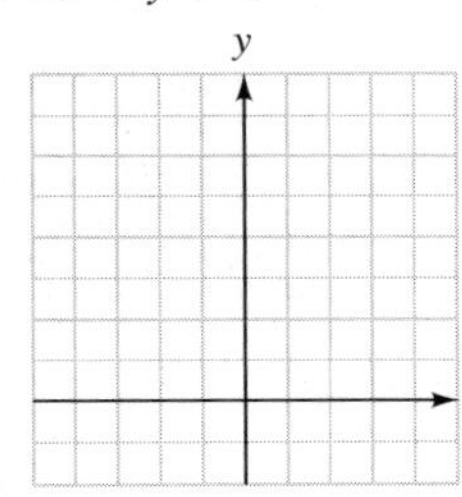

36. $x - 2y \geq 4$

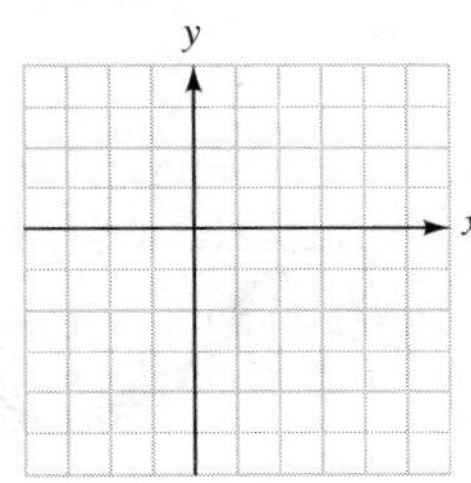

37. $3x \geq -y + 3$

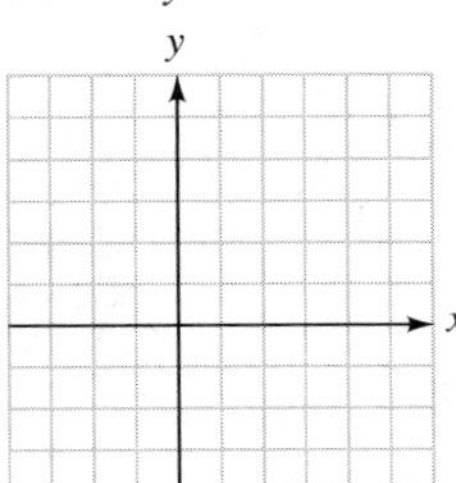

38. $2x \leq -3y - 12$

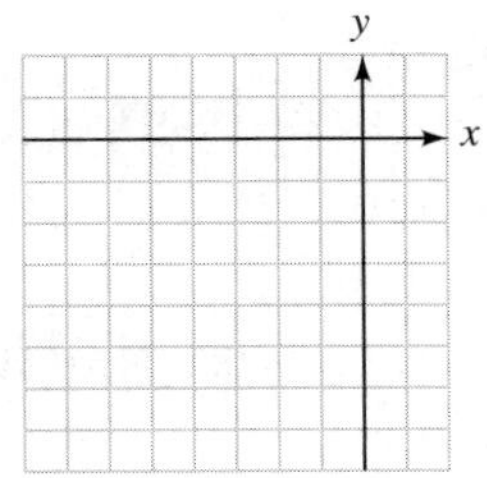

Use a graphing calculator to graph each inequality.

39. $y < 0.27x - 1$

40. $y > -3.5x + 2.7$

41. $y \geq -2.37x + 1.5$

42. $y \leq 3.37x - 1.7$

Find the equation of the boundary line or lines. Then write the inequality whose graph is shown.

43.

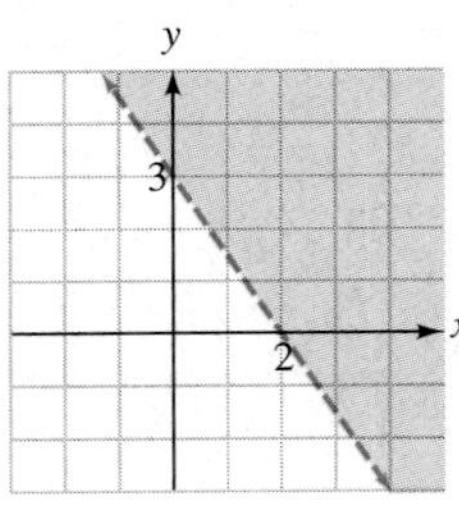

44.

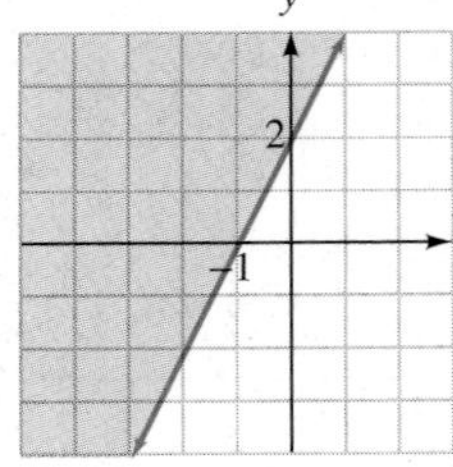

45.

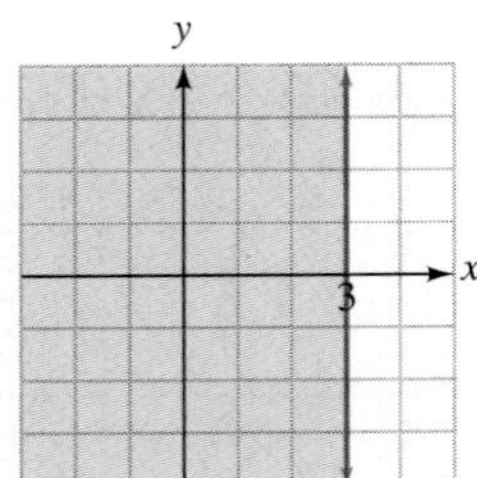

46.

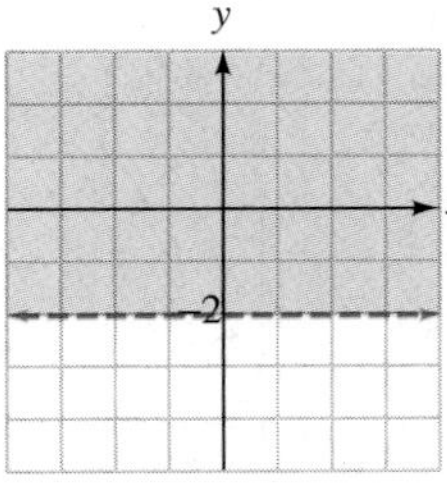

47.

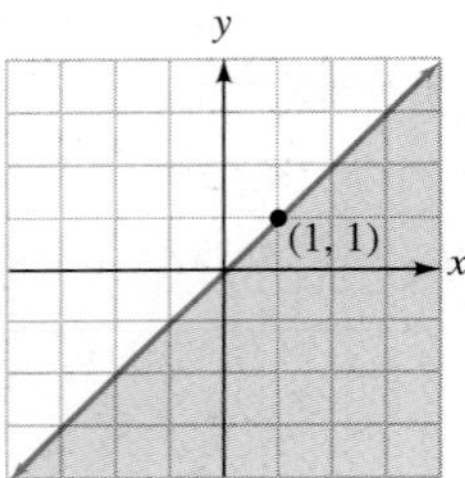

48.

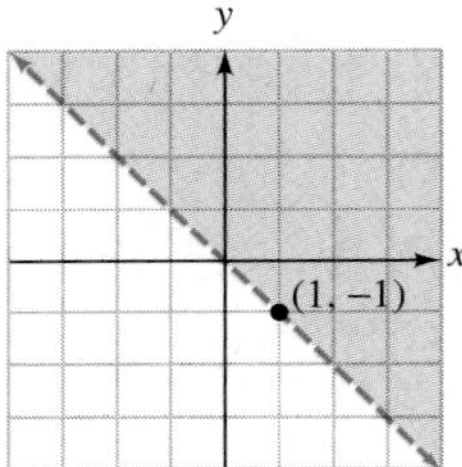

49.

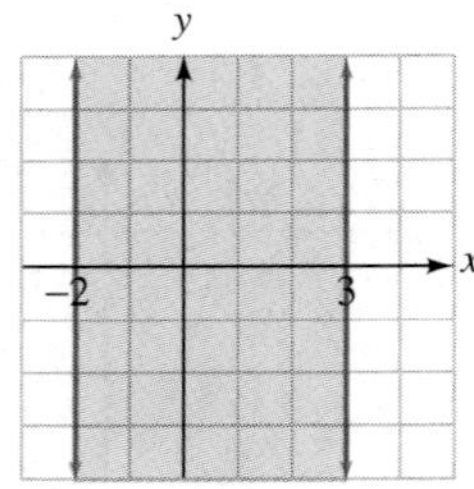

50.

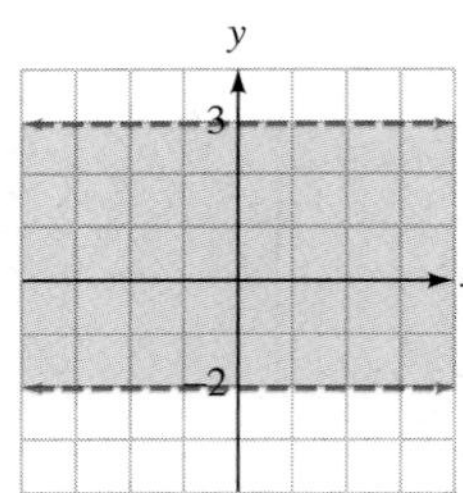

51.

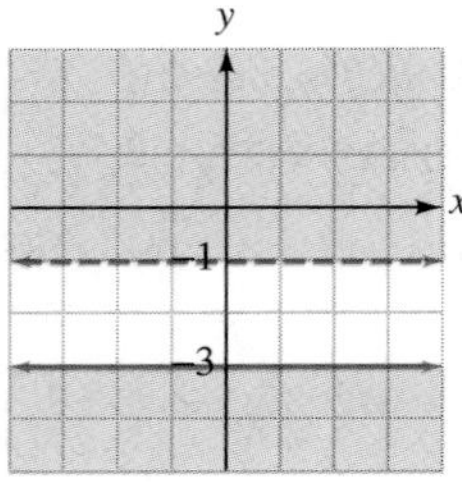

52.

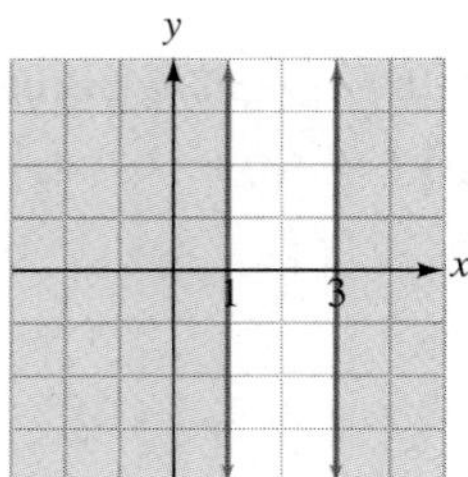

APPLICATIONS ***Graph each inequality for nonnegative values of x and y. Then give some ordered pairs that satisfy the inequality.*** **See Example 4. (Objective 3)**

53. Figuring taxes On average, it takes an accountant 1 hour to complete a simple tax return and 3 hours to complete a complicated return. If the accountant wants to work no more than 9 hours per day, use the illustration to graph an inequality that shows the possible ways that simple returns (x) and complicated returns (y) can be completed each day.

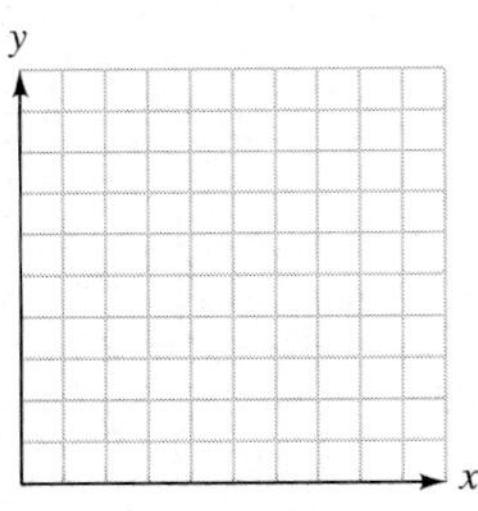

54. Selling trees During a sale, a garden store sold more than \$2,000 worth of trees. If a 6-foot maple costs \$100 and a 5-foot pine costs \$125, use the illustration to graph an inequality that shows the possible ways that maple trees (x) and pine trees (y) were sold.

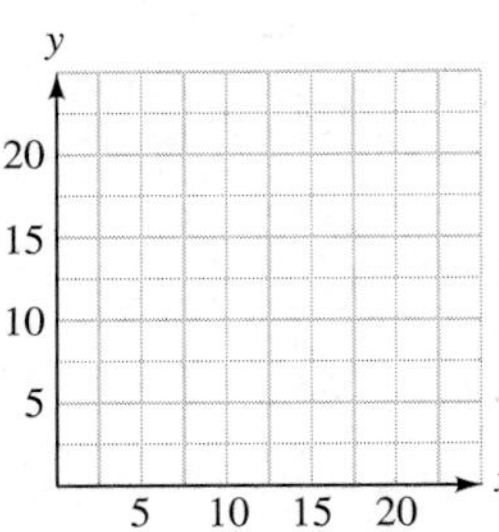

55. Choosing housekeepers One housekeeper charges \$6 per hour, and another charges \$7 per hour. If Sarah can afford no more than \$42 per week to clean her house, use the illustration to graph an inequality that shows the possible ways that she can hire the first housekeeper (x) and the second housekeeper (y).

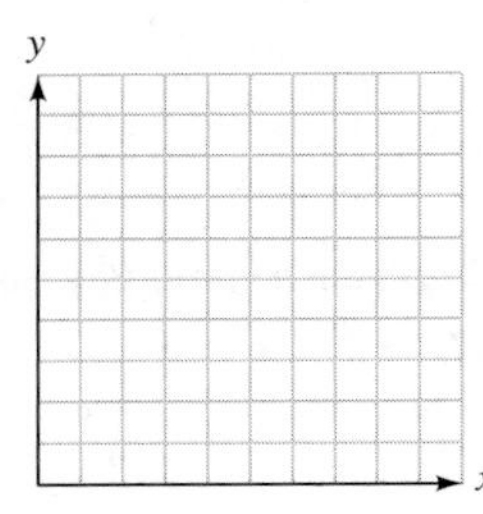

56. Making sporting goods A sporting goods manufacturer allocates at least 1,200 units of time per day to make fishing rods and reels. If it takes 10 units of time to make a rod and 15 units of time to make a reel, use the illustration to graph an inequality that shows the possible ways to schedule the time to make rods (x) and reels (y).

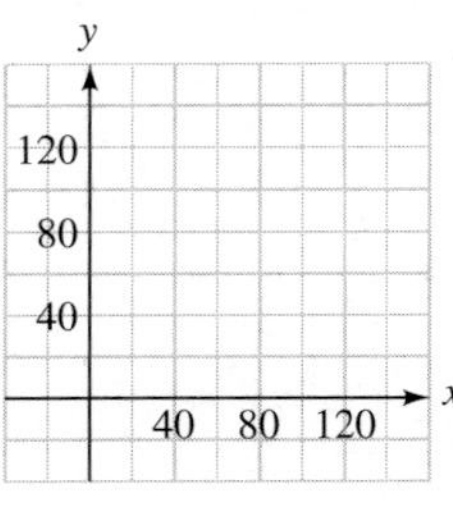

57. Investing A woman has up to \$6,000 to invest. If stock in Traffico sells for \$50 per share and stock in Cleanco sells for \$60 per share, use the illustration to graph an inequality that shows the possible ways that she can buy shares of Traffico (x) and Cleanco (y).

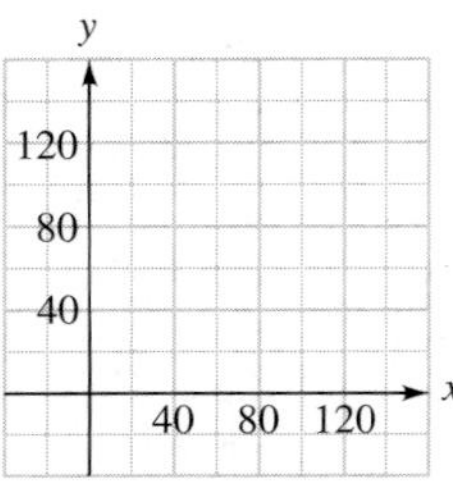

58. Buying concert tickets Tickets to a concert cost \$6 for reserved seats and \$4 for general admission. If receipts must be at least \$10,200 to meet expenses, use the illustration to graph an inequality that shows the possible ways that the box office can sell reserved seats (x) and general admission tickets (y) to meet expenses.

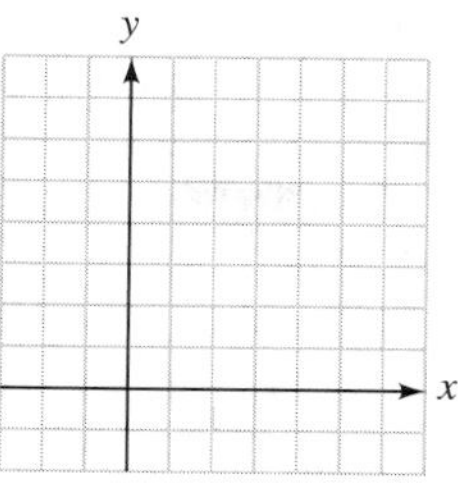

WRITING ABOUT MATH

59. Explain how to decide where to draw the boundary of the graph of a linear inequality, and whether to draw it as a solid or a broken line.

60. Explain how to decide which side of the boundary of the graph of a linear inequality should be shaded.

SOMETHING TO THINK ABOUT

61. Can an inequality be an identity, one that is satisfied by all (x, y) pairs? Illustrate.

62. Can an inequality have no solutions? Illustrate.

SECTION 4.4 Solving Systems of Linear and Quadratic Inequalities in Two Variables

Objectives

1. Solve a system of linear inequalities in two variables by graphing.
2. Solve a system of quadratic inequalities in two variables by graphing.
3. Solve an application problem requiring the use of a system of inequalities in two variables.

Getting Ready

Do the coordinates satisfy the inequalities $y > 2x - 3$ and $x - y > -1$?

1. $(0, 0)$ **2.** $(1, 1)$ **3.** $(-2, 0)$ **4.** $(0, 4)$

In the previous section, we learned how to graph linear inequalities. In this section, we will learn how to solve systems of linear inequalities.

1 Solve a system of linear inequalities in two variables by graphing.

We now consider the graphs of systems of inequalities in the variables x and y. These graphs will usually be the intersection of half-planes.

EXAMPLE 1 Graph the solution set of $\begin{cases} x + y \leq 1 \\ 2x - y > 2 \end{cases}$.

Solution On one set of coordinate axes, we graph each inequality as shown in Figure 4-30 on the next page.

The graph of $x + y \leq 1$ includes the line graph of the equation $x + y = 1$ and all points below it. Since the edge is included, we draw it as a solid line.

The graph of $2x - y > 2$ contains the points below the graph of the equation $2x - y = 2$. Since the edge is not included, we draw it as a dashed line.

The area where the half-planes intersect represents the solution of the system of inequalities, because any point in that region has coordinates that will satisfy both inequalities.

$x + y = 1$

x	y	(x, y)
0	1	(0, 1)
1	0	(1, 0)

$2x - y = 2$

x	y	(x, y)
0	−2	(0, −2)
1	0	(1, 0)

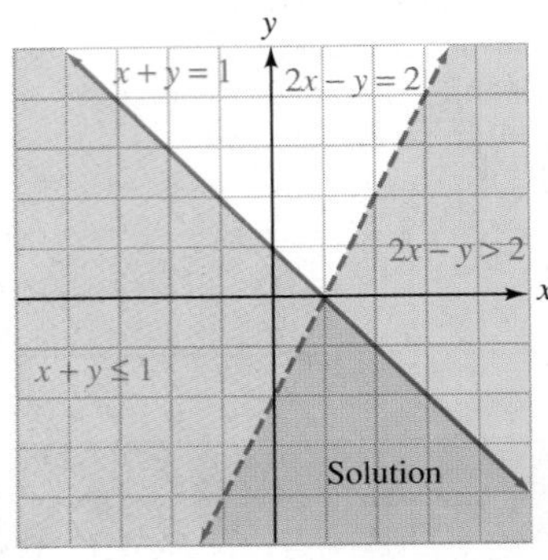

Figure 4-30

SELF CHECK 1 Graph the solution set of $\begin{cases} x + y \geq 1 \\ 2x - y < 2 \end{cases}$.

EXAMPLE 2 Graph the solution set of $\begin{cases} x \geq 1 \\ y \geq x \\ 4x + 5y < 20 \end{cases}$.

Solution The graph of $x \geq 1$ includes the points that lie on the graph of $x = 1$ and to the right, as shown in Figure 4-31(a).

The graph of $y \geq x$ includes the points that lie on the graph of $y = x$ and above it, as shown in Figure 4-31(b).

The graph of $4x + 5y < 20$ includes the points that lie below the graph of $4x + 5y = 20$, as shown in Figure 4-31(c).

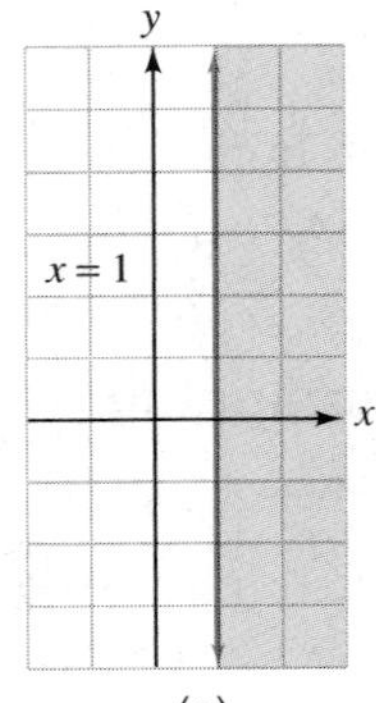

(a)

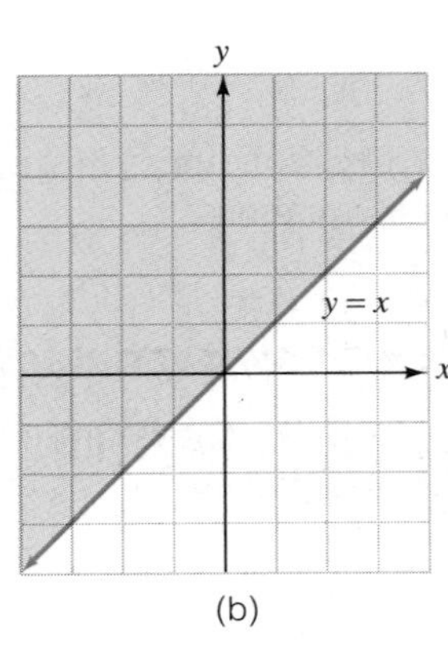

(b)

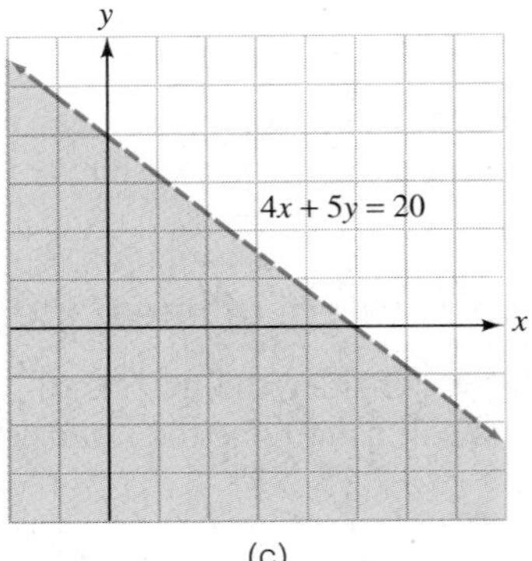

(c)

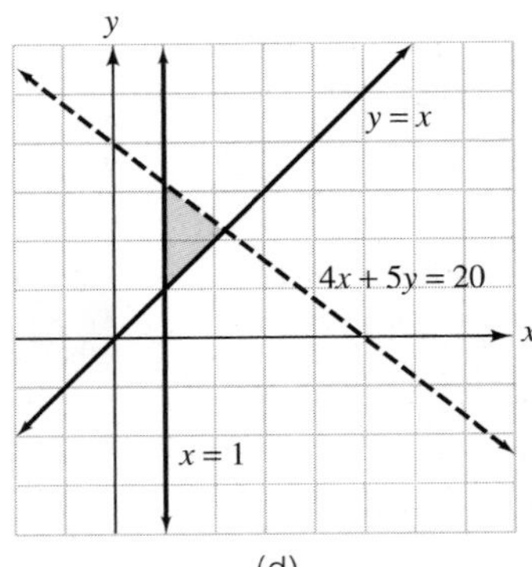

(d)

Figure 4-31

If we merge these graphs onto one set of coordinate axes, we see that the graph of the system includes the points that lie within a shaded triangle, together with the points on two of the three sides of the triangle, as shown in Figure 4-31(d).

SELF CHECK 2 Graph the solution set of $\begin{cases} x \geq 0 \\ y \leq 0 \\ y \geq -2 \end{cases}$.

2 Solve a system of quadratic inequalities in two variables by graphing.

EXAMPLE 3 Graph the solution set of $\begin{cases} y < x^2 \\ y > \frac{x^2}{4} - 2 \end{cases}$.

Solution The graph of $y = x^2$ is the red, dashed parabola shown in Figure 4-32. The points with coordinates that satisfy $y < x^2$ are the points that lie below the parabola.

The graph of $y = \frac{x^2}{4} - 2$ is the blue, dashed parabola. This time, the points that lie above the parabola satisfy the inequality. Thus, the solution of the system is the area between the parabolas.

$y = x^2$

x	y	(x, y)
0	0	(0, 0)
1	1	(1, 1)
−1	1	(−1, 1)
2	4	(2, 4)
−2	4	(−2, 4)

$y = \frac{x^2}{4} - 2$

x	y	(x, y)
0	−2	(0, −2)
2	−1	(2, −1)
−2	−1	(−2, −1)
4	2	(4, 2)
−4	2	(−4, 2)

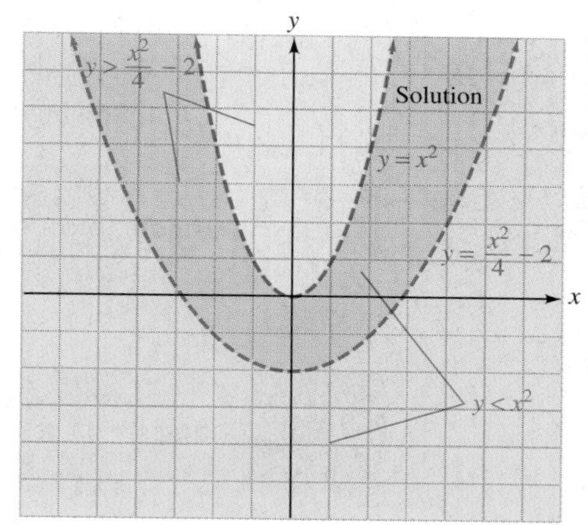

Figure 4-32

SELF CHECK 3 Graph the solution set of $\begin{cases} y \geq x^2 \\ y \leq x + 2 \end{cases}$.

ACCENT ON TECHNOLOGY

Solving Systems of Inequalities

To solve the system of Example 1, we use window settings of $x = [-10, 10]$ and $y = [-10, 10]$. To graph $x + y \leq 1$, we enter the equation $x + y = 1$ $(y = -x + 1)$ and change the graphing-style icon to below (◣). To graph $2x - y > 2$, we enter the equation $2x - y = 2$ $(y = 2x - 2)$ and change the graphing-style icon to below (◣). Finally, we press **GRAPH** to obtain Figure 4-33(a).

To solve the system of Example 3, we enter the equation $y = x^2$ and change the graphing-style icon to below (◣). We then enter the equation $y = \frac{x^2}{4} - 2$ and change the graphing-style icon to above (◥). Finally, we press **GRAPH** to obtain Figure 4-33(b).

(continued)

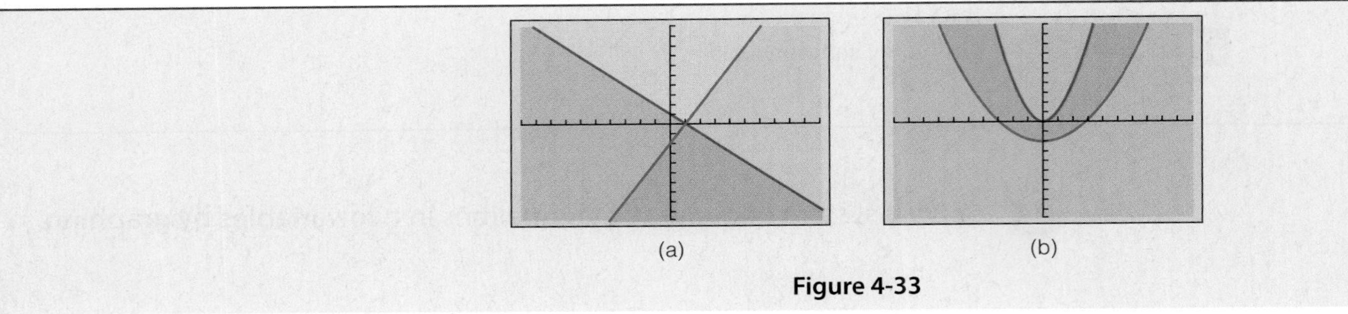
(a) (b)

Figure 4-33

3 Solve an application problem requiring the use of a system of inequalities in two variables.

EXAMPLE 4 **LANDSCAPING** A homeowner budgets from \$300 to \$600 for trees and bushes to landscape his yard. After shopping around, he finds that good trees cost \$150 and mature bushes cost \$75. What combinations of trees and bushes can he afford?

Analyze the problem: If x represents the number of trees purchased, \$150x will be the cost of the trees. If y represents the number of bushes purchased, \$75y will be the cost of the bushes.

Form two inequalities: We know that the sum of these costs must be from \$300 to \$600. We can then form the following system of inequalities:

The cost of a tree	times	the number of trees purchased	plus	the cost of a bush	times	the number of bushes purchased	is greater than or equal to	\$300.
\$150	$\cdot$	x	+	\$75	$\cdot$	y	$\geq$	\$300

The cost of a tree	times	the number of trees purchased	plus	the cost of a bush	times	the number of bushes purchased	is less than or equal to	\$600.
\$150	$\cdot$	x	+	\$75	$\cdot$	y	$\leq$	\$600

Solve the system: We graph the system

$$\begin{cases} 150x + 75y \geq 300 \\ 150x + 75y \leq 600 \end{cases}$$

as in Figure 4-34. The coordinates of each point shown in the graph give a possible combination of trees (x) and bushes (y) that can be purchased.

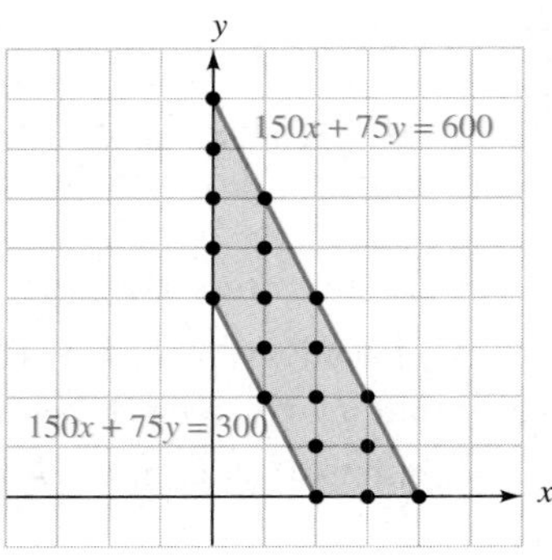

Figure 4-34

State the conclusion These possibilities are

(0, 4), (0, 5), (0, 6), (0, 7), (0, 8)
(1, 2), (1, 3), (1, 4), (1, 5), (1, 6)
(2, 0), (2, 1), (2, 2), (2, 3), (2, 4)
(3, 0), (3, 1), (3, 2), (4, 0)

Only these points can be used, because the homeowner cannot buy a portion of a tree.

Check the result Check some of the ordered pairs to verify that they satisfy both inequalities.

SELF CHECK ANSWERS

1. 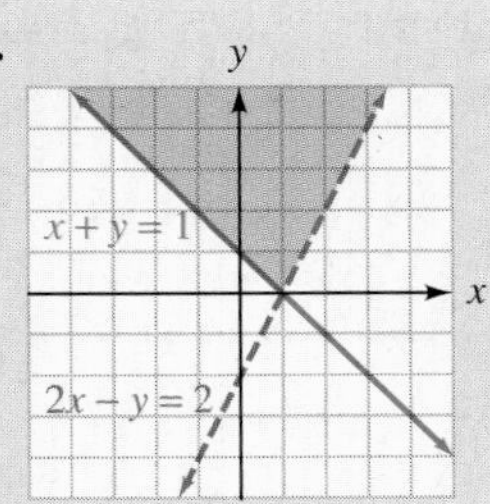

2. 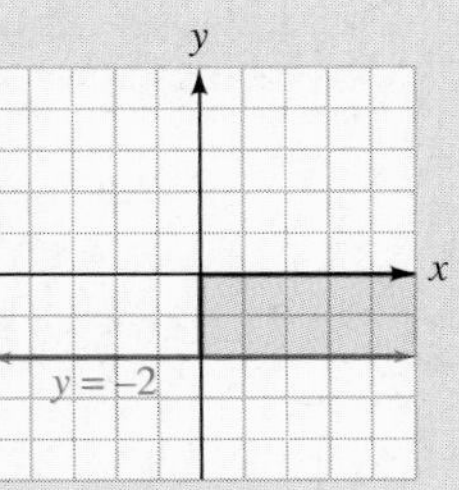

3. 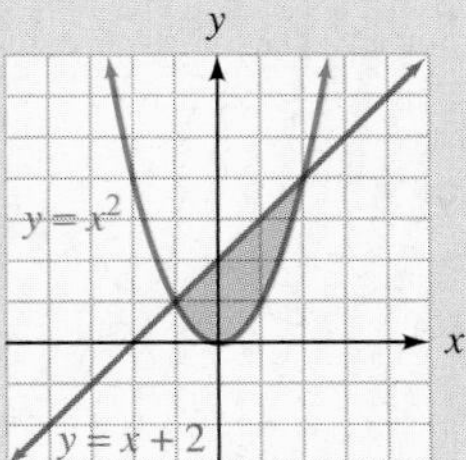

NOW TRY THIS

1. $\begin{cases} x - y > 0 \\ y \le x + 2 \end{cases}$

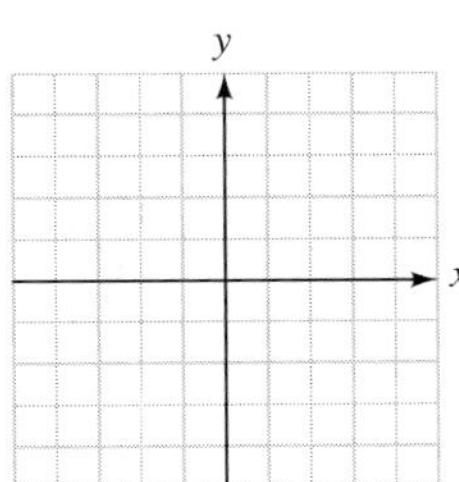

2. $\begin{cases} 4x - 3y \le -4 \\ y < \frac{4}{3}x - \frac{5}{2} \end{cases}$

3. $\begin{cases} y \le x^2 - 6x + 7 \\ x - y < 3 \end{cases}$

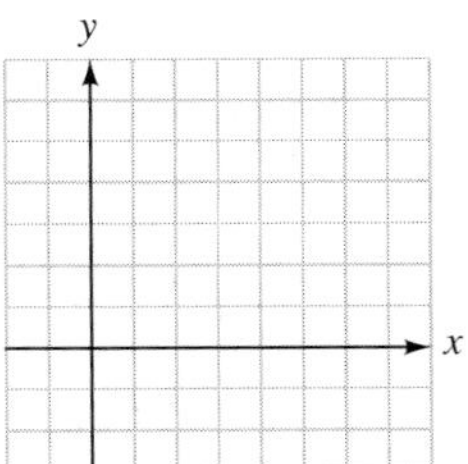

4.4 EXERCISES

WARM-UPS *Do the coordinates* (1, 1) *satisfy both inequalities?*

1. $\begin{cases} y < x + 1 \\ y > x - 1 \end{cases}$

2. $\begin{cases} y > 2x - 3 \\ y < -x + 3 \end{cases}$

REVIEW *Solve each formula for the given variable.*

3. $A = p + prt$ for r

4. $C = \frac{5}{9}(F - 32)$ for F

5. $z = \frac{x - \mu}{\sigma}$ for x

6. $P = 2l + 2w$ for w

7. $l = a + (n - 1)d$ for d

8. $z = \frac{x - \mu}{\sigma}$ for μ

VOCABULARY AND CONCEPTS *Fill in the blanks.*

9. To solve a system of inequalities by graphing, we graph each inequality. The solution is the region where the graphs ________.

10. If an edge is included in the graph of an inequality, we draw it as a ______ line and if not, it will be ________.

GUIDED PRACTICE

Graph the solution set of each system. See Example 1. (Objective 1)

11. $\begin{cases} y < 3x + 2 \\ y < -2x + 3 \end{cases}$

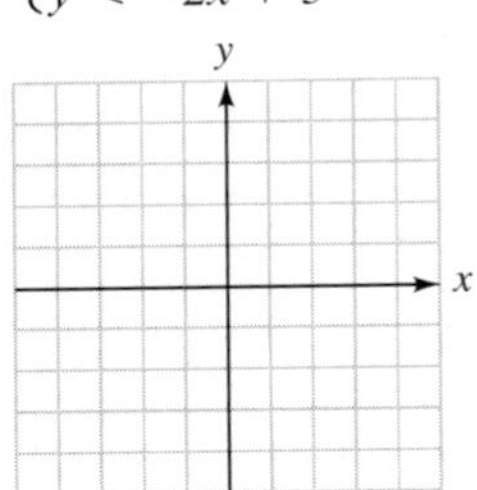

12. $\begin{cases} y \leq x - 2 \\ y \geq 2x + 1 \end{cases}$

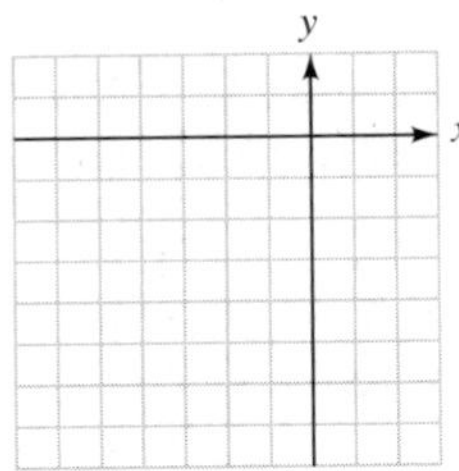

13. $\begin{cases} 3x + 2y > 6 \\ x + 3y \leq 2 \end{cases}$

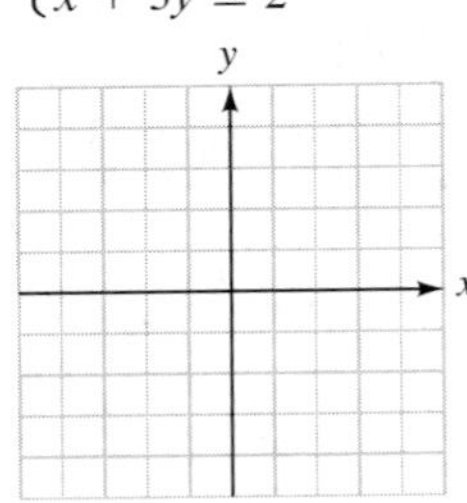

14. $\begin{cases} x + y < 2 \\ x + y \leq 1 \end{cases}$

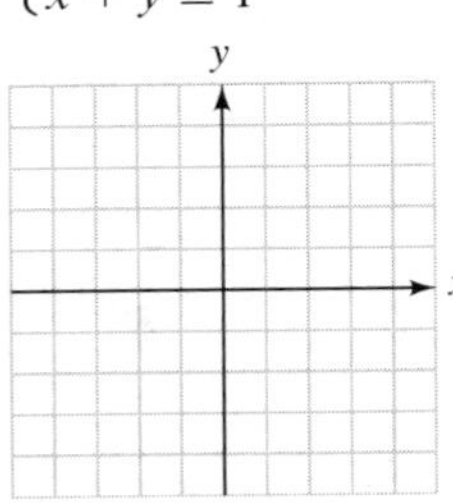

Graph the solution set of each system. See Example 2. (Objective 1)

15. $\begin{cases} 2x + 3y \leq 6 \\ 3x + y \leq 1 \\ x \leq 0 \end{cases}$

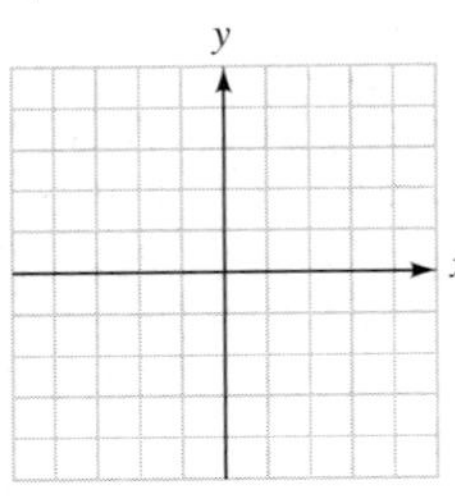

16. 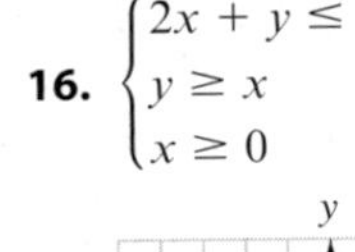

$\begin{cases} 2x + y \leq 2 \\ y \geq x \\ x \geq 0 \end{cases}$

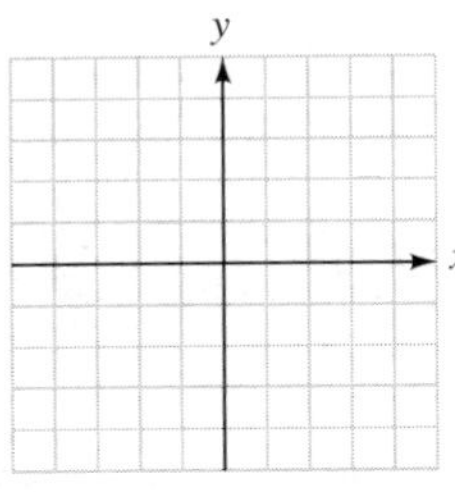

17. $\begin{cases} x - y < 4 \\ y \leq 0 \\ x \geq 0 \end{cases}$

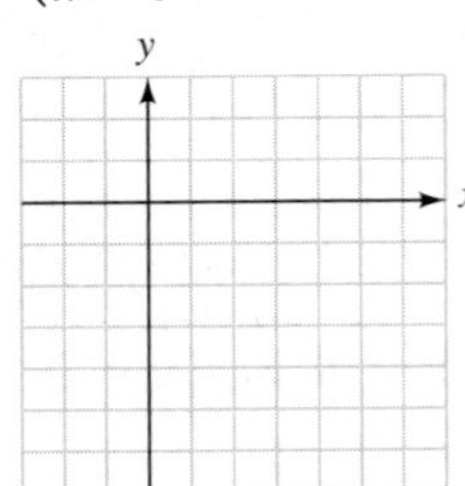

18. $\begin{cases} x + y \leq 4 \\ x \geq 0 \\ y \geq 0 \end{cases}$

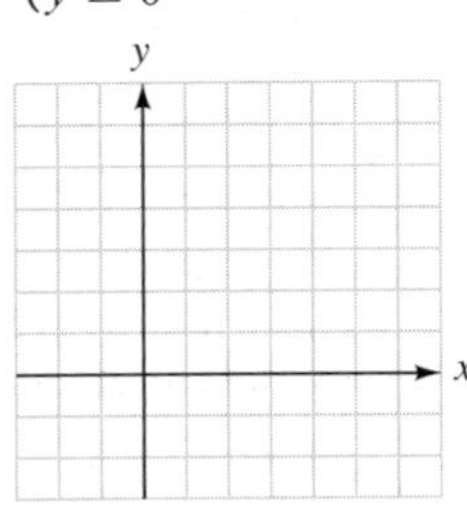

Graph the solution set of each system. See Example 3. (Objective 2)

19. $\begin{cases} 2x - y > 4 \\ y < -x^2 + 2 \end{cases}$

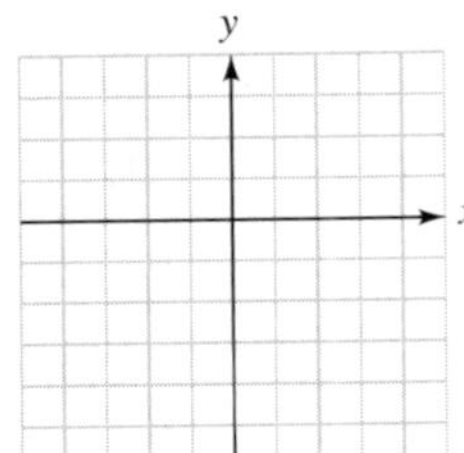

20. 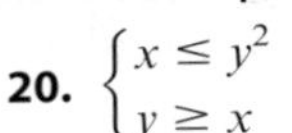

$\begin{cases} x \leq y^2 \\ y \geq x \end{cases}$

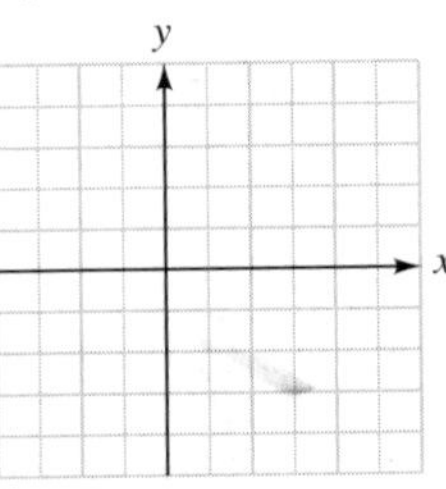

21. $\begin{cases} y > x^2 - 4 \\ y < -x^2 + 4 \end{cases}$

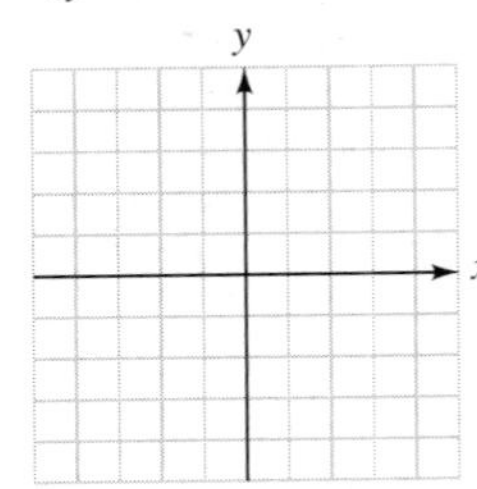

22. $\begin{cases} x \geq y^2 \\ y \geq x^2 \end{cases}$

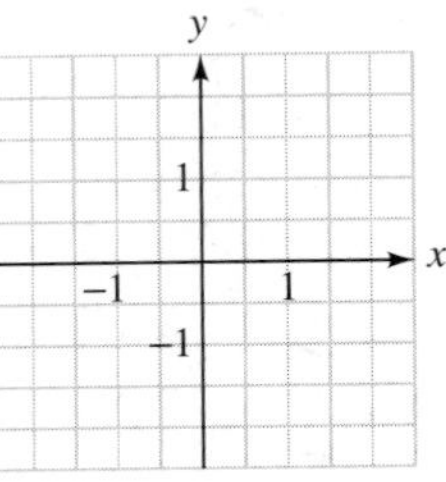

ADDITIONAL PRACTICE

Graph the solution set of each system.

23. $\begin{cases} 3x + y \leq 1 \\ -x + 2y \geq 6 \end{cases}$

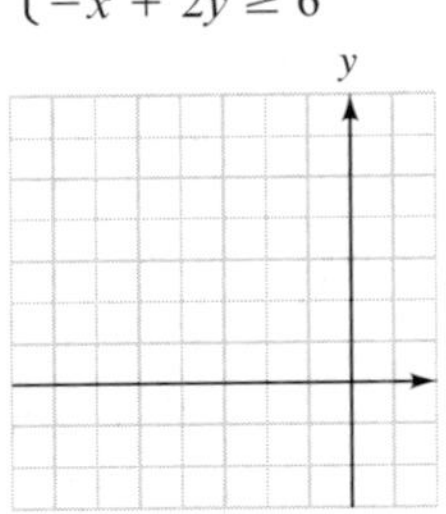

24. $\begin{cases} x + 2y < 3 \\ 2x + 4y < 8 \end{cases}$

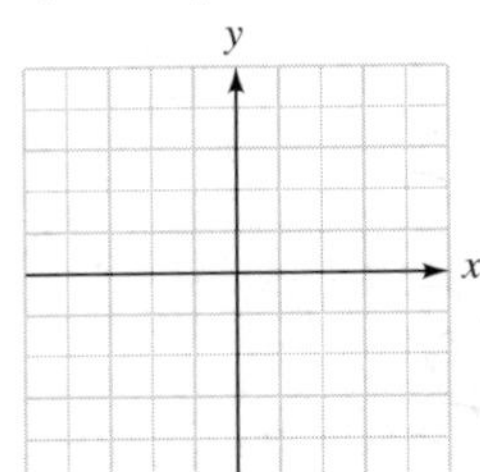

25. $\begin{cases} x \geq 0 \\ y \geq 0 \\ 9x + 3y \leq 18 \\ 3x + 6y \leq 18 \end{cases}$

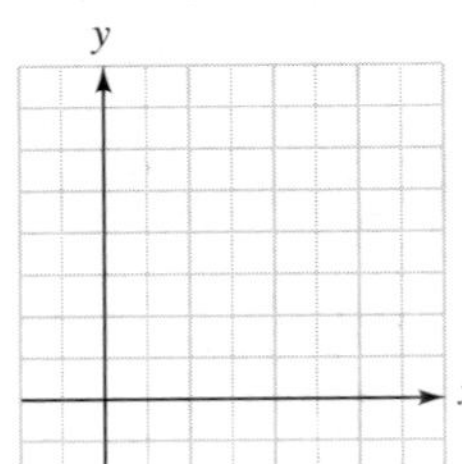

26. $\begin{cases} x + y \geq 1 \\ x - y \leq 1 \\ x - y \geq 0 \\ x \leq 2 \end{cases}$

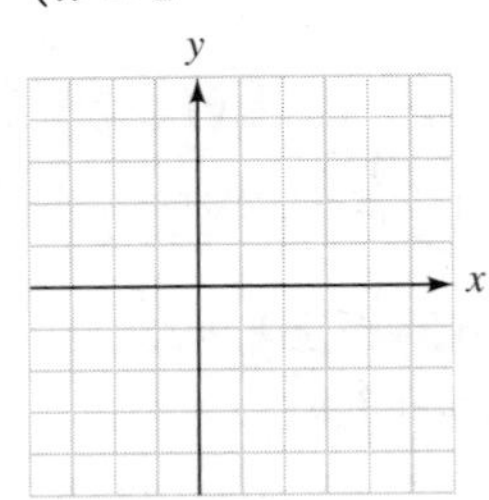

Use a graphing calculator to solve each system. See Example 3. (Objective 2)

27. $\begin{cases} y < 3x + 2 \\ y < -2x + 3 \end{cases}$
(See Exercise 11)

28. $\begin{cases} y > x^2 - 4 \\ y < -x^2 + 4 \end{cases}$
(See Exercise 21)

APPLICATIONS

Set up a system of inequalities and graph the solution. See Example 4. (Objective 3)

29. Football In 2003, the Green Bay Packers scored either a touchdown or a field goal 65.4% of the time when their offense was in the red zone. This was the best record in the NFL! If x represents the yard line the football is on, a team's *red zone* is an area on their opponent's half of the field that can be described by the system

$$\begin{cases} x > 0 \\ x \le 20 \end{cases}$$

Shade the red zone on the field shown below.

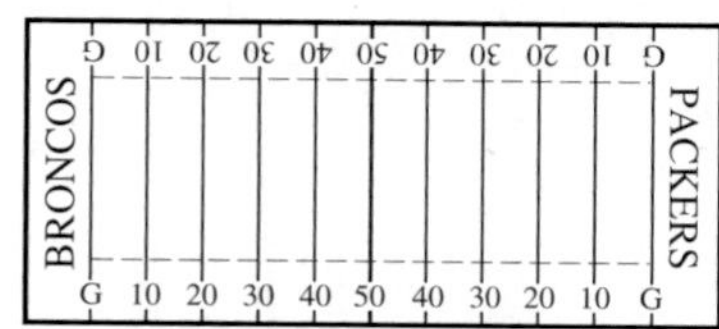

30. Track and field In the shot put, the solid metal ball must land in a marked sector for it to be a fair throw. In the illustration, graph the system of inequalities that describes the region in which a shot must land.

$$\begin{cases} y \le \frac{3}{8}x \\ y \ge -\frac{3}{8}x \\ x \ge 1 \end{cases}$$

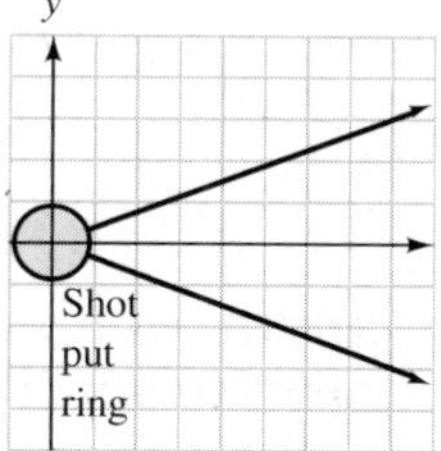

Graph a system of inequalities and give two possible solutions to each problem.

31. Buying compact discs Melodic Music has compact discs on sale for either \$10 or \$15. If a customer wants to spend at least \$30 but no more than \$60 on CDs, use the illustration to graph a system of inequalities that will show the possible ways a customer can buy \$10 CDs (x) and \$15 CDs (y).

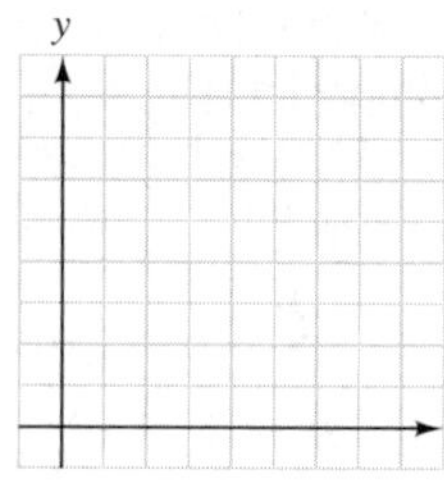

32. Buying boats Dry Boat Works wholesales aluminum boats for \$800 and fiberglass boats for \$600. Northland Marina wants to order at least \$2,400 worth, but no more than \$4,800 worth of boats. Use the illustration to graph a system of inequalities that will show the possible combinations of aluminum boats (x) and fiberglass boats (y) that can be ordered.

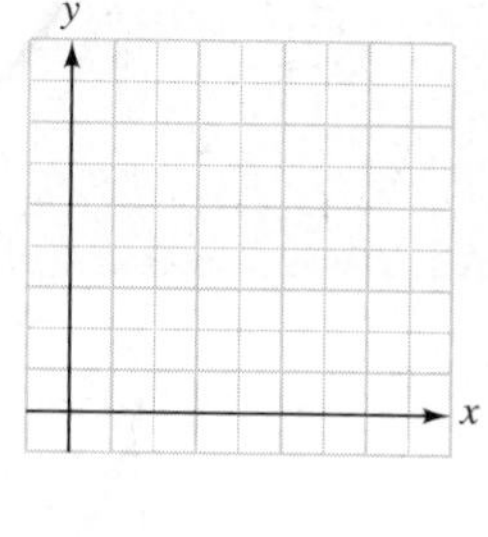

33. Buying furniture A distributor wholesales desk chairs for \$150 and side chairs for \$100. Best Furniture wants to order no more than \$900 worth of chairs, including more side chairs than desk chairs. Use the illustration to graph a system of inequalities that will show the possible combinations of desk chairs (x) and side chairs (y) that can be ordered.

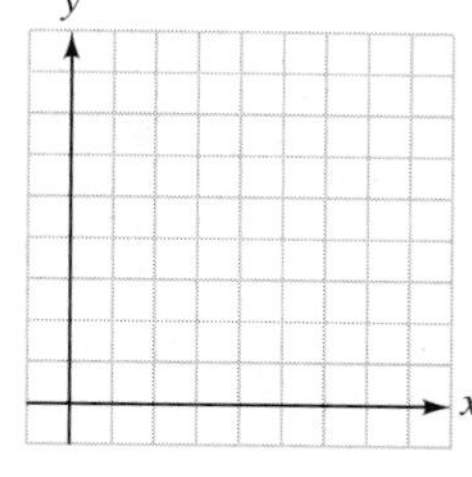

34. Ordering furnace equipment Bolden Heating Company wants to order no more than \$2,000 worth of electronic air cleaners and humidifiers from a wholesaler that charges \$500 for air cleaners and \$200 for humidifiers. If Bolden wants more humidifiers than air cleaners, use the illustration to graph a system of inequalities that will show the possible combinations of air cleaners (x) and humidifiers (y) that can be ordered.

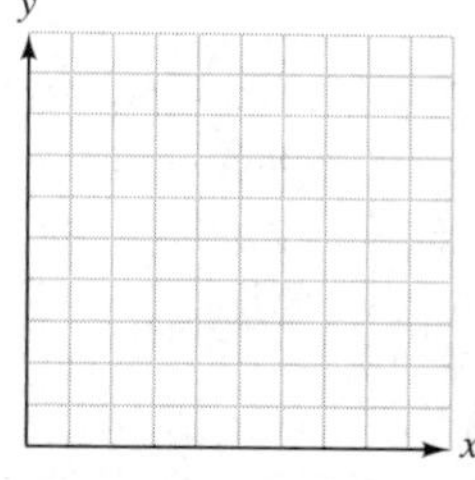

WRITING ABOUT MATH

35. When graphing a system of linear inequalities, explain how to decide which region to shade.

36. Explain how a system of two linear inequalities might have no solution.

SOMETHING TO THINK ABOUT

37. The solution of a system of inequalities in two variables is *bounded* if it is possible to draw a circle around it. Can a nonempty solution of two linear inequalities be bounded?

38. The solution of $\begin{cases} y \ge |x| \\ y \le k \end{cases}$ has an area of 25. Find k.

SECTION 4.5 Solving Systems Using Linear Programming

Objectives

1. Find the maximum and minimum value of an equation in the form $P = ax + by$, subject to specific constraints.
2. Solve an application problem using linear programming.

Vocabulary

linear programming
objective function
constraints
feasibility region

Getting Ready

Evaluate $2x + 3y$ *for each pair of coordinates.*

1. (0, 0) **2.** (3, 0) **3.** (2, 2) **4.** (0, 4)

We now use our knowledge of solving systems of inequalities to solve linear programming problems.

1 Find the maximum and minimum value of an equation in the form $P = ax + by$, subject to specific constraints.

Linear programming is a mathematical technique used to find the optimal allocation of resources in the military, business, telecommunications, and other fields. It got its start during World War II when it became necessary to move huge quantities of people, materials, and supplies as efficiently and economically as possible.

To solve a linear program, we maximize (or minimize) a function (called the **objective function**) subject to given conditions on its variables. These conditions (called **constraints**) are usually given as a system of linear inequalities. For example, suppose that the annual profit (in millions of dollars) earned by a business is given by the equation $P = y + 2x$, where the profit is determined by the sale of two different items and x and y are subject to the following constraints:

$$\begin{cases} 3x + y \leq 120 \\ x + y \leq 60 \\ x \geq 0 \\ y \geq 0 \end{cases}$$

To find the maximum profit P that can be earned by the business, we solve the system of inequalities as shown in Figure 4-35(a) and find the coordinates of each *corner point* of the region R, called a **feasibility region.** We can then write the profit equation

$$P = y + 2x \qquad \text{in the form} \qquad y = -2x + P$$

The equation $y = -2x + P$ is the equation of a set of parallel lines, each with a slope of -2 and a y-intercept of P. To find the line that passes through region R and provides the maximum value of P, we refer to Figure 4-35(b) and locate the line with the greatest y-intercept. Since line l has the greatest y-intercept and intersects region R at the corner point (30, 30), the maximum value of P (subject to the given constraints) is

$$\begin{aligned} P &= y + 2x \\ &= 30 + 2(30) \\ &= 90 \end{aligned}$$

Thus, the maximum profit P that can be earned is \$90 million. This profit occurs when $x = 30$ and $y = 30$.

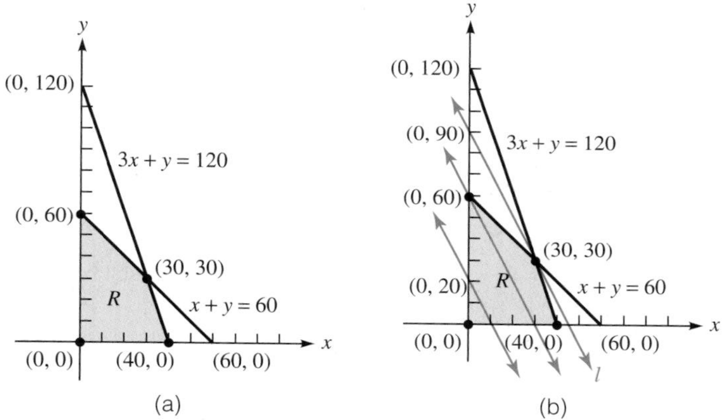

Figure 4-35

PERSPECTIVE How to Solve It

As a young student, George Polya (1888–1985) enjoyed mathematics and understood the solutions presented by his teachers. However, Polya had questions still asked by mathematics students today: "Yes, the solution works, but how is it possible to come up with such a solution? How could I discover such things by myself?" These questions still concerned him years later when, as Professor of Mathematics at Stanford University, he developed an approach to teaching mathematics that was very popular with faculty and students. His book, *How to Solve It,* became a bestseller.

George Polya (1888–1985)

Polya's problem-solving approach involves four steps.

- *Understand the problem.* What is the unknown? What information is known? What are the conditions?
- *Devise a plan.* Have you seen anything like it before? Do you know any related problems you have solved before? If you can't solve the proposed problem, can you solve a similar but easier problem?
- *Carry out the plan.* Check each step. Can you explain why each step is correct?
- *Look back.* Examine the solution. Can you check the result? Can you use the result, or the method, to solve any other problem?

The preceding discussion illustrates the following important fact.

Maximum or Minimum of an Objective Function

If a linear function, subject to the constraints of a system of linear inequalities in two variables, attains a maximum or a minimum value, that value will occur at a corner point or along an entire edge of the region R that represents the solution of the system.

EXAMPLE 1 If $P = 2x + 3y$, find the maximum value of P subject to the following constraints:

$$\begin{cases} x + y \leq 4 \\ 2x + y \leq 6 \\ x \geq 0 \\ y \geq 0 \end{cases}$$

Solution We solve the system of inequalities to find the feasibility region R shown in Figure 4-36. The coordinates of its corner points are (0, 0), (3, 0), (0, 4), and (2, 2).

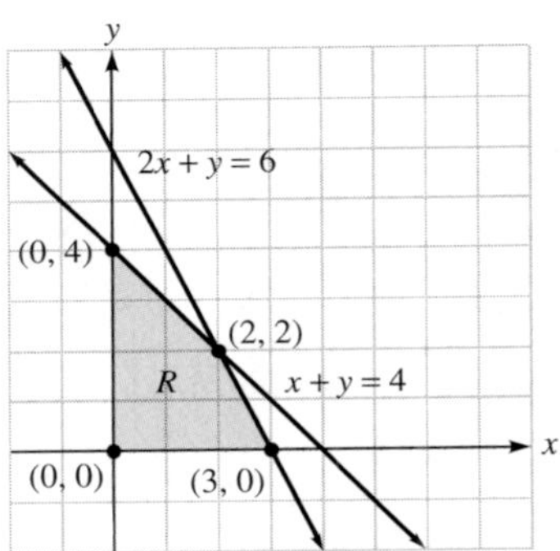

Figure 4-36

Since the maximum value of P will occur at a corner of R, we substitute the coordinates of each corner point into the objective function $P = 2x + 3y$ and find the one that gives the maximum value of P.

Point	$P = 2x + 3y$
(0, 0)	$P = 2(0) + 3(0) = 0$
(3, 0)	$P = 2(3) + 3(0) = 6$
(2, 2)	$P = 2(2) + 3(2) = 10$
(0, 4)	$P = 2(0) + 3(4) = 12$

The maximum value $P = 12$ occurs when $x = 0$ and $y = 4$.

SELF CHECK 1 Find the maximum value of $P = 4x + 3y$, subject to the constraints of Example 1.

EXAMPLE 2 If $P = 3x + 2y$, find the minimum value of P subject to the following constraints:

$$\begin{cases} x + y \geq 1 \\ x - y \leq 1 \\ x - y \geq 0 \\ x \leq 2 \end{cases}$$

Solution We refer to the feasibility region shown in Figure 4-37 with corner points at $\left(\frac{1}{2}, \frac{1}{2}\right)$, (2, 2), (2, 1), and (1, 0).

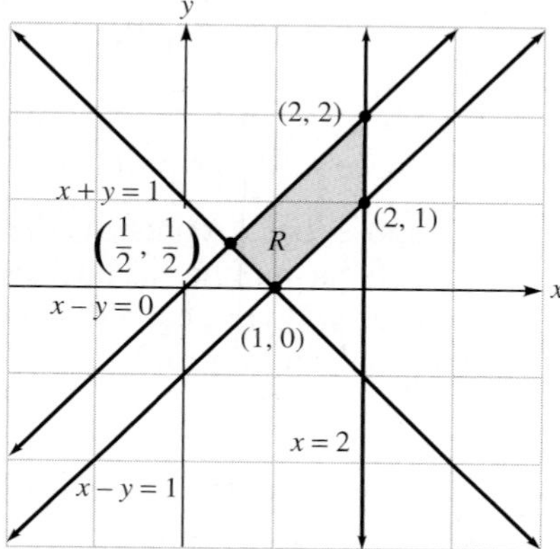

Figure 4-37

Since the minimum value of P occurs at a corner point of region R, we substitute the coordinates of each corner point into the objective function $P = 3x + 2y$ and find the one that gives the minimum value of P.

Point	$P = 3x + 2y$
$\left(\frac{1}{2}, \frac{1}{2}\right)$	$P = 3\left(\frac{1}{2}\right) + 2\left(\frac{1}{2}\right) = \frac{5}{2}$
(2, 2)	$P = 3(2) + 2(2) = 10$
(2, 1)	$P = 3(2) + 2(1) = 8$
(1, 0)	$P = 3(1) + 2(0) = 3$

The minimum value $P = \frac{5}{2}$ occurs when $x = \frac{1}{2}$ and $y = \frac{1}{2}$.

SELF CHECK 2 Find the minimum value of $P = 2x + y$, subject to the constraints of Example 2.

2 Solve an application problem using linear programming.

Linear programming problems can be complex and involve hundreds of variables. In this section, we will consider a few simple problems. Since they involve only two variables, we can solve them using graphical methods.

EXAMPLE 3 **MAXIMIZING INCOME** An accountant prepares tax returns for individuals and for small businesses. On average, each individual return requires 3 hours of her time and 1 hour of computer time. Each business return requires 4 hours of her time and 2 hours of computer time. Because of other business considerations, her time is limited to 240 hours, and the computer time is limited to 100 hours. If she earns a profit of \$80 on each individual return and a profit of \$150 on each business return, how many returns of each type should she prepare to maximize her profit?

Solution First, we organize the given information into a table.

	Individual tax return	Business tax return	Time available
Accountant's time	3	4	240 hours
Computer time	1	2	100 hours
Profit	\$80	\$150	

Then we solve the problem using the following steps.

Find the objective function

Suppose that x represents the number of individual returns to be completed and y represents the number of business returns to be completed. Since each of the x individual returns will earn an \$80 profit, and each of the y business returns will earn a \$150 profit, the total profit is given by the equation

$$P = 80x + 150y$$

Find the feasibility region

Since the number of individual returns and business returns cannot be negative, we know that $x \geq 0$ and $y \geq 0$.

Since each of the x individual returns will take 3 hours of her time, and each of the y business returns will take 4 hours of her time, the total number of hours she will work will be $(3x + 4y)$ hours. This amount must be less than or equal to her available time, which is 240 hours. Thus, the inequality $3x + 4y \leq 240$ is a constraint on the accountant's time.

Since each of the x individual returns will take 1 hour of computer time, and each of the y business returns will take 2 hours of computer time, the total number of hours of computer time will be $(x + 2y)$ hours. This amount must be less than or equal to the available computer time, which is 100 hours. Thus, the inequality $x + 2y \leq 100$ is a constraint on the computer time.

We have the following constraints on the values of x and y.

$$\begin{cases} x \geq 0 & \text{The number of individual returns is nonnegative.} \\ y \geq 0 & \text{The number of business returns is nonnegative.} \\ 3x + 4y \leq 240 & \text{The accountant's time must be less than or equal to 240 hours.} \\ x + 2y \leq 100 & \text{The computer time must be less than or equal to 100 hours.} \end{cases}$$

To find the feasibility region, we graph each of the constraints to find region R, as in Figure 4-38. The four corner points of this region have coordinates of (0, 0), (80, 0), (40, 30), and (0, 50).

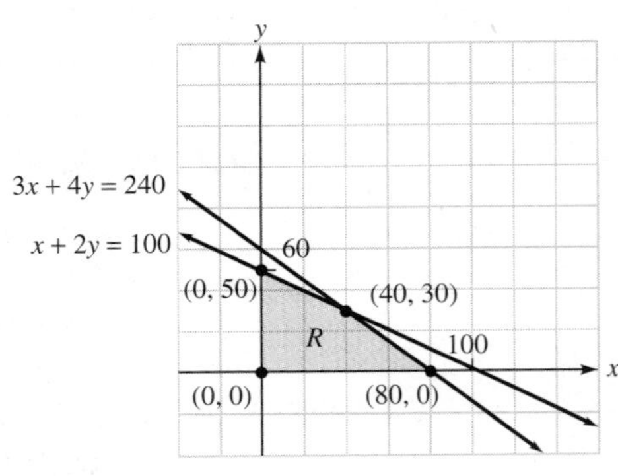

Figure 4-38

Find the maximum profit

To find the maximum profit, we substitute the coordinates of each corner point into the objective function $P = 80x + 150y$.

Point	$P = 80x + 150y$
(0, 0)	$P = 80(0) + 150(0) = 0$
(80, 0)	$P = 80(80) + 150(0) = 6{,}400$
(40, 30)	$P = 80(40) + 150(30) = 7{,}700$
(0, 50)	$P = 80(0) + 150(50) = 7{,}500$

From the table, we can see that the accountant will earn a maximum profit of $7,700 if she prepares 40 individual returns and 30 business returns.

EXAMPLE 4 **DIETS** Vigortab and Robust are two diet supplements. Each Vigortab tablet costs 50¢ and contains 3 units of calcium, 20 units of Vitamin C, and 40 units of iron. Each Robust tablet costs 60¢ and contains 4 units of calcium, 40 units of Vitamin C, and 30 units of iron. At least 24 units of calcium, 200 units of Vitamin C, and 120 units of iron are required for the daily needs of one patient. How many tablets of each supplement should be taken daily for a minimum cost? Find the daily minimum cost.

Solution First, we organize the given information into a table.

	Vigortab	Robust	Amount required
Calcium	3	4	24
Vitamin C	20	40	200
Iron	40	30	120
Cost	50¢	60¢	

Find the objective function

We can let x represent the number of Vigortab tablets to be taken daily and y the corresponding number of Robust tablets. Because each of the x Vigortab tablets will cost 50¢, and each of the y Robust tablets will cost 60¢, the total cost will be given by the equation

$$C = 0.50x + 0.60y \quad \text{50¢ = \$0.50 and 60¢ = \$0.60.}$$

Find the feasibility region

Since there are requirements for calcium, Vitamin C, and iron, there is a constraint for each. Note that neither x nor y can be negative.

$$\begin{cases} 3x + 4y \geq 24 & \text{The amount of calcium must be greater than or equal to 24 units.} \\ 20x + 40y \geq 200 & \text{The amount of Vitamin C must be greater than or equal to 200 units.} \\ 40x + 30y \geq 120 & \text{The amount of iron must be greater than or equal to 120 units.} \\ x \geq 0,\ y \geq 0 & \text{The number of tablets taken must be greater than or equal to 0.} \end{cases}$$

We graph the inequalities to find the feasibility region and the coordinates of its corner points, as in Figure 4-39 on the next page.

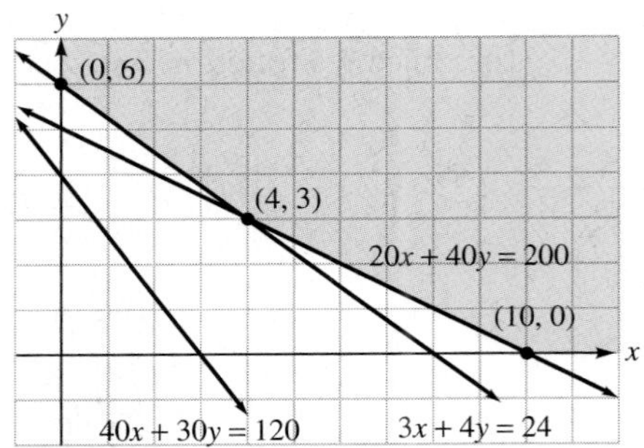

Figure 4-39

Find the minimum cost

In this case, the feasibility region is not bounded on all sides. The coordinates of the corner points are (0, 6), (4, 3), and (10, 0). To find the minimum cost, we substitute each pair of coordinates into the objective function.

Point	$C = 0.50x + 0.60y$
(0, 6)	$C = 0.50(0) + 0.60(6) = 3.60$
(4, 3)	$C = 0.50(4) + 0.60(3) = 3.80$
(10, 0)	$C = 0.50(10) + 0.60(0) = 5.00$

A minimum cost will occur if no Vigortab and 6 Robust tablets are taken daily. The minimum daily cost is \$3.60.

EXAMPLE 5 **PRODUCTION SCHEDULES** A television program director must schedule comedy skits and musical numbers for prime-time variety shows. Each comedy skit requires 2 hours of rehearsal time, costs \$3,000, and brings in \$20,000 from the show's sponsors. Each musical number requires 1 hour of rehearsal time, costs \$6,000, and generates \$12,000. If 250 hours are available for rehearsal, and \$600,000 is budgeted for comedy and music, how many segments of each type should be produced to maximize income? Find the maximum income.

EVERYDAY CONNECTIONS Fuel Oil Production

Crude Oil, Gasoline, and Natural Gas Futures	
Heating Oil NY Harbor	\$1.9805
Gasoline NY Harbor	\$2.1368

Source: http://www.wtrg.com/#Crude

Suppose that the refining process at United States refineries requires the production of at least three gallons of gasoline for each gallon of fuel oil. To meet the anticipated demands of winter, at least 2.10 million gallons of fuel oil will need to be produced per day. The demand for gasoline is not more than 8 million gallons a day. The wholesale price of gasoline is \$2.14 per gallon and the wholesale price of fuel oil is \$1.98/gal.

1. How much of each should be produced in order to maximize revenue?
2. What is the maximum revenue?

Solution First, we organize the given information into a table.

	Comedy	Musical	Available
Rehearsal time (hours)	2	1	250
Cost (in $1,000s)	3	6	600
Generated income (in $1,000s)	20	12	

Find the objective function

We can let x represent the number of comedy skits and y the number of musical numbers to be scheduled. Since each of the x comedy skits generates $20 thousand, the income generated by the comedy skits is $$20x$ thousand. The musical numbers produce $$12y$ thousand. The objective function to be maximized is

$$V = 20x + 12y$$

Find the feasibility region

Since there are limits on rehearsal time and budget, there is a constraint for each. Note that neither x nor y can be negative.

$$\begin{cases} 2x + y \le 250 & \text{The total rehearsal time must be less than or equal to 250 hours.} \\ 3x + 6y \le 600 & \text{The total cost must be less than or equal to \$600 thousand.} \\ x \ge 0,\ y \ge 0 & \text{The numbers of skits and musical numbers must be greater than or equal to 0.} \end{cases}$$

We graph the inequalities to find the feasibility region shown in Figure 4-40 and find the coordinates of each corner point.

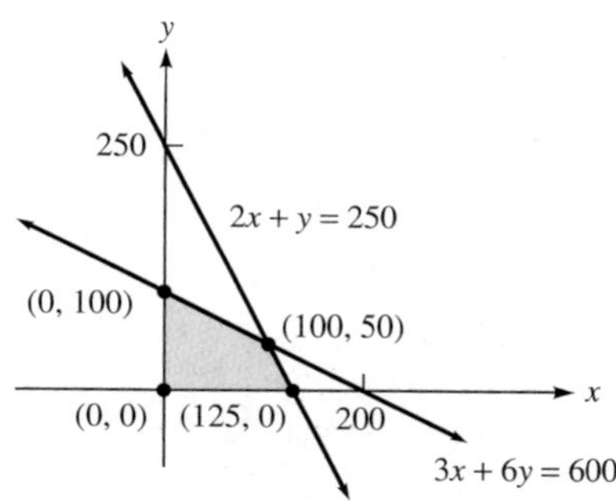

Figure 4-40

Find the maximum income

The coordinates of the corner points of the feasible region are (0, 0), (0, 100), (100, 50), and (125, 0). To find the maximum income, we substitute each pair of coordinates into the objective function.

Corner point	$V = 20x + 12y$
(0, 0)	$V = 20(0) + 12(0) = 0$
(0, 100)	$V = 20(0) + 12(100) = 1{,}200$
(100, 50)	$V = 20(100) + 12(50) = 2{,}600$
(125, 0)	$V = 20(125) + 12(0) = 2{,}500$

Maximum income will occur if 100 comedy skits and 50 musical numbers are scheduled. The maximum income will be 2,600 thousand dollars, or $2,600,000.

SELF CHECK ANSWERS **1.** 14 **2.** $\frac{3}{2}$

NOW TRY THIS

1. The FDA set the standard that fat calories should be limited to between 20% and 35% of the total daily intake, carbohydrates between 45% and 65%, and protein between 10% and 35%. Due to diabetes, Dale must limit his carbohydrate intake to the minimum (45%) and consume between 1,600 and 2,000 calories per day. Let x represent the number of fat calories and y the number of protein calories.
   a. Between what numbers of calories can Dale consume daily in carbohydrates?
   b. Between what numbers of calories does this leave for fats (x) and protein (y) combined?
   c. Between what numbers of calories can he allot to fats?
   d. Between what numbers of calories can be allotted to proteins?
   e. What are the corner points in this graph?
   f. Which one will minimize fat but maximize protein?

4.5 EXERCISES

WARM-UPS

Evaluate $P = 2x + 5y$ when

1. $x = 0, y = 5$
2. $x = 2, y = 1$

Find the corner points of the region determined by

3. $\begin{cases} x \geq 0 \\ y \geq 0 \\ x + y \leq 3 \end{cases}$
4. $\begin{cases} x \geq 0 \\ y \geq 0 \\ x + 2y \leq 4 \end{cases}$

REVIEW *Consider the line passing through $(-2, 4)$ and $(5, 7)$.*

5. Find the slope of the line.
6. Write the equation of the line in general form.
7. Write the equation of the line in slope-intercept form.
8. Write the equation of the line that passes through the origin and is parallel to the line.

VOCABULARY AND CONCEPTS *Fill in the blanks.*

9. In a linear program, the inequalities are called ________.
10. Ordered pairs that satisfy the constraints of a linear program are called ________ solutions.
11. The function to be maximized (or minimized) in a linear program is called the ________ function.
12. The objective function of a linear program attains a maximum (or minimum), subject to the constraints, at a ______ or along an ______ of the feasibility region.

GUIDED PRACTICE

Maximize P subject to the following constraints. See Example 1. (Objective 1)

13. $P = 2x + 3y$
$\begin{cases} x \geq 0 \\ y \geq 0 \\ x + y \leq 4 \end{cases}$
14. $P = 3x + 2y$
$\begin{cases} x \geq 0 \\ y \geq 0 \\ x + y \leq 4 \end{cases}$
15. $P = y + \frac{1}{2}x$
$\begin{cases} x \geq 0 \\ y \geq 0 \\ 2y - x \leq 1 \\ y - 2x \geq -2 \end{cases}$
16. $P = 4y - x$
$\begin{cases} x \leq 2 \\ y \geq 0 \\ x + y \geq 1 \\ 2y - x \leq 1 \end{cases}$

Minimize P subject to the following constraints. See Example 2. (Objective 1)

17. $P = 5x + 12y$
$\begin{cases} x \geq 0 \\ y \geq 0 \\ x + y \leq 4 \end{cases}$
18. $P = 3x + 6y$
$\begin{cases} x \geq 0 \\ y \geq 0 \\ x + y \leq 4 \end{cases}$
19. $P = 3y + x$
$\begin{cases} x \geq 0 \\ y \geq 0 \\ 2y - x \leq 1 \\ y - 2x \geq -2 \end{cases}$
20. $P = 5y + x$
$\begin{cases} x \leq 2 \\ y \geq 0 \\ x + y \geq 1 \\ 2y - x \leq 1 \end{cases}$

ADDITIONAL PRACTICE

Maximize P subject to the following constraints.

21. $P = 2x + y$

$$\begin{cases} y \geq 0 \\ y - x \leq 2 \\ 2x + 3y \leq 6 \\ 3x + y \leq 3 \end{cases}$$

22. $P = x - 2y$

$$\begin{cases} x + y \leq 5 \\ y \leq 3 \\ x \leq 2 \\ x \geq 0 \\ y \geq 0 \end{cases}$$

23. $P = 3x - 2y$

$$\begin{cases} x \leq 1 \\ x \geq -1 \\ y - x \leq 1 \\ x - y \leq 1 \end{cases}$$

24. $P = x - y$

$$\begin{cases} 5x + 4y \leq 20 \\ y \leq 5 \\ x \geq 0 \\ y \geq 0 \end{cases}$$

Minimize P subject to the following constraints.

25. $P = 6x + 2y$

$$\begin{cases} y \geq 0 \\ y - x \leq 2 \\ 2x + 3y \leq 6 \\ 3x + y \leq 3 \end{cases}$$

26. $P = 2y - x$

$$\begin{cases} x \geq 0 \\ y \geq 0 \\ x + y \leq 5 \\ x + 2y \geq 2 \end{cases}$$

27. $P = 2x - 2y$

$$\begin{cases} x \leq 1 \\ x \geq -1 \\ y - x \leq 1 \\ x - y \leq 1 \end{cases}$$

28. $P = y - 2x$

$$\begin{cases} x + 2y \leq 4 \\ 2x + y \leq 4 \\ x + 2y \geq 2 \\ 2x + y \geq 2 \end{cases}$$

APPLICATIONS *Write the objective function and the inequalities that describe the constraints in each problem. Graph the feasibility region, showing the corner points. Then find the maximum or minimum value of the objective function.*

See Examples 3–5. (Objective 2)

29. Making furniture Two woodworkers, Tom and Carlos, bring in \$100 for making a table and \$80 for making a chair. On average, Tom must work 3 hours and Carlos 2 hours to make a chair. Tom must work 2 hours and Carlos 6 hours to make a table. If neither wants to work more than 42 hours per week, how many tables and how many chairs should they make each week to maximize their income? Find the maximum income.

	Table	Chair	Time available
Income (\$)	100	80	
Tom's time (hr)	2	3	42
Carlos's time (hr)	6	2	42

30. Making crafts Two artists, Nina and Rob, make yard ornaments. They bring in \$80 for each wooden snowman they make and \$64 for each wooden Santa Claus. On average, Nina must work 4 hours and Rob 2 hours to make a snowman. Nina must work 3 hours and Rob 4 hours to make a Santa Claus. If neither wants to work more than 20 hours per week, how many of each ornament should they make each week to maximize their income? Find the maximum income.

	Snowman	Santa Claus	Time available
Income (\$)	80	64	
Nina's time (hr)	4	3	20
Rob's time (hr)	2	4	20

31. Inventories An electronics store manager stocks from 20 to 30 IBM-compatible computers and from 30 to 50 Macintosh computers. There is room in the store to stock up to 60 computers. The manager receives a commission of \$50 on the sale of each IBM-compatible computer and \$40 on the sale of each Macintosh computer. If the manager can sell all of the computers, how many should she stock to maximize her commissions? Find the maximum commission.

Inventory	IBM	Macintosh
Minimum	20	30
Maximum	30	50
Commission	\$50	\$40

32. Diet problem A diet requires at least 16 units of Vitamin C and at least 34 units of Vitamin B complex. Two food supplements are available that provide these nutrients in the amounts and costs shown in the table. How much of each should be used to minimize the cost?

Supplement	Vitamin C	Vitamin B	Cost
A	3 units/g	2 units/g	3¢/g
B	2 units/g	6 units/g	4¢/g

33. Production Manufacturing DVD players and TVs requires the use of the electronics, assembly, and finishing departments of a factory, according to the following schedule:

	Hours for DVD player	Hours for TV	Hours available per week
Electronics	3	4	180
Assembly	2	3	120
Finishing	2	1	60

(continued)

Each DVD player has a profit of \$40, and each TV has a profit of \$32. How many DVD players and TVs should be manufactured weekly to maximize profit? Find the maximum profit.

34. **Production problem** A company manufactures one type of computer chip that runs at 1.66 GHz and another that runs at 2.66 GHz. The company can make a maximum of 50 fast chips per day and a maximum of 100 slow chips per day. It takes 6 hours to make a fast chip and 3 hours to make a slow chip, and the company's employees can provide up to 360 hours of labor per day. If the company makes a profit of \$20 on each 2.66-GHz chip and \$27 on each 1.66-GHz chip, how many of each type should be manufactured to earn the maximum profit?

35. **Financial planning** A stockbroker has \$200,000 to invest in stocks and bonds. She wants to invest at least \$100,000 in stocks and at least \$50,000 in bonds. If stocks have an annual yield of 9% and bonds have an annual yield of 7%, how much should she invest in each to maximize her income? Find the maximum return.

36. **Production** A small country exports soybeans and flowers. Soybeans require 8 workers per acre, flowers require 12 workers per acre, and 100,000 workers are available. Government contracts require that there be at least 3 times as many acres of soybeans as flowers planted. It costs \$250 per acre to plant soybeans and \$300 per acre to plant flowers, and there is a budget of \$3 million. If the profit from soybeans is \$1,600 per acre and the profit from flowers is \$2,000 per acre, how many acres of each crop should be planted to maximize profit? Find the maximum profit.

WRITING ABOUT MATH

37. What is meant by the constraints of a linear program?
38. What is meant by a feasible solution of a linear program?

SOMETHING TO THINK ABOUT

39. Try to construct a linear programming problem. What difficulties do you encounter?
40. Try to construct a linear programming problem that will have a maximum at every point along an edge of the feasibility region.

PROJECTS

Project 1

A farmer is building a machine shed onto his barn, as shown in the illustration. It is to be 12 feet wide, 20 feet long, and h_2 must be no more than 20 feet. In order for all of the shed to be useful for storing machinery, h_1 must be at least 6 feet. For the roof to shed rain and melting snow adequately, the slope of the roof must be at least $\frac{1}{2}$, but to be easily shingled, it must have a slope that is no greater than 1.

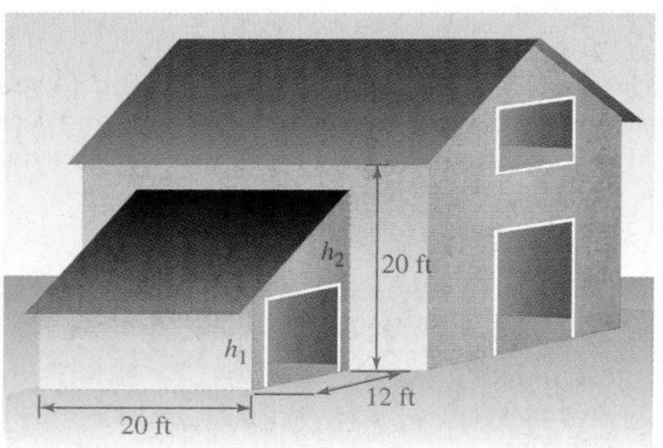

a. Represent on a graph all of the possible values for h_1 and h_2, subject to the constraints listed above.

b. The farmer wants to minimize the construction costs while still making sure that the shed is large enough for his purposes. He does this by setting a lower bound on the volume of the shed (3,000 cubic feet) and then minimizing the surface area of the walls that must be built. The volume of the shed can be expressed in a formula that contains h_1 and h_2. Derive this formula, and include the volume restriction in your design constraints. Then find the dimensions that will minimize the total area of the two ends of the shed and the outside wall. (The inner wall is already present as a wall of the barn and therefore involves no new cost.)

Project 2

Knowing any three points on the graph of a parabolic function is enough to determine the equation of that parabola. It follows that for any three points that could possibly lie on a parabola, there is exactly one parabola that passes through those points, and this parabola will have the equation $y = ax^2 + bx + c$ for appropriate a, b, and c.

a. In order for a set of three points to lie on the graph of a parabolic function, no two of the points can have the same x-coordinate, and not all three can have the same y-coordinate. Explain why we need these restrictions.

b. Suppose that the points $(1, 3)$ and $(2, 6)$ are on the graph of a parabola. What restrictions would have to be placed on a and b to guarantee that the y-intercept of the parabola has an absolute value of 4 or less? Can $(-1, 8)$ be a third point on such a parabola?

Chapter 4 REVIEW

SECTION 4.1 Solving Linear Inequalities in One Variable

DEFINITIONS AND CONCEPTS	EXAMPLES
Trichotomy property: $a < b, a = b$, or $a > b$	Either $x < 3$, $x = 3$, or $x > 3$.
Transitive properties: If $a < b$ and $b < c$, then $a < c$. If $a > b$ and $b > c$, then $a > c$.	If $-2 < 5$ and $5 < 10$, then $-2 < 10$. If $20 > 7$ and $7 > -5$, then $20 > -5$.
Properties of inequality: If a and b are real numbers and $a < c$, then $a + c < b + c$ $a - c < b - c$ $ac < bc \quad (c > 0)$ $ac > bc \quad (c < 0)$ $\frac{a}{c} < \frac{b}{c} \quad (c > 0)$ $\frac{a}{c} > \frac{b}{c} \quad (c < 0)$	To solve the linear inequality $3(2x + 6) < 18$, we use the same steps as for solving equations. $3(2x + 6) < 18$ $6x + 18 < 18$ Remove parentheses. $6x < 0$ Subtract 18 from both sides. $x < 0$ Divide both sides by 6. The solution set is $\{x \mid x < 0\}$. The graph is shown below. The parenthesis at 0 indicates that 0 is not included in the solution set. 0 To solve $-4(3x - 4) \le -8$, we proceed as follows: $-4(3x - 4) \le -8$ $-12x + 16 \le -8$ Remove parentheses. $-12x \le -24$ Subtract 16 from both sides. $x \ge 2$ Divide both sides by -12 and reverse inequality symbol. The solution set is $\{x \mid x \ge 2\}$. The graph is shown below. The bracket at 2 indicates that 2 is included in the solution set. 2

Compound inequalities:

$c < x < d$ is equivalent to $c < x$ and $x < d$.

To solve the inequality $-7 \le 2x - 5 < 3$, we isolate x between the inequality symbols.

$$-7 \le 2x - 5 < 3$$

$$-2 \le 2x < 8 \quad \text{Add 5 to all three parts.}$$

$$-1 \le x < 4 \quad \text{Divide all three parts by 2.}$$

The solution set is $\{x \mid -1 \le x < 4\}$, which is the interval $[-1, 4)$. The graph is shown below.

The word *or* between two inequalities indicates that only one of the inequalities needs to be true to make the entire statement true.

To solve $x \le 2$ or $x > 7$, we must find all numbers that make either inequality true. The following graph shows those numbers. The solution set is $\{x \mid x \le 2 \text{ or } x > 7\}$, which is the union of two intervals $(-\infty, 2] \cup (7, \infty)$.

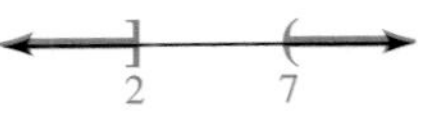

REVIEW EXERCISES

Graph each inequality and give each solution set in interval notation.

1. $5(x - 2) \le 5$

2. $3x + 4 > 10$

3. $\frac{1}{3}x - 2 \ge \frac{1}{2}x + 2$

4. $\frac{7}{4}(x + 3) < \frac{3}{8}(x - 3)$

5. $3 < 3x + 4 < 10$

6. $4x > 3x + 2 > x - 3$

7. $-5 \le 2x - 3 < 5$

8. Investing A woman invests \$10,000 at 6% annual interest. How much more must she invest at 7% so that her annual income is at least \$2,000?

SECTION 4.2 Solving Absolute-Value Equations and Inequalities in One Variable

DEFINITIONS AND CONCEPTS	EXAMPLES

If $x \ge 0$, $|x| = x$.

If $x < 0$, $|x| = -x$.

If $k > 0$, $|x| = k$ is equivalent to $x = k$ or $x = -k$.

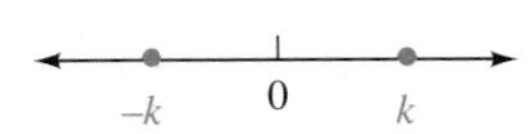

$|4| = 4$

$|-4| = -(-4) = 4$

To solve the equation $|2x - 3| = 1$, we can write it as

$$2x - 3 = 1 \quad \text{or} \quad 2x - 3 = -1$$

and solve each equation for x:

$$\begin{aligned} 2x - 3 &= 1 & \text{or} \quad 2x - 3 &= -1 \\ 2x &= 4 & 2x &= 2 \\ x &= 2 & x &= 1 \end{aligned}$$

Verify that both solutions check.

$|a| = |b|$ is equivalent to $a = b$ or $a = -b$.

To solve the equation $|4x - 3| = |2x + 15|$, we note that the equation is true when $4x - 3 = 2x + 15$ or when $4x - 3 = -(2x + 15)$. We then solve each equation for x.

$$\begin{array}{rcl|rcl} 4x - 3 &=& 2x + 15 \quad \text{or} & 4x - 3 &=& -(2x + 15) \\ 4x &=& 2x + 18 & 4x - 3 &=& -2x - 15 \\ 2x &=& 18 & 6x &=& -12 \\ x &=& 9 & x &=& -2 \end{array}$$

Verify that both solutions check.

If $k > 0$, $|x| < k$ is equivalent to $-k < x < k$.

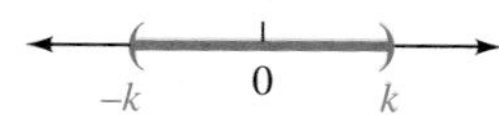

To solve the inequality $|2x - 3| < 1$, we note that $|2x - 3| < 1$ is equivalent to $-1 < 2x - 3 < 1$, an inequality that we can solve.

$$-1 < 2x - 3 < 1$$

$$2 < 2x < 4 \quad \text{Add 3 to each part.}$$

$$1 < x < 2 \quad \text{Divide each part by 2.}$$

The solution set contains all numbers between 1 and 2, not including either 1 or 2. This is the interval (1, 2), whose graph is shown below.

$|x| > k$ is equivalent to $x < -k$ or $x > k$.

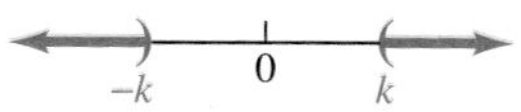

To solve the inequality $|5x - 5| > 15$, we write the inequality as two separate inequalities and solve each one for x. Since $|5x - 5| > 15$ is equivalent to

$$5x - 5 < -15 \quad \text{or} \quad 5x - 5 > 15$$

we have

$$\begin{array}{rcl|rcll} 5x - 5 &<& -15 \quad \text{or} & 5x - 5 &>& 15 & \\ 5x &<& -10 & 5x &>& 20 & \text{Add 5 to both sides.} \\ x &<& -2 & x &>& 4 & \text{Divide both sides by 5.} \end{array}$$

Thus,

$$x < -2 \quad \text{or} \quad x > 4$$

This is the interval $(-\infty, -2) \cup (4, \infty)$, whose graph appears below.

REVIEW EXERCISES

Find each absolute value.

9. $|-7|$
10. $|8|$
11. $-|7|$
12. $-|-12|$

Graph each function.

13. $f(x) = |x + 1| - 3$

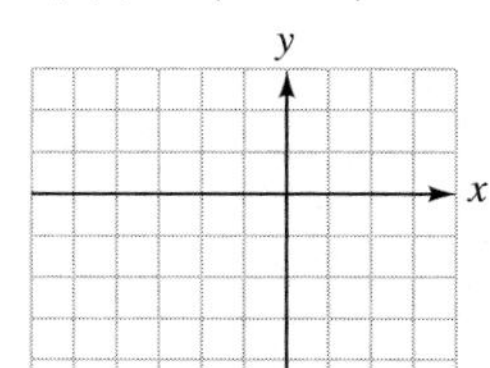

14. $f(x) = |x - 2| + 1$

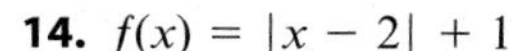

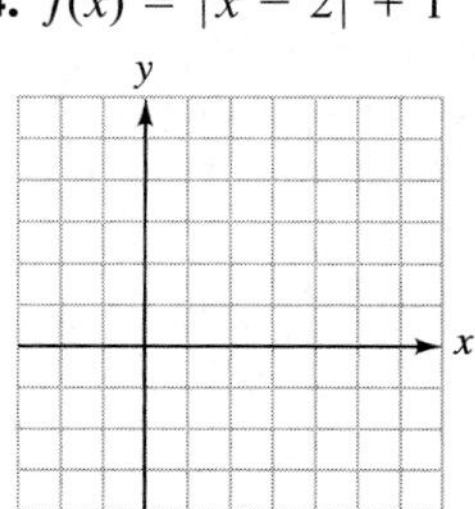

Solve and check each equation.

15. $|3x + 1| = 10$

16. $\left|\frac{3}{2}x - 4\right| + 2 = 11$

17. $\left|\frac{2 - x}{3}\right| = 4$

18. $|3x + 2| = |2x - 3|$

19. $|5x - 4| = |4x - 5|$

20. $\left|\frac{3 - 2x}{2}\right| = \left|\frac{3x - 2}{3}\right|$

Solve each inequality. Give the solution in interval notation and graph it.

21. $|2x + 7| < 3$

22. $|5 - 3x| \le 14$

23. $\left|\frac{2}{3}x + 14\right| < 0$

24. $\left|\frac{1 - 5x}{3}\right| > 7$

25. $|3x - 8| - 3 \ge 1$

26. $\left|\frac{3}{2}x - 14\right| \ge 0$

SECTION 4.3 Solving Linear Inequalities in One or Two Variables

DEFINITIONS AND CONCEPTS	EXAMPLES
To graph a linear inequality in x and y, graph the boundary line, and then use a test point to determine which side of the boundary should be shaded. If the inequality is $<$ or $>$, the boundary line will be dashed because the points that lie on the line are not included in the solution. If the inequality is $\le$ or $\ge$, the boundary line will be solid.	To graph the inequality $3x - 2y > 6$, we start by graphing $3x - 2y = 6$ to find the boundary line. We draw a dashed line because equality is not included. To determine which half-plane represents $3x - 2y > 6$, we check to see whether the coordinates of the origin satisfy the inequality. $3x - 2y > 6$ $3(0) - 2(0) \overset{?}{>} 6$ Substitute 0 for x and 0 for y. $0 > 6$ Since the coordinates do not satisfy the inequality, the origin is not in the half-plane that is the graph of $3x - 2y > 6$. The graph is shown below. 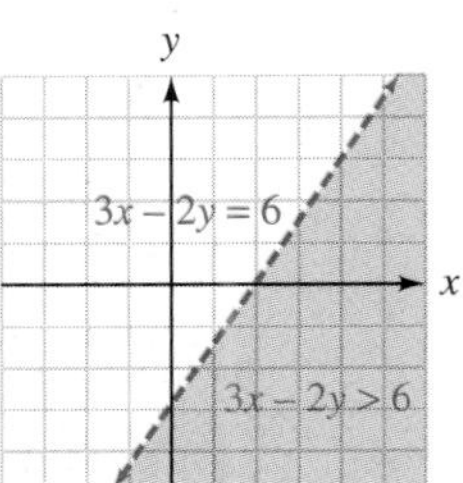

REVIEW EXERCISES

Graph each inequality on the coordinate plane.

27. $2x + 3y > 6$

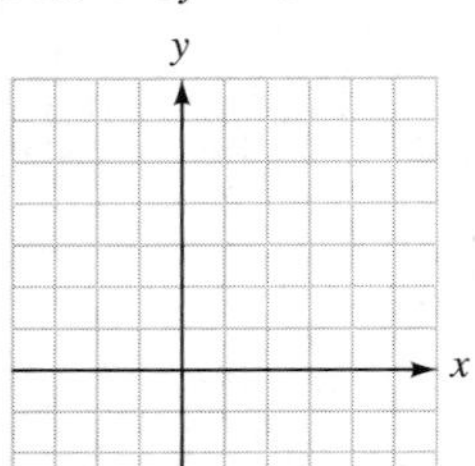

28. $y \le 4 - x$

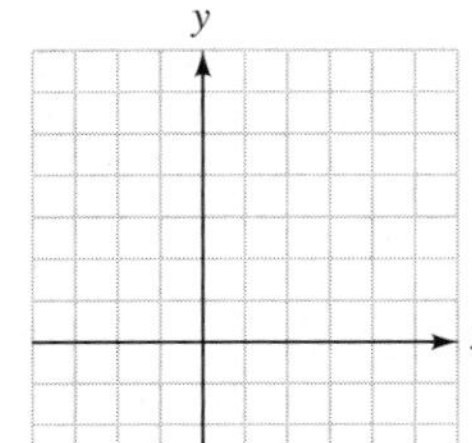

29. $-2 < x < 4$

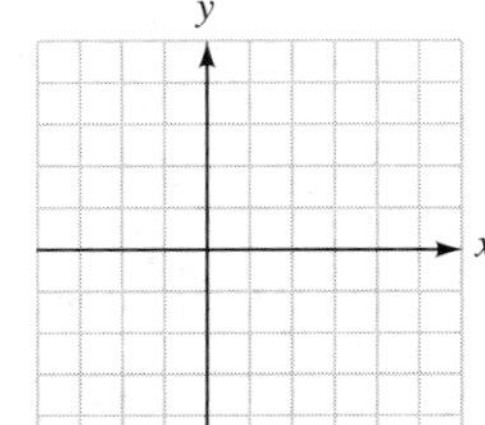

30. $y \le -2$ or $y > 1$

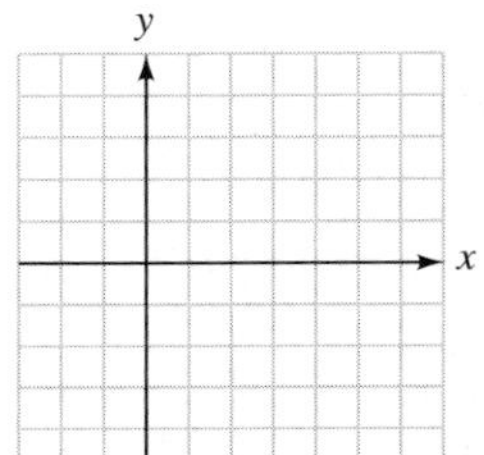

SECTION 4.4 Solving Systems of Linear and Quadratic Inequalities in Two Variables

DEFINITIONS AND CONCEPTS	EXAMPLES
To graph a system of linear inequalities in x and y, graph each boundary and determine whether the graph will be solid or dashed based on the inequality symbol. Use a test point to determine which side of each of the boundaries should be shaded. The answer will be the region shaded by both inequalities.	To graph the solution set of $\begin{cases} x - 2y \le 2 \\ 2x + y > 4 \end{cases}$, we graph each inequality on one set of coordinate axes as shown below. The graph of $x - 2y \le 2$ includes the line graph of the equation $x - 2y = 2$ and all points above it. Since the edge is included, we draw it as a solid line. The graph of $2x + y > 4$ contains the points to the right of the graph of the equation $2x + y = 4$. Since the edge is not included, we draw it as a dashed line. The area where the half-planes intersect represents the solution of the system of inequalities, because any point in that region has coordinates that will satisfy both inequalities. 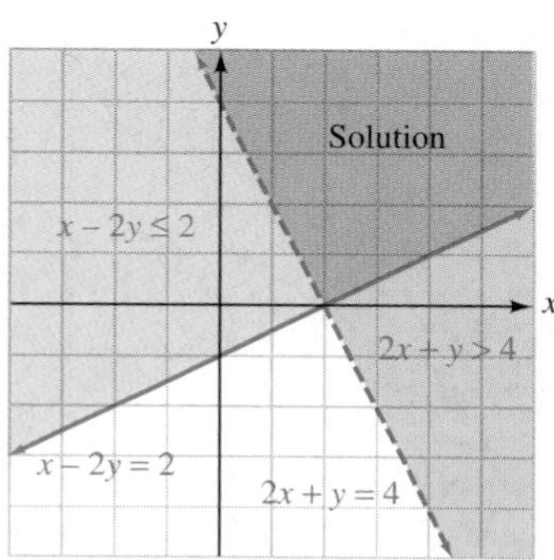

REVIEW EXERCISES

Graph the solution set of each system of inequalities.

31. $\begin{cases} y \ge x + 1 \\ 3x + 2y < 6 \end{cases}$

32. $\begin{cases} y \ge x^2 - 4 \\ y < x + 3 \end{cases}$

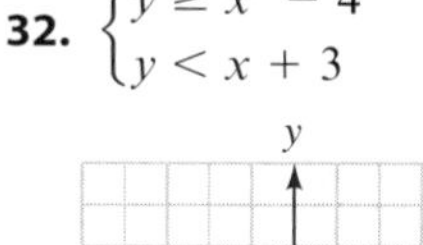

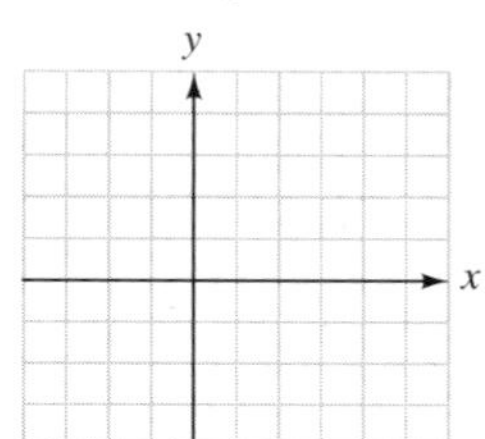

SECTION 4.5 Solving Systems Using Linear Programming

DEFINITIONS AND CONCEPTS

If a linear function, subject to the constraints of a system of linear inequalities in two variables, attains a maximum or a minimum value, that value will occur at a corner or along an entire edge of the region R that represents the solution of the system.

EXAMPLES

To solve a linear program, we maximize (or minimize) a function (called the **objective function**) subject to given conditions on its variables, called **constraints.**

To find the maximum value P of the function $P = 3x + 2y$ subject to the constraints

$$\begin{cases} x - y \leq 4 \\ x + y \leq 6 \\ x \geq 0 \\ y \geq 0 \end{cases}$$

we solve the system of inequalities to find the feasibility region R shown in the figure below. The coordinates of its corner points are (0, 0), (4, 0), (0, 6), and (5, 1).

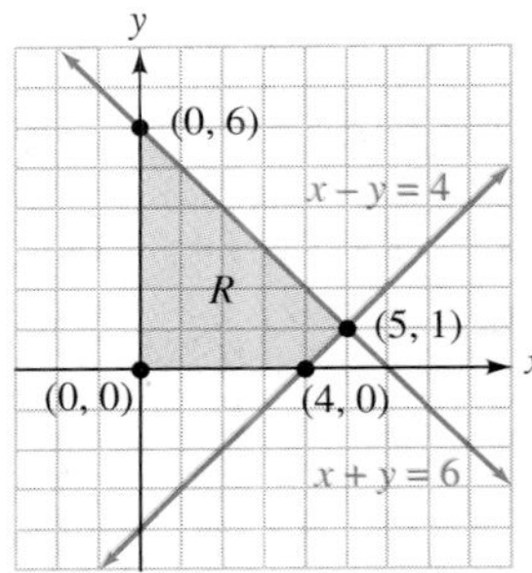

Since the maximum value of P will occur at a corner of R, we substitute the coordinates of each corner point into the objective function $P = 3x + 2y$ and find the one that gives the maximum value of P.

Point	$P = 3x + 2y$
(0, 0)	$P = 3(0) + 2(0) = 0$
(4, 0)	$P = 3(4) + 2(0) = 12$
(5, 1)	$P = 3(5) + 2(1) = 17$
(0, 6)	$P = 3(0) + 2(6) = 12$

The maximum value $P = 17$ occurs when $x = 5$ and $y = 1$.

REVIEW EXERCISES

33. Maximize $P = 2x + y$ subject to $\begin{cases} x \geq 0 \\ y \geq 0 \\ x + y \leq 3 \end{cases}$.

34. Making fertilizer A company manufactures fertilizers X and Y. Each 50-pound bag requires three ingredients, which are available in the limited quantities shown in the table.

Ingredient	Number of pounds in fertilizer X	Number of pounds in fertilizer Y	Total number of pounds available
Nitrogen	6	10	20,000
Phosphorus	8	6	16,400
Potash	6	4	12,000

The profit on each bag of fertilizer X is \$6, and on each bag of Y, \$5. How many bags of each should be produced to maximize profit?

Chapter 4 TEST

Graph the solution of each inequality and give the solution in interval notation.

1. $-2(2x + 3) \geq 14$
2. $-2 < \dfrac{x - 4}{3} < 4$

Write each expression without absolute value symbols.

3. $|5 - 8|$
4. $|4\pi - 4|$

Graph each function.

5. $f(x) = |x + 1| - 4$
6. $f(x) = |x - 2| + 3$

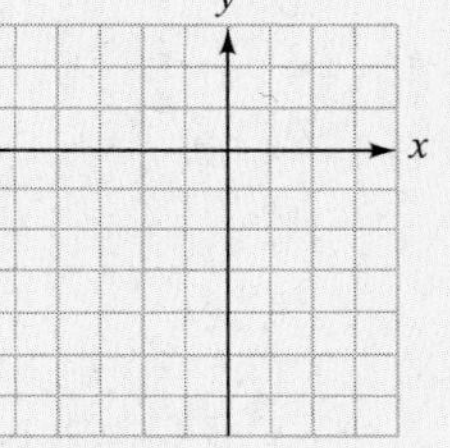

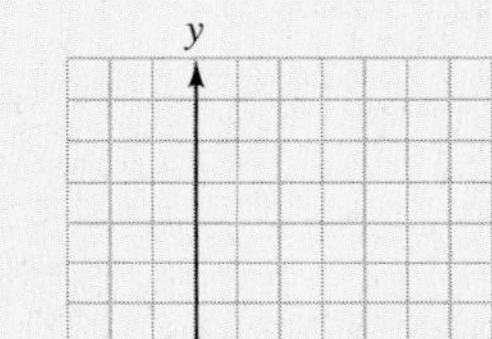

Solve each equation.

7. $|2x + 3| = 11$
8. $|4 - 3x| - 2 = 17$
9. $|3x + 4| = |x + 12|$
10. $|3 - 2x| = |2x + 3|$

Graph the solution of each inequality and give the solution in interval notation.

11. $|x + 3| \leq 4$
12. $|2x - 4| > 22$
13. $|4 - 2x| > 2$
14. $|2x - 4| \leq 2$

Graph each inequality.

15. $3x + 2y \geq 6$
16. $-2 \leq y < 5$

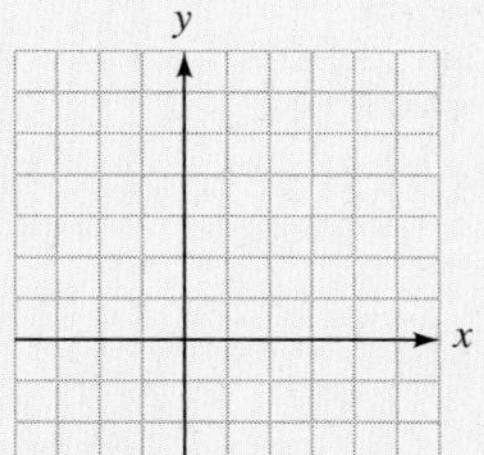

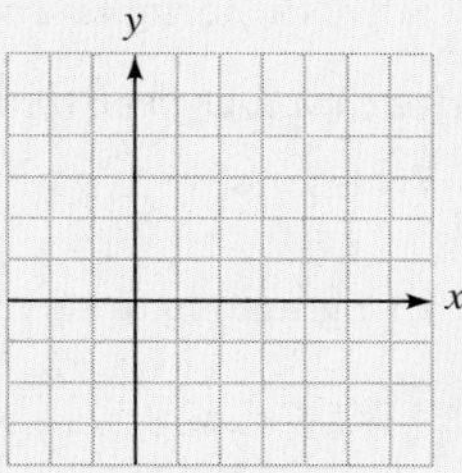

Use graphing to solve each system.

17. $\begin{cases} 2x - 3y \geq 6 \\ y \leq -x + 1 \end{cases}$
18. $\begin{cases} y \geq x^2 \\ y < x + 3 \end{cases}$

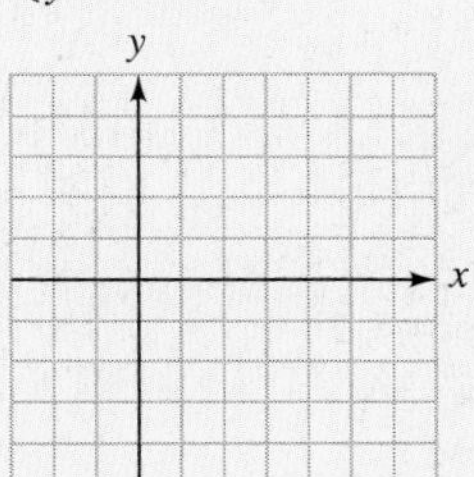

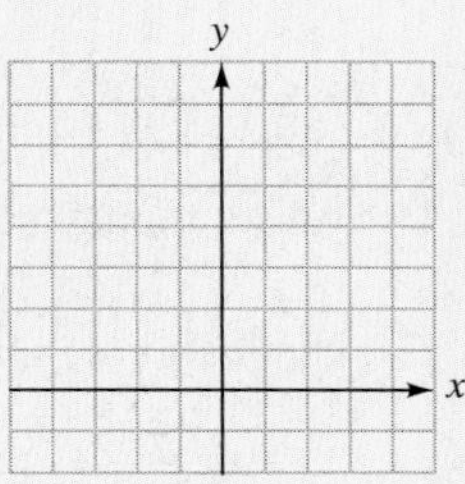

19. Maximize $P = 3x - y$ subject to $\begin{cases} y \geq 1 \\ y \leq 2 \\ y \leq 3x + 1 \\ x \leq 1 \end{cases}$.

Cumulative Review Exercises

1. Draw a number line and graph the prime numbers from 50 to 60.
2. Find the additive inverse of -5.

Evaluate each expression when $x = 2$ and $y = -4$.

3. $x - xy$
4. $\dfrac{x^2 - y^2}{3x + y}$

Simplify each expression. Assume no variable is zero.

5. $(x^2x^3)^2$
6. $(x^2)^3(x^4)^2$
7. $\left(\dfrac{x^3}{x^5}\right)^{-2}$
8. $\dfrac{a^2b^n}{a^nb^2}$
9. Write 32,600,000 in scientific notation.
10. Write 0.000012 in scientific notation.

Solve each equation.

11. $3x - 6 = 20$
12. $6(x - 1) = 2(x + 3)$

13. $\frac{5b}{2} - 10 = \frac{b}{3} + 3$

14. $2a - 5 = -2a + 4(a - 2) + 1$

Determine whether the lines represented by the equations are parallel, perpendicular, or neither.

15. $3x + 2y = 12, 2x - 3y = 5$

16. $3x = y + 4, y = 3(x - 4) - 1$

17. Write the equation of the line passing through $(-2, 3)$ and perpendicular to the graph of $3x + y = 8$.

18. Solve the formula $A = \frac{1}{2}h(b_1 + b_2)$ for h.

Find each value, given that $f(x) = 3x^2 - x$.

19. $f(2)$

20. $f(-2)$

21. Use graphing to solve $\begin{cases} 2x + y = 5 \\ x - 2y = 0 \end{cases}$.

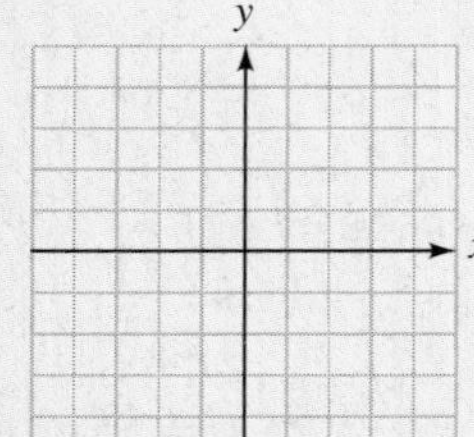

22. Use substitution to solve $\begin{cases} 3x + y = 4 \\ 2x - 3y = -1 \end{cases}$.

23. Use elimination to solve $\begin{cases} x + 2y = -2 \\ 2x - y = 6 \end{cases}$.

24. Solve: $\begin{cases} \frac{x}{10} + \frac{y}{5} = \frac{1}{2} \\ \frac{x}{2} - \frac{y}{5} = \frac{13}{10} \end{cases}$.

25. Solve: $\begin{cases} x + y + z = 1 \\ 2x - y - z = -4. \\ x - 2y + z = 4 \end{cases}$

26. Solve: $\begin{cases} x + 2y + 3z = 1 \\ 3x + 2y + z = -1. \\ 2x + 3y + z = -2 \end{cases}$

27. Evaluate: $\begin{vmatrix} 3 & -2 \\ 1 & -1 \end{vmatrix}$.

28. Evaluate: $\begin{vmatrix} 2 & 3 & -1 \\ -1 & -1 & 2 \\ 4 & 1 & -1 \end{vmatrix}$.

Use Cramer's rule to solve each system.

29. $\begin{cases} 4x - 3y = -1 \\ 3x + 4y = -7 \end{cases}$

30. $\begin{cases} x - 2y - z = -2 \\ 3x + y - z = 6 \\ 2x - y + z = -1 \end{cases}$

Solve each inequality.

31. $-3(x - 4) \geq x - 32$

32. $-8 < -3x + 1 < 10$

Solve each equation.

33. $|4x - 3| = 9$

34. $|2x - 1| = |3x + 4|$

Solve each inequality.

35. $|3x - 2| \leq 4$

36. $|2x + 3| - 1 > 4$

Use graphing to solve each inequality.

37. $2x - 3y \leq 12$

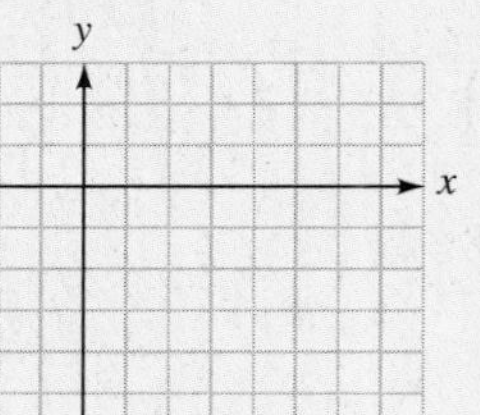

38. $3 > x \geq -2$

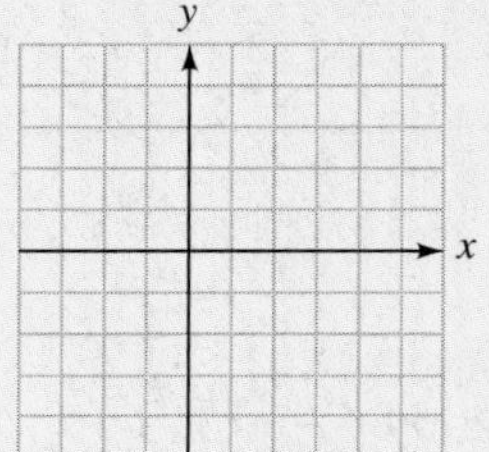

Use graphing to solve each system.

39. $\begin{cases} 3x - 2y < 6 \\ y < -x + 2 \end{cases}$

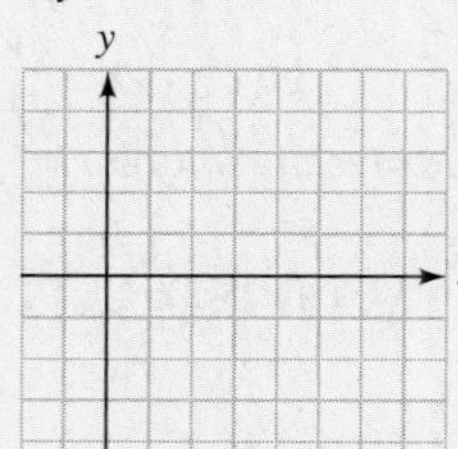

40. $\begin{cases} y < x + 2 \\ 3x + y \leq 6 \end{cases}$

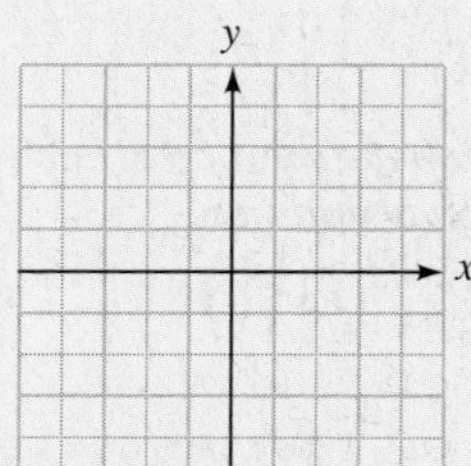

41. **Research** To conduct an experiment with mice and rats, a researcher will place the animals into one of two mazes for the number of minutes shown in the table. Find the greatest number of animals that can be used in this experiment.

	Time per mouse	Time per rat	Time available
Maze 1	12 min	8 min	240 min
Maze 2	10 min	15 min	300 min

©Shutterstock.com/Sebastien Burel

CHAPTER 5

Polynomials and Polynomial Functions

5.1 Polynomials and Polynomial Functions
5.2 Adding and Subtracting Polynomials
5.3 Multiplying Polynomials
5.4 The Greatest Common Factor and Factoring by Grouping
5.5 The Difference of Two Squares; the Sum and Difference of Two Cubes
5.6 Factoring Trinomials
5.7 Summary of Factoring Techniques
5.8 Solving Equations by Factoring
■ *Projects*
CHAPTER REVIEW
CHAPTER TEST

Careers and Mathematics

ARTISTS

Artists create art to communicate ideas, thoughts, or feelings. They use a variety of methods—painting, sculpting, or illustrating.

Artists held about 218,000 jobs in 2006. More than half were self-employed. Training requirements for artists vary by specialty. However, it is very difficult to become skilled enough to make a living without some formal training.

Job Outlook:
Employment of artists and related workers is expected to grow 16% through 2016. However, competition for salaried jobs and freelance work in some areas is expected to be keen.

Annual Earnings:
$28,500–$58,550

For More Information:
http://www.bls.gov/oco/ocos092.htm

For a Sample Application:
See Problem 85 in Section 5.8.

In this chapter ▶

Polynomials can be considered to be the numbers of algebra. In this chapter, we will learn how to add, subtract, and multiply them. Then we will learn how to reverse the operation of multiplication and find the factors of a known product. We conclude by solving equations using factoring.

SECTION 5.1 Polynomials and Polynomial Functions

Objectives

1. Classify a polynomial as a monomial, binomial, trinomial, or none of these.
2. Determine the degree of a polynomial.
3. Evaluate a polynomial function.
4. Graph a polynomial function.

Vocabulary

algebraic term	binomial	descending order
polynomial	trinomial	ascending order
monomial	degree of a polynomial	polynomial function

Getting Ready

Write each expression using exponents.

1. $3aabb$ **2.** $-5xxxy$
3. $4pp + 7qq$ **4.** $aaa - bb$

In this section, we discuss important algebraic expressions called *polynomials.* We then examine several polynomial functions.

Classify a polynomial as a monomial, binomial, trinomial, or none of these.

Algebraic terms are expressions that contain constants and/or variables. Some examples are

$$17, \quad 9x, \quad 15y^2, \quad \text{and} \quad -24x^4y^5$$

The *numerical coefficient* of 17 is 17. The numerical coefficients of the remaining terms are 9, 15, and -24, respectively.

A **polynomial** is the sum of one or more algebraic terms whose variables have whole-number exponents.

Polynomial in One Variable

A **polynomial in one variable** (say, x) is the sum of one or more terms of the form ax^n, where a is a real number and n is a whole number.

The following expressions are polynomials in x. Note that $17 = 17x^0$.

$$3x^2 + 2x, \qquad \frac{3}{2}x^5 - \frac{7}{3}x^4 - \frac{8}{3}x^3, \qquad \text{and} \qquad 19x^{20} + \sqrt{3}x^{14} + 4.5x^{11} - 17$$

The following expressions are not polynomials.

$$\frac{2x}{x^2 + 1}, \qquad x^{1/2} - 1, \qquad \text{and} \qquad x^{-3} + 2x$$

The first expression is the quotient of two polynomials, and the last two have exponents that are not whole numbers.

Polynomial in More than One Variable

A **polynomial in several variables** (say x, y, and z) is the sum of one or more terms of the form $ax^my^nz^p$; where a is a real number and m, n, and p are whole numbers.

The following expressions are polynomials in more than one variable.

$$3xy, \qquad 5x^2y + 2yz^3 - 3xz, \qquad \text{and} \qquad u^2v^2w^2 + x^3y^3 + 1$$

A polynomial with one term is called a **monomial,** a polynomial with two terms is called a **binomial,** and a polynomial with three terms is called a **trinomial:**

Monomials (*one term*)	***Binomials*** (*two terms*)	***Trinomials*** (*three terms*)
$2x^3$	$2x^4 + 5$	$2x^3 + 4x^2 + 3$
a^2b	$-17t^{45} - 3xy$	$3mn^3 - m^2n^3 + 7n$
$3x^3y^5z^2$	$32x^{13}y^5 + 47x^3yz$	$-12x^5y^2 + 13x^4y^3 - 7x^3y^3$

2 Determine the degree of a polynomial.

Because the variable x occurs three times as a factor in the monomial $2x^3$, the monomial is called a *third-degree monomial* or a *monomial of degree 3.* The monomial $3x^3y^5z^2$ is called a *monomial of degree 10,* because the variables x, y, and z occur as factors a total of ten times. These examples illustrate the following definition.

Degree of a Monomial

If $a \neq 0$, the **degree of ax^n** is n. The degree of a monomial containing several variables is the sum of the exponents on those variables.

EXAMPLE 1 Find the degree of **a.** $3x^4$ **b.** $-18x^3y^2z^{12}$ **c.** $4^7x^2y^3$ **d.** 3.

Solution

a. $3x^4$ is a monomial of degree 4, because the variable x occurs as a factor four times.

b. $-18x^3y^2z^{12}$ is a monomial of degree 17, because the sum of the exponents on the variables is 17.

c. $4^7x^2y^3$ is a monomial of degree 5, because the sum of the exponents on the variables is 5.

d. 3 is a monomial of degree 0, because $3 = 3x^0$.

COMMENT Since $a \neq 0$ in the previous definition, 0 has no defined degree.

SELF CHECK 1 Find the degree of: **a.** $-12a^5$ **b.** $8a^3b^2$.

We define the degree of a polynomial by considering the degrees of each of its terms.

Degree of a Polynomial The **degree of a polynomial** is the same as the degree of the term in the polynomial with largest degree.

EXAMPLE 2 Find the degree of each polynomial: **a.** $3x^5 + 4x^2 + 7$ **b.** $7x^2y^8 - 3xy$ **c.** $3x + 2y - xy$ **d.** $18x^2y^3 - 12x^7y^2 + 3x^9y^3 - 3$.

Solution

a. $3x^5 + 4x^2 + 7$ is a trinomial of degree 5, because the largest degree of the three monomials is 5.

b. $7x^2y^8 - 3xy$ is a binomial of degree 10.

c. $3x + 2y - xy$ is a trinomial of degree 2.

d. $18x^2y^3 - 12x^7y^2 + 3x^9y^3 - 3$ is a polynomial of degree 12.

SELF CHECK 2 Find the degree of $7a^3b^2 - 14a^2b^4$.

If the terms of a polynomial in one variable are written so that the exponents decrease as we move from left to right, we say that the terms are written with their exponents in **descending order.** If the terms are written so that the exponents increase as we move from left to right, we say that the terms are written with their exponents in **ascending order.**

EXAMPLE 3 Write the terms of $7x^2 - 5x^4 + 3x + 2x^3 - 1$ in: **a.** descending order **b.** ascending order.

Solution

a. $-5x^4 + 2x^3 + 7x^2 + 3x - 1$

b. $-1 + 3x + 7x^2 + 2x^3 - 5x^4$

SELF CHECK 3 Write the terms of $3x^2 + 2x^4 - 5x - 3$ in descending order.

In the following polynomial, the exponents on x are in descending order, and the exponents on y are in ascending order.

$$7x^4 - 2x^3y + 4x^2y^2 - 8xy^3 + 12y^4$$

3 Evaluate a polynomial function.

A function of the form $y = P(x)$, where $P(x)$ is a polynomial, is called a **polynomial function.** Some examples are

$$P(x) = -5x - 10 \quad \text{and} \quad Q(t) = 2t^2 + 5t - 2$$

Read $P(x)$ as "P of x" and $Q(t)$ as "Q of t."

Polynomial Functions A **polynomial function in one variable (say, x)** is defined by an equation of the form $y = f(x) = P(x)$, where $P(x)$ is a polynomial in the variable x.

The **degree of the polynomial function** $y = f(x) = P(x)$ is the same as the degree of the polynomial.

To evaluate a polynomial function at a specific value of its variable, we substitute the value of the variable and simplify. For example, to evaluate $P(x)$ at $x = 1$, we substitute 1 for x and simplify.

$$\begin{aligned} P(x) &= x^6 + 4x^5 - 3x^2 + x - 2 \\ P(1) &= (1)^6 + 4(1)^5 - 3(1)^2 + 1 - 2 \quad \text{Substitute 1 for } x. \\ &= 1 + 4 - 3 + 1 - 2 \\ &= 1 \end{aligned}$$

Thus, $P(1) = 1$.

EXAMPLE 4 Given the polynomial function $P(x) = 3x^2 - 2x + 7$, find:
a. $P(-2)$ **b.** $P(a)$ **c.** $P(-2t)$.

Solution **a.**

$$\begin{aligned} P(x) &= 3x^2 - 2x + 7 \\ P(-2) &= 3(-2)^2 - 2(-2) + 7 \\ &= 3(4) + 4 + 7 \\ &= 23 \end{aligned}$$

b.

$$\begin{aligned} P(x) &= 3x^2 - 2x + 7 \\ P(a) &= 3(a)^2 - 2(a) + 7 \\ &= 3a^2 - 2a + 7 \end{aligned}$$

c.

$$\begin{aligned} P(x) &= 3x^2 - 2x + 7 \\ P(-2t) &= 3(-2t)^2 - 2(-2t) + 7 \\ &= 12t^2 + 4t + 7 \end{aligned}$$

SELF CHECK 4 Find $P(2t)$.

To evaluate polynomials with more than one variable, we substitute values for the variables into the polynomial and simplify.

EXAMPLE 5 Evaluate $4x^2y - 5xy^3$ at $x = 3$ and $y = -2$.

Solution We substitute 3 for x and -2 for y and simplify.

$$\begin{aligned} 4x^2y - 5xy^3 &= 4(3)^2(-2) - 5(3)(-2)^3 \\ &= 4(9)(-2) - 5(3)(-8) \\ &= -72 + 120 \\ &= 48 \end{aligned}$$

SELF CHECK 5 Evaluate $3ab^2 - 2a^2b$ at $a = 2$ and $b = -3$.

EXAMPLE 6 **HEIGHT OF A ROCKET** If a toy rocket is launched straight up with an initial velocity of 128 feet per second, its height h (in feet) above the ground after t seconds is given by the polynomial function

$$P(t) = -16t^2 + 128t \quad \text{The height } h \text{ is the value } P(t).$$

Find the height of the rocket at **a.** 0 second **b.** 3 seconds **c.** 7.9 seconds.

Solution **a.** To find the height at 0 second, we find $P(0)$ by substituting 0 for t and simplifying.

$$P(t) = -16t^2 + 128t$$
$$P(0) = -16(0)^2 + 128(0)$$
$$= 0$$

At 0 second, the rocket is on the ground waiting to be launched.

b. To find the height at 3 seconds, we find $P(3)$ by substituting 3 for t and simplifying.

$$P(3) = -16(3)^2 + 128(3)$$
$$= -16(9) + 384$$
$$= -144 + 384$$
$$= 240$$

At 3 seconds, the height of the rocket is 240 feet.

c. To find the height at 7.9 seconds, we find $P(7.9)$ by substituting 7.9 for t and simplifying.

$$P(7.9) = -16(7.9)^2 + 128(7.9)$$
$$= -16(62.41) + 1{,}011.2$$
$$= -998.56 + 1{,}011.2$$
$$= 12.64$$

At 7.9 seconds, the height is 12.64 feet. It has fallen nearly back to Earth.

SELF CHECK 6 Find the height of the rocket at 4 seconds.

4 Graph a polynomial function.

In earlier sections we discussed three basic polynomial functions. The graph of the linear function $f(x) = 3x + 1$, the graph of the squaring function $f(x) = x^2$, and the graph of the cubing function $f(x) = x^3$ are shown in Figure 5-1.

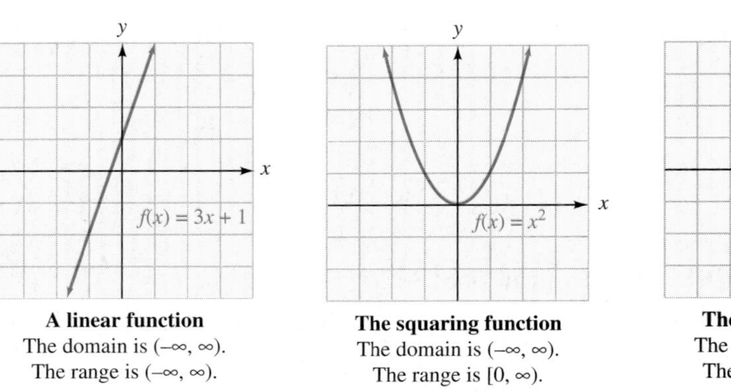

A linear function
The domain is $(-\infty, \infty)$.
The range is $(-\infty, \infty)$.
(a)

The squaring function
The domain is $(-\infty, \infty)$.
The range is $[0, \infty)$.
(b)

The cubing function
The domain is $(-\infty, \infty)$.
The range is $(-\infty, \infty)$.
(c)

Figure 5-1

We have seen that the graphs of many polynomial functions are translations or reflections of these graphs. For example, the graph of $f(x) = x^2 - 2$ is the graph of

Srinivasa Ramanujan (1887–1920)

Ramanujan was one of India's most prominent mathematicians. When in high school, he read *Synopsis of Elementary Results in Pure Mathematics* and from this book taught himself mathematics. He entered college but failed several times because he would study only mathematics. He went on to teach at Cambridge University in England.

$f(x) = x^2$ translated 2 units downward, as shown in Figure 5-2(a). The graph of $f(x) = -x^2$ is the graph of $f(x) = x^2$ reflected about the x-axis, as shown in Figure 5-2(b).

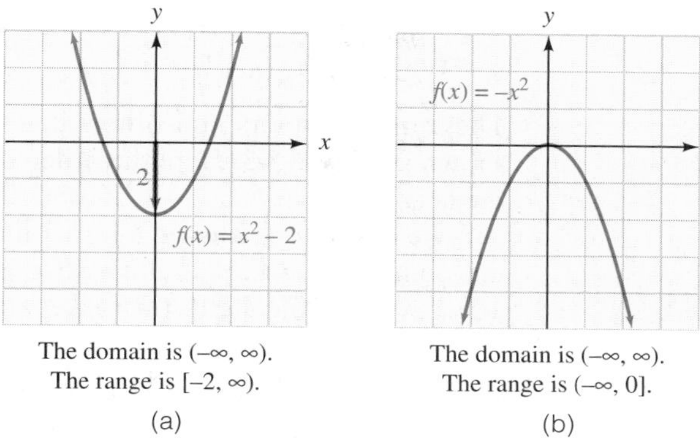

The domain is $(-\infty, \infty)$.
The range is $[-2, \infty)$.
(a)

The domain is $(-\infty, \infty)$.
The range is $(-\infty, 0]$.
(b)

Figure 5-2

In Example 6, we saw that a polynomial function can describe (or model) the height of a rocket. Since the height (h) of the rocket depends on time (t), we say that the height is a function of time, and we can write $h = f(t)$. To graph this function, we can make a table of values, plot the points, and join them with a smooth curve, as shown in Example 7.

EVERYDAY CONNECTIONS Stretching Rubber Bands

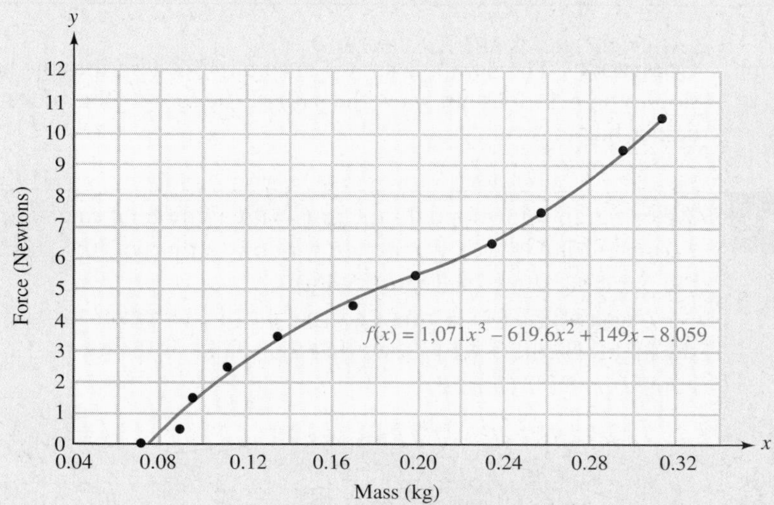

Source: http://www.madphysics.com/exp/hysteresis_and_rubber_bands.htm

The polynomial function shown above models the stretching force on a rubber band when a mass is attached to one end.

Use the graph of the function to answer the following questions.

1. What mass stretched the rubber band with a force of approximately 5.5 newtons?
2. What was the stretching force caused by a mass of 0.28 kg?
3. What is the predicted stretching force caused by a mass of 0.32 kg?

EXAMPLE 7 Graph $h = f(t) = -16t^2 + 128t$.

Solution From Example 6, we saw that:

When $t = 3$, then $h = 240$.
When $t = 0$, then $h = 0$.

These ordered pairs and others that satisfy $h = -16t^2 + 128t$ are given in the table shown in Figure 5-3. Here the ordered pairs are pairs (t, h), where t is the time and h is the height.

We plot these pairs on a horizontal t-axis and a vertical h-axis and join the resulting points to get the parabola shown in the figure. From the graph, we can see that 4 seconds into the flight, the rocket attains a maximum height of 256 feet.

$h = f(t) = -16t^2 + 128t$

t	$f(t)$	$(t, f(t))$
0	0	(0, 0)
1	112	(1, 112)
2	192	(2, 192)
3	240	(3, 240)
4	256	(4, 256)
5	240	(5, 240)
6	192	(6, 192)
7	112	(7, 112)
8	0	(8, 0)

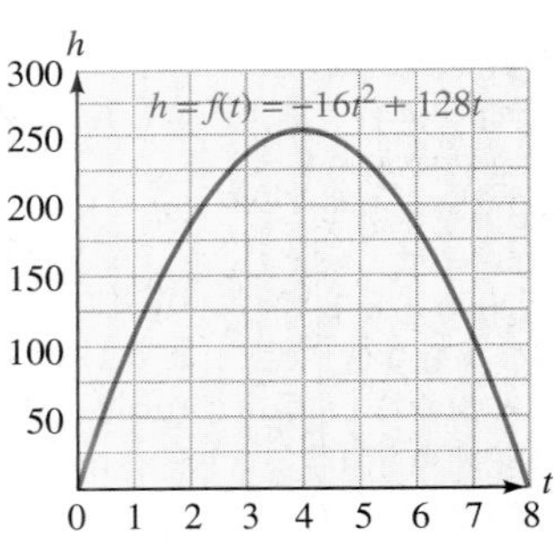

Figure 5-3

SELF CHECK 7 Use Figure 5-3 to estimate the height of the rocket at 6.5 seconds.

COMMENT The parabola shown in Figure 5-3 describes the height of the rocket in relation to time. It does not show the path of the rocket. The rocket goes straight up and then comes back down.

ACCENT ON TECHNOLOGY

Graphing Polynomial Functions

We can graph polynomial functions with a graphing calculator. For example, to graph $P(t) = -16t^2 + 128t$, we can use window settings of [0, 8] for x and [0, 260] for y to get the parabola shown in Figure 5-4(a).

We can trace to estimate the height of the rocket for any number of seconds into the flight. Figure 5-4(b) shows that the height of the rocket 1.6 seconds into the flight is approximately 165 feet.

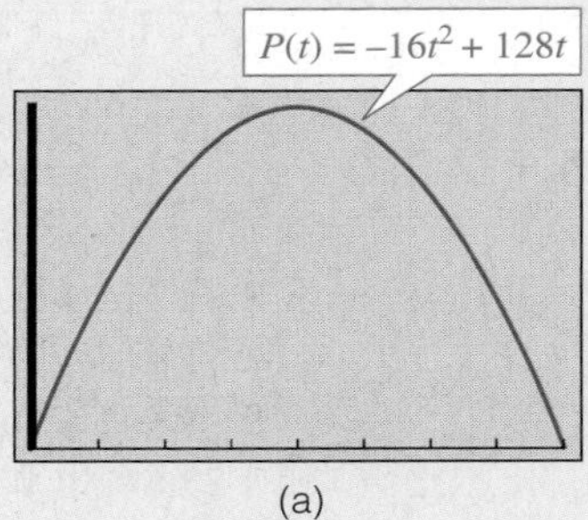

(a)

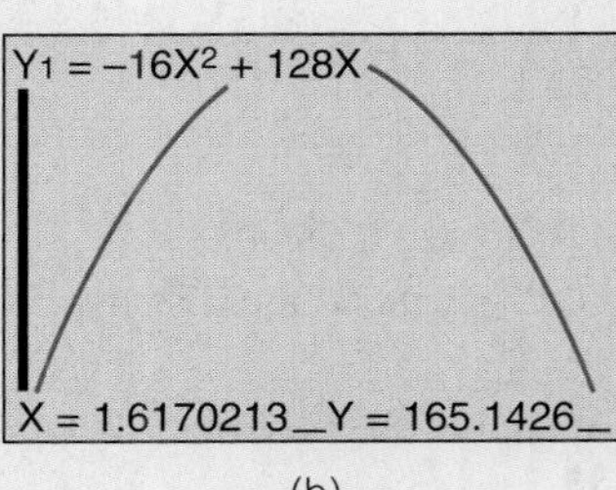

(b)

Figure 5-4

SELF CHECK ANSWERS **1. a.** 5 **b.** 5 **2.** 6 **3.** $2x^4 + 3x^2 - 5x - 3$ **4.** $12t^2 - 4t + 7$ **5.** 78 **6.** 256 ft **7.** 150 ft

NOW TRY THIS

1. Graph the polynomial function $f(x) = x^3 - 4x$ and find the x-intercepts.

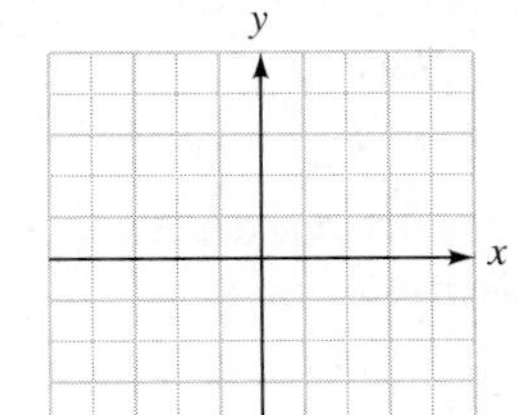

2. Given $P(x) = 5x^2 + 8x - 4$, find $P(-2x)$.

3. Given $P(x, y) = 3x^2y - 4xy^2$, find $P(-1, 5)$.

5.1 EXERCISES

WARM-UPS

Give the degree of each polynomial. Assume no variable is zero.

1. $8x^3$ **2.** $-4x^3y$

3. $4x^2y + 2x$ **4.** $3x + 4xyz$

If $P(x) = 2x + 1$, find each value.

5. $P(0)$ **6.** $P(2)$

7. $P(-1)$ **8.** $P(-2)$

REVIEW *Write each expression with a single exponent.*

9. a^3a^2 **10.** $\frac{b^3b^3}{b^4}$

11. $\frac{3(y^3)^{10}}{y^3y^4}$ **12.** $\frac{4x^{-4}x^5}{2x^{-6}}$

13. Astronomy The distance from Mars to the Sun is about 114,000,000 miles. Express this number in scientific notation.

14. Science One angstrom is about 0.0000001 millimeter. Express this number in scientific notation.

VOCABULARY AND CONCEPTS *Fill in the blanks.*

15. A polynomial is the ____ of one or more algebraic terms whose variables have ______ number exponents.

16. A monomial is a polynomial with ____ term.

17. A ________ is a polynomial with two terms.

18. A ________ is a polynomial with three terms.

19. The polynomial $3x^5 + 3x^3 - x^2 - 7x$ is written in ____________ of powers of x.

20. The polynomial $-2 + 3y^2 + y^3 + 8y^4$ is written in ____________ of powers of y.

21. The equation $y = P(x)$, where $P(x)$ is a polynomial, defines a function, because each input value x determines ____ output value y.

22. The degree of the polynomial function $y = P(x)$ is the same as the degree of the __________.

GUIDED PRACTICE

Classify each polynomial as a monomial, a binomial, a trinomial, or not a polynomial. (Objective 1)

23. $3x^2$ **24.** $2y^3 + 4y^2$

25. $3x^2y - 2x + 3y$ **26.** $a^2 + b^2$

27. $x^2 - y^2$ **28.** $\frac{17}{2}x^3 + 3x^{-2} - x - 4$

29. 5 **30.** $8x^3y^5$

Find the degree of each polynomial. See Examples 1–2. (Objective 2)

31. $-5x^7$ **32.** $7x$

33. 121 **34.** $3^2x^2y^3$

35. $3x^2 + 2$ **36.** x^{17}

37. $4x^8 + 3x^2y^4$ **38.** $19x^2y^4 - y^{10}$

39. $4x^2 - 5y^3z^3t^4$ **40.** $x^2y^3z^4 + z^{12}$

Write each polynomial with the exponents on x in descending order and y in ascending order. See Example 3. (Objective 2)

41. $3x - 2x^4 + 7 - 5x^2$

42. $-x^2 + 3x^5 - 7x + 3x^3$

43. $4y^2 - 2y^5 + 7y - 5y^3$

44. $y^3 + 3y^2 + 8y^4 - 2$

Consider the polynomial function $P(x) = 2x^2 + x + 2$ and find each value. See Example 4. (Objective 3)

45. $P(0)$
46. $P(1)$
47. $P(-2)$
48. $P(-3)$
49. $P(3v)$
50. $P(2t)$
51. $P(-3x)$
52. $P(-2x)$

Find each value when $x = 2$ and $y = -3$. See Example 5. (Objective 3)

53. $x^2 + y^2$
54. $x^3 + y^3$
55. $x^3 - y^3$
56. $x^2 - y^2$
57. $3x^2y + xy^3$
58. $8xy - xy^2$
59. $-2xy^2 + x^2y$
60. $-x^3y - x^2y^2$

The height h, in feet, of a ball shot straight up with an initial velocity of 64 feet per second is given by the polynomial function $h = f(t) = -16t^2 + 64t$. Find the height of the ball after the given number of seconds. See Example 6. (Objective 3)

61. 0 second
62. 1 second
63. 2 seconds
64. 4 seconds

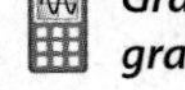
Graph each polynomial function. Check your work with a graphing calculator. See Example 7. (Objective 4)

65. $f(x) = x^2 + 2$

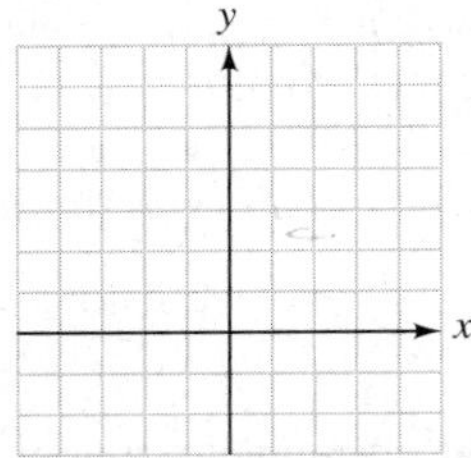

66. $f(x) = x^3 - 2$

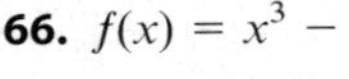
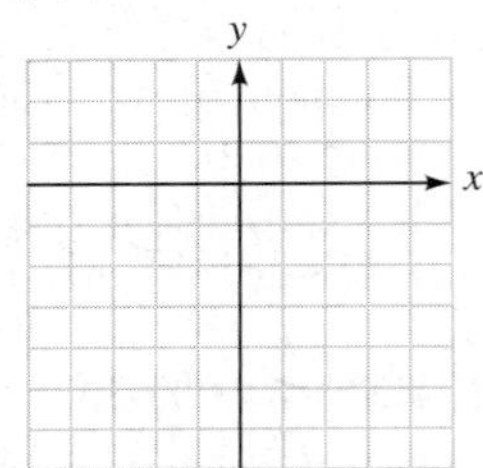

67. $f(x) = -x^3$

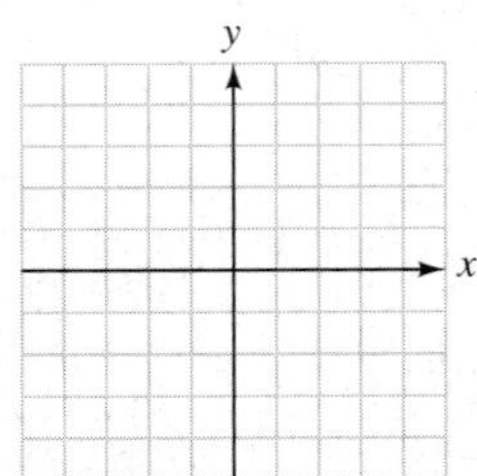

68. $f(x) = -x^2 + 1$

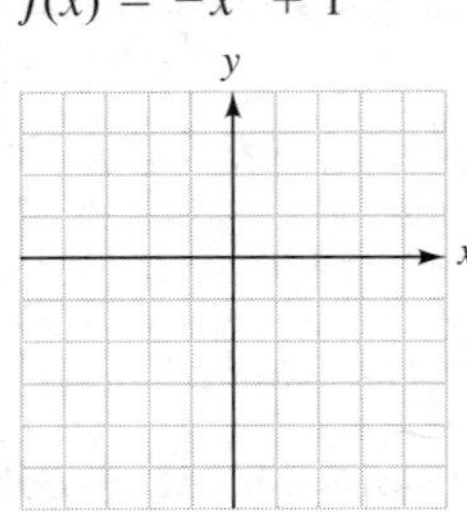

69. $f(x) = -x^3 + x$

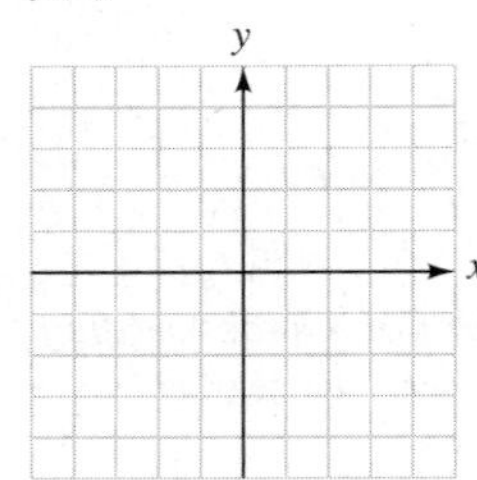

70. $f(x) = x^3 - x$

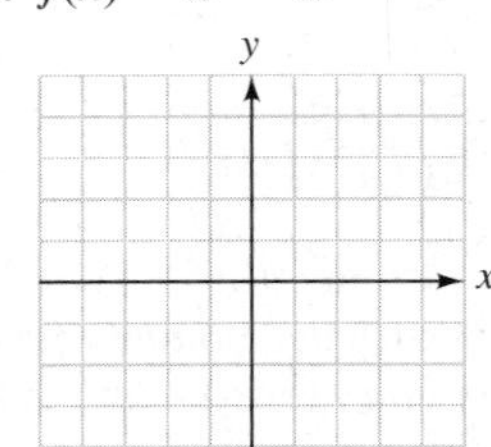

71. $f(x) = x^2 - 2x + 1$

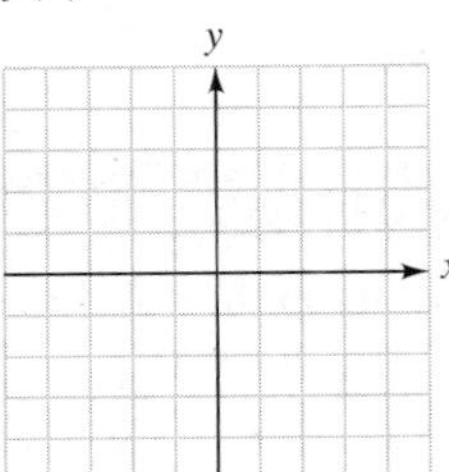

72. $f(x) = x^2 - 2x - 3$

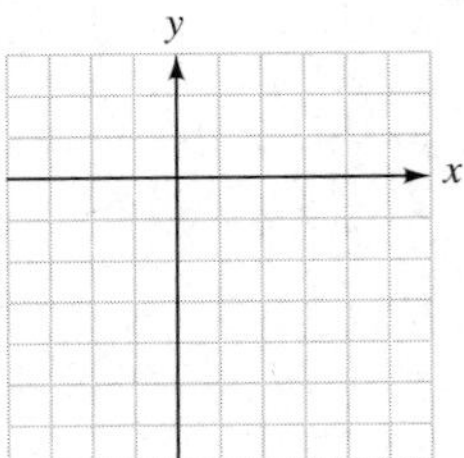

ADDITIONAL PRACTICE

Write each polynomial with the exponents on x in descending order or y in ascending order. Then give the degree of each polynomial.

73. $a^2x - ax^3 + 7a^3x^5 - 5a^3x^2$

74. $4x^2y^7 - 3x^5y^2 + 4x^3y^3 - 2x^4y^6 + 5x^6$

75. $5x^3y^6 + 2x^4y - 5x^3y^3 + x^5y^7 - 2y^4$

76. $-x^3y^2 + x^2y^3 - 2x^3y + x^7y^6 - 3x^6$

Use a calculator to find each value when $x = 3.7$, $y = -2.5$, and $z = 8.9$.

77. x^2y
78. xyz^2
79. $\frac{x^2}{z^2}$
80. $\frac{z^3}{y^2}$
81. $\frac{x + y + z}{xyz}$
82. $\frac{x + yz}{xy + z}$

Use a graphing calculator to graph each polynomial function. Use window settings of [−4, 6] for x and [−5, 5] for y.

83. $f(x) = 2.75x^2 - 4.7x + 1.5$

84. $f(x) = -2.5x^2 + 1.7x + 3.2$

85. $f(x) = 0.25x^2 - 0.5x - 2.5$

86. $f(x) = 0.37x^2 - 1.4x - 1.5$

APPLICATIONS

87. Juggling During a performance, a juggler tosses one ball straight upward while continuing to juggle three others. The height $f(t)$, in feet, of the ball is given by the polynomial function $f(t) = -16t^2 + 32t + 4$, where t is the time in seconds since the ball was thrown. Find the height of the ball 1 second after it is tossed upward.

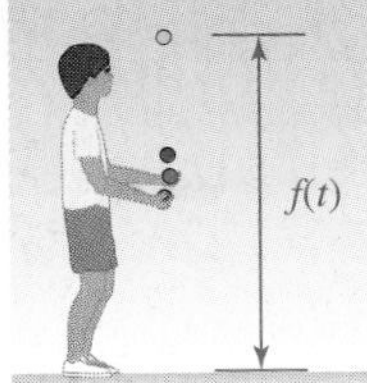

88. Juggling Find the height of the ball discussed in Exercise 87, 2 seconds after it is tossed upward.

Braking distance *The number of feet that a car travels before stopping depends on the driver's reaction time and the braking distance. See the illustration. For one driver, the stopping distance d is given by the polynomial function $d(v) = 0.04v^2 + 0.9v$ where v is the speed of the car. Find the stopping distance for each of the following speeds.*

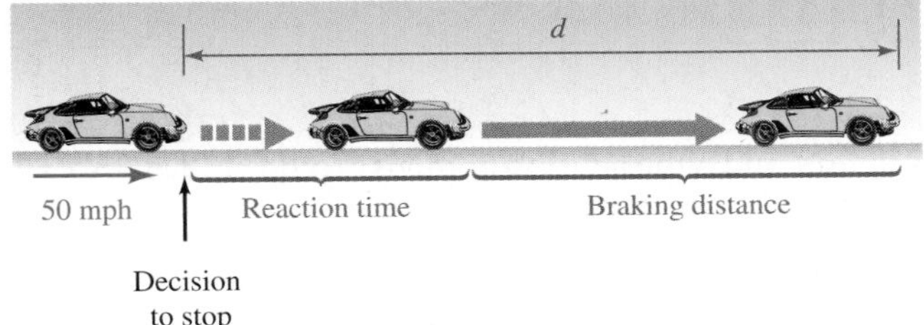

89. 30 mph

90. 50 mph

91. 60 mph

92. 70 mph

Making gutters *A rectangular sheet of metal will be used to make a rain gutter by bending up its sides, as shown in the illustration. Since a cross section is a rectangle, the cross-sectional area is the product of its length and width, and the capacity c of the gutter is a polynomial function of x: $f(x) = -2x^2 + 12x$. Find the capacity for each value of x.*

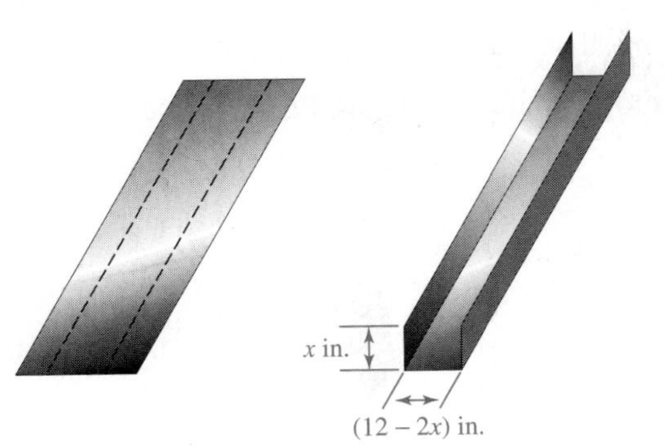

93. 1 inch

94. 2 inches

95. 3 inches

96. 4 inches

97. Roller coasters The polynomial function $f(x) = 0.001x^3 - 0.12x^2 + 3.6x + 10$ models the path of a portion of the track of a roller coaster. Find the height of the track for $x = 0, 20, 40$, and 60 meters.

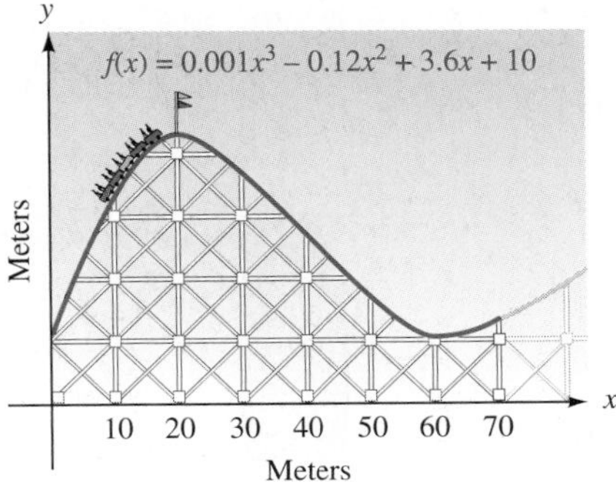

98. Customer service A technical help service of a computer software company has found that on Mondays, the polynomial function $C(t) = -0.0625t^4 + t^3 + 16t$ approximates the number of calls received at any one time. If t represents the time (in hours) since the service opened at 8:00 A.M., how many calls are expected at noon?

WRITING ABOUT MATH

99. Explain how to find the degree of a polynomial.

100. Explain why $P(x) = x^6 + 4x^5 - 3x^2 + x - 2$ defines a function.

SOMETHING TO THINK ABOUT

101. If $P(x) = x^2 - 5x$, is $P(2) + P(3) = P(2 + 3)$?

102. If $P(x) = x^3 - 3x$, is $P(2) - P(3) = P(2 - 3)$?

103. If $P(x) = x^2 - 2x - 3$, find $P(P(0))$.

104. If $P(x) = 2x^2 - x - 5$, find $P(P(-1))$.

105. Graph $f(x) = x^2$, $f(x) = 2x^2$, and $f(x) = 4x^2$. What do you discover?

106. Graph $f(x) = x^2$, $f(x) = \frac{1}{2}x^2$, and $f(x) = \frac{1}{4}x^2$. What do you discover?

SECTION 5.2 Adding and Subtracting Polynomials

Objectives

1. Combine like terms.
2. Add two or more polynomials.
3. Subtract two polynomials.

Getting Ready

Use the distributive property to remove parentheses.

1. $2(x + 3)$ **2.** $-4(2x - 5)$

3. $\frac{5}{2}(2x + 6)$ **4.** $-\frac{2}{3}(3x - 9)$

In this section, we will see how to add and subtract polynomials. To do so, we first will learn how to combine like terms.

1 Combine like terms.

Recall that when terms have the same variables with the same exponents, they are called *like* or *similar terms.*

- $3x^2$, $5x^2$, and $7x^2$ are like terms, because they have the same variables with the same exponents.
- $5x^3y^2$, $17x^3y^2$, and $103x^3y^2$ are like terms, because they have the same variables with the same exponents.
- $4x^4y^2$, $12xy^5$, and $98x^7y^9$ are unlike terms. They have the same variables, but with different exponents.
- $3x^4y$ and $5x^4z^2$ are unlike terms. They have different variables.

The distributive property enables us to combine like terms. For example,

$$3x + 7x = (3 + 7)x \quad \text{Use the distributive property.}$$
$$= 10x \quad 3 + 7 = 10$$
$$5x^2y^3 + 22x^2y^3 = (5 + 22)x^2y^3 \quad \text{Use the distributive property.}$$
$$= 27x^2y^3 \quad 5 + 22 = 27$$
$$9xy^4 + 6xy^4 + xy^4 = 9xy^4 + 6xy^4 + 1xy^4 \quad xy^4 = 1xy^4$$
$$= (9 + 6 + 1)xy^4 \quad \text{Use the distributive property.}$$
$$= 16xy^4 \quad 9 + 6 + 1 = 16$$

The results of the previous example suggest that to add like terms, *we add their numerical coefficients and keep the same variables with the same exponents.*

Hypatia
(370 A.D.–415 A.D.)
Hypatia is the earliest known woman in the history of mathematics. She was a professor at the University of Alexandria. Because of her scientific beliefs, she was considered to be a heretic. At the age of 45, she was attacked by a mob and murdered for her beliefs.

COMMENT The terms in the following binomials cannot be combined, because they are not like terms.

$$3x^2 - 5y^2, \qquad -2a^2 + 3a^3, \qquad \text{and} \qquad 5y^2 + 17xy$$

2 Add two or more polynomials.

To add polynomials, we use the distributive property to remove parentheses and combine like terms, whenever possible.

EXAMPLE 1 Add $3x^2 - 2x + 4$ and $2x^2 + 4x - 3$.

Solution

$(3x^2 - 2x + 4) + (2x^2 + 4x - 3)$

$= 1(3x^2 - 2x + 4) + 1(2x^2 + 4x - 3)$ — Each polynomial has an understood coefficient of 1.

$= 3x^2 - 2x + 4 + 2x^2 + 4x - 3$ — Use the distributive property to remove parentheses.

$= 3x^2 + 2x^2 - 2x + 4x + 4 - 3$ — Use the commutative property of addition to get the terms involving x^2 together and the terms involving x together.

$= 5x^2 + 2x + 1$ — Combine like terms.

SELF CHECK 1 Add $2a^2 - 3a + 5$ and $5a^2 - 4a - 2$.

EXAMPLE 2 Add $5x^3y^2 + 4x^2y^3$ and $-2x^3y^2 + 5x^2y^3$.

Solution

$$\begin{aligned}
&(5x^3y^2 + 4x^2y^3) + (-2x^3y^2 + 5x^2y^3)\\
&= 1(5x^3y^2 + 4x^2y^3) + 1(-2x^3y^2 + 5x^2y^3)\\
&= 5x^3y^2 + 4x^2y^3 - 2x^3y^2 + 5x^2y^3\\
&= 5x^3y^2 - 2x^3y^2 + 4x^2y^3 + 5x^2y^3\\
&= 3x^3y^2 + 9x^2y^3
\end{aligned}$$

SELF CHECK 2 Add $6a^2b^3 + 5a^3b^2$ and $-3a^2b^3 - 2a^3b^2$.

The additions in Examples 1 and 2 also can be done by aligning the terms vertically.

$$\begin{array}{r} 3x^2 - 2x + 4\\ \underline{2x^2 + 4x - 3}\\ 5x^2 + 2x + 1 \end{array} \qquad \begin{array}{r} 5x^3y^2 + 4x^2y^3\\ \underline{-2x^3y^2 + 5x^2y^3}\\ 3x^3y^2 + 9x^2y^3 \end{array}$$

3 Subtract two polynomials.

To subtract one monomial from another, we use the rule $a - b = a + (-b)$. In words, we add the negative (or opposite) of the monomial that is to be subtracted.

EXAMPLE 3 Perform each subtraction.

a. $8x^2 - 3x^2 = 8x^2 + (-3x^2)$
$= 5x^2$

b. $3x^2y - 9x^2y = 3x^2y + (-9x^2y)$
$= -6x^2y$

c. $-5x^5y^3z^2 - 3x^5y^3z^2 = -5x^5y^3z^2 + (-3x^5y^3z^2)$
$= -8x^5y^3z^2$

SELF CHECK 3 Subtract: **a.** $-2a^2b^3 - 5a^2b^3$ **b.** $-2a^2b^3 - (-5a^2b^3)$.

To subtract polynomials, we use the distributive property to remove parentheses and combine like terms, whenever possible.

EXAMPLE 4 Perform each subtraction.

a. $(8x^3y + 2x^2y) - (2x^3y - 3x^2y)$

$= \mathbf{1}(8x^3y + 2x^2y) - \mathbf{1}(2x^3y - 3x^2y)$ Insert the understood coefficients of 1.

$= 8x^3y + 2x^2y - 2x^3y + 3x^2y$ Use the distributive property to remove parentheses.

$= 6x^3y + 5x^2y$ Combine like terms.

b. $(3rt^2 + 4r^2t^2) - (8rt^2 - 4r^2t^2 + r^3t^2)$

$= \mathbf{1}(3rt^2 + 4r^2t^2) - \mathbf{1}(8rt^2 - 4r^2t^2 + r^3t^2)$ Insert the understood coefficients of 1.

$= 3rt^2 + 4r^2t^2 - 8rt^2 + 4r^2t^2 - r^3t^2$ Use the distributive property to remove parentheses.

$= -5rt^2 + 8r^2t^2 - r^3t^2$ Combine like terms.

SELF CHECK 4 Subtract: $(6a^2b^3 - 2a^2b^2) - (-2a^2b^3 + a^2b^2)$.

To subtract polynomials in vertical form, we add the negative (or opposite) of the polynomial that is being subtracted.

$$- \begin{array}{r} 8x^3y + 2x^2y \\ \underline{2x^3y - 3x^2y} \end{array} \Rightarrow + \begin{array}{r} 8x^3y + 2x^2y \\ \underline{-2x^3y + 3x^2y} \\ 6x^3y + 5x^2y \end{array}$$

To add or subtract multiples of one polynomial and another, we use the distributive property to remove parentheses and then combine like terms.

EXAMPLE 5 Simplify: $3(2x^2 + 4x - 7) - 2(3x^2 - 4x - 5)$.

Solution

$3(2x^2 + 4x - 7) - 2(3x^2 - 4x - 5)$

$= 6x^2 + 12x - 21 - 6x^2 + 8x + 10$ Remove parentheses.

$= 20x - 11$ Combine like terms.

SELF CHECK 5 Simplify: $-2(3a^3 + a^2) - 5(2a^3 - a^2 + a)$.

SELF CHECK ANSWERS 1. $7a^2 - 7a + 3$ 2. $3a^2b^3 + 3a^3b^2$ 3. a. $-7a^2b^3$ b. $3a^2b^3$ 4. $8a^2b^3 - 3a^2b^2$ 5. $-16a^3 + 3a^2 - 5a$

NOW TRY THIS

1. $x^{2n-1}x^{3n-2}$
2. $\frac{x^{2n-1}}{x^{3n-2}}$

5.2 EXERCISES

WARM-UPS

Combine like terms.

1. $4x^2 + 5x^2$
2. $3y^2 - 5y^2$

Perform the operations.

3. $(x^2 + 2x + 1) + (2x^2 - 2x + 1)$
4. $(x^2 + 2x + 1) - (2x^2 - 2x + 1)$
5. $(2x^2 - x - 3) + (x^2 - 3x - 1)$
6. $(2x^2 - x - 3) - (x^2 - 3x - 1)$

REVIEW *Solve each inequality. Give the result in interval notation.*

7. $2x + 3 \le 11$
8. $\frac{2}{3}x + 5 > 11$
9. $|x - 4| < 5$
10. $|2x + 1| \ge 7$

VOCABULARY AND CONCEPTS

Fill in the blanks.

11. If two algebraic terms have the same variables each with the same __________, they are called like terms.
12. $-6a^3b^2$ and $7a^2b^3$ are ______ terms.
13. To add two like monomials, we add their numerical __________ and keep the same variables with the same exponents.
14. To subtract one like monomial from another, we add the __________________ of the monomial that is to be subtracted.

Determine whether the terms are like or unlike terms. If they are like terms, combine them.

15. $3x, 7x$
16. $-8x, 3y$
17. $7x, 7y$
18. $3mn, 5mn$
19. $3r^2t^3, -8r^2t^3$
20. $9u^2v, 10u^2v$
21. $9x^2y^3, 3x^2y^2$
22. $27x^6y^4z, 9x^6y^4z^2$

GUIDED PRACTICE

Simplify each expression. (Objective 1)

23. $8x + 4x$
24. $-2y + 16y$
25. $5x^3y^2z - 3x^3y^2z$
26. $8wxy - 12wxy$
27. $-2x^2y^3 + 3xy^4 - 5x^2y^3$
28. $3ab^4 - 4a^2b^2 - 2ab^4 + 2a^2b^2$
29. $(3x^2y)^2 + 2x^4y^2 - x^4y^2$
30. $(5x^2y^4)^3 - (5x^3y^6)^2$

Perform each operation. See Examples 1–2. (Objective 2)

31. $(2a + 3) + (3a + 5)$
32. $(5y - 2) + (-2y + 3)$
33. $(3x^2 + 2x + 1) + (-2x^2 - 7x + 5)$
34. $(-2a^2 - 5a - 7) + (-3a^2 + 7a + 1)$
35. $(7y^3 + 4y^2 + y + 3) + (-8y^3 - y + 3)$
36. $(3x^2 + 4x - 3) + (2x^2 - 3x - 1) + (x^2 + x + 7)$
37. $(6a^2b + 1) + (4a^2b + 3)$
38. $(-3p^3q^2 - 2pq) + (7p^3q^2 - 3pq)$
39. $(-4a^2b^4 + 2a^2b^2) + (8a^2b^4 - 2a^2b^2)$
40. $(8m^3n^3 - 3mn) + (10m^3n^3 - 2mn)$

Perform each operation. See Examples 3–4. (Objective 3)

41. $(4t + 3) - (2t + 5)$
42. $(6z - 5) - (-2z + 1)$
43. $(-a^2 + 2a + 3) - (4a^2 - 2a - 1)$
44. $(x^2 - 3x + 8) - (3x^2 + x + 3)$

45. $(7pq^2 + 3pq) - (2pq^2 - 2pq)$

46. $(-2a^2b^2 - 2ab) - (5a^2b^2 + 3ab)$

47. $(10xy - 3x^2y^2) - (-3xy - 5x^2y^2)$

48. $(5c^2d + 3cd) - (7c^2d + 5cd)$

Simplify each expression. **See Example 5. (Objectives 2–3)**

49. $3(x + 2) + 2(x - 5)$

50. $-2(x - 4) + 5(x + 1)$

51. $-6(t - 4) - 5(t - 1)$

52. $4(a + 5) - 3(a - 1)$

53. $2(x^3 + x^2) + 3(2x^3 - x^2)$

54. $-3(4x^3 - 2x^2 + 4) - 4(3x^3 + 4x^2 + 3x) + 5(3x - 4)$

55. $5(2a^2 + 4a - 2) - 2(-3a^2 - a + 12) - 2(a^2 + 3a - 5)$

56. $-2(2b^2 - 3b + 3) + 3(3b^2 + 2b - 8) - (3b^2 - b + 4)$

ADDITIONAL PRACTICE

Simplify each expression.

57. $(3x^3 - 2x + 3) + (4x^3 + 3x^2 - 2) + (-4x^3 - 3x^2 + x + 12)$

58. $(x^4 - 3x^2 + 4) + (-2x^4 - x^3 + 3x^2) + (3x^2 + 2x + 1)$

59. $(6x^3 + 3x - 2) - (2x^3 + 3x^2 + 5)$

60. $(-2x^2 + 6x + 5) - (-4x^2 - 7x + 2) - (4x^2 + 10x + 5)$

61. $(3y^2 - 2y + 4) + [(2y^2 - 3y + 2) - (y^2 + 4y + 3)]$

62. $(-t^2 - t - 1) - [(t^2 + 3t - 1) - (-2t^2 + 4)]$

63. $3(y^2 + 2y) - 4(y^2 - 4)$

64. $-3(2m - n) + 2(m - 3n)$

65. $5(p - 2q) - 4(2p + q)$

66. $-5(2x^3 + 7x^2 + 4x) - 2(3x^3 - 4x^2 - 4x)$

67. $-3(3a^2 + 4b^3 + 7) + 4(5a^2 - 2b^3 + 3)$

68. $4(3z^2 - 4z + 5) + 6(-2z^2 - 3z + 4) - 2(4z^2 + 3z - 5)$

Add the polynomials.

69.
$$\begin{array}{r} 3x^3 - 2x^2 + 4x - 3 \\ -2x^3 + 3x^2 + 3x - 2 \\ \underline{5x^3 - 7x^2 + 7x - 12} \end{array}$$

70.
$$\begin{array}{r} 7a^3 \qquad\quad + 3a + 7 \\ -2a^3 + 4a^2 \qquad\quad - 13 \\ \underline{3a^3 - 3a^2 + 4a + 5} \end{array}$$

71.
$$\begin{array}{r} -2y^4 - 2y^3 + 4y^2 - 3y + 10 \\ -3y^4 + 7y^3 - y^2 + 14y - 3 \\ -3y^3 - 5y^2 - 5y + 7 \\ \underline{-4y^4 + y^3 - 13y^2 + 14y - 2} \end{array}$$

72.
$$\begin{array}{r} 17t^4 + 3t^3 - 2t^2 - 3t + 4 \\ -12t^4 - 2t^3 + 3t^2 - 5t - 17 \\ -2t^4 - 7t^3 + 4t^2 + 12t - 5 \\ \underline{5t^4 + t^3 + 5t^2 - 13t + 12} \end{array}$$

Subtract the bottom polynomial from the top polynomial.

73.
$$\begin{array}{r} 3x^2 - 4x + 17 \\ \underline{2x^2 + 4x - 5} \end{array}$$

74.
$$\begin{array}{r} -2y^2 - 4y + 3 \\ \underline{3y^2 + 10y - 5} \end{array}$$

75.
$$\begin{array}{r} -5y^3 + 4y^2 - 11y + 3 \\ \underline{-2y^3 - 14y^2 + 17y - 32} \end{array}$$

76.
$$\begin{array}{r} 17x^4 - 3x^2 - 65x - 12 \\ \underline{23x^4 + 14x^2 + 3x - 23} \end{array}$$

APPLICATIONS

77. Geometry Write a polynomial that represents the perimeter of the triangle.

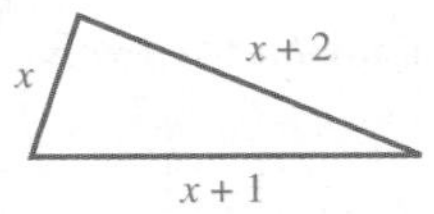

78. Geometry Write a polynomial that represents the perimeter of the rectangle.

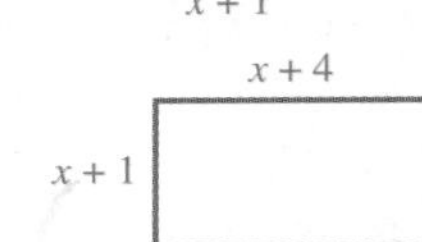

79. Carpentry The wooden plank is $(2x^2 + 5x + 1)$ meters long. If the shorter end is cut off, write a polynomial to express the length of the remaining piece.

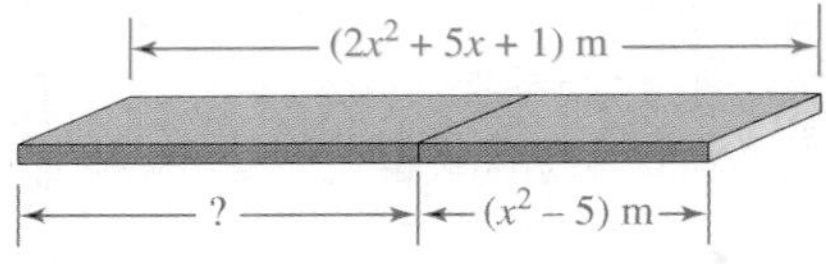

80. Gardening Write a polynomial that expresses how long the hose will be if the shorter pieces are joined together.

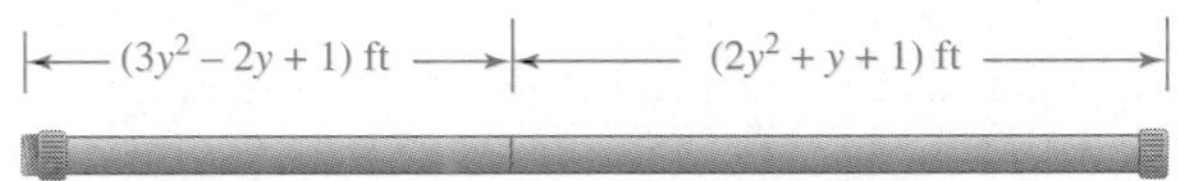

Consider the following information. If a house is purchased for $125,000 and is expected to appreciate $1,100 per year, its value y after x years is given by the polynomial function **$f(x) = 1{,}100x + 125{,}000$.**

81. Value of a house Find the expected value of the house in 10 years.

82. Value of a house A second house is purchased for \$150,000 and is expected to appreciate \$1,400 per year.

a. Find a polynomial function that will give the value y of the house in x years.

b. Find the value of the second house after 12 years.

83. Value of two houses Find one polynomial function that will give the combined value y of both houses after x years.

84. Value of two houses Find the value of the two houses after 20 years by

a. substituting 20 into the polynomial functions

$$f(x) = 1{,}100x + 125{,}000 \text{ and}$$
$$f(x) = 1{,}400x + 150{,}000$$

and adding.

b. substituting 20 into the result of Exercise 83.

Consider the following information. A business bought two cars, one for \$16,600 and the other for \$19,200. The first car is expected to depreciate \$2,100 per year and the second \$2,700 per year.

85. Value of a car Find a polynomial function that will give the value of the first car after x years.

86. Value of a car Find a polynomial function that will give the value of the second car after x years.

87. Value of two cars Find one polynomial function that will give the value of both cars after x years.

88. Value of two cars In two ways, find the total value of the cars after 3 years.

WRITING ABOUT MATH

89. Explain why the terms x^2y and xy^2 are not like terms.

90. Explain how to recognize like terms, and how to add them.

SOMETHING TO THINK ABOUT

91. Find the difference when $3x^2 + 4x - 3$ is subtracted from the sum of $-2x^2 - x + 7$ and $5x^2 + 3x - 1$.

92. Find the difference when $8x^3 + 2x^2 - 1$ is subtracted from the sum of $x^2 + x + 2$ and $2x^3 - x + 9$.

93. Find the sum when $2x^2 - 4x + 3$ minus $8x^2 + 5x - 3$ is added to $-2x^2 + 7x - 4$.

94. Find the sum when $7x^3 - 4x$ minus $x^2 + 2$ is added to $5 + 3x$.

SECTION 5.3 Multiplying Polynomials

Objectives

1. Multiply two monomials.
2. Multiply a polynomial by a monomial.
3. Multiply a polynomial by a polynomial.
4. Multiply two binomials using the FOIL method.
5. Square a binomial and multiply two conjugate binomials (special products).
6. Solve an application problem requiring multiplication of polynomials.
7. Multiply two expressions that are not polynomials.

Vocabulary

conjugate binomials   difference of two squares

Getting Ready

Simplify.

1. $(3a)(4)$
2. $4aaa(a)$
3. $-7a^3 \cdot a$
4. $12a^3a^2$

Use the distributive property to remove parentheses.

5. $2(a + 4)$
6. $a(a - 3)$
7. $-4(a - 3)$
8. $-2b(b + 2)$

We now discuss how to multiply polynomials.

1 Multiply two monomials.

In an earlier section, we saw that to multiply one monomial by another, *we multiply the numerical factors (coefficients) and then multiply the variable factors.*

EXAMPLE 1 Perform each multiplication.
a. $(3x^2)(6x^3)$ **b.** $(-8x)(2y)(xy)$ **c.** $(2a^3b)(-7b^2c)(-12ac^4)$

Solution We can use the commutative and associative properties of multiplication to rearrange the factors and regroup the numbers.

a.
$$\begin{aligned}(3x^2)(6x^3) &= 3 \cdot x^2 \cdot 6 \cdot x^3 \\ &= (3 \cdot 6)(x^2 \cdot x^3) \\ &= 18x^5\end{aligned}$$

b.
$$\begin{aligned}(-8x)(2y)(xy) &= -8 \cdot x \cdot 2 \cdot y \cdot x \cdot y \\ &= (-8 \cdot 2) \cdot x \cdot x \cdot y \cdot y \\ &= -16x^2y^2\end{aligned}$$

c.
$$\begin{aligned}(2a^3b)(-7b^2c)(-12ac^4) &= 2 \cdot a^3 \cdot b \cdot (-7) \cdot b^2 \cdot c \cdot (-12) \cdot a \cdot c^4 \\ &= 2(-7)(-12) \cdot a^3 \cdot a \cdot b \cdot b^2 \cdot c \cdot c^4 \\ &= 168a^4b^3c^5\end{aligned}$$

SELF CHECK 1 Multiply: **a.** $(-2a^3)(4a^2)$ **b.** $(-5b^3)(-3a)(a^2b)$.

2 Multiply a polynomial by a monomial.

To multiply a polynomial by a monomial, *we use the distributive property and multiply each term of the polynomial by the monomial.*

EXAMPLE 2 Perform each multiplication. **a.** $3x^2(6xy + 3y^2)$ **b.** $5x^3y^2(xy^3 - 2x^2y)$
c. $-2ab^2(3bz - 2az + 4z^3)$

Solution We can use the distributive property to remove parentheses.

a.
$$\begin{aligned}3x^2(6xy + 3y^2) &= 3x^2 \cdot 6xy + 3x^2 \cdot 3y^2 \\ &= 18x^3y + 9x^2y^2\end{aligned}$$

b.
$$\begin{aligned}5x^3y^2(xy^3 - 2x^2y) &= 5x^3y^2 \cdot xy^3 - 5x^3y^2 \cdot 2x^2y \\ &= 5x^4y^5 - 10x^5y^3\end{aligned}$$

c. $-2ab^2(3bz - 2az + 4z^3) = -2ab^2 \cdot 3bz - (-2ab^2) \cdot 2az + (-2ab^2) \cdot 4z^3$
$= -6ab^3z + 4a^2b^2z - 8ab^2z^3$

SELF CHECK 2 Multiply: $-2a^2(a^2 - a + 3)$.

3 Multiply a polynomial by a polynomial.

To multiply a polynomial by a polynomial, we use the distributive property more than once.

EXAMPLE 3 Perform each multiplication. **a.** $(3x + 2)(4x + 9)$ **b.** $(2a - b)(3a^2 - 4ab + b^2)$

Solution We can use the distributive property twice to remove parentheses.

a. $(3x + 2)(4x + 9) = (3x + 2) \cdot 4x + (3x + 2) \cdot 9$
$= 12x^2 + 8x + 27x + 18$
$= 12x^2 + 35x + 18$ Combine like terms.

b. $(2a - b)(3a^2 - 4ab + b^2)$
$= (2a - b)3a^2 - (2a - b)4ab + (2a - b)b^2$
$= 6a^3 - 3a^2b - 8a^2b + 4ab^2 + 2ab^2 - b^3$
$= 6a^3 - 11a^2b + 6ab^2 - b^3$

SELF CHECK 3 Multiply: $(2a + b)(3a + 2ab - b)$.

The results of Example 3 suggest that to multiply one polynomial by another, *we multiply each term of one polynomial by each term of the other polynomial and combine like terms, when possible.*

In the next example, we organize the work done in Example 3 vertically.

EXAMPLE 4 Perform each multiplication.

a.

$$\begin{array}{rrrl} 3x & +2 & & \\ 4x & +9 & & \\ \hline 12x^2 & +8x & & \leftarrow 4x(3x+2) \\ & +27x & +18 & \leftarrow 9(3x+2) \\ \hline 12x^2 & +35x & +18 & \end{array}$$

b.

$$\begin{array}{rrrrl} 3a^2 & -\ 4ab & +b^2 & & \\ & 2a & -b & & \\ \hline 6a^3 & -\ 8a^2b & +2ab^2 & & \leftarrow 2a(3a^2-4ab+b^2) \\ & -\ 3a^2b & +4ab^2 & -b^3 & \leftarrow -b(3a^2-4ab+b^2) \\ \hline 6a^3 & -\ 11a^2b & +6ab^2 & -b^3 & \end{array}$$

SELF CHECK 4 Multiply:

$$\begin{array}{r} 3x^2+2x-5. \\ 2x+1 \\ \hline \end{array}$$

4 Multiply two binomials using the FOIL method.

When multiplying two binomials, the distributive property requires that each term of one binomial be multiplied by each term of the other binomial. This fact can be emphasized by drawing lines to show the indicated products. For example, to multiply $3x + 2$ and $x + 4$, we can write

First terms Last terms

$$(3x+2)(x+4) = 3x \cdot x + 3x \cdot 4 + 2 \cdot x + 2 \cdot 4$$

Inner terms

$$= 3x^2 + 12x + 2x + 8$$

Outer terms

$$= 3x^2 + 14x + 8 \quad \text{Combine like terms.}$$

We note that

- the product of the **F**irst terms is $3x^2$,
- the product of the **O**uter terms is $12x$,
- the product of the **I**nner terms is $2x$, and
- the product of the **L**ast terms is 8.

This scheme is called the **FOIL** method of multiplying two binomials. FOIL is an acronym for **F**irst terms, **O**uter terms, **I**nner terms, and **L**ast terms. Of course, the resulting terms of the products must be combined, if possible.

To multiply binomials by sight, we find the product of the first terms, then find the products of the outer terms and the inner terms and add them (when possible), and then find the product of the last terms.

EXAMPLE 5 Find each product.

a. $(2x-3)(3x+2) = 6x^2 - 5x - 6$

The middle term of $-5x$ in the result comes from combining the outer and inner products of $4x$ and $-9x$.

$$4x + (-9x) = -5x$$

b. $(3x + 1)(3x + 4) = 9x^2 + 15x + 4$

The middle term of $+15x$ in the result comes from combining the products $12x$ and $3x$.

$$12x + 3x = 15x$$

c. $(4x - y)(2x + 3y) = 8x^2 + 10xy - 3y^2$

The middle term of $+10xy$ in the result comes from combining the products $12xy$ and $-2xy$.

$$12xy - 2xy = 10xy$$

SELF CHECK 5 Multiply: $(3a + 4b)(2a - b)$.

5 Square a binomial and multiply two conjugate binomials (special products).

We can square a binomial by using it as a factor in a multiplication two times.

EXAMPLE 6 Find each square: **a.** $(x + y)^2$ **b.** $(x - y)^2$.

Solution We note that $(x + y)^2 = (x + y)(x + y)$ and find the product $(x + y)(x + y)$ by multiplying each term of the first binomial by each term of the second binomial. Then we can combine like terms.

a. $(x + y)^2 = (x + y)(x + y)$

$= x^2 + xy + xy + y^2$ Use the FOIL method.

$= x^2 + 2xy + y^2$ Combine like terms.

We see that the square of the binomial is the square of the first term, plus twice the product of the terms, plus the square of the last term. This product can be illustrated graphically as shown in Figure 5-5.

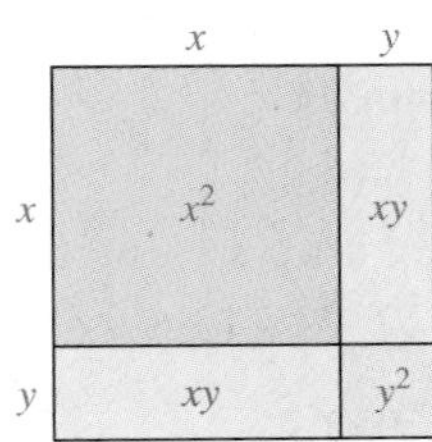

Figure 5-5

The area of the large square is the product of its length and width: $(x + y)(x + y) = (x + y)^2$.

The area of the large square is also the sum of its four pieces: $x^2 + xy + xy + y^2 = x^2 + 2xy + y^2$.

Thus, $(x + y)^2 = x^2 + 2xy + y^2$.

b. $(x - y)^2 = (x - y)(x - y)$

$= x^2 - xy - xy + y^2$ Use the FOIL method.

$= x^2 - 2xy + y^2$ Combine like terms.

We see that the square of the binomial is the square of the first term, minus twice the product of the terms, plus the square of the last term. For a geometric interpretation, see Exercise 125.

SELF CHECK 6 Find the squares: **a.** $(a + 2)^2$ **b.** $(a - 4)^2$.

Binomials with the same terms, but with opposite signs between their terms are called **conjugate binomials.** In the next example, we will show that the product of two conjugate binomials is also a binomial.

EXAMPLE 7 Multiply: $(x + y)(x - y)$.

Solution

$(x + y)(x - y) = x^2 - xy + xy - y^2$ Use the FOIL method.

$= x^2 - y^2$ Combine like terms.

From this example, we see that the product of the sum of two quantities and the difference of the same two quantities is the square of the first quantity minus the square of the second quantity. Such a product is called the **difference of two squares.** For a geometric interpretation, see Exercise 126.

SELF CHECK 7 Multiply: $(a + 3)(a - 3)$.

The products discussed in Examples 6 and 7 are called *special products.* Since they occur so often, it is useful to learn their forms.

Special Product Formulas

$(x + y)^2 = (x + y)(x + y) = x^2 + 2xy + y^2$

$(x - y)^2 = (x - y)(x - y) = x^2 - 2xy + y^2$

$(x + y)(x - y) = x^2 - y^2$

Because $x^2 + 2xy + y^2 = (x + y)^2$ and $x^2 - 2xy + y^2 = (x - y)^2$, the two trinomials are called *perfect square trinomials.*

COMMENT The squares $(x + y)^2$ and $(x - y)^2$ have trinomials for their products. Don't forget to write the middle terms in these products. Remember that

$$(x + y)^2 \neq x^2 + y^2 \quad \text{and that} \quad (x - y)^2 \neq x^2 - y^2$$

Also remember that the product $(x + y)(x - y)$ is the binomial $x^2 - y^2$.

At first, the expression $3[x^2 - 2(x + 3)]$ doesn't look like a polynomial. However, if we remove the parentheses and the brackets, it takes on the form of a polynomial.

$$\begin{aligned} 3[x^2 - 2(x + 3)] &= 3[x^2 - 2x - 6] \\ &= 3x^2 - 6x - 18 \end{aligned}$$

If an expression has one set of grouping symbols that is enclosed within another set, we always eliminate the inner set first.

EXAMPLE 8 Find the product of $-2[y^3 + 3(y^2 - 2)]$ and $5[y^2 - 2(y + 1)]$.

Solution We change each expression into polynomial form:

$$\begin{array}{ll} -2[y^3 + 3(y^2 - 2)] & 5[y^2 - 2(y + 1)] \\ -2(y^3 + 3y^2 - 6) & 5(y^2 - 2y - 2) \\ -2y^3 - 6y^2 + 12 & 5y^2 - 10y - 10 \end{array}$$

Then we do the multiplication:

$$\begin{array}{rrrrrr} & -2y^3 & -6y^2 & & +12 & \\ & 5y^2 & -10y & & -10 & \\ \hline -10y^5 & -30y^4 & & +60y^2 & & \\ & +20y^4 & +60y^3 & + & -120y & \\ & & +20y^3 & +60y^2 & & -120 \\ \hline -10y^5 & -10y^4 & +80y^3 & +120y^2 & -120y & -120 \end{array}$$

SELF CHECK 8 Find the product of $2[a^2 + 3(a - 2)]$ and $3[a^2 + 3(a - 1)]$.

EXAMPLE 9 Multiply: $(p + q)(2p - q)(p + 2q)$.

Solution First we find the product of $p + q$ and $2p - q$. Then we multiply that product by $p + 2q$.

$$\begin{aligned} (p + q)(2p - q)(p + 2q) &= (2p^2 - pq + 2pq - q^2)(p + 2q) \\ &= (2p^2 + pq - q^2)(p + 2q) && \text{Combine like terms.} \\ &= (2p^2 + pq - q^2)(p) + (2p^2 + pq - q^2)(2q) && \text{Distribute the multiplication by } 2p^2 + pq - q^2. \\ &= 2p^3 + p^2q - pq^2 + 4p^2q + 2pq^2 - 2q^3 && \text{Use the distributive property to remove parentheses.} \\ &= 2p^3 + 5p^2q + pq^2 - 2q^3 && \text{Combine like terms.} \end{aligned}$$

COMMENT Because of the associative property, we can find the product $(2p - q)(p + 2q)$ first and then multiply the result by $p + q$. We will get the same result.

SELF CHECK 9 Multiply: $(x + y)(x - y)(2x + 3y)$.

PERSPECTIVE

François Vieta (1540–1603) was one of the first to use a mathematical notation that is close to the notation we use today. Trained as a lawyer, Vieta served in the parliament of Brittany and as the personal lawyer of Henry of Navarre. If he had continued as a successful lawyer, Vieta might now be forgotten. However, he lost his job.

When political opposition forced him out of office in 1584, Vieta had time to devote himself entirely to his hobby, mathematics. He studied the work of earlier mathematicians and adapted and improved their ideas.

Vieta was the first to use letters to represent unknown numbers, but he did not use modern notation for exponents. To us, his notation seems awkward. For example, what we would write as

$$(x + 1)^3 = x^3 + 3x^2 + 3x + 1$$

Vieta would have written as

$$\overline{x + 1} \text{ cubus aequalis } x \text{ cubus} + x \text{ quad. } 3 + x \text{ in } 3 + 1$$

6 Solve an application problem requiring multiplication of polynomials.

The profit p earned on the sale of one or more items is given by the formula

$$p = r - c$$

where r is the *revenue* (amount taken in) and c is the *wholesale cost* (amount going out).

If a salesperson has 12 vacuum cleaners and sells them for \$225 each, the revenue will be $r = \$(12 \cdot 225) = \$2{,}700$. This illustrates the following formula for finding the revenue r:

$$r = \boxed{\text{number of items sold } (x)} \cdot \boxed{\text{selling price of each item } (p)} = xp = px$$

EXAMPLE 10 **SELLING VACUUM CLEANERS** Over the years, a saleswoman has found that the number of vacuum cleaners she can sell depends on price. The lower the price, the more she can sell. She has determined that the number of vacuums (x) that she can sell at a price (p) is related by the equation $x = -\frac{2}{25}p + 28$.

a. Find a formula for the revenue r.
b. How much revenue will be taken in if the vacuums are priced at \$250?

Solution **a.** To find a formula for revenue, we substitute $-\frac{2}{25}p + 28$ for x in the formula $r = px$ and solve for r.

$$r = px \qquad \text{This is the formula for revenue.}$$

$$r = p\left(-\frac{2}{25}p + 28\right) \qquad \text{Substitute } -\tfrac{2}{25}p + 28 \text{ for } x.$$

$$r = -\frac{2}{25}p^2 + 28p \qquad \text{Multiply the polynomials.}$$

b. To find how much revenue will be taken in if the vacuums are priced at \$250, we substitute 250 for p in the formula for revenue.

$$r = -\frac{2}{25}p^2 + 28p \qquad \text{This is the formula for revenue.}$$

$$r = -\frac{2}{25}(250)^2 + 28(250) \qquad \text{Substitute 250 for } p.$$

$$= -5{,}000 + 7{,}000$$
$$= 2{,}000$$

The revenue will be \$2,000.

7 Multiply two expressions that are not polynomials.

The following examples show that the process of multiplying two polynomials can be applied to expressions that are not polynomials.

EXAMPLE 11 Find the product of $x^{-2} + y$ and $x^2 - y^{-2}$.

Solution We multiply each term of the second expression by each term of the first expression and then simplify.

$$\begin{aligned}(x^{-2} + y)(x^2 - y^{-2}) &= x^{-2}x^2 - x^{-2}y^{-2} + yx^2 - yy^{-2}\\ &= x^{-2+2} - x^{-2}y^{-2} + yx^2 - y^{1+(-2)}\\ &= x^0 - \frac{1}{x^2y^2} + x^2y - y^{-1} \quad \text{Recall that } x^{-2} = \tfrac{1}{x^2} \text{ and } y^{-2} = \tfrac{1}{y^2}.\\ &= 1 - \frac{1}{x^2y^2} + x^2y - \frac{1}{y}\end{aligned}$$

SELF CHECK 11 Multiply: $(a^{-3} + b)(a^2 - b^{-1})$.

EXAMPLE 12 Find the product of $x^n + 2x$ and $x^n + 3x^{-n}$.

Solution We multiply each term of the second expression by each term of the first expression and simplify:

$$\begin{aligned}(x^n + 2x)(x^n + 3x^{-n}) &= x^nx^n + x^n(3x^{-n}) + 2x(x^n) + 2x(3x^{-n})\\ &= x^{n+n} + 3x^{n+(-n)} + 2x^{1+n} + 6xx^{-n}\\ &= x^{2n} + 3x^0 + 2x^{n+1} + 6x^{1+(-n)}\\ &= x^{2n} + 3 + 2x^{n+1} + 6x^{1-n}\end{aligned}$$

SELF CHECK 12 Multiply: $(a^n + b)(a^n + b^n)$.

SELF CHECK ANSWERS

1. a. $-8a^5$ **b.** $15a^3b^4$ **2.** $-2a^4 + 2a^3 - 6a^2$ **3.** $6a^2 + 4a^2b + ab + 2ab^2 - b^2$ **4.** $6x^3 + 7x^2 - 8x - 5$ **5.** $6a^2 + 5ab - 4b^2$ **6. a.** $a^2 + 4a + 4$ **b.** $a^2 - 8a + 16$ **7.** $a^2 - 9$ **8.** $6a^4 + 36a^3 - 162a + 108$ **9.** $2x^3 + 3x^2y - 2xy^2 - 3y^3$ **11.** $\frac{1}{a} - \frac{1}{a^3b} + a^2b - 1$ **12.** $a^{2n} + a^nb^n + a^nb + b^{n+1}$

NOW TRY THIS

1. Given $f(x) = x^2 - 5x + 1$, find $f(x + 2)$.
2. Given the figure below, find the volume.

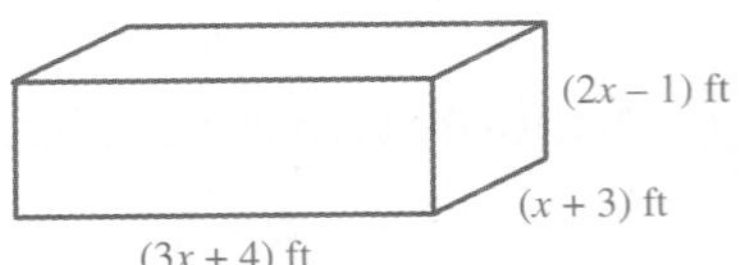

3. $[(x + 2) - y][(x + 2) + y]$

5.3 EXERCISES

WARM-UPS *Find each product.*

1. $(-2a^2b)(3ab^2)$
2. $(4xy^2)(-2xy)$
3. $3a^2(2a - 1)$
4. $-4n^2(4m - n)$
5. $(x + 1)(2x + 1)$
6. $(3y - 2)(2y + 1)$

REVIEW *Let a = −2 and b = 4 and find the absolute value of each expression.*

7. $|3a - b|$
8. $|ab - b^2|$
9. $-|a^2b - b^0|$
10. $\left|\dfrac{a^3b^2 + ab}{2(ab)^2 - a^3}\right|$
11. **Investing** A woman owns 200 shares of ABC Company, valued at \$125 per share, and 350 shares of WD Company, valued at \$75 per share. One day, ABC rose 1.5 points and WD fell 1.5 points. Find the current value of her portfolio.
12. **Astronomy** One light year is approximately 5,870,000,000,000 miles. Write this number in scientific notation.

VOCABULARY AND CONCEPTS *Fill in the blanks.*

13. To multiply a monomial by a monomial, we multiply the numerical factors and then multiply the ________ factors.
14. To multiply a polynomial by a monomial, we multiply each term of the polynomial by the ________.
15. To multiply a polynomial by a polynomial, we multiply each _____ of one polynomial by each term of the other polynomial.
16. FOIL is an acronym for ______ terms, ______ terms, ______ terms, and _____ terms.
17. $(x + y)^2 = (x + y)(x + y) =$ ____________
18. $(x - y)^2 = (x - y)(x - y) =$ ____________
19. $(x + y)(x - y) =$ _______. The binomials $x + y$ and $x - y$ are called ________ binomials, and the product $x^2 - y^2$ is called the ________ of two squares.
20. $x^2 + 2xy + y^2$ and $x^2 - 2xy + y^2$ are called ____________ trinomials.

GUIDED PRACTICE

Find each product. See Example 1. (Objective 1)

21. $(2a^2)(-3ab)$
22. $(-3x^2y)(3xy)$
23. $(-3ab^2c)(5ac^2)$
24. $(-2m^2n)(-4mn^3)$
25. $(4a^2b)(-5a^3b^2)(6a^4)$
26. $(2x^2y^3)(4xy^5)(-5y^6)$
27. $(5x^3y^2)^4\left(\frac{1}{5}x^2\right)^2$
28. $(4a^2b)^2(2a^3b^4)^4$

Find each product. See Example 2. (Objective 2)

29. $3(x + 2)$
30. $-5(a + b)$
31. $-a(a - b)$
32. $y^2(y - 1)$
33. $3x(x^2 + 3x)$
34. $-2x(3x^2 - 2)$
35. $-2x(3x^2 - 3x + 2)$
36. $3a(4a^2 + 3a - 4)$

37. $5a^2b^3(2a^4b - 5a^0b^3)$

38. $-2a^3b(3a^0b^4 - 2a^2b^3)$

39. $7rst(r^2 + s^2 - t^2)$

40. $3x^2yz(x^2 - 2y + 3z^2)$

Find each product. See Examples 3–5. (Objectives 3–4)

41. $(x + 2)(x + 3)$

42. $(y - 3)(y + 4)$

43. $(z - 7)(z - 2)$

44. $(x + 3)(x - 5)$

45. $(2a + 1)(a - 2)$

46. $(3b - 1)(2b - 1)$

47. $(3t - 2)(2t + 3)$

48. $(p + 3)(3p - 4)$

49. $(3y - z)(2y - z)$

50. $(2m + n)(3m + n)$

51. $(2x - 3y)(x + 2y)$

52. $(3y + 2z)(y - 3z)$

53. $(3x + y)(3x - 3y)$

54. $(2x - y)(3x + 2y)$

55. $(4a - 3b)(2a + 5b)$

56. $(3a + 2b)(2a - 7b)$

Find each product. See Examples 6–7. (Objective 5)

57. $(x + 2)^2$

58. $(x - 3)^2$

59. $(a - 4)^2$

60. $(y + 5)^2$

61. $(2a + b)^2$

62. $(a - 2b)^2$

63. $(2x - y)^2$

64. $(3m + 4n)(3m + 4n)$

65. $(x + 2)(x - 2)$

66. $(z + 3)(z - 3)$

67. $(a + b)(a - b)$

68. $(p + q)(p - q)$

69. $(2x + 3y)(2x - 3y)$

70. $(3a + 4b)(3a - 4b)$

71. $(x - y)(x^2 + xy + y^2)$

72. $(x + y)(x^2 - xy + y^2)$

Find each product. See Example 8. (Objective 3)

73. $3x(2x + 4) - 3x^2$

74. $2y - 3y(y^2 + 4)$

75. $3pq - p(p - q)$

76. $-4rs(r - 2) + 4rs$

77. $-3y(2y + z) + (2y - z)(3y + 2z)$

78. $(x + 3)(x - 3) + (2x - 1)(x + 2)$

79. $(3y + 1)^2 + (2y - 4)^2$

80. $3(x - 3y)^2 + 2(3x + y)^2$

Find each product. See Example 9. (Objective 3)

81. $(a + b)(a - b)(a - 3b)$

82. $(x - y)(x + 2y)(x - 2y)$

83. $(a + b)^3$

84. $(2m - n)^3$

Find each product. Write all answers without using negative exponents. See Examples 11–12. (Objective 7)

85. $x^3(2x^2 + x^{-2})$

86. $x^{-4}(2x^{-3} - 5x^2)$

87. $x^3y^{-6}z^{-2}(3x^{-2}y^2z - x^3y^{-4})$

88. $ab^{-2}c^{-3}(a^{-4}bc^3 + a^{-3}b^4c^3)$

89. $(x^{-1} + y)(x^{-1} - y)$

90. $(x^{-1} - y)(x^{-1} - y)$

91. $(2x^{-3} + y^3)(2x^3 - y^{-3})$

92. $(5x^{-4} - 4y^2)(5x^2 - 4y^{-4})$

93. $x^n(x^{2n} - x^n)$

94. $a^{2n}(a^n + a^{2n})$

95. $(x^n + 1)(x^n - 1)$

96. $(x^n - a^n)(x^n + a^n)$

ADDITIONAL PRACTICE

Simplify each expression. Write any expression with positive exponents only.

97. $(-5xx^2)(-3xy)^4$

98. $(-2a^2ab^2)^3(-3ab^2b^2)$

99. $4m^2n(-3mn)(m + n)$

100. $-3a^2b^3(2b)(3a + b)$

101. $(3y + 1)(2y^2 + 3y + 2)$

102. $(a + 2)(3a^2 + 4a - 2)$

103. $(2a - b)(4a^2 + 2ab + b^2)$

104. $(x - 3y)(x^2 + 3xy + 9y^2)$

105. $(2x - 1)[2x^2 - 3(x + 2)]$

106. $(x + 1)^2[x^2 - 2(x + 2)]$

107. $(2p + q)(p - 2q)^2$

108. $(a - b)(2a + b)^2$

109. $2m(m - n) - (m + n)(m - 2n)$

110. $(2b + 3)(b - 1) - (b + 2)(3b - 1)$

111. $(3x - 4)^2 - (2x + 3)^2$

112. $2(x - y^2)^2 - 3(y^2 + 2x)^2$

113. $5(2y - z)^2 + 4(y + 2z)^2$

114. $3(x + 2z)^2 - 2(2x - z)^2$

115. $(x^n - y^n)(x^n - y^{-n})$

116. $(x^n + y^n)(x^n + y^{-n})$

117. $(x^{2n} + y^{2n})(x^{2n} - y^{2n})$

118. $(a^{3n} - b^{3n})(3^{3n} + b^{3n})$

119. $(x^n + y^n)(x^n + 1)$

120. $(1 - x^n)(x^{-n} - 1)$

Use a calculator to find each product.

121. $(3.21x - 7.85)(2.87x + 4.59)$

122. $(7.44y + 56.7)(-2.1y - 67.3)$

123. $(-17.3y + 4.35)^2$

124. $(-0.31x + 29.3)(-81x - 0.2)$

125. Refer to the illustration and answer the following questions.

a. Find the area of the large square.
b. Find the area of square I.
c. Find the area of rectangle II.
d. Find the area of rectangle III.
e. Find the area of square IV.

Use the answers to the preceding questions to show that $(x - y)^2 = x^2 - 2xy + y^2$.

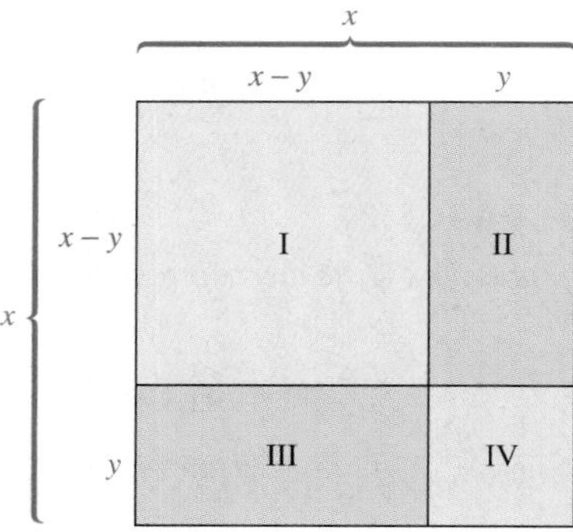

126. Refer to the illustration and answer the following questions.

a. Find the area of rectangle $ABCD$.
b. Find the area of rectangle I.
c. Find the area of rectangle II.

Use the answers to the preceding questions to show that $(x + y)(x - y) = x^2 - y^2$.

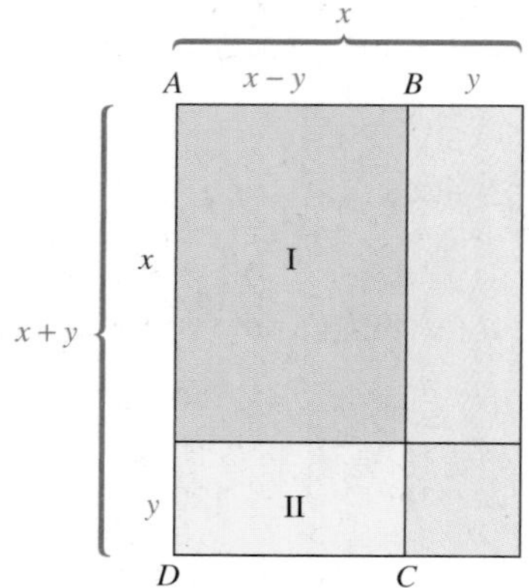

APPLICATIONS *Solve each application problem.*

Selling TVs *Assume that the number (x) of televisions a salesman can sell at a certain price (p) is given by the equation* $x = -\frac{1}{5}p + 90$. **See Example 10. (Objective 6)**

127. Find the number of TVs he will sell if the price is \$375.

128. If he sold 20 TVs, what was the price?

129. Write a formula for the revenue when x TVs are sold.

130. Find the revenue generated by the sales of the TVs if they are priced at \$400 each.

131. Geometry Write a polynomial that represents the area of the square garden.

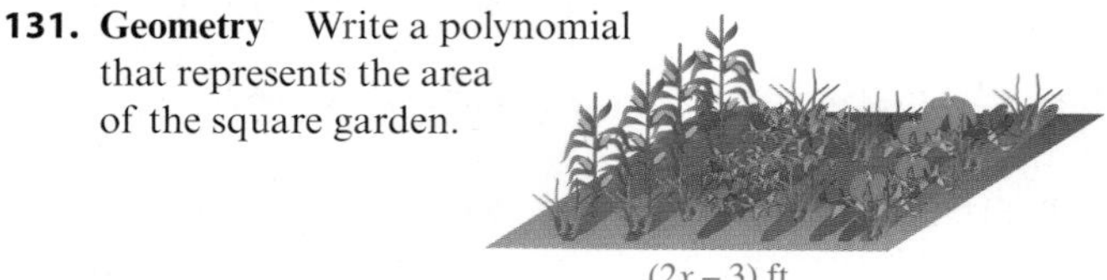

132. Geometry Write a polynomial that represents the area of the rectangular field.

133. Geometry Write a polynomial that represents the area of the triangle.

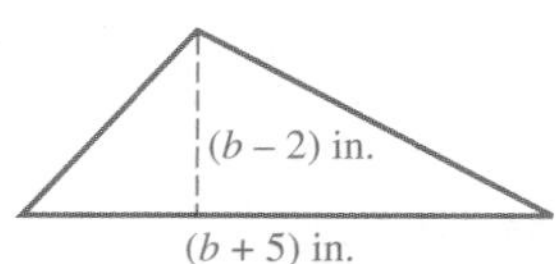

134. Geometry Write a polynomial that represents the volume of the cube.

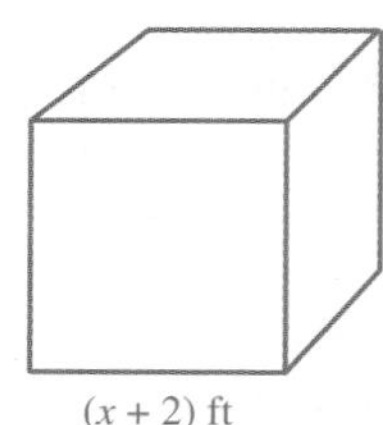

WRITING ABOUT MATH

135. Explain how to use the FOIL method.

136. Explain how to multiply two trinomials.

SOMETHING TO THINK ABOUT

137. The numbers 0.35×10^7 and 1.96×10^7 both involve the same power of 10. Find their sum.

138. Without converting to standard notation, find the sum: $1.435 \times 10^8 + 2.11 \times 10^7$. (*Hint:* The first number in the previous exercise is not in scientific notation.)

SECTION 5.4 The Greatest Common Factor and Factoring by Grouping

Objectives

1. Find the prime-factored form of a natural number.
2. Find the greatest common factor (GCF) of two or more monomials.
3. Factor the greatest common factor (GCF) from a polynomial.
4. Factor a polynomial with four or six terms by grouping.
5. Solve a formula for a specified variable by factoring.

Vocabulary

factoring
prime-factored form
greatest common factor (GCF)
prime polynomial

Getting Ready

Simplify each expression by removing parentheses.

1. $4(a + 2)$ **2.** $-5(b - 5)$
3. $a(a + 5)$ **4.** $-b(b - 3)$
5. $6a(a + 2b)$ **6.** $-2p^2(3p + 5)$

In this section, we will reverse the operation of multiplication and show how to find the factors of a known product. The process of finding the individual factors of a product is called **factoring.**

1 Find the prime-factored form of a natural number.

If one number a divides a second number b, then a is called a factor of b. For example, because 3 divides 24, it is a factor of 24. Each number in the following list is a factor of 24, because each number divides 24.

1, 2, 3, 4, 6, 8, 12, and 24

To factor a natural number means to write it as a product of other natural numbers. If each factor is a prime number, the natural number is said to be written in **prime-factored form.** Example 1 shows how to find the prime-factored forms of 60, 84, and 180, respectively.

EXAMPLE 1 Find the prime factorization of each number.

a. $60 = 6 \cdot 10$
$= \mathbf{2 \cdot 3 \cdot 2} \cdot 5$
$= \mathbf{2^2 \cdot 3} \cdot 5$

b. $84 = 4 \cdot 21$
$= \mathbf{2 \cdot 2 \cdot 3} \cdot 7$
$= \mathbf{2^2 \cdot 3} \cdot 7$

c. $180 = 10 \cdot 18$
$= 2 \cdot 5 \cdot 3 \cdot 6$
$= 2 \cdot 5 \cdot \mathbf{3 \cdot 3 \cdot 2}$
$= \mathbf{2^2 \cdot 3^2} \cdot 5$

SELF CHECK 1 Find the prime factorization of 120.

2 Find the greatest common factor (GCF) of two or more monomials.

The largest natural number that divides 60, 84, and 180 is called the **greatest common factor (GCF)** of the numbers. Because 60, 84, and 180 all have two factors of 2 and one factor of 3, the GCF of these three numbers is $2^2 \cdot 3 = 12$. We note that

$$\frac{60}{12} = 5, \qquad \frac{84}{12} = 7, \qquad \text{and} \qquad \frac{180}{12} = 15$$

There is no natural number greater than 12 that divides 60, 84, and 180 exactly.

Algebraic monomials also can have greatest common factors.

EXAMPLE 2 Find the GCF of $6a^2b^3c$, $9a^3b^2c$, and $18a^4c^3$.

Solution We begin by finding the prime factorization of each monomial:

$$\begin{aligned} 6a^2b^3c &= 3 \cdot 2 \cdot a \cdot a \cdot b \cdot b \cdot b \cdot c \\ 9a^3b^2c &= 3 \cdot 3 \cdot a \cdot a \cdot a \cdot b \cdot b \cdot c \\ 18a^4c^3 &= 2 \cdot 3 \cdot 3 \cdot a \cdot a \cdot a \cdot a \cdot c \cdot c \cdot c \end{aligned}$$

Since each monomial has one factor of 3, two factors of a, and one factor of c in common, their GCF is

$$3^1 \cdot a^2 \cdot c^1 = 3a^2c$$

SELF CHECK 2 Find the GCF of $-24x^3y^3$, $3x^3y$, and $18x^2y^2$.

To find the GCF of several monomials, we follow these steps.

Steps for Finding the GCF

1. Find the prime-factored form of each monomial.
2. Identify the prime factors and variable factors that are common to each monomial.
3. Find the product of the factors found in Step 2 with each factor raised to the smallest power that occurs in any one monomial.

3 Factor the greatest common factor (GCF) from a polynomial.

We have seen that the distributive property provides a method for multiplying a polynomial by a monomial. For example,

$$\begin{aligned} 2x^3y^3(3x^2 - 4y^3) &= 2x^3y^3 \cdot 3x^2 - 2x^3y^3 \cdot 4y^3 \\ &= 6x^5y^3 - 8x^3y^6 \end{aligned}$$

If the product of a multiplication is $6x^5y^3 - 8x^3y^6$, we can use the distributive property to find the individual factors.

$$6x^5y^3 - 8x^3y^6 = \mathbf{2x^3y^3} \cdot 3x^2 - \mathbf{2x^3y^3} \cdot 4y^3$$
$$= \mathbf{2x^3y^3}(3x^2 - 4y^3)$$

Since $2x^3y^3$ is the GCF of the terms of $6x^5y^3 - 8x^3y^6$, this process is called *factoring out the greatest common factor.*

COMMENT Although it is true that

$$6x^5y^3 - 8x^3y^6 = 2x^3(3x^2y^3 - 4y^6)$$

the polynomial $6x^5y^3 - 8x^3y^6$ is not completely factored, because y^3 can be factored from $3x^2y^3 - 4y^6$. Whenever you factor a polynomial, be sure to factor it completely.

EXAMPLE 3 Factor: $25a^3b + 15ab^3$.

Solution We begin by finding the prime factorization of each monomial:

$$25a^3b = \mathbf{5} \cdot 5 \cdot \mathbf{a} \cdot a \cdot a \cdot \mathbf{b}$$
$$15ab^3 = \mathbf{5} \cdot 3 \cdot \mathbf{a} \cdot \mathbf{b} \cdot b \cdot b$$

Since each term has at least one factor of 5, one factor of a, and one factor of b in common and there are no other common factors, $5ab$ is the GCF of the two terms. We can use the distributive property to factor it out.

$$25a^3b + 15ab^3 = \mathbf{5ab} \cdot 5a^2 + \mathbf{5ab} \cdot 3b^2$$
$$= \mathbf{5ab}(5a^2 + 3b^2)$$

SELF CHECK 3 Factor: $9x^4y^2 - 12x^3y^3$.

EXAMPLE 4 Factor: $3xy^2z^3 + 6xyz^3 - 3xz^2$.

Solution We begin by finding the prime factorization of each monomial:

$$3xy^2z^3 = \mathbf{3} \cdot \mathbf{x} \cdot y \cdot y \cdot \mathbf{z} \cdot \mathbf{z} \cdot z$$
$$6xyz^3 = \mathbf{3} \cdot 2 \cdot \mathbf{x} \cdot y \cdot \mathbf{z} \cdot \mathbf{z} \cdot z$$
$$-3xz^2 = \mathbf{-3} \cdot \mathbf{x} \cdot \mathbf{z} \cdot \mathbf{z}$$

Since each term has one factor of 3, one factor of x, and two factors of z in common, and because there are no other common factors, $3xz^2$ is the GCF of the three terms. We can use the distributive property to factor it out.

COMMENT The last term $-3xz^2$ of the given trinomial has a coefficient of -3. When the $3xz^2$ is factored out, remember to write the -1.

$$3xy^2z^3 + 6xyz^3 - 3xz^2 = \mathbf{3xz^2} \cdot y^2z + \mathbf{3xz^2} \cdot 2yz - \mathbf{3xz^2} \cdot 1$$
$$= \mathbf{3xz^2}\,(y^2z + 2yz - 1)$$

SELF CHECK 4 Factor: $2a^4b^2 + 6a^3b^2 - 2a^2b$.

A polynomial that cannot be factored using only rational numbers is called a **prime polynomial** or an **irreducible polynomial** over the rational numbers.

EXAMPLE 5 Factor $3x^2 + 4y + 7$, if possible.

Solution We find the prime factorization of each monomial:

$$3x^2 = 3 \cdot x \cdot x \qquad 4y = 2 \cdot 2 \cdot y \qquad 7 = 7$$

There are no common factors other than 1. This polynomial is an example of a prime polynomial.

SELF CHECK 5 Factor $6a^3 + 7b^2 + 5$, if possible.

EXAMPLE 6 Factor the negative of the GCF from $-6u^2v^3 + 8u^3v^2$.

Solution Because the GCF of the two terms is $2u^2v^2$, the negative of the GCF is $-2u^2v^2$. To factor out $-2u^2v^2$, we proceed as follows:

$$\begin{aligned} -6u^2v^3 + 8u^3v^2 &= -2u^2v^2 \cdot 3v + 2u^2v^2 \cdot 4u \\ &= -2u^2v^2 \cdot 3v - (-2u^2v^2)4u \\ &= -2u^2v^2(3v - 4u) \end{aligned}$$

SELF CHECK 6 Factor the negative of the GCF from $-8a^2b^2 - 12ab^3$.

We also can factor out common factors with variable exponents.

EXAMPLE 7 Factor x^{2n} from $x^{4n} + x^{3n} + x^{2n}$.

Solution We write the trinomial as $x^{2n} \cdot x^{2n} + x^{2n} \cdot x^n + x^{2n} \cdot 1$ and factor out x^{2n}.

$$\begin{aligned} x^{4n} + x^{3n} + x^{2n} &= x^{2n} \cdot x^{2n} + x^{2n} \cdot x^n + x^{2n} \cdot 1 \\ &= x^{2n}(x^{2n} + x^n + 1) \end{aligned}$$

SELF CHECK 7 Factor a^n from $a^{6n} + a^{3n} - a^n$.

EXAMPLE 8 Factor $a^{-2}b^{-2}$ from $a^{-2}b - a^3b^{-2}$.

Solution We write $a^{-2}b - a^3b^{-2}$ as $a^{-2}b^{-2} \cdot b^3 - a^{-2}b^{-2} \cdot a^5$ and factor out $a^{-2}b^{-2}$.

$$\begin{aligned} a^{-2}b - a^3b^{-2} &= a^{-2}b^{-2} \cdot b^3 - a^{-2}b^{-2} \cdot a^5 \\ &= a^{-2}b^{-2}(b^3 - a^5) \end{aligned}$$

SELF CHECK 8 Factor $x^{-3}y^{-2}$ from $x^{-3}y^{-2} - x^6y^{-2}$.

A common factor can have more than one term. For example, in the expression

$$x(a + b) + y(a + b)$$

the binomial $a + b$ is a factor of both terms. We can factor it out to get

$$x(a + b) + y(a + b) = (a + b)x + (a + b)y \quad \text{Use the commutative property of multiplication.}$$
$$= (a + b)(x + y)$$

EXAMPLE 9 Factor: $a(x - y + z) - b(x - y + z) + 3(x - y + z)$.

Solution We can factor out the GCF of $(x - y + z)$.

$$a(x - y + z) - b(x - y + z) + 3(x - y + z)$$
$$= (x - y + z)a - (x - y + z)b + (x - y + z)3$$
$$= (x - y + z)(a - b + 3)$$

SELF CHECK 9 Factor: $x(a + b - c) - y(a + b - c)$.

4 Factor a polynomial with four or six terms by grouping.

Suppose that we want to factor

$$ac + ad + bc + bd$$

Although no factor is common to all four terms, there is a common factor of a in the first two terms and a common factor of b in the last two terms. We can factor out these common factors to get

$$ac + ad + bc + bd = a(c + d) + b(c + d)$$

We can now factor out the common factor of $c + d$ on the right side:

$$ac + ad + bc + bd = (c + d)(a + b)$$

The grouping in this type of problem is not always unique, but the factorization is. For example, if we write the expression $ac + ad + bc + bd$ in the form

$$ac + bc + ad + bd$$

and factor c from the first two terms and d from the last two terms, we obtain

$$ac + bc + ad + bd = c(a + b) + d(a + b)$$
$$= (a + b)(c + d)$$

COMMENT Because of the commutative property of multiplication, the previous result can be written in the equivalent form $(c + d)(a + b)$.

The method used in the previous examples is called *factoring by grouping.*

EXAMPLE 10 Factor: $3ax^2 + 3bx^2 + a + 5bx + 5ax + b$.

Solution Although no factor is common to all six terms, $3x^2$ can be factored out of the first two terms, and $5x$ can be factored out of the fourth and fifth terms to get

$$3ax^2 + 3bx^2 + a + 5bx + 5ax + b = 3x^2(a + b) + a + 5x(b + a) + b$$

This result can be written in the form

$$3ax^2 + 3bx^2 + a + 5bx + 5ax + b = 3x^2(a + b) + 5x(a + b) + (a + b)$$

Since $a + b$ is common to all three terms, it can be factored out to get

$$3ax^2 + 3bx^2 + a + 5bx + 5ax + b = (a + b)(3x^2 + 5x + 1)$$

SELF CHECK 10 Factor: $4x^3 + 3x^2 + x + 4x^2y + 3xy + y$.

To factor an expression, it is often necessary to factor more than once, as the following example illustrates.

EXAMPLE 11 Factor: $3x^3y - 4x^2y^2 - 6x^2y + 8xy^2$.

Solution We begin by factoring out the common factor of xy.

$$3x^3y - 4x^2y^2 - 6x^2y + 8xy^2 = xy(3x^2 - 4xy - 6x + 8y)$$

We can now factor $3x^2 - 4xy - 6x + 8y$ by grouping:

$$3x^3y - 4x^2y^2 - 6x^2y + 8xy^2$$
$$= xy(3x^2 - 4xy - 6x + 8y)$$
$$= xy[x(\mathbf{3x - 4y}) - 2(\mathbf{3x - 4y})]$$ Factor x from $3x^2 - 4xy$ and -2 from $-6x + 8y$.
$$= xy(\mathbf{3x - 4y})(x - 2)$$ Factor out $3x - 4y$.

Because no more factoring can be done, the factorization is complete.

COMMENT Whenever you factor an expression, always factor it completely. Each factor of a completely factored expression will be prime over the rational numbers except possibly for a monomial GCF.

SELF CHECK 11 Factor: $3a^3b + 3a^2b - 2a^2b^2 - 2ab^2$.

5 Solve a formula for a specified variable by factoring.

Factoring often is required to solve a literal equation for one of its variables.

EXAMPLE 12 **ELECTRONICS** The formula $r_1r_2 = rr_2 + rr_1$ is used in electronics to relate the combined resistance r of two resistors wired in parallel. The variable r_1 represents the resistance of the first resistor, and the variable r_2 represents the resistance of the second. Solve for r_2.

Solution To isolate r_2 on one side of the equation, we get all terms involving r_2 on the left side and all terms not involving r_2 on the right side. We then proceed as follows:

$$r_1r_2 = rr_2 + rr_1$$
$$r_1r_2 - rr_2 = rr_1$$ Subtract rr_2 from both sides.
$$r_2(r_1 - r) = rr_1$$ Factor out r_2 on the left side.
$$r_2 = \frac{rr_1}{r_1 - r}$$ Divide both sides by $r_1 - r$.

SELF CHECK 12 Solve $A = p + prt$ for p.

SELF CHECK ANSWERS

1. $2^3 \cdot 3 \cdot 5$ **2.** $3x^2y$ **3.** $3x^3y^2(3x - 4y)$ **4.** $2a^2b(a^2b + 3ab - 1)$ **5.** a prime polynomial **6.** $-4ab^2(2a + 3b)$ **7.** $a^n(a^{5n} + a^{2n} - 1)$ **8.** $x^{-3}y^{-2}(1 - x^9)$ **9.** $(a + b - c)(x - y)$ **10.** $(x + y)(4x^2 + 3x + 1)$ **11.** $ab(3a - 2b)(a + 1)$ **12.** $p = \frac{A}{1 + rt}$

NOW TRY THIS

Factor out the smallest power of x:

1. $x^{5n} + x^{2n} - 3x^n$

2. $x^{4/3} + x - 3x^{1/3}$

3. $x^{-2} + x^{-1} - 3x$

Solve for y.

4. $3xy + 2x = 5y - 4$

5.4 EXERCISES

WARM-UPS *Factor each expression.*

1. $3x^2 - x$

2. $7t^3 + 14t^2$

3. $-3a^2 - 6a$

4. $-4x^2 + 12x$

5. $3(a + b) + x(a + b)$

6. $a(m - n) - b(m - n)$

REVIEW *Perform each multiplication.*

7. $(a + 4)(a - 4)$

8. $(2b + 3)(2b - 3)$

9. $(4r^2 + 3s)(4r^2 - 3s)$

10. $(5a + 2b^3)(5a - 2b^3)$

11. $(m + 4)(m^2 - 4m + 16)$

12. $(p - q)(p^2 + pq + q^2)$

VOCABULARY AND CONCEPTS *Fill in the blanks.*

13. The process of finding the individual factors of a known product is called ________.

14. If a natural number is written as the product of prime factors, we say that it is written in ____________ form.

15. The abbreviation GCF stands for ________________.

16. If a polynomial cannot be factored, it is called a ______ polynomial or an _________ polynomial.

GUIDED PRACTICE

Find the prime-factored form of each number. See Example 1. (Objective 1)

17. 6

18. 10

19. 135

20. 98

21. 128

22. 357

23. 325

24. 288

Find the GCF of each set of monomials. See Example 2. (Objective 2)

25. 36, 48

26. 45, 75

27. 42, 36, 98

28. 16, 40, 60

29. $4a^2b$, $8a^3c$

30. $6x^3y^2z$, $9xyz^2$

31. $18x^4y^3z^2$, $-12xy^2z^3$

32. $6x^2y^3$, $24xy^3$, $40x^2y^2z^3$

Complete each factorization. (Objective 3)

33. $3a - 12 = 3(a - \square)$

34. $5t + 25 = 5(t + \square)$

35. $8z^2 + 2z = 2z(4z + \square)$

36. $9t^3 - 3t^2 = 3t^2(3t - \square)$

Factor each expression. See Example 3. (Objective 3)

37. $2x + 8$

38. $3y - 9$

39. $2x^2 - 6x$

40. $3y^3 + 3y^2$

41. $7x^2 + 14x$

42. $15x^2y - 10x^2y^2$

43. $63x^3y^2 + 81x^2y^4$

44. $13ab^2c^3 - 26a^3b^2c$

Factor each expression. **See Example 4. (Objective 3)**

45. $24s^3 - 12s^2t + 6st^2$
46. $18y^2z^2 + 12y^2z^3 - 24y^4z^3$
47. $45x^{10}y^3 - 63x^7y^7 + 81x^{10}y^{10}$
48. $48u^6v^6 - 16u^4v^4 - 3u^6v^3$

Factor each expression, if possible. **See Example 5. (Objective 3)**

49. $5xy + 12ab^2$
50. $14r^2s^3 + 15t^6$
51. $11m^3n^2 - 12x^2y$
52. $33a^3b^4c - 16xyz$

Factor out the negative of the greatest common factor. **See Example 6. (Objective 3)**

53. $-3a - 6$
54. $-6b + 12$
55. $-3x^2 - x$
56. $-4a^3 + a^2$
57. $-6x^2 - 3xy$
58. $-15y^3 + 25y^2$
59. $-18a^2b - 12ab^2$
60. $-21t^5 + 28t^3$

Factor out the designated factor. **See Examples 7–8. (Objective 3)**

61. x^2 from $x^{n+2} + x^{n+3}$
62. y^3 from $y^{n+3} + y^{n+5}$
63. y^n from $2y^{n+2} - 3y^{n+3}$
64. x^n from $4x^{n+3} - 5x^{n-5}$
65. x^{-2} from $x^4 - 5x^6$
66. y^{-4} from $7y^4 + y$
67. t^{-3} from $t^5 + 4t^{-6}$
68. p^{-5} from $6p^3 - p^{-2}$

Factor each expression. **See Examples 9–11. (Objective 4)**

69. $4(x + y) + t(x + y)$
70. $5(a - b) - t(a - b)$
71. $(a - b)r - (a - b)s$
72. $(x + y)u + (x + y)v$
73. $(u + v)^2 - (u + v)$
74. $a(x - y) - (x - y)^2$
75. $-a(x + y) + b(x + y)$
76. $-bx(a - b) - cx(a - b)$
77. $3c - cd + 3d - c^2$
78. $x^2 + 4y - xy - 4x$
79. $a^2 - 4b + ab - 4a$
80. $7u + v^2 - 7v - uv$
81. $ax + bx - a - b$
82. $x^2y - ax - xy + a$
83. $x^2 + xy + xz + xy + y^2 + zy$
84. $ab - b^2 - bc + ac - bc - c^2$

Solve for the indicated variable. **See Example 12. (Objective 5)**

85. $r_1r_2 = rr_2 + rr_1$ for r_1
86. $r_1r_2 = rr_2 + rr_1$ for r
87. $d_1d_2 = fd_2 + fd_1$ for f
88. $d_1d_2 = fd_2 + fd_1$ for d_1
89. $b^2x^2 + a^2y^2 = a^2b^2$ for a^2
90. $b^2x^2 + a^2y^2 = a^2b^2$ for b^2
91. $S(1 - r) = a - lr$ for r
92. $Sn = (n - 2)180°$ for n

ADDITIONAL PRACTICE

Factor completely.

93. $27z^3 + 12z^2 + 3z$
94. $25t^6 - 10t^3 + 5t^2$
95. $25x^3 - 14y^3 + 36x^3y^3$
96. $9m^4n^3p^2 + 18m^2n^3p^4 - 27m^3n^4p$
97. $-63u^3v^6z^9 + 28u^2v^7z^2 - 21u^3v^3z^4$
98. $-56x^4y^3z^2 - 72x^3y^4z^5 + 80xy^2z^3$
99. $4y^{-2n}$ from $8y^{2n} + 12 + 16y^{-2n}$
100. $7x^{-3n}$ from $21x^{6n} + 7x^{3n} + 14$
101. $3(m + n + p) + x(m + n + p)$
102. $x(x - y - z) + y(x - y - z)$
103. $(x + y)(x + y) + z(x + y)$
104. $(a - b)^2 + (a - b)$
105. $ax + bx + ay + by$
106. $ar - br + as - bs$
107. $x^2 + yx + 2x + 2y$
108. $2c + 2d - cd - d^2$
109. $mpx + mqx + npx + nqx$
110. $abd - abe + acd - ace$
111. $x^2y + xy^2 + 2xyz + xy^2 + y^3 + 2y^2z$
112. $a^3 - 2a^2b + a^2c - a^2b + 2ab^2 - abc$
113. $2n^4p - 2n^2 - n^3p^2 + np + 2mn^3p - 2mn$
114. $a^2c^3 + ac^2 + a^3c^2 - 2a^2bc^2 - 2bc^2 + c^3$

Solve for the indicated variable.

115. $H(a + b) = 2ab$ for a
116. $H(a + b) = 2ab$ for b
117. $3xy - x = 2y + 3$ for y
118. $x(5y + 3) = y - 1$ for y

APPLICATIONS

119. Framing a picture The dimensions of a family portrait and the frame in which it is mounted are given in the illustration. Write an algebraic expression that describes

a. the area of the picture frame.

b. the area of the portrait.

c. the area of the mat used in the framing. Express the result in factored form.

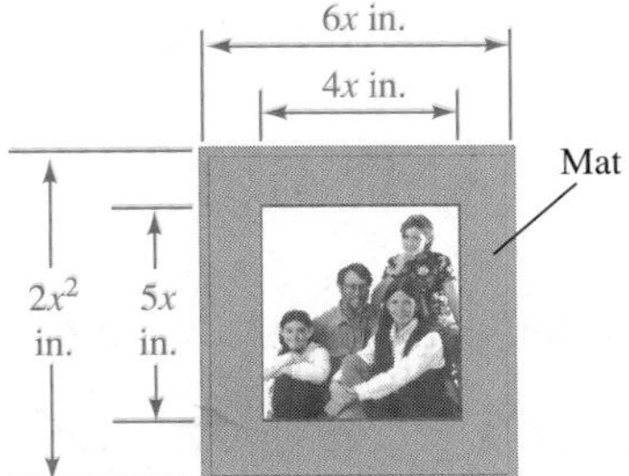

120. Aircraft carriers The illustration shows the deck of the aircraft carrier *Enterprise*. The rectangular-shaped landing area of $(x^3 + 4x^2 + 5x + 20)$ ft^2 is shaded. If the dimensions can be determined by factoring the expression, write an expression that represents the width of the landing area.

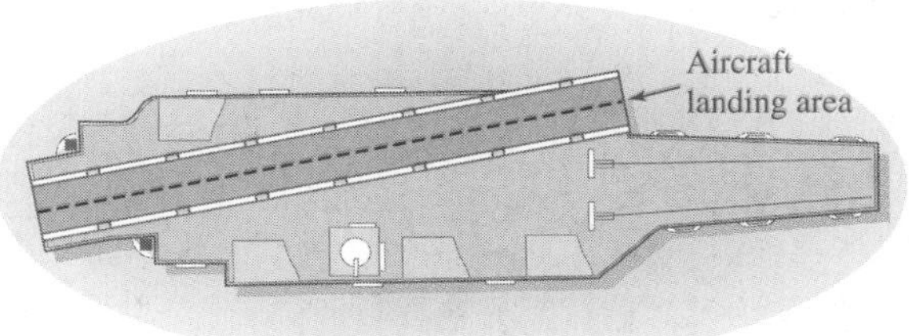

WRITING ABOUT MATH

121. Explain how to find the greatest common factor of two natural numbers.

122. Explain how to recognize when a number is prime.

SOMETHING TO THINK ABOUT

123. Pick two natural numbers. Divide their product by their greatest common factor. The result is called the **least common multiple** of the two numbers you picked. Why?

124. The number 6 is called a **perfect number,** because the sum of all the divisors of 6 is twice 6: $1 + 2 + 3 + 6 = 12$. Verify that 28 is also a perfect number.

If the greatest common factor of several terms is 1, the terms are called* relatively prime. *Determine whether the terms in each set are relatively prime.

125. 14, 45

126. 24, 63, 112

127. 60, 28, 36

128. 55, 49, 78

129. $12x^2y$, $5ab^3$, $35x^2b^3$

130. $18uv$, $25rs$, $12rsuv$

SECTION 5.5 The Difference of Two Squares; the Sum and Difference of Two Cubes

Objectives

1. Factor the difference of two squares.
2. Factor the sum and difference of two cubes.
3. Factor a polynomial by grouping.

Vocabulary

perfect square
difference of squares
sum of squares
perfect cube
sum of cubes
difference of cubes

Getting Ready

Multiply:

1. $(a + b)(a - b)$
2. $(5p + q)(5p - q)$
3. $(3m + 2n)(3m - 2n)$
4. $(2a^2 + b^2)(2a^2 - b^2)$
5. $(a - 3)(a^2 + 3a + 9)$
6. $(p + 2)(p^2 - 2p + 4)$

In this section, we will discuss three special types of factoring problems. These types can be factored using the appropriate factoring formula.

1 Factor the difference of two squares.

To factor the difference of two squares, it is helpful to know the first 20 integers that are **perfect squares.**

1, 4, 9, 16, 25, 36, 49, 64, 81, 100, 121, 144, 169, 196, 225, 256, 289, 324, 361, 400

Expressions like $x^6y^4z^2$ are also perfect squares, because they can be written as the square of another quantity:

$$x^6y^4z^2 = (x^3y^2z)^2$$

All exponential expressions that have even-numbered exponents are perfect squares.

In Section 5.3, we developed the special product formula

$$(1)\qquad (x + y)(x - y) = x^2 - y^2$$

The binomial $x^2 - y^2$ is called the **difference of two squares,** because x^2 represents the square of x, y^2 represents the square of y, and $x^2 - y^2$ represents the difference of these squares.

Equation 1 can be written in reverse order to give a formula for factoring the difference of two squares.

Factoring the Difference of Two Squares

$$x^2 - y^2 = (x + y)(x - y)$$

The formula above shows that the difference of the squares of two quantities factors into a product of two conjugate binomials.

If we think of the difference of two squares as the square of a **F**irst quantity minus the square of a **L**ast quantity, we have the formula

$$\text{F}^2 - \text{L}^2 = (\text{F} + \text{L})(\text{F} - \text{L})$$

and we say: *To factor the square of a* **F**irst *quantity minus the square of a* **L**ast *quantity, we multiply the* **F**irst *plus the* **L**ast *by the* **F**irst *minus the* **L**ast.

EXAMPLE 1 Factor: $49x^2 - 16$.

Solution We can write $49x^2 - 16$ in the form $(7x)^2 - (4)^2$ and use the formula for factoring the difference of two squares:

$$\begin{array}{l} F^2 - L^2 = (F + L)(F - L) \\ \downarrow \quad \downarrow \quad \downarrow \quad \downarrow \quad \downarrow \quad \downarrow \\ (7x)^2 - 4^2 = (7x + 4)(7x - 4) \end{array}$$ Substitute $7x$ for F and 4 for L.

We can verify this result by multiplication.

$$\begin{aligned} (7x + 4)(7x - 4) &= 49x^2 - 28x + 28x - 16 \\ &= 49x^2 - 16 \end{aligned}$$

SELF CHECK 1 Factor: $81p^2 - 25$.

COMMENT Expressions such as $(7x)^2 + 4^2$, which represent the **sum of two squares,** cannot be factored in the real-number system unless there are common factors. Thus, the binomial $49x^2 + 16$ is a prime binomial.

EXAMPLE 2 Factor: $64a^4 - 25b^2$.

Solution We can write $64a^4 - 25b^2$ in the form $(8a^2)^2 - (5b)^2$ and use the formula for factoring the difference of two squares.

$$\begin{array}{l} F^2 - L^2 = (F + L)(F - L) \\ \downarrow \quad \downarrow \quad \downarrow \quad \downarrow \quad \downarrow \quad \downarrow \\ (8a^2)^2 - (5b)^2 = (8a^2 + 5b)(8a^2 - 5b) \end{array}$$

Verify by multiplication.

SELF CHECK 2 Factor: $36r^4 - s^2$.

EXAMPLE 3 Factor: $x^4 - 1$.

Solution Because the binomial is the difference of the squares of x^2 and 1, it factors into the sum of x^2 and 1, and the difference of x^2 and 1.

$$\begin{aligned} x^4 - 1 &= (x^2)^2 - (1)^2 \\ &= (x^2 + 1)(x^2 - 1) \end{aligned}$$

The factor $x^2 + 1$ is the sum of two squares and is prime. However, the factor $x^2 - 1$ is the difference of two squares and can be factored as $(x + 1)(x - 1)$. Thus,

$$\begin{aligned} x^4 - 1 &= (x^2 + 1)(x^2 - 1) \\ &= (x^2 + 1)(x + 1)(x - 1) \end{aligned}$$

SELF CHECK 3 Factor: $a^4 - 16$.

EXAMPLE 4 Factor: $(x + y)^4 - z^4$.

Solution This expression is the difference of two squares that can be factored:

$$\begin{aligned}(x + y)^4 - z^4 &= [(x + y)^2]^2 - (z^2)^2 \\ &= [(x + y)^2 + z^2][(x + y)^2 - z^2]\end{aligned}$$

The factor $(x + y)^2 + z^2$ is the sum of two squares and is prime. However, the factor $(x + y)^2 - z^2$ is the difference of two squares and can be factored as $(x + y + z)(x + y - z)$. Thus,

$$\begin{aligned}(x + y)^4 - z^4 &= [(x + y)^2 + z^2][(x + y)^2 - z^2] \\ &= [(x + y)^2 + z^2](x + y + z)(x + y - z)\end{aligned}$$

SELF CHECK 4 Factor: $(a - b)^2 - c^2$.

When possible, we will always factor out a common factor before factoring the difference of two squares. The factoring process is easier when all common factors are factored out first.

EXAMPLE 5 Factor: $2x^4y - 32y$.

Solution

$$\begin{aligned}2x^4y - 32y &= 2y(x^4 - 16) && \text{Factor out } 2y. \\ &= 2y(x^2 + 4)(x^2 - 4) && \text{Factor } x^4 - 16. \\ &= 2y(x^2 + 4)(x + 2)(x - 2) && \text{Factor } x^2 - 4.\end{aligned}$$

SELF CHECK 5 Factor: $3a^4 - 3$.

2 Factor the sum and difference of two cubes.

The number 64 is called a **perfect cube,** because $4^3 = 64$. To factor the sum or difference of two cubes, it is helpful to know the first ten positive perfect cubes:

1, 8, 27, 64, 125, 216, 343, 512, 729, 1,000

Expressions like $x^9y^6z^3$ are also perfect cubes, because they can be written as the cube of another quantity:

$$x^9y^6z^3 = (x^3y^2z)^3$$

To find formulas for factoring the **sum or difference of two cubes,** we use the following product formulas:

(2) $(x + y)(x^2 - xy + y^2) = x^3 + y^3$

(3) $(x - y)(x^2 + xy + y^2) = x^3 - y^3$

To verify Equation 2, we multiply $x^2 - xy + y^2$ by $x + y$ and see that the product is $x^3 + y^3$.

$$\begin{aligned}(x + y)(x^2 - xy + y^2) &= (x + y)x^2 - (x + y)xy + (x + y)y^2 \\ &= x \cdot x^2 + y \cdot x^2 - x \cdot xy - y \cdot xy + x \cdot y^2 + y \cdot y^2\end{aligned}$$

$$= x^3 + x^2y - x^2y - xy^2 + xy^2 + y^3$$
$$= x^3 + y^3 \quad \text{Combine like terms.}$$

Equation 3 also can be verified by multiplication.

If we write Equations 2 and 3 in reverse order, we have the formulas for factoring the sum and difference of two cubes.

Sum and Difference of Two Cubes

$$x^3 + y^3 = (x + y)(x^2 - xy + y^2)$$
$$x^3 - y^3 = (x - y)(x^2 + xy + y^2)$$

If we think of the sum of two cubes as the sum of the cube of a **F**irst quantity plus the cube of a **L**ast quantity, we have the formula

$$\text{F}^3 + \text{L}^3 = (\text{F} + \text{L})(\text{F}^2 - \text{FL} + \text{L}^2)$$

and we say to factor the cube of a First quantity plus the cube of a Last quantity, we multiply the sum of the First and Last by

- *the First squared*
- *minus the First times the Last*
- *plus the Last squared.*

The formula for the difference of two cubes is

$$\text{F}^3 - \text{L}^3 = (\text{F} - \text{L})(\text{F}^2 + \text{FL} + \text{L}^2)$$

and we say to factor the cube of a First quantity minus the cube of a Last quantity, we multiply the difference of the First and Last by

- *the First squared*
- *plus the First times the Last*
- *plus the Last squared.*

EXAMPLE 6 Factor: $a^3 + 8$.

Solution Since $a^3 + 8$ can be written as $a^3 + 2^3$, we have the sum of two cubes, which factors as follows:

$$\begin{aligned} \text{F}^3 + \text{L}^3 &= (\text{F} + \text{L})(\text{F}^2 - \text{F}\ \ \text{L} + \text{L}^2) \\ \downarrow \quad \downarrow \quad & \quad \ \downarrow \quad \downarrow \quad \downarrow \quad \ \ \downarrow \ \ \downarrow \quad \downarrow \\ a^3 + 2^3 &= (a + 2)(a^2 - a\ \ 2 + 2^2) \\ &= (a + 2)(a^2 - 2a + 4) \end{aligned}$$

Thus, $a^3 + 8 = (a + 2)(a^2 - 2a + 4)$. Check by multiplication.

SELF CHECK 6 Factor: $p^3 + 27$.

EXAMPLE 7 Factor: $27a^3 - 64b^3$.

Solution Since $27a^3 - 64b^3$ can be written as $(3a)^3 - (4b)^3$, we have the difference of two cubes, which factors as follows:

$$\begin{array}{ccccccccccccc} F^3 & - & L^3 & = (& F & - & L &)(& F^2 & + & F\ L & + & L^2) \\ \downarrow & & \downarrow & & \downarrow & & \downarrow & & \downarrow & & \downarrow\ \downarrow & & \downarrow \end{array}$$

$$\begin{aligned}(3a)^3 - (4b)^3 &= (3a - 4b)[(3a)^2 + (3a)(4b) + (4b)^2] \\ &= (3a - 4b)(9a^2 + 12ab + 16b^2)\end{aligned}$$

Thus, $27a^3 - 64b^3 = (3a - 4b)(9a^2 + 12ab + 16b^2)$. Check by multiplication.

SELF CHECK 7 Factor: $8p^3 - 27q^3$.

EXAMPLE 8 Factor: $2a^5 + 128a^2$.

Solution We first factor out the common factor of $2a^2$ to obtain

$$2a^5 + 128a^2 = 2a^2(a^3 + 64)$$

Then we factor $a^3 + 64$ as the sum of two cubes to obtain

$$2a^5 + 128a^2 = 2a^2(a + 4)(a^2 - 4a + 16)$$

SELF CHECK 8 Factor: $3x^5 + 24x^2$.

EXAMPLE 9 Factor: $x^6 - 64$.

Solution This expression can be considered as the difference of two squares or the difference of two cubes. If we treat it as the difference of two squares, it factors into the product of a sum and a difference.

$$\begin{aligned}x^6 - 64 &= (x^3)^2 - 8^2 \\ &= (x^3 + 8)(x^3 - 8)\end{aligned}$$

Each of these factors further, however, for one is the sum of two cubes and the other is the difference of two cubes:

$$x^6 - 64 = (x + 2)(x^2 - 2x + 4)(x - 2)(x^2 + 2x + 4)$$

SELF CHECK 9 Factor: $x^6 - 1$.

EXAMPLE 10 Factor: $a^3 - (c + d)^3$.

Solution

$$\begin{aligned}a^3 - (c + d)^3 &= [a - (c + d)][a^2 + a(c + d) + (c + d)^2] \\ &= (a - c - d)(a^2 + ac + ad + c^2 + 2cd + d^2)\end{aligned}$$

SELF CHECK 10 Factor: $(p + q)^3 - r^3$.

EXAMPLE 11 Factor: $16r^{6m} - 54t^{3n}$.

Solution

$$\begin{aligned} 16r^{6m} - 54t^{3n} &= 2(8r^{6m} - 27t^{3n}) && \text{Factor out 2.} \\ &= 2[(2r^{2m})^3 - (3t^n)^3] && \text{Write } 8r^{6m} \text{ as } (2r^{2m})^3 \text{ and } 27t^{3n} \text{ as } (3t^n)^3. \\ &= 2[(2r^{2m} - 3t^n)(4r^{4m} + 6r^{2m}t^n + 9t^{2n})] && \text{Factor } (2r^{2m})^3 - (3t^n)^3. \end{aligned}$$

SELF CHECK 11 Factor: $2a^{3m} - 16b^{3n}$.

3 Factor polynomials by grouping.

EXAMPLE 12 Factor: $x^2 - y^2 + x - y$.

Solution If we group the first two terms and factor the difference of two squares, we have

$$\begin{aligned} x^2 - y^2 + x - y &= (x + y)(x - y) + (x - y) && \text{Factor } x^2 - y^2. \\ &= (x - y)(x + y + 1) && \text{Factor out } x - y. \end{aligned}$$

SELF CHECK 12 Factor: $a^2 - b^2 + a + b$.

SELF CHECK ANSWERS

1. $(9p + 5)(9p - 5)$ **2.** $(6r^2 + s)(6r^2 - s)$ **3.** $(a^2 + 4)(a + 2)(a - 2)$ **4.** $(a - b + c)(a - b - c)$ **5.** $3(a^2 + 1)(a + 1)(a - 1)$ **6.** $(p + 3)(p^2 - 3p + 9)$ **7.** $(2p - 3q)(4p^2 + 6pq + 9q^2)$ **8.** $3x^2(x + 2)(x^2 - 2x + 4)$ **9.** $(x + 1)(x^2 - x + 1)(x - 1)(x^2 + x + 1)$ **10.** $(p + q - r)(p^2 + 2pq + q^2 + pr + qr + r^2)$ **11.** $2(a^m - 2b^n)(a^{2m} + 2a^mb^n + 4b^{2n})$ **12.** $(a + b)(a - b + 1)$

NOW TRY THIS

Factor.

1. $\frac{1}{9}x^2 - \frac{1}{4}$

2. $x^{2/3} - 1$

3. $x^{3n} - 1$

5.5 EXERCISES

WARM-UPS *Factor each expression, if possible.*

1. $x^2 - 1$

2. $a^4 - 16$

3. $x^3 + 1$

4. $a^3 - 8$

5. $2x^2 - 8$

6. $x^4 + 25$

REVIEW *Perform each multiplication.*

7. $(x + 1)(x + 1)$
8. $(2m - 3)(m - 2)$
9. $(2m + n)(2m + n)$
10. $(3m - 2n)(3m - 2n)$
11. $(a + 4)(a + 3)$
12. $(3b + 2)(2b - 5)$
13. $(4r - 3s)(2r - s)$
14. $(5a - 2b)(3a + 4b)$

VOCABULARY AND CONCEPTS *Fill in the blanks.*

15. Write the first ten perfect-square natural numbers.
 ____
16. The difference of two squares is factored as $p^2 - q^2 = (p + q)$______
17. The sum of two squares ______ be factored over the real numbers except for common factors.
18. Write the first ten perfect-cube natural numbers.
 ____
19. The sum of two cubes is factored as $p^3 + q^3 = (p + q)$ ____________.
20. The difference of two cubes is factored as $p^3 - q^3 = (p - q)$____________.

GUIDED PRACTICE

Factor each expression, if possible. See Example 1. (Objective 1)

21. $x^2 - 4$
22. $y^2 - 9$
23. $t^2 - 225$
24. $p^2 - 400$
25. $b^2 - 81$
26. $a^2 - 64$

Factor each expression, if possible. See Examples 2–3. (Objective 1)

27. $x^2 + 25$
28. $y^2 + 49$
29. $9y^2 - 64$
30. $16x^4 - 81y^2$
31. $144a^2 - b^4$
32. $625a^2 - 169b^4$
33. $4y^2 + 9z^4$
34. $81a^4 - 49b^2$
35. $x^4 - y^4$
36. $64r^6 - 121s^2$

Factor each expression. See Example 4. (Objective 1)

37. $(x + y)^2 - z^2$
38. $a^2 - (b - c)^2$
39. $(a - b)^2 - c^2$
40. $(m + n)^2 - p^4$

Factor completely. See Example 5. (Objective 1)

41. $2x^2 - 288$
42. $8x^2 - 72$
43. $2x^3 - 32x$
44. $3x^3 - 243x$
45. $5x^3 - 125x$
46. $6x^4 - 216x^2$
47. $r^2s^2t^2 - t^2x^4y^2$
48. $16a^4b^3c^4 - 64a^2bc^6$

Factor each expression, if possible. See Examples 6–7. (Objective 2)

49. $r^3 + s^3$
50. $t^3 - v^3$
51. $p^3 - q^3$
52. $m^3 + n^3$
53. $x^3 - 8y^3$
54. $27a^3 + b^3$
55. $81x^3 + y^3$
56. $64a^3 - 125b^6$

Factor each expression completely. See Example 8. (Objective 2)

57. $5x^3 + 625$
58. $2x^3 - 128$
59. $4x^5 - 256x^2$
60. $2x^6 + 54x^3$
61. $128u^2v^3 - 2t^3u^2$
62. $56rs^2t^3 + 7rs^2v^6$
63. $(a + b)x^3 + 27(a + b)$
64. $(c - d)r^3 - (c - d)s^3$

Factor each expression. See Examples 9–10. (Objective 2)

65. $x^6 - y^6$
66. $x^9 + y^9$
67. $x^3 + (y - z)^3$
68. $x^3 - (y + 2)^3$

Factor each expression. Assume m and n are natural numbers. See Example 11. (Objective 2).

69. $x^{2m} - y^{4n}$
70. $a^{4m} - b^{8n}$
71. $100a^{4m} - 81b^{2n}$
72. $25x^{8m} - 36y^{4n}$

Factor each expression by grouping. See Example 12. (Objective 3)

73. $a^2 - b^2 + a + b$
74. $x^2 - y^2 - x - y$
75. $a^2 - b^2 + 2a - 2b$
76. $m^2 - n^2 + 3m + 3n$

ADDITIONAL PRACTICE *Factor each expression completely, if possible. Assume variable exponents are natural numbers.*

77. $4a^2b^4c^6 - 9d^8$

78. $36x^4y^2 - 49z^4$

79. $8x^6 + 125y^3$

80. $16a^4 - 81b^4$

81. $256x^4y^4 - z^8$

82. $125x^3y^6 + 216z^9$

83. $1{,}000a^6 - 343b^3c^6$

84. $225a^4 - 16b^8c^{12}$

85. $2x + y + 4x^2 - y^2$

86. $m - 2n + m^2 - 4n^2$

87. $x^{3n} - 8$

88. $a^{3m} + 64$

89. $a^{3m} + b^{3n}$

90. $x^{6m} - y^{3n}$

91. $2x^{6m} + 16y^{3m}$

92. $24 + 3c^{3m}$

APPLICATIONS

93. Physics The illustration shows a time-sequence picture of a falling apple. Factor the expression, which gives the difference in the distance fallen by the apple during the time interval from t_1 to t_2 seconds.

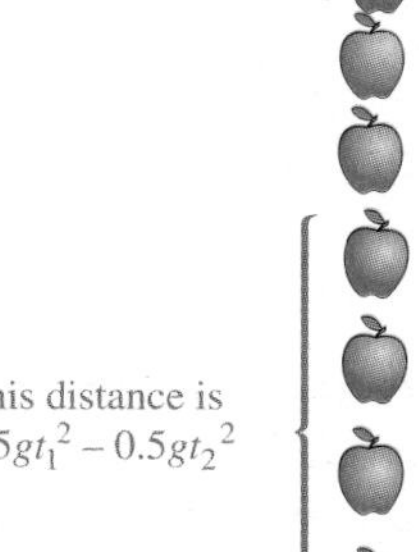

This distance is $0.5gt_1^2 - 0.5gt_2^2$

94. Darts A circular dart board has a series of rings around a solid center, called the bullseye. To find the area of the outer white ring, we can use the formula

$$A = \pi R^2 - \pi r^2$$

Factor the expression on the right side of the equation.

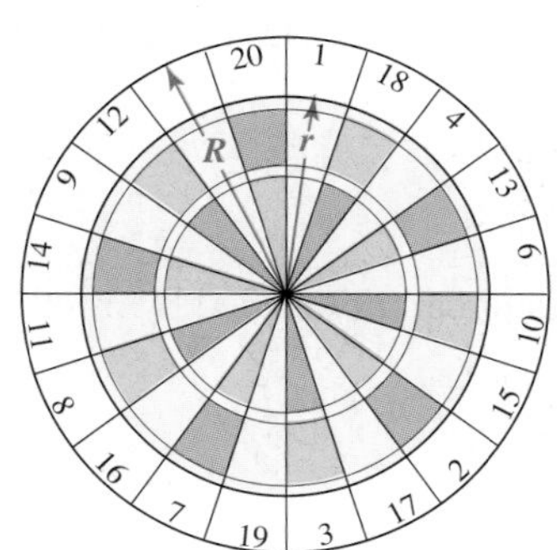

95. Candy To find the amount of chocolate in the outer coating of a malted-milk ball, we can use the formula for the volume V of the chocolate shell:

$$V = \frac{4}{3}\pi r_1^3 - \frac{4}{3}\pi r_2^3$$

Factor the expression on the right side.

96. Movie stunts The distance that a stuntman is above the ground t seconds after he jumps from the top of the building shown in the illustration is given by the formula

$$h = 144 - 16t^2$$

Factor the right side of the equation.

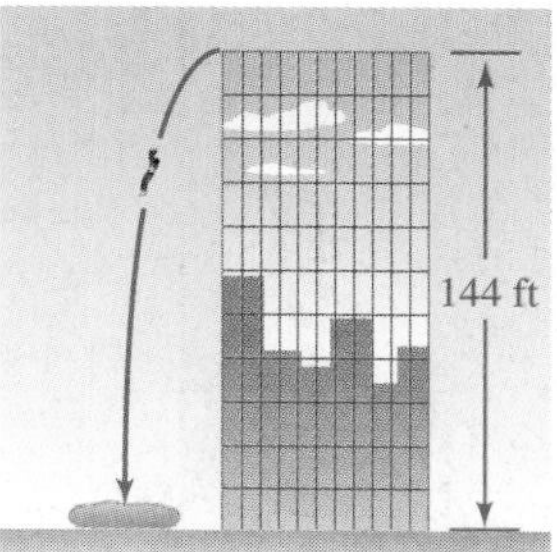

WRITING ABOUT MATH

97. Describe the pattern used to factor the difference of two squares.

98. Describe the patterns used to factor the sum and the difference of two cubes.

SOMETHING TO THINK ABOUT

99. Factor $x^{32} - y^{32}$.

100. Find the error in this proof that $2 = 1$.

$$\begin{aligned} x &= y \\ x^2 &= xy \\ x^2 - y^2 &= xy - y^2 \\ (x + y)(x - y) &= y(x - y) \\ \frac{(x + y)(x - y)}{x - y} &= \frac{y(x - y)}{x - y} \\ x + y &= y \\ y + y &= y \\ 2y &= y \\ \frac{2y}{y} &= \frac{y}{y} \\ 2 &= 1 \end{aligned}$$

SECTION 5.6 Factoring Trinomials

Objectives

1. Factor a perfect-square trinomial.
2. Factor a trinomial with a leading coefficient of 1.
3. Factor a trinomial with a leading coefficient other than 1.
4. Determine whether a trinomial is factorable over the integers.
5. Factor a trinomial using substitution.
6. Factor a trinomial by grouping (*ac* method).
7. Factor a polynomial with four terms by grouping a trinomial.

Vocabulary

key number

Getting Ready

Multiply.

1. $(a + 3)(a - 4)$
2. $(3a + 5)(2a - 1)$
3. $(a + b)(a + 2b)$
4. $(2a + b)(a - b)$
5. $(3a - 2b)(2a - 3b)$
6. $(4a - 3b)^2$

In this section, we will learn how to factor trinomials. We will consider two methods: a trial-and-error method and a factoring-by-grouping method.

1 Factor a perfect-square trinomial.

Perfect-square trinomials can be factored by using the following special product formulas.

(1) $(x + y)(x + y) = x^2 + 2xy + y^2$

(2) $(x - y)(x - y) = x^2 - 2xy + y^2$

To factor the perfect-square trinomial $x^2 + 6x + 9$, we note that it can be written in the form $x^2 + 2(3)x + 3^2$. If $y = 3$, this form matches the right side of Equation 1. Thus, $x^2 + 6x + 9$ factors as

$$\begin{aligned} x^2 + 6x + 9 &= x^2 + 2(3)x + 3^2 \\ &= (x + 3)(x + 3) \end{aligned}$$

which is generally written as $(x + 3)^2$. This result can be verified by multiplication:

$$\begin{aligned} (x + 3)(x + 3) &= x^2 + 3x + 3x + 9 \\ &= x^2 + 6x + 9 \end{aligned}$$

To factor the perfect-square trinomial $x^2 - 4xz + 4z^2$, we note that it can be written in the form $x^2 - 2x(2z) + (2z)^2$. If $y = 2z$, this form matches the right side of Equation 2. Thus, $x^2 - 4xz + 4z^2$ factors as

$$\begin{aligned} x^2 - 4xz + 4z^2 &= x^2 - 2x(2z) + (2z)^2 \\ &= (x - 2z)(x - 2z) \\ &= (x - 2z)^2 \end{aligned}$$

This result also can be verified by multiplication.

2 Factor a trinomial with a leading coefficient of 1.

Since the product of two binomials is often a trinomial, we expect that many trinomials will factor as two binomials. For example, to factor $x^2 + 7x + 12$, we must find two binomials $x + a$ and $x + b$ such that

$$x^2 + 7x + 12 = (x + a)(x + b)$$

where $ab = 12$ and $ax + bx = 7x$.

To find the numbers a and b, we list the possible factorizations of 12 and find the one where the sum of the factors is 7.

The one to choose (↓ 4(3))

$12(1) \quad 6(2) \quad 4(3) \quad -12(-1) \quad -6(-2) \quad -4(-3)$

Thus, $a = 4$, $b = 3$, and

$$x^2 + 7x + 12 = (x + a)(x + b)$$

$$x^2 + 7x + 12 = (x + 4)(x + 3) \qquad (3)$$ Substitute 4 for a and 3 for b.

This factorization can be verified by multiplying $x + 4$ and $x + 3$ and observing that the product is $x^2 + 7x + 12$.

Because of the commutative property of multiplication, the order of the factors in Equation 3 is not important.

To factor trinomials with leading coefficients of 1, we follow these steps.

Factoring Trinomials with Leading Coefficients of 1

1. Write the trinomial in descending powers of one variable.
2. List the factorizations of the third term of the trinomial.
3. Pick the factorization where the sum of the factors is the coefficient of the middle term.

EXAMPLE 1 Factor: $x^2 - 6x + 8$.

Solution Since the trinomial is written in descending powers of x, we can move to Step 2 and list the possible factorizations of the third term, which is 8.

The one to choose (↓ −4(−2))

$8(1) \quad 4(2) \quad -8(-1) \quad -4(-2)$

In the trinomial, the coefficient of the middle term is -6. The only factorization where the sum of the factors is -6 is $-4(-2)$. Thus, $a = -4$, $b = -2$, and

$$x^2 - 6x + 8 = (x + a)(x + b)$$
$$= (x - 4)(x - 2) \quad \text{Substitute } -4 \text{ for } a \text{ and } -2 \text{ for } b.$$

We can verify this result by multiplication:

$$(x - 4)(x - 2) = x^2 - 2x - 4x + 8$$
$$= x^2 - 6x + 8$$

SELF CHECK 1 Factor: $a^2 - 7a + 12$.

EXAMPLE 2 Factor: $-x + x^2 - 12$.

Solution We begin by writing the trinomial in descending powers of x:

$$-x + x^2 - 12 = x^2 - x - 12$$

The possible factorizations of the third term are

The one to choose ↓

$12(-1) \quad 6(-2) \quad 4(-3) \quad 1(-12) \quad 2(-6) \quad 3(-4)$

In the trinomial, the coefficient of the middle term is -1. The only factorization where the sum of the factors is -1 is $3(-4)$. Thus, $a = 3$, $b = -4$, and

$$-x + x^2 - 12 = x^2 - x - 12$$
$$= (x + a)(x + b)$$
$$= (x + 3)(x - 4) \quad \text{Substitute 3 for } a \text{ and } -4 \text{ for } b.$$

SELF CHECK 2 Factor: $-3a + a^2 - 10$.

EXAMPLE 3 Factor completely: $30x - 4xy - 2xy^2$.

Solution We begin by writing the trinomial in descending powers of y:

$$30x - 4xy - 2xy^2 = -2xy^2 - 4xy + 30x$$

Each term in this trinomial has a common monomial factor of $-2x$, which we will factor out.

$$30x - 4xy - 2xy^2 = -2x(y^2 + 2y - 15)$$

To factor $y^2 + 2y - 15$, we list the factors of -15 and find the pair whose sum is 2.

COMMENT In Example 3, be sure to include all factors in the final result! Remember that you must be able to multiply all factors and obtain the original polynomial.

The one to choose ↓

$15(-1) \quad 5(-3) \quad 1(-15) \quad 3(-5)$

The only factorization where the sum of the factors is 2 (the coefficient of the middle term of $y^2 + 2y - 15$) is $5(-3)$. Thus, $a = 5$, $b = -3$, and

$$\begin{aligned} 30x - 4xy - 2xy^2 &= -2x(y^2 + 2y - 15) \\ &= -2x(y + 5)(y - 3) \end{aligned}$$

SELF CHECK 3 Factor: $18a + 3ab - 3ab^2$.

3 Factor a trinomial with a leading coefficient other than 1.

There are more combinations of factors to consider when factoring trinomials with leading coefficients other than 1. To factor $5x^2 + 7x + 2$, for example, we must find two binomials of the form $ax + b$ and $cx + d$ such that

$$5x^2 + 7x + 2 = (ax + b)(cx + d)$$

Since the first term of the trinomial $5x^2 + 7x + 2$ is $5x^2$, the first terms of the binomial factors must be $5x$ and x.

$$5x^2 + 7x + 2 = (\overbrace{5x + b)(x}^{5x^2} + d)$$

Since the product of the last terms must be 2, and the sum of the products of the outer and inner terms must be $7x$, we must find two numbers whose product is 2 that will give a middle term of $7x$.

$$5x^2 + 7x + 2 = (\underbrace{5x + \overbrace{b)(x + d}^{2}}_{\text{O} + \text{I} = 7x})$$

Since $2(1)$ and $(-2)(-1)$ give a product of 2, there are four possible combinations to consider:

$$\begin{array}{ll} \mathbf{(5x + 2)(x + 1)} & (5x - 2)(x - 1) \\ (5x + 1)(x + 2) & (5x - 1)(x - 2) \end{array}$$

Of these possibilities, only the one printed in red gives the correct middle term of $7x$. Thus,

$$5x^2 + 7x + 2 = (5x + 2)(x + 1)$$

We can verify this result by multiplication:

$$\begin{aligned} (5x + 2)(x + 1) &= 5x^2 + 5x + 2x + 2 \\ &= 5x^2 + 7x + 2 \end{aligned}$$

4 Determine whether a trinomial is factorable over the integers.

If a trinomial has the form $ax^2 + bx + c$, with integer coefficients and $a \neq 0$, we can test to see whether it is factorable over the integers.

- If the value of $b^2 - 4ac$ is a perfect square, the trinomial can be factored using only integers.
- If the value is not a perfect square, the trinomial cannot be factored using only integers.

For example, $5x^2 + 7x + 2$ is a trinomial in the form $ax^2 + bx + c$ with

$$a = 5, \qquad b = 7, \qquad \text{and} \qquad c = 2$$

For this trinomial, the value of $b^2 - 4ac$ is

$$\begin{aligned} b^2 - 4ac &= 7^2 - 4(5)(2) && \text{Substitute 7 for } b, \text{ 5 for } a, \text{ and 2 for } c. \\ &= 49 - 40 \\ &= 9 \end{aligned}$$

Since 9 is a perfect square, the trinomial is factorable. Its factorization is $(5x + 2)(x + 1)$, as discussed above.

Test for Factorability

A trinomial of the form $ax^2 + bx + c$, with integer coefficients and $a \neq 0$, will factor into two binomials with integer coefficients if the value of $b^2 - 4ac$ is a perfect square. If $b^2 - 4ac = 0$, the factors will be the same.

EXAMPLE 4 Determine whether the polynomial $3p^2 - 4p - 4$ is factorable. If it is, factor it.

Solution In the trinomial, $a = 3$, $b = -4$, and $c = -4$. To see whether it factors, we evaluate $b^2 - 4ac$.

$$\begin{aligned} b^2 - 4ac &= (-4)^2 - 4(3)(-4) && \text{Substitute } -4 \text{ for } b, \text{ 3 for } a, \text{ and } -4 \text{ for } c. \\ &= 16 + 48 \\ &= 64 \end{aligned}$$

Since 64 is a perfect square, it is factorable.

To factor the trinomial, we note that the first terms of the binomial factors must be $3p$ and p to give the first term of $3p^2$.

$$3p^2 - 4p - 4 = (3p + ?)(p + ?) \quad [3p^2]$$

The product of the last terms must be -4, and the sum of the products of the outer terms and the inner terms must be $-4p$.

$$3p^2 - 4p - 4 = (3p + ?)(p + ?) \quad [-4]$$

$$\text{O} + \text{I} = -4p$$

Because $1(-4)$, $-1(4)$, and $-2(2)$ all give a product of -4, there are six possible combinations to consider:

$$\begin{aligned} &(3p + 1)(p - 4) && (3p - 4)(p + 1) \\ &(3p - 1)(p + 4) && (3p + 4)(p - 1) \\ &(3p - 2)(p + 2) && (3p + 2)(p - 2) \end{aligned}$$

Of these possibilities, only the one printed in red gives the correct middle term of $-4p$. Thus,

$$3p^2 - 4p - 4 = (3p + 2)(p - 2)$$

SELF CHECK 4 Determine whether the polynomial $4q^2 - 9q - 9$ is factorable.

EXAMPLE 5 Factor $4t^2 - 3t - 5$, if possible.

Solution In the trinomial, $a = 4$, $b = -3$, and $c = -5$. To see whether it is factorable, we evaluate $b^2 - 4ac$ by substituting the values of a, b, and c.

$$\begin{aligned} b^2 - 4ac &= (-3)^2 - 4(4)(-5) \\ &= 9 + 80 \\ &= 89 \end{aligned}$$

Since 89 is not a perfect square, the trinomial is not factorable using only integer coefficients. It is prime.

SELF CHECK 5 Factor: $5a^2 - 8a + 2$, if possible.

To factor trinomials, the following hints are helpful.

Factoring a Trinomial

1. Write the trinomial in descending powers of one variable.
2. Factor out any greatest common factor (including -1 if that is necessary to make the coefficient of the first term positive).
3. Test the trinomial for factorability.
4. When the sign of the first term of a trinomial is $+$ and the sign of the third term is $+$, the signs between the terms of each binomial factor are the same as the sign of the middle term of the trinomial.
   When the sign of the first term is $+$ and the sign of the third term is $-$, the signs between the terms of the binomials are opposites.
5. Try various combinations of first terms and last terms until you find the one that works.
6. Check the factorization by multiplication.

EXAMPLE 6 Factor completely: $24y + 10xy - 6x^2y$.

Solution We write the trinomial in descending powers of x and factor out the common factor of $-2y$:

$$\begin{aligned} 24y + 10xy - 6x^2y &= -6x^2y + 10xy + 24y \\ &= -2y(3x^2 - 5x - 12) \end{aligned}$$

In the trinomial $3x^2 - 5x - 12$, $a = 3$, $b = -5$, and $c = -12$.

$$\begin{aligned} b^2 - 4ac &= (-5)^2 - 4(3)(-12) \\ &= 25 + 144 \\ &= 169 \end{aligned}$$

Since 169 is a perfect square, the trinomial will factor.

Since the sign of the third term of $3x^2 - 5x - 12$ is $-$, the signs between the binomial factors will be opposite. Because the first term is $3x^2$, the first terms of the binomial factors must be $3x$ and x.

$$-2y(3x^2 - 5x - 12) = -2y(\overset{3x^2}{\overbrace{3x \quad)(x}} \quad)$$

The product of the last terms must be -12, and the sum of the product of the outer terms and the product of the inner terms must be $-5x$.

$$-2y(3x^2 - 5x - 12) = -2y(3x \quad ?)(x \quad ?)$$

(last terms: -12; $\text{O} + \text{I} = -5x$)

Since $1(-12)$, $2(-6)$, $3(-4)$, $12(-1)$, $6(-2)$, and $4(-3)$ all give a product of -12, there are 12 possible combinations to consider.

	$(3x + 1)(x - 12)$	$(3x - 12)(x + 1)$
	$(3x + 2)(x - 6)$	$(3x - 6)(x + 2)$
	$(3x + 3)(x - 4)$	$(3x - 4)(x + 3)$
	$(3x + 12)(x - 1)$	$(3x - 1)(x + 12)$
	$(3x + 6)(x - 2)$	$(3x - 2)(x + 6)$
The one to choose →	$(3x + 4)(x - 3)$	$(3x - 3)(x + 4)$

The combinations marked in blue cannot work, because one of the binomial factors has a common factor. This implies that $3x^2 - 5x - 12$ would have a common factor, which it doesn't.

After mentally trying the remaining combinations, we find that only $(3x + 4)(x - 3)$ gives the correct middle term of $-5x$.

$$\begin{aligned} 24y + 10xy - 6x^2y &= -2y(3x^2 - 5x - 12) \\ &= -2y(3x + 4)(x - 3) \end{aligned}$$

Verify this result by multiplication.

SELF CHECK 6 Factor: $9b - 6a^2b - 3ab$.

EXAMPLE 7 Factor: $6y + 13x^2y + 6x^4y$.

Solution We write the trinomial in descending powers of x and factor out the common factor of y to obtain

$$\begin{aligned} 6y + 13x^2y + 6x^4y &= 6x^4y + 13x^2y + 6y \\ &= y(6x^4 + 13x^2 + 6) \end{aligned}$$

A test of the trinomial $6x^4 + 13x^2 + 6$ will show that it does factor. Since the coefficients of the first and last terms are positive, the signs between the terms in each binomial will be +.

Since the first term of the trinomial is $6x^4$, the first terms of the binomial factors must be either $2x^2$ and $3x^2$ or x^2 and $6x^2$.

Since the product of the last terms of the binomial factors must be 6, we must find two numbers whose product is 6 that will lead to a middle term of $13x^2$. After trying some combinations, we find the one that works.

$$\begin{aligned} 6y + 13x^2y + 6x^4y &= y(6x^4 + 13x^2 + 6) \\ &= y(2x^2 + 3)(3x^2 + 2) \end{aligned}$$

Verify this result by multiplication.

SELF CHECK 7 Factor: $4b + 11a^2b + 6a^4b$.

EXAMPLE 8 Factor: $x^{2n} + x^n - 2$.

Solution Since the first term is x^{2n}, the first terms of the binomial factors must be x^n and x^n.

$$x^{2n} + x^n - 2 = (\overset{x^{2n}}{x^n \quad)(x^n} \quad)$$

Since the third term of the trinomial is -2, the last terms of the binomial factors must have opposite signs, have a product of -2, and lead to a middle term of x^n. The only combination that works is

$$x^{2n} + x^n - 2 = (x^n + 2)(x^n - 1)$$

Verify this result by multiplication.

SELF CHECK 8 Factor: $a^{2n} + a^n - 6$.

5 Factor a trinomial using substitution.

EXAMPLE 9 Factor: $(x + y)^2 + 7(x + y) + 12$.

Solution We can write the trinomial $(x + y)^2 + 7(x + y) + 12$ as $z^2 + 7z + 12$, where $z = (x + y)$. The trinomial $z^2 + 7z + 12$ factors as $(z + 4)(z + 3)$.

To find the factorization of $(x + y)^2 + 7(x + y) + 12$, we substitute $x + y$ for z in the expression $(z + 4)(z + 3)$ to obtain

$$z^2 + 7z + 12 = (z + 4)(z + 3)$$
$$(x + y)^2 + 7(x + y) + 12 = (x + y + 4)(x + y + 3)$$

SELF CHECK 9 Factor: $(a + b)^2 - 3(a + b) - 10$.

6 Factor a trinomial by grouping (*ac* method).

Factoring by grouping can be used to help factor trinomials of the form $ax^2 + bx + c$. For example, to factor $6x^2 + 7x - 3$, we proceed as follows:

1. We find the product ac: $6(-3) = -18$. This is called the **key number.**
2. We find two factors of the key number -18 whose sum is $b = 7$:

$$9(-2) = -18 \qquad \text{and} \qquad 9 + (-2) = 7$$

3. We use the factors 9 and -2 as coefficients of two terms to be placed between $6x^2$ and -3:

$$6x^2 + 7x - 3 = 6x^2 + 9x - 2x - 3$$

4. We factor by grouping:

$$6x^2 + 9x - 2x - 3 = 3x(2x + 3) - 1(2x + 3)$$
$$= (2x + 3)(3x - 1) \quad \text{Factor out } 2x + 3.$$

We can verify this factorization by multiplication.

EXAMPLE 10 Factor: $10x^2 + 13x - 3$.

Solution Since $a = 10$ and $c = -3$ in the trinomial, $ac = -30$. We must find two factors of -30 whose sum is $+13$. Two such factors are 15 and -2. We use these factors as coefficients of two terms to be placed between $10x^2$ and -3:

$$10x^2 + 15x - 2x - 3$$

Finally, we factor by grouping.

$$10x^2 + 15x - 2x - 3 = 5x(2x + 3) - 1(2x + 3)$$
$$= (2x + 3)(5x - 1)$$

Thus, $10x^2 + 13x - 3 = (2x + 3)(5x - 1)$.

SELF CHECK 10 Factor: $15a^2 + 17a - 4$.

7 Factor a polynomial with four terms by grouping a trinomial.

In the next example, we will group three terms together and factor a trinomial.

EXAMPLE 11 Factor: $x^2 + 6x + 9 - z^2$.

Solution We group the first three terms together and factor the trinomial to get

$$x^2 + 6x + 9 - z^2 = (x + 3)(x + 3) - z^2$$
$$= (x + 3)^2 - z^2$$

We now can factor the difference of two squares to get

$$x^2 + 6x + 9 - z^2 = (x + 3 + z)(z + 3 - z)$$

SELF CHECK 11 Factor: $a^2 + 4a + 4 - b^2$.

SELF CHECK ANSWERS **1.** $(a - 4)(a - 3)$ **2.** $(a + 2)(a - 5)$ **3.** $-3a(b + 2)(b - 3)$ **4.** $(4q + 3)(q - 3)$ **5.** a prime polynomial **6.** $-3b(2a + 3)(a - 1)$ **7.** $b(2a^2 + 1)(3a^2 + 4)$ **8.** $(a^n + 3)(a^n - 2)$ **9.** $(a + b + 2)(a + b - 5)$ **10.** $(3a + 4)(5a - 1)$ **11.** $(a + 2 + b)(a + 2 - b)$

NOW TRY THIS

Factor completely.

1. $x^2 - x + \frac{1}{4}$

2. $3x^{2/3} - x^{1/3} - 2$

3. $3(\tan x)^2 - \tan x \cos x - 2(\cos x)^2$

5.6 EXERCISES

WARM-UPS *Factor each expression.*

1. $x^2 + 3x + 2$
2. $x^2 + 5x + 4$
3. $x^2 - 5x + 6$
4. $x^2 - 3x - 4$
5. $2x^2 + 3x + 1$
6. $3x^2 + 4x + 1$

REVIEW *Solve each equation.*

7. $\frac{2 + x}{11} = 3$
8. $\frac{3y - 12}{2} = 9$
9. $\frac{2}{3}(5t - 3) = 38$
10. $3(p + 2) = 4p$
11. $11r + 6(3 - r) = 3$
12. $2q^2 - 9 = q(q + 3) + q^2$

VOCABULARY AND CONCEPTS

Fill in the blanks.

13. $x^2 +$ ________ $= (x + y)(x + y)$,
 $x^2 -$ ________ $= (x - y)(x - y)$
14. $(x + y)(x - y) =$ ________
15. To factor a trinomial of the form $ax^2 + bx + c$ using the ac method, we must find the product ac. This product is called the ________.
16. If ________ is a perfect square, the trinomial $ax^2 + bx + c$ will factor using ________ coefficients.

Complete each factorization.

17. $x^2 + 5x + 6 = (x + 3)($____$)$
18. $x^2 - 6x + 8 = (x - 4)($____$)$
19. $x^2 + 2x - 15 = (x + 5)($____$)$
20. $x^2 - 3x - 18 = (x - 6)($____$)$
21. $2a^2 + 9a + 4 = ($____$)(a + 4)$
22. $6p^2 - 5p - 4 = ($____$)(2p + 1)$
23. $4m^2 + 8mn + 3n^2 = ($____$)(2m + n)$
24. $12r^2 + 5rs - 2s^2 = ($____$)(3r + 2s)$

GUIDED PRACTICE

Factor each perfect-square trinomial. (Objective 1)

25. $x^2 + 2x + 1$
26. $y^2 - 2y + 1$
27. $a^2 - 18a + 81$
28. $b^2 + 12b + 36$
29. $4y^2 + 4y + 1$
30. $9x^2 + 6x + 1$
31. $9b^2 - 12b + 4$
32. $4a^2 - 12a + 9$

Factor, if possible. See Example 1. (Objective 2)

33. $x^2 + 9x + 8$
34. $y^2 + 7y + 6$
35. $x^2 - 7x + 10$
36. $c^2 - 7c + 12$
37. $b^2 + 8b + 18$
38. $x^2 - 12x + 35$
39. $x^2 - x - 30$
40. $a^2 + 4a - 45$

Factor out all common monomials first (including −1 if the leading coefficient is negative) and then factor the remaining trinomial. See Examples 2–3. (Objective 2)

41. $-a^2 + 4a + 32$
42. $-x^2 - 2x + 15$

43. $-3x^2 + 15x - 18$

44. $-2y^2 - 16y + 40$

45. $3x^2 + 12x - 63$

46. $2y^2 + 4y - 48$

47. $a^2b^2 - 13ab^2 + 22b^2$

48. $a^2b^2x^2 - 18a^2b^2x + 81a^2b^2$

49. $b^2x^2 - 12bx^2 + 35x^2$

50. $c^3x^2 + 11c^3x - 42c^3$

51. $-4x^2 + 4x + 80$

52. $-5a^2 + 40a - 75$

Test each trinomial for factorability and factor it, if possible. See Examples 4–5. (Objective 4)

53. $6y^2 + 7y + 2$

54. $6x^2 - 11x + 3$

55. $8a^2 + 6a - 9$

56. $15b^2 + 4b - 4$

57. $6x^2 - 5x - 4$

58. $18y^2 - 3y - 10$

59. $5x^2 + 4x + 1$

60. $6z^2 + 17z + 12$

61. $3x^3 - 10x^2 + 3x$

62. $3t^3 - 3t^2 + t$

63. $-4x^2 - 9 + 12x$

64. $6x + 4 + 9x^2$

Test each trinomial for factorability and factor it, if possible. See Examples 6–7. (Objective 4)

65. $2y^2 + yt - 6t^2$

66. $3x^2 - 10xy - 8y^2$

67. $-3a^2 + ab + 2b^2$

68. $-2x^2 + 3xy + 5y^2$

69. $5a^2 + 45b^2 - 30ab$

70. $-90x^2 + 2 - 8x$

71. $8x^2z + 6xyz + 9y^2z$

72. $x^3 - 60xy^2 + 7x^2y$

Factor each expression. Assume any variable exponent is a natural number. See Example 8. (Objectives 2–3)

73. $x^{2n} + 2x^n + 1$

74. $x^{4n} - 2x^{2n} + 1$

75. $2a^{6n} - 3a^{3n} - 2$

76. $b^{2n} - b^n - 6$

77. $x^{4n} + 2x^{2n}y^{2n} + y^{4n}$

78. $y^{6n} + 2y^{3n}z + z^2$

79. $6x^{2n} + 7x^n - 3$

80. $12y^{4n} + 10y^{2n} + 2$

Factor each expression by substitution. Assume any variable exponent is a natural number. See Example 9. (Objective 5)

81. $(x + 1)^2 + 2(x + 1) + 1$

82. $(a + b)^2 - 2(a + b) + 1$

83. $(a + b)^2 - 2(a + b) - 24$

84. $(x - y)^2 + 3(x - y) - 10$

Use grouping (ac method) to factor each trinomial. See Example 10. (Objective 6)

85. $a^2 - 17a + 16$

86. $b^2 - 4b - 21$

87. $2u^2 + 5u + 3$

88. $6y^2 + 5y - 6$

89. $20r^2 - 7rs - 6s^2$

90. $6s^2 + st - 12t^2$

91. $20u^2 + 19uv + 3v^2$

92. $12m^2 + mn - 6n^2$

Factor each expression by first grouping a trinomial. See Example 11. (Objective 7)

93. $x^2 + 4x + 4 - y^2$

94. $x^2 - 6x + 9 - 4y^2$

95. $x^2 + 2x + 1 - 9z^2$

96. $x^2 + 10x + 25 - 16z^2$

97. $c^2 - 4a^2 + 4ab - b^2$

98. $4c^2 - a^2 - 6ab - 9b^2$

99. $a^2 - b^2 + 8a + 16$

100. $a^2 + 14a - 25b^2 + 49$

ADDITIONAL PRACTICE *Factor each polynomial completely, if possible.*

101. $a^2 - 3ab - 4b^2$

102. $b^2 + 2bc - 80c^2$

103. $9z^2 + 24z + 16$

104. $16z^2 - 24z + 9$

105. $a^2 + 5a - 50$

106. $b^2 + 9b - 36$

107. $y^2 - 4y - 21$

108. $x^2 + 4x - 28$

109. $4a^2 + 20a + 3$

110. $8x^2 - 10x + 3$

111. $21x^4 - 10x^3 - 16x^2$

112. $16x^3 - 50x^2 + 36x$

113. $x^4 + 8x^2 + 15$

114. $x^4 + 11x^2 + 24$

115. $y^4 - 13y^2 + 30$

116. $y^4 - 13y^2 + 42$

117. $a^4 - 13a^2 + 36$

118. $b^4 - 17b^2 + 16$

119. $z^4 - z^2 - 12$

120. $c^4 - 8c^2 - 9$

121. $4x^3 + x^6 + 3$

122. $a^6 - 2 + a^3$

123. $6(x + y)^2 - 7(x + y) - 20$

124. $2(x - z)^2 + 9(x - z) + 4$

125. $4x^2 - z^2 + 4xy + y^2$

126. $x^2 - 4xy - 4z^2 + 4y^2$

APPLICATIONS

127. Ice cubes The surface area of one face of an ice cube is given by $x^2 + 6x + 9$. Write a binomial that expresses the length of an edge of the cube.

128. Checkers The area of a square checkerboard is represented by $25x^2 - 40x + 16$. Write a binomial that expresses the length of one side of the board.

129. Office furniture The area of the desktop is given by the expression $(4x^2 + 20x - 11)$ in.^{2}. Factor this expression to find the expressions that can represent its length and width. Then determine the difference in the length and width of the desktop.

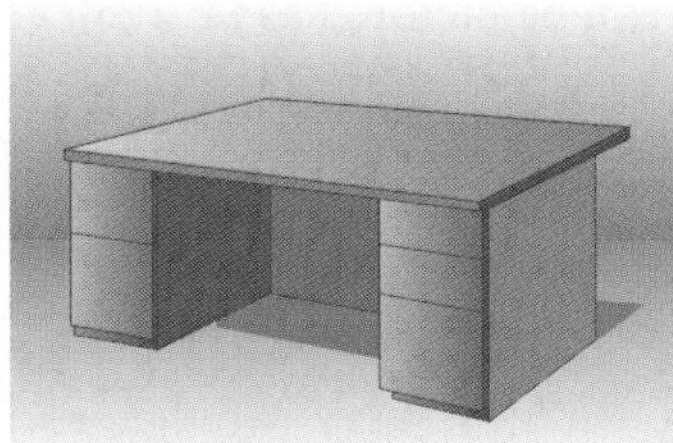

130. Storage The volume of the 8-foot-wide portable storage container is given by the expression $(72x^2 + 120x - 400)$ ft^3. If its dimensions can be determined by factoring the expression, find the height and the length of the container.

WRITING ABOUT MATH

131. Explain how you would factor -1 from a trinomial.

132. Explain how you would test the polynomial $ax^2 + bx + c$ for factorability.

SOMETHING TO THINK ABOUT

133. Because it is the difference of two squares, $x^2 - q^2$ always factors. Does the test for factorability predict this?

134. The polynomial $ax^2 + ax + a$ factors, because a is a common factor. Does the test for factorability predict this? If not, is there something wrong with the test? Explain.

SECTION 5.7 Summary of Factoring Techniques

Objectives

1 Factor a polynomial by applying an appropriate technique.

Getting Ready

Factor each polynomial.

1. $3p^2q - 3pq^2$
2. $4p^2 - 9q^2$
3. $p^2 + 5p - 6$
4. $6p^2 - 13pq + 6q^2$

In this section, we will discuss ways to approach a randomly chosen factoring problem.

1 Factor a polynomial by applying an appropriate technique.

Suppose we want to factor the trinomial

$$x^2y^2z^3 + 7xy^2z^3 + 6y^2z^3$$

We begin by attempting to identify the problem type. The first type to look for is one containing a common monomial. Because the trinomial has a common monomial factor of y^2z^3, we factor it out:

$$x^2y^2z^3 + 7xy^2z^3 + 6y^2z^3 = y^2z^3(x^2 + 7x + 6)$$

We note that $x^2 + 7x + 6$ is a trinomial that can be factored as $(x + 6)(x + 1)$. Thus,

$$\begin{aligned} x^2y^2z^3 + 7xy^2z^3 + 6y^2z^3 &= y^2z^3(x^2 + 7x + 6) \\ &= y^2z^3(x + 6)(x + 1) \end{aligned}$$

To identify the type of factoring problem, we follow these steps.

Identifying Factoring Problem Types

1. Factor out all common monomial factors.
2. If an expression has two terms, check to see if the problem type is
   a. **The difference of two squares:** $x^2 - y^2 = (x + y)(x - y)$
   b. **The sum of two cubes:** $x^3 + y^3 = (x + y)(x^2 - xy + y^2)$
   c. **The difference of two cubes:** $x^3 - y^3 = (x - y)(x^2 + xy + y^2)$
3. If an expression has three terms, attempt to factor it as a **trinomial.**
4. If an expression has four or more terms, try factoring by **grouping.**
5. Continue until each individual factor is prime, except possibly a monomial GCF.
6. Check the results by multiplying.

EXAMPLE 1 Factor: $48a^4c^3 - 3b^4c^3$.

Solution We begin by factoring out the common monomial factor of $3c^3$:

$$48a^4c^3 - 3b^4c^3 = 3c^3(16a^4 - b^4)$$

Since the expression $16a^4 - b^4$ has two terms, we check to see whether it is the difference of two squares, which it is. As the difference of two squares, it factors as $(4a^2 + b^2)(4a^2 - b^2)$.

$$\begin{aligned} 48a^4c^3 - 3b^4c^3 &= 3c^3(\mathbf{16a^4 - b^4}) \\ &= 3c^3(\mathbf{4a^2 + b^2})(\mathbf{4a^2 - b^2}) \end{aligned}$$

The binomial $4a^2 + b^2$ is the sum of two squares and is prime. However, $4a^2 - b^2$ is the difference of two squares and factors as $(2a + b)(2a - b)$.

$$\begin{aligned} 48a^4c^3 - 3b^4c^3 &= 3c^3(16a^4 - b^4) \\ &= 3c^3(4a^2 + b^2)(\mathbf{4a^2 - b^2}) \\ &= 3c^3(4a^2 + b^2)(\mathbf{2a + b})(\mathbf{2a - b}) \end{aligned}$$

Since each of the individual factors is prime, the factorization is complete.

SELF CHECK 1 Factor: $3p^4r^3 - 3q^4r^3$.

EXAMPLE 2 Factor: $x^5y + x^2y^4 - x^3y^3 - y^6$.

Solution We begin by factoring out the common monomial factor of y:

$$x^5y + x^2y^4 - x^3y^3 - y^6 = y(x^5 + x^2y^3 - x^3y^2 - y^5)$$

Because the expression $x^5 + x^2y^3 - x^3y^2 - y^5$ has four terms, we try factoring by grouping to obtain

$$\begin{aligned} x^5y + x^2y^4 - x^3y^3 - y^6 &= y(x^5 + x^2y^3 - x^3y^2 - y^5) && \text{Factor out } y. \\ &= y[x^2(x^3 + y^3) - y^2(x^3 + y^3)] && \text{Factor by grouping.} \\ &= y(x^3 + y^3)(x^2 - y^2) && \text{Factor out } x^3 + y^3. \end{aligned}$$

Finally, we factor $x^3 + y^3$ (the sum of two cubes) and $x^2 - y^2$ (the difference of two squares) to obtain

$$x^5y + x^2y^4 - x^3y^3 - y^6 = y(x + y)(x^2 - xy + y^2)(x + y)(x - y)$$

Because each of the individual factors is prime, the factorization is complete.

SELF CHECK 2 Factor: $a^5p - a^3b^2p + a^2b^3p - b^5p$.

EXAMPLE 3 Factor: $p^{-4} - p^{-2} - 6$ and write the result without using negative exponents.

Solution We factor this expression as if it were a trinomial:

$$\begin{aligned} p^{-4} - p^{-2} - 6 &= (p^{-2} - 3)(p^{-2} + 2) \\ &= \left(\frac{1}{p^2} - 3\right)\left(\frac{1}{p^2} + 2\right) \end{aligned}$$

SELF CHECK 3 Factor: $x^{-4} - x^{-2} - 20$.

EXAMPLE 4 Factor: $x^3 + 5x^2 + 6x + x^2y + 5xy + 6y$.

Solution Since there are more than three terms, we try factoring by grouping. We can factor x from the first three terms and y from the last three terms and proceed as follows:

$$\begin{aligned} &x^3 + 5x^2 + 6x + x^2y + 5xy + 6y \\ &\quad = x(x^2 + 5x + 6) + y(x^2 + 5x + 6) \\ &\quad = (x^2 + 5x + 6)(x + y) && \text{Factor out } x^2 + 5x + 6. \\ &\quad = (x + 3)(x + 2)(x + y) && \text{Factor } x^2 + 5x + 6. \end{aligned}$$

SELF CHECK 4 Factor: $a^3 - 5a^2 + 6a + a^2b - 5ab + 6b$.

EXAMPLE 5 Factor: $x^4 + 2x^3 + x^2 + x + 1$.

Solution Since there are more than three terms, we try factoring by grouping. We can factor x^2 from the first three terms and proceed as follows:

$$\begin{aligned} x^4 + 2x^3 + x^2 + x + 1 &= x^2(x^2 + 2x + 1) + (x + 1) \\ &= x^2(x + 1)(x + 1) + (x + 1) && \text{Factor } x^2 + 2x + 1. \\ &= (x + 1)[x^2(x + 1) + 1] && \text{Factor out } x + 1. \\ &= (x + 1)(x^3 + x^2 + 1) && \text{Remove inner parentheses.} \end{aligned}$$

SELF CHECK 5 Factor: $a^4 - a^3 - 2a^2 + a - 2$.

SELF CHECK ANSWERS **1.** $3r^3(p^2 + q^2)(p + q)(p - q)$ **2.** $p(a + b)(a^2 - ab + b^2)(a + b)(a - b)$ **3.** $\left(\frac{1}{x^2} + 4\right)\left(\frac{1}{x^2} - 5\right)$ **4.** $(a - 2)(a - 3)(a + b)$ **5.** $(a - 2)(a^3 + a^2 + 1)$

NOW TRY THIS

Factor completely.

1. $64x^2 + 64$

2. $64x^2 - 1$

3. $64x^3 - 1$

4. $64x^2 + 1$

5.7 EXERCISES

WARM-UPS *Factor each expression.*

1. $x^2 - y^2$

2. $2x^3 - 4x^4$

3. $x^2 + 4x + 4$

4. $x^2 - 5x + 6$

5. $x^3 - 8$

6. $x^3 + 8$

REVIEW *Perform the operations.*

7. $(3a^2 + 4a - 2) + (4a^2 - 3a - 5)$

8. $(-4b^2 - 3b - 2) - (3b^2 - 2b + 5)$

9. $5(2y^2 - 3y + 3) - 2(3y^2 - 2y + 6)$

10. $4(3x^2 + 3x + 3) + 3(x^2 - 3x - 4)$

11. $(m + 4)(m - 2)$

12. $(3p + 4q)(2p - 3q)$

VOCABULARY AND CONCEPTS *Fill in the blanks.*

13. In any factoring problem, always factor out any ______________ first.

14. If an expression has two terms, check to see whether the problem type is the __________ of two squares, the sum of two ______, or the __________ of two cubes.

15. If an expression has three terms, try to factor it as a _________.

16. If an expression has four or more terms, try factoring it by _________.

GUIDED PRACTICE *Factor each polynomial completely, if possible.* See Examples 1–5. (Objective 1)

17. $x^2 + 8x + 16$

18. $20 + 11x - 3x^2$

19. $8x^3y^3 - 27$

20. $3x^2y + 6xy^2 - 12xy$

21. $xy - ty + xs - ts$

22. $bc + b + cd + d$
23. $25x^2 - 16y^2$
24. $27x^9 - y^3$
25. $12x^2 + 52x + 35$
26. $12x^2 + 14x - 6$
27. $6x^2 - 14x + 8$
28. $12x^2 - 12$
29. $9x^2 + 25$
30. $56x^2 - 15x + 1$
31. $7x^2 - 57x + 8$
32. $4x^2y^2 + 4xy^2 + y^2$
33. $100z^2 - 81t^2$
34. $x^3 + (a^2y)^3$
35. $4x^2y^2z^2 - 26x^2y^2z^3$
36. $2x^3 - 54$
37. $12x^2 - 9x + 8$
38. $4(xy)^3 + 256$
39. $ae + bf + af + be$
40. $a^2x^2 + b^2y^2 + b^2x^2 + a^2y^2$
41. $2(x + y)^2 + (x + y) - 3$
42. $(x - y)^3 + 125$
43. $625x^4 - 256y^4$
44. $2(a - b)^2 + 5(a - b) + 3$
45. $8x^4 - 6x^3 + 12x^2 + 24x$
46. $36x^4 - 36$
47. $6x^2 - 63 - 13x$
48. $2x^6 + 2y^6$
49. $x^4 - x^4y^4$
50. $a^4 - 13a^2 + 36$
51. $x^4 - 17x^2 + 16$
52. $x^2 + 6x + 9 - y^2$
53. $x^2 + 10x + 25 - y^8$
54. $4x^2 + 4x + 1 - 4y^2$
55. $9x^2 - 6x + 1 - 25y^2$
56. $x^2 - y^2 - 2y - 1$
57. $64x^3 + 81$
58. $a^2 - b^2 + 4b - 4$
59. $x^5 + x^2 - x^3 - 1$
60. $x^5 - x^2 - 4x^3 + 4$
61. $x^5 - 9x^3 + 8x^2 - 72$
62. $x^5 - 4x^3 - 8x^2 + 32$
63. $2x^5z - 2x^2y^3z - 2x^3y^2z + 2y^5z$
64. $x^2y^3 - 4x^2y - 9y^3 + 36y$
65. $x^{2m} - x^m - 6$
66. $a^{2n} - b^{2n}$
67. $a^{3n} - b^{3n}$
68. $x^{3m} + y^{3m}$
69. $x^{-2} + 2x^{-1} + 1$
70. $4a^{-2} - 12a^{-1} + 9$
71. $6x^{-2} - 5x^{-1} - 6$
72. $x^{-4} - y^{-4}$

WRITING ABOUT MATH

73. What is your strategy for factoring a polynomial?
74. Explain how you can know that your factorization is correct.

SOMETHING TO THINK ABOUT

75. If you have the choice of factoring a polynomial as the difference of two squares or as the difference of two cubes, which do you do first? Why?
76. Can several polynomials have a greatest common factor? Find the GCF of $2x^2 + 7x + 3$ and $x^2 - 2x - 15$.
77. Factor: $x^4 + x^2 + 1$. (*Hint:* Add and subtract x^2.)
78. Factor: $x^4 + 7x^2 + 16$. (*Hint:* Add and subtract x^2.)

SECTION 5.8 Solving Equations by Factoring

Objectives

1. Solve a quadratic equation by factoring.
2. Solve a higher-degree polynomial equation by factoring.
3. Solve an application problem using a quadratic equation.

Vocabulary

quadratic equation — zero factor property

Getting Ready

Factor each polynomial.

1. $2a^2 - 4a$
2. $a^2 - 25$
3. $6a^2 + 5a - 6$
4. $6a^3 - a^2 - 2a$

In this section, we will learn how to solve quadratic equations by factoring. Then we will use this skill to solve application problems.

Solve a quadratic equation by factoring.

An equation such as $3x^2 + 4x - 7 = 0$ or $-5y^2 + 3y + 8 = 0$ is called a *quadratic* (or *second-degree*) equation.

Quadratic Equations

A **quadratic equation** is any equation that can be written in the form

$$ax^2 + bx + c = 0$$

where a, b, and c are real numbers and $a \neq 0$.

Many quadratic equations can be solved by factoring and then using the **zero-factor property.**

Zero-Factor Property

If a and b are real numbers, then

If $ab = 0$, then $a = 0$ or $b = 0$.

The zero-factor property and its extensions state that *if the product of two or more numbers is* 0, *then at least one of the numbers must be* 0.

To solve $x^2 + 5x + 6 = 0$, we factor its left side to obtain

$$(x + 3)(x + 2) = 0$$

Since the product of $x + 3$ and $x + 2$ is 0, then at least one of the factors is 0. Thus, we can set factor equal to 0 and solve each resulting linear equation for x:

$$x + 3 = 0 \quad \text{or} \quad x + 2 = 0$$
$$x = -3 \quad \bigg| \quad x = -2$$

To check these solutions, we first substitute -3 and then -2 for x in the equation and verify that each number satisfies the equation.

$$\begin{aligned} x^2 + 5x + 6 &= 0 \\ (-3)^2 + 5(-3) + 6 &\stackrel{?}{=} 0 \\ 9 - 15 + 6 &\stackrel{?}{=} 0 \\ 0 &= 0 \end{aligned} \qquad \begin{aligned} x^2 + 5x + 6 &= 0 \\ (-2)^2 + 5(-2) + 6 &\stackrel{?}{=} 0 \\ 4 - 10 + 6 &\stackrel{?}{=} 0 \\ 0 &= 0 \end{aligned}$$

Both -3 and -2 are solutions, because both satisfy the equation.

EXAMPLE 1 Solve: $3x^2 + 6x = 0$.

Solution We factor the left side, set each factor equal to 0, and solve each resulting equation for x.

$$3x^2 + 6x = 0$$

$$3x(x + 2) = 0 \quad \text{Factor out the common factor of } 3x.$$

$$3x = 0 \quad \text{or} \quad x + 2 = 0$$

$$x = 0 \quad \big| \quad x = -2$$

Verify that both solutions check.

SELF CHECK 1 Solve: $4p^2 - 12p = 0$.

EXAMPLE 2 Solve: $x^2 - 16 = 0$.

Solution We factor the difference of two squares on the left side, set each factor equal to 0, and solve each resulting equation.

$$x^2 - 16 = 0$$

$$(x + 4)(x - 4) = 0$$

$$x + 4 = 0 \quad \text{or} \quad x - 4 = 0$$

$$x = -4 \quad \big| \quad x = 4$$

Verify that both solutions check.

SELF CHECK 2 Solve: $a^2 - 81 = 0$.

Many equations that do not appear to be quadratic can be put into quadratic form $(ax^2 + bx + c = 0)$ and then solved by factoring.

EXAMPLE 3 Solve: $x = \frac{6}{5} - \frac{6}{5}x^2$.

Solution We write the equation in quadratic form and then solve by factoring.

$$x = \frac{6}{5} - \frac{6}{5}x^2$$

$$5x = 6 - 6x^2 \quad \text{Multiply both sides by 5.}$$

$$6x^2 + 5x - 6 = 0 \quad \text{Add } 6x^2 \text{ to both sides and subtract 6 from both sides.}$$
$$(3x - 2)(2x + 3) = 0 \quad \text{Factor the trinomial.}$$
$$3x - 2 = 0 \quad \text{or} \quad 2x + 3 = 0 \quad \text{Set each factor equal to 0.}$$
$$3x = 2 \quad\quad 2x = -3$$
$$x = \frac{2}{3} \quad\quad x = -\frac{3}{2}$$

Verify that both solutions check.

SELF CHECK 3 Solve: $x = \frac{6}{7}x^2 - \frac{3}{7}$.

COMMENT To solve a quadratic equation by factoring, be sure to set the quadratic polynomial equal to 0 before factoring and using the zero-factor property. Do not make the following error:

$$6x^2 + 5x = 6$$
$$x(6x + 5) = 6 \quad \text{If the product of two numbers is 6, neither number need be 6. For example, } 2 \cdot 3 = 6.$$
$$x = 6 \quad \text{or} \quad 6x + 5 = 6$$
$$x = \frac{1}{6}$$

Neither solution checks.

ACCENT ON TECHNOLOGY

Solving Quadratic Equations

To solve a quadratic equation such as $x^2 + 4x - 5 = 0$ with a graphing calculator, we can use standard window settings of $[-10, 10]$ for x and $[-10, 10]$ for y and graph the quadratic function $y = x^2 + 4x - 5$, as shown in Figure 5-6(a). Then we can use trace to find the x-coordinates of the x-intercepts of the parabola. See Figures 5-6(b) and 5-6(c). For better results, we can zoom in and see that x-values of -5 and 1 make $y = 0$. These numbers are the solutions of the equation.

We also can find the solutions by using the ZERO command found in the CALC menu.

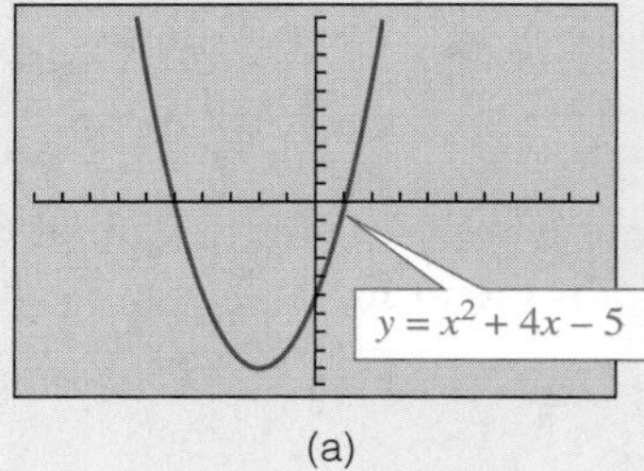

(a)

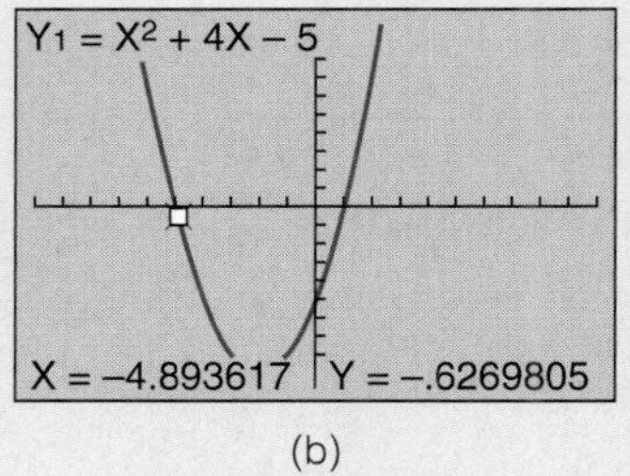

(b)

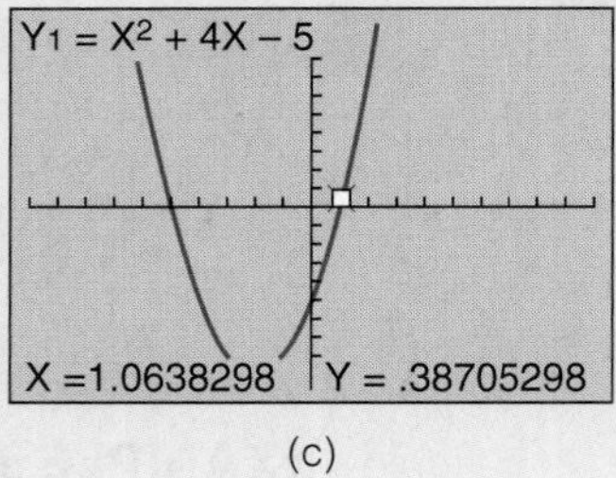

(c)

Figure 5-6

2 Solve a higher-degree polynomial equation by factoring.

We can solve many polynomial equations with degrees greater than 2 by factoring.

EXAMPLE 4 Solve: $6x^3 - x^2 - 2x = 0$.

Solution We factor x from the polynomial on the left side and proceed as follows:

$$6x^3 - x^2 - 2x = 0$$

$$x(6x^2 - x - 2) = 0 \quad \text{Factor out } x.$$

$$x(3x - 2)(2x + 1) = 0 \quad \text{Factor } 6x^2 - x - 2.$$

$$x = 0 \quad \text{or} \quad 3x - 2 = 0 \quad \text{or} \quad 2x + 1 = 0 \quad \text{Set each factor equal to 0.}$$

$$x = \frac{2}{3} \qquad x = -\frac{1}{2}$$

Verify that the solutions check.

SELF CHECK 4 Solve: $5x^3 + 13x^2 - 6x = 0$.

EXAMPLE 5 Solve: $x^4 - 5x^2 + 4 = 0$.

Solution We factor the trinomial on the left side and proceed as follows:

$$x^4 - 5x^2 + 4 = 0$$

$$(x^2 - 1)(x^2 - 4) = 0$$

$$(x + 1)(x - 1)(x + 2)(x - 2) = 0 \quad \text{Factor } x^2 - 1 \text{ and } x^2 - 4.$$

$$x + 1 = 0 \quad \text{or} \quad x - 1 = 0 \quad \text{or} \quad x + 2 = 0 \quad \text{or} \quad x - 2 = 0$$

$$x = -1 \qquad x = 1 \qquad x = -2 \qquad x = 2$$

Verify that each solution checks.

SELF CHECK 5 Solve: $a^4 - 13a^2 + 36 = 0$.

EXAMPLE 6 Given that $f(x) = 6x^3 + x^2 - 2x$, find all x such that $f(x) = 0$.

Solution We first set $f(x)$ equal to 0.

$$f(x) = 6x^3 + x^2 - 2x$$

$$0 = 6x^3 + x^2 - 2x \quad \text{Substitute 0 for } f(x).$$

Then we solve for x.

$$6x^3 + x^2 - 2x = 0$$

$$x(6x^2 + x - 2) = 0 \quad \text{Factor out } x.$$

$$x(2x - 1)(3x + 2) = 0 \quad \text{Factor } 6x^2 + x - 2.$$

$$x = 0 \quad \text{or} \quad 2x - 1 = 0 \quad \text{or} \quad 3x + 2 = 0 \quad \text{Set each factor equal to 0.}$$

$$x = \frac{1}{2} \qquad x = -\frac{2}{3}$$

Verify that $f(0) = 0$, $f\left(\frac{1}{2}\right) = 0$, and $f\left(-\frac{2}{3}\right) = 0$.

SELF CHECK 6 Given that $f(x) = x^4 - 4x^2$, find all x such that $f(x) = 0$.

ACCENT ON TECHNOLOGY

Solving Equations

To solve the equation $x^4 - 5x^2 + 4 = 0$ with a graphing calculator, we can use window settings of $[-6, 6]$ for x and $[-5, 10]$ for y and graph the polynomial function $y = x^4 - 5x^2 + 4$ as shown in Figure 5-7. Then we can read the values of x that make $y = 0$. They are $x = -2, -1, 1$, and 2. If the x-coordinates of the x-intercepts were not obvious, we could get their values by using the trace and zoom features. We also can use the ZERO or INTERSECT command to find the solutions.

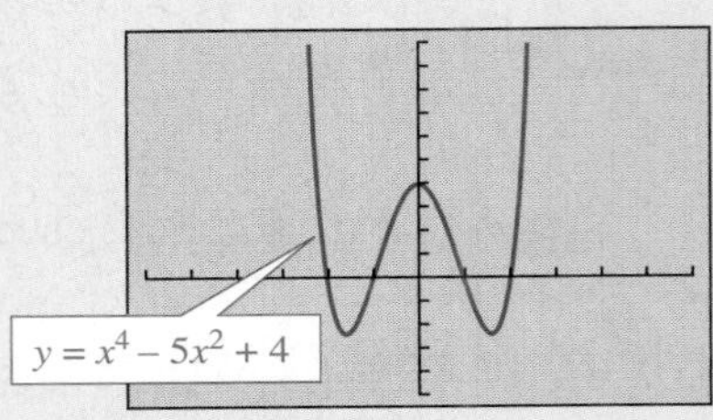

Figure 5-7

3 Solve an application problem using a quadratic equation.

EXAMPLE 7 **FINDING THE DIMENSIONS OF A TRUSS** The width of the truss shown in Figure 5-8 is 3 times its height. The area of the triangle is 96 square feet. Find its width and height.

Analyze the problem

We can let x be the positive number that represents the height of the truss. Then $3x$ represents its width.

Form and solve an equation

We can substitute x for h, $3x$ for b, and 96 for A in the formula for the area of a triangle $A = \frac{1}{2}bh$ and solve for x.

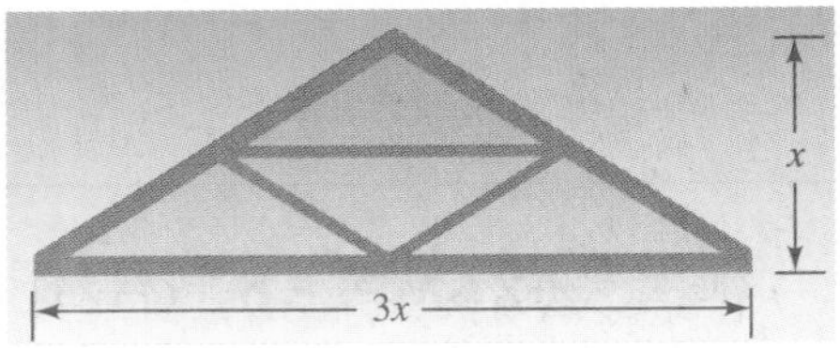

Figure 5-8

$$A = \frac{1}{2}bh$$

$$96 = \frac{1}{2}(3x)x$$

$$192 = 3x^2 \quad \text{Multiply both sides by 2.}$$

$$64 = x^2 \quad \text{Divide both sides by 3.}$$

$$0 = x^2 - 64 \quad \text{Subtract 64 from both sides.}$$

$$0 = (x + 8)(x - 8) \quad \text{Factor the difference of two squares.}$$

$$x + 8 = 0 \quad \text{or} \quad x - 8 = 0$$

$$x = -8 \quad \Big| \quad x = 8$$

State the conclusion

Since the height of a triangle cannot be negative, we must discard the negative solution. Thus, the height of the truss is 8 feet, and its width is 3(8), or 24 feet.

Check the result

The area of a triangle with a width (base) of 24 feet and a height of 8 feet is 96 square feet:

$$A = \frac{1}{2}bh = \frac{1}{2}(24)(8) = 12(8) = 96$$

EXAMPLE 8 **BALLISTICS** If the initial velocity of an object thrown straight up into the air is 176 feet per second, when will it hit the ground?

Analyze the problem: Recall the formula for the height of an object thrown or shot straight up into the air is $h = vt - 16t^2$, where h represents height in feet, v represents initial velocity in feet per second, and t represents the number of seconds since the object was released.

Form and solve an equation: When the object hits the ground, its height will be 0. Thus, we set h equal to 0, set v equal to 176, and solve for t.

$$h = vt - 16t^2$$
$$0 = 176t - 16t^2$$
$$0 = 16t(11 - t) \quad \text{Factor out } 16t.$$
$$16t = 0 \quad \text{or} \quad 11 - t = 0 \quad \text{Set each factor equal to 0.}$$
$$t = 0 \quad \bigg| \quad t = 11$$

State the conclusion: When $t = 0$, the object's height above the ground is 0 feet, because it has not been released. When $t = 11$, the height is again 0 feet, because the object has hit the ground. The solution is 11 seconds.

Check the result: Verify that $h = 0$ when $t = 11$.

SELF CHECK ANSWERS **1.** 0, 3 **2.** 9, −9 **3.** $\frac{3}{2}, -\frac{1}{3}$ **4.** $0, \frac{2}{5}, -3$ **5.** 2, −2, 3, −3 **6.** 0, 0, 2, −2

NOW TRY THIS

Factor or solve as indicated.

1. $6x^2 + 5x - 6$

2. $25x^2 = 25x$

3. $6x^2 - 6 = 5x$

4. $25x^3 - 25x$

5.8 EXERCISES

WARM-UPS *Solve each equation.*

1. $(x - 2)(x - 3) = 0$

2. $(x + 4)(x - 2) = 0$

3. $(x - 2)(x - 3)(x + 1) = 0$

4. $(x + 3)(x + 2)(x - 5)(x - 6) = 0$

REVIEW

5. List the prime numbers less than 10.

6. List the composite numbers between 7 and 17.

7. The formula for the volume of a sphere is $V = \frac{4}{3}\pi r^3$. Find the volume when $r = 21.23$ centimeters. Round the answer to the nearest hundredth.

8. The formula for the volume of a cone is $V = \frac{1}{3}\pi r^2 h$. Find the volume when $r = 12.33$ meters and $h = 14.7$ meters. Round the answer to the nearest hundredth.

VOCABULARY AND CONCEPTS *Fill in the blanks.*

9. A quadratic equation is any equation that can be written in the form ______________ $(a \neq 0)$.

10. The zero-factor property: If a and b are real numbers, and if ______, then $a = 0$ or $b = 0$.

GUIDED PRACTICE

Solve each equation. (Objective 1)

11. $z^2 + 8z + 15 = 0$

12. $w^2 + 7w + 12 = 0$

13. $y^2 - 7y + 12 = 0$

14. $x^2 - 3x + 2 = 0$

15. $y^2 - 7y + 6 = 0$

16. $n^2 - 5n + 6 = 0$

17. $3m^2 + 10m + 3 = 0$

18. $2r^2 + 5r + 3 = 0$

Solve each equation. See Examples 1–2. (Objective 1)

19. $4x^2 + 8x = 0$

20. $x^2 - 9 = 0$

21. $y^2 - 16 = 0$

22. $5y^2 - 10y = 0$

23. $x^2 + x = 0$

24. $x^2 - 3x = 0$

25. $p^2 - 49 = 0$

26. $p^2 - 49p$

Solve each equation. See Example 3. (Objective 1)

27. $\frac{3a^2}{2} = \frac{1}{2} - a$

28. $x^2 = \frac{1}{2}(x + 1)$

29. $\frac{1}{2}x^2 - \frac{5}{4}x = -\frac{1}{2}$

30. $\frac{1}{4}x^2 + \frac{3}{4}x = 1$

31. $b(6b - 7) = 10$

32. $2y(4y + 3) = 9$

33. $3a(a^2 + 5a) = -18a$

34. $7t^3 = 2t\left(t + \frac{5}{2}\right)$

Solve each equation. See Examples 4–5. (Objective 2)

35. $x^3 + x^2 = 0$

36. $2x^4 + 8x^3 = 0$

37. $y^3 - 49y = 0$

38. $2z^3 - 200z = 0$

39. $x^3 - 4x^2 - 21x = 0$

40. $x^3 + 8x^2 - 9x = 0$

41. $z^4 - 13z^2 + 36 = 0$

42. $y^4 - 10y^2 + 9 = 0$

43. $x^3 + 3x^2 - x - 3 = 0$

44. $x^3 - x^2 - 4x + 4 = 0$

45. $2r^3 + 3r^2 - 18r - 27 = 0$

46. $3s^3 - 2s^2 - 3s + 2 = 0$

Find all x that will make f(x) = 0. See Example 6. (Objectives 1–2.)

47. $f(x) = x^2 - 49$

48. $f(x) = x^2 + 11x$

49. $f(x) = 2x^2 + 5x - 3$

50. $f(x) = 3x^2 - x - 2$

ADDITIONAL PRACTICE

Solve each equation.

51. $2y^2 - 5y + 2 = 0$

52. $2x^2 - 3x + 1 = 0$

53. $2x^2 - x - 1 = 0$

54. $2x^2 - 3x - 5 = 0$

55. $3s^2 - 5s - 2 = 0$

56. $8t^2 + 10t - 3 = 0$

57. $8a^2 = 3 - 10a$

58. $5z^2 = 6 - 13z$

59. $5y^2 - 25y = 0$

60. $y^2 - 36 = 0$

61. $x(x - 6) + 9 = 0$

62. $x^2 + 8(x + 2) = 0$

63. $x^2 + 6x + 8 = 0$

64. $x^2 + 9x + 20 = 0$

65. $x\left(3x + \frac{22}{5}\right) = 1$

66. $x\left(\frac{x}{11} - \frac{1}{7}\right) = \frac{6}{77}$

67. $\frac{x^2(6x + 37)}{35} = x$

68. $x^2 = -\frac{4x^3(3x + 5)}{3}$

69. $3y^3 + y^2 = 4(3y + 1)$

70. $w^3 + 16 = w(w + 16)$

Find all x that will make f(x) = 0.

71. $f(x) = 5x^3 + 3x^2 - 2x$

72. $f(x) = x^4 - 26x^2 + 25$

Use a graphing calculator to find the real solutions of each equation, if any exist. If an answer is not exact, give the answer to the nearest hundredth.

73. $x^2 - 4x + 7 = 0$

74. $2x^2 - 7x + 4 = 0$

75. $-3x^3 - 2x^2 + 5 = 0$

76. $-2x^3 - 3x - 5 = 0$

77. Integer problem The product of two consecutive even integers is 288. Find the integers.

78. Integer problem The product of two consecutive odd integers is 143. Find the integers.

79. Integer problem The sum of the squares of two consecutive positive integers is 85. Find the integers.

80. Integer problem The sum of the squares of three consecutive positive integers is 77. Find the integers.

APPLICATIONS *Set up an equation and solve.*

See Examples 7–8. (Objective 3)

81. Geometry One side of a rectangle is three times longer than another. If its area is 147 square centimeters, find its dimensions.

82. Geometry Find the dimensions of the rectangle, given that its area is 375 square feet.

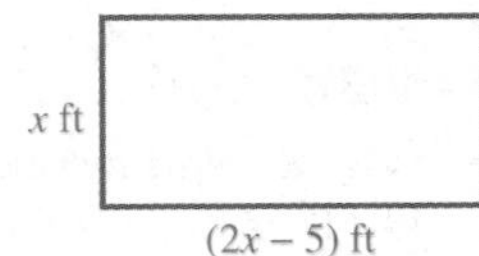

83. Geometry Find the height of the triangle, given that its area is 162 square centimeters.

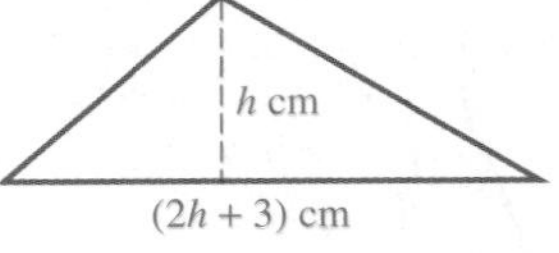

84. Geometry Find the perimeter of the rectangle.

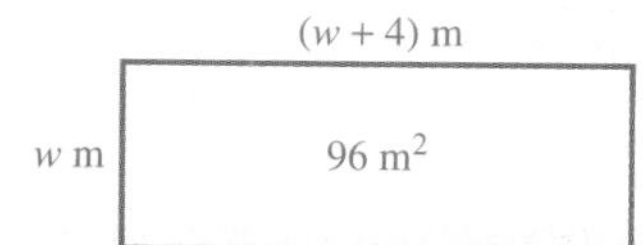

85. Fine arts An artist intends to paint a 60-square-foot mural on a large wall as shown in the illustration. Find the dimensions of the mural if the artist leaves a border of uniform width around it.

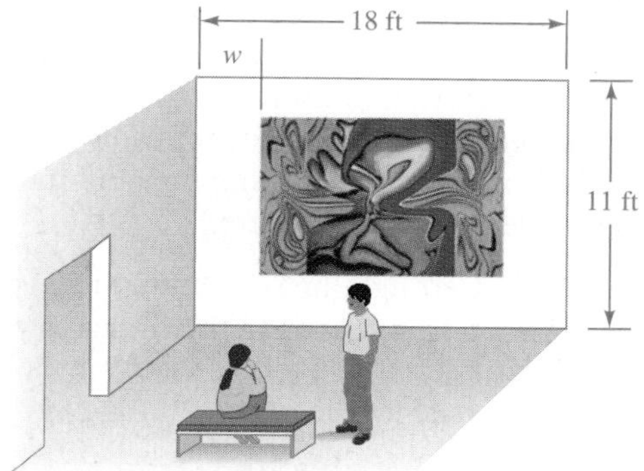

86. Gardening A woman plans to use one-fourth of her 48-foot-by-100-foot rectangular backyard to plant a garden. Find the perimeter of the garden if the length is to be 40 feet greater than the width.

87. Architecture The rectangular room shown in the illustration is twice as long as it is wide. It is divided into two rectangular parts by a partition, positioned as shown. If the larger part of the room contains 560 square feet, find the dimensions of the entire room.

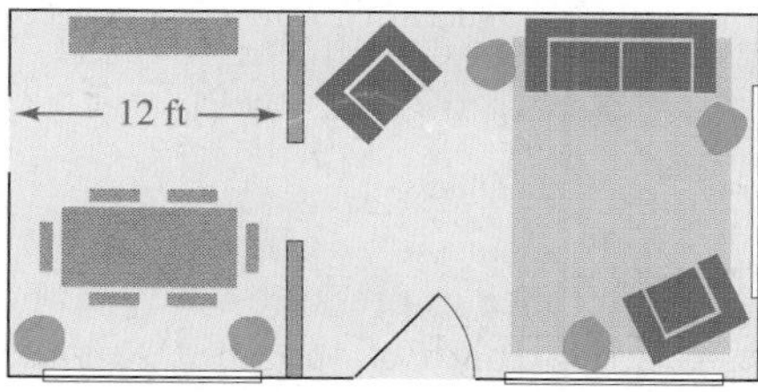

88. Perimeter of a square If the length of each side of a square is increased by 4 inches, the area of the square becomes 9 times greater. Find the perimeter of the original square.

89. Time of flight After how many seconds will an object hit the ground if it is thrown upward with an initial velocity of 160 feet per second?

90. Time of flight After how many seconds will an object hit the ground if it is thrown upward with an initial velocity of 208 feet per second?

91. Ballistics The muzzle velocity of a cannon is 480 feet per second. If a cannonball is fired vertically, at what times will it be at a height of 3,344 feet?

92. Ballistics A slingshot can provide an initial velocity of 128 feet per second. At what times will a stone, shot vertically upward, be 192 feet above the ground?

93. Winter recreation The length of the rectangular ice-skating rink is 20 meters greater than twice its width. Find the width.

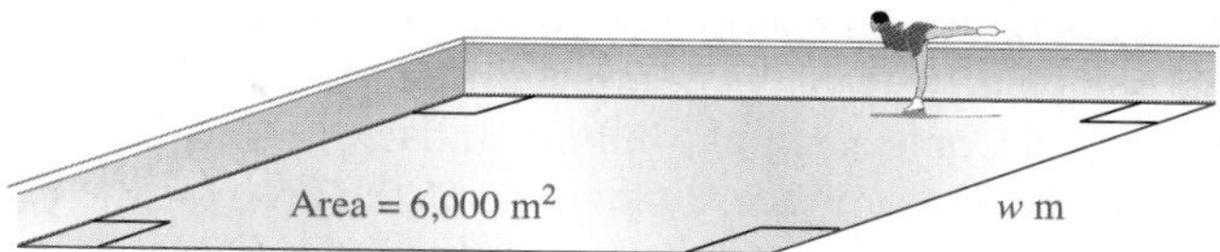

94. Carpentry A 285-square-foot room is 4 feet longer than it is wide. What length of crown molding is needed to trim the perimeter of the ceiling?

95. Designing a swimming pool Building codes require that the rectangular swimming pool in the illustration be surrounded by a uniform-width walkway of at least 516 square feet. If the length of the pool is 10 feet less than twice the width, how wide should the border be?

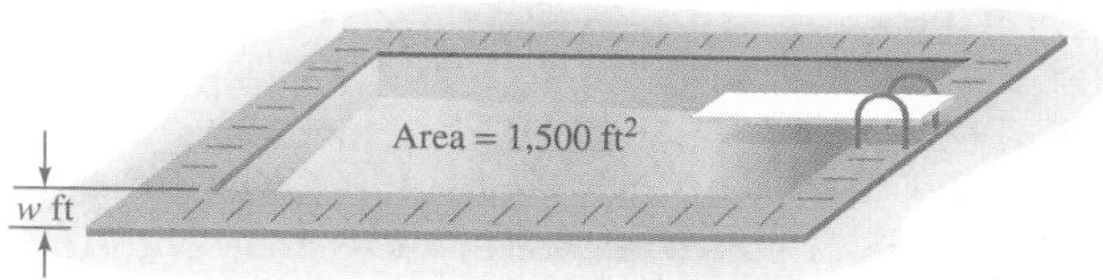

96. House construction The formula for the area of a trapezoid is $A = \frac{h(B + b)}{2}$. The area of the truss is 44 square feet. Find the height of the truss if the shorter base is the same as the height.

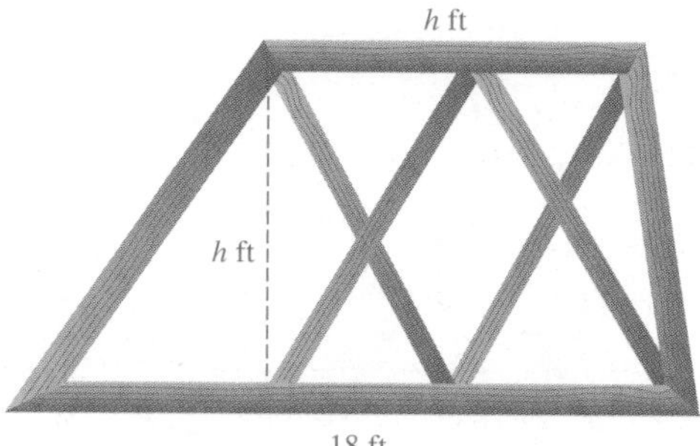

WRITING ABOUT MATH

97. Describe the steps for solving an application problem.

98. Explain how to check the solution of an application problem.

SOMETHING TO THINK ABOUT *Find a quadratic equation with the given solutions.*

99. 3, 5

100. -2, 6

101. 0, -5

102. $\frac{1}{2}, \frac{1}{3}$

PROJECTS

Project 1

Mac operates a bait shop located between two sharp bends on the Snarly River. The area around Mac's shop has recently become a popular hiking and camping area. Mac has decided to produce some maps of the area for the use of the visitors. Although he knows the region well, he has very little idea of the actual distances from one location to another. What he knows is:

1. Big Falls, a beautiful waterfall on Snarly River, is due east of Mac's.
2. Grandview Heights, a fabulous rock climbing site, is due west of Mac's, right on the river.
3. Foster's General Store, the only sizable camping and climbing outfitter in the area, is located on the river some distance west and north of Mac's.

Mac hires an aerial photographer to take pictures of the area, with some surprising results. If Mac's bait shop is treated as the origin of a coordinate system, with the y-axis running north–south and the x-axis running east–west, then on the domain $-4 \leq x \leq 4$ (where the units are miles), the river follows the curve

$$P(x) = \frac{1}{4}(x^3 - x^2 - 6x)$$

The aerial photograph is shown in the illustration.

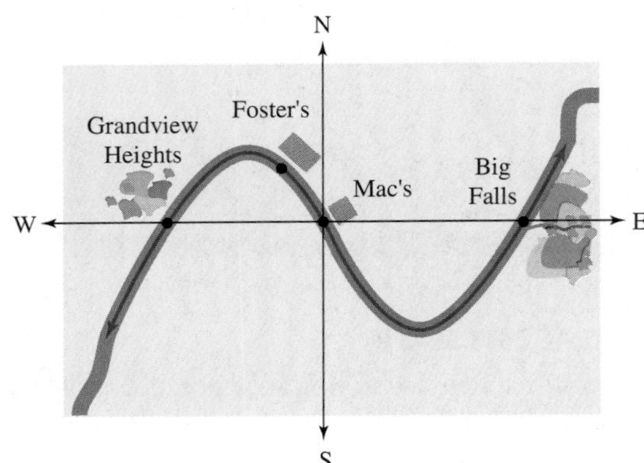

a. Mac would like to include on his maps the exact locations (relative to Mac's) of Big Falls and Grandview Heights. Find these for him, and explain to Mac why your answers must be correct.

b. Mac and Foster have determined that Foster's General Store is 0.7 miles west of the bait shop. Since it is on the river, it is a bit north as well. They decide that to promote business, they will join together to clear a few campsites in the region bordered by the straight-line paths that run between Mac's and Foster's, Mac's and Grandview Heights, and Foster's and Grandview Heights. If they clear 1 campsite for each 40 full acres of area, how many campsites can they put in? (*Hint:* A square mile contains 640 acres.)

c. A path runs in a straight line directly southeast (along the line $y = -x$) from Mac's to the river. How far east and how far south has a hiker on this trail traveled when he or she reaches the river?

Project 2

The rate at which fluid flows through a cylindrical pipe, or any cylinder-shaped tube (an artery, for instance), is

$$\text{Velocity of fluid flow} = V = \frac{p}{nL}(R^2 - r^2)$$

Cross section of tube

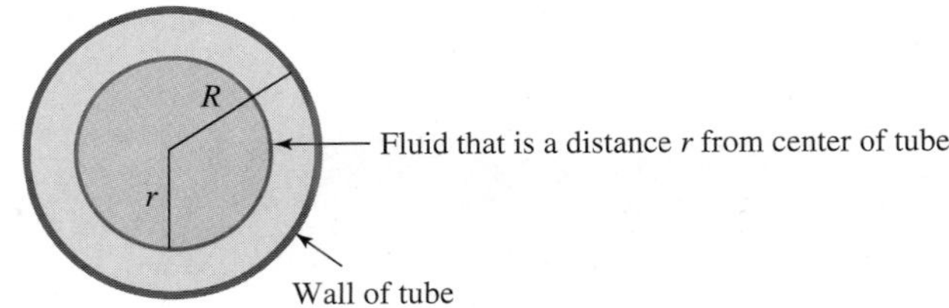

where p is the difference in pressure between the two ends of the tube, L is the length of the tube, R is the radius of the tube, and n is the *viscosity constant,* a measure of the thickness of a fluid. Since the variable r represents the distance from the center of the tube, $0 \leq r \leq R$. In most situations, p, L, R, and n are constants, so V is a function of r.

$$V(r) = \frac{p}{nL}(R^2 - r^2)$$

It can be shown that the velocity of a fluid moving in the tube depends on how far from the center (or how close to the wall of the tube) the fluid is.

a. Consider a pipe with radius 5 centimeters and length 60 centimeters. Suppose that $p = 15$ and $n = 0.001$

(water has a viscosity of approximately 0.001). Find the velocity of the fluid at the center of the pipe. The units of measurement for V will be centimeters per second.

Find the velocity of the fluid when it is halfway between the center and the wall of the pipe. What percent of the velocity at the center of the pipe does this represent? Where in the pipe is the fluid flowing with a velocity of 4,000 centimeters per second?

b. Suppose that the situation is the same, but the fluid is now machinery oil, with a viscosity of 0.15. Answer the same questions as in part **a,** except find where in the pipe the oil flows with a velocity of 15 cm per second. Note that the oil travels at a much slower speed than water.

c. Medical doctors use various methods to increase the rate of blood flow through an artery. The patient may take a drug that "thins the blood" (lowers its viscosity) or a drug that dilates the artery, or the patient may undergo angioplasty, a surgical procedure that widens the canal through which the blood passes. Explain why each of these increases the velocity V at a given distance r from the center of the artery.

Chapter 5 REVIEW

SECTION 5.1 Polynomials and Polynomial Functions

DEFINITIONS AND CONCEPTS	EXAMPLES
The **degree of a polynomial** is the degree of the term with highest degree contained within the polynomial.	The degree of $2x^2 - 14x^5 + 8x + 5x^3 - 7$ is 5 because 5 is the highest degree of any term.
If $P(x)$ is a polynomial in x then $P(r)$ is the value of the polynomial at $x = r$.	Given $P(x) = 2x^2 - 6x + 1$, find $P(-1)$. $P(-1) = 2(-1)^2 - 6(-1) + 1$ $= 2(1) + 6 + 1$ $= 9$

REVIEW EXERCISES

1. Find the degree of $P(x) = 3x^5 + 4x^3 + 2$.
2. Find the degree of $9x^2y + 13x^3y^2 + 8x^4y^4$.

Find each value when $P(x) = -x^2 + 4x + 6$.

3. $P(0)$
4. $P(1)$
5. $P(-t)$
6. $P(z)$

Graph each function.

7. $f(x) = x^3 - 1$

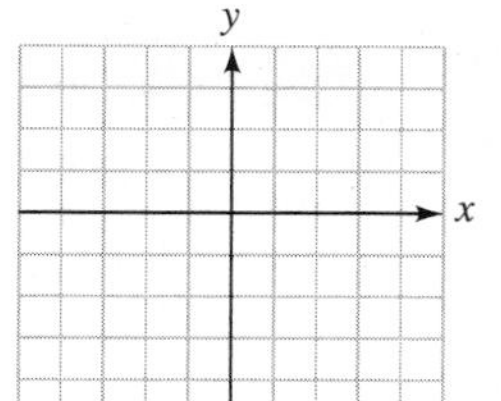

8. $f(x) = x^2 - 2x$

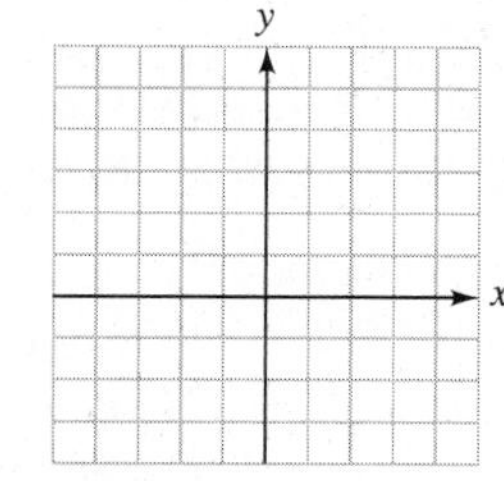

SECTION 5.2 Adding and Subtracting Polynomials

DEFINITIONS AND CONCEPTS	EXAMPLES
To *add* polynomials, add their like terms.	To add $(2x^2 - 2x - 3)$ and $(3x^2 + 4x + 4)$, remove parentheses and combine like terms. $(2x^2 - 2x - 3) + (3x^2 + 4x + 4)$ $= 2x^2 - 2x - 3 + 3x^2 + 4x + 4$ $= 5x^2 + 2x + 1$

To *subtract* polynomials, add the opposite of the polynomial that is to be subtracted from the other polynomial.	To subtract $(3x^2 + 4x - 4)$ from $(2x^2 - 2x - 3)$, write the expression as a subtraction, remove the parentheses to get the opposite of each term in the second polynomial, and combine like terms. $(2x^2 - 2x - 3) - (3x^2 + 4x - 4)$ $= 2x^2 - 2x - 3 - 3x^2 - 4x + 4$ $= -x^2 - 6x + 1$

REVIEW EXERCISES

Simplify each expression.

9. $(3x^2 + 4x + 9) + (2x^2 - 2x + 7)$

10. $(4x^3 + 4x^2 + 7) - (-2x^3 - x - 2)$

11. $(2x^2 - 5x + 9) - (x^2 - 3) - (-3x^2 + 4x - 7)$

12. $2(7x^3 - 6x^2 + 4x - 3) - 3(7x^3 + 6x^2 + 4x - 3)$

SECTION 5.3 Multiplying Polynomials

DEFINITIONS AND CONCEPTS	EXAMPLES
To multiply monomials, multiply their numerical factors and multiply their variable factors.	$(-5a^2b)(6ab^3) = -5 \cdot 6 \cdot a^2 \cdot a \cdot b \cdot b^3$ $= -30a^3b^4$
To multiply a polynomial by a monomial, multiply each term of the polynomial by the monomial.	$4x^2(3x^2 + 6x - 1) = 4x^2 \cdot 3x^2 + 4x^2 \cdot 6x - 4x^2 \cdot 1$ $= 12x^4 + 24x^3 - 4x^2$
To multiply polynomials, multiply each term of one polynomial by each term of the other polynomial. Then combine like terms, if possible.	$(6x - 2)(5x^3 - 8x + 9)$ $= 30x^4 - 48x^2 + 54x - 10x^3 + 16x - 18$ $= 30x^4 - 10x^3 - 48x^2 + 70x - 18$

REVIEW EXERCISES

Find each product.

13. $(8a^2b^2)(-2abc)$ **14.** $(-3xy^2z)(2xz^3)$

Find each product.

15. $2xy^2(x^3y - 4xy^5)$ **16.** $a^2b(a^2 + 2ab + b^2)$

Find each product.

17. $(8x - 5)(2x + 3)$ **18.** $(3x + 2)(2x - 4)$

19. $(2x - 5y)^2$ **20.** $(5x - 4)(3x - 2)$

21. $(3x^2 - 2)(x^2 - x + 2)$

22. $(2a - b)(a + b)(a - 2b)$

SECTION 5.4 The Greatest Common Factor and Factoring by Grouping

DEFINITIONS AND CONCEPTS	EXAMPLES
Always factor out common monomial factors as the first step in a factoring problem.	To factor $36x^3 - 6x^2 + 12x$, use the distributive property to factor out the common factor of $6x$. $6x(6x^2 - x + 2)$ Factor the common term.
If an expression has four or more terms, try to factor the expression by grouping.	$25x^3 - 10x^2 + 20x - 8$ $= (25x^3 - 10x^2) + (20x - 8)$ Group the first two terms and the last two terms. $= 5x^2(5x - 2) + 4(5x - 2)$ Factor each grouping. $= (5x - 2)(5x^2 + 4)$ Factor out $5x - 2$.

REVIEW EXERCISES

Factor each expression.

23. $4x + 8$ **24.** $3x^2 - 6x$

25. $5x^2y^3 - 10xy^2$ **26.** $7a^4b^2 + 49a^3b$

27. $-8x^2y^3z^4 - 12x^4y^3z^2$

28. $12a^6b^4c^2 + 15a^2b^4c^6$

29. $27x^3y^3z^3 + 81x^4y^5z^2 - 90x^2y^3z^7$

30. $-36a^5b^4c^2 + 60a^7b^5c^3 - 24a^2b^3c^7$

31. Factor x^n from $x^{2n} + x^n$.
32. Factor y^{2n} from $y^{2n} - y^{4n}$.
33. Factor x^{-2} from $x^{-4} - x^{-2}$.
34. Factor a^{-3} from $a^6 + 1$.
35. $5x^2(x + y)^3 - 15x^3(x + y)^4$
36. $-49a^3b^2(a - b)^4 + 63a^2b^4(a - b)^3$

Factor each expression completely.

37. $xy + 2y + 4x + 8$
38. $ac + bc + 3a + 3b$
39. $x^4 + 4y + 4x^2 + x^2y$
40. $a^5 + b^2c + a^2c + a^3b^2$

Solve for the indicated variable.

41. $S = 2wh + 2wl + 2lh$ for h
42. $S = 2wh + 2wl + 2lh$ for l

SECTION 5.5 The Difference of Two Squares; the Sum and Difference of Two Cubes

DEFINITIONS AND CONCEPTS	EXAMPLES
Difference of two squares: $x^2 - y^2 = (x + y)(x - y)$	$64x^2 - 25$ $= (8x)^2 - (5^2)$ Write each factor as a square. $= (8x + 5)(8x - 5)$ Factor.
Sum of two cubes: $x^3 + y^3 = (x + y)(x^2 - xy + y^2)$	$64x^3 + 125$ $= (4x)^3 + (5)^3$ Write each factor as a cube. $= (4x + 5)[(4x)^2 - (4x)(5) + (5)^2]$ Factor. $= (4x + 5)(16x^2 - 20x + 25)$ Simplify.
Difference of two cubes: $x^3 - y^3 = (x - y)(x^2 + xy + y^2)$	$27a^3 - 1$ $= (3a)^3 - (1)^3$ Write each factor as a cube. $= (3a - 1)[(3a)^2 + (3a)(1) + (1)^2]$ Factor. $= (3a - 1)(9a^2 + 3a + 1)$ Simplify.

REVIEW EXERCISES

Factor each expression completely, if possible.

43. $z^2 - 16$
44. $y^2 - 121$
45. $x^2y^4 - 64z^6$
46. $a^2b^2 + c^2$
47. $(x + z)^2 - t^2$
48. $c^2 - (a + b)^2$
49. $2x^4 - 98$
50. $3x^6 - 300x^2$
51. $x^3 + 343$
52. $a^3 - 125$
53. $8y^3 - 512$
54. $4x^3y + 108yz^3$

SECTION 5.6 Factoring Trinomials

DEFINITIONS AND CONCEPTS	EXAMPLES
Special-product formulas: $x^2 + 2xy + y^2 = (x + y)(x + y)$ $x^2 - 2xy + y^2 = (x - y)(x - y)$	The trinomial $9a^2 + 30a + 25$ can be written in the form $(3a)^2 - 2(3a)(5) + (5)^2$. This can be factored as $(3a + 5)(3a + 5)$ or $(3a + 5)^2$ The trinomial $4a^2 - 12a + 9$ can be written in the form $(2a)^2 - 2(2a)(3) + (3)^2$. This can be factored as $(2a - 3)(2a - 3)$ or $(2a - 3)^2$

Factoring a trinomial with a leading coefficient of 1: **1.** Write the trinomial in descending powers of one variable. **2.** List the factorizations of the third term of the trinomial. **3.** Pick the factorization in which the sum of the factors is the coefficient of the middle term.	To factor $x^2 - 8x + 12$, first note that the factors of 12 are 1 and 12 -1 and -12 2 and 6 -2 and -6 3 and 4 -3 and -4 Since -2 and -6 are the only pair whose sum is -8, the factorization is $(x - 2)(x - 6)$.
Test for factorability: A trinomial of the form $ax^2 + bx + c \quad (a \neq 0)$ will factor with integer coefficients if $b^2 - 4ac$ is a perfect square.	To determine whether $5x^2 - 8x + 3$ is factorable with integer coefficients, note that $a = 5$, $b = -8$, and $c = 3$. Then calculate $b^2 - 4ac$. $(-8)^2 - 4(5)(3) = 64 - 60 = 4$ Since 4 is a perfect square, $5x^2 - 8x + 3$ is factorable with integer coefficients.
Factoring a trinomial with a leading coefficient other than 1:	To factor $5x^2 - 8x + 3$, follow the steps shown on the left.
1. Write the trinomial in descending powers of one variable.	The polynomial is already written in descending powers of x.
2. Factor out any greatest common factor (including -1 if that is necessary to make the coefficient of the first term positive).	In this polynomial, there are no common factors.
3. Test the trinomial for factorability.	We previously have determined that $5x^2 - 8x + 3$ is factorable.
4. When the sign of the first term of a trinomial is + and the sign of the third term is +, the signs between the terms of each binomial factor are the same as the sign of the middle term of the trinomial. When the sign of the first term is + and the sign of the third term is −, the signs between the terms of the binomials are opposites.	Since the last term is + and the middle term −, the signs in both sets of parentheses will be −. The factors of the first term 5 are 5 and 1. The factors of the last term 3 are 3 and 1.
5. Try various combinations of first terms and last terms until you find the one that works.	The possible combinations are $(5x - 3)(x - 1)$ and $(5x - 1)(x - 3)$
6. Check the factorization by multiplication.	The pair that will give a middle term of $-8x$ is $(5x - 3)(x - 1)$.

REVIEW EXERCISES

Factor each expression completely, if possible.

55. $x^2 + 10x + 25$
56. $a^2 - 14a + 49$
57. $y^2 + 21y + 20$
58. $z^2 - 11z + 30$
59. $-x^2 - 3x + 28$
60. $y^2 - 5y - 24$
61. $4a^2 - 5a + 1$
62. $3b^2 + 2b + 1$
63. $7x^2 + x + 2$
64. $-15x^2 + 14x + 8$
65. $y^3 + y^2 - 2y$
66. $2a^4 + 4a^3 - 6a^2$
67. $-3x^2 - 9x - 6$
68. $8x^2 - 4x - 24$
69. $15x^2 - 57xy - 12y^2$
70. $30x^2 + 65xy + 10y^2$
71. $24x^2 - 23xy - 12y^2$
72. $14x^2 + 13xy - 12y^2$

SECTION 5.7 Summary of Factoring Techniques

DEFINITIONS AND CONCEPTS	EXAMPLES
Identifying factoring problem types: 1. Factor out all common monomial factors. 2. If an expression has two terms, check to see whether it is a. **The difference of two squares:** $x^2 - y^2 = (x + y)(x - y)$ b. **The sum of two cubes:** $x^3 + y^3 = (x + y)(x^2 - xy + y^2)$ c. **The difference of two cubes:** $x^3 - y^3 = (x - y)(x^2 + xy + y^2)$ 3. If an expression has three terms, attempt to factor it as a **trinomial.** 4. If an expression has four or more terms, try factoring by **grouping.** 5. Continue until each individual factor (except possibly a numerical coefficient) is prime. 6. Check the results.	To factor $ax^3 - bx^3 - axy^2 + bxy^2$, first factor out the common factor of x to get $x(ax^2 - bx^2 - ay^2 + by^2)$ Because the expression in the parentheses has four terms, factor it by grouping. $x(ax^2 - bx^2 - ay^2 + by^2) = x[x^2(a - b) - y^2(a - b)]$ Then factor out $a - b$ to get $= x(a - b)(x^2 - y^2)$ and finally factor the difference of two squares to get $= x(a - b)(x + y)(x - y)$ Thus, $ax^3 - bx^3 - axy^2 + bxy^2 = x(a - b)(x + y)(x - y)$ To check, verify that $x(a - b)(x + y)(x - y) = ax^3 - bx^3 - axy^2 + bxy^2$

REVIEW EXERCISES

Factor each expression, if possible.

73. $x^3 + 5x^2 - 6x$

74. $3x^2y - 12xy - 63y$

75. $z^2 - 4 + zx - 2x$

76. $x^2 + 2x + 1 - p^2$

77. $x^2 + 4x + 4 - 4p^4$

78. $y^2 + 3y + 2 + 2x + xy$

79. $x^{2m} + 2x^m - 3$

80. $x^{-2} - x^{-1} - 2$

SECTION 5.8 Solving Equations by Factoring

DEFINITIONS AND CONCEPTS	EXAMPLES
Zero-factor property: If $xy = 0$, then $x = 0$ or $y = 0$.	To solve $7x^3 = 2x\left(x + \frac{5}{2}\right)$, proceed as follows: $7x^3 = 2x^2 + 5x$ — Remove parentheses. $7x^3 - 2x^2 - 5x = 0$ — Subtract $2x^2$ and $5x$ from both sides. $x(7x^2 - 2x - 5) = 0$ — Factor out x. $x(7x + 5)(x - 1) = 0$ — Factor the trinomial. $x = 0$ or $7x + 5 = 0$ or $x - 1 = 0$ — Use the zero-factor property. $x = 0$ | $x = -\frac{5}{7}$ | $x = 1$ — Solve each equation. Verify that the solutions are 0, $-\frac{5}{7}$, and 1.

REVIEW EXERCISES

Solve each equation.

81. $4x^2 - 3x = 0$

82. $x^2 - 36 = 0$

83. $12x^2 + 4x - 5 = 0$

84. $7y^2 - 37y + 10 = 0$

85. $t^2(15t - 2) = 8t$

86. $3u^3 = u(19u + 14)$

87. Volume The volume V of the rectangular solid is given by the formula $V = lwh$, where l is its length, w is its width, and h is its height. If the volume is 840 cubic centimeters, the length is 12 centimeters, and the width exceeds the height by 3 centimeters, find the height.

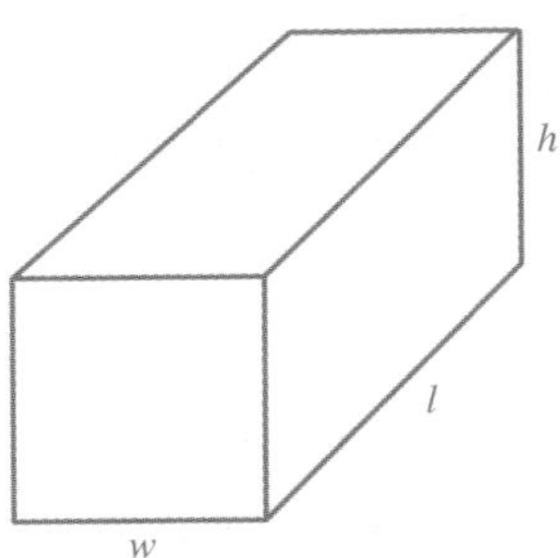

88. Volume of a pyramid The volume of the pyramid is given by the formula $V = \frac{Bh}{3}$, where B is the area of its base and h is its height. The volume of the pyramid is 1,020 cubic meters. Find the dimensions of its rectangular base if one edge of the base is 3 meters longer than the other and the height of the pyramid is 9 meters.

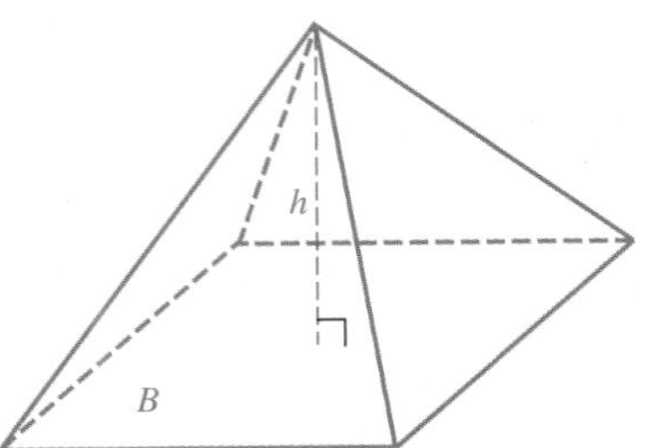

Chapter 5 TEST

Find the degree of each polynomial.

1. $3x^3 - 4x^5 - 3x^2 - 5$
2. $3x^5y^3 - x^8y^2 + 2x^9y^4 - 3x^2y^5 + 4$

Let $P(x) = -3x^2 + 2x - 1$ and find each value.

3. $P(2)$
4. $P(-1)$
5. Graph the function $f(x) = x^2 + 2x$.

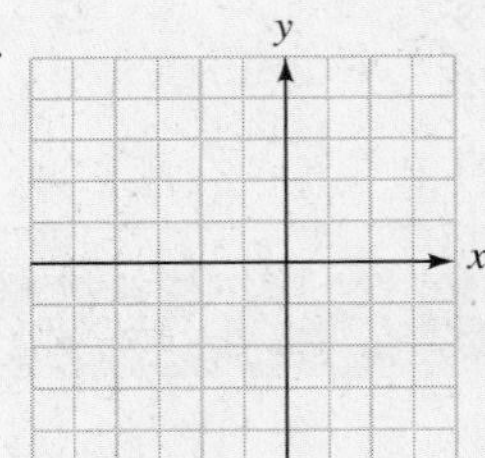

Perform the operations.

6. $(2y^2 + 4y + 3) + (3y^2 - 3y - 4)$
7. $(-3u^2 + 2u - 7) - (u^2 + 7)$
8. $3(2a^2 - 4a + 2) - 4(-a^2 - 3a - 4)$
9. $-2(2x^2 - 2) + 3(x^2 + 5x - 2)$
10. $(3x^3y^2z)(-2xy^{-1}z^3)$
11. $-5a^2b(3ab^3 - 2ab^4)$
12. $(z + 4)(z - 4)$
13. $(3x - 2)(4x + 3)$
14. $(u - v)^2(2u + v)$

Factor each expression completely.

15. $3xy^2 + 6x^2y$
16. $12a^3b^2c - 3a^2b^2c^2 + 6abc^3$
17. Factor y^n from $x^2y^{n+2} + y^n$.
18. Factor b^n from $a^nb^n - ab^{-n}$.
19. $(u - v)r + (u - v)s$
20. $ax - xy + ay - y^2$
21. $x^2 - 49$
22. $2x^2 - 32$
23. $4y^4 - 64$
24. $b^3 + 125$
25. $b^3 - 27$
26. $3u^3 - 24$
27. $a^2 - 5a - 6$
28. $6b^2 + b - 2$
29. $6u^2 + 9u - 6$
30. $20r^2 - 15r - 5$
31. $x^{2n} + 2x^n + 1$
32. $x^2 + 6x + 9 - y^2$

33. Solve for r: $r_1r_2 - r_2r = r_1r$.
34. Solve for x: $x^2 - 5x - 6 = 0$.
35. **Integer problem** The product of two consecutive positive integers is 156. Find their sum.
36. **Preformed concrete** The slab of concrete in the illustration is twice as long as it is wide. The area in which it is placed includes a 1-foot-wide border of 70 square feet. Find the dimensions of the slab.

©Shutterstock.com/Keith A Frith

CHAPTER 6

Rational Expressions

6.1 Finding the Domain of Rational Functions and Simplifying Rational Expressions
6.2 Multiplying and Dividing Rational Expressions
6.3 Adding and Subtracting Rational Expressions
6.4 Simplifying Complex Fractions
6.5 Solving Equations Containing Rational Expressions
6.6 Dividing Polynomials
6.7 Synthetic Division
6.8 Proportion and Variation
- ***Projects***
CHAPTER REVIEW
CHAPTER TEST
CUMULATIVE REVIEW EXERCISES

Careers and Mathematics

ENVIRONMENTAL ENGINEER

Using the principles of biology and chemistry, environmental engineers develop solutions to environmental problems. They are involved in water- and air-pollution control, recycling, waste disposal, and public health issues. Environmental engineers held about 54,000 jobs in 2006. Almost half worked in professional, scientific, and technical services and about 15,000 were employed in federal, state, and local government agencies. A bachelor's degree is required for almost all entry-level jobs.

Job Outlook:
Employment of environmental engineers is expected to increase much faster than the average for all occupations through 2016.

Annual Earnings:
$54,150–$88,480

For More Information:
http://www.bls.gov/oco/ocos027.htm

For a Sample Application:
See Problem 109 in Section 6.1.

In this chapter ▶

Rational expressions are the fractions of algebra. In this chapter, we will learn how to simplify, add, subtract, multiply, and divide them. Then we will solve equations containing rational expressions.

SECTION 6.1 Finding the Domain of Rational Functions and Simplifying Rational Expressions

Objectives

1. Find all values of a variable for which a rational expression is undefined.
2. Find the domain of a rational function.
3. Simplify a rational expression.
4. Simplify a rational expression containing factors that are negatives.
5. Evaluate a rational function that models an application problem.

Vocabulary

rational function   rational expression   asymptote

Getting Ready

Simplify each fraction.

1. $\frac{6}{8}$ **2.** $\frac{12}{15}$ **3.** $\frac{-25}{65}$ **4.** $\frac{-49}{-63}$

We begin our discussion of rational expressions by considering **rational functions.** Then, we will discuss how to simplify rational expressions.

1 Find all values of a variable for which a rational expression is undefined.

Rational expressions are fractions that indicate the quotient of two polynomials. Some examples of rational expressions are

$$\frac{3x}{x-7}, \quad \frac{5m+n}{8m+16}, \quad \text{and} \quad \frac{a^3+2a^2+7}{2a^2-5a+4}$$

Since division by 0 is undefined, the value of a polynomial occurring in the denominator cannot be 0. For example, x cannot be 7 in the rational expression $\frac{3x}{x-7}$, because the denominator would be 0. In $\frac{5m+n}{8m+16}$, m cannot be -2, because that would make the denominator equal to 0.

To evaluate a rational expression for a given value, we replace the variable in the expression with the given value and simplify. For example, to evaluate $\frac{x+1}{x-3}$ for $x = -1$, $x = 0$, and $x = 3$, we proceed as follows:

$$\frac{x+1}{x-3} = \frac{-1+1}{-1-3} = \frac{0}{-4} = 0 \qquad \frac{x+1}{x-3} = \frac{0+1}{0-3} = \frac{1}{-3} = -\frac{1}{3} \qquad \frac{x+1}{x-3} = \frac{3+1}{3-3} = \frac{4}{0} \text{ (undefined)}$$

In the expression $\frac{x+1}{x-3}$, x cannot be 3 because 3 would make the denominator 0.

EXAMPLE 1 Find all values of x for which each expression is undefined.

a. $\dfrac{9x}{5x-4}$ **b.** $\dfrac{3x+2}{x^2+x-6}$

Solution **a.** Because the denominator of the rational expression cannot be 0, the denominator $5x - 4$ cannot be 0. So we must exclude all values of x such that $5x - 4 = 0$. To find these values we will solve the equation.

$$5x - 4 = 0$$
$$5x = 4 \quad \text{Add 4 to both sides.}$$
$$x = \frac{4}{5} \quad \text{Divide both sides by 5.}$$

Since substituting $\frac{4}{5}$ for x will make the denominator 0, the expression will be undefined when $x = \frac{4}{5}$.

b. Because the denominator of the rational expression cannot be 0, the denominator $x^2 + x - 6$ cannot be 0. So we must exclude all values of x such that $x^2 + x - 6 = 0$. To find these values, we will solve the equation.

$$x^2 + x - 6 = 0$$
$$(x+3)(x-2) = 0 \quad \text{Factor the trinomial.}$$
$$x + 3 = 0 \quad \text{or} \quad x - 2 = 0 \quad \text{Set each factor equal to 0.}$$
$$x = -3 \quad \Big| \quad x = 2 \quad \text{Solve each linear equation.}$$

Since substituting 2 or -3 for x will make the denominator 0, the expression will be undefined when $x = 2$ or $x = -3$.

SELF CHECK 1 Find all values of x for which each expression is undefined.

a. $\dfrac{5x+2}{6x-9}$ **b.** $\dfrac{3x}{5x^2+3x-2}$

2 Find the domain of a rational function.

Rational expressions often define functions. For example, the equation $f(x) = \frac{9x}{5x-4}$ defines a function because it sets up a correspondence in which every number x determines exactly one value of y. Since division by 0 is undefined, we must exclude any values that make the denominator 0 from the domain of the function. From Example 1(a), we saw that x cannot be $\frac{4}{5}$. Thus, the domain of $f(x) = \frac{9x}{5x-4}$ is the set of all real numbers except $\frac{4}{5}$. In interval notation, the domain is $\left(-\infty, \frac{4}{5}\right) \cup \left(\frac{4}{5}, \infty\right)$.

EXAMPLE 2 Find the domain of $f(x) = \dfrac{3x + 2}{x^2 + x - 6}$.

Solution From Example 1(b), we saw that $x = 2$ and $x = -3$ make the denominator of the fraction $\frac{3x + 2}{x^2 + x - 6}$ equal to 0. Thus, the domain of the function is the set of all real numbers except -3 and 2. In interval notation, the domain is $(-\infty, -3) \cup (-3, 2) \cup (2, \infty)$.

SELF CHECK 2 Find the domain of $f(x) = \frac{x^2 + 1}{x - 2}$.

ACCENT ON TECHNOLOGY

Finding the Domain and Range of a Function

We can find the domain and range of the function in Example 2 by looking at its graph. If we use window settings of $[-10, 10]$ for x and $[-10, 10]$ for y and graph the function

$$f(x) = \frac{3x + 2}{x^2 + x - 6}$$

we will obtain the graph in Figure 6-1(a).

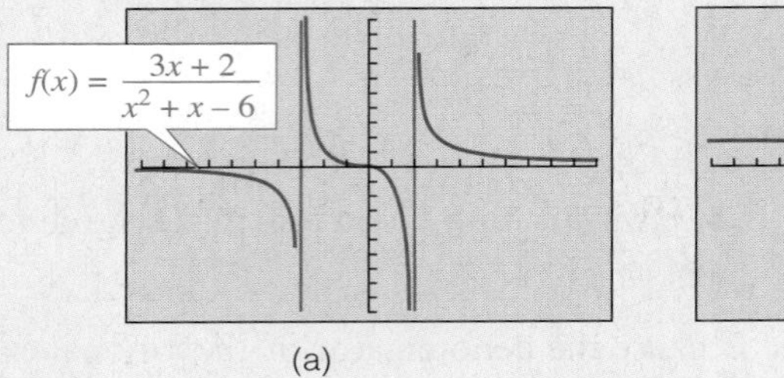

(a)

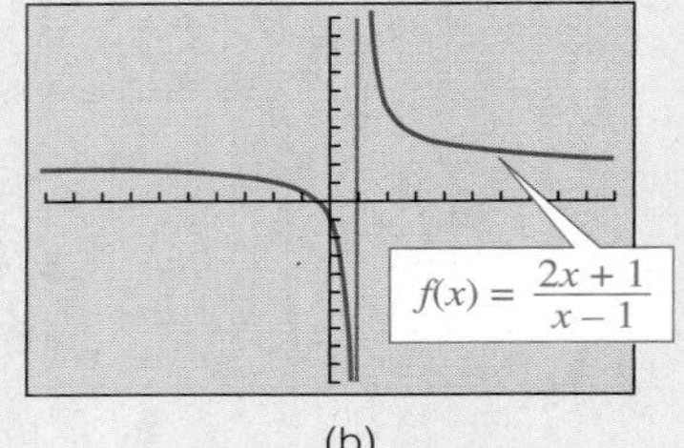

(b)

Figure 6-1

From the figure, we can see that

- As x approaches -3 from the left, the values of y decrease, and the graph approaches the vertical line $x = -3$.
- As x approaches -3 from the right, the values of y increase, and the graph approaches the vertical line $x = -3$.

We also can see that

- As x approaches 2 from the left, the values of y decrease, and the graph approaches the vertical line $x = 2$.
- As x approaches 2 from the right, the values of y increase, and the graph approaches the vertical line $x = 2$.

The lines $x = -3$ and $x = 2$ are *vertical asymptotes.* Although the vertical lines in the graph appear to be the graphs of $x = -3$ and $x = 2$, they are not. Graphing calculators draw graphs by connecting dots whose x-coordinates are close together. Often, when two such points straddle a vertical asymptote and their y-coordinates are far apart, the calculator draws a line between them anyway, producing what appears to be a vertical asymptote. If you set your calculator to dot mode instead of connected mode, the vertical lines will not appear.

From Figure 6-1(a), we also can see that

- As x increases to the right of 2, the values of y decrease and approach the value $y = 0$.

- As x decreases to the left of -3, the values of y increase and approach the value $y = 0$.

The line $y = 0$ (the x-axis) is a *horizontal asymptote.* Graphing calculators do not draw lines that appear to be horizontal asymptotes.

From the graph, we can see that all real numbers x, except -3 and 2, give a value of y. This confirms that the domain of the function is $(-\infty, -3) \cup (-3, 2) \cup (2, \infty)$. We also can see that y can be any value. Thus, the range is $(-\infty, \infty)$.

To find the domain and range of the function $f(x) = \frac{2x+1}{x-1}$, we use a calculator to draw the graph shown in Figure 6-1(b). From this graph, we can see that the line $x = 1$ is a vertical asymptote and that the line $y = 2$ is a horizontal asymptote. Since x can be any real number except 1, the domain is the interval $(-\infty, 1) \cup (1, \infty)$. Since y can be any value except 2, the range is $(-\infty, 2) \cup (2, \infty)$.

3 Simplify a rational expression.

Since rational expressions are the fractions of algebra, the rules for arithmetic fractions apply.

Properties of Rational Expressions

If there are no divisions by 0, then

1. $\frac{a}{b} = \frac{c}{d}$ if and only if $ad = bc$ **2.** $\frac{a}{1} = a$ and $\frac{a}{a} = 1$

3. $\frac{ak}{bk} = \frac{a}{b}$ **4.** $-\frac{a}{b} = \frac{-a}{b} = \frac{a}{-b}$

To simplify rational expressions, we will use Property 3, which enables us to divide out factors that are common to the numerator and the denominator.

EXAMPLE 3 Simplify: **a.** $\frac{10k}{25k^2}$ **b.** $\frac{-8y^3z^5}{6y^4z^3}$. Assume no variable is zero.

Solution We find the prime factorization of each numerator and denominator and divide out the common factors:

a.
$$\frac{10k}{25k^2} = \frac{5 \cdot 2 \cdot k}{5 \cdot 5 \cdot k \cdot k} = \frac{\overset{1}{\cancel{5}} \cdot 2 \cdot \overset{1}{\cancel{k}}}{\underset{1}{\cancel{5}} \cdot 5 \cdot \underset{1}{\cancel{k}} \cdot k} = \frac{2}{5k}$$

b.
$$\frac{-8y^3z^5}{6y^4z^3} = \frac{-2 \cdot 4 \cdot y \cdot y \cdot y \cdot z \cdot z \cdot z \cdot z \cdot z}{2 \cdot 3 \cdot y \cdot y \cdot y \cdot y \cdot z \cdot z \cdot z} = \frac{-\overset{1}{\cancel{2}} \cdot 4 \cdot \overset{1}{\cancel{y}} \cdot \overset{1}{\cancel{y}} \cdot \overset{1}{\cancel{y}} \cdot \overset{1}{\cancel{z}} \cdot \overset{1}{\cancel{z}} \cdot \overset{1}{\cancel{z}} \cdot z \cdot z}{\underset{1}{\cancel{2}} \cdot 3 \cdot \underset{1}{\cancel{y}} \cdot \underset{1}{\cancel{y}} \cdot \underset{1}{\cancel{y}} \cdot y \cdot \underset{1}{\cancel{z}} \cdot \underset{1}{\cancel{z}} \cdot \underset{1}{\cancel{z}}} = -\frac{4z^2}{3y}$$

SELF CHECK 3 Simplify: $\frac{-12a^4b^2}{-3ab^4}$. Assume no variable is zero.

The rational expressions in Example 3 also can be simplified by using the rules of exponents:

$$\frac{10k}{25k^2} = \frac{5 \cdot 2}{5 \cdot 5}k^{1-2} \qquad \frac{-8y^3z^5}{6y^4z^3} = \frac{-2 \cdot 4}{2 \cdot 3}y^{3-4}z^{5-3}$$

$$= \frac{2}{5} \cdot k^{-1} \qquad = \frac{-4}{3} \cdot y^{-1}z^2$$

$$= \frac{2}{5} \cdot \frac{1}{k} \qquad = -\frac{4}{3} \cdot \frac{1}{y} \cdot \frac{z^2}{1}$$

$$= \frac{2}{5k} \qquad = -\frac{4z^2}{3y}$$

EXAMPLE 4 Simplify: $\frac{x^2 - 16}{x + 4}$. Assume no denominator of zero.

Solution We factor $x^2 - 16$ and use the fact that $\frac{x + 4}{x + 4} = 1$.

$$\frac{x^2 - 16}{x + 4} = \frac{\overset{1}{\cancel{(x + 4)}}(x - 4)}{1\underset{1}{\cancel{(x + 4)}}}$$

$$= \frac{x - 4}{1}$$

$$= x - 4$$

SELF CHECK 4 Simplify: $\frac{x^2 - 9}{x - 3}$. Assume no denominator of zero.

ACCENT ON TECHNOLOGY

Checking an Algebraic Simplification

To show that the simplification in Example 4 is correct, we can graph the following functions and show that the graphs are the same.

$$f(x) = \frac{x^2 - 16}{x + 4} \quad \text{and} \quad g(x) = x - 4$$

(See Figure 6-2.) Except for the point where $x = -4$, the graphs are the same.

The point where $x = -4$ is excluded from the graph of $f(x) = \frac{x^2 - 16}{x + 4}$, because -4 is not in the domain of f. However, graphing calculators do not show that this point is excluded. The point where $x = -4$ is included in the graph of $g(x) = x - 4$, because -4 is in the domain of g.

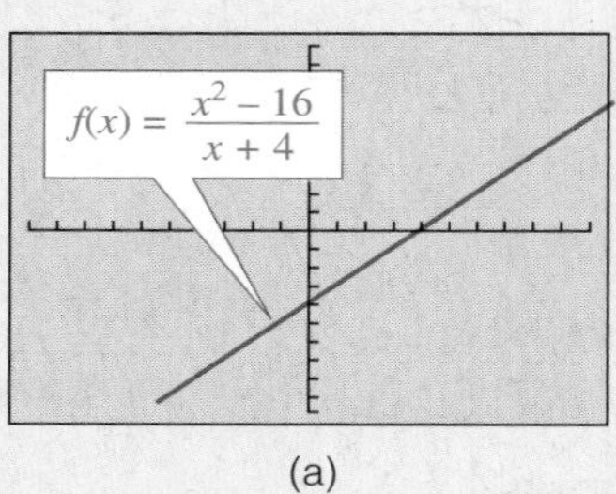

(a)

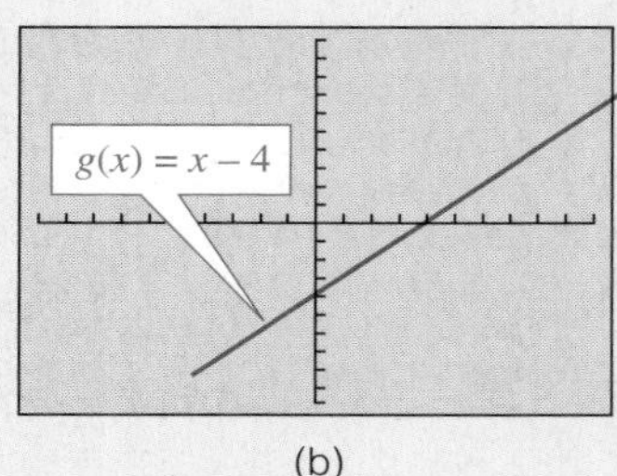

(b)

Figure 6-2

EXAMPLE 5 Simplify: $\dfrac{2x^2 + 11x + 12}{3x^2 + 11x - 4}$. Assume no denominator of zero.

Solution We factor the numerator and denominator and use the fact that $\frac{x+4}{x+4} = 1$.

$$\frac{2x^2 + 11x + 12}{3x^2 + 11x - 4} = \frac{(2x + 3)\overset{1}{\cancel{(x + 4)}}}{(3x - 1)\underset{1}{\cancel{(x + 4)}}} \qquad \frac{x+4}{x+4} = 1$$

$$= \frac{2x + 3}{3x - 1}$$

COMMENT Only factors that are common to the entire numerator and the entire denominator can be divided out. Terms common to both the numerator and denominator cannot be divided out. For example, it is incorrect to divide out the common term of 3 in the following simplification, because doing so gives a wrong answer.

$$\frac{3 + 7}{3} = \frac{\overset{1}{\cancel{3}} + 7}{\underset{1}{\cancel{3}}} = \frac{1 + 7}{1} = 8 \qquad \text{The correct simplification is } \frac{3+7}{3} = \frac{10}{3}.$$

The 3s in the fraction $\frac{2x+3}{3x-1}$ cannot be divided out either, because the 3 in the numerator is a term and the 3 in the denominator is a factor of the first term only. To be divided out, the 3 must be a factor of the entire numerator and the entire denominator.

SELF CHECK 5 Simplify: $\dfrac{2x^2 + 7x - 15}{2x^2 + 13x + 15}$. Assume no denominator of zero.

EXAMPLE 6 Simplify: $\dfrac{(x^2 + 2x)(x^2 + 2x - 3)}{(x^2 + x - 2)(x^2 + 3x)}$. Assume no denominator of zero.

Solution We factor the numerator and denominator and divide out all common factors:

$$\frac{(x^2 + 2x)(x^2 + 2x - 3)}{(x^2 + x - 2)(x^2 + 3x)} = \frac{\overset{1}{\cancel{x}}\overset{1}{\cancel{(x + 2)}}\overset{1}{\cancel{(x + 3)}}\overset{1}{\cancel{(x - 1)}}}{\underset{1}{\cancel{(x + 2)}}\underset{1}{\cancel{(x - 1)}}\underset{1}{\cancel{x}}\underset{1}{\cancel{(x + 3)}}} \qquad \frac{x}{x} = 1, \frac{x+2}{x+2} = 1, \frac{x+3}{x+3} = 1, \frac{x-1}{x-1} = 1$$

$$= 1$$

SELF CHECK 6 Simplify: $\dfrac{(a^2 + 4a)(a^2 - a - 2)}{a(a^2 + 2a - 8)}$. Assume no denominator of zero.

Many rational expressions we shall encounter are already in simplified form. For example, to attempt to simplify

$$\frac{x^2 + xa + 2x + 2a}{x^2 + x - 6}$$

we factor the numerator and denominator and divide out any common factors:

$$\frac{x^2 + xa + 2x + 2a}{x^2 + x - 6} = \frac{x(x + a) + 2(x + a)}{(x - 2)(x + 3)} = \frac{(x + a)(x + 2)}{(x - 2)(x + 3)}$$

Since there are no common factors in the numerator and denominator, the result cannot be simplified.

4 Simplify a rational expression containing factors that are negatives.

To simplify $\frac{b - a}{a - b}$ $(a \neq b)$, we factor -1 from the numerator and divide out any factors common to both the numerator and the denominator:

$$\frac{b - a}{a - b} = \frac{-a + b}{a - b} \quad (a \neq b)$$

$$= \frac{-\overset{1}{\cancel{(a - b)}}}{\underset{1}{\cancel{(a - b)}}} \qquad \frac{a - b}{a - b} = 1$$

$$= \frac{-1}{1}$$

$$= -1$$

In general, we have the following principle.

Quotient of a Quantity and Its Opposite

The quotient of any nonzero quantity and its negative (or opposite) is -1.

EXAMPLE 7 Simplify: $\frac{3x^2 - 10xy - 8y^2}{4y^2 - xy}$. Assume no denominator of zero.

Solution We factor the numerator and denominator and note that because $x - 4y$ and $4y - x$ are negatives, their quotient is -1.

$$\frac{3x^2 - 10xy - 8y^2}{4y^2 - xy} = \frac{(3x + 2y)\overset{-1}{\cancel{(x - 4y)}}}{y\underset{1}{\cancel{(4y - x)}}} \qquad \frac{x - 4y}{4y - x} = \frac{-1(4y - x)}{4y - x} = -1$$

$$= \frac{-(3x + 2y)}{y}$$

$$= \frac{-3x - 2y}{y}$$

SELF CHECK 7 Simplify: $\frac{2a^2 - 5ab - 3b^2}{3b^2 - ab}$. Assume no denominator of zero.

5 Evaluate a rational function that models an application problem.

Rational expressions often define functions. For example, if the cost of subscribing to an online research service is \$6 per month plus \$1.50 per hour of access time, the average (mean) hourly cost of the service is the total monthly cost, divided by the number of hours of access time.

$$\bar{c} = \frac{c}{n} = \frac{1.50n + 6}{n}$$ $\bar{c}$ is the mean hourly cost, c is the total monthly cost, and n is the number of hours the service is used.

The function

$$(1) \qquad \bar{c} = f(n) = \frac{1.50n + 6}{n} \quad (n > 0)$$

gives the mean hourly cost of using the research service for n hours per month.

Figure 6-3 shows the graph of the rational function $\bar{c} = f(n) = \frac{1.50n + 6}{n}$ $(n > 0)$. Since $n > 0$, the domain of this function is the interval $(0, \infty)$.

Sonya Kovalevskaya (1850–1891)

This talented young Russian woman hoped to study mathematics at the University of Berlin, but strict rules prohibited women from attending lectures. Undaunted, Sonya studied privately with the great mathematician Karl Weierstrauss and published several important papers.

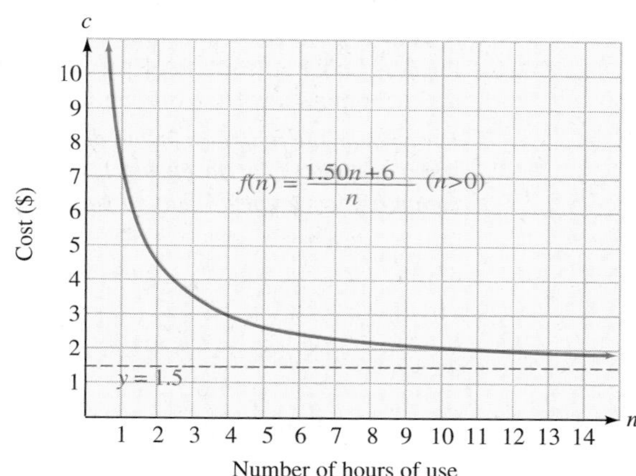

Figure 6-3

From the graph, we can see that the mean hourly cost decreases as the number of hours of access time increases. Since the cost of each extra hour of access time is \$1.50, the mean hourly cost can approach \$1.50 but never drop below it. Thus, the graph of the function approaches the line $y = 1.5$ as n increases without bound. When a graph approaches a line as the independent variable gets large, we call the line an **asymptote.** The line $y = 1.5$ is a *horizontal asymptote* of the graph.

As n gets smaller and approaches 0, the graph approaches the y-axis but never touches it. The y-axis is a *vertical asymptote* of the graph.

EXAMPLE 8 Find the mean hourly cost when the service described above is used for

a. 3 hours **b.** 70.4 hours.

Solution **a.** To find the mean hourly cost for 3 hours of access time, we substitute 3 for n in Equation 1 and simplify:

$$\bar{c} = f(3) = \frac{1.50(3) + 6}{3} = 3.5$$

The mean hourly cost for 3 hours of access time is \$3.50.

b. To find the mean hourly cost for 70.4 hours of access time, we substitute 70.4 for n in Equation 1 and simplify:

$$\bar{c} = f(70.4) = \frac{1.50(70.4) + 6}{70.4} = 1.585227273$$

The mean hourly cost for 70.4 hours of access time is approximately \$1.59.

SELF CHECK 8 Find the mean hourly cost when the service is used for 5 hours.

SELF CHECK ANSWERS

1. a. $x = \frac{3}{2}$ **b.** $x = \frac{2}{5}, -1$ **2.** $(-\infty, 2) \cup (2, \infty)$ **3.** $\frac{4a^3}{b^2}$ **4.** $x + 3$ **5.** $\frac{2x - 3}{2x + 3}$ **6.** $a + 1$ **7.** $-\frac{2a + b}{b}$ **8.** \$2.70

NOW TRY THIS

A vertical asymptote appears at any value for which a simplified rational function is undefined. A hole in the graph of the function occurs at any value that is common to the numerator and denominator of the original function.

For each of the functions below, determine where the vertical asymptote(s) occur (if any) and where the hole(s) occur (if any). Graph the original function and any simplified function and compare their graphs. What do you notice?

1. $f(x) = \dfrac{x^2 + 4}{x^2 - 9}$

2. $g(x) = \dfrac{x^3 - 3x^2}{15x - 5x^2}$

3. $h(x) = \dfrac{x^2 - 6x + 9}{x^2 - 5x + 6}$

Note: Graph the functions using the window settings of $x = [-4.7, 4.7]$ and $y = [-5, 5]$ in order to see the holes. Use the TRACE feature to verify the location of the holes. (There will be no value for y.)

6.1 EXERCISES

WARM-UPS

Evaluate $f(x) = \frac{2x - 3}{x - 1}$ *when*

1. $x = 1$

2. $x = 0$

Give the domain of each function.

3. $f(x) = \dfrac{x + 7}{2x - 4}$

4. $f(x) = \dfrac{3x - 4}{x^2 - 9}$

Simplify each expression.

5. $\dfrac{25}{30}$

6. $\dfrac{x^2}{xy}$

7. $\dfrac{2x - 4}{x - 2}$

8. $\dfrac{x - 2}{2 - x}$

REVIEW *Factor each expression.*

9. $3x^2 - 9x$

10. $6t^2 - 5t - 6$

11. $27x^6 + 64y^3$

12. $x^2 + ax + 2x + 2a$

VOCABULARY AND CONCEPTS *Fill in the blanks.*

13. If a fraction is the quotient of two polynomials, it is called a ________ expression.

14. The denominator of a fraction can never be __.

15. If a graph approaches a line, the line is called an ________.

16. $\frac{a}{b} = \frac{c}{d}$ if and only if ________.

17. $\frac{a}{1} =$ __ and $\frac{a}{a} =$ __, provided that $a \neq$ __.

18. Rational expressions can often define a ________ function.

19. $\frac{ak}{bk} = \frac{\quad}{\_\_}$, provided that $b \neq$ __ and $k \neq 0$.

20. The quotient of any nonzero quantity and its negative is ____.

GUIDED PRACTICE

Find all values of the variable for which the rational expression is undefined. **See Example 1. (Objective 1)**

21. $\frac{3x - 8}{x + 6}$

22. $\frac{3a^2 + 5a}{2a - 3}$

23. $\frac{12x - 7}{6x + 5}$

24. $\frac{3x - 13}{x^2 + x - 2}$

Give the domain of each function in interval notation. **See Example 2. (Objective 2)**

25. $\frac{3a^2 + 5a}{2a - 3}$

26. $\frac{12x - 7}{6x + 5}$

27. $\frac{3x - 13}{x^2 + x - 2}$

28. $\frac{3x - 8}{x + 6}$

Use a graphing calculator to graph each function. From the graph, determine its domain. **(Objective 2)**

29. $f(x) = \frac{x}{x - 2}$

30. $f(x) = \frac{x + 2}{x}$

31. $f(x) = \frac{x + 1}{x^2 - 4}$

32. $f(x) = \frac{x - 2}{x^2 - 3x - 4}$

Simplify each expression when possible. Assume no denominator of zero. **See Example 3. (Objective 3)**

33. $\frac{12}{18}$

34. $\frac{25}{55}$

35. $-\frac{112}{36}$

36. $-\frac{49}{21}$

37. $\frac{12x^3}{3x}$

38. $-\frac{15a^2}{25a^3}$

39. $\frac{-24x^3y^4}{18x^4y^3}$

40. $\frac{15a^5b^4}{21b^3c^2}$

41. $-\frac{11x(x - y)}{22(x - y)}$

42. $\frac{x(x - 2)^2}{(x - 2)^3}$

43. $\frac{9y^2(y - z)}{21y(y - z)^2}$

44. $\frac{(a - b)(c - d)}{(c - d)(a - b)}$

Simplify each expression when possible. Assume no denominator of zero. **See Examples 4–5. (Objective 3)**

45. $\frac{x^2 - 25}{x + 5}$

46. $\frac{y - 7}{y^2 - 49}$

47. $\frac{x^2 + 2x - 15}{x^2 - 25}$

48. $\frac{5x - 10}{x^2 - 4x + 4}$

49. $\frac{2a^2 + 5a + 3}{6a + 9}$

50. $\frac{2x^2 + 2x - 12}{x^3 + 3x^2 - 4x - 12}$

51. $\frac{x^2 + 2x + 1}{x^2 + 4x + 3}$

52. $\frac{6x^2 + x - 2}{8x^2 + 2x - 3}$

53. $\frac{4x^2 + 24x + 32}{16x^2 + 8x - 48}$

54. $\frac{-4x - 4 + 3x^2}{4x^2 - 2 - 7x}$

55. $\frac{x^3 + 8}{x^2 - 2x + 4}$

56. $\frac{a^2 - 4}{a^3 - 8}$

57. $\frac{x^2 + 3x + 9}{x^3 - 27}$

58. $\frac{x^2 - x - 2}{2x + 4}$

59. $\frac{x + y}{x^2 - y^2}$

60. $\frac{x - y}{x^2 - y^2}$

61. $\frac{3x + 6y}{x + 2y}$

62. $\frac{y - xy}{xy - x}$

63. $\frac{4a^2 - 9b^2}{2a^2 - ab - 6b^2}$

64. $\frac{x^2 + 2xy}{x + 2y + x^2 - 4y^2}$

65. $\frac{x - y}{x^3 - y^3 - x + y}$

66. $\frac{(m + n)^3}{m^2 + 2mn + n^2}$

Simplify each expression when possible. Assume no denominator of zero. **See Example 6. (Objective 3)**

67. $\frac{(x^2 - 1)(x + 1)}{(x^2 - 2x + 1)^2}$

68. $\frac{(x^2 + 2x + 1)(x^2 - 2x + 1)}{(x^2 - 1)^2}$

69. $\frac{(2x^2 + 3xy + y^2)(3a + b)}{(x + y)(2xy + 2bx + y^2 + by)}$

70. $\frac{(x - 1)(6ax + 9x + 4a + 6)}{(3x + 2)(2ax - 2a + 3x - 3)}$

Simplify each expression when possible. Assume no denominator of zero. See Example 7. (Objective 4)

71. $\frac{(p + q)(p - r)}{(r - p)(p + q)}$

72. $\frac{a - b}{b^2 - a^2}$

73. $\frac{-3ab^2(a - b)}{9ab(b - a)}$

74. $\frac{12 - 3x^2}{x^2 - x - 2}$

Each of the following applications models a rational function. Evaluate each one for the given values. See Example 8. (Objective 5)

The time, t, it takes to travel 600 miles is a function of the mean rate of speed, r: $t = f(r) = \frac{600}{r}$. Find t for each value of r.

75. 30 mph

76. 40 mph

77. 50 mph

78. 60 mph

Suppose the cost (in dollars) of removing p% of the pollution in a river is given by the function $c = f(p) = \frac{50{,}000p}{100 - p}$ $(0 \le p < 100)$. Find the cost of removing each percent of pollution.

79. 10%

80. 30%

81. 50%

82. 80%

ADDITIONAL PRACTICE

Simplify each expression when possible. Assume no denominator of zero.

83. $\frac{288}{312}$

84. $\frac{144}{72}$

85. $-\frac{244}{74}$

86. $-\frac{512}{236}$

87. $\frac{(3x^3)^2}{9x^4}$

88. $\frac{8(x^2y^3)^3}{2(xy^2)^2}$

89. $\frac{4x^2 + 8x + 3}{6 + x - 2x^2}$

90. $\frac{6x^2 + 13x + 6}{6 - 5x - 6x^2}$

91. $\frac{a^3 + 27}{4a^2 - 36}$

92. $\frac{x^3 - 27}{3x^2 - 8x - 3}$

93. $\frac{x^2 + x - 30}{x^2 - x - 20}$

94. $\frac{6x^2 - 7x - 5}{2x^2 + 5x + 2}$

95. $\frac{x^2 + y^2}{x + y}$

96. $\frac{2x^2 - 3x - 9}{2x^2 + 3x - 9}$

97. $\frac{3m - 6n}{3n - 6m}$

98. $\frac{x^4 - y^4}{(x^2 + 2xy + y^2)(x^2 + y^2)}$

99. $\frac{px - py + qx - qy}{px + qx + py + qy}$

100. $\frac{6xy - 4x - 9y + 6}{6y^2 - 13y + 6}$

101. $\frac{m^3 - mn^2}{mn^2 + m^2n - 2m^3}$

102. $\frac{p^3 + p^2q - 2pq^2}{pq^2 + p^2q - 2p^3}$

103. $\frac{3x^2 - 3y^2}{x^2 + 2y + 2x + yx}$

104. $\frac{ax + by + ay + bx}{a^2 - b^2}$

The function $f(t) = \frac{t^2 + 2t}{2t + 2}$ gives the number of days it would take two construction crews, working together, to frame a house that crew 1 (working alone) could complete in t days and crew 2 (working alone) could complete in t + 2 days.

105. If crew 1 can frame a certain house in 15 days, how long would it take both crews working together?

106. If crew 2 can frame a certain house in 20 days, how long would it take both crews working together?

The function $f(t) = \frac{t^2 + 3t}{2t + 3}$ gives the number of hours it would take two pipes to fill a pool that the larger pipe (working alone) could fill in t hours and the smaller pipe (working alone) could fill in t + 3 hours.

107. If the smaller pipe can fill a pool in 7 hours, how long would it take both pipes to fill the pool?

108. If the larger pipe can fill a pool in 8 hours, how long would it take both pipes to fill the pool?

APPLICATIONS

109. **Environmental cleanup** The cost (in dollars) of removing p% of the pollution in a river is approximated by the rational function

$$f(p) = \frac{50{,}000p}{100 - p} \quad (0 \le p < 100)$$

Find the cost of removing each percent of pollution.

a. 40%

b. 70%

110. **Directory costs** The average (mean) cost for a service club to publish a directory of its members is given by the rational function

$$f(x) = \frac{1.25x + 800}{x}$$

where x is the number of directories printed. Find the average cost per directory if

a. 700 directories are printed.

b. 2,500 directories are printed.

A service club wants to publish a directory of its members. Some investigation shows that the cost of typesetting and photography will be \$700 and the cost of printing each directory will be \$1.25.

111. Find a function that gives the total cost c of printing x directories.

112. Find a function that gives the mean cost per directory, $\overline{c}$, of printing x directories.

113. Find the total cost of printing 500 directories.

114. Find the mean cost per directory if 500 directories are printed.

115. Find the mean cost per directory if 1,000 directories are printed.

116. Find the mean cost per directory if 2,000 directories are printed.

An electric company charges \$7.50 per month plus 9¢ for each kilowatt hour (kwh) of electricity used.

117. Find a function that gives the total cost c of n kwh of electricity.

118. Find a function that gives the mean cost per kwh, $\overline{c}$, when using n kwh.

119. Find the total cost for using 775 kwh.

120. Find the mean cost per kwh when 775 kwh are used.

121. Find the mean cost per kwh when 1,000 kwh are used.

122. Find the mean cost per kwh when 1,200 kwh are used.

WRITING ABOUT MATH

123. Explain how to simplify a rational expression.

124. Explain how to recognize that a rational expression is in lowest terms.

SOMETHING TO THINK ABOUT

125. A student compares an answer of $\frac{a-3b}{2b-a}$ to an answer of $\frac{3b-a}{a-2b}$. Are the two answers the same?

126. Is this work correct? Explain.

$$\frac{3x^2+6}{3y} = \frac{\not{3}x^2+6}{\not{3}y} = \frac{x^2+6}{y}$$

127. In which parts can you divide out the 4s?

a. $\frac{4x}{4y}$ **b.** $\frac{4x}{x+4}$

c. $\frac{4+x}{4+y}$ **d.** $\frac{4x}{4+4y}$

128. In which parts can you divide out the 3s?

a. $\frac{3x+3y}{3z}$ **b.** $\frac{3(x+y)}{3x+y}$

c. $\frac{3x+3y}{3a-3b}$ **d.** $\frac{x+3}{3y}$

SECTION 6.2 Multiplying and Dividing Rational Expressions

Objectives

1. Multiply two rational expressions and write the result in simplest form.
2. Find the square of a rational expression.
3. Divide two rational expressions and write the result in simplest form.
4. Perform combined operations on three or more rational expressions.

Getting Ready

Simplify each fraction.

1. $\frac{45}{30}$ **2.** $\frac{72}{180}$ **3.** $\frac{600}{450}$ **4.** $\frac{210}{45}$

In this section, we will learn how to multiply and divide rational expressions. After mastering these skills, we will consider mixed operations.

1 Multiply two rational expressions and write the result in simplest form.

In Section 6.1, we introduced four basic properties of fractions. We now provide the rule for multiplying fractions.

Multiplying Fractions

If no denominators are 0, then

$$\frac{a}{b} \cdot \frac{c}{d} = \frac{a \cdot c}{b \cdot d} = \frac{ac}{bd}$$

Thus, *to multiply two fractions, we multiply the numerators and multiply the denominators.*

$$\frac{3}{5} \cdot \frac{2}{7} = \frac{3 \cdot 2}{5 \cdot 7} = \frac{6}{35}$$

$$\frac{4}{7} \cdot \frac{5}{8} = \frac{4 \cdot 5}{7 \cdot 8} = \frac{\overset{1}{\cancel{2}} \cdot \overset{1}{\cancel{2}} \cdot 5}{7 \cdot \underset{1}{\cancel{2}} \cdot \underset{1}{\cancel{2}} \cdot 2} = \frac{5}{14} \qquad \frac{2}{2} = 1$$

The same rule applies to rational expressions. If $t \neq 0$, then

$$\frac{x^2y}{t} \cdot \frac{xy^3}{t^3} = \frac{x^2y \cdot xy^3}{tt^3} = \frac{x^2x \cdot yy^3}{t^4} = \frac{x^3y^4}{t^4}$$

EXAMPLE 1 Find the product of $\frac{x^2 - 6x + 9}{x}$ and $\frac{x^2}{x - 3}$ $(x \neq 0, 3)$.

Solution We multiply the numerators and multiply the denominators and then simplify.

$$\frac{x^2 - 6x + 9}{x} \cdot \frac{x^2}{x - 3} = \frac{(x^2 - 6x + 9)x^2}{x(x - 3)}$$ Multiply the numerators and multiply the denominators.

$$= \frac{(x - 3)(x - 3)xx}{x(x - 3)}$$ Factor the numerator.

$$= \frac{\overset{1}{\cancel{(x - 3)}}(x - 3)\overset{1}{\cancel{x}}x}{\underset{1}{\cancel{x}}\underset{1}{\cancel{(x - 3)}}}$$ $\frac{x-3}{x-3} = 1$ and $\frac{x}{x} = 1$.

$$= x(x - 3)$$

SELF CHECK 1 Multiply: $\frac{a^2 + 6a + 9}{a} \cdot \frac{a^3}{a + 3}$ $(a \neq 0, -3)$.

ACCENT ON TECHNOLOGY

Checking an Algebraic Simplification

We can check the simplification in Example 1 by graphing the rational functions $f(x) = \left(\frac{x^2 - 6x + 9}{x}\right)\left(\frac{x^2}{x - 3}\right)$ and $g(x) = x(x - 3)$ and observing that the graphs are the same. Note that 0 and 3 are not included in the domain of the first function.

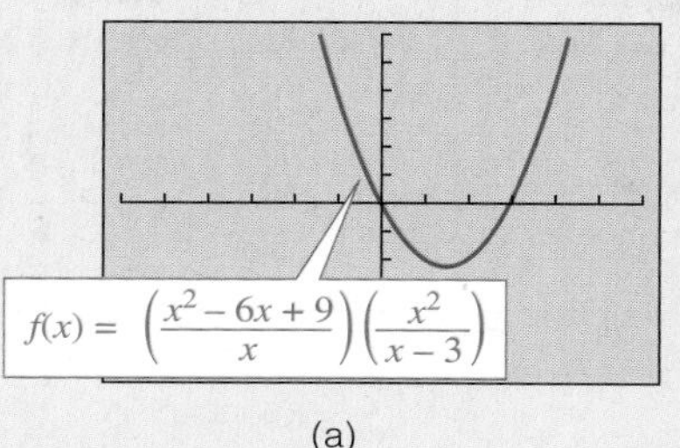

(a)

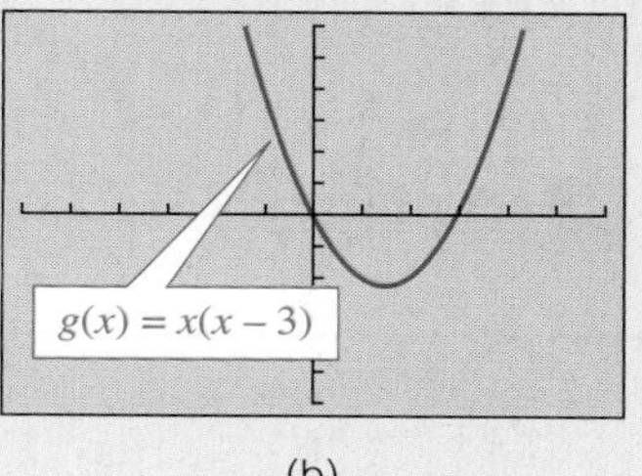

(b)

Figure 6-4

EXAMPLE 2 Multiply: $\frac{x^2 - x - 6}{x^2 - 4} \cdot \frac{x^2 + x - 6}{x^2 - 9}$ $(x \neq 2, -2, 3, -3)$.

Solution

$$\frac{x^2 - x - 6}{x^2 - 4} \cdot \frac{x^2 + x - 6}{x^2 - 9}$$

$$= \frac{(x^2 - x - 6)(x^2 + x - 6)}{(x^2 - 4)(x^2 - 9)}$$ Multiply the numerators and multiply the denominators.

$$= \frac{(x - 3)(x + 2)(x + 3)(x - 2)}{(x + 2)(x - 2)(x + 3)(x - 3)}$$ Factor the polynomials.

$$= \frac{\overset{1}{\cancel{(x - 3)}}\overset{1}{\cancel{(x + 2)}}\overset{1}{\cancel{(x + 3)}}\overset{1}{\cancel{(x - 2)}}}{\underset{1}{\cancel{(x + 2)}}\underset{1}{\cancel{(x - 2)}}\underset{1}{\cancel{(x + 3)}}\underset{1}{\cancel{(x - 3)}}}$$ $\frac{x - 3}{x - 3} = 1, \frac{x + 2}{x + 2} = 1, \frac{x + 3}{x + 3} = 1$, and $\frac{x - 2}{x - 2} = 1$.

$$= 1$$

COMMENT Note that when all factors divide out, the result is 1 and not 0.

SELF CHECK 2 Multiply: $\frac{a^2 + a - 6}{a^2 - 9} \cdot \frac{a^2 - a - 6}{a^2 - 4}$ $(a \neq 2, -2, 3, -3)$.

EXAMPLE 3 Multiply: $\frac{6x^2 + 5x - 4}{2x^2 + 5x + 3} \cdot \frac{8x^2 + 6x - 9}{12x^2 + 7x - 12}$ $\left(x \neq -\frac{3}{2}, -1, -\frac{4}{3}, \frac{3}{4}\right)$.

Solution

$$\frac{6x^2 + 5x - 4}{2x^2 + 5x + 3} \cdot \frac{8x^2 + 6x - 9}{12x^2 + 7x - 12}$$

$$= \frac{(6x^2 + 5x - 4)(8x^2 + 6x - 9)}{(2x^2 + 5x + 3)(12x^2 + 7x - 12)}$$ Multiply the numerators and multiply the denominators.

$$= \frac{(3x + 4)(2x - 1)(4x - 3)(2x + 3)}{(2x + 3)(x + 1)(3x + 4)(4x - 3)}$$ Factor the polynomials.

$$= \frac{\overset{1}{\cancel{(3x+4)}}(2x-1)\overset{1}{\cancel{(4x-3)}}\overset{1}{\cancel{(2x+3)}}}{\underset{1}{\cancel{(2x+3)}}(x+1)\underset{1}{\cancel{(3x+4)}}\underset{1}{\cancel{(4x-3)}}}$$ $\frac{3x+4}{3x+4} = 1$, $\frac{4x-3}{4x-3} = 1$, and $\frac{2x+3}{2x+3} = 1$.

$$= \frac{2x-1}{x+1}$$

SELF CHECK 3 Multiply: $\frac{2a^2+5a-12}{2a^2+11a+12} \cdot \frac{2a^2-3a-9}{2a^2-a-3}$ $\left(a \neq -\frac{3}{2}, -4, \frac{3}{2}, -1\right)$.

EXAMPLE 4 Multiply: $(2x - x^2) \cdot \frac{x}{x^2 - 5x + 6}$ $(x \neq 3, 2)$.

Solution

$$(2x - x^2) \cdot \frac{x}{x^2 - 5x + 6}$$

$$= \frac{2x - x^2}{1} \cdot \frac{x}{x^2 - 5x + 6}$$ Write $2x - x^2$ as $\frac{2x - x^2}{1}$.

$$= \frac{(2x - x^2)x}{1(x^2 - 5x + 6)}$$ Multiply the fractions.

$$= \frac{x\overset{-1}{\cancel{(2-x)}}x}{\underset{1}{\cancel{(x-2)}}(x-3)}$$ Factor out x and note that the quotient of any nonzero quantity and its negative is -1.

$$= \frac{-x^2}{x-3}$$

COMMENT In Examples 1–4, we can obtain the same answers if we factor first and divide out the common factors before we multiply.

Since $\frac{-a}{b} = -\frac{a}{b}$, the $-$ sign can be written in front of the fraction. For this reason, the final result can be written as

$$-\frac{x^2}{x-3}$$

SELF CHECK 4 Multiply: $\frac{x^2 + 5x + 6}{(x^2 + 4x)(x + 2)}(x^3 + 4x^2)$ $(x \neq 0, -4, -2)$.

2 Find the square of a rational expression.

To square a rational expression, we can write the expression without an exponent and then multiply as we have in the previous examples.

EXAMPLE 5 Find: $\left(\frac{x^2 + x - 1}{2x + 3}\right)^2$ $\left(x \neq -\frac{3}{2}\right)$.

Solution To square the expression, we write it as a factor twice and perform the multiplication.

$$\left(\frac{x^2 + x - 1}{2x + 3}\right)^2 = \left(\frac{x^2 + x - 1}{2x + 3}\right)\left(\frac{x^2 + x - 1}{2x + 3}\right)$$

$$= \frac{(x^2 + x - 1)(x^2 + x - 1)}{(2x + 3)(2x + 3)}$$ Multiply the numerators and multiply the denominators.

$$= \frac{x^4 + 2x^3 - x^2 - 2x + 1}{4x^2 + 12x + 9}$$ Do the multiplications.

SELF CHECK 5 Find: $\left(\frac{x + 5}{x^2 - 6x}\right)^2$ $(x \neq 0, 6)$.

3 Divide two rational expressions and write the result in simplest form.

Recall the rule for dividing fractions.

Dividing Fractions If no denominators are 0, then

$$\frac{a}{b} \div \frac{c}{d} = \frac{a}{b} \cdot \frac{d}{c} = \frac{ad}{bc}$$

We can prove this rule as follows:

$$\frac{a}{b} \div \frac{c}{d} = \frac{\frac{a}{b}}{\frac{c}{d}} \cdot 1 = \frac{\frac{a}{b}}{\frac{c}{d}} \cdot \frac{\frac{d}{c}}{\frac{d}{c}} = \frac{\frac{a}{b} \cdot \frac{d}{c}}{\frac{c}{d} \cdot \frac{d}{c}} = \frac{\frac{a}{b} \cdot \frac{d}{c}}{\frac{cd}{cd}} = \frac{\frac{a}{b} \cdot \frac{d}{c}}{1} = \frac{a}{b} \cdot \frac{d}{c}$$

Thus, *to divide two fractions, we can invert the divisor and multiply.*

$$\frac{3}{5} \div \frac{2}{7} = \frac{3}{5} \cdot \frac{7}{2} = \frac{3 \cdot 7}{5 \cdot 2} = \frac{21}{10}$$

$$\frac{4}{7} \div \frac{2}{21} = \frac{4}{7} \cdot \frac{21}{2} = \frac{4 \cdot 21}{7 \cdot 2} = \frac{\overset{1}{\cancel{2}} \cdot 2 \cdot 3 \cdot \overset{1}{\cancel{7}}}{\underset{1}{\cancel{7}} \cdot \underset{1}{\cancel{2}}} = 6$$

The same rule applies to rational expressions.

$$\frac{x^2}{y^3z^2} \div \frac{x^2}{yz^3} = \frac{x^2}{y^3z^2} \cdot \frac{yz^3}{x^2}$$ Invert the divisor and multiply.

$$= \frac{x^2yz^3}{x^2y^3z^2}$$ Multiply the numerators and the denominators.

$$= x^{2-2}y^{1-3}z^{3-2}$$ To divide exponential expressions with the same base, keep the base and subtract the exponents.

$$= x^0y^{-2}z^1$$

$$= 1 \cdot y^{-2} \cdot z \qquad x^0 = 1$$

$$= \frac{z}{y^2} \qquad y^{-2} = \frac{1}{y^2}$$

EXAMPLE 6 Divide: $\dfrac{x^3 + 8}{x + 1} \div \dfrac{x^2 - 2x + 4}{2x^2 - 2}$. Assume no division by zero.

Solution We invert the divisor and multiply.

$$\frac{x^3 + 8}{x + 1} \div \frac{x^2 - 2x + 4}{2x^2 - 2}$$

$$= \frac{x^3 + 8}{x + 1} \cdot \frac{2x^2 - 2}{x^2 - 2x + 4}$$

$$= \frac{(x^3 + 8)(2x^2 - 2)}{(x + 1)(x^2 - 2x + 4)}$$

$$= \frac{(x + 2)\overset{1}{\cancel{(x^2 - 2x + 4)}}2\overset{1}{\cancel{(x + 1)}}(x - 1)}{\underset{1}{\cancel{(x + 1)}}\underset{1}{\cancel{(x^2 - 2x + 4)}}} \qquad \frac{x^2 - 2x + 4}{x^2 - 2x + 4} = 1, \frac{x + 1}{x + 1} = 1$$

$$= 2(x + 2)(x - 1)$$

SELF CHECK 6 Divide: $\dfrac{x^3 - 8}{x - 1} \div \dfrac{x^2 + 2x + 4}{3x^2 - 3x}$. Assume no division by zero.

EXAMPLE 7 Divide: $\dfrac{x^2 - 4}{x - 1} \div (x - 2)$. Assume no division by zero.

Solution

$$\frac{x^2 - 4}{x - 1} \div (x - 2)$$

$$= \frac{x^2 - 4}{x - 1} \div \frac{x - 2}{1} \qquad \text{Write } x - 2 \text{ as a fraction with a denominator of 1.}$$

$$= \frac{x^2 - 4}{x - 1} \cdot \frac{1}{x - 2} \qquad \text{Invert the divisor and multiply.}$$

$$= \frac{x^2 - 4}{(x - 1)(x - 2)} \qquad \text{Multiply the numerators and the denominators.}$$

$$= \frac{(x + 2)\overset{1}{\cancel{(x - 2)}}}{(x - 1)\underset{1}{\cancel{(x - 2)}}} \qquad \text{Factor } x^2 - 4 \text{ and divide out } x - 2\text{: } \frac{x - 2}{x - 2} = 1.$$

$$= \frac{x + 2}{x - 1}$$

SELF CHECK 7 Divide: $\dfrac{a^3 - 9a}{a - 3} \div (a^2 + 3a)$. Assume no division by zero.

4 Perform combined operations on three or more rational expressions.

The following example involves both a division and a multiplication.

EXAMPLE 8 Simplify: $\frac{x^2 + 2x - 3}{6x^2 + 5x + 1} \div \frac{2x^2 - 2}{2x^2 - 5x - 3} \cdot \frac{6x^2 + 4x - 2}{x^2 - 2x - 3}$. Assume no division by zero.

Solution Since multiplications and divisions are done in order from left to right, we change the division to a multiplication

$$\left(\frac{x^2 + 2x - 3}{6x^2 + 5x + 1} \div \frac{2x^2 - 2}{2x^2 - 5x - 3}\right)\frac{6x^2 + 4x - 2}{x^2 - 2x - 3}$$

$$= \left(\frac{x^2 + 2x - 3}{6x^2 + 5x + 1} \cdot \frac{2x^2 - 5x - 3}{2x^2 - 2}\right)\frac{6x^2 + 4x - 2}{x^2 - 2x - 3}$$

and then multiply the rational expressions and simplify the result.

$$= \frac{(x^2 + 2x - 3)(2x^2 - 5x - 3)(6x^2 + 4x - 2)}{(6x^2 + 5x + 1)(2x^2 - 2)(x^2 - 2x - 3)}$$

$$= \frac{(x + 3)\overset{1}{\cancel{(x - 1)}}\overset{1}{\cancel{(2x + 1)}}\overset{1}{\cancel{(x - 3)}}\overset{1}{\cancel{2}}(3x - 1)\overset{1}{\cancel{(x + 1)}}}{(3x + 1)\underset{1}{\cancel{(2x + 1)}}\underset{1}{\cancel{2}}(x + 1)\underset{1}{\cancel{(x - 1)}}\underset{1}{\cancel{(x - 3)}}\underset{1}{\cancel{(x + 1)}}}$$

$\frac{x - 1}{x - 1} = 1, \frac{2x + 1}{2x + 1} = 1, \frac{x - 3}{x - 3} = 1, \frac{2}{2} = 1, \frac{x + 1}{x + 1} = 1$

$$= \frac{(x + 3)(3x - 1)}{(3x + 1)(x + 1)}$$

SELF CHECK 8 Simplify: $\frac{2x^2 - x - 1}{x^2 + 5x + 4} \div \frac{x^2 + x - 2}{2x^2 + 5x + 2} \cdot \frac{x^2 + 5x + 4}{4x^2 + 4x + 1}$. Assume no division by zero.

SELF CHECK ANSWERS **1.** $a^2(a + 3)$ **2.** 1 **3.** $\frac{a - 3}{a + 1}$ **4.** $x(x + 3)$ **5.** $\frac{x^2 + 10x + 25}{x^4 - 12x^3 + 36x^2}$ **6.** $3x(x - 2)$ **7.** 1 **8.** 1

NOW TRY THIS

Simplify:

1. $\frac{6x + 11}{6} \div \frac{5x}{7}$
2. $\frac{3x^2 - x - 2}{x^2 - 1} \div \frac{x^2 - 3x + 2}{3x^2 - 4x - 4}$
3. $\frac{3 - 2x}{5x - 2} \cdot \frac{x}{2x^2 - 3x}$

6.2 EXERCISES

Assume no divisions by zero.

WARM-UPS *Perform the operations and simplify, if possible.*

1. $\frac{3}{2} \cdot \frac{3}{4}$
2. $\frac{3x}{7} \cdot \frac{7}{6x}$
3. $\frac{x-2}{y} \cdot \frac{y}{x+2}$
4. $\frac{3}{4} \div \frac{4}{3}$
5. $\frac{5a}{b} \div \frac{a}{b}$
6. $\frac{x^2y}{ab} \div \frac{2x^2}{ba}$

REVIEW *Perform each operation.*

7. $-2a^2(3a^3 - a^2)$
8. $(2t - 1)^2$
9. $(m^n + 2)(m^n - 2)$
10. $(3b^{-n} + c)(b^{-n} - c)$

VOCABULARY AND CONCEPTS *Fill in the blanks.*

11. $\frac{a}{b} \cdot \frac{c}{d} =$ ___ $(b \neq 0, d \neq 0)$
12. $\frac{a}{b} \div \frac{c}{d} =$ ___ $(b \neq 0, c \neq 0, d \neq 0)$
13. The denominator of a fraction cannot be ___.
14. $\frac{a+1}{a+1} =$ ___ $(a \neq -1)$

GUIDED PRACTICE

Perform the operations and simplify. (Objective 1)

15. $\frac{3}{4} \cdot \frac{5}{3} \cdot \frac{8}{7}$
16. $-\frac{5}{6} \cdot \frac{3}{7} \cdot \frac{14}{25}$
17. $-\frac{6}{11} \div \frac{36}{55}$
18. $\frac{17}{12} \div \frac{34}{3}$
19. $\frac{x^2y^2}{cd} \cdot \frac{c^{-2}d^2}{x}$
20. $\frac{a^{-2}b^2}{x^{-1}y} \cdot \frac{a^4b^4}{x^2y^3}$
21. $\frac{-x^2y^{-2}}{x^{-1}y^{-3}} \div \frac{x^{-3}y^2}{x^4y^{-1}}$
22. $\frac{(a^3)^2}{b^{-1}} \div \frac{(a^3)^{-2}}{b^{-1}}$

Perform the operations and simplify. See Examples 1–2. (Objective 1)

23. $\frac{x^2 + 2x + 1}{x} \cdot \frac{x^2 - x}{x^2 - 1}$
24. $\frac{a + 6}{a^2 - 16} \cdot \frac{3a - 12}{3a + 18}$
25. $\frac{x^2 - x - 6}{x^2 - 4} \cdot \frac{x^2 - x - 2}{9 - x^2}$
26. $\frac{x^4 - 3x^2 - 4}{x^4 - 1} \cdot \frac{x^2 + 3x + 2}{x^2 + 4x + 4}$
27. $\frac{x^3 + 2x^2 + 4x + 8}{y^2 - 1} \cdot \frac{y^2 + 2y + 1}{x^4 - 16}$
28. $\frac{x^2 + 3x + yx + 3y}{x^2 - 9} \cdot \frac{x - 3}{x + 3}$
29. $\frac{m^2 - n^2}{2x^2 + 3x - 2} \cdot \frac{2x^2 + 5x - 3}{n^2 - m^2}$
30. $\frac{p^3 - q^3}{q^2 - p^2} \cdot \frac{q^2 + pq}{p^3 + p^2q + pq^2}$

Perform the operations and simplify. See Examples 3. (Objective 1)

31. $\frac{2x^2 - x - 3}{x^2 - 1} \cdot \frac{x^2 + x - 2}{2x^2 + x - 6}$
32. $\frac{3t^2 - t - 2}{6t^2 - 5t - 6} \cdot \frac{4t^2 - 9}{2t^2 + 5t + 3}$
33. $\frac{2p^2 - 5p - 3}{p^2 - 9} \cdot \frac{2p^2 + 5p - 3}{2p^2 + 5p + 2}$
34. $\frac{9x^2 + 3x - 20}{3x^2 - 7x + 4} \cdot \frac{3x^2 - 5x + 2}{9x^2 + 18x + 5}$

Perform the operations and simplify. See Example 4. (Objective 1)

35. $(x + 1) \cdot \frac{1}{x^2 + 2x + 1}$
36. $(x^2 - x - 2) \cdot \frac{x^2 + 3x + 2}{x^2 - 4}$
37. $\frac{x}{2x^2 + x} \cdot (2x^2 - 9x - 5)$
38. $\frac{2x^2 - x - 1}{x - 2} \cdot (2x^2 - 3x - 2)$

Find the square of each rational expression. See Example 5. (Objective 2)

39. $\left(\frac{x - 3}{x^3 + 4}\right)^2$
40. $\left(\frac{2t^2 + t}{t - 1}\right)^2$
41. $\left(\frac{2m^2 - m - 3}{x^2 - 1}\right)^2$
42. $\left(\frac{-k - 3}{x^2 - x + 1}\right)^2$

Perform the operations and simplify. See Example 6. (Objective 3)

43. $\frac{x^2 - 16}{x^2 - 25} \div \frac{x + 4}{x - 5}$
44. $\frac{a^2 - 9}{a^2 - 49} \div \frac{a + 3}{a + 7}$
45. $\frac{3n^2 + 5n - 2}{12n^2 - 13n + 3} \div \frac{n^2 + 3n + 2}{4n^2 + 5n - 6}$

46. $\frac{8y^2 - 14y - 15}{6y^2 - 11y - 10} \div \frac{4y^2 - 9y - 9}{3y^2 - 7y - 6}$

47. $\frac{a^2 + 2a - 35}{12x} \div \frac{ax - 3x}{a^2 + 4a - 21}$

48. $\frac{x^2 - 4}{2b - bx} \div \frac{x^2 + 4x + 4}{2b + bx}$

49. $\frac{x^2 - 6x + 9}{4 - x^2} \div \frac{x^2 - 9}{x^2 - 8x + 12}$

50. $\frac{2x^2 - 7x - 4}{20 - x - x^2} \div \frac{2x^2 - 9x - 5}{x^2 - 25}$

Perform the operations and simplify. **See Example 7. (Objective 3)**

51. $\frac{x^2 - 4}{x} \div (x + 2)$

52. $(2x^2 - 15x + 25) \div \frac{2x^2 - 3x - 5}{x + 1}$

53. $(x^2 - 6x + 9) \div \frac{x^2 - 9}{x + 3}$

54. $\frac{2x^2 - x - 1}{x - 2} \div (x - 1)$

Perform the operations and simplify. **See Example 8. (Objective 4)**

55. $\frac{3x^2y^2}{6x^3y} \cdot \frac{-4x^7y^{-2}}{18x^{-2}y} \div \frac{36x}{18y^{-2}}$

56. $\frac{9ab^3}{7xy} \cdot \frac{14xy^2}{27z^3} \div \frac{18a^2b^2x}{3z^2}$

57. $(4x + 12) \cdot \frac{x^2}{2x - 6} \div \frac{2}{x - 3}$

58. $(4x^2 - 9) \div \frac{2x^2 + 5x + 3}{x + 2} \div (2x - 3)$

59. $\frac{2x^2 - 2x - 4}{x^2 + 2x - 8} \cdot \frac{3x^2 + 15x}{x + 1} \div \frac{4x^2 - 100}{x^2 - x - 20}$

60. $\frac{6a^2 - 7a - 3}{a^2 - 1} \div \frac{4a^2 - 12a + 9}{a^2 - 1} \cdot \frac{2a^2 - a - 3}{3a^2 - 2a - 1}$

61. $\frac{2t^2 + 5t + 2}{t^2 - 4t + 16} \div \frac{t + 2}{t^3 + 64} \div \frac{2t^3 + 9t^2 + 4t}{t + 1}$

62. $\frac{a^6 - b^6}{a^4 - a^3b} \cdot \frac{a^3}{a^4 + a^2b^2 + b^4} \div \frac{1}{a}$

ADDITIONAL PRACTICE *Simplify each expression.*

63. $\frac{x^2 - y^2}{2x^2 + 2xy + x + y} \cdot \frac{2x^2 - 5x - 3}{yx - 3y - x^2 + 3x}$

64. $\frac{ax + ay + bx + by}{x^3 - 27} \cdot \frac{x^2 + 3x + 9}{xc + xd + yc + yd}$

65. $\frac{x^3 + y^3}{x^3 - y^3} \div \frac{x^2 - xy + y^2}{x^2 + xy + y^2}$

66. $\frac{2x^2 + 3xy + y^2}{y^2 - x^2} \div \frac{6x^2 + 5xy + y^2}{2x^2 - xy - y^2}$

67. $(x^2 - x - 6) \div (x - 3) \div (x - 2)$

68. $(x^2 - x - 6) \div [(x - 3) \div (x - 2)]$

69. $\frac{3x^2 - 2x}{3x + 2} \div (3x - 2) \div \frac{3x}{3x - 3}$

70. $(2x^2 - 3x - 2) \div \frac{2x^2 - x - 1}{x - 2} \div (x - 1)$

71. $\frac{2x^2 + 5x - 3}{x^2 + 2x - 3} \div \left(\frac{x^2 + 2x - 35}{x^2 - 6x + 5} \div \frac{x^2 - 9x + 14}{2x^2 - 5x + 2}\right)$

72. $\frac{x^2 - 4}{x^2 - x - 6} \div \left(\frac{x^2 - x - 2}{x^2 - 8x + 15} \cdot \frac{x^2 - 3x - 10}{x^2 + 3x + 2}\right)$

73. $\frac{x^2 - x - 12}{x^2 + x - 2} \div \frac{x^2 - 6x + 8}{x^2 - 3x - 10} \cdot \frac{x^2 - 3x + 2}{x^2 - 2x - 15}$

74. $\frac{4x^2 - 10x + 6}{x^4 - 3x^3} \div \frac{2x - 3}{2x^3} \cdot \frac{x - 3}{2x - 2}$

APPLICATIONS

75. Area of a triangle Find an expression that represents the area of the triangle.

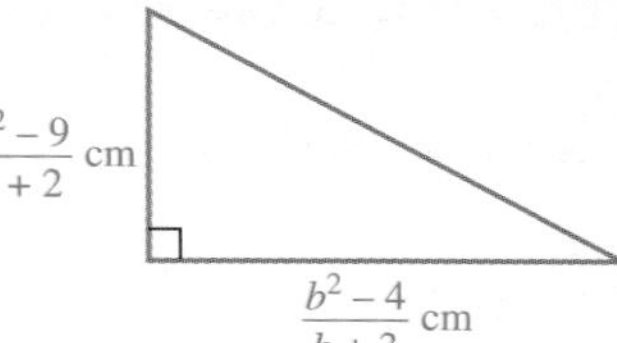

76. Storage capacity Find the capacity of the trunk shown in the illustration.

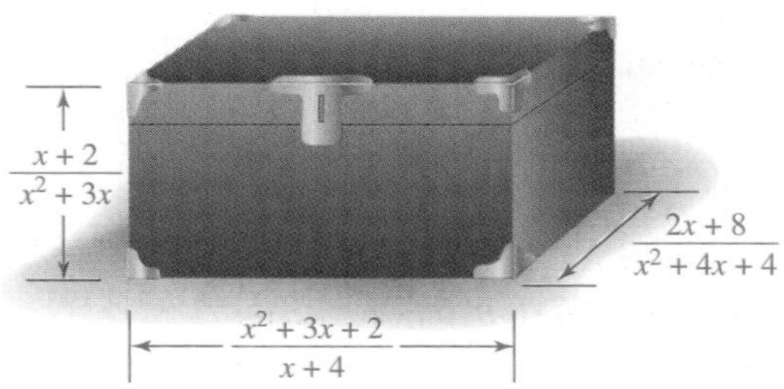

77. Distance, rate, and time Complete the following table.

Rate (mph)	Time (hr)	Distance (m)
$\frac{k^2 - k - 6}{k - 4}$	$\frac{k^2 - 16}{k^2 - 2k - 3}$	

78. Lab experiments The following table shows data obtained from a physics experiment in which k and k_1 are constants. Complete the table.

Trial	Rate (m/sec)	Time (sec)	Distance (m)
1	$\frac{k_1^2 + 3k_1 + 2}{k_1 - 3}$	$\frac{k_1^2 - 3k_1}{k_1 + 1}$	
2	$\frac{k_2^2 + 6k_2 + 5}{k_2 + 1}$	$k_2 + 6$	$k_2^2 + 11k_2 + 30$

WRITING ABOUT MATH

79. Explain how to multiply two rational expressions.

80. Explain how to divide one rational expression by another.

SOMETHING TO THINK ABOUT *Insert either a multiplication or a division symbol in each box to make a true statement.*

81. $\frac{x^2}{y} \blacksquare \frac{x}{y^2} \blacksquare \frac{x^2}{y^2} = \frac{x^3}{y}$

82. $\frac{x^2}{y} \blacksquare \frac{x}{y^2} \blacksquare \frac{x^2}{y^2} = \frac{y^3}{x}$

SECTION 6.3 Adding and Subtracting Rational Expressions

Objectives

1. Add and subtract two rational expressions with like denominators.
2. Find the least common denominator of two or more rational expressions.
3. Add and subtract two rational expressions with unlike denominators.
4. Perform combined operations on three or more rational expressions.

Getting Ready

Determine whether the fractions are equal.

1. $\frac{3}{5}, \frac{3}{5}$ **2.** $\frac{3}{7}, \frac{18}{40}$ **3.** $\frac{8}{13}, \frac{40}{65}$

4. $\frac{7}{25}, \frac{42}{150}$ **5.** $\frac{23}{32}, \frac{46}{66}$ **6.** $\frac{8}{9}, \frac{64}{72}$

We now discuss how to add and subtract rational expressions. After mastering these skills, we will simplify expressions that involve more than one operation.

1 Add and subtract two rational expressions with like denominators.

Rational expressions with like denominators are added and subtracted according to the following rules.

Adding and Subtracting Rational Expressions

If there are no divisions by 0, then

$$\frac{a}{b} + \frac{c}{b} = \frac{a+c}{b} \quad \text{and} \quad \frac{a}{b} - \frac{c}{b} = \frac{a-c}{b}$$

In words, *we add (or subtract) rational expressions with like denominators by adding (or subtracting) the numerators and keeping the common denominator.* Whenever possible, we should simplify the result.

EXAMPLE 1 Simplify: **a.** $\frac{17}{22} + \frac{13}{22}$ **b.** $\frac{3}{2x} + \frac{7}{2x}$ **c.** $\frac{4x}{x+2} - \frac{7x}{x+2}$.

Assume no denominator is zero.

Solution **a.**

$$\frac{17}{22} + \frac{13}{22} = \frac{17+13}{22} = \frac{30}{22} = \frac{15 \cdot \cancel{2}}{11 \cdot \cancel{2}} = \frac{15}{11}$$

b.

$$\frac{3}{2x} + \frac{7}{2x} = \frac{3+7}{2x} = \frac{10}{2x} = \frac{\cancel{2} \cdot 5}{\cancel{2} \cdot x} = \frac{5}{x}$$

c.

$$\frac{4x}{x+2} - \frac{7x}{x+2} = \frac{4x - 7x}{x+2} = \frac{-3x}{x+2}$$

SELF CHECK 1 Simplify: **a.** $\frac{5}{3a} - \frac{2}{3a}$ **b.** $\frac{3a}{a-2} + \frac{2a}{a-2}$.

Assume no denominator is zero.

ACCENT ON TECHNOLOGY

Checking Algebra

We can check the subtraction in part c of Example 1 by graphing the rational functions $f(x) = \frac{4x}{x+2} - \frac{7x}{x+2}$, shown in Figure 6-5(a), and $g(x) = \frac{-3x}{x+2}$, shown in Figure 6-5(b), and observing that the graphs are the same. Note that -2 is not in the domain of either function.

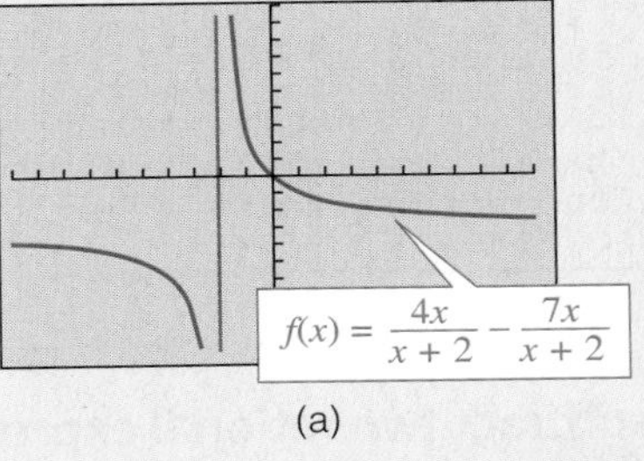

(a)

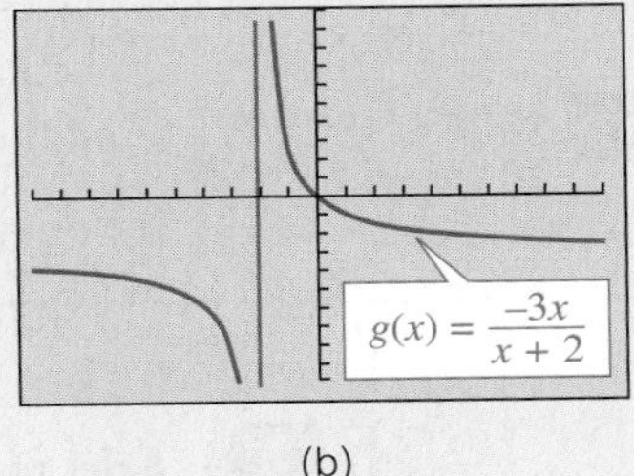

(b)

Figure 6-5

2 Find the least common denominator of two or more rational expressions.

When adding fractions with unlike denominators, we change the fractions into fractions having the smallest common denominator possible, called the *least* (or lowest) *common denominator (LCD)*.

Suppose we have the fractions $\frac{1}{12}$, $\frac{1}{20}$, and $\frac{1}{35}$. To find the LCD of these fractions, we first find the prime factorizations of each denominator.

$$12 = 4 \cdot 3 = 2^2 \cdot 3$$
$$20 = 4 \cdot 5 = 2^2 \cdot 5$$
$$35 = 5 \cdot 7$$

Since the LCD is the smallest number that can be exactly divided by 12, 20, and 35, it must contain factors of 2^2, 3, 5, and 7, and we have

$$\text{LCD} = 2^2 \cdot 3 \cdot 5 \cdot 7 = 420$$

We can find the LCD of several rational expressions in a similar way by following these steps.

Finding the LCD

1. Prime-factor the denominator of each rational expression.
2. List the different factors of each denominator.
3. Write each factor found in Step 2 to the highest power that occurs in any one factorization.
4. The LCD is the product of the factors to their highest powers found in Step 3.

EXAMPLE 2 Find the LCD of $\dfrac{1}{x^2 + 7x + 6}$, $\dfrac{3}{x^2 - 36}$, and $\dfrac{5}{x^2 + 12x + 36}$. Assume no denominator is zero.

Solution We prime-factor each denominator:

$$x^2 + 7x + 6 = (x + 6)(x + 1)$$
$$x^2 - 36 = (x + 6)(x - 6)$$
$$x^2 + 12x + 36 = (x + 6)(x + 6) = (x + 6)^2$$

and list the individual factors:

$$x + 6, \qquad x + 1, \qquad \text{and} \qquad x - 6$$

To find the LCD, we use the highest power of each of these factors:

$$\text{LCD} = (x + 6)^2(x + 1)(x - 6)$$

SELF CHECK 2 Find the LCD of $\dfrac{1}{a^2 - 25}$ and $\dfrac{7}{a^2 + 7a + 10}$. Assume no denominator is zero.

3 Add and subtract two rational expressions with unlike denominators.

To add or subtract rational expressions with unlike denominators, we change them into expressions with a common denominator. When the denominators are negatives (opposites), we can multiply one of them by 1, written in the form $\frac{-1}{-1}$, to get a common denominator.

EXAMPLE 3 Add: $\dfrac{x}{x - y} + \dfrac{y}{y - x}$. Assume no denominator is zero.

Solution

$$\frac{x}{x - y} + \frac{y}{y - x} = \frac{x}{x - y} + \left(\frac{-1}{-1}\right)\frac{y}{y - x}$$ $\frac{-1}{-1} = 1$; multiplying a fraction by 1 does not change its value.

$$= \frac{x}{x - y} + \frac{-y}{-y + x}$$ Multiply.

$$= \frac{x}{x-y} + \frac{-y}{x-y}$$ $-y + x = x - y$

$$= \frac{x-y}{x-y}$$ Add the numerators and keep the common denominator.

$$= 1$$ Simplify.

SELF CHECK 3 Add: $\frac{2a}{a-b} + \frac{b}{b-a}$. Assume no denominator is zero.

When the denominators of two or more rational expressions are different, we often have to multiply one or more of them by 1, written in some appropriate form, to get a common denominator.

EXAMPLE 4 Simplify: **a.** $\frac{2}{3} - \frac{3}{2}$ **b.** $\frac{3}{x} + \frac{4}{y}$. Assume no denominator is zero.

Solution **a.** $\frac{2}{3} - \frac{3}{2} = \frac{2}{3} \cdot 1 - \frac{3}{2} \cdot 1$ Multiply each fraction by 1.

$$= \frac{2}{3} \cdot \frac{2}{2} - \frac{3}{2} \cdot \frac{3}{3}$$ $\frac{2}{2} = 1$ and $\frac{3}{3} = 1$. The LCD is 6.

$$= \frac{2 \cdot 2}{3 \cdot 2} - \frac{3 \cdot 3}{2 \cdot 3}$$ Multiply the fractions by multiplying their numerators and their denominators.

$$= \frac{4}{6} - \frac{9}{6}$$ Simplify.

$$= \frac{-5}{6}$$ Subtract the numerators and keep the common denominator.

$$= -\frac{5}{6}$$

Charles Babbage (1792–1871)
In 1823, Babbage built a steam-powered digital calculator, which he called a difference engine. Thought to be a crack-pot by his London neighbors, Babbage was a visionary. His machine embodied principles still used in modern computers.

b. $\frac{3}{x} + \frac{4}{y} = \frac{3}{x} \cdot 1 + \frac{4}{y} \cdot 1$ Multiply each fraction by 1.

$$= \frac{3}{x} \cdot \frac{y}{y} + \frac{4}{y} \cdot \frac{x}{x}$$ $\frac{y}{y} = 1 \quad (y \neq 0)$ and $\frac{x}{x} = 1 \quad (x \neq 0)$. The LCD is xy.

$$= \frac{3y}{xy} + \frac{4x}{xy}$$ Multiply the fractions by multiplying their numerators and their denominators.

$$= \frac{3y + 4x}{xy}$$ Add the numerators and keep the common denominator.

SELF CHECK 4 Simplify: $\frac{5}{a} - \frac{7}{b}$. Assume no denominator is zero.

EXAMPLE 5 Simplify: **a.** $3 + \frac{7}{x-2}$ **b.** $\frac{4x}{x+2} - \frac{7x}{x-2}$. Assume no denominator is zero.

Solution **a.** $3 + \frac{7}{x-2} = \frac{3}{1} + \frac{7}{x-2}$ $3 = \frac{3}{1}$

$$= \frac{3(x-2)}{1(x-2)} + \frac{7}{x-2}$$ $\frac{x-2}{x-2} = 1$. The LCD is $x - 2$.

$$= \frac{3x-6}{x-2} + \frac{7}{x-2}$$ Remove parentheses.

$$= \frac{3x-6+7}{x-2}$$ Add the numerators and keep the common denominator.

$$= \frac{3x+1}{x-2}$$ Simplify.

b. $\frac{4x}{x+2} - \frac{7x}{x-2}$

$$= \frac{4x(x-2)}{(x+2)(x-2)} - \frac{(x+2)7x}{(x+2)(x-2)}$$ $\frac{x-2}{x-2} = 1$ and $\frac{x+2}{x+2} = 1$. The LCD is $(x-2)(x+2)$.

$$= \frac{(4x^2-8x)-(7x^2+14x)}{(x+2)(x-2)}$$ Multiply in the numerators, subtract the numerators, and keep the common denominator.

COMMENT The $-$ sign between the fractions in Step 1 of part b applies to both terms of $7x^2 + 14x$.

$$= \frac{4x^2-8x-7x^2-14x}{(x+2)(x-2)}$$ Use the distributive property to remove parentheses in the numerator.

$$= \frac{-3x^2-22x}{(x+2)(x-2)}$$ Simplify the numerator.

$$= \frac{-x(3x+22)}{(x+2)(x-2)}$$ Factor the numerator to see whether the fraction can be simplified.

SELF CHECK 5 Simplify: $\frac{3a}{a+3} - \frac{5a}{a-3}$. Assume no denominator is zero.

EXAMPLE 6 Simplify: $\frac{x}{x^2-2x+1} + \frac{3}{x^2-1}$. Assume no denominator is zero.

Solution We prime-factor each denominator to find the LCD:

$$x^2 - 2x + 1 = (x-1)(x-1) = (x-1)^2$$
$$x^2 - 1 = (x+1)(x-1)$$

The LCD is $(x-1)^2(x+1)$.

We now write the rational expressions with their denominators in factored form and change them into expressions with an LCD of $(x-1)^2(x+1)$. Then we add.

$$\frac{x}{x^2-2x+1} + \frac{3}{x^2-1}$$

$$= \frac{x}{(x-1)(x-1)} + \frac{3}{(x+1)(x-1)}$$ Factor each denominator.

$$= \frac{x(x+1)}{(x-1)(x-1)(x+1)} + \frac{3(x-1)}{(x+1)(x-1)(x-1)}$$ $\frac{x+1}{x+1} = 1, \frac{x-1}{x-1} = 1$

$$= \frac{x^2+x+3x-3}{(x-1)(x-1)(x+1)}$$ Add the numerators and keep the denominators.

$$= \frac{x^2+4x-3}{(x-1)^2(x+1)}$$ This result does not simplify.

SELF CHECK 6 Simplify: $\dfrac{a}{a^2 - 4a + 4} - \dfrac{2}{a^2 - 4}$. Assume no denominator is zero.

EXAMPLE 7 Simplify: $\dfrac{3x}{x - 1} - \dfrac{2x^2 + 3x - 2}{(x + 1)(x - 1)}$. Assume no denominator is zero.

Solution We write each rational expression in a form having the LCD of $(x + 1)(x - 1)$, remove the resulting parentheses in the first numerator, do the subtraction, and simplify.

(1)
$$\frac{3x}{x - 1} - \frac{2x^2 + 3x - 2}{(x + 1)(x - 1)} = \frac{(x + 1)3x}{(x + 1)(x - 1)} - \frac{2x^2 + 3x - 2}{(x + 1)(x - 1)} \qquad \frac{x+1}{x+1} = 1$$

$$= \frac{3x^2 + 3x}{(x + 1)(x - 1)} - \frac{2x^2 + 3x - 2}{(x + 1)(x - 1)} \qquad \text{Multiply.}$$

$$= \frac{3x^2 + 3x - (2x^2 + 3x - 2)}{(x + 1)(x - 1)} \qquad \text{Add the numerators and keep the denominators.}$$

$$= \frac{3x^2 + 3x - 2x^2 - 3x + 2}{(x + 1)(x - 1)} \qquad \text{Subtract.}$$

$$= \frac{x^2 + 2}{(x + 1)(x - 1)} \qquad \text{Simplify.}$$

COMMENT The $-$ sign between the rational expressions in Equation 1 affects every term of the numerator of $2x^2 + 3x - 2$. Whenever we subtract one rational expression from another, we must remember to subtract each term of the numerator in the second expression.

SELF CHECK 7 Simplify: $\dfrac{2a}{a + 2} - \dfrac{a^2 - 4a + 4}{a^2 + a - 2}$. Assume no denominator is zero.

4 Perform combined operations on three or more rational expressions.

We can use the previous techniques to simplify expressions that contain both an addition and a subtraction.

EXAMPLE 8 Simplify: $\dfrac{2x}{x^2 - 4} - \dfrac{1}{x^2 - 3x + 2} + \dfrac{x + 1}{x^2 + x - 2}$. Assume no denominator is zero.

Solution We factor each denominator to find the LCD:

$$\text{LCD} = (x + 2)(x - 2)(x - 1)$$

We then write each rational expression as one with the LCD as its denominator and do the subtraction and addition.

$$\frac{2x}{x^2 - 4} - \frac{1}{x^2 - 3x + 2} + \frac{x + 1}{x^2 + x - 2}$$

$$= \frac{2x}{(x - 2)(x + 2)} - \frac{1}{(x - 2)(x - 1)} + \frac{x + 1}{(x - 1)(x + 2)}$$

$$= \frac{2x(x-1)}{(x-2)(x+2)(x-1)} - \frac{1(x+2)}{(x-2)(x-1)(x+2)} + \frac{(x+1)(x-2)}{(x-1)(x+2)(x-2)}$$

$$= \frac{2x(x-1) - 1(x+2) + (x+1)(x-2)}{(x+2)(x-2)(x-1)}$$ Add the numerators and keep the denominators.

$$= \frac{2x^2 - 2x - x - 2 + x^2 - x - 2}{(x+2)(x-2)(x-1)}$$ Add and subtract.

$$= \frac{3x^2 - 4x - 4}{(x+2)(x-2)(x-1)}$$ Simplify.

Since the final result simplifies, we have

$$\frac{2x}{x^2-4} - \frac{1}{x^2-3x+2} + \frac{x+1}{x^2+x-2}$$

$$= \frac{3x^2 - 4x - 4}{(x+2)(x-2)(x-1)}$$

$$= \frac{(3x+2)\cancel{(x-2)}}{(x+2)\cancel{(x-2)}(x-1)}$$ Factor the numerator and divide out the $x-2$; $\frac{x-2}{x-2} = 1$.

$$= \frac{3x+2}{(x+2)(x-1)}$$

SELF CHECK 8 Simplify: $\frac{y}{y^2+y} - \frac{1}{y^2-1} + \frac{2}{y}$. Assume no denominator is zero.

EXAMPLE 9 Simplify: $\left(\frac{x^2}{x-2} + \frac{4}{2-x}\right)^2$. Assume no denominator is zero.

Solution We can do the addition within the parentheses. Since the denominators are negatives (opposites) of each other, we can write the rational expressions with a common denominator by multiplying both the numerator and denominator of $\frac{4}{2-x}$ by -1. Then we can add, simplify, and square the result.

$$\left(\frac{x^2}{x-2} + \frac{4}{2-x}\right)^2 = \left[\frac{x^2}{x-2} + \frac{(-1)4}{(-1)(2-x)}\right]^2$$ $\frac{-1}{-1} = 1$

$$= \left[\frac{x^2}{x-2} + \frac{-4}{x-2}\right]^2$$ Simplify.

$$= \left[\frac{x^2-4}{x-2}\right]^2$$ Add.

$$= \left[\frac{(x+2)(x-2)}{x-2}\right]^2$$ Factor the numerator.

$$= (x+2)^2$$ Simplify the fraction.

$$= x^2 + 4x + 4$$

SELF CHECK 9 Simplify: $\left(\frac{a^2}{a-3} - \frac{9-6a}{3-a}\right)^2$. Assume no denominator is zero.

SELF CHECK ANSWERS 1. a. $\frac{1}{a}$ b. $\frac{5a}{a-2}$ 2. $(a+5)(a+2)(a-5)$ 3. $\frac{2a-b}{a-b}$ 4. $\frac{5b-7a}{ab}$ 5. $\frac{-2a(a+12)}{(a+3)(a-3)}$ 6. $\frac{a^2+4}{(a-2)^2(a+2)}$ 7. $\frac{a^2+2a-4}{(a+2)(a-1)}$ 8. $\frac{3y^2-2y-2}{y(y+1)(y-1)}$ 9. a^2-6a+9

NOW TRY THIS

1. $5x(x-2)^{-1} - 6(x+3)^{-1}$
2. $8(x-2)^{-2} - 30(x-2)^{-1} + 7$
3. $\left(\frac{2}{5} + \frac{2}{x}\right) \div \left(\frac{x-2}{x} - \frac{3}{5}\right)$

6.3 EXERCISES

WARM-UPS *Add or subtract the rational expressions and simplify the result, if possible.*

1. $\frac{x}{2} + \frac{x}{2}$
2. $\frac{3a}{4} - \frac{a}{4}$
3. $\frac{x}{x+2} + \frac{2}{x+2}$
4. $\frac{2a}{a+4} - \frac{a-4}{a+4}$
5. $\frac{2x}{3} + \frac{x}{2}$
6. $\frac{5}{x} - \frac{3}{y}$

REVIEW *Graph each interval on a number line.*

7. $(-1, 4]$
8. $(-\infty, -5] \cup [4, \infty)$

Solve each formula for the indicated letter.

9. $P = 2l + 2w$; for w
10. $S = \frac{a - lr}{1 - r}$; for a

VOCABULARY AND CONCEPTS *Fill in the blanks.*

11. $\frac{a}{b} + \frac{c}{b} =$ ______ $b \neq 0$
12. $\frac{a}{b} - \frac{c}{b} =$ ______ $b \neq 0$
13. To subtract fractions with like denominators, we ______ the numerators and ______ the common denominator.
14. To add fractions with like denominators, we ______ the numerators and keep the ______ denominator.
15. The abbreviation for the least common denominator is ______.
16. To find the LCD of several fractions, we ______ each denominator and use each factor to the ______ power that it appears in any factorization.

GUIDED PRACTICE

Perform the operations and simplify. Assume no denominator is zero. See Example 1. (Objective 1)

17. $\frac{3}{4} + \frac{7}{4}$
18. $\frac{5}{11} + \frac{2}{11}$
19. $\frac{10}{33} - \frac{21}{33}$
20. $\frac{8}{15} - \frac{2}{15}$
21. $\frac{3}{4y} + \frac{8}{4y}$
22. $\frac{5}{3z^2} - \frac{6}{3z^2}$
23. $\frac{3}{a+b} - \frac{a}{a+b}$
24. $\frac{x}{x+4} + \frac{5}{x+4}$
25. $\frac{3x}{2x+2} + \frac{x+4}{2x+2}$
26. $\frac{4y}{y-4} - \frac{16}{y-4}$
27. $\frac{3x}{x-3} - \frac{9}{x-3}$
28. $\frac{9x}{x-y} - \frac{9y}{x-y}$
29. $\frac{5x}{x+1} + \frac{3}{x+1} - \frac{2x}{x+1}$
30. $\frac{4}{a+4} - \frac{2a}{a+4} + \frac{3a}{a+4}$
31. $\frac{3(x^2+x)}{x^2-5x+6} + \frac{-3(x^2-x)}{x^2-5x+6}$
32. $\frac{2x+4}{x^2+13x+12} - \frac{x+3}{x^2+13x+12}$

The denominators of several fractions are given. Find the LCD. See Example 2. (Objective 2)

33. 8, 12, 18
34. 10, 15, 28

35. $x^2 + 3x, x^2 - 9$

36. $3y^2 - 6y, 3y(y - 4)$

37. $x^3 + 27, x^2 + 6x + 9$

38. $x^3 - 8, x^2 - 4x + 4$

39. $2x^2 + 5x + 3, 4x^2 + 12x + 9, x^2 + 2x + 1$

40. $2x^2 + 5x + 3, 4x^2 + 12x + 9, 4x + 6$

Perform the operations and simplify. Assume no denominator is zero. **See Example 3. (Objective 3)**

41. $\dfrac{x+8}{x-3} - \dfrac{x-14}{3-x}$

42. $\dfrac{3-x}{2-x} + \dfrac{x-1}{x-2}$

43. $\dfrac{2a+1}{3a-2} - \dfrac{a-4}{2-3a}$

44. $\dfrac{4}{x-2} + \dfrac{5}{4-x^2}$

Perform the operations and simplify. Assume no denominator is zero. **See Example 4. (Objective 3)**

45. $\dfrac{1}{2} + \dfrac{1}{3}$

46. $\dfrac{5}{6} + \dfrac{2}{7}$

47. $\dfrac{7}{15} - \dfrac{17}{25}$

48. $\dfrac{8}{9} - \dfrac{5}{12}$

49. $\dfrac{a}{2} + \dfrac{2a}{5}$

50. $\dfrac{b}{6} + \dfrac{3a}{4}$

51. $\dfrac{3a}{2} - \dfrac{4b}{7}$

52. $\dfrac{2m}{3} - \dfrac{4n}{5}$

53. $\dfrac{3}{4x} + \dfrac{2}{3x}$

54. $\dfrac{2}{5a} + \dfrac{3}{2b}$

55. $\dfrac{3a}{2b} - \dfrac{2b}{3a}$

56. $\dfrac{5m}{2n} - \dfrac{3n}{4m}$

Perform the operations and simplify. Assume no denominator is zero. **See Examples 5–6. (Objective 3)**

57. $2 + \dfrac{5}{x-1}$

58. $\dfrac{4}{b-2} - 3$

59. $\dfrac{3}{x+2} + \dfrac{5}{x-4}$

60. $\dfrac{2}{a+4} - \dfrac{6}{a+3}$

61. $\dfrac{x+2}{x+5} - \dfrac{x-3}{x+7}$

62. $\dfrac{7}{x+3} + \dfrac{4x}{x+6}$

63. $\dfrac{x}{x^2+5x+6} + \dfrac{x}{x^2-4}$

64. $\dfrac{x}{3x^2-2x-1} + \dfrac{4}{3x^2+10x+3}$

65. $\dfrac{4}{x^2-2x-3} - \dfrac{x}{3x^2-7x-6}$

66. $\dfrac{2a}{a^2-2a-8} + \dfrac{3}{a^2-5a+4}$

Perform the operations and simplify. Assume no denominator is zero. **See Example 7. (Objective 3)**

67. $\dfrac{2x+y}{y-2x} + \dfrac{3x^2+y}{y^2-4xy+4x^2}$

68. $\dfrac{2x}{x^2-1} - \dfrac{x}{x^2+2x+1}$

69. $\dfrac{x}{1-x} - \dfrac{2x^2-1}{x^2-2x+1}$

70. $\dfrac{5}{x} + \dfrac{x+1}{x^3-5x^2}$

Perform the operations and simplify. Assume no denominator is zero. **See Example 8. (Objective 4)**

71. $\dfrac{8}{x^2-9} + \dfrac{2}{x-3} - \dfrac{6}{x}$

72. $\dfrac{x}{x^2-4} - \dfrac{x}{x+2} + \dfrac{2}{x}$

73. $\dfrac{x}{x+1} - \dfrac{x}{1-x^2} + \dfrac{1}{x}$

74. $\dfrac{y}{y-2} - \dfrac{2}{y+2} - \dfrac{-8}{4-y^2}$

75. $2x + 3 + \dfrac{1}{x+1}$

76. $x + 1 + \dfrac{1}{x-1}$

77. $\dfrac{y+4}{y^2+7y+12} - \dfrac{y-4}{y+3} + \dfrac{47}{y+4}$

78. $\dfrac{3}{x+1} - \dfrac{2}{x-1} + \dfrac{x+3}{x^2-1}$

79. $\dfrac{2}{x-2} + \dfrac{3}{x+2} - \dfrac{x-1}{x^2-4}$

80. $\dfrac{x-2}{x^2-3x} + \dfrac{2x-1}{x^2+3x} - \dfrac{2}{x^2-9}$

81. $\dfrac{2}{x-1} - \dfrac{2x}{x^2-1} - \dfrac{x}{x^2+2x+1}$

82. $\dfrac{5}{x^2-25} - \dfrac{3}{2x^2-9x-5} + 1$

Raise each rational expression to the stated power. Assume no denominator is zero. **See Example 9. (Objective 3)**

83. $\left(\dfrac{1}{x-1} + \dfrac{1}{1-x}\right)^2$

84. $\left(\dfrac{1}{a-1} - \dfrac{1}{1-a}\right)^2$

85. $\left(\dfrac{x}{x-3} + \dfrac{3}{3-x}\right)^3$

86. $\left(\dfrac{2y}{y+4} + \dfrac{8}{y+4}\right)^3$

ADDITIONAL PRACTICE ***Perform the operations and simplify. Assume no denominator is zero.***

87. $x + \dfrac{1}{x}$

88. $2 - \dfrac{1}{x+1}$

89. $\dfrac{x+3}{2x^2-5x+2}-\dfrac{3x-1}{x^2-x-2}$

90. $\dfrac{a+b}{3}+\dfrac{a-b}{7}$

91. $\dfrac{x-y}{2}+\dfrac{x+y}{3}$

92. $1+x-\dfrac{x}{x-5}$

93. $2-x+\dfrac{3}{x-9}$

94. $\dfrac{3x}{x-1}-2x-x^2$

95. $\dfrac{23}{x-1}+4x-5x^2$

96. $\dfrac{3x}{2x-1}+\dfrac{x+1}{3x+2}+\dfrac{2x}{6x^3+x^2-2x}$

97. $\dfrac{3x}{x-3}+\dfrac{4}{x-2}-\dfrac{5x}{x^3-5x^2+6x}$

98. $\dfrac{2x-1}{x^2+x-6}-\dfrac{3x-5}{x^2-2x-15}+\dfrac{2x-3}{x^2-7x+10}$

99. $2+\dfrac{4a}{a^2-1}-\dfrac{2}{a+1}$

100. $\dfrac{a}{a-1}-\dfrac{a+1}{2a-2}+a$

101. $\dfrac{x+5}{2x^2-2}+\dfrac{x}{2x+2}-\dfrac{3}{x-1}$

102. $\dfrac{a}{2-a}+\dfrac{3}{a-2}-\dfrac{3a-2}{a^2-4}$

103. $\dfrac{a}{a-b}+\dfrac{b}{a+b}+\dfrac{a^2+b^2}{b^2-a^2}$

104. $\dfrac{1}{x+y}-\dfrac{1}{x-y}-\dfrac{2y}{y^2-x^2}$

105. $\dfrac{7n^2}{m-n}+\dfrac{3m}{n-m}-\dfrac{3m^2-n}{m^2-2mn+n^2}$

106. $\dfrac{3b}{2a-b}+\dfrac{2a-1}{b-2a}-\dfrac{3a^2+b}{b^2-4ab+4a^2}$

107. $\dfrac{m+1}{m^2+2m+1}+\dfrac{m-1}{m^2-2m+1}+\dfrac{2}{m^2-1}$
(*Hint:* Think about this before finding the LCD.)

108. $\dfrac{a+2}{a^2+3a+2}+\dfrac{a-1}{a^2-1}+\dfrac{3}{a+1}$

109. Show that $\dfrac{a}{b}+\dfrac{c}{d}=\dfrac{ad+bc}{bd}$.

110. Show that $\dfrac{a}{b}-\dfrac{c}{d}=\dfrac{ad-bc}{bd}$.

APPLICATIONS

111. Monuments Find an expression that represents the total height of the monument.

112. Perimeter of a rectangle Write an expression that represents the perimeter of the rectangle.

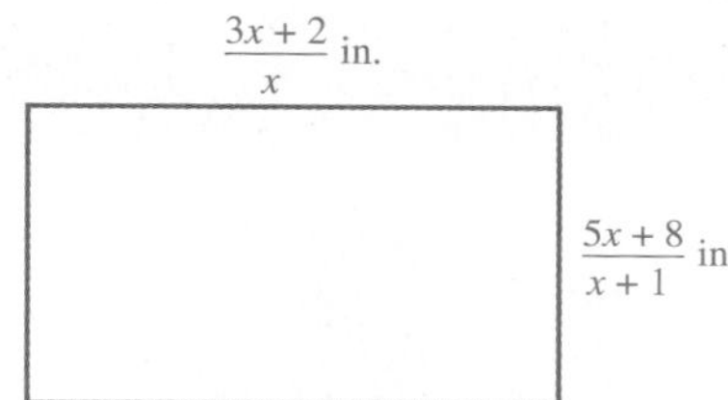

113. Width of a ring Find an expression that represents the width of the ring.

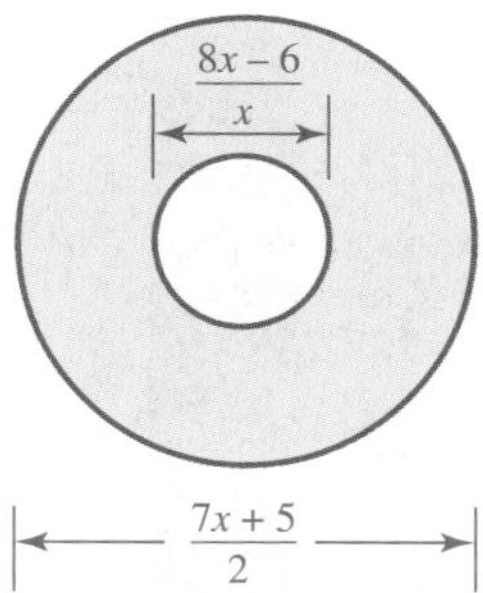

114. Work rates After working together on a common job, one worker completes $\frac{x+2}{4}$ of the work and another completes $\frac{x+3}{6}$ of the work. How much more did the first worker accomplish than the second?

WRITING ABOUT MATH

115. Explain how to find the least common denominator.

116. Explain how to add two fractions.

SOMETHING TO THINK ABOUT

117. Find the error:

$$\frac{8x+2}{5} - \frac{3x+8}{5} = \frac{8x+2-3x+8}{5} = \frac{5x+10}{5} = x+2$$

118. Find the error:

$$\frac{(x+y)^2}{2} + \frac{(x-y)^2}{3} = \frac{3 \cdot (x+y)^2}{3 \cdot 2} + \frac{2 \cdot (x-y)^2}{2 \cdot 3} = \frac{3x^2+3y^2+2x^2-2y^2}{6} = \frac{5x^2+y^2}{6}$$

SECTION 6.4 Simplifying Complex Fractions

Objectives

1 Simplify a complex fraction.

Vocabulary

complex fraction

Getting Ready

Use the distributive property to remove parentheses and simplify.

1. $2\left(1+\frac{1}{2}\right)$ **2.** $12\left(\frac{1}{6}-3\right)$ **3.** $a\left(\frac{3}{a}+2\right)$ **4.** $b\left(\frac{5}{b}-2\right)$

In this section, we will consider **complex fractions,** fractions that have a fraction in their numerator or their denominator. Examples of complex fractions are

$$\frac{\frac{3a}{b}}{\frac{6ac}{b^2}}, \quad \frac{\frac{2}{x}+1}{3+x}, \quad \text{and} \quad \frac{\frac{1}{x}+\frac{1}{y}}{\frac{1}{x}-\frac{1}{y}}$$

1 Simplify a complex fraction.

We can use two methods to simplify the complex fraction

$$\frac{\frac{3a}{b}}{\frac{6ac}{b^2}}$$

In Method 1, we eliminate the rational expressions in the numerator and denominator by writing the complex fraction as a division and using the division rule for fractions:

$$\frac{\frac{3a}{b}}{\frac{6ac}{b^2}} = \frac{3a}{b} \div \frac{6ac}{b^2}$$

$$= \frac{3a}{b} \cdot \frac{b^2}{6ac} \quad \text{Invert the divisor and multiply.}$$

$$= \frac{b}{2c} \quad \text{Multiply and simplify.}$$

In Method 2, we eliminate the rational expressions in the numerator and denominator by multiplying the expression by 1, written in the form $\frac{b^2}{b^2}$. We use $\frac{b^2}{b^2}$, because b^2 is the LCD of $\frac{3a}{b}$ and $\frac{6ac}{b^2}$.

$$\frac{\frac{3a}{b}}{\frac{6ac}{b^2}} = \frac{b^2 \cdot \frac{3a}{b}}{b^2 \cdot \frac{6ac}{b^2}} \quad \frac{b^2}{b^2} = 1$$

$$= \frac{\frac{3ab^2}{b}}{\frac{6acb^2}{b^2}} \quad \text{Multiply.}$$

$$= \frac{3ab}{6ac} \quad \text{Simplify the fractions in the numerator and denominator.}$$

$$= \frac{b}{2c} \quad \text{Divide out the common factor of } 3a.$$

With either method, the result is the same.

EXAMPLE 1 Simplify: $\dfrac{\frac{2}{x} + 1}{3 + x}$. Assume no denominator is zero.

Solution **Method 1:** We can add in the numerator and proceed as follows:

$$\frac{\frac{2}{x} + 1}{3 + x} = \frac{\frac{2}{x} + \frac{x}{x}}{\frac{3 + x}{1}} \quad \text{Write 1 as } \tfrac{x}{x} \text{ and } 3 + x \text{ as } \tfrac{3 + x}{1}.$$

$$= \frac{\frac{2 + x}{x}}{\frac{3 + x}{1}}$$ Add $\frac{2}{x}$ and $\frac{x}{x}$ to get $\frac{2 + x}{x}$.

$$= \frac{2 + x}{x} \div \frac{3 + x}{1}$$ Write the complex fraction as a division.

$$= \frac{2 + x}{x} \cdot \frac{1}{3 + x}$$ Invert the divisor and multiply.

$$= \frac{2 + x}{x^2 + 3x}$$ After noting that there are no common factors, multiply the numerators and multiply the denominators.

Method 2: To eliminate the denominator of x, we can multiply the numerator and the denominator by x, the LCD of the rational expressions in the complex fraction.

$$\frac{\frac{2}{x} + 1}{3 + x} = \frac{x\left(\frac{2}{x} + 1\right)}{x(3 + x)}$$ $\frac{x}{x} = 1$

$$= \frac{x \cdot \frac{2}{x} + x \cdot 1}{x \cdot 3 + x \cdot x}$$ Use the distributive property to remove parentheses.

$$= \frac{2 + x}{x^2 + 3x}$$ Simplify.

SELF CHECK 1 Simplify: $\frac{\frac{3}{a} + 2}{2 + a}$. Assume no denominator is zero.

ACCENT ON TECHNOLOGY

Checking Algebra

We can check the simplification in Example 1 by graphing the functions

$$f(x) = \frac{\frac{2}{x} + 1}{3 + x}$$

shown in Figure 6-6(a), and

$$g(x) = \frac{2 + x}{x^2 + 3x}$$

shown in Figure 6-6(b), and observing that the graphs are the same. Each graph has window settings of $[-5, 3]$ for x and $[-6, 6]$ for y.

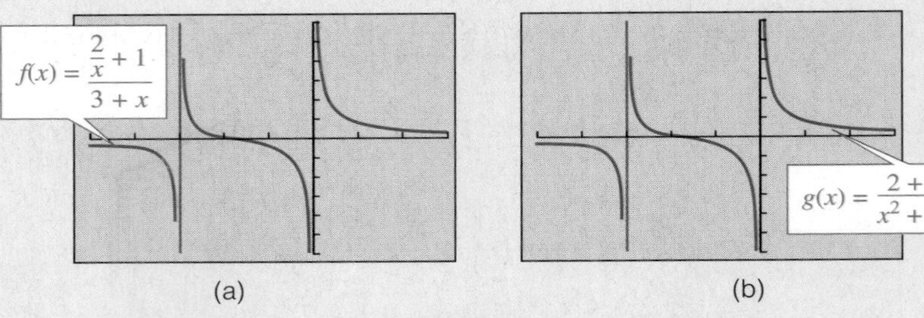

Figure 6-6

EXAMPLE 2 Simplify: $\dfrac{\frac{1}{x}+\frac{1}{y}}{\frac{1}{x}-\frac{1}{y}}$. Assume no denominator is zero.

Solution **Method 1:** We can add in the numerator and in the denominator and proceed as follows:

$$\frac{\frac{1}{x}+\frac{1}{y}}{\frac{1}{x}-\frac{1}{y}} = \frac{\frac{1y}{xy}+\frac{x1}{xy}}{\frac{1y}{xy}-\frac{x1}{xy}}$$ $\frac{y}{y}=1, \frac{x}{x}=1$

$$= \frac{\frac{y+x}{xy}}{\frac{y-x}{xy}}$$ Add the fractions in the numerator and subtract the fractions in the denominator.

$$= \frac{y+x}{xy} \div \frac{y-x}{xy}$$ Write the complex fraction as a division.

$$= \frac{y+x}{xy} \cdot \frac{xy}{y-x}$$ Invert the divisor and multiply.

$$= \frac{y+x}{y-x}$$ Multiply the numerators and multiply the denominators and simplify.

Method 2: We can multiply the numerator and denominator by xy (the LCD of the rational expressions appearing in the complex fraction) and simplify.

$$\frac{\frac{1}{x}+\frac{1}{y}}{\frac{1}{x}-\frac{1}{y}} = \frac{xy\left(\frac{1}{x}+\frac{1}{y}\right)}{xy\left(\frac{1}{x}-\frac{1}{y}\right)}$$

$$= \frac{xy \cdot \frac{1}{x} + xy \cdot \frac{1}{y}}{xy \cdot \frac{1}{x} - xy \cdot \frac{1}{y}}$$ Use the distributive property to remove parentheses.

$$= \frac{y+x}{y-x}$$ Simplify.

SELF CHECK 2 Simplify: $\dfrac{\frac{1}{a}-\frac{1}{b}}{\frac{1}{a}+\frac{1}{b}}$. Assume no denominator is zero.

EXAMPLE 3 Simplify: $\dfrac{x^{-1}+y^{-1}}{x^{-2}-y^{-2}}$. Assume no denominator is zero.

Solution **Method 1:** We can use the rule $n^{-1} = \frac{1}{n}$ and proceed as follows:

$$\frac{x^{-1}+y^{-1}}{x^{-2}-y^{-2}} = \frac{\frac{1}{x}+\frac{1}{y}}{\frac{1}{x^2}-\frac{1}{y^2}}$$ Write the fraction without using negative exponents.

$$= \frac{\frac{y}{xy} + \frac{x}{xy}}{\frac{y^2}{x^2y^2} - \frac{x^2}{x^2y^2}}$$ Get a common denominator in the numerator and denominator.

$$= \frac{\frac{y + x}{xy}}{\frac{y^2 - x^2}{x^2y^2}}$$ Add the fractions in the numerator and denominator.

$$= \frac{y + x}{xy} \div \frac{y^2 - x^2}{x^2y^2}$$ Write the fraction as a division.

$$= \frac{y + x}{xy} \cdot \frac{xxyy}{(y - x)(y + x)}$$ Invert and factor $y^2 - x^2$.

$$= \frac{(y + x)xxyy}{xy(y - x)(y + x)}$$ Multiply the numerators and the denominators and divide out the common factors of x, y, and $y + x$.

$$= \frac{xy}{y - x}$$ Simplify.

Method 2: We can multiply both numerator and denominator by x^2y^2, the LCD of the rational expressions, and proceed as follows:

$$\frac{x^{-1} + y^{-1}}{x^{-2} - y^{-2}} = \frac{\frac{1}{x} + \frac{1}{y}}{\frac{1}{x^2} - \frac{1}{y^2}}$$ Write the fraction without negative exponents.

$$= \frac{x^2y^2\left(\frac{1}{x} + \frac{1}{y}\right)}{x^2y^2\left(\frac{1}{x^2} - \frac{1}{y^2}\right)}$$ Multiply numerator and denominator by x^2y^2.

$$= \frac{x^2y^2 \cdot \frac{1}{x} + x^2y^2 \cdot \frac{1}{y}}{x^2y^2 \cdot \frac{1}{x^2} - x^2y^2 \cdot \frac{1}{y^2}}$$ Use the distributive property to remove parentheses.

$$= \frac{xy^2 + x^2y}{y^2 - x^2}$$ Simplify.

$$= \frac{xy(y + x)}{(y + x)(y - x)}$$ Factor the numerator and denominator.

$$= \frac{xy}{y - x}$$ $\frac{y + x}{y + x} = 1$.

COMMENT $x^{-1} + y^{-1}$ means $\frac{1}{x} + \frac{1}{y}$, and $(x + y)^{-1}$ means $\frac{1}{x + y}$. Thus, $x^{-1} + y^{-1} \neq (x + y)^{-1}$.

SELF CHECK 3 Simplify: $\frac{a^{-2} + b^{-2}}{a^{-1} - b^{-1}}$. Assume no denominator is zero.

PERSPECTIVE

Each of the complex fractions in the list

$$1 + \frac{1}{2},\quad 1 + \frac{1}{1 + \frac{1}{2}},\quad 1 + \frac{1}{1 + \frac{1}{1 + \frac{1}{2}}},\ \ldots$$

can be simplified by using the value of the expression preceding it. For example, to simplify the second expression in the list, replace $1 + \frac{1}{2}$ with $\frac{3}{2}$:

$$1 + \frac{1}{1 + \frac{1}{2}} = 1 + \frac{1}{\frac{3}{2}} = 1 + \frac{2}{3} = \frac{5}{3}$$

To simplify the third expression, we replace $1 + \frac{1}{1 + \frac{1}{2}}$ with $\frac{5}{3}$:

$$1 + \frac{1}{1 + \frac{1}{1 + \frac{1}{2}}} = 1 + \frac{1}{\frac{5}{3}} = 1 + \frac{3}{5} = \frac{8}{5}$$

The complex fractions in the list simplify to the fractions $\frac{3}{2}$, $\frac{5}{3}$, and $\frac{8}{5}$. The decimal values of these fractions get closer and closer to the irrational number 1.61803398875 . . . , known as the **golden ratio.** This number often appears in the architecture of the ancient Greeks and Egyptians.

For example, the width of the stairs in front of the Greek Parthenon (Illustration 1), divided by the building's height, is the golden ratio. The height of the triangular face of the Great Pyramid of Cheops (Illustration 2), divided by the pyramid's width, is also the golden ratio.

Illustration 1

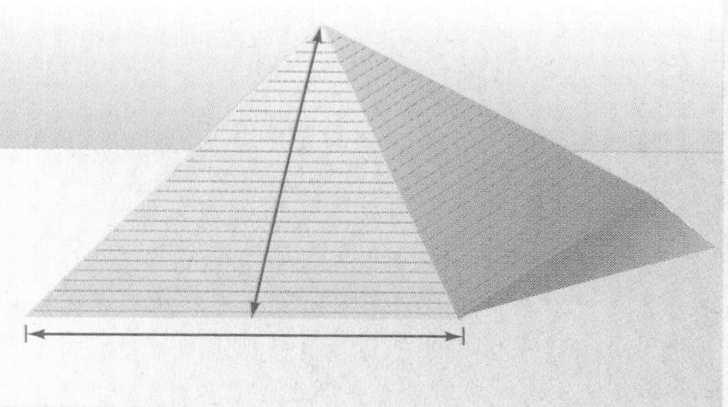

Illustration 2

EXAMPLE 4 Simplify: $\dfrac{\dfrac{2x}{1 - \frac{1}{x}} + 3}{3 - \frac{2}{x}}$. Assume no denominator is zero.

Solution In more complicated complex fractions such as this one, it is usually easiest to use Method 2. To do so, we begin by multiplying the numerator and denominator of

$$\frac{2x}{1 - \frac{1}{x}}$$

by x to eliminate the complex fraction in the numerator.

$$\frac{\dfrac{2x}{1 - \frac{1}{x}} + 3}{3 - \frac{2}{x}} = \frac{\dfrac{x2x}{x\left(1 - \frac{1}{x}\right)} + 3}{3 - \frac{2}{x}} \qquad \frac{x}{x} = 1$$

$$= \frac{\dfrac{2x^2}{x \cdot 1 - x \cdot \dfrac{1}{x}} + 3}{3 - \dfrac{2}{x}}$$ Use the distributive property to remove parentheses.

$$= \frac{\dfrac{2x^2}{x - 1} + 3}{3 - \dfrac{2}{x}}$$ Simplify: $x \cdot 1 = x$ and $x \cdot \frac{1}{x} = 1$.

We then multiply the numerator and denominator of the previous fraction by $x(x - 1)$, the LCD of $\frac{2x^2}{x - 1}$, 3, and $\frac{2}{x}$, and simplify:

$$\frac{\dfrac{2x}{1 - \dfrac{1}{x}} + 3}{3 - \dfrac{2}{x}} = \frac{x(x-1)\left(\dfrac{2x^2}{x - 1} + 3\right)}{x(x-1)\left(3 - \dfrac{2}{x}\right)}$$ $\frac{x}{x} = 1, \frac{x-1}{x-1} = 1$

$$= \frac{x(x-1) \cdot \dfrac{2x^2}{x-1} + x(x-1) \cdot 3}{x(x-1) \cdot 3 - x(x-1) \cdot \dfrac{2}{x}}$$ Use the distributive property.

$$= \frac{2x^3 + 3x(x - 1)}{3x(x - 1) - 2(x - 1)}$$ Simplify.

$$= \frac{2x^3 + 3x^2 - 3x}{3x^2 - 5x + 2}$$

This result does not simplify.

SELF CHECK 4 Simplify: $\dfrac{3 + \dfrac{2}{a}}{\dfrac{2a}{1 + \dfrac{1}{a}} + 3}$. Assume no denominator is zero.

SELF CHECK ANSWERS **1.** $\frac{2a + 3}{a^2 + 2a}$ **2.** $\frac{b - a}{b + a}$ **3.** $\frac{b^2 + a^2}{ab(b - a)}$ **4.** $\frac{(3a + 2)(a + 1)}{a(2a^2 + 3a + 3)}$

NOW TRY THIS

1. Simplify: $\left(\frac{x + 11}{3x - 3} - \frac{2}{x}\right) \div \left(\frac{3}{2x} + \frac{x - 9}{6x - 6}\right)$.

2. Simplify: $\dfrac{\frac{2}{x} - \frac{x + 11}{3x - 3}}{\frac{9 - x}{6x - 6} - \frac{3}{2x}}$.

3. Did the form of the problem influence your choice of methods? Do you find one method easier than the other? Discuss your choice with a classmate.

6.4 EXERCISES

WARM-UPS *Simplify each complex fraction.*

1. $\dfrac{\frac{3}{4}}{\frac{5}{4}}$

2. $\dfrac{\frac{a}{b}}{\frac{d}{b}}$

3. $\dfrac{\frac{3}{4}}{\frac{3}{8}}$

4. $\dfrac{\frac{x+y}{x}}{\frac{x-y}{x}}$

5. $\dfrac{\frac{x}{y}-1}{\frac{x}{y}}$

6. $\dfrac{1+\frac{a}{b}}{\frac{a}{b}}$

REVIEW *Solve each equation.*

7. $\dfrac{8(a-5)}{3} = 2(a-4)$

8. $\dfrac{3t^2}{5} + \dfrac{7t}{10} = \dfrac{3t+6}{5}$

9. $a^4 - 13a^2 + 36 = 0$

10. $|2x - 1| = 9$

VOCABULARY AND CONCEPTS *Fill in the blanks.*

11. A ________ fraction is a fraction that has fractions in its numerator or denominator.

12. The fraction $\dfrac{\frac{a}{b}}{\frac{c}{d}}$ is equivalent to $\frac{a}{b}$ ___ $\frac{c}{d}$.

GUIDED PRACTICE

Simplify using any method. Assume no denominator is zero. **(Objective 1)**

13. $\dfrac{\frac{1}{2}}{\frac{3}{4}}$

14. $-\dfrac{\frac{3}{4}}{\frac{1}{2}}$

15. $\dfrac{\frac{4x}{y}}{\frac{6xz}{y^2}}$

16. $\dfrac{\frac{5t^4}{9x}}{\frac{2t}{18x}}$

17. $\dfrac{5ab^2}{\frac{ab}{25}}$

18. $\dfrac{\frac{6a^2b}{4t}}{3a^2b^2}$

19. $\dfrac{\frac{1}{2}-\frac{2}{3}}{\frac{2}{3}+\frac{1}{2}}$

20. $\dfrac{\frac{2}{3}+\frac{4}{5}}{\frac{2}{5}-\frac{1}{3}}$

Simplify using any method. Assume no denominator is zero. **See Example 1. (Objective 1)**

21. $\dfrac{\frac{x-y}{xy}}{\frac{y-x}{x}}$

22. $\dfrac{\frac{x^2+5x+6}{3xy}}{\frac{x^2-9}{6xy}}$

23. $\dfrac{\frac{1}{x}-\frac{1}{y}}{xy}$

24. $\dfrac{xy}{\frac{1}{x}-\frac{1}{y}}$

25. $\dfrac{\frac{1}{a}+\frac{1}{b}}{\frac{1}{a}}$

26. $\dfrac{\frac{1}{b}}{\frac{1}{a}-\frac{1}{b}}$

27. $\dfrac{1+\frac{x}{y}}{1-\frac{x}{y}}$

28. $\dfrac{\frac{x}{y}+1}{1-\frac{x}{y}}$

Simplify using any method. Assume no denominator is zero. **See Example 2. (Objective 1)**

29. $\dfrac{\frac{y}{x}-\frac{x}{y}}{\frac{1}{x}+\frac{1}{y}}$

30. $\dfrac{\frac{y}{x}-\frac{x}{y}}{\frac{1}{y}-\frac{1}{x}}$

31. $\dfrac{\frac{1}{a}-\frac{1}{b}}{\frac{a}{b}-\frac{b}{a}}$

32. $\dfrac{\frac{1}{a}+\frac{1}{b}}{\frac{a}{b}-\frac{b}{a}}$

33. $\dfrac{x+1-\frac{6}{x}}{\frac{1}{x}}$

34. $\dfrac{x-1-\frac{2}{x}}{\frac{x}{3}}$

35. $\dfrac{5xy}{1+\frac{1}{xy}}$

36. $\dfrac{3a}{a+\frac{1}{a}}$

37. $\dfrac{1+\dfrac{6}{x}+\dfrac{8}{x^2}}{1+\dfrac{1}{x}-\dfrac{12}{x^2}}$

38. $\dfrac{1-x-\dfrac{2}{x}}{\dfrac{6}{x^2}+\dfrac{1}{x}-1}$

39. $\dfrac{\dfrac{1}{a+1}+1}{\dfrac{3}{a-1}+1}$

40. $\dfrac{2+\dfrac{3}{x+1}}{\dfrac{1}{x}+x+x^2}$

Simplify using any method. Assume no denominator is zero.
See Example 3. (Objective 1)

41. $\dfrac{x^{-1}+y^{-1}}{x}$

42. $\dfrac{x^{-1}-y^{-1}}{y}$

43. $\dfrac{y}{x^{-1}-y^{-1}}$

44. $\dfrac{x^{-1}+y^{-1}}{(x+y)^{-1}}$

45. $\dfrac{x^{-1}+y^{-1}}{x^{-1}-y^{-1}}$

46. $\dfrac{(x+y)^{-1}}{x^{-1}+y^{-1}}$

47. $\dfrac{x-y^{-2}}{y-x^{-2}}$

48. $\dfrac{x^{-2}-y^{-2}}{x^{-1}-y^{-1}}$

Simplify using any method. Assume no denominator is zero.
See Example 4. (Objective 1)

49. $\dfrac{1+\dfrac{a}{b}}{1-\dfrac{a}{1-\frac{a}{b}}}$

50. $\dfrac{1+\dfrac{2}{1+\frac{a}{b}}}{1-\frac{a}{b}}$

51. $\dfrac{x-\dfrac{1}{x}}{1+\dfrac{1}{\frac{1}{x}}}$

52. $\dfrac{\dfrac{a^2+3a+4}{ab}}{2+\dfrac{3+a}{\frac{2}{a}}}$

53. $\dfrac{b}{b+\dfrac{2}{2+\frac{1}{2}}}$

54. $\dfrac{2y}{y-\dfrac{y}{3-\frac{1}{2}}}$

55. $a+\dfrac{a}{1+\dfrac{a}{a+1}}$

56. $b+\dfrac{b}{1-\dfrac{b+1}{b}}$

ADDITIONAL PRACTICE *Simplify each complex fraction.*

57. $\dfrac{-\dfrac{2}{3}}{\dfrac{6}{9}}$

58. $\dfrac{\dfrac{11}{18}}{\dfrac{22}{27}}$

59. $\dfrac{\dfrac{1}{2}+\dfrac{1}{3}}{\dfrac{1}{4}}$

60. $\dfrac{\dfrac{1}{4}-\dfrac{1}{5}}{\dfrac{1}{3}}$

61. $\dfrac{x+y}{x^{-1}+y^{-1}}$

62. $\dfrac{x-y}{x^{-1}-y^{-1}}$

63. $\dfrac{x-\dfrac{1}{1-\frac{x}{2}}}{\dfrac{3}{x+\frac{2}{3}}-x}$

64. $\dfrac{\dfrac{2x}{x-\frac{1}{x}}-\dfrac{1}{x}}{2x+\dfrac{2x}{1-\frac{1}{x}}}$

65. $\dfrac{2x+\dfrac{1}{2-\frac{x}{2}}}{\dfrac{4}{\frac{x}{2}-2}-x}$

66. $\dfrac{3x-\dfrac{1}{3-\frac{x}{2}}}{\dfrac{3}{\frac{x}{2}-3}+x}$

67. $\dfrac{\dfrac{1}{x^2+3x+2}+\dfrac{1}{x^2+x-2}}{\dfrac{3x}{x^2-1}-\dfrac{x}{x+2}}$

68. $\dfrac{\dfrac{1}{x^2-1}-\dfrac{2}{x^2+4x+3}}{\dfrac{2}{x^2+2x-3}+\dfrac{1}{x+3}}$

APPLICATIONS

69. **Engineering** The stiffness k of the shaft is given by the formula

$$k=\frac{1}{\dfrac{1}{k_1}+\dfrac{1}{k_2}}$$

where k_1 and k_2 are the individual stiffnesses of each section. Simplify the complex fraction.

70. **Electronics** In electronic circuits, resistors oppose the flow of an electric current. To find the total resistance of a parallel combination of the two resistors shown in the illustration, we can use the formula

$$\text{Total resistance}=\frac{1}{\dfrac{1}{R_1}+\dfrac{1}{R_2}}$$

where R_1 is the resistance of the first resistor and R_2 is the resistance of the second. Simplify the complex fraction on the right-hand side of the formula.

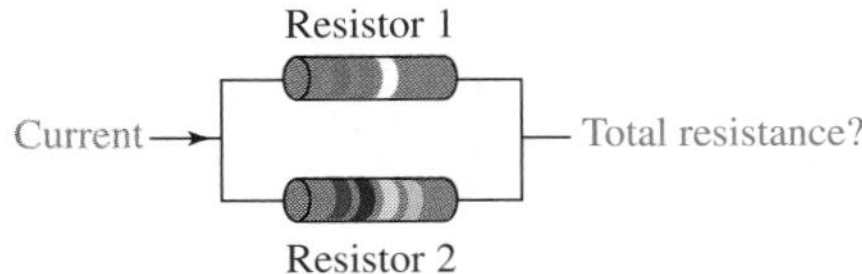

71. Data analysis Use the data in the table to find the average measurement for the three-trial experiment.

	Trial 1	Trial 2	Trial 3
Measurement	$\frac{k}{2}$	$\frac{k}{3}$	$\frac{k}{2}$

72. Transportation If a car travels a distance d_1 at a speed s_1, and then travels a distance d_2 at a speed s_2, the average (mean) speed is given by the formula

$$\bar{s} = \frac{d_1 + d_2}{\frac{d_1}{s_1} + \frac{d_2}{s_2}}$$

Simplify the complex fraction.

WRITING ABOUT MATH

73. There are two methods used to simplify a complex fraction. Explain one of them.

74. Explain the other method of simplifying a complex fraction.

SOMETHING TO THINK ABOUT

75. Simplify: $(x^{-1}y^{-1})(x^{-1} + y^{-1})^{-1}$.

76. Simplify: $[(x^{-1} + 1)^{-1} + 1]^{-1}$.

SECTION 6.5 Solving Equations Containing Rational Expressions

Objectives

1. Solve a rational equation.
2. Solve a formula involving a rational equation for a specified variable.
3. Solve an application problem involving a rational equation.

Vocabulary

rational equation   extraneous solution

Getting Ready

Solve each equation.

1. $\frac{3}{x} = \frac{6}{9}$ **2.** $\frac{x}{8} = \frac{2}{x}$

We now show how to solve equations that contain rational expressions. Then we will use these skills to solve application problems.

1 Solve a rational equation.

If an equation contains one or more rational expressions, it is called a **rational equation.** Some examples are

$$\frac{3}{5}+\frac{7}{x+2}=2, \qquad \frac{x+3}{x-3}=\frac{2}{x^2-4}, \qquad \text{and} \qquad \frac{-x^2+10}{x^2-1}+\frac{3x}{x-1}=\frac{2x}{x+1}$$

To solve a rational equation, we can multiply both sides of the equation by a nonzero expression to clear the equation of fractions.

EXAMPLE 1 Solve: $\frac{3}{5}+\frac{7}{x+2}=2.$

Solution We first note that x cannot be -2, because this would give a 0 in the denominator of $\frac{7}{x+2}$. If $x \neq -2$, we can multiply both sides of the equation by the LCD of the fractions, which is $5(x + 2)$.

$$5(x+2)\left(\frac{3}{5}+\frac{7}{x+2}\right)=5(x+2)2$$

$$5(x+2)\left(\frac{3}{5}\right)+5(x+2)\left(\frac{7}{x+2}\right)=5(x+2)2 \qquad \text{Use the distributive property on the left side.}$$

$$3(x+2)+5(7)=10(x+2) \qquad \text{Simplify.}$$

$$3x+6+35=10x+20 \qquad \text{Use the distributive property and simplify.}$$

$$3x+41=10x+20 \qquad \text{Simplify.}$$

$$-7x=-21 \qquad \text{Add } -10x \text{ and } -41 \text{ to both sides.}$$

$$x=3 \qquad \text{Divide both sides by } -7.$$

The solution is 3.

Check: To check, we substitute 3 for x in the original equation and simplify:

$$\frac{3}{5}+\frac{7}{x+2}=2$$

$$\frac{3}{5}+\frac{7}{3+2}\stackrel{?}{=}2$$

$$\frac{3}{5}+\frac{7}{5}\stackrel{?}{=}2$$

$$2=2 \qquad \text{The result checks.}$$

SELF CHECK 1 Solve: $\frac{2}{5}+\frac{5}{x-2}=\frac{29}{10}$.

ACCENT ON TECHNOLOGY

Solving Equations

To use a graphing calculator to approximate the solution of $\frac{3}{5}+\frac{7}{x+2}=2$, we graph the functions $f(x)=\frac{3}{5}+\frac{7}{x+2}$ and $g(x) = 2$. If we use window settings of $[-10, 10]$ for x and $[-10, 10]$ for y, we will obtain the graph shown in Figure 6-7(a).

If we trace and move the cursor close to the intersection point of the two graphs, we will get the approximate value of x shown in Figure 6-7(b). If we zoom twice and trace again, we get the results shown in Figure 6-7(c). Algebra will show that the exact solution is 3, as shown in Example 1.

We can find the intersection point exactly by using the INTERSECT command found in the CALC menu.

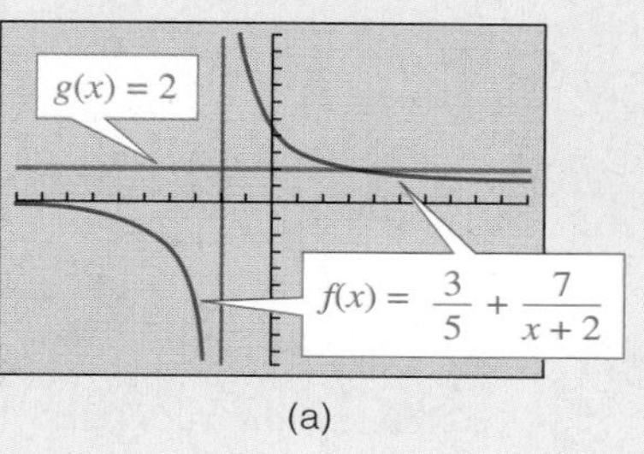

(a)

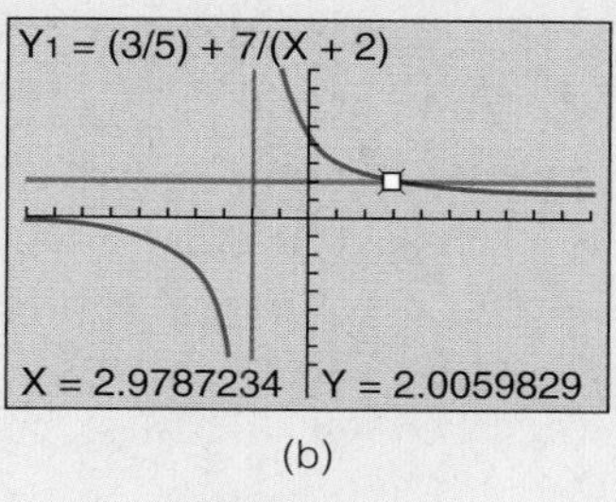

(b)

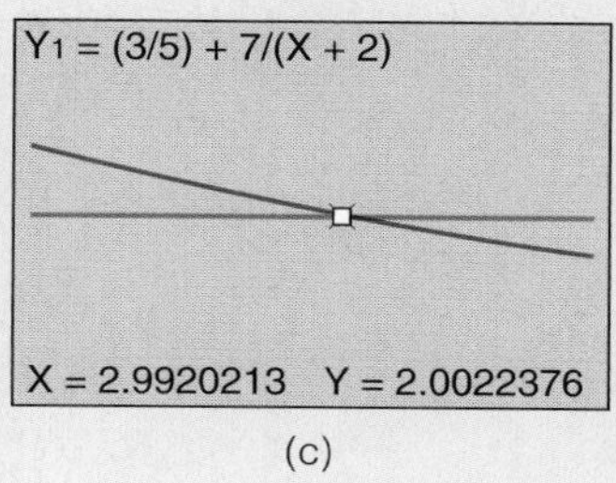

(c)

Figure 6-7

EXAMPLE 2 Solve: $\frac{-x^2 + 10}{x^2 - 1} + \frac{3x}{x - 1} = \frac{2x}{x + 1}$.

Solution We first note that x cannot be 1 or -1, because this would give a 0 in a denominator. If $x \neq 1$ and $x \neq -1$, we can clear the equation of denominators by multiplying both sides by the LCD of the three rational expressions, $(x + 1)(x - 1)$, and proceed as follows:

$$\frac{-x^2 + 10}{x^2 - 1} + \frac{3x}{x - 1} = \frac{2x}{x + 1}$$

$$\frac{-x^2 + 10}{(x + 1)(x - 1)} + \frac{3x}{x - 1} = \frac{2x}{x + 1} \quad \text{Factor } x^2 - 1.$$

$$\frac{\mathbf{(x + 1)(x - 1)}(-x^2 + 10)}{(x + 1)(x - 1)} + \frac{3x\mathbf{(x + 1)(x - 1)}}{x - 1} = \frac{2x\mathbf{(x + 1)(x - 1)}}{x + 1} \quad \text{Multiply both sides by } (x + 1)(x - 1).$$

$$-x^2 + 10 + 3x(x + 1) = 2x(x - 1) \quad \text{Divide out common factors.}$$

$$-x^2 + 10 + 3x^2 + 3x = 2x^2 - 2x \quad \text{Remove parentheses.}$$

$$2x^2 + 10 + 3x = 2x^2 - 2x \quad \text{Combine like terms.}$$

$$10 + 3x = -2x \quad \text{Subtract } 2x^2 \text{ from both sides.}$$

$$10 + 5x = 0 \quad \text{Add } 2x \text{ to both sides.}$$

$$5x = -10 \quad \text{Subtract 10 from both sides.}$$

$$x = -2 \quad \text{Divide both sides by 5.}$$

The solution is -2. Verify that -2 is a solution of the original equation.

SELF CHECK 2 Solve: $\frac{2x^2}{x^2 - 4} = \frac{3}{x + 2} + \frac{2x}{x - 2}$.

When we multiply both sides of an equation by a quantity that contains a variable, we can get false solutions, called **extraneous solutions.** We must exclude extraneous solutions from the solution set of an equation.

EXAMPLE 3 Solve: $\dfrac{2(x+1)}{x-3} = \dfrac{x+5}{x-3}$.

Solution We start by noting that x cannot be 3, because this would give a 0 in a denominator. If $x \neq 3$, we can clear the equation of denominators by multiplying both sides by $x - 3$.

$$\frac{2(x+1)}{x-3} = \frac{x+5}{x-3}$$

$$(x-3)\frac{2(x+1)}{x-3} = (x-3)\frac{x+5}{x-3} \quad \text{Multiply both sides by } x - 3.$$

$$2(x+1) = x+5 \quad \text{Simplify.}$$

$$2x+2 = x+5 \quad \text{Remove parentheses.}$$

$$x+2 = 5 \quad \text{Subtract } x \text{ from both sides.}$$

$$x = 3 \quad \text{Subtract 2 from both sides.}$$

Since x cannot be 3, the 3 must be discarded. This equation has no solution. Its solution set is the empty set, $\varnothing$.

SELF CHECK 3 Solve: $\frac{6}{x-6} + 3 = \frac{x}{x-6}$.

EXAMPLE 4 Solve: $\dfrac{x+1}{5} - 2 = -\dfrac{4}{x}$.

Solution We note that x cannot be 0, because this would give a 0 in a denominator. If $x \neq 0$, we can clear the equation of denominators by multiplying both sides by $5x$.

$$\frac{x+1}{5} - 2 = -\frac{4}{x}$$

$$5x\left(\frac{x+1}{5} - 2\right) = 5x\left(-\frac{4}{x}\right) \quad \text{Multiply both sides by } 5x.$$

$$x(x+1) - 10x = -20 \quad \text{Remove parentheses and simplify.}$$

$$x^2 + x - 10x = -20 \quad \text{Remove parentheses.}$$

$$x^2 - 9x + 20 = 0 \quad \text{Combine like terms and add 20 to both sides.}$$

$$(x-5)(x-4) = 0 \quad \text{Factor } x^2 - 9x + 20.$$

$$x - 5 = 0 \quad \text{or} \quad x - 4 = 0 \quad \text{Set each factor equal to 0.}$$

$$x = 5 \quad \Big| \quad x = 4$$

Since 4 and 5 both satisfy the original equation, the solutions are 4 and 5.

SELF CHECK 4 Solve: $a + \frac{2}{3} = \frac{2a-12}{3(a-3)}$.

2 Solve a formula involving a rational equation for a specified variable.

Many formulas must be cleared of denominators before we can solve them for a specific variable.

EXAMPLE 5 Solve: $\frac{1}{r} = \frac{1}{r_1} + \frac{1}{r_2}$ for r.

Solution This formula is from electronics. Since r, r_1, and r_2 represent resistances, they will not be zero.

$$\frac{1}{r} = \frac{1}{r_1} + \frac{1}{r_2}$$

$$\frac{rr_1r_2}{r} = \frac{rr_1r_2}{r_1} + \frac{rr_1r_2}{r_2} \quad \text{Multiply both sides by } rr_1r_2 \text{, the LCD.}$$

$$r_1r_2 = rr_2 + rr_1 \quad \text{Simplify each fraction.}$$

$$r_1r_2 = r(r_2 + r_1) \quad \text{Factor out } r \text{ on the right side.}$$

$$\frac{r_1r_2}{r_2 + r_1} = r \quad \text{Divide both sides by } r_2 + r_1.$$

$$r = \frac{r_1r_2}{r_2 + r_1}$$

SELF CHECK 5 Solve: $\frac{1}{a} - \frac{1}{b} = 1$ for b $(a, b \neq 0)$.

3 Solve an application problem involving a rational equation.

EXAMPLE 6 **DRYWALLING A HOUSE** A contractor knows that one crew can drywall a house in 4 days and that another crew can drywall the same house in 5 days. One day must be allowed for the plaster coat to dry. If the contractor uses both crews, can the house be ready for painting in 4 days?

Analyze the problem: We will let x represent the number of days it takes for both crews to drywall the house.

Form an equation: Since the first crew can drywall the house in 4 days, it can do $\frac{1}{4}$ of the job in 1 day. Since the second crew can drywall the house in 5 days, it can do $\frac{1}{5}$ of the job in 1 day. If it takes x days for both crews to finish the house, together they can do $\frac{1}{x}$ of the job in 1 day. The amount of work the first crew can do in 1 day plus the amount of work the second crew can do in 1 day equals the amount of work both crews can do in 1 day. This gives the following equation:

What crew 1 can do in one day	plus	what crew 2 can do in one day	equals	what they can do together in one day.
$\frac{1}{4}$	$+$	$\frac{1}{5}$	$=$	$\frac{1}{x}$

Solve the equation: We can solve this equation as follows.

$$20x\left(\frac{1}{4} + \frac{1}{5}\right) = 20x\left(\frac{1}{x}\right) \quad \text{Multiply both sides by } 20x \text{, the LCD of } \tfrac{1}{4}, \tfrac{1}{5} \text{ and } \tfrac{1}{x}.$$

$$5x + 4x = 20 \quad \text{Remove parentheses and simplify.}$$

$$9x = 20 \quad \text{Combine like terms.}$$

$$x = \frac{20}{9} \quad \text{Divide both sides by 9.}$$

State the conclusion

Because 1 day is necessary for drying, the drywallers must complete their work in 3 days. Since it will take $2\frac{2}{9}$ days for both crews to drywall the house, it will be ready for painting in less than 4 days.

Check the result

$$\frac{1}{4} + \frac{1}{5} \stackrel{?}{=} \frac{1}{\frac{20}{9}}$$

$$\frac{1}{4} + \frac{1}{5} \stackrel{?}{=} \frac{9}{20}$$

$$\frac{9}{20} = \frac{9}{20} \qquad \text{The result checks.}$$

EXAMPLE 7 **DRIVING TO A CONVENTION** A man drove 200 miles to a convention. Because of road construction, his average speed on the return trip was 10 mph less than his average speed going to the convention. If the return trip took 1 hour longer, how fast did he drive in each direction?

Analyze the problem

Let r represent the average rate of speed going to the meeting. Then $r - 10$ represents the average rate of speed on the return trip. We will need the formula $d = rt$ for this problem. (d is distance, r is the rate of speed, and t is time.)

Form an equation

We know the distance and the rate for each part of the trip, and the formula for time is $t = \frac{d}{r}$. We can organize the given information in the table shown in Figure 6-8.

	Rate	· Time	= Distance
Going	r	$\frac{200}{r}$	200
Returning	$r - 10$	$\frac{200}{r - 10}$	200

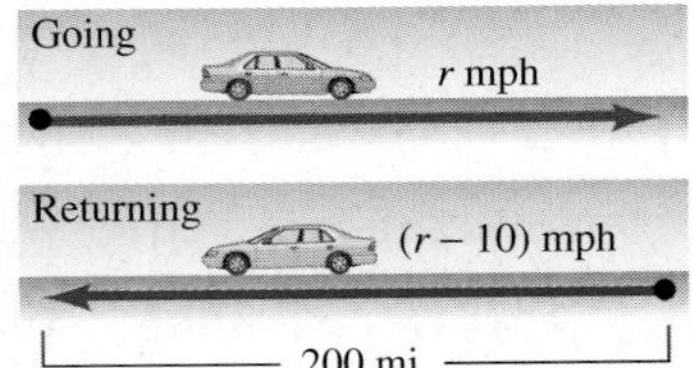

Figure 6-8

Because the return trip took 1 hour longer, we can form the following equation:

The time it took to travel to the convention	plus	1	equals	the time it took to return.
$\frac{200}{r}$	$+$	1	$=$	$\frac{200}{r - 10}$

Solve the equation

We can solve the equation as follows:

$$r(r - 10)\left(\frac{200}{r} + 1\right) = r(r - 10)\left(\frac{200}{r - 10}\right) \qquad \text{Multiply both sides by } r(r - 10).$$

$$200(r - 10) + r(r - 10) = 200r \qquad \text{Remove parentheses and simplify.}$$

$$200r - 2{,}000 + r^2 - 10r = 200r \qquad \text{Remove parentheses.}$$

$$r^2 - 10r - 2{,}000 = 0 \qquad \text{Subtract } 200r \text{ from both sides.}$$

$$(r - 50)(r + 40) = 0 \qquad \text{Factor } r^2 - 10r - 2{,}000.$$

$$r - 50 = 0 \quad \text{or} \quad r + 40 = 0 \qquad \text{Set each factor equal to 0.}$$

$$r = 50 \quad \Big| \quad r = -40$$

State the conclusion We must exclude the solution of -40, because a speed cannot be negative. Thus, the man averaged 50 mph going to the convention, and he averaged $50 - 10$, or 40 mph, returning.

Check the result At 50 mph, the 200-mile trip took 4 hours. At 40 mph, the return trip took 5 hours, which is 1 hour longer.

EXAMPLE 8 **A RIVER CRUISE** The Forest City Queen can make a 9-mile trip down the Rock River and return in a total of 1.6 hours. If the riverboat travels 12 mph in still water, find the speed of the current in the Rock River.

Analyze the problem We can let c represent the speed of the current. Since the boat travels 12 mph and a current of c mph pushes the boat while it is going downstream, the speed of the boat going downstream is $(12 + c)$ mph. On the return trip, the current pushes against the boat, and its speed is $(12 - c)$ mph.

Form an equation Since $t = \frac{d}{r}\left(\text{time} = \frac{\text{distance}}{\text{rate}}\right)$, the time required for the downstream leg of the trip is $\frac{9}{12 + c}$ hours, and the time required for the upstream leg of the trip is $\frac{9}{12 - c}$ hours.

We can organize this information in the table shown in Figure 6-9. Furthermore, we know that the total time required for the round trip is 1.6 or $\frac{8}{5}$ hours.

	Rate	·	Time	=	Distance
Going downstream	$12 + c$		$\frac{9}{12 + c}$		9
Going upstream	$12 - c$		$\frac{9}{12 - c}$		9

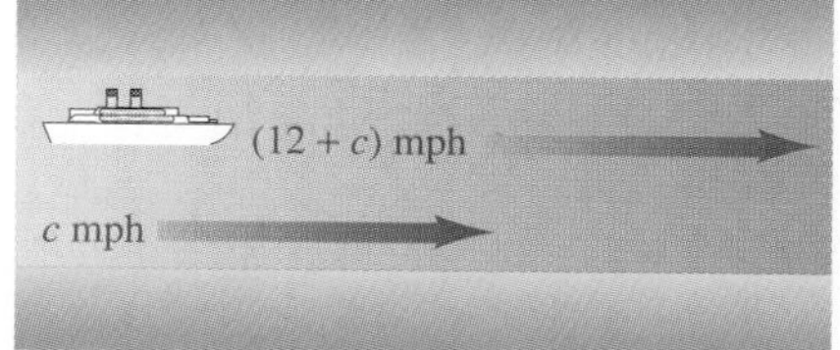

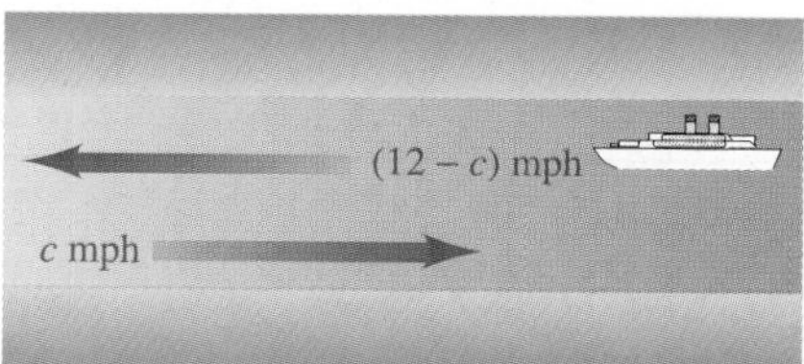

Figure 6-9

This gives the following equation.

The time it takes to travel downstream	plus	the time it takes to travel upstream	equals	the total time for the round trip.
$\frac{9}{12 + c}$	$+$	$\frac{9}{12 - c}$	$=$	$\frac{8}{5}$

Solve the equation We can multiply both sides of this equation by $5(12 + c)(12 - c)$ to clear it of denominators and proceed as follows:

$$\frac{5(12 + c)(12 - c)9}{12 + c} + \frac{5(12 + c)(12 - c)9}{12 - c} = \frac{5(12 + c)(12 - c)8}{5}$$

$$45(12 - c) + 45(12 + c) = 8(12 + c)(12 - c) \quad \frac{12 + c}{12 + c} = 1, \frac{12 - c}{12 - c} = 1, \text{ and } \frac{5}{5} = 1.$$

$$540 - 45c + 540 + 45c = 8(144 - c^2) \quad \text{Multiply.}$$

$$1{,}080 = 1{,}152 - 8c^2$$ Combine like terms and multiply.

$$8c^2 - 72 = 0$$ Add $8c^2$ and $-1{,}152$ to both sides.

$$c^2 - 9 = 0$$ Divide both sides by 8.

$$(c + 3)(c - 3) = 0$$ Factor $c^2 - 9$.

$$c + 3 = 0 \quad \text{or} \quad c - 3 = 0$$ Set each factor equal to 0.

$$c = -3 \quad | \quad c = 3$$

State the conclusion Since the current cannot be a negative rate, the apparent solution of -3 must be discarded. Thus, the speed of the current in the Rock River is 3 mph.

Check the result The boat travels 9 miles downstream at a speed of 15 mph $(12 + 3)$ and 9 miles back upstream at a speed of 9 mph $(12 - 3)$. It takes $\frac{3}{5}$ hr to go downstream and 1 hour to return for a total trip of $1\frac{3}{5}$ or 1.6 hours.

SELF CHECK ANSWERS **1.** 4 **2.** $\frac{6}{7}$ **3.** 6 is extraneous, $\varnothing$ **4.** 1, 2 **5.** $b = \frac{a}{1 - a}$

NOW TRY THIS

Solve:

1. $\frac{3x - 5}{x^2 - x - 6} = 0$

2. $\frac{2}{(x - 2)^2} - \frac{5}{x - 2} + 3 = 0$

3. $\frac{1}{x^2 - 5x + 6} + \frac{x - 2}{x^2 - x - 6} = \frac{x}{x^2 - 4}$

6.5 EXERCISES

WARM-UPS *Solve each equation.*

1. $\frac{4}{x} = 2$

2. $\frac{9}{y} = 3$

3. $\frac{4}{p} + \frac{5}{p} = 9$

4. $\frac{5}{r} - \frac{2}{r} = 1$

5. $\frac{4}{y} - \frac{1}{y} = 3$

6. $\frac{8}{t} - \frac{2}{t} = 3$

REVIEW *Simplify each expression. Write each answer without using negative exponents.*

7. $(m^2n^{-3})^{-2}$

8. $\frac{a^{-1}}{a^{-1} + 1}$

9. $\frac{a^0 + 2a^0 - 3a^0}{(a - b)^0}$

10. $(4x^{-2} + 3)(2x - 4)$

VOCABULARY AND CONCEPTS *Fill in the blanks.*

11. If an equation contains a rational expression, it is called a ________ equation.

12. A false solution to an equation is called an ________ solution.

GUIDED PRACTICE

Solve each equation. **See Example 1. (Objective 1)**

13. $\frac{1}{4} + \frac{9}{x} = 1$

14. $\frac{1}{3} - \frac{10}{x} = -3$

15. $\frac{34}{x} - \frac{3}{2} = -\frac{13}{20}$

16. $\frac{1}{2} + \frac{7}{x} = 2 + \frac{1}{x}$

17. $\frac{3}{y} + \frac{7}{2y} = 13$

18. $\frac{2}{x} + \frac{1}{2} = \frac{7}{2x}$

19. $\frac{7}{5x} - \frac{1}{2} = \frac{5}{6x} + \frac{1}{3}$

20. $\frac{2}{x} + \frac{1}{2} = \frac{9}{4x} - \frac{1}{2x}$

Solve each equation. See Example 2. (Objective 1)

21. $\frac{x}{x+2} = 1 - \frac{3x+2}{x^2+4x+4}$

22. $\frac{3+2a}{a^2+6+5a} + \frac{2-5a}{a^2-4} = \frac{2-3a}{a^2-6+a}$

23. $\frac{2}{x-2} + \frac{1}{x+1} = \frac{1}{x^2-x-2}$

24. $\frac{5}{2z^2+z-3} - \frac{2}{2z+3} = \frac{z+1}{z-1} - 1$

Solve each equation. If a solution is extraneous, so indicate. See Example 3. (Objective 1)

25. $\frac{x+1}{x} - \frac{x-1}{x} = 0$

26. $\frac{x-3}{x-1} - \frac{2x-4}{x-1} = 0$

27. $\frac{y-3}{y+2} = 3 - \frac{1-2y}{y+2}$

28. $\frac{3a-5}{a-1} - 2 = \frac{2a}{1-a}$

Solve each equation. If a solution is extraneous, so indicate. See Example 4. (Objective 1)

29. $\frac{5}{x+4} + \frac{1}{x+4} = x - 1$

30. $\frac{2}{x-1} + \frac{x-2}{3} = \frac{4}{x-1}$

31. $\frac{3}{x+1} - \frac{x-2}{2} = \frac{x-2}{x+1}$

32. $\frac{2}{x-3} + \frac{3}{4} = \frac{17}{2x}$

33. $\frac{x-4}{x-3} + \frac{x-2}{x-3} = x - 3$

34. $1 = \frac{3}{x-2} - \frac{12}{x^2-4}$

35. $\frac{30}{y-2} + \frac{24}{y-5} = 13$

36. $\frac{x+4}{x+7} - \frac{x}{x+3} = \frac{3}{8}$

Solve each formula for the indicated variable. Assume no denominator is zero. See Example 5. (Objective 2)

37. $\frac{1}{R} = \frac{1}{r_1} + \frac{1}{r_2}$ for R

38. $S = \frac{a}{1-r}$ for r

39. $\frac{1}{p} + \frac{1}{q} = \frac{1}{f}$ for f

40. $\frac{1}{p} + \frac{1}{q} = \frac{1}{f}$ for p

ADDITIONAL PRACTICE

Solve each equation. If a solution is extraneous, so indicate.

41. $\frac{3-5y}{2+y} = \frac{3+5y}{2-y}$

42. $\frac{x}{x-2} = 1 + \frac{1}{x-3}$

43. $\frac{a+2}{a+1} = \frac{a-4}{a-3}$

44. $\frac{z+2}{z+8} - \frac{z-3}{z-2} = 0$

45. $\frac{x+2}{x+3} - 1 = \frac{1}{3-2x-x^2}$

46. $\frac{x-3}{x-2} - \frac{1}{x} = \frac{x-3}{x}$

47. $\frac{5}{x+6} - \frac{3}{x-4} = \frac{2}{x+3}$

48. $\frac{a-1}{a+3} - \frac{1-2a}{3-a} = \frac{2-a}{a-3}$

49. $\frac{y^2}{y+2} + 4 = \frac{y+6}{y+2} + 3$

50. $\frac{5}{y-1} + \frac{3}{y-3} = \frac{8}{y-2}$

51. $\frac{5}{x+4} - \frac{1}{3} = \frac{x-1}{x}$

52. $\frac{3}{a-2} - \frac{1}{a-1} = \frac{7}{(a-2)(a-1)}$

Solve each formula for the indicated variable. Assume no denominator is zero.

53. $S = \frac{a-lr}{1-r}$ for r

54. $H = \frac{2ab}{a+b}$ for a

55. $\frac{1}{R} = \frac{1}{r_1} + \frac{1}{r_2} + \frac{1}{r_3}$ for R

56. $\frac{1}{R} = \frac{1}{r_1} + \frac{1}{r_2} + \frac{1}{r_3}$ for r_1

APPLICATIONS

Solve each application problem. See Examples 6–8. (Objective 3)

57. House painting If one painter can paint a house in 5 days and another painter can paint the same house in 3 days, how long will it take them to paint the house working together?

58. Reading proof A proofreader can read 250 pages in 8 hours, and a second proofreader can read 250 pages in 10 hours. If they both work on a 250-page book, can they meet a five-hour deadline?

59. Storing corn In 10 minutes, a conveyor belt can move 1,000 bushels of corn into the storage bin shown in the illustration. A smaller belt can move 1,000 bushels to the storage bin in 14 minutes. If both belts are used, how long will it take to move 1,000 bushels to the storage bin?

60. Roofing a house One roofing crew can finish a 2,800-square-foot roof in 12 hours, and another crew can do the job in 10 hours. If they work together, can they finish before a predicted rain in 5 hours?

61. Draining a pool A drain can empty the swimming pool shown in the illustration in 3 days. A second drain can empty the pool in 2 days. How long will it take to empty the pool if both drains are used?

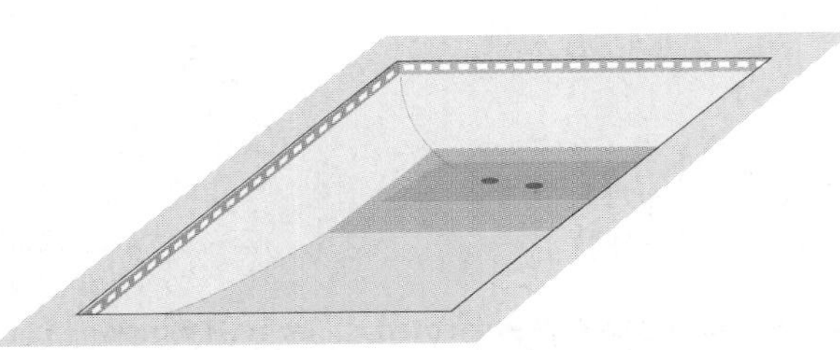

62. Filling a pool A pipe can fill a pool in 9 hours. If a second pipe is also used, the pool can be filled in 3 hours. How long would it take the second pipe alone to fill the pool?

63. Filling a pond One pipe can fill a pond in 3 weeks, and a second pipe can fill the pond in 5 weeks. However, evaporation and seepage can empty the pond in 10 weeks. If both pipes are used, how long will it take to fill the pond?

64. Housecleaning Sally can clean the house in 6 hours, and her father can clean the house in 4 hours. Sally's younger brother, Dennis, can completely mess up the house in 8 hours. If Sally and her father clean and Dennis plays, how long will it take to clean the house?

65. Walking for charity A man bicycles 5 mph faster than he can walk. He bicycles 24 miles and walks back along the same route for a charity race in 11 hours. How fast does he walk?

66. Finding rates Two trains made the same 315-mile run. Since one train traveled 10 mph faster than the other, it arrived 2 hours earlier. Find the speed of each train.

67. Train travel A train traveled 120 miles from Freeport to Chicago and returned the same distance in a total time of 5 hours. If the train traveled 20 mph slower on the return trip, how fast did the train travel in each direction?

68. Time on the road A car traveled from Rockford to Chicago in 3 hours less time than it took a second car to travel from Rockford to St. Louis. If the cars traveled at the same average speed, how long was the first driver on the road?

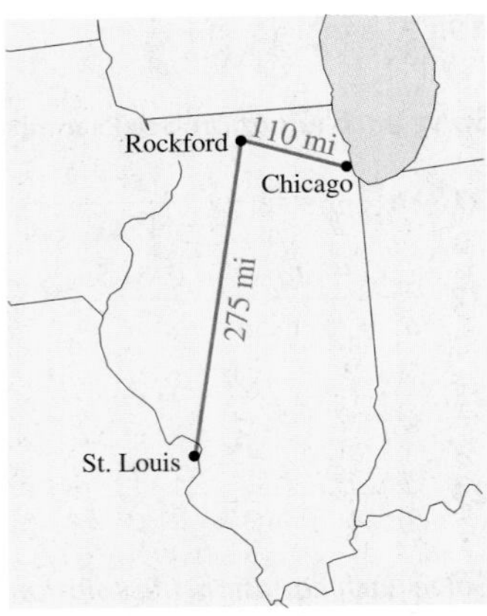

69. Boating A man can drive a motorboat 45 miles down the Rock River in the same amount of time that he can drive it 27 miles upstream. Find the speed of the current if the speed of the boat is 12 mph in still water.

70. Rowing a boat A woman who can row 3 mph in still water rows 10 miles downstream on the Eagle River and returns upstream in a total of 12 hours. Find the speed of the current.

71. Aviation A plane that can fly 340 mph in still air can fly 200 miles downwind in the same amount of time that it can fly 140 miles upwind. Find the velocity of the wind.

72. Aviation An airplane can fly 650 miles with the wind in the same amount of time as it can fly 475 miles against the wind. If the wind speed is 40 mph, find the speed of the plane in still air.

73. Discount buying A repairman purchased some washing-machine motors for a total of $224. When the unit cost decreased by $4, he was able to buy one extra motor for the same total price. How many motors did he buy originally?

74. Price increases An appliance store manager bought several microwave ovens for a total of $1,800. When her unit cost increased by $25, she was able to buy one less oven for the same total price. How many ovens did she buy originally?

75. Stretching a vacation A student saved $1,200 for a trip to Europe. By cutting $20 from her daily expenses, she was able to stay three extra days. How long had she originally planned to be gone?

76. Focal length The design of a camera lens uses the equation

$$\frac{1}{f} = \frac{1}{s_1} + \frac{1}{s_2}$$

which relates the focal length f of a lens to the image distance s_1 and the object distance s_2. Find the focal length of the lens shown in the illustration. (*Hint:* Convert feet to inches.)

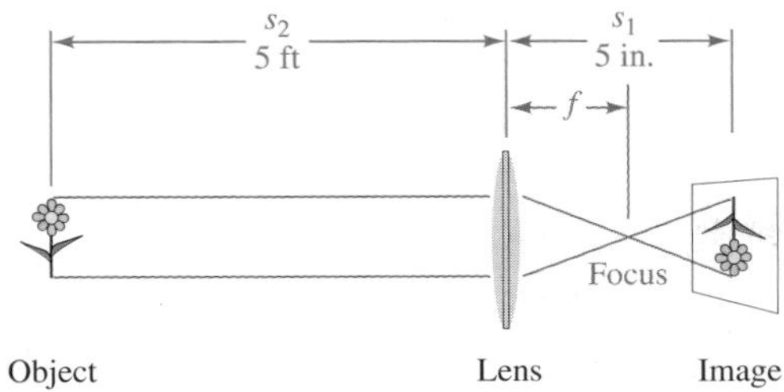

77. Lens maker's formula The focal length f of a lens is given by the lens maker's formula

$$\frac{1}{f} = 0.6\left(\frac{1}{r_1} + \frac{1}{r_2}\right)$$

where f is the focal length of the lens and r_1 and r_2 are the radii of the two circular surfaces. Find the focal length of the lens shown in the illustration.

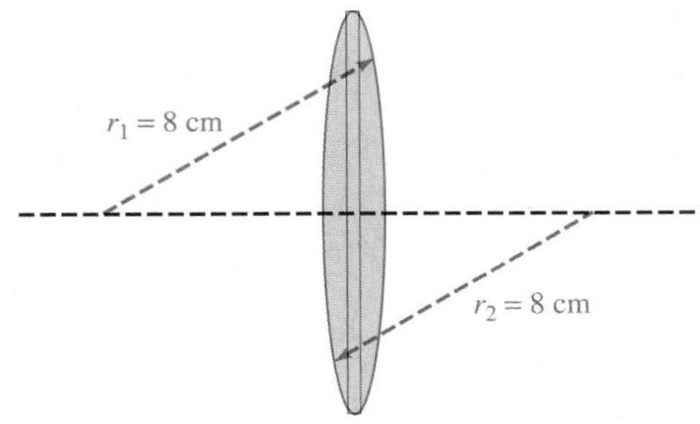

78. Engineering The stiffness of the shaft shown is given by the formula

$$k = \frac{1}{\frac{1}{k_1} + \frac{1}{k_2}}$$

where k_1 and k_2 represent the respective stiffness of each section. If the stiffness k_2 of Section 2 is 4,200,000 in. lb/rad, and the design specifications require that the stiffness k of the entire shaft be 1,900,000 in. lb/rad, what must the stiffness k_1 of Section 1 be?

WRITING ABOUT MATH

79. Why is it necessary to check the solutions of a rational equation?

80. Explain the steps you would use to solve a rational equation.

SOMETHING TO THINK ABOUT

81. Invent a rational equation that has an extraneous solution of 3.

82. Solve: $(x - 1)^{-1} - x^{-1} = 6^{-1}$.

SECTION 6.6 Dividing Polynomials

Objectives

1. Divide a monomial by a monomial.
2. Divide a polynomial by a monomial.
3. Divide a polynomial by a polynomial.
4. Divide a polynomial with one or more missing terms by a polynomial.

Vocabulary

algorithm

Getting Ready

Simplify the fraction or perform the division.

1. $\frac{16}{24}$ **2.** $-\frac{45}{81}$ **3.** $23\overline{)115}$ **4.** $32\overline{)512}$

We have previously seen how to add, subtract, and multiply polynomials. We now consider how to divide them.

1 Divide a monomial by a monomial.

In Example 1, we review how to divide a monomial by a monomial.

EXAMPLE 1 Simplify: $(3a^2b^3) \div (2a^3b) \quad (a, b \neq 0)$.

Solution We can use the rules of exponents or divide out all common factors.

Method 1

$$\frac{3a^2b^3}{2a^3b} = \frac{3}{2}a^{2-3}b^{3-1}$$
$$= \frac{3}{2}a^{-1}b^2$$
$$= \frac{3}{2}\left(\frac{1}{a}\right)\frac{b^2}{1}$$
$$= \frac{3b^2}{2a}$$

Method 2

$$\frac{3a^2b^3}{2a^3b} = \frac{3aabbb}{2aaab}$$
$$= \frac{3\cancel{a}\cancel{a}\cancel{b}bb}{2\cancel{a}\cancel{a}a\cancel{b}}$$
$$= \frac{3b^2}{2a}$$

SELF CHECK 1 Simplify: $\frac{6x^3y^2}{8x^2y^3} \quad (x, y \neq 0)$.

2 Divide a polynomial by a monomial

In Example 2, we use the fact that $\frac{a+b}{c} = \frac{a}{c} + \frac{b}{c} \quad (c \neq 0)$ to divide a polynomial by a monomial.

EXAMPLE 2 Divide: $(4x^3y^2 + 3xy^5 - 12xy)$ by $3x^2y^3 \quad (x, y \neq 0)$.

Solution We can write the division in the form $\frac{4x^3y^2 + 3xy^5 - 12xy}{3x^2y^3}$ and proceed as follows:

$$\frac{4x^3y^2 + 3xy^5 - 12xy}{3x^2y^3} = \frac{4x^3y^2}{3x^2y^3} + \frac{3xy^5}{3x^2y^3} - \frac{12xy}{3x^2y^3}$$ Write the fraction as three separate fractions.

$$= \frac{4x}{3y} + \frac{y^2}{x} - \frac{4}{xy^2}$$ Simplify each fraction.

SELF CHECK 2 Simplify: $\dfrac{8a^3b^4 - 4a^4b^2 + a^2b^2}{4a^2b^2}$ $(a, b \neq 0)$.

3 Divide a polynomial by a polynomial.

There is an **algorithm** (a repeating series of steps) to use when the divisor is not a monomial. It is very similar to the division algorithm we used in arithmetic. To use the division algorithm to divide $x^2 + 7x + 12$ by $x + 4$, we write the division in long division form and proceed as follows:

$$\begin{array}{r} x \\ x + 4\overline{)x^2 + 7x + 12} \end{array}$$

How many times does x divide x^2? $\frac{x^2}{x} = x$. Place the x in the quotient.

$$\begin{array}{r} x \\ x + 4\overline{)x^2 + 7x + 12} \\ \underline{x^2 + 4x} \\ 3x + 12 \end{array}$$

Multiply each term in the divisor by x to get $x^2 + 4x$, subtract $x^2 + 4x$ from $x^2 + 7x$, and bring down the 12.

$$\begin{array}{r} x + 3 \\ x + 4\overline{)x^2 + 7x + 12} \\ \underline{x^2 + 4x} \\ 3x + 12 \end{array}$$

How many times does x divide $3x$? $\frac{3x}{x}$. Place the $+3$ in the quotient.

$$\begin{array}{r} x + 3 \\ x + 4\overline{)x^2 + 7x + 12} \\ \underline{x^2 + 4x} \\ 3x + 12 \\ \underline{3x + 12} \\ 0 \end{array}$$

Multiply each term in the divisor by 3 to get $3x + 12$, and subtract $3x + 12$ from $3x + 12$ to get 0.

The division process stops when the result of the subtraction is a constant or a polynomial with degree less than the degree of the divisor. Here, the quotient is $x + 3$ and the remainder is 0.

We can check the quotient by multiplying the divisor by the quotient. The product should be the dividend.

$$\underbrace{(x + 4)}_{\text{Divisor}} \cdot \underbrace{(x + 3)}_{\text{quotient}} + \underbrace{0}_{\text{remainder}} = \underbrace{x^2 + 7x + 12}_{\text{dividend}}$$

The quotient checks.

EXAMPLE 3 Divide $(2a^3 + 9a^2 + 5a - 6)$ by $(2a + 3)$ $\left(a \neq -\frac{3}{2}\right)$.

Solution

$$\begin{array}{r} a^2 \\ 2a + 3\overline{)2a^3 + 9a^2 + 5a - 6} \end{array}$$

How many times does $2a$ divide $2a^3$? $\frac{2a^3}{2a} = a^2$. Place a^2 in the quotient.

$$\begin{array}{r} a^2 \\ 2a + 3\overline{)2a^3 + 9a^2 + 5a - 6} \\ \underline{2a^3 + 3a^2} \\ 6a^2 + 5a \end{array}$$

Multiply each term in the divisor by a^2 to get $2a^3 + 3a^2$, subtract $2a^3 + 3a^2$ from $2a^3 + 9a^2$, and bring down the $5a$.

$$\begin{array}{r l} & a^2 + 3a \\ 2a + 3 \overline{)\,2a^3 + 9a^2 + 5a - 6} & \\ \underline{2a^3 + 3a^2} & \\ 6a^2 + 5a & \end{array}$$

How many times does $2a$ divide $6a^2$? $\frac{6a^2}{2a} = 3a$. Place the $+3a$ in the quotient.

$$\begin{array}{r} a^2 + 3a \\ 2a + 3 \overline{)\,2a^3 + 9a^2 + 5a - 6} \\ \underline{2a^3 + 3a^2} \\ 6a^2 + 5a \\ \underline{6a^2 + 9a} \\ -4a - 6 \end{array}$$

Multiply each term in the divisor by $3a$ to get $6a^2 + 9a$, subtract $6a^2 + 9a$ from $6a^2 + 5a$, and bring down -6.

$$\begin{array}{r} a^2 + 3a - 2 \\ 2a + 3 \overline{)\,2a^3 + 9a^2 + 5a - 6} \\ \underline{2a^3 + 3a^2} \\ 6a^2 + 5a \\ \underline{6a^2 + 9a} \\ -4a - 6 \end{array}$$

How many times does $2a$ divide $-4a$? $-\frac{4a}{2a} = -2$. Place the -2 in the quotient.

$$\begin{array}{r} a^2 + 3a - 2 \\ 2a + 3 \overline{)\,2a^3 + 9a^2 + 5a - 6} \\ \underline{2a^3 + 3a^2} \\ 6a^2 + 5a \\ \underline{6a^2 + 9a} \\ -4a - 6 \\ \underline{-4a - 6} \\ 0 \end{array}$$

Multiply each term in the divisor by -2 to get $-4a - 6$, and subtract $-4a - 6$ from $-4a - 6$ to get 0.

Since the remainder is 0, the quotient is $a^2 + 3a - 2$. We can check the quotient by verifying that

$$\overbrace{(2a + 3)}^{\text{Divisor}} \cdot \overbrace{(a^2 + 3a - 2)}^{\text{quotient}} + \overbrace{0}^{\text{remainder}} = \overbrace{2a^3 + 9a^2 + 5a - 6}^{\text{dividend}}$$

SELF CHECK 3 Divide: $2p + 1\overline{)\,2p^3 - 5p^2 - p + 1}$ $\left(p \neq -\frac{1}{2}\right)$.

EXAMPLE 4 Divide: $\dfrac{3x^3 + 2x^2 - 3x + 8}{x - 2}$ $(x \neq 2)$.

Solution

$$\begin{array}{r} 3x^2 + 8x + 13 \\ x - 2 \overline{)\,3x^3 + 2x^2 - 3x + 8} \\ \underline{3x^3 - 6x^2} \\ 8x^2 - 3x \\ \underline{8x^2 - 16x} \\ 13x + 8 \\ \underline{13x - 26} \\ 34 \end{array}$$

This division gives a quotient of $3x^2 + 8x + 13$ and a remainder of 34. It is common to form a fraction with the remainder as the numerator and the divisor as the denominator and to write the result as

$$3x^2 + 8x + 13 + \frac{34}{x - 2}$$

To check, we verify that

$$\underbrace{(x-2)}_{\text{Divisor}} \cdot \underbrace{(3x^2 + 8x + 13)}_{\text{quotient}} + \underbrace{34}_{\text{remainder}} = \underbrace{3x^3 + 2x^2 - 3x + 8}_{\text{dividend}}$$

SELF CHECK 4 Divide: $a - 3\overline{)2a^3 + 3a^2 - a + 2}$ $(a \neq 3)$.

EXAMPLE 5 Divide: $\dfrac{-9x + 8x^3 + 10x^2 - 9}{3 + 2x}$ $\left(x \neq -\dfrac{3}{2}\right)$.

Solution The division algorithm works best when the polynomials in the dividend and the divisor are written in descending powers of x. We can use the commutative property of addition to rearrange the terms. Then the division is routine:

$$\begin{array}{r} 4x^2 - x - 3 \\ 2x + 3\overline{)8x^3 + 10x^2 - 9x - 9} \\ \underline{8x^3 + 12x^2} \\ -2x^2 - 9x \\ \underline{-2x^2 - 3x} \\ -6x - 9 \\ \underline{-6x - 9} \\ 0 \end{array}$$

Thus,

$$\frac{-9x + 8x^3 + 10x^2 - 9}{3 + 2x} = 4x^2 - x - 3$$

SELF CHECK 5 Divide: $2 + 3a\overline{)-4a + 15a^2 + 18a^3 - 4}$ $\left(a \neq -\frac{2}{3}\right)$.

4 Divide a polynomial with one or more missing terms by a polynomial.

Sometimes the divisor or the dividend will have a missing term.

EXAMPLE 6 Divide: $\dfrac{8x^3 + 1}{2x + 1}$ $\left(x \neq -\dfrac{1}{2}\right)$.

Solution When we write the terms in the dividend in descending powers of x, we see that the terms involving x^2 and x are missing in the numerator. We must include the terms $0x^2$ and $0x$ in the dividend or leave spaces for them. Then the division is routine.

$$\begin{array}{r} 4x^2 - 2x + 1 \\ 2x + 1\overline{)8x^3 + 0x^2 + 0x + 1} \\ \underline{8x^3 + 4x^2} \qquad\qquad \\ -4x^2 + 0x \qquad \\ \underline{-4x^2 - 2x} \qquad \\ 2x + 1 \\ \underline{2x + 1} \\ 0 \end{array}$$

Thus,

$$\frac{8x^3 + 1}{2x + 1} = 4x^2 - 2x + 1$$

SELF CHECK 6 Divide: $3a - 1\overline{)27a^3 - 1}$ $\left(a \neq \frac{1}{3}\right)$.

EXAMPLE 7 Divide: $\dfrac{-17x^2 + 5x + x^4 + 2}{x^2 - 1 + 4x}$. Assume the denominator is not 0.

Solution We write the problem with the divisor and the dividend in descending powers of x. After leaving space for the missing term in the dividend, we proceed as follows:

$$\begin{array}{r} x^2 - 4x \\ x^2 + 4x - 1\overline{)x^4 \qquad\quad - 17x^2 + 5x + 2} \\ \underline{x^4 + 4x^3 - \quad x^2} \qquad\qquad\quad \\ -4x^3 - 16x^2 + 5x \qquad \\ \underline{-4x^3 - 16x^2 + 4x} \qquad \\ x + 2 \end{array}$$

This division gives a quotient of $x^2 - 4x$ and a remainder of $x + 2$.

$$\frac{-17x^2 + 5x + x^4 + 2}{x^2 - 1 + 4x} = x^2 - 4x + \frac{x + 2}{x^2 + 4x - 1}$$

SELF CHECK 7 Divide: $\dfrac{2a^2 + 3a^3 + a^4 - 7 + a}{a^2 - 2a + 1}$ $(a \neq 1)$.

SELF CHECK ANSWERS **1.** $\frac{3x}{4y}$ **2.** $2ab^2 - a^2 + \frac{1}{4}$ **3.** $p^2 - 3p + 1$ **4.** $2a^2 + 9a + 26 + \frac{80}{a - 3}$ **5.** $6a^2 + a - 2$ **6.** $9a^2 + 3a + 1$ **7.** $a^2 + 5a + 11 + \frac{18a - 18}{a^2 - 2a + 1}$

NOW TRY THIS

1. Divide: $\frac{10x^3 + 25x^2 + 5x}{5x}$.

2. The product of three binomials with integer coefficients is $12x^3 - 32x^2 - 19x + 60$. If one of the binomials is $2x - 3$, find the other two.

3. Divide: $\frac{x^4 - 3x^3 + 2x^2 - 7}{x^2 - 2}$.

6.6 EXERCISES

WARM-UPS *Divide.*

1. $\frac{6x^2y^2}{2xy}$

2. $\frac{4ab^2 + 8a^2b}{2ab}$

3. $\frac{x^2 + 2x + 1}{x + 1}$

4. $\frac{x^2 - 4}{x - 2}$

REVIEW *Remove parentheses and simplify.*

5. $2(x^2 + 4x - 1) + 3(2x^2 - 2x + 2)$
6. $3(2a^2 - 3a + 2) - 4(2a^2 + 4a - 7)$
7. $-2(3y^3 - 2y + 7) - 3(y^2 + 2y - 4) + 4(y^3 + 2y - 1)$
8. $3(4y^3 + 3y - 2) + 2(3y^2 - y + 3) - 5(2y^3 - y^2 - 2)$

VOCABULARY AND CONCEPTS *Fill in the blanks.*

9. $\frac{a}{b} = \_\_ \cdot a \quad (b \neq 0)$
10. The division __________ is a repeating series of steps used to do a long division.
11. To check a division problem, the divisor · ________ + remainder = dividend.
12. If a polynomial is divided by $3a - 2$ and the quotient is $3a^2 + 5$ with a remainder of 6, we usually write the result as $3a^2 + 5 +$ _____.

GUIDED PRACTICE

Perform each division. Write each answer without using negative exponents and assume no denominator is 0. See Example 1. (Objective 1)

13. $(4x^2y^3) \div (8x^5y^2)$

14. $(25x^4y^7) \div (5xy^9)$

15. $\frac{33a^{-2}b^2}{44a^2b^{-2}}$

16. $\frac{-63a^4b^{-3}}{81a^{-3}b^3}$

17. $\frac{45x^{-2}y^{-3}t^0}{-63x^{-1}y^4t^2}$

18. $\frac{112a^0b^2c^{-3}}{48a^4b^0c^4}$

19. $\frac{-65a^{2n}b^nc^{3n}}{-15a^nb^{-n}c}$

20. $\frac{-32x^{-3n}y^{-2n}z}{40x^{-2}y^{-n}z^{n+1}}$

Perform each division. Write each answer without using negative exponents and assume no denominator is 0. See Example 2. (Objective 2)

21. $(4x^2 - x^3) \div (6x)$

22. $(5y^4 + 45y^3) \div (15y^2)$

23. $\frac{4x^2y^3 + x^3y^2}{6xy}$

24. $\frac{3a^3y^2 - 18a^4y^3}{27a^2y^2}$

25. $\frac{28x^2y^3 + 4xy - 20x^3y^2}{4xy}$

26. $\frac{45a^5b^3 - 63a^4b^4 - 9a^2b^3}{-9a^2b^3}$

27. $\frac{24x^6y^7 - 12x^5y^{12} + 36xy}{48x^2y^3}$

28. $\frac{9x^4y^3 + 18x^2y - 27xy^4}{9x^3y^3}$

Divide. Assume no divisor is 0. See Example 3. (Objective 3)

29. $(x^2 + 5x + 6) \div (x + 3)$

30. $(x^2 - 5x + 6) \div (x - 3)$

31. $x + 3\overline{)x^2 + 10x + 21}$

32. $x + 7\overline{)x^2 + 10x + 21}$

33. $\frac{6x^2 - x - 12}{2x - 3}$

34. $\frac{6x^2 - 7x + 2}{3x - 2}$

35. $\dfrac{6x^3 - x^2 - 6x - 9}{2x - 3}$

36. $\dfrac{16x^3 + 16x^2 - 9x - 5}{4x + 5}$

Divide. Assume no divisor is 0. See Example 4. (Objective 3)

37. $\dfrac{6x^2 - x - 12}{2x + 3}$

38. $\dfrac{4a^3 + a^2 - 3a + 7}{a + 1}$

39. $\dfrac{3x^3 - 2x^2 + x + 6}{x - 1}$

40. $\dfrac{6x^3 + 11x^2 - x - 2}{3x - 2}$

41. $\dfrac{6x^3 + 11x^2 - x + 10}{2x + 3}$

42. $\dfrac{4x^3 - 12x^2 + 17x - 12}{2x - 3}$

43. $\dfrac{x^4 + 2x^3 + 4x^2 + 3x + 2}{x^2 + x + 2}$

44. $\dfrac{2x^4 + 3x^3 + 3x^2 - 5x - 3}{2x^2 - x - 1}$

Divide. Assume no divisor is 0. See Example 5. (Objective 3)

45. $\dfrac{2a + 1 + a^2}{a + 1}$

46. $\dfrac{a - 15 + 6a^2}{2a - 3}$

47. $\dfrac{6y - 4 + 10y^2}{5y - 2}$

48. $\dfrac{-18x + 12 + 6x^2}{x - 1}$

49. $\dfrac{-9x^2 + 8x + 9x^3 - 4}{3x - 2}$

50. $\dfrac{6x^2 + 8x^3 - 13x + 3}{4x - 3}$

51. $\dfrac{x^3 + 3x + 5x^2 + 6 + x^4}{x^2 + 3}$

52. $\dfrac{13x + 16x^4 + 3x^2 + 3}{4x + 3}$

Divide. Assume no divisor is 0. See Examples 6–7. (Objective 4)

53. $\dfrac{a^3 + 1}{a - 1}$

54. $\dfrac{27a^3 - 8b^3}{3a - 2b}$

55. $x^2 - 2\overline{)x^6 - x^4 + 2x^2 - 8}$

56. $x^2 + 3\overline{)x^6 + 2x^4 - 6x^2 - 9}$

ADDITIONAL PRACTICE

Divide. Assume no divisor is 0.

57. $\dfrac{x^ny^n - 3x^{2n}y^{2n} + 6x^{3n}y^{3n}}{x^ny^n}$

58. $\dfrac{2a^n - 3a^nb^{2n} - 6b^{4n}}{a^nb^{n-1}}$

59. $\dfrac{3a^{-2}b^3 - 6a^2b^{-3} + 9a^{-2}}{12a^{-1}b}$

60. $\dfrac{4x^3y^{-2} + 8x^{-2}y^2 - 12y^4}{12x^{-1}y^{-1}}$

61. $\dfrac{3x^2 + 9x^3 + 4x + 4}{3x + 2}$

62. $\dfrac{27x + 23x^2 + 6x^3}{2x + 3}$

63. $\dfrac{15a^3 - 29a^2 + 16}{3a - 4}$

64. $\dfrac{x^5 + 3x + 2}{x^3 + 1 + 2x}$

65. $\dfrac{-10xy + x^2 + 16y^2}{x - 2y}$

66. $y - 2\overline{)-24y + 24 + 6y^2}$

67. $3 - a\overline{)21a - a^2 - 54}$

68. $2x + y\overline{)32x^5 + y^5}$

69. $3x - y\overline{)81x^4 - y^4}$

Use a calculator to help find each quotient.

70. $x - 2\overline{)9.8x^2 - 3.2x - 69.3}$

71. $2.5x - 3.7\overline{)-22.25x^2 - 38.9x - 16.65}$

APPLICATIONS

72. **Geometry** Find an expression for the length of the longer sides of the rectangle.

$3x + 2$ | Area = $9x^2 + 21x + 10$

73. **Geometry** Find an expression for the height of the triangle.

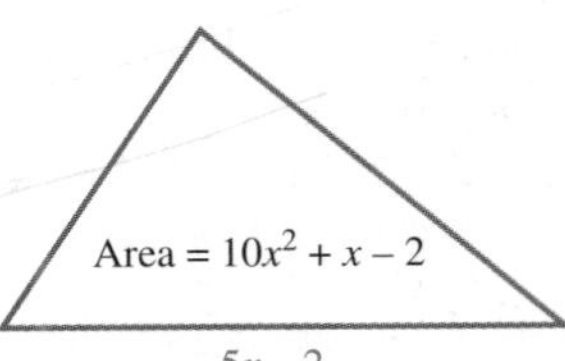

74. **Winter travel** Complete the following table, which lists the rate (mph), time traveled (hr), and distance traveled (mi) by an Alaskan trail guide using two different means of travel.

	r ·	t =	d
Dog sled		$4x + 7$	$12x^2 + 13x - 14$
Snowshoes	$3x + 4$		$3x^2 + 19x + 20$

75. Pricing Complete the table for two items sold at a produce store.

	Price per lb	·	Number of lb	=	Value
Cashews	$x^2 + 2x + 4$				$x^4 + 4x^2 + 16$
Sunflower seeds			$x^2 + 6$		$x^4 - x^2 - 42$

WRITING ABOUT MATH

76. Explain how to divide a monomial by a monomial.

77. Explain how to check the result of a division problem.

SOMETHING TO THINK ABOUT

78. Since 6 is a factor of 24, 6 divides 24 with no remainder. Decide whether $2x - 3$ is a factor of $10x^2 - x - 21$.

79. Is $x - 1$ a factor of $x^5 - 1$?

SECTION 6.7 Synthetic Division

Objectives

1. Divide a polynomial by a binomial of the form $x - r$ using synthetic division.
2. Apply the remainder theorem to evaluate a polynomial.
3. Apply the factor theorem to determine whether a specific value is a zero of a polynomial.

Vocabulary

synthetic division   remainder theorem   factor theorem

Getting Ready

Divide each polynomial $P(x)$ by $x - 2$ and find $P(2)$.

1. $P(x) = x^2 - x - 1$

2. $P(x) = x^2 + x + 3$

1 Divide a polynomial by a binomial of the form $x - r$ using synthetic division.

There is a method, called **synthetic division,** that we can use to divide a polynomial by a binomial of the form $x - r$. To see how it works, we consider the division of $4x^3 - 5x^2 - 11x + 20$ by $x - 2$.

$$
\begin{array}{r|l}
 & 4x^2 + 3x - 5 \\ \hline
x - 2 & 4x^3 - 5x^2 - 11x + 20 \\
 & \underline{4x^3 - 8x^2} \\
 & \quad 3x^2 - 11x \\
 & \quad \underline{3x^2 - 6x} \\
 & \qquad - 5x + 20 \\
 & \qquad \underline{- 5x + 10} \\
 & \qquad\qquad 10 \quad \text{(remainder)}
\end{array}
\qquad
\begin{array}{r|l}
 & 4 \quad 3 - 5 \\ \hline
1 - 2 & 4 - 5 - 11 \quad 20 \\
 & \underline{4 - 8} \\
 & \quad 3 - 11 \\
 & \quad \underline{3 - 6} \\
 & \qquad - 5 \quad 20 \\
 & \qquad \underline{-5 \quad 10} \\
 & \qquad\qquad 10 \quad \text{(remainder)}
\end{array}
$$

On the left is the long division, and on the right is the same division with the variables and their exponents removed and the numbers in the quotient moved to the left. The various powers of x can be remembered without actually writing them, because the exponents of the terms in the divisor, dividend, and quotient were written in descending order.

We can further shorten the version on the right. The numbers printed in color need not be written, because they are duplicates of the numbers above them. Thus, we can write the division in the following form:

$$
\begin{array}{r|l}
 & 4 \quad 3 - 5 \\ \hline
1 - 2 & 4 - 5 - 11 \quad 20 \\
 & \quad \underline{- 8} \\
 & \quad\; 3 \\
 & \qquad \underline{- 6} \\
 & \qquad - 5 \\
 & \qquad\qquad \underline{10} \\
 & \qquad\qquad 10
\end{array}
$$

We can shorten the process further by compressing the work vertically and eliminating the 1 (the coefficient of x in the divisor):

$$
\begin{array}{r|rrrr}
 & 4 & 3 & -5 & \\ \hline
-2 & 4 & -5 & -11 & 20 \\
 & & -8 & -6 & 10 \\ \cline{2-5}
 & & 3 & -5 & 10
\end{array}
$$

If we write the 4 in the quotient on the bottom line, the bottom line gives the coefficients of the quotient and the remainder. If we eliminate the top line, the division appears as follows:

$$
\begin{array}{r|rrrr}
-2 & 4 & -5 & -11 & 20 \\
 & & -8 & -6 & 10 \\ \hline
 & 4 & 3 & -5 & 10
\end{array}
$$

The bottom line was obtained by subtracting the middle line from the top line. If we replace the -2 in the divisor by $+2$, the division process will reverse the signs of every entry in the middle line, and then the bottom line can be obtained by addition. This gives the final form of the synthetic division.

$$
\begin{array}{r|rrrrl}
+2 & 4 & -5 & -11 & 20 & \text{The coefficients of the dividend.} \\
 & & 8 & 6 & -10 & \\ \hline
 & 4 & 3 & -5 & 10 & \text{The coefficients of the quotient and the remainder.}
\end{array}
$$

Thus,

$$\frac{4x^3 - 5x^2 - 11x + 20}{x - 2} = 4x^2 + 3x - 5 + \frac{10}{x - 2}.$$

EXAMPLE 1 Use synthetic division to divide $(6x^2 + 5x - 2)$ by $(x - 5)$ $(x \neq 5)$.

Solution We write the coefficients in the dividend and the 5 in the divisor in the following form:

$$\begin{array}{r|rrr} 5 & 6 & 5 & -2 \\ & & & \\ \hline \end{array}$$

Then we follow these steps:

$$\begin{array}{r|rrr} 5 & 6 & 5 & -2 \\ & & & \\ \hline & 6 & & \end{array}$$ Begin by bringing down the 6.

$$\begin{array}{r|rrr} 5 & 6 & 5 & -2 \\ & & 30 & \\ \hline & 6 & & \end{array}$$ Multiply 5 by 6 to get 30.

$$\begin{array}{r|rrr} 5 & 6 & 5 & -2 \\ & & 30 & \\ \hline & 6 & 35 & \end{array}$$ Add 5 and 30 to get 35.

$$\begin{array}{r|rrr} 5 & 6 & 5 & -2 \\ & & 30 & 175 \\ \hline & 6 & 35 & \end{array}$$ Multiply 35 by 5 to get 175.

$$\begin{array}{r|rrr} 5 & 6 & 5 & -2 \\ & & 30 & 175 \\ \hline & 6 & 35 & 173 \end{array}$$ Add −2 and 175 to get 173.

The numbers 6 and 35 represent the quotient $6x + 35$, and 173 is the remainder. Thus,

$$\frac{6x^2 + 5x - 2}{x - 5} = 6x + 35 + \frac{173}{x - 5}$$

SELF CHECK 1 Use synthetic division to divide $(6x^2 - 5x + 2)$ by $(x - 5)$ $(x \neq 5)$.

EXAMPLE 2 Use synthetic division to divide $(5x^3 + x^2 - 3)$ by $(x - 2)$ $(x \neq 2)$.

Solution We begin by writing

$$\begin{array}{r|rrrr} 2 & 5 & 1 & 0 & -3 \end{array}$$ Write 0 for the coefficient of x, the missing term.

and complete the division as follows:

$$\begin{array}{r|rrrr} 2 & 5 & 1 & 0 & -3 \\ & & 10 & & \\ \hline & 5 & 11 & & \end{array} \quad \begin{array}{r|rrrr} 2 & 5 & 1 & 0 & -3 \\ & & 10 & 22 & \\ \hline & 5 & 11 & 22 & \end{array} \quad \begin{array}{r|rrrr} 2 & 5 & 1 & 0 & -3 \\ & & 10 & 22 & 44 \\ \hline & 5 & 11 & 22 & 41 \end{array}$$

Thus,

$$\frac{5x^3 + x^2 - 3}{x - 2} = 5x^2 + 11x + 22 + \frac{41}{x - 2}$$

SELF CHECK 2 Use synthetic division to divide $(5x^3 - x^2 + 3)$ by $(x - 2)$ $(x \neq 2)$.

EXAMPLE 3 Use synthetic division to divide $(5x^2 + 6x^3 + 2 - 4x)$ by $(x + 2)$ $(x \neq -2)$.

Solution First, we write the dividend with the exponents in descending order.

$$6x^3 + 5x^2 - 4x + 2$$

Then we write the divisor in $x - r$ form: $x - (-2)$. Using synthetic division, we begin by writing

$$\begin{array}{r|rrrr} -2 & 6 & 5 & -4 & 2 \\ & & & & \\ \hline \end{array}$$

and complete the division.

$$\begin{array}{r|rrrr} -2 & 6 & 5 & -4 & 2 \\ & & -12 & 14 & -20 \\ \hline & 6 & -7 & 10 & -18 \end{array}$$

Thus,

$$\frac{5x^2 + 6x^3 + 2 - 4x}{x + 2} = 6x^2 - 7x + 10 + \frac{-18}{x + 2}$$

SELF CHECK 3 Divide $2x - 4x^2 + 3x^3 - 3$ by $x - 1$ $(x \neq 1)$.

2 Apply the remainder theorem to evaluate a polynomial.

Synthetic division is important because of the **remainder theorem.**

Remainder Theorem If a polynomial $P(x)$ is divided by $x - r$, the remainder is $P(r)$.

We illustrate the remainder theorem in the next example.

EXAMPLE 4 Let $P(x) = 2x^3 - 3x^2 - 2x + 1$. Find
a. $P(3)$ **b.** the remainder when $P(x)$ is divided by $x - 3$.

Solution **a.** $P(3) = 2(3)^2 - 3(3)^3 - 2(3) + 1$ Substitute 3 for x.
$= 2(27) - 3(9) - 6 + 1$
$= 54 - 27 - 6 + 1$
$= \mathbf{22}$

b. We can use the following synthetic division to find the remainder when $P(x) = 2x^3 - 3x^2 - 2x + 1$ is divided by $x - 3$.

$$\begin{array}{r|rrrr} 3 & 2 & -3 & -2 & 1 \\ & & 6 & 9 & 21 \\ \hline & 2 & 3 & 7 & \mathbf{22} \end{array}$$ The remainder is 22.

The results of parts **a** and **b** show that when $P(x)$ is divided by $x - 3$, the remainder is $P(3)$.

SELF CHECK 4 Use the polynomial of Example 4 and find:
a. $P(2)$ **b.** the remainder when the polynomial is divided by $x - 2$.

COMMENT It is often easier to find $P(r)$ by using synthetic division than by substituting r for x in $P(x)$. This is especially true if r is a fraction or decimal.

3 Apply the factor theorem to determine whether a specific value is a zero of a polynomial.

Recall that if two quantities are multiplied, each is called a *factor* of the product. Thus, $x - 2$ is one factor of $6x - 12$, because $6(x - 2) = 6x - 12$. A theorem, called the **factor theorem,** enables us to find one factor of a polynomial if the remainder of a certain division is 0.

Factor Theorem

If $P(x)$ is a polynomial in x, then

$$P(r) = 0 \quad \text{if and only if} \quad x - r \text{ is a factor of } P(x)$$

If $P(x)$ is a polynomial in x and if $P(r) = 0$, r is called a *zero of the polynomial.*

EXAMPLE 5 Let $P(x) = 3x^3 - 5x^2 + 3x - 10$. Show that:
a. $P(2) = 0$ **b.** $x - 2$ is a factor of $P(x)$.

Solution **a.** Use the remainder theorem to evaluate $P(2)$ by dividing $P(x)$ by $(x - 2)$.

$$\begin{array}{r|rrrr} 2 & 3 & -5 & 3 & -10 \\ & & 6 & 2 & 10 \\ \hline & 3 & 1 & 5 & 0 \end{array}$$

The remainder in this division is 0. By the remainder theorem, the remainder is $P(2)$. Thus, $P(2) = 0$, and 2 is a zero of the polynomial.

b. Because the remainder is 0, the numbers 3, 1, and 5 in the synthetic division in part **a** represent the quotient $3x^2 + x + 5$. Thus,

$$\underbrace{(x - 2)}_{\text{Divisor}} \cdot \underbrace{(3x^2 + x + 5)}_{\text{quotient}} + \underbrace{0}_{\text{remainder}} = \underbrace{3x^3 - 5x^2 + 3x - 10}_{\text{the dividend, } P(x)}$$

or

$$(x - 2)(3x^2 + x + 5) = 3x^3 - 5x^2 + 3x - 10$$

Thus, $x - 2$ is a factor of $P(x)$.

SELF CHECK 5 Use the polynomial $P(x) = x^2 - 5x + 6$ and show that $P(2) = 0$ and that $x - 2$ is a factor of $P(x)$.

The result in Example 5 is true, because the remainder, $P(2)$, is 0. If the remainder had not been 0, then $x - 2$ would not have been a factor of $P(x)$.

ACCENT ON TECHNOLOGY

Approximating Zeros of Polynomials

We can use a graphing calculator to approximate the real zeros of a polynomial function $f(x)$. For example, to find the real zeros of $f(x) = 2x^3 - 6x^2 + 7x - 21$, we graph the function as in Figure 6-10.

It is clear from the figure that the function f has a zero at $x = 3$.

$$
\begin{aligned}
f(3) &= 2(3)^3 - 6(3)^2 + 7(3) - 21 \quad \text{Substitute 3 for } x. \\
&= 2(27) - 6(9) + 21 - 21 \\
&= 0
\end{aligned}
$$

From the factor theorem, we know that $x - 3$ is a factor of the polynomial. To find the other factor, we can synthetically divide by 3.

$$
\begin{array}{r|rrrr}
3 & 2 & -6 & 7 & -21 \\
  &   & 6 & 0 & 21 \\
\hline
  & 2 & 0 & 7 & 0
\end{array}
$$

Thus, $f(x) = (x - 3)(2x^2 + 7)$. Since $2x^2 + 7$ cannot be factored over the real numbers, we can conclude that 3 is the only real zero of the polynomial function.

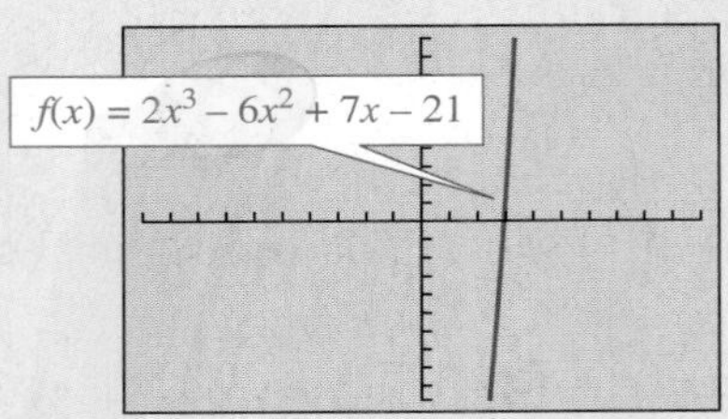

Figure 6-10

SELF CHECK ANSWERS

1. $6x + 25 + \frac{127}{x - 5}$ **2.** $5x^2 + 9x + 18 + \frac{39}{x - 2}$ **3.** $3x^2 - x + 1 + \frac{-2}{x - 1}$ **4.** both are 1 **5.** $(x - 2)(x - 3) = x^2 - 5x + 6$

NOW TRY THIS

1. Given that $x + 2$ is a factor of $2x^3 - 3x^2 - 11x + 6$, find the remaining factors.
2. Solve $2x^3 - 3x^2 - 11x + 6 = 0$ without a calculator.
3. Find the x-intercepts for $f(x) = 2x^3 - 3x^2 - 11x + 6$.

6.7 EXERCISES

Assume no division by zero.

WARM-UPS

Find the remainder in each division.

1. $(x^2 + 2x + 1) \div (x - 2)$

2. $(x^2 - 4) \div (x + 1)$

Determine whether $x - 2$ is a factor of each polynomial.

3. $x^3 - 2x^2 + x - 2$

4. $x^3 + 4x^2 - 1$

REVIEW

Let $f(x) = 3x^2 + 2x - 1$ and find each value.

5. $f(1)$
6. $f(-2)$
7. $f(2a)$
8. $f(-t)$

Remove parentheses and simplify.

9. $2(x^2 + 4x - 1) + 3(2x^2 - 2x + 2)$
10. $-2(3y^3 - 2y + 7) - 3(y^2 + 2y - 4) + 4(y^3 + 2y - 1)$

VOCABULARY AND CONCEPTS *Fill in the blanks.*

11. In order to use synthetic division, the divisor must be in the form ______.
12. If the power of the leading term of the dividend is x^5 and the divisor is $x - r$, the power of the leading term of the quotient will be ___.
13. The remainder theorem states that if a polynomial $P(x)$ is divided by $x - r$, the remainder is _____.
14. The factor theorem states that if $P(x)$ is a polynomial in x, then $P(r) = 0$ if and only if ______ is a factor of $P(x)$.

GUIDED PRACTICE

Use synthetic division to perform each division. See Example 1. (Objective 1)

15. $(x^2 - 7x + 12) \div (x - 4)$
16. $(x^2 - 6x + 5) \div (x - 5)$
17. $x - 1\overline{)x^2 + x - 2}$
18. $x - 2\overline{)x^2 + x - 6}$
19. $(3x^3 - 10x^2 + 5x - 6) \div (x - 3)$
20. $(2x^3 - 9x^2 + 10x - 3) \div (x - 3)$
21. $(x^2 - 5x + 14) \div (x + 2)$
22. $(5x^2 - 8x - 2) \div (x - 2)$

Use synthetic division to perform each division. See Examples 2–3. (Objective 1)

23. $(x^2 + 8 + 6x) \div (x + 4)$
24. $(x^2 - 15 - 2x) \div (x + 3)$
25. $(5x^2 + 6x^3 + 4) \div (x + 1)$
26. $(2x^3 - 5x - 6) \div (x - 2)$
27. $(-6x^2 + 4x^3 + 2 - 5x) \div (x - 2)$
28. $(4x^2 + 3x^3 + 8) \div (x + 2)$

Let $P(x) = 2x^3 - 4x^2 + 2x - 1$. Evaluate $P(x)$ for the given value by using the remainder theorem. Then evaluate by substituting the value of x into the polynomial and simplifying. See Example 4. (Objective 2)

29. $P(1)$
30. $P(2)$
31. $P(-2)$
32. $P(-1)$
33. $P(3)$
34. $P(-4)$
35. $P(0)$
36. $P(4)$

Use the remainder theorem and synthetic division to find P(r). (Objective 2)

37. $P(x) = x^3 - 4x^2 + x - 2; r = 2$
38. $P(x) = x^3 - 3x^2 + x + 1; r = 1$
39. $P(x) = x^4 - 2x^3 + x^2 - 3x + 2; r = -2$
40. $P(x) = 2x^3 + x + 2; r = 3$
41. $P(x) = x^3 + x^2 + 1; r = -2$
42. $P(x) = x^5 + 3x^4 - x^2 + 1; r = -1$
43. $P(x) = 2x^3 - x^2 + 4x - 5; r = \frac{1}{2}$
44. $P(x) = 3x^3 + x^2 - 8x + 4; r = \frac{2}{3}$

Use the factor theorem and determine whether the first expression is a factor of P(x). See Example 5. (Objective 3)

45. $x - 3; P(x) = x^3 - 3x^2 + 5x - 15$
46. $x + 1; P(x) = x^3 + 2x^2 - 2x - 3$ (*Hint:* Write $x + 1$ as $x - (-1)$.)
47. $x - 6; P(x) = 2x^3 - 8x^2 + x - 36$
48. $x + 2; P(x) = 3x^2 - 7x + 4$ (*Hint:* Write $x + 2$ as $x - (-2)$.)
49. $x; P(x) = 7x^3 - 5x^2 - 8x$ (*Hint:* $x = x - 0$.)
50. $x; P(x) = 5x^3 + 6x + 2$
51. $x - 5; P(x) = x^3 - 125$
52. $x + 4; P(x) = x^3 - 64$

ADDITIONAL PRACTICE

Divide using synthetic division.

53. $(4 - 3x^2 + x) \div (x - 4)$
54. $(x^2 + 13x + 42) \div (x + 6)$
55. $(4x^3 + 5x^2 - 1) \div (x + 2)$
56. $(x^3 - x + 6) \div (x - 2)$

Evaluate P(x) for the given value using the remainder theorem.

57. $P(x) = 3x^5 + 1; c = -\frac{1}{2}$
58. $P(x) = 5x^7 - 7x^4 + x^2 + 1; c = 2$

Let $Q(x) = x^4 - 3x^3 + 2x^2 + x - 3$. Evaluate Q(x) for the given value by using the remainder theorem. Then evaluate by substituting the value of x into the polynomial and simplifying. See Example 4. (Objective 2)

59. $Q(-1)$
60. $Q(1)$
61. $Q(2)$
62. $Q(-2)$
63. $Q(3)$
64. $Q(0)$
65. $Q(-3)$
66. $Q(-4)$

Use a calculator and synthetic division to perform each division.

67. $(7.2x^2 - 2.1x + 0.5) \div (x - 0.2)$
68. $(8.1x^2 + 3.2x - 5.7) \div (x - 0.4)$

69. $(2.7x^2 + x - 5.2) \div (x + 1.7)$

70. $(1.3x^2 - 0.5x - 2.3) \div (x + 2.5)$

71. $(9x^3 - 25) \div (x + 57)$

72. $(0.5x^3 + x) \div (x - 2.3)$

Use a calculator to work each problem.

73. Find 2^6 by using synthetic division to evaluate the polynomial $P(x) = x^6$ at $x = 2$. Then check the answer by evaluating 2^6 with a calculator.

74. Find $(-3)^5$ by using synthetic division to evaluate the polynomial $P(x) = x^5$ at $x = -3$. Then check the answer by evaluating $(-3)^5$ with a calculator.

WRITING ABOUT MATH

75. If you are given $P(x)$, explain how to use synthetic division to calculate $P(a)$.

76. Explain the factor theorem.

SOMETHING TO THINK ABOUT ***Suppose that*** $P(x) = x^{100} - x^{99} + x^{98} - x^{97} + \cdots + x^2 - x + 1$.

77. Find the remainder when $P(x)$ is divided by $x - 1$.

78. Find the remainder when $P(x)$ is divided by $x + 1$.

SECTION 6.8 Proportion and Variation

Objectives

1. Solve a proportion.
2. Solve an application problem involving a proportion.
3. Use similar triangles to determine a missing length of one of the sides.
4. Solve an application problem involving direct variation.
5. Solve an application problem involving inverse variation.
6. Solve an application problem involving joint variation.
7. Solve an application problem involving combined variation.

Vocabulary

ratio	means	inverse variation
unit cost	similar triangles	joint variation
rates	direct variation	combined variation
proportion	constant of proportionality (constant of variation)	
extremes		

Getting Ready

Solve each equation.

1. $30k = 70$

2. $\dfrac{k}{4{,}000^2} = 90$

Classify each function as a linear function or a rational function.

3. $f(x) = 3x$

4. $f(x) = \dfrac{3}{x} \quad (x > 0)$

In this section, we will discuss ratio and proportion. Then we will use these skills to solve variation problems.

1 Solve a proportion.

The quotient of two numbers is often called a **ratio.** For example, the fraction $\frac{2}{3}$ can be read as "the ratio of 2 to 3." Some more examples of ratios are

$$\frac{4x}{7y} \quad \text{(the ratio of } 4x \text{ to } 7y) \qquad \text{and} \qquad \frac{x-2}{3x} \quad \text{(the ratio of } x-2 \text{ to } 3x)$$

Ratios often are used to express **unit costs,** such as the cost per pound of ground round steak.

$$\begin{array}{r} \text{The cost of a package of ground round} \rightarrow \\ \text{The weight of the package} \rightarrow \end{array} \frac{\$18.75}{5 \text{ lb}} = \$3.75 \text{ per lb} \leftarrow \text{The cost per pound.}$$

Ratios also are used to express **rates,** such as an average rate of speed.

$$\begin{array}{r} \text{A distance traveled} \rightarrow \\ \text{In a period of time} \rightarrow \end{array} \frac{372 \text{ miles}}{6 \text{ hours}} = 62 \text{ mph} \leftarrow \text{The average rate of speed.}$$

An equation indicating that two ratios are equal is called a **proportion.** Two examples of proportions are

$$\frac{1}{4} = \frac{2}{8} \qquad \text{and} \qquad \frac{4}{7} = \frac{12}{21}$$

In the proportion $\frac{a}{b} = \frac{c}{d}$, a and d are called the **extremes** and b and c are called the **means.**

To develop a fundamental property of proportions, we suppose that

$$\frac{a}{b} = \frac{c}{d}$$

is a proportion and multiply both sides by bd to obtain

$$bd\left(\frac{a}{b}\right) = bd\left(\frac{c}{d}\right)$$

$$\frac{\cancel{b}da}{\cancel{b}} = \frac{b\cancel{d}c}{\cancel{d}}$$

$$ad = bc$$

Thus, if $\frac{a}{b} = \frac{c}{d}$, then $ad = bc$. This shows that in a proportion, the *product of the extremes equals the product of the means.*

EXAMPLE 1 Solve: $\dfrac{x+1}{x} = \dfrac{x}{x+2} \quad (x \neq 0, -2)$.

Solution We will use the fact that *in a proportion, the product of the extremes is equal to the product of the means.*

$$\frac{x+1}{x} = \frac{x}{x+2}$$

$(x+1)(x+2) = x \cdot x$ — In a proportion, the product of the extremes equals the product of the means.

$x^2 + 3x + 2 = x^2$ — Multiply.

$3x + 2 = 0$ — Subtract x^2 from both sides.

$x = -\frac{2}{3}$ — Subtract 2 from both sides and divide by 3.

Thus, $x = -\frac{2}{3}$.

SELF CHECK 1 Solve: $\frac{x-1}{x} = \frac{x}{x+3}$ $(x \neq 0, -3)$.

EXAMPLE 2 Solve: $\frac{5a+2}{2a} = \frac{18}{a+4}$ $(a \neq 0, -4)$.

Solution We will use the fact that *in a proportion, the product of the extremes is equal to the product of the means.*

$$\frac{5a+2}{2a} = \frac{18}{a+4}$$

$(5a+2)(a+4) = 2a(18)$ In a proportion, the product of the extremes equals the product of the means.

$5a^2 + 22a + 8 = 36a$ Multiply.

$5a^2 - 14a + 8 = 0$ Subtract $36a$ from both sides.

$(5a-4)(a-2) = 0$ Factor.

$5a - 4 = 0$ or $a - 2 = 0$ Set each factor equal to 0.

$5a = 4$ | $a = 2$ Solve each linear equation.

$a = \frac{4}{5}$

Thus, $a = \frac{4}{5}$ or $a = 2$.

SELF CHECK 2 Solve: $\frac{3x+1}{12} = \frac{x}{x+2}$ $(x \neq -2)$.

2 Solve an application problem involving a proportion.

EXAMPLE 3 **GROCERY SHOPPING** If 7 pears cost \$2.73, how much will 11 pears cost?

Solution We can let c represent the cost of 11 pears. The price per pear of 7 pears is $\frac{\$2.73}{7}$, and the price per pear of 11 pears is $\frac{\$c}{11}$. Since these costs are equal, we can set up and solve the following proportion.

$$\frac{2.73}{7} = \frac{c}{11}$$

$11(2.73) = 7c$ In a proportion, the product of the extremes is equal to the product of the means.

$30.03 = 7c$ Multiply.

$\frac{30.03}{7} = c$ Divide both sides by 7.

$c = 4.29$ Simplify.

Eleven pears will cost \$4.29.

SELF CHECK 3 How much will 28 pears cost?

EVERYDAY CONNECTIONS Video Action Games

The Sims 3

Game Genres in Order of Popularity

Genre (examples)	% teens who report playing
Racing (NASCAR, Mario Kart, Burnout)	74%
Puzzle (Bejeweled, Tetris, Solitaire)	72
Sports (Madden, FIFA, Tony Hawk)	68
Action (Grand Theft Auto, Devil May Cry, Ratchet and Clank)	67
Adventure (Legend of Zelda, Tomb Raider)	66
Rhythm (Guitar Hero, Dance Dance Revolution, Lumines)	61
Strategy (Civilization IV, StarCraft, Command and Conquer)	59
Simulation (The Sims, Rollercoaster Tycoon, Ace Combat)	49
Fighting (Tekken, Super Smash Bros., Mortal Kombat)	49
First-Person Shooters (Halo, Counter-Strike, Half-Life)	47
Role-Playing (Final Fantasy, Blue Dragon, Knights of the Old Republic)	36
Survival Horror (Resident Evil, Silent Hill, Condemned)	32
MMOGs (World of Warcraft)	21
Virtual Worlds (Second Life, Gaia, Habbo Hotel)	10

Source: Pew Internet & American Life Project. Gaming and Civic Engagement Survey of Teens/Parents, Nov. 2007–Feb. 2008.

1. Given the data in the table, determine how many of the 1,064 teens surveyed reported playing action games.

3 Use similar triangles to determine a missing length of one of the sides.

If two angles of one triangle have the same measure as two angles of a second triangle, the triangles will have the same shape. In this case, we call the triangles **similar triangles.** Here are some facts about similar triangles.

Similar Triangles

If two triangles are similar, then

1. The three angles of the first triangle have the same measures, respectively, as the three angles of the second triangle.
2. The lengths of all corresponding sides are in proportion.

Since the triangles shown in Figure 6-11 are similar, their corresponding sides are in proportion.

$$\frac{2}{4} = \frac{x}{2x}, \quad \frac{x}{2x} = \frac{1}{2}, \quad \frac{1}{2} = \frac{2}{4}$$

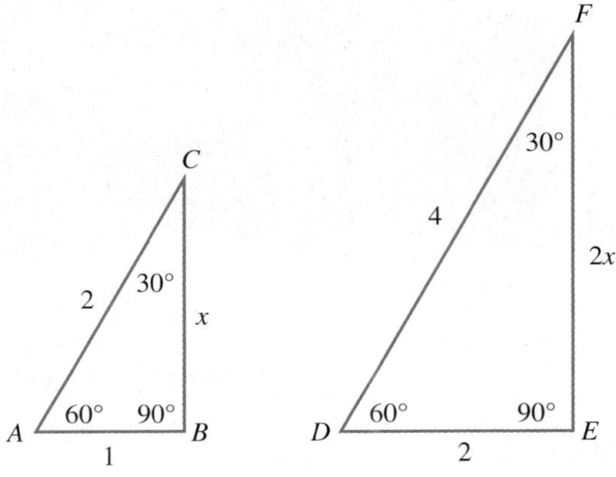

Figure 6-11

The properties of similar triangles often enable us to find the lengths of the sides of a triangle indirectly. For example, on a sunny day we can find the height of a tree and stay safely on the ground.

EXAMPLE 4 **HEIGHT OF A TREE** A tree casts a shadow of 29 feet at the same time as a vertical yardstick casts a shadow of 2.5 feet. Find the height of the tree.

Solution We will let h represent the height of the tree. Refer to Figure 6-12, which shows the triangles determined by the tree and its shadow, and the yardstick and its shadow. Because the triangles have the same shape, they are similar, and the measures of their corresponding sides are in proportion. We can find h by setting up and solving the following proportion.

$$\frac{h}{3} = \frac{39}{2.5}$$

$2.5h = 3(29)$ In a proportion, the product of the extremes is equal to the product of the means.

$2.5h = 87$ Simplify.

$h = 34.8$ Divide both sides by 2.5.

The tree is about 35 feet tall.

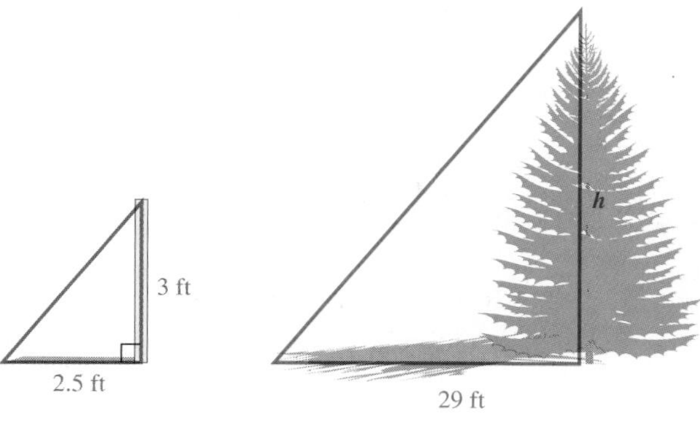

Figure 6-12

SELF CHECK 4 Find the height of the tree if its shadow is 25 feet long.

4 Solve an application problem involving direct variation.

To introduce **direct variation,** we consider the formula

$$C = \pi D$$

for the circumference of a circle, where C is the circumference, D is the diameter, and $\pi \approx 3.14159$. If we double the diameter of a circle, we determine another circle with a larger circumference C_1 such that

$$C_1 = \pi(2D) = 2\pi D = 2C$$

Thus, doubling the diameter results in doubling the circumference. Likewise, if we triple the diameter, we triple the circumference.

In this formula, we say that the variables C and D *vary directly,* or that they are *directly proportional.* This is because as one variable gets larger, so does the other, and in a predictable way. In this example, the constant π is called the *constant of variation* or the *constant of proportionality.*

COMMENT In this section, we will assume that k is a positive number.

Direct Variation

The words "y varies directly with x" or "y is directly proportional to k" mean that $y = kx$ for some nonzero constant k. The constant k is called the **constant of variation** or the **constant of proportionality.**

Since the formula for direct variation ($y = kx$) defines a linear function, its graph is always a line with a y-intercept at the origin. The graph of $y = kx$ appears in Figure 6-13 for three positive values of k.

One example of direct variation is Hooke's law from physics. Hooke's law states that the distance a spring will stretch varies directly with the force that is applied to it.

If d represents a distance and f represents a force, Hooke's law is expressed mathematically as

$$d = kf$$

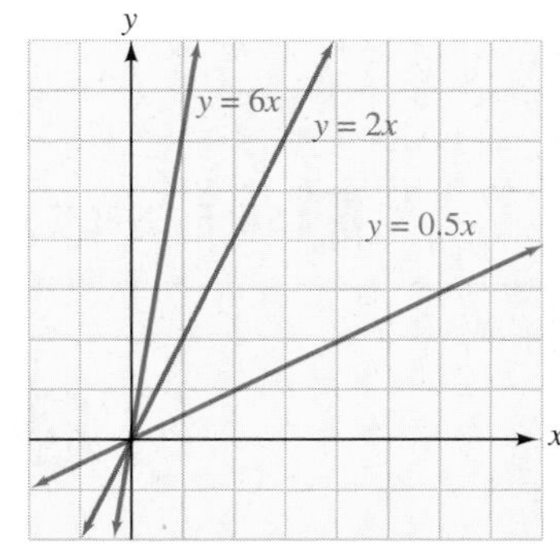

Figure 6-13

where k is the constant of variation. If the spring stretches 10 inches when a weight of 6 pounds is attached, k can be found as follows:

$$d = kf$$
$$10 = k(6) \quad \text{Substitute 10 for } d \text{ and 6 for } f.$$
$$\frac{5}{3} = k \quad \text{Divide both sides by 6 and simplify.}$$

To find the force required to stretch the spring a distance of 35 inches, we can solve the equation $d = kf$ for f, with $d = 35$ and $k = \frac{5}{3}$.

$$d = kf$$
$$35 = \frac{5}{3}f \quad \text{Substitute 35 for } d \text{ and } \tfrac{5}{3} \text{ for } k.$$
$$105 = 5f \quad \text{Multiply both sides by 3.}$$
$$21 = f \quad \text{Divide both sides by 5.}$$

The force required to stretch the spring a distance of 35 inches is 21 pounds.

EXAMPLE 5 **DIRECT VARIATION** The distance traveled in a given time is directly proportional to the speed. If a car goes 70 miles at 30 mph, how far will it go in the same time at 45 mph?

Solution The words *distance is directly proportional to speed* can be expressed by the equation

$$(1) \qquad d = ks$$

where d is distance, k is the constant of variation, and s is the speed. To find k, we substitute 70 for d and 30 for s, and solve for k.

$$d = ks$$
$$70 = k(30)$$
$$k = \frac{7}{3}$$

To find the distance traveled at 45 mph, we substitute $\frac{7}{3}$ for k and 45 for s in Equation 1 and simplify.

$$d = ks$$
$$d = \frac{7}{3}(45)$$
$$= 105$$

In the time it took to go 70 miles at 30 mph, the car could go 105 miles at 45 mph.

SELF CHECK 5 How far will the car go in the same time at 60 mph?

5 Solve an application problem involving inverse variation.

In the formula $w = \frac{12}{l}$, w gets smaller as l gets larger, and w gets larger as l gets smaller. Since these variables vary in opposite directions in a predictable way, we say that the variables *vary inversely,* or that they are *inversely proportional.* The constant 12 is the constant of variation.

Inverse Variation The words "y varies inversely with x" or "y is inversely proportional to x" mean that $y = \frac{k}{x}$ for some nonzero constant k. The constant k is called the **constant of variation.**

The formula for inverse variation $\left(y = \frac{k}{x}\right)$ defines a rational function. The graph of $y = \frac{k}{x}$ appears in Figure 6-14 for three positive values of k.

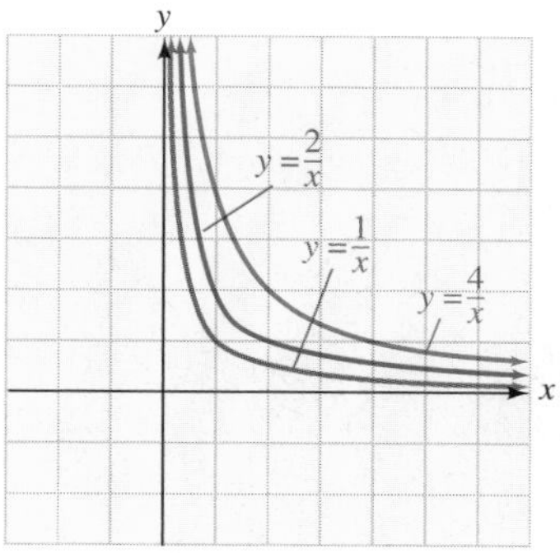

Because of gravity, an object in space is attracted to Earth. The force of this attraction varies inversely with the square of the distance from the object to Earth's center.

If f represents the force and d represents the distance, this information can be expressed by the equation

$$f = \frac{k}{d^2}$$

Figure 6-14

If we know that an object 4,000 miles from Earth's center is attracted to Earth with a force of 90 pounds, we can find k.

$$f = \frac{k}{d^2}$$

$$90 = \frac{k}{4{,}000^2} \quad \text{Substitute 90 for } f \text{ and 4,000 for } d.$$

$$k = 90(4{,}000^2)$$

$$= 1.44 \times 10^9$$

To find the force of attraction when the object is 5,000 miles from Earth's center, we proceed as follows:

$$f = \frac{k}{d^2}$$

$$f = \frac{1.44 \times 10^9}{5{,}000^2} \quad \text{Substitute } 1.44 \times 10^9 \text{ for } k \text{ and 5,000 for } d.$$

$$= 57.6$$

The object will be attracted to Earth with a force of 57.6 pounds when it is 5,000 miles from Earth's center.

EXAMPLE 6 **LIGHT INTENSITY** The intensity I of light received from a light source varies inversely with the square of the distance from the source. If the intensity of a light source 4 feet from an object is 8 candelas, find the intensity at a distance of 2 feet.

Solution The words *intensity varies inversely with the square of the distance* d can be expressed by the equation

$$I = \frac{k}{d^2}$$

To find k, we substitute 8 for I and 4 for d and solve for k.

$$I = \frac{k}{d^2}$$

$$8 = \frac{k}{4^2}$$

$$128 = k$$

To find the intensity when the object is 2 feet from the light source, we substitute 2 for d and 128 for k and simplify.

$$I = \frac{k}{d^2}$$

$$I = \frac{128}{2^2}$$

$$= 32$$

The intensity at 2 feet is 32 candelas.

SELF CHECK 6 Find the intensity at a distance of 8 feet.

6 Solve an application problem involving joint variation.

There are times when one variable varies with the product of several variables. For example, the area of a triangle varies directly with the product of its base and height:

$$A = \frac{1}{2}bh$$

Such variation is called *joint variation.*

Joint Variation

If one variable varies directly with the product of two or more variables, the relationship is called **joint variation.** If y varies jointly with x and z, then $y = kxz$. The nonzero constant k is called the **constant of variation.**

EXAMPLE 7 **VOLUME OF A CONE** The volume V of a cone varies jointly with its height h and the area of its base b. If $V = 6\text{ cm}^3$ when $h = 3$ cm and $B = 6\text{ cm}^2$, find V when $h = 2$ cm and $B = 8\text{ cm}^2$.

Solution The words *V varies jointly with h and B* can be expressed by the equation

$V = khB$ The relationship also can be read as "V is directly proportional to the product of h and B."

We can find k by substituting 6 for V, 3 for h, and 6 for B.

$$V = khB$$

$$6 = k(3)(6)$$

$$6 = k(18)$$

$$\frac{1}{3} = k$$ Divide both sides by 18; $\frac{6}{18} = \frac{1}{3}$.

To find V when $h = 2$ and $B = 8$, we substitute these values into the formula $V = \frac{1}{3}hB$.

$$V = \frac{1}{3}hB$$

$$V = \left(\frac{1}{3}\right)(2)(8)$$

$$= \frac{16}{3}$$

The volume is $5\frac{1}{3}$ cm^3.

SELF CHECK 7 Find V when $B = 10$ cm.

Solve an application problem involving combined variation.

Many applied problems involve a combination of direct and inverse variation. Such variation is called **combined variation.**

EXAMPLE 8 **BUILDING HIGHWAYS** The time it takes to build a highway varies directly with the length of the road, but inversely with the number of workers. If it takes 100 workers 4 weeks to build 2 miles of highway, how long will it take 80 workers to build 10 miles of highway?

Solution We can let t represent the time in weeks, l represent the length in miles, and w represent the number of workers. The relationship between these variables can be expressed by the equation

$$t = \frac{kl}{w}$$

We substitute 4 for t, 100 for w, and 2 for l to find k:

$$4 = \frac{k(2)}{100}$$

$400 = 2k$ Multiply both sides by 100.

$200 = k$ Divide both sides by 2.

We now substitute 80 for w, 10 for l, and 200 for k in the equation $t = \frac{kl}{w}$ and simplify:

$$t = \frac{kl}{w}$$

$$t = \frac{200(10)}{80}$$

$$= 25$$

It will take 25 weeks for 80 workers to build 10 miles of highway.

SELF CHECK 8 How long will it take 60 workers to build 6 miles of highway?

SELF CHECK ANSWERS 1. $\frac{3}{2}$ 2. $\frac{2}{3}$, 1 3. $10.92 4. 30 ft 5. 140 mi 6. 2 candelas 7. $6\frac{2}{3}$ cm^3 8. 20 weeks

NOW TRY THIS

1. Solve $\frac{5x+2}{6x+3} = \frac{2x}{2x+1}$ as a proportion.
2. Solve $\frac{5x+2}{6x+3} = \frac{2x}{2x+1}$ by multiplying by the LCD.

   Explain to a classmate why you did not get the same answer as in problem 1.

The time T, in hours, required for a satellite to complete an orbit around Earth varies directly as the radius r of the orbit measured from the center of Earth and inversely as the velocity v, in miles per hour. The radius of Earth is approximately 4,000 miles.

3. The space shuttle makes one orbit around Earth in 1.5 hours at a rate of 17,000 mph. Its altitude above Earth's surface is about 200 miles. Find the constant of variation rounded to 2 decimal places.
4. The Pentagon's Defense Satellite Communications System (DSCS) orbits at 23,500 miles above Earth's surface at a speed of approximately 6,955 mph. Using the same constant of variation, how long is its orbit? What can you conclude about the satellite?

6.8 EXERCISES

WARM-UPS

Solve each proportion.

1. $\frac{x}{2} = \frac{3}{6}$
2. $\frac{3}{x} = \frac{4}{12}$

Express each sentence with a formula.

3. a varies directly with b.
4. a varies inversely with b.
5. a varies jointly with b and c.
6. a varies directly with b but inversely with c.

REVIEW *Simplify each expression.*

7. $(x^2x^3)^2$
8. $\left(\frac{a^3a^5}{a^{-2}}\right)^3$
9. $\frac{b^0 - 2b^0}{b^0}$
10. $\left(\frac{2r^{-2}r^{-3}}{4r^{-5}}\right)^{-3}$
11. Write 35,000 in scientific notation.
12. Write 0.00035 in scientific notation.
13. Write 2.5×10^{-3} in standard notation.
14. Write 2.5×10^4 in standard notation.

VOCABULARY AND CONCEPTS

Fill in the blanks.

15. Ratios are used to express ________ and ______.
16. An equation stating that two ratios are equal is called a ________.
17. In a proportion, the product of the ________ is equal to the product of the _______.
18. If two angles of one triangle have the same measures as two angles of a second triangle, the triangles are _______.
19. The equation $y = kx$ indicates ______ variation.
20. The equation $y = \frac{k}{x}$ indicates _______ variation.
21. Inverse variation is represented by a ________ function.
22. Direct variation is represented by a ______ function through the origin.
23. The equation $y = kxz$ indicates ______ variation and k represents the ____________________.
24. The equation $y = \frac{kx}{z}$ indicates _________ variation.

Determine whether the graph represents direct variation, inverse variation, or neither.

25.

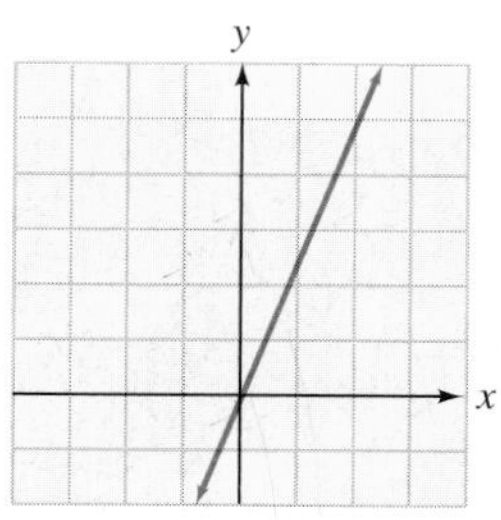

26.

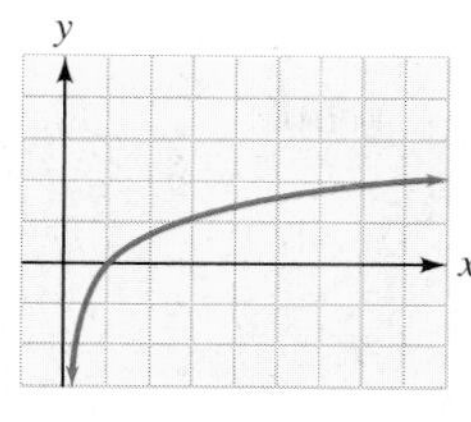

27.

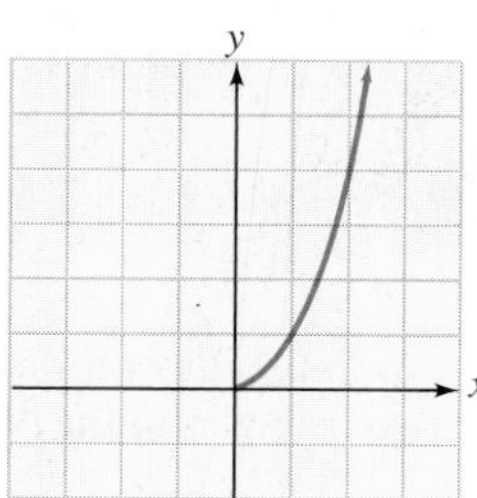

28.

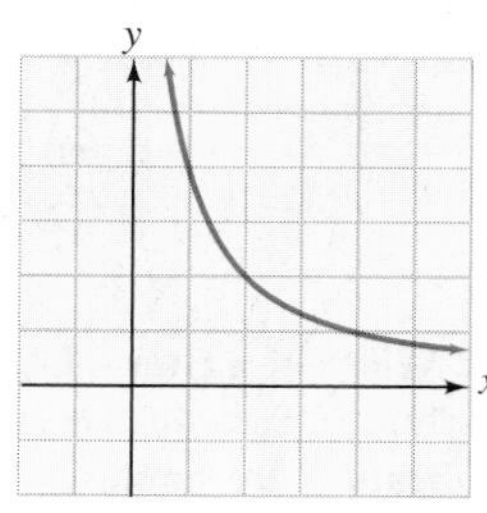

GUIDED PRACTICE

Solve each proportion for the variable, if possible. See Examples 1–2. (Objective 1)

29. $\frac{x}{5} = \frac{15}{25}$

30. $\frac{4}{y} = \frac{6}{27}$

31. $\frac{r-2}{3} = \frac{r}{5}$

32. $\frac{x+1}{x-1} = \frac{6}{4}$

33. $\frac{3}{n} = \frac{2}{n+1}$

34. $\frac{4}{x+3} = \frac{3}{5}$

35. $\frac{5}{5z+3} = \frac{2z}{2z^2+6}$

36. $\frac{9t+6}{t(t+3)} = \frac{7}{t+3}$

37. $\frac{1}{x+3} = \frac{-2x}{x+5}$

38. $\frac{x-1}{x+1} = \frac{2}{3x}$

39. $\frac{a-4}{a+2} = \frac{a-5}{a+1}$

40. $\frac{z+2}{z+6} = \frac{z-4}{z-2}$

Express each sentence as a formula. See Examples 5–8. (Objectives 4–7)

41. A varies directly with the square of p.

42. r varies directly with the square of s.

43. z varies inversely with the cube of t.

44. v varies inversely with the cube of r.

45. B varies jointly with m and n.

46. C varies jointly with x, y, and z.

47. X varies directly with w and inversely with q.

48. T varies directly with P and inversely with the square of v.

ADDITIONAL PRACTICE

Solve each proportion.

49. $\frac{2}{c} = \frac{c-3}{2}$

50. $\frac{y}{4} = \frac{4}{y}$

51. $\frac{2}{3x} = \frac{6x}{36}$

52. $\frac{2}{x+6} = \frac{-2x}{5}$

53. $\frac{2(x+3)}{3} = \frac{4(x-4)}{5}$

54. $\frac{x+4}{5} = \frac{3(x-2)}{3}$

Express each sentence as a formula.

55. P varies directly with the square of a, and inversely with the cube of j.

56. M varies inversely with the cube of n, and jointly with x and the square of z.

Express each formula in words. In each formula, k is the constant of variation.

57. $L = kmn$

58. $P = \frac{km}{n}$

59. $E = kab^2$

60. $U = krs^2t$

61. $X = \frac{kx^2}{y^2}$

62. $Z = \frac{kw}{xy}$

63. $R = \frac{kL}{d^2}$

64. $e = \frac{kPL}{A}$

APPLICATIONS

Set up and solve the required proportion. See Example 3. (Objective 2)

65. Selling shirts Consider the following ad. How much will 5 shirts cost?

66. **Cooking** A recipe requires four 16-ounce bottles of catsup to make two gallons of spaghetti sauce. How many bottles are needed to make ten gallons of sauce?

67. **Gas consumption** A car gets 42 mpg. How much gas will be needed to go 315 miles?

68. **Model railroading** An HO-scale model railroad engine is 9 inches long. The HO scale is 87 feet to 1 foot. How long is a real engine?

69. **Hobbies** Standard dollhouse scale is 1 inch to 1 foot. Heidi's dollhouse is 32 inches wide. How wide would it be if it were a real house?

70. **Staffing** A school board has determined that there should be 3 teachers for every 50 students. How many teachers are needed for an enrollment of 2,700 students?

71. **Drafting** In a scale drawing, a 280-foot antenna tower is drawn 7 inches high. The building next to it is drawn 2 inches high. How tall is the actual building?

72. **Mixing fuel** The instructions on a can of oil intended to be added to lawnmower gasoline read:

Recommended	Gasoline	Oil
50 to 1	6 gal	16 oz

Are these instructions correct? (*Hint:* There are 128 ounces in 1 gallon.)

73. **Recommended dosage** The recommended child's dose of the sedative hydroxine is 0.006 gram per kilogram of body mass. Find the dosage for a 30-kg child.

74. **Body mass** The proper dose of the antibiotic cephalexin in children is 0.025 gram per kilogram of body mass. Find the mass of a child receiving a $1\frac{1}{8}$-gram dose.

Use similar triangles to help solve each problem. **See Example 4. (Objective 3)**

75. **Height of a tree** A tree casts a shadow of 28 feet at the same time as a 6-foot man casts a shadow of 4 feet. Find the height of the tree.

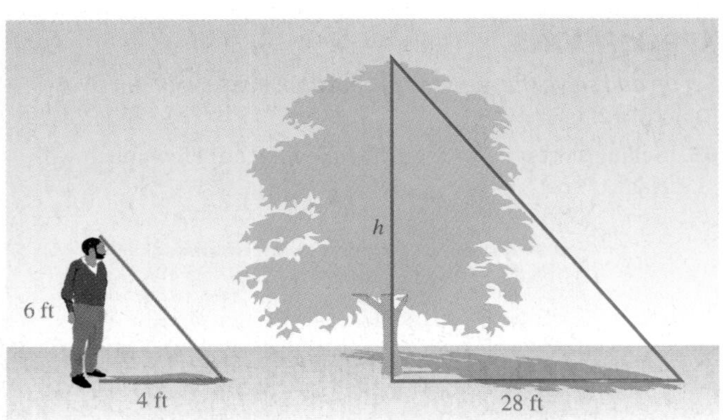

76. **Height of a flagpole** A man places a mirror on the ground and sees the reflection of the top of a flagpole. The two triangles in the illustration are similar. Find the height, h, of the flagpole.

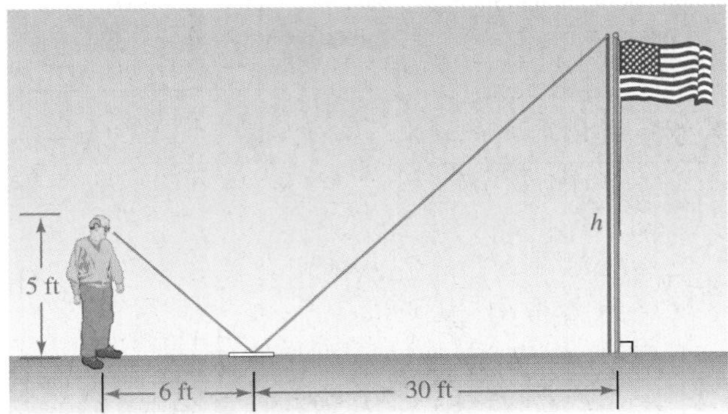

77. **Width of a river** Use the dimensions in the illustration to find w, the width of the river. The two triangles in the illustration are similar.

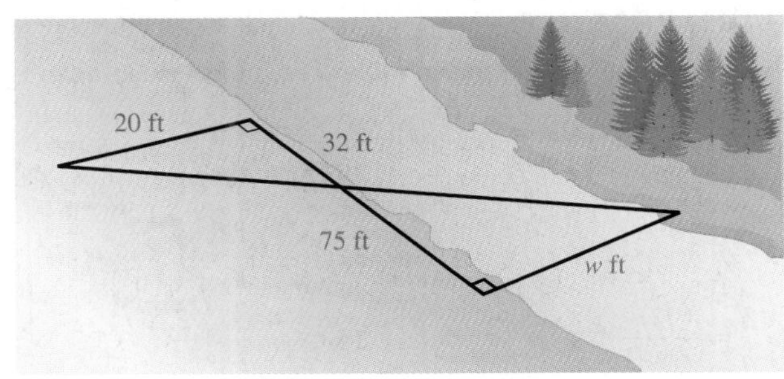

78. **Flight paths** An airplane ascends 150 feet as it flies a horizontal distance of 1,000 feet. How much altitude will it gain as it flies a horizontal distance of 1 mile? (*Hint:* 5,280 feet = 1 mile.)

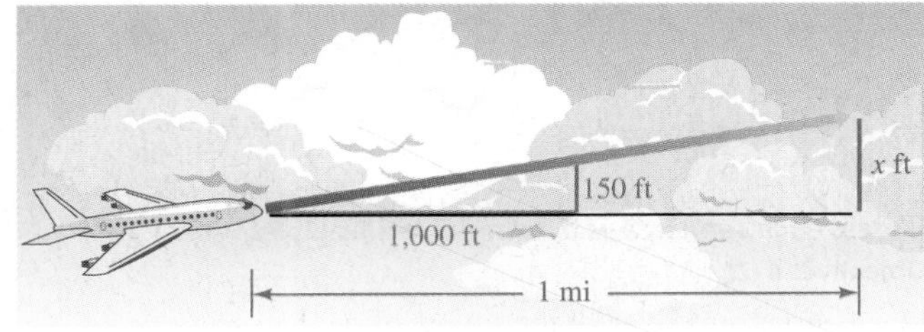

79. **Flight paths** An airplane descends 1,350 feet as it flies a horizontal distance of 1 mile. How much altitude is lost as it flies a horizontal distance of 5 miles?

80. **Ski runs** A ski course with $\frac{1}{2}$ mile of horizontal run falls 100 feet in every 300 feet of run. Find the height of the hill.

Solve each variation problem. See Examples 5–8. (Objectives 4–7)

81. Area of a circle The area of a circle varies directly with the square of its radius. The constant of variation is π. Find the area of a circle with a radius of 6 inches.

82. Falling objects An object in free fall travels a distance s that is directly proportional to the square of the time t. If an object falls 1,024 feet in 8 seconds, how far will it fall in 10 seconds?

83. Finding distance The distance that a car can go is directly proportional to the number of gallons of gasoline it consumes. If a car can go 288 miles on 12 gallons of gasoline, how far can it go on a full tank of 18 gallons?

84. Farming A farmer's harvest in bushels varies directly with the number of acres planted. If 8 acres can produce 144 bushels, how many acres are required to produce 1,152 bushes?

85. Farming The length of time that a given number of bushels of corn will last when feeding cattle varies inversely with the number of animals. If x bushels will feed 25 cows for 10 days, how long will the feed last for 10 cows?

86. Geometry For a fixed area, the length of a rectangle is inversely proportional to its width. A rectangle has a width of 18 feet and a length of 12 feet. If the length is increased to 16 feet, find the width.

87. Gas pressure Under constant temperature, the volume occupied by a gas is inversely proportional to the pressure applied. If the gas occupies a volume of 20 cubic inches under a pressure of 6 pounds per square inch, find the volume when the gas is subjected to a pressure of 10 pounds per square inch.

88. Value of a boat The value of a boat usually varies inversely with its age. If a boat is worth \$7,000 when it is 3 years old, how much will it be worth when it is 7 years old?

89. Organ pipes The frequency of vibration of air in an organ pipe is inversely proportional to the length l of the pipe. If a pipe 2 feet long vibrates 256 times per second, how many times per second will a 6-foot pipe vibrate?

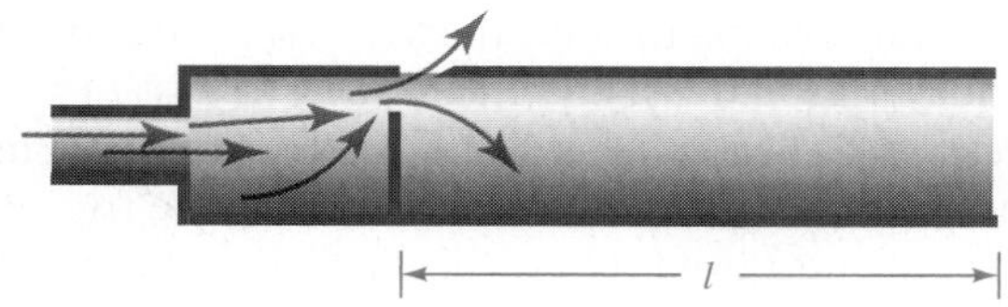

90. Geometry The area of a rectangle varies jointly with its length and width. If both the length and the width are tripled, by what factor is the area multiplied?

91. Geometry The volume of a rectangular solid varies jointly with its length, width, and height. If the length is doubled, the width is tripled, and the height is doubled, by what factor is the volume multiplied?

92. Costs of a trucking company The costs incurred by a trucking company vary jointly with the number of trucks in service and the number of hours each is used. When 4 trucks are used for 6 hours each, the costs are \$1,800. Find the costs of using 10 trucks, each for 12 hours.

93. Storing oil The number of gallons of oil that can be stored in a cylindrical tank varies jointly with the height of the tank and the square of the radius of its base. The constant of proportionality is 23.5. Find the number of gallons that can be stored in the cylindrical tank shown in the illustration.

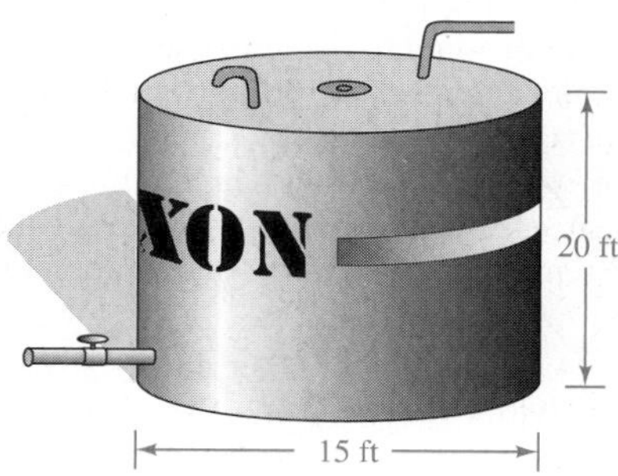

94. Finding the constant of variation A quantity l varies jointly with x and y and inversely with z. If the value of l is 30 when $x = 15$, $y = 5$, and $z = 10$, find k.

95. Electronics The voltage (in volts) measured across a resistor is directly proportional to the current (in amperes) flowing through the resistor. The constant of variation is the **resistance** (in ohms). If 6 volts is measured across a resistor carrying a current of 2 amperes, find the resistance.

96. Electronics The power (in watts) lost in a resistor (in the form of heat) is directly proportional to the square of the current (in amperes) passing through it. The constant of proportionality is the resistance (in ohms). What power is lost in a 5-ohm resistor carrying a 3-ampere current?

97. Building construction The deflection of a beam is inversely proportional to its width and the cube of its depth. If the deflection of a 4-inch-by-4-inch beam is 1.1 inches, find the deflection of a 2-inch-by-8-inch beam positioned as in the illustration.

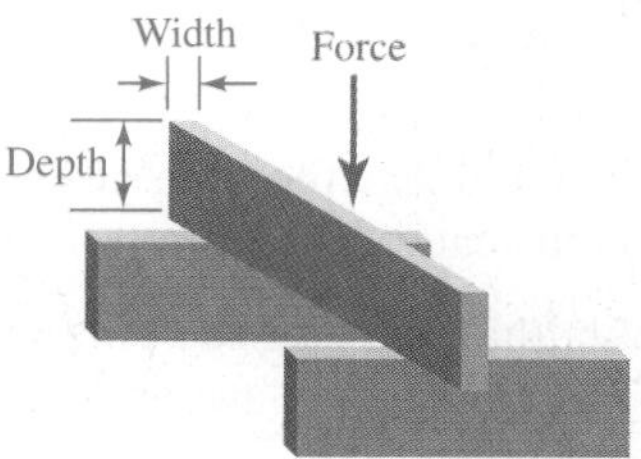

98. Building construction Find the deflection of the beam in Exercise 97 when the beam is positioned as in the illustration.

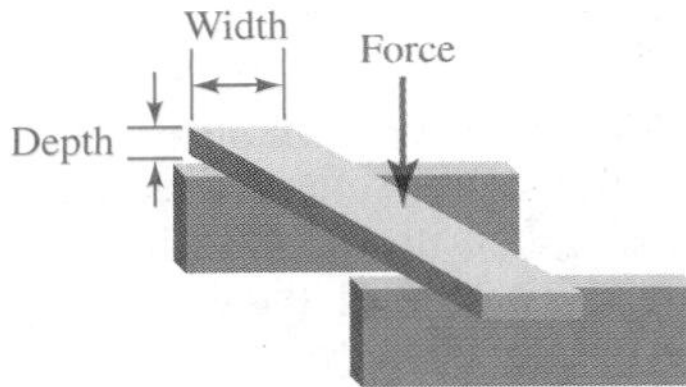

99. Gas pressure The pressure of a certain amount of gas is directly proportional to the temperature (measured on the Kelvin scale) and inversely proportional to the volume. A sample of gas at a pressure of 1 atmosphere occupies a volume of 1 cubic meter at a temperature of 273 Kelvin. When heated, the gas expands to twice its volume, but the pressure remains constant. To what temperature is it heated?

100. Tension A yo-yo, twirled at the end of a string, is kept in its circular path by the tension of the string. The tension T is directly proportional to the square of the speed s and inversely proportional to the radius r of the circle. In the illustration, the tension is 32 pounds when the speed is 8 feet/second and the radius is 6 feet. Find the tension when the speed is 4 feet/second and the radius is 3 feet.

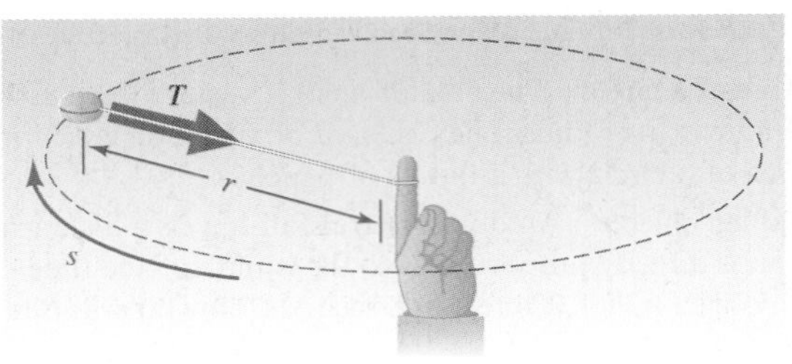

WRITING ABOUT MATH

101. Explain the terms *means* and *extremes.*

102. Distinguish between a *ratio* and a *proportion.*

103. Explain the term *joint variation.*

104. Explain why the equation $\frac{y}{x} = k$ indicates that y varies directly with x.

SOMETHING TO THINK ABOUT

105. As temperature increases on the Fahrenheit scale, it also increases on the Celsius scale. Is this direct variation? Explain.

106. As the cost of a purchase (less than $5) increases, the amount of change received from a five-dollar bill decreases. Is this inverse variation? Explain.

PROJECTS

Project 1

Suppose that a motorist usually makes an 80-mile trip at an average speed of 50 mph.

a. How much time does the driver save by increasing her rate by 5 mph? By 10 mph? By 15 mph? Give all answers to the nearest minute.

b. Consider your work in part a, and find a formula that will tell how much time is saved if the motorist travels x mph faster than 50 mph. That is, find an expression involving x that will give the time saved by traveling $(50 + x)$ mph instead of 50 mph.

c. Find a formula that will give the time saved on a trip of d miles by traveling at an average rate that is x mph faster than a usual speed of y mph. When you have this formula, simplify it into a nice, compact form.

d. Test your formula by doing part a over again using the formula. The answers you get should agree with those found earlier.

e. Use your formula to solve the following problem. Every holiday season, Kurt and Ellen travel to visit their relatives, a distance of 980 miles. Under normal circumstances, they can average 60 mph during the trip. However, improved roads will enable them to travel 4 mph faster this year. How much time (to the nearest minute) will they save on this year's trip? How much faster than normal (to the nearest tenth of a mph) would Kurt and Ellen have to travel to save two hours?

Project 2

By cross-fertilization among a number of species of corn, Professor Greenthumb has succeeded in creating a miracle hybrid corn plant. Under ideal conditions, the plant will produce twice as much corn as an ordinary hybrid. However, the new hybrid is very sensitive to the amount of sun and water it receives. If conditions are not ideal, the corn yield diminishes.

To determine what yield the plants will have, daily measurements record the amount of sun the plants receive (the sun index, S) and the amount of water they receive (W, measured in millimeters). Then the daily yield diminution is calculated using the formula

$$\text{Daily yield diminution} = \frac{S^3 + W^3 - (S + W)^2}{(S + W)^3}$$

At the end of a 100-day growing season, the daily yield diminutions are added together to make the total diminution (D). The yield for the year is then determined using the formula

$$\text{Annual yield} = 1 - 0.01D$$

The resulting decimal is the percent of maximum yield that the corn plants will achieve. Remember that maximum yield for these plants is twice the yield of an ordinary corn plant.

As Greenthumb's research assistant, you have been asked to handle a few questions that have come up regarding her research.

a. First, show that if $S = W = 2$, the daily yield diminution is 0. These would be the optimal conditions for Greenthumb's plants.

b. Now suppose that for each day of the 100-day growing season, the sun index is 8, and 6 millimeters of rain fall on the plants. Find the daily yield diminution. To the nearest tenth of a percent, what percentage of the yield of normal plants will the special hybrid plants produce?

c. Show that when the daily yield diminution is 0.5 for each of the 100 days in the growing season, the annual yield will be 0.5 (exactly the yield of ordinary corn plants). Now suppose that through the use of an irrigation system, you arrange for the corn to receive 10 millimeters of rain each day. What would the sun index have to be each day to give an annual yield of 0.5? To answer this question, first simplify the daily yield diminution formula. Do this *before* you substitute any numbers into the formula.

d. Another assistant is also working with Greenthumb's formula, but he is getting different results. Something is wrong; for the situation given in part b, he finds that the daily yield diminution for one day is 34. He simplified Greenthumb's formula first, then substituted in the values for S and W. He shows you the following work. Find all of his mistakes and explain what he did wrong in making each of them.

$$\begin{aligned}
\text{Daily yield diminution} &= \frac{S^3 + W^3 - (S + W)^2}{(S + W)^3} \\
&= \frac{S^3 + W^3 - S^2 + W^2}{(S + W)^3} \\
&= \frac{(S^3 + W^3) - (S^2 - W^2)}{(S + W)^3} \\
&= \frac{(S + W)\cdot(S^2 + SW + W^2) - (S + W)\cdot(S - W)}{(S + W)^3} \\
&= \frac{(S^2 + SW + W^2) - S - W}{(S + W)^2} \\
&= \frac{S^2 + SW + W^2 - S - W}{S^2 + W^2} \\
&= SW - S - W
\end{aligned}$$

So for $S = 8$ and $W = 6$, he gets 34 as the daily yield diminution.

Chapter 6 REVIEW

SECTION 6.1 Finding the Domain of Rational Functions and Simplifying Rational Expressions

DEFINITIONS AND CONCEPTS	EXAMPLES
Division by 0 is undefined. Any value in a rational expression that will cause the denominator to be 0 must be excluded.	To find any restricted values for the variable in the expression $\frac{6x+7}{x^2-5x-6}$, factor the denominator to get $$\frac{6x+7}{(x-6)(x+1)}$$ The restricted values are 6 and -1 because both of these will create a 0 in the denominator.
Finding the domain of a rational function: The **domain** of a rational function is all values of the variable for which the function is defined.	From the work above, we have seen that 6 and -1 are not permissible replacements for x in the expression $\frac{6x+7}{x^2-5x-6}$. Therefore, the domain is the set of real numbers except 6 and -1. In interval notation, the domain is $(-\infty, -1) \cup (-1, 6) \cup (6, \infty)$.
Simplifying a rational expression: To simplify a rational expression, factor the numerator and denominator and divide out all factors common to the numerator and denominator. $\frac{ak}{bk} = \frac{a}{b}$ $(b \neq 0 \text{ and } k \neq 0)$	To simplify $\frac{x^2-5x+6}{x^2-9}$ $(x \neq 3, -3)$, factor the numerator and the denominator and divide out the resulting common factor: $\frac{(x-3)(x-2)}{(x+3)(x-3)}$ Factor the numerator and the denominator. $\frac{x-2}{x+3}$ Divide out $x-3$.

REVIEW EXERCISES

1. Evaluate $f(x) = \frac{3x+2}{x}$ when $x = 3$.

2. Evaluate $f(x) = \frac{3x+2}{x}$ when $x = 100$.

3. Graph the rational function $f(x) = \frac{3x+2}{x}$ $(x > 0)$. Find the equation of the horizontal and vertical asymptotes.

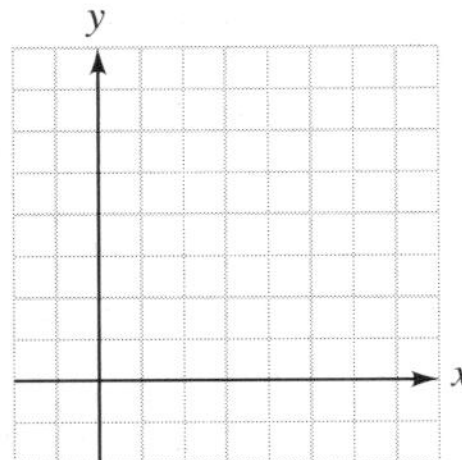

Simplify each rational expression. Assume no division by zero.

4. $\frac{248x^2y}{576xy^2}$

5. $\frac{212m^3n}{588m^2n^3}$

6. $\frac{x^2-49}{x^2+14x+49}$

7. $\frac{x^2+6x+36}{x^3-216}$

8. $\frac{x^2-2x+4}{2x^3+16}$

9. $\frac{x-y}{y-x}$

10. $\frac{2m-2n}{n-m}$

11. $\frac{m^3+m^2n-2mn^2}{2m^3-mn^2-m^2n}$

12. $\frac{ac-ad+bc-bd}{d^2-c^2}$

SECTION 6.2 Multiplying and Dividing Rational Expressions

DEFINITIONS AND CONCEPTS	EXAMPLES
Multiplying rational expressions: To multiply two rational expressions, follow the same procedure as multiplying two numerical fractions. $\frac{a}{b} \cdot \frac{c}{d} = \frac{ac}{bd}$ $(b, d \neq 0)$	$\frac{a+6}{(ax+4a-4x-16)} \cdot \frac{3a-12}{3a+18}$ $= \frac{(a+6)(3a-12)}{(ax+4a-4x-16)(3a+18)}$ Multiply the numerators and multiply the denominators. $= \frac{(a+6)3(a-4)}{(a-4)(x+4)3(a+6)}$ Factor. $= \frac{1}{x+4}$ Divide out all common factors.
Dividing rational expressions: To divide two rational expressions, follow the same procedure as dividing two numerical fractions. $\frac{a}{b} \div \frac{c}{d} = \frac{a}{b} \cdot \frac{d}{c}$ $(b, d, c \neq 0)$	$\frac{x^2-4}{2b-bx} \div \frac{x^2+4x+4}{2b+bx}$ $= \frac{x^2-4}{2b-bx} \cdot \frac{2b+bx}{x^2+4x+4}$ Multiply by the reciprocal of the second expression. $= \frac{(x^2-4)(2b+bx)}{(2b-bx)(x^2+4x+4)}$ Multiply the numerators and multiply the denominators. $= \frac{(x-2)(x+2)b(2+x)}{b(2-x)(x+2)(x+2)}$ Factor. $= \frac{x-2}{2-x}$ Divide out all common factors. $= -1$ $x - 2$ and $2 - x$ are opposites.

REVIEW EXERCISES

Perform the operations and simplify.

13. $\frac{x^2+4x+4}{x^2-x-6} \cdot \frac{x^2-9}{x^2+5x+6}$

14. $\frac{x^3-64}{x^2+4x+16} \div \frac{x^2-16}{x+4}$

15. $\frac{x^2+3x+2}{x^2-x-6} \cdot \frac{3x^2-3x}{x^2-3x-4} \div \frac{x^2+3x+2}{x^2-2x-8}$

16. $\frac{x^2-x-6}{x^2-3x-10} \div \frac{x^2-x}{x^2-5x} \cdot \frac{x^2-4x+3}{x^2-6x+9}$

SECTION 6.3 Adding and Subtracting Rational Expressions

DEFINITIONS AND CONCEPTS	EXAMPLES
Adding and subtracting rational expressions with the same denominator: To add or subtract two rational expressions with like denominators, add the numerators and keep the denominator. Simplify, if necessary. $\frac{a}{b} + \frac{c}{b} = \frac{a+c}{b}$ $(b \neq 0)$ $\frac{a}{b} - \frac{c}{b} = \frac{a-c}{b}$ $(b \neq 0)$	$\frac{2x+4}{x^2+13x+12} - \frac{x+3}{x^2+13x+12}$ $= \frac{2x+4-x-3}{x^2+13x+12}$ Subtract the numerators and keep the common denominator. $= \frac{x+1}{x^2+13x+12}$ Combine like terms. $= \frac{x+1}{(x+12)(x+1)}$ Factor the denominator. $= \frac{1}{x+12}$ Divide out $x + 1$.

Finding the LCD of two rational expressions:

To find the LCD of two fractions, factor each denominator and use each factor the greatest number of times that it appears in any one denominator. The product of these factors is the LCD.

To find the LCD of $\frac{2a}{a^2 - 2a - 8}$ and $\frac{3}{a^2 - 5a + 4}$, factor each denominator to get

$$\frac{2a}{(a-4)(a+2)} \qquad \frac{3}{(a-4)(a-1)}$$

The LCD is $(a-4)(a+2)(a-1)$.

Adding and subtracting rational expressions with different denominators.

To add or subtract two rational expressions with different denominators, find the LCD of the two expressions and write each fraction as an equivalent fraction with the new denominator. Add the numerators and keep the denominator. Simplify if necessary.

$$\frac{2a}{a^2 - 2a - 8} + \frac{3}{a^2 - 5a + 4}$$

$$= \frac{2a}{(a-4)(a+2)} + \frac{3}{(a-4)(a-1)} \quad \text{Factor each denominator.}$$

Write each fraction with the LCD of $(a-4)(a+2)(a-1)$, found above.

$$= \frac{2a(a-1)}{(a-4)(a+2)(a-1)} + \frac{3(a+2)}{(a-4)(a-1)(a+2)}$$

$$= \frac{2a^2 - 2a}{(a-4)(a+2)(a-1)} + \frac{3a+6}{(a-4)(a-1)(a+2)}$$

$$= \frac{2a^2 + a + 6}{(a-4)(a+2)(a-1)}$$

REVIEW EXERCISES

Perform the operations and simplify.

17. $\frac{5y}{x-y} - \frac{3}{x-y}$

18. $\frac{3x-1}{x^2+2} + \frac{3(x-2)}{x^2+2}$

19. $\frac{3}{x+2} + \frac{2}{x+3}$

20. $\frac{4x}{x-4} - \frac{3}{x+3}$

21. $\frac{2x}{x+1} + \frac{3x}{x+2} + \frac{4x}{x^2+3x+2}$

22. $\frac{5x}{x-3} + \frac{5}{x^2-5x+6} + \frac{x+3}{x-2}$

23. $\frac{3(x+2)}{x^2-1} - \frac{2}{x+1} + \frac{4(x+3)}{x^2-2x+1}$

24. $\frac{-2(3+x)}{x^2+6x+9} + \frac{3(x+2)}{x^2-6x+9} - \frac{1}{x^2-9}$

SECTION 6.4 Simplifying Complex Fractions

DEFINITIONS AND CONCEPTS | **EXAMPLES**

Simplifying complex fractions:

A fraction that has a fraction in its numerator or its denominator is called a **complex fraction.**

There are two ways to simplify a complex fraction.

Method 1: Eliminate the rational expressions in the numerator and denominator by writing the complex fraction as a division and using the division rule for fractions.

Method 2: Eliminate the rational expressions in the numerator and denominator by multiplying the expression by 1, written in the form $\frac{\text{LCD}}{\text{LCD}}$.

To use Method 2 to simplify $\frac{\frac{1}{a} + \frac{1}{b}}{\frac{a}{b} - \frac{b}{a}}$, proceed as follows:

$$\frac{\frac{1}{a} + \frac{1}{b}}{\frac{a}{b} - \frac{b}{a}} \cdot \frac{\mathbf{ab}}{\mathbf{ab}} \quad \text{Multiply by } \frac{ab}{ab} = 1.$$

$$= \frac{b+a}{a^2 - b^2} \quad \text{Simplify.}$$

$$= \frac{b+a}{(a+b)(a-b)} \quad \text{Factor.}$$

$$= \frac{1}{a-b} \quad \text{Simplify.}$$

REVIEW EXERCISES

Simplify each complex fraction.

25. $\dfrac{\frac{3}{x} - \frac{2}{y}}{xy}$

26. $\dfrac{\frac{1}{x} + \frac{2}{y}}{\frac{2}{x} - \frac{1}{y}}$

29. $\dfrac{1 - \frac{1}{x} - \frac{2}{x^2}}{1 + \frac{4}{x} + \frac{3}{x^2}}$

30. $\dfrac{x^{-1} + 1}{x + 1}$

27. $\dfrac{2x + 3 + \frac{1}{x}}{x + 2 + \frac{1}{x}}$

28. $\dfrac{6x + 13 + \frac{6}{x}}{6x + 5 - \frac{6}{x}}$

31. $\dfrac{x^{-1} - y^{-1}}{x^{-1} + y^{-1}}$

32. $\dfrac{(x - y)^{-2}}{x^{-2} - y^{-2}}$

SECTION 6.5 Solving Equations Containing Rational Expressions

DEFINITIONS AND CONCEPTS	EXAMPLES
Solving rational equations: To solve a rational equation, we can multiply both sides of the equation by the LCD of the rational expressions in the equation to clear it of fractions. Because this method can give *extraneous solutions,* it is necessary to check all possible solutions.	To solve $1 = \frac{3}{x-2} - \frac{12}{x^2-4}$, first determine that x cannot be 2 or -2. Then multiply both sides by the LCD of $(x + 2)(x - 2)$. $(x + 2)(x - 2)(1) = \left(\frac{3}{x-2} - \frac{12}{(x+2)(x-2)}\right)(x + 2)(x - 2)$ $(x + 2)(x - 2) = 3(x + 2) - 12$ $x^2 - 4 = 3x + 6 - 12$ Multiply. $x^2 - 4 = 3x - 6$ Combine terms. $x^2 - 3x + 2 = 0$ Add $-3x$ and 6 to both sides. $(x - 2)(x - 1) = 0$ Factor. $x - 2 = 0$ or $x - 1 = 0$ Use the zero-factor property. $x = 2$ | $x = 1$ Solve each linear equation. Since 2 is a restricted value, it is an extraneous solution. The only solution is 1.
Solving a formula for a specified variable: Clear the formula of any denominators and isolate the identified variable.	To solve $S = \frac{a}{1-r}$ for r, proceed as follows: $(1 - r)S = \frac{a}{1-r}(1 - r)$ Multiply both sides by $1 - r$. $S - rS = a$ Remove parentheses. $S = a + rS$ Add rS to both sides. $S - a = rS$ Subtract a from both sides. $\frac{S-a}{S} = r$ Divide by S. The result can also be written as $1 - \frac{a}{S} = r$.

REVIEW EXERCISES

Solve each equation, if possible.

33. $\frac{4}{x} - \frac{1}{10} = \frac{7}{2x}$

34. $\frac{2}{x+5} - \frac{1}{6} = \frac{1}{x+4}$

35. $\frac{2(x-5)}{x-2} = \frac{6x+12}{4-x^2}$

36. $\frac{7}{x+9} - \frac{x+2}{2} = \frac{x+4}{x+9}$

Solve each formula for the indicated variable.

37. $\frac{x^2}{a^2} - \frac{y^2}{b^2} = 1$ for y^2

38. $H = \frac{2ab}{a+b}$ for b

39. **Trip length** Traffic reduced Jim's usual speed by 10 mph, which lengthened his 200-mile trip by 1 hour. Find his usual average speed.

40. **Flying speed** On a 600-mile trip, a pilot can save 30 minutes by increasing her usual speed by 40 mph. Find her usual speed.

41. **Draining a tank** If one outlet pipe can drain a tank in 24 hours and another pipe can drain the tank in 36 hours, how long will it take for both pipes to drain the tank?

42. **Siding a house** Two men have estimated that they can side a house in 8 days. If one of them, who could have sided the house alone in 14 days, gets sick, how long will it take the other man to side the house alone?

SECTION 6.6 Dividing Polynomials

DEFINITIONS AND CONCEPTS

Dividing a monomial by a monomial:

To find the quotient of two monomials, express the quotient as a fraction and use the rules of exponents to simplify.

EXAMPLES

To simplify $\frac{25x^4y^7}{5xy^9}$ by using the rule of exponents, proceed as follows:

$$\frac{25x^4y^7}{5xy^9} = 5 \cdot x^{4-1} \cdot y^{7-9} \qquad \frac{x^4}{x} = x^{4-1}, \frac{y^7}{y^9} = y^{7-9}$$

$$= 5 \cdot x^3 \cdot y^{-2} \qquad \text{Simplify.}$$

$$= 5x^3 \cdot \frac{1}{y^2} \qquad \text{Write with positive exponents.}$$

$$= \frac{5x^3}{y^2} \qquad \text{Write as a single fraction.}$$

Dividing a polynomial by a monomial:

To find the quotient of a polynomial and a monomial, divide each term of the polynomial by the monomial and simplify each resulting rational expression.

To divide $\frac{45a^5b^3 - 63a^4b^4 - 9a^2b^3}{-9a^2b^3}$, write the fraction as three separate fractions and simplify each one.

$$\frac{45a^5b^3}{-9a^2b^3} - \frac{63a^4b^4}{-9a^2b^3} - \frac{9a^2b^3}{-9a^2b^3} = -5a^3 + 7a^2b + 1$$

Dividing a polynomial by a polynomial:

To find the quotient of two polynomials, use long division.

To divide $\frac{4a^3 + a^2 - 3a + 7}{a+1}$, use long division.

$$\begin{array}{r} 4a^2 - 3a + 0 \\ a + 1\overline{)4a^3 + a^2 - 3a + 7} \\ \underline{4a^3 + 4a^2} \qquad\qquad \\ -3a^2 - 3a \qquad \\ \underline{-3a^2 - 3a} \qquad \\ 0a + 7 \end{array}$$

Thus, $\frac{4a^3 + a^2 - 3a + 7}{a+1} = 4a^2 - 3a + \frac{7}{a+1}$.

REVIEW EXERCISES

Perform each division.

43. $\frac{-5x^6y^3}{10x^3y^6}$

44. $\frac{30x^3y^2 - 15x^2y - 10xy^2}{-10xy}$

45. $(3x^2 + 13xy - 10y^2) \div (3x - 2y)$

46. $(2x^3 + 7x^2 + 3 + 4x) \div (2x + 3)$

SECTION 6.7 Synthetic Division

DEFINITIONS AND CONCEPTS	EXAMPLES
Dividing a polynomial by a binomial of the form $x - r$ using synthetic division: Write the coefficients of the polynomial in the dividend and r in the divisor. Use multiplication and addition to complete the division.	Use synthetic division to perform the division $(5x^2 - 8x - 2) \div (x - 2)$ $\begin{array}{r\|rrr} 2 & 5 & -8 & -2 \\ & & 10 & 4 \\ \hline & 5 & 2 & 2 \end{array}$ Thus, $(5x^2 - 8x - 2) \div (x - 2) = 5x + 2 + \frac{2}{x-2}$.
Using the remainder theorem to evaluate a polynomial: **Remainder theorem:** If a polynomial $P(x)$ is divided by $x - r$, then the remainder is $P(r)$.	Given $P(x) = 8x^3 - 2x^2 - 9$, find $P(-2)$. $\begin{array}{r\|rrrr} -2 & 8 & -2 & 0 & -9 \\ & & -16 & 36 & -72 \\ \hline & 8 & -18 & 36 & -81 \end{array}$ Insert a 0 for the missing term. Thus, $P(-2) = -81$.
Using the factor theorem to determine if a binomial is a factor of a polynomial: **Factor theorem:** If $P(x)$ is divided by $x - r$, then $P(r) = 0$, if and only if $x - r$ is a factor of $P(x)$.	Determine whether $x + 4$ is a factor of $x^4 + 4x^3 + 9x^2 + 37x + 4$. $\begin{array}{r\|rrrrr} -4 & 1 & 4 & 9 & 37 & 4 \\ & & -4 & 0 & -36 & -4 \\ \hline & 1 & 0 & 9 & 1 & 0 \end{array}$ Since the remainder is 0, $x + 4$ is a factor of $x^4 + 4x^3 + 9x^2 + 37x + 4$.

REVIEW EXERCISES

Use the factor theorem to determine whether the first expression is a factor of $P(x)$.

47. $x - 5$; $P(x) = x^3 - 3x^2 - 8x - 10$

48. $x + 5$; $P(x) = x^3 + 4x^2 - 5x + 5$ (*Hint:* Write $x + 5$ as $x - (-5)$.)

SECTION 6.8 Proportion and Variation

DEFINITIONS AND CONCEPTS	EXAMPLES
Solving a proportion: In a proportion, the product of the extremes is equal to the product of the means.	To solve $\frac{x-1}{x+1} = \frac{2}{3x}$, proceed as follows: $3x(x - 1) = 2(x + 1)$ The product of the extremes is equal to the product of the means. $3x^2 - 3x = 2x + 2$ Remove parentheses. $3x^2 - 5x - 2 = 0$ Subtract $2x$ and 2 from both sides. $(3x + 1)(x - 2) = 0$ Factor. $3x + 1 = 0$ or $x - 2 = 0$ Use the zero-factor property. $x = -\frac{1}{3}$ | $x = 2$ Solve each linear equation. The solutions are $-\frac{1}{3}, 2$.

Using similar triangles to find the length of a missing side:

If two angles of one triangle have the same measure as two angles of a second triangle, the triangles are similar.

The corresponding sides of similar triangles are in proportion.

The two triangles below are similar. Find x.

$\frac{x}{5} = \frac{8}{20}$ Corresponding sides are in proportion.

$20x = 40$ The product of the extremes is equal to the product of the means.

$x = 2$ Solve for x.

8 ft

x ft

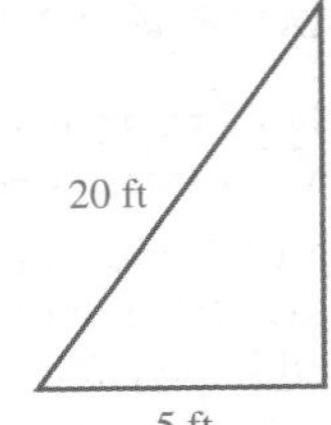

The length of the missing side is 2 ft.

Solving variation problems

Direct variation:

$y = kx$ (k is a constant)

Express each sentence as a formula:

The distance, d, a car travels is directly proportional to the time, t, it has been traveling.

$d = kt$

Inverse variation:

$y = \frac{k}{x}$ (k is a constant)

The temperature, T, of the coffee in the mug varies inversely to the time, t, it has been sitting on the counter.

$T = \frac{k}{t}$

Joint variation:

$y = kxz$ (k is a constant)

The interest, I, on the money in a bank account is jointly proportional to the principle, P, and the interest rate, r.

$I = kPr$

Combined variation:

$y = \frac{kx}{z}$ (k is a constant)

The pressure, P, of the gas varies directly as the temperature, T, and inversely as the volume, V.

$P = \frac{kT}{V}$

REVIEW EXERCISES

Solve each proportion.

49. $\frac{x+1}{8} = \frac{4x-2}{24}$

50. $\frac{1}{x+6} = \frac{x+10}{12}$

51. Find the height of a tree if it casts a 44-foot shadow when a 4-foot tree casts a $2\frac{1}{2}$-foot shadow.

52. Assume that x varies directly with y. If $x = 12$ when $y = 2$, find x when $y = 12$.

53. Assume that x varies inversely with y. If $x = 24$ when $y = 3$, find y when $x = 12$.

54. Assume that x varies jointly with y and z. Find the constant of variation if $x = 24$ when $y = 3$ and $z = 4$.

55. Assume that x varies directly with t and inversely with y. Find the constant of variation if $x = 2$ when $t = 8$ and $y = 64$.

56. **Taxes** The property taxes in a city vary directly with the assessed valuation. If a tax of \$1,575 is levied on a house assessed at \$90,000, find the tax on a building assessed at \$312,000.

Chapter 6 TEST

Simplify each rational expression. Assume no division by zero.

1. $\frac{-12x^2y^3z^2}{18x^3y^4z^2}$

2. $\frac{2x+4}{x^2-4}$

3. $\frac{3y-6z}{2z-y}$

4. $\frac{2x^2+7x+3}{4x+12}$

5. Evaluate $f(x) = \frac{5x - 2}{x}$ when $x = 10$.

Perform the operations and simplify, if necessary. Write all answers without negative exponents.

6. $\frac{x^2y^{-2}}{x^3z^2} \cdot \frac{x^2z^4}{y^2z}$

7. $\frac{(x + 1)(x + 2)}{10} \cdot \frac{5}{x + 2}$

8. $\frac{u^2 + 5u + 6}{u^2 - 4} \cdot \frac{u^2 - 5u + 6}{u^2 - 9}$

9. $\frac{x^3 + y^3}{4} \div \frac{x^2 - xy + y^2}{2x + 2y}$

10. $\frac{xu + 2u + 3x + 6}{u^2 - 9} \cdot \frac{2u - 6}{x^2 + 3x + 2}$

11. $\frac{a^2 + 7a + 12}{a + 3} \div \frac{16 - a^2}{a - 4}$

12. $\frac{3t}{t + 3} + \frac{9}{t + 3}$

13. $\frac{3w}{w - 5} + \frac{w + 10}{5 - w}$

14. $\frac{2}{r} + \frac{r}{s}$

15. $\frac{x + 2}{x + 1} - \frac{x + 1}{x + 2}$

Simplify each complex fraction.

16. $\frac{\frac{2u^2w^3}{v^2}}{\frac{4uw^4}{uv}}$

17. $\frac{\frac{x}{y} + \frac{1}{2}}{\frac{x}{2} - \frac{1}{y}}$

Solve each equation.

18. $\frac{2}{x - 1} + \frac{5}{x + 2} = \frac{11}{x + 2}$

19. $\frac{u - 2}{u - 3} + 3 = u + \frac{u - 4}{3 - u}$

Solve each formula for the indicated variable.

20. $\frac{x^2}{a^2} + \frac{y^2}{b^2} = 1$ for a^2

21. $\frac{1}{r} = \frac{1}{r_1} + \frac{1}{r_2}$ for r_2

22. **Sailing time** A boat sails a distance of 440 nautical miles. If the boat had averaged 11 nautical miles more each day, the trip would have required 2 fewer days. How long did the trip take?

23. **Investing** A student can earn \$300 interest annually by investing in a certificate of deposit at a certain interest rate. If she were to receive an annual interest rate that is 4% higher, she could receive the same annual interest by investing \$2,000 less. How much would she invest at each rate?

24. Divide: $\frac{18x^2y^3 - 12x^3y^2 + 9xy}{-3xy^4}$.

25. Divide: $(6x^3 + 5x^2 - 2) \div (2x - 1)$.

26. Find the remainder: $\frac{x^3 - 4x^2 + 5x + 3}{x + 1}$.

27. Use synthetic division to find the remainder when $4x^3 + 3x^2 + 2x - 1$ is divided by $x - 2$.

28. Find the height of a tree that casts a shadow of 12 feet when a vertical yardstick casts a shadow of 2 feet.

29. Solve the proportion: $\frac{3}{x - 2} = \frac{x + 3}{2x}$.

30. V varies inversely with t. If $V = 55$ when $t = 20$, find t when $V = 75$.

Cumulative Review Exercises

Simplify each expression. Assume no division by zero.

1. $a^3b^2a^5b^2$

2. $\frac{a^3b^6}{a^7b^2}$

3. $\left(\frac{2a^2}{3b^4}\right)^{-4}$

4. $\left(\frac{x^{-2}y^3}{x^2x^3y^4}\right)^{-3}$

Write each number in standard notation.

5. 4.25×10^4

6. 7.12×10^{-4}

Solve each equation.

7. $\frac{a + 2}{5} - \frac{8}{5} = 4a - \frac{a + 9}{2}$

8. $\frac{3x - 4}{6} - \frac{x - 2}{2} = \frac{-2x - 3}{3}$

Find the slope of the line with the given properties.

9. Passing through $(-2, 5)$ and $(4, 10)$

10. Has an equation of $3x + 4y = 13$

11. Parallel to a line with equation of $y = 3x + 2$

12. Perpendicular to a line with equation of $y = 3x + 2$

Let $f(x) = x^2 - 2x$ and find each value.

13. $f(0)$

14. $f(-2)$

15. $f\left(\frac{2}{5}\right)$

16. $f(t - 1)$

17. Express as a formula: *y varies directly with the product of x and z, but inversely with r.*

18. Does the graph below represent a function?

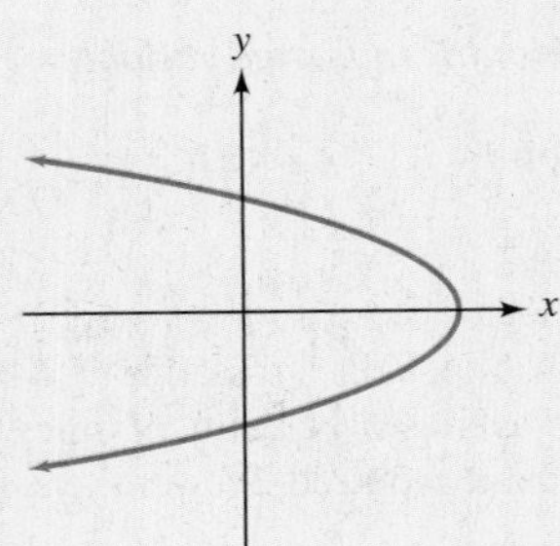

Solve each inequality, graph the solution set, and write the solution in interval notation.

19. $x - 2 \leq 3x + 1 \leq 5x - 4$

20. $\left|\frac{3a}{5} - 2\right| + 1 \geq \frac{6}{5}$

21. Is $3 + x + x^2$ a monomial, a binomial, or a trinomial?

22. Find the degree of $3 + x^2y + 17x^3y^4$.

23. If $f(x) = -3x^3 + x - 4$, find $f(-2)$.

24. Graph: $y = f(x) = 2x^2 - 3$.

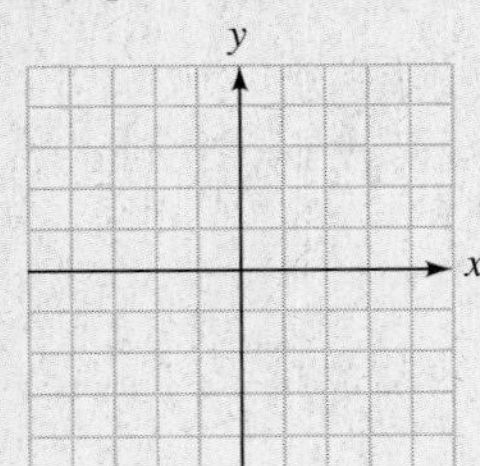

Perform the operations and simplify.

25. $(3x^2 - 2x + 7) + (-2x^2 + 2x + 5) + (3x^2 - 4x + 2)$

26. $(-5x^2 + 3x + 4) - (-2x^2 + 3x + 7)$

27. $(3x + 4)(2x - 5)$

28. $(2x^n - 1)(x^n + 2)$

Factor each expression.

29. $3r^2s^3 - 6rs^4$

30. $5(x - y) - a(x - y)$

31. $xu + yv + xv + yu$

32. $81x^4 - 16y^4$

33. $8x^3 - 27y^6$

34. $6x^2 + 5x - 6$

35. $9x^2 - 30x + 25$

36. $15x^2 - x - 6$

37. $27a^3 + 8b^3$

38. $6x^2 + x - 35$

39. $x^2 + 10x + 25 - y^4$

40. $y^2 - x^2 + 4x - 4$

Solve each equation.

41. $x^3 - 4x = 0$

42. $6x^2 + 7 = -23x$

Simplify each expression.

43. $\dfrac{2x^2y + xy - 6y}{3x^2y + 5xy - 2y}$

44. $\dfrac{x^2 - 4}{x^2 + 9x + 20} \div \dfrac{x^2 + 5x + 6}{x^2 + 4x - 5} \cdot \dfrac{x^2 + 3x - 4}{(x - 1)^2}$

45. $\dfrac{2}{x + y} + \dfrac{3}{x - y} - \dfrac{x - 3y}{x^2 - y^2}$

46. $\dfrac{\frac{a}{b} + b}{a - \frac{b}{a}}$

Solve each equation.

47. $\dfrac{5x - 3}{x + 2} = \dfrac{5x + 3}{x - 2}$

48. $\dfrac{3}{x - 2} + \dfrac{x^2}{(x + 3)(x - 2)} = \dfrac{x + 4}{x + 3}$

Perform the operation.

49. Divide: $(x^2 + 9x + 20) \div (x + 5)$.

50. Divide: $(2x^2 + 4x - x^3 + 3) \div (x - 1)$.

CHAPTER 7

Radicals and Rational Exponents

7.1 Radical Expressions

7.2 Applications of the Pythagorean Theorem and the Distance Formula

7.3 Rational Exponents

7.4 Simplifying and Combining Radical Expressions

7.5 Multiplying and Dividing Radical Expressions

7.6 Radical Equations

7.7 Complex Numbers

■ ***Projects***

CHAPTER REVIEW

CHAPTER TEST

Careers and Mathematics

PHOTOGRAPHERS

Photographers produce and preserve images that paint a picture, tell a story, or record an event. They use either a traditional camera that records images on silver halide film that is developed into prints or a digital camera that electronically records images. Photographers held about 129,000 jobs in 2004. More than half were self-employed. Employers usually seek applicants with a "good eye," imagination, and creativity. Entry-level positions in photojournalism generally require a college degree in journalism or photography.

Job Outlook:
Employment of photographers is expected to increase about as fast as the average for all occupations through 2014. However, photographers can expect keen competition for job openings because the work is attractive to many people.

Annual Earnings:
$18,380–$37,370

For More Information:
http://www.bls.gov/oco/ocos264.htm

For a Sample Application:
See Problem 107 in Section 7.5.

In this chapter ▶

In this chapter, we will reverse the squaring process and learn how to find square roots of numbers. We also will learn how to find other roots of numbers, solve radical equations, and discuss complex numbers.

SECTION 7.1 Radical Expressions

Objectives

1. Simplify a perfect-square root.
2. Simplify a perfect-square root expression.
3. Simplify a perfect-cube root.
4. Simplify a perfect *n*th root.
5. Find the domain of a square-root function and a cube-root function.
6. Use a square root to find the standard deviation of a set of data.

Vocabulary

square root	integer squares	index
radical sign	cube root	square-root function
radicand	odd root	cube-root function
principal square root	even root	standard deviation

Getting Ready

Find each power.

1. 0^2 **2.** 4^2 **3.** $(-4)^2$ **4.** -4^2

5. $\left(\frac{2}{5}\right)^3$ **6.** $\left(-\frac{3}{4}\right)^4$ **7.** $(7xy)^2$ **8.** $(7xy)^3$

In this section, we will discuss square roots and other roots of algebraic expressions. We also will consider their related functions.

1 Simplify a perfect-square root.

When solving problems, we often must find what number must be squared to obtain a second number a. If such a number can be found, it is called a **square root** of a. For example,

- 0 is a square root of 0, because $0^2 = 0$.
- 4 is a square root of 16, because $4^2 = 16$.
- -4 is a square root of 16, because $(-4)^2 = 16$.
- $7xy$ is a square root of $49x^2y^2$, because $(7xy)^2 = 49x^2y^2$.
- $-7xy$ is a square root of $49x^2y^2$, because $(-7xy)^2 = 49x^2y^2$.

All positive numbers have two real-number square roots: one that is positive and one that is negative.

EXAMPLE 1 Find the two square roots of 121.

Solution The two square roots of 121 are 11 and -11, because

$$11^2 = 121 \qquad \text{and} \qquad (-11)^2 = 121$$

SELF CHECK 1 Find the square roots of 144.

To express square roots, we use the symbol $\sqrt{}$, called a **radical sign.** For example,

$\sqrt{121} = 11$ Read as "The positive square root of 121 is 11."

$-\sqrt{121} = -11$ Read as "The negative square root of 121 is -11."

The number under the radical sign is called a **radicand.**

Square Root of *a*

If $a > 0$, $\sqrt{a}$ is the positive number whose square is a. In symbols,

$$\left(\sqrt{a}\right)^2 = a$$

The positive number $\sqrt{a}$ is called the **principal square root of *a*.**

If $a = 0$, $\sqrt{a} = \sqrt{0} = 0$. The principal square root of 0 is 0.

If $a < 0$, $\sqrt{a}$ is not a real number.

COMMENT The principal square root of a positive number is always positive. Although 5 and -5 are both square roots of 25, only 5 is the principal square root. The radical expression $\sqrt{25}$ represents 5. The radical expression $-\sqrt{25}$ represents -5.

COMMENT These examples suggest that if any number a can be factored into two equal factors, either of those factors is a square root of a.

Because of the previous definition, the square root of any number squared is that number. For example,

$$\left(\sqrt{10}\right)^2 = \sqrt{10} \cdot \sqrt{10} = 10 \qquad \left(\sqrt{a}\right)^2 = \sqrt{a} \cdot \sqrt{a} = a$$

EXAMPLE 2 Simplify each radical.

a. $\sqrt{1} = 1$

b. $\sqrt{81} = 9$

c. $-\sqrt{81} = -9$

d. $-\sqrt{225} = -15$

e. $\sqrt{\frac{1}{4}} = \frac{1}{2}$

f. $-\sqrt{\frac{16}{121}} = -\frac{4}{11}$

g. $\sqrt{0.04} = 0.2$

h. $-\sqrt{0.0009} = -0.03$

SELF CHECK 2 Simplify: **a.** $-\sqrt{49}$ **b.** $\sqrt{\frac{25}{49}}$.

Numbers such as 1, 4, 9, 16, 49, and 1,600 are called **integer squares,** because each one is the square of an integer. The square root of every integer square is an integer.

$$\sqrt{1} = 1 \qquad \sqrt{4} = 2 \qquad \sqrt{9} = 3 \qquad \sqrt{16} = 4 \qquad \sqrt{49} = 7 \qquad \sqrt{1{,}600} = 40$$

PERSPECTIVE Calculating Square Roots

The Bakhshali manuscript is an early mathematical manuscript that was discovered in India in the late 19th century. Mathematical historians estimate that the manuscript was written sometime around 400 A.D. One section of the manuscript presents a procedure for calculating square roots using basic arithmetic. Specifically, we can use the formula

$$\sqrt{Q} = A + \frac{b}{2A} - \left(\frac{b^2}{4A(2A^2 + b)}\right)$$

where A^2 = a perfect square close to the number Q, and $b = Q - A^2$.

Source: http://www.gap-system.org/~history/HistTopics/Bakhshali_manuscript.html

For example, if we want to compute an approximation of $\sqrt{21}$, we can choose $A^2 = 16$. Thus, $A = 4$ and $b = 21 - 16 = 5$. So we get

$$\begin{aligned}\sqrt{21} &= 4 + \frac{5}{(2)(4)} - \left(\frac{5^2}{(4)(4)((2)(4)^2 + 5)}\right)\\ &= 4 + \frac{5}{8} - \left(\frac{25}{(16)(37)}\right)\\ &= 4 + \frac{5}{8} - \frac{25}{592} \approx 4.58277027\end{aligned}$$

Using the square root key on a calculator, we see that, to nine decimal places, $\sqrt{21} = 4.582575695$. Therefore, the formula gives an answer that is correct to three decimal places.

1. Use the formula to approximate $\sqrt{105}$. How accurate is your answer?

2. Use the formula to approximate $\sqrt{627}$. How accurate is your answer?

The square roots of many positive integers are not rational numbers. For example, $\sqrt{11}$ is an *irrational number.* To find an approximate value of $\sqrt{11}$ with a calculator, we enter these numbers and press these keys.

11 **2ND** **√**	Using a scientific calculator
2ND **√** 11 **ENTER**	Using a graphing calculator

Either way, we will see that

$$\sqrt{11} \approx 3.31662479$$

Square roots of negative numbers are not real numbers. For example, $\sqrt{-9}$ is not a real number, because no real number squared equals -9. Square roots of negative numbers come from a set called *imaginary numbers,* which we will discuss later in this chapter.

2 Simplify a perfect-square root expression.

If $x \neq 0$, the positive number x^2 has x and $-x$ for its two square roots. To denote the positive square root of $\sqrt{x^2}$, we must know whether x is positive or negative.

If $x > 0$, we can write

$\sqrt{x^2} = x$ $\sqrt{x^2}$ represents the positive square root of x^2, which is x.

If x is negative, then $-x > 0$, and we can write

$\sqrt{x^2} = -x$ $\sqrt{x^2}$ represents the positive square root of x^2, which is $-x$.

If we don't know whether x is positive or negative, we must use absolute value symbols to guarantee that $\sqrt{x^2}$ is positive.

Definition of $\sqrt{x^2}$

If x can be any real number, then

$$\sqrt{x^2} = |x|$$

EXAMPLE 3 Simplify each expression. Assume that x can be any real number.

a. $\sqrt{16x^2} = \sqrt{(4x)^2}$ Write $16x^2$ as $(4x)^2$.

$= |4x|$ Because $(|4x|)^2 = 16x^2$. Since x could be negative, absolute value symbols are needed.

$= 4|x|$ Since 4 is a positive constant in the product $4x$, we can write it outside the absolute value symbols.

b. $\sqrt{x^2 + 2x + 1}$

$= \sqrt{(x + 1)^2}$ Factor $x^2 + 2x + 1$.

$= |x + 1|$ Because $(x + 1)^2 = x^2 + 2x + 1$. Since $x + 1$ can be negative (for example, when $x = -5$), absolute value symbols are needed.

c. $\sqrt{x^4} = x^2$ Because $(x^2)^2 = x^4$. Since $x^2 \geq 0$, no absolute value symbols are needed.

SELF CHECK 3 Simplify: **a.** $\sqrt{25a^2}$ **b.** $\sqrt{x^2 + 4x + 4}$ **c.** $\sqrt{16a^4}$.

3 Simplify a perfect-cube root.

The **cube root of x** is any number whose cube is x. For example,

4 is a cube root of 64, because $4^3 = 64$.
$3x^2y$ is a cube root of $27x^6y^3$, because $(3x^2y)^3 = 27x^6y^3$.
$-2y$ is a cube root of $-8y^3$, because $(-2y)^3 = -8y^3$.

Cube Root of a

The cube root of a is denoted as $\sqrt[3]{a}$ and is the number whose cube is a. In symbols,

$$\left(\sqrt[3]{a}\right)^3 = a$$

If a is any real number, then

$$\sqrt[3]{a^3} = a$$

We note that 64 has two real-number square roots, 8 and -8. However, 64 has only one real-number cube root, 4, because 4 is the only real number whose cube is 64. Since every real number has exactly one real cube root, it is unnecessary to use absolute value symbols when simplifying cube roots.

EXAMPLE 4 Simplify each radical.

a. $\sqrt[3]{125} = 5$ Because $5^3 = 5 \cdot 5 \cdot 5 = 125$

b. $\sqrt[3]{\dfrac{1}{8}} = \dfrac{1}{2}$ Because $\left(\frac{1}{2}\right)^3 = \frac{1}{2} \cdot \frac{1}{2} \cdot \frac{1}{2} = \frac{1}{8}$

c. $\sqrt[3]{-27x^3} = -3x$ Because $(-3x)^3 = (-3x)(-3x)(-3x) = -27x^3$

d. $\sqrt[3]{-\frac{8a^3}{27b^3}} = -\frac{2a}{3b}$ Because $\left(-\frac{2a}{3b}\right)^3 = \left(-\frac{2a}{3b}\right)\left(-\frac{2a}{3b}\right)\left(-\frac{2a}{3b}\right) = -\frac{8a^3}{27b^3}$

e. $\sqrt[3]{0.216x^3y^6} = 0.6xy^2$ Because $(0.6xy^2)^3 = (0.6xy^2)(0.6xy^2)(0.6xy^2) = 0.216x^3y^6$

SELF CHECK 4 Simplify: **a.** $\sqrt[3]{1,000}$ **b.** $\sqrt[3]{\frac{1}{27}}$ **c.** $\sqrt[3]{125a^3}$.

COMMENT The previous examples suggest that if a can be factored into three equal factors, any one of those factors is a cube root of a.

4 Simplify a perfect *n*th root.

Just as there are square roots and cube roots, there are fourth roots, fifth roots, sixth roots, and so on.

When n is an odd natural number greater than 1, $\sqrt[n]{x}$ represents an **odd root.** Since every real number has only one real nth root when n is odd, we don't need to use absolute value symbols when finding odd roots. For example,

$\sqrt[5]{243} = 3$ because $3^5 = 243$

$\sqrt[7]{-128x^7} = -2x$ because $(-2x)^7 = -128x^7$

When n is an even natural number greater than 1, $\sqrt[n]{x}$ represents an **even root.** In this case, there will be one positive and one negative real nth root. For example, the two real sixth roots of 729 are 3 and -3, because $3^6 = 729$ and $(-3)^6 = 729$. When finding even roots, we use absolute value symbols to guarantee that the principal nth root is positive.

$\sqrt[4]{(-3)^4} = |-3| = 3$ Because $3^4 = (-3)^4$. We also could simplify this as follows: $\sqrt[4]{(-3)^4} = \sqrt[4]{81} = 3$.

$\sqrt[6]{729x^6} = |3x| = 3|x|$ Because $(3|x|)^6 = 729x^6$. The absolute value symbols guarantee that the sixth root is positive.

In the radical $\sqrt[n]{x}$, n is called the **index** (or **order**) of the radical. When the index is 2, the radical is a square root, and we usually do not write the index.

$$\sqrt[2]{x} = \sqrt{x}$$

COMMENT When n is an even number greater than 1 and $x < 0$, $\sqrt[n]{x}$ is not a real number. For example, $\sqrt[4]{-81}$ is not a real number, because no real number raised to the 4th power is -81. However, when n is odd, $\sqrt[n]{x}$ is a real number.

EXAMPLE 5 Simplify each radical.

a. $\sqrt[4]{625} = 5$, because $5^4 = 625$ Read $\sqrt[4]{625}$ as "the fourth root of 625."

b. $\sqrt[5]{-32} = -2$, because $(-2)^5 = -32$ Read $\sqrt[5]{-32}$ as "the fifth root of -32."

c. $\sqrt[6]{\frac{1}{64}} = \frac{1}{2}$, because $\left(\frac{1}{2}\right)^6 = \frac{1}{64}$ Read $\sqrt[6]{\frac{1}{64}}$ as "the sixth root of $\frac{1}{64}$."

d. $\sqrt[7]{10^7} = 10$, because $10^7 = 10^7$ Read $\sqrt[7]{10^7}$ as "the seventh root of 10^7."

SELF CHECK 5 Simplify: **a.** $\sqrt[4]{\frac{1}{81}}$ **b.** $\sqrt[5]{10^5}$.

When finding the nth root of an nth power, we can use the following rules.

Definition of $\sqrt[n]{a^n}$

If n is an odd natural number greater than 1, then $\sqrt[n]{a^n} = a$.

If n is an even natural number, then $\sqrt[n]{a^n} = |a|$.

EXAMPLE 6 Simplify each radical. Assume that x can be any real number.

Solution

a. $\sqrt[5]{x^5} = x$ — Since n is odd, absolute value symbols aren't needed.

b. $\sqrt[4]{16x^4} = |2x| = 2|x|$ — Since n is even and x can be negative, absolute value symbols are needed to guarantee that the result is positive.

c. $\sqrt[6]{(x + 4)^6} = |x + 4|$ — Absolute value symbols are needed to guarantee that the result is positive.

d. $\sqrt[3]{(x + 1)^3} = x + 1$ — Since n is odd, absolute value symbols aren't needed.

e. $\sqrt{(x^2 + 6x + 9)^2} = \sqrt{[(x + 3)^2]^2}$ — Factor $x^2 + 6x + 9$.

$= \sqrt{(x + 3)^4}$

$= (x + 3)^2$ — Since $(x + 3)^2$ is always positive, absolute value symbols aren't needed.

SELF CHECK 6 Simplify: **a.** $\sqrt[4]{16a^4}$ **b.** $\sqrt[5]{(a + 5)^5}$.

We summarize the possibilities for $\sqrt[n]{x}$ as follows:

Definition for $\sqrt[n]{x}$

If n is a natural number greater than 1 and x is a real number, then

If $x > 0$, then $\sqrt[n]{x}$ is the positive number such that $\left(\sqrt[n]{x}\right)^n = x$.

If $x = 0$, then $\sqrt[n]{x} = 0$.

If $x < 0$ $\begin{cases} \text{and } n \text{ is odd, then } \sqrt[n]{x} \text{ is the real number such that } \left(\sqrt[n]{x}\right)^n = x. \\ \text{and } n \text{ is even, then } \sqrt[n]{x} \text{ is not a real number.} \end{cases}$

5 Find the domain of a square-root function and a cube-root function.

Since there is one principal square root for every nonnegative real number x, the equation $f(x) = \sqrt{x}$ determines a function, called the **square-root function.**

EXAMPLE 7 Consider the function $f(x) = \sqrt{x}$.

a. Find its domain. **b.** Graph the function. **c.** Find its range.

Solution

a. To find the domain, we note that $x \geq 0$ in the function because the radicand must be nonnegative. Thus, the domain is the set of nonnegative real numbers. In interval notation, the domain is the interval $[0, \infty)$.

b. We can make a table of values and plot points to get the graph shown in Figure 7-1(a). If we use a graphing calculator, we can choose window settings of $[-1, 9]$ for x and $[-2, 5]$ for y to get the graph shown in Figure 7-1(b). Since the equation defines a function, its graph passes the vertical line test.

$f(x) = \sqrt{x}$

x	$f(x)$	$(x, f(x))$
0	0	(0, 0)
1	1	(1, 1)
4	2	(4, 2)
9	3	(9, 3)

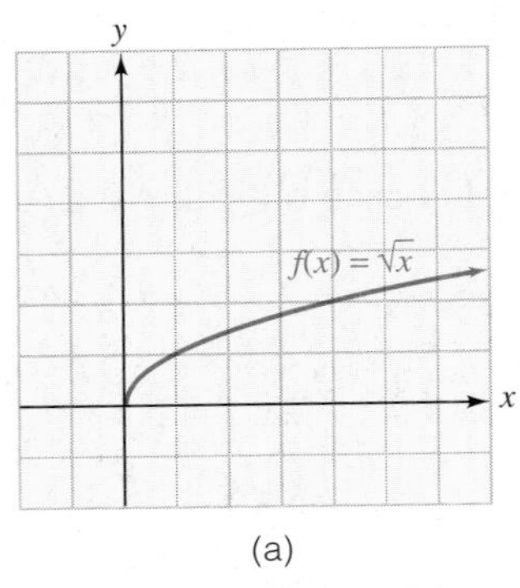

(a)

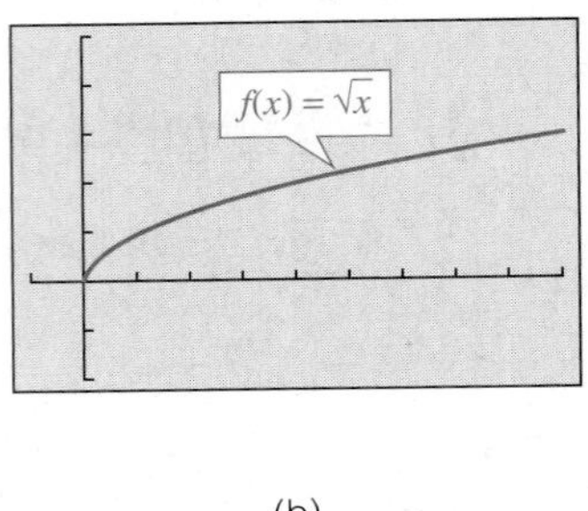

(b)

Figure 7-1

c. From either graph, we can see that the range of the function is the set of nonnegative real numbers, which is the interval $[0, \infty)$. The graph also confirms that the domain is the interval $[0, \infty)$.

SELF CHECK 7 Graph $f(x) = \sqrt{x} + 2$ and compare it to the graph of $f(x) = \sqrt{x}$. Find the domain and the range.

The graphs of many functions are translations or reflections of the square-root function. For example, if $k > 0$,

- The graph of $f(x) = \sqrt{x} + k$ is the graph of $f(x) = \sqrt{x}$ translated k units up.
- The graph of $f(x) = \sqrt{x} - k$ is the graph of $f(x) = \sqrt{x}$ translated k units down.
- The graph of $f(x) = \sqrt{x + k}$ is the graph of $f(x) = \sqrt{x}$ translated k units to the left.
- The graph of $f(x) = \sqrt{x - k}$ is the graph of $f(x) = \sqrt{x}$ translated k units to the right.
- The graph of $f(x) = -\sqrt{x}$ is the graph of $f(x) = \sqrt{x}$ reflected about the x-axis.

EXAMPLE 8 Consider the function $f(x) = -\sqrt{x + 4} - 2$.
a. Find its domain. **b.** Graph the function. **c.** Find its range.

Solution **a.** To find the domain, we note that the radicand must be nonnegative and solve the following inequality:

$$x + 4 \geq 0$$
$$x \geq -4$$

In interval notation, the domain is the interval $[-4, \infty)$.

b. This graph will be the reflection of $f(x) = \sqrt{x}$ about the x-axis, translated 4 units to the left and 2 units down. (See Figure 7-2(a).) We can confirm this graph by using a graphing calculator. If we choose window settings of $[-5, 6]$ for x and $[-6, 2]$ for y, we will get the graph shown in Figure 7-2(b).

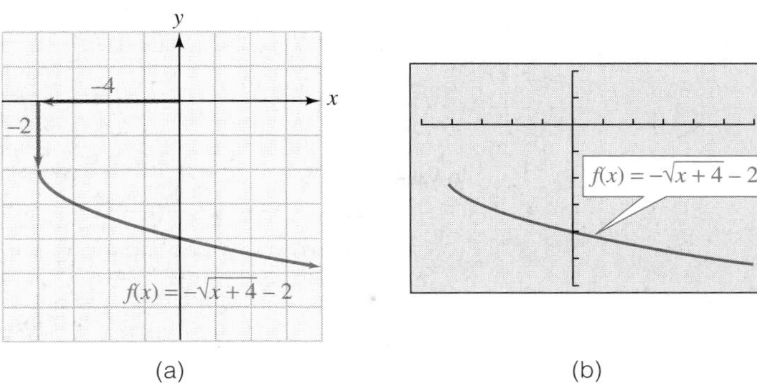

(a) (b)

Figure 7-2

c. From either graph, we can see that the range is the interval $(-\infty, -2]$. The graph also confirms that the domain is the interval $[-4, \infty)$.

SELF CHECK 8 Graph $f(x) = \sqrt{x-2} - 4$ and find the domain and range.

EXAMPLE 9 **PERIOD OF A PENDULUM** The *period of a pendulum* is the time required for the pendulum to swing back and forth to complete one cycle. (See Figure 7-3.) The period t (in seconds) is a function of the pendulum's length l, which is defined by the formula

$$t = f(l) = 2\pi\sqrt{\frac{l}{32}}$$

Find the period of a pendulum that is 5 feet long.

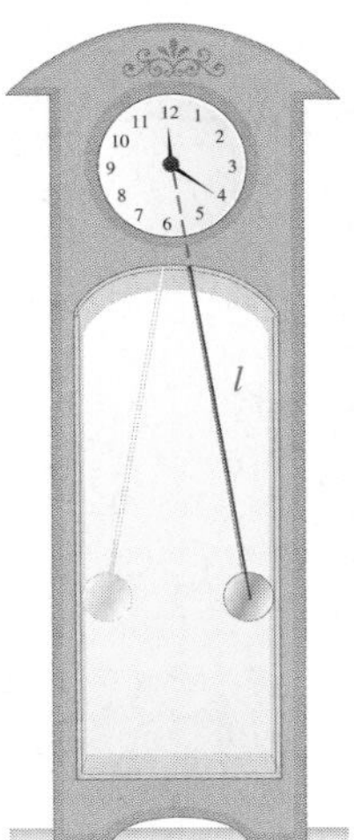

Figure 7-3

Solution We substitute 5 for l in the formula and simplify.

$$t = 2\pi\sqrt{\frac{l}{32}}$$

$$t = 2\pi\sqrt{\frac{5}{32}} \quad \text{Substitute.}$$

$$\approx 2.483647066 \quad \text{Use a calculator.}$$

To the nearest tenth, the period is 2.5 seconds.

SELF CHECK 9 To the nearest hundredth, find the period of a pendulum that is 3 feet long.

ACCENT ON TECHNOLOGY

Finding the Period of a Pendulum

To solve Example 9 with a graphing calculator with window settings of $[-2, 10]$ for x and $[-2, 10]$ for y, we graph the function $f(x) = 2\pi\sqrt{\frac{x}{32}}$, as in Figure 7-4(a) on the next page. We then trace and move the cursor toward an x-value of 5 until we see the coordinates shown in Figure 7-4(b). The period is given by the y-value shown on the screen. By zooming in, we can get better results.

(continued)

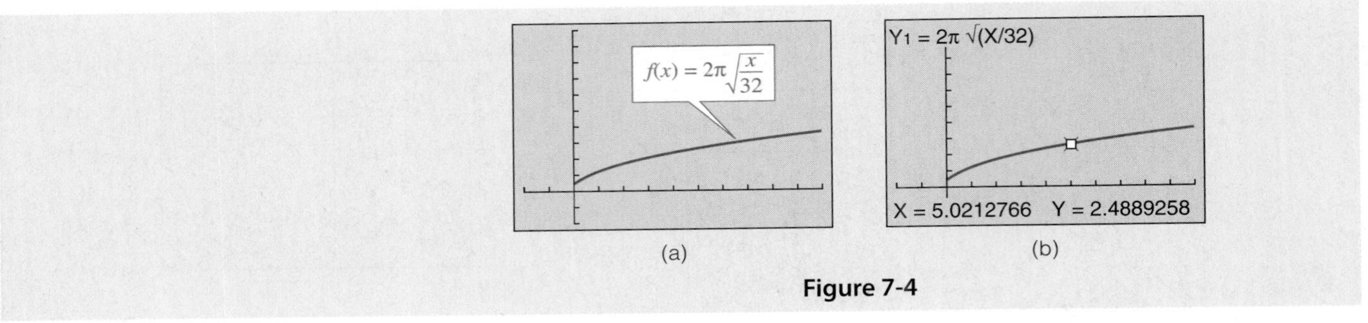

(a) (b)

Figure 7-4

The equation $f(x) = \sqrt[3]{x}$ defines the **cube-root function.** From the graph shown in Figure 7-5(a), we can see that the domain and range of the function $f(x) = \sqrt[3]{x}$ are the set of real numbers. Note that the graph of $f(x) = \sqrt[3]{x}$ passes the vertical line test. Figures 7-5(b) and 7-5(c) show several translations of the cube-root function.

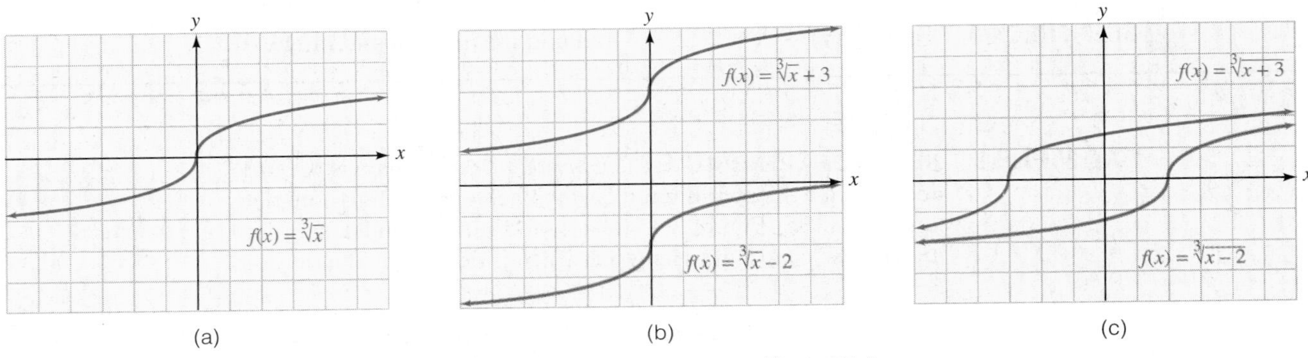

(a) (b) (c)

Figure 7-5

6 Use a square root to find the standard deviation of a set of data.

In statistics, the **standard deviation** of a data set is a measure of how tightly the data points are grouped around the mean (average) of the data set.

To see how to compute the standard deviation of a distribution, we consider the distribution 4, 5, 5, 8, 13 and construct the following table.

Original terms	Mean of the distribution	Differences (original term minus mean)	Squares of the differences from the mean
4	7	−3	9
5	7	−2	4
5	7	−2	4
8	7	1	1
13	7	6	36

The population *standard deviation* of the distribution is the positive square root of the mean of the numbers shown in column 4 of the table.

$$\text{Standard deviation} = \sqrt{\frac{\text{sum of the squares of the differences from the mean}}{\text{number of differences}}}$$

$$= \sqrt{\frac{9 + 4 + 4 + 1 + 36}{5}}$$

$$= \sqrt{\frac{54}{5}}$$

$$\approx 3.286335345 \quad \text{Use a calculator.}$$

To the nearest hundredth, the standard deviation of the given distribution is 3.29.

The symbol for the population standard deviation is σ, the lowercase Greek letter *sigma.*

EXAMPLE 10 Which of the following distributions has the most variability?

a. 3, 5, 7, 8, 12 **b.** 1, 4, 6, 11

Solution We compute the standard deviation of each distribution.

a.

Original terms	Mean of the distribution	Differences (original term minus mean)	Squares of the differences from the mean
3	7	−4	16
5	7	−2	4
7	7	0	0
8	7	1	1
12	7	5	25

$$\sigma = \sqrt{\frac{16 + 4 + 0 + 1 + 25}{5}} = \sqrt{\frac{46}{5}} \approx 3.03$$

b.

Original terms	Mean of the distribution	Differences (original term minus mean)	Squares of the differences from the mean
1	5.5	−4.5	20.25
4	5.5	−1.5	2.25
6	5.5	0.5	0.25
11	5.5	5.5	30.25

$$\sigma = \sqrt{\frac{20.25 + 2.25 + 0.25 + 30.25}{4}} = \sqrt{\frac{53}{4}} \approx 3.64$$

Since the standard deviation for the second distribution is greater than the standard deviation for the first distribution, the second distribution has the greater variability.

SELF CHECK ANSWERS

1. 12, −12 **2. a.** −7 **b.** $\frac{5}{7}$ **3. a.** $5|a|$ **b.** $|x + 2|$ **c.** $4a^2$ **4. a.** 10 **b.** $\frac{1}{3}$ **c.** $5a$ **5. a.** $\frac{1}{3}$ **b.** 10 **6. a.** $2|a|$ **b.** $a + 5$ **7.** It is 2 units higher.

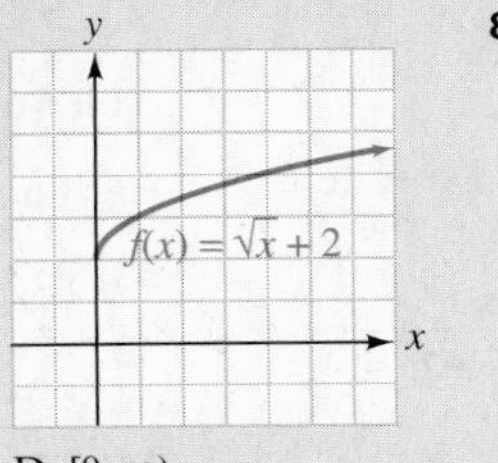

D: $[0, \infty)$
R: $[2, \infty)$

8.

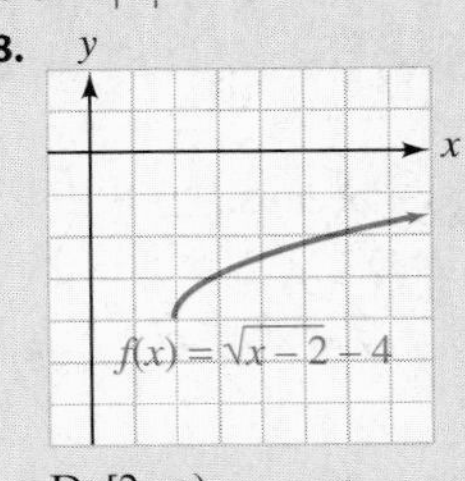

D: $[2, \infty)$
R: $[-4, \infty)$

9. 1.92 sec

NOW TRY THIS

1. Simplify: $\sqrt{100x^{100}}$.

2. Without using a calculator, between which two integers will the value of each expression be found?

   a. $\sqrt{8}$ b. $\sqrt{54}$ c. $\sqrt[3]{54}$

3. Given that $2^{-1} = \frac{1}{2}$, $2^0 = 1$, $2^1 = 2$, $2^2 = 4$, and $2^3 = 8$, between which two integers would you expect the value of each expression to be found?

   a. $2^{1/2}$ b. $2^{3/2}$ c. $2^{5/2}$

7.1 EXERCISES

WARM-UPS *Simplify each radical, if possible.*

1. $\sqrt{9}$
2. $-\sqrt{16}$
3. $\sqrt[3]{-8}$
4. $\sqrt[5]{32}$
5. $\sqrt{64x^2}$
6. $\sqrt[3]{-27x^3}$
7. $\sqrt{-3}$
8. $\sqrt[4]{(x+1)^8}$

REVIEW

Simplify each rational expression. Assume no division by zero.

9. $\dfrac{x^2+7x+12}{x^2-16}$
10. $\dfrac{a^3-b^3}{b^2-a^2}$

Perform the operations. Assume no division by zero.

11. $\dfrac{x^2-x-6}{x^2-2x-3} \cdot \dfrac{x^2-1}{x^2+x-2}$
12. $\dfrac{x^2-3x-4}{x^2-5x+6} \div \dfrac{x^2-2x-3}{x^2-x-2}$
13. $\dfrac{3}{m+1} + \dfrac{3m}{m-1}$
14. $\dfrac{2x+3}{3x-1} - \dfrac{x-4}{2x+1}$

VOCABULARY AND CONCEPTS

Fill in the blanks.

15. $5x^2$ is the square root of $25x^4$, because ______ $= 25x^4$ and 6 is a square root of 36 because ______.
16. The numbers 1, 4, 9, 16, 25, . . . are called ______.
17. The principal square root of x ($x > 0$) is the ______ square root of x.
18. The graph of $f(x) = \sqrt{x} + 3$ is the graph of $f(x) = \sqrt{x}$ translated __ units ___.
19. The graph of $f(x) = \sqrt{x+5}$ is the graph of $f(x) = \sqrt{x}$ translated __ units to the ___.
20. When n is an odd number greater than 1, $\sqrt[n]{x}$ represents an ______ root.
21. Given the radical $\sqrt[a]{b}$, the symbol $\sqrt{}$ is the ______ sign, a is the ______, and b is the ______.
22. The square-root function has the domain ______, while the cube-root function has the domain ______.
23. $\sqrt{x^2} =$ ___
24. $\left(\sqrt[3]{x}\right)^3 =$ __
25. $\sqrt[3]{x^3} =$ __
26. $\sqrt{0} =$ __
27. When n is a positive ______ number, $\sqrt[n]{x}$ represents an even root.
28. The ______ deviation of a set of numbers is the positive square root of the mean of the squares of the differences of the numbers from the mean.

Identify the radicand in each expression.

29. $\sqrt{3x^2}$
30. $5\sqrt{x}$
31. $ab^2\sqrt{a^2+b^3}$
32. $\frac{1}{2}x\sqrt{\frac{x}{y}}$

GUIDED PRACTICE

Find each square root, if possible. See Examples 1–2. (Objective 1)

33. $\sqrt{121}$
34. $\sqrt{144}$
35. $-\sqrt{64}$
36. $-\sqrt{1}$
37. $\sqrt{\dfrac{1}{9}}$
38. $-\sqrt{\dfrac{4}{25}}$

39. $-\sqrt{\frac{25}{49}}$ **40.** $\sqrt{\frac{49}{81}}$

41. $\sqrt{-25}$ **42.** $\sqrt{0.25}$

43. $\sqrt{0.16}$ **44.** $\sqrt{-49}$

Find each square root. Assume that all variables are unrestricted, and use absolute value symbols when necessary. See Example 3. (Objective 2)

45. $\sqrt{4x^2}$ **46.** $\sqrt{16y^4}$

47. $\sqrt{9a^4}$ **48.** $\sqrt{16b^2}$

49. $\sqrt{(t + 5)^2}$ **50.** $\sqrt{(a + 6)^2}$

51. $\sqrt{a^2 + 6a + 9}$ **52.** $\sqrt{x^2 + 10x + 25}$

Simplify each cube root. See Example 4. (Objective 3)

53. $\sqrt[3]{1}$ **54.** $\sqrt[3]{-8}$

55. $\sqrt[3]{-125}$ **56.** $\sqrt[3]{512}$

57. $\sqrt[3]{-\frac{8}{27}}$ **58.** $\sqrt[3]{\frac{125}{216}}$

59. $\sqrt[3]{0.064}$ **60.** $\sqrt[3]{0.001}$

61. $\sqrt[3]{8a^3}$ **62.** $\sqrt[3]{-27x^6}$

63. $\sqrt[3]{-1{,}000p^3q^3}$ **64.** $\sqrt[3]{343a^6b^3}$

Simplify each radical, if possible. See Example 5. (Objective 4)

65. $\sqrt[4]{81}$ **66.** $\sqrt[6]{64}$

67. $-\sqrt[5]{243}$ **68.** $-\sqrt[4]{625}$

69. $\sqrt[5]{-32}$ **70.** $\sqrt[6]{729}$

71. $\sqrt[4]{\frac{16}{625}}$ **72.** $\sqrt[5]{-\frac{243}{32}}$

73. $-\sqrt[5]{-\frac{1}{32}}$ **74.** $\sqrt[6]{-729}$

75. $\sqrt[4]{-256}$ **76.** $-\sqrt[4]{\frac{81}{256}}$

Simplify each radical. Assume that all variables are unrestricted, and use absolute value symbols where necessary. See Example 6. (Objective 4)

77. $\sqrt[4]{16x^4}$ **78.** $\sqrt[5]{32a^5}$

79. $\sqrt[3]{8a^3}$ **80.** $\sqrt[6]{64x^6}$

81. $\sqrt[4]{\frac{1}{16}x^4}$ **82.** $\sqrt[4]{\frac{1}{81}x^8}$

83. $\sqrt[4]{x^{12}}$ **84.** $\sqrt[8]{x^{24}}$

85. $\sqrt[5]{-x^5}$ **86.** $\sqrt[3]{-x^6}$

87. $\sqrt[3]{-27a^6}$ **88.** $\sqrt[5]{-32x^5}$

Find each value given that f(x) = $\sqrt{x - 4}$. (Objective 5)

89. $f(4)$ **90.** $f(8)$

91. $f(20)$ **92.** $f(29)$

Find each value given that g(x) = $\sqrt{x - 8}$. (Objective 5)

93. $g(9)$ **94.** $g(17)$

95. $g(8.25)$ **96.** $g(8.64)$

Graph each function and find its domain and range. See Examples 7–8. (Objective 5)

97. $f(x) = \sqrt{x + 4}$ **98.** $f(x) = -\sqrt{x - 2}$

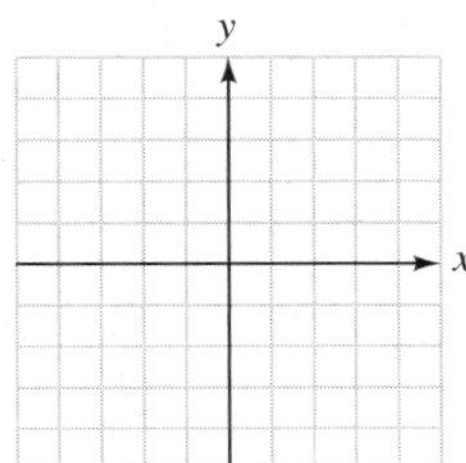

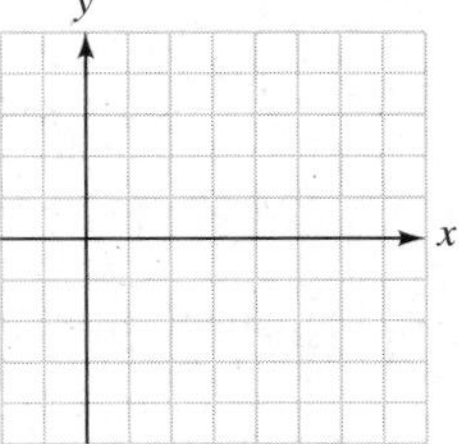

99. $f(x) = \sqrt[3]{x} - 1$ **100.** $f(x) = -\sqrt{x} - 3$

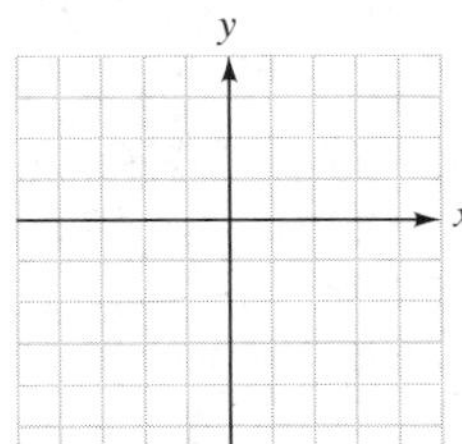

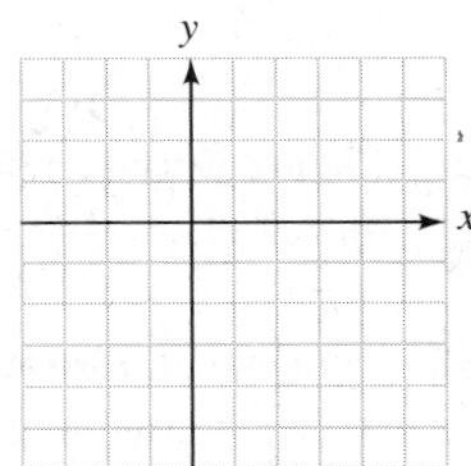

ADDITIONAL PRACTICE

Simplify each radical. Assume that all variables are unrestricted, and use absolute value symbols where necessary.

101. $\sqrt{(-4)^2}$ **102.** $\sqrt{(-9)^2}$

103. $\sqrt{-36}$ **104.** $-\sqrt{-4}$

105. $\sqrt{(-5b)^2}$ **106.** $\sqrt{(-8c)^2}$

107. $\sqrt{t^2 + 24t + 144}$ **108.** $\sqrt{m^2 + 30m + 225}$

109. $\sqrt[3]{-\frac{1}{8}m^6n^3}$ **110.** $\sqrt[3]{\frac{27}{1{,}000}a^6b^6}$

111. $\sqrt[3]{0.008z^9}$ **112.** $\sqrt[3]{0.064s^9t^6}$

113. $\sqrt[25]{(x + 2)^{25}}$ **114.** $\sqrt[44]{(x + 4)^{44}}$

115. $\sqrt[8]{0.00000001x^{16}y^8}$ **116.** $\sqrt[5]{0.00032x^{10}y^5}$

Use a calculator to find each square root. Give the answer to four decimal places.

117. $\sqrt{12}$ **118.** $\sqrt{340}$

119. $\sqrt{679.25}$ **120.** $\sqrt{0.0063}$

Find each value given that f(x) = $\sqrt{x^2 + 1}$. Give each answer to four decimal places.

121. $f(4)$ **122.** $f(6)$

123. $f(2.35)$ **124.** $f(21.57)$

APPLICATIONS *Use a calculator to solve each problem.*

See Examples 9–10. (Objectives 2 and 6)

125. Find the standard deviation of the following distribution to the nearest hundredth: 2, 5, 5, 6, 7.

126. Find the standard deviation of the following distribution to the nearest hundredth: 3, 6, 7, 9, 11, 12.

127. Statistics In statistics, the formula

$$s_{\bar{x}} = \frac{s}{\sqrt{N}}$$

gives an estimate of the standard error of the mean. Find $s_{\bar{x}}$ to four decimal places when $s = 65$ and $N = 30$.

128. Statistics In statistics, the formula

$$\sigma_{\bar{x}} = \frac{\sigma}{\sqrt{N}}$$

gives the standard deviation of means of samples of size N. Find $\sigma_{\bar{x}}$ to four decimal places when $\sigma = 12.7$ and $N = 32$.

129. Radius of a circle The radius r of a circle is given by the formula $r = \sqrt{\frac{A}{\pi}}$, where A is its area. Find the radius of a circle whose area is 9π square units.

130. Diagonal of a baseball diamond The diagonal d of a square is given by the formula $d = \sqrt{2s^2}$, where s is the length of each side. Find the diagonal of the baseball diamond.

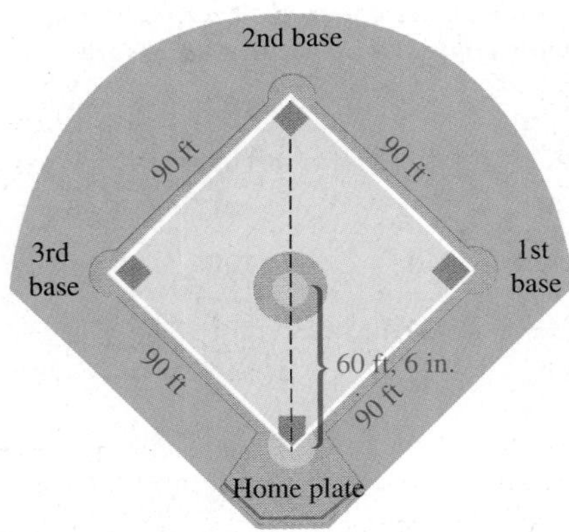

131. Falling objects The time t (in seconds) that it will take for an object to fall a distance of s feet is given by the formula

$$t = \frac{\sqrt{s}}{4}$$

If a stone is dropped down a 256-foot well, how long will it take it to hit bottom?

132. Law enforcement Police sometimes use the formula $s = k\sqrt{l}$ to estimate the speed s (in mph) of a car involved in an accident. In this formula, l is the length of the skid in feet, and k is a constant depending on the condition of the pavement. For wet pavement, $k \approx 3.24$. How fast was a car going if its skid was 400 feet on wet pavement?

133. Electronics When the resistance in a circuit is 18 ohms, the current I (measured in amperes) and the power P (measured in watts) are related by the formula

$$I = \sqrt{\frac{P}{18}}$$

Find the current used by an electrical appliance that is rated at 980 watts.

134. Medicine The approximate pulse rate p (in beats per minute) of an adult who is t inches tall is given by the formula

$$p = \frac{590}{\sqrt{t}}$$

Find the approximate pulse rate of an adult who is 71 inches tall.

WRITING ABOUT MATH

135. If x is any real number, then $\sqrt{x^2} = x$ is not correct. Explain.

136. If x is any real number, then $\sqrt[3]{x^3} = |x|$ is not correct. Explain.

SOMETHING TO THINK ABOUT

137. Is $\sqrt{x^2 - 4x + 4} = x - 2$? What are the exceptions?

138. When is $\sqrt{x^2} \neq x$?

SECTION 7.2 Applications of the Pythagorean Theorem and the Distance Formula

Objectives

1. Apply the Pythagorean theorem to find the missing length of one side of a right triangle.
2. Find the distance between two points on the coordinate plane.

Vocabulary

hypotenuse   Pythagorean theorem

Getting Ready

Evaluate each expression.

1. $3^2 + 4^2$
2. $5^2 + 12^2$
3. $(5 - 2)^2 + (2 + 1)^2$
4. $(111 - 21)^2 + (60 - 4)^2$

In this section, we will discuss the Pythagorean theorem, a theorem that shows the relationship of the sides of a right triangle. We will then use this theorem to develop a formula that gives the distance between two points on the coordinate plane.

1 Apply the Pythagorean theorem to find the missing length of one side of a right triangle.

If we know the lengths of two legs of a right triangle, we can find the length of the **hypotenuse** (the side opposite the 90° angle) by using the **Pythagorean theorem.** In fact, if we know the lengths of any two sides of a right triangle, we can find the length of the third side.

Pythagorean Theorem

If a and b are the lengths of two legs of a right triangle and c is the length of the hypotenuse, then

$$a^2 + b^2 = c^2$$

In words, the Pythagorean theorem says,

In any right triangle, the square of the length of the hypotenuse is equal to the sum of the squares of the lengths of the two legs.

Figure 7-6

Suppose the right triangle shown in Figure 7-6 has legs of length 3 and 4 units. To find the length of the hypotenuse, we can use the Pythagorean theorem.

$$a^2 + b^2 = c^2$$
$$3^2 + 4^2 = c^2 \quad \text{Substitute.}$$
$$9 + 16 = c^2 \quad \text{Simplify.}$$
$$25 = c^2 \quad \text{Add.}$$

To find c, we ask "what number, when squared is equal to 25?" There are two such numbers: the positive square root of 25 and the negative square root of 25. Since c represents the length of the hypotenuse and cannot be negative, it follows that c is the positive square root of 25.

$$\sqrt{25} = c \quad \text{Recall that the radical symbol } \sqrt{} \text{ represents the positive, or principal, square root of a number.}$$
$$5 = c$$

The length of the hypotenuse is 5 units.

EXAMPLE 1 **FIGHTING FIRES** To fight a forest fire, the forestry department plans to clear a rectangular fire break around the fire, as shown in Figure 7-7. Crews are equipped with mobile communications with a 3,000-yard range. Can crews at points A and B remain in radio contact?

Solution Points A, B, and C form a right triangle. The lengths of its sides are represented as a, b, and c where a is opposite point A, b is opposite point B, and c is opposite point C. To find the distance c, we can use the Pythagorean theorem, substituting 2,400 for a and 1,000 for b and solving for c.

$$a^2 + b^2 = c^2$$
$$\mathbf{2{,}400}^2 + \mathbf{1{,}000}^2 = c^2 \quad \text{Substitute.}$$
$$5{,}760{,}000 + 1{,}000{,}000 = c^2 \quad \text{Square each value.}$$
$$6{,}760{,}000 = c^2 \quad \text{Add.}$$
$$\sqrt{6{,}760{,}000} = c \quad \text{Since } c \text{ represents a length, it must be the positive square root of 6,760,000.}$$
$$2{,}600 = c \quad \text{Use a calculator to find the square root.}$$

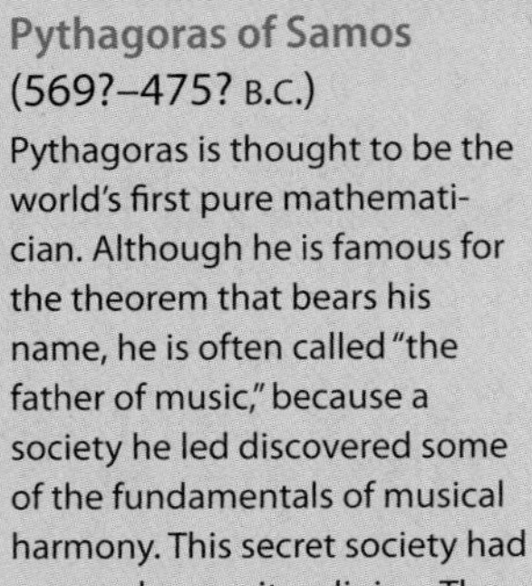

Pythagoras of Samos (569?–475? B.C.)

Pythagoras is thought to be the world's first pure mathematician. Although he is famous for the theorem that bears his name, he is often called "the father of music," because a society he led discovered some of the fundamentals of musical harmony. This secret society had numerology as its religion. The society is also credited with the discovery of irrational numbers.

Figure 7-7

The two crews are 2,600 yards apart. Because this distance is less than the range of the radios, they can communicate.

SELF CHECK 1 Can the crews communicate if $b = 1{,}500$ yards?

PERSPECTIVE

Pythagoras was a teacher. Although it was unusual at that time, his classes were coeducational. He and his followers formed a secret society with two rules: Membership was for life, and members could not reveal the secrets they knew.

Much of their teaching was good mathematics, but some ideas were strange. To them, numbers were sacred. Because beans were used as counters to represent numbers, Pythagoreans refused to eat beans. They also believed that the *only* numbers were the whole numbers. To them, fractions were not numbers; $\frac{2}{3}$ was just a way of comparing the whole numbers 2 and 3. They believed that whole numbers were the building blocks of the universe. The basic Pythagorean doctrine was, "All things are numbers," and they meant *whole* numbers.

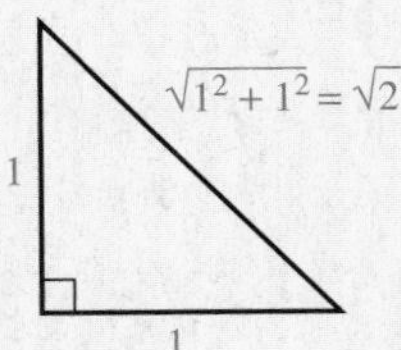

The Pythagorean theorem was an important discovery of the Pythagorean school, yet it caused some controversy. The right triangle in the illustration has two legs of length 1. By the Pythagorean theorem, the length of the hypotenuse is $\sqrt{2}$. One of their own group, Hippasus of Metapontum, discovered that $\sqrt{2}$ is an irrational number: There are *no* whole numbers a and b that make the fraction $\frac{a}{b}$ exactly equal to $\sqrt{2}$. This discovery was not appreciated by the other Pythagoreans. How could everything in the universe be described with whole numbers, when the side of this simple triangle couldn't? The Pythagoreans had a choice. Either expand their beliefs, or cling to the old. According to legend, the group was at sea at the time of the discovery. Rather than upset the system, they threw Hippasus overboard.

2 Find the distance between two points on the coordinate plane.

We can use the Pythagorean theorem to develop a formula to find the distance between any two points that are graphed on a rectangular coordinate system.

To find the distance d between points P and Q shown in Figure 7-8, we construct the right triangle PRQ. Because line segment RQ is vertical, point R will have the same x-coordinate as point Q. Because line segment PR is horizontal, point R will have same y-coordinate as point P. The distance between P and R is $|x_2 - x_1|$, and the distance between R and Q is $|y_2 - y_1|$. We apply the Pythagorean theorem to the right triangle PRQ to get

$$(PQ)^2 = (PR)^2 + (RQ)^2$$ Read PQ as "the length of segment PQ."

$$d^2 = |x_2 - x_1|^2 + |y_2 - y_1|^2$$ Substitute the value of each expression.

$$d^2 = (x_2 - x_1)^2 + (y_2 - y_1)^2$$ Because $|x_2 - x_1|^2 = (x_2 - x_1)^2$ and $|y_2 - y_1|^2 = (y_2 - y_1)^2$

(1) $$d = \sqrt{(x_2 - x_1)^2 + (y_2 - y_1)^2}$$ Since d represents a length, it must be the positive square root of $(x_2 - x_1)^2 + (y_2 - y_1)^2$.

Equation 1 is called the **distance formula.**

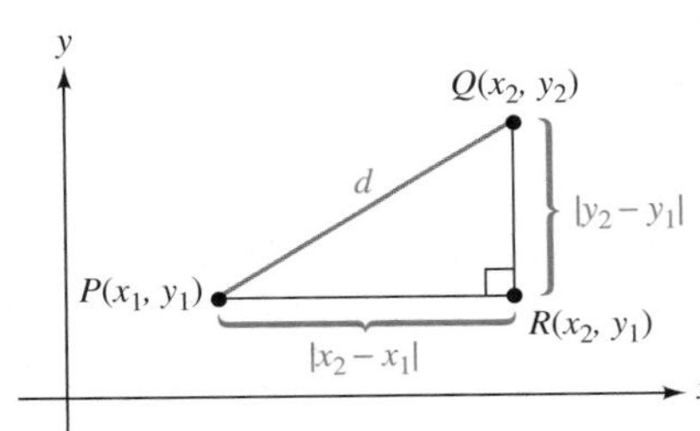

Figure 7-8

Distance Formula The distance d between two points (x_1, y_1) and (x_2, y_2) is given by the formula

$$d = \sqrt{(x_2 - x_1)^2 + (y_2 - y_1)^2}$$

EXAMPLE 2 Find the distance between the points $(-2, 3)$ and $(4, -5)$.

Solution To find the distance, we can use the distance formula by substituting 4 for x_2, -2 for x_1, -5 for y_2, and 3 for y_1.

COMMENT Recall that $\sqrt{a^2 + b^2} \neq \sqrt{a} + \sqrt{b}$.

$$\begin{aligned} d &= \sqrt{(x_2 - x_1)^2 + (y_2 - y_1)^2} \\ &= \sqrt{[4 - (-2)]^2 + (-5 - 3)^2} && \text{Substitute.} \\ &= \sqrt{(4 + 2)^2 + (-5 - 3)^2} && \text{Simplify.} \\ &= \sqrt{6^2 + (-8)^2} && \text{Simplify.} \\ &= \sqrt{36 + 64} && \text{Square each value.} \\ &= \sqrt{100} && \text{Add.} \\ &= 10 && \text{Take the square root.} \end{aligned}$$

The distance between the two points is 10 units.

SELF CHECK 2 Find the distance between $P(-2, -2)$ and $Q(3, 10)$.

EXAMPLE 3 **BUILDING A FREEWAY** In a city, streets run north and south, and avenues run east and west. Streets are 850 feet apart and avenues are 850 feet apart. The city plans to construct a straight freeway from the intersection of 25th Street and 8th Avenue to the intersection of 115th Street and 64th Avenue. How long will the freeway be?

Solution We can represent the roads by the coordinate system in Figure 7-9, where the units on each axis represent 850 feet. We represent the end of the freeway at 25th Street and 8th Avenue by the point $(x_1, y_1) = (25, 8)$. The other end is $(x_2, y_2) = (115, 64)$.

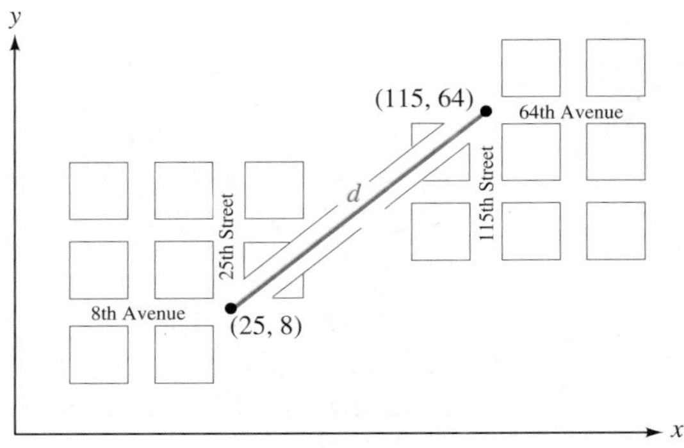

Figure 7-9

We can use the distance formula to find the number of units between the two designated points.

$d = \sqrt{(x_2 - x_1)^2 + (y_2 - y_1)^2}$

$d = \sqrt{(\mathbf{115} - \mathbf{25})^2 + (\mathbf{64} - \mathbf{8})^2}$ Substitute.

$= \sqrt{90^2 + 56^2}$ Remove parentheses.

$= \sqrt{8{,}100 + 3{,}136}$ Square each value.

$= \sqrt{11{,}236}$ Add.

$= 106$ Use a calculator to find the square root.

There are approximately 106 units between the two designated points, and because each unit is 850 feet, the length of the freeway is $106(850) = 90{,}100$ feet. Since 5,280 feet $=$ 1 mile, we can divide 90,100 by 5,280 to convert 90,100 feet to 17.064394 miles. Thus, the freeway will be about 17 miles long.

SELF CHECK 3 Find the diagonal distance in feet from the intersection of 25th Street and 8th Avenue to the intersection of 28th Street and 12th Avenue.

EXAMPLE 4 **BOWLING** The velocity, v, of an object after it has fallen d feet is given by the equation $v^2 = 64d$. If an inexperienced bowler lofts the ball 4 feet, with what velocity does it strike the alley?

Solution We find the velocity by substituting 4 for d in the equation $v^2 = 64d$ and solving for v.

$v^2 = 64\mathbf{d}$

$v^2 = 64(\mathbf{4})$ Substitute.

$v^2 = 256$ Multiply.

$v = \sqrt{256}$ Since v represents a velocity, it must be the positive square root of 256.

$= 16$ Simplify.

The ball strikes the alley with a velocity of 16 feet per second.

SELF CHECK 4 Find the velocity if the bowler lofts the ball 3 feet. Round to the nearest tenth.

SELF CHECK ANSWERS **1.** yes **2.** 13 units **3.** 4,250 ft **4.** 13.9 ft/sec

NOW TRY THIS

1. Determine whether the points $(4, -2)$, $(-2, -4)$, and $(-4, 2)$ are the vertices of an *isosceles triangle* (two equal sides), an *equilateral triangle* (three equal sides), or neither. Is the triangle a right triangle?
2. Find the distance between points with coordinates of $(2x + 1, x + 1)$ and $(2x - 3, x - 2)$.

7.2 EXERCISES

WARM-UPS *Evaluate each expression.*

1. $\sqrt{25}$
2. $\sqrt{169}$
3. $\sqrt{3^2 + 4^2}$
4. $\sqrt{5^2 + 12^2}$
5. $\sqrt{5^2 - 3^2}$
6. $\sqrt{5^2 - 4^2}$

REVIEW *Find each product.*

7. $(4x + 2)(3x - 5)$
8. $(3y - 5)(2y + 3)$
9. $(5t + 4s)(3t - 2s)$
10. $(4r - 3)(2r^2 + 3r - 4)$

VOCABULARY AND CONCEPTS *Fill in the blanks.*

11. In a right triangle, the side opposite the 90° angle is called the ________.

12. In a right triangle, the two shorter sides are called ____.

13. If a and b are the lengths of two legs of a right triangle and c is the length of the hypotenuse, then ________.

14. In any right triangle, the square of the length of the hypotenuse is equal to the ____ of the squares of the lengths of the two ____. This fact is known as the ____________.

15. If $x^2 = 25$ and x is positive, we can conclude that x is the ______ square root of 25. Thus, $x = 5$.

16. The formula for finding the distance between two points on a rectangular coordinate system is

$d =$ ________________.

GUIDED PRACTICE

The lengths of two sides of the right triangle ABC shown in the illustration are given. Find the length of the missing side. See Example 1. (Objective 1)

17. $a = 6$ ft and $b = 8$ ft
18. $a = 10$ cm and $c = 26$ cm
19. $b = 18$ m and $c = 82$ m
20. $b = 7$ ft and $c = 25$ ft
21. $a = 14$ in. and $c = 50$ in.
22. $a = 8$ cm and $b = 15$ cm
23. $a = \sqrt{6}$ mi and $b = \sqrt{3}$ mi
24. $a = \sqrt{15}$ ft and $b = \sqrt{21}$ ft

Find the distance between the given points. If an answer is not exact, use a calculator and give an approximation to the nearest tenth. See Example 2. (Objective 2)

25. $(0, 0), (3, -4)$
26. $(0, 0), (-6, 8)$
27. $(2, 4), (5, 8)$
28. $(5, 9), (8, 13)$
29. $(-2, -8), (3, 4)$
30. $(-5, -2), (7, 3)$
31. $(6, 8), (12, 16)$
32. $(10, 4), (2, -2)$
33. $(-3, 5), (-5, -5)$
34. $(2, -3), (4, -8)$
35. $(-9, 3), (4, 7)$
36. $(-1, -3), (-5, 8)$

ADDITIONAL PRACTICE

In Exercises 37–40, use a calculator to approximate each value to the nearest foot. The baseball diamond is a square, 90 feet on a side.

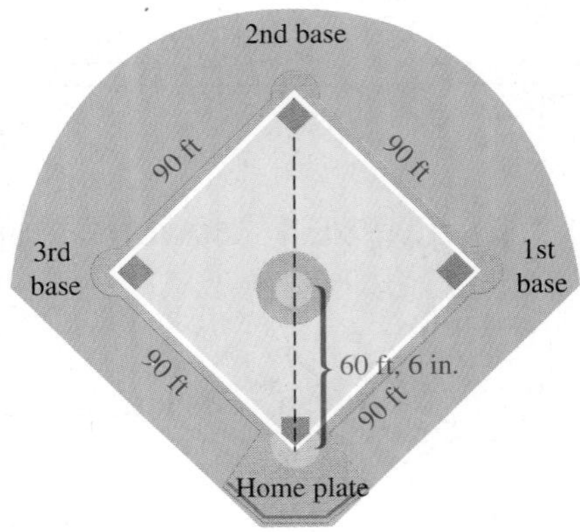

37. Baseball How far must a catcher throw the ball to throw out a runner stealing second base?

38. Baseball In baseball, the pitcher's mound is 60 feet, 6 inches from home plate. How far from the mound is second base?

39. Baseball If the third baseman fields a ground ball 10 feet directly behind third base, how far must he throw the ball to throw a runner out at first base?

40. Baseball The shortstop fields a grounder at a point one-third of the way from second base to third base. How far will he have to throw the ball to make an out at first base?

For Exercises 41–42, approximate each answer to the nearest tenth.

41. Geometry Find the length of the diagonal of one of the faces of the cube.

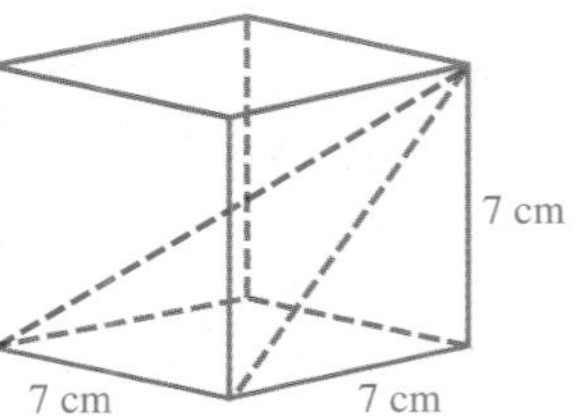

42. Geometry Find the length of the diagonal of the cube shown in the illustration in Exercise 41.

43. Geometry Show that the point (5, 1) is equidistant from points (7, 0) and (3, 0).

44. Geometry Show that a triangle with vertices at (2, 3), (−3, 4), and (1, −2) is a right triangle. (*Hint:* If the Pythagorean theorem holds, the triangle is a right triangle.)

45. Geometry Show that a triangle with vertices at (−2, 4), (2, 8), and (6, 4) is isosceles.

46. Geometry Show that a triangle with vertices at (−2, 13), (−8, 9), and (−2, 5) is isosceles.

APPLICATIONS

See Examples 1, 3, and 4. (Objectives 1–2)

47. Sailing Refer to the sailboat in the illustration. How long must a rope be to fasten the top of the mast to the bow?

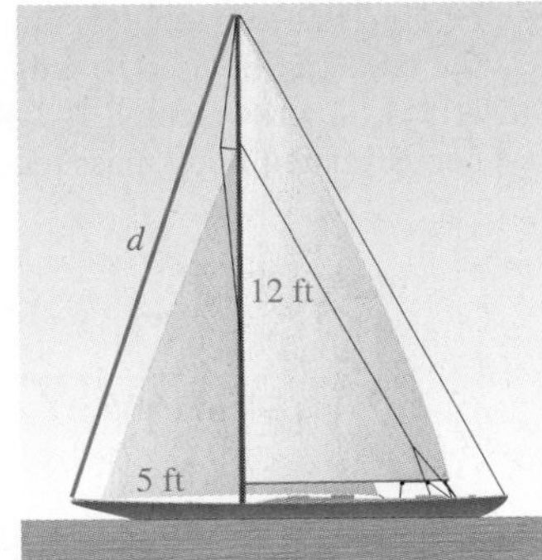

48. Carpentry The gable end of the roof shown is divided in half by a vertical brace. Find the distance from an eave to the peak.

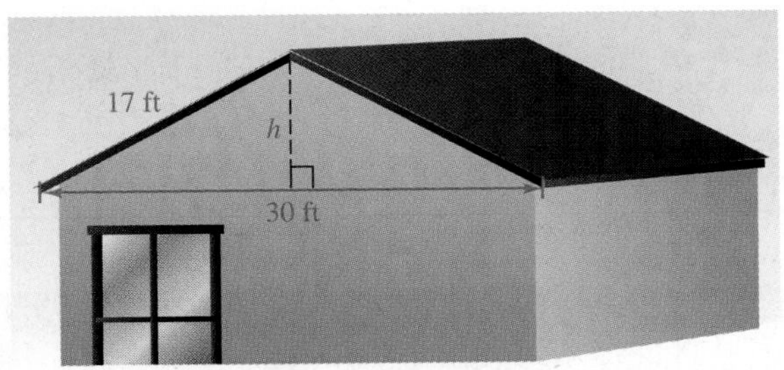

49. Reach of a ladder The base of the 37-foot ladder in the illustration is 9 feet from the wall. Will the top reach a window ledge that is 35 feet above the ground?

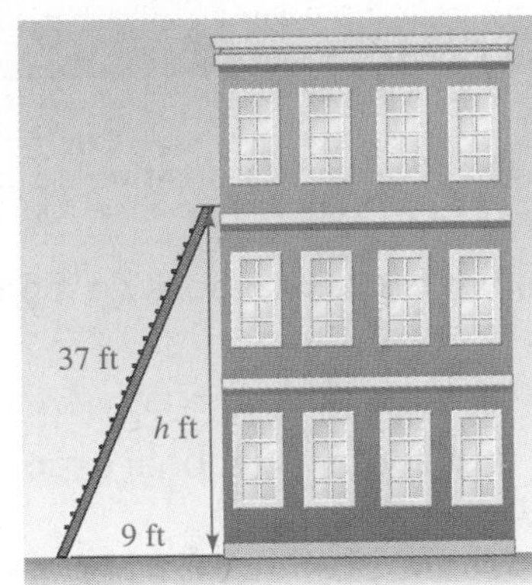

50. Telephone service The telephone cable in the illustration currently runs from A to B to C to D. How much cable is required to run from A to D directly?

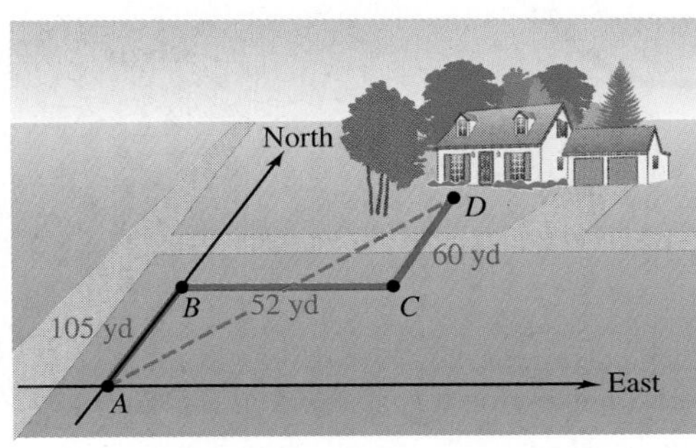

51. Electric service The power company routes its lines as shown in the illustration. How much wire could be saved by going directly from A to E?

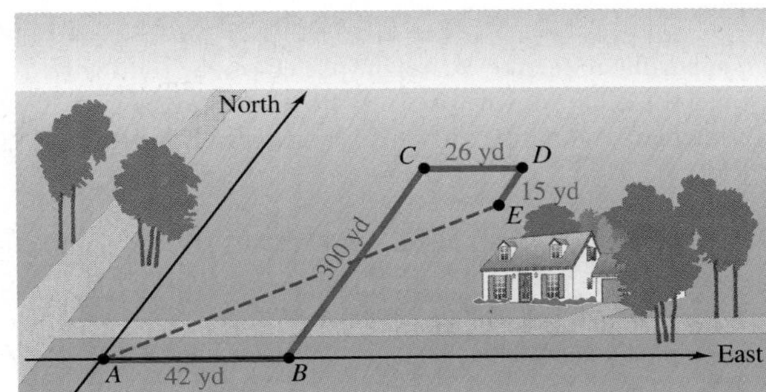

52. Geometry The side, s, of a square with area A square feet is given by the formula $s = \sqrt{A}$. Find the perimeter of a square with an area of 49 square feet.

53. Surface area of a cube The total surface area, A, of a cube is related to its volume, V, by the formula $A = 6\sqrt[3]{V^2}$. Find the surface area of a cube with a volume of 8 cubic centimeters.

54. Area of many cubes A grain of table salt is a cube with a volume of approximately 6×10^{-6} cubic in., and there are about 1.5 million grains of salt in one cup. Find the total surface area of the salt in one cup. (See Exercise 53.)

Use a calculator to help answer each question.

55. Packing a tennis racket The diagonal d of a rectangular box with dimensions $a \times b \times c$ is given by

$$d = \sqrt{a^2 + b^2 + c^2}$$

Will the racket shown on the next page fit in the shipping carton?

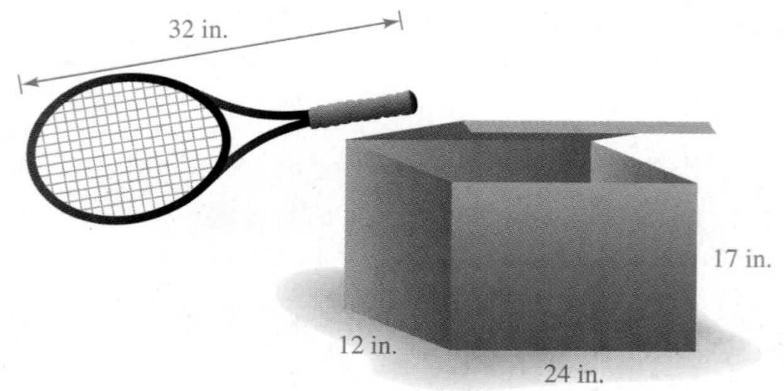

56. Packing a tennis racket Will the racket in Exercise 55 fit in a carton with dimensions that are 15 in. · 15 in. · 17 in.?

57. Shipping packages A delivery service won't accept a package for shipping if any dimension exceeds 21 inches. An archaeologist wants to ship a 36-inch femur bone. Will it fit in a 3-inch-tall box that has a 21-inch-square base?

58. Shipping packages Can the archaeologist in Exercise 57 ship the femur bone in a cubical box 21 inches on an edge?

59. Supporting a weight A weight placed on the tight wire pulls the center down 1 foot. By how much is the wire stretched? Round the answer to the nearest hundredth of a foot.

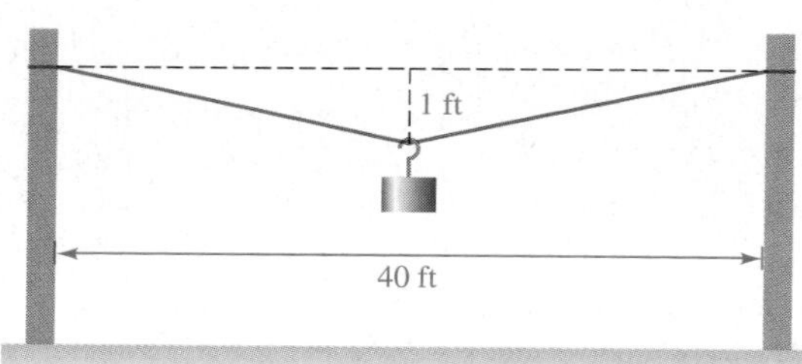

60. Supporting a weight If the weight in Exercise 59 pulls the center down 2 feet, by how much would the wire stretch? Round the answer to the nearest tenth of a foot.

WRITING ABOUT MATH

61. State the Pythagorean theorem.

62. Explain the distance formula.

SOMETHING TO THINK ABOUT

63. Body mass The formula

$$I = \frac{703w}{h^2}$$

(where w is weight in pounds and h is height in inches) can be used to estimate body mass index, I. The scale shown in the table can be used to judge a person's risk of heart attack. A girl weighing 104 pounds is 54.1 inches tall. Find her estimated body mass index.

20–26	normal
27–29	higher risk
30 and above	very high risk

64. What is the risk of a heart attack for a man who is 6 feet tall and weighs 220 pounds?

SECTION 7.3 Rational Exponents

Objectives

1. Simplify an expression that contains a positive rational exponent with a numerator of 1.
2. Simplify an expression that contains a positive rational exponent with a numerator other than 1.
3. Simplify an expression that contains a negative rational exponent.
4. Simplify an expression that contains rational exponents by applying the properties of exponents.
5. Simplify a radical expression by first writing it as an expression with a rational exponent.

Getting Ready

Simplify each expression. Assume no variable is zero.

1. x^3x^4 **2.** $(a^3)^4$ **3.** $\frac{a^8}{a^4}$ **4.** a^0

5. x^{-4} **6.** $\frac{1}{x^{-3}}$ **7.** $\left(\frac{b^2}{c^3}\right)^3$ **8.** $(a^2a^3)^2$

1 Simplify an expression that contains a positive rational exponent with a numerator of 1.

We have seen that positive integer exponents indicate the number of times that a base is to be used as a factor in a product. For example, x^5 means that x is to be used as a factor five times.

$$x^5 = \overbrace{x \cdot x \cdot x \cdot x \cdot x}^{\text{5 factors of } x}$$

Furthermore, we recall the following properties of exponents.

Rules of Exponents

If there are no divisions by 0, then for all integers m and n,

1. $x^mx^n = x^{m+n}$ **2.** $(x^m)^n = x^{mn}$ **3.** $(xy)^n = x^ny^n$

4. $\left(\frac{x}{y}\right)^n = \frac{x^n}{y^n}$ **5.** $x^0 = 1 \quad (x \neq 0)$ **6.** $x^{-n} = \frac{1}{x^n}$

7. $\frac{x^m}{x^n} = x^{m-n}$ **8.** $\left(\frac{x}{y}\right)^{-n} = \left(\frac{y}{x}\right)^n$ **9.** $\frac{1}{x^{-n}} = x^n$

To show how to raise bases to rational powers, we consider the expression $10^{1/2}$. Since rational exponents must obey the same rules as integer exponents, the square of $10^{1/2}$ is equal to 10.

$$\begin{aligned} (10^{1/2})^2 &= 10^{(1/2)2} && \text{Keep the base and multiply the exponents.} \\ &= 10^1 && \tfrac{1}{2} \cdot 2 = 1 \\ &= 10 && 10^1 = 10 \end{aligned}$$

However, we have seen that

$$\left(\sqrt{10}\right)^2 = 10$$

Since $(10^{1/2})^2$ and $\left(\sqrt{10}\right)^2$ both equal 10, we define $10^{1/2}$ to be $\sqrt{10}$. Likewise, we define

$10^{1/3}$ to be $\sqrt[3]{10}$ and $10^{1/4}$ to be $\sqrt[4]{10}$

Rational Exponents

If n is a natural number greater than 1, and $\sqrt[n]{x}$ is a real number, then

$$x^{1/n} = \sqrt[n]{x}$$

EXAMPLE 1 Simplify each expression. Assume all variables represent nonnegative values.

a. $9^{1/2} = \sqrt{9} = 3$

b. $-\left(\frac{16}{9}\right)^{1/2} = -\sqrt{\frac{16}{9}} = -\frac{4}{3}$

c. $(-64)^{1/3} = \sqrt[3]{-64} = -4$

d. $16^{1/4} = \sqrt[4]{16} = 2$

e. $\left(\frac{1}{32}\right)^{1/5} = \sqrt[5]{\frac{1}{32}} = \frac{1}{2}$

f. $0^{1/8} = \sqrt[8]{0} = 0$

g. $-(32x^5)^{1/5} = -\sqrt[5]{32x^5} = -2x$

h. $(xyz)^{1/4} = \sqrt[4]{xyz}$

SELF CHECK 1 Assume that $x > 0$. Simplify:
a. $16^{1/2}$ **b.** $\left(\frac{27}{8}\right)^{1/3}$ **c.** $-(16x^4)^{1/4}$.

EXAMPLE 2 Write each radical using a rational exponent: **a.** $\sqrt[4]{5xyz}$ **b.** $\sqrt[5]{\frac{xy^2}{15}}$.

Solution **a.** $\sqrt[4]{5xyz} = (5xyz)^{1/4}$ **b.** $\sqrt[5]{\frac{xy^2}{15}} = \left(\frac{xy^2}{15}\right)^{1/5}$

SELF CHECK 2 Write the radical using a rational exponent: $\sqrt[6]{4ab}$.

As with radicals, when n is even in the expression $x^{1/n}$ $(n > 1)$, there are two real nth roots and we must use absolute value symbols to guarantee that the simplified result is positive.

When n is odd, there is only one real nth root, and we don't need to use absolute value symbols.

When n is even and x is negative, the expression $x^{1/n}$ is not a real number.

EXAMPLE 3 Assume that all variables can be any real number, and simplify each expression using absolute value symbols when necessary.

a. $(-27x^3)^{1/3} = -3x$ Because $(-3x)^3 = -27x^3$. Since n is odd, no absolute value symbols are needed.

b. $(49x^2)^{1/2} = |7x| = 7|x|$ Because $(|7x|)^2 = 49x^2$. Since $7x$ can be negative, absolute value symbols are needed.

c. $(256a^8)^{1/8} = 2|a|$ Because $(2|a|)^8 = 256a^8$. Since a can be any real number, $2a$ can be negative. Thus, absolute value symbols are needed.

d. $[(y + 1)^2]^{1/2} = |y + 1|$ Because $|y + 1|^2 = (y + 1)^2$. Since y can be any real number, $y + 1$ can be negative, and the absolute value symbols are needed.

e. $(25b^4)^{1/2} = 5b^2$ Because $(5b^2)^2 = 25b^4$. Since $b^2 \geq 0$, no absolute value symbols are needed.

f. $(-256x^4)^{1/4}$ is not a real number. Because no real number raised to the 4th power is $-256x^4$

SELF CHECK 3 Simplify each expression using absolute value symbols when necessary.

a. $(625a^4)^{1/4}$ **b.** $(b^4)^{1/2}$

We summarize the cases as follows.

Summary of the Definitions of $x^{1/n}$

If n is a natural number greater than 1 and x is a real number, then

If $x > 0$, then $x^{1/n}$ is the positive number such that $(x^{1/n})^n = x$.

If $x = 0$, then $x^{1/n} = 0$.

$$\text{If } x < 0 \begin{cases} \text{and } n \text{ is odd, then } x^{1/n} \text{ is the real number such that } (x^{1/n})^n = x. \\ \text{and } n \text{ is even, then } x^{1/n} \text{ is not a real number.} \end{cases}$$

2 Simplify an expression that contains a positive rational exponent with a numerator other than 1.

We can extend the definition of $x^{1/n}$ to include rational exponents with numerators other than 1. For example, since $4^{3/2}$ can be written as $(4^{1/2})^3$, we have

$$4^{3/2} = (4^{1/2})^3 = \left(\sqrt{4}\right)^3 = 2^3 = 8$$

Thus, we can simplify $4^{3/2}$ by cubing the square root of 4. We can also simplify $4^{3/2}$ by taking the square root of 4 cubed.

$$4^{3/2} = (4^3)^{1/2} = 64^{1/2} = \sqrt{64} = 8$$

In general, we have the following rule.

Changing from Rational Exponents to Radicals

If m and n are positive integers, $x \geq 0$, and $\frac{m}{n}$ is in simplified form, then

$$x^{m/n} = \left(\sqrt[n]{x}\right)^m = \sqrt[n]{x^m}$$

Because of the previous definition, we can interpret $x^{m/n}$ in two ways:

1. $x^{m/n}$ means the mth power of the nth root of x.
2. $x^{m/n}$ means the nth root of the mth power of x.

EXAMPLE 4 Simplify each expression.

a. $27^{2/3} = \left(\sqrt[3]{27}\right)^2 = 3^2 = 9$ or $27^{2/3} = \sqrt[3]{27^2} = \sqrt[3]{729} = 9$

b. $\left(\frac{1}{16}\right)^{3/4} = \left(\sqrt[4]{\frac{1}{16}}\right)^3 = \left(\frac{1}{2}\right)^3 = \frac{1}{8}$ or $\left(\frac{1}{16}\right)^{3/4} = \sqrt[4]{\left(\frac{1}{16}\right)^3} = \sqrt[4]{\frac{1}{4{,}096}} = \frac{1}{8}$

COMMENT To avoid large numbers, it is usually better to find the root of the base first, as shown in Example 4.

c. $(-8x^3)^{4/3} = \left(\sqrt[3]{-8x^3}\right)^4 = (-2x)^4 = 16x^4$ or $(-8x^3)^{4/3} = \sqrt[3]{(-8x^3)^4} = \sqrt[3]{4{,}096x^{12}} = 16x^4$

SELF CHECK 4 Simplify: **a.** $16^{3/2}$ **b.** $(-27x^6)^{2/3}$.

ACCENT ON TECHNOLOGY

Rational Exponents

We can evaluate expressions containing rational exponents using the exponential key y^x or x^y on a scientific calculator. For example, to evaluate $10^{2/3}$, we enter

10 y^x (2 ÷ 3) = **4.641588834**

Note that parentheses were used when entering the power. Without them, the calculator would interpret the entry as $10^2 \div 3$.

To evaluate the exponential expression using a graphing calculator, we use the ∧ key, which raises a base to a power. Again, we use parentheses when entering the power.

10 ∧ (2 ÷ 3) ENTER

10 ∧ (2/3)
4.641588834

To the nearest hundredth, $10^{2/3} \approx 4.64$.

3 Simplify an expression that contains a negative rational exponent.

To be consistent with the definition of negative integer exponents, we define $x^{-m/n}$ as follows.

Definition of $x^{-m/n}$

If m and n are positive integers, $\frac{m}{n}$ is in simplified form, and $x^{1/n}$ is a real number ($x \neq 0$), then

$$x^{-m/n} = \frac{1}{x^{m/n}} \quad \text{and} \quad \frac{1}{x^{-m/n}} = x^{m/n}$$

EXAMPLE 5 Write each expression without negative exponents, if possible.

a. $64^{-1/2} = \frac{1}{64^{1/2}} = \frac{1}{8}$

b. $16^{-3/2} = \frac{1}{16^{3/2}} = \frac{1}{(16^{1/2})^3} = \frac{1}{64}$ $\quad (16^{1/2})^3 = 4^3 = 64$

c. $(-32x^5)^{-2/5} = \frac{1}{(-32x^5)^{2/5}} \quad (x \neq 0)$
$= \frac{1}{[(-32x^5)^{1/5}]^2}$

d. $(-16)^{-3/4}$ is not a real number, because $(-16)^{-1/4}$ is not a real number.

$$= \frac{1}{(-2x)^2}$$

$$= \frac{1}{4x^2}$$

SELF CHECK 5 Write each expression without negative exponents.

a. $25^{-3/2}$ **b.** $(-27a^3)^{-2/3}$

COMMENT By definition, 0^0 is undefined. A base of 0 raised to a negative power is also undefined, because 0^{-2} would equal $\frac{1}{0^2}$, which is undefined since we cannot divide by 0.

4 Simplify an expression that contains rational exponents by applying the properties of exponents.

We can use the properties of exponents to simplify many expressions with rational exponents.

EXAMPLE 6 Write all answers without negative exponents. Assume that all variables represent positive numbers. Thus, no absolute value symbols are necessary.

a. $5^{2/7}5^{3/7} = 5^{2/7+3/7}$ Use the rule $x^mx^n = x^{m+n}$.

$= 5^{5/7}$ Add: $\frac{2}{7} + \frac{3}{7} = \frac{5}{7}$.

b. $(5^{2/7})^3 = 5^{(2/7)(3)}$ Use the rule $(x^m)^n = x^{mn}$.

$= 5^{6/7}$ Multiply: $\frac{2}{7}(3) = \frac{2}{7}\left(\frac{3}{1}\right) = \frac{6}{7}$.

c. $(a^{2/3}b^{1/2})^6 = (a^{2/3})^6(b^{1/2})^6$ Use the rule $(xy)^n = x^ny^n$.

$= a^{12/3}b^{6/2}$ Use the rule $(x^m)^n = x^{mn}$ twice.

$= a^4b^3$ Simplify the exponents.

d. $\dfrac{a^{8/3}a^{1/3}}{a^2} = a^{8/3+1/3-2}$ Use the rules $x^mx^n = x^{m+n}$ and $\frac{x^m}{x^n} = x^{m-n}$.

$= a^{8/3+1/3-6/3}$ $2 = \frac{6}{3}$

$= a^{3/3}$ $\frac{8}{3} + \frac{1}{3} - \frac{6}{3} = \frac{3}{3}$

$= a$ $\frac{3}{3} = 1$

SELF CHECK 6 Simplify: **a.** $(x^{1/3}y^{3/2})^6$ **b.** $\dfrac{x^{5/3}x^{2/3}}{x^{1/3}}$.

EXAMPLE 7 Assume that all variables represent positive numbers and perform the operations.

COMMENT Note that $a + a^{7/5} \neq a^{1+7/5}$. The expression $a + a^{7/5}$ cannot be simplified, because a and $a^{7/5}$ are not like terms.

a. $a^{4/5}(a^{1/5} + a^{3/5}) = a^{4/5}a^{1/5} + a^{4/5}a^{3/5}$ Use the distributive property.

$= a^{4/5+1/5} + a^{4/5+3/5}$ Use the rule $x^mx^n = x^{m+n}$.

$= a^{5/5} + a^{7/5}$ Simplify the exponents.

$= a + a^{7/5}$

Carl Friedrich Gauss (1777–1855)
Many people consider Gauss to be the greatest mathematician of all time. He made contributions in the areas of number theory, solutions of equations, geometry of curved surfaces, and statistics. For his efforts, he earned the title "Prince of the Mathematicians."

b. $x^{1/2}(x^{-1/2} + x^{1/2}) = x^{1/2}x^{-1/2} + x^{1/2}x^{1/2}$ Use the distributive property.
$= x^{1/2-1/2} + x^{1/2+1/2}$ Use the rule $x^m x^n = x^{m+n}$.
$= x^0 + x^1$ Simplify.
$= 1 + x$ $x^0 = 1$

c. $(x^{2/3} + 1)(x^{2/3} - 1) = x^{4/3} - x^{2/3} + x^{2/3} - 1$ Use the FOIL method.
$= x^{4/3} - 1$ Combine like terms.

d. $(x^{1/2} + y^{1/2})^2 = (x^{1/2} + y^{1/2})(x^{1/2} + y^{1/2})$ Use the FOIL method.
$= x + 2x^{1/2}y^{1/2} + y$

SELF CHECK 7 Assume that all variables represent positive numbers and perform the operations.
a. $p^{1/5}(p^{4/5} + p^{2/5})$ **b.** $(p^{2/3} + q^{1/3})(p^{2/3} - q^{1/3})$

5 Simplify a radical expression by first writing it as an expression with a rational exponent.

We can simplify many radical expressions by using the following steps.

Using Fractional Exponents to Simplify Radicals

1. Change the radical expression into an exponential expression with rational exponents.
2. Simplify the rational exponents.
3. Change the exponential expression back into a radical.

EXAMPLE 8 Simplify. Assume variables represent positive values.
a. $\sqrt[4]{3^2}$ **b.** $\sqrt[8]{x^6}$ **c.** $\sqrt[9]{27x^6y^3}$

Solution **a.** $\sqrt[4]{3^2} = 3^{2/4}$ Use the rule $\sqrt[n]{x^m} = x^{m/n}$.
$= 3^{1/2}$ $\frac{2}{4} = \frac{1}{2}$
$= \sqrt{3}$ Change back to radical notation.

b. $\sqrt[8]{x^6} = x^{6/8}$ Use the rule $\sqrt[n]{x^m} = x^{m/n}$.
$= x^{3/4}$ $\frac{6}{8} = \frac{3}{4}$
$= (x^3)^{1/4}$ $\frac{3}{4} = 3\left(\frac{1}{4}\right)$
$= \sqrt[4]{x^3}$ Change back to radical notation.

c. $\sqrt[9]{27x^6y^3} = (3^3x^6y^3)^{1/9}$ Write 27 as 3^3 and change the radical to an exponential expression.

$= 3^{3/9}x^{6/9}y^{3/9}$ Raise each factor to the $\frac{1}{9}$ power by multiplying the fractional exponents.

$= 3^{1/3}x^{2/3}y^{1/3}$ Simplify each fractional exponent.

$= (3x^2y)^{1/3}$ Use the rule $(xy)^n = x^ny^n$.

$= \sqrt[3]{3x^2y}$ Change back to radical notation.

SELF CHECK 8 Simplify. Assume variables represent positive values.
a. $\sqrt[6]{3^3}$ **b.** $\sqrt[4]{49x^2y^2}$

SELF CHECK ANSWERS **1. a.** 4 **b.** $\frac{3}{2}$ **c.** $-2x$ **2.** $(4ab)^{1/6}$ **3. a.** $5|a|$ **b.** b^2 **4. a.** 64 **b.** $9x^4$ **5. a.** $\frac{1}{125}$ **b.** $\frac{1}{9a^2}$ **6. a.** x^2y^9 **b.** x^2 **7. a.** $p + p^{3/5}$ **b.** $p^{4/3} - q^{2/3}$ **8. a.** $\sqrt{3}$ **b.** $\sqrt{7xy}$

NOW TRY THIS

1. Evaluate each exponential expression, if possible.
   a. $-64^{-2/3}$ **b.** $-64^{-3/2}$ **c.** $(-64)^{-2/3}$
   d. $(-64)^{-3/2}$
2. Simplify each expression and then write the answer using radical notation. Assume all variables represent positive numbers.
   a. $x^{a/4} \cdot x^{a/2}$ **b.** $\frac{x^{n/m}}{x^{(n-1)/m}}$
3. Simplify $\frac{\sqrt{5}}{\sqrt[3]{5}}$ and write the answer using radical notation.

7.3 EXERCISES

Assume no division by zero.

WARM-UPS *Simplify each expression.*

1. $4^{1/2}$ **2.** $9^{1/2}$
3. $27^{1/3}$ **4.** $1^{1/4}$
5. $4^{3/2}$ **6.** $8^{2/3}$
7. $\left(\frac{1}{4}\right)^{1/2}$ **8.** $\left(\frac{1}{4}\right)^{-1/2}$
9. $(8x^3)^{1/3}$ **10.** $(16x^8)^{1/4}$

REVIEW *Solve each inequality.*

11. $5x - 4 < 11$ **12.** $2(3t - 5) \geq 8$
13. $\frac{4}{5}(r - 3) > \frac{2}{3}(r + 2)$ **14.** $-4 < 2x - 4 \leq 8$

15. Mixing solutions How much water must be added to 5 pints of a 20% alcohol solution to dilute it to a 15% solution?

16. Selling apples A grocer bought some boxes of apples for \$70. However, 4 boxes were spoiled. The grocer sold the remaining boxes at a profit of \$2 each. How many boxes did the grocer sell if she managed to break even?

VOCABULARY AND CONCEPTS *Fill in the blanks.*

17. $a^4 =$ ______ **18.** $a^ma^n =$ ______
19. $(a^m)^n =$ ______ **20.** $(ab)^n =$ ______

21. $\left(\frac{a}{b}\right)^n = \frac{\quad}{\quad}$

22. $a^0 =$ ___, provided $a \neq$ ___.

23. $a^{-n} = \frac{\quad}{\quad}$, provided $a \neq$ ___.

24. $\frac{a^m}{a^n} =$ _____, provided $a \neq 0$.

25. $\left(\frac{a}{b}\right)^{-n} = \frac{\quad}{\quad}$

26. $x^{1/n} =$ ____

27. $(x^n)^{1/n} =$ ___, provided n is even.

28. $x^{m/n} = \sqrt[n]{x^m} =$ ______

GUIDED PRACTICE

Change each expression into radical notation. (Objective 1)

29. $7^{1/3}$

30. $26^{1/2}$

31. $8^{1/5}$

32. $13^{1/7}$

33. $(3x)^{1/4}$

34. $(4ab)^{1/6}$

35. $\left(\frac{1}{2}x^3y\right)^{1/4}$

36. $\left(\frac{3}{4}a^2b^2\right)^{1/5}$

37. $(4a^2b^3)^{1/5}$

38. $(5pq^2)^{1/3}$

39. $(x^2 + y^2)^{1/2}$

40. $(x^3 + y^3)^{1/3}$

Simplify each expression, if possible. See Example 1. (Objective 1)

41. $4^{1/2}$

42. $64^{1/2}$

43. $27^{1/3}$

44. $125^{1/3}$

45. $\left(\frac{1}{4}\right)^{1/2}$

46. $\left(\frac{1}{16}\right)^{1/2}$

47. $\left(\frac{1}{8}\right)^{1/3}$

48. $\left(\frac{1}{16}\right)^{1/4}$

49. $-16^{1/4}$

50. $-125^{1/3}$

51. $(-64)^{1/2}$

52. $(-216)^{1/2}$

Change each radical to an exponential expression. See Example 2. (Objective 1)

53. $\sqrt{11}$

54. $\sqrt[3]{12}$

55. $\sqrt[4]{3a}$

56. $\sqrt[7]{12xy}$

57. $3\sqrt[5]{a}$

58. $4\sqrt[3]{p}$

59. $\sqrt[6]{\frac{1}{7}abc}$

60. $\sqrt[7]{\frac{3}{8}p^2q}$

61. $\sqrt[5]{\frac{1}{2}mn}$

62. $\sqrt[8]{\frac{2}{7}p^2q}$

63. $\sqrt[3]{a^2 - b^2}$

64. $\sqrt{x^2 + y^2}$

Simplify each expression. Assume that all variables can be any real number, and use absolute value symbols if necessary. See Example 3. (Objective 1)

65. $(25y^2)^{1/2}$

66. $(16x^4)^{1/4}$

67. $(243x^5)^{1/5}$

68. $(-27x^3)^{1/3}$

69. $[(x + 1)^4]^{1/4}$

70. $[(x + 5)^3]^{1/3}$

71. $(-64x^8)^{1/4}$

72. $(-16x^4)^{1/2}$

Simplify each expression, if possible. Assume that all variables are unrestricted, and use absolute value symbols when necessary. See Example 4. (Objective 2)

73. $36^{3/2}$

74. $27^{2/3}$

75. $81^{3/4}$

76. $100^{3/2}$

77. $144^{3/2}$

78. $1{,}000^{2/3}$

79. $\left(\frac{1}{8}\right)^{2/3}$

80. $\left(\frac{4}{9}\right)^{3/2}$

Write each expression without using negative exponents. Assume that all variables represent positive numbers. See Example 5. (Objective 3)

81. $4^{-1/2}$

82. $8^{-1/3}$

83. $(4)^{-3/2}$

84. $25^{-5/2}$

85. $(16x^2)^{-3/2}$

86. $(81c^4)^{-3/2}$

87. $(-27y^3)^{-2/3}$

88. $(-8z^9)^{-2/3}$

89. $\left(\frac{1}{4}\right)^{-3/2}$

90. $\left(\frac{4}{25}\right)^{-3/2}$

91. $\left(\frac{27}{8}\right)^{-4/3}$

92. $\left(\frac{25}{49}\right)^{-3/2}$

Perform the operations. Write answers without negative exponents. Assume that all variables represent positive numbers. See Example 6. (Objective 4)

93. $5^{4/9}5^{4/9}$

94. $4^{2/5}4^{2/5}$

95. $(4^{1/5})^3$

96. $(3^{1/3})^5$

97. $6^{-2/3}6^{-4/3}$

98. $5^{1/3}5^{-5/3}$

99. $\frac{9^{4/5}}{9^{3/5}}$

100. $\frac{7^{2/3}}{7^{1/2}}$

101. $\frac{7^{1/2}}{7^0}$

102. $\frac{3^{4/3}3^{1/3}}{3^{2/3}}$

103. $\frac{2^{5/6}2^{1/3}}{2^{1/2}}$

104. $\frac{5^{1/3}5^{1/2}}{5^{1/3}}$

105. $a^{2/3}a^{1/3}$

106. $b^{3/5}b^{1/5}$

107. $(a^{2/3})^{1/3}$

108. $(t^{4/5})^{10}$

Perform the operations. Write answers without negative exponents. Assume that all variables represent positive numbers. See Example 7. (Objective 4)

109. $y^{1/3}(y^{2/3} + y^{5/3})$

110. $y^{2/5}(y^{-2/5} + y^{3/5})$

111. $x^{3/5}(x^{7/5} - x^{2/5} + 1)$

112. $(x^{1/2} + 2)(x^{1/2} - 2)$

113. $(x^{1/2} + y^{1/2})(x^{1/2} - y^{1/2})$

114. $(x^{2/3} - x)(x^{2/3} + x)$

115. $(x^{2/3} + y^{2/3})^2$

116. $(a^{3/2} - b^{3/2})^2$

Use rational exponents to simplify each radical. Assume that all variables represent positive numbers. See Example 8. (Objective 5)

117. $\sqrt[6]{p^3}$
118. $\sqrt[8]{q^2}$
119. $\sqrt[4]{25b^2}$
120. $\sqrt[9]{-8x^6}$

ADDITIONAL PRACTICE

Simplify each expression, if possible. Assume all variables represent positive numbers. Write answers without negative exponents.

121. $16^{1/4}$
122. $625^{1/4}$
123. $32^{1/5}$
124. $0^{1/5}$
125. $0^{1/3}$
126. $(-243)^{1/5}$
127. $(-27)^{1/3}$
128. $(-125)^{1/3}$
129. $(25x^4)^{3/2}$
130. $(27a^3b^3)^{2/3}$
131. $\left(\frac{8x^3}{27}\right)^{2/3}$
132. $\left(\frac{27}{64y^6}\right)^{2/3}$
133. $(-32p^5)^{-2/5}$
134. $(16q^6)^{-5/2}$
135. $\left(-\frac{8x^3}{27}\right)^{-1/3}$
136. $\left(\frac{16}{81y^4}\right)^{-3/4}$
137. $(a^{1/2}b^{1/3})^{3/2}$
138. $(a^{3/5}b^{3/2})^{2/3}$
139. $(mn^{-2/3})^{-3/5}$
140. $(r^{-2}s^3)^{1/3}$
141. $\frac{(4x^3y)^{1/2}}{(9xy)^{1/2}}$
142. $\frac{(27x^3y)^{1/3}}{(8xy^2)^{2/3}}$
143. $(27x^{-3})^{-1/3}$
144. $(16a^{-2})^{-1/2}$
145. $x^{4/3}(x^{2/3} + 3x^{5/3} - 4)$
146. $(x^{1/3} + x^2)(x^{1/3} - x^2)$
147. $(x^{-1/2} - x^{1/2})^2$
148. $(a^{1/2} - b^{2/3})^2$

Use a calculator to evaluate each expression. Round to the nearest hundredth.

149. $15^{1/3}$
150. $50.5^{1/4}$
151. $1.045^{1/5}$
152. $(-1{,}000)^{2/5}$

Use a calculator to evaluate each expression. Round to the nearest hundredth.

153. $17^{-1/2}$
154. $2.45^{-2/3}$
155. $(-0.25)^{-1/5}$
156. $(-17.1)^{-3/7}$

WRITING ABOUT MATH

157. Explain how you would decide whether $a^{1/n}$ is a real number.

158. The expression $(a^{1/2} + b^{1/2})^2$ is not equal to $a + b$. Explain.

SOMETHING TO THINK ABOUT

159. The fraction $\frac{2}{4}$ is equal to $\frac{1}{2}$. Is $16^{2/4}$ equal to $16^{1/2}$? Explain.

160. How would you evaluate an expression with a mixed-number exponent? For example, what is $8^{1\frac{1}{3}}$? What is $25^{2\frac{1}{2}}$? Explain.

SECTION 7.4 Simplifying and Combining Radical Expressions

Objectives

1. Simplify a radical expression by applying the properties of radicals.
2. Add and subtract two or more radical expressions.
3. Find the length of a side of a 30°–60°–90° triangle and a 45°–45°–90° triangle.

Vocabulary

like (similar) radicals
altitude

Getting Ready

Simplify each radical. Assume that all variables represent positive numbers.

1. $\sqrt{225}$ **2.** $\sqrt{576}$ **3.** $\sqrt[3]{125}$ **4.** $\sqrt[3]{343}$

5. $\sqrt{16x^4}$ **6.** $\sqrt{\frac{64}{121}x^6}$ **7.** $\sqrt[3]{27a^3b^9}$ **8.** $\sqrt[3]{-8a^{12}}$

In this section, we will introduce several properties of radicals and use them to simplify radical expressions. Then we will add and subtract radical expressions.

1 Simplify a radical expression by applying the properties of radicals.

Many properties of exponents have counterparts in radical notation. For example, because $a^{1/n}b^{1/n} = (ab)^{1/n}$, we have

(1) $\sqrt[n]{a}\sqrt[n]{b} = \sqrt[n]{ab}$

For example,

$$\sqrt{5}\sqrt{5} = \sqrt{5 \cdot 5} = \sqrt{5^2} = 5$$

$$\sqrt[3]{7x}\sqrt[3]{49x^2} = \sqrt[3]{7x \cdot 7^2x^2} = \sqrt[3]{7^3 \cdot x^3} = 7x$$

$$\sqrt[4]{2x^3}\sqrt[4]{8x} = \sqrt[4]{2x^3 \cdot 2^3x} = \sqrt[4]{2^4 \cdot x^4} = 2x \quad (x > 0)$$

If we rewrite Equation 1, we have the following rule.

Multiplication Property of Radicals

If $\sqrt[n]{a}$ and $\sqrt[n]{b}$ are real numbers, then

$$\sqrt[n]{ab} = \sqrt[n]{a}\sqrt[n]{b}$$

As long as all radicals represent real numbers, *the nth root of the product of two numbers is equal to the product of their nth roots.*

COMMENT The multiplication property of radicals applies to the nth root of the product of two numbers. There is no such property for sums or differences. For example,

$$\sqrt{9 + 4} \neq \sqrt{9} + \sqrt{4} \qquad \sqrt{9 - 4} \neq \sqrt{9} - \sqrt{4}$$

$$\sqrt{13} \neq 3 + 2 \qquad \sqrt{5} \neq 3 - 2$$

$$\sqrt{13} \neq 5 \qquad \sqrt{5} \neq 1$$

Thus, $\sqrt{a + b} \neq \sqrt{a} + \sqrt{b}$ and $\sqrt{a - b} \neq \sqrt{a} - \sqrt{b}$.

A second property of radicals involves quotients. Because

$$\frac{a^{1/n}}{b^{1/n}} = \left(\frac{a}{b}\right)^{1/n}$$

it follows that

(2) $\dfrac{\sqrt[n]{a}}{\sqrt[n]{b}} = \sqrt[n]{\dfrac{a}{b}} \quad (b \neq 0)$

For example,

$$\frac{\sqrt{8x^3}}{\sqrt{2x}} = \sqrt{\frac{8x^3}{2x}} = \sqrt{4x^2} = 2x \quad (x > 0)$$

$$\frac{\sqrt[3]{54x^5}}{\sqrt[3]{2x^2}} = \sqrt[3]{\frac{54x^5}{2x^2}} = \sqrt[3]{27x^3} = 3x \quad (x \neq 0)$$

If we rewrite Equation 2, we have the following rule.

Division Property of Radicals

If $\sqrt[n]{a}$ and $\sqrt[n]{b}$ are real numbers, then

$$\sqrt[n]{\frac{a}{b}} = \frac{\sqrt[n]{a}}{\sqrt[n]{b}} \quad (b \neq 0)$$

As long as all radicals represent real numbers, *the nth root of the quotient of two numbers is equal to the quotient of their nth roots.*

A radical expression is said to be in simplified form when each of the following statements is true.

Simplified Form of a Radical Expression

A radical expression is in simplified form when

1. Each prime and variable factor in the radicand appears to a power that is less than the index of the radical.
2. The radicand contains no fractions or negative numbers.
3. No radicals appear in the denominator of a fraction.

EXAMPLE 1 Simplify: **a.** $\sqrt{12}$ **b.** $\sqrt{98}$ **c.** $\sqrt[3]{54}$.

Solution **a.** Recall that squares of integers, such as 1, 4, 9, 16, 25, and 36, are *perfect squares.* To simplify $\sqrt{12}$, we factor 12 so that one factor is the largest perfect square that divides 12. Since 4 is the largest perfect-square factor of 12, we write 12 as $4 \cdot 3$, use the multiplication property of radicals, and simplify.

$$\begin{aligned} \sqrt{12} &= \sqrt{4 \cdot 3} && \text{Write 12 as } 4 \cdot 3. \\ &= \sqrt{4}\sqrt{3} && \sqrt{4 \cdot 3} = \sqrt{4}\sqrt{3} \\ &= 2\sqrt{3} && \sqrt{4} = 2 \end{aligned}$$

b. Since the largest perfect-square factor of 98 is 49, we have

$$\begin{aligned} \sqrt{98} &= \sqrt{49 \cdot 2} && \text{Write 98 as } 49 \cdot 2. \\ &= \sqrt{49}\sqrt{2} && \sqrt{49 \cdot 2} = \sqrt{49}\sqrt{2} \\ &= 7\sqrt{2} && \sqrt{49} = 7 \end{aligned}$$

c. Numbers that are cubes of integers, such as 1, 8, 27, 64, 125, and 216, are called *perfect cubes.* Since the largest perfect-cube factor of 54 is 27, we have

$$\sqrt[3]{54} = \sqrt[3]{27 \cdot 2} \qquad \text{Write 54 as } 27 \cdot 2.$$
$$= \sqrt[3]{27}\sqrt[3]{2} \qquad \sqrt[3]{27 \cdot 2} = \sqrt[3]{27}\sqrt[3]{2}$$
$$= 3\sqrt[3]{2} \qquad \sqrt[3]{27} = 3$$

SELF CHECK 1 Simplify: **a.** $\sqrt{20}$ **b.** $\sqrt[3]{24}$.

EXAMPLE 2 Simplify: **a.** $\sqrt{\dfrac{15}{49x^2}}$ $(x > 0)$ **b.** $\sqrt[3]{\dfrac{10x^2}{27y^6}}$ $(y \neq 0)$.

Solution **a.** We can write the square root of the quotient as the quotient of the square roots and simplify the denominator. Since $x > 0$, we have

$$\sqrt{\frac{15}{49x^2}} = \frac{\sqrt{15}}{\sqrt{49x^2}}$$
$$= \frac{\sqrt{15}}{7x}$$

b. We can write the cube root of the quotient as the quotient of two cube roots. Since $y \neq 0$, we have

$$\sqrt[3]{\frac{10x^2}{27y^6}} = \frac{\sqrt[3]{10x^2}}{\sqrt[3]{27y^6}}$$
$$= \frac{\sqrt[3]{10x^2}}{3y^2}$$

SELF CHECK 2 Simplify: **a.** $\sqrt{\dfrac{11}{36a^2}}$ $(a > 0)$ **b.** $\sqrt[3]{\dfrac{8a^2}{125y^3}}$ $(y \neq 0)$.

EXAMPLE 3 Simplify each expression. Assume that all variables represent positive numbers.

a. $\sqrt{128a^5}$ **b.** $\sqrt[3]{24x^5}$ **c.** $\dfrac{\sqrt{45xy^2}}{\sqrt{5x}}$ **d.** $\dfrac{\sqrt[3]{-432x^5}}{\sqrt[3]{8x}}$

Solution **a.** We can write $128a^5$ as $64a^4 \cdot 2a$ and use the multiplication property of radicals.

$$\sqrt{128a^5} = \sqrt{64a^4 \cdot 2a} \qquad 64a^4 \text{ is the largest perfect square that divides } 128a^5.$$
$$= \sqrt{64a^4}\sqrt{2a} \qquad \text{Use the multiplication property of radicals.}$$
$$= 8a^2\sqrt{2a} \qquad \sqrt{64a^4} = 8a^2$$

b. We can write $24x^5$ as $8x^3 \cdot 3x^2$ and use the multiplication property of radicals.

$$\sqrt[3]{24x^5} = \sqrt[3]{8x^3 \cdot 3x^2} \qquad 8x^3 \text{ is the largest perfect cube that divides } 24x^5.$$
$$= \sqrt[3]{8x^3}\sqrt[3]{3x^2} \qquad \text{Use the multiplication property of radicals.}$$
$$= 2x\sqrt[3]{3x^2} \qquad \sqrt[3]{8x^3} = 2x$$

c. We can write the quotient of the square roots as the square root of a quotient.

$$\frac{\sqrt{45xy^2}}{\sqrt{5x}} = \sqrt{\frac{45xy^2}{5x}} \quad \text{Use the quotient property of radicals.}$$

$$= \sqrt{9y^2} \quad \text{Simplify the fraction.}$$

$$= 3y$$

d. We can write the quotient of the cube roots as the cube root of a quotient.

$$\frac{\sqrt[3]{-432x^5}}{\sqrt[3]{8x}} = \sqrt[3]{\frac{-432x^5}{8x}} \quad \text{Use the quotient property of radicals.}$$

$$= \sqrt[3]{-54x^4} \quad \text{Simplify the fraction.}$$

$$= \sqrt[3]{-27x^3 \cdot 2x} \quad -27x^3 \text{ is the largest perfect cube that divides } -54x^4.$$

$$= \sqrt[3]{-27x^3}\sqrt[3]{2x} \quad \text{Use the multiplication property of radicals.}$$

$$= -3x\sqrt[3]{2x}$$

SELF CHECK 3 Simplify each expression. Assume that all variables represent positive numbers.

a. $\sqrt{98b^3}$ **b.** $\sqrt[3]{54y^5}$ **c.** $\frac{\sqrt{50ab^2}}{\sqrt{2a}}$

To simplify more complicated radicals, we can use the prime factorization of the radicand to find its perfect-square factors. For example, to simplify $\sqrt{3{,}168x^5y^7}$, we first find the prime factorization of $3{,}168x^5y^7$.

$$3{,}168x^5y^7 = 2^5 \cdot 3^2 \cdot 11 \cdot x^5 \cdot y^7$$

Then we have

$$\sqrt{3{,}168x^5y^7} = \sqrt{2^4 \cdot 3^2 \cdot x^4 \cdot y^6 \cdot 2 \cdot 11 \cdot x \cdot y}$$

$$= \sqrt{2^4 \cdot 3^2 \cdot x^4 \cdot y^6}\sqrt{2 \cdot 11 \cdot x \cdot y}$$

Write each perfect square under the left radical and each nonperfect square under the right radical.

$$= 2^2 \cdot 3x^2y^3\sqrt{22xy}$$

$$= 12x^2y^3\sqrt{22xy}$$

2 Add and subtract two or more radical expressions.

Radical expressions with the same index and the same radicand are called **like** or **similar radicals.** For example, $3\sqrt{2}$ and $5\sqrt{2}$ are like radicals. However,

$3\sqrt{5}$ and $5\sqrt{2}$ are not like radicals, because the radicands are different.

$3\sqrt{5}$ and $2\sqrt[3]{5}$ are not like radicals, because the indexes are different.

We often can combine like terms. For example, to simplify the expression $3\sqrt{2} + 2\sqrt{2}$, we use the distributive property to factor out $\sqrt{2}$ and simplify.

$$3\sqrt{2} + 2\sqrt{2} = (3 + 2)\sqrt{2}$$

$$= 5\sqrt{2}$$

Radicals with the same index but different radicands can often be written as like radicals. For example, to simplify the expression $\sqrt{27} - \sqrt{12}$, we simplify both radicals and combine the resulting like radicals.

$$\begin{aligned}\sqrt{27} - \sqrt{12} &= \sqrt{9 \cdot 3} - \sqrt{4 \cdot 3} \\ &= \sqrt{9}\sqrt{3} - \sqrt{4}\sqrt{3} && \sqrt{ab} = \sqrt{a}\sqrt{b} \\ &= 3\sqrt{3} - 2\sqrt{3} && \sqrt{9} = 3 \text{ and } \sqrt{4} = 2. \\ &= (3 - 2)\sqrt{3} && \text{Factor out } \sqrt{3}. \\ &= \sqrt{3}\end{aligned}$$

As the previous examples suggest, we can use the following rule to add or subtract radicals.

Adding and Subtracting Radicals To add or subtract radicals, simplify each radical and combine all like radicals. To combine like radicals, add the coefficients and keep the common radical.

EXAMPLE 4 Simplify: $2\sqrt{12} - 3\sqrt{48} + 3\sqrt{3}$.

Solution We simplify each radical separately and combine like radicals.

$$\begin{aligned}2\sqrt{12} - 3\sqrt{48} + 3\sqrt{3} &= 2\sqrt{4 \cdot 3} - 3\sqrt{16 \cdot 3} + 3\sqrt{3} \\ &= 2\sqrt{4}\sqrt{3} - 3\sqrt{16}\sqrt{3} + 3\sqrt{3} \\ &= 2(2)\sqrt{3} - 3(4)\sqrt{3} + 3\sqrt{3} \\ &= 4\sqrt{3} - 12\sqrt{3} + 3\sqrt{3} \\ &= (4 - 12 + 3)\sqrt{3} \\ &= -5\sqrt{3}\end{aligned}$$

SELF CHECK 4 Simplify: $3\sqrt{75} - 2\sqrt{12} + 2\sqrt{48}$.

EXAMPLE 5 Simplify: $\sqrt[3]{16} - \sqrt[3]{54} + \sqrt[3]{24}$.

Solution We simplify each radical separately and combine like radicals.

COMMENT We cannot combine $-\sqrt[3]{2}$ and $2\sqrt[3]{3}$, because the radicals have different radicands.

$$\begin{aligned}\sqrt[3]{16} - \sqrt[3]{54} + \sqrt[3]{24} &= \sqrt[3]{8 \cdot 2} - \sqrt[3]{27 \cdot 2} + \sqrt[3]{8 \cdot 3} \\ &= \sqrt[3]{8}\sqrt[3]{2} - \sqrt[3]{27}\sqrt[3]{2} + \sqrt[3]{8}\sqrt[3]{3} \\ &= 2\sqrt[3]{2} - 3\sqrt[3]{2} + 2\sqrt[3]{3} \\ &= -\sqrt[3]{2} + 2\sqrt[3]{3}\end{aligned}$$

SELF CHECK 5 Simplify: $\sqrt[3]{24} - \sqrt[3]{16} + \sqrt[3]{54}$.

EXAMPLE 6 Simplify: $\sqrt[3]{16x^4} + \sqrt[3]{54x^4} - \sqrt[3]{-128x^4}$.

Solution We simplify each radical separately, factor out $\sqrt[3]{2x}$, and combine like radicals.

$$\begin{aligned}
&\sqrt[3]{16x^4} + \sqrt[3]{54x^4} - \sqrt[3]{-128x^4} \\
&\quad = \sqrt[3]{8x^3 \cdot 2x} + \sqrt[3]{27x^3 \cdot 2x} - \sqrt[3]{-64x^3 \cdot 2x} \\
&\quad = \sqrt[3]{8x^3}\sqrt[3]{2x} + \sqrt[3]{27x^3}\sqrt[3]{2x} - \sqrt[3]{-64x^3}\sqrt[3]{2x} \\
&\quad = \mathbf{2x}\sqrt[3]{2x} + \mathbf{3x}\sqrt[3]{2x} + \mathbf{4x}\sqrt[3]{2x} \\
&\quad = (\mathbf{2x + 3x + 4x})\sqrt[3]{2x} \\
&\quad = 9x\sqrt[3]{2x}
\end{aligned}$$

SELF CHECK 6 Simplify: $\sqrt{32x^3} + \sqrt{50x^3} - \sqrt{18x^3}$ $(x > 0)$.

3 Find the length of a side of a 30°–60°–90° triangle and a 45°–45°–90° triangle.

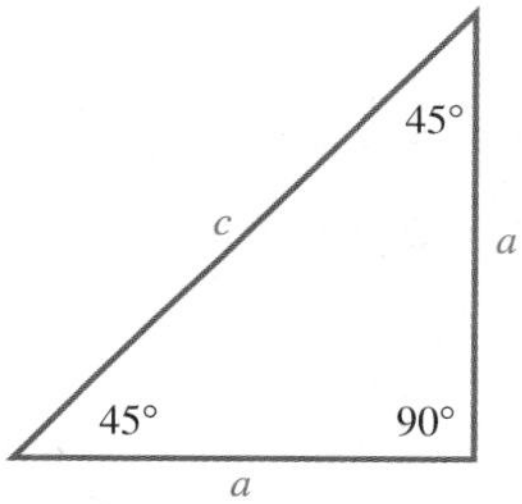

An isosceles right triangle is a right triangle with two legs of equal length. If we know the length of one leg of an isosceles right triangle, we can use the Pythagorean theorem to find the length of the hypotenuse. Since the triangle shown in Figure 7-10 is a right triangle, we have

45°
c
a
45°
90°
a

Figure 7-10

$c^2 = a^2 + a^2$	Use the Pythagorean theorem.
$c^2 = 2a^2$	Combine like terms.
$c = \sqrt{2a^2}$	Since c represents a length, take the positive square root.
$c = a\sqrt{2}$	$\sqrt{2a^2} = \sqrt{2}\sqrt{a^2} = \sqrt{2}a = a\sqrt{2}$. No absolute value symbols are needed, because a is positive.

Thus, *in an isosceles right triangle, the length of the hypotenuse is the length of one leg times* $\sqrt{2}$.

EXAMPLE 7 **GEOMETRY** If one leg of the isosceles right triangle shown in Figure 7-10 is 10 feet long, find the length of the hypotenuse.

Solution Since the length of the hypotenuse is the length of a leg times $\sqrt{2}$, we have

$$c = 10\sqrt{2}$$

The length of the hypotenuse is $10\sqrt{2}$ feet. To two decimal places, the length is 14.14 feet.

SELF CHECK 7 Find the length of the hypotenuse of an isosceles right triangle if one leg is 12 meters long.

If the length of the hypotenuse of an isosceles right triangle is known, we can use the Pythagorean theorem to find the length of each leg.

EXAMPLE 8 **GEOMETRY** Find the length of each leg of the isosceles right triangle shown in Figure 7-11 on the next page.

Solution We use the Pythagorean theorem.

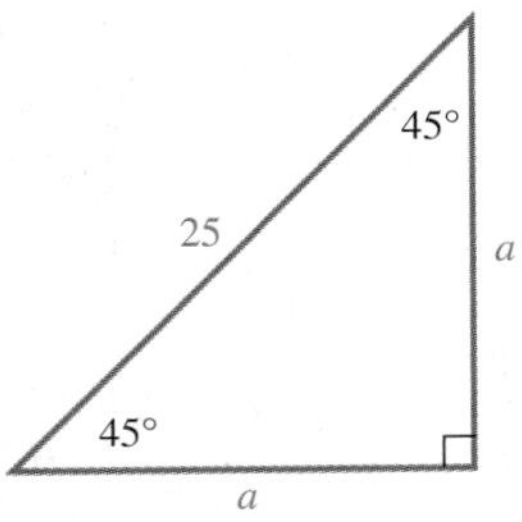

Figure 7-11

$$c^2 = a^2 + a^2$$
$$25^2 = 2a^2 \quad \text{Substitute 25 for } c \text{ and combine like terms.}$$
$$\frac{625}{2} = a^2 \quad \text{Square 25 and divide both sides by 2.}$$
$$\sqrt{\frac{625}{2}} = a \quad \text{Since } a \text{ represents a length, take the positive square root.}$$
$$a \approx 17.67766953 \quad \text{Use a calculator.}$$

To two decimal places, the length is 17.68 units.

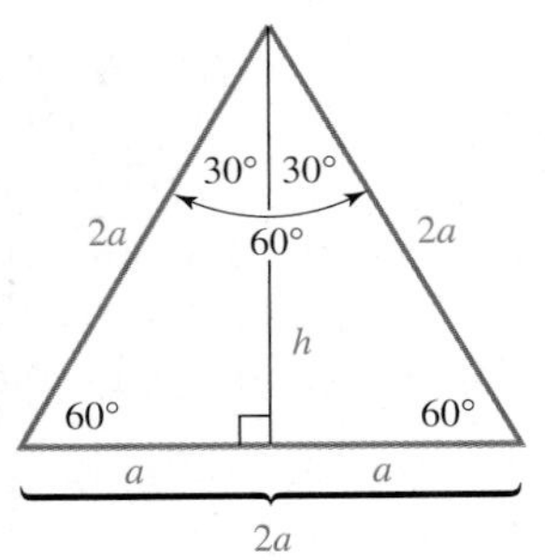

Figure 7-12

From geometry, we know that an *equilateral triangle* is a triangle with three sides of equal length and three 60° angles. If an **altitude** is drawn upon the base of an equilateral triangle, as shown in Figure 7-12, it bisects the base and divides the triangle into two 30°–60°–90° triangles. We can see that the shortest leg of each 30°–60°–90° triangle is a units long. Thus,

The shorter leg of a 30°–60°–90° triangle is half as long as its hypotenuse.

We can find the length of the altitude, h, by using the Pythagorean theorem.

$$a^2 + h^2 = (2a)^2$$
$$a^2 + h^2 = 4a^2 \quad (2a)^2 = (2a)(2a) = 4a^2$$
$$h^2 = 3a^2 \quad \text{Subtract } a^2 \text{ from both sides.}$$
$$h = \sqrt{3a^2} \quad \text{Since } h \text{ represents a length, take the positive square root.}$$
$$h = a\sqrt{3} \quad \sqrt{3a^2} = \sqrt{3}\sqrt{a^2} = a\sqrt{3}. \text{ No absolute value symbols are needed, because } a \text{ is positive.}$$

Thus,

The length of the longer leg is the length of the shorter side times $\sqrt{3}$.

EXAMPLE 9 **GEOMETRY** Find the length of the hypotenuse and the longer leg of the right triangle shown in Figure 7-13.

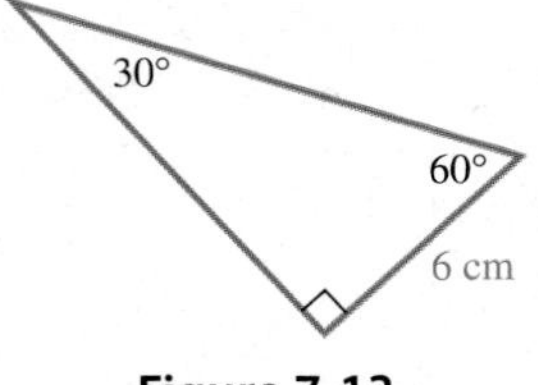

Figure 7-13

Solution Since the shorter leg of a 30°–60°–90° triangle is half as long as its hypotenuse, the hypotenuse is 12 centimeters long.

Since the length of the longer leg is the length of the shorter leg times $\sqrt{3}$, the longer leg is $6\sqrt{3}$ (about 10.39) centimeters long.

SELF CHECK 9 Find the length of the hypotenuse and the longer leg of a 30°–60°–90° right triangle if the shorter leg is 8 centimeters long.

EXAMPLE 10 **GEOMETRY** Find the length of each leg of the triangle shown in Figure 7-14.

Solution Since the shorter leg of a 30°–60°–90° triangle is half as long as its hypotenuse, the shorter leg is $\frac{9}{2}$ centimeters long.

Since the length of the longer leg is the length of the shorter leg times $\sqrt{3}$, the longer leg is $\frac{9}{2}\sqrt{3}$ (or about 7.79) centimeters long.

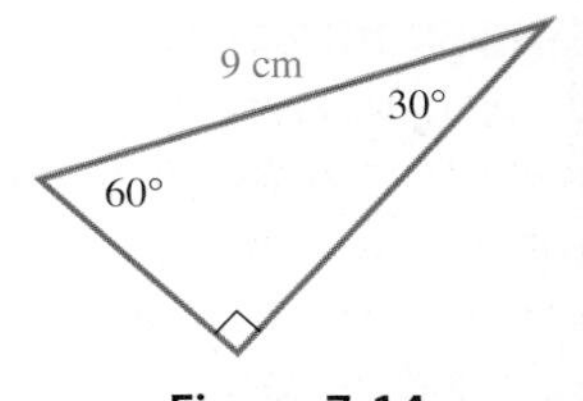

Figure 7-14

SELF CHECK 10 Find the length of the longer leg of the right triangle shown in Figure 7-14 if its hypotenuse is 10 cm long.

SELF CHECK ANSWERS **1. a.** $2\sqrt{5}$ **b.** $2\sqrt[3]{3}$ **2. a.** $\frac{\sqrt{11}}{6a}$ **b.** $\frac{2\sqrt[3]{a^2}}{5y}$ **3. a.** $7b\sqrt{2b}$ **b.** $3y\sqrt[3]{2y^2}$ **c.** $5b$ **4.** $19\sqrt{3}$ **5.** $2\sqrt[3]{3} + \sqrt[3]{2}$ **6.** $6x\sqrt{2x}$ **7.** $12\sqrt{2}$ m **9.** 16 cm, $8\sqrt{3}$ cm **10.** $5\sqrt{3}$ cm

NOW TRY THIS

Simplify.

1. $\sqrt[3]{8x - 8} + \sqrt[3]{x - 1}$

2. $\sqrt{\dfrac{75x^7y}{3xy}}$

3. $\dfrac{5\sqrt[3]{32}}{2}$

7.4 EXERCISES

WARM-UPS

Simplify.

1. $\sqrt{7}\sqrt{7}$

2. $\sqrt[3]{4^2}\sqrt[3]{4}$

3. $\dfrac{\sqrt[3]{54}}{\sqrt[3]{2}}$

Simplify each expression. Assume that $b \neq 0$.

4. $\sqrt{18}$

5. $\sqrt[3]{16}$

6. $\sqrt[3]{\dfrac{3x^2}{64b^6}}$

Combine like terms.

7. $3\sqrt{3} + 4\sqrt{3}$

8. $5\sqrt{7} - 2\sqrt{7}$

9. $2\sqrt[3]{9} + 3\sqrt[3]{9}$

10. $10\sqrt[5]{4} - 2\sqrt[5]{4}$

REVIEW *Perform each operation.*

11. $3x^2y^2(-5x^3y^{-3})$

12. $-2a^2b^{-2}(4a^{-2}b^4 - 2a^2b + 3a^3b^2)$

13. $(3t + 2)^2$

14. $(5r - 3s)(5r + 2s)$

15. $2p - 5\overline{)6p^2 - 7p - 25}$

16. $3m + n\overline{)6m^3 - m^2n + 2mn^2 + n^3}$

VOCABULARY AND CONCEPTS *Fill in the blanks.*

17. $\sqrt[n]{ab} =$ ______

18. $\sqrt[n]{\dfrac{a}{b}} =$ ______

19. If two radicals have the same index and the same radicand, they are called ______ radicals.

20. The perpendicular distance from a vertex of a triangle to the opposite side is called an ________.

GUIDED PRACTICE

Simplify each expression. Assume that all variables represent positive numbers. **(Objective 1)**

21. $\sqrt{6}\sqrt{6}$
22. $\sqrt{11}\sqrt{11}$
23. $\sqrt{t}\sqrt{t}$
24. $-\sqrt{z}\sqrt{z}$
25. $\sqrt[3]{5x^2}\sqrt[3]{25x}$
26. $\sqrt[4]{25a}\sqrt[4]{25a^3}$
27. $\dfrac{\sqrt{500}}{\sqrt{5}}$
28. $\dfrac{\sqrt{128}}{\sqrt{2}}$
29. $\dfrac{\sqrt{98x^3}}{\sqrt{2x}}$
30. $\dfrac{\sqrt{75y^5}}{\sqrt{3y}}$
31. $\dfrac{\sqrt{180ab^4}}{\sqrt{5ab^2}}$
32. $\dfrac{\sqrt{112ab^3}}{\sqrt{7ab}}$
33. $\dfrac{\sqrt[3]{48}}{\sqrt[3]{6}}$
34. $\dfrac{\sqrt[3]{64}}{\sqrt[3]{8}}$
35. $\dfrac{\sqrt[3]{189a^4}}{\sqrt[3]{7a}}$
36. $\dfrac{\sqrt[3]{243x^7}}{\sqrt[3]{9x}}$

Simplify each radical. **See Example 1. (Objective 1)**

37. $\sqrt{20}$
38. $\sqrt{8}$
39. $-\sqrt{200}$
40. $-\sqrt{250}$
41. $\sqrt[3]{80}$
42. $\sqrt[3]{270}$
43. $\sqrt[3]{-81}$
44. $\sqrt[3]{-72}$
45. $\sqrt[4]{32}$
46. $\sqrt[4]{48}$
47. $\sqrt[5]{96}$
48. $\sqrt[7]{256}$

Simplify each radical. Assume no divisions by 0. **See Example 2. (Objective 1)**

49. $\sqrt{\dfrac{7}{9x^2}}$
50. $\sqrt{\dfrac{3}{4y^2}}$
51. $\sqrt[3]{\dfrac{7a^3}{64}}$
52. $\sqrt[3]{\dfrac{4b^3}{125}}$
53. $\sqrt[4]{\dfrac{3p^4}{10{,}000q^4}}$
54. $\sqrt[5]{\dfrac{4r^5}{243s^{10}}}$
55. $\sqrt[5]{\dfrac{3m^{15}}{32n^{10}}}$
56. $\sqrt[6]{\dfrac{5a^6}{64b^{12}}}$

Simplify each radical. Assume that all variables represent positive numbers. **See Example 3. (Objective 1)**

57. $\sqrt{50x^2}$
58. $\sqrt{75a^2}$
59. $\sqrt{32b}$
60. $\sqrt{80c}$
61. $-\sqrt{112a^3}$
62. $\sqrt{147a^5}$
63. $\sqrt{175a^2b^3}$
64. $\sqrt{128a^3b^5}$
65. $-\sqrt{300xy}$
66. $\sqrt{200x^2y}$
67. $\sqrt[3]{-54x^6}$
68. $-\sqrt[3]{-81a^3}$
69. $\sqrt[3]{16x^{12}y^3}$
70. $\sqrt[3]{40a^3b^6}$
71. $\sqrt{\dfrac{z^2}{16x^2}}$
72. $\sqrt{\dfrac{b^4}{64a^8}}$

Simplify and combine like radicals. **See Example 4. (Objective 2)**

73. $\sqrt{3} + \sqrt{27}$
74. $\sqrt{8} + \sqrt{32}$
75. $\sqrt{2} - \sqrt{8}$
76. $\sqrt{20} - \sqrt{125}$
77. $\sqrt{98} - \sqrt{50}$
78. $\sqrt{72} - \sqrt{200}$
79. $3\sqrt{24} + \sqrt{54}$
80. $\sqrt{18} + 2\sqrt{50}$
81. $\sqrt{18} + \sqrt{300} - \sqrt{243}$
82. $\sqrt{80} - \sqrt{128} + \sqrt{288}$

Simplify and combine like radicals. **See Example 5. (Objective 2)**

83. $\sqrt[3]{24} + \sqrt[3]{3}$
84. $\sqrt[3]{16} + \sqrt[3]{128}$
85. $\sqrt[3]{32} - \sqrt[3]{108}$
86. $\sqrt[3]{80} - \sqrt[3]{10{,}000}$
87. $2\sqrt[3]{125} - 5\sqrt[3]{64}$
88. $\sqrt[3]{81} - \sqrt[3]{24}$
89. $2\sqrt[3]{16} - \sqrt[3]{54} - 3\sqrt[3]{128}$
90. $\sqrt[3]{250} - 4\sqrt[3]{5} + \sqrt[3]{16}$

Simplify and combine like radicals. All variables represent positive numbers. **See Example 6. (Objective 2)**

91. $\sqrt[3]{3x^5} - \sqrt[3]{24x^5}$
92. $\sqrt[3]{16x^4} - \sqrt[3]{54x^4}$
93. $\sqrt{25yz^2} + \sqrt{9yz^2}$
94. $\sqrt{36xy^2} + \sqrt{49xy^2}$
95. $\sqrt{y^5} - \sqrt{9y^5} - \sqrt{25y^5}$
96. $\sqrt{8y^7} + \sqrt{32y^7} - \sqrt{2y^7}$
97. $3\sqrt[3]{2x} - \sqrt[3]{54x}$
98. $2\sqrt[3]{64a} + 2\sqrt[3]{8a}$

For Exercises 99–106, find the lengths of the remaining sides of the triangle. **See Examples 7–8. (Objective 3)**

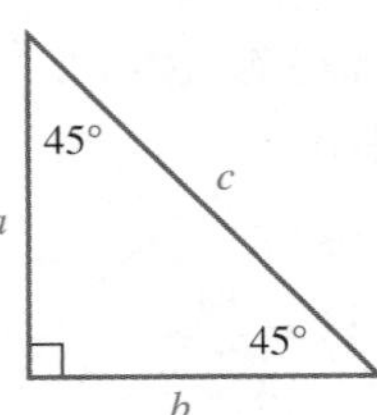

99. $a = 3$
100. $a = 8$
101. $b = \dfrac{2}{3}$
102. $b = \dfrac{3}{10}$
103. $a = 5\sqrt{2}$
104. $a = 12\sqrt{2}$
105. $c = 7\sqrt{2}$
106. $c = 16\sqrt{2}$

For Exercises 107–114, find the lengths of the remaining sides of the triangle. **See Examples 9–10. (Objective 3)**

107. $a = 5$

108. $a = 8$

109. $b = 9\sqrt{3}$

110. $b = 18\sqrt{3}$

111. $c = 24$

112. $c = 8$

113. $c = 15$

114. $c = 25$

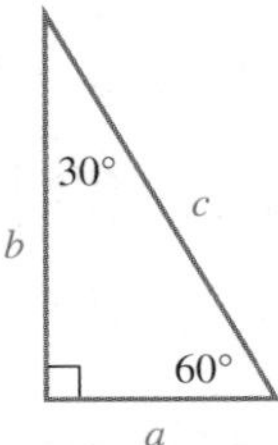

ADDITIONAL PRACTICE

Simplify. Assume that all variables represent positive numbers.

115. $4\sqrt{2x} + 6\sqrt{2x}$

116. $\sqrt{25y^2z} - \sqrt{16y^2z}$

117. $\sqrt[4]{32x^{12}y^4}$

118. $\sqrt[5]{64x^{10}y^5}$

119. $\sqrt[4]{\dfrac{5x}{16z^4}}$

120. $\sqrt{\dfrac{11a^2}{125b^6}}$

121. $\sqrt{98} - \sqrt{50} - \sqrt{72}$

122. $\sqrt{20} + \sqrt{125} - \sqrt{80}$

123. $3\sqrt[3]{27} + 12\sqrt[3]{216}$

124. $14\sqrt[4]{32} - 15\sqrt[4]{162}$

125. $23\sqrt[4]{768} + \sqrt[4]{48}$

126. $3\sqrt[4]{512} + 2\sqrt[4]{32}$

127. $4\sqrt[4]{243} - \sqrt[4]{48}$

128. $\sqrt[4]{48} - \sqrt[4]{243} - \sqrt[4]{768}$

129. $6\sqrt[3]{5y} + 3\sqrt[3]{5y}$

130. $8\sqrt[5]{7a^2} - 7\sqrt[5]{7a^2}$

131. $10\sqrt[6]{12xyz} - \sqrt[6]{12xyz}$

132. $3\sqrt[4]{x^4y} - 2\sqrt[4]{x^4y}$

133. $\sqrt[5]{x^6y^2} + \sqrt[5]{32x^6y^2} + \sqrt[5]{x^6y^2}$

134. $\sqrt[3]{xy^4} + \sqrt[3]{8xy^4} - \sqrt[3]{27xy^4}$

135. $\sqrt{x^2 + 2x + 1} + \sqrt{x^2 + 2x + 1}$

136. $\sqrt{4x^2 + 12x + 9} + \sqrt{9x^2 + 6x + 1}$

Find the missing lengths in each triangle. Give each answer to two decimal places.

137.

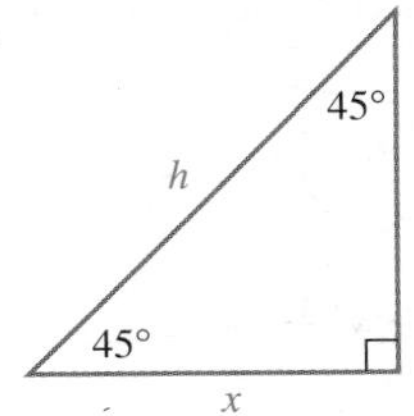

138.

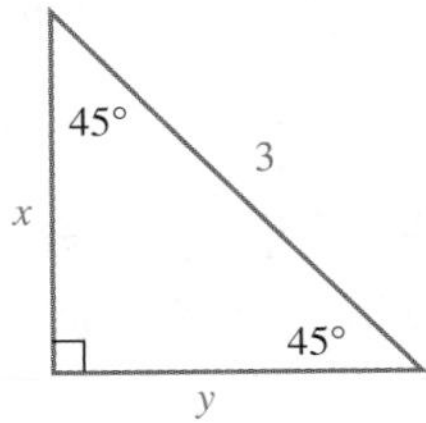

139.

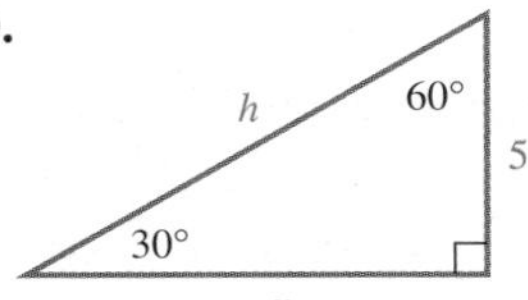

140.

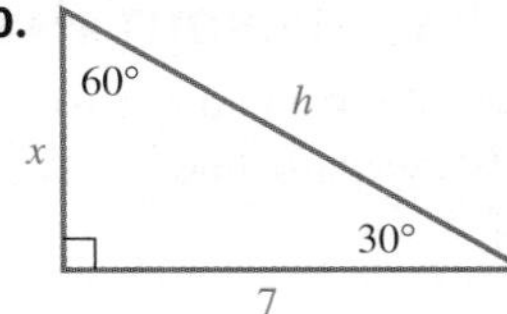

141.

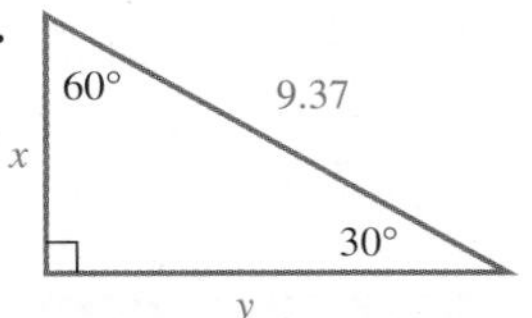

142.

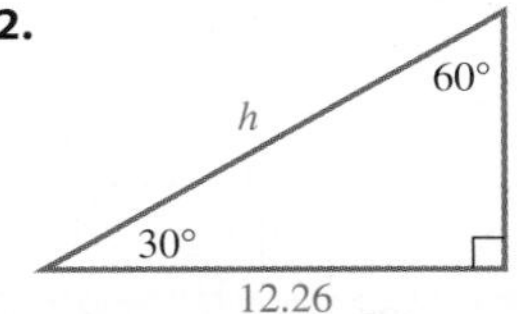

143.

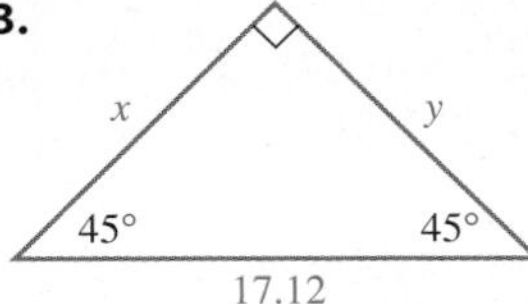

144.

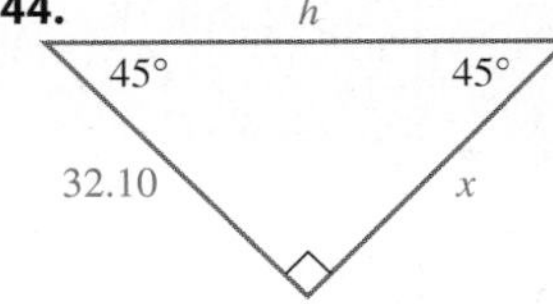

APPLICATIONS *Find the exact answer and then give an approximation to the nearest hundredth.*

145. Hardware The sides of a regular hexagonal nut are 10 millimeters long. Find the height h of the nut.

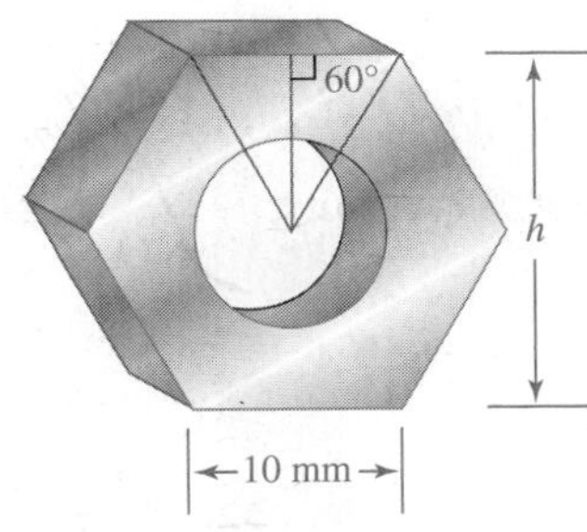

146. Ironing boards Find the height h of the ironing board shown in the illustration.

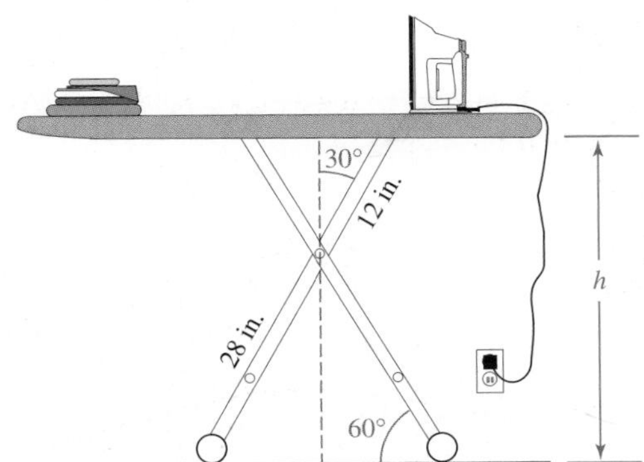

WRITING ABOUT MATH

147. Explain how to recognize like radicals.

148. Explain how to combine like radicals.

SOMETHING TO THINK ABOUT

149. Can you find any numbers a and b such that

$$\sqrt{a+b} = \sqrt{a} + \sqrt{b}$$

150. Find the sum.

$$\sqrt{3} + \sqrt{3^2} + \sqrt{3^3} + \sqrt{3^4} + \sqrt{3^5}$$

SECTION 7.5 Multiplying and Dividing Radical Expressions

Objectives

1. Multiply two radical expressions.
2. Rationalize the denominator of a fraction that contains a radical expression.
3. Rationalize the numerator of a fraction that contains a radical expression.
4. Solve an application problem containing a radical expression.

Vocabulary

rationalize denominators conjugates rationalize numerators

Getting Ready

Perform each operation and simplify, if possible.

1. a^3a^4 **2.** $\frac{b^5}{b^2}$ **3.** $a(a-2)$ **4.** $3b^2(2b+3)$

5. $(a+2)(a-5)$ **6.** $(2a+3b)(2a-3b)$

We now learn how to multiply and divide radical expressions. Then we will use these skills to solve application problems.

1 Multiply two radical expressions.

Radical expressions with the same index can be multiplied and divided.

EXAMPLE 1 Multiply: **a.** $(3\sqrt{6})(2\sqrt{3})$ **b.** $(\sqrt[3]{3a})(\sqrt[3]{9a^4})$.

Solution We use the commutative and associative properties of multiplication to multiply the coefficients and the radicals separately. Then we simplify any radicals in the product, if possible.

a. $3\sqrt{6} \cdot 2\sqrt{3} = 3(2)\sqrt{6}\sqrt{3}$ Multiply the coefficients and multiply the radicals.

$= 6\sqrt{18}$ $3(2) = 6$ and $\sqrt{6}\sqrt{3} = \sqrt{18}$.

$= 6\sqrt{9}\sqrt{2}$ $\sqrt{18} = \sqrt{9 \cdot 2} = \sqrt{9}\sqrt{2}$

$= 6(3)\sqrt{2}$ $\sqrt{9} = 3$

$= 18\sqrt{2}$

b. $\left(\sqrt[3]{3a}\right)\left(\sqrt[3]{9a^4}\right) = \sqrt[3]{27a^5}$ Multiply the radicals.

$= \sqrt[3]{27a^3 \cdot a^2}$ Factor $27a^5$.

$= \sqrt[3]{27a^3}\sqrt[3]{a^2}$ $\sqrt[3]{ab} = \sqrt[3]{a}\sqrt[3]{b}$

$= 3a\sqrt[3]{a^2}$ $\sqrt[3]{27a^3} = 3a$

SELF CHECK 1 Multiply $-2\sqrt{7}$ by $5\sqrt{2}$.

To multiply a radical expression with two or more terms by a radical expression, we use the distributive property to remove parentheses and then simplify each resulting term, if possible.

EXAMPLE 2 Multiply: $3\sqrt{3}\left(4\sqrt{8} - 5\sqrt{10}\right)$.

Solution $3\sqrt{3}\left(4\sqrt{8} - 5\sqrt{10}\right)$

$= 3\sqrt{3} \cdot 4\sqrt{8} - 3\sqrt{3} \cdot 5\sqrt{10}$ Use the distributive property.

$= 12\sqrt{24} - 15\sqrt{30}$ Multiply the coefficients and multiply the radicals.

$= 12\sqrt{4}\sqrt{6} - 15\sqrt{30}$

$= 12(2)\sqrt{6} - 15\sqrt{30}$

$= 24\sqrt{6} - 15\sqrt{30}$

SELF CHECK 2 Multiply: $4\sqrt{2}\left(3\sqrt{5} - 2\sqrt{8}\right)$.

To multiply two radical expressions, each with two or more terms, we use the distributive property as we did when we multiplied two polynomials. Then we simplify each resulting term, if possible.

EXAMPLE 3 Multiply: $\left(\sqrt{7} + \sqrt{2}\right)\left(\sqrt{7} - 3\sqrt{2}\right)$.

Solution $\left(\sqrt{7} + \sqrt{2}\right)\left(\sqrt{7} - 3\sqrt{2}\right)$

$= \left(\sqrt{7}\right)^2 - 3\sqrt{7}\sqrt{2} + \sqrt{2}\sqrt{7} - 3\sqrt{2}\sqrt{2}$

$= 7 - 3\sqrt{14} + \sqrt{14} - 3(2)$

$$= 7 - 2\sqrt{14} - 6$$
$$= 1 - 2\sqrt{14}$$

SELF CHECK 3 Multiply: $(\sqrt{5} + 2\sqrt{3})(\sqrt{5} - \sqrt{3})$.

EXAMPLE 4 Multiply: $(\sqrt{3x} - \sqrt{5})(\sqrt{2x} + \sqrt{10})$.

Solution

$$(\sqrt{3x} - \sqrt{5})(\sqrt{2x} + \sqrt{10})$$
$$= \sqrt{3x}\sqrt{2x} + \sqrt{3x}\sqrt{10} - \sqrt{5}\sqrt{2x} - \sqrt{5}\sqrt{10}$$
$$= \sqrt{6x^2} + \sqrt{30x} - \sqrt{10x} - \sqrt{50}$$
$$= \sqrt{6}\sqrt{x^2} + \sqrt{30x} - \sqrt{10x} - \sqrt{25}\sqrt{2}$$
$$= \sqrt{6}x + \sqrt{30x} - \sqrt{10x} - 5\sqrt{2}$$

COMMENT Note that x is not under the radical in the first term, but it is under the radical in the second and third terms.

SELF CHECK 4 Multiply: $(\sqrt{x} + 1)(\sqrt{x} - 3)$.

COMMENT It is important to draw radical signs so they completely cover the radicand, but no more than the radicand. To avoid confusion, we can use the commutative property of multiplication and write an expression such as $\sqrt{6}x$ in the form $x\sqrt{6}$.

2 Rationalize the denominator of a fraction that contains a radical expression.

To divide radical expressions, we **rationalize the denominator** of a fraction to replace the denominator with a rational number. For example, to divide $\sqrt{70}$ by $\sqrt{3}$ we write the division as the fraction

$$\frac{\sqrt{70}}{\sqrt{3}}$$

To eliminate the radical in the denominator, we multiply the numerator and the denominator by a number that will give a perfect square under the radical in the denominator. Because $3 \cdot 3 = 9$ and 9 is a perfect square, $\sqrt{3}$ is such a number.

$$\frac{\sqrt{70}}{\sqrt{3}} = \frac{\sqrt{70} \cdot \sqrt{3}}{\sqrt{3} \cdot \sqrt{3}} \quad \text{Multiply numerator and denominator by } \sqrt{3}.$$
$$= \frac{\sqrt{210}}{3} \quad \text{Multiply the radicals.}$$

Since there is no radical in the denominator and $\sqrt{210}$ cannot be simplified, the expression $\frac{\sqrt{210}}{3}$ is in simplest form, and the division is complete.

EXAMPLE 5 Rationalize the denominator.

a. $\sqrt{\frac{20}{7}}$ **b.** $\frac{4}{\sqrt[3]{2}}$

Solution **a.** We first write the square root of the quotient as the quotient of two square roots.

$$\sqrt{\frac{20}{7}} = \frac{\sqrt{20}}{\sqrt{7}}$$

Because the denominator is a square root, we must then multiply the numerator and the denominator by a number that will give a rational number in the denominator. Such a number is $\sqrt{7}$.

$$\frac{\sqrt{20}}{\sqrt{7}} = \frac{\sqrt{20} \cdot \sqrt{7}}{\sqrt{7} \cdot \sqrt{7}} \quad \text{Multiply numerator and denominator by } \sqrt{7}.$$

$$= \frac{\sqrt{140}}{7} \quad \text{Multiply the radicals.}$$

$$= \frac{2\sqrt{35}}{7} \quad \text{Simplify } \sqrt{140}\text{: } \sqrt{140} = \sqrt{4 \cdot 35} = \sqrt{4}\sqrt{35} = 2\sqrt{35}.$$

b. Since the denominator is a cube root, we multiply the numerator and the denominator by a number that will give a perfect cube under the radical sign. Since $2 \cdot 4 = 8$ is a perfect cube, $\sqrt[3]{4}$ is such a number.

$$\frac{4}{\sqrt[3]{2}} = \frac{4 \cdot \sqrt[3]{4}}{\sqrt[3]{2} \cdot \sqrt[3]{4}} \quad \text{Multiply numerator and denominator by } \sqrt[3]{4}.$$

$$= \frac{4\sqrt[3]{4}}{\sqrt[3]{8}} \quad \text{Multiply the radicals in the denominator.}$$

$$= \frac{4\sqrt[3]{4}}{2} \quad \sqrt[3]{8} = 2$$

$$= 2\sqrt[3]{4} \quad \text{Simplify.}$$

SELF CHECK 5 Rationalize the denominator: $\frac{5}{\sqrt[4]{3}}$.

EXAMPLE 6 Rationalize the denominator: $\frac{\sqrt[3]{5}}{\sqrt[3]{18}}$.

Solution We multiply the numerator and the denominator by a number that will result in a perfect cube under the radical sign in the denominator.

Since 216 is the smallest perfect cube that is divisible by 18 ($216 \div 18 = 12$) multiplying the numerator and the denominator by $\sqrt[3]{12}$ will give the smallest possible perfect cube under the radical in the denominator.

$$\frac{\sqrt[3]{5}}{\sqrt[3]{18}} = \frac{\sqrt[3]{5} \cdot \sqrt[3]{12}}{\sqrt[3]{18} \cdot \sqrt[3]{12}} \quad \text{Multiply numerator and denominator by } \sqrt[3]{12}.$$

$$= \frac{\sqrt[3]{60}}{\sqrt[3]{216}} \quad \text{Multiply the radicals.}$$

$$= \frac{\sqrt[3]{60}}{6} \qquad \sqrt[3]{216} = 6$$

SELF CHECK 6 Rationalize the denominator: $\dfrac{\sqrt[3]{2}}{\sqrt[3]{9}}$.

EXAMPLE 7 Rationalize the denominator of $\dfrac{\sqrt{5xy^2}}{\sqrt{xy^3}}$ (x and y are positive numbers).

Solution

Method 1

$$\frac{\sqrt{5xy^2}}{\sqrt{xy^3}} = \sqrt{\frac{5xy^2}{xy^3}} = \sqrt{\frac{5}{y}} = \frac{\sqrt{5}}{\sqrt{y}} = \frac{\sqrt{5}\sqrt{y}}{\sqrt{y}\sqrt{y}} = \frac{\sqrt{5y}}{y}$$

Method 2

$$\frac{\sqrt{5xy^2}}{\sqrt{xy^3}} = \sqrt{\frac{5xy^2}{xy^3}} = \sqrt{\frac{5}{y}} = \sqrt{\frac{5 \cdot y}{y \cdot y}} = \frac{\sqrt{5y}}{\sqrt{y^2}} = \frac{\sqrt{5y}}{y}$$

SELF CHECK 7 Rationalize the denominator: $\dfrac{\sqrt{4ab^3}}{\sqrt{2a^2b^2}}$ $(a > 0, b > 0)$.

To rationalize the denominator of a fraction with square roots in a binomial denominator, we can multiply the numerator and denominator by the **conjugate** of the denominator. Conjugate binomials are binomials with the same terms but with opposite signs between their terms.

Conjugate Binomials The conjugate of $a + b$ is $a - b$, and the conjugate of $a - b$ is $a + b$.

If we multiply an expression such as $5 + \sqrt{2}$ by its conjugate $5 - \sqrt{2}$, we will obtain an expression without any radical terms.

$$\begin{aligned}(5 + \sqrt{2})(5 - \sqrt{2}) &= 25 - 5\sqrt{2} + 5\sqrt{2} - 2 \\ &= 23\end{aligned}$$

EXAMPLE 8 Rationalize the denominator: $\dfrac{1}{\sqrt{2} + 1}$.

Solution We multiply the numerator and denominator of the fraction by $\sqrt{2} - 1$, which is the conjugate of the denominator.

$$\frac{1}{\sqrt{2}+1} = \frac{1(\sqrt{2}-1)}{(\sqrt{2}+1)(\sqrt{2}-1)} \qquad \frac{\sqrt{2}-1}{\sqrt{2}-1} = 1$$

$$= \frac{\sqrt{2}-1}{(\sqrt{2})^2 - 1} \qquad (\sqrt{2}+1)(\sqrt{2}-1) = (\sqrt{2})^2 - 1$$

$$= \frac{\sqrt{2}-1}{2-1} \qquad (\sqrt{2})^2 = 2$$

$$= \sqrt{2} - 1 \qquad \frac{\sqrt{2}-1}{2-1} = \frac{\sqrt{2}-1}{1} = \sqrt{2}-1$$

SELF CHECK 8 Rationalize the denominator: $\frac{2}{\sqrt{3}+1}$.

EXAMPLE 9 Rationalize the denominator: $\frac{\sqrt{x}+\sqrt{2}}{\sqrt{x}-\sqrt{2}}$ $(x > 0, x \neq 2)$.

Solution We multiply the numerator and denominator by $\sqrt{x}+\sqrt{2}$, which is the conjugate of the denominator, and simplify.

$$\frac{\sqrt{x}+\sqrt{2}}{\sqrt{x}-\sqrt{2}} = \frac{(\sqrt{x}+\sqrt{2})(\sqrt{x}+\sqrt{2})}{(\sqrt{x}-\sqrt{2})(\sqrt{x}+\sqrt{2})}$$

$$= \frac{x + \sqrt{2x} + \sqrt{2x} + 2}{x-2} \qquad \text{Use the FOIL method.}$$

$$= \frac{x + 2\sqrt{2x} + 2}{x-2}$$

SELF CHECK 9 Rationalize the denominator: $\frac{\sqrt{x}-\sqrt{2}}{\sqrt{x}+\sqrt{2}}$ $(x > 0)$.

3 Rationalize the numerator of a fraction that contains a radical expression.

In calculus, we sometimes have to **rationalize a numerator** by multiplying the numerator and denominator of the fraction by the conjugate of the numerator.

EXAMPLE 10 Rationalize the numerator: $\frac{\sqrt{x}-3}{\sqrt{x}}$ $(x > 0)$.

Solution We multiply the numerator and denominator by $\sqrt{x}+3$, which is the conjugate of the numerator.

$$\frac{\sqrt{x}-3}{\sqrt{x}} = \frac{(\sqrt{x}-3)(\sqrt{x}+3)}{\sqrt{x}(\sqrt{x}+3)}$$

$$= \frac{x + 3\sqrt{x} - 3\sqrt{x} - 9}{x + 3\sqrt{x}}$$

$$= \frac{x-9}{x+3\sqrt{x}}$$

The final expression is not in simplified form. However, this nonsimplified form is sometimes desirable in calculus.

SELF CHECK 10 Rationalize the numerator: $\frac{\sqrt{x}+3}{\sqrt{x}}$ $(x > 0)$.

4 Solve an application problem containing a radical expression.

EXAMPLE 11 **PHOTOGRAPHY** Many camera lenses (see Figure 7-15) have an adjustable opening called the *aperture,* which controls the amount of light passing through the lens. The *f-number* of a lens is its *focal length* divided by the diameter of its circular aperture.

Figure 7-15

$$f\text{-number} = \frac{f}{d} \quad f \text{ is the focal length, and } d \text{ is the diameter of the aperture.}$$

A lens with a focal length of 12 centimeters and an aperture with a diameter of 6 centimeters has an f-number of $\frac{12}{6}$ and is an $f/2$ lens. If the area of the aperture is reduced to admit half as much light, the f-number of the lens will change. Find the new f-number.

Solution We first find the area of the aperture when its diameter is 6 centimeters.

$A = \pi r^2$ The formula for the area of a circle.

$A = \pi(3)^2$ Since a radius is half the diameter, substitute 3 for r.

$A = 9\pi$

When the size of the aperture is reduced to admit half as much light, the area of the aperture will be $\frac{9\pi}{2}$ square centimeters. To find the diameter of a circle with this area, we proceed as follows:

$A = \pi r^2$ This is the formula for the area of a circle.

$\frac{9\pi}{2} = \pi\left(\frac{d}{2}\right)^2$ Substitute $\frac{9\pi}{2}$ for A and $\frac{d}{2}$ for r.

$\frac{9\pi}{2} = \frac{\pi d^2}{4}$ $\left(\frac{d}{2}\right)^2 = \frac{d^2}{4}$

$18 = d^2$ Multiply both sides by 4, and divide both sides by π.

$d = 3\sqrt{2}$ $\sqrt{18} = \sqrt{9}\sqrt{2} = 3\sqrt{2}$

Since the focal length of the lens is still 12 centimeters and the diameter is now $3\sqrt{2}$ centimeters, the new f-number of the lens is

$f\text{-number} = \frac{f}{d} = \frac{12}{3\sqrt{2}}$ Substitute 12 for f and $3\sqrt{2}$ for d.

$= \frac{4}{\sqrt{2}}$ Simplify.

$= \frac{4\sqrt{2}}{2}$ Rationalize the denominator.

$= 2\sqrt{2}$ Simplify.

≈ 2.828427125 Use a calculator.

The lens is now an $f/2.8$ lens.

EVERYDAY CONNECTIONS Traveling Through Water

©Shutterstock.com/Steven Lee

The Froude number, named after William Froude, measures the resistance of an object moving through water. For a ship, the Froude number is calculated by the formula:

$$Fr = \frac{V}{\sqrt{gL}}$$

where V is the ship's velocity, L is the ship's length, and g is the acceleration due to gravity.

Source:
http://www.solarnavigator.net/froude_number_speed_length_ratio.htm

1. Use the formula to compute the Froude number for a ship that is 100 feet long, traveling with a velocity of 74 feet per second. Note that the acceleration due to gravity is 32 ft/s^2.
2. Use the formula to compute the Froude number for a ship that is 25 meters long, traveling with a velocity of 22 meters per second. Note that the acceleration due to gravity is 9.8 m/s^2.

SELF CHECK ANSWERS

1. $-10\sqrt{14}$ **2.** $12\sqrt{10} - 32$ **3.** $-1 + \sqrt{15}$ **4.** $x - 2\sqrt{x} - 3$ **5.** $\frac{5\sqrt[4]{27}}{3}$ **6.** $\frac{\sqrt[3]{6}}{3}$ **7.** $\frac{\sqrt{2ab}}{a}$ **8.** $\sqrt{3} - 1$ **9.** $\frac{x - 2\sqrt{2x} + 2}{x - 2}$ **10.** $\frac{x - 9}{x - 3\sqrt{x}}$

NOW TRY THIS

Find the domain of each of the following. Give your answer in interval notation.

1. $f(x) = \dfrac{3x - 2}{\sqrt{x} + 1}$

2. $g(x) = \dfrac{3x - 2}{\sqrt{x + 1}}$

3. $h(x) = \dfrac{\sqrt{x - 2}}{\sqrt{x} + 1}$

7.5 EXERCISES

WARM-UPS *Simplify.*

1. $\sqrt{3}\sqrt{3}$

2. $\sqrt[3]{2}\sqrt[3]{2}\sqrt[3]{2}$

3. $\sqrt{3}\sqrt{9}$

4. $\sqrt{a^3b}\sqrt{ab}$ $(a > 0, b > 0)$

5. $3\sqrt{2}(\sqrt{2} + 1)$

6. $(\sqrt{2} + 1)(\sqrt{2} - 1)$

7. $\frac{1}{\sqrt{2}}$

8. $\frac{1}{\sqrt{3} - 1}$

REVIEW *Solve each equation.*

9. $\frac{2}{3 - a} = 1$

10. $5(s - 4) = -5(s - 4)$

11. $\frac{8}{b - 2} + \frac{3}{2 - b} = -\frac{1}{b}$

12. $\frac{2}{x - 2} + \frac{1}{x + 1} = \frac{1}{(x + 1)(x - 2)}$

VOCABULARY AND CONCEPTS *Fill in the blanks.*

13. To multiply $2\sqrt{7}$ by $3\sqrt{5}$, we multiply __ by 3 and then multiply ____ by ____.

14. To multiply $2\sqrt{5}(3\sqrt{8} + \sqrt{3})$, we use the __________ property to remove parentheses and simplify each resulting term.

15. The conjugate of $\sqrt{x} + 1$ is _______.

16. To multiply $(\sqrt{3} + \sqrt{2})(\sqrt{3} - 2\sqrt{2})$, we can use the ______ method.

17. To rationalize the denominator of $\frac{1}{\sqrt{3} - 1}$, multiply both the numerator and denominator by the ________ of the denominator.

18. To rationalize the numerator of $\frac{\sqrt{5} + 2}{\sqrt{5} - 2}$, multiply both the numerator and denominator by _______.

GUIDED PRACTICE

Simplify. All variables represent positive values. See Example 1. (Objective 1)

19. $\sqrt{2}\sqrt{8}$

20. $\sqrt{3}\sqrt{27}$

21. $\sqrt{5}\sqrt{10}$

22. $\sqrt{7}\sqrt{35}$

23. $2\sqrt{3}\sqrt{6}$

24. $3\sqrt{11}\sqrt{33}$

25. $\sqrt[3]{5}\sqrt[3]{25}$

26. $\sqrt[3]{7}\sqrt[3]{49}$

27. $\sqrt[3]{2}\sqrt[3]{12}$

28. $\sqrt[3]{3}\sqrt[3]{18}$

29. $\sqrt{ab^3}\sqrt{ab}$

30. $\sqrt{8x}\sqrt{2x^3y}$

31. $\sqrt[3]{5r^2s}\sqrt[3]{2r}$

32. $\sqrt[3]{3xy^2}\sqrt[3]{9x^3}$

33. $\sqrt{x(x + 3)}\sqrt{x^3(x + 3)}$

34. $\sqrt{y^2(x + y)}\sqrt{(x + y)^3}$

Simplify. See Example 2. (Objective 1)

35. $3\sqrt{5}(4 - \sqrt{5})$

36. $2\sqrt{7}(3\sqrt{7} - 1)$

37. $3\sqrt{2}(4\sqrt{3} + 2\sqrt{7})$

38. $-\sqrt{3}(\sqrt{7} - \sqrt{5})$

Simplify. All variables represent positive values. See Examples 3–4. (Objective 1)

39. $(\sqrt{2} + 1)(\sqrt{2} - 3)$

40. $(2\sqrt{3} + 1)(\sqrt{3} - 1)$

41. $(4\sqrt{x} + 3)(2\sqrt{x} - 5)$

42. $(7\sqrt{y} + 2)(3\sqrt{y} - 5)$

43. $(\sqrt{5z} + \sqrt{3})(\sqrt{5z} + \sqrt{3})$

44. $(\sqrt{3p} - \sqrt{2})(\sqrt{3p} + \sqrt{2})$

45. $(2\sqrt{3a} - \sqrt{b})(\sqrt{3a} + 3\sqrt{b})$

46. $(5\sqrt{p} - \sqrt{3q})(\sqrt{p} + 2\sqrt{3q})$

47. $(3\sqrt{2r} - 2)^2$

48. $(2\sqrt{3t} + 5)^2$

49. $-2(\sqrt{3x} + \sqrt{3})^2$

50. $3(\sqrt{5x} - \sqrt{3})^2$

Rationalize each denominator. See Examples 5–6. (Objective 2)

51. $\sqrt{\frac{1}{7}}$

52. $\sqrt{\frac{5}{3}}$

53. $\sqrt{\frac{2}{3}}$

54. $\sqrt{\frac{3}{2}}$

55. $\frac{\sqrt{5}}{\sqrt{8}}$

56. $\frac{\sqrt{3}}{\sqrt{50}}$

57. $\frac{\sqrt{8}}{\sqrt{2}}$

58. $\frac{\sqrt{27}}{\sqrt{3}}$

59. $\frac{1}{\sqrt[3]{2}}$

60. $\frac{2}{\sqrt[3]{6}}$

61. $\frac{\sqrt[3]{2}}{\sqrt[3]{9}}$

62. $\frac{\sqrt[3]{9}}{\sqrt[3]{54}}$

Rationalize each denominator. All variables represent positive values. See Example 7. (Objective 2)

63. $\frac{\sqrt{8x^2y}}{\sqrt{xy}}$

64. $\frac{\sqrt{9xy}}{\sqrt{3x^2y}}$

65. $\frac{\sqrt{10xy^2}}{\sqrt{2xy^3}}$

66. $\frac{\sqrt{5ab^2c}}{\sqrt{10abc}}$

Rationalize each denominator. See Example 8. (Objective 2)

67. $\frac{1}{\sqrt{2} - 1}$

68. $\frac{3}{\sqrt{3} - 1}$

69. $\frac{\sqrt{2}}{\sqrt{5} + 3}$

70. $\frac{\sqrt{3}}{\sqrt{3} - 2}$

71. $\frac{\sqrt{3} + 1}{\sqrt{3} - 1}$

72. $\frac{\sqrt{2} - 1}{\sqrt{2} + 1}$

73. $\frac{\sqrt{7} - \sqrt{2}}{\sqrt{2} + \sqrt{7}}$

74. $\frac{\sqrt{3} + \sqrt{2}}{\sqrt{3} - \sqrt{2}}$

Rationalize each denominator. All variables represent positive values. See Example 9. (Objective 2)

75. $\dfrac{2}{\sqrt{x}+1}$

76. $\dfrac{3}{\sqrt{x}-2}$

77. $\dfrac{x}{\sqrt{x}-4}$

78. $\dfrac{2x}{\sqrt{x}+1}$

79. $\dfrac{\sqrt{x}-\sqrt{y}}{\sqrt{x}+\sqrt{y}}$

80. $\dfrac{\sqrt{x}+\sqrt{y}}{\sqrt{x}-\sqrt{y}}$

Rationalize each numerator. All variables represent positive values. See Example 10. (Objective 3)

81. $\dfrac{\sqrt{3}+1}{2}$

82. $\dfrac{\sqrt{5}-1}{2}$

83. $\dfrac{\sqrt{x}+3}{x}$

84. $\dfrac{2+\sqrt{x}}{5x}$

ADDITIONAL PRACTICE

Simplify each radical expression. All variables represent positive values.

85. $(3\sqrt[3]{9})(2\sqrt[3]{3})$

86. $(2\sqrt[3]{16})(-\sqrt[3]{4})$

87. $\sqrt{5ab}\sqrt{5a}$

88. $\sqrt{15rs^2}\sqrt{10r}$

89. $\sqrt[3]{a^5b}\sqrt[3]{16ab^5}$

90. $\sqrt[3]{3x^4y}\sqrt[3]{18x}$

91. $\sqrt[3]{6x^2(y+z)^2}\sqrt[3]{18x(y+z)}$

92. $\sqrt[3]{9x^2y(z+1)^2}\sqrt[3]{6xy^2(z+1)}$

93. $-2\sqrt{5x}(4\sqrt{2x}-3\sqrt{3})$

94. $3\sqrt{7t}(2\sqrt{7t}+3\sqrt{3t^2})$

Rationalize each denominator. All variables represent positive values.

95. $\dfrac{3}{\sqrt[3]{9}}$

96. $\dfrac{2}{\sqrt[3]{a}}$

97. $\dfrac{1}{\sqrt[4]{4}}$

98. $\dfrac{4}{\sqrt[4]{32}}$

99. $\dfrac{1}{\sqrt[5]{16}}$

100. $\dfrac{1}{\sqrt[5]{2}}$

101. $\dfrac{\sqrt[3]{4a^2}}{\sqrt[3]{2ab}}$

102. $\dfrac{\sqrt[3]{9x}}{\sqrt[3]{3xy}}$

103. $\dfrac{2z-1}{\sqrt{2z}-1}$

104. $\dfrac{3t-1}{\sqrt{3t}+1}$

Rationalize each numerator. All variables represent positive values.

105. $\dfrac{\sqrt{x}+\sqrt{y}}{\sqrt{x}}$

106. $\dfrac{\sqrt{x}-\sqrt{y}}{\sqrt{x}+\sqrt{y}}$

APPLICATIONS *Solve each application problem.*

See Example 11. (Objective 4)

107. Photography We have seen that a lens with a focal length of 12 centimeters and an aperture $3\sqrt{2}$ centimeters in diameter is an $f/2.8$ lens. Find the f-number if the area of the aperture is again cut in half.

108. Photography A lens with a focal length of 12 centimeters and an aperture 3 centimeters in diameter is an $f/4$ lens. Find the f-number if the area of the aperture is cut in half.

109. Targets The radius r of the target is given by the formula

$$r = \sqrt{\frac{A}{\pi}}$$

where A is the area. Write the formula in a form in which the denominator is not part of the radicand.

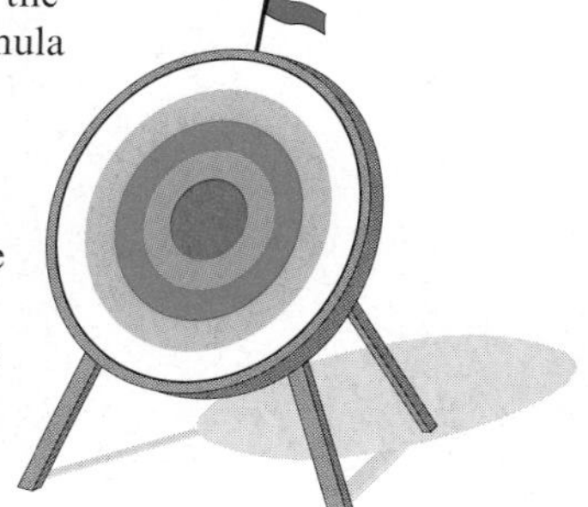

110. Pulse rates The approximate pulse rate (in beats per minute) of an adult who is t inches tall is given by the function

$$p(t) = \frac{590}{\sqrt{t}}$$

Write the formula in a form in which the denominator is a rational expression.

111. If the hypotenuse of an isosceles right triangle is 8 cm, find the length of each leg.

112. The hypotenuse of a 45°–45°–90° triangle is 14 m. Find the length of each leg.

113. The longer leg of a 30°–60°–90° triangle is 6 ft. Find the length of remaining sides.

114. The altitude of an equilateral triangle is 24 mm. Find the lengths of the sides of the triangle.

WRITING ABOUT MATH

115. Explain how to simplify a fraction with the monomial denominator $\sqrt[3]{3}$.

116. Explain how to simplify a fraction with the monomial denominator $\sqrt[3]{9}$.

SOMETHING TO THINK ABOUT *Assume that x is a rational number.*

117. Change the numerator of $\dfrac{\sqrt{x}-3}{4}$ to a rational number.

118. Rationalize the numerator: $\dfrac{2\sqrt{3x}+4}{\sqrt{3x}-1}$.

SECTION 7.6 Radical Equations

Objectives

1. Solve a radical equation containing one radical.
2. Solve a radical equation containing two radicals.
3. Solve a radical equation containing three radicals.
4. Solve a formula containing a radical for a specified variable.

Vocabulary

power rule

Getting Ready

Find each power.

1. $\left(\sqrt{a}\right)^2$ **2.** $\left(\sqrt{5x}\right)^2$ **3.** $\left(\sqrt{x+4}\right)^2$ **4.** $\left(\sqrt[4]{y-3}\right)^4$

In this section, we will solve equations that contain radicals. To do so, we will use the **power rule.**

1 Solve a radical equation containing one radical.

The Power Rule

If x, y, and n are real numbers and $x = y$, then

$$x^n = y^n$$

If we raise both sides of an equation to the same power, the resulting equation might not be equivalent to the original equation. For example, if we square both sides of the equation

(1) $x = 3$ With a solution set of $\{3\}$

we obtain the equation

(2) $x^2 = 9$ With a solution set of $\{3, -3\}$

Equations 1 and 2 are not equivalent, because they have different solution sets, and the solution -3 of Equation 2 does not satisfy Equation 1. Since raising both sides of an equation to the same power can produce an equation with roots that don't satisfy the original equation, we must check each suspected solution in the original equation.

EXAMPLE 1 Solve: $\sqrt{x + 3} = 4$.

Solution To eliminate the radical, we apply the power rule by squaring both sides of the equation, and proceed as follows:

$$\sqrt{x + 3} = 4$$

$$\left(\sqrt{x + 3}\right)^2 = (4)^2 \quad \text{Square both sides.}$$

$$x + 3 = 16$$

$$x = 13 \quad \text{Subtract 3 from both sides.}$$

To check the apparent solution of 13, we can substitute 13 for x and see whether it satisfies the original equation.

$$\sqrt{x + 3} = 4$$

$$\sqrt{13 + 3} \stackrel{?}{=} 4 \quad \text{Substitute 13 for } x.$$

$$\sqrt{16} \stackrel{?}{=} 4$$

$$4 = 4$$

Since 13 satisfies the original equation, it is a solution.

SELF CHECK 1 Solve: $\sqrt{a - 2} = 3$.

To solve an equation with radicals, we follow these steps.

Solving an Equation Containing Radicals

1. Isolate one radical expression on one side of the equation.
2. Raise both sides of the equation to the power that is the same as the index of the radical.
3. Solve the resulting equation. If it still contains a radical, go back to Step 1.
4. Check the possible solutions to eliminate the ones that do not satisfy the original equation.

EXAMPLE 2 **HEIGHT OF A BRIDGE** The distance d (in feet) that an object will fall in t seconds is given by the formula

$$t = \sqrt{\frac{d}{16}}$$

To find the height of a bridge, a man drops a stone into the water. (See Figure 7-16.) If it takes the stone 3 seconds to hit the water, how far above the river is the bridge?

Figure 7-16

Solution We substitute 3 for t in the formula and solve for d.

$$t = \sqrt{\frac{d}{16}}$$

$$3 = \sqrt{\frac{d}{16}}$$

$$9 = \frac{d}{16} \quad \text{Square both sides.}$$

$$144 = d \quad \text{Multiply both sides by 16.}$$

The bridge is 144 feet above the river.

SELF CHECK 2 How high is the bridge if it takes 4 seconds for the stone to hit the water?

EXAMPLE 3 Solve: $\sqrt{3x + 1} + 1 = x$.

Solution We first subtract 1 from both sides to isolate the radical. Then, to eliminate the radical, we square both sides of the equation and proceed as follows:

$$\sqrt{3x + 1} + 1 = x$$

$$\sqrt{3x + 1} = x - 1 \quad \text{Subtract 1 from both sides.}$$

$$\left(\sqrt{3x + 1}\right)^2 = (x - 1)^2 \quad \text{Square both sides to eliminate the square root.}$$

$$3x + 1 = x^2 - 2x + 1 \quad (x - 1)^2 = (x - 1)(x - 1) = x^2 - x - x + 1 = x^2 - 2x + 1$$

$$0 = x^2 - 5x \quad \text{Subtract } 3x \text{ and 1 from both sides.}$$

$$0 = x(x - 5) \quad \text{Factor } x^2 - 5x.$$

$$x = 0 \quad \text{or} \quad x - 5 = 0 \quad \text{Set each factor equal to 0.}$$

$$x = 0 \quad | \quad x = 5$$

We must check each apparent solution to see whether it satisfies the original equation.

Check:

$$\sqrt{3x + 1} + 1 = x \qquad \sqrt{3x + 1} + 1 = x$$

$$\sqrt{3(0) + 1} + 1 \stackrel{?}{=} 0 \qquad \sqrt{3(5) + 1} + 1 \stackrel{?}{=} 5$$

$$\sqrt{1} + 1 \stackrel{?}{=} 0 \qquad \sqrt{16} + 1 \stackrel{?}{=} 5$$

$$2 \neq 0 \qquad 5 = 5$$

Since 0 does not check, it is extraneous and must be discarded. The only solution of the original equation is 5.

SELF CHECK 3 Solve: $\sqrt{4x + 1} + 1 = x$.

ACCENT ON TECHNOLOGY

Solving Equations Containing Radicals

To find approximate solutions for $\sqrt{3x + 1} + 1 = x$ with a graphing calculator, we graph the functions $f(x) = \sqrt{3x + 1} + 1$ and $g(x) = x$, and then adjust the window settings to $[-5, 10]$ for x and $[-2, 8]$ for y as in Figure 7-17(a). We then trace to find the approximate x-coordinate of their intersection point, as in Figure 7-17(b). After repeated zooms, we will see that $x = 5$.

We can also find the exact x-coordinate of the intersection point by using the INTERSECT command found in the CALC menu.

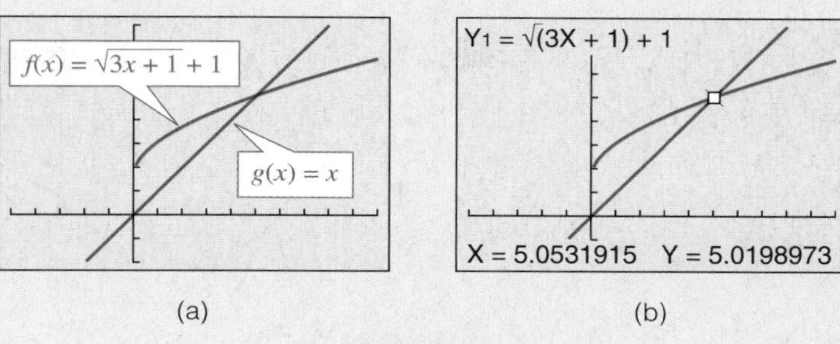

(a) (b)

Figure 7-17

EXAMPLE 4 Solve: $\sqrt[3]{x^3 + 7} = x + 1$.

Solution To eliminate the radical, we cube both sides of the equation and proceed as follows:

$$\sqrt[3]{x^3 + 7} = x + 1$$
$$\left(\sqrt[3]{x^3 + 7}\right)^3 = (x + 1)^3 \quad \text{Cube both sides to eliminate the cube root.}$$
$$x^3 + 7 = x^3 + 3x^2 + 3x + 1$$
$$0 = 3x^2 + 3x - 6 \quad \text{Subtract } x^3 \text{ and 7 from both sides.}$$
$$0 = x^2 + x - 2 \quad \text{Divide both sides by 3.}$$
$$0 = (x + 2)(x - 1) \quad \text{Factor the trinomial.}$$
$$x + 2 = 0 \quad \text{or} \quad x - 1 = 0 \quad \text{Set each factor equal to 0.}$$
$$x = -2 \quad | \quad x = 1 \quad \text{Solve each linear equation.}$$

We check each apparent solution to see whether it satisfies the original equation.

Check:

$$\sqrt[3]{x^3 + 7} = x + 1 \qquad \sqrt[3]{x^3 + 7} = x + 1$$
$$\sqrt[3]{(-2)^3 + 7} \stackrel{?}{=} -2 + 1 \qquad \sqrt[3]{1 + 7} \stackrel{?}{=} 1 + 1$$
$$\sqrt[3]{-8 + 7} \stackrel{?}{=} -1 \qquad \sqrt[3]{8} \stackrel{?}{=} 2$$
$$\sqrt[3]{-1} \stackrel{?}{=} -1 \qquad 2 = 2$$
$$-1 = -1$$

Both solutions satisfy the original equation.

SELF CHECK 4 Solve: $\sqrt[3]{x^3 + 8} = x + 2$.

2 Solve a radical equation containing two radicals.

When more than one radical appears in an equation, it is often necessary to apply the power rule more than once.

EXAMPLE 5 Solve: $\sqrt{x} + \sqrt{x + 2} = 2$.

Solution To remove the radicals, we square both sides of the equation. Since this is easier to do if one radical is on each side of the equation, we subtract $\sqrt{x}$ from both sides to isolate one radical on one side of the equation.

$$\sqrt{x} + \sqrt{x+2} = 2$$

$$\sqrt{x+2} = 2 - \sqrt{x}$$ Subtract $\sqrt{x}$ from both sides.

$$\left(\sqrt{x+2}\right)^2 = \left(2 - \sqrt{x}\right)^2$$ Square both sides to eliminate the square root.

$$x + 2 = 4 - 4\sqrt{x} + x$$ $\left(2 - \sqrt{x}\right)\left(2 - \sqrt{x}\right) = 4 - 2\sqrt{x} - 2\sqrt{x} + x = 4 - 4\sqrt{x} + x$

$$2 = 4 - 4\sqrt{x}$$ Subtract x from both sides.

$$-2 = -4\sqrt{x}$$ Subtract 4 from both sides.

$$\frac{1}{2} = \sqrt{x}$$ Divide both sides by -4.

$$\frac{1}{4} = x$$ Square both sides.

Check: $\sqrt{x} + \sqrt{x+2} = 2$

$$\sqrt{\frac{1}{4}} + \sqrt{\frac{1}{4} + 2} \stackrel{?}{=} 2$$

$$\frac{1}{2} + \sqrt{\frac{9}{4}} \stackrel{?}{=} 2$$

$$\frac{1}{2} + \frac{3}{2} \stackrel{?}{=} 2$$

$$2 = 2$$

The solution checks.

SELF CHECK 5 Solve: $\sqrt{a} + \sqrt{a+3} = 3$.

ACCENT ON TECHNOLOGY

Solving Equations Containing Radicals

To find approximate solutions for $\sqrt{x} + \sqrt{x+2} = 5$ with a graphing calculator, we graph the functions $f(x) = \sqrt{x} + \sqrt{x+2}$ and $g(x) = 5$, and adjust the window settings to [2, 10] for x and [−2, 8] for y as in Figure 7-18(a). We then trace to find an approximation of the x-coordinate of their intersection point, as in Figure 7-18(b). From the figure, we can see that $x \approx 5.15$. We can zoom to get better results.

We also can find the exact x-coordinate of the intersection point by using the INTERSECT command found in the CALC menu.

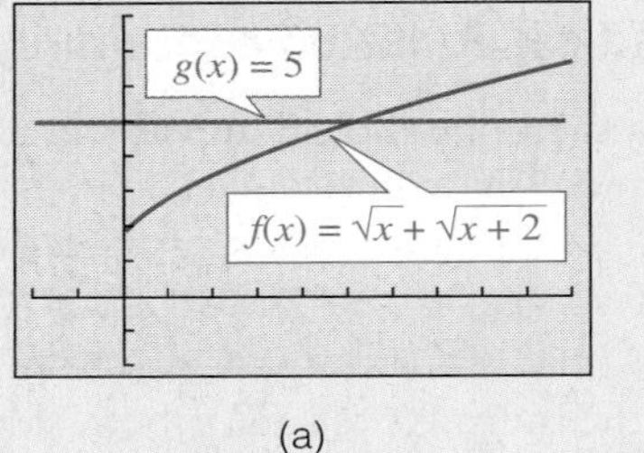

(a)

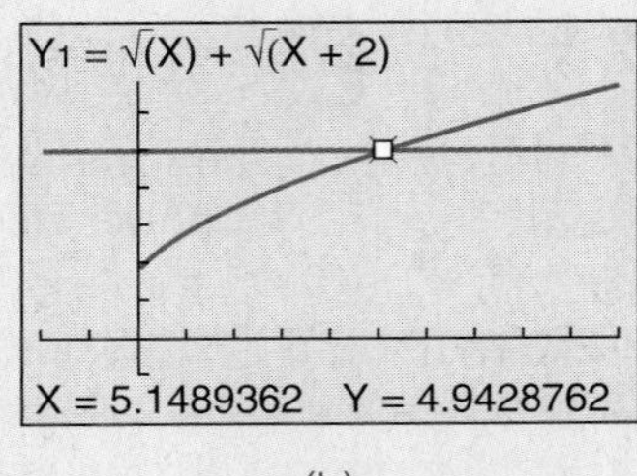

(b)

Figure 7-18

3 Solve a radical equation containing three radicals.

EXAMPLE 6 Solve: $\sqrt{x+2} + \sqrt{2x} = \sqrt{18-x}$.

Solution In this case, it is impossible to isolate one radical on each side of the equation, so we begin by squaring both sides. Then we proceed as follows.

$$\sqrt{x+2} + \sqrt{2x} = \sqrt{18-x}$$

$$\left(\sqrt{x+2} + \sqrt{2x}\right)^2 = \left(\sqrt{18-x}\right)^2 \quad \text{Square both sides to eliminate one square root.}$$

$$x + 2 + 2\sqrt{x+2}\sqrt{2x} + 2x = 18 - x$$

$$2\sqrt{x+2}\sqrt{2x} = 16 - 4x \quad \text{Subtract } 3x \text{ and 2 from both sides.}$$

$$\sqrt{x+2}\sqrt{2x} = 8 - 2x \quad \text{Divide both sides by 2.}$$

$$\left(\sqrt{x+2}\sqrt{2x}\right)^2 = (8-2x)^2 \quad \text{Square both sides to eliminate the other square roots.}$$

$$(x+2)2x = 64 - 32x + 4x^2$$

$$2x^2 + 4x = 64 - 32x + 4x^2$$

$$0 = 2x^2 - 36x + 64 \quad \text{Write the equation in quadratic form.}$$

$$0 = x^2 - 18x + 32 \quad \text{Divide both sides by 2.}$$

$$0 = (x-16)(x-2) \quad \text{Factor the trinomial.}$$

$$x - 16 = 0 \quad \text{or} \quad x - 2 = 0 \quad \text{Set each factor equal to 0.}$$

$$x = 16 \quad | \quad x = 2$$

Verify that 2 satisfies the equation, but 16 does not. Thus, the only solution is 2.

SELF CHECK 6 Solve: $\sqrt{3x+4} + \sqrt{x+9} = \sqrt{x+25}$.

4 Solve a formula containing a radical for a specified variable.

To *solve a formula for a variable* means to isolate that variable on one side of the equation, with all other quantities on the other side.

EXAMPLE 7 **DEPRECIATION RATES** Some office equipment that is now worth V dollars originally cost C dollars 3 years ago. The rate r at which it has depreciated is given by

$$r = 1 - \sqrt[3]{\frac{V}{C}}$$

Solve the formula for C.

Solution We begin by isolating the cube root on the right side of the equation.

$$r = 1 - \sqrt[3]{\frac{V}{C}}$$

$$r - 1 = -\sqrt[3]{\frac{V}{C}} \quad \text{Subtract 1 from both sides.}$$

$$(r - 1)^3 = \left(-\sqrt[3]{\frac{V}{C}}\right)^3$$ To eliminate the radical, cube both sides.

$$(r - 1)^3 = -\frac{V}{C}$$ Simplify the right side.

$$C(r - 1)^3 = -V$$ Multiply both sides by C.

$$C = -\frac{V}{(r - 1)^3}$$ Divide both sides by $(r - 1)^3$.

SELF CHECK 7 A formula used in statistics to determine the size of a sample to obtain a desired degree of accuracy is

$$E = z_0\sqrt{\frac{pq}{n}}$$

Solve the formula for n.

SELF CHECK ANSWERS **1.** 11 **2.** 256 ft **3.** 6; 0 is extraneous **4.** 0, −2 **5.** 1 **6.** 0 **7.** $n = \frac{z_0^2 pq}{E^2}$

NOW TRY THIS

Solve.

1. $x^{1/3} = 2$
2. $x^{2/3} = 4$
3. $(x + 1)^{-1/2} = 3$

7.6 EXERCISES

WARM-UPS *Solve each equation.*

1. $\sqrt{x + 2} = 3$
2. $\sqrt{x - 2} = 1$
3. $\sqrt[3]{x + 1} = 1$
4. $\sqrt[3]{x - 1} = 2$
5. $\sqrt[4]{x - 1} = 2$
6. $\sqrt[5]{x + 1} = 2$

REVIEW *If $f(x) = 3x^2 - 4x + 2$, find each quantity.*

7. $f(0)$
8. $f(-3)$
9. $f(2)$
10. $f\left(\frac{1}{2}\right)$

VOCABULARY AND CONCEPTS *Fill in the blanks.*

11. If x, y, and n are real numbers and $x = y$, then ________, called the __________.
12. When solving equations containing radicals, try to ______ one radical expression on one side of the equation.
13. To solve the equation $\sqrt{x + 4} = 5$, we first ______ both sides.
14. To solve the equation $\sqrt[3]{x + 4} = 2$, we first _____ both sides.
15. Squaring both sides of an equation can introduce __________ solutions.
16. Always remember to _____ the solutions of an equation containing radicals to eliminate any __________ solutions.

GUIDED PRACTICE

Solve each equation. See Example 1. (Objective 1)

17. $\sqrt{5x-6}=2$

18. $\sqrt{7x-10}=12$

19. $\sqrt{6x+1}+2=7$

20. $\sqrt{6x+13}-2=5$

21. $\sqrt[3]{7n-1}=3$

22. $\sqrt[3]{12m+4}=4$

23. $x=\frac{\sqrt{12x-5}}{2}$

24. $x=\frac{\sqrt{16x-12}}{2}$

Solve each equation. Identify any extraneous solution. See Examples 3–4. (Objective 1)

25. $r-9=\sqrt{2r-3}$

26. $-s-3=2\sqrt{5-s}$

27. $\sqrt{-5x+24}=6-x$

28. $\sqrt{-x+2}=x-2$

29. $\sqrt{y+2}=4-y$

30. $\sqrt{22y+86}=y+9$

31. $\sqrt[3]{x^3-7}=x-1$

32. $\sqrt[3]{x^3+56}-2=x$

Solve each equation. Identify any extraneous solution. See Example 5. (Objective 2)

33. $2\sqrt{4x+1}=\sqrt{x+4}$

34. $\sqrt{3(x+4)}=\sqrt{5x-12}$

35. $\sqrt{x+2}=\sqrt{4-x}$

36. $\sqrt{6-x}=\sqrt{2x+3}$

37. $2\sqrt{x}=\sqrt{5x-16}$

38. $3\sqrt{x}=\sqrt{3x+12}$

39. $\sqrt{2y+1}=1-2\sqrt{y}$

40. $\sqrt{u}+3=\sqrt{u-3}$

41. $1+\sqrt{z}=\sqrt{z+3}$

42. $\sqrt{x}+2=\sqrt{x+4}$

43. $\sqrt{4s+1}-\sqrt{6s}=-1$

44. $\sqrt{y+7}+3=\sqrt{y+4}$

45. $\sqrt{2x+5}+\sqrt{x+2}=5$

46. $\sqrt{2x+5}+\sqrt{2x+1}+4=0$

Solve each equation. Identify any extraneous solution. See Example 6. (Objective 3)

47. $\sqrt{v}+\sqrt{3}=\sqrt{v+3}$

48. $\sqrt{x+1}+\sqrt{3x}=\sqrt{5x+1}$

49. $\sqrt{3x}-\sqrt{x+1}=\sqrt{x-2}$

50. $\sqrt{x+2}+\sqrt{2x-3}=\sqrt{11-x}$

Solve each formula for the indicated variable. See Example 7. (Objective 4)

51. $v=\sqrt{2gh}$ for h

52. $d=1.4\sqrt{h}$ for h

53. $T=2\pi\sqrt{\frac{l}{32}}$ for l

54. $d=\sqrt[3]{\frac{12V}{\pi}}$ for V

55. $r=\sqrt[3]{\frac{A}{P}}-1$ for A

56. $r=\sqrt[3]{\frac{A}{P}}-1$ for P

57. $L_A=L_B\sqrt{1-\frac{v^2}{c^2}}$ for v^2

58. $R_1=\sqrt{\frac{A}{\pi}-R_2^{\,2}}$ for A

ADDITIONAL PRACTICE *Solve each equation.*

59. $5r+4=\sqrt{5r+20}+4r$

60. $\sqrt{x}\sqrt{x+16}=15$

61. $\sqrt{x}\sqrt{x+6}=4$

62. $\frac{6}{\sqrt{x+5}}=\sqrt{x}$

63. $\sqrt[4]{x^4+4x^2-4}=-x$

64. $\sqrt[4]{8x-8}+2=0$

65. $2+\sqrt{u}=\sqrt{2u+7}$

66. $\sqrt[4]{12t+4}+2=0$

67. $u=\sqrt[4]{u^4-6u^2+24}$

68. $\sqrt{6t+1}-3\sqrt{t}=-1$

69. $\sqrt{x-5}-\sqrt{x+3}=4$

70. $\sqrt[4]{10p+1}=\sqrt[4]{11p-7}$

71. $\sqrt{x+8}-\sqrt{x-4}=-2$

72. $\sqrt[4]{10y+2}=2\sqrt[4]{2}$

73. $\sqrt{z-1}+\sqrt{z+2}=3$

74. $\sqrt{16v+1}+\sqrt{8v+1}=12$

75. $\sqrt{\sqrt{a}+\sqrt{a+8}}=2$

76. $\sqrt{\sqrt{2y}-\sqrt{y-1}}=1$

77. $\frac{\sqrt{2x}}{\sqrt{x+2}}=\sqrt{x-1}$

78. $\sqrt{8-x}-\sqrt{3x-8}=\sqrt{x-4}$

APPLICATIONS

Solve each application problem. See Example 2. (Objective 1)

79. Highway design A curve banked at 8° will accommodate traffic traveling s mph if the radius of the curve is r feet, according to the formula $s=1.45\sqrt{r}$. If engineers expect 65-mph traffic, what radius should they specify? See the illustration on the next page.

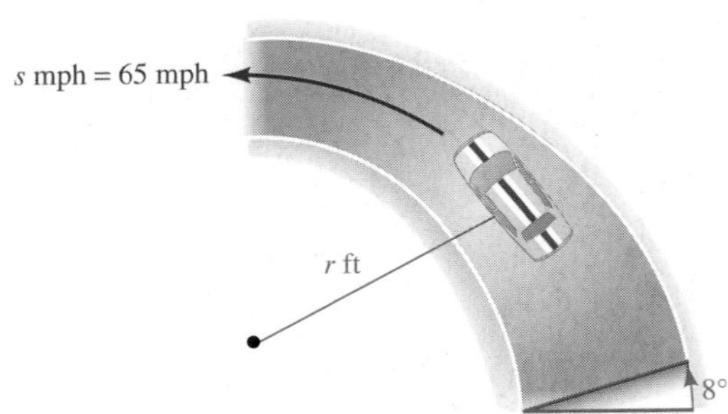

80. Horizon distance The higher a lookout tower is built, the farther an observer can see. That distance d (called the *horizon distance*, measured in miles) is related to the height h of the observer (measured in feet) by the formula $d = 1.4\sqrt{h}$. How tall must a lookout tower be to see the edge of the forest, 25 miles away?

81. Generating power The power generated by a windmill is related to the velocity of the wind by the formula

$$v = \sqrt[3]{\frac{P}{0.02}}$$

where P is the power (in watts) and v is the velocity of the wind (in mph). Find the speed of the wind when the windmill is generating 500 watts of power.

82. Carpentry During construction, carpenters often brace walls as shown in the illustration, where the length of the brace is given by the formula

$$l = \sqrt{f^2 + h^2}$$

If a carpenter nails a 10-ft brace to the wall 6 feet above the floor, how far from the base of the wall should he nail the brace to the floor?

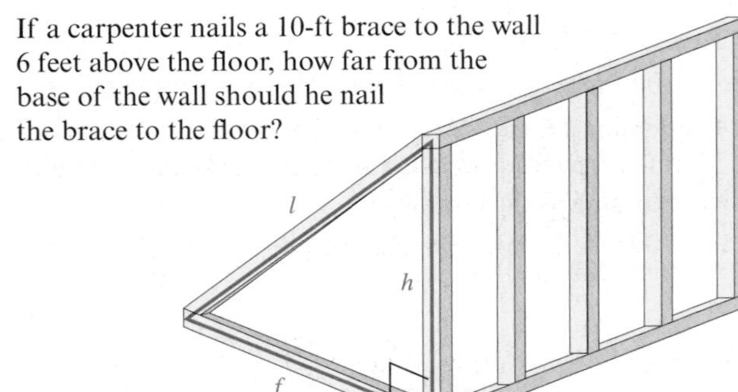

Use a graphing calculator.

83. Depreciation The formula

$$r = 1 - \sqrt[n]{\frac{T}{C}}$$

gives the annual depreciation rate r of a car that had an original cost of C dollars, a useful life of n years, and a trade-in value of T dollars. Find the annual depreciation rate of a car that cost \$22,000 and was sold 5 years later for \$9,000. Give the result to the nearest percent.

84. Savings accounts The interest rate r earned by a savings account after n compoundings is given by the formula

$$\sqrt[n]{\frac{V}{P}} - 1 = r$$

where V is the current value and P is the original principal. What interest rate r was paid on an account in which a deposit of \$1,000 grew to \$1,338.23 after 5 compoundings?

85. Marketing The number of wrenches that will be produced at a given price can be predicted by the formula $s = \sqrt{5x}$, where s is the supply (in thousands) and x is the price (in dollars). If the demand, d, for wrenches can be predicted by the formula $d = \sqrt{100 - 3x^2}$, find the equilibrium price.

86. Marketing The number of footballs that will be produced at a given price can be predicted by the formula $s = \sqrt{23x}$, where s is the supply (in thousands) and x is the price (in dollars). If the demand, d, for footballs can be predicted by the formula $d = \sqrt{312 - 2x^2}$, find the equilibrium price.

87. Medicine The resistance R to blood flow through an artery can be found using the formula

$$r = \sqrt[4]{\frac{8kl}{\pi R}}$$

where r is the radius of the artery, k is the viscosity of blood, and l is the length of the artery. Solve the formula for R.

88. Generating power The power P generated by a windmill is given by the formula

$$s = \sqrt[3]{\frac{P}{0.02}}$$

where s is the speed of the wind. Solve the formula for P.

WRITING ABOUT MATH

89. If both sides of an equation are raised to the same power, the resulting equation might not be equivalent to the original equation. Explain.

90. Explain why you must check each apparent solution of a radical equation.

SOMETHING TO THINK ABOUT

91. Solve: $\sqrt[3]{2x} = \sqrt{x}$.

92. Solve: $\sqrt[4]{x} = \sqrt{\frac{x}{4}}$.

SECTION 7.7 Complex Numbers

Objectives

1. Simplify an imaginary number.
2. Simplify an expression containing complex numbers.
3. Rationalize the denominator of a fraction that contains a complex number.
4. Find a specified power of *i*.
5. Find the absolute value of a complex number.

Vocabulary

imaginary number complex number

Getting Ready

Perform the following operations.

1. $(3x + 5) + (4x - 5)$
2. $(3x + 5) - (4x - 5)$
3. $(3x + 5)(4x - 5)$
4. $(3x + 5)(3x - 5)$

We have seen that square roots of negative numbers are not real numbers. However, there is a broader set of numbers, called the *complex numbers*, in which negative numbers do have square roots. In this section, we will discuss this broader set of numbers.

1 Simplify an imaginary number.

Consider the number $\sqrt{-3}$. Since no real number squared is -3, $\sqrt{-3}$ is not a real number. For years, people believed that numbers such as

$$\sqrt{-1}, \quad \sqrt{-3}, \quad \sqrt{-4}, \quad \text{and} \quad \sqrt{-9}$$

were nonsense. In the 17th century, René Descartes (1596–1650) called them **imaginary numbers.** Today, imaginary numbers have many important uses, such as describing the behavior of alternating current in electronics.

The imaginary number $\sqrt{-1}$ often is denoted by the letter i:

$$i = \sqrt{-1}$$

Because i represents the square root of -1, it follows that

$$i^2 = -1$$

PERSPECTIVE

The Pythagoreans (ca. 500 B.C.) understood the universe as a harmony of whole numbers. They did not classify fractions as numbers, and were upset that $\sqrt{2}$ was not the ratio of whole numbers. For 2,000 years, little progress was made in the understanding of the various kinds of numbers.

The father of algebra, François Vieta (1540–1603), understood the whole numbers, fractions, and certain irrational numbers. But he was unable to accept negative numbers, and certainly not imaginary numbers.

René Descartes (1596–1650) thought these numbers to be nothing more than figments of his imagination, so he called them *imaginary numbers.* Leonhard Euler (1707–1783) used the letter i for $\sqrt{-1}$; Augustin Cauchy (1789–1857) used the term *conjugate;* and Carl Gauss (1777–1855) first used the word *complex.*

Today, we accept complex numbers without question, but it took many centuries and the work of many mathematicians to make them respectable.

If we assume that multiplication of imaginary numbers is commutative and associative, then

$$\begin{aligned}(2i)^2 &= 2^2i^2 \\ &= 4(-1) \qquad i^2 = -1 \\ &= -4\end{aligned}$$

Since $(2i)^2 = -4$, $2i$ is a square root of -4, and we can write

$$\sqrt{-4} = 2i$$

This result also can be obtained by using the multiplication property of radicals:

$$\sqrt{-4} = \sqrt{4(-1)} = \sqrt{4}\sqrt{-1} = 2i$$

We can use the multiplication property of radicals to simplify any imaginary number. For example,

$$\sqrt{-25} = \sqrt{25(-1)} = \sqrt{25}\sqrt{-1} = 5i$$

$$\sqrt{\frac{-100}{49}} = \sqrt{\frac{100}{49}(-1)} = \frac{\sqrt{100}}{\sqrt{49}}\sqrt{-1} = \frac{10}{7}i$$

These examples illustrate the following rule.

Properties of Radicals

If at least one of a and b is a nonnegative real number, then

$$\sqrt{ab} = \sqrt{a}\sqrt{b} \qquad \text{and} \qquad \sqrt{\frac{a}{b}} = \frac{\sqrt{a}}{\sqrt{b}} \quad (b \neq 0)$$

COMMENT If a and b are negative, then $\sqrt{ab} \neq \sqrt{a}\sqrt{b}$. For example, if $a = -16$ and $b = -4$, we have

$$\sqrt{(-16)}\sqrt{(-4)} = (4i)(2i) = 8i^2 = 8(-1) = -8$$

Note that $\sqrt{(-16)}\sqrt{(-4)}$ does not simplify as $\sqrt{(-16)(-4)} = \sqrt{64} = 8$.

2 Simplify an expression containing complex numbers.

The imaginary numbers are a subset of a set of numbers called the *complex numbers.*

Complex Numbers

A **complex number** is any number that can be written in the standard form $a + bi$, where a and b are real numbers and $i = \sqrt{-1}$.

In the complex number $a + bi$, a is called the *real part,* and b is called the *imaginary part.*

If $b = 0$, the complex number $a + bi$ is a real number. If $b \neq 0$ and $a = 0$, the complex number $0 + bi$ (or just bi) is an imaginary number.

Any imaginary number can be expressed in bi form. For example,

COMMENT The expression $\sqrt{3}i$ is often written as $i\sqrt{3}$ to make it clear that i is not part of the radicand. Don't confuse $\sqrt{3}i$ with $\sqrt{3i}$.

$$\sqrt{-1} = i$$

$$\sqrt{-9} = \sqrt{9(-1)} = \sqrt{9}\sqrt{-1} = 3i$$

$$\sqrt{-3} = \sqrt{3(-1)} = \sqrt{3}\sqrt{-1} = \sqrt{3}i$$

The relationship between the real numbers, the imaginary numbers, and the complex numbers is shown in Figure 7-19.

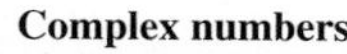

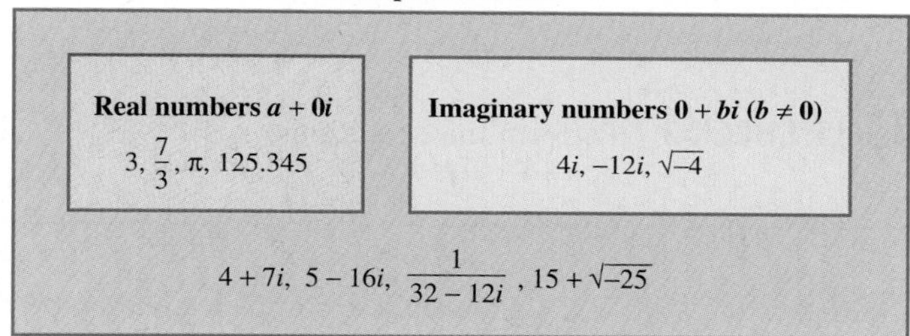

Figure 7-19

Equality of Complex Numbers

The complex numbers $a + bi$ and $c + di$ are equal if and only if

$$a = c \quad \text{and} \quad b = d$$

Because of the previous definition, complex numbers are equal when their real parts are equal and their imaginary parts are equal.

EXAMPLE 1

a. $2 + 3i = \sqrt{4} + \frac{6}{2}i$ because $2 = \sqrt{4}$ and $3 = \frac{6}{2}$.

b. $4 - 5i = \frac{12}{3} - \sqrt{25}i$ because $4 = \frac{12}{3}$ and $-5 = -\sqrt{25}$.

c. $x + yi = 4 + 7i$ if and only if $x = 4$ and $y = 7$.

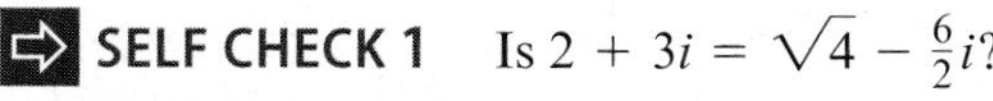

SELF CHECK 1 Is $2 + 3i = \sqrt{4} - \frac{6}{2}i$?

Addition and Subtraction of Complex Numbers

Complex numbers are added and subtracted as if they were binomials:

$$(a + bi) + (c + di) = (a + c) + (b + d)i$$
$$(a + bi) - (c + di) = (a + bi) + (-c - di) = (a - c) + (b - d)i$$

The previous definition suggests that when adding or subtracting two complex numbers, we add or subtract the real parts and then add or subtract the imaginary parts.

EXAMPLE 2 Perform the operations.

a. $(8 + 4i) + (12 + 8i) = 8 + 4i + 12 + 8i$
$= 20 + 12i$

b. $(7 - 4i) + (9 + 2i) = 7 - 4i + 9 + 2i$
$= 16 - 2i$

c. $(-6 + i) - (3 - 4i) = -6 + i - 3 + 4i$
$= -9 + 5i$

d. $(2 - 4i) - (-4 + 3i) = 2 - 4i + 4 - 3i$
$= 6 - 7i$

SELF CHECK 2 Perform the operations.

a. $(3 - 5i) + (-2 + 7i)$ **b.** $(3 - 5i) - (-2 + 7i)$

To multiply a complex number by an imaginary number, we use the distributive property to remove parentheses and simplify. For example,

$-5i(4 - 8i) = -5i(4) - (-5i)8i$	Use the distributive property.
$= -20i + 40i^2$	Simplify.
$= -20i + 40(-1)$	Remember that $i^2 = -1$.
$= -40 - 20i$	

To multiply two complex numbers, we use the following definition.

Multiplying Complex Numbers

Complex numbers are multiplied as if they were binomials, with $i^2 = -1$:

$$(a + bi)(c + di) = ac + adi + bci + bdi^2$$
$$= ac + adi + bci + bd(-1)$$
$$= (ac - bd) + (ad + bc)i$$

EXAMPLE 3 Multiply the complex numbers.

a. $(2 + 3i)(3 - 2i) = 6 - 4i + 9i - 6i^2$	Use the FOIL method.
$= 6 + 5i + 6$	$i^2 = -1$, combine $-4i$ and $9i$.
$= 12 + 5i$	

b. $(3 + i)(1 + 2i) = 3 + 6i + i + 2i^2$ Use the FOIL method.

$= 3 + 7i - 2$ $i^2 = -1$, combine $6i$ and i.

$= 1 + 7i$

c. $(-4 + 2i)(2 + i) = -8 - 4i + 4i + 2i^2$ Use the FOIL method.

$= -8 - 2$ $i^2 = -1$, combine $-4i$ and $4i$.

$= -10$

SELF CHECK 3 Multiply: $(-2 + 3i)(3 - 2i)$.

The next two examples show how to write complex numbers in $a + bi$ form. It is common to use $a - bi$ as a substitute for $a + (-b)i$.

EXAMPLE 4 Write each number in $a + bi$ form.

a. $7 = 7 + 0i$

b. $3i = 0 + 3i$

c. $4 - \sqrt{-16} = 4 - \sqrt{-1(16)}$

$= 4 - \sqrt{16}\sqrt{-1}$

$= 4 - 4i$

d. $5 + \sqrt{-11} = 5 + \sqrt{-1(11)}$

$= 5 + \sqrt{11}\sqrt{-1}$

$= 5 + \sqrt{11}i$

SELF CHECK 4 Write $3 - \sqrt{-25}$ in $a + bi$ form.

Complex Conjugates

The complex numbers $a + bi$ and $a - bi$ are called **complex conjugates.**

For example,

$3 + 4i$ and $3 - 4i$ are complex conjugates.

$5 - 7i$ and $5 + 7i$ are complex conjugates.

EXAMPLE 5 Find the product of $3 + i$ and its complex conjugate.

Solution The complex conjugate of $3 + i$ is $3 - i$. We can find the product as follows:

$(3 + i)(3 - i) = 9 - 3i + 3i - i^2$ Use the FOIL method.

$= 9 - i^2$ Combine like terms.

$= 9 - (-1)$ $i^2 = -1$

$= 10$

SELF CHECK 5 Multiply: $(2 + 3i)(2 - 3i)$.

The product of the complex number $a + bi$ and its complex conjugate $a - bi$ is the real number $a^2 + b^2$, as the following work shows:

$$\begin{aligned}(a + bi)(a - bi) &= a^2 - abi + abi - b^2i^2 && \text{Use the FOIL method.}\\ &= a^2 - b^2(-1) && i^2 = -1\\ &= a^2 + b^2\end{aligned}$$

3 Rationalize the denominator of a fraction that contains a complex number.

If $b \neq 0$, the complex number $a + bi$ contains the square root $i = \sqrt{-1}$. Since a square root cannot remain in the denominator of a fraction, we often have to rationalize a denominator when dividing complex numbers.

EXAMPLE 6 Divide and write the result in $a + bi$ form: $\dfrac{1}{3 + i}$.

Solution We can rationalize the denominator by multiplying the numerator and the denominator by the complex conjugate of the denominator.

$$\begin{aligned}\frac{1}{3 + i} &= \frac{1}{3 + i} \cdot \frac{3 - i}{3 - i} && \frac{3 - i}{3 - i} = 1\\ &= \frac{3 - i}{9 - 3i + 3i - i^2} && \text{Multiply the numerators and multiply the denominators.}\\ &= \frac{3 - i}{9 - (-1)} && i^2 = -1\\ &= \frac{3 - i}{10}\\ &= \frac{3}{10} - \frac{1}{10}i && \text{Write the result in } a + bi \text{ form.}\end{aligned}$$

SELF CHECK 6 Rationalize the denominator: $\dfrac{1}{5 - i}$.

EXAMPLE 7 Write $\dfrac{3 - i}{2 + i}$ in $a + bi$ form.

Solution We multiply the numerator and the denominator of the fraction by the complex conjugate of the denominator.

$$\begin{aligned}\frac{3 - i}{2 + i} &= \frac{3 - i}{2 + i} \cdot \frac{2 - i}{2 - i} && \frac{2 - i}{2 - i} = 1\\ &= \frac{6 - 3i - 2i + i^2}{4 - 2i + 2i - i^2} && \text{Multiply the numerators and multiply the denominators.}\\ &= \frac{5 - 5i}{4 - (-1)} && 6 + i^2 = 6 - 1 = 5\end{aligned}$$

$$= \frac{5(1 - i)}{5}$$ Factor out 5 in the numerator.

$$= 1 - i$$ Simplify.

SELF CHECK 7 Rationalize the denominator: $\frac{2 + i}{5 - i}$.

EXAMPLE 8 Write $\frac{4 + \sqrt{-16}}{2 + \sqrt{-4}}$ in $a + bi$ form.

Solution

$$\frac{4 + \sqrt{-16}}{2 + \sqrt{-4}} = \frac{4 + 4i}{2 + 2i}$$ Write each number in $a + bi$ form.

$$= \frac{2(2 + 2i)}{2 + 2i}$$ Factor out 2 in the numerator and simplify.

$$= 2 + 0i$$

SELF CHECK 8 Divide: $\frac{3 + \sqrt{-25}}{2 + \sqrt{-9}}$.

COMMENT To avoid mistakes, always put complex numbers in $a + bi$ form before doing any operations with complex numbers.

4 Find a specified power of *i*.

The powers of i produce an interesting pattern:

$$i = \sqrt{-1} = i \qquad i^5 = i^4 i = 1i = i$$
$$i^2 = \left(\sqrt{-1}\right)^2 = \mathbf{-1} \qquad i^6 = i^4 i^2 = 1(-1) = \mathbf{-1}$$
$$i^3 = i^2 i = -1i = \mathbf{-i} \qquad i^7 = i^4 i^3 = 1(-i) = \mathbf{-i}$$
$$i^4 = i^2 i^2 = (-1)(-1) = \mathbf{1} \qquad i^8 = i^4 i^4 = (1)(1) = \mathbf{1}$$

The pattern continues: $i, -1, -i, 1, \ldots$.

EXAMPLE 9 Simplify: i^{29}.

Solution We note that 29 divided by 4 gives a quotient of 7 and a remainder of 1. Thus, $29 = 4 \cdot 7 + 1$, and

$$i^{29} = i^{4 \cdot 7 + 1}$$ $29 = 4 \cdot 7 + 1$

$$= (i^4)^7 \cdot i$$ $i^{4 \cdot 7 + 1} = i^{4 \cdot 7} \cdot i^1 = (i^4)^7 \cdot i$

$$= \mathbf{1}^7 \cdot i$$ $i^4 = 1$

$$= i$$

SELF CHECK 9 Simplify: i^{31}.

The results of Example 9 illustrate the following fact.

Powers of i

If n is a natural number that has a remainder of r when divided by 4, then

$$i^n = i^r$$

When n is divisible by 4, the remainder r is 0 and $i^0 = 1$.

EXAMPLE 10 Simplify: i^{55}.

Solution We divide 55 by 4 and get a remainder of 3. Therefore,

$$i^{55} = i^3 = -i$$

SELF CHECK 10 Simplify: i^{62}.

EXAMPLE 11 Simplify each expression. If a denominator has a factor of i, multiply the expression by $\frac{i}{i}$.

a. $2i^2 + 4i^3 = 2(-1) + 4(-i)$
$= -2 - 4i$

b. $\frac{3}{2i} = \frac{3}{2i} \cdot \frac{i}{i}$ $\quad \frac{i}{i} = 1$
$= \frac{3i}{2i^2}$
$= \frac{3i}{2(-1)}$
$= \frac{3i}{-2}$
$= 0 - \frac{3}{2}i$

c. $-\frac{5}{i} = -\frac{5}{i} \cdot \frac{i}{i}$ $\quad \frac{i}{i} = 1$
$= -\frac{5(i)}{i^2}$
$= -\frac{5i}{-1}$
$= 5i$
$= 0 + 5i$

d. $\frac{6}{i^3} = \frac{6i}{i^3 i}$ $\quad \frac{i}{i} = 1$
$= \frac{6i}{i^4}$
$= \frac{6i}{1}$
$= 6i$
$= 0 + 6i$

SELF CHECK 11 Simplify. **a.** $3i^3 - 2i^2$ **b.** $\frac{2}{3i}$

5 Find the absolute value of a complex number.

Absolute Value of a Complex Number

The **absolute value** of the complex number $a + bi$ is $\sqrt{a^2 + b^2}$. In symbols,

$$|a + bi| = \sqrt{a^2 + b^2}$$

EXAMPLE 12 Find each absolute value.

a. $|3 + 4i| = \sqrt{3^2 + 4^2}$
$= \sqrt{9 + 16}$
$= \sqrt{25}$
$= 5$

b. $|3 - 4i| = \sqrt{3^2 + (-4)^2}$
$= \sqrt{9 + 16}$
$= \sqrt{25}$
$= 5$

c. $|-5 - 12i| = \sqrt{(-5)^2 + (-12)^2}$
$= \sqrt{25 + 144}$
$= \sqrt{169}$
$= 13$

d. $|a + 0i| = \sqrt{a^2 + 0^2}$
$= \sqrt{a^2}$
$= |a|$

SELF CHECK 12 Evaluate: $|5 + 12i|$.

SELF CHECK ANSWERS **1.** no **2. a.** $1 + 2i$ **b.** $5 - 12i$ **3.** $13i$ **4.** $3 - 5i$ **5.** 13 **6.** $\frac{5}{26} + \frac{1}{26}i$ **7.** $\frac{9}{26} + \frac{7}{26}i$ **8.** $\frac{21}{13} + \frac{1}{13}i$ **9.** $-i$ **10.** -1 **11. a.** $2 - 3i$ **b.** $0 - \frac{2}{3}i$ **12.** 13

NOW TRY THIS

1. Simplify: $-\sqrt{-8}\sqrt{-2}$.
2. Evaluate $3x^2 - 2x - 4$ for $x = 2 - 3i$.

7.7 EXERCISES

WARM-UPS

Write each imaginary number in bi form.

1. $\sqrt{-49}$
2. $\sqrt{-64}$
3. $\sqrt{-100}$
4. $\sqrt{-81}$

Simplify each power of i.

5. i^3
6. i^2
7. i^4
8. i^5

Find each absolute value.

9. $|-3 + 4i|$
10. $|5 - 12i|$

REVIEW *Perform each operation.*

11. $\frac{x^2 - x - 6}{9 - x^2} \cdot \frac{x^2 + x - 6}{x^2 - 4}$
12. $\frac{3x + 4}{x - 2} + \frac{x - 4}{x + 2}$
13. **Wind speed** A plane that can fly 200 mph in still air makes a 330-mile flight with a tail wind and returns, flying into the same wind. Find the speed of the wind if the total flying time is $3\frac{1}{3}$ hours.
14. **Finding rates** A student drove a distance of 135 miles at an average speed of 50 mph. How much faster would he have to drive on the return trip to save 30 minutes of driving time?

VOCABULARY AND CONCEPTS *Fill in the blanks.*

15. $\sqrt{-1} =$ _
16. $i^6 =$ ____
17. $i^7 =$ ____
18. $i^8 =$ __
19. $\sqrt{-1}$, $\sqrt{-3}$, $\sqrt{-4}$ are examples of __________ numbers.
20. $\sqrt{ab} =$ ________, provided a and b are not both negative.
21. $\sqrt{\frac{a}{b}} = \frac{\_\_\_\_}{\_\_\_\_}$ $(b \neq 0)$, provided a and b are not both negative.
22. $3 + 5i$, $2 - 7i$, and $5 - \frac{1}{2}i$ are examples of _________ numbers.
23. The real part of $5 + 7i$ is __. The imaginary part is __.
24. $a + bi = c + di$ if and only if $a =$ __ and $b =$ __.
25. $a + bi$ and $a - bi$ are called complex __________.
26. $|a + bi| =$ _________

GUIDED PRACTICE

Write each imaginary number in simplified form. (Objective 1)

27. $\sqrt{-9}$
28. $\sqrt{-16}$
29. $\sqrt{-36}$
30. $\sqrt{-81}$
31. $\sqrt{-7}$
32. $\sqrt{-11}$
33. $\sqrt{-8}$
34. $\sqrt{-24}$

Determine whether the complex numbers are equal. See Example 1. (Objective 1)

35. $3 + 7i$, $\sqrt{9} + (5 + 2)i$
36. $\sqrt{4} + \sqrt{25}i$, $2 - (-5)i$
37. $\sqrt{4} + \sqrt{-4}$, $2 - 2i$
38. $\sqrt{-9} - i$, $4i$

Simplify each expression. Write all answers in standard form. See Example 2. (Objective 2)

39. $(3 + 4i) + (5 - 6i)$
40. $(5 + 3i) - (6 - 9i)$
41. $(7 - 3i) - (4 + 2i)$
42. $(8 + 3i) + (-7 - 2i)$
43. $(8 + 5i) + (7 + 2i)$
44. $(-7 + 9i) - (-2 - 8i)$
45. $(1 + i) - 2i + (5 - 7i)$
46. $(-9 + i) - 5i + (2 + 7i)$

Simplify each expression. Write all answers in standard form. (Objective 2)

47. $3i(2 - i)$
48. $-4i(3 + 4i)$
49. $-5i(5 - 5i)$
50. $2i(7 + 2i)$

Simplify each expression. Write all answers in standard form. See Example 3. (Objective 2)

51. $(2 + i)(3 - i)$
52. $(4 - i)(2 + i)$
53. $(2 - 4i)(3 + 2i)$
54. $(3 - 2i)(4 - 3i)$
55. $(2 + \sqrt{2}i)(3 - \sqrt{2}i)$
56. $(5 + \sqrt{3}i)(2 - \sqrt{3}i)$
57. $(2 + i)^2$
58. $(3 - 2i)^2$
59. $(2 + 3i)^2$
60. $(1 - 3i)^2$
61. $i(5 + i)(3 - 2i)$
62. $i(-3 - 2i)(1 - 2i)$

Simplify each expression. Write in standard form. See Examples 4–5. (Objective 2)

63. $(8 - \sqrt{-1})(-2 - \sqrt{-16})$
64. $(-1 + \sqrt{-4})(2 + \sqrt{-9})$
65. $(6 - 5i)(6 + 5i)$
66. $(7 + 2i)(7 - 2i)$

Divide and write each expression in standard form. See Example 6. (Objective 3)

67. $\frac{5}{2 - i}$
68. $\frac{26}{3 - 2i}$
69. $\frac{13i}{5 + i}$
70. $\frac{2i}{5 + 3i}$
71. $\frac{-12}{7 - \sqrt{-1}}$
72. $\frac{4}{3 + \sqrt{-1}}$
73. $\frac{5i}{6 + 2i}$
74. $\frac{-4i}{2 - 6i}$

Divide and write each expression in standard form. See Examples 7–8. (Objective 3)

75. $\frac{3 - 2i}{3 + 2i}$
76. $\frac{2 + 3i}{2 - 3i}$
77. $\frac{3 + 2i}{3 + i}$
78. $\frac{2 - 5i}{2 + 5i}$
79. $\frac{\sqrt{5} - \sqrt{3}i}{\sqrt{5} + \sqrt{3}i}$
80. $\frac{\sqrt{3} + \sqrt{2}i}{\sqrt{3} - \sqrt{2}i}$
81. $\left(\frac{i}{3 + 2i}\right)^2$
82. $\left(\frac{5 + i}{2 + i}\right)^2$

Simplify each expression. See Examples 9–10. (Objective 4)

83. i^{21}
84. i^{19}
85. i^{27}
86. i^{22}
87. i^{100}
88. i^{42}
89. i^{97}
90. i^{200}

Simplify each expression. See Example 11. (Objectives 3–4)

91. $3i^3 + i^2$
92. $4i^2 - 3i^3$
93. $\frac{1}{i}$
94. $\frac{1}{i^3}$
95. $\frac{4}{5i^3}$
96. $\frac{3}{2i}$
97. $\frac{3i}{8\sqrt{-9}}$
98. $\frac{5i^3}{2\sqrt{-4}}$

99. $\dfrac{-3}{5i^5}$

100. $\dfrac{-4}{6i^7}$

Find each value. See Example 12. (Objective 5)

101. $|6 + 8i|$

102. $|12 + 5i|$

103. $|12 - 5i|$

104. $|3 - 4i|$

105. $|5 + 7i|$

106. $|6 - 5i|$

107. $\left|\dfrac{3}{5} - \dfrac{4}{5}i\right|$

108. $\left|\dfrac{5}{13} + \dfrac{12}{13}i\right|$

ADDITIONAL PRACTICE

Are the two numbers equal?

109. $8 + 5i,\ 2^3 + \sqrt{25}i^3$

110. $4 - 7i,\ -4i^2 + 7i^3$

Simplify each expression. Write the answer in standard form.

111. $(5 + 3i) - (3 - 5i) + \sqrt{-1}$

112. $(8 + 7i) - \left(-7 - \sqrt{-64}\right) + (3 - i)$

113. $\left(-8 - \sqrt{3}i\right) - \left(7 - 3\sqrt{3}i\right)$

114. $\left(2 + 2\sqrt{2}i\right) + \left(-3 - \sqrt{2}i\right)$

115. $(2 + i)(2 - i)(1 + i)$

116. $(3 + 2i)(3 - 2i)(i + 1)$

117. $(3 + i)[(3 - 2i) + (2 + i)]$

118. $(2 - 3i)[(5 - 2i) - (2i + 1)]$

119. $\dfrac{i(3 - i)}{3 + i}$

120. $\dfrac{5 + 3i}{i(3 - 5i)}$

121. $\dfrac{(2 - 5i) - (5 - 2i)}{5 - i}$

122. $\dfrac{5i}{(5 + 2i) + (2 + i)}$

123. Show that $1 - 5i$ is a solution of $x^2 - 2x + 26 = 0$.

124. Show that $3 - 2i$ is a solution of $x^2 - 6x + 13 = 0$.

125. Show that i is a solution of $x^4 - 3x^2 - 4 = 0$.

126. Show that $2 + i$ is *not* a solution of $x^2 + x + 1 = 0$.

APPLICATIONS

In electronics, the formula $V = IR$ is called Ohm's Law. It gives the relationship in a circuit between the voltage V (in volts), the current I (in amperes), and the resistance R (in ohms).

127. Electronics Find V when $I = 2 - 3i$ amperes and $R = 2 + i$ ohms.

128. Electronics Find R when $I = 3 - 2i$ amperes and $V = 18 + i$ volts.

In electronics, the formula $Z = \frac{V}{I}$ is used to find the impedance Z of a circuit, where V is the voltage and I is the current.

129. Electronics Find the impedance of a circuit when the voltage is $1.7 + 0.5i$ and the current is $0.5i$.

130. Electronics Find the impedance of a circuit when the voltage is $1.6 - 0.4i$ and the current is $-0.2i$.

WRITING ABOUT MATH

131. Determine how to decide whether two complex numbers are equal.

132. Define the complex conjugate of a complex number.

SOMETHING TO THINK ABOUT

133. Rationalize the numerator: $\dfrac{3 - i}{2}$.

134. Rationalize the numerator: $\dfrac{2 + 3i}{2 - 3i}$.

PROJECTS

Project 1

The size of a television screen is measured along the diagonal of its screen, as shown in the illustrations. The screen of a traditional TV has an aspect ratio of 4:3. This means that the ratio of the width of the screen to its height is $\frac{4}{3}$. The screen of a wide-screen set has an aspect ratio of 16:9. This means that the ratio of the width of the screen to its height is $\frac{16}{9}$.

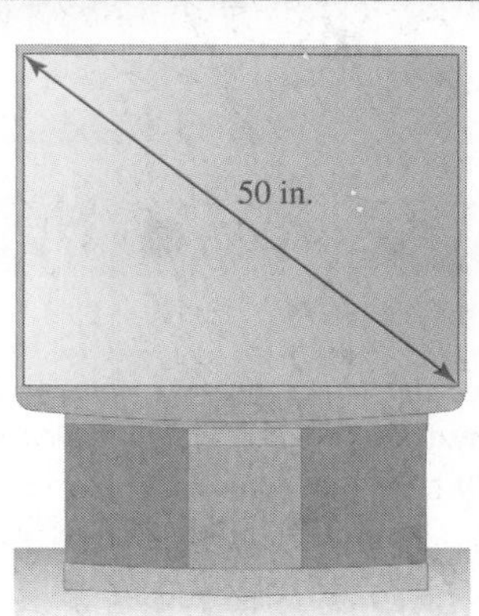

a. Find the width and height of the traditional-screen set shown in the illustration on the previous page. $\left(\textit{Hint: } \frac{4}{3} = \frac{4x}{3x}.\right)$

b. Find the viewing area of the traditional-screen set in square inches.

c. Find the width and height of the wide-screen set shown in the illustration above.

d. Find the viewing area of the wide-screen set in square inches.

e. Which set has the larger viewing area? Give the answer as a percent.

Project 2

Tom and Brian arrange to have a bicycle race. Each leaves his own house at the same time and rides to the other's house, whereupon the winner of the race calls his own house and leaves a message for the loser. A map of the race is shown in the illustration. Brian stays on the highway, averaging 21 mph. Tom knows that he and Brian are evenly matched when biking on the highway, so he cuts across country for the first part of his trip, averaging 15 mph. When Tom reaches the highway at point A, he turns right and follows the highway, averaging 21 mph.

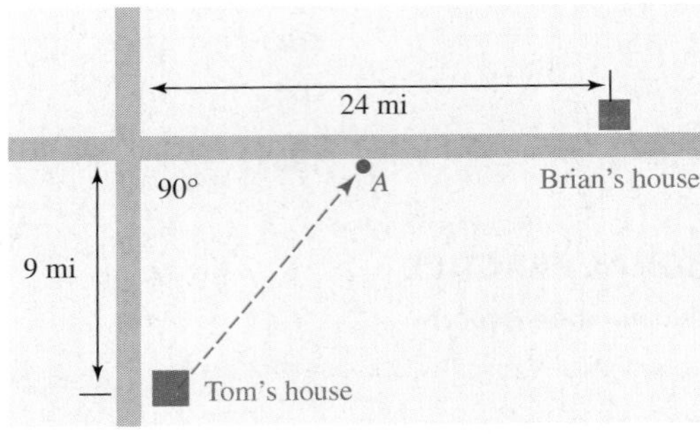

Tom and Brian never meet during the race and, amazingly, the race is a tie. Each of them calls the other at exactly the same moment!

a. How long (to the nearest second) did it take each person to complete the race?

b. How far from the intersection of the two highways is point A? (*Hint:* Set the travel times for Brian and Tom equal to each other. You may find two answers, but only one of them matches all of the information.)

c. Show that if Tom had started straight across country for Brian's house (in order to minimize the distance he had to travel), he would have lost the race. By how much time (to the nearest second) would he have lost? Then show that if Tom had biked across country to a point 9 miles from the intersection of the two highways, he would have won the race. By how much time (to the nearest second) would he have won?

Chapter 7 REVIEW

SECTION 7.1 Radical Expressions

DEFINITIONS AND CONCEPTS	EXAMPLES
Simplifying radicals: If n is a natural number greater than 1 and x is a real number, then	
If $x > 0$, then $\sqrt[n]{x}$ is the positive number such that $\left(\sqrt[n]{x}\right)^n = x$.	$\sqrt{25} = 5$ because $5^2 = 25$.
If $x = 0$, then $\sqrt[n]{x} = 0$.	$\sqrt[4]{0} = 0$ because $0^4 = 0$.

If $x < 0$, and n is odd, $\sqrt[n]{x}$ is the real number such that $\left(\sqrt[n]{x}\right)^n = x$. If $x < 0$, and n is even, $\sqrt[n]{x}$ is not a real number.	$\sqrt[3]{-64} = -4$ because $(-4)^3 = -64$. $\sqrt[4]{-8} =$ is not a real number.
If n is an even natural number, $\sqrt[n]{a^n} = \|a\|$ If n is an odd natural number, greater than 1, $\sqrt[n]{a^n} = a$	$\sqrt{64x^2} = 8\|x\|$ Absolute value bars are necessary because x could be a negative number. $\sqrt[3]{-8} = -2$ because $(-2)^3 = -8$.
Finding the domain of a radical function: If $f(x) = \sqrt[n]{x}$, then the domain of $f(x)$ will be $[0, \infty)$ if n is even $(-\infty, \infty)$ if n is odd	To find the domain of $f(x) = \sqrt{x + 4}$, set the radicand to be greater than or equal to 0, and solve for x. $x + 4 \geq 0$ The radicand must be ≥ 0. $x \geq -4$ Subtract 4 from each side. The domain is $[-4, \infty)$. To find the domain of $g(x) = \sqrt[3]{x - 9}$, we note that in a cube root the radicand can be any real number. Therefore, x can be any real number, and the domain is $(-\infty, \infty)$.
Standard deviation of a data set: Standard deviation $= \sqrt{\dfrac{\text{sum of the squares of the differences from the mean}}{\text{number of differences}}}$	Find the standard deviation of the data set 1, 3, 4, 8. (see table below) The sum of the squares of the differences is 26 and there are 4 values in the data set. The standard deviation is $\sqrt{\dfrac{26}{4}} \approx 2.549509757$. To the nearest hundredth, the standard deviation is 2.55.

Original terms	Mean	Difference	Square of the differences
1	4	−3	9
3	4	−1	1
4	4	0	0
8	4	4	16

REVIEW EXERCISES

Simplify each radical. Assume that x can be any number.

1. $\sqrt{49}$

2. $-\sqrt{121}$

3. $-\sqrt{36}$

4. $\sqrt{225}$

5. $\sqrt[3]{-27}$

6. $-\sqrt[3]{216}$

7. $\sqrt[4]{625}$

8. $\sqrt[5]{-32}$

9. $\sqrt{25x^2}$

10. $\sqrt{x^2 + 4x + 4}$

11. $\sqrt[3]{27a^6b^3}$

12. $\sqrt[4]{256x^8y^4}$

Graph each function.

13. $f(x) = \sqrt{x + 2}$

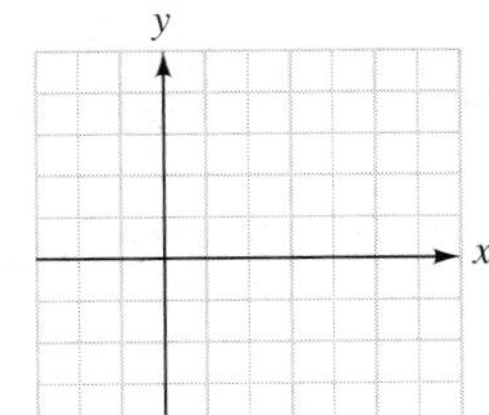

14. $f(x) = -\sqrt{x - 1}$

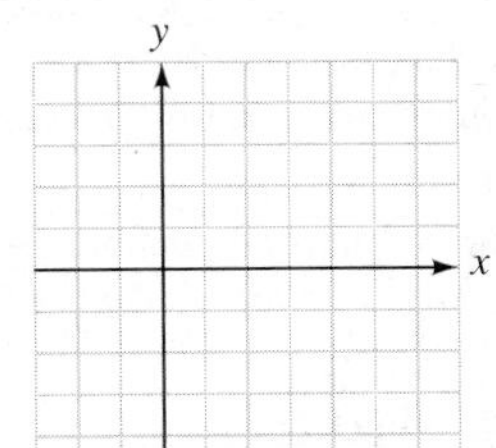

15. $f(x) = -\sqrt{x} + 2$

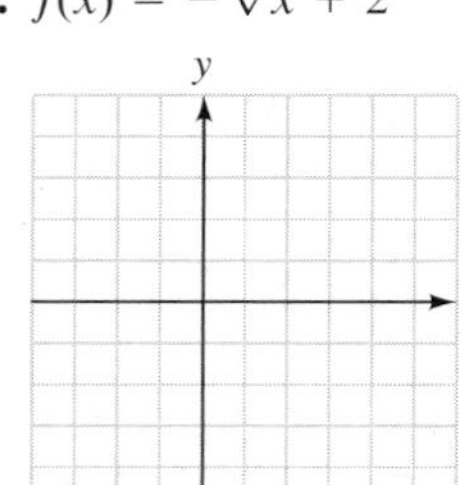

16. $f(x) = -\sqrt[3]{x} + 3$

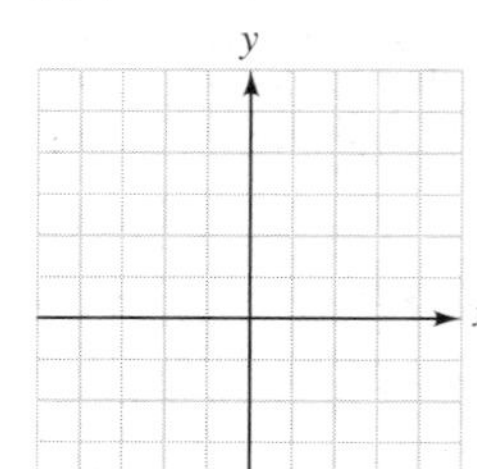

Consider the distribution 4, 8, 12, 16, 20.

17. Find the mean of the distribution.

18. Find the standard deviation.

SECTION 7.2 Applications of the Pythagorean Theorem and the Distance Formula

DEFINITIONS AND CONCEPTS	EXAMPLES
The Pythagorean theorem: If a and b are the lengths of the legs of a right triangle and c is the length of the hypotenuse, then $a^2 + b^2 = c^2$	To find the length of the hypotenuse of a right triangle with legs of length 9 ft and 12 ft, proceed as follows: $a^2 + b^2 = c^2$ The Pythagorean theorem. $(9)^2 + (12)^2 = c^2$ Substitute the values. $81 + 144 = c^2$ Square each value. $225 = c^2$ Add. $\sqrt{225} = c$ Since c is a length, take the positive square root. $15 = c$ $\sqrt{225} = 15$ The hypotenuse is 15 ft.
The distance formula: The distance between two points, (x_1, y_1) and (x_2, y_2), on a coordinate plane is $d = \sqrt{(x_2 - x_1)^2 + (y_2 - y_1)^2}$	To find the distance between $(6, -1)$ and $(3, 3)$, use the distance formula: $d = \sqrt{(x_2 - x_1)^2 + (y_2 - y_1)^2}$ The distance formula. $= \sqrt{(3 - 6)^2 + [3 - (-1)]^2}$ Substitute values. $= \sqrt{(-3)^2 + 4^2}$ Remove parentheses $= \sqrt{9 + 16}$ Square each value. $= \sqrt{25}$ Add. $= 5$ Simplify. The distance between the points is 5 units.

REVIEW EXERCISES

In Exercises 19–20, the horizon distance d (measured in miles) is related to the height h (measured in feet) of the observer by the formula $d = 1.4\sqrt{h}$.

19. View from a submarine A submarine's periscope extends 4.7 feet above the surface. How far away is the horizon?

20. View from a submarine How far out of the water must a submarine periscope extend to provide a 4-mile horizon?

21. Sailing A technique called *tacking* allows a sailboat to make progress into the wind. A sailboat follows the course in the illustration. Find d, the distance the boat advances into the wind.

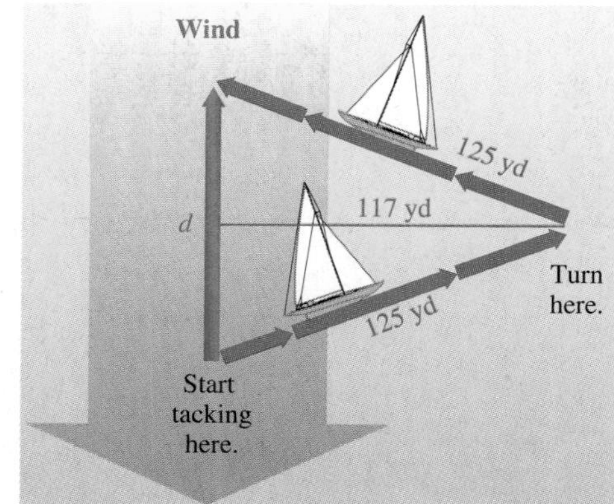

22. Communications Some campers 3,900 yards from a highway are talking to truckers on a citizen's band radio with an 8,900-yard range. Over what length of highway can these conversations take place?

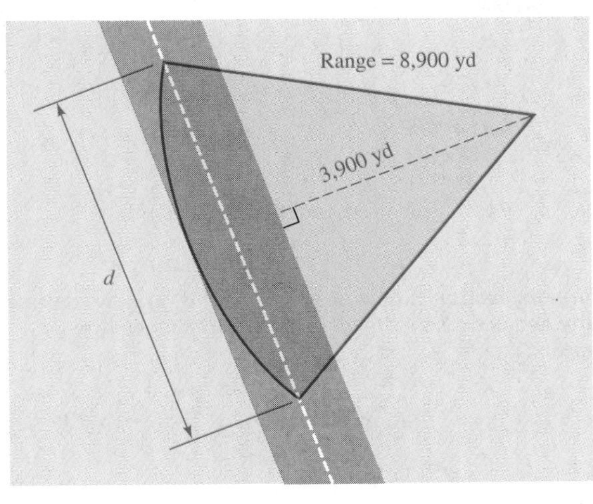

23. Find the distance between points (0, 0) and (5, −12).

24. Find the distance between points (−4, 6) and (−2, 8). Give the result to the nearest hundredth.

SECTION 7.3 Rational Exponents

DEFINITIONS AND CONCEPTS	EXAMPLES
Rational exponents with numerator of 1: If n $(n > 1)$ is a natural number and $\sqrt[n]{x}$ is a real number, then $x^{1/n} = \sqrt[n]{x}$.	$16^{1/2} = \sqrt{16} = 4$
If n is even, $(x^n)^{1/n} = \lvert x \rvert$.	$(25x^2)^{1/2} = \sqrt{25x^2} = 5\lvert x \rvert$
If n is a natural number greater than 1 and x is a real number, then	
If $x > 0$, then $x^{1/n}$ is the positive number such that $(x^{1/n})^n = x$.	$64^{1/2} = \sqrt{64} = 8$
If $x = 0$, then $x^{1/n} = 0$.	$0^{1/4} = \sqrt[4]{0} = 0$
If $x < 0$ and n is odd, then $x^{1/n}$ is the real number such that $(x^{1/n})^n = x$.	$(-27)^{1/3} = \sqrt[3]{-27} = -3$
If $x < 0$ and n is even, then $x^{1/n}$ is not a real number.	$(-64)^{1/2} = \sqrt{-64}$ is not a real number.
Rational exponents with numerator other than 1: If m and n are positive integers and $x > 0$, $x^{m/n} = \sqrt[n]{x^m} = \left(\sqrt[n]{x}\right)^m$	$64^{2/3} = \sqrt[3]{(64)^2} = \sqrt[3]{4{,}096} = 16$ $64^{2/3} = \left(\sqrt[3]{64}\right)^2 = 4^2 = 16$
Negative rational exponents: $x^{-m/n} = \dfrac{1}{x^{m/n}}$ $\dfrac{1}{x^{-m/n}} = x^{m/n}$ $(x \neq 0)$	$25^{-1/2} = \dfrac{1}{25^{1/2}} = \dfrac{1}{\sqrt{25}} = \dfrac{1}{5}$ $\dfrac{1}{16^{-3/2}} = 16^{3/2} = \left(\sqrt{16}\right)^3 = 4^3 = 64$

Simplifying expressions with rational exponents: Apply the properties of exponents.	$\frac{x^{3/4} \cdot x^{2/3}}{x^{7/6}} = x^{3/4+2/3-7/6}$ Use the rules $x^m \cdot x^n = x^{m+n}$ and $\frac{x^m}{x^n} = x^{m-n}$. $= x^{1/4}$ $\frac{3}{4} + \frac{2}{3} - \frac{7}{6} = \frac{1}{4}$ $= \sqrt[4]{x}$ Change to radical notation.
Simplifying radical expressions: **1.** Change the radical expression into an exponential expression with rational exponents. **2.** Simplify the rational exponents. **3.** Change the exponential expression back into a radical.	$\sqrt[4]{81x^2} = (81x^2)^{1/4}$ Use the rule $\sqrt[n]{x} = x^{1/n}$. $= (3^4x^2)^{1/4}$ $81 = 3^4$ $= 3x^{1/2}$ Use the rule $(xy)^m = x^m \cdot y^m$. $= 3\sqrt{x}$ Change to radical notation.

REVIEW EXERCISES

Simplify each expression, if possible. Assume that all variables represent positive numbers.

25. $25^{1/2}$
26. $-36^{1/2}$
27. $9^{3/2}$
28. $16^{3/2}$
29. $(-8)^{1/3}$
30. $-8^{2/3}$
31. $8^{-2/3}$
32. $8^{-1/3}$
33. $-49^{5/2}$
34. $\frac{1}{25^{5/2}}$
35. $\left(\frac{1}{4}\right)^{-3/2}$
36. $\left(\frac{4}{9}\right)^{-3/2}$
37. $(27x^3y)^{1/3}$
38. $(81x^4y^2)^{1/4}$
39. $(25x^3y^4)^{3/2}$
40. $(8u^2v^3)^{-2/3}$

Perform the multiplications. Assume that all variables represent positive numbers and write all answers without negative exponents.

41. $5^{1/4}5^{1/2}$
42. $a^{3/7}a^{2/7}$
43. $u^{1/2}(u^{1/2} - u^{-1/2})$
44. $v^{2/3}(v^{1/3} + v^{4/3})$
45. $(x^{1/2} + y^{1/2})^2$
46. $(a^{2/3} + b^{2/3})(a^{2/3} - b^{2/3})$

Simplify each expression. Assume that all variables are positive.

47. $\sqrt[6]{5^2}$
48. $\sqrt[8]{x^4}$
49. $\sqrt[9]{27a^3b^6}$
50. $\sqrt[4]{25a^2b^2}$

SECTION 7.4 Simplifying and Combining Radical Expressions

DEFINITIONS AND CONCEPTS	EXAMPLES
Properties of radicals: $\sqrt[n]{ab} = \sqrt[n]{a}\sqrt[n]{b}$	$\sqrt{24} = \sqrt{4 \cdot 6} = \sqrt{4}\sqrt{6} = 2\sqrt{6}$ $\sqrt[3]{24} = \sqrt[3]{8 \cdot 3} = \sqrt[3]{8}\sqrt[3]{3} = 2\sqrt[3]{3}$
$\sqrt[n]{\frac{a}{b}} = \frac{\sqrt[n]{a}}{\sqrt[n]{b}}$ $(b \neq 0)$	$\sqrt{\frac{21}{64x^6}} = \frac{\sqrt{21}}{\sqrt{64x^6}} = \frac{\sqrt{21}}{8x^3}$ $\sqrt[3]{\frac{21}{64x^6}} = \frac{\sqrt[3]{21}}{\sqrt[3]{64x^6}} = \frac{\sqrt[3]{21}}{4x^2}$
Adding and subtracting radical expressions: Like radicals can be combined by addition and subtraction. Radicals that are not similar often can be simplified to radicals that are similar and then combined.	$9\sqrt{2} + 4\sqrt{2} = (9 + 4)\sqrt{2} = 13\sqrt{2}$ $\sqrt{2} + \sqrt{18} = \sqrt{2} + \sqrt{9}\sqrt{2}$ $= \sqrt{2} + 3\sqrt{2}$ $= 4\sqrt{2}$

Special right triangles: In a 45°–45°–90° triangle, the length of the hypotenuse is the length of one leg times $\sqrt{2}$.	If each leg of an isosceles triangle is 17 cm, the hypotenuse measures $17\sqrt{2}$ cm.
The shorter leg of a 30°–60°–90° triangle is half as long as the hypotenuse. The longer leg is the length of the shorter leg times $\sqrt{3}$.	If the shorter leg of a 30°–60°–90° triangle is 12 units, the hypotenuse is 24 cm and the longer leg is $12\sqrt{3}$ units.

REVIEW EXERCISES

Simplify each expression. Assume that all variables represent positive numbers.

51. $\sqrt{240}$ **52.** $\sqrt[3]{54}$

53. $\sqrt[4]{32}$ **54.** $\sqrt[5]{96}$

55. $\sqrt{8x^3}$ **56.** $\sqrt{18x^4y^3}$

57. $\sqrt[3]{16x^5y^4}$ **58.** $\sqrt[3]{54x^7y^3}$

59. $\dfrac{\sqrt{32x^3}}{\sqrt{2x}}$ **60.** $\dfrac{\sqrt[3]{16x^5}}{\sqrt[3]{2x^2}}$

61. $\sqrt[3]{\dfrac{2a^2b}{27x^3}}$ **62.** $\sqrt{\dfrac{17xy}{64a^4}}$

Simplify and combine like radicals. Assume that all variables represent positive numbers.

63. $\sqrt{2} + \sqrt{8}$ **64.** $\sqrt{20} - \sqrt{5}$

65. $2\sqrt[3]{3} - \sqrt[3]{24}$ **66.** $\sqrt[4]{32} + 2\sqrt[4]{162}$

67. $2x\sqrt{8} + 2\sqrt{200x^2} + \sqrt{50x^2}$

68. $3\sqrt{27a^3} - 2a\sqrt{3a} + 5\sqrt{75a^3}$

69. $\sqrt[3]{54} - 3\sqrt[3]{16} + 4\sqrt[3]{128}$

70. $2\sqrt[4]{32x^5} + 4\sqrt[4]{162x^5} - 5x\sqrt[4]{512x}$

71. Geometry Find the length of the hypotenuse of an isosceles right triangle whose legs measure 7 meters.

72. Geometry The hypotenuse of a 30°–60°–90° triangle measures $12\sqrt{3}$ centimeters. Find the length of each leg.

Find x to two decimal places.

73.

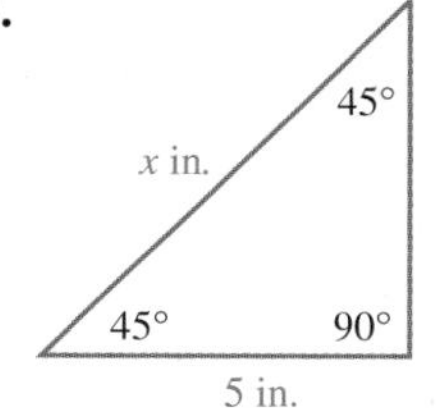

74.

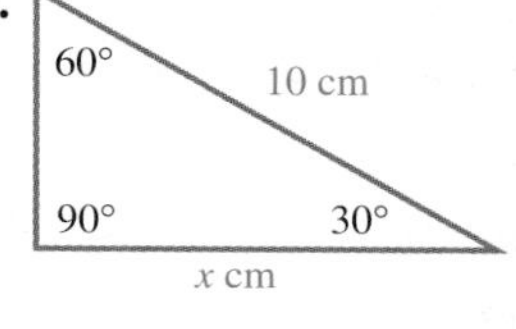

SECTION 7.5 Multiplying and Dividing Radical Expressions

DEFINITIONS AND CONCEPTS	EXAMPLES
Multiplying radical expressions: If two radicals have the same index, they can be multiplied:	$\sqrt{3x}\sqrt{6x} = \sqrt{18x^2} \quad (x > 0)$ $= \sqrt{9x^2}\sqrt{2}$ $= 3x\sqrt{2}$ $(5 - 2\sqrt{3})(7 + 6\sqrt{3})$ $= 35 + 30\sqrt{3} - 14\sqrt{3} - 12\sqrt{9}$ Distribute. $= 35 + 16\sqrt{3} - 12(3)$ Combine like terms. $= 35 + 16\sqrt{3} - 36$ Multiply. $= -1 + 16\sqrt{3}$ Simplify.
Rationalizing the denominator: To eliminate a single radical in the denominator, we multiply the numerator and the denominator by a number that will give a perfect square (or cube, or 4th power, etc.) under the radical in the denominator.	$\sqrt{\dfrac{5}{8}} = \dfrac{\sqrt{5}}{\sqrt{8}}$ $\quad \sqrt{\dfrac{a}{b}} = \dfrac{\sqrt{a}}{\sqrt{b}}$ $= \dfrac{\sqrt{5}}{\sqrt{8}} \cdot \dfrac{\sqrt{2}}{\sqrt{2}}$ Multiply by 1: $\dfrac{\sqrt{2}}{\sqrt{2}} = 1$. $= \dfrac{\sqrt{10}}{\sqrt{16}}$ Multiply radicals. $= \dfrac{\sqrt{10}}{4}$ Simplify.

To rationalize a fraction whose denominator has two terms with one or both containing square roots, we multiply its numerator and denominator by the *conjugate* of its denominator.	$\frac{5}{4-\sqrt{2}} \cdot \frac{4+\sqrt{2}}{4+\sqrt{2}}$	Multiply the numerator and the denominator by the conjugate of the denominator.
	$= \frac{20+5\sqrt{2}}{16-2}$	Multiply the numerators and denominators.
	$= \frac{20+5\sqrt{2}}{14}$	Simplify.

REVIEW EXERCISES

Simplify each expression. Assume that all variables represent positive numbers.

75. $(2\sqrt{5})(3\sqrt{2})$
76. $2\sqrt{6}\sqrt{216}$
77. $\sqrt{9x}\sqrt{x}$
78. $\sqrt[3]{3}\sqrt[3]{9}$
79. $-\sqrt[3]{2x^2}\sqrt[3]{4x}$
80. $-\sqrt[4]{256x^5y^{11}}\sqrt[4]{625x^9y^3}$
81. $\sqrt{2}(\sqrt{8}-3)$
82. $\sqrt{2}(\sqrt{2}+3)$
83. $\sqrt{5}(\sqrt{2}-1)$
84. $\sqrt{3}(\sqrt{3}+\sqrt{2})$
85. $(\sqrt{2}+1)(\sqrt{2}-1)$
86. $(\sqrt{3}+\sqrt{2})(\sqrt{3}+\sqrt{2})$
87. $(\sqrt{x}+\sqrt{y})(\sqrt{x}-\sqrt{y})$
88. $(2\sqrt{u}+3)(3\sqrt{u}-4)$

Rationalize each denominator.

89. $\frac{1}{\sqrt{3}}$
90. $\frac{\sqrt{3}}{\sqrt{5}}$
91. $\frac{x}{\sqrt{xy}}$
92. $\frac{\sqrt[3]{uv}}{\sqrt[3]{u^5v^7}}$
93. $\frac{2}{\sqrt{2}-1}$
94. $\frac{\sqrt{2}}{\sqrt{3}-1}$
95. $\frac{2x-32}{\sqrt{x}+4}$
96. $\frac{\sqrt{a}+1}{\sqrt{a}-1}$

Rationalize each numerator. All variables represent positive numbers.

97. $\frac{\sqrt{3}}{5}$
98. $\frac{\sqrt[3]{9}}{3}$
99. $\frac{3-\sqrt{x}}{2}$
100. $\frac{\sqrt{a}-\sqrt{b}}{\sqrt{a}}$

SECTION 7.6 Radical Equations

DEFINITIONS AND CONCEPTS	EXAMPLES
The power rule: If x, y, and n are real numbers, If $x = y$, then $x^n = y^n$. Raising both sides of an equation to the same power can lead to extraneous solutions. Be sure to check all suspected solutions.	To solve $\sqrt{x+4} = x-2$, proceed as follows:

$(\sqrt{x+4})^2 = (x-2)^2$ Square both sides to eliminate the square root.

$x + 4 = x^2 - 4x + 4$ Square the binomial.

$0 = x^2 - 5x$ Subtract x and 4 from both sides.

$0 = x(x-5)$ Factor $x^2 - 5x$.

$x = 0$ or $x - 5 = 0$ Set each factor equal to 0.

$x = 0$ | $x = 5$

Check:

$\sqrt{x+4} = x-2$	$\sqrt{x+4} = x-2$
$\sqrt{0+4} \stackrel{?}{=} 0-2$	$\sqrt{5+4} \stackrel{?}{=} 5-2$
$\sqrt{4} \stackrel{?}{=} -2$	$\sqrt{9} \stackrel{?}{=} 3$
$2 \neq -2$	$3 = 3$

Since 0 does not check, it is extraneous and must be discarded. The only solution of the original equation is 5.

REVIEW EXERCISES

Solve each equation.

101. $\sqrt{y+3} = \sqrt{2y-19}$

102. $u = \sqrt{25u - 144}$

103. $r = \sqrt{12r - 27}$

104. $\sqrt{z+1} + \sqrt{z} = 2$

105. $\sqrt{2x+5} - \sqrt{2x} = 1$

106. $\sqrt[3]{x^3+8} = x + 2$

SECTION 7.7 Complex Numbers

DEFINITIONS AND CONCEPTS	EXAMPLES
Simplifying imaginary numbers: $\sqrt{-1}$ is defined as the **imaginary number *i*.**	$\sqrt{-12} = \sqrt{-4}\sqrt{3}$ Write -12 as $-4(3)$. $= \sqrt{-1}\sqrt{4}\sqrt{3}$ Write -4 as the product of -1 and 4. $= i(2)\sqrt{3}$ $\sqrt{-1} = i$ $= 2\sqrt{3}i$
Operations with complex numbers: If a, b, c, and d are real numbers and $i^2 = -1$, $a + bi = c + di$ if and only if $a = c$ and $b = d$ $(a + bi) + (c + di) = (a + c) + (b + d)i$ $(a + bi) - (c + di) = (a - c) + (b - d)i$ $(a + bi)(c + di) = (ac - bd) + (ad + bc)i$	$3 + 5i = \sqrt{9} + \frac{10}{2}i$ because $\sqrt{9} = 3$ and $\frac{10}{2} = 5$. $(6 + 5i) + (2 + 9i) = (6 + 2) + (5 + 9)i = 8 + 14i$ $(6 + 5i) - (2 + 9i) = (6 - 2) + (5 - 9)i = 4 - 4i$ $(6 - 5i)(2 + 9i) = 12 + 54i - 10i - 45i^2$ $= 12 + 44i - 45(-1)$ $= 12 + 44i + 45$ $= 57 + 44i$
Dividing complex numbers: To divide complex numbers, write the division as a fraction, and rationalize the denominator.	$\frac{6}{3+i} \cdot \frac{3-i}{3-i} = \frac{18-6i}{9-i^2}$ Multiply the numerator and the denominator by the conjugate of the denominator. Multiply the fractions. $= \frac{18-6i}{9+1}$ $i^2 = -1$ $= \frac{18-6i}{10}$ Add. $= \frac{18}{10} - \frac{6}{10}i$ Write in $a + bi$ form. $= \frac{9}{5} - \frac{3}{5}i$ Simplify each fraction.
Absolute value of a complex number: $\lvert a + bi \rvert = \sqrt{a^2 + b^2}$	$\lvert 4 + 8i \rvert = \sqrt{4^2 + 8^2}$ $\lvert a + bi \rvert = \sqrt{a^2 + b^2}$ $= \sqrt{16 + 64}$ Square each value. $= \sqrt{80}$ Add. $= \sqrt{16 \cdot 5}$ Factor: $80 = 16(5)$. $= 4\sqrt{5}$ Simplify: $\sqrt{16 \cdot 5} = \sqrt{16}\sqrt{5} = 4\sqrt{5}$.

REVIEW EXERCISES

Perform the operations and give all answers in $a + bi$ form.

107. $(5 + 4i) + (7 - 12i)$

108. $(-6 - 40i) - (-8 + 28i)$

109. $\left(-32 + \sqrt{-144}\right) - \left(64 + \sqrt{-81}\right)$

110. $\left(-8 + \sqrt{-8}\right) + \left(6 - \sqrt{-32}\right)$

111. $(2 - 7i)(-3 + 4i)$

112. $(-5 + 6i)(2 + i)$

113. $\left(5 - \sqrt{-27}\right)\left(-6 + \sqrt{-12}\right)$

114. $\left(2 + \sqrt{-128}\right)\left(3 - \sqrt{-98}\right)$

115. $\dfrac{3}{4i}$

116. $\dfrac{-2}{5i^3}$

117. $\dfrac{6}{2 + i}$

118. $\dfrac{7}{3 - i}$

119. $\dfrac{4 + i}{4 - i}$

120. $\dfrac{3 - i}{3 + i}$

121. $\dfrac{3}{5 + \sqrt{-4}}$

122. $\dfrac{2}{3 - \sqrt{-9}}$

Simplify.

123. $|9 + 12i|$

124. $|24 - 10i|$

125. i^{12}

126. i^{583}

Chapter 7 TEST

Find each root.

1. $\sqrt{49}$

2. $\sqrt[3]{64}$

3. $\sqrt{4x^2}$

4. $\sqrt[3]{8x^3}$

Graph each function and find its domain and range.

5. $f(x) = \sqrt{x - 2}$

6. $f(x) = \sqrt[3]{x} + 3$

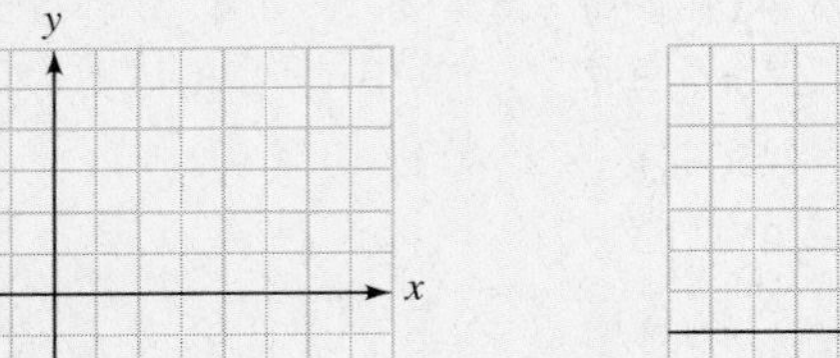

Use a calculator.

7. Shipping crates The diagonal brace on the shipping crate shown in the illustration is 53 inches. Find the height, h, of the crate.

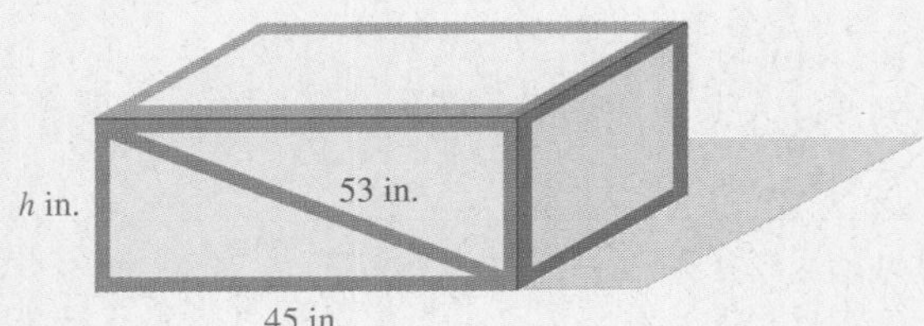

8. Pendulums The 2-meter pendulum rises 0.1 meter at the extremes of its swing. Find the width w of the swing.

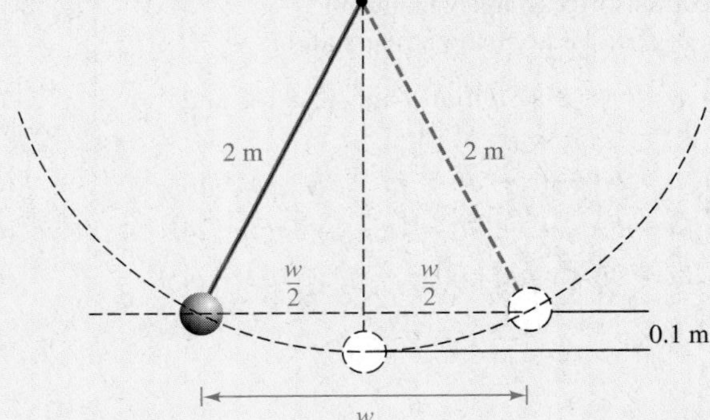

Find the distance between the points.

9. $(6, 8), (0, 0)$

10. $(-2, 5), (22, 12)$

Simplify each expression. Assume that all variables represent positive numbers, and write answers without using negative exponents.

11. $16^{1/4}$

12. $27^{2/3}$

13. $36^{-3/2}$

14. $\left(-\dfrac{8}{27}\right)^{-2/3}$

15. $\dfrac{2^{5/3}2^{1/6}}{2^{1/2}}$

16. $\dfrac{(8x^3y)^{1/2}(8xy^5)^{1/2}}{(x^3y^6)^{1/3}}$

Simplify each expression. Assume that all variables represent positive numbers.

17. $\sqrt{48}$

18. $\sqrt{250x^3y^5}$

19. $\dfrac{\sqrt[3]{24x^{15}y^4}}{\sqrt[3]{y}}$

20. $\sqrt{\dfrac{3a^5}{48a^7}}$

Simplify each expression. Assume that the variables are unrestricted.

21. $\sqrt{12x^2}$

22. $\sqrt{8x^6}$

23. $\sqrt[3]{81x^3}$

24. $\sqrt{18x^4y^9}$

Simplify and combine like radicals. Assume that all variables represent positive numbers.

25. $\sqrt{12} - \sqrt{27}$

26. $2\sqrt[3]{40} - \sqrt[3]{5{,}000} + 4\sqrt[3]{625}$

27. $2\sqrt{48y^5} - 3y\sqrt{12y^3}$

28. $\sqrt[4]{768z^5} + z\sqrt[4]{48z}$

Perform each operation and simplify, if possible. All variables represent positive numbers.

29. $-2\sqrt{xy}\left(3\sqrt{x} + \sqrt{xy^3}\right)$

30. $\left(3\sqrt{2} + \sqrt{3}\right)\left(2\sqrt{2} - 3\sqrt{3}\right)$

Rationalize each denominator.

31. $\dfrac{1}{\sqrt{5}}$

32. $\dfrac{3t - 1}{\sqrt{3t} - 1}$

Rationalize each numerator.

33. $\dfrac{\sqrt{3}}{\sqrt{7}}$

34. $\dfrac{\sqrt{a} + \sqrt{b}}{\sqrt{a} - \sqrt{b}}$

Solve and check each equation.

35. $\sqrt[3]{6n + 4} - 4 = 0$

36. $1 - \sqrt{u} = \sqrt{u - 3}$

Perform the operations. Give all answers in standard form.

37. $(2 + 4i) + (-3 + 7i)$

38. $\left(3 - \sqrt{-9}\right) - \left(-1 + \sqrt{-16}\right)$

39. $2i(3 - 4i)$

40. $(3 + 2i)(-4 - i)$

41. $\dfrac{1}{i\sqrt{2}}$

42. $\dfrac{2 + i}{3 - i}$

CHAPTER **8**

Quadratic Functions, Inequalities, and Algebra of Functions

8.1 Solving Quadratic Equations Using the Square-Root Property and by Completing the Square
8.2 Solving Quadratic Equations by the Quadratic Formula
8.3 The Discriminant and Equations That Can Be Written in Quadratic Form
8.4 Graphs of Quadratic Functions
8.5 Quadratic and Other Nonlinear Inequalities
8.6 Algebra and Composition of Functions
8.7 Inverses of Functions
■ ***Projects***
CHAPTER REVIEW
CHAPTER TEST
CUMULATIVE REVIEW EXERCISES

Careers and Mathematics

POLICE OFFICERS AND DETECTIVES

People depend on police officers and detectives to protect their lives and property. Law enforcement officers, some of whom are state or federal special agents, perform these duties in a variety of ways. Uniformed police officers maintain regular patrols and respond to calls for service. They may direct traffic, investigate a burglary, or give first aid to an accident victim. Police work can be very dangerous and stressful.

Police officers and detectives held about 861,000 jobs in 2006.

Job Outlook:
Employment of police officers and detectives is expected to grow as fast as the average for all occupations through 2016.

Annual Earnings:
$35,600–$59,880

For More Information:
http://www.bls.gov/oco/ocos160.htm

For a Sample Application:
See Problem 83 in Section 8.1.

In this chapter ▶

We have discussed how to solve linear equations and certain quadratic equations in which the quadratic expression is factorable. In this chapter, we will discuss more general methods for solving quadratic equations and quadratic inequalities, and we will consider the graphs of quadratic functions. Finally, we will discuss operations on fuctions.

SECTION 8.1 Solving Quadratic Equations Using the Square-Root Property and by Completing the Square

Objectives

1. Solve a quadratic equation by factoring.
2. Solve a quadratic equation by applying the square-root property.
3. Solve a quadratic equation by completing the square.
4. Solve an application problem requiring the use of the square-root property.

Vocabulary

square-root property   completing the square

Getting Ready

Factor each expression.

1. $x^2 - 25$
2. $b^2 - 81$
3. $6x^2 + x - 2$
4. $4x^2 - 4x - 3$

We begin this section by reviewing how to solve quadratic equations by factoring. We will then discuss how to solve these equations by applying the **square-root property** and completing the square and use these skills to solve application problems.

1 Solve a quadratic equation by factoring.

Recall that a *quadratic equation* is an equation of the form $ax^2 + bx + c = 0 \quad (a \neq 0)$, where a, b, and c are real numbers. We will solve the first two examples by factoring.

EXAMPLE 1 Solve: $x^2 = 9$.

Solution To solve this quadratic equation by factoring, we proceed as follows:

$$x^2 = 9$$

$$x^2 - 9 = 0 \qquad \text{Subtract 9 from both sides.}$$

$$(x + 3)(x - 3) = 0 \qquad \text{Factor the binomial.}$$

$$x + 3 = 0 \quad \text{or} \quad x - 3 = 0 \qquad \text{Set each factor equal to 0.}$$

$$x = -3 \quad | \quad x = 3 \qquad \text{Solve each linear equation.}$$

Check: ***For x = −3*** ***For x = 3***

$$x^2 = 9 \qquad x^2 = 9$$
$$(-3)^2 \stackrel{?}{=} 9 \qquad (3)^2 \stackrel{?}{=} 9$$
$$9 = 9 \qquad 9 = 9$$

Since both results check, the solutions are 3 and −3.

SELF CHECK 1 Solve: $p^2 = 64$.

EXAMPLE 2 Solve: $6x^2 - 7x - 3 = 0$.

Solution To solve this quadratic equation by factoring, we proceed as follows:

$$6x^2 - 7x - 3 = 0$$
$$(2x - 3)(3x + 1) = 0 \quad \text{Factor.}$$
$$2x - 3 = 0 \quad \text{or} \quad 3x + 1 = 0 \quad \text{Set each factor equal to 0.}$$
$$x = \frac{3}{2} \quad \Big| \quad x = -\frac{1}{3} \quad \text{Solve each linear equation.}$$

Check: ***For*** $x = \frac{3}{2}$ ***For*** $x = -\frac{1}{3}$

$$6x^2 - 7x - 3 = 0 \qquad 6x^2 - 7x - 3 = 0$$
$$6\left(\frac{3}{2}\right)^2 - 7\left(\frac{3}{2}\right) - 3 \stackrel{?}{=} 0 \qquad 6\left(-\frac{1}{3}\right)^2 - 7\left(-\frac{1}{3}\right) - 3 \stackrel{?}{=} 0$$
$$6\left(\frac{9}{4}\right) - 7\left(\frac{3}{2}\right) - 3 \stackrel{?}{=} 0 \qquad 6\left(\frac{1}{9}\right) - 7\left(-\frac{1}{3}\right) - 3 \stackrel{?}{=} 0$$
$$\frac{27}{2} - \frac{21}{2} - \frac{6}{2} \stackrel{?}{=} 0 \qquad \frac{2}{3} + \frac{7}{3} - \frac{9}{3} \stackrel{?}{=} 0$$
$$0 = 0 \qquad 0 = 0$$

Since both results check, the solutions are $\frac{3}{2}$ and $-\frac{1}{3}$.

SELF CHECK 2 Solve: $6m^2 - 5m + 1 = 0$.

Unfortunately, many quadratic expressions do not factor easily. For example, it would be difficult to solve $2x^2 + 4x + 1 = 0$ by factoring, because $2x^2 + 4x + 1$ cannot be factored by using only integers.

2 Solve a quadratic equation by applying the square-root property.

To develop general methods for solving all quadratic equations, we first solve $x^2 = c$ by a method similar to the one used in Example 1.

$$x^2 = c$$
$$x^2 - c = 0 \quad \text{Subtract } c \text{ from both sides.}$$
$$x^2 - \left(\sqrt{c}\right)^2 = 0 \quad c = \left(\sqrt{c}\right)^2$$
$$\left(x + \sqrt{c}\right)\left(x - \sqrt{c}\right) = 0 \quad \text{Factor the difference of two squares.}$$

$x + \sqrt{c} = 0$ or $x - \sqrt{c} = 0$ Set each factor equal to 0.

$x = -\sqrt{c}$ | $x = \sqrt{c}$ Solve each linear equation.

The two solutions of $x^2 = c$ are $x = \sqrt{c}$ and $x = -\sqrt{c}$.

The Square-Root Property

The equation $x^2 = c$ has two solutions. They are

$$x = \sqrt{c} \quad \text{or} \quad x = -\sqrt{c}$$

We often use the symbol $\pm\sqrt{c}$ to represent the two solutions $\sqrt{c}$ and $-\sqrt{c}$. The symbol $\pm\sqrt{c}$ is read as "the positive or negative square root of c."

EXAMPLE 3 Use the square-root property to solve $x^2 - 12 = 0$.

Solution We can write the equation as $x^2 = 12$ and use the square-root property.

$x^2 - 12 = 0$

$x^2 = 12$ Add 12 to both sides.

$x = \sqrt{12}$ or $x = -\sqrt{12}$ Use the square-root property.

$x = 2\sqrt{3}$ | $x = -2\sqrt{3}$ $\sqrt{12} = \sqrt{4}\sqrt{3} = 2\sqrt{3}$

The solutions can be written as $\pm 2\sqrt{3}$. Verify that each one satisfies the equation.

SELF CHECK 3 Use the square-root property to solve $x^2 - 18 = 0$.

EXAMPLE 4 Use the square-root property to solve $(x - 3)^2 = 16$.

Solution We can use the square-root property.

$(x - 3)^2 = 16$

$x - 3 = \sqrt{16}$ or $x - 3 = -\sqrt{16}$ Use the square-root property.

$x - 3 = 4$ | $x - 3 = -4$ $\sqrt{16} = 4$ and $-\sqrt{16} = -4$.

$x = 3 + 4$ | $x = 3 - 4$ Add 3 to both sides.

$x = 7$ | $x = -1$ Simplify.

Verify that each solution satisfies the equation.

SELF CHECK 4 Use the square-root property to solve $(x + 2)^2 = 9$.

In the following example, the solutions are imaginary numbers.

EXAMPLE 5 Use the square-root property to solve $9x^2 + 25 = 0$.

Solution We can write the equation as $x^2 = -\frac{25}{9}$ and use the square-root property.

$$9x^2 + 25 = 0$$

$$x^2 = -\frac{25}{9}$$ Subtract 25 from both sides and divide both sides by 9.

$$x = \sqrt{-\frac{25}{9}} \quad \text{or} \quad x = -\sqrt{-\frac{25}{9}}$$ Use the square-root property.

$$x = \sqrt{\frac{25}{9}}\sqrt{-1} \quad\Big|\quad x = -\sqrt{\frac{25}{9}}\sqrt{-1}$$ $\sqrt{-\frac{25}{9}} = \sqrt{\frac{25}{9}(-1)} = \sqrt{\frac{25}{9}}\sqrt{-1}$

$$x = \frac{5}{3}i \quad\Big|\quad x = -\frac{5}{3}i$$ $\sqrt{\frac{25}{9}} = \frac{5}{3}$; $\sqrt{-1} = i$

Check:

$9x^2 + 25 = 0$	$9x^2 + 25 = 0$
$9\left(\frac{5}{3}i\right)^2 + 25 \stackrel{?}{=} 0$	$9\left(-\frac{5}{3}i\right)^2 + 25 \stackrel{?}{=} 0$
$9\left(\frac{25}{9}\right)i^2 + 25 \stackrel{?}{=} 0$	$9\left(\frac{25}{9}\right)i^2 + 25 \stackrel{?}{=} 0$
$25(-1) + 25 \stackrel{?}{=} 0$	$25(-1) + 25 \stackrel{?}{=} 0$
$0 = 0$	$0 = 0$

Since both results check, the solutions are $\pm\frac{5}{3}i$.

SELF CHECK 5 Use the square-root property to solve $4x^2 + 36 = 0$.

3 Solve a quadratic equation by completing the square.

All quadratic equations can be solved by a method called **completing the square.** This method is based on the special products

$$x^2 + 2ax + a^2 = (x + a)^2 \quad \text{and} \quad x^2 - 2ax + a^2 = (x - a)^2$$

Recall that the trinomials $x^2 + \mathbf{2}ax + a^2$ and $x^2 - \mathbf{2}ax + a^2$ are both *perfect-square trinomials,* because both factor as the square of a binomial. In each case, the coefficient of the first term is 1 and if we take one-half of the coefficient of x in the middle term and square it, we obtain the third term.

$$\left[\frac{1}{2}(2a)\right]^2 = a^2 \qquad \left[\tfrac{1}{2}(2a)\right]^2 = (a)^2 = a^2$$

$$\left[\frac{1}{2}(-2a)\right]^2 = (-a)^2 = a^2 \qquad \left[\tfrac{1}{2}(-2a)\right]^2 = (-a)^2 = a^2$$

EXAMPLE 6 Find the number that when added to each binomial results in a perfect-square trinomial: **a.** $x^2 + 10x$ **b.** $x^2 - 6x$ **c.** $x^2 - 11x$.

Solution **a.** To make $x^2 + 10x$ a perfect-square trinomial, we first find one-half of 10 to get 5 and square 5 to get 25.

$$\left[\frac{1}{2}(10)\right]^2 = (5)^2 = 25$$

Then we add 25 to $x^2 + 10x$ and obtain $x^2 + 10x + 25$. This is a perfect-square trinomial because $x^2 + 10x + 25 = (x + 5)^2$.

b. To make $x^2 - 6x$ a perfect-square trinomial, we first find one-half of -6 to get -3 and square -3 to get 9.

$$\left[\frac{1}{2}(-6)\right]^2 = (-3)^2 = 9$$

Then we add 9 to $x^2 - 6x$ and obtain $x^2 - 6x + 9$. This is a perfect-square trinomial because $x^2 - 6x + 9 = (x - 3)^2$.

c. To make $x^2 - 11x$ a perfect-square trinomial, we first find one-half of -11 to get $-\frac{11}{2}$ and square $-\frac{11}{2}$ to get $\frac{121}{4}$.

$$\left[\frac{1}{2}(-11)\right]^2 = \left(-\frac{11}{2}\right)^2 = \frac{121}{4}$$

Then we add $\frac{121}{4}$ to $x^2 - 11x$ and obtain $x^2 - 11x + \frac{121}{4}$. This is a perfect-square trinomial because $x^2 - 11x + \frac{121}{4} = \left(x - \frac{11}{2}\right)^2$.

SELF CHECK 6 Find the number that when added to $a^2 - 5a$ results in a perfect-square trinomial.

To see geometrically why completing the square works on $x^2 + 10x$, we refer to Figure 8-1(a), which shows a polygon with an area of $x^2 + 10x$. To turn the polygon into a square, we can divide the area of $10x$ into two areas of $5x$ and then reassemble the polygon as shown in Figure 8-1(b). To fill in the missing corner, we must add a square with an area of $5^2 = 25$. Thus, we complete the square.

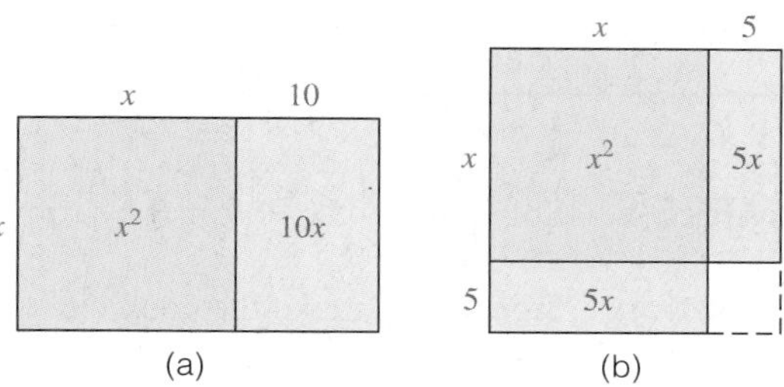

Figure 8-1

To solve an equation of the form $ax^2 + bx + c = 0 \quad (a \neq 0)$ by completing the square, we use the following steps.

Completing the Square

1. Make sure that the coefficient of x^2 is 1. If it isn't, make it 1 by dividing both sides of the equation by the coefficient of x^2.
2. If necessary, add a number to both sides of the equation to place the constant term on the right side of the equal sign.
3. Complete the square:
   - **a.** Find one-half of the coefficient of x and square it.
   - **b.** Add the square to both sides of the equation.
4. Factor the trinomial square on one side of the equation and combine like terms on the other side.
5. Solve the resulting equation by using the square-root property.

EXAMPLE 7 Solve $x^2 + 8x + 7 = 0$ by completing the square.

Solution **Step 1** In this example, the coefficient of x^2 is already 1.

Step 2 We add -7 to both sides to place the constant on the right side of the equal sign:

$$x^2 + 8x + 7 = 0$$
$$x^2 + 8x = -7$$

Step 3 The coefficient of x is 8, one-half of 8 is 4, and $4^2 = 16$. To complete the square, we add 16 to both sides.

$$x^2 + 8x + \mathbf{16} = -7 + \mathbf{16}$$

Step 4 Since the left side of the above equation is a perfect-square trinomial, we can factor it to get $(x + 4)^2$ and simplify on the right side to obtain

$$(x + 4)^2 = 9$$

Step 5 We can then solve the resulting equation by using the square-root property.

$$(x + 4)^2 = 9$$

$$x + 4 = \sqrt{9} \quad \text{or} \quad x + 4 = -\sqrt{9}$$
$$x + 4 = 3 \qquad\qquad x + 4 = -3$$
$$x = -1 \qquad\qquad x = -7$$

After checking both results, we see that the solutions are -1 and -7. Note that this equation could be solved by factoring.

SELF CHECK 7 Solve $a^2 + 5a + 4 = 0$ by completing the square.

EXAMPLE 8 Solve $6x^2 + 5x - 6 = 0$ by completing the square.

Solution **Step 1** To make the coefficient of x^2 equal to 1, we divide both sides by 6.

$$6x^2 + 5x - 6 = 0$$

$$\frac{6x^2}{6} + \frac{5}{6}x - \frac{6}{6} = \frac{0}{6} \quad \text{Divide both sides by 6.}$$

$$x^2 + \frac{5}{6}x - 1 = 0 \quad \text{Simplify.}$$

Step 2 We add 1 to both sides to place the constant on the right side.

$$x^2 + \frac{5}{6}x = 1$$

Step 3 The coefficient of x is $\frac{5}{6}$, one-half of $\frac{5}{6}$ is $\frac{5}{12}$, and $\left(\frac{5}{12}\right)^2 = \frac{25}{144}$. To complete the square, we add $\frac{25}{144}$ to both sides.

$$x^2 + \frac{5}{6}x + \frac{\mathbf{25}}{\mathbf{144}} = 1 + \frac{\mathbf{25}}{\mathbf{144}}$$

Step 4 Since the left side of the above equation is a perfect-square trinomial, we can factor it to get $\left(x + \frac{5}{12}\right)^2$ and simplify the right side to obtain

$$\left(x + \frac{5}{12}\right)^2 = \frac{169}{144} \qquad 1 + \frac{25}{144} = \frac{144}{144} + \frac{25}{144} = \frac{169}{144}$$

Step 5 We can solve this equation by using the square-root property.

$$x + \frac{5}{12} = \sqrt{\frac{169}{144}} \quad \text{or} \quad x + \frac{5}{12} = -\sqrt{\frac{169}{144}} \qquad \text{Apply the square-root property.}$$

$$x + \frac{5}{12} = \frac{13}{12} \qquad x + \frac{5}{12} = -\frac{13}{12} \qquad \sqrt{\frac{169}{144}} = \frac{13}{12}$$

$$x = -\frac{5}{12} + \frac{13}{12} \qquad x = -\frac{5}{12} - \frac{13}{12} \qquad \text{Subtract } \tfrac{5}{12} \text{ from both sides.}$$

$$x = \frac{8}{12} \qquad x = -\frac{18}{12} \qquad \text{Add.}$$

$$x = \frac{2}{3} \qquad x = -\frac{3}{2} \qquad \text{Simplify each fraction.}$$

After checking both results, we see that the solutions are $\frac{2}{3}$ and $-\frac{3}{2}$. Note that this equation could be solved by factoring.

SELF CHECK 8 Solve $6p^2 - 5p - 6 = 0$ by completing the square.

EXAMPLE 9 Solve $2x^2 + 4x + 1 = 0$ by completing the square.

Solution

$$2x^2 + 4x + 1 = 0$$

$$x^2 + 2x + \frac{1}{2} = \frac{0}{2} \qquad \text{Divide both sides by 2 to make the coefficient of } x^2 \text{ equal to 1.}$$

$$x^2 + 2x = -\frac{1}{2} \qquad \text{Subtract } \tfrac{1}{2} \text{ from both sides.}$$

$$x^2 + 2x + \mathbf{1} = -\frac{1}{2} + \mathbf{1} \qquad \text{Square half the coefficient of } x \text{ and add it to both sides.}$$

$$(x + 1)^2 = \frac{1}{2} \qquad \text{Factor on one side and combine like terms on the other.}$$

$$x + 1 = \sqrt{\frac{1}{2}} \quad \text{or} \quad x + 1 = -\sqrt{\frac{1}{2}} \qquad \text{Use the square-root property.}$$

$$x + 1 = \frac{\sqrt{2}}{2} \qquad x + 1 = -\frac{\sqrt{2}}{2} \qquad \sqrt{\frac{1}{2}} = \frac{1}{\sqrt{2}} = \frac{1 \cdot \sqrt{2}}{\sqrt{2}\sqrt{2}} = \frac{\sqrt{2}}{2}$$

$$x = -1 + \frac{\sqrt{2}}{2} \qquad x = -1 - \frac{\sqrt{2}}{2}$$

These solutions can be written as $-1 \pm \frac{\sqrt{2}}{2}$.

COMMENT Note that $-1 \pm \frac{\sqrt{2}}{2}$ can be written as

$$-1 \pm \frac{\sqrt{2}}{2} = \frac{-2}{2} \pm \frac{\sqrt{2}}{2} = \frac{-2 \pm \sqrt{2}}{2}$$

SELF CHECK 9 Solve $3x^2 + 6x + 1 = 0$ by completing the square.

ACCENT ON TECHNOLOGY

Checking Solutions of Quadratic Equations

We can use a graphing calculator to check the solutions of the equation $2x^2 + 4x + 1 = 0$ found in Example 9. Using a TI-84 Plus calculator, we first find the decimal value of $-1 + \frac{\sqrt{2}}{2}$ by pressing these keys:

(−1 + 2ND √ 2) ÷ 2) ENTER

We will obtain the screen shown in Figure 8-2(a). We can now store this decimal value in the calculator by pressing these keys:

STO X, T, θ, n ENTER   X, T, θ, n is one key.

Finally, we enter $2x^2 + 4x + 1$ by pressing

2 X, T, θ, n ∧ 2 + 4 X, T, θ, n + 1

After pressing ENTER one more time, we will obtain the screen shown in Figure 8-2(b). The 0 on the screen confirms that $-1 + \frac{\sqrt{2}}{2}$ satisfies the equation $2x^2 + 4x + 1 = 0$ and is a solution.

We can check the other solution in a similar way.

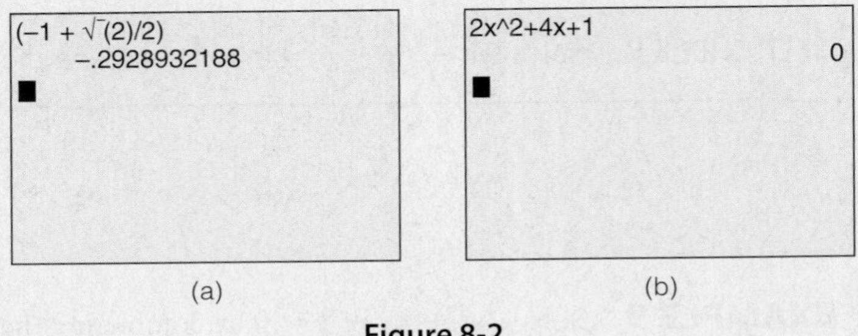

(a) (b)

Figure 8-2

In the next example, the solutions are complex numbers.

EXAMPLE 10 Solve $3x^2 + 2x + 2 = 0$ by completing the square.

Solution

$$3x^2 + 2x + 2 = 0$$

$$x^2 + \frac{2}{3}x + \frac{2}{3} = \frac{0}{3}$$ Divide both sides by 3 to make the coefficient of x^2 equal to 1.

$$x^2 + \frac{2}{3}x = -\frac{2}{3}$$ Subtract $\frac{2}{3}$ from both sides.

$$x^2 + \frac{2}{3}x + \mathbf{\frac{1}{9}} = -\frac{2}{3} + \mathbf{\frac{1}{9}}$$ Square half the coefficient of x and add it to both sides.

$$\left(x + \frac{1}{3}\right)^2 = -\frac{5}{9}$$ Factor on one side and combine terms on the other: $\frac{1}{9} - \frac{2}{3} = \frac{1}{9} - \frac{6}{9} = -\frac{5}{9}$.

$$x + \frac{1}{3} = \sqrt{-\frac{5}{9}} \quad \text{or} \quad x + \frac{1}{3} = -\sqrt{-\frac{5}{9}}$$ Use the square-root property.

$$x + \frac{1}{3} = \sqrt{\frac{5}{9}}\sqrt{-1} \quad \Big| \quad x + \frac{1}{3} = -\sqrt{\frac{5}{9}}\sqrt{-1}$$ $\sqrt{-\frac{5}{9}} = \sqrt{\frac{5}{9}(-1)} = \sqrt{\frac{5}{9}}\sqrt{-1}$

$$x + \frac{1}{3} = \frac{\sqrt{5}}{3}i \qquad\qquad x + \frac{1}{3} = -\frac{\sqrt{5}}{3}i \qquad \sqrt{\frac{5}{9}} = \frac{\sqrt{5}}{\sqrt{9}} = \frac{\sqrt{5}}{3}$$

$$x = -\frac{1}{3} + \frac{\sqrt{5}}{3}i \qquad\qquad x = -\frac{1}{3} - \frac{\sqrt{5}}{3}i \qquad \text{Subtract } \tfrac{1}{3} \text{ from both sides.}$$

These solutions can be written as $x = -\frac{1}{3} \pm \frac{\sqrt{5}}{3}i$.

SELF CHECK 10 Solve $x^2 + 4x + 6 = 0$ by completing the square.

4 Solve an application problem requiring the use of the square-root property.

Many application problems involving equations containing squared terms can be solved using the square-root property.

EXAMPLE 11 **DVDs** A DVD used for recording movies has a surface area of 17.72 square inches on one side. Find the radius of a disc.

Solution The formula for the area of a circular disc is $A = \pi r^2$. We can find the radius of a disc by substituting 17.72 for A and solving for r.

$$A = \pi r^2$$

$$\mathbf{17.72} = \pi r^2 \qquad \text{Substitute 17.72 for } A.$$

$$\frac{17.72}{\pi} = r^2 \qquad \text{Divide both sides by } \pi.$$

$$r = \sqrt{\frac{17.72}{\pi}} \quad \text{or} \quad r = -\sqrt{\frac{17.72}{\pi}} \qquad \text{Use the square-root property.}$$

Since the radius of a disc cannot be negative, we will discard the negative result. Thus, the radius of a disc is $\sqrt{\frac{17.72}{\pi}}$ inches or, to the nearest hundredth, 2.37 inches.

When you deposit money in a bank account, it earns interest. If you leave the money in the account, the earned interest is deposited back into the account and also earns interest. When this is the case, the account is earning *compound interest.* There is a formula we can use to compute the amount in an account at any time t.

Formula for Compound Interest

If P dollars is deposited in an account and interest is paid once a year at an annual rate r, the amount A in the account after t years is given by the formula

$$A = P(1 + r)^t$$

EXAMPLE 12 **SAVING MONEY** A woman invests \$10,000 in an account. Find the annual interest rate if the account will be worth \$11,025 in 2 years.

Solution We substitute 11,025 for A, 10,000 for P, and 2 for t in the compound interest formula and solve for r.

$$A = P(1 + r)^t$$

$$\mathbf{11{,}025} = \mathbf{10{,}000}(1 + r)^2 \quad \text{Substitute.}$$

$$\frac{11{,}025}{10{,}000} = (1 + r)^2 \quad \text{Divide both sides by 10,000.}$$

$$1.1025 = (1 + r)^2 \quad \frac{11{,}025}{10{,}000} = 1.1025$$

$$1 + r = 1.05 \quad \text{or} \quad 1 + r = -1.05 \quad \text{Use the square-root property: } \sqrt{1.1025} = 1.05.$$

$$r = 0.05 \quad \Big| \quad r = -2.05 \quad \text{Subtract 1 from both sides.}$$

Since an interest rate cannot be negative, we must discard the result of -2.05. Thus, the annual interest rate is 0.05, or 5%.

We can check this result by substituting 0.05 for r, 10,000 for P, and 2 for t in the formula and confirming that the deposit of \$10,000 will grow to \$11,025 in 2 years.

$$A = P(1 + r)^t = \mathbf{10{,}000}(1 + \mathbf{0.05})^2 = 10{,}000(1.1025) = 11{,}025$$

SELF CHECK ANSWERS **1.** 8, −8 **2.** $\frac{1}{3}, \frac{1}{2}$ **3.** $\pm 3\sqrt{2}$ **4.** 1, −5 **5.** $\pm 3i$ **6.** $\frac{25}{4}$ **7.** −1, −4 **8.** $-\frac{2}{3}, \frac{3}{2}$ **9.** $-1 \pm \frac{\sqrt{6}}{3}$ **10.** $-2 \pm i\sqrt{2}$

NOW TRY THIS

1. Solve using the square-root property.
   a. $(3x + 5)^2 = 18$
   b. $(x + 6)^2 = 0$
2. Solve $x^2 - 2\sqrt{2}x + 1 = 0$ by completing the square.

8.1 EXERCISES

WARM-UPS

Solve each equation.

1. $x^2 = 49$ **2.** $x^2 = 10$

Find the number that when added to the binomial will make it a perfect-square trinomial.

3. $x^2 + 4x$ **4.** $x^2 - 6x$

5. $x^2 - 3x$ **6.** $x^2 + 5x$

REVIEW *Solve each equation or inequality.*

7. $\frac{t+9}{2} + \frac{t+2}{5} = \frac{8}{5} + 4t$ **8.** $\frac{1-5x}{2x} + 4 = \frac{x+3}{x}$

9. $3(t - 3) + 3t \le 2(t + 1) + t + 1$

10. $-2(y + 4) - 3y + 8 \ge 3(2y - 3) - y$

VOCABULARY AND CONCEPTS *Fill in the blanks.*

11. The square-root property states that the solutions of $x^2 = c$ are ________ and ________.

12. To complete the square on x in $x^2 + 6x = 17$, find one-half of __, square it to get __, and add __ to both sides of the equation.

13. The symbol ± is read as ________________.

14. The formula for annual compound interest is ____________.

GUIDED PRACTICE

Use factoring to solve each equation. **See Examples 1–2. (Objective 1)**

15. $2y^2 - 50 = 0$
16. $4y^2 - 64 = 0$
17. $6x^2 + 12x = 0$
18. $5x^2 + 11x = 0$
19. $r^2 + 6r + 8 = 0$
20. $x^2 + 9x + 20 = 0$
21. $6s^2 + 11s - 10 = 0$
22. $3x^2 + 10x - 8 = 0$

Use the square-root property to solve each equation. **See Example 3. (Objective 2)**

23. $x^2 = 36$
24. $x^2 = 144$
25. $z^2 = 5$
26. $u^2 = 24$

Use the square-root property to solve each equation. **See Example 4. (Objective 2)**

27. $(y + 1)^2 = 1$
28. $(y - 1)^2 = 4$
29. $(x - 2)^2 - 5 = 0$
30. $(x - 5)^2 - 11 = 0$

Use the square-root property to solve each equation. **See Example 5. (Objective 2)**

31. $p^2 + 16 = 0$
32. $q^2 + 25 = 0$
33. $4m^2 + 81 = 0$
34. $9n^2 + 121 = 0$

Use completing the square to solve each equation. **See Examples 6–7. (Objective 3)**

35. $x^2 + 2x - 8 = 0$
36. $x^2 + 6x + 5 = 0$
37. $x^2 - 6x + 8 = 0$
38. $x^2 + 8x + 15 = 0$
39. $x^2 + 5x + 4 = 0$
40. $x^2 - 11x + 30 = 0$
41. $x^2 - 9x - 10 = 0$
42. $x^2 - 3x + 2 = 0$

Use completing the square to solve each equation. **See Example 8. (Objective 3)**

43. $6x^2 + 11x + 3 = 0$
44. $6x^2 + x - 2 = 0$
45. $6x^2 - 7x - 5 = 0$
46. $4x^2 - x - 3 = 0$
47. $9 - 6r = 8r^2$
48. $11m - 10 = 3m^2$
49. $x + 1 = 2x^2$
50. $-2 = 2x^2 - 5x$

Use completing the square to solve each equation. **See Examples 9–10. (Objective 3)**

51. $\frac{7x + 1}{5} = -x^2$
52. $\frac{3x^2}{8} = \frac{1}{8} - x$
53. $p^2 + 2p + 2 = 0$
54. $x^2 - 6x + 10 = 0$
55. $y^2 + 8y + 18 = 0$
56. $t^2 + t + 3 = 0$
57. $3m^2 - 2m + 3 = 0$
58. $4p^2 + 2p + 3 = 0$

ADDITIONAL PRACTICE

Solve using any method.

59. $7x - 6 = x^2$
60. $5t - 6 = t^2$
61. $3x^2 - 16 = 0$
62. $5x^2 - 49 = 0$
63. $(s - 7)^2 - 9 = 0$
64. $(t + 4)^2 = 16$
65. $(x + 5)^2 - 3 = 0$
66. $(x + 3)^2 - 7 = 0$
67. $2z^2 - 5z + 2 = 0$
68. $2x^2 - x - 1 = 0$
69. $3x^2 - 6x + 1 = 0$
70. $3x^2 + 9x + 5 = 0$
71. $2x^2 - x + 8 = 0$
72. $4x^2 + 2x + 5 = 0$

Solve for the indicated variable. Assume that all variables represent positive numbers. Express all radicals in simplified form. **See Example 11.**

73. $2d^2 = 3h$ for d
74. $2x^2 = d^2$ for d
75. $E = mc^2$ for c
76. $S = \frac{1}{2}gt^2$ for t

Find all values of x that will make f(x) = 0.

77. $f(x) = 2x^2 + x - 5$
78. $f(x) = 3x^2 - 2x - 4$
79. $f(x) = x^2 + x - 3$
80. $f(x) = x^2 + 2x - 4$

APPLICATIONS ***Solve each application problem.***

See Examples 11–12. (Objective 4)

81. Falling objects The distance s (in feet) that an object will fall in t seconds is given by the formula $s = 16t^2$. How long will it take an object to fall 256 feet?

82. Pendulums The time (in seconds) it takes a pendulum to swing back and forth to complete one cycle is related to its length l (in feet) by the formula:

$$l = \frac{32t^2}{4\pi^2}$$

How long will it take a 5-foot pendulum to swing through one cycle? Give the result to the nearest hundredth.

83. Law enforcement To estimate the speed s (in mph) of a car involved in an accident, police often use the formula $s^2 = 10.5l$, where l is the length of any skid mark. Approximately how fast was a car going that was involved in an accident and left skid marks of 495 feet?

84. Medicine The approximate pulse rate (in beats per minute) of an adult who is t inches tall is given by the formula

$$p^2 = \frac{348{,}100}{t}$$

Find the pulse rate of an adult who is 64 inches tall.

85. Saving money A student invests \$8,500 in a savings account drawing interest that is compounded annually. Find the annual rate if the money grows to \$9,193.60 in 2 years.

86. Saving money A woman invests \$12,500 in a savings account drawing interest that is compounded annually. Find the annual rate if the money grows to \$14,045 in 2 years.

87. Flags In 1912, an order by President Taft fixed the width and length of the U.S. flag in the ratio 1 to 1.9. If 100 square feet of cloth are to be used to make a U.S. flag, estimate its dimensions to the nearest $\frac{1}{4}$ foot.

88. Accidents The height h (in feet) of an object that is dropped from a height of s feet is given by the formula $h = s - 16t^2$, where t is the time the object has been falling. A 5-foot-tall woman on a sidewalk looks directly overhead and sees a window washer drop a bottle from 4 stories up. How long does she have to get out of the way? Round to the nearest tenth. (A story is 10 feet.)

WRITING ABOUT MATH

89. Explain how to complete the square.

90. Explain why a cannot be 0 in the quadratic equation $ax^2 + bx + c = 0$.

SOMETHING TO THINK ABOUT

91. What number must be added to $x^2 + \sqrt{3}x$ to make it a perfect-square trinomial?

92. Solve $x^2 + \sqrt{3}x - \frac{1}{4} = 0$ by completing the square.

SECTION 8.2 Solving Quadratic Equations by the Quadratic Formula

Objectives

1. Solve a quadratic equation using the quadratic formula.
2. Solve a formula for a specified variable using the quadratic formula.
3. Solve an application problem involving a quadratic equation.

Vocabulary

quadratic formula

Getting Ready

Add a number to each binomial to complete the square. Then write the resulting trinomial as the square of a binomial.

1. $x^2 + 12x$

2. $x^2 - 7x$

Evaluate $\sqrt{b^2 - 4ac}$ *for the following values.*

3. $a = 6, b = 1, c = -2$

4. $a = 4, b = -4, c = -3$

Solving quadratic equations by completing the square is often tedious. Fortunately, there is an easier way. In this section, we will develop a formula, called the **quadratic formula,** that we can use to solve quadratic equations with a minimum of effort. To develop this formula, we will use the skills we learned in the last section and complete the square.

1 Solve a quadratic equation using the quadratic formula.

To develop a formula to solve quadratic equations, we will solve the general quadratic equation $ax^2 + bx + c = 0 \quad (a \neq 0)$ by completing the square.

$$ax^2 + bx + c = 0$$

$$\frac{ax^2}{a} + \frac{bx}{a} + \frac{c}{a} = \frac{0}{a}$$ To make the coefficient of x^2 equal to 1, we divide both sides by a.

$$x^2 + \frac{bx}{a} = -\frac{c}{a}$$ $\frac{0}{a} = 0$; subtract $\frac{c}{a}$ from both sides.

$$x^2 + \frac{bx}{a} + \left(\frac{b}{2a}\right)^2 = \left(\frac{b}{2a}\right)^2 - \frac{c}{a}$$ Complete the square on x by adding $\left(\frac{b}{2a}\right)^2$ to both sides.

$$x^2 + \frac{b}{a}x + \frac{b^2}{4a^2} = \frac{b^2}{4a^2} - \frac{4ac}{4aa}$$ Remove parentheses and get a common denominator on the right side.

(1) $$\left(x + \frac{b}{2a}\right)^2 = \frac{b^2 - 4ac}{4a^2}$$ Factor the left side and add the fractions on the right side.

We can solve Equation 1 using the square-root property.

$$x + \frac{b}{2a} = \sqrt{\frac{b^2 - 4ac}{4a^2}} \quad \text{or} \quad x + \frac{b}{2a} = -\sqrt{\frac{b^2 - 4ac}{4a^2}}$$

$$x + \frac{b}{2a} = \frac{\sqrt{b^2 - 4ac}}{2a} \qquad x + \frac{b}{2a} = -\frac{\sqrt{b^2 - 4ac}}{2a}$$

$$x = -\frac{b}{2a} + \frac{\sqrt{b^2 - 4ac}}{2a} \qquad x = -\frac{b}{2a} - \frac{\sqrt{b^2 - 4ac}}{2a}$$

$$= \frac{-b + \sqrt{b^2 - 4ac}}{2a} \qquad = \frac{-b - \sqrt{b^2 - 4ac}}{2a}$$

These two solutions give the *quadratic formula.*

The Quadratic Formula

The solutions of $ax^2 + bx + c = 0 \quad (a \neq 0)$ are given by the formula

$$x = \frac{-b \pm \sqrt{b^2 - 4ac}}{2a}$$

COMMENT Be sure to draw the fraction bar under both parts of the numerator, and be sure to draw the radical sign exactly over $b^2 - 4ac$. Don't write the quadratic formula as

$$x = -b \pm \frac{\sqrt{b^2 - 4ac}}{2a} \quad \text{or as} \quad x = -b \pm \sqrt{\frac{b^2 - 4ac}{2a}}$$

EXAMPLE 1 Use the quadratic formula to solve $2x^2 - 3x - 5 = 0$.

Solution In this equation $a = 2$, $b = -3$, and $c = -5$.

$$x = \frac{-b \pm \sqrt{b^2 - 4ac}}{2a}$$

$$= \frac{-(-3) \pm \sqrt{(-3)^2 - 4(2)(-5)}}{2(2)}$$ Substitute 2 for a, -3 for b, and -5 for c.

$$= \frac{3 \pm \sqrt{9 + 40}}{4}$$ Simplify.

$$= \frac{3 \pm \sqrt{49}}{4}$$ Add.

$$= \frac{3 \pm 7}{4}$$ Simplify the radical.

$$x = \frac{3 + 7}{4} \quad \text{or} \quad x = \frac{3 - 7}{4}$$

$$x = \frac{10}{4} \quad \Big| \quad x = \frac{-4}{4}$$

$$x = \frac{5}{2} \quad \Big| \quad x = -1$$

After checking the results, we see that the solutions are $\frac{5}{2}$ and -1. Note that this equation can be solved by factoring.

SELF CHECK 1 Use the quadratic formula to solve $3x^2 - 5x - 2 = 0$.

EXAMPLE 2 Use the quadratic formula to solve $2x^2 + 1 = -4x$.

Solution We begin by writing the equation in $ax^2 + bx + c = 0$ form (called *standard form*) before identifying a, b, and c.

$$2x^2 + 4x + 1 = 0$$

In this equation, $a = 2$, $b = 4$, and $c = 1$.

$$x = \frac{-b \pm \sqrt{b^2 - 4ac}}{2a}$$

$$= \frac{-4 \pm \sqrt{4^2 - 4(2)(1)}}{2(2)}$$ Substitute 2 for a, 4 for b, and 1 for c.

$$= \frac{-4 \pm \sqrt{16 - 8}}{4}$$ Simplify.

$$= \frac{-4 \pm \sqrt{8}}{4}$$ Subtract.

$$= \frac{-4 \pm 2\sqrt{2}}{4}$$ $\sqrt{8} = \sqrt{4 \cdot 2} = \sqrt{4}\sqrt{2} = 2\sqrt{2}$

$$= \frac{-2 \pm \sqrt{2}}{2}$$ $\frac{-4 \pm 2\sqrt{2}}{4} = \frac{2(-2 \pm \sqrt{2})}{4} = \frac{-2 \pm \sqrt{2}}{2}$

Note that these solutions can be written as $-1 \pm \frac{\sqrt{2}}{2}$.

SELF CHECK 2 Use the quadratic formula to solve $3x^2 - 2x - 3 = 0$.

In the next example, the solutions are complex numbers.

EXAMPLE 3 Use the quadratic formula to solve $x^2 + x = -1$.

Solution We begin by writing the equation in standard form before identifying a, b, and c.

$$x^2 + x + 1 = 0$$

In this equation, $a = 1$, $b = 1$, and $c = 1$:

$$x = \frac{-b \pm \sqrt{b^2 - 4ac}}{2a}$$

$$= \frac{-1 \pm \sqrt{1^2 - 4(1)(1)}}{2(1)}$$ Substitute 1 for a, 1 for b, and 1 for c.

$$= \frac{-1 \pm \sqrt{1 - 4}}{2}$$ Simplify the expression under the radical.

PERSPECTIVE The Fibonacci Sequence and the Golden Ratio

Perhaps one of the most intriguing examples of how a mathematical idea can represent natural phenomena is the *Fibonacci Sequence,* a list of whole numbers that is generated by a very simple rule. This sequence was first developed by the Italian mathematician Leonardo da Pisa, more commonly known as Fibonacci. The Fibonacci Sequence is the following list of numbers

1, 1, 2, 3, 5, 8, 13, 21, . . .

where each successive number in the list is obtained by adding the two preceding numbers. Although Fibonacci originally developed this sequence to solve a mathematical puzzle, subsequent study of the numbers in this sequence has uncovered many examples in the natural world in which this sequence emerges. For example, the arrangement of the seeds on the face of a sunflower, the hibernation periods of certain insects, and the branching patterns of many plants all give rise to Fibonacci numbers.

Among the many special properties of these numbers is the fact that, as we generate more and more numbers in the list, the ratio of successive numbers approaches a constant value. This value is designated by the symbol ϕ and often is referred to as the "Golden Ratio." One way to calculate the value of ϕ is to solve the quadratic equation $\phi^2 - \phi - 1 = 0$.

1. Using the quadratic formula, find the exact value of ϕ.

2. Using a calculator, find a decimal approximation of ϕ, correct to three decimal places.

$$= \frac{-1 \pm \sqrt{-3}}{2}$$ Subtract.

$$= \frac{-1 \pm \sqrt{3}i}{2}$$ Simplify the radical expression.

Note that these solutions can be written as $-\frac{1}{2} \pm \frac{\sqrt{3}}{2}i$.

SELF CHECK 3 Use the quadratic formula to solve $a^2 + 2a + 3 = 0$.

2 Solve a formula for a specified variable using the quadratic formula.

EXAMPLE 4 An object thrown straight up with an initial velocity of v_0 feet per second will reach a height of s feet in t seconds according to the formula $s = -16t^2 + v_0t$. Solve the formula for t.

Solution We begin by writing the equation in standard form:

$$s = -16t^2 + v_0t$$

$$16t^2 - v_0t + s = 0$$

In this equation, $a = 6$, $b = -v_0$, and $c = s$. Then we can use the quadratic formula to solve for t.

$$t = \frac{-b \pm \sqrt{b^2 - 4ac}}{2a}$$

$$t = \frac{-(-v_0) \pm \sqrt{(-v_0)^2 - 4(16)(s)}}{2(16)}$$ Substitute into the quadratic formula.

$$t = \frac{v_0 \pm \sqrt{v_0{}^2 - 64s}}{32}$$ Simplify.

Thus, $t = \frac{v_0 \pm \sqrt{v_0{}^2 - 64s}}{32}$.

3 Solve an application problem involving a quadratic equation.

EXAMPLE 5 **DIMENSIONS OF A RECTANGLE** Find the dimensions of the rectangle shown in Figure 8-3, given that its area is 253 cm^2.

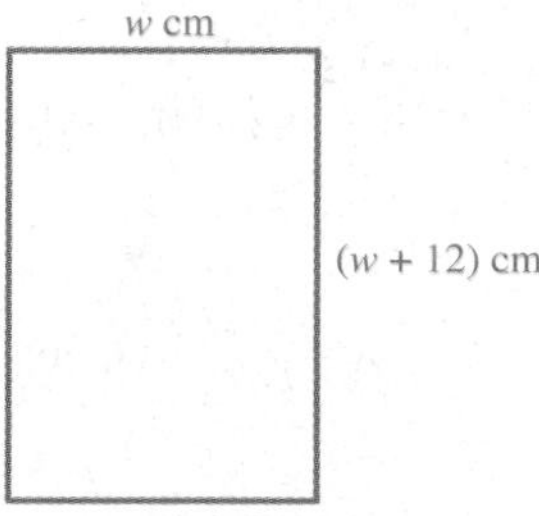

Figure 8-3

Solution If we let w represent the width of the rectangle, then $w + 12$ represents its length. Since the area of the rectangle is 253 square centimeters, we can form the equation

$w(w + 12) = 253$ Area of a rectangle = width · length.

and solve it as follows:

$$w(w + 12) = 253$$

$$w^2 + 12w = 253 \quad \text{Use the distributive property to remove parentheses.}$$

$$w^2 + 12w - 253 = 0 \quad \text{Subtract 253 from both sides.}$$

Solution by factoring

$$(w - 11)(w + 23) = 0$$

$$w - 11 = 0 \quad \text{or} \quad w + 23 = 0$$

$$w = 11 \quad\quad w = -23$$

Solution by formula

$$w = \frac{-12 \pm \sqrt{12^2 - 4(1)(-253)}}{2(1)}$$

$$= \frac{-12 \pm \sqrt{144 + 1{,}012}}{2}$$

$$= \frac{-12 \pm \sqrt{1{,}156}}{2}$$

$$= \frac{-12 \pm 34}{2}$$

$$w = 11 \quad \text{or} \quad w = -23$$

Since the rectangle cannot have a negative width, we discard the solution of -23. Thus, the only solution is $w = 11$. Since the rectangle is 11 centimeters wide and $(11 + 12)$ centimeters long, its dimensions are 11 centimeters by 23 centimeters.

Check: 23 is 12 more than 11, and the area of a rectangle with dimensions of 23 centimeters by 11 centimeters is 253 square centimeters.

SELF CHECK ANSWERS

1. $2, -\frac{1}{3}$ **2.** $\frac{1}{3} \pm \frac{\sqrt{10}}{3}$ **3.** $-1 \pm i\sqrt{2}$

NOW TRY THIS

1. The length of a rectangular garden is 1 ft less than 3 times the width. If the area is 44 sq ft, find the length of the garden.

2. The product of 2 consecutive integers is 90. Find the numbers. (Consecutive means one after another.)

3. Solve $x^3 - 8 = 0$. (*Hint:* Recall how to factor the difference of cubes.)

4. Graph $y = x^3 - 8$ and identify the x-intercept(s).

8.2 EXERCISES

WARM-UPS *Identify a, b, and c in each quadratic equation.*

1. $3x^2 - 4x + 7 = 0$

2. $-2x^2 + x = 5$

REVIEW

Solve for the indicated variable.

3. $Ax + By = C$ for y

4. $R = \dfrac{kL}{d^2}$ for L

Simplify each radical.

5. $\sqrt{24}$

6. $\sqrt{288}$

7. $\dfrac{3}{\sqrt{3}}$

8. $\dfrac{1}{2 - \sqrt{3}}$

VOCABULARY AND CONCEPTS *Fill in the blanks.*

9. In the quadratic equation $3x^2 - 2x + 6 = 0$, a = __, b = ____, and c = __.

10. The solutions of $ax^2 + bx + c = 0$ $(a \neq 0)$ are given by the quadratic formula, which is x = ____________.

GUIDED PRACTICE

Solve each equation using the quadratic formula. See Example 1. (Objective 1)

11. $x^2 + 3x + 2 = 0$

12. $x^2 - 3x + 2 = 0$

13. $x^2 - 2x - 15 = 0$

14. $x^2 - 2x - 35 = 0$

15. $x^2 + 12x = -36$

16. $y^2 - 18y = -81$

17. $2x^2 - x - 3 = 0$

18. $3x^2 - 10x + 8 = 0$

Solve each equation using the quadratic formula. See Example 2. (Objective 1)

19. $15x^2 - 14x = 8$

20. $4x^2 = -5x + 6$

21. $8u = -4u^2 - 3$

22. $4t + 3 = 4t^2$

23. $16y^2 + 8y - 3 = 0$

24. $16x^2 + 16x + 3 = 0$

25. $5x^2 + 5x + 1 = 0$

26. $4w^2 + 6w + 1 = 0$

Solve each equation using the quadratic formula. See Example 3. (Objective 1)

27. $x^2 + 2x + 2 = 0$

28. $x^2 + 3x + 3 = 0$

29. $2x^2 + x + 1 = 0$

30. $3x^2 + 2x + 1 = 0$

31. $3x^2 - 4x = -2$

32. $2x^2 + 3x = -3$

33. $3x^2 - 2x = -3$

34. $5x^2 = 2x - 1$

Solve each formula for the indicated variable. See Example 4. (Objective 2)

35. $C = \dfrac{N^2 - N}{2}$, for n

(The formula for a selection sort in data processing)

36. $A = 2\pi r^2 + 2\pi hr$, for r

(The formula for the surface area of a right circular-cylinder)

37. $x^2 - kx = -ay$ for x

38. $xy^2 + 3xy + 7 = 0$ for y

ADDITIONAL PRACTICE

Solve each equation using any method.

39. $6x^2 - x - 1 = 0$

40. $2x^2 + 5x - 3 = 0$

41. $\dfrac{x^2}{2} + \dfrac{5}{2}x = -1$

42. $-3x = \dfrac{x^2}{2} + 2$

43. $2x^2 - 1 = 3x$

44. $-9x = 2 - 3x^2$

Find all x-values that will make f(x) = 0.

45. $f(x) = 4x^2 + 4x - 19$

46. $f(x) = 9x^2 + 12x - 8$

47. $f(x) = 3x^2 + 2x + 2$

48. $f(x) = 4x^2 + x + 1$

Use the quadratic formula and a calculator to solve each equation. Give all answers to the nearest hundredth.

49. $0.7x^2 - 3.5x - 25 = 0$

50. $-4.5x^2 + 0.2x + 3.75 = 0$

Note that a and b are the solutions to the equation $(x - a)(x - b) = 0$.

51. Find a quadratic equation that has a solution set of $\{3, 5\}$.

52. Find a quadratic equation that has a solution set of $\{-4, 6\}$.

53. Find a third-degree equation that has a solution set of $\{2, 3, -4\}$.

54. Find a fourth-degree equation that has a solution set of $\{3, -3, 4, -4\}$.

APPLICATIONS

Solve each problem. See Example 5. (Objective 3)

55. **Dimensions of a rectangle** The rectangle has an area of 96 square feet. Find its dimensions.

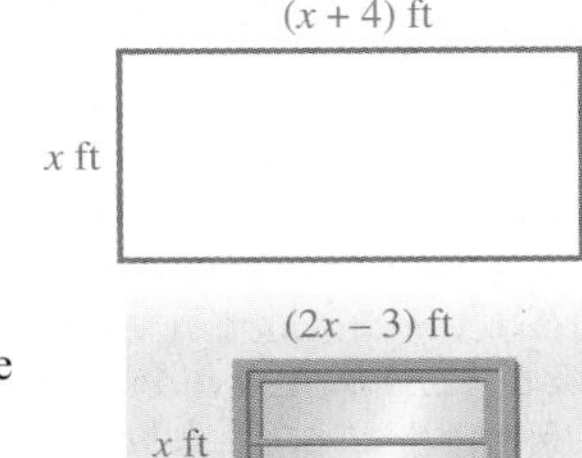

56. **Dimensions of a window** The area of the window is 77 square feet. Find its dimensions.

(2x – 3) ft

x ft

57. **Side of a square** The area of a square is numerically equal to its perimeter. Find the length of each side of the square.

58. **Perimeter of a rectangle** A rectangle is 2 inches longer than it is wide. Numerically, its area exceeds its perimeter by 11. Find the perimeter.

Solve each problem.

59. **Base of a triangle** The height of a triangle is 5 centimeters longer than three times its base. Find the base of the triangle if its area is 6 square centimeters.

60. **Height of a triangle** The height of a triangle is 4 meters longer than twice its base. Find the height if the area of the triangle is 15 square meters.

61. **Integer problem** The product of two consecutive even integers is 288. Find the integers. (*Hint:* If one even integer is x, the next consecutive even integer is $x + 2$.)

62. **Integer problem** The product of two consecutive odd integers is 143. Find the integers. (*Hint:* If one odd integer is x, the next consecutive odd integer is $x + 2$.)

63. **Integer problem** The sum of the squares of two consecutive integers is 85. Find the integers. (*Hint:* If one integer is x, the next consecutive positive integer is $x + 1$.)

64. **Integer problem** The sum of the squares of three consecutive integers is 77. Find the integers. (*Hint:* If one integer is x, the next consecutive positive integer is $x + 1$, and the third is $x + 2$.)

65. **Finding rates** A woman drives her snowmobile 150 miles at the rate of r mph. She could have gone the same distance in 2 hours less time if she had increased her speed by 20 mph. Find r.

66. **Finding rates** Jeff bicycles 160 miles at the rate of r mph. The same trip would have taken 2 hours longer if he had decreased his speed by 4 mph. Find r.

67. **Pricing concert tickets** Tickets to a concert cost \$4, and the projected attendance is 300 people. It is further projected that for every 10¢ increase in ticket price, the average attendance will decrease by 5. At what ticket price will the nightly receipts be \$1,248?

68. **Setting bus fares** A bus company has 3,000 passengers daily, paying a \$1.25 fare. For each 25¢ increase in fare, the company estimates that it will lose 80 passengers. What is the smallest increase in fare that will produce a \$4,970 daily revenue?

69. **Computing profit** The *Gazette's* profit is \$20 per year for each of its 3,000 subscribers. Management estimates that the profit per subscriber will increase by 1¢ for each additional subscriber over the current 3,000. How many subscribers will bring a total profit of \$120,000?

70. **Finding interest rates** A woman invests \$1,000 in a mutual fund for which interest is compounded annually at a rate r. After one year, she deposits an additional \$2,000. After two years, the balance A in the account is

$$A = \$1{,}000(1 + r)^2 + \$2{,}000(1 + r)$$

If this amount is \$3,368.10, find r.

71. **Framing a picture** The frame around the picture in the illustration has a constant width. To the nearest hundredth, how wide is the frame if its area equals the area of the picture?

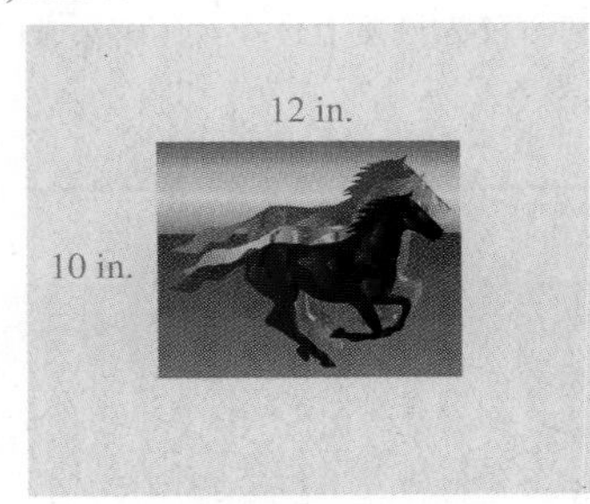

72. **Metal fabrication** A box with no top is to be made by cutting a 2-inch square from each corner of the square sheet of metal shown in the illustration. After bending up the sides, the volume of the box is to be 200 cubic inches. How large should the piece of metal be?

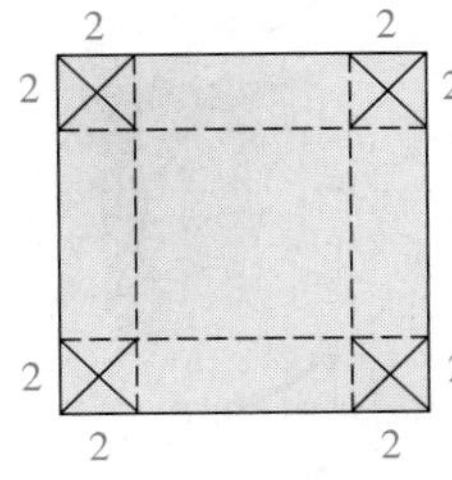

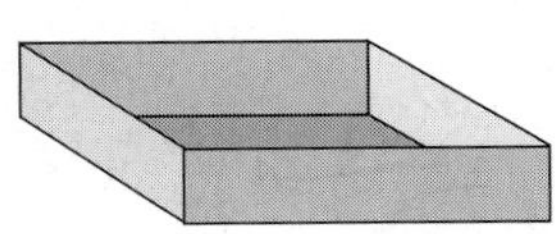

Use a calculator.

73. Labor force The labor force participation rate P (in percent) for workers ages 16 and older from 1966 to 2008 is approximated by the quadratic equation

$$P = -0.0072x^2 + 0.4904x + 58.2714$$

where $x = 0$ corresponds to the year 1966, $x = 1$ corresponds to 1967, and so on. (Thus, $0 \leq x \leq 42$.) In what year in this range were 65% of the workers ages 16 and older part of the workforce?

74. Space program The yearly budget B (in billions of dollars) for the National Aeronautics and Space Administration (NASA) is approximated by the quadratic equation

$$B = 0.0518x^2 - 0.2122x + 14.1112$$

where x is the number of years since 1997 and $0 \leq x \leq 12$. In what year does the model indicate that NASA's budget was about \$17 billion?

75. Chemistry A weak acid (0.1 M concentration) breaks down into free cations (the hydrogen ion, H^+) and anions (A^-). When this acid dissociates, the following equilibrium equation is established:

$$\frac{[H^+][A^-]}{[HA]} = 4 \times 10^{-4}$$

where $[H^+]$, the hydrogen ion concentration, is equal to $[A^-]$, the anion concentration. $[HA]$ is the concentration of the undissociated acid itself. Find $[H^+]$ at equilibrium. (*Hint:* If $[H^+] = x$, then $[HA] = 0.1 - x$.)

76. Chemistry A saturated solution of hydrogen sulfide (0.1 M concentration) dissociates into cation $[H^+]$ and anion $[HS^-]$, where $[H^+] = [HS^-]$. When this solution dissociates, the following equilibrium equation is established:

$$\frac{[H^+][HS^-]}{[HHS]} = 1.0 \times 10^{-7}$$

Find $[H^+]$. (*Hint:* If $[H^+] = x$, then $[HHS] = 0.1 - x$.)

WRITING ABOUT MATH

77. Explain why $x = -b \pm \frac{\sqrt{b^2 - 4ac}}{2a}$ is not a correct statement of the quadratic formula.

78. Explain why $x = \frac{b \pm \sqrt{b^2 - 4ac}}{2a}$ is not a correct statement of the quadratic formula.

SOMETHING TO THINK ABOUT ***All of the equations we have solved so far have had rational-number coefficients. However, the quadratic formula can be used to solve quadratic equations with irrational or even imaginary coefficients. Try solving each of the following equations.***

79. $x^2 + 2\sqrt{2}x - 6 = 0$

80. $\sqrt{2}x^2 + x - \sqrt{2} = 0$

81. $x^2 - 3ix - 2 = 0$

82. $ix^2 + 3x - 2i = 0$

SECTION 8.3 The Discriminant and Equations That Can Be Written in Quadratic Form

Objectives

1. Use the discriminant to determine the type of solutions to a given quadratic equation.
2. Solve an equation that can be written in quadratic form.
3. Verify the solutions of a quadratic equation by showing that the sum of the solutions is $-\frac{b}{a}$ and the product is $\frac{c}{a}$.

Vocabulary

discriminant

Getting Ready

Evaluate $b^2 - 4ac$ for the following values.

1. $a = 2, b = 3$, and $c = -1$

2. $a = -2, b = 4$, and $c = -3$

We can use part of the quadratic formula to predict the type of solutions, if any, that a quadratic equation will have. We don't even have to solve the equation.

1 Use the discriminant to determine the type of solutions to a given quadratic equation.

Suppose that the coefficients a, b, and c in the equation $ax^2 + bx + c = 0 \quad (a \neq 0)$ are real numbers. Then the solutions of the equation are given by the quadratic formula

$$x = \frac{-b \pm \sqrt{b^2 - 4ac}}{2a} \quad (a \neq 0)$$

If $b^2 - 4ac \geq 0$, the solutions are real numbers. If $b^2 - 4ac < 0$, the solutions are nonreal complex numbers. Thus, the value of $b^2 - 4ac$, called the **discriminant,** determines the type of solutions for a particular quadratic equation.

The Discriminant

If $ax^2 + bx + c = 0 \quad (a \neq 0)$ and if a, b, and c are real numbers, then

$b^2 - 4ac > 0$, there are two unequal real solutions.
$b^2 - 4ac = 0$, there are two equal real solutions (called a *double root*).
$b^2 - 4ac < 0$, the solutions are complex conjugates.

If a, b, and c are rational numbers and the discriminant

is a perfect square greater than 0, there are two unequal rational solutions.
is positive but not a perfect square, the solutions are irrational and unequal.

EXAMPLE 1 Determine the type of solutions for the equations.
a. $x^2 + x + 1 = 0$ **b.** $3x^2 + 5x + 2 = 0$

Solution **a.** We calculate the discriminant for $x^2 + x + 1 = 0$.

$$\begin{aligned} b^2 - 4ac &= 1^2 - 4(1)(1) \quad && a = 1, b = 1, \text{ and } c = 1. \\ &= -3 \end{aligned}$$

Since $b^2 - 4ac < 0$, the solutions will be complex conjugates.

b. We calculate the discriminant for $3x^2 + 5x + 2 = 0$.

$$\begin{aligned} b^2 - 4ac &= 5^2 - 4(3)(2) \quad && a = 3, b = 5, \text{ and } c = 2. \\ &= 25 - 24 \\ &= 1 \end{aligned}$$

Since $b^2 - 4ac > 0$ and $b^2 - 4ac$ is a perfect square, there will be two unequal rational solutions.

SELF CHECK 1 Determine the type of solutions.

a. $x^2 + x - 1 = 0$

b. $4x^2 - 10x + 25 = 0$

EXAMPLE 2 What value of k will make the solutions of the equation $kx^2 - 12x + 9 = 0$ equal?

Solution We calculate the discriminant:

$$\begin{aligned} \boldsymbol{b^2 - 4ac} &= \boldsymbol{(-12)^2 - 4(k)(9)} && a = k, b = -12, \text{ and } c = 9. \\ &= 144 - 36k \\ &= -36k + 144 \end{aligned}$$

Since the solutions are to be equal, we let $-36k + 144 = 0$ and solve for k.

$$\begin{aligned} -36k + 144 &= 0 \\ -36k &= -144 && \text{Subtract 144 from both sides.} \\ k &= 4 && \text{Divide both sides by } -36. \end{aligned}$$

If $k = 4$, the solutions will be equal. Verify this by solving $4x^2 - 12x + 9 = 0$ and showing that the solutions are equal.

SELF CHECK 2 Find the value of k will make the solutions of $kx^2 - 20x + 25 = 0$ equal.

2 Solve an equation that can be written in quadratic form.

Many equations that are not quadratic can be written in quadratic form ($ax^2 + bx + c = 0$) and then solved using the techniques discussed in previous sections. For example, an inspection of the equation $x^4 - 5x^2 + 4 = 0$ shows that

The leading term x^4 is the square of x^2, the variable part of the middle term: $x^4 = (x^2)^2$

$$x^4 - 5x^2 + 4 = 0$$

The last term is a constant.

To solve the equation $x^4 - 5x^2 + 4 = 0$, we can write the equation in a different form and proceed as follows:

$$\begin{aligned} x^4 - 5x^2 + 4 &= 0 \\ (x^2)^2 - 5(x^2) + 4 &= 0 \end{aligned}$$

If we replace each x^2 with u, we will obtain a quadratic equation with the variable u that we can solve by factoring.

$$\begin{aligned} u^2 - 5u + 4 &= 0 && \text{Let } x^2 = u. \\ (u - 4)(u - 1) &= 0 && \text{Factor } u^2 - 5u + 4. \\ u - 4 = 0 \quad \text{or} \quad u - 1 &= 0 && \text{Set each factor equal to 0.} \\ u = 4 \quad \big| \quad u &= 1 \end{aligned}$$

Since $u = x^2$, it follows that $x^2 = 4$ or $x^2 = 1$. Thus,

$$x^2 = 4 \quad \text{or} \quad x^2 = 1$$
$$x = 2 \quad \text{or} \quad x = -2 \quad | \quad x = 1 \quad \text{or} \quad x = -1$$

This equation has four solutions: 1, -1, 2, and -2. Verify that each one satisfies the original equation. Note that this equation can be solved by factoring.

EXAMPLE 3 Solve: $x - 7\sqrt{x} + 12 = 0$.

Solution We examine the leading term and middle term.

The leading term x is the square of $\sqrt{x}$, the variable part of the middle term: $x - 7\sqrt{x} + 12 = 0$

$x = (\sqrt{x})^2$

If we write x as $(\sqrt{x})^2$, the equation takes the form

$$(\sqrt{x})^2 - 7\sqrt{x} + 12 = 0$$

and it is said to be *quadratic in* $\sqrt{x}$. We can solve this equation by letting $\sqrt{x} = u$ and factoring.

$$u^2 - 7u + 12 = 0 \qquad \text{Replace each } \sqrt{x} \text{ with } u.$$
$$(u - 3)(u - 4) = 0 \qquad \text{Factor } u^2 - 7u + 12.$$
$$u - 3 = 0 \quad \text{or} \quad u - 4 = 0 \qquad \text{Set each factor equal to 0.}$$
$$u = 3 \quad | \quad u = 4$$

To find x, we undo the substitutions by replacing each u with $\sqrt{x}$. Then we solve the radical equations by squaring both sides.

$$\sqrt{x} = 3 \quad \text{or} \quad \sqrt{x} = 4$$
$$x = 9 \quad | \quad x = 16$$

The solutions are 9 and 16. Verify that both satisfy the original equation.

SELF CHECK 3 Solve: $x + \sqrt{x} - 6 = 0$.

EXAMPLE 4 Solve: $2m^{2/3} - 2 = 3m^{1/3}$.

Solution First we subtract $3m^{1/3}$ from both sides to write the equation in the form

$$2m^{2/3} - 3m^{1/3} - 2 = 0$$

Then we write the equation in the form

$$2(m^{1/3})^2 - 3m^{1/3} - 2 = 0 \qquad (m^{1/3})^2 = m^{2/3}$$

If we substitute u for $m^{1/3}$, this equation can be written in a form that can be solved by factoring.

$$2u^2 - 3u - 2 = 0 \qquad \text{Replace each } m^{1/3} \text{ with } u.$$
$$(2u + 1)(u - 2) = 0 \qquad \text{Factor } 2u^2 - 3u - 2.$$

$$2u + 1 = 0 \quad \text{or} \quad u - 2 = 0 \qquad \text{Set each factor equal to 0.}$$
$$u = -\frac{1}{2} \qquad\qquad u = 2$$

To find m, we undo the substitutions by replacing each u with $m^{1/3}$ and solve each resulting equation by cubing both sides.

$$m^{1/3} = -\frac{1}{2} \quad \text{or} \quad m^{1/3} = 2$$
$$(m^{1/3})^3 = \left(-\frac{1}{2}\right)^3 \qquad (m^{1/3})^3 = (2)^3 \qquad \text{Cube both sides.}$$
$$m = -\frac{1}{8} \qquad m = 8 \qquad \text{Simplify.}$$

The solutions are $-\frac{1}{8}$ and 8. Verify that both satisfy the original equation.

SELF CHECK 4 Solve: $a^{2/3} = -3a^{1/3} + 10$.

EXAMPLE 5 Solve: $\dfrac{24}{x} + \dfrac{12}{x+1} = 11$.

Solution Since the denominator cannot be 0, x cannot be 0 or -1. If either 0 or -1 appears as a suspected solution, it is extraneous and must be discarded.

$$\frac{24}{x} + \frac{12}{x+1} = 11$$
$$x(x+1)\left(\frac{24}{x} + \frac{12}{x+1}\right) = x(x+1)11 \qquad \text{Multiply both sides by } x(x+1).$$
$$24(x+1) + 12x = (x^2 + x)11 \qquad \text{Simplify.}$$
$$24x + 24 + 12x = 11x^2 + 11x \qquad \text{Use the distributive property to remove parentheses.}$$
$$36x + 24 = 11x^2 + 11x \qquad \text{Combine like terms.}$$
$$0 = 11x^2 - 25x - 24 \qquad \text{Subtract } 36x \text{ and 24 from both sides.}$$
$$0 = (11x + 8)(x - 3) \qquad \text{Factor } 11x^2 - 25x - 24.$$
$$11x + 8 = 0 \quad \text{or} \quad x - 3 = 0 \qquad \text{Set each factor equal to 0.}$$
$$x = -\frac{8}{11} \qquad\qquad x = 3$$

Verify that $-\frac{8}{11}$ and 3 satisfy the original equation.

SELF CHECK 5 Solve: $\frac{12}{x} + \frac{6}{x+3} = 5$.

EXAMPLE 6 Solve: $15a^{-2} - 8a^{-1} + 1 = 0$.

Solution First we write the equation in the form

$$15(a^{-1})^2 - 8a^{-1} + 1 = 0 \qquad (a^{-1})^2 = a^{-2}$$

If we substitute u for a^{-1}, this equation can be written in a form that can be solved by factoring.

$$
\begin{aligned}
15u^2 - 8u + 1 &= 0 && \text{Replace each } a^{-1} \text{ with } u.\\
(5u - 1)(3u - 1) &= 0 && \text{Factor } 15u^2 - 8u + 1.\\
5u - 1 = 0 \quad \text{or} \quad 3u - 1 &= 0 && \text{Set each factor equal to 0.}\\
u = \frac{1}{5} \quad \Big| \quad u &= \frac{1}{3}
\end{aligned}
$$

To find a, we undo the substitutions by replacing each u with a^{-1} and solve each resulting equation.

$$
\begin{aligned}
a^{-1} &= \frac{1}{5} & \Big| \quad a^{-1} &= \frac{1}{3} &&\\
\frac{1}{a} &= \frac{1}{5} & \Big| \quad \frac{1}{a} &= \frac{1}{3} && a^{-1} = \frac{1}{a}\\
5 &= a & \Big| \quad 3 &= a && \text{Solve the proportions.}
\end{aligned}
$$

The solutions are 5 and 3. Verify that both satisfy the original equation.

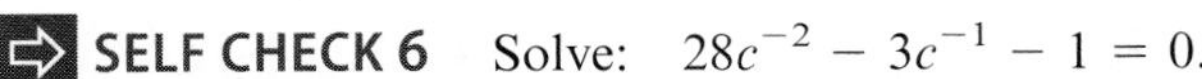

SELF CHECK 6 Solve: $28c^{-2} - 3c^{-1} - 1 = 0$.

EXAMPLE 7 Solve the formula $s = 16t^2 - 32$ for t.

Solution We proceed as follows:

$$
\begin{aligned}
s &= 16t^2 - 32 &&\\
s + 32 &= 16t^2 && \text{Add 32 to both sides.}\\
\frac{s + 32}{16} &= t^2 && \text{Divide both sides by 16.}\\
t^2 &= \frac{s + 32}{16} && \text{Write } t^2 \text{ on the left side.}\\
t &= \pm\sqrt{\frac{s + 32}{16}} && \text{Apply the square-root property.}\\
t &= \pm\frac{\sqrt{s + 32}}{\sqrt{16}} && \sqrt{\tfrac{a}{b}} = \tfrac{\sqrt{a}}{\sqrt{b}}\\
t &= \pm\frac{\sqrt{s + 32}}{4} &&
\end{aligned}
$$

SELF CHECK 7 Solve $a^2 + b^2 = c^2$ for a.

3 Verify the solutions of a quadratic equation by showing that the sum of the solutions is $-\frac{b}{a}$ and the product is $\frac{c}{a}$.

Solutions of a Quadratic Equation

If r_1 and r_2 are the solutions of the quadratic equation $ax^2 + bx + c = 0$, with $a \neq 0$, then

$$r_1 + r_2 = -\frac{b}{a} \quad \text{and} \quad r_1 r_2 = \frac{c}{a}$$

Proof We note that the solutions to the equation are given by the quadratic formula

$$r_1 = \frac{-b + \sqrt{b^2 - 4ac}}{2a} \quad \text{and} \quad r_2 = \frac{-b - \sqrt{b^2 - 4ac}}{2a}$$

Thus,

$$\begin{aligned} r_1 + r_2 &= \frac{-b + \sqrt{b^2 - 4ac}}{2a} + \frac{-b - \sqrt{b^2 - 4ac}}{2a} \\ &= \frac{-b + \sqrt{b^2 - 4ac} - b - \sqrt{b^2 - 4ac}}{2a} && \text{Keep the denominator and add the numerators.} \\ &= -\frac{2b}{2a} \\ &= -\frac{b}{a} \end{aligned}$$

and

$$\begin{aligned} r_1 r_2 &= \frac{-b + \sqrt{b^2 - 4ac}}{2a} \cdot \frac{-b - \sqrt{b^2 - 4ac}}{2a} \\ &= \frac{b^2 - (b^2 - 4ac)}{4a^2} && \text{Multiply the numerators and multiply the denominators.} \\ &= \frac{b^2 - b^2 + 4ac}{4a^2} \\ &= \frac{4ac}{4a^2} && b^2 - b^2 = 0 \\ &= \frac{c}{a} \end{aligned}$$

It can also be shown that if

$$r_1 + r_2 = -\frac{b}{a} \quad \text{and} \quad r_1 r_2 = \frac{c}{a}$$

then r_1 and r_2 are solutions of $ax^2 + bx + c = 0$. We can use this fact to check the solutions of quadratic equations.

EXAMPLE 8 Show that $\frac{3}{2}$ and $-\frac{1}{3}$ are solutions of $6x^2 - 7x - 3 = 0$.

Solution Since $a = 6$, $b = -7$, and $c = -3$, we have

$$-\frac{b}{a} = -\frac{-7}{6} = \frac{7}{6} \quad \text{and} \quad \frac{c}{a} = \frac{-3}{6} = -\frac{1}{2}$$

Since $\frac{3}{2} + \left(-\frac{1}{3}\right) = \frac{7}{6}$ and $\left(\frac{3}{2}\right)\left(-\frac{1}{3}\right) = -\frac{1}{2}$, these numbers are solutions. Solve the equation to verify that the roots are $\frac{3}{2}$ and $-\frac{1}{3}$.

SELF CHECK 8 Are $-\frac{3}{2}$ and $\frac{1}{3}$ solutions of $6x^2 + 7x - 3 = 0$?

SELF CHECK ANSWERS

1. a. irrational and unequal **b.** complex conjugates **2.** 4 **3.** 4 **4.** 8, −125 **5.** 3, $-\frac{12}{5}$ **6.** −7, 4 **7.** $a = \pm\sqrt{c^2 - b^2}$ **8.** yes

NOW TRY THIS

1. Solve: $x - 3\sqrt{x} - 4 = 0$.

2. Without substituting, show that $3 + 5i$ and $3 - 5i$ are solutions of $x^2 - 6x + 34 = 0$.

3. Find the discriminant of $\sqrt{2}x^2 - \sqrt{65}x - 2\sqrt{2} = 0$.

8.3 EXERCISES

WARM-UPS

Find $b^2 - 4ac$ when

1. $a = 1, b = 1, c = 1$

2. $a = 2, b = 1, c = 1$

Determine the type of solutions for

3. $x^2 - 4x + 1 = 0$

4. $8x^2 - x + 2 = 0$

Are the following numbers solutions of $x^2 - 7x + 6 = 0$?

5. 1, 5

6. 1, 6

REVIEW *Solve each equation. Assume no division by 0.*

7. $\frac{1}{4} + \frac{1}{t} = \frac{1}{2t}$

8. $\frac{p-3}{3p} + \frac{1}{2p} = \frac{1}{4}$

9. Find the slope of the line passing through (−2, −4) and (3, 5).

10. Write an equation of the line passing through (−2, −4) and (3, 5) in general form.

VOCABULARY AND CONCEPTS *Consider the equation $ax^2 + bx + c = 0$ $(a \neq 0)$, and fill in the blanks.*

11. The discriminant is ________.

12. If $b^2 - 4ac < 0$, the solutions of the equation are complex ________.

13. If $b^2 - 4ac$ is a nonzero perfect square, the solutions are ________ numbers and ________.

14. If r_1 and r_2 are the solutions of the equation, then $r_1 + r_2 =$ __ and $r_1r_2 =$ __.

GUIDED PRACTICE

Use the discriminant to determine what type of solutions exist for each quadratic equation. Do not solve the equation. See Example 1. (Objective 1)

15. $4x^2 - 4x + 1 = 0$

16. $6x^2 - 5x - 6 = 0$

17. $5x^2 + x + 2 = 0$

18. $3x^2 + 10x - 2 = 0$

19. $2x^2 = 4x - 1$

20. $9x^2 = 12x - 4$

21. $x(2x - 3) = 20$

22. $x(x - 3) = -10$

Find the values of k that will make the solutions of each given quadratic equation equal. See Example 2. (Objective 1)

23. $x^2 + kx + 9 = 0$

24. $kx^2 - 12x + 4 = 0$

25. $9x^2 + 4 = -kx$

26. $9x^2 - kx + 25 = 0$

Solve each equation. (Objective 2)

27. $x^4 - 17x^2 + 16 = 0$

28. $x^4 - 10x^2 + 9 = 0$

29. $x^4 - 3x^2 = -2$

30. $x^4 - 29x^2 = -100$

31. $x^4 = 6x^2 - 5$

32. $x^4 = 8x^2 - 7$

33. $2x^4 - 10x^2 = -8$

34. $3x^4 + 12 = 15x^2$

Solve each equation. **See Example 3. (Objective 2)**

35. $x - 6\sqrt{x} + 8 = 0$

36. $x - 5\sqrt{x} + 4 = 0$

37. $2x - \sqrt{x} = 3$

38. $3x - 4 = -4\sqrt{x}$

39. $2x + x^{1/2} - 3 = 0$

40. $2x - x^{1/2} - 1 = 0$

41. $3x + 5x^{1/2} + 2 = 0$

42. $3x - 4x^{1/2} + 1 = 0$

Solve each equation. **See Example 4. (Objective 2)**

43. $x^{2/3} + 5x^{1/3} + 6 = 0$

44. $x^{2/3} - 7x^{1/3} + 12 = 0$

45. $x^{2/3} - 2x^{1/3} - 3 = 0$

46. $x^{2/3} + 4x^{1/3} - 5 = 0$

Solve each equation. **See Example 5. (Objective 2)**

47. $x + 5 + \frac{4}{x} = 0$

48. $x - 4 + \frac{3}{x} = 0$

49. $x + 1 = \frac{20}{x}$

50. $x + \frac{15}{x} = 8$

51. $\frac{1}{x-1} + \frac{3}{x+1} = 2$

52. $\frac{6}{x-2} - \frac{12}{x-1} = -1$

53. $\frac{1}{x+2} + \frac{24}{x+3} = 13$

54. $\frac{3}{x} + \frac{4}{x+1} = 2$

Solve each equation. **See Example 6. (Objective 2)**

55. $x^{-4} - 2x^{-2} + 1 = 0$

56. $4x^{-4} + 1 = 5x^{-2}$

57. $8a^{-2} - 10a^{-1} - 3 = 0$

58. $2y^{-2} - 5y^{-1} = 3$

Solve each equation for the indicated variable. **See Example 7. (Objective 2)**

59. $x^2 + y^2 = r^2$ for x

60. $x^2 + y^2 = r^2$ for y

61. $xy^2 + 3xy + 7 = 0$ for y

62. $kx = ay - x^2$ for x

Solve each equation and verify that the sum of the solutions is $-\frac{b}{a}$ and that the product of the solutions is $\frac{c}{a}$. **See Example 8. (Objective 3)**

63. $12x^2 - 5x - 2 = 0$

64. $8x^2 - 2x - 3 = 0$

65. $2x^2 + 5x + 1 = 0$

66. $3x^2 + 9x + 1 = 0$

67. $3x^2 - 2x + 4 = 0$

68. $2x^2 - x + 4 = 0$

69. $x^2 + 2x + 5 = 0$

70. $x^2 - 4x + 13 = 0$

ADDITIONAL PRACTICE

71. Use the discriminant to determine whether the solutions of $1{,}492x^2 + 1{,}776x - 1{,}984 = 0$ are real numbers.

72. Use the discriminant to determine whether the solutions of $1{,}776x^2 - 1{,}492x + 1{,}984 = 0$ are real numbers.

Solve using any method.

73. $4x - 5\sqrt{x} - 9 = 0$

74. $9x - 5\sqrt{x} = 4$

75. $3x^{2/3} - x^{1/3} - 2 = 0$

76. $4x^{2/3} + 4x^{1/3} + 1 = 0$

77. $2x^4 + 24 = 26x^2$

78. $4x^4 = -9 + 13x^2$

79. $t^4 + 3t^2 = 28$

80. $t^4 + 4t^2 - 5 = 0$

81. $4(2x - 1)^2 - 3(2x - 1) - 1 = 0$

82. $4(x^2 - 1)^2 + 13(x^2 - 1) + 9 = 0$

83. $x^{-2/3} - 2x^{-1/3} - 3 = 0$

84. $4x^{-1} - 5x^{-1/2} - 9 = 0$

85. $x + \frac{2}{x-2} = 0$

86. $x + \frac{x+5}{x-3} = 0$

87. $8(m + 1)^{-2} - 30(m + 1)^{-1} + 7 = 0$

88. $2(p - 2)^{-2} + 3(p - 2)^{-1} - 5 = 0$

89. $I = \frac{k}{d^2}$ for d

90. $V = \frac{1}{3}\pi r^2 h$ for r

91. $\sigma = \sqrt{\frac{\Sigma x^2}{N} - \mu^2}$ for μ^2

92. $\sigma = \sqrt{\frac{\Sigma x^2}{N} - \mu^2}$ for N

Find the values of k that will make the solutions of each given quadratic equation equal.

93. $(k - 1)x^2 + (k - 1)x + 1 = 0$

94. $(k + 3)x^2 + 2kx + 4 = 0$

95. $(k + 4)x^2 + 2kx + 9 = 0$

96. $(k + 15)x^2 + (k - 30)x + 4 = 0$

97. Determine k such that the solutions of $3x^2 + 4x = k$ are complex numbers.

98. Determine k such that the solutions of $kx^2 - 4x = 7$ are complex numbers.

WRITING ABOUT MATH

99. Describe how to predict what type of solutions the equation $3x^2 - 4x + 5 = 0$ will have.

100. How is the discriminant related to the quadratic formula?

SOMETHING TO THINK ABOUT

101. Can a quadratic equation with integer coefficients have one real and one complex solution? Why?

102. Can a quadratic equation with complex coefficients have one real and one complex solution? Why?

SECTION 8.4 Graphs of Quadratic Functions

Objectives

1. Graph a quadratic function of the form $f(x) = ax^2$.
2. Use a vertical translation of $f(x) = ax^2$ to graph $f(x) = ax^2 + c$.
3. Use a horizontal translation of $f(x) = ax^2$ to graph $f(x) = a(x - h)^2$.
4. Use both a vertical and horizontal translation of $f(x) = ax^2$ to graph $f(x) = a(x - h)^2 + k$.
5. Graph a quadratic function in standard form by writing it in the form $f(x) = a(x - h)^2 + k$.
6. Find the vertex of a parabola using $\left(-\frac{b}{2a}, f\left(-\frac{b}{2a}\right)\right)$.
7. Graph a quadratic function in standard form by finding the vertex, axis of symmetry, and the x- and y-intercepts.
8. Solve an application problem using a quadratic function.

Vocabulary

quadratic function	vertex	variance
parabola	axis of symmetry	

Getting Ready

If $f(x) = 3x^2 + x - 2$, find each value.

1. $f(0)$ **2.** $f(1)$ **3.** $f(-1)$ **4.** $f(-2)$

If $x = -\frac{b}{2a}$, find x when a and b have the following values.

5. $a = 3$ and $b = -6$ **6.** $a = 5$ and $b = -40$

In this section, we consider graphs of second-degree polynomial functions, called *quadratic functions.*

The graph shown in Figure 8-4 on the next page shows the height (in relation to time) of a toy rocket launched straight up into the air.

COMMENT Note that the graph describes the height of the rocket, not the path of the rocket. The rocket goes straight up and comes straight down.

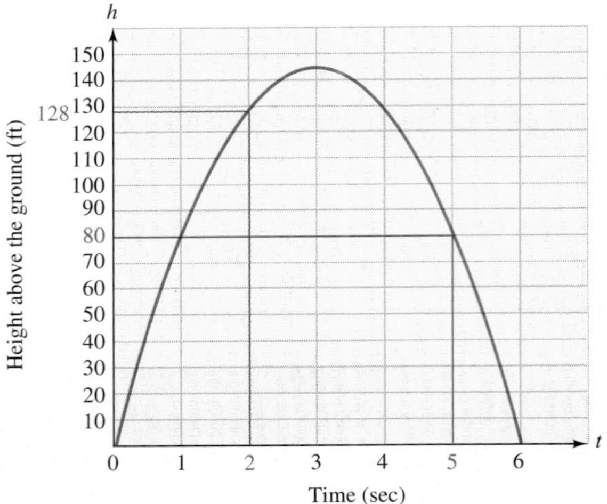

Figure 8-4

From the graph, we can see that the height of the rocket 2 seconds after it was launched is about 128 feet and that the height of the rocket 5 seconds after it was launched is 80 feet.

The parabola shown in Figure 8-4 is the graph of a *quadratic function.*

Quadratic Functions

A **quadratic function** is a second-degree polynomial function of the form

$$f(x) = ax^2 + bx + c \quad (a \neq 0)$$

where a, b, and c are real numbers.

We begin the discussion of graphing quadratic functions by considering the graph of $f(x) = ax^2 + bx + c$, where $b = 0$ and $c = 0$.

1 Graph a quadratic function of the form $f(x) = ax^2$.

EXAMPLE 1 Graph: **a.** $f(x) = x^2$ **b.** $g(x) = 3x^2$ **c.** $h(x) = \frac{1}{3}x^2$.

Solution We can make a table of ordered pairs that satisfy each equation, plot each point, and join them with a smooth curve, as in Figure 8-5. We note that the graph of $h(x) = \frac{1}{3}x^2$ is wider than the graph of $f(x) = x^2$, and that the graph of $g(x) = 3x^2$ is narrower than the graph of $f(x) = x^2$. In the function $f(x) = ax^2$, the smaller the value of $|a|$, the wider the graph.

$f(x) = x^2$

x	$f(x)$	$(x, f(x))$
−2	4	(−2, 4)
−1	1	(−1, 1)
0	0	(0, 0)
1	1	(1, 1)
2	4	(2, 4)

$g(x) = 3x^2$

x	$g(x)$	$(x, g(x))$
−2	12	(−2, 12)
−1	3	(−1, 3)
0	0	(0, 0)
1	3	(1, 3)
2	12	(2, 12)

$h(x) = \frac{1}{3}x^2$

x	$h(x)$	$(x, h(x))$
−2	$\frac{4}{3}$	$\left(-2, \frac{4}{3}\right)$
−1	$\frac{1}{3}$	$\left(-1, \frac{1}{3}\right)$
0	0	(0, 0)
1	$\frac{1}{3}$	$\left(1, \frac{1}{3}\right)$
2	$\frac{4}{3}$	$\left(2, \frac{4}{3}\right)$

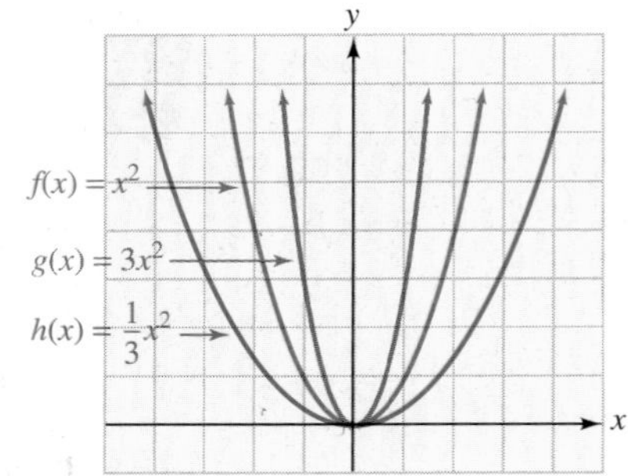

Figure 8-5

SELF CHECK 1 On the same set of coordinate axes, graph each function.
a. $f(x) = 2x^2$ **b.** $f(x) = \frac{1}{2}x^2$

If we consider the graph of $f(x) = -3x^2$, we will see that it opens downward and has the same shape as the graph of $g(x) = 3x^2$.

EXAMPLE 2 Graph: $f(x) = -3x^2$.

Solution We make a table of ordered pairs that satisfy the equation, plot each point, and join them with a smooth curve, as in Figure 8-6.

$f(x) = -3x^2$

x	$f(x)$	$(x, f(x))$
−2	−12	(−2, −12)
−1	−3	(−1, −3)
0	0	(0, 0)
1	−3	(1, −3)
2	−12	(2, −12)

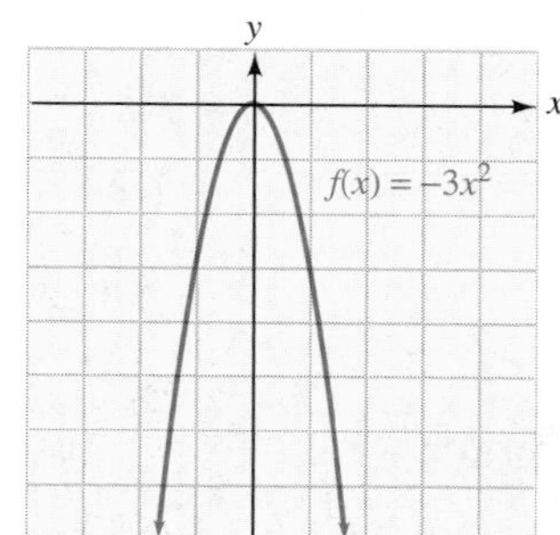

Figure 8-6

SELF CHECK 2 Graph: $f(x) = -\frac{1}{3}x^2$.

The graphs of quadratic functions are called **parabolas.** They open upward when $a > 0$ and downward when $a < 0$. The lowest point (*minimum*) of a parabola that opens upward, or the highest point (*maximum*) of a parabola that opens downward, is called the **vertex** of the parabola. The vertex of the parabola shown in Figure 8-6 is the point (0, 0).

The vertical line, called an **axis of symmetry,** that passes through the vertex divides the parabola into two congruent halves. The axis of symmetry of the parabola shown in Figure 8-6 is the y-axis, written as the equation $x = 0$.

2 Use a vertical translation of $f(x) = ax^2$ to graph $f(x) = ax^2 + c$.

EXAMPLE 3 Graph: **a.** $f(x) = 2x^2$ **b.** $g(x) = 2x^2 + 3$ **c.** $h(x) = 2x^2 - 3$.

Solution We make a table of ordered pairs that satisfy each equation, plot each point, and join them with a smooth curve, as in Figure 8-7 on the next page. We note that the graph of $g(x) = 2x^2 + 3$ is identical to the graph of $f(x) = 2x^2$, except that it has been translated 3 units upward. The graph of $h(x) = 2x^2 - 3$ is identical to the graph of $f(x) = 2x^2$, except that it has been translated 3 units downward.

$f(x) = 2x^2$

x	$f(x)$	$(x, f(x))$
−2	8	(−2, 8)
−1	2	(−1, 2)
0	0	(0, 0)
1	2	(1, 2)
2	8	(2, 8)

$g(x) = 2x^2 + 3$

x	$g(x)$	$(x, g(x))$
−2	11	(−2, 11)
−1	5	(−1, 5)
0	3	(0, 3)
1	5	(1, 5)
2	11	(2, 11)

$h(x) = 2x^2 - 3$

x	$h(x)$	$(x, h(x))$
−2	5	(−2, 5)
−1	−1	(−1, −1)
0	−3	(0, −3)
1	−1	(1, −1)
2	5	(2, 5)

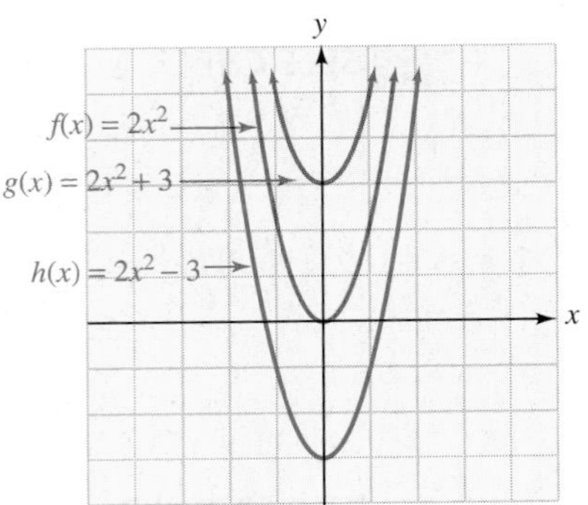

Figure 8-7

SELF CHECK 3 On the same set of coordinate axes, graph each function and tell how it differs from the graph of $f(x) = x^2$.

a. $f(x) = x^2 + 1$
b. $f(x) = x^2 - 5$

The results of Example 3 confirm the following facts.

Vertical Translations of Graphs

If $y = f(x)$ is a function and k is a positive number, then

- The graph of $y = f(x) + k$ is identical to the graph of $y = f(x)$, except that it is translated k units upward.
- The graph of $y = f(x) - k$ is identical to the graph of $y = f(x)$, except that it is translated k units downward.

3 Use a horizontal translation of $f(x) = ax^2$ to graph $f(x) = a(x - h)^2$.

EXAMPLE 4 Graph: **a.** $f(x) = 2x^2$ **b.** $g(x) = 2(x - 3)^2$ **c.** $h(x) = 2(x + 3)^2$.

Solution We make a table of ordered pairs that satisfy each equation, plot each point, and join them with a smooth curve, as in Figure 8-8. We note that the graph of $g(x) = 2(x - 3)^2$ is identical to the graph of $f(x) = 2x^2$, except that it has been translated 3 units to the right. The graph of $h(x) = 2(x + 3)^2$ is identical to the graph of $f(x) = 2x^2$, except that it has been translated 3 units to the left.

$f(x) = 2x^2$

x	$f(x)$	$(x, f(x))$
−2	8	(−2, 8)
−1	2	(−1, 2)
0	0	(0, 0)
1	2	(1, 2)
2	8	(2, 8)

$g(x) = 2(x - 3)^2$

x	$g(x)$	$(x, g(x))$
1	8	(1, 8)
2	2	(2, 2)
3	0	(3, 0)
4	2	(4, 2)
5	8	(5, 8)

$h(x) = 2(x + 3)^2$

x	$h(x)$	$(x, h(x))$
−5	8	(−5, 8)
−4	2	(−4, 2)
−3	0	(−3, 0)
−2	2	(−2, 2)
−1	8	(−1, 8)

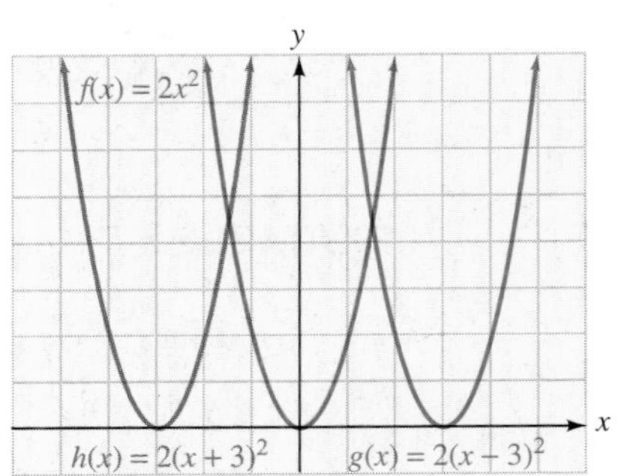

Figure 8-8

SELF CHECK 4 On the same set of coordinate axes, graph each function and tell how it differs from the graph of $f(x) = x^2$.

a. $f(x) = (x - 2)^2$
b. $f(x) = (x + 5)^2$

The results of Example 4 confirm the following facts.

Horizontal Translations of Graphs

If $y = f(x)$ is a function and h is a positive number, then

- The graph of $y = f(x - h)$ is identical to the graph of $y = f(x)$, except that it is translated h units to the right.
- The graph of $y = f(x + h)$ is identical to the graph of $y = f(x)$, except that it is translated h units to the left.

4 Use both a vertical and horizontal translation of $f(x) = ax^2$ to graph $f(x) = a(x - h)^2 + k$.

EXAMPLE 5 Graph: $f(x) = 2(x - 3)^2 - 4$.

Solution The graph of $f(x) = 2(x - 3)^2 - 4$ is identical to the graph of $g(x) = 2(x - 3)^2$, except that it has been translated 4 units downward. The graph of $g(x) = 2(x - 3)^2$ is identical to the graph of $h(x) = 2x^2$, except that it has been translated 3 units to the right. Thus, to graph $f(x) = 2(x - 3)^2 - 4$, we can graph $h(x) = 2x^2$ and shift it 3 units to the right and then 4 units downward, as shown in Figure 8-9.

The vertex of the graph is the point $(3, -4)$, and the axis of symmetry is the line $x = 3$.

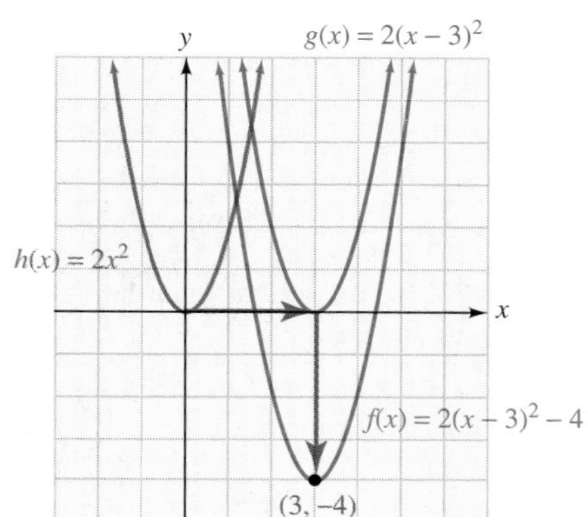

Figure 8-9

SELF CHECK 5 Graph: $f(x) = 2(x + 3)^2 + 1$.

The results of Example 5 confirm the following facts.

Vertex and Axis of Symmetry of a Parabola

The graph of the function

$$f(x) = a(x - h)^2 + k \quad (a \neq 0)$$

is a parabola with vertex at (h, k). (See Figure 8-10.)

The parabola opens upward when $a > 0$ and downward when $a < 0$. The axis of symmetry is the line $x = h$.

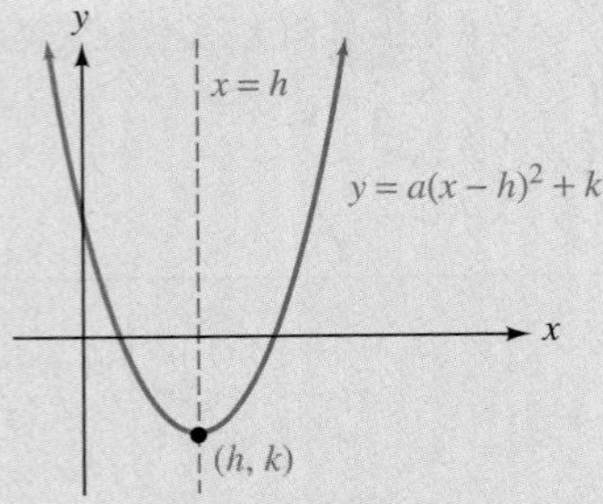

Figure 8-10

5 Graph a quadratic function in standard form by writing it in the form $f(x) = a(x - h)^2 + k$.

To graph functions of the form $f(x) = ax^2 + bx + c$, we can complete the square to write the function in the form $f(x) = a(x - h)^2 + k$.

EXAMPLE 6 Graph: $f(x) = 2x^2 - 4x - 1$.

Solution We complete the square on x to write the function in the form $f(x) = a(x - h)^2 + k$.

$f(x) = 2x^2 - 4x - 1$

$f(x) = 2(x^2 - 2x) - 1$ — Factor 2 from $2x^2 - 4x$.

$f(x) = 2(x^2 - 2x + \mathbf{1}) - 1 - \mathbf{2}$ — Complete the square on x. Since this adds 2 to the right side, we also subtract 2 from the right side.

(1) $f(x) = 2(x - 1)^2 - 3$ — Factor $x^2 - 2x + 1$ and combine like terms.

From Equation 1, we can see that the vertex will be at the point $(1, -3)$. We can plot the vertex and a few points on either side of the vertex and draw the graph, which appears in Figure 8-11.

COMMENT Note that this is the graph of $f(x) = 2x^2$ translated one unit to the right and three units down.

$f(x) = 2x^2 - 4x - 1$

x	$f(x)$	$(x, f(x))$
−1	5	(−1, 5)
0	−1	(0, −1)
1	−3	(1, −3)
2	−1	(2, −1)
3	5	(3, 5)

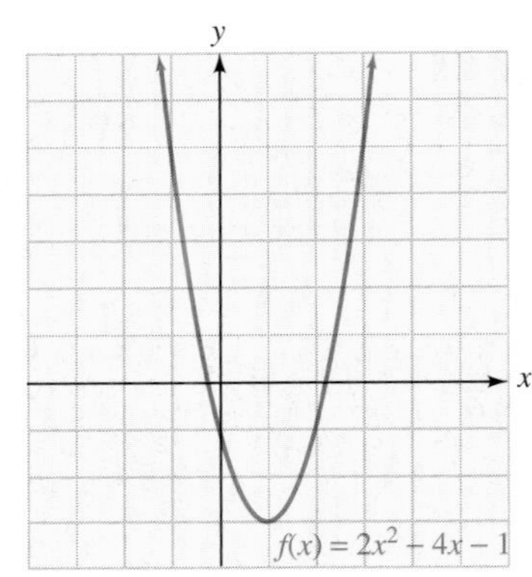

Figure 8-11

SELF CHECK 6 Graph: $f(x) = 2x^2 - 4x + 1$.

6 Find the vertex of a parabola using $\left(-\frac{b}{2a}, f\left(-\frac{b}{2a}\right)\right)$.

We can derive a formula for the vertex of the graph of $f(x) = ax^2 + bx + c$ by completing the square in the same manner as we did in Example 6. After using similar steps, the result is

$$f(x) = a\left[x - \underset{\substack{\uparrow \\ h}}{\left(-\frac{b}{2a}\right)}\right]^2 + \underset{\substack{\uparrow \\ k}}{\frac{4ac - b^2}{4a}}$$

The x-coordinate of the vertex is $-\frac{b}{2a}$. The y-coordinate of the vertex is $\frac{4ac - b^2}{4a}$. We can also find the y-coordinate of the vertex by substituting the x-coordinate, $-\frac{b}{2a}$, for x in the quadratic function and simplifying.

Formula for the Vertex of a Parabola

The vertex of the graph of the quadratic function $f(x) = ax^2 + bx + c$ is

$$\left(-\frac{b}{2a}, f\left(-\frac{b}{2a}\right)\right)$$

and the axis of symmetry of the parabola is the line $x = -\frac{b}{2a}$.

EXAMPLE 7 Find the vertex of the graph of $f(x) = 2x^2 - 4x - 1$.

Solution The function is written in $f(x) = ax^2 + bx + c$ form, where $a = 2$, $b = -4$, and $c = -1$. We can find the x-coordinate of the vertex by evaluating $-\frac{b}{2a}$.

$$-\frac{b}{2a} = -\frac{-4}{2(2)} = -\frac{-4}{4} = 1$$

COMMENT This is the same as finding $f(1)$.

We can find the y-coordinate by evaluating $f\left(-\frac{b}{2a}\right)$.

$$f\left(-\frac{b}{2a}\right) = f(1) = 2(1)^2 - 4(1) - 1 = -3$$

The vertex is the point $(1, -3)$. This agrees with the result we obtained in Example 6 by completing the square.

SELF CHECK 7 Find the vertex of the graph of $f(x) = 3x^2 - 12x + 8$.

7 Graph a quadratic function in standard form by finding the vertex, axis of symmetry, and the *x*- and *y*-intercepts.

Much can be determined about the graph of $f(x) = ax^2 + bx + c$ from the coefficients a, b, and c. This information is summarized as follows:

Graphing a Quadratic Function $f(x) = ax^2 + bx + c$

Determine whether the parabola opens upward or downward:

If $a > 0$, the parabola opens upward.
If $a < 0$, the parabola opens downward.

The x-coordinate of the vertex of the parabola is $x = -\frac{b}{2a}$.

To find the y-coordinate of the vertex, substitute $-\frac{b}{2a}$ for x and find $f\left(-\frac{b}{2a}\right)$.

The axis of symmetry is the vertical line passing through the vertex. The axis of symmetry is $x = -\frac{b}{2a}$.

The y-intercept is determined by the value of $f(x)$ when $x = 0$. The y-intercept is $(0, c)$.

The x-intercepts (if any) are determined by the values of x that make $f(x) = 0$. To find them, solve the quadratic equation $ax^2 + bx + c = 0$.

EXAMPLE 8 Graph: $f(x) = -2x^2 - 8x - 8$.

Solution **Step 1** *Determine whether the parabola opens upward or downward.* The function is in the form $f(x) = ax^2 + bx + c$, with $a = -2$, $b = -8$, and $c = -8$. Since $a < 0$, the parabola opens downward.

Step 2 *Find the vertex and draw the axis of symmetry.* To find the coordinates of the vertex, we evaluate $-\frac{b}{2a}$ by substituting -2 for a and -8 for b.

$$x = -\frac{b}{2a} \quad x = -\frac{-8}{2(-2)} = -2$$

We then find $f(-2)$.

$$f\left(-\frac{b}{2a}\right) = f(-2) = -2(-2)^2 - 8(-2) - 8 = -8 + 16 - 8 = 0$$

The vertex of the parabola is the point $(-2, 0)$. The axis of symmetry is the line $x = -2$.

Step 3 *Find the x- and y-intercepts.* Since $c = -8$, the y-intercept of the parabola is $(0, -8)$. The point $(-4, -8)$, two units to the left of the axis of symmetry, must also be on the graph. We plot both points in black on the graph.

To find the x-intercepts, we set $f(x)$ equal to 0 and solve the resulting quadratic equation.

$$\begin{aligned} f(x) &= -2x^2 - 8x - 8 & \\ 0 &= -2x^2 - 8x - 8 & \text{Set } f(x) = 0. \\ 0 &= x^2 + 4x + 4 & \text{Divide both sides by } -2. \\ 0 &= (x + 2)(x + 2) & \text{Find the trinomial.} \end{aligned}$$

$$x + 2 = 0 \quad \text{or} \quad x + 2 = 0 \qquad \text{Set each factor equal to 0.}$$
$$x = -2 \quad \Big| \quad x = -2$$

Since the solutions are the same, the graph has only one x-intercept: $(-2, 0)$. This point is the vertex of the parabola and has already been plotted.

Step 4 *Plot another point.* Finally, we find another point on the parabola. If $x = -3$, then $f(-3) = -2$. We plot $(-3, -2)$ and use symmetry to determine that $(-1, -2)$ is also on the graph. Both points are in black.

Step 5 *Draw a smooth curve through the points,* as shown.

$f(x) = -2x^2 - 8x - 8$

x	$f(x)$	$(x, f(x))$
-3	-2	$(-3, -2)$

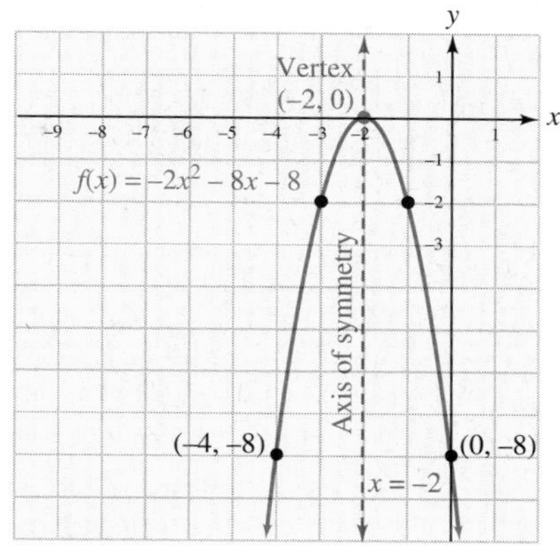

Figure 8-12

ACCENT ON TECHNOLOGY

Graphing Quadratic Functions

To use a graphing calculator to graph $f(x) = 0.7x^2 + 2x - 3.5$, we can use window settings of $[-10, 10]$ for x and $[-10, 10]$ for y, enter the function, and press **GRAPH** to obtain Figure 8-13(a).

To find approximate coordinates of the vertex of the graph, we trace to move the cursor near the lowest point of the graph as shown in Figure 8-13(b). By zooming in twice and tracing as in Figure 8-13(c), we can see that the vertex is a point whose coordinates are approximately $(-1.422872, -4.928549)$.

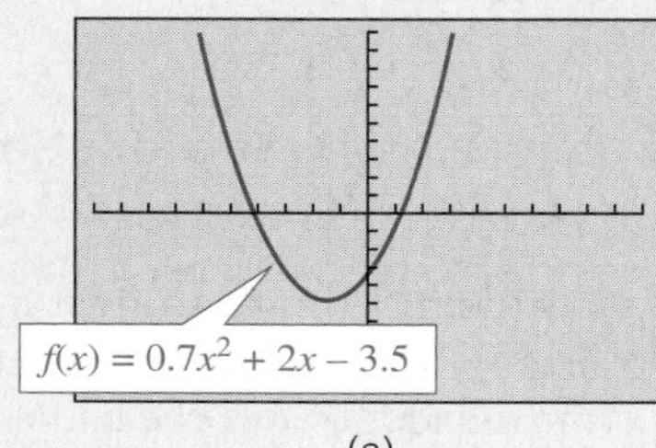

(a)

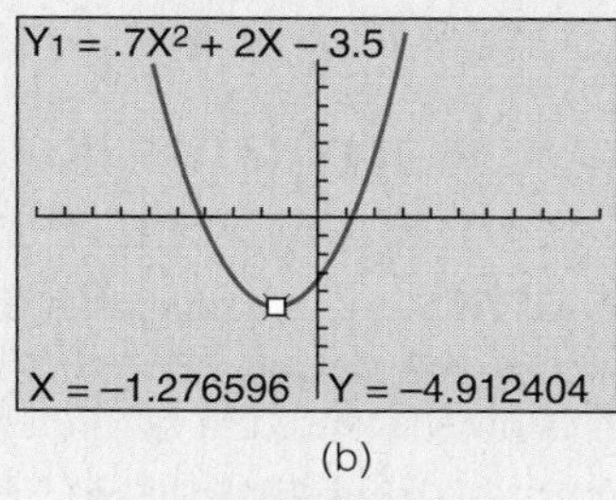

(b)

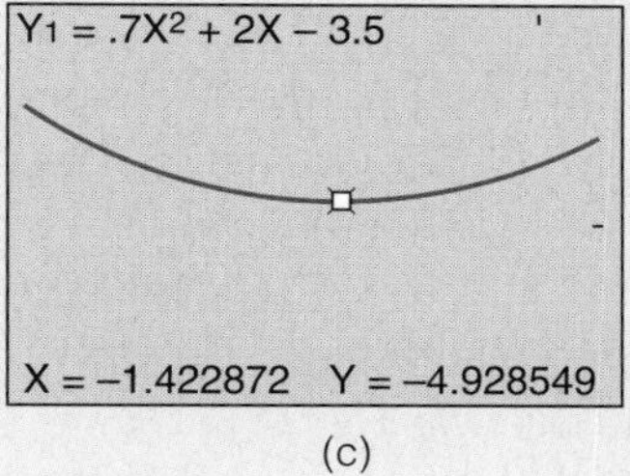

(c)

Figure 8-13

We can find the vertex more efficiently by using the MINIMUM command found in the CALC menu. We first graph the function $f(x) = 0.7x^2 + 2x - 3.5$ as in Figure 8-13(a). We then select 3 in the CALC menu, enter -3 for a left guess, and press **ENTER**. We then enter 0 for a right guess and press **ENTER**. After pressing **ENTER** again, we will obtain the minimum value $(-1.42857, -4.928571)$.

The solutions of the quadratic equation $0.7x^2 + 2x - 3.5 = 0$ are the numbers x that will make $f(x) = 0$ in the function $f(x) = 0.7x^2 + 2x - 3.5$. To approximate these numbers, we graph the function as shown in Figure 8-14(a) and find the x-intercepts by

(continued)

tracing to move the cursor near each x-intercept, as in Figures 8-14(b) and 8-14(c). From the graphs, we can read the approximate value of the x-coordinate of each x-intercept. For better results, we can zoom in.

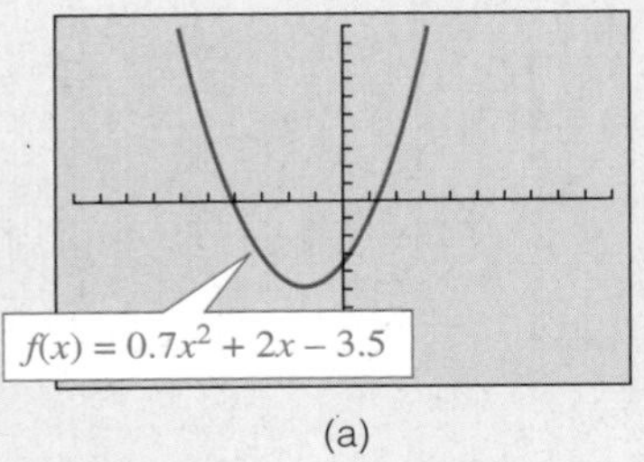

(a)

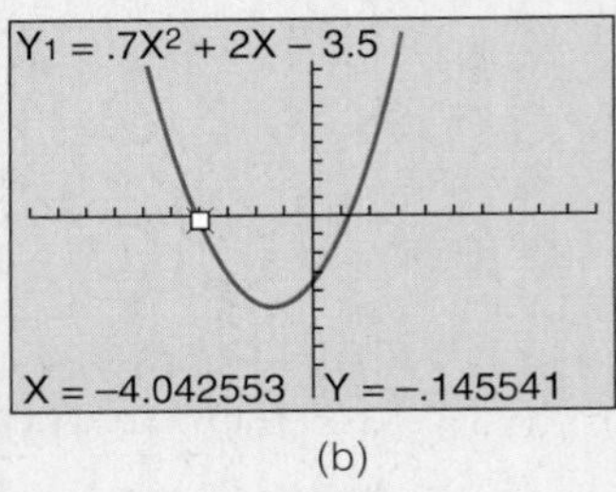

(b)

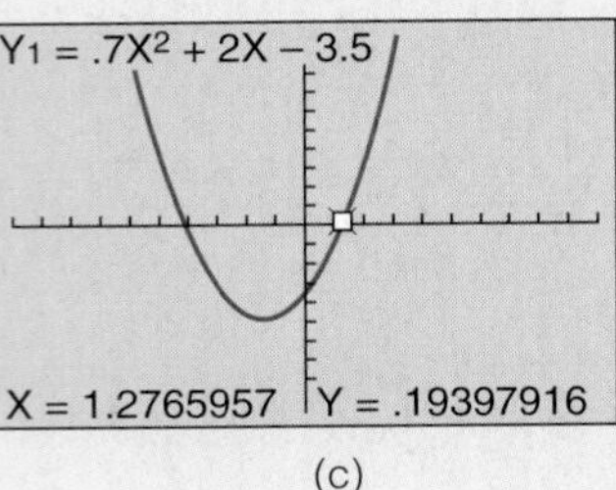

(c)

Figure 8-14

We can also solve the equation by using the ZERO command found in the CALC menu. We first graph the function $f(x) = 0.7x^2 + 2x - 3.5$ as in Figure 8-15(a). We then select 2 in the CALC menu to get Figure 8-15(b). We enter -5 for a left guess and press **ENTER**. We then enter -2 for a right guess and press **ENTER**. After pressing **ENTER** again, we will obtain Figure 8-15(c). We can find the second solution in a similar way.

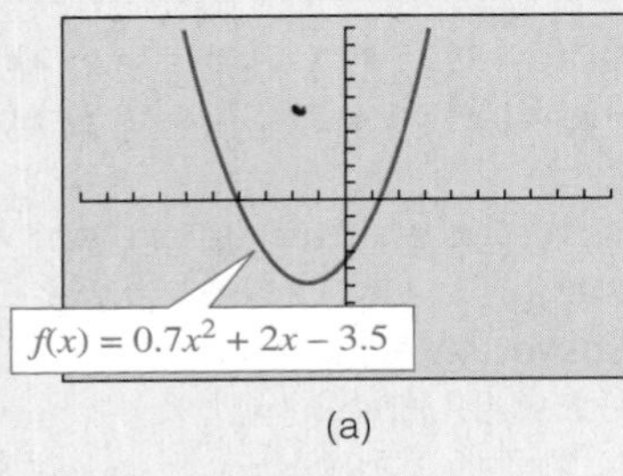

(a)

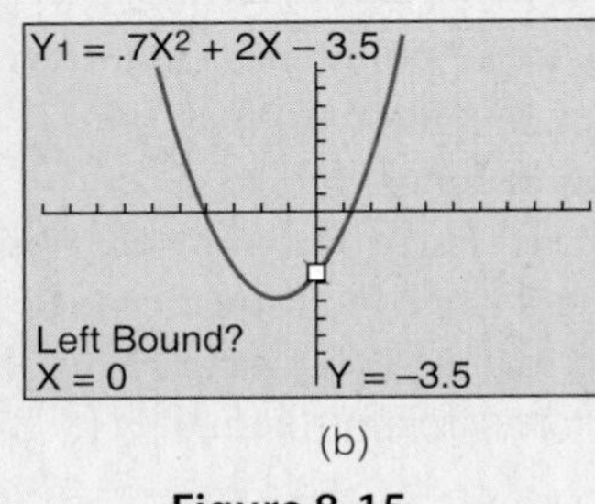

(b)

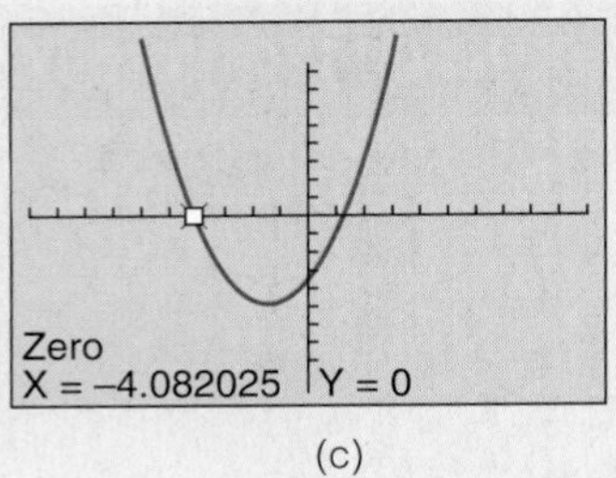

(c)

Figure 8-15

8 Solve an application problem using a quadratic function.

EXAMPLE 9 **BALLISTICS** The ball shown in Figure 8-16(a) is thrown straight up with a velocity of 128 feet per second. The function $s = h(t) = -16t^2 + 128t$ gives the relation between t (the time measured in seconds) and s (the number of feet the ball is above the ground). How long will it take the ball to reach its maximum height, and what is that height?

Solution The graph of $s = h(t) = -16t^2 + 128t$ is a parabola. Since the coefficient of t^2 is negative, it opens downward. The time it takes the ball to reach its maximum height is given by the t-coordinate of its vertex, and the maximum height of the ball is given by the s-coordinate of the vertex. To find the vertex, we find its t-coordinate and s-coordinate. To find the t-coordinate, we compute

$$-\frac{b}{2a} = -\frac{128}{2(-16)} \quad b = 128 \text{ and } a = -16.$$

$$= -\frac{128}{-32}$$

$$= 4$$

To find the s-coordinate, we substitute 4 for t in $h(t) = -16t^2 + 128t$.

$$\begin{aligned} h(\mathbf{4}) &= -16(\mathbf{4})^2 + 128(\mathbf{4}) \\ &= -256 + 512 \\ &= 256 \end{aligned}$$

Since $t = 4$ and $s = 256$ are the coordinates of the vertex, the ball will reach a maximum height in 4 seconds and that maximum height will be 256 feet.

To solve this problem with a graphing calculator with window settings of [0, 10] for x and [0, 300] for y, we graph the function $h(t) = -16t^2 + 128t$ to get the graph in Figure 8-16(b). By using the MAXIMUM command found under the CALC menu in the same way as the MINIMUM command, we can determine that the ball reaches a height of 256 feet in 4 seconds.

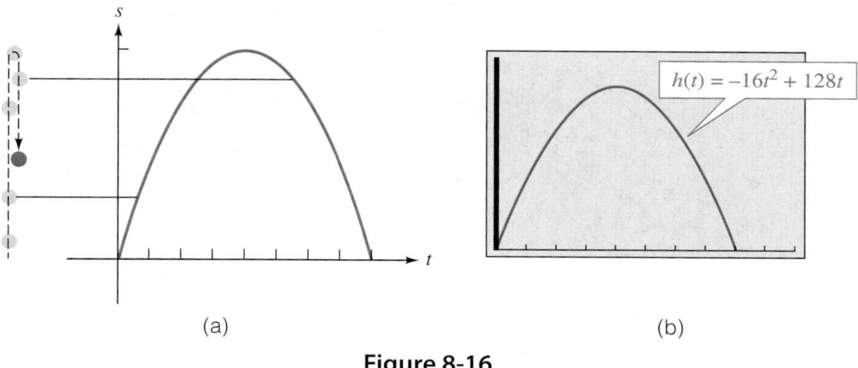

(a) (b)

Figure 8-16

EXAMPLE 10 **MAXIMIZING AREA** A man wants to build the rectangular pen shown in Figure 8-17(a) on the next page to house his dog. If he uses one side of his barn, find the maximum area that he can enclose with 80 feet of fencing.

Solution If we use w to represent the width of the pen, the length is represented by $80 - 2w$. Since the area A of the pen is the product of its length and width, we have

$$\begin{aligned} A &= (80 - 2w)w \\ &= 80w - 2w^2 \\ &= -2w^2 + 80w \end{aligned}$$

Since the graph of $A = -2w^2 + 80w$ is a parabola opening downward, the maximum area will be given by the A-coordinate of the vertex of the graph. To find the vertex, we first find its w-coordinate by letting $b = 80$ and $a = -2$ and computing $-\frac{b}{2a}$.

$$-\frac{\mathbf{b}}{\mathbf{2a}} = -\frac{\mathbf{80}}{2(\mathbf{-2})} = 20$$

We can then find the A-coordinate of the vertex by substituting 20 into the function $A = -2w^2 + 80w$.

$$\begin{aligned} A &= -2w^2 + 80w \\ &= -2(\mathbf{20})^2 + 80(\mathbf{20}) \\ &= -2(400) + 1{,}600 \end{aligned}$$

$= -800 + 1{,}600$

$= 800$

Thus, the coordinates of the vertex of the graph of the quadratic function are (20, 800), and the maximum area is 800 square feet. This occurs when the width is 20 feet.

To solve this problem using a graphing calculator with window settings of [0, 50] for x and [0, 1,000] for y, we graph the function $A = -2w^2 + 80w$ to get the graph in Figure 8-17(b). By using the MAXIMUM command, we can determine that the maximum area is 800 square feet when the width is 20 feet.

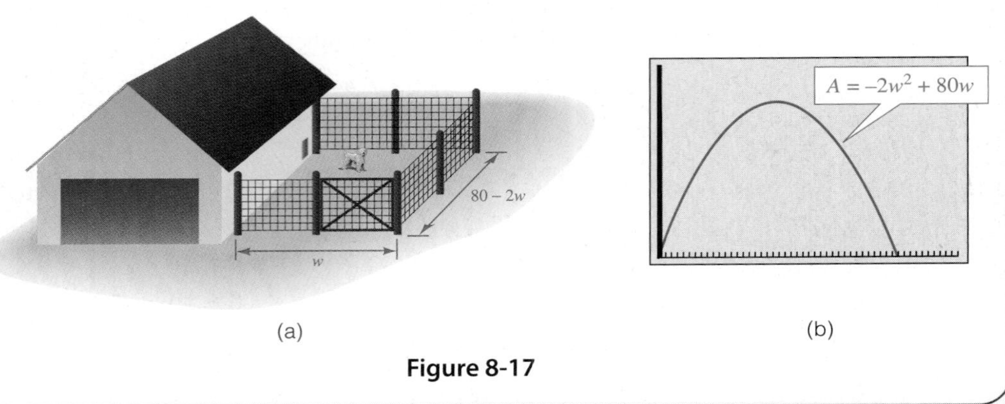

(a) (b)

Figure 8-17

In statistics, the square of the standard deviation is called the **variance.**

EXAMPLE 11 **VARIANCE** If p is the chance that a person selected at random has the HIV virus, then $1 - p$ is the chance that the person does not have the HIV virus. If 100 people are randomly sampled, we know from statistics that the variance of this type of sample distribution will be $100p(1 - p)$. Find the value of p that will maximize the variance.

Solution The variance is given by the function

$$v(p) = 100p(1 - p) \qquad \text{or} \qquad v(p) = -100p^2 + 100p$$

In this setting, all values of p are between 0 and 1, including 0 and 1. We use window settings of [0, 1] for x when graphing the function $v(p) = -100p^2 + 100p$ on a graphing calculator. If we also use window settings of [0, 30] for y, we will obtain the graph shown in Figure 8-18(a). After using the MAXIMUM command to obtain Figure 8-18(b), we can see that a value of 0.5 will give the maximum variance.

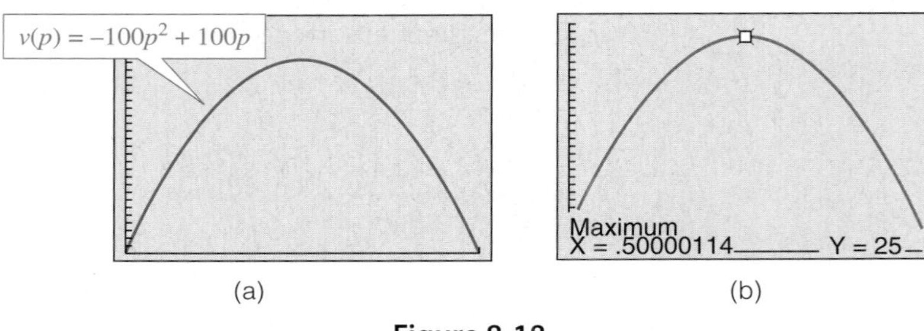

(a) (b)

Figure 8-18

SELF CHECK ANSWERS

1. 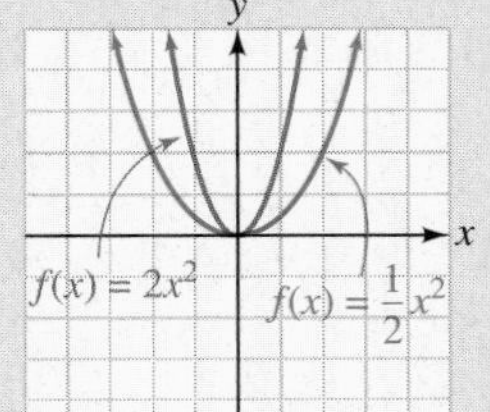

2. 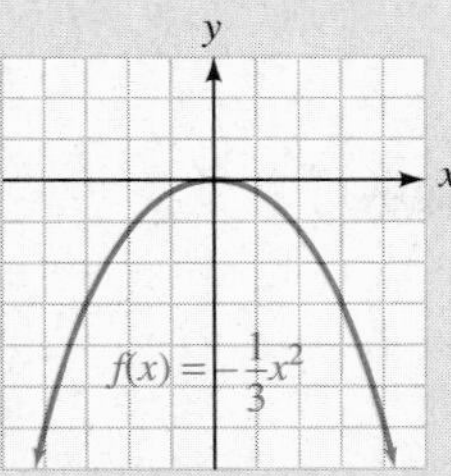

3. 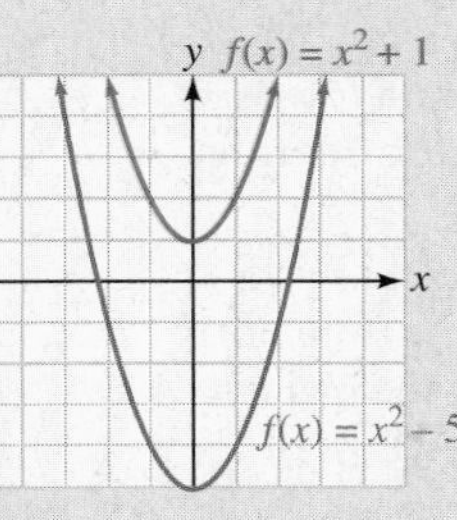

a. shifted 1 unit up
b. shifted 5 units down

4. 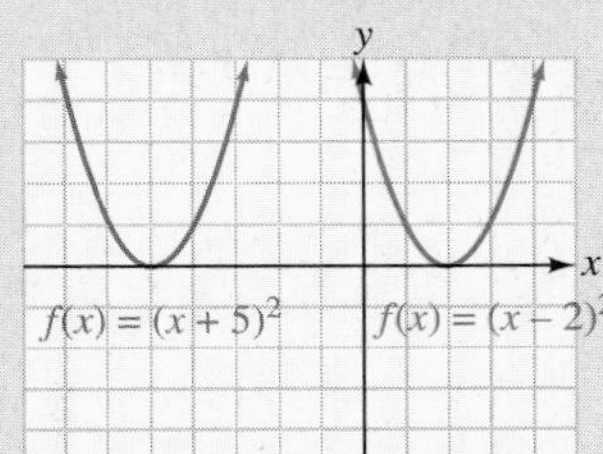

a. shifted 2 units to the right
b. shifted 5 units to the left

5. 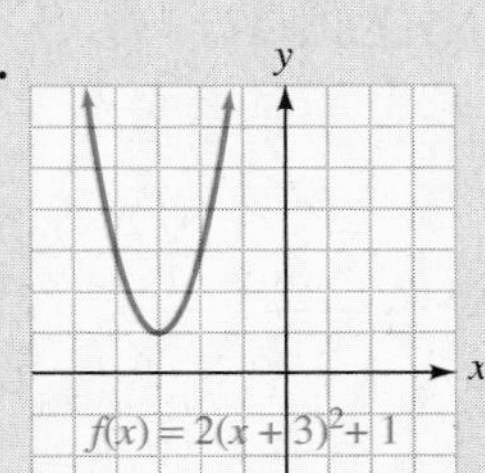

6. 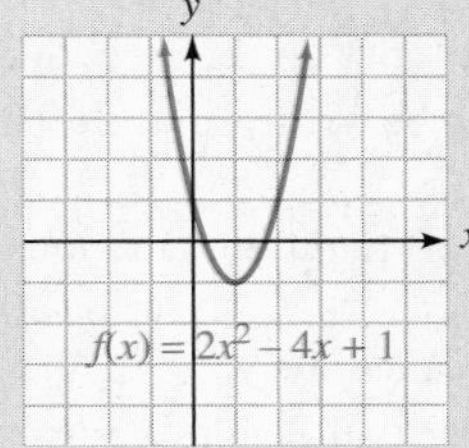

7. $(2, -4)$

NOW TRY THIS

Answer the following questions about the graph of $f(x)$.

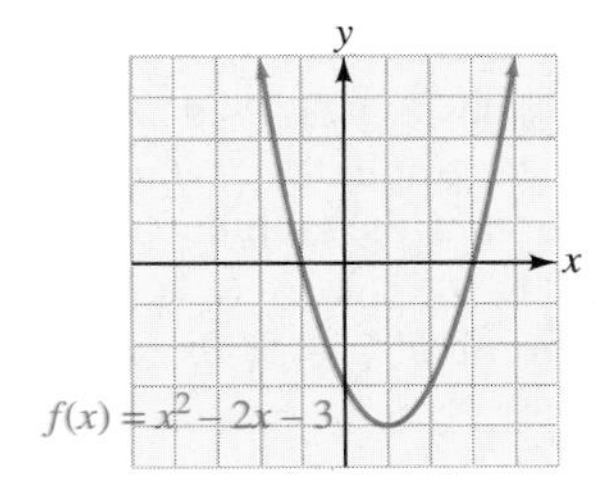

1. What is the vertex?
2. What is the axis of symmetry?
3. What is the domain?
4. What is the range?
5. For what values of x will $f(x) = 0$?
6. For what values of x will y be positive? ($f(x) > 0$)
7. For what values of x will y be negative? ($f(x) < 0$)

8.4 EXERCISES

WARM-UPS

Determine whether the graph of each equation opens up or down.

1. $y = -3x^2 + x - 5$
2. $y = 4x^2 + 2x - 3$
3. $y = 2(x - 3)^2 - 1$
4. $y = -3(x + 2)^2 + 2$

Find the vertex of the parabola determined by each equation.

5. $y = 2(x - 3)^2 - 1$
6. $y = -3(x + 2)^2 + 2$

REVIEW *Find the value of x.*

7. 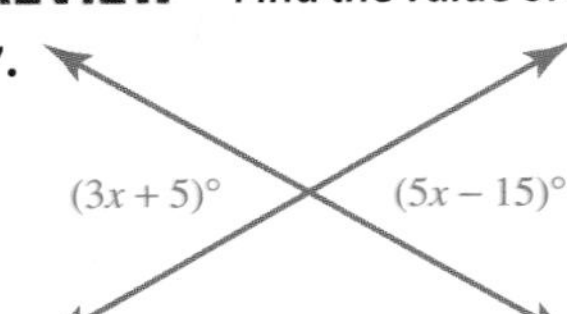

8. Lines r and s are parallel.

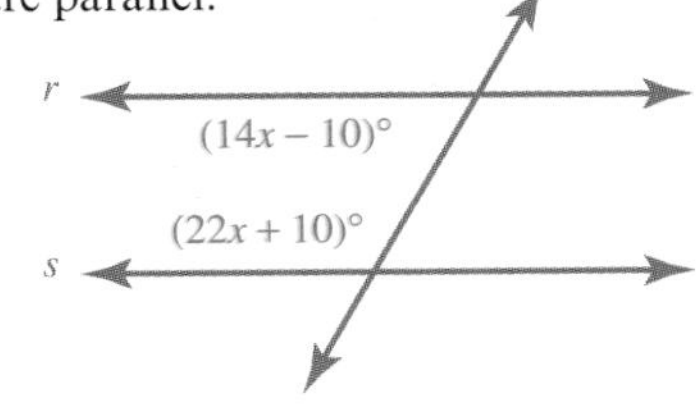

9. **Travel** Madison and St. Louis are 385 miles apart. One train leaves Madison and heads toward St. Louis at the rate of 30 mph. Three hours later, a second train leaves Madison, bound for St. Louis. If the second train travels at the rate of 55 mph, in how many hours will the faster train overtake the slower train?

10. **Investing** A woman invests \$25,000, some at 7% annual interest and the rest at 8%. If the annual income from both investments is \$1,900, how much is invested at the higher rate?

VOCABULARY AND CONCEPTS *Fill in the blanks.*

11. A quadratic function is a second-degree polynomial function that can be written in the form ________________, where ______.
12. The graphs of quadratic functions are called ________.
13. The highest (________) or lowest (________) point on a parabola is called the ______.
14. A vertical line that divides a parabola into two halves is called an ____ of symmetry.
15. The graph of $y = f(x) + k$ $(k > 0)$ is identical to the graph of $y = f(x)$, except that it is translated k units ________.
16. The graph of $y = f(x) - k$ $(k > 0)$ is a vertical translation of the graph of $y = f(x)$, k units ________.
17. The graph of $y = f(x - h)$ $(h > 0)$ is a horizontal translation of the graph of $y = f(x)$, h units ________.
18. The graph of $y = f(x + h)$ $(h > 0)$ is identical to the graph of $y = f(x)$, except that it is translated h units ________.
19. The graph of $y = f(x) = ax^2 + bx + c$ $(a \neq 0)$ opens ________ when $a > 0$.
20. In statistics, the square of the standard deviation is called the ________.

GUIDED PRACTICE

Graph each function. See Examples 1–3. (Objectives 1–2)

21. $f(x) = x^2$

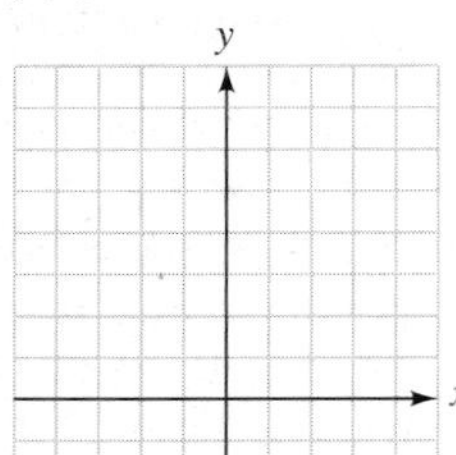

22. $f(x) = -x^2$

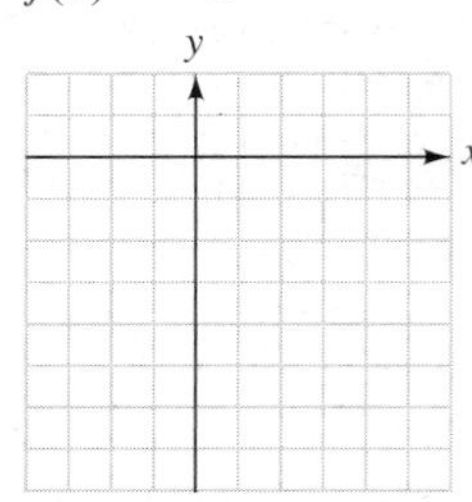

23. $f(x) = x^2 + 2$

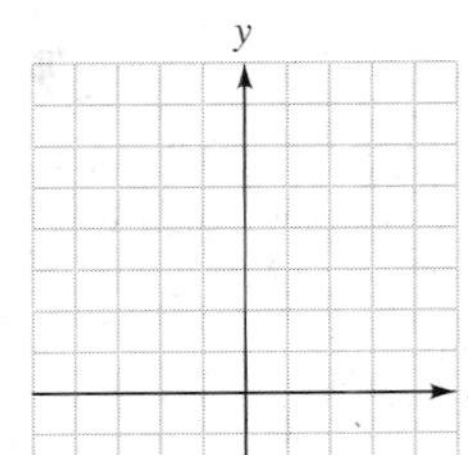

24. $f(x) = x^2 - 3$

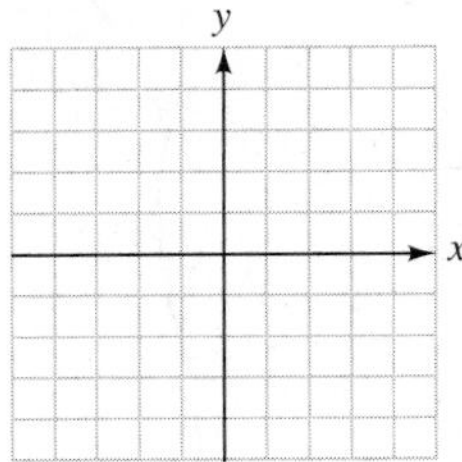

Graph each function. See Examples 4–5. (Objectives 3–4)

25. $f(x) = -(x - 2)^2$

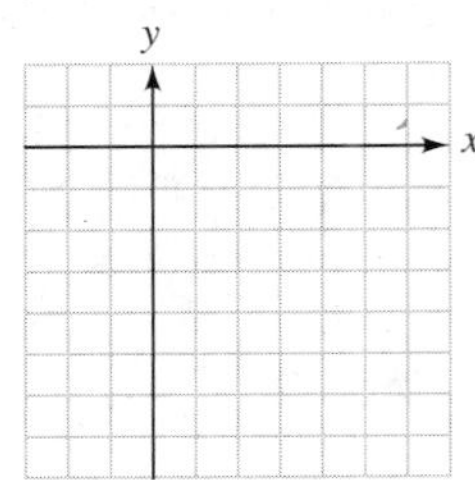

26. $f(x) = (x + 2)^2$

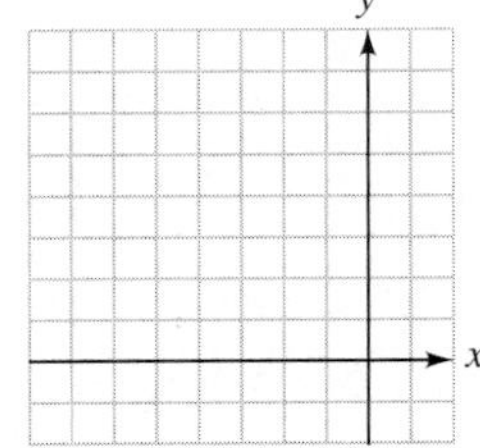

27. $f(x) = (x - 3)^2 + 2$

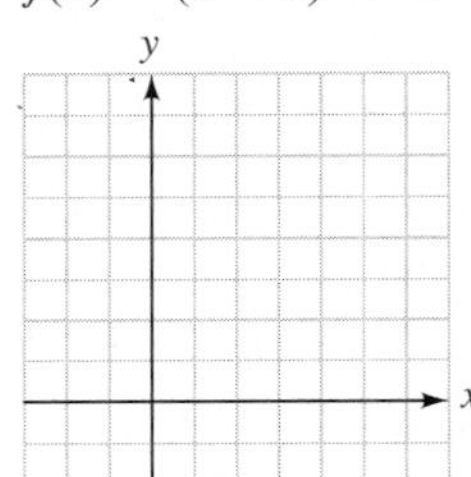

28. $f(x) = (x + 1)^2 - 2$

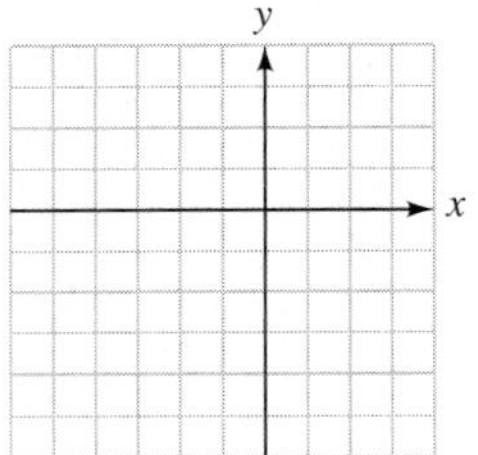

Find the coordinates of the vertex and the axis of symmetry of the graph of each equation. If necessary, complete the square on x to write the equation in the form $y = a(x - h)^2 + k$. Do not graph the equation. See Example 6. (Objective 5)

29. $y = (x - 1)^2 + 2$
30. $y = 2(x - 2)^2 - 1$
31. $y = 2(x + 3)^2 - 4$
32. $y = -3(x + 1)^2 + 3$
33. $y = -3x^2$
34. $y = 3x^2 - 3$
35. $y = 2x^2 - 4x$
36. $y = 3x^2 + 6x$

Find the coordinates of the vertex and the axis of symmetry of the graph of each equation. Use $\left(-\frac{b}{2a}, f\left(-\frac{b}{2a}\right)\right)$ to find the vertex and graph the equation. See Examples 7–8. (Objectives 6–7)

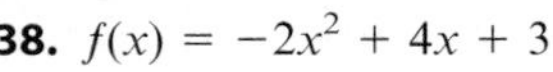

37. $f(x) = -2x^2 + 4x + 1$

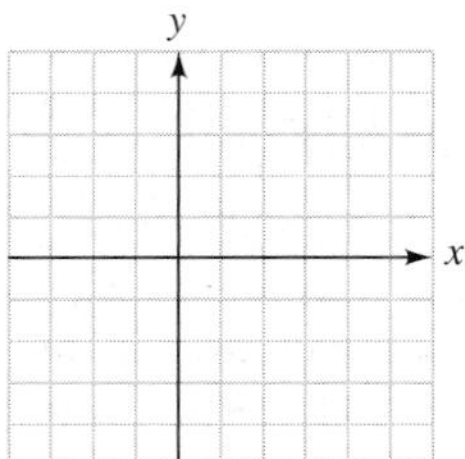

38. $f(x) = -2x^2 + 4x + 3$

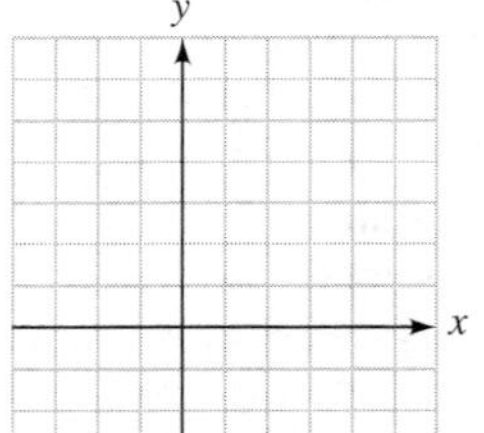

39. $f(x) = 3x^2 - 12x + 10$

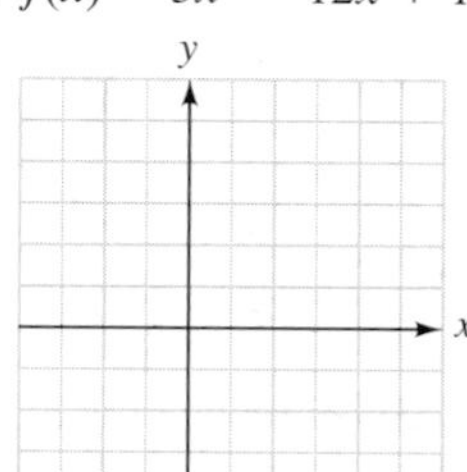

40. $f(x) = 3x^2 - 12x + 9$

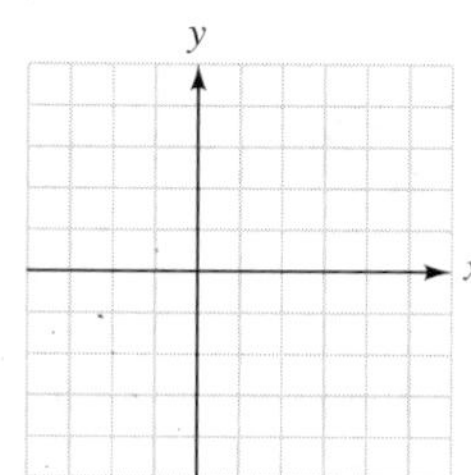

ADDITIONAL PRACTICE

Find the vertex and axis of symmetry using any method. Do not graph.

41. $y = -4x^2 + 16x + 5$

42. $y = 5x^2 + 20x + 25$

43. $y - 7 = 6x^2 - 5x$

44. $y - 2 = 3x^2 + 4x$

Graph each function.

45. $f(x) = x^2 + x - 6$

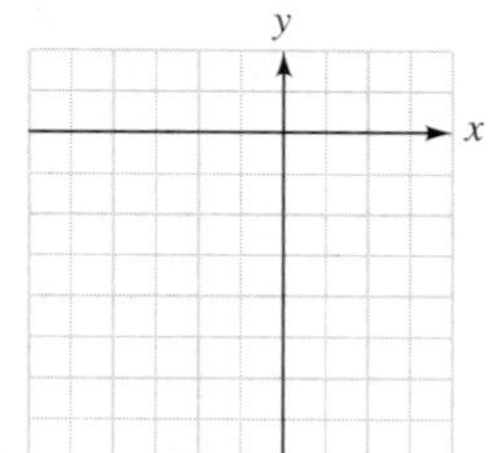

46. $f(x) = x^2 - x - 6$

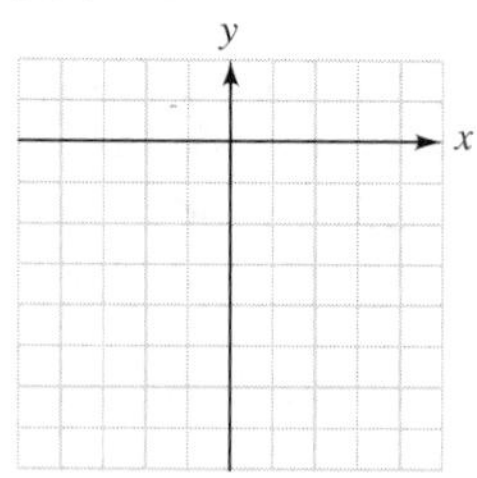

Use a graphing calculator to find the coordinates of the vertex of the graph of each quadratic function. Give results to the nearest hundredth.

47. $y = 2x^2 - x + 1$

48. $y = x^2 + 5x - 6$

49. $y = 7 + x - x^2$

50. $y = 2x^2 - 3x + 2$

Use a graphing calculator to solve each equation. If a result is not exact, give the result to the nearest hundredth.

51. $x^2 + x - 6 = 0$

52. $2x^2 - 5x - 3 = 0$

53. $0.5x^2 - 0.7x - 3 = 0$

54. $2x^2 - 0.5x - 2 = 0$

55. The equation $y - 2 = (x - 5)^2$ represents a quadratic function whose graph is a parabola. Find its vertex.

56. Show that $y = ax^2$, where $a \neq 0$, represents a quadratic function whose vertex is at the origin.

APPLICATIONS *Solve each application problem. Use a graphing calculator if necessary.* See Examples 9–11. (Objective 8)

57. Ballistics If a ball is thrown straight up with an initial velocity of 48 feet per second, its height s after t seconds is given by the equation $s = 48t - 16t^2$. Find the maximum height attained by the ball and the time it takes for the ball to reach that height.

58. Ballistics From the top of the building, a ball is thrown straight up with an initial velocity of 32 feet per second. The equation $s = -16t^2 + 32t + 48$ gives the height s of the ball t seconds after it is thrown. Find the maximum height reached by the ball and the time it takes for the ball to hit the ground. (*Hint:* Let $s = 0$ and solve for t.)

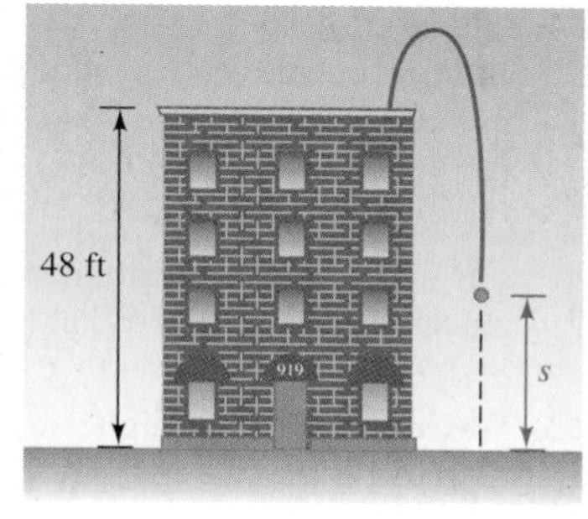

59. Maximizing area Find the dimensions of the rectangle of maximum area that can be constructed with 200 feet of fencing. Find the maximum area.

60. Fencing a field A farmer wants to fence in three sides of a rectangular field with 1,000 feet of fencing. The other side of the rectangle will be a river. If the enclosed area is to be maximum, find the dimensions of the field.

61. Finding the variance If p is the chance that a person sampled at random has high blood pressure, $1 - p$ is the chance that the person doesn't. If 50 people are sampled at random, the variance of the sample will be $50p(1 - p)$. What two values of p will give a variance of 9.375?

62. Finding the variance If p is the chance that a person sampled at random smokes, then $1 - p$ is the chance that the person doesn't. If 75 people are sampled at random, the variance of the sample will be $75p(1 - p)$. What two values of p will give a variance of 12?

63. Police investigations A police officer seals off the scene of a car collision using a roll of yellow police tape that is 300 feet long. What dimensions should be used to seal off the maximum rectangular area around the collision? What is the maximum area?

64. Operating costs The cost C in dollars of operating a certain concrete-cutting machine is related to the number of minutes n the machine is run by the function

$$C(n) = 2.2n^2 - 66n + 655$$

For what number of minutes is the cost of running the machine a minimum? What is the minimum cost?

65. Water usage The height (in feet) of the water level in a reservoir over a 1-year period is modeled by the function

$$H(t) = 3.3t^2 - 59.4t + 281.3$$

How low did the water level get that year?

66. School enrollment The total annual enrollment (in millions) in U.S. elementary and secondary schools for the years 1975–1996 is given by the function

$$E(x) = 0.058x^2 - 1.162x + 50.604$$

For this period, what was the lowest enrollment?

67. Maximizing revenue The revenue R received for selling x stereos is given by the equation

$$R = -\frac{x^2}{1{,}000} + 10x$$

Find the number of stereos that must be sold to obtain the maximum revenue.

68. Maximizing revenue In Exercise 67, find the maximum revenue.

69. Maximizing revenue The revenue received for selling x radios is given by the formula

$$R = -\frac{x^2}{728} + 9x$$

How many radios must be sold to obtain the maximum revenue? Find the maximum revenue.

70. Maximizing revenue The revenue received for selling x stereos is given by the formula

$$R = -\frac{x^2}{5} + 80x - 1{,}000$$

How many stereos must be sold to obtain the maximum revenue? Find the maximum revenue.

71. Maximizing revenue When priced at \$30 each, a toy has annual sales of 4,000 units. The manufacturer estimates that each \$1 increase in cost will decrease sales by 100 units. Find the unit price that will maximize total revenue. (*Hint:* Total revenue = price · the number of units sold.)

72. Maximizing revenue When priced at \$57, one type of camera has annual sales of 525 units. For each \$1 the camera is reduced in price, management expects to sell an additional 75 cameras. Find the unit price that will maximize total revenue. (*Hint:* Total revenue = price · the number of units sold.)

WRITING ABOUT MATH

73. The graph of $y = ax^2 + bx + c \quad (a \neq 0)$ passes the vertical line test. Explain why this shows that the equation defines a function.

74. The graph of $x = y^2 - 2y$ is a parabola. Explain why its graph does not represent a function.

SOMETHING TO THINK ABOUT

75. Can you use a graphing calculator to find solutions of the equation $x^2 + x + 1 = 0$? What is the problem? How do you interpret the result?

76. Complete the square on x in the equation $y = ax^2 + bx + c$ and show that the vertex of the parabolic graph is the point with coordinates of

$$\left(-\frac{b}{2a}, \frac{4ac - b^2}{4a}\right)$$

SECTION 8.5 Quadratic and Other Nonlinear Inequalities

Objectives

1. Solve a quadratic inequality.
2. Solve a rational inequality.
3. Graph a nonlinear inequality in two variables.

Vocabulary

quadratic inequality | critical values | critical points

Getting Ready

Factor each trinomial.

1. $x^2 + 2x - 15$

2. $x^2 - 3x + 2$

We have previously solved linear inequalities. We will now discuss how to solve quadratic and rational inequalities.

1 Solve a quadratic inequality.

Quadratic inequalities in one variable, say x, are inequalities that can be written in one of the following forms, where $a \neq 0$:

$$ax^2 + bx + c < 0 \qquad ax^2 + bx + c > 0$$
$$ax^2 + bx + c \leq 0 \qquad ax^2 + bx + c \geq 0$$

To solve one of these inequalities, we must find its solution set. For example, to solve

$$x^2 + x - 6 < 0$$

we must find the values of x that make the inequality true. To find these values, we can factor the trinomial to obtain

$$(x + 3)(x - 2) < 0$$

Since the product of $x + 3$ and $x - 2$ is to be less than 0, the values of the factors must be opposite in sign. This will happen when one of the factors is positive and the other is negative.

To keep track of the sign of $x + 3$, we can construct the following graph.

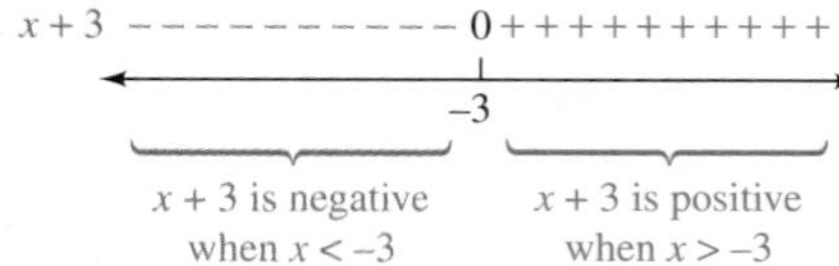

To keep track of the sign of $x - 2$, we can construct the following graph.

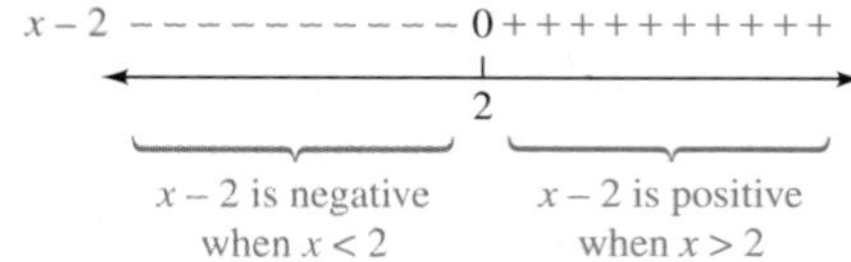

We can merge these graphs as shown in Figure 8-19 and note where the signs of the factors are opposite. This occurs in the interval $(-3, 2)$. Therefore, the product $(x + 3)(x - 2)$ will be less than 0 when

$$-3 < x < 2$$

The graph of the solution set is shown on the number line in the figure.

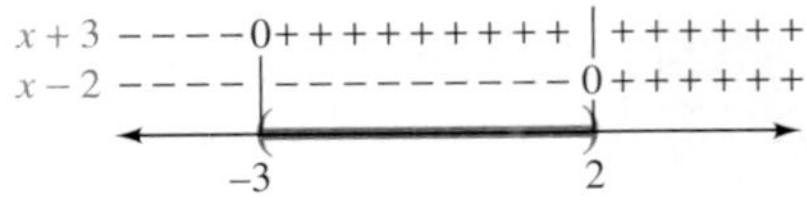

Figure 8-19

Another way to solve the inequality $x^2 + x - 6 < 0$ is to solve its related quadratic equation $x^2 + x - 6 = 0$. The solutions to this equation are sometimes called **critical values** and they establish points on the number line, called **critical points.**

$$x^2 + x - 6 = 0$$

$$(x + 3)(x - 2) = 0$$

$$x + 3 = 0 \quad \text{or} \quad x - 2 = 0$$

$$x = -3 \quad | \quad x = 2$$

The graphs of these critical values establish the three intervals shown on the number line. To determine which intervals are solutions, we test a number in each interval and see whether it satisfies the inequality.

Interval	Test value	Inequality $x^2 + x - 6 < 0$	Result
$(-\infty, -3)$	-6	$(-6)^2 - (-6) - 6 \overset{?}{<} 0$ $36 < 0$ false	The numbers in this interval are not solutions.
$(-3, 2)$	0	$(0)^2 - (0) - 6 \overset{?}{<} 0$ $-6 < 0$ true	The numbers in this interval are solutions.
$(2, \infty)$	5	$(5)^2 - (5) - 6 \overset{?}{<} 0$ $14 < 0$ false	The numbers in this interval are not solutions.

Table 8-1

Figure 8-20

The solution set is the interval $(-3, 2)$, as shown in Figure 8-20.

EXAMPLE 1 Solve: $x^2 + 2x - 3 \geq 0$.

Solution **Method 1** We can factor the trinomial to get $(x - 1)(x + 3)$ and construct the sign chart shown in Figure 8-21.

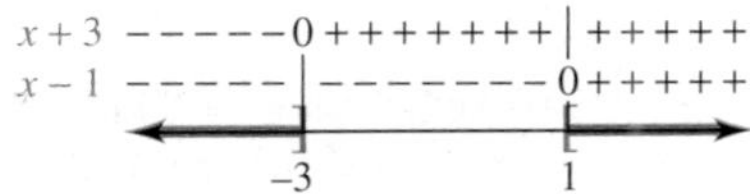

Figure 8-21

- $x - 1$ is 0 when $x = 1$, is positive when $x > 1$, and is negative when $x < 1$.
- $x + 3$ is 0 when $x = -3$, is positive when $x > -3$, and is negative when $x < -3$.

The product of $x - 1$ and $x + 3$ will be greater than 0 when the signs of the binomial factors are the same. This occurs in the intervals $(-\infty, -3)$ and $(1, \infty)$. The numbers -3 and 1 are also included, because they make the product equal to 0. Thus, the solution set is the union of two intervals

$$(-\infty, -3] \cup [1, \infty)$$

The graph of the solution set is shown on the number line in Figure 8-21.

Method 2 We can obtain the same result by solving the related quadratic equation $x^2 + 2x - 3 = 0$ and establishing critical points on the number line.

$$x^2 + 2x - 3 = 0$$
$$(x + 3)(x - 1) = 0$$
$$x + 3 = 0 \quad \text{or} \quad x - 1 = 0$$
$$x = -3 \quad | \quad x = 1$$

The graphs of these critical values establish the three intervals shown on the number line. To determine which intervals are solutions, we test a number in each interval and see whether it satisfies the inequality.

Interval	Test value	Inequality $x^2 + 2x - 3 \geq 0$	Result
$(-\infty, -3)$	-5	$(-5)^2 + 2(-5) - 3 \overset{?}{\geq} 0$ $12 \geq 0$ true	The numbers in this interval are solutions.
$(-3, 1)$	0	$(0)^2 + 2(0) - 3 \overset{?}{\geq} 0$ $-3 \geq 0$ false	The numbers in this interval are not solutions.
$(1, \infty)$	4	$(4)^2 + 2(4) - 3 \overset{?}{\geq} 0$ $21 \geq 0$ true	The numbers in this interval are solutions.

Table 8-2

From the table, we see that numbers in the intervals $(-\infty, -3)$ and $(1, \infty)$ satisfy the inequality. Because the quadratic inequality contains an $\geq$ symbol, the critical values of $x = -3$ and $x = 1$ also satisfy the inequality. Thus, the solution set is the union of two intervals: $(-\infty, -3] \cup [1, \infty)$, as shown in Figure 8-22.

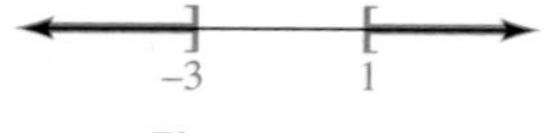

Figure 8-22

SELF CHECK 1 Solve $x^2 + 2x - 15 > 0$ and graph its solution set.

2 Solve a rational inequality.

Making a sign chart is useful for solving many inequalities that are neither linear nor quadratic.

EXAMPLE 2 Solve: $\frac{1}{x} < 6$.

Solution We subtract 6 from both sides to make the right side equal to 0, find a common denominator, and add the fractions:

$$\frac{1}{x} < 6$$

$$\frac{1}{x} - 6 < 0 \quad \text{Subtract 6 from both sides.}$$

$$\frac{1}{x} - \frac{6x}{x} < 0$$ Write each fraction with the same denominator.

$$\frac{1 - 6x}{x} < 0$$ Subtract the numerators and keep the common denominator.

To use Method 1, we make a sign chart, as in Figure 8-23.

- The denominator x is 0 when $x = 0$, is positive when $x > 0$, and is negative when $x < 0$.
- The numerator $1 - 6x$ is 0 when $x = \frac{1}{6}$, is positive when $x < \frac{1}{6}$, and is negative when $x > \frac{1}{6}$.

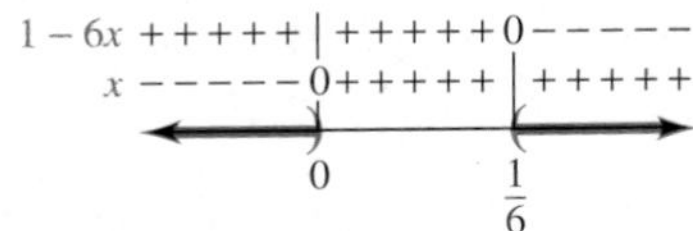

Figure 8-23

The fraction $\frac{1 - 6x}{x}$ will be less than 0 when the numerator and denominator are opposite in sign. This occurs in the union of two intervals:

$$(-\infty, 0) \cup \left(\frac{1}{6}, \infty\right)$$

The graph of this union is shown in Figure 8-23.

To solve this inequality by Method 2, we can find the critical values by finding the values of x that make the numerator of $\frac{1 - 6x}{x}$ equal to 0 and the values of x that make the denominator equal to 0. The critical values are the numbers $\frac{1}{6}$ and 0 and the solution is $(-\infty, 0) \cup \left(\frac{1}{6}, \infty\right)$.

COMMENT Since we don't know whether x is positive, 0, or negative, multiplying both sides of the inequality $\frac{1}{x} < 6$ by x is a three-case situation:

- If $x > 0$, then $1 < 6x$.
- If $x = 0$, then the fraction $\frac{1}{x}$ is undefined.
- If $x < 0$, then $1 > 6x$.

If you multiply both sides by x and solve $1 < 6x$, you are only considering one case and will get only part of the answer.

SELF CHECK 2 Solve: $\frac{3}{x} > 5$.

EXAMPLE 3 Solve: $\frac{x^2 - 3x + 2}{x - 3} \geq 0$.

Solution We write the fraction with the numerator in factored form.

$$\frac{(x - 2)(x - 1)}{x - 3} \geq 0$$

To keep track of the signs of the binomials, we construct the sign chart shown in Figure 8-24. The fraction will be positive in the intervals where all factors are positive, or where two factors are negative. The numbers 1 and 2 are included, because they make the numerator (and thus the fraction) equal to 0. The number 3 is not included, because it gives a 0 in the denominator.

The solution is the union of two intervals $[1, 2] \cup (3, \infty)$. The graph appears in Figure 8-24.

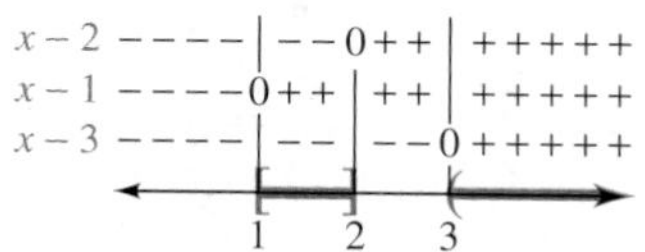

Figure 8-24

To solve this inequality by Method 2, we can find the critical values by finding the values of x that make the numerator of $\frac{(x-2)(x-1)}{x-3}$ equal to 0 and the values of x that make the denominator equal to 0. The critical values are the numbers 1, 2, and 3 and the solution set is $[1, 2] \cup (3, \infty)$.

SELF CHECK 3 Solve: $\frac{x+2}{x^2-2x-3} > 0$ and graph the solution set.

EXAMPLE 4 Solve: $\frac{3}{x-1} < \frac{2}{x}$.

Solution We subtract $\frac{2}{x}$ from both sides to get 0 on the right side and proceed as follows:

$$\frac{3}{x-1} < \frac{2}{x}$$

$$\frac{3}{x-1} - \frac{2}{x} < 0 \quad \text{Subtract } \tfrac{2}{x} \text{ from both sides.}$$

$$\frac{3x}{(x-1)x} - \frac{2(x-1)}{x(x-1)} < 0 \quad \text{Write each fraction with the same denominator.}$$

$$\frac{3x-2x+2}{x(x-1)} < 0 \quad \text{Keep the denominator and subtract the numerators.}$$

$$\frac{x+2}{x(x-1)} < 0 \quad \text{Combine like terms.}$$

We can keep track of the signs of the three factors with the sign chart shown in Figure 8-25. The fraction will be negative in the intervals with either one or three negative factors. The numbers 0 and 1 are not included, because they give a 0 in the denominator, and the number -2 is not included, because it does not satisfy the inequality.

The solution is the union of two intervals $(-\infty, -2) \cup (0, 1)$, as shown in Figure 8-25.

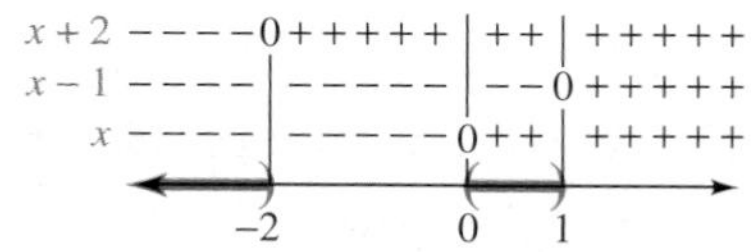

Figure 8-25

To solve this inequality by Method 2, we can find the critical values by finding the values of x that make the numerator of $\frac{x+2}{x(x-1)}$ equal to 0 and the values of x that make the denominator equal to 0. The critical values are the numbers -2, 0, and 1 and the solution set is $(-\infty, -2) \cup (0, 1)$.

SELF CHECK 4 Solve $\frac{2}{x+1} > \frac{1}{x}$ and graph the solution set.

ACCENT ON TECHNOLOGY

Solving Inequalities

To approximate the solutions of $x^2 + 2x - 3 \geq 0$ (Example 1) by graphing, we can use window settings of $[-10, 10]$ for x and $[-10, 10]$ for y and graph the quadratic function $y = x^2 + 2x - 3$, as in Figure 8-26. The solutions of the inequality will be those numbers x for which the graph of $y = x^2 + 2x - 3$ lies above or on the x-axis. We can trace to find that this interval is $(-\infty, -3] \cup [1, \infty)$.

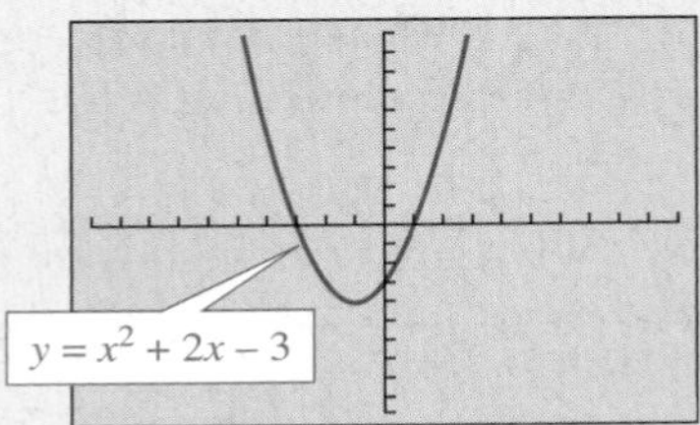

Figure 8-26

To approximate the solutions of $\frac{3}{x-1} < \frac{2}{x}$ (Example 4), we first write the inequality in the form

$$\frac{3}{x-1} - \frac{2}{x} < 0$$

Then we use window settings of $[-5, 5]$ for x and $[-3, 3]$ for y and graph the function $y = \frac{3}{x-1} - \frac{2}{x}$, as in Figure 8-27(a). The solutions of the inequality will be those numbers x for which the graph lies below the x-axis.

We can trace to see that the graph is below the x-axis when x is less than -2. Since we cannot see the graph in the interval $0 < x < 1$, we redraw the graph using window settings of $[-1, 2]$ for x and $[-25, 10]$ for y. See Figure 8-27(b).

We can now see that the graph is below the x-axis in the interval $(0, 1)$. Thus, the solution of the inequality is the union of two intervals:

$$(-\infty, -2) \cup (0, 1)$$

COMMENT Graphing calculators cannot determine whether a critical value is included in a solution set. You must make that determination yourself.

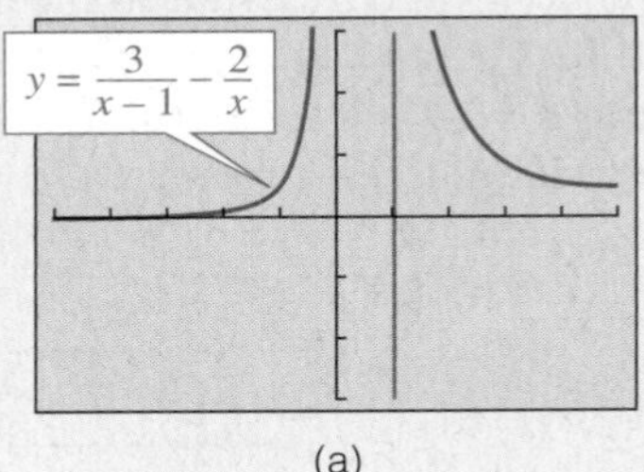

(a)

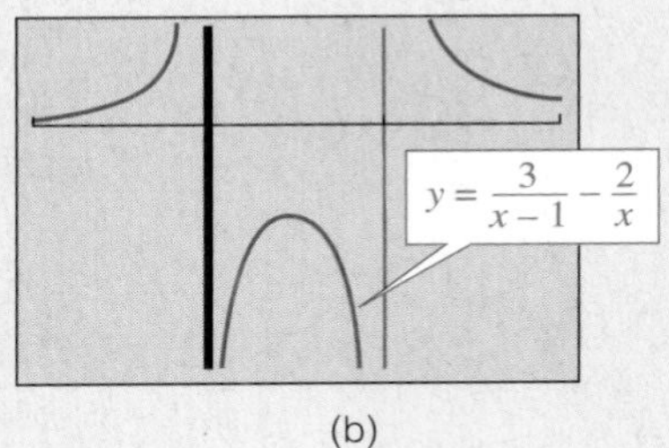

(b)

Figure 8-27

3 Graph a nonlinear inequality in two variables.

We now consider the graphs of nonlinear inequalities in two variables.

EXAMPLE 5 Graph: $y < -x^2 + 4$.

Solution The graph of $y = -x^2 + 4$ is the parabolic boundary separating the region representing $y < -x^2 + 4$ and the region representing $y > -x^2 + 4$.

We graph $y = -x^2 + 4$ as a dashed parabola, because there is no equality symbol in the original inequality. Since the coordinates of the origin satisfy the inequality $y < -x^2 + 4$, the point (0, 0) is in the graph. The complete graph is shown in Figure 8-28.

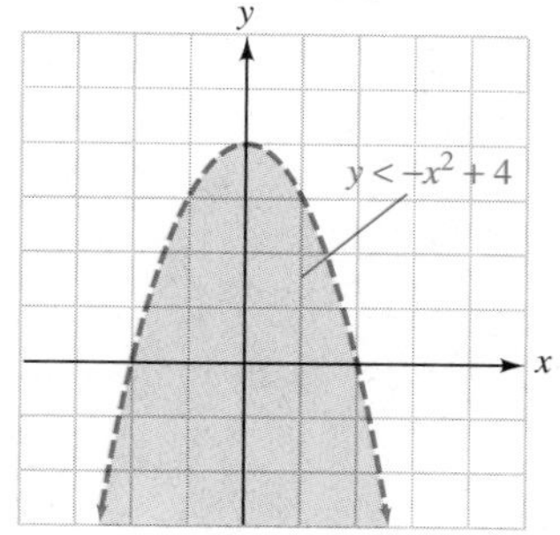

Figure 8-28

SELF CHECK 5 Graph: $y \geq -x^2 + 4$.

EXAMPLE 6 Graph: $x \leq |y|$.

Solution We first graph $x = |y|$ as in Figure 8-29(a), using a solid line because the symbol in the inequality is $\leq$. Since the origin is on the graph, we cannot use it as a test point. However, another point, such as (1, 0), will do. We substitute 1 for x and 0 for y into the inequality to get

$$x \leq |y|$$
$$\mathbf{1} \leq |\mathbf{0}|$$
$$1 \leq 0$$

Since $1 \leq 0$ is a false statement, the point (1, 0) does not satisfy the inequality and is not part of the graph. Thus, the graph of $x \leq |y|$ is to the left of the boundary.

The complete graph is shown in Figure 8-29(b).

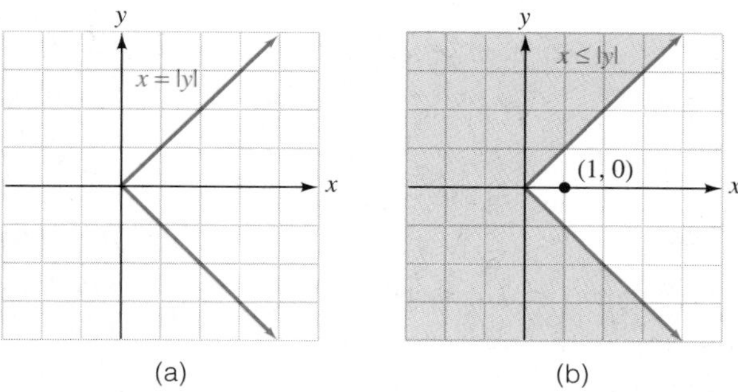

Figure 8-29

SELF CHECK 6 Graph: $x \geq -|y|$.

SELF CHECK ANSWERS

1. $(-\infty, -5) \cup (3, \infty)$

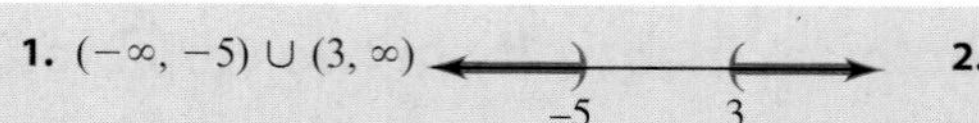

2. $\left(0, \frac{3}{5}\right)$ 0 3/5

3. $(-2, -1) \cup (3, \infty)$ −2 −1 3

4. $(-1, 0) \cup (1, \infty)$ −1 0 1

5.

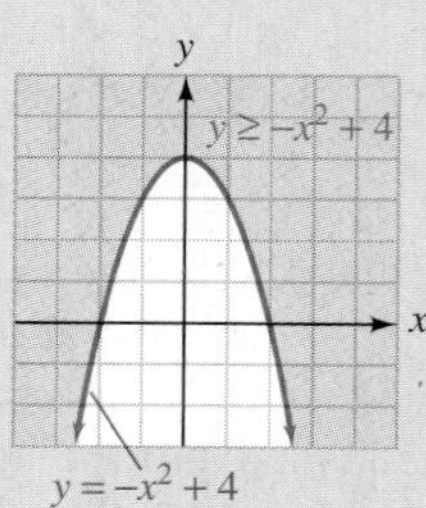

6.

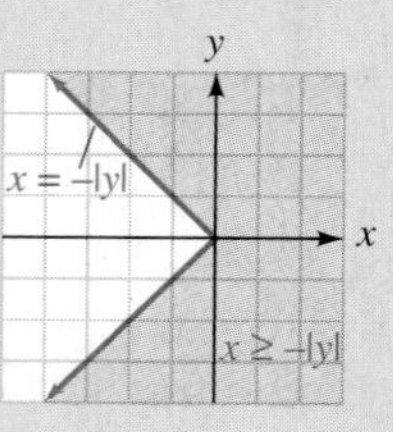

NOW TRY THIS

Find the domain of each of the following.

1. $f(x) = \sqrt{x - 6}$
2. $y \geq x^2 - 5x - 6$
3. $h(x) = \sqrt{x^2 - 5x - 6}$
4. $k(x) = \dfrac{3x - 5}{\sqrt{x^2 - 5x - 6}}$

8.5 EXERCISES

WARM-UPS

Determine where $x - 2$ is

1. 0
2. positive
3. negative

Determine where $x + 3$ is

4. 0
5. positive
6. negative

Multiply both sides of the equation $\frac{1}{x} < 2$ by x when x is

7. positive
8. negative

REVIEW

Write each expression as an equation.

9. y varies directly with x.
10. y varies inversely with t.
11. t varies jointly with x and y.
12. d varies directly with t but inversely with u^2.

Find the slope of the graph of each equation.

13. $y = 3x - 4$
14. $\dfrac{2x - y}{5} = 8$

VOCABULARY AND CONCEPTS *Fill in the blanks.*

15. When $x > 3$, the binomial $x - 3$ is ________ than zero.
16. When $x < 3$, the binomial $x - 3$ is _____ than zero.
17. The expression $x^2 + 9x + 18 \leq 0$ is an example of a __________ inequality.
18. The expression $\frac{x + 3}{x - 2} > 0$ is an example of a ________ inequality.
19. If $x = 0$, the fraction $\frac{1}{x}$ is __________.

20. To solve $x^2 + 2x - 3 < 0$, we can find the solutions of the related equation $x^2 + 2x - 3 = 0$. The solutions are called ____________. They establish points on a number line that separate the line into ________.

21. To keep track of the signs of factors in a product or quotient, we can use a _____ chart.

22. The inequality $|x + 3| < 0$ will be graphed with a _______ line.

GUIDED PRACTICE

Solve each inequality. Give each result in interval notation and graph the solution set. See Example 1. (Objective 1)

23. $x^2 - 5x + 4 < 0$

24. $x^2 - 3x - 4 > 0$

25. $x^2 - 8x + 15 > 0$

26. $x^2 + 2x - 8 < 0$

27. $x^2 + x - 12 \le 0$

28. $x^2 + 7x + 12 \ge 0$

29. $x^2 + 2x \ge 15$

30. $x^2 - 8x \le -15$

31. $x^2 + 8x < -16$

32. $x^2 + 6x \ge -9$

33. $x^2 \ge 9$

34. $x^2 \ge 16$

Solve each inequality. Give each result in interval notation and graph the solution set. See Example 2. (Objective 2)

35. $\frac{1}{x} < 2$

36. $\frac{1}{x} > 3$

37. $\frac{4}{x} \ge 2$

38. $-\frac{6}{x} < 12$

Solve each inequality. Give each result in interval notation and graph the solution set. See Example 3. (Objective 2)

39. $\frac{x^2 - x - 12}{x - 1} < 0$

40. $\frac{x^2 + x - 6}{x - 4} \ge 0$

41. $\frac{x^2 + x - 20}{x + 2} \ge 0$

42. $\frac{x^2 - 10x + 25}{x + 5} < 0$

43. $\frac{x^2 - 4x + 4}{x + 4} < 0$

44. $\frac{2x^2 - 5x + 2}{x + 2} > 0$

45. $\frac{6x^2 - 5x + 1}{2x + 1} > 0$

46. $\frac{6x^2 + 11x + 3}{3x - 1} < 0$

Solve each inequality. Give each result in interval notation and graph the solution set. See Example 4. (Objective 2)

47. $\frac{3}{x - 2} < \frac{4}{x}$

48. $\frac{-6}{x + 1} \ge \frac{1}{x}$

49. $\frac{-5}{x + 2} \ge \frac{4}{2 - x}$

50. $\frac{-6}{x - 3} < \frac{5}{3 - x}$

51. $\frac{7}{x - 3} \ge \frac{2}{x + 4}$

52. $\frac{-5}{x - 4} < \frac{3}{x + 1}$

53. $(x + 2)^2 > 0$

54. $(x - 3)^2 < 0$

Graph each inequality. See Example 5. (Objective 3)

55. $y < x^2 + 1$

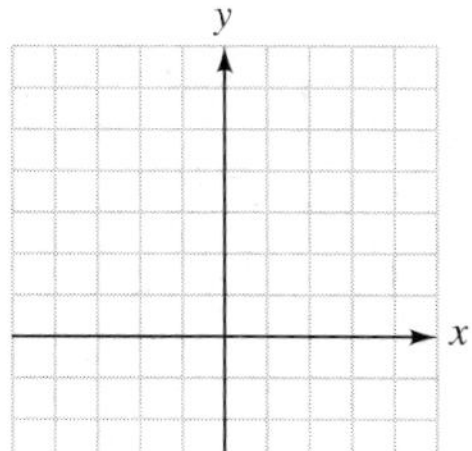

56. $y > x^2 - 3$

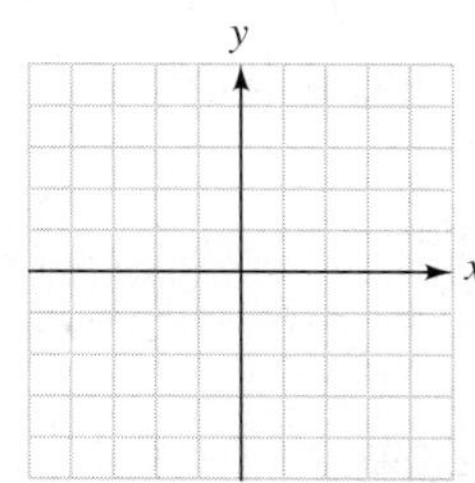

57. $y \le x^2 + 5x + 6$

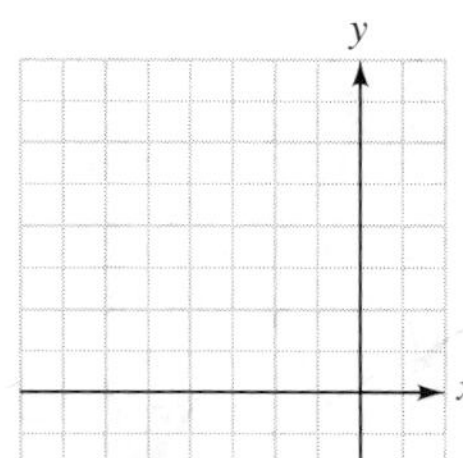

58. $y \ge x^2 + 5x + 4$

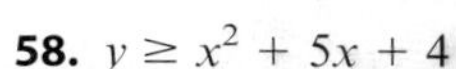

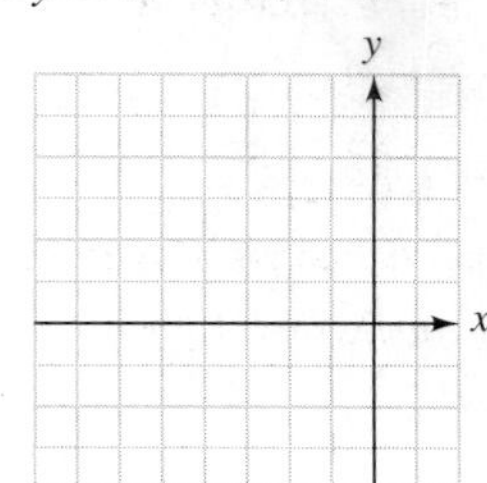

59. $y \geq (x - 1)^2$

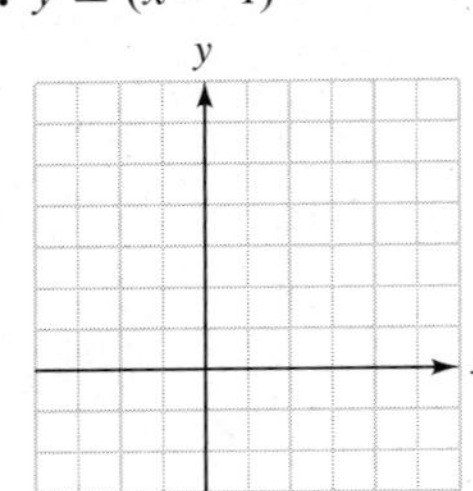

60. $y \leq (x + 2)^2$

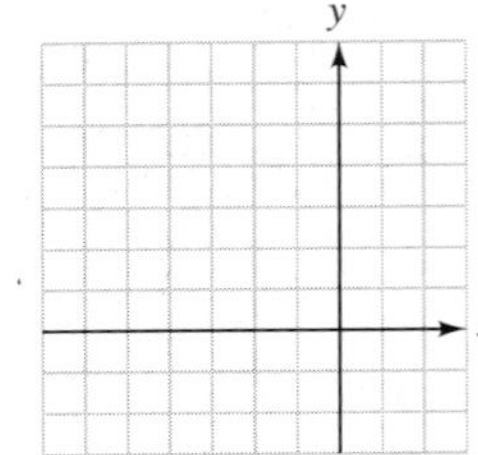

61. $-x^2 - y + 6 > -x$

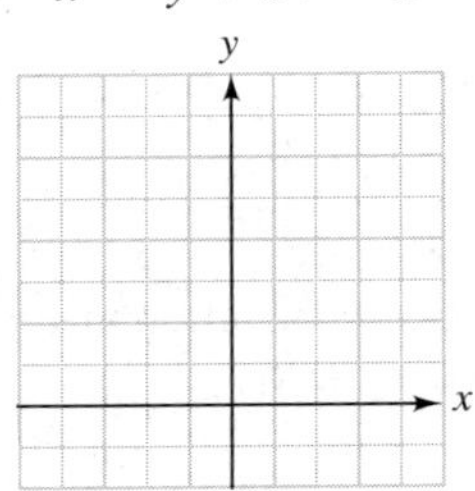

62. $y > (x + 3)(x - 2)$

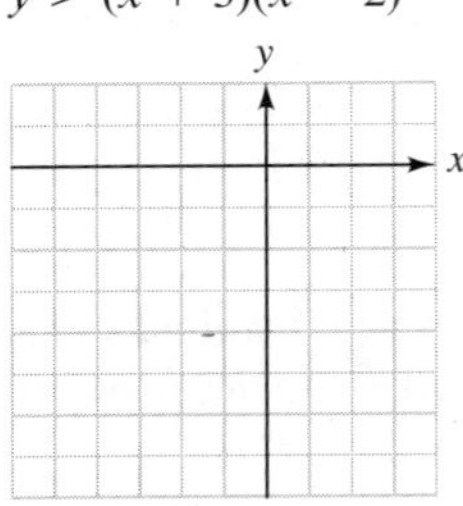

Graph each inequality. See Example 6. (Objective 3)

63. $y < |x + 4|$

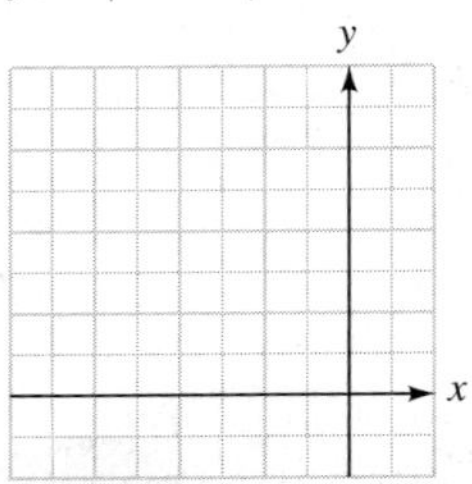

64. $y \geq |x - 3|$

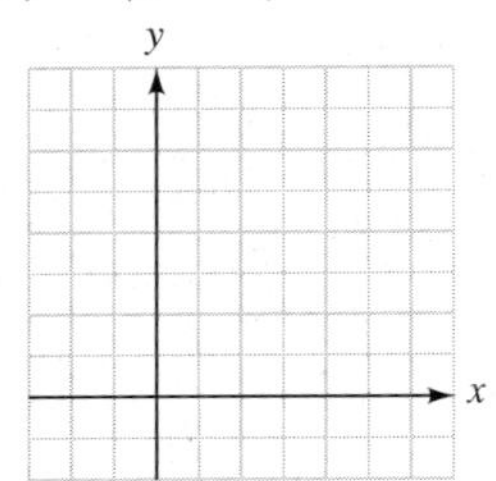

65. $y \leq -|x| + 2$

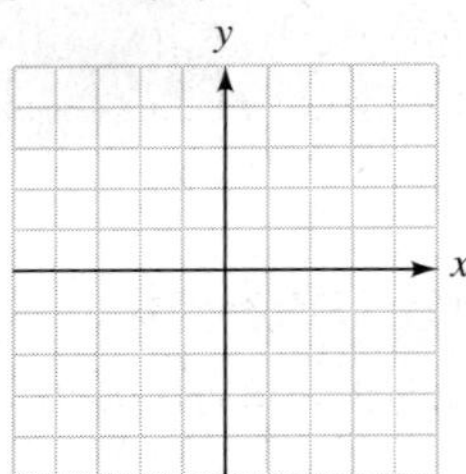

66. $y > |x| - 2$

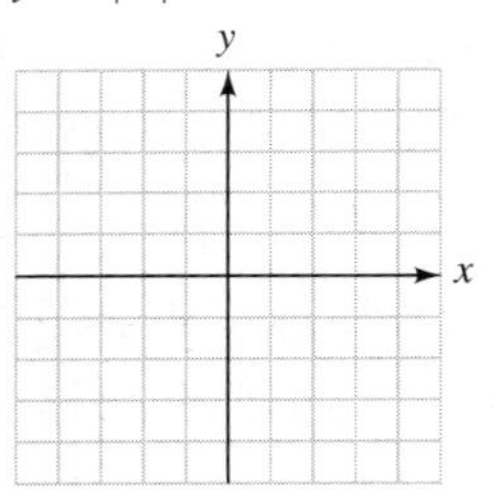

ADDITIONAL PRACTICE

Solve each inequality. Give each result in interval notation and graph the solution set.

67. $2x^2 - 50 < 0$

68. $3x^2 - 243 < 0$

69. $-\frac{5}{x} < 3$

70. $\frac{4}{x} \geq 8$

71. $\frac{x}{x + 4} \leq \frac{1}{x + 1}$

72. $\frac{x}{x + 9} \geq \frac{1}{x + 1}$

73. $\frac{x}{x + 16} > \frac{1}{x + 1}$

74. $\frac{x}{x + 25} < \frac{1}{x + 1}$

Use a graphing calculator to solve each inequality. Give the answer in interval notation.

75. $x^2 - 2x - 3 < 0$

76. $x^2 + x - 6 > 0$

77. $\frac{x + 3}{x - 2} > 0$

78. $\frac{3}{x} < 2$

WRITING ABOUT MATH

79. Explain why $(x - 4)(x + 5)$ will be positive only when the signs of $x - 4$ and $x + 5$ are the same.

80. Explain how to find the graph of $y \geq x^2$.

SOMETHING TO THINK ABOUT

81. Under what conditions will the fraction $\frac{(x - 1)(x + 4)}{(x + 2)(x + 1)}$ be positive?

82. Under what conditions will the fraction $\frac{(x - 1)(x + 4)}{(x + 2)(x + 1)}$ be negative?

SECTION 8.6 Algebra and Composition of Functions

Objectives

1. Find the sum, difference, product, and quotient of two functions.
2. Find the composition of two functions.
3. Find the difference quotient of a function.
4. Solve an application problem requiring the composition of two functions.

Vocabulary

composition
composite functions
identity function
difference quotient

Getting Ready

Assume that $P(x) = 2x + 1$ and $Q(x) = x - 2$. Find each expression.

1. $P(x) + Q(x)$

2. $P(x) - Q(x)$

3. $P(x) \cdot Q(x)$

4. $\dfrac{P(x)}{Q(x)}$

Throughout the text, we have talked about functions. In this section, we will show how to add, subtract, multiply, and divide them. We also will show how to find the composition of two functions.

Find the sum, difference, product, and quotient of two functions.

We now consider how functions can be added, subtracted, multiplied, and divided.

Operations on Functions

If the domains and ranges of functions f and g are subsets of the real numbers,

The *sum* of f and g, denoted as $f + g$, is defined by

$$(f + g)(x) = f(x) + g(x)$$

The *difference* of f and g, denoted as $f - g$, is defined by

$$(f - g)(x) = f(x) - g(x)$$

The *product* of f and g, denoted as $f \cdot g$, is defined by

$$(f \cdot g)(x) = f(x)g(x)$$

The *quotient* of f and g, denoted as f/g, is defined by

$$(f/g)(x) = \frac{f(x)}{g(x)} \quad (g(x) \neq 0)$$

The domain of each of these functions is the set of real numbers x that are in the domain of both f and g. In the case of the quotient, there is the further restriction that $g(x) \neq 0$.

EXAMPLE 1 Let $f(x) = 2x^2 + 1$ and $g(x) = 5x - 3$. Find each function and its domain.
a. $f + g$ **b.** $f - g$

Solution **a.**
$$\begin{aligned}(f + g)(x) &= f(x) + g(x) \\ &= (2x^2 + 1) + (5x - 3) \\ &= 2x^2 + 5x - 2\end{aligned}$$

The domain of $f + g$ is the set of real numbers that are in the domain of both f and g. Since the domain of both f and g is the interval $(-\infty, \infty)$, the domain of $f + g$ is also the interval $(-\infty,\infty)$.

b.
$$\begin{aligned}(f - g)(x) &= f(x) - g(x) \\ &= (2x^2 + 1) - (5x - 3) \\ &= 2x^2 + 1 - 5x + 3 && \text{Remove parentheses.} \\ &= 2x^2 - 5x + 4 && \text{Combine like terms.}\end{aligned}$$

Since the domain of both f and g is $(-\infty, \infty)$, the domain of $f - g$ is also the interval $(-\infty, \infty)$.

SELF CHECK 1 Let $f(x) = 3x - 2$ and $g(x) = 2x^2 + 3x$. Find each function and its domain.
a. $f + g$ **b.** $f - g$

EXAMPLE 2 Let $f(x) = 2x^2 + 1$ and $g(x) = 5x - 3$. Find each function and its domain.
a. $f \cdot g$ **b.** f/g

Solution **a.**
$$\begin{aligned}(f \cdot g)(x) &= f(x)g(x) \\ &= (2x^2 + 1)(5x - 3) \\ &= 10x^3 - 6x^2 + 5x - 3 && \text{Multiply.}\end{aligned}$$

The domain of $f \cdot g$ is the set of real numbers that are in the domain of both f and g. Since the domain of both f and g is the interval $(-\infty, \infty)$, the domain of $f \cdot g$ is also the interval $(-\infty, \infty)$.

b.
$$\begin{aligned}(f/g)(x) &= \frac{f(x)}{g(x)} \\ &= \frac{2x^2 + 1}{5x - 3}\end{aligned}$$

Since the denominator of the fraction cannot be 0, $x \neq \frac{3}{5}$. The domain of f/g is the union of two intervals $\left(-\infty, \frac{3}{5}\right) \cup \left(\frac{3}{5}, \infty\right)$.

SELF CHECK 2 Let $f(x) = 2x^2 - 3$ and $g(x) = x - 1$. Find each function and its domain.
a. $f \cdot g$ **b.** f/g

2 Find the composition of two functions.

We have seen that a function can be represented by a machine: We put in a number from the domain, and a number from the range comes out. For example, if we put the number

2 into the machine shown in Figure 8-30(a), the number $f(2) = 5(2) - 2 = 8$ comes out. In general, if we put x into the machine shown in Figure 8-30(b), the value $f(x)$ comes out.

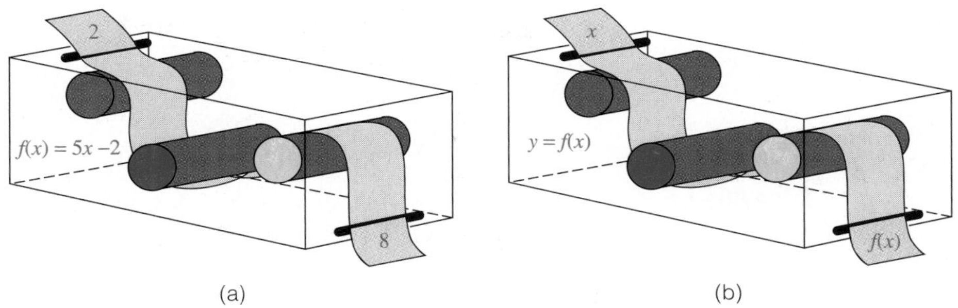

(a) (b)

Figure 8-30

Often one quantity is a function of a second quantity that depends, in turn, on a third quantity. For example, the cost of a car trip is a function of the gasoline consumed. The amount of gasoline consumed, in turn, is a function of the number of miles driven. Such chains of dependence can be analyzed mathematically as **compositions of functions.**

The function machines shown in Figure 8-31 illustrate the composition of functions f and g. When we put a number x into the function g, $g(x)$ comes out. The value $g(x)$ goes into function f, which transforms $g(x)$ into $f(g(x))$. This two-step process defines a new function, called a **composite function.** If the function machines for g and f were connected to make a single machine, that machine would be named $f \circ g$, read as "f composition g."

To be in the domain of the composite function $f \circ g$, a number x has to be in the domain of g. Also, the output of g must be in the domain of f. Thus, the domain of $f \circ g$ consists of those numbers x that are in the domain of g, and for which $g(x)$ is in the domain of f.

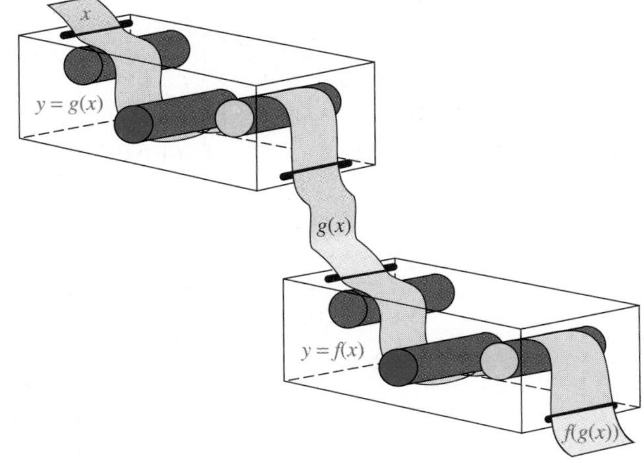

Figure 8-31

Composite Functions

The **composite function** $f \circ g$ is defined by

$$(f \circ g)(x) = f(g(x))$$

COMMENT Note that in this example, $(f \circ g)(x) \neq (g \circ f)(x)$. This shows that the composition of functions is not commutative.

For example, if $f(x) = 4x - 5$ and $g(x) = 3x + 2$, then

$$\begin{aligned}(f \circ g)(x) &= f(g(x)) \\ &= f(3x + 2) \\ &= 4(3x + 2) - 5 \\ &= 12x + 8 - 5 \\ &= 12x + 3\end{aligned} \qquad \begin{aligned}(g \circ f)(x) &= g(f(x)) \\ &= g(4x - 5) \\ &= 3(4x - 5) + 2 \\ &= 12x - 15 + 2 \\ &= 12x - 13\end{aligned}$$

EXAMPLE 3 Let $f(x) = 2x + 1$ and $g(x) = x - 4$. Find:
a. $(f \circ g)(9)$ **b.** $(f \circ g)(x)$ **c.** $(g \circ f)(-2)$.

Solution **a.** $(f \circ g)(9)$ means $f(g(9))$. In Figure 8-32(a), function g receives the number 9, subtracts 4, and releases the number $g(9) = 5$. The 5 then goes into the f function, which doubles 5 and adds 1. The final result, 11, is the output of the composite function $f \circ g$:

$$(f \circ g)(9) = f(g(9)) = f(5) = 2(5) + 1 = 11$$

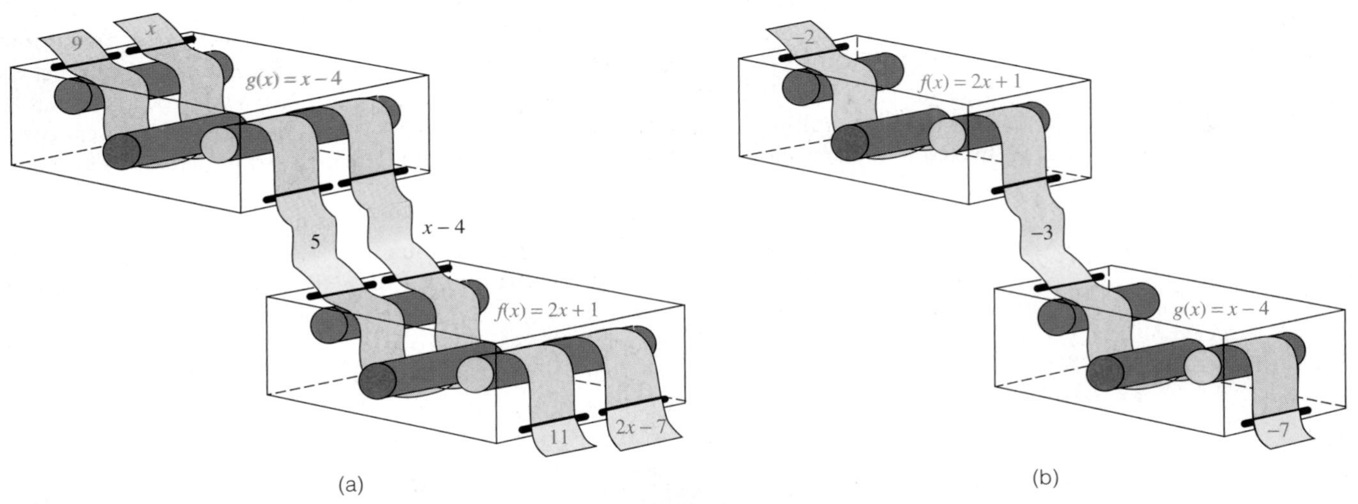

Figure 8-32

b. $(f \circ g)(x)$ means $f(g(x))$. In Figure 8-32(a), function g receives the number x, subtracts 4, and releases the number $x - 4$. The $x - 4$ then goes into the f function, which doubles $x - 4$ and adds 1. The final result, $2x - 7$, is the output of the composite function $f \circ g$.

$$(f \circ g)(x) = f(g(x)) = f(x - 4) = 2(x - 4) + 1 = 2x - 7$$

c. $(g \circ f)(-2)$ means $g(f(-2))$. In Figure 8-32(b), function f receives the number -2, doubles it and adds 1, and releases -3 into the g function. Function g subtracts 4 from -3 and releases a final result of -7. Thus,

$$(g \circ f)(-2) = g(f(-2)) = g(-3) = -3 - 4 = -7$$

SELF CHECK 3 Let $f(x) = 3x + 2$ and $g(x) = 9x - 5$. Find
a. $(f \circ g)(2)$ **b.** $(g \circ f)(x)$ **c.** $(g \circ f)(-4)$.

Recall that in the real number system, 0 is called the *additive identity* because $0 + x = x$ and 1 is called the *multiplicative identity* because $1 \cdot x = x$. There is an identity for functions as well. The **identity function** is defined by the equation $I(x) = x$. Under this function, the value that corresponds to any real number x is x itself. If f is any function, the composition of f with the identity function is the function f:

$$(f \circ I)(x) = (I \circ f)(x) = f(x)$$

We can show this as follows:

$(f \circ I)(x)$ means $f(I(x))$. Because $I(x) = x$, we have

$$(f \circ I)(x) = f(I(x)) = f(x)$$

$(I \circ f)(x)$ means $I(f(x))$. Because I passes any number through unchanged, we have $I(f(x)) = f(x)$ and

$$(I \circ f)(x) = I(f(x)) = f(x)$$

3 Find the difference quotient of a function.

An important function in calculus, called the **difference quotient,** represents the slope of a line that passes through two given points on the graph of a function. The difference quotient is defined as follows:

$$\frac{f(x + h) - f(x)}{h}$$

EXAMPLE 4 If $f(x) = x^2 - 4$, evaluate the difference quotient.

Solution First, we evaluate $f(x + h)$.

$$f(x) = x^2 - 4$$

$$f(x + h) = (x + h)^2 - 4 \qquad \text{Substitute } x + h \text{ for } h.$$

$$= x^2 + 2xh + h^2 - 4 \qquad (x + h)^2 = x^2 + 2hx + h^2$$

Then we note that $f(x) = x^2 - 4$. We can now substitute the values of $f(x + h)$ and $f(x)$ into the difference quotient and simplify.

$$\frac{f(x + h) - f(x)}{h} = \frac{(x^2 + 2xh + h^2 - 4) - (x^2 - 4)}{h}$$

$$= \frac{x^2 + 2xh + h^2 - 4 - x^2 + 4}{h} \qquad \text{Remove parentheses.}$$

$$= \frac{2xh + h^2}{h} \qquad \text{Combine like terms.}$$

$$= \frac{h(2x + h)}{h} \qquad \text{Factor out } h \text{ in the numerator.}$$

$$= 2x + h \qquad \text{Divide out } h; \tfrac{h}{h} = 1.$$

The difference quotient for this function simplifies as $2x + h$.

SELF CHECK 4 If $f(x) = 3x^2 - 4$, evaluate the difference quotient.

4 Solve an application problem requiring the composition of two functions.

EXAMPLE 5 **TEMPERATURE CHANGE** A laboratory sample is removed from a cooler at a temperature of 15° Fahrenheit. Technicians are warming the sample at a controlled rate of 3° F per hour. Express the sample's Celsius temperature as a function of the time, t (in hours), since it was removed from refrigeration.

Solution The temperature of the sample is 15° F when $t = 0$. Because it warms at 3° F per hour, it warms $3t°$ after t hours. The Fahrenheit temperature after t hours is given by the function

$$F(t) = 3t + 15$$

The Celsius temperature is a function of the Fahrenheit temperature, given by the formula

$$C(F) = \frac{5}{9}(F - 32)$$

To express the sample's Celsius temperature as a function of time, we find the composition function $C \circ F$.

$$\begin{aligned}
(C \circ F)(t) &= C(F(t)) \\
&= \frac{5}{9}(F(t) - 32) \\
&= \frac{5}{9}[(3t + 15) - 32] && \text{Substitute } 3t + 15 \text{ for } F(t). \\
&= \frac{5}{9}(3t - 17) && \text{Simplify.} \\
&= \frac{15}{9}t - \frac{85}{9} \\
&= \frac{5}{3}t - \frac{85}{9}
\end{aligned}$$

SELF CHECK ANSWERS

1. a. $2x^2 + 6x - 2, (-\infty, \infty)$ **b.** $-2x^2 - 2, (-\infty, \infty)$ **2. a.** $2x^3 - 2x^2 - 3x + 3, (-\infty, \infty)$ **b.** $\frac{2x^2 - 3}{x - 1}, (-\infty, 1) \cup (1, \infty)$ **3. a.** 41 **b.** $27x + 13$ **c.** −95 **4.** $6x + 3h$

NOW TRY THIS

Given $f(x) = 3x - 2$ and $g(x) = x^2 - 5x + 1$, find:

1. $(g \circ f)(-1)$
2. $(g \circ f)(x)$
3. $(f \circ f)(x)$

8.6 EXERCISES

Assume no denominators are 0.

WARM-UPS *If f(x) = 2x, g(x) = 3x, and h(x) = 4x, find:*

1. $f + g$
2. $h - g$
3. $f \cdot h$
4. g/f
5. h/f
6. $g \cdot h$
7. $(f \circ h)(x)$
8. $(f \circ g)(x)$

REVIEW *Simplify each expression.*

9. $\dfrac{3x^2 + x - 14}{4 - x^2}$
10. $\dfrac{2x^3 + 14x^2}{3 + 2x - x^2} \cdot \dfrac{x^2 - 3x}{x}$
11. $\dfrac{8 + 2x - x^2}{12 + x - 3x^2} \div \dfrac{3x^2 + 5x - 2}{3x - 1}$
12. $\dfrac{x - 1}{1 + \dfrac{x}{x - 2}}$

VOCABULARY AND CONCEPTS *Fill in the blanks.*

13. $(f + g)(x) =$ ______
14. $(f - g)(x) =$ ______
15. $(f \cdot g)(x) =$ ______
16. $(f/g)(x) = \dfrac{\text{\_\_\_\_}}{\text{\_\_\_\_}}$ $(g(x) \neq 0)$
17. In Exercises 13–15, the domain of each function is the set of real numbers x that are in the ______ of both f and g.
18. The ______ of functions f and g is denoted by $(f \circ g)(x)$ or $f \circ g$.
19. $(f \circ g)(x) =$ ______
20. If I is the identity function, then $(f \circ I)(x) =$ ______.
21. If I is the identity function, then $(I \circ f)(x) =$ ______.
22. The difference quotient is defined as ______.

GUIDED PRACTICE

Let f(x) = 3x and g(x) = 4x. Find each function and its domain. See Examples 1–2. (Objective 1)

23. $f + g$
24. $f - g$
25. $f \cdot g$
26. f/g
27. $g - f$
28. $g + f$
29. g/f
30. $g \cdot f$

Let f(x) = 2x + 1 and g(x) = x − 3. Find each function and its domain. See Examples 1–2. (Objective 1)

31. $f + g$
32. $f - g$
33. $f \cdot g$
34. f/g
35. $g - f$
36. $g + f$
37. g/f
38. $g \cdot f$

Let f(x) = 2x + 1 and g(x) = x² − 1. Find each value. See Example 3. (Objective 2)

39. $(f \circ g)(2)$
40. $(g \circ f)(2)$
41. $(g \circ f)(-3)$
42. $(f \circ g)(-3)$
43. $(f \circ g)(0)$
44. $(g \circ f)(0)$
45. $(f \circ g)\left(\frac{1}{2}\right)$
46. $(g \circ f)\left(\frac{1}{3}\right)$
47. $(f \circ g)(x)$
48. $(g \circ f)(x)$
49. $(g \circ f)(2x)$
50. $(f \circ g)(2x)$

Find $\frac{f(x + h) - f(x)}{h}$. See Example 4. (Objective 3)

51. $f(x) = 2x + 3$
52. $f(x) = 3x - 5$
53. $f(x) = x^2$
54. $f(x) = x^2 - 1$
55. $f(x) = 2x^2 - 1$
56. $f(x) = 3x^2$
57. $f(x) = x^2 + x$
58. $f(x) = x^2 - x$
59. $f(x) = x^2 + 3x - 4$
60. $f(x) = x^2 - 4x + 3$
61. $f(x) = 2x^2 + 3x - 7$
62. $f(x) = 3x^2 - 2x + 4$

ADDITIONAL PRACTICE

Let f(x) = 3x − 2 and g(x) = 2x² + 1. Find each function and its domain.

63. $f - g$
64. $f + g$
65. f/g
66. $f \cdot g$

Let f(x) = x² − 1 and g(x) = x² − 4. Find each function and its domain.

67. $f - g$
68. $f + g$
69. g/f
70. $g \cdot f$

Let f(x) = 3x − 2 and g(x) = x² + x. Find each value.

71. $(f \circ g)(4)$
72. $(g \circ f)(4)$

73. $(g \circ f)(-3)$

74. $(f \circ g)(-3)$

75. $(g \circ f)(0)$

76. $(f \circ g)(0)$

77. $(g \circ f)(x)$

78. $(f \circ g)(x)$

Find $\frac{f(x) - f(a)}{x - a}$ $(x \neq a)$.

79. $f(x) = 2x + 3$

80. $f(x) = 3x - 5$

81. $f(x) = x^2$

82. $f(x) = x^2 - 1$

83. $f(x) = 2x^2 - 1$

84. $f(x) = 3x^2$

85. $f(x) = x^2 + x$

86. $f(x) = x^2 - x$

87. $f(x) = x^2 + 3x - 4$

88. $f(x) = x^2 - 4x + 3$

89. $f(x) = 2x^2 + 3x - 7$

90. $f(x) = 3x^2 - 2x + 4$

91. If $f(x) = x + 1$ and $g(x) = 2x - 5$, show that $(f \circ g)(x) \neq (g \circ f)(x)$.

92. If $f(x) = x^2 + 1$ and $g(x) = 3x^2 - 2$, show that $(f \circ g)(x) \neq (g \circ f)(x)$.

93. If $f(x) = x^2 + 2x - 3$, find $f(a)$, $f(h)$, and $f(a + h)$. Then show that $f(a + h) \neq f(a) + f(h)$.

94. If $g(x) = 2x^2 + 10$, find $g(a)$, $g(h)$, and $g(a + h)$. Then show that $g(a + h) \neq g(a) + g(h)$.

95. If $f(x) = x^3 - 1$, find $\frac{f(x + h) - f(x)}{h}$.

96. If $f(x) = x^3 + 2$, find $\frac{f(x + h) - f(x)}{h}$.

APPLICATIONS *Solve each application problem.*

See Example 5. (Objective 4)

97. Alloys A molten alloy must be cooled slowly to control crystallization. When removed from the furnace, its temperature is 2,700° F, and it will be cooled at 200° F per hour. Express the Celsius temperature as a function of the number of hours t since cooling began.

98. Weather forecasting A high-pressure area promises increasingly warmer weather for the next 48 hours. The temperature is now 34° Celsius and will rise 1° every 6 hours. Express the Fahrenheit temperature as a function of the number of hours from now.

WRITING ABOUT MATH

99. Explain how to find the domain of f/g.

100. Explain why the difference quotient represents the slope of a line passing through $(x, f(x))$ and $(x + h, f(x + h))$.

SOMETHING TO THINK ABOUT

101. Is composition of functions associative? Choose functions f, g, and h and determine whether $[f \circ (g \circ h)](x) = [(f \circ g) \circ h](x)$.

102. Choose functions f, g, and h and determine whether $f \circ (g + h) = f \circ g + f \circ h$.

SECTION 8.7 Inverses of Functions

Objectives

1. Determine whether a function is one-to-one.
2. Apply the horizontal line test to determine if the graph of a function is one-to-one.
3. Find the inverse of a function.

Vocabulary

one-to-one function   horizontal line test   inverse function

Getting Ready

Solve each equation for y.

1. $x = 3y + 2$

2. $x = \frac{3}{2}y + 5$

We already know that real numbers have inverses. For example, the additive inverse of 3 is -3, because $3 + (-3) = 0$. The multiplicative inverse of 3 is $\frac{1}{3}$, because $3\left(\frac{1}{3}\right) = 1$. In a similar way, functions have inverses. After discussing one-to-one functions, we will learn how to find the inverse of a function.

1 Determine whether a function is one-to-one.

Recall that for each input into a function, there is a single output. For some functions, different inputs have the same output, as shown in Figure 8-33(a). For other functions, different inputs have different outputs, as shown in Figure 8-33(b).

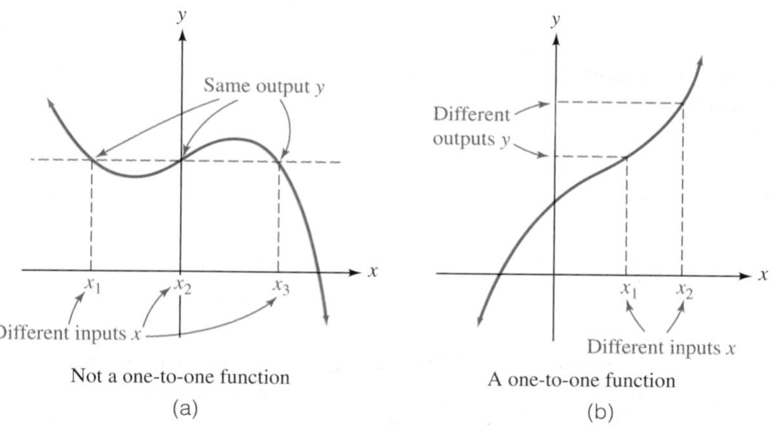

Figure 8-33

When every output of a function corresponds to exactly one input, we say that the function is *one-to-one.*

One-to-One Functions A function is called **one-to-one** if each input value of x in the domain determines a different output value of y in the range.

EXAMPLE 1 Determine whether **a.** $f(x) = x^2$ and **b.** $f(x) = x^3$ are one-to-one.

Solution **a.** The function $f(x) = x^2$ is not one-to-one, because different input values x can determine the same output value y. For example, inputs of 3 and -3 produce the same output value of 9.

$$f(3) = 3^2 = 9 \quad \text{and} \quad f(-3) = (-3)^2 = 9$$

b. The function $f(x) = x^3$ is one-to-one, because different input values x determine different output values of y for all x. This is because different numbers have different cubes.

SELF CHECK 1 Determine whether $f(x) = 2x + 3$ is one-to-one.

2 Apply the horizontal line test to determine if the graph of a function is one-to-one.

A **horizontal line test** can be used to determine whether the graph of a function represents a one-to-one function. If every horizontal line that intersects the graph of a function does so only once, the function is one-to-one. Otherwise, the function is not one-to-one. See Figure 8-34.

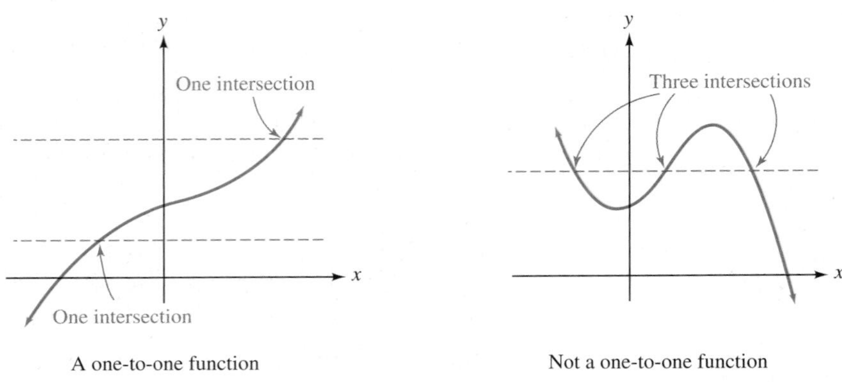

Figure 8-34

EXAMPLE 2 The graphs in Figure 8-35 represent functions. Use the horizontal line test to determine whether the graphs represent one-to-one functions.

Solution **a.** Because many horizontal lines intersect the graph shown in Figure 8-35(a) twice, the graph does not represent a one-to-one function.

b. Because each horizontal line that intersects the graph in Figure 8-35(b) does so exactly once, the graph does represent a one-to-one function.

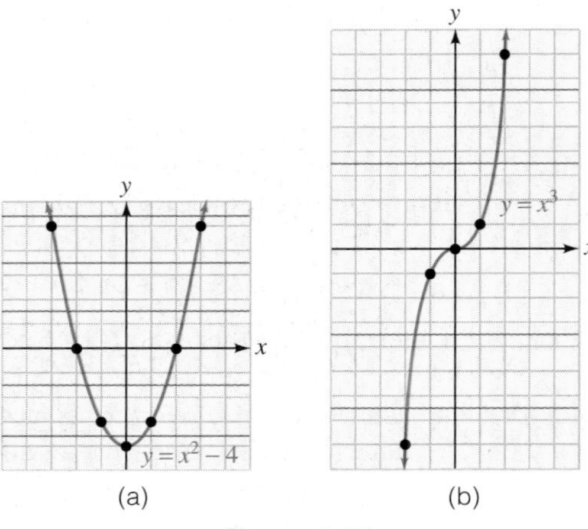

Figure 8-35

SELF CHECK 2 Does the following graph represent a one-to-one function?

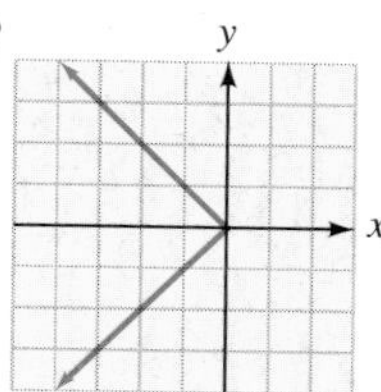

COMMENT Use the vertical line test to determine whether a graph represents a function. If it does, use the horizontal line test to determine whether the function is one-to-one.

3 Find the inverse of a function.

The function defined by $C = \frac{5}{9}(F - 32)$ is the formula that we use to convert degrees Fahrenheit to degrees Celsius. If we substitute a Fahrenheit reading into the formula, a Celsius reading comes out. For example, if we substitute 41° for F we obtain a Celsius reading of 5°:

$$C = \frac{5}{9}(F - 32)$$

$$= \frac{5}{9}(\mathbf{41} - 32) \quad \text{Substitute 41 for } F.$$

$$= \frac{5}{9}(9)$$

$$= 5$$

If we want to find a Fahrenheit reading from a Celsius reading, we need a formula into which we can substitute a Celsius reading and have a Fahrenheit reading come out. Such a formula is $F = \frac{9}{5}C + 32$, which takes the Celsius reading of 5° and turns it back into a Fahrenheit reading of 41°.

$$F = \frac{9}{5}C + 32$$

$$= \frac{9}{5}(\mathbf{5}) + 32 \quad \text{Substitute 5 for } C.$$

$$= 41$$

The functions defined by these two formulas do opposite things. The first turns 41° F into 5° Celsius, and the second turns 5° Celsius back into 41° F. For this reason, we say that the functions are *inverses* of each other.

If f is the function determined by the table shown in Figure 8-36(a), it turns the number 1 into 10, 2 into 20, and 3 into 30. Since the inverse of f must turn 10 back into 1, 20 back into 2, and 30 back into 3, it consists of the ordered pairs shown in Figure 8-36(b) on the next page.

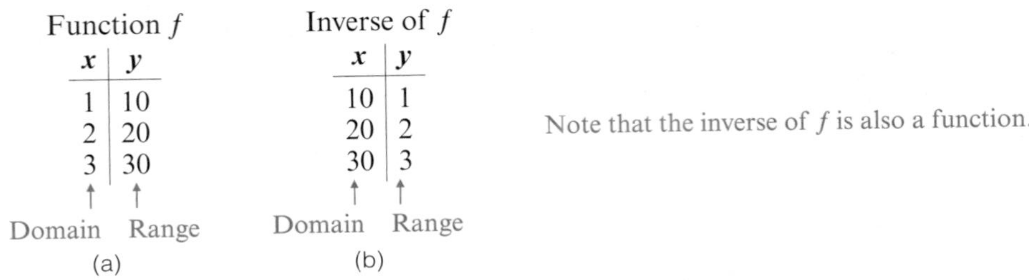

Function f

x	y
1	10
2	20
3	30
↑ Domain	↑ Range

(a)

Inverse of f

x	y
10	1
20	2
30	3
↑ Domain	↑ Range

(b)

Note that the inverse of f is also a function.

Figure 8-36

We note that the domain of f and the range of its inverse is $\{1, 2, 3\}$. The range of f and the domain of its inverse is $\{10, 20, 30\}$.

This example suggests that to form the inverse of a function f, we simply interchange the coordinates of each ordered pair that determines f. When the inverse of a function is also a function, we call it f *inverse* and denote it with the symbol f^{-1}.

COMMENT The symbol $f^{-1}(x)$ is read as "the inverse of $f(x)$" or just "f inverse." The -1 in the notation $f^{-1}(x)$ is not an exponent. Remember that $f^{-1}(x) \neq \frac{1}{f(x)}$.

Finding the Inverse of a One-to-One Function

If a function is one-to-one, we find its inverse as follows:

1. Replace $f(x)$ with y, if necessary.
2. Interchange the variables x and y.
3. Solve the resulting equation for y.
4. This equation is $y = f^{-1}(x)$.

EXAMPLE 3 If $f(x) = 4x + 2$, find the inverse of f and determine whether it is a function.

Solution To find the inverse, we replace $f(x)$ with y and interchange the positions of x and y.

$$f(x) = 4x + 2$$

$$y = 4x + 2 \quad \text{Replace } f(x) \text{ with } y.$$

$$x = 4y + 2 \quad \text{Interchange the variables } x \text{ and } y.$$

Then we solve the equation for y.

$$x = 4y + 2$$

$$x - 2 = 4y \quad \text{Subtract 2 from both sides.}$$

$$y = \frac{x - 2}{4} \quad \text{Divide both sides by 4 and write } y \text{ on the left side.}$$

Because each input x that is substituted into the resulting equation gives one output y, the inverse of f is a function. Expressing the inverse in function notation, we have

$$f^{-1}(x) = \frac{x - 2}{4}$$

SELF CHECK 3 If $f(x) = -5x - 3$, find the inverse of f and determine whether it is a function.

COMMENT If the inverse of a function is also a function, we usually write the inverse function using function notation.

To emphasize an important relationship between a function and its inverse, we substitute some number x, such as $x = 3$, into the function $f(x) = 4x + 2$ of Example 3. The corresponding value of y produced is

$$f(3) = 4(3) + 2 = 14$$

If we substitute 14 into the inverse function, f^{-1}, the corresponding value of y that is produced is

$$f^{-1}(14) = \frac{14 - 2}{4} = 3$$

Thus, the function f turns 3 into 14, and the inverse function f^{-1} turns 14 back into 3. In general, *the composition of a function and its inverse is the identity function.*

To prove that $f(x) = 4x + 2$ and $f^{-1}(x) = \frac{x-2}{4}$ are inverse functions, we must show that their composition (in both directions) is the identity function:

$$\begin{aligned}(f \circ f^{-1})(x) &= f(f^{-1}(x)) \\ &= f\left(\frac{x-2}{4}\right) \\ &= 4\left(\frac{x-2}{4}\right) + 2 \\ &= x - 2 + 2 \\ &= x\end{aligned} \qquad \begin{aligned}(f^{-1} \circ f)(x) &= f^{-1}(f(x)) \\ &= f^{-1}(4x + 2) \\ &= \frac{4x + 2 - 2}{4} \\ &= \frac{4x}{4} \\ &= x\end{aligned}$$

Thus, $(f \circ f^{-1})(x) = (f^{-1} \circ f)(x) = x$, which is the identity function $I(x)$.

EXAMPLE 4 The set of all pairs (x, y) determined by $3x + 2y = 6$ is a function. Find its inverse function, and graph the function and its inverse on one coordinate system.

Solution To find the inverse function of $3x + 2y = 6$, we interchange x and y to obtain

$$3y + 2x = 6$$

and then solve the equation for y.

$$\begin{aligned}3y + 2x &= 6 \\ 3y &= -2x + 6 && \text{Subtract } 2x \text{ from both sides.} \\ y &= -\frac{2}{3}x + 2 && \text{Divide both sides by 3.}\end{aligned}$$

Since the resulting equation represents a function, we can write it in inverse function notation as

$$f^{-1}(x) = -\frac{2}{3}x + 2$$

The graphs of $3x + 2y = 6$ and $f^{-1}(x) = -\frac{2}{3}x + 2$ appear in Figure 8-37.

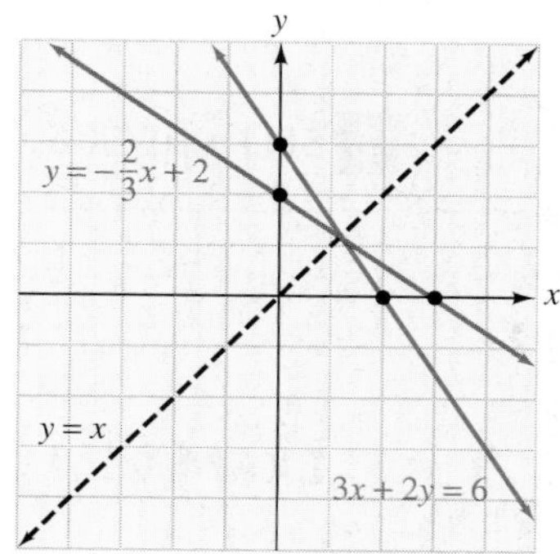

Figure 8-37

SELF CHECK 4 Find the inverse of the function defined by $2x - 3y = 6$. Graph the function and its inverse on one coordinate system.

In Example 4, the graph of $3x + 2y = 6$ and $f^{-1}(x) = -\frac{2}{3}x + 2$ are symmetric about the line $y = x$. In general, *any function and its inverse are symmetric about the line* $y = x$, because when the coordinates (a, b) satisfy an equation, the coordinates (b, a) will satisfy its inverse.

In each example so far, the inverse of a function has been another function. This is not always true, as the following example will show.

EXAMPLE 5 Find the inverse of the function determined by $f(x) = x^2$.

Solution

$y = x^2$ Replace $f(x)$ with y.

$x = y^2$ Interchange x and y.

$y = \pm\sqrt{x}$ Use the square-root property and write y on the left side.

When the inverse $y = \pm\sqrt{x}$ is graphed as in Figure 8-38, we see that the graph does not pass the vertical line test. Thus, it is not a function.

The graph of $y = x^2$ is also shown in the figure. As expected, the graphs of $y = x^2$ and $y = \pm\sqrt{x}$ are symmetric about the line $y = x$.

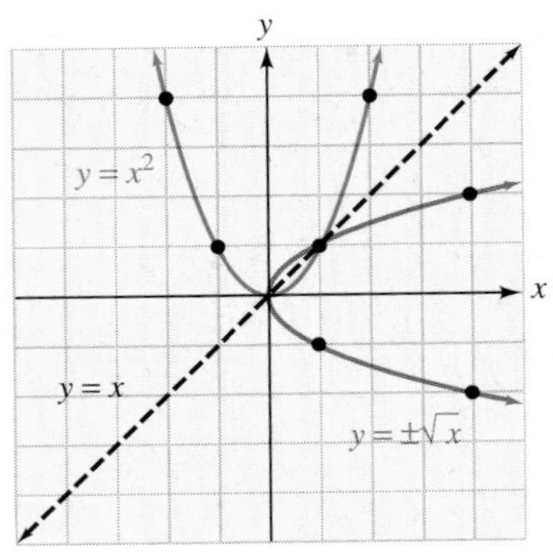

Figure 8-38

SELF CHECK 5 Find the inverse of the function determined by $f(x) = 4x^2$.

EXAMPLE 6 Find the inverse of $f(x) = x^3$.

Solution To find the inverse, we proceed as follows:

$y = x^3$ Replace $f(x)$ with y.

$x = y^3$ Interchange the variables x and y.

$\sqrt[3]{x} = y$ Take the cube root of both sides.

We note that to each number x there corresponds one real cube root. Thus, $y = \sqrt[3]{x}$ represents a function. In function notation, we have

$$f^{-1}(x) = \sqrt[3]{x}$$

SELF CHECK 6 Find the inverse of $f(x) = x^5$.

If a function is not one-to-one, we often can make it a one-to-one function by restricting its domain.

EXAMPLE 7 Find the inverse of $f(x) = x^2$ $(x \geq 0)$. Then determine whether the inverse is a function. Graph the function and its inverse.

Solution The inverse of the function $f(x) = x^2$ with $x \geq 0$ is

$y = x^2$ with $x \geq 0$ Replace $f(x)$ with y.

$x = y^2$ with $y \geq 0$ Interchange the variables x and y.

$y = \pm\sqrt{x}$ with $y \geq 0$ Solve for y and write y on the left side.

Considering the restriction $y \geq 0$, the equation can be written more simply as

$$y = \sqrt{x} \quad \text{This is the inverse of } f(x) = x^2.$$

In this equation, each number x gives only one value of y. Thus, the inverse is a function, which we can write as

$$f^{-1}(x) = \sqrt{x}$$

The graphs of the two functions appear in Figure 8-39. The line $y = x$ is included so that we can see that the graphs are symmetric about the line $y = x$.

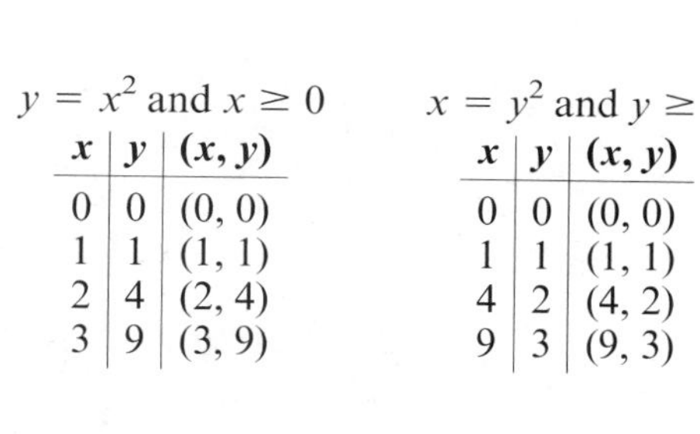

$y = x^2$ and $x \geq 0$

x	y	(x, y)
0	0	(0, 0)
1	1	(1, 1)
2	4	(2, 4)
3	9	(3, 9)

$x = y^2$ and $y \geq 0$

x	y	(x, y)
0	0	(0, 0)
1	1	(1, 1)
4	2	(4, 2)
9	3	(9, 3)

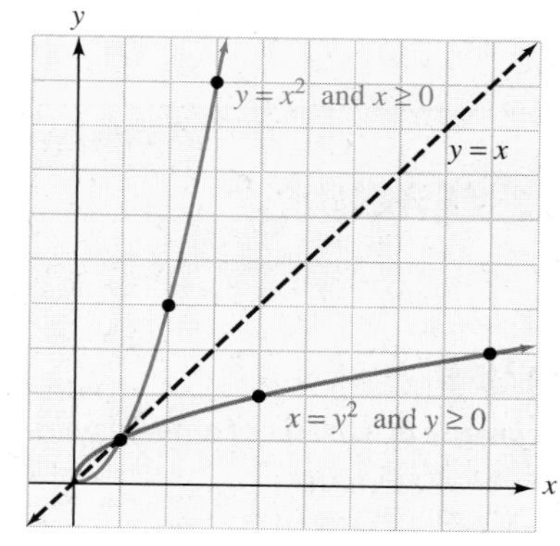

Figure 8-39

SELF CHECK 7 Find the inverse of $f(x) = x^2 - 8 \quad (x \geq 0)$. Graph the function and its inverse.

SELF CHECK ANSWERS

1. yes **2.** no **3.** $f^{-1}(x) = -\frac{1}{5}x - \frac{3}{5}$; yes **4.** $f^{-1}(x) = \frac{3}{2}x + 3$

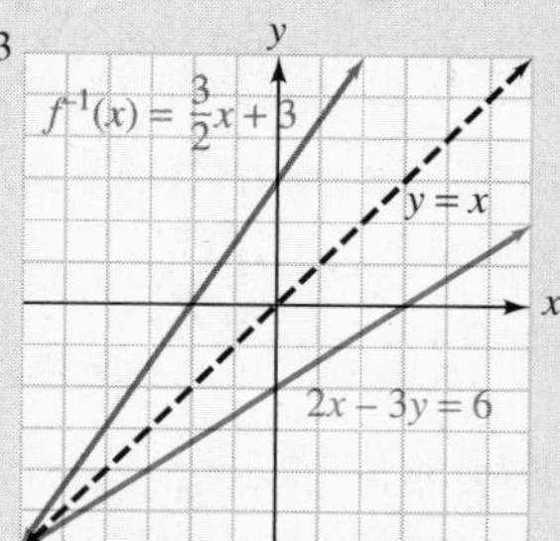

5. $y = \pm\frac{\sqrt{x}}{2}$ **6.** $f^{-1}(x) = \sqrt[5]{x}$

7. $f^{-1}(x) = \sqrt{x + 8}$

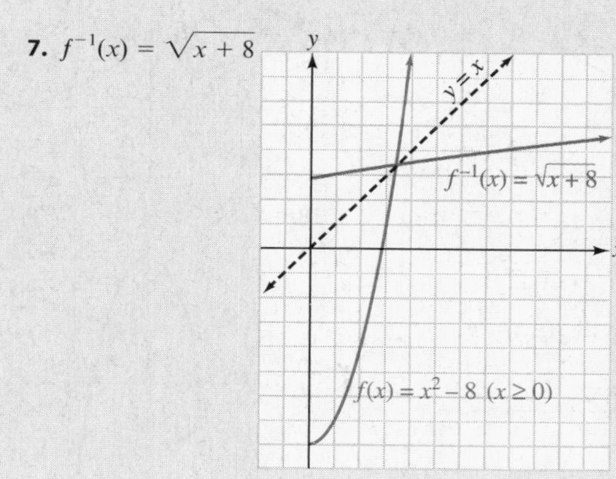

NOW TRY THIS

1. Find the inverse of $f(x) = \frac{1}{x}$.
2. Given $f(x) = \frac{5x + 1}{3x - 2}$, find $f^{-1}(x)$.

8.7 EXERCISES

WARM-UPS

Find the inverse of each set of ordered pairs.

1. $\{(1, 2), (2, 3), (5, 10)\}$
2. $\{(1, 1), (2, 8), (4, 64)\}$

Find the inverse function of each linear function.

3. $y = \frac{1}{2}x$
4. $y = 2x$

Determine whether each function is one-to-one.

5. $y = x^2 - 2$
6. $y = x^3$

REVIEW *Write each complex number in a + bi form or find each value.*

7. $3 - \sqrt{-64}$
8. $(2 - 3i) + (4 + 5i)$
9. $(3 + 4i)(2 - 3i)$
10. $\frac{6 + 7i}{3 - 4i}$
11. $|6 - 8i|$
12. $\left|\frac{2 + i}{3 - i}\right|$

VOCABULARY AND CONCEPTS *Fill in the blanks.*

13. A function is called __________ if each input determines a different output.
14. If every __________ line that intersects the graph of a function does so only once, the function is one-to-one.
15. If a one-to-one function turns an input of 2 into an output of 5, the inverse function will turn 5 into __.
16. The symbol $f^{-1}(x)$ is read as ____________ or ________.
17. $(f \circ f^{-1})(x) = (f^{-1} \circ f)(x) =$ __.
18. The graphs of a function and its inverse are symmetrical about the line ______.

GUIDED PRACTICE

Determine whether each function is one-to-one. See Example 1. (Objective 1)

19. $f(x) = 2x$
20. $f(x) = |x|$
21. $f(x) = x^4$
22. $f(x) = x^3 + 1$

Each graph represents a function. Use the horizontal line test to decide whether the function is one-to-one. See Example 2. (Objective 2)

23.
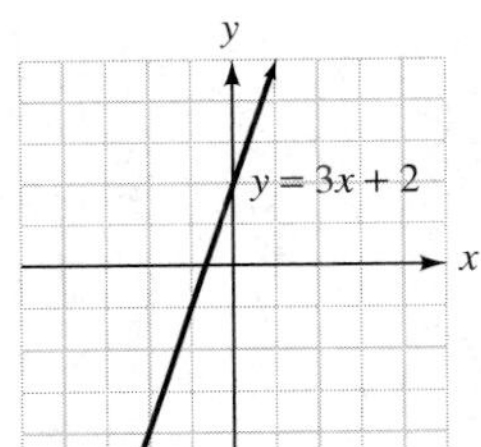

24.
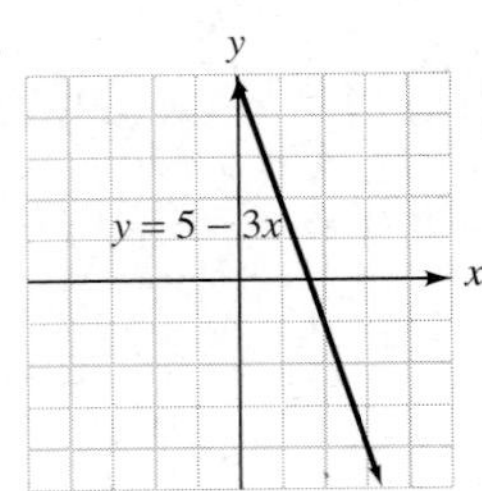

25.
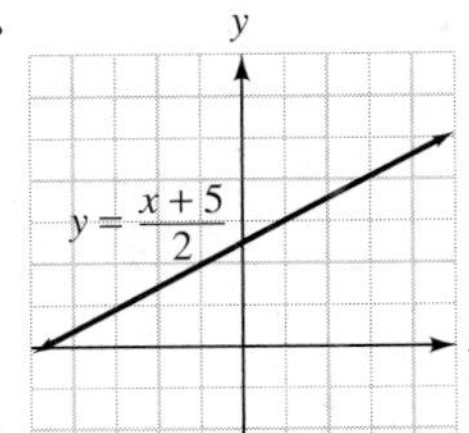

26.
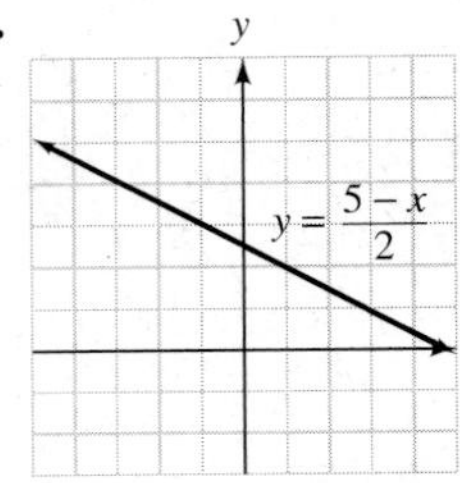

27.
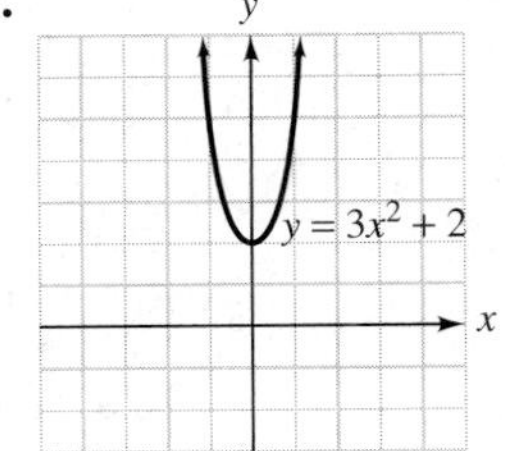

28.
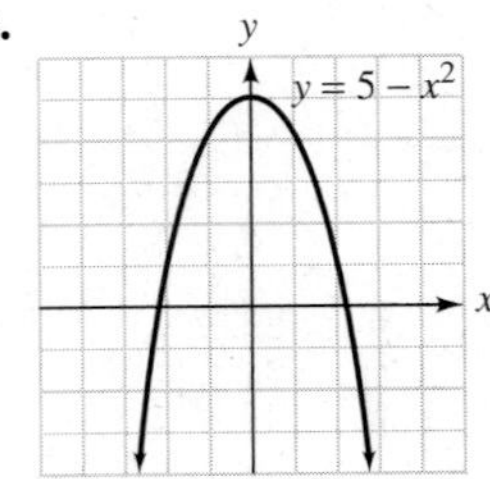

29.

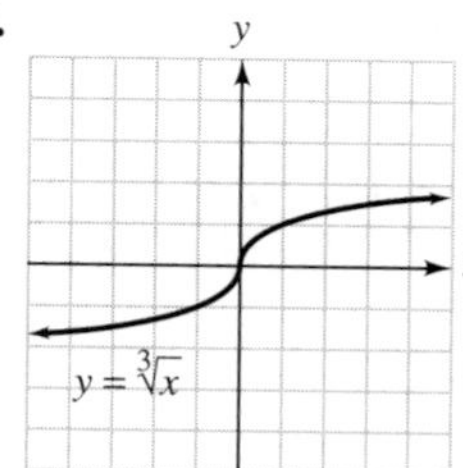

30.

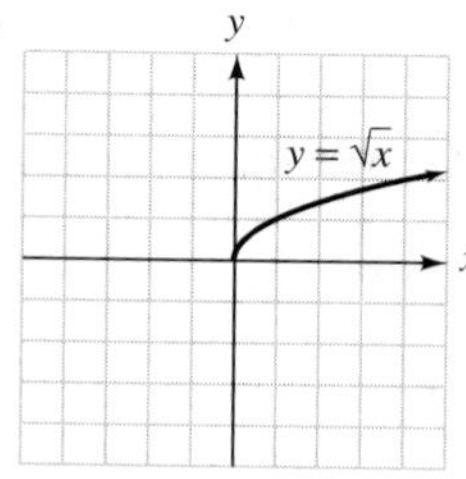

Find the inverse of each set of ordered pairs (x, y) and determine whether the inverse is a function. **(Objective 3)**

31. $\{(3, 2), (2, 1), (1, 0)\}$

32. $\{(4, 1), (5, 1), (6, 1), (7, 1)\}$

33. $\{(1, 2), (2, 3), (1, 3), (1, 5)\}$

34. $\{(-1, -1), (0, 0), (1, 1), (2, 2)\}$

Find the inverse of each function and express it in the form $y = f^{-1}(x)$. Verify each result by showing that $(f \circ f^{-1})(x) = (f^{-1} \circ f)(x) = (x)$. **See Example 3. (Objective 3)**

35. $f(x) = 3x + 1$

36. $y + 1 = 5x$

37. $x + 4 = 5y$

38. $x = 3y + 1$

39. $f(x) = \dfrac{x - 4}{5}$

40. $f(x) = \dfrac{2x + 6}{3}$

41. $4x - 5y = 20$

42. $3x + 5y = 15$

Find the inverse of each function. Then graph the function and its inverse on one coordinate system. Draw the line of symmetry on the graph. **See Example 4. (Objective 3)**

43. $y = 4x + 3$

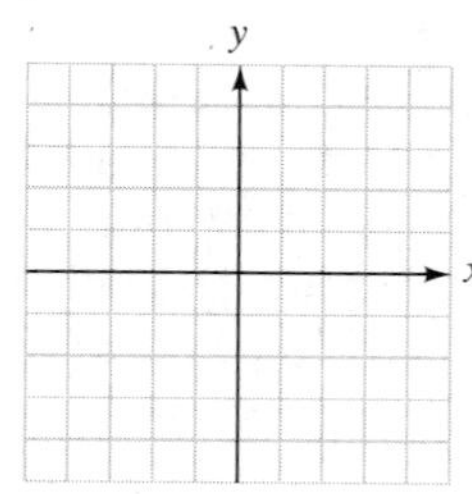

44. $x = 3y - 1$

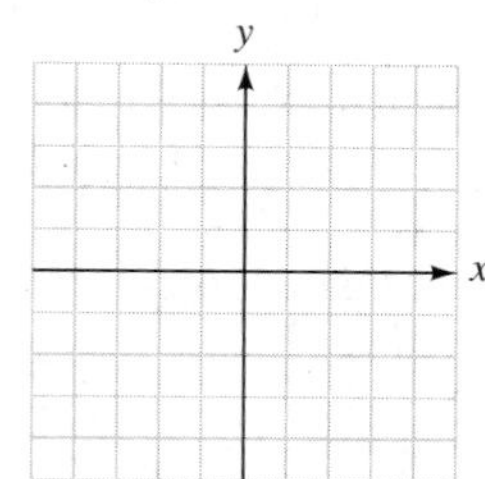

45. $x = \dfrac{y - 2}{3}$

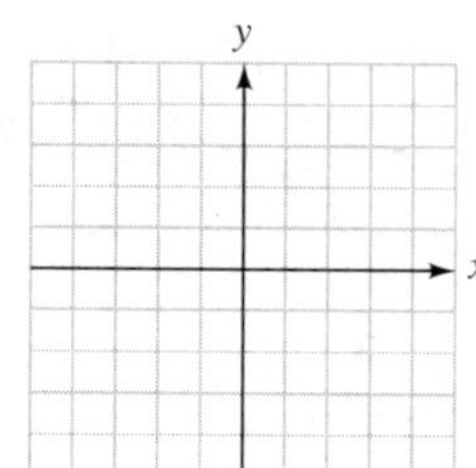

46. $y = \dfrac{x + 3}{4}$

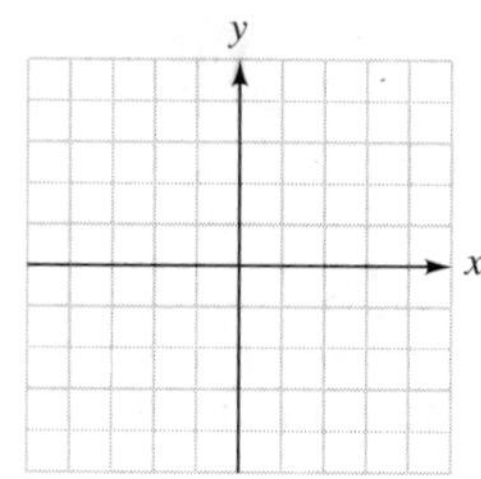

47. $3x - y = 5$

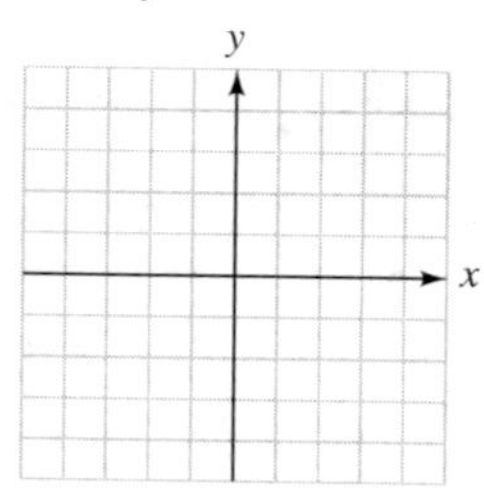

48. $2x + 3y = 9$

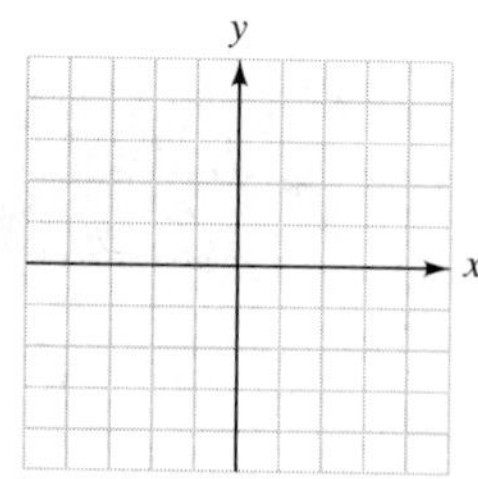

49. $3(x + y) = 2x + 4$

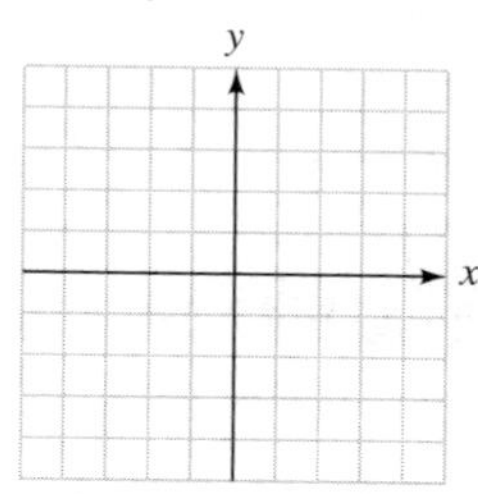

50. $-4(y - 1) + x = 2$

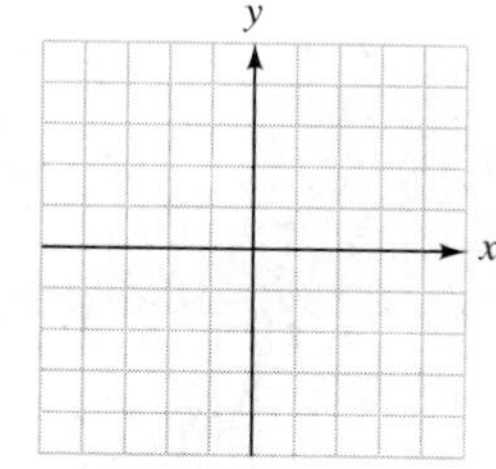

Find the inverse of each function and determine whether it is a function. If it is a function, express it in function notation. **See Examples 5–6. (Objective 3)**

51. $y = x^2 + 4$

52. $y = x^2 + 5$

53. $y = x^3$

54. $xy = 4$

Graph each equation and its inverse on one set of coordinate axes. Find the axis of symmetry. **See Example 7. (Objective 3)**

55. $y = x^2 + 1$

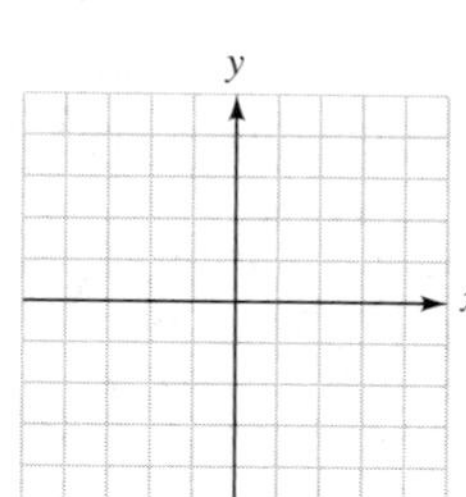

56. $y = \dfrac{1}{4}x^2 - 3$

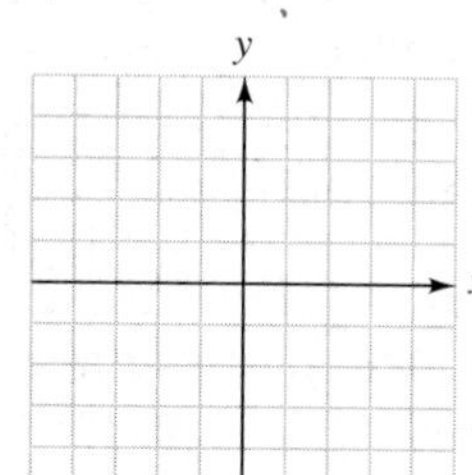

57. $y = \sqrt{x}$

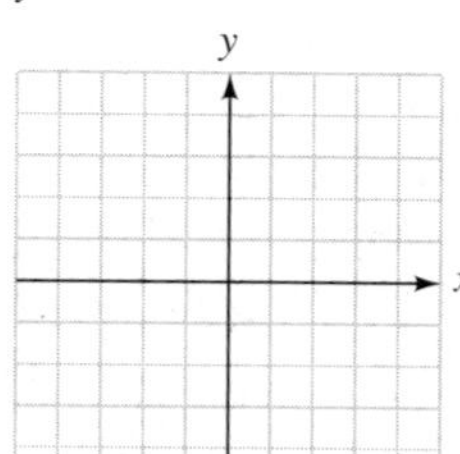

58. $y = |x|$

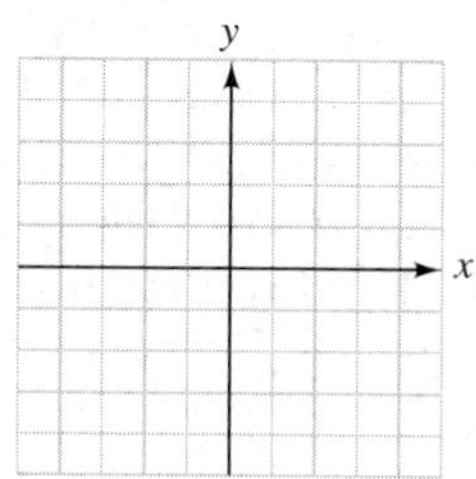

ADDITIONAL PRACTICE ***Find the inverse of each function. Do not rationalize denominators.***

59. $\{(1, 1), (2, 4), (3, 9), (4, 16)\}$

60. $\{(1, 1), (2, 1), (3, 1), (4, 1)\}$

61. $y = |x|$

62. $y = \sqrt[3]{x}$

63. $f(x) = 2x^3 - 3$

64. $f(x) = \dfrac{3}{x^3} - 1$

WRITING ABOUT MATH

65. Explain the purpose of the vertical line test.

66. Explain the purpose of the horizontal line test.

SOMETHING TO THINK ABOUT

67. Find the inverse of $y = \dfrac{x + 1}{x - 1}$.

68. Using the functions of Exercise 67, show that $(f \circ f^{-1})(x) = x$.

PROJECTS

Project 1

Ballistics is the study of how projectiles fly. The general formula for the height above the ground of an object thrown straight up or down is given by the function

$$h(t) = -16t^2 + v_0t + h_0$$

where h is the object's height (in feet) above the ground t seconds after it is thrown. The initial velocity v_0 is the velocity with which the object is thrown, measured in feet per second. The initial height h_0 is the object's height (in feet) above the ground when it is thrown. (If $v_0 > 0$, the object is thrown upward; if $v_0 < 0$, the object is thrown downward.)

This formula takes into account the force of gravity, but disregards the force of air resistance. It is much more accurate for a smooth, dense ball than for a crumpled piece of paper.

One act in the Bungling Brothers Circus is Amazing Glendo's cannonball-catching act. A cannon fires a ball vertically into the air; Glendo, standing on a platform above the cannon, uses his catlike reflexes to catch the ball as it passes by on its way toward the roof of the big top. As the balls fly past, they are within Glendo's reach only during a two-foot interval of their upward path.

As an investigator for the company that insures the circus, you have been asked to find answers to the following questions. The answers will determine whether or not Bungling Brothers' insurance policy will be renewed.

a. In the first part of the act, cannonballs are fired from the end of a six-foot cannon with an initial velocity of 80 feet per second. Glendo catches one ball between 40 and 42 feet above the ground. Then he lowers his platform and catches another ball between 25 and 27 feet above the ground.

i. Show that if Glendo missed a cannonball, it would hit the roof of the 56-foot-tall big top. How long would it take for a ball to hit the big top? To prevent this from happening, a special net near the roof catches and holds any missed cannonballs.

ii. Find (to the nearest thousandth of a second) how long the cannonballs are within Glendo's reach for each of his catches. Which catch is easier? Why does your answer make sense? Your company is willing to insure against injuries to Glendo if he has at least 0.025 second to make each catch. Should the insurance be offered?

b. For Glendo's grand finale, the special net at the roof of the big top is removed, making Glendo's catch more significant to the people in the audience, who worry that if Glendo misses, the tent will collapse around them. To make it even more dramatic, Glendo's arms are tied to restrict his reach to a one-foot interval of the ball's flight, and he stands on a platform just under the peak of the big top, so that his catch is made at the very last instant (between 54 and 55 feet above the ground). For this part of the act, however, Glendo has the cannon charged with less gunpowder, so that the muzzle velocity of the cannon is 56 feet per second. Show work to prove that Glendo's big finale is in fact his easiest catch, and that even if he misses, the big top is never in any danger of collapsing, so insurance should be offered against injury to the audience.

Project 2

The center of Sterlington is the intersection of Main Street (running east–west) and Due North Road (running north–south). The recreation area for the townspeople is Robin Park, a few blocks from there. The park is bounded on the south by Main Street and on every other side by Parabolic Boulevard, named for its distinctive shape. In fact, if Main Street and Due North Road were used as the axes of a rectangular coordinate system, Parabolic

Boulevard would have the equation $y = -(x - 4)^2 + 5$, where each unit on the axes is 100 yards.

The city council has recently begun to consider whether or not to put two walkways through the park. (See Illustration 1.) The walkways would run from two points on Main Street and converge at the northernmost point of the park, dividing the area of the park exactly into thirds.

The city council is pleased with the esthetics of this arrangement but needs to know two important facts.

a. For planning purposes, they need to know exactly where on Main Street the walkways would begin.

b. In order to budget for the construction, they need to know how long the walkways will be.

Provide answers for the city council, along with explanations and work to show that your answers are correct. You will need to use the formula shown in Illustration 2, due to Archimedes (287–212 B.C.), for the area under a parabola but above a line perpendicular to the axis of symmetry of the parabola.

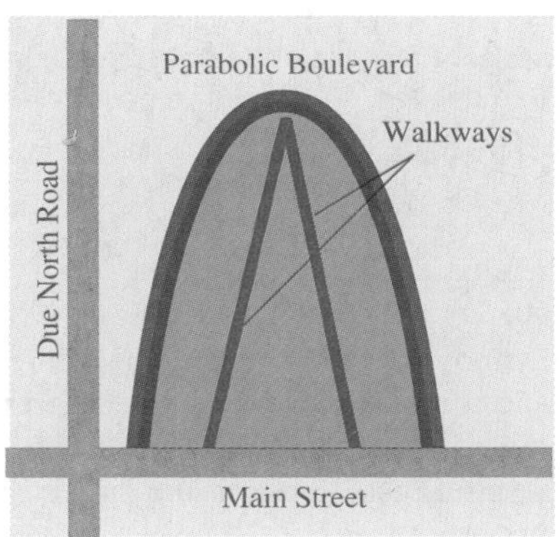

Illustration 1

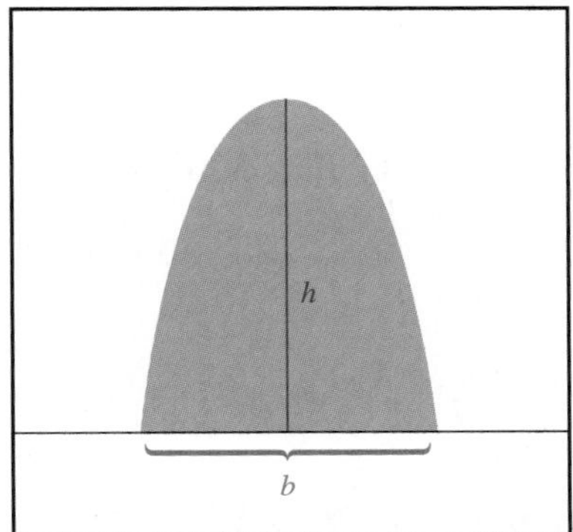

Shaded area = $\frac{2}{3} \cdot b \cdot h$

Illustration 2

Chapter 8 REVIEW

SECTION 8.1 Solving Quadratic Equations Using the Square-Root Property and by Completing the Square

DEFINITIONS AND CONCEPTS	EXAMPLES
Square-root property: The equation $x^2 = c$ has two solutions: $x = \sqrt{c}$ and $x = -\sqrt{c}$	To solve $x^2 - 28 = 0$, proceed as follows: $x^2 = 28$ Add 28 to both sides. $x = \sqrt{28}$ or $x = -\sqrt{28}$ Use the square-root property. $x = 2\sqrt{7}$ | $x = -2\sqrt{7}$ $\sqrt{28} = \sqrt{4}\sqrt{7} = 2\sqrt{7}$ The solutions are $2\sqrt{7}$ and $-2\sqrt{7}$, or $\pm 2\sqrt{7}$.
Completing the square: **1.** Make sure that the coefficient of x^2 is 1. If not, make it 1 by dividing both sides of the equation by the coefficient of x^2.	To solve $2x^2 - 12x + 24 = 0$, we first make the coefficient of x^2 equal to 1 by dividing both sides by 2. **1.** $\frac{2x^2}{2} - \frac{12x}{2} + \frac{24}{2} = \frac{0}{2}$ Divide both sides by 2. $x^2 - 6x + 12 = 0$ Simplify.

2. If necessary, get the constant term on the right side of the equal sign.	**2.** $x^2 - 6x = -12$ Subtract 12 from both sides.
3. Complete the square: **a.** Find one-half of the coefficient of x and square it. **b.** Add the square to both sides of the equation.	**3.** Since $\left[\frac{1}{2}(-6)\right]^2 = (-3)^2 = 9$, we complete the square by adding 9 to both sides. $x^2 - 6x + 9 = -12 + 9$
4. Factor the trinomial square on one side of the equation and combine like terms on the other side.	**4.** $(x - 3)^2 = -3$ Factor the left side and combine terms on the right side.
5. Solve the resulting equation by using the square-root property.	**5.** $x - 3 = \sqrt{-3}$ or $x - 3 = -\sqrt{-3}$ Use the square-root property. $x - 3 = \sqrt{3}i$ | $x - 3 = -\sqrt{3}i$ Simplify $\sqrt{-3}$. $x = 3 + \sqrt{3}i$ | $x = 3 - \sqrt{3}i$ Add 3 to both sides. The solutions are $3 \pm \sqrt{3}i$.

REVIEW EXERCISES

Solve each equation by factoring or by using the square-root property.

1. $12x^2 + x - 6 = 0$
2. $6x^2 + 17x + 5 = 0$
3. $15x^2 + 2x - 8 = 0$
4. $(x + 2)^2 = 36$

Solve each equation by completing the square.

5. $x^2 + 6x + 8 = 0$
6. $2x^2 - 9x + 7 = 0$
7. $2x^2 - x - 5 = 0$

SECTION 8.2 Solving Quadratic Equations by the Quadratic Formula

DEFINITIONS AND CONCEPTS	EXAMPLES
Quadratic formula: The solutions of $ax^2 + bx + c = 0 \quad (a \neq 0)$ are given by the formula $x = \dfrac{-b \pm \sqrt{b^2 - 4ac}}{2a}$	To solve $2x^2 - 5x + 4 = 0$, note that in the equation $a = 2$, $b = -5$, and $c = 4$, and substitute these values into the quadratic formula. $x = \dfrac{-b \pm \sqrt{b^2 - 4ac}}{2a}$ $= \dfrac{-(-5) \pm \sqrt{(-5)^2 - 4(2)(4)}}{2(2)}$ Substitute the values. $= \dfrac{5 \pm \sqrt{25 - 32}}{4}$ Simplify. $= \dfrac{5 \pm \sqrt{-7}}{4}$ Add. $= \dfrac{5 \pm i\sqrt{7}}{4}$ $\sqrt{-7} = i\sqrt{7}$ The solutions are $\frac{5}{4} \pm \frac{\sqrt{7}}{4}i$.

REVIEW EXERCISES

Solve each equation by using the quadratic formula.

8. $x^2 - 8x - 9 = 0$
9. $x^2 - 10x = 0$
10. $2x^2 + 13x - 7 = 0$
11. $3x^2 + 20x - 7 = 0$
12. $2x^2 - x - 2 = 0$
13. $x^2 + x + 2 = 0$

14. Dimensions of a rectangle A rectangle is 2 centimeters longer than it is wide. If both the length and width are doubled, its area is increased by 72 square centimeters. Find the dimensions of the original rectangle.

15. Dimensions of a rectangle A rectangle is 1 foot longer than it is wide. If the length is tripled and the width is doubled, its area is increased by 30 square feet. Find the dimensions of the original rectangle.

16. Ballistics If a rocket is launched straight up into the air with an initial velocity of 112 feet per second, its height after t seconds is given by the formula $h = 112t - 16t^2$, where h represents the height of the rocket in feet. After launch, how long will it be before it hits the ground?

17. Ballistics What is the maximum height of the rocket discussed in Exercise 16?

SECTION 8.3 The Discriminant and Equations That Can Be Written in Quadratic Form

DEFINITIONS AND CONCEPTS | **EXAMPLES**

DEFINITIONS AND CONCEPTS

The discriminant: If a, b, and c are real numbers and a is not 0,

If $b^2 - 4ac > 0$, the equation $ax^2 + bx + c = 0$ has two unequal solutions.

If $b^2 - 4ac = 0$, the equation $ax^2 + bx + c = 0$ has two equal solutions (called a *double root*).

If $b^2 - 4ac < 0$, the equation $ax^2 + bx + c = 0$ has two solutions that are complex conjugates.

If a, b, and c are rational numbers and the discriminant:

is a perfect square greater than 0, there are two unequal rational solutions.

is positive but not a perfect square, the solutions are irrational and unequal.

EXAMPLES

To determine the type of solutions for the equation $x^2 - x + 9 = 0$, we calculate the discriminant.

$$\begin{aligned} b^2 - 4ac &= (-1)^2 - 4(1)(9) \qquad a = 1, b = -1, \text{ and } c = 9. \\ &= 1 - 36 \\ &= -35 \end{aligned}$$

Since $b^2 - 4ac < 0$, the solutions are complex conjugates.

To determine the type of solutions for the equation $2x^2 + 4x - 5 = 0$, we calculate the discriminant.

$$\begin{aligned} b^2 - 4ac &= 4^2 - 4(2)(-5) \qquad a = 2, b = 4, \text{ and } c = -5. \\ &= 16 + 40 \\ &= 56 \end{aligned}$$

Since $b^2 - 4ac > 0$, the solutions are unequal and irrational.

DEFINITIONS AND CONCEPTS

Solving equations quadratic in form:
Use u substitution when necessary.

EXAMPLES

To solve the equation $x^4 - 3x^2 - 4 = 0$, we can proceed as follows:

$$\begin{aligned} x^4 - 3x^2 - 4 &= 0 \\ (x^2)^2 - 3(x^2) - 4 &= 0 \\ u^2 - 3u - 4 &= 0 \qquad \text{Substitute } u \text{ for } x^2. \\ (u - 4)(u + 1) &= 0 \qquad \text{Factor } u^2 - 5u + 4. \end{aligned}$$

$$u - 4 = 0 \quad \text{or} \quad u + 1 = 0 \qquad \text{Set each factor equal to 0.}$$

$$u = 4 \quad \Big| \quad u = -1$$

Since $x^2 = u$, it follows that $x^2 = 4$ or $x^2 = -1$. Thus,

$$x^2 = 4 \quad \text{or} \quad x^2 = -1$$

$$x = 2 \quad \text{or} \quad x = -2 \quad \Big| \quad x = i \quad \text{or} \quad x = -i$$

DEFINITIONS AND CONCEPTS

Verifying solutions:
If r_1 and r_2 are solutions of $ax^2 + bx + c = 0$ $(a = 0)$, then

$$r_1 + r_2 = -\frac{b}{a} \quad \text{and} \quad r_1r_2 = \frac{c}{a}$$

EXAMPLES

To verify that $\frac{5}{3}$ and 2 are the solutions of $3x^2 - 11x + 10 = 0$, make the following calculations:

$$-\frac{b}{a} = -\frac{-11}{3} = \frac{11}{3} \quad \text{and} \quad \frac{c}{a} = \frac{10}{3}$$

Since $\frac{5}{3} + 2 = \frac{5}{3} + \frac{6}{3} = \frac{11}{3}$ and $\frac{5}{3}(2) = \frac{5}{3} \cdot \frac{2}{1} = \frac{10}{3}$, $\frac{5}{3}$ and 2 are the solutions of $3x^2 - 11x + 10 = 0$.

REVIEW EXERCISES

Use the discriminant to determine what types of solutions exist for each equation.

18. $3x^2 + 4x - 3 = 0$

19. $4x^2 - 5x + 7 = 0$

20. Find the values of k that will make the solutions of $(k - 8)x^2 + (k + 16)x = -49$ equal.

21. Find the values of k such that the solutions of $3x^2 + 4x = k + 1$ will be real numbers.

Solve each equation.

22. $x - 13x^{1/2} + 12 = 0$

23. $a^{2/3} + a^{1/3} - 6 = 0$

24. $\dfrac{1}{x+1} - \dfrac{1}{x} = -\dfrac{1}{x+1}$

25. $\dfrac{6}{x+2} + \dfrac{6}{x+1} = 5$

26. Find the sum of the solutions of the equation $3x^2 - 14x + 3 = 0$.

27. Find the product of the solutions of the equation $3x^2 - 14x + 3 = 0$.

SECTION 8.4 Graphs of Quadratic Functions

DEFINITIONS AND CONCEPTS | **EXAMPLES**

Graphing quadratic functions:

If f is a function and k and h positive numbers, then:

The graph of $f(x) + k$ is identical to the graph of $f(x)$, except that it is translated k units upward.

The graph of $f(x) - k$ is identical to the graph of $f(x)$, except that it is translated k units downward.

The graph of $f(x - h)$ is identical to the graph of $f(x)$, except that it is translated h units to the right.

The graph of $f(x + h)$ is identical to the graph of $f(x)$, except that it is translated h units to the left.

Graph each of the following:

a. $f(x) = 4x^2$

b. $f(x) = 4x^2 - 3$

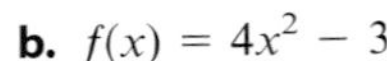

c. $f(x) = 4(x - 3)^2$

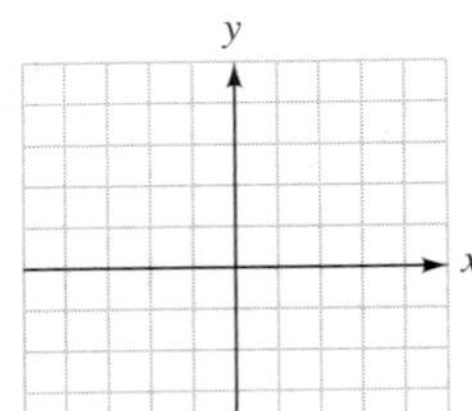

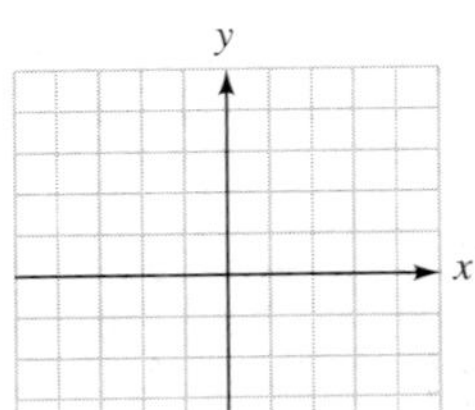

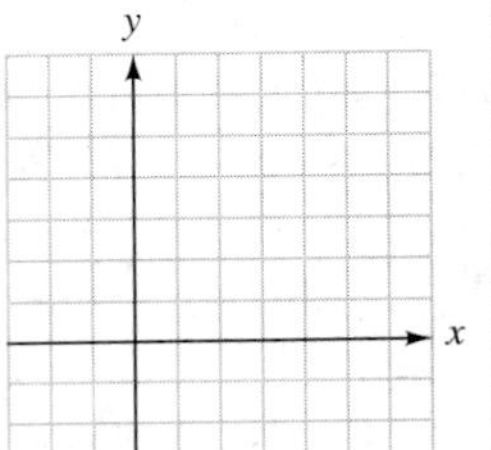

Finding the vertex of a parabola:

If $a \neq 0$, the graph of $y = a(x - h)^2 + k$ is a parabola with vertex at (h, k). It opens upward when $a > 0$ and downward when $a < 0$.

The vertex of the graph of $y = 2(x - 3)^2 - 5$ is $(3, -5)$. Since $2 > 0$, the graph will open upward.

The coordinates of the vertex of the graph of

$$f(x) = ax^2 + bc + c \quad (a \neq 0)$$

are $\left(-\frac{b}{2a}, f\left(-\frac{b}{2a}\right)\right)$.

The axis of symmetry is $x = -\frac{b}{2a}$.

Find the axis of symmetry and the vertex of the graph of $f(x) = 2x^2 - 4x - 5$.

$$x\text{-coordinate} = -\frac{b}{2a} = -\frac{-4}{2(2)} = -\frac{-4}{4} = \frac{4}{4} = 1$$

The axis of symmetry is $x = 1$.

To find the y-coordinate of the vertex, substitute 1 for x in $f(x) = 2x^2 - 4x - 5$.

$$\begin{aligned} f(x) &= 2x^2 - 4x - 5 \\ f(1) &= 2(1)^2 - 4(1) - 5 \\ &= 2 - 4 - 5 \\ &= -7 \end{aligned}$$

The vertex is $(1, -7)$.

REVIEW EXERCISES

Graph each function and give the coordinates of the vertex of the resulting parabola.

28. $y = 2x^2 - 3$

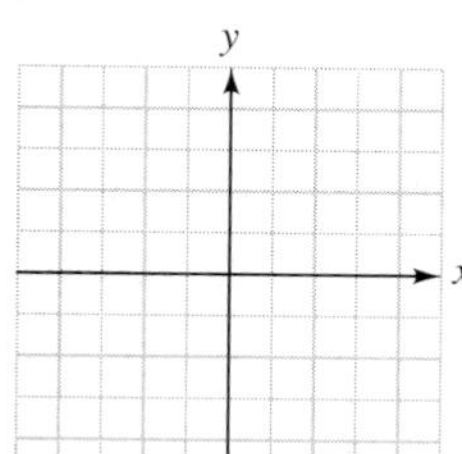

29. $y = -2x^2 - 1$

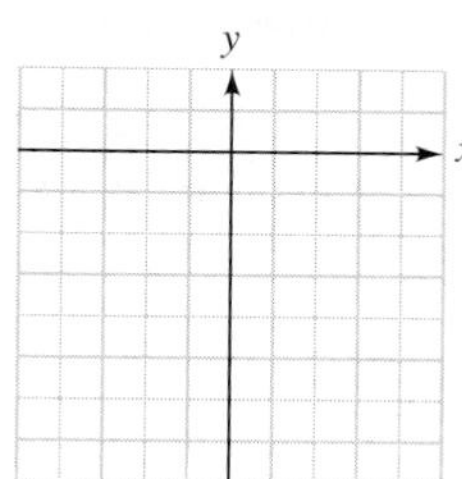

30. $y = -4(x - 2)^2 + 1$

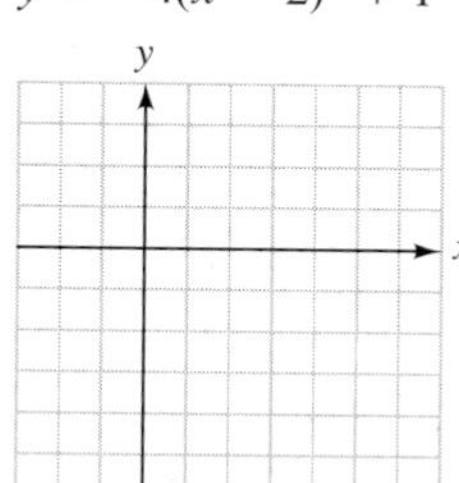

31. $y = 5x^2 + 10x - 1$

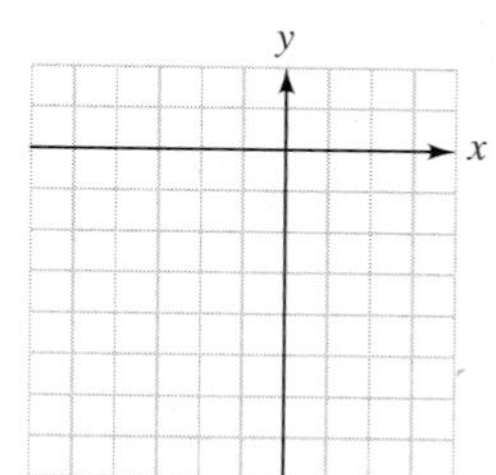

32. Find the vertex of the graph of $f(x) = 3x^2 - 12x - 5$.

SECTION 8.5 Quadratic and Other Nonlinear Inequalities

DEFINITIONS AND CONCEPTS	EXAMPLES
Solving quadratic inequalities: To solve a quadratic inequality in one variable, make a sign chart and determine which intervals are solutions.	To solve $x^2 - 2x - 8 > 0$, factor the trinomial to get $(x - 4)(x + 2)$ and construct the chart shown below. The critical values will occur at 4 and -2. 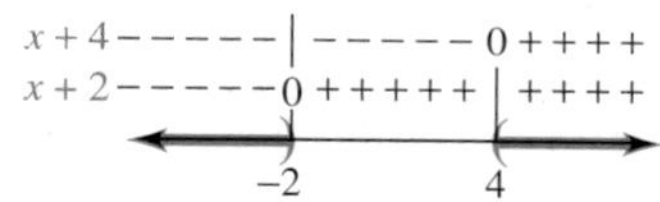 $x - 4$ is positive when $x > 4$, and is negative when $x < 4$. $x + 2$ is positive when $x > -2$, and is negative when $x < -2$. The product of $x - 4$ and $x + 2$ will be greater than 0 when the signs of the binomial factors are the same. This occurs in the intervals $(-\infty, -2)$ and $(4, \infty)$. The numbers -2 and 4 are not included, because equality is not indicated in the original inequality. Thus, the solution set in interval notation is $(-\infty, -2) \cup (4, \infty)$ The graph of the solution set is shown on the number line below. 
Graphing rational inequalities: To solve inequalities with rational expressions, get 0 on the right side, add the fractions, and then factor the numerator and denominator. Use a sign chart to determine the solution.	To solve $\frac{1}{x} \geq -4$, add 4 to both sides to make the right side equal to 0 and proceed as follows: $\frac{1}{x} + 4 \geq 0$ $\frac{1}{x} + \frac{4x}{x} \geq 0$ Write each fraction with a common denominator. $\frac{1 + 4x}{x} \geq 0$ Add. Finally, make a sign chart, as shown. 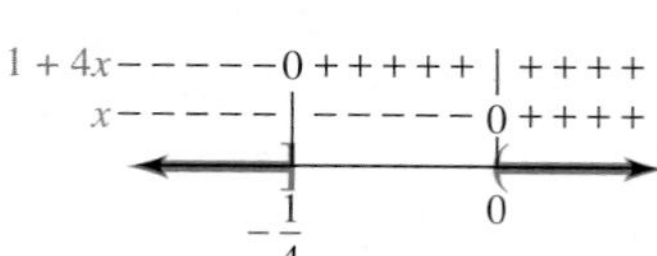

The denominator x is undefined when $x = 0$, positive when $x > 0$, and negative when $x < 0$. The numerator $1 + 4x$ is 0 when $x = -\frac{1}{4}$, positive when $x > -\frac{1}{4}$, and negative when $x < -\frac{1}{4}$. The critical values occur when the numerator or denominator is 0: 0 and $-\frac{1}{4}$.

The fraction $\frac{1 + 4x}{x}$ will be greater than or equal to 0 when the numerator and denominator are the same sign and when the numerator is 0. This occurs in the interval

$$\left(-\infty, -\frac{1}{4}\right] \cup (0, \infty)$$

The graph of this interval is shown.

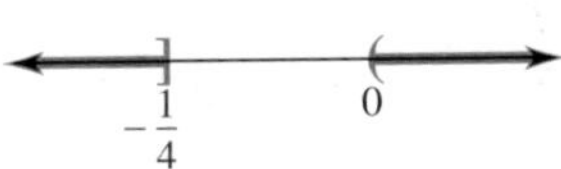

Graphing nonlinear inequalities in two variables: To graph a nonlinear inequality, first graph the equation. Then determine which region represents the graph of the inequality.

To graph $x > |y|$, first graph $x = |y|$ as in the illustration below using a dashed line, because equality is not indicated in the original inequality.

Since the origin is on the graph, we cannot use it as a test point. We select another point, such as (1, 0). We substitute 1 for x and 0 for y into the inequality to get

$$x > |y|$$
$$1 > |\mathbf{0}|$$
$$1 > 0$$

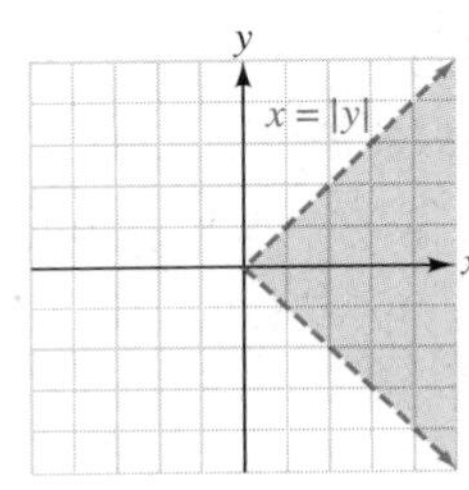

Since $1 > 0$ is a true statement, the point (1, 0) satisfies the inequality and is part of the graph. Thus, the graph of $x > |y|$ is to the right of the boundary.

The complete graph is shown.

REVIEW EXERCISES

Solve each inequality. Give each result in interval notation and graph the solution set.

33. $x^2 + 2x - 35 > 0$

34. $x^2 + 7x - 18 < 0$

35. $\frac{3}{x} \leq 5$

36. $\frac{2x^2 - x - 28}{x - 1} > 0$

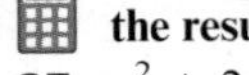

Use a graphing calculator to solve each inequality. Compare the results with Review Exercises 33–36.

37. $x^2 + 2x - 35 > 0$

38. $x^2 + 7x - 18 < 0$

39. $\frac{3}{x} \leq 5$

40. $\frac{2x^2 - x - 28}{x - 1} > 0$

Graph each inequality.

41. $y < \frac{1}{2}x^2 - 1$

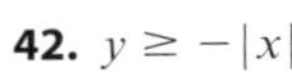

42. $y \geq -|x|$

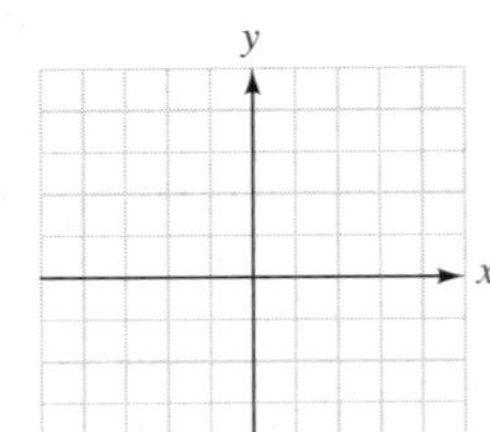

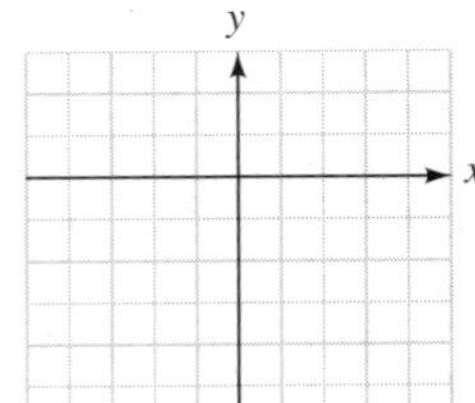

SECTION 8.6 Algebra and Composition of Functions

DEFINITIONS AND CONCEPTS	EXAMPLES
Operations with functions: $(f + g)(x) = f(x) + g(x)$ $(f - g)(x) = f(x) - g(x)$ $(f \cdot g)(x) = f(x)g(x)$ $(f/g)(x) = \dfrac{f(x)}{g(x)} \quad (g(x) \neq 0)$	Given $f(x) = 6x - 2$ and $g(x) = x^2 + 4$, find **a.** $f + g$ **b.** $f - g$ **c.** $f \cdot g$ **d.** f/g **a.** $f + g = f(x) + g(x)$ $= (6x - 2) + (x^2 + 4)$ $= x^2 + 6x + 2$ Combine like terms. **b.** $f - g = (6x - 2) - (x^2 + 4)$ $= 6x - 2 - x^2 - 4$ Remove parentheses. $= -x^2 + 6x - 6$ Combine like terms. **c.** $f \cdot g = (6x - 2)(x^2 + 4)$ $= 6x^3 + 24x - 2x^2 - 8$ Multiply. $= 6x^3 - 2x^2 + 24x - 8$ Write exponents in descending order. **d.** $f/g = \dfrac{6x - 2}{x^2 + 4}$
Composition of functions: $(f \circ g)(x) = f(g(x))$	Let $f(x) = 5x - 4$ and $g(x) = x^2 + 2$. Find **a.** $(f \circ g)(x)$ **b.** $(g \circ f)(x)$ **a.** $(f \circ g)(x)$ means $f(g(x))$. $f(g(x)) = f(x^2 + 2)$ $= 5(x^2 + 2) - 4$ Substitute $x^2 + x$ for x. $= 5x^2 + 10 - 4$ Multiply. $= 5x^2 + 6$ Add. **b.** $(g \circ f)(x)$ means $g(f(x))$. $g(f(x)) = g(5x - 4)$ $= (5x - 4)^2 + 2$ Substitute $5x - 4$ for x. $= 25x^2 - 40x + 16 + 2$ Square the binomial. $= 25x^2 - 40x + 18$ Add.
The difference quotient: The difference quotient is defined as follows: $\dfrac{f(x + h) - f(x)}{h}$	To evaluate the difference quotient for $f(x) = x^2 + 2x - 6$, first evaluate $f(x + h)$. $f(x) = x^2 + 2x - 6$ $f(x + h) = (x + h)^2 + 2(x + h) - 6$ Substitute $x + h$ for h. $= x^2 + 2xh + h^2 + 2x + 2h - 6$ $(x + h)^2 = x^2 + 2hx + h^2$

Now substitute the values of $f(x + h)$ and $f(x)$ into the difference quotient and simplify.

$$\frac{f(x+h) - f(x)}{h}$$

$$= \frac{(x^2 + 2xh + h^2 + 2x + 2h - 6) - (x^2 + 2x - 6)}{h}$$

$$= \frac{x^2 + 2xh + h^2 + 2x + 2h - 6 - x^2 - 2x + 6}{h}$$ Remove parentheses.

$$= \frac{2xh + h^2 + 2h}{h}$$ Combine like terms.

$$= \frac{h(2x + h + 2)}{h}$$ Factor out h in the numerator.

$$= 2x + h + 2$$ Divide out h.

The difference quotient for this function is $2x + h + 2$.

REVIEW EXERCISES

Let $f(x) = 2x$ and $g(x) = x + 1$. Find each function or value.

43. $f + g$

44. $f - g$

45. $f \cdot g$

46. f/g

47. $(f \circ g)(2)$

48. $(g \circ f)(-1)$

49. $(f \circ g)(x)$

50. $(g \circ f)(x)$

SECTION 8.7 Inverses of Functions

DEFINITIONS AND CONCEPTS	EXAMPLES

Horizontal line test:
If every horizontal line that intersects the graph of a function does so only once, the function is one-to-one.

Determine if the graph of each function is one-to-one.

a.

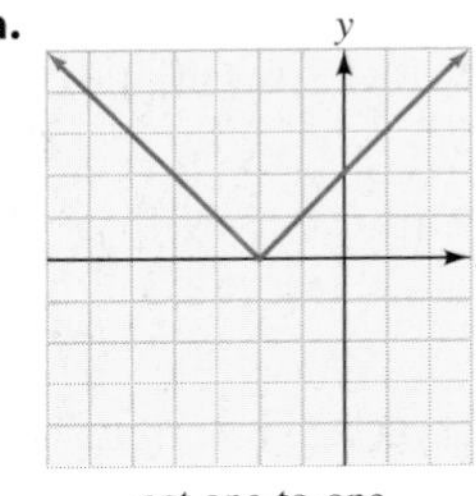

not one-to-one

b.

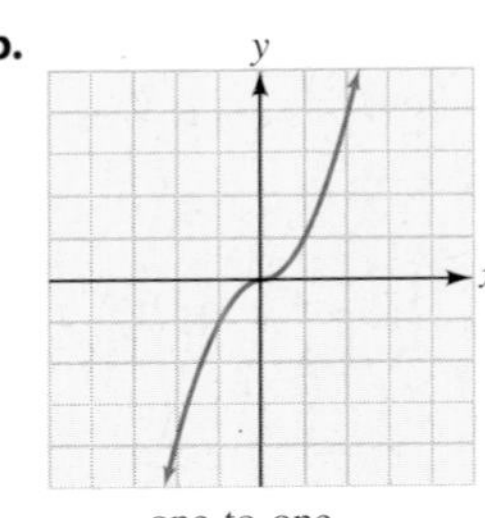

one-to-one

a. The function is not one-to-one because a horizontal line will cross its graph more than once.

b. The function is one-to-one because any horizontal line will cross its graph no more than once.

Finding the inverse of a one-to-one function:
If a function is one-to-one, we find its inverse as follows:

1. Replace $f(x)$ with y, if necessary.
2. Interchange the variables x and y.
3. Solve the resulting equation for y.
4. If the equation is a function, write the equation in function notation.

To find the inverse of $f(x) = x^3 - 5$, we proceed as follows:

$y = x^3 - 5$ Replace $f(x)$ with y.

$x = y^3 - 5$ Interchange the variables x and y.

$x + 5 = y^3$ Add 5 to both sides.

$\sqrt[3]{x + 5} = y$ Take the cube root of both sides.

Because there corresponds one real cube root for each x, $y = \sqrt[3]{x + 5}$ represents a function. In function notation, we describe the inverse as

$$f^{-1}(x) = \sqrt[3]{x + 5}$$

REVIEW EXERCISES

Graph each function and use the horizontal line test to decide whether the function is one-to-one.

51. $f(x) = 2(x - 3)$

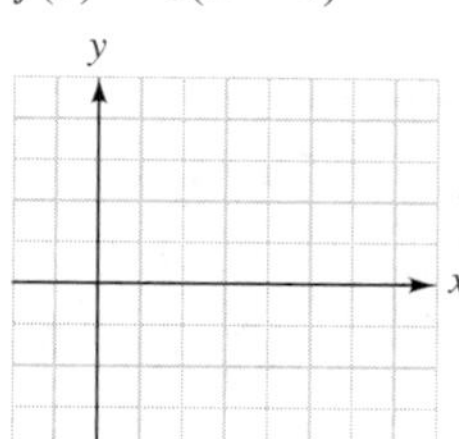

52. $f(x) = x(2x - 3)$

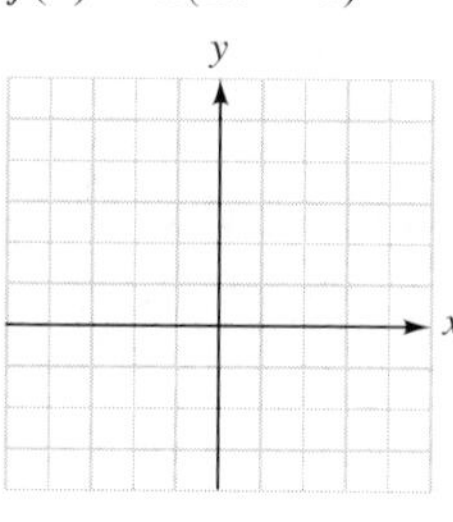

53. $f(x) = -3(x - 2)^2 + 5$

54. $f(x) = |x|$

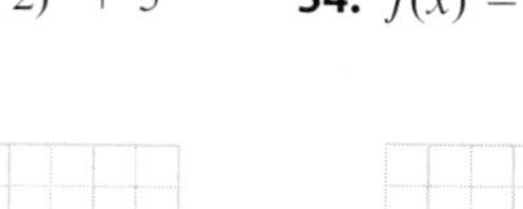

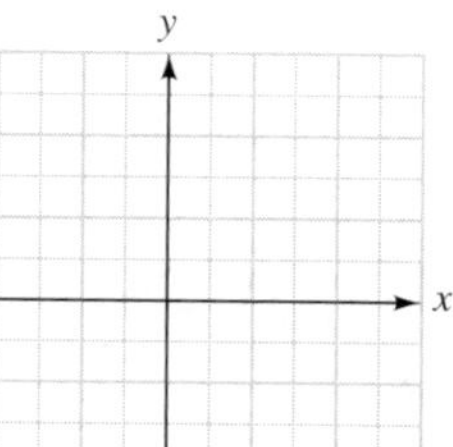

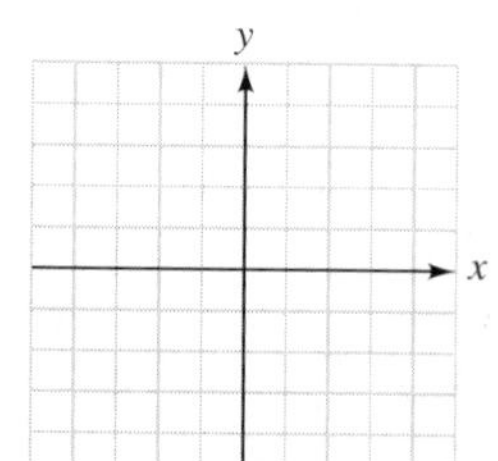

Find the inverse of each function. Do not rationalize the result.

55. $f(x) = 6x - 3$

56. $f(x) = 4x + 5$

57. $y = 2x^2 - 1 \quad (x \geq 0)$

58. $y = |x|$

Chapter 8 TEST

Solve each equation by factoring.

1. $x^2 + 3x - 18 = 0$

2. $x(6x + 19) = -15$

Determine what number must be added to each binomial to make it a perfect square.

3. $x^2 + 24x$

4. $x^2 - 50x$

Solve each equation by completing the square.

5. $x^2 + 4x + 1 = 0$

6. $x^2 - 5x - 3 = 0$

Solve each equation by the quadratic formula.

7. $2x^2 + 5x + 1 = 0$

8. $x^2 - x + 3 = 0$

9. Determine whether the solutions of $3x^2 + 5x + 17 = 0$ are real or nonreal numbers.

10. For what value(s) of k are the solutions of $4x^2 - 2kx + k - 1 = 0$ equal?

11. One leg of a right triangle is 14 inches longer than the other, and the hypotenuse is 26 inches. Find the length of the shorter leg.

12. Solve: $2y - 3y^{1/2} + 1 = 0$.

13. Graph $f(x) = \frac{1}{2}x^2 - 4$ and give the coordinates of its vertex.

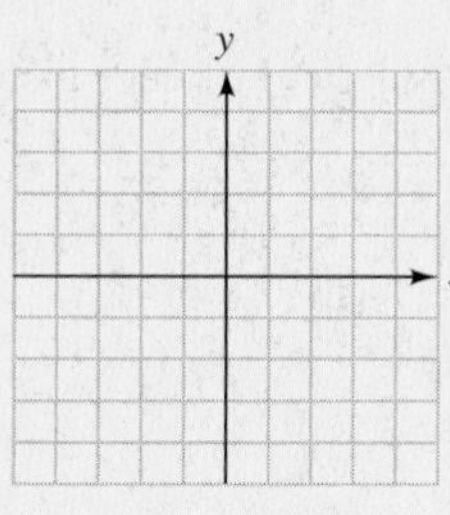

14. Find the vertex of the graph of $f(x) = -2x^2 + 8x - 7$.

15. Graph: $y \leq -x^2 + 3$.

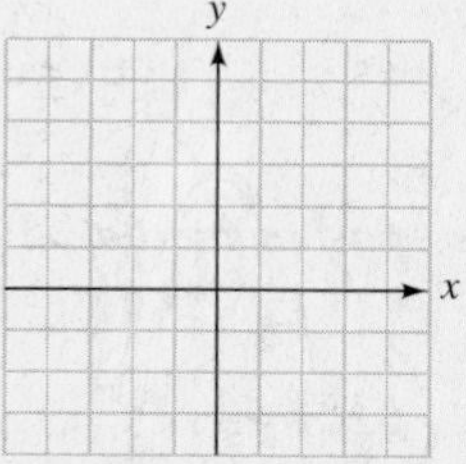

Solve each inequality and graph the solution set.

16. $x^2 - 2x - 8 > 0$

17. $\dfrac{x - 2}{x + 3} \leq 0$

Let f(x) = 4x and g(x) = x − 1. Find each function.

18. $g + f$

19. $f - g$

20. $g \cdot f$

21. g/f

Let f(x) = 4x and g(x) = x − 1. Find each value.

22. $(g \circ f)(1)$

23. $(f \circ g)(0)$

24. $(f \circ g)(-1)$

25. $(g \circ f)(-2)$

Let f(x) = 4x and g(x) = x − 1. Find each function.

26. $(f \circ g)(x)$

27. $(g \circ f)(x)$

Find the inverse of each function. Do not rationalize the result.

28. $3x + 2y = 12$

29. $y = 3x^2 + 4 \quad (x \le 0)$

Cumulative Review Exercises

Find the domain and range of each function.

1. $f(x) = 2x^2 - 3$

2. $f(x) = -|x - 4|$

Write the equation of the line with the given properties.

3. $m = 3$, passing through $(-2, -4)$

4. parallel to the graph of $2x + 3y = 6$ and passing through $(0, -2)$

Perform each operation.

5. $(2a^2 + 4a - 7) - 2(3a^2 - 4a)$

6. $(3x + 2)(2x - 3)$

Factor each expression completely using only integers.

7. $x^4 - 16y^4$

8. $15x^2 - 2x - 8$

Solve each equation.

9. $x^2 - 5x - 6 = 0$

10. $6a^3 - 2a = a^2$

Simplify each expression. Assume that all variables represent positive numbers.

11. $\sqrt{25x^4}$

12. $\sqrt{48t^3}$

13. $\sqrt[3]{-27x^3}$

14. $\sqrt[3]{\dfrac{128x^4}{2x}}$

15. $8^{-1/3}$

16. $64^{2/3}$

17. $\dfrac{y^{2/3}y^{5/3}}{y^{1/3}}$

18. $\dfrac{x^{5/3}x^{1/2}}{x^{3/4}}$

Graph each function and give the domain and the range.

19. $f(x) = \sqrt{x - 2}$

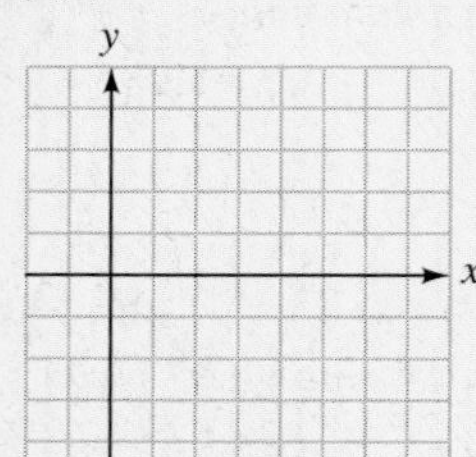

20. $f(x) = -\sqrt{x + 2}$

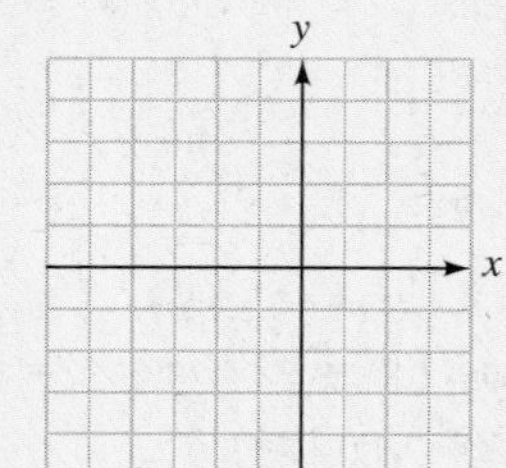

Perform the operations.

21. $(x^{2/3} - x^{1/3})(x^{2/3} + x^{1/3})$

22. $(x^{-1/2} + x^{1/2})^2$

Simplify each statement. Assume no division by 0.

23. $\sqrt{50} - \sqrt{8} + \sqrt{32}$

24. $-3\sqrt[4]{32} - 2\sqrt[4]{162} + 5\sqrt[4]{48}$

25. $3\sqrt{2}(2\sqrt{3} - 4\sqrt{12})$

26. $\dfrac{5}{\sqrt[3]{x}}$

27. $\dfrac{\sqrt{x} + 2}{\sqrt{x} - 1}$

28. $\sqrt[6]{x^3y^3}$

Solve each equation.

29. $5\sqrt{x + 2} = x + 8$

30. $\sqrt{x} + \sqrt{x + 2} = 2$

31. Find the length of the hypotenuse of the right triangle shown in Illustration 1.

32. Find the length of the hypotenuse of the right triangle shown in Illustration 2.

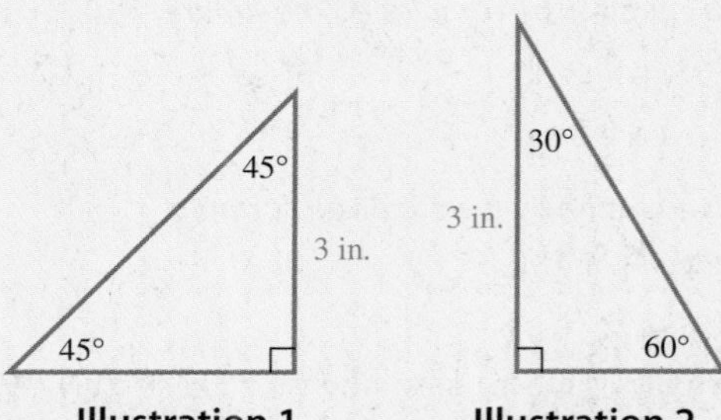

Illustration 1 **Illustration 2**

33. Find the distance between $(-2, 6)$ and $(4, 14)$.

34. What number must be added to $x^2 + 6x$ to make a perfect trinomial square?

35. Use the method of completing the square to solve $2x^2 + x - 3 = 0$.

36. Use the quadratic formula to solve $3x^2 + 4x - 1 = 0$.

37. Graph $f(x) = \frac{1}{2}x^2 + 5$ and find the coordinates of its vertex.

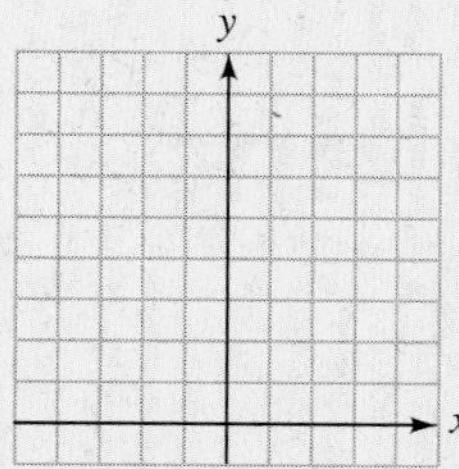

38. Graph $y \leq -x^2 + 3$ and find the coordinates of its vertex.

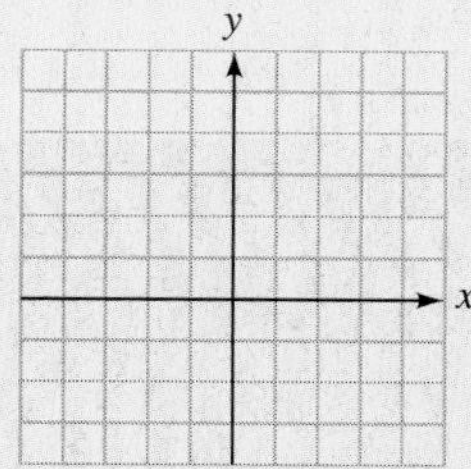

Write each expression as a real number or as a complex number in a + bi form.

39. $(3 + 5i) + (4 - 3i)$

40. $(7 - 4i) - (12 + 3i)$

41. $(2 - 3i)(2 + 3i)$

42. $(3 + i)(3 - 3i)$

43. $(3 - 2i) - (4 + i)^2$

44. $\dfrac{5}{3 - i}$

45. $|3 + 2i|$

46. $|5 - 6i|$

47. For what values of k will the solutions of $2x^2 + 4x = k$ be equal?

48. Solve: $a - 7a^{1/2} + 12 = 0$.

Solve each inequality and graph the solution set on the number line.

49. $x^2 - x - 6 > 0$

50. $x^2 - x - 6 \leq 0$

Let f(x) = 3x² + 2 and g(x) = 2x − 1. Find each value or composite function.

51. $f(-1)$

52. $(g \circ f)(2)$

53. $(f \circ g)(x)$

54. $(g \circ f)(x)$

Find the inverse of each function.

55. $f(x) = 3x + 2$

56. $f(x) = x^3 + 4$

CHAPTER 9

Exponential and Logarithmic Functions

9.1 Exponential Functions
9.2 Base-e Exponential Functions
9.3 Logarithmic Functions
9.4 Natural Logarithms
9.5 Properties of Logarithms
9.6 Exponential and Logarithmic Equations
■ *Projects*
CHAPTER REVIEW
CHAPTER TEST

Careers and Mathematics

PEST-CONTROL WORKERS

Few people welcome roaches, rats, mice, spiders, termites, fleas, ants, and bees into their homes. It is the job of pest-control workers to locate, identify, destroy, control, and repel these pests.

Of the 70,000 pest-control workers that held jobs in 2006, 85 percent were employed in services related to the building industry, primarily in states with warmer climates. About 9 percent are self-employed.

Both federal and state laws require pest-control workers to be certified.

Job Outlook:
Employment of pest-control workers is expected to grow 15 percent between 2006 and 2016, which is faster than the average for all occupations.

Hourly Earnings:
\$10.79–\$16.76.

For More Information:
http://www.bls.gov/oco/ocos254.htm

For a Sample Application:
See Problem 87 in Section 9.6.

In this chapter ▶

In this chapter, we will discuss two functions that are important in many applications of mathematics. Exponential functions are used to compute compound interest, find radioactive decay, and model population growth. Logarithmic functions are used to measure acidity of solutions, drug dosage, gain of an amplifier, magnitude of earthquakes, and safe noise levels in factories.

SECTION 9.1 Exponential Functions

Objectives

1. Simplify an expression containing irrational exponents.
2. Graph an exponential function.
3. Graph a translation of an exponential function.
4. Evaluate an application problem containing an exponential function.

Vocabulary

exponential function
increasing function
decreasing function
compound interest
future value
periodic interest rate
compounding period

Getting Ready

Find each value.

1. 2^3 **2.** $25^{1/2}$ **3.** 5^{-2} **4.** $\left(\frac{3}{2}\right)^{-3}$

The graph in Figure 9-1 shows the balance in an investment account in which \$10,000 was invested in 2000 at 9% annual interest, compounded monthly. The graph shows that in the year 2010, the value of the account will be approximately \$25,000, and in the year 2030, the value will be approximately \$147,000. The curve shown in Figure 9-1 is the graph of a function called an *exponential function,* the topic of this section.

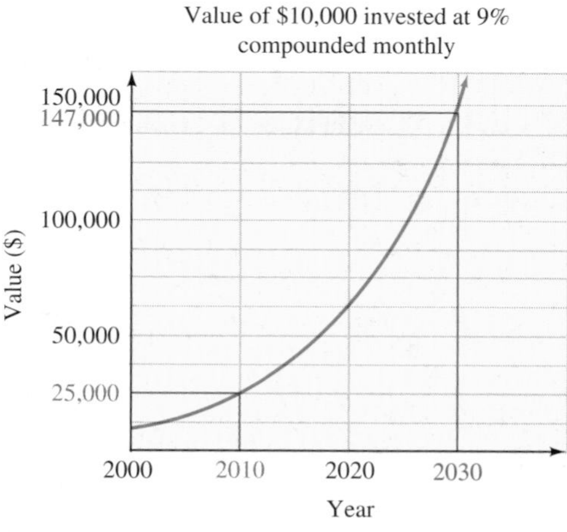

Figure 9-1

1 Simplify an expression containing irrational exponents.

We have discussed expressions of the form b^x, where x is a rational number.

$8^{1/2}$ means "the square root of 8."

$5^{1/3}$ means "the cube root of 5."

$3^{-2/5} = \frac{1}{3^{2/5}}$ means "the reciprocal of the fifth root of 3^2."

To give meaning to b^x when x is an irrational number, we consider the expression

$5^{\sqrt{2}}$ where $\sqrt{2}$ is the irrational number 1.414213562. . .

Each number in the following list is defined, because each exponent is a rational number.

$$5^{1.4},\ 5^{1.41},\ 5^{1.414},\ 5^{1.4142},\ 5^{1.41421},\ \ldots$$

Since the exponents are getting closer to $\sqrt{2}$, the numbers in this list are successively better approximations of $5^{\sqrt{2}}$. We can use a calculator to obtain a very good approximation.

ACCENT ON TECHNOLOGY

Evaluating Exponential Expressions

To approximate the value of $5^{\sqrt{2}}$ with a scientific calculator, we enter these numbers and press these keys:

5 y^x 2 $\sqrt{\ }$ =

The display will read **9.738517742**.

With a graphing calculator, we enter these numbers and press these keys:

5 ^ $\sqrt{\ }$ 2) ENTER

The display will read 5^√(2) 9.738517742.

In general, if b is positive and x is a real number, b^x represents a positive number. It can be shown that all of the rules of exponents hold true for irrational exponents.

EXAMPLE 1 Use the rules of exponents to simplify: **a.** $(5^{\sqrt{2}})^{\sqrt{2}}$ **b.** $b^{\sqrt{3}} \cdot b^{\sqrt{12}}$.

Solution

a.
$$\begin{aligned}(5^{\sqrt{2}})^{\sqrt{2}} &= 5^{\sqrt{2}\sqrt{2}} && \text{Keep the base and multiply the exponents.}\\ &= 5^2 && \sqrt{2}\sqrt{2} = \sqrt{4} = 2\\ &= 25\end{aligned}$$

b.
$$\begin{aligned}b^{\sqrt{3}} \cdot b^{\sqrt{12}} &= b^{\sqrt{3}+\sqrt{12}} && \text{Keep the base and add the exponents.}\\ &= b^{\sqrt{3}+2\sqrt{3}} && \sqrt{12} = \sqrt{4}\sqrt{3} = 2\sqrt{3}\\ &= b^{3\sqrt{3}} && \sqrt{3} + 2\sqrt{3} = 3\sqrt{3}\end{aligned}$$

SELF CHECK 1 Simplify: **a.** $(3^{\sqrt{2}})^{\sqrt{8}}$ **b.** $b^{\sqrt{2}} \cdot b^{\sqrt{18}}$.

2 Graph an exponential function.

If $b > 0$ and $b \neq 1$, the function $f(x) = b^x$ is called an *exponential function.* Since x can be any real number, its domain is the set of real numbers. This is the interval $(-\infty, \infty)$. Since b is positive, the value of $f(x)$ is positive and the range is the set of positive numbers. This is the interval $(0, \infty)$.

Since $b \neq 1$, an exponential function cannot be the constant function $f(x) = 1^x$, in which $f(x) = 1$ for every real number x.

Exponential Functions

An **exponential function with base *b*** is defined by the equation

$$f(x) = b^x \quad (b > 0, b \neq 1, \text{ and } x \text{ is a real number})$$

The **domain of any exponential function** is the interval $(-\infty, \infty)$. The **range** is the interval $(0, \infty)$.

Since the domain and range of $f(x) = b^x$ are subsets of the real numbers, we can graph exponential functions on a rectangular coordinate system.

EXAMPLE 2 Graph: $f(x) = 2^x$.

Solution To graph $f(x) = 2^x$, we find several points (x, y) whose coordinates satisfy the equation, plot the points, and join them with a smooth curve, as shown in Figure 9-2.

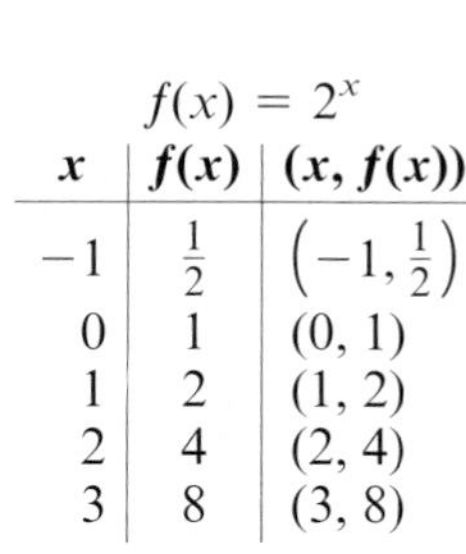

$f(x) = 2^x$

x	$f(x)$	$(x, f(x))$
-1	$\frac{1}{2}$	$\left(-1, \frac{1}{2}\right)$
0	1	(0, 1)
1	2	(1, 2)
2	4	(2, 4)
3	8	(3, 8)

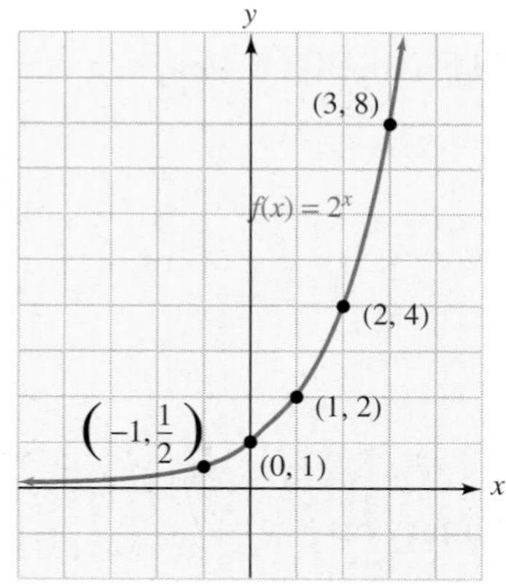

Figure 9-2

By looking at the graph, we can verify that the domain is the interval $(-\infty, \infty)$ and that the range is the interval $(0, \infty)$.

Note that as x decreases, the values of $f(x)$ decrease and approach 0, but will never be 0. Thus, the x-axis is the horizontal asymptote of the graph.

Also note that the graph of $f(x) = 2^x$ passes through the points (0, 1) and (1, 2).

SELF CHECK 2 Graph: $f(x) = 4^x$.

In Example 2, the values of $f(x)$ increase as the values of x increase. When the graph of a function rises as we move to the right, we call the function an **increasing function.** When $b > 1$, the larger the value of b, the steeper the curve.

EXAMPLE 3 Graph: $f(x) = \left(\frac{1}{2}\right)^x$.

Solution We find and plot pairs (x, y) that satisfy the equation. The graph of $f(x) = \left(\frac{1}{2}\right)^x$ appears in Figure 9-3.

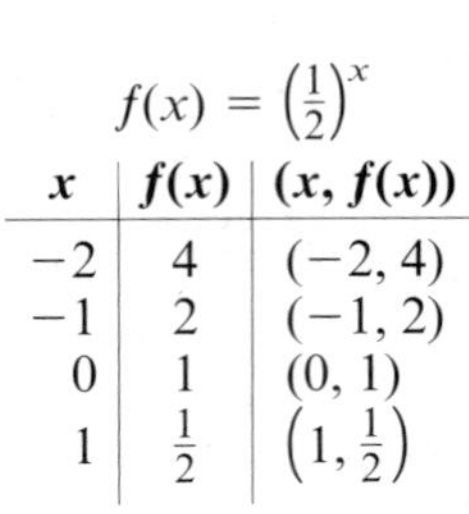

$f(x) = \left(\frac{1}{2}\right)^x$

x	$f(x)$	$(x, f(x))$
-2	4	$(-2, 4)$
-1	2	$(-1, 2)$
0	1	$(0, 1)$
1	$\frac{1}{2}$	$\left(1, \frac{1}{2}\right)$

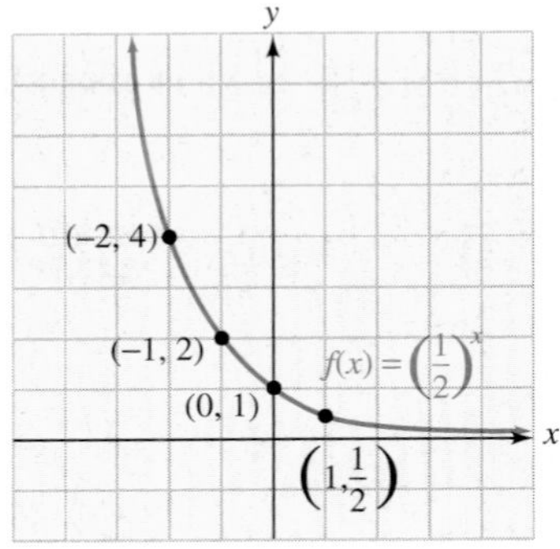

Figure 9-3

By looking at the graph, we can see that the domain is the interval $(-\infty, \infty)$ and that the range is the interval $(0, \infty)$.

In this case, as x increases, the values of $f(x)$ decrease and approach 0. The x-axis is a horizontal asymptote. Note that the graph of $f(x) = \left(\frac{1}{2}\right)^x$ passes through the points $(0, 1)$ and $\left(1, \frac{1}{2}\right)$.

SELF CHECK 3 Graph: $f(x) = \left(\frac{1}{4}\right)^x$.

In Example 3, the values of $f(x)$ decrease as the values of x increase. When the graph of a function drops as we move to the right, we call the function a **decreasing function.** When $0 < b < 1$, the smaller the value of b, the steeper the curve.

Examples 2 and 3 illustrate the following properties of exponential functions.

Properties of Exponential Functions

The **domain** of the exponential function $y = f(x) = b^x$ is the interval $(-\infty, \infty)$.

The **range** is the interval $(0, \infty)$.

The graph has a y-intercept of $(0, 1)$.

The x-axis is an asymptote of the graph.

The graph of $f(x) = b^x$ passes through the point $(1, b)$.

If $b > 1$, then $f(x) = b^x$ is an **increasing function.**

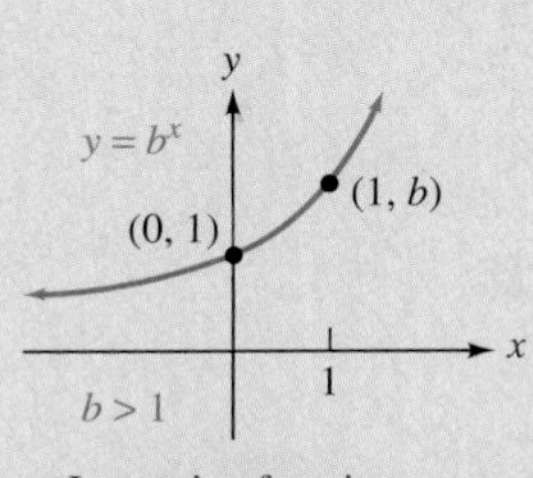

Increasing function

If $0 < b < 1$, then $f(x) = b^x$ is a **decreasing function.**

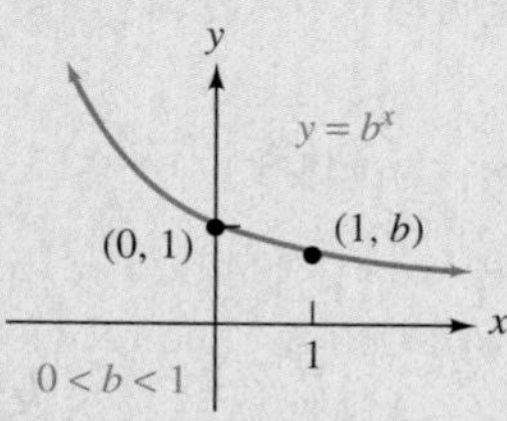

Decreasing function

EXAMPLE 4 From the graph of $f(x) = b^x$ shown in Figure 9-4, find the value of b.

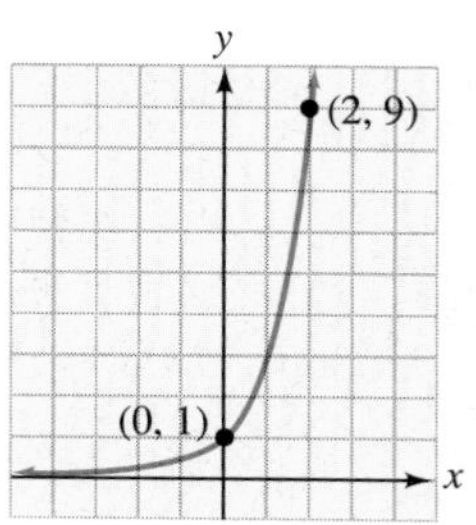

Figure 9-4

Solution We first note that the graph passes through (0, 1). Since the point (2, 9) is on the graph, we substitute 9 for y and 2 for x in the equation $y = b^x$ to get

$$y = b^x$$
$$9 = b^2$$
$$3 = b \quad \text{Because 3 is the positive number whose square is 9.}$$

The base b is 3.

SELF CHECK 4 From the graph of $f(x) = b^x$ shown to the right, find the value of b.

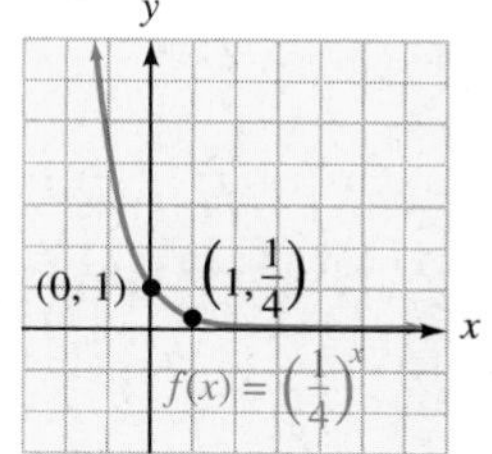

An exponential function with base b is either increasing (for $b > 1$) or decreasing ($0 < b < 1$). Since different real numbers x determine different values of b^x, exponential functions are one-to-one.

The exponential function defined by

$$f(x) = b^x \quad \text{where } b > 0 \text{ and } b \neq 1$$

is one-to-one. Thus,

1. If $b^r = b^s$, then $r = s$.
2. If $r \neq s$, then $b^r \neq b^s$.

ACCENT ON TECHNOLOGY

Graphing Exponential Functions

To use a graphing calculator to graph $f(x) = \left(\frac{2}{3}\right)^x$ and $f(x) = \left(\frac{3}{2}\right)^x$, we enter the right sides of the equations. The screen will show the following equations.

$\backslash Y_1 = (2/3)^\wedge X$
$\backslash Y_2 = (3/2)^\wedge X$

If we use window settings of $[-10, 10]$ for x and $[-2, 10]$ for y and press **GRAPH**, we will obtain the graph shown in Figure 9-5.

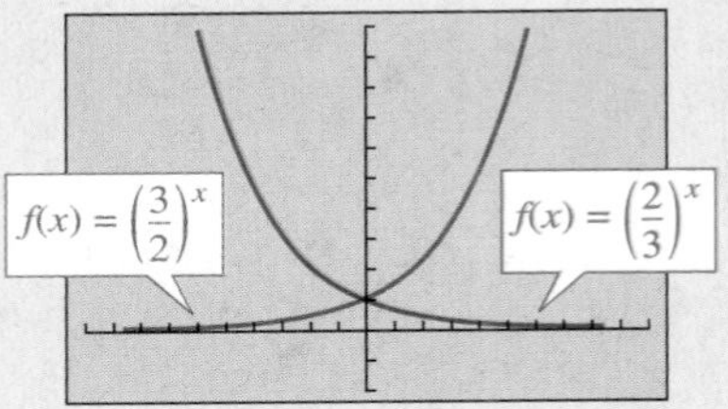

Figure 9-5

The graph of $f(x) = \left(\frac{2}{3}\right)^x$ passes through the points $(0, 1)$ and $\left(1, \frac{2}{3}\right)$. Since $\frac{2}{3} < 1$, the function is decreasing.

The graph of $f(x) = \left(\frac{3}{2}\right)^x$ passes through the points $(0, 1)$ and $\left(1, \frac{3}{2}\right)$. Since $\frac{3}{2} > 1$, the function is increasing.

Since both graphs pass the horizontal line test, each function is one-to-one.

3 Graph a translation of an exponential function.

We have seen that when $k > 0$ the graph of

$y = f(x) + k$ is the graph of $y = f(x)$ translated k units upward.
$y = f(x) - k$ is the graph of $y = f(x)$ translated k units downward.
$y = f(x - k)$ is the graph of $y = f(x)$ translated k units to the right.
$y = f(x + k)$ is the graph of $y = f(x)$ translated k units to the left.

EXAMPLE 5 On one set of axes, graph $f(x) = 2^x$ and $f(x) = 2^x + 3$.

Solution The graph of $f(x) = 2^x + 3$ is identical to the graph of $f(x) = 2^x$, except that it is translated 3 units upward. (See Figure 9-6.)

$f(x) = 2^x$

x	$f(x)$	$(x, f(x))$
-4	$\frac{1}{16}$	$\left(-4, \frac{1}{16}\right)$
0	1	$(0, 1)$
2	4	$(2, 4)$

$f(x) = 2^x + 3$

x	$f(x)$	$(x, f(x))$
-4	$3\frac{1}{16}$	$\left(-4, 3\frac{1}{16}\right)$
0	4	$(0, 4)$
2	7	$(2, 7)$

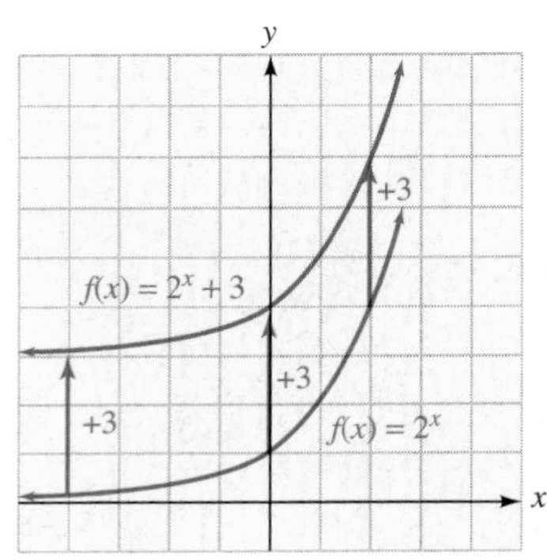

Figure 9-6

SELF CHECK 5 On one set of axes, graph $f(x) = 4^x$ and $f(x) = 4^x - 3$.

EXAMPLE 6 On one set of axes, graph $f(x) = 2^x$ and $f(x) = 2^{x+3}$.

Solution The graph of $f(x) = 2^{x+3}$ is identical to the graph of $f(x) = 2^x$, except that it is translated 3 units to the left. (See Figure 9-7.)

$f(x) = 2^x$

x	$f(x)$	$(x, f(x))$
0	1	(0, 1)
2	4	(2, 4)
4	16	(4, 16)

$f(x) = 2^{x+3}$

x	$f(x)$	$(x, f(x))$
0	8	(0, 8)
−1	4	(−1, 4)
−2	2	(−2, 2)

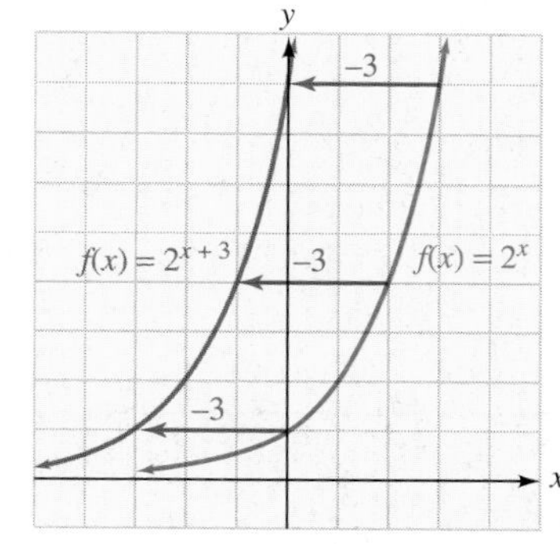

Figure 9-7

SELF CHECK 6 On one set of axes, graph $f(x) = 4^x$ and $f(x) = 4^{x-3}$.

The graphs of $f(x) = kb^x$ and $f(x) = b^{kx}$ are vertical and horizontal stretchings, respectively, of the graph of $f(x) = b^x$. To graph these functions, we can plot several points and join them with a smooth curve or use a graphing calculator.

ACCENT ON TECHNOLOGY

Graphing Exponential Functions

To use a graphing calculator to graph the exponential function $f(x) = 3(2^{x/3})$, we enter the right side of the equation. The display will show the equation

\Y$_1$ = 3(2^(X/3))

If we use window settings of $[-10, 10]$ for x and $[-2, 18]$ for y and press **GRAPH**, we will obtain the graph shown in Figure 9-8.

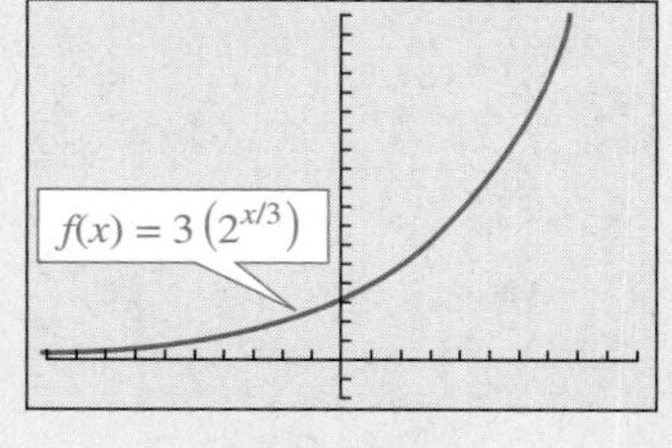

Figure 9-8

4 Evaluate an application problem containing an exponential function.

EXAMPLE 7 **CELL PHONE GROWTH** In the decade from 1990 to 2000, the U.S. cellular telephone industry experienced exponential growth. The exponential function $S(n) = 5.74(1.39)^n$ approximates the number of cellular telephone subscribers in millions from 1990 to 2000, where n is the number of years since 1990.

a. How many subscribers were there at the beginning of the decade?

b. How many subscribers were there at the end of the decade?

Solution **a.** To find the number of subscribers in 1990, we substitute 0 for n in the function and find $S(0)$.

$$\begin{aligned} S(n) &= 5.74(1.39)^n \\ S(0) &= 5.74(1.39)^0 && \text{Substitute 0 for } n. \\ &= 5.74 \cdot 1 && (1.39)^0 = 1 \\ &= 5.74 \end{aligned}$$

In 1990, there were approximately 5.74 million cellular telephone subscribers in the U.S.

b. To find the number of subscribers in 2000, we substitute 10 for n in the function and find $S(10)$.

$$\begin{aligned} S(n) &= 5.74(1.39)^n \\ S(10) &= 5.74(1.39)^{10} && \text{Substitute 10 for } n. \\ &\approx 154.5467565 && \text{Use a calculator to find an approximation.} \end{aligned}$$

In 2000, there were approximately 154.55 million cellular telephone subscribers.

If we deposit $\$P$ in an account paying an annual simple interest rate r, we can find the amount A in the account at the end of t years by using the formula $A = P + Prt$ or $A = P(1 + rt)$.

Suppose that we deposit \$500 in an account that pays interest every six months. Then $P = 500$, and after six months $\left(\frac{1}{2}\text{ year}\right)$, the amount in the account will be

$$\begin{aligned} A &= 500(1 + rt) \\ &= 500\left(1 + r \cdot \frac{1}{2}\right) && \text{Substitute } \tfrac{1}{2} \text{ for } t. \\ &= 500\left(1 + \frac{r}{2}\right) \end{aligned}$$

The account will begin the second six-month period with a value of $\$500\left(1 + \frac{r}{2}\right)$. After the second six-month period, the amount will be

$$\begin{aligned} A &= P(1 + rt) \\ A &= \left[500\left(1 + \frac{r}{2}\right)\right]\left(1 + r \cdot \frac{1}{2}\right) && \text{Substitute } 500\left(1 + \tfrac{r}{2}\right) \text{ for } P \text{ and } \tfrac{1}{2} \text{ for } t. \\ &= 500\left(1 + \frac{r}{2}\right)\left(1 + \frac{r}{2}\right) \\ &= 500\left(1 + \frac{r}{2}\right)^2 \end{aligned}$$

At the end of the third six-month period, the amount in the account will be

$$A = 500\left(1 + \frac{r}{2}\right)^3$$

In this discussion, the earned interest is deposited back in the account and also earns interest. When this is the case, we say that the account is earning **compound interest.**

Formula for Compound Interest

If $\$P$ is deposited in an account and interest is paid k times a year at an annual rate r, the amount A in the account after t years is given by

$$A = P\left(1 + \frac{r}{k}\right)^{kt}$$

EXAMPLE 8 **SAVING FOR COLLEGE** To save for college, parents invest \$12,000 for their newborn child in a mutual fund that should average a 10% annual return. If the quarterly interest is reinvested, how much will be available in 18 years?

Solution We substitute 12,000 for P, 0.10 for r, and 18 for t into the formula for compound interest and find A. Since interest is paid quarterly, $k = 4$.

$$A = P\left(1 + \frac{r}{k}\right)^{kt}$$

$$\begin{aligned} A &= \mathbf{12{,}000}\left(1 + \frac{\mathbf{0.10}}{\mathbf{4}}\right)^{\mathbf{4(18)}} \\ &= 12{,}000(1 + 0.025)^{72} \\ &= 12{,}000(1.025)^{72} \\ &\approx 71{,}006.74 && \text{Use a calculator.} \end{aligned}$$

In 18 years, the account will be worth \$71,006.74.

SELF CHECK 8 How much would be available if the parents invest \$20,000?

In business applications, the initial amount of money deposited is the **present value** (PV). The amount to which the money will grow is called the **future value** (FV). The interest rate used for each compounding period is the **periodic interest rate** (i), and the number of times interest is compounded is the number of **compounding periods** (n). Using these definitions, we have an alternate formula for compound interest.

Formula for Future Value

$$FV = PV(1 + i)^n \quad \text{where } i = \frac{r}{k} \text{ and } n = kt$$

This alternate formula appears on business calculators. To use this formula to solve Example 8, we proceed as follows:

$$\begin{aligned} FV &= PV(1 + i)^n \\ FV &= \mathbf{12{,}000}(1 + \mathbf{0.025})^{\mathbf{72}} && i = \tfrac{0.10}{4} = 0.025 \text{ and } n = 4(18) = 72. \\ &\approx 71{,}006.74 \end{aligned}$$

ACCENT ON TECHNOLOGY

Solving Investment Problems

Suppose \$1 is deposited in an account earning 6% annual interest, compounded monthly. To use a graphing calculator to estimate how much money will be in the account in 100 years, we can substitute 1 for P, 0.06 for r, and 12 for k into the formula

(continued)

$$A = P\left(1 + \frac{r}{k}\right)^{kt}$$

$$A = 1\left(1 + \frac{0.06}{12}\right)^{12t}$$

and simplify to get

$$A = (1.005)^{12t}$$

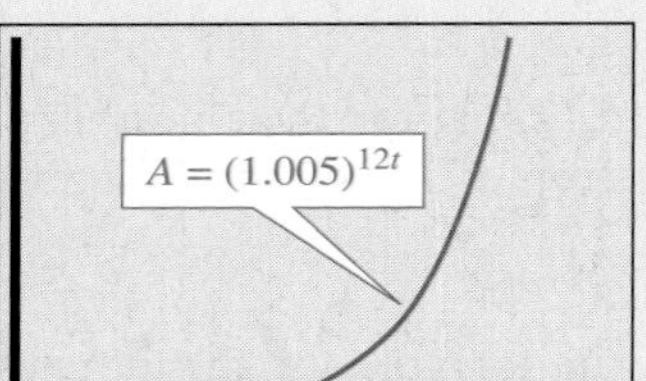

Figure 9-9

We now graph $A = (1.005)^{12t}$ using window settings of [0, 120] for t and [0, 400] for A to obtain the graph shown in Figure 9-9. We can then trace and zoom to estimate that \$1 grows to be approximately \$397 in 100 years. From the graph, we can see that the money grows slowly in the early years and rapidly in the later years.

We can use the VALUE feature under the CALC menu to determine that exactly \$397.44 will be in the account in 100 years.

SELF CHECK ANSWERS

1. a. 81 **b.** $b^{4\sqrt{2}}$

2.

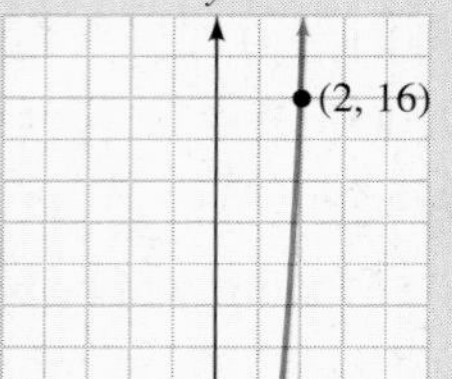

3.

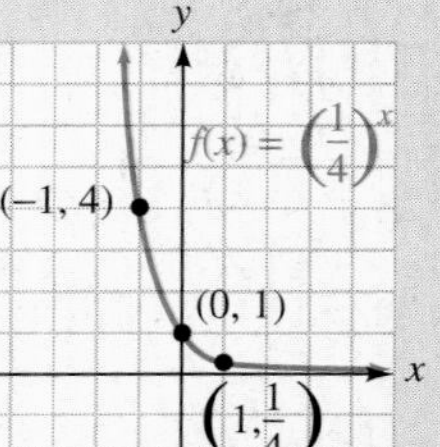

4. $\frac{1}{4}$

5.

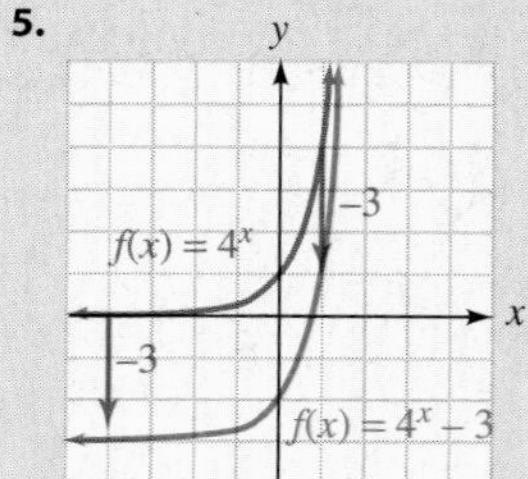

6.

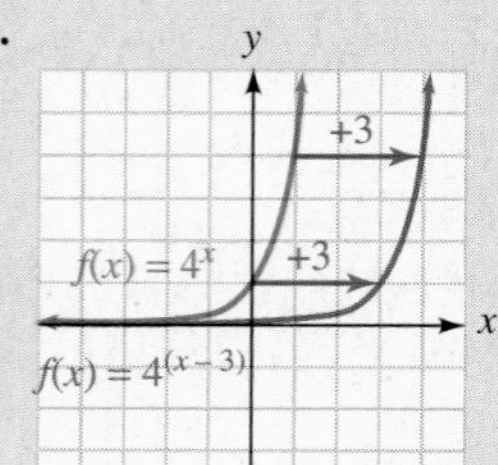

8. \$118,344.56

NOW TRY THIS

1. If you were given \$1 on May 1, \$2 on May 2, \$4 on May 3, \$8 on May 4, and double the previous day's earnings each day after, how much would you earn on

a. May 10?

b. May 15?

c. May 21?

Try to find an equation to model this problem.

9.1 EXERCISES

WARM-UPS

If $x = 2$, evaluate each expression.

1. 2^x **2.** 5^x

3. $2(3^x)$ **4.** 3^{x-1}

If $x = -2$, evaluate each expression.

5. 2^x **6.** 5^x

7. $2(3^x)$ **8.** 3^{x-1}

REVIEW *In the illustration, lines r and s are parallel.*

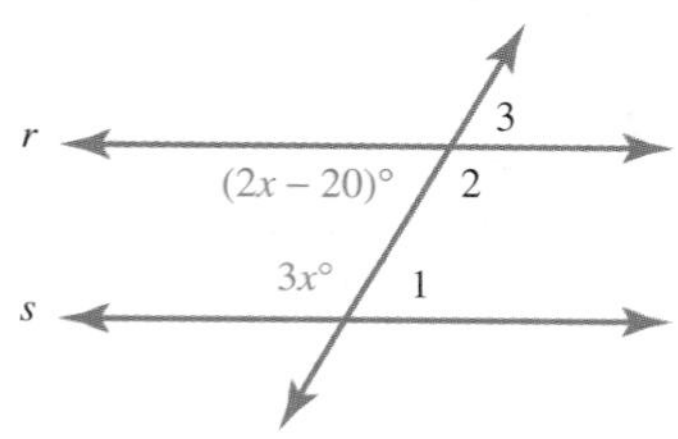

9. Find the value x.

10. Find the measure of $\angle 1$.

11. Find the measure of $\angle 2$.

12. Find the measure of $\angle 3$.

VOCABULARY AND CONCEPTS *Fill in the blanks.*

13. If $b > 0$ and $b \neq 1$, $y = f(x) = b^x$ is called an ________ function.

14. The ______ of an exponential function is $(-\infty, \infty)$.

15. The range of an exponential function is the interval ______.

16. The graph of $y = f(x) = 3^x$ passes through the points $(0, \_\_)$ and $(1, \_\_)$.

17. If $b > 1$, then $y = f(x) = b^x$ is an ________ function.

18. If $0 < b < 1$, then $y = f(x) = b^x$ is a ________ function.

19. The formula for compound interest is $A =$ ________.

20. An alternate formula for compound interest is $FV =$ ________, where PV stands for present value, i stands for the ________________, and n stands for the number of ________________.

GUIDED PRACTICE

Find each value to four decimal places. **(Objective 1)**

21. $2^{\sqrt{2}}$ **22.** $7^{\sqrt{2}}$

23. $5^{\sqrt{5}}$ **24.** $6^{\sqrt{3}}$

Simplify and give the exact value of each expression. **See Example 1. (Objective 1)**

25. $\left(2^{\sqrt{3}}\right)^{\sqrt{3}}$ **26.** $3^{\sqrt{2}}3^{\sqrt{18}}$

27. $7^{\sqrt{3}}7^{\sqrt{12}}$ **28.** $\left(3^{\sqrt{5}}\right)^{\sqrt{5}}$

Graph each exponential function. Check your work with a graphing calculator. **See Examples 2–3. (Objective 2)**

29. $f(x) = 3^x$ **30.** $f(x) = 5^x$

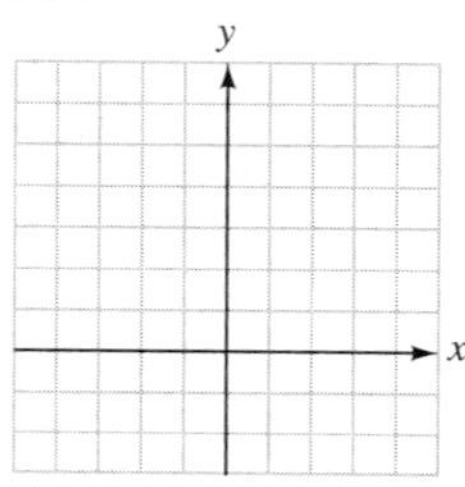

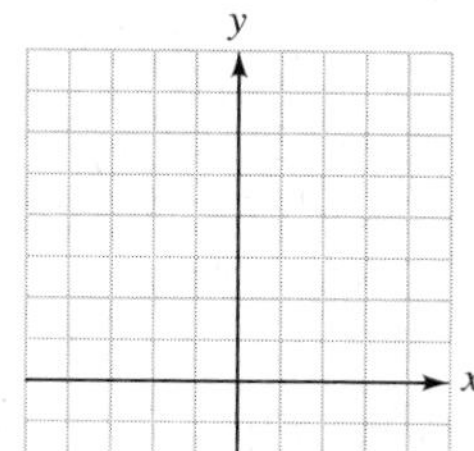

31. $f(x) = \left(\frac{1}{3}\right)^x$ **32.** $f(x) = \left(\frac{1}{5}\right)^x$

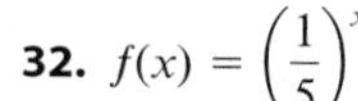

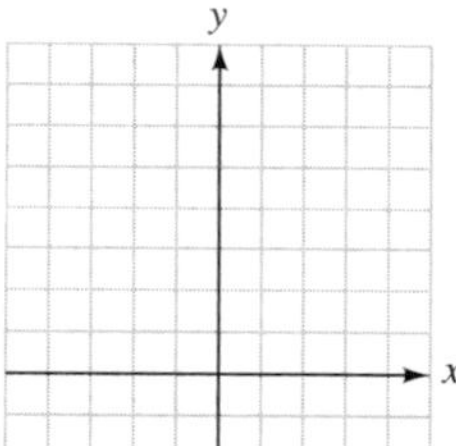

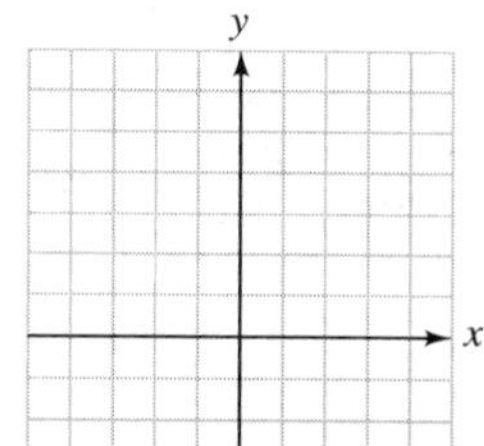

Find the value of b that would cause the graph of $y = b^x$ to look like the graph indicated. **See Example 4. (Objective 2)**

33.

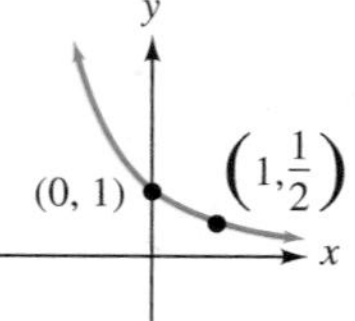

34.

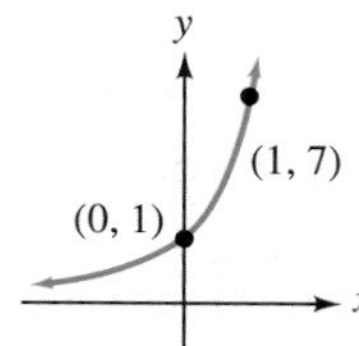

35.

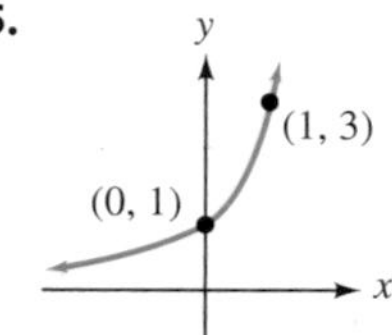

36.

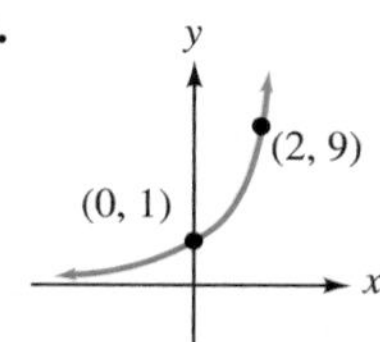

Graph each exponential function. Check your work with a graphing calculator. See Examples 5–6. (Objective 3)

37. $f(x) = 3^x - 2$

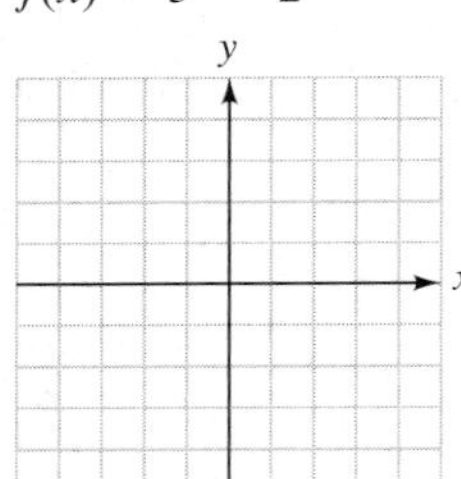

38. $f(x) = 2^x + 1$

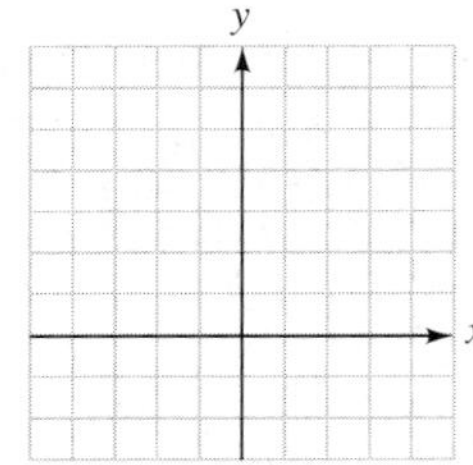

39. $f(x) = 3^{x-1}$

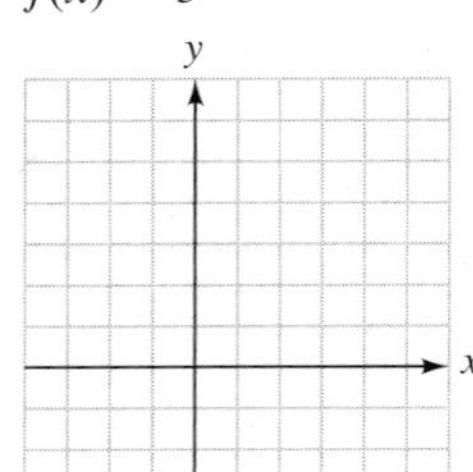

40. $f(x) = 2^{x+1}$

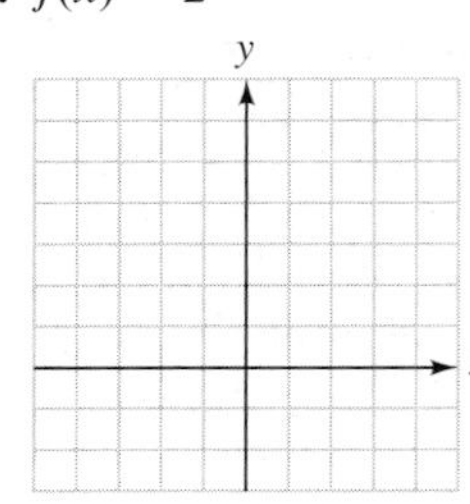

ADDITIONAL PRACTICE

Find the value of b that would cause the graph of $y = b^x$ to look like the graph indicated.

41.

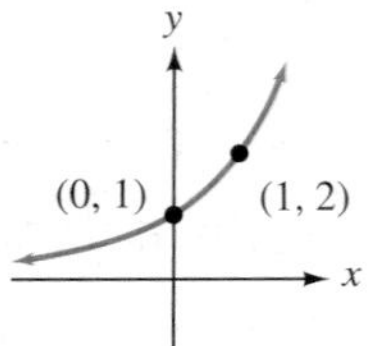

42.

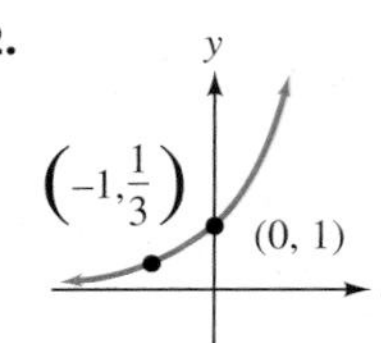

Use a graphing calculator to graph each function. Determine whether the function is an increasing or a decreasing function.

43. $f(x) = \frac{1}{2}(3^{x/2})$

44. $f(x) = -3(2^{x/3})$

45. $f(x) = 2(3^{-x/2})$

46. $f(x) = -\frac{1}{4}(2^{-x/2})$

APPLICATIONS

Evaluate each application. See Examples 7–8. (Objective 4)

47. Cell phone usage Refer to Example 7 and find the number of cell phone users in 1995.

48. Cell phone usage Refer to Example 7 and find the number of cell phone users in 1998.

49. Radioactive decay A radioactive material decays according to the formula $A = A_0\left(\frac{2}{3}\right)^t$, where A_0 is the initial amount present and t is measured in years. Find an expression for the amount present in 5 years.

50. Bacteria cultures A colony of 6 million bacteria is growing in a culture medium. (See the illustration.) The population P after t hours is given by the formula $P = (6 \times 10^6)(2.3)^t$. Find the population after 4 hours.

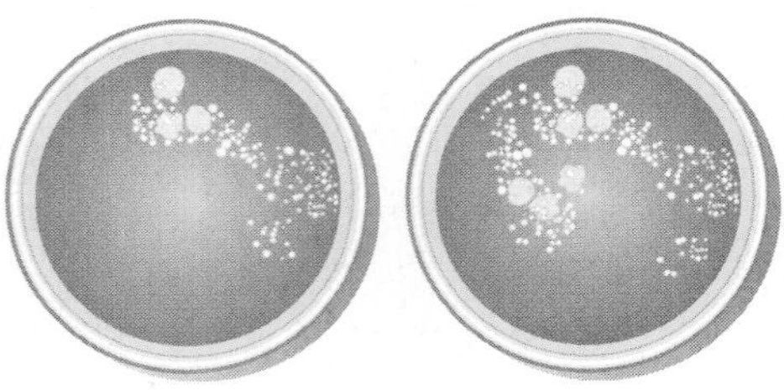

Assume that there are no deposits or withdrawals.

51. Compound interest An initial deposit of \$10,000 earns 8% interest, compounded quarterly. How much will be in the account after 10 years?

52. Compound interest An initial deposit of \$10,000 earns 8% interest, compounded monthly. How much will be in the account after 10 years?

53. Comparing interest rates How much more interest could \$1,000 earn in 5 years, compounded quarterly, if the annual interest rate were $5\frac{1}{2}\%$ instead of 5%?

54. Comparing savings plans Which institution in the two ads provides the better investment?

Fidelity Savings & Loan
Earn 5.25%
compounded monthly

Union Trust
Money Market Account
paying 5.35%
compounded annually

55. Compound interest If \$1 had been invested on July 4, 1776, at 5% interest, compounded annually, what would it be worth on July 4, 2076?

56. Frequency of compounding \$10,000 is invested in each of two accounts, both paying 6% annual interest. In the first account, interest compounds quarterly, and in the second account, interest compounds daily. Find the difference between the accounts after 20 years.

57. **Discharging a battery** The charge remaining in a battery decreases as the battery discharges. The charge C (in coulombs) after t days is given by the formula $C = (3 \times 10^{-4})(0.7)^t$. Find the charge after 5 days.

58. **Town population** The population of North Rivers is decreasing exponentially according to the formula $P = 3{,}745(0.93)^t$, where t is measured in years from the present date. Find the population in 6 years, 9 months.

59. **Salvage value** A small business purchases a computer for \$4,700. It is expected that its value each year will be 75% of its value in the preceding year. If the business disposes of the computer after 5 years, find its salvage value (the value after 5 years).

60. **Louisiana Purchase** In 1803, the United States acquired territory from France in the Louisiana Purchase. The country doubled its territory by adding 827,000 square miles of land for \$15 million. If the value of the land has appreciated at the rate of 6% each year, what would one square mile of land be worth in 2023?

WRITING ABOUT MATH

61. If the world population is increasing exponentially, why is there cause for concern?
62. How do the graphs of $y = b^x$ differ when $b > 1$ and $0 < b < 1$?

SOMETHING TO THINK ABOUT

63. In the definition of the exponential function, b could not equal 0. Why not?
64. In the definition of the exponential function, b could not be negative. Why not?

SECTION 9.2 Base-*e* Exponential Functions

Objectives

1. Compute the continuously compounded interest of an investment given the principal, rate, and duration.
2. Graph an exponential function.
3. Solve an application problem involving exponential growth or decay.

Vocabulary

compound interest | annual growth rate | exponential function

Getting Ready

Evaluate $\left(1 + \frac{1}{n}\right)^n$ *for the following values. Round each answer to the nearest hundredth.*

1. $n = 1$
2. $n = 2$
3. $n = 4$
4. $n = 10$

In this section, we will discuss special exponential functions with a base of e, an important number with a value of approximately 2.72.

 Compute the continuously compounded interest of an investment given the principal, rate, and duration.

If a bank pays interest twice a year, we say that interest is *compounded semiannually.* If it pays interest four times a year, we say that interest is *compounded quarterly.* If it pays interest continuously (infinitely many times in a year), we say that interest is *compounded continuously.*

To develop the formula for continuous **compound interest,** we start with the formula

$$A = P\left(1 + \frac{r}{k}\right)^{kt} \quad \text{This is the formula for compound interest.}$$

and substitute rn for k. Since r and k are positive numbers, so is n.

$$A = P\left(1 + \frac{r}{rn}\right)^{rnt}$$

We can then simplify the fraction $\frac{r}{rn}$ and use the commutative property of multiplication to change the order of the exponents.

$$A = P\left(1 + \frac{1}{n}\right)^{nrt}$$

Finally, we can use a property of exponents to write this formula as

$$(1) \quad A = P\left[\left(1 + \frac{1}{n}\right)^{n}\right]^{rt} \quad \text{Use the property } a^{mn} = (a^m)^n.$$

To find the value of $\left(1 + \frac{1}{n}\right)^n$, we use a calculator to evaluate it for several values of n , as shown in Table 9-1.

n	$\left(1 + \frac{1}{n}\right)^n$
1	2
2	2.25
4	2.44140625 . . .
12	2.61303529 . . .
365	2.71456748 . . .
1,000	2.71692393 . . .
100,000	2.71826823 . . .
1,000,000	2.71828046 . . .

Table 9-1

Leonhard Euler (1707–1783)

Euler first used the letter *i* to represent $\sqrt{-1}$, the letter *e* for the base of natural logarithms, and the symbol Σ for summation. Euler was one of the most prolific mathematicians of all time, contributing to almost all areas of mathematics. Much of his work was accomplished after he became blind.

The results suggest that as n gets larger, the value of $\left(1 + \frac{1}{n}\right)^n$ approaches an irrational number with a value of 2.71828. . . . This number is called e, which has the following approximate value.

$$e \approx \mathbf{2.718281828459}$$

In continuous compound interest, k (the number of compoundings) is infinitely large. Since k, r, and n are all positive and $k = rn$, as k gets very large (approaches infinity), then so does n. Therefore, we can replace $\left(1 + \frac{1}{n}\right)^n$ in Equation 1 with e to get

$$A = Pe^{rt}$$

Formula for Exponential Growth

If a quantity P increases or decreases at an annual rate r, compounded continuously, then the amount A after t years is given by

$$A = Pe^{rt}$$

If time is measured in years, then r is called the **annual growth rate.** If r is negative, the "growth" represents a decrease, commonly referred to as *decay.*

To compute the amount to which \$12,000 will grow if invested for 18 years at 10% annual interest, compounded continuously, we substitute 12,000 for P, 0.10 for r, and 18 for t in the formula for exponential growth:

$$\begin{aligned} A &= Pe^{rt} \\ &= \mathbf{12{,}000}e^{\mathbf{0.10(18)}} \\ &= 12{,}000e^{1.8} \\ &\approx 72{,}595.76957 \quad \text{Use a calculator.} \end{aligned}$$

After 18 years, the account will contain \$72,595.77. This is \$1,589.03 more than the result in Example 8 in the previous section, where interest was compounded quarterly.

EXAMPLE 1 **CONTINUOUS COMPOUND INTEREST** If \$25,000 accumulates interest at an annual rate of 8%, compounded continuously, find the balance in the account in 50 years.

Solution We substitute 25,000 for P, 0.08 for r, and 50 for t.

$$\begin{aligned} A &= Pe^{rt} \\ A &= \mathbf{25{,}000}e^{\mathbf{(0.08)(50)}} \\ &= 25{,}000e^{4} \\ &\approx 1{,}364{,}953.751 \quad \text{Use a calculator.} \end{aligned}$$

In 50 years, the balance will be \$1,364,953.75—over one million dollars.

SELF CHECK 1 Find the balance in 60 years.

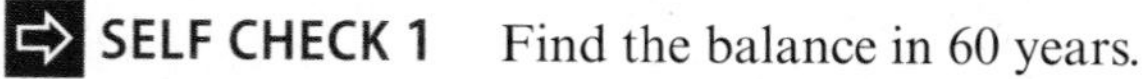

The exponential function $f(x) = e^x$ is so important that it is often called the **exponential function.**

2 Graph an exponential function.

To graph the exponential function $f(x) = e^x$, we plot several points and join them with a smooth curve, as shown in Figure 9-10.

$f(x) = e^x$

x	$f(x)$	$(x, f(x))$
−2	0.1	(−2, 0.1)
−1	0.4	(−1, 0.4)
0	1	(0, 1)
1	2.7	(1, 2.7)
2	7.4	(2, 7.4)

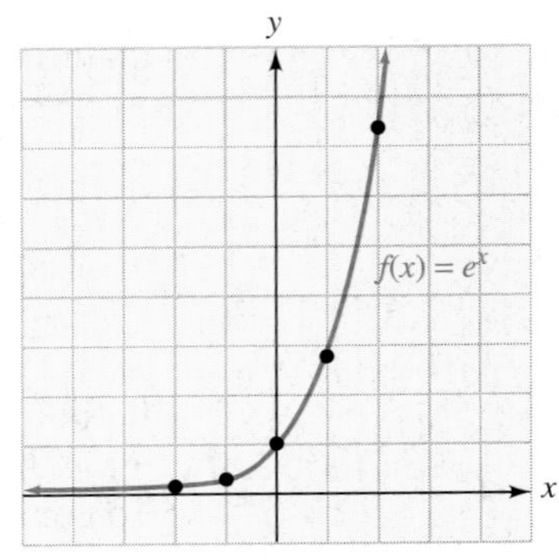

Figure 9-10

ACCENT ON TECHNOLOGY

Translations of the Exponential Function

Figure 9-11(a) shows the graphs of $f(x) = e^x$, $f(x) = e^x + 5$, and $f(x) = e^x - 3$. To graph these functions with window settings of [−3, 6] for x and [−5, 15] for y, we enter the right sides of the equations after the symbols $Y_1 =$, $Y_2 =$, and $Y_3 =$. The display will show

$$Y_1 = e^\wedge(x)$$
$$Y_2 = e^\wedge(x) + 5$$
$$Y_3 = e^\wedge(x) - 3$$

After graphing these functions, we can see that the graph of $f(x) = e^x + 5$ is 5 units above the graph of $f(x) = e^x$, and that the graph of $f(x) = e^x - 3$ is 3 units below the graph of $f(x) = e^x$.

Figure 9-11(b) shows the calculator graphs of $f(x) = e^x$, $f(x) = e^{x+5}$, and $f(x) = e^{x-3}$. To graph these functions with window settings of [−7, 10] for x and [−5, 15] for y, we enter the right sides of the equations after the symbols $Y_1 =$, $Y_2 =$, $Y_3 =$. The display will show

$$Y_1 = e^\wedge(x)$$
$$Y_2 = e^\wedge(x + 5)$$
$$Y_3 = e^\wedge(x - 3)$$

After graphing these functions, we can see that the graph of $f(x) = e^{x+5}$ is 5 units to the left of the graph of $f(x) = e^x$, and that the graph of $f(x) = e^{x-3}$ is 3 units to the right of the graph of $f(x) = e^x$.

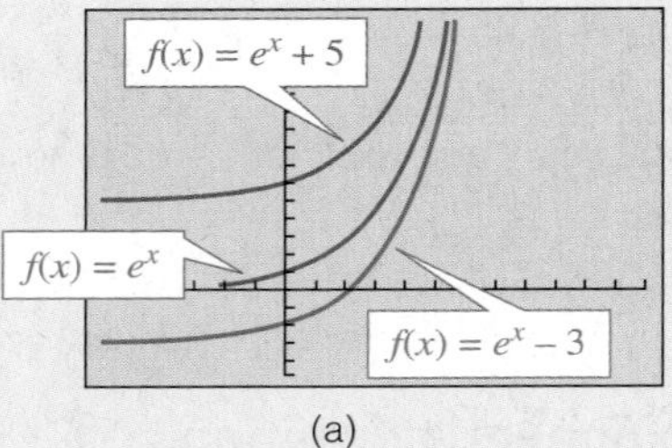

(a)

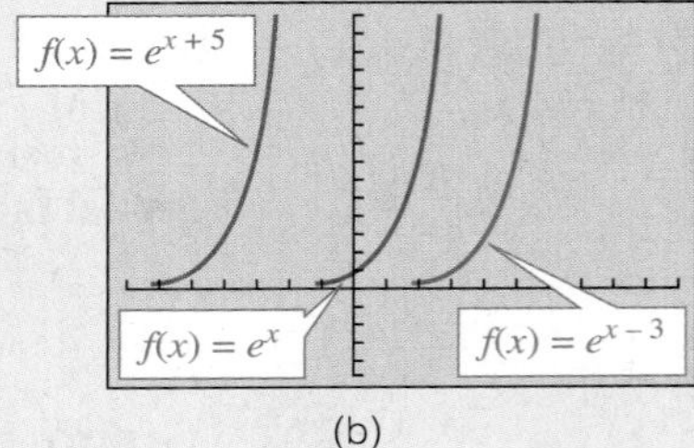

(b)

Figure 9-11

ACCENT ON TECHNOLOGY

Graphing Exponential Functions

Figure 9-12 shows the calculator graph of $f(x) = 3e^{-x/2}$. To graph this function with window settings of $[-7, 10]$ for x and $[-5, 15]$ for y, we enter the right side of the equation after the symbol $Y_1 =$. The display will show the equation

$$Y_1 = 3(e\text{^}(-x/2))$$

Explain why the graph has a y-intercept of $(0, 3)$.

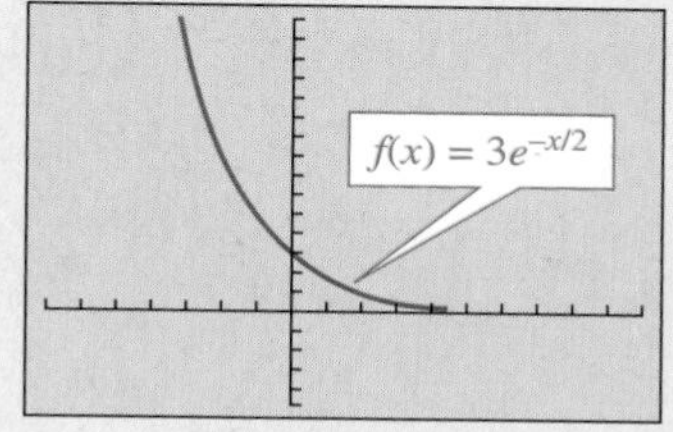

Figure 9-12

3 Solve an application problem involving exponential growth or decay.

An equation based on the exponential function provides a model for population growth. In the Malthusian model for population growth, the future or past population of a colony is related to the present population by the formula $A = Pe^{rt}$.

EXAMPLE 2 **CITY PLANNING** The population of a city is currently 15,000, but changing economic conditions are causing the population to decrease by 2% each year. If this trend continues, find the population in 30 years.

Solution Since the population is decreasing by 2% each year, the annual growth rate is −2%, or −0.02. We can substitute −0.02 for r, 30 for t, and 15,000 for P in the formula for exponential growth and find A.

$$A = Pe^{rt}$$
$$A = 15{,}000e^{-0.02(30)}$$
$$= 15{,}000e^{-0.6}$$
$$\approx 8{,}232.174541$$

In 30 years, city planners expect a population of approximately 8,232.

SELF CHECK 2 Find the population in 50 years.

The English economist Thomas Robert Malthus (1766–1834) pioneered in population study. He believed that poverty and starvation were unavoidable, because the human population tends to grow exponentially, but the food supply tends to grow linearly.

EXAMPLE 3 **POPULATION GROWTH** Suppose that a country with a population of 1,000 people is growing exponentially according to the formula

$$P = 1{,}000e^{0.02t}$$

where t is in years. Furthermore, assume that the food supply measured in adequate food per day per person is growing linearly according to the formula

$$y = 30.625x + 2{,}000$$

where x is in years. In how many years will the population outstrip the food supply?

Solution We can use a graphing calculator, with window settings of [0, 100] for x and [0, 10,000] for y. After graphing the functions, we obtain Figure 9-13(a). If we trace, as in Figure 9-13(b), we can find the point where the two graphs intersect. From the graph, we can see that the food supply will be adequate for about 71 years. At that time, the population of approximately 4,160 people will have problems.

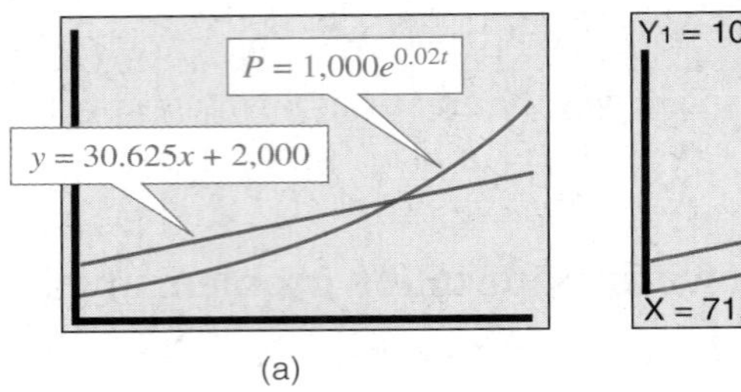

(a) (b)

Figure 9-13

If we use the INTERSECT feature in the CALC menu, we will find that the food supply will be adequate for about 71.7 years when the population will be about 4,196.

SELF CHECK 3 Suppose that the population grows at a 3% rate. For how many years will the food supply be adequate?

The atomic structure of a radioactive material changes as the material emits radiation. The amount of radioactive material that is present decays exponentially according to the following formula.

Radioactive Decay Formula

The amount A of radioactive material present at a time t is given by the formula

$$A = A_0 e^{kt}$$

where A_0 is the amount that was present at $t = 0$ and k is a negative number.

EXAMPLE 4 **RADIOACTIVE DECAY** The radioactive material radon-22 decays according to the formula $A = A_0 e^{-0.181t}$, where t is expressed in days. How much radon-22 will be left if a sample of 100 grams decays for 20 days?

Solution To find the number of grams of radon-22 that will be left, we substitute 100 for A_0 and 20 for t and simplify.

COMMENT Note that the radioactive decay formula is the same as the formula for exponential growth, except for the variables.

$$A = A_0 e^{-0.181t}$$

$$A = \mathbf{100}e^{-0.181(\mathbf{20})}$$

$$= 100e^{-3.62} \qquad -0.181 \cdot 20 = -3.62$$

$$= 100(0.0267826765) \qquad e^{-3.62} \approx 0.0267826765$$

$$\approx 2.678267649 \qquad \text{Multiply.}$$

To the nearest hundredth, 2.68 grams of radon-22 will be left in 20 days.

 **SELF CHECK 4** To the nearest hundredth, how much radon-22 will be left in 30 days?

SELF CHECK ANSWERS **1.** \$3,037,760.44 **2.** approximately 5,518 **3.** about 38 years **4.** 0.44 gram

NOW TRY THIS

Give all answers to the nearest milligran (mg).

The half-life of Ibuprofen is approximately 1.8 hours. This translates to $k \approx -0.39$. (We will learn how to compute this value in Section 9.6.)

1. If a woman takes two 200mg tablets, how many mg of Ibuprofen are in her system 4 hours after taking the medication?
2. If she takes 2 more tablets 4 hours after taking the first dose, how many mg are in her system 6 hours after the first dose?
3. If she doesn't take any more medication, how many mg are in her system 12 hours after the first dose? 18 hours?

9.2 EXERCISES

WARM-UPS

Use a calculator to find each value to the nearest hundredth.

1. e^0 **2.** e^1

3. e^2 **4.** e^3

Fill in the blanks.

5. The graph of $f(x) = e^x + 2$ is __ units above the graph of $f(x) = e^x$.

6. The graph of $f(x) = e^{(x-2)}$ is __ units to the right of the graph of $f(x) = e^x$.

REVIEW *Simplify each expression. Assume that all variables represent positive numbers.*

7. $\sqrt{240x^5}$ **8.** $\sqrt[3]{-125x^5y^4}$

9. $4\sqrt{48y^3} - 3y\sqrt{12y}$ **10.** $\sqrt[4]{48z^5} + \sqrt[4]{768z^5}$

VOCABULARY AND CONCEPTS *Fill in the blanks.*

11. To two decimal places, the value of e is _____.

12. The formula for continuous compound interest is $A =$ _____.

13. Since $e > 1$, the base-e exponential function is a(n) __________ function.

14. The graph of the exponential function $y = e^x$ passes through the points (0, 1) and _____.

15. The Malthusian population growth formula is ________.

16. The Malthusian theory is pessimistic, because __________ grows exponentially, but food supplies grow ________.

GUIDED PRACTICE

Graph each function. Check your work with a graphing calculator. Compare each graph to the graph of $f(x) = e^x$. (Objective 2)

17. $f(x) = e^x + 1$

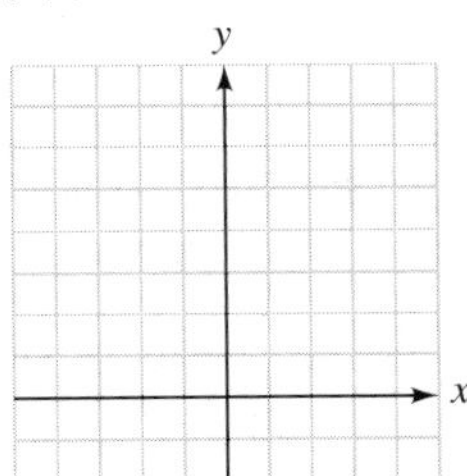

18. $f(x) = e^x - 2$

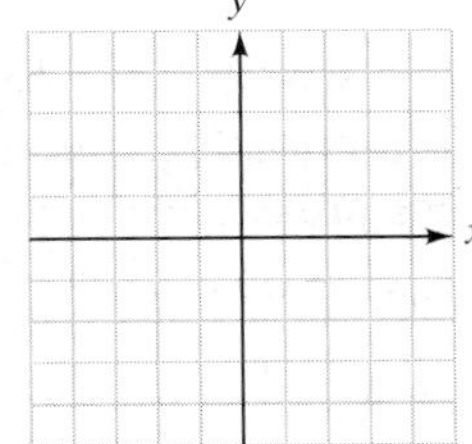

19. $f(x) = e^{x+3}$

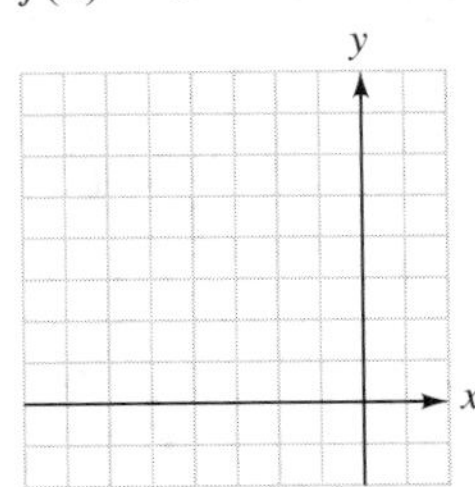

20. $f(x) = e^{x-5}$

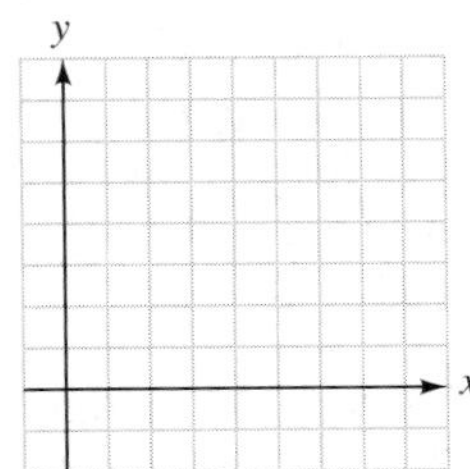

21. $f(x) = -e^x$

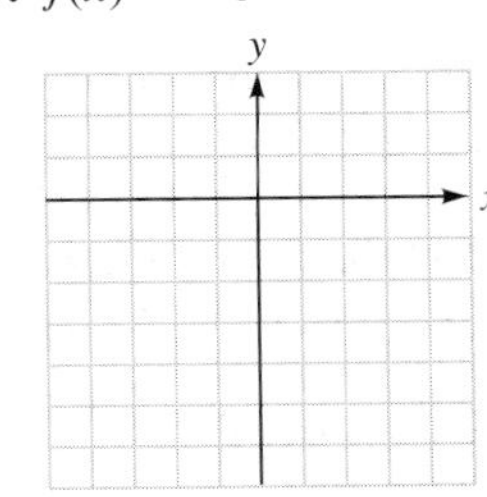

22. $f(x) = -e^x + 1$

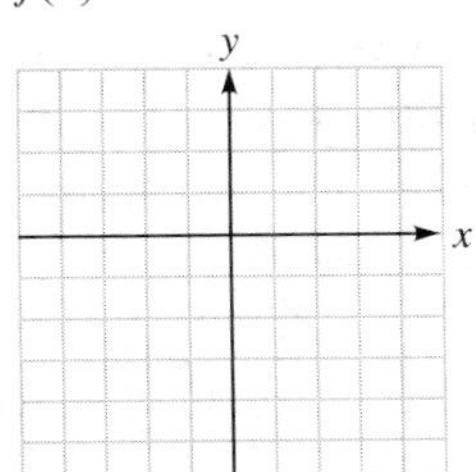

23. $f(x) = 2e^x$

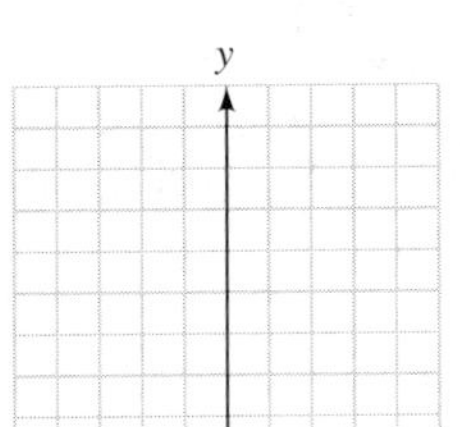

24. $f(x) = \frac{1}{2}e^x$

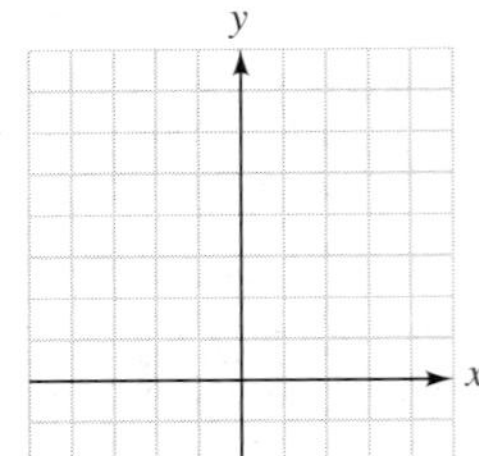

Determine whether the graph of f(x) = e^x could look like the graph shown here. **(Objective 2)**

25.

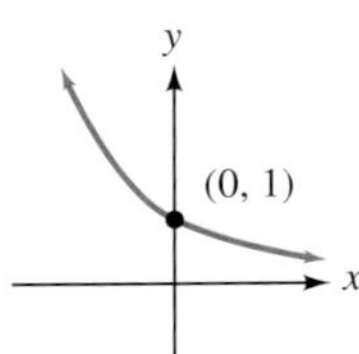

26.

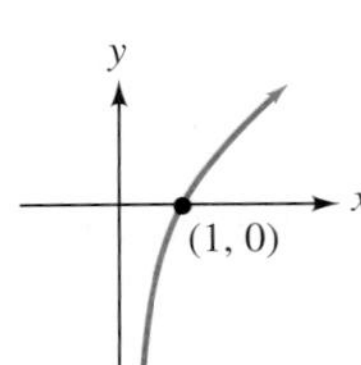

27.

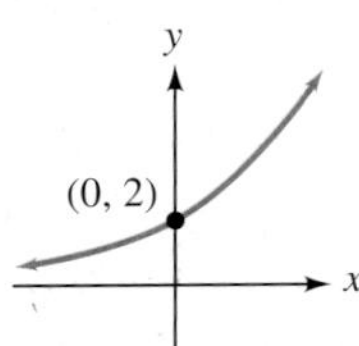

28.

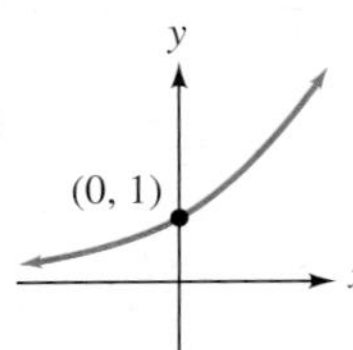

APPLICATIONS

Use a calculator to help solve each problem. Assume that there are no deposits or withdrawals. **See Example 1. (Objective 1)**

29. Continuous compound interest An investment of $5,000 earns 6% interest, compounded continuously. What will the investment be worth in 12 years?

30. Continuous compound interest An investment of $6,000 earns 7% interest, compounded continuously. What will the investment be worth in 35 years?

31. Determining the initial deposit An account now contains $12,000 and has been accumulating 7% annual interest, compounded continuously, for 9 years. Find the initial deposit.

32. Determining a previous balance An account now contains $8,000 and has been accumulating 8% annual interest, compounded continuously. How much was in the account 6 years ago?

Use a calculator to help solve each problem. **See Example 2. (Objective 3)**

33. World population growth Earth's population is approximately 6 billion people and is growing at an annual rate of 1.9%. Assuming a Malthusian growth model, find the world population in 30 years.

34. World population growth Earth's population is approximately 6 billion people and is growing at an annual rate of 1.9%. Assuming a Malthusian growth model, find the world population in 40 years.

35. World population growth Assuming a Malthusian growth model and an annual growth rate of 1.9%, by what factor will Earth's current population increase in 50 years? (See Exercise 33.)

36. Population growth The growth of a population is modeled by

$$P = 173e^{0.03t}$$

How large will the population be when $t = 30$?

Use a calculator to help solve each problem. **See Example 3. (Objective 3)**

37. In Example 3, suppose that better farming methods change the formula for food growth to $y = 31x + 2{,}000$. How long will the food supply be adequate?

38. In Example 3, suppose that a birth-control program changed the formula for population growth to $P = 1{,}000e^{0.01t}$. How long will the food supply be adequate?

Use a calculator to help solve each problem. Round each answer to the nearest hundredth. **See Example 4. (Objective 3)**

39. Radioactive decay The radioactive material iodine-131 decays according to the formula $A = A_0e^{-0.087t}$, where t is expressed in days. To the nearest hundredth, how much iodine-131 will be left if a sample of 50 grams decays for 30 days?

40. Radioactive decay The radioactive material strontium-89 decays according to the formula $A = A_0e^{-0.013t}$, where t is expressed in days. To the nearest hundredth, how much strontium-89 will be left if a sample of 25 grams decays for 45 days?

41. Radioactive decay The radioactive material tin-126 decays according to the formula $A = A_0e^{-0.00000693t}$, where t is expressed in years. To the nearest hundredth, how much tin-126 will be left if a sample of 2,500 grams decays for 100 years?

42. Radioactive decay The radioactive material plutonium-239 decays according to the formula $A = A_0e^{-0.0000284t}$, where t is expressed in years. To the nearest hundredth, how much plutonium-239 will be left if a sample of 75 grams decays for 50,000 years?

Use a graphing calculator to solve each problem.

43. Comparison of compounding methods An initial deposit of $5,000 grows at an annual rate of 8.5% for 5 years. Compare the final balances resulting from continuous compounding and annual compounding.

44. **Comparison of compounding methods** An initial deposit of \$30,000 grows at an annual rate of 8% for 20 years. Compare the final balances resulting from continuous compounding and annual compounding.

45. **Population decline** The decline of a population is modeled by

$$P = 8{,}000e^{-0.008t}$$

How large will the population be when $t = 20$?

46. **Epidemics** The spread of foot and mouth disease through a herd of cattle can be modeled by the formula

$$P = P_0e^{0.27t} \quad (t \text{ is in days})$$

If a rancher does not act quickly to treat two cases, how many cattle will have the disease in 10 days?

47. **Medicine** The concentration, x, of a certain drug in an organ after t minutes is given by $x = 0.08(1 - e^{-0.1t})$. Find the concentration of the drug after 30 minutes.

48. **Medicine** Refer to Exercise 47. Find the initial concentration of the drug.

49. **Skydiving** Before her parachute opens, a skydiver's velocity v in meters per second is given by $v = 50(1 - e^{-0.2t})$. Find the initial velocity.

50. **Skydiving** Refer to Exercise 49 and find the velocity after 20 seconds.

51. **Free-falling objects** After t seconds, a certain falling object has a velocity v given by $v = 50(1 - e^{-0.3t})$. Which is falling faster after 2 seconds, this object or the skydiver in Exercise 49?

52. **Alcohol absorption** In one individual, the blood alcohol level t minutes after drinking two shots of whiskey is given by $P = 0.3(1 - e^{-0.05t})$. Find the blood alcohol level after 15 minutes.

53. **Depreciation** A camping trailer originally purchased for \$4,570 is continuously losing value at the rate of 6% per year. Find its value when it is $6\frac{1}{2}$ years old.

54. **Depreciation** A boat purchased for \$7,500 has been continuously decreasing in value at the rate of 2% each year. It is now 8 years, 3 months old. Find its value.

WRITING ABOUT MATH

55. Explain why the graph of $f(x) = e^x - 5$ is 5 units below the graph of $f(x) = e^x$.

56. Explain why the graph of $f(x) = e^{(x+5)}$ is 5 units to the left of the graph of $f(x) = e^x$.

SOMETHING TO THINK ABOUT

57. The value of e can be calculated to any degree of accuracy by adding the first several terms of the following list:

$$1, 1, \frac{1}{2}, \frac{1}{2 \cdot 3}, \frac{1}{2 \cdot 3 \cdot 4}, \frac{1}{2 \cdot 3 \cdot 4 \cdot 5}, \ldots$$

The more terms that are added, the closer the sum will be to e. Add the first six numbers in the preceding list. To how many decimal places is the sum accurate?

58. Graph the function defined by the equation

$$y = f(x) = \frac{e^x + e^{-x}}{2}$$

from $x = -2$ to $x = 2$. The graph will look like a parabola, but it is not. The graph, called a **catenary,** is important in the design of power distribution networks, because it represents the shape of a uniform flexible cable whose ends are suspended from the same height. The function is called the **hyperbolic cosine function.**

59. If $e^{t+5} = ke^t$, find k.

60. If $e^{5t} = k^t$, find k.

SECTION 9.3 Logarithmic Functions

Objectives

1. Write a logarithmic function as an exponential function and write an exponential function as a logarithmic function.
2. Graph a logarithmic function.
3. Graph a vertical and horizontal translation of a logarithmic function.
4. Evaluate a common logarithm.
5. Solve an application problem involving a logarithm.

Vocabulary

logarithmic function
logarithm
common logarithm
decibel
Richter scale

Getting Ready

Find each value.

1. 7^0 **2.** 5^2 **3.** 5^{-2} **4.** $16^{1/2}$

In this section, we consider the inverse function of an exponential function $f(x) = b^x$. The inverse function is called a *logarithmic function.* These functions can be used to solve application problems from fields such as electronics, seismology, and business.

1 Write a logarithmic function as an exponential function and write an exponential function as a logarithmic function.

Since the exponential function $y = b^x$ is one-to-one, it has an inverse function defined by the equation $x = b^y$. To express this inverse function in the form $y = f^{-1}(x)$, we must solve the equation $x = b^y$ for y. To do this, we need the following definition.

Logarithmic Functions

If $b > 0$ and $b \neq 1$, the **logarithmic function with base *b*** is defined by

$$y = \log_b x \quad \text{if and only if} \quad x = b^y$$

The **domain of the logarithmic function** is the interval $(0, \infty)$. The **range** is the interval $(-\infty, \infty)$.

COMMENT Since the domain of the logarithmic function is the set of positive numbers, the logarithm of 0 or the logarithm of a negative number is not defined in the real-number system.

COMMENT Since $b^y = x$ is equivalent to $y = \log_b x$, then $b^{\log_b x} = x$ by substitution.

Since the function $y = \log_b x$ is the inverse of the one-to-one exponential function $y = b^x$, the logarithmic function is also one-to-one.

The previous definition guarantees that any pair (x, y) that satisfies the equation $y = \log_b x$ also satisfies the equation $x = b^y$.

$\log_4 1 = \mathbf{0}$ because $1 = 4^{\mathbf{0}}$

$\log_5 25 = \mathbf{2}$ because $25 = 5^{\mathbf{2}}$

$\log_5 \frac{1}{25} = \mathbf{-2}$ because $\frac{1}{25} = 5^{\mathbf{-2}}$

$\log_{16} 4 = \mathbf{\frac{1}{2}}$ because $4 = 16^{\mathbf{1/2}}$

$\log_2 \frac{1}{8} = \mathbf{-3}$ because $\frac{1}{8} = 2^{\mathbf{-3}}$

$\log_b x = y$ because $x = b^y$

In each of these examples, the **logarithm** of a number is an exponent. In fact,

$\log_b x$ is the exponent to which b is raised to get x.

In equation form, we write

$$b^{\log_b x} = x$$

EXAMPLE 1 Find y in each equation: **a.** $\log_6 1 = y$ **b.** $\log_3 27 = y$ **c.** $\log_5 \frac{1}{5} = y$.

Solution **a.** We can change the equation $\log_6 1 = y$ into the equivalent exponential equation $6^y = 1$. Since $6^0 = 1$, it follows that $y = 0$. Thus,

$$\log_6 1 = 0$$

b. $\log_3 27 = y$ is equivalent to $3^y = 27$. Since $3^3 = 27$, it follows that $3^y = 3^3$, and $y = 3$. Thus,

$$\log_3 27 = 3$$

c. $\log_5 \frac{1}{5} = y$ is equivalent to $5^y = \frac{1}{5}$. Since $5^{-1} = \frac{1}{5}$, it follows that $5^y = 5^{-1}$, and $y = -1$. Thus,

$$\log_5 \frac{1}{5} = -1$$

SELF CHECK 1 Find y: **a.** $\log_3 9 = y$ **b.** $\log_2 64 = y$ **c.** $\log_5 \frac{1}{125} = y$.

EXAMPLE 2 Find x in each equation: **a.** $\log_3 81 = x$ **b.** $\log_x 125 = 3$ **c.** $\log_4 x = 3$.

Solution **a.** $\log_3 81 = x$ is equivalent to $3^x = 81$. Because $3^4 = 81$, it follows that $3^x = 3^4$. Thus, $x = 4$.

b. $\log_x 125 = 3$ is equivalent to $x^3 = 125$. Because $5^3 = 125$, it follows that $x^3 = 5^3$. Thus, $x = 5$.

c. $\log_4 x = 3$ is equivalent to $4^3 = x$. Because $4^3 = 64$, it follows that $x = 64$.

SELF CHECK 2 Find x: **a.** $\log_2 32 = x$ **b.** $\log_x 8 = 3$ **c.** $\log_5 x = 2$.

EXAMPLE 3 Find x in each equation.
a. $\log_{1/3} x = 2$ **b.** $\log_{1/3} x = -2$ **c.** $\log_{1/3} \frac{1}{27} = x$

Solution **a.** $\log_{1/3} x = 2$ is equivalent to $\left(\frac{1}{3}\right)^2 = x$. Thus, $x = \frac{1}{9}$.

b. $\log_{1/3} x = -2$ is equivalent to $\left(\frac{1}{3}\right)^{-2} = x$. Thus,

$$x = \left(\frac{1}{3}\right)^{-2} = 3^2 = 9$$

c. $\log_{1/3} \frac{1}{27} = x$ is equivalent to $\left(\frac{1}{3}\right)^x = \frac{1}{27}$. Because $\left(\frac{1}{3}\right)^3 = \frac{1}{27}$, it follows that $x = 3$.

SELF CHECK 3 Find x: **a.** $\log_{1/4} x = 3$ **b.** $\log_{1/4} x = -2$.

2 Graph a logarithmic function.

To graph the logarithmic function $y = \log_2 x$, we calculate and plot several points with coordinates (x, y) that satisfy the equation $x = 2^y$. After joining these points with a smooth curve, we have the graph shown in Figure 9-14(a).

To graph $y = \log_{1/2} x$, we calculate and plot several points with coordinates (x, y) that satisfy the equation $x = \left(\frac{1}{2}\right)^y$. After joining these points with a smooth curve, we have the graph shown in Figure 9-14(b).

$y = \log_2 x$

x	y	(x, y)
$\frac{1}{4}$	-2	$\left(\frac{1}{4}, -2\right)$
$\frac{1}{2}$	-1	$\left(\frac{1}{2}, -1\right)$
1	0	(1, 0)
2	1	(2, 1)
4	2	(4, 2)
8	3	(8, 3)

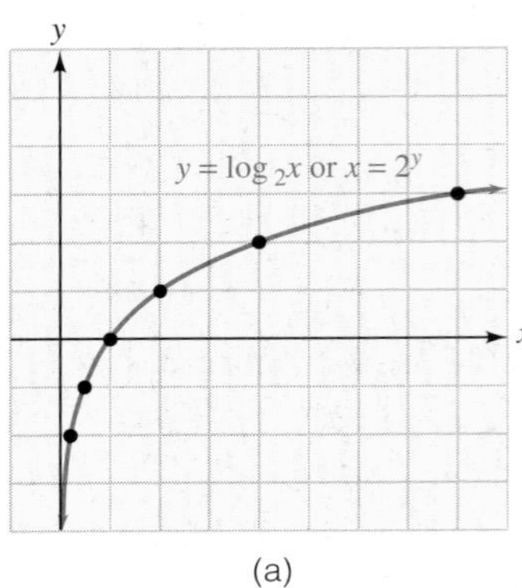

(a)

$y = \log_{1/2} x$

x	y	(x, y)
$\frac{1}{4}$	2	$\left(\frac{1}{4}, 2\right)$
$\frac{1}{2}$	1	$\left(\frac{1}{2}, 1\right)$
1	0	(1, 0)
2	-1	$(2, -1)$
4	-2	$(4, -2)$
8	-3	$(8, -3)$

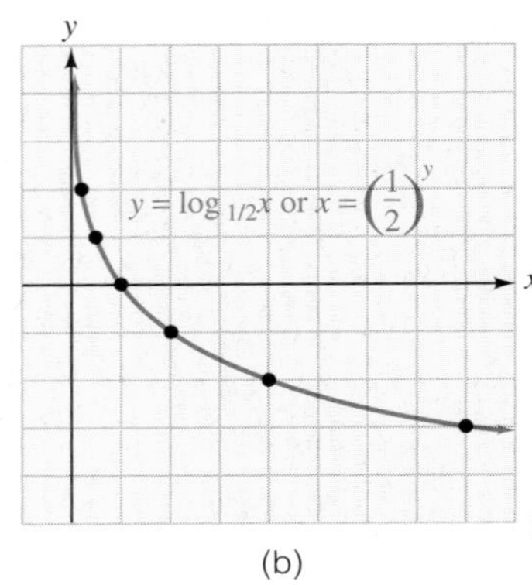

(b)

Figure 9-14

The graphs of all logarithmic functions are similar to those in Figure 9-15. If $b > 1$, the logarithmic function is increasing, as in Figure 9-15(a). If $0 < b < 1$, the logarithmic function is decreasing, as in Figure 9-15(b).

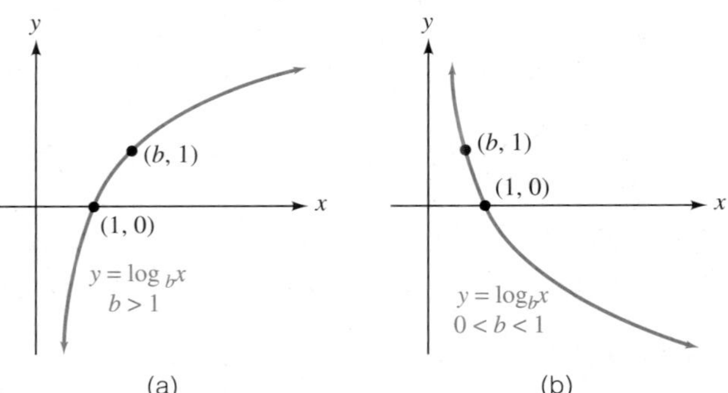

(a) (b)

Figure 9-15

The graph of $f(x) = \log_b x$ has the following properties.

1. It passes through the point (1, 0).
2. It passes through the point $(b, 1)$.
3. The y-axis is an asymptote.
4. The domain is $(0, \infty)$ and the range is $(-\infty, \infty)$.

The exponential and logarithmic functions are inverses of each other and, therefore, have symmetry about the line $y = x$. The graphs $y = \log_b x$ and $y = b^x$ are shown in Figure 9-16(a) when $b > 1$, and in Figure 9-16(b) when $0 < b < 1$.

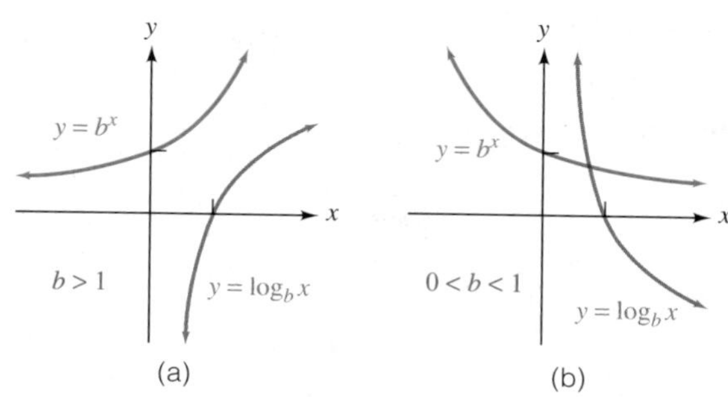

Figure 9-16

3 Graph a vertical and horizontal translation of a logarithmic function.

The graphs of many functions involving logarithms are translations of the basic logarithmic graphs.

EXAMPLE 4 Graph the function defined by $f(x) = 3 + \log_2 x$.

Solution The graph of $f(x) = 3 + \log_2 x$ is identical to the graph of $f(x) = \log_2 x$, except that it is translated 3 units upward. (See Figure 9-17.)

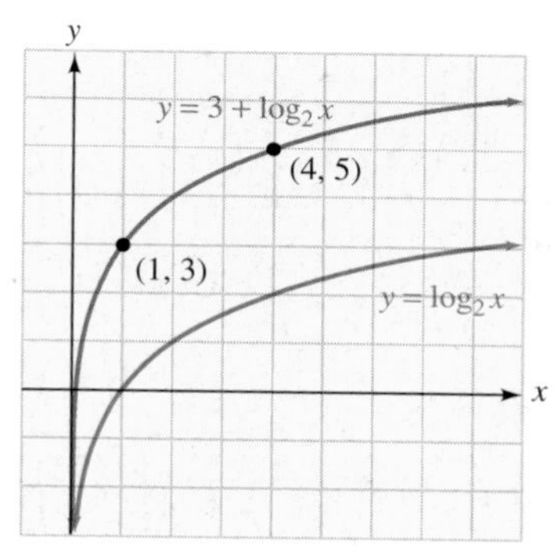

Figure 9-17

SELF CHECK 4 Graph: $f(x) = \log_3 x - 2$.

EXAMPLE 5 Graph: $f(x) = \log_{1/2} (x - 1)$.

Solution The graph of $f(x) = \log_{1/2} (x - 1)$ is identical to the graph of $f(x) = \log_{1/2} x$, except that it is translated 1 unit to the right. (See Figure 9-18.)

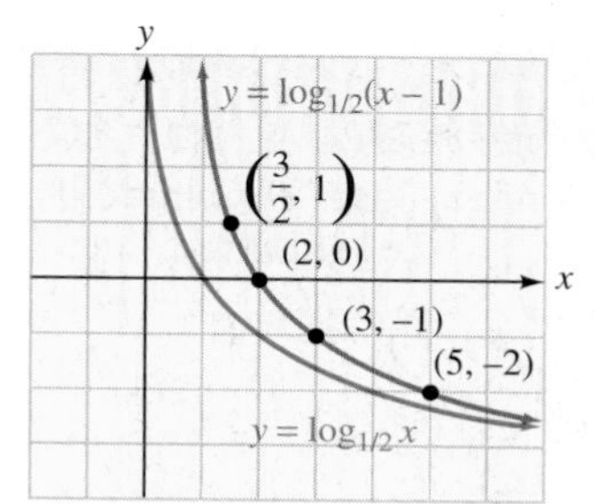

Figure 9-18

SELF CHECK 5 Graph: $f(x) = \log_{1/3} (x + 2)$.

ACCENT ON TECHNOLOGY

Graphing Logarithmic Functions

Graphing calculators can graph logarithmic functions directly only if the base of the logarithmic function is 10 or e. To use a calculator to graph $f(x) = -2 + \log_{10}\left(\frac{1}{2}x\right)$, we enter the right side of the equation after the symbol $Y_1 =$. The display will show the equation

$$Y_1 = -2 + \log(1/2*x)$$

If we use window settings of $[-1, 5]$ for x and $[-4, 1]$ for y and press **GRAPH**, we will obtain the graph shown in Figure 9-19.

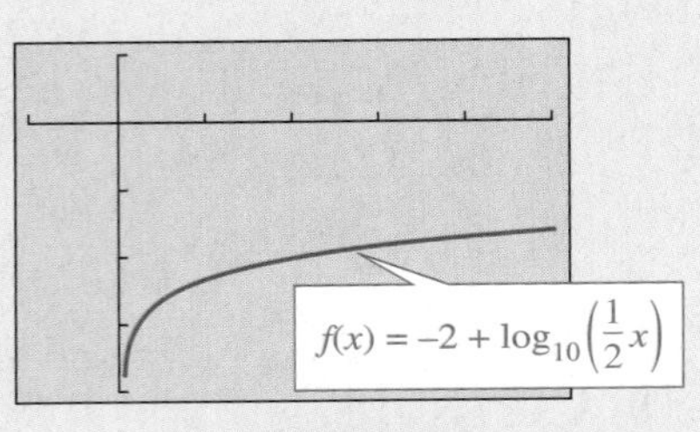

Figure 9-19

4 Evaluate a common logarithm.

For computational purposes and in many applications, we will use base-10 logarithms (also called **common logarithms**). When the base b is not indicated in the notation $\log x$, we assume that $b = 10$:

$$\log x \quad \textbf{means} \quad \log_{10} x$$

Because base-10 logarithms appear so often, it is a good idea to become familiar with the following base-10 logarithms:

$\log_{10} \frac{1}{100} = -2$ because $10^{-2} = \frac{1}{100}$

$\log_{10} \frac{1}{10} = -1$ because $10^{-1} = \frac{1}{10}$

$\log_{10} 1 = 0$ because $10^0 = 1$

$\log_{10} 10 = 1$ because $10^1 = 10$

$\log_{10} 100 = 2$ because $10^2 = 100$

$\log_{10} 1{,}000 = 3$ because $10^3 = 1{,}000$

In general, we have

$$\log_{10} 10^x = x$$

ACCENT ON TECHNOLOGY

Finding Base-10 (Common) Logarithms

Before calculators, extensive tables provided logarithms of numbers. Today, logarithms are easy to find with a calculator. For example, to find log 32.58 with a scientific calculator, we enter these numbers and press these keys:

32.58 **LOG**

The display will read **1.51295108**. To four decimal places, log 32.58 = 1.5130.

To use a graphing calculator, we enter these numbers and press these keys:

LOG 32.58 **ENTER**

The display will read **LOG (32.58)** **1.51295108**.

To four decimal places, log 32.58 = 1.5130.

EXAMPLE 6 Find x in the equation $\log x = 0.3568$. Round to four decimal places.

Solution The equation $\log x = 0.3568$ is equivalent to $10^{0.3568} = x$. To find x with a calculator, we can enter these numbers and press these keys:

Scientific Calculator	*Graphing Calculator*
10 y^x 0.3568 =	10 ^ 0.3568 ENTER

Either way, the display will read 2.274049951. To four decimal places,

$$x = 2.2740$$

SELF CHECK 6 Solve: $\log x = 2.7$. Round to four decimal places.

5 Solve an application problem involving a logarithm.

Common logarithms are used in electrical engineering to express the voltage gain (or loss) of an electronic device such as an amplifier. The unit of gain (or loss), called the **decibel,** is defined by a logarithmic relation.

Decibel Voltage Gain

If E_O is the output voltage of a device and E_I is the input voltage, the decibel voltage gain is given by

$$\text{dB gain} = 20 \log \frac{E_O}{E_I}$$

EXAMPLE 7 **FINDING dB GAIN** If the input to an amplifier is 0.4 volt and the output is 50 volts, find the decibel voltage gain of the amplifier.

Solution We can find the decibel voltage gain by substituting 0.4 for E_I and 50 for E_O into the formula for dB gain:

$$\text{dB gain} = 20 \log \frac{E_O}{E_I}$$

$$\text{dB gain} = 20 \log \frac{50}{0.4}$$

$$= 20 \log 125$$

$$\approx 42 \quad \text{Use a calculator.}$$

The amplifier provides a 42-decibel voltage gain.

In seismology, the study of earthquakes, common logarithms are used to measure the magnitude (ground motion) of earthquakes on the **Richter scale.** The magnitude of an earthquake is given by the following logarithmic function.

Richter Scale

If R is the magnitude of an earthquake, A is the amplitude (measured in micrometers), and P is the period (the time of one oscillation of Earth's surface, measured in seconds), then

$$R = \log \frac{A}{P}$$

EXAMPLE 8 **MEASURING EARTHQUAKES** Find the measure on the Richter scale of an earthquake with an amplitude of 10,000 micrometers (1 centimeter) and a period of 0.1 second.

Solution We substitute 10,000 for A and 0.1 for P in the Richter scale formula and simplify:

$$R = \log \frac{A}{P}$$

$$\begin{aligned} R &= \log \frac{\mathbf{10{,}000}}{\mathbf{0.1}} \\ &= \log 100{,}000 \\ &= 5 \end{aligned}$$

The earthquake measures 5 on the Richter scale.

SELF CHECK ANSWERS

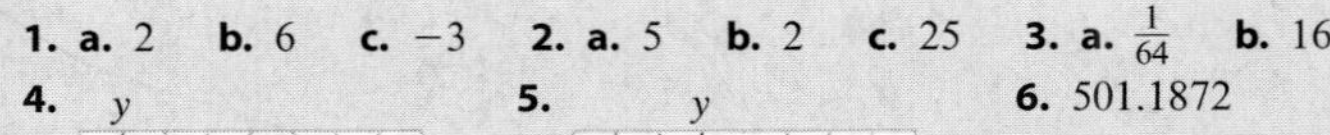

1. a. 2 **b.** 6 **c.** −3 **2. a.** 5 **b.** 2 **c.** 25 **3. a.** $\frac{1}{64}$ **b.** 16

4.

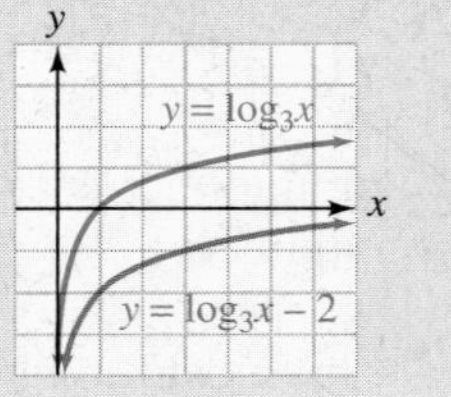

5.

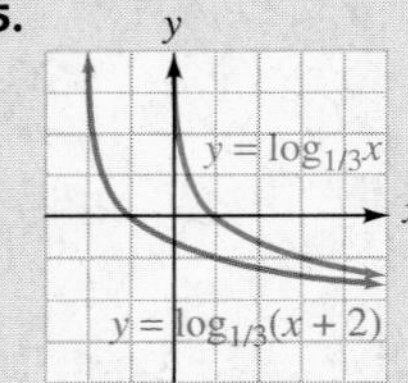

6. 501.1872

NOW TRY THIS

In 1989, an earthquake measuring 7.1 on the Richter scale rocked San Francisco.

1. Write a logarithmic equation to describe the earthquake.

2. Write the equation in exponential form.

In 1964, an earthquake measuring 9.1 on the Richter scale devastated Juneau, Alaska.

3. Write the equation in exponential form.

4. Given that each earthquake had the same period, the amplitude in Juneau was how many times greater than that in San Francisco?

Source: http://pubs.usgs.gov/gip/earthq1/measure.html

9.3 EXERCISES

WARM-UPS *Find the value of x in each equation.*

1. $\log_2 8 = x$
2. $\log_3 9 = x$
3. $\log_x 125 = 3$
4. $\log_x 8 = 3$
5. $\log_4 16 = x$
6. $\log_x 32 = 5$
7. $\log_{1/2} x = 2$
8. $\log_9 3 = x$
9. $\log_x \frac{1}{4} = -2$

REVIEW *Solve each equation.*

10. $\sqrt[3]{6x + 4} = 4$
11. $\sqrt{3x - 4} = \sqrt{-7x + 2}$
12. $\sqrt{a + 1} - 1 = 3a$
13. $3 - \sqrt{t - 3} = \sqrt{t}$

VOCABULARY AND CONCEPTS *Fill in the blanks.*

14. The equation $y = \log_b x$ is equivalent to ______.
15. The domain of the logarithmic function is the interval ______.
16. The ______ of the logarithmic function is the interval $(-\infty, \infty)$.
17. $b^{\log_b x} =$ ___.
18. Because an exponential function is one-to-one, it has an ______ function that is called a ______ function.
19. $\text{Log}_b\, x$ is the ______ to which b is raised to get x.
20. The y-axis is an ______ to the graph of $f(x) = \log_b x$.
21. The graph of $f(x) = \log_b x$ passes through the points ______ and ______.
22. A logarithm with a base of 10 is called a ______ logarithm and $\log_{10} 10^x =$ ___.
23. The decibel voltage gain is found using the equation dB gain = ______.
24. The magnitude of an earthquake is measured by the formula $R =$ ______.

GUIDED PRACTICE

Write each equation in exponential form. (Objective 1)

25. $\log_3 27 = 3$
26. $\log_8 8 = 1$
27. $\log_{1/2} \frac{1}{4} = 2$
28. $\log_{1/5} 1 = 0$
29. $\log_4 \frac{1}{64} = -3$
30. $\log_6 \frac{1}{36} = -2$
31. $\log_{1/2} \frac{1}{8} = 3$
32. $\log_{1/5} 25 = -2$

Write each equation in logarithmic form. (Objective 1)

33. $6^2 = 36$
34. $10^3 = 1{,}000$
35. $5^{-2} = \frac{1}{25}$
36. $3^{-3} = \frac{1}{27}$
37. $\left(\frac{1}{2}\right)^{-5} = 32$
38. $\left(\frac{1}{3}\right)^{-3} = 27$
39. $x^y = z$
40. $m^n = p$

Find each value of x. See Example 1. (Objective 1)

41. $\log_7 x = 2$
42. $\log_5 x = 0$
43. $\log_6 x = 1$
44. $\log_2 x = 4$
45. $\log_{25} x = \frac{1}{2}$
46. $\log_4 x = \frac{1}{2}$
47. $\log_5 x = -2$
48. $\log_{27} x = -\frac{1}{3}$

Find each value of x. See Example 2. (Objective 1)

49. $\log_x 5^3 = 3$
50. $\log_x 5 = 1$
51. $\log_x \frac{9}{4} = 2$
52. $\log_x \frac{\sqrt{3}}{3} = \frac{1}{2}$

Find each value of x. See Example 3. (Objective 1)

53. $\log_2 16 = x$
54. $\log_3 9 = x$
55. $\log_4 16 = x$
56. $\log_6 216 = x$
57. $\log_{1/2} \frac{1}{8} = x$
58. $\log_{1/3} \frac{1}{81} = x$
59. $\log_9 3 = x$
60. $\log_{125} 5 = x$

Graph each function. Determine whether each function is an increasing or decreasing function. (Objective 2)

61. $f(x) = \log_3 x$

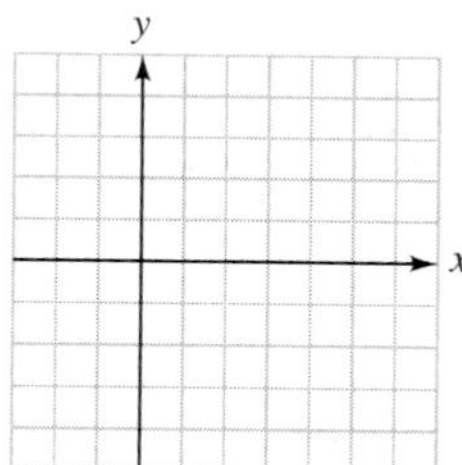

62. $f(x) = \log_{1/3} x$

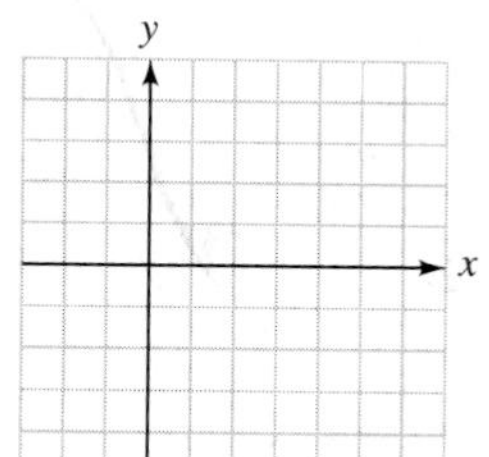

63. $f(x) = \log_{1/2} x$

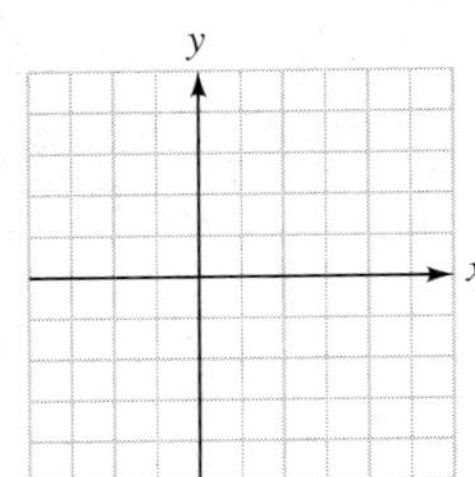

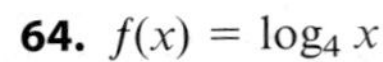

64. $f(x) = \log_4 x$

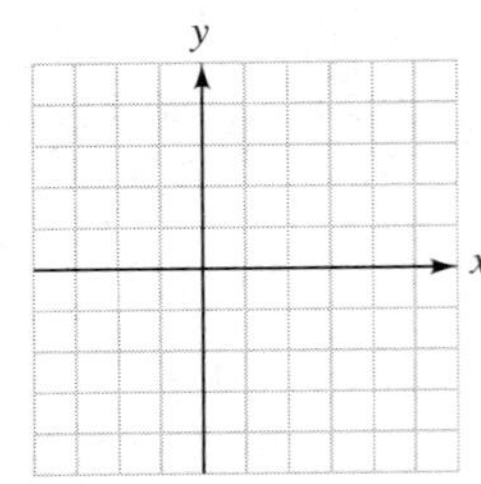

Graph each pair of inverse functions on a single coordinate system. **(Objective 2)**

65. $f(x) = 2^x$

$g(x) = \log_2 x$

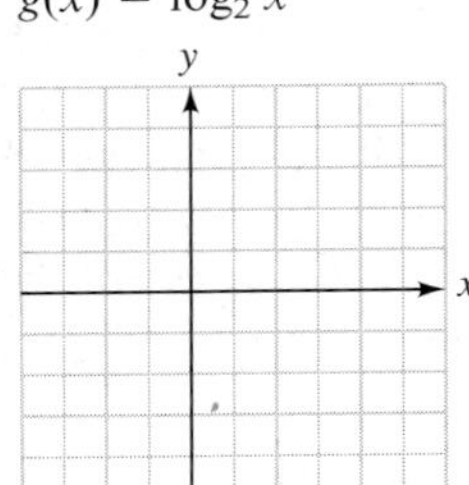

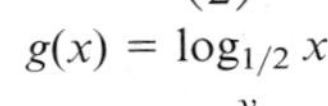

66. $f(x) = \left(\frac{1}{2}\right)^x$

$g(x) = \log_{1/2} x$

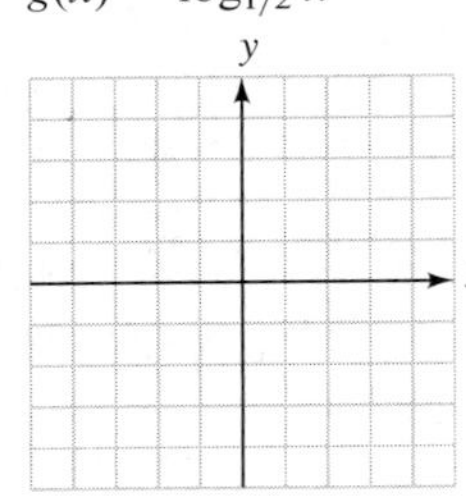

67. $f(x) = \left(\frac{1}{4}\right)^x$

$g(x) = \log_{1/4} x$

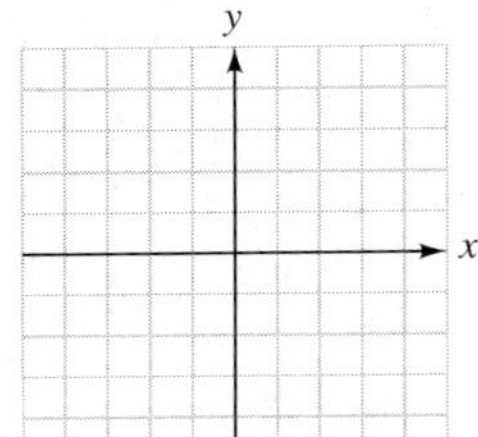

68. $f(x) = 4^x$

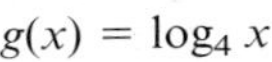

$g(x) = \log_4 x$

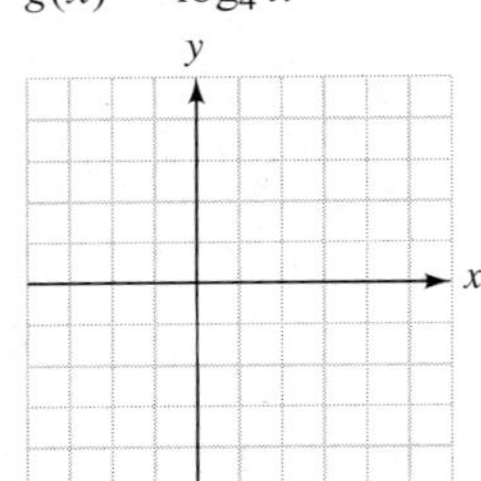

Graph each function. **See Examples 4–5. (Objective 3)**

69. $f(x) = 3 + \log_3 x$

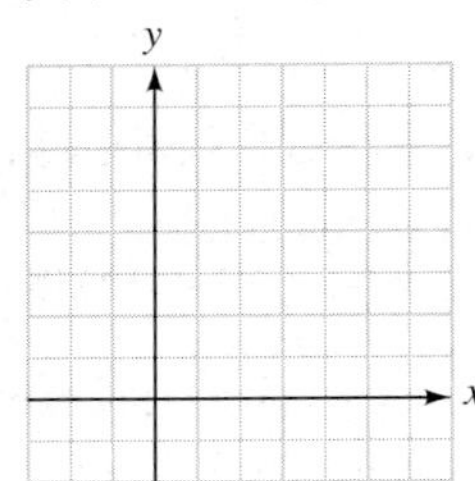

70. $f(x) = \log_{1/3} x - 1$

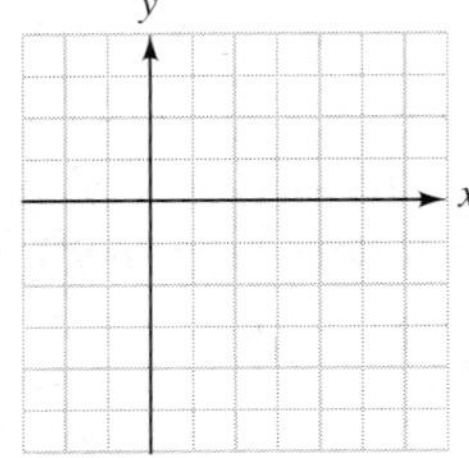

71. $f(x) = \log_{1/2}(x - 2)$

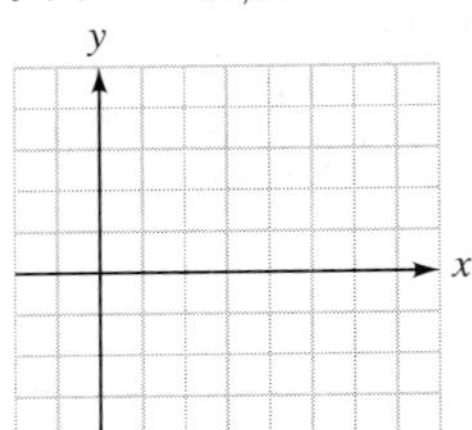

72. $f(x) = \log_4(x + 2)$

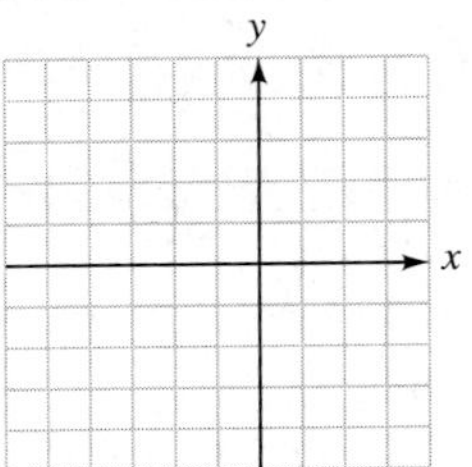

Use a calculator to find each value. Give answers to four decimal places. **(Objective 4)**

73. $\log 8.25$ **74.** $\log 0.77$

75. $\log 0.00867$ **76.** $\log 375.876$

Use a calculator to find each value of y. If an answer is not exact, give the answer to two decimal places. **(Objective 4)**

77. $\log y = 4.24$ **78.** $\log y = 0.926$

79. $\log y = -3.71$ **80.** $\log y = -0.28$

ADDITIONAL PRACTICE

Find each value of x.

81. $\log_{36} x = -\frac{1}{2}$ **82.** $\log_{27} x = -\frac{1}{3}$

83. $\log_{1/2} 8 = x$ **84.** $\log_{1/2} 16 = x$

85. $\log_{100} \frac{1}{1{,}000} = x$ **86.** $\log_{5/2} \frac{4}{25} = x$

87. $\log_{27} 9 = x$ **88.** $\log_{12} x = 0$

89. $\log_{2\sqrt{2}} x = 2$ **90.** $\log_4 8 = x$

91. $\log_x \frac{1}{64} = -3$ **92.** $\log_x \frac{1}{100} = -2$

93. $2^{\log_2 4} = x$ **94.** $3^{\log_3 5} = x$

95. $x^{\log_4 6} = 6$ **96.** $x^{\log_3 8} = 8$

97. $\log 10^3 = x$ **98.** $\log 10^{-2} = x$

99. $10^{\log x} = 100$ **100.** $10^{\log x} = \frac{1}{10}$

Use a calculator to find each value of y. If an answer is not exact, give the answer to two decimal places.

101. $\log y = 1.4023$ **102.** $\log y = 2.6490$

103. $\log y = \log 8$ **104.** $\log y = \log 7$

Find the value of b, if any, that would cause the graph of $f(x) = \log_b x$ to look like the graph indicated.

105.

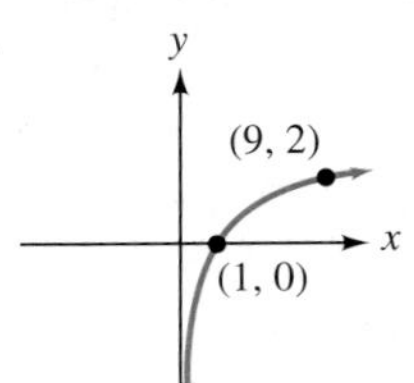

106.

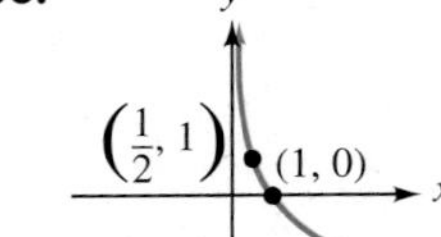

107.

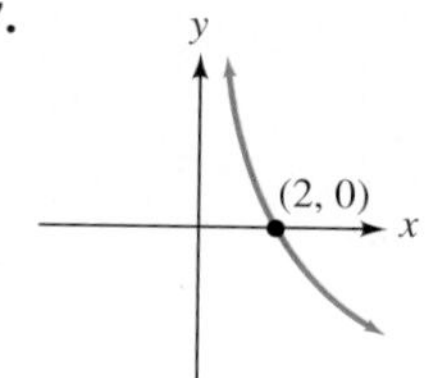

108.

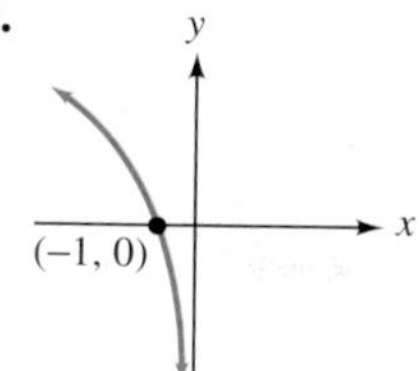

APPLICATIONS

Use a calculator to help solve each problem. If an answer is not exact, round to the nearest tenth. **See Example 6.** (Objective 5)

109. Finding the gain of an amplifier Find the dB gain of an amplifier if the input voltage is 0.71 volt when the output voltage is 20 volts.

110. Finding the gain of an amplifier Find the dB gain of an amplifier if the output voltage is 2.8 volts when the input voltage is 0.05 volt.

111. dB gain of an amplifier Find the dB gain of the amplifier.

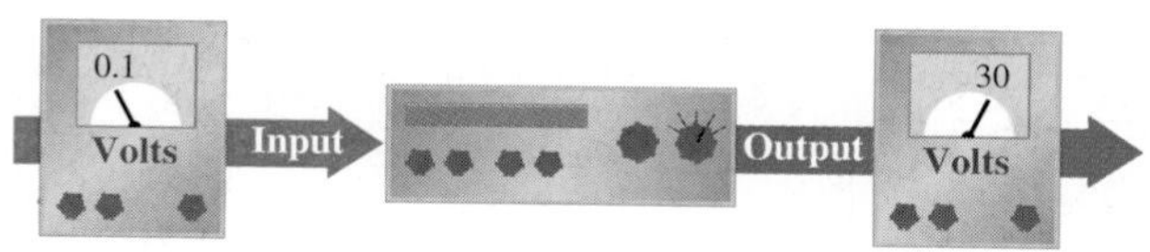

112. dB gain of an amplifier An amplifier produces an output of 80 volts when driven by an input of 0.12 volts. Find the amplifier's dB gain.

Use a calculator to help solve each problem. If an answer is not exact, round to the nearest tenth. **See Example 7.** (Objective 6)

113. Earthquakes An earthquake has an amplitude of 5,000 micrometers and a period of 0.2 second. Find its measure on the Richter scale.

114. Earthquakes The period of an earthquake with amplitude of 80,000 micrometers is 0.08 second. Find its measure on the Richter scale.

115. Earthquakes An earthquake has a period of $\frac{1}{4}$ second and an amplitude of 2,500 micrometers. Find its measure on the Richter scale.

116. Earthquakes By what factor must the amplitude of an earthquake change to increase its magnitude by 1 point on the Richter scale? Assume that the period remains constant.

Use a calculator to help solve each problem. If an answer is not exact, round to the nearest tenth.

117. Depreciation Business equipment is often depreciated using the double declining-balance method. In this method, a piece of equipment with a life expectancy of N years, costing $\$C$, will depreciate to a value of $\$V$ in n years, where n is given by the formula

$$n = \frac{\log V - \log C}{\log\left(1 - \frac{2}{N}\right)}$$

A computer that cost \$17,000 has a life expectancy of 5 years. If it has depreciated to a value of \$2,000, how old is it?

118. Depreciation See Exercise 117. A printer worth \$470 when new had a life expectancy of 12 years. If it is now worth \$189, how old is it?

119. Time for money to grow If $\$P$ is invested at the end of each year in an annuity earning annual interest at a rate r, the amount in the account will be $\$A$ after n years, where

$$n = \frac{\log\left(\frac{Ar}{P} + 1\right)}{\log(1 + r)}$$

If \$1,000 is invested each year in an annuity earning 12% annual interest, how long will it take for the account to be worth \$20,000?

120. Time for money to grow If \$5,000 is invested each year in an annuity earning 8% annual interest, how long will it take for the account to be worth \$50,000? (See Exercise 119.)

WRITING ABOUT MATH

121. Describe the appearance of the graph of $y = f(x) = \log_b x$ when $0 < b < 1$ and when $b > 1$.

122. Explain why it is impossible to find the logarithm of a negative number.

SOMETHING TO THINK ABOUT

123. Graph $f(x) = -\log_3 x$. How does the graph compare to the graph of $f(x) = \log_3 x$?

124. Find a logarithmic function that passes through the points (1, 0) and (5, 1).

125. Explain why an earthquake measuring 7 on the Richter scale is much worse than an earthquake measuring 6.

SECTION 9.4 Natural Logarithms

Objectives

1. Evaluate a natural logarithm.
2. Solve a logarithmic equation.
3. Graph a natural logarithmic function.
4. Solve an application problem involving a natural logarithm.

Vocabulary

natural logarithm

Getting Ready

Evaluate each expression.

1. $\log_4 16$ **2.** $\log_2 \frac{1}{8}$ **3.** $\log_5 5$ **4.** $\log_7 1$

In this section, we will discuss special logarithmic functions with a base of e. They play an important role in advanced mathematics courses.

1 Evaluate a natural logarithm.

We have seen the importance of base-e exponential functions in mathematical models of events in nature. Base-e logarithms are just as important. They are called **natural logarithms** or **Napierian logarithms,** after John Napier (1550–1617), and usually are written as $\ln x$, rather than $\log_e x$:

$$\ln x \quad \text{means} \quad \log_e x$$

As with all logarithmic functions, the domain of $f(x) = \ln x$ is the interval $(0, \infty)$, and the range is the interval $(-\infty, \infty)$.

We have seen that the logarithm of a number is an exponent. For natural logarithms,

$\ln x$ is the exponent to which e is raised to get x.

In equation form, we write

$$e^{\ln x} = x$$

To find the base-e logarithms of numbers, we can use a calculator.

ACCENT ON TECHNOLOGY

Evaluating Logarithms

To use a scientific calculator to find the value of ln 9.87, we enter these numbers and press these keys:

9.87 **LN**

The display will read **2.289499853**. To four decimal places, ln 9.87 = 2.2895.

To use a graphing calculator, we enter

LN 9.87 **)** **ENTER**

The display will read **ln (9.87)** **2.289499853**.

EXAMPLE 1 Use a calculator to find each value: **a.** ln 17.32 **b.** ln (log 0.05).

Solution **a.** We can enter these numbers and press these keys:

Scientific Calculator	*Graphing Calculator*
17.32 **LN**	**LN** 17.32 **)** **ENTER**

Either way, the result is 2.851861903.

b. We can enter these numbers and press these keys:

Scientific Calculator	*Graphing Calculator*
0.05 **LOG** **LN**	**LN** **(** **LOG** 0.05 **)** **)** **ENTER**

Either way, we obtain an error, because log 0.05 is a negative number. Because the domain of $\ln x$ is $(0, \infty)$, we cannot take the logarithm of a negative number.

SELF CHECK 1 Find each value to four decimal places.

a. $\ln \pi$ **b.** $\ln\left(\log \frac{1}{2}\right)$

2 Solve a logarithmic equation.

EXAMPLE 2 Find the value of x to four decimal places.

a. $\ln x = 1.335$ **b.** $\ln x = \log 5.5$

Solution **a.** The equation $\ln x = 1.335$ is equivalent to $e^{1.335} = x$. To use a scientific calculator to find x, we enter these numbers and press these keys:

1.335 e^x

The display will read 3.799995946. To four decimal places,

$$x = 3.8000$$

b. The equation $\ln x = \log 5.5$ is equivalent to $e^{\log 5.5} = x$. To use a scientific calculator to find x, we press these keys:

5.5 **LOG** e^x

The display will read 2.096695826. To four decimal places,

$$x = 2.0967$$

SELF CHECK 2 Find the value of x to four decimal places.
a. $\ln x = 2.5437$ **b.** $\log x = \ln 5$

3 Graph a natural logarithmic function.

John Napier (1550–1617)

Napier is famous for his work with natural logarithms. In fact, natural logarithms are often called *Napierian logarithms.* He also invented a device, called *Napier's rods,* that did multiplications mechanically. His device was a forerunner of modern-day computers.

The equation $y = \ln x$ is equivalent to the equation $x = e^y$. To graph $f(x) = \ln x$, we can plot points that satisfy the equation $x = e^y$ and join them with a smooth curve, as shown in Figure 9-20(a). Figure 9-20(b) shows the calculator graph.

$f(x) = \ln x$

x	y	$(x, f(x))$
$\frac{1}{e} \approx 0.4$	-1	$(0.4, -1)$
1	0	$(1, 0)$
$e \approx 2.7$	1	$(2.7, 1)$
$e^2 \approx 7.4$	2	$(7.4, 2)$

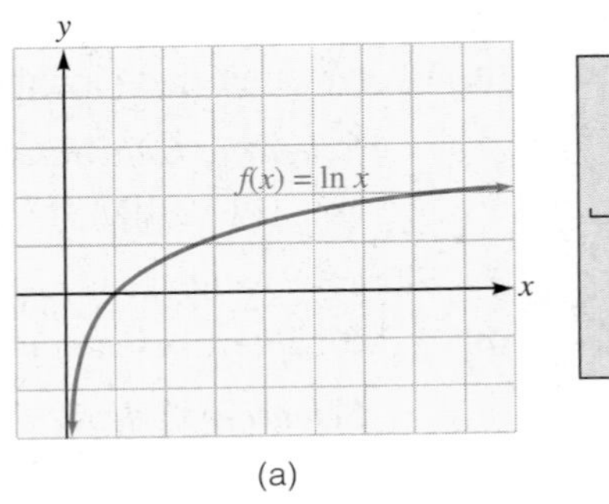

(a)

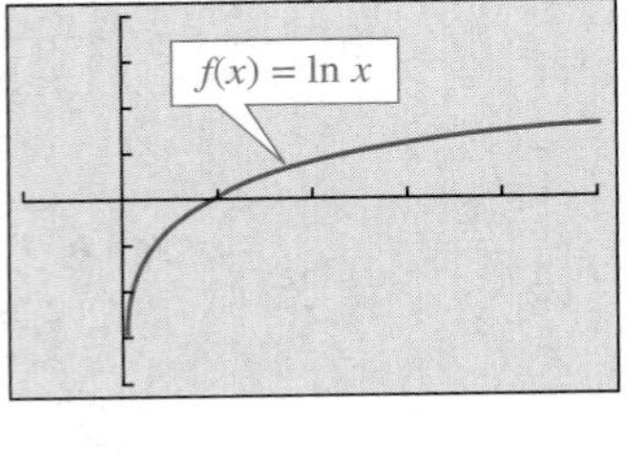

(b)

Figure 9-20

ACCENT ON TECHNOLOGY

Graphing Logarithmic Functions

Many graphs of logarithmic functions involve translations of the graph of $f(x) = \ln x$. For example, Figure 9-21 shows calculator graphs of the functions $f(x) = \ln x$, $f(x) = \ln x + 2$, and $f(x) = \ln x - 3$.

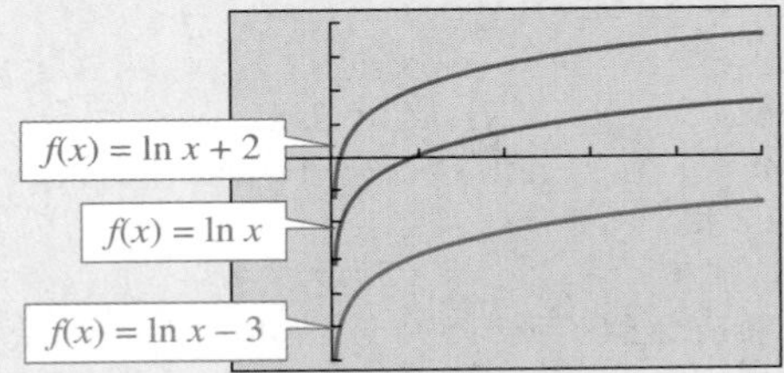

The graph of $f(x) = \ln x + 2$ is 2 units above the graph of $f(x) = \ln x$.

The graph of $f(x) = \ln x - 3$ is 3 units below the graph of $f(x) = \ln x$.

Figure 9-21

Figure 9-22 shows the calculator graphs of the functions $f(x) = \ln x$, $f(x) = \ln (x - 2)$, and $f(x) = \ln (x + 2)$.

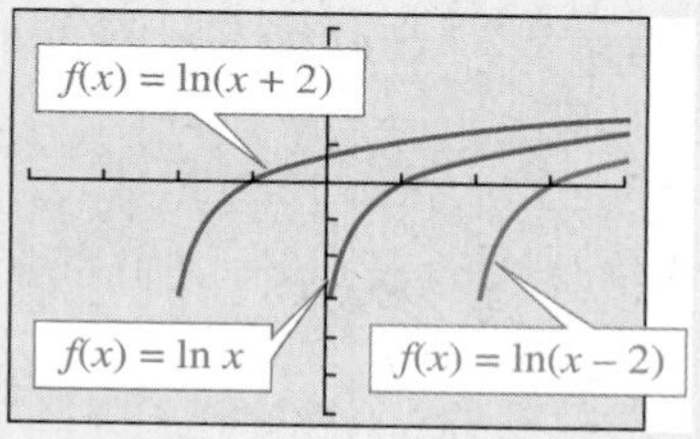

The graph of $f(x) = \ln(x - 2)$ is 2 units to the right of the graph of $f(x) = \ln x$.

The graph of $f(x) = \ln(x + 2)$ is 2 units to the left of the graph of $f(x) = \ln x$.

Figure 9-22

Natural logarithms have many applications.

4 Solve an application problem involving a natural logarithm.

If a population grows exponentially at a certain annual rate, the time required for the population to double is called the *doubling time.*

EXAMPLE 3 **DOUBLING TIME** If the Earth's population continues to grow at the approximate rate of 2% per year, how long will it take for its population to double?

Solution In Section 9.2, we learned that the formula for population growth is $A = Pe^{rt}$, where P is the population at time $t = 0$, A is the population after t years, and r is the annual rate of growth, compounded continuously. Since we want to find out how long it takes for the population to double, we can substitute $2P$ for A and 0.02 for r in the formula and proceed as follows.

$$A = Pe^{rt}$$

$$2P = Pe^{0.02t} \quad \text{Substitute } 2P \text{ for } A \text{ and } 0.02 \text{ for } r.$$

$$2 = e^{0.02t} \quad \text{Divide both sides by } P.$$

$$\ln 2 = 0.02t \quad \ln 2 \text{ is the exponent to which } e \text{ is raised to get 2.}$$

$$34.65735903 \approx t \quad \text{Divide both sides by 0.02 and simplify.}$$

The population will double in about 35 years.

SELF CHECK 3 If the world population's annual growth rate could be reduced to 1.5% per year, what would be the doubling time? Give the result to the nearest year.

By solving the formula $A = Pe^{rt}$ for t, we can obtain a simpler formula for finding the doubling time.

$$A = Pe^{rt}$$

$$2P = Pe^{rt} \quad \text{Substitute } 2P \text{ for } A.$$

$$2 = e^{rt} \quad \text{Divide both sides by } P.$$

$$\ln 2 = rt \quad \ln 2 \text{ is the exponent to which } e \text{ is raised to get 2.}$$

$$\frac{\ln 2}{r} = t \quad \text{Divide both sides by } r.$$

This result gives a specific formula for finding the doubling time.

Formula for Doubling Time

If r is the annual rate (compounded continuously) and t is the time required for a population to double, then

$$t = \frac{\ln 2}{r}$$

EXAMPLE 4 **DOUBLING TIME** How long will it take \$1,000 to double at an annual rate of 8%, compounded continuously?

Solution We can substitute 0.08 for r and simplify:

$$t = \frac{\ln 2}{r}$$

$$t = \frac{\ln 2}{\mathbf{0.08}}$$

$$\approx 8.664339757$$

It will take about $8\frac{2}{3}$ years for the money to double.

COMMENT To find the doubling time, you can use either the method in Example 3 or the method in Example 4.

SELF CHECK 4 How long will it take at 9%, compounded continuously?

SELF CHECK ANSWERS **1. a.** 1.1447 **b.** no value **2. a.** 12.7267 **b.** 40.6853 **3.** 46 years **4.** about 7.7 years

NOW TRY THIS

Between 2000 and 2007, McKinney, TX, was ranked overall as the fastest growing city in the United States.

1. The population more than doubled from 54,369 to 115,620 during this period. To the nearest tenth, what was the average growth rate?
2. If the growth rate slowed to half that of problem 1 and was estimated to remain there for the next 10 years, project the population in 2015.
3. Discuss with another student (or in a group) the difficulties a city might face with such a growth rate.

9.4 EXERCISES

WARM-UPS

1. Write $y = \ln x$ as an exponential equation.
2. Write $e^a = b$ as a logarithmic equation.
3. Write the formula for doubling time.

REVIEW

Write the equation of the required line.

4. Parallel to $y = 5x + 8$ and passing through the origin
5. Having a slope of 9 and a y-intercept of (0, 5)

6. Passing through the point (3, 2) and perpendicular to the line $y = \frac{2}{3}x - 12$
7. Parallel to the line $3x + 2y = 9$ and passing through the point $(-3, 5)$
8. Vertical line through the point (5, 3)
9. Horizontal line through the point (2, 5)

Simplify each expression. Assume no denominators are 0.

10. $\dfrac{2x + 3}{4x^2 - 9}$
11. $\dfrac{x + 1}{x} + \dfrac{x - 1}{x + 1}$
12. $\dfrac{x^2 + 3x + 2}{3x + 12} \cdot \dfrac{x + 4}{x^2 - 4}$
13. $\dfrac{1 + \frac{y}{x}}{\frac{y}{x} - 1}$

VOCABULARY AND CONCEPTS *Fill in the blanks.*

14. The expression $\ln x$ means ______ and is called a ________ logarithm.
15. The domain of the function $f(x) = \ln x$ is the interval ______ and the range is the interval ________.
16. The graph of $y = f(x) = \ln x$ has the ______ as an asymptote.
17. In the expression $\log x$, the base is understood to be ___.
18. In the expression $\ln x$, the base is understood to be __.
19. If a population grows exponentially at a rate r, the time it will take the population to double is given by the formula $t = \frac{\_\_\_}{\_\_\_}$.
20. The logarithm of a negative number is _________.

GUIDED PRACTICE

Use a calculator to find each value, if possible. Express all answers to four decimal places. See Example 1. (Objective 1)

21. ln 25.25
22. ln 0.523
23. ln 9.89
24. ln 0.00725
25. log (ln 2)
26. ln (log 28.8)
27. ln (log 0.5)
28. log (ln 0.2)

Use a calculator to find y, if possible. Express all answers to four decimal places. See Example 2. (Objective 2)

29. $\ln y = 2.3015$
30. $\ln y = 1.548$
31. $\ln y = 3.17$
32. $\ln y = 0.837$
33. $\ln y = -4.72$
34. $\ln y = -0.48$
35. $\log y = \ln 6$
36. $\ln y = \log 5$

Use a graphing calculator to graph each function. (Objective 3)

37. $y = -\ln x$
38. $y = \ln x^2$
39. $y = \ln(-x)$
40. $y = \ln\left(\frac{1}{2}x\right)$

Determine whether the graph could represent the graph of $y = \ln x$.

41.

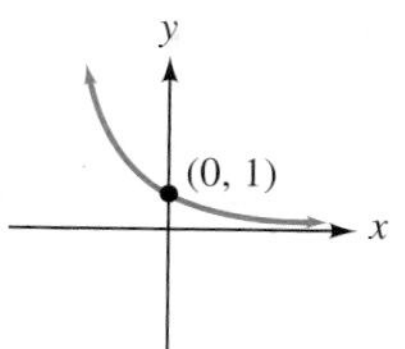

42.

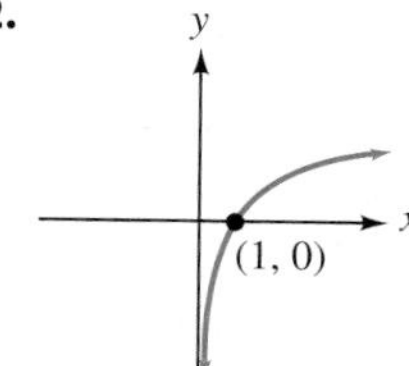

43.

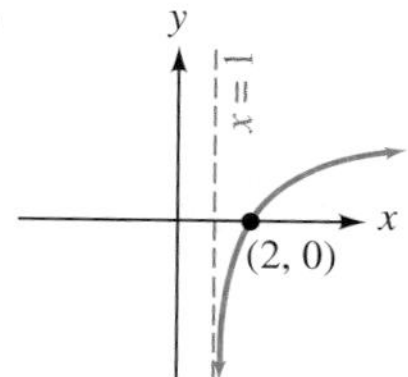

44.

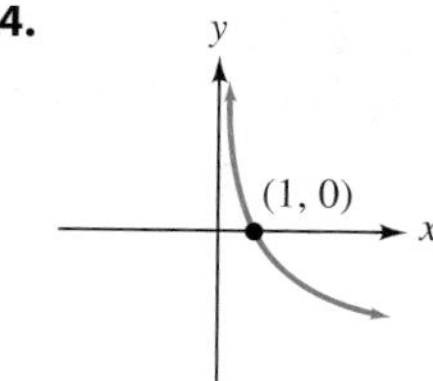

APPLICATIONS

Use a calculator to solve each problem. Round each answer to the nearest tenth. See Examples 3–4. (Objective 4)

45. **Population growth** See the ad. How long will it take the population of River City to double?

46. **Population growth** A population growing at an annual rate r will triple in a time t given by the formula

$$t = \frac{\ln 3}{r}$$

How long will it take the population of a town growing at the rate of 12% per year to triple?

47. **Doubling money** How long will it take \$1,000 to double if it is invested at an annual rate of 5%, compounded continuously?
48. **Tripling money** Find the length of time for \$25,000 to triple if invested at 6% annual interest, compounded continuously. (See Exercise 46.)

49. Making Jell-O After the contents of a package of Jell-O are combined with boiling water, the mixture is placed in a refrigerator whose temperature remains a constant 38° F. Estimate the number of hours t that it will take for the Jell-O to cool to 50° F using the formula

$$t = -\frac{1}{0.9}\ln\frac{50 - T_r}{200 - T_r}$$

where T_r is the temperature of the refrigerator.

50. Forensic medicine To estimate the number of hours t that a murder victim had been dead, a coroner used the formula

$$t = \frac{1}{0.25}\ln\frac{98.6 - T_s}{82 - T_s}$$

where T_s is the temperature of the surroundings where the body was found. If the crime took place in an apartment where the thermostat was set at 72° F, approximately how long ago did the murder occur?

WRITING ABOUT MATH

51. The time it takes money to double at an annual rate r, compounded continuously, is given by the formula $t = (\ln 2)/r$. Explain why money doubles more quickly as the rate increases.

52. The time it takes money to triple at an annual rate r, compounded continuously, is given by the formula $t = (\ln 3)/r$. Explain why money triples less quickly as the rate decreases.

SOMETHING TO THINK ABOUT

53. Use the formula $P = P_0e^{rt}$ to verify that P will be three times as large as P_0 when $t = \frac{\ln 3}{r}$.

54. Use the formula $P = P_0e^{rt}$ to verify that P will be four times as large as P_0 when $t = \frac{\ln 4}{r}$.

55. Find a formula to find how long it will take a sum of money to become five times as large.

56. Use a graphing calculator to graph

$$f(x) = \frac{1}{1 + e^{-2x}}$$

and discuss the graph.

SECTION 9.5 Properties of Logarithms

Objectives

1. Simplify a logarithmic expression by applying a property of logarithms.
2. Expand a logarithmic expression.
3. Write a logarithmic expression as a single logarithm.
4. Evaluate a logarithm by applying properties of logarithms.
5. Apply the change-of-base formula.
6. Solve an application problem using one or more logarithmic properties.

Vocabulary

change-of-base formula

Getting Ready

Simplify each expression. Assume $x \neq 0$.

1. $x^m x^n$ **2.** x^0 **3.** $(x^m)^n$ **4.** $\frac{x^m}{x^n}$

In this section, we will consider many properties of logarithms. We will then use these properties to solve application problems.

1 Simplify a logarithmic expression by applying a property of logarithms.

Since logarithms are exponents, the properties of exponents have counterparts in the theory of logarithms. We begin with four basic properties.

Properties of Logarithms

If b is a positive number and $b \neq 1$, then

1. $\log_b 1 = 0$ **2.** $\log_b b = 1$

3. $\log_b b^x = x$ **4.** $b^{\log_b x} = x \quad (x > 0)$

Properties 1 through 4 follow directly from the definition of a logarithm.

1. $\log_b 1 = 0$, because $b^0 = 1$.

2. $\log_b b = 1$, because $b^1 = b$.

3. $\log_b b^x = x$, because $b^x = b^x$.

4. $b^{\log_b x} = x$, because $\log_b x$ is the exponent to which b is raised to get x.

Properties 3 and 4 also indicate that the composition of the exponential and logarithmic functions with the same base (in both directions) is the identity function. This is expected, because the exponential and logarithmic functions with the same base are inverse functions.

EXAMPLE 1 Simplify each expression: **a.** $\log_5 1$ **b.** $\log_3 3$ **c.** $\log_7 7^3$ **d.** $b^{\log_b 7}$.

Solution **a.** By Property 1, $\log_5 1 = 0$, because $5^0 = 1$.

b. By Property 2, $\log_3 3 = 1$, because $3^1 = 3$.

c. By Property 3, $\log_7 7^3 = 3$, because $7^3 = 7^3$.

d. By Property 4, $b^{\log_b 7} = 7$, because $\log_b 7$ is the power to which b is raised to get 7.

SELF CHECK 1 Simplify: **a.** $\log_4 1$ **b.** $\log_5 5$ **c.** $\log_2 2^4$ **d.** $5^{\log_5 2}$.

The next two properties state that

The logarithm of a product is the sum of the logarithms.

The logarithm of a quotient is the difference of the logarithms.

The Product and Quotient Properties of Logarithms

If M, N, and b are positive numbers and $b \neq 1$, then

5. $\log_b MN = \log_b M + \log_b N$ **6.** $\log_b \frac{M}{N} = \log_b M - \log_b N$

Proof To prove the product property of logarithms, we let $x = \log_b M$ and $y = \log_b N$. We use the definition of logarithms to write each equation in exponential form.

$$M = b^x \quad \text{and} \quad N = b^y$$

Then $MN = b^x b^y$ and a property of exponents gives

$$MN = b^{x+y} \quad b^x b^y = b^{x+y}\text{: Keep the base and add the exponents.}$$

We write this exponential equation in logarithmic form as

$$\log_b MN = x + y$$

Substituting the values of x and y completes the proof.

$$\log_b MN = \log_b M + \log_b N$$

The proof of the quotient property of logarithms is similar.

COMMENT By the product property of logarithms, the logarithm of a *product* is equal to the *sum* of the logarithms. The logarithm of a sum or a difference usually does not simplify. In general,

$$\log_b(M + N) = \log_b M + \log_b N \quad \text{(crossed out)}$$
$$\log_b(M - N) = \log_b M - \log_b N \quad \text{(crossed out)}$$

By the quotient property of logarithms, the logarithm of a *quotient* is equal to the *difference* of the logarithms. The logarithm of a quotient is not the quotient of the logarithms:

$$\log_b \frac{M}{N} = \frac{\log_b M}{\log_b N} \quad \text{(crossed out)}$$

ACCENT ON TECHNOLOGY

Verifying Properties of Logarithms

We can use a calculator to illustrate the product property of logarithms by showing that

$$\ln[(3.7)(15.9)] = \ln 3.7 + \ln 15.9$$

We calculate the left and right sides of the equation separately and compare the results. To use a calculator to find ln [(3.7)(15.9)], we enter these numbers and press these keys:

3.7 × 15.9 = **LN** Using a scientific calculator
LN 3.7 × 15.9) **ENTER** Using a graphing calculator

The display will read **4.074651929**.

To find $\ln 3.7 + \ln 15.9$, we enter these numbers and press these keys:

3.7 **LN** + 15.9 **LN** = Using a scientific calculator
LN 3.7) + **LN** 15.9) **ENTER** Using a graphing calculator

The display will read **4.074651929**. Since the left and right sides are equal, the equation is true.

The power rule of logarithms states that

The logarithm of an expression to a power is the power times the logarithm of the expression.

The Power Rule of Logarithms

If M, p, and b are positive numbers and $b \neq 1$, then

7. $\log_b M^p = p \log_b M$

Proof To prove the power rule, we let $x = \log_b M$, write the expression in exponential form, and raise both sides to the pth power:

$$M = b^x$$

$$(M)^p = (b^x)^p \quad \text{Raise both sides to the } p\text{th power.}$$

$$M^p = b^{px} \quad \text{Keep the base and multiply the exponents.}$$

Using the definition of logarithms gives

$$\log_b M^p = px$$

Substituting the value for x completes the proof.

$$\log_b M^p = p \log_b M$$

The logarithmic property of equality states that

If the logarithms of two numbers are equal, the numbers are equal.

The Logarithmic Property of Equality

If M, p, and b are positive numbers and $b \neq 1$, then

8. If $\log_b x = \log_b y$, then $x = y$.

The logarithmic property of equality follows from the fact that the logarithmic function is a one-to-one function. It will be important in the next section when we solve logarithmic equations.

2 Expand a logarithmic expression.

We can use the properties of logarithms to write a logarithm as the sum or difference of several logarithms.

EXAMPLE 2 Assume that b, x, y, and z are positive numbers and $b \neq 1$. Write each expression in terms of the logarithms of x, y, and z.

a. $\log_b xyz$ **b.** $\log_b \frac{xy}{z}$

Solution **a.** $\log_b xyz = \log_b(xy)z$

$$= \log_b(xy) + \log_b z \quad \text{The log of a product is the sum of the logs.}$$

$$= \log_b x + \log_b y + \log_b z \quad \text{The log of a product is the sum of the logs.}$$

b. $\log_b \dfrac{xy}{z} = \log_b(xy) - \log_b z$ The log of a quotient is the difference of the logs.

$= (\log_b x + \log_b y) - \log_b z$ The log of a product is the sum of the logs.

$= \log_b x + \log_b y - \log_b z$ Remove parentheses.

SELF CHECK 2 Write $\log_b \frac{x}{yz}$ in terms of the logarithms of x, y, and z.

EXAMPLE 3 Assume that b, x, y, and z are positive numbers and $b \neq 1$. Write each expression in terms of the logarithms of x, y, and z.

a. $\log_b(x^2y^3z)$ **b.** $\log_b \dfrac{\sqrt{x}}{y^3z}$

Solution **a.** $\log_b(x^2y^3z) = \log_b x^2 + \log_b y^3 + \log_b z$ The log of a product is the sum of the logs.

$= 2\log_b x + 3\log_b y + \log_b z$ The log of an expression to a power is the power times the log of the expression.

b. $\log_b \dfrac{\sqrt{x}}{y^3z} = \log_b \sqrt{x} - \log_b(y^3z)$ The log of a quotient is the difference of the logs.

$= \log_b x^{1/2} - (\log_b y^3 + \log_b z)$ $\sqrt{x} = x^{1/2}$. The log of a product is the sum of the logs.

$= \dfrac{1}{2}\log_b x - (3\log_b y + \log_b z)$ The log of a power is the power times the log.

$= \dfrac{1}{2}\log_b x - 3\log_b y - \log_b z$ Use the distributive property to remove parentheses.

SELF CHECK 3 Write $\log_b \sqrt[4]{\dfrac{x^3y}{z}}$ in terms of the logarithms of x, y, and z.

3 Write a logarithmic expression as a single logarithm.

We can use the properties of logarithms to combine several logarithms into one logarithm.

EXAMPLE 4 Assume that b, x, y, and z are positive numbers and $b \neq 1$. Write each expression as one logarithm.

a. $3\log_b x + \frac{1}{2}\log_b y$ **b.** $\frac{1}{2}\log_b(x-2) - \log_b y + 3\log_b z$

Solution **a.** $3\log_b x + \dfrac{1}{2}\log_b y = \log_b x^3 + \log_b y^{1/2}$ A power times a log is the log of the power.

$= \log_b(x^3y^{1/2})$ The sum of two logs is the log of a product.

$= \log_b\left(x^3\sqrt{y}\right)$ $y^{1/2} = \sqrt{y}$

b. $\dfrac{1}{2}\log_b(x-2) - \log_b y + 3\log_b z$

$= \log_b(x-2)^{1/2} - \log_b y + \log_b z^3$ A power times a log is the log of the power.

$$= \log_b \frac{(x-2)^{1/2}}{y} + \log_b z^3$$ The difference of two logs is the log of the quotient.

$$= \log_b \frac{z^3\sqrt{x-2}}{y}$$ The sum of two logs is the log of a product.

SELF CHECK 4 Write the expression as one logarithm: $2 \log_b x + \frac{1}{2}\log_b y - 2\log_b(x-y)$.

We summarize the properties of logarithms as follows.

Properties of Logarithms

If b, M, and N are positive numbers and $b \neq 1$, then

1. $\log_b 1 = 0$

2. $\log_b b = 1$

3. $\log_b b^x = x$

4. $b^{\log_b x} = x$

5. $\log_b MN = \log_b M + \log_b N$

6. $\log_b \frac{M}{N} = \log_b M - \log_b N$

7. $\log_b M^p = p \log_b M$

8. If $\log_b x = \log_b y$, then $x = y$.

4 Evaluate a logarithm by applying properties of logarithms.

EXAMPLE 5 Given that $\log 2 \approx 0.3010$ and $\log 3 \approx 0.4771$, find approximations for **a.** $\log 6$ **b.** $\log 9$ **c.** $\log 18$ **d.** $\log 2.5$ without using a calculator.

Solution

a.
$$\log 6 = \log(2 \cdot 3)$$
$$= \log 2 + \log 3$$ The log of a product is the sum of the logs.
$$\approx 0.3010 + 0.4771$$ Substitute the value of each logarithm.
$$\approx 0.7781$$

b.
$$\log 9 = \log(3^2)$$
$$= 2\log 3$$ The log of a power is the power times the log.
$$\approx 2(0.4771)$$ Substitute the value of log 3.
$$\approx 0.9542$$

c.
$$\log 18 = \log(2 \cdot 3^2)$$
$$= \log 2 + \log 3^2$$ The log of a product is the sum of the logs.
$$= \log 2 + 2\log 3$$ The log of a power is the power times the log.
$$\approx 0.3010 + 2(0.4771)$$
$$\approx 1.2552$$

d.
$$\log 2.5 = \log\left(\frac{5}{2}\right)$$
$$= \log 5 - \log 2$$ The log of a quotient is the difference of the logs.
$$= \log \frac{10}{2} - \log 2$$ Write 5 as $\frac{10}{2}$.

$= \log 10 - \log 2 - \log 2$ The log of a quotient is the difference of the logs.
$= 1 - 2 \log 2$ $\log_{10} 10 = 1$
$\approx 1 - 2(0.3010)$
≈ 0.3980

SELF CHECK 5 Use the values given in Example 5 and approximate
a. $\log 1.5$ **b.** $\log 0.2$.

5 Apply the change-of-base formula.

If we know the base-a logarithm of a number, we can find its logarithm to some other base b with a formula called the **change-of-base formula.**

Change-of-Base Formula

If a, b, and x are real numbers, $b > 0$, and $b \neq 1$, then

$$\log_b x = \frac{\log_a x}{\log_a b}$$

Proof To prove this formula, we begin with the equation $\log_b x = y$.

$y = \log_b x$

$x = b^y$ Change the equation from logarithmic to exponential form.

$\log_a x = \log_a b^y$ Take the base-a logarithm of both sides.

$\log_a x = y \log_a b$ The log of a power is the power times the log.

$y = \dfrac{\log_a x}{\log_a b}$ Divide both sides by $\log_a b$.

$\log_b x = \dfrac{\log_a x}{\log_a b}$ Refer to the first equation and substitute $\log_b x$ for y.

If we know logarithms to base a (for example, $a = 10$), we can find the logarithm of x to a new base b. We simply divide the base-a logarithm of x by the base-a logarithm of b.

EXAMPLE 6 Find $\log_4 9$ using base-10 logarithms.

Solution We can substitute 4 for b, 10 for a, and 9 for x into the change-of-base formula:

$$\log_b x = \frac{\log_a x}{\log_a b}$$

$$\log_4 9 = \frac{\log_{10} 9}{\log_{10} 4}$$

$$\approx 1.584962501$$

To four decimal places, $\log_4 9 = 1.5850$.

COMMENT $\frac{\log_a x}{\log_a b}$ means that one logarithm is to be divided by the other. They are not to be subtracted.

SELF CHECK 6 Find $\log_5 3$ to four decimal places using base-10 logarithms.

COMMENT It does not matter what base you choose when applying the change-of-base formula. You could use base-e (natural logarithm) and obtain the same result. In the example above, $\log_4 9 = \frac{\ln 9}{\ln 4} \approx 1.584962501$.

6 Solve an application problem using one or more logarithmic properties.

Common logarithms are used to express the acidity of solutions. The more acidic a solution, the greater the concentration of hydrogen ions. This concentration is indicated indirectly by the **pH scale,** or **hydrogen ion index.** The pH of a solution is defined by the following equation.

pH of a Solution

If $[H^+]$ is the hydrogen ion concentration in gram-ions per liter, then

$$\text{pH} = -\log [H^+]$$

EXAMPLE 7 **FINDING THE pH OF A SOLUTION** Find the pH of pure water, which has a hydrogen ion concentration of 10^{-7} gram-ions per liter.

Solution Since pure water has approximately 10^{-7} gram-ions per liter, its pH is

$$\begin{aligned} \text{pH} &= -\log [H^+] \\ \text{pH} &= -\log 10^{-7} \\ &= -(-7)\log 10 && \text{The log of a power is the power times the log.} \\ &= -(-7)(1) && \log 10 = 1 \\ &= 7 \end{aligned}$$

EXAMPLE 8 **FINDING THE HYDROGEN-ION CONCENTRATION** Find the hydrogen-ion concentration of seawater if its pH is 8.5.

Solution To find its hydrogen-ion concentration, we substitute 8.5 for the pH and find $[H^+]$.

$$\begin{aligned} 8.5 &= -\log [H^+] \\ -8.5 &= \log [H^+] && \text{Multiply both sides by } -1. \\ [H^+] &= 10^{-8.5} && \text{Change the equation to exponential form.} \end{aligned}$$

We can use a calculator to find that

$$[H^+] \approx 3.2 \times 10^{-9} \text{ gram-ions per liter}$$

In physiology, experiments suggest that the relationship between the loudness and the intensity of sound is a logarithmic one known as the **Weber–Fechner law.**

Weber–Fechner Law

If L is the apparent loudness of a sound, I is the actual intensity, and k is a constant, then

$$L = k \ln I$$

EXAMPLE 9 **WEBER–FECHNER LAW** Find the increase in intensity that will cause the apparent loudness of a sound to double.

Solution If the original loudness L_O is caused by an actual intensity I_O, then

(1) $L_O = k \ln I_O$

To double the apparent loudness, we multiply both sides of Equation 1 by 2 and use the power rule of logarithms:

$$\begin{aligned} 2\,L_O &= 2\,k \ln I_O \\ &= k \ln (I_O)^2 \end{aligned}$$

To double the loudness of a sound, the intensity must be squared.

SELF CHECK 9 What decrease in intensity will cause a sound to be half as loud?

SELF CHECK ANSWERS **1. a.** 0 **b.** 1 **c.** 4 **d.** 2 **2.** $\log_b x - \log_b y - \log_b z$ **3.** $\frac{1}{4}(3 \log_b x + \log_b y - \log_b z)$ **4.** $\log_b \frac{x^2\sqrt{y}}{(x-y)^2}$ **5. a.** 0.1761 **b.** −0.6990 **6.** 0.6826 **9.** the square root of the intensity

NOW TRY THIS

Evaluate.

1. $\log 5 + \log 20$

2. $\log_3 24 - \log_3 8$

3. $7 \log_2 \frac{1}{2} - \log_2 \frac{1}{8}$

9.5 EXERCISES

WARM-UPS *Find the value of x in each equation.*

1. $\log_3 9 = x$

2. $\log_x 5 = 1$

3. $\log_7 x = 3$

4. $\log_2 x = -2$

5. $\log_4 x = \frac{1}{2}$

6. $\log_x 4 = 2$

7. $\log_{1/2} x = 2$

8. $\log_9 3 = x$

9. $\log_x \frac{1}{4} = -2$

REVIEW *Consider the line that passes through* **(−2, 3)** *and* **(4, −4)**.

10. Find the slope of the line.

11. Find the distance between the points.

12. Find the midpoint of the segment.

13. Write an equation of the line.

VOCABULARY AND CONCEPTS *Fill in the blanks.*

14. $\log_b 1 =$ __

15. $\log_b b =$ __

16. $\log_b MN = \log_b \_\_ + \log_b \_\_$

17. $b^{\log_b x} = \_\_$

18. If $\log_b x = \log_b y$, then $\_\_ = \_\_$.

19. $\log_b \dfrac{M}{N} = \log_b M \_\_ \log_b N$

20. $\log_b x^p = p \cdot \log_b \_\_$

21. $\log_b b^x = \_\_$

22. $\log_b (A + B) \_\_ \log_b A + \log_b B$

23. $\log_b A + \log_b B \_\_ \log_b AB$

24. The change-of-base formula states that $\log_b x = \_\_\_\_$.

GUIDED PRACTICE

Simplify each expression. See Example 1. (Objective 1)

25. $\log_4 1 = \_\_$

26. $\log_4 4 = \_\_$

27. $\log_4 4^7 = \_\_$

28. $4^{\log_4 8} = \_\_$

29. $5^{\log_5 10} = \_\_$

30. $\log_5 5^2 = \_\_$

31. $\log_5 5 = \_\_$

32. $\log_5 1 = \_\_$

33. $\log_7 1 = \_\_$

34. $\log_9 9 = \_\_$

35. $\log_3 3^7 = \_\_$

36. $5^{\log_5 8} = \_\_$

37. $8^{\log_8 10} = \_\_$

38. $\log_4 4^2 = \_\_$

39. $\log_9 9 = \_\_$

40. $\log_3 1 = \_\_$

Use a calculator to verify each equation. (Objective 1)

41. $\log[(2.5)(3.7)] = \log 2.5 + \log 3.7$

42. $\ln \dfrac{11.3}{6.1} = \ln 11.3 - \ln 6.1$

43. $\ln (2.25)^4 = 4 \ln 2.25$

44. $\log 45.37 = \dfrac{\ln 45.37}{\ln 10}$

Assume that x, y, z, and b are positive numbers ($b \neq 1$). Use the properties of logarithms to write each expression in terms of the logarithms of x, y, and z. See Examples 2–3. (Objective 2)

45. $\log_b xyz$

46. $\log_b 4xz$

47. $\log_b \dfrac{2x}{y}$

48. $\log_b \dfrac{x}{yz}$

49. $\log_b x^3y^2$

50. $\log_b xy^2z^3$

51. $\log_b (xy)^{1/2}$

52. $\log_b x^3y^{1/2}$

53. $\log_b x\sqrt{z}$

54. $\log_b \sqrt{xy}$

55. $\log_b \dfrac{\sqrt[3]{x}}{\sqrt[4]{yz}}$

56. $\log_b \sqrt[4]{\dfrac{x^3y^2}{z^4}}$

Assume that x, y, z, and b are positive numbers ($b \neq 1$). Use the properties of logarithms to write each expression as the logarithm of a single quantity. See Example 4. (Objective 3)

57. $\log_b (x + 1) - \log_b x$

58. $\log_b x + \log_b (x + 2) - \log_b 8$

59. $2 \log_b x + \dfrac{1}{2} \log_b y$

60. $-2 \log_b x - 3 \log_b y + \log_b z$

61. $-3 \log_b x - 2 \log_b y + \dfrac{1}{2} \log_b z$

62. $3 \log_b (x + 1) - 2 \log_b (x + 2) + \log_b x$

63. $\log_b \left(\dfrac{x}{z} + x\right) - \log_b \left(\dfrac{y}{z} + y\right)$

64. $\log_b (xy + y^2) - \log_b (xz + yz) + \log_b z$

Assume that $\log 4 \approx 0.6021$, $\log 7 \approx 0.8451$, and $\log 9 \approx 0.9542$. Use these values and the properties of logarithms to approximate each value. Do not use a calculator. See Example 5. (Objective 4)

65. $\log 28$

66. $\log \dfrac{7}{4}$

67. $\log 2.25$

68. $\log 36$

69. $\log \dfrac{63}{4}$

70. $\log \dfrac{4}{63}$

71. $\log 252$

72. $\log 49$

73. $\log 112$

74. $\log 324$

75. $\log \dfrac{144}{49}$

76. $\log \dfrac{324}{63}$

Use a calculator and the change-of-base formula to find each logarithm to four decimal places. See Example 6. (Objective 5)

77. $\log_3 7$

78. $\log_7 3$

79. $\log_{1/3} 3$

80. $\log_{1/2} 6$

81. $\log_3 8$

82. $\log_5 10$

83. $\log_{\sqrt{2}} \sqrt{5}$

84. $\log_\pi e$

ADDITIONAL PRACTICE

Use a calculator to verify each equation.

85. $\log \sqrt{24.3} = \dfrac{1}{2} \log 24.3$

86. $\ln 8.75 = \dfrac{\log 8.75}{\log e}$

Determine whether each statement is true. If a statement is false, explain why.

87. $\log_b 0 = 1$

88. $\log_b (x + y) \neq \log_b x + \log_b y$

89. $\log_b xy = (\log_b x)(\log_b y)$

90. $\log_b ab = \log_b a + 1$

91. $\log_7 7^7 = 7$

92. $7^{\log_7 7} = 7$

93. $\dfrac{\log_b A}{\log_b B} = \log_b A - \log_b B$

94. $\log_b (A - B) = \dfrac{\log_b A}{\log_b B}$

95. $3 \log_b \sqrt[3]{a} = \log_b a$

96. $\dfrac{1}{3} \log_b a^3 = \log_b a$

97. $\log_b \dfrac{1}{a} = -\log_b a$

98. $\log_b 2 = \log_2 b$

APPLICATIONS *Use a calculator to find each value.*

See Examples 7–9. (Objective 6)

99. pH of a solution Find the pH of a solution with a hydrogen ion concentration of 1.7×10^{-5} gram-ions per liter.

100. Hydrogen ion concentration Find the hydrogen ion concentration of a saturated solution of calcium hydroxide whose pH is 13.2.

101. Aquariums To test for safe pH levels in a fresh-water aquarium, a test strip is compared with the scale shown in the illustration. Find the corresponding range in the hydrogen ion concentration.

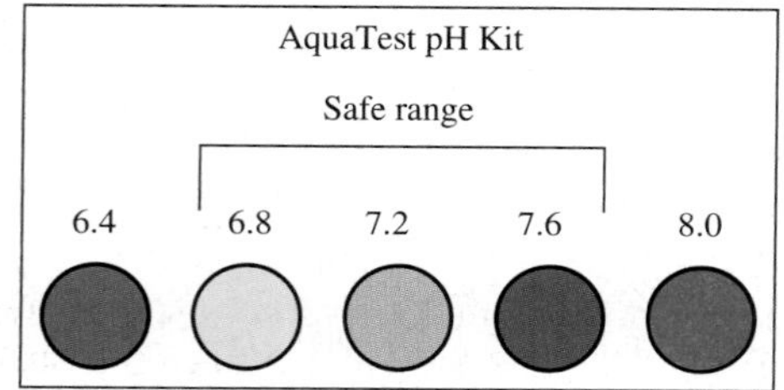

102. pH of pickles The hydrogen ion concentration of sour pickles is 6.31×10^{-4}. Find the pH.

103. Change in loudness If the intensity of a sound is doubled, find the apparent change in loudness.

104. Change in loudness If the intensity of a sound is tripled, find the apparent change in loudness.

105. Change in intensity What change in intensity of sound will cause an apparent tripling of the loudness?

106. Change in intensity What increase in the intensity of a sound will cause the apparent loudness to be multiplied by 4?

WRITING ABOUT MATH

107. Explain why $\ln (\log 0.9)$ is undefined in the real-number system.

108. Explain why $\log_b (\ln 1)$ is undefined in the real-number system.

SOMETHING TO THINK ABOUT

109. Show that $\ln(e^x) = x$.

110. If $\log_b 3x = 1 + \log_b x$, find the value of x.

111. Show that $\log_{b^2} x = \frac{1}{2} \log_b x$.

112. Show that $e^{x \ln a} = a^x$.

SECTION 9.6 Exponential and Logarithmic Equations

Objectives

1. Solve an exponential equation.
2. Solve a logarithmic equation.
3. Solve an application problem involving an exponential or logarithmic equation.

Vocabulary

exponential equation   logarithmic equation   half-life

Getting Ready

Write each expression without using exponents.

1. $\log x^2$

2. $\log x^{1/2}$

3. $\log x^0$

4. $\log a^b + b \log a$

An **exponential equation** is an equation that contains a variable in one of its exponents. Some examples of exponential equations are

$$3^x = 5, \qquad 6^{x-3} = 2^x, \qquad \text{and} \qquad 3^{2x+1} - 10(3^x) + 3 = 0$$

A **logarithmic equation** is an equation with logarithmic expressions that contain a variable. Some examples of logarithmic equations are

$$\log(2x) = 25, \qquad \ln x - \ln(x - 12) = 24, \qquad \text{and} \qquad \log x = \log\frac{1}{x} + 4$$

In this section, we will learn how to solve many of these equations.

1 Solve an exponential equation.

EXAMPLE 1 Solve: $4^x = 7$.

Solution Since logarithms of equal numbers are equal, we can take the common or natural logarithm of both sides of the equation and obtain a new equation. For the sake of discussion, we will use the common logarithm. The power rule of logarithms then provides a way of moving the variable x from its position as an exponent to a position as a coefficient.

$$4^x = 7$$

$$\log(4^x) = \log(7) \qquad \text{Take the common logarithm of both sides.}$$

$$x \log 4 = \log 7 \qquad \text{The log of a power is the power times the log.}$$

$$(1) \qquad x = \frac{\log 7}{\log 4} \qquad \text{Divide both sides by log 4.}$$

$$\approx 1.403677461 \qquad \text{Use a calculator.}$$

To four decimal places, $x = 1.4037$.

SELF CHECK 1 Solve: $5^x = 4$. Give the result to four decimal places.

COMMENT A careless reading of Equation 1 leads to a common error. The right side of Equation 1 calls for a division, not a subtraction.

$$\frac{\log 7}{\log 4} \quad \text{means} \quad (\log 7) \div (\log 4)$$

It is the expression $\log\left(\frac{7}{4}\right)$ that means $\log 7 - \log 4$.

EXAMPLE 2 Solve: $73 = 1.6(1.03)^t$.

Solution We divide both sides by 1.6 to obtain

$$\frac{73}{1.6} = 1.03^t$$

and solve the equation as in Example 1.

$$1.03^t = \frac{73}{1.6}$$

$$\log(1.03^t) = \log\left(\frac{73}{1.6}\right)$$ Take the common logarithm of both sides.

$$t \log 1.03 = \log \frac{73}{1.6}$$ The logarithm of a power is the power times the logarithm.

$$t = \frac{\log \frac{73}{1.6}}{\log 1.03}$$ Divide both sides by log 1.03.

$$t \approx 129.2493444$$ Use a calculator.

To four decimal places, $t = 129.2493$.

SELF CHECK 2 Solve: $47 = 2.5(1.05)^t$. Give the result to the nearest tenth.

EXAMPLE 3 Solve: $6^{x-3} = 2^x$.

Solution

$$6^{x-3} = 2^x$$

$$\log(6^{x-3}) = \log(2^x)$$ Take the common logarithm of both sides.

$$(x - 3)\log 6 = x \log 2$$ The log of a power is the power times the log.

$$x \log 6 - 3 \log 6 = x \log 2$$ Use the distributive property.

$$x \log 6 - x \log 2 = 3 \log 6$$ Add 3 log 6 and subtract x log 2 from both sides.

$$x(\log 6 - \log 2) = 3 \log 6$$ Factor out x on the left side.

$$x = \frac{3 \log 6}{\log 6 - \log 2}$$ Divide both sides by log 6 − log 2.

$$x \approx 4.892789261$$ Use a calculator.

To four decimal places, $x = 4.8928$.

SELF CHECK 3 Solve to four decimal places: $5^{x-2} = 3^x$.

EXAMPLE 4 Solve: $2^{x^2+2x} = \frac{1}{2}$.

Solution Since $\frac{1}{2} = 2^{-1}$, we can write the equation in the form

$$2^{x^2+2x} = 2^{-1}$$

Since equal quantities with equal bases have equal exponents, we have

$$
\begin{aligned}
x^2 + 2x &= -1 \\
x^2 + 2x + 1 &= 0 && \text{Add 1 to both sides.} \\
(x + 1)(x + 1) &= 0 && \text{Factor the trinomial.} \\
x + 1 = 0 \quad &\text{or} \quad x + 1 = 0 && \text{Set each factor equal to 0.} \\
x = -1 \quad &| \quad x = -1 && -1 \text{ is a double root.}
\end{aligned}
$$

Verify that -1 satisfies the equation.

SELF CHECK 4 Solve: $3^{x^2-2x} = \frac{1}{3}$.

ACCENT ON TECHNOLOGY

Solving Exponential Equations

To use a graphing calculator to approximate the solutions of $2^{x^2+2x} = \frac{1}{2}$ (see Example 4), we can subtract $\frac{1}{2}$ from both sides of the equation to get

$$2^{x^2+2x} - \frac{1}{2} = 0$$

and graph the corresponding function

$$f(x) = 2^{x^2+2x} - \frac{1}{2}$$

If we use window settings of $[-4, 4]$ for x and $[-2, 6]$ for y, we obtain the graph shown in Figure 9-23(a).

Since the solutions of the equation are its x-intercepts, we can approximate the solutions by zooming in on the values of the x-intercepts, as in Figure 9-23(b). Since $x = -1$ is the only x-intercept, -1 is the only solution. In this case, we have found an exact solution.

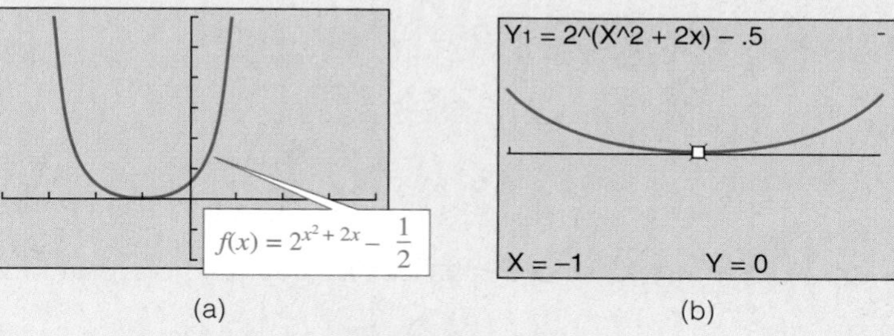

Figure 9-23

We also can find the solution by using the ZERO feature in the CALC menu.

2 Solve a logarithmic equation.

In each of the following examples, we use the properties of logarithms to change a logarithmic equation into an algebraic equation.

EXAMPLE 5 Solve: $\log_b(3x + 2) - \log_b(2x - 3) = 0$.

Solution

$$
\begin{aligned}
\log_b(3x + 2) - \log_b(2x - 3) &= 0 \\
\log_b(3x + 2) &= \log_b(2x - 3) && \text{Add } \log_b(2x - 3) \text{ to both sides.}
\end{aligned}
$$

$$3x + 2 = 2x - 3 \qquad \text{If } \log_b r = \log_b s, \text{ then } r = s.$$

$$x = -5 \qquad \text{Subtract } 2x \text{ and } 2 \text{ from both sides.}$$

Check:

$$\log_b(3x + 2) - \log_b(2x - 3) = 0$$

$$\log_b[3(-5) + 2] - \log_b[2(-5) - 3] \stackrel{?}{=} 0$$

$$\log_b(-13) - \log_b(-13) \stackrel{?}{=} 0$$

COMMENT Example 5 illustrates that you must check the solutions of all logarithmic equations.

Since the logarithm of a negative number does not exist, the apparent solution of -5 must be discarded. Since this equation has no solution, its solution set is $\varnothing$.

SELF CHECK 5 Solve: $\log_b(5x + 2) - \log_b(7x - 2) = 0$.

EXAMPLE 6 Solve: $\log x + \log(x - 3) = 1$.

Solution

$$\log x + \log(x - 3) = 1$$

$$\log[x(x - 3)] = 1 \qquad \text{The sum of two logs is the log of a product.}$$

$$x(x - 3) = 10^1 \qquad \text{Use the definition of logarithms to change the equation to exponential form.}$$

$$x^2 - 3x - 10 = 0 \qquad \text{Remove parentheses and subtract 10 from both sides.}$$

$$(x + 2)(x - 5) = 0 \qquad \text{Factor the trinomial.}$$

$$x + 2 = 0 \quad \text{or} \quad x - 5 = 0 \qquad \text{Set each factor equal to 0.}$$

$$x = -2 \quad \mid \quad x = 5$$

Check: The number -2 is not a solution, because it does not satisfy the equation. (A negative number does not have a logarithm.) We will check the remaining number, 5.

$$\log x + \log(x - 3) = 1$$

$$\log 5 + \log(5 - 3) \stackrel{?}{=} 1 \qquad \text{Substitute 5 for } x.$$

$$\log 5 + \log 2 \stackrel{?}{=} 1$$

$$\log 10 \stackrel{?}{=} 1 \qquad \text{The sum of two logs is the log of a product.}$$

$$1 = 1 \qquad \log 10 = 1$$

Since 5 satisfies the equation, it is a solution.

SELF CHECK 6 Solve: $\log x + \log(x + 3) = 1$.

EXAMPLE 7 Solve: $\dfrac{\log(5x - 6)}{\log x} = 2$.

Solution We can multiply both sides of the equation by $\log x$ to get

$$\log(5x - 6) = 2 \log x$$

and apply the power rule of logarithms to get

$$\log(5x - 6) = \log x^2$$

By Property 8 of logarithms, $5x - 6 = x^2$. Thus,

$$5x - 6 = x^2$$
$$0 = x^2 - 5x + 6$$
$$0 = (x - 3)(x - 2)$$
$$x - 3 = 0 \quad \text{or} \quad x - 2 = 0$$
$$x = 3 \quad | \quad x = 2$$

Verify that both 2 and 3 satisfy the equation.

SELF CHECK 7 Solve: $\dfrac{\log(5x + 6)}{\log x} = 2.$

ACCENT ON TECHNOLOGY

Solving Logarithmic Equations

To use a graphing calculator to approximate the solutions of $\log x + \log(x - 3) = 1$ (see Example 6), we can subtract 1 from both sides of the equation to get

$$\log x + \log(x - 3) - 1 = 0$$

and graph the corresponding function

$$f(x) = \log x + \log(x - 3) - 1$$

If we use window settings of [0, 20] for x and [−2, 2] for y, we obtain the graph shown in Figure 9-24. Since the solution of the equation is the x-intercept, we can find the solution by zooming in on the value of the x-intercept or by using the ZERO command. The solution is $x = 5$.

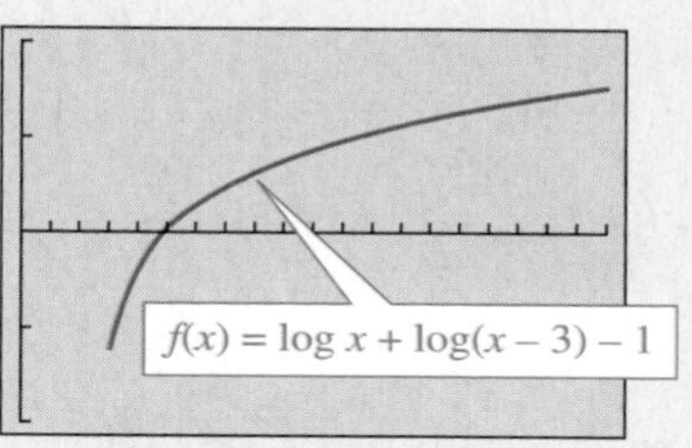

Figure 9-24

3 Solve an application problem involving an exponential or logarithmic equation.

We have learned that the amount A of radiation present in a radioactive material decays exponentially according to the formula $A = A_0e^{kt}$, where A_0 is the amount of radioactive material present at time $t = 0$, and k is a negative number.

Experiments have determined the time it takes for one-half of a sample of a radioactive element to decompose. That time is a constant, called the given material's **half-life.**

EXAMPLE 8 **HALF-LIFE OF RADON-22** In Example 4 of Section 9.2, we learned that radon-22 decays according to the formula $A = A_0e^{-0.181t}$, where t is expressed in days. Find the material's half-life.

Solution At time $t = 0$, the amount of radioactive material is A_0. At the end of one half-life, the amount present will be $\frac{1}{2}A_0$. To find the material's half-life, we can substitute $\frac{1}{2}A_0$ for A in the formula and solve for t.

$$A = A_0 e^{-0.181t}$$

$$\frac{1}{2}A_0 = A_0 e^{-0.181t} \quad \text{Substitute } \tfrac{1}{2}A_0 \text{ for } A.$$

$$\frac{1}{2} = e^{-0.181t} \quad \text{Divide both sides by } A_0.$$

$$\ln\left(\frac{1}{2}\right) = \ln(e^{-0.181t}) \quad \text{Take the natural logarithm of both sides.}$$

$$-0.6931471806 \approx -0.181t \ln e \quad \text{Find } \ln \tfrac{1}{2}, \text{ and use the property } \ln M^p = p \ln M.$$

$$3.829542434 \approx t \quad \text{Divide both sides by } -0.181 \text{ and note that } \ln e = 1.$$

To the nearest hundredth, the half-life of radon-22 is 3.83 days.

SELF CHECK 8 To the nearest hundredth, find the half-life of iodine-131 given that it decays according to the formula $A = A_0 e^{-0.087t}$ where t is expressed in days.

When a living organism dies, the oxygen/carbon dioxide cycle common to all living things stops and carbon-14, a radioactive isotope with a half-life of 5,700 years, is no longer absorbed. By measuring the amount of carbon-14 present in an ancient object, archaeologists can estimate the object's age.

EXAMPLE 9 **CARBON-14 DATING** How old is a wooden statue that retains one-third of its original carbon-14 content?

Solution The formula for radioactive decay is $A = A_0 e^{kt}$, where A_0 is the original amount of carbon-14 present, t is the age of the object, and k is a negative number. We can find k by using the fact that after 5,730 years, half of the original amount of carbon-14 will remain.

$$A = A_0 e^{kt}$$

$$\frac{1}{2}A_0 = A_0 e^{k(5{,}730)} \quad \text{Substitute } \tfrac{1}{2}A_0 \text{ for } A \text{ and 5,730 for } t.$$

$$\frac{1}{2} = e^{5{,}730k} \quad \text{Divide both sides by } A_0.$$

$$\ln\left(\frac{1}{2}\right) = \ln(e^{5{,}730k}) \quad \text{Take the natural logarithm of both sides.}$$

$$-0.6931471806 \approx 5{,}730k \ln e \quad \text{Find } \ln \tfrac{1}{2}, \text{ and use the property } \ln M^p = p \ln M.$$

$$-0.000120968094 \approx k \quad \text{Divide both sides by 5,730 and note that } \ln e = 1.$$

Thus, the formula for radioactive decay for carbon-14 can be written as

$$A \approx A_0 e^{-0.000120968094t}$$

Since $\frac{1}{3}$ of the original carbon-14 still remains, we can proceed as follows:

$$A \approx A_0 e^{-0.000120968094t}$$

$$\frac{1}{3}A_0 \approx A_0 e^{-0.000120968094t} \quad \text{Substitute } \tfrac{1}{3}A_0 \text{ for } A.$$

$$\frac{1}{3} \approx e^{-0.000120968094t} \quad \text{Divide both sides by } A_0.$$

$$\ln\left(\frac{1}{3}\right) \approx \ln(e^{-0.000120968094t})$$ Take the natural logarithm of both sides.

$$-1.098612289 \approx (-0.000120968094t)\ln e$$ Find $\ln \frac{1}{3}$, and use the property $\ln M^p = p \ln M$.

$$9081.835155 \approx t$$ Divide both sides by -0.000120968094 and note that $\ln e = 1$.

To the nearest one hundred years, the statue is 9,100 years old.

SELF CHECK 9 To the nearest hundred years, find the age of a statue that retains 25% of its original carbon-14 content.

Recall that when there is sufficient food and space, populations of living organisms tend to increase exponentially according to the Malthusian growth model.

Malthusian Growth Model

If P is the population at some time t, P_0 is the initial population at $t = 0$, and k is the rate of growth, then

$$P = P_0e^{kt}$$

COMMENT Note that this formula is the same as all exponential growth formulas, except for the variables.

EXAMPLE 10

POPULATION GROWTH The bacteria in a laboratory culture increased from an initial population of 500 to 1,500 in 3 hours. How long will it take for the population to reach 10,000?

Solution We substitute 500 for P_0, 1,500 for P, and 3 for t and simplify to find k:

$$P = P_0e^{kt}$$

$$1{,}500 = 500(e^{k3})$$ Substitute 1,500 for P, 500 for P_0, and 3 for t.

$$3 = e^{3k}$$ Divide both sides by 500.

$$3k = \ln 3$$ Change the equation from exponential to logarithmic form.

$$k = \frac{\ln 3}{3}$$ Divide both sides by 3.

To find when the population will reach 10,000, we substitute 10,000 for P, 500 for P_0, and $\frac{\ln 3}{3}$ for k in the equation $P = P_0e^{kt}$ and solve for t:

$$P = P_0e^{kt}$$

$$10{,}000 = 500e^{[(\ln 3)/3]t}$$

$$20 = e^{[(\ln 3)/3]t}$$ Divide both sides by 500.

$$\left(\frac{\ln 3}{3}\right)t = \ln 20$$ Change the equation to logarithmic form.

$$t = \frac{3 \ln 20}{\ln 3}$$ Multiply both sides by $\frac{3}{\ln 3}$.

$$\approx 8.180499084$$ Use a calculator.

The culture will reach 10,000 bacteria in about 8 hours.

SELF CHECK 10 How long will it take to reach 20,000?

EXAMPLE 11 **GENERATION TIME** If a medium is inoculated with a bacterial culture that contains 1,000 cells per milliliter, how many generations will pass by the time the culture has grown to a population of 1 million cells per milliliter?

Solution During bacterial reproduction, the time required for a population to double is called the *generation time.* If b bacteria are introduced into a medium, then after the generation time of the organism has elapsed, there are $2b$ cells. After another generation, there are $2(2b)$, or $4b$ cells, and so on. After n generations, the number of cells present will be

$$(1) \qquad B = b \cdot 2^n$$

To find the number of generations that have passed while the population grows from b bacteria to B bacteria, we solve Equation 1 for n.

EVERYDAY CONNECTIONS U.S. Population Growth

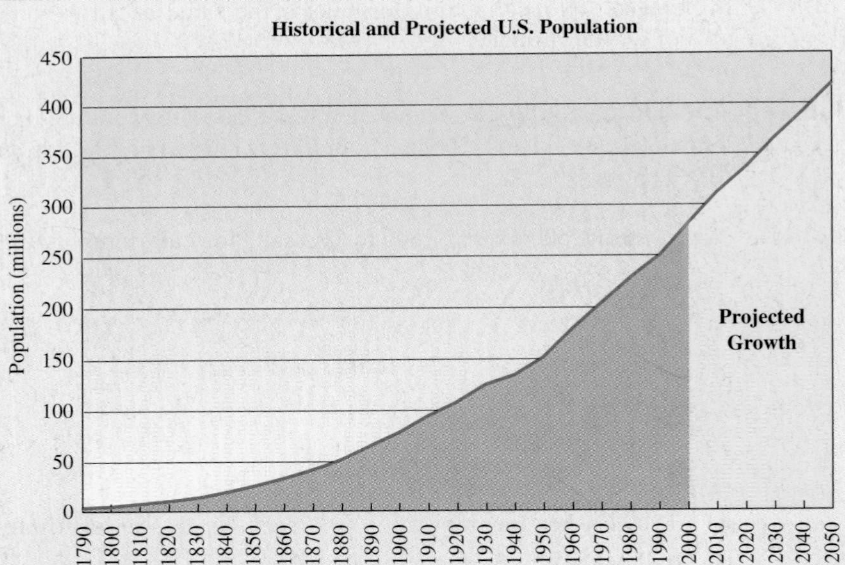

Source: http://txsdc.utsa.edu/txdata/apport/hist_a.php

Population growth in the United States can be modeled by an exponential function of the form $P(t) = P_0 \cdot e^{rt}$, where P_0 = the initial population during a given time interval, and t represents the number of years in that time interval.

1. Given that the United States population was approximately 150 million in 1950 and approximately 200 million in 1970, determine the growth rate of the population during this time period.
2. Given that the United States population was 200 million in 1970 and approximately 280 million in 2000, determine the growth rate of the population during this time period.
3. Using the population growth rate from question 2, to the nearest year how long would it take for the United States population to double?

$$\log B = \log(b \cdot 2^n) \quad \text{Take the common logarithm of both sides.}$$

$$\log B = \log b + n \log 2 \quad \text{Apply the product and power rules of logarithms.}$$

$$\log B - \log b = n \log 2 \quad \text{Subtract } \log b \text{ from both sides.}$$

$$n = \frac{1}{\log 2}(\log B - \log b) \quad \text{Multiply both sides by } \tfrac{1}{\log 2}.$$

$$(2) \qquad n = \frac{1}{\log 2}\left(\log \frac{B}{b}\right) \quad \text{Use the quotient rule of logarithms.}$$

Equation 2 is a formula that gives the number of generations that will pass as the population grows from b bacteria to B bacteria.

To find the number of generations that have passed while a population of 1,000 cells per milliliter has grown to a population of 1 million cells per milliliter, we substitute 1,000 for b and 1,000,000 for B in Equation 2 and solve for n.

$$n = \frac{1}{\log 2} \log \frac{1{,}000{,}000}{1{,}000}$$

$$= \frac{1}{\log 2} \log 1{,}000 \quad \text{Simplify.}$$

$$\approx 3.321928095(3) \quad \tfrac{1}{\log 2} \approx 3.321928095 \text{ and } \log 1{,}000 = 3.$$

$$\approx 9.965784285$$

Approximately 10 generations will have passed.

SELF CHECK ANSWERS

1. 0.8614 **2.** 60.1 **3.** 6.3013 **4.** 1, 1 **5.** 2 **6.** 2; -5 is extraneous **7.** 6; -1 is extraneous **8.** 8.00 days **9.** about 11,500 years **10.** about 10 hours

NOW TRY THIS

Given $f(x) = \log_3(x + 2) + \log_3 x$,

1. solve $f(x) = 1$.

2. solve $f(x) = 2$.

9.6 EXERCISES

WARM-UPS *Solve each equation for x.* Do not simplify answers.

1. $3^x = 5$

2. $5^x = 3$

3. $2^{-x} = 7$

4. $6^{-x} = 1$

5. $\log 2x = \log(x + 2)$

6. $\log 2x = 0$

7. $\log x^4 = 4$

8. $\log \sqrt{x} = \frac{1}{2}$

REVIEW *Solve each equation.*

9. $5x^2 - 25x = 0$

10. $4y^2 - 25 = 0$

11. $3p^2 + 10p = 8$

12. $4t^2 + 1 = -6t$

VOCABULARY AND CONCEPTS *Fill in the blanks.*

13. An equation with a variable as an exponent is called a(n) __________ equation.

14. An equation with a logarithmic expression that contains a variable is a(n) __________ equation.

15. The formula for carbon dating is $A =$ ______.

16. The ______ of a radioactive element is determined by how long it takes for half of a sample to decompose.

GUIDED PRACTICE

Solve each exponential equation. If an answer is not exact, give the answer to four decimal places. **See Examples 1–2. (Objective 1)**

17. $4^x = 5$

18. $7^x = 12$

19. $e^t = 50$

20. $e^{-t} = 0.25$

21. $2^x = 3^x$

22. $3^{2x} = 4^x$

23. $5 = 2.1(1.04)^t$

24. $61 = 1.5(1.02)^t$

Solve each exponential equation. If an answer is not exact, give the answer to four decimal places. **See Example 3. (Objective 1)**

25. $13^{x-1} = 2$

26. $5^{x+1} = 3$

27. $2^{x+1} = 3^x$

28. $5^{x-3} = 3^{2x}$

Solve each exponential equation. If an answer is not exact, give the answer to four decimal places. **See Example 4. (Objective 1)**

29. $2^{x^2-3x} = 16$

30. $3^{x^2-3x} = 81$

31. $3^{x^2+4x} = \frac{1}{81}$

32. $7^{x^2+3x} = \frac{1}{49}$

33. $7^{x^2} = 10$

34. $8^{x^2} = 11$

35. $8^{x^2} = 9^x$

36. $5^{x^2} = 2^{5x}$

Use a calculator to solve each equation, if possible. Give all answers to the nearest tenth. **(Objective 1)**

37. $2^{x+1} = 7$

38. $3^{x-1} = 2^x$

39. $2^{x^2-2x} - 8 = 0$

40. $3^x - 10 = 3^{-x}$

Solve each logarithmic equation. Check all solutions. **See Example 5. (Objective 2)**

41. $\log 2x = \log 4$

42. $\log 3x = \log 9$

43. $\log (3x + 1) = \log (x + 7)$

44. $\log (x^2 + 4x) = \log (x^2 + 16)$

45. $\log (3 - 2x) - \log (x + 24) = 0$

46. $\log (3x + 5) - \log (2x + 6) = 0$

47. $\log x^2 = 2$

48. $\log x^3 = 3$

Solve each logarithmic equation. Check all solutions. **See Example 6. (Objective 2)**

49. $\log x + \log (x - 48) = 2$

50. $\log x + \log (x + 9) = 1$

51. $\log x + \log (x - 15) = 2$

52. $\log x + \log (x + 21) = 2$

53. $\log (x + 90) = 3 - \log x$

54. $\log (x - 90) = 3 - \log x$

55. $\log (x - 6) - \log (x - 2) = \log \frac{5}{x}$

56. $\log (3 - 2x) - \log (x + 9) = 0$

Solve each logarithmic equation. Check all solutions. **See Example 7. (Objective 2)**

57. $\dfrac{\log (2x + 1)}{\log (x - 1)} = 2$

58. $\dfrac{\log (4x + 9)}{\log (2x - 3)} = 2$

59. $\dfrac{\log (3x + 4)}{\log x} = 2$

60. $\dfrac{\log (8x - 7)}{\log x} = 2$

Use a graphing calculator to solve each equation. If an answer is not exact, give all answers to the nearest tenth. **(Objective 2)**

61. $\log x + \log (x - 15) = 2$

62. $\log x + \log (x + 3) = 1$

63. $\ln(2x + 5) - \ln 3 = \ln(x - 1)$

64. $2 \log (x^2 + 4x) = 1$

ADDITIONAL PRACTICE *Solve each equation. If the answer is not exact, round to 4 decimal places.*

65. $4^{x+2} - 4^x = 15$ (*Hint:* $4^{x+2} = 4^x4^2$.)

66. $3^{x+3} + 3^x = 84$ (*Hint:* $3^{x+3} = 3^x3^3$.)

67. $\dfrac{\log (5x + 6)}{2} = \log x$

68. $\frac{1}{2}\log (4x + 5) = \log x$

69. $\log_3 x = \log_3 \left(\frac{1}{x}\right) + 4$

70. $\log_5(7 + x) + \log_5(8 - x) - \log_5 2 = 2$

71. $2(3^x) = 6^{2x}$

72. $2(3^{x+1}) = 3(2^{x-1})$

73. $\log x^2 = (\log x)^2$

74. $\log (\log x) = 1$

75. $2 \log_2 x = 3 + \log_2(x - 2)$

76. $2 \log_3 x - \log_3(x - 4) = 2 + \log_3 2$

77. $\log (7y + 1) = 2 \log (y + 3) - \log 2$

78. $2 \log (y + 2) = \log (y + 2) - \log 12$

79. $\log \dfrac{4x + 1}{2x + 9} = 0$

80. $\log \dfrac{2 - 5x}{2(x + 8)} = 0$

APPLICATIONS

Solve each application problem. See Examples 8–11. (Objective 3)

81. Half-life To the nearest day, find the half-life of strontium-89, given that it decays according to the formula $A = A_0e^{-0.013t}$.

82. Half-life To the nearest thousand years, find the half-life of plutonium-239, given that it decays according to the formula $A = A_0e^{-0.0000284t}$.

83. Radioactive decay In two years, 20% of a radioactive element decays. Find its half-life.

84. Tritium decay The half-life of tritium is 12.4 years. How long will it take for 25% of a sample of tritium to decompose?

85. Carbon-14 dating The bone fragment shown in the illustration contains 60% of the carbon-14 that it is assumed to have had initially. How old is it?

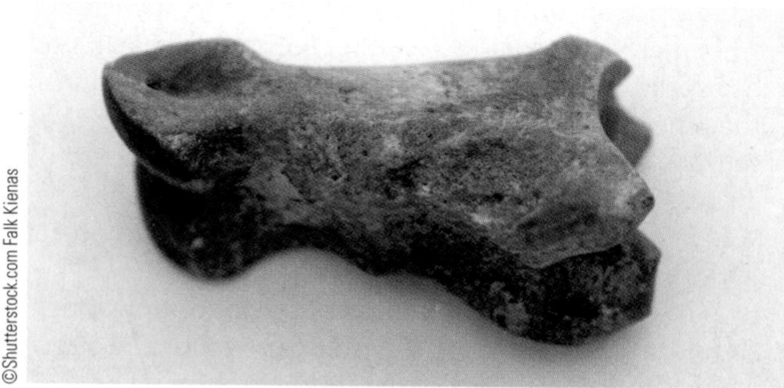

86. Carbon-14 dating Only 10% of the carbon-14 in a small wooden bowl remains. How old is the bowl?

87. Rodent control The rodent population in a city is currently estimated at 30,000. If it is expected to double every 5 years, when will the population reach 1 million?

88. Population growth The population of a city is expected to triple every 15 years. When can the city planners expect the present population of 140 persons to double?

89. Bacterial culture A bacterial culture doubles in size every 24 hours. By how much will it have increased in 36 hours?

90. Bacterial growth A bacterial culture grows according to the formula

$$P = P_0a^t$$

If it takes 5 days for the culture to triple in size, how long will it take to double in size?

91. Medicine If a medium is inoculated with a bacterial culture containing 500 cells per milliliter, how many generations will have passed by the time the culture contains 5×10^6 cells per milliliter?

92. Medicine If a medium is inoculated with a bacterial culture containing 800 cells per milliliter, how many generations will have passed by the time the culture contains 6×10^7 cells per milliliter?

Solve each application problem.

93. Thorium decay An isotope of thorium, ^{227}Th, has a half-life of 18.4 days. How long will it take for 80% of the sample to decompose?

94. Lead decay An isotope of lead, ^{201}Pb, has a half-life of 8.4 hours. How many hours ago was there 30% more of the substance?

95. Compound interest If \$500 is deposited in an account paying 8.5% annual interest, compounded semiannually, how long will it take for the account to increase to \$800?

96. Continuous compound interest In Exercise 95, how long will it take if the interest is compounded continuously?

97. Compound interest If \$1,300 is deposited in a savings account paying 9% interest, compounded quarterly, how long will it take the account to increase to \$2,100?

98. Compound interest A sum of \$5,000 deposited in an account grows to \$7,000 in 5 years. Assuming annual compounding, what interest rate is being paid?

99. Rule of seventy A rule of thumb for finding how long it takes an investment to double is called the **rule of seventy.** To apply the rule, divide 70 by the interest rate written as a percent. At 5%, it takes $\frac{70}{5} = 14$ years to double an investment. At 7%, it takes $\frac{70}{7} = 10$ years. Explain why this formula works.

100. Oceanography The intensity I of a light a distance x meters beneath the surface of a lake decreases exponentially. From the illustration, find the depth at which the intensity will be 20%.

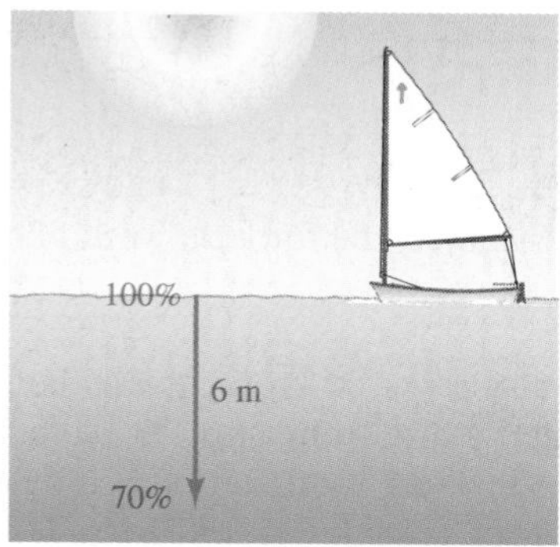

WRITING ABOUT MATH

101. Explain how to solve the equation $2^x = 7$.

102. Explain how to solve the equation $x^2 = 7$.

SOMETHING TO THINK ABOUT

103. Without solving the following equation, find the values of x that cannot be a solution:

$$\log(x - 3) - \log(x^2 + 2) = 0$$

104. Solve the equation $x^{\log x} = 10{,}000$.

PROJECTS

Project 1

When an object moves through air, it encounters air resistance. So far, all ballistics problems in this text have ignored air resistance. We now consider the case where an object's fall is affected by air resistance.

At relatively low velocities ($v < 200$ feet per second), the force resisting an object's motion is a constant multiple of the object's velocity:

$$\text{Resisting force} = f_r = bv$$

where b is a constant that depends on the size, shape, and texture of the object, and has units of kilograms per second. This is known as **Stokes' law of resistance.**

In a vacuum, the downward velocity of an object dropped with an initial velocity of 0 feet per second is

$$v(t) = 32t \quad \text{(no air resistance)}$$

t seconds after it is released. However, with air resistance, the velocity is given by the formula

$$v(t) = \frac{32m}{b}(1 - e^{-(b/m)t})$$

where m is the object's mass (in kilograms). There is also a formula for the distance an object falls (in feet) during the first t seconds after release, taking into account air resistance:

$$d(t) = \frac{32m}{b}t - \frac{32m^2}{b^2}(1 - e^{-(b/m)t})$$

Without air resistance, the formula would be

$$d(t) = 16t^2$$

a. Fearless Freda, a renowned skydiving daredevil, performs a practice dive from a hot-air balloon with an altitude of 5,000 feet. With her parachute on, Freda has a mass of 75 kg, so that $b = 15$ kg/sec. How far (to the nearest foot) will Freda fall in 5 seconds? Compare this with the answer you get by disregarding air resistance.

b. What downward velocity (to the nearest ft/sec) does Freda have after she has fallen for 2 seconds? For 5 seconds? Compare these answers with the answers you get by disregarding air resistance.

c. Find Freda's downward velocity after falling for 20, 22, and 25 seconds. (Without air resistance, Freda would hit the ground in less than 18 seconds.) Note that Freda's velocity increases only slightly. This is because for a large enough velocity, the force of air resistance almost counteracts the force of gravity; after Freda has been falling for a few seconds, her velocity becomes nearly constant. The constant velocity that a falling object approaches is called the *terminal velocity.*

$$\text{Terminal velocity} = \frac{32m}{b}$$

Find Freda's terminal velocity for her practice dive.

d. In Freda's show, she dives from a hot-air balloon with an altitude of only 550 feet, and pulls her ripcord when her velocity is 100 feet per second. (She can't tell her speed, but she knows how long it takes to reach that speed.) It takes a fall of 80 more feet for the chute to open fully, but then the chute increases the force of air resistance, making $b = 80$. After that, Freda's velocity approaches the terminal velocity of an object with this new b-value.

To the nearest hundredth of a second, how long should Freda fall before she pulls the ripcord? To the nearest foot, how close is she to the ground when she pulls the ripcord? How close to the ground is she when the chute takes full effect? At what velocity will Freda hit the ground?

Project 2

If an object at temperature T_0 is surrounded by a constant temperature T_s (for instance, an oven or a large amount of fluid that has a constant temperature), the temperature of the object will change with time t according to the formula

$$T(t) = T_s + (T_0 - T_s)e^{-kt}$$

This is **Newton's law of cooling and warming.** The number k is a constant that depends on how well the object absorbs and dispels heat.

In the course of brewing "yo ho! grog," the dread pirates of Hancock Isle have learned that it is important that their rather disgusting, soupy mash be heated slowly to allow all of the ingredients a chance to add their particular offensiveness to the mixture. However, after the mixture has simmered for several hours, it is equally important that the grog be cooled very quickly, so that it retains its potency. The kegs of grog are then stored in a cool spring.

By trial and error, the pirates have learned that by placing the mash pot into a tub of boiling water (100° C), they can heat the mash in the correct amount of time. They have also learned that they can cool the grog to the temperature of the spring by placing it in ice caves for 1 hour.

With a thermometer, you find that the pirates heat the mash from 20° C to 95° C and then cool the grog from 95° C to 7° C. Calculate how long the pirates cook the mash, and how cold the ice caves are. Assume that $k = 0.5$, and t is measured in hours.

Chapter 9 REVIEW

SECTION 9.1 Exponential Functions

DEFINITIONS AND CONCEPTS	EXAMPLES
An exponential function with base b is defined by the equation $f(x) = b^x \quad (b > 0, b \neq 1)$	Graph $f(x) = 4^x$ and $f(x) = \left(\frac{1}{4}\right)^x$. 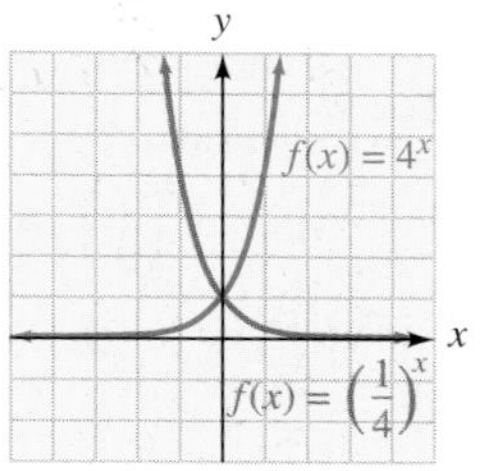 y; $f(x) = 4^x$; x; $f(x) = \left(\frac{1}{4}\right)^x$
Increasing/decreasing functions: The graph of $f(x) = b^x$ is an increasing function if $b > 1$ and decreasing if $0 < b < 1$.	The graph of $f(x) = 4^x$ is an increasing function because as x gets larger, $f(x)$ also gets larger. The graph of $f(x) = \left(\frac{1}{4}\right)^x$ is a decreasing function because as x gets larger, $f(x)$ gets smaller.
Graphing translations: Graphing a translation of an exponential function follows the same rules as for polynomial functions.	Graph $f(x) = 4^x - 2$ and $f(x) = 4^{x-2}$. The graph of $f(x) = 4^x - 2$ is the same as the graph of $f(x) = 4^x$ but shifted 2 units down. The graph of $f(x) = 4^{x-2}$ is the same as the graph of $f(x) = 4^x$ but shifted 2 units to the right. 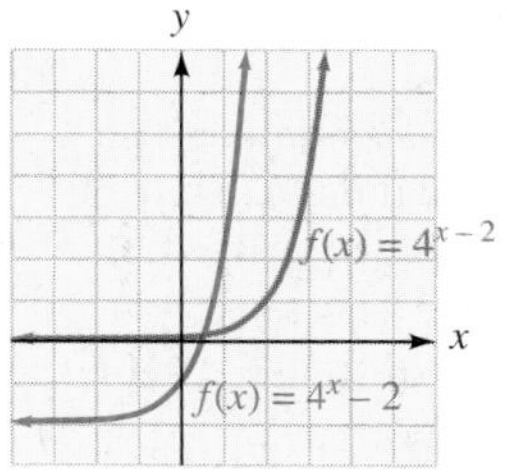 y; $f(x) = 4^{x-2}$; x; $f(x) = 4^x - 2$

Compound interest:
The amount of money in an account with an initial deposit of $\$P$ at an annual interest rate of $r\%$, compounded k times a year for t years, can be found using the formula $A = P\left(1 + \frac{r}{k}\right)^{kt}$.

To find the balance in an account after 3 years when \$5,000 is deposited at 6% interest, compounded quarterly, we note that

$$P = 5{,}000 \qquad r = 0.06 \qquad k = 4 \qquad t = 3$$

We can substitute these values into the formula for compound interest and proceed as follows:

$$A = P\left(1 + \frac{r}{k}\right)^{kt}$$

$$A = 5{,}000\left(1 + \frac{0.06}{4}\right)^{4(3)}$$

$$A = 5{,}000(1 + 0.015)^{12}$$

$$A = 5{,}000(1.015)^{12}$$

$$A \approx 5{,}978.090857 \qquad \text{Use a calculator.}$$

There will be \$5,978.09 in the account after 3 years.

REVIEW EXERCISES

Use properties of exponents to simplify.

1. $5^{\sqrt{2}} \cdot 5^{\sqrt{2}}$
2. $\left(2^{\sqrt{5}}\right)^{\sqrt{2}}$

Graph the function defined by each equation.

3. $y = 3^x$
4. $y = \left(\frac{1}{3}\right)^x$

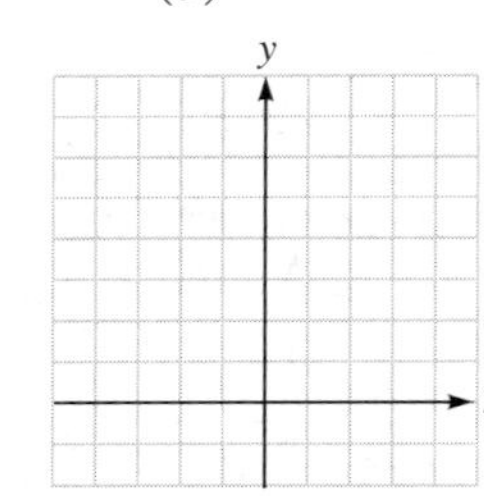

5. The graph of $f(x) = 6^x$ will pass through the points $(0, x)$ and $(1, y)$. Find x and y.

6. Give the domain and range of the function $f(x) = b^x$.

Graph each function by using a translation.

7. $f(x) = \left(\frac{1}{2}\right)^x - 2$

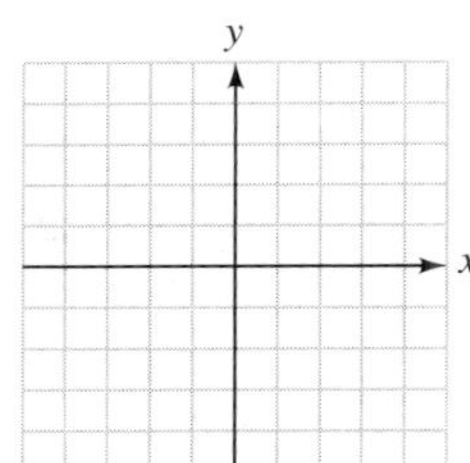

8. 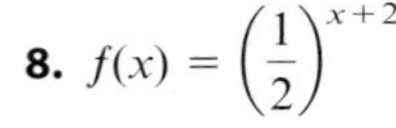 $f(x) = \left(\frac{1}{2}\right)^{x+2}$

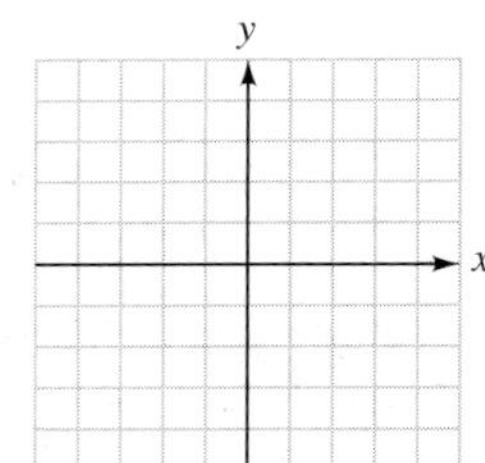

9. **Savings** How much will \$10,500 become if it earns 9% per year for 60 years, compounded quarterly?

SECTION 9.2 Base-*e* Exponential Functions

DEFINITIONS AND CONCEPTS

$e \approx 2.71828182845904$

Continuous compound interest:
The amount of money A in an account with an initial deposit of $\$P$ at an annual interest rate of $r\%$, compounded continuously for t years, can be found using the formula

$$A = Pe^{rt}$$

EXAMPLES

To find the balance in an account after 3 years when \$5,000 is deposited at 6% interest, compounded continuously, we note that

$$P = 5{,}000 \qquad r = 0.06 \qquad t = 3$$

We can substitute these values into the formula $A = Pe^{rt}$ and simplify to obtain

$$A = Pe^{rt}$$

$$A = 5{,}000e^{0.06(3)}$$

$$A = 5{,}000e^{0.18}$$

$$A \approx 5{,}986.09 \qquad \text{Use a calculator.}$$

There will be \$5,986.08 in the account.

Malthusian population growth: The same formula for continuous compound interest is used for population growth, where A is the new population, P is the previous population, r is the growth rate, and t is the number of years. $A = Pe^{rt}$	To find the population of a town in 5 years with a current population of 4,000 and a growth rate of 1.5%, we note that $P = 4{,}000 \quad r = 0.015 \quad t = 5$. We can substitute these values into the formula $A = Pe^{rt}$ to obtain $A = Pe^{rt}$; $A = 4{,}000e^{0.015(5)}$; $A = 4{,}000e^{0.075}$; $A \approx 4{,}311.54$ Use a calculator. There will be about 4,312 people in 5 years.
Radioactive decay: The formula $A = A_0e^{kt}$ can be used to determine the amount of material left (A) when an initial amount (A_0) has been decaying for t years at a rate of k. The value of k in this type of problem will be negative.	The radioactive material radon-22 decays with a rate of $k = -0.181t$, where t is in days. To find how much radon-22 will be left if a sample of 100 grams decays for 30 days, we can substitute the values into the radioactive decay formula and proceed as follows: $A = A_0e^{kt}$; $A = 100e^{-0.181(30)}$; $A = 100e^{-5.43}$; $A \approx .4383095803$ Use a calculator. There will be approximately 0.44 grams remaining.

REVIEW EXERCISES

10. If \$10,500 accumulates interest at an annual rate of 9%, compounded continuously, how much will be in the account in 60 years?

Graph each function.

11. $f(x) = e^x + 1$

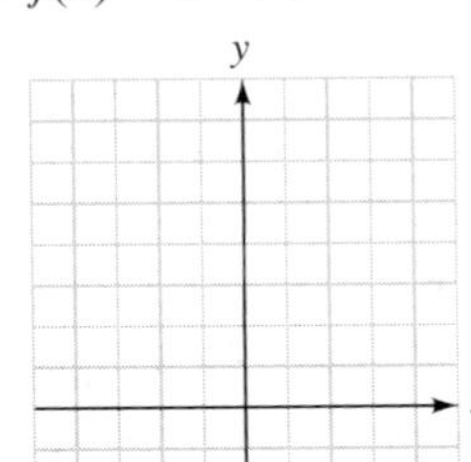

12. $f(x) = e^{x-3}$

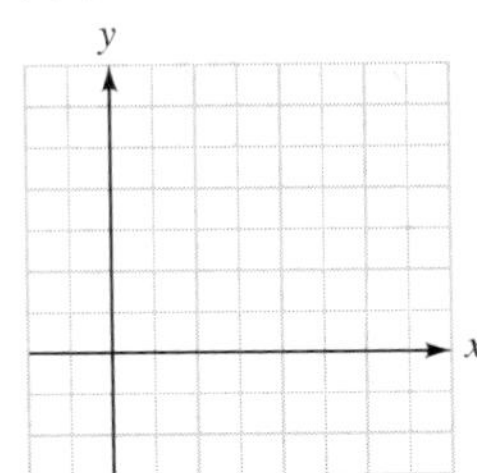

13. **U.S. population** The population of the United States is approximately 275,000,000 people. Find the population in 50 years if $k = 0.015$.

14. **Radioactive decay** The radioactive material strontium-90 decays according to the formula $A = A_0e^{-0.0244t}$, where t is expressed in years. To the nearest hundredth, how much of the material will remain if a sample of 50 grams decays for 20 years?

SECTION 9.3 Logarithmic Functions

DEFINITIONS AND CONCEPTS	EXAMPLES
If $b > 0$ and $b \neq 1$, then $y = \log_b x$ means $x = b^y$	$y = \log_3 27$ is equivalent to $27 = 3^y$. Since $\log_4 x = 2$ is equivalent to $x = 4^2$, we have $x = 16$. Since $\log_2 32 = x$ is equivalent to $32 = 2^x$ or $2^5 = 2^x$, we have $x = 5$. Since $\log_x 64 = 3$ is equivalent to $64 = x^3$ or $4^3 = x^3$, we have $x = 4$.

REVIEW EXERCISES

15. Give the domain and range of the logarithmic function $y = \log_b x$.

16. Explain why the functions $y = b^x$ and $y = \log_b x$ are called inverse functions.

Find each value.

17. $\log_3 9$

18. $\log_9 \frac{1}{3}$

19. $\log_\pi 1$

20. $\log_5 0.04$

21. $\log_a \sqrt{a}$

22. $\log_a \sqrt[3]{a}$

Find the value of x.

23. $\log_2 x = 5$

24. $\log_{\sqrt{3}} x = 4$

25. $\log_{\sqrt{3}} x = 6$

26. $\log_{0.1} 10 = x$

27. $\log_x 2 = -\frac{1}{3}$

28. $\log_x 32 = 5$

29. $\log_{0.25} x = -1$

30. $\log_{0.125} x = -\frac{1}{3}$

31. $\log_{\sqrt{2}} 32 = x$

32. $\log_{\sqrt{5}} x = -4$

33. $\log_{\sqrt{3}} 9\sqrt{3} = x$

34. $\log_{\sqrt{5}} 5\sqrt{5} = x$

Graph each function.

35. $f(x) = \log(x - 2)$

36. $f(x) = 3 + \log x$

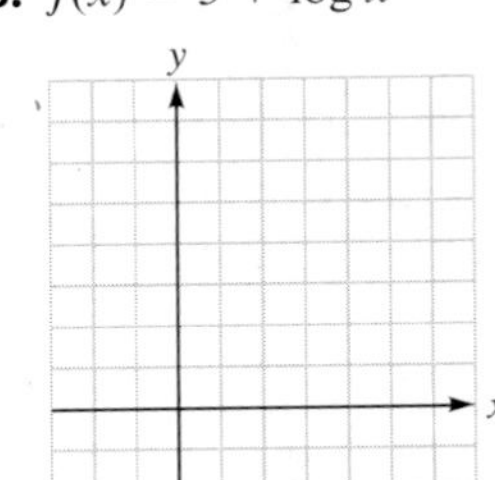

Graph each pair of equations on one set of coordinate axes.

37. $y = 4^x$ and $y = \log_4 x$

38. $y = \left(\frac{1}{3}\right)^x$ and $y = \log_{1/3} x$

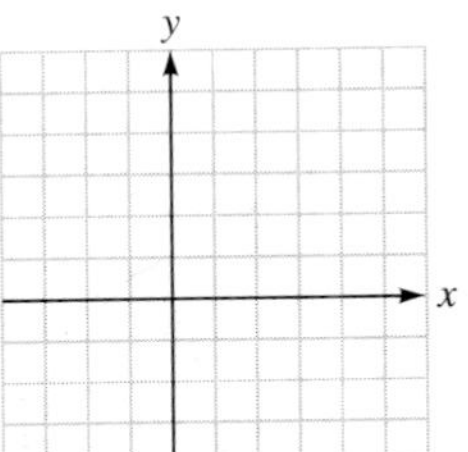

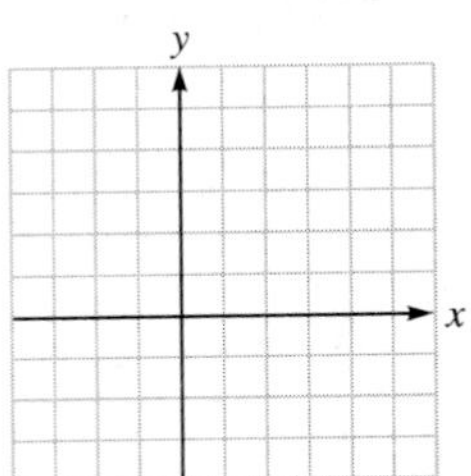

39. dB gain An amplifier has an output of 18 volts when the input is 0.04 volt. Find the dB gain.

40. Earthquakes An earthquake had a period of 0.3 second and an amplitude of 7,500 micrometers. Find its measure on the Richter scale.

SECTION 9.4 Natural Logarithms

DEFINITIONS AND CONCEPTS	EXAMPLES
$\ln x$ means $\log_e x$.	Use a calculator to evaluate ln 5.89. $\ln 5.89 \approx 1.7732$
Population doubling time: $A = A_0e^{kt}$ or $t = \frac{\ln 2}{r}$	To find the time it takes for the population of a colony to double if the growth rate is 1.5% per year, substitute 0.015 into the formula for doubling time: $t = \frac{\ln 2}{r}$ $t = \frac{\ln 2}{0.015}$ $t \approx 46.2098$ Use a calculator. The colony will double in population in about 46 years.

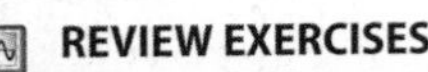

REVIEW EXERCISES

Use a calculator to find each value to four decimal places.

41. $\ln 452$

42. $\ln(\log 7.85)$

Find x.

43. $\ln x = 2.336$

44. $\ln x = \log 8.8$

Graph each function.

45. $f(x) = 1 + \ln x$

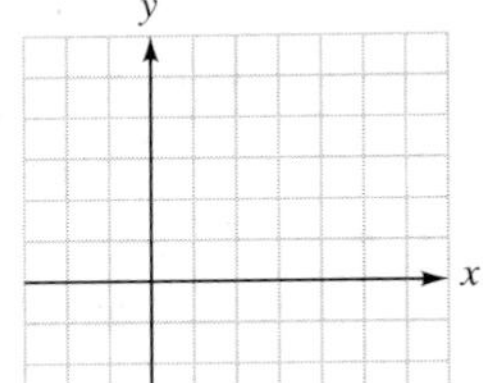

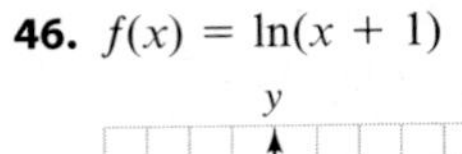

46. $f(x) = \ln(x + 1)$

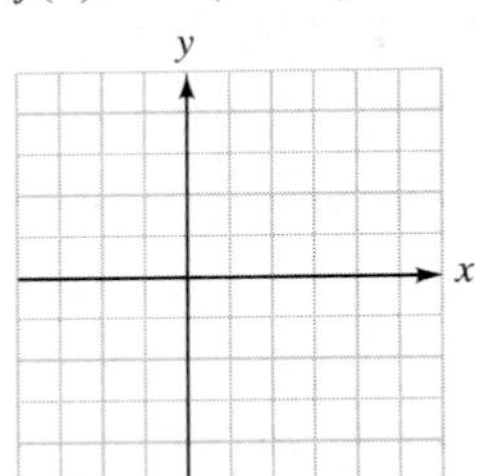

47. U.S. population How long will it take the population of the United States to double if the growth rate is 3% per year?

SECTION 9.5 Properties of Logarithms

DEFINITIONS AND CONCEPTS	EXAMPLES
Properties of logarithms: If b is a positive number and $b \neq 1$, **1.** $\log_b 1 = 0$ **2.** $\log_b b = 1$ **3.** $\log_b b^x = x$ **4.** $b^{\log_b x} = x$ **5.** $\log_b MN = \log_b M + \log_b N$ **6.** $\log_b \frac{M}{N} = \log_b M - \log_b N$ **7.** $\log_b M^p = p \log_b M$ **8.** If $\log_b x = \log_b y$, then $x = y$.	**1.** $\log_b 1 = 0$ because $b^0 = 1$. **2.** $\log_b b = 1$ because $b^1 = b$. **3.** $\log_b b^x = x$ because $b^x = b^x$. **4.** $b^{\log_b x} = x$ because $b^y = x$ and $y = \log_b x$. **5.** $\log_5 xy = \log_5 x + \log_5 y$ **6.** $\log_5 \frac{x}{y} = \log_5 x - \log_5 y$ **7.** $\log_5 x^3 = 3 \log_5 x$ **8.** If $\log_3 x = \log_3 9$, then $x = 9$.
Change-of-base formula: $\log_b y = \frac{\log_a y}{\log_a b}$	To evaluate $\log_7 16$, substitute 7 and 16 into the change-of-base formula. $\log_7 \mathbf{16} = \frac{\log_{10} \mathbf{16}}{\log_{10} \mathbf{7}} \approx 1.4248$ Use a calculator. or $\log_7 \mathbf{16} = \frac{\ln \mathbf{16}}{\ln \mathbf{7}} \approx 1.4248$ Use a calculator.

REVIEW EXERCISES

Simplify each expression.

48. $\log_7 1$

49. $\log_7 7$

50. $\log_7 7^3$

51. $7^{\log_7 4}$

Simplify each expression.

52. $\ln e^4$

53. $\ln 1$

54. $10^{\log_{10} 7}$

55. $e^{\ln 3}$

56. $\log_b b^4$

57. $\ln e^9$

Write each expression in terms of the logarithms of x, y, and z.

58. $\log_b \frac{x^2y^3}{z^4}$

59. $\log_b \sqrt{\frac{x}{yz^2}}$

Write each expression as the logarithm of one quantity.

60. $3 \log_b x - 5 \log_b y + 7 \log_b z$

61. $\frac{1}{2} \log_b x + 3 \log_b y - 7 \log_b z$

Assume that $\log a = 0.6$, $\log b = 0.36$, and $\log c = 2.4$. Find the value of each expression.

62. $\log abc$

63. $\log a^2b$

64. $\log \frac{ac}{b}$

65. $\log \frac{a^2}{c^3b^2}$

66. To four decimal places, find $\log_5 17$.

67. pH of grapefruit The pH of grapefruit juice is about 3.1. Find its hydrogen ion concentration.

68. Find the decrease in loudness if the intensity is cut in half.

SECTION 9.6 Exponential and Logarithmic Equations

DEFINITIONS AND CONCEPTS

Solving exponential and logarithmic equations:
If the bases of an exponential equation are the same, set the exponents equal.

EXAMPLES

1. Solve each equation.

a.

$$9^x = 3^{x-1}$$
$$(3^2)^x = 3^{x-1}$$
$$3^{2x} = 3^{x-1}$$
$$2x = x - 1$$
$$x = -1$$

If the bases of an exponential equation are not the same, take the logarithms of both sides. Then use the properties of logarithms to solve for the variable.

b.

$9^x = 5$

$\ln(9^x) = \ln 5$ Take the natural logarithm of both sides.

$x \ln 9 = \ln 5$ The logarithm of a power is the power times the logarithm.

$x = \dfrac{\ln 5}{\ln 9}$ Divide both sides by ln 9.

$x \approx .7325$ Use a calculator.

Use the properties of logarithms to combine multiple logarithms into a single logarithm.

c. $\log_2(x + 1) + \log_2(x - 1) = 3$

$\log_2[(x + 1)(x - 1)] = 3$ Write as a single logarithm.

$\log_2(x^2 - 1) = 3$ Multiply inside the brackets.

$2^3 = x^2 - 1$ Write as an exponential expression.

$8 = x^2 - 1$ Simplify.

$9 = x^2$ Add 1 to both sides.

$x = 3, -3$ Take the square root of both sides.

Since 3 checks, it is a solution. Since −3 does not check, it is extraneous.

REVIEW EXERCISES

Solve each equation for x. If an answer is not exact, round to four decimal places.

69. $3^x = 7$

70. $5^{x+2} = 625$

71. $25 = 5.5(1.05)^t$

72. $4^{2t-1} = 64$

73. $2^x = 3^{x-1}$

74. $2^{x^2+4x} = \dfrac{1}{8}$

Solve each equation for x.

75. $\log x + \log(29 - x) = 2$

76. $\log_2 x + \log_2(x - 2) = 3$

77. $\log_2(x + 2) + \log_2(x - 1) = 2$

78. $\dfrac{\log(7x - 12)}{\log x} = 2$

79. $\log x + \log(x - 5) = \log 6$

80. $\log 3 - \log(x - 1) = -1$

81. $e^{x \ln 2} = 9$

82. $\ln x = \ln(x - 1)$

83. $\ln x = \ln(x - 1) + 1$

84. $\ln x = \log_{10} x$ (*Hint:* Use the change-of-base formula.)

85. Carbon-14 dating A wooden statue found in Egypt has a carbon-14 content that is two-thirds of that found in living wood. If the half-life of carbon-14 is 5,730 years, how old is the statue?

Chapter 9 TEST

Graph each function.

1. $f(x) = 2^x + 1$

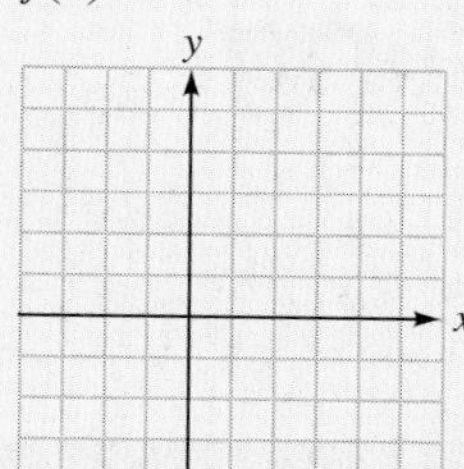

2. $f(x) = 2^{-x}$

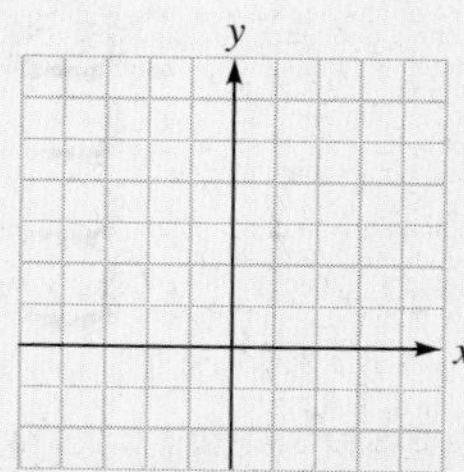

Solve each equation.

3. **Radioactive decay** A radioactive material decays according to the formula $A = A_0(2)^{-t}$. How much of a 3-gram sample will be left in 6 years?
4. **Investing** An initial deposit of \$1,000 earns 6% interest, compounded twice a year. How much will be in the account in one year?
5. Graph the function $f(x) = e^x$.

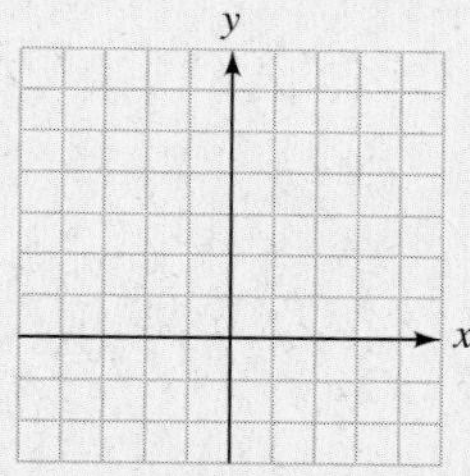

6. **Investing** An account contains \$2,000 and has been earning 8% interest, compounded continuously. How much will be in the account in 10 years?

Find the value of x.

7. $\log_4 16 = x$
8. $\log_x 81 = 4$
9. $\log_3 x = -3$
10. $\log_x 100 = 2$
11. $\log_{3/2} \frac{9}{4} = x$
12. $\log_{2/3} x = -3$

Graph each function.

13. $f(x) = -\log_3 x$

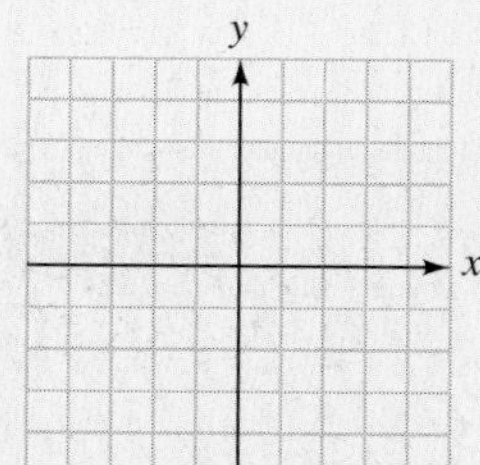

14. $f(x) = \ln x$

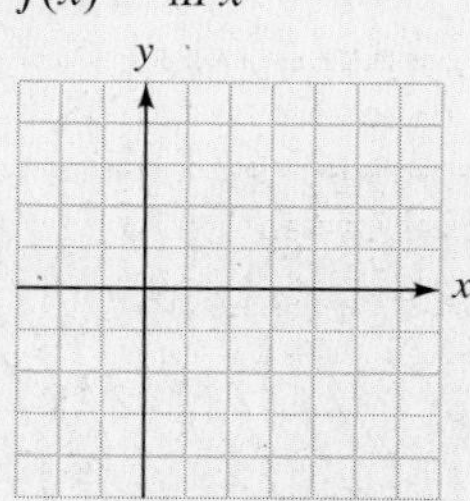

Write each expression in terms of the logarithms of a, b, and c.

15. $\log a^2bc^3$
16. $\ln\sqrt{\frac{a}{b^2c}}$

Write each expression as a logarithm of a single quantity.

17. $\frac{1}{2}\log(a + 2) + \log b - 3 \log c$
18. $\frac{1}{3}(\log a - 2 \log b) - \log c$

Assume that $\log 2 \approx 0.3010$ and $\log 3 \approx 0.4771$. Find each value. Do not use a calculator.

19. $\log 24$
20. $\log \frac{8}{3}$

Use the change-of-base formula to find each logarithm. Do not simplify the answer.

21. $\log_7 3$
22. $\log_\pi e$

Determine whether each statement is true. If a statement is not true, explain why.

23. $\log_a ab = 1 + \log_a b$
24. $\frac{\log a}{\log b} = \log a - \log b$
25. $\log a^{-3} = \frac{1}{3 \log a}$
26. $\ln(-x) = -\ln x$
27. Find the pH of a solution with a hydrogen ion concentration of 3.7×10^{-7}. (*Hint:* $\text{pH} = -\log[\text{H}^+]$.)
28. Find the dB gain of an amplifier when $E_O = 60$ volts and $E_I = 0.3$ volt. (*Hint:* dB gain $= 20 \log(E_O/E_I)$.)

Solve each equation. Do not simplify the logarithms.

29. $5^x = 3$
30. $3^{x-1} = 100^x$

Solve each equation.

31. $\log(5x + 2) = \log(2x + 5)$
32. $\log x + \log(x - 9) = 1$

CHAPTER 10

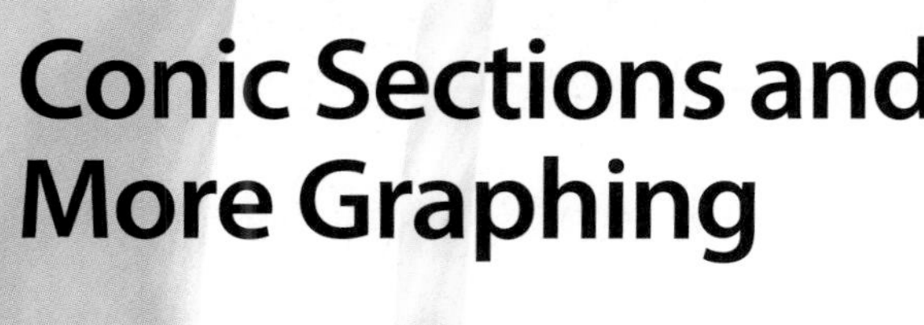

Conic Sections and More Graphing

10.1 The Circle and the Parabola

10.2 The Ellipse

10.3 The Hyperbola

10.4 Systems Containing Second-Degree Equations

10.5 Piecewise-Defined Functions and the Greatest Integer Function

■ ***Projects***

CHAPTER REVIEW

CHAPTER TEST

CUMULATIVE REVIEW EXERCISES

Careers and Mathematics

WATER TRANSPORTATION OCCUPATIONS

The movement of cargo and passengers between nations depends on workers in the water transportation occupations, also known on commercial ships as merchant mariners. They operate and maintain deep-sea merchant ships, tugboats, ferries, and excursion vessels.

Water transportation workers held more than 84,000 jobs in 2006. About 17 percent worked in inland waters, primarily the Mississippi River system, 23 percent worked on the deep seas and the Great Lakes, and another 24 percent were employed in harbor operations.

Job Outlook:
Employment in this industry is projected to grow 16 percent over the 2006–2016 decade. This is faster than average for all occupations.

Annual Earnings:
\$30,630–\$54,280

For More Information:
http://www.bls.gov/oco/ocos247.htm

For a Sample Application:
See Problem 38 in Section 10.3.

In this chapter ▶

We have seen that the graphs of linear functions are straight lines, and that the graphs of quadratic functions are parabolas. In this chapter, we will discuss some special curves, called conic sections. Then we will discuss piecewise-defined functions and step functions.

SECTION 10.1 The Circle and the Parabola

Objectives

1. Find the center and radius of a circle given an equation in standard and general form.
2. Write an equation of a circle in general form given the center and the radius.
3. Solve an application problem involving a circle.
4. Graph a parabola of the form $x = (y - k)^2 + h$.
5. Solve an application problem involving a parabola.

Vocabulary

conic section
circle
center of a circle
radius of a circle
standard form of the equation of a circle
point circle
general form of the equation of a circle

Getting Ready

Square each binomial.

1. $(x - 2)^2$ **2.** $(x + 4)^2$

What number must be added to each binomial to make it a perfect square trinomial?

3. $x^2 + 9x$ **4.** $x^2 - 12x$

In this chapter, we will introduce a group of curves, called *conic sections.*

The graphs of second-degree equations in x and y represent figures that were investigated in the 17th century by René Descartes (1596–1650) and Blaise Pascal (1623–1662). Descartes discovered that graphs of second-degree equations fall into one of several categories: a pair of lines, a point, a circle, a parabola, an ellipse, a hyperbola, or no graph at all. Because all of these graphs can be formed by the intersection of a plane and a right-circular cone, they are called **conic sections.** See Figure 10-1 on the next page.

In this first section, we will discuss the circle and extend the discussion of parabolas. In Section 10.2, we will discuss the ellipse, and in Section 10.3, we will discuss the hyperbola.

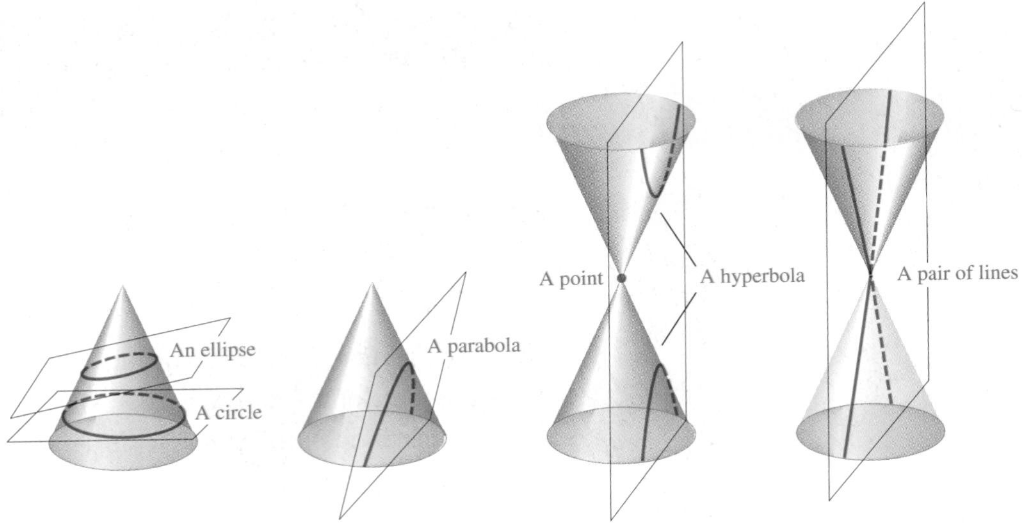

Figure 10-1

1 Find the center and radius of a circle given an equation in standard and general form.

A *circle* is one of the most common of the conic sections. Everyone knows about circular wheels and gears, pizza cutters, and Ferris wheels. Because of their importance, we will begin the study of conic sections with the circle.

Every conic section can be represented by a second-degree equation in x and y. To find the form of an equation of a circle, we use the following definition.

The Circle

A **circle** is the set of all points in a plane that are a fixed distance from a point, called its **center.**

The fixed distance is the **radius** of the circle.

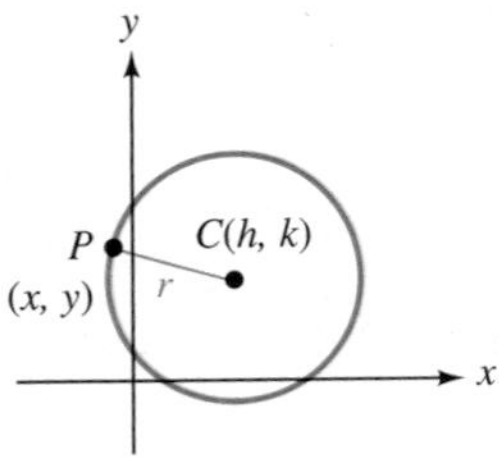

Figure 10-2

To develop the general equation of a circle, we must write the equation of a circle with a radius of r and with center at some point $C(h, k)$, as in Figure 10-2. This task is equivalent to finding all points $P(x, y)$ such that the length of line segment CP is r. We can use the distance formula to find r.

$$r = \sqrt{(x - h)^2 + (y - k)^2}$$

We then square both sides to obtain

$$(1) \qquad r^2 = (x - h)^2 + (y - k)^2$$

Equation 1 is called the **standard form of the equation of a circle** with radius r and center at the point with coordinates (h, k).

Standard Form of the Equation of a Circle with Center at (h, k)

Any equation that can be written in the form

$$(x - h)^2 + (y - k)^2 = r^2$$

has a graph that is a circle with radius r and center at point (h, k).

If $r = 0$, the graph reduces to a single point called a **point circle.** If $r^2 < 0$, a circle does not exist. If both coordinates of the center are 0, the center of the circle is the origin.

Standard Form of the Equation of a Circle with Center at (0, 0)

Any equation that can be written in the form

$$x^2 + y^2 = r^2$$

has a graph that is a circle with radius r and center at the origin.

EXAMPLE 1 Find the center and the radius of each circle and then graph it.

a. $(x - 4)^2 + (y - 1)^2 = 9$ **b.** $x^2 + y^2 = 25$ **c.** $(x + 3)^2 + y^2 = 12$

Solution **a.** It is easy to determine the center and the radius of a circle when its equation is written in standard form.

$$(x - 4)^2 + (y - 1)^2 = 9$$
$$\uparrow \quad \uparrow \quad \uparrow$$
$$(x - h)^2 + (y - k)^2 = r^2$$

$h = 4$, $k = 1$, and $r^2 = 9$. Since the radius of a circle must be positive, $r = 3$.

The center of the circle is $(h, k) = (4, 1)$ and the radius is 3.

To plot four points on the circle, we move up, down, left, and right 3 units from the center, as shown in Figure 10-3(a). Then we draw a circle through the points to get the graph of $(x - 4)^2 + (y - 1)^2 = 9$, as shown in Figure 10-3(b).

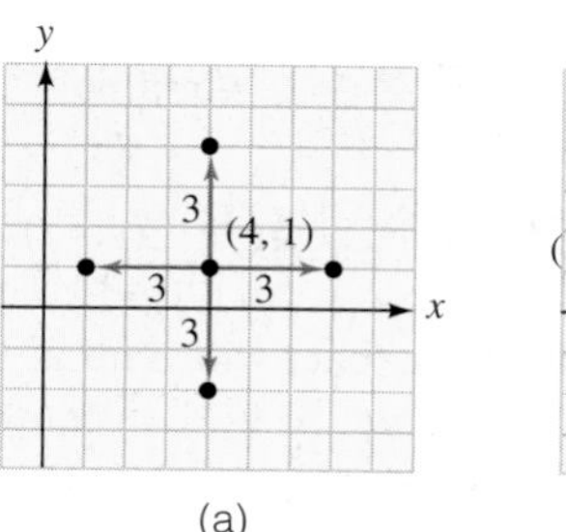

(a)

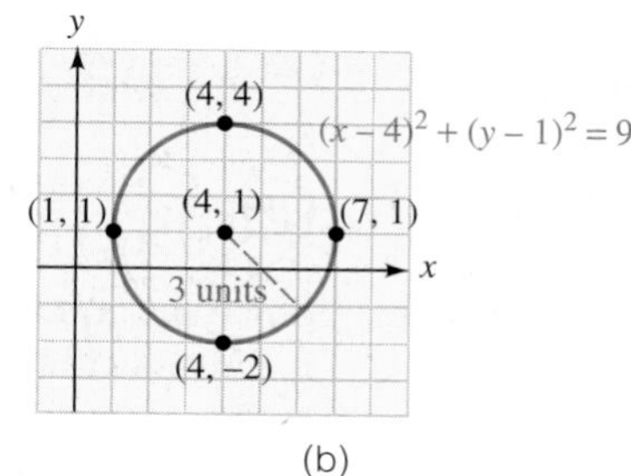

(b)

Figure 10-3

b. To verify that the center of the circle is the origin, we can write $x^2 + y^2 = 25$ in the following way:

$$(x - 0)^2 + (y - 0)^2 = 25$$
$$\uparrow \quad \uparrow \quad \uparrow$$
$$h \quad k \quad r^2$$

$h = 0$, $k = 0$, and $r^2 = 25$. Since the radius of a circle must be positive, $r = 5$.

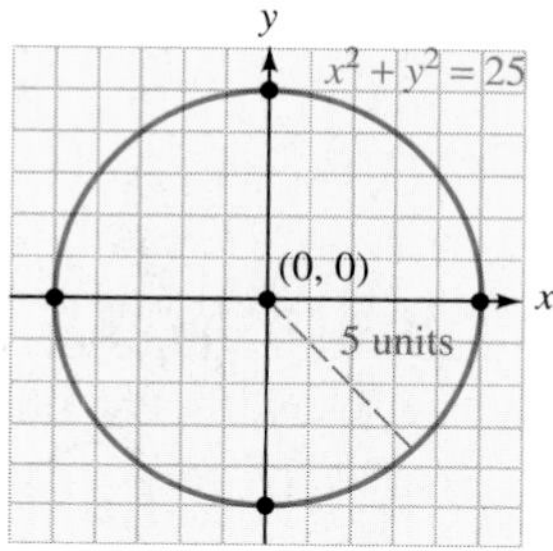

Figure 10-4

This shows that the center of the circle is at (0, 0) and the radius is 5.

To plot four points on the circle, we move up, down, left, and right 5 units from the center. Then we draw a circle through the points to get the graph of $x^2 + y^2 = 25$, as shown in Figure 10-4.

c. To determine h in the equation $(x + 3)^2 + y^2 = 12$, it is helpful to write $x + 3$ as $x - (-3)$.

Standard form requires a minus symbol here.

$$[x - (-3)]^2 + (y - 0)^2 = 12$$

h k r^2 $\quad h = -3, k = 0, \text{ and } r^2 = 12.$

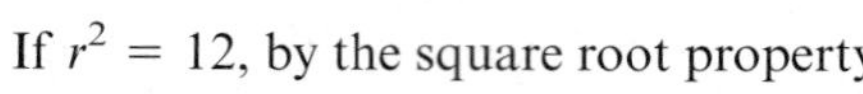

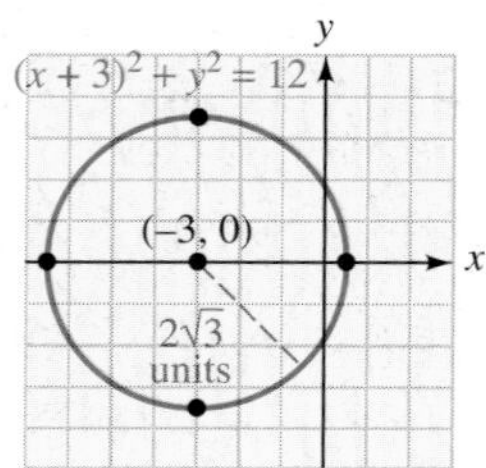

Figure 10-5

If $r^2 = 12$, by the square root property

$$r = \pm\sqrt{12} = \pm 2\sqrt{3}$$

Since the radius can't be negative, we get $r = 2\sqrt{3}$. The center of the circle is at $(-3, 0)$ and the radius is $2\sqrt{3}$.

To plot four points on the circle, we move up, down, left, and right $2\sqrt{3} \approx 3.5$ units from the center. Then we draw a circle through the points to get the graph of $(x + 3)^2 + y^2 = 12$, as shown in Figure 10-5.

SELF CHECK 1 Find the center and the radius of each circle and then graph it.

a. $(x - 3)^2 + (y + 4)^2 = 4$ **b.** $x^2 + y^2 = 8$

Another important form of the equation of a circle is called the *general form of the equation of a circle.*

General Form of the Equation of a Circle

The equation of any circle can be written in the form

$$x^2 + y^2 + Dx + Ey + F = 0$$

EXAMPLE 2 Graph: $x^2 + y^2 - 4x + 2y - 20 = 0$.

Solution Since the equation matches the general form of the equation of a circle, we know that its graph will be a circle. To find its center and radius, we must complete the square on both x and y and write the equation in standard form.

$$x^2 + y^2 - 4x + 2y = 20 \quad \text{Add 20 to both sides.}$$
$$x^2 - 4x + y^2 + 2y = 20$$

To complete the square on x and y, add 4 and 1 to both sides.

$$x^2 - 4x + 4 + y^2 + 2y + 1 = 20 + 4 + 1$$
$$(x - 2)^2 + (y + 1)^2 = 25 \quad \text{Factor } x^2 - 4x + 4 \text{ and } y^2 + 2y + 1.$$
$$(x - 2)^2 + [y - (-1)]^2 = 5^2$$

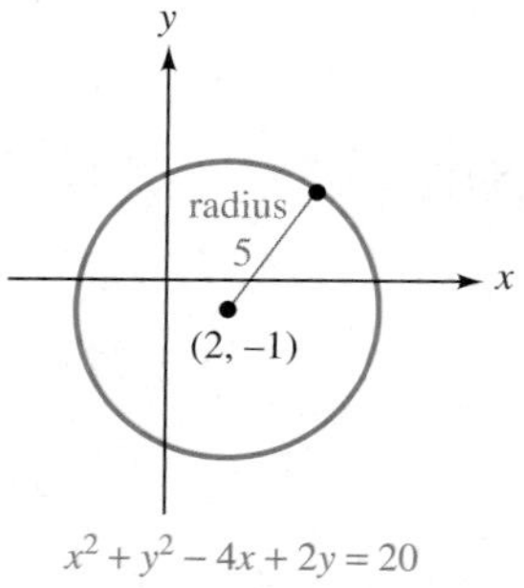

We can now see that this result is the standard equation of a circle with a radius of 5 and center at $h = 2$ and $k = -1$. If we plot the center and draw a circle with a radius of 5 units, we will obtain the circle shown in Figure 10-6.

y
x
radius 5
(2, –1)
$x^2 + y^2 - 4x + 2y = 20$

Figure 10-6

SELF CHECK 2 Write the equation $x^2 + y^2 + 2x - 4y - 11 = 0$ in standard form and graph it.

2 Write an equation of a circle in general form given the center and the radius.

EXAMPLE 3 Find the general form of the equation of the circle with radius 5 and center at (3, 2).

Solution We substitute 5 for r, 3 for h, and 2 for k in the standard form of a circle and proceed as follows:

$$(x - h)^2 + (y - k)^2 = r^2$$
$$(x - 3)^2 + (y - 2)^2 = 5^2$$
$$x^2 - 6x + 9 + y^2 - 4y + 4 = 25 \quad (x - 3)^2 = x^2 - 6x + 9;\ (y - 2)^2 = y^2 - 4y + 4$$
$$x^2 + y^2 - 6x - 4y - 12 = 0 \quad \text{Subtract 25 from both sides and simplify.}$$

The general form of the equation is $x^2 + y^2 - 6x - 4y - 12 = 0$.

SELF CHECK 3 Find the general form of the equation of the circle with radius 6 and center at (2, 3).

ACCENT ON TECHNOLOGY

Graphing Circles

Since the graphs of circles fail the vertical line test, their equations do not represent functions. It is somewhat more difficult to use a graphing calculator to graph equations that are not functions. For example, to graph the circle described by $(x - 1)^2 + (y - 2)^2 = 4$, we must split the equation into two functions and graph each one separately. We begin by solving the equation for y.

$$(x - 1)^2 + (y - 2)^2 = 4$$
$$(y - 2)^2 = 4 - (x - 1)^2 \quad \text{Subtract } (x - 1)^2 \text{ from both sides.}$$
$$y - 2 = \pm\sqrt{4 - (x - 1)^2} \quad \text{Use the square-root property.}$$
$$y = 2 \pm \sqrt{4 - (x - 1)^2} \quad \text{Add 2 to both sides.}$$

This equation defines two functions. If we use window settings of $[-3, 5]$ for x and $[-3, 5]$ for y and graph the functions

$$y = 2 + \sqrt{4 - (x - 1)^2} \quad \text{and} \quad y = 2 - \sqrt{4 - (x - 1)^2}$$

we get the distorted circle shown in Figure 10-7(a). To get a better circle, graphing calculators have a squaring feature, ZSquare, that gives an equal unit distance on both the x- and y-axes. After using this feature, we get the circle shown in Figure 10-7(b).

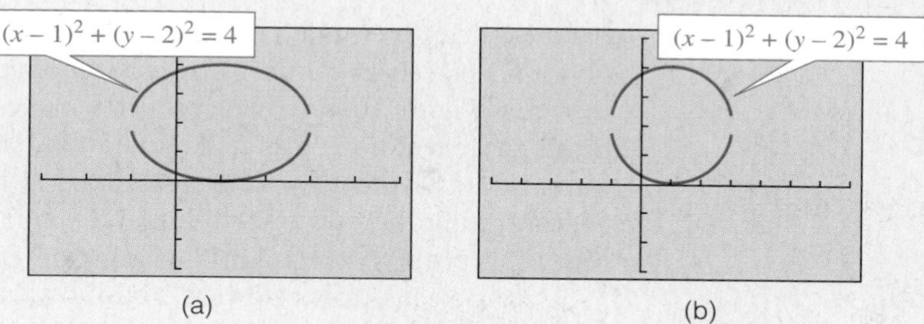

Figure 10-7

3 Solve an application problem involving a circle.

EXAMPLE 4 **TELEVISION TRANSLATORS** The broadcast area of a television station is bounded by the circle $x^2 + y^2 = 3{,}600$, where x and y are measured in miles. A translator station picks up the signal and retransmits it from the center of a circular area bounded by $(x + 30)^2 + (y - 40)^2 = 1{,}600$. Find the location of the translator and the greatest distance from the main transmitter that the signal can be received.

Solution The coverage of the television station is bounded by $x^2 + y^2 = 60^2$, a circle centered at the origin with a radius of 60 miles, as shown in Figure 10-8. Because the translator is at the center of the circle $(x + 30)^2 + (y - 40)^2 = 1{,}600$, it is located at $(-30, 40)$, a point 30 miles west and 40 miles north of the television station. The radius of the translator's coverage is $\sqrt{1{,}600}$, or 40 miles.

As shown in the figure, the greatest distance of reception is the sum of A, the distance from the translator to the TV station, and 40 miles, the radius of the translator's coverage.

To find A, we use the distance formula to find the distance between $(x_1, y_1) = (-30, 40)$ and the origin, $(x_2, y_2) = (0, 0)$.

$$A = \sqrt{(x_1 - x_2)^2 + (y_1 - y_2)^2}$$
$$A = \sqrt{(-30 - 0)^2 + (40 - 0)^2}$$
$$= \sqrt{(-30)^2 + 40^2}$$
$$= \sqrt{2{,}500}$$
$$= 50$$

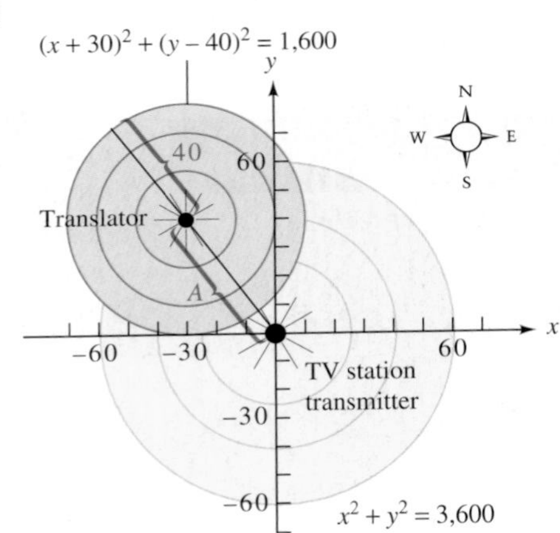

Figure 10-8

Since the translator is 50 miles from the TV station and it broadcasts the signal an additional 40 miles, the greatest reception distance is 50 + 40, or 90 miles.

4 Graph a parabola of the form $x = (y - k)^2 + h$.

Parabolas can be rotated to generate dish-shaped surfaces called *paraboloids.* Any light or sound placed at the *focus* of a paraboloid is reflected outward in parallel paths, as shown in Figure 10-9(a). This property makes parabolic surfaces ideal for flashlight and headlight reflectors. It also makes parabolic surfaces good antennas, because signals captured by such antennas are concentrated at the focus. Parabolic mirrors are capable of concentrating the rays of the Sun at a single point and thereby generating tremendous heat. This property is used in the design of solar furnaces.

Any object thrown upward and outward travels in a parabolic path, as shown in Figure 10-9(b). In architecture, many arches are parabolic in shape, because this gives strength. Cables that support suspension bridges hang in the form of a parabola. (See Figure 10-9(c).)

Parabolas

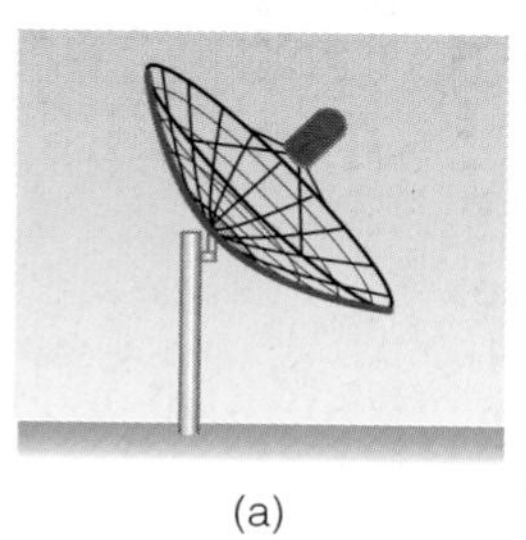

(a)

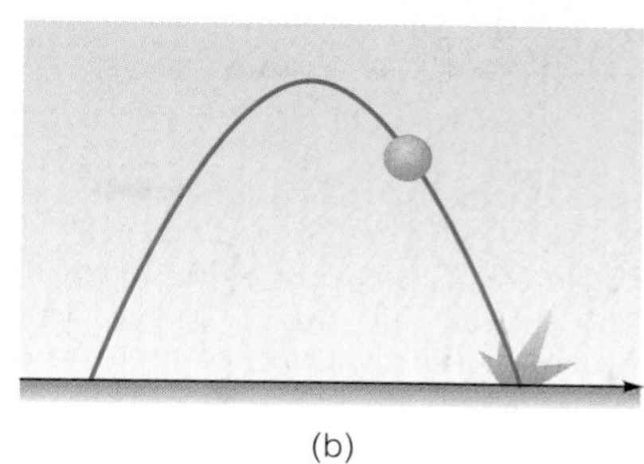

(b)

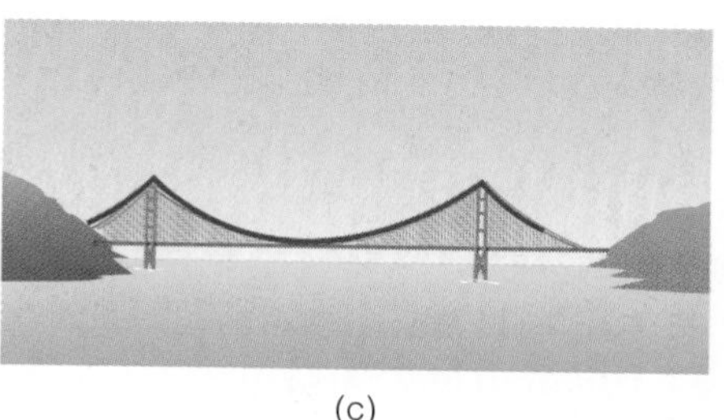

(c)

Figure 10-9

We have seen that equations of the form $y = a(x - h)^2 + k$, with $a \neq 0$, represent parabolas with the vertex at the point (h, k). They open upward when $a > 0$ and downward when $a < 0$.

Equations of the form $x = a(y - k)^2 + h \quad (a \neq 0)$, also represent parabolas with vertex at point (h, k). However, they open to the right when $a > 0$ and to the left when $a < 0$. Parabolas that open to the right or left do not represent functions, because their graphs fail the vertical line test.

Standard equations of many parabolas are summarized in the following table.

Equations of Parabolas

Parabola opening	*Vertex at origin*	*Vertex at (h, k)*
Up	$y = ax^2 \quad (a > 0)$	$y = a(x - h)^2 + k \quad (a > 0)$
Down	$y = ax^2 \quad (a < 0)$	$y = a(x - h)^2 + k \quad (a < 0)$
Right	$x = ay^2 \quad (a > 0)$	$x = a(y - k)^2 + h \quad (a > 0)$
Left	$x = ay^2 \quad (a < 0)$	$x = a(y - k)^2 + h \quad (a < 0)$

EXAMPLE 5 Graph: **a.** $x = \frac{1}{2}y^2$ **b.** $x = -2(y - 2)^2 + 3$.

Solution **a.** We can make a table of ordered pairs that satisfy the equation, plot each pair, and draw the parabola, as in Figure 10-10(a). Because the equation is of the form $x = ay^2$ with $a > 0$, the parabola opens to the right and has its vertex at the origin.

b. We can make a table of ordered pairs that satisfy the equation, plot each pair, and draw the parabola, as in Figure 10-10(b). Because the equation is of the form $x = a(y - k)^2 + h \quad (a < 0)$, the parabola opens to the left and has its vertex at the point with coordinates (3, 2).

$x = \frac{1}{2}y^2$

x	y	(x, y)
0	0	(0, 0)
2	2	(2, 2)
2	−2	(2, −2)
8	4	(8, 4)
8	−4	(8, −4)

(a)

$x = -2(y - 2)^2 + 3$

x	y	(x, y)
−5	0	(−5, 0)
1	1	(1, 1)
3	2	(3, 2)
1	3	(1, 3)
−5	4	(−5, 4)

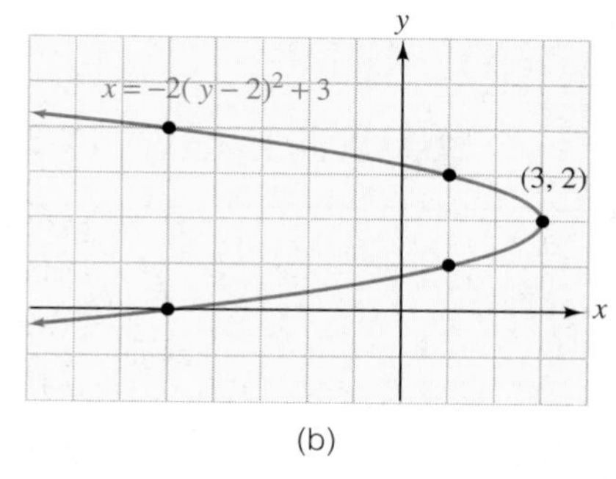

(b)

Figure 10-10

SELF CHECK 5 Graph: $x = \frac{1}{2}(y - 1)^2 - 2$.

The general forms of the equations of a parabola are as follows:

General Form of the Equation of a Parabola That Opens Upward or Downward

The general form of the equation of a parabola that opens upward or downward is

$$y = ax^2 + bx + c \quad (a \neq 0)$$

If $a > 0$, the parabola opens upward. If $a < 0$, the parabola opens downward.

General Form of the Equation of a Parabola That Opens Left or Right

The general form of the equation of a parabola that opens to the left or to the right is

$$x = ay^2 + by + c \quad (a \neq 0)$$

If $a > 0$, the parabola opens to the right. If $a < 0$, the parabola opens to the left.

EXAMPLE 6 Graph: $x = -2y^2 + 12y - 15$.

Solution This equation is the general form of a parabola that opens left or right. Since $a = -2$ and $-2 < 0$, the parabola opens to the left. To find the coordinates of its vertex, we write the equation in standard form by completing the square on y.

$$\begin{aligned} x &= -2y^2 + 12y - 15 \\ &= -2(y^2 - 6y) - 15 && \text{Factor out } -2 \text{ from } -2y^2 + 12y. \\ &= \mathbf{-2(y^2 - 6y + 9) - 15 + 18} && \text{Subtract and add 18; } -2(9) = -18. \\ &= -2(y - 3)^2 + 3 \end{aligned}$$

Because the equation is written in the form $x = a(y - k)^2 + h$, we can see that the parabola has its vertex at (3, 3). The graph is shown in Figure 10-11.

$x = -2y^2 + 12y - 15$

x	y	(x, y)
−5	1	(−5, 1)
1	2	(1, 2)
3	3	(3, 3)
1	4	(1, 4)
−5	5	(−5, 5)

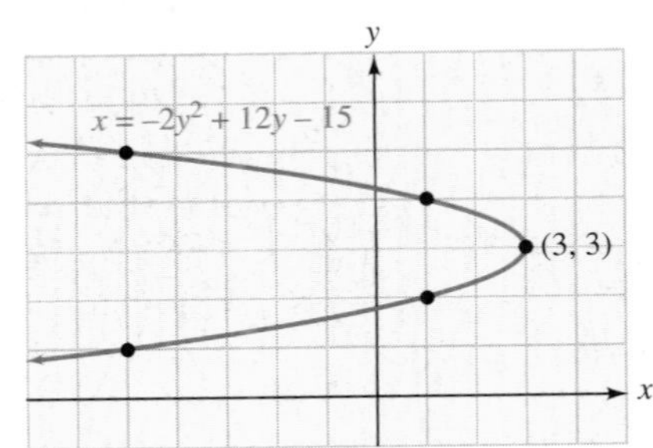

Figure 10-11

SELF CHECK 6 Graph: $x = 0.5y^2 - y - 1$.

COMMENT To find the y-coordinate of the vertex for this horizontal parabola, we could just as well compute $-\frac{b}{2a}$. To find the x-coordinate, we could substitute the value of $-\frac{b}{2a}$ for y and find x.

5 Solve an application problem involving a parabola.

EXAMPLE 7 **GATEWAY ARCH** The shape of the Gateway Arch in St. Louis is approximately a parabola 630 feet high and 630 feet wide, as shown in Figure 10-12(a). How high is the arch 100 feet from its foundation?

Solution We place the parabola in a coordinate system as in Figure 10-12(b), with ground level on the x-axis and the vertex of the parabola at the point $(h, k) = (0, 630)$.

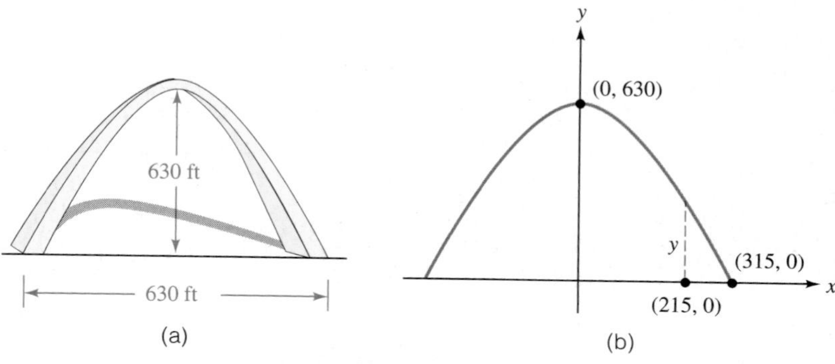

Figure 10-12

The equation of this downward-opening parabola has the form

$$\begin{aligned} y &= a(x - h)^2 + k && \text{With } a < 0. \\ &= a(x - \mathbf{0})^2 + \mathbf{630} && \text{Substitute } h = 0 \text{ and } k = 630. \\ &= ax^2 + 630 && \text{Simplify.} \end{aligned}$$

Because the Gateway Arch is 630 feet wide at its base, the parabola passes through the point $\left(\frac{630}{2}, 0\right)$, or (315, 0). To find a in the equation of the parabola, we proceed as follows:

$$\begin{aligned} y &= ax^2 + 630 \\ \mathbf{0} &= a(\mathbf{315})^2 + 630 && \text{Substitute 315 for } x \text{ and 0 for } y. \\ \frac{-630}{315^2} &= a && \text{Subtract 630 from both sides and divide both sides by } 315^2. \\ -\frac{2}{315} &= a && \text{Simplify; } \tfrac{-630}{315^2} = \tfrac{-2}{315}. \end{aligned}$$

The equation of the parabola that approximates the shape of the Gateway Arch is

$$y = -\frac{2}{315}x^2 + 630$$

To find the height of the arch at a point 100 feet from its foundation, we substitute 315 − 100, or 215, for x in the equation of the parabola and solve for y.

$$\begin{aligned} y &= -\frac{2}{315}x^2 + 630 \\ &= -\frac{2}{315}(\mathbf{215})^2 + 630 \\ &= 336.5079365 \end{aligned}$$

At a point 100 feet from the foundation, the height of the arch is about 337 feet.

SELF CHECK ANSWERS

1. a. (3, −4), 2

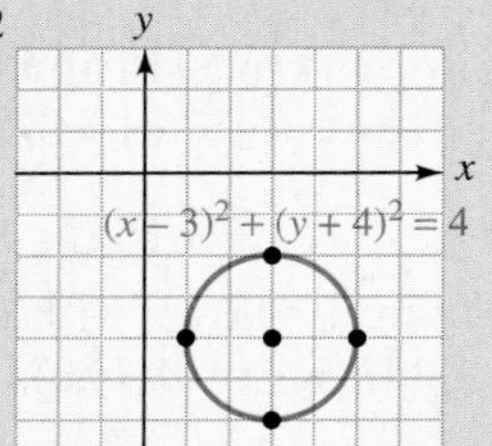

b. (0, 0), $2\sqrt{2}$

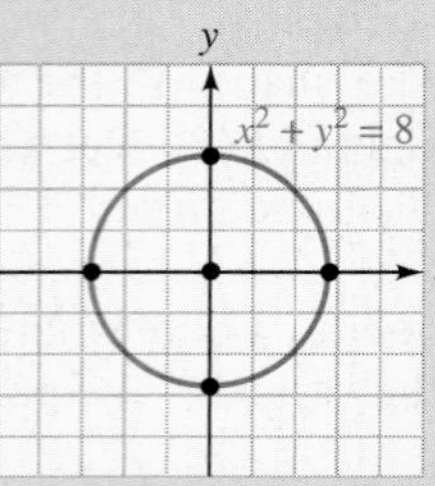

2. $(x+1)^2+(y-2)^2=16$

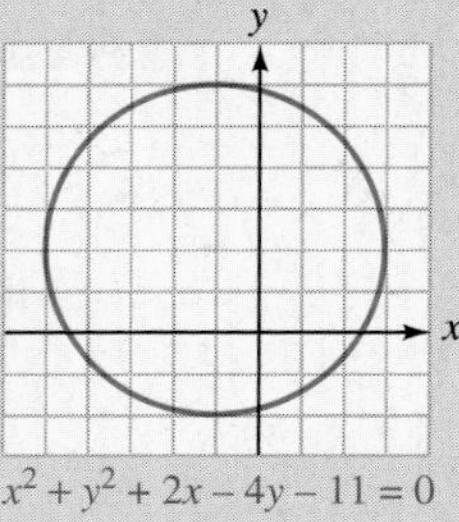

3. $x^2+y^2-4x-6y-23=0$

5.

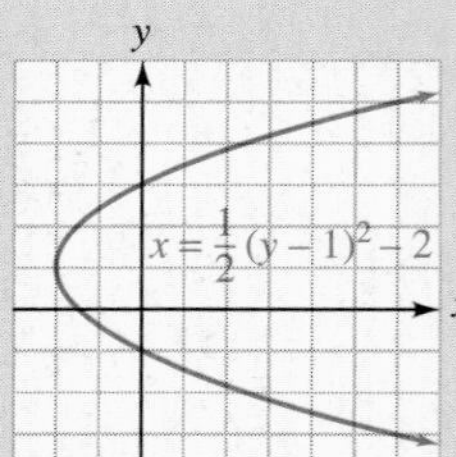

6.

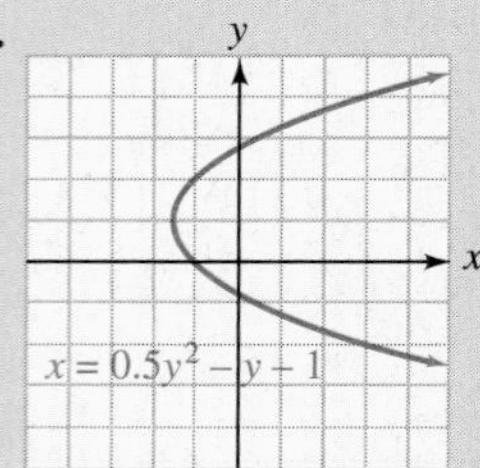

NOW TRY THIS

1. The equation $4y^2+(4x-1)y+x^2-5x-3=0$ is an equation for a parabola that is neither vertical nor horizontal. Use the quadratic formula to solve for y and then graph the equation with a calculator. (*Hint:* $a=4$, $b=4x-1$, and $c=x^2-5x-3$.)

10.1 EXERCISES

WARM-UPS

Find the center and the radius of each circle.

1. $x^2+y^2=144$

2. $x^2+y^2=121$

3. $(x-2)^2+y^2=16$

4. $x^2+(y+1)^2=9$

Determine whether the graph of each parabola opens up or down or left or right.

5. $y=-3x^2-2$

6. $y=7x^2-5$

7. $x=-3y^2$

8. $x=(y-3)^2$

REVIEW *Solve each equation.*

9. $|3x-4|=11$

10. $\left|\frac{4-3x}{5}\right|=12$

11. $|3x+4|=|5x-2|$

12. $|6-4x|=|x+2|$

VOCABULARY AND CONCEPTS *Fill in the blanks.*

13. A ______ section is determined by the intersection of a plane and a right-circular cone.

14. A ______ is the set of all points in a ______ that are a fixed distance from a given point. The fixed distance is called the ______ and the point is called the ______.

15. The equation of the circle $x^2+(y-3)^2=16$ is in ________ form with the center at ______ and a radius __.

16. The graph of the equation $x^2+y^2=0$ is a __________.

17. The equation $x^2+y^2-10x-8y-8=0$ is an equation of a ______ written in _______ form.

18. The graph of $y=ax^2$ $(a>0)$ is a ________ with vertex at the ______ that opens _______.

19. The graph of $x = a(y - 2)^2 + 3 \quad (a > 0)$ is a ________ with vertex at ______ that opens to the ______.

20. The graph of $x = a(y - 1)^2 - 3 \quad (a < 0)$ is a ________ with vertex at ________ that opens to the ______.

GUIDED PRACTICE

Graph each equation and find the center and radius of the resulting circle. **See Example 1. (Objective 1)**

21. $x^2 + y^2 = 9$

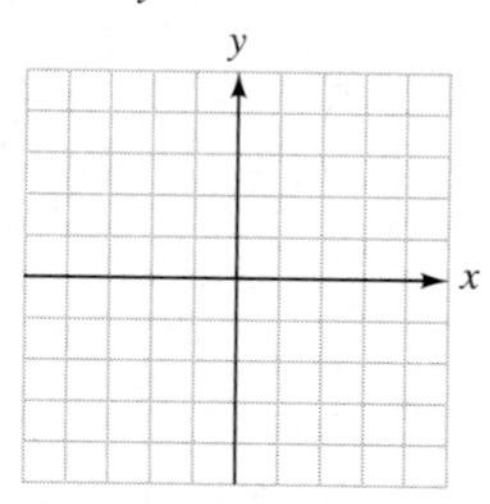

22. $x^2 + y^2 = 16$

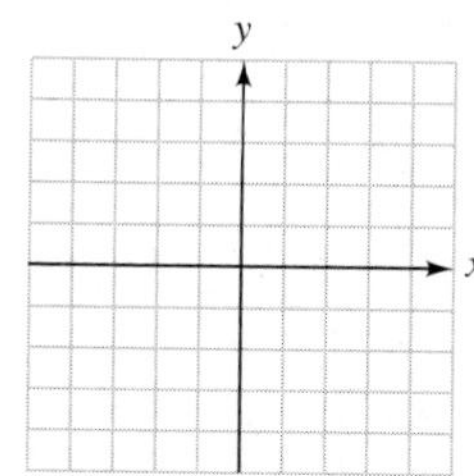

23. $(x - 2)^2 + y^2 = 9$

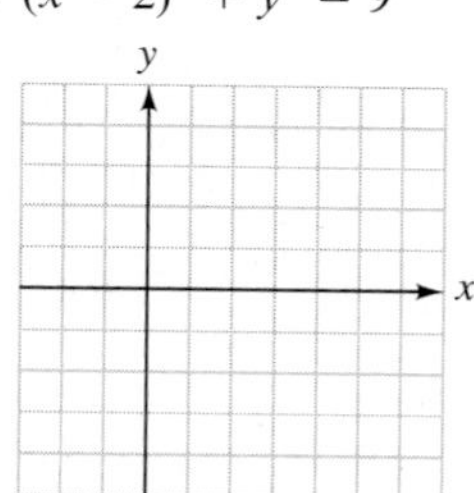

24. $x^2 + (y - 3)^2 = 4$

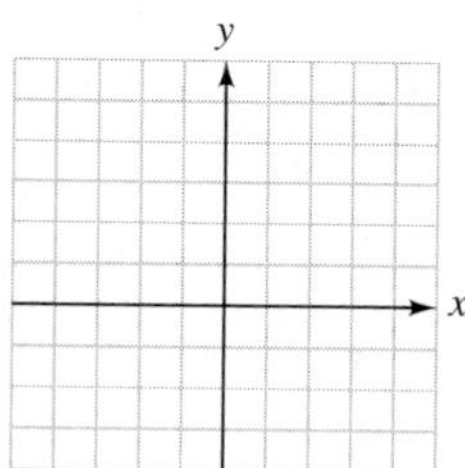

25. $(x - 2)^2 + (y - 4)^2 = 4$

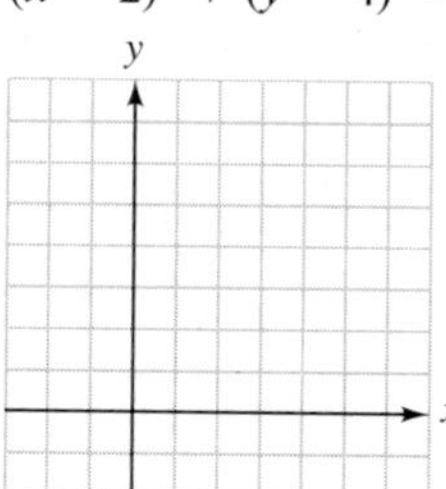

26. $(x - 3)^2 + (y - 2)^2 = 4$

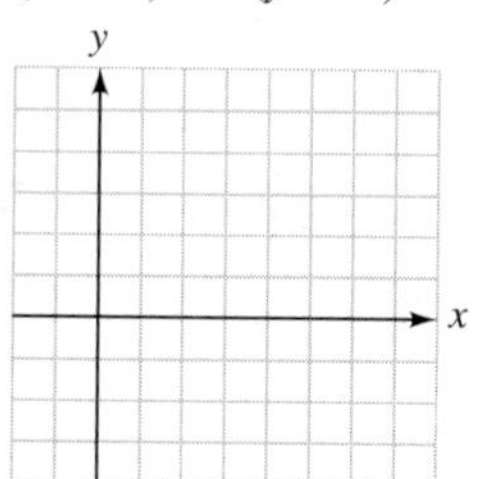

27. $(x + 3)^2 + (y - 1)^2 = 16$

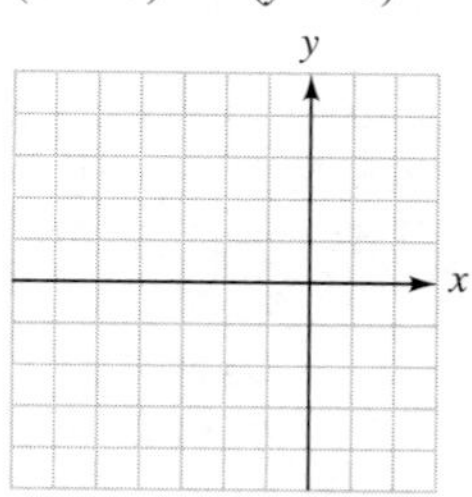

28. $(x - 1)^2 + (y + 4)^2 = 9$

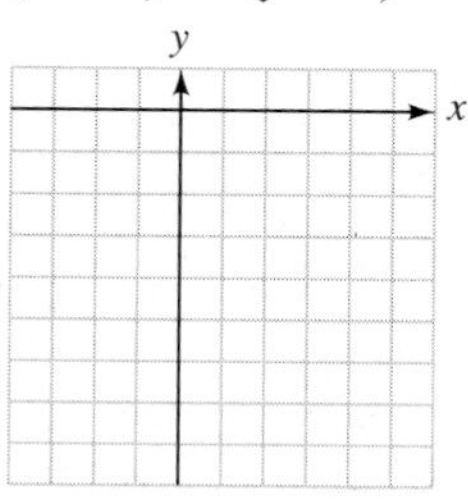

Graph each circle. Give the coordinates of the center and find the radius. **See Example 2. (Objective 1)**

29. $x^2 + y^2 + 2x - 8 = 0$

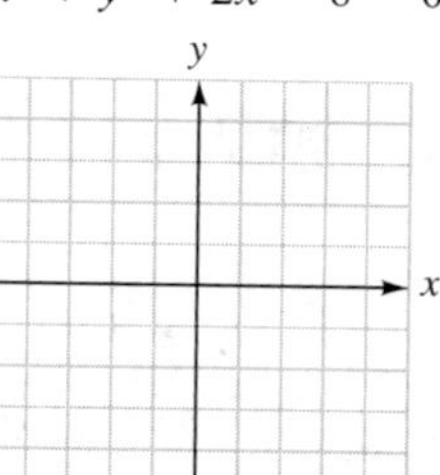

30. $x^2 + y^2 - 4y = 12$

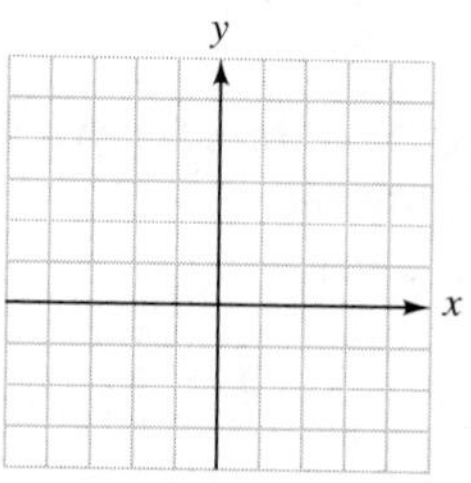

31. $9x^2 + 9y^2 - 12y = 5$

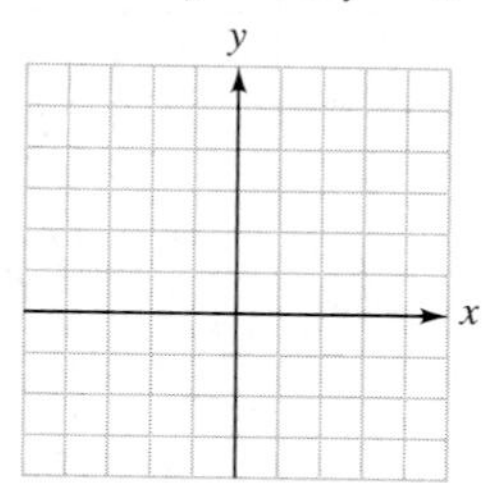

32. $4x^2 + 4y^2 + 4y = 15$

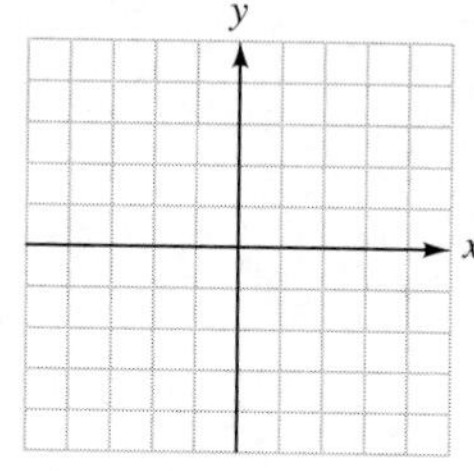

33. $x^2 + y^2 - 2x + 4y = -1$

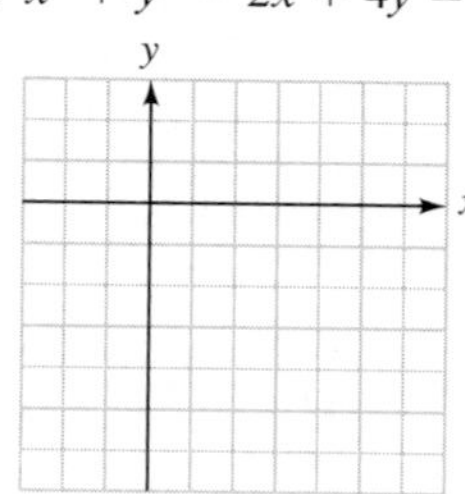

34. $x^2 + y^2 + 4x + 2y = 4$

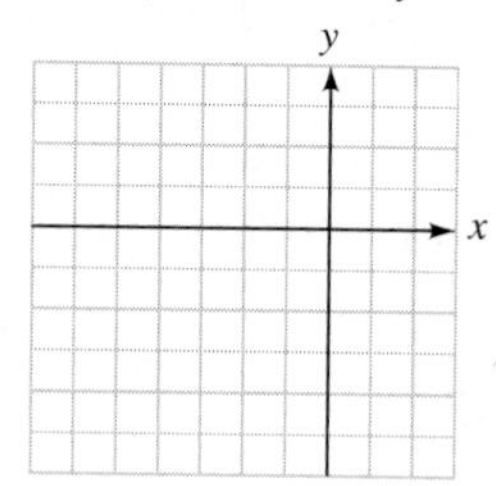

35. $x^2 + y^2 + 6x - 4y = -12$

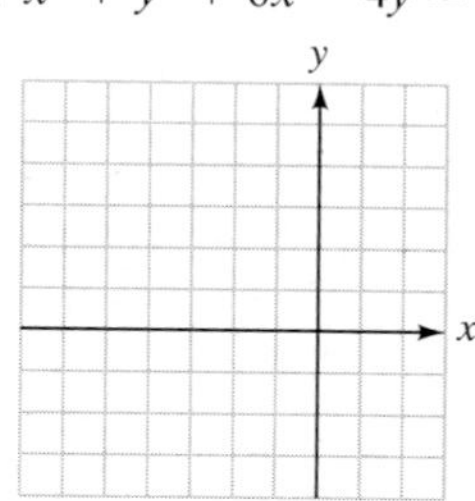

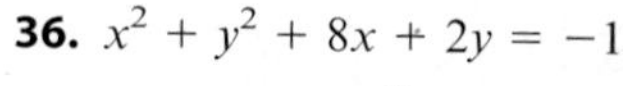

36. $x^2 + y^2 + 8x + 2y = -13$

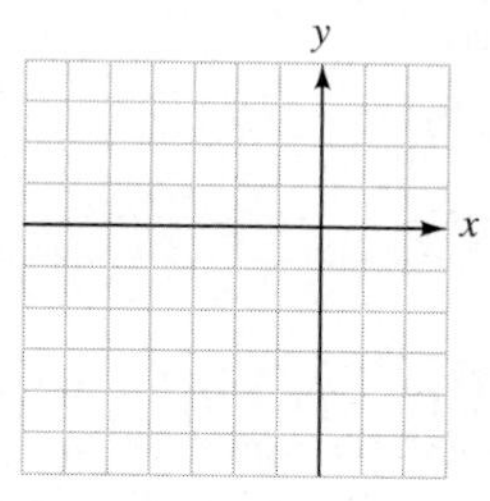

Write the equation of the circle with the following properties in standard form and in general form. **See Example 3. (Objective 2)**

37. Center at origin; radius 1

38. Center at origin; radius 4

39. Center at (6, 8); radius 5

40. Center at (5, 3); radius 2

41. Center at (−2, 6); radius 12

42. Center at (5, −4); radius 6

43. Center at the origin; diameter $2\sqrt{2}$

44. Center at the origin; diameter $8\sqrt{3}$

Find the vertex of each parabola and graph it. **See Examples 5–6. (Objective 4)**

45. $x = y^2$

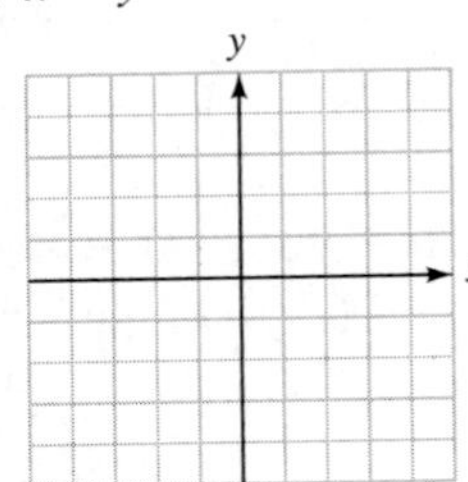

46. $x = -y^2 + 1$

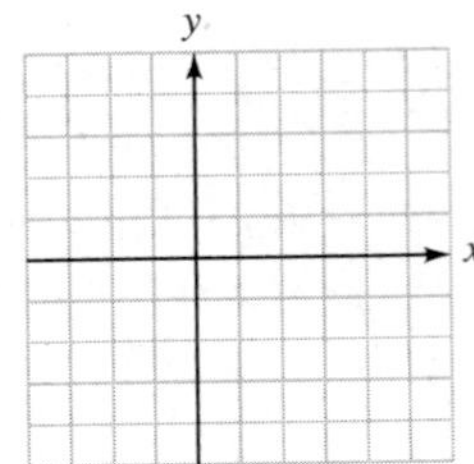

47. $x = -\frac{1}{4}y^2$

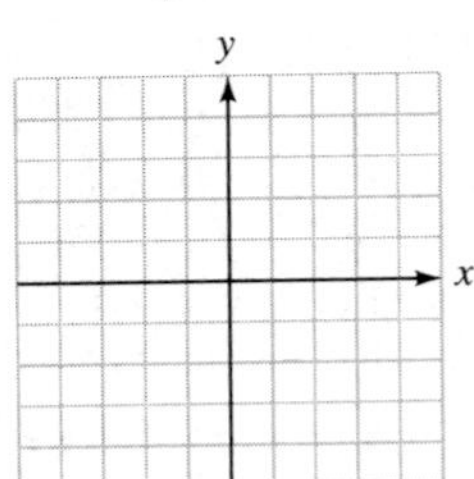

48. $x = 4y^2$

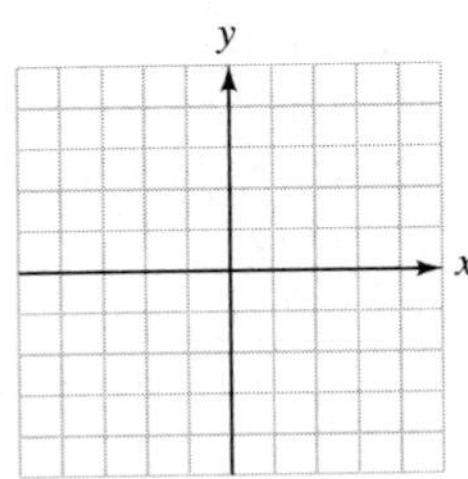

49. $y^2 + 4x - 6y = -1$

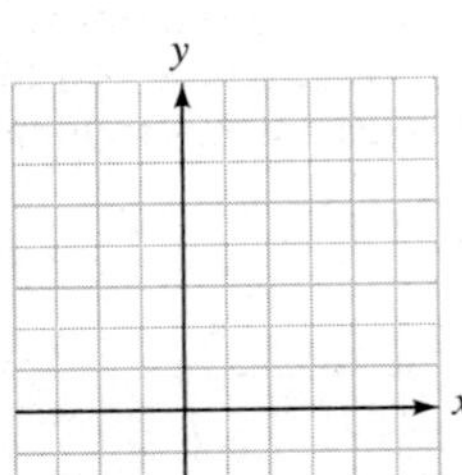

50. $x = \frac{1}{2}y^2 + 2y$

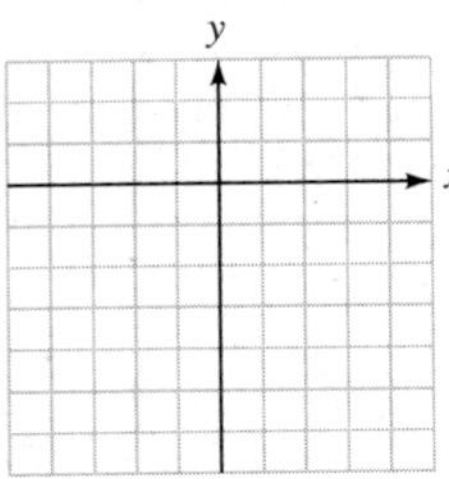

51. $y = 2(x - 1)^2 + 3$

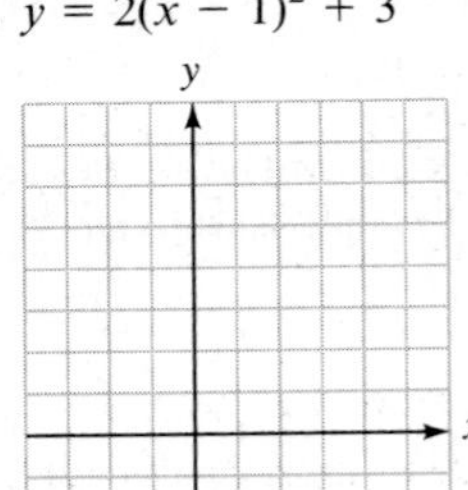

52. $y = -2(x + 1)^2 + 2$

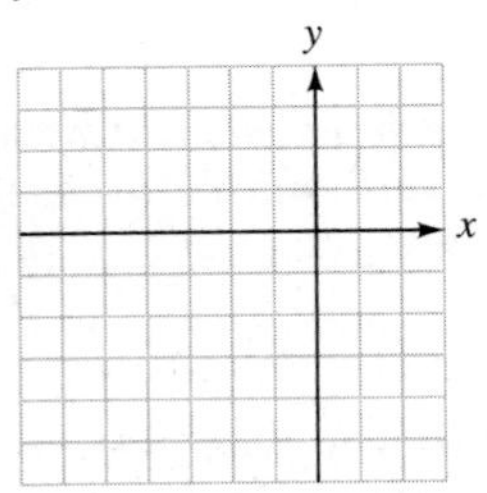

ADDITIONAL PRACTICE

Graph each equation.

53. $x^2 + (y + 3)^2 = 1$

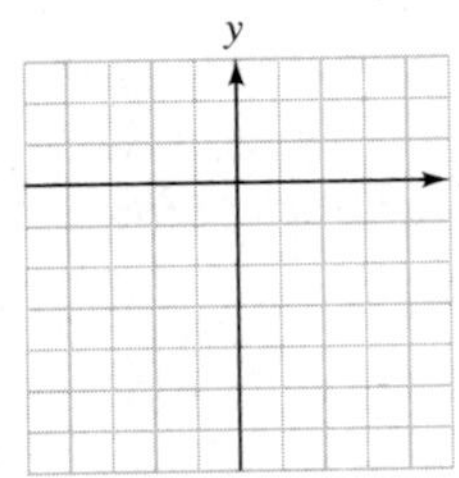

54. $(x + 4)^2 + y^2 = 1$

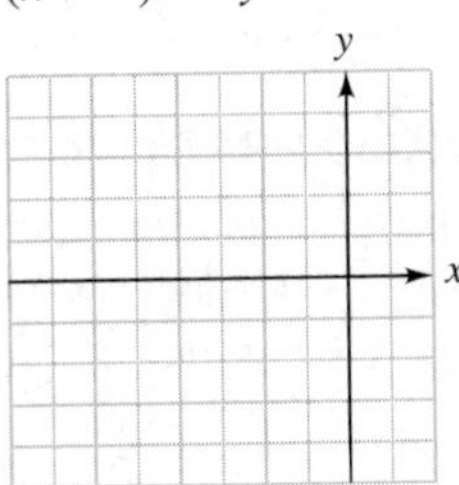

55. $y = x^2 + 4x + 5$

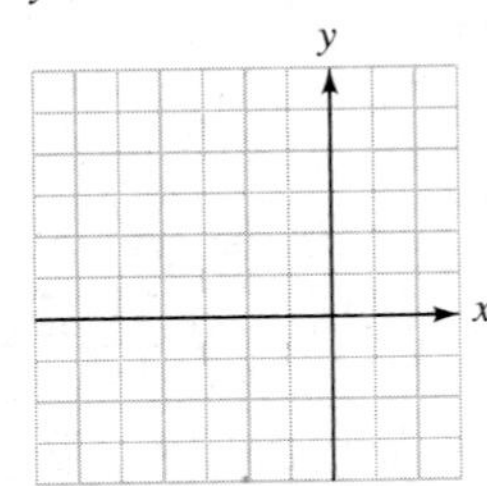

56. $y = -x^2 - 2x + 3$

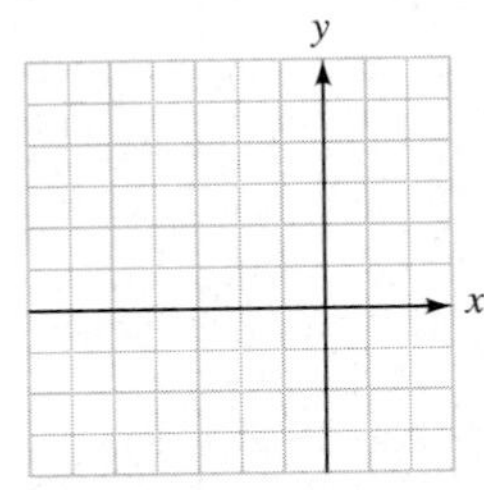

57. $y = -x^2 - x + 1$

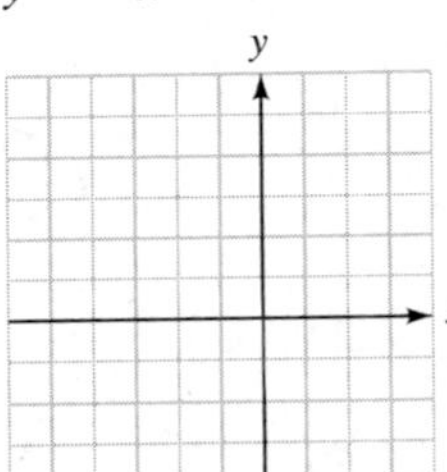

58. $x^2 - 2y - 2x = -7$

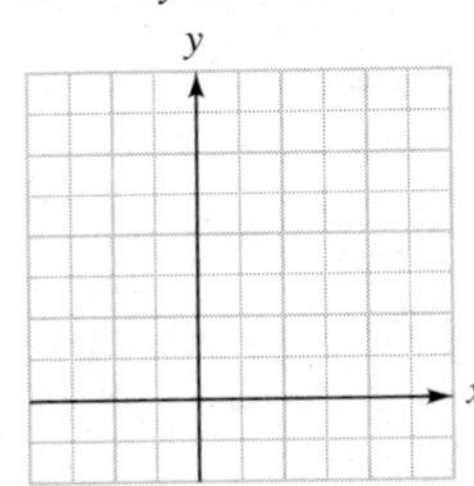

Use a graphing calculator to graph each equation.

59. $3x^2 + 3y^2 = 16$

60. $2x^2 + 2y^2 = 9$

61. $(x + 1)^2 + y^2 = 16$

62. $x^2 + (y - 2)^2 = 4$

63. $x = 2y^2$

64. $x = y^2 - 4$

65. $x^2 - 2x + y = 6$

66. $x = -2(y - 1)^2 + 2$

APPLICATIONS *Solve each application.* See Examples 4 and 7. (Objectives 3 and 5)

67. Meshing gears For design purposes, the large gear is the circle $x^2 + y^2 = 16$. The smaller gear is a circle centered at (7, 0) and tangent to the larger circle. Find the equation of the smaller gear.

68. Width of a walkway The following walkway is bounded by the two circles $x^2 + y^2 = 2{,}500$ and $(x - 10)^2 + y^2 = 900$, measured in feet. Find the largest and the smallest width of the walkway.

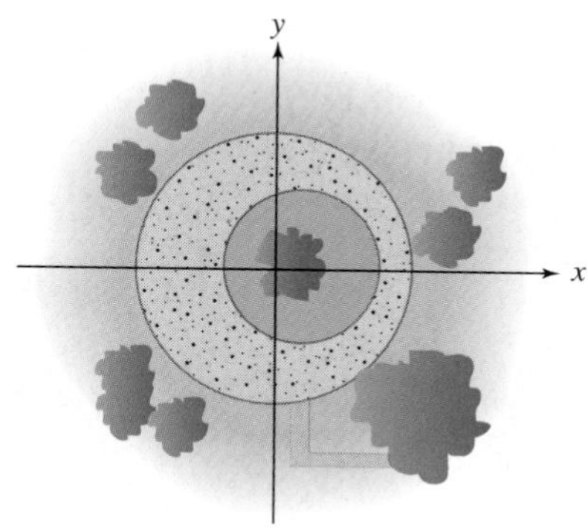

69. Broadcast ranges Radio stations applying for licensing may not use the same frequency if their broadcast areas overlap. One station's coverage is bounded by $x^2 + y^2 - 8x - 20y + 16 = 0$, and the other's by $x^2 + y^2 + 2x + 4y - 11 = 0$. May they be licensed for the same frequency?

70. Highway design Engineers want to join two sections of highway with a curve that is one-quarter of a circle as shown in the illustration. The equation of the circle is $x^2 + y^2 - 16x - 20y + 155 = 0$, where distances are measured in kilometers. Find the locations (relative to the center of town) of the intersections of the highway with State and with Main.

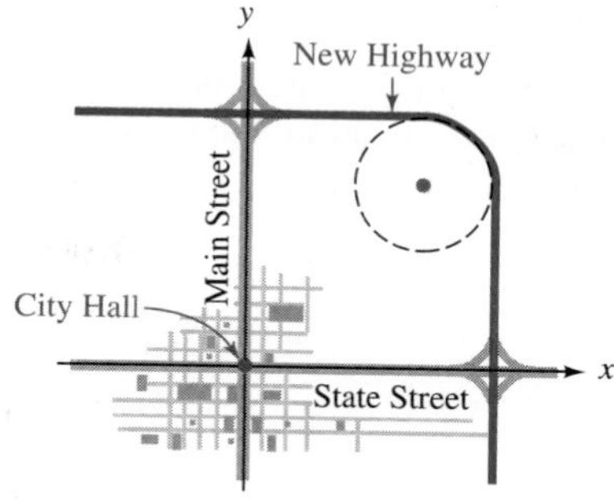

71. Projectiles The cannonball in the illustration follows the parabolic trajectory $y = 30x - x^2$. Where does it land?

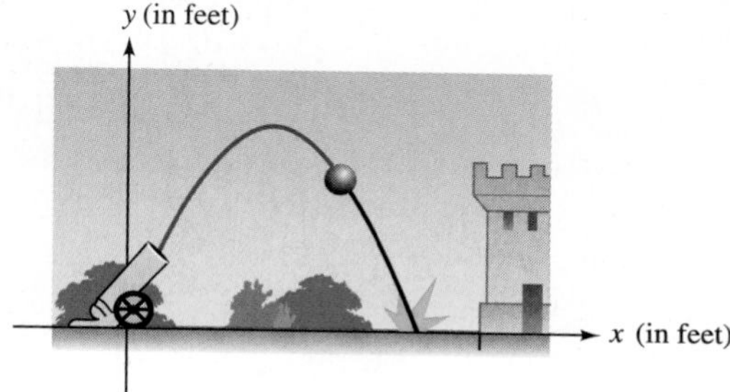

72. Projectiles In Exercise 71, how high does the cannonball rise?

73. Path of a comet If the path of a comet is given by the equation $2y^2 - 9x = 18$, how far is it from the Sun at the vertex of the orbit? Distances are measured in astronomical units (AU).

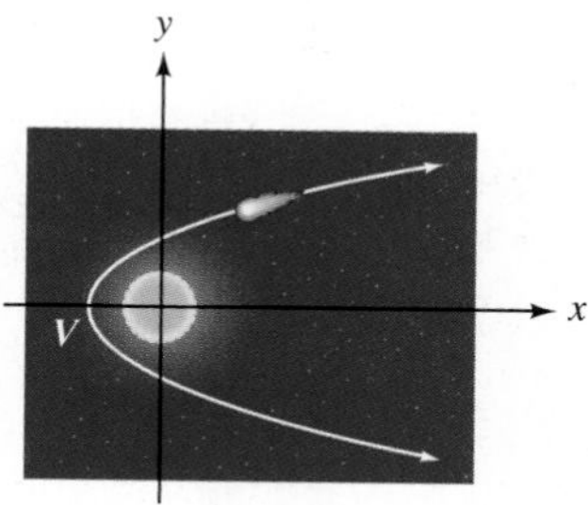

74. Satellite antennas The cross section of the satellite antenna is a parabola given by the equation $y = \frac{1}{16}x^2$, with distances measured in feet. If the dish is 8 feet wide, how deep is it?

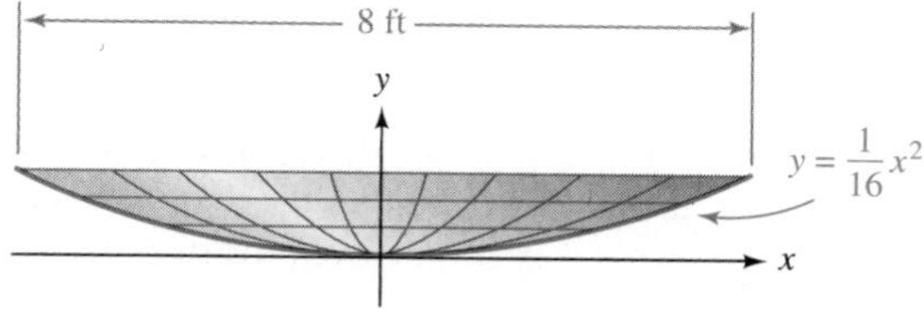

WRITING ABOUT MATH

75. Explain how to decide from its equation whether the graph of a parabola opens up, down, right, or left.

76. From the equation of a circle, explain how to determine the radius and the coordinates of the center.

SOMETHING TO THINK ABOUT

77. From the values of a, h, and k, explain how to determine the number of x-intercepts of the graph of $y = a(x - h)^2 + k$.

78. Under what conditions will the graph of $x = a(y - k)^2 + h$ have no y-intercepts?

SECTION 10.2 The Ellipse

Objectives

1. Graph an ellipse given an equation in standard form.
2. Graph an ellipse given an equation in general form.
3. Solve an application problem involving an ellipse.

Vocabulary

ellipse	eccentricity	major axis
focus	vertices	minor axis
foci		

Getting Ready

Solve each equation for the indicated variable $(a \neq 0, b \neq 0)$.

1. $\dfrac{y^2}{b^2} = 1$ for y

2. $\dfrac{x^2}{a^2} = 1$ for x

A third conic section is an oval-shaped curve called an *ellipse.* Ellipses can be nearly round or they can be long and narrow. In this section, we will learn how to construct ellipses and how to graph equations that represent them.

Ellipses have optical and acoustical properties that are useful in architecture and engineering. For example, many arches are portions of an ellipse, because the shape is pleasing to the eye. (See Figure 10-13(a).) The planets and many comets have elliptical orbits. (See Figure 10-13(b).) Gears are often cut into elliptical shapes to provide nonuniform motion. (See Figure 10-13(c).)

Ellipses

Arches
(a)

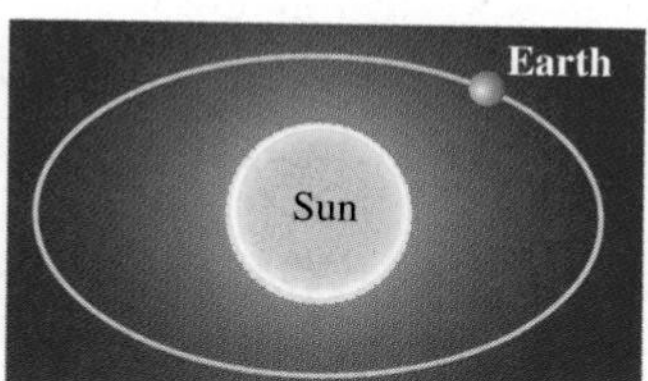

Earth's orbit
(b)

Gears
(c)

Figure 10-13

1 Graph an ellipse given an equation in standard form.

The Ellipse

An **ellipse** is the set of all points P in the plane the sum of whose distances from two fixed points is a constant. See Figure 10-14, in which $d_1 + d_2$ is a constant.

Each of the two points is called a **focus.** Midway between the **foci** is the *center* of the ellipse.

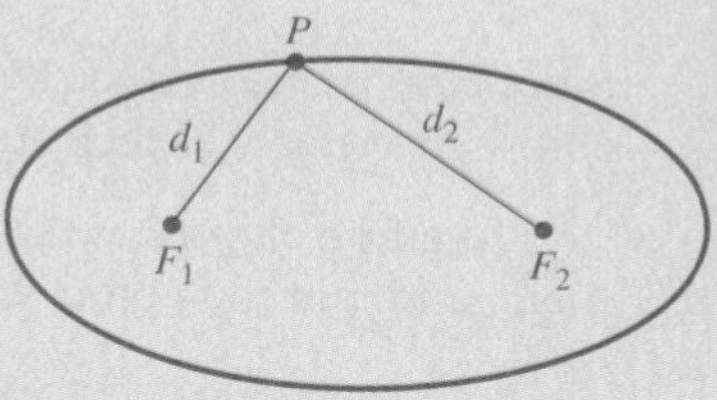

Figure 10-14

We can construct an ellipse by placing two thumbtacks fairly close together, as in Figure 10-15. We then tie each end of a piece of string to a thumbtack, catch the loop with the point of a pencil, and, while keeping the string taut, draw the ellipse.

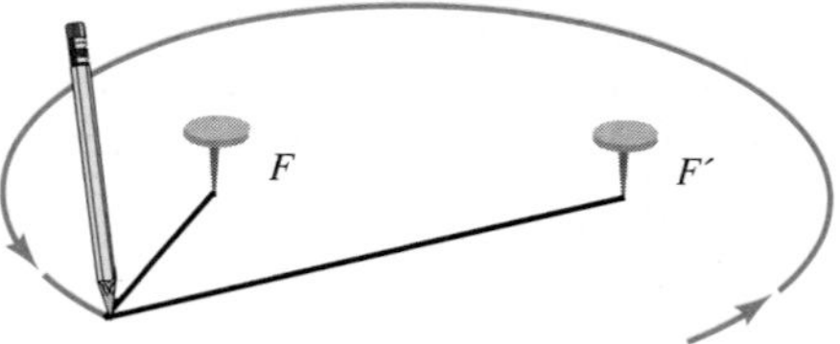

Figure 10-15

Using this method, we can construct an ellipse of any specific size. For example, to construct an ellipse that is 10 inches wide and 6 inches high, we must find the length of string to use and the distance between thumbtacks.

To do this, we will let a represent the distance between the center and vertex V, as shown in Figure 10-16(a) on the next page. We will also let c represent the distance between the center of the ellipse and either focus. When the pencil is at vertex V, the length of the string is $c + a + (a - c)$, or just $2a$. Because $2a$ is the 10 inch width of the ellipse, the string needs to be 10 inches long. The distance $2a$ is constant for any point on the ellipse, including point B shown in Figure 10-16(b).

Sir Isaac Newton
(1642–1727)

Newton was an English scientist and mathematician. Because he was not a good farmer, he went to Cambridge University to become a preacher. When he had to leave Cambridge because of the plague, he made some of his most important discoveries. He is best known in mathematics for developing calculus and in physics for discovering the laws of motion. Newton probably contributed more to science and mathematics than anyone else in history.

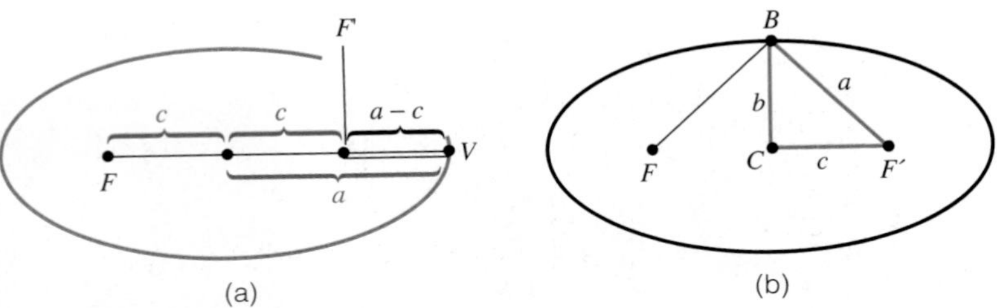

Figure 10-16

From right triangle BCF' in Figure 10-16(b) and the Pythagorean theorem, we can find c as follows:

$$a^2 = b^2 + c^2 \quad \text{or} \quad c = \sqrt{a^2 - b^2}$$

Since distance b is one-half of the height of the ellipse, $b = 3$. Since $2a = 10$, $a = 5$. We can now substitute $a = 5$ and $b = 3$ into the formula to find c:

$$\begin{aligned} c &= \sqrt{5^2 - 3^2} \\ &= \sqrt{25 - 9} \\ &= \sqrt{16} \\ &= 4 \end{aligned}$$

Since $c = 4$, the distance between the thumbtacks must be 8 inches. We can construct the ellipse by tying a 10 inch string to thumbtacks that are 8 inches apart.

To graph ellipses, we can make a table of ordered pairs that satisfy the equation, plot them, and join the points with a smooth curve.

EXAMPLE 1 Graph: $\dfrac{x^2}{36} + \dfrac{y^2}{9} = 1$.

Solution We note that the equation can be written in the form

$$\frac{x^2}{6^2} + \frac{y^2}{3^2} = 1 \qquad 36 = 6^2 \text{ and } 9 = 3^2.$$

After making a table of ordered pairs that satisfy the equation, plotting each of them, and joining the points with a curve, we obtain the ellipse shown in Figure 10-17.

We note that the center of the ellipse is the origin, the ellipse intersects the x-axis at points $(6, 0)$ and $(-6, 0)$, and the ellipse intersects the y-axis at points $(0, 3)$ and $(0, -3)$.

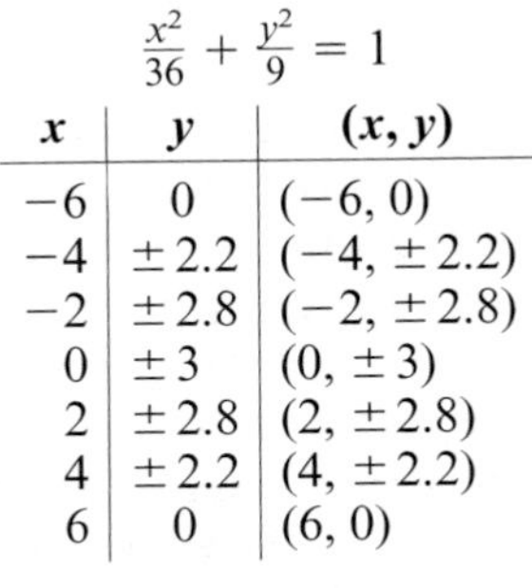

$\dfrac{x^2}{36} + \dfrac{y^2}{9} = 1$

x	y	(x, y)
−6	0	(−6, 0)
−4	±2.2	(−4, ±2.2)
−2	±2.8	(−2, ±2.8)
0	±3	(0, ±3)
2	±2.8	(2, ±2.8)
4	±2.2	(4, ±2.2)
6	0	(6, 0)

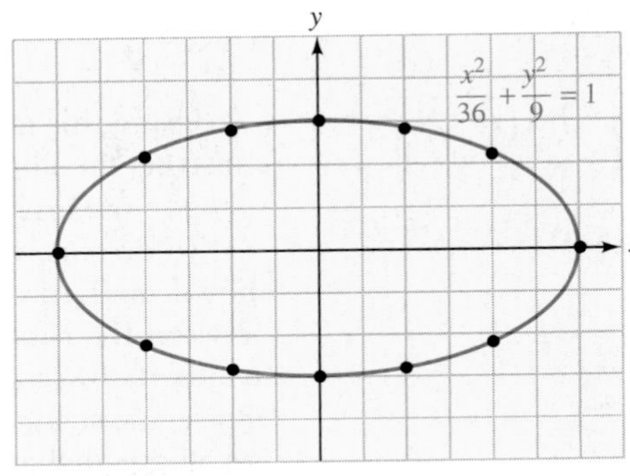

Figure 10-17

SELF CHECK 1 Graph: $\frac{x^2}{4} + \frac{y^2}{16} = 1.$

Example 1 illustrates that the graph of

$$\frac{x^2}{a^2} + \frac{y^2}{b^2} = 1$$

is an ellipse centered at the origin. To find the x-intercepts of the graph, we can let $y = 0$ and solve for x.

$$\frac{x^2}{a^2} + \frac{0^2}{b^2} = 1$$

$$\frac{x^2}{a^2} + 0 = 1$$

$$x^2 = a^2$$

$$x = a \quad \text{or} \quad x = -a$$

The x-intercepts are $(a, 0)$ and $(-a, 0)$.

To find the y-intercepts, we let $x = 0$ and solve for y.

$$\frac{0^2}{a^2} + \frac{y^2}{b^2} = 1$$

$$0 + \frac{y^2}{b^2} = 1$$

$$y^2 = b^2$$

$$y = b \quad \text{or} \quad y = -b$$

The y-intercepts are $(0, b)$ and $(0, -b)$.

In general, we have the following results.

Equations of an Ellipse Centered at the Origin

The equation of an ellipse centered at the origin, with x-intercepts at $V_1(a, 0)$ and $V_2(-a, 0)$ and with y-intercepts of $(0, b)$ and $(0, -b)$, is

$$\frac{x^2}{a^2} + \frac{y^2}{b^2} = 1 \quad (a > b > 0)$$ See Figure 10-18(a) on the next page.

The equation of an ellipse centered at the origin, with y-intercepts at $V_1(0, a)$ and $V_2(0, -a)$ and x-intercepts of $(b, 0)$ and $(-b, 0)$, is

$$\frac{x^2}{b^2} + \frac{y^2}{a^2} = 1 \quad (a > b > 0)$$ See Figure 10-18(b).

In Figure 10-18, the points V_1 and V_2 are the **vertices** of the ellipse, the midpoint of segment V_1V_2 is the *center* of the ellipse, and the distance between the center and either vertex is a. The segment V_1V_2 is called the **major axis,** and the segment joining either $(0, b)$ and $(0, -b)$ or $(b, 0)$ and $(-b, 0)$ is called the **minor axis.**

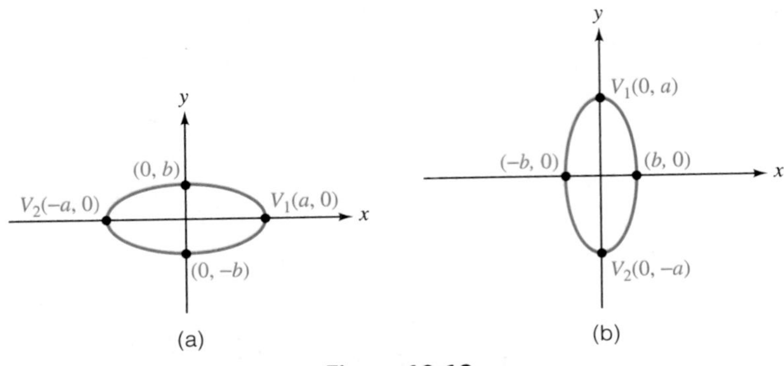

Figure 10-18

The equations for ellipses centered at (h, k) are as follows.

Standard Equation of a Horizontal Ellipse Centered at (*h*, *k*)

The equation of a horizontal ellipse centered at (h, k), with major axis parallel to the x-axis, is

$$(1)\quad \frac{(x-h)^2}{a^2} + \frac{(y-k)^2}{b^2} = 1 \quad (a > b > 0)$$

Standard Equation of a Vertical Ellipse Centered at (*h*, *k*)

The equation of a vertical ellipse centered at (h, k), with major axis parallel to the y-axis, is

$$(2)\quad \frac{(x-h)^2}{b^2} + \frac{(y-k)^2}{a^2} = 1 \quad (a > b > 0)$$

EVERYDAY CONNECTIONS Eccentricity of an Ellipse

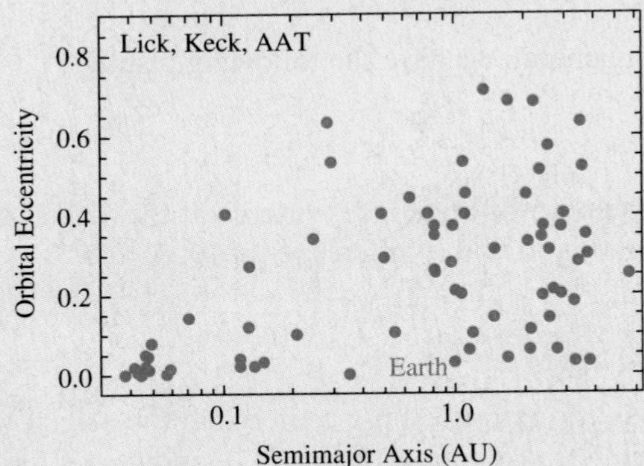

Source: Eccentricity vs. semimajor axis for extrasolar planets. The 75 planets shown were found in a Doppler survey of 1,300 FGKM main sequence stars using the Lick, Keck, and AAT telescopes. The survey was carried out by the California-Carnegie planet search team. http://exoplanets.org/newsframe.html

In Figure 10-16(a) on page 702, the ratio of c to a is called the **eccentricity** of the ellipse. We can use the eccentricity of an ellipse to judge its shape. If the eccentricity is close to 1, the ellipse is relatively flat. If it is close to 0, the ellipse is more circular. Specifically, the eccentricity of a true circle equals 0.

Use the data plot above to estimate how many of the 75 planets shown follow orbits that are true circles.

COMMENT To determine whether an ellipse is horizontal or vertical, look at the denominators in its standard equation. If the largest denominator is associated with the x-term, the ellipse will be horizontal. If the largest denominator is associated with the y-term, the ellipse will be vertical.

EXAMPLE 2 Graph: $25(x-2)^2 + 16(y+3)^2 = 400$.

Solution We first write the equation in standard form.

$$25(x-2)^2 + 16(y+3)^2 = 400$$

$$\frac{25(x-2)^2}{400} + \frac{16(y+3)^2}{400} = \frac{400}{400} \qquad \text{Divide both sides by 400.}$$

$$\frac{(x-2)^2}{16} + \frac{(y+3)^2}{25} = 1 \qquad \text{Simplify each fraction.}$$

$$\frac{(x-2)^2}{4^2} + \frac{[y-(-3)]^2}{5^2} = 1 \quad (5 > 4)$$

This is the equation of a vertical ellipse centered at $(h, k) = (2, -3)$ with major axis parallel to the y-axis and with $b = 4$ and $a = 5$. We first plot the center, as shown in Figure 10-19. Since a is the distance from the center to a vertex, we can locate the vertices by counting 5 units above and 5 units below the center. The vertices are at points $(2, 2)$ and $(2, -8)$.

Since $b = 4$, we can locate two more points on the ellipse by counting 4 units to the left and 4 units to the right of the center. The points $(-2, -3)$ and $(6, -3)$ are also on the graph.

Using these four points as guides, we can draw the ellipse.

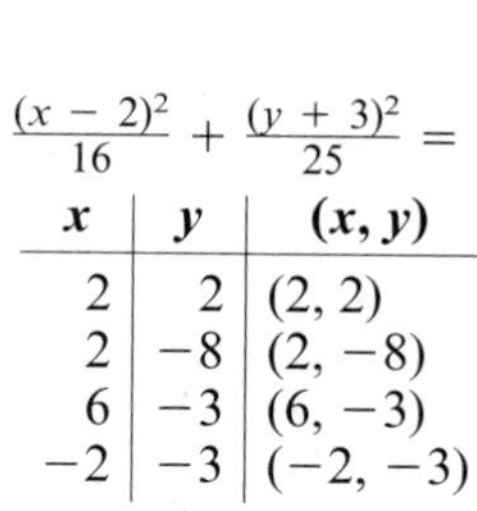

$$\frac{(x-2)^2}{16} + \frac{(y+3)^2}{25} = 1$$

x	y	(x, y)
2	2	(2, 2)
2	−8	(2, −8)
6	−3	(6, −3)
−2	−3	(−2, −3)

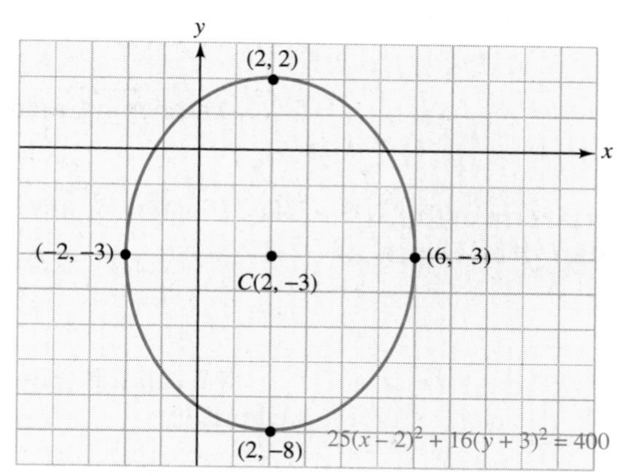

Figure 10-19

SELF CHECK 2 Graph: $16(x-1)^2 + 9(y+2)^2 = 144$.

ACCENT ON TECHNOLOGY

Graphing Ellipses

To use a graphing calculator to graph

$$\frac{(x+2)^2}{4} + \frac{(y-1)^2}{25} = 1$$

(continued)

we first clear the fractions by multiplying both sides by 100 and solving for y.

$$25(x+2)^2 + 4(y-1)^2 = 100 \quad \text{Multiply both sides by 100.}$$
$$4(y-1)^2 = 100 - 25(x+2)^2 \quad \text{Subtract } 25(x+2)^2 \text{ from both sides.}$$
$$(y-1)^2 = \frac{100 - 25(x+2)^2}{4} \quad \text{Divide both sides by 4.}$$
$$y - 1 = \pm\frac{\sqrt{100 - 25(x+2)^2}}{2} \quad \text{Use the square-root property.}$$
$$y = 1 \pm \frac{\sqrt{100 - 25(x+2)^2}}{2} \quad \text{Add 1 to both sides.}$$

If we use window settings $[-6, 6]$ for x and $[-6, 6]$ for y and graph the functions

$$y = 1 + \frac{\sqrt{100 - 25(x+2)^2}}{2} \quad \text{and} \quad y = 1 - \frac{\sqrt{100 - 25(x+2)^2}}{2}$$

we will obtain the ellipse shown in Figure 10-20.

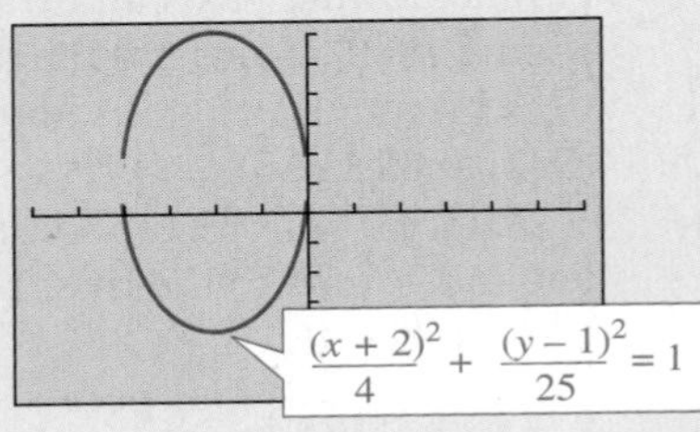

Figure 10-20

Another important form of the equation of an ellipse is called the *general form.*

General Form of the Equation of an Ellipse

The equation of any ellipse can be written in the form

$$Ax^2 + Cy^2 + Dx + Ey + F = 0$$

We can use completing the square to write the general form of the equation of an ellipse.

2 Graph an ellipse given an equation in general form.

EXAMPLE 3 Write $4x^2 + 9y^2 - 16x - 18y - 11 = 0$ in standard form to show that the equation represents an ellipse. Then graph the equation.

Solution We write the equation in standard form by completing the square on x and y:

$$4x^2 + 9y^2 - 16x - 18y - 11 = 0$$
$$4x^2 + 9y^2 - 16x - 18y = 11 \quad \text{Add 11 to both sides.}$$
$$4x^2 - 16x + 9y^2 - 18y = 11 \quad \text{Use the commutative property to rearrange terms.}$$

$$4(x^2 - 4x) + 9(y^2 - 2y) = 11$$ Factor 4 from $4x^2 - 16x$ and factor 9 from $9y^2 - 18y$ to get coefficients of 1 for the squared terms.

$$4(x^2 - 4x + 4) + 9(y^2 - 2y + 1) = 11 + 16 + 9$$ Complete the square to make $x^2 - 4x$ and $y^2 - 2y$ perfect trinomial squares. Since $16 + 9$ is added to the left side, add $16 + 9$ to the right side.

$$4(x - 2)^2 + 9(y - 1)^2 = 36$$ Factor $x^2 - 4x + 4$ and $y^2 - 2y + 1$.

$$\frac{(x - 2)^2}{9} + \frac{(y - 1)^2}{4} = 1$$ Divide both sides by 36.

Since this equation matches Equation 1 on page 704, it represents an ellipse with $h = 2$, $k = 1$, $a = 3$, and $b = 2$. Its graph is shown in Figure 10-21.

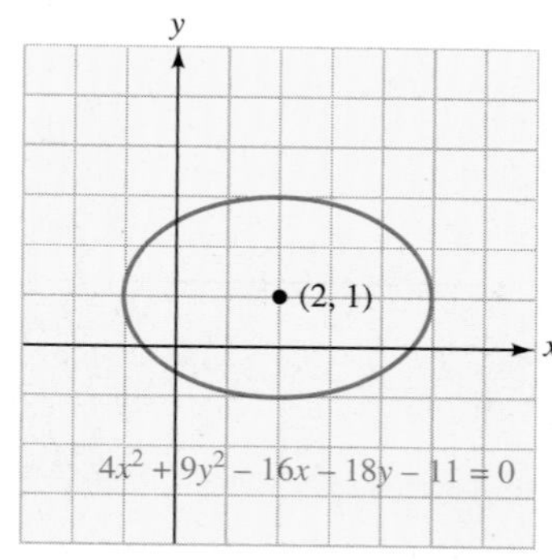

Figure 10-21

 **SELF CHECK 3** Graph: $4x^2 - 8x + 9y^2 - 36y = -4$.

COMMENT To distinguish between an equation of an ellipse and an equation of a circle, look at the coefficients of the squared terms. If the coefficients are the same, the equation is the equation of a circle. If the coefficients are different, but both positive, the equation is the equation of an ellipse.

Solve an application problem involving an ellipse.

EXAMPLE 4 **LANDSCAPE DESIGN** A landscape architect is designing an elliptical pool that will fit in the center of a 20-by-30-foot rectangular garden, leaving at least 5 feet of space on all sides. Find the equation of the ellipse.

Solution We place the rectangular garden in a coordinate system, as in Figure 10-22 on the next page. To maintain 5 feet of clearance at the ends of the ellipse, the vertices must be the points $V_1(10, 0)$ and $V_2(-10, 0)$. Similarly, the y-intercepts are the points $(0, 5)$ and $(0, -5)$.

The equation of the ellipse has the form

$$\frac{x^2}{a^2} + \frac{y^2}{b^2} = 1$$

with $a = 10$ and $b = 5$. Thus, the equation of the boundary of the pool is

$$\frac{x^2}{100} + \frac{y^2}{25} = 1$$

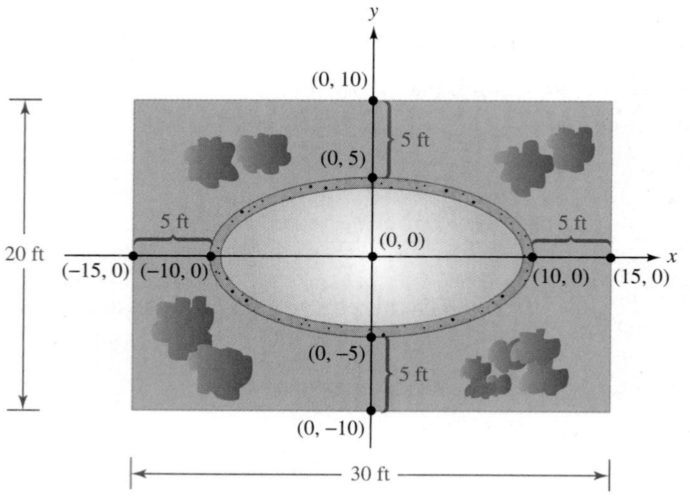

Figure 10-22

SELF CHECK ANSWERS

1.

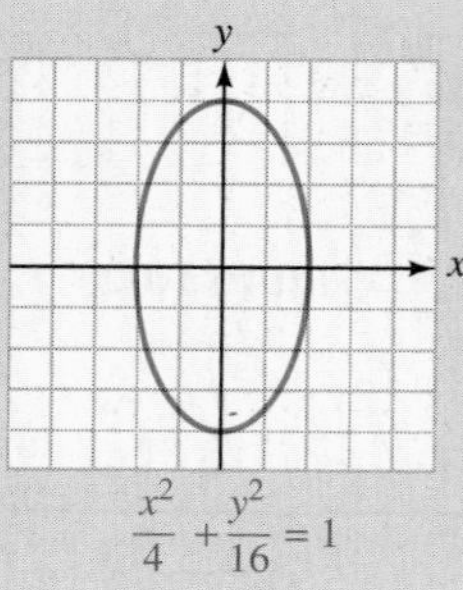

$\frac{x^2}{4} + \frac{y^2}{16} = 1$

2.

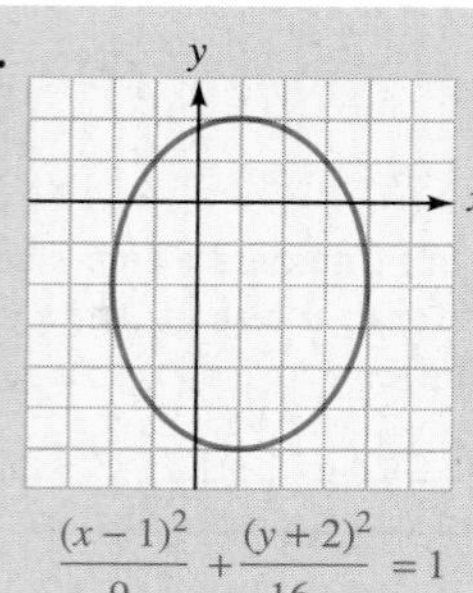

$\frac{(x-1)^2}{9} + \frac{(y+2)^2}{16} = 1$

3.

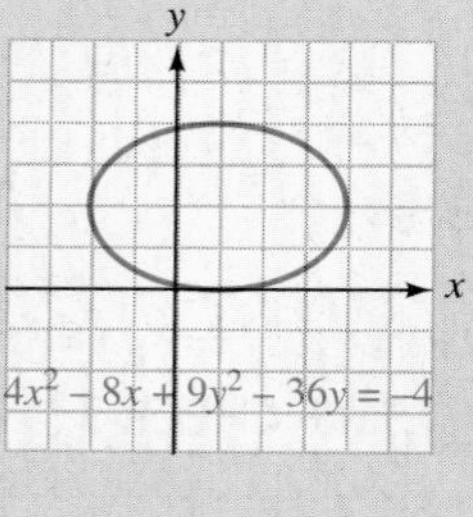

NOW TRY THIS

1. If the vertices of an ellipse are $(5, -1)$ and $(5, 5)$ and the endpoints of the minor axis are $(3, 2)$ and $(7, 2)$, find the equation.
2. The eccentricity of an ellipse is $\frac{c}{a}$. If the eccentricity is $\frac{2}{3}$, $a = 3$, and the center is located at $(4, -1)$, find the equation of the horizontal ellipse.

10.2 EXERCISES

WARM-UPS

Find the x- and y-intercepts of each ellipse.

1. $\frac{x^2}{9} + \frac{y^2}{16} = 1$
2. $\frac{x^2}{25} + \frac{y^2}{36} = 1$

Find the center of each ellipse.

3. $\frac{(x-2)^2}{9} + \frac{y^2}{16} = 1$
4. $\frac{x^2}{25} + \frac{(y+1)^2}{36} = 1$

REVIEW

Find each product.

5. $3x^{-2}y^2(4x^2 + 3y^{-2})$

6. $(2a^{-2} - b^{-2})(2a^{-2} + b^{-2})$

Write each expression without using negative exponents.

7. $\dfrac{x^{-2} + y^{-2}}{x^{-2} - y^{-2}}$

8. $\dfrac{2x^{-3} - 2y^{-3}}{4x^{-3} + 4y^{-3}}$

VOCABULARY AND CONCEPTS *Fill in the blanks.*

9. An ______ is the set of all points in a plane the _____ of whose distances from two fixed points is a constant.

10. The fixed points in Exercise 9 are the _____ of the ellipse.

11. The midpoint of the line segment joining the foci of an ellipse is called the _______ of the ellipse.

12. The graph of $\frac{x^2}{a^2} + \frac{y^2}{b^2} = 1 \quad (a > b > 0)$ has vertices at ________, y-intercepts at ________, and eccentricity of __.

13. The center of the ellipse with an equation of $\frac{x^2}{a^2} + \frac{y^2}{b^2} = 1$ is ______ with the __________ having a length of $2a$ and the minor axis having a length of ___.

14. The center of the ellipse with an equation of $\frac{(x-h)^2}{a^2} + \frac{(y-k)^2}{b^2} = 1$ is the point ______.

GUIDED PRACTICE

Graph each equation. See Example 1. (Objective 1)

15. $\dfrac{x^2}{4} + \dfrac{y^2}{9} = 1$

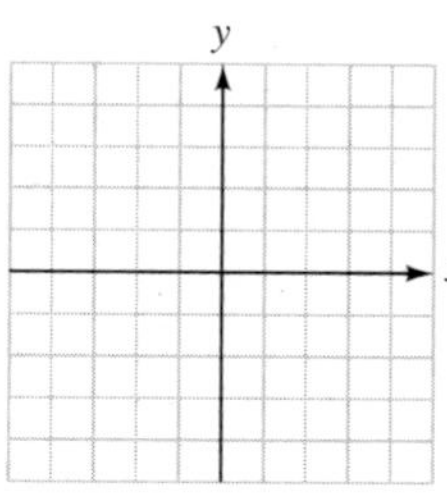

16. $x^2 + \dfrac{y^2}{9} = 1$

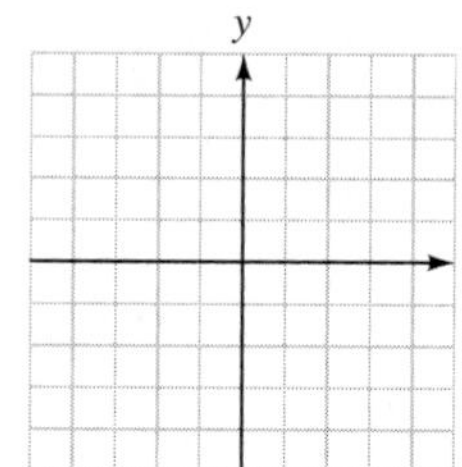

17. $\dfrac{x^2}{9} + \dfrac{y^2}{16} = 1$

18. $\dfrac{x^2}{25} + \dfrac{y^2}{36} = 1$

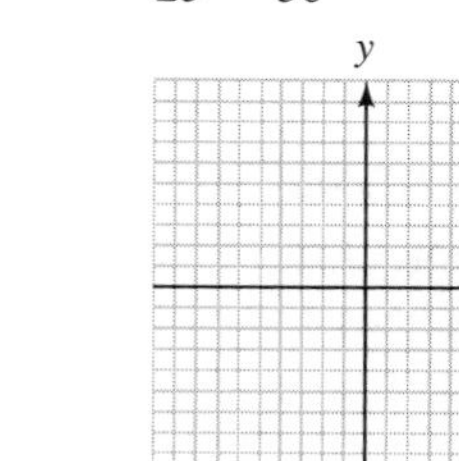

19. $\dfrac{(x-2)^2}{16} + \dfrac{y^2}{25} = 1$

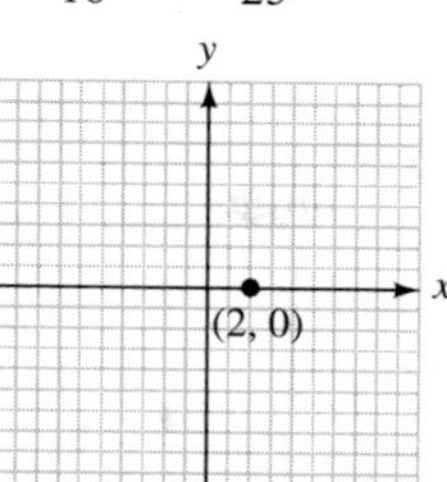

20. $\dfrac{x^2}{25} + \dfrac{(y+1)^2}{36} = 1$

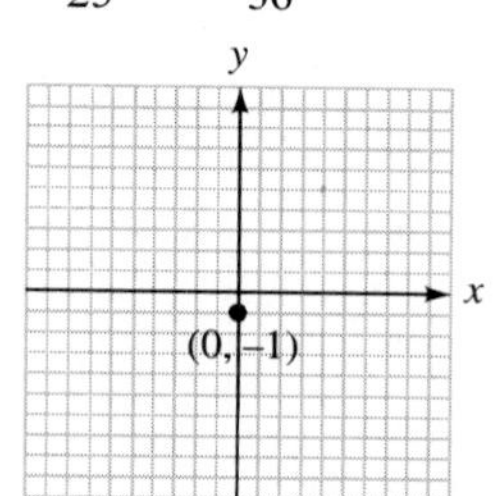

21. $\dfrac{(x-2)^2}{9} + \dfrac{(y-1)^2}{4} = 1$

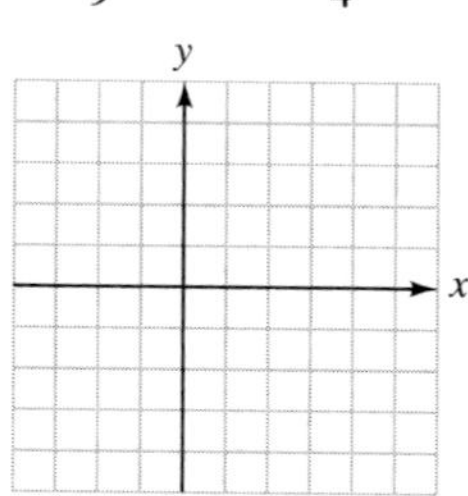

22. $\dfrac{(x-1)^2}{9} + \dfrac{(y-3)^2}{4} = 1$

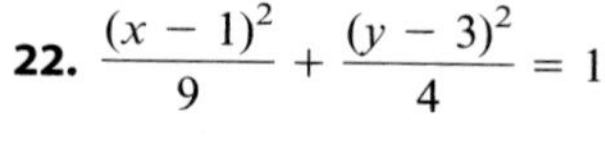

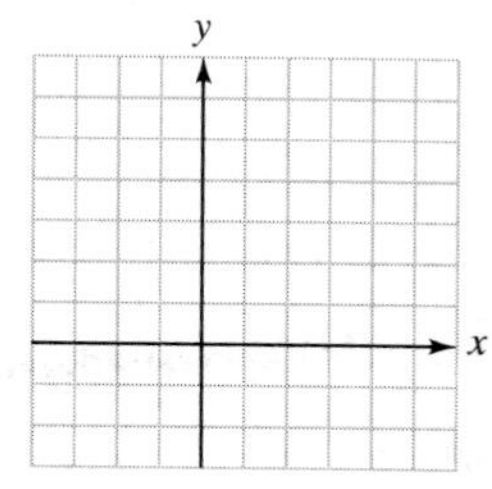

Graph each equation. See Example 2. (Objective 1)

23. $x^2 + 9y^2 = 9$

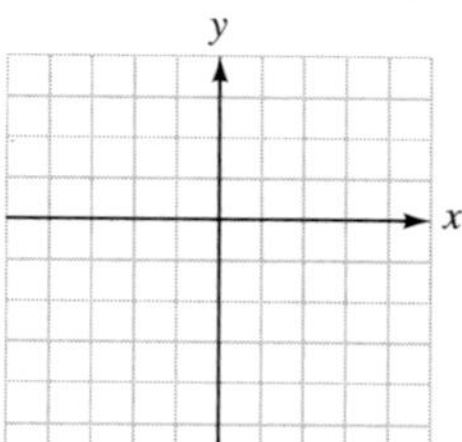

24. $25x^2 + 9y^2 = 225$

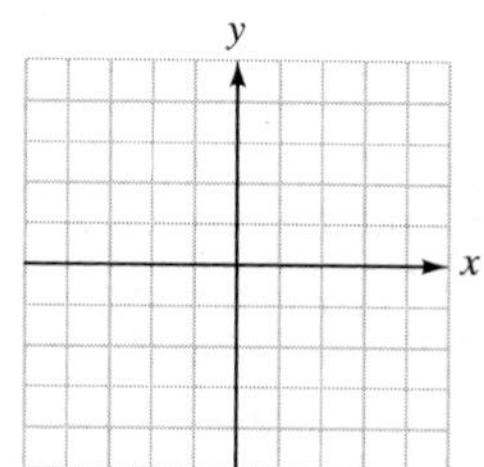

25. $16x^2 + 4y^2 = 64$

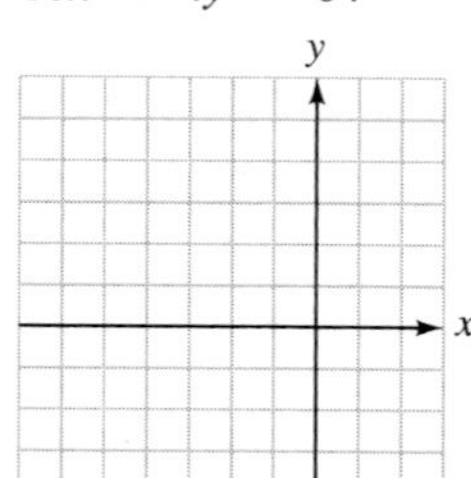

26. $4x^2 + 9y^2 = 36$

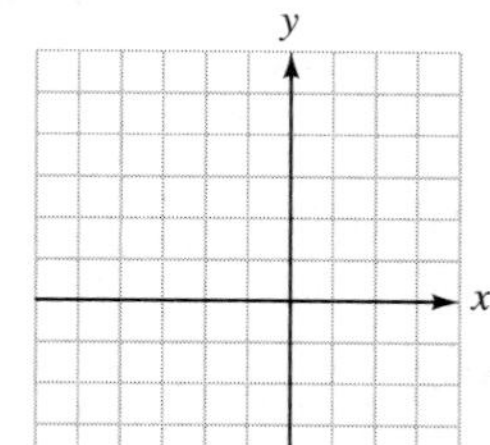

27. $(x + 1)^2 + 4(y + 2)^2 = 4$

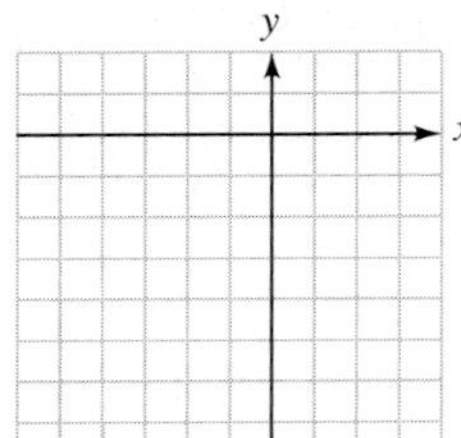

28. $9(x - 5)^2 + (y + 2)^2 = 9$

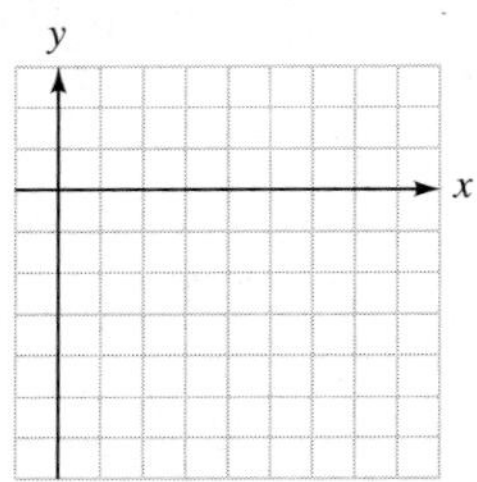

29. $25(x + 1)^2 + 9y^2 = 225$

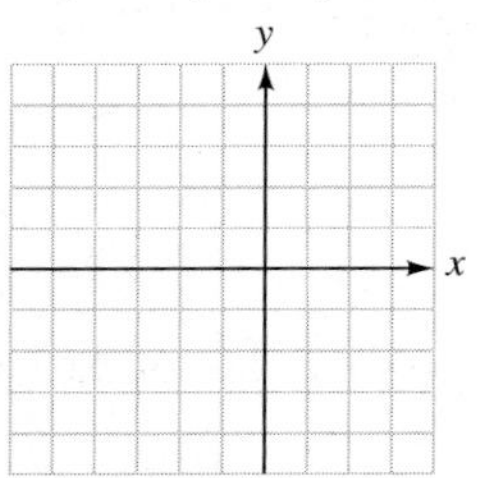

30. $4(x - 6)^2 + 25(y - 3)^2 = 100$

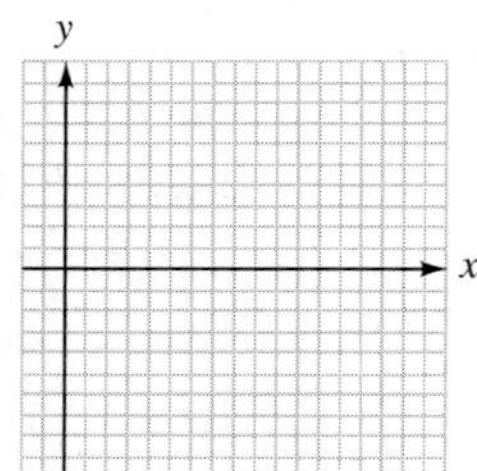

Write each equation in standard form and graph it. See Example 3. (Objective 2)

31. $x^2 + 4y^2 - 4x + 8y + 4 = 0$

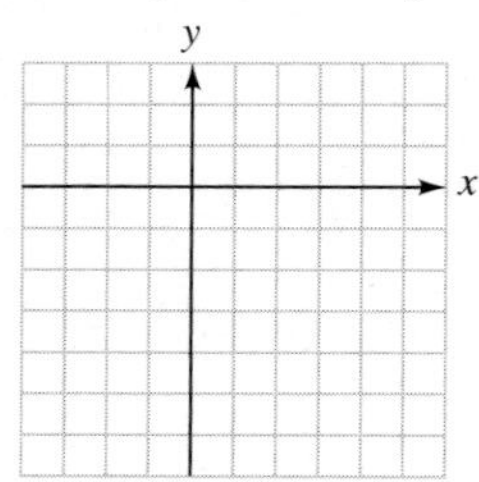

32. $x^2 + 4y^2 - 2x - 16y = -13$

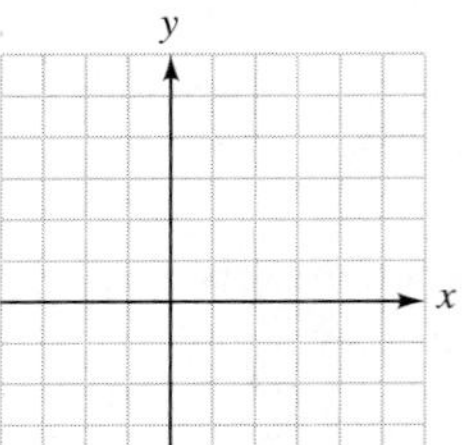

33. $9x^2 + 4y^2 - 18x + 16y = 11$

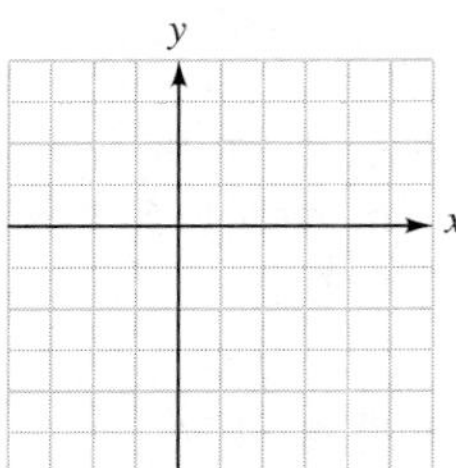

34. $16x^2 + 25y^2 - 160x - 200y + 400 = 0$

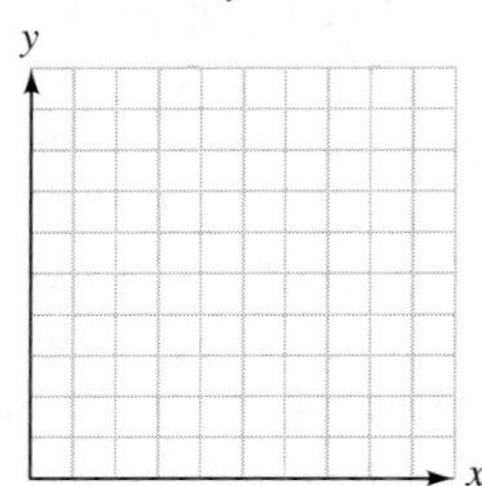

ADDITIONAL PRACTICE *Use a graphing calculator to graph each equation.*

35. $\frac{x^2}{9} + \frac{y^2}{4} = 1$

36. $x^2 + 16y^2 = 16$

37. $\frac{x^2}{4} + \frac{(y - 1)^2}{9} = 1$

38. $\frac{(x + 1)^2}{9} + \frac{(y - 2)^2}{4} = 1$

APPLICATIONS See Example 4. (Objective 3)

39. Fitness equipment With elliptical cross-training equipment, the feet move through the natural elliptical pattern that one experiences when walking, jogging, or running. Write the equation of the elliptical pattern shown on the next page.

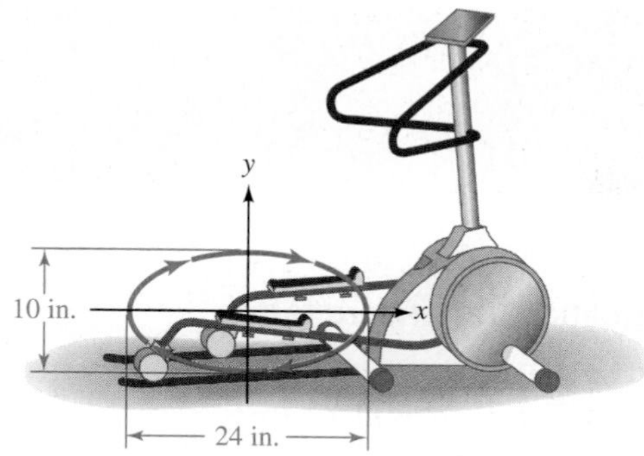

40. **Pool tables** Find the equation of the outer edge of the elliptical pool table shown below. Assume the red ball is at the focus.

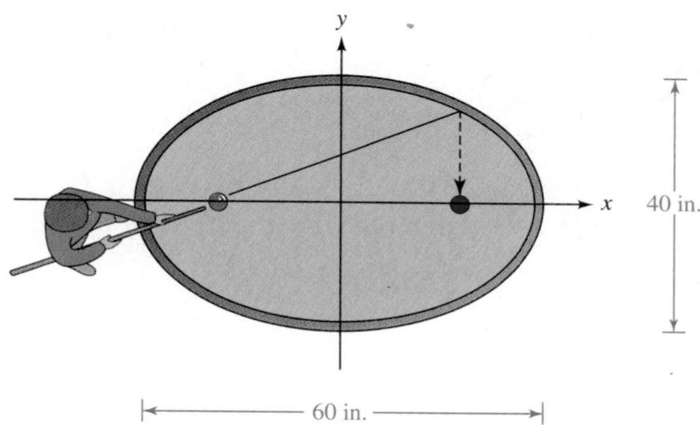

41. **Designing an underpass** The arch of the underpass is half of an ellipse. Find the equation of the arch.

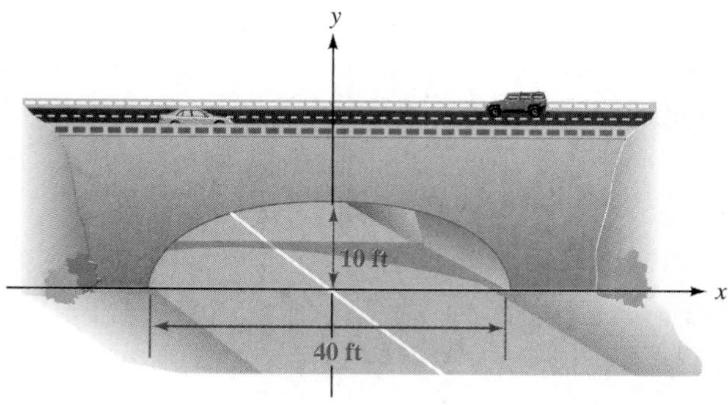

42. **Calculating clearance** Find the height of the elliptical arch in Exercise 41 at a point 10 feet from the center of the roadway.

43. **Area of an ellipse** The area A of the ellipse

$$\frac{x^2}{a^2} + \frac{y^2}{b^2} = 1$$

is given by $A = \pi ab$. Find the area of the ellipse $9x^2 + 16y^2 = 144$.

44. **Area of a track** The elliptical track is bounded by the ellipses $4x^2 + 9y^2 = 576$ and $9x^2 + 25y^2 = 900$. Find the area of the track. (See Exercise 43.)

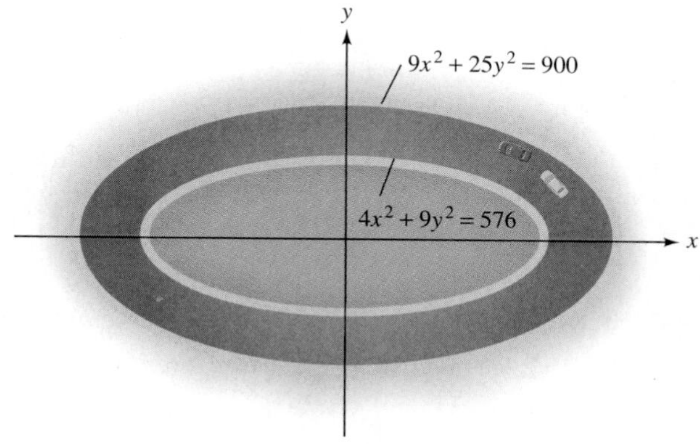

WRITING ABOUT MATH

45. Explain how to find the x- and the y-intercepts of the graph of the ellipse

$$\frac{x^2}{a^2} + \frac{y^2}{b^2} = 1$$

46. Explain the relationship between the center, focus, and vertex of an ellipse.

SOMETHING TO THINK ABOUT

47. What happens to the graph of $\frac{x^2}{a^2} + \frac{y^2}{b^2} = 1$ when $a = b$?

48. Explain why the graph of $x^2 - 2x + y^2 + 4y + 20 = 0$ does not exist.

SECTION 10.3 The Hyperbola

Objectives

1. Graph a hyperbola given an equation in standard form.
2. Graph a hyperbola given an equation in general form.
3. Graph a hyperbola of the form $xy = k$.
4. Solve an application problem involving a hyperbola.

Vocabulary

hyperbola	center	fundamental rectangle
focus (foci)	vertices	

Getting Ready

Find the value of y when $\frac{x^2}{25} - \frac{y^2}{9} = 1$ and x is the given value. Give each result to the nearest tenth.

1. $x = 6$

2. $x = -7$

The last conic section, the *hyperbola,* is a curve with two branches. In this section, we will see how to graph equations that represent hyperbolas.

Hyperbolas are the basis of a navigational system known as LORAN (LOng RAnge Navigation). (See Figure 10-23.) They are also used to find the source of a distress signal, are the basis for the design of hypoid gears, and describe the paths of some comets.

Hyperbola

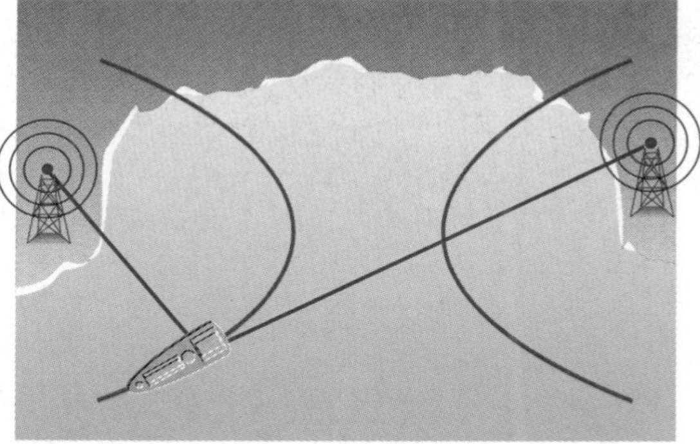

Figure 10-23

1 Graph a hyperbola given an equation in standard form.

The Hyperbola

A **hyperbola** is the set of all points P in the plane for which the difference of the distances of each point from two fixed points is a constant. See Figure 10-24, in which $d_1 - d_2$ is a constant.

Each of the two points is called a **focus.** Midway between the foci is the **center** of the hyperbola.

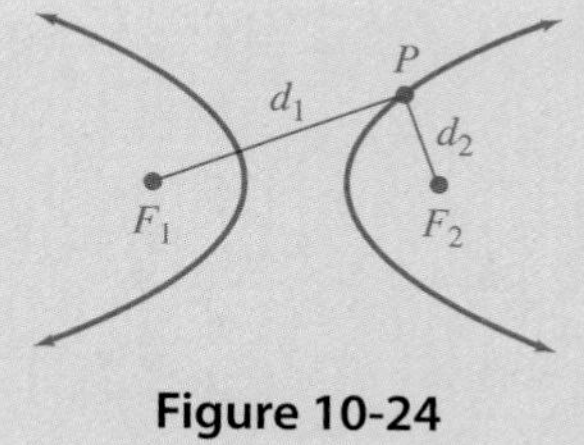

Figure 10-24

The graph of the equation

$$\frac{x^2}{25} - \frac{y^2}{9} = 1$$

is a hyperbola. To graph the equation, we make a table of ordered pairs that satisfy the equation, plot each pair, and join the points with a smooth curve as in Figure 10-25.

$$\frac{x^2}{25} - \frac{y^2}{9} = 1$$

x	y	(x, y)
-7	± 2.9	$(-7, \pm 2.9)$
-6	± 2.0	$(-6, \pm 2.0)$
-5	0	$(-5, 0)$
5	0	$(5, 0)$
6	± 2.0	$(6, \pm 2.0)$
7	± 2.9	$(7, \pm 2.9)$

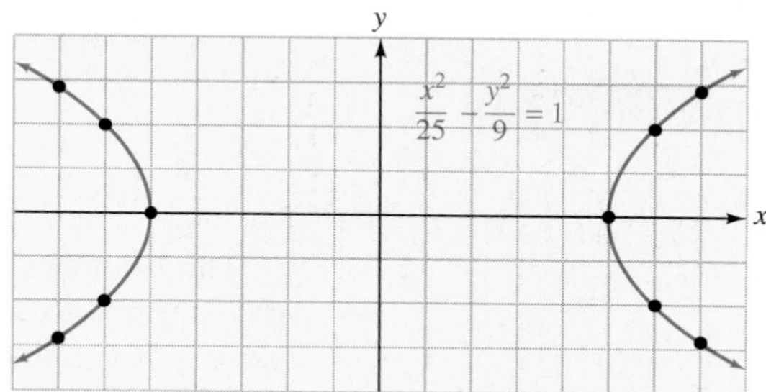

Figure 10-25

This graph is centered at the origin and intersects the x-axis at $(5, 0)$ and $(-5, 0)$. We also note that the graph does not intersect the y-axis.

It is possible to draw a hyperbola without plotting points. For example, if we want to graph the hyperbola with an equation of

$$\frac{x^2}{a^2} - \frac{y^2}{b^2} = 1$$

we first look at the x- and y-intercepts. To find the x-intercepts, we let $y = 0$ and solve for x:

$$\frac{x^2}{a^2} - \frac{0^2}{b^2} = 1$$

$$x^2 = a^2$$

$$x = \pm a$$

Thus, the hyperbola crosses the x-axis at the points $V_1(a, 0)$ and $V_2(-a, 0)$, called the **vertices** of the hyperbola. See Figure 10-26.

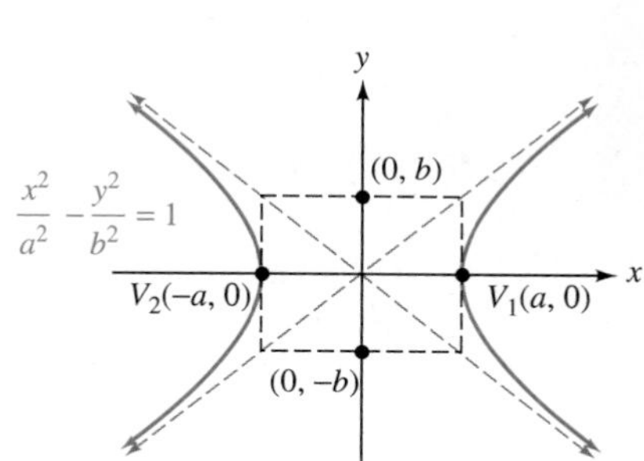

Figure 10-26

To attempt to find the y-intercepts, we let $x = 0$ and solve for y:

$$\frac{0^2}{a^2} - \frac{y^2}{b^2} = 1$$
$$y^2 = -b^2$$
$$y = \pm\sqrt{-b^2}$$

Since b^2 is always positive, $\sqrt{-b^2}$ is an imaginary number. This means that the hyperbola does not cross the y-axis.

If we construct a rectangle, called the **fundamental rectangle,** whose sides pass horizontally through $\pm b$ on the y-axis and vertically through $\pm a$ on the x-axis, the extended diagonals of the rectangle will be asymptotes of the hyperbola.

Equation of a Hyperbola Centered at the Origin with Vertices on the *x*-Axis

Any equation that can be written in the form

$$\frac{x^2}{a^2} - \frac{y^2}{b^2} = 1$$

has a graph that is a hyperbola centered at the origin, as in Figure 10-27. The x-intercepts are the vertices $V_1(a, 0)$ and $V_2(-a, 0)$. There are no y-intercepts.

The asymptotes of the hyperbola are the extended diagonals of the rectangle shown in the figure.

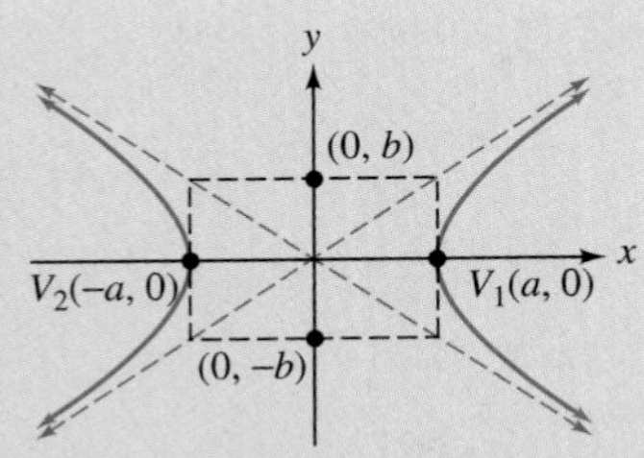

Figure 10-27

The branches of the hyperbola in previous discussions open to the left and to the right. It is possible for hyperbolas to have different orientations with respect to the x- and y-axes. For example, the branches of a hyperbola can open upward and downward. In that case, the following equation applies.

Equation of a Hyperbola Centered at the Origin with Vertices on the *y*-Axis

Any equation that can be written in the form

$$\frac{y^2}{a^2} - \frac{x^2}{b^2} = 1$$

has a graph that is a hyperbola centered at the origin, as in Figure 10-28. The y-intercepts are the vertices $V_1(0, a)$ and $V_2(0, -a)$. There are no x-intercepts.

The asymptotes of the hyperbola are the extended diagonals of the rectangle shown in the figure.

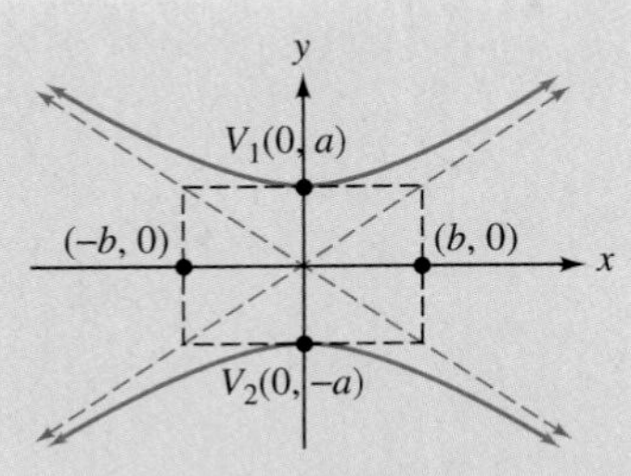

Figure 10-28

COMMENT To determine whether a hyperbola opens horizontally, as in Figure 10-27, or vertically, as in Figure 10-28, we can look at the signs of the terms. If the term containing x^2 is positive, the hyperbola will open horizontally. If the term containing y^2 is positive, the hyperbola will open vertically.

EXAMPLE 1 Graph: $9y^2 - 4x^2 = 36$.

Solution To write the equation in standard form, we divide both sides by 36 to obtain

$$\frac{9y^2}{36} - \frac{4x^2}{36} = 1$$

$$\frac{y^2}{4} - \frac{x^2}{9} = 1 \quad \text{Simplify each fraction.}$$

Because the term containing y^2 is positive, the hyperbola will open vertically. We can find the y-intercepts of the graph by letting $x = 0$ and solving for y:

$$\frac{y^2}{4} - \frac{0^2}{9} = 1$$

$$y^2 = 4$$

Thus, $y = \pm 2$, and the vertices of the hyperbola are $V_1(0, 2)$ and $V_2(0, -2)$. (See Figure 10-29.)

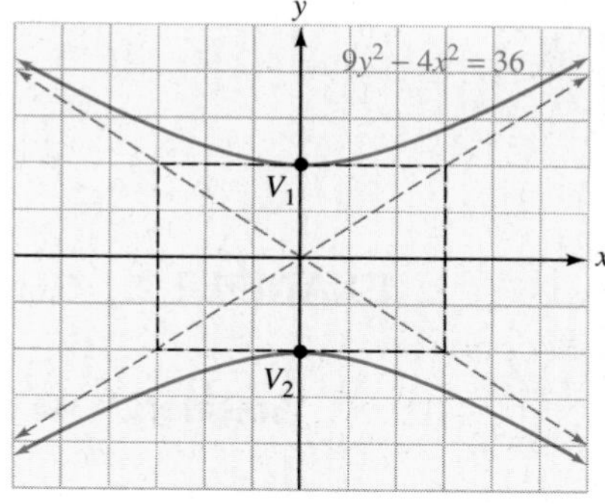

Figure 10-29

Since $\pm\sqrt{9} = \pm 3$, we can use the points (3, 0) and (−3, 0) on the x-axis to help draw the fundamental rectangle. We then draw its extended diagonals and sketch the hyperbola.

SELF CHECK 1 Graph: $9x^2 - 4y^2 = 36$.

ACCENT ON TECHNOLOGY

Graphing Hyperbolas

To graph $\frac{x^2}{9} - \frac{y^2}{16} = 1$ using a graphing calculator, we follow the same procedure that we used for circles and ellipses. To write the equation as two functions, we solve for y to get $y = \pm\frac{\sqrt{16x^2 - 144}}{3}$. Then we graph the following two functions in a square window setting to get the graph of the hyperbola shown in Figure 10-30.

$$y = \frac{\sqrt{16x^2 - 144}}{3} \quad \text{and} \quad y = -\frac{\sqrt{16x^2 - 144}}{3}$$

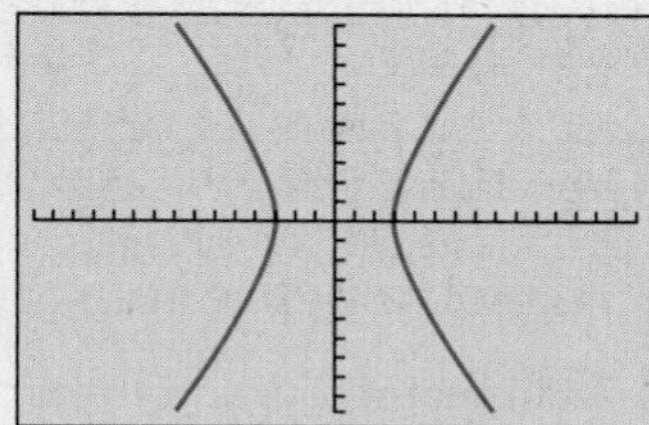
Figure 10-30

If a hyperbola is centered at a point with coordinates (h, k), the equations on the next page apply.

Equations of Hyperbolas Centered at (h, k)

Any equation that can be written in the form

$$\frac{(x-h)^2}{a^2} - \frac{(y-k)^2}{b^2} = 1$$

is a hyperbola centered at (h, k) that opens left and right.

Any equation of the form

$$\frac{(y-k)^2}{a^2} - \frac{(x-h)^2}{b^2} = 1$$

is a hyperbola centered at (h, k) that opens up and down.

EXAMPLE 2 Graph: $\frac{(x-3)^2}{16} - \frac{(y+1)^2}{4} = 1$.

Solution We write the equation in the form

$$\frac{(x-3)^2}{16} - \frac{[y-(-1)]^2}{4} = 1$$

to see that its graph will be a hyperbola centered at the point $(h, k) = (3, -1)$. Its vertices are located at $a = 4$ units to the right and left of center, at $(7, -1)$ and $(-1, -1)$. Since $b = 2$, we can count 2 units above and below center to locate points $(3, 1)$ and $(3, -3)$. With these points, we can draw the fundamental rectangle along with its extended diagonals. We can then sketch the hyperbola, as shown in Figure 10-31.

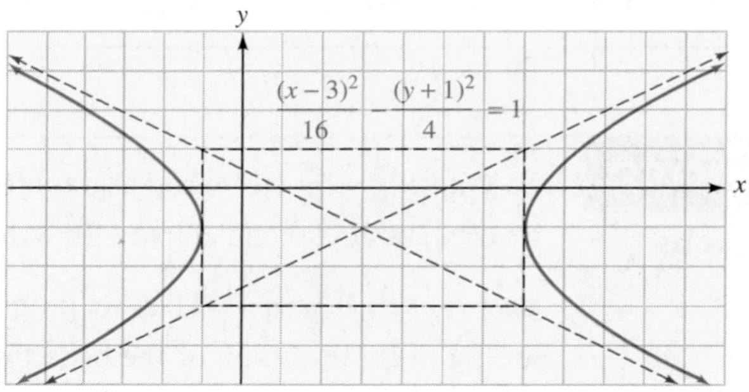

Figure 10-31

SELF CHECK 2 Graph: $\frac{(x+2)^2}{9} - \frac{(y-1)^2}{4} = 1$.

2 Graph a hyperbola given an equation in general form.

Another important form of the equation of a hyperbola is called the *general form.*

General Form of the Equation of a Hyperbola

The equation of any hyperbola can be written in the form

$$Ax^2 - Cy^2 + Dx + Ey + F = 0$$

EXAMPLE 3 Write the equation $x^2 - y^2 - 2x + 4y - 12 = 0$ in standard form to show that the equation represents a hyperbola. Then graph it.

Solution We proceed as follows.

$$x^2 - y^2 - 2x + 4y - 12 = 0$$

$$x^2 - y^2 - 2x + 4y = 12 \quad \text{Add 12 to both sides.}$$

$$x^2 - 2x - y^2 + 4y = 12 \quad \text{Use the commutative property to group the } x \text{ terms and } y \text{ terms.}$$

$$x^2 - 2x - (y^2 - 4y) = 12 \quad \text{Factor } -1 \text{ from } -y^2 + 4y.$$

We then complete the square on x and y to make $x^2 - 2x$ and $y^2 - 4y$ perfect trinomial squares.

$$x^2 - 2x + \mathbf{1} - (y^2 - 4y + \mathbf{4}) = 12 + \mathbf{1} - \mathbf{4}$$

We then factor $x^2 - 2x + 1$ and $y^2 - 4y + 4$ to get

$$(x - 1)^2 - (y - 2)^2 = 9$$

$$\frac{(x - 1)^2}{9} - \frac{(y - 2)^2}{9} = 1 \quad \text{Divide both sides by 9.}$$

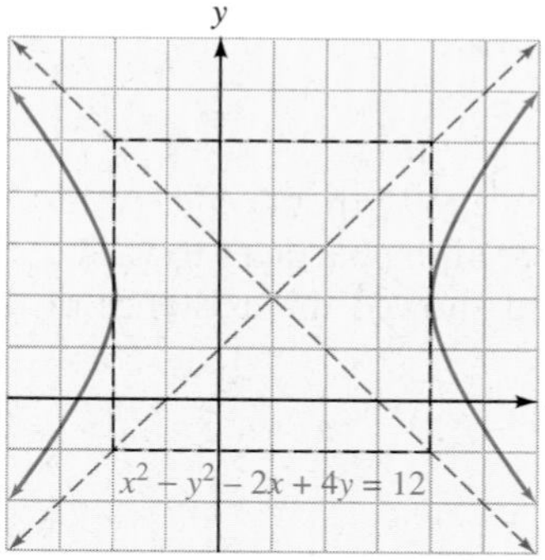

Figure 10-32

This is the equation of a hyperbola centered at (1, 2). Its graph is shown in Figure 10-32.

SELF CHECK 3 Graph: $x^2 - 4y^2 + 2x - 8y = 7$.

3 Graph a hyperbola of the form $xy = k$.

There is a special type of hyperbola (also centered at the origin) that does not intersect either the x- or the y-axis. These hyperbolas have equations of the form $xy = k$, where $k \neq 0$.

EXAMPLE 4 Graph: $xy = -8$.

Solution We make a table of ordered pairs, plot each pair, and join the points with a smooth curve to obtain the hyperbola in Figure 10-33.

$xy = -8$

x	y	(x, y)
1	−8	(1, −8)
2	−4	(2, −4)
4	−2	(4, −2)
8	−1	(8, −1)
−1	8	(−1, 8)
−2	4	(−2, 4)
−4	2	(−4, 2)
−8	1	(−8, 1)

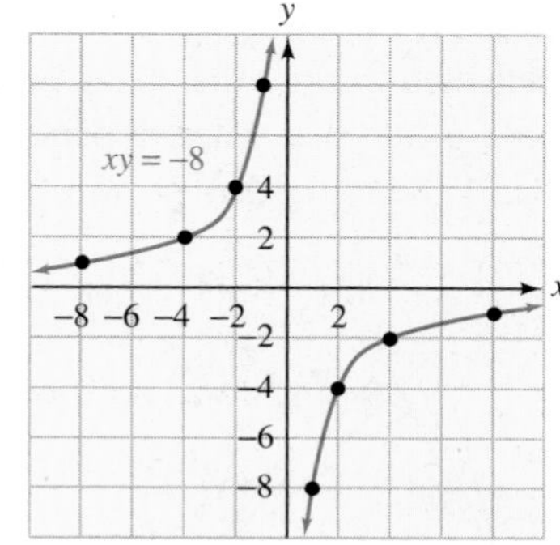

Figure 10-33

SELF CHECK 4 Graph: $xy = 6$.

The result in Example 4 illustrates the following general equation.

Equations of Hyperbolas of the Form $xy = k$

Any equation of the form $xy = k$, where $k \neq 0$, has a graph that is a hyperbola that does not intersect either the x-axis or the y-axis.

Solve an application problem involving a hyperbola.

EXAMPLE 5 **ATOMIC STRUCTURE** In an experiment that led to the discovery of the atomic structure of matter, Lord Rutherford (1871–1937) shot high-energy alpha particles toward a thin sheet of gold. Because many were reflected, Rutherford showed the existence of

EVERYDAY CONNECTIONS Focus on Conics

Satellite dishes and flashlight reflectors are familiar examples of a conic's ability to reflect a beam of light or to concentrate incoming satellite signals at one point. That property is shown in the following illustration.

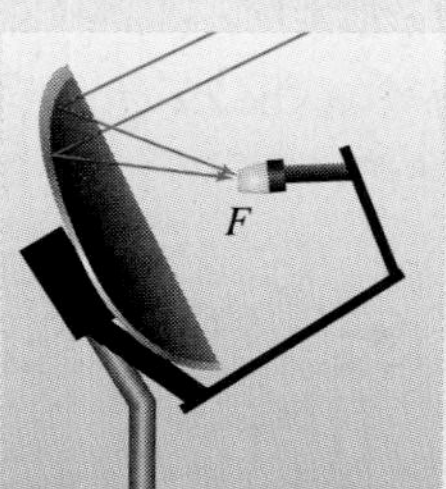

An ellipse has two foci, the points marked in the following illustration. Any light or signal that starts at one focus will be reflected to the other. This property is the basis of whispering galleries, where a person standing at one focus can clearly hear another person speaking at the other focus.

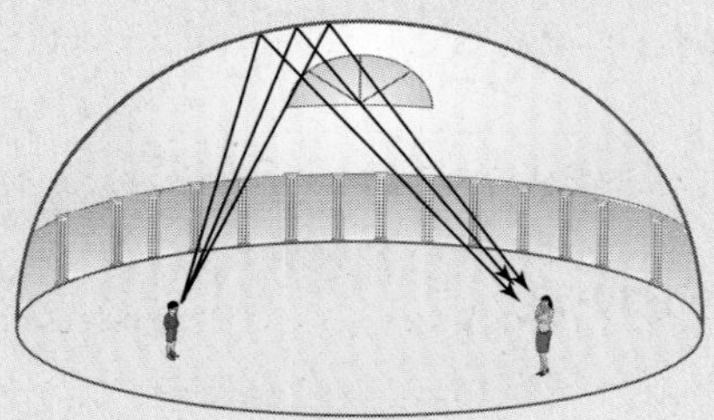

The focal property of the ellipse is also used in *lithotripsy,* a medical procedure for treating kidney stones. The patient is placed in an elliptical tank of water with the kidney stone at one focus. Shock waves from a small controlled explosion at the other focus are concentrated on the stone, pulverizing it.

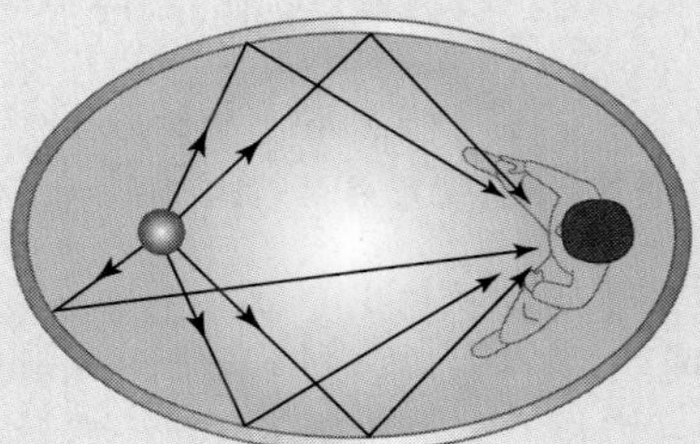

The hyperbola also has two foci, the two points labeled F in the following illustration. As in the ellipse, light aimed at one focus is reflected toward the other. Hyperbolic mirrors are used in some reflecting telescopes.

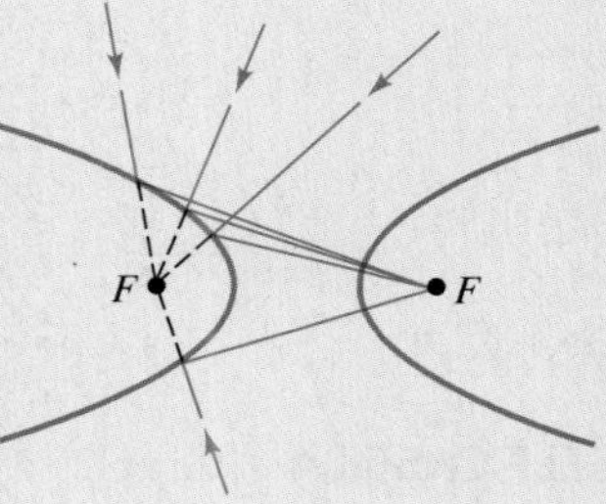

the nucleus of a gold atom. The alpha particle in Figure 10-34 is repelled by the nucleus at the origin; it travels along the hyperbolic path given by $4x^2 - y^2 = 16$. How close does the particle come to the nucleus?

Solution To find the distance from the nucleus at the origin, we must find the coordinates of the vertex V. To do so, we write the equation of the particle's path in standard form:

$$4x^2 - y^2 = 16$$

$$\frac{4x^2}{16} - \frac{y^2}{16} = \frac{16}{16} \quad \text{Divide both sides by 16.}$$

$$\frac{x^2}{4} - \frac{y^2}{16} = 1 \quad \text{Simplify.}$$

$$\frac{x^2}{2^2} - \frac{y^2}{4^2} = 1 \quad \text{Write 4 as } 2^2 \text{ and 16 as } 4^2.$$

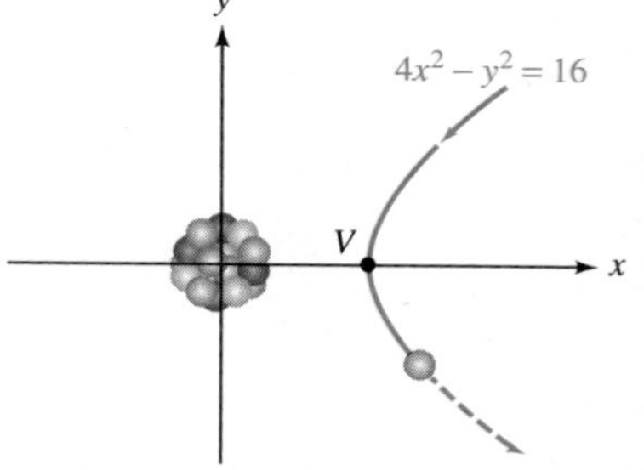

Figure 10-34

This equation is in the form

$$\frac{x^2}{a^2} - \frac{y^2}{b^2} = 1$$

with $a = 2$. Thus, the vertex of the path is (2, 0). The particle is never closer than 2 units from the nucleus.

SELF CHECK ANSWERS

1. 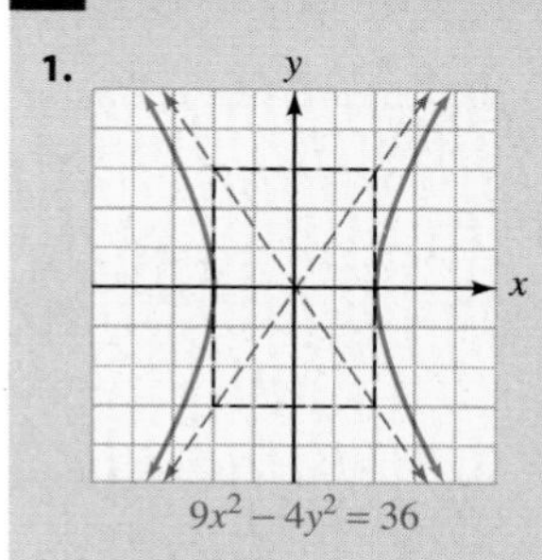

$9x^2 - 4y^2 = 36$

2. 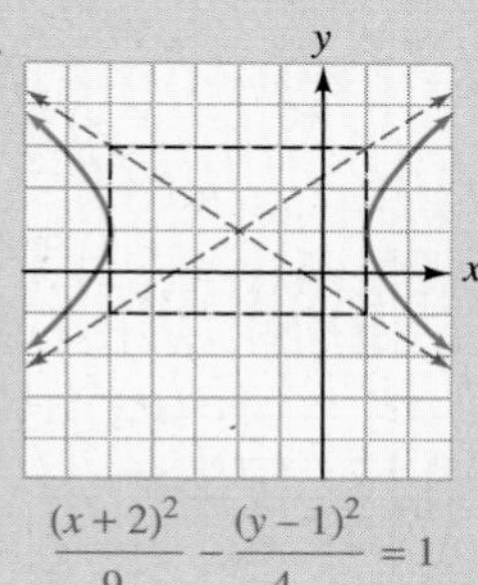

$\frac{(x+2)^2}{9} - \frac{(y-1)^2}{4} = 1$

3. 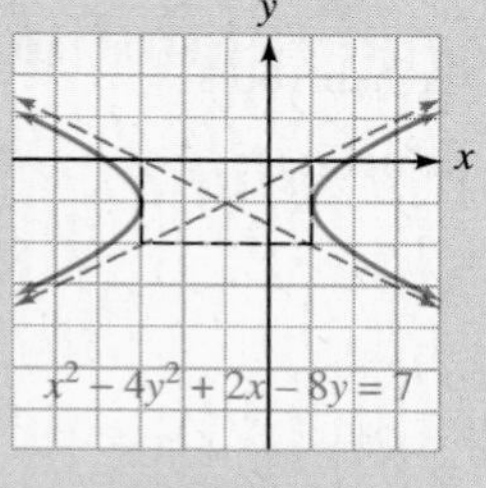

4. 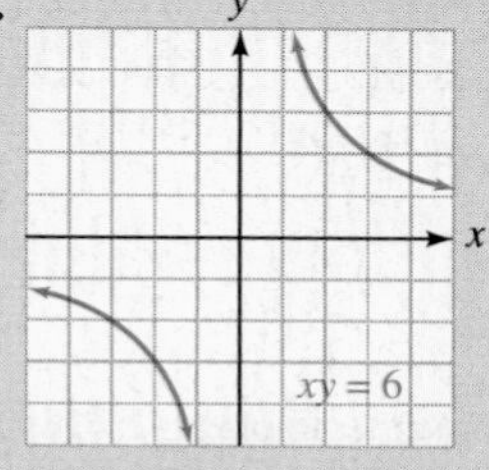

NOW TRY THIS

Given the equation $\frac{x^2}{a^2} - \frac{y^2}{b^2} = 1$, the equations for the asymptotes are $y = \frac{b}{a}x$, and $y = -\frac{b}{a}x$.

1. Find the equations for the asymptotes for the hyperbola described by the equation $\frac{x^2}{25} - \frac{y^2}{9} = 1$.
2. Find the equations for the asymptotes for the hyperbola described by the equation $\frac{y^2}{4} - \frac{x^2}{9} = 1$.
3. Find the equations for the asymptotes for the hyperbola described by the equation $\frac{(x + 2)^2}{9} - \frac{(y - 1)^2}{4} = 1$.

10.3 EXERCISES

WARM-UPS *Find the x- or y-intercepts of each hyperbola.*

1. $\dfrac{x^2}{9} - \dfrac{y^2}{16} = 1$

2. $\dfrac{x^2}{25} - \dfrac{y^2}{36} = 1$

REVIEW *Factor each expression.*

3. $-6x^4 + 9x^3 - 6x^2$

4. $4a^2 - b^2$

5. $15a^2 - 4ab - 4b^2$

6. $8p^3 - 27q^3$

VOCABULARY AND CONCEPTS *Fill in the blanks.*

7. A __________ is the set of all points in a plane for which the __________ of the distances from two fixed points is a constant.

8. The fixed points in Exercise 7 are the _____ of the hyperbola.

9. The midpoint of the line segment joining the foci of a hyperbola is called the ______ of the hyperbola.

10. To graph a hyperbola, we locate its center and vertices, sketch the __________________, and sketch the asymptotes.

11. The hyperbolic graph of $\frac{x^2}{a^2} - \frac{y^2}{b^2} = 1$ has x-intercepts of _______. There are no ___________.

12. The center of the hyperbola with an equation of $\frac{x^2}{a^2} - \frac{y^2}{b^2} = 1$ is the point ______. The center of the hyperbola with an equation of $\frac{(x-h)^2}{a^2} - \frac{(y-k)^2}{b^2} = 1$ is the point ______.

GUIDED PRACTICE

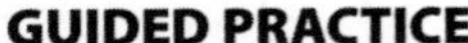

Graph each hyperbola. **See Examples 1–2. (Objective 1)**

13. $\dfrac{x^2}{9} - \dfrac{y^2}{4} = 1$

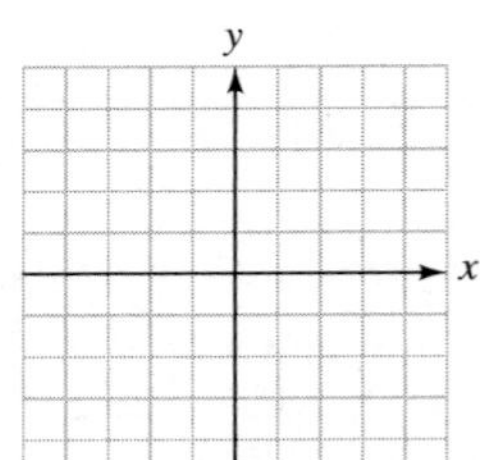

14. $\dfrac{x^2}{4} - \dfrac{y^2}{4} = 1$

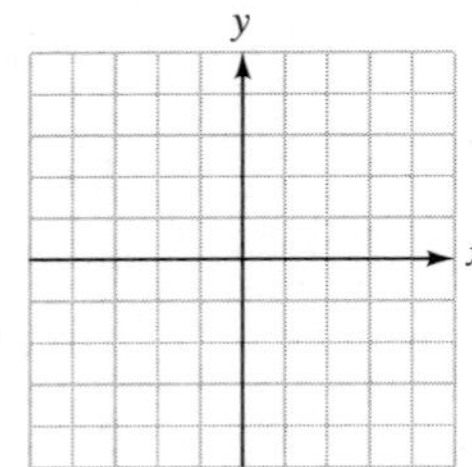

15. $\dfrac{y^2}{4} - \dfrac{x^2}{9} = 1$

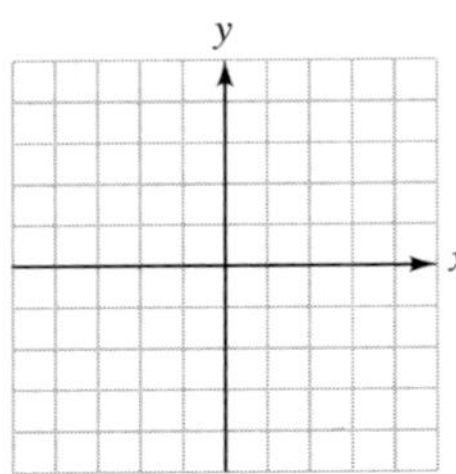

16. $\dfrac{y^2}{4} - \dfrac{x^2}{64} = 1$

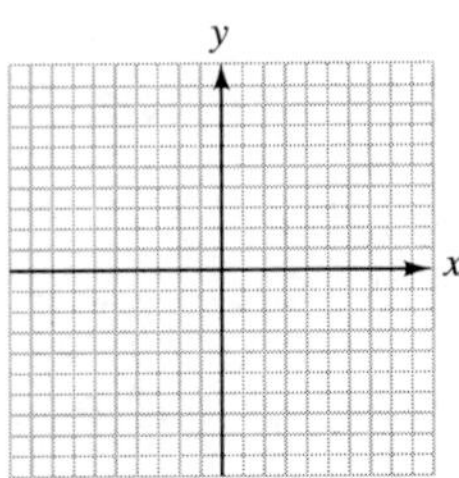

17. $25x^2 - y^2 = 25$

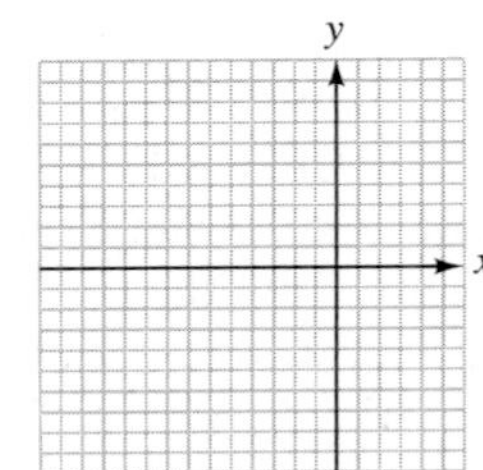

18. $9x^2 - 4y^2 = 36$

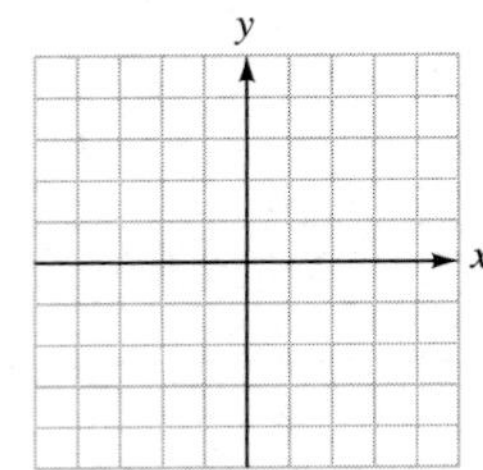

19. $\dfrac{(x-2)^2}{9} - \dfrac{y^2}{16} = 1$

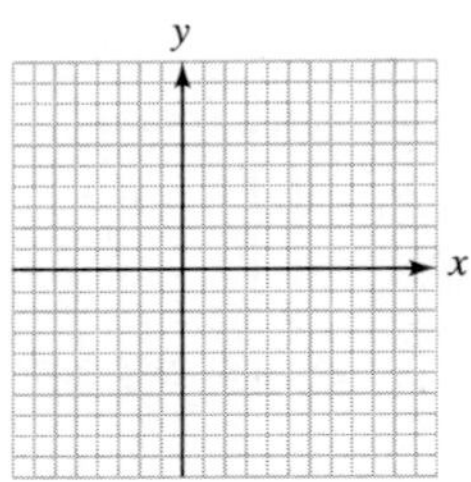

20. $\dfrac{(x+2)^2}{16} - \dfrac{(y-3)^2}{25} = 1$

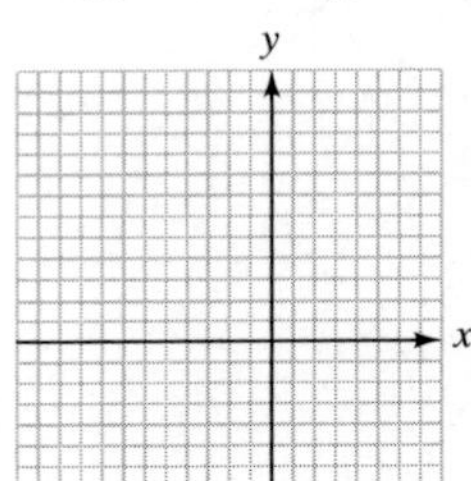

21. $\dfrac{(y+1)^2}{1} - \dfrac{(x-2)^2}{4} = 1$

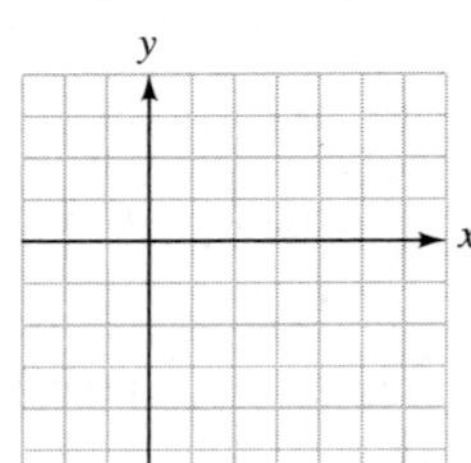

22. $\dfrac{(y-2)^2}{4} - \dfrac{(x+1)^2}{1} = 1$

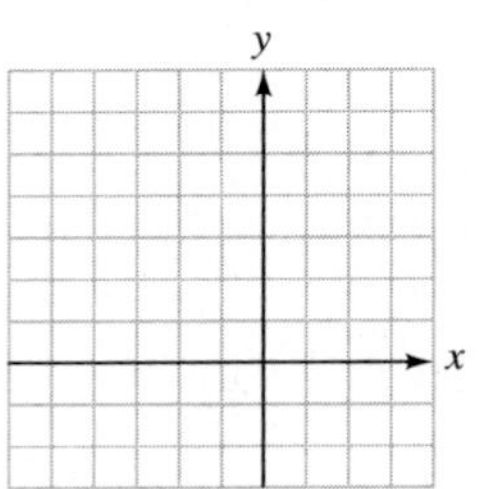

23. $4(x + 3)^2 - (y - 1)^2 = 4$

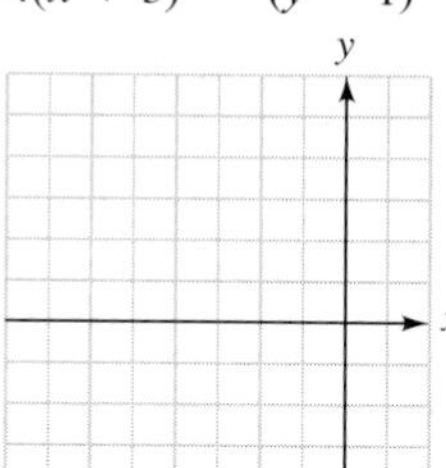

24. $(x + 5)^2 - 16y^2 = 16$

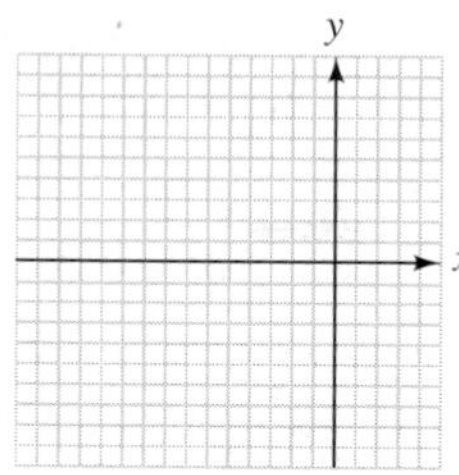

Write each equation in standard form and graph it. **See Example 3. (Objective 2)**

25. $4x^2 - y^2 + 8x - 4y = 4$

26. $x^2 - 9y^2 - 4x - 54y = 86$

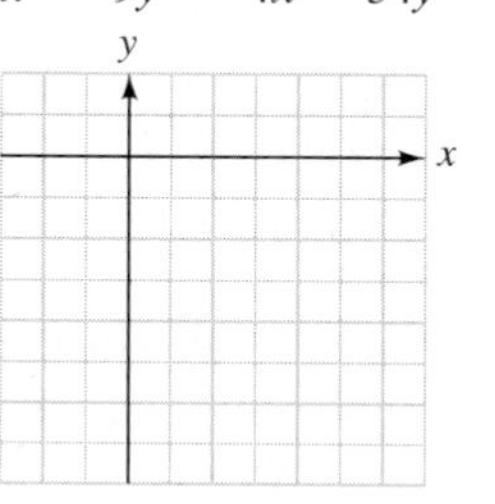

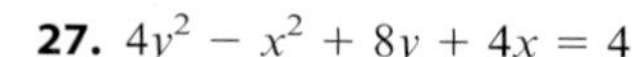

27. $4y^2 - x^2 + 8y + 4x = 4$

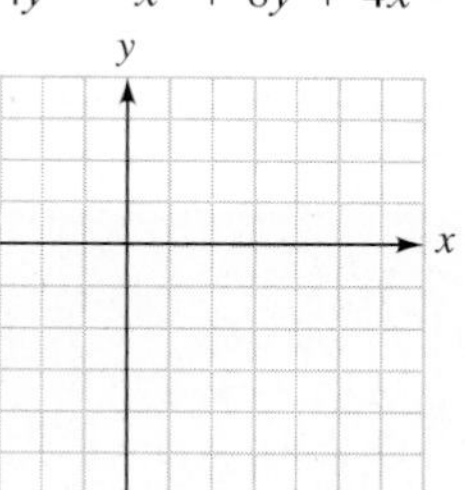

28. $y^2 - 4x^2 - 4y - 8x = 4$

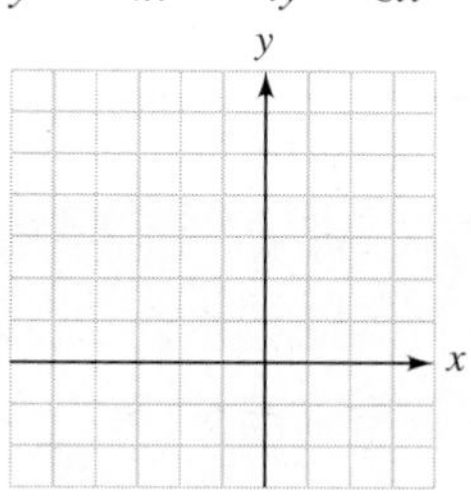

Graph each hyperbola. **See Example 4. (Objective 3)**

29. $xy = 8$

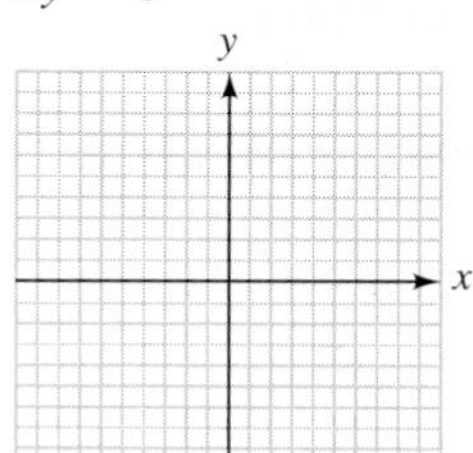

30. $xy = -10$

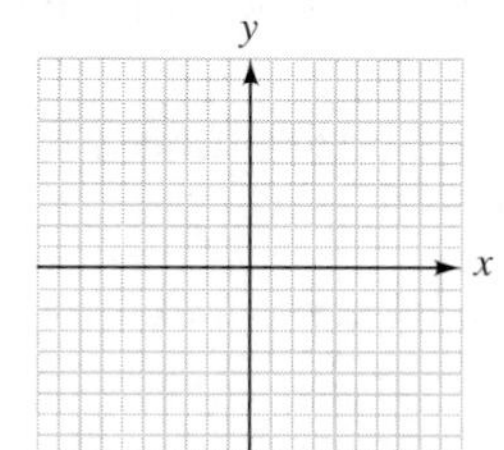

31. $xy = -12$

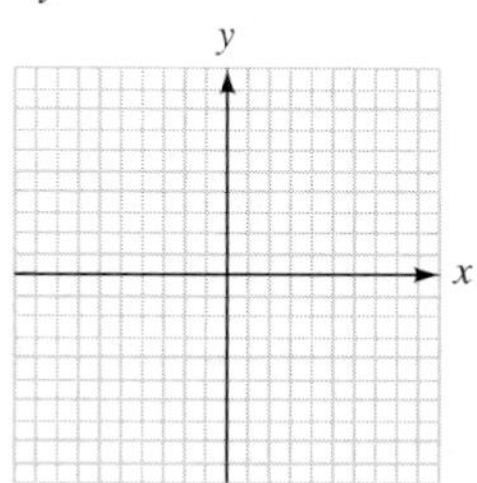

32. $xy = 6$

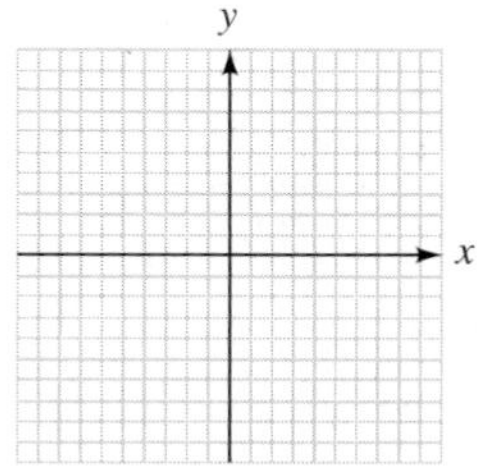

ADDITIONAL PRACTICE *Use a graphing calculator to graph each equation.*

33. $\dfrac{x^2}{9} - \dfrac{y^2}{4} = 1$

34. $y^2 - 16x^2 = 16$

35. $\dfrac{x^2}{4} - \dfrac{(y - 1)^2}{9} = 1$

36. $\dfrac{(y + 1)^2}{9} - \dfrac{(x - 2)^2}{4} = 1$

APPLICATIONS See Example 5. (Objective 4)

37. Alpha particles The particle in the illustration approaches the nucleus at the origin along the path $9y^2 - x^2 = 81$. How close does the particle come to the nucleus?

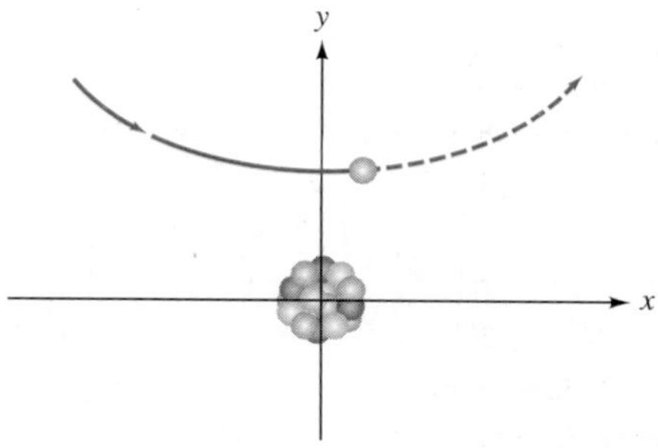

38. LORAN By determining the difference of the distances between the ship and two land-based radio transmitters, the LORAN system places the ship on the hyperbola $x^2 - 4y^2 = 576$. If the ship is also 5 miles out to sea, find its coordinates.

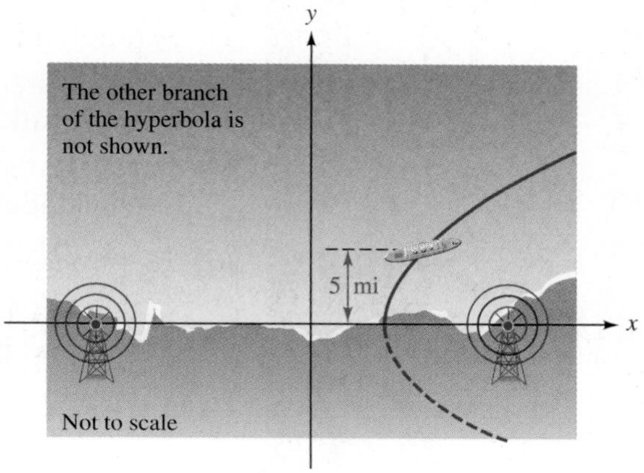

39. Sonic boom The position of the sonic boom caused by faster-than-sound aircraft is the hyperbola $y^2 - x^2 = 25$ in the coordinate system shown on the next page. How wide is the hyperbola 5 units from its vertex?

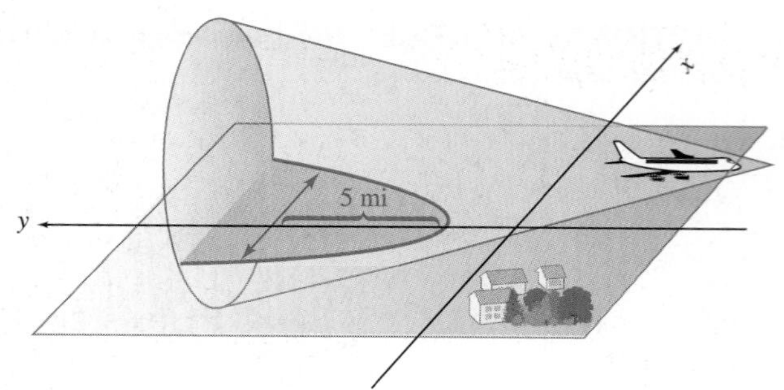

40. **Electrostatic repulsion** Two similarly charged particles are shot together for an almost head-on collision, as shown in the illustration. They repel each other and travel the two branches of the hyperbola given by $x^2 - 4y^2 = 4$. How close do they get?

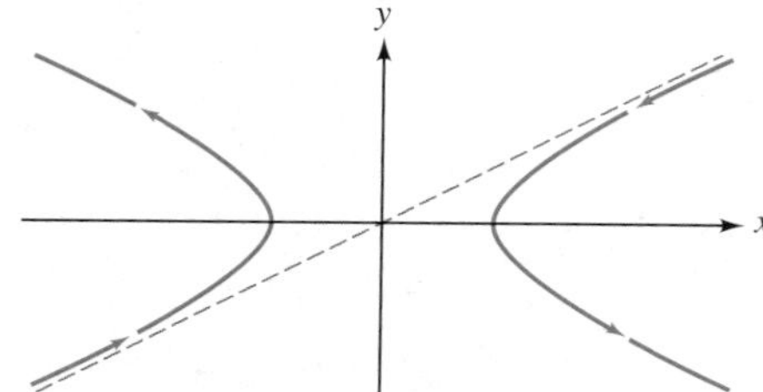

WRITING ABOUT MATH

41. Explain how to find the x- and the y-intercepts of the graph of the hyperbola

$$\frac{x^2}{a^2} - \frac{y^2}{b^2} = 1$$

42. Explain why the graph of the hyperbola

$$\frac{x^2}{a^2} - \frac{y^2}{b^2} = 1$$

has no y-intercept.

SOMETHING TO THINK ABOUT

43. Describe the fundamental rectangle of

$$\frac{x^2}{a^2} - \frac{y^2}{b^2} = 1$$

when $a = b$.

44. The hyperbolas $x^2 - y^2 = 1$ and $y^2 - x^2 = 1$ are called **conjugate hyperbolas.** Graph both on the same axes. What do they have in common?

SECTION 10.4 Systems Containing Second-Degree Equations

Objectives

1. Solve a system of equations containing a second-degree equation by graphing.
2. Solve a system of equations containing a second-degree equation by substitution.
3. Solve a system of equations containing a second-degree equation by elimination (addition).

Getting Ready

Add the left sides and the right sides of the following equations.

1. $\begin{array}{r} 3x^2 + 2y^2 = 12 \\ \underline{-3x^2 - 3y^2 = 32} \end{array}$

2. $\begin{array}{r} -7x^2 - 5y^2 = -17 \\ \underline{12x^2 + 5y^2 = 25} \end{array}$

We now discuss ways to solve systems of two equations in two variables where at least one of the equations is of second degree.

1 Solve a system of equations containing a second-degree equation by graphing.

EXAMPLE 1 Solve $\begin{cases} x^2 + y^2 = 25 \\ 2x + y = 10 \end{cases}$ by graphing.

Solution The graph of $x^2 + y^2 = 25$ is a circle with center at the origin and radius of 5. The graph of $2x + y = 10$ is a line. Depending on whether the line is a secant (intersecting the circle at two points) or a tangent (intersecting the circle at one point) or does not intersect the circle at all, there are two, one, or no solutions to the system, respectively.

After graphing the circle and the line, as shown in Figure 10-35, we see that there are two intersection points (3, 4) and (5, 0). These are the solutions of the system.

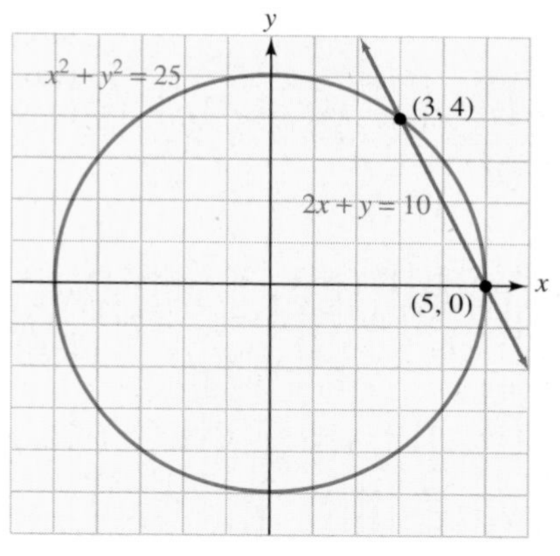

Figure 10-35

SELF CHECK 1 Solve by graphing: $\begin{cases} x^2 + y^2 = 13 \\ y = -\frac{1}{5}x + \frac{13}{5}. \end{cases}$

ACCENT ON TECHNOLOGY

Solving Systems of Equations

To solve Example 1 with a graphing calculator, we graph the circle and the line on one set of coordinate axes (see Figure 10-36(a)). We then find approximations of the coordinates of the intersection points of the graphs (see Figure 10-36(b) and Figure 10-36(c)).

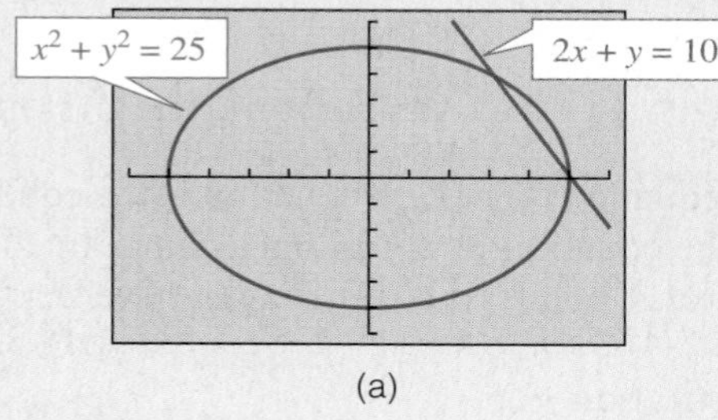

(a)

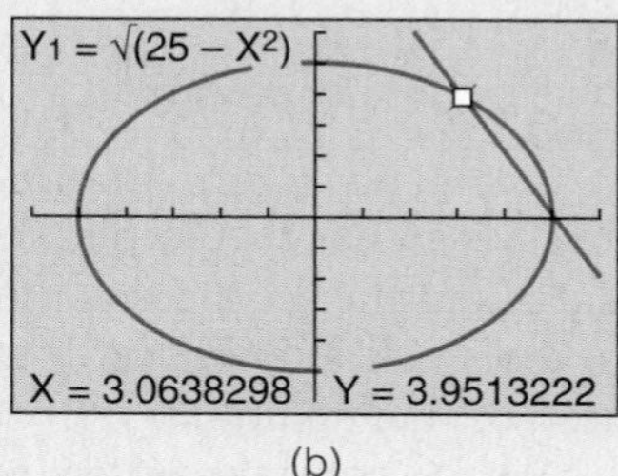

(b)

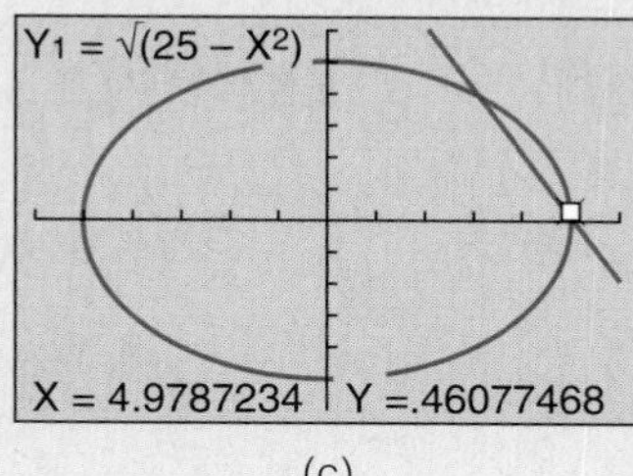

(c)

Figure 10-36

We can find the exact answers by using the INTERSECT feature found in the CALC menu.

2 Solve a system of equations containing a second-degree equation by substitution.

Algebraic methods also can be used to solve systems of equations. In the next three examples, we will solve a system using the substitution method.

EXAMPLE 2 Solve using substitution: $\begin{cases} x^2 + y^2 = 25 \\ 2x + y = 10 \end{cases}$.

Solution This system has one second-degree equation and one first-degree equation. The first equation is the equation of a circle and the second equation is the equation of a line. Since a line can intersect a circle in 0, 1, or 2 points, this system can have zero, one, or two solutions.

We can solve the system by substitution. Solving the linear equation for y gives

$$2x + y = 10$$

$$(1) \qquad y = -2x + 10$$

We can substitute $-2x + 10$ for y in the second-degree equation and solve the resulting quadratic equation for x:

$$x^2 + y^2 = 25$$

$$x^2 + (\mathbf{-2x + 10})^2 = 25$$

$$x^2 + 4x^2 - 40x + 100 = 25 \qquad (-2x + 10)(-2x + 10) = 4x^2 - 40x + 100$$

$$5x^2 - 40x + 75 = 0 \qquad \text{Combine like terms and subtract 25 from both sides.}$$

$$x^2 - 8x + 15 = 0 \qquad \text{Divide both sides by 5.}$$

$$(x - 5)(x - 3) = 0 \qquad \text{Factor } x^2 - 8x + 15.$$

$$x - 5 = 0 \quad \text{or} \quad x - 3 = 0 \qquad \text{Set each factor equal to 0.}$$

$$x = 5 \quad \Big| \quad x = 3$$

COMMENT Note that if we substitute 5 and 3 into the equation $x^2 + y^2 = 25$, we will get ordered pairs that do not satisfy the equation $2x + y = 10$.

If we substitute 5 for x in Equation 1, we get $y = 0$. If we substitute 3 for x in Equation 1, we get $y = 4$. The two solutions of this system are (5, 0) and (3, 4).

SELF CHECK 2 Solve using substitution: $\begin{cases} x^2 + y^2 = 13 \\ y = -\frac{1}{5}x + \frac{13}{5} \end{cases}$.

EXAMPLE 3 Solve using substitution: $\begin{cases} 4x^2 + 9y^2 = 5 \\ y = x^2 \end{cases}$.

Solution This system has two second-degree equations. The first equation is the equation of an ellipse and the second equation is the equation of a parabola. Since an ellipse and a parabola can intersect in 0, 1, 2, 3, or 4 points, this system can have zero, one, two, three, or four solutions.

We can solve the system by substitution.

$$4x^2 + 9y^2 = 5$$

$$4y + 9y^2 = 5 \qquad \text{Substitute } y \text{ for } x^2.$$

$$9y^2 + 4y - 5 = 0 \qquad \text{Subtract 5 from both sides.}$$

$$(9y - 5)(y + 1) = 0 \qquad \text{Factor } 9y^2 + 4y - 5.$$

$$9y - 5 = 0 \quad \text{or} \quad y + 1 = 0 \qquad \text{Set each factor equal to 0.}$$

$$y = \frac{5}{9} \quad \Big| \quad y = -1$$

Since $y = x^2$, the values of x are found by solving the equations

$$x^2 = \frac{5}{9} \qquad \text{and} \qquad x^2 = -1$$

Because $x^2 = -1$ has no real solutions, this possibility is discarded. The solutions of $x^2 = \frac{5}{9}$ are

$$x = \frac{\sqrt{5}}{3} \quad \text{or} \quad x = -\frac{\sqrt{5}}{3}$$

The two solutions of the system are

$$\left(\frac{\sqrt{5}}{3}, \frac{5}{9}\right) \quad \text{and} \quad \left(-\frac{\sqrt{5}}{3}, \frac{5}{9}\right)$$

SELF CHECK 3 Solve using substitution: $\begin{cases} x^2 + y^2 = 20 \\ y = x^2 \end{cases}$.

3 Solve a system of equations containing a second-degree equation by elimination (addition).

In the next example, we will solve a system by elimination.

EXAMPLE 4 Solve by elimination: $\begin{cases} 3x^2 + 2y^2 = 36 \\ 4x^2 - y^2 = 4 \end{cases}$.

Solution This system has two second-degree equations. The first equation is the equation of an ellipse and the second equation is the equation of a hyperbola. Since an ellipse and a hyperbola can intersect in 0, 1, 2, 3, or 4 points, this system can have 0, 1, 2, 3, or 4 solutions.

Since both equations are in the form $ax^2 + by^2 = c$, we can solve the system by elimination.

We can keep the first equation and multiply the second equation by 2 to obtain the equivalent system

$$\begin{cases} 3x^2 + 2y^2 = 36 \\ 8x^2 - 2y^2 = 8 \end{cases}$$

We add the equations to eliminate y and solve the resulting equation for x:

$$11x^2 = 44$$
$$x^2 = 4$$
$$x = 2 \quad \text{or} \quad x = -2$$

To find y, we substitute 2 for x and then -2 for x in the first equation and proceed as follows:

For x = 2

$$\begin{aligned} 3x^2 + 2y^2 &= 36 \\ 3(2)^2 + 2y^2 &= 36 \\ 12 + 2y^2 &= 36 \\ 2y^2 &= 24 \\ y^2 &= 12 \end{aligned}$$
$$y = +\sqrt{12} \quad \text{or} \quad y = -\sqrt{12}$$
$$y = 2\sqrt{3} \quad \Big| \quad y = -2\sqrt{3}$$

For x = −2

$$\begin{aligned} 3x^2 + 2y^2 &= 36 \\ 3(-2)^2 + 2y^2 &= 36 \\ 12 + 2y^2 &= 36 \\ 2y^2 &= 24 \\ y^2 &= 12 \end{aligned}$$
$$y = +\sqrt{12} \quad \text{or} \quad y = -\sqrt{12}$$
$$y = 2\sqrt{3} \quad \Big| \quad y = -2\sqrt{3}$$

The four solutions of this system are

$$\left(2, 2\sqrt{3}\right), \left(2, -2\sqrt{3}\right), \left(-2, 2\sqrt{3}\right), \text{ and } \left(-2, -2\sqrt{3}\right)$$

SELF CHECK 4 Solve by elimination: $\begin{cases} x^2 + 4y^2 = 16 \\ x^2 - y^2 = 1 \end{cases}$.

SELF CHECK ANSWERS **1.** (3, 2), (−2, 3)

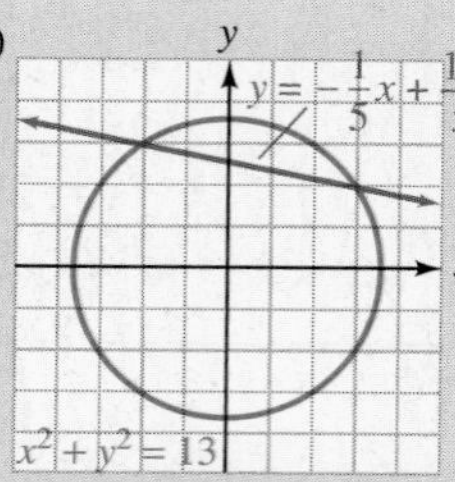

2. (3, 2), (−2, 3) **3.** (2, 4), (−2, 4) **4.** $\left(2, \sqrt{3}\right), \left(2, -\sqrt{3}\right), \left(-2, \sqrt{3}\right), \left(-2, -\sqrt{3}\right)$

NOW TRY THIS

1. Solve by substitution.

$$\begin{cases} x^2 - y^2 = 4 \\ 9x^2 + 16y^2 = 144 \end{cases}$$

10.4 EXERCISES

WARM-UPS *Give the possible number of solutions of a system when the graphs of the equations are*

1. A line and a parabola

2. A line and a hyperbola

3. A circle and a parabola

4. A circle and a hyperbola

REVIEW *Simplify each radical expression. Assume that all variables represent positive numbers.*

5. $\sqrt{200x^2} - 3\sqrt{98x^2}$

6. $a\sqrt{112a} - 5\sqrt{175a^3}$

7. $\dfrac{3t\sqrt{2t} - 2\sqrt{2t^3}}{\sqrt{18t} - \sqrt{2t}}$

8. $\sqrt[3]{\dfrac{x}{4}} + \sqrt[3]{\dfrac{x}{32}} - \sqrt[3]{\dfrac{x}{500}}$

VOCABULARY AND CONCEPTS *Fill in the blanks.*

9. We can solve systems of equations by ________, elimination (addition), or ________.

10. A line can intersect an ellipse in at most ____ points.

11. A parabola can intersect an ellipse in at most _____ points.

12. An ellipse can intersect a hyperbola in at most _____ points.

GUIDED PRACTICE

Solve each system of equations by graphing. See Example 1. (Objective 1)

13. $\begin{cases} 8x^2 + 32y^2 = 256 \\ x = 2y \end{cases}$

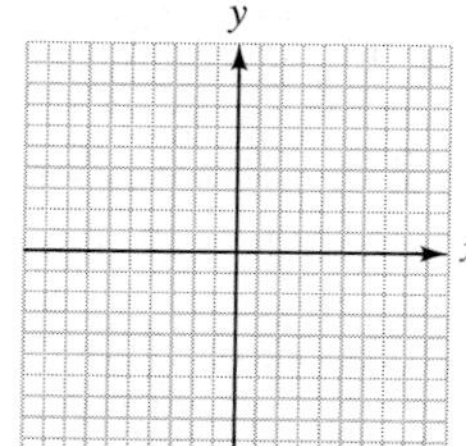

14. $\begin{cases} x^2 + y^2 = 2 \\ x + y = 2 \end{cases}$

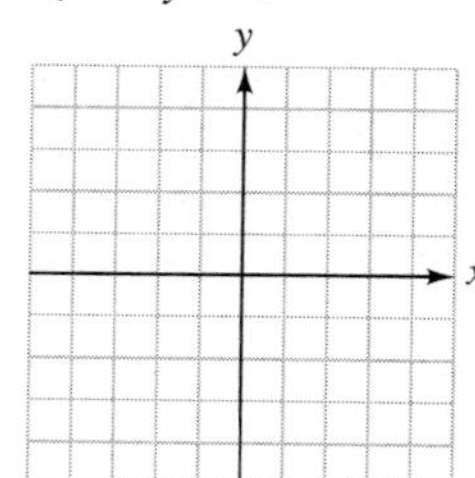

15. $\begin{cases} x^2 + y^2 = 10 \\ y = 3x^2 \end{cases}$

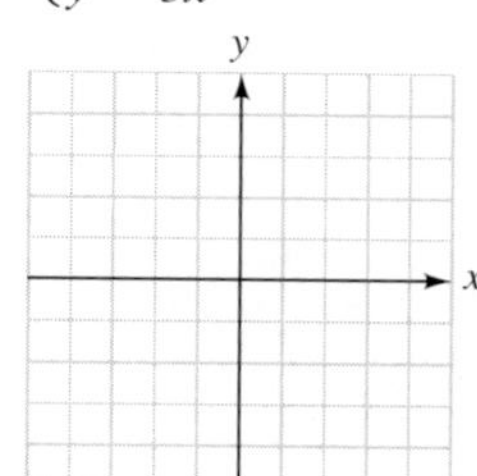

16. $\begin{cases} x^2 + y^2 = 5 \\ x + y = 3 \end{cases}$

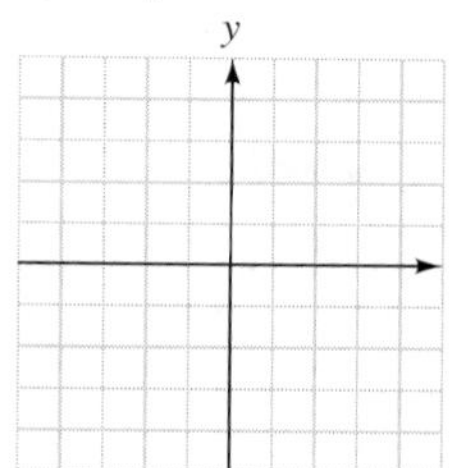

17. $\begin{cases} x^2 + y^2 = 25 \\ 12x^2 + 64y^2 = 768 \end{cases}$

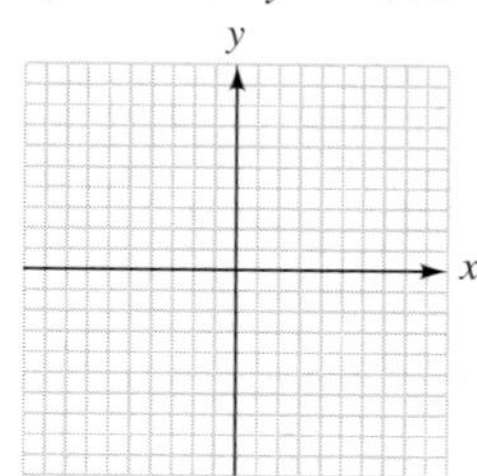

18. $\begin{cases} x^2 + y^2 = 13 \\ y = x^2 - 1 \end{cases}$

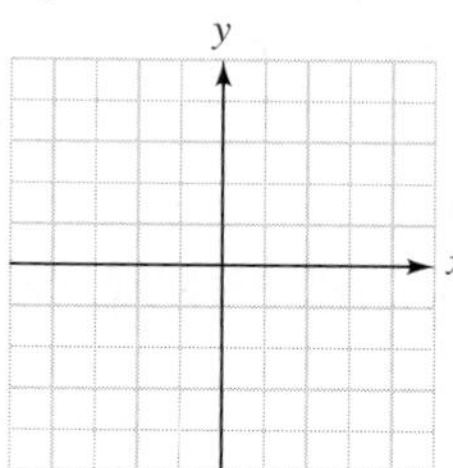

19. $\begin{cases} x^2 - 13 = -y^2 \\ y = 2x - 4 \end{cases}$

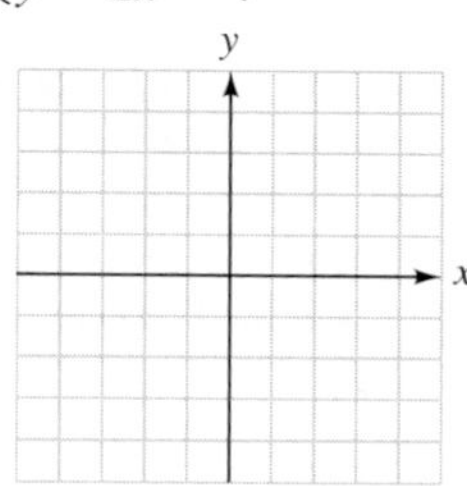

20. $\begin{cases} x^2 + y^2 = 20 \\ y = x^2 \end{cases}$

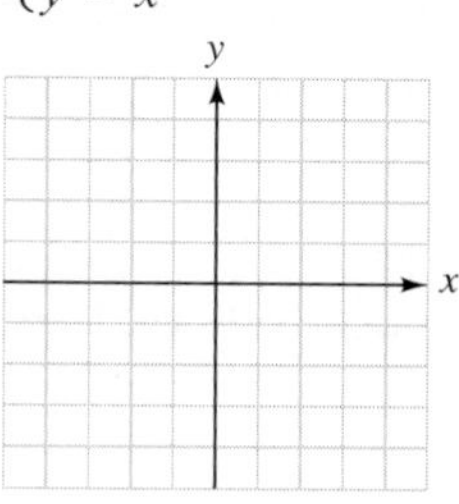

Solve each system by substitution. See Examples 2–3. (Objective 2)

21. $\begin{cases} 25x^2 + 9y^2 = 225 \\ 5x + 3y = 15 \end{cases}$

22. $\begin{cases} x^2 + y^2 = 20 \\ y = x^2 \end{cases}$

23. $\begin{cases} x^2 + y^2 = 2 \\ x + y = 2 \end{cases}$

24. $\begin{cases} x^2 + y^2 = 36 \\ 49x^2 + 36y^2 = 1{,}764 \end{cases}$

25. $\begin{cases} x^2 + y^2 = 5 \\ x + y = 3 \end{cases}$

26. $\begin{cases} x^2 - x - y = 2 \\ 4x - 3y = 0 \end{cases}$

27. $\begin{cases} x^2 + y^2 = 13 \\ y = x^2 - 1 \end{cases}$

28. $\begin{cases} x^2 + y^2 = 25 \\ 2x^2 - 3y^2 = 5 \end{cases}$

Solve each system by elimination. See Example 4. (Objective 3)

29. $\begin{cases} x^2 + y^2 = 13 \\ x^2 - y^2 = 5 \end{cases}$

30. $\begin{cases} 2x^2 + y^2 = 6 \\ x^2 - y^2 = 3 \end{cases}$

31. $\begin{cases} x^2 + y^2 = 20 \\ x^2 - y^2 = -12 \end{cases}$

32. $\begin{cases} x^2 - 6x - y = -5 \\ x^2 - 6x + y = -5 \end{cases}$

33. $\begin{cases} y = x^2 - 4 \\ x^2 - y^2 = -16 \end{cases}$

34. $\begin{cases} 6x^2 + 8y^2 = 182 \\ 8x^2 - 3y^2 = 24 \end{cases}$

35. $\begin{cases} y^2 = 40 - x^2 \\ y = x^2 - 10 \end{cases}$

36. $\begin{cases} x^2 + y^2 = 10 \\ 2x^2 - 3y^2 = 5 \end{cases}$

ADDITIONAL PRACTICE

Solve each system using any method.

37. $\begin{cases} x^2 + y^2 = 30 \\ y = x^2 \end{cases}$

38. $\begin{cases} 9x^2 - 7y^2 = 81 \\ x^2 + y^2 = 9 \end{cases}$

39. $\begin{cases} x^2 - y^2 = -5 \\ 3x^2 + 2y^2 = 30 \end{cases}$

40. $\begin{cases} \frac{1}{x} + \frac{1}{y} = 5 \\ \frac{1}{x} - \frac{1}{y} = -3 \end{cases}$

41. $\begin{cases} \frac{1}{x} + \frac{2}{y} = 1 \\ \frac{2}{x} - \frac{1}{y} = \frac{1}{3} \end{cases}$

42. $\begin{cases} \frac{1}{x} + \frac{3}{y} = 4 \\ \frac{2}{x} - \frac{1}{y} = 7 \end{cases}$

43. $\begin{cases} xy = -\frac{9}{2} \\ 3x + 2y = 6 \end{cases}$

44. $\begin{cases} 3y^2 = xy \\ 2x^2 + xy - 84 = 0 \end{cases}$

45. $\begin{cases} xy = \frac{1}{6} \\ y + x = 5xy \end{cases}$

46. $\begin{cases} xy = \frac{1}{12} \\ y + x = 7xy \end{cases}$

Use a graphing calculator to solve each system.

47. $\begin{cases} x^2 - 6x - y = -5 \\ x^2 - 6x + y = -5 \end{cases}$

48. $\begin{cases} x^2 - y^2 = -5 \\ 3x^2 + 2y^2 = 30 \end{cases}$

APPLICATIONS

49. Integer problem The product of two integers is 32, and their sum is 12. Find the integers.

50. Number problem The sum of the squares of two numbers is 221, and the sum of the numbers is 9. Find the numbers.

51. Geometry problem The area of a rectangle is 63 square centimeters, and its perimeter is 32 centimeters. Find the dimensions of the rectangle.

52. Investing money Ignacio receives $225 annual income from one investment. Carol invested $500 more than Ignacio, but at an annual rate of 1% less. Carol's annual income is $240. What is the amount and rate of Ignacio's investment?

53. Investing money Rania receives $67.50 annual income from one investment. Jerome invested $150 more than Rania at an annual rate of $1\frac{1}{2}$% more. Jerome's annual income is $94.50. What is the amount and rate of Rania's investment? (*Hint:* There are two answers.)

54. Artillery The shell fired from the base of the hill in the illustration follows the parabolic path $y = -\frac{1}{6}x^2 + 2x$ with distances measured in miles. The hill has a slope of $\frac{1}{3}$. How far from the gun is the point of impact? (*Hint:* Find the coordinates of the point and then the distance.)

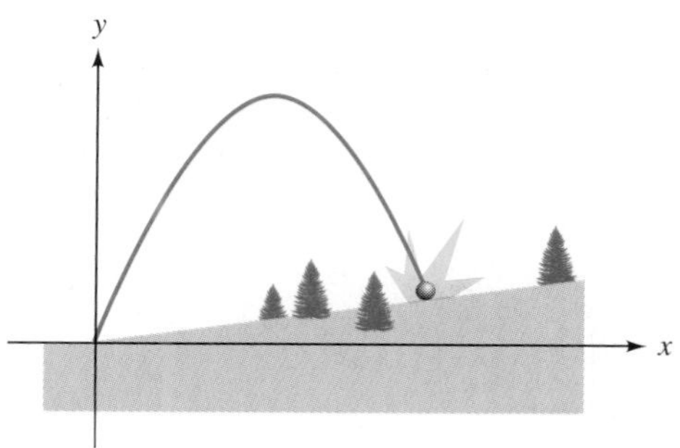

55. Driving rates Jim drove 306 miles. Jim's brother made the same trip at a speed 17 mph slower than Jim did and required an extra $1\frac{1}{2}$ hours. What was Jim's rate and time?

WRITING ABOUT MATH

56. Describe the benefits of the graphical method for solving a system of equations.

57. Describe the drawbacks of the graphical method.

SOMETHING TO THINK ABOUT

58. The graphs of the two independent equations of a system are parabolas. How many solutions might the system have?

59. The graphs of the two independent equations of a system are hyperbolas. How many solutions might the system have?

SECTION 10.5 Piecewise-Defined Functions and the Greatest Integer Function

Objectives

1. Graph a piecewise-defined function and determine the open intervals over which the function is increasing, decreasing, and constant.
2. Graph the greatest integer function.

Vocabulary

piecewise-defined function   greatest integer function   step function

Getting Ready

1. Is $f(x) = x^2$ positive or negative when $x > 0$?
2. Is $f(x) = -x^2$ positive or negative when $x > 0$?
3. What is the largest integer that is less than 98.6?
4. What is the largest integer that is less than -2.7?

Some functions are defined by using different equations for different parts of their domains. Such functions are called **piecewise-defined functions.**

1 Graph a piecewise-defined function and determine the open intervals over which the function is increasing, decreasing, and constant.

A simple piecewise-defined function is the absolute value function, $f(x) = |x|$, which can be written in the form

$$f(x) = \begin{cases} x \text{ when } x \geq 0 \\ -x \text{ when } x < 0 \end{cases}$$

When x is in the interval $[0, \infty)$, we use the function $f(x) = x$ to evaluate $|x|$. However, when x is in the interval $(-\infty, 0)$, we use the function $f(x) = -x$ to evaluate $|x|$. The graph of the absolute value function is shown in Figure 10-37.

For $x \geq 0$

x	y	(x, y)
0	0	(0, 0)
1	1	(1, 1)
2	2	(2, 2)
3	3	(3, 3)

For $x < 0$

x	y	(x, y)
−4	4	(−4, 4)
−3	3	(−3, 3)
−2	2	(−2, 2)
−1	1	(−1, 1)

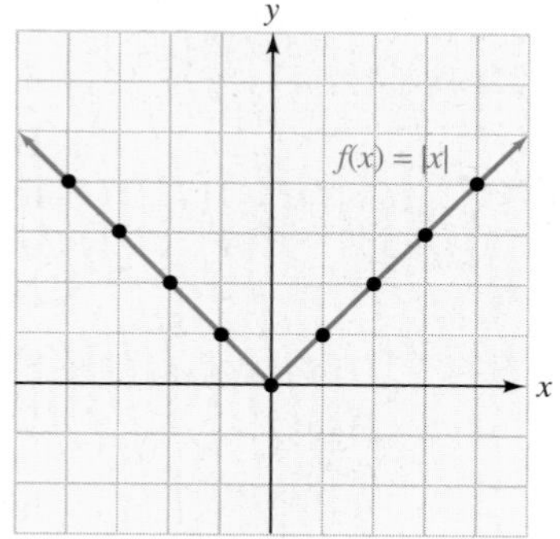

Figure 10-37

If the values of $f(x)$ increase as x increases on an open interval, we say that the function is *increasing on the interval* (see Figure 10-38(a)) on the next page. If the values of $f(x)$ decrease as x increases on an open interval, we say that the function is *decreasing on the interval* (see Figure 10-38(b)). If the values of $f(x)$ remain constant as x increases on an open interval, we say that the function is *constant on the interval* (see Figure 10-38(c)).

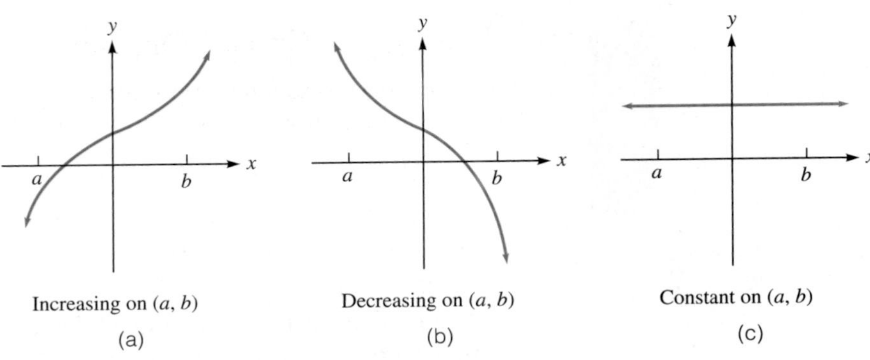

Figure 10-38

The absolute value function, shown in Figure 10-37, is decreasing on the interval $(-\infty, 0)$ and is increasing on the interval $(0, \infty)$.

EXAMPLE 1 Graph the piecewise-defined function given by

$$f(x) = \begin{cases} x^2 \text{ when } x \le 0 \\ x \text{ when } 0 < x < 2 \\ -1 \text{ when } x \ge 2 \end{cases}$$

and determine where the function is increasing, decreasing, and constant.

Solution For each number x, we decide which of the three equations will be used to find the corresponding value of y:

- For numbers $x \le 0$, $f(x)$ is determined by $f(x) = x^2$, and the graph is the left half of a parabola. See Figure 10-39. Since the values of $f(x)$ decrease on this graph as x increases, the function is decreasing on the interval $(-\infty, 0)$.
- For numbers $0 < x < 2$, $f(x)$ is determined by $f(x) = x$, and the graph is part of a line. Since the values of $f(x)$ increase on this graph as x increases, the function is increasing on the interval $(0, 2)$.
- For numbers $x \ge 2$, $f(x)$ is the constant -1, and the graph is part of a horizontal line. Since the values of $f(x)$ remain constant on this line, the function is constant on the interval $(2, \infty)$.

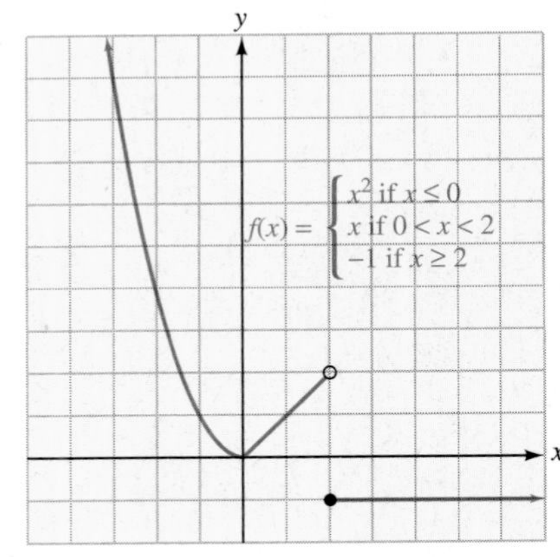

Figure 10-39

The use of solid and open circles on the graph indicates that $f(x) = -1$ when $x = 2$.

Since every number x determines one value y, the domain of this function is the interval $(-\infty, \infty)$. The range is $\{-1\} \cup [0, \infty)$.

In set-builder notation, the range is $\{y \mid y = -1 \text{ and } y \ge 0\}$.

SELF CHECK 1 Graph: $f(x) = \begin{cases} 2x \text{ when } x \le 0 \\ \frac{1}{2}x \text{ when } x > 0 \end{cases}$.

2 Graph the greatest integer function.

The **greatest integer function** is important in computer applications. It is a function determined by the equation

$f(x) = [\![x]\!]$ Read as "y equals the greatest integer in x."

where the value of y that corresponds to x is the greatest integer that is less than or equal to x. For example,

$$[\![4.7]\!] = 4, \quad \left[\!\!\left[2\frac{1}{2}\right]\!\!\right] = 2, \quad [\![\pi]\!] = 3, \quad [\![-3.7]\!] = -4, \quad [\![-5.7]\!] = -6$$

COMMENT One way to help determine the greatest integer is to visualize the number on a number line. The integer directly to the left of the number is the greatest integer.

EXAMPLE 2 Graph: $f(x) = [\![x]\!]$.

Solution We list several intervals and the corresponding values of the greatest integer function:

$[0, 1)$ $y = [\![x]\!] = 0$ For numbers from 0 to 1, not including 1, the greatest integer in the interval is 0.

$[1, 2)$ $y = [\![x]\!] = 1$ For numbers from 1 to 2, not including 2, the greatest integer in the interval is 1.

$[2, 3)$ $y = [\![x]\!] = 2$ For numbers from 2 to 3, not including 3, the greatest integer in the interval is 2.

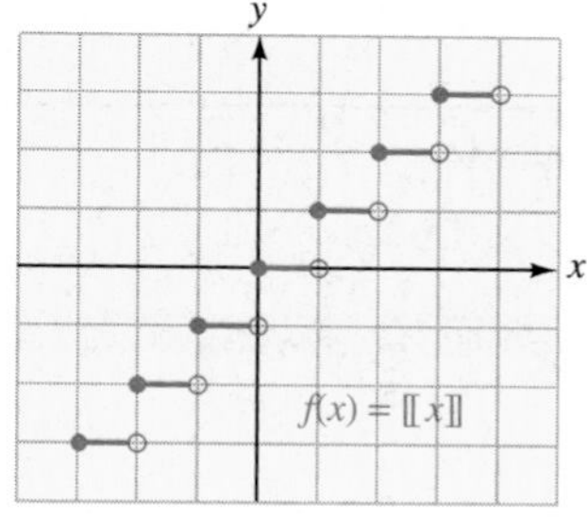

Figure 10-40

In each interval, the values of y are constant, but they jump by 1 at integer values of x. The graph is shown in Figure 10-40. From the graph, we see that the domain is $(-\infty, \infty)$, and the range is the set of integers $\{\ldots, -3, -2, -1, 0, 1, 2, 3, \ldots\}$.

Since the greatest integer function is made up of a series of horizontal line segments, it is an example of a group of functions called **step functions.**

SELF CHECK 2 Graph: $f(x) = [\![x]\!] + 1$.

EXAMPLE 3 **PRINTING STATIONERY** To print stationery, a printer charges \$10 for setup charges, plus \$20 for each box. Any portion of a box counts as a full box. Graph this step function.

Solution If we order stationery and cancel the order before it is printed, the cost will be \$10. Thus, the ordered pair (0, 10) will be on the graph.

If we purchase 1 box, the cost will be \$10 for setup plus \$20 for printing, for a total cost of \$30. Thus, the ordered pair (1, 30) will be on the graph.

The cost of $1\frac{1}{2}$ boxes will be the same as the cost of 2 boxes, or \$50. Thus, the ordered pairs (1.5, 50) and (2, 50) will be on the graph. The complete graph is shown in Figure 10-41.

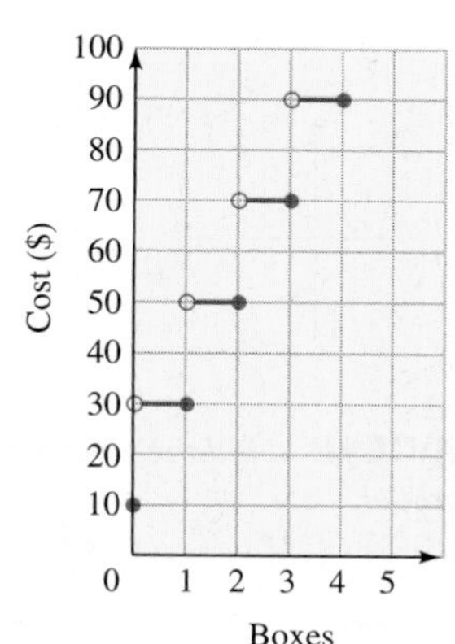

Figure 10-41

SELF CHECK 3 How much will $3\frac{1}{2}$ boxes cost?

SELF CHECK ANSWERS

1. 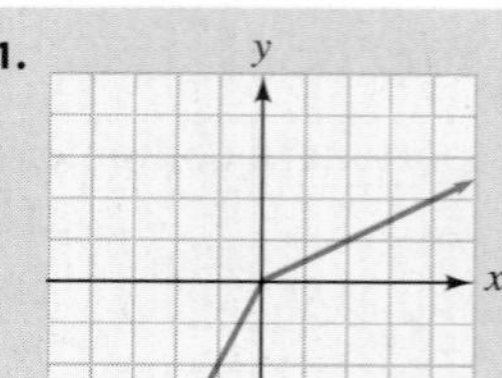

2. 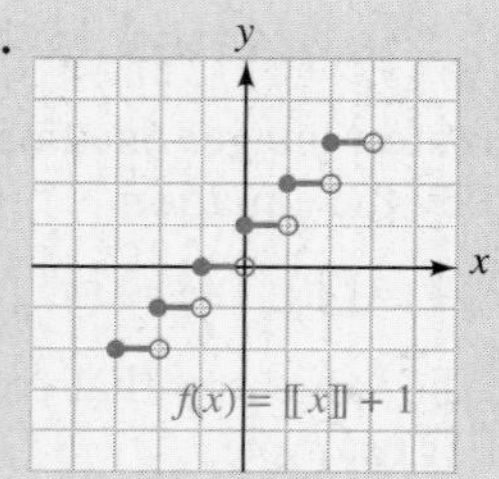

3. $90

NOW TRY THIS

1. Graph the piecewise-defined function.

$$f(x) = \begin{cases} -x^2 \text{ if } x < 1 \\ [x] \text{ if } 1 \le x < 4 \\ x - 1 \text{ if } x \ge 4 \end{cases}$$

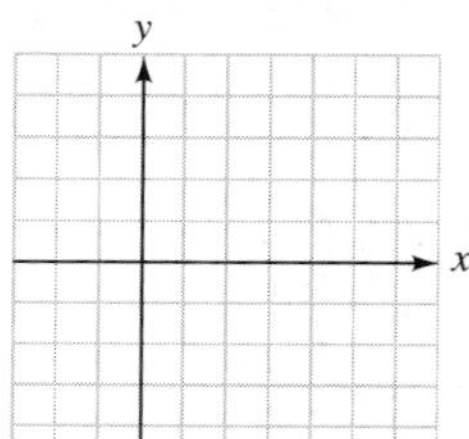

10.5 EXERCISES

WARM-UPS *Determine whether each function is increasing, decreasing, or constant on the interval (−2, 3).*

1. 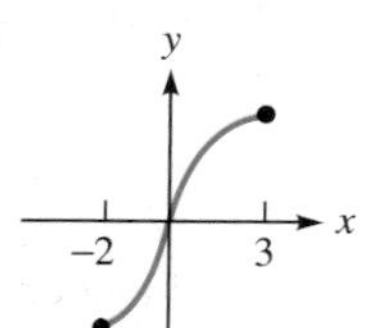

2. 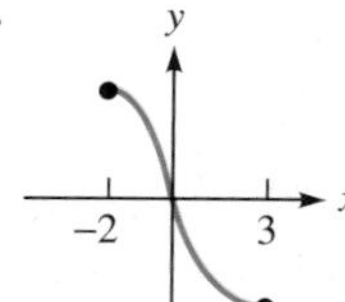

3. 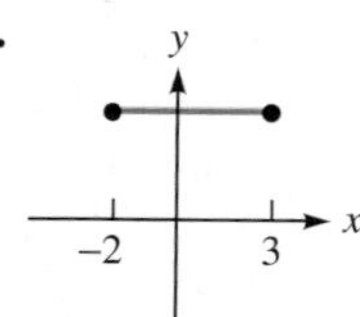

4. 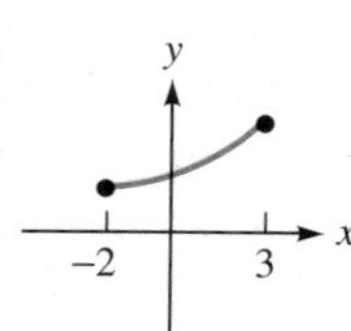

REVIEW *Find the value of x. Assume that lines r and s are parallel.*

5. 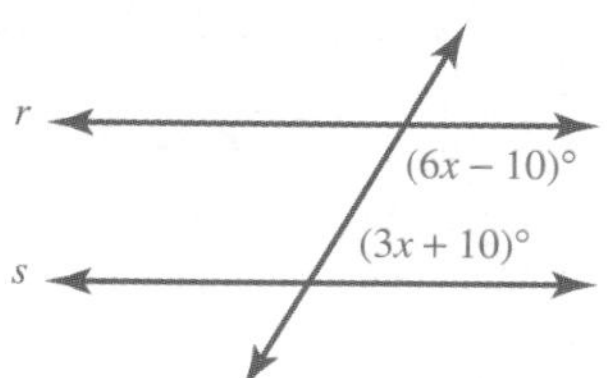

6. 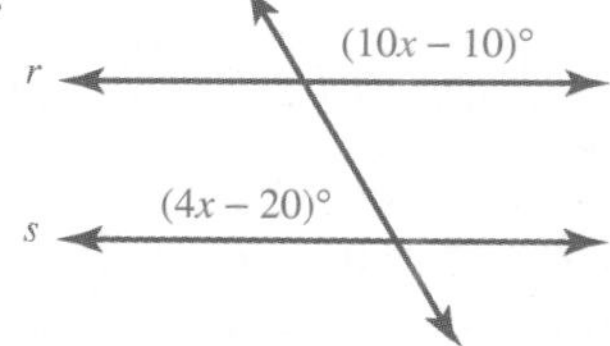

VOCABULARY AND CONCEPTS *Fill in the blanks.*

7. Piecewise-defined functions are defined by using different functions for different parts of their ________.
8. When the values $f(x)$ increase as the values of x increase over an interval, we say that the function is an ________ function over that interval.
9. In a ________ function, the values of ____ are the same.
10. When the values of $f(x)$ decrease as the values of x ________ over an interval, we say that the function is a decreasing function over that interval.
11. When the graph of a function contains a series of horizontal line segments, the function is called a ____ function.
12. The function that gives the largest integer that is less than or equal to a number x is called the ____________ function.

GUIDED PRACTICE

Give the intervals on which each function is increasing, decreasing, or constant. **(Objective 1)**

13.

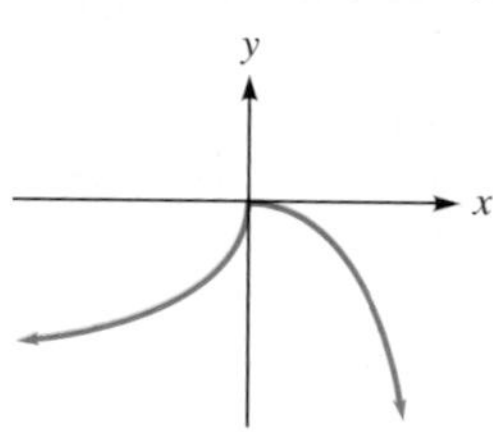

14.

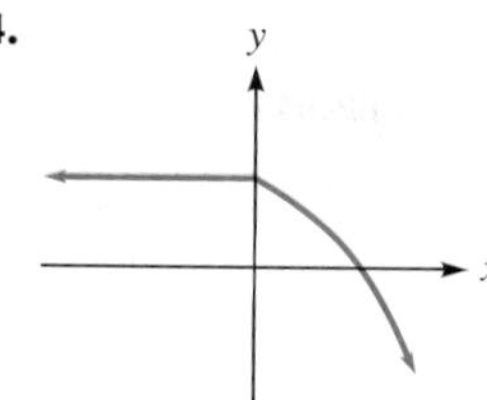

15.

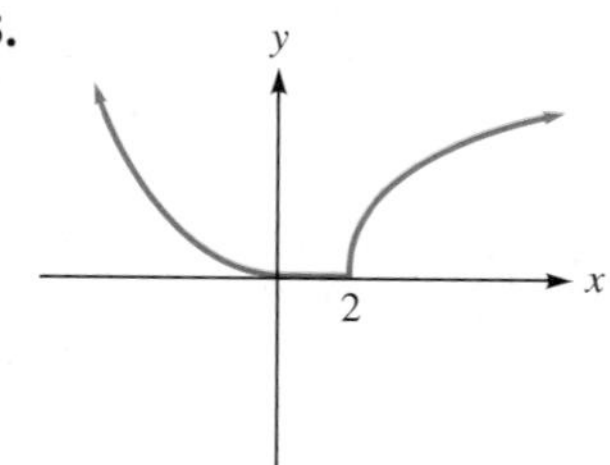

16

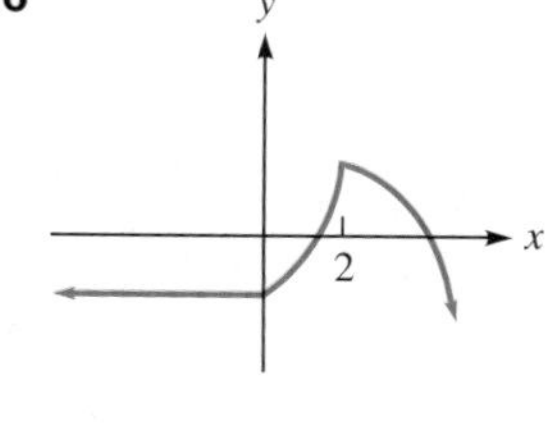

Graph each function and give the intervals on which f is increasing, decreasing, or constant. **See Example 1. (Objective 1)**

17. $f(x) = \begin{cases} -1 \text{ when } x \leq 0 \\ x \text{ when } x > 0 \end{cases}$

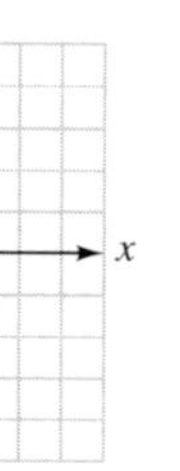

18. $f(x) = \begin{cases} -2 \text{ if } x \leq 0 \\ x^2 \text{ if } x > 0 \end{cases}$

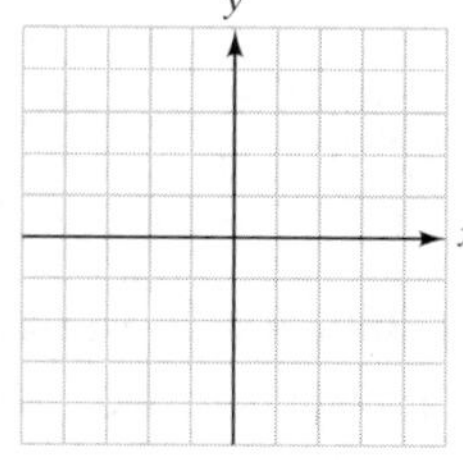

19. $f(x) = \begin{cases} -x \text{ if } x \leq 0 \\ x \text{ if } 0 < x < 2 \\ -x \text{ if } x \geq 2 \end{cases}$

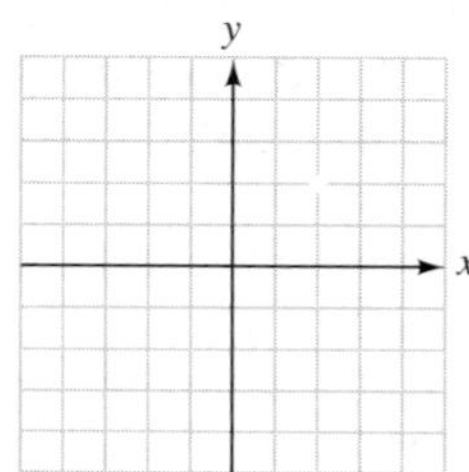

20. $f(x) = \begin{cases} -x \text{ if } x < 0 \\ x^2 \text{ if } 0 \leq x \leq 1 \\ 1 \text{ if } x > 1 \end{cases}$

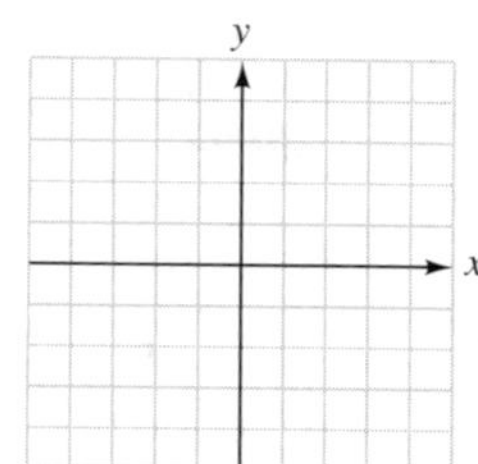

Graph each function. **See Example 2. (Objective 1)**

21. $f(x) = -[\![x]\!]$

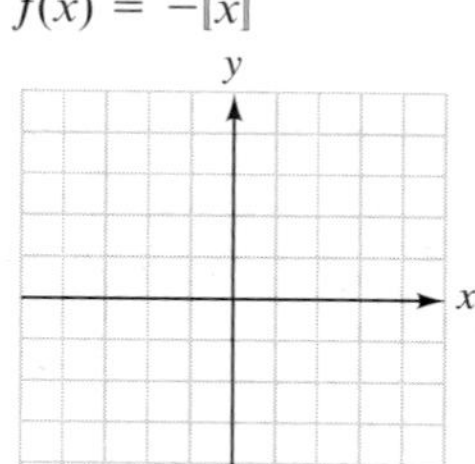

22. $f(x) = [\![x]\!] + 2$

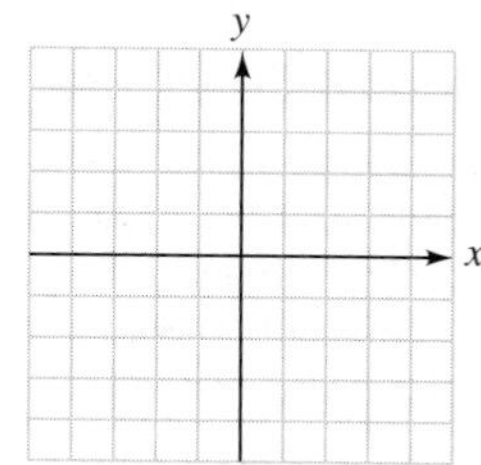

23. $f(x) = 2[\![x]\!]$

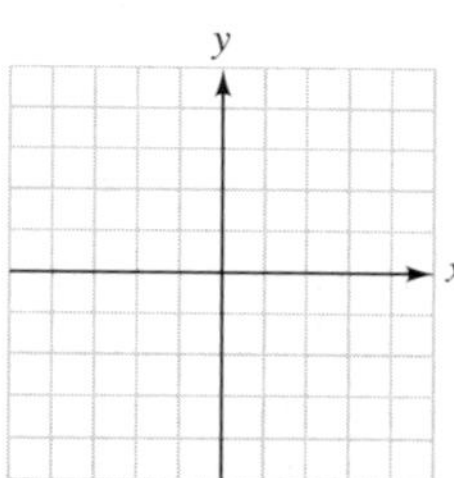

24. $f(x) = \left[\!\!\left[\frac{1}{2}x\right]\!\!\right]$

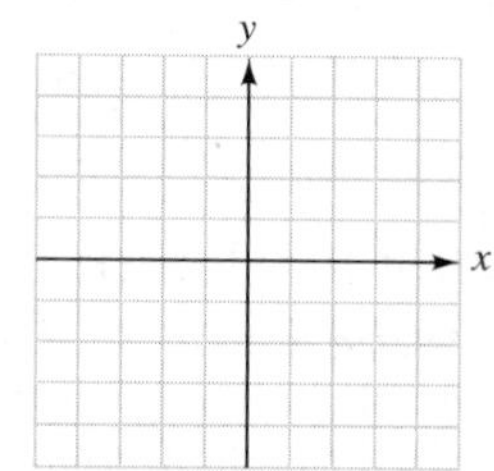

ADDITIONAL PRACTICE

25. Signum function Computer programmers use a function denoted by $f(x) = \text{sgn } x$ that is defined in the following way:

$$f(x) = \begin{cases} -1 \text{ if } x < 0 \\ 0 \text{ if } x = 0 \\ 1 \text{ if } x > 0 \end{cases}$$

Graph this function.

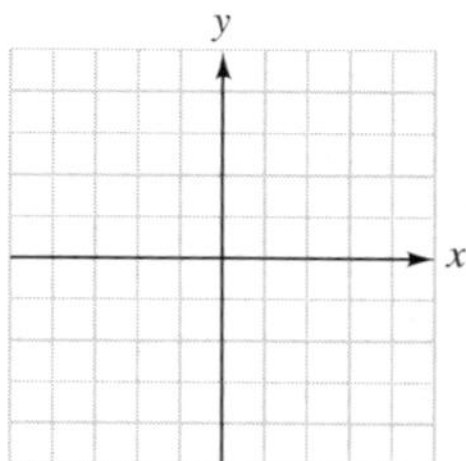

26. Heaviside unit step function This function, used in calculus, is defined by

$$f(x) = \begin{cases} 1 \text{ if } x > 0 \\ 0 \text{ if } x < 0 \end{cases}$$

Graph this function.

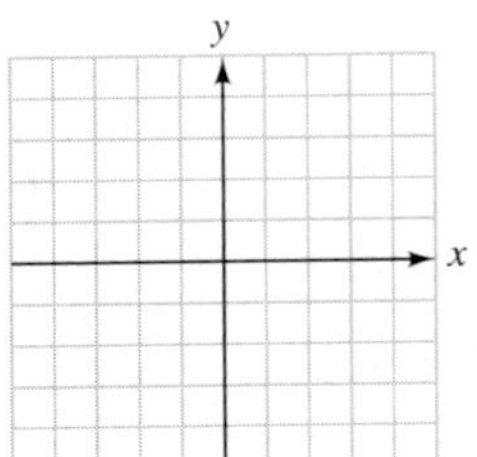

APPLICATIONS See Example 3. (Objective 2)

27. Renting a jet ski A marina charges \$20 to rent a jet ski for 1 hour, plus \$5 for every extra hour (or portion of an hour). In the illustration, graph the ordered pairs (h, c), where h represents the number of hours and c represents the cost. Find the cost if the ski is used for 2.5 hours.

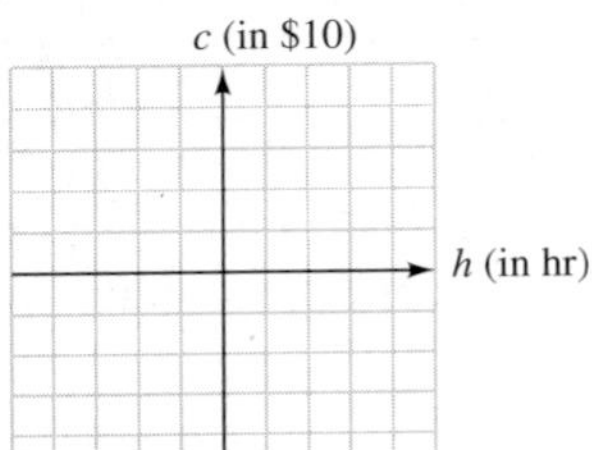

28. Riding in a taxi A cab company charges \$3 for a trip up to 1 mile, and \$2 for every extra mile (or portion of a mile). In the illustration, graph the ordered pairs (m, c), where m represents the number of miles traveled and c represents the cost. Find the cost to ride $10\frac{1}{4}$ miles.

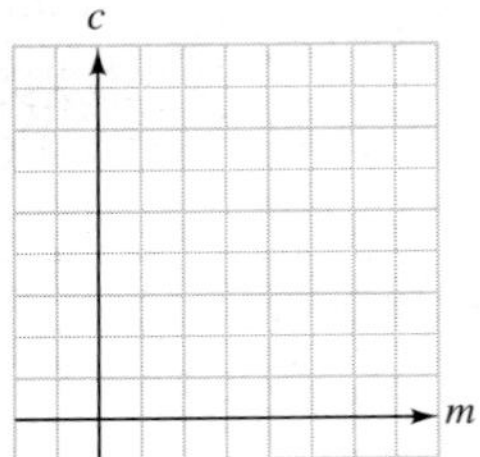

29. Information access Computer access to international data network A costs \$10 per day plus \$8 per hour or fraction of an hour. Network B charges \$15 per day, but only \$6 per hour or fraction of an hour. For each network, graph the ordered pairs (t, C), where t represents the connect time and C represents the total cost. Find the minimum daily usage at which it would be more economical to use network B.

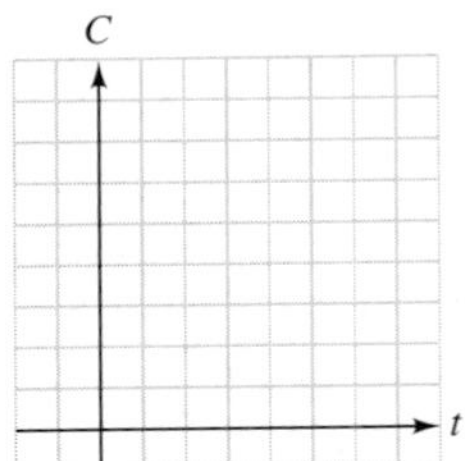

30. Royalties A publisher has agreed to pay the author of a novel 7% royalties on sales of the first 50,000 copies and 10% on sales thereafter. If the book sells for \$10, express the royalty income I as a function of s, the number of copies sold, and graph the function. (*Hint:* When sales are into the second 50,000 copies, how much was earned on the first 50,000?)

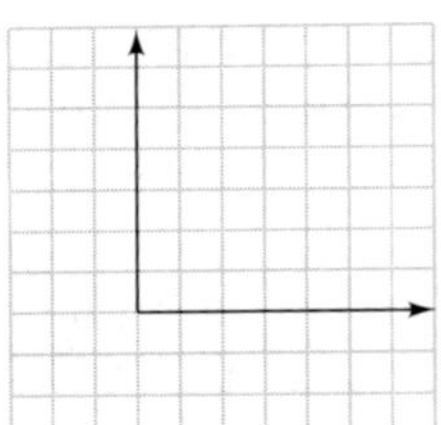

WRITING ABOUT MATH

31. Explain how to decide whether a function is increasing on the interval (a, b).
32. Describe the greatest integer function.

SOMETHING TO THINK ABOUT

33. Find a piecewise-defined function that is increasing on the interval $(-\infty, -2)$ and decreasing on the interval $(-2, \infty)$.
34. Find a piecewise-defined function that is constant on the interval $(-\infty, 0)$, increasing on the interval $(0, 5)$, and decreasing on the interval $(5, \infty)$.

PROJECTS

The zillionaire G. I. Luvmoney is known for his love of flowers. On his estate, he recently set aside a circular plot of land with a radius of 100 yards to be made into a flower garden. He has hired your landscape design firm to do the job. If Luvmoney is satisfied, he will hire your firm to do more lucrative jobs. Here is Luvmoney's plan.

The center of the circular plot of land is to be the origin of a rectangular coordinate system. You are to make 100 circles, all centered at the origin, with radii of 1 yard, 2 yards, 3 yards, and so on up to the outermost circle, which will have a radius of 100 yards. Inside the innermost circle, he wants a fountain with a circular walkway around it. In the ring between the first and second circle, he wants to plant his favorite kind of flower, in the next ring his second favorite, and so on, until you reach the edge of the circular plot. Luvmoney provides you with a list ranking his 99 favorite flowers.

The first thing he wants to know is the area of each ring, so that he will know how many of each plant to order. Then he wants a simple formula that will give the area of any ring just by substituting in the number of the ring.

He also wants a walkway to go through the garden in the form of a hyperbolic path, following the equation

$$x^2 - \frac{y^2}{9} = 1$$

Luvmoney wants to know the x- and y-coordinates of the points where the path will intersect the circles, so that those points can be marked with stakes to keep gardeners from planting flowers where the walkway will later be built. He wants a formula (or two) that will enable him to put in the number of a circle and get out the intersection points.

Finally, although cost has no importance for Luvmoney, his accountants will want an estimate of the total cost of all of the flowers.

You go back to your office with Luvmoney's list. You find that because the areas of the rings grow from the inside of the garden to the outside, and because of Luvmoney's ranking of flowers, a strange thing happens. The first ring of flowers will cost \$360, and the flowers in every ring after that will cost 110% as much as the flowers in the previous ring. That is, the second ring of flowers will cost \$360(1.1) = \$396, the third will cost \$435.60, and so on.

Answer all of Luvmoney's questions, and show work that will convince him that you are right.

Chapter 10 REVIEW

SECTION 10.1 The Circle and the Parabola

DEFINITIONS AND CONCEPTS	EXAMPLES
Equations of a circle: **Standard forms:** $x^2 + y^2 = r^2$ center (0, 0), radius r	To graph a circle, we need to know the center and the radius. Graph each circle. **a.** $x^2 + y^2 = 25$ $C(0, 0)$ From the formula $r^2 = 25$ From the formula $r = 5$ The center is (0, 0) and the radius is 5 units.

$(x - h)^2 + (y - k)^2 = r^2$
center (h, k), radius r

b. $(x - 3)^2 + (y + 4)^2 = 16$

$(x - 3)^2 + [y - (-4)]^2 = 16$	Write in standard form.
$C(3, -4)$	Comparing to the formula, $h = 3$ and $k = -4$.
$r^2 = 16$	Comparing to the formula
$r = 4$	Take the positive square root of 16.

The center is at $(3, -4)$ and the radius is 4 units.

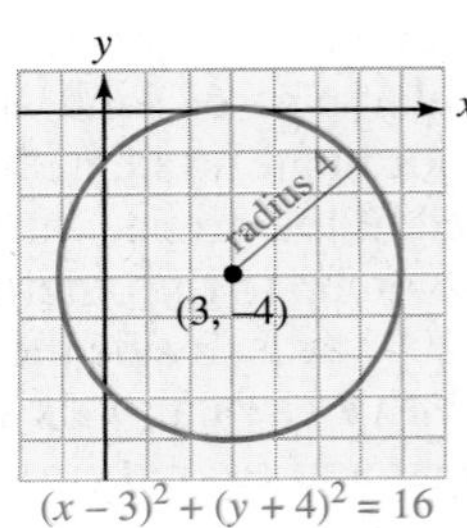

General form:

$x^2 + y^2 + Dx + Ey + F = 0$

c. $x^2 + y^2 + 2x + 6y + 6 = 0$
To write the equation in standard form, we can complete the square on both x and y.

$x^2 + y^2 + 2x + 6y = -6$	Subtract 6 from both sides.
$x^2 + 2x + y^2 + 6y = -6$	Rearrange the terms.
$x^2 + 2x + 1 + y^2 + 6y + 9 = -6$	Complete the square on x and y.
$(x + 1)^2 + (y + 3)^2 = 4$	Factor and simplify.
$[x - (-1)]^2 + [y - (-3)]^2 = 4$	Write in standard form.

Comparing to the standard form, we see that $C(-1, -3)$, $r^2 = 4$, and $r = 2$.

The center is at $(-1, -3)$ with a radius of 2 units.

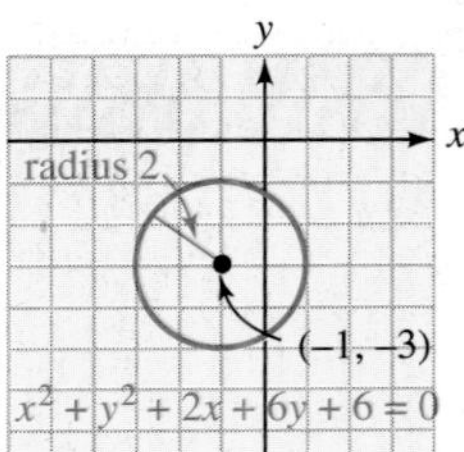

Equations of parabolas:

Parabola opening	Vertex at origin	
Up	$y = ax^2$	$(a > 0)$
Down	$y = ax^2$	$(a < 0)$
Right	$x = ay^2$	$(a > 0)$
Left	$x = ay^2$	$(a < 0)$

Graph each parabola.

a. $x = y^2$

The parabola is horizontal and opens to the right because $a > 0$. To obtain the graph, we can plot several points and connect them with a smooth curve.

$x = y^2$

x	y	(x, y)
0	0	(0, 0)
4	2	(4, 2)
4	−2	(4, −2)
9	3	(9, 3)
9	−3	(9, −3)

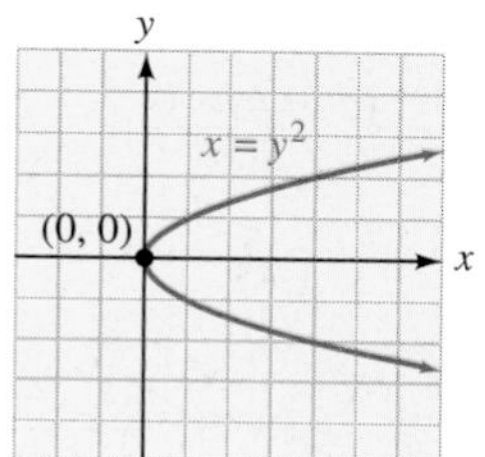

Parabola opening	Vertex at (h, k)	
Up	$y = a(x - h)^2 + k$	$(a > 0)$
Down	$y = a(x - h)^2 + k$	$(a < 0)$
Right	$x = a(y - k)^2 + h$	$(a > 0)$
Left	$x = a(y - k)^2 + h$	$(a < 0)$

b. $x = -2(y - 1)^2 + 3$

The parabola is horizontal and opens to the left because $a < 0$. To obtain the graph, we can plot several points and connect them with a smooth curve.

$x = -2(y - 1)^2 + 3$

x	y	(x, y)
1	0	(1, 0)
1	2	(1, 2)
3	1	(3, 1)
−5	−1	(−5, −1)
−5	3	(3, 1)

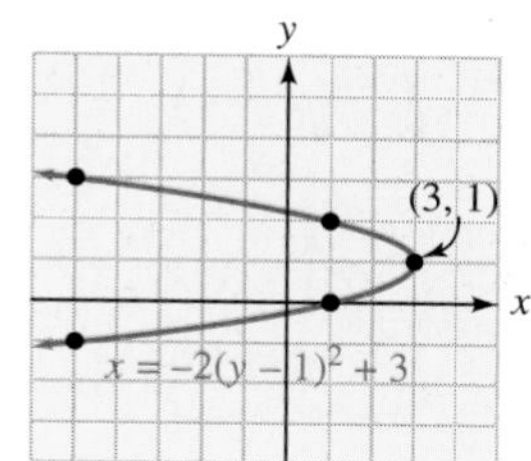

REVIEW EXERCISES

Graph each equation.

1. $(x - 1)^2 + (y + 2)^2 = 9$

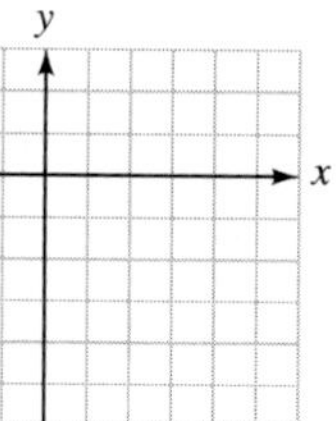

2. $x^2 + y^2 = 16$

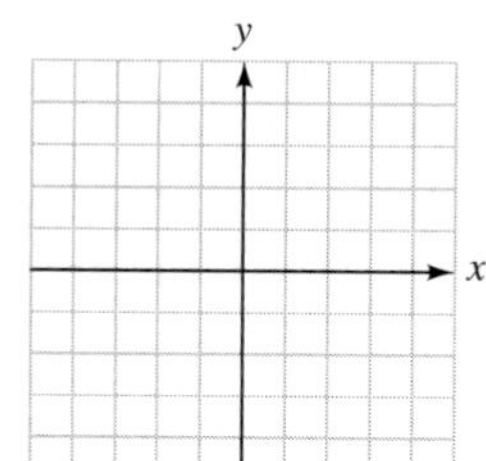

3. Write the equation in standard form and graph it.

$x^2 + y^2 + 4x - 2y = 4$

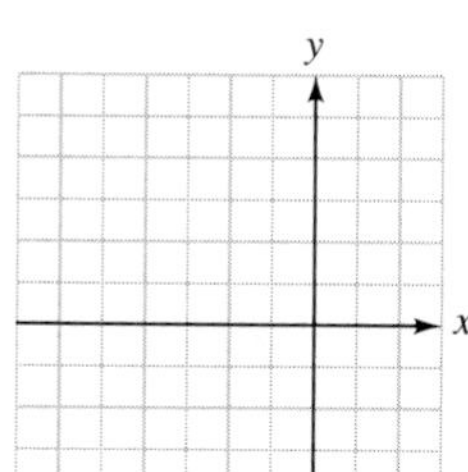

Graph each equation.

4. $x = -3(y - 2)^2 + 5$

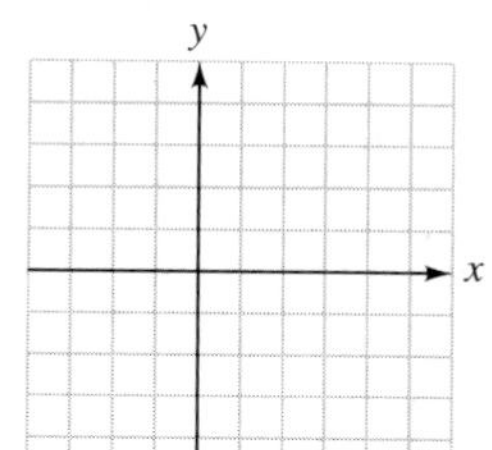

5. $x = 2(y + 1)^2 - 2$

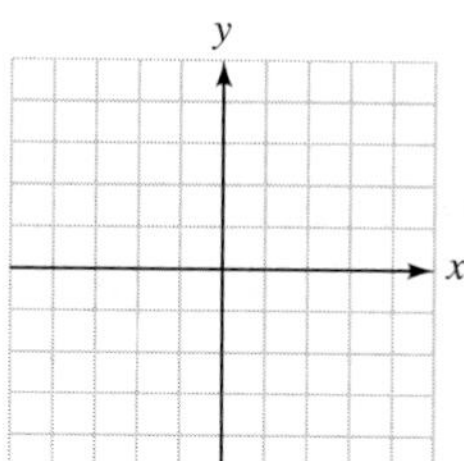

SECTION 10.2 The Ellipse

DEFINITIONS AND CONCEPTS	EXAMPLES
Equations of an ellipse: **Standard forms:** Center at $(0, 0)$ $\frac{x^2}{a^2} + \frac{y^2}{b^2} = 1 \quad (a > b > 0)$ $\frac{x^2}{b^2} + \frac{y^2}{a^2} = 1 \quad (a > b > 0)$	To graph an ellipse, we need to know the center and the endpoints of the major and minor axes. Graph each ellipse: **a.** $\frac{x^2}{9} + \frac{y^2}{16} = 1$ Comparing to the formula, the center is $(0, 0)$ and $a^2 = 16 \quad b^2 = 9$ $a = 4 \quad b = 3$ The ellipse will be vertical because the larger denominator is associated with the y-term. The endpoints of the major axis will be 4 units above and below the center, $(0, 4)$ and $(0, -4)$. The endpoints of the minor axis will be 3 units to the left and right of center, $(3, 0)$ and $(-3, 0)$. 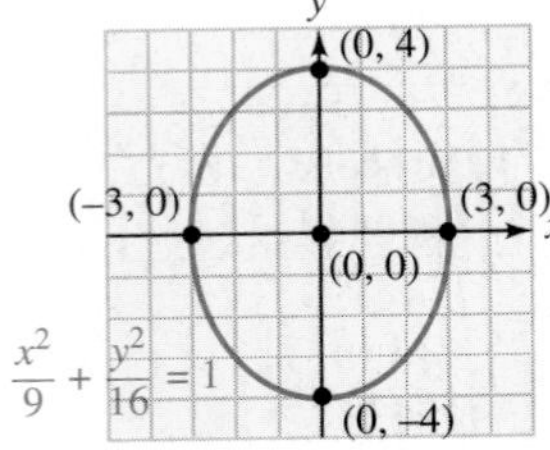
Center at (h, k) $\frac{(x - h)^2}{a^2} + \frac{(y - k)^2}{b^2} = 1$ $\frac{(x - h)^2}{b^2} + \frac{(y - k)^2}{a^2} = 1$ In either case, The length of the major axis is $2a$. The length of the minor axis is $2b$.	**b.** $\frac{(x - 3)^2}{9} + \frac{(y - 4)^2}{4} = 1$ From the formula, the center is $(3, 4)$ and $a^2 = 9 \quad b^2 = 4$ $a = 3 \quad b = 2$ The ellipse will be horizontal because the larger denominator is associated with the x-term. The endpoints of the major axis will be 3 units to the left and right of the center, $(6, 4)$ and $(0, 4)$. The endpoints of the minor axis will be 2 units above and below the center, $(3, 6)$ and $(3, 2)$. 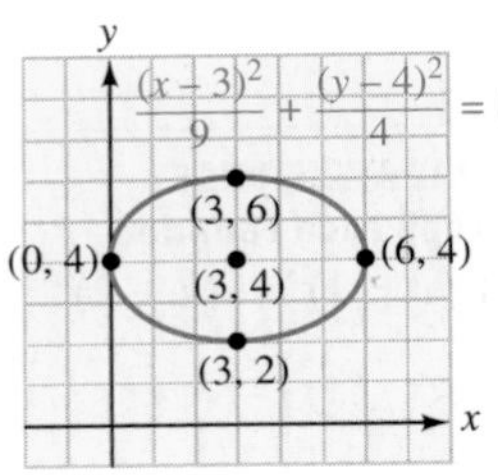
General form: $Ax^2 + Cy^2 + Dx + Ey + F = 0$	To find the standard equation of the ellipse with equation $4x^2 + y^2 - 8x - 2y - 11 = 0$, proceed as follows: $4x^2 + y^2 - 8x - 2y = 11$ — Add 11 to both sides. $4x^2 - 8x + y^2 - 2y = 11$ — Rearrange terms. $4(x^2 - 2x) + y^2 - 2y = 11$ — Factor to get a coefficient of 1 for the term involving x-squared. $4(x^2 - 2x + \mathbf{1}) + (y^2 - 2y + \mathbf{1}) = 11 + \mathbf{4} + \mathbf{1}$ — Complete the square on both x and y. $4(x - 1)^2 + (y - 1)^2 = 16$ — Factor and simplify. $\frac{(x - 1)^2}{4} + \frac{(y - 1)^2}{16} = 1$ — Divide both sides by 16.

REVIEW EXERCISES
Graph each ellipse.

6. $9x^2 + 16y^2 = 144$

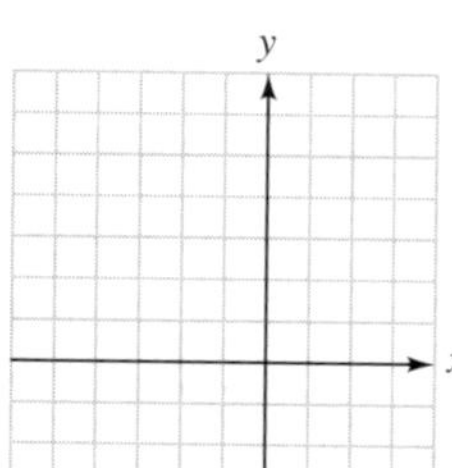

7. $\frac{(x-2)^2}{4} + \frac{(y-1)^2}{9} = 1$

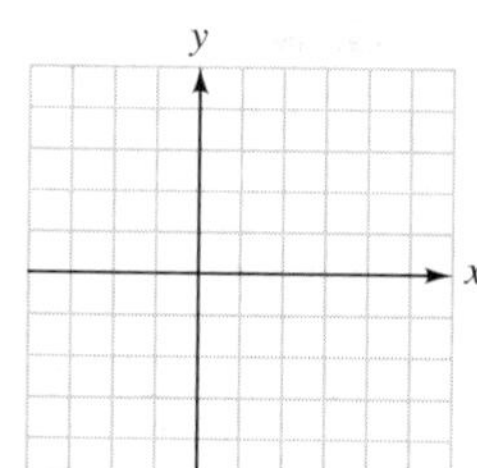

8. Write the equation in standard form and graph it.

$4x^2 + 9y^2 + 8x - 18y = 23$

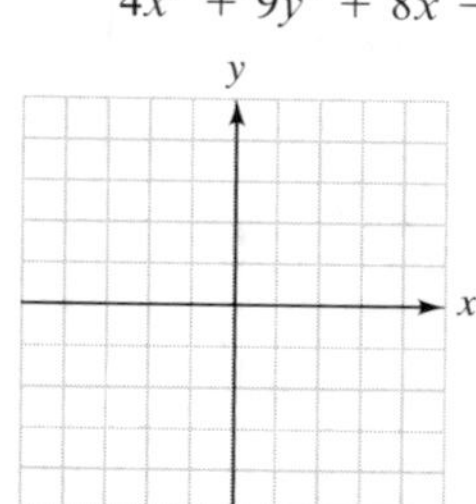

SECTION 10.3 The Hyperbola

DEFINITIONS AND CONCEPTS

Equations of a hyperbola:

Standard forms:

Center at (0, 0)

$$\frac{x^2}{a^2} - \frac{y^2}{b^2} = 1$$

$$\frac{y^2}{a^2} - \frac{x^2}{b^2} = 1$$

EXAMPLES

To graph a hyperbola, we need to know where it is centered, the coordinates of the vertices, and the location of the asymptotes.

Graph $9y^2 - 4x^2 = 36$.

$$\frac{9y^2}{36} - \frac{4x^2}{36} = 1 \quad \text{Write the equation in standard form.}$$

$$\frac{y^2}{4} - \frac{x^2}{9} = 1 \quad \text{Simplify each fraction.}$$

From the previous equation, we can determine that $a = 2$ and $b = 3$. Because the y-term is positive, the hyperbola will be vertical and the vertices of the hyperbola are $V_1(0, 2)$ and $V_2(0, -2)$.

Since $b = 3$, we can use the points (3, 0) and $(-3, 0)$ on the x-axis to help draw the fundamental rectangle. We then draw its extended diagonals and sketch the hyperbola.

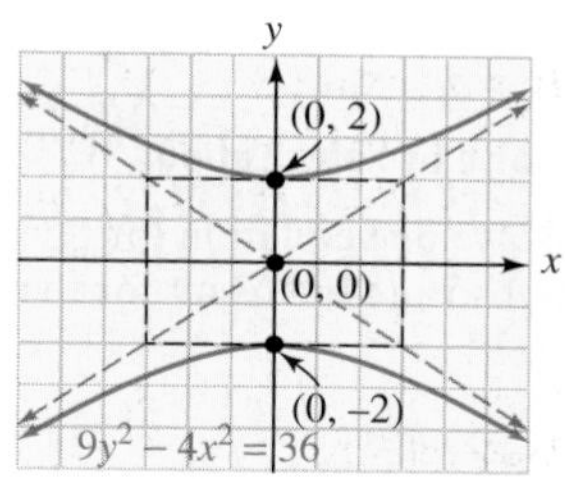

DEFINITIONS AND CONCEPTS

Center at (h, k)

$$\frac{(x-k)^2}{a^2} - \frac{(y-h)^2}{b^2} = 1 \quad \text{opens left or right}$$

$$\frac{(y-k)^2}{a^2} - \frac{(x-h)^2}{b^2} = 1 \quad \text{opens up or down}$$

EXAMPLES

Graph: $\frac{(x-3)^2}{4} - \frac{(y+1)^2}{4} = 1$.

From the equation, we see that the hyperbola is centered at $(3, -1)$. Its vertices are located 2 units to the right and left of center, at $(5, -1)$ and $(1, -1)$. Since $b = 2$, we can count 2 units above and below center to locate points $(3, -3)$ and $(3, 1)$. With these points, we can draw the fundamental rectangle along with its extended diagonals. Then we can sketch the hyperbola.

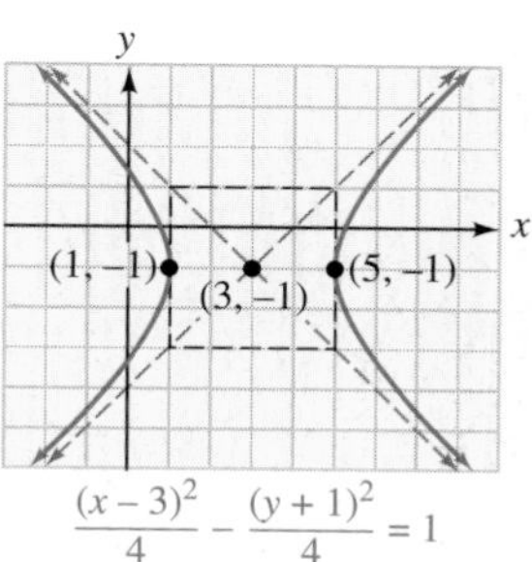

REVIEW EXERCISES

Graph each hyperbola.

9. $9x^2 - y^2 = -9$

10. $xy = 9$

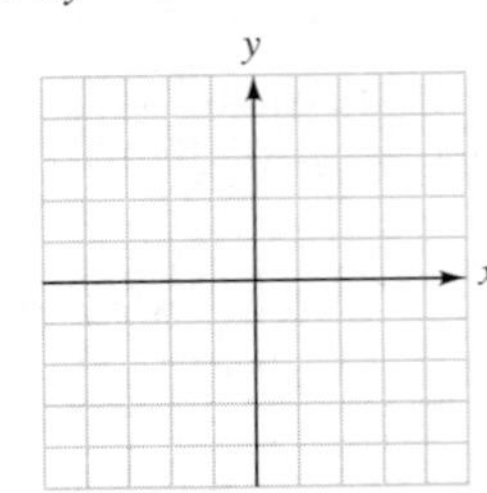

11. Write the equation $4x^2 - 2y^2 + 8x - 8y = 8$ in standard form and determine whether its graph will be an ellipse or a hyperbola.

12. Write the equation in standard form and graph it.

$$9x^2 - 4y^2 - 18x - 8y = 31$$

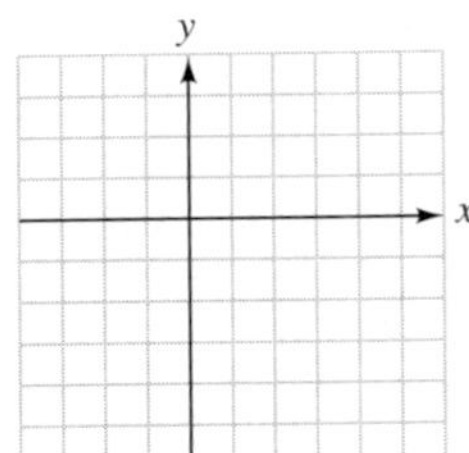

SECTION 10.4 Solving Systems Containing Second-Degree Equations

DEFINITIONS AND CONCEPTS	EXAMPLES
Solve by graphing: To solve a system of equations by graphing, graph both equations. The coordinates of the intersection points of the graphs will be the solutions of the system.	Solve the system by graphing. $\begin{cases} x^2 + y^2 = 4 \\ x - y = 2 \end{cases}$ 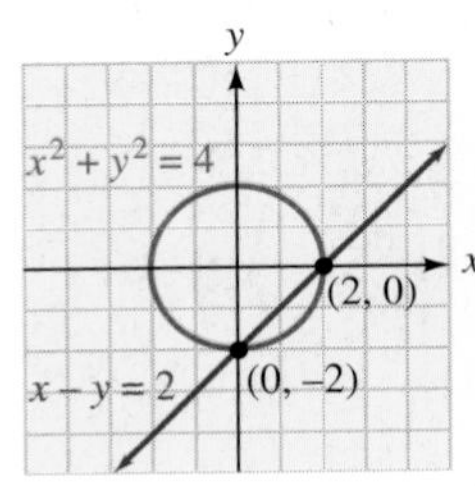 The equation $x^2 + y^2 = 4$ is that of a circle and $x - y = 2$ is that of a line. There is a possibility of 0, 1, or 2 solutions. After graphing the equations, we will find that the two graphs intersect at the points (2, 0) and (0, −2). These are the solutions.
Solve by substitution: Solve one equation for one variable, say y, and substitute the result for the other variable.	Solve the system by substitution. $\begin{cases} x^2 + 9y^2 = 10 \\ y = x^2 \end{cases}$ The equation $x^2 + 9y^2 = 10$ is that of an ellipse and $y = x^2$ is that of a parabola. We have the possibility of 0, 1, 2, 3, or 4 solutions. To find the solutions, proceed as follows: $x^2 + 9y^2 = 10$ $y + 9y^2 = 10$ Substitute y for x^2. $9y^2 + y - 10 = 0$ Subtract 10 from both sides. $(9y + 10)(y - 1) = 0$ Factor $9y^2 + y - 10$. $9y + 10 = 0$ or $y - 1 = 0$ Set each factor equal to 0. $y = -\frac{10}{9}$ | $y = 1$ Since $y = x^2$, the values of x can be found by solving the equations $x^2 = -\frac{10}{9}$ and $x^2 = 1$ Because $x^2 = -\frac{10}{9}$ has no real solutions, this possibility is discarded. The solutions of $x^2 = 1$ are $x = 1$ or $x = -1$. The solutions of the system are (1, 1) and (−1,1).

Solve by elimination (addition): To solve a system of equations by elimination, add the equations to eliminate one of the variables, say y. Then solve the resulting equation for the other variable. Then substitute this value into one of the equations to find y.	Solve the system by elimination (addition). $\begin{cases} 4x^2 - y^2 = 1 \\ 4x^2 + y^2 = 1 \end{cases}$ The equation $4x^2 - y^2 = 1$ is that of a hyperbola and the equation $4x^2 + y^2 = 1$ is that of an ellipse. We have the possibility of 0, 1, 2, 3, or 4 solutions. If we add the equations $\begin{cases} 4x^2 - y^2 = 1 \\ 4x^2 + y^2 = 1 \end{cases}$, we have $8x^2 = 2$ $x^2 = \frac{1}{4}$ $x = \frac{1}{2}, -\frac{1}{2}$ After substituting each value of x for y in the first equation, we have $4\left(\frac{1}{2}\right)^2 - y^2 = 1 \qquad 4\left(-\frac{1}{2}\right)^2 - y^2 = 1$ $4\left(\frac{1}{4}\right) - y^2 = 1 \qquad 4\left(\frac{1}{4}\right) - y^2 = 1$ $1 - y^2 = 1 \qquad 1 - y^2 = 1$ $-y^2 = 0 \qquad -y^2 = 0$ $y = 0 \qquad y = 0$ The solutions are $\left(\frac{1}{2}, 0\right)$ and $\left(-\frac{1}{2}, 0\right)$.

REVIEW EXERCISES

Solve each system.

13. $\begin{cases} 3x^2 + y^2 = 52 \\ x^2 - y^2 = 12 \end{cases}$

14. $\begin{cases} \frac{x^2}{16} + \frac{y^2}{12} = 1 \\ x^2 - \frac{y^2}{3} = 1 \end{cases}$

SECTION 10.5 Piecewise-Defined Functions and the Greatest Integer Function

DEFINITIONS AND CONCEPTS	EXAMPLES
Piecewise-defined functions: A piecewise-defined function is a function that has different rules for different intervals of x. **Increasing and decreasing functions:** A function is increasing on the interval (a, b) if the values of $f(x)$ increase as x increases from a to b. A function is decreasing on the interval (a, b) if the values of $f(x)$ decrease as x increases from a to b. A function is constant on the interval (a, b) if the value of $f(x)$ is constant as x increases from a to b.	Graph the function $f(x) = \begin{cases} -x^2 & \text{if } x < 0 \\ -x & \text{if } 0 \le x < 3 \\ -1 & \text{if } x \ge 3 \end{cases}$ and determine where the function is increasing, decreasing, and constant. 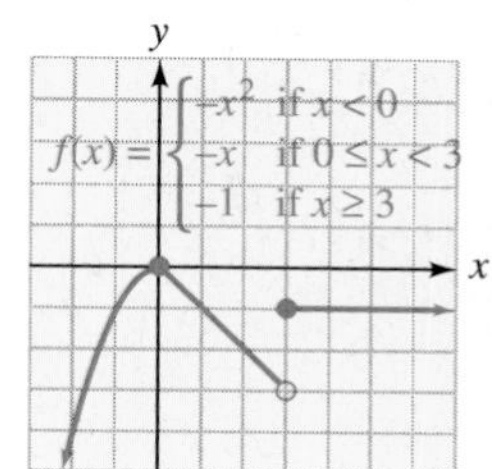

For each number x, we decide which of the three equations will be used to find the corresponding value of y:

- For numbers $x < 0$, $f(x)$ is determined by $f(x) = -x^2$, and the graph is the left half of a parabola. Since the values of $f(x)$ increase on this graph as x increases, the function is increasing on the interval $(-\infty, 0)$.
- For numbers $0 < x < 3$, $f(x)$ is determined by $f(x) = -x$, and the graph is part of a line. Since the values of $f(x)$ decrease on this graph as x increases, the function is decreasing on the interval $(0, 3)$.
- For numbers $x \geq 3$, $f(x)$ is the constant -1, and the graph is part of a horizontal line. Since the values of $f(x)$ remain constant on this line, the function is constant on the interval $(3, \infty)$.

Greatest integer function:

The function $f(x) = \lfloor x \rfloor$ describes the greatest integer function. To find the greatest integer, visualize the number on a number line and the integer directly to the left is the greatest integer.

Graph: $f(x) = \lfloor x \rfloor - 1$.

We list several intervals and the corresponding values of the greatest integer function:

$[0, 1)$	$f(x) = \lfloor x \rfloor - 1$	For numbers from 0 to 1, not including 1, the greatest integer in the interval is 0 and then we subtract 1 and graph -1 in the interval.
$[1, 2)$	$f(x) = \lfloor x \rfloor - 1$	For numbers from 1 to 2, not including 2, the greatest integer in the interval is 1 and then we subtract 1 and graph 0 in the interval.
$[2, 3)$	$f(x) = \lfloor x \rfloor - 1$	For numbers from 2 to 3, not including 3, the greatest integer in the interval is 0 and then we subtract 1 and graph 1 in the interval.

REVIEW EXERCISES

15. Determine when the function is increasing, decreasing, or constant.

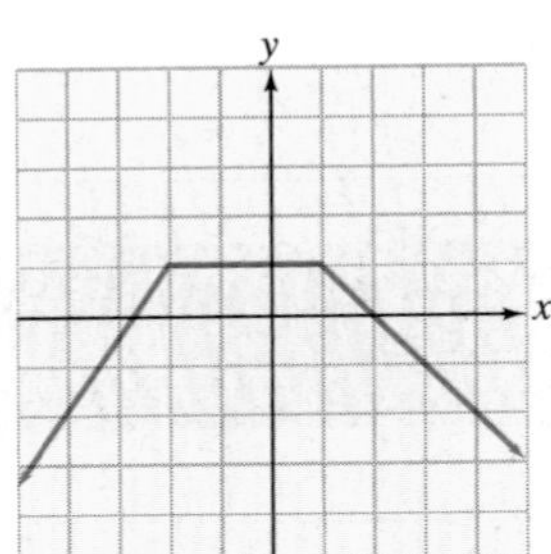

Graph each function.

16. $f(x) = \begin{cases} x \text{ if } x \leq 1 \\ -x^2 \text{ if } x > 1 \end{cases}$

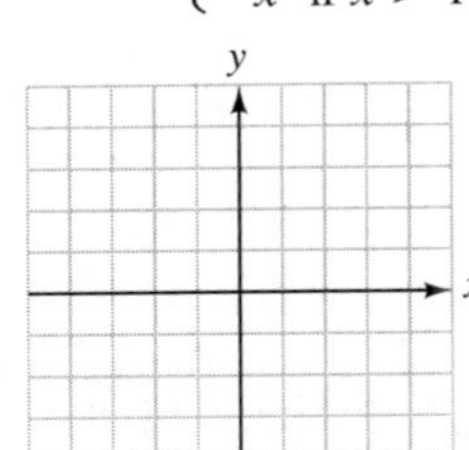

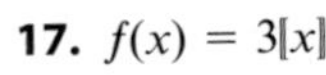
17. $f(x) = 3\lfloor x \rfloor$

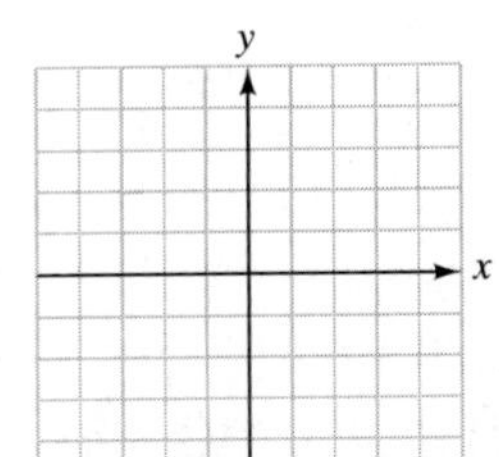

Chapter 10 TEST

1. Find the center and the radius of the circle $(x - 2)^2 + (y + 3)^2 = 4$.
2. Find the center and the radius of the circle $x^2 + y^2 + 4x - 6y = 3$.

Graph each equation.

3. $(x + 1)^2 + (y - 2)^2 = 9$

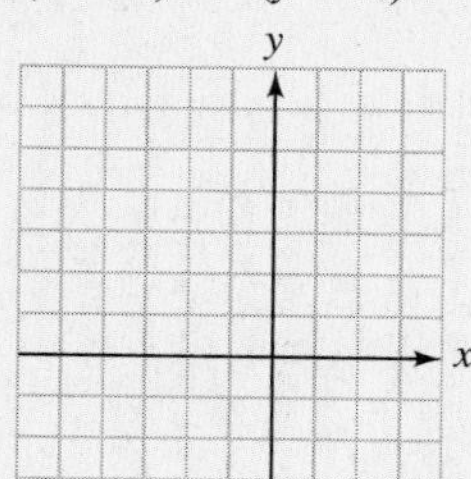

4. $x = (y - 2)^2 - 1$

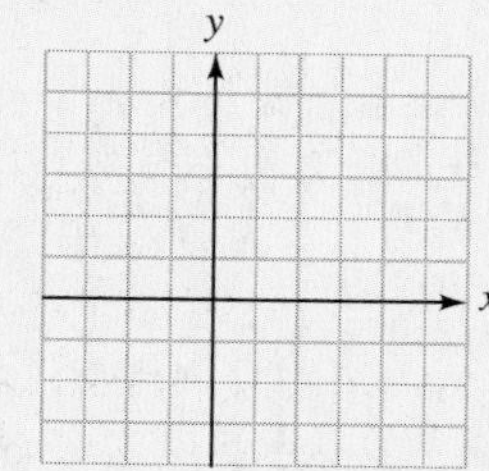

5. $9x^2 + 4y^2 = 36$

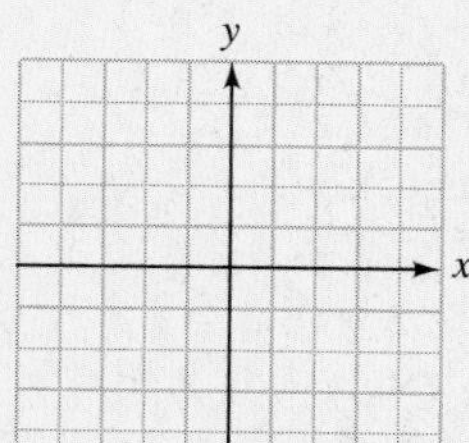

6. $\dfrac{(x - 2)^2}{9} - y^2 = 1$

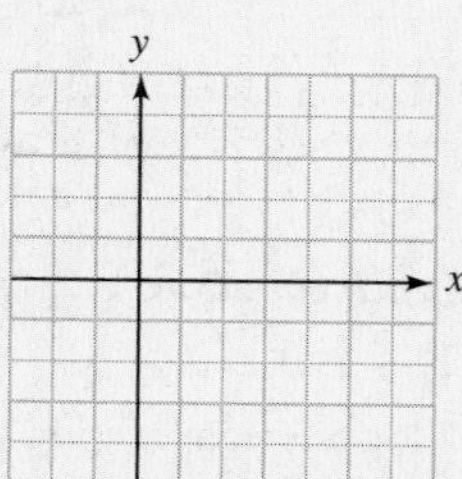

Write each equation in standard form and graph the equation.

7. $4x^2 + y^2 - 24x + 2y = -33$

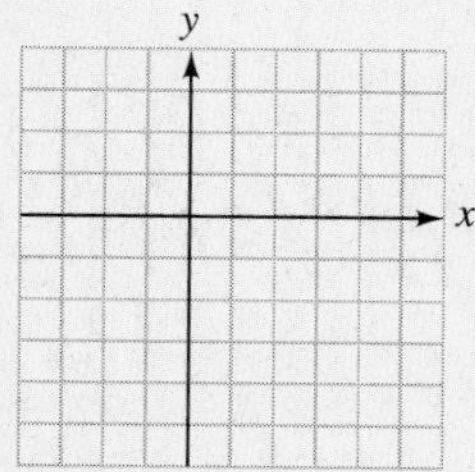

8. $x^2 - 9y^2 + 2x + 36y = 44$

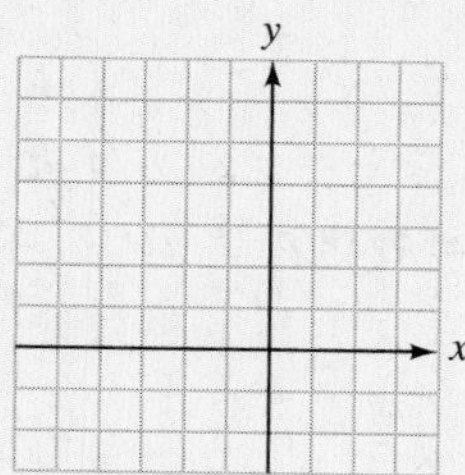

Solve each system.

9. $\begin{cases} 2x - y = -2 \\ x^2 + y^2 = 16 + 4y \end{cases}$

10. $\begin{cases} x^2 + y^2 = 25 \\ 4x^2 - 9y = 0 \end{cases}$

11. Determine where the function is increasing, decreasing, or constant.

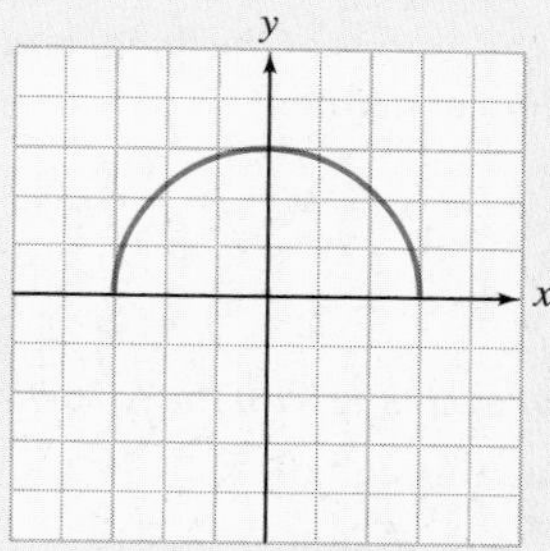

12. Graph: $f(x) = \begin{cases} -x^2, \text{ when } x < 0 \\ -x, \text{ when } x \geq 0 \end{cases}$.

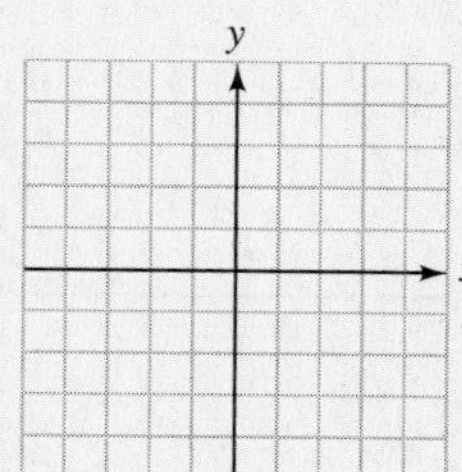

Cumulative Review Exercises

Perform the operations.

1. $(4x - 3y)(3x + y)$
2. $(a^n + 1)(a^n - 3)$

Simplify each fraction. Assume no division by zero.

3. $\dfrac{5a - 10}{a^2 - 4a + 4}$
4. $\dfrac{a^4 - 5a^2 + 4}{a^2 + 3a + 2}$

Perform the operations and simplify the result. Assume no division by zero.

5. $\dfrac{a^2 - a - 6}{a^2 - 4} \div \dfrac{a^2 - 9}{a^2 + a - 6}$
6. $\dfrac{2}{a - 2} + \dfrac{3}{a + 2} - \dfrac{a - 1}{a^2 - 4}$

Determine whether the graphs of the linear equations are parallel, perpendicular, or neither.

7. $3x - 4y = 12$, $y = \dfrac{3}{4}x - 5$
8. $y = 3x + 4$, $x = -3y + 4$

Write the equation of each line with the following properties.

9. $m = -2$, passing through $(0, 5)$
10. Passing through $(8, -5)$ and $(-5, 4)$

Graph each inequality.

11. $2x - 3y < 6$

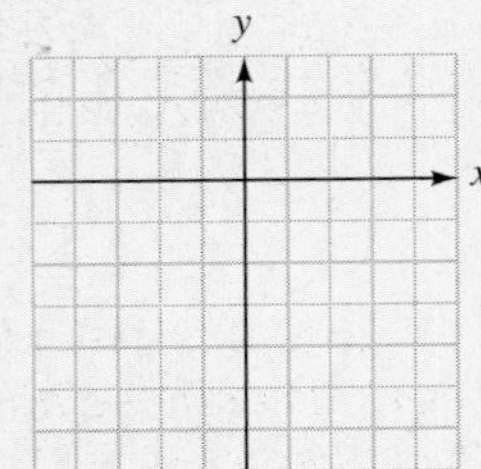

12. $y \geq x^2 - 4$

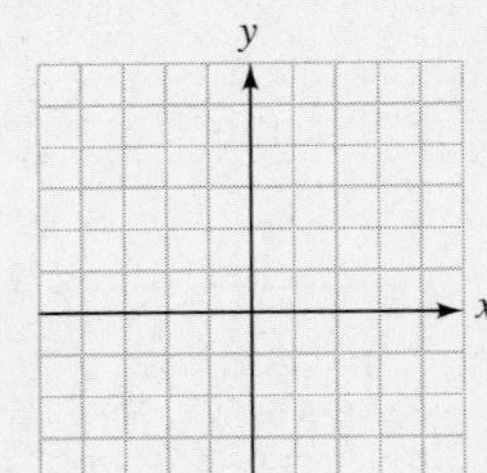

Simplify each expression.

13. $\sqrt{98} + \sqrt{8} - \sqrt{32}$
14. $12\sqrt[3]{648x^4} + 3\sqrt[3]{81x^4}$

Solve each equation.

15. $\sqrt{3a + 1} = a - 1$
16. $\sqrt{x + 3} - \sqrt{3} = \sqrt{x}$
17. $6a^2 + 5a - 6 = 0$
18. $3x^2 + 8x - 1 = 0$
19. If $f(x) = x^2 - 2$ and $g(x) = 2x + 1$, find $(f \circ g)(x)$.
20. Find the inverse function of $y = 2x^3 - 1$.
21. Graph $y = \left(\dfrac{1}{2}\right)^x$.

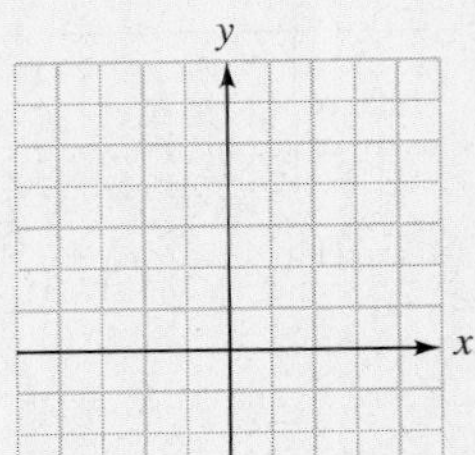

22. Write $y = \log_2 x$ as an exponential equation.

Solve each equation.

23. $2^{x+2} = 3^x$
24. $2 \log 5 + \log x - \log 4 = 2$

Graph each equation.

25. $x^2 + (y + 1)^2 = 9$

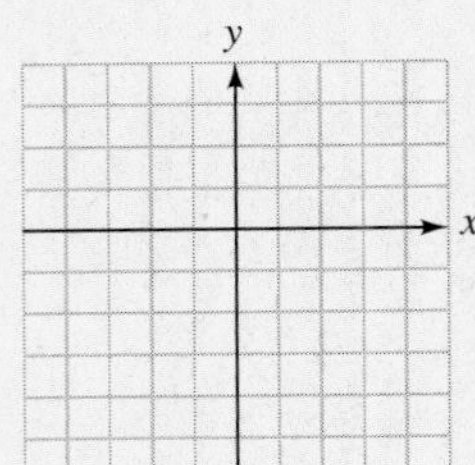

26. $x^2 - 9(y + 1)^2 = 9$

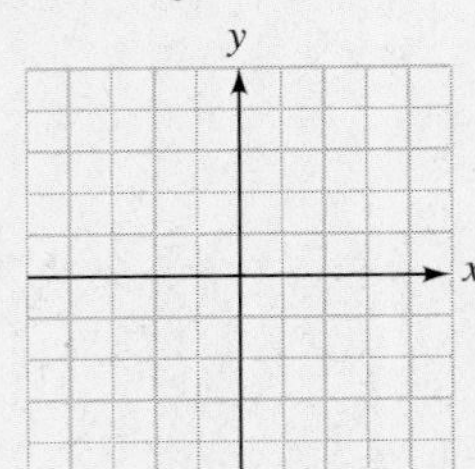

©Shutterstock.com/Rob Marmion

CHAPTER 11

Miscellaneous Topics

11.1 The Binomial Theorem
11.2 The *n*th Term of a Binomial Expansion
11.3 Arithmetic Sequences
11.4 Geometric Sequences
11.5 Infinite Geometric Sequences
11.6 Permutations and Combinations
11.7 Probability
■ ***Projects***
CHAPTER REVIEW
CHAPTER TEST
CUMULATIVE REVIEW EXERCISES

Careers and Mathematics

FINANCIAL ANALYSTS AND FINANCIAL PLANNERS

Financial analysts and personal financial planners provide analysis and guidance to businesses and individuals to help them with their investment decisions. Both types of specialists gather financial information, analyze it, and make recommendations to their clients.

Financial analysts and personal financial planners held 397,000 jobs in 2006, of which financial analysts held 221,000. Approximately 30 percent of personal financial planners are self-employed.

A bachelor's or graduate degree is required for financial analysts.

Job Outlook:
Employment of financial analysts and advisors is expected to grow by 37 percent during the 2006–2016 decade. This is much faster than the average for all occupations.

Annual Earnings:
$44,130–$114,260

For More Information:
http://www.bls.gov/oco/ocos259.htm

For a Sample Application:
See Example 7 in Section 11.4.

In this chapter ▶

In this chapter, we introduce several topics that have applications in advanced mathematics and in many occupational areas. The binomial theorem, permutations, and combinations are used in statistics. Arithmetic and geometric sequences are used in the mathematics of finance.

SECTION 11.1

The Binomial Theorem

Objectives

1. Raise a binomial to a power.
2. Complete a specified number of rows of Pascal's triangle.
3. Simplify an expression involving factorial notation.
4. Apply the binomial theorem to expand a binomial.

Vocabulary

Pascal's triangle | factorial notation | binomial theorem

Getting Ready

Raise each binomial to the indicated power.

1. $(x + 2)^2$ **2.** $(x - 3)^2$

3. $(x + 1)^3$ **4.** $(x - 2)^3$

We have seen how to square and cube binomials. In this section, we will learn how to raise binomials to higher powers.

1 Raise a binomial to a power.

We have discussed how to raise binomials to positive integral powers. For example, we know that

$$(a + b)^2 = a^2 + 2ab + b^2$$

and that

$$\begin{aligned}(a + b)^3 &= (a + b)(a + b)^2 \\ &= (a + b)(a^2 + 2ab + b^2) \\ &= a^3 + 2a^2b + ab^2 + a^2b + 2ab^2 + b^3 \\ &= a^3 + 3a^2b + 3ab^2 + b^3\end{aligned}$$

To show how to raise binomials to positive-integer powers without doing the actual multiplications, we consider the following binomial expansions:

$(a + b)^0 = 1$ 1 term

$(a + b)^1 = a + b$ 2 terms

$(a + b)^2 = a^2 + 2ab + b^2$ 3 terms

Blaise Pascal (1623–1662)

Pascal was torn between the fields of religion and mathematics. Each surfaced at times in his life to dominate his interest. In mathematics, Pascal made contributions to the study of conic sections, probability, and differential calculus. At the age of 19, he invented a calculating machine. He is best known for a triangular array of numbers that bears his name.

$$(a+b)^3 = a^3 + 3a^2b + 3ab^2 + b^3 \quad \text{4 terms}$$
$$(a+b)^4 = a^4 + 4a^3b + 6a^2b^2 + 4ab^3 + b^4 \quad \text{5 terms}$$
$$(a+b)^5 = a^5 + 5a^4b + 10a^3b^2 + 10a^2b^3 + 5ab^4 + b^5 \quad \text{6 terms}$$
$$(a+b)^6 = a^6 + 6a^5b + 15a^4b^2 + 20a^3b^3 + 15a^2b^4 + 6ab^5 + b^6 \quad \text{7 terms}$$

Several patterns appear in these expansions:

1. Each expansion has one more term than the power of the binomial.
2. The degree of each term in each expansion is equal to the exponent of the binomial that is being expanded. For example, in the expansion of $(a+b)^5$, the sum of the exponents in each term is 5:

$$(a+b)^5 = a^5 + 5\underbrace{a^4b}_{4+1=5} + 10\underbrace{a^3b^2}_{3+2=5} + 10\underbrace{a^2b^3}_{2+3=5} + 5\underbrace{ab^4}_{1+4=5} + b^5$$

3. The first term in each expansion is a raised to the power of the binomial, and the last term in each expansion is b raised to the power of the binomial.
4. The exponents on a decrease by 1 in each successive term. The exponents of b, beginning with $b^0 = 1$ in the first term, increase by 1 in each successive term. For example, the expansion of $(a+b)^4$ is

$$a^4b^0 + 4a^3b^1 + 6a^2b^2 + 4a^1b^3 + a^0b^4$$

Thus, the variables have the pattern

$$a^n, \quad a^{n-1}b, \quad a^{n-2}b^2, \quad \ldots, \quad ab^{n-1}, \quad b^n$$

2 Complete a specified number of rows of Pascal's triangle.

To see another pattern, we write the coefficients of each expansion in the following triangular array:

```
                  1                     Row 0
               1     1                  Row 1
            1     2     1               Row 2
         1     3     3     1            Row 3
      1     4     6     4     1         Row 4
   1    (5)  (10)   10    5     1       Row 5
1     6    15    20    15    6     1    Row 6
```

In this array, called **Pascal's triangle,** each entry between the 1's is the sum of the closest pair of numbers in the line immediately above it. For example, the first 15 in the bottom row is the sum of the 5 and 10 immediately above it. Pascal's triangle continues with the same pattern forever. The next two lines are

```
   1    7    21    35    35    21    7    1       Row 7
1    8    28    56    70    56    28    8    1    Row 8
```

EXAMPLE 1 Expand: $(x+y)^5$.

Solution The first term in the expansion is x^5, and the exponents of x decrease by 1 in each successive term. A y first appears in the second term, and the exponents on y increase by 1

in each successive term, concluding when the term y^5 is reached. Thus, the variables in the expansion are

$$x^5, \quad x^4y, \quad x^3y^2, \quad xy^4, \quad y^5$$

The coefficients of these variables are given in row 5 of Pascal's triangle.

1 5 10 10 5 1

Combining this information gives the following expansion:

$$(x + y)^5 = x^5 + 5x^4y + 10x^3y^2 + 10x^2y^3 + 5xy^4 + y^5$$

SELF CHECK 1 Expand: $(x + y)^4$.

EXAMPLE 2 Expand: $(u - v)^4$.

Solution We note that $(u - v)^4$ can be written in the form $[u + (-v)]^4$. The variables in this expansion are

$$u^4, \quad u^3(-v), \quad u^2(-v)^2, \quad u(-v)^3, \quad (-v)^4$$

and the coefficients are given in row 4 of Pascal's triangle.

1 4 6 4 1

Thus, the required expansion is

$$\begin{aligned}(u - v)^4 &= u^4 + 4u^3(-v) + 6u^2(-v)^2 + 4u(-v)^3 + (-v)^4 \\ &= u^4 - 4u^3v + 6u^2v^2 - 4uv^3 + v^4\end{aligned}$$

SELF CHECK 2 Expand: $(x - y)^5$.

3 Simplify an expression involving factorial notation.

Although Pascal's triangle gives the coefficients of the terms in a binomial expansion, it can be a tedious way to expand a binomial for large powers. To develop a more efficient way, we introduce **factorial notation.**

Factorial Notation

If n is a natural number, the symbol $n!$ (read as **"n factorial"** or as **"factorial n"**) is defined as

$$n! = n(n - 1)(n - 2)(n - 3) \cdots (3)(2)(1)$$

Zero factorial is defined as

$$0! = 1$$

EXAMPLE 3 Write each expression without using factorial notation.
a. $2!$ **b.** $5!$ **c.** $-9!$ **d.** $(n - 2)!$ **e.** $4! \cdot 0!$

Solution
a. $2! = 2 \cdot 1 = 2$
b. $5! = 5 \cdot 4 \cdot 3 \cdot 2 \cdot 1 = 120$

c. $-9! = -9 \cdot 8 \cdot 7 \cdot 6 \cdot 5 \cdot 4 \cdot 3 \cdot 2 \cdot 1 = -362{,}880$

d. $(n - 2)! = (n - 2)(n - 3)(n - 4) \cdot \cdots \cdot 3 \cdot 2 \cdot 1$

e. $4! \cdot 0! = (4 \cdot 3 \cdot 2 \cdot 1) \cdot 1 = 24$

COMMENT According to the previous definition, part d is meaningful only if $n - 2$ is a natural number.

SELF CHECK 3 Write each expression without using factorial notation:
a. $6!$ **b.** $x!$

ACCENT ON TECHNOLOGY

Factorials

We can find factorials using a calculator. For example, to find 12! with a scientific calculator, we enter

12 **x!** (You may have to use a **2ND** or **SHIFT** key first.) **479001600**

To find 12! on a TI-84 Plus graphing calculator, we enter

12 **MATH** ► ► ► **4** **ENTER**

12!
479001600

To discover an important property of factorials, we note that

$$5 \cdot 4! = 5 \cdot 4 \cdot 3 \cdot 2 \cdot 1 = 5!$$
$$7 \cdot 6! = 7 \cdot 6 \cdot 5 \cdot 4 \cdot 3 \cdot 2 \cdot 1 = 7!$$
$$10 \cdot 9! = 10 \cdot 9 \cdot 8 \cdot 7 \cdot 6 \cdot 5 \cdot 4 \cdot 3 \cdot 2 \cdot 1 = 10!$$

These examples suggest the following property.

Property of Factorials

If n is a positive integer, then

$$n(n - 1)! = n!$$

EXAMPLE 4 Simplify each expression. **a.** $\dfrac{6!}{5!}$ **b.** $\dfrac{10!}{8!(10-8)!}$

Solution **a.** If we write $6!$ as $6 \cdot 5!$, we can simplify the fraction by removing the common factor $5!$ in the numerator and denominator.

$$\frac{6!}{5!} = \frac{6 \cdot 5!}{5!} = \frac{6 \cdot \cancel{5!}}{\cancel{5!}} = 6 \qquad \text{Simplify: } \tfrac{5!}{5!} = 1.$$

b. First, we subtract within the parentheses. Then we write $10!$ as $10 \cdot 9 \cdot 8!$ and simplify.

$$\frac{10!}{8!(10-8)!} = \frac{10!}{8! \cdot 2!} = \frac{10 \cdot 9 \cdot \cancel{8!}}{\cancel{8!} \cdot 2!} = \frac{5 \cdot \cancel{2} \cdot 9}{\cancel{2} \cdot 1} = 45 \qquad \text{Simplify: } \tfrac{8!}{8!} = 1. \text{ Factor 10 as } 5 \cdot 2 \text{ and simplify: } \tfrac{2}{2} = 1.$$

SELF CHECK 4 Simplify: **a.** $\frac{4!}{3!}$ **b.** $\frac{7!}{5!(7-5)!}$.

4 Apply the binomial theorem to expand a binomial.

We now state the **binomial theorem.**

The Binomial Theorem

If n is any positive integer, then

$$(a + b)^n = a^n + \frac{n!}{1!(n-1)!}a^{n-1}b + \frac{n!}{2!(n-2)!}a^{n-2}b^2 + \frac{n!}{3!(n-3)!}a^{n-3}b^3 + \cdots + \frac{n!}{r!(n-r)!}a^{n-r}b^r + \cdots + b^n$$

In the binomial theorem, the exponents of the variables follow the familiar pattern:

- The sum of the exponents on a and b in each term is n,
- the exponents on a decrease in each subsequent term, and
- the exponents on b increase in each subsequent term.

Only the method of finding the coefficients is different. Except for the first and last terms, the numerator of each coefficient is $n!$. If the exponent of b in a particular term is r, the denominator of the coefficient of that term is $r!(n - r)!$.

EXAMPLE 5 Use the binomial theorem to expand $(a + b)^3$.

Solution We can substitute directly into the binomial theorem and simplify:

$$\begin{aligned}(a + b)^3 &= a^3 + \frac{3!}{1!(3-1)!}a^2b + \frac{3!}{2!(3-2)!}ab^2 + b^3 \\ &= a^3 + \frac{3!}{1! \cdot 2!}a^2b + \frac{3!}{2! \cdot 1!}ab^2 + b^3 \\ &= a^3 + \frac{3 \cdot 2 \cdot 1}{1 \cdot 2 \cdot 1}a^2b + \frac{3 \cdot 2 \cdot 1}{2 \cdot 1 \cdot 1}ab^2 + b^3 \\ &= a^3 + 3a^2b + 3ab^2 + b^3\end{aligned}$$

SELF CHECK 5 Use the binomial theorem to expand $(a + b)^4$.

EXAMPLE 6 Use the binomial theorem to expand $(x - y)^4$.

Solution We can write $(x - y)^4$ in the form $[x + (-y)]^4$, substitute directly into the binomial theorem, and simplify:

$$\begin{aligned}(x - y)^4 &= [x + (-y)]^4 \\ &= x^4 + \frac{4!}{1!(4-1)!}x^3(-y) + \frac{4!}{2!(4-2)!}x^2(-y)^2 + \frac{4!}{3!(4-3)!}x(-y)^3 + (-y)^4 \\ &= x^4 - \frac{4 \cdot 3!}{1! \cdot 3!}x^3y + \frac{4 \cdot 3 \cdot 2!}{2! \cdot 2!}x^2y^2 - \frac{4 \cdot 3!}{3! \cdot 1!}xy^3 + y^4 \\ &= x^4 - 4x^3y + 6x^2y^2 - 4xy^3 + y^4 \quad \text{Note the alternating signs.}\end{aligned}$$

SELF CHECK 6 Use the binomial theorem to expand $(x - y)^3$.

EXAMPLE 7 Use the binomial theorem to expand $(3u - 2v)^4$.

Solution We write $(3u - 2v)^4$ in the form $[3u + (-2v)]^4$ and let $a = 3u$ and $b = -2v$. Then we can use the binomial theorem to expand $(a + b)^4$.

$$(a + b)^4 = a^4 + \frac{4!}{1!(4-1)!}a^3b + \frac{4!}{2!(4-2)!}a^2b^2 + \frac{4!}{3!(4-3)!}ab^3 + b^4$$
$$= a^4 + 4a^3b + 6a^2b^2 + 4ab^3 + b^4$$

Now we can substitute $3u$ for a and $-2v$ for b and simplify:

$$(3u - 2v)^4 = (3u)^4 + 4(3u)^3(-2v) + 6(3u)^2(-2v)^2 + 4(3u)(-2v)^3 + (-2v)^4$$
$$= 81u^4 - 216u^3v + 216u^2v^2 - 96uv^3 + 16v^4$$

SELF CHECK 7 Use the binomial theorem to expand $(2a - 3b)^3$.

SELF CHECK ANSWERS **1.** $x^4 + 4x^3y + 6x^2y^2 + 4xy^3 + y^4$ **2.** $x^5 - 5x^4y + 10x^3y^2 - 10x^2y^3 + 5xy^4 - y^5$ **3. a.** 720 **b.** $x(x-1)(x-2) \cdot \cdots \cdot 3 \cdot 2 \cdot 1$ **4. a.** 4 **b.** 21 **5.** $a^4 + 4a^3b + 6a^2b^2 + 4ab^3 + b^4$ **6.** $x^3 - 3x^2y + 3xy^2 - y^3$ **7.** $8a^3 - 36a^2b + 54ab^2 - 27b^3$

NOW TRY THIS

Expand $(1 + 2i)^7$ using any method.

11.1 EXERCISES

WARM-UPS

Find each value.

1. $1!$

2. $4!$

3. $0!$

4. $5!$

Expand each binomial.

5. $(m + n)^2$

6. $(m - n)^2$

7. $(p + 2q)^2$

8. $(2p - q)^2$

REVIEW *Find each value of x.*

9. $\log_4 16 = x$

10. $\log_x 49 = 2$

11. $\log_{25} x = \frac{1}{2}$

12. $\log_{1/2} \frac{1}{8} = x$

VOCABULARY AND CONCEPTS *Fill in the blanks.*

13. Every binomial expansion has ____ more term than the power of the binomial.

14. The first term in the expansion of $(a + b)^{20}$ is ____.

15. The triangular array that can be used to find the coefficients of a binomial expansion is called ________ triangle.

16. The symbol 5! is read as "____________."

17. $6 \cdot 5 \cdot 4 \cdot 3 \cdot 2 \cdot 1 =$ ___ (Write your answer in factorial notation.)

18. $8! = 8 \cdot$ ___

19. $0! =$ ___

20. According to the binomial theorem, the third term of the expansion of $(a + b)^n$ is ____________.

GUIDED PRACTICE

Expand each expression using Pascal's triangle. See Examples 1–2. (Objectives 1–2)

21. $(a + b)^3$

22. $(a + b)^4$

23. $(a - b)^4$

24. $(a - b)^3$

Evaluate each expression. See Example 3. (Objective 3)

25. $3!$

26. $7!$

27. $-5!$

28. $-6!$

29. $3! + 4!$

30. $2!(3!)$

31. $3!(4!)$

32. $4! + 4!$

Evaluate each expression. See Example 4. (Objective 3)

33. $\dfrac{9!}{11!}$

34. $\dfrac{13!}{10!}$

35. $\dfrac{49!}{47!}$

36. $\dfrac{101!}{100!}$

37. $\dfrac{9!}{7! \cdot 0!}$

38. $\dfrac{7!}{5! \cdot 0!}$

39. $\dfrac{5!}{3!(5 - 3)!}$

40. $\dfrac{6!}{4!(6 - 4)!}$

Use the binomial theorem to expand each expression. See Example 5. (Objective 4)

41. $(x + y)^3$

42. $(x + y)^4$

43. $(x - y)^4$

44. $(x - y)^3$

Use the binomial theorem to expand each expression. See Example 6. (Objective 4)

45. $(2x + y)^3$

46. $(x + 2y)^3$

47. $(x - 2y)^3$

48. $(2x - y)^3$

Use the binomial theorem to expand each expression. See Example 7. (Objective 4)

49. $(2x + 3y)^3$

50. $(3x - 2y)^3$

51. $\left(\dfrac{x}{2} - \dfrac{y}{3}\right)^3$

52. $\left(\dfrac{x}{3} + \dfrac{y}{2}\right)^3$

ADDITIONAL PRACTICE

Evaluate each expression.

53. $8(7!)$

54. $4!(5)$

55. $\dfrac{7!}{5!(7 - 5)!}$

56. $\dfrac{8!}{6!(8 - 6)!}$

57. $\dfrac{5!(8 - 5)!}{4! \cdot 7!}$

58. $\dfrac{6! \cdot 7!}{(8 - 3)!(7 - 4)!}$

Use a calculator to find each factorial.

59. $11!$

60. $13!$

61. $20!$

62. $55!$

Expand using any method.

63. $(3 + 2y)^4$

64. $(2x + 3)^4$

65. $\left(\dfrac{x}{3} - \dfrac{y}{2}\right)^4$

66. $\left(\dfrac{x}{2} + \dfrac{y}{3}\right)^4$

67. Without referring to the text, write the first ten rows of Pascal's triangle.

68. Find the sum of the numbers in each row of the first ten rows of Pascal's triangle. What is the pattern?

WRITING ABOUT MATH

69. Explain how to construct Pascal's triangle.

70. Explain how to find the variables of the terms in the expansion of $(r + s)^4$.

SOMETHING TO THINK ABOUT

71. If we apply the pattern of the coefficients to the coefficient of the first term in a binomial expansion, the coefficient would be $\frac{n!}{0!(n - 0)!}$. Show that this expression is 1.

72. If we apply the pattern of the coefficients to the coefficient of the last term in a binomial expansion, the coefficient would be $\frac{n!}{n!(n - n)!}$. Show that this expression is 1.

73. Find the sum of the numbers in the designated diagonal rows of Pascal's triangle shown in the illustration. What is the pattern?

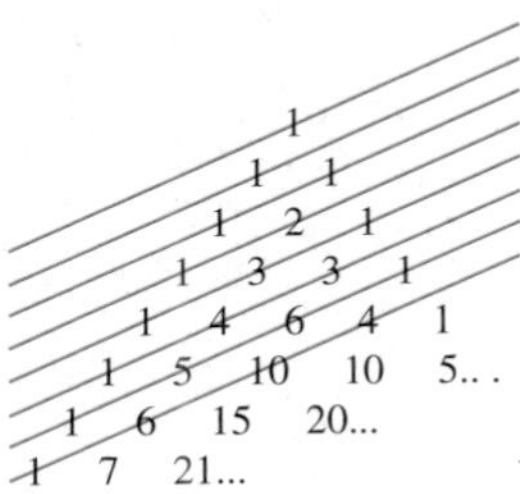

SECTION 11.2 The *n*th Term of a Binomial Expansion

Objectives

1 Find a particular term of a binomial expansion.

Getting Ready

Expand the binomial and find the coefficient of the third term in each expansion.

1. $(r + s)^4$ **2.** $(r + s)^5$

Expand the binomial and find the coefficient of the fourth term in each expansion.

3. $(p - q)^6$ **4.** $(p - q)^4$

In this section, we will consider a way to find a particular term of a binomial expansion.

1 Find a particular term of a binomial expansion.

To find the fourth term of the expansion of $(a + b)^9$, we could raise the binomial $a + b$ to the 9th power and look at the fourth term. However, this task would be tedious. By using the binomial theorem, we can construct the fourth term without finding the complete expansion of $(a + b)^9$.

EXAMPLE 1 Find the fourth term in the expansion of $(a + b)^9$.

Solution Since b^1 appears in the second term, b^2 appears in the third term, and so on, the exponent on b in the fourth term is 3. Since the exponent on b added to the exponent on a must equal 9, the exponent on a must be 6. Thus, the variables of the fourth term are

a^6b^3 The sum of the exponents must be 9.

We can find the coefficient of a^6b^3 by using the formula $\frac{n!}{r!(n-r)!}$, where n is the power of the expansion and r is the exponent of the second variable b.

To find the coefficient of the fourth term, we substitute 9 for n and 3 for r and simplify.

$$\frac{n!}{r!(n-r)!} = \frac{9!}{3!(9-3)!}$$

The complete fourth term is

$$\frac{9!}{3!(9-3)!}a^6b^3 = \frac{9 \cdot 8 \cdot 7 \cdot 6!}{3 \cdot 2 \cdot 1 \cdot 6!}a^6b^3$$
$$= 84a^6b^3$$

SELF CHECK 1 Find the third term of the expansion of $(a + b)^9$.

EXAMPLE 2 Find the sixth term in the expansion of $(x - y)^7$.

Solution We first find the sixth term of $[x + (-y)]^7$. In the sixth term, the exponent on $(-y)$ is 5. Thus, the variables in the sixth term are

$x^2(-y)^5$ The sum of the exponents must be 7.

The coefficient of these variables is

$$\frac{n!}{r!(n - r)!} = \frac{7!}{5!(7 - 5)!}$$

The complete sixth term is

$$\begin{aligned}\frac{7!}{5!(7 - 5)!}x^2(-y)^5 &= -\frac{7 \cdot 6 \cdot 5!}{5! \cdot 2 \cdot 1}x^2y^5 \\ &= -21x^2y^5\end{aligned}$$

SELF CHECK 2 Find the fifth term of the expansion of $(a - b)^7$.

EXAMPLE 3 Find the fourth term of the expansion of $(2x - 3y)^6$.

Solution We can let $a = 2x$ and $b = -3y$ and find the fourth term of the expansion of $(a + b)^6$:

$$\begin{aligned}\frac{6!}{3!(6 - 3)!}a^3b^3 &= \frac{6 \cdot 5 \cdot 4 \cdot 3!}{3! \cdot 3 \cdot 2 \cdot 1}a^3b^3 \\ &= 20a^3b^3\end{aligned}$$

We can now substitute $2x$ for a and $-3y$ for b and simplify:

$$\begin{aligned}20a^3b^3 &= 20(2x)^3(-3y)^3 \\ &= -4{,}320x^3y^3\end{aligned}$$

The fourth term is $-4{,}320x^3y^3$.

SELF CHECK 3 Find the third term of the expansion of $(2a - 3b)^6$.

SELF CHECK ANSWERS **1.** $36a^7b^2$ **2.** $35a^3b^4$ **3.** $2{,}160a^4b^2$

NOW TRY THIS

Find the specified term of each expansion.

1. 4th term of $\left(5x + \frac{1}{4}\right)^6$

2. 5th term of $(3 - 2i)^8$.

11.2 EXERCISES

WARM-UPS

In the expansion of $(x + y)^8$, find the exponent on y in the

1. 3rd term
2. 4th term
3. 7th term

In the expansion of $(x + y)^8$, find the exponent on x in the

4. 3rd term
5. 4th term
6. 7th term

In the expansion of $(x + y)^8$, find the coefficient of the

7. 1st term
8. 2nd term

REVIEW

Solve each system of equations.

9. $\begin{cases} 3x + 2y = 12 \\ 2x - y = 1 \end{cases}$

10. $\begin{cases} a + b + c = 6 \\ 2a + b + 3c = 11 \\ 3a - b - c = 6 \end{cases}$

Evaluate each determinant.

11. $\begin{vmatrix} 2 & -3 \\ 4 & -2 \end{vmatrix}$

12. $\begin{vmatrix} 1 & 2 & 3 \\ 4 & 5 & 0 \\ -1 & -2 & 1 \end{vmatrix}$

VOCABULARY AND CONCEPTS *Fill in the blanks.*

13. The exponent on b in the fourth term of the expansion of $(a + b)^6$ is __.

14. The exponent on b in the fifth term of the expansion of $(a + b)^6$ is __.

15. In the expansion of $(a + b)^7$, the sum of the exponents on a and b is __.

16. The coefficient of the fourth term of the expansion of $(a + b)^9$ is ______.

GUIDED PRACTICE

Find the specified term of each expansion. See Example 1. (Objective 1)

17. $(a + b)^3$; second term
18. $(a + b)^3$; third term
19. $(x + y)^6$; fifth term
20. $(x + y)^7$; fifth term

Find the specified term of each expansion. See Example 2. (Objective 1)

21. $(x - y)^4$; fourth term
22. $(x - y)^5$; second term
23. $(x - y)^8$; third term
24. $(x - y)^9$; seventh term

Find the specified term of each expansion. See Example 3. (Objective 1)

25. $(4x + y)^5$; third term
26. $(x + 4y)^5$; fourth term
27. $(x - 3y)^4$; second term
28. $(3x - y)^5$; third term
29. $(2x - 5)^7$; fourth term
30. $(2x + 3)^6$; sixth term
31. $(2x - 3y)^5$; fifth term
32. $(3x - 2y)^4$; second term
33. $\left(\sqrt{2}x + \sqrt{3}y\right)^6$; third term
34. $\left(\sqrt{3}x + \sqrt{2}y\right)^5$; second term
35. $\left(\frac{x}{2} - \frac{y}{3}\right)^4$; second term
36. $\left(\frac{x}{3} + \frac{y}{2}\right)^5$; fourth term

ADDITIONAL PRACTICE *Find the specified term of each expansion.*

37. $(x + 3)^5$; third term
38. $(x - 2)^4$; second term
39. $(a + b)^n$; fourth term
40. $(a + b)^n$; third term
41. $(a - b)^n$; fifth term
42. $(a - b)^n$; sixth term
43. $(a + b)^n$; rth term
44. $(a + b)^n$; $(r + 1)$th term

WRITING ABOUT MATH

45. Explain how to find the coefficients in the expansion of $(x + y)^5$.

46. Explain why the signs alternate in the expansion of $(x - y)^9$.

SOMETHING TO THINK ABOUT

47. Find the constant term in the expansion of $\left(x + \frac{1}{x}\right)^{10}$.

48. Find the coefficient of a^5 in the expansion of $\left(a - \frac{1}{a}\right)^9$.

SECTION 11.3 Arithmetic Sequences

Objectives

1. Find a specified term of a sequence given the general term.
2. Find a specified term of an arithmetic sequence given the first term and the common difference.
3. Find a specified term of an arithmetic sequence given the first term and one other term.
4. Insert one or more arithmetic means between two numbers.
5. Find the sum of the first n terms of an arithmetic sequence.
6. Expand and find the sum of a series written in summation notation.

Vocabulary

sequence
Fibonnaci sequence
finite sequence
infinite sequence
series
arithmetic sequence
common difference
arithmetic means
arithmetic series
summation notation
index of summation

Getting Ready

Complete each table.

1.

n	$2n + 1$
1	
2	
3	
4	

2.

n	$3n - 5$
3	
4	
5	
6	

We will now discuss ordered lists of numbers called *sequences.* In this section, we will examine a special type of sequence called an *arithmetic sequence.*

1 Find a specified term of a sequence given the general term.

A **sequence** is a function whose domain is the set of natural numbers. For example, the function $f(n) = 3n + 2$, where n is a natural number, is a sequence. Because a sequence is a function whose domain is the set of natural numbers, we can write its values as a list. If the natural numbers are substituted for n, the function $f(n) = 3n + 2$ generates the list

$$5, 8, 11, 14, 17, \ldots$$

It is common to call the list, as well as the function, a sequence. Each number in the list is called a *term* of the sequence. Other examples of sequences are

$1^3, 2^3, 3^3, 4^3, \ldots$ The ordered list of the cubes of the natural numbers

$4, 8, 12, 16, \ldots$ The ordered list of the positive multiples of 4

2, 3, 5, 7, 11, . . .   The ordered list of prime numbers

1, 1, 2, 3, 5, 8, 13, 21, . . .   The Fibonacci sequence

The **Fibonacci sequence** is named after the 12th-century mathematician Leonardo of Pisa—also known as Fibonacci. Beginning with the 2, each term of the sequence is the sum of the two preceding terms.

Finite sequences contain a finite number of terms and **infinite sequences** contain infinitely many terms. One example of each type of sequence is:

Finite sequence: 1, 5, 9, 13, 17, 21, 25

Infinite sequence: 3, 6, 9, 12, 15, . . .   The ellipsis, . . . , indicates that the sequence goes on forever.

In this section, we will use a_n (read as "a sub n") to denote the nth term of a sequence. For example, in the sequence 3, 6, 9, 12, 15, . . . , we have

1st term	2nd term	3rd term	4th term	5th term	
3,	6,	9,	12,	15,	. . .
↑	↑	↑	↑	↑	
a_1	a_2	a_3	a_4	a_5	

To describe all the terms of a sequence, we can write a formula for a_n, called the *general term* of the sequence. For the sequence 3, 6, 9, 12, 15, . . . , we note that $a_1 = 3 \cdot 1$, $a_2 = 3 \cdot 2$, $a_3 = 3 \cdot 3$, and so on. In general, the nth term of the sequence is found by multiplying n by 3.

$$a_n = 3n$$

We can use this formula to find any term of the sequence. For example, to find the 12th term, we substitute 12 for n.

$$a_{12} = 3(12) = 36$$

EXAMPLE 1 Given an infinite sequence with $a_n = 2n - 3$, find:

a. the first four terms **b.** a_{50}.

Solution **a.** To find the first four terms of the sequence, we substitute 1, 2, 3, and 4 for n in $a_n = 2n - 3$ and simplify.

$a_1 = 2(1) - 3 = -1$ Substitute 1 for n.

$a_2 = 2(2) - 3 = 1$ Substitute 2 for n.

$a_3 = 2(3) - 3 = 3$ Substitute 3 for n.

$a_4 = 2(4) - 3 = 5$ Substitute 4 for n.

The first four terms of the sequence are -1, 1, 3, and 5.

b. To find a_{50}, the 50th term of the sequence, we let $n = 50$:

$$a_{50} = 2(50) - 3 = 97$$

SELF CHECK 1 Given an infinite sequence with $a_n = 3n + 5$, find:

a. the first three terms **b.** a_{100}.

2 Find a specified term of an arithmetic sequence given the first term and the common difference.

One common type of sequence is the **arithmetic sequence.**

Arithmetic Sequence

An **arithmetic sequence** is a sequence of the form

$$a_1, \quad a_1 + d, \quad a_1 + 2d, \quad \ldots, \quad a_1 + (n-1)d, \ldots$$

where a_1 is the *first term* and d is the **common difference.** The *nth term* is given by

$$a_n = a_1 + (n-1)d$$

We note that the second term of an arithmetic sequence has an addend of $1d$, the third term has an addend of $2d$, the fourth term has an addend of $3d$, and the nth term has an addend of $(n-1)d$. We also note that the difference between any two consecutive terms in an arithmetic sequence is d.

EXAMPLE 2 An arithmetic sequence has a first term of 5 and a common difference of 4.

a. Write the first six terms of the sequence.

b. Write the 25th term of the sequence.

Solution **a.** Since the first term is $a = 5$ and the common difference is $d = 4$, the first six terms are

$$\underset{a_1}{5}, \quad \underset{a_2}{5+4}, \quad \underset{a_3}{5+2(4)}, \quad \underset{a_4}{5+3(4)}, \quad \underset{a_5}{5+4(4)}, \quad \underset{a_6}{5+5(4)}$$

or

$$5, 9, 13, 17, 21, 25$$

b. The nth term is $a_n = a + (n-1)d$. Since we want the 25th term, we let $n = 25$:

$$\begin{aligned} a_n &= a_1 + (n-1)d \\ a_{25} &= 5 + (25-1)4 \quad \text{Remember that } a_1 = 5 \text{ and } d = 4. \\ &= 5 + 24(4) \\ &= 5 + 96 \\ &= 101 \end{aligned}$$

SELF CHECK 2 **a.** Write the seventh term of the sequence in Example 2.

b. Write the 30th term of the sequence in Example 2.

3 Find a specified term of an arithmetic sequence given the first term and one other term.

EXAMPLE 3 The first three terms of an arithmetic sequence are 3, 8, and 13. Find:

a. the 67th term **b.** the 100th term.

Solution We first find d, the common difference. It is the difference between two successive terms:

$$d = 8 - 3 = 13 - 8 = 5$$

a. We substitute 3 for a_1, 67 for n, and 5 for d in the formula for the nth term and simplify:

$$\begin{aligned} a_n &= a_1 + (n - 1)\boldsymbol{d} \\ a_{67} &= \mathbf{3} + (\mathbf{67} - 1)\mathbf{5} \\ &= 3 + 66(5) \\ &= 333 \end{aligned}$$

b. We substitute 3 for a_1, 100 for n, and 5 for d in the formula for the nth term, and simplify:

$$\begin{aligned} a_n &= a_1 + (n - 1)\boldsymbol{d} \\ a_{100} &= \mathbf{3} + (\mathbf{100} - 1)\mathbf{5} \\ &= 3 + 99(5) \\ &= 498 \end{aligned}$$

SELF CHECK 3 Find the 50th term of the sequence in Example 3.

EXAMPLE 4 The first term of an arithmetic sequence is 12, and the 50th term is 3,099. Write the first six terms of the sequence.

Solution The key is to find the common difference. Because the 50th term of this sequence is 3,099, we can let $n = 50$ and solve the following equation for d:

$$\begin{aligned} a_n &= a_1 + (n - 1)\boldsymbol{d} \\ a_{50} &= \mathbf{12} + (\mathbf{50} - 1)\boldsymbol{d} \\ \mathbf{3{,}099} &= 12 + 49d && \text{Substitute 3,099 for } a_{50} \text{ and simplify.} \\ 3{,}087 &= 49d && \text{Subtract 12 from both sides.} \\ 63 &= d && \text{Divide both sides by 49.} \end{aligned}$$

Since the first term of the sequence is 12, and the common difference is 63, its first six terms are

12, 75, 138, 201, 264, 327 Add 63 to a term to get the next term.

SELF CHECK 4 The first term of an arithmetic sequence is 15, and the 12th term is 92. Write the first four terms of the sequence.

4 Insert one or more arithmetic means between two numbers.

If numbers are inserted between two numbers a and b to form an arithmetic sequence, the inserted numbers are called **arithmetic means** between a and b.

If a single number is inserted between the numbers a and b to form an arithmetic sequence, that number is called the *arithmetic mean* between a and b.

EXAMPLE 5 Insert two arithmetic means between 6 and 27.

Solution Here the first term is $a_1 = 6$, and the fourth term is $a_4 = 27$. We must find the common difference so that the terms

$$\underset{a_1}{\underset{\uparrow}{6,}} \quad \underset{a_2}{\underset{\uparrow}{6 + d,}} \quad \underset{a_3}{\underset{\uparrow}{6 + 2d,}} \quad \underset{a_4}{\underset{\uparrow}{27}}$$

form an arithmetic sequence. To find d, we substitute 6 for a_1 and 4 for n into the formula for the nth term:

$$\begin{aligned} a_n &= a_1 + (n - 1)d \\ a_4 &= 6 + (4 - 1)d \\ 27 &= 6 + 3d && \text{Substitute 27 for } a_4 \text{ and simplify.} \\ 21 &= 3d && \text{Subtract 6 from both sides.} \\ 7 &= d && \text{Divide both sides by 3.} \end{aligned}$$

The two arithmetic means between 6 and 27 are

$$\begin{aligned} 6 + d &= 6 + 7 \\ &= 13 \end{aligned} \quad \text{or} \quad \begin{aligned} 6 + 2d &= 6 + 2(7) \\ &= 6 + 14 \\ &= 20 \end{aligned}$$

The numbers 6, 13, 20, and 27 are the first four terms of an arithmetic sequence.

SELF CHECK 5 Insert two arithmetic means between 8 and 44.

5 Find the sum of the first *n* terms of an arithmetic sequence.

We now consider a formula that gives the sum of the first n terms of an arithmetic sequence. To develop this formula, we let S_n represent the sum of the first n terms of an arithmetic sequence:

$$S_n = a_1 + [a_1 + d] + [a_1 + 2d] + \cdots + [a_1 + (n - 1)d]$$

We write the same sum again, but in reverse order:

$$S_n = [a_1 + (n - 1)d] + [a_1 + (n - 2)d] + [a_1 + (n - 3)d] + \cdots + a_1$$

We add these two equations together, term by term, to get

$$2S_n = [2a_1 + (n - 1)d] + [2a_1 + (n - 1)d] + [2a_1 + (n - 1)d] + \cdots + [2a_1 + (n - 1)d]$$

Because there are n equal terms on the right side of the preceding equation, we can write

$$\begin{aligned} 2S_n &= n[2a_1 + (n - 1)d] \\ 2S_n &= n[a_1 + a_1 + (n - 1)d] && 2a_1 = a_1 + a_1 \\ 2S_n &= n[a_1 + a_n] && \text{Substitute } a_n \text{ for } a_1 + (n - 1)d. \\ S_n &= \frac{n(a_1 + a_n)}{2} \end{aligned}$$

This reasoning establishes the following theorem.

Sum of the First n Terms of an Arithmetic Sequence

The sum of the first n terms of an arithmetic sequence is given by the formula

$$S_n = \frac{n(a_1 + a_n)}{2} \quad \text{with } a_n = a_1 + (n - 1)d$$

where a_1 is the first term, a_n is the nth term, and n is the number of terms in the sequence.

EXAMPLE 6 Find the sum of the first 40 terms of the arithmetic sequence 4, 10, 16, ...

Solution In this example, we let $a_1 = 4$, $n = 40$, $d = 6$, and $a_{40} = 4 + (40 - 16)6 = 238$ and substitute these values into the formula for S_n:

$$S_n = \frac{n(a_1 + a_{40})}{2}$$

$$S_{40} = \frac{40(4 + 238)}{2}$$

$$= 20(242)$$

$$= 4{,}840$$

The sum of the first 40 terms is 4,840.

SELF CHECK 6 Find the sum of the first 50 terms of the arithmetic sequence 3, 8, 13, ...

When the commas between the terms of a sequence are replaced by + signs, we call the indicated sum a **series.** The sum of the terms of an arithmetic sequence is called an **arithmetic series.** Some examples are

$4 + 8 + 12 + 16 + 20 + 24$ Since this series has a limited number of terms, it is a finite arithmetic series.

$5 + 8 + 11 + 14 + 17 + \cdots$ Since this series has infinitely many terms, it is an infinite arithmetic series.

6 Expand and find the sum of a series written in summation notation.

We can use a shorthand notation for indicating the sum of a finite number of consecutive terms in a series. This notation, called **summation notation,** involves the Greek letter Σ (sigma). The expression

$$\sum_{k=2}^{5} 3k$$ Read as "the summation of $3k$ as k runs from 2 to 5."

designates the sum of all terms obtained if we successively substitute the numbers 2, 3, 4, and 5 for k, called the **index of the summation.** Thus, we have

$k = 2$ $k = 3$ $k = 4$ $k = 5$

↓ ↓ ↓ ↓

$$\sum_{k=2}^{5} 3k = 3(2) + 3(3) + 3(4) + 3(5)$$

$$= 6 + 9 + 12 + 15$$

$$= 42$$

EXAMPLE 7 Write the series associated with each summation.

a. $\sum_{k=1}^{4} k$ **b.** $\sum_{k=2}^{5} (k-1)^3$

Solution **a.** $\sum_{k=1}^{4} k = \mathbf{1} + \mathbf{2} + \mathbf{3} + \mathbf{4}$

b. $\sum_{k=2}^{5} (k-1)^3 = (\mathbf{2}-1)^3 + (\mathbf{3}-1)^3 + (\mathbf{4}-1)^3 + (\mathbf{5}-1)^3$

$= 1^3 + 2^3 + 3^3 + 4^3$

$= 1 + 8 + 27 + 64$

SELF CHECK 7 Write the series associated with the summation $\sum_{t=3}^{5} t^2$.

EXAMPLE 8 Find each sum. **a.** $\sum_{k=3}^{5} (2k+1)$ **b.** $\sum_{k=2}^{5} k^2$ **c.** $\sum_{k=1}^{3} (3k^2+3)$

Solution **a.** $\sum_{k=3}^{5} (2k+1) = [2(\mathbf{3})+1] + [2(\mathbf{4})+1] + [2(\mathbf{5})+1]$

$= 7 + 9 + 11$

$= 27$

b. $\sum_{k=2}^{5} k^2 = \mathbf{2}^2 + \mathbf{3}^2 + \mathbf{4}^2 + \mathbf{5}^2$

$= 4 + 9 + 16 + 25$

$= 54$

c. $\sum_{k=1}^{3} (3k^2+3) = [3(\mathbf{1})^2 + 3] + [3(\mathbf{2})^2 + 3] + [3(\mathbf{3})^2 + 3]$

$= 6 + 15 + 30$

$= 51$

SELF CHECK 8 Evaluate: $\sum_{k=1}^{4} (2k^2 - 3)$.

SELF CHECK ANSWERS

1. a. 8, 11, 14 **b.** 305 **2. a.** 29 **b.** 121 **3.** 248 **4.** 15, 22, 29, 36 **5.** 20, 32 **6.** 6,275 **7.** 9 + 16 + 25 **8.** 48

NOW TRY THIS

Given $a_5 = 3a - 2$ and $a_9 = 11a + 10$ in an arithmetic sequence, find

1. the first 5 terms.

2. the sum of the first 10 terms.

11.3 EXERCISES

WARM-UPS

Find the next term in each arithmetic sequence.

1. 2, 6, 10, . . .

2. 10, 7, 4, . . .

Find the common difference in each arithmetic sequence.

3. −2, 3, 8, . . .

4. 5, −1, −7, . . .

Find each sum.

5. $\sum_{k=1}^{2} k$

6. $\sum_{k=2}^{3} k$

REVIEW *Perform the operations and simplify, if possible. Assume no division by 0.*

7. $3(2x^2 - 4x + 7) + 4(3x^2 + 5x - 6)$

8. $(2p + q)(3p^2 + 4pq - 3q^2)$

9. $\frac{3a + 4}{a - 2} + \frac{3a - 4}{a + 2}$

10. $2t - 3\overline{)8t^4 - 12t^3 + 8t^2 - 16t + 6}$

VOCABULARY AND CONCEPTS *Fill in the blanks.*

11. A ________ is a function whose domain is the set of natural numbers. If a sequence has a limited number of terms, it is called a ______ sequence. If it has infinitely many terms, it is an _______ sequence.

12. The sequence 1, 1, 2, 3, 5, 8, 13, 21, . . . is called the __________ sequence.

13. The sequence 3, 9, 15, 21, . . . is an example of an __________ sequence with a common __________ of 6 and a first term of 3.

14. The last term (or *n*th term) of an arithmetic sequence is given by the formula ________________.

15. If a number *c* is inserted between two numbers *a* and *b* to form an arithmetic sequence, then *c* is called the ______________ between *a* and *b*.

16. The sum of the first *n* terms of an arithmetic sequence is given by the formula $S_n =$ ________.

17. The indicated sum of the terms of an arithmetic sequence is called an arithmetic ______.

18. The symbol Σ is the Greek letter ______.

19. $\sum_{k=1}^{5} k$ means ________________.

20. In the expression $\sum_{k=1}^{5} (2k - 5)$, *k* is called the ______ of the summation.

GUIDED PRACTICE

Given the infinite sequence $a_n = 3n - 2$, find each value. See Example 1. (Objective 1)

21. a_1

22. a_3

23. a_{25}

24. a_{50}

Write the first five terms of each arithmetic sequence with the given properties. See Example 2. (Objective 2)

25. $a_1 = 3, d = 2$

26. $a_1 = -2, d = 3$

27. $a_1 = -5, d = -3$

28. $a_1 = 8, d = -5$

Write the first five terms of each arithmetic sequence with the given properties. See Examples 3–4. (Objective 3)

29. $a_1 = 5$, fifth term is 29

30. $a_1 = 4$, sixth term is 39

31. $a_1 = -4$, sixth term is −39

32. $a_1 = -5$, fifth term is −37

Insert the specified number of arithmetic means. See Example 5. (Objective 4)

33. Insert three arithmetic means between 2 and 11.

34. Insert four arithmetic means between 5 and 25.

35. Insert four arithmetic means between 10 and 20.

36. Insert three arithmetic means between 20 and 30.

37. Find the arithmetic mean between 10 and 19.

38. Find the arithmetic mean between 5 and 23.

39. Find the arithmetic mean between −4.5 and 7.

40. Find the arithmetic mean between −6.3 and −5.2.

Find the sum of the first n terms of each arithmetic sequence. See Example 6. (Objective 5)

41. 1, 4, 7, . . . ; $n = 30$

42. 2, 6, 10, . . . ; $n = 28$

43. −5, −1, 3, . . . ; $n = 17$

44. −7, −1, 5, . . . ; $n = 15$

Write the series associated with each summation. See Example 7. (Objective 6)

45. $\sum_{k=1}^{4} (3k)$

46. $\sum_{k=1}^{3} (k - 9)$

47. $\sum_{k=4}^{6} k^2$

48. $\sum_{k=3}^{5} (-2k)$

Find each sum. See Example 8. (Objective 6)

49. $\sum_{k=1}^{4} 6k$

50. $\sum_{k=2}^{5} 3k$

51. $\sum_{k=3}^{4} (k^2 + 3)$

52. $\sum_{k=2}^{6} (k^2 + 1)$

ADDITIONAL PRACTICE

Write the first five terms of each arithmetic sequence with the given properties.

53. $d = 7$, sixth term is -83

54. $d = 3$, seventh term is 12

55. $d = -3$, seventh term is 16

56. $d = -5$, seventh term is -12

57. The 19th term is 131 and the 20th term is 138.

58. The 16th term is 70 and the 18th term is 78.

Find the specified value.

59. Find the 30th term of the arithmetic sequence with $a_1 = 7$ and $d = 12$.

60. Find the 55th term of the arithmetic sequence with $a_1 = -5$ and $d = 4$.

61. Find the 37th term of the arithmetic sequence with a second term of -4 and a third term of -9.

62. Find the 40th term of the arithmetic sequence with a second term of 6 and a fourth term of 16.

63. Find the first term of the arithmetic sequence with a common difference of 11 and whose 27th term is 263.

64. Find the common difference of the arithmetic sequence with a first term of -164 if its 36th term is -24.

65. Find the common difference of the arithmetic sequence with a first term of 40 if its 44th term is 556.

66. Find the first term of the arithmetic sequence with a common difference of -5 and whose 23rd term is -625.

Find the sum of the first n terms.

67. Second term is 7, third term is 12; $n = 12$

68. Second term is 5, fourth term is 9; $n = 16$

69. $f(n) = 2n + 1$, nth term is 31; n is a natural number

70. $f(n) = 4n + 3$, nth term is 23; n is a natural number

71. Find the sum of the first 50 natural numbers.

72. Find the sum of the first 100 natural numbers.

73. Find the sum of the first 50 odd natural numbers.

74. Find the sum of the first 50 even natural numbers.

Find each sum.

75. $\sum_{k=4}^{4} (2k + 4)$

76. $\sum_{k=3}^{5} (3k^2 - 7)$

APPLICATIONS

77. Saving money Yasmeen puts \$60 into a safety deposit box. Each month, she puts \$50 more in the box. Write the first six terms of an arithmetic sequence that gives the monthly amounts in her savings, and find her savings after 10 years.

78. Installment loans Maria borrowed \$10,000, interest-free, from her mother. She agreed to pay back the loan in monthly installments of \$275. Write the first six terms of an arithmetic sequence that shows the balance due after each month, and find the balance due after 17 months.

79. Designing a patio Each row of bricks in the triangular patio is to have one more brick than the previous row, ending with the longest row of 150 bricks. How many bricks will be needed?

80. Falling objects The equation $s = 16t^2$ represents the distance s in feet that an object will fall in t seconds. After 1 second, the object has fallen 16 feet. After 2 seconds, the object has fallen 64 feet, and so on. Find the distance that the object will fall during the second and third seconds.

81. Falling objects Refer to Exercise 80. How far will the object fall during the 12th second?

82. Interior angles The sums of the angles of several polygons are given in the table. Assuming that the pattern continues, complete the table.

Figure	Number of sides	Sum of angles
Triangle	3	180°
Quadrilateral	4	360°
Pentagon	5	540°
Hexagon	6	720°
Octagon	8	
Dodecagon	12	

WRITING ABOUT MATH

83. Define an arithmetic sequence.

84. Develop the formula for finding the sum of the first n terms of an arithmetic sequence.

SOMETHING TO THINK ABOUT

85. Write the addends of the sum given by

$$\sum_{n=1}^{6}\left(\frac{1}{2}n + 1\right)$$

86. Find the sum of the sequence given in Exercise 85.

87. Show that the arithmetic mean between a and b is the average of a and b: $\frac{a+b}{2}$.

88. Show that the sum of the two arithmetic means between a and b is $a + b$.

89. Show that $\sum_{k=1}^{5} 5k = 5\sum_{k=1}^{5} k$.

90. Show that $\sum_{k=3}^{6} (k^2 + 3k) = \sum_{k=3}^{6} k^2 + \sum_{k=3}^{6} 3k$.

91. Show that $\sum_{k=1}^{n} 3 = 3n$. (*Hint:* Consider 3 to be $3k^0$.)

92. Show that $\sum_{k=1}^{3} \frac{k^2}{k} \neq \frac{\sum_{k=1}^{3} k^2}{\sum_{k=1}^{3} k}$.

SECTION 11.4 Geometric Sequences

Objectives

1. Find a specified term of a geometric sequence given the first term and the common ratio.
2. Find a specified term of a geometric sequence given the first term and one other term.
3. Find one or more geometric means given two terms of a sequence.
4. Find the sum of the first n terms of a geometric sequence.
5. Solve an application problem involving a geometric sequence.

Vocabulary

geometric sequence | common ratio | geometric mean

Getting Ready

Complete each table.

1.

n	$5(2^n)$
1	
2	
3	

2.

n	$6(3^n)$
1	
2	
3	

Another common type of sequence is called a *geometric sequence.*

1 Find a specified term of a geometric sequence given the first term and the common ratio.

Each term of a geometric sequence is found by multiplying the previous term by the same number.

Geometric Sequence

A **geometric sequence** is a sequence of the form

$$a_1, \quad a_1r, \quad a_1r^2, \quad a_1r^3, \quad \ldots \quad a_1r^{n-1}, \quad \ldots$$

where a_1 is the *first term* and r is the **common ratio.** The *nth term* is given by

$$a_n = a_1r^{n-1}$$

We note that the second term of a geometric sequence has a factor of r^1, the third term has a factor of r^2, the fourth term has a factor of r^3, and the nth term has a factor of r^{n-1}. We also note that the quotient obtained when any term is divided by the previous term is r.

EXAMPLE 1 A geometric sequence has a first term of 5 and a common ratio of 3.

a. Write the first five terms of the sequence.

b. Find the ninth term.

Solution **a.** Since the first term is $a = 5$ and the common ratio is $r = 3$, the first five terms are

$$\underset{a_1}{\overset{5,}{\uparrow}} \quad \underset{a_2}{\overset{5(3),}{\uparrow}} \quad \underset{a_3}{\overset{5(3^2),}{\uparrow}} \quad \underset{a_4}{\overset{5(3^3),}{\uparrow}} \quad \underset{a_5}{\overset{5(3^4)}{\uparrow}}$$

Each term is found by multiplying the previous term by 3.

or

$$5, 15, 45, 135, 405$$

b. The nth term is a_1r^{n-1} where $a_1 = 5$ and $r = 3$. Because we want the ninth term, we let $n = 9$:

$$\begin{aligned} a_n &= a_1r^{n-1} \\ a_9 &= 5(3)^{9-1} \\ &= 5(3)^8 \\ &= 5(6{,}561) \\ &= 32{,}805 \end{aligned}$$

SELF CHECK 1 A geometric sequence has a first term of 3 and a common ratio of 4.

a. Write the first four terms.

b. Find the eighth term.

2 Find a specified term of a geometric sequence given the first term and one other term.

EXAMPLE 2 The first three terms of a geometric sequence are 16, 4, and 1. Find the seventh term.

Solution We note that the first term a is 16 and that

$$r = \frac{a_2}{a_1} = \frac{4}{16} = \frac{1}{4}$$

Also note that $\frac{a_3}{a_2} = \frac{1}{4}$.

We substitute 16 for a_1, $\frac{1}{4}$ for r, and 7 for n in the formula for the nth term and simplify:

$$a_n = a_1 r^{n-1}$$
$$a_7 = 16\left(\frac{1}{4}\right)^{7-1}$$
$$= 16\left(\frac{1}{4}\right)^{6}$$
$$= 16\left(\frac{1}{4{,}096}\right)$$
$$= \frac{1}{256}$$

SELF CHECK 2 Find the tenth term of the sequence in Example 2.

3 Find one or more geometric means given two terms of a sequence.

If numbers are inserted between two numbers a and b to form a geometric sequence, the inserted numbers are called **geometric means** between a and b.

If a single number is inserted between the numbers a and b to form a geometric sequence, that number is called the **geometric mean** between a and b.

EXAMPLE 3 Insert two geometric means between 7 and 1,512.

Solution Here the first term is $a_1 = 7$. Because we are inserting two geometric means between 7 and 1,512, the fourth term is $a_4 = 1{,}512$. To find the common ratio r so that the terms

$$\begin{array}{cccc} 7, & 7r, & 7r^2, & 1{,}512 \\ \uparrow & \uparrow & \uparrow & \uparrow \\ a_1 & a_2 & a_3 & a_4 \end{array}$$

form a geometric sequence, we substitute 4 for n and 7 for a_1 into the formula for the nth term of a geometric sequence and solve for r.

$$a_n = a_1 r^{n-1}$$
$$a_4 = 7r^{4-1}$$
$$1{,}512 = 7r^3$$
$$216 = r^3 \quad \text{Divide both sides by 7.}$$
$$6 = r \quad \text{Take the cube root of both sides.}$$

The two geometric means between 7 and 1,512 are

$$7r = 7(6) = \mathbf{42} \qquad \text{and} \qquad 7r^2 = 7(6)^2 = 7(36) = \mathbf{252}$$

The numbers 7, 42, 252, and 1,512 are the first four terms of a geometric sequence.

SELF CHECK 3 Insert three positive geometric means between 1 and 16.

EXAMPLE 4 Find a geometric mean between 2 and 20.

Solution We want to find the middle term of the three-term geometric sequence

$$\underset{a_1}{\overset{\uparrow}{2,}} \quad \underset{a_2}{\overset{\uparrow}{2r,}} \quad \underset{a_3}{\overset{\uparrow}{20}}$$

with $a_1 = 2$, $a_3 = 20$, and $n = 3$. To find r, we substitute these values into the formula for the nth term of a geometric sequence:

$$\begin{aligned} a_n &= ar^{n-1} \\ a_3 &= 2r^{3-1} \\ 20 &= 2r^2 \\ 10 &= r^2 && \text{Divide both sides by 2.} \\ \pm\sqrt{10} &= r && \text{Take the square root of both sides.} \end{aligned}$$

Because r can be either $\sqrt{10}$ or $-\sqrt{10}$, there are two values for the geometric mean. They are

$$2r = 2\sqrt{10} \qquad \text{and} \qquad 2r = -2\sqrt{10}$$

The numbers 2, $2\sqrt{10}$, 20 and 2, $-2\sqrt{10}$, 20 both form geometric sequences. The common ratio of the first sequence is $\sqrt{10}$, and the common ratio of the second sequence is $-\sqrt{10}$.

SELF CHECK 4 Find the positive geometric mean between 2 and 200.

4 Find the sum of the first *n* terms of a geometric sequence.

There is a formula that gives the sum of the first n terms of a geometric sequence. To develop this formula, we let S_n represent the sum of the first n terms of a geometric sequence.

(1) $$S_n = a_1 + a_1r + a_1r^2 + a_1r^3 + \cdots + a_1r^{n-1}$$

We multiply both sides of Equation 1 by r to get

(2) $$S_nr = a_1r + a_1r^2 + a_1r^3 + \cdots + a_1r^{n-1} + a_1r^n$$

We now subtract Equation 2 from Equation 1 and solve for S_n:

$$\begin{aligned} S_n - S_nr &= a_1 - a_1r^n \\ S_n(1 - r) &= a_1 - a_1r^n && \text{Factor out } S_n \text{ from the left side.} \\ S_n &= \frac{a_1 - a_1r^n}{1 - r} && \text{Divide both sides by } 1 - r. \end{aligned}$$

This reasoning establishes the following formula.

Sum of the First *n* Terms of a Geometric Sequence

The sum of the first n terms of a geometric sequence is given by the formula

$$S_n = \frac{a_1 - a_1r^n}{1 - r} \quad (r \neq 1)$$

where S_n is the sum, a_1 is the first term, r is the common ratio, and n is the number of terms.

EXAMPLE 5 Find the sum of the first six terms of the geometric sequence 250, 50, 10, ...

Solution Here $a_1 = 250$, $r = \frac{1}{5}$, and $n = 6$. We substitute these values into the formula for the sum of the first n terms of a geometric sequence and simplify:

$$S_n = \frac{a_1 - a_1 r^n}{1 - r}$$

$$S_n = \frac{250 - 250\left(\frac{1}{5}\right)^6}{1 - \frac{1}{5}}$$

$$= \frac{250 - 250\left(\frac{1}{15{,}625}\right)}{\frac{4}{5}}$$

$$= \frac{5}{4}\left(250 - \frac{250}{15{,}625}\right)$$

$$= \frac{5}{4}\left(\frac{3{,}906{,}000}{15{,}625}\right)$$

$$= 312.48$$

The sum of the first six terms is 312.48.

SELF CHECK 5 Find the sum of the first five terms of the geometric sequence 100, 20, 4, ...

5 Solve an application problem involving a geometric sequence.

EXAMPLE 6 **GROWTH OF A TOWN** The mayor of Eagle River (population 1,500) predicts a growth rate of 4% each year for the next 10 years. Find the expected population of Eagle River 10 years from now.

Solution Let P_0 be the initial population of Eagle River. After 1 year, there will be a different population, P_1. The initial population (P_0) plus the growth (the product of P_0 and the rate of growth, r) will equal this new population P_1:

$$P_1 = P_0 + P_0 r = P_0(1 + r)$$

The population after 2 years will be P_2, and

$$\begin{aligned} P_2 &= P_1 + P_1 r \\ &= P_1(1 + r) && \text{Factor out } P_1. \\ &= P_0(1 + r)(1 + r) && \text{Remember that } P_1 = P_0(1 + r). \\ &= P_0(1 + r)^2 \end{aligned}$$

The population after 3 years will be P_3, and

$$\begin{aligned} P_3 &= P_2 + P_2 r \\ &= P_2(1 + r) && \text{Factor out } P_2. \end{aligned}$$

$= P_0(1 + r)^2(1 + r)$ Remember that $P_2 = P_0(1 + r)^2$.

$= P_0(1 + r)^3$

The yearly population figures

$$P_0, \quad P_1, \quad P_2, \quad P_3, \quad \ldots$$

or

$$P_0, \quad P_0(1 + r), \quad P_0(1 + r)^2, \quad P_0(1 + r)^3, \quad \ldots$$

form a geometric sequence with a first term of P_0 and a common ratio of $1 + r$. The population of Eagle River after 10 years is P_{10}, which is the 11th term of this sequence:

$$\begin{aligned} a_n &= ar^{n-1} \\ P_{10} = a_{11} &= P_0(1 + r)^{10} \\ &= 1{,}500(1 + 0.04)^{10} \\ &= 1{,}500(1.04)^{10} \\ &\approx 1{,}500(1.480244285) \\ &\approx 2{,}220 \end{aligned}$$

The expected population 10 years from now is 2,220 people.

EXAMPLE 7 **AMOUNT OF AN ANNUITY** An *annuity* is a sequence of equal payments made periodically over a length of time. The sum of the payments and the interest earned during the *term* of the annuity is called the *amount* of the annuity.

After a sales clerk works six months, her employer will begin an annuity for her and will contribute \$500 every six months to a fund that pays 8% annual interest. After she has been employed for two years, what will be the amount of the annuity?

Solution Because the payments are to be made semiannually, there will be four payments of \$500, each earning a rate of 4% per six-month period. These payments will occur at the end of 6 months, 12 months, 18 months, and 24 months. The first payment, to be made after 6 months, will earn interest for three interest periods. Thus, the amount of the first payment is $\$500(1.04)^3$. The amounts of each of the four payments after two years are shown in Table 11-1.

The amount of the annuity is the sum of the amounts of the individual payments, a sum of \$2,123.23.

Payment (at the end of period)	Amount of payment at the end of 2 years
1	$\$500(1.04)^3 = \$562.43$
2	$\$500(1.04)^2 = \$540.80$
3	$\$500(1.04)^1 = \$520.00$
4	$\$500 = \$500.00$
	$A_n = \$2{,}123.23$

Table 11-1

SELF CHECK ANSWERS **1. a.** 3, 12, 48, 192 **b.** 49,152 **2.** $\frac{1}{16{,}384}$ **3.** 2, 4, 8 **4.** 20 **5.** 124.96

NOW TRY THIS

Given $a_1 = 8x - 12$ and $r = \frac{1}{2}$ in a geometric sequence:

1. Find the first 4 terms.
2. Find the sum of the first 4 terms by adding the terms.
3. Set up the formula for finding the sum of the first 4 terms and simplify it to show that the result is the same as the result obtained in Problem 2.

11.4 EXERCISES

WARM-UPS

Find the next term in each geometric sequence.

1. $1, 3, 9, \ldots$

2. $1, \frac{1}{3}, \frac{1}{9}, \ldots$

Find the common ratio in each geometric sequence.

3. $0.2, 0.5, 1.25, \ldots$

4. $\sqrt{3}, 3, 3\sqrt{3}, \ldots$

Find the value of x in each geometric sequence.

5. $2, x, 18, 54, \ldots$

6. $3, x, \frac{1}{3}, \frac{1}{9}, \ldots$

REVIEW *Solve each inequality. Assume no division by 0.*

7. $x^2 - 5x - 6 \leq 0$

8. $a^2 - 7a + 12 \geq 0$

9. $\frac{x - 4}{x + 3} \geq 0$

10. $\frac{t^2 + t - 20}{t + 2} < 0$

VOCABULARY AND CONCEPTS *Fill in the blanks.*

11. A sequence of the form $a_1, a_1r, a_1r^2, \ldots$ is called a ________ sequence.

12. The formula for the nth term of a geometric sequence is ________.

13. In a geometric sequence, r is called the ________.

14. A number inserted between two numbers a and b to form a geometric sequence is called a geometric ______ between a and b.

15. The sum of the first n terms of a geometric sequence is given by the formula ________.

16. In the formula for Exercise 15, a_1 is the _____ term of the sequence.

GUIDED PRACTICE

Write the first five terms of each geometric sequence and find the eighth term. See Example 1. (Objective 1)

17. $a_1 = 3, r = 2$

18. $a_1 = -2, r = 2$

19. $a_1 = -5, r = \frac{1}{5}$

20. $a_1 = 8, r = \frac{1}{2}$

Write the first five terms of each geometric sequence. See Example 2. (Objective 2)

21. $a_1 = 2, r > 0$, third term is 32

22. $a_1 = 3$, fourth term is 24

23. $a_1 = -3$, fourth term is -192

24. $a_1 = 2, r < 0$, third term is 50

25. $a_1 = -64, r < 0$, fifth term is -4

26. $a_1 = -64, r > 0$, fifth term is -4

27. $a_1 = -64$, sixth term is -2

28. $a_1 = -81$, sixth term is $\frac{1}{3}$

Insert the specified number of geometric means. See Examples 3–4. (Objective 3)

29. Insert three positive geometric means between 2 and 162.

30. Insert four geometric means between 3 and 96.

31. Insert four geometric means between -4 and $-12{,}500$.

32. Insert three geometric means (two positive and one negative) between -64 and $-1{,}024$.

33. Find the negative geometric mean between 2 and 128.

34. Find the positive geometric mean between 3 and 243.

35. Find the positive geometric mean between 10 and 20.

36. Find the negative geometric mean between 5 and 15.

Find the sum of the first n terms of each geometric sequence. See Example 5. (Objective 4)

37. $2, 6, 18, \ldots; n = 6$

38. $2, -6, 18, \ldots; n = 6$

39. $2, -6, 18, \ldots; n = 5$

40. $3, -6, 12, \ldots; n = 5$

41. The second term is 1, and the third term is $\frac{1}{5}$; $n = 4$.

42. The second term is 1, and the third term is 4; $n = 5$.

43. The third term is -2, and the fourth term is 1; $n = 6$.

44. The third term is -3, and the fourth term is 1; $n = 5$.

ADDITIONAL PRACTICE

Find the first five terms of the geometric sequence.

45. The second term is 10, and the third term is 50.

46. The third term is -27, and the fourth term is 81.

Find the indicated quantity.

47. Find the tenth term of the geometric sequence with $a_1 = 7$ and $r = 2$.

48. Find the 12th term of the geometric sequence with $a_1 = 64$ and $r = \frac{1}{2}$.

49. Find the first term of the geometric sequence with a common ratio of -3 and an eighth term of -81.

50. Find the first term of the geometric sequence with a common ratio of 2 and a tenth term of 384.

51. Find the common ratio of the geometric sequence with a first term of -8 and a sixth term of $-1{,}944$.

52. Find the common ratio of the geometric sequence with a first term of 12 and a sixth term of $\frac{3}{8}$.

53. Find a geometric mean, if possible, between -50 and 10.

54. Find a negative geometric mean, if possible, between -25 and -5.

Find the sum of the first n terms of each geometric sequence.

55. $3, -6, 12, \ldots; n = 8$

56. $3, 6, 12, \ldots; n = 8$

57. $3, 6, 12, \ldots; n = 7$

58. $3, -6, 12, \ldots; n = 7$

APPLICATIONS *Use a calculator to help solve each problem.* See Examples 6–7. (Objective 5)

59. Population growth The population of Union is predicted to increase by 6% each year. What will be the population of Union 5 years from now if its current population is 500?

60. Population decline The population of Bensonville is decreasing by 10% each year. If its current population is 98, what will be the population 8 years from now?

61. Declining savings John has \$10,000 in a safety deposit box. Each year he spends 12% of what is left in the box. How much will be in the box after 15 years?

62. Savings growth Lu Ling has \$5,000 in a savings account earning 12% annual interest. How much will be in her account 10 years from now? (Assume that Lu Ling makes no deposits or withdrawals.)

63. House appreciation A house appreciates by 6% each year. If the house is worth \$70,000 today, how much will it be worth 12 years from now?

64. Motorboat depreciation A motorboat that cost \$5,000 when new depreciates at a rate of 9% per year. How much will the boat be worth in 5 years?

65. Inscribed squares Each inscribed square in the illustration joins the midpoints of the next larger square. The area of the first square, the largest, is 1. Find the area of the 12th square.

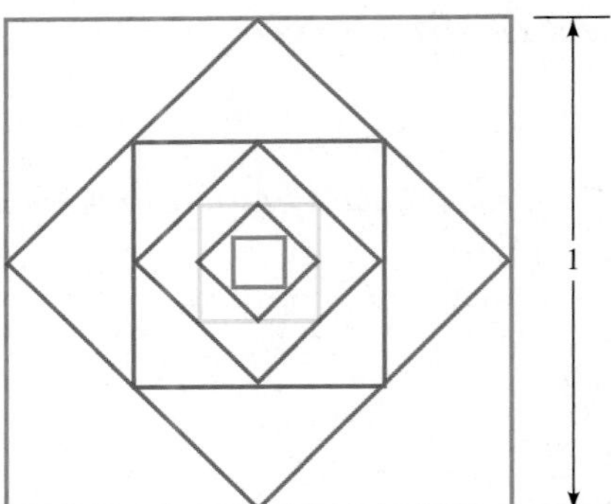

66. Genealogy The following family tree spans 3 generations and lists 7 people. How many names would be listed in a family tree that spans 10 generations?

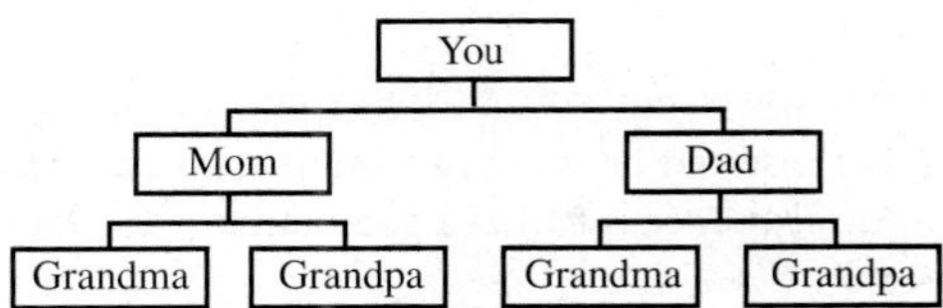

67. Annuities Find the amount of an annuity if \$1,000 is paid semiannually for two years at 6% annual interest. Assume that the first of the four payments is made immediately.

68. **Annuities** Note that the amounts shown in Table 11-1 form a geometric sequence. Verify the answer for Example 7 by using the formula for the sum of a geometric sequence.

WRITING ABOUT MATH

69. Define a geometric sequence.

70. Develop the formula for finding the sum of the first n terms of a geometric sequence.

SOMETHING TO THINK ABOUT

71. Show that the formula for the sum of the first n terms of a geometric sequence can be found by using the formula $S = \frac{a_1 - a_n r}{1 - r}$.

72. Show that the geometric mean between a and b is $\sqrt{ab}$.

73. If $a > b > 0$, which is larger: the arithmetic mean between a and b or the geometric mean between a and b?

74. Is there a geometric mean between -5 and 5?

75. Show that the formula for the sum of the first n terms of a geometric sequence can be written in the form

$$S_n = \frac{a_1 - a_n r}{1 - r} \quad \text{where } a_n = a_1 r^{n-1}$$

76. Show that the formula for the sum of the first n terms of a geometric sequence can be written in the form

$$S_n = \frac{a_1(1 - r^n)}{1 - r}$$

SECTION 11.5 Infinite Geometric Sequences

Objectives

1. Find the sum of an infinite geometric series, if possible.
2. Convert a repeating decimal to a fraction.

Vocabulary

infinite geometric series   partial sum

Getting Ready

Evaluate each expression.

1. $\frac{2}{1 - \frac{1}{2}}$ **2.** $\frac{3}{1 - \frac{1}{3}}$ **3.** $\frac{\frac{3}{2}}{1 - \frac{1}{2}}$ **4.** $\frac{\frac{5}{3}}{1 - \frac{1}{3}}$

In this section, we will consider geometric sequences with infinitely many terms.

1 Find the sum of an infinite geometric series, if possible.

If we form the sum of the terms of an infinite geometric sequence, we get a series called an **infinite geometric series.** For example, if the common ratio r is 3, we have

Infinite geometric sequence	*Infinite geometric series*
2, 6, 18, 54, 162, . . .	$2 + 6 + 18 + 54 + 162 + \cdots$

Under certain conditions, we can find the sum of an infinite geometric series. To define this sum, we consider the geometric series

$$a_1 + a_1r + a_1r^2 + a_1r^3 + \cdots + a_1r^{n-1}$$

- The first **partial sum,** S_1, of the sequence is $S_1 = a_1$.
- The second partial sum, S_2, of the sequence is $S_2 = a_1 + a_1r$.
- The third partial sum, S_3, of the sequence is $S_3 = a_1 + a_1r + a_1r^2$.
- The nth partial sum, S_n, of the sequence is $S_n = a_1 + a_1r + a_1r^2 + \cdots + a_1r^{n-1}$.

If the nth *partial sum,* S_n, approaches some number S_∞ as n approaches infinity, then S_∞ is called the *sum of the infinite geometric series.*

To develop a formula for finding the sum (if it exists) of an infinite geometric series, we consider the formula that gives the sum of the first n terms of a geometric sequence.

$$S_n = \frac{a_1 - a_1r^n}{1 - r} \quad (r \neq 1)$$

If $|r| < 1$ and a_1 is constant, then the term a_1r^n in the above formula approaches 0 as n becomes very large. For example,

$$a_1\left(\frac{1}{4}\right)^1 = \frac{1}{4}a_1, \quad a_1\left(\frac{1}{4}\right)^2 = \frac{1}{16}a_1, \quad a_1\left(\frac{1}{4}\right)^3 = \frac{1}{64}a_1$$

and so on. When n is very large, the value of a_1r^n is negligible, and the term a_1r^n in the above formula can be ignored. This reasoning justifies the following theorem.

Sum of an Infinite Geometric Series

If a_1 is the first term and r is the common ratio of an infinite geometric sequence, and if $|r| < 1$, the sum of the related geometric series is given by the formula

$$S_\infty = \frac{a_1}{1 - r}$$

Comment Recall that $|r| < 1$ is equivalent to the inequality $-1 < r < 1$. This implies that an infinite series will have a sum if r is between -1 and 1.

EXAMPLE 1 Find the sum of the infinite geometric series $125 + 25 + 5 + \ldots$ . .

Solution Here $a = 125$ and $r = \frac{1}{5}$. Since $|r| = \left|\frac{1}{5}\right| = \frac{1}{5} < 1$, we can find the sum of the series. We do this by substituting 125 for a_1 and $\frac{1}{5}$ for r in the formula $S_\infty = \frac{a_1}{1 - r}$ and simplifying:

$$S_\infty = \frac{a_1}{1 - r} = \frac{125}{1 - \frac{1}{5}} = \frac{125}{\frac{4}{5}} = \frac{5}{4}(125) = \frac{625}{4}$$

The sum of the series is $\frac{625}{4}$ or 156.25.

SELF CHECK 1 Find the sum of the infinite geometric series $100 + 20 + 4 + \cdots$.

EXAMPLE 2 Find the sum of the infinite geometric series $64 + (-4) + \frac{1}{4} + \cdots$.

Solution Here $a_1 = 64$ and $r = -\frac{1}{16}$. Since $|r| = \left|-\frac{1}{16}\right| = \frac{1}{16} < 1$, we can find the sum of the series. We substitute 64 for a_1 and $-\frac{1}{16}$ for r in the formula $S_\infty = \frac{a_1}{1-r}$ and simplify:

$$S_\infty = \frac{a_1}{1-r} = \frac{64}{1-\left(-\frac{1}{16}\right)} = \frac{64}{\frac{17}{16}} = \frac{16}{17}(64) = \frac{1{,}024}{17}$$

The sum of the terms of the infinite geometric sequence $64, -4, \frac{1}{4}, \ldots$ is $\frac{1{,}024}{17}$.

SELF CHECK 2 Find the sum of the infinite geometric series $81 + 27 + 9 + \cdots$.

2 Convert a repeating decimal to a fraction.

We can use the sum of an infinite geometric series to convert a repeating decimal to a fraction.

EXAMPLE 3 Change $0.\overline{8}$ to a common fraction.

Solution The decimal $0.\overline{8}$ can be written as an infinite geometric series:

$$0.\overline{8} = 0.888\cdots = \frac{8}{10} + \frac{8}{100} + \frac{8}{1{,}000} + \cdots$$

where $a_1 = \frac{8}{10}$ and $r = \frac{1}{10}$. Because $|r| = \left|\frac{1}{10}\right| = \frac{1}{10} < 1$, we can find the sum as follows:

$$S_\infty = \frac{a_1}{1-r} = \frac{\frac{8}{10}}{1-\frac{1}{10}} = \frac{\frac{8}{10}}{\frac{9}{10}} = \frac{8}{9}$$

Thus, $0.\overline{8} = \frac{8}{9}$. Long division will verify that $\frac{8}{9} = 0.888. . . .$

SELF CHECK 3 Change $0.\overline{6}$ to a common fraction.

EXAMPLE 4 Change $0.\overline{25}$ to a common fraction.

Solution The decimal $0.\overline{25}$ can be written as an infinite geometric series:

$$0.\overline{25} = 0.252525\cdots = \frac{25}{100} + \frac{25}{10{,}000} + \frac{25}{1{,}000{,}000} + \cdots$$

where $a_1 = \frac{25}{100}$ and $r = \frac{1}{100}$. Since $|r| = \left|\frac{1}{100}\right| = \frac{1}{100} < 1$, we can find the sum as follows:

$$S_\infty = \frac{a_1}{1 - r} = \frac{\frac{25}{100}}{1 - \frac{1}{100}} = \frac{\frac{25}{100}}{\frac{99}{100}} = \frac{25}{99}$$

Thus, $0.\overline{25} = \frac{25}{99}$. Long division will verify that this is true.

SELF CHECK 4 Change $0.\overline{15}$ to a common fraction.

SELF CHECK ANSWERS **1.** 125 **2.** $\frac{243}{2}$ **3.** $\frac{2}{3}$ **4.** $\frac{5}{33}$

NOW TRY THIS

Find each sum, if possible.

1. $\sum_{k=1}^{\infty} 3\left(\frac{2}{3}\right)^k$

2. $\sum_{k=1}^{\infty} \frac{1}{4}\left(\frac{3}{2}\right)^k$

3. $\sum_{k=1}^{10} 5\left(\frac{1}{2}k\right)$

11.5 EXERCISES

WARM-UPS

Find the common ratio in each infinite geometric sequence.

1. $\frac{1}{64}, \frac{1}{8}, 1, \ldots$

2. $1, \frac{1}{8}, \frac{1}{64}, \ldots$

3. $\frac{2}{3}, \frac{1}{3}, \frac{1}{6}, \ldots$

4. $64, 8, 1, \ldots$

Find the sum of each infinite geometric series.

5. $18 + 6 + 2 + \cdots$

6. $12 + 3 + \frac{3}{4} + \cdots$

REVIEW *Determine whether each equation determines y to be a function of x.*

7. $y = 3x^3 - 4$

8. $xy = 12$

9. $3x = y^2 + 4$

10. $x = |y|$

VOCABULARY AND CONCEPTS *Fill in the blanks.*

11. If a geometric sequence has infinitely many terms, it is called an _______ geometric sequence.

12. The third partial sum of the series $2 + 6 + 18 + 54 + \cdots$ is __________ = 26.

13. The formula for the sum of an infinite geometric series with $|r| < 1$ is ________.

14. Write $0.\overline{75}$ as an infinite geometric series.

____

GUIDED PRACTICE

Find the sum of each infinite geometric series, if possible.
See Examples 1–2. (Objective 1)

15. $8 + 4 + 2 + \cdots$

16. $12 + 6 + 3 + \cdots$

17. $54 + 18 + 6 + \cdots$

18. $45 + 15 + 5 + \cdots$

19. $12 + (-6) + 3 + \cdots$

20. $8 + (-4) + 2 + \cdots$

21. $\frac{9}{2} + 6 + 8 + \cdots$

22. $-45 + 15 + (-5) + \cdots$

Change each decimal to a common fraction. Then check the answer by using long division. See Examples 3–4. (Objective 2)

23. $0.\overline{1}$

24. $0.\overline{2}$

25. $-0.\overline{3}$

26. $-0.\overline{4}$

27. $0.\overline{12}$

28. $0.\overline{21}$

29. $0.\overline{75}$

30. $0.\overline{57}$

ADDITIONAL PRACTICE *Find the sum of each infinite geometric series, if possible.*

31. $-54 + 18 + (-6) + \cdots$

32. $-112 + (-28) + (-7) + \cdots$

33. $-\frac{27}{2} + (-9) + (-6) + \cdots$

34. $\frac{18}{25} + \frac{6}{5} + 2 + \cdots$

APPLICATIONS

35. Bouncing balls On each bounce, the rubber ball in the illustration rebounds to a height one-half of that from which it fell. Find the total distance the ball travels.

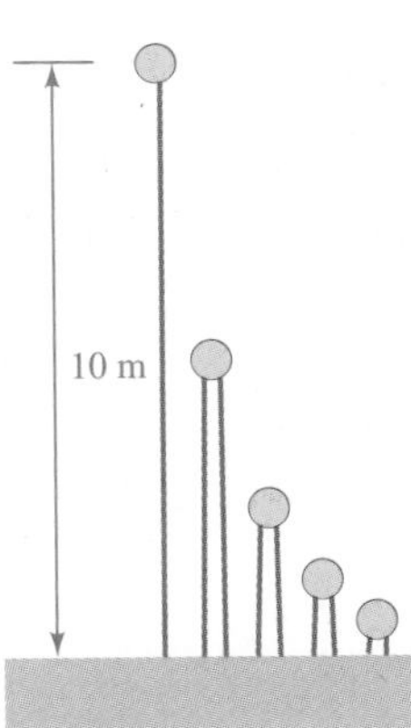

36. Bouncing balls A golf ball is dropped from a height of 12 feet. On each bounce, it returns to a height two-thirds of that from which it fell. Find the total distance the ball travels.

37. Controlling moths To reduce the population of a destructive moth, biologists release 1,000 sterilized male moths each day into the environment. If 80% of these moths alive one day survive until the next, then after a long time, the population of sterile males is the sum of the infinite geometric sequence

$$1{,}000 + 1{,}000(0.8) + 1{,}000(0.8)^2 + 1{,}000(0.8)^3 + \cdots$$

Find the long-term population.

38. Controlling moths If mild weather increases the day-to-day survival rate of the sterile male moths in Exercise 37 to 90%, find the long-term population.

WRITING ABOUT MATH

39. Why must the absolute value of the common ratio be less than 1 before an infinite geometric series can have a sum?

40. Can an infinite arithmetic series have a sum?

SOMETHING TO THINK ABOUT

41. An infinite geometric series has a sum of 5 and a first term of 1. Find the common ratio.

42. An infinite geometric series has a common ratio of $-\frac{2}{3}$ and a sum of 9. Find the first term.

43. Show that $0.\overline{9} = 1$.

44. Show that $1.\overline{9} = 2$.

45. Does $0.999999 = 1$? Explain.

46. If $f(x) = 1 + x + x^2 + x^3 + x^4 + \cdots$, find $f\left(\frac{1}{2}\right)$ and $f\left(-\frac{1}{2}\right)$.

SECTION 11.6 Permutations and Combinations

Objectives

1. Use the multiplication principle to determine the number of ways one event can be followed by another.
2. Use permutations to find the number of n things taken r at a time.
3. Use combinations to find the number of n things taken r at a time.
4. Use combinations to find the coefficients of the terms of a binomial expansion.

Vocabulary

tree diagram
event
multiplication principle for events
permutation
combination

Getting Ready

Evaluate each expression:

1. $4 \cdot 3 \cdot 2 \cdot 1$

2. $5 \cdot 4 \cdot 3 \cdot 2 \cdot 1$

3. $\dfrac{6 \cdot 5 \cdot 4 \cdot 3 \cdot 2 \cdot 1}{4 \cdot 3 \cdot 2 \cdot 1}$

4. $\dfrac{8 \cdot 7 \cdot 6 \cdot 5 \cdot 4 \cdot 3 \cdot 2 \cdot 1}{2(5 \cdot 4 \cdot 3 \cdot 2 \cdot 1)}$

In this section, we will discuss methods of counting the different ways we can do something like arranging books on a shelf or selecting a committee. These kinds of problems are important in statistics, insurance, telecommunications, and other fields.

1 Use the multiplication principle to determine the number of ways one event can be followed by another.

Steven goes to the cafeteria for lunch. He has a choice of three different sandwiches (hamburger, hot dog, or ham and cheese) and four different beverages (cola, root beer, orange, or milk). How many different lunches can he choose?

He has three choices of sandwich, and for any one of these choices, he has four choices of drink. The different options are shown in the **tree diagram** in Figure 11-1.

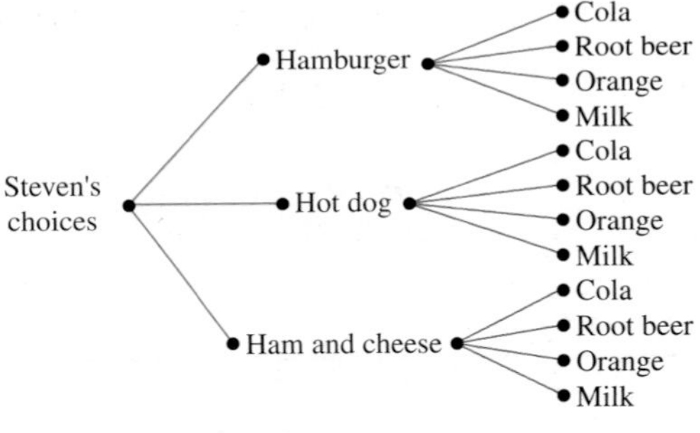

Figure 11-1

The tree diagram shows that he has a total of 12 different lunches to choose from. One of the possibilities is a hamburger with a cola, and another is a hot dog with milk.

A situation that can have several different outcomes—such as choosing a sandwich—is called an **event.** Choosing a sandwich and choosing a beverage can be thought of as two events. The preceding example illustrates the **multiplication principle for events.**

Multiplication Principle for Events

Let E_1 and E_2 be two events. If E_1 can be done in a_1 ways, and if—after E_1 has occurred—E_2 can be done in a_2 ways, the event "E_1 followed by E_2" can be done in $a_1 \cdot a_2$ ways.

EXAMPLE 1 After dinner, Heidi plans to watch the evening news and then a situation comedy on television. If there are choices of four news broadcasts and two comedies, in how many ways can she choose to watch television?

Solution Let E_1 be the event "watching the news" and E_2 be the event "watching a comedy." Because there are four ways to accomplish E_1 and two ways to accomplish E_2, the number of choices that she has is $4 \cdot 2 = 8$.

SELF CHECK 1 If Jose has 7 shirts and 5 pairs of pants, how many outfits could be created?

The multiplication principle can be extended to any number of events. In Example 2, we use it to complete the number of ways that we can arrange objects in a row.

EXAMPLE 2 In how many ways can we arrange five books on a shelf?

Solution We can fill the first space with any of the 5 books, the second space with any of the remaining 4 books, the third space with any of the remaining 3 books, the fourth space with any of the remaining 2 books, and the fifth space with the remaining 1 (or last) book. By the multiplication principle for events, the number of ways in which the books can be arranged is

$$5 \cdot 4 \cdot 3 \cdot 2 \cdot 1 = 120$$

SELF CHECK 2 In how many ways can 4 men line up in a row?

EXAMPLE 3 **SENDING SIGNALS** If a sailor has six flags, each of a different color, to hang on a flagpole, how many different signals can the sailor send by using four flags?

Solution The sailor must find the number of arrangements of 4 flags when there are 6 flags to choose from. The sailor can hang any one of the 6 flags in the top position, any one of the remaining 5 flags in the second position, any one of the remaining 4 flags in the third position, and any one of the remaining 3 flags in the lowest position. By the multiplication principle for events, the total number of signals that can be sent is

$$6 \cdot 5 \cdot 4 \cdot 3 = 360$$

SELF CHECK 3 How many different signals can the sailor send if each signal uses three flags?

2 Use permutations to find the number of *n* things taken *r* at a time.

When computing the number of possible arrangements of objects such as books on a shelf or flags on a pole, we are finding the number of **permutations** of those objects. In these cases, order is important. A blue flag followed by a yellow flag has a different meaning than a yellow flag followed by a blue flag.

In Example 2, we found that the number of permutations of five books, using all five of them, is 120. In Example 3, we found that the number of permutations of six flags, using four of them, is 360.

The symbol $P(n, r)$, read as "the number of permutations of n things r at a time," is often used to express permutation problems. In Example 2, we found that $P(5, 5) = 120$. In Example 3, we found that $P(6, 4) = 360$.

EXAMPLE 4 **SENDING SIGNALS** If Sarah has seven flags, each of a different color, to hang on a flagpole, how many different signals can she send by using three flags?

Solution She must find $P(7, 3)$ (the number of permutations of 7 things 3 at a time). In the top position Sarah can hang any of the 7 flags, in the middle position any one of the remaining 6 flags, and in the bottom position any one of the remaining 5 flags. According to the multiplication principle for events,

$$P(7, 3) = 7 \cdot 6 \cdot 5 = 210$$

She can send 210 signals using only three of the seven flags.

SELF CHECK 4 How many different signals can Sarah send using four flags?

Although it is correct to write $P(7, 3) = 7 \cdot 6 \cdot 5$, there is an advantage in changing the form of this answer to obtain a formula for computing $P(7, 3)$:

$$\begin{aligned} P(7, 3) &= 7 \cdot 6 \cdot 5 \\ &= \frac{7 \cdot 6 \cdot 5 \cdot \mathbf{4 \cdot 3 \cdot 2 \cdot 1}}{\mathbf{4 \cdot 3 \cdot 2 \cdot 1}} && \text{Multiply both the numerator and denominator by } 4 \cdot 3 \cdot 2 \cdot 1. \\ &= \frac{7!}{4!} \\ &= \frac{7!}{(7-3)!} \end{aligned}$$

The generalization of this idea gives the following formula.

Finding $P(n, r)$

The number of permutations of n things r at a time is given by the formula

$$P(n, r) = \frac{n!}{(n-r)!}$$

EXAMPLE 5 Compute: **a.** $P(8, 2)$ **b.** $P(7, 5)$ **c.** $P(n, n)$ **d.** $P(n, 0)$.

Solution We substitute into the permutation formula $P(n, r) = \frac{n!}{(n-r)!}$.

a. $$\begin{aligned} P(8, 2) &= \frac{8!}{(8-2)!} \\ &= \frac{8 \cdot 7 \cdot 6!}{6!} \\ &= 8 \cdot 7 \\ &= 56 \end{aligned}$$

b. $$\begin{aligned} P(7, 5) &= \frac{7!}{(7-5)!} \\ &= \frac{7 \cdot 6 \cdot 5 \cdot 4 \cdot 3 \cdot 2!}{2!} \\ &= 7 \cdot 6 \cdot 5 \cdot 4 \cdot 3 \\ &= 2{,}520 \end{aligned}$$

c. $P(n, n) = \frac{n!}{(n - n)!}$
$= \frac{n!}{0!}$
$= \frac{n!}{1}$
$= n!$

d. $P(n, 0) = \frac{n!}{(n - 0)!}$
$= \frac{n!}{n!}$
$= 1$

SELF CHECK 5 Compute: **a.** $P(10, 6)$ **b.** $P(10, 0)$.

Parts c and d of Example 5 establish the following formulas.

Finding $P(n, n)$ and $P(n, 0)$

The number of permutations of n things n at a time and n things 0 at a time are given by the formulas

$$P(n, n) = n! \quad \text{and} \quad P(n,0) = 1$$

EXAMPLE 6 **TV PROGRAMMING** **a.** In how many ways can a television executive arrange the Saturday night lineup of 6 programs if there are 15 programs to choose from?
b. If there are only 6 programs to choose from?

Solution **a.** To find the number of permutations of 15 programs 6 at a time, we will use the formula $P(n, r) = \frac{n!}{(n - r)!}$ with $n = 15$ and $r = 6$.

$$\begin{aligned} P(15, 6) &= \frac{15!}{(15 - 6)!} \\ &= \frac{15 \cdot 14 \cdot 13 \cdot 12 \cdot 11 \cdot 10 \cdot 9!}{9!} \\ &= 15 \cdot 14 \cdot 13 \cdot 12 \cdot 11 \cdot 10 \\ &= 3{,}603{,}600 \end{aligned}$$

b. To find the number of permutations of 6 programs 6 at a time, we use the formula $P(n, n) = n!$ with $n = 6$.

$$P(6, 6) = 6! = 720$$

SELF CHECK 6 How many ways are there if the executive has 20 programs to choose from?

3 Use combinations to find the number of n things taken r at a time.

Suppose that a student must read 4 books from a reading list of 10 books. The order in which he reads them is not important. For the moment, however, let's assume that order is important and find the number of permutations of 10 things 4 at a time:

$$
\begin{aligned}
P(10, 4) &= \frac{10!}{(10 - 4)!} \\
&= \frac{10 \cdot 9 \cdot 8 \cdot 7 \cdot 6!}{6!} \\
&= 10 \cdot 9 \cdot 8 \cdot 7 \\
&= 5{,}040
\end{aligned}
$$

If order is important, there are 5,040 ways of choosing 4 books when there are 10 books to choose from. However, because the order in which the student reads the books does not matter, the previous result of 5,040 is too big. Since there are 24 (or 4!) ways of ordering the 4 books that are chosen, the result of 5,040 is exactly 24 (or 4!) times too big. Therefore, the number of choices that the student has is the number of permutations of 10 things 4 at a time, divided by 24:

$$\frac{P(10, 4)}{24} = \frac{5{,}040}{24} = 210$$

The student has 210 ways of choosing 4 books to read from the list of 10 books.

In situations where order is not important, we are interested in **combinations,** not permutations. The symbols $C(n, r)$ and $\binom{n}{r}$ both mean the number of combinations of n things r at a time.

If a selection of r books is chosen from a total of n books, the number of possible selections is $C(n, r)$ and there are $r!$ arrangements of the r books in each selection. If we consider the selected books as an ordered grouping, the number of orderings is $P(n, r)$. Therefore, we have

(1) $\quad r! \cdot C(n, r) = P(n, r)$

We can divide both sides of Equation 1 by $r!$ to get the formula for finding $C(n, r)$:

$$C(n, r) = \binom{n}{r} = \frac{P(n, r)}{r!} = \frac{n!}{r!(n - r)!}$$

Finding $C(n, r)$

The number of combinations of n things r at a time is given by

$$C(n, r) = \frac{n!}{r!(n - r)!}$$

EXAMPLE 7 Compute: **a.** $C(8, 5)$ **b.** $\binom{7}{2}$ **c.** $C(n, n)$ **d.** $C(n, 0)$.

Solution We will substitute into the combination formula $C(n, r) = \frac{n!}{r!(n - r)!}$.

a.
$$
\begin{aligned}
C(8, 5) &= \frac{8!}{5!(8 - 5)!} \\
&= \frac{8 \cdot 7 \cdot 6 \cdot 5!}{5! \cdot 3!} \\
&= 8 \cdot 7 \\
&= 56
\end{aligned}
$$

b.
$$
\begin{aligned}
\binom{7}{2} &= \frac{7!}{2!(7 - 2)!} \\
&= \frac{7 \cdot 6 \cdot 5!}{2 \cdot 1 \cdot 5!} \\
&= 21
\end{aligned}
$$

c. $C(n, n) = \dfrac{n!}{n!(n - n)!}$

$= \dfrac{n!}{n!(0!)}$

$= \dfrac{n!}{n!(1)}$

$= 1$

d. $C(n, 0) = \dfrac{n!}{0!(n - 0)!}$

$= \dfrac{n!}{0! \cdot n!}$

$= \dfrac{1}{0!}$

$= \dfrac{1}{1}$

$= 1$

The symbol $C(n, 0)$ indicates that we choose 0 things from the available n things.

SELF CHECK 7 Compute: **a.** $C(9, 6)$ **b.** $C(10, 10)$.

Parts c and d of Example 7 establish the following formulas.

Finding $C(n, n)$ and $C(n, 0)$

The number of combinations of n things n at a time is 1. The number of combinations of n things 0 at a time is 1.

$$C(n, n) = 1 \quad \text{and} \quad C(n, 0) = 1$$

EXAMPLE 8 **PICKING COMMITTEES** If 15 students want to pick a committee of 4 students to plan a party, how many different committees are possible?

Solution Since the ordering of people on each possible committee is not important, we find the number of combinations of 15 people 4 at a time:

$$C(15, 4) = \frac{15!}{4!(15 - 4)!}$$

$$= \frac{15 \cdot 14 \cdot 13 \cdot 12 \cdot 11!}{4 \cdot 3 \cdot 2 \cdot 1 \cdot 11!}$$

$$= \frac{15 \cdot 14 \cdot 13 \cdot 12}{4 \cdot 3 \cdot 2 \cdot 1}$$

$$= 1{,}365$$

There are 1,365 possible committees.

SELF CHECK 8 In how many ways can 20 students pick a committee of 5 students to plan a party?

EXAMPLE 9 **CONGRESS** A committee in Congress consists of ten Democrats and eight Republicans. In how many ways can a subcommittee be chosen if it is to contain five Democrats and four Republicans?

Solution There are $C(10, 5)$ ways of choosing the 5 Democrats and $C(8, 4)$ ways of choosing the 4 Republicans. By the multiplication principle for events, there are $C(10, 5) \cdot C(8, 4)$ ways of choosing the subcommittee:

$$\begin{aligned} C(10, 5) \cdot C(8, 4) &= \frac{10!}{5!(10 - 5)!} \cdot \frac{8!}{4!(8 - 4)!} \\ &= \frac{10 \cdot 9 \cdot 8 \cdot 7 \cdot 6 \cdot 5!}{120 \cdot 5!} \cdot \frac{8 \cdot 7 \cdot 6 \cdot 5 \cdot 4!}{24 \cdot 4!} \\ &= \frac{10 \cdot 9 \cdot 8 \cdot 7 \cdot 6}{120} \cdot \frac{8 \cdot 7 \cdot 6 \cdot 5}{24} \\ &= 17{,}640 \end{aligned}$$

There are 17,640 possible subcommittees.

SELF CHECK 9 In how many ways can a subcommittee be chosen if it is to contain four members from each party?

EVERYDAY CONNECTIONS State Lotteries

Gambling is a diversion for some and an obsession for others. Whether it is horse racing, slot machines, or state lotteries, the lure of instant riches is hard to resist.

Many states conduct lotteries. One scheme is typical: For \$1, you have two chances to match 6 numbers chosen from 55 numbers and win a grand prize of about \$5 million. How likely are you to win? Is it worth \$1 to play the game?

To match 6 numbers chosen from 55, you must choose the one winning combination out of $C(55, 6)$ possibilities:

$$C(n, r) = \frac{n!}{r!(n - r)!}$$

$$\begin{aligned} C(55, 6) &= \frac{55!}{6!(55 - 6)!} \\ &= \frac{55 \cdot 54 \cdot 53 \cdot 52 \cdot 51 \cdot 50}{6 \cdot 5 \cdot 4 \cdot 3 \cdot 2} \\ &= 28{,}989{,}675 \end{aligned}$$

In this game, you have two chances in about 29 million of winning \$5 million. Over the long haul, you will win $\frac{2}{29{,}000{,}000}$ of the time, so each ticket is worth $\frac{2}{29{,}000{,}000}$ of \$5,000,000, or about 35¢, if you don't have to share the prize with another winner. For every dollar spent to play the game, you can expect to throw away 65¢. This state lottery is a poor bet. Casinos pay better than 50¢ on the dollar, with some slot machines returning 90¢. "You can't win if you're not in!" is the claim of the lottery promoters. A better claim would be "You won't regret if you don't bet!"

4 Use combinations to find the coefficients of the terms of a binomial expansion.

We have seen that the expansion of $(x + y)^3$ is

$$(x + y)^3 = 1x^3 + 3x^2y + 3xy^2 + 1y^3$$

and that

$$\binom{3}{0} = 1, \quad \binom{3}{1} = 3, \quad \binom{3}{2} = 3, \quad \text{and} \quad \binom{3}{3} = 1$$

Putting these facts together gives the following way of writing the expansion of $(x + y)^3$:

$$(x + y)^3 = \binom{3}{0}x^3 + \binom{3}{1}x^2y + \binom{3}{2}xy^2 + \binom{3}{3}y^3$$

Likewise, we have

$$(x + y)^4 = \binom{4}{0}x^4 + \binom{4}{1}x^3y + \binom{4}{2}x^2y^2 + \binom{4}{3}xy^3 + \binom{4}{4}y^4$$

The generalization of this idea allows us to state the binomial theorem using combinatorial notation.

The Binomial Theorem

If n is any positive integer, then

$$(a + b)^n = \binom{n}{0}a^n + \binom{n}{1}a^{n-1}b + \binom{n}{2}a^{n-2}b^2 + \cdots + \binom{n}{r}a^{n-r}b^r + \cdots + \binom{n}{n}b^n$$

EXAMPLE 10 Use the combinatorial form of the binomial theorem to expand $(x + y)^6$.

Solution

$$\begin{aligned}(x + y)^6 &= \binom{6}{0}x^6 + \binom{6}{1}x^5y + \binom{6}{2}x^4y^2 + \binom{6}{3}x^3y^3 + \binom{6}{4}x^2y^4 + \binom{6}{5}xy^5 + \binom{6}{6}y^6 \\ &= x^6 + 6x^5y + 15x^4y^2 + 20x^3y^3 + 15x^2y^4 + 6xy^5 + y^6\end{aligned}$$

SELF CHECK 10 Use the combinatorial form of the binomial theorem to expand $(a + b)^2$.

EXAMPLE 11 Use the combinatorial form of the binomial theorem to expand $(2x - y)^3$.

Solution

$$\begin{aligned}(2x - y)^3 &= [2x + (-y)]^3 \\ &= \binom{3}{0}(2x)^3 + \binom{3}{1}(2x)^2(-y) + \binom{3}{2}(2x)(-y)^2 + \binom{3}{3}(-y)^3 \\ &= 1(2x)^3 + 3(4x^2)(-y) + 3(2x)(y^2) + (-y)^3 \\ &= 8x^3 - 12x^2y + 6xy^2 - y^3\end{aligned}$$

SELF CHECK 11 Use the combinatorial form of the binomial theorem to expand $(3a + b)^3$.

SELF CHECK ANSWERS

1. 35 **2.** 24 **3.** 120 **4.** 840 **5. a.** 151,200 **b.** 1 **6.** 27,907,200 **7. a.** 84 **b.** 1 **8.** 15,504 **9.** 14,700 **10.** $a^2 + 2ab + b^2$ **11.** $27a^3 + 27a^2b + 9ab^2 + b^3$

NOW TRY THIS

1. In Texas, license plate numbering has often been changed to accommodate a growing population. Find the greatest number of license plates that could be issued during each of these periods. A or B represents a letter and 0 or 1 represents a number.

1975–1982	AAA-000	(I or O is not used)
1982–1990	000-AAA	(I or O is not used)
1990–1998	BBB-00B	(no vowels or Q used)
1998–2004	B00-BBB	(no vowels or Q used)
2004–2007	000-BBB	(no vowels or Q used)
2007–mid-2009	BBB-000	(no vowels or Q used)

2. Find the greatest number of license plates that could be distributed from 1975–mid-2009.

3. Beginning in mid-2009, Texas began issuing 7-digit license numbers of the form BB1-B001 (no vowels or Q is used and no 0 is used in the third or last place). How many license plates can be issued using this form?

Source: http://www.licenseplateinfo.com/txchart/pass-tables.html

11.6 EXERCISES

WARM-UPS

1. If there are 3 books and 5 magazines, in how many ways can you pick 1 book and 1 magazine?
2. In how many ways can 5 soldiers stand in line?
3. Find $P(3, 1)$.
4. Find $P(3, 3)$.
5. Find $C(3, 0)$.
6. Find $C(3, 3)$.

REVIEW *Find each value of x. Assume no division by 0.*

7. $|2x - 3| = 9$
8. $2x^2 - x = 15$
9. $\frac{3}{x - 5} = \frac{8}{x}$
10. $\frac{3}{x} = \frac{x - 2}{8}$

VOCABULARY AND CONCEPTS *Fill in the blanks.*

11. If an event E_1 can be done in p ways and, after it occurs, a second event E_2 can be done in q ways, the event E_1 followed by E_2 can be done in _____ ways.
12. We can use a _____ diagram to illustrate the multiplication principle.
13. A ___________ is an arrangement of objects in which order matters.
14. The symbol _______ means the number of permutations of n things taken r at a time.
15. The formula for the number of permutations of n things taken r at a time is ______________.
16. $P(n, n) =$ __ and $P(n, 0) =$ __.
17. A ___________ is an arrangement of objects in which order does not matter.
18. The symbol $C(n, r)$ or ___ means the number of ____________ of n things taken r at a time.
19. The formula for the number of combinations of n things taken r at a time is ______________.
20. $C(n, n) =$ __ and $C(n, 0) =$ __.

GUIDED PRACTICE

Evaluate using permutations. See Example 5. (Objective 2)

21. $P(3, 3)$
22. $P(4, 4)$
23. $P(5, 3)$
24. $P(3, 2)$
25. $P(2, 2) \cdot P(3, 3)$
26. $P(3, 2) \cdot P(3, 3)$
27. $\frac{P(5, 3)}{P(4, 2)}$
28. $\frac{P(6, 2)}{P(5, 4)}$

Evaluate using combinations. See Example 7. (Objective 3)

29. $C(5, 3)$
30. $C(5, 4)$

31. $\binom{6}{3}$

32. $\binom{6}{4}$

33. $\binom{5}{4}\binom{5}{3}$

34. $\binom{6}{5}\binom{6}{4}$

35. $\dfrac{C(38, 37)}{C(19, 18)}$

36. $\dfrac{C(25, 23)}{C(40, 39)}$

Use the combinatorial form of the binomial theorem to expand each binomial. See Examples 10–11. (Objective 4)

37. $(x + y)^4$

38. $(x - y)^2$

39. $(2x + y)^3$

40. $(2x + 1)^4$

ADDITIONAL PRACTICE

Evaluate.

41. $C(12, 0) \cdot C(12, 12)$

42. $\dfrac{C(8, 0)}{C(8, 1)}$

43. $\dfrac{P(6, 2) \cdot P(7, 3)}{P(5, 1)}$

44. $\dfrac{P(8, 3)}{P(5, 3) \cdot P(4, 3)}$

45. $C(n, 2)$

46. $C(n, 3)$

Expand the binomial.

47. $(3x - 2)^4$

48. $(3 - x^2)^3$

Find the indicated term of the binomial expansion.

49. $(x - 5y)^5$; fourth term

50. $(2x - y)^5$; third term

51. $(x^2 - y^3)^4$; second term

52. $(x^3 - y^2)^4$; fourth term

APPLICATIONS

See Examples 1–3. (Objective 1)

53. Arranging an evening Kyoro plans to go to dinner and see a movie. In how many ways can she arrange her evening if she has a choice of five movies and seven restaurants?

54. Travel choices Paula has five ways to travel from New York to Chicago, three ways to travel from Chicago to Denver, and four ways to travel from Denver to Los Angeles. How many choices are available to Paula if she travels from New York to Los Angeles?

55. Arranging books In how many ways can seven books be placed on a shelf?

56. Lining up In how many ways can the people shown be placed in a line?

See Examples 4 and 6. (Objective 2)

57. Making license plates How many six-digit license plates can be manufactured? Note that there are ten choices—0, 1, 2, 3, 4, 5, 6, 7, 8, 9—for each digit.

58. Making license plates How many six-digit license plates can be manufactured if no digit can be repeated?

59. Making license plates How many six-digit license plates can be manufactured if no license can begin with 0 and if no digit can be repeated?

60. Making license plates How many license plates can be manufactured with two letters followed by four digits?

61. Phone numbers How many seven-digit phone numbers are available in area code 815 if no phone number can begin with 0 or 1?

62. Phone numbers How many ten-digit phone numbers are available if area codes 000 and 911 cannot be used and if no local number can begin with 0 or 1?

63. Arranging books In how many ways can four novels and five biographies be arranged on a shelf if the novels are placed on the left?

64. Making a ballot In how many ways can six candidates for mayor and four candidates for the county board be arranged on a ballot if all of the candidates for mayor must be placed on top?

65. Combination locks How many permutations does a combination lock have if each combination has three numbers, no two numbers of any combination are equal, and the lock has 25 numbers?

66. Combination locks How many permutations does a combination lock have if each combination has three numbers, no two numbers of any combination are equal, and the lock has 50 numbers?

67. Arranging appointments The receptionist at a dental office has only three appointment times available before Tuesday, and ten patients have toothaches. In how many ways can the receptionist fill those appointments?

68. Computers In many computers, a word consists of 32 *bits*—a string of thirty-two 1's and 0's. How many different words are possible?

69. Palindromes A palindrome is any word, such as *madam* or *radar,* that reads the same backward and forward. How many five-digit numerical palindromes (such as 13531) are there? (*Hint:* A leading 0 would be dropped.)

70. Call letters The call letters of U.S. commercial radio stations have 3 or 4 letters, and the first is always a W or a K. How many radio stations could this system support?

See Examples 8–9. (Objective 3)

71. Planning a picnic A class of 14 students wants to pick a committee of 3 students to plan a picnic. How many committees are possible?

72. Choosing books Jeff must read 3 books from a reading list of 15 books. How many choices does he have?

73. **Forming committees** The number of three-person committees that can be formed from a group of people is ten. How many people are in the group?

74. **Forming committees** The number of three-person committees that can be formed from a group of people is 20. How many people are in the group?

75. **Winning a lottery** In one state lottery, anyone who picks the correct 6 numbers (in any order) wins. With the numbers 0 through 99 available, how many choices are possible?

76. **Taking a test** The instructions on a test read: "Answer any ten of the following fifteen questions. Then choose one of the remaining questions for homework and turn in its solution tomorrow." In how many ways can the questions be chosen?

77. **Forming a committee** In how many ways can we select a committee of two men and two women from a group containing three men and four women?

78. **Forming a committee** In how many ways can we select a committee of three men and two women from a group containing five men and three women?

79. **Choosing clothes** In how many ways can we select 2 shirts and 3 neckties from a group of 12 shirts and 10 neckties?

80. **Choosing clothes** In how many ways can we select five dresses and two coats from a wardrobe containing nine dresses and three coats?

WRITING ABOUT MATH

81. State the multiplication principle for events.

82. Explain why *permutation lock* would be a better name for a combination lock.

SOMETHING TO THINK ABOUT

83. How many ways could five people stand in line if two people insist on standing together?

84. How many ways could five people stand in line if two people refuse to stand next to each other?

SECTION 11.7 Probability

Objectives

1. Find a sample space for an experiment.
2. Find the probability of an event.

Vocabulary

probability sample space event
experiment

Getting Ready

Answer each question.

1. What are the possible outcomes on one roll of one die?
2. What are the possible outcomes after flipping a single coin two times?

The **probability** that an event will occur is a measure of the likelihood of that event. A tossed coin, for example, can land in two ways, either heads or tails. Because one of these two equally likely outcomes is heads, we expect that out of several tosses, about half will be heads. We say that the probability of obtaining heads in a single toss of the coin is $\frac{1}{2}$.

If records show that out of 100 days with weather conditions like today's, 30 have received rain, the weather service will report, "There is a $\frac{30}{100}$ or 30% probability of rain today."

1 Find a sample space for an experiment.

Activities such as tossing a coin, rolling a die, drawing a card, and predicting rain are called **experiments.** For any experiment, a list of all possible outcomes is called a **sample space.** For example, the sample space S for the experiment of tossing two coins is the set

$S = \{(H, H), (H, T), (T, H), (T, T)\}$ There are four possible outcomes.

where the ordered pair (H, T) represents the outcome "heads on the first coin and tails on the second coin."

EXAMPLE 1 List the sample space of the experiment "rolling two dice a single time."

Solution We can list ordered pairs and let the first number be the result on the first die and the second number the result on the second die. The sample space S is the following set of ordered pairs:

(1, 1) (1, 2) (1, 3) (1, 4) (1, 5) (1, 6)
(2, 1) (2, 2) (2, 3) (2, 4) (2, 5) (2, 6)
(3, 1) (3, 2) (3, 3) (3, 4) (3, 5) (3, 6)
(4, 1) (4, 2) (4, 3) (4, 4) (4, 5) (4, 6)
(5, 1) (5, 2) (5, 3) (5, 4) (5, 5) (5, 6)
(6, 1) (6, 2) (6, 3) (6, 4) (6, 5) (6, 6)

By counting, we see that the experiment has 36 equally likely possible outcomes.

SELF CHECK 1 How many pairs in the sample space have a sum of 4?

An **event** is a subset of the sample space of an experiment. For example, if E is the event "getting at least one heads" in the experiment of tossing two coins, then

$E = \{(H, H), (H, T), (T, H)\}$ There are 3 ways of getting at least one heads.

Because the outcome of getting at least one heads can occur in 3 out of 4 possible ways, we say that the *probability* of E is $\frac{3}{4}$, and we write

$$P(E) = P(\text{at least one heads}) = \frac{3}{4}$$

Probability of an Event If a sample space of an experiment has n distinct and equally likely outcomes and E is an event that occurs in s of those ways, the *probability of E* is

$$P(E) = \frac{s}{n}$$

Since $0 \le s \le n$, it follows that $0 \le \frac{s}{n} \le 1$. This implies that all probabilities have a value from 0 to 1. If an event cannot happen, its probability is 0. If an event is certain to happen, its probability is 1.

2 Find the probability of an event.

EXAMPLE 2 Find the probability of the event "rolling a sum of 7 on one roll of two dice."

Solution In the sample space listed in Example 1, the following ordered pairs give a sum of 7:

$$\{(1, 6), (2, 5), (3, 4), (4, 3), (5, 2), (6, 1)\}$$

Since there are 6 ordered pairs whose numbers give a sum of 7 out of a total of 36 equally likely outcomes, we have

$$P(E) = P(\text{rolling a sum of 7}) = \frac{s}{n} = \frac{6}{36} = \frac{1}{6}$$

SELF CHECK 2 Find the probability of rolling a sum of 4.

A standard playing deck of 52 cards has two red suits, hearts and diamonds, and two black suits, clubs and spades. Each suit has 13 cards, including the ace, king, queen, jack, and cards numbered from 2 to 10. We will refer to a standard deck of cards in many examples and exercises.

EXAMPLE 3 Find the probability of drawing an ace on one draw from a standard card deck.

Solution Since there are 4 aces in the deck, the number of favorable outcomes is $s = 4$. Since there are 52 cards in the deck, the total number of possible outcomes is $n = 52$. The probability of drawing an ace is the ratio of the number of favorable outcomes to the number of possible outcomes.

$$P(\text{an ace}) = \frac{s}{n} = \frac{4}{52} = \frac{1}{13}$$

The probability of drawing an ace is $\frac{1}{13}$.

SELF CHECK 3 Find the probability of drawing a red ace on one draw from a standard card deck.

EVERYDAY CONNECTIONS Winning the Lottery

©Shutterstock.com/rj lerich

Suppose a certain state lottery has the following Mega Millions design.

- A spinning machine contains 49 different balls. Each ball is labeled with exactly one number from the list $\{1, 2, 3, \ldots, 49\}$. The spinning machine ensures that each ball has an equal chance of being selected.
- Six different balls are drawn, without replacement, from the spinning machine.
- Each lottery ticket has six different numbers from the list $\{1, 2, 3, \ldots, 49\}$.
- The last number must match the sixth drawn number.

Suppose that a grand-prize winning ticket consists of all six numbers drawn from the machine. Find the probability associated with the winning ticket.

EXAMPLE 4 Find the probability of drawing 5 cards, all hearts, from a standard card deck.

Solution The number of ways we can draw 5 hearts from the 13 hearts is $C(13, 5)$, the number of combinations of 13 things taken 5 at a time. The number of ways to draw 5 cards from the deck is $C(52, 5)$, the number of combinations of 52 things taken 5 at a time. The probability of drawing 5 hearts is the ratio of the number of favorable outcomes to the number of possible outcomes.

$$P(5 \text{ hearts}) = \frac{s}{n} = \frac{C(13,5)}{C(52,5)}$$

$$P(5 \text{ hearts}) = \frac{\dfrac{13!}{5!8!}}{\dfrac{52!}{5!47!}}$$

$$= \frac{13!}{\cancel{5!}8!} \cdot \frac{\cancel{5!}47!}{52!} \qquad \frac{5!}{5!} = 1$$

$$= \frac{13 \cdot 12 \cdot 11 \cdot 10 \cdot 9 \cdot 8!}{8!} \cdot \frac{47!}{52 \cdot 51 \cdot 50 \cdot 49 \cdot 48 \cdot 47!}$$

$$= \frac{13 \cdot 12 \cdot 11 \cdot 10 \cdot 9}{52 \cdot 51 \cdot 50 \cdot 49 \cdot 48}$$

$$= \frac{33}{66{,}640}$$

The probability of drawing 5 hearts is $\frac{33}{66{,}640}$ or $4.951980792 \times 10^{-4}$.

SELF CHECK 4 Find the probability of drawing 6 cards, all spades, from a standard card deck.

SELF CHECK ANSWERS **1.** 3 **2.** $\frac{1}{12}$ **3.** $\frac{1}{26}$ **4.** $\frac{33}{391{,}510}$ or $8.428903476 \times 10^{-5}$

NOW TRY THIS

The notation $P(E)$ represents the probability that event E will occur. The notation $P(\overline{E})$ represents the probability that event E will not occur. Because either event E or event $\overline{E}$ must occur, $P(E) + P(\overline{E}) = 1$. Thus, $P(E) = 1 - P(\overline{E})$.

Sometimes it is easier to determine $P(E)$ by finding $P(\overline{E})$ and subtracting the result from 1.

1. In a family of 5 children, find the probability that at least one child is a girl.

2. In a clinical trial, the probability of experiencing a serious side effect is 0.15 while the probability of a minor side effect is 0.45, with no overlapping side effects. What is the probability that a person in the trial picked at random experiences no side effects?

11.7 EXERCISES

WARM-UPS

1. Find the probability of rolling a 3 on one roll of an ordinary die.

2. Find the probability of drawing the ace of spades with one draw from an ordinary card deck.

REVIEW *Solve each equation.*

3. $5^{4x} = \frac{1}{125}$

4. $8^{-x+1} = \frac{1}{64}$

5. $2^{x^2-2x} = 8$

6. $3^{x^2-3x} = 81$

7. $3^{x^2+4x} = \frac{1}{81}$

8. $7^{x^2+3x} = \frac{1}{49}$

VOCABULARY AND CONCEPTS *Fill in the blanks.*

9. An __________ is any activity for which the outcome is uncertain.

10. A list of all possible outcomes for an experiment is called a ________ ______.

11. The probability of an event E is defined as $P(E) = \frac{\_}{\_}$.

12. If an event is certain to happen, its probability is __.

13. If an event cannot happen, its probability is __.

14. All probability values are between __ and __, inclusive.

15. Fill in the blanks to find the probability of drawing a black face card from a standard deck.

a. The number of black face cards is __.

b. The number of cards in the deck is ___.

c. The probability is $\frac{\_}{\_}$ or $\frac{\_}{\_}$.

16. Fill in the blanks to find the probability of drawing 4 aces from a standard card deck.

a. The number of ways to draw 4 aces from 4 aces is $C(4, 4) =$ __.

b. The number of ways to draw 4 cards from 52 cards is $C(52, 4) =$ ________.

c. The probability is $\frac{\_}{\_\_\_\_}$.

GUIDED PRACTICE

List the sample space of each experiment. See Example 1. (Objective 1)

17. Rolling one die and tossing one coin

18. Tossing three coins

19. Selecting a letter of the alphabet

20. Picking a one-digit number

An ordinary die is rolled once. Find the probability of each event. **See Example 2. (Objective 2)**

21. Rolling a 2

22. Rolling a number greater than 4

23. Rolling a number larger than 1 but less than 6

24. Rolling an odd number

Balls numbered from 1 to 42 are placed in a container and stirred. If one is drawn at random, find the probability of each result. **See Example 3. (Objective 2)**

25. The number is less than 20.

26. The number is less than 50.

27. The number is a prime number.

28. The number is less than 10 or greater than 40.

Refer to the following spinner. If the spinner is spun, find the probability of each event. Assume that the spinner never stops on a line. **See Example 3. (Objective 2)**

29. The spinner stops on red.

30. The spinner stops on green.

31. The spinner stops on brown.

32. The spinner stops on yellow.

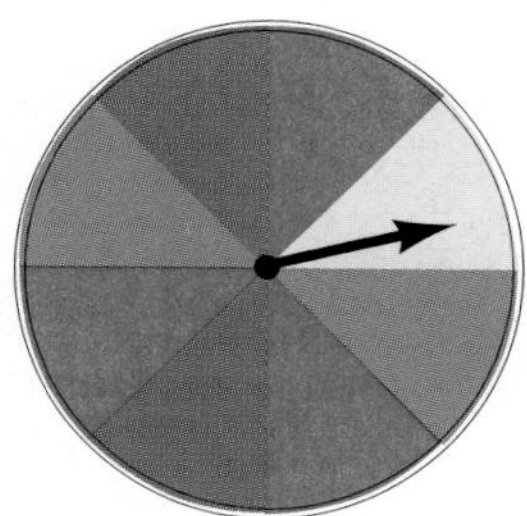

Find the probability of each event. **See Example 4. (Objective 2)**

33. Drawing a diamond on one draw from a standard card deck

34. Drawing a face card from a standard card deck

35. Drawing a red face card from a standard card deck

36. Drawing a face card from a standard deck followed by a 10 after replacing the first card

37. Drawing an ace from a standard deck followed by a 10 after replacing the first card

38. Drawing 6 diamonds from a standard card deck without replacing the cards after each draw

39. Drawing 5 aces from a standard card deck without replacing the cards after each draw

40. Drawing 5 clubs from the black cards in a standard card deck

ADDITIONAL PRACTICE

Find the probability of each event.

41. Rolling a sum of 4 on one roll of two dice

42. Drawing a red egg from a basket containing 5 red eggs and 7 blue eggs

43. Drawing a yellow egg from a basket containing 5 red eggs and 7 yellow eggs

Assume that the probability that a backup generator will fail a test is $\frac{1}{2}$ and that the college in question has 4 backup generators. In Exercises 45–49, find each probability.

44. Construct a sample space for the test.

45. All generators will fail the test.

46. Exactly 1 generator will fail.

47. Exactly 2 generators will fail.

48. Exactly 3 generators will fail.

49. No generator will fail.

50. Find the sum of the probabilities in Exercises 45–49.

APPLICATIONS

51. Quality control In a batch of 10 tires, 2 are known to be defective. If 4 tires are chosen at random, find the probability that all 4 tires are good.

52. Medicine Out of a group of 9 patients treated with a new drug, 4 suffered a relapse. If 3 patients are selected at random from the group of 9, find the probability that none of the 3 patients suffered a relapse.

A survey of 282 people is taken to determine the opinions of doctors, teachers, and lawyers on a proposed piece of legislation, with the results shown in the table. A person is chosen at random from those surveyed. Refer to the table to find each probability.

53. The person favors the legislation.

54. A doctor opposes the legislation.

55. A person who opposes the legislation is a lawyer.

	Number that favor	Number that oppose	Number with no opinion	Total
Doctors	70	32	17	119
Teachers	83	24	10	117
Lawyers	23	15	8	46
Total	176	71	35	282

WRITING ABOUT MATH

56. Explain why all probability values range from 0 to 1.

57. Explain the concept of probability.

SOMETHING TO THINK ABOUT *If P(A) represents the probability of event A, and P(B | A) represents the probability that event B will occur after event A, then*

$$P(A \text{ and } B) = P(A) \cdot P(B \mid A)$$

58. In a school, 30% of the students are gifted in math and 10% are gifted in art and math. If a student is gifted in math, find the probability that the student is also gifted in art.

59. The probability that a person owns a luxury car is 0.2, and the probability that the owner of such a car also owns a second car is 0.7. Find the probability that a person chosen at random owns both a luxury car and a second car.

PROJECTS

Project 1

Baytown is building an auditorium. The city council has already decided on the layout shown in the illustration. Each of the sections A, B, C, D, E is to be 60 feet in length from front to back. The aisle widths cannot be changed due to fire regulations. The one thing left to decide is how many rows of seats to put in each section. Based on the following information regarding each section of the auditorium, help the council decide on a final plan.

Sections A and C each have four seats in the front row, five seats in the second row, six seats in the third row and so on, adding one seat per row as we count from front to back.

Section B has eight seats in the front row and adds one seat per row as we count from front to back.

Sections D and E each have 28 seats in the front row and add two seats per row as we count from front to back.

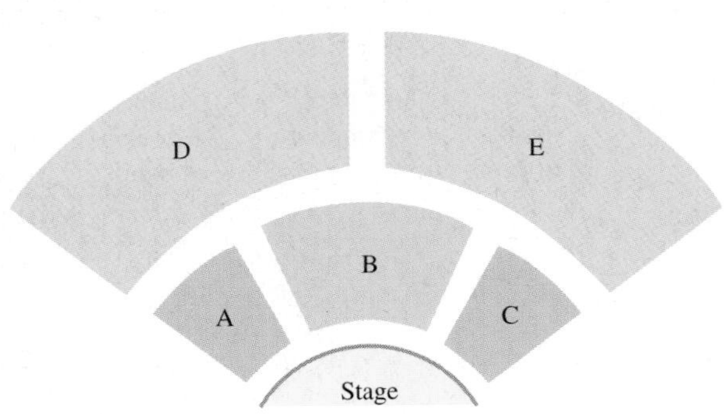

a. One plan calls for a distance of 36 inches, front to back, for each row of seats. Another plan allows for 40 inches (an extra four inches of legroom) for each row. How many seats will the auditorium have under each of these plans?

b. Another plan calls for the higher-priced seats (Sections A, B, and C) to have the extra room afforded by 40-inch rows, but for Sections D and E to have enough rows to make sure that the auditorium holds at least 2,700 seats. Determine how many rows Section D and E would have to contain for this to work. (This answer should be an integer.) How much space (to the nearest tenth of an inch) would be allotted for each row in Sections D and E?

Project 2

Pascal's triangle contains a wealth of interesting patterns. You have seen two in Exercises 68 and 73 of Section 11.1. Here are a few more.

a. Find the hockey-stick pattern in the numbers in the illustration. What would be the missing number in the rightmost hockey stick? Does this pattern work for larger hockey sticks? Experiment.

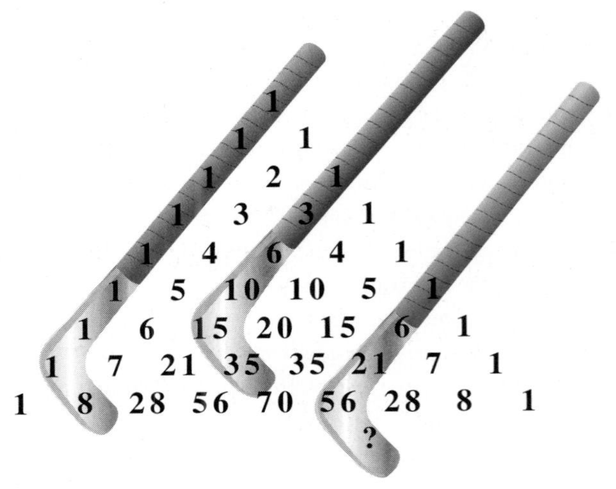

b. In Illustration 1, find the pattern in the sums of increasingly larger portions of Pascal's triangle. Find the sum of all of the numbers up to and including the row that begins 1 10 45 . . . .

c. In Illustration 2, find the pattern in the sums of the squares of the numbers in each row of the triangle. What is the sum of the squares of the numbers in the row that begins 1 10 45 . . . ? (*Hint:* Calculate $P(2, 1)$, $P(4, 2)$, $P(6, 3)$, ... Do these numbers appear elsewhere in the triangle?)

d. In 1653, Pascal described the triangle in *Treatise on the Arithmetic Triangle,* writing, "I have left out many more properties than I have included. It is remarkable how fertile in properties this triangle is. *Everyone can try his hand.*" Accept Pascal's invitation. Find some of the triangle's patterns for yourself and share your discoveries with your class. Illustration 3 is an idea to get you started.

$1^2 = 1$

$1^2 + 1^2 = 2$

$1^2 + 2^2 + 1^2 = 6$

$1^2 + 3^2 + 3^2 + 1^2 = 20$

$1^2 + 4^2 + 6^2 + 4^2 + 1^2 = 70$

$1^2 + 5^2 + 10^2 + 10^2 + 5^2 + 1^2 = ?$

$1^2 + 6^2 + 15^2 + 20^2 + 15^2 + 6^2 + 1^2 = ?$

ILLUSTRATION 2

1
1 1
1 2 1
1 3 3 1
1 4 6 4 1
1 5 10 10 5 1
1 6 15 20 15 6 1
1 7 21 35 35 21 7 1
1 8 28 56 70 56 28 8 1

ILLUSTRATION 3

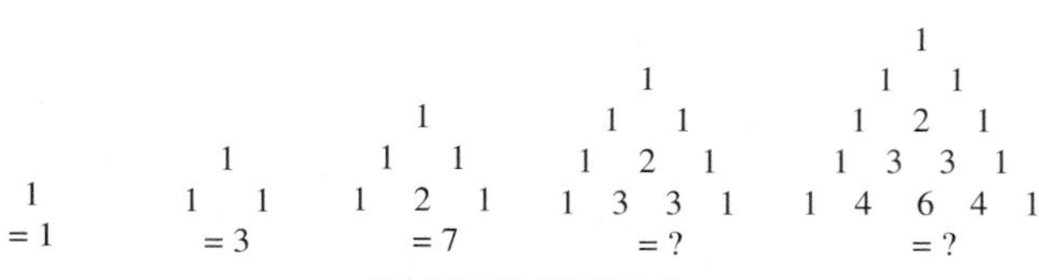

ILLUSTRATION 1

Chapter 11 REVIEW

SECTION 11.1 The Binomial Theorem

DEFINITIONS AND CONCEPTS	EXAMPLES
The symbol $n!$ **(factorial)** is defined as $n! = n(n-1)(n-2)\cdot\cdots\cdot(3)(2)(1)$ where n is a natural number. $0! = 1$ $n(n-1)! = n!$ (n is a natural number)	$7! = 7\cdot6\cdot5\cdot4\cdot3\cdot2\cdot1 = 5{,}040$ $\frac{5!}{2!3!} = \frac{5\cdot4\cdot\cancel{3!}}{2\cdot1\cdot\cancel{3!}} = \frac{20}{2} = 10$ $7\cdot6! = 7\cdot6\cdot5\cdot4\cdot3\cdot2\cdot1 = 7!$
The binomial theorem: $(a+b)^n = a^n + \frac{n!}{1!(n-1)!}a^{n-1}b + \frac{n!}{2!(n-2)!}a^{n-2}b^2 + \cdots + b^n$	$(x-2y)^4$ $= [x + (-2y)]^4$ $= x^4 + \frac{4!}{1!3!}x^3(-2y) + \frac{4!}{2!2!}x^2(-2y)^2 + \frac{4!}{3!1!}x(-2y)^3 + (-2y)^4$ $= x^4 + 4x^3(-2y) + 6x^2(4y)^2 + 4(-8y^3) + 16y^4$ $= x^4 - 8x^3y + 24x^2y^2 - 32xy^3 + 16y^4$

REVIEW EXERCISES

Evaluate each expression.

1. $(4!)(3!)$
2. $\frac{5!}{3!}$
3. $\frac{6!}{2!(6-2)!}$
4. $\frac{12!}{3!(12-3)!}$
5. $(n-n)!$
6. $\frac{8!}{7!}$

Use the binomial theorem to find each expansion.

7. $(x+y)^5$
8. $(x-y)^4$
9. $(4x-y)^3$
10. $(x+4y)^3$

SECTION 11.2 The *n*th Term of a Binomial Expansion

DEFINITIONS AND CONCEPTS	EXAMPLES
Specific term of a binomial expansion: The binomial theorem can be used to find a specific term of a binomial expansion. The coefficient of the variables can be found using the formula $\frac{n!}{r!(n-r)!}$ where n is the power of the expansion and r is the exponent of the second variable.	To find the third term of $(x+y)^5$, note that the exponent on y will be 2, because the exponent on y is 1 less than the number of the term. The power on x will be 3 because the sum of the powers must be 5. Thus, the variables will be x^3y^2. The coefficient of the variables can be found with the formula. $\frac{5!}{2!(5-2)!}x^3y^2$ Substitute the values into the formula. $\frac{5 \cdot 4 \cdot \cancel{3!}}{2 \cdot 1 \cdot \cancel{3!}}x^3y^2$ Expand the factorials. $10x^3y^2$ Simplify.

REVIEW EXERCISES

Find the specified term in each expansion.

11. $(x+y)^4$; third term
12. $(x-y)^5$; fourth term
13. $(3x-4y)^3$; second term
14. $(4x+3y)^4$; third term

SECTION 11.3 Arithmetic Sequences

DEFINITIONS AND CONCEPTS	EXAMPLES
An **arithmetic sequence** is a sequence of the form $a_1, a_1+d, a_1+2d, \ldots, a_1+(n-1)d$ where a_1 is the first term the nth term is $a_n = a_1+(n-1)d$ and d is the common difference.	Find the first 5 terms of the sequence with $a_1 = 15$ and $d = -2$. $a_1 = 15$ $a_2 = 15 + (-2) = 13$ $a_3 = 15 + 2(-2) = 11$ $a_4 = 15 + 3(-2) = 9$ $a_5 = 15 + 4(-2) = 7$
Arithmetic means: If numbers are inserted between two given numbers a and b to form an arithmetic sequence, the inserted numbers are **arithmetic means** between a and b.	To insert two arithmetic means between 7 and 25, note that $a_1 = 7$ and $a_4 = 25$. Then substitute these values into the formula for the nth term and solve for d. $a_n = a_1 + (n-1)d$ $25 = 7 + (4-1)d$ Substitute values. $25 = 7 + 3d$ Simplify. $18 = 3d$ Subtract 7 from both sides. $6 = d$ Divide both sides by 3. The two arithmetic means are $7 + 6 = 13$ and $7 + 2(6) = 19$.

Sum of terms of arithmetic sequences: The sum of the first n terms of an arithmetic sequence is given by $$S_n = \frac{n(a_1 + a_n)}{2} \text{ with } a_n = a_1 + (n-1)d$$ where a_1 is the first term, a_n is the nth term, and n is the number of terms in the sequence.	To find the sum of the first 12 terms of the sequence 4, 12, 20, 28, . . . , note that the common difference is 8 and substitute into the formula for the nth term to find the twelfth term: $$a_{12} = 4 + (12 - 1)8 = 4 + 88 = 92$$ Then substitute into the formula for the sum of the first n terms: $$S_{12} = \frac{12(4 + 92)}{2} = \frac{12(96)}{2} = \frac{1{,}152}{2} = 576$$
Summation notation: $$\sum_{k=1}^{n} f(k) = f(1) + f(2) + \cdots + f(n)$$	$$\begin{aligned}\sum_{k=1}^{3}(3k - 1) &= [3(1) - 1] + [3(2) - 1] + [3(3) - 1] \\ &= (3 - 1) + (6 - 1) + (9 - 1) \\ &= 2 + 5 + 8 \\ &= 15\end{aligned}$$

REVIEW EXERCISES

15. Find the eighth term of an arithmetic sequence whose first term is 7 and whose common difference is 5.

16. Write the first five terms of the arithmetic sequence whose ninth term is 242 and whose seventh term is 212.

17. Find two arithmetic means between 8 and 25.

18. Find the sum of the first 20 terms of the sequence 11, 18, 25, ...

19. Find the sum of the first ten terms of the sequence 9, $6\frac{1}{2}$, 4, ...

Find each sum.

20. $\sum_{k=4}^{6} \frac{1}{2}k$

21. $\sum_{k=2}^{5} 7k^2$

22. $\sum_{k=1}^{4} (3k - 4)$

23. $\sum_{k=10}^{10} 36k$

SECTION 11.4 Geometric Sequences

DEFINITIONS AND CONCEPTS	EXAMPLES
Geometric sequences: A geometric sequence is a sequence of the form $$a_1, a_1r, a_1r^2, a_1r^3, \ldots, a_1r^{n-1}$$ where a_1 is the first term; $a_n = a_1r^{n-1}$ is the nth term and r is the common ratio.	The first four terms of the geometric sequence with $a_1 = 8$ and $r = 2$ are as follows: $$\begin{aligned}a_1 &= 8 \\ a_2 &= 8(2) = 16 \\ a_3 &= 8(2)^2 = 32 \\ a_4 &= 8(2)^3 = 64\end{aligned}$$
Geometric means: If numbers are inserted between a and b to form a geometric sequence, the inserted numbers are geometric means between a and b.	To insert the positive geometric mean between 4 and 1, note that $a_1 = 4$ and $a_3 = 1$. Then substitute these values into the formula for the nth term and solve for d. $a_n = a_1r^{n-1}$ Formula for the nth term; $a_3 = 4r^{3-1}$ Formula for the 3rd term; $1 = 4r^2$ Substitute 1 for a_3.; $\frac{1}{4} = r^2$ Divide both sides by 4.; $\frac{1}{2} = r$ Take the positive square root of both sides. Since $r = \frac{1}{2}$, $a_2 = a_1r = 4\left(\frac{1}{2}\right) = 2$. The geometric mean between 4 and 1 is 2.

Sum of terms of a geometric sequence: The sum of the first n terms of a geometric sequence is given by $$S_n = \frac{a_1 - a_1 r^n}{1 - r} \quad (r \neq 1)$$ where S_n is the sum, a_1 is the first term, r is the common ratio, and n is the number of terms in the sequence.	To find the sum of the first 4 terms of the geometric sequence defined by $a_n = 2(3)^n$, note that the first term and the common ratio are $$a_1 = 2(3)^1 = 2(3) = 6 \quad \text{and} \quad r = 3$$ Then substitute into the formula for the sum of the first n terms and simplify: $$S_n = \frac{a_1 - a_1 r^n}{1 - r}$$ $$S_4 = \frac{6 - 6(3)^4}{1 - 3} \quad \text{Substitute values into the formula.}$$ Thus, $$S_4 = \frac{6 - 6(81)}{-2} = \frac{6 - 486}{-2} = 240$$ The sum of the first 4 terms is 240.

REVIEW EXERCISES

24. Write the first five terms of the geometric sequence whose fourth term is 3 and whose fifth term is $\frac{3}{2}$.

25. Find the sixth term of a geometric sequence with a first term of $\frac{1}{8}$ and a common ratio of 2.

26. Find two geometric means between -6 and 384.

27. Find the sum of the first seven terms of the sequence 162, 54, 18, ...

28. Find the sum of the first eight terms of the sequence $\frac{1}{8}, -\frac{1}{4}, \frac{1}{2}, \ldots$.

29. Car depreciation A $5,000 car depreciates at the rate of 20% of the previous year's value. How much is the car worth after 5 years?

30. Stock appreciation The value of Mia's stock portfolio is expected to appreciate at the rate of 18% per year. How much will the portfolio be worth in 10 years if its current value is $25,700?

31. Planting corn A farmer planted 300 acres in corn this year. He intends to plant an additional 75 acres in corn in each successive year until he has 1,200 acres in corn. In how many years will that be?

32. Falling objects If an object is in free fall, the sequence 16, 48, 80, . . . represents the distance in feet that the object falls during the first second, during the second second, during the third second, and so on. How far will the object fall during the first 10 seconds?

SECTION 11.5 Infinite Geometric Series

DEFINITIONS AND CONCEPTS	EXAMPLES
Sum of an infinite geometric series: If r is the common ratio of an infinite geometric series and if $\|r\| < 1$, the sum of the series is given by $$S_\infty = \frac{a_1}{1 - r}$$ where a_1 is the first term and r is the common ratio.	To find the sum of the geometric series $8 + 4 + 2 + \cdots$, note that the first term is 8 and the common ratio is $r = \frac{1}{2}$. Since $\left\|\frac{1}{2}\right\| < 1$, the sum exists. To find it, proceed as follows: $$S_\infty = \frac{a_1}{1 - r}$$ $$S_\infty = \frac{8}{1 - \frac{1}{2}} \quad \text{Substitute values into the formula.}$$ Thus, $$S_\infty = \frac{8}{\frac{1}{2}} = 8 \cdot 2 = 16$$ The sum is 16.

REVIEW EXERCISES

33. Find the sum of the infinite geometric series $25 + 20 + 16 + \cdots$.

34. Change the decimal $0.\overline{05}$ to a common fraction.

SECTION 11.6 Permutations and Combinations

DEFINITIONS AND CONCEPTS	EXAMPLES
Multiplication principle for events: If E_1 and E_2 are two events and if E_1 can be done in a_1 ways and E_2 can be done in a_2 ways, then the event "E_1 followed by E_2" can be done in $a_1 \cdot a_2$ ways.	If Danielle has 5 shirts, 4 pairs of jeans, and 3 pairs of shoes, she has a choice of $5 \cdot 4 \cdot 3 = 60$ different outfits to wear.
Formulas for permutations: $P(n, r) = \dfrac{n!}{(n - r)!}$ $P(n, n) = n!$ $P(n, 0) = 1$ Order matters in a permutation.	$P(6, 4) = \dfrac{6!}{(6 - 4)!} = \dfrac{6 \cdot 5 \cdot 4 \cdot 3 \cdot 2 \cdot 1}{(2)!} = \dfrac{720}{2 \cdot 1} = 360$ $P(5, 5) = 5! = 5 \cdot 4 \cdot 3 \cdot 2 \cdot 1 = 120$ $P(5, 0) = 1$
Formulas for combinations: $C(n, r) = \dbinom{n}{r} = \dfrac{n!}{r!(n - r)!}$ $C(n, n) = \dbinom{n}{n} = 1 \quad C(n, 0) = \dbinom{n}{0} = 1$ Order does not matter in a combination.	$C(12, 5) = \dfrac{12!}{5!(12 - 5)!} = \dfrac{12!}{5!(7)!}$ $= \dfrac{12 \cdot 11 \cdot 10 \cdot 9 \cdot 8 \cdot \cancel{7!}}{5 \cdot 4 \cdot 3 \cdot 2 \cdot 1 \cdot \cancel{7!}} = \dfrac{12 \cdot 11 \cdot 10 \cdot 9 \cdot 8}{5 \cdot 4 \cdot 3 \cdot 2 \cdot 1} = 792$ $C(5, 5) = 1 \quad C(5, 0) = 1$

REVIEW EXERCISES

35. Planning a trip If there are 17 flights from New York to Chicago, and 8 flights from Chicago to San Francisco, in how many different ways could a passenger plan a trip from New York to San Francisco?

Evaluate each expression.

36. $P(7, 7)$

37. $P(7, 0)$

38. $P(8, 6)$

39. $\dfrac{P(9, 6)}{P(10, 7)}$

Evaluate each expression.

40. $C(7, 7)$

41. $C(7, 0)$

42. $\dbinom{8}{6}$

43. $\dbinom{9}{6}$

44. $C(6, 3) \cdot C(7, 3)$

45. $\dfrac{C(7, 3)}{C(6, 3)}$

46. Lining up In how many ways can five people be arranged in a line?

47. Lining up In how many ways can three men and five women be arranged in a line if the women are placed ahead of the men?

48. Choosing people In how many ways can we pick three people from a group of ten?

49. Forming committees In how many ways can we pick a committee of two Democrats and two Republicans from a group containing five Democrats and six Republicans?

SECTION 11.7 Probability

DEFINITIONS AND CONCEPTS	EXAMPLES
An event that cannot occur has a probability of 0. An event that is certain to occur has a probability of 1. All other events have probabilities between 0 and 1.	The probability of rolling a 7 with a single standard die is 0 because there is no 7 on a standard die. The probability of drawing a red card from a stack of red cards is 1 because all the cards are red.
Sample space: The sample space consists of all possibilities of outcomes for an experiment.	The sample space of rolling one standard die is the set of all possible outcomes: $\{1, 2, 3, 4, 5, 6\}$
Probability of an event: If S is the *sample space* of an experiment with n distinct and equally likely outcomes, and E is an event that occurs in s of those ways, then the probability of E is $P(E) = \frac{s}{n}$	To find the probability of rolling a prime number with one roll of a standard die, first determine that there are 6 possible outcomes in the sample space: The set of possible outcomes is $\{1, 2, 3, 4, 5, 6\}$. Then determine that there are 3 possible favorable outcomes: The prime numbers on a face of a die are 2, 3, and 5. Thus, $P(\text{prime number}) = \frac{3}{6} = \frac{1}{2}$ The probability of rolling a prime number is $\frac{1}{2}$.

REVIEW EXERCISES

In Exercises 50–52, assume that a dart is randomly thrown at the colored chart.

1	2	3	4
5	6	7	8
9	10	11	12
13	14	15	16

50. What is the probability that the dart lands in a blue area?
51. What is the probability that the dart lands in an even-numbered area?
52. What is the probability that the dart lands in an area whose number is greater than 2?
53. Find the probability of rolling an 11 on one roll of two dice.
54. Find the probability of drawing a black card from a stack of red cards.
55. Find the probability of drawing a 10 from a standard deck of cards.
56. Find the probability of drawing a 5-card poker hand that has exactly 3 aces.
57. Find the probability of drawing 5 cards, all spades, from a standard card deck.

Chapter 11 TEST

1. Evaluate: $\frac{7!}{4!}$.
2. Evaluate: $0!$.
3. Find the second term in the expansion of $(x - y)^5$.
4. Find the third term in the expansion of $(x + 2y)^4$.
5. Find the tenth term of an arithmetic sequence with the first three terms of 3, 10, and 17.
6. Find the sum of the first 12 terms of the sequence $-2, 3, 8, \ldots$.

7. Find two arithmetic means between 2 and 98.

8. Evaluate: $\sum_{k=1}^{3}(2k - 3)$.

9. Find the seventh term of the geometric sequence whose first three terms are $-\frac{1}{9}$, $-\frac{1}{3}$, and -1.

10. Find the sum of the first six terms of the sequence $\frac{1}{27}, \frac{1}{9}, \frac{1}{3}, \ldots$.

11. Find two geometric means between 3 and 648.

12. Find the sum of all of the terms of the infinite geometric sequence 9, 3, 1, ...

Find the value of each expression.

13. $P(5, 4)$

14. $P(8, 8)$

15. $C(6, 4)$

16. $C(8, 3)$

17. $C(6, 0) \cdot P(6, 5)$

18. $P(8, 7) \cdot C(8, 7)$

19. $\dfrac{P(6, 4)}{C(6, 4)}$

20. $\dfrac{C(9, 6)}{P(6, 4)}$

21. **Choosing people** In how many ways can we pick three people from a group of seven?

22. **Choosing committees** From a group of five men and four women, how many three-person committees can be chosen that will include two women?

Find each probability.

23. Rolling a 5 on one roll of a die

24. Drawing a jack or a queen from a standard card deck

25. Receiving 5 hearts for a 5-card poker hand

26. Tossing 2 heads in 5 tosses of a fair coin

Cumulative Review Exercises

1. Use graphing to solve $\begin{cases} 2x + y = 5 \\ x - 2y = 0 \end{cases}$.

2. Use substitution to solve $\begin{cases} 3x + y = 4 \\ 2x - 3y = -1 \end{cases}$.

3. Use elimination to solve $\begin{cases} x + 2y = -2 \\ 2x - y = 6 \end{cases}$.

4. Use any method to solve $\begin{cases} \frac{x}{10} + \frac{y}{5} = \frac{1}{2} \\ \frac{x}{2} - \frac{y}{5} = \frac{13}{10} \end{cases}$.

5. Evaluate: $\begin{vmatrix} 3 & -2 \\ 1 & -1 \end{vmatrix}$.

6. Use Cramer's rule and solve for y only: $\begin{cases} 4x - 3y = -1 \\ 3x + 4y = -7 \end{cases}$.

7. Solve: $\begin{cases} x + y + z = 1 \\ 2x - y - z = -4 \\ x - 2y + z = 4 \end{cases}$.

8. Solve for z only: $\begin{cases} x + 2y + 3z = 6 \\ 3x + 2y + z = 6 \\ 2x + 3y + z = 6 \end{cases}$.

9. Solve: $\begin{cases} 3x - 2y < 6 \\ y < -x + 2 \end{cases}$.

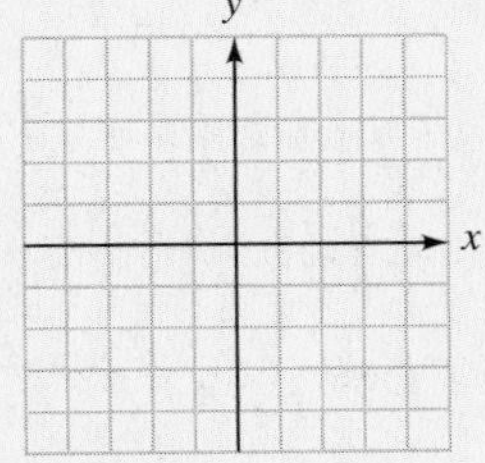

10. Solve: $\begin{cases} y < x + 2 \\ 3x + y \le 6 \end{cases}$.

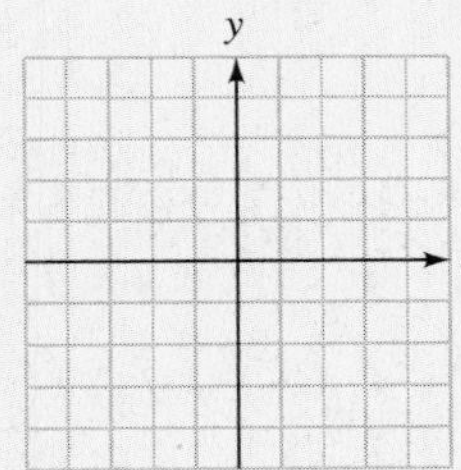

11. Graph: $f(x) = \left(\frac{1}{2}\right)^x$.

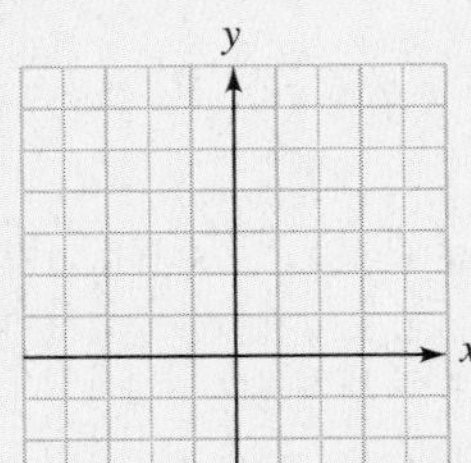

12. Write $y = \log_2 x$ as an exponential equation.

Find the value of x.

13. $\log_x 25 = 2$

14. $\log_5 125 = x$

15. $\log_3 x = -3$

16. $\log_5 x = 0$

17. Find the inverse of $y = \log_2 x$.

18. If $\log_{10} 10^x = y$, then y equals what quantity?

If log 7 = 0.8451 and log 14 = 1.1461, evaluate each expression without using a calculator or tables.

19. $\log 98$

20. $\log 2$

21. $\log 49$

22. $\log \frac{7}{5}$ (*Hint:* $\log 10 = 1$.)

23. Solve: $2^{x+5} = 3^x$.

24. Solve: $\log 5 + \log x - \log 4 = 1$.

Use a calculator.

25. Boat depreciation How much will a \$9,000 boat be worth after 9 years if it depreciates 12% per year?

26. Find $\log_6 8$.

27. Evaluate: $\frac{6!7!}{5!}$.

28. Use the binomial theorem to expand $(3a - b)^4$.

29. Find the seventh term in the expansion of $(2x - y)^8$.

30. Find the 20th term of an arithmetic sequence with a first term of -11 and a common difference of 6.

31. Find the sum of the first 20 terms of an arithmetic sequence with a first term of 6 and a common difference of 3.

32. Insert two arithmetic means between -3 and 30.

33. Evaluate: $\sum_{k=1}^{3} 3k^2$.

34. Evaluate: $\sum_{k=3}^{5} (2k + 1)$.

35. Find the seventh term of a geometric sequence with a first term of $\frac{1}{27}$ and a common ratio of 3.

36. Find the sum of the first ten terms of the sequence $\frac{1}{64}, \frac{1}{32}, \frac{1}{16}, \ldots$.

37. Insert two geometric means between -3 and 192.

38. Find the sum of all the terms of the sequence $9, 3, 1, \ldots$.

39. Evaluate: $P(9, 3)$.

40. Evaluate: $C(7, 4)$.

41. Evaluate: $\frac{C(8, 4) \cdot C(8, 0)}{P(6, 2)}$.

42. If $n > 1$, which is smaller: $P(n, n)$ or $C(n, n)$?

43. Lining up In how many ways can seven people stand in a line?

44. Forming a committee In how many ways can a committee of three people be chosen from a group containing nine people?

45. Cards Find the probability of drawing a black face card from a standard deck of cards.

Glossary

absolute value The distance between a given number and 0 on a number line, usually denoted with | |

absolute value function The function $f(x) = |x|$

acute angle An angle with a measure between 0° and 90°

additive identity The number zero, because for all real numbers a, $a + 0 = 0 + a = a$

additive inverses A pair of numbers, a and b, are additive inverses if $a + b = 0$; also called **negatives** or **opposites**

algebraic term An expression that is a constant, variable, or a product of constants and variables; for example, 37, xyz, and $32t$ are terms.

algorithm A repeating series of steps or processes

altitude The length of the perpendicular line from a vertex of a triangle to its opposite side

annual growth rate The rate r that a quantity P increases or decreases within one year; the variable r in the exponential growth formula $A = Pe^{rt}$

area The amount of surface enclosed by a two-dimensional geometric figure

arithmetic means Numbers that are inserted between two elements in a sequence to form a new arithmetic sequence

arithmetic sequence A sequence in which each successive term is equal to the sum of the preceding term and a constant; written $a, a + d, a + 2d, a + 3d, \ldots$

arithmetic series The indicated sum of the terms of an arithmetic sequence

ascending order An ordering of the terms of a polynomial such that a given variable's exponents occur in increasing order

associative properties The properties that state for real numbers a, b, and c, $(a + b) + c = a + (b + c)$ and $(ab)c = a(bc)$ *Note:* The associative property does not hold for subtraction or division.

asymptote A straight line that is approached by a graph

augmented matrix A matrix representing a system of linear equations written in standard form, with columns consisting of the coefficients and a column of the constant terms of the equations

axis of symmetry The line that passes through the vertex of a parabola and divides it into two congruent halves

back substitution The process of finding the value of a second (or third, etc.) variable after finding the first by substituting the first into an equation of two variables and, if necessary, substituting both values into an equation of three values to find the third, etc.

base In the expression x^y, x is the base and y is the exponent. The base, x, will be used as a factor y times.

base angles of an isosceles triangle The two angles of an isosceles triangle opposite the two sides that are of the same length

binomial A polynomial with exactly two terms

binomial theorem The theorem used to expand a binomial

boundary line A line that divides the coordinate plane into two half-planes

center of a circle The point that is equidistant from all points on a circle

center of a hyperbola The midpoint of the segment joining the foci of a hyperbola

change-of-base formula A formula used to convert a logarithm from one base a to some other base b

circle The set of all points in a plane that are the same distance from a fixed point (center)

coefficient matrix A matrix consisting of the coefficients of the variables in a system of linear equations

combinations The number of ways to choose r things from a set of n things if order is not important

common difference The constant difference between successive terms in an arithmetic sequence, that is, the number d in the formula $a_n = a_1 + (n - 1)d$

common logarithm A logarithm with a base of 10

common ratio The constant quotient of two successive terms in a geometric sequence

commutative properties The properties that state for real numbers a and b, $a + b = b + a$ and $ab = ba$. *Note:* The commutative property does not hold for subtraction and division.

complementary angles Two angles whose measures sum to 90°

completing the square The process of forming a perfect-square trinomial from a binomial of the form $x^2 + bx$

complex fraction A fraction that contains a fraction in its numerator and/or denominator

complex number The sum of a real number and an imaginary number

composite function A new function formed when one function is evaluated in terms of another, denoted by $(f \circ g)(x)$

composite number A natural number greater than 1 with factors other than 1 and itself

composition The process of using the output of one function as the input of another function

compound inequality A single statement representing the intersection or union of two inequalities

compounding periods The number of times per year interest is compounded

compound interest The total interest earned on money deposited in an account where the interest for each compounding period is deposited back into the account, which increases the principle for the next compounding period

conditional equation A linear equation in one variable that has exactly one solution

cone A three-dimensional surface with a circular base whose cross sections (parallel to the base) are circles that decrease in diameter until it comes to a point

conic section A graph that can be formed by the intersection of a plane and a right-circular cone

conjugate An expression that contains the same two terms as another but with opposite signs between them; for example, $a + \sqrt{b}$ and $a - \sqrt{b}$.

consistent system A system of linear equations with at least one solution

constant A term whose variable factor(s) have an exponent of 0

constant of proportionality If y varies directly with x, that is, $y = kx$, we say k is the constant of proportionality.

constraints Inequalities that limit the possible values of the variables of the objective function

contradiction An equation that is false for all values of its variables; its solution set is $\varnothing$.

coordinate The number that corresponds to a given point on a number line

coordinate plane The plane that contains the x-axis and y-axis

Cramer's rule A method using determinants to solve systems of linear equations

critical points Given a quadratic or rational inequality, the points on the number line corresponding to its critical values

critical values Given a quadratic inequality in standard form, the solutions to the quadratic equation $ax^2 + bx + c = 0$; given a rational inequality in standard form, the values where the denominator is equal to zero and the numerator is equal to zero

cube A rectangular solid whose faces are all congruent squares

cube-root function The function $f(x) = \sqrt[3]{x}$

cubing function The function $f(x) = x^3$

cylinder A three-dimensional surface whose cross sections (parallel to the base) are the same size and shape as the base of the cylinder

decibel A unit of measure that expresses the voltage gain (or loss) of an electronic device such as an amplifier

decreasing function A function whose value decreases as the independent value increases (graph drops as we move to the right)

degree of a polynomial The largest degree of a polynomial's terms

dependent equations A system of equations that has infinitely many solutions

dependent variable The variable in an equation of two variables whose value is determined by the independent variable (usually y in an equation involving x and y)

descending order An ordering of the terms of a polynomial, such that a given variable's exponents occur in decreasing order

determinant A calculated value from the elements in a square matrix. For a two-by-two matrix $\begin{bmatrix} a & b \\ c & d \end{bmatrix}$, the determinant is $ad - bc$. For a three-by-three matrix or larger, we use the method of expanding by minors.

difference The result of subtracting two expressions

difference of two cubes An expression of the form $a^3 - b^3$

difference of two squares An expression of the form $a^2 - b^2$

difference quotient The function defined by $\dfrac{f(x + h) - f(x)}{h}$

direct variation A relationship between two variables x and y of the form $y = kx$, where k is the constant of proportionality

discriminant The part of the quadratic formula, $b^2 - 4ac$, used to determine the number and nature of solutions to the quadratic equation

distributive property The property that states for real numbers a, b, and c, $c(a + b) = ca + cb$ and $(a + b)c = ac + bc$

domain of a function The set of all permissible input values of a function

domain of a relation The set of all first elements (components) of a relation

eccentricity A measure of the flatness of an ellipse

element of a matrix One of the entries in a matrix

elements of a set The objects in a given set

ellipse The set of all points in a plane the sum of whose distances from two fixed points (foci) is a constant

equation A statement indicating that two quantities are equal

equilateral triangle A triangle with all sides of equal length and all angles of equal measure

equivalent equations Two equations that have the same solution set

equivalent systems Two systems of equations that have the same solution set

even integers The set of integers that are exactly divisible by 2 (*Note:* 0 is an even integer.)

even root If $b = \sqrt[n]{a}$, b is an even root if n is even.

event A situation that can have several different outcomes or a subset of the sample space of an experiment

exponent In the expression x^y, x is the base and y is the exponent. The exponent y states the number of times that the base x will be used as a factor.

exponential equation An equation that contains a variable in one of its exponents

exponential function A function of the form $f(x) = (a \cdot b)^x$

exponential function (natural) A function of the form $f(t) = Pe^{kt}$

extraneous solution A solution to an equation that does not result in a true statement when substituted for the variable in the original equation

extremes In the proportion $\frac{a}{b} = \frac{c}{d}$, the numbers a and d are called the extremes.

factorial For a natural number n, the product of all the natural numbers less than or equal to n, denoted as $n!$ (*Exception:* 0! is defined as 1.)

factoring The process of finding the individual factors of a product

factors of a number The natural numbers that divide a given number; for example, the factors of 12 are 1, 2, 3, 4, 6, and 12.

factor theorem The theorem that states if $P(x)$ is a polynomial, then $P(r) = 0$ if and only if $(x - r)$ is a factor of $P(r)$

feasibility region The set of points that satisfy all of the constraints of a system of inequalities

Fibonacci sequence The sequence 1, 1, 2, 3, 5, 8, ... with every successive term the sum of the two previous terms

finite sequence A sequence with a finite number of terms

foci (of an ellipse) The two fixed points used in the definition of the ellipse

focus (foci) of a hyperbola The two fixed points used in the definition of the hyperbola

fulcrum The point on which a lever pivots

function A relation in which to each first element there corresponds exactly one second element, usually denoted by $f(x)$

future value The amount to which an invested amount of money will grow

Gaussian elimination A method of solving a system of linear equations by working with its associated augmented matrix

general form of the equation of a circle The equation of a circle written in the form $x^2 + y^2 + Dx + Ey + F = 0$

general form of a linear equation A linear equation written in the form $Ax + By = C$

geometric means Numbers that are inserted between two elements in a sequence to form a new geometric sequence

geometric sequence A sequence in which each successive term is equal to the product of the preceding term and a constant, written $a_1, a_1r, \ldots, a_1r^2, a_1r^3, \ldots$

graph of a point A point's location in the rectangular or Cartesian plane

greatest integer function The function whose output for a given x is the greatest integer that is less than or equal to x, denoted by $[\ ]$

grouping symbol A symbol (such as a radical sign or a fraction bar) or pair of symbols (such as parentheses, braces, or brackets) that indicate that these operations should be computed before other operations

half-life The time it takes for one-half of a sample of a radioactive element to decompose

half-plane A subset of the coordinate plane consisting of all points on a given side of a boundary line

horizontal line A line parallel to the x-axis; a line with the equation $y = b$

horizontal-line test A test used to determine if a given function is one-to-one: If every horizontal line that intersects the graph of the function does so exactly once, then the graph is the graph of a one-to-one function.

horizontal translation A graph that is the same shape as a given graph, except that it is shifted horizontally

hyperbola The set of all points in a plane the difference of whose distances from two fixed points (foci) is a constant

hypotenuse The longest side of a right triangle, the side opposite the 90° angle

identity An equation that is true for all values of its variables; its solution set is $\mathbb{R}$.

identity function The function whose rule is $f(x) = x$

imaginary number The square root of a negative number; the square root of -1 is denoted by i

inconsistent system A system of equations that has no solution; its solution set is $\varnothing$.

increasing function A function whose value increases as the independent variable increases (graph rises as we move to the right)

independent equations of a two-by-two system A system of two equations with two variables for which each equation's graph is different

independent variable The variable in an equation of two variables to which we assign input values (usually x in an equation involving x and y)

index If $b = \sqrt[n]{a}$, we say n is the index of the radical.

inequality A mathematical statement indicating that two quantities are not necessarily equal; most commonly a statement involving the symbols $<, \leq, >$, or $\geq$

infinite geometric series The indicated sum of the terms of a geometric sequence for which $|r| < 1$

infinite sequence A sequence with infinitely many terms

input value A value substituted for the independent variable

integers The set of numbers given by $\{\ldots, -4, -3, -2, -1, 0, 1, 2, 3, 4, \ldots\}$

integer square An integer that is the square of another integer

intercept method The method of graphing a linear equation by first graphing the intercepts, and then connecting them by a straight line

intersection of two intervals A combination of two intervals that contains any point in the first interval that is also in the second interval

interval A set of real numbers between two given real numbers a and b; it must be specified whether or not a and b are included in the interval.

interval notation A way of writing intervals using brackets or parentheses to distinguish between endpoints that are included in a given interval, and endpoints that are not included

inverse function The function, denoted by f^{-1}, obtained by reversing the coordinates of each ordered pair $(x, f(x))$; for example, if $f(2) = 3$, then $f^{-1}(3) = 2$.

irrational numbers The set of numbers that cannot be put in the form of a fraction with integer numerator and nonzero integer denominator; can be expressed only as non-terminating, non-repeating decimals

isosceles triangle A triangle that has two sides of the same length

like radicals Radicals that have the same index and the same radicand

like terms Terms containing the same variables each with the same exponents

linear equation in one variable An equation of the form $ax + b = c$, where a, b, and c are real numbers and $a \neq 0$

linear equation in two variables An equation of the form $Ax + By = C$, where A, B, and C are real numbers; A and B cannot both be 0.

linear inequality in one variable An inequality containing an expression of the form $Ax + B \quad (A \neq 0)$

linear inequality in two variables An inequality containing an expression of the form $Ax + By \quad (A$ and B not both equal to zero)

linear programming A mathematical technique used to find the optimal allocation of resources

logarithm The exponent to which b is raised to get x, denoted by $\log_b x$

logarithmic equation An equation with logarithmic expressions that contain a variable

logarithmic function The function $f(x) = \log_b x$; $y = \log_b x$ if and only if $x = b^y$

major axis of an ellipse The line segment passing through the foci joining the vertices of an ellipse

matrix A rectangular array of numbers

mean The sum of a collection of value divided by the number of values

means In the proportion $\frac{a}{b} = \frac{c}{d}$, the numbers b and c are called the means.

median The middle value of a collection of values, obtained by arranging them in increasing order, and either choosing the middle value, or the average of the two middle values

midpoint The point half way between two given points

minor A number associated with an element of a square matrix used to find a determinant of next lower order, by crossing out the elements of the matrix that are in the same row and column as the element

minor axis of an ellipse The line segment passing through the center of an ellipse perpendicular to the major axis

mode The value that occurs most often in a collection of values; there can be more than one mode for a collection of values.

monomial A polynomial with exactly one term

multiplication principle of events The method of determining the number of ways that one event can be followed by another

multiplicative identity The number 1, because, for all real numbers a, $a \cdot 1 = 1 \cdot a = a$

multiplicative inverse Two numbers whose product is 1; also called **reciprocals**

natural logarithm A logarithm with a base of e

natural numbers The set of numbers given by $\{1, 2, 3, 4, 5, \ldots\}$; also called **positive integers**

numerical coefficient The numerical factor of a term; for example, the numerical coefficient of $8ab$ is 8.

objective function A function whose value is maximized or minimized

odd integers The set of integers that cannot be divided exactly by 2

odd root If $b = \sqrt[n]{a}$, b is an odd root if n is odd.

one-to-one function A function for which every output corresponds to exactly one input

ordered pair A pair of real numbers, written as (x, y), that describes a unique point in the Cartesian plane

origin The point on a number line that represents the number zero; the point on the rectangular coordinate system that represents the point $(0, 0)$

output value The value of the dependent variable, determined by the choice of input value

parabola The graph of a quadratic function

partial sum The sum of finitely many consecutive terms of a series, starting at the first term

Pascal's triangle A triangular array of numbers in which 1's begin and end each row, and each entry between the 1's is the sum of the closest pair of numbers in the line immediately above it; the first 4 rows of Pascal's triangle are

$$\begin{array}{ccccccc} & & & 1 & & & \\ & & 1 & & 1 & & \\ & 1 & & 2 & & 1 & \\ 1 & & 3 & & 3 & & 1 \end{array}$$

perfect cube An integer that can be written as the cube of another integer or, more generally, a polynomial expression that can be written as the cube of a polynomial expression

perfect square An integer that can be written as the square of another integer or, more generally, a polynomial expression that can be written as the square of a polynomial expression

periodic interest rate The interest rate used for each compounding period

permutations The number of ways to choose r things from a set of n things if order is important

piecewise-defined function A function defined by using different equations for different parts of its domain

plane The graph of an equation of the form $ax + by + cz = d$

point circle A circle with radius zero; a single point

point-slope form The equation of a line in the form $y - y_1 = m(x - x_1)$, where m is the slope and (x_1, y_1) is a point on the line

polynomial An algebraic expression that is a single term or the sum of several terms containing whole-number exponents on the variables

polynomial function A function whose equation is a polynomial, for example, $f(x) = x^2 - 3x + 6$

power of *x* An expression of the form x^y; also called an **exponential expression**

power rule The rule that states if x, y, and n are real numbers and $x = y$, then $x^n = y^n$

prime number A natural number greater than 1 that can be divided exactly only by 1 and itself

prime polynomial A polynomial that does not factor over the rational numbers

principal square root The positive square root of a number

probability A measure of the likelihood of an event

product The result of multiplying two or more expressions

proportion A statement that two ratios are equal

pyramid A three-dimensional surface whose cross sections (parallel to the base) are polygons the same shape as the base but decrease in size until it comes to a point

Pythagorean theorem If the length of the hypotenuse of a right triangle is c and the lengths of the two legs are a and b, then $a^2 + b^2 = c^2$.

quadrants The four regions in the Cartesian plane, formed by the x- and y-axes

quadratic equation An equation that can be written in the form $ax^2 + bx + c = 0$ where $a \neq 0$

quadratic formula A formula that gives the solutions to the general quadratic equation; $x = \dfrac{-b \pm \sqrt{b^2 - 4ac}}{2a}$

quadratic function A polynomial function whose equation is of the form $f(x) = ax^2 + bx + c \quad (a \neq 0)$

quadratic inequality An inequality involving a quadratic polynomial in one variable

quotient The result of dividing two expressions

radical sign The symbol used to represent the root of a number

radicand The expression under a radical sign

radius A segment drawn from the center of a circle to a point on the circle; the length of such a line segment

range of a function The set of all output values of a function

range of a relation The set of all second elements of a relation

rate A ratio used to compare two quantities of different units

ratio The comparison of two numbers by their indicated quotient

rational equation An equation that contains one or more rational expressions

rational expression The quotient of two polynomials, where the polynomial in the denominator cannot be equal to 0

rational function A function of the form $f(x) = \dfrac{p(x)}{q(x)}$ where $p(x)$ and $q(x)$ are polynomial functions and $q(x) \neq 0$

rationalizing the denominator The process of simplifying a fraction so that there are no radicals in the denominator

rationalizing the numerator The process of simplifying a fraction so that there are no radicals in the numerator

rational numbers The set of fractions that have an integer numerator and a nonzero integer denominator; alternately, any terminating or repeating decimal

real numbers The set of decimals that contains the rational numbers and the irrational numbers

rectangular coordinate system A grid that allows us to identify each point in a plane with a unique pair of numbers; also called the **Cartesian coordinate system**

rectangular solid A three-dimensional surface whose cross sections (parallel to the base) are the same size and shape as the rectangular base

reflection A graph that is identical in shape to a given graph, except that it is flipped across a line (usually the x- or y-axis)

relation A set of ordered pairs

remainder theorem The theorem that states if a polynomial function $P(x)$ is divided by $(x - r)$, the remainder is $P(r)$

Richter scale A unit of measure that expresses magnitude (ground motion) of an earthquake

right angle An angle whose measure is 90°

right triangle A triangle that has an angle whose measure is 90°

root A number that makes an equation true when substituted for its variable; if there are several variables, then the set of numbers that make the equation true; also called a **solution**

roster method A method of describing a set by listing its elements within braces

sample space The set of all possible outcomes for an experiment

scientific notation The representation of a number as the product of a number between 1 and 10 (1 included), and an integer power of ten, for example, 6.02×10^{23}

sequence An ordered set of numbers that are defined by the position (1st, 2nd, 3rd, etc.) they hold

set A collection of objects whose members are listed or defined within braces

set-builder notation A method of describing a set that uses a variable (or variables) to represent the elements and a rule to determine the possible values of the variable; for example, we can describe the natural numbers as $\{x \mid x \text{ is an integer and } x > 0\}$.

similar triangles Two triangles that have the same shape, that is, two triangles whose corresponding angles have the same measure

slope-intercept form The equation of the line in the form $y = mx + b$ where m is the slope and b is the y-coordinate of the y-intercept

solution/solution set A number that makes an equation true when substituted for its variable; if there are several variables, then the numbers that make the equation true; also called a **root**

sphere A three-dimensional surface in which all points are equidistant from a fixed point (center); a ball shape

square matrix A matrix with the same number of rows as columns

square root If $a = b^2$, then b is a square root of a.

square-root function The function $f(x) = \sqrt{x}$

square-root property The property that states the equation $x^2 = c$ has two solutions: $\sqrt{c}$ and $-\sqrt{c}$

squaring function The function $f(x) = x^2$

standard deviation A measure of how closely a set of data is grouped about its mean

standard form of the equation of a circle The equation of a circle written in the form $(x - h)^2 + (y - k)^2 = r^2$, where (h, k) is the center and r is the radius

standard notation The representation of a given number as an integer part, followed by a decimal, for example, 212.3337012

step function A function whose graph is a series of horizontal line segments

straight angle An angle whose measure is 180°

subset A set, all of whose elements are included in a different set; for example, the set $\{1, 3, 5\}$ is a subset of the set $\{1, 2, 3, 4, 5\}$.

sum The result of adding two expressions

summation notation A shorthand notation for indicating the sum of a number of consecutive terms in a series, denoted by Σ

sum of two cubes An expression of the form $a^3 + b^3$

sum of two squares An expression of the form $a^2 + b^2$

supplementary angles Two angles whose measures sum to 180°

system of equations Two or more equations that are solved simultaneously

synthetic division A method used to divide a polynomial by a binomial of the form $x - r$

tree diagram A way of visually representing all the possible outcomes of an event

triangular form of a matrix A matrix with all zeros below its main diagonal

trinomial A polynomial with exactly three terms

union of two intervals A combination of two intervals that contains all points in the first interval combined with any additional points in the second interval

unit cost The ratio of an item's cost to its quantity

variables Letters that are used to represent real numbers

vertex The lowest point of a parabola that opens up, or the highest point of a parabola that opens down

vertex angle of an isosceles triangle The angle in an isosceles triangle formed by the two sides that are of the same length

vertical line A line parallel to the y-axis; a line with the equation $x = b$

vertical-line test A method of testing whether the graph of an equation represents a function

vertical translation A graph that is the same shape as a given graph, only shifted vertically

vertices of a hyperbola The endpoints of the axis that passes through the foci

vertices of an ellipse The endpoints of the major axis

volume The amount of space enclosed by a three-dimensional geometric figure

whole numbers The set of numbers given by $\{0, 1, 2, 3, 4, 5, \ldots\}$, that is, the set of integers that are greater than or equal to zero

***x*-axis** The horizontal number line in the rectangular coordinate system

x-coordinate The first number in the rectangular coordinates of a point

x-intercept The point $(a, 0)$ where a graph intersects the x-axis

y-axis The vertical number line in the rectangular coordinate system

y-coordinate The second number in the rectangular coordinates of a point

y-intercept The point $(0, b)$ where a graph intersects the y-axis

zero-factor property The property of real numbers that if the product of two quantities is 0, then at least one of those quantities must be equal to 0

Appendix I Symmetries of Graphs

There are several ways that a graph can exhibit symmetry about the coordinate axes and the origin. It is often easier to draw graphs of equations if we first find the x- and y-intercepts and find any of the following symmetries of the graph:

1. **y-axis symmetry:** If the point $(-x, y)$ lies on a graph whenever the point (x, y) does, as in Figure I-1(a), we say that the graph is **symmetric about the y-axis.**
2. **Symmetry about the origin:** If the point $(-x, -y)$ lies on the graph whenever the point (x, y) does, as in Figure I-1(b), we say that the graph is **symmetric about the origin.**
3. **x-axis symmetry:** If the point $(x, -y)$ lies on the graph whenever the point (x, y) does, as in Figure I-1(c), we say that the graph is **symmetric about the x-axis.**

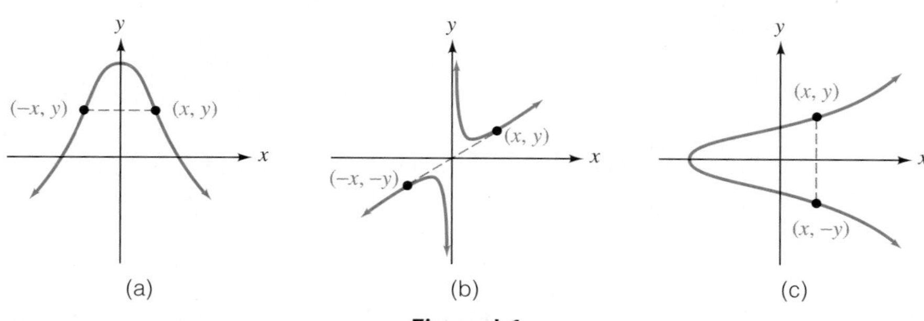

Figure I-1

Tests for Symmetry for Graphs in x and y

- To test a graph for y-axis symmetry, replace x with $-x$. If the new equation is equivalent to the original equation, the graph is symmetric about the y-axis. Symmetry about the y-axis will occur whenever x appears with only even exponents.
- To test a graph for symmetry about the origin, replace x with $-x$ and y with $-y$. If the resulting equation is equivalent to the original equation, the graph is symmetric about the origin.
- To test a graph for x-axis symmetry, replace y with $-y$. If the resulting equation is equivalent to the original equation, the graph is symmetric about the x-axis. The only function that is symmetric about the x-axis is $f(x) = 0$.

EXAMPLE 1 Find the intercepts and the symmetries of the graph of $y = f(x) = x^3 - 9x$. Then graph the function.

Solution ***x*-intercepts:** To find the x-intercepts, we let $y = 0$ and solve for x:

$$y = x^3 - 9x$$

$$0 = x^3 - 9x \quad \text{Substitute 0 for } y.$$

$$0 = x(x^2 - 9) \quad \text{Factor out } x.$$

$$0 = x(x + 3)(x - 3) \quad \text{Factor } x^2 - 9.$$

$$x = 0 \quad \text{or} \quad x + 3 = 0 \quad \text{or} \quad x - 3 = 0 \quad \text{Set each factor equal to 0.}$$

$$x = -3 \qquad x = 3$$

Since the x-coordinates of the x-intercepts are 0, -3, and 3, the graph intersects the x-axis at (0, 0), (-3, 0), and (3, 0).

***y*-intercepts:** To find the y-intercepts, we let $x = 0$ and solve for y.

$$y = x^3 - 9x$$

$$y = 0^3 - 9(0) \quad \text{Substitute 0 for } x.$$

$$y = 0$$

Since the y-coordinate of the y-intercept is 0, the graph intersects the y-axis at (0, 0).

Symmetry: We test for symmetry about the y-axis by replacing x with $-x$, simplifying, and comparing the result to the original equation.

(1) $y = x^3 - 9x$ This is the original equation.

$y = (-x)^3 - 9(-x)$ Replace x with $-x$.

(2) $y = -x^3 + 9x$ Simplify.

Because Equation 2 is not equivalent to Equation 1, the graph is not symmetric about the y-axis.

We test for symmetry about the origin by replacing x with $-x$ and y with $-y$, respectively, and comparing the result to the original equation.

(1) $y = x^3 - 9x$ This is the original equation.

$-y = (-x)^3 - 9(-x)$ Replace x with $-x$, and y with $-y$.

$-y = -x^3 + 9x$ Simplify.

(3) $y = x^3 - 9x$ Multiply both sides by -1 to solve for y.

Because Equation 3 is equivalent to Equation 1, the graph is symmetric about the origin. Because the equation is the equation of a nonzero function, there is no symmetry about the x-axis.

To graph the equation, we plot the x-intercepts of (-3, 0), (0, 0), and (3, 0) and the y-intercept of (0, 0). We also plot other points for positive values of x and use the symmetry about the origin to draw the rest of the graph, as in Figure I-2(a). (Note that the scale on the x-axis is different from the scale on the y-axis.)

If we graph the equation with a graphing calculator, with window settings of $[-10, 10]$ for x and $[-10, 10]$ for y, we will obtain the graph shown in Figure I-2(b).

From the graph, we can see that the domain is the interval $(-\infty, \infty)$, and the range is the interval $(-\infty, \infty)$.

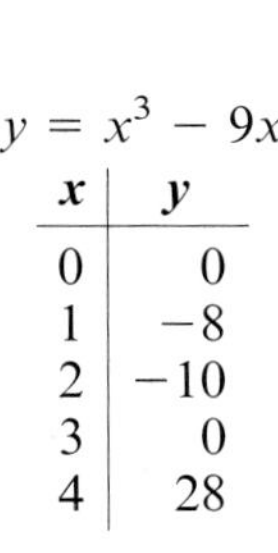

$y = x^3 - 9x$

x	y
0	0
1	−8
2	−10
3	0
4	28

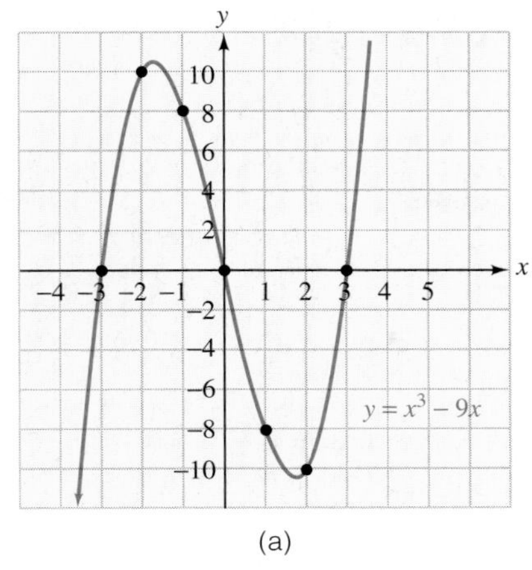

(a)

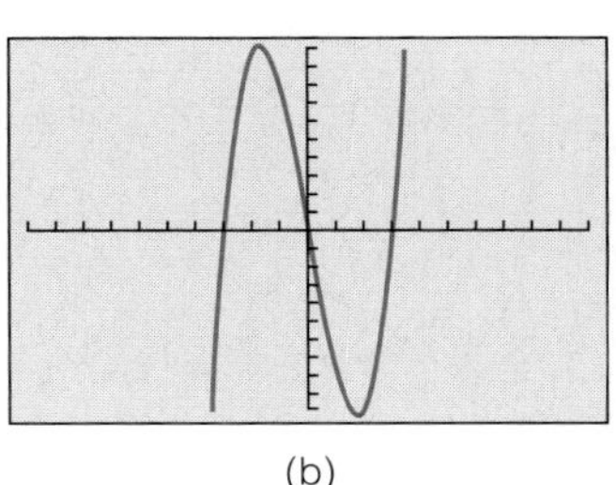

(b)

Figure I-2

EXAMPLE 2 Graph the function $y = f(x) = |x| - 2$.

Solution ***x*-intercepts:** To find the x-intercepts, we let $y = 0$ and solve for x:

$$y = |x| - 2$$
$$\mathbf{0} = |x| - 2$$
$$2 = |x|$$
$$x = -2 \quad \text{or} \quad x = 2$$

Since -2 and 2 are solutions, the points $(-2, 0)$ and $(2, 0)$ are the x-intercepts, and the graph passes through $(-2, 0)$ and $(2, 0)$.

***y*-intercepts:** To find the y-intercepts, we let $x = 0$ and solve for y:

$$y = |x| - 2$$
$$y = |\mathbf{0}| - 2$$
$$y = -2$$

Since $y = -2$, $(0, -2)$ is the y-intercept, and the graph passes through the point $(0, -2)$.

Symmetry: To test for y-axis symmetry, we replace x with $-x$.

(4) $y = |x| - 2$ This is the original equation.

$y = |-x| - 2$ Replace x with $-x$.

(5) $y = |x| - 2$ $|-x| = |x|$

Since Equation 5 is equivalent to Equation 4, the graph is symmetric about the y-axis. The graph has no other symmetries.

We plot the x- and y-intercepts and several other points (x, y), and use the y-axis symmetry to obtain the graph shown in Figure I-3(a) on the next page.

If we graph the equation with a graphing calculator, with window settings of $[-10, 10]$ for x and $[-10, 10]$ for y, we will obtain the graph shown in Figure I-3(b).

From the graph, we see that the domain is the interval $(-\infty, \infty)$, and the range is the interval $[-2, \infty)$.

$y = |x| - 2$

x	y
0	−2
1	−1
2	0
3	1
4	2

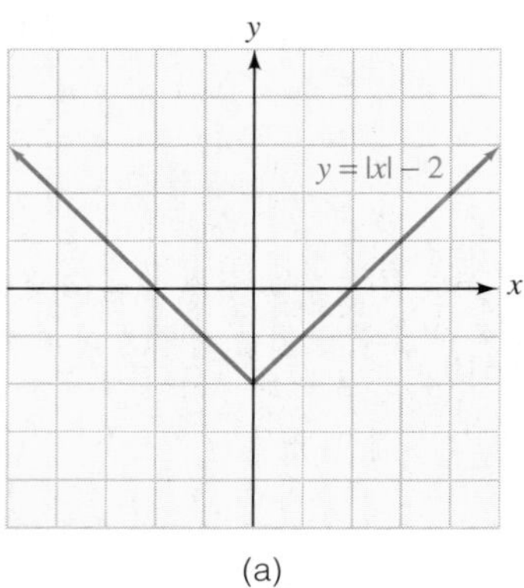

(a)

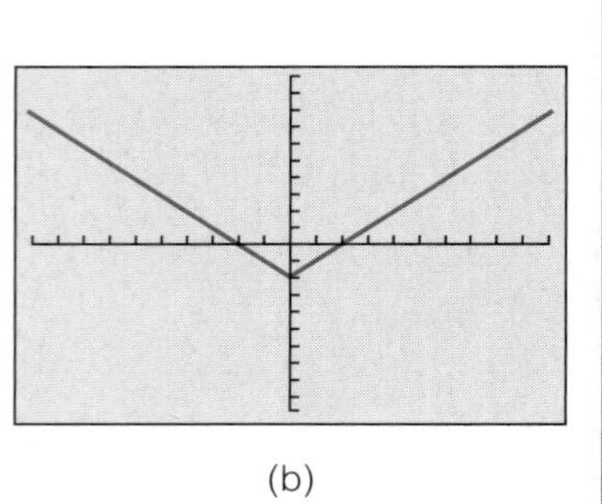

(b)

Figure I-3

I.1 EXERCISES

Find the symmetries of the graph of each equation. **Do not draw the graph.**

1. $y = x^2 - 1$

2. $y = x^3$

3. $y = x^5$

4. $y = x^4$

5. $y = -x^2 + 2$

6. $y = x^3 + 1$

7. $y = x^2 - x$

8. $y^2 = x + 7$

9. $y = -|x + 2|$

10. $y = |x| - 3$

11. $|y| = x$

12. $y = 2\sqrt{x}$

Graph each function and give its domain and range. Check each graph with a graphing calculator.

13. $f(x) = x^4 - 4$

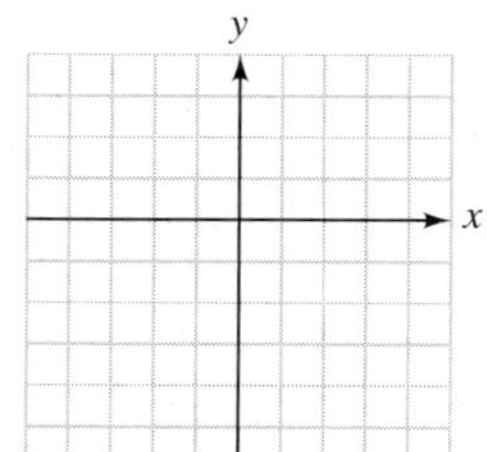

14. $f(x) = \frac{1}{2}x^4 - 1$

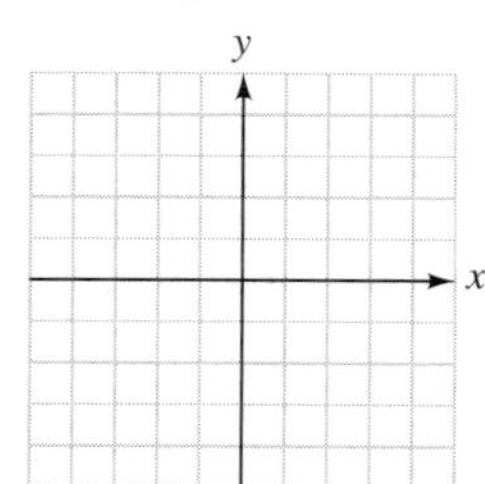

15. $f(x) = -x^3$

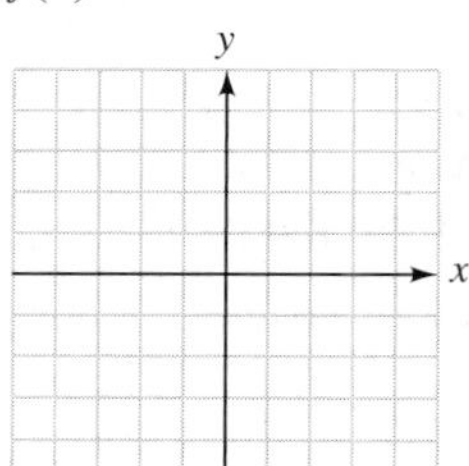

16. $f(x) = x^3 + 2$

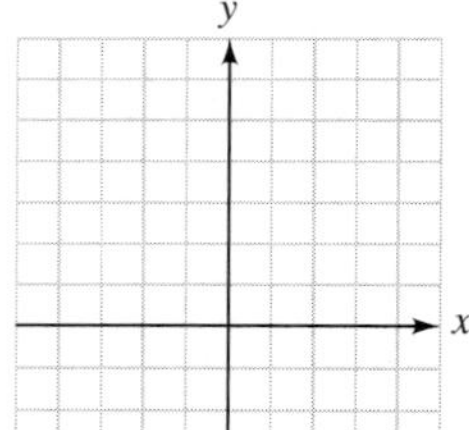

17. $f(x) = x^4 + x^2$

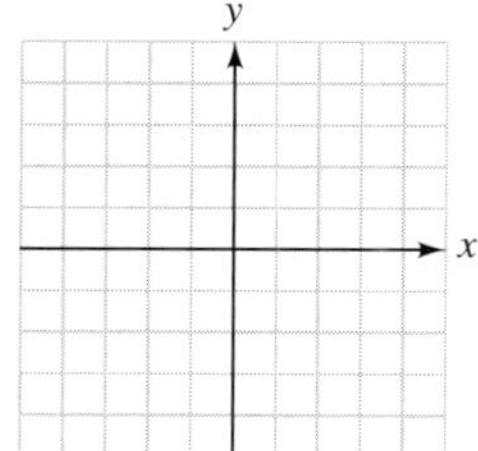

18. $f(x) = 3 - x^4$

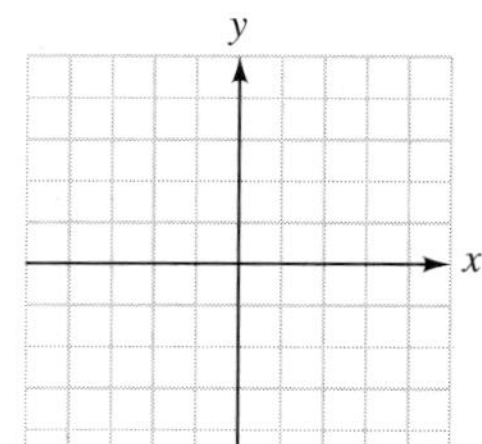

19. $f(x) = x^3 - x$

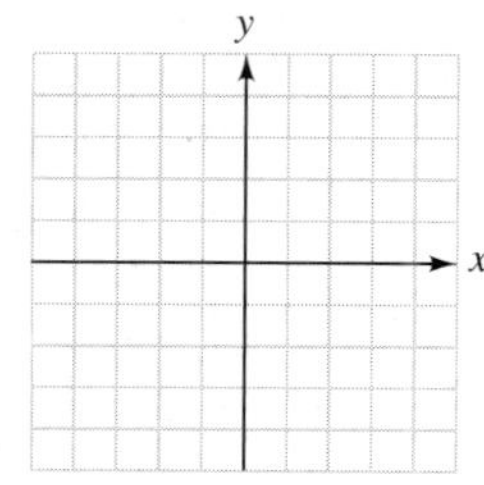

20. $f(x) = x^3 + x$

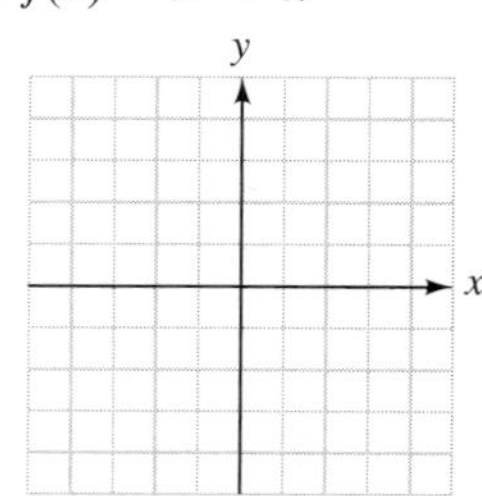

21. $f(x) = \frac{1}{2}|x| - 1$

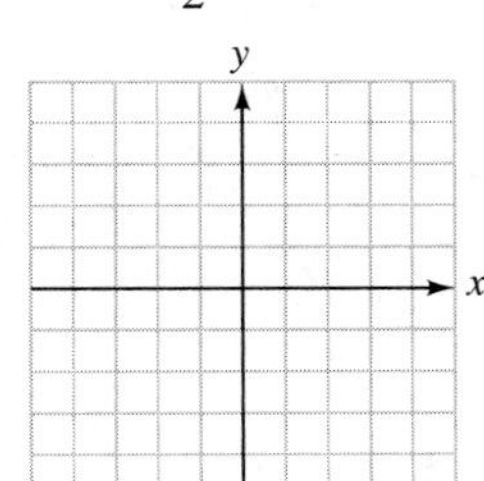

22. $f(x) = -|x| + 1$

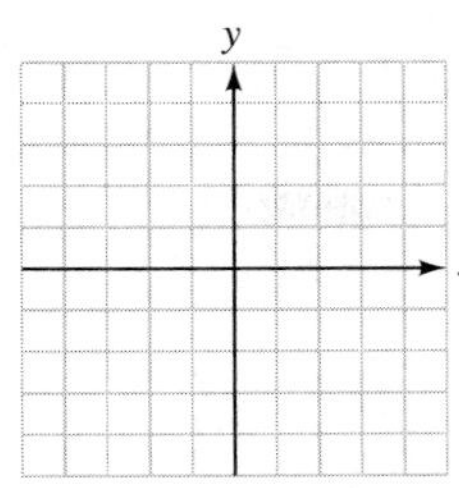

23. $f(x) = -|x + 2|$

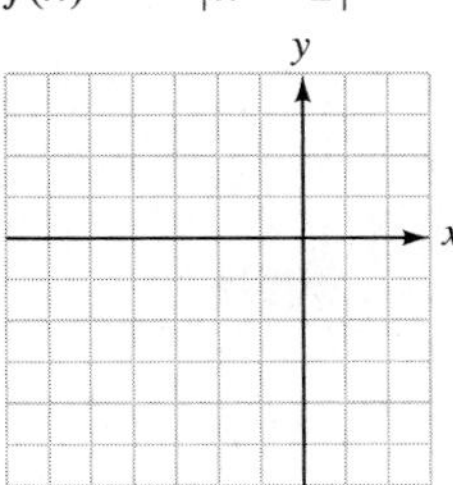

24. $f(x) = |x - 2|$

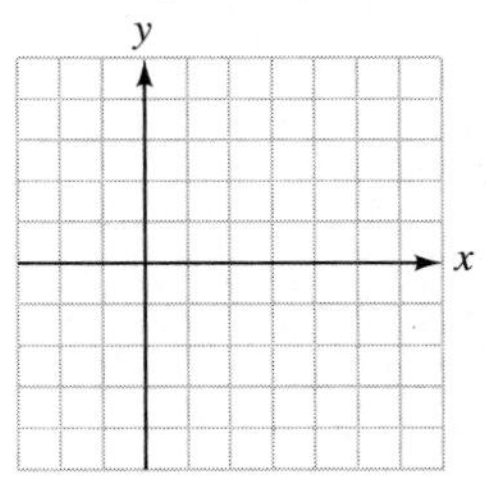

Appendix II Sample Final Examination

1. How many prime numbers are there in the interval from 40 to 50?

   a. 3 **b.** 4 **c.** 5 **d.** none of the above

2. The commutative property of multiplication is written symbolically as

   a. $ab = ba$ **b.** $(ab)c = a(bc)$
   c. If $a = b$, then $b = a$. **d.** none of the above

3. If $a = 3$, $b = -2$, and $c = 6$, the value of $\frac{c - ab}{bc}$ is

   a. 1 **b.** 0 **c.** -1 **d.** none of the above

4. Evaluate: $\frac{1}{2} + \frac{3}{4} \div \frac{5}{6}$.

   a. $\frac{3}{2}$ **b.** $\frac{7}{2}$ **c.** $\frac{9}{8}$ **d.** none of the above

5. The expression $\left(\frac{a^2}{a^5}\right)^{-5}$ can be written as

   a. 15 **b.** a^{-15} **c.** a^{15} **d.** none of the above

6. Write 0.0000234 in scientific notation.

   a. 2.34×10^{-5} **b.** 2.34×10^{5} **c.** $234. \times 10^{-7}$
   d. none of the above

7. If $P(x) = 2x^2 - x - 1$, find $P(-1)$.

   a. 4 **b.** 0 **c.** 2 **d.** none of the above

8. Simplify: $(3x + 2) - (2x - 1) + (x - 3)$.

   a. $2x - 2$ **b.** $2x$ **c.** 2 **d.** none of the above

9. Multiply: $(3x - 2)(2x + 3)$.

   a. $6x^2 + 5x - 6$ **b.** $6x^2 - 5x - 6$
   c. $6x^2 + 5x + 6$ **d.** none of the above

10. Divide: $2x + 1\overline{)2x^2 - 3x - 2}$.

    a. $x + 2$ **b.** $x - 1$ **c.** $x - 2$
    d. none of the above

11. Solve for x: $5x - 3 = -2x + 10$.

    a. $\frac{11}{7}$ **b.** $\frac{13}{3}$ **c.** 1 **d.** none of the above

12. The sum of two consecutive odd integers is 44. The product of the integers is

    a. an even integer **b.** 463 **c.** 483
    d. none of the above

13. Solve for x: $2ax - a = b + x$.

    a. $\frac{a + b}{2a}$ **b.** $\frac{a + b}{2a - 1}$ **c.** $a + b - 2a$
    d. $-a + b + 1$

14. The sum of the solutions of $|2x + 5| = 13$ is

    a. 4 **b.** 8 **c.** 12 **d.** none of the above

15. Solve for x: $-2x + 5 > 9$.

    a. $x > 7$ **b.** $x < 7$ **c.** $x < -2$ **d.** $x > -2$

16. Solve for x: $|2x - 5| \leq 9$.

    a. $2 \leq x \leq 7$ **b.** $-2 \geq x$ and $x \leq 7$ **c.** $x \leq 7$
    d. $-2 \leq x \leq 7$

17. Factor completely: $3ax^2 + 6a^2x$.

    a. $3(ax^2 + 2a^2x)$ **b.** $3a(x^2 + 2ax)$
    c. $3x(ax + 2a^2)$ **d.** none of the above

18. The sum of the prime factors of $x^4 - 16$ is

    a. $2x^2$ **b.** $x^2 + 2x + 4$
    c. $4x + 4$ **d.** none of the above

19. The sum of the factors of $8x^2 - 2x - 3$ is

    a. $6x - 2$ **b.** $8x - 3$ **c.** $6x + 2$
    d. none of the above

20. One of the factors of $27a^3 + 8$ is

    a. $3a - 2$ **b.** $9a^2 + 12a + 4$ **c.** $9a^2 + 6a + 4$
    d. none of the above

21. The smallest solution of the equation $6x^2 - 5x - 6 = 0$ is

    a. $-\frac{3}{2}$ **b.** $\frac{2}{3}$ **c.** $-\frac{2}{3}$ **d.** none of the above

22. Simplify: $\frac{x^2 + 5x + 6}{x^2 - 9}$.

    a. $-\frac{2}{3}$ **b.** $\frac{x + 2}{x - 3}$ **c.** $\frac{-5x + 6}{9}$
    d. none of the above

23. Simplify: $\frac{3x + 6}{x + 3} - \frac{x^2 - 4}{x^2 + x - 6}$.

    a. $\frac{3(x + 2)(x + 2)}{(x + 3)(x + 3)}$ **b.** 3 **c.** $\frac{1}{3}$
    d. none of these

24. The numerator of the sum $\dfrac{y}{x+y} + \dfrac{x}{x-y}$ is

a. $x^2 + 2xy - y^2$ **b.** $y + x$ **c.** $y - x$
d. none of the above

25. Simplify: $\dfrac{\frac{1}{x} + \frac{1}{y}}{\frac{1}{y}}$.

a. $\dfrac{1}{xy}$ **b.** 1 **c.** $\dfrac{y+x}{x}$ **d.** none of the above

26. Solve for y: $\dfrac{2}{y+1} = \dfrac{1}{y+1} - \dfrac{1}{3}$.

a. 4 **b.** 3 **c.** -4 **d.** none of the above

27. The sum of the x- and y-coordinatesof the x- and y-intercepts of the graph of $2x + 3y = 6$ is

a. $-\dfrac{2}{3}$ **b.** 0 **c.** 5 **d.** none of the above

28. The slope of the line passing through $(3, -2)$ and $(5, -1)$ is

a. $-\dfrac{1}{2}$ **b.** 2 **c.** -2 **d.** $\dfrac{1}{2}$

29. The graphs of the equations $\begin{cases} 2x - 3y = 4 \\ 3x + 2y = 1 \end{cases}$

a. are parallel **b.** are perpendicular
c. do not intersect **d.** are the same line

30. An equation of the line passing through $(-2, 5)$ and $(6, 7)$ is

a. $y = -\dfrac{1}{4}x - \dfrac{11}{2}$ **b.** $y = \dfrac{1}{4}x + \dfrac{11}{2}$

c. $y = \dfrac{1}{4}x - \dfrac{11}{2}$ **d.** $y = -\dfrac{1}{4}x + \dfrac{11}{2}$

31. If $g(x) = x^2 - 3$, then $g(t + 1)$ is

a. $t^2 - 2$ **b.** -2 **c.** $t - 2$ **d.** $t^2 + 2t - 2$

32. Assume that d varies directly with t. Find the constant of variation if $d = 12$ when $t = 3$.

a. 36 **b.** 4 **c.** -36 **d.** none of the above

33. The expression $x^{a/2}x^{a/5}$ can be expressed as

a. $x^{a/7}$ **b.** $x^{7a/10}$ **c.** $x^{a/10}$ **d.** $x^{a/5}$

34. The product of $(x^{1/2} + 2)$ and $(x^{-1/2} - 2)$ is

a. -3 **b.** $-3 + 2x$ **c.** $-3 - 2x^{1/2} + 2x^{-1/2}$
d. none of the above

35. Completely simplify $\sqrt{112a^3}$ $(a \geq 0)$.

a. $a\sqrt{112a}$ **b.** $4a^2\sqrt{7a}$ **c.** $4\sqrt{7a^2}$

d. none of the above

36. Simplify and combine terms: $\sqrt{50} - \sqrt{98} + \sqrt{128}$.

a. $-2\sqrt{2} + \sqrt{128}$ **b.** $6\sqrt{2}$ **c.** $20\sqrt{2}$
d. $-4\sqrt{2}$

37. Rationalize the denominator and simplify: $\dfrac{3}{2 - \sqrt{3}}$.

a. $-6 - \sqrt{3}$ **b.** $6 + 3\sqrt{3}$
c. $2(2 + \sqrt{3})$ **d.** $6(2 + \sqrt{3})$

38. The distance between the points $(-2, 3)$ and $(6, -8)$ is

a. $\sqrt{185}$ **b.** $\sqrt{41}$ **c.** $\sqrt{57}$
d. none of the above

39. Solve for x: $\sqrt{x + 7} - 2x = -1$.

a. 2 **b.** $-\dfrac{3}{4}$ **c.** $\varnothing$ **d.** $2, -\dfrac{3}{4}$

40. The graph of $y > 3x + 2$ contains no points in

a. quadrant I **b.** quadrant II **c.** quadrant III
d. quadrant IV

41. The quadratic formula is

a. $x = \dfrac{b \pm \sqrt{b^2 - 4ac}}{2a}$ **b.** $x = \dfrac{-b \pm \sqrt{b^2 - 4ac}}{a}$

c. $x = \dfrac{-b \pm \sqrt{b - 4ac}}{a}$ **d.** $x = \dfrac{-b \pm \sqrt{b^2 - 4ac}}{2a}$

42. Write the complex number $(2 + 3i)^2$ in $a + bi$ form.

a. $-5 + 12i$ **b.** -5 **c.** $13 + 12i$ **d.** 13

43. Write the complex number $\dfrac{i}{3 + i}$ in $a + bi$ form.

a. $\dfrac{1}{3} + 0i$ **b.** $\dfrac{1}{10} + \dfrac{3}{10}i$ **c.** $\dfrac{1}{8} + \dfrac{3}{8}i$
d. none of the above

44. The vertex of the parabola determined by $y = 2x^2 + 4x - 3$ is at point

a. $(0, -3)$ **b.** $(12, 13)$ **c.** $(-1, -5)$ **d.** $(1, 3)$

45. Solve for x: $\dfrac{2}{x} < 3$.

a. $0 < x < \dfrac{2}{3}$ **b.** $x > \dfrac{2}{3}$ **c.** $x < 0$ or $x > \dfrac{2}{3}$

d. $x < 0$ and $x > \dfrac{2}{3}$

46. If $f(x) = 2x^2 + 1$, then $f(3) =$

a. 7 **b.** 19 **c.** 17 **d.** 37

47. The inverse function of $y = 3x + 2$ is

a. $y = \frac{x + 2}{3}$ **b.** $x = 3y - 2$ **c.** $y = \frac{x - 2}{3}$

d. $x = \frac{y - 2}{3}$

48. The equation of a circle with center at $(-2, 4)$ and radius of 4 units is

a. $(x + 2)^2 + (y - 4)^2 = 16$

b. $(x + 2)^2 + (y - 4)^2 = 4$

c. $(x + 2)^2 + (y - 4)^2 = 2$

d. $(x - 2)^2 + (y + 4)^2 = 16$

49. Find the sum of the solutions of the system

$$\begin{cases} \frac{4}{x} + \frac{2}{y} = 2 \\ \frac{2}{x} - \frac{3}{y} = -1 \end{cases}$$

a. 6 **b.** -6 **c.** 2 **d.** -4

50. Find the y-value of the solution of the system

$$\begin{cases} 4x + 6y = 5 \\ 8x - 9y = 3 \end{cases}$$

a. $\frac{3}{4}$ **b.** $\frac{1}{3}$ **c.** $\frac{1}{2}$ **d.** $\frac{2}{3}$

51. The value of the determinant $\begin{vmatrix} 2 & -3 \\ 4 & 4 \end{vmatrix}$ is

a. 0 **b.** 20 **c.** -4 **d.** -20

52. The value of z in the system $\begin{cases} x + y + z = 4 \\ 2x + y + z = 6 \\ 3x + y + 2z = 8 \end{cases}$ is

a. 0 **b.** 1 **c.** 2 **d.** 3

53. $\log_a N = x$ means

a. $a^x = N$ **b.** $a^N = x$ **c.** $x^N = a$ **d.** $N^a = x$

54. Find $\log_2 \frac{1}{32}$.

a. 5 **b.** $\frac{1}{5}$ **c.** -5 **d.** $-\frac{1}{5}$

55. Find $\log 7 + \log 5$.

a. $\log 12$ **b.** $\log \frac{7}{12}$ **c.** $\log 2$ **d.** $\log 35$

56. Find $b^{\log_b x}$.

a. b **b.** x **c.** 10 **d.** 0

57. Solve for y: $\log y + \log (y + 3) = 1$.

a. $-5, 2$ **b.** 2 **c.** 5 **d.** none of the above

58. The coefficient of the third term in the expansion of $(a + b)^6$ is

a. 30 **b.** 2! **c.** 15 **d.** 120

59. Compute: $P(7, 3)$.

a. 35 **b.** 210 **c.** 21 **d.** none of the above

60. Compute: $C(7, 3)$.

a. 35 **b.** 210 **c.** $7! \cdot 3!$ **d.** none of the above

61. Find the 100th term of 2, 5, 8, 11, ...

a. 301 **b.** 299 **c.** 297 **d.** 295

62. Find the sum of all the terms of $1, \frac{1}{3}, \frac{1}{9}, \frac{1}{27}, \ldots$.

a. 2 **b.** $\frac{2}{3}$ **c.** $\frac{3}{2}$ **d.** $\frac{5}{3}$

63. Find the probability of rolling a 5 or 6 on one roll of a fair die.

a. $\frac{1}{6}$ **b.** $\frac{1}{3}$ **c.** $\frac{1}{2}$ **d.** none of the above

Table A Powers and Roots

n	n^2	$\sqrt{n}$	n^3	$\sqrt[3]{n}$	n	n^2	$\sqrt{n}$	n^3	$\sqrt[3]{n}$
1	1	1.000	1	1.000	51	2,601	7.141	132,651	3.708
2	4	1.414	8	1.260	52	2,704	7.211	140,608	3.733
3	9	1.732	27	1.442	53	2,809	7.280	148,877	3.756
4	16	2.000	64	1.587	54	2,916	7.348	157,464	3.780
5	25	2.236	125	1.710	55	3,025	7.416	166,375	3.803
6	36	2.449	216	1.817	56	3,136	7.483	175,616	3.826
7	49	2.646	343	1.913	57	3,249	7.550	185,193	3.849
8	64	2.828	512	2.000	58	3,364	7.616	195,112	3.871
9	81	3.000	729	2.080	59	3,481	7.681	205,379	3.893
10	100	3.162	1,000	2.154	60	3,600	7.746	216,000	3.915
11	121	3.317	1,331	2.224	61	3,721	7.810	226,981	3.936
12	144	3.464	1,728	2.289	62	3,844	7.874	238,328	3.958
13	169	3.606	2,197	2.351	63	3,969	7.937	250,047	3.979
14	196	3.742	2,744	2.410	64	4,096	8.000	262,144	4.000
15	225	3.873	3,375	2.466	65	4,225	8.062	274,625	4.021
16	256	4.000	4,096	2.520	66	4,356	8.124	287,496	4.041
17	289	4.123	4,913	2.571	67	4,489	8.185	300,763	4.062
18	324	4.243	5,832	2.621	68	4,624	8.246	314,432	4.082
19	361	4.359	6,859	2.668	69	4,761	8.307	328,509	4.102
20	400	4.472	8,000	2.714	70	4,900	8.367	343,000	4.121
21	441	4.583	9,261	2.759	71	5,041	8.426	357,911	4.141
22	484	4.690	10,648	2.802	72	5,184	8.485	373,248	4.160
23	529	4.796	12,167	2.844	73	5,329	8.544	389,017	4.179
24	576	4.899	13,824	2.884	74	5,476	8.602	405,224	4.198
25	625	5.000	15,625	2.924	75	5,625	8.660	421,875	4.217
26	676	5.099	17,576	2.962	76	5,776	8.718	438,976	4.236
27	729	5.196	19,683	3.000	77	5,929	8.775	456,533	4.254
28	784	5.292	21,952	3.037	78	6,084	8.832	474,552	4.273
29	841	5.385	24,389	3.072	79	6,241	8.888	493,039	4.291
30	900	5.477	27,000	3.107	80	6,400	8.944	512,000	4.309
31	961	5.568	29,791	3.141	81	6,561	9.000	531,441	4.327
32	1,024	5.657	32,768	3.175	82	6,724	9.055	551,368	4.344
33	1,089	5.745	35,937	3.208	83	6,889	9.110	571,787	4.362
34	1,156	5.831	39,304	3.240	84	7,056	9.165	592,704	4.380
35	1,225	5.916	42,875	3.271	85	7,225	9.220	614,125	4.397
36	1,296	6.000	46,656	3.302	86	7,396	9.274	636,056	4.414
37	1,369	6.083	50,653	3.332	87	7,569	9.327	658,503	4.431
38	1,444	6.164	54,872	3.362	88	7,744	9.381	681,472	4.448
39	1,521	6.245	59,319	3.391	89	7,921	9.434	704,969	4.465
40	1,600	6.325	64,000	3.420	90	8,100	9.487	729,000	4.481
41	1,681	6.403	68,921	3.448	91	8,281	9.539	753,571	4.498
42	1,764	6.481	74,088	3.476	92	8,464	9.592	778,688	4.514
43	1,849	6.557	79,507	3.503	93	8,649	9.644	804,357	4.531
44	1,936	6.633	85,184	3.530	94	8,836	9.695	830,584	4.547
45	2,025	6.708	91,125	3.557	95	9,025	9.747	857,375	4.563
46	2,116	6.782	97,336	3.583	96	9,216	9.798	884,736	4.579
47	2,209	6.856	103,823	3.609	97	9,409	9.849	912,673	4.595
48	2,304	6.928	110,592	3.634	98	9,604	9.899	941,192	4.610
49	2,401	7.000	117,649	3.659	99	9,801	9.950	970,299	4.626
50	2,500	7.071	125,000	3.684	100	10,000	10.000	1,000,000	4.642

Table B Base-10 Logarithms

N	0	1	2	3	4	5	6	7	8	9
1.0	.0000	.0043	.0086	.0128	.0170	.0212	.0253	.0294	.0334	.0374
1.1	.0414	.0453	.0492	.0531	.0569	.0607	.0645	.0682	.0719	.0755
1.2	.0792	.0828	.0864	.0899	.0934	.0969	.1004	.1038	.1072	.1106
1.3	.1139	.1173	.1206	.1239	.1271	.1303	.1335	.1367	.1399	.1430
1.4	.1461	.1492	.1523	.1553	.1584	.1614	.1644	.1673	.1703	.1732
1.5	.1761	.1790	.1818	.1847	.1875	.1903	.1931	.1959	.1987	.2014
1.6	.2041	.2068	.2095	.2122	.2148	.2175	.2201	.2227	.2253	.2279
1.7	.2304	.2330	.2355	.2380	.2405	.2430	.2455	.2480	.2504	.2529
1.8	.2553	.2577	.2601	.2625	.2648	.2672	.2695	.2718	.2742	.2765
1.9	.2788	.2810	.2833	.2856	.2878	.2900	.2923	.2945	.2967	.2989
2.0	.3010	.3032	.3054	.3075	.3096	.3118	.3139	.3160	.3181	.3201
2.1	.3222	.3243	.3263	.3284	.3304	.3324	.3345	.3365	.3385	.3404
2.2	.3424	.3444	.3464	.3483	.3502	.3522	.3541	.3560	.3579	.3598
2.3	.3617	.3636	.3655	.3674	.3692	.3711	.3729	.3747	.3766	.3784
2.4	.3802	.3820	.3838	.3856	.3874	.3892	.3909	.3927	.3945	.3962
2.5	.3979	.3997	.4014	.4031	.4048	.4065	.4082	.4099	.4116	.4133
2.6	.4150	.4166	.4183	.4200	.4216	.4232	.4249	.4265	.4281	.4298
2.7	.4314	.4330	.4346	.4362	.4378	.4393	.4409	.4425	.4440	.4456
2.8	.4472	.4487	.4502	.4518	.4533	.4548	.4564	.4579	.4594	.4609
2.9	.4624	.4639	.4654	.4669	.4683	.4698	.4713	.4728	.4742	.4757
3.0	.4771	.4786	.4800	.4814	.4829	.4843	.4857	.4871	.4886	.4900
3.1	.4914	.4928	.4942	.4955	.4969	.4983	.4997	.5011	.5024	.5038
3.2	.5051	.5065	.5079	.5092	.5105	.5119	.5132	.5145	.5159	.5172
3.3	.5185	.5198	.5211	.5224	.5237	.5250	.5263	.5276	.5289	.5302
3.4	.5315	.5328	.5340	.5353	.5366	.5378	.5391	.5403	.5416	.5428
3.5	.5441	.5453	.5465	.5478	.5490	.5502	.5514	.5527	.5539	.5551
3.6	.5563	.5575	.5587	.5599	.5611	.5623	.5635	.5647	.5658	.5670
3.7	.5682	.5694	.5705	.5717	.5729	.5740	.5752	.5763	.5775	.5786
3.8	.5798	.5809	.5821	.5832	.5843	.5855	.5866	.5877	.5888	.5899
3.9	.5911	.5922	.5933	.5944	.5955	.5966	.5977	.5988	.5999	.6010
4.0	.6021	.6031	.6042	.6053	.6064	.6075	.6085	.6096	.6107	.6117
4.1	.6128	.6138	.6149	.6160	.6170	.6180	.6191	.6201	.6212	.6222
4.2	.6232	.6243	.6253	.6263	.6274	.6284	.6294	.6304	.6314	.6325
4.3	.6335	.6345	.6355	.6365	.6375	.6385	.6395	.6405	.6415	.6425
4.4	.6435	.6444	.6454	.6464	.6474	.6484	.6493	.6503	.6513	.6522
4.5	.6532	.6542	.6551	.6561	.6571	.6580	.6590	.6599	.6609	.6618
4.6	.6628	.6637	.6646	.6656	.6665	.6675	.6684	.6693	.6702	.6712
4.7	.6721	.6730	.6739	.6749	.6758	.6767	.6776	.6785	.6794	.6803
4.8	.6812	.6821	.6830	.6839	.6848	.6857	.6866	.6875	.6884	.6893
4.9	.6902	.6911	.6920	.6928	.6937	.6946	.6955	.6964	.6972	.6981
5.0	.6990	.6998	.7007	.7016	.7024	.7033	.7042	.7050	.7059	.7067
5.1	.7076	.7084	.7093	.7101	.7110	.7118	.7126	.7135	.7143	.7152
5.2	.7160	.7168	.7177	.7185	.7193	.7202	.7210	.7218	.7226	.7235
5.3	.7243	.7251	.7259	.7267	.7275	.7284	.7292	.7300	.7308	.7316
5.4	.7324	.7332	.7340	.7348	.7356	.7364	.7372	.7380	.7388	.7396

Table B (continued)

N	0	1	2	3	4	5	6	7	8	9
5.5	.7404	.7412	.7419	.7427	.7435	.7443	.7451	.7459	.7466	.7474
5.6	.7482	.7490	.7497	.7505	.7513	.7520	.7528	.7536	.7543	.7551
5.7	.7559	.7566	.7574	.7582	.7589	.7597	.7604	.7612	.7619	.7627
5.8	.7634	.7642	.7649	.7657	.7664	.7672	.7679	.7686	.7694	.7701
5.9	.7709	.7716	.7723	.7731	.7738	.7745	.7752	.7760	.7767	.7774
6.0	.7782	.7789	.7796	.7803	.7810	.7818	.7825	.7832	.7839	.7846
6.1	.7853	.7860	.7868	.7875	.7882	.7889	.7896	.7903	.7910	.7917
6.2	.7924	.7931	.7938	.7945	.7952	.7959	.7966	.7973	.7980	.7987
6.3	.7993	.8000	.8007	.8014	.8021	.8028	.8035	.8041	.8048	.8055
6.4	.8062	.8069	.8075	.8082	.8089	.8096	.8102	.8109	.8116	.8122
6.5	.8129	.8136	.8142	.8149	.8156	.8162	.8169	.8176	.8182	.8189
6.6	.8195	.8202	.8209	.8215	.8222	.8228	.8235	.8241	.8248	.8254
6.7	.8261	.8267	.8274	.8280	.8287	.8293	.8299	.8306	.8312	.8319
6.8	.8325	.8331	.8338	.8344	.8351	.8357	.8363	.8370	.8376	.8382
6.9	.8388	.8395	.8401	.8407	.8414	.8420	.8426	.8432	.8439	.8445
7.0	.8451	.8457	.8463	.8470	.8476	.8482	.8488	.8494	.8500	.8506
7.1	.8513	.8519	.8525	.8531	.8537	.8543	.8549	.8555	.8561	.8567
7.2	.8573	.8579	.8585	.8591	.8597	.8603	.8609	.8615	.8621	.8627
7.3	.8633	.8639	.8645	.8651	.8657	.8663	.8669	.8675	.8681	.8686
7.4	.8692	.8698	.8704	.8710	.8716	.8722	.8727	.8733	.8739	.8745
7.5	.8751	.8756	.8762	.8768	.8774	.8779	.8785	.8791	.8797	.8802
7.6	.8808	.8814	.8820	.8825	.8831	.8837	.8842	.8848	.8854	.8859
7.7	.8865	.8871	.8876	.8882	.8887	.8893	.8899	.8904	.8910	.8915
7.8	.8921	.8927	.8932	.8938	.8943	.8949	.8954	.8960	.8965	.8971
7.9	.8976	.8982	.8987	.8993	.8998	.9004	.9009	.9015	.9020	.9025
8.0	.9031	.9036	.9042	.9047	.9053	.9058	.9063	.9069	.9074	.9079
8.1	.9085	.9090	.9096	.9101	.9106	.9112	.9117	.9122	.9128	.9133
8.2	.9138	.9143	.9149	.9154	.9159	.9165	.9170	.9175	.9180	.9186
8.3	.9191	.9196	.9201	.9206	.9212	.9217	.9222	.9227	.9232	.9238
8.4	.9243	.9248	.9253	.9258	.9263	.9269	.9274	.9279	.9284	.9289
8.5	.9294	.9299	.9304	.9309	.9315	.9320	.9325	.9330	.9335	.9340
8.6	.9345	.9350	.9355	.9360	.9365	.9370	.9375	.9380	.9385	.9390
8.7	.9395	.9400	.9405	.9410	.9415	.9420	.9425	.9430	.9435	.9440
8.8	.9445	.9450	.9455	.9460	.9465	.9469	.9474	.9479	.9484	.9489
8.9	.9494	.9499	.9504	.9509	.9513	.9518	.9523	.9528	.9533	.9538
9.0	.9542	.9547	.9552	.9557	.9562	.9566	.9571	.9576	.9581	.9586
9.1	.9590	.9595	.9600	.9605	.9609	.9614	.9619	.9624	.9628	.9633
9.2	.9638	.9643	.9647	.9652	.9657	.9661	.9666	.9671	.9675	.9680
9.3	.9685	.9689	.9694	.9699	.9703	.9708	.9713	.9717	.9722	.9727
9.4	.9731	.9736	.9741	.9745	.9750	.9754	.9759	.9763	.9768	.9773
9.5	.9777	.9782	.9786	.9791	.9795	.9800	.9805	.9809	.9814	.9818
9.6	.9823	.9827	.9832	.9836	.9841	.9845	.9850	.9854	.9859	.9863
9.7	.9868	.9872	.9877	.9881	.9886	.9890	.9894	.9899	.9903	.9908
9.8	.9912	.9917	.9921	.9926	.9930	.9934	.9939	.9943	.9948	.9952
9.9	.9956	.9961	.9965	.9969	.9974	.9978	.9983	.9987	.9991	.9996

Table C Base-*e* Logarithms

N	0	1	2	3	4	5	6	7	8	9
1.0	.0000	.0100	.0198	.0296	.0392	.0488	.0583	.0677	.0770	.0862
1.1	.0953	.1044	.1133	.1222	.1310	.1398	.1484	.1570	.1655	.1740
1.2	.1823	.1906	.1989	.2070	.2151	.2231	.2311	.2390	.2469	.2546
1.3	.2624	.2700	.2776	.2852	.2927	.3001	.3075	.3148	.3221	.3293
1.4	.3365	.3436	.3507	.3577	.3646	.3716	.3784	.3853	.3920	.3988
1.5	.4055	.4121	.4187	.4253	.4318	.4383	.4447	.4511	.4574	.4637
1.6	.4700	.4762	.4824	.4886	.4947	.5008	.5068	.5128	.5188	.5247
1.7	.5306	.5365	.5423	.5481	.5539	.5596	.5653	.5710	.5766	.5822
1.8	.5878	.5933	.5988	.6043	.6098	.6152	.6206	.6259	.6313	.6366
1.9	.6419	.6471	.6523	.6575	.6627	.6678	.6729	.6780	.6831	.6881
2.0	.6931	.6981	.7031	.7080	.7129	.7178	.7227	.7275	.7324	.7372
2.1	.7419	.7467	.7514	.7561	.7608	.7655	.7701	.7747	.7793	.7839
2.2	.7885	.7930	.7975	.8020	.8065	.8109	.8154	.8198	.8242	.8286
2.3	.8329	.8372	.8416	.8459	.8502	.8544	.8587	.8629	.8671	.8713
2.4	.8755	.8796	.8838	.8879	.8920	.8961	.9002	.9042	.9083	.9123
2.5	.9163	.9203	.9243	.9282	.9322	.9361	.9400	.9439	.9478	.9517
2.6	.9555	.9594	.9632	.9670	.9708	.9746	.9783	.9821	.9858	.9895
2.7	.9933	.9969	1.0006	.0043	.0080	.0116	.0152	.0188	.0225	.0260
2.8	1.0296	.0332	.0367	.0403	.0438	.0473	.0508	.0543	.0578	.0613
2.9	.0647	.0682	.0716	.0750	.0784	.0818	.0852	.0886	.0919	.0953
3.0	1.0986	.1019	.1053	.1086	.1119	.1151	.1184	.1217	.1249	.1282
3.1	.1314	.1346	.1378	.1410	.1442	.1474	.1506	.1537	.1569	.1600
3.2	.1632	.1663	.1694	.1725	.1756	.1787	.1817	.1848	.1878	.1909
3.3	.1939	.1969	.2000	.2030	.2060	.2090	.2119	.2149	.2179	.2208
3.4	.2238	.2267	.2296	.2326	.2355	.2384	.2413	.2442	.2470	.2499
3.5	1.2528	.2556	.2585	.2613	.2641	.2669	.2698	.2726	.2754	.2782
3.6	.2809	.2837	.2865	.2892	.2920	.2947	.2975	.3002	.3029	.3056
3.7	.3083	.3110	.3137	.3164	.3191	.3218	.3244	.3271	.3297	.3324
3.8	.3350	.3376	.3403	.3429	.3455	.3481	.3507	.3533	.3558	.3584
3.9	.3610	.3635	.3661	.3686	.3712	.3737	.3762	.3788	.3813	.3838
4.0	1.3863	.3888	.3913	.3938	.3962	.3987	.4012	.4036	.4061	.4085
4.1	.4110	.4134	.4159	.4183	.4207	.4231	.4255	.4279	.4303	.4327
4.2	.4351	.4375	.4398	.4422	.4446	.4469	.4493	.4516	.4540	.4563
4.3	.4586	.4609	.4633	.4656	.4679	.4702	.4725	.4748	.4770	.4793
4.4	.4816	.4839	.4861	.4884	.4907	.4929	.4951	.4974	.4996	.5019
4.5	1.5041	.5063	.5085	.5107	.5129	.5151	.5173	.5195	.5217	.5239
4.6	.5261	.5282	.5304	.5326	.5347	.5369	.5390	.5412	.5433	.5454
4.7	.5476	.5497	.5518	.5539	.5560	.5581	.5602	.5623	.5644	.5665
4.8	.5686	.5707	.5728	.5748	.5769	.5790	.5810	.5831	.5851	.5872
4.9	.5892	.5913	.5933	.5953	.5974	.5994	.6014	.6034	.6054	.6074
5.0	1.6094	.6114	.6134	.6154	.6174	.6194	.6214	.6233	.6253	.6273
5.1	.6292	.6312	.6332	.6351	.6371	.6390	.6409	.6429	.6448	.6467
5.2	.6487	.6506	.6525	.6544	.6563	.6582	.6601	.6620	.6639	.6658
5.3	.6677	.6696	.6715	.6734	.6752	.6771	.6790	.6808	.6827	.6845
5.4	.6864	.6882	.6901	.6919	.6938	.6956	.6974	.6993	.7011	.7029

Use the properties of logarithms and $\ln 10 \approx 2.3026$ to find logarithms of numbers less than 1 or greater than 10.

Table C (*continued*)

N	0	1	2	3	4	5	6	7	8	9
5.5	1.7047	.7066	.7084	.7102	.7120	.7138	.7156	.7174	.7192	.7210
5.6	.7228	.7246	.7263	.7281	.7299	.7317	.7334	.7352	.7370	.7387
5.7	.7405	.7422	.7440	.7457	.7475	.7492	.7509	.7527	.7544	.7561
5.8	.7579	.7596	.7613	.7630	.7647	.7664	.7681	.7699	.7716	.7733
5.9	.7750	.7766	.7783	.7800	.7817	.7834	.7851	.7867	.7884	.7901
6.0	1.7918	.7934	.7951	.7967	.7984	.8001	.8017	.8034	.8050	.8066
6.1	.8083	.8099	.8116	.8132	.8148	.8165	.8181	.8197	.8213	.8229
6.2	.8245	.8262	.8278	.8294	.8310	.8326	.8342	.8358	.8374	.8390
6.3	.8405	.8421	.8437	.8453	.8469	.8485	.8500	.8516	.8532	.8547
6.4	.8563	.8579	.8594	.8610	.8625	.8641	.8656	.8672	.8687	.8703
6.5	1.8718	.8733	.8749	.8764	.8779	.8795	.8810	.8825	.8840	.8856
6.6	.8871	.8886	.8901	.8916	.8931	.8946	.8961	.8976	.8991	.9006
6.7	.9021	.9036	.9051	.9066	.9081	.9095	.9110	.9125	.9140	.9155
6.8	.9169	.9184	.9199	.9213	.9228	.9242	.9257	.9272	.9286	.9301
6.9	.9315	.9330	.9344	.9359	.9373	.9387	.9402	.9416	.9430	.9445
7.0	1.9459	.9473	.9488	.9502	.9516	.9530	.9544	.9559	.9573	.9587
7.1	.9601	.9615	.9629	.9643	.9657	.9671	.9685	.9699	.9713	.9727
7.2	.9741	.9755	.9769	.9782	.9796	.9810	.9824	.9838	.9851	.9865
7.3	.9879	.9892	.9906	.9920	.9933	.9947	.9961	.9974	.9988	2.0001
7.4	2.0015	.0028	.0042	.0055	.0069	.0082	.0096	.0109	.0122	.0136
7.5	2.0149	.0162	.0176	.0189	.0202	.0215	.0229	.0242	.0255	.0268
7.6	.0281	.0295	.0308	.0321	.0334	.0347	.0360	.0373	.0386	.0399
7.7	.0412	.0425	.0438	.0451	.0464	.0477	.0490	.0503	.0516	.0528
7.8	.0541	.0554	.0567	.0580	.0592	.0605	.0618	.0631	.0643	.0656
7.9	.0669	.0681	.0694	.0707	.0719	.0732	.0744	.0757	.0769	.0782
8.0	2.0794	.0807	.0819	.0832	.0844	.0857	.0869	.0882	.0894	.0906
8.1	.0919	.0931	.0943	.0956	.0968	.0980	.0992	.1005	.1017	.1029
8.2	.1041	.1054	.1066	.1078	.1090	.1102	.1114	.1126	.1138	.1150
8.3	.1163	.1175	.1187	.1199	.1211	.1223	.1235	.1247	.1258	.1270
8.4	.1282	.1294	.1306	.1318	.1330	.1342	.1353	.1365	.1377	.1389
8.5	2.1401	.1412	.1424	.1436	.1448	.1459	.1471	.1483	.1494	.1506
8.6	.1518	.1529	.1541	.1552	.1564	.1576	.1587	.1599	.1610	.1622
8.7	.1633	.1645	.1656	.1668	.1679	.1691	.1702	.1713	.1725	.1736
8.8	.1748	.1759	.1770	.1782	.1793	.1804	.1815	.1827	.1838	.1849
8.9	.1861	.1872	.1883	.1894	.1905	.1917	.1928	.1939	.1950	.1961
9.0	2.1972	.1983	.1994	.2006	.2017	.2028	.2039	.2050	.2061	.2072
9.1	.2083	.2094	.2105	.2116	.2127	.2138	.2148	.2159	.2170	.2181
9.2	.2192	.2203	.2214	.2225	.2235	.2246	.2257	.2268	.2279	.2289
9.3	.2300	.2311	.2322	.2332	.2343	.2354	.2364	.2375	.2386	.2396
9.4	.2407	.2418	.2428	.2439	.2450	.2460	.2471	.2481	.2492	.2502
9.5	2.2513	.2523	.2534	.2544	.2555	.2565	.2576	.2586	.2597	.2607
9.6	.2618	.2628	.2638	.2649	.2659	.2670	.2680	.2690	.2701	.2711
9.7	.2721	.2732	.2742	.2752	.2762	.2773	.2783	.2793	.2803	.2814
9.8	.2824	.2834	.2844	.2854	.2865	.2875	.2885	.2895	.2905	.2915
9.9	.2925	.2935	.2946	.2956	.2966	.2976	.2986	.2996	.3006	.3016

Exercises 1.1 (page 12)

1. 2, 3, 5, 7 **3.** 6 **5.** $\frac{3}{4}$ **7.** $\frac{4}{5}$ **9.** $\frac{3}{20}$ **11.** $\frac{14}{9}$ **13.** 1 **15.** $\frac{22}{15}$ **17.** variable **19.** natural **21.** integers **23.** even **25.** natural, 1, itself **27.** 0 **29.** rational **31.** $<$ **33.** $\approx$ **35.** inequality, compound inequality **37.** union, intersection **39.** 0.875, terminating **41.** $-0.7\overline{3}$, repeating **43.** 1, 2, 9 **45.** $-3, 0, 1, 2, 9$ **47.** $\sqrt{3}$ **49.** 2 **51.** 2 **53.** 9 **55.** $<$ **57.** $>$ **59.** $<$ **61.** $>$ **63.** $12 < 19$ **65.** $-5 \geq -6$ **67.** $-3 \leq 5$ **69.** $0 > -10$

71. 0 1 2 3 4 5 6 7 8

73. 10 11 12 13 14 15 16 17 18

75. 3 **77.** -5

79. -3 3 **81.** -6 5

83. 6 10 **85.** -1 3

87. 20 **89.** -6 **91.** 7 **93.** 20 **95.** 3 or -3 **97.** $x \geq 0$ **99.** 7

101. 0 5 **103.** 6

105. 20 18 16 14 12 10 8 6 4 2 0 −2 −4 −6 −8 −10 **111.** 99

Getting Ready (page 14)

1. 9 **2.** 9 **3.** 12 **4.** 12 **5.** 5 **6.** 5 **7.** 3 **8.** 6

Exercises 1.2 (page 25)

1. -2 **3.** -28 **5.** -3

7. 4 **9.** 2 10

11. \$77.53 **13.** absolute, common **15.** change, add **17.** negative **19.** product, quotient **21.** 0 **23.** $C = \pi D$; $C = 2\pi r$ **25.** $a + b = b + a$ **27.** reciprocals **29.** $5, \frac{1}{7}$ **31.** -8 **33.** -5 **35.** $\frac{1}{6}$ **37.** $\frac{7}{36}$ **39.** -7 **41.** 0 **43.** $\frac{11}{10}$ **45.** $-\frac{1}{6}$ **47.** -12 **49.** 21 **51.** $-\frac{6}{7}$ **53.** $-\frac{27}{28}$ **55.** -2 **57.** 4 **59.** -2 **61.** $\frac{24}{25}$ **63.** 23 **65.** -13 **67.** 0 **69.** 2 **71.** 1 **73.** 4 **75.** 8 **77.** 9 **79.** 12 **81.** -8 **83.** -9 **85.** $-\frac{5}{4}$ **87.** $-\frac{1}{8}$ **89.** 19,900 **91.** 100.4 ft **93.** comm. prop. of add. **95.** distrib. prop. **97.** additive identity prop. **99.** mult. inverse prop. **101.** assoc. prop. of add. **103.** comm. prop. of mult. **105.** assoc. prop. of add. **107.** distrib. prop. **109.** 70 **111.** 24 **113.** 168 **115.** 4 **117.** -1 **119.** -4 **121.** -12 **123.** -20 **125.** 11 **127.** 1 **129.** \$62 **131.** $+4°$ **133.** $12°$ **135.** 6,900 gal **137.** +1,325 m **139.** \$421.88 **141.** \$1,211 **143.** 80 **145.** yes **147.** 30 cm

Getting Ready (page 28)

1. 4 **2.** 27 **3.** -64 **4.** 81 **5.** $\frac{1}{27}$ **6.** $-\frac{16}{625}$

Exercises 1.3 (page 37)

1. 16 **3.** x^5 **5.** 1 **7.** a^6b^3 **9.** $\frac{1}{25}$ **11.** x^3 **13.** 7 **15.** 1 **17.** base, exponent, power **19.** x^{m+n} **21.** x^ny^n **23.** 1 **25.** x^{m-n} **27.** $A = s^2$ **29.** $A = \frac{1}{2}bh$ **31.** $A = \pi r^2$ **33.** $V = lwh$ **35.** $V = Bh$ **37.** $V = \frac{1}{3}Bh$ **39.** base is 5, exponent is 3 **41.** base is x, exponent is 5 **43.** base is b, exponent is 6 **45.** base is $-mn^2$, exponent is 3 **47.** 9 **49.** -9 **51.** 9 **53.** $-32x^5$ **55.** x^5 **57.** k^7 **59.** x^{10} **61.** p^{10} **63.** x^5y^4 **65.** x^{28} **67.** a^{20} **69.** y^8 **71.** $64x^6$ **73.** $x^{12}y^8$ **75.** $\frac{a^{15}}{b^{10}}$ **77.** $\frac{1}{16}a^8b^{20}$ **79.** 1 **81.** 1 **83.** 5 **85.** 50 **87.** $\frac{1}{25}$ **89.** $-\frac{1}{25}$ **91.** $\frac{1}{25}$ **93.** a^4 **95.** a^5 **97.** c^7 **99.** $3m$ **101.** m **103.** $\frac{a^6}{b^4}$ **105.** $\frac{16}{y^{16}}$ **107.** a^{n-1} **109.** b^{3n-9} **111.** 108 **113.** $-\frac{1}{216}$ **115.** $\frac{1}{324}$ **117.** $\frac{27}{8}$ **119.** 15 m^2 **121.** 113 cm^2 **123.** 45 cm^2 **125.** 300 cm^2 **127.** 343 m^3 **129.** 360 ft^3 **131.** 168 ft^3 **133.** 2,714 m^3 **135.** a^4b^5 **137.** $-243p^{10}q^{15}$ **139.** x^{12} **141.** $\frac{s^3}{r^9}$ **143.** $\frac{1}{b^{72}}$ **145.** $-\frac{1}{d^3}$ **147.** 1 **149.** $\frac{1}{9x^3}$ **151.** $\frac{1}{a^{n+1}}$ **153.** a^{2-n} **155.** $\frac{64b^{12}}{27a^9}$ **157.** $\frac{-b^3}{8a^{21}}$ **159.** 3.462825992 **161.** -244.140625 **171.** \$922,824.13 **177.** $\frac{7}{12}$

Getting Ready (page 40)

1. 10 **2.** 100 **3.** 1,000 **4.** 10,000 **5.** $\frac{1}{100}$ **6.** $\frac{1}{10,000}$ **7.** 4,000 **8.** $\frac{7}{10,000}$

Exercises 1.4 (page 46)

1. 3.52×10^2 **3.** 2×10^{-3} **5.** 350 **7.** 0.27 **9.** $0.\overline{75}$ **11.** $1.\overline{4}$ **13.** 89 **15.** scientific, 10^n **17.** left **19.** 3.9×10^3 **21.** 1.76×10^7 **23.** 7.8×10^{-3} **25.** 9.6×10^{-6} **27.** -4.5×10^4 **29.** -2.1×10^{-4} **31.** 270 **33.** 796,000 **35.** 0.00323 **37.** 0.00037 **39.** 23,650,000 **41.** 32,300,000 **43.** 2×10^{11} **45.** 1.44×10^7 **47.** 4×10^{-2} **49.** 6×10^3 **51.** 6.0×10^{-4} **53.** 5.27×10^3 **55.** 3.17×10^{-4} **57.** 5.23×10^1 **59.** 1.2874×10^{13} **61.** 5.671×10^{10} **63.** 0.64 **65.** 3.6×10^{25} **67.** g, x, v, i, r **69.** 1.19×10^8 cm/hr **71.** 1.67248×10^{-18} g **73.** 1.51×10^{10} in. **75.** 5.9×10^4 mi **77.** 2.54×10^8 **79.** almost 23 years **81.** 2.5×10^{13} mi **85.** 332

Getting Ready (page 49)

1. 2 **2.** 4 **3.** 3 **4.** 6

Exercises 1.5 (page 58)

1. $9x$ **3.** no **5.** no **7.** 3 **9.** 5 **11.** −64 **13.** 1 **15.** x^8 **17.** $\frac{1}{8x^3}$ **19.** equation **21.** equivalent **23.** c, c **25.** term **27.** numerical coefficients, variables **29.** yes **31.** yes **33.** 2 **35.** 25 **37.** 3 **39.** 28 **41.** $\frac{2}{3}$ **43.** 6 **45.** −8 **47.** $\frac{8}{3}$ **49.** yes, $8x$ **51.** no **53.** yes, $-2x^2$ **55.** no **57.** $15x - 27$ **59.** $43x + 44$ **61.** $-4x + 9$ **63.** $21x - 19$ **65.** −11 **67.** 13 **69.** −8 **71.** $-\frac{7}{3}$ **73.** 6 **75.** 4 **77.** 3 **79.** 0 **81.** −2 **83.** 24 **85.** 6 **87.** −6 **89.** 3 **91.** 2,500 **93.** identity, ℝ **95.** contradiction, ∅ **97.** $w = \frac{A}{l}$ **99.** $B = \frac{3V}{h}$ **101.** $t = \frac{I}{pr}$ **103.** $w = \frac{p - 2l}{2}$ **105.** $B = \frac{2A}{h} - b$ **107.** $x = \frac{y - b}{m}$ **109.** $l = \frac{a - S + Sr}{r}$ **111.** $l = \frac{2S - na}{n}$ **113.** 3 **115.** 9 **117.** −4 **119.** −6 **121.** 13 **123.** 1.7 **125.** ∅, contradiction **127.** $-\frac{5}{2}$ **129.** $n = \frac{l - a + d}{d}$ **131.** $m = \frac{Fd^2}{(GM)}$ **133.** $C = \frac{5}{9}(F - 32)$; 0°, 21.1°, 100° **135.** $n = \frac{(100 \times (C - 6.50))}{7}$; 621, 1,000, 1,692.9 kwh **137.** $R = \frac{E}{I}$; $R = 8$ ohms **139.** $n = \frac{360°}{(180° - a)}$; 8

Getting Ready (page 61)

1. 20 **2.** 72 **3.** 41 **4.** 7

Exercises 1.6 (page 67)

1. 100 **3.** \$270 **5.** 24 m^2 **7.** $\frac{256x^{20}}{81}$ **9.** a^{m+1} **11.** $5x + 4$ **13.** $40x$ **15.** acute **17.** 180° **19.** isosceles **21.** equilateral **23.** 32 ft **25.** 7 ft, 15 ft **27.** 20% **29.** $233\frac{1}{3}\%$ **31.** 300 shares of BB, 200 shares of SS **33.** 35 \$15 calculators, 50 \$67 calculators **35.** 60° **37.** 10 **39.** 72.5° **41.** 56° **43.** 12 m by 24 m **45.** 156 ft by 312 ft **47.** 5 ft **49.** 90 lb **51.** 13 roses **53.** 60 mi, 120 mi **55.** \$355 **57.** 12 in. **59.** 50 **61.** 10 ft **63.** 4 ft **65.** −40°

Getting Ready (page 72)

1. \$40 **2.** $\$0.08x$ **3.** \$6,000 **4.** $\$(15,000 - x)$ **5.** 200 mi **6.** \$20

Exercises 1.7 (page 76)

1. \$90 **3.** $\$(30,000 - x)$ **5.** 0.08 gal **7.** 1 **9.** 8 **11.** principal, rate, time **13.** value, price, number **15.** \$2,000 at 4%, \$10,000 at 6% **17.** 10.8% **19.** \$8,250 **21.** 25 **23.** $3\frac{1}{2}$ hr **25.** $\frac{2}{3}$ hr **27.** $\frac{1}{8}$ hr **29.** 12 mi **31.** 20 lb of \$2.95 candy; 10 lb of \$3.10 candy **33.** 15 **35.** 2 gal **37.** 30 **39.** \$50 **41.** $1\frac{1}{2}$ hr **43.** 10 oz **45.** 6 in.

Chapter Review (page 80)

1. 0, 1, 2, 4 **2.** 1, 2, 4 **3.** $-4, -\frac{2}{3}, 0, 1, 2, 4$ **4.** −4, 0, 1, 2, 4 **5.** π **6.** $-4, -\frac{2}{3}, 0, 1, 2, \pi, 4$ **7.** $-4, -\frac{2}{3}$ **8.** $1, 2, \pi, 4$ **9.** 2 **10.** 4 **11.** −4, 0, 2, 4 **12.** 1

13. 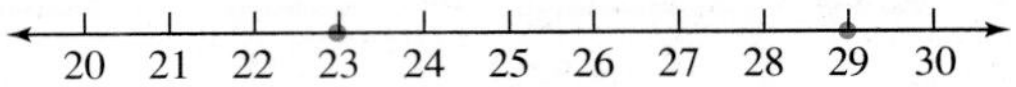

14. 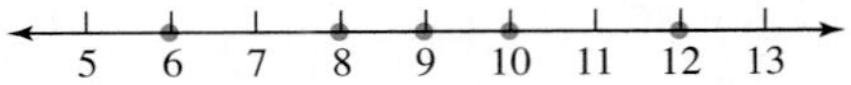

15. 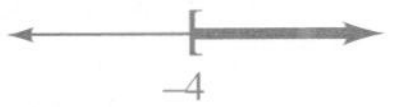

16. 

17. 

18. 

19. 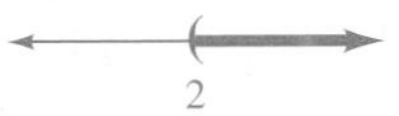

20. 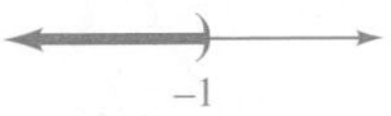

21. 

22. 0 **23.** 1 **24.** 8 **25.** −8 **26.** 8 **27.** −9 **28.** −28 **29.** 57 **30.** 3 **31.** −9 **32.** 5 **33.** −2 **34.** −13 **35.** 5 **36.** −13 **37.** −5 **38.** 39 **39.** 53 **40.** 35 **41.** 42 **42.** 4 **43.** 5 **44.** −12 **45.** −24 **46.** −4 **47.** −2 **48.** 12 **49.** 27 **50.** 0 **51.** −60 **52.** −1 **53.** 4 **54.** 15.4 **55.** 15 **56.** 15 **57.** yes **58.** $-\frac{19}{6}$ **59.** $\frac{2}{15}$ **60.** $-\frac{4}{3}$ **61.** $\frac{11}{5}$ **62.** distrib. prop. **63.** comm. prop. of add. **64.** assoc. prop. of add. **65.** add. identity prop. **66.** add. inverse prop. **67.** comm. prop. of mult. **68.** assoc. prop. of mult. **69.** mult. identity prop. **70.** mult. inverse prop. **71.** double negative rule **72.** 729 **73.** −64 **74.** −64 **75.** $-\frac{1}{625}$ **76.** $-6x^6$ **77.** $-3x^8$ **78.** $\frac{1}{x}$ **79.** x^2 **80.** $27x^6$ **81.** $256x^{16}$ **82.** $-32x^{10}$ **83.** $243x^{15}$ **84.** $\frac{1}{x^{10}}$ **85.** x^{20} **86.** $\frac{x^6}{9}$ **87.** $\frac{16}{x^{16}}$ **88.** x^2 **89.** x^5 **90.** $\frac{1}{a^5}$ **91.** $\frac{1}{a^3}$ **92.** $\frac{1}{y^7}$ **93.** y^9 **94.** $\frac{1}{x}$ **95.** x^3 **96.** $9x^4y^6$ **97.** $\frac{1}{81a^{12}b^8}$ **98.** $\frac{64y^9}{27x^6}$ **99.** $\frac{64y^3}{125}$ **100.** 1.93×10^{10} **101.** 2.73×10^{-8} **102.** 72,000,000 **103.** 0.0000000083 **104.** 5.02×10^{11} **105.** 5 **106.** −9 **107.** 8 **108.** 7

109. 19 **110.** 8 **111.** 12 **112.** 5 **113.** 2
114. 18,000 **115.** $r^3 = \frac{3V}{4\pi}$ **116.** $h = \frac{3V}{\pi r^2}$
117. $x = \frac{6v}{ab} - y$ or $x = \frac{6v - aby}{ab}$ **118.** $r = \frac{V}{\pi h^2} + \frac{h}{3}$
119. 5 ft from one end **120.** 45 m^2 **121.** $2\frac{2}{3}$ ft
122. \$18,000 at 10%, \$7,000 at 9% **123.** 10 liters
124. $\frac{1}{3}$ hr

Chapter 1 Test (page 89)

1. 1, 2, 5 **2.** $\sqrt{7}$

3. **4.** **5.** **6.**

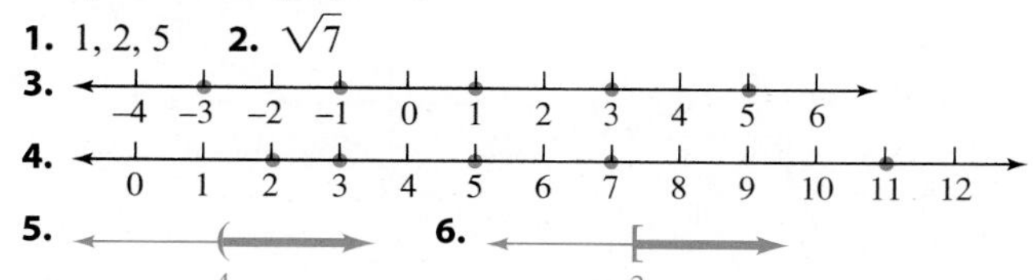

7. **8.**

9. −8 **10.** 5 **11.** 2 **12.** 20 **13.** −4 **14.** 1
15. 0.3 **16.** 0.5 **17.** −6 **18.** −10 **19.** 6 **20.** 1
21. comm. prop. of add. **22.** distrib. prop. **23.** x^8
24. $8x^6y^9$ **25.** $\frac{1}{m^8}$ **26.** $\frac{m^4}{n^{10}}$ **27.** 4.7×10^6
28. 2.3×10^{-7} **29.** 653,000 **30.** 0.0245 **31.** −12
32. −2 **33.** 6 **34.** 400 **35.** $i = \frac{f(P - L)}{s}$ **36.** 36 cm^2
37. \$4,000 **38.** 80 liters

Getting Ready (page 91)

1. **2.** **3.**

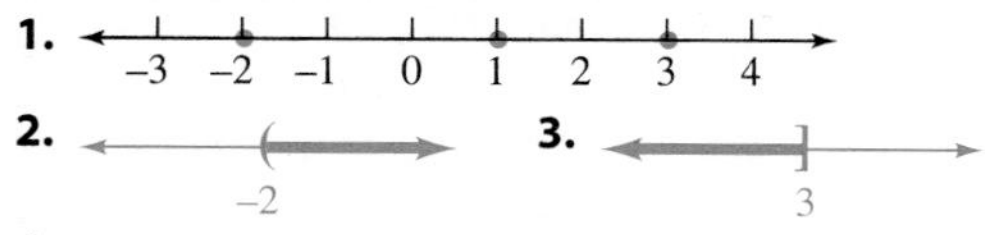

4.

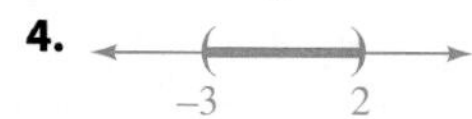

Exercises 2.1 (page 102)

1. (3, 0), (0, 3) **3.** (8, 0), (0, 2) **5.** vertical **7.** (4, 6)
9. 12 **11.** $\frac{1}{2}$ **13.** 7 **15.** ordered pair **17.** origin
19. rectangular or Cartesian coordinate **21.** linear equation, general form **23.** x-axis **25.** vertical **27.** sub 1

29-36.

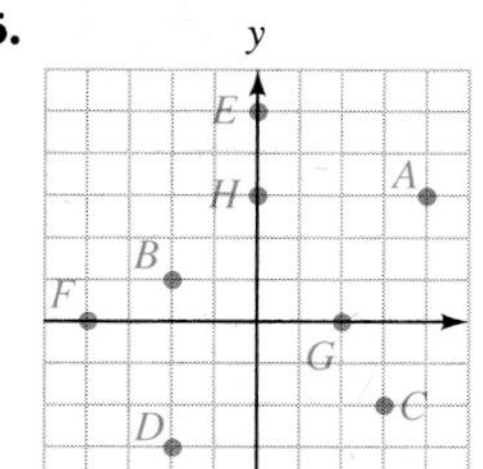
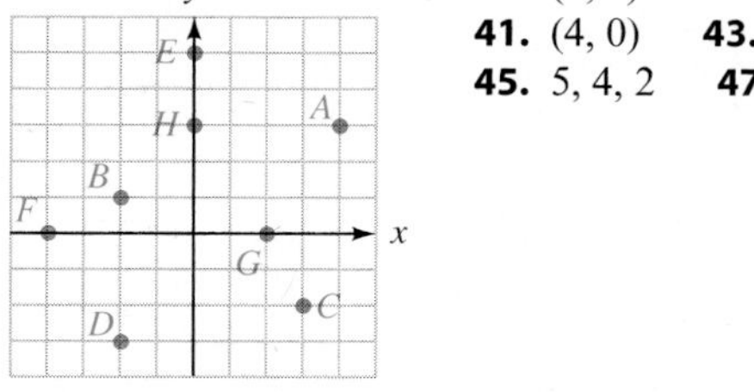

37. (2, 4) **39.** (−2, −1)
41. (4, 0) **43.** (0, 0)
45. 5, 4, 2 **47.** −5, −3, 3

49.

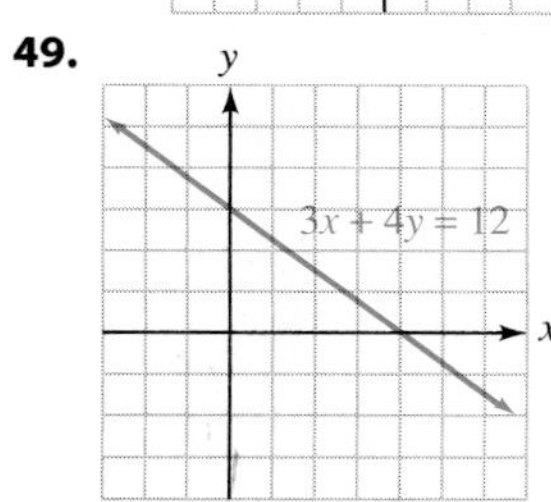

51.

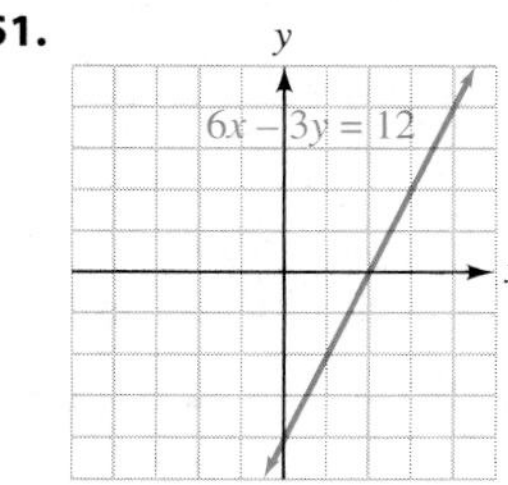

53.

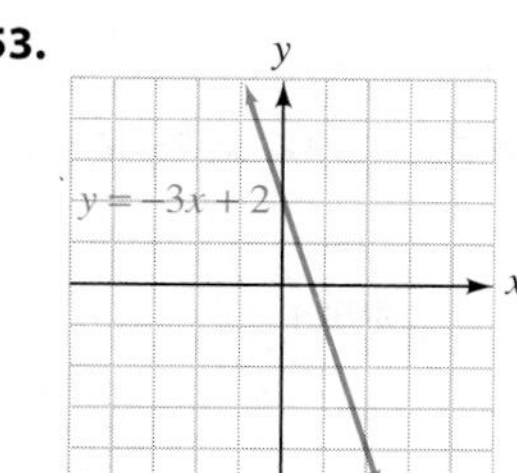

55.

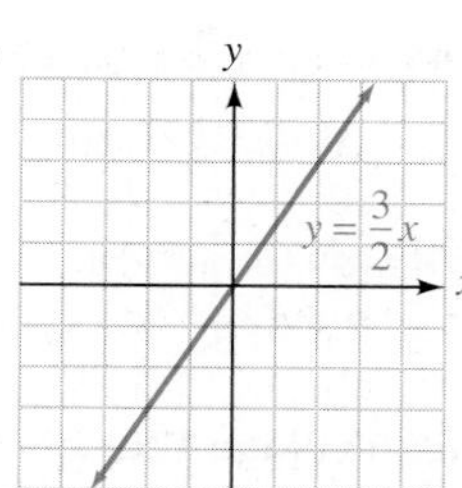

57.

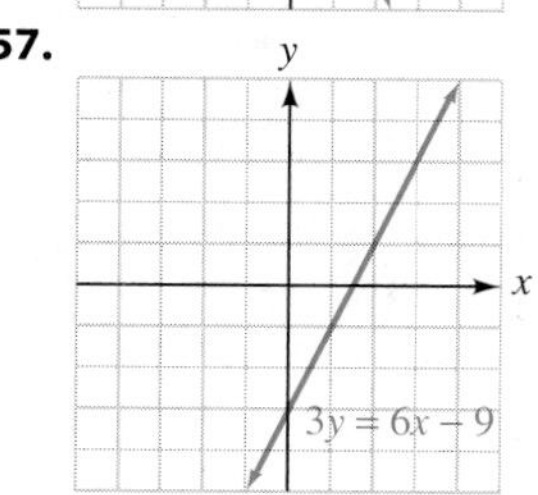

59.

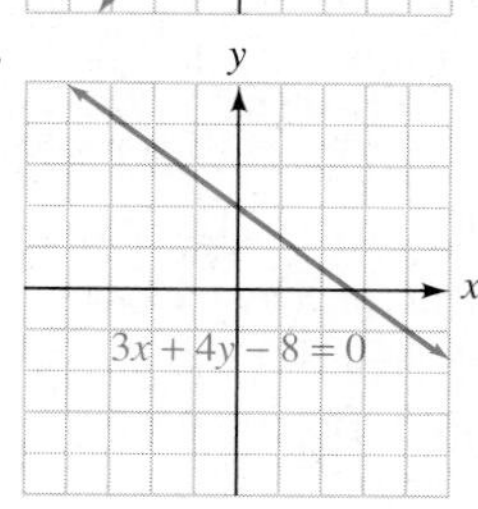

61.

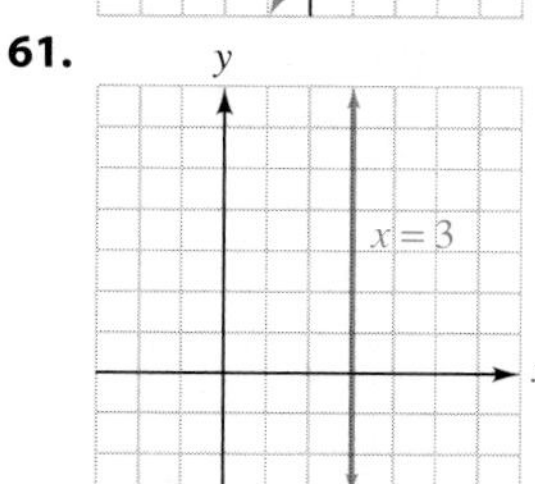

63.

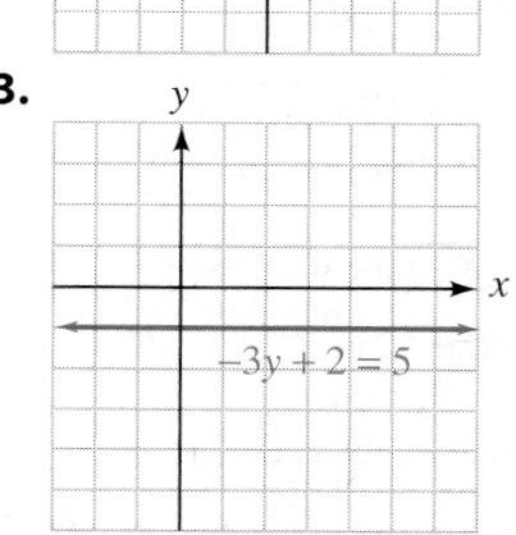

65. \$48 **67.** \$3,000 **69.** (3, 4) **71.** (9, 12) **73.** $\left(\frac{7}{2}, 6\right)$
75. $\left(\frac{1}{2}, -2\right)$ **77.** (−4, 0) **79.** (4, 1) **81.** −1
83. −3.933 **85.** 1.22 **87.** 4.67 **89.** \$162,500 **91.** 200
93. 12 mi **95. a.** $y = 0.25x + 5$ **b.** 6, 7, 8, 9 **c.** \$10

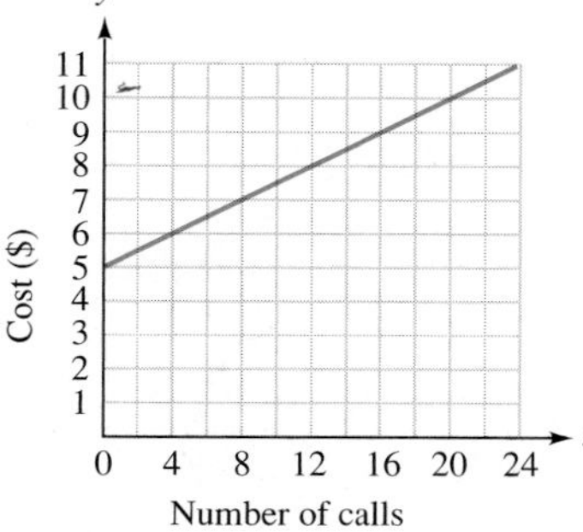

97. a. In 1990, there were 65.5 million swimmers.
b. 45.7 million
101. $a = 0, b > 0$

Getting Ready (page 107)

1. 1 **2.** −1 **3.** $\frac{13}{14}$ **4.** $\frac{3}{14}$

Exercises 2.2 (page 115)

1. 3 **3.** yes **5.** yes **7.** x^6 **9.** $x^{24}y^{12}$ **11.** $x^{16}y^2$
13. change **15.** rise **17.** horizontal **19.** positive
21. perpendicular, reciprocals **23.** 2 **25.** 3 **27.** −1
29. $-\frac{1}{3}$ **31.** 0 **33.** not defined **35.** $-\frac{3}{2}$ **37.** $\frac{3}{4}$
39. $\frac{1}{2}$ **41.** 0 **43.** negative **45.** positive
47. not defined **49.** 0 **51.** perpendicular **53.** neither
55. parallel **57.** perpendicular **59.** perpendicular
61. neither **63.** parallel **65.** same line **67.** same line
69. not the same line **71.** $x = 0$, no defined slope **79.** $\frac{1}{4}$

81. $\frac{1}{25}, \frac{1}{10}, \frac{4}{25}$ **83.** $\frac{7}{500}$ of a degree increase per year
85. a. $\frac{2}{25}$ **b.** $\frac{1}{20}, \frac{1}{20}$ **87.** \$525.13 per yr **93.** -4

Getting Ready (page 118)

1. 14 **2.** $-\frac{5}{3}$ **3.** $y = 3x - 4$ **4.** $x = \frac{-By - 3}{A}$

Exercises 2.3 (page 128)

1. $y - 3 = 2(x - 2)$ **3.** $y = -3x + 5$ **5.** parallel **7.** 6
9. -1 **11.** 20 oz **13.** $y - y_1 = m(x - x_1)$
15. $Ax + By = C$ **17.** perpendicular **19.** $y - 7 = 5x$
21. $y = -3(x - 2)$ **23.** $y = x$ **25.** $y = \frac{7}{3}x - 3$
27. $y = 3x + 17$ **29.** $y = -7x + 54$ **31.** $y = -4$
33. $y = -\frac{1}{2}x + 11$ **35.** $\frac{3}{2}, (0, -4)$ **37.** $-\frac{1}{3}, \left(0, -\frac{5}{6}\right)$
39. $1, (0, -1)$ **41.** $\frac{2}{3}, (0, 2)$

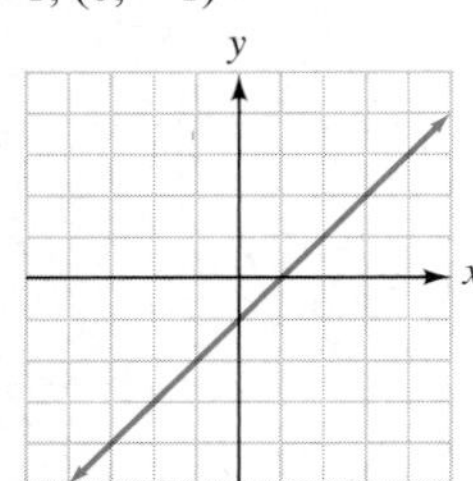

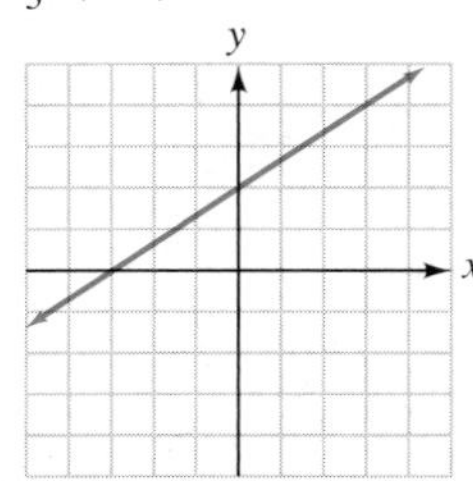

43. parallel **45.** perpendicular **47.** parallel
49. perpendicular **51.** perpendicular **53.** perpendicular
55. $y = 4x$ **57.** $y = 4x - 3$ **59.** $y = -\frac{1}{4}x$
61. $y = -\frac{1}{4}x + \frac{11}{2}$ **63.** $y = \frac{4}{5}x - \frac{26}{5}$ **65.** $y = -\frac{5}{4}x + 3$
67. perpendicular **69.** parallel
71. $\frac{7}{2}, (0, 2)$ **73.** $-\frac{2}{3}, (0, 6)$

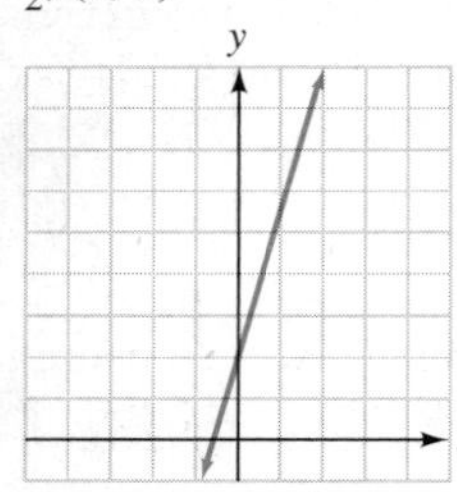

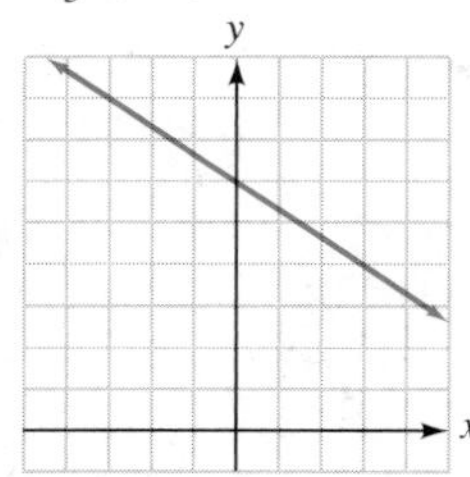

75. $y = \frac{2}{3}x + \frac{11}{3}$ **77.** $y = -\frac{7}{4}x + \frac{1}{2}$ **79.** $x = -2$
81. $x = 5$ **83.** $y = -\frac{A}{B}x + \frac{C}{B}$ **85.** $y = -2{,}298x + 19{,}984$
87. $y = 1{,}811{,}250x + 36{,}225{,}000$
89. $y = 50{,}000x + 250{,}000$ **91.** $y = -\frac{950}{3}x + 1{,}750$
93. \$490 **95.** \$154,000 **97.** \$180 **107.** $a < 0, b > 0$

Getting Ready (page 132)

1. 1 **2.** 7 **3.** -20 **4.** $-\frac{11}{4}$

Exercises 2.4 (page 141)

1. yes **3.** no **5.** 3 **7.** -2 **9.** 2 **11.** relation
13. domain **15.** 0 **17.** cannot **19.** slope; y-intercept
21. y **23.** y **25.** D: $\{3, 5, -4, 0\}$; R: $\{-2, 0, -5\}$; yes
27. D: $\{-2, 6, 5\}$; R: $\{3, 8, 5, 4\}$; no
29. D: $(-\infty, 1]$; R: $(-\infty, \infty)$; not a function
31. D: $(-\infty, \infty)$; R: $(-\infty, \infty)$; a function **33.** $9, -3, 0$
35. $3, -5, \frac{3}{2}$ **37.** 4, 9 **39.** 7, 26 **41.** 4, 4 **43.** 2, 2
45. $2w, 2w + 2$ **47.** $3w - 5, 3w - 2$ **49.** 12
51. $2b - 2a$ **53.** yes **55.** yes **57.** yes **59.** no
61. D: $\{-2, 4, 6\}$ **63.** D: $(-\infty, 4) \cup (4, \infty)$
65. $(-\infty, -3) \cup (-3, \infty)$ **67.** $(-\infty, \infty)$
69. D: $(-\infty, \infty)$; R: $(-\infty, \infty)$ **71.** D: $(-\infty, \infty)$; R: $(-\infty, \infty)$

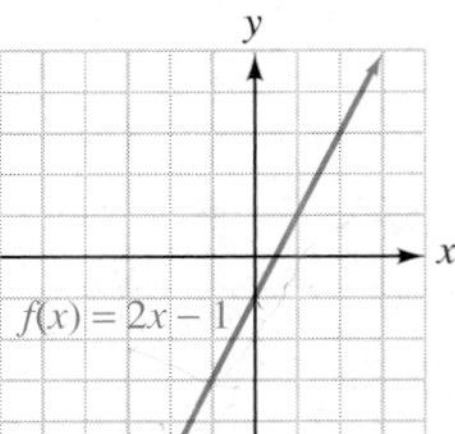

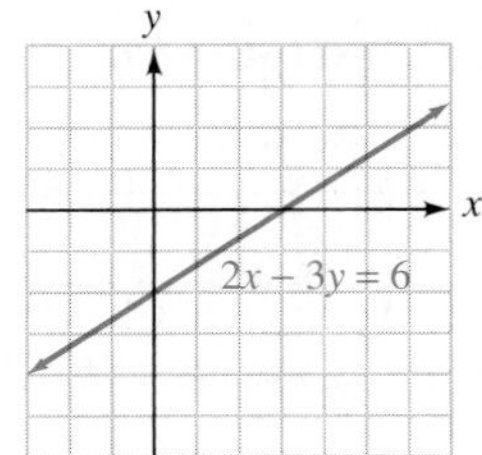

73. no **74.** yes **77.** 9, 16 **79.** 6, 15 **81.** 22, 2
83. 3, 11 **85.** $2b$ **87.** 1 **89.** 192
91. a. $I(h) = 0.10h + 50$ **b.** \$61.50 **93.** 624 ft
95. 12 ft **97.** 77°F **101.** yes

Getting Ready (page 144)

1. $2, (0, -3)$ **2.** $-3, (0, 4)$ **3.** $6, -9$ **4.** $4, \frac{5}{2}$

Exercises 2.5 (page 155)

9. 41, 43, 47 **11.** $a \cdot b = b \cdot a$ **13.** 1 **15.** squaring, parabola **17.** absolute value **19.** horizontal
21. 2, down **23.** 4, to the left **25.** reflected
27.

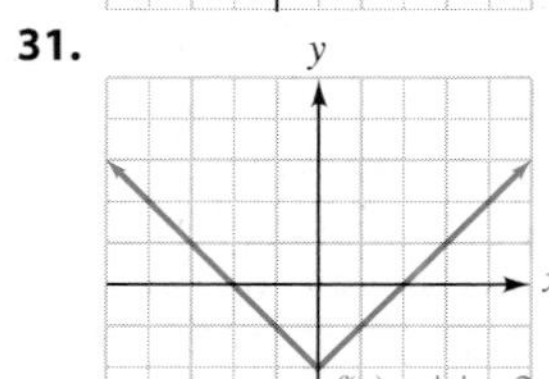

29.

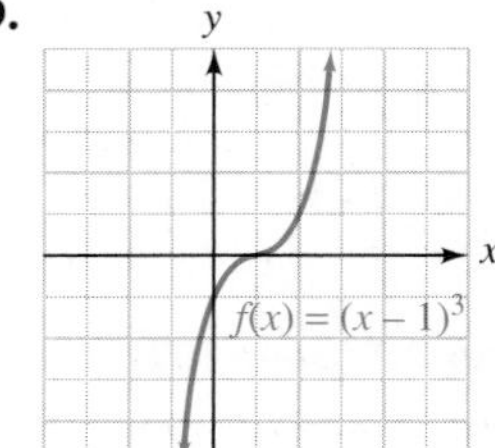

31.

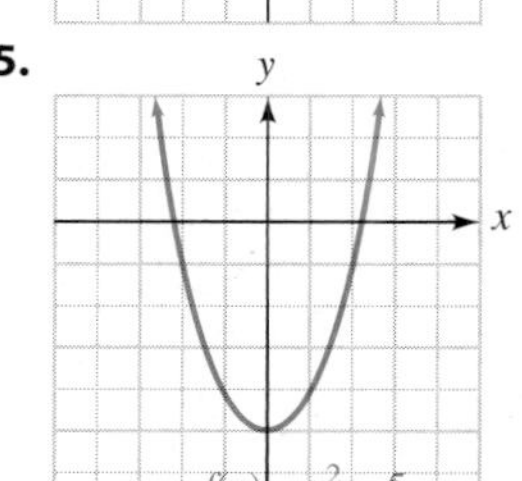

33.

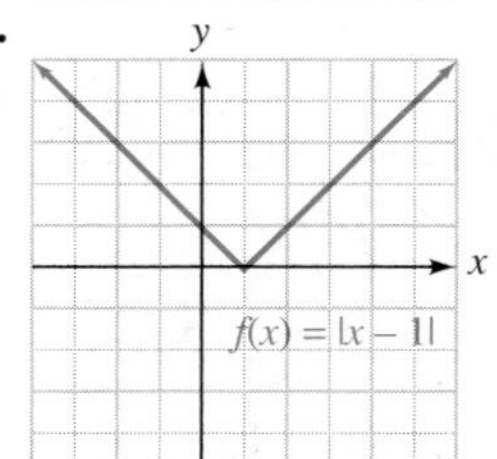

35.

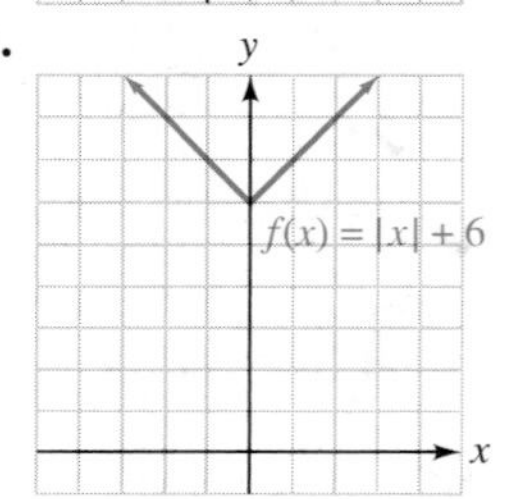

37.

39.

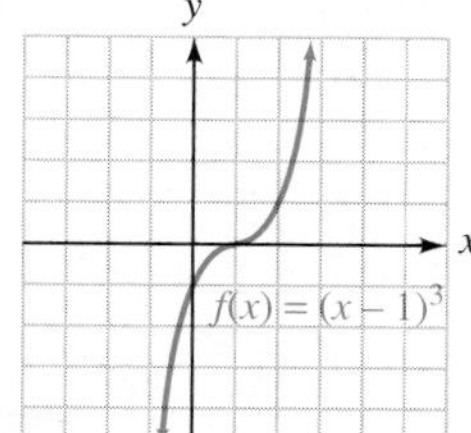

41.

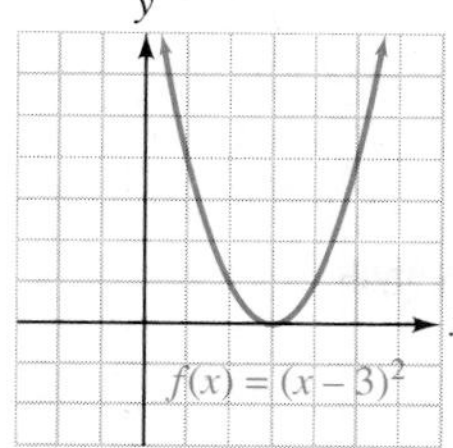

43.

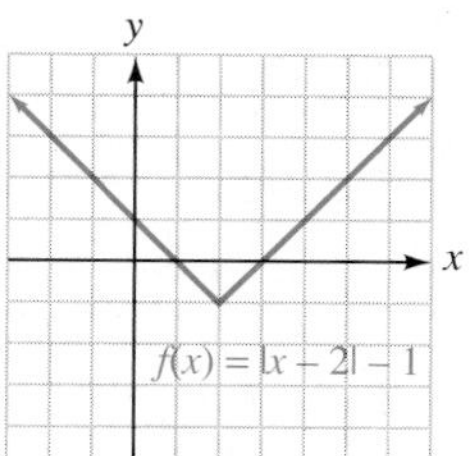

45.

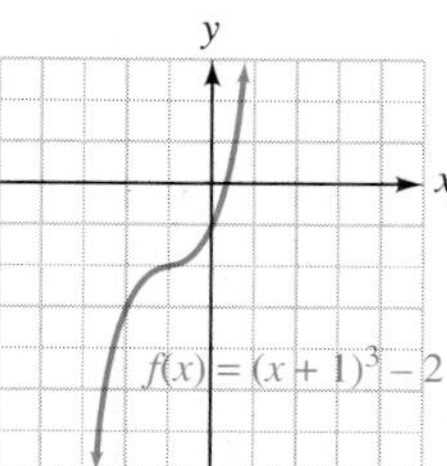

47.

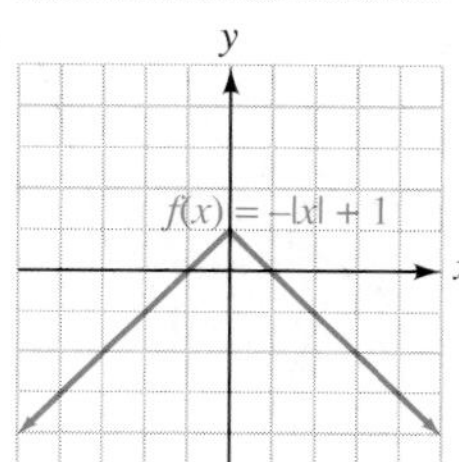

49.

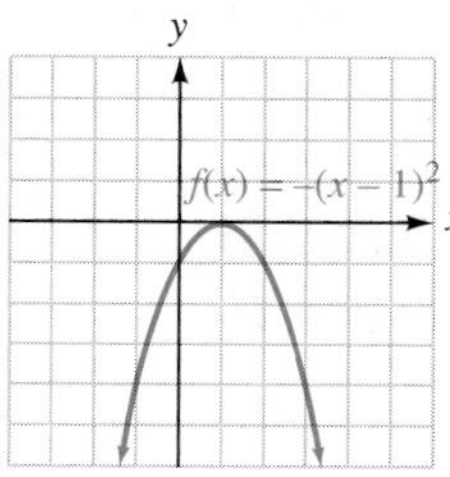

51.

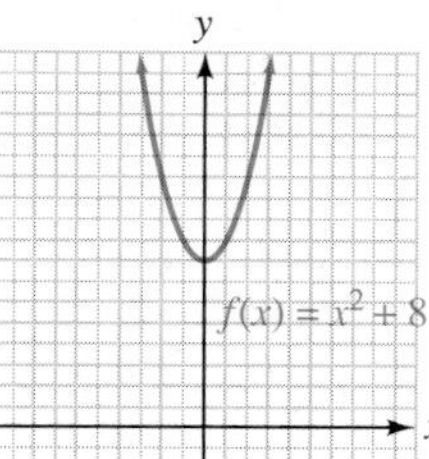

53.

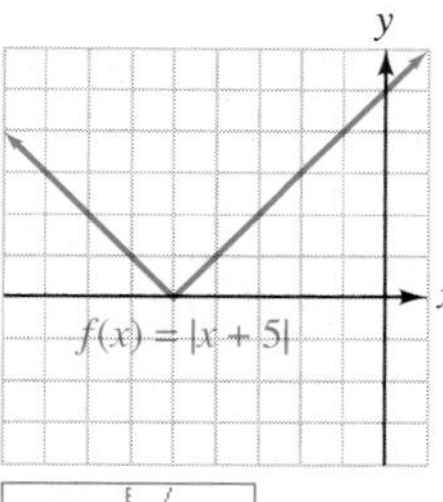

55.

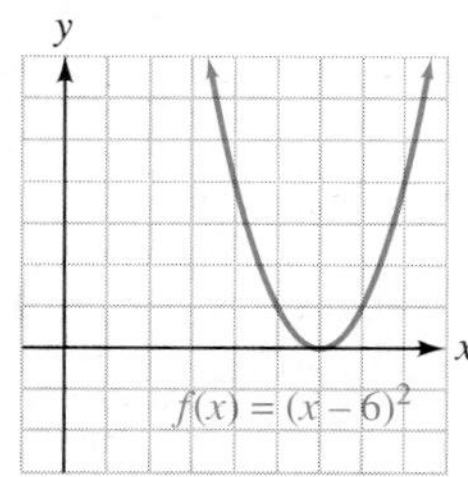

57.

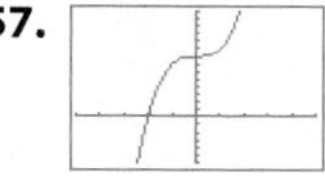

59. -2 **61.** 4 **63.** -3

Chapter Review (page 157)

1. $-10, -6, -6, 0, -2, 6, 9$

2.

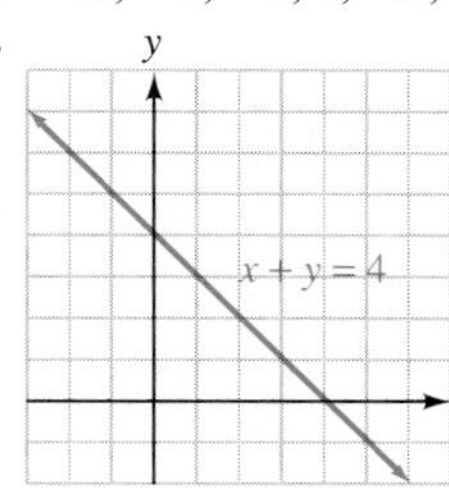

3.

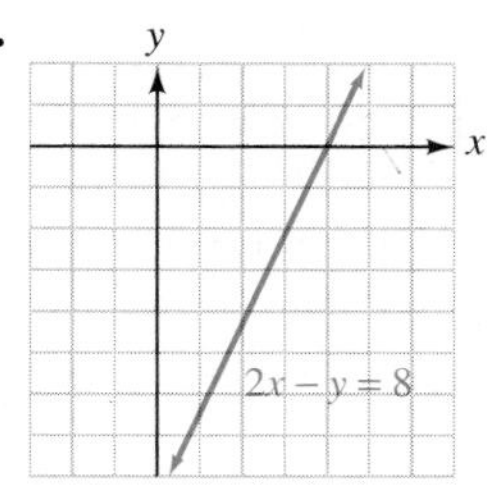

4.

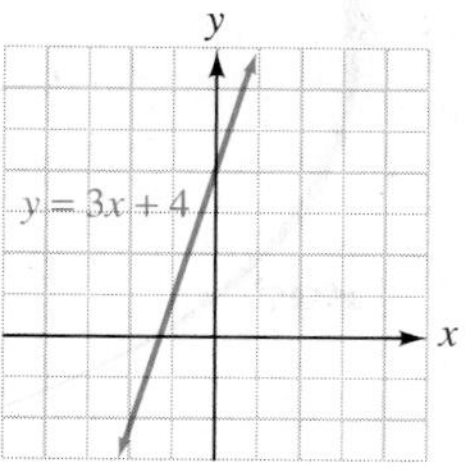

5.

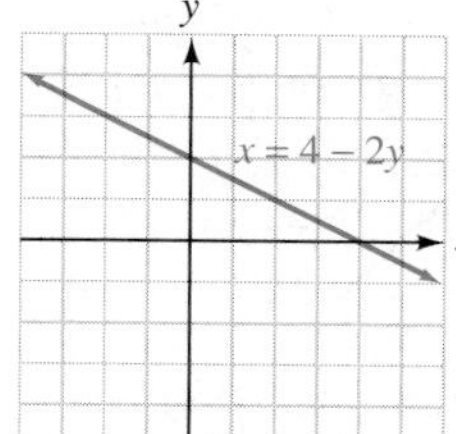

6.

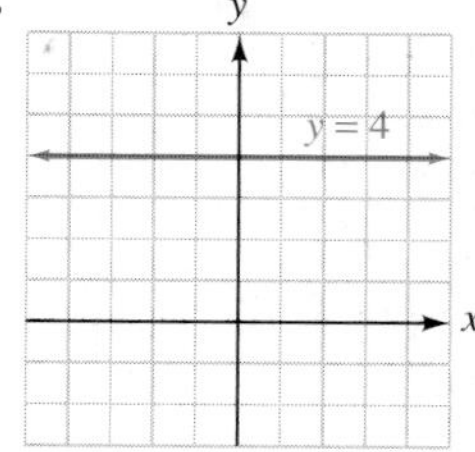

7.

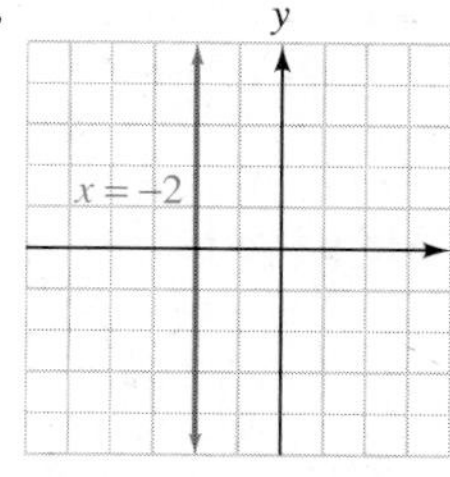

8.

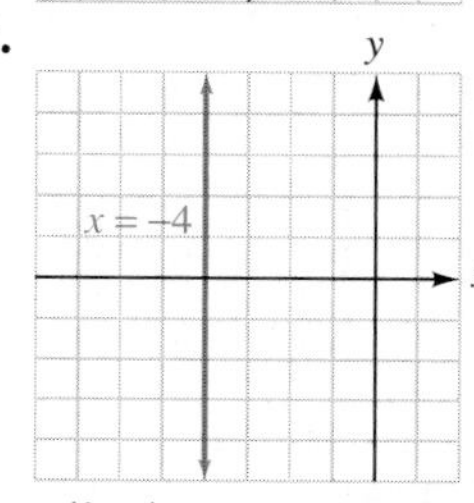

9.

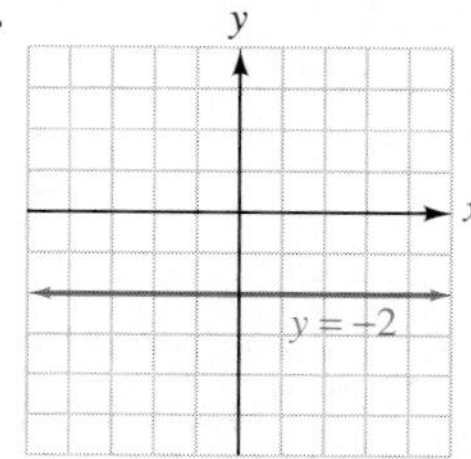

10. $\left(\frac{3}{2}, 8\right)$ **11.** 1 **12.** $\frac{14}{9}$ **13.** 5 **14.** $\frac{5}{11}$ **15.** 0 **16.** no defined slope **17.** $\frac{2}{3}$ **18.** -2 **19.** no defined slope **20.** 0 **21.** perpendicular **22.** parallel **23.** neither **24.** perpendicular **25.** \$21,666.67 **26.** $3x - y = -29$ **27.** $13x + 8y = 6$ **28.** $3x - 2y = 1$ **29.** $2x + 3y = -21$ **30.** $y = -1{,}720x + 8{,}700$ **31.** yes **32.** yes **33.** no **34.** no **35.** -7 **36.** 60 **37.** 0 **38.** 17 **39.** D: $(-\infty, \infty)$ **40.** D: $(-\infty, \infty)$ **41.** D: $(-\infty, \infty)$ **42.** D: $(-\infty, 2) \cup (2, \infty)$ **43.** D: $(-\infty, 3) \cup (3, \infty)$ **44.** D: $(-\infty, \infty)$ **45.** a function; D: $(-\infty, \infty)$, R: $(-\infty, \infty)$ **46.** not a function **47.** not a function **48.** a function; D: $(-\infty, \infty)$, R: $[-3, \infty)$ **49.** yes **50.** yes **51.** yes **52.** no

53.

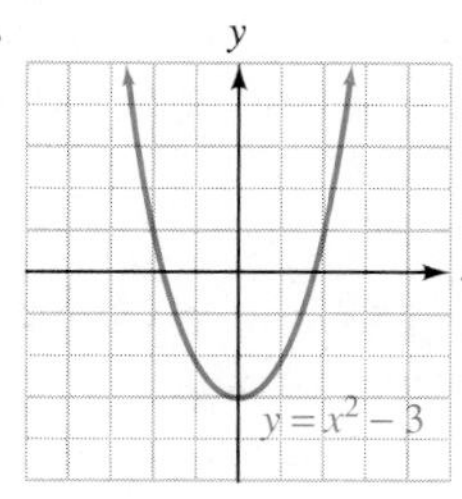

54.

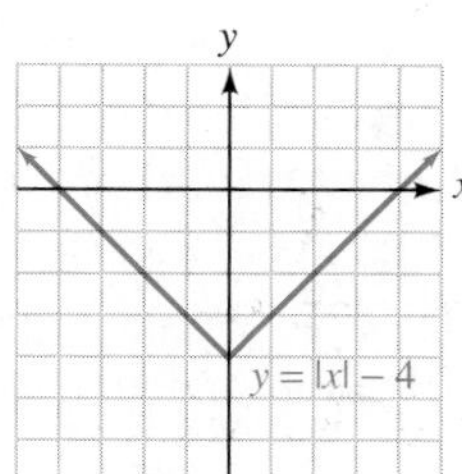

55.

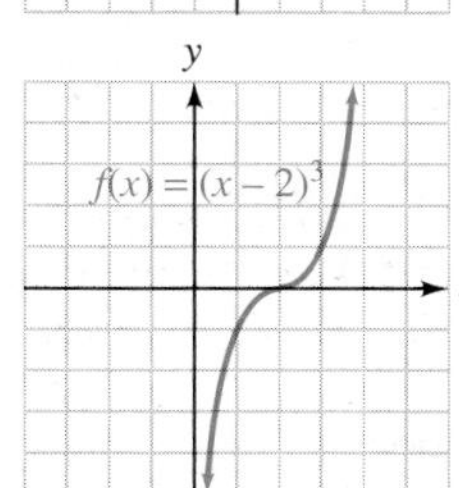

56.

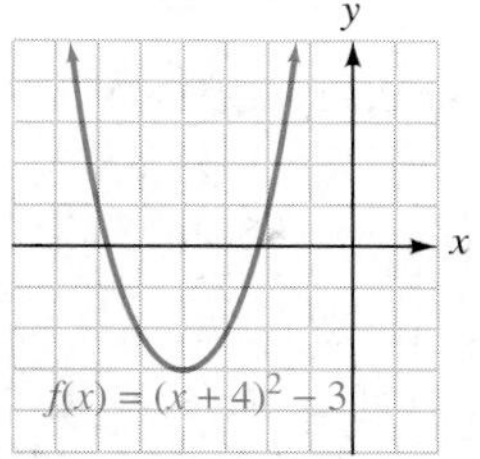

57. 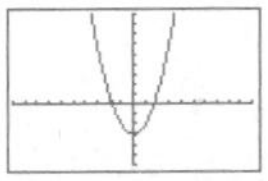
58. 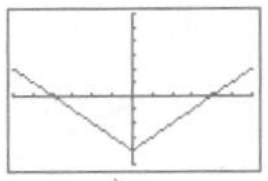
59. 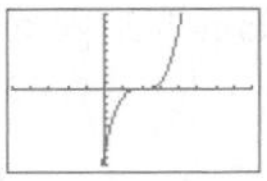
60. 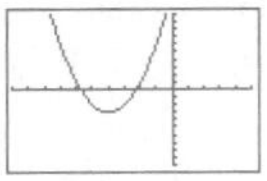
61. 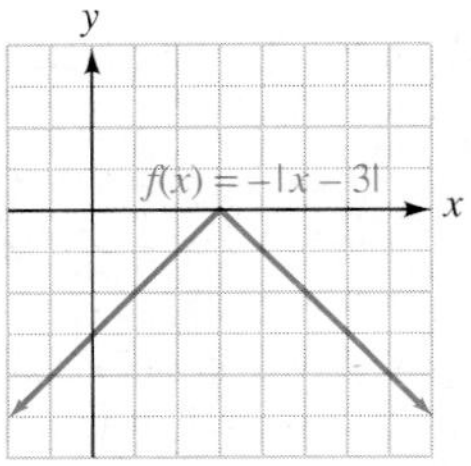

Chapter 2 Test (page 163)

1. 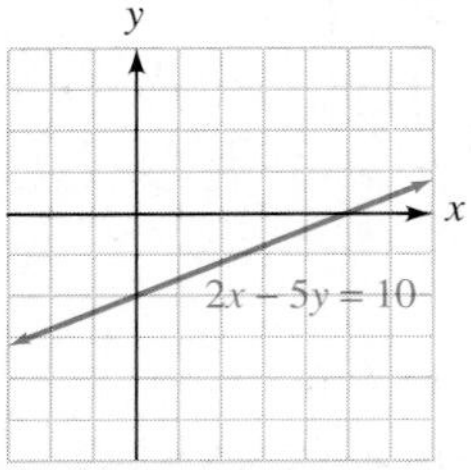

2. x-intercept: (3, 0), y-intercept: $\left(0, -\frac{3}{5}\right)$
3. $\left(\frac{1}{2}, \frac{1}{2}\right)$ **4.** $\frac{1}{2}$ **5.** $\frac{2}{3}$
6. no defined slope **7.** 0
8. $y = \frac{2}{3}x - \frac{23}{3}$
9. $8x - y = -22$
10. $m = -\frac{1}{3}, \left(0, -\frac{3}{2}\right)$
11. neither **12.** perpendicular **13.** $y = \frac{3}{2}x$
14. $y = \frac{3}{2}x + \frac{21}{2}$ **15.** no **16.** D: $(-\infty, \infty)$; R: $[0, \infty)$
17. D: $(-\infty, \infty)$; R: $(-\infty, \infty)$ **18.** 10 **19.** -2
20. $3a + 1$ **21.** $x^2 - 2$ **22.** yes **23.** no
24. 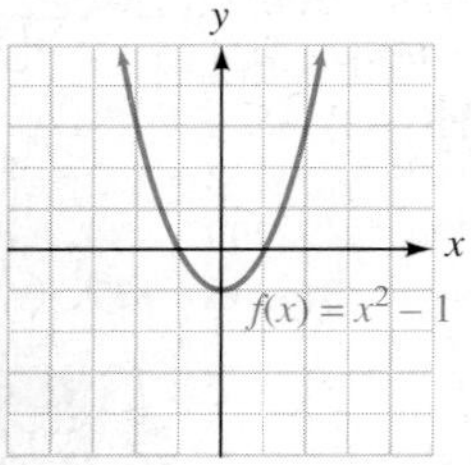

25. 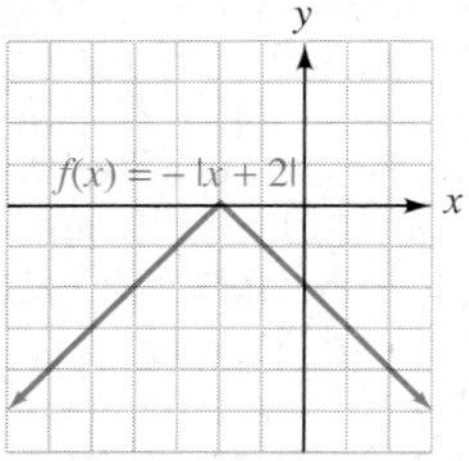

Cumulative Review Exercises (page 164)

1. 1, 2, 6, 7 **2.** 0, 1, 2, 6, 7 **3.** $-2, 0, 1, 2, \frac{13}{12}, 6, 7$
4. $\sqrt{5}, \pi$ **5.** -2 **6.** $-2, 0, 1, 2, \frac{13}{12}, 6, 7, \sqrt{5}, \pi$ **7.** 2, 7
8. 6 **9.** $-2, 0, 2, 6$ **10.** 1, 7 **11.** 

12. 

13. -2 **14.** -2
15. 22 **16.** -2 **17.** -4 **18.** -3 **19.** 4 **20.** -5
21. assoc. prop. of add. **22.** distrib. prop.
23. comm. prop. of add. **24.** assoc. prop. of mult.
25. x^8y^{12} **26.** c^2 **27.** $-\frac{b^3}{a^2}$ **28.** 1 **29.** 4.97×10^{-6}
30. 932,000,000 **31.** 8 **32.** -27 **33.** -1 **34.** 6
35. $a = \frac{2S}{n} - l$ **36.** $h = \frac{2A}{b_1 + b_2}$ **37.** 28, 30, 32
38. 14 cm by 42 cm **39.** yes **40.** $-\frac{7}{5}$ **41.** $y = -\frac{7}{5}x + \frac{11}{5}$
42. $y = -3x - 3$ **43.** 5 **44.** -1 **45.** $2t - 1$
46. $3r^2 + 2$

47. yes; D: $(-\infty, \infty)$; R: $(-\infty, 1]$
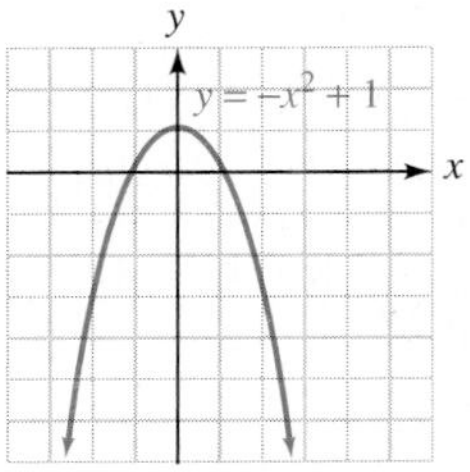

48. yes; D: $(-\infty, \infty)$; R: $[0, \infty)$
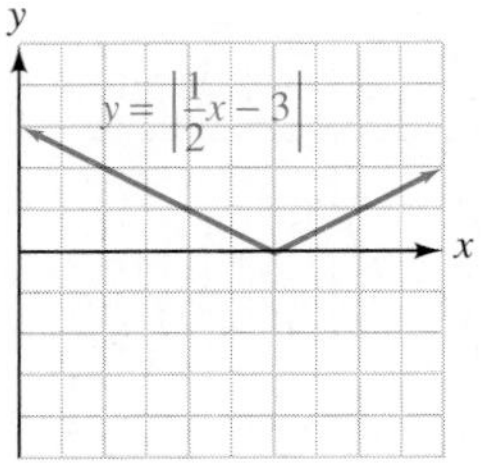

Getting Ready (page 166)

1. 2 **2.** -7 **3.** 11 **4.** 3

Exercises 3.1 (page 172)

1. no solution **3.** one solution **5.** 9.3×10^7
7. 3.45×10^4 **9.** system **11.** inconsistent
13. dependent **15.** yes **17.** no
19. 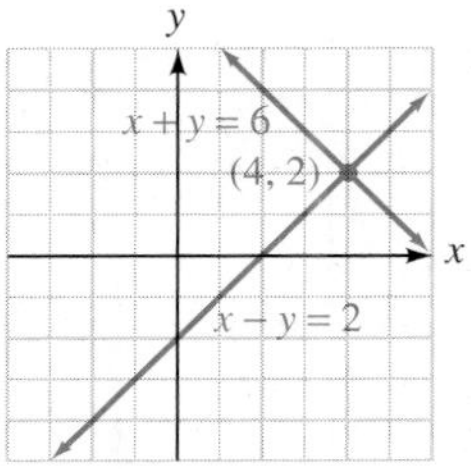

21. 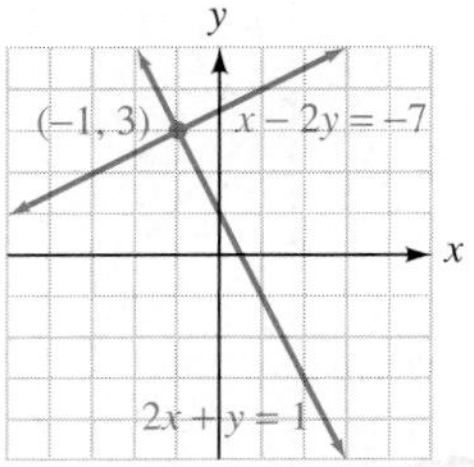

23. 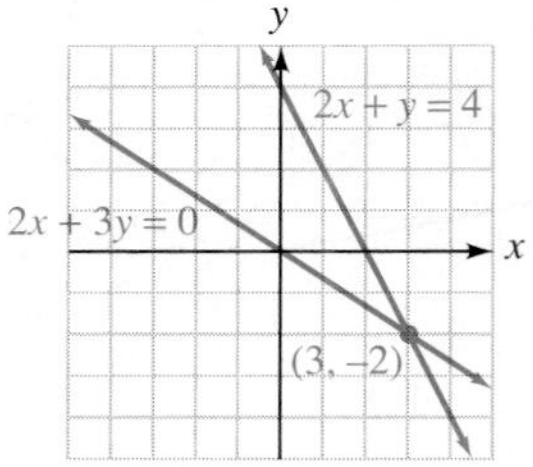

25. 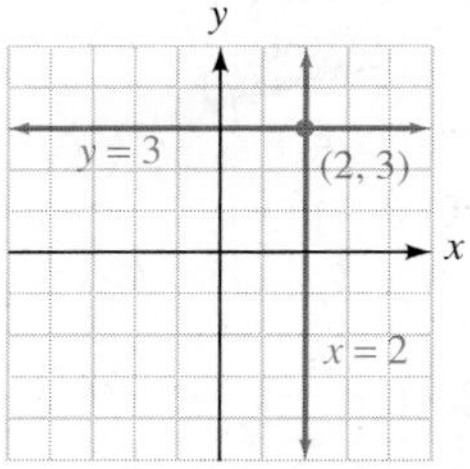

27. 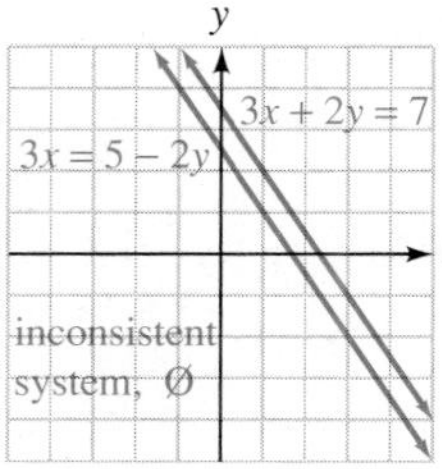

29. 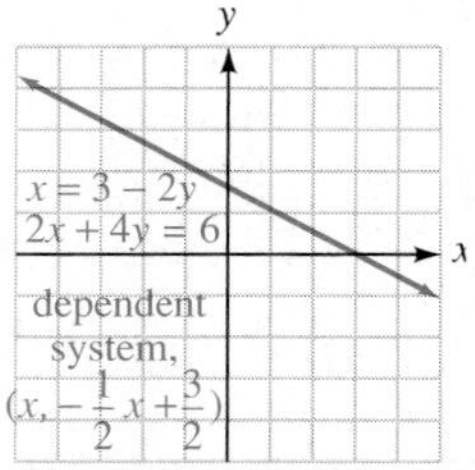

31. 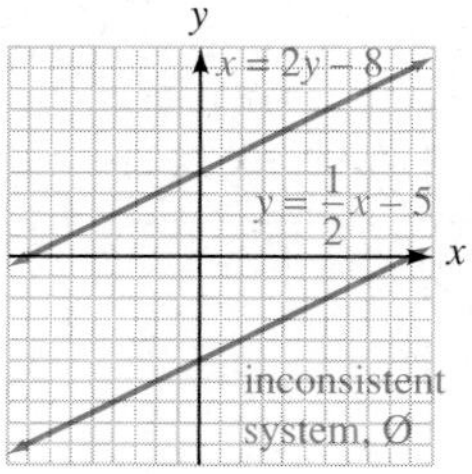

33. 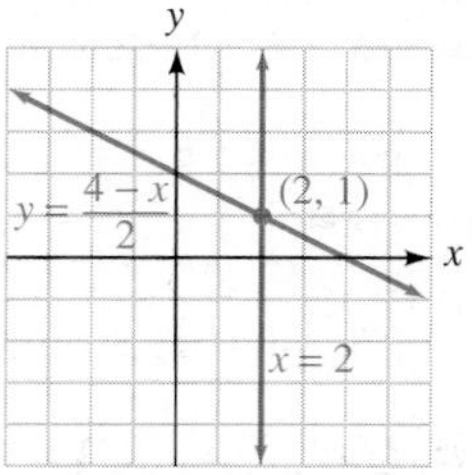

35.

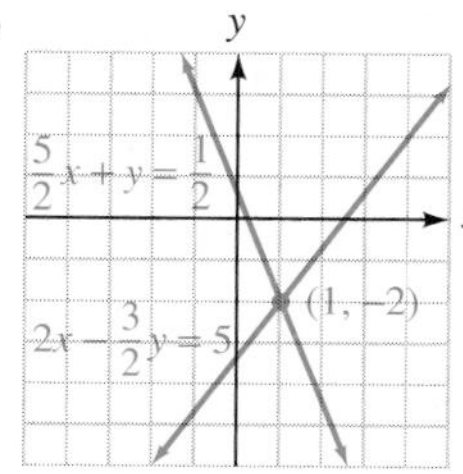

37. −18 **39.** 13

41.

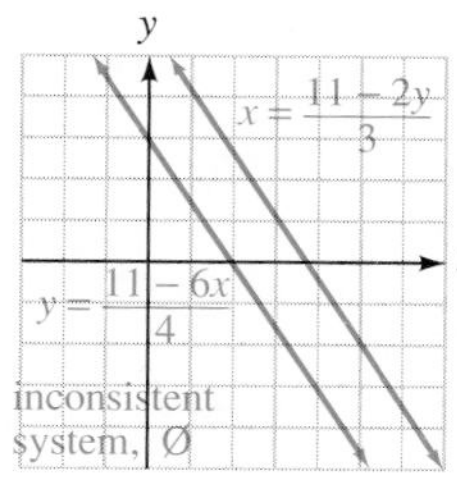

43.

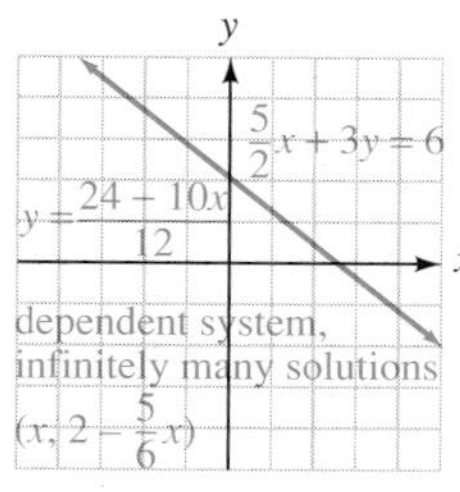

45.

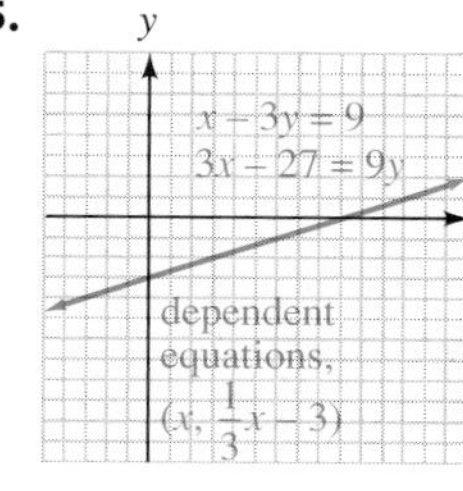

47. $\left(3, \frac{5}{2}\right)$ **49.** $\left(-1, \frac{2}{3}\right)$ **51.** (−0.37, −2.69) **53.** (−7.64, 7.04) **55. a.** \$2 million **b.** \$3 million **c.** 10,000 cameras **57. a.** yes **b.** (3.75, −0.5) **c.** no **59.** 3 hr **61.** (5, 1,400)

65. One possible answer is $\begin{cases} x + y = -3 \\ x - y = -7 \end{cases}$.

Getting Ready (page 176)

1. $6x - 21$ **2.** $-12x - 20$ **3.** $3x - 9$ **4.** $-2x + 2$ **5.** $7x = 15$ **6.** $-9b = 27$

Exercises 3.2 (page 186)

1. 2 **3.** 4 **5.** a^{22} **7.** $\frac{1}{81x^{32}y^4}$ **9.** setup, unit **11.** parallelogram **13.** opposite **15.** (2, 2) **17.** (5, 3) **19.** (−2, 4) **21.** (4, 8) **23.** (5, 2) **25.** (−4, −2) **27.** (3, −2) **29.** $\left(\frac{3}{4}, \frac{1}{3}\right)$ **31.** (1, 2) **33.** $\left(\frac{2}{3}, \frac{3}{2}\right)$ **35.** ∅ **37.** $(x, 2x - 4)$ **39.** $(x, 2x + 4)$ **41.** ∅ **43.** $\frac{1}{3}$ **45.** $-\frac{691}{1,980}$ **47.** $\left(5, \frac{3}{2}\right)$ **49.** $\left(-2, \frac{3}{2}\right)$ **51.** (−2, 9) **53.** (20, −12) **55.** (2, 3) **57.** $\left(-\frac{1}{3}, 1\right)$ **59.** \$57 **61.** 625Ω, 750Ω **63.** 16 m by 20 m **65.** \$3,000 at 10%, \$5,000 at 12% **67.** 40 oz of 8% solution, 60 oz of 15% solution **69.** 55 mph **71.** 85 racing bikes, 120 mountain bikes **73.** 200 plates **75.** 21 **77.** 750 **79.** 6,500 gal per month **81.** A (smaller loss) **83.** 590 units per month **85.** A (smaller loss) **87.** A **89.** 35°, 145° **91.** $x = 22.5, y = 67.5$ **93.** 72° **95.** $f^2 = \frac{1}{4\pi^2 LC}$

Getting Ready (page 191)

1. yes **2.** yes **3.** no **4.** yes

Exercises 3.3 (page 197)

1. yes **3.** $\frac{9}{5}$ **5.** 1 **7.** $2s^2 + 1$ **9.** plane **11.** infinitely **13.** yes **15.** (1, 1, 2) **17.** (0, 2, 2) **19.** $\left(\frac{3}{4}, \frac{1}{2}, \frac{1}{3}\right)$ **21.** ∅ **23.** infinitely many solutions of the form $(19 - 9z, 5 - 3z, z)$ **25.** (3, 2, 1) **27.** (2, 6, 9) **29.** ∅ **31.** infinitely many solutions of the form $\left(\frac{1}{2}, \frac{5}{3} - \frac{4}{3}z, z\right)$ **33.** −2, 4, 16 **35.** $A = 40°, B = 60°, C = 80°$ **37.** 1, 2, 3 **39.** 30 expensive, 50 middle-priced, 100 inexpensive **41.** 250 \$5 tickets, 375 \$3 tickets, 125 \$2 tickets **43.** 3 poles, 2 bears, 4 deer **45.** 78%, 21%, 1% **47.** $y = x^2 - 4x$ **49.** $x^2 + y^2 - 2x - 2y - 2 = 0$ **53.** (1, 1, 0, 1)

Getting Ready (page 200)

1. 5 8 13 **2.** 0 3 7 **3.** −1 −1 −2 **4.** 3 3 −5

Exercises 3.4 (page 207)

1. $\begin{bmatrix} 3 & 2 \\ 4 & -3 \end{bmatrix}$ **3.** yes **5.** 9.3×10^7 **7.** 6.3×10^4 **9.** matrix **11.** 3, columns **13.** augmented, coefficient **15.** type 1 **17.** nonzero **19.** 0 **21.** 8 **23.** (1, 1) **25.** (2, −3) **27.** (1, 2, 3) **29.** (−1, −1, 2) **31.** ∅ **33.** (1, 2) **35.** (2, 0) **37.** ∅ **39.** $(-6 - z, 2 - z, z)$ **41.** $(2 - z, 1 - z, z)$ **43.** (0, −3) **45.** (8, 8) **47.** (4, 5, 4) **49.** (1, 2) **51.** (2, 3) **53.** ∅ **55.** (2, 1, 0) **57.** $(x, 3x - 9)$ **59.** $(4 - z, 2, z)$ **61.** $(x, 0, 1 - x)$ **63.** 22°, 68° **65.** 40°, 65°, 75° **67.** $y = 2x^2 - x + 1$ **69.** 76°, 104° **71.** 20, 40, 4 **75.** $k \neq 0$

Getting Ready (page 209)

1. −22 **2.** 22 **3.** −13 **4.** −13

Exercises 3.5 (page 217)

1. 1 **3.** 0 **5.** $\begin{vmatrix} 5 & 2 \\ 4 & -1 \end{vmatrix}$ **7.** −3 **9.** 0 **11.** number, square **13.** $\begin{vmatrix} a_2 & c_2 \\ a_3 & c_3 \end{vmatrix}$ **15.** Cramer's rule **17.** consistent, independent **19.** 8 **21.** −2 **23.** 0 **25.** −13 **27.** 26 **29.** 0 **31.** (−1, 3) **33.** (4, 2) **35.** ∅ **37.** dependent equations, $\left(x, -\frac{2}{3}x + 3\right)$ **39.** (1, 1, 2) **41.** (3, 2, 1) **43.** ∅ **45.** dependent equations, $\left(\frac{1}{2}, \frac{5}{3} - \frac{4}{3}z, z\right)$ **47.** $x^2 - y^2$ **49.** $10a$ **51.** 0 **53.** −23 **55.** 26 **57.** $\left(-\frac{1}{2}, \frac{1}{3}\right)$ **59.** (2, −1) **61.** $\left(5, \frac{14}{5}\right)$ **63.** (3, −2, 1) **65.** $\left(\frac{3}{4}, \frac{1}{2}, \frac{1}{3}\right)$ **67.** (−2, 3, 1) **69.** 2 **71.** 2 **73.** 50°, 80° **75.** \$5,000 in HiTech, \$8,000 in SaveTel, \$7,000 in HiGas **81.** −4

Chapter Review (page 221)

1. 

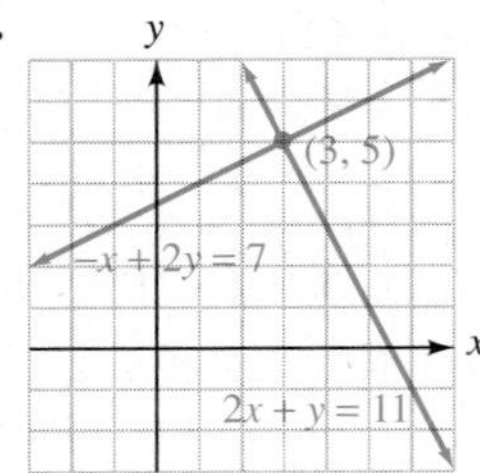

2.

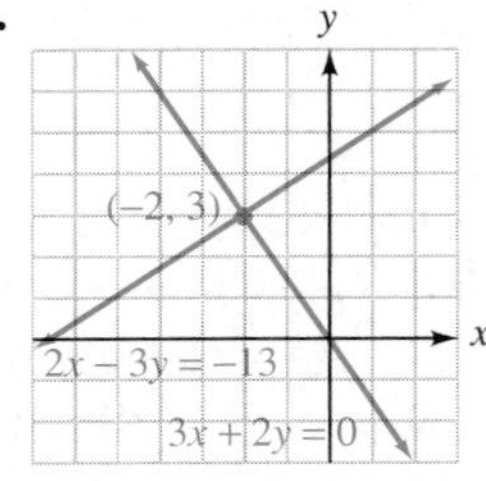

3. 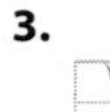

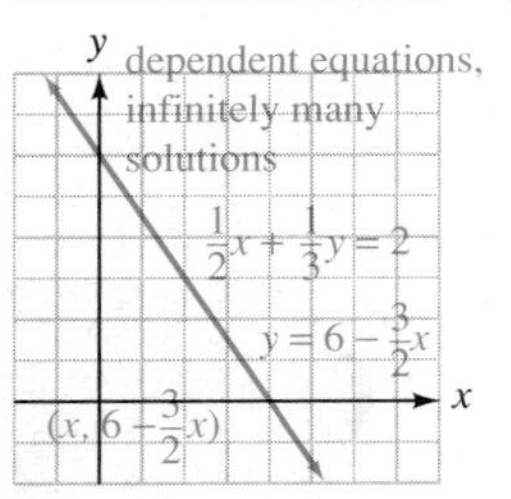

4. 

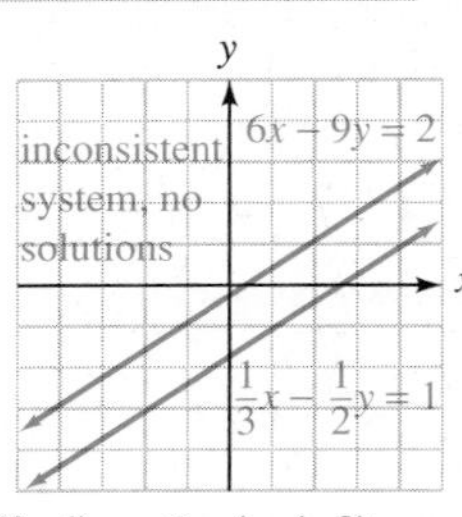

5. $(-1, 3)$ **6.** $(-3, -1)$ **7.** $(3, 4)$ **8.** $(-4, 2)$
9. $(-3, 1)$ **10.** $(1, -1)$ **11.** $(9, -4)$ **12.** $\left(4, \frac{1}{2}\right)$
13. $(1, 2, 3)$ **14.** $\varnothing$ **15.** $(2, 1)$ **16.** $(1, 3, 2)$ **17.** $(1, 2)$
18. $(3z, 1 - 2z, z)$ **19.** 18 **20.** 38 **21.** −3 **22.** 28
23. $(2, 1)$ **24.** $(-1, 3)$ **25.** $(1, -2, 3)$ **26.** $(-3, 2, 2)$

Chapter 3 Test (page 226)

1.

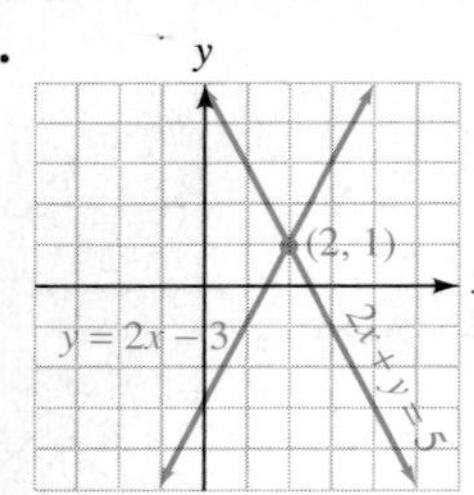

2. $(7, 0)$ **3.** $(2, -3)$
4. $(-6, 4)$ **5.** dependent
6. consistent **7.** 6 **8.** −8

9. $\left[\begin{array}{ccc|c} 1 & 1 & 1 & 4 \\ 1 & 1 & -1 & 6 \\ 2 & -3 & 1 & -1 \end{array}\right]$ **10.** $\begin{bmatrix} 1 & 1 & 1 \\ 1 & 1 & -1 \\ 2 & -3 & 1 \end{bmatrix}$

11. $(2, 2)$ **12.** $(-1, 3)$ **13.** 22 **14.** −17 **15.** 4
16. 13 **17.** $\begin{vmatrix} -6 & -1 \\ -6 & 1 \end{vmatrix}$ **18.** $\begin{vmatrix} 1 & -1 \\ 3 & 1 \end{vmatrix}$ **19.** −3
20. 3 **21.** 3 **22.** −1

Getting Ready (page 229)

1. −3 **2.** 4 **3.** 5
4. −1

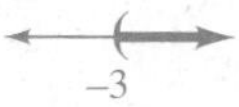

Exercises 4.1 (page 237)

1. $\{x \mid x < 2\}, (-\infty, 2)$ **3.** $\{x \mid x < -4\}, (-\infty, -4)$ **5.** $\frac{1}{t^{12}}$
7. 471 or more **9.** $\neq$ **11.** $<$ **13.** $\geq$
15. $a < c$ **17.** reversed **19.** $c < x, x < d$ **21.** open
23. $\{x \mid x < 1\}$ 1 **25.** $\{x \mid x \geq -5\}$ −5 **27.** $\{x \mid x < 3\}$ 3

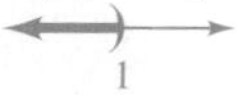

29. $\{x \mid x \geq -2\}$ −2 **31.** $(-36, \infty)$ −36 **33.** $(-\infty, 45/7]$ 45/7
35. $\{x \mid -2 < x < 4\}$ −2 4 **37.** $(-2, 5)$ −2 5
39. $(8, 11)$ 8 11 **41.** $[-4, 6)$ −4 6
43. $(2, 3)$ 2 3  **45.** no solution **47.** $(-\infty, -15)$ −15
49. $[2, \infty)$ 2 **51.** $(-\infty, 2) \cup (7, \infty)$ 2 7 **53.** $(-\infty, 1)$ 1
55. $\{x \mid x > 2\}$ 2 **57.** $(-3, \infty)$ −3  **59.** $(-\infty, -2]$ −2 
61. $(-\infty, -2)$ −2 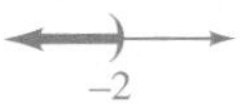 **63.** $(-\infty, 20]$ 20 **65.** $(-\infty, 10)$ 10
67. $(-\infty, 1.5]$ 1.5  **69.** no solution **71.** $[-2, 4]$ −2 4 
73. $x < 1$ **75.** $x \geq -4$ **77.** 5 hr **79.** 4 **81.** more than \$5,000 **83.** 18 **85.** 88 or higher **87.** 13
89. anything over \$900 **91.** 139 **95.** no **97.** a, b, c

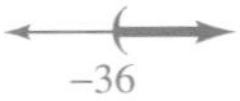

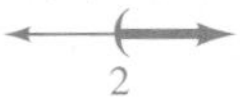

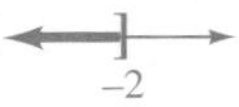

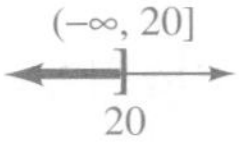

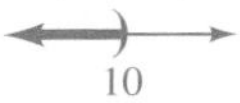

Getting Ready (page 240)

1. 5 **2.** 0 **3.** 10 **4.** −5 **5.** $(-5, 4)$
6. $(-\infty, -3) \cup (3, \infty)$

Exercises 4.2 (page 250)

1. 5 **3.** −6 **5.** 8 or −8 **7.** $-8 < x < 8$ **9.** $x \leq -4$ or $x \geq 4$ **11.** $\frac{3}{4}$ **13.** 6 **15.** $t = \frac{A - p}{pr}$ **17.** x **19.** 0
21. reflected **23.** $a = b$ or $a = -b$ **25.** $x \leq -k$ or $x \geq k$
27. 8 **29.** −2 **31.** −30 **33.** $4 - \pi$
35.

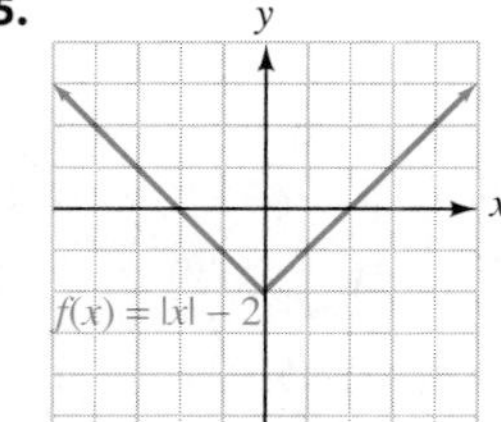

37.

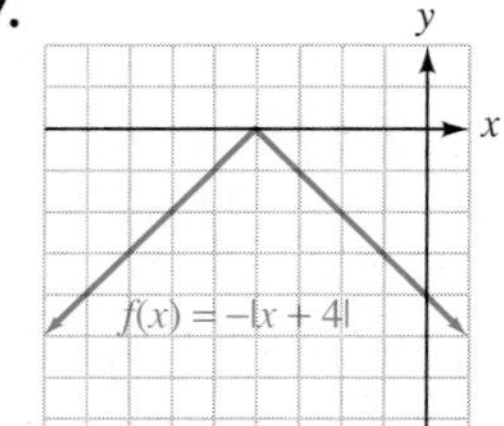

39. 4, −4 **41.** 9, −3 **43.** 4, −1 **45.** 8, −4 **47.** $2, -\frac{1}{2}$
49. −8 **51.** −4, −28 **53.** 7, −8 **55.** $-2, -\frac{4}{5}$
57. 0, −2 **59.** 0 **61.** $\varnothing$
63. $(-4, 4)$ −4 4 **65.** $[-21, 3]$ −21 3  **67.** $\varnothing$
69. $(-6, 18)$ −6 18 **71.** $[5, 5]$ 5 
73. $[-7, -7]$ −7 **75.** $(-\infty, -1) \cup (1, \infty)$ −1 1

77. $(-\infty, -16/3) \cup (4, \infty)$

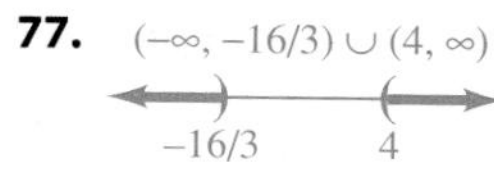

79. $(-\infty, \infty)$

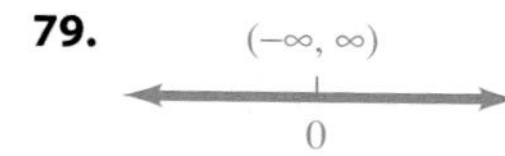

81. $(-\infty, -2] \cup [10/3, \infty)$

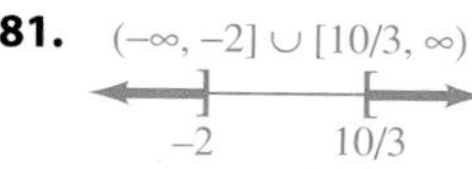

83. $(-\infty, 3/8) \cup (3/8, \infty)$ (3/8)

85. $[-10, 14]$ (−10, 14) **87.** $0, -6$ **89.** $(-\infty, -4] \cup [-1, \infty)$ (−4, −1)

91. $[-3/2, 2]$ (−3/2, 2) **93.** $(-2, 5)$ (−2, 5) **95.** $3, -1$

97. $(-\infty, -2) \cup (5, \infty)$ (−2, 5) **99.** $\frac{4}{3}$ **101.** $\varnothing$

103. $(-5/3, 1)$ (−5/3, 1) **105.** $(-\infty, -24) \cup (-18, \infty)$ (−24, −18)

107. $\frac{14}{3}, -6$ **109.** $(-\infty, -12) \cup (36, \infty)$ (−12, 36) **111.** $\varnothing$ **113.** $\frac{20}{3}$

115. $|x| < 4$ **117.** $|x + 3| > 6$ **119.** $4 \text{ ft} \le d \le 6 \text{ ft}$
121. $70° \le t \le 86°$ **123.** $|c - 0.6°| \le 0.5°$ **129.** $k < 0$
131. x and y must have different signs.

Getting Ready (page 253)

1. yes **2.** yes **3.** no **4.** yes

Exercises 4.3 (page 258)

1. yes **3.** no **5.** no **7.** yes **9.** $(3, 1)$ **11.** $(2, -3)$
13. linear **15.** edge **17.** dashed

19.

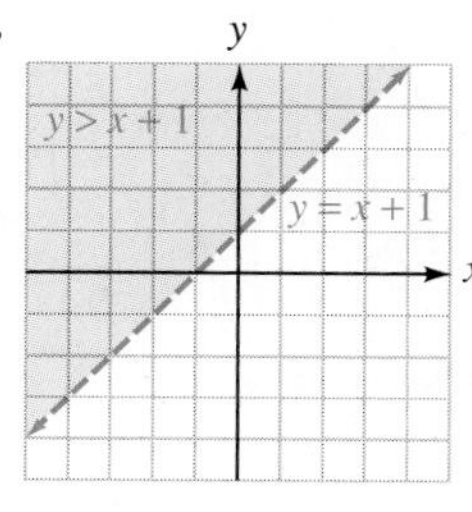

21.

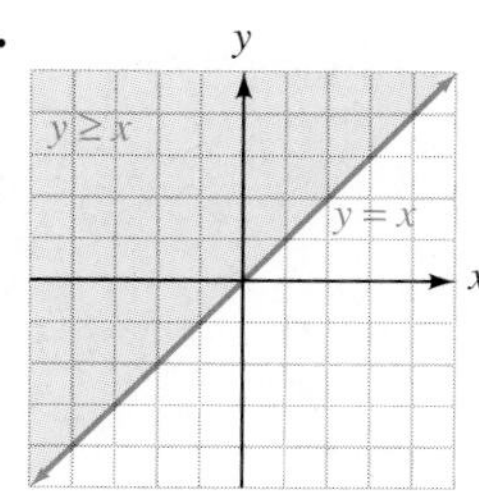

23.

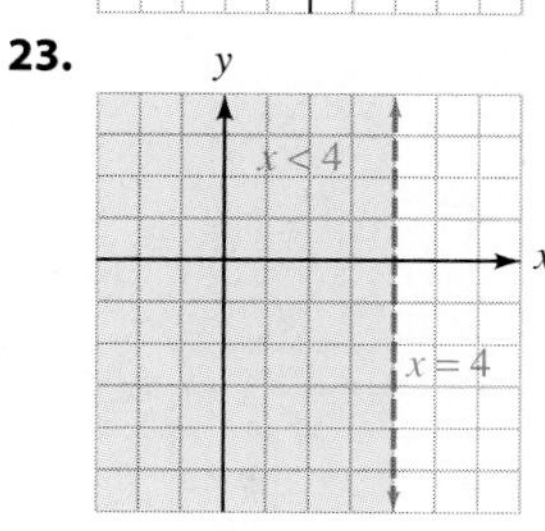

25.

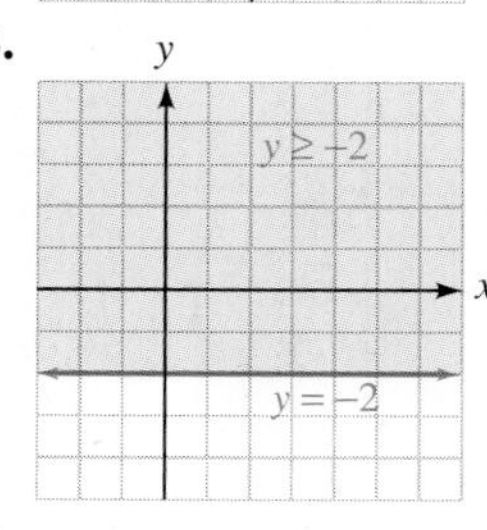

27.

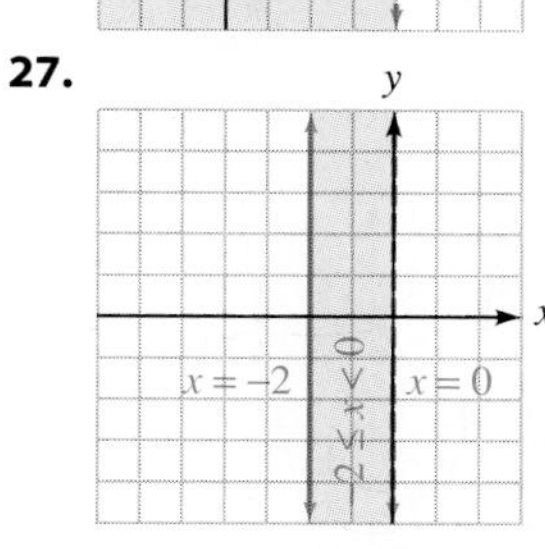

29.

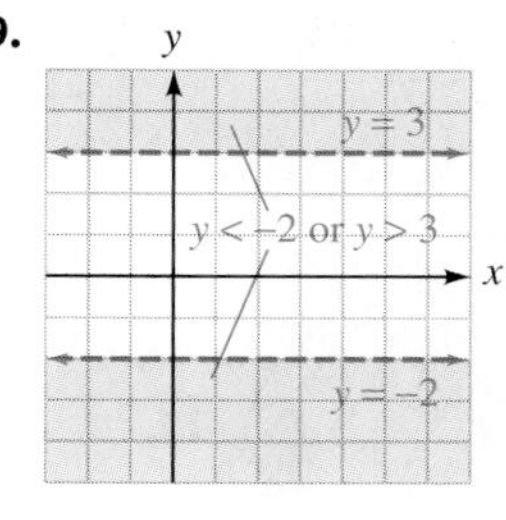

31.

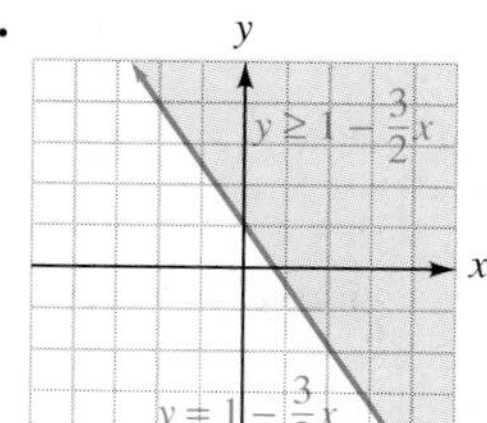

33.

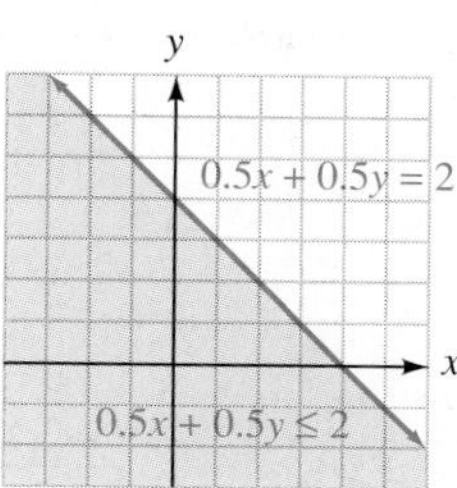

35.

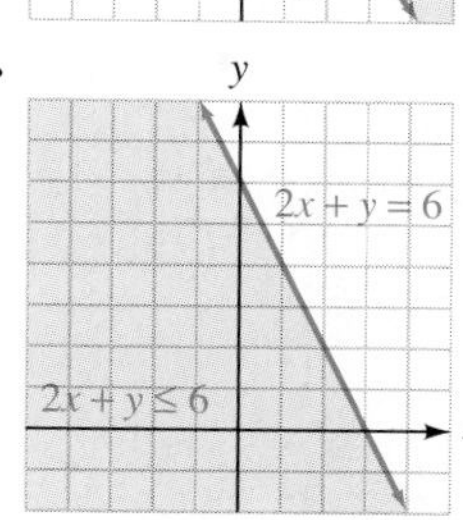

37.

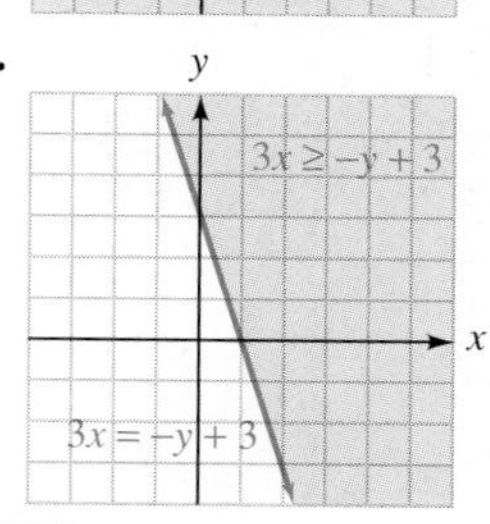

39.

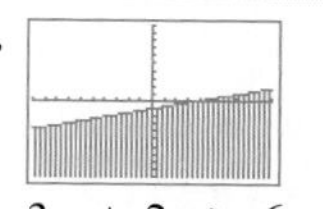

41.

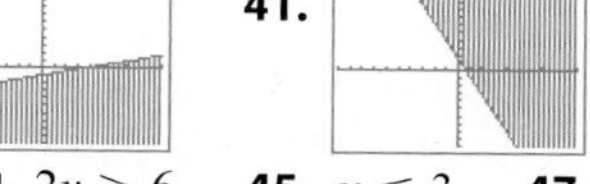

43. $3x + 2y > 6$ **45.** $x \le 3$ **47.** $y \le x$
49. $-2 \le x \le 3$ **51.** $y > -1$ or $y \le -3$
53. $(1, 1), (2, 1), (2, 2)$

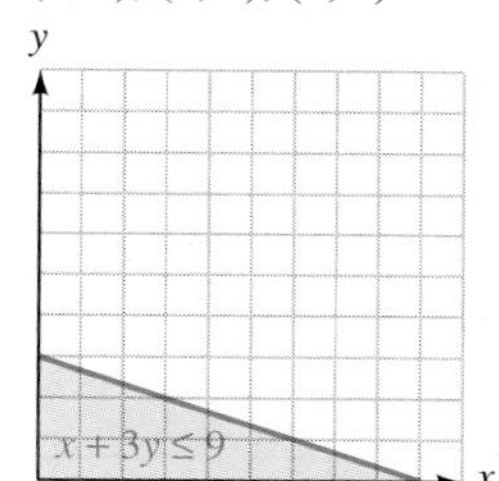

55. $(2, 2), (3, 3), (5, 1)$

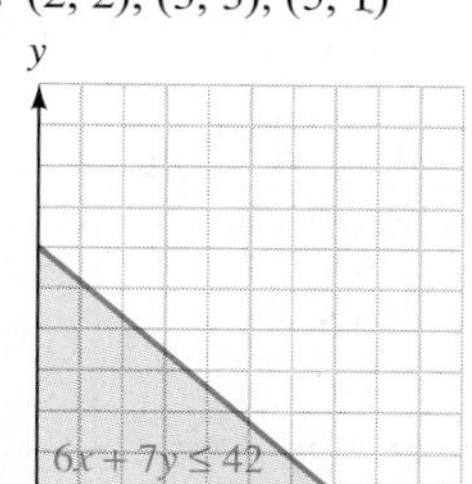

57. $(40, 20), (60, 40), (80, 20)$

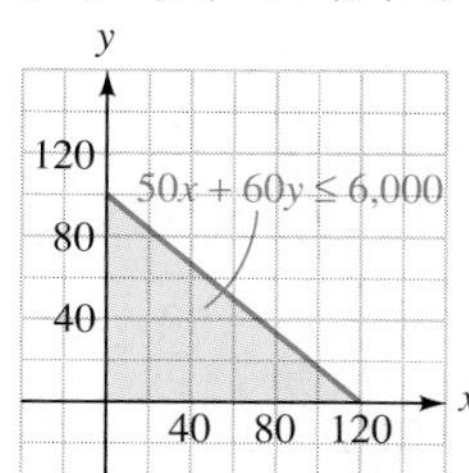

Getting Ready (page 261)

1. yes **2.** yes **3.** no **4.** no

Exercises 4.4 (page 265)

1. yes **3.** $r = \frac{A - p}{pt}$ **5.** $x = z\sigma + \mu$ **7.** $d = \frac{l - a}{n - 1}$
9. intersect

11.

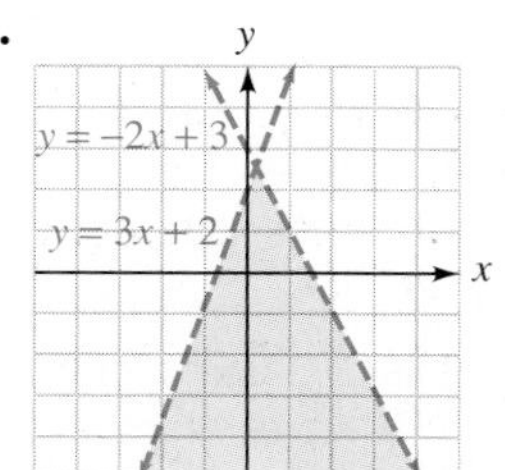

13.

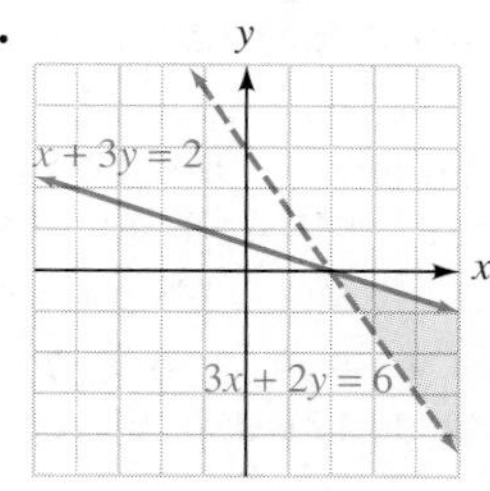

15.
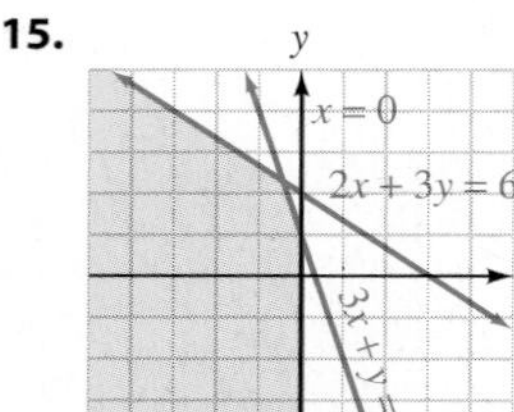

17.
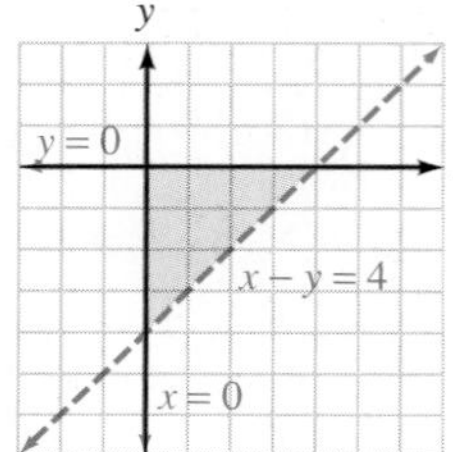

19.
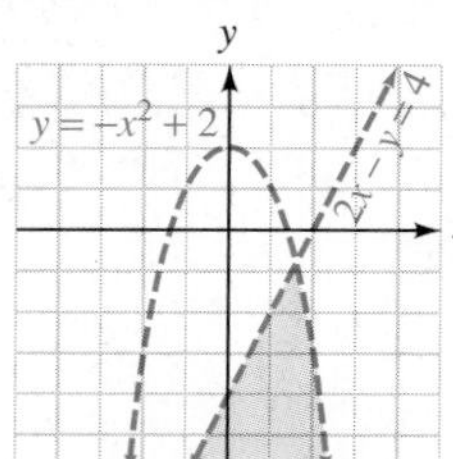

21.
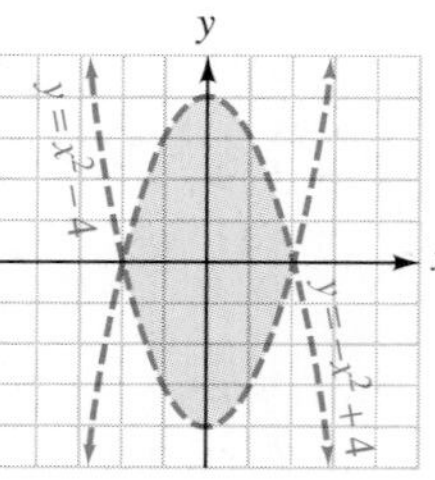

23.
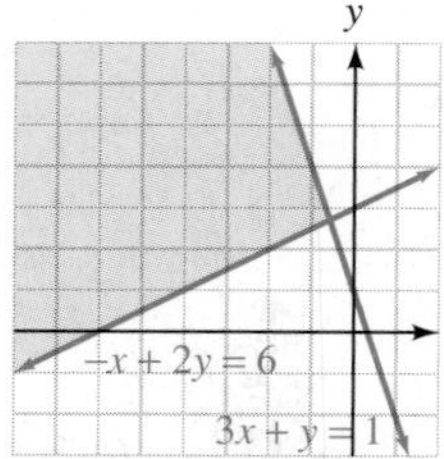

25.
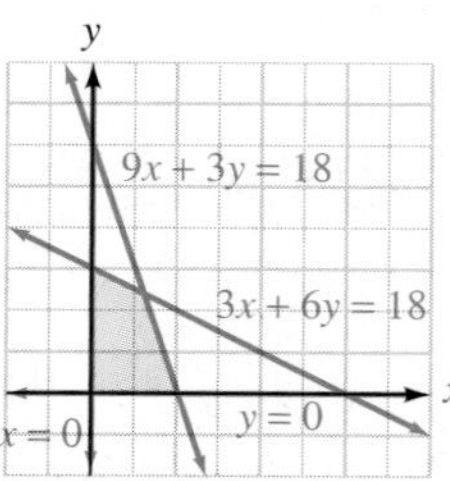

27.
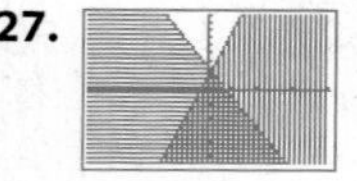

29.
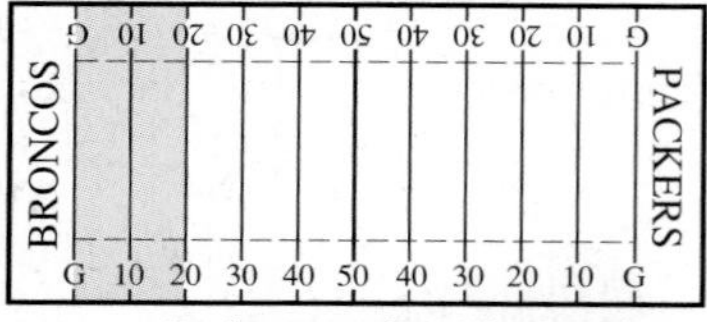

31. 1 \$10 CD and 2 \$15 CDs, 4 \$10 CDs and 1 \$15 CD

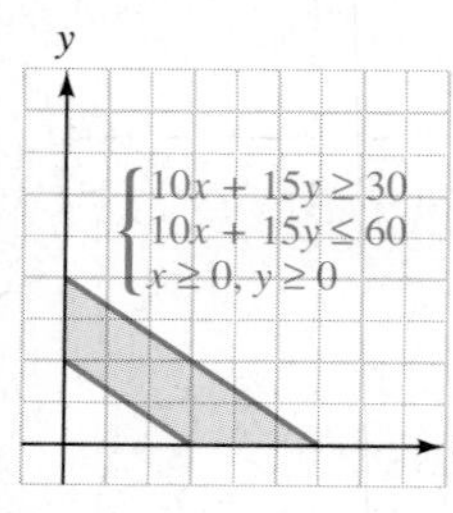

33. 2 desk chairs and 4 side chairs, 1 desk chair and 5 side chairs

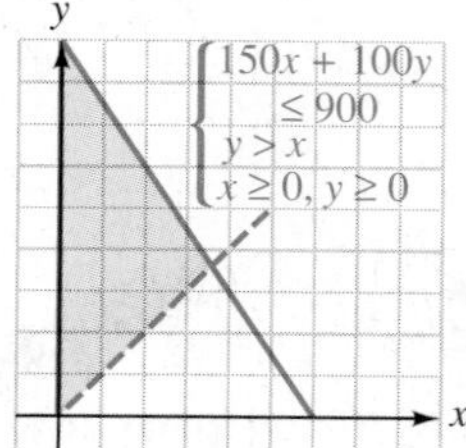

37. no

Getting Ready (page 268)

1. 0 **2.** 6 **3.** 10 **4.** 12

Exercises 4.5 (page 276)

1. 25 **3.** (0, 0), (0, 3), (3, 0) **5.** $\frac{3}{7}$ **7.** $y = \frac{3}{7}x + \frac{34}{7}$
9. constraints **11.** objective **13.** $P = 12$ at (0, 4)
15. $P = \frac{13}{6}$ at $\left(\frac{5}{3}, \frac{4}{3}\right)$ **17.** $P = 0$ at (0, 0) **19.** $P = 0$ at (0, 0)
21. $P = \frac{18}{7}$ at $\left(\frac{3}{7}, \frac{12}{7}\right)$ **23.** $P = 3$ at (1, 0)
25. $P = -12$ at (−2, 0) **27.** $P = -2$ at (1, 2) and (−1, 0)
29. 3 tables, 12 chairs, \$1,260 **31.** 30 IBMs, 30 Macs, \$2,700
33. 15 DVD players, 30 TVs, \$1,560 **35.** \$150,000 in stocks, \$50,000 in bonds; \$17,000

Chapter Review (page 279)

1. (−∞, 3]
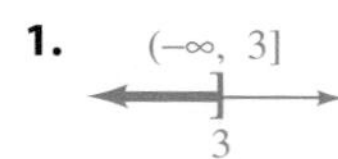

2. (2, ∞)

3. (−∞, −24]
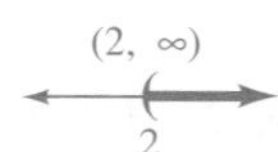

4. (−∞, −51/11)

5. (−1/3, 2)

6. (2, ∞)

7. [−1, 4)

8. \$20,000 or more **9.** 7 **10.** 8
11. −7 **12.** −12

13.
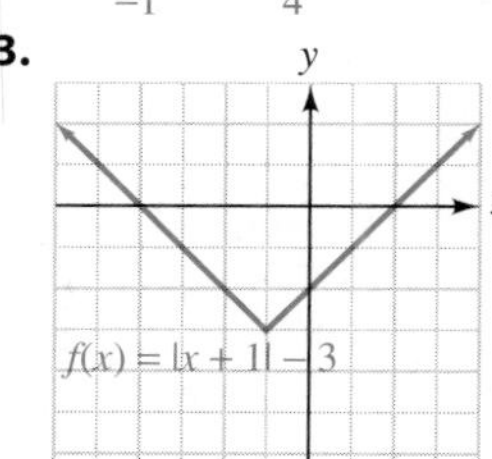

14.
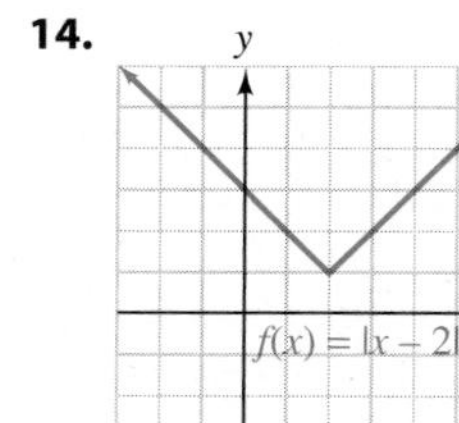

15. 3, $-\frac{11}{3}$ **16.** $\frac{26}{3}$, $-\frac{10}{3}$ **17.** 14, −10 **18.** $\frac{1}{5}$, −5
19. −1, 1 **20.** $\frac{13}{12}$

21. (−5, −2)
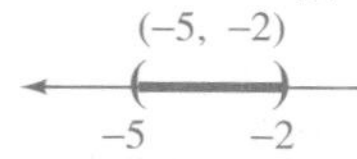

22. [−3, 19/3]
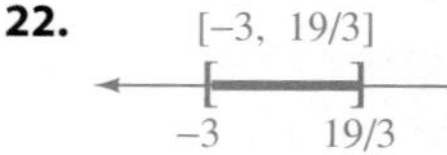

23. ∅ **24.** (−∞, −4) ∪ (22/5, ∞)

25. (−∞, 4/3] ∪ [4, ∞)

26. (−∞, ∞)
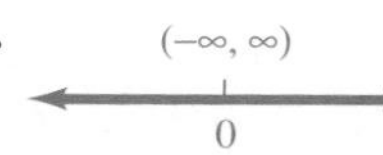

27.
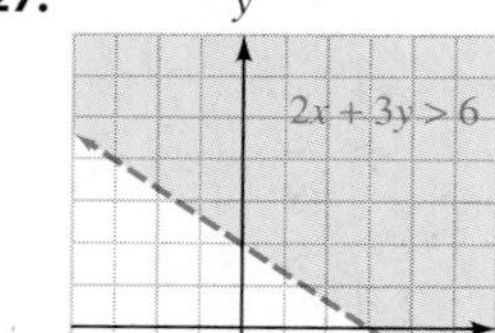

28.
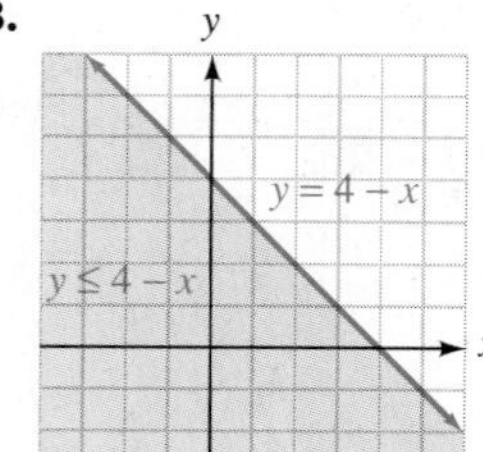

29.
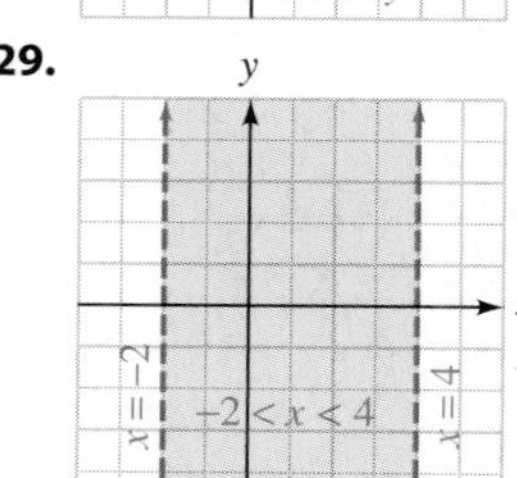

30.
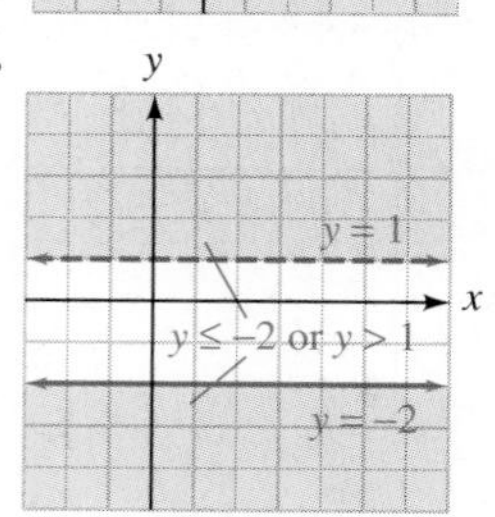

31.
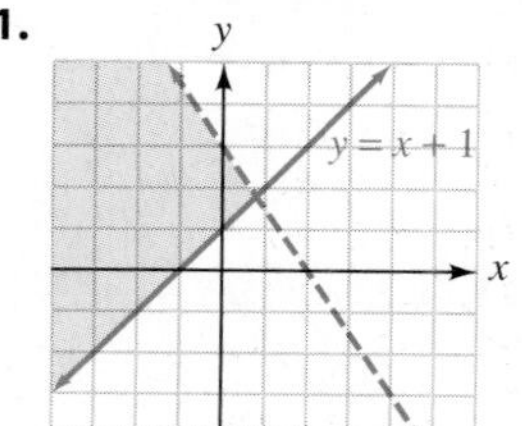

32.
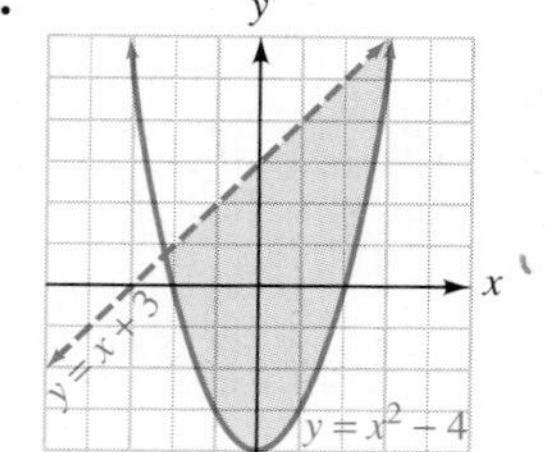

33. max. of 6 at (3, 0) **34.** 1,000 bags of X, 1,400 bags of Y

Chapter 4 Test (page 285)

1. 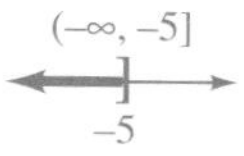

2. 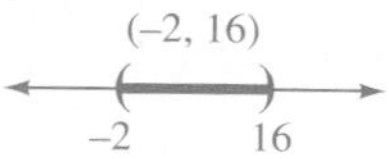

3. 3 **4.** $4\pi - 4$

5. 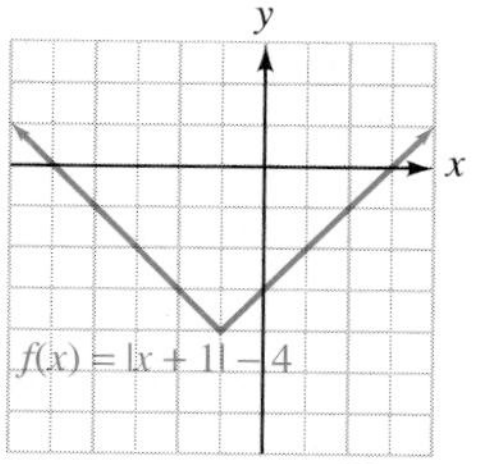

6. 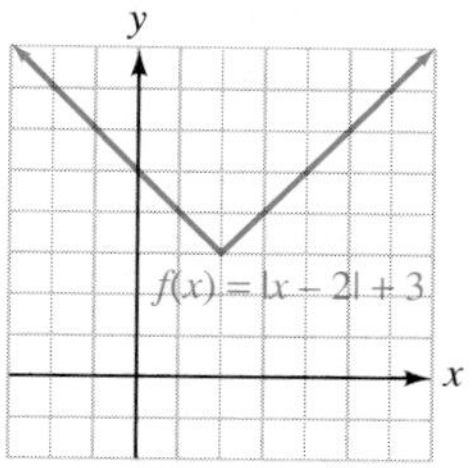

7. 4, −7 **8.** $-5, \frac{23}{3}$ **9.** 4, −4 **10.** 0

11. 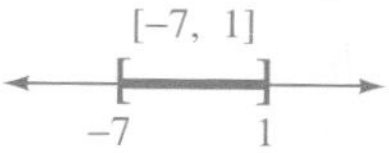

12. 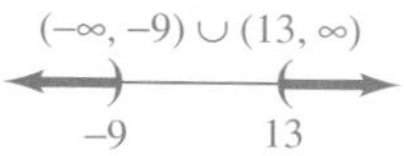

13. 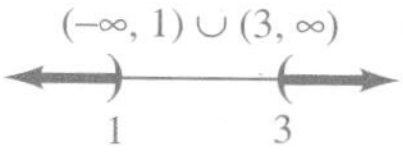

14. 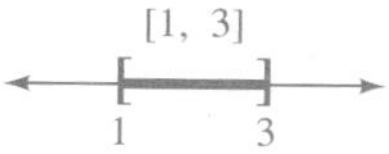

15. 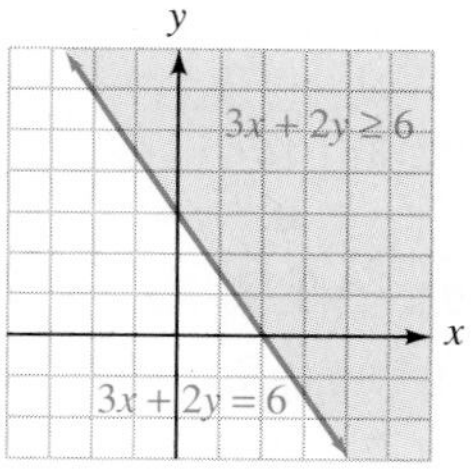

16. 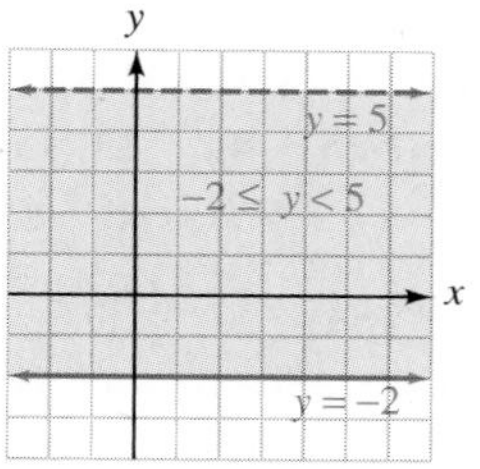

17. 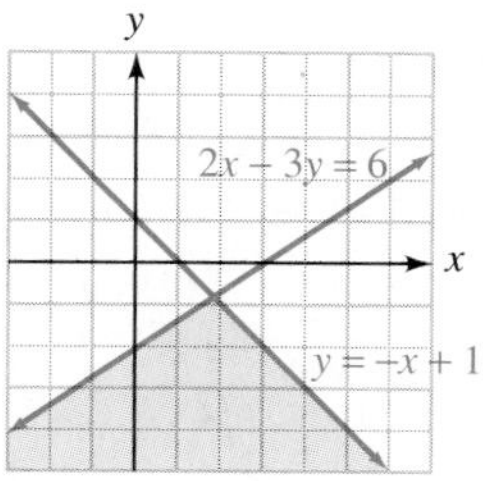

18. 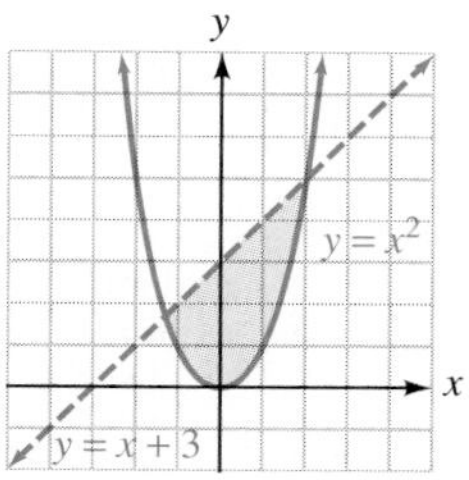

19. $P = 2$ at (1, 1)

Cumulative Review Exercises (page 285)

1. 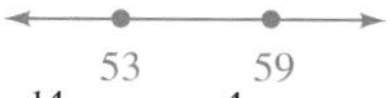

2. 5 **3.** 10 **4.** −6 **5.** x^{10}
6. x^{14} **7.** x^4 **8.** $a^{2-n}b^{n-2}$ **9.** 3.26×10^7
10. 1.2×10^{-5} **11.** $\frac{26}{3}$ **12.** 3 **13.** 6 **14.** ∅
15. perpendicular **16.** parallel **17.** $y = \frac{1}{3}x + \frac{11}{3}$
18. $h = \frac{2A}{b_1 + b_2}$ **19.** 10 **20.** 14
21. (2, 1)

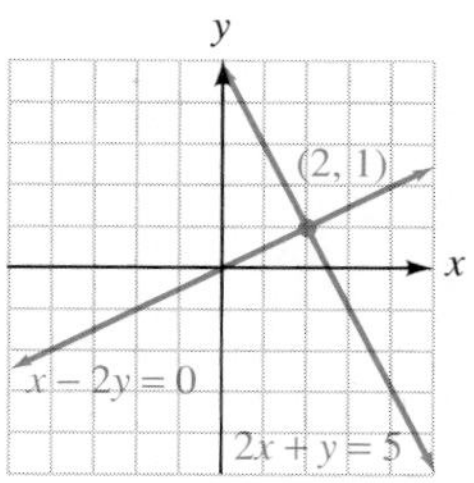

22. (1, 1) **23.** (2, −2)
24. (3, 1)
25. (−1, −1, 3)
26. (0, −1, 1) **27.** −1
28. 16 **29.** (−1, −1)
30. (1, 2, −1)
31. $x \le 11$
32. $-3 < x < 3$
33. $3, -\frac{3}{2}$ **34.** $-5, -\frac{3}{5}$

35. $-\frac{2}{3} \le x \le 2$ **36.** $x < -4$ or $x > 1$

37. 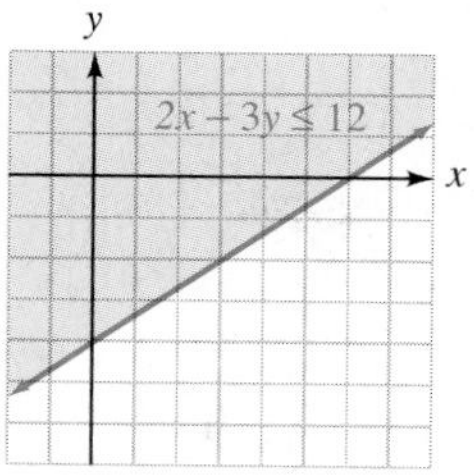

38. 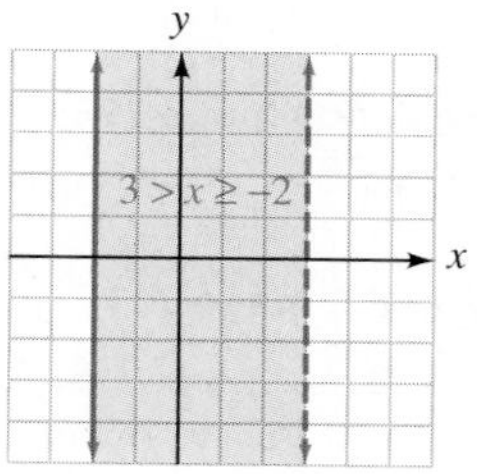

39. 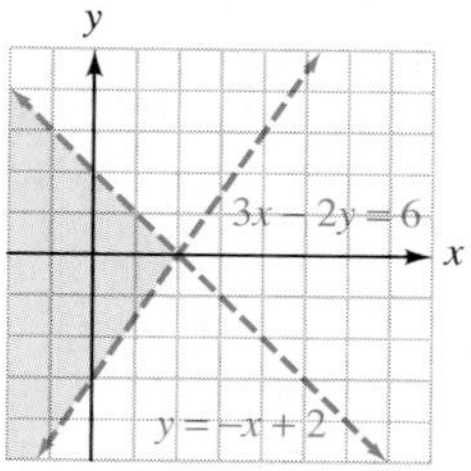

40. 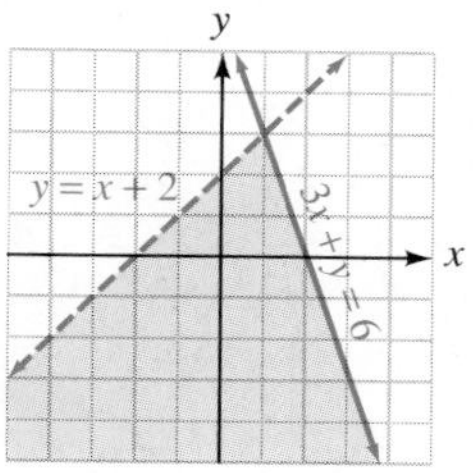

41. 24

Getting Ready (page 288)

1. $3a^2b^2$ **2.** $-5x^3y$ **3.** $4p^2 + 7q^2$ **4.** $a^3 - b^2$

Exercises 5.1 (page 295)

1. 3 **3.** 3 **5.** 1 **7.** −1 **9.** a^5 **11.** $3y^{23}$
13. 1.14×10^8 **15.** sum, whole **17.** binomial
19. descending order **21.** one **23.** monomial
25. trinomial **27.** binomial **29.** monomial **31.** 7
33. 0 **35.** 2 **37.** 8 **39.** 10 **41.** $-2x^4 - 5x^2 + 3x + 7$
43. $7y + 4y^2 - 5y^3 - 2y^5$ **45.** 2 **47.** 8
49. $18v^2 + 3v + 2$ **51.** $18x^2 - 3x + 2$ **53.** 13 **55.** 35
57. −90 **59.** −48 **61.** 0 ft **63.** 64 ft

65. 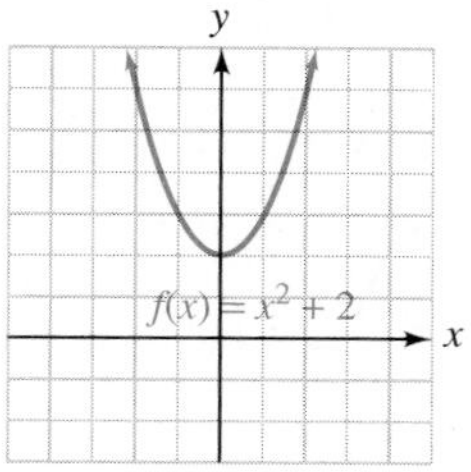

67. 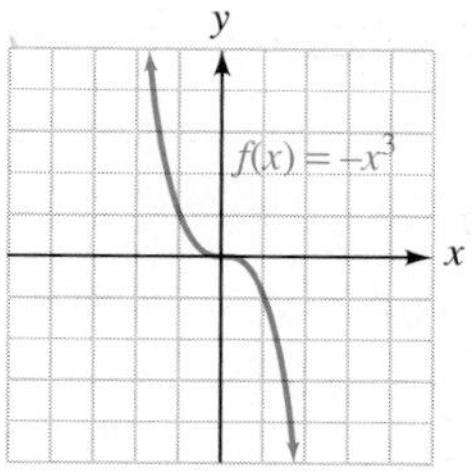

69. 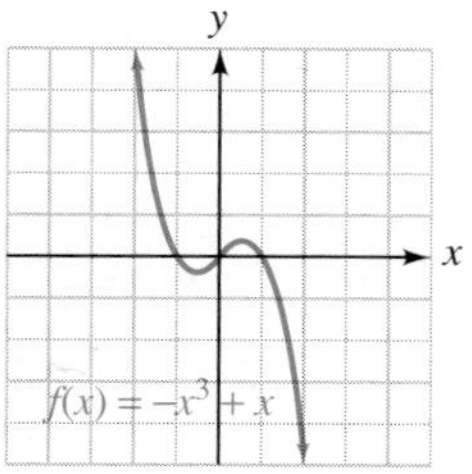

71. 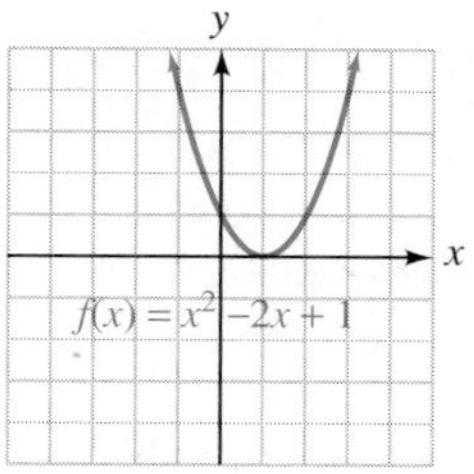

73. $7a^3x^5 - ax^3 - 5a^3x^2 + a^2x$; 8
75. $2x^4y - 5x^3y^3 - 2y^4 + 5x^3y^6 + x^5y^7$; 12 **77.** −34.225
79. 0.1728317132 **81.** −0.1226844822

83. 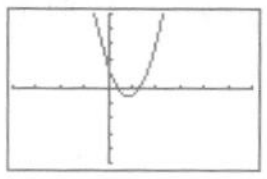

85. 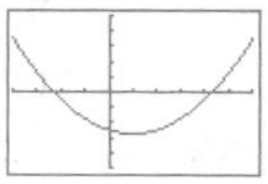

87. 20 ft **89.** 63 ft **91.** 198 ft **93.** 10 in.^{2} **95.** 18 in.^{2}
97. 10 m, 42 m, 26 m, 10 m **101.** no **103.** 12

Getting Ready (page 298)

1. $2x + 6$ **2.** $-8x + 20$ **3.** $5x + 15$ **4.** $-2x + 6$

Exercises 5.2 (page 301)

1. $9x^2$ **3.** $3x^2 + 2$ **5.** $3x^2 - 4x - 4$ **7.** $(-\infty, 4]$
9. $(-1, 9)$ **11.** exponents **13.** coefficients
15. like terms, $10x$ **17.** unlike terms
19. like terms, $-5r^2t^3$ **21.** unlike terms **23.** $12x$
25. $2x^3y^2z$ **27.** $-7x^2y^3 + 3xy^4$ **29.** $10x^4y^2$ **31.** $5a + 8$
33. $x^2 - 5x + 6$ **35.** $-y^3 + 4y^2 + 6$ **37.** $10a^2b + 4$
39. $4a^2b^4$ **41.** $2t - 2$ **43.** $-5a^2 + 4a + 4$
45. $5pq^2 + 5pq$ **47.** $13xy + 2x^2y^2$ **49.** $5x - 4$
51. $-11t + 29$ **53.** $8x^3 - x^2$ **55.** $14a^2 + 16a - 24$
57. $3x^3 - x + 13$ **59.** $4x^3 - 3x^2 + 3x - 7$
61. $4y^2 - 9y + 3$ **63.** $-y^2 + 6y + 16$ **65.** $-3p - 14q$
67. $11a^2 - 20b^3 - 9$ **69.** $6x^3 - 6x^2 + 14x - 17$
71. $-9y^4 + 3y^3 - 15y^2 + 20y + 12$ **73.** $x^2 - 8x + 22$
75. $-3y^3 + 18y^2 - 28y + 35$ **77.** $3x + 3$
79. $(x^2 + 5x + 6)$ m **81.** \$136,000
83. $f(x) = 2{,}500x + 275{,}000$ **85.** $f(x) = -2{,}100x + 16{,}600$
87. $f(x) = -4{,}800x + 35{,}800$ **91.** $-2x + 9$
93. $-8x^2 - 2x + 2$

Getting Ready (page 304)

1. $12a$ **2.** $4a^4$ **3.** $-7a^4$ **4.** $12a^5$ **5.** $2a + 8$
6. $a^2 - 3a$ **7.** $-4a + 12$ **8.** $-2b^2 - 4b$

Exercises 5.3 (page 312)

1. $-6a^3b^3$ **3.** $6a^3 - 3a^2$ **5.** $2x^2 + 3x + 1$ **7.** 10
9. -15 **11.** \$51,025 **13.** variable **15.** term
17. $x^2 + 2xy + y^2$ **19.** $x^2 - y^2$; conjugate; difference
21. $-6a^3b$ **23.** $-15a^2b^2c^3$ **25.** $-120a^9b^3$ **27.** $25x^{16}y^8$
29. $3x + 6$ **31.** $-a^2 + ab$ **33.** $3x^3 + 9x^2$
35. $-6x^3 + 6x^2 - 4x$ **37.** $10a^6b^4 - 25a^2b^6$
39. $7r^3st + 7rs^3t - 7rst^3$ **41.** $x^2 + 5x + 6$
43. $z^2 - 9z + 14$ **45.** $2a^2 - 3a - 2$ **47.** $6t^2 + 5t - 6$
49. $6y^2 - 5yz + z^2$ **51.** $2x^2 + xy - 6y^2$
53. $9x^2 - 6xy - 3y^2$ **55.** $8a^2 + 14ab - 15b^2$
57. $x^2 + 4x + 4$ **59.** $a^2 - 8a + 16$ **61.** $4a^2 + 4ab + b^2$
63. $4x^2 - 4xy + y^2$ **65.** $x^2 - 4$ **67.** $a^2 - b^2$
69. $4x^2 - 9y^2$ **71.** $x^3 - y^3$ **73.** $3x^2 + 12x$
75. $-p^2 + 4pq$ **77.** $-2yz - 2z^2$ **79.** $13y^2 - 10y + 17$
81. $a^3 - 3a^2b - ab^2 + 3b^3$ **83.** $a^3 + 3a^2b + 3ab^2 + b^3$
85. $2x^5 + x$ **87.** $\frac{3x}{y^4z} - \frac{x^6}{y^{10}z^2}$ **89.** $\frac{1}{x^2} - y^2$
91. $2x^3y^3 - \frac{2}{x^3y^3} + 3$ **93.** $x^{3n} - x^{2n}$ **95.** $x^{2n} - 1$
97. $-405x^7y^4$ **99.** $-12m^4n^2 - 12m^3n^3$
101. $6y^3 + 11y^2 + 9y + 2$ **103.** $8a^3 - b^3$
105. $4x^3 - 8x^2 - 9x + 6$ **107.** $2p^3 - 7p^2q + 4pq^2 + 4q^3$
109. $m^2 - mn + 2n^2$ **111.** $5x^2 - 36x + 7$
113. $24y^2 - 4yz + 21z^2$ **115.** $x^{2n} - \frac{x^n}{y^n} - x^ny^n + 1$
117. $x^{4n} - y^{4n}$ **119.** $x^{2n} + x^n + x^ny^n + y^n$
121. $9.2127x^2 - 7.7956x - 36.0315$
123. $299.29y^2 - 150.51y + 18.9225$ **127.** 15
129. $r = -\frac{1}{5}p^2 + 90p$ **131.** $(4x^2 - 12x + 9)$ ft^2
133. $\frac{1}{2}(b^2 + 3b - 10)$ in.^{2} **137.** 2.31×10^7

Getting Ready (page 315)

1. $4a + 8$ **2.** $-5b + 25$ **3.** $a^2 + 5a$ **4.** $-b^2 + 3b$
5. $6a^2 + 12ab$ **6.** $-6p^3 - 10p^2$

Exercises 5.4 (page 321)

1. $x(3x - 1)$ **3.** $-3a(a + 2)$ **5.** $(a + b)(3 + x)$
7. $a^2 - 16$ **9.** $16r^4 - 9s^2$ **11.** $m^3 + 64$ **13.** factoring
15. greatest common factor **17.** $2 \cdot 3$ **19.** $3^3 \cdot 5$ **21.** 2^7
23. $5^2 \cdot 13$ **25.** 12 **27.** 2 **29.** $4a^2$ **31.** $6xy^2z^2$
33. 4 **35.** 1 **37.** $2(x + 4)$ **39.** $2x(x - 3)$
41. $7x(x + 2)$ **43.** $9x^2y^2(7x + 9y^2)$ **45.** $6s(4s^2 - 2st + t^2)$
47. $9x^7y^3(5x^3 - 7y^4 + 9x^3y^7)$ **49.** prime **51.** prime
53. $-3(a + 2)$ **55.** $-x(3x + 1)$ **57.** $-3x(2x + y)$
59. $-6ab(3a + 2b)$ **61.** $x^2(x^n + x^{n+1})$ **63.** $y^n(2y^2 - 3y^3)$
65. $x^{-2}(x^6 - 5x^8)$ **67.** $t^{-3}(t^8 + 4t^{-3})$ **69.** $(x + y)(4 + t)$
71. $(a - b)(r - s)$ **73.** $(u + v)(u + v - 1)$
75. $-(x + y)(a - b)$ **77.** $(3 - c)(c + d)$
79. $(a + b)(a - 4)$ **81.** $(a + b)(x - 1)$
83. $(x + y)(x + y + z)$ **85.** $r_1 = \frac{rr_2}{r_2 - r}$ **87.** $f = \frac{d_1d_2}{d_2 + d_1}$
89. $a^2 = \frac{b^2x^2}{b^2 - y^2}$ **91.** $r = \frac{S - a}{S - l}$ **93.** $3z(9z^2 + 4z + 1)$
95. prime **97.** $-7u^2v^3z^2(9uv^3z^7 - 4v^4 + 3uz^2)$
99. $4y^{-2n}(2y^{4n} + 3y^{2n} + 4)$ **101.** $(m + n + p)(3 + x)$
103. $(x + y)(x + y + z)$ **105.** $(x + y)(a + b)$
107. $(x + 2)(x + y)$ **109.** $x(m + n)(p + q)$
111. $y(x + y)(x + y + 2z)$ **113.** $n(2n - p + 2m)(n^2p - 1)$
115. $a = \frac{Hb}{2b - H}$ **117.** $y = \frac{x + 3}{3x - 2}$ **119. a.** $12x^3$ in.^{2}
b. $20x^2$ in.^{2} **c.** $4x^2(3x - 5)$ in.^{2} **125.** yes **127.** no
129. yes

Getting Ready (page 324)

1. $a^2 - b^2$ **2.** $25p^2 - q^2$ **3.** $9m^2 - 4n^2$ **4.** $4a^4 - b^4$
5. $a^3 - 27$ **6.** $p^3 + 8$

Exercises 5.5 (page 329)

1. $(x + 1)(x - 1)$ **3.** $(x + 1)(x^2 - x + 1)$
5. $2(x + 2)(x - 2)$ **7.** $x^2 + 2x + 1$ **9.** $4m^2 + 4mn + n^2$
11. $a^2 + 7a + 12$ **13.** $8r^2 - 10rs + 3s^2$
15. 1, 4, 9, 16, 25, 36, 49, 64, 81, 100 **17.** cannot
19. $(p^2 - pq + q^2)$ **21.** $(x + 2)(x - 2)$
23. $(t + 15)(t - 15)$ **25.** $(b + 9)(b - 9)$ **27.** prime
29. $(3y + 8)(3y - 8)$ **31.** $(12a + b^2)(12a - b^2)$ **33.** prime
35. $(x^2 + y^2)(x + y)(x - y)$ **37.** $(x + y + z)(x + y - z)$
39. $(a - b + c)(a - b - c)$ **41.** $2(x + 12)(x - 12)$
43. $2x(x + 4)(x - 4)$ **45.** $5x(x + 5)(x - 5)$
47. $t^2(rs + x^2y)(rs - x^2y)$ **49.** $(r + s)(r^2 - rs + s^2)$
51. $(p - q)(p^2 + pq + q^2)$ **53.** $(x - 2y)(x^2 + 2xy + 4y^2)$
55. prime **57.** $5(x + 5)(x^2 - 5x + 25)$

59. $4x^2(x - 4)(x^2 + 4x + 16)$
61. $2u^2(4v - t)(16v^2 + 4vt + t^2)$
63. $(a + b)(x + 3)(x^2 - 3x + 9)$
65. $(x + y)(x^2 - xy + y^2)(x - y)(x^2 + xy + y^2)$
67. $(x + y - z)(x^2 - xy + xz + y^2 - 2yz + z^2)$
69. $(x^m + y^{2n})(x^m - y^{2n})$ **71.** $(10a^{2m} + 9b^n)(10a^{2m} - 9b^n)$
73. $(a + b)(a - b + 1)$ **75.** $(a - b)(a + b + 2)$
77. $(2ab^2c^3 + 3d^4)(2ab^2c^3 - 3d^4)$
79. $(2x^2 + 5y)(4x^4 - 10x^2y + 25y^2)$
81. $(16x^2y^2 + z^4)(4xy + z^2)(4xy - z^2)$
83. $(10a^2 - 7bc^2)(100a^4 + 70a^2bc^2 + 49b^2c^4)$
85. $(2x + y)(1 + 2x - y)$ **87.** $(x^n - 2)(x^{2n} + 2x^n + 4)$
89. $(a^m + b^n)(a^{2m} - a^mb^n + b^{2n})$
91. $2(x^{2m} + 2y^m)(x^{4m} - 2x^{2m}y^m + 4y^{2m})$
93. $0.5g(t_1 + t_2)(t_1 - t_2)$ **95.** $\frac{4}{3}\pi(r_1 - r_2)(r_1^2 + r_1r_2 + r_2^2)$
99. $(x^{16} + y^{16})(x^8 + y^8)(x^4 + y^4)(x^2 + y^2)(x + y)(x - y)$

Getting Ready (page 332)

1. $a^2 - a - 12$ **2.** $6a^2 + 7a - 5$ **3.** $a^2 + 3ab + 2b^2$
4. $2a^2 - ab - b^2$ **5.** $6a^2 - 13ab + 6b^2$
6. $16a^2 - 24ab + 9b^2$

Exercises 5.6 (page 341)

1. $(x + 2)(x + 1)$ **3.** $(x - 3)(x - 2)$ **5.** $(2x + 1)(x + 1)$
7. 31 **9.** 12 **11.** −3 **13.** $2xy + y^2$; $2xy + y^2$
15. key number **17.** $x + 2$ **19.** $x - 3$ **21.** $2a + 1$
23. $2m + 3n$ **25.** $(x + 1)^2$ **27.** $(a - 9)^2$ **29.** $(2y + 1)^2$
31. $(3b - 2)^2$ **33.** $(x + 8)(x + 1)$ **35.** $(x - 2)(x - 5)$
37. prime **39.** $(x + 5)(x - 6)$ **41.** $-(a - 8)(a + 4)$
43. $-3(x - 3)(x - 2)$ **45.** $3(x + 7)(x - 3)$
47. $b^2(a - 11)(a - 2)$ **49.** $x^2(b - 7)(b - 5)$
51. $-4(x - 5)(x + 4)$ **53.** $(3y + 2)(2y + 1)$
55. $(4a - 3)(2a + 3)$ **57.** $(3x - 4)(2x + 1)$ **59.** prime
61. $x(3x - 1)(x - 3)$ **63.** $-(2x - 3)^2$
65. $(2y - 3t)(y + 2t)$ **67.** $-(3a + 2b)(a - b)$
69. $5(a - 3b)^2$ **71.** $z(8x^2 + 6xy + 9y^2)$ **73.** $(x^n + 1)^2$
75. $(2a^{3n} + 1)(a^{3n} - 2)$ **77.** $(x^{2n} + y^{2n})^2$
79. $(3x^n - 1)(2x^n + 3)$ **81.** $(x + 2)^2$
83. $(a + b + 4)(a + b - 6)$ **85.** $(a - 16)(a - 1)$
87. $(2u + 3)(u + 1)$ **89.** $(5r + 2s)(4r - 3s)$
91. $(5u + v)(4u + 3v)$ **93.** $(x + 2 + y)(x + 2 - y)$
95. $(x + 1 + 3z)(x + 1 - 3z)$ **97.** $(c + 2a - b)(c - 2a + b)$
99. $(a + 4 + b)(a + 4 - b)$ **101.** $(a + b)(a - 4b)$
103. $(3z + 4)^2$ **105.** $(a + 10)(a - 5)$ **107.** $(y - 7)(y + 3)$
109. prime **111.** $x^2(7x - 8)(3x + 2)$
113. $(x^2 + 5)(x^2 + 3)$ **115.** $(y^2 - 10)(y^2 - 3)$
117. $(a + 3)(a - 3)(a + 2)(a - 2)$
119. $(z^2 + 3)(z + 2)(z - 2)$
121. $(x^3 + 3)(x + 1)(x^2 - x + 1)$
123. $(3x + 3y + 4)(2x + 2y - 5)$
125. $(2x + y + z)(2x + y - z)$ **127.** $x + 3$
129. $(2x + 11)$ in.; $(2x - 1)$ in.; 12 in. **133.** yes

Getting Ready (page 343)

1. $3pq(p - q)$ **2.** $(2p + 3q)(2p - 3q)$ **3.** $(p + 6)(p - 1)$
4. $(2p - 3q)(3p - 2q)$

Exercises 5.7 (page 346)

1. $(x + y)(x - y)$ **3.** $(x + 2)^2$ **5.** $(x - 2)(x^2 + 2x + 4)$
7. $7a^2 + a - 7$ **9.** $4y^2 - 11y + 3$ **11.** $m^2 + 2m - 8$
13. common factors **15.** trinomial **17.** $(x + 4)^2$
19. $(2xy - 3)(4x^2y^2 + 6xy + 9)$ **21.** $(x - t)(y + s)$
23. $(5x + 4y)(5x - 4y)$ **25.** $(6x + 5)(2x + 7)$
27. $2(3x - 4)(x - 1)$ **29.** prime **31.** $(7x - 1)(x - 8)$
33. $(10z + 9t)(10z - 9t)$ **35.** $2x^2y^2z^2(2 - 13z)$ **37.** prime
39. $(a + b)(f + e)$ **41.** $(2x + 2y + 3)(x + y - 1)$
43. $(25x^2 + 16y^2)(5x + 4y)(5x - 4y)$
45. $2x(4x^3 - 3x^2 + 6x + 12)$ **47.** $(2x - 9)(3x + 7)$
49. $-x^4(y^2 + 1)(y + 1)(y - 1)$
51. $(x + 1)(x - 1)(x + 4)(x - 4)$
53. $(x + 5 + y^4)(x + 5 - y^4)$
55. $(3x - 1 + 5y)(3x - 1 - 5y)$ **57.** prime
59. $(x + 1)(x^2 - x + 1)(x + 1)(x - 1)$
61. $(x + 3)(x - 3)(x + 2)(x^2 - 2x + 4)$
63. $2z(x + y)(x - y)^2(x^2 + xy + y^2)$ **65.** $(x^m - 3)(x^m + 2)$
67. $(a^n - b^n)(a^{2n} + a^nb^n + b^{2n})$ **69.** $\left(\frac{1}{x} + 1\right)^2$
71. $\left(\frac{3}{x} + 2\right)\left(\frac{2}{x} - 3\right)$ **77.** $(x^2 + x + 1)(x^2 - x + 1)$

Getting Ready (page 348)

1. $2a(a - 2)$ **2.** $(a + 5)(a - 5)$ **3.** $(3a - 2)(2a + 3)$
4. $a(3a - 2)(2a + 1)$

Exercises 5.8 (page 353)

1. 2, 3 **3.** 2, 3, −1 **5.** 2, 3, 5, 7 **7.** 40,081.00 cm^3
9. $ax^2 + bx + c = 0$ **11.** −3, −5 **13.** 3, 4 **15.** 1, 6
17. $-3, -\frac{1}{3}$ **19.** 0, −2 **21.** 4, −4 **23.** 0, −1
25. −7, 7 **27.** $\frac{1}{3}, -1$ **29.** $2, \frac{1}{2}$ **31.** $2, -\frac{5}{6}$
33. 0, −2, −3 **35.** 0, 0, −1 **37.** 0, 7, −7 **39.** 0, 7, −3
41. 3, −3, 2, −2 **43.** 1, −1, −3 **45.** $3, -3, -\frac{3}{2}$
47. 7, −7 **49.** $\frac{1}{2}, -3$ **51.** $\frac{1}{2}, 2$ **53.** $1, -\frac{1}{2}$ **55.** $2, -\frac{1}{3}$
57. $\frac{1}{4}, -\frac{3}{2}$ **59.** 0, 5 **61.** 3, 3 **63.** −2, −4 **65.** $\frac{1}{5}, -\frac{5}{3}$
67. $0, \frac{5}{6}, -7$ **69.** $2, -2, -\frac{1}{3}$ **71.** $0, -1, \frac{2}{5}$
73. no solution **75.** 1 **77.** 16, 18, or −18, −16 **79.** 6, 7
81. 7 cm by 21 cm **83.** 12 cm **85.** 5 ft by 12 ft
87. 20 ft by 40 ft **89.** 10 sec **91.** 11 sec and 19 sec
93. 50 m **95.** 3 ft **99.** $x^2 - 8x + 15 = 0$
101. $x^2 + 5x = 0$

Chapter Review (page 357)

1. 5 **2.** 8 **3.** 6 **4.** 9 **5.** $-t^2 - 4t + 6$
6. $-z^2 + 4z + 6$

7.

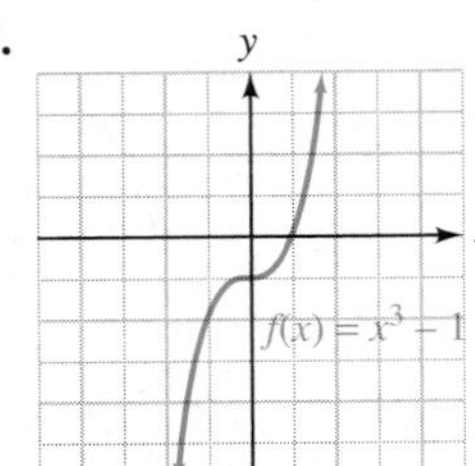

8.

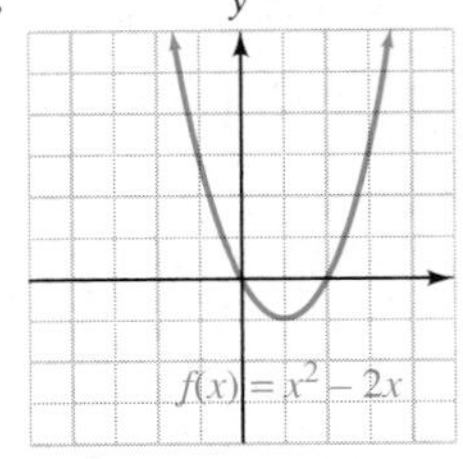

9. $5x^2 + 2x + 16$ **10.** $6x^3 + 4x^2 + x + 9$
11. $4x^2 - 9x + 19$ **12.** $-7x^3 - 30x^2 - 4x + 3$
13. $-16a^3b^3c$ **14.** $-6x^2y^2z^4$ **15.** $2x^4y^3 - 8x^2y^7$
16. $a^4b + 2a^3b^2 + a^2b^3$ **17.** $16x^2 + 14x - 15$
18. $6x^2 - 8x - 8$ **19.** $4x^2 - 20xy + 25y^2$
20. $15x^2 - 22x + 8$ **21.** $3x^4 - 3x^3 + 4x^2 + 2x - 4$
22. $2a^3 - 3a^2b - 3ab^2 + 2b^3$ **23.** $4(x + 2)$ **24.** $3x(x - 2)$
25. $5xy^2(xy - 2)$ **26.** $7a^3b(ab + 7)$
27. $-4x^2y^3z^2(2z^2 + 3x^2)$ **28.** $3a^2b^4c^2(4a^4 + 5c^4)$
29. $9x^2y^3z^2(3xz + 9x^2y^2 - 10z^5)$
30. $-12a^2b^3c^2(3a^3b - 5a^5b^2c + 2c^5)$ **31.** $x^n(x^n + 1)$
32. $y^{2n}(1 - y^{2n})$ **33.** $x^{-2}(x^{-2} - 1)$ **34.** $a^{-3}(a^9 + a^3)$
35. $5x^2(x + y)^3(1 - 3x^2 - 3xy)$
36. $-7a^2b^2(a - b)^3(7a^2 - 7ab - 9b^2)$ **37.** $(x + 2)(y + 4)$
38. $(a + b)(c + 3)$ **39.** $(x^2 + 4)(x^2 + y)$
40. $(a^3 + c)(a^2 + b^2)$ **41.** $h = \frac{S - 2wl}{2w + 2l}$ **42.** $l = \frac{S - 2wh}{2w + 2h}$
43. $(z + 4)(z - 4)$ **44.** $(y + 11)(y - 11)$
45. $(xy^2 + 8z^3)(xy^2 - 8z^3)$ **46.** prime
47. $(x + z + t)(x + z - t)$ **48.** $(c + a + b)(c - a - b)$
49. $2(x^2 + 7)(x^2 - 7)$ **50.** $3x^2(x^2 + 10)(x^2 - 10)$
51. $(x + 7)(x^2 - 7x + 49)$ **52.** $(a - 5)(a^2 + 5a + 25)$
53. $8(y - 4)(y^2 + 4y + 16)$ **54.** $4y(x + 3z)(x^2 - 3xz + 9z^2)$
55. $(x + 5)(x + 5)$ **56.** $(a - 7)(a - 7)$
57. $(y + 20)(y + 1)$ **58.** $(z - 5)(z - 6)$
59. $-(x + 7)(x - 4)$ **60.** $(y - 8)(y + 3)$
61. $(4a - 1)(a - 1)$ **62.** prime **63.** prime
64. $-(5x + 2)(3x - 4)$ **65.** $y(y + 2)(y - 1)$
66. $2a^2(a + 3)(a - 1)$ **67.** $-3(x + 2)(x + 1)$
68. $4(2x + 3)(x - 2)$ **69.** $3(5x + y)(x - 4y)$
70. $5(6x + y)(x + 2y)$ **71.** $(8x + 3y)(3x - 4y)$
72. $(2x + 3y)(7x - 4y)$ **73.** $x(x - 1)(x + 6)$
74. $3y(x + 3)(x - 7)$ **75.** $(z - 2)(z + x + 2)$
76. $(x + 1 + p)(x + 1 - p)$ **77.** $(x + 2 + 2p^2)(x + 2 - 2p^2)$
78. $(y + 2)(y + 1 + x)$ **79.** $(x^m + 3)(x^m - 1)$
80. $\left(\frac{1}{x} - 2\right)\left(\frac{1}{x} + 1\right)$ **81.** $0, \frac{3}{4}$ **82.** $6, -6$ **83.** $\frac{1}{2}, -\frac{5}{6}$
84. $\frac{2}{7}, 5$ **85.** $0, -\frac{2}{3}, \frac{4}{5}$ **86.** $-\frac{2}{3}, 7, 0$ **87.** 7 cm
88. 17 m by 20 m

Chapter 5 Test (page 362)

1. 5 **2.** 13 **3.** -9 **4.** -6

5.

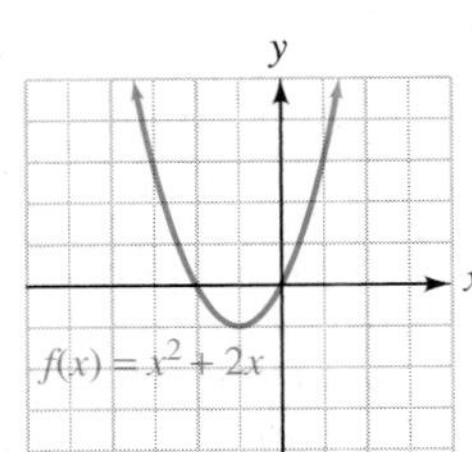

6. $5y^2 + y - 1$
7. $-4u^2 + 2u - 14$
8. $10a^2 + 22$
9. $-x^2 + 15x - 2$
10. $-6x^4yz^4$
11. $-15a^3b^4 + 10a^3b^5$
12. $z^2 - 16$ **13.** $12x^2 + x - 6$
14. $2u^3 - 3u^2v + v^3$
15. $3xy(y + 2x)$ **16.** $3abc(4a^2b - abc + 2c^2)$
17. $y^n(x^2y^2 + 1)$ **18.** $b^n(a^n - ab^{-2n})$ **19.** $(u - v)(r + s)$
20. $(a - y)(x + y)$ **21.** $(x + 7)(x - 7)$
22. $2(x + 4)(x - 4)$ **23.** $4(y^2 + 4)(y + 2)(y - 2)$
24. $(b + 5)(b^2 - 5b + 25)$ **25.** $(b - 3)(b^2 + 3b + 9)$
26. $3(u - 2)(u^2 + 2u + 4)$ **27.** $(a - 6)(a + 1)$
28. $(3b + 2)(2b - 1)$ **29.** $3(u + 2)(2u - 1)$
30. $5(4r + 1)(r - 1)$ **31.** $(x^n + 1)^2$
32. $(x + 3 + y)(x + 3 - y)$ **33.** $r = \frac{r_1r_2}{r_2 + r_1}$ **34.** $6, -1$
35. 25 **36.** 11 ft by 22 ft

Getting Ready (page 364)

1. $\frac{3}{4}$ **2.** $\frac{4}{5}$ **3.** $-\frac{5}{13}$ **4.** $\frac{7}{9}$

Exercises 6.1 (page 372)

1. undefined **3.** $(-\infty, 2) \cup (2, \infty)$ **5.** $\frac{5}{6}$ **7.** 2
9. $3x(x - 3)$ **11.** $(3x^2 + 4y)(9x^4 - 12x^2y + 16y^2)$
13. rational **15.** asymptote **17.** a, 1, 0 **19.** $\frac{a}{b}$, 0
21. -6 **23.** $-\frac{5}{6}$ **25.** $\left(-\infty, \frac{3}{2}\right) \cup \left(\frac{3}{2}, \infty\right)$
27. $(-\infty, -2) \cup (-2, 1) \cup (1, \infty)$
29. $(-\infty, 2) \cup (2, \infty)$ **31.** $(-\infty, -2) \cup (-2, 2) \cup (2,\infty)$

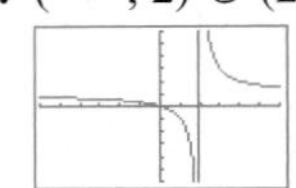

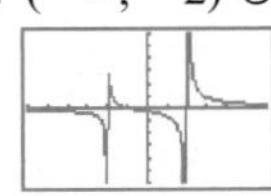

33. $\frac{2}{3}$ **35.** $-\frac{28}{9}$ **37.** $4x^2$ **39.** $-\frac{4y}{3x}$ **41.** $-\frac{x}{2}$
43. $\frac{3y}{7(y - z)}$ **45.** $x - 5$ **47.** $\frac{x - 3}{x - 5}$ **49.** $\frac{a + 1}{3}$ **51.** $\frac{x + 1}{x + 3}$
53. $\frac{x + 4}{2(2x - 3)}$ **55.** $x + 2$ **57.** $\frac{1}{x - 3}$ **59.** $\frac{1}{x - y}$ **61.** 3
63. $\frac{2a - 3b}{a - 2b}$ **65.** $\frac{1}{x^2 + xy + y^2 - 1}$ **67.** $\frac{(x + 1)^2}{(x - 1)^3}$ **69.** $\frac{3a + b}{y + b}$
71. -1 **73.** $\frac{b}{3}$ **75.** 20 hr **77.** 12 hr **79.** \$5,555.56
81. \$50,000 **83.** $\frac{12}{13}$ **85.** $-\frac{122}{37}$ **87.** x^2 **89.** $\frac{2x + 1}{2 - x}$
91. $\frac{a^2 - 3a + 9}{4(a - 3)}$ **93.** $\frac{x + 6}{x + 4}$ **95.** in lowest terms
97. $\frac{m - 2n}{n - 2m}$ **99.** $\frac{x - y}{x + y}$ **101.** $-\frac{m + n}{2m + n}$ **103.** $\frac{3(x - y)}{x + 2}$
105. almost 8 days **107.** about 2.55 hr
109. **a.** \$33,333.33 **b.** \$116,666.67
111. $c = f(x) = 1.25x + 700$ **113.** \$1,325 **115.** \$1.95
117. $c = f(n) = 0.09n + 7.50$ **119.** \$77.25 **121.** 9.75¢
125. yes **127.** a, d

Getting Ready (page 375)

1. $\frac{3}{2}$ **2.** $\frac{2}{5}$ **3.** $\frac{4}{3}$ **4.** $\frac{14}{3}$

Exercises 6.2 (page 382)

1. $\frac{9}{8}$ **3.** $\frac{x-2}{x+2}$ **5.** 5 **7.** $-6a^5+2a^4$ **9.** $m^{2n}-4$
11. $\frac{ac}{bd}$ **13.** 0 **15.** $\frac{10}{7}$ **17.** $-\frac{5}{6}$ **19.** $\frac{xy^2d}{c^3}$ **21.** $-\frac{x^{10}}{y^2}$
23. $x+1$ **25.** $-\frac{x+1}{x+3}$ **27.** $\frac{y+1}{(y-1)(x-2)}$ **29.** $-\frac{x+3}{x+2}$
31. 1 **33.** $\frac{2p-1}{p+2}$ **35.** $\frac{1}{x+1}$ **37.** $x-5$
39. $\frac{x^2-6x+9}{x^6+8x^3+16}$ **41.** $\frac{4m^4-4m^3-11m^2+6m+9}{x^4-2x^2+1}$ **43.** $\frac{x-4}{x+5}$
45. $\frac{n+2}{n+1}$ **47.** $\frac{(a+7)^2(a-5)}{12x^2}$ **49.** $\frac{-(x-3)(x-6)}{(x+2)(x+3)}$
51. $\frac{x-2}{x}$ **53.** $x-3$ **55.** $-\frac{x^7}{18y^4}$ **57.** $x^2(x+3)$ **59.** $\frac{3x}{2}$
61. $\frac{t+1}{t}$ **63.** -1 **65.** $\frac{x+y}{x-y}$ **67.** $\frac{x+2}{x-2}$ **69.** $\frac{x-1}{3x+2}$
71. $\frac{x-7}{x+7}$ **73.** 1 **75.** $\frac{(b-2)(b-3)}{2}$ cm^2 **77.** $\frac{(k+2)(k+4)}{k+1}$
81. $\div, \cdot$

Getting Ready (page 384)

1. yes **2.** no **3.** yes **4.** yes **5.** no **6.** yes

Exercises 6.3 (page 391)

1. x **3.** 1 **5.** $\frac{7x}{6}$ **7.** $(-1, 4]$

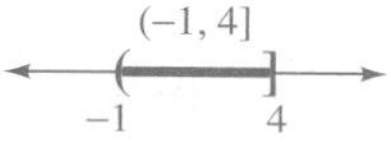

9. $w=\frac{P-2l}{2}$
11. $\frac{a+c}{b}$ **13.** subtract, keep **15.** LCD **17.** $\frac{5}{2}$ **19.** $-\frac{1}{3}$
21. $\frac{11}{4y}$ **23.** $\frac{3-a}{a+b}$ **25.** 2 **27.** 3 **29.** 3
31. $\frac{6x}{(x-3)(x-2)}$ **33.** 72 **35.** $x(x+3)(x-3)$
37. $(x+3)^2(x^2-3x+9)$ **39.** $(2x+3)^2(x+1)^2$ **41.** 2
43. $\frac{3(a-1)}{3a-2}$ **45.** $\frac{5}{6}$ **47.** $-\frac{16}{75}$ **49.** $\frac{9a}{10}$
51. $\frac{21a-8b}{14}$ **53.** $\frac{17}{12x}$ **55.** $\frac{9a^2-4b^2}{6ab}$ **57.** $\frac{2x+3}{x-1}$
59. $\frac{8x-2}{(x+2)(x-4)}$ **61.** $\frac{7x+29}{(x+5)(x+7)}$ **63.** $\frac{2x^2+x}{(x+3)(x+2)(x-2)}$
65. $\frac{-x^2+11x+8}{(3x+2)(x+1)(x-3)}$ **67.** $\frac{y^2+y-x^2}{(y-2x)^2}$ **69.** $\frac{-3x^2+x+1}{(x-1)^2}$
71. $\frac{-4x^2+14x+54}{x(x+3)(x-3)}$ **73.** $\frac{x^3+x^2-1}{x(x+1)(x-1)}$ **75.** $\frac{2x^2+5x+4}{x+1}$
77. $\frac{-y^2+48y+161}{(y+4)(y+3)}$ **79.** $\frac{4x-1}{(x+2)(x-2)}$ **81.** $\frac{-x^2+3x+2}{(x-1)(x+1)^2}$
83. 0 **85.** 1 **87.** $\frac{x^2+1}{x}$ **89.** $-\frac{5x+1}{(2x-1)(x+1)}$
91. $\frac{5x-y}{6}$ **93.** $\frac{-x^2+11x-15}{x-9}$ **95.** $\frac{-5x^3+9x^2-4x+23}{x-1}$
97. $\frac{3x^2-2x-17}{(x-3)(x-2)}$ **99.** $\frac{2a}{a-1}$ **101.** $\frac{x^2-6x-1}{2(x+1)(x-1)}$
103. $\frac{2b}{a+b}$ **105.** $\frac{7mn^2-7n^3-6m^2+3mn+n}{(m-n)^2}$ **107.** $\frac{2}{m-1}$
111. $\frac{11x^2+4x+12}{2x}$ ft **113.** $\frac{7x^2-11x+12}{4x}$

Getting Ready (page 394)

1. 3 **2.** -34 **3.** $3+2a$ **4.** $5-2b$

Exercises 6.4 (page 401)

1. $\frac{3}{5}$ **3.** 2 **5.** $\frac{x-y}{x}$ **7.** 8 **9.** 2, -2, 3, -3
11. complex **13.** $\frac{2}{3}$ **15.** $\frac{2y}{3z}$ **17.** $125b$ **19.** $-\frac{1}{7}$
21. $-\frac{1}{y}$ **23.** $\frac{y-x}{x^2y^2}$ **25.** $\frac{b+a}{b}$ **27.** $\frac{y+x}{y-x}$ **29.** $y-x$
31. $\frac{-1}{a+b}$ **33.** x^2+x-6 **35.** $\frac{5x^2y^2}{xy+1}$ **37.** $\frac{x+2}{x-3}$
39. $\frac{a-1}{a+1}$ **41.** $\frac{y+x}{x^2y}$ **43.** $\frac{xy^2}{y-x}$ **45.** $\frac{y+x}{y-x}$
47. $\frac{x^2(xy^2-1)}{y^2(x^2y-1)}$ **49.** $\frac{(b+a)(b-a)}{b(b-a-ab)}$ **51.** $\frac{x-1}{x}$ **53.** $\frac{5b}{5b+4}$
55. $\frac{3a^2+2a}{2a+1}$ **57.** -1 **59.** $\frac{10}{3}$ **61.** xy
63. $\frac{(-x^2+2x-2)(3x+2)}{(2-x)(-3x^2-2x+9)}$ **65.** $\frac{2(x^2-4x-1)}{-x^2+4x+8}$
67. $\frac{-2}{x^2-3x-7}$ **69.** $\frac{k_1k_2}{k_2+k_1}$ **71.** $\frac{4k}{9}$ **75.** $\frac{1}{y+x}$

Getting Ready (page 403)

1. $\frac{9}{2}$ **2.** 4, -4

Exercises 6.5 (page 410)

1. 2 **3.** 1 **5.** 1 **7.** $\frac{n^6}{m^4}$ **9.** 0 **11.** rational **13.** 12
15. 40 **17.** $\frac{1}{2}$ **19.** $\frac{17}{25}$ **21.** 2 **23.** $\frac{1}{3}$ **25.** $\varnothing$
27. $\varnothing$, -2 is extraneous **29.** 2, -5 **31.** -4, 3
33. 5, 3 is extraneous **35.** 8, $\frac{41}{13}$ **37.** $R=\frac{r_1r_2}{r_2+r_1}$
39. $f=\frac{pq}{q+p}$ **41.** 0 **43.** 1 **45.** 2 **47.** $-\frac{11}{6}$
49. 2, -2 is extraneous **51.** 2, $-\frac{3}{2}$ **53.** $r=\frac{S-a}{S-l}$
55. $R=\frac{r_1r_2r_3}{r_1r_3+r_1r_2+r_2r_3}$ **57.** $1\frac{7}{8}$ days **59.** $5\frac{5}{6}$ min
61. $1\frac{1}{5}$ days **63.** $2\frac{4}{13}$ weeks **65.** 3 mph **67.** 60 mph and 40 mph **69.** 3 mph **71.** 60 mph **73.** 7 **75.** 12 days
77. $\frac{20}{3}$ cm

Getting Ready (page 414)

1. $\frac{2}{3}$ **2.** $-\frac{5}{9}$ **3.** 5 **4.** 16

Exercises 6.6 (page 419)

1. $3xy$ **3.** $x+1$ **5.** $8x^2+2x+4$
7. $-2y^3-3y^2+6y-6$ **9.** $\frac{1}{b}$ **11.** quotient **13.** $\frac{y}{2x^3}$
15. $\frac{3b^4}{4a^4}$ **17.** $\frac{-5}{7xy^7t^2}$ **19.** $\frac{13a^nb^{2n}c^{3n-1}}{3}$ **21.** $\frac{2x}{3}-\frac{x^2}{6}$
23. $\frac{2xy^2}{3}+\frac{x^2y}{6}$ **25.** $7xy^2+1-5x^2y$ **27.** $\frac{x^4y^4}{2}-\frac{x^3y^9}{4}+\frac{3}{4xy^2}$
29. $x+2$ **31.** $x+7$ **33.** $3x+4$ **35.** $3x^2+4x+3$
37. $3x-5+\frac{3}{2x+3}$ **39.** $3x^2+x+2+\frac{8}{x-1}$
41. $3x^2+x-2+\frac{16}{2x+3}$ **43.** x^2+x+1 **45.** $a+1$
47. $2y+2$ **49.** $3x^2-x+2$ **51.** x^2+x+2
53. $a^2+a+1+\frac{2}{a-1}$ **55.** x^4+x^2+4
57. $1-3x^ny^n+6x^{2n}y^{2n}$ **59.** $\frac{b^2}{4a}-\frac{a^3}{2b^4}+\frac{3}{4ab}$
61. $3x^2-x+2$ **63.** $5a^2-3a-4$ **65.** $x-8y$
67. $-18+a$ or $a-18$ **69.** $27x^3+9x^2y+3xy^2+y^3$
71. $-8.9x-28.732+\frac{-122.9584}{2.5x-3.7}$ **73.** $4x+2$
75. x^2-7; x^2-2x+4 **79.** yes

Getting Ready (page 421)

1. $x+1$ with a remainder of 1, 1 **2.** $x+3$ with a remainder of 9, 9

Exercises 6.7 (page 426)

1. 9 **3.** yes **5.** 4 **7.** $12a^2+4a-1$ **9.** $8x^2+2x+4$
11. $x-r$ **13.** $P(r)$ **15.** $x-3$ **17.** $x+2$

19. $3x^2 - x + 2$ **21.** $x - 7 + \frac{28}{x + 2}$ **23.** $x + 2$
25. $6x^2 - x + 1 + \frac{3}{x + 1}$ **27.** $4x^2 + 2x - 1$ **29.** -1
31. -37 **33.** 23 **35.** -1 **37.** -8 **39.** 44 **41.** -3
43. -3 **45.** yes **47.** no **49.** yes **51.** yes
53. $-3x - 11 + \frac{-40}{x - 4}$ **55.** $4x^2 - 3x + 6 + \frac{-13}{x + 2}$ **57.** $\frac{29}{32}$
59. 2 **61.** -1 **63.** 18 **65.** 174
67. $7.2x - 0.66 + \frac{0.368}{x - 0.2}$ **69.** $2.7x - 3.59 + \frac{0.903}{x + 1.7}$
71. $9x^2 - 513x + 29,241 + \frac{-1,666,762}{x + 57}$ **73.** 64 **77.** 1

Getting Ready (page 428)

1. $\frac{7}{3}$ **2.** 1.44×10^9 **3.** linear function
4. rational function

Exercises 6.8 (page 438)

1. 1 **3.** $a = kb$ **5.** $a = kbc$ **7.** x^{10} **9.** -1
11. 3.5×10^4 **13.** 0.0025 **15.** unit costs, rates
17. extremes, means **19.** direct **21.** rational **23.** joint, constant of proportionality **25.** direct **27.** neither
29. 3 **31.** 5 **33.** -3 **35.** 5 **37.** $-\frac{5}{2}, -1$
39. no solution **41.** $A = kp^2$ **43.** $z = \frac{k}{t^3}$ **45.** $B = kmn$
47. $X = \frac{kw}{q}$ **49.** 4, -1 **51.** 2, -2 **53.** 39 **55.** $P = \frac{ka^2}{j^3}$
57. L varies jointly with m and n. **59.** E varies jointly with a and the square of b. **61.** X varies directly with x^2 and inversely with y^2. **63.** R varies directly with L and inversely with d^2. **65.** \$62.50 **67.** $7\frac{1}{2}$ gal **69.** 32 ft **71.** 80 ft
73. 0.18 g **75.** 42 ft **77.** $46\frac{7}{8}$ ft **79.** 6,750 ft
81. 36π in.^{2} **83.** 432 mi **85.** 25 days **87.** 12 in.^{3}
89. 85.3 **91.** 12 **93.** 26,437.5 gal **95.** 3 ohms
97. 0.275 in. **99.** 546 Kelvin

Chapter Review (page 444)

1. $\frac{11}{3}$ **2.** $\frac{302}{100}$
3.

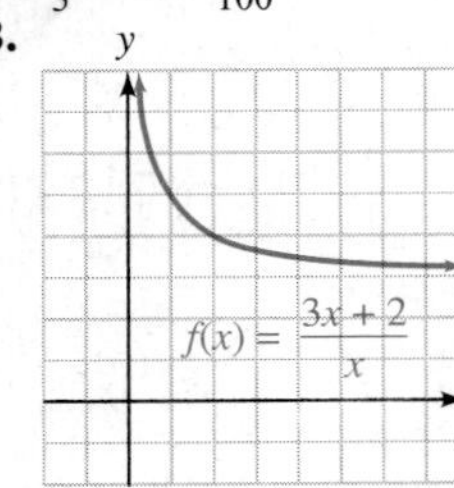

horizontal asymptote: $y = 3$; vertical asymptote: $x = 0$ **4.** $\frac{31x}{72y}$
5. $\frac{53m}{147n^2}$ **6.** $\frac{x - 7}{x + 7}$ **7.** $\frac{1}{x - 6}$
8. $\frac{1}{2x + 4}$ **9.** -1 **10.** -2
11. $\frac{m + 2n}{2m + n}$ **12.** $\frac{-a - b}{c + d}$ **13.** 1
14. 1 **15.** $\frac{3x(x - 1)}{(x - 3)(x + 1)}$ **16.** 1
17. $\frac{5y - 3}{x - y}$ **18.** $\frac{6x - 7}{x^2 + 2}$
19. $\frac{5x + 13}{(x + 2)(x + 3)}$ **20.** $\frac{4x^2 + 9x + 12}{(x - 4)(x + 3)}$ **21.** $\frac{5x^2 + 11x}{(x + 1)(x + 2)}$
22. $\frac{2(3x + 1)}{x - 3}$ **23.** $\frac{5x^2 + 23x + 4}{(x + 1)(x - 1)^2}$ **24.** $\frac{x^2 + 26x + 3}{(x + 3)(x - 3)^2}$
25. $\frac{3y - 2x}{x^2y^2}$ **26.** $\frac{y + 2x}{2y - x}$ **27.** $\frac{2x + 1}{x + 1}$ **28.** $\frac{3x + 2}{3x - 2}$
29. $\frac{x - 2}{x + 3}$ **30.** $\frac{1}{x}$ **31.** $\frac{y - x}{y + x}$ **32.** $\frac{x^2y^2}{(x - y)^2(y^2 - x^2)}$ **33.** 5
34. $-1, -2$ **35.** $\varnothing$; 2 and -2 are extraneous.
36. $-1, -12$ **37.** $y^2 = \frac{x^2b^2 - a^2b^2}{a^2}$ **38.** $b = \frac{Ha}{2a - H}$
39. 50 mph **40.** 200 mph **41.** $14\frac{2}{5}$ hr **42.** $18\frac{2}{3}$ days
43. $-\frac{x^3}{2y^3}$ **44.** $-3x^2y + \frac{3x}{2} + y$ **45.** $x + 5y$
46. $x^2 + 2x - 1 + \frac{6}{2x + 3}$ **47.** yes **48.** no **49.** 5
50. $-4, -12$ **51.** 70.4 ft **52.** 72 **53.** 6 **54.** 2
55. 16 **56.** \$5,460

Chapter 6 Test (page 450)

1. $\frac{-2}{3xy}$ **2.** $\frac{2}{x - 2}$ **3.** -3 **4.** $\frac{2x + 1}{4}$ **5.** $\frac{24}{5}$ **6.** $\frac{xz}{y^4}$
7. $\frac{x + 1}{2}$ **8.** 1 **9.** $\frac{(x + y)^2}{2}$ **10.** $\frac{2}{x + 1}$ **11.** -1 **12.** 3
13. 2 **14.** $\frac{2s + r^2}{rs}$ **15.** $\frac{2x + 3}{(x + 1)(x + 2)}$ **16.** $\frac{u^2}{2vw}$
17. $\frac{2x + y}{xy - 2}$ **18.** $\frac{5}{2}$ **19.** 5; 3 is extraneous **20.** $a^2 = \frac{x^2b^2}{b^2 - y^2}$
21. $r_2 = \frac{rr_1}{r_1 - r}$ **22.** 10 days **23.** \$5,000 at 6% and \$3,000 at 10% **24.** $\frac{-6x}{y} + \frac{4x^2}{y^2} - \frac{3}{y^3}$ **25.** $3x^2 + 4x + 2$ **26.** -7
27. 47 **28.** 18 ft **29.** 6, -1 **30.** $\frac{44}{3}$

Cumulative Review Exercises (page 451)

1. a^8b^4 **2.** $\frac{b^4}{a^4}$ **3.** $\frac{81b^{16}}{16a^8}$ **4.** $x^{21}y^3$ **5.** 42,500
6. 0.000712 **7.** 1 **8.** -2 **9.** $\frac{5}{6}$ **10.** $-\frac{3}{4}$ **11.** 3
12. $-\frac{1}{3}$ **13.** 0 **14.** 8 **15.** $-\frac{16}{25}$ **16.** $t^2 - 4t + 3$
17. $y = \frac{kxz}{r}$ **18.** no **19.** $\left[\frac{5}{2}, \infty\right)$ 5/2
20. $(-\infty, 3] \cup \left[\frac{11}{3}, \infty\right)$ 3 11/3
21. trinomial **22.** 7 **23.** 18
24. $y = 2x^2 - 3$
25. $4x^2 - 4x + 14$
26. $-3x^2 - 3$
27. $6x^2 - 7x - 20$
28. $2x^{2n} + 3x^n - 2$
29. $3rs^3(r - 2s)$
30. $(x - y)(5 - a)$
31. $(x + y)(u + v)$
32. $(9x^2 + 4y^2)(3x + 2y)(3x - 2y)$
33. $(2x - 3y^2)(4x^2 + 6xy^2 + 9y^4)$ **34.** $(2x + 3)(3x - 2)$
35. $(3x - 5)^2$ **36.** $(5x + 3)(3x - 2)$
37. $(3a + 2b)(9a^2 - 6ab + 4b^2)$ **38.** $(2x + 5)(3x - 7)$
39. $(x + 5 + y^2)(x + 5 - y^2)$ **40.** $(y + x - 2)(y - x + 2)$
41. 0, 2, -2 **42.** $-\frac{1}{3}, -\frac{7}{2}$ **43.** $\frac{2x - 3}{3x - 1}$ **44.** $\frac{x - 2}{x + 3}$
45. $\frac{4}{x - y}$ **46.** $\frac{a^2 + ab^2}{a^2b - b^2}$ **47.** 0 **48.** -17 **49.** $x + 4$
50. $-x^2 + x + 5 + \frac{8}{x - 1}$

Getting Ready (page 454)

1. 0 **2.** 16 **3.** 16 **4.** -16 **5.** $\frac{8}{125}$ **6.** $\frac{81}{256}$
7. $49x^2y^2$ **8.** $343x^3y^3$

Exercises 7.1 (page 464)

1. 3 **3.** -2 **5.** $8|x|$ **7.** not real **9.** $\frac{x + 3}{x - 4}$ **11.** 1
13. $\frac{3(m^2 + 2m - 1)}{(m + 1)(m - 1)}$ **15.** $(5x^2)^2$, $6^2 = 36$ **17.** positive
19. 5, left **21.** radical, index, radicand **23.** $|x|$ **25.** x
27. even **29.** $3x^2$ **31.** $a^2 + b^3$ **33.** 11 **35.** -8

37. $\frac{1}{3}$ **39.** $-\frac{5}{7}$ **41.** not real **43.** 0.4 **45.** $2|x|$
47. $3a^2$ **49.** $|t + 5|$ **51.** $|a + 3|$ **53.** 1 **55.** -5
57. $-\frac{2}{3}$ **59.** 0.4 **61.** $2a$ **63.** $-10pq$ **65.** 3 **67.** -3
69. -2 **71.** $\frac{2}{5}$ **73.** $\frac{1}{2}$ **75.** not real **77.** $2|x|$ **79.** $2a$
81. $\frac{1}{2}|x|$ **83.** $|x^3|$ **85.** $-x$ **87.** $-3a^2$ **89.** 0 **91.** 4
93. 1 **95.** 0.5
97. D: $[-4, \infty)$, R: $[0, \infty)$ **99.** D: $(-\infty, \infty)$, R: $(-\infty, \infty)$

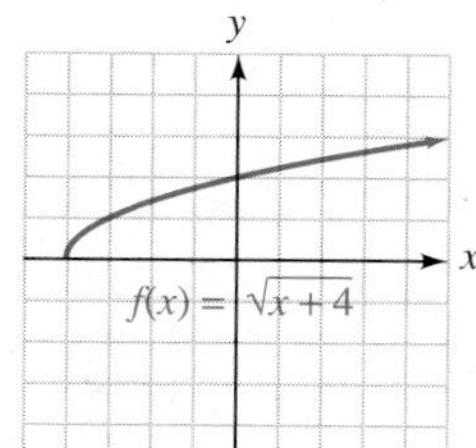

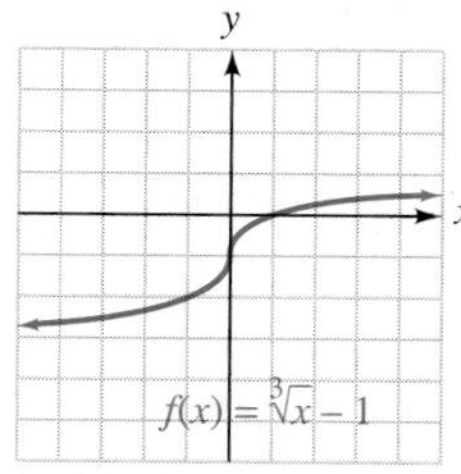

101. 4 **103.** not real **105.** $5|b|$ **107.** $|t + 12|$
109. $-\frac{1}{2}m^2n$ **111.** $0.2z^3$ **113.** $x + 2$ **115.** $0.1x^2|y|$
117. 3.4641 **119.** 26.0624 **121.** 4.1231 **123.** 2.5539
125. 1.67 **127.** 11.8673 **129.** 3 units **131.** 4 sec
133. about 7.4 amperes

Getting Ready (page 467)

1. 25 **2.** 169 **3.** 18 **4.** 11,236

Exercises 7.2 (page 472)

1. 5 **3.** 5 **5.** 4 **7.** $12x^2 - 14x - 10$
9. $15t^2 + 2ts - 8s^2$ **11.** hypotenuse **13.** $a^2 + b^2 = c^2$
15. positive **17.** 10 ft **19.** 80 m **21.** 48 in. **23.** 3 mi
25. 5 **27.** 5 **29.** 13 **31.** 10 **33.** 10.2 **35.** 13.6
37. about 127 ft **39.** about 135 ft **41.** 9.9 cm **47.** 13 ft
49. yes **51.** 90 yd **53.** 24 cm^2 **55.** yes **57.** no
59. 0.05 ft **63.** about 25

Getting Ready (page 475)

1. x^7 **2.** a^{12} **3.** a^4 **4.** 1 **5.** $\frac{1}{x^4}$ **6.** x^3 **7.** $\frac{b^6}{c^9}$
8. a^{10}

Exercises 7.3 (page 481)

1. 2 **3.** 3 **5.** 8 **7.** $\frac{1}{2}$ **9.** $2x$ **11.** $x < 3$
13. $r > 28$ **15.** $1\frac{2}{3}$ pints **17.** $a \cdot a \cdot a \cdot a$ **19.** a^{mn}
21. $\frac{a^n}{b^n}$ **23.** $\frac{1}{a^n}$, 0 **25.** $\left(\frac{b}{a}\right)^n$ **27.** $|x|$ **29.** $\sqrt[3]{7}$ **31.** $\sqrt[5]{8}$
33. $\sqrt[4]{3x}$ **35.** $\sqrt[4]{\frac{1}{2}x^3y}$ **37.** $\sqrt[5]{4a^2b^3}$ **39.** $\sqrt{x^2 + y^2}$
41. 2 **43.** 3 **45.** $\frac{1}{2}$ **47.** $\frac{1}{2}$ **49.** -2 **51.** not real
53. $11^{1/2}$ **55.** $(3a)^{1/4}$ **57.** $3a^{1/5}$ **59.** $\left(\frac{1}{7}abc\right)^{1/6}$
61. $\left(\frac{1}{2}mn\right)^{1/5}$ **63.** $(a^2 - b^2)^{1/3}$ **65.** $5|y|$ **67.** $3x$
69. $|x + 1|$ **71.** not real **73.** 216 **75.** 27 **77.** 1,728
79. $\frac{1}{4}$ **81.** $\frac{1}{2}$ **83.** $\frac{1}{8}$ **85.** $\frac{1}{64x^3}$ **87.** $\frac{1}{9y^2}$ **89.** 8
91. $\frac{16}{81}$ **93.** $5^{8/9}$ **95.** $4^{3/5}$ **97.** $\frac{1}{36}$ **99.** $9^{1/5}$ **101.** $7^{1/2}$
103. $2^{2/3}$ **105.** a **107.** $a^{2/9}$ **109.** $y + y^2$
111. $x^2 - x + x^{3/5}$ **113.** $x - y$
115. $x^{4/3} + 2x^{2/3}y^{2/3} + y^{4/3}$ **117.** $\sqrt{p}$ **119.** $\sqrt{5b}$
121. 2 **123.** 2 **125.** 0 **127.** -3 **129.** $125x^6$
131. $\frac{4x^2}{9}$ **133.** $\frac{1}{4p^2}$ **135.** $-\frac{3}{2x}$ **137.** $a^{3/4}b^{1/2}$ **139.** $\frac{n^{2/5}}{m^{3/5}}$
141. $\frac{2x}{3}$ **143.** $\frac{1}{3}x$ **145.** $x^2 + 3x^3 - 4x^{4/3}$
147. $\frac{1}{x} - 2 + x$ **149.** 2.47 **151.** 1.01 **153.** 0.24
155. -1.32 **159.** yes

Getting Ready (page 484)

1. 15 **2.** 24 **3.** 5 **4.** 7 **5.** $4x^2$ **6.** $\frac{8}{11}x^3$ **7.** $3ab^3$
8. $-2a^4$

Exercises 7.4 (page 491)

1. 7 **3.** 3 **5.** $2\sqrt[3]{2}$ **7.** $7\sqrt{3}$ **9.** $5\sqrt[3]{9}$ **11.** $\frac{-15x^5}{y}$
13. $9t^2 + 12t + 4$ **15.** $3p + 4 + \frac{-5}{2p - 5}$ **17.** $\sqrt[n]{a}\sqrt[n]{b}$
19. like **21.** 6 **23.** t **25.** $5x$ **27.** 10 **29.** $7x$
31. $6b$ **33.** 2 **35.** $3a$ **37.** $2\sqrt{5}$ **39.** $-10\sqrt{2}$
41. $2\sqrt[3]{10}$ **43.** $-3\sqrt[3]{3}$ **45.** $2\sqrt[4]{2}$ **47.** $2\sqrt[5]{3}$ **49.** $\frac{\sqrt{7}}{3x}$
51. $\frac{a\sqrt[3]{7}}{4}$ **53.** $\frac{p\sqrt[4]{3}}{10q}$ **55.** $\frac{m^3\sqrt[5]{3}}{2n^2}$ **57.** $5x\sqrt{2}$ **59.** $4\sqrt{2b}$
61. $-4a\sqrt{7a}$ **63.** $5ab\sqrt{7b}$ **65.** $-10\sqrt{3xy}$
67. $-3x^2\sqrt[3]{2}$ **69.** $2x^4y\sqrt[3]{2}$ **71.** $\frac{z}{4x}$ **73.** $4\sqrt{3}$
75. $-\sqrt{2}$ **77.** $2\sqrt{2}$ **79.** $9\sqrt{6}$ **81.** $3\sqrt{2} + \sqrt{3}$
83. $3\sqrt[3]{3}$ **85.** $-\sqrt[3]{4}$ **87.** -10 **89.** $-11\sqrt[3]{2}$
91. $-x\sqrt[3]{3x^2}$ **93.** $8z\sqrt{y}$ **95.** $-7y^2\sqrt{y}$ **97.** 0
99. $b = 3, c = 3\sqrt{2}$ **101.** $a = \frac{2}{3}, c = \frac{2}{3}\sqrt{2}$
103. $b = 5\sqrt{2}, c = 10$ **105.** $a = 7, b = 7$
107. $b = 5\sqrt{3}, c = 10$ **109.** $a = 9, c = 18$
111. $a = 12, b = 12\sqrt{3}$ **113.** $a = \frac{15}{2}, b = \frac{15}{2}\sqrt{3}$
115. $10\sqrt{2x}$ **117.** $2x^3y\sqrt[4]{2}$ **119.** $\frac{\sqrt[4]{5x}}{2z}$ **121.** $-4\sqrt{2}$
123. 81 **125.** $94\sqrt[4]{3}$ **127.** $10\sqrt[4]{3}$ **129.** $9\sqrt[3]{5y}$
131. $9\sqrt[6]{12xyz}$ **133.** $4x\sqrt[5]{xy^2}$ **135.** $2x + 2$
137. $h = 2.83, x = 2.00$ **139.** $x = 8.66, h = 10.00$
141. $x = 4.69, y = 8.11$ **143.** $x = 12.11, y = 12.11$
145. $10\sqrt{3}$ mm, 17.32 mm **149.** If $a = 0$, then b can be any nonnegative real number. If $b = 0$, then a can be any nonnegative real number.

Getting Ready (page 494)

1. a^7 **2.** b^3 **3.** $a^2 - 2a$ **4.** $6b^3 + 9b^2$
5. $a^2 - 3a - 10$ **6.** $4a^2 - 9b^2$

Exercises 7.5 (page 501)

1. 3 **3.** $3\sqrt{3}$ **5.** $6 + 3\sqrt{2}$ **7.** $\frac{\sqrt{2}}{2}$ **9.** 1 **11.** $\frac{1}{3}$
13. 2, $\sqrt{7}$, $\sqrt{5}$ **15.** $\sqrt{x} - 1$ **17.** conjugate **19.** 4
21. $5\sqrt{2}$ **23.** $6\sqrt{2}$ **25.** 5 **27.** $2\sqrt[3]{3}$ **29.** ab^2
31. $r\sqrt[3]{10s}$ **33.** $x^2(x + 3)$ **35.** $12\sqrt{5} - 15$
37. $12\sqrt{6} + 6\sqrt{14}$ **39.** $-1 - 2\sqrt{2}$
41. $8x - 14\sqrt{x} - 15$ **43.** $5z + 2\sqrt{15z} + 3$
45. $6a + 5\sqrt{3ab} - 3b$ **47.** $18r - 12\sqrt{2r} + 4$

49. $-6x - 12\sqrt{x} - 6$ **51.** $\frac{\sqrt{7}}{7}$ **53.** $\frac{\sqrt{6}}{3}$ **55.** $\frac{\sqrt{10}}{4}$
57. 2 **59.** $\frac{\sqrt[3]{4}}{2}$ **61.** $\frac{\sqrt[3]{6}}{3}$ **63.** $2\sqrt{2x}$ **65.** $\frac{\sqrt{5y}}{y}$
67. $\sqrt{2} + 1$ **69.** $\frac{3\sqrt{2} - \sqrt{10}}{4}$ **71.** $2 + \sqrt{3}$
73. $\frac{9 - 2\sqrt{14}}{5}$ **75.** $\frac{2(\sqrt{x} - 1)}{x - 1}$ **77.** $\frac{x(\sqrt{x} + 4)}{x - 16}$
79. $\frac{x - 2\sqrt{xy} + y}{x - y}$ **81.** $\frac{1}{\sqrt{3} - 1}$ **83.** $\frac{x - 9}{x(\sqrt{x} - 3)}$ **85.** 18
87. $5a\sqrt{b}$ **89.** $2a^2b^2\sqrt[3]{2}$ **91.** $3x(y + z)\sqrt[3]{4}$
93. $-8x\sqrt{10} + 6\sqrt{15x}$ **95.** $\sqrt[3]{3}$ **97.** $\frac{\sqrt[4]{4}}{2}$ **99.** $\frac{\sqrt[5]{2}}{2}$
101. $\frac{\sqrt[3]{2ab^2}}{b}$ **103.** $\sqrt{2z} + 1$ **105.** $\frac{x - y}{\sqrt{x}(\sqrt{x} - \sqrt{y})}$
107. $f/4$ **109.** $r = \frac{\sqrt{\pi A}}{\pi}$ **111.** $4\sqrt{2}$ cm
113. $2\sqrt{3}$ ft, $4\sqrt{3}$ ft **117.** $\frac{x - 9}{4(\sqrt{x} + 3)}$

Getting Ready (page 504)

1. a **2.** $5x$ **3.** $x + 4$ **4.** $y - 3$

Exercises 7.6 (page 510)

1. 7 **3.** 0 **5.** 17 **7.** 2 **9.** 6 **11.** $x^n = y^n$, power rule
13. square **15.** extraneous **17.** 2 **19.** 4 **21.** 4
23. $\frac{5}{2}, \frac{1}{2}$ **25.** 14; 6 is extraneous **27.** 4, 3
29. 2; 7 is extraneous **31.** 2, -1 **33.** 0 **35.** 1 **37.** 16
39. 0; 4 is extraneous **41.** 1 **43.** 6; 0 is extraneous
45. 2; 142 is extraneous **47.** 0 **49.** 3; -1 is extraneous
51. $h = \frac{v^2}{2g}$ **53.** $l = \frac{8T^2}{\pi^2}$ **55.** $A = P(r + 1)^3$
57. $v^2 = c^2\left(1 - \frac{L_A^2}{L_B^2}\right)$ **59.** 1, -4 **61.** 2; -8 is extraneous
63. -1; 1 is extraneous **65.** 1, 9 **67.** 2; -2 is extraneous
69. $\varnothing$; 6 is extraneous **71.** $\varnothing$; 8 is extraneous **73.** 2
75. 1 **77.** 2, -1 is extraneous **79.** 2,010 ft
81. about 29 mph **83.** 16% **85.** \$5 **87.** $R = \frac{8kl}{\pi r^4}$
91. 0, 4

Getting Ready (page 513)

1. $7x$ **2.** $-x + 10$ **3.** $12x^2 + 5x - 25$ **4.** $9x^2 - 25$

Exercises 7.7 (page 521)

1. $7i$ **3.** $10i$ **5.** $-i$ **7.** 1 **9.** 5 **11.** -1
13. 20 mph **15.** i **17.** $-i$ **19.** imaginary **21.** $\frac{\sqrt{a}}{\sqrt{b}}$
23. 5, 7 **25.** conjugates **27.** $3i$ **29.** $6i$ **31.** $i\sqrt{7}$
33. $2i\sqrt{2}$ **35.** yes **37.** no **39.** $8 - 2i$ **41.** $3 - 5i$
43. $15 + 7i$ **45.** $6 - 8i$ **47.** $3 + 6i$ **49.** $-25 - 25i$
51. $7 + i$ **53.** $14 - 8i$ **55.** $8 + \sqrt{2}i$ **57.** $3 + 4i$
59. $-5 + 12i$ **61.** $7 + 17i$ **63.** $-20 - 30i$ **65.** 61
67. $2 + i$ **69.** $\frac{1}{2} + \frac{5}{2}i$ **71.** $-\frac{42}{25} - \frac{6}{25}i$ **73.** $\frac{1}{4} + \frac{3}{4}i$
75. $\frac{5}{13} - \frac{12}{13}i$ **77.** $\frac{11}{10} + \frac{3}{10}i$ **79.** $\frac{1}{4} - \frac{\sqrt{15}}{4}i$
81. $-\frac{5}{169} + \frac{12}{169}i$ **83.** i **85.** $-i$ **87.** 1 **89.** i
91. $-1 - 3i$ **93.** $0 - i$ **95.** $0 + \frac{4}{5}i$ **97.** $\frac{1}{8} - 0i$
99. $0 + \frac{3}{5}i$ **101.** 10 **103.** 13 **105.** $\sqrt{74}$ **107.** 1
109. no **111.** $2 + 9i$ **113.** $-15 + 2\sqrt{3}i$ **115.** $5 + 5i$
117. $16 + 2i$ **119.** $\frac{3}{5} + \frac{4}{5}i$ **121.** $-\frac{6}{13} - \frac{9}{13}i$
127. $7 - 4i$ volts **129.** $1 - 3.4i$ **133.** $\frac{5}{3 + i}$

Chapter Review (page 524)

1. 7 **2.** -11 **3.** -6 **4.** 15 **5.** -3 **6.** -6 **7.** 5
8. -2 **9.** $5|x|$ **10.** $|x + 2|$ **11.** $3a^2b$ **12.** $4x^2|y|$

13.

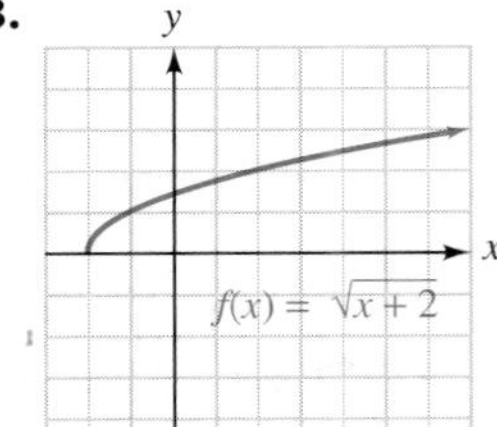

14.

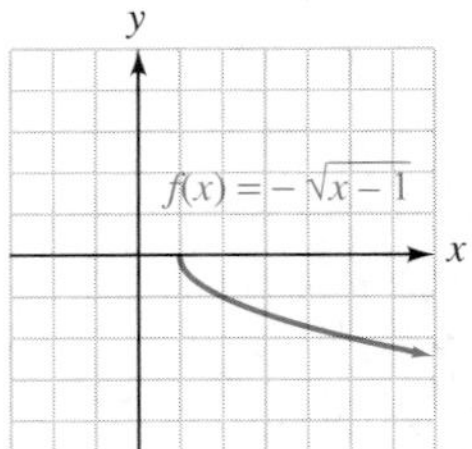

15.

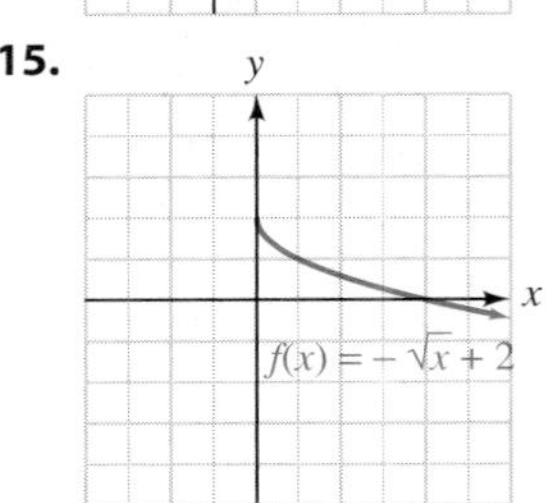

16.

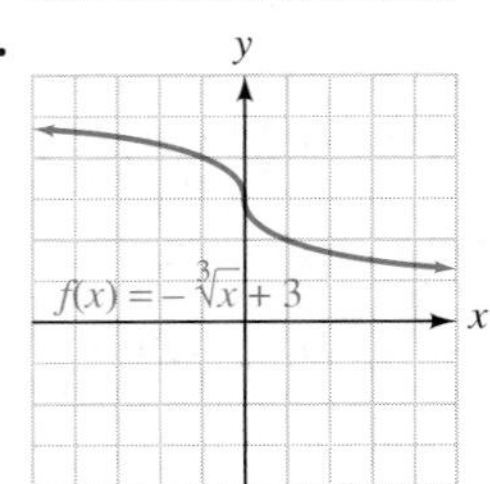

17. 12 **18.** about 5.7 **19.** 3 mi **20.** 8.2 ft **21.** 88 yd
22. 16,000 yd, or about 9 mi **23.** 13 **24.** 2.83 units
25. 5 **26.** -6 **27.** 27 **28.** 64 **29.** -2 **30.** -4
31. $\frac{1}{4}$ **32.** $\frac{1}{2}$ **33.** $-16{,}807$ **34.** $\frac{1}{3{,}125}$ **35.** 8 **36.** $\frac{27}{8}$
37. $3xy^{1/3}$ **38.** $3xy^{1/2}$ **39.** $125x^{9/2}y^6$ **40.** $\frac{1}{4u^{4/3}v^2}$
41. $5^{3/4}$ **42.** $a^{5/7}$ **43.** $u - 1$ **44.** $v + v^2$
45. $x + 2x^{1/2}y^{1/2} + y$ **46.** $a^{4/3} - b^{4/3}$ **47.** $\sqrt[3]{5}$ **48.** $\sqrt{x}$
49. $\sqrt[3]{3ab^2}$ **50.** $\sqrt{5ab}$ **51.** $4\sqrt{15}$ **52.** $3\sqrt[3]{2}$
53. $2\sqrt[4]{2}$ **54.** $2\sqrt[5]{3}$ **55.** $2x\sqrt{2x}$ **56.** $3x^2y\sqrt{2y}$
57. $2xy\sqrt[3]{2x^2y}$ **58.** $3x^2y\sqrt[3]{2x}$ **59.** $4x$ **60.** $2x$
61. $\frac{\sqrt[3]{2a^2b}}{3x}$ **62.** $\frac{\sqrt{17xy}}{8a^2}$ **63.** $3\sqrt{2}$ **64.** $\sqrt{5}$ **65.** 0
66. $8\sqrt[4]{2}$ **67.** $29x\sqrt{2}$ **68.** $32a\sqrt{3a}$ **69.** $13\sqrt[3]{2}$
70. $-4x\sqrt[4]{2x}$ **71.** $7\sqrt{2}$ m **72.** $6\sqrt{3}$ cm, 18 cm
73. 7.07 in. **74.** 8.66 cm **75.** $6\sqrt{10}$ **76.** 72 **77.** $3x$
78. 3 **79.** $-2x$ **80.** $-20x^3y^3\sqrt{xy}$ **81.** $4 - 3\sqrt{2}$
82. $2 + 3\sqrt{2}$ **83.** $\sqrt{10} - \sqrt{5}$ **84.** $3 + \sqrt{6}$ **85.** 1
86. $5 + 2\sqrt{6}$ **87.** $x - y$ **88.** $6u + \sqrt{u} - 12$ **89.** $\frac{\sqrt{3}}{3}$
90. $\frac{\sqrt{15}}{5}$ **91.** $\frac{\sqrt{xy}}{y}$ **92.** $\frac{\sqrt[3]{u^2}}{u^2v^2}$ **93.** $2(\sqrt{2} + 1)$
94. $\frac{\sqrt{6} + \sqrt{2}}{2}$ **95.** $2(\sqrt{x} - 4)$ **96.** $\frac{a + 2\sqrt{a} + 1}{a - 1}$
97. $\frac{3}{5\sqrt{3}}$ **98.** $\frac{1}{\sqrt[3]{3}}$ **99.** $\frac{9 - x}{2(3 + \sqrt{x})}$ **100.** $\frac{a - b}{a + \sqrt{ab}}$
101. 22 **102.** 16, 9 **103.** 3, 9 **104.** $\frac{9}{16}$ **105.** 2
106. 0, -2 **107.** $12 - 8i$ **108.** $2 - 68i$ **109.** $-96 + 3i$
110. $-2 - 2\sqrt{2}i$ **111.** $22 + 29i$ **112.** $-16 + 7i$
113. $-12 + 28\sqrt{3}i$ **114.** $118 + 10\sqrt{2}i$ **115.** $0 - \frac{3}{4}i$

116. $0 - \frac{2}{5}i$ **117.** $\frac{12}{5} - \frac{6}{5}i$ **118.** $\frac{21}{10} + \frac{7}{10}i$ **119.** $\frac{15}{17} + \frac{8}{17}i$
120. $\frac{4}{5} - \frac{3}{5}i$ **121.** $\frac{15}{29} - \frac{6}{29}i$ **122.** $\frac{1}{3} + \frac{1}{3}i$ **123.** 15
124. 26 **125.** 1 **126.** $-i$

Chapter 7 Test (page 532)

1. 7 **2.** 4 **3.** $2|x|$ **4.** $2x$
5. D: $[2, \infty)$, R: $[0, \infty)$ **6.** D: $(-\infty, \infty)$, R: $(-\infty, \infty)$

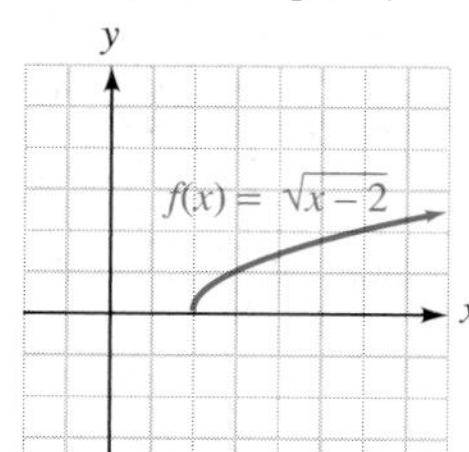

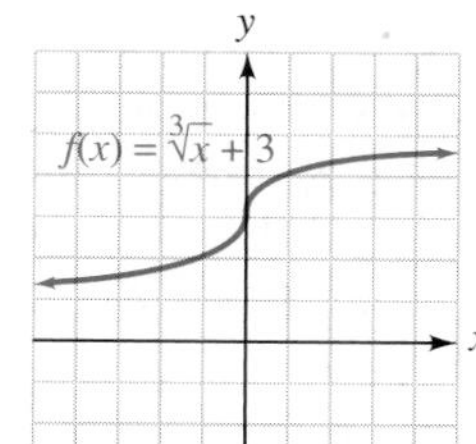

7. 28 in. **8.** 1.25 m **9.** 10 **10.** 25 **11.** 2 **12.** 9
13. $\frac{1}{216}$ **14.** $\frac{9}{4}$ **15.** $2^{4/3}$ **16.** $8xy$ **17.** $4\sqrt{3}$
18. $5xy^2\sqrt{10xy}$ **19.** $2x^5y\sqrt[3]{3}$ **20.** $\frac{1}{4a}$ **21.** $2|x|\sqrt{3}$
22. $2|x^3|\sqrt{2}$ **23.** $3x\sqrt[3]{3}$ **24.** $3x^2y^4\sqrt{2y}$ **25.** $-\sqrt[3]{3}$
26. $14\sqrt[3]{5}$ **27.** $2y^2\sqrt{3y}$ **28.** $6z\sqrt[4]{3z}$
29. $-6x\sqrt{y} - 2xy^2$ **30.** $3 - 7\sqrt{6}$ **31.** $\frac{\sqrt{5}}{5}$
32. $\sqrt{3t} + 1$ **33.** $\frac{3}{\sqrt{21}}$ **34.** $\frac{a - b}{a - 2\sqrt{ab} + b}$ **35.** 10
36. $\varnothing$; 4 is extraneous **37.** $-1 + 11i$ **38.** $4 - 7i$
39. $8 + 6i$ **40.** $-10 - 11i$ **41.** $0 - \frac{\sqrt{2}}{2}i$ **42.** $\frac{1}{2} + \frac{1}{2}i$

Getting Ready (page 535)

1. $(x + 5)(x - 5)$ **2.** $(b + 9)(b - 9)$ **3.** $(3x + 2)(2x - 1)$
4. $(2x - 3)(2x + 1)$

Exercises 8.1 (page 544)

1. ± 7 **3.** 4 **5.** $\frac{9}{4}$ **7.** 1 **9.** $t \le 4$
11. $x = \sqrt{c}, x = -\sqrt{c}$ **13.** positive or negative
15. $5, -5$ **17.** $0, -2$ **19.** $-2, -4$ **21.** $\frac{2}{3}, -\frac{5}{2}$ **23.** ± 6
25. $\pm\sqrt{5}$ **27.** $0, -2$ **29.** $2 \pm \sqrt{5}$ **31.** $\pm 4i$ **33.** $\pm\frac{9}{2}i$
35. $2, -4$ **37.** $2, 4$ **39.** $-1, -4$ **41.** $10, -1$
43. $-\frac{1}{3}, -\frac{3}{2}$ **45.** $\frac{5}{3}, -\frac{1}{2}$ **47.** $\frac{3}{4}, -\frac{3}{2}$ **49.** $1, -\frac{1}{2}$
51. $-\frac{7}{10} \pm \frac{\sqrt{29}}{10}$ **53.** $-1 \pm i$ **55.** $-4 \pm i\sqrt{2}$
57. $\frac{1}{3} \pm \frac{2\sqrt{2}}{3}i$ **59.** 6, 1 **61.** $\pm\frac{4\sqrt{3}}{3}$ **63.** 4, 10
65. $-5 \pm \sqrt{3}$ **67.** $2, \frac{1}{2}$ **69.** $1 \pm \frac{\sqrt{6}}{3}$ **71.** $\frac{1}{4} + \frac{3\sqrt{7}}{4}i$
73. $d = \frac{\sqrt{6h}}{2}$ **75.** $c = \frac{\sqrt{Em}}{m}$ **77.** $-\frac{1}{4} \pm \frac{\sqrt{41}}{4}$
79. $-\frac{1}{2} \pm \frac{\sqrt{13}}{2}$ **81.** 4 sec **83.** 72 mph **85.** 4%
87. width: $7\frac{1}{4}$ ft.; length: $13\frac{3}{4}$ ft **91.** $\frac{3}{4}$

Getting Ready (page 547)

1. $x^2 + 12x + 36, (x + 6)^2$ **2.** $x^2 - 7x + \frac{49}{4}, \left(x - \frac{7}{2}\right)^2$
3. 7 **4.** 8

Exercises 8.2 (page 552)

1. $3, -4, 7$ **3.** $y = \frac{-Ax + C}{B}$ **5.** $2\sqrt{6}$ **7.** $\sqrt{3}$
9. $3, -2, 6$ **11.** $-1, -2$ **13.** $-3, 5$ **15.** $-6, -6$
17. $-1, \frac{3}{2}$ **19.** $\frac{4}{3}, -\frac{2}{5}$ **21.** $-\frac{3}{2}, -\frac{1}{2}$ **23.** $\frac{1}{4}, -\frac{3}{4}$
25. $-\frac{1}{2} \pm \frac{\sqrt{5}}{10}$ **27.** $-1 \pm i$ **29.** $-\frac{1}{4} \pm \frac{\sqrt{7}}{4}i$
31. $\frac{2}{3} \pm \frac{\sqrt{2}}{3}i$ **33.** $\frac{1}{3} \pm \frac{2\sqrt{2}}{3}i$ **35.** $N = \frac{1 \pm \sqrt{1 + 8C}}{2}$
37. $x = \frac{k}{2} \pm \frac{\sqrt{k^2 - 4ay}}{2}$ **39.** $\frac{1}{2}, -\frac{1}{3}$ **41.** $-\frac{5}{2} \pm \frac{\sqrt{17}}{2}$
43. $\frac{3}{4} \pm \frac{\sqrt{17}}{4}$ **45.** $-\frac{1}{2} \pm \sqrt{5}$ **47.** $-\frac{1}{3} \pm \frac{\sqrt{5}}{3}i$
49. $8.98, -3.98$ **51.** $x^2 - 8x + 15 = 0$
53. $x^3 - x^2 - 14x + 24 = 0$ **55.** 8 ft by 12 ft **57.** 4 units
59. $\frac{4}{3}$ cm **61.** 16, 18 or $-16, -18$ **63.** 6, 7 or $-6, -7$
65. 30 mph **67.** \$4.80 or \$5.20 **69.** 4,000
71. 2.26 in. **73.** 1985 **75.** about 6.13×10^{-3} M
79. $\sqrt{2}, -3\sqrt{2}$ **81.** $i, 2i$

Getting Ready (page 555)

1. 17 **2.** -8

Exercises 8.3 (page 561)

1. -3 **3.** irrational and unequal **5.** no **7.** -2 **9.** $\frac{9}{5}$
11. $b^2 - 4ac$ **13.** rational, unequal **15.** rational, equal
17. complex conjugates **19.** irrational, unequal
21. rational, unequal **23.** $6, -6$ **25.** $12, -12$
27. $1, -1, 4, -4$ **29.** $1, -1, \sqrt{2}, -\sqrt{2}$
31. $1, -1, \sqrt{5}, -\sqrt{5}$ **33.** $1, -1, 2, -2$ **35.** 16, 4
37. $\frac{9}{4}$; 1 is extraneous **39.** 1; $\frac{9}{4}$ is extraneous
41. $\varnothing$; $\frac{4}{9}$ and 1 are both extraneous **43.** $-8, -27$
45. $-1, 27$ **47.** $-1, -4$ **49.** $4, -5$ **51.** 0, 2
53. $-1, -\frac{27}{13}$ **55.** $1, 1, -1, -1$ **57.** $-4, \frac{2}{3}$
59. $x = \pm\sqrt{r^2 - y^2}$ **61.** $y = \frac{-3x \pm \sqrt{9x^2 - 28x}}{2x}$
63. $\frac{2}{3}, -\frac{1}{4}$ **65.** $-\frac{5}{4} \pm \frac{\sqrt{17}}{4}$ **67.** $\frac{1}{3} \pm \frac{\sqrt{11}}{3}i$ **69.** $-1 \pm 2i$
71. yes **73.** $\frac{81}{16}$; 1 is extraneous **75.** $-\frac{8}{27}, 1$
77. $1, -1, 2\sqrt{3}, -2\sqrt{3}$ **79.** $2, -2, i\sqrt{7}, -i\sqrt{7}$ **81.** $\frac{3}{8}, 1$
83. $\frac{1}{27}, -1$ **85.** $1 \pm i$ **87.** $-\frac{5}{7}, 3$ **89.** $d = \pm\frac{\sqrt{kI}}{I}$
91. $\mu^2 = \frac{\Sigma x^2}{N} - \sigma^2$ **93.** 5 **95.** $12, -3$ **97.** $k < -\frac{4}{3}$
101. no

Getting Ready (page 563)

1. -2 **2.** 2 **3.** 0 **4.** 8 **5.** 1 **6.** 4

Exercises 8.4 (page 575)

1. down **3.** up **5.** $(3, -1)$ **7.** 10 **9.** $3\frac{3}{5}$ hr
11. $f(x) = ax^2 + bx + c, a \neq 0$ **13.** maximum, minimum, vertex **15.** upward **17.** to the right **19.** upward

21.

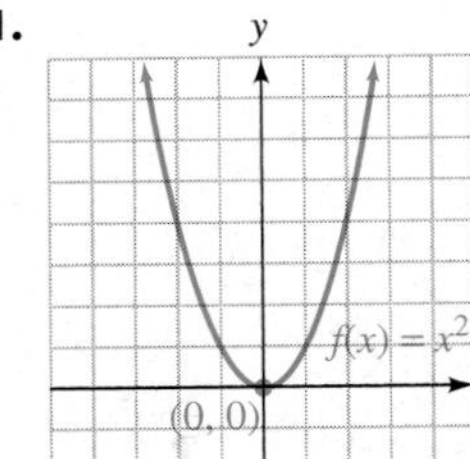

23.

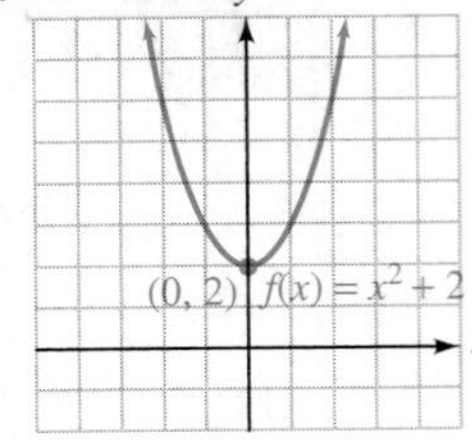

25.

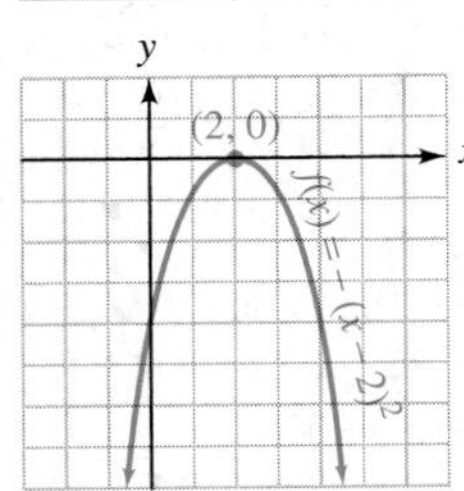

27.

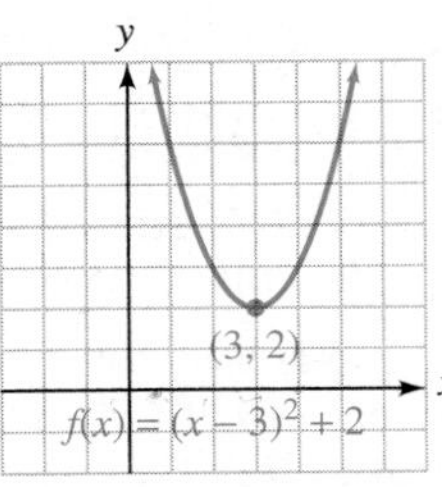

29. $(1, 2), x = 1$ **31.** $(-3, -4), x = -3$ **33.** $(0, 0), x = 0$
35. $(1, -2), x = 1$

37.

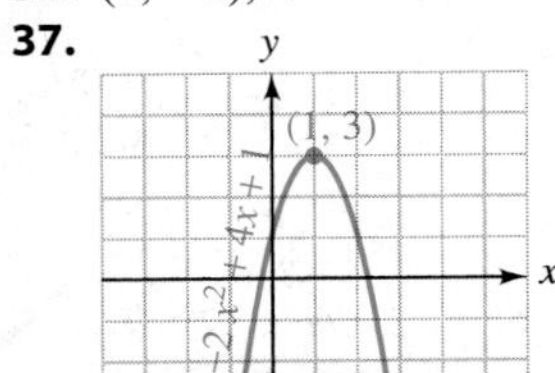

39.

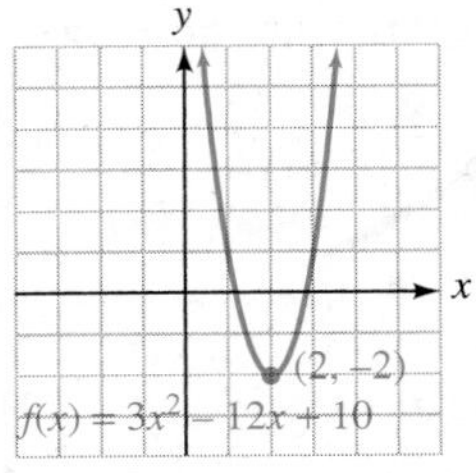

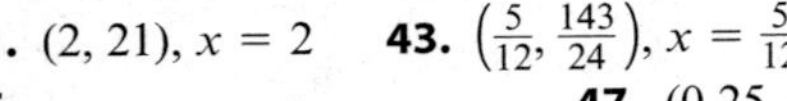

41. $(2, 21), x = 2$ **43.** $\left(\frac{5}{12}, \frac{143}{24}\right), x = \frac{5}{12}$

45.

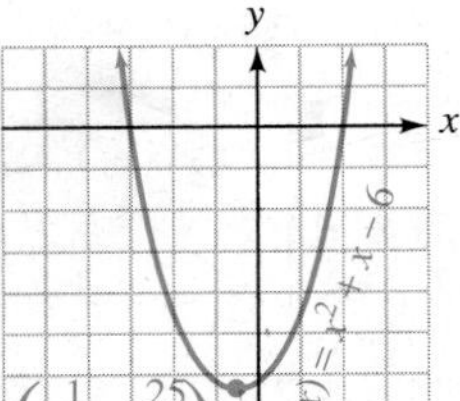

47. $(0.25, 0.88)$ **49.** $(0.5, 7.25)$
51. $2, -3$ **53.** $-1.85, 3.25$
55. $(5, 2)$ **57.** 36 ft, 1.5 sec
59. 50 ft by 50 ft, 2,500 ft^2
61. 0.25 and 0.75 **63.** 75 ft by 75 ft, 5,625 ft^2 **65.** 14 ft
67. 5,000 **69.** 3,276, \$14,742
71. \$35

Getting Ready (page 579)

1. $(x + 5)(x - 3)$ **2.** $(x - 2)(x - 1)$

Exercises 8.5 (page 586)

1. $x = 2$ **3.** $x < 2$ **5.** $x > -3$ **7.** $1 < 2x$ **9.** $y = kx$
11. $t = kxy$ **13.** 3 **15.** greater **17.** quadratic
19. undefined **21.** sign

23.

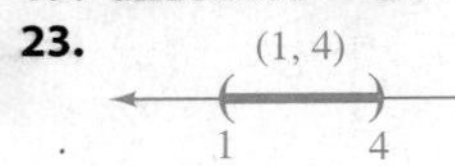

25.

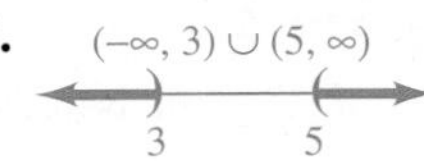

27. [−4, 3]

−4 3

29.

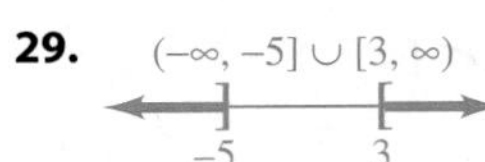

31. ∅

33.

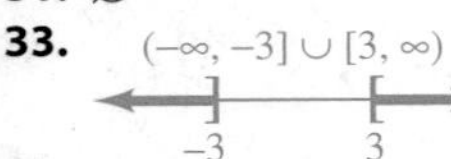

35.

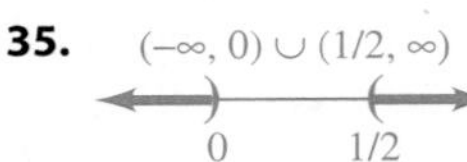

37.

39. (−∞, −3) ∪ (1, 4)

−3 1 4

41. [−5, −2) ∪ [4, ∞)

−5 −2 4

43.

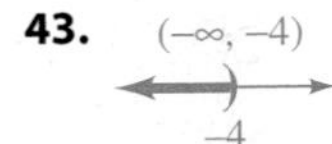

45. (−1/2, 1/3) ∪ (1/2, ∞)

−1/2 1/3 1/2

47. (0, 2) ∪ (8, ∞)

0 2 8

49. (−∞, −2) ∪ (2, 18]

−2 2 18

51. [−34/5, −4) ∪ (3, ∞)

−34/5 −4 3

53. (−∞, −2) ∪ (−2, ∞)

−2

55.

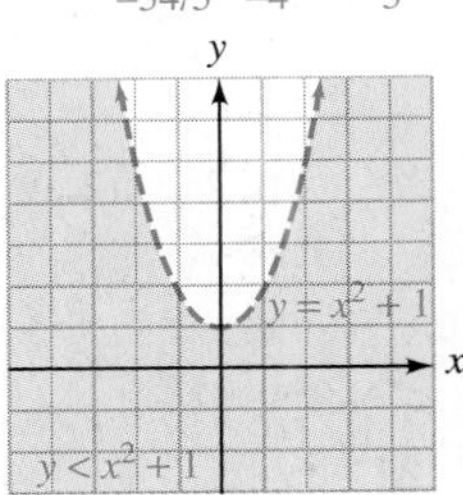

57. y = x² + 5x + 6; y ≤ x² + 5x + 6

59.

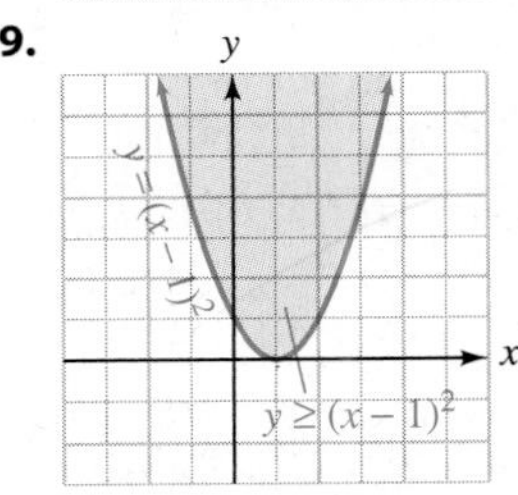

61. −x² − y + 6 > −x; −x² − y + 6 = −x

63.

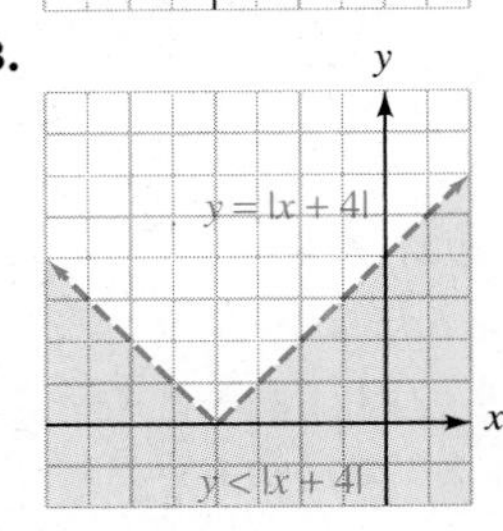

65.

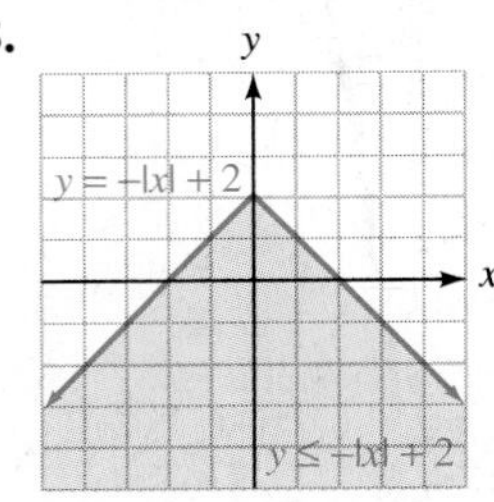

67.

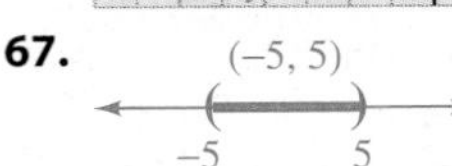

69.

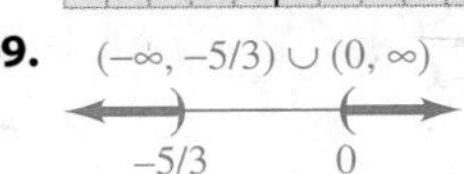

71.

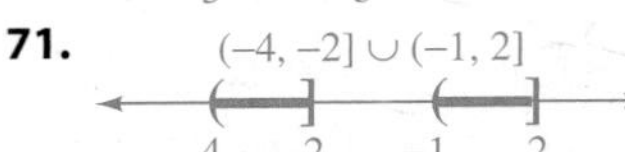

73.

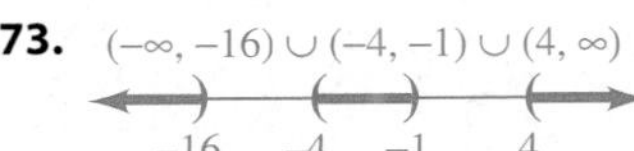

75. $(-1, 3)$ **77.** $(-\infty, -3) \cup (2, \infty)$
81. when 4 factors are negative, 2 factors are negative, or no factors are negative

Getting Ready (page 589)

1. $3x - 1$ **2.** $x + 3$ **3.** $2x^2 - 3x - 2$ **4.** $\frac{2x + 1}{x - 2}$ $(x \neq 2)$

Exercises 8.6 (page 595)

1. $5x$ **3.** $8x^2$ **5.** 2 **7.** $8x$ **9.** $-\frac{3x+7}{x+2}$
11. $\frac{x-4}{3x^2-x-12}$ **13.** $f(x)+g(x)$ **15.** $f(x)g(x)$
17. domain **19.** $f(g(x))$ **21.** $f(x)$ **23.** $7x, (-\infty, \infty)$
25. $12x^2, (-\infty, \infty)$ **27.** $x, (-\infty, \infty)$
29. $\frac{4}{3}, (-\infty, 0) \cup (0, \infty)$ **31.** $3x-2, (-\infty, \infty)$
33. $2x^2-5x-3, (-\infty, \infty)$ **35.** $-x-4, (-\infty, \infty)$
37. $\frac{x-3}{2x+1}, \left(-\infty, -\frac{1}{2}\right) \cup \left(-\frac{1}{2}, \infty\right)$ **39.** 7 **41.** 24 **43.** -1
45. $-\frac{1}{2}$ **47.** $2x^2-1$ **49.** $16x^2+8x$ **51.** 2
53. $2x+h$ **55.** $4x+2h$ **57.** $2x+h+1$
59. $2x+h+3$ **61.** $4x+2h+3$ **63.** $-2x^2+3x-3, (-\infty, \infty)$ **65.** $(3x-2)/(2x^2+1), (-\infty, \infty)$ **67.** $3, (-\infty, \infty)$
69. $(x^2-4)/(x^2-1), (-\infty, -1) \cup (-1, 1) \cup (1, \infty)$ **71.** 58
73. 110 **75.** 2 **77.** $9x^2-9x+2$ **79.** 2 **81.** $x+a$
83. $2x+2a$ **85.** $x+a+1$ **87.** $x+a+3$
89. $2x+2a+3$ **95.** $3x^2+3xh+h^2$
97. $C(t) = \frac{5}{9}(2{,}668 - 200t)$

Getting Ready (page 597)

1. $y = \frac{x-2}{3}$ **2.** $y = \frac{2x-10}{3}$

Exercises 8.7 (page 604)

1. $\{(2, 1), (3, 2), (10, 5)\}$ **3.** $f^{-1}(x) = 2x$ **5.** no
7. $3-8i$ **9.** $18-i$ **11.** 10 **13.** one-to-one **15.** 2
17. x **19.** yes **21.** no

23.

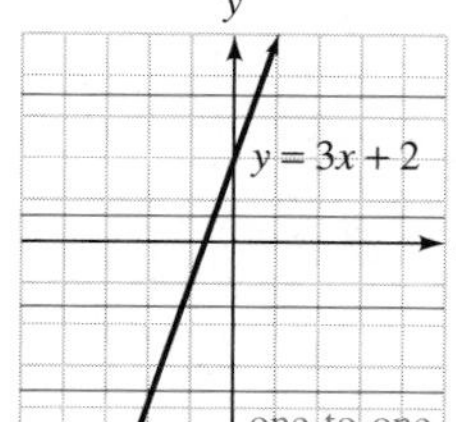

25.

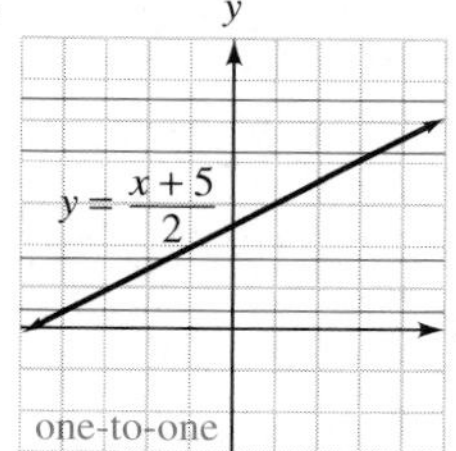

27.

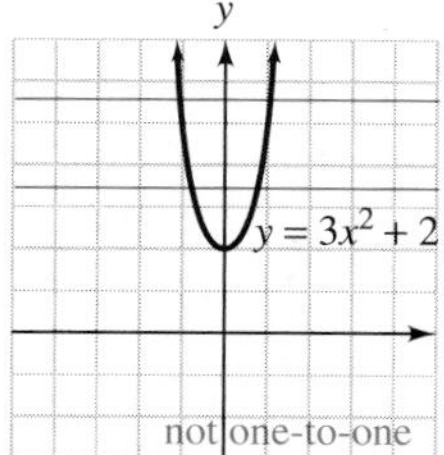

29.

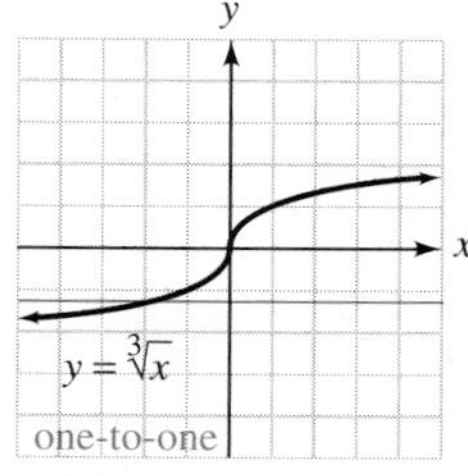

31. $\{(2, 3), (1, 2), (0, 1)\}$; yes **33.** $\{(2, 1), (3, 2), (3, 1), (5, 1)\}$; no **35.** $f^{-1}(x) = \frac{1}{3}x - \frac{1}{3}$ **37.** $f^{-1}(x) = 5x - 4$
39. $f^{-1}(x) = 5x + 4$ **41.** $f^{-1}(x) = \frac{5}{4}x + 5$

43.

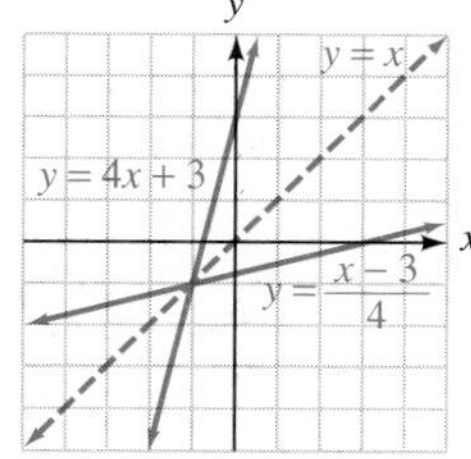

45.

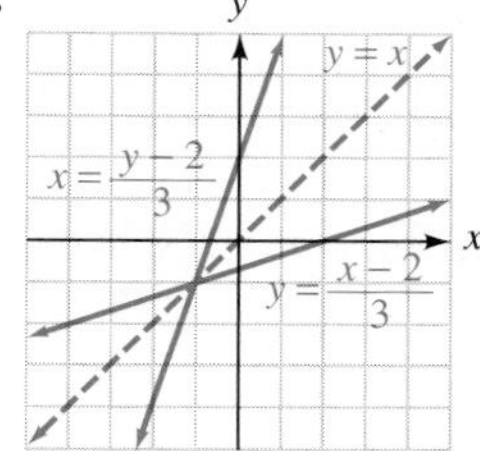

47.

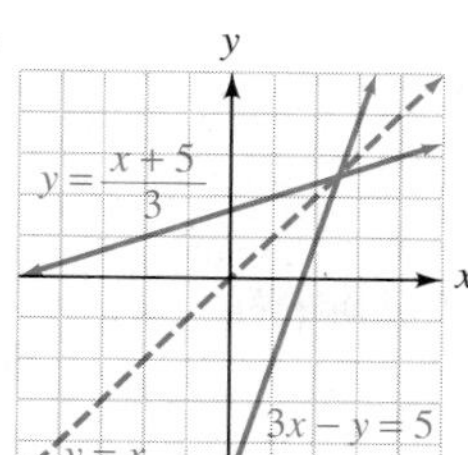

49.

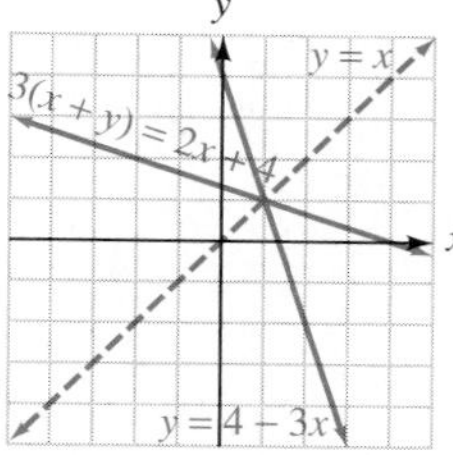

51. $y = \pm\sqrt{x-4}$, no **53.** $f^{-1}x = \sqrt[3]{x}$, yes

55.

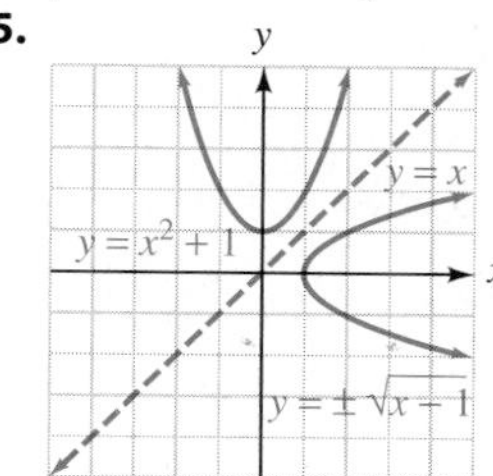

57. y, x, $y = x^2$, $x \geq 0$, $y = \sqrt{x}$, $y = x$

59. $\{(1, 1), (4, 2), (9, 3), (16, 4)\}$ **61.** $x = |y|$
63. $f^{-1}(x) = \sqrt[3]{\frac{x+3}{2}}$ **67.** $f^{-1}(x) = \frac{x+1}{x-1}$

Chapter Review (page 607)

1. $\frac{2}{3}, -\frac{3}{4}$ **2.** $-\frac{1}{3}, -\frac{5}{2}$ **3.** $\frac{2}{3}, -\frac{4}{5}$ **4.** $4, -8$ **5.** $-4, -2$
6. $\frac{7}{2}, 1$ **7.** $\frac{1}{4} \pm \frac{\sqrt{41}}{4}$ **8.** $9, -1$ **9.** $0, 10$ **10.** $\frac{1}{2}, -7$
11. $-7, \frac{1}{3}$ **12.** $\frac{1}{4} \pm \frac{\sqrt{17}}{4}$ **13.** $-\frac{1}{2} \pm \frac{\sqrt{7}}{2}i$
14. 4 cm by 6 cm **15.** 2 ft by 3 ft **16.** 7 sec **17.** 196 ft
18. irrational, unequal **19.** complex conjugates
20. 12, 152 **21.** $k \geq -\frac{7}{3}$ **22.** 1, 144 **23.** $8, -27$
24. 1 **25.** $1, -\frac{8}{5}$ **26.** $\frac{14}{3}$ **27.** 1

28.

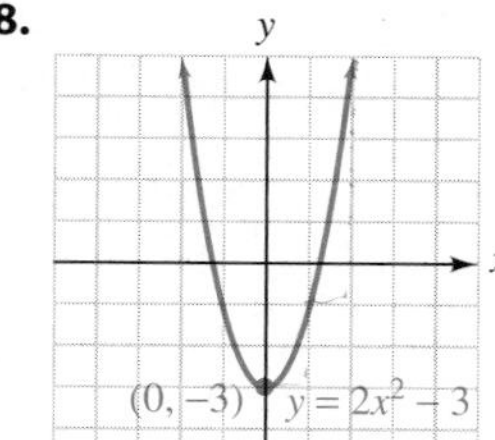

29.

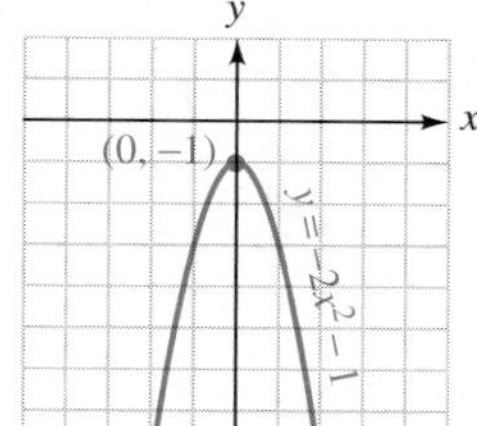

30. y, x, (2, 1), $y = -4(x-2)^2 + 1$

31. y, x, (−1, −6), $y = 5x^2 + 10x - 1$

32. $(2, -17)$

33. $(-\infty, -7) \cup (5, \infty)$; −7, 5

34. $(-9, 2)$; −9, 2

35. $(-\infty, 0) \cup [3/5, \infty)$; 0, 3/5

36. $(-7/2, 1) \cup (4, \infty)$; −7/2, 1, 4

37. $x < -7$ or $x > 5$ **38.** $-9 < x < 2$

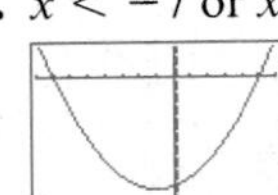

39. $x < 0$ or $x \geq \frac{3}{5}$ **40.** $-\frac{7}{2} < x < 1$ or $x > 4$

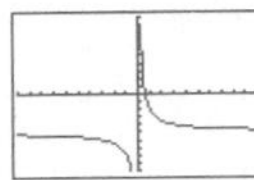

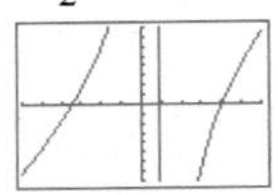

41.

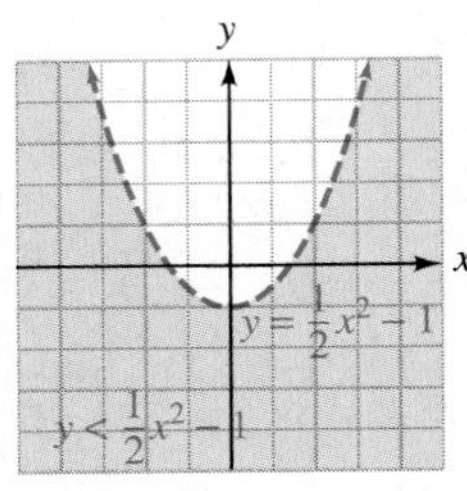

42.

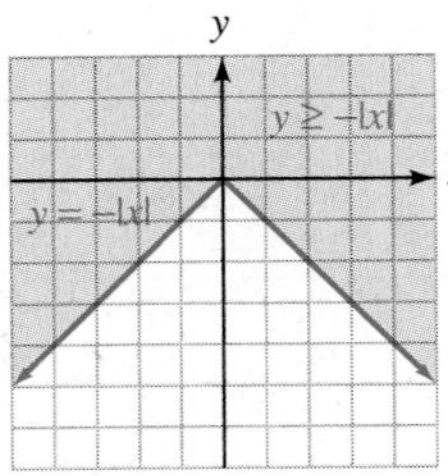

43. $(f + g)(x) = 3x + 1$ **44.** $(f - g)(x) = x - 1$

45. $(f \cdot g)(x) = 2x^2 + 2x$ **46.** $(f/g)(x) = \frac{2x}{x + 1}$ $(x \neq -1)$

47. 6 **48.** -1 **49.** $2(x + 1)$ **50.** $2x + 1$

51. yes

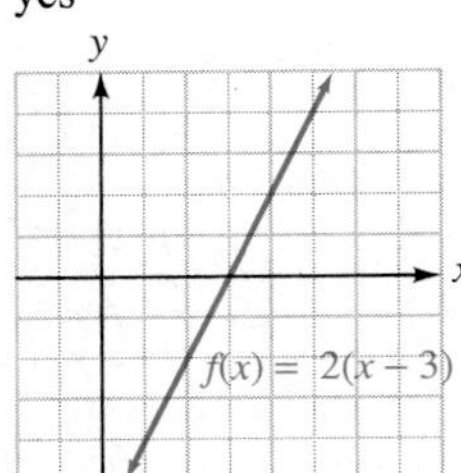

52. no

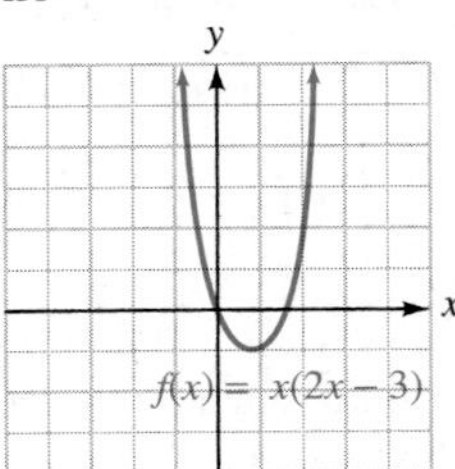

53. no

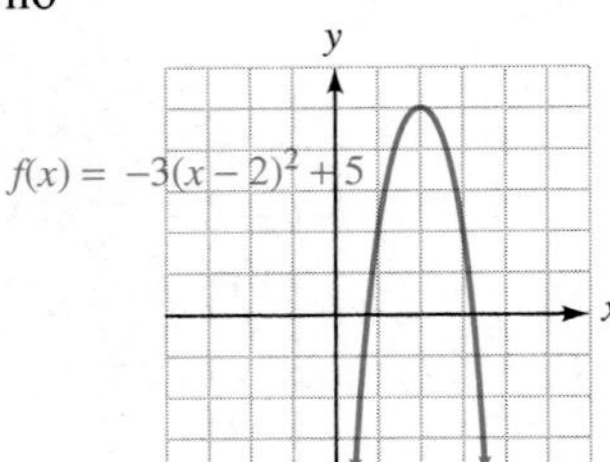

54. no

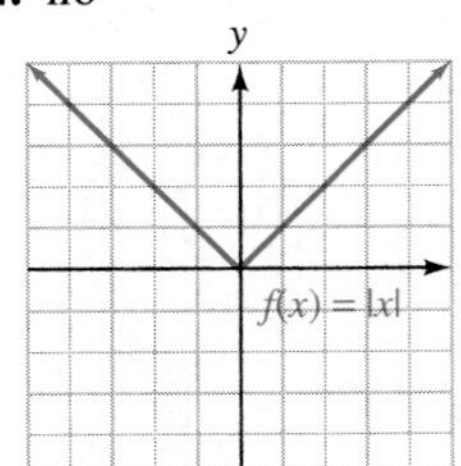

55. $f^{-1}(x) = \frac{x + 3}{6}$ **56.** $f^{-1}(x) = \frac{x - 5}{4}$

57. $y = \sqrt{\frac{x + 1}{2}}$ **58.** $x = |y|$

Chapter 8 Test (page 615)

1. 3, -6 **2.** $-\frac{3}{2}, -\frac{5}{3}$ **3.** 144 **4.** 625 **5.** $-2 \pm \sqrt{3}$

6. $\frac{5}{2} \pm \frac{\sqrt{37}}{2}$ **7.** $-\frac{5}{4} \pm \frac{\sqrt{17}}{4}$ **8.** $\frac{1}{2} \pm \frac{\sqrt{11}}{2}i$ **9.** nonreal

10. 2 **11.** 10 in. **12.** 1, $\frac{1}{4}$

13.

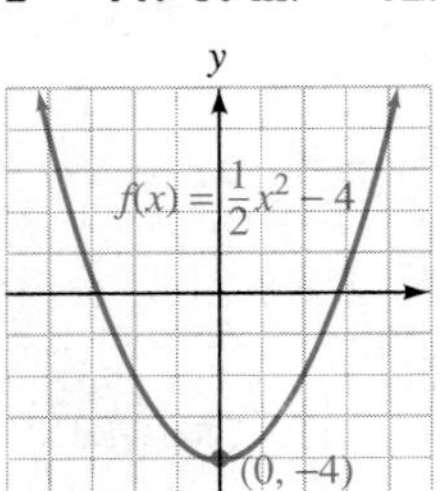

14. (2, 1)

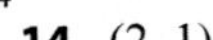

15.

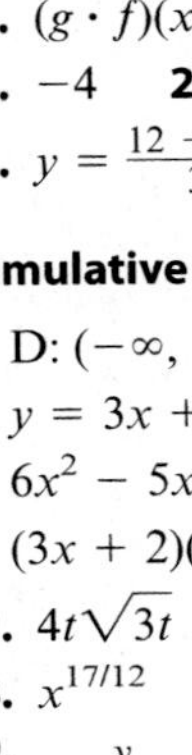

16. $(-\infty, -2) \cup (4, \infty)$

−2 4

17. $(-3, 2]$

−3 2

18. $(g + f)(x) = 5x - 1$

19. $(f - g)(x) = 3x + 1$

20. $(g \cdot f)(x) = 4x^2 - 4x$ **21.** $(g/f)(x) = \frac{x - 1}{4x}$ **22.** 3

23. -4 **24.** -8 **25.** -9 **26.** $4(x - 1)$ **27.** $4x - 1$

28. $y = \frac{12 - 2x}{3}$ **29.** $y = -\sqrt{\frac{x - 4}{3}}$

Cumulative Review Exercises (page 616)

1. D: $(-\infty, \infty)$; R: $[-3, \infty)$ **2.** D: $(-\infty, \infty)$; R: $(-\infty, 0]$

3. $y = 3x + 2$ **4.** $y = -\frac{2}{3}x - 2$ **5.** $-4a^2 + 12a - 7$

6. $6x^2 - 5x - 6$ **7.** $(x^2 + 4y^2)(x + 2y)(x - 2y)$

8. $(3x + 2)(5x - 4)$ **9.** 6, -1 **10.** 0, $\frac{2}{3}$, $-\frac{1}{2}$ **11.** $5x^2$

12. $4t\sqrt{3t}$ **13.** $-3x$ **14.** $4x$ **15.** $\frac{1}{2}$ **16.** 16 **17.** y^2

18. $x^{17/12}$

19.

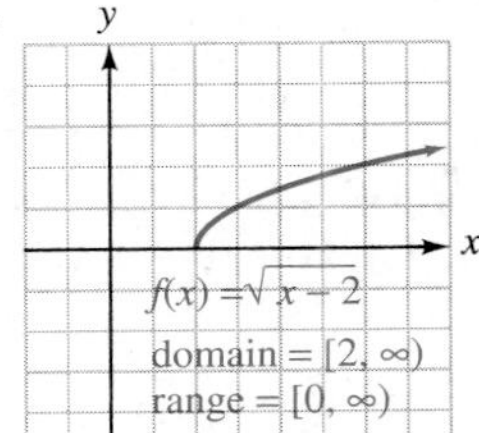

20.

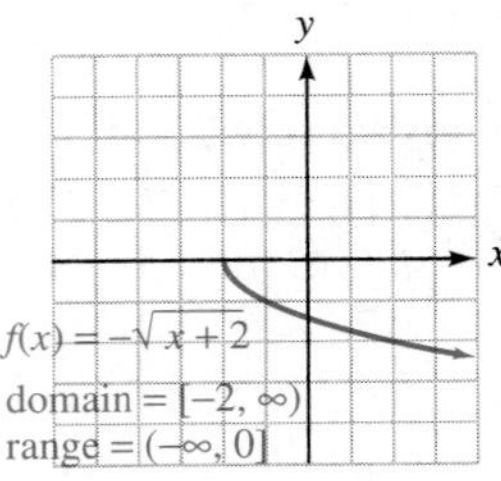

21. $x^{4/3} - x^{2/3}$ **22.** $\frac{1}{x} + 2 + x$ **23.** $7\sqrt{2}$

24. $-12\sqrt[4]{2} + 10\sqrt[4]{3}$ **25.** $-18\sqrt{6}$ **26.** $\frac{5\sqrt[3]{x^2}}{x}$

27. $\frac{x + 3\sqrt{x} + 2}{x - 1}$ **28.** $\sqrt{xy}$ **29.** 2, 7 **30.** $\frac{1}{4}$

31. $3\sqrt{2}$ in. **32.** $2\sqrt{3}$ in. **33.** 10 **34.** 9

35. 1, $-\frac{3}{2}$ **36.** $-\frac{2}{3} \pm \frac{\sqrt{7}}{3}$

37.

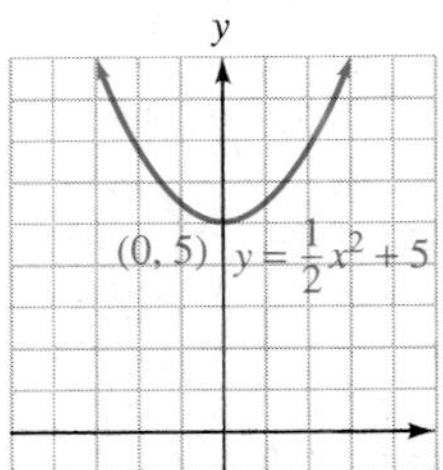

38.

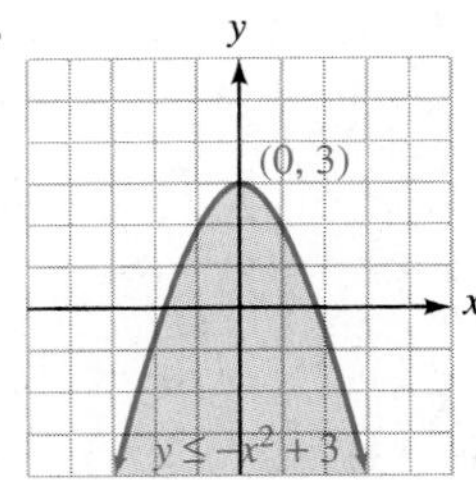

39. $7 + 2i$ **40.** $-5 - 7i$ **41.** 13 **42.** $12 - 6i$

43. $-12 - 10i$ **44.** $\frac{3}{2} + \frac{1}{2}i$ **45.** $\sqrt{13}$ **46.** $\sqrt{61}$

47. -2 **48.** 9, 16

49. $(-\infty, -2) \cup (3, \infty)$

−2 3

50. $[-2, 3]$

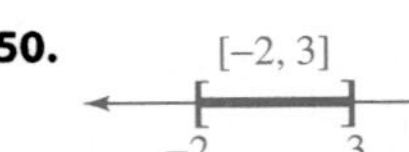

51. 5 **52.** 27 **53.** $12x^2 - 12x + 5$ **54.** $6x^2 + 3$

55. $f^{-1}(x) = \frac{x - 2}{3}$ **56.** $f^{-1}(x) = \sqrt[3]{x - 4}$

Getting Ready (page 619)

1. 8 **2.** 5 **3.** $\frac{1}{25}$ **4.** $\frac{8}{27}$

Exercises 9.1 (page 629)

1. 4 **3.** 18 **5.** $\frac{1}{4}$ **7.** $\frac{2}{9}$ **9.** 40 **11.** 120°
13. exponential **15.** $(0, \infty)$ **17.** increasing
19. $P\left(1 + \frac{r}{k}\right)^{kt}$ **21.** 2.6651 **23.** 36.5548 **25.** 8
27. $7^{3\sqrt{3}}$

29.

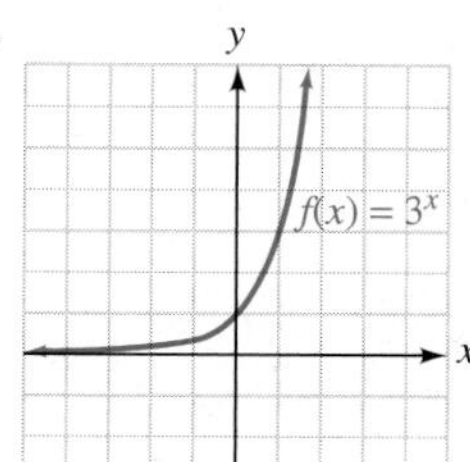

31.

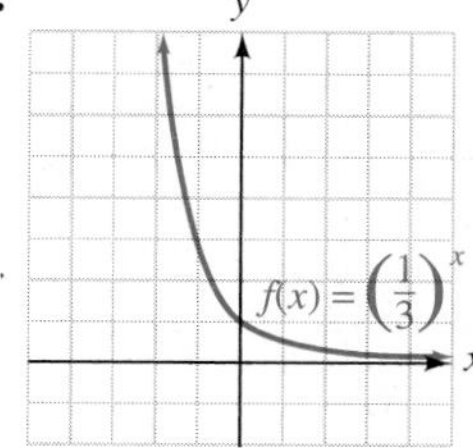

33. $b = \frac{1}{2}$ **35.** $b = 3$

37.

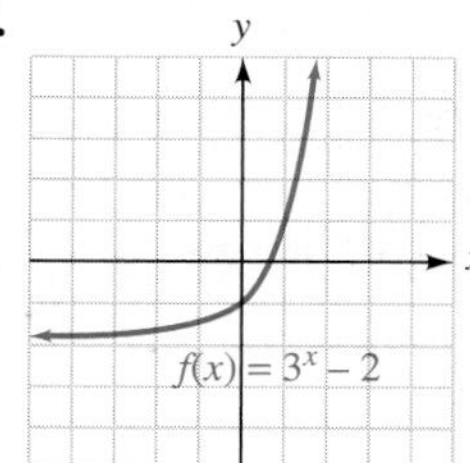

39.

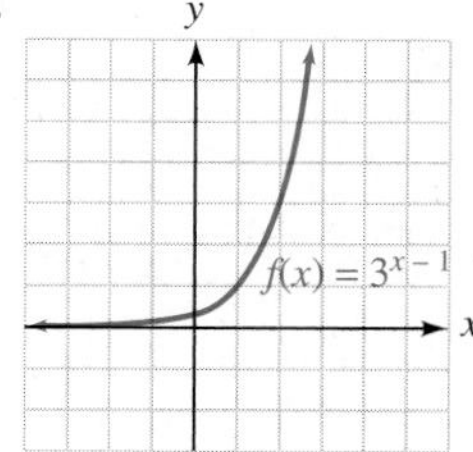

41. $b = 2$
43. increasing function **45.** decreasing function

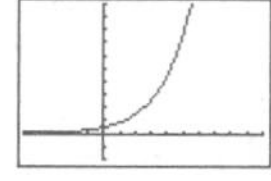

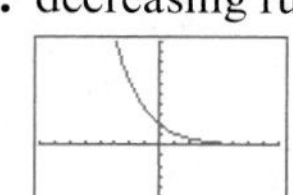

47. approximately 29.8 million
49. $\frac{32}{243}A_0$ **51.** \$22,080.40 **53.** \$32.03
55. \$2,273,996.13 **57.** 5.0421×10^{-5} coulombs
59. \$1,115.33

Getting Ready (page 631)

1. 2 **2.** 2.25 **3.** 2.44 **4.** 2.59

Exercises 9.2 (page 637)

1. 1 **3.** 7.39 **5.** 2 **7.** $4x^2\sqrt{15x}$ **9.** $10y\sqrt{3y}$
11. 2.72 **13.** increasing **15.** $A = Pe^{rt}$

17.

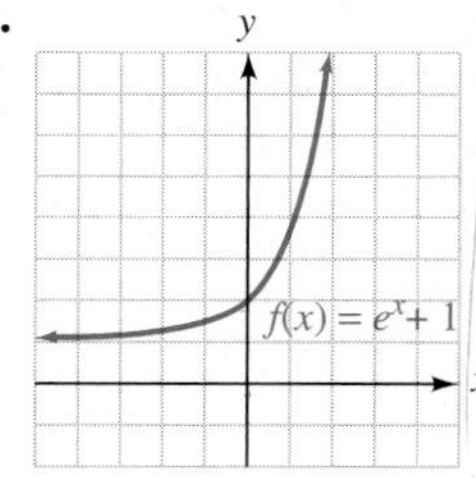

19.

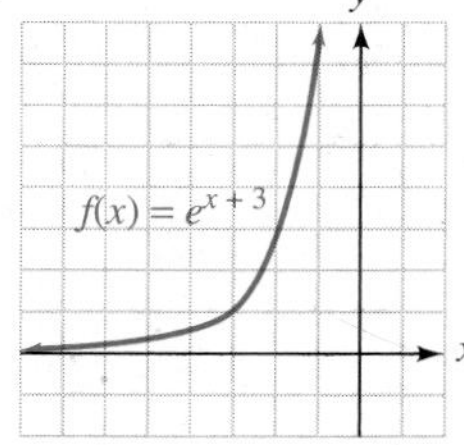

21.

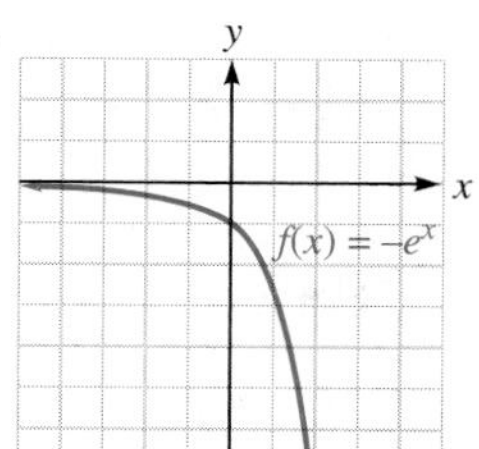

23.

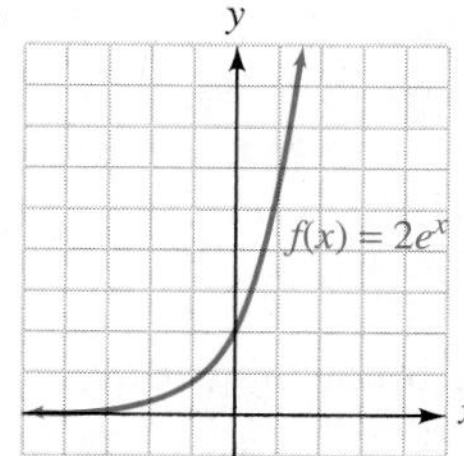

25. no **27.** no **29.** \$10,272.17 **31.** \$6,391.10
33. 10.6 billion **35.** 2.6 **37.** 72 yr **39.** 3.68 grams
41. 2,498.27grams **43.** \$7,518.28 from annual compounding; \$7,647.95 from continuous compounding **45.** 6,817
47. 0.076 **49.** 0 mps **51.** this object **53.** \$3,094.15
57. 2 **59.** $k = e^5$

Getting Ready (page 640)

1. 1 **2.** 25 **3.** $\frac{1}{25}$ **4.** 4

Exercises 9.3 (page 647)

1. 3 **3.** 5 **5.** 2 **7.** $\frac{1}{4}$ **9.** 2 **11.** $\varnothing$; $\frac{3}{5}$ is extraneous
13. 4 **15.** $(0, \infty)$ **17.** x **19.** exponent
21. $(b, 1), (1, 0)$ **23.** $20 \log \frac{E_O}{E_I}$ **25.** $3^3 = 27$
27. $\left(\frac{1}{2}\right)^2 = \frac{1}{4}$ **29.** $4^{-3} = \frac{1}{64}$ **31.** $\left(\frac{1}{2}\right)^3 = \frac{1}{8}$
33. $\log_6 36 = 2$ **35.** $\log_5 \frac{1}{25} = -2$ **37.** $\log_{1/2} 32 = -5$
39. $\log_x z = y$ **41.** 49 **43.** 6 **45.** 5 **47.** $\frac{1}{25}$ **49.** 5
51. $\frac{3}{2}$ **53.** 4 **55.** 2 **57.** 3 **59.** $\frac{1}{2}$
61. increasing **63.** decreasing

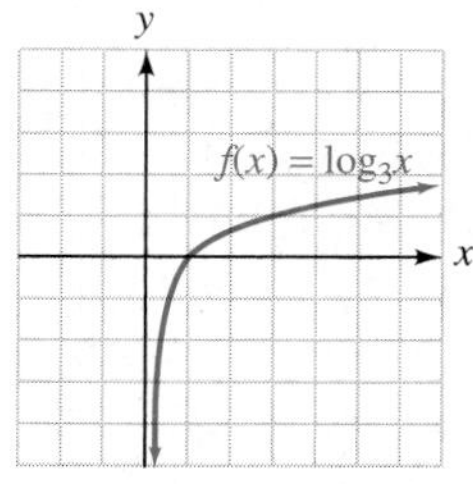

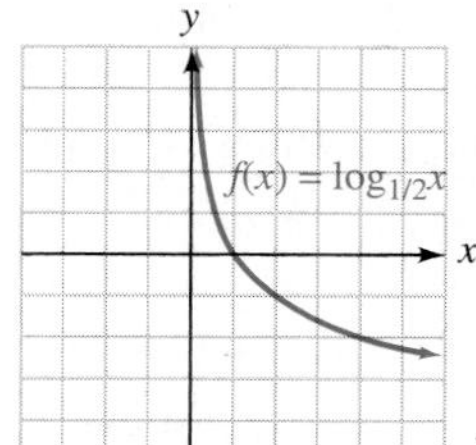

65.

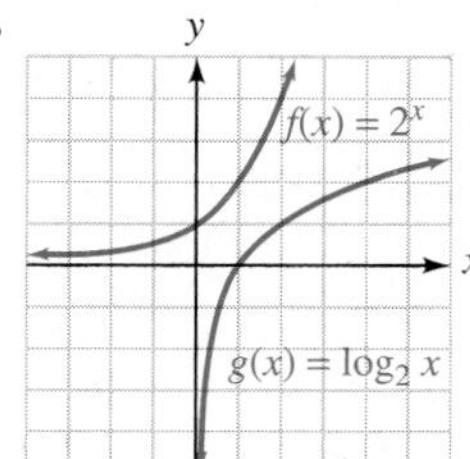

67.

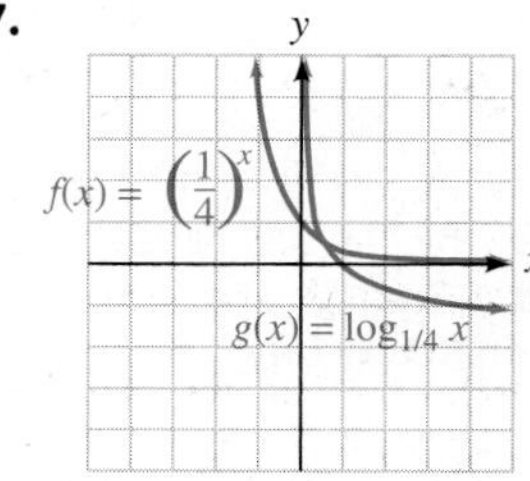

69.

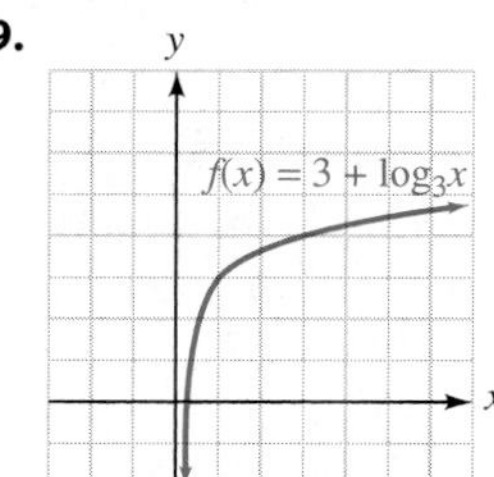

71.

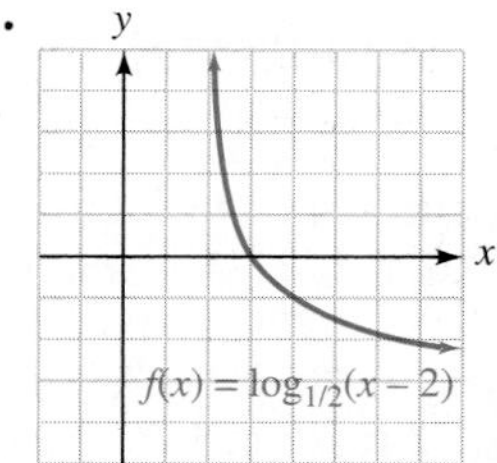

73. 0.9165 **75.** −2.0620 **77.** 17,378.01 **79.** 0.00

81. $\frac{1}{6}$ **83.** -3 **85.** $-\frac{3}{2}$ **87.** $\frac{2}{3}$ **89.** 8 **91.** 4 **93.** 4
95. 4 **97.** 3 **99.** 100 **101.** 25.25 **103.** 8
105. $b = 3$ **107.** no value of b **109.** 29.0 dB
111. 49.5 dB **113.** 4.4 **115.** 4 **117.** 4.2 yr old
119. 10.8 yr

Getting Ready (page 650)

1. 2 **2.** -3 **3.** 1 **4.** 0

Exercises 9.4 (page 654)

1. $e^y = x$ **3.** $t = \frac{\ln 2}{r}$ **5.** $y = 9x + 5$ **7.** $y = -\frac{3}{2}x + \frac{1}{2}$
9. $y = 5$ **11.** $\frac{2x^2 + x + 1}{x(x + 1)}$ **13.** $\frac{x + y}{y - x}$ **15.** $(0, \infty), (-\infty, \infty)$
17. 10 **19.** $\frac{\ln 2}{r}$ **21.** 3.2288 **23.** 2.2915 **25.** -0.1592
27. no real value **29.** 9.9892 **31.** 23.8075 **33.** 0.0089
35. 61.9098 **37.**

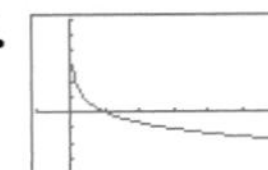

39.

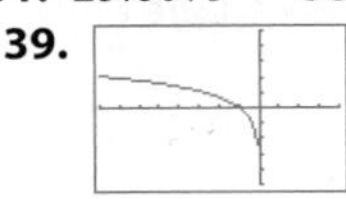

41. no **43.** no **45.** 5.8 yr **47.** 13.9 yr
49. about 2.9 hr

Getting Ready (page 657)

1. x^{m+n} **2.** 1 **3.** x^{mn} **4.** x^{m-n}

Exercises 9.5 (page 664)

1. 2 **3.** 343 **5.** 2 **7.** $\frac{1}{4}$ **9.** 2 **11.** $\sqrt{85}$
13. $y = -\frac{7}{6}x + \frac{2}{3}$ **15.** 1 **17.** x **19.** $-$ **21.** x
23. $=$ **25.** 0 **27.** 7 **29.** 10 **31.** 1 **33.** 0 **35.** 7
37. 10 **39.** 1 **45.** $\log_b x + \log_b y + \log_b z$
47. $\log_b 2 + \log_b x - \log_b y$ **49.** $3 \log_b x + 2 \log_b y$
51. $\frac{1}{2}(\log_b x + \log_b y)$ **53.** $\log_b x + \frac{1}{2} \log_b z$
55. $\frac{1}{3} \log_b x - \frac{1}{4} \log_b y - \frac{1}{4} \log_b z$ **57.** $\log_b \frac{x + 1}{x}$
59. $\log_b x^2 y^{1/2}$ **61.** $\log_b \frac{z^{1/2}}{x^3 y^2}$ **63.** $\log_b \frac{\frac{x}{z} + x}{\frac{y}{z} + y} = \log_b \frac{x}{y}$
65. 1.4472 **67.** 0.3521 **69.** 1.1972 **71.** 2.4014
73. 2.0493 **75.** 0.4682 **77.** 1.7712 **79.** -1.0000
81. 1.8928 **83.** 2.3219 **87.** false **89.** false **91.** true
93. false **95.** true **97.** true **99.** 4.77
101. from 2.5119×10^{-8} to 1.585×10^{-7} **103.** It will increase by $k \ln 2$. **105.** The intensity must be cubed.

Getting Ready (page 667)

1. $2 \log x$ **2.** $\frac{1}{2} \log x$ **3.** 0 **4.** $2b \log a$

Exercises 9.6 (page 675)

1. $\frac{\log 5}{\log 3}$ **3.** $-\frac{\log 7}{\log 2}$ **5.** 2 **7.** 10 **9.** 0, 5 **11.** $\frac{2}{3}, -4$
13. exponential **15.** $A_0 e^{-kt}$ **17.** 1.1610 **19.** 3.9120
21. 0 **23.** 22.1184 **25.** 1.2702 **27.** 1.7095 **29.** 4, -1
31. $-2, -2$ **33.** ± 1.0878 **35.** 0, 1.0566 **37.** 1.8
39. 3, -1 **41.** 2 **43.** 3 **45.** -7 **47.** 10, -10
49. 50; -2 is extraneous **51.** 20; -5 is extraneous
53. 10; -100 is extraneous **55.** 10; 1 is extraneous
57. 4; 0 is extraneous **59.** 4; 1 is extraneous **61.** 20
63. 8 **65.** 0 **67.** 6; -1 is extraneous
69. 9; -9 is extraneous **71.** 0.2789 **73.** 100, 1 **75.** 4
77. 1, 7 **79.** 4 **81.** 53 days **83.** 6.2 yr **85.** about 4,200 yr **87.** 25.3 yr **89.** 2.828 times larger **91.** 13.3
93. 42.7 days **95.** 5.6 yr **97.** 5.4 yr
99. because $\ln 2 \approx 0.7$ **103.** $x \leq 3$

Chapter Review (page 679)

1. $5^{2\sqrt{2}}$ **2.** $2^{\sqrt{10}}$
3.

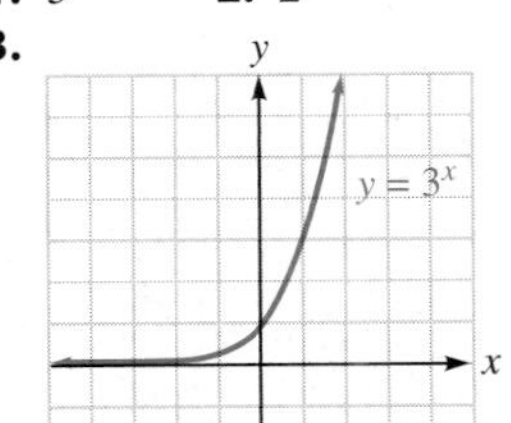

4.

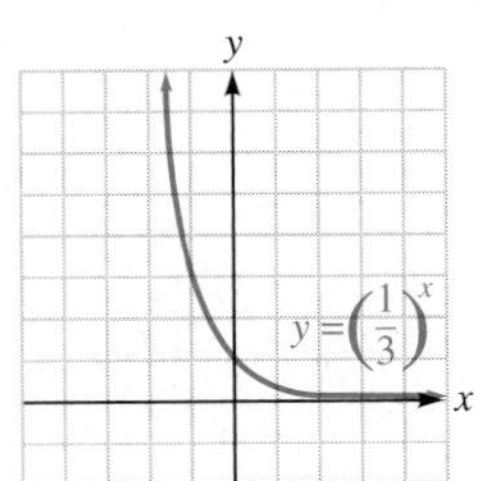

5. $x = 1, y = 6$ **6.** D: $(-\infty, \infty)$, R: $(0, \infty)$
7.

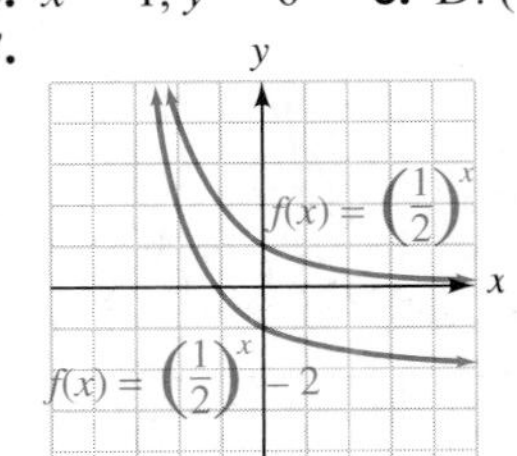

8.

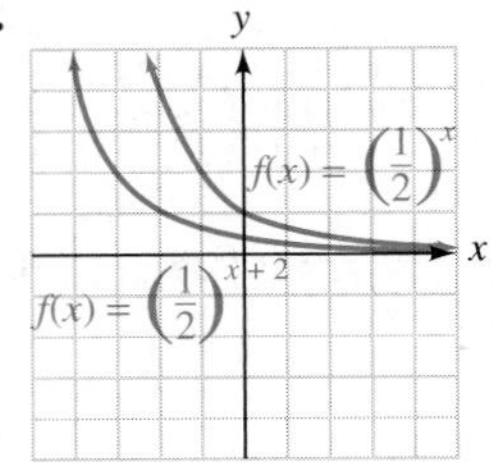

9. \$2,189,703.45 **10.** \$2,324,767.37
11.

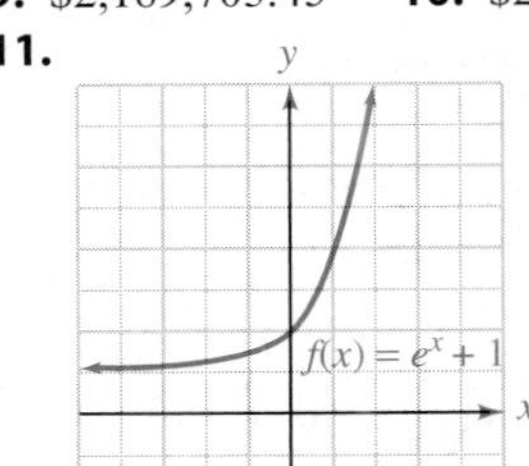

12.

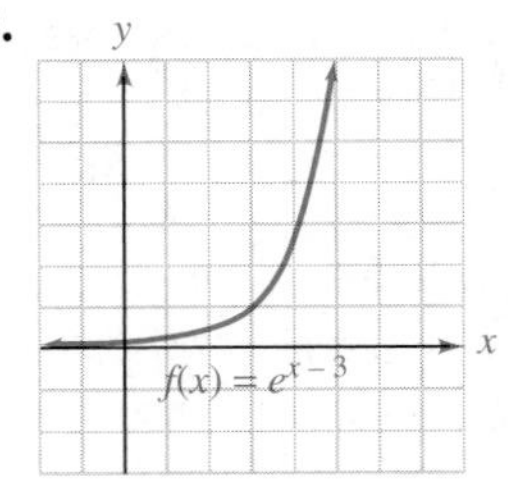

13. about 582,000,000 **14.** 30.69 grams
15. D: $(0, \infty)$, R: $(-\infty, \infty)$ **17.** 2 **18.** $-\frac{1}{2}$ **19.** 0
20. -2 **21.** $\frac{1}{2}$ **22.** $\frac{1}{3}$ **23.** 32 **24.** 9 **25.** 27
26. -1 **27.** $\frac{1}{8}$ **28.** 2 **29.** 4 **30.** 2 **31.** 10 **32.** $\frac{1}{25}$
33. 5 **34.** 3
35.

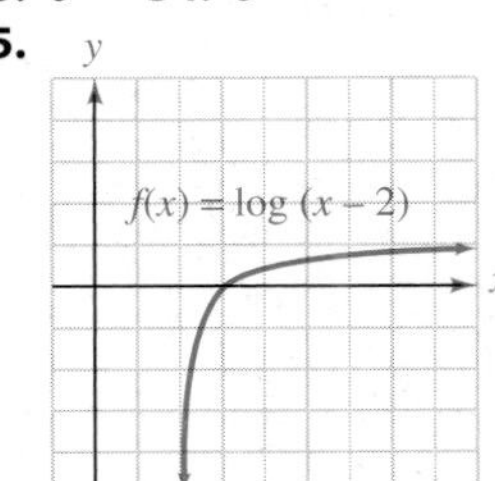

36.

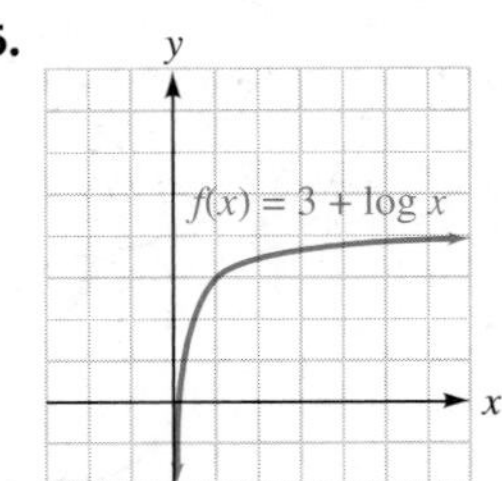

37.

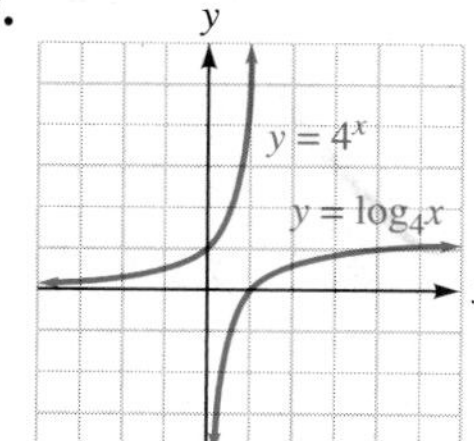

38.

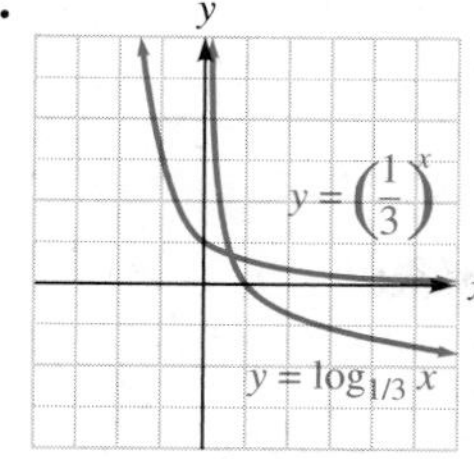

39. 53 dB **40.** 4.4 **41.** 6.1137 **42.** -0.1111
43. 10.3398 **44.** 2.5715
45.

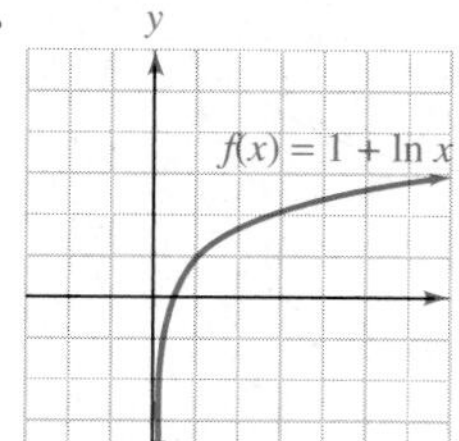

46.

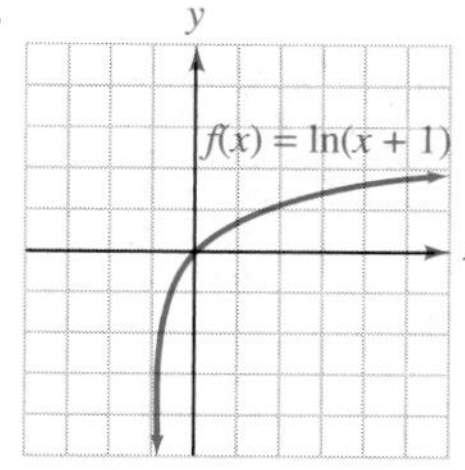

47. 23 yr **48.** 0 **49.** 1 **50.** 3 **51.** 4 **52.** 4 **53.** 0
54. 7 **55.** 3 **56.** 4 **57.** 9
58. $2 \log_b x + 3 \log_b y - 4 \log_b z$
59. $\frac{1}{2}(\log_b x - \log_b y - 2 \log_b z)$ **60.** $\log_b \frac{x^3z^7}{y^5}$
61. $\log_b \frac{y^3\sqrt{x}}{z^7}$ **62.** 3.36 **63.** 1.56 **64.** 2.64
65. -6.72 **66.** 1.7604 **67.** about 7.94×10^{-4} gram-ions per liter **68.** $k \ln 2$ less **69.** $\frac{\log 7}{\log 3} \approx 1.7712$ **70.** 2
71. 31.0335 **72.** 2 **73.** $\frac{\log 3}{\log 3 - \log 2} \approx 2.7095$
74. $-1, -3$ **75.** 25, 4 **76.** 4; -2 is extraneous
77. 2; -3 is extraneous **78.** 4, 3 **79.** 6; -1 is extraneous
80. 31 **81.** $\frac{\ln 9}{\ln 2} \approx 3.1699$ **82.** $\varnothing$ **83.** $\frac{e}{e - 1} \approx 1.5820$
84. 1 **85.** about 3,400 yr

Chapter 9 Test (page 685)

1.

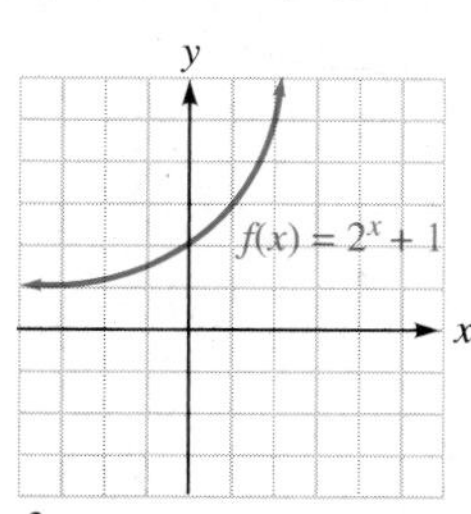

2.

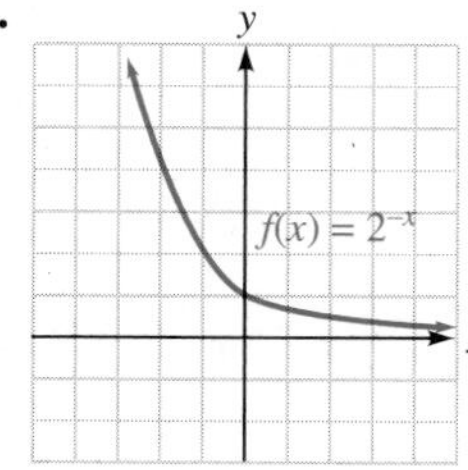

3. $\frac{3}{64}$ gram **4.** $1,060.90 **5.**

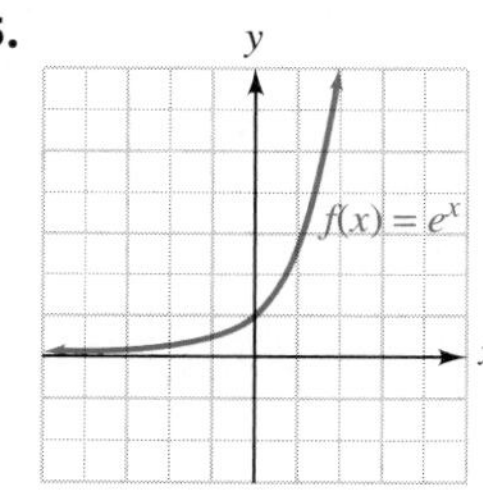

6. $4,451.08 **7.** 2 **8.** 3 **9.** $\frac{1}{27}$
10. 10 **11.** 2 **12.** $\frac{27}{8}$

13.

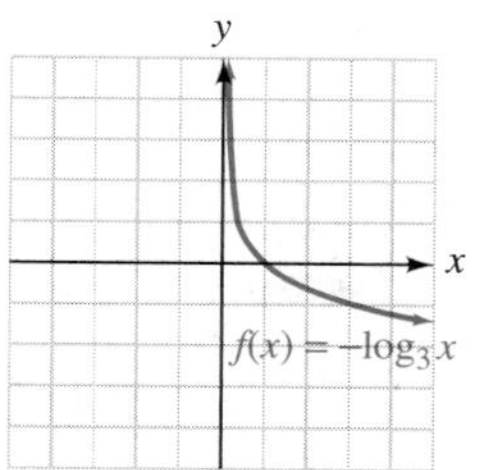

14.

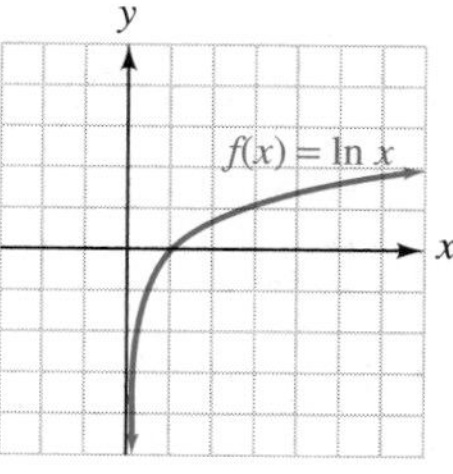

15. $2 \log a + \log b + 3 \log c$ **16.** $\frac{1}{2}(\ln a - 2 \ln b - \ln c)$
17. $\log \frac{b\sqrt{a + 2}}{c^3}$ **18.** $\log \frac{\sqrt[3]{a}}{c\sqrt[3]{b^2}}$ **19.** 1.3801
20. 0.4259 **21.** $\frac{\log 3}{\log 7}$ or $\frac{\ln 3}{\ln 7}$ **22.** $\frac{\log e}{\log \pi}$ or $\frac{\ln e}{\ln \pi}$ **23.** true
24. false **25.** false **26.** false **27.** 6.4 **28.** 46
29. $\frac{\log 3}{\log 5}$ **30.** $\frac{\log 3}{(\log 3) - 2}$ **31.** 1 **32.** 10; -1 is extraneous

Getting Ready (page 687)

1. $x^2 - 4x + 4$ **2.** $x^2 + 8x + 16$ **3.** $\frac{81}{4}$ **4.** 36

Exercises 10.1 (page 696)

1. (0, 0), 12 **3.** (2, 0), 4 **5.** down **7.** left **9.** 5, $-\frac{7}{3}$
11. 3, $-\frac{1}{4}$ **13.** conic **15.** standard, (0, 3), 4
17. circle, general **19.** parabola, (3, 2), right
21.

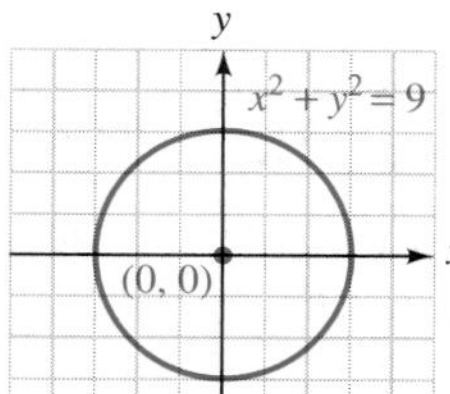

23.

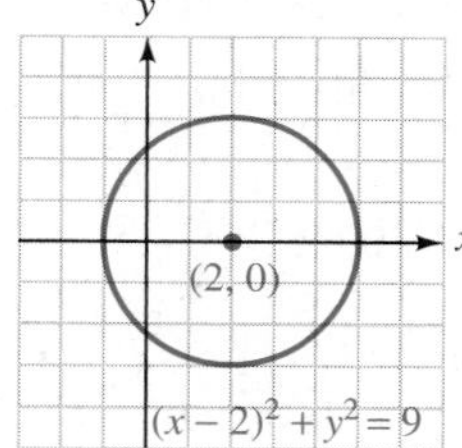

25.

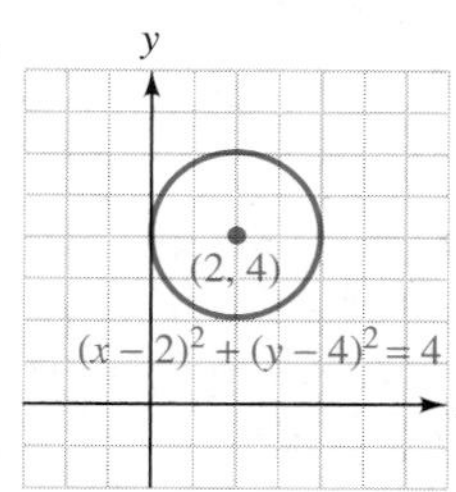

27.

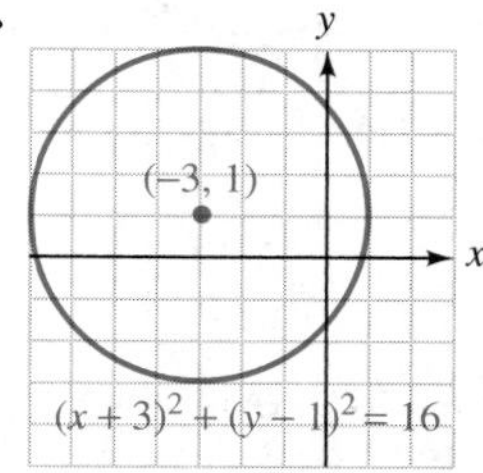

29.

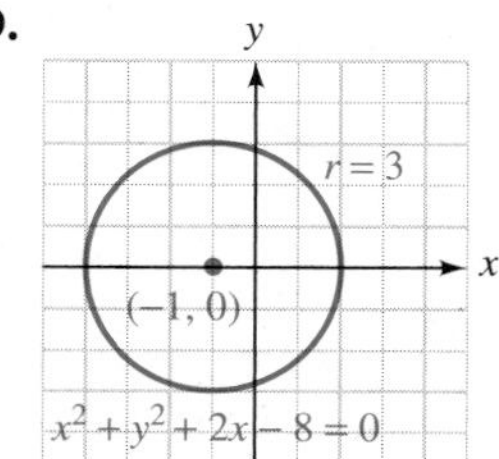

31.

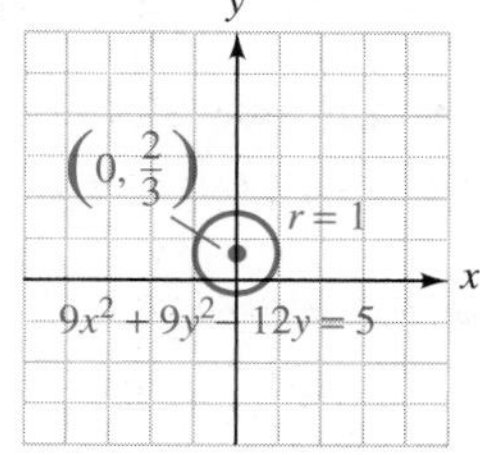

33.

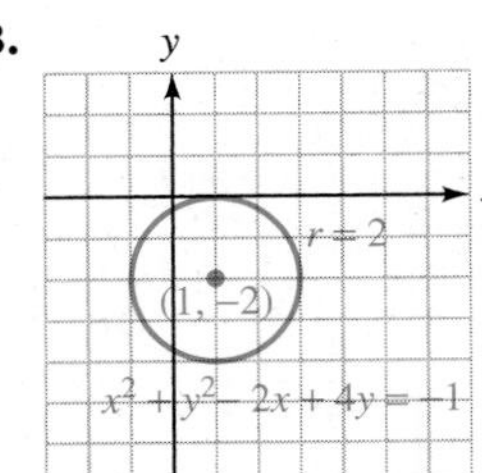

35.

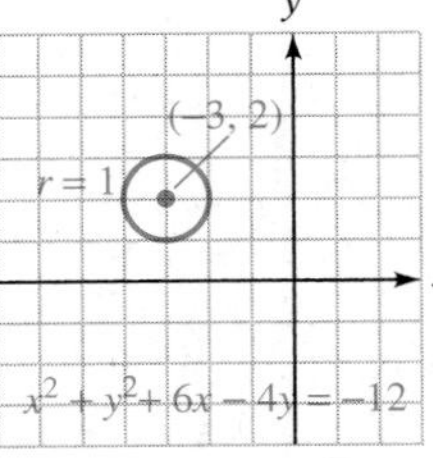

37. $x^2 + y^2 = 1, x^2 + y^2 - 1 = 0$
39. $(x - 6)^2 + (y - 8)^2 = 25, x^2 + y^2 - 12x - 16y + 75 = 0$
41. $(x + 2)^2 + (y - 6)^2 = 144, x^2 + y^2 + 4x - 12y - 104 = 0$
43. $x^2 + y^2 = 2, x^2 + y^2 - 2 = 0$

45.

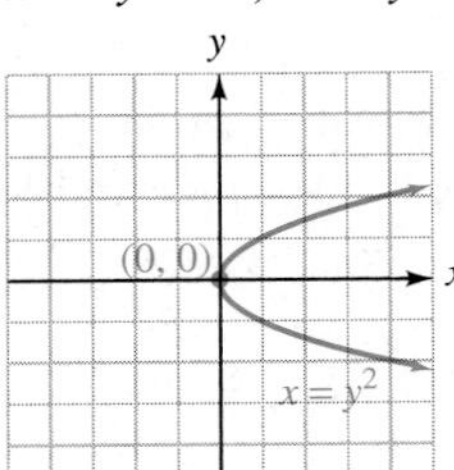

47.

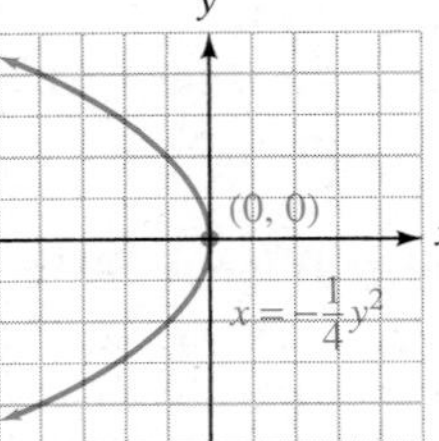

49.

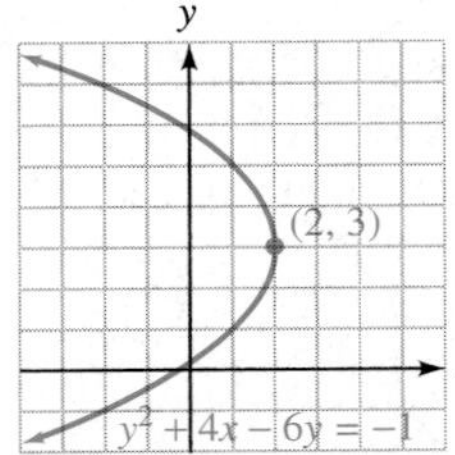

51.

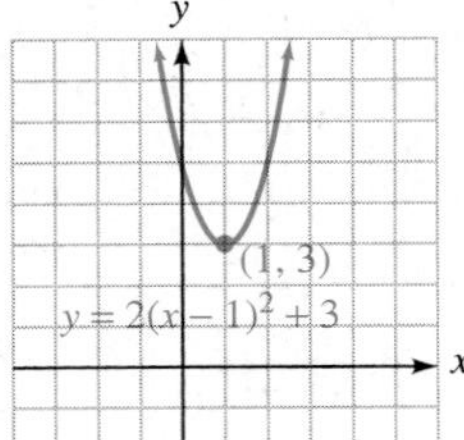

53.

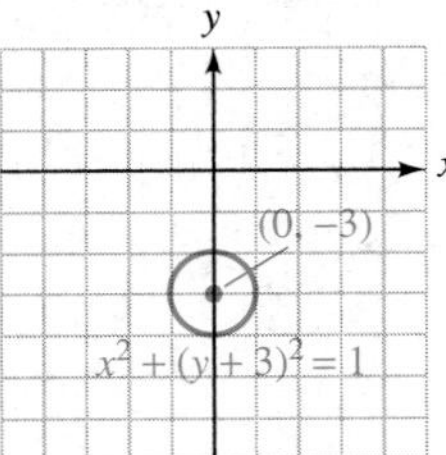

55.

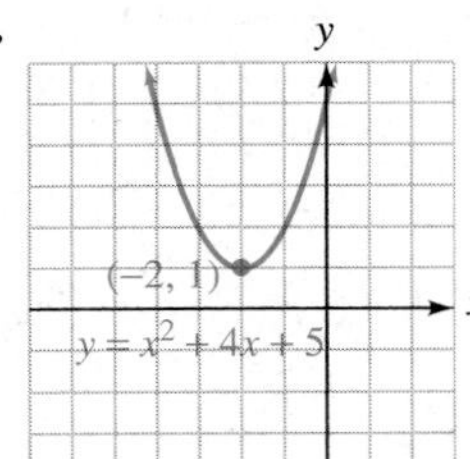

57.

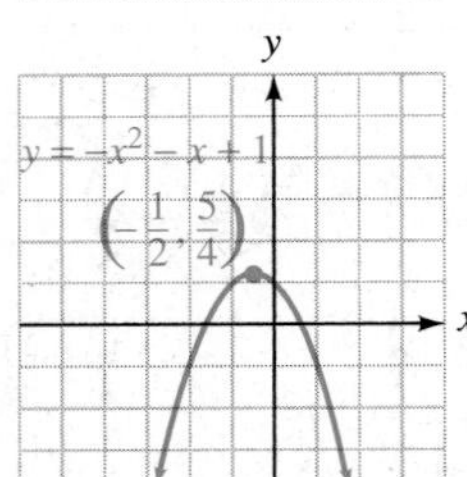

59.

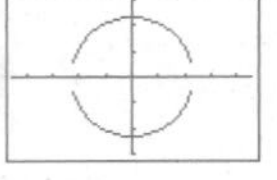

61.

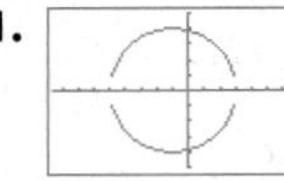

63.

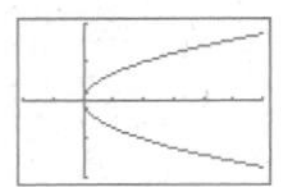

65.

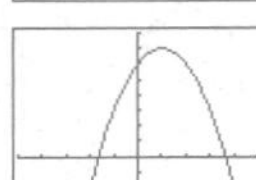

67. $(x - 7)^2 + y^2 = 9$ **69.** no **71.** 30 ft away
73. 2 AU

Getting Ready (page 700)

1. $y = \pm b$ **2.** $x = \pm a$

Exercises 10.2 (page 708)

1. $(\pm 3, 0), (0, \pm 4)$ **3.** $(2, 0)$ **5.** $12y^2 + \frac{9}{x^2}$ **7.** $\frac{y^2 + x^2}{y^2 - x^2}$
9. ellipse, sum **11.** center **13.** $(0, 0)$, major axis, $2b$

15.

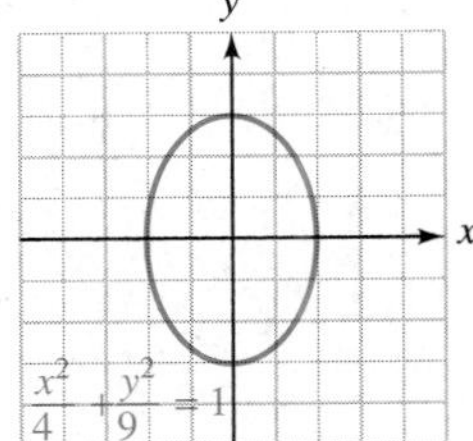

17.

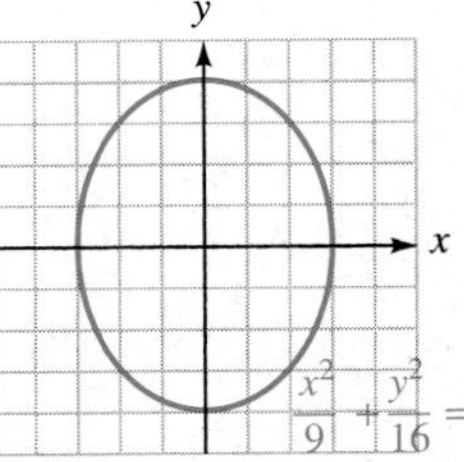

19.

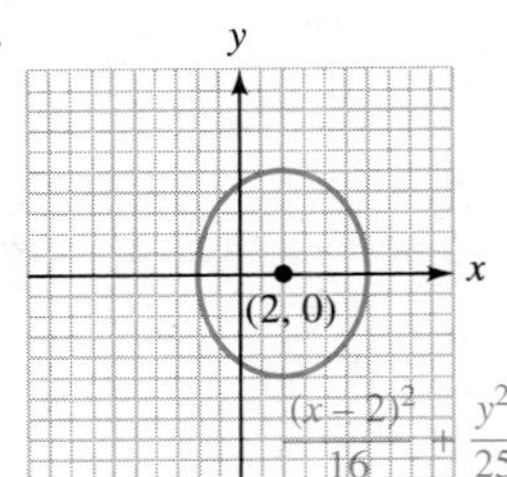

21.

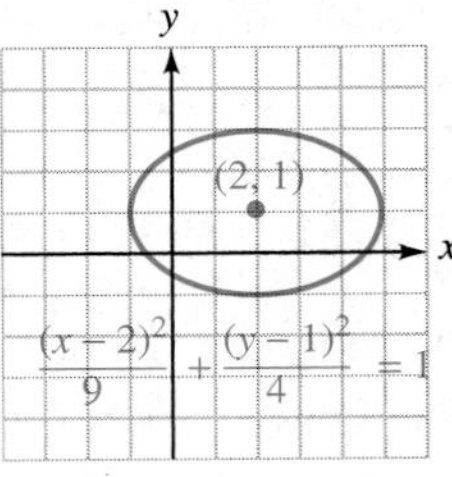

23.

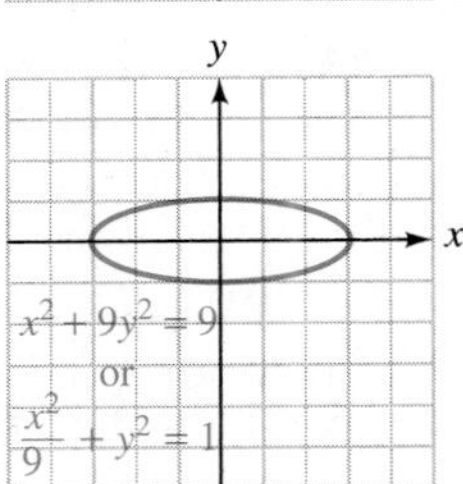

25.

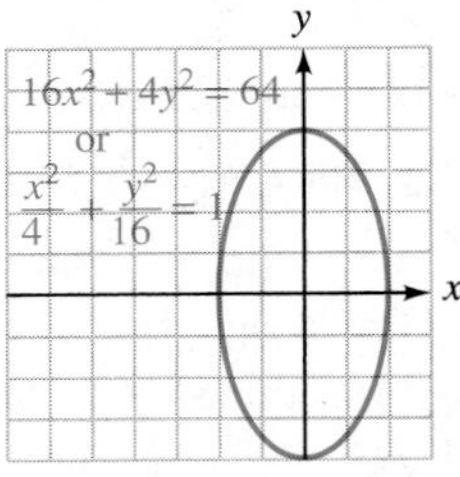

27.

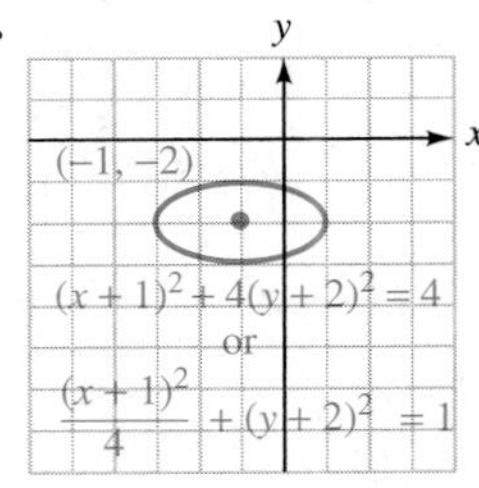

29.

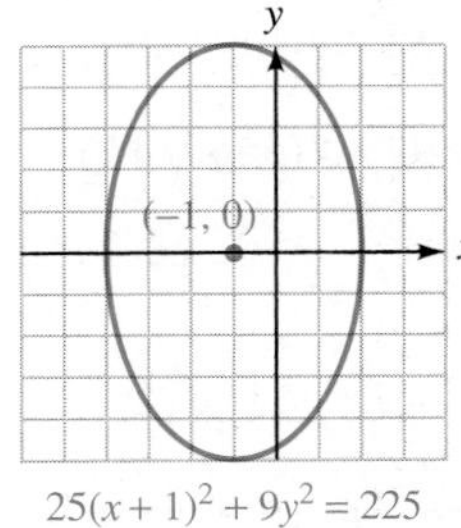

$25(x + 1)^2 + 9y^2 = 225$
or
$\frac{(x + 1)^2}{9} + \frac{y^2}{25} = 1$

31.

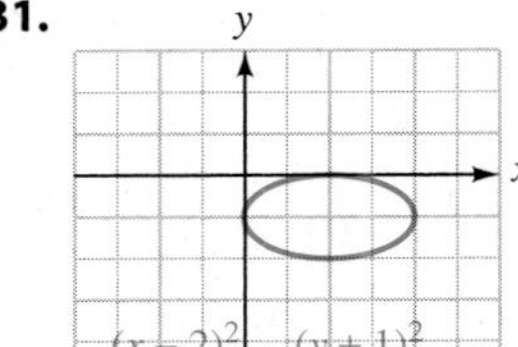

33.

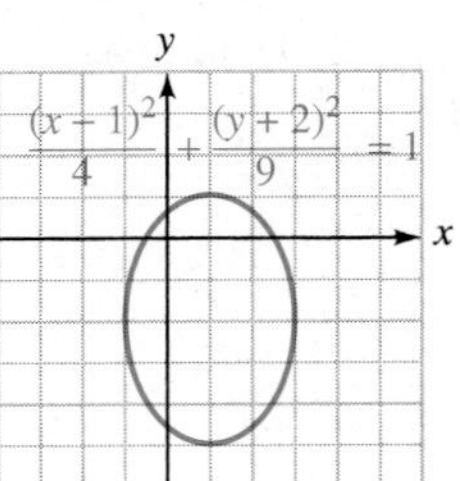

35.

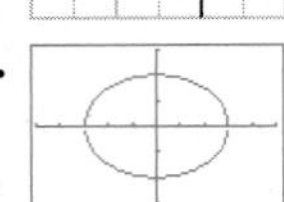

37.

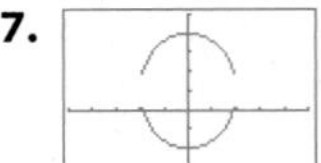

39. $\frac{x^2}{144} + \frac{y^2}{25} = 1$ **41.** $y = \frac{1}{2}\sqrt{400 - x^2}$ **43.** 12π sq. units

Getting Ready (page 712)

1. $y = \pm 2.0$ **2.** $y = \pm 2.9$

Exercises 10.3 (page 720)

1. $(\pm 3, 0)$ **3.** $-3x^2(2x^2 - 3x + 2)$ **5.** $(5a + 2b)(3a - 2b)$ **7.** hyperbola, difference **9.** center **11.** $(\pm a, 0)$, y-intercepts

13.

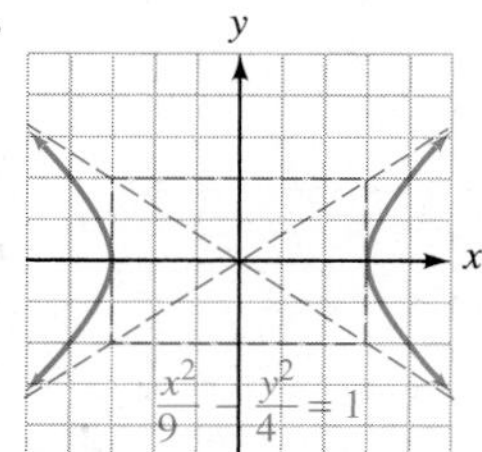

15.

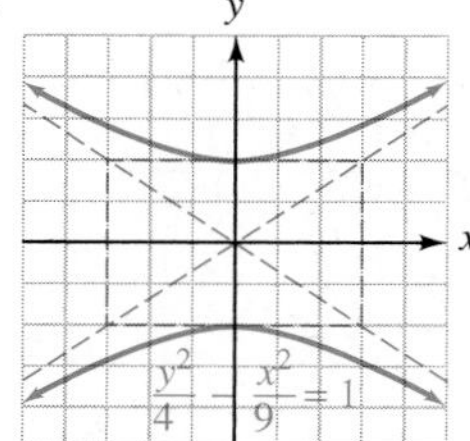

17.

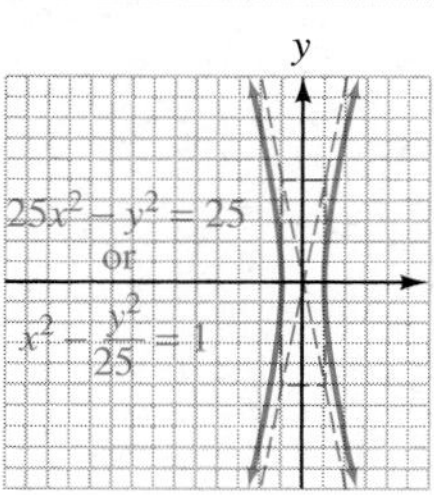

19.

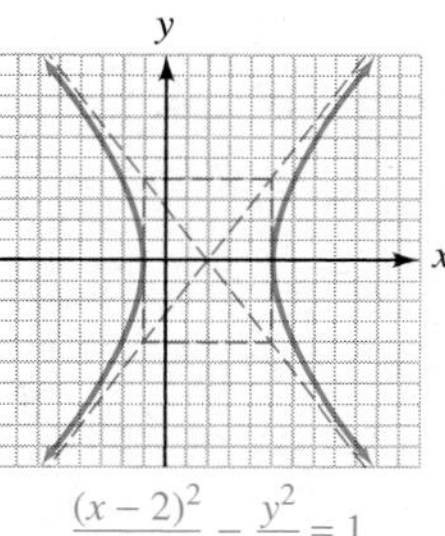

$\frac{(x-2)^2}{9} - \frac{y^2}{16} = 1$

21.

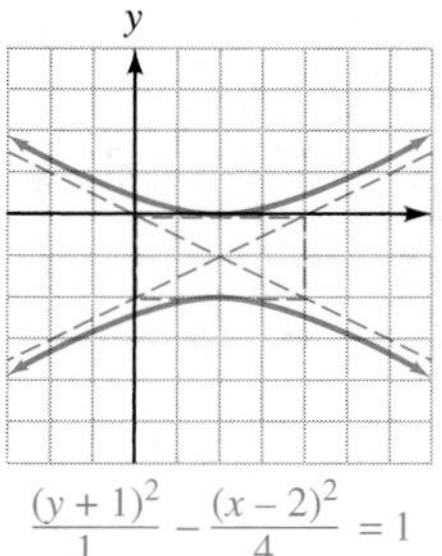

$\frac{(y+1)^2}{1} - \frac{(x-2)^2}{4} = 1$

23.

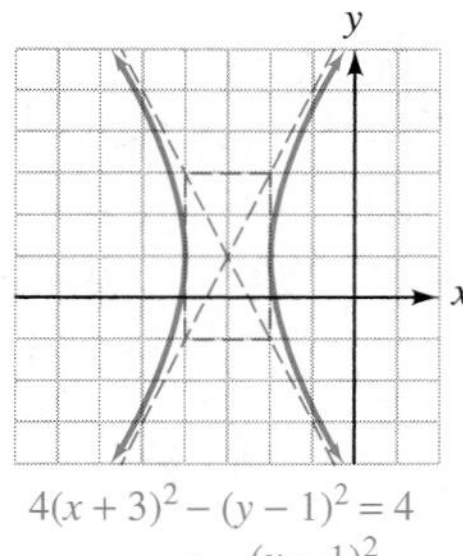

$4(x+3)^2 - (y-1)^2 = 4$

or $(x+3)^2 - \frac{(y-1)^2}{4} = 1$

25.

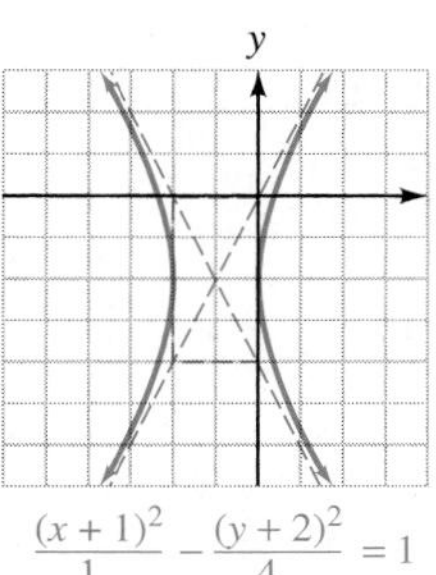

$\frac{(x+1)^2}{1} - \frac{(y+2)^2}{4} = 1$

27.

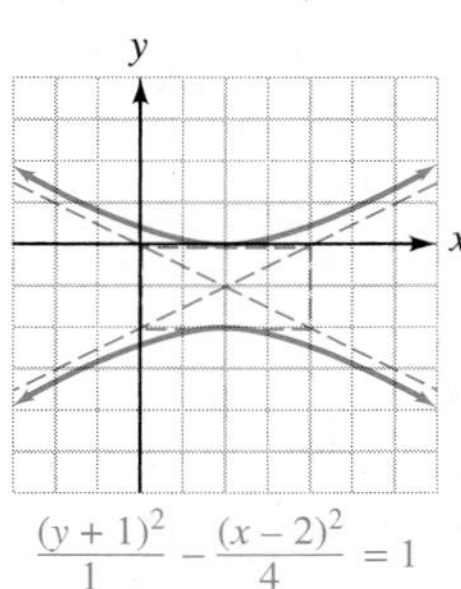

$\frac{(y+1)^2}{1} - \frac{(x-2)^2}{4} = 1$

29.

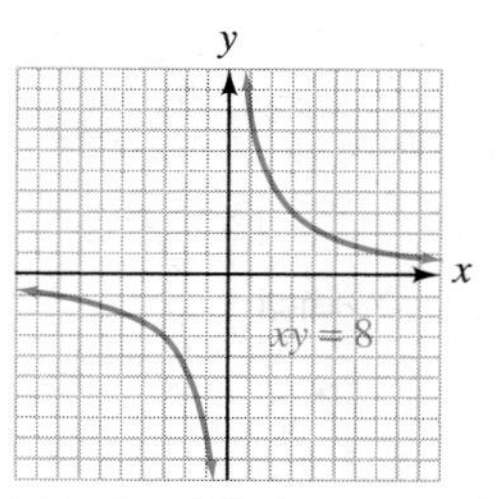

31.

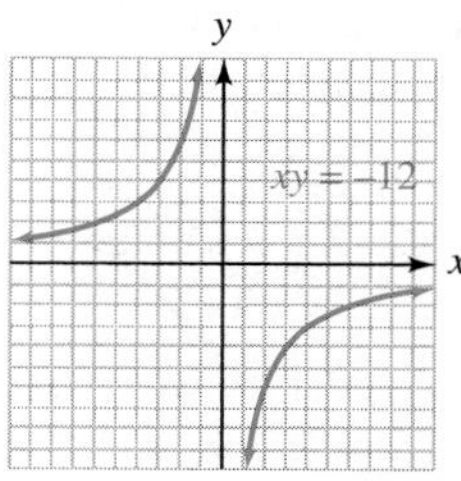

33.

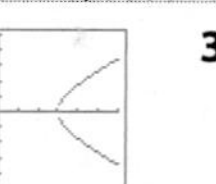

35.

37. 3 units **39.** $10\sqrt{3}$ units

Getting Ready (page 722)

1. $-y^2 = 44$ **2.** $5x^2 = 8$

Exercises 10.4 (page 726)

1. 0, 1, 2 **3.** 0, 1, 2, 3, 4 **5.** $-11x\sqrt{2}$ **7.** $\frac{t}{2}$ **9.** graphing, substitution **11.** four

13.

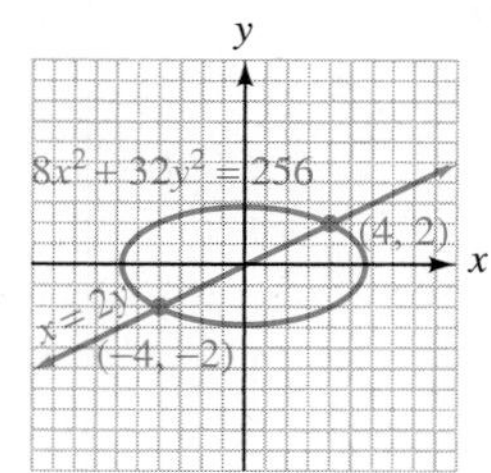

15.

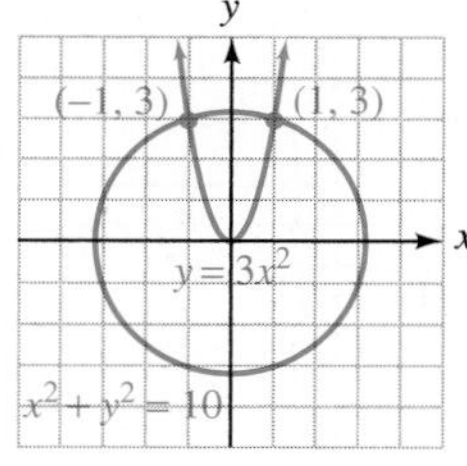

17.

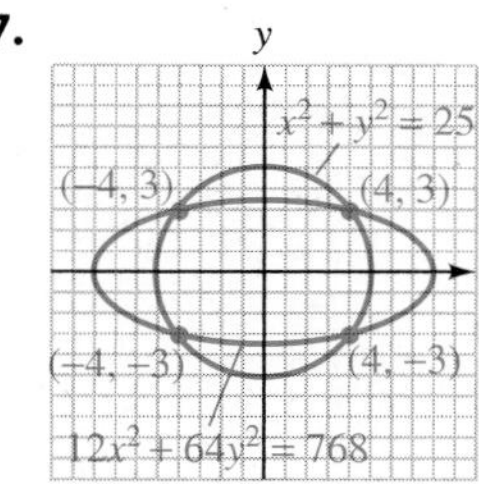

19.

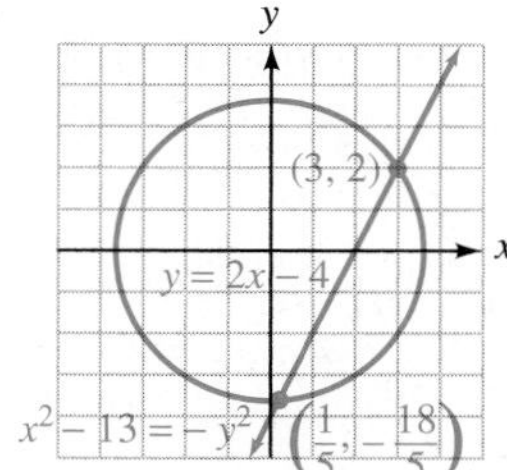

21. (3, 0), (0, 5) **23.** (1, 1) **25.** (1, 2), (2, 1) **27.** (−2, 3), (2, 3) **29.** (3, 2), (3, −2), (−3, 2), (−3, −2) **31.** (2, 4), (2, −4), (−2, 4), (−2, −4) **33.** (0, −4), (−3, 5), (3, 5) **35.** $\left(-\sqrt{15}, 5\right)$, $\left(\sqrt{15}, 5\right)$, (−2, −6),(2, −6) **37.** $\left(\sqrt{5}, 5\right)$, $\left(-\sqrt{5}, 5\right)$ **39.** (−2, 3), (2, 3), (−2, −3), (2, −3) **41.** (3, 3) **43.** $\left(-1, \frac{9}{2}\right)$, $\left(3, -\frac{3}{2}\right)$ **45.** $\left(\frac{1}{2}, \frac{1}{3}\right)$, $\left(\frac{1}{3}, \frac{1}{2}\right)$ **47.** (1, 0), (5, 0) **49.** 4 and 8 **51.** 7 cm by 9 cm **53.** either \$750 at 9% or \$900 at 7.5% **55.** 68 mph, 4.5 hr **59.** 0, 1, 2, 3, 4

Getting Ready (page 729)

1. positive **2.** negative **3.** 98 **4.** −3

Exercises 10.5 (page 732)

1. increasing **3.** constant **5.** 20 **7.** domains **9.** constant, $f(x)$ **11.** step **13.** increasing on $(-\infty, 0)$, decreasing on $(0, \infty)$ **15.** decreasing on $(-\infty, 0)$, constant on (0, 2), increasing on $(2, \infty)$

17. constant on $(-\infty, 0)$, increasing on $(0, \infty)$

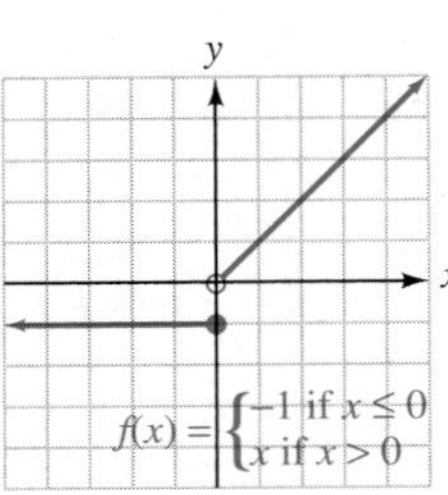

19. decreasing on $(-\infty, 0)$, increasing on $(0, 2)$, decreasing on $(2, \infty)$

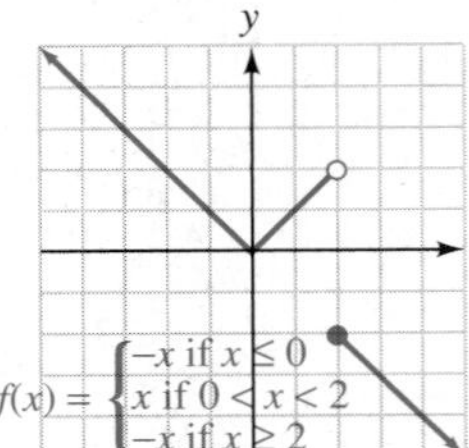

21.

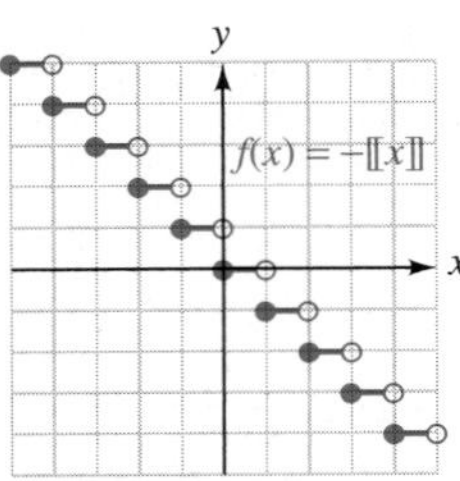

23.

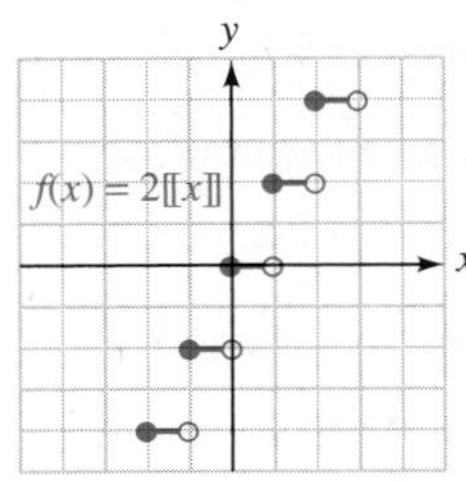

25.

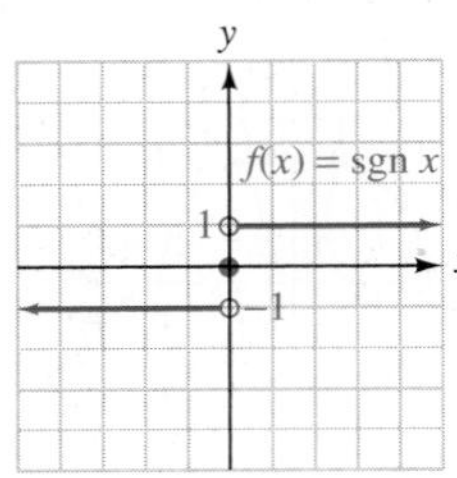

27. \$30

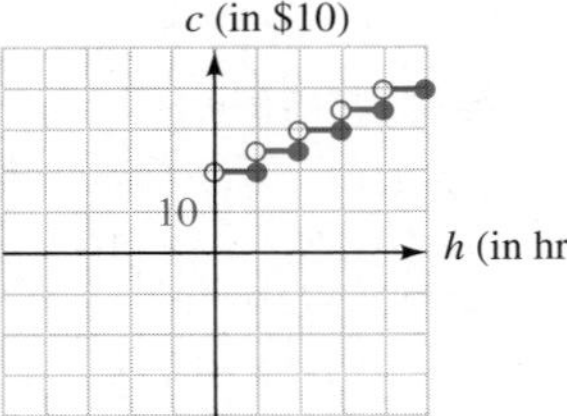

29. After 2 hours, network B is cheaper.

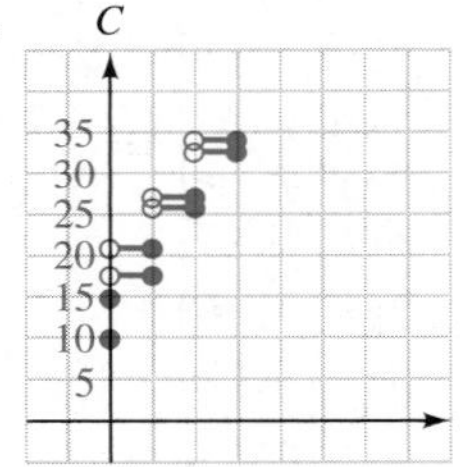

Chapter Review (page 735)

1.

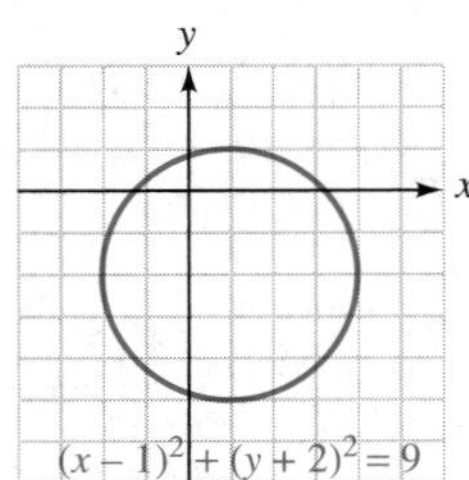

2.

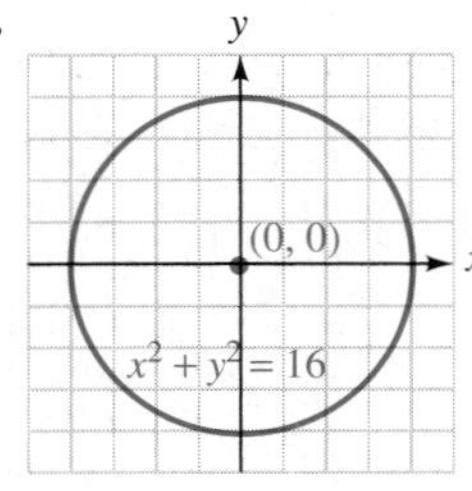

3.

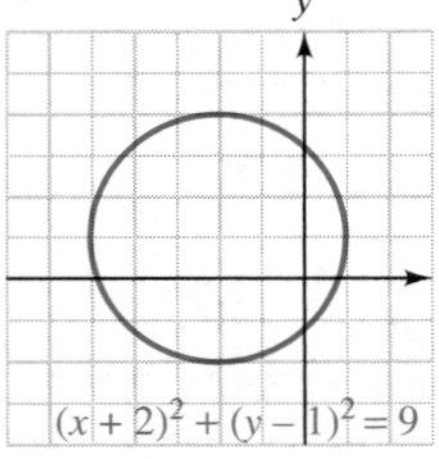

4.

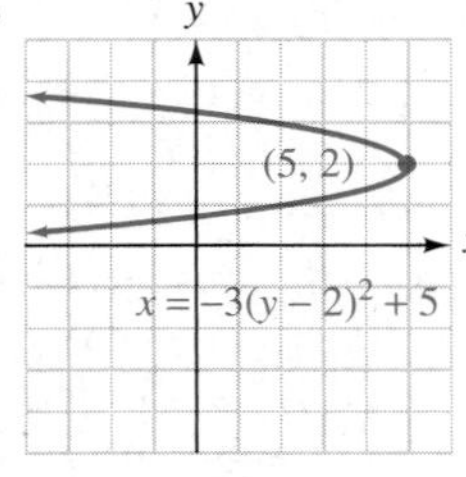

5.

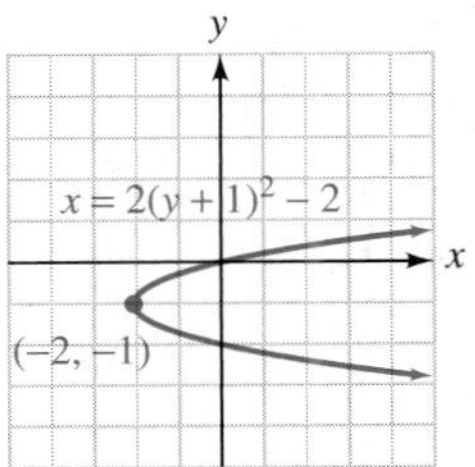

6.

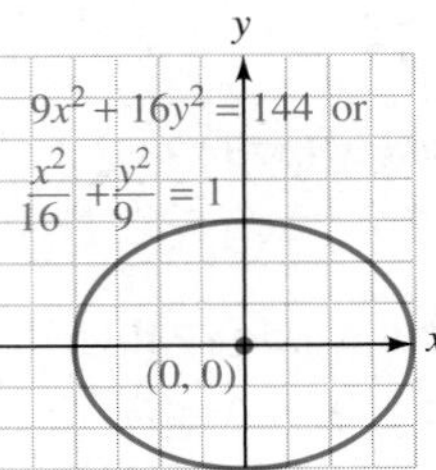

7.

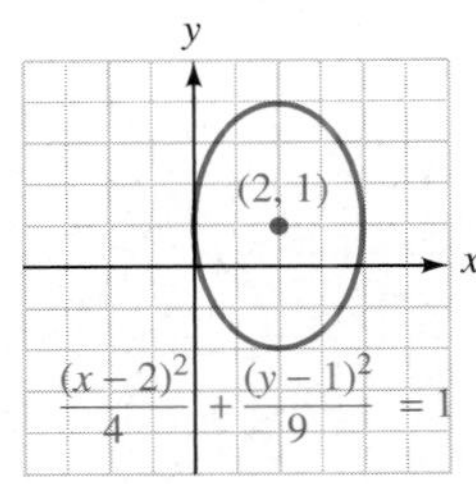

8.

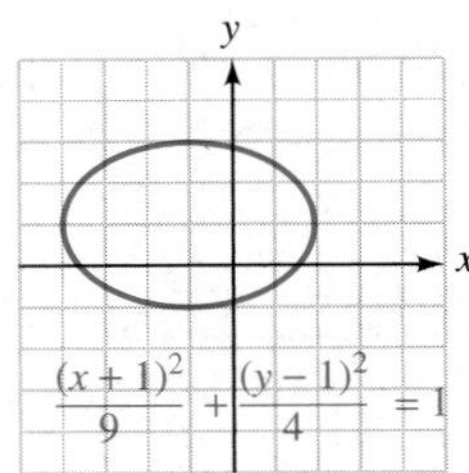

9.

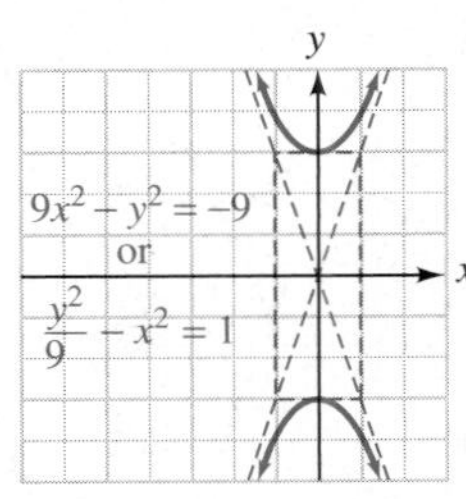

10.

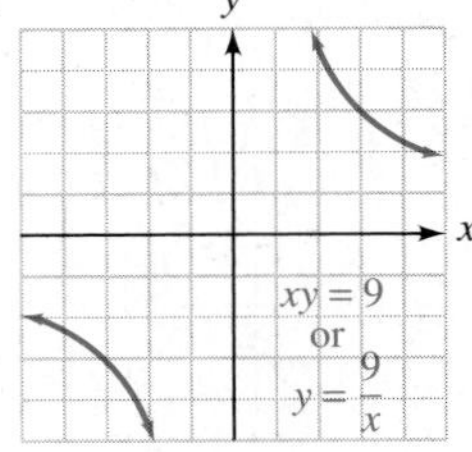

11. hyperbola **12.**

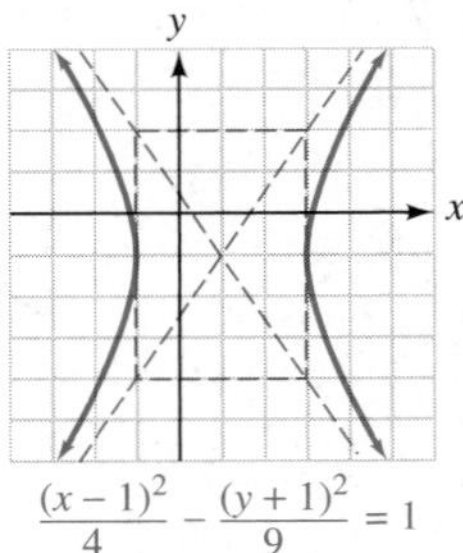

13. $(4, 2)$, $(4, -2)$, $(-4, 2)$, $(-4, -2)$ **14.** $(2, 3)$, $(2, -3)$, $(-2, 3)$, $(-2, -3)$ **15.** increasing on $(-\infty, -2)$, constant on $(-2, 1)$, decreasing on $(1, \infty)$

16.

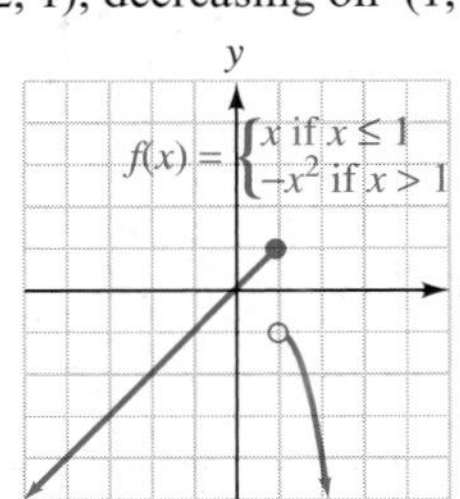

17.

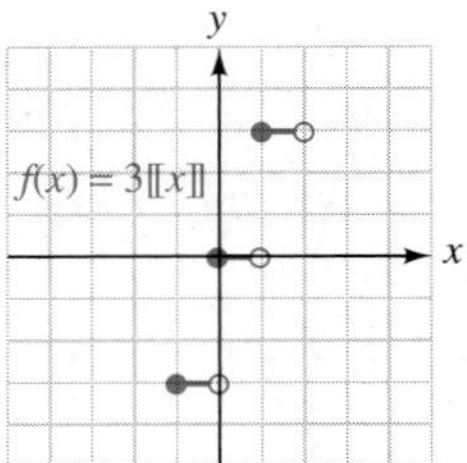

Chapter 10 Test (page 743)

1. $(2, -3)$, 2 **2.** $(-2, 3)$, 4 **3.**

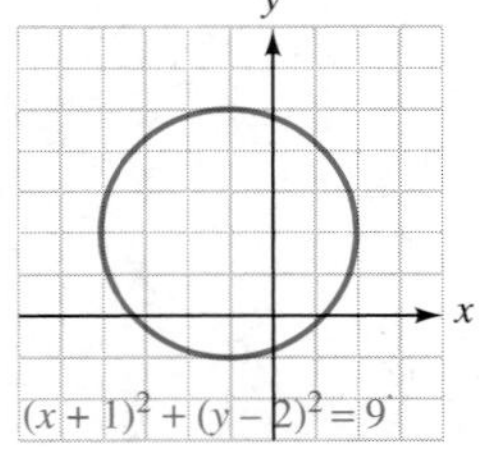

4.

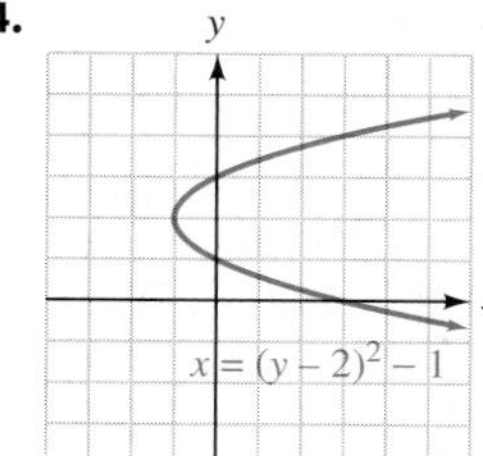

5.

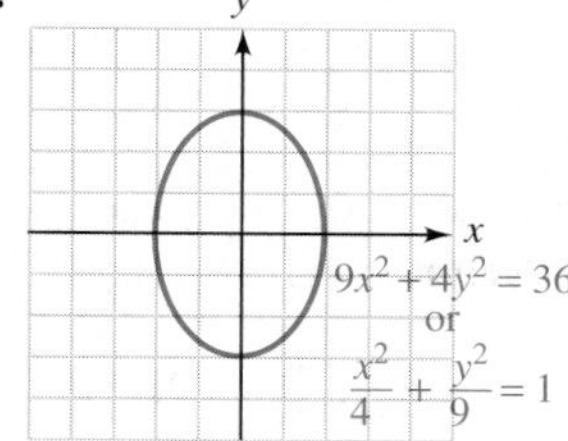

6.

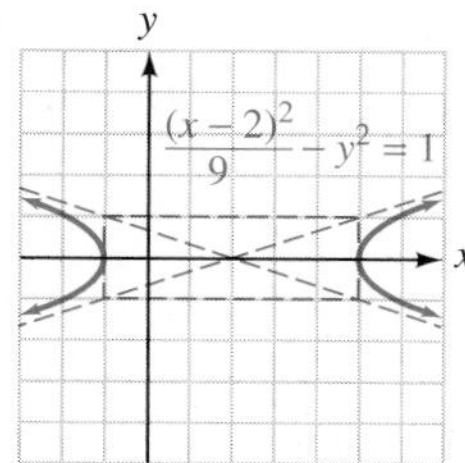

7.

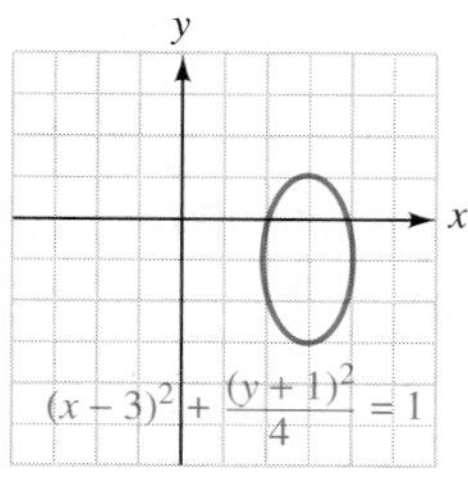

8.

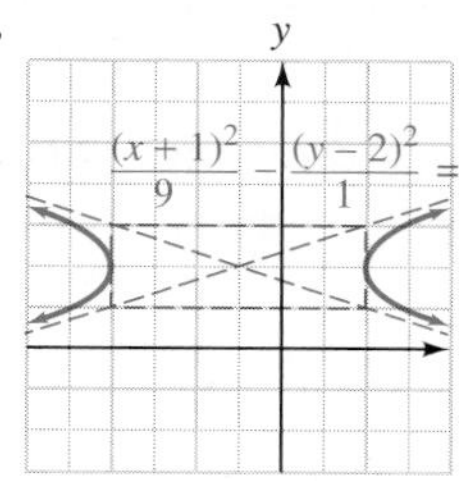

9. $(2, 6), (-2, -2)$
10. $(3, 4), (-3, 4)$
11. increasing on $(-3, 0)$, decreasing on $(0, 3)$

12.

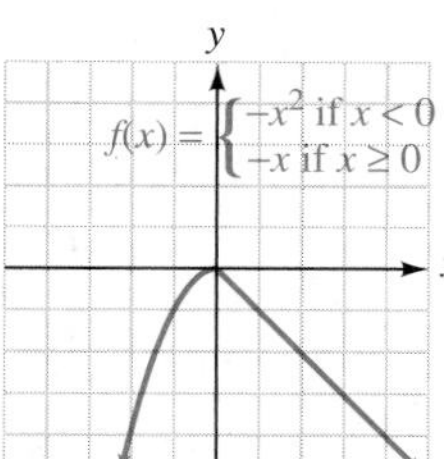

Cumulative Review Exercises (page 744)

1. $12x^2 - 5xy - 3y^2$ **2.** $a^{2n} - 2a^n - 3$ **3.** $\frac{5}{a-2}$
4. $a^2 - 3a + 2$ **5.** 1 **6.** $\frac{4a-1}{(a+2)(a-2)}$ **7.** parallel
8. perpendicular **9.** $y = -2x + 5$ **10.** $y = -\frac{9}{13}x + \frac{7}{13}$
11.

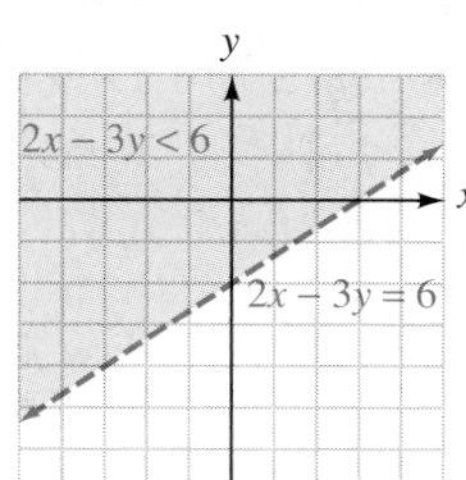

12.

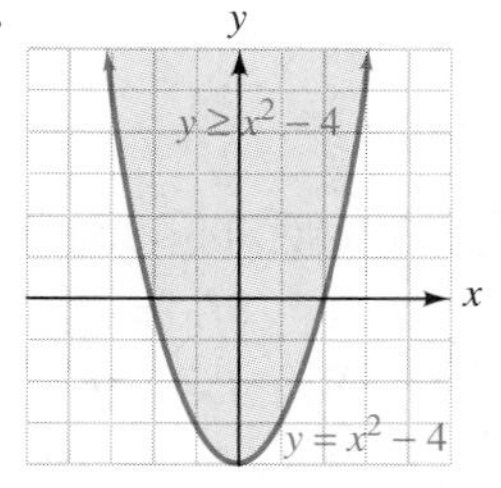

13. $5\sqrt{2}$ **14.** $81x\sqrt[3]{3x}$ **15.** 5; 0 is extraneous **16.** 0
17. $\frac{2}{3}, -\frac{3}{2}$ **18.** $-\frac{4}{3} \pm \frac{\sqrt{19}}{3}$ **19.** $4x^2 + 4x - 1$
20. $f^{-1}(x) = \sqrt[3]{\frac{x+1}{2}} = \frac{\sqrt[3]{4x+4}}{2}$

21.

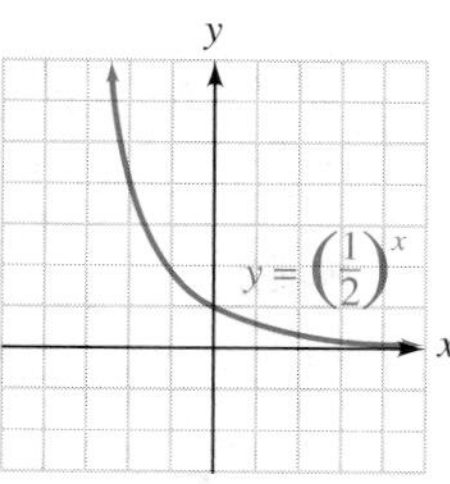

22. $2^y = x$ **23.** $\frac{2 \log 2}{\log 3 - \log 2}$
24. 16

25.

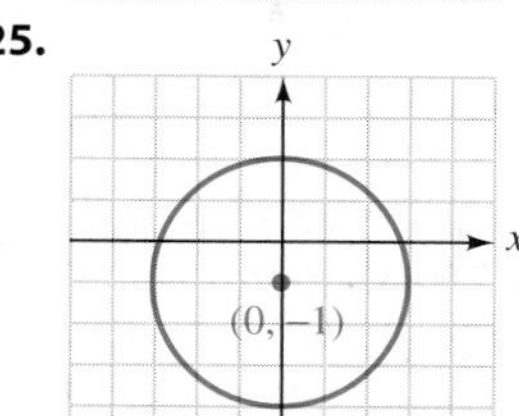

26.

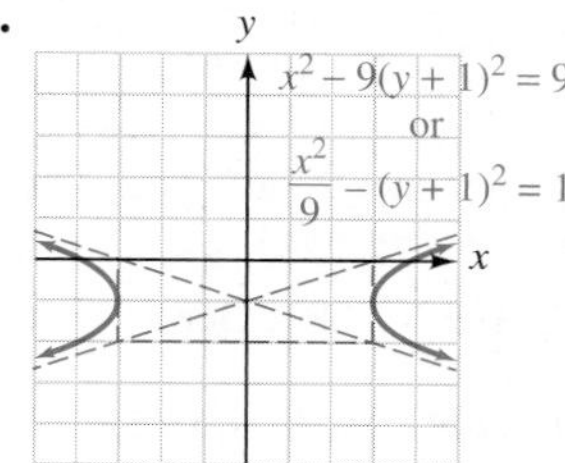

Getting Ready (page 746)

1. $x^2 + 4x + 4$ **2.** $x^2 - 6x + 9$ **3.** $x^3 + 3x^2 + 3x + 1$
4. $x^3 - 6x^2 + 12x - 8$

Exercises 11.1 (page 751)

1. 1 **3.** 1 **5.** $m^2 + 2mn + n^2$ **7.** $p^2 + 4pq + 4q^2$
9. 2 **11.** 5 **13.** one **15.** Pascal's **17.** 6! **19.** 1
21. $a^3 + 3a^2b + 3ab^2 + b^3$
23. $a^4 - 4a^3b + 6a^2b^2 - 4ab^3 + b^4$ **25.** 6 **27.** -120
29. 30 **31.** 144 **33.** $\frac{1}{110}$ **35.** 2,352 **37.** 72 **39.** 10
41. $x^3 + 3x^2y + 3xy^2 + y^3$
43. $x^4 - 4x^3y + 6x^2y^2 - 4xy^3 + y^4$
45. $8x^3 + 12x^2y + 6xy^2 + y^3$ **47.** $x^3 - 6x^2y + 12xy^2 - 8y^3$
49. $8x^3 + 36x^2y + 54xy^2 + 27y^3$ **51.** $\frac{x^3}{8} - \frac{x^2y}{4} + \frac{xy^2}{6} - \frac{y^3}{27}$
53. 40,320 **55.** 21 **57.** $\frac{1}{168}$
59. 39,916,800 **61.** $2.432902008 \times 10^{18}$
63. $81 + 216y + 216y^2 + 96y^3 + 16y^4$
65. $\frac{x^4}{81} - \frac{2x^3y}{27} + \frac{x^2y^2}{6} - \frac{xy^3}{6} + \frac{y^4}{16}$ **73.** 1, 1, 2, 3, 5, 8, 13, . . . ; beginning with 2, each number is the sum of the previous two numbers.

Getting Ready (page 753)

1. 6 **2.** 10 **3.** -20 **4.** -4

Exercises 11.2 (page 755)

1. 2 **3.** 6 **5.** 5 **7.** 1 **9.** (2, 3) **11.** 8 **13.** 3
15. 7 **17.** $3a^2b$ **19.** $15x^2y^4$ **21.** $-4xy^3$ **23.** $28x^6y^2$
25. $640x^3y^2$ **27.** $-12x^3y$ **29.** $-70{,}000x^4$ **31.** $810xy^4$
33. $180x^4y^2$ **35.** $-\frac{1}{6}x^3y$ **37.** $90x^3$ **39.** $\frac{n!}{3!(n-3)!}a^{n-3}b^3$
41. $\frac{n!}{4!(n-4)!}a^{n-4}b^4$ **43.** $\frac{n!}{(r-1)!(n-r+1)!}a^{n-r+1}b^{r-1}$
47. 252

Getting Ready (page 756)

1. 3, 5, 7, 9 **2.** 4, 7, 10, 13

Exercises 11.3 (page 763)

1. 14 **3.** 5 **5.** 3 **7.** $18x^2 + 8x - 3$ **9.** $\frac{6a^2 + 16}{(a + 2)(a - 2)}$
11. sequence, finite, infinite **13.** arithmetic, difference
15. arithmetic mean **17.** series **19.** $1 + 2 + 3 + 4 + 5$
21. 1 **23.** 73 **25.** 3, 5, 7, 9, 11
27. $-5, -8, -11, -14, -17$ **29.** 5, 11, 17, 23, 29
31. $-4, -11, -18, -25, -32$ **33.** $\frac{17}{4}, \frac{13}{2}, \frac{35}{4}$
35. 12, 14, 16, 18 **37.** $\frac{29}{2}$ **39.** $\frac{5}{4}$ **41.** 1,335 **43.** 459
45. $3 + 6 + 9 + 12$ **47.** $16 + 25 + 36$ **49.** 60 **51.** 31
53. $-118, -111, -104, -97, -90$ **55.** 34, 31, 28, 25, 22
57. 5, 12, 19, 26, 33 **59.** 355 **61.** -179 **63.** -23
65. 12 **67.** 354 **69.** 255 **71.** 1,275 **73.** 2,500
75. 12 **77.** \$60, \$110, \$160, \$210, \$260, \$310; \$6,060
79. 11,325 **81.** 368 ft **85.** $\frac{3}{2}, 2, \frac{5}{2}, 3, \frac{7}{2}, 4$

Getting Ready (page 765)

1. 10, 20, 40 **2.** 18, 54, 162

Exercises 11.4 (page 771)

1. 27 **3.** 2.5 **5.** 6 **7.** $[-1, 6]$ **9.** $(-\infty, -3) \cup [4, \infty)$
11. geometric **13.** common ratio **15.** $S_n = \frac{a_1 - a_1r^n}{1 - r}$
17. 3, 6, 12, 24, 48; 384 **19.** $-5, -1, -\frac{1}{5}, -\frac{1}{25}, -\frac{1}{125}; -\frac{1}{15{,}625}$
21. 2, 8, 32, 128, 512 **23.** $-3, -12, -48, -192, -768$
25. $-64, 32, -16, 8, -4$ **27.** $-64, -32, -16, -8, -4$
29. 6, 18, 54 **31.** $-20, -100, -500, -2{,}500$ **33.** -16
35. $10\sqrt{2}$ **37.** 728 **39.** 122 **41.** $\frac{156}{25}$ **43.** $-\frac{21}{4}$
45. 2, 10, 50, 250, 1,250 **47.** 3,584 **49.** $\frac{1}{27}$ **51.** 3
53. No geometric mean exists. **55.** -255 **57.** 381
59. about 669 people **61.** \$1,469.74 **63.** \$140,853.75
65. $\left(\frac{1}{2}\right)^{11} \approx 0.0005$ **67.** \$4,309.14 **73.** arithmetic mean

Getting Ready (page 773)

1. 4 **2.** 4.5 **3.** 3 **4.** 2.5

Exercises 11.5 (page 776)

1. 8 **3.** $\frac{1}{2}$ **5.** 27 **7.** yes **9.** no **11.** infinite
13. $S_\infty = \frac{a_1}{1 - r}$ **15.** 16 **17.** 81 **19.** 8
21. no sum because $r > 1$ **23.** $\frac{1}{9}$ **25.** $-\frac{1}{3}$ **27.** $\frac{4}{33}$
29. $\frac{25}{33}$ **31.** $-\frac{81}{2}$ **33.** $-\frac{81}{2}$ **35.** 30 m **37.** 5,000
41. $\frac{4}{5}$ **45.** no; $0.999999 = \frac{999{,}999}{1{,}000{,}000} < 1$

Getting Ready (page 778)

1. 24 **2.** 120 **3.** 30 **4.** 168

Exercises 11.6 (page 786)

1. 15 **3.** 3 **5.** 1 **7.** 6, -3 **9.** 8 **11.** $p \cdot q$
13. permutation **15.** $P(n, r) = \frac{n!}{(n - r)!}$ **17.** combination
19. $C(n, r) = \frac{n!}{r!(n - r)!}$ **21.** 6 **23.** 60 **25.** 12 **27.** 5
29. 10 **31.** 20 **33.** 50 **35.** 2
37. $x^4 + 4x^3y + 6x^2y^2 + 4xy^3 + y^4$
39. $8x^3 + 12x^2y + 6xy^2 + y^3$ **41.** 1 **43.** 1,260
45. $C(n, 2) = \frac{n!}{2!(n - 2)!}$
47. $81x^4 - 216x^3 + 216x^2 - 96x + 16$ **49.** $-1{,}250x^2y^3$
51. $-4x^6y^3$ **53.** 35 **55.** 5,040 **57.** 1,000,000
59. 136,080 **61.** 8,000,000 **63.** 2,880 **65.** 13,800
67. 720 **69.** 900 **71.** 364 **73.** 5 **75.** 1,192,052,400
77. 18 **79.** 7,920 **83.** 48

Getting Ready (page 788)

1. $\{1, 2, 3, 4, 5, 6\}$ **2.** $\{(H, H), (H, T), (T, H), (T, T)\}$

Exercises 11.7 (page 792)

1. $\frac{1}{6}$ **3.** $-\frac{3}{4}$ **5.** 3, -1 **7.** $-2, -2$ **9.** experiment
11. $\frac{s}{n}$ **13.** 0 **15. a.** 6 **b.** 52 **c.** $\frac{6}{52}, \frac{3}{26}$
17. {(1, H), (2, H), (3, H), (4, H), (5, H), (6, H), (1, T), (2, T), (3, T), (4, T), (5, T), (6, T)} **19.** {a, b, c, d, e, f, g, h, i, j, k, l, m, n, o, p, q, r, s, t, u, v, w, x, y, z} **21.** $\frac{1}{6}$ **23.** $\frac{2}{3}$ **25.** $\frac{19}{42}$
27. $\frac{13}{42}$ **29.** $\frac{3}{8}$ **31.** 0 **33.** $\frac{1}{4}$ **35.** $\frac{3}{26}$ **37.** $\frac{1}{169}$ **39.** 0
41. $\frac{1}{12}$ **43.** $\frac{7}{12}$ **45.** $\frac{1}{16}$ **47.** $\frac{3}{8}$ **49.** $\frac{1}{16}$ **51.** $\frac{1}{3}$
53. $\frac{88}{141}$ **55.** $\frac{15}{71}$ **59.** 0.14

Chapter Review (page 795)

1. 144 **2.** 20 **3.** 15 **4.** 220 **5.** 1 **6.** 8
7. $x^5 + 5x^4y + 10x^3y^2 + 10x^2y^3 + 5xy^4 + y^5$
8. $x^4 - 4x^3y + 6x^2y^2 - 4xy^3 + y^4$
9. $64x^3 - 48x^2y + 12xy^2 - y^3$
10. $x^3 + 12x^2y + 48xy^2 + 64y^3$ **11.** $6x^2y^2$ **12.** $-10x^2y^3$
13. $-108x^2y$ **14.** $864x^2y^2$ **15.** 42
16. 122, 137, 152, 167, 182 **17.** $\frac{41}{3}, \frac{58}{3}$ **18.** 1,550
19. $-\frac{45}{2}$ **20.** $\frac{15}{2}$ **21.** 378 **22.** 14 **23.** 360
24. 24, 12, 6, 3, $\frac{3}{2}$ **25.** 4 **26.** 24, -96 **27.** $\frac{2{,}186}{9}$
28. $-\frac{85}{8}$ **29.** \$1,638.40 **30.** \$134,509.57 **31.** 12 yr
32. 1,600 ft **33.** 125 **34.** $\frac{5}{99}$ **35.** 136 **36.** 5,040
37. 1 **38.** 20,160 **39.** $\frac{1}{10}$ **40.** 1 **41.** 1 **42.** 28
43. 84 **44.** 700 **45.** $\frac{7}{4}$ **46.** 120 **47.** 720 **48.** 120
49. 150 **50.** $\frac{3}{8}$ **51.** $\frac{1}{2}$ **52.** $\frac{7}{8}$ **53.** $\frac{1}{18}$ **54.** 0 **55.** $\frac{1}{13}$
56. $\frac{94}{54{,}145}$ **57.** $\frac{33}{66{,}640}$

Chapter 11 Test (page 800)

1. 210 **2.** 1 **3.** $-5x^4y$ **4.** $24x^2y^2$ **5.** 66 **6.** 306
7. 34, 66 **8.** 3 **9.** -81 **10.** $\frac{364}{27}$ **11.** 18, 108 **12.** $\frac{27}{2}$
13. 120 **14.** 40,320 **15.** 15 **16.** 56 **17.** 720
18. 322,560 **19.** 24 **20.** $\frac{7}{30}$ **21.** 35 **22.** 30 **23.** $\frac{1}{6}$
24. $\frac{2}{13}$ **25.** $\frac{33}{66{,}640}$ **26.** $\frac{5}{16}$

Cumulative Review Exercises (page 801)

1. (2, 1) **2.** (1, 1) **3.** (2, -2) **4.** (3, 1) **5.** -1
6. -1 **7.** (-1, -1, 3) **8.** 1

9.

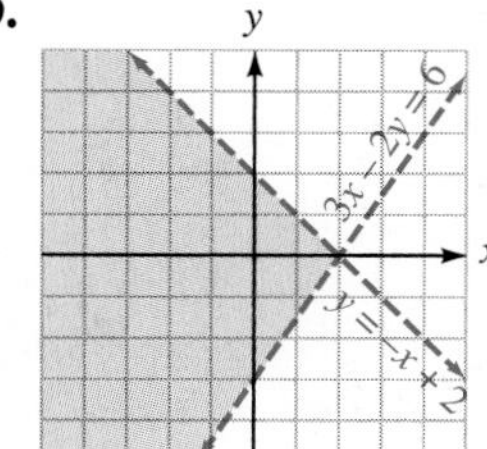

10.

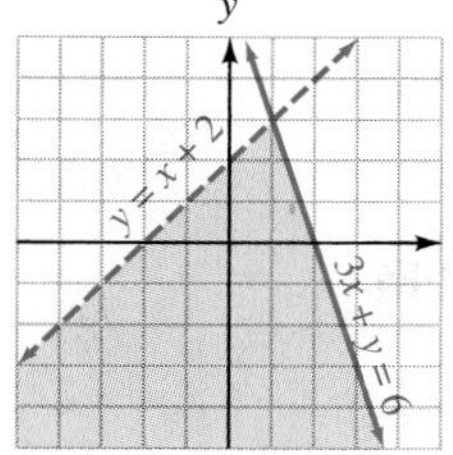

11.

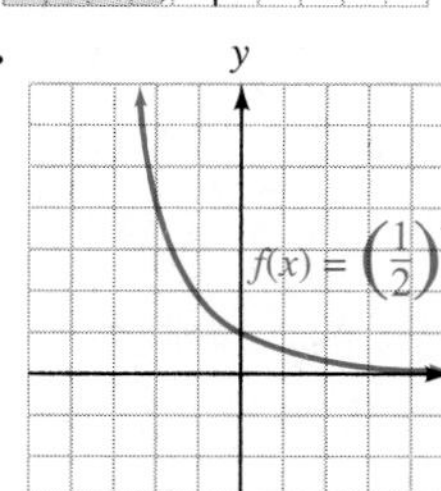

12. $2^y = x$ **13.** 5 **14.** 3
15. $\frac{1}{27}$ **16.** 1 **17.** $y = 2^x$
18. x **19.** 1.9912
20. 0.3010 **21.** 1.6902
22. 0.1461 **23.** $\frac{5 \log 2}{\log 3 - \log 2}$
24. 8 **25.** $2,848.31
26. 1.16056 **27.** 30,240

28. $81a^4 - 108a^3b + 54a^2b^2 - 12ab^3 + b^4$ **29.** $112x^2y^6$
30. 103 **31.** 690 **32.** 8 and 19 **33.** 42 **34.** 27
35. 27 **36.** $\frac{1,023}{64}$ **37.** 12, −48 **38.** $\frac{27}{2}$ **39.** 504
40. 35 **41.** $\frac{7}{3}$ **42.** $C(n, n)$ **43.** 5,040 **44.** 84 **45.** $\frac{3}{26}$

Exercises I.1 (page A-4)

1. y-axis **3.** origin **5.** y-axis **7.** none **9.** none
11. x-axis

13.

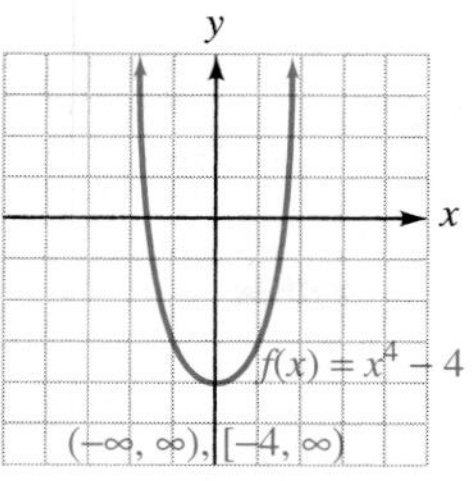

15.

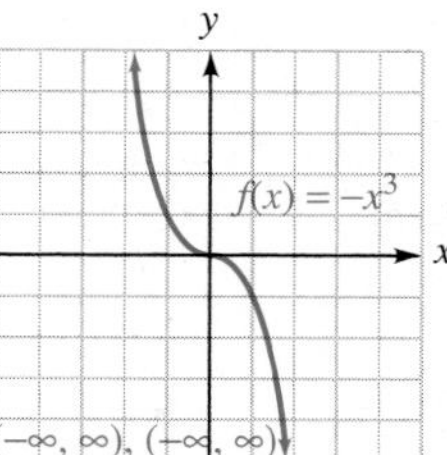

17.

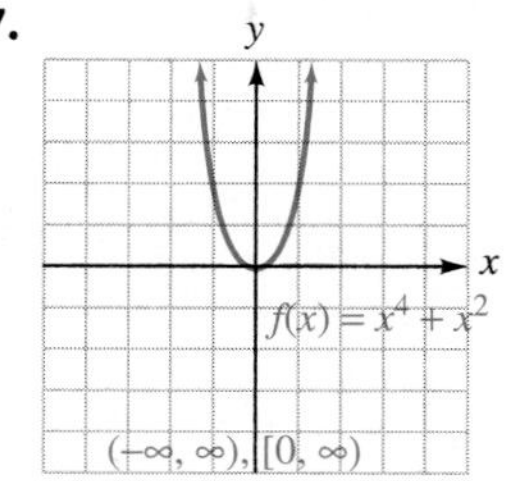

19.

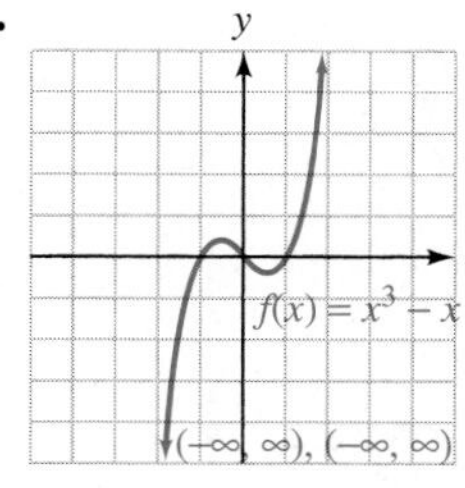

21.

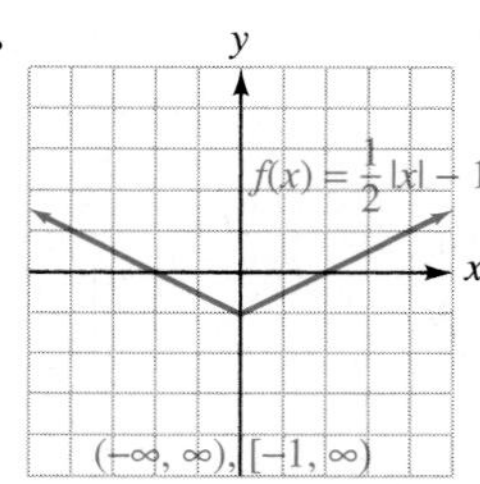

23.

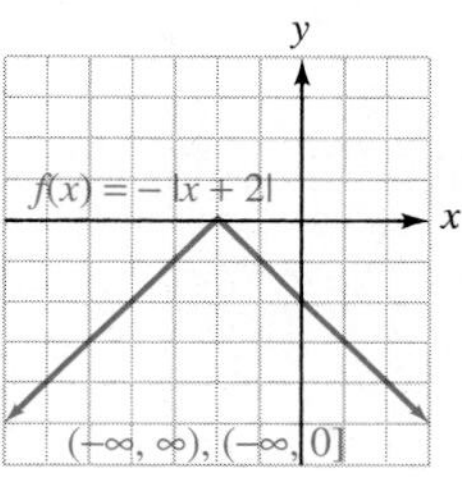

Index

A

Absolute value
  of complex number, 520–521
  defined, 241
  of real numbers, 11
Absolute value equations, 243, 245
Absolute value functions, 146–147, 148, 149–151, 241–242
Absolute value inequalities, 246, 247
Accent on Technology
  approximating zeros of polynomials, 426
  checking algebra, 385, 396
  checking algebraic simplification, 368, 377
  checking solutions of quadratic equations, 542
  evaluating determinants, 213
  evaluating exponential expressions, 620
  evaluating logarithms, 651
  factorials, 749
  finding base-10 (common) logarithms, 644
  finding powers, 35
  finding the domain and range of function, 366–367
  finding the period of pendulum, 461–462
  generating tables of solutions, 97
  graphing circles, 691
  graphing ellipses, 705–706
  graphing exponential functions, 624, 625, 635
  graphing functions, 147
  graphing hyperbolas, 715
  graphing inequalities, 257–258
  graphing lines, 100–102
  graphing logarithmic functions, 644, 652–653
  graphing polynomial functions, 294
  graphing quadratic functions, 571–572
  rational exponents, 478
  solving absolute-value inequalities, 249–250
  solving equations, 352, 404–405
  solving equations containing radicals, 506–507, 508
  solving equations graphically, 172
  solving equations with graphing calculators, 152–153
  solving exponential equations, 669
  solving inequalities, 584
  solving investment problems, 627–628
  solving linear inequalities, 236
  solving logarithmic equations, 671
  solving quadratic equations, 350
  solving systems by graphing, 170–171
  solving systems of equations, 723
  solving systems of inequalities, 263–264
  translations of the exponential function, 634
  using scientific notation, 43–44
  verifying properties of logarithms, 658
Acute angles, 64
Adding
  complex numbers, 516
  polynomials, 299
  radical expressions, 487–489
  radicals, 488
  rational expressions, 384–389
  real numbers, 15
Addition method, 178–180, 725–726
Addition property
  of equality, 50
  of inequalities, 230
Additive identity, 23
Additive inverse property, 24
Additive inverses, 23
Adjacent angles, 69
Ahmes papyrus, 42
Algebraic expressions
  English phrases converted to, 61–62
  evaluating, 21–22
Algebraic terms, 51, 288
Algorithms, defined, 415
Al-jabr, 1
Alternate interior angles, 185
Altitude, of equilateral triangle, 490
Amount of an annuity, 770
Angles
  acute, 64
  adjacent, 69
  alternate interior, 185
  complementary, 65
  right, 64
  straight, 64
  supplementary, 65
  vertical, 69
Angstrom, 46, 48
Annual depreciation rate, 125
Annual growth rate, 633
Annuity
  amount of, 770
  term of, 770
Areas
  of circles, 36
  of rectangles, 36
  of squares, 36
  of trapezoids, 36
  of triangles, 36
Arithmetic means, 759–760
Arithmetic sequences
  common difference, 758
  defined, 758
  first term of, 758–759
  *n*th term of, 758
  summation notation, 761–762
  sum of the first *n* terms, 760–761
Arithmetic series, 761
Array of signs for 3×3 determinant, 211
Ascending order, of exponents in polynomials, 290
Associative property
  for addition, 22–23
  for multiplication, 22–23
Astronomical unit, 48
Asymptotes
  horizontal, 371
  vertical, 371
Augmented matrix, 201
Axes
  x-, 92
  y-, 92
Axis of symmetry, of parabolas, 565, 568, 569–571

B

Babbage, Charles, 387
Back substitution, 202
Base-10 logarithms, 644–645
Base angles of isosceles triangles, 65
Base-*e* exponential functions, 631–636
Base-*e* logarithms, 650–654

Base of an exponential expression, 28–29
Bimodal distribution, 21
Binomial expansion
  binomial theorem, 750–751
  combinations for finding terms of, 784–785
  factorial notation, 748–749
  finding particular term of, 753–754
  Pascal's triangle, 747–748
  raising binomial to power, 746–747
Binomials
  conjugate binomials, 307–309, 498
  defined, 289
  multiplying by FOIL method, 306–307
  raise to power, 746–747
Binomial theorem, 750–751, 785
Boundary line, 254
Break point, 184–185
Briggs, Henry, 644

C
Carbon-14 dating, 672
Careers and Mathematics
  artists, 287
  atmospheric scientists–weather forecasters, 90
  automotive service technicians and mechanics, 228
  cosmetologists, 165
  environmental engineers, 363
  farmers, ranchers, and agricultural managers, 1
  financial analysts and financial planners, 745
  pest control workers, 618
  photographers, 453
  police officers and detectives, 534
  water transportation occupations, 686
Cartesian coordinate system, 92
Catenary, 639, 695
Cauchy, Augustin, 514
Cayley, Arthur, 201
Center
  of a circle, 688
  of an ellipse, 701
  of a hyperbola, 713
Change-of-base formula, 662–663
Circles
  area of, 36
  center of, 688
  circumference of, 21
  defined, 688
  general form of the equation of, 690, 691
  radius, 688
  standard equations of, 688, 689
Circumference, 21
Clavius, Christopher, 16
Closed intervals, 232, 233
Coefficient matrix, 201
Coefficients, numerical, 51, 288
Column of constants, 201
Combinations
  $C(n, 0)$, 783
  $C(n, n)$, 783
  $C(n, r)$, 781–784
  terms of binomial expansion, 784–785
Combined variation, 437
Common difference of arithmetic sequence, 758
Common logarithms, 644–645
Common ratio of geometric sequence, 766
Commutative property
  of addition, 22
  of multiplication, 22
Complementary angles, 65
Complement of an angle, 65
Completing the square
  solving quadratic equations by, 538–543
  steps in, 539
Complex conjugates, 517
Complex fractions, 394–400
Complex numbers
  absolute value of, 520–521
  adding, 516
  defined, 515
  dividing, 518–519
  equality of, 515
  imaginary numbers and, 515
  multiplying, 516
  real numbers and, 515
  simplifying expressions containing, 514–518
  simplifying imaginary numbers, 513–514
  subtracting, 516
Composite functions, 591
Composite numbers, 3–4
Composition of functions, 590–593
Compound inequalities
  defined, 8, 234
  graphing, 8–10, 256
  solving in one variable, 234–235
Compounding periods, 627
Compound interest, 543, 626, 632–633
Compound interest formula, 543, 627
Conditional equations, 55–56
Cone, volume of, 36
Congruent triangles, 185
Conic sections, 687–688
Conjugate binomials, 307–309, 498
Conjugate hyperbolas, 722
Consistent systems, 167
Constant functions, 140
Constant of proportionality, 433
Constant of variation, 433, 435, 436
Constants
  column of, 201
  defined, 51
Constraints, 268
Contradiction, defined, 55–56
Coordinate plane, 92–94, 254, 469–471
Coordinates
  of a point, 93
  defined, 3
  x-, 93
  y-, 93
Coordinate systems
  Cartesian, 92
  rectangular, 92
Cramer, Gabriel, 212, 213
Cramer's rule, 212–217
Critical points, 580
Critical values, 580
Cube, volume of, 36
Cube-root function, 462
Cube roots
  defined, 457
  simplifying perfect cube root, 457–458
Cubing functions, 145–146, 148, 149–151, 292
Cylinder, volume of, 36

D
Decibels, 645
Decibel voltage gain, 645
Decimals
  repeating, 5, 181, 775–776
  terminating, 5
Decreasing functions, 622, 623
Degree
  of a monomial, 289
  of a polynomial, 289–290
  of a polynomial function, 290
Demand equations, 105
Dependent equations, 168–170, 180–181, 194–195
Dependent variables, 137
Descartes, René, 92, 513, 514, 687
Descending order, of exponents in polynomials, 290
Determinants
  defined, 210
  solving systems of linear equations with, 210–217
  value of 2×2, 210
  value of 3×3, 210–211
Difference
  defined, 15

of two cubes, 326–329, 344
of two functions, 589–590
of two squares, 308, 324–326, 344
Difference quotient, of functions, 593
Direct variation, 433–434
Discriminant, defined, 555
Distance formula, 469–471
Distributive property
extended, 23
of multiplication over addition, 22–23
Dividing
complex numbers, 518–519
fractions, 379
monomials, 414
polynomials, 414–418, 421–424
radicals, 485
rational expressions, 379–380
real numbers, 17–18
Division, synthetic, 421–426
Division by 0, 17
Division property of radicals, 485
Domain
of an exponential function, 621, 622
of functions, 133–134, 138–139
of a logarithmic function, 640
of a rational function, 365–366
of a relation, 132
of a square-root function, 459–462
Double negative rule, 24
Doubling time, 653, 654

E

e, 632
Eccentricity, of ellipses, 704
Edge, 254
Elementary row operations, 202, 203–205
Elements
of a matrix, 201
of sets, 2, 4
Elimination (addition) method, 178–180, 725–726
Ellipses
center of, 701
defined, 701
eccentricity of, 704
equations centered at origin, 703
focus of, 701, 718
general form of the equation of, 706
graphing in general form of equation, 706–707
graphing in standard form of equations, 701–706
major axis, 703
minor axis, 703
standard equations of, 704
vertex, 703
Ellipsis, 3, 757
Empty set, 56, 245
English phrases, converted to algebraic expressions, 61–62
Equality
of complex numbers, 515
logarithmic property of, 659
properties of, 49–51
Equations. *See also* Linear equations; Quadratic equations; Systems of equations; Systems of linear equations
absolute value, 243, 245
conditional, 55–56
contradictions, 55–56
defined, 49
demand, 105
dependent, 168–170, 180–181, 194–195
of ellipses, 703, 704, 706
equivalent, 50
exponential, 667–669
finding slope of line with, 109
functions represented by, 137–138
general form of line, 95, 123–124
of hyperbolas, 714, 716, 718
independent, 168–170
logarithmic, 651–652, 667, 669–671
of parabolas, 196–197, 693, 694
point-slope form of line, 118–120
prediction, 126–127
in quadratic form, 556–559
radical, 504–509
rational, 404–410
regression, 126–127
roots of, 49
second-degree, 722–726
slope-intercept form of line, 120–127
solution set of, 49
solutions of, 49
solving three in three variables, 191–197
supply, 105
with two absolute values, 245
Equilateral triangles, 65, 490
Equivalent equations, 50
Equivalent systems, 169
Euler, Leonhard, 514, 551, 632
Evaluating
algebraic expressions, 21–22
logarithms, 661–662
natural logarithms, 650–651
polynomial functions, 290–292
rational expressions, 371–372
Even integers, defined, 4
Even root, 458
Events
defined, 789
multiplication principle for, 778–779
probability of, 789–792
Everyday Connections
CO_2 emissions in Asia, 45
eccentricity of an ellipse, 704
focus on conics, 718
fuel oil production, 274
labor force rate of growth, 114
staffing, 206
state lotteries, 784
stretching rubber bands, 293
traveling through water, 501
U.S. population growth, 674
video action games, 431
winning the lottery, 791
Experiments, 789
Exponential equations, solving, 667–669
Exponential functions
base-*e*, 631–636
defined, 621
domain of, 621, 622
graphing, 621–624, 625, 633–634
graphing translation of, 624–625
graphs of, 619
one-to-one, 623
properties of, 622
range of, 621, 622
writing logarithmic functions as, 640–641
Exponential growth, formula for, 633, 673
Exponents
defined, 28–29
descending or ascending order of, 290
fractions to negative powers, 34
irrational, 620
natural-number, 29
negative, 32
power rules of, 31
product rule of, 30
properties of, 29–35
quotient rule of, 33
rational, 475–481
rules of, 29–35, 475
zero exponents, 32
Expressions. *See also* Radical expressions; Rational expressions
algebraic, 21–22, 61–62
rules for order of operations, 18–19, 35, 51–53
Extended distributive property, 23
Extraneous solutions, 405
Extremes of proportion, 429

F

Factorability, test for, 336
Factorial notation, 748–749
Factorials, property of, 749

Factoring
  ac method, 339–340
  defined, 315
  difference of two cubes, 326–329, 344
  difference of two squares, 324–326, 344
  greatest common factor, 316–319
  by grouping, 319–320, 329, 339–340, 344
  hints for trinomials, 337
  perfect-square trinomials, 332–333
  problem types, 344
  solving polynomial equations by, 350–351
  solving quadratic equations by, 348–350, 352–353, 535–536
  sum of two cubes, 326–329, 344
  sum of two squares, 325
  trinomials, 332–341, 344
  trinomials by grouping, 339–340
  trinomials using integers, 335–339
  trinomials using substitution, 339
  trinomials with leading coefficient of 1, 333–335
  trinomials with leading coefficient other than 1, 335
Factors, defined, 29
Factor theorem, 425
Feasibility region, 269
Fibonacci, Leonardo, 549, 757
Fibonacci sequence, 549, 757
Finite sequences, 757
First term of arithmetic sequences, 758–759
First term of geometric sequences, 765–766
Focus
  of an ellipse, 701, 718
  of a hyperbola, 713, 718
  of a parabola, 692
FOIL method, 306–307
Formulas
  area of circle, 36
  area of rectangle, 36
  area of square, 36
  area of trapezoid, 36
  area of triangle, 36
  change-of-base, 662–663
  circumference of circle, 21
  $C(n, 0)$, 783
  $C(n, n)$, 783
  $C(n, r)$, 782
  compound interest, 543, 627
  distance, 469–471
  doubling time, 653, 654
  exponential growth, 633, 673
  factoring difference of two squares, 324
  factoring sum or difference of two cubes, 326–327
  Malthusian growth model, 635, 673
  midpoint, 100
  Ohm's Law, 60
  perimeter of rectangle, 21
  perimeter of square, 21
  perimeter of trapezoid, 21
  perimeter of triangle, 21
  pH of solution, 663
  $P(n, 0)$, 781
  $P(n, n)$, 781
  $P(n, r)$, 780
  quadratic, 547–548
  radioactive decay, 636
  Richter scale, 646
  rule of seventy, 677
  slope of a nonvertical line, 108
  special product, 308
  sum of first n terms of arithmetic sequence, 761
  sum of first n terms of geometric sequence, 768
  sum of infinite geometric series, 774
  vertex of parabola, 569
  volume of cone, 36
  volume of cube, 36
  volume of cylinder, 36
  volume of pyramid, 36
  volume of rectangular solid, 36
  volume of sphere, 36
  Weber-Fechner law, 663
Fractions
  adding, 384–389
  complex, 394–400
  dividing, 379
  multiplying, 376
  to negative powers, 34
  rationalizing denominator of fractions with complex numbers, 518–519
  rationalizing denominator of fractions with radical expression, 496–499
  rationalizing numerator of fractions with radical expression, 499–500
  repeating decimals in fractional form, 181, 775–776
  subtracting, 384–389
Froude, William, 501
Function notation, 134–136
Functions
  Absolute-value, 146–147, 148, 149–151, 241–242
  composite, 591
  composition of, 590–593, 594
  constant, 140
  cube-root, 462
  cubing, 145–146, 148, 149–151, 292
  decreasing, 622, 623
  defined, 133
  difference quotient of, 593
  domain of, 133–134, 138–139
  equations representing, 137–138
  exponential, 619, 621–625, 631–636
  greatest integer, 731
  hyperbolic cosine, 639
  identity, 593
  increasing, 622
  inverse of, 599–603
  linear, 139–141, 292
  logarithmic, 640–643, 652–653
  objective, 268, 270
  one-to-one, 597–600, 623
  operations on, 589–590
  piecewise-defined, 729–730
  polynomial, 290–294
  quadratic, 145, 564–571
  range of, 133–134
  rational, 364, 365–366
  square-root, 459–462
  squaring, 145, 148, 149–151, 292
  step, 731
Fundamental rectangle, 714
Future value, 627

G
Gauss, Carl Friedrich, 480, 514
Gaussian elimination, 202
GCF (greatest common factor), 316–319
General form
  of equation of circle, 690, 691
  of equation of ellipse, 706
  of equation of hyperbola, 716
  of equation of line, 95, 123–124
  of equations of parabolas, 694
Generation time, 674
Geometric means, 767–768
Geometric sequences
  common ratio, 766
  first term, 765–766
  nth term, 766
  specified term of, 766–767
  sum of the first n terms, 768–769
Golden ratio, 399, 549
Graphing
  compound inequalities, 8–10, 256
  exponential functions, 621–624, 625, 633–634
  functions, 139–141
  linear equations, 91–102, 121
  logarithmic functions, 642–643, 652–653
  nonlinear inequalities, 585
  a point, 93
  polynomial functions, 292–294
  quadratic functions, 564–571
  solving linear inequalities in two variables by, 261–263
  solving systems of equations by, 166–172, 723

solving systems of quadratic inequalities in two variables by, 263
Graphing method, 166–167, 171, 723
Graphs
of absolute-value functions, 241–242
defined, 93
of equations, 94–95
of exponential functions, 619
horizontal translations of, 567
of intervals, 10, 233
of linear functions, 139–141
of linear inequalities, 231–232, 233, 253–256
of ordered pairs, 93
of real numbers, 7–8
reflections of, 150–151
slope of line, 107–108
symmetries of, A-1–A-4
vertical translations of, 566
Greatest common factor (GCF)
of polynomials, 316–319
steps for finding, 316
of two or more monomials, 316
Greatest integer function, 731
Grouping symbols, 18–19

H
Half-life, 671
Half-open intervals, 232, 233
Half planes, 254
Hippasus of Metapontum, 469
Hooke's law, 433–434
Hopper, Grace Murray, 107
Horizontal asymptotes, 371
Horizontal lines
defined, 98
slopes of, 109–110, 124
Horizontal line test, 598–599
Horizontal translations, 149–150, 566–568, 643
Hydrogen-ion index, 663
Hypatia, 298
Hyperbolas
center of, 713
defined, 713
equations of, 714, 716, 718
focus of, 713, 718
fundamental rectangle of, 714
graphing, 713–718
LORAN, 712
vertices, 713
Hyperbolic cosine function, 639
Hypotenuse, 467

I
i, powers of, 519–520
Identity, defined, 55, 56
Identity function, 593
Imaginary numbers, 456, 513–514, 515
Inconsistent systems, 167–168, 180, 193–194
Increasing functions, 622
Increment value, 97
Independent equations, 168–170
Independent variables, 137
Index
hydrogen-ion index, 663
of radicals, 458, 487–488
of summation, 761
Inequalities. *See also* Linear inequalities
with absolute values, 246, 247
compound, 8–10, 234–235, 256
defined, 229
properties of, 230–231
quadratic, 263, 579–581
rational, 581–584
symbols, 7, 229
systems of, 261–265
Infinite geometric sequences, 773–776
Infinite geometric series, 774–776
Infinite sequence, 757
Infinity symbol, 8
Input value, 137–138
Integers
defined, 3
even, 4
factoring trinomials using, 335–339
odd, 4
as subset of rational numbers, 5
Integer squares, 455
Intercept method of graphing a line, 96–97
Intercepts
x-, 96–97, 569–571
y-, 96–97, 121, 124, 569–571
Intersection, defined, 9
Interval notation
of compound inequalities, 9
defined, 8
of intervals, 10, 233
Intervals
closed, 232, 233
defined, 7
graphs of, 10, 233
half-open, 232, 233
interval notation, 10, 233
open, 232, 233
set notation, 10, 233
unbounded, 232, 233
Inverse functions, 599–603
Inverse variation, 434–436
Irrational exponents, 620
Irrational numbers
defined, 5–6
square roots as, 456
Irreducible polynomials, 317
Isosceles right triangles, 489–491
Isosceles triangles
base angles of, 65
vertex angles of, 65

J
Joint variation, 436–437

K
Key numbers, 339
Kovalevskaya, Sonya, 371

L
Least common denominator (LCD), 385–386
Least common multiple (LCM), 323
Leibniz, Gottfried Wilhelm, 43, 93
Light year, 48
Like radicals, 487
Like signs
adding real numbers with, 15
dividing real numbers with, 17
multiplying real numbers with, 17
Like terms, 51–53, 298–299
Linear equations. *See also* Systems of linear equations
defined, 50, 95
graphing, 91–102, 121
simplifying one or both sides, 53–55
solving in one variable, 48–57, 62–67, 71–75
solving with one absolute-value term, 243–245
solving with two absolute-value terms, 245–246
systems of, 182–186
Linear functions
defined, 140
graphs of, 139–141, 292
Linear inequalities
defined, 231, 254
graphs of, 231–232, 233, 253–256
solving in one or two variables, 253–258
solving in one variable, 231–237
solving with one absolute-value term, 246–250
systems of, 261–265
Linear programming, 268, 271–275
Lithotripsy, 718
Logarithmic equations
defined, 667
solving, 651–652, 669–671
Logarithmic functions
defined, 640
domain of, 640
graphing, 642–643, 652–653
range of, 640

vertical and horizontal translation of, 643
writing as exponential function, 640–641
Logarithmic property of equality, 659
Logarithms
base-10, 644–645
change-of-base formula, 662–663
common, 644–645
evaluating, 661–662
as exponents, 641
Napierian, 650, 652
natural, 650–654
power rule of, 659
product property of, 657–658
properties of, 657–664
quotient property of, 657–658
LORAN, 712
Lowest common denominator (LCD), 385–386

M
Major axis of ellipse, 703
Malthus, Thomas Robert, 635
Malthusian growth model, 635, 673
Markdown, 63
Mathematical models, 61
Matrices
augmented, 201
coefficient, 201
defined, 201
determinants of, 210–217
elementary row operations, 202, 203–205
elements of, 201
square, 201
triangular form, 202
Maximum of objective function, 268, 270
Means
arithmetic, 759–760
defined, 20–21
geometric, 767–768
of proportion, 429
Median, defined, 20–21
Midpoint, 100
Midpoint formula, 100
Minimum of objective function, 268, 270
Minor axis of ellipse, 703
Minors of determinant, 210
Mode, defined, 20–21
Monomials
defined, 289
dividing, 414
greatest common factor, 316
multiplying, 304
multiplying polynomial by, 304–305
Multiplication principle for events, 778–779
Multiplication property
of inequalities, 230–231
of radicals, 484
of 0, 17, 23
Multiplicative identity, 23
Multiplicative inverse property, 24
Multiplicative inverses, 24
Multiplying
binomials by FOIL method, 306–307
complex numbers, 516
fractions, 376
monomials, 304
polynomial by monomial, 304–305
polynomial by polynomial, 305–306
radical expressions, 494–496
radicals, 494–496
rational expressions, 376–378
real numbers, 16–17

N
Napier, John, 652
Napierian logarithms, 650, 652
Napier's rods, 652
Natural logarithms, 650–654
Natural-number exponents, 29
Natural numbers
defined, 3
prime-factored form of, 315–316
Negative exponents, 32
Negative numbers, 3, 231
Negative powers of fractions, 34
Negative reciprocals, 111–112
Negatives, 23
Newton, Isaac, 702
Newton's law of cooling and warming, 678
Noether, Amalie, 232
Nonlinear inequalities, graphing, 585
Nonvertical line, slope of, 108
Notation
factorial, 748–749
function, 134–136
interval, 8, 9, 10, 233
scientific, 40–45
set-builder, 4, 180–181
standard, 40–42
summation, 761–762
*n*th roots, simplifying perfect *n*th root, 458–459
*n*th terms
of an arithmetic sequence, 758
of a geometric sequence, 766
Numbers. *See also* Real numbers
complex, 513–521
composite, 3–4
imaginary, 456, 513–514, 515
integers, 3, 4, 5
interrelationship of, 6
irrational, 5–6, 456
natural, 3, 315–316
negative, 3, 231
perfect, 323
positive, 3
prime, 3–4
rational, 4–5
whole, 3
Numerical coefficients, 51, 288

O
Objective function
maximum of, 268, 270
minimum of, 268, 270
Odd integers, defined, 4
Odd root, 458
Ohm's Law, 60
One-to-one functions, 597–600, 623
Open intervals, 232, 233
Operations on functions, 589–590
Opposites, 23
Ordered pairs
defined, 93
functions as, 133
plotting on coordinate plane, 92–94
relations as, 132
Order of operations, 18–19, 35, 51–53
Order of radical, 458
Oresme, Nicole, 475
Origin, defined, 92
Oughtred, William, 16
Output value, 137–138

P
Parabolas
axis of symmetry, 565, 568, 569–571
defined, 145, 565
equations of, 196–197, 693, 694
graphing, 692–694
vertex of, 565, 568, 569–571
Paraboloids, 151, 692
Parallel lines
general form of equation, 123–124
slope-intercept form of equation, 122–123
slopes of, 110–111
Parallelograms, 185
Parsec, 48
Partial sum, 774
Pascal, Blaise, 687, 747
Pascal's triangle, 747–748
Path of a comet, 699
Perfect cubes, 326, 485
Perfect numbers, 323

Perfect squares, 324, 485
Perfect-square trinomials, 308, 332–333, 538
Periodic interest rate, 627
Period of pendulum, 461–462
Permutations
  $P(n, 0)$, 781
  $P(n, n)$, 781
  $P(n, r)$, 779–781
Perpendicular lines
  general form of equation, 123–124
  slope-intercept form of equation, 122–123
  slopes of, 112
Perspectives
  Ahmes papyrus, 42
  calculating square roots, 456
  complex numbers, 514
  Fibonacci sequence, 549
  François Vieta, 310
  golden ratio, 399
  graphs in space, 151
  infinite series, 551
  Lewis Carroll, 212
  Pythagorean theorem, 469
pH of solution, 663
pH scale, 663
Piecewise-defined functions, 729–730
Plane
  coordinate plane, 92–94, 254, 469–471
  defined, 191
Plotting a point, 94
Point circle, 689
Point-slope form of equation of line, 118–120, 124
Polya, George, 269
Polynomial equations, solving by factoring, 350–351
Polynomial functions
  evaluating, 290–292
  graphing, 292–294
Polynomials
  adding, 299
  defined, 288
  degree of, 289–290
  dividing by monomial, 414–415
  dividing by polynomial, 415–417
  dividing polynomial with missing terms, 417–418
  factoring by grouping, 329
  factoring with four terms by grouping trinomial, 340
  greatest common factor from, 316–319
  irreducible, 317
  multiplying binomials, 306–307
  multiplying by monomial, 304–305
  multiplying by polynomials, 305–306
  multiplying non-polynomial expressions, 311
  in one variable, 288
  prime, 317
  in several variables, 289
  subtracting, 299–300
  zeros of, 425, 426
Population study, 635
Positive numbers, 3
Power of x, 29
Power rule, 504
Power rule of logarithms, 659
Power rules of exponents, 31
Powers of i, 519–520
Prediction equations, 126–127
Present value, 627
Prime-factored form of natural number, 315–316
Prime numbers, defined, 3–4
Prime polynomials, 317
Principal square root, 455
Probability, of events, 789–792
Problem solving. *See also* Solving
  linear equations used for, 62–67, 71–75
  mathematical models, 61
  Polya's approach, 269
  steps in, 62
  systems of linear equations, 182–186
Product
  defined, 16
  of two functions, 589–590
Product rule of exponents, 30
Projects
  Chapter 1, 79
  Chapter 2, 156–157
  Chapter 3, 220–221
  Chapter 4, 278–279
  Chapter 5, 356–357
  Chapter 6, 442–443
  Chapter 7, 523–524
  Chapter 8, 606–607
  Chapter 9, 678–679
  Chapter 10, 735
  Chapter 11, 794–795
Properties
  addition of inequalities, 230
  additive inverse, 24
  associative of addition, 22–23
  associative of multiplication, 22–23
  commutative of addition, 22
  commutative of multiplication, 22
  distributive of multiplication over addition, 22–23
  division of radicals, 485
  of equality, 49–51
  of exponential functions, 622
  of exponents, 29–35
  of factorials, 749
  of inequalities, 230–231
  logarithmic of equality, 659
  of logarithms, 657–664
  multiplication of inequalities, 230–231
  multiplication of radicals, 484
  multiplicative inverse, 24
  of parallelograms, 185
  power rule of logarithms, 659
  of radicals, 514
  of real numbers, 22–24
  square-root, 537
  transitive, 230
  trichotomy, 14, 230
  of 0 and 1, 23
  zero-factor, 348
Proportions
  combined variation, 437
  defined, 429
  direct variation, 433–434
  extremes of, 429
  inverse variation, 434–436
  joint variation, 436–437
  means of, 429
  similar triangles, 431–433
  solving, 429–431
Pyramid, volume of, 36
Pythagoras of Samos, 468, 469, 514
Pythagorean theorem, 467–468, 469

Q

Quadrants, defined, 92
Quadratic equations
  defined, 348
  discriminant determining types of solutions, 555–556
  equations written in quadratic form, 556–559
  solutions of, 559–560
  solving by completing the square, 538–543
  solving by factoring, 348–350, 352–353, 535–536
  solving with quadratic formula, 547–550
  solving with square-root property, 536–538
  standard form of, 348
Quadratic formula
  defined, 547–548
  solving formula for specified variable using, 550
  solving quadratic equations using, 547–550

Quadratic functions
defined, 145, 564
graphing, 564–571
Quadratic inequalities
defined, 579
solving, 579–581
systems of, 263
Quadrilateral, 69
Quotient
defined, 17
difference quotient, 593
of a quantity and its opposite, 370
of two functions, 589–590
Quotient rule of exponents, 33

R

Radical equations
solving with one radical, 504–507
solving with three radicals, 509
solving with two radicals, 507–508
Radical expressions
adding, 487–489
multiplying, 494–496
rationalizing denominators of fractions with, 496–499
rationalizing numerators of fractions with, 499–500
simplified form of, 485
simplifying perfect cube root, 457–458
simplifying perfect *n*th root, 458–459
simplifying perfect square root, 454–456
simplifying perfect square-root expressions, 456–457
simplifying with properties of radicals, 484–487
simplifying with rational exponents, 480–481
subtracting, 487–489
Radicals
adding, 488
changing from rational exponents to, 477
division property of, 485
like, 487
multiplication property of, 484
multiplying, 494–496
properties of, 514
similar, 487
solving equations containing, 505, 506–507, 508
for specified variables, 509–510
subtracting, 488
Radical signs, 455, 496
Radicand, 455, 487
Radioactive decay formula, 636
Radius of circle, 688
Rahn, Johann Heinrich, 17
Ramanujan, Srinivasa, 293
Range
of an exponential function, 621, 622
of a function, 133–134
of a logarithmic function, 640
of a relation, 132
Rate of growth (or decrease), 114
Rates, 429
Rational equations
defined, 404
solving, 404–410
solving formulas for specified variable, 406–407
Rational exponents
changing to radicals, 477
defined, 475, 477
simplifying expression with negative rational exponent, 478–479
simplifying expression with positive rational exponent, 475–478
simplifying expression with properties of exponents, 479–480
simplifying radical expression with, 480–481
Rational expressions
adding with like denominators, 384–385
adding with unlike denominators, 386–389
combined operations on, 381, 389–390
defined, 364
dividing, 379–380
evaluating, 371–372
finding domain of rational function, 365–366
finding least common denominator, 385–386
finding square of, 378–379
multiplying, 376–378
properties of, 367
simplifying, 367–370
simplifying containing factors that are negatives, 370
subtracting with like denominators, 384–385
subtracting with unlike denominators, 386–389
values of variables for undefined rational expression, 364–365
Rational functions, 364, 365–366
Rational inequalities, solving, 581–584
Rationalizing denominators, 496–499, 518–519
Rationalizing numerators, 499–500
Rational numbers, defined, 4–5
Ratios
defined, 429
golden, 399
Real numbers
absolute value of, 11
adding, 15
complex numbers and, 515
defined, 6
dividing, 17–18
graphs of, 7–8
multiplying, 16–17
properties of, 22–24
subtracting, 15–16
Reciprocals
defined, 24
negative, 111–112
Rectangles
area of, 36
perimeter of, 21
Rectangular coordinate system, 92
Rectangular solid, volume of, 36
Reflections of graph, 150–151
Regression equations, 126–127
Relations, defined, 132
Remainder theorem, 424–425
Repeating decimals, 5, 181, 775–776
Resistance, 441
Rhind, Henry, 42
Richter scale, 645, 646
Right angles, 64
Right triangles, 65, 467–468
Rise, 108
Roots of equations, 49
Roster method, 4
Rudolff, Christoff, 455
Rule of seventy, 677
Rules
Cramer's, 212–217
double negative rule, 24
of exponents, 29–35, 475
for order of operations, 18–19, 35
power rule, 504
Run, 108
Rutherford, Ernest, 718–719

S

Salvage value, 125
Sample space, 789
Scattergrams, 126
Scientific notation
converting from standard notation, 40–41
converting to standard notation, 41–42
defined, 40
rounding to significant digit, 44–45
simplifying computations with, 43
Second-degree equations, 722–726

Sequences
arithmetic, 758–762
Fibonacci, 549, 757
finite, 757
general term, 756–757
geometric, 765–770
infinite, 757
infinite geometric, 773–776
term of, 757
Series, 761
Set-builder notation, 4, 180–181
Set notation, 10
Sets
defined, 2
elements of, 2, 4
empty set, 56
solution sets, 49
subsets of, 3
Setup costs, 184–185
Significant digits, finding, 44–45
Similar radicals, 487
Similar terms, 51–53, 298–299
Similar triangles, 431–433
Simplifying
complex fractions, 394–400
computations with scientific notation, 43
expressions containing complex numbers, 514–518
imaginary numbers, 513–514
one or both sides of linear equations, 53–55
radical expressions, 454–459, 480–481, 484–487
rational expressions, 367–370
Slope-intercept form of equation of line, 120–127
Slope of the line, 107
Slopes
finding with equations, 109
finding with graph, 107–108
of horizontal lines, 109–110, 124
of nonvertical line, 108
of parallel lines, 110–111
passing through two points, 108
of perpendicular lines, 112
problem solving, 112–114
of vertical line, 109–110, 124
Solutions
of equations, 169
extraneous, 405
of quadratic equations, 559–560
Solution set, 49
Solving. *See also* Problem solving
equations containing radicals, 505, 506–507, 508
formulas containing radicals, 509–510
formulas for indicated variable, 56–57
higher-degree polynomial equations, 350–351
linear equations in one variable, 48–57, 62–67, 71–75
linear equations with absolute-value terms, 243–246
linear inequalities in one or two variables, 253–258
linear inequalities in one variable, 231–237
linear inequalities with one absolute-value term, 246–250
logarithmic equations, 651–652, 669–671
proportions, 429–431
quadratic equations by factoring, 348–350, 352–353, 535–536
quadratic inequalities, 579–581
radical equations, 504–509
rational equations, 404–410
rational inequalities, 581–584
systems of equations, 166–172, 176–180, 722–726
systems of linear equations, 182–186, 191–197, 201–206, 210–217, 261–263
three equations in three variables, 192–193
Special product formulas, 308
Special products, 308
Spheres, volume of, 36
Square matrix, 201, 210
Square-root function, domain of, 459–462
Square root property
defined, 537
solving quadratic equations with, 536–538
Square roots
defined, 454
principal, 455
simplifying perfect square root, 454–456
simplifying perfect square-root expression, 456–457
standard deviation, 462–463
Squares
area of, 36
perimeter of, 21
Squaring function, 145, 148, 149–151, 292
Standard deviation, 462–463
Standard equations
of circles, 688, 689
of ellipses, 704
of hyperbolas, 713
Standard form of quadratic equation, 548
Standard notation
converting from scientific notation, 41–42
converting to scientific notation, 40–41
defined, 41
Step functions, 731
Stokes' law of resistance, 678
Straight angles, 64
Subsets, defined, 3
Substitution method, 176–178, 339, 723–725
Subtracting
complex numbers, 516
polynomials, 299–300
radical expressions, 487–489
radicals, 488
rational expressions, 384–389
real numbers, 15–16
Sum
defined, 15
of first n terms of an arithmetic sequence, 760–761
of first n terms of a geometric sequence, 768–769
of infinite a geometric series, 774–776
of two cubes, 326–329, 344
of two functions, 589–590
of two squares, 325
Summation, index of, 761
Summation notation, 761–762
Supplementary angles, 65
Supplement of an angle, 65
Supply equation, 105
Symbols
grouping, 18–19
inequality, 7, 229
infinity, 8
Symmetries, of graphs, A-1–A-4
Symmetry, axis of, 565
Synthetic division, 421–426
Systems of equations. *See also* Systems of linear equations
consistent, 167
equivalent, 169
inconsistent, 167–168, 180, 193–194
solving by elimination (addition), 178–180, 725–726
solving by graphing, 166–172, 723
solving by substitution, 176–178, 723–725
solving containing second-degree equations, 722–726
Systems of inequalities, 261–265
Systems of linear equations
solving in two variables by graphing, 261–263

solving three in three variables, 191–197
solving two, 182–186
solving using Cramer's rule, 212–217
solving using determinants, 210–217
solving using matrices, 201–206

T

Television translators, 692
Terminating decimals, 5
Terms
algebraic, 51, 288
of annuity, 770
like, 51–53, 298–299
of sequences, 757
similar, 51–53, 298–299
unlike, 51, 298
Tests
for factorability, 336
horizontal line, 598–599
vertical line, 133–134, 599
Theorems
binomial, 750–751, 785
factor, 425
Pythagorean, 467–468, 469
remainder, 424–425
Transitive property, 230
Translations
of exponential functions, 624–625
horizontal, 149–150, 566–568, 643
vertical, 148, 565–568, 643
Trapezoids
area of, 36
perimeter of, 21
Tree diagrams, 778
Triangles
area of, 36
congruent, 185
equilateral, 65, 490
isosceles, 65
isosceles right, 489–491
Pascal's, 747–748
perimeter of, 21
right, 65, 467–468
similar, 431–433
Triangular form of matrix, 202
Trichotomoy property, 14, 230
Trinomials
defined, 289
factoring, 332–341, 344
with leading coefficient of 1, 333–335
with leading coefficient other than 1, 335
perfect-square, 308, 332–333, 538

U

Unbounded intervals, 232, 233
Union, defined, 8
Unit costs, 184–185, 429
Unlike signs
adding real numbers with, 15
dividing real numbers with, 17
multiplying real numbers with, 17
Unlike terms, 51, 298

V

Value
of 2 × 2 determinant, 210
of 3 × 3 determinant, 210–211
Variables
defined, 2
dependent, 137
independent, 137
Variance, 574
Variation
combined, 437
constant of, 433, 435, 436
direct, 433–434
inverse, 434–436
joint, 436–437
Vertex
of an ellipse, 703
of a hyperbola, 713
of a parabola, 565, 568, 569–571
Vertex angles of isosceles triangles, 65
Vertical angles, 69
Vertical asymptotes, 371
Vertical lines
defined, 98
as parallel lines, 111
slope of, 109–110, 124
Vertical line test, 133–134, 599
Vertical translations, 148, 565–568, 643
Vieta, François, 6, 310, 514
Volume
of cones, 36
of cubes, 36
of cylinders, 36
of pyramids, 36
of rectangular solids, 36
of spheres, 36

W

Weber-Fechner law, 663
Weierstrauss, Karl, 241
Whole numbers, defined, 3
Widmann, John, 15

X

x-axis, 92
x-coordinates, 93
x-intercepts, 96–97, 569–571

Y

y-axis, 92
y-coordinates, 93
y-intercepts, 96–97, 121, 124, 569–571
y is a function of x, 138

Z

Zero exponents, 32
Zero-factor property, 348
Zeros of polynomials, 425, 426

Basic Calculator Keystroke Guide

TI-83/84 Families of Calculators

Words in RED are calculator keys
For details on these and other functions, refer to the owner's manual.

BASIC SETUP	MODE All values down the left side should be highlighted. To return to the home screen at any time: 2ND MODE Normal Sci Eng Float 0123456789 Radian Degree Func Par Pol Seq Connected Dot Sequential Simul Real a+bi re^θ*i* Full Horiz G–T
Entering a rational expression:	(numerator)/(denominator) ENTER
Raising a value (or variable) to a power:	For x^2: value (or variable) x^2 For other powers: value (or variable) ^ power
Converting a decimal to a fraction:	decimal MATH ENTER ENTER
Entering an absolute value:	MATH ► ENTER value or expression)
Storing a value for x:	value STO X,T,θ,n ENTER
Storing a value for a variable other than x:	value STO ALPHA choose a variable from letters above keys on the right, ENTER
Accessing π:	2ND ^
Graphing an equation:	Solve for y if needed: Y= right side of equation GRAPH
Changing the viewing window for a graph:	WINDOW enter values and desired scales
Tracing along a graph: (An equation must be entered.)	From the graph: TRACE ► or ◄ as desired
Finding a y-value from a graph: (An equation must be entered.)	From the graph: 2ND TRACE ENTER value ENTER *Note:* Your x-value must be within your viewing window. You do not need to repeat the process for an additional value; enter a new one.
Finding the point of intersection of two graphs: (Two equations must be entered.)	From the graph: 2ND TRACE scroll to INTERSECT ENTER Adjust window if necessary to see the intersection. Using ► ◄ move cursor to approximate intersection. ENTER ENTER ENTER If two graphs are tangent, this process may not identify the point of intersection. *Note:* Your x-value must be within your viewing window. You will need to repeat the entire process for any additional intersection point.
Solving an equation by graphing: (Two equations must be entered.)	Y= left side of equation (Y_1) ENTER right side of equation (Y_2). GRAPH Locate point(s) of intersection as directed above.
Finding the x-intercept(s): (An equation must be entered.)	From the graph: 2ND TRACE scroll to ZERO ENTER numerical value that lies to the LEFT of the point ENTER numerical value that lies to the RIGHT of the point ENTER ENTER
Converting a decimal value obtained in a graph to a fraction:	From the graph, for x-value: 2ND MODE X,T,θ,n MATH ENTER ENTER From the graph, for y-value: 2ND MODE ALPHA 1 MATH ENTER ENTER *Note:* If the decimal value is irrational, there will be no equivalent fraction. It is possible that a rational number may not be converted to a fraction.

BASIC SETUP	MODE All values down the left side should be highlighted. To return to the home screen at any time: 2ND MODE Normal Sci Eng Float 0123456789 Radian Degree Func Par Pol Seq Connected Dot Sequential Simul Real a+bi re^θ*i* Full Horiz G–T
Accessing a table: (One or more equations must be entered.)	2ND GRAPH ▲ or ▼ as desired
Adjusting a table: (One or more equations must be entered.)	2ND WINDOW Set start value (TblStart=) and increment (ΔTbl=). AUTO should be highlighted for both Indpnt and Depend.
Finding a square root ($\sqrt{\ }$):	2ND x^2 value or expression)
Finding a cube root ($\sqrt[3]{\ }$):	MATH scroll to $\sqrt[3]{\ }$ ENTER value or expression)
Finding an *x*th root ($\sqrt[x]{y}$):	root (index) MATH scroll to $\sqrt[x]{\ }$ (value or expression) ENTER
Finding the maximum (minimum) point: (An equation must be entered.)	2ND TRACE scroll to MAXIMUM (MINIMUM) ENTER numerical value that lies to the left of the point ENTER numerical value that lies to the right of the point ENTER ENTER
Solving an inequality in two variables: (shading)	Y= right side of equation ◄ to the left of Y_1 For $y >$ ENTER ENTER (◥) GRAPH For $y <$ ENTER ENTER ENTER (◣) GRAPH If solving a system of inequalities, repeat for Y_2. *Note:* The calculator cannot distinguish between solid and dashed boundaries.
Evaluating a function: (An equation must be entered for $Y_{\#}$.)	VARS ► ENTER select $Y_{\#}$ ENTER (value) ENTER Or: Use the VALUE feature from the graph.
Accessing *i*:	2ND •
Accessing a function from the catalog:	2ND 0 Press first letter of desired function (ALPHA letter) or scroll (up for symbols).

NOTES:

CHAPTER 8 QUADRATIC FUNCTIONS, INEQUALITIES, AND ALGEBRA OF FUNCTIONS

The quadratic formula:
If $a \neq 0$, the solutions of $ax^2 + bx + c = 0$ are given by

$$x = \frac{-b \pm \sqrt{b^2 - 4ac}}{2a}$$

The discriminant:

If	then the roots of $ax^2 + bx + c = 0$ are
$b^2 - 4ac > 0$	real and unequal
$b^2 - 4ac = 0$	rational and equal
$b^2 - 4ac < 0$	complex conjugates

If r_1 and r_2 are the roots of $ax^2 + bx + c = 0$,

$$r_1 + r_2 = -\frac{b}{a} \qquad r_1 r_2 = \frac{c}{a}$$

Graphs of quadratic functions:
The graph of $y = ax^2 + bx + c \quad (a \neq 0)$ is a parabola. It opens

$$\begin{Bmatrix}\text{upward}\\ \text{downward}\end{Bmatrix} \text{ when } \begin{Bmatrix}a > 0\\ a < 0\end{Bmatrix}$$

The graph of $y = a(x - h)^2 + k \quad (a \neq 0)$ is a parabola with vertex at (h, k). It opens

$$\begin{Bmatrix}\text{upward}\\ \text{downward}\end{Bmatrix} \text{ when } \begin{Bmatrix}a > 0\\ a < 0\end{Bmatrix}$$

Vertex of a parabola:
$f(x) = ax^2 + bx + c \quad (a \neq 0)$

$$\left(-\frac{b}{2a}, f\left(-\frac{b}{2a}\right)\right)$$

Algebra of functions:

$$(f + g)(x) = f(x) + g(x)$$
$$(f - g)(x) = f(x) - g(x)$$
$$(f \cdot g)(x) = f(x)g(x)$$
$$(f/g)(x) = \frac{f(x)}{g(x)} \quad (g(x) \neq 0)$$
$$(f \circ g)(x) = f(g(x))$$

CHAPTER 9 EXPONENTIAL AND LOGARITHMIC FUNCTIONS

$y = \log_b x$ if and only if $x = b^y \quad (b > 0, b \neq 1)$

$y = b^x$ and $y = \log_b x$ are inverse functions.

Compound interest: $A = P\left(1 + \frac{r}{k}\right)^{kt}$

Continuous compound interest: $A = Pe^{rt}$

Malthusian population growth: $A = Pe^{rt}$

Decibel voltage gain: db gain $= 20 \log \frac{E_o}{E_I}$

Richter scale: $R = \log \frac{A}{P}$

Doubling time: $t = \frac{\ln 2}{r}$

Properties of logarithms: If M, N, and b are positive numbers and $b \neq 1$, then

$$\log_b 1 = 0$$
$$\log_b b = 1$$
$$\log_b b^x = x$$
$$b^{\log_b x} = x$$
$$\log_b MN = \log_b M + \log_b N$$
$$\log_b \frac{M}{N} = \log_b M - \log_b N$$
$$\log_b M^p = p \log_b M$$

If $\log_b x = \log_b y$, then $x = y$.

Change-of-base formula: $\log_b y = \frac{\log_a y}{\log_a b}$

pH scale: $\text{pH} = -\log[\text{H}^+]$

Weber–Fechner law: $L = k \ln I$

Carbon dating: $A = A_0 e^{kt}$

CHAPTER 10 CONIC SECTIONS AND MORE GRAPHING

Equations of a circle with radius r:

$(x - h)^2 + (y - k)^2 = r^2$ center at (h, k)

$x^2 + y^2 = r^2$ center at $(0, 0)$

Equations of a parabola:

Parabola opening	Vertex at origin	Vertex at (h, k)
Up	$y = ax^2 \quad (a > 0)$	$y = a(x - h)^2 + k \quad (a > 0)$
Down	$y = ax^2 \quad (a < 0)$	$y = a(x - h)^2 + k \quad (a < 0)$
Right	$x = ay^2 \quad (a > 0)$	$x = a(y - k)^2 + h \quad (a > 0)$
Left	$x = ay^2 \quad (a < 0)$	$x = a(y - k)^2 + h \quad (a < 0)$